HANDBOOK OF SPIN TRANSPORT AND MAGNETISM

EDITED BY
EVGENY Y. TSYMBAL
IGOR ŽUTIĆ

CRC Press is an imprint of the
Taylor & Francis Group, an **informa** business

CRC Press
Taylor & Francis Group
6000 Broken Sound Parkway NW, Suite 300
Boca Raton, FL 33487-2742

First issued in paperback 2020

Version Date: 20110831

ISBN-13: 978-0-367-57689-9 (pbk)
ISBN-13: 978-1-4398-0377-6 (hbk)

Library of Congress Cataloging-in-Publication Data

Handbook of spin transport and magnetism / editors, Evgeny Y. Tsymbal, Igor Zutic.
p. cm.
Includes bibliographical references and index.
ISBN 978-1-4398-0377-6 (hardcover : alk. paper)
1. Spintronics. 2. Magnetism. 3. Magnetoresistance. I. Tsymbal, E. Y. (Evgeny Y.) II. Zutic, Igor.

TK7874.887.H36 2012
538--dc23 2011030270

Visit the Taylor & Francis Web site at
http://www.taylorandfrancis.com

and the CRC Press Web site at
http://www.crcpress.com

Contents

PART VII Applications

Preface

This book addresses two intimately related subjects—spin transport and magnetism. The field of spin transport has emerged from studying the properties of magnetic materials and formed a relatively new research area known as spin electronics, or spintronics. In the past two to three decades, the research on spin transport and magnetism has led to remarkable scientific and technological breakthroughs. A prominent example is the discovery of giant magnetoresistance in magnetic metallic multilayers that was recognized by the 2007 Nobel Prize in Physics awarded to Albert Fert and Peter Grünberg.

The roots of spin transport and spintronics lie in the quantum description of solids, which predicts the existence of the spin, an intrinsic quantity characterizing electrons. The importance of the electron's spin in controlling transport properties was already emphasized by Sir Nevill Mott in 1936 to understand electrical conductivities of the transition metals and their alloys. His two-current model, in which the conductivity is expressed as the sum of two unequal contributions from two different spin projections, explained the high resistivity of ferromagnetic metals relative to their nonmagnetic counterparts. This model introduced the concept of spin transport and showed the importance of magnetism in controlling transport properties. The spin nature of electron transport lies at the heart of modern magnetic data storage technologies, such as computer hard drives and magnetic random access memories (MRAMs). Their operation employs a magnetoresistive device, a metallic spin valve or a magnetic tunnel junction, that consists of two ferromagnetic electrodes separated by a thin spacer layer of nonmagnetic material. The flow of carriers through the device is determined by the direction of their spin relative to the magnetization of the device's electrodes. When the magnetization of the two electrodes is parallel, current flow across the device is permitted; whereas when the magnetization is antiparallel, the current flow is restricted. The corresponding difference in resistance between these two configurations is known as giant magnetoresistance (GMR) if the spacer is metallic and as tunneling magnetoresistance (TMR) if the spacer is insulating.

Early experimental demonstration of Mott's ideas, that the electrical current in magnetic materials is spin polarized, can be inferred from two classes of experiments pertaining to tunneling processes. In 1970, Paul Tedrow and Robert Meservey demonstrated that the tunneling current, originating from a ferromagnetic metal electrode and passing through an insulating alumina barrier, remains spin polarized even outside of the ferromagnetic region. The detection of the spin-polarized current was performed using Zeeman-split density of states in a superconductor, allowing them to quantify the degree of tunneling spin polarization for various ferromagnetic metals. Their studies of the corresponding ferromagnetic/insulator/superconductor junctions have directly guided the discovery of TMR in magnetic tunnel junctions, a key functioning element in modern MRAMs. In 1967, Leo Esaki, Philip Stiles, and Stephan von Molnar employed a metal/magnetic insulator/metal tunnel junction to demonstrate what perhaps can be viewed as the first prototype of a spintronic device. They found that after a large applied bias, the tunneling current depended directly on the magnetization state of the insulator. The height of the insulating barrier layer made of Eu-based chalcogenides could be altered by an applied magnetic field, and the device effectively acted as a spin filter, allowing predominantly the transfer of spins of one direction determined by the magnetization orientation in the insulating film. Doping Eu-chalcogenides with carriers revealed that their ferromagnetic transition temperature can be enhanced, paving a path for versatile magnetic semiconductors that have been intensely studied in the past two decades in which the ferromagnetism may be electrically or optically tuned.

These pioneering contributions to what has subsequently become known as spintronics have foreshadowed the need to study spin transport and magnetism in a wide class of different materials and structures. As this book reveals, some of the key phenomena that were first discovered in one class of materials may have been revisited decades later in other materials systems. A prominent example is the demonstration of spin injection in metals, which, two decades later, was explored in entirely different materials such as silicon, organic semiconductors, carbon nanotubes, and graphene, as well as carefully engineered nanostructures, often with surprising twists and unexpected revelations.

To provide a balanced account of the state-of-the-art standing of the field, we employ a comprehensive approach that covers all the most active research areas in spintronics. Consequently, the content of the book is divided into six parts mostly focusing on specific materials and structures with extensive cross-references between chapters to reflect the fact that the discoveries in one class of materials may be directly relevant to the breakthroughs in seemingly unrelated systems.

Part I contains a historical and personal perspective of the field written by the Nobel Prize laureate Albert Fert. In addition to his pioneering work on GMR in metallic magnetic multilayers, it contains his first-hand vision of several other topics and materials systems currently actively studied in spintronics.

Part II addresses diverse physical phenomena discovered in hybrid structures of ferromagnetic and normal metals. The part starts from the relevant background in magnetism of magnetic nanostructures where the concepts of magnetic anisotropy, interlayer exchange coupling, exchange bias, and magnetostatic interactions are introduced and discussed. The phenomenon of GMR continues the part. It was discovered that the resistance of a magnetic metallic multilayer drops dramatically when the magnetization alignment of the magnetic layers changes from antiparallel to parallel due to the application of an external magnetic field. A sizable GMR effect persisted at room temperature that opened the way to present-day spintronic applications. Advances in magnetic data storage have enabled nearly a 1000-fold increase in the capacity of computer hard drives using metal-based GMR (and later TMR) spin valves over the last decade. Experiments on electrical spin injection, first envisioned three decades ago, have provided an effective way to generate a nonequilibrium spin population and use it as a sensitive spectroscopic tool to probe fundamental properties such as spin–orbit coupling, spin–Hall effect, hyperfine interactions, and pairing symmetry of unconventional superconductors. Specific nonlocal geometries of lateral spin valves for the spin detection and demonstration of pure spin currents flowing without any accompanying charge currents are now increasingly important in a wide range of materials. The discovery of the spin-torque effect has provided a powerful alternative to the conventional magnetization switching using an external magnetic field, thus offering a wealth of new opportunities for spin transport. Spin-polarized currents interact with ferromagnets, so that a fraction of the spin angular momentum carried by the electrons can be transferred to the ferromagnet. This induces spin-torque-driven magnetic excitations and may be sufficient to reverse the magnetization of the ferromagnet.

Part III is devoted to impressive breakthroughs in spin-dependent tunneling. For more than two decades, after early experiments on spin-dependent tunneling, this area of research has been reserved for only a handful of experts. However, several key developments have dramatically altered this situation. The demonstration of a large and reproducible room temperature TMR in magnetic tunnel junctions in the mid-1990s has shown the practical side of these structures. The latter became especially apparent when a replacement of the amorphous Al_2O_3 tunnel barrier by crystalline MgO led to a more than 10-fold increase in the room temperature TMR. The underlying principle of atomically controlled interfaces for desirable symmetry selection in spin transport, predicted from the first-principles quantum-mechanical calculations, has already led to the development of advanced magnetic fields sensors used in modern hard drives and magnetic data storage elements employed in MRAMs. The tunneling structures are now among the most promising platforms to provide superior spin-based devices and reveal novel fundamental phenomena. For example, using a ferroelectric barrier layer in a magnetic tunnel junction offers a versatile electrical control of TMR through ferroelectric polarization of the barrier.

Traditionally, both spin transport and magnetism have been primarily studied in metallic materials. Part IV reviews the challenges and opportunities for controlling spin and magnetism in semiconductors. Elucidating the underlying mechanisms for spin relaxation and decoherence in semiconductors provides predictive tools to identify the materials and structures that could provide desirably long time- and length-scales for the transfer of spin information. Recent scientific breakthroughs demonstrate that silicon is not only central to conventional electronics, but also offers desirable spin-dependent properties. In silicon, spin can be injected and transferred over more than a millimeter at room temperature. The prospect of silicon spintronics offers an intriguing possibility to combine well-established semiconductor technology with novel spin-related functionalities. Two classes of magnetic semiconductors have been the subject of the intense research effort in the past 15 years. However, the origin of their magnetic ordering remains to be fully understood and continues to be a theoretical challenge. One class is dilute magnetic semiconductors, typically III–V or II–VI compounds doped with Mn of low density. They are remarkably tunable in nature: the ferromagnetic Curie temperature can be controlled externally by light or applied bias. Another class is magnetic oxides, where the ferromagnetic ordering has been reported even in the absence of d-electrons and magnetic impurities. Magnetic semiconductors also offer new possibilities for spin-torque effects, including lower critical currents for magnetization reversal. Spin–orbit coupling arising due to bulk and structure inversion asymmetries is both friend and foe. While often it is desirable to have the spin–orbit coupling reduced, for example, in Si or organic semiconductors to have a longer spin relaxation time, it also is the origin of the fascinating phenomena ranging from the family of spin-dependent Hall effects to generating spin polarization simply by a dc charge current. In addition to the longer-standing tradition of optically injecting and detecting spins in semiconductors, electrical spin injection including related nonlocal effects by now is also well established in semiconductors. Using a tunnel barrier between a ferromagnetic metal and a semiconductor provides robust room temperature spin injection into semiconductors.

Driven by the technological challenge to increase data storage density and enhance the speed of data processing, there is a continuing effort to miniaturize electronic and magnetoresistive devices. A versatile fabrication of various nanostructures can offer favorable scaling properties such as enhanced magnetoresistive effects and room temperature operation otherwise limited to cryogenic temperatures. Part V focuses on the phenomena characteristic for such nanostructures made of a wide range of materials, including metals, superconductors, molecular magnets, carbon nanotubes, or graphene. Spin-dependent tunneling in nanoscale structures can be used to implement a spin-polarized tunneling microscope that can be used not only to provide atomically resolved information about magnetism, but also to place individual magnetic impurities with atomic-scale precision. Transport measurements, employing Andreev reflection in junctions with a superconducting tip and a ferromagnet, have provided a sensitive probe for the degree of transport spin polarization in a wide range of materials. Spin injection in nanoscale structures can be more pronounced than in their bulk-like counterparts. Efficient magnetization reversal can be implemented using pure spin currents in nonlocal geometries of lateral spin valves while the family of spin-dependent Hall effects can be observed even at room temperature. Semiconductor quantum dots allow for a versatile control of the number of carriers, spin, and the effects of quantum confinement, which leads to improved optical, transport, and magnetic properties as compared to their bulk counterparts. Adding a single carrier in magnetic quantum dots can both strongly change the carrier's spin and the temperature of the magnetization onset.

Part VI addresses the existing and potential spin-based applications that closely rely on the fundamental properties of spin transport and magnetism, discussed in the preceding chapters of the book. This part illustrates methods for improving the operation of metallic spin valves and magnetic tunnel junctions desirable for particular applications and explores paths to novel spin-based applications that are not limited to magnetoresistive effects or conventional metallic structures. The successful use of spin valves and magnetic tunnel junctions is discussed in the chapters devoted to magnetic field sensing and biosensing as well as to nonvolatile MRAMs. A vision for the novel class of spintronic memories explores the possibility to utilize the domain wall motion and implement a three-dimensional integration. Semiconductor spin lasers, in which spin-polarized carriers are injected electrically or optically, have already been demonstrated and have shown advantages over conventional lasers. While the push for ferromagnetic semiconductors has often been motivated by the desire to seamlessly integrate charge-based logic and spin-based information storage, there is also an alternative scheme relying on electrical spin injection in semiconductor/ferromagnetic metal junctions. The envisioned performance in terms of the speed or energy consumption could exceed the ultimate limits possible within conventional electronics.

Overall, this book offers a balanced and thorough treatment of the core principles, theoretical models, experimental approaches, and state-of-the-art applications of spin transport and magnetism, making it suitable for a broad readership. We hope that this book will be an important reference both for graduate students and experienced researchers specializing in the fields of condensed matter physics, materials science, device engineering, and nanoscience.

None of this would have been possible without the dedicated effort of the authors who abandoned, for extended periods of time, their own research to complete their respective chapters, while dealing with the editorial requests for reshaping their style and content. We are thankful to the reviewers who have carefully examined all the chapters and made many helpful suggestions to improve this book. We are also grateful to Verona Skomski for providing invaluable assistance in preparing the book for submission to the publisher. Finally, we acknowledge the support of NSF-DMR, NSF-MRSEC, NSF-ECCS, NSF-EPSCoR, US ONR, AFOSR-DCT, DOE-BES, and SRC, which has enabled our editorial work.

Evgeny Y. Tsymbal
Department of Physics and Astronomy
Nebraska Center for Materials and Nanoscience
University of Nebraska
Lincoln, Nebraska

Igor Žutić
Department of Physics
University at Buffalo
State University of New York
Buffalo, New York

Editors

Evgeny Y. Tsymbal is a Charles Bessey Professor in the Department of Physics and Astronomy at the University of Nebraska-Lincoln (UNL) and the director of the UNL's Materials Research Science and Engineering Center. He joined UNL in 2002 as an associate professor, was promoted to a full professor with tenure in 2005, and named a Charles Bessey Professor of Physics in 2009. Prior to his appointment at UNL, he was a research scientist at the University of Oxford, United Kingdom; a research fellow of the Alexander von Humboldt Foundation at the Research Center-Jülich, Germany; and a research scientist at the Russian Research Center "Kurchatov Institute," Moscow.

Evgeny Tsymbal's research is focused on computational materials science, aiming at the understanding of fundamental properties of advanced ferromagnetic and ferroelectric nanostructures and materials relevant to nanoelectronics and spintronics. He has published over 150 papers, review articles, and book chapters and has presented over 100 invited talks in the areas of spin-dependent transport, magnetoresistive phenomena, nanoscale magnetism, complex oxide heterostructures, interface magnetoelectric phenomena, two-dimensional electron gases, and ferroelectric tunnel junctions. His research has been supported by the National Science Foundation, Semiconductor Research Corporation, the Office of Naval Research, the Department of Energy, Seagate Technology, and the W. M. Keck Foundation.

Evgeny Tsymbal is a fellow of the American Physical Society; a fellow of the Institute of Physics, United Kingdom; and a recipient of the UNL's College of Arts & Sciences Outstanding Research and Creativity Award.

Igor Žutić received his PhD in theoretical physics from the University of Minnesota, after undergraduate studies at the University of Zagreb, Croatia. He was a postdoc at the University of Maryland and the Naval Research Lab. In 2005, he joined the State University of New York at Buffalo as an assistant professor of physics and got promoted to an associate professor in 2009.

Interested to find out if the emerging area of spintronics is only hype or has some future, he proposed and chaired *Spintronics 2001: International Conference on Novel Aspects of Spin-Polarized Transport and Spin Dynamics*, at Washington, DC. The conference appeared to be a success (spintronics was not all hype) and he was invited to write a comprehensive review, Spintronics: Fundamentals and Applications, for the *Reviews of Modern Physics*—currently among the most cited articles on spin transport and magnetism. Work with his collaborators spans a range of topics from high-temperature superconductors and ferromagnetism that can get stronger as the temperature is increased, to prediction of various spin-based devices (some of which were actually realized). He has published over 50 refereed articles and given about 80 invited presentations on spin transport, magnetism, spintronics, and superconductivity.

Igor Žutić is a recipient of the 2006 National Science Foundation CAREER Award, 2005 National Research Council/American Society for Engineering Education Postdoctoral Research Award, and the National Research Council Fellowship (2003–2005). His research is supported by the National Science Foundation, the Office of Naval Research, the Air Force Office of Scientific Research, and the Department of Energy.

Contributors

Ramon Aguado
Department of Teoria y Simulación de Materiales
Instituto de Ciencia de Materiales de Madrid (ICMM), CSIC
Madrid, Spain

Johan Åkerman
Department of Physics
University of Gothenburg
Gothenburg, Sweden

Ian Appelbaum
Department of Physics
University of Maryland
College Park, Maryland

Anthony S. Arrott
Department of Physics
Simon Fraser University
Burnaby, British Columbia, Canada

Agnès Barthélémy
Unité Mixte de Physique
Centre National de la Recherche Scientifique-Thales
Université Paris-Sud
Palaiseau, France

Jack Bass
Department of Physics and Astronomy
Michigan State University
East Lansing, Michigan

Kirill D. Belashchenko
Department of Physics and Astronomy
Nebraska Center for Materials and Nanoscience
University of Nebraska
Lincoln, Nebraska

Ilaria Bergenti
Institute for Nanostructured Materials Studies
Bologna, Italy

Manuel Bibes
Unité Mixte de Physique
Centre National de la Recherche Scientifique-Thales
Université Paris-Sud
Palaiseau, France

Matthias Bode
Department of Physics
University of Würzburg
Würzburg, Germany

J.M.D. Coey
Centre for Research on Adaptive Nanostructures and Nanodevices
School of Physics
Trinity College Dublin
Dublin, Ireland

Scott A. Crooker
National High Magnetic Field Laboratory
Los Alamos National Laboratory
Los Alamos, New Mexico

Paul A. Crowell
School of Physics and Astronomy
University of Minnesota
Minneapolis, Minnesota

Valentin Dediu
Institute for Nanostructured Materials Studies
Bologna, Italy

Hanan Dery
Department of Electrical & Computer Engineering
University of Rochester
Rochester, New York

Bernard Doudin
Institut de Physique et de Chimie des Matériaux de Strasbourg
Centre National de la Recherche Scientifique
University Louis Pasteur
Strasbourg, France

Jaroslav Fabian
Department of Physics
University of Regensburg
Regensburg, Germany

Joaquín Fernández-Rossier
Departamento de Física Aplicada
Universidad de Alicante
Alicante, Spain

Albert Fert
Unité Mixte de Physique
Centre National de la Recherche Scientifique-Thales
Université Paris-Sud
Palaiseau, France

William Gallagher
IBM Thomas J. Watson Research Center
Yorktown Heights, New York

Sergey D. Ganichev
Department of Physics
University of Regensburg
Regensburg, Germany

Richard S. Gaster
Department of Bioengineering
Stanford University
Stanford, California

Jean-Marie George
Unité Mixte de Physique
Centre National de la Recherche Scientifique-Thales
Université Paris-Sud
Palaiseau, France

Christian Gøthgen
Department of Physics
University at Buffalo
State University of New York
Buffalo, New York

Drew A. Hall
Department of Electrical Engineering
Stanford University
Stanford, California

Masamitsu Hayashi
National Institute for Materials Science
Tsukuba, Japan

Bretislav Heinrich
Department of Physics
Simon Fraser University
Burnaby, British Columbia, Canada

Luis E. Hueso
CIC nanoGUNE Consolider
Basque Foundation for Science
Bilbao, Spain

Henri Jaffrès
Unité Mixte de Physique
Centre National de la Recherche Scientifique-Thales
Université Paris-Sud
Palaiseau, France

Ron Jansen
National Institute of Advanced Industrial Science and Technology
Tsukuba, Japan

Xin Jiang
IBM Almaden Research Center
San Jose, California

Mark Johnson
Materials Physics Division
Naval Research Laboratory
Washington, DC

Berend T. Jonker
Magnetoelectronic Materials and Devices Branch
Materials Science and Technology Division
Naval Research Laboratory
Washington, DC

Csaba Józsa
Physics of Nanodevices
Zernike Institute for Advanced Materials
University of Groningen
Groningen, the Netherlands

Tomás Jungwirth
Institute of Physics
Academy of Sciences of the Czech Republic
Prague, Czech Republic

N.T. Kemp
Institut de Physique et de Chimie des Matériaux de Strasbourg
Centre National de la Recherche Scientifique
University Louis Pasteur
Strasbourg, France

Takashi Kimura
Department of Electronic Materials and Devices
INAMORI Frontier Research Center
Kyushu University
Fukuoka, Japan

Hitoshi Kubota
National Institute of Advanced Industrial Science and Technology
Spintronics Research Center
Tsukuba, Japan

Patrick R. LeClair
Department of Physics
and
Astronomy Center for Materials for Information Technology
University of Alabama
Tuscaloosa, Alabama

Jeongsu Lee
Department of Physics
University at Buffalo
State University of New York
Buffalo, New York

Sadamichi Maekawa
Advanced Science Research Center
Japan Atomic Energy Agency
Tokai, Japan

A. Manchon
Department of Physics
University of Arizona
Tucson, Arizona

and

Physical Science and Engineering
King Abdullah University of Science and Technology
Thuwal, Saudi Arabia

Jagadeesh S. Moodera
Francis Bitter Magnet Laboratory
Massachusetts Institute of Technology
Cambridge, Massachusetts

Rai Moriya
Institute of Industrial Science
University of Tokyo
Tokyo, Japan

Boris E. Nadgorny
Department of Physics & Astronomy
Wayne State University
Detroit, Michigan

Rafał Oszwałdowski
Department of Physics
University at Buffalo
State University of New York
Buffalo, New York

Yoshichika Otani
The Institute for Solid State Physics
The University of Tokyo
Chiba, Japan

and

RIKEN ASI Quantum Nano-Scale Magnetics Laboratory
Saitama, Japan

Stuart Parkin
IBM Almaden Research Center
San Jose, California

D.G. Pettifor
Department of Materials
University of Oxford
Oxford, United Kingdom

Tiffany S. Santos
Center for Nanoscale Materials
Argonne National Laboratory
Argonne, Illinois

Stefano Sanvito
Centre for Research on Adaptive Nanostructures and Nanodevices
School of Physics
Trinity College Dublin
Dublin, Ireland

John Schliemann
Department of Physics
University of Regensburg
Regensburg, Germany

Jairo Sinova
Department of Physics
Texas A&M University
College Station, Texas

Yoshishige Suzuki
Department of Materials Engineering Science
Osaka University
Osaka, Japan

Saburo Takahashi
Institute for Materials Research
Tohoku University
Sendai, Japan

Luc Thomas
IBM Almaden Research Center
San Jose, California

Carsten Timm
Institute for Theoretical Physics
Technische Universität Dresden
Dresden, Germany

Maxim Trushin
Department of Physics
University of Regensburg
Regensburg, Germany

Maxim Tsoi
Department of Physics
The University of Texas at Austin
Austin, Texas

Evgeny Y. Tsymbal
Department of Physics and Astronomy
Nebraska Center for Materials and Nanoscience
University of Nebraska
Lincoln, Nebraska

Bart J. van Wees
Physics of Nanodevices
Zernike Institute for Advanced Materials
University of Groningen
Groningen, the Netherlands

Shan X. Wang
Departments of Materials Science and Electrical Engineering
Stanford University
Stanford, California

M.W. Wu
Hefei National Laboratory for Physical Sciences at Microscale
and
Department of Physics
University of Science and Technology of China
Anhui, China

Jörg Wunderlich
Hitachi Cambridge Laboratory
Cambridge, United Kingdom

Gang Xiao
Department of Physics
Brown University
Providence, Rhode Island

Shinji Yuasa
National Institute of Advanced Industrial Science and Technology
Tsukuba, Japan

Shufeng Zhang
Department of Physics
University of Arizona
Tucson, Arizona

Igor Žutić
Department of Physics
University at Buffalo
State University of New York
Buffalo, New York

I

Introduction

1

Historical Overview: From Electron Transport in Magnetic Materials to Spintronics

Albert Fert
Université Paris-Sud

1.1 Introduction

Spintronics is now an important field of research with major applications in several technologies. Its development has been triggered by the discovery [1,2] of the giant magnetoresistance (GMR) in 1988. The basic concept of spintronics is the manipulation of spin-polarized currents, in contrast to mainstream electronics in which the spin of the electron is ignored. Adding the spin degree of freedom provides new effects, new capabilities, and new functionalities. Spin-polarized currents can be generated by exploiting the influence of the spin on the transport properties of the electrons in ferromagnetic conductors. This influence, first suggested by Mott [3], had been experimentally demonstrated and theoretically described in early works [4,5] more than 10 years before the discovery of the GMR. The GMR was the first step on the road of the utilization of the spin degree of freedom in magnetic nanostructures. Its application to the read heads of hard discs greatly contributed to the fast rise in the density of stored information and led to the extension of the hard disk technology to consumer's electronics. Then, the development of an intensive research revealed many other phenomena related to the control and manipulation of spin-polarized currents. Today, the field of spintronics is expanding considerably, with very promising new axes like the manipulation of magnetic moments and the generation of microwaves by spin transfer, spintronics with semiconductors, molecular spintronics, the spin Hall effect (SHE), the quantum spin Hall effect (QSHE), and single-electron spintronics for quantum computing. In this chapter, I will tell the story of this development from the early experiments on spin-dependent conduction in ferromagnets to the emerging directions of today.

1.2 Spin-Dependent Conduction in Ferromagnets and Early Examples of Spin Transport Experiments

GMR and spintronics take their roots in previous researches on the influence of the spin on the electrical conduction in ferromagnetic metals [3–5]. The spin dependence of the conduction can be understood from the typical band structure of a ferromagnetic metal, which is shown in Figure 1.1a. Due to the splitting between the energies of the "majority spin" and "minority spin" directions (spin up and spin down in the usual notation), the electrons at the Fermi level, which carry the electrical current, are in different states and exhibit different conduction properties for opposite spin directions. This spin-dependent conduction was proposed by Mott [3] in 1936 to explain some features of the resistivity of ferromagnetic metals at the Curie temperature. However, in 1966, when I started my PhD thesis, the subject was still almost completely unexplored. My supervisor, Ian Campbell, and I investigated the transport properties of

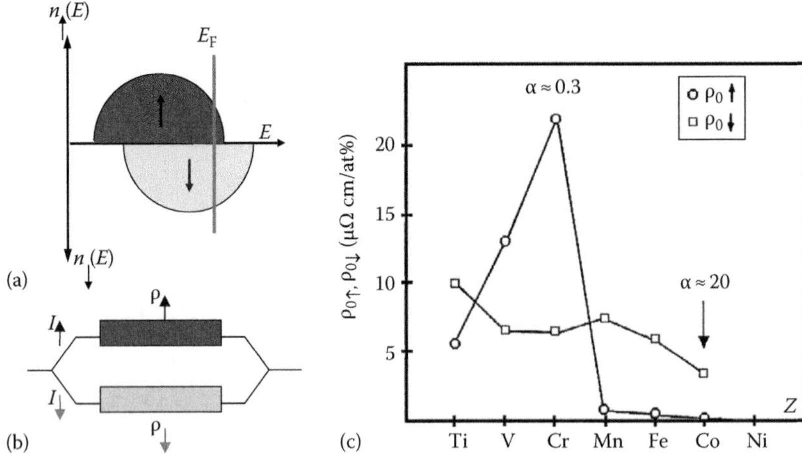

FIGURE 1.1 Basics of spintronics. (a) Schematic band structure of a ferromagnetic metal. (b) Schematic for spin-dependent conduction through independent spin-up and spin-down channels in the limit of negligible spin mixing ($\rho_{\uparrow\downarrow} = 0$ in the formalism of Ref. [4]). (c) Resistivities of the spin-up and spin-down conduction channels for nickel doped with 1% of several types of impurity (measurements at 4.2 K) [4]. The ratio α between the resistivities $\rho_{0\downarrow}$ and $\rho_{0\uparrow}$ can be as large as 20 (Co impurities) or, as well, smaller than 1 (Cr or V impurities). (From Fert, A. and Campbell, I.A., *J. Phys. F Met. Phys.*, 6, 849, 1976. With permission.)

Ni- and Fe-based alloys and, from the analysis of the temperature dependence of the resistivity and also from experiments on ternary alloys I describe in Section 1.3, we demonstrated the spin dependence of conduction in various metals and alloys. In particular, we showed that the resistivities of the two channels can be very different in metals doped with impurities presenting a strongly spin-dependent scattering cross-section [4]. In Figure 1.1c, I show the example of the spin-up and spin-down resistivities of nickel doped with 1% of different impurities. It can be seen that the ratio α of the spin-down resistivity to the spin-up one can be as large as 20 for Co impurities or, as well, smaller than the 1 for Cr or V impurities. This was consistent with the theoretical models developed at this time by Jacques Friedel for the electronic structures of impurities in metals. The two-current conduction was rapidly confirmed in other groups and, for example, extended to Co-based alloys by Loegel and Gautier [5].

The so-called two-current model [4] for the conduction in ferromagnetic metals was worked out for the quantitative interpretation of the experiments described in the preceding paragraph. This model is based on a picture of spin-up and spin-down currents coupled by spin mixing, that is, by exchange of current between the spin-up and spin-down channels. Spin mixing comes from spin-flip scattering, mainly from electron-magnon scattering, which conserves the total spin of the electronic system but is a mechanism of current transfer between the two channels. It increases with temperature and tends to equalize partly the spin-up and spin-down currents at room temperature in most ferromagnetic metals [4]. The two-current model is the basis of spintronics today, but, except in very few publications [6,7] discussing the temperature dependence of the GMR, the interpretation of the spintronics phenomena is surprisingly based on a simplified version of the model neglecting spin mixing and assuming that the conduction is by two independent channels in parallel, as in the sketch of Figure 1.1c. It should be certainly useful to revisit the interpretation of many recent experiments by taking into account the spin mixing contributions (note that the spin mixing mechanism by spin-flips should not be confused with the spin relaxation mechanism transferring spin accumulation to the lattice and due mainly to spin–orbit scattering).

The research on spin transport developed before the discovery of GMR has not only worked out the basic physics of spin transport in ferromagnetic conductors but also explored some topics that came into fashion only recently in spintronics. I take the example of the SHE. When an electrical current flows in a nonmagnetic conductor, the electrons of opposite spins are deflected to opposite transverse directions by spin–orbit interactions. With equal spin-up and spin-down currents, there is no charge accumulation and consequently no Hall voltage between edge contacts, but the deflections induce opposite spin accumulations on the edges of the conductor, as illustrated in Figure 1.2a. This is the SHE, already described in 1971 by D'yakonov and Perel' [8].

As the spin accumulation by SHE can be used to generate spin currents, the SHE is nowadays presented as a possible way for spintronics without magnetic materials, which explains the intense current research on the topic. In most experiments of the recent years, the SHE is detected by breaking the symmetry between the opposite spin directions to obtain a transverse voltage, either by using a ferromagnetic metal for one of the Hall probes or by injecting locally a spin-polarized current (or even a pure spin current) from a ferromagnetic contact, the so-called inverse SHE detection. This can be done only in nano-devices fabricated by lithographic techniques [9–11]. Thirty years ago, it was not technically possible to fabricate such nano-devices,

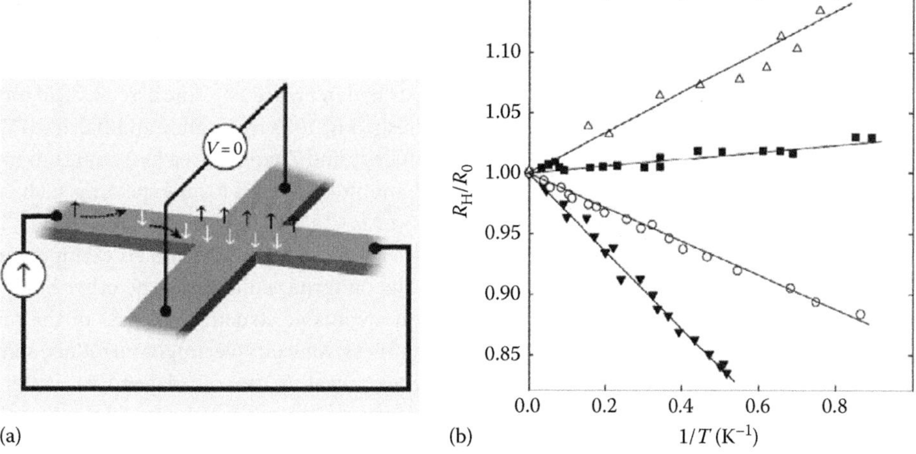

FIGURE 1.2 (a) Sketch illustrating the SHE in a nonmagnetic conductor. (b) Experimental results [12] on the SHE induced by resonant scattering on 5d levels (split by spin–orbit coupling) of nonmagnetic impurities in Cu (open triangles = Lu 0.013%, circles = Ta 0.023%, black triangle = Ir 0.19%). The inverse SHE of the Cu alloys is revealed by adding ≈ 0.01% of Mn impurities and applying a field H to spin-polarize the current by exchange scattering on the spin-polarized Mn impurities. With a paramagnetic-like spin polarization proportional to H/T, the SHE is revealed by a contribution to the Hall constant proportional to T^{-1} with different slopes for different impurities (the squares represent measurement on Cu with only 0.015% of Mn and without 5d impurities, which shows the quasi-absence of $1/T$ contribution to the Hall effect from Mn alone). Note the characteristic change of sign of the SHE between the beginning (Lu) and the end (Ir) of the 5d series.

but, nevertheless, a precise determination of the SHE could be performed in other ways. In Figure 1.2b, we show the results published in 1981 [12] on the SHE induced by 5d impurities (Lu, Ta, Ir, Au) in Cu. The SHE is detected by adding 0.01% of Mn impurities and applying a magnetic field. It had previously been shown that applying a field with only Mn impurities in Cu does not give any significant contribution to the Hall effect but induces (by exchange scattering) a spin polarization of the current, which can be known from the associated negative magnetoresistance (GMR-like effect). In brief, the spin polarization induced by exchange with dilute impurities of Mn replaces the spin injection of modern experiments. As the current polarization follows the polarization of the paramagnetic Mn impurities and is inversely proportional to the temperature (T), one obtains the variations of the Hall constant as $1/T$ seen in Figure 1.2b for different types of nonmagnetic impurities. The amplitude of the SHE can be characterized by the Hall angle, $\Phi_H = \rho_{xy}/\rho_{xx}$. From the results of Figure 1.2b, Fert et al. [12] have derived Hall angles varying from -2.4×10^{-2} for CuLu to $+5.2 \times 10^{-2}$ for CuIr, with the typical change of sign between the beginning and the end of the 5d series predicted by a model of resonant scattering on impurity 5d states split by spin–orbit interaction (−1.2% and +2.6% if we divide by 2 to take into account the different definition of the Hall angle today).

The mechanism of resonant skew scattering by impurity levels explains that the SHE Hall angle in Ref. [12] is larger by one or two orders of magnitude than that found in modern experiments for most pure metals [9–11] or semiconductors [13]. Actually a larger angle was recently reported for Au spin-injected from FePt [11], but more recent experiments in the same group seem to show this large SHE comes from Pt impurities and probably from the same mechanism as in the alloys of Figure 1.2b.

1.3 Concept of GMR in Experiments on Ternary Magnetic Alloys

Twenty years before the discovery of GMR, some experiments with ferromagnetic metals doped with two types of impurities [4] were already anticipating the GMR. This is illustrated in Figure 1.3. Suppose, for example, that nickel is doped with impurities A (Co, for example), which scatter strongly the electrons of the spin-down channel and with impurities B (Rh, for example), which scatter strongly the spin-up electrons. In the ternary alloy Ni(Co + Rh), that I call of type 1, the electrons of both channels are strongly scattered either by Co in one of the channels or by Rh in the other, so that there is no shorting by one of the channels and the resistivity is strongly enhanced, as illustrated in Figure 1.3a. In contrast, there is no such enhancement in alloys of type 2 doped with impurities A and B (Co and Au, for example) scattering strongly the electrons in the same channel and leaving the second channel open, as in Figure 1.3b.

The idea of GMR is the replacement of the impurities A and B of the ternary alloy by the successive layers A and B of the same magnetic metal in a multilayer. If the magnetizations of the layers A and B are antiparallel, this corresponds to the situation of strong scattering in both channels in alloys of type 1, while the configuration with parallel magnetizations corresponds to the situation with a relatively free channel in alloys of type 2. What is new with respect to the ternary alloys is the possibility of switching between high and low resistivity states applying a magnetic field and by simply changing the relative orientation of the magnetizations of layers A and B from antiparallel to parallel. However, the transport equations tell us that the relative

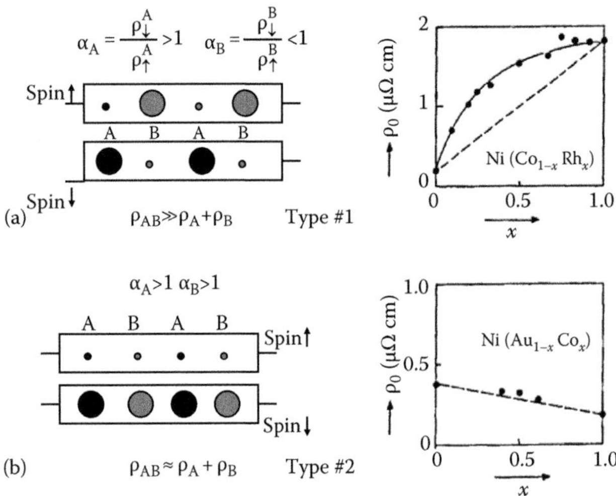

FIGURE 1.3 Experiments on ternary alloys based on the same concept as that of the GMR [4]. On the sketches illustrating the conduction by two channels in a ferromagnet doped with impurities A (black) and B (grey), the circles are at the scale of the scattering cross-sections of impurities A and B. (a) Schematic for the spin-dependent conduction in alloys with impurities of opposite scattering spin asymmetries ($\alpha_A = \rho_{A\downarrow}/\rho_{A\uparrow} > 1$, $\alpha_B = \rho_{B\downarrow}/\rho_{B\uparrow} < 1$, $\rho_{AB} \gg \rho_A + \rho_B$) and experimental results for Ni($Co_{1-x}Rh_x$) alloys. (b) Same for alloys with impurities of similar scattering spin asymmetries ($\alpha_A = \rho_{A\downarrow}/\rho_{A\uparrow} > 1$, $\alpha_B = \rho_{B\downarrow}/\rho_{B\uparrow} > 1$, $\rho_{AB} \approx \rho_A + \rho_B$) and experimental results for Ni($Au_{1-x}Co_x$) alloys. In GMR, the impurities A and B are replaced by multilayers, the situation of (a) corresponding to the antiparallel magnetic configurations of adjacent magnetic layers and (b) corresponding to parallel.

orientation of layers A and B can be felt by the electrons only if their distance is smaller than the electron mean free path, that is, practically, if they are spaced by only a few nanometers. Unfortunately, in the 1970s, it was not technically possible to make multilayers with layers as thin as a few nanometers, and the discovery of the GMR waited until the development of sophisticated deposition techniques.

1.4 Discovery of the GMR

In the mid-1980s, with the development of techniques like the molecular beam epitaxy (MBE), it became possible to fabricate multilayers composed of very thin individual layers and one could consider trying to extend the experiments on ternary alloys to multilayers. In addition, in 1986, I saw the beautiful Brillouin scattering experiments of Grünberg and coworkers [14] revealing the existence of antiferromagnetic interlayer exchange couplings in Fe/Cr multilayers. Fe/Cr appeared as a magnetic multilayered system in which it was possible to switch the relative orientation of the magnetization in adjacent magnetic layers from antiparallel to parallel by applying a magnetic field. In collaboration with the group of Alain Friederich at the Thomson-CSF company, I started the fabrication and investigation of Fe/Cr multilayers. This led us in 1988 to the discovery [1] of very large magnetoresistance effects that we called GMR (Figure 1.4a). Effects of the same type in Fe/Cr/Fe trilayers were obtained practically at the same time by Grünberg at Jülich [2] (Figure 1.4b). The interpretation of the GMR is similar to that described above for the ternary alloys and is illustrated in Figure 1.4c. The first classical model of the GMR was published in 1989 by Camley and Barnas [15], and I collaborated with Levy and Zhang for the first quantum model [16] in 1991.

I am often asked if I was expecting such large MR effects. My answer is yes and no: on one hand, a very large magnetoresistance could be expected from an extrapolation of the preceding results on ternary alloys, on the other hand one could fear that the unavoidable structural defects of the multilayers, interface roughness, for example, might introduce spin-independent scatterings canceling the spin-dependent scattering inside the magnetic layers. The good luck was finally that the scattering by the roughness of the interfaces is also spin dependent and adds its contribution to the "bulk" one (the "bulk" and interface contributions can be separately derived from GMR experiments with the current perpendicular to the layers).

1.5 Golden Age of GMR

Rapidly, the papers reporting the discovery of GMR attracted attention for their fundamental interest as well as for the many possibilities of applications, and the research on magnetic multilayers and GMR became a very hot topic. In my team, as well as in the small but rapidly increasing community working in the field, we had the exalting impression of exploring a wide virgin country with so many amazing surprises in store. On the experimental side, two important results were published in 1990. Parkin et al. [17] demonstrated the existence of GMR in multilayers made by the simpler and faster technique of sputtering (Fe/Cr, Co/Ru, and Co/Cr) and found the oscillatory behavior of the GMR due to the oscillations of the interlayer exchange as a function of the thickness of the nonmagnetic layers (see Figure 1.5). Also in 1990, Shinjo and Yamamoto [18], as well as Dupas et al. [19], demonstrated that GMR effects can be found in multilayers without antiferromagnetic interlayer coupling but composed of magnetic layers of different coercivities. Another important result, in 1991, was the observation of large and oscillatory GMR effects in Co/Cu, which became an archetypical GMR system. The first observations were obtained at Orsay [20] with multilayers prepared by sputtering at Michigan State University and at about the same time at IBM [21].

Also in 1991, Dieny et al. [22] reported the first observation of GMR in spin-valves, that is, trilayered structures in which the magnetization of one of the two magnetic layers is pinned by coupling with an antiferromagnetic layer while the magnetization of the second one is free. The magnetization of the free layer can be reversed by very small magnetic fields, so that the concept is now used in many devices. The various applications of the GMR are described in other chapters of the book. Its application to the read heads of the hard discs is certainly the most important [23,24]. The GMR, by providing a sensitive and scalable read technique, has led to an increase of the areal recording density by more than two orders of magnitude (from ≈1 to ≈600 Gbit/in.^{2}

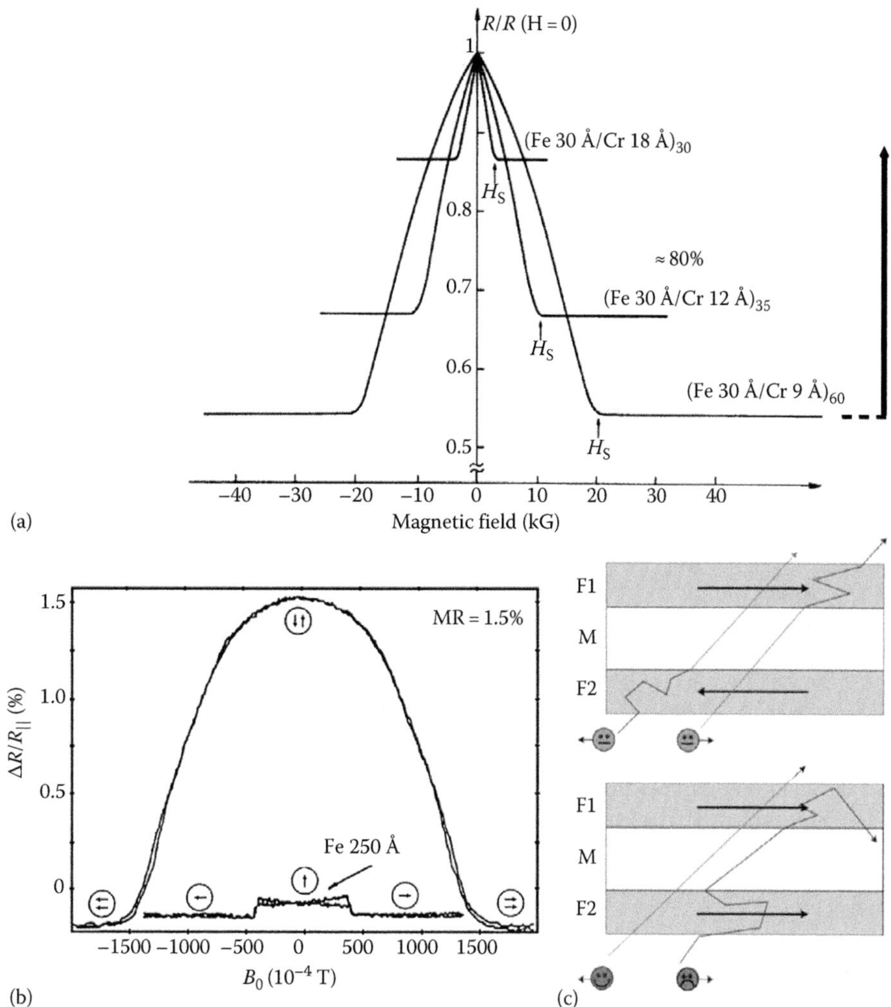

FIGURE 1.4 First observations of giant magnetoresistance. (a) Fe/Cr(001) multilayers [1] (with the current definition of the magnetoresistance ratio, MR= $100[R_{AP}–R_P]/R_P$, MR = 85% for the Fe 3 nm/Cr 0.9 nm multilayer). (b) Fe/Cr/Fe trilayers [2]. (c) Schematic of the mechanism of the GMR. In the parallel magnetic configuration (bottom), the electrons of one of the spin directions can go easily through all the magnetic layers and the short circuit through this channel leads to a small resistance. In the antiparallel configuration (top), the electrons of each channel are slowed down every second magnetic layer and the resistance is high. (From Chappert, C. et al., *Nat. Mater.*, 6, 813, 2007. With permission.)

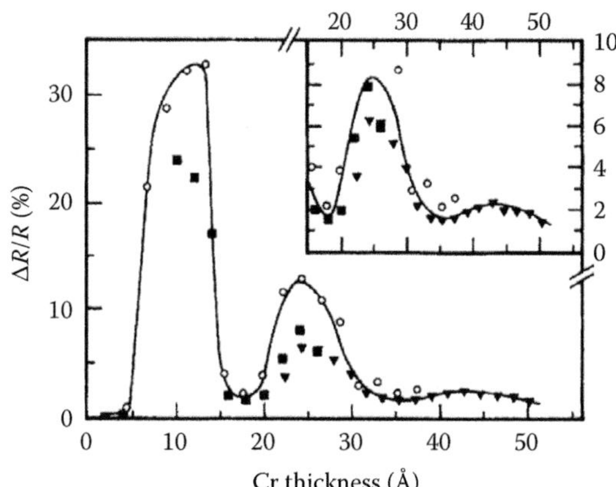

FIGURE 1.5 Oscillatory variation of the GMR ratio of Fe/Cr multilayers as a function of the thickness of the Cr layers. (From Parkin, S.S.P. et al., *Phys. Rev. Lett.*, 64, 2304, 1990. With permission.)

in 2009). This increase opened the way both to unprecedented drive capacities (up to 1 terabyte) for video recording or backup and to smaller hard disc drive (HDD) sizes (down to .85 in. disk diameter) for mobile appliances like ultra-light laptops or portable multimedia players. GMR sensors are also used in many other types of application, mainly in automotive industry and biomedical technology [25].

1.6 TMR Relays GMR

An important stage in the development of spintronics has been the research on the tunneling magnetoresistance (TMR) of the magnetic tunnel junctions (MTJs). The MTJs are tunnel junctions with ferromagnetic electrodes, and their resistance is different for the parallel and antiparallel magnetic configurations of their electrodes. Some early observations of small TMR effects had been already reported by Jullière [26] in 1975 and Maekawa

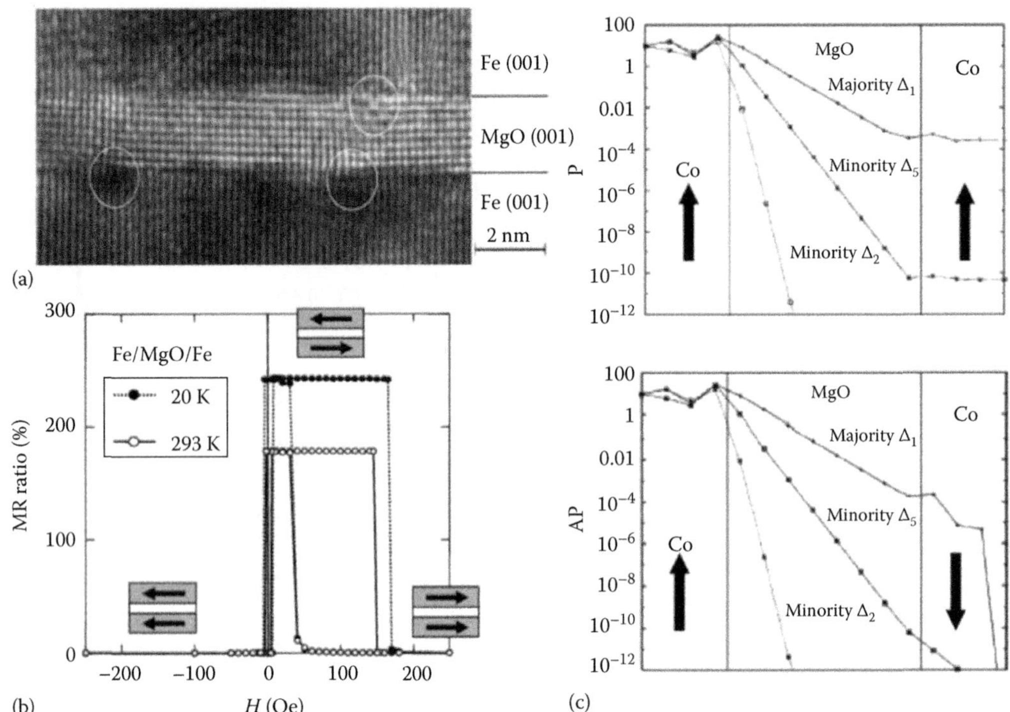

FIGURE 1.6 TMR of MTJ with MgO barrier. (a) Electron microscopy image of an Fe(001)/MgO(001)/Fe(001) MTJ [30]. (b) $MR = (R_{max} - R_{min})/R_{min}$ measured by Yuasa et al. [30] for a Fe(001)/MgO(001)/Fe(001) MTJ. (c) Physics of the TMR illustrated by the decay of evanescent electronic waves of different symmetries in an MgO(001) layer between cobalt electrodes calculated by Zhang and Butler [37]. The Δ_1 symmetry of the slowly decay tunneling channel is well represented at the Fermi level of the spin conduction band of cobalt for the majority spin direction and not for the minority spin one, so that a spin conserving connection through the channel Δ_1 is possible in the parallel configuration. This explains the very high TMR of this type of junction.

and Gäfvert [27] in 1982, but they were found to be hardly reproducible and actually could not be really reproduced until 1995. It was at this time only that large (≈20%) and reproducible effects were obtained by Moodera's and Miyasaki's groups on MTJ with a tunnel barrier of amorphous alumina [28].

After 1995, the research on TMR became very active, and the most important step was the transition from MTJ with amorphous tunnel barrier (alumina) to single-crystal MTJ and especially MTJ with MgO barrier. In the first results with MgO, the TMR ratio was only slightly larger than that found with alumina barriers and similar electrodes [29]. The important breakthrough came in 2004 at Tsukuba [30] and IBM [31] where it was found that very large TMR ratios, up to 200% at room temperature, could be obtained with MgO MTJ of very high structural quality, as illustrated in Figure 1.6. Since 2004, these results have been progressively improved [32], and TMR ratios up to 1000% have been now reached [33].

The large TMR of MTJ with single-crystal tunnel barriers like MgO(001) comes from symmetry selection [34–37]. This is illustrated in Figure 1.6c where one sees the calculated density of states (DOS) of evanescent wave functions of different symmetries, Δ_1, Δ_5, etc., and the much slower decay of the symmetry Δ_1 in a MgO(001) barrier between Co electrodes [37]. The key point is that, at least for interfaces of high quality, an evanescent wave function of a given symmetry is connected to the Bloch functions of the same symmetry and same spin direction at the Fermi level of the electrodes. For Co electrodes, the Δ_1 symmetry is well represented at the Fermi level in the majority spin direction sub-band and not in the minority one. Consequently, a good connection between the majority spin direction sub-bands of the Co electrodes by the slowly decaying channel Δ_1 can be obtained only in their parallel magnetic configuration, which explains the very high TMR. Other types of barriers can select symmetries other than the symmetry Δ_1 selected by MgO(001). For example, an $SrTiO_3$ barrier predominantly selects evanescent wave functions of Δ_5 symmetry, which are well connected to minority spin states of cobalt [38]. This explains the negative effective spin polarization of cobalt observed in $SrTiO_3$-based MTJ [39].

The high spin polarization obtained by symmetry selection gives a very good illustration of what is under the word "spin polarization" in a spintronic experiment. There is no intrinsic spin polarization of a magnetic conductor. In an MTJ, the effective polarization is related to the symmetry selected by the barrier and, depending on the barrier, can be positive or negative, large or small. In the same way, as we have seen in Section 1.2, the spin polarization of metallic conduction depends strongly on the spin dependence of the scattering by impurities, as shown in Figure 1.1b.

The MTJ with MgO barrier are today the most efficient "spin polarizers" and are used in many experiments and devices.

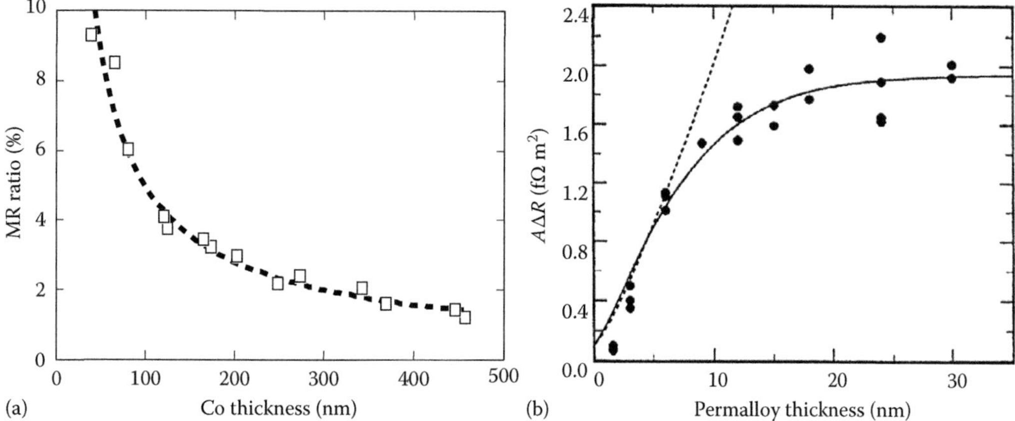

FIGURE 1.7 Experimental results illustrating the thickness dependence of the CPP-GMR and its relation with the SDL. (a) CPP-GMR of Co/Cu/Co multilayers electrodeposited into pores of membranes [44]. With the long SDL of Co (60 nm), the GMR can be seen up to Co thicknesses in the μm range. (b) CPP-GMR of Py/Cu/Py multilayers [43]. With an SDL of only 5 nm, the resistance change between P and AP configuration flattens out, for permalloy layers thicker than 10 nm and, as the resistance still increase, the GMR ratio decreases to zero.

However, other directions are also explored to obtain even larger TMR. One of these directions is the research of half-metallic materials, in other words materials of metallic character for one of the spin directions and insulating for the other, which means a 100% polarization. Some oxides and Heusler alloys close to half-metallicity have been found [40,41]. Spin filtering by a ferromagnetic barrier is another interesting concept [42].

From a technological point of view, the interest of the MTJ with respect to the metallic spin valves comes from the vertical direction of the current and from the resulting possibility of a reduction of the lateral size to a submicronic scale by lithographic techniques. The MTJs are at the basis of a new concept of magnetic memory called magnetic random access memory (MRAM), which is expected to combine the short access time of the semiconductor-based RAMs and the nonvolatile character of the magnetic memories [24]. In the first MRAMs, put onto the market in 2006, the memory cells are MTJs with an alumina barrier. The magnetic fields generated by "word" and "bit" lines are used to switch their magnetic configuration. The next generation of MRAM, based on MgO tunnel junctions and switching by spin transfer, is expected to have a much stronger impact on the technology of the computers.

1.7 Spin Accumulation and Spin Currents

During the first years of the research on GMR, the experiments were performed only with currents flowing along the layer planes—in the geometry called CIP (current in plane). It is only in 1991 that experiments of GMR with the current perpendicular to the layer planes (CPP-GMR) begun to be performed. This was done first by sandwiching a magnetic multilayer between superconducting electrodes [43], then by electrodepositing multilayers into the pores of a polycarbonate membrane [44], and, more recently, in vertical nanostructures fabricated by e-beam lithographic techniques (pillars). In the CPP geometry, the GMR not only is definitely higher than that in CIP but also subsists in multilayers with relatively thick layers, up to the micron range in Figure 1.7a, for example. The Valet–Fert model [45] explains that, owing to spin accumulation effects occurring in the diffusive regime of transport, the length scale of the CPP-GMR becomes the long spin diffusion length (SDL) in place of the short mean free path for the CIP geometry.

Actually, the CPP-GMR revealed clearly the spin accumulation effects, which govern the propagation of a spin-polarized current through a succession of magnetic and nonmagnetic materials and play an important role in all the current developments of spintronics. The diffusion current induced by the accumulation of spins at the magnetic/nonmagnetic interface is the mechanism driving a spin-polarized current in a nonmagnetic conductor to a long distance from the interface, well beyond the ballistic range (i.e., well beyond the mean free path) up to the distance of the SDL. In carbon molecules, for example, the SDL exceeds largely the micron range and, as we see in Section 1.11, strongly spin-polarized currents can be transported throughout long carbon nanotubes (CNTs) [46].

The physics of the spin accumulation occurring when an electron flux crosses an interface between a ferromagnetic and a nonmagnetic material is explained in Figure 1.8 for a simple situation (single interface, no interface resistance, no band bending, and single polarity). To sum up, there is a broad zone of spin accumulation, which extends on both sides of an F/N interface and in which the spin polarization of the injected current subsists in N to a distance of the order of the SDL (or start at this distance for the opposite current direction). The spin polarization just at the interface depends on the proportion of the depolarizing (polarizing) spin-flips induced by the spin accumulation in F and N. If the DOS and the spin relaxation times are similar on both sides, there is a balanced proportion of spin-flips on the N and F sides, and the current is significantly spin-polarized in

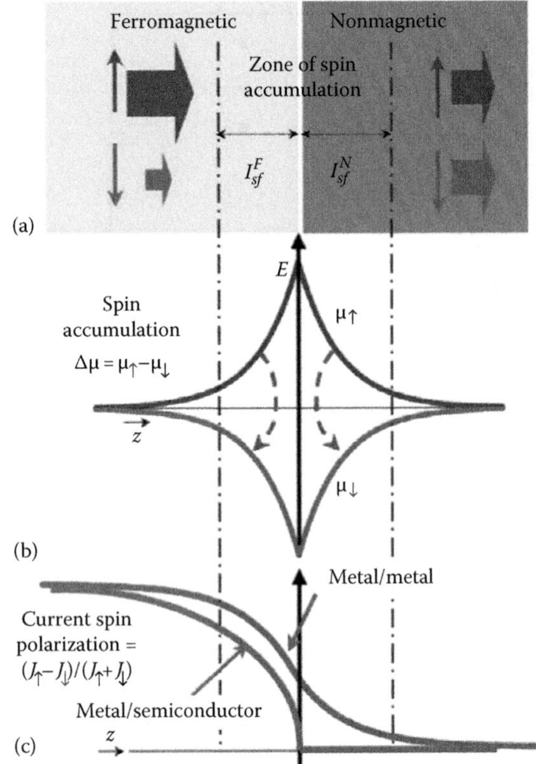

FIGURE 1.8 (See color insert.) Schematic representation of the spin accumulation at an interface between a ferromagnetic metal and a nonmagnetic layer. (a) Incoming and outgoing spin-up and spin-down currents. (b) Splitting of the chemical potentials, $E_{F\uparrow}$ and $E_{F\downarrow}$, in the interface region with arrows symbolizing the spin-flips induced by the spin accumulation (out of equilibrium distribution), spin-flips controlling the progressive depolarization of the current. With an opposite direction of the current, the inversion of the spin accumulation and the opposite spin-flips polarize progressively the current. (c) Variation of the current spin polarization when there is an approximate balance between the spin-flips on both sides (metal/metal curve) and when the spin-flips on the right side are predominant (metal/semiconductor curve in the situation without spin-dependent interface resistance). (From Chappert, C. et al., *Nat. Mater.*, 6, 813, 2007. With permission.)

N (as for the metal/metal curve in Figure 1.8c). In contrast, if the DOS is much smaller or the spin relaxation time much shorter in N, the depolarization (or polarization for spin extraction) occurs mainly in F and the current is negligibly spin-polarized in N. This corresponds to the situation for the metal/semiconductor curve in Figure 1.8c. This difficulty when F is a metal and N a semiconductor has been first described by Schmidt et al. [47] and called "conductivity mismatch." The problem can be solved by inserting a spin-dependent interface resistance (tunnel junction) to induce a discontinuity of the spin accumulation, which increases the Fermi energy spitting and the proportion of spin-flips on the side N of the interface [48,49].

The physics of spin accumulation and spin-polarized currents, which plays an important role in most fields of spintronics today, can be described by new types of transport equation [45,49,50], often called drift/diffusion equations, in which the electrical potential is replaced by a spin- and position-dependent electrochemical potential. These equations can also be extended to take into account band bending and high current density effects [51]. They have been frequently applied to the general situation of multi-contact systems with interplay of the spin accumulation effects at different contacts [49,52,53]. A standard structure is a nonmagnetic lateral channel, metal, semiconductor, or carbon-based conductor, between a spin-polarized source F1 and a spin-polarized drain F2. The output "spin signal" is the resistance or voltage difference between different magnetic configurations of the contacts. Very generally, one distinguishes local geometries in which the output signal is measured between the source and the drain (Figure 1.9a), and nonlocal geometries in which the output signal is between two other contacts (Figure 1.9b). In the nonlocal geometry of Figure 1.9b, where the current flows to the left with a spin accumulation spreading to the right (Figure 1.9c), the output voltage V reflects the spin accumulation splitting at the magnetic contact F2 and is inverted when one reverses the magnetization of F2. Note that, in the right part of the channel, the opposite gradients of spin-up and spin-down chemical potentials generate opposite spin-up and spin-down currents, or, in other words, a pure spin current without charge current.

In Figure 1.9e, we show an example of the application of the drift-diffusion equations to the problem of spin transport in a nonmagnetic lateral channel between spin-polarized source and drains connected to the channel by a spin-dependent interface resistance (MTJ, for example). The calculation is performed for the local and nonlocal detections of Figure 1.9a and b and also for the local detection of Figure 1.9d in which the channel does not extend outside the length L between the contacts (confined spin relaxation). The output signal, $\Delta R = (V_{AP} - V_P)/I$, is plotted as a function of the ratio of the mean interface resistance R_T^* (defined in the caption and supposed to be the same for the contacts F1 and F2) to the spin resistance of the channel (R_N = product of the resistivity by the SDL λ_N divided by the section; the spin relaxation per unit length of N is inversely proportional to R_N) for the three geometries of Figure 1.9a–d (other details on the parameters are in the caption of the figure). The calculation is performed in a situation of "conductivity mismatch," that is, $R_F \ll R_N$, so that, for the three curves, the spin signal turns out when R_T^* exceeds R_N. Then, the difference between the three curves illustrates an important rule: the scale of ΔR is the "resistance" controlling the spin relaxation. For a and b with $R_T^* \gg R_N$, the relaxation in N between − and $+\lambda_N$ is larger ($\propto 1/R_N$) than the spin escape through R_T^* ($\propto 1 R_T^*$) and ΔR flattens out at a level of the order of $R_N \ll R_T^*$ ($\approx 2\gamma^2 R_N \exp(-L/\lambda_N)$ for local detection and $\approx \gamma^2 R_N \exp(-L/\lambda_N)$ for nonlocal). In contrast for d with $R_N \ll R_T^* \ll R_N\lambda_N/L$, the relaxation is controlled by the spin escape through R_T^* (rather than by the relaxation inside L) and ΔR goes on increasing in proportion of R_T^* well above the saturation value $\approx R_N$ of the open structures a and b. This regime, also described by a dwell time τ_n shorter than the spin lifetime τ_{sf}, is illustrated by the large MR observed in CNTs (60%–70%, ΔR in the MΩ range well above the spin resistance of the nanotube)

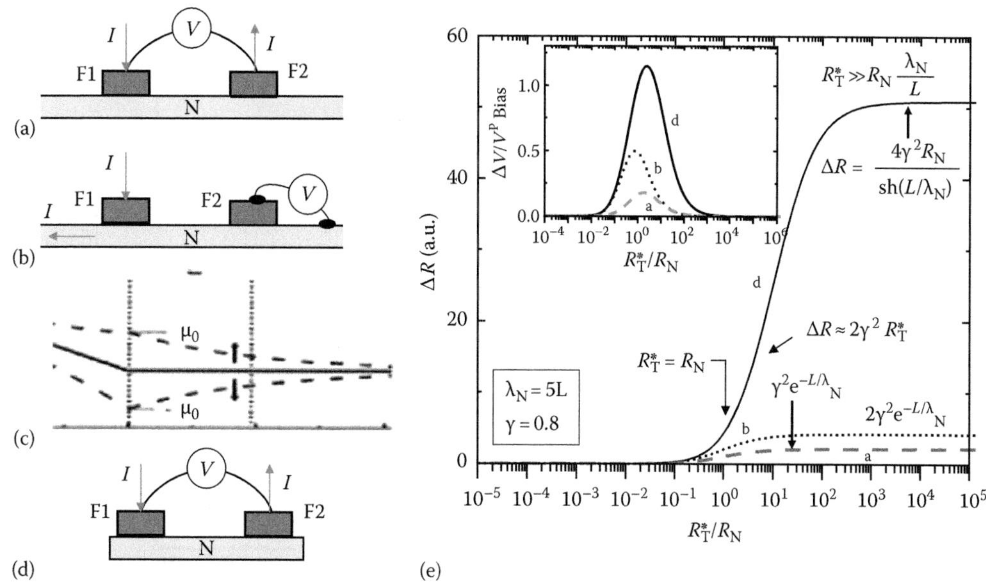

FIGURE 1.9 (a) and (b) Geometries for lateral spin transport between spin-polarized sources and drains with, respectively, local and nonlocal detection of the "spin signal." (c) Spin-up and spin-down electrochemical potential profiles for the nonlocal geometry of (b). The nonlocal signal V is related to the splitting (spin accumulation) under the nonlocal contact. (d) Geometry similar to (a) but with a lateral channel not extending outside the contacts. (e) Output spin signal, $\Delta R = [V_{AP} - V_P]/I$, for geometries of (a), (b), (d) calculated from drift/diffusion equations as a function of R_T^*/R_N with the following definitions: L = length between F1 and F2, $R_{T\uparrow(\downarrow)} = 2\,[1 - (+)\gamma]\,R_T^*$ = tunnel resistances between the channel and F1 or F2, $R_N = \rho_N\lambda_N/S_N$ = channel spin resistance (ρ_N = resistivity, λ_N = SDL, S_N = section), $R_F = \rho_F^*\lambda_F/S_F \ll R_N$ ($\gamma = 0.8$, $\lambda_N = 5L$ in the calculation). ΔR flattens out at a value of the order of R_N for $R_T^* \gg R_N$ in the open structures of (a) and (b) (dotted and dashed lines). For the geometry of (d), ΔR does not flatten out for $R_T^* > R_N$ and goes on increasing proportionally to R_T^* before reaching saturation at the level $4\gamma^2 R_N/\mathrm{sh}(L/\lambda_N)$, well above R_N, for $R_T^* \gg R_N\lambda_N/L$ (solid line).

[46]. Then, for $R_T^* \gg R_N\lambda_N/L$, the relaxation inside L becomes predominant and ΔR flattens out at the level $4\gamma^2 R_N/\mathrm{sh}(\lambda_N/L) \approx 4\gamma^2 R_N\lambda_N/L$ (well above the saturation for curves a and b in Figure 1.9e). For confined relaxation with nonlocal detection $4\gamma^2 R_N/\mathrm{sh}(L/\lambda_N)$ becomes $3\gamma^2 R_N/\mathrm{sh}(L/\lambda_N)$, as shown by Jaffrès et al. [49]. The different rules described above can explain the main features of the results found in various structures with metals, semiconductor, or carbon-based materials.

In Figure 1.10, we show examples of lateral spin transport in the simple situation of a metallic channel and ohmic contacts with metallic electrodes [50,52,54,55]. The output signal ΔR, at the typical scale of the channel spin resistance, is typically of the order of a few mΩ, which corresponds to local or nonlocal output voltages in the μV range for current densities in the 10^7 A/cm^2 range (the ratio to the bias voltage can reach 10%). With a metallic channel but with tunnel contacts, most experiments have been performed in the geometry of Figure 1.9a and b, so that the output signals are larger but do not exceed significantly the spin resistance of a metallic channel ($\Delta R \approx 10^{-3} - 10^{-4}\,R_{bias}$). For the experiments with a semiconducting channel, generally performed with tunnel or Schottky contacts to solve the "conductivity mismatch," the review article of Jonker and Flatté [56] concludes that the contrast between the conductance of the parallel and antiparallel configuration never exceeds the range 0.1%–1%. With CNTs and tunnel contacts in the range of 100 MΩ with $La_{2/3}Sr_{1/3}MnO_3$ (LSMO) electrodes [46], some experiments in the geometry of Figure 1.9d have led to output signals $\Delta R \approx 90$ MΩ (60% of the total resistance of the two contacts), which appears to correspond to the situation of the corresponding curve in Figure 1.9e with $\Delta R \approx R_T \gg R_{CNT}$. This shows the possibility of obtaining large spin signals at the level of tunnel resistances by preventing the spin relaxation in branches of small spin resistance. This also reflects the potential of carbon-based materials like CNT [46,57] or graphene [58], which can combine a long spin lifetime, due to the small spin–orbit of carbon, with the advantage of their very high electron velocity for short dwell times.

1.8 Spin Transfer

Spin transfer gives means to manipulate the magnetic moment of a ferromagnetic body without applying any magnetic field but only by transfer of spin angular momentum from a spin-polarized current. The concept that has been introduced by Slonczewski [59] and appears also in papers of Berger [60] is illustrated in Figure 1.11. As described in the caption of the figure, the transfer of the transverse component of a spin current to the "free" magnetic layer F2 can be described by a torque acting on the magnetic moment of F2. This torque can induce a switching or reorientation of this magnetic moment, which can be described as a transport of nonvolatile magnetization by an electrical wire. In a second regime, generally in the presence of an applied field, the spin transfer torque generates periodic motions of the magnetic moment and voltage oscillations in the microwave frequency

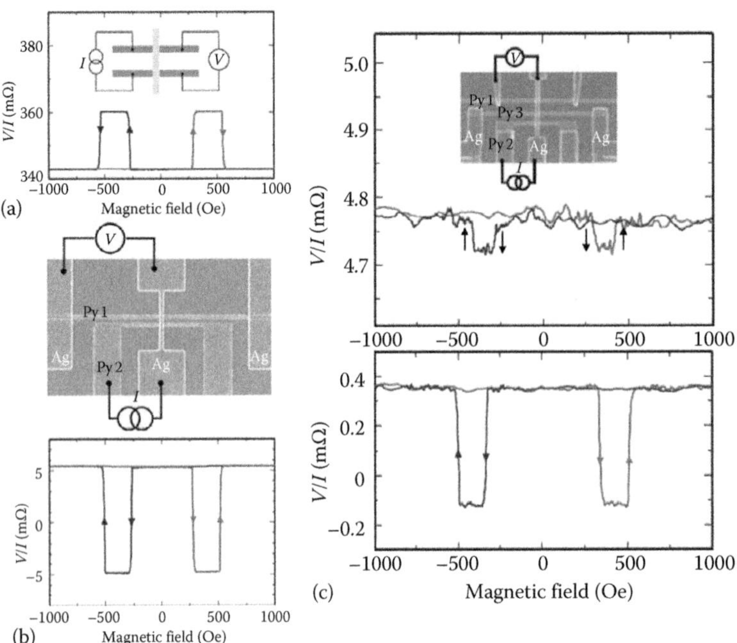

FIGURE 1.10 Experimental examples of lateral spin transport in permalloy/silver nanodevices [55]. (a) Local spin-valve measurements at 77 K. A sketch of the nanodevice is shown above the experimental curves. (b) Nonlocal spin-valve measurements at 77 K. An image of the nanodevice is shown on the top. (c) Nonlocal spin-valve measurements at 77 K (bottom) and room temperature (middle) on a nanodevices similar to that of (b) but with an additional permalloy contact (Py3 on the image) between the injection and detection contacts. The signal reduction with respect to devices without additional contact is due to the spin current escaping through this contact and relaxing in the additional channel. The reduction is inversely proportional to the spin resistance of the additional channel and, for example, becomes larger and larger as one goes from Cu to Au and Py. (From Kimura, T. et al., *Phys. Rev. B*, 72, 014461-1–6, 2005. With permission.)

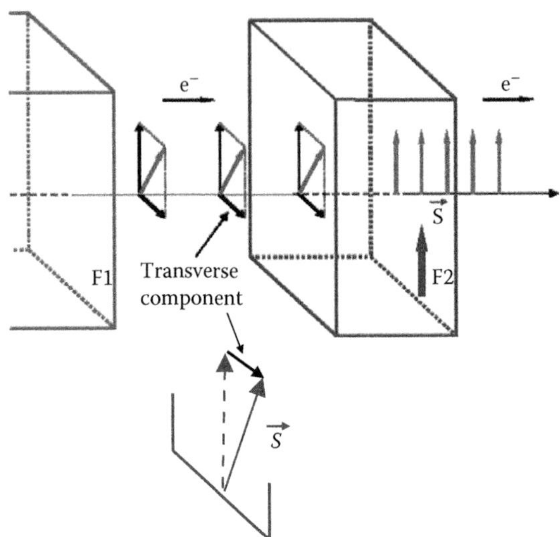

FIGURE 1.11 Illustration of the spin transfer concept introduced by Slonczewski [59]. A spin-polarized current is prepared by a first magnetic layer F with an obliquely oriented spin polarization with respect to the magnetization axis of a second layer F2. When it goes through F2, the exchange interaction aligns its spin polarization along the magnetization axis. The exchange interaction being spin conserving, the transverse spin polarization lost by the current is transferred to the total spin of F2, which can also be described by a spin-transfer torque acting on F2. (From Chappert, C. et al., *Nat. Mater.*, 6, 813, 2007. With permission.)

range. Actually the concept of spin transfer has opened a new face of spintronics: GMR or TMR allowed us to detect electrically a magnetic configuration, while spin transfer now enables us to create a magnetic configuration with an electrical current.

The first evidence that spin transfer can work was indicated by experiments of spin injection through point contacts by Tsoi et al. [61], but a clear understanding came later from measurements [63,64] performed on pillar-shaped metallic trilayers and tunnel junctions (Figure 1.12a), or by injecting a current into a magnetic trilayer through the type of nanocontact represented in Figure 1.12b [64]. In Figure 1.12c, we show an example of magnetic switching detected by GMR in a metallic pillar. Figure 1.12d represents the microwave power spectrum of the voltage oscillations generated by current-induced magnetic precessions in a structure of nanocontact type. Actually the spin transfer torque can be used not only to induce switchings and precessions of magnetic moments but also more sophisticated magnetic excitations. Figure 1.12e shows the microwave power spectrum generated by the gyration of the core of a magnetic vortex in one of the electrodes of an MTJ [65]. Alternatively, the spin transfer torque can also be used to move domain walls in nanowires. These many very different spin transfer experiments raise a great variety of interesting physical problems mixing spin transport, spin dynamics, and micromagnetism.

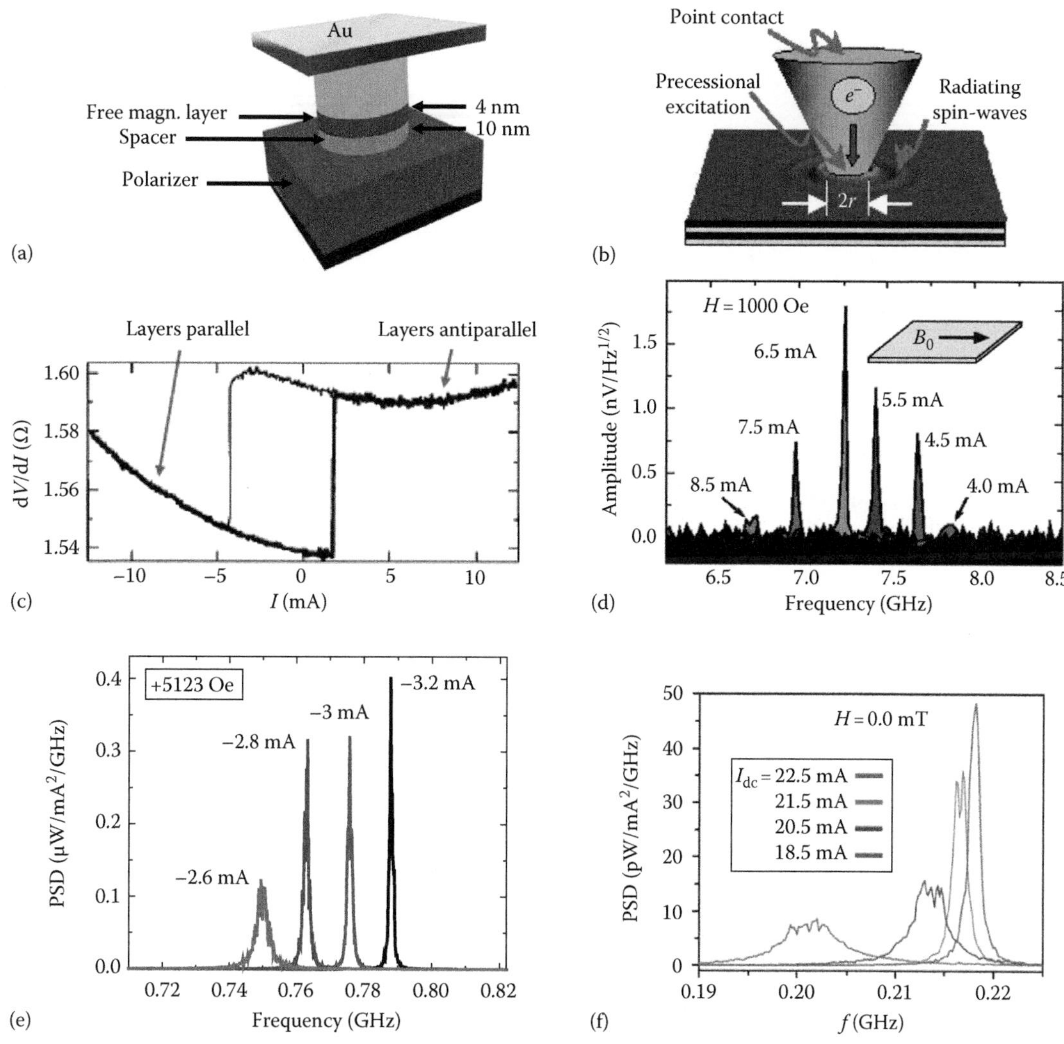

FIGURE 1.12 (See color insert.) (a) and (b) Nanodevices for spin transfer experiments ("pillar" and nanocontacts [64]). (c) Differential resistance vs. current for a Co/Cu/Co "pillar" with current-induced switchings between parallel and antiparallel magnetic configurations [62]. (d) Microwave spectra obtained by current injection into a CoFe/Cu/NiFe trilayer through a nanocontact at different values of current [64]. (e) Microwave power spectra obtained at different values of current from vortex gyrations in the permalloy layer of a CoFeB/MgO/Permalloy MTJ [65]. (f) Progressive synchronization of the gyration of four vortices appearing at increasing current in the microwave spectrum of a device for current injection into a Co/Cu/permalloy trilayer through four nanocontacts. (From Ruotolo, A. et al., *Nat. Nanotechnol.*, 4, 528, 2009. With permission.)

The spin transfer phenomena will have certainly important applications. Switching by spin transfer will be used in the next generation of MRAMs (called STT-RAMs) for a more precise addressing and a smaller energy consumption. The generation of oscillations in the microwave frequency range will lead to the design of spin transfer oscillators (STOs). One of the main interests of the STOs is their tunability, that is the possibility of changing rapidly their frequency by tuning a DC current. Their disadvantage is the very small microwave power of an individual STO. The microwave power obtained with MTJs is not too far from what is needed for applied devices but the linewidth of their emission is still much too broad. Today it seems that the excitation of magnetic vortices in MTJs heads the race on the way to application as they combine power and narrow linewidth see Figure 1.12e [65]. An interesting challenge, to increase the emitted power and reduce the linewidth, is the synchronization of a set of STOs. Figure 1.12f shows an example of synchronization of the gyration of a square of four vortices [66].

1.9 Spintronics Today and Tomorrow

Nowadays spintronics is expanding in so many directions that we will only list and summarize the main axes.

The very active research [67] on the axis "spintronics with semiconductors" aims at combining the potential of semiconductors (control of current by gate, coupling with optics, etc.) with the potential of the magnetic materials (control of current by spin manipulation, non-volatility, etc.). It should be possible,

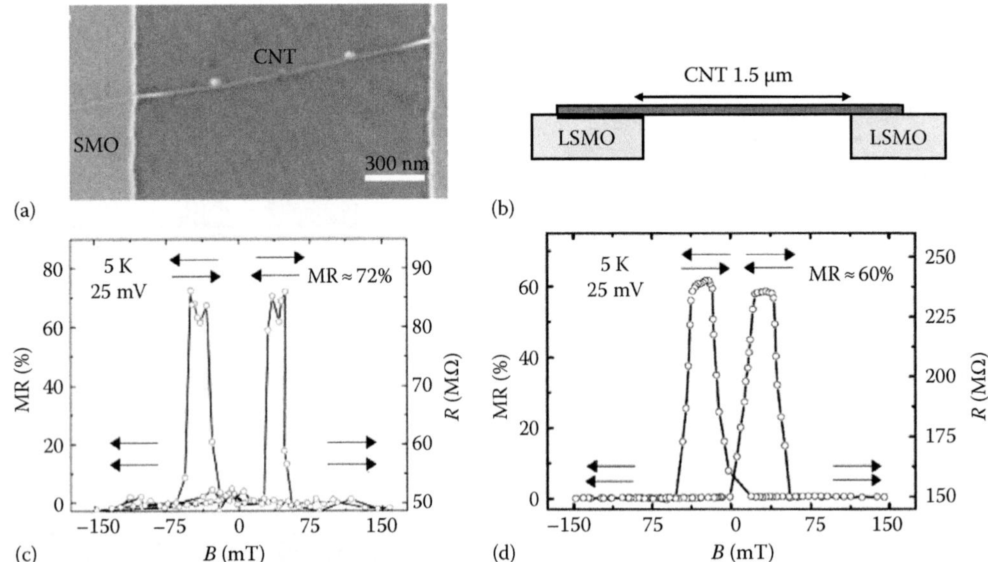

FIGURE 1.13 Spin transport through carbon nanotubes (CNTs) [46]. (a) Electron microscope image of a CNT between LSMO electrodes. (b) Schematic side view of the device. (c), (d) Examples of MR experimental results. The resistance R of the device is predominantly due to the tunnel resistance at the LSMO/CNT interfaces, and its change between the parallel and antiparallel magnetic configuration is about 72% (c) or 60% (d).

for example, to gather storage, detection, logic, and communication capabilities on a single chip that could replace several components. One of the ways is with ferromagnetic semiconductors (FSs) based on the well-known semiconductors of today. Excellent results have been obtained with GaAs doped wit Mn [68], but it has not been possible to increase its Curie temperature above 170 K and this rules out most possibilities of applied devices. Some announcements of FS at room temperature have been made, but the situation is still rather confused on this point today. The second way is the fabrication of hybrid structures associating ferromagnetic metals and conventional semiconductors. In a spin LED [69–72], for example, the injection of spin-polarized current into an LED leads to the emission of a circularly polarized light. In other experiments, it has been now possible to detect electrically the spin accumulation injected into a semiconductor [72–75]. An intense research has been devoted to the implementation of the concept of spin transistor proposed by Datta and Das [76], but only very modest advances [77] have been made on this topic up to now. Finally, a hot topic of today is the SHE and the QSHE [78,79].

Spintronics with multiferroics is also a promising axis. Multiferroic materials combine ferromagnetic and ferroelectric properties, and the possible coupling between the ferromagnetic and ferroelectric orders could give means to control a magnetic configuration with an electric field [80–82]. The recent results [82] with $BiFeO_3$ are promising, and the research on several other types of multiferroic oxides is now very active. Note that an alternative way for an electric field control of the magnetization direction turns out to be through direct effects on the band structure by spin–orbit interactions in FSs [83] or even in ultrathin layers of ferromagnetic metals [84].

Other interesting new properties of oxides have been revealed recently. For example, the existence of 2DEGs of high mobility at the interface between two insulating oxides [85–87] opens the way to spintronic devices exploiting these 2DEGs in "all oxide" nanostructures. Also the extension of the technologies developed for MTJ to tunnel junctions with ferroelectric barriers has led to the observation of huge electroresistance effects [88] and to interesting prospects for the design of ferroelectric memories.

Another important field of research today is spintronics with carbon-based materials. Their general advantage comes from the long spin lifetime related to the small spin–orbit coupling of carbon. For metallic CNT and graphene, another advantage comes from their linear dispersion curves and very large velocities even for small carrier concentrations. In lateral spin-valve structures of CNT between LSMO electrodes (Figure 1.13), the MR can reach 60%–70% ($V_{AP} - V_P \approx 60$ mV for $V_P \approx 100$ mV for the sample of Figure 1.13d), well above the output electrical signals that can be obtained with semiconductor lateral channels. These results can be explained not only by the long spin lifetime in CNT but also by the short dwell time resulting from the high velocity of the electrons inside CNTs [46]. Promising experimental results with graphene and various molecules have been also recently published [58,89,90].

Finally, with qbits based on the spin state of a quantum box, nanospintronics is one of the ways to quantum computing. This is certainly one of the most exiting perspectives, out of the scope of this chapter, and, on this point, we refer to a recent review by Trauzettel et al. [91].

1.10 Conclusion

In about 20 years, we have seen spintronics increasing considerably the capacity of our hard discs, extending the hard disc technology to mobile appliances like cameras or portable multimedia players, entering the automotive industry and the biomedical technology and, with TMR and spin transfer, getting ready to enter the RAM of our computers or the microwave emitters of our cell phones. The researches of today on the spin transfer phenomena, on multiferroic materials, on spintronics with semiconductors and molecular spintronics, open fascinating new fields and are also very promising in the "beyond CMOS" perspective. Spintronics is still a young baby in the science family, but this baby is growing remarkably well.

References

1. M. N. Baibich, J. M. Broto, A. Fert et al., Giant magnetoresistance of (001) Fe/(001) Cr magnetic superlattices, *Phys. Rev. Lett.* **61**, 2472–2475 (1988).
2. G. Binash, P. Grünberg, F. Saurenbach et al., Enhanced magnetoresistance in layered magnetic structures with antiferromagnetic interlayer exchange, *Phys. Rev. B* **39**, 4828–4830 (1989).
3. N. F. Mott, The electrical conductivity of transition metals, *Proc. Roy. Soc. A* **153**, 699–718 (1936).
4. A. Fert and I. A. Campbell, Two current conduction in nickel, *Phys. Rev. Lett.* **21**, 1190–1192 (1968); A. Fert and I. A. Campbell, Transport properties of ferromagnetic transition metals, *J. Phys.* **32**, C1-46-50 (1971); A. Fert and I. A. Campbell, Electrical resistivity of ferromagnetic nickel and iron based alloys, *J. Phys. F Met. Phys.* **6**, 849–871 (1976).
5. B. Loegel and F. Gautier, Origine de la résistivité dans le cobalt et ses alliages dilués, *J. Phys. Chem. Sol.* **32**, 2723–2735 (1971).
6. J. Barnas, O. Baksalary, and Y. Bruynseraede, Effect of interchannel transitions in the current-in-plane giant magnetoresistance, *J. Phys.: Condens. Matter* **7**, 6437–6448 (1995).
7. A. Fert, A. Barthélémy, and F. Petroff, Spin transport in magnetic multilayers and tunnel junctions, in *Nanomagnetism: Ultrathin Films and Nanostructures*, F. Mills and J. A. C. Bland (Eds.), Elsevier, Amsterdam, the Netherlands, pp. 153–226, 2006.
8. M. I. D'yakonov and V. I. Perel', Possibility of orienting electron spins with current, *JETP Lett.* **13**, 467–469 (1971).
9. L. Vila, T. Kjimura, and Y. Otani, Evolution of the spin Hall effect in Pt nanowires: Size and temperature effects, *Phys. Rev. Lett.* **99**, 226604-1–4 (2007).
10. S. O. Valenzuela and M. Tinkham, Direct measurement of the spin Hall effect, *Nature* **442**, 176–179 (2006).
11. T. Seki, Y. Hasegawa, S. Mitani et al., Giant spin Hall effect in perpendicularly spin-polarized FePt/Au devices, *Nat. Mater.* **7**, 125–128 (2008).
12. A. Fert, A. Friederich, and A. Hamzic, Hall effect in dilute magnetic alloys, *J. Magn. Magn. Mater.* **24**, 231–257 (1981).
13. Y. K. Kato, R. C. Myers, A. C. Gossard et al., Observation of the spin Hall effect in Semiconductors, *Science* **306**, 1910–1913 (2004).
14. P. Grünberg, R. Schreiber, Y. Young et al., Layered magnetic structures: Evidence for antiferromagnetic coupling of Fe layers across Cr interlayers, *Phys. Rev. Lett.* **57**, 2442–2445 (1986).
15. R. E. Camley and J. Barnas, Theory of giant magnetoresistance effects in magnetic layered structures with antiferromagnetic coupling, *Phys. Rev. Lett.* **63**, 664–667 (1989).
16. P. M. Levy, S. Zhang, and A. Fert, Electrical conductivity of magnetic multilayered structures, *Phys. Rev. Lett.* **65**, 1643–1646 (1990).
17. S. S. P. Parkin, N. More, and K. P. Roche, Oscillations in exchange coupling and magnetoresistance in metallic superlattice structure: Co/Ru, Co/Cr, and Fe/Cr, *Phys. Rev. Lett.* **64**, 2304–2307 (1990).
18. T. Shinjo and H. Yamamoto, Large magnetoresistance of field-induced giant ferromagnetic multilayers, *J. Phys. Soc. Jpn.* **59**, 3061–3064 (1990).
19. C. Dupas, P. Beauvillain, C. Chappert et al., Very large magnetoresistance effects induced by antiparallel magnetization in two ultrathin cobalt films, *J. Appl. Phys.* **67**, 5680–5682 (1990).
20. D. H. Mosca, F. Petroff, A. Fert et al., Oscillatory interlayer coupling and giant magnetoresistance in Co/Cu multilayers, *J. Magn. Magn. Mater.* **94**, L1–L5 (1991).
21. S. S. P. Parkin, R. Bhadra, and K. P. Roche, Oscillatory magnetic exchange coupling through thin copper layers, *Phys. Rev. Lett.* **66**, 2152–2155 (1991).
22. B. Dieny, V. S. Speriosu, S. S. P. Parkin et al., Giant magnetoresistance in soft ferromagnetic multilayers, *Phys. Rev. B* **43**, 1297–1300 (1991).
23. S. S. P. Parkin, Applications of magnetic nanostructures, in *Spin Dependent Transport in Magnetic Nanostructures*, S. Maekawa and T. Shinjo (Eds.), Taylor & Francis, London, pp. 237–279, 2002.
24. C. Chappert, A. Fert, and F. Nguyen Van Dau, The emergence of spin electronics in data storage, *Nat. Mater.* **6**, 813–823 (2007).
25. P. P. Freitas, H. Ferreira, D. Graham et al., Magnetoresistive biochips, *Europhys. News* **34/6**, 224–226 (2003).
26. M. Jullière, Tunneling between ferromagnetic films, *Phys. Lett. A* **54**, 225–226 (1975).
27. S. Maekawa and O. Gäfvert, Electron tunneling between ferromagnetic films, *IEEE Trans. Magn. MAG* **18**, 707–708 (1982).
28. J. S. Moodera, L. R. Kinder, T. M. Wong et al., Large magnetoresistance at room temperature in ferromagnetic thin film tunnel junctions, *Phys. Rev. Lett.* 74, 3273–3276 (1995); T. Miyazaki and N. Tezuka, Giant magnetic tunneling effect in Fe/Al_2O_3/Fe junction, *J. Magn. Magn. Mater.* **139**, L231–L234 (1995).
29. M. Bowen, V. Cros, F. Petroff et al., Large magnetoresistance in Fe/MgO/FeCo(001) epitaxial tunnel junctions on GaAs(001), *Appl. Phys. Lett.* **79**, 1655–1657 (2001).
30. S. Yuasa, T. Nagahama, K. Fukushima et al., Giant room-temperature magnetoresistance in single-crystal Fe/MgO/Fe magnetic tunnel junctions, *Nat. Mater.* **3**, 868–871 (2004).

31. S. S. P. Parkin, C. Kaiser, A. Panchula et al., Giant tunneling magnetoresistance at room temperature with MgO(100) tunnel barriers, *Nat. Mater.* **3**, 862–867 (2004).
32. Y. M. Lee, J. Hayakawa, S. Ikeda et al., Effect of electrode composition on the tunnel magnetoresistance of pseudo-spin-valve magnetic tunnel junction with a MgO tunnel barrier, *Appl. Phys. Lett.* **90**, 212507-1-3 (2007).
33. L. Jiang, H. Naganuma, M. Oogane et al., Large tunnel magnetoresistance of 1056% at room temperature in MgO based double barrier magnetic tunnel junctions, *Appl. Phys. Express* **2**, 083002-1-3 (2009).
34. I. I.Oleinik, E. Y. Tsymbal, and D. G. Pettifor, Structural and electronic properties of $CoAl_2O_3$/Co magnetic tunnel junction from first principles, *Phys. Rev. B* **62**, 3952–3959 (2000); I. I. Oleinik, E. Y. Tsymbal, and D. G. Pettifor, Atomic and electronic structure of Co/$SrTiO_3$/Co magnetic tunnel junctions, *Phys. Rev. B* **65**, 020401-1-4 (2002).
35. J. Mathon and A. Umerski, Theory of tunneling magnetoresistance in a junction with a nonmagnetic metallic interlayer, *Phys. Rev. B* **60**, 1117–1121 (1999).
36. Ph. Mavropoulos, N. Papanikolaou, and P. H. Dederichs, Complex band structure and tunneling through ferromagnet/insulator/ferromagnet junctions, *Phys. Rev. Lett.* **85**, 1088–1091 (2000).
37. X.-G. Zhang and W. H. Butler, Large magnetoresistance in bcc Co/MgO/Co and FeCo/MgO/FeCo tunnel junctions, *Phys. Rev. B* **70**, 173407-1-4 (2004).
38. J. P. Velev, K. D. Belashchenko, D. A. Stewart et al., Negative spin polarization and large tunneling magnetoresistance in epitaxial Co/$SrTiO_3$/Co magnetic tunnel junctions, *Phys. Rev. Lett.* **95**, 216601-1-4 (2005).
39. J. M. De Teresa, A. Barthélémy, A. Fert et al., Role of metal-oxide interface in determining the spin polarization of magnetic tunnel junctions, *Science* **286**, 507–509 (1999).
40. M. Bowen, M. Bibes, A. Barthélémy et al., Nearly total spin polarization in $La_{2/3}Sr_{1/3}MnO_3$ from tunneling experiments, *Appl. Phys. Lett.* **82**, 233–235 (2003).
41. T. Ishikawa, T. Marukame, H. Kijima et al., Spin-dependent tunneling characteristics of fully epitaxial magnetic tunneling junctions with a full-Heusler alloy Co_2MnSi thin film and a MgO tunnel barrier, *Appl. Phys. Lett.* **89**, 192505-1-3 (2006).
42. P. LeClair, J. K. Ha, J. M. Swagten et al., Large magnetoresistance using hybrid spin filter devices, *Appl. Phys. Lett.* **80**, 625–627 (2002).
43. W. P. Pratt, S. F. Lee, J. M. Slaughter et al., Perpendicular giant magnetoresistances of Ag/Co multilayers, *Phys. Rev. Lett.* **66**, 3060–3063 (1991); J. Bass and W. P. Pratt, Jr., Current-perpendicular (CPP) magnetoresistance in magnetic metallic multilayers, *J. Magn. Magn. Mater.* **200**, 274–289 (1999).
44. L. Piraux, J.-M. George, C. Leroy et al., Giant magnetoresistance in magnetic multilayered nanowires, *Appl. Phys. Lett.* **65**, 2484–2486 (1994); A. Fert and L. Piraux, Magnetic nanowires, *J. Magn. Magn. Mater.* **200**, 338–358 (1999).
45. T. Valet and A. Fert, Theory of the perpendicular magnetoresistance in magnetic multilayers, *Phys. Rev. B* **48**, 7099–7013 (1993).
46. L. E. Hueso, J. M. Pruneda, V. Ferrari et al., Transformation of spin information into large electrical signals using carbon nanotubes, *Nature* **445**, 410–413 (2007).
47. G. Schmidt, D. Ferrand, L. W. Molenkamp et al., Fundamental obstacle for electrical spin injection from a ferromagnetic metal into a diffusive semiconductor, *Phys. Rev. B* **62**, 4790–4793 (2000).
48. E. I. Rashba, Theory of electrical spin injection: Tunnel contacts as a solution of the conductivity mismatch problem, *Phys. Rev. B* **62**, R16267–R1670 (2000).
49. A. Fert and H. Jaffrès, Conditions for efficient spin injection from a ferromagnetic metal into a semiconductor, *Phys. Rev. B* **64**, 184420-1-9 (2001); A. Fert, J.-M. George, H. Jaffrès, and R. Mattana, Semiconductors between spin-polarized sources and drains, *IEEE Trans. Electron Dev.* **54**, 921–932 (2007); H. Jaffrès, A. Fert, and J.-M. George, *Phys. Rev. B.* 67, 140408(R), 8–11 (2010).
50. M. Johnson and R. H. Silsbee, Thermodynamic analysis of interfacial transport and of the thermomagnetoelectric system, *Phys. Rev. B* **35**, 4959–4972 (1987).
51. Z. G. Yu and M. E. Flatté, Electric-field dependent spin diffusion and spin injection into semiconductors, *Phys. Rev. B* **66**, 201202-1-4 (2002).
52. T. Kimura, J. Hamrle, and Y. Otani, Estimation of spin-diffusion length from the magnitude of spin-current absorption: Multi-terminal ferromagnetic/non-ferromagnetic hybrid structures, *Phys. Rev. B* **72**, 014461-1-6 (2005).
53. S. Takahashi and S. Maekawa, Spin currents in metals and superconductors, *J. Phys. Soc. Jpn.* **77**, 031009-1-14 (2008).
54. F. J. Jedema, M. S. Nijboer, A. T. Filip et al., Spin injection and spin accumulation in all-metal mesoscopic spin valves, *Phys. Rev. B* **67**, 085319-1-16 (2003).
55. T. Kimura and Y. Otani, Large spin accumulation in a permalloy-silver lateral spin valve, *Phys. Rev. Lett.* **99**, 196604 (2007).
56. B. T. Jonker and M. E. Flatté, Electrical spin injection and transport in semiconductors, in *Nanomagnetism: Ultrathin Films, Multilayers and Nanostructures*, D. L. Mills and J. A. C. Bland (Eds.), Elsevier, Amsterdam, pp. 227–272, 2006.
57. A. Cottet, T. Kontos, S. Sahooo et al., Nanospintronics with carbon nanotubes, *Semicond. Sci. Technol.* **21**, S78–S95 (2006).
58. N. Tombros, C. Jozsa, M. Popinciuc et al., Electronic transport and spin precession in single graphene layers at room temperature. *Nature* **448**, 571–574 (2006); M. Ohishi, M. Shiraishi, R. Nouchi et al., Spin injection into graphene at room temperature, *Jpn. J. Appl. Phys.* **46**, 25–28 (2007); W. Han, K. Pi, K. McCreary et al., Tunneling spin injection into single layer graphene, *Phys. Rev. Lett.* **105**, 16702–16704 (2010).
59. J. C. Slonczewski, Current-driven excitation of magnetic multilayers, *J. Magn. Magn. Mater.* **159**, L1–L7 (1996).
60. L. Berger, Emission of spin waves by a magnetic multilayer traversed by a current, *Phys. Rev. B* **54**, 9353–9358 (1996).

61. M. Tsoi, A. G. M. Jansen, J. Bass et al., Excitation of a magnetic multilayer by an electric current, *Phys. Rev. Lett.* **80**, 4281–4284 (1998).
62. F. J. Albert, J. A. Katine, F. J. Albert et al., Spin-polarized current switching of a Co thin film nanomagnet, *Appl. Phys. Lett.* **77**, 3809–3811 (2000).
63. J. Grollier, V. Cros, A. Hamzic et al., Spin-polarized current induced switching in Co/Cu/Co pillars, *Appl. Phys. Lett.* **78**, 3663–3665 (2001).
64. W. H. Rippart, M. R. Pufall, S. Kaka et al., Direct-current induced dynamics in $Co_{90}Fe_{10}/Ni_{80}Fe_{20}$ point contacts, *Phys. Rev. Lett.* **92**, 027201-1–4 (2004).
65. A. Dussaux, B. Georges, J. Grollier et al., Large microwave generation from current-driven magnetic vortex oscillators in magnetic tunnel junctions, *Nat. Commun.* (2010) DOI:10.1038/ncommas1006.
66. A. Ruotolo V. Cros, B. Georges et al., Spin transfer induced dynamics of an array of interacting vortices in a periodic potential, *Nat. Nanotechnol.* **4**, 528–532 (2009).
67. I. Žutić, J. Fabian, and S. Das Sarma, Spintronics: Fundamentals and applications, *Rev. Mod. Phys.* **76**, 323–410 (2004).
68. H. Ohno, A. Shen, and F. Matsukura, (Ga,Mn)As: A new diluted magnetic semiconductor based on GaAs, *Appl. Phys. Lett.* **69**, 363–365 (1996).
69. V. F. Motsnyi, P. Van Dorpe, W. Van Roy et al., Optical investigation of spin injection into semiconductors, *Phys. Rev. B* **68**, 245319 (2003).
70. A. T. Hanbicki, M. J. van't Erve, and R. Magno, Analysis of the transport process providing spin injection through an Fe/AlGaAS Schottky barrier, *Appl. Phys. Lett.* **82**, 4092–4094 (2003).
71. Y. Lu, V. G. Truong, P. Renucci et al., MgO thickness dependence of spin injection efficiency in spin-light emitting diodes, *Appl. Phys. Lett.* **93**, 152102-1–3 (2008).
72. X. Lou, C. Adelmann, S. A. Crooker et al., Electrical detection of spin transport in lateral ferromagnet–semiconductor devices, *Nat. Phys.* **3**, 197–202 (2007).
73. O. Van't Erve, A. T. Hanbicki, M. Holub et al., Electrical injection and detection of spin-polarized carriers in silicon in a lateral transport geometry, *Appl. Phys. Lett.* **91**, 212109-1-3 (2007).
74. M. Tran, H. Jaffrès, C. Deranlot et al., Enhancement of the spin accumulation at the interface between a spin-polarized tunnel junction and a semiconductor, *Phys. Rev. Lett.* **102**, 036601-1–4 (2009).
75. S. P. Dash, S. Sharma, R. S. Pald et al., Electrical creation of spin polarization in silicon at room temperature, *Nature* **462**, 491–494 (2009).
76. S. Datta and B. Das, Electronic analog of the electro-optic modulator, *Appl. Phys. Lett.* **56**, 665–667 (1990).
77. H.-C. Koo, J.-H. Kwon, J. Eom et al., Control of spin precession in a spin-injected field effect transistor, *Science* **325**, 1515–1518 (2009).
78. M. König, S. Wiedmann, and C. Brüne, Quantum spin Hall insulator state in HgTe quantum wells, *Science* **318**, 766–770 (2007).
79. D. Hsieh, Y. Xia, D. Qian et al., Observation of time-reversal-protected single-Dirac-Cone topological-insulator states in Bi_2Te_3 and Sb_2Te_3, *Phys. Rev. Lett.* **103**, 146401-1–4 (2009).
80. T. Zhao, A. Scholl, F. Zavaliche et al., Electrical control of antiferromagnetic domains in multiferroic $BiFeO_3$ films at room temperature, *Nat. Mater.* **5**, 823–829 (2006); H. Béa, M. Bibes, F. Ott et al., Mechanisms of exchange bias with multiferroic $BiFeO_3$ epitaxial films, *Phys. Rev. Lett.* **100**, 017204 (2008); J. D. Burton and E. Y. Tsymbal, Prediction of electrically induced magnetic reconstruction at the manganite/ferroelectric interface, *Phys. Rev. B* **80**, 1744061–1744065 (2009).
81. M. Bibes and A. Barthélémy, Oxide spintronics, *IEEE Trans. Electron. Dev.* **54**, 1003–1023 (2007).
82. Y.-H. Chu, L. W. Martin, M. B. Holcomb et al., Electric-field control of local ferromagnetism using a magnetoelectric multiferroic, *Nat. Mater.* **7**, 478–482 (2008).
83. D. Chiba, M. Sawicki, Y. Nishitami et al., Magnetization vector manipulation by electric field, *Nature* **455**, 515–518 (2008).
84. M. Weisheit, S. Fähler, A. Marty et al., Electric field-induced modification of magnetization in thin-film ferromagnets, *Science* **315**, 349 (2007).
85. A. Ohtomo and H. Y. Hwang, A high-mobility electron gas at the $LaAlO_3/SrTiO_3$ heterointerface, *Nature* **427**, 423–426 (2004).
86. M. Basletic, J.-L. Maurice, C. Carrétéro et al., Mapping the spatial distribution of charge carriers in $LaAlO_3/SrTiO_3$ heterostructures, *Nat. Mater.* **7**, 621–25 (2008); C. Cen, S. Thiel, J. Mannhart et al., Oxide nanoelectronics on demand, *Science* **323**, 1026–1028 (2009).
87. M. K. Niranjan, Y. Wang, S. S. Jaswal et al., Prediction of a switchable two-dimensional electron gas at ferroelectric oxide interfaces, *Phys. Rev. Lett.* **103**, 016804-1-4 (2009).
88. V. Garcia, S. Fusil, K. Bouzehouane et al., Giant tunnel electroresistance for non-destructive readout of ferroelectric states, *Nature* **460**, 81–84 (2009).
89. Z. H. Xiong, D. Wu, V. Vardeny et al., Giant magnetoresistance in organic spin-valves, *Nature* **427**, 821–824 (2004).
90. W. J. M. Naber, S. Faez, and W. G. Van der Wiel, Organic spintronics, *J. Phys. D: Appl. Phys.* **40**, R205–R228 (2007).
91. B. Trauzettel, M. Borhani, M. Trif et al., Theory of spin qubits in nanostructures, *J. Phys. Soc. Jpn.* **77**, 031012 (2008).

II

Spin Transport and Magnetism in Magnetic Metallic Multilayers

II

2

Basics of Nano-Thin Film Magnetism

Bretislav Heinrich
Simon Fraser University

2.1 Introduction

Ultrathin magnetic films have been studied and used in low-dimensional systems and materials employed in spintronics. In these systems, the interfaces play a crucial role. The magnetic properties of interface atoms are different from those in the bulk. First-principle calculations indicate that the bulk properties are almost fully acquired just a few atomic layer from the interface. Ultrathin films behave like giant magnetic molecules having their magnetic properties given by an admixture of the interface and bulk properties. The ability to admix the interface and bulk magnetic properties and combine the magnetic films together by nonmagnetic interlayers allows one to engineer unique magnetic materials. In the following three sections, the emphases will be put on the basic magnetic properties: magnetic anisotropy, interlayer exchange coupling, and exchange bias (EB) coupling. Each section starts with a general theoretical introduction, followed by most common experimental techniques, and a brief summary of important experimental studies, which played a crucial role in the development of magnetic nanostructures. I apologize to those who feel that their work and contributions were either insufficiently covered or perhaps even omitted. This is certainly not intentional. The goal of these sections is not to provide a comprehensive account of this field. The page limitation allows one to provide only basic ideas that are needed for specialists and graduate students entering the field of magnetic nanostructures and spintronics applications. There are many references within this chapter to important books and review articles where further details and references can be found.

Since a large range of problems is covered, the following provides a brief guided tour through these sections. In Section 2.2, the concept of free energy of magnetic anisotropy is described, starting from a phenomenological description satisfying the lattice symmetry requirement. It is shown that the magnetic anisotropy exerts a torque on magnetic moment by an effective field given by Equation 2.6. It is shown that the interfaces due to their decreased lattice symmetry play a crucial role in magnetic anisotropies. In ultrathin films, the overall magnetic properties are then given by adding the bulk and interface properties using scaling with the film thickness shown in Equation 2.10 allowing one to use a concept of giant magnetic molecules. Theory of magnetic anisotropy in 3D transition elements is described in Section 2.2.1. It is shown that it is caused by the spin orbit contribution to the energy of electron states occupying the valence bands. Néel's model of magnetic anisotropy based on a magnetic atom pair interaction is described in Section 2.2.2. This allows one to include the magnetoelastic contributions, which, in a strained lattice, results in perpendicular and in-plane uniaxial fields. MacLaren and Victora's extension of Néel's model includes in addition to a pair interaction between two ferromagnetic atoms also a pair interaction between a ferromagnetic atom and nonmagnetic nearest neighbor substrate atoms allowing one to include the dependence of interface anisotropy on the type of nonmagnetic substrate. Dzyaloshinski–Moriya exchange anisotropy energy and noncollinear magnetism in ultrathin films are discussed in Section 2.2.3. The measurements of magnetic anisotropies by ferromagnetic resonance (FMR) are described in Section 2.2.4. The interpretation of the data in this section is based on a detailed solution of Landau–Lifshitz–Gilbert (LLG) equations of motion. It requires evaluating the effective fields arising from the magnetic anisotropies including the demagnetizing energy and magnetoelastic contributions. Expressions for the effective fields and corresponding torque equations are shown in details. The results of magnetic anisotropies in seminal ultrathin film systems are presented in Section 2.2.5.

Interlayer exchange coupling is covered in Section 2.3. Phenomenology of the bilinear and biquadratic interface exchange coupling is discussed in Section 2.3.1. Theory of bilinear exchange coupling using quantum well states (QWS) in a normal metal (NM) spacer is discussed in Section 2.3.2. The mechanism of biquadratic exchange coupling based on interface roughness is discussed in Section 2.3.3. It is also shown that in any system that exhibits a lateral inhomogeneity, one must expect additional energy terms. They originate from intrinsic magnetic energy terms that fluctuate in strength across the sample interface. These additional terms have the next higher angular power compared with that of the intrinsic term, and they have only one sign. The static and dynamic techniques employed in measurements of the interlayer exchange coupling are discussed in Section 2.3.4. The seminal studies of interlayer exchange coupling are summarized in Section 2.3.5. It covers the Cr spacers with the long-range spin density waves (SDWs) and NMs where the interlayer exchange coupling is generated by QWS passing through the Fermi surface.

The EB coupling is covered in Section 2.4. The EB mechanism is an important part of spintronics devices. This mechanism originates in very thin region between the antiferromagnetic (AFM) and ferromagnetic materials, where the magnetism, structure, and the role of interface morphology abruptly change and acquire unique features, which are strongly dependent on sample preparation. In this respect, there is no unified theory applicable to these systems. However, some basic principles governing the EB mechanism have been found and are useful in designing and fabricating spin valve and spin torque devices where it provides pinning mechanism for hard magnetic layers. In this section, a brief review of basic models for EB is presented, starting from the Meiklejohn–Beam and then followed by Malozemoff and Mauri models, respectively.

2.2 Magnetic Anisotropy

The energy of a ferromagnet (FM) depends on the orientation of the magnetic moment with respect to the crystalline axes. This dependence is called magnetic anisotropy. The free energy contribution due to anisotropy in a cubic crystal has to satisfy the lattice symmetry and can be written to lowest order in the magnetization components,

$$\mathcal{F}_a = -\frac{K_1}{2}\left[\left(\frac{M_x}{M_s}\right)^4 + \left(\frac{M_y}{M_s}\right)^4 + \left(\frac{M_z}{M_s}\right)^4\right], \tag{2.1}$$

where

K_1 is the cubic anisotropy parameter (erg/cm^3)
M_x, M_y, and M_z are the components of $\vec{M}_s$ along the three cube axes

$\mathcal{F}_a$ can also be rewritten in a more common form using the directional cosines of the magnetization components ($\alpha_{x,y,z} = M_{x,y,z}/M_s$),

$$\mathcal{F}_a = K_1\left[\alpha_x^2\alpha_y^2 + \alpha_x^2\alpha_z^2 + \alpha_y^2\alpha_z^2\right]. \tag{2.2}$$

The equivalence of Equations 2.1 and 2.2 follows from the relation $\alpha_x^2 + \alpha_y^2 + \alpha_z^2 = 1$.

For hexagonal crystal symmetry, the anisotropic free energy can be expressed in lowest orders by

$$\mathcal{F}_a = K_2 \sin^2(\phi) + K_4 \sin^4(\phi), \tag{2.3}$$

where

$$\sin^2(\phi) = 1 - \left(\frac{M_z}{M_s}\right)^2,$$

where

ϕ is the angle between $\vec{M}_s$ and the *c*-axis of the crystal
K_2 and K_4 are hexagonal anisotropy parameters

The first term in Equation 2.3 represents an example of the magnetic energy density in systems having a single axis of symmetry. This symmetry leads in general to a uniaxial anisotropy

$$\mathcal{F}_u = -\frac{K_u}{M_s^2}\left(\vec{M}_s \cdot \vec{u}\right)^2, \tag{2.4}$$

where

K_u is a uniaxial anisotropy parameter
$\vec{u}$ is a unit vector that specifies the orientation of the twofold axis

The variation of the free energy density with magnetization direction in the crystal results in a torque density that acts so as to align the magnetization along a direction that minimizes the free energy density. This torque density can be written as

$$\vec{L} = \vec{M}_s \times \vec{H}_a, \tag{2.5}$$

where

$$\vec{H}_a = -\left(\frac{\partial \mathcal{F}_a}{\partial M_x}\right)\hat{u}_x - \left(\frac{\partial \mathcal{F}_a}{\partial M_y}\right)\hat{u}_y - \left(\frac{\partial \mathcal{F}_a}{\partial M_z}\right)\hat{u}_z \tag{2.6}$$

is an effective magnetic field that exerts the same torque on the magnetization as does a real magnetic field [1]. $\hat{u}_{x,y,z}$ are unit vectors along the x-, y-, and z-axes.

The description of magnetic coupling will be mostly restricted to ultrathin magnetic films. In ultrathin films, magnetic variations across the thickness of the film are mainly suppressed. The magnetization in such thin films is uniform for internal exchange fields that are significantly larger than typical anisotropy fields. The exchange interaction fields between spins lead to a ferromagnetic state and, at the same time, make any spatial variation of the magnetization density relatively costly in free energy [1] in ultrathin films. This means that the magnetic moments on lattice sites across the film thickness are nearly parallel to each other. This is not exactly correct, but greatly simplifies the treatment of magnetic properties. The limits of this concept are described in Ref. [2]. The film can be usefully considered to be ultrathin when its thickness does not exceed the exchange length δ, see Ref. [2],

$$\delta = \left(\frac{A}{2\pi M_s^2}\right)^{0.5}, \tag{2.7}$$

where

A is the exchange stiffness coefficient (for Fe $A=2\times10^{-6}$ erg/cm)

M_s is the saturation magnetization

The exchange length in Fe is 3.2 nm.

It is clear that the free energy density of a thin film is likely to be quite different when the magnetization is directed along the film normal as compared with the case in which the magnetization lies in the film plane. The source of this free energy difference is the magnetic fields generated by the shape of the film. When the magnetization lies in the plane of a film, a few millimeters in lateral dimensions but a few nanometers thick, the magnetic field generated by the magnetization density discontinuity at the film edges can be ignored (the approximation of infinite lateral dimensions). However, when this same uniform magnetization has a component, M_z, directed along the film normal, the discontinuity in magnetization at the film surfaces generates an internal demagnetizing field $H_d=-4\pi M_z$. The interaction between this demagnetizing field and the magnetization density contributes a term to the free energy density having the form of a uniaxial anisotropy

$$\mathcal{F}_d = 2\pi M_z^2. \tag{2.8}$$

This uniform demagnetizing energy is true only in magnetic continuum. The treatment of demagnetizing energy in atomic lattices with magnetic moments localized around the lattice sites is described in Section 2.2.4, see Equation 2.39.

Another important source of difference between a bulk magnetic crystal and a magnetic thin film comes about because the thin film is grown on a substrate. If the crystal structure of the thin film is not exactly matched to the crystal structure of the substrate, the bonding between film and substrate must result in the deformation of the film. In general, the film structure becomes distorted both in-plane and out-of-plane. The simplest case is that in which the film retains fourfold symmetry in plane, but the lattice spacing in the perpendicular direction changes to keep a constant atomic volume, and the thin film adopts a tetragonal symmetry. For tetragonal symmetry, the magnetocrystalline free energy density can be written as

$$\mathcal{F}_a = -\frac{K_1^{\parallel}}{2}\left[\left(\frac{M_x}{M_s}\right)^4 + \left(\frac{M_y}{M_s}\right)^4\right] - \frac{K_1^{\perp}}{2}\left(\frac{M_z}{M_s}\right)^4 - K_{\perp}\left(\frac{M_z}{M_s}\right)^2, \tag{2.9}$$

where $K_1^{\parallel}$, $K_1^{\perp}$, and $K_{\perp}$ are anisotropy parameters for magnetization components parallel and perpendicular to the film surface, respectively. An example of this case is the growth of an Fe(001) film on the (001) surface of a silver substrate [2]. The atoms in the fcc Ag(001) surface net form a square array 2.889 Å on a side. The atoms on the bcc Fe(001) surface planes form a square array 2.866 Å on a side. Thus, the fourfold hollows on the silver surface can accommodate the Fe(001) surface atoms if the iron net is expanded in plane by 0.8%. At the same time, the spacing between iron atomic layers is reduced by 1.5%. This lattice mismatch then leads to a tetragonal distortion, which results in the uniaxial anisotropy energy described by the last term in Equation 2.9.

The atomic layers in the vicinity of an interface generally have different magnetic properties from those that are farther away from the interface, i.e., in the bulk, see Figure 2.1. The total free energy functional can be written as the sum of a surface term plus a volume term [2]

$$\mathcal{F}_{tot} = \frac{\mathcal{F}_{int}(\vec{M})}{d} + \mathcal{F}_b(\vec{M}), \tag{2.10}$$

where

$\mathcal{F}_{int}(\vec{M})$ includes interface energies in erg/cm^2

d is the film thickness

It has been observed that each of the parameters in Equations 2.9 and 2.4 usually exhibits a thickness dependence, such that

$$K_i = K_{i,b} + \left(\frac{K_{i,int}}{d}\right), \tag{2.11}$$

where

$K_{i,b}$ is a thickness independent volume anisotropy with the units of erg/cm^3

$K_{i,int}$ is an interface energy in erg/cm^2

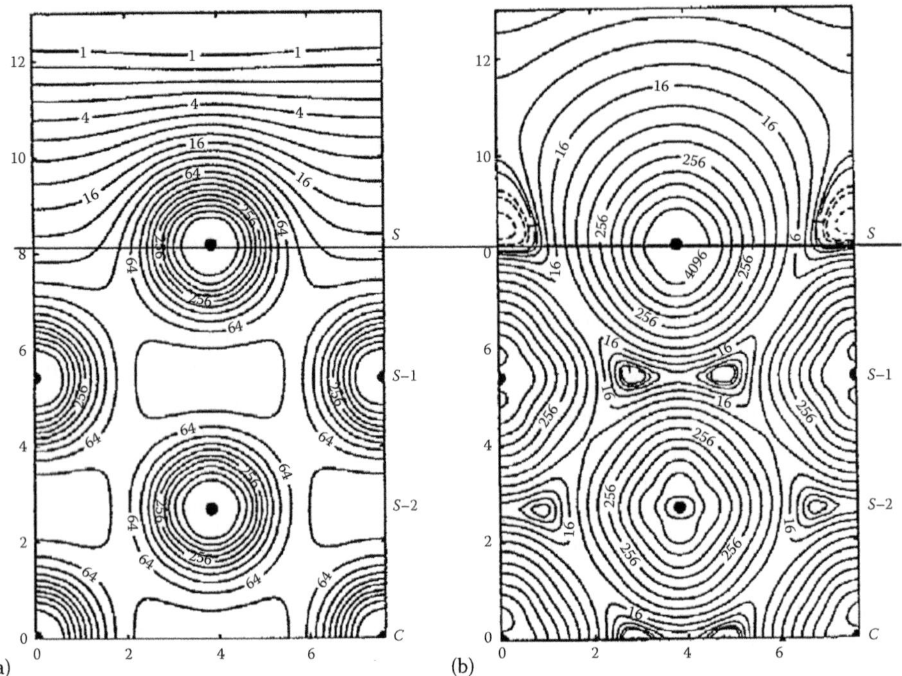

FIGURE 2.1 The charge density (a) and spin density (b) distributions around Ni atoms in the Ni(011) plane. The continuous solid horizontal line marks the interface between the Ni and empty space. (Courtesy of Freeman, A., Northwestern University; From Heinrich, B., *Can. J. Phys.*, 2000. With permission.)

In fact, the interface anisotropy $K_{i,int}$ is found to be sensitive to the structure and chemical composition of the film surfaces and their interfaces. An example is shown in Figure 2.2 of the perpendicular, in-plane uniaxial $2K_u^{\|}/M_s$, fourfold (cubic) $2K_1^{\|}/M_s$, and $4\pi M_{s,eff} - 2K_u^{\perp}/M_s$ anisotropy fields in Au/Fe/GaAs(001) films as a function of $1/d$, where d is the film thickness. Detailed discussion of $4\pi M_{s,eff}$ can be found in text after Equation 2.20 in Section 2.2.2.

2.2.1 Magnetic Anisotropies: Theory

2.2.1.1 First-Principle Electronic Band Calculations

First-principle calculations of magnetic anisotropies are based upon the solution of Schrödinger's equation for interacting electrons with spin. The form of the equations to be solved is discussed by Gay and Richter [3]. The magnetic anisotropies are caused by the spin–orbit interaction. A simple argument can be used to describe the spin–orbit interaction term. In the rest coordinate system of moving electrons, the electric field $\vec{E}$ appears as an effective field $\vec{B} \sim \vec{p} \times \vec{E}$ due to the Lorentz transformation. This leads to a relativistic contribution to the electron Hamiltonian of the form $\mathcal{H}_{so} \sim -\vec{B} \cdot \vec{S}$, where $\vec{S}$ is the spin momentum. For an electron passing an atomic potential having radial symmetry $\vec{E} = -(dV/dr)(\vec{r}/r)$, this results in $\mathcal{H}_{so} \sim (1/r)(dV(r)/dr)\vec{S} \cdot (\vec{r} \times \vec{p})$, which can be rewritten as $((1/r)(dV/dr))\vec{S} \times \vec{L}$, where $\vec{L}$ is the orbital momentum. Therefore, for a radially symmetric potential, the Hamiltonian for the spin–orbit interaction is given by

$$\mathcal{H}_{so} = \xi(r)\vec{S} \cdot \vec{L}, \quad (2.12)$$

where ξ is the spin-interaction parameter, which is proportional to the radial derivative of the atomic potential. In lattices exhibiting a high degree of symmetry, the crystal field hybridizes the d-orbitals and results in zero-orbital momentum. The spin–orbit interaction introduces orbital momentum. In second-order perturbation theory, the magnetic orbital momentum is given by Equation 2.34 in Ref. [4]

$$\overrightarrow{m_{so}} = \mu_B \sum_{i,j} \frac{\langle \phi_i | \vec{L} | \phi_j \rangle \langle \phi_i | H_{so} | \phi_j \rangle}{E_i - E_j}, \quad (2.13)$$

where the state i is occupied and the state j is unoccupied. The wave functions ϕ_i belong to all bands crossing the Fermi level E_F. When the majority and minority spin bands are well separated, it was shown by Bruno [5] that the magnetic anisotropy in a second-order perturbation calculation for a uniaxial system can be expressed as

$$\Delta E_{so} \simeq -\frac{\xi}{4\mu_B}[m_{orb}^{\perp} - m_{orb}^{\|}], \quad (2.14)$$

where $m_{orb}^{\perp,\|}$ is the orbital moment for directions perpendicular and parallel to the film. This means that the uniaxial magnetic anisotropy is proportional to the difference in the components

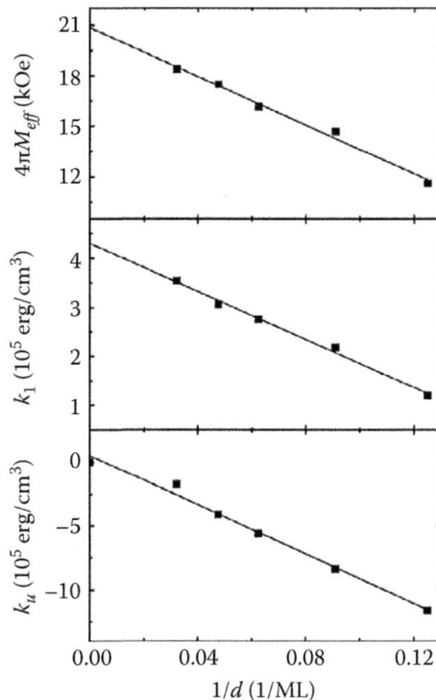

FIGURE 2.2 The effective demagnetizing field perpendicular to the film surface $4\pi M_{eff} = 4\pi M_{s,eff} - 2K^{\perp}_{u,int}/dM_s$, the in-plane fourfold (cubic) anisotropy parameter, K_1, and the uniaxial anisotropy parameter, K_u, are plotted as functions of $1/d$, where d is the Fe film thickness. The Fe films were grown on GaAs(001) and covered by 20 ML of Au(001). The in-plane uniaxial axis is directed along the $[\bar{1}0]$ crystallographic direction. The in-plane anisotropies comprise bulk and interface contributions: $K^{\parallel}_{1,b} = 4.3 \times 10^5$ erg/cm^3, $K^{\parallel}_{1,int} = -0.036$ erg/cm; $K^{\parallel}_{u,int} = -0.13$ erg/cm^2; and $4\pi M_{s,eff} = 20.3$ kOe, $K^{\perp}_{u,int} = 0.88$ erg/cm, where b and int stand for the bulk and interface parts of the corresponding anisotropies.

of orbital momentum along the uniaxial axis and the plane perpendicular to that axis. This is a particularly useful concept for the systems with reduced dimensionality and symmetry. At the surface, the atom dimensionality is effectively reduced from 3 to 2, and the lattice symmetry is also reduced. This reduced lattice symmetry results in an appreciable interface anisotropy. Weller et al. [6] using x-ray magnetic circular dichroism (XMCD) on Au/Co/Au(111) textured wedge samples measured an anisotropy in the orbital moment components parallel and perpendicular to the film surface. This anisotropy in the orbital moment components resulted in $m^{\perp}_{orb} - m^{\parallel}_{orb} < 0$, and, therefore, the easy axis was oriented perpendicular to the surface. Equation 2.13 leads to a uniaxial anisotropy, which is only a factor 2 larger than that obtained from the magnetic measurements on Au/Co/Ag(111) textured samples. This was good agreement considering the simplicity of Equation 2.13.

In order to calculate anisotropies accurately, it is necessary to calculate the ground state energy of the system under investigation using relativistic local density approximation (LDA) band calculations for various orientations of the magnetization vector with respect to the crystalline axes. This is a formidable task since the energy differences in question are of the order of meV per atom and are very small compared with the total ground state energy, which is of the order of eV per atom.

The precision with which the ground state energies could be calculated improved with time until it is now claimed that anisotropy energies for Fe, Ni, and Co monolayers (ML), which are of the order of 1 meV per atom, can be calculated with an uncertainty of the order of 0.001 meV per atom (1 meV/atom corresponds to a surface energy of 2.45 erg/cm^2 for atoms arranged on a Cu(001) surface net—a square net 2.56 Å on a side). See the review article by Wu and Freeman [7].

Magnetic anisotropy calculations are usually carried out for smooth and perfect surface interface planes whereas in real specimens these interface planes are usually rough on a scale of at least ±1 atomic layer. Therefore, one cannot expect perfect agreement between calculated and observed anisotropy coefficients no matter how exactly one evaluates the theoretical model.

2.2.2 Magneto-Elasticity: Néel Model

In 1954, Néel introduced a phenomenological model for the ferromagnetic state based upon the sum of pairwise interactions between ferromagnetic atoms [8]. This model was meant to help understand the origin of surface anisotropies as well as to elucidate the physics of magnetoelastic phenomena. The pair interaction was written as the sum of Legendre polynomials:

$$w(r,\theta) = g(r) + L(r)P_2(\cos\theta) + g(r)P_4(\cos\theta) + \cdots, \quad (2.15)$$

where

- r is the separation of the atom pair
- $L(r)$ and $g(r)$ are phenomenological parameters
- θ is the angle between the magnetization direction and the line that joins the atom pair

Usually, only the first two terms of the above series are used to discuss magnetic anisotropies and magnetoelastic effects. Thus, ignoring a constant term, the pair interaction can be written as

$$w(r,\theta) = L_f(r)(\vec{u}_r \cdot \vec{m})^2, \quad (2.16)$$

where

- $L_f(r)$ is an atom pair interaction parameter between the magnetic sites
- $\vec{u}_r$ is a unit vector directed along the line joining the atom pair
- $\vec{m}$ is a unit vector in the direction of the magnetization vector

The first term $g(r)$ is isotropic and results in a volume magnetostriction [9]. This simple model can be used to calculate the free energy density for an ultrathin film in terms of the parameter L_f. Such a free energy function will consist of the sum of a term proportional to the film thickness, a volume term, plus a term independent of film thickness, a surface term. It is clear that the free energy density must depend upon any strains introduced by film growth on a lattice mismatched substrate because it depends upon the angles between the magnetization direction and the orientation of the lines joining nearest neighbor atoms. In Chapter 8 of Ref. [9], Chikazumi and Charap demonstrate the use of the Néel model to deduce the form of the magnetoelastic coupling energy for cubic crystals. The presence of a magnetic moment will change the lattice. Using the strain tensor $e_{i,j}(i,j=x,y,z)$, one can write the bulk contribution to the magnetoelastic energy as

$$\mathcal{E}_{mag-el} = B_1\left[e_{xx}\left(\alpha_1^2-\frac{1}{3}\right)+e_{yy}\left(\alpha_2^2-\frac{1}{3}\right)+e_{zz}\left(\alpha_3^2-\frac{1}{3}\right)\right] + B_2(e_{xy}\alpha_1\alpha_2+e_{yz}\alpha_2\alpha_3+e_{zx}\alpha_3\alpha_1), \quad (2.17)$$

where

α_1, α_2, and α_3 are the directional cosines of the magnetic moment with respect to x, y, and z (usually major crystallographic axes)

B_1 and B_2 are the basic mangetoelastic parameters

The elastic energy is given by

$$\mathcal{E}_{el} = \frac{1}{2}C_{11}(e_{xx}^2+e_{yy}^2+e_{zz}^2)+\frac{1}{2}C_{44}(e_{xy}^2+e_{yz}^2+e_{zx}^2) + C_{12}(e_{xx}e_{yy}+e_{xx}e_{zz}+e_{yy}e_{zz}), \quad (2.18)$$

where C_{11}, C_{12}, and C_{44} are the elastic moduli.

For a (001) oriented cubic lattice whose in-plane lattice parameter has been altered by $\varepsilon_1=\Delta a/a$, and whose lattice parameter perpendicular to the plane has been altered by $\varepsilon_2=\Delta c/a$, then the condition that the film must be stress free along the film normal leads to

$$\frac{\varepsilon_2}{\varepsilon_1} = -\frac{2C_{12}}{C_{11}}, \quad (2.19)$$

see also Sander [10]. From Equations 2.17 and 2.19 and the condition $M_x^2+M_y^2+M_z^2=M_s^2$, the strain contribution to the free energy density can be written as

$$\mathcal{F}_{me} = const. + B_1[\varepsilon_2-\varepsilon_1]\left(\frac{M_z}{M_s}\right)^2 = const. - B_1\left(1+\frac{2C_{12}}{C_{11}}\right)\varepsilon_1\left(\frac{M_z}{M_s}\right)^2. \quad (2.20)$$

This is a typical expression for a uniaxial anisotropy perpendicular to the film surface with the coefficient of uniaxial anisotropy equal to $K_\perp=B_1(1+(2C_{12}/C_{11}))$. For iron grown on GaAs(001), the in-plane strain $e_\parallel=-0.0136$ and Equation 2.17 lead to $e_\perp/e_\parallel=-1.212$; in addition, $C_{11}=2.41\times10^{12}$ erg/cc, $C_{12}=1.46\times10^{12}$ erg/cc, and $B_1=-34.3\times10^6$ erg/cc. Therefore, the magnetoelastic field perpendicular to the surface using Equation 2.20 is 1.2 kOe. Since for the bulk Fe at RT $4\pi M_s=21.5$ kG, the perpendicular uniaxial magnetoelastic field leads to $4\pi M_{s,eff}=4\pi M_s-1.2=20.3$ kG, see Figure 2.2.

The above Néel model results in an anisotropic free energy density that is independent of the substrate or overlayer materials. In order to correct this defect, Victora and MacLaren [11] have proposed an extension of the Néel theory that includes in addition to a pair interaction between two ferromagnetic atoms also a pair interaction between a ferromagnetic atom and nonmagnetic nearest neighbor atoms having the form

$$W(r,\theta) = L_m(r)(\vec{u}_r\cdot\vec{m})^2, \quad (2.21)$$

where

$\vec{u}_r$ is a unit vector in the direction of the line joining a ferromagnetic atom with a nonmagnetic atom

$\vec{m}$ is a unit vector parallel with the magnetization vector

In this version of the theory, at least two interaction parameters are required, L_f and L_m. MacLaren and Victora have determined these two parameters for the Co(001)/Pd system by comparison of the resulting anisotropic free energy expression with the free energy obtained from first principles calculations using the experimentally observed strained lattice spacings as measured by Engel et al. [12]. The results of their first principles calculations were in good agreement with experiment. They used the values of L_f, L_m so determined to calculate the anisotropy coefficients for (111) and (110) oriented Co/Pd surfaces. The results were in reasonable agreement with the experimental data of Engel et al. [12].

2.2.3 Dzyaloshinski Moriya Interaction and Noncollinear Magnetic Configurations

The low dimentionality of magnetic nanostructures can lead to additional potential gradients, which then result in additional electric fields. A typical example occurs at the interface between magnetic and nonmagnetic materials. The lack of inversion symmetry leads to a gradient of the lattice potential, which will, just as for the spin–orbit interaction, result in an additional energy term due to the spins at the sites i and j. It was shown by Dzyaloshinski [13] and Moriya [14] that this results in the energy interaction term

$$\mathcal{H}_{DM} = \sum_{i,j}\vec{D}_{i,j}\cdot(\vec{S}_i\times\vec{S}_j), \quad (2.22)$$

where $\vec{D}$ is a vector dependent on the lattice symmetry and is proportional to the spin interaction parameter ξ. Notice that this term acts to tilt the magnetic moments away from collinearity. Since this term arises in the first-order perturbation expansion of the spin interaction, its contribution to the total Hamiltonian can be important and result in noncollinear magnetic moments in magnetic nanostructures. $\mathcal{H}_{DM}$ is well

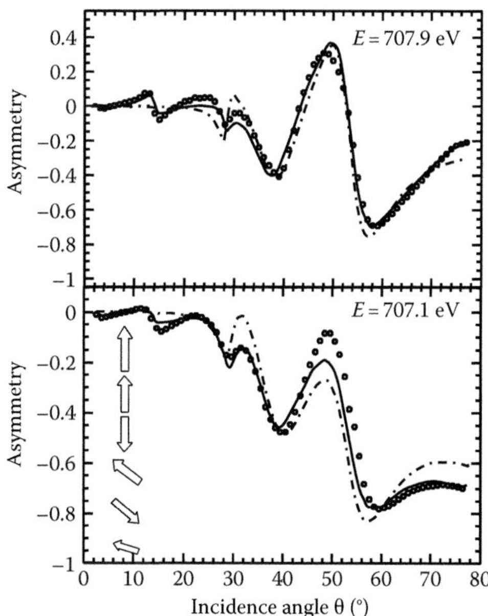

FIGURE 2.3 fcc Fe films 6 atomic layers thick were grown on a Cu(001) substrate using pulsed laser deposition (PLD). The measured x-ray reflectivity θ–2θ asymmetry parameter (using the right and left hand polarized x-ray beams) is shown by the open circles for two different x-ray energies, which straddle the Fe-L_3 absorption edge. The data were fit using the two magnetic configurations. The noncollinear configuration is shown in the inset. Computer fits using the noncollinear configuration is shown by the solid lines. The dashed lines are the best fits one can get for a fully collinear configuration. See details in Ref. [15]. (From Ohldag, H. et al., *Phys. Rev. Lett.*, 87, 247201, 2001. Copyright 2001 by the American Physical Society. With permission.)

known in weak magnetic materials where a weak magnetic moment arises from an AFM configuration in systems, which lack inversion symmetry. Recently, noncollinear configurations were found in ultrathin metastable metallic films. Meyerheim et al. [15] used soft x-ray resonant magnetic scattering in fcc Fe on Cu(001) to demonstrate the existence of a noncollinear magnetic structure in 6(fccFe)/Cu(001). The integer represents the number of atomic layers. Resonant magnetic scattering of a circularly polarized light beam using high scattering angles combined with first principles density functional theory (DFT) allows one to determine the magnetic moments in the individual atomic layers. The solid lines in Figure 2.3 show that the magnetic moments at the Fe/Cu interface are clearly noncollinear, and their orientation requires the presence of both the collinear exchange coupling and the Dzyaloahinski–Moriya interaction.

2.2.4 Measurement of Ultrathin Film Anisotropies: Ferromagnetic Resonance

The effective fields arising from the density of magnetization energy exert torques on the magnetization Equations 2.5 and 2.6. Thus, any experimental technique that permits one to measure torques exerted on the magnetization can be used to obtain the anisotropy coefficients of Equations 2.9 and 2.4. There are two main methods for the investigation of these torques: (1) FMR experiments and (2) experiments designed to measure the orientation of the magnetization vector as a function of the strength and orientation of an applied magnetic field. In this section, only FMR will be described. A more general description of the techniques employed in the study of magnetic anisotropies in ultrathin films can be found in Ref. [16]. FMR is a dynamic technique in which the frequency is measured when the magnetization, having been perturbed from equilibrium, precesses around its equilibrium orientation. The precessional frequency depends upon all of the magnetic fields to which the magnetization is subject, including the dipolar and anisotropy effective fields.

The directional cosines with respect to the lattice coordinate system (*X*, *Y*, *Z*), see Figure 2.4, are then given by simple trigonometry:

$$\alpha_x = \frac{M_x}{M_s}\cos\varphi_M \sin\Theta_M - \frac{M_y}{M_s}\sin\varphi_M - \frac{M_z}{M_s}\cos\varphi_M \cos\Theta_M \quad (2.23)$$

$$\alpha_y = \frac{M_x}{M_s}\sin\varphi_M \sin\Theta_M + \frac{M_y}{M_s}\cos\varphi_M - \frac{M_z}{M_s}\sin\varphi_M \cos\Theta_M \quad (2.24)$$

$$\alpha_z = \frac{M_x}{M_s}\cos\Theta_M + \frac{M_z}{M_s}\sin\Theta_M. \quad (2.25)$$

In this section, a simple example will be worked out at some length. The configuration of the magnetic field and the applied dc field is shown in Figure 2.4. For simplicity, let us assume that the contributions of the anisotropy to the free energy is given by the in-plane and perpendicular uniaxial terms and that the

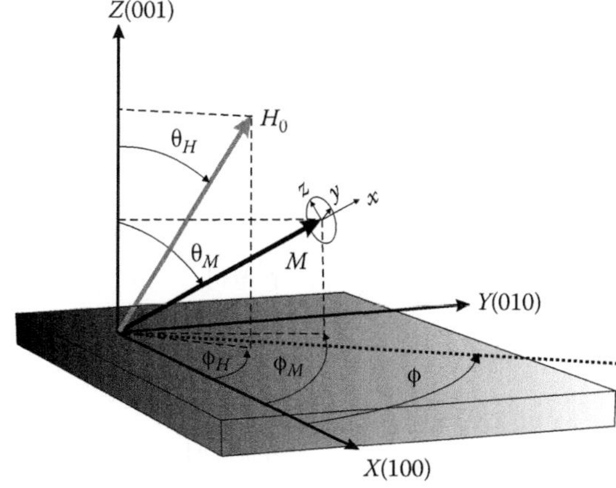

FIGURE 2.4 The laboratory system of coordinates is described by *X*, *Y*, *Z*. $\vec{M}$ and $\vec{H}$ are the magnetization and applied magnetic field, respectively. The *x*, *y*, *z* coordinates are used to describe the magnetization dynamics. The *x*-axis lies parallel with the static magnetization vector $\vec{M}_s$. (With kind permission from Springer Science+Business Media: *Magnetic Heterostructures: Advances and Perspectives in Spinstructures and Spintransport*, Exchange coupling in magnetic multilayers, vol. 227, 2008, 183, Heinrich, B.)

applied field and the static magnetization lie in the plane of the film, $\theta_{M,H}=0$. The Gibbs energy density is then given by

$$\mathcal{G} = -\frac{K_u^{\perp}}{2}\alpha_Z^2 - K_u^{\parallel}\frac{(\vec{u}\cdot\vec{M})^2}{M_s^2} - \vec{M}\cdot\vec{H}. \tag{2.26}$$

Using Equations 2.23 through 2.25 in Equation 2.26, one obtains the energy in terms of the variables $M_{x,y,z}$. The effective anisotropy fields are given by Equation 2.6.

$$H_x^{\parallel,u} = \frac{K_u^{\parallel}}{2M_s^2}\left[M_x(1+\cos 2(\varphi_M-\varphi_u)) - M_y\sin 2(\varphi_M-\varphi_u)\right], \tag{2.27}$$

$$H_y^{\parallel,u} = -\frac{K_u^{\parallel}}{M_s^2}\left[M_x\sin 2(\varphi_M-\varphi_u) - M_y(1-\cos 2(\varphi_M-\varphi_u)\right], \tag{2.28}$$

$$H_z^{\perp,u} = -4\pi D M_z + \frac{2K_u^{\perp}}{M_s^2}M_z, \tag{2.29}$$

where φ_u is the angle of the in-plane uniaxial anisotropy with respect to the [100] axis. The dc applied field ($\vec{H_0}$) is given by

$$\vec{H}_0 = H_0\left[\cos(\varphi_M - \cos\varphi_H)\vec{x} - \sin(\varphi_M - \cos\varphi_H)\vec{y}\right]. \tag{2.30}$$

We shall employ these equations to describe FMR. In FMR, one usually uses a small transverse field $\vec{h}$ in addition to the dc applied field $\vec{H_0}$. In this case, the torque on the magnetic moment leads to a precessional motion around the effective field having the transverse rf components $m_x, m_y > M_s$. The total effective field $\vec{H_{eff}}$ is given by the sum of the above fields

$$\vec{H}_{eff} = \vec{H}^u + \vec{H}_0 + h\vec{y}. \tag{2.31}$$

The response of the magnetization is usually described by the LLG equations of motions

$$\frac{1}{\gamma}\frac{\partial\vec{M}}{\partial t} = -\left[\vec{M}\times H_{eff}\right] + \frac{G}{\gamma^2 M_S}\left[\vec{M}\times\frac{\partial\hat{n}}{\partial t}\right] \tag{2.32}$$

$$= -\left[\vec{M}\times\left(\vec{H}_{eff} - \frac{\alpha}{\gamma}\frac{\partial\hat{n}}{\partial t}\right)\right], \tag{2.33}$$

where the first term on the right represents the torque due to the effective field, and the second term is the Gilbert damping torque. G is the Gilbert damping parameter, $\alpha = G/\gamma M_s$, $\vec{n}$ is the unit vector in the direction of the instantaneous magnetization $\vec{M}$, and γ is the absolute value of the gyromagnetic ratio, see further details in Ref. [2]. Solutions for $\vec{m}$ were found using the ansatz $\vec{m}, \vec{h} \sim \exp^{i\omega t}$, where ω is the angular microwave frequency.

Inserting the effective field $\vec{H_{eff}}$ in Equation 2.33 and keeping only terms linear in m_x, m_y, h, one gets

$$i\frac{\omega}{\gamma}m_y + \left(\boldsymbol{B_{eff}} + i\alpha\frac{\omega}{\gamma}\right)m_z = 0, \tag{2.34}$$

$$\left(\boldsymbol{H_{eff}} + i\alpha\frac{\omega}{\gamma}\right)m_y - i\frac{\omega}{\gamma}m_z = hM_s. \tag{2.35}$$

The y component of the LLG equation of motion also includes a large-term proportional only to the large effective field components. This term is zero after minimizing the total Gibbs potential in Equation 2.26. It states that the dc torque is zero in equilibrium. The effective fields $\boldsymbol{B_{eff}}$ and $\boldsymbol{H_{eff}}$ are

$$\boldsymbol{B_{eff}} = H_{dc}\cos(\varphi_M - \varphi_H) + 4\pi M_{eff} + \frac{K_1^{\parallel}}{2M_s}(3+\cos 4\varphi_M) + \frac{K_u^{\parallel}}{M_s}(1+\cos 2(\varphi_M-\varphi_u)), \tag{2.36}$$

$$\boldsymbol{H_{eff}} = H_{dc}\cos(\varphi_M - \varphi_H) + \frac{2K_1^{\parallel}}{M_s}\cos 4\varphi_M + \frac{2K_u^{\parallel}}{M_s}\cos 2(\varphi_M - \varphi_u), \tag{2.37}$$

where φ_u is the angle of the in-plane uniaxial anisotropy with respect to the [100] crystallographic axis. The contributions from the fourfold anisotropy were added in Equations 2.36 and 2.37. Here, the demagnetizing factor $4\pi M_s$ is grouped together with the perpendicular uniaxial anisotropy contribution:

$$4\pi M_{eff} = 4\pi M_s - \frac{2K_u^{\perp}}{M_s}. \tag{2.38}$$

The demagnetizing field is usually described by a magnetic continuum in which the demagnetizing field is $4\pi M_s$. This approach is incorrect in ultrathin films where the atomic magnetic moments are localized around the atomic sites. The discreteness of atomic moments results in a demagnetizing field that changes across the film thickness. The demagnetizing field from a given atomic layer decreases exponentially away from its surface with a decay length corresponding to the in-plane lattice spacing. Consequently, demagnetizing field decreases when approaching the film surface. This results in a perpendicular surface dipole–dipole uniaxial energy term

$$E_{d-d,s} = -c_{str}(2\pi M_s^2)a_0\alpha_z^2, \tag{2.39}$$

where a_0 is the interplanar spacing (in cm). The structural factor $c_{str}=0.425$ and 0.234 for the bcc and fcc lattice, respectively. For bcc Fe(001), this interface dipole–dipole interface term gives $K_{u,int}^{\perp} = 0.06$ erg/cm^2. See further details in Ref. [2].

Notice that only the in-plane fourfold effective field affects FMR. The effective fourfold field perpendicular to the surface includes only a third-order term in the perpendicular component m_z, see Equation 2.1. The transverse rf susceptibility χ_y can be found from the system of Equations 2.34 and 2.35:

$$\chi_y \equiv \chi'_y + i\chi''_y \equiv \frac{m_y}{h} = \frac{M_s\left(\boldsymbol{B}_{eff} - i\alpha\frac{\omega}{\gamma}\right)}{\left(\boldsymbol{B}_{eff} - i\alpha\frac{\omega}{\gamma}\right)\left(\boldsymbol{H}_{eff} - i\alpha\frac{\omega}{\gamma}\right) - \left(\frac{\omega}{\gamma}\right)^2}, \tag{2.40}$$

where χ'_y and χ''_y are dispersive and absorbtive parts of the rf susceptibility, respectively. FMR occurs when the denominator of the susceptibility function χ_y is at a minimum. Neglecting a small damping contribution to the resonance condition,

$$\left(\frac{\omega}{\gamma}\right)^2 = \boldsymbol{B}_{eff}\boldsymbol{H}_{eff}\,|_{H_{FMR}}. \tag{2.41}$$

Equation 2.40 shows that the dependence of χ''_y on the externally applied field can be mathematically represented by a Lorentzian function centered around the FMR field H_{FMR} with a half width at half maximum given by $\Delta H = \alpha\omega/\gamma$

$$\chi''_y = \left(\frac{M_s\boldsymbol{B}_{eff}}{\boldsymbol{B}_{eff} + \boldsymbol{H}_{eff}}\right)_{H=H_{FMR}} \times \frac{\Delta H}{(\Delta H)^2 + (H_{dc} - H_{FMR})^2}. \tag{2.42}$$

A simple calculation can be carried out for perpendicular FMR, where $\vec{M_s}$ and $\vec{H}$ are perpendicular to the film surface. In this case, the effective in-plane fourfold anisotropy field is proportional to the third power of the rf magnetization components and can be neglected. It is convenient to use a coordinate system with the x-axis parallel to the in-plane uniaxial anisotropy and the z axis perpendicular to the film surface. The perpendicular anisotropy $K_1^{\perp}$ leads to an effective field $2K_1^{\perp}/M_s$ in the perpendicular direction and results in an effective perpendicular dc field of $4\pi M_{eff} - 2K_1^{\perp}/M_s$. The effective field due to the in-plane uniaxial anisotropy can be easily obtained using the second term in Equation 2.26 and partial derivatives (Equation 2.6) with respect to the magnetization components $m_{x,y}$; $h_{x,u}^{\parallel} = (2K_u^{\parallel}/M_s^2)m_x$. Using LLG equation, one obtains

$$\left(\frac{\omega}{\gamma}\right)^2 = \left(H_{\perp} - 4\pi M_{eff} + \frac{2K_1^{\perp}}{M_s}\right)\left(H_{\perp} - 4\pi M_{eff} + \frac{2K_1^{\perp}}{M_s} - \frac{2K_u^{\parallel}}{M_s}\right). \tag{2.43}$$

Notice that that the in-plane uniaxial anisotropy leads to an elliptically polarized rf magnetization precession even in the perpendicular configuration.

The in-plane FMR measurements allow one to determine the in-plane anisotropy fields. The perpendicular FMR allows one to determine the perpendicular fourfold anisotropy field $2K_1^{\perp}/M_s$. In ultrathin films, the in-plane and perpendicular fourfold anisotropy fields are often dramatically different (even in sign) for a film thickness less than 10 ML. This is caused by an additional anisotropy arising from an inhomogeneous lateral distribution of the interface perpendicular uniaxial anisotropy, see Section 2.3.3. The gyromagnetic ratio ($\gamma \sim g$ factor) can be determined either by using out of plane FMR measurements in which the external field is rotated from the in-plane to the perpendicular FMR configuration or by using FMR measurements at two or more different microwave frequencies.

In FMR experiments using microwave standard waveguides, one usually maintains a constant microwave frequency and sweeps the applied field. In microstrip and coplanar waveguides, FMR measurements can be also carried out by sweeping the microwave frequency at constant applied magnetic field. The resonance field corresponds to a maximum in $\chi^{\parallel}$; see the review articles by Heinrich [17], Farle [18], and Poulopoulos and Baberschke [19]. An example of the use of FMR to determine magnetic anisotropies is shown in Figure 2.5. Note that FMR measurements yield effective fields. It is therefore necessary to know M_s in order to obtain the anisotropy coefficients K_i. Usually, the saturation magnetization must be measured in a separate experiment using a sensitive magnetometer such as a superconducting quantum interference device (SQUID) or a vibrating sample magnetometer (VSM). However, the strength of the FMR absorption signal is proportional to M_s, and it proves to be possible to obtain M_s with an accuracy of a few percentage by means of very careful absorption measurements [20].

2.2.5 Anisotropy Data

More detailed data and discussion are provided in the review article by Heinrich and Cochran [16]: in particular, refer to their Table 1 for anisotropy parameters measured for interfaces between Fe(001) and vacuum, Ag, Cu, Au, and Pd. The review article by Farle [18] contains an extended discussion of the Ni/Cu(001) and the Gd/W(001) systems. The case of Fe grown on GaAs(001) has become very interesting because the small mismatch between the Fe(001) and the GaAs(001) interface makes this combination an obvious choice for devices based on a hybrid ferromagnetic/semiconductor system. The Fe/GaAs(001) system is exhaustively discussed in the monumental review article by Wastlbauer and Bland [21]: this article includes a compilation of measured anisotropy coefficients plus relevant references. Bayreuther and coworkers extensively studied the in-plane uniaxial anisotropy [22,23].

Seki et al. [24] have shown that thin films of the alloy $Fe_{38}Pt_{62}$ having a uniaxial perpendicular anisotropy parameter as large as $K_u^{\perp} = 1.8 \times 10^7$ erg/cc can be grown on an MgO(001) substrate at substrate temperatures as low as 300°C. These films, 18 nm thick, were grown by means of co-deposition of Fe and Pt by

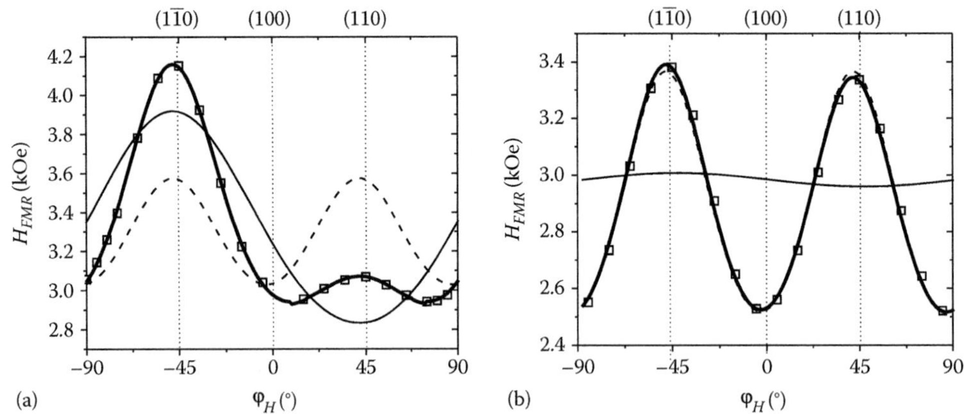

FIGURE 2.5 The angular dependence of the resonant magnetic field H_{res} at 24 GHz. (a) 20Au/16Fe/GaAs(001); (b) 20Au/70Fe/GaAs(001). The integers represent the number of atomic layers. Both specimens were grown using molecular beam epitaxy (MBE). The angle of the applied magnetic field, φ_H, was rotated in the plane of the film and measured with respect to the $[100]_{Fe}$ crystallographic direction. The data are plotted using open squares. The contributions of the fourfold $K_1^{\|}$ and uniaxial $K_u^{\|}$ magnetic anisotropies are shown using the dashed and thin solid lines, respectively. The thick solid line is a computer fit to the data using the following magnetic parameters: (a) 20Au/16Fe/GaAs – $K_1^{\|} = 2.61 \times 10^5$ erg/cm^3, $K_u^{\|} = -5.01 \times 10^5$ erg/cm^3, $4\pi M_{eff} = 16.8$ kOe; (b) 20Au/70Fe/GaAs – $K_1^{\|} = 3.77 \times 10^5$ erg/cm^3, $K_u^{\|} = -0.2 \times 10^5$ erg/cm^3, $4\pi M_{eff} = 19.6$ kOe. The g factor was fixed at $g = 2.09$, and the hard axis of the in-plane uniaxial anisotropy lies along the [1$\bar{1}$0] direction. Clearly, the 70 ML thick Fe film and the 16 ML thick Fe film exhibit different magnetic anisotropies. The significantly smaller in-plane uniaxial anisotropy in the 70Fe film compared to the 16Fe film reveals a complex interface origin.

sputtering on a (1 nm Fe + 40 nm Pt) buffer layer deposited on the MgO(001) crystal at room temperature. The films exhibited long-range intermetallic order having the $L1_0$ structure. Such films may be useful for ultrahigh-density recording media.

2.2.5.1 Vicinal Templates

A special case of decreased symmetry is provided by film growth on a template having a self-assembled network of oriented atomic steps. If a crystal is cut so that the surface plane makes a small angle with respect to a principle crystallographic plane, the result is a surface containing many monatomic steps. Such a surface is called a vicinal surface. In a series of experiments in which Fe films and Fe films covered by Ni films were grown on vicinal Ag substrates, it was discovered that the specimens exhibited an in-plane twofold magnetic anisotropy in which the direction of the uniaxial anisotropy axis was correlated with the orientation of the vicinal surface steps [25]. A few years later, systematic studies by Krams et al. [26] have shown that the in-plane uniaxial anisotropy due to a stepped surface is a consequence of the lattice mismatch between the magnetic thin film and the substrate upon which it has been grown. See further the discussion of magnetic anisotropy on vicinal surfaces [16]. Recently, Li et al. [27] studied the in-plane uniaxial anisotropy for the structure Fe/Ag(1,1,10). For temperatures less than 200 K, they observed strong oscillations of the uniaxial magnetic anisotropy as a function of the Fe thickness with a period of 5.7 atomic layers, see Figure 2.6. For Au-covered samples, the uniaxial anisotropy axis oscillated between the perpendicular and parallel (P) directions to the atomic steps. This behavior is attributed to QWS in the Fe film.

2.3 Interlayer Exchange Coupling

2.3.1 Phenomenology of Magnetic Coupling

The simplest form of magnetic coupling per unit area of an ultrathin film trilayer structure consisting of two ferromagnetic films

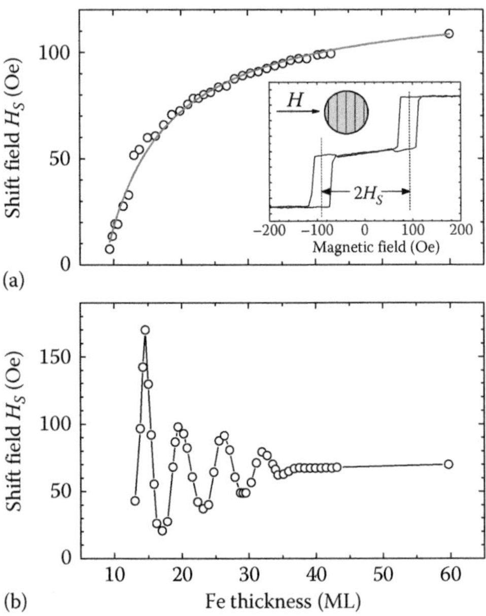

FIGURE 2.6 The shift field (saturation field) measured as a function of the Fe film thickness for an Fe film grown on a vicinal surface [Fe/Ag(1,1,10)] at (a) RT and (b) 5 K. The inset shows a typical MOKE signal. (From Unguris, J. et al., *Phys. Rev. Lett.*, 67, 140, 1991. Copyright 1991 by the American Physical Society. With permission.)

(F) separated by a NM spacer, F1/NM/F2, can be described by an interface bilinear form

$$\mathcal{E}_1 = -J_1 \vec{n}_1 \cdot \vec{n}_2, \quad (2.44)$$

where
- $\mathcal{E}_1$ is the exchange coupling coefficient in erg/cm^2
- $\vec{n_1}$ and $\vec{n_2}$ are unit vectors along the magnetic moments in layers 1 and 2, respectively

Another common coupling equation has a biquadratic form

$$\mathcal{E}_1 = -J_2 (\vec{n}_1 \cdot \vec{n}_2)^2, \quad (2.45)$$

where J_2 describes the strength of biquadratic coupling. Cases of the bilinear coupling J_1 are found having either a positive sign (+) or a negative sign (−). For a + sign, the minimum of the bilinear energy term is reached for a parallel (P) orientation of the magnetic moments; for a − sign, the minimum energy corresponds to antiparallel (AP) magnetic moments. J_2 is almost exclusively found to be positive, and the minimum of the biquadratic energy term is reached when the magnetic moments are oriented perpendicularly to each other. A detailed description of bilinear and biquadratic coupling terms will be carried out in Section 2.3.3.

The equilibrium of the magnetic moments and their dynamic response can be found by using the LLG equations of motion (2.32)

$$\frac{1}{\gamma_{1,2}} \frac{\partial \vec{M}_{1,2}}{\partial t} = -\left[\vec{M}_{1,2} \times \vec{H}_{eff,1,2}\right] + \frac{G_{1,2}}{\gamma_{1,2}^2 M_{s,1,2}} \left[\vec{M}_{1,2} \times \frac{\partial \vec{n}_{1,2}}{\partial t}\right], \quad (2.46)$$

where $\gamma_{1,2}$, $\vec{M_{1,2}}$, and $G_{1,2}$ are the absolute values of the electron gyromagnetic ratios, magnetic moments, and Gilbert damping parameters of layers 1 and 2, respectively. The damping parameter is often expressed in the form of a dimensionless parameter $\alpha = G/\gamma M_s$. The first term on the right-hand side of Equation 2.46 represents the precessional torque per unit area, and the second term represents the well-known Gilbert damping torque per unit area. The effective fields $\vec{H_{eff,1,2}}$ are given by the derivatives of the magnetic Gibbs energy density, $\mathcal{G}$, with respect to the components $(\mathcal{M}_{x,1,2}, \mathcal{M}_{y,1,2}, \mathcal{M}_{z,1,2})$ of the magnetization vector densities $\vec{\mathcal{M}}_{1,2}$, see Equation 2.6,

$$\vec{H}_{eff} = -\frac{\partial \mathcal{G}}{\partial \vec{\mathcal{M}}}. \quad (2.47)$$

$\mathcal{G}$ includes the Zeeman energy of the dc applied magnetic field, demagnetizing fields $\vec{H_{dem}}$, rf magnetic field $\vec{h}$, free energy $\mathcal{F}_a$ given by Equation 2.9, and the interlayer magnetic coupling energy divided by an appropriate film thickness, see Equations 2.10 and 2.66. In order to carry out appropriate partial derivatives in Equation 2.47, the magnetic bilinear and biquadratic coupling terms Equations 2.44 and 2.45 are rewritten by replacing $\vec{n_{1,2}}$ by $\vec{\mathcal{M}}_{1,2}/\mathcal{M}_{s,1,2}$, and they have to be divided by the F film thickness, see Equation 2.10. In ultrathin films with a symmetric rf driving field, any variations of the magnetization across the film thickness can be neglected.

2.3.2 Quantum Well States and Interlayer Exchange Coupling

The first successful model of interlayer exchange coupling was introduced by Mathon et al. [28]. They correctly pointed out that exchange coupling is primarily a property of the NM spacer and is related to the confinement of Fermi surface electrons in the NM. This model was quickly extended to include the spin-dependent electron reflectivity at the F/NM interfaces [29,30]. One has to include the itinerant nature of the 3d, 4sp electrons in the Ferromagnetic (F) layers. The interlayer bilinear coupling, J_1, is given by the difference in energy between the AP and P alignment of the magnetic moments in the F/NM/F structure [31],

$$J_1 = \frac{1}{2A}(E_{\uparrow\downarrow} - E_{\uparrow\uparrow}), \quad (2.48)$$

where A is the area of the film. Calculations of energy differences are simplified by using the *force theorem*. The main problem is how to treat electron correlations self-consistently. The force theorem says that the energy difference between the two configurations is well accounted for by taking the difference in single particle energies. It is adequate to take an approximate spin-dependent potential and to calculate the single particle energies in the P and AP configurations. This difference in energy is very close to that obtained from self-consistent calculations, see the further discussion in Ref. [31]. In fact, this section closely follows Stiles's Section 4.3 in Ref. [31]. This procedure based on the force theorem significantly simplifies the calculation of exchange coupling and interface magnetic anisotropies. In calculations of the interlayer exchange coupling energies, one does not create a large error by neglecting spin–orbit interactions, while in calculations of the interface anisotropies spin–orbit coupling is the crucial ingredient. Single particle energy calculations require one to evaluate the electron energy for four QWS, see Figure 2.7. For thick F layers, one finds large energy contributions. However, these large contributions cancel out in the difference Equation 2.48. In order to avoid mistakes in this procedure, it is better to calculate the cohesive energy of the QWS by subtracting the bulk contributions,

$$\Delta E_{QWS} = E_{tot} - E_F V_F - E_{NM} V_{NM}, \quad (2.49)$$

where $V_{F,NM}$ and $E_{F,NM}$ are the total volumes and bulk energies for F and NM layers, respectively.

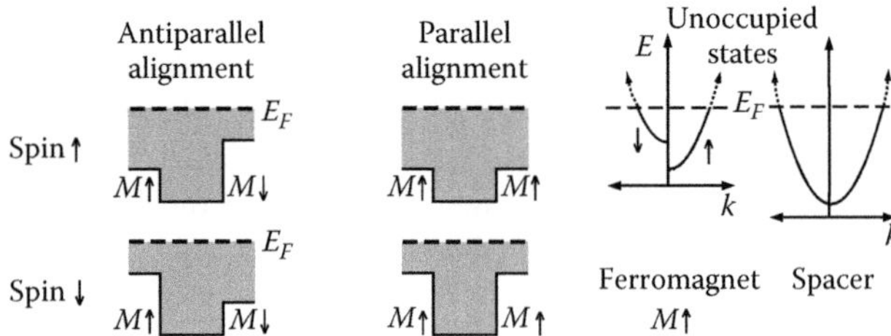

FIGURE 2.7 Quantum wells employed in the calculation of the interlayer exchange coupling energy. These spin-dependent potentials correspond reasonably well to a Co/Cu/Co(001) system. On the left side, the four panels show quantum wells for spin-up and spin-down electrons and for P and AP alignment of the magnetic moments. The gray regions show the occupied states. (With kind permission from Springer Science+Business Media: *Ultrathin Magnetic Structures*, Interlayer exchange coupling, vol. III, 2005, 99, Stiles, M.D.)

2.3.2.1 Quantum Interference

Let us consider a simple 1D model in which an electron having a wave vector $k^{\perp}$ travels inside the NM spacer and is partially reflected at the interfaces, F/NM (interface A) and NM/F (interface B). The reflection coefficients are $R_{A,B}=r_{A,B}\exp(i\phi)_{A,B}$. After multiple interference, the electron density of states (EDS) changes. The phase of the wavefunction after a round trip changes by

$$\Delta\phi = 2k_{\perp}d + \phi_A + \phi_B. \tag{2.50}$$

The amplitude after multiple reflections is given by a sum of round trips

$$\sum_1^{\infty} r_A r_B e^{i\Delta\phi} = \frac{r_A r_B e^{i\Delta\phi}}{1 - r_A r_B e^{i\Delta\phi}}. \tag{2.51}$$

The denominator becomes small when one obtains a constructive interference, $\Delta\phi = 2n\pi$. For energies less than the potential barrier at the interface $r_A = r_B = 1$ and one gets perfect QWS. For energies greater than the barrier energy, the QWS become broader resonances due to an extension of the wave amplitude into the surrounding FM layers. By changing the NM spacer thickness, these states can be made to pass through the Fermi energy, see Figure 2.8, which leads to an oscillatory behavior of the cohesive energy and consequently to an oscillatory interlayer exchange coupling energy. The first clear experimental observation of QWS was presented by Himpsel and Ortega [32,33] using photoemission and inverse photoemission using a nonmagnetic layer on top of a magnetic layer.

In the first approximation, the change in the density of states (DOS) due to interference, $\Delta n(\varepsilon)$, should be proportional to $r_A r_B$ cos ($\Delta\phi$) and to the spacer width d and the DOS per unit length, $(2/\pi)(dk_{\perp}/d\varepsilon)$ [30]. Therefore, $\Delta n(\varepsilon)$ per spin can be written as

$$\Delta n(\varepsilon) \simeq \frac{2d}{\pi}\frac{dk_{\perp}}{d\varepsilon} r_A r_B \cos(\Delta\phi). \tag{2.52}$$

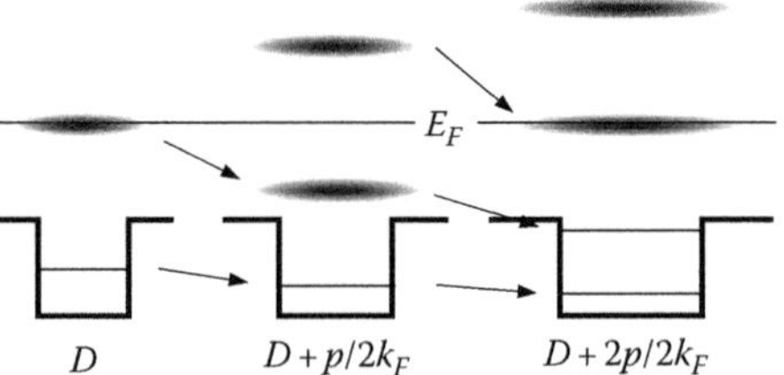

FIGURE 2.8 Evolution of QWS as a function of the film thickness. The solid lines represent bound states (localized in the QW), and resonance states are shown by the "fuzzy ellipses." E_F is the Fermi energy. (With kind permission from Springer Science+Business Media: *Ultrathin Magnetic Structures*, Interlayer exchange coupling, vol. III, 2005, 99, Stiles, M.D.)

For multiple scattering, one has to use the expression in Equation 2.51. It is relatively easy to show that Equation 2.52 can be generalized to [34]

$$\Delta n(\varepsilon) = -\frac{1}{\pi} Im \frac{d}{d\varepsilon}\left[ln(1 - r_A r_B e^{i\Delta\phi})\right]. \tag{2.53}$$

Note that Equation 2.53 equals to Equation 2.52 for small reflection coefficients.

The cohesive energy is then given by

$$E_{coh} = -\int_{-\infty}^{E_F} d\varepsilon(\varepsilon - E_F)\Delta n(\varepsilon). \tag{2.54}$$

Using integration by parts, one gets

$$E_{coh} = \frac{1}{\pi} Im \int_{-\infty}^{E_F} d\varepsilon ln(1 - r_A r_B e^{i\Delta\phi}). \tag{2.55}$$

For fixed thickness d, the integral oscillates rapidly as a function of $k_{\perp}$. Only energies close to the Fermi level will generate nonzero contributions. It can be shown that in these regions for large d [31]

$$E_{coh} = \frac{\hbar\upsilon_F}{2\pi d}\sum_n \frac{1}{n} Re\left((r_A r_B)^n e^{in\Delta\phi(k_F)}\right). \tag{2.56}$$

For small reflection coefficients

$$E_{coh} \simeq \frac{\hbar\upsilon_F}{2\pi d} r_A r_B \cos(2k_F d + \phi_A + \phi_B). \tag{2.57}$$

The interlayer exchange energy, J_1, is then given by adding all cohesive energies (cohesive energy is defined in Equation 2.49 in Figure 2.8, assuming the same reflection coefficients at the A and B interfaces. In the limit of large d

$$J_1 \simeq \frac{\hbar v_F}{4\pi d} Re(R_\uparrow R_\downarrow + R_\downarrow R_\uparrow - R_\uparrow^2 - R_\uparrow^2)e^{i2k_F d}$$
$$= -\frac{\hbar v_F}{4\pi d} Re(R_\uparrow - R_\downarrow)^2 e^{i2k_F d}. \quad (2.58)$$

The exchange coupling in this simple 1D limit is inversely proportional to the film thickness, d, and its associated oscillatory spatial period in d is given by the Fermi spanning vector $2k_F$. In 3D space, one has to take into account all k-vectors parallel to the surface. These k-vectors, resulting from the lattice periodicity, are conserved in going from F to NM regions. In this 3D case, the 1D QWS have additional $\vec{K}$-wave-vectors parallel to the interface. The total cohesive energy per unit area involves the integration of the QWS over the interface Brillouin zone. The integrand for E_{coh} oscillates rapidly with the $\vec{K}$-wave-vectors except on areas of the Fermi surface where opposite sheets of the Fermi surface are nearly parallel. The vector connecting these parts of the Fermi surface are called *critical spanning vectors.* The spanning k-vectors for (001) interfaces for simple metals such as Cu are shown in Figure 2.9.

The exchange coupling involves the difference in cohesive energies (see Equation 2.49) for P and AP configurations of the magnetic moments. In its asymptotic form, this coupling can be written as

$$J \simeq \sum_i \frac{\hbar v_\perp^i \kappa^i}{16\pi^2 d^2} Re\left((R_\uparrow^i - R_\downarrow^i)^2 e^{iq_\perp^i d} e^{i\chi^i}\right), \quad (2.59)$$

where

$v_\perp^i$ are Fermi velocities at the spanning vectors
$q_\perp^i$ is the length of a critical spanning wave-vector
κ^i is the phase associated with the type of critical point
$R_\uparrow^i$ and $R_\downarrow^i$ are corresponding reflectivities

The periods of the observed exchange coupling oscillations as the film thickness is varied are in good agreement with those obtained in de Haas–van Alphen measurements, see Table 4.1 in Ref. [31]. A detailed discussion of calculations of exchange coupling for Co/Cu/Co(001), Fe/Au/Fe(001), and Fe/Ag/Fe(001) systems can be found in Ref. [31]. The quantitative agreement for the exchange coupling between theory and experiment is far from being good. The main reason is that the interfaces in real samples are far from being ideal, and measurements are often not carried out in the asymptotic thickness limit.

Comprehensive studies of exchange coupling and its relationship to QWS were carried out by the Qiu group at the University of California at Berkeley [35] (see references within) using a wedged Cu spacer in Co/Cu/Co(001) structures grown on Cu(001) single crystal substrates. This system was particularly convenient for such studies because Cu has a simple Fermi surface whose sp bands can be easily separated from the other energy bands, see the Fermi surface of Cu in Figure 2.9. Cu and Co can be grown in the (001) orientation with atomically flat interfaces. Angular resolved photoemission spectroscopy (ARPES) of QWS was carried out at the advanced light source (ALS) of the Lawrence Berkeley National Lab. The DOS is significantly increased at energies corresponding to the QWS. This allows one to follow the QWS as a function of energy for different Cu thicknesses. In Figure 2.10, ARPES measurements show the formation of QWS corresponding to the belly direction of the fcc Cu Fermi surface, see Figure 2.9. The study was carried out for 20 ML thick Co grown on a Cu(001) substrate and with a Cu wedge grown on top of the Co layer. The ARPES oscillations have clearly shown the QWS corresponding to the sp electrons in the Cu layer. The

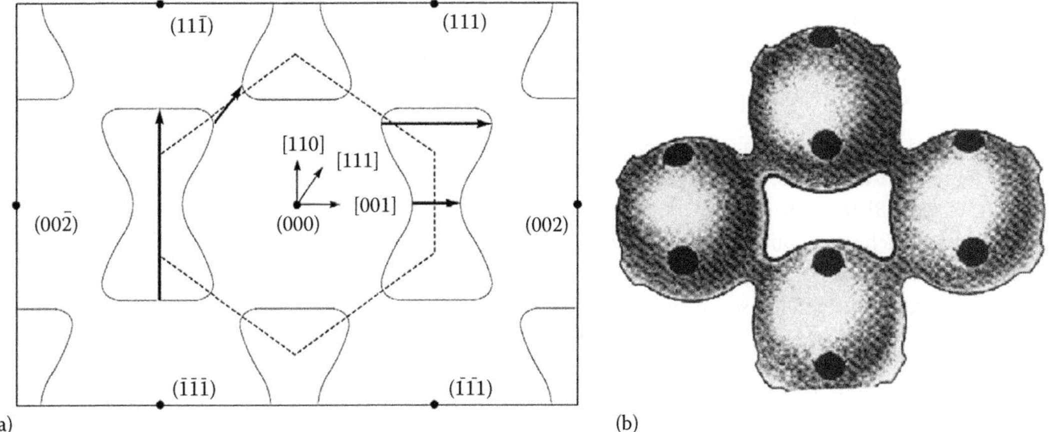

FIGURE 2.9 (a) The [1$\bar{1}$0] cross-section of the fcc Cu Fermi surface. The hexagon of straight lines outlines the first Brillouin zone. The solid dots represent reciprocal k-vectors. All three important orientations are present. The critical spanning vectors in the extended Brillouin zone are denoted by the solid arrows. Along the [001] direction, the two critical spanning vectors are located at the belly and neck of the Fermi surface. (b) A part of the extended Brillouin zone in 3D. (With kind permission from Springer Science+Business Media: *Magnetic Heterostructures: Advances and Perspectives in Spinstructures and Spintransport*, Exchange coupling in magnetic multilayers, vol. 227, 2008, 183, Heinrich, B; Heinrich, B., *Can. J. Phys.*, 2000.)

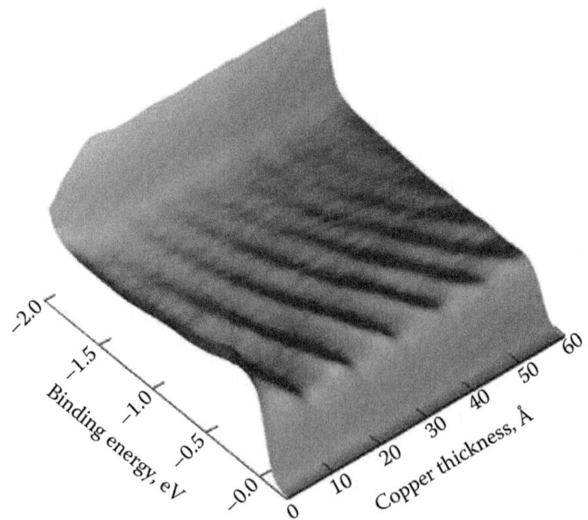

FIGURE 2.10 Photoemission spectra obtained along the surface normal corresponding to the belly direction of the fcc Cu Fermi surface [35]. Oscillations in intensity as a function of the Cu layer thickness and electron energy demonstrate the formation of QWS. (From Qiu, Z.Q. and Smith, N.V., *J. Phys.: Cond. Matter*, 14, R169, 2002. With permission.)

periodicity of the oscillations corresponding the belly direction (photoemission spectra were obtained along the surface normal) was found to be 5.88 atomic layers at the Fermi level, and this is exactly the periodicity of the long period exchange coupling oscillations observed for Co/Cu/Co(001) systems.

The interlayer exchange coupling rarely behaves as outlined in the above theory. The interface roughness often smears out the presence of short wavelength oscillations and even significantly decreases the strength of long wavelength interlayer coupling oscillations. Various contributions to magnetic interlayer coupling and the role of interface roughness and interface chemistry are extensively described in Ref. [36]. Here, the discussion will be limited to biquadratic exchange coupling.

2.3.3 Biquadratic Exchange Coupling

The presence of biquadratic exchange coupling was observed at the same time by Heinrich et al. [37] for Co/Cu/Co(001) trilayers and by Ruehrig et al. [38] for Fe/Cr/Fe(001) trilayers. The evidence for biquadratic exchange coupling in Ref. [37] was obtained using magnetization loops. In order to properly explain the observed critical fields, one needed to use an angular-dependent bilinear exchange coupling parameter in the form of

$$J(\theta) = J_1 - J_2\cos(\theta), \tag{2.60}$$

where θ is the angle between the magnetic moments. Consequently, the corresponding exchange energy was given by

$$E(\theta) = -J(\theta)\cos(\theta) = -J_1\cos(\theta) + J_2\cos^2(\theta). \tag{2.61}$$

Ruehrig et al. observed a perpendicular orientation of the magnetic moments in an Fe/wedge Cr/Fe(001) sample in which the Cr interlayer was grown with a linearly variable thickness. They explained the observed perpendicular configuration by using

$$E(\theta) = -J(\theta)\cos(\theta) = -J_1\cos(\theta) - J_2\sin^2(\theta). \tag{2.62}$$

Clearly, these two concepts are identical. Slonczewski soon after that proposed a theoretical interpretation [39]. He realized that fluctuations in the NM interlayer thickness could result in an additional coupling term. The nonmagnetic layers at different parts of the sample have different thicknesses and consequently generate different strengths of magnetic coupling between the two ferromagnetic layers. Short-wavelength oscillations can even result in changing the coupling from ferromagnetic to AFM. His model is applicable when lateral variations in the bilinear interlayer exchange field are suppressed by corresponding lateral variations of the bulk exchange field. This means that local angular variations from the average direction of the magnetic moments are small so that, in this case, the problem can be treated by perturbation theory. The magnetic moments are frustrated across the film surface by a variable interlayer coupling. Consequently, there is an additional energy term that prefers to orient the magnetic moments in the F layers perpendicularly to each other. This additional coupling has then an angular dependence given by $\cos^2(\theta)$, for which the name "biquadratic exchange coupling" was coined. Its strength is given by the competition between variations in the interlayer exchange coupling field, $\Delta J_1/M_s d$, and the in-plane intralayer exchange field, $2Ak^2/M_s$. The length of the k-vector, k, is given by the average size of atomic terraces, L; $k = \pi/L$. Slonczewski has shown that the interlayer exchange coupling fluctuations are decreased due to exchange averaging. In the simplest form, the strength of the biquadratic exchange coupling can be expressed by

$$J_2 = \frac{4}{\pi}\Delta J_1 \frac{\Delta J_1/M_s d}{2Ak^2/M_s}. \tag{2.63}$$

Notice that the large fraction describes the exchange averaging effect. A more general description can be found in Ref. [39]. The above expression shows that biquadratic coupling has only a positive sign, and therefore it always encourages a perpendicular orientation of the magnetic moments. The angle between the magnetic moments is given by a competition between the bilinear, biquadratic magnetic couplings, and the magnetic anisotropies, see Section 2.3.4. In zero field, this angle can range continuously from 0 to π.

It is often believed that biquadratic exchange coupling occurs only from short wavelength exchange coupling oscillations where the exchange coupling changes its sign between two subsequent atomic terraces. This is not correct. Any lateral variations in magnetic coupling strength (including ferromagnetic coupling) will result in biquadratic exchange coupling. Once the magnetic moments are arranged in a noncollinear state, the magnetic frustration due to an inhomogeneous magnetic coupling strength results in biquadratic magnetic coupling.

The Slonczewski idea of additional energy terms due to imperfect interfaces is more general. It has been shown [40] that "in any system that exhibits a lateral inhomogeneity, one must expect additional energy terms. They originate from intrinsic magnetic energy terms that fluctuate in strength across the sample interface. These additional terms have the next higher angular power compared with that of the intrinsic term, and they have only one sign." The power of the higher order angular term has to satisfy the requirements of the sample symmetry including time inversion symmetry. Here are several examples: (a) Variations of the interlayer exchange coupling results in a $\cos^2(\theta)$ angular term; (b) variations in a uniaxial interface perpendicular anisotropy results in an angular dependent term having the form $\cos^4(\vartheta)$, where ϑ is the angle between the magnetic moment and the film normal: this term has an easy axis at 45° from the surface normal; (c) variation in the in-plane uniaxial anisotropy results in an in-plane fourfold anisotropy with the easy axis rotated by 45° with respect to the uniaxial anisotropy axis; (d) variation in the EB field results in a uniaxial anisotropy with the easy axis perpendicular to the EB field direction.

2.3.4 Measuring Techniques

2.3.4.1 Static Measurements

In SQUID and VSMs, the total magnetic moment is measured, usually along the direction of the applied field. In the magneto optical Kerr effect (MOKE), the magnetic moment is commonly measured in the polar and longitudinal configurations. In the longitudinal configuration, the dc field is applied parallel to the film surface, and it is usually applied in the plane of incidence of the laser beam. In the polar configuration, the incident laser beam impinges nearly perpendicularly on the film surface, and the MOKE signal is mainly sensitive to the magnetic moment component that is perpendicular to the film surface.

The discussion in this Section will be limited to the micromagnetics of trilayer structures consisting of a crystalline ultrathin film F1/NM/F2 structure. For simplicity, these calculations will be limited to cubic materials with the film surface oriented in the (001) plane. Films of 3d transition group elements are often, but not exclusively, grown on (001) templates. This is a particular case but it involves all the ingredients needed to formulate calculations for any other configurations involving an arbitrary direction of the applied field and crystalline orientation. The discussion below should be viewed as an example taken from a user's manual.

The total free energy per unit area for the in-plane configuration (the magnetic moments $\vec{M_{s1}}$ and $\vec{M_{s2}}$, and the field $\vec{H}$ applied parallel to the film surface) is given by

$$\mathcal{F} = \sum_{i=1,2}\left[-\frac{K_{1,i}^{\parallel}}{2}\left(\alpha_{X,i}^4+\alpha_{Y,i}^4\right)-K_{u,i}^{\parallel}\frac{\left(\vec{n}_{u,i}\cdot\vec{M}_i\right)^2}{M_{s,i}^2}-K_{u,i}^{\perp}\alpha_{Z,i}^2-\vec{M}_i\cdot\vec{H}\right]d_i - J_1\left(\vec{m}_1\cdot\vec{m}_2\right)+J_2\left(\vec{m}_1\cdot\vec{m}_2\right)^2, \tag{2.64}$$

where

$\alpha_{X,i}$, $\alpha_{Y,i}$, and $\alpha_{Z,i}$ are the directional cosines between the magnetization vector M_i and the crystallographic axes [100], [010], and [001], respectively

$K_{1,i}^{\parallel}$, $K_{u,i}^{\parallel}$, and $K_{u,i}^{\perp}$ are the parameters that describe the in-plane effective fourfold (cubic) anisotropy, the in-plane uniaxial anisotropy, and the perpendicular uniaxial anisotropy

$\vec{n_{u,i}}$ are the directions of the in-plane uniaxial axes

d_i are the film thicknesses

$\vec{m_1}$ and $\vec{m_2}$ are unit vectors directed along the magnetizations of the coupled films

J_1 and J_2 are the bilinear and biquadratic coupling coefficients

The indices $i=1$ and 2 describe the properties of the layers F1 and F2, respectively. In ultrathin films, the magnetic moments across the film are locked together by intralayer exchange coupling, and they can be considered to be giant magnetic molecules [2]. For ultrathin films, the role of the interface anisotropies associated with a uniformly magnetized sample can be included in the effective anisotropy parameter, see Equation 2.11. The energy expression in Equation 2.64 is valid for a wide range of magnetic ultrathin films such as Fe on Ag(001) [2] and GaAs(001) [41,42] templates. One can easily generalize it by using the appropriate film symmetry.

The static equilibrium is found by minimizing the total Gibbs energy with respect to the angles φ_{M1} and φ_{M2} for a given angle φ_H, see Figure 2.4. For the in-plane configuration, $\theta_{M1,2}=\theta_H=\pi/2$.

There are a number of minimization procedures available, and they are usually implemented by individual groups according to their liking. One should realize that minimum energy solutions can exhibit metastable states. Usually, one looks for the lowest energy state. This means that no hysteresis is present in the magnetization measurements. Experimentally, this is not often the case, especially when using films that were grown on GaAs(001) substrates [41]. Therefore, it is imperative to carry out MOKE measurements with the specimen in the lowest energy state. The lowest energy state for a given magnetic field can be achieved by cycling the magnetic state at the given applied field using a transverse ac magnetic field, which increases in amplitude to some preselected maximum, and then the amplitude is gradually decreased to zero. This has to be repeated for each applied dc field. An example of such a procedure can be found in Ref. [41] using exchange coupled GaAs/Fe/Au/Fe/Au(001) structures.

Several examples of hysteresis loops using exchange coupled magnetic bilayers F1/NM/F2 are shown in Figure 2.11. The external field was applied along the magnetic easy axis of the thicker film. In that case, the thicker film remains oriented close to the easy axis. It is the thinner film that undergoes a full angular dependence due to the presence of interlayer AFM coupling, see the insets in Figure 2.11. Since the two magnetic moments are different it is easy to identify the contributions from each individual layer.

The family of curves in Figure 2.11 shows the role of biquadratic magnetic coupling. When the biquadratic magnetic coupling becomes comparable to the bilinear coupling, one is not

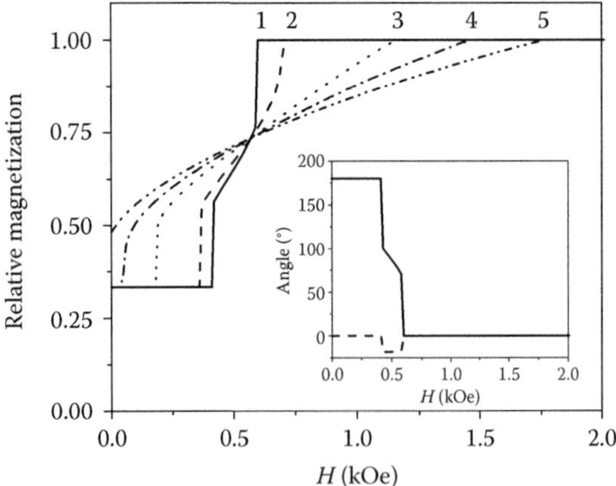

FIGURE 2.11 Simulations of magnetization curves using a set of exchange coupling parameters. The lines 1, 2, 3, 4, and 5 correspond to $J_2 = 0.0$, 0.01, 0.04, 0.06, and 0.08 erg/cm², respectively. The bilinear coupling parameter was kept constant, $J_1 = -0.1$ erg/cm². The calculations were carried out for the magnetic parameters obtained using GaAs/8Fe/Au/16Fe/Au(001) structures [42], where the integers represent the number of atomic layers. 16Fe: $K_1^{\|} = 3.1 \times 10^5$ erg/cm³; 8Fe: $K_1^{\|} = 1.33 \times 10^5$ erg/cm³. $4\pi M_s = 21.5$ kOe. In-plane uniaxial anisotropies were omitted in order to keep the easy magnetic axis in both films directed along the [100] crystallographic axis. The applied field was oriented along the magnetic easy axis [100]. The inset shows the field dependence of the magnetization angle for $J_1 = -0.1$ erg/cm² and $J_2 = 0$. The dashed line corresponds to the 16Fe film, and the solid line corresponds to the 8Fe film. Note that the first jump brings the magnetic moment of the thinner 8Fe film from the P orientation over the first hard axis, [110], close to the second easy axis, [010]. The second jump corresponds to pulling the magnetic moment over the second hard axis, [1$\bar{1}$0], resulting in an AP configuration of the magnetic moments. (With kind permission from Springer Science+Business Media: *Magnetic Heterostructures: Advances and Perspectives in Spinstructures and Spintransport*, Exchange coupling in magnetic multilayers, vol. 227, 2008, 183, Heinrich, B.)

able to achieve an AP configuration of magnetic moments in small magnetic fields. In zero field, the biquadratic coupling can lead to an angle between the two magnetic moments ranging from −180° to 90°. The biquadratic magnetic coupling can also affect the approach to saturation. For even small J_2, the saturation field is reached without a jump in the magnetic moment, see Figure 2.11. The saturation field can be easily determined when there is no discontinuity in the magnetic moment. The deviation from saturation can be treated using a small angle expansion. It is easy to show that the saturation field along the easy direction can be described by the simple expression

$$H_{sat} + \frac{2K_{eff}}{M_s} = -\frac{J_1 - 2J_2}{M_s}\left(\frac{1}{d_1} + \frac{1}{d_2}\right), \tag{2.65}$$

where it has been assumed that the saturation magnetization M_s is the same for both films. K_{eff} is an effective anisotropy field obtained from the minimization of Equation 2.64.

In F1/NM/F2 structures, one can measure only negative bilinear J_1 (AFM coupling) using MOKE. One is able to measure a positive (ferromagnetic) bilinear coupling using spin "engineered structures." An additional magnetic F0 layer plus a nonmagnetic metal spacer that creates a large AFM coupling between F0 and F1 is needed [43], e.g., F0/NM/F1/NM/F2. In these structures, the magnetic moments in F1 and F2 in zero-applied field are oriented AP to the magnetization in F0. One needs to apply a dc field along the direction of the magnetic moment of the F0 layer to overcome the ferromagnetic coupling and thus to orient the moments in F1 and F2 in an AP configuration. Strong AFM coupling can be achieved by using ultrathin Ru spacers [43].

2.3.4.2 Ferromagnetic Resonance

The magnetic coupling can be measured by means of rf techniques, see Refs. [2,17]. In FMR, one usually fixes the microwave frequency and sweeps the field. However, with network analyzers (NAs), this is not a limitation; one can set the field and sweep the frequency. When the field is held constant, the angle between the magnetic moments is fixed. This is a simpler situation compared to a regular FMR measurement (holding the frequency constant and sweeping the field) where the angle between the magnetic moments changes in noncollinear configurations for samples, which are not fully saturated (usual in measurements at low microwave frequencies). For a saturated sample, the difference between constant field and constant frequency sweeps is minimal. Interpretation of FMR results can be carried out using rf solutions of Equation 2.46. The basic principles of FMR including the rf susceptibility are outlined in Section 2.2.4. Solutions of the LLG equations of motion 2.32 require the internal effective fields given by Equation 2.6 using the free energy density Equation 2.10 along with the Zeeman energy. The Gibbs energy density for the individual layers i (the in-plane configuration) is

$$\mathcal{G}_i = -\frac{K_{1,i}^{\|}}{2}\left(\alpha_{X,i}^4 + \alpha_{Y,i}^4\right) - K_{u,i}^{\|}\frac{\left(\vec{n}_{u,i}\cdot\vec{M}_i\right)^2}{M_{s,i}^2} - K_{u,i}^{\perp}\alpha_{Z,i}^2 - \vec{M}_i\cdot\vec{H}$$
$$-\frac{\left[J_1\vec{m}_1\cdot\vec{m}_2 + J_2\left(\vec{m}_1\cdot\vec{m}_2\right)^2\right]}{d_i}, \tag{2.66}$$

where $\vec{M}_i$ is the magnetization densities. The partial derivatives for $H_{i,eff}$ are taken with respect to $M_{x,i}$, $M_{y,i}$, and $M_{z,i}$. In FMR, the rf transversal components are usually negligible compared to the magnetization component parallel to the x-axis (static effective fields). The coupled equations for the rf magnetization components lead to two solutions. The precessional motions are coupled and result in an acoustic mode in which the magnetic moments in the two layers precess in phase and in an optical mode in which the magnetic moments precess in antiphase. The simplest interpretation of the coupling can be obtained in the saturated case when the dc magnetic moments are parallel to

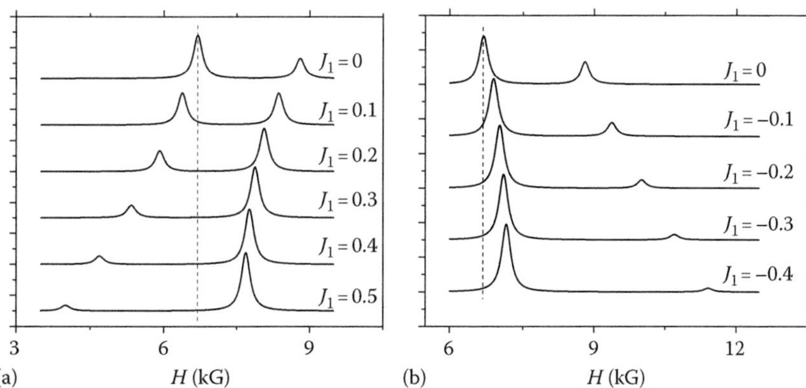

FIGURE 2.12 Simulations of acoustic and optical resonance absorption peaks at $f=36\,\text{GHz}$ as a function of the bilayer exchange coupling strength in an FM1/NM/FM2 structure. In panel (a), $J_1=0.0$, 0.1, 0.2, 0.3, 0.4, and $0.5\,\text{erg/cm}^2$. In panel (b), $J_1=0.0$, -0.1, -0.2, -0.3, and $-0.4\,\text{erg/cm}^2$. The following magnetic parameters were used: 16Fe: $K_1^{\parallel}=3.1\times10^5\,\text{erg/cm}^3$, $K_{u,s}^{\perp}=0.88\,\text{erg/cm}^2$, and $K_u^{\parallel}=3.3\times10^4\,\text{erg/cm}^3$; 8Fe: $K_1^{\parallel}=1.33\times10^5\,\text{erg/cm}^3$, $K_{\perp,s}=0.82\,\text{erg/cm}^2$, and $K_{u,eff}^{\parallel}=-1.44\times10^6\,\text{erg/cm}^3$. $4\pi M_s=21.5\,\text{kG}$, $g=2.09$, and $\alpha=0.009$. The in-plane uniaxial easy axes for the 16Fe and 8Fe films were oriented along the [110] and [1$\bar{1}$0] directions, respectively. The applied field was oriented along the [1$\bar{1}$0] crystallographic axis. (With kind permission from Springer Science+Business Media: *Magnetic Heterostructures: Advances and Perspectives in Spinstructures and Spintransport*, Exchange coupling in magnetic multilayers, vol. 227, 2008, 183, Heinrich, B.)

the applied field. The isolated films F1 and F2 must have different resonant frequencies (fields) in order to be able to observe acoustic and optical modes. If the resonance frequencies (fields) of the two films are exactly the same, then the strength of the optical mode is zero. Different resonant fields are easy to establish by choosing different film thicknesses for the films F1 and F2. The interface anisotropies scale with $1/d$ and in consequence the isolated films exhibit two separate resonance frequencies (fields), and the optical mode becomes observable. The sign of the coupling can be determined from the relative positions of the acoustic and optical modes. Calculated spectra for a magnetic double layer F1/NM/F2 for P and AP coupling are shown in Figure 2.12. For AP coupling, the acoustic and optical peaks move to higher magnetic fields at a fixed FMR frequency. The acoustic peaks keep increasing their intensity with increasing AP coupling whereas the optic peaks get weaker with increasing coupling. The acoustic peaks gradually approach a fixed magnetic field point, which is located between the resonance peaks for the uncoupled films. For P coupling, this trend is exactly opposite. The resonance fields decrease with an increasing coupling. It is the film with a higher resonance field, which approaches the fixed point.

In the saturated state (collinear magnetic moments), the overall strength of the interlayer coupling, J_{eff}, is given by the superposition of bilinear and biquadratic interlayer couplings,

$$J_{eff}=J_1-2J_2. \tag{2.67}$$

2.3.5 Experimental Results

Only several examples will be shown to demonstrate the basic behavior and role of interfaces in the interlayer exchange coupling in magnetic ultrathin film structures. No attempt is made to account for the complete work done in this field. A more detailed presentation containing a large number of references can be found in Ref. [36].

2.3.5.1 Simple Normal Metal Spacers: Cu and Ag

2.3.5.1.1 Cu

It was pointed out in Section 2.3.2 that comprehensive studies of exchange coupling and its relationship to QWS were carried out by the Qiu group at the University of California, Berkeley [35] (see references within). They used a wedged Cu spacer in Co/Cu/Co(001) structures grown on Cu(001) single crystal substrates. The copper spacer was oriented along the [001] crystallographic direction and provided two critical spanning vectors at the belly and at the neck of the Fermi surface, see Figure 2.9. The belly, 5.88, and neck, 2.67, atomic layer periodicities can be understood by means of the extended Brillouin zone picture, see the solid arrows in Figure 2.9. In this case, one subtracts from the regular spanning vector inside the first Brillouin zone (from belly to belly, from neck to neck) a k-vector having the atomic layer periodicity ($4\pi/a$). The oscillatory period for the exchange coupling through Cu is given by

$$2k^e d_{Cu}-\phi_A-\phi_B=2\pi n, \tag{2.68}$$

where

- $k^e=k_{BZ}-k$, $k_{BZ}=2\pi/a$ is a Brillouin-zone vector
- n is an integer
- $\phi_{A,B}$ are the phase shifts of the electron wavefunctions upon reflection at the two boundaries of the potential well formed by the Cu layer surrounded by Co and vacuum
- a is the lattice spacing of Cu

Equation 2.68 explains the long and short wavelength coupling oscillation periods as due to the belly and neck spanning

k-vectors, respectively. They are caused because the strength of the exchange coupling is evaluated at discrete atomic layer separations. This is often called the *alising effect*. Oscillations in ARPES intensity as a function of the Cu layer thickness clearly demonstrated the formation of QWS along the belly and neck of the Fermi surface, e.g., see Figure 2.10, showing the presence of long wavelength oscillations across the belly of the Fermi surface.

The QWS form the underlying basis for the presence of the interlayer exchange coupling. To insure the direct comparison of the exchange coupling periodicity with the QWS variation as a function of the Cu spacer thickness, half of the wedged sample was covered by a Co film 3 ML thick. XMCD measurements are only surface sensitive, and consequently FM and AFM coupling can be determined by monitoring the XMCD signal coming from the 3 ML thick Co. Images of the DOS (using photoemission measurements) at the belly and neck of the Fermi surface were obtained by scanning the photon beam across the Cu wedge on the Co/Cu side of the wedged Cu film. Figure 2.13c shows the observed XMCD signal with maxima and minima intensities corresponding to AP and P couplings, respectively. Clearly long and short wavelength oscillations are easily visible, see Figure 2.13. The coupling between the ferromagnetic layers is determined by the energy difference between the P and AP alignment of the magnetic moments

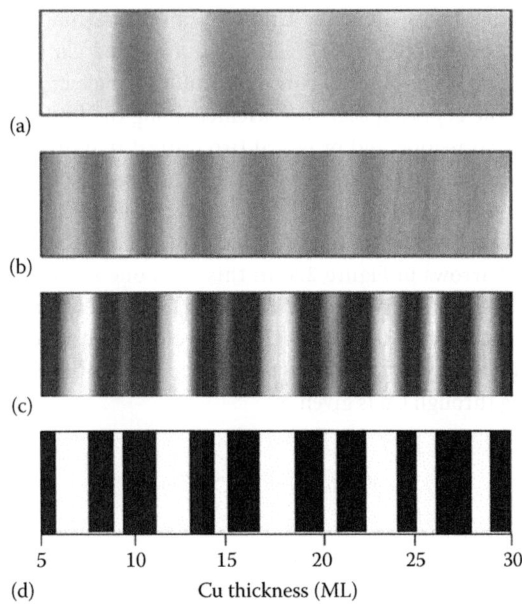

FIGURE 2.13 (a) QWS at the belly of the Cu Fermi surface. (b) QWS at the neck of the Cu Fermi surface. (c) XMCD from the top three atomic layers of Co evaporated over the Cu-wedged spacer. See further details in Ref. [35]. The dark and light regions correspond to ferromagnetic and antiferromagnetic coupling. (d) Calculated interlayer coupling. Notice the remarkable agreement between the theoretical predictions and experiment for the sign of the exchange coupling. (From Qiu, Z.Q. and Smith, N.V., *J. Phys.: Cond. Matter*, 14, R169, 2002. With permission.)

$$2J \sim E_{AP} - E_P = \int_{-\infty}^{E_F} E\Delta D dE, \tag{2.69}$$

where $\Delta D = D_{AP} - D_P$ is the difference of the DOS between P and AP alignment of the magnetic moments. For the P configuration of the magnetic moments, the minority spins are confined and form well-defined QWS. At the neck of the Fermi surface, the minority spins are completely confined by the spin potential of the Co. At the belly of the Fermi surface, they are only partially confined. Whenever the energy of a QWS crosses the Fermi level, it adds energy to E_P making the P configuration of the magnetic moments unfavorable. Fitting the MXCD data using

$$J = -\frac{A_1}{d^2}\sin\left(\frac{2\pi}{\Lambda_1} + \Phi_1\right) - \frac{A_2}{d^2}\sin\left(\frac{2\pi}{\Lambda_2} + \Phi_2\right), \tag{2.70}$$

resulted in $\Lambda_1 = 5.88$ ML and $\Lambda_2 = 2.67$ ML. MXLD is not able to determine the strength of the coupling.

2.3.5.2 Spin Density Wave in Cr

Fe-whisker/Cr/Fe(001) wedge structures, see Figure 2.14, have played a crucial role in the study of interlayer exchange coupling. The Cr spacers are expected to contain SDWs with short and long wavelength periods, see the spanning k-wavevectors in Figure 2.15.

Unguris et al. [44] used scanning electron microscopy with polarization analysis (SEMPA) to study Fe-whisker/Cr/Fe(001) samples. They showed that the exchange coupling oscillates with a short-wavelength period of 2 ML. The SEMPA images revealed in a very explicit way the presence of both short-wavelength and long-wavelength oscillations in the thickness range from 5 to 80 ML, see Figure 2.16.

The period of the short-wavelength oscillations, $\lambda = 2.11$ ML, was found to be slightly incommensurate with the Cr lattice

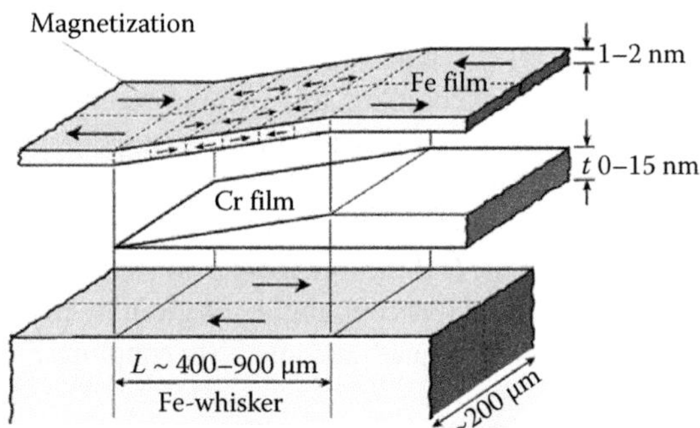

FIGURE 2.14 A schematic view of an Fe-whisker/Cr/Fe(001) sample containing a wedged Cr spacer as employed by the NIST group for their SEMPA measurements. The arrows show the directions of the magnetization. The Fe-whisker is demagnetized by a single 180° domain wall. The direction of the magnetization in the Fe film follows the sign of the local interlayer exchange coupling. (From Li, J. et al., *Phys. Rev. Lett.*, 102, 207206, 2009. Copyright 2009 by the American Physical Society. With permission.)

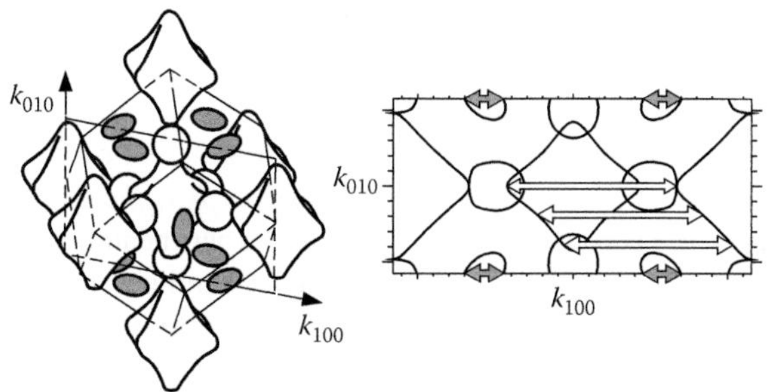

FIGURE 2.15 The left panel shows a representation of the Cr Fermi surface in the paramagnetic state. The gray-shaded areas are ellipsoids centered at the *N* points in the bcc reciprocal lattice. The right panel shows a slice through the Fermi surface as indicated in the left panel. The gray-shaded arrows are the critical spanning vectors at the *N*-centered ellipsoids, and the white arrows indicate the nested parts of the Fermi surface that give rise to SDW antiferromagnetism. (With kind permission from Springer Science+Business Media: *Ultrathin Magnetic Structures*, Interlayer exchange coupling, vol. III, 2005, 99, Stiles, M.D.)

spacing; the period of the long wavelength oscillations was found to be 12 ML. These two basic periods were expected from the complex Cr Fermi surface, see Figure 2.15. The incommensurate nature of the short wavelength oscillations in Cr, $\lambda = 2.11$ ML, results in phase slips of exchange coupling at 24 and 44 atomic layers of Cr (see Figure 2.16 and further details in Ref. [45]).

Heinrich et al. (Simon Fraser University [SFU] group) carried out quantitative exchange coupling measurements on Fe-whisker/Cr/Fe(001) samples [2,46]. The objective of the SFU group was to grow samples having the best available interfaces to measure quantitatively the strength of the exchange coupling and to compare these coupling strengths with ab initio calculations that explicitly include the presence of spin-density waves in the Cr. The requirement of smooth interfaces limited our study to samples that were grown on Fe-whisker templates with the Cr spacers terminated at an integral number of Cr atomic layers. It was found that the strength of the exchange coupling through the Cr(001) spacer even for these smooth Cr layers was extremely sensitive to small variations in the growth conditions. Cr ultrathin films do not possess a robust spin-density wave; the spatial variation of the spin moments is extremely sensitive to the interface structures. In our studies, we concentrated on samples for which the Cr thickness ranged from 4 to 13 atomic layers where the role of interfaces was most pronounced. The measured exchange coupling was found to be reproducible only in those samples that exhibited true layer by layer growth, see Ref. [46]. Quantitative Brillouin Light Scattering (BLS) studies of the exchange coupling in Fe-whisker/Cr/Fe(001) have been discussed in Refs. [2,46,47]: the main results are shown in Figure 2.17. These studies showed that the exchange coupling through Cr(001) contains both oscillatory bilinear J_1 and positive biquadratic J_2 exchange coupling terms. The exchange coupling first becomes AFM at 4 ML of Cr. For Cr spacer thicknesses $d_{Cr} < 8$ ML, the strength of the short-wavelength oscillations is quite weak, ~0.1 erg/cm^2. The exchange coupling in this range is AFM. This is due to the presence of an AFM long-wavelength bias. This AFM bias is peaked in the range between 6 and 7 ML. It is interesting to note that the strength of the long-wavelength AFM bias is very nearly the same as that observed in Fe/Cr/Fe(001) epitaxial multilayers prepared by sputtering where a relatively large interface roughness annihilates the presence of the short wavelength oscillations [48]. Exchange coupling in

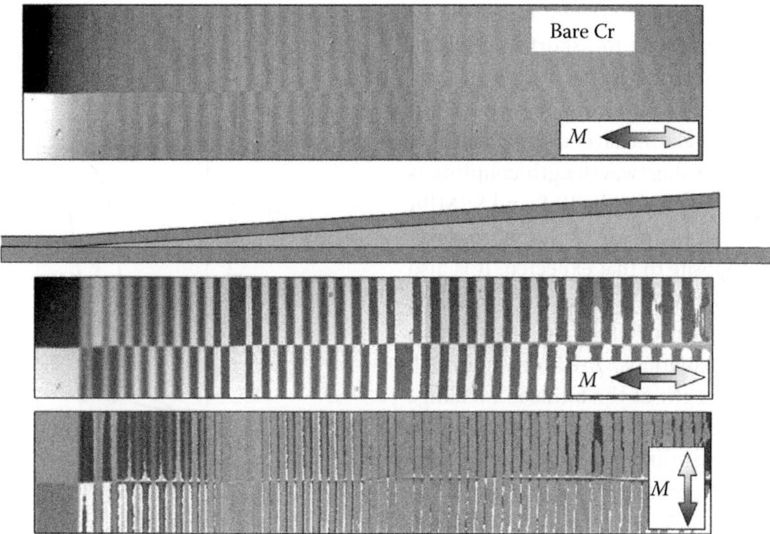

FIGURE 2.16 The direction of the magnetic moment in an Fe film grown over a Cr wedge is shown in the second SEMPA image. The top SEMPA image shows the presence of a magnetic moment in the Cr wedge grown on an Fe-whisker substrate. White and black contrast indicate P and AP orientations of the magnetic moment with respect to the Fe-whisker magnetization. Note in the bottom SEMPA image that at the boundaries between the P and AP orientation of the magnetic moments a perpendicular component of the magnetic moment is present. This is caused by magnetic frustration due to a partial completion of the top Cr/Fe interface, which results in a strong biquadratic exchange coupling.

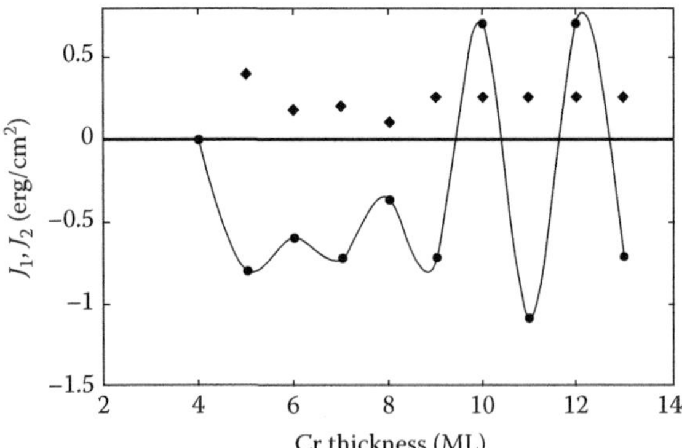

FIGURE 2.17 The thickness dependence of the bilinear J_1 (•) and biquadratic J_2 (♦) exchange coupling. The biquadratic coupling can be measured only for AFM coupled samples. The values of J_2 for FM-coupled samples (10 and 12 ML) were assumed to be the same as those for the AFM-coupled samples whose spacer thicknesses were 9, 11, and 13 ML of Cr. Note that the coupling becomes AFM for thicknesses greater than 4 ML, and the thickness dependence of J_1 has a broad maximum around 7 ML of Cr. This behavior is caused by long wavelength oscillations in the coupling strength. Strong short wavelength oscillations appear for Cr thicknesses greater than 9 ML.

the sputtered films exhibited long-wavelength oscillations with a rapidly decreasing coupling strength for thicknesses greater than 10 ML of Cr. For Cr spacers thicker than 8 ML, $d_{Cr} > 8$ ML, the exchange coupling in films grown on a whisker substrate is dominated by the short-wavelength oscillations. In this thickness range, the samples are AFM coupled for an odd number of Cr atomic layers and ferromagnetically (FM) coupled for an even number of Cr atomic layers.

The coupling between the Fe and Cr atoms at the Fe/Cr interface is expected to be strongly AFM [49] and, consequently, the spin-density wave in Cr is locked to the orientation of the Fe magnetic moments. Since the period of the short-wavelength oscillations is close to 2 ML, one would expect antiferromagnet (AF) coupling for an even number of Cr atomic layers and FM coupling for an odd number of Cr atomic layers. For the period 2.11 ML, the first-phase slip in the shortwavelength coupling is predicted to occur at 20 ML. Surprisingly, the BLS and SEMPA measurements showed clearly that the phase of the short-wavelength oscillations is exactly opposite to that expected. It is also important to note that the strength of the exchange coupling was found to be much less than that obtained by Stoeffler and Gautier using first-principle calculations, $J_1 \sim 30\,\text{erg/cm}^2$ [50]. Our studies showed that the strength of the bilinear exchange coupling J_1 is very sensitive to the initial growth conditions: a lower initial substrate temperature resulted in a larger exchange coupling strength. The bilinear exchange coupling could be changed by as much as a factor of 5 by varying the substrate temperature during the growth of the first Cr atomic layer [51]. This behavior led us to believe that the atomic formation of the Cr layer was more complex than had been previously acknowledged. Angular-resolved Auger spectroscopy (ARAES) [52], STM [53], and proton-induced ARAES [54] have shown that the formation of the Fe/Cr(001) interface is strongly affected by an interface atom exchange mechanism (interface alloying). Freyss et al. [55] investigated the phase of the exchange coupling for intermixed Fe/Cr interfaces. The calculations were carried out using two mixed atomic layers Fe/$Fe_{1-x}Cr_x$/$Cr_{1-x}Fe_x$/Cr. They were able to account for two important experimental observations. First, the crossover to AFM coupling and onset of short-wavelength oscillations was predicted to occur at 4–5 ML of Cr, in good agreement with our observations, see Figure 2.17, and in agreement with NIST studies using the SEMPA imaging technique. Second, the phase of the coupling was reversed for a concentration $x \gtrsim 0.2$. AFM and FM coupling was obtained at an odd and an even number of Cr atomic layers, respectively, in perfect agreement with experiment.

2.4 Exchange Bias

A metallic film in contact with an AF can display a hysteresis loop shifted along the field axis after cooling in an applied field. A typical example is shown in Figure 2.18. In this measurement, Jungblut et al. [56] used a ferromagnetic (F) permalloy (py) film, $Ni_{79}Fe_{21}$, and an AF $Fe_{43}Mn_{57}$ film in a crystalline sandwich structure grown on a crystalline Cu(111) substrate.

The hysteresis loop shows two clear features. The loop is shifted by $H_{eb} \sim 275$ Oe, and the hysteresis loop is 180 Oe wide. The unidirectional shift of the magnetization reversal loop is called EB.

2.4.1 Meiklejohn and Bean Model

EB was discovered by Meiklejohn and Bean [57] in 1956 in their study of oxidized Co particles. They explained EB by means of the simplified picture shown in Figure 2.19.

The Meikeljohn and Bean model (MB) can be easily quantified using the total energy for the F/AF structure

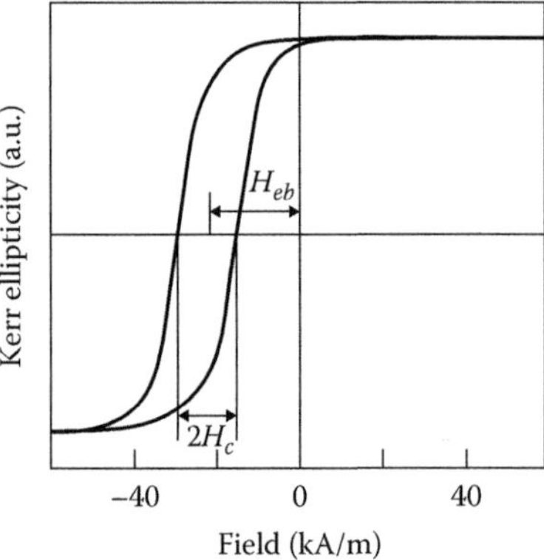

FIGURE 2.18 The magnetization reversal loop for a py film exchange coupled to an antiferromagnetic (AFM) $Fe_{43}Mn_{57}$ film in a crystalline structure grown on Cu(111). Note that the magnetization reversal loop is shifted by 275 Oe with respect to zero field. This is called the EB effect. (Courtesy of Coehoorn, R., Phillips Research Laboratory; From Jungblut, R. et al., *J. Appl. Phys.*, 75, 6659, 1994. With permission.)

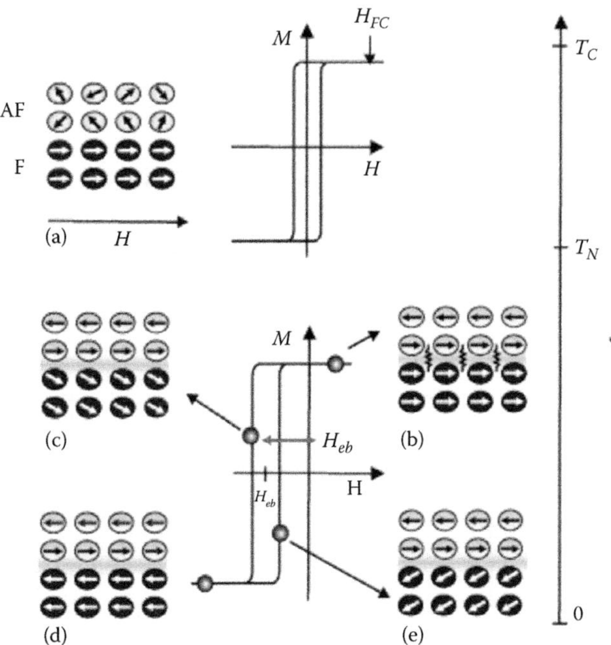

FIGURE 2.19 A simplified view of the EB mechanism. (a) For temperatures greater than the ordering Neel temperature (T_N) of AF, the exchange field between AF and F is zero. (b) After cooling in a positive magnetic field $T < T_N$, the AF moment becomes ordered, and, consequently, an exchange coupling develops between the magnetic moments of AF and F at the AFM/F interface. An EB is established when the magnetic moments of AF are held sufficiently firmly along the easy axis of the AF internal anisotropy: usually, the sample must be cooled to a temperature less than T_N. This required temperature is called the blocking temperature (T_B). (c) and (e) In ultrathin magnetic films, the reversal of the magnetization proceeds by a rotation of the magnetic moment. (d) Reversal of the FM moment also reverses the sign of the coupling at the AF/F interface. This means that the exchange field due to AF acts as an effective field (unidirectional anisotropy) resulting in a shift of the hysteresis loop. In this figure, the coupling between the AF and FM moments at the interface was assumed to be positive, and, consequently, the center of the hysteresis loop moved to a negative field. This is a case of positive EB. (With kind permission from Springer Science+Business Media: *Magnetic Heterostructures*, EB effect of ferro-/antiferromagnetic structures, vol. 227, 2008, 95, Zabel, D. and Bader, S.D.)

$$\varepsilon = -M_F H t_F \cos(\theta - \beta) + K_F t_F \sin^2\beta + K_{AF} t_A \sin^2\alpha - J_{eb}\cos(\beta - \alpha), \quad (2.71)$$

where

- M_F is the saturation magnetization of F
- $K_{F,AF}$ are the anisotropy energy densities for F and AF, respectively
- J_{eb} is the interface F/AF exchange energy (erg/cm^2)
- t_F and t_{AF} are the thicknesses of F and AF layers, respectively
- α, β, and θ are the angles between the axis of the AF moments, the magnetic moment of F, and the external field H and the reference direction given by the AF uniaxial anisotropy axis, respectively

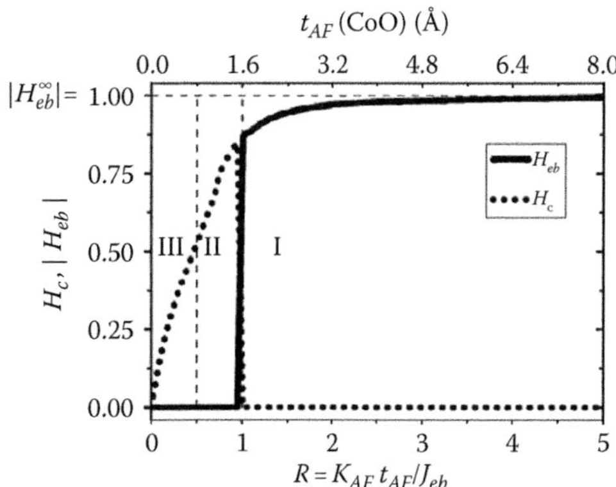

FIGURE 2.20 Meikeljohn and Bean model: The phase diagram of the EB field $|H_{eb}|$ (solid line) and coercive field H_c (dotted line) as a function of the parameter R. Notice that for $R < 1$, there is no EB. (With kind permission from Springer Science+Business Media: *Magnetic Heterostructures*, EB effect of ferro-/antiferromagnetic structures, vol. 227, 2008, 95, Zabel, D. and Bader, S.D.)

The minimum of energy is given by $\partial\mathcal{E}/\partial\alpha = \partial\mathcal{E}/\partial\beta = 0$. Solutions can be found numerically by introducing two parameters: $H_{eb}^{\infty} = -J_{eb}/(M_F t_F)$ and $R = K_{AF} t_{AF}/J_{eb}$. Solutions are shown in Figures 2.20 and 2.21.

The MB model is obviously a gross simplification. In fact, a completely uncompensated moment in AF at the interface would lead to a bias field several orders of magnitude larger than is observed. For a fully compensated AF interface plane, no EB develops: a uniaxial anisotropy is induced in F with the magnetic axis oriented perpendicular to the spin moment in AF [58]. This implies that a realistic F/AF interface is far from being ideal. It is affected by surface roughness and a complex interface chemistry. This creates an interface with a complex magnetic structure, which cannot in general be described. However, there are some common features that are shared by a number of EB systems: (a) In the saturated state, the F and AF spins are collinear at the F/AF interface; (b) A small fraction of the uncompensated magnetic moment is pinned to AF and is exchange coupled to F; (c) The sign of the EB is determined by the sign of the exchange coupling between the pinned interfacial and F spins. A schematic picture of an F/AF interface is shown in Figure 2.22.

Direct imaging of interfacial magnetic spins in a Co/NiO EB structure was demonstrated by Ohldag et al. [59]. Using XMCD (determining the magnetic moment in F) and x-ray magnetic linear dichroism (XMLD) (determining the AFM axis in AF), they were able to identify the magnetic state of a Co/NiO structure, see Figure 2.23. X-ray absorption spectra measured on Co/NiO revealed that the Co/NiO interface went through a slight oxidation of Co and a slight reduction of NiO [60]. This clearly demonstrates an appreciable degree of interface chemistry. The oxygen-reduced NiO showed a XMCD signal indicating that the Ni at the interface is magnetic. The second and third layers show that the AF axis of the NiO was collinear with the magnetic moment of the interfacial Ni. This clearly indicates

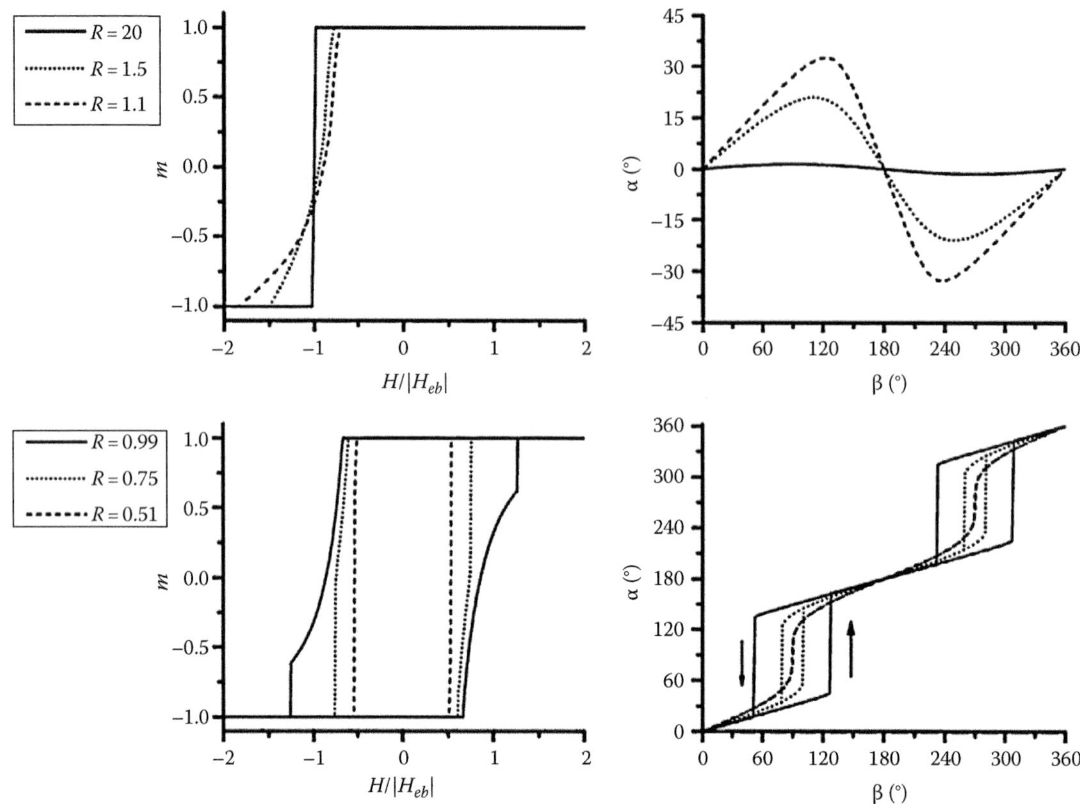

FIGURE 2.21 Simulations of several hysteresis loops and the AF spin orientation during the magnetization reversal. (With kind permission from Springer Science+Business Media: *Magnetic Heterostructures*, EB effect of ferro-/antiferromagnetic structures, vol. 227, 2008, 95, Zabel, D. and Bader, S.D.)

that the interfacial Ni provides uncompensated spins at the Co/NiO interface. The EB in their samples was very weak indicating that most of the uncompensated spins followed the magnetic moment of Co, and only a small fraction of the uncompensated spins was pinned and therefore contributed to the EB. This is strongly supported by the XMCD images in the first and second layers of Figure 2.23: the magnetic moment of the interface Ni is mainly parallel to Co magnetization. The third layer in Figure 2.23 shows another important feature of AF in exchange-biased systems; NiO does not support a single AF domain.

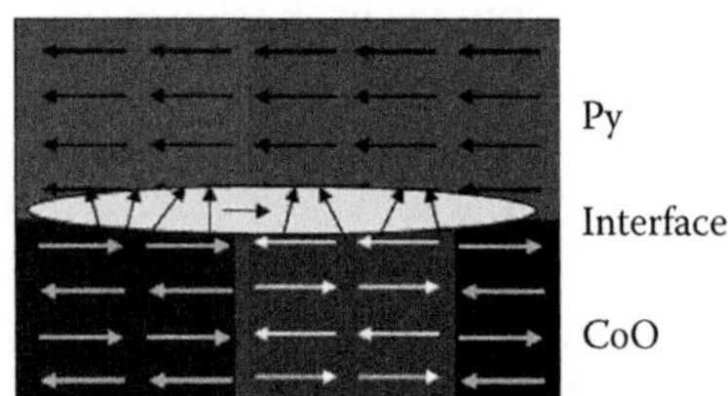

FIGURE 2.22 A schematic diagram of the interface between AF domains in CoO and F (Py). The white background region shows an uncompensated magnetic moment at the F/AF interface, which originates in AF. A part of the spins rotate (shown by the arrows randomly oriented) during the magnetization reversal, and a part is pinned to the AF. The pinned spins lead to the EB while the rotatable spins create an asymmetric hysteresis loop having a larger saturated magnetic moment with the magnetic field oriented parallel to the pinned spins (for ferromagnetic coupling).

2.4.2 Malozemoff Model

Malozemoff [61] recognized that randomness of the interface exchange interaction between F and AF leads to the breaking of

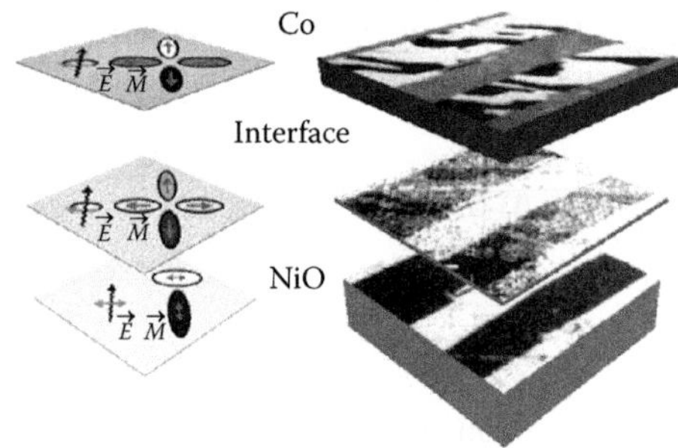

FIGURE 2.23 XMCD and XMLD images of a cleaved NiO(100) sample covered by eight atomic layers of Co. The two top layers show ferromagnetic XMCD images recorded at the top Co and middle Ni L absorption edges. The bottom layer shows an XMLD image of NiO. The icons on the left side show the contrast for different domain orientations and the electric polarization of the x-ray beam. (Courtesy of Jo Stoehr; From Meyerheim, H.L. et al., *Phys. Rev. Lett.*, 103, 267202, 2009. Copyright 2009 by the American Physical Society. With permission.)

AF into domains. Malozemoff has shown that if it is favorable to keep a single magnetic domain in F, then the interface roughness in AF prefers to break AF into magnetic domains with the domain walls perpendicular to the interface. Malozemoff found that H_{eb} is proportional to the domain wall energy in AF and inversely proportional to the total magnetic moment in F,

$$H_{eb} \simeq \frac{2(A_{AF}K_{AF})^{1/2}}{\pi^2 M_F t_F}, \tag{2.72}$$

where

$A_{AF}=J_{AF}/a_{AF}$ is the exchange stiffness coefficient for AF
J_{AF} is the exchange coupling energy for AF
a_{AF} is the lattice constant of AF

2.4.3 Mauri Model

A simple model that includes a domain wall in AF was introduced by Mauri et al. [62]. A sketch describing this geometry is shown in Figure 2.24. The total magnetic energy is

$$\begin{aligned}\varepsilon = &-M_F H t_F \cos(\theta-\beta) + K_F t_F \sin^2\beta + K_{AF} t_A \sin^2\alpha \\ &- J_{eb}\cos(\beta-\alpha) - 2(A_{AF}K_{AF})^{1/2}(1-\cos\alpha),\end{aligned} \tag{2.73}$$

where the last term represents the energy of an AF domain. The shift H_{eb} is given by

$$H_{eb} = -H_{eb}^{\infty}\frac{P}{(1+P^2)^{1/2}}, \tag{2.74}$$

where the parameter $P=(A_{AF}K_{AF})^{1/2}/J_{eb}$ describes the stiffness of the AF EB structure. Notice that in the Malozemoff and the Mauri models, the onset of EB has no threshold: this is opposite to the MB model of EB, compare Figures 2.22 and 2.24. For large P, the EB shift H_{eb} approaches a maxim value H_{eb}° asymptotically, see Figure 2.24. In both models, H_{eb} is always inversely proportional to the F layer thickness. This is expected because the F layer was treated in the ultrathin film limit.

Experimentally, it is found that the onset of H_{eb} occurs at a temperature lower than the Néel temperature T_N of AF. This lower temperature is called the *blocking temperature* T_B. For thick AF films, $T_B \lesssim T_N$ whereas for thin films $T_B \ll T_N$ [63]. This is obviously associated with the onset of a sufficiently large wall energy for AF, see Equations 2.72 and 2.74.

AF materials are available both as metallic oxides and as metallic alloys. Among the oxides, the monoxides NiO, CoO, and $Ni_xCo_{1-x}O$ are the most commonly used. NiO and CoO have Neel temperatures (T_N) of 525 and 293 K, respectively. In spintronic device applications, the AF materials are mainly metallic alloys of Fe–Mn, Ni–Mn, Ir–Mn, Pt–Mn, and Rh–Mn. For the AFM γ phase of the Fe–Mn alloy (with 30 and 35 atomic % of Mn), T_N lies in the range of 425–525 K; and J_{eb} ~ 0.1 erg/cm² at the $Fe_{46}Mn_{54}$/py interface. In the NiMn/py interface, J_{eb} ~ 0.27 erg/cm². A detailed discussion and further references

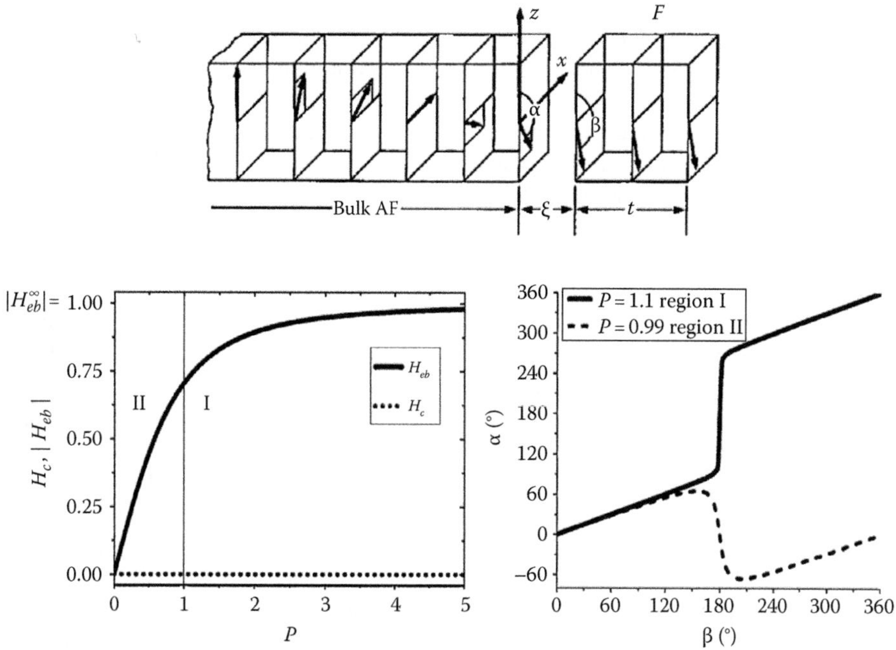

FIGURE 2.24 The Mauri model is shown in the upper diagram. In the Mauri model, the AF domain wall is parallel to the interface. Notice that the spin directions of only one AF sublattice are shown, and the other AF sublattice having AP spin moments is omitted for simplicity. The left figure in the lower set shows the dependence of H_{eb} on the parameter $P=(A_{AF}K_{AF})^{1/2}/J_{eb}$ describing the magnetic stiffness of the AF; the right figure shows the dependence of the angle α on the angle β of the applied field H. (With kind permission from Springer Science+Business Media: *Magnetic Heterostructures*, EB effect of ferro-/antiferromagnetic structures, vol. 227, 2008, 95, Zabel, D. and Bader, S.D.)

to work on AF materials used in EB systems can be found in the article by Berkowitz and Takano [64], in Chapter 13.4.3 in Magnetism by Stoehr and Siegmann [65], and in Chapter 3 in Magnetic Heterostructures by Radu and Zabel [66]. These chapters greatly motivated the author in writing this chapter on exchange bias. Figures 2.19 through 2.22 and 2.24 were provided through the courtesy of Zabel [66].

Acknowledgments

I thank my colleagues Dr. B. Kardasz, Dr. O. Mosendz, Dr. G. Wolterdorf, and Prof. J.C. Cochran and Prof. A.S. Arrott for stimulating discussions and invaluable help in preparation of this manuscript. I have also held extensive discussions with Prof. H. Zabel, Prof. J. Stöhr, Prof. Z.Q. Qiu and Dr. M. D. Stiles, Dr. J. Unguris, and Dr. Y.Z. Wu. Several figures presented in these sections were obtained through their courtesy. I thank especially the Natural Science Engineering Research Council (NSERC) of Canada and Canadian Institute for Advanced Research (CIFAR) for continued research fundings, which make this work possible. I also express my thanks to Professor J. Kirschner, Max Planck Institute in Halle, for providing me with a generous support during my summer research semesters in Germany where this manuscript was partly prepared.

References

1. W. Brown, *Micromagnetics*, Robert E. Krieger Publishing Co., Huntington, New York, 1978.
2. B. Heinrich and J. F. Cochran, Ultrathin metallic magnetic films: Magnetic anisotropies and exchange interaction, *Adv. Phys.* **42**, 523–639 (1993).
3. J. G. Gay and R. Richter, Electronic structure of magnetic thin films, in *Ultrathin Magnetic Structures*, J. A. C. Bland and B. Heinrich (Eds.), vol. I, Chapter 2.1, Springer Verlag, Berlin, Heidelberg, p. 22, 2004.
4. S. Bluegel and G. Bihlmayer, Magnetism of low-dimensional systems: Theory, in *Handbook of Magnetism and Advanced Magnetic Materials*, H. Kronmueller and S. Parkin (Eds.), vol. 1, Wiley, New York, pp. 598–639, 2007.
5. P. Bruno, Tight-binding approach to the orbital magnetic moment and magnetocrystalline anisotropy of transition monolayers, *Phys. Rev. B* **39**, 865 (1989).
6. D. Weller, J. Stoehr, R. Nakajima et al., Macroscopic origin of magnetic anisotropy in Au/Co/Au probed by x-ray magnetic circular dichroism, *Phys. Rev. Lett.* **75**, 3752 (1995).
7. R. Wu and A. J. Freeman, Spin-orbit induced magnetic phenomena in bulk metals and their surfaces and interfaces, *J. Magn. Magn. Mater.* **200**, 498 (1999).
8. L. Néel, Anisotropie magnetique superficielle et surstructures d'orientation, *J. Phys. Radium* **15**, 225 (1954).
9. S. Chikazumi and S. H. Charap, *Physics of Magnetism*, John Wiley & Sons Inc., New York, 1964.
10. D. Sander, The correlation between the mechanical stress and magnetic anisotropy in ultrathin films, *Rep. Prog. Phys.* **62**, 809 (1999).
11. R. H. Victora and J. M. MacLaren, Theory of magnetic interface anisotropy, *Phys. Rev. B* **47**, 11583 (1993).
12. B. N. Engel, C. D. England, R. A. van Leeuwen, M. H. Wiedman, and C. Falco, Magnetocrystalline and magnetoelastic anisotropy in epitaxial co/pd superlattices, *J. Appl. Phys.* **70**, 5873 (1991).
13. I. Dzyaloshinski, A thermodynamic theory of "weak" ferromagnetism of antiferromagnets, *J. Phys. Chem. Solids* **4**, 241 (1958).
14. T. Moriya, Anisotropic superexchange interaction and weak ferromagnetism, *Phys. Rev.* **120**, 91 (1960).
15. H. L. Meyerheim, J.-M. Tonnerre, L. Sandratskii et al., A new model for magnetism in ultrathin fcc Fe on Cu(001), *Phys. Rev. Lett.* **103**, 267202 (2009).
16. B. Heinrich and J. F. Cochran, Magnetic ultrathin films, in *Handbook of Magnetism and Advanced Magnetic Materials*, H. Kroenmueller and S. Parkin (Eds.), vol. 4, Wiley, New York, pp. 2285–2305, 2007.
17. B. Heinrich, Ferromagnetic resonance in ultrathin film structures, in *Ultrathin Magnetic Structures*, B. Heinrich and J. A. C. Bland (Eds.), vol. II, Springer Verlag, Berlin, Heidelberg, Germany, pp. 195–222, 1994.
18. M. Farle, Ferromagnetic resonance of ultrathin metallic layers, *Rep. Prog. Phys.* **61**, 755 (1998).
19. P. Poulopoulos and K. Baberschke, Magnetism in thin films, *J. Phys.: Condens. Matter* **11**, 9495 (1999).
20. Z. Celinski, K. B. Urquhart, and B. Heinrich, Using ferromagnetic resonance to measure magnetic moments of ultrathin films, *J. Magn. Magn. Mater.* **166**, 6–26 (1997).
21. G. Wastlbauer and J. A. C. Bland, Structural and magnetic properties of ultrathin epitaxial Fe films on GaAs(001) and related semiconductor substrates, *Adv. Phys.* **54**, 137–219 (2005).
22. R. Moosbuhler, F. Bensch, M. Dumm, and G. Bayreuther, Epitaxial Fe films on GaAs(001): Does the substrate surface reconstruction affect the uniaxial magnetic anisotropy? *J. Appl. Phys.* **91**, 8757 (2002).
23. M. Brockmann, M. Zolfl, S. Miethaner, and G. Bayreuther, In-plane volume and interface magnetic anisotropies in epitaxial Fe films on GaAs(001), *J. Magn. Magn. Mater.* **198–199**, 384–386 (1999).
24. T. Seki, T. Shima, K. Takanashi, Y. Takahashi, E. Matsubara, and K. Hono, L1-0 ordering off-stoichiometric FePt(001) thin films at reduced temperature, *Appl. Phys. Lett.* **82**, 2461 (2003).
25. B. Heinrich, S. T. Purcell, J. R. Dutcher, K. B. Urquhart, J. F. Cochran, and A. S. Arrott, Structural and magnetic properties of ultrathin Ni/Fe bilayers grown epitaxially on Ag(001), *Phys. Rev. B* **64**, 5334–5336 (1988).
26. P. Krams, F. Lauks, R. L. Stamps, B. Hillebrands, G. Güntherodt, and H. P. Oepen, Magnetic anisotropies of ultrathin Co films on Cu(001) and Cu(1113) substrates, *J. Magn. Magn. Mater.* **121**, 479F (1993).

27. J. Li, M. Przybylski, F. Yildiz, X. D. Ma, and Y. Z. Wu, Oscillatory magnetic anisotropy originating from quantum well states in Fe films, *Phys. Rev. Lett.* **102**, 207206 (2009).
28. J. Mathon, M. Villeret, and D. M. Edwards, Exchange coupling in magnetic multilayers: Effect of partial confinement of carriers, *J. Phys. Condens. Matter* **4**, 9873 (1992).
29. M. D. Stiles, Exchange coupling in magnetic heterostructures, *Phys. Rev. B* **48**, 7238 (1993).
30. P. Bruno, Theory of interlayer magnetic coupling, *Phys. Rev. B* **52**, 411 (1995).
31. M. D. Stiles, Interlayer exchange coupling, in *Ultrathin Magnetic Structures*, J. A. C. Bland and B. Heinrich (Eds.), vol. III, Springer Verlag, Berlin, Heidelberg, Germany, pp. 99–142, 2005.
32. F. J. Himpsel, Fe on Au(001): Quantum-well states down to a monolayer, *Phys. Rev. B* **44**, 5966 (1991).
33. J. E. Ortega and F. J. Himpsel, Quantum well states as mediators of magnetic coupling in superlattices, *Phys. Rev. Lett.* **69**, 844 (1992).
34. P. Bruno, Theory of interlayer exchange interactions in magnetic multilayers, *J. Phys. Condens. Matter* **11**, 9403 (1999).
35. Z. Q. Qiu and N. V. Smith, Topical review: Quantum well states and oscillatory magnetic interlayer coupling, *J. Phys.: Condens. Matter* **14**, R169 (2002).
36. B. Heinrich, Exchange coupling in magnetic multilayers, in *Magnetic Heterostructures: Advances and Perspectives in Spinstructures and Spintransport*, H. Zabel and S. D. Bader (Eds.), vol. 227, Springer Tracts in Modern Physics, pp. 183–250, 2008.
37. B. Heinrich, J. F. Cochran, M. Kowalewski et al., Magnetic anisotropies and exchange coupling in ultrathin fcc Co(001) structures, *Phys. Rev. B* **44**, 9348 (1991).
38. M. Ruehrig, R. Schaefer, A. Hubert et al., Domain observations on Fe/Cr/Fe layered structures: Evidence for a biquadratic coupling effect, *Phys. Stat. Sol. A* **125**, 635 (1991).
39. J. C. Slonczewski, Fluctuation mechanism for biquadratic exchange coupling in magnetic multilayers, *Phys. Rev. Lett.* **67**, 3127 (1991).
40. B. Heinrich, T. Monchesky, and R. Urban, Role of interfaces in higher order angular terms of magnetic anisotropies: Ultrathin film structures, *J. Magn. Magn. Mater.* **236**, 339 (2001).
41. T. L. Monchesky, B. Heinrich, R. Urban, K. Myrtle, M. Klaua, and J. Kirschner, Magnetoresistance and magnetic properties of Fe/Cu/Fe/GaAs(100), *Phys. Rev. B* **60**, 10242–10251 (1999).
42. R. Urban, G. Woltersdorf, and B. Heinrich, Gilbert damping in single and multilayer ultrathin films: Role of interfaces in nonlocal spin dynamics, *Phys. Rev. Lett.* **87**, 217204 (2001).
43. S. S. P. Parkin and D. Mauri, Spin engineering: Direct determination of the rkkr far-field range function in ruthenium, *Phys. Rev. B* **44**, 7131 (1991).
44. J. Unguris, R. J. Cellota, and D. T. Pierce, Observation of two different oscillation periods in the exchange coupling of Fe/Cr/Fe(001), *Phys. Rev. Lett.* **67**, 140 (1991).
45. D. T. Pierce, J. Unguris, and R. J. Celotta, Investigation of exchange coupled magnetic layers by SEMPA, in *Ultrathin Magnetic Structures*, B. Heinrich and J. A. C. Bland (Eds.), vol. II, Chapter 2.3, Springer Verlag, Berlin, Heidelberg, Germany, pp. 117–147, 2005.
46. B. Heinrich, J. F. Cochran, T. Monchesky, and R. Urban, Exchange coupling through spin-density waves in Cr(001) structures: Fe-whisker/Cr/Fe(001) studies, *Phys. Rev. B* **59**, 14520 (1999).
47. J. F. Cochran, Brillouin light scattering intensities for a film exchange coupled to a bulk substrate, *J. Magn. Magn. Mater.* **169**, 1 (1997).
48. E. Fullerton, M. J. Conover, J. E. Mattson, C. H. Sowers, and S. D. Bader, Oscillatory interlayer coupling and giant magnetoresistance in epitaxial Fe/Cr(211) and (100) superlattices, *Phys. Rev. B* **48**, 15755 (1993).
49. C. Carbone and S. F. Alvarado, Antiparallel coupling between Fe layers separated by a Cr interlayer: Dependence of the magnetization on the film thickness, *Phys. Rev. B* **36**, 2433 (1987).
50. D. Stoeffler and F. Gautier, Magnetic properties of ferro/antiferromagnetic-based sandwiches, in *Magnetism and Structure in Systems of Reduced Dimensions*, R. F. C. Farrow, B. Dienny, M. Donath, A. Fert, and B. D. Hersmeier (Eds.), NATO-ASI B, vol. 309, Plenum, New York, p. 411, 1993.
51. B. Heinrich, J. F. Cochran, D. Venus, K. Totland, C. Schneider, and K. Myrtle, Role of interface alloying in Fe-whisker/Cr/Fe(001) structures, angular resolved auger electron and MOKE studies, *J. Magn. Magn. Mater.* **156**, 215 (1996).
52. D. Venus and B. Heinrich, Interfacial mixing of ultrathin Cr films grown on an Fe whisker, *Phys. Rev. B* **53**, R1733 (1996).
53. A. Davis, J. A. Strocio, D. T. Pierce, and R. J. Celotta, Atomic-scale observations of alloying at the Cr-Fe(001) interface, *Phys. Rev. Lett.* **76**, 4175 (1996).
54. R. Pfandzelter, T. Igel, and H. Winter, Intermixing during growth of Cr on Fe(001) studied by proton- and electron-induced Auger-electron spectroscopy, *Phys. Rev. B* **54**, 4496 (1996).
55. M. Freyss, D. Stoeffler, and H. Dreysee, Interfacial alloying and interfacial coupling in Cr/Fe(001), *Phys. Rev. B* **56**, 6047 (1997).
56. R. Jungblut, R. Coehoorn, M. T. Johnson, J. Aan de Stege, and A. Reinders, Orientational dependence of the exchange biasing in MBE grown Ni80Fe20/Fe50Mn50 bilayers, *J. Appl. Phys.* **75**, 6659 (1994).
57. R. H. Meiklejohn and C. P. Bean, New magnetic anisotropy, *Phys. Rev.* **102**, 1413 (1956).
58. T. C. Schulthess and W. H. Butler, Consequence of spin-flop coupling in exchange biased films, *Phys. Rev. Lett.* **81**, 4516 (1998).
59. H. Ohldag, T. J. Regan, J. Stoehr et al., Spectroscopic identification and direct imaging of interfacial magnetic spins, *Phys. Rev. Lett.* **87**, 247201 (2001).

60. T. J. Regan, H. Ohldag, C. Stamm et al., Chemical effects at metal/oxide interfaces studied by x-ray-absorption spectroscopy, *Phys. Rev. B* **64**, 214422 (2001).
61. A. P. Malozemoff, Random-field model of exchange anisotropy at rough ferromagnetic-antiferromagnetic interfaces, *Phys. Rev. B* **35**, 3679 (1987).
62. D. Mauri, H. C. Siegmann, P. S. Bagus, and E. Kay, Simple model for thin ferromagnetic film exchange coupled to an antiferromagnetic substrate, *J. Appl. Phys.* **62**, 3047 (1987).
63. J. Nogues and I. Schuller, Exchange bias, *J. Magn. Magn. Mater.* **192**, 203 (1999).
64. A. E. Berkowitz and K. Takano, Exchange anisotropy—A review, *J. Magn. Magn. Mater.* **200**, 552 (1999).
65. J. Stoehr and H. Ch. Siegmann, Surfaces and interfaces of ferromagnetic metals, *Magnetism: From Fundamentals to Nanoscale Dynamics*, Springer Series in Solid State Sciences, vol. 152, Chapter 13.4.3, Springer, Berlin, Heidelberg, Germany, pp. 617–629, 2006.
66. H. Zabel and S. D. Bader, Exchange bias effect of ferro-/antiferromagnetic structures, in *Magnetic Heterostructures*, H. Zabel and S. D. Bader (Eds.), Springer Tracks in Modern Physics, vol. 227, Springer, Berlin, Heidelberg, Germany, pp. 95–184, 2008.
67. B. Heinrich, Magnetic nanostructures: From physical principles to spintronics, *Can. J. Phys.* **78**, 161 (2000).

3

Micromagnetism as a Prototype for Complexity

Anthony S. Arrott
Simon Fraser University

3.1 Introduction

Magnetostatic dipole–dipole interactions play an important role in determining properties of nanomagnetic systems along with the exchange coupling and magnetocrystalline anisotropy, which were discussed in Chapter 2. The integral–differential equations of micromagnetics lead to complexity. These equations are nonlinear because the magnetization is a vector of constant magnitude. The nature of the dipole–dipole interactions favors patterns that are divergence free in volume and induce magnetic charges on the surfaces that create demagnetizing fields opposed to applied fields. The emerging field of spintronics with spin-polarized currents in semiconductors proceeds with the magnetic moment of the electron as just the tail of the dog. Much of the work on giant magnetoresistance (see Chapters 4 and 5), tunneling magnetoresistance (see Chapters 10–12), and magnetic random access memories (see Chapter 35) is performed by using intelligent ways to avoid the complexities of dipole–dipole interactions. The movement to ultrathin films was motivated by reducing the role of dipole–dipole interactions in directing the process of magnetization, while at the same time eliminating the variation of magnetization in the z-direction, perpendicular to the plane of the ultrathin film. Ewing understood the importance of the dipole–dipole interaction in ferromagnetism as early as in 1890. Ewing gave the name to "hysteresis" [1].* In 1935, Landau gave the first 3D model of a structure that minimized the effects of the dipole–dipole interaction (see Figure 3.1). In the 1970s, a beautiful new technology was created using magnetic bubbles† for which understanding of the dipole–dipole interactions was critical. This chapter is intended as an ode to complexity in anticipation that emerging technologies may yet benefit if the current view of dipole–dipole interactions is not far from sight. The concept of splay saving is a recurrent theme in this ode. That concept underlies the recent work of Riccardo Hertel's group on the propagation of domain walls along a nanotube of permalloy [2].

* The past introduced J. A. Ewing to those of the new millennium. The present was dedicated to Alex Hubert and contains a description of the critical role of magnetoelasticity in a current carrying iron whisker, a subject beyond the scope of this chapter.

† The most convenient reference finder today is Google, which includes Wikipedia, that points to additional references.

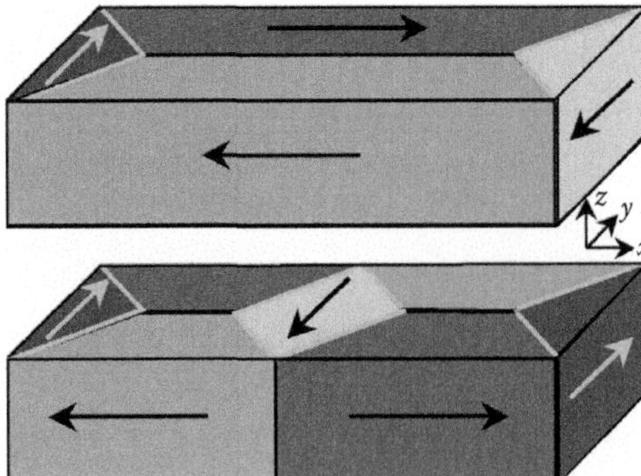

FIGURE 3.1 The Landau structure as presented in domain theory is shown in the top panel with the four domains separated by four 90° walls and one 180° Bloch wall. This is a vortex structure with winding number +1. It does not have inversion symmetry. The structure with a diamond in the center has inversion symmetry and a winding number of +2.

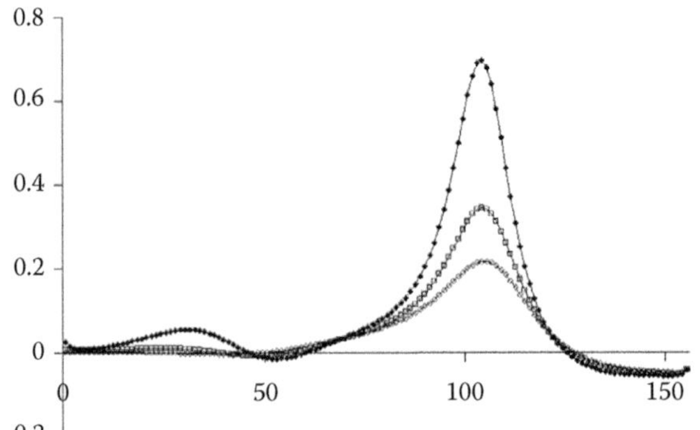

FIGURE 3.2 External fields for a $156 \times 96 \times 60\,\text{nm}^3$ nanobrick using 1.25 nm cubes for calculation. The scan is on a line of constant y through the center of the swirl at positions 1, 10, and 20 nm above the centers of the uppermost calculation cells. The large peak in the H_z field is from one swirl on the top surface. The swirl on the bottom surface is 60 nm away and creates a negative contribution to H_z that is negligible on the top surface. There are also large H_z fields at the four vertical corners of the nanobrick.

This chapter is about the 3D patterns of magnetization in a nanobrick of iron with dimensions at the current limit of lithography. This nanobrick is a parallelepiped with typical dimensions $X = 130\,\text{nm}$, $Y = 80\,\text{nm}$, and $Z = 50\,\text{nm}$. The iron nanobrick is compared with an ellipsoid of comparable dimensions to show that the main actor in this drama is a vortex, capable of playing many roles. It is anticipated that some of these roles will aid nanoscience and nanotechnology. This chapter presents calculations of the behavior of the nanobrick and provides insight into the consequences of magnetic dipole–dipole interactions and their role in computational micromagnetics. The vortex states discussed here have minimal problems from surfaces and can be manipulated by sufficiently small fields and currents to be attractive for nanoscale devices. Hubert has given the name "swirl" for the circulating pattern of magnetization in the region where a vortex meets a surface [3].* The external magnetic fields from the vortices intersecting the opposing surfaces in ultrathin films are small because the magnetic charge on one end of the vortex cancels the effects of the charge on the other end. In a nanobrick, the swirls on the two ends of the vortex can be moved far apart so that the external fields can be greater than one quarter of the saturation induction of iron (see Figure 3.2). The swirls can be moved easily and rapidly over large distances. They become highly nonlinear oscillators that serve as prototypes for the complexity of bifurcations and chaos. The calculations for the nanobrick also serve as a primer on the effective use of modern codes for micromagnetism.

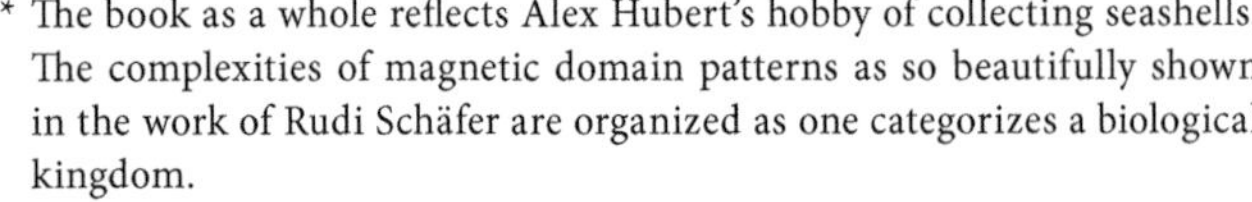

* The book as a whole reflects Alex Hubert's hobby of collecting seashells. The complexities of magnetic domain patterns as so beautifully shown in the work of Rudi Schäfer are organized as one categorizes a biological kingdom.

3.2 Symmetry Breaking in a Cycle of Magnetization

For an ellipsoid in a mathematical model with continuous variation of magnetization, the application of a large applied magnetic field **H** along the ***k***-direction, parallel to the z-axis, can result in a uniform magnetization. For an iron nanobrick in a high field, this is not the case. The magnetization is along ***k*** at the z-axis, but deviates from ***k*** for finite x and y, for coordinates with the origin at the geometric center of the brick. At high fields, that deviation is outward at the top surface and inward at the bottom surface. This is called the flower state. The flower state is discussed in the caption of Figure 3.3. Discussion of the visual patterns and details of the calculations are to be found in the figure captions throughout this chapter.

As the field is lowered, the magnetization pattern changes continuously, starting at a critical field, from the flower state to the curling state in which the magnetization has components that circulate about the central axis (see Figure 3.4). This is explained below as a result of the dipole–dipole interaction and the concept of splay saving. The sense of the circulation, clockwise (*cw*) or counterclockwise (*ccw*), is determined while breaking the reflection symmetry of the flower state. The energy contours present a three-tined fork [4].† The paths to both the right and the left of the fork become lower in energy than the path along the central tine. Yogi Berra, catcher for the New York Yankees, said famously, "When you come to a fork in the road, take it." To help the computer "take it," a current i_z is passed up the z-axis.

† References given to the mathematics of bifurcations.

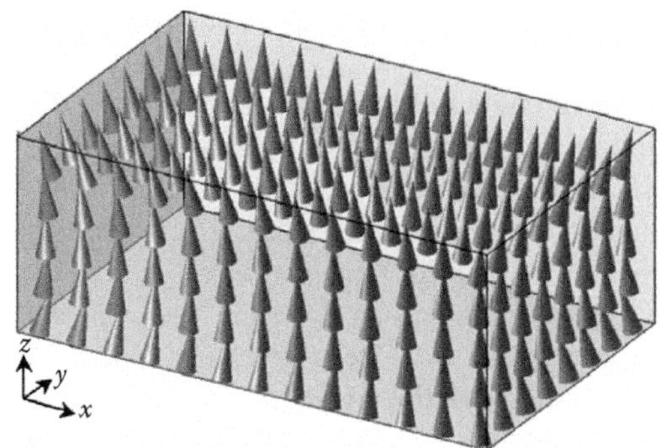

FIGURE 3.3 An iron nanobrick with dimensions $156 \times 96 \times 60\,\text{nm}^3$ in the flower state at a field $\mu_0 H_z = 2.2\,\text{T}$, showing the directions of the magnetization as hollow cones on three sides of the brick. Each cone represents the average components of the magnetization in a volume of $288\,\text{nm}^3$ combining 36 computational cells in a plane. The computational cells are cubes with 2 nm edges. The magnetic surface charge density on the top surface is $\sigma_m \approx M_s$, where M_s is the saturation magnetization of iron at room temperature, taken here to be 1714 emu or $\mu_0 M_s = 2.154\,\text{T}$. This charge produces a magnetic field that is in the plane of the top surface and points outward. The charge density on the bottom surface is $\sigma_m \approx -M_s$. The field from this is inward in the plane of the bottom surface. There is also magnetic charge density on the side surfaces that varies almost proportional to z, measured from the center of the brick. The flower state appears in high fields for nanobricks of all dimensions, including a cluster of nine iron atoms at $T = 0\,\text{K}$ in a cube with one of the atoms in the body-centered position. The flower state has inversion symmetry at equilibrium. Changing the uniformly applied field will cause the magnetic moments to precess, yet the inversion symmetry is maintained. Changing the field nonuniformly by passing a current along the z-axis will break the inversion symmetry.

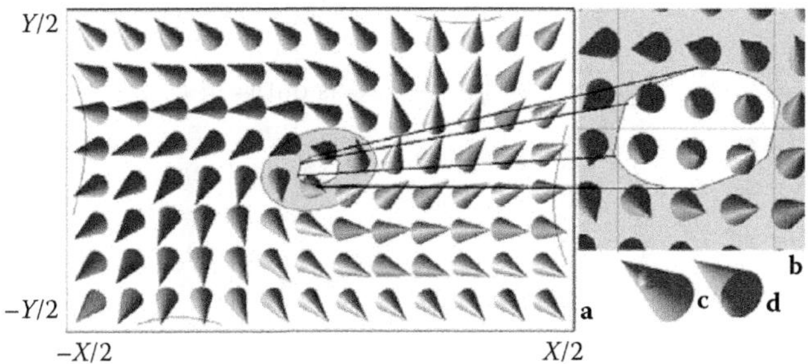

FIGURE 3.4 An iron nanobrick with dimensions $156 \times 96 \times 60\,\text{nm}^3$ in the curling state at a field $\mu_0 H_z = 0.8\,\text{T}$. The contour lines are for direction cosines $m_z = 0.95$ and $m_z = 0.8$. The area between these is shaded gray. The cones on the left at **a** are the average of the magnetization for 9 cells of 4 nm cubes in the topmost plane centered at $z = 28\,\text{nm}$. In the blow up on the right at **b**, the cones are the magnetization in each cell. As the field is lowered below $\mu_0 H_z = 0.9\,\text{T}$, each cone in the top surface rotates *ccw* from its position in the flower state as it would if a current were passed up the z-axis. This configuration at and near the surface is called a *swirl*. Each cone is treated as if it were centered on a position one-third of the distance from the apex, denoted, in detail at **c** in the lower right, as a sphere superimposed on the cone that points out of the plane of the page. The cones are hollow, so that when viewed from below, light is usually not reflected. The hollow cone at **d** points into the page. Using cones to indicate the direction of magnetization creates artificial waves in the visual pattern, when the base of the cone moves from one side of a grid line to the other, for example, the line of cones at $x = 0$.

3.2.1 Breaking Inversion Symmetry and Choosing a Handedness

The inversion symmetry of the flower state may or may not be maintained in the curling state. Note that the inversion symmetry for a magnetic dipole is opposite to that of an electric dipole. Consider the moments at two points equidistant from the center of an inversion symmetry on opposite ends of a line through that center. For electric dipoles, the moments point in opposite directions. For magnetic dipoles they point in the same direction. If the inversion symmetry were maintained on the transition to the curling state, the circulation at the bottom surface would be opposite to that on the top surface (see Figure 3.5, where the "top" and "bottom" are along the x-axis). This leads to complex magnetization patterns, as the field is lowered [1] with the formation of two vortices, one of which exits the nanobrick leaving the other to arrive at the same final state, which is achieved by lowering the H_z field.

Here the discussion is focused on reaching that state by lowering H_z. To have the circulation in the same sense on both the surfaces, it is necessary to break the inversion symmetry. A controlled way of achieving this is to pass a small current along the z-axis while lowering H_z.

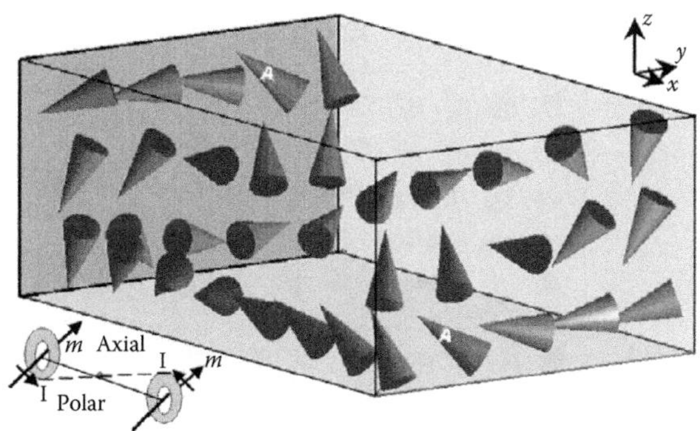

FIGURE 3.5 Curling with inversion symmetry for an iron nanobrick with dimensions $112 \times 70 \times 42\,\text{nm}^3$. The moments near the front surface at $x = X/2$ show a *cw* circulation. The moments near the back surface at $x = -X/2$ show a *ccw* circulation. The line of moments joining the front and back surfaces shows a Néel wall in the form of an end-over-end helix. Inversion symmetry is maintained. Note that the two cones labeled ***A*** are connected by a line through the center of the brick and have the same orientation. The rendering of the magnetization as cones shows an object that does not have inversion symmetry (a line joining the tips of the cones does not go through the center of the brick). It is the magnetization at the grid points that has the inversion symmetry. The field along the x-axis, where the two surfaces are separated by 112 nm, was used for this illustration of inversion symmetry. When the field is along the z-axis, the two surfaces are separated by 42 nm. Then the exchange energy presents a sufficiently large barrier that the *cw*/*ccw* structure is suppressed (see Figure 3.6). The distinction in the meaning of inversion symmetry between a polar vector and an axial vector is illustrated in the lower left corner using the polar vector ***i*** for the current in a ring to produce an axial vector ***m***. The two current vectors, equidistant from the center of inversion symmetry, point in opposite directions while the magnetizations they produce point in the same direction.

3.2.2 Response of Isosurfaces to Changing Fields

In the presence of the field, from a small current i_z, the flower pattern develops a slight swirl even at the highest fields. This is analogous to the magnetization of a ferromagnetic material in a small field above its Curie temperature. When the temperature is lowered below T_c, the response to the small field increases rapidly as the spontaneous magnetization develops. In the nanobrick, below a critical field H_{zc}, there is a spontaneous contribution to the angle of rotation of the moments away from their direction in the flower state. The swirls are centered on the z-axis (see Figure 3.6). As the swirls develop with decreasing H_z, the action is first concentrated at the core of the swirl. As the field is decreased to zero, the moments in the corners are the last to fully participate in the curling pattern, turning from pointing out along a line at 45° to pointing perpendicular to that line.

The direction cosines of the magnetization $\boldsymbol{M}$ are denoted by m_x, m_y, and m_z. Starting from 1, m_z decreases with distance from the z-axis. Tubes of constant m_z, called m_z-isosurfaces, connect the top and bottom surfaces (see Figure 3.7). The cross-sections of the tubes in planes of constant z are elliptical and centered on the z-axis, reflecting the geometry of the brick. The ellipses are largest in the midplane $z=0$. The magnetization does not lie in the m_z-isosurfaces except in the midplane. Elsewhere it has a radial component as well as a circulating component. The sense of the circulating component has been forced to be the same at all z by the small current along z while decreasing H_z. The radial component is inward for $z<0$ and outward for $z>0$ as a result of the dipole–dipole interaction. The m_z-isosurfaces define a vortex structure. The role of vortices in magnetism has often been overlooked.

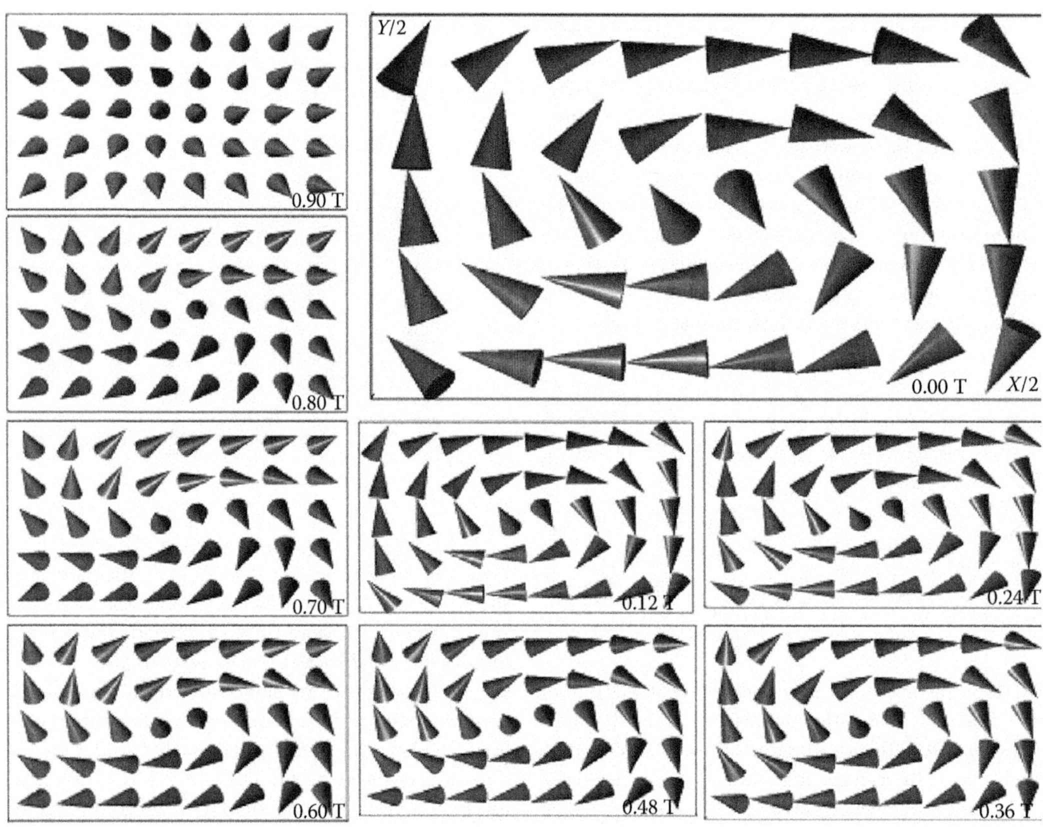

FIGURE 3.6 The curling pattern that develops on reducing $\mu_0 H_z$ from 0.9 T for the nanobrick in Figure 3.5 is called the *I-vortex state*. The circulation is cw (or ccw) in all planes of constant z. The *I-vortex state* persists to zero field for the dimensions $102\times70\times42\,\text{nm}^3$, but, if $Z>42\,\text{nm}$, the *I-vortex state* becomes unstable at low field. When there is a computational cell centered at the origin, the cells along the z-axis all have $m_z=1$. Here the center of the vortex is midway between two cells on the y-axis displaced in x by 1 nm on either side of the vortex center. When the cells are grouped in bundles of 49 cells in a plane to produce the cones shown here, the center of the bundles nearest the core of the swirl are displaced by 7 nm from the axis. This is far enough from the center of the swirl that the magnetization lies close to the surface when H goes to zero. The development of the *I-vortex state* is followed here with the 42 nm high brick in the presence of a small current i_z. Note the progress of the corner moments. These change little at high fields. At lower fields, they rotate so that they are parallel to one of the two corner edges. At the lowest fields, they lose their radial component and at the same time turn down in response to the dipole field from the central core of the vortex structure. Note that the hollow cone in the lower left corner of the large panel for 0.00 T appears darker at the bottom of the cone where the light enters. This cone points into the iron nanobrick as is the case for all corners. It is more evident in the lower left.

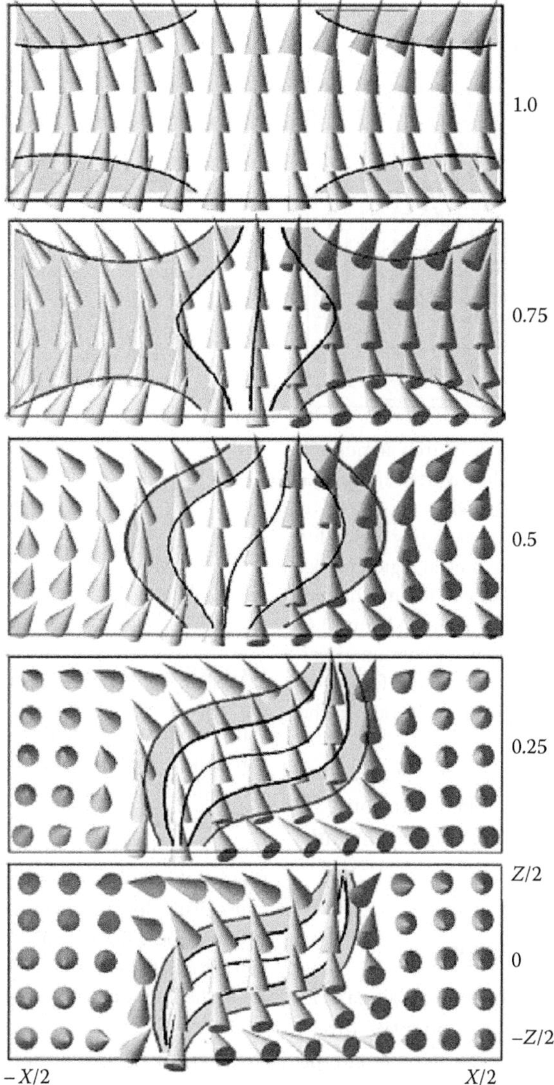

FIGURE 3.7 The *S-vortex state* develops from the *I-vortex state* with decreasing H_z for the iron brick with dimensions $156\times100\times60\,\text{nm}^3$. Each cone represents 9 cubes with 4 nm edges in a plane of constant *y*. The coarse grid is used in this calculation for visual purposes, even though it is too crude for accurate assessment of critical fields. The width of the iron nanobrick was increased by one computational cell compared to Figures 3.3 and 3.4 in order to show the central cross-section, $y=0$, of the *I-vortex state*. The formation of the *S-vortex state* is followed using m_z-isosurfaces that connect the top and bottom swirls of the *I-vortex state*. The intersections with the midplane at $y=0$ for $\mu_0H_z>0.4\,\text{T}$ or at $y=4\,\text{nm}$ for $\mu_0H_z<0.4\,\text{T}$ are shown for the contour lines in that plane for $m_z=0.95$ and $m_z=0.80$ with the central white regions corresponding to $m_z>0.95$ and the gray regions to $0.8<m_z<0.95$. A small bias current in the *x*-direction creates a field H_y that is positive at the top and negative at the bottom, tilting the magnetization in the +*x*-direction at the top and in the −*x*-direction at the bottom. To make this clearer, a contour for $m_y=0$ is shown in the middle of the central white region. The tilt of the $m_y=0$ contour in the second panel at 0.75 T is the result of the bias current. The larger displacement of the swirls in the panel at 0.50 T is almost all the result of a spontaneous displacement of the swirls in opposite directions to form the *S-vortex state*. In zero field, the swirls reach their maximum displacement and appear near $x=(X-Y)/2$, $y=y_1$, $z=Z/2$ and $x=-(X-Y)/2$, $y=y_1$, $z=-Z/2$, where y_1 is a small displacement (see Figure 3.9). At $Z=60\,\text{nm}$, the *I-state vortex* can persist as a structure in unstable equilibrium all the way to $H_z=0$. The width of the contours at $z=0$ is essentially the same for the unstable *I-state vortex* and the *S-state vortex*.

3.2.3 Displacing the Swirls

The vortex structure can be moved off the center by bias fields perpendicular to the *z*-axis. The vortex moves in the direction perpendicular to the bias field with the swirls remaining above one another, but the ellipsoidal cross-sections distort and need not be centered with respect to the swirls—this is for equilibrium configurations. In dynamic responses the swirls move with respect to each other. The m_z-isosurface takes on a life of its own. The tubes can bulge, twist, and bend. If the bias fields are derived from a current i_x along the *x*-axis in the +*x*-direction, the upper swirl moves in the +*x*-direction and the lower swirl moves in the −*x*-direction as a result of fields in the +*y*-direction at the top and in the −*y*-direction at the bottom. The H_y fields from i_x

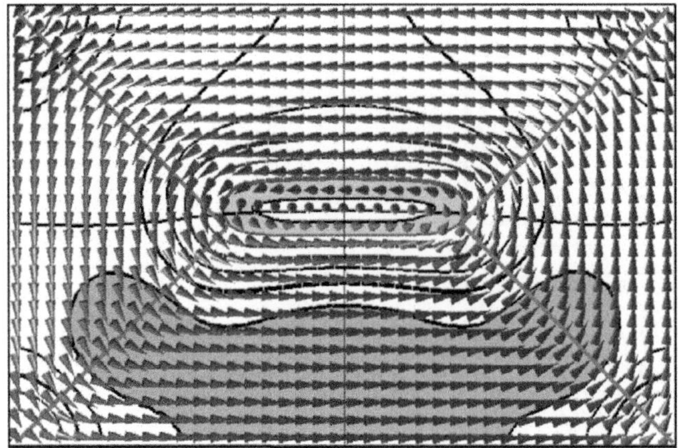

FIGURE 3.8 Central cross-section for $z=0$ for an iron nanobrick with dimensions $156 \times 100 \times 60\,nm^3$. The core of the S-vortex and center of the 180° Bloch wall are the central white region, where $m_z > 0.95$. In the surrounding gray region $0.95 > m_x > 0.8$. In the dark gray region for $y<0$, $-0.4<m_z<-0.2$. The open contour about the central region marks $m_z=0$. In the corners $m_z \approx 0.8$. A negative H_z is required to turn the corners down for the $Z=60\,nm$ structure, while for $Z=42\,nm$ in Figure 3.6, the corners are already down at $H_z=0$. The slightly wavy horizontal line is the contour for $m_x=0$. The vertical thin line is the contour for $m_y=0$. The gray lines from the four corners show the positions of walls in the divergence-free van den Berg construction.

are largest at the top and bottom surfaces. When a uniform H_y field is superimposed on the field from i_x, one can independently manipulate the two swirls in any given H_z.

3.2.4 Onset of the Landau Structure

If i_x is maintained while H_z is lowered, there is a critical H_z, below which the displacements of the swirls in opposite directions increase rapidly with decreasing H_z. The development of the S-vortex state from the I-vortex state is shown in Figure 3.7. The correspondence of the S-vortex state with the well-known Landau configuration is not at first obvious, if the dimensions are at the threshold for the instability of the I-vortex state. When viewed in the central cross-section, $z=0$, the pattern more closely reflects the Landau configuration (see Figure 3.8). When X and Y are much larger than the values used here, the swirls sit at the ends of the 180° Bloch wall of the Landau configuration at $x=(X-Y)/2$, $y=y_1$, $z=Z/2$ and $x=-(X-Y)/2, y=y_1, z=-Z/2$, where y_1 is a small displacement (see Figure 3.9 for a portrait of the S-vortex, which is the heart of the Landau configuration). If the current i_x were in the $-x$-direction while H_z was lowered through the critical H_z, the swirls would be positioned at $x=-(X-Y)/2$, $y=-y_1$, $z=Z/2$ and $x=(X-Y)/2, y=-y_1, z=-Z/2$. The statics and dynamics of switching between these two configurations, called the S^+-vortex and the S^--vortex, are the principal focus of this chapter.

3.2.5 Landau Structure

Landau structure, as shown in Figure 3.1, was postulated for a large iron brick, where the anisotropy causes a clearer distinction between the domains and the domain walls. The Landau structure does not have inversion symmetry, but the structure with a diamond in the center, often observed in the 1950s by researchers at General Electric and General Motors, does have inversion symmetry. Landau put the walls at 90° to avoid a discontinuity in the component of $\boldsymbol{m}$ perpendicular to the wall. The magnetization was assumed to be in the z-plane everywhere except in the 180° Bloch wall. The Bloch wall creates surface charges, which Néel eliminated by having the magnetization lie in the surface as it turns through 180°. These are the Néel caps on the Bloch wall that were first calculated by LaBonte in the 1960s in his treatment of a never-ending Bloch wall (see Figure 3.10). Later, it was pointed out that the Néel cap was an extension of one or the other of the end-closure domains with that extension, terminating in a swirl at the opposite end [5]. The swirls satisfy the topological necessity of the magnetization pointing out of the surface in at least two places. One of the first scanning electron microscopy with polarization analysis (SEMPA) experiments was to show that this is the case with iron whiskers [6]. Structures where both the swirls are on the same surface are possible [7], but not considered here. Which end of the Bloch wall claims the swirl depends on the sense of rotation

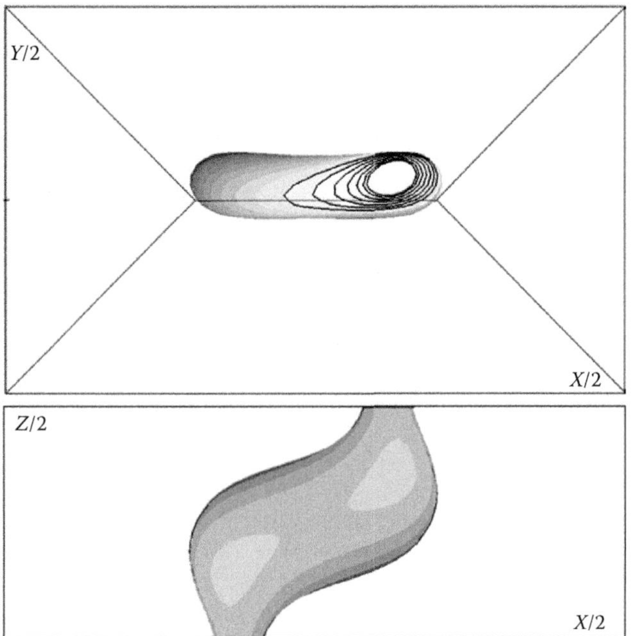

FIGURE 3.9 Portrait of an S-vortex state core showing contours of $m_z=0.95$ in successive planes in y or z spaced by 1.5 nm. The S-vortex state is a 3D structure that is not fully characterized by a single slice through the m_z-isosurfaces. The upper panel shows that the swirl is offset from the core of the 180° wall, which itself is offset from the plane $y=0$. The silhouette of the structure in the lower panel has contributions from several slices in y. The lines in the upper panel show the positions of the 180° Bloch wall and the four 90° walls delineating the two end-closure domains in the classic structure, first postulated by Landau in 1935. These lines are also the positions of the walls obtained by van den Berg in his solution to the problem of ideally soft magnetic materials in the limit of ultrathin films.

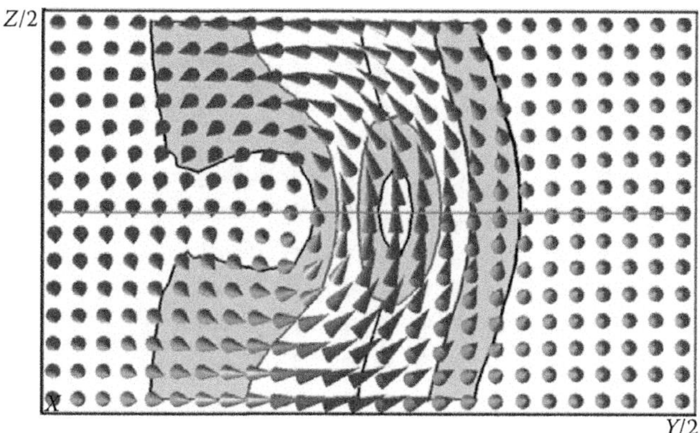

FIGURE 3.10 Cross-section in the plane $x=0$ for the $156\times100\times60\,\text{nm}^3$ iron nanobrick. Each cone represents a 4 nm cubic computational cell. The almost vertical bowed line in the center is the contour for $m_x=0$. It is centered along the top and bottom surfaces, but bows to the right by 4 nm at the center of the nanobrick. In the white region about $y=4\,\text{nm}$ and $z=0$, $m_z>0.95$. This is the core of the *S*-vortex and the center of the 180° Bloch wall of the Landau structure. The light line at $z=0$ is also the contour for $m_y=0$. Along that line the cones rotate through an angle greater than 180° from $y=-Y/2$ to $y=Y/2$. This is the Bloch wall separating the two principal domains of the Landau structure, where, in the white region on the far left, $m_x<-0.95$, and in the white region on the far right, $m_x>0.95$. In the gray region on the left, $-0.95<m_x<-0.8$. In the gray region on the right, $0.95>m_x>0.8$. In the gray region in the center, $0.95>m_z>0.8$. This figure is in one-to-one correspondence with the LaBonte's calculation of the cross-section of a never-ending-Bloch wall from the 1960s, which showed clearly the existence of the Néel caps to the left of center at both surfaces. The magnetization in the Néel caps and the core of the *S*-vortex circulate about the white bulge on the left, which forms the core of a partial vortex in the $-x$-direction. This partial vortex accounts for the component $\langle m_x\rangle$ that accompanies the transformation of the *I*-vortex to the *S*-vortex, as shown in Figure 3.14.

in the Néel cap, which itself depends on the sense of rotation of the magnetization about the *y*-axis, on transversing the Bloch wall in the *y*-direction. The diamond structure, also shown in Figure 3.1, requires additional vortices as there are now two swirls and two half antiswirls on the top and bottom surfaces. But this discussion will be taken up another time.

The Bloch wall with the Néel caps is seen in Figure 3.10 for an *S*-vortex in the iron nanobrick for a cross-section in the plane $x=0$. This structure is essentially the same as that calculated by LaBonte in the 1960s for a never-ending 180° Bloch wall [8]. The vortex structure about a line in the *x*-direction has long been noted, without recognizing that the vertical section of such circulation is actually the core of an *S*-vortex connecting the upper and lower surfaces. The swirls do not appear in LaBonte's calculation, for they are displaced to infinity in the never-ending Bloch wall. They do not appear in Figure 3.10, because the cross-section is midway between the two swirls. But the displacement of the core of the *S*-vortex does appear in Figures 3.8 and 3.9 as well as in LaBonte's calculation. The reason for the displacement has long been understood. It is to make room for the Néel caps that not only remove magnetic charge from the surfaces, but also minimize volume charge by curling about the line that is parallel to the *x*-axis.

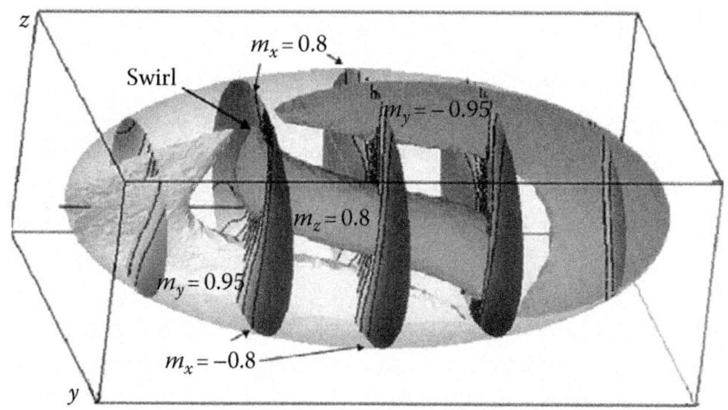

FIGURE 3.11 The Landau structure in an ellipsoid ($260\times161\times100\,\text{nm}^3$) is calculated using Hertel's finite element program TetraMag. The isosurfaces at $m_y=+0.95$ and $m_y=-0.95$ delineate the combined closure domains and Néel walls at each end of the core of the Bloch wall, outlined by the isosurface $m_z=0.8$. The core of the Bloch wall terminates at the surfaces in swirls. The Bloch and Néel walls separate the regions of high magnetization in the $+x$ and $-x$ directions, indicated by the darker regions of the five planes perpendicular to the *x*-axis. This pattern is topologically equivalent to the structure of Figures 3.7 through 3.9. It is generated by the same sequence of fields used in Figure 3.7 to produce the Landau structure starting from the high-field flower state. But, the flower state exists only to the extent that the finite elements are not sufficiently effective in producing an exact ellipsoid. The ellipsoid is simpler than the brick because there are no corners or edges. For the ellipsoid, this configuration is one of the eight ground states, which differ in the choice of polarization with respect to the *z*-axis, the handedness of the circulation about the *z*-axis and whether the upper swirl is on the left in an S^- state as shown here or on the right in an S^+ state. The two swirls and the core of the Bloch wall are displaced from the $y=Y/2$ plane as shown for the brick in Figure 3.9. The swirls can be displaced in opposite directions by the field from a current in the *x*-direction as shown in Figure 3.7 for the brick. A uniform H_y will move both the swirls toward one end or the other of the ellipsoid. A large H_y will drive an S^+-pattern into an I^*-pattern (see Figure 3.16) near the end of the ellipsoid. Then, on decreasing H_y, the I^*-pattern becomes an S^+-pattern again, if biased by a field in H_x. If the bias field is in $-H_x$, the I^*-pattern becomes an S^--pattern. In a high enough H_y field, the swirls move together out from one end of the ellipsoid.

The *S*-vortex state for a nanoellipsoid is shown in Figure 3.11. The nanoellipsoid avoids the discussion of what happens in the corners for the Landau structure in the nanobrick. The central core of the vortex in the nanoellipsoid is the Bloch wall terminating at the displaced swirls illustrated by an m_z-isosurface for $m_z=0.8$, which corresponds to the 3-4-5 triangle with acos $m_z\approx37°$. (The m_z-isosurfaces in Figure 3.7 are also for $m_z=0.8$. In Figure 3.9, the portrait of the *S*-vortex, the m_z-isosurfaces are for $m_z=0.95$.) The Néel caps and the closure domains are represented by the m_y-isosurfaces for $m_y=\pm0.95$ in Figure 3.11. The contour lines on the circular slices at various values of *x*, show the m_x components. The principal domains of the Landau structure in the ellipsoid are suggested by contours with $m_x>0.8$ and $m_x<-0.8$.

3.2.6 Manipulating the Landau Structure

Near the critical field, it is easy to drive the swirls back and forth between their small offset positions. At $H_z = 0$ it is harder. But it can still be done. It is easier to do this dynamically with the right time-sequence of fields. The moving swirls form a highly nonlinear oscillator. They can move over distances of 50 nm in 200 ps. The swirls carry with them localized external magnetic fields with $\mu_0 H_z \approx 0.5\,\mathrm{T}$ from the surface magnetic charges. At resonance, they will oscillate as long as the energy is supplied to compensate for the damping losses. As the swirls oscillate back and forth in the x-direction, they also make excursions in the y-direction, as they follow paths of almost constant energy. This is in contrast with the motion of the swirls in varying fields, where the paths are perpendicular to the paths of constant energy. It is easier to go around a barrier than to climb over it. The ability to easily and quickly move well-localized sources of large external magnetic fields is a phenomenon waiting to be exploited in the world of nanoscience and nanotechnology.

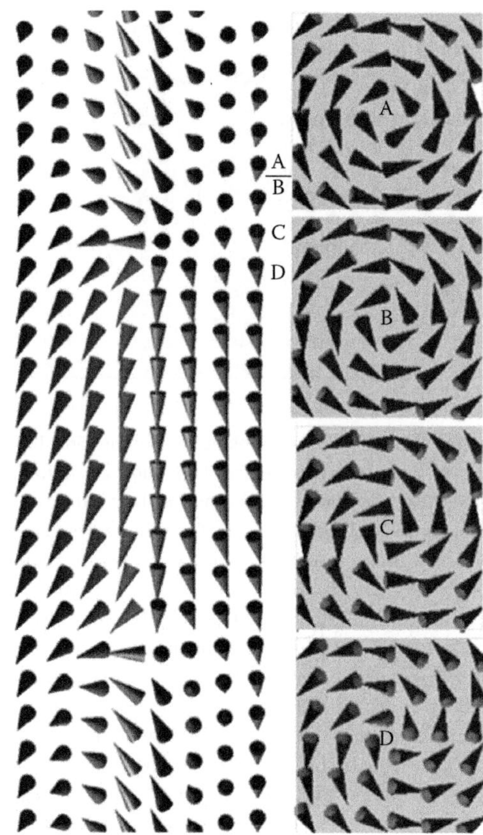

FIGURE 3.12 Pair creation in a reversed applied magnetic field $\mu_0 H_z = -0.8\,\mathrm{T}$. The centers of the propagating singularities are in the middle of the four central cones in planes of constant z in each of the panels A, B, C, D on the right. The panel on the left is a cut in a plane of constant y through the two cones just above that center. By symmetry, the two central columns of the cones on the left are two views of each cone of the four central cones. In panels A, B, C, D, the four central cones all rotate in unison as one views this as a sequence of layers one above the other. It is also the time sequence for any given layer as the singularity passes up the axis. For the upper singularity, $\partial m_z/\partial z$ is positive and in the lower singularity, it is negative. For splay canceling, a negative m_r is required for the first case and a positive m_r for the lower case, as $\nabla \cdot \boldsymbol{m} \approx m_r/r + \partial m_z/\partial z$. As m_r is positive in both regions, there is splay saving in the case of the lower singularity. The lower one propagates more quickly as the splay saving lowers the energy barrier.

3.2.7 Core Reversal by Pair Creation

When H_z is increased in the $-\boldsymbol{k}$ direction, the swirls move back toward the z-axis, reaching the axis at a critical field that is lower in magnitude than the critical field for forming the displaced swirls on the decreasing field in the $\boldsymbol{k}$-direction. During all these changes, the $m_z = 1$-line continues to go from the bottom surface to the top surface. But at high enough field in the $-\boldsymbol{k}$ direction, a pair of singularities are created that propagate from near the mid-plane outward toward the surfaces, reversing the core of the vortex as they propagate (see Figure 3.12).

The two singularities are singular in a continuum model. On a finite grid, the core of the vortex centers itself on a position between the grid points, so that the singularity itself is a mathematical point between the grid points. This should be the case for a real lattice, where the center of the vortex would lie between the atoms. But that is a classical description for which there is no quantum mechanical calculation to support the concept of the atoms maintaining a rigid magnetic moment during the reversal. The magnetic moment density can vary in direction as well as magnitude across an atom [9].

3.3 Landau–Lifshitz–Gilbert Equations and Their Numerical Solutions

The above description is for a vortex that intersects the top and bottom surfaces. Starting with fields in the x- or y-directions, vortices can form with swirls on the end and/or side surfaces. In either case, the removal or reversal of these fields returns the system to the state where the swirls are on the top and bottom surfaces. This process can be quite complex with the swirls moving from one surface to another or by the vortices leaving the brick and then reentering. Starting with the Landau structure in zero field, there is a rich landscape of the responses, steady and dynamic, to fields and field gradients applied in the x–y plane.

All of the above can be seen by solving the Landau–Lifshitz–Gilbert (LLG) equations of motion. A modern review of micromagnetics is given by 33 authors in 500 pages of Vol. 2 of the *Handbook of Magnetism*. The chapters on numerical methods are quite detailed with hundreds of equations [10].

There are few cases where these equations can be treated by analytical techniques. The LLG equations are simultaneous integral–differential equations for the direction cosines of the components of the magnetization. The differential part comes from the exchange energy responsible in the first place for ferromagnetism. Any variation of the moments from parallel alignment locally increases the negative exchange energy. The integral part comes from the dipole–dipole interactions in which, if there are N moments, there are $N(N-1)/2$ pairs to consider.

3.3.1 Zeeman Energy

The independent variable is the applied magnetic field $\boldsymbol{H}_{\text{ext}}$. The Zeeman energy is $E_{\text{ext}} = -\boldsymbol{\mu B}$ of a magnetic moment $\boldsymbol{\mu}$ in an externally applied magnetic induction, $\boldsymbol{B}$, defined by the force on a moving charge. In free space, $\boldsymbol{B} = \mu_0 \boldsymbol{H}_{\text{ext}}$, where $\mu_0 \equiv 4\pi \times 10^{-7}\,\text{H/m}$, is an arbitrary constant called the permeability of free space in SI units, which were enacted by a one-vote margin at a conference in 1931 dominated by electrical engineers with little appreciation of magnetism. Magneticians have resisted the adoption of these units in favor of Gaussian units for reasons of good physical insight. In Gaussian units $\mu_0 = 1$. The magnetic moment of a nanobrick is $\boldsymbol{\mu} = \langle \boldsymbol{M} \rangle V$ where $\langle \boldsymbol{M} \rangle$ is the average magnetization in the volume V. In SI units, the magnetization $\boldsymbol{M}$ is replaced by the magnetic polarization $\boldsymbol{J}$, where $\boldsymbol{\mu} = \langle \boldsymbol{J} \rangle V/\mu_0$. In terms of the unit vector $\boldsymbol{m}$ representing the direction of magnetization or magnetic polarization, the Zeeman energy in SI units is $E_{\text{ext}} = -J_s \boldsymbol{m} \boldsymbol{H}_{\text{ext}} V = -M_s \boldsymbol{m} \cdot (\mu_0 \boldsymbol{H}_{\text{ext}}) V$, where J_s and M_s are the spontaneous polarization and the spontaneous magnetization, respectively, of the iron nanobrick. For iron, $J_s = 4\pi \times 1714/10^4\,\text{T} = 2.154\,\text{T}$, where the 1714 comes from M_s for iron in Gaussian units and the 10^4 comes from the conversion of Gauss to Tesla. In Gaussian units the polarization has the same units as the magnetization but differs by the factor 4π; that is, $J_s = 4\pi M_s$. In SI units, $J_s = \mu_0 M_s$. This means that one cannot convert from SI to Gaussian just by setting $\mu_0 = 1$. The magnetic susceptibility, which is dimensionless in both systems, differs by that factor of 4π. The engineers removed the 4π's in Maxwell's equations, from where they properly belong, in front of the source terms. In Maxwell's equations $\boldsymbol{M}$ appears as a current source given by $\boldsymbol{j}_{\text{m}} = \nabla \times \boldsymbol{M}$. The 4π will reappear when the dipole–dipole interaction is considered. The anisotropy energy and the exchange energies depend only on the directions of magnetization, so the question of units does not affect their energy expressions.

3.3.2 Anisotropy

When a magnetic moment is rotated with respect to the atomic lattice, there is a change in energy because of spin–orbit coupling to the lattice. The derivatives of the anisotropy energy with respect to the magnetization components produce effective fields acting on the magnetization. These are not real fields, and do not necessarily obey the Rabi–Schwinger theorem [11], that a magnetic moment in a real field can be treated classically, but they are treated as real fields in the LLG equations. As the anisotropy fields play very minor roles in the behavior of an iron nanobrick, quantum mechanical subtleties will be ignored here. The magnetization patterns for an iron nanobrick and those for a permalloy nanobrick are indistinguishable in zero field at the level of visually comparing graphs of lines of constant components of the reduced magnetization. To see where anisotropy is important in spintronics, visit Chapter 2 of this book.

3.3.3 Exchange Field

The effective field from the exchange interaction can be written using the Laplacian of the direction cosines of the magnetization $\nabla^2 \boldsymbol{m}$, leading to nine terms in the torque equations representing curvatures in each of the components in each of the three directions. The coefficient A of the Laplacian ∇^2 is called the exchange stiffness constant. Due to $|\boldsymbol{m}|^2 = 1$ and hence $\boldsymbol{m} \cdot \partial \boldsymbol{m}/\partial x = 0$, the exchange energy density e_{ex} can be written as

$$e_{\text{ex}} = A(\nabla \boldsymbol{m})^2 = -A\boldsymbol{m} \cdot \nabla^2 \boldsymbol{m}, \tag{3.1}$$

where the Laplacian can also be written as

$$\nabla^2 \boldsymbol{m} = \nabla(\nabla \cdot \boldsymbol{m}) - \nabla \times (\nabla \times \boldsymbol{m}). \tag{3.2}$$

In a divergence-free pattern of magnetization, the only contributions to the exchange field come from $\nabla \times \boldsymbol{m}$. In polar coordinates, there are 15 terms in the Laplacian. Even so, it is easier to think about divergence-free patterns using the Laplacian rather than $(\nabla \boldsymbol{m})^2$. The simplest magnetization pattern for a vortex is $m_r = 0$, $m_\phi = \tanh(ar)$, and $m_z = \text{sech}(ar)$, where a is a constant. There are no derivatives with respect to ϕ or z, there is no m_r component, and $\nabla \cdot \boldsymbol{m} = 0$. In this case there remain five terms in the Laplacian. When the dot product of the Laplacian and the magnetization is considered, two of the terms cancel and two of the terms combine, leaving the local exchange energy density $e_{\text{ex}}(r)$ as

$$e_{\text{ex}}(r) = A\left(\frac{\text{sech}^2(ar)}{a^2} + \frac{\tanh^2(ar)}{r^2}\right). \tag{3.3}$$

Even after these simplifications, the integration of Equation 3.3 requires approximations [12].* It would be helpful to have the total exchange energy in terms of $\langle m_z \rangle$, but that requires the evaluation of $\equiv r\,\text{sech}(ar)\,\mathrm{d}r$, which also does not have an analytic expression. Progress in micromagnetics is difficult without numerical methods. The exchange fields calculated from the simple case using $m_z = \text{sech}(ar)$ and $m_r = \tanh(ar)$ do not quite point in the same direction as the magnetization, indicating that these are not self-consistent solutions of the torque equations. They are, however, quite good approximations for the magnetization around a vortex in an ultrathin circular disc (one for which there is no significant dependence on z), when the full micromagnetic

* See p. 131 for the approximation to the integral of $\tanh^2(\rho/\lambda)/\rho$. This reference follows micromagnetics from a single iron atom imbedded in a lattice to a 4 nm thick circular disk of 96 nm diameter. The present chapter continues that development to thicknesses where the patterns of magnetization are fully 3D.

calculation includes all the dipole–dipole interactions [12]. In such situation, the parameter a is close to $1/\lambda$, where λ is the exchange length discussed below.

3.3.4 Dipole–Dipole Interactions

The treatment of the dipole–dipole interaction is different in the finite element program *TetraMag* by Riccardo Hertel.* and the finite grid program *LLG Micromagnetic Simulator* by Scheinfein.† In *TetraMag*, the moments in each element are used to solve Poisson's equation for the magnetic potential, from which the fields are derived. In the LLG Micromagnetic Simulator the dipoles on a uniform grid are summed using fast Fourier transforms, which require that every grid point be treated similarly so that the interactions depend only on the vector, connecting any two grid points. Using fast Fourier transforms reduces the calculation from the order of N^2 to the order of $N \ln(N)$. Fortunately, the equations of micromagnetics have attracted mathematicians who have brought sophisticated methods to bear on the problem of treating the dipole–dipole problem, sufficiently sophisticated to be beyond the intent of this article.

There is a problem in the finite grid approach in treating boundaries that are not aligned with the grid. This is avoided here by choosing the parallelepiped as the object of interest. One solution to the jaggy-edge problem is to treat the local regions on the order of $n(n-1)/2$ and farther regions at the level of $N \ln(N)$ for all N points, adding $Nn(n-1)/2$ calculations, where n is the number of points in the local region. It takes but a small n for $n(n-1)/2$ to be bigger than $\ln(N)$.

The dipole–dipole energy is written using the demagnetizing field. This is taken from the fact that the sources of $\nabla \times \boldsymbol{B}$ are currents and there are no sources for $\nabla \cdot \boldsymbol{B}$. The vector $\boldsymbol{H}$ is a mixed vector combining the source vector $\boldsymbol{M}$ with the field vector $\boldsymbol{B}$. In Gaussian units the combination defines $\boldsymbol{H}$ as $\boldsymbol{H} \equiv \boldsymbol{B} - 4\pi \boldsymbol{M}$, where the 4π belongs in front of the source vector. In SI units, $\boldsymbol{H}$ is defined as $\mu_0 \boldsymbol{H} \equiv \boldsymbol{B} - \boldsymbol{J}$. The sources of $\nabla \times \boldsymbol{H}$ are real currents. The sources of $\nabla \cdot \boldsymbol{H}$ are magnetic charges given by $\nabla \cdot \boldsymbol{m}$. The part of $\boldsymbol{H}$ that derives from these charges is called the demagnetizing field $\boldsymbol{H}_D$. In SI units, $\nabla \cdot \boldsymbol{H}_D = -(J_s/\mu_0)\, \nabla \cdot \boldsymbol{m}$. In Gaussian units $\nabla \cdot \boldsymbol{H}_D = -4\pi M_s\, \nabla \cdot \boldsymbol{m}$. The demagnetizing field energy is the integral over the nanobrick to get $E_{dem} = -(1/2) \iiint \boldsymbol{H}_D \cdot \boldsymbol{J}\, dV$ in SI units and $E_{dem} = -(1/2) \iiint \boldsymbol{H}_D \cdot \boldsymbol{M}\, dV$ in Gaussian units. The factor (1/2) comes from these being self-energies. For a sphere uniformly magnetized in the z-direction, in Gaussian units $\boldsymbol{H}_D = 4\pi(1/3) M_s \mathbf{k}$ and $\boldsymbol{M} = M_s \mathbf{k}$ for which $E_{dem} = -(1/2)\ 4\pi(1/3) M_s^2$. In SI units, $\boldsymbol{J} = J_s \mathbf{k}$ and $\boldsymbol{H}_D = (1/3) J_s/\mu_0 \boldsymbol{k}$ for which $E_{dem} = -(1/2)\ (1/3)\ J_s^2\ V/\mu_0$ or $E_{dem} = -(1/2)\ (1/3) \mu_0 M_s^2\ V$. In all cases, the (1/3) is the demagnetizing coefficient N for a sphere. In the general ellipse, $N_x + N_y + N_z = 1$. In Gaussian units, $4\pi N$ is called the demagnetizing factor.

3.3.5 Torque Equation

The basic equation of micromagnetics is the torque equation for the precession of an electron in a magnetic induction $\boldsymbol{B}$. The electron has a magnetic moment $\mu = -g\mu_B S$ and an angular momentum $S\hbar$, where $\hbar$ is the reduced Planck's constant, g is very close to 2 and $S = 1/2$. The minus sign appears because the spin and the moment are in opposite direction for the negatively charged electron. The ratio of the angular momentum to the magnetic moment of the electron is $\gamma_e = -g\mu_B/\hbar$. These are used in a classical equation of motion, where the angular momentum of the magnetic moment is $\boldsymbol{L}_\mu = \boldsymbol{\mu}/\gamma_e$ and the torque acting is $\boldsymbol{\mu} \times \boldsymbol{B}$; that is,

$$\left(\frac{1}{\gamma_e}\right)\frac{d\mathrm{m}}{dt} = \mathrm{m} \times \boldsymbol{B}. \tag{3.4}$$

$\boldsymbol{B}$ can be replaced by $\mu_0 \boldsymbol{H}$ in the torque. Dividing both sides by a small volume and letting $\boldsymbol{\mu}$ stand for the moment in that volume, $\boldsymbol{\mu}$ can be replaced by $\boldsymbol{M}$ to give

$$\left(\frac{1}{\gamma_e}\right)\frac{d\boldsymbol{M}}{dt} = \mu_0 \boldsymbol{M} \times \boldsymbol{H}, \tag{3.5}$$

which is in SI units, but produces the torque equation in Gaussian units by replacing $\mu_0 = 4\pi \times 10^{-7}$ H/m by a dimensionless 1. It is common practice to absorb the μ_0 into the gyromagnetic ratio to write

$$\left(\frac{1}{\gamma_0}\right)\frac{d\boldsymbol{M}}{dt} = \boldsymbol{M} \times \boldsymbol{H}, \tag{3.6}$$

where $\gamma_0 \equiv \mu_0 \gamma_e$. As $\boldsymbol{M}$, the magnetization, appears on both sides of the equation, it can be replaced by $\boldsymbol{J}$, the magnetic polarization ($\boldsymbol{J} = \mu_0 \boldsymbol{M}$), to obtain

$$\left(\frac{1}{\gamma_0}\right)\frac{d\boldsymbol{J}}{dt} = \boldsymbol{J} \times \boldsymbol{H} = \boldsymbol{J} \times \{-\nabla_J(e_{ext})\}. \tag{3.7}$$

The reason this is done is to treat the $\boldsymbol{H}$ in Equation 3.7 as an effective field in an expression for the free-energy density in which the Zeeman term is $e_{ext} = -\boldsymbol{M} \cdot \boldsymbol{B} = -\boldsymbol{J} \cdot \boldsymbol{H}$. The price for using the magnetic polarization as the variable in the torque equations is to put some μ_0's into expressions for the anisotropy and exchange energies where there is no physical reason for them to exist there. The other terms are derived from expressions

* A concise summary of TetraMag is found in TetraMag—A general-purpose finite-element micromagnetic simulation package and high resolution large-scale micromagnetic simulations with hierarchical matrices by Riccardo Hertel and Attila Kakay. These are available by inserting TetraMag Hertel into the Google search box.

† A full description of Michael R. Scheinfein's LLG Micromagnetic Simulator is available by inserting Scheinfein micromagnetic simulator into the Google search box.

where the variable $\boldsymbol{m}$ is the direction of the polarization (and the magnetization).

$$\nabla_J(e_{\text{tot}}) = \frac{\nabla_m(e_{\text{tot}})}{J_s} = \frac{\nabla_m(e_{\text{tot}})}{(\mu_0 M_s)}. \tag{3.8}$$

The $\boldsymbol{H}$ in Equation 3.7 becomes an effective field. It includes the applied field, the demagnetizing field from all the other magnetic moments, the exchange field from the variation of the moment direction with position, the anisotropy field, and the damping field. The damping field was formulated by Gilbert to be proportional to the rate of change of the components of the magnetization [13].* The damping in iron comes from the repopulation of the Fermi surfaces of spin-up and spin-down electrons as the magnetization direction is rotated locally. These spin currents dissipate energy. Spin currents can be created externally and used to cause the magnetization to rotate by forced repopulation of the Fermi surfaces. The two processes differ in the sign of the coefficient of the contribution of $\partial \boldsymbol{m}/\partial t$ to the effective field. There is much more about the subject of damping in Bretislav Heinrich's chapter in this book (Chapter 2). Spin currents are not discussed in this chapter because the author has not applied them to the nanobrick, yet.

3.4 Applying the Micromagnetic Equations of Motion

This chapter describes in some detail the results of calculations for the iron nanobrick. The parameters are those of iron except for the calculation of the equilibrium configurations, where the damping coefficient α is greatly increased from the low value of iron to the value that gives critical damping. The dynamic calculations use $\alpha = 0.02$ and the equilibrium calculations use $\alpha = 1$. For $\alpha = 0.02$, the time constant for the approach to equilibrium is typically 10's of ns. For $\alpha = 1$ the time constants can be <1 ns, unless the system is near a critical point for which the torques vanish. Then, the time can be too long to compute and the usual approach to equilibrium, as an exponential, changes to a $1/t$ approach. As critical points are of interest in describing magnetic configurations, it is necessary to have techniques to obtain the answers more quickly.

The time steps used in micromagnetic calculations are small compared to the time resolution adequate to describe the fastest of dynamic responses. The torque equations have mathematical instabilities that often require that the time steps be as small as a fraction of a femtosecond. The time step must be smaller when the grid size is smaller. The grid size itself should be smaller by at least a factor of two than the exchange length λ given by $\lambda^2 = A/K$, where A is exchange-stiffness constant of Equation 3.1 and K is the magnetostatic-energy density; $K = \mu_0 M_s^2/2$. In classical micromagnetics [14],† the language is that of Gaussian units, where M, H, and B all have the same dimensions; there, μ_0 is replaced by 4π in the magnetostatic energy used in the definition of the exchange length.

The use of a grid spacing that is too large can lead to results that are completely misleading. In a treatment of the magnetization processes in a nanobox with square sides, $X = Y > Z$, it was shown that the moments in the I_z-vortex have $m_z = 0$, when the grid is greater than 3λ [15]. Critical fields are sensitive to the grid size even for grid size $<\lambda/10$.

3.4.1 Bias Fields

The fields from currents used to break the symmetry at critical points also provide a means of avoiding the prohibitively long computation times. Such bias fields are called anticipatory fields [16], as they select among three prongs of a three-tined fork. Without a proper bias field, one can stay on the central tine, even though the other tines lead to lower energies. If an anticipatory field is not used, the middle tine will be abandoned after numerical round-off errors propagate exponentially with time for more than 10-time constants. Away from critical points, the total energy comes to equilibrium exponentially, while the components of the magnetization and individual terms in the energy come to equilibrium as damped oscillators. The convergence of a calculation is achieved when the extrapolation of the exponential or the damped oscillators to infinite time no longer changes significantly with time. To find a critical field, it is convenient to use a bias field that takes the magnetization from one configuration to another quickly and then analyze the results of that calculation to obtain what would have happened in the absence of that bias field. This is called the path method [17].

3.4.2 Internal Energy

The path method relies on the insensitivity the internal energy E_{int} to the values of the external fields needed to reach configurations that have the same values of the average components of the magnetization $\langle m_x \rangle$, $\langle m_y \rangle$, $\langle m_z \rangle$. E_{int} is the sum of all the energy terms excluding the Zeeman term. The energy, divided by the volume V, along an equilibrium path can be written as

$$\frac{E}{V} = -\left\langle \boldsymbol{m}^* \right\rangle \left(J_s \boldsymbol{H} - \nabla_{\langle \mathbf{m} \rangle} \left(E_{\text{int}} \right) \big|_{\langle \mathbf{m} \rangle = \langle \mathbf{m}^* \rangle} \right), \tag{3.9}$$

where $\langle \boldsymbol{m}^* \rangle$ is $\langle \boldsymbol{m} \rangle$ at some point along the path. Knowing the gradient of the internal energy E_{int} with respect to the components of the average magnetization as a function of $\langle \boldsymbol{m} \rangle$ along an equilibrium path, one can calculate the field necessary to reach equilibrium (not necessarily stable) for a given $\langle \boldsymbol{m}^* \rangle$ along that path. The path method assumes that knowledge of E_{int} along a path that is

* The theoretical foundation is of current interest as described in the chapter by Bret Heinrich and in the interesting exchange in *Physical Review Letters* that can be accessed by entering Hickey Replies in the Google search box.

† These are still worth reading today, not only for the physics but also for the style. Brown was also an English teacher.

close to a given equilibrium path will give the same result; that is, $\nabla_{\langle \mathbf{m}\rangle}(E_{\text{int}})|_{\langle \mathbf{m}\rangle=\langle \mathbf{m}^*\rangle}$ is assumed to be insensitive to small bias fields. The calculations of the magnetic response to applied fields are carried out for sufficient number of fields to determine the functional form of E_{int}, but no more than necessary. Once one has an analytic expression for $E_{\text{int}}(\langle \boldsymbol{m}\rangle)$ of a particular type of pattern, one can produce the entire dependence of the magnetization on field. When this works, it can save many orders of magnitude in computation time. To determine whether it works can take time, but that need only be done once for a given type of system.

3.4.3 Using Anticipatory Fields

An example of the use of the path method and anticipatory fields is the calculation of the field for a vortex parallel to the z-axis to reenter an iron nanobox [15] after being driven out through the $y = Y/2$ surface by a field H_x. After the vortex leaves, the magnetization is in a *C-state*, which can be viewed as a virtual I_z-vortex just outside the brick. When H_x is reduced sufficiently, the vortex should reenter the surface through which it exited. The three-tined fork in this case is the fact that the virtual vortex must choose a direction of the magnetization for the core in order to reenter. The virtual vortex does not have a polarization before it enters unless the *C-state* itself has a bias in the direction of the vortex that exited, which can happen for particular geometries. If a bias field along the z-axis is not applied, the system stays on the central tine and the vortex does not reenter until long after round-off error provides an initial bias. If a bias field is applied along the z-axis, it changes slightly the H_x at which the vortex enters, but it changes greatly, by many orders of magnitude, how long one would have to wait for that to happen.

3.4.4 Using the Path Method

A hypothetical example of the path method is given in Figure 3.13 where it is supposed that the dependence of the internal energy E_{int} upon $\langle M_x\rangle$ is shown in the left panel as the solid curve for an arbitrarily constructed

$$E_{\text{int}} = \frac{\left(E_2\langle m_x\rangle^2 + E_4\langle m_x\rangle^4 + E_6\langle m_x\rangle^6\right)}{\left(1-\langle m_x\rangle^{12}\right)} \tag{3.10}$$

to produce four inflection points and also mimic the approach to saturation. The system is in unstable equilibrium for the region between the inflection points at **a** and **b**. The dashed line shows $\mathrm{d}E_{\text{int}}/\mathrm{d}\langle m_x\rangle$, which is shown again in the right panel as an independent variable to produce a magnetization curve with hysteresis. For equilibrium the applied field must be equal and opposite to the internal field given by $-\mathrm{d}E_{\text{int}}/\mathrm{d}\langle m_x\rangle$, which explains why there are no minus signs in this illustration of the path method.

E_{int} can be constructed by the analysis of the calculations in regions of stable equilibrium to interpolate the regions of unstable equilibrium. E_{int} can be calculated using the regions of unstable equilibrium if the time scale of changes is appropriate for obtaining close to equilibrium configurations while the configuration as a whole is moving in time.

The path method works only for large damping. For small damping, the magnetization moves on a path of almost constant energy, while the path method has the magnetization moving in the direction of maximum gradient of the energy.

3.5 Around the $\langle M_z\rangle$–H_z Hysteresis Loops

The computations are designed to obtain E_{int} as a function of $\langle \boldsymbol{M}\rangle$ and $\langle \boldsymbol{M}\rangle$ as a function of H_z for each configuration using appropriate bias fields when necessary. Hysteresis loops are shown for $\langle m_z\rangle$ and $\langle m_x\rangle$ in Figure 3.14 with and without a small bias field from a current $i_x = 0.1\,\text{mA}$. The iron brick has dimensions $50 \times 80 \times 130\,\text{nm}^3$. The sequence of configurations starting from high H_z includes the following.

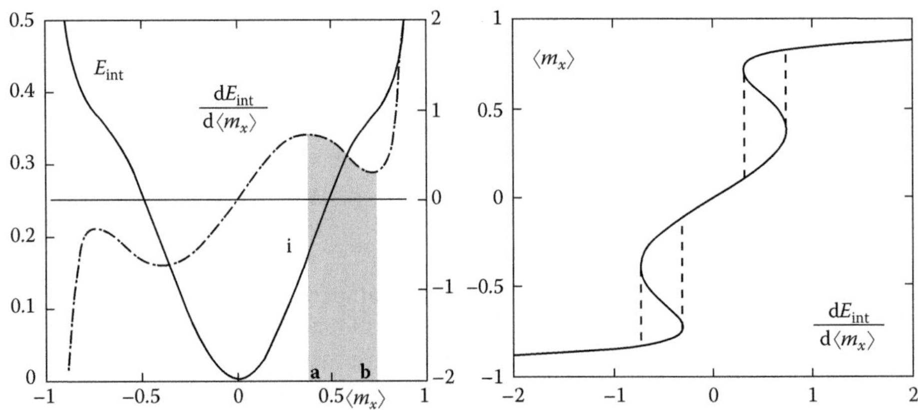

FIGURE 3.13 Illustration of the *path method*. The dependence of the internal energy E_{int} upon $\langle M_x\rangle$ is shown in the left panel as the solid curve for an arbitrarily constructed $E_{\text{int}} = E_2\langle M_x\rangle^2 + E_4\langle M_x\rangle^4 + E_6\langle M_x\rangle^6)/(1-\langle M_x\rangle^{12})$ to produce four inflection points and also mimic the approach to saturation. The system is in unstable equilibrium for the region between the inflection points at **a** and **b**. The dashed line shows $\mathrm{d}E_{\text{int}}/\mathrm{d}\langle M_x\rangle$, which is shown again in the right panel as the independent variable ($\sim H_x$) to produce a magnetization curve with hysteresis.

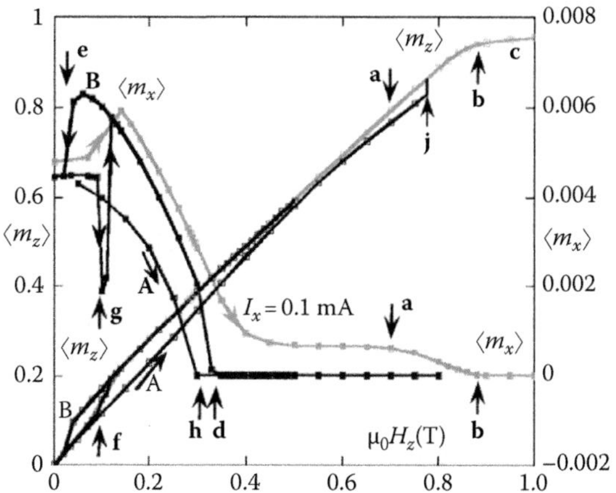

FIGURE 3.14 Hysteresis loops for an iron nanobrick with dimensions $130 \times 80 \times 50\,\text{nm}^3$. The first quadrant of the major hysteresis loop is shown for both $\langle m_z \rangle$ and $\langle m_x \rangle$, the latter with much magnification (its ordinate is on the right.) The gray curves for $\langle m_z \rangle$ and $\langle m_x \rangle$ were calculated with a bias current $i_x = 0.1\,\text{mA}$ to anticipate the transitions to and from the *I* state. The bias current has little effect on $\langle m_z \rangle$, but facilitates following the transitions from the effects it has on $\langle m_x \rangle$. The gray curve for $\langle m_x \rangle$ measures the degree of curling between points **a** and **b** and measures the susceptibility of the *I* state to form the *S* state in the field region between 0.3 and 0.4 T. The two black curves for $\langle m_x \rangle$ show the sharp transition between the *S* and *I* state near 0.3 T that occur in the absence of the bias current. The curve for $\langle m_x \rangle$ labeled **A** is for the vortex core m_z in the $-z$ direction after coming from saturation at high negative field. The curve for $\langle m_x \rangle$ labeled **B** is for the vortex core m_z in the $+z$ direction with the transition occurring at a slightly higher field, **d** compared to **h**. All the hysteresis at low fields is from the switching of the magnetization directions in the four corners of the nanobrick. For $\langle m_z \rangle$, the black curves, one of which hides the gray curve, are for no bias current. The curve for $\langle m_z \rangle$ labeled A is for the core of the vortex in the $-z$-direction with the transition for the state with the core m_z in the $+z$ direction taking place by pair creation and propagation at **j**. There are two different states with the same components of the magnetization in zero field. One of these has the core of the vortex in the $+z$-direction and the other has the core in the $-z$-direction. This would give different values for the magnetization in zero field if it were not for the almost complete compensation of the net magnetization by the moments in the four corners, which are opposed to the core magnetization in both cases.

3.5.1 Flower State

The magnetization splays out from the center on the top surface and inward toward the axis on the bottom surface with most of the magnetization along the $+z$-axis (see Figure 3.3 and the section of Figure 3.14 labeled **c**, $\mu_0 H_z \sim 1T$).

3.5.2 Curling State with Inversion Symmetry

The circulation is cw on one half of the nanobrick and ccw on the other (see Figure 3.5). This is not shown in Figure 3.14. (The reader will be spared the complexities of magnetization processes proceeding from this state.)

3.5.3 I_z-Vortex State

The same handedness throughout the nanobrick was achieved by applying a bias field from a current along the z-axis, which can be removed once the handedness is chosen.

The gray curves in Figure 3.14 were obtained in the presence of a bias field from a current along the x-axis, $i_x = 0.1\,\text{mA}$, which anticipates the transformation from the I_z-vortex state to the S_z-vortex state as H_z is lowered. The degree of displacement of the swirls is tracked by $\langle m_x \rangle$. The inflexion point on the gray curve at $\mu_0 H_z = 0.32\,\text{T}$ corresponds to the field at which the transition takes place in the absence of the bias field, labeled **d**. The bias current i_x does not produce any $\langle m_x \rangle$ in the flower state but does in the curling state, so that $\langle m_x \rangle$ tracks the onset of curling at **b** and the approach to saturation of that effect at **a** on the gray curves.

3.5.4 Landau Type Curling State

The I_z-vortex distorts spontaneously into either the S_z^+-vortex or the S_z^--vortex, depending on the bias field applied from a current along the x-axis (see Figures 3.5 through 3.9). The presence of the S_z-vortex is signaled by the presence of $\langle m_x \rangle$ in the absence of bias fields. The presence of the S_z-vortex has a minor effect on $\langle m_z \rangle$ that can be noticed after subtracting the demagnetizing field, the dominant effect of the magnetostatic energy.

3.5.5 Corner States

The Landau type state has four choices of polarization, plus (p) or minus (m) for the virtual vortices along the four vertical corners. Without additional bias fields, the four corners for the S_z^+-vortex state are either *pppp*, *ppmm*, or *mmmm*, labeled *cw* from the corner ($-X/2$, $Y/2$). For the S_z^--vortex state the sequence is *pppp*, *mmpp*, *mmmm*. Each of these states has its range of stability with hysteresis in the minor loops for the switching of the corners. The minor hysteresis loops at low fields are shown in Figure 3.14 and in more detail in Figure 3.15, where the mean effect of the demagnetizing field has been subtracted. All of this is avoided in the ellipsoid, which has no corners.

3.5.6 S-Vortex in Reversed Field

The *S*-vortex is at its maximum extension for $H_z = 0$. For $H_z < 0$ the core of the *S*-vortex is opposite to the field, but most of the magnetization follows H_z because the energetics is a competition between the magnetostatic energy and the Zeeman energy, both of which are reduced slightly by the core magnetization remaining in the positive z-direction. The *S*-vortex changes back to the *I*-vortex in negative field. This is seen in the curve labeled **A** for the $\langle m_x \rangle$ component, which goes to zero for a smaller magnitude of H_z when the fields are in the direction opposite to the core magnetization; compare points **d** and **h** in Figure 3.14. The curves labeled **A** are calculated for negative fields and then replotted in the first quadrant for direct comparison with the positive fields.

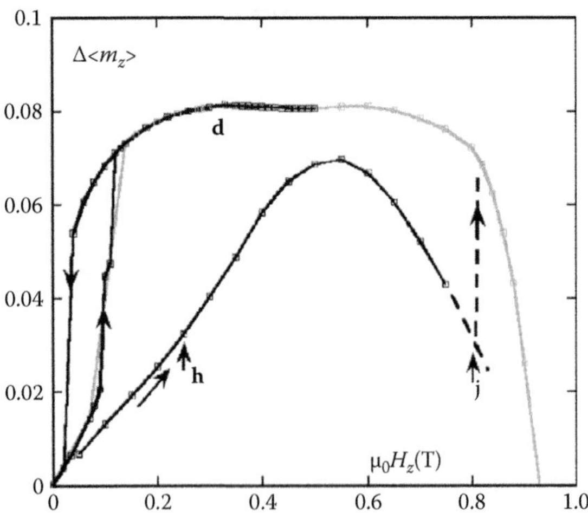

FIGURE 3.15 The hysteresis loops for $\langle m_z \rangle$ versus $\mu_0 H_z$ after subtracting H_z/H_s from each $\langle m_z \rangle$ in Figure 3.14. The hysteresis in low fields accompanies the reversal of the magnetization in the four corners. All four corners flipped in a single large step in field on decreasing field, but on increasing field in smaller steps, two flip back first and, then, after a second step in field, the other two flip back. When resolved in time, the four flip independently from one another. The onset of the S-vortex is signaled by $\Delta\langle m_z \rangle$, which begin to decrease below **d**. The lower curve is for the magnetization of the core opposite to the applied field. The S-vortex goes back to the I-vortex at **h**. The higher slope for the reversed field reflects the smaller demagnetizing field for the trapped vortex. Point **j** is where the core reverses by pair creation and propagation.

Those curves would be obtained directly in the first quadrant if the magnetization process started with large negative fields and then proceeded to positive fields.

3.5.7 Trapped I_z-Vortex State

When the field is large in the reversed direction, the magnetization is negative everywhere except in the core of the vortex. The demagnetizing field becomes smaller for the trapped state leading to a higher slope of m_z vs. H_z.

3.5.8 Transient Core Reversal State

Starting with the trapped $+I_z$-vortex, as the field becomes more negative, near $-\mu_0 H_z = 0.8$ T, a pair of point singularities is created near the origin that propagate up and down the z-axis reversing the magnetization of the core to produce the $-I_z$-vortex with the same circulation as maintained in all these processes (see Figure 3.12 and point **j** in Figure 3.14).

3.5.9 Saturation

At a high negative H_z, the $-I_z$-vortex state goes to the flower state with the moments splaying outward from the center of the bottom surface and inward toward the center of the top surface, with most of the magnetization along the $-z$-axis.

3.6 Discussion of Magnetization Processes

The programs for micromagnetic calculations keep track of the components of the magnetization for each grid point for every-so-many iterations as well as the net magnetizations and each term in the energy of the nanobrick as a whole. These are analyzed to gain insight into the competition among the energy terms and how they lead to the various magnetization patterns.

3.6.1 Contributing Factors

The leading energy term is the Zeeman energy. Even when the external fields are all zero, the configuration in a real nanobrick is one that reflects the past history of the magnetization. On the computer, one can arbitrarily assign a configuration A and calculate the configuration B that minimizes the energy starting from A. Then, if $H_z = 0$, the dominant term in the energy of the iron nanobrick is the dipole–dipole energy. At all other fields the competition is between the Zeeman energy and the dipole–dipole energy with the exchange playing a supporting role and the anisotropy almost no role at all.

3.6.1.1 Demagnetizing Factor

To a good approximation the dipole–dipole energy is quadratic in $\langle m_z \rangle$ for all the above states of the nanobrick. The Zeeman energy is, of course, linear in $\langle m_z \rangle$. The magnetization curve is then approximately linear in H_z until the flower state is approached at high fields. The magnetization is given to a good approximation by $H_z - H_s\langle m_z \rangle = 0$, where $H_s\langle m_z \rangle$ is called the effective demagnetizing field. To emphasize the role of the supporting actors in this drama and to show the nonquadratic terms in E_{dem}, a magnetization $\langle m_z \rangle_D \equiv H_z/H_s$ is subtracted from each point in Figure 3.14 to produce Figure 3.15, where the value of H_s has been chosen to make the deviation of $\Delta\langle m_z \rangle = \langle m_z \rangle - \langle m_z \rangle_D$ independent of H_z over much of the range of H_z.

3.6.1.2 Vortex Contributions to the Fields

To some extent the iron nanobrick behaves like an ideally soft magnetic material, but the core of the vortex has its own life. It builds up during the transition from the flower state to the curling state. It costs exchange energy to do this. During the build up, the exchange energy is proportional to the deviation of $\langle m_z \rangle$ from unity. The derivative of the exchange energy with respect to $\langle m_z \rangle$ is constant during this process, giving rise to a constant effective-exchange field that aids the Zeeman field in maintaining the magnetization. Once the vortex breaks free of confinement to the symmetry axis, it is a moving wall in which changes in exchange energy are slight and compensated by changes in magnetostatic self-energy. In the dynamic response in constant applied field, the vortex can moves back and forth between the S^+ and the S^- state on paths of constant internal energy in which there are oscillations in the exchange energy and the magnetostatic energy, which are equal and opposite to one another. When the vortex is free to move as it is in the *S* state, the exchange energy has little effect on the magnetization loop.

3.6.1.3 Vortex Contributions to the Magnetization

The core of the vortex does have an effect on the magnetization loop because the core is magnetized. It is a separate permanent magnet that has its own magnetization loop, reversing only in fields of the order of $\mu_0 H_z = 1\,\mathrm{T}$. The volume of that permanent magnet changes somewhat with field, shrinking as the field is lowered from saturation and continuing to shrink as the field is increased in the reverse direction.

3.6.1.4 Corners as Partial Antivortices

The four corners of the cube are each one-quarter of a virtual vortex that is just outside the corners. These partial virtual vortices are antivortices with a winding number of −1. Each of the corner partial-virtual antivortices can be described using Preisach diagrams (with sloping sides). When a corner reverses, there is a change in the exchange energy in the region between the corners and the core of the *S*- or *I*-vortex in the center. The change in exchange energy with $\langle m_z \rangle$ during the reversal of a corner is an exchange field that adds to or subtracts from the applied field at the same time that the magnetization of the corner changes its contribution to the magnetization. The flipping of a corner shifts the sloping line of the demagnetizing field competition with the Zeeman field both sideways from the exchange field effect and also up or down from the change in magnetic moment.

The composite picture is then of five Preisach diagrams added to the ideal soft magnet, plus some exchange energy to be provided for the buildup of the five Preisach regions.

3.6.2 Transition from the Flower State to the Curling State

The flower state has reflection symmetry in the planes $x = 0$ and $y = 0$. It has inversion symmetry with respect to the origin at the center of the nanobrick or the nanoellipsoid. There are states such as the diamond structure that maintain inversion symmetry while breaking the reflection symmetries; the states, considered here, all have broken the inversion symmetry as well as the reflection symmetry. A current breaks both symmetries to produce curling states.

3.6.2.1 Magnetic Charge Density

The magnetic surface charge density $\sigma_m = \boldsymbol{n} \cdot \boldsymbol{M}$ is positive M_s at the top surface and negative M_s at the bottom. The fields from these charges are radially outward from the positive charge and radially in toward the negative charge. This produces the flower state at high fields for the flat top of the brick where the surface charge is in the planar surface. The transition to the curling state results from minimizing the volume magnetic charge density $\rho_m = -\nabla \cdot \boldsymbol{m}$ that arises from the change in m_z in the *z*-direction, as the surface is approached.

3.6.2.2 Magnetic Conductors

Both the applied field and the exchange energy favor parallel alignment of the magnetic dipoles, but the dipole–dipole interactions favor minimizing the surface magnetic charge density and at the same time minimizing the volume magnetic charge density. The surface charge density creates a demagnetizing field that opposes the applied field. If there were no anisotropy and the system were large enough, the magnetic surface charge density would create a demagnetizing field that is equal and opposite to the applied field for any shape, just as an electrical conductor produces surface charges to cancel applied electric fields within the conductor. The fact that the magnetic conductor has its charge limited by M_s produces major differences in the response to external fields.

3.6.2.3 Ideally Soft Magnetic Materials

For a magnetic system to act like a conductor, the magnetization pattern has to adjust itself to produce the required surface charges while remaining divergence free within the volume. It can do this if there is no anisotropy and the system is large enough that the increase in exchange energy required by the divergence free pattern is very small. A large enough ferromagnetic body without anisotropy behaves as a magnetic conductor with no net field inside the body, except that in a singly connected body, the magnetic conductor cannot topologically escape the need for two swirls. For a large enough body, the magnetization pattern is divergence free everywhere except in the vicinity of the swirls. This is called the ideally soft magnetic material. This is realized experimentally in iron whiskers with $X = Y \sim 0.1\,\mathrm{mm}$ and $Z \sim 10\,\mathrm{mm}$ just below the Curie temperature, where the magnetic anisotropy goes to zero much faster than the spontaneous magnetization [18]. Although this work was inspired by measurements at high temperature, the calculations are all for low temperature, where thermal agitation is completely neglected, except for its effect on the material constants that are those of ambient temperature.

The electrical charge on the surface of an electrical conductor is a very small fraction of the charge on a surface atom. The magnetic charge on the surface of a magnetic conductor is limited by the finite moment of the surface atom. As the charge necessary to cancel an applied field at an edge or corner of a brick goes to infinity, the corners become saturated (in the direction of the net field), as the external field penetrates the surface.

In high fields in the curling state the iron nanobrick also mimics a magnetically soft material as long as the high field is not so high as to force the flower state.

3.6.2.4 Splay Saving

For an ideally soft ferromagnet in not too high fields, the pattern is set by the surface charge density distribution that produces a field equal and opposite to the applied field. For a cylinder in an axial field, that charge density is, to a crude approximation, linear along the cylindrical surface and constant across the top and bottom surfaces. Once there is a component of the magnetization in the plane of the top surface and the magnetization is saturated in the midplane, there is a variation in M_z with z away from the surface. This would produce magnetic volume charge density $\rho_m = -\nabla \cdot \boldsymbol{m}$, if $\partial M_z/\partial z$ were not compensated by an equal and opposite contribution to the divergence from $(1/r)\ \partial(rM_r)/\partial r$. If near the surface $z = Z/2$, the magnetization is approximately given by

$$M_z = M_s \cos\psi(r,z) \equiv \frac{M_s(Z/2 - z)}{\text{sqrt}[b^2r^2 + (Z/2 - z)^2]}, \quad (3.11)$$

where b is a constant then on the surface

$$M_r = M_s \sin\psi\left(r, \frac{Z}{2}\right)\cos\chi(r) \quad (3.12)$$

and

$$M_\phi = M_s \sin\psi\left(r, \frac{Z}{2}\right)\sin\chi(r), \quad (3.13)$$

where $\chi(r)$ is the angle that the magnetic moment in the plane makes with the radial vector from the axis. The pattern is more complicated than this simple expression, but it is a good local approximation to the configuration in any small region. The value of b changes slowly with distance from the center of the swirl. For each b there is a $\chi(r)$ for which the contribution to $\nabla \cdot \boldsymbol{m}$ from $(1/r)\ \partial(rM_r)/\partial r$ cancels the contribution from $\partial M_z/\partial z$. For $\cos\chi = b$ the divergence vanishes. When $b = 1$, the magnetization is completely radial. For $b > 1$, the maximum amount of splay canceling accompanies the radial pattern. As H_z is increased, the position where $b = 1$ moves in toward the center of the swirl. When it reaches the swirl the entire pattern becomes radial. To a first approximation the exchange energy is independent of χ. It is the gradual change in χ with distance from the core that contributes to the linear increase in exchange energy proportional to $1-\langle m_z\rangle$. The transition from the flower state to the curling state with lowering of H_z starts with χ increasing from zero in the core of the swirl. If one artificially decreases the exchange energy in the iron nanobrick by a factor of 10 from its value in iron, the position, where $\chi = 0$ moves outward from the center as H_z decreases. In the iron, the exchange is strong enough to couple the rotations of χ in all distances from the center of the swirl. So, once the core of the swirl has an increase in χ, all regions have increases in χ, but smaller, depending on the distance from the core. The rotation of the center of the swirl is continuous starting at $\chi = 0$ at the threshold field. If the exchange energy is artificially decreased, that threshold field increases.

3.6.2.5 Absence of the Flower State in Ellipsoids

When the calculations are carried out for an ellipsoid, the flower state does not occur. Splay saving works right up to saturation. The swirl in the ellipsoid is not at a flat surface. It is the curvature that forestalls the breakdown of splay saving when $b = 1$ in the above argument. The ellipsoid goes directly from saturation to the swirl with the handedness chosen by a symmetry-breaking field. As the swirl forms, the exchange energy increases directly with $(M_s - \langle M_z\rangle)$. The linear increase in exchange energy with decreasing $\langle M_z\rangle$ is a constant exchange field that adds to the applied field. When H_z increases below saturation, the exchange field brings the ellipsoid to saturation at a lower H_z than one would obtain for a paramagnet with infinite susceptibility to reach M_s. The magnetization is linear in the applied field with the slope determined completely by the demagnetizing field. The demagnetizing field line is offset by the constant exchange field. For the ellipsoid, this is true for a field along any of the three principal axes. The slope is different for each axis because of the change in demagnetizing factor. The constant offset is different because the curvatures of the surface change the contribution of the exchange energy to the energy of formation of the swirl.

3.6.3 Curling States

The curling states include all the structures that occur once a current i_z is applied while lowering H_z from the flower state.

3.6.3.1 Curling in an Ellipsoid

To predict the line of $\langle M_z\rangle$ versus H_z for an ellipsoid, all one needs is Osborn's formulae [19] for the demagnetizing factors of the ellipsoid and a single number for $\partial E_{ex}/\partial\langle M_z\rangle$ for the chosen axis. The latter can be obtained from a micromagnetic calculation of $\langle M_z\rangle$ at a single field below saturation. Precise agreement with the analytic formulae has been found using TetraMag to calculate the properties of the mathematical ellipsoid with a triangular mesh on the boundaries shown in Figure 3.11. A full micromagnetic calculation of $\langle M_z\rangle$ versus H_z for the approach to saturation for an ellipsoid would require very long computational times because the torques become very low as the swirl saturates. There is an important message to workers in the field of micromagnetics.

3.6.3.2 Message

Modeling the results of the calculations can lead to insights that greatly shorten the computational time for a given problem. The ellipsoid at high fields is an extreme example in which one calculation in a single field, where the torques are large and the relaxation time short, produces the entire magnetization "curve" in the region where swirl remains along the axis and $\partial E_{ex}/\partial\langle M_z\rangle$ remains constant with change in H_z. In this case, the field for saturation is determined precisely. The instability field at which the I_z-vortex moves off the axis cannot be determined without calculating the pattern changes when the swirls move off the axis.

3.6.3.3 Reversal of a "Stoner–Wohlfarth Particle"

Below a second critical field, the centered position of the swirls on the ends of the principal axis of the ellipsoid becomes an energy maximum and the swirls move off the center, if the dimensions of the ellipsoid are sufficiently large. At small enough dimensions the ellipsoid remains "uniformly magnetized" at all fields. The magnetization process is limited to rotations in the Stoner–Wohlfarth model. It is assumed that the exchange energy does not change in the process. This model has served for 60 years as the starting point for understanding magnetization processes as a competition between the Zeeman energy and the anisotropy energy, where the anisotropy energy includes the dipole-dipole interactions and the crystalline anisotropy. Variations of the exchange energy in "uniform" rotation would also appear as an addition to the anisotropy. Even an ellipsoid in the Stoner–Wohlfarth model requires bias fields for the uniform rotation of the magnetization away from a principal axis.

3.6.3.4 Propagating Singularities

In a micromagnetic calculation, one can eliminate all the geometrical biases and cause the small ellipsoid to reverse its magnetization by a nonuniform distortion in which a singularity propagates along the principal axis starting at the swirl. The field must be applied fast enough such that the round-off errors in the numerical calculation do not have sufficient time to nucleate the uniform rotation by the displacement of the swirls. Even then, one needs a symmetry-breaking field to choose the handedness of the swirls. Here again, the numerical round-off error can provide the handedness. The field must be larger than necessary and must be applied fast enough such that the round-off error favors the reversal by singularity propagation rather than by uniform rotation.

3.6.3.5 Pair Creation and Propagation in the I_z-Vortex

A rule for the formation of a pair of singularities in micromagnetics has been given by Sebastian Gliga who worked in high energy physics before specializing in magnetism [20]. If enough energy is provided to create each of the singularities in a given region, the program for solving the micromagnetic equations will find that solution in which the pair is created.

As the field is increased in the negative direction for a $+I_z$-vortex state, the magnetization turns to the $-z$-direction everywhere except in the immediate vicinity of the axis of the vortex. There is a wall in which exchange energy becomes higher as the field in the $-z$-direction increases. In a bcc lattice, the singularity appears at the center of a tetrahedron of iron atoms. The four moments can no longer sustain the wall when their m_z decreases to a critical value

3.6.3.6 Bias Fields in the Stoner–Wohlfarth Model

The importance of bias fields for reversal of magnetization was first pointed out by Smith at the second MMM conference in Boston, in 1956. The subject of the session was the failure of experiments to show switching with the time constant predicted by the Landau–Lifshitz equations. The experimentalists were asking what was wrong with the Landau–Lifshitz equations. Smith showed that if the experiments and the theory are done using bias fields, they agree.

3.6.3.7 I_z-Vortex State

In the nanobrick the curling pattern with the swirls centered on the z-axis is called the I_z-vortex state. The m_z-isosurfaces are elliptical in cross-section. The central bulge along the x-axis corresponds to the 180° wall of the Landau structure. As the field H_z is lowered, the bulge extends toward the positions $x = \pm(X/2 - Y/2)$. When the I_z-vortex state is maintained to $H_z = 0$, for $|x| > (X/2 - Y/2)$ and $y = 0$, magnetization lies almost in the midplane with $|m_y| \sim 1$ corresponding to the closure-domain pattern of the Landau structure. For Z above a critical thickness that depends somewhat on X and Y, the I_z-vortex is not stable for $H_z = 0$, but it is always an equilibrium state that persists as long as there is no symmetry-breaking field or the inevitable effect of computational round-off error has not yet developed. Once one sees the correspondence between the Landau structure and the I_z-vortex state, one can view the I_z-vortex state as the Landau structure with its two swirls centered on the z-axis. Or one can view the Landau structure as a vortex with its two swirls moving off the z-axis. The two swirls can be manipulated to move along the top and bottom surfaces distorting the 180° Bloch wall as they move.

3.6.3.8 S-Vortex States

The swirls of the I_z-vortex state can be manipulated. In response to a current i_x in the $+x$-direction, the I_z-vortex state takes the S_z^+-vortex configuration. For the dimensions of the iron nanobricks chosen for this chapter, the S_z^+-vortex configuration appears spontaneously below a critical magnitude of H_z. In the absence of bias fields in the x- or y-directions, the two swirls of the spontaneous S_z^+ and S_z^--vortex states have coordinates $(x_s, y_s, Z/2)$ and $(-x_s, y_s, -Z/2)$, respectively, where x_s and y_s increase with decreasing H_z reaching a maximum at $H_z = 0$. As x_s increases it is accompanied by an increase in $\langle m_x \rangle$, as shown in Figure 3.14. This occurs because there is a displacement of the core in the $-y$-direction increasing the volume in which the magnetization in the $+x$-direction is dominant. In the midplane, where $x = 0$, the core of the vortex with $m_z = 1$ and the two Néel caps with $m_y = 1$ at the top and $m_y = -1$ at the bottom form a circulating magnetization pattern on one side of the z-vortex. The circulation is about an x-axis displaced from the midplane in the $-y$-direction, as originally calculated by LaBonte for a never-ending Bloch wall (see Figure 3.10).

The lowest Z for the spontaneous appearance of an S_z-vortex structure is $Z_{crit} = 25\,nm$ for $Y = 35\,nm$ with X varying from 120 to 126 nm. For $X = 119\,nm$ the S_z-vortex configuration goes to the I_z-vortex state. For $X = 127\,nm$, an S_z-vortex structure is unstable with respect to the formation of an I_x-vortex along the x-axis. For $X = 130\,nm$ and $Y = 80\,nm$, the spontaneous S_z-vortex structure occurs for $Z > 42\,nm$.

There are also limits on the sizes of the nanobrick for which the I_z-vortex state is stable in the absence of a magnetic field. If the nanobrick is too small, the I_z-vortex state moves away from the axis and disappears out of the nearby Y face, as H_z is reduced. The range of dimensions (X, Y, Z) and applied fields

(H_x, H_y, or H_z) for which the S_z^+ and S_z^- and I_z-vortex states are stable has been studied by Templeton [21], who includes the effects of the configurations in the four corners in his elaborate phase diagrams.

3.6.3.9 I_z^*-Vortex State

The spontaneous S_z^+ state is distorted by applying a uniform field H_y. The swirl on the left moves toward the swirl on the right, which moves only slightly to the right. There is a critical field H_y ($\mu_0 H_y \sim 0.1$ T) at which the left swirl catches up to the right swirl. The I_z^*-vortex state is the Landau structure with both swirls on the same end of the 180° Bloch wall. The centers of the two swirls are at the same x-position, but the small displacements of the swirls in the y-direction are in opposite y-directions. The m_z-isosurfaces are far from symmetric. The 180° Bloch wall is attached to one side of the I_z^*-vortex. The central cross-section of the m_z-isosurfaces bulges to include the Bloch wall (see Figure 3.16). On lowering H_y, there is a critical field for re-nucleating the S_z-vortex that restores the Landau state with the swirls on opposite ends as, H_y goes to zero. The transitions from S_z to I_z^* in H_y are not quite continuous and hysteresis occurs. The transitions from I_z^* to S_z depends on a bias to select between S_z^+ and S_z^-. From the view of the swirls, one of the swirls traveled further to form the I_z^*-vortex state in the large H_y. The choice of which swirl travels back along the nanobrick when H_y is reduced is, again, a three-tined fork. If no decision is made, the I_z^*-vortex state persists for a long time in a region where an S_z-vortex would be more stable. A small bias field in H_x would make the selection. For a finite step in H_y, the dynamics makes the selection. In experimental studies at Grenoble of faceted nanogems of iron, it is the facets on the top surface that chooses the upper swirl as the one that propagates further on reducing H_y [22].

The transition to the I_z^*-vortex state from the spontaneous S_z-vortex state can also be made by applying a large current, i_y (~20 mA). The latter produces a large gradient field in the z-direction ($\mu_0 H_z = 0.06$ T at the end surface), which stabilizes

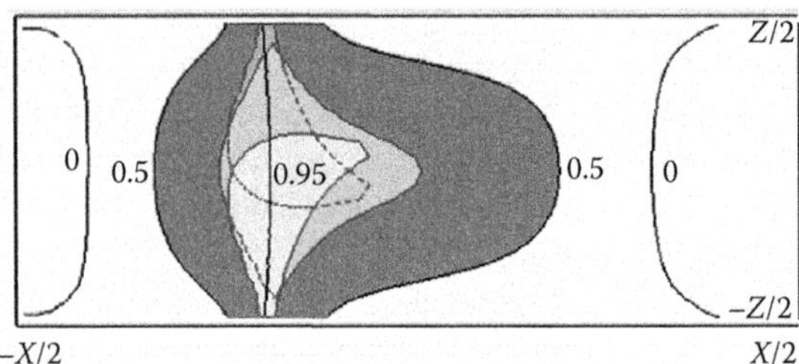

FIGURE 3.16 The I^*-vortex formed by displacing both swirls to the same end of the Bloch wall, using a field H_y. The two swirls have the same x-coordinate but are displaced in y by −4 nm on the bottom surface and +4 nm on the top. The regions of light gray ($m_z > 0.8$) and dark gray ($0.8 > m_z > 0.5$) are for m_z-isosurfaces intersecting the $y = 0$ plane. The slightly curved and almost-vertical line is the contour for $m_y = 0$ in that plane. The I^*-vortex maintains a memory of the Bloch wall as it bulges to the right. The area in white labeled 0.95 is for the plane $y = -4$ nm and the dotted contour is in the plane $y = 4$ nm. The I^*-vortex has a twofold symmetry on rotation about the x-axis.

the I_z^*-vortex state in a first-order jump, with the complication of turning the magnetization in the corners into the direction of the gradient field, that is *mmmm* goes to *mppm* for the four corners ordered cw from (−X/2, Y/2). Removing the large i_y does not restore the corners to their original configurations.

3.6.3.10 Switching

The most dramatic result of the attack on the micromagnetics of the nanobrick was the discovery by computation that the two swirls of the Landau structure when viewed as the ends of the S-vortex could be switched back and forth, using modest driving fields, over long distances in short times, carrying with them large external fields [23]. The switching of the S-vortex had already been observed experimentally in 2004, but not specifically identified with the reversal of the positions of the two swirls [24]. The switching between two stable states is discussed here in terms of the energy landscape correlated with the positions of the two swirls. This is a gross simplification, but in equilibrium and for heavily damped dynamics, there is some usefulness in thinking about the internal energy along and near the equilibrium path.

The combination of the field from a current i_x, producing a field that is $+H_y$ at the top surface and $-H_y$ at the bottom, and a uniform H_y, permits the independent manipulation of the positions of the two swirls in any given H_z. The internal energy at equilibrium in the combined fields changes with the positions of the two swirls. At constant H_z, one has an energy landscape with minima at the symmetric positions that the two swirls for S^+ or S^- take in the absence of bias fields. If the motion of the two swirls between S^+ and S^- is determined by a slowly varying current oscillating between $+i_x$ and $-i_x$, the internal energy along the path goes through the minima when the current goes through zero. The displacement in x from that equilibrium has the separation between the two swirls first increasing, reaching a maximum for the highest current, returning to equilibrium as the current goes through zero, and then having the displacement in x go toward zero as the swirls approach each other. But before they reach each other, the swirls have reached an energy position where it is all down hill toward a stable position that lies beyond the far equilibrium position in zero current. That position will be reached if the current is maintained at the critical current for switching ($i_x = 1.2$ mA producing a maximum field $\mu_0 H_z = 0.06$ T at the surfaces for the nanobrick $156 \times 96 \times 60\ \text{nm}^3$).

3.6.3.11 Double-Well Potential

Along the path to the critical current for switching from the S^+ to S^- configuration, the internal energy contour is one-half of a double-well potential. That potential is completely determined at each current up to the critical current. The inflection point on that curve is reached at the critical current. If the motion is calculated using large damping, a path beyond the inflection point can be determined from the damped dynamic response and the full double-well potential can be determined. This path leads through the position where the separation in x of the swirls goes through zero; but when the separation in x is zero, the separation in y is not zero. The paths of the two swirls on the opposite faces of the nanobrick are narrow ellipses tilted in the x–y planes. The double-well potential is defined along the ellipses. The swirls do not pass over the maximum in the potential where the displacements in x and y are both zero.

3.6.3.12 Forced Oscillations

An example of the nonlinear oscillations of the S-vortex with large amplitude is found using a 1.28 GHz driving current i_x = 0.6 mA with a period of 780 ps. This current is one-half of that necessary to reverse the S-vortex with slowly varying currents. The swirls move in elliptical paths on the top and bottom surfaces that avoid the region of the local maximum in the internal energy at the center of the faces as they follow contours of almost constant energy over the saddle point on the $+y$ and $-y$ sides of the origin. The path on the top surface is shown in Figure 3.17 for one period of oscillation. The path on the bottom surface is the mirror image in either the $x=0$ or $y=0$ plane with the two ends of the m_z-isosurface moving ccw along the paths. When the swirls pass one another at $x=0$, the two ends are displaced in $y=\pm 8$ nm. At their greatest separation $x=\pm 36$ nm at $y=16$ nm.

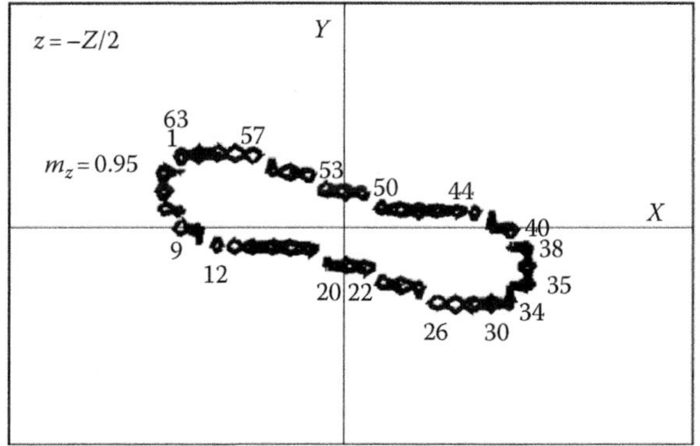

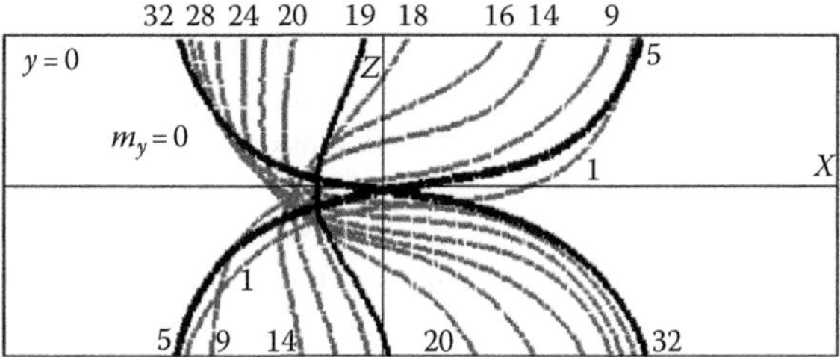

FIGURE 3.17 The dynamic response of a nanobrick with dimensions $156\times 96\times 60\,\text{nm}^3$ to an ac driving current $i_x=0.6$ mA with a period $\tau=780$ ps. The path of the swirl on the bottom surface is followed in the upper panel by tracing the contours $m_z=0.95$ in steps of 12.8 ps. The calculation was carried out with a grid of 4 nm cubes, which exaggerates the interaction of the swirl with the grid resulting in the steps of 4 nm in both the x- and y-directions. A much smoother ellipse is obtained for a grid of 1.25 nm. At no time does the central core of the S-vortex lie in a plane, let alone in the $y=0$ plane, but the contours with $m_y=0$ in the plane $y=0$ are used in the bottom panel to reflect the distortions of the m_z-isosurfaces, which, to be fully appreciated, require 3D movies.

3.6.3.13 Mired at the Central Maximum

In the dynamic response with low damping, the path of either swirl can go through the origin, but generally not at the same time. But in one dynamical calculation, the two swirls came through the origin at the same time during 1 cycle of a damped oscillation. The two swirls then stayed there for a time equal to the period of the nonlinear oscillator. During this time, a higher harmonic of the dynamic response corresponding to a wave propagating up and down the z-axis provided the driving force to allow the two vortices to move away from the central energy maximum.

3.6.4 Motion Pictures

Most of the switching in Figure 3.17 takes place in 200 ps in each direction. The time between contours 16 and 24 is 100 ps. There is a crack-the-whip effect as the swirls rush past each other, followed by a dwell time before they make an assault on the barrier between the two stable S-states. For a lower driving current, swirls make it over the barrier in some attempts and then not in others, only to regroup and try again on the next cycle. Whether a swirl makes it over the barrier depends on the phase of oscillations along the vortex core. Such motions are not easily envisioned even with movies of the magnetization patterns. One can view the tubes of constant M_z with 3D glasses or one can have a changing camera angle to add rotation of the observer to follow the 3D shape of the tubes and the oscillations that propagate along the tube, as the tube itself wanders through space. No DVD is included with this chapter, but movies are sometimes submitted as supplemental material that is available online; an example is the visual display of the small amplitude oscillations in a nanobrick [25].

3.6.5 What Causes the S-Vortex to Form Spontaneously?

If the exchange energy in the iron nanobrick is reduced by a factor of 10, the S_z-vortex is stable to a field that is about twice that of the iron nanobrick with the correct exchange energy. The lowering of the magnetostatic dipole–dipole interaction energy is the driving force for the spontaneous formation of the S_z-vortex state. One gets the sign correctly if one compares the demagnetizing energy for the magnetization along the z-axis, where the demagnetizing factor is largest, to along the x-axis, where it is smallest. But a vortex is far from a state where the magnetization is uniform. There is magnetic charge in the center of the swirls at the surfaces. The m_z-isosurfaces move inward toward the center of the swirl as the swirls move off the center. This decreases the amount of charge within any given radius from the center of the swirl. This decreases the self-energy of those charges. What causes the isosurfaces to move inward is the distance to the surfaces at $x=\pm X/2$. For the I-vortex, the isosurfaces move outward, if X is increased. If the swirl moves toward the $X/2$ end, it slightly increases the outward displacement of the isosurface toward the receding $-X/2$ end, but increases more the inward displacement of the isosurface toward the approaching $X/2$ surface. Thus, there now is a partial understanding of the full Landau structure some 65 years after it was first drawn without the details of what happens at the point where the walls meet the surfaces and each other.

Acknowledgments

Bretislav Heinrich and the author started their studies of the magnetic response of iron whiskers over 40 years ago. Dan S. Bloomberg, Murray J. Press, Scott D. Hanham, Amikam Aharoni, T.L. Templeton, and J.-G. Lee all contributed to the partial understanding of a long list of phenomena observed in whiskers. The micromagnetics of a nanobrick is a continuation of that work. The nanobrick and the nanoellipsoid provide simple examples of 3D magnetic configurations in a singly connected body. Only the configurations with a single vortex have been described above. The larger the nanobrick, the more room there is for more vortices and the more the need for more computing power to fully appreciate the complexity they bring with them. In over 1000 pictures in their treatise *Magnetic Domains*, Rudi Schafer and the late Alex Hubert have shown how complex and beautiful the patterns of magnetization can be. Riccardo Hertel and Mike Scheinfein have provided tools for and participated in the attack on this complexity. Bill Gates has removed his 2 Gbyte memory barrier for the personal computer. Attila Kakay has now implemented micromagnetism for parallel processing using the hundreds of central processing units on the graphic processor unit [26]. My work has just begun.

References

1. A. S. Arrott, The past, present and future of soft magnetic materials, *J. Magn. Magn. Mater.* **215–216**, 6–10 (2000).
2. M. Yan, A. Kákay, S. Gliga, and R. Hertel, Beating the walker limit with massless domain walls in cylindrical nanowires, *Phys. Rev. Lett.* **104**, 057201(1–4) (2010).
3. A. Hubert and R. Schäfer, *Magnetic Domains. The Analysis of Magnetic Microstructures*, Springer, Berlin, Germany 1998, Chapter 3 of Magnetic domains remains, after more than a decade, the key reference in the magnetism of soft magnetic materials.
4. A. J. Newell and R. T. Merrill, The curling nucleation mode in a ferromagnetic cube, *J. Appl. Phys.* **84**, 4394 (1998).
5. A. S. Arrott, B. Heinrich, and A. Aharoni, Point singularities and magnetization reversal in ideally soft ferromagnetic cylinders, *IEEE Trans. Magn.* **15**, 128 (1979).
6. M. R. Scheinfein, J. Unguris, M. H. Kelley, D. T. Pierce, and R. J. Celotta, Scanning electron microscopy with polarization analysis (SEMPA)—Studies of domains, domain walls and magnetic singularities at surfaces and in thin films, *J. Magn. Magn. Mater.* **93**, 109–115 (1991); *Phys. Rev. B* **43**, 3395 (1991).
7. A. S. Arrott and R. Hertel, Formation and transformation of vortex structures in soft ferromagnetic ellipsoids, *J. Appl. Phys.* **103**, 07E39 (2008).
8. A. E. LaBonte, Two dimensional Bloch-type domain walls in ferromagnetic thin films, *J. Appl. Phys.* **40**, 2450–2458 (1969).
9. A. Arrott, Antiferromagnetism in metals, in *Magnetism*, H. Suhl and G. T. Rado (Eds.), vol. III B, Academic Press, New York, 1966.
10. H. Kronmüller and S. Parkin, in *Handbook of Magnetism and Advance Magnetic Materials*, vol. 2, Micromagnetics, John Wiley & Sons, Chichester, 2007; see in particular chapters by J. E. Miltat and J. Donahoe, *Numerical Micromagnetics: Finite Difference Methods*, and by T. Schrefl et al., *Numerical Methods in Micromagnetics* (Finite Element Method).
11. I. Rabi, N. E. Ramsey, and J. Schwinger, Use of rotating coordinates in magnetic resonance problems, *Rev. Mod. Phys.* **26**, 167–171 (1954).
12. A. S. Arrott, Introduction to micromagnetics, in *Ultrathin Magnetic Structures IV*, B. Heinrich and J. A. C. Bland (Eds.), Springer-Verlag, Berlin, Germany, 2005.
13. The story of Gilbert damping starts with an abstract by T. Gilbert in 1955. Gilbert's idea was formalized by R. Kikuchi. 1956. On the minimum of magnetization reversal time. *J. Appl. Phys.* **27**, 1352–1357. Fifty years later Gilbert commented on his work in 2004. *IEEE Trans. Magn.* **40**, 3443. The basis for Gilbert's phenomenology is firmly established in experiments by R. Urban, G. Woltersdorf, and B. Heinrich, *Phys. Rev. Lett.* **87**, 2173 (2001).
14. Classical micromagnetics refers to two monographs by William Fuller Brown, Jr. published in 1962 and 1963 by Interscience, New York: *Magnetostatic Principles in Ferromagnetism* and *Micromagnetism*.
15. D. Dotze and A. S. Arrott, Micromagnetic studies of vortices leaving and entering square nanoboxes, *J. Appl. Phys.* **97**, 10E307 (2005).
16. A. S. Arrott and R. Hertel, Mode anticipation fields for symmetry breaking, *IEEE Trans. Magn.* **43**, 2911 (2007).
17. M. R. Scheinfein and A. S. Arrott, Increased efficiency and accuracy in micromagnetic calculations of switching asteroids, *J. Appl. Phys.* **93**, 6802 (2003).
18. For the years 1971–1975 the *Proceedings of the MMM Conference* were published in books by the American Institute of Physics. These are known in magnetism as the "lost years" because the work was so infrequently cited and still are not referenced in the Web of Science. (Now AIP now has its conference proceedings online at $28 per paper.) In those years Heinrich and Arrott studied iron whiskers at the Curie temperature. The conjectured curling pattern just below T_c appeared in the seldom-cited and hard-to-find paper by B. Heinrich and A. S. Arrott, in *Proceedings of the International Conference of Magnetism ICM-73*, vol. IV, Publishing House NAUKA, Moscow, pp. 556–561, 1974.

19. J. A. Osborn, Demagnetizing factors of the general ellipsoid, *Phys. Rev.* **67**, 351 (1945).
20. S. Gliga, Private communication, to be published.
21. T. L. Templeton, Private communication, to be published.
22. F. Cheynis, A. Masseboeuf, O. Fruchart et al., Controlled switching of Néel caps in flux-closure dots, *Phys. Rev. Lett.* **102**, 107201 (2009).
23. A. S. Arrott and R. Hertel, Large amplitude oscillations (switching) of bi-stable vortex structures in zero field, *J. Magn. Magn. Mater.* **322**, 1389–1391 (2010).
24. S. B. Choe, Y. Acremann, A. Scholl et al., Vortex core driven magnetic dynamics, *Science* **304**, 420–422 (2004). Their Fig. 1 panel IV shows the oscillation of a swirl in the Landau structure. See also the discussion by Xiaowei Yu. 2009. In Time-resolved x-ray imaging of spin-torque-induced magnetic vortex oscillation, PhD thesis, Applied Physics, Stanford.
25. M. Yan, R. Hertel, and C. M. Schneider, Calculations of three-dimensional magnetic normal modes in mesoscopic permalloy prisms with vortex structure, *Phys. Rev. B* **76**, 094407 (2007).
26. A. Kákay, E. Westphal, and R. Hertel, Speedup of FEM micromagnetic simulations with graphical processing units, *IEEE Trans. Magn.* **46**, 2303 (2010).

4

Giant Magnetoresistance: Experiment

Jack Bass
Michigan State University

4.1 Introduction and Overview

This chapter reviews experimental data on giant magnetoresistance (GMR) in magnetic multilayers composed of ferromagnetic (*F*) and nonmagnetic (*N*) metals. Since Chapter 5 reviews the theory underlying GMR [1], theoretical issues are treated only as needed to motivate and understand experimental data. As background, the review begins with a brief overview of the MR for a single magnetic film—called anisotropic magnetoresistance (AMR). It then covers, in more detail, the traditional current-in-plane (CIP) MR and the current-perpendicular-to-plane (CPP) MR in multilayers. Results on current-at-an-angle-to-the-plane (CAP) MR in multilayers [2,3] and MR in granular films (gMR) [4–6] are treated only briefly. Figure 4.1a shows the different current directions for CIP and CPP. In a real CIP film, the aspect ratio in Figure 4.1a is inverted—the length of the CIP film in the direction of the current flow is much greater than the total thickness of its layers.

Since the discovery and early history of GMR are covered in Chapter 1 [7], this chapter touches only briefly upon those topics. This review is also not encyclopedic but rather focuses on a number of questions of special physical interest that include the following: (1) How strong is scattering at *F*/*N* interfaces and how important is it in GMR? Such scattering is a new transport phenomenon arising with multilayers. When GMR was first discovered, almost nothing was known about such scattering. In the CIP geometry on thin films, earlier analyses of electron scattering at the metal/air surfaces had assumed a single parameter that characterized the probability of diffuse (as opposed to specular) reflection at the interface [8]. Understanding CIP- and CPP-GMRs, in contrast, requires ascertaining both the strength of the scattering of electrons at an *F*/*N* interface and the asymmetry of such scattering for electrons with magnetic moments oriented along (↑) or opposite (↓) the moment of the *F*-metal. In CIP-MR, current flow is parallel to the layers and interfaces, whereas in CPP-MR, it is perpendicular to both. Given this difference, are interface parameters the same in the two cases? Are the relative weightings of scattering in the bulk *F*-metals and at the *F*/*N* interfaces similar? Especially for the CPP-MR, there is an interest also in the scattering at the *N*1/*N*2 and *F*/*S* (*S*, superconductor) interfaces. (2) Are the bulk and interfacial contributions to CIP- and CPP-MRs dominated by band structure effects, or by scattering from impurities? (3) Can GMR be adequately understood with semiclassical models, or are quantum coherence effects needed?

This chapter is organized as follows: Section 4.2 contains definitions of AMR, GMR, and various parameters that are needed later. Section 4.3 compares CIP- and CPP-MRs and

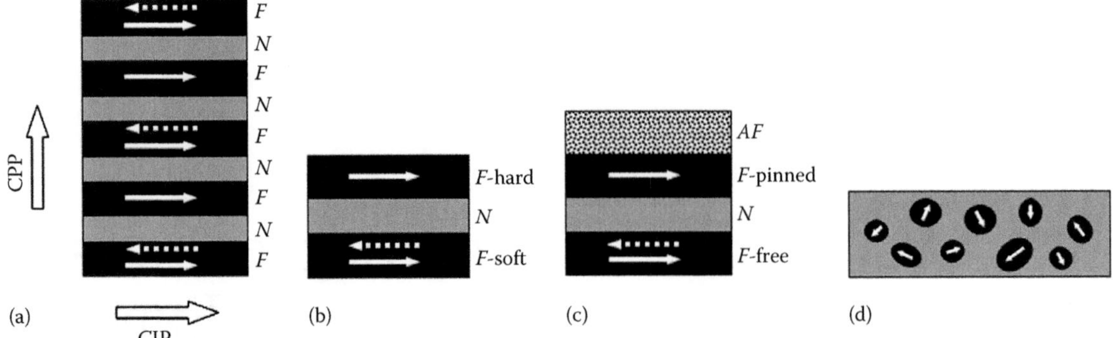

FIGURE 4.1 Sample geometries: The magnetic field is applied horizontally (in-plane). (a) A multilayer composed of alternating ferromagnetic (*F*) and nonmagnetic (*N*) metals, and *F*-layer moment orientations for the *P*-state (all solid arrows pointing to the right) and *AP*-state (all dotted arrows pointing to the left). The arrows outside the multilayer show the current directions for CIP and CPP geometries. (b) A trilayer composed of a soft *F*-layer, the moment of which reverses at low field, and a hard *F*-layer, the moment of which stays fixed until a much higher field. (c) An EBSV, where an antiferromagnet (*AF*) pins the moment of an adjacent *F*-layer, leaving the other "free" *F*-layer to reverse its magnetization at low fields. (d) A granular *N* alloy, containing *F*-inclusions with initially randomly aligned moments that align in a large enough magnetic field.

gives examples of CAP and granular MR data. Section 4.4 more fully covers CIP-MR data and what we have learned from them. Section 4.5 covers CPP-MR data and what we have learned from them. Section 4.6 contains a summary and conclusions.

For more information about GMR, and additional references, we recommend several earlier reviews [9–14].

4.2 Definitions and Background

4.2.1 Anisotropic Magnetoresistance

The resistivity ρ of a thin film of an *F*-metal to which a magnetic field *H* is applied in the plane of the film was found long ago [15] to vary with the angle θ between the current *I* and the magnetization *M* as [1,16]

$$\rho(\theta) = \rho_{\perp} + \left(\rho_{\parallel} - \rho_{\perp}\right)\cos^2\theta \tag{4.1}$$

Here, the normally larger $\rho_{\parallel}$ and smaller $\rho_{\perp}$ are the limiting resistivities when the current is parallel or perpendicular to the *F*-layer magnetization. AMR arises from the asymmetry of scattering by the spin–orbit interaction, which is standardly stronger for electrons moving parallel to the magnetization [1]. AMR in the alloy permalloy, $Ni_{1-x}Fe_x$ with $x \sim 0.2$, was the source of the magnetoresistance in the magnetoresistive read heads in computers prior to the use of GMR [7]. When the CIP-MR is small, account must be taken of the MR contribution from AMR.

4.2.2 Giant Magnetoresistance

GMR is the change in the electrical resistance *R* of a magnetic multilayer composed of alternating *F* and *N* layers, upon application of *H*, usually in the layer plane. Figure 4.1 shows examples of the different types of multilayers that give GMRs: (a) an $[F/N]_n$ multilayer repeating *n* identical *F* and *N* layers, (b) an [*F*/*N*/*F*] multilayer containing a magnetically soft layer that reverses in a small field and a magnetically hard one that requires a larger field, (c) an [*AF*/*F*/*N*/*F*] exchange-biased spin valve (EBSV) containing a free *F*-layer that reverses in a small field and an *F*-layer that is "pinned" by an adjacent antiferromagnet (AF) to reverse only at a larger field, and (d) a granular alloy containing *F*-inclusions in an *N*-matrix. In all four cases, so long as the *F*-layers are not coupled ferromagnetically, the sample will show a magnetoresistance as *H* is swept from a large positive value to a large negative one. As described in Chapter 1 [7], GMR was discovered in samples in which Fe layers were coupled antiferromagnetically across Cr layers [17,18], so that at $H = 0$, the magnetizations of the *F*-layers were oriented antiparallel (*AP*) to each other. This state gave the highest resistance, R_{AP}. Application of *H* large enough to align all of the Fe layer moments parallel (*P*) to each other gave a lower resistance, R_P. The resulting MR is shown in Chapter 1.

It was quickly discovered that large MRs did not require such *AF* coupling. As an example, Figure 4.2 shows the MR for a magnetically uncoupled, simple $[Co/Ag]_n$ multilayer (*n* is the number of repeats). The largest value of *R* in Figure 4.2, R_0, occurs in the as-prepared (virgin) state where $M = 0$. A later study [20] showed that this state most closely approaches the *AP* state (but probably does not quite reach it—see Figure 4.3), because the micron-sized domains in the adjacent *F*-layers are initially oriented close to *AP*. In contrast, the peaks in *R* in Figure 4.2 after *H* has been taken to saturation do not occur at $H = 0$, but rather at the coercive field, H_c. Although the total sample magnetization at H_c is again $M = 0$, now the relative orientations between moments of domains in the adjacent layers are close to random. Thus, these peaks are not *AP* states, and R_{peak} is much less than the resistance of the virgin state, R_0.

For *F*-inclusions in *N*-hosts, gMR gives qualitatively similar curves to Figure 4.2, except that R_0 is barely larger than R_{peak} at the "peaks" at H_c after saturation, and R_{peak} occurs at much higher magnetic fields ($H_c \sim 1$ kG for gMR vs $H_c \sim 0.1$ kG for GMR) [4]. Again, at these peaks, the moments of the individual

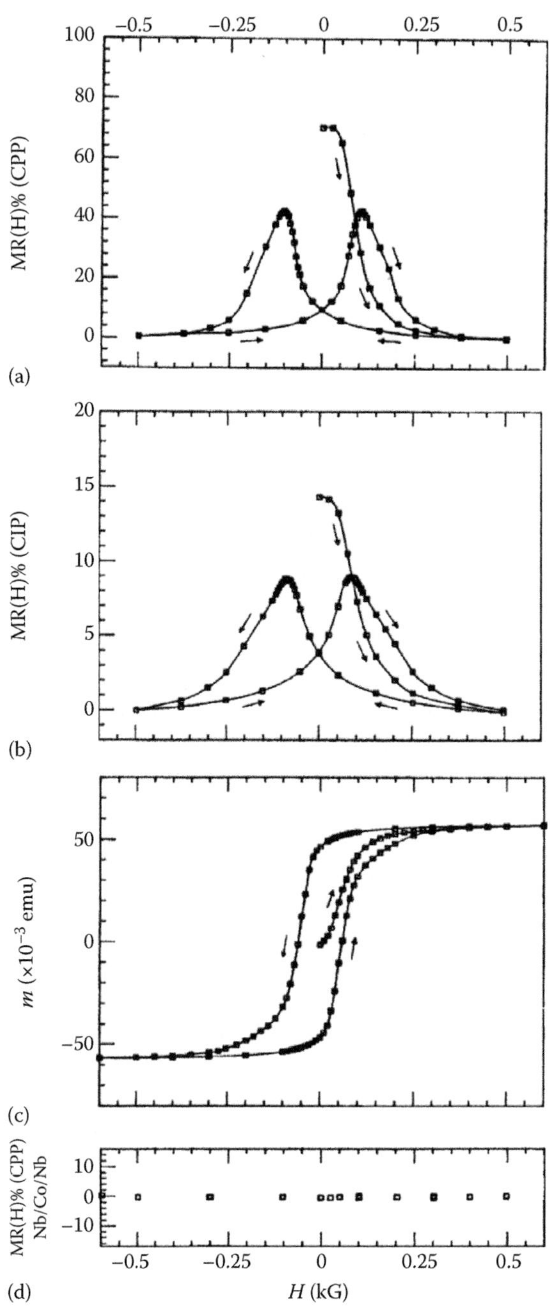

FIGURE 4.2 (a) CPP-MR, (b) CIP-MR, and (c) magnetization, $M(H)$, for $[\text{Co}(6\,\text{nm})/\text{Ag}(6\,\text{nm})]_{60}$ multilayers, and (d) negligible CPP-MR for a single Co(9 nm) film. In the text, the maximum CPP-MR in the "virgin" state is labeled R_0, the maximum after saturation at large H is labeled R_{peak}, and the minimum is labeled R_P. (After Pratt, W.P. et al., *Phys. Rev. Lett.*, 66, 3060, 1991. With permission.)

inclusions are presumably aligned randomly, rather than *AP*. The large values of H_c for granular samples have limited their use in devices.

As just noted, the largest MR (GMR) occurs when the order of the magnetic moments of the two *F*-layers changes from *P* (usually at high *H*) to *AP* (usually at low *H*). Assuming that these two states can be achieved, we formally define the GMR as

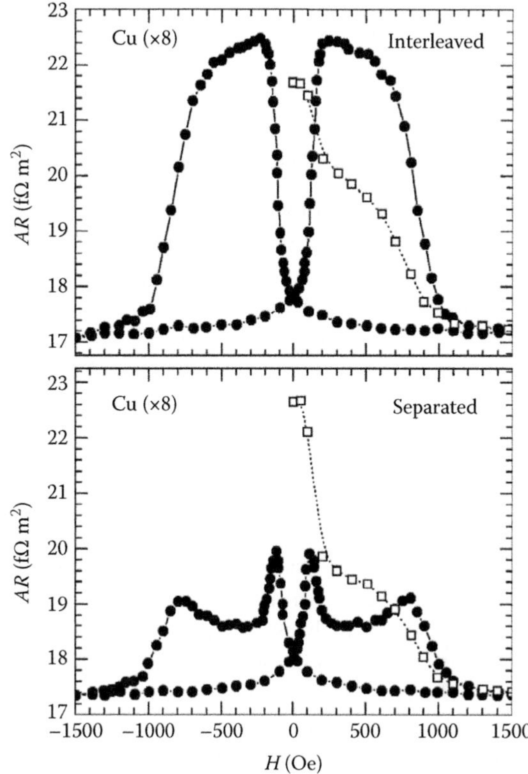

FIGURE 4.3 Comparison of CPP data for $A\Delta R$ versus H(Oe) for (top) "interleaved" $[\text{Co}(6)/\text{Cu}(20)/\text{Co}(1)/\text{Cu}(20)]_8$ and (bottom) "separated" $[\text{Co}(6)/\text{Cu}(20)]_8$ $[\text{Co}(1)/\text{Cu}(20)]_8$ samples. The use of two different Co thicknesses leads to an *AP* state of maximum $A\Delta R$ after saturation that is larger than the virgin state value of $A\Delta R$ at $H=0$ (contrast with Figure 4.2). The behavior of the "separated" sample will be discussed in Section 4.5.9. (After Eid, K. et al., *Phys. Rev. B*, 65, 054424, 2002. With permission.)

$$\text{MR} = \frac{R_{AP} - R_P}{R_P} = \frac{\Delta R}{R_P} \tag{4.2}$$

As explained in Chapter 1, for multilayers containing only a single pair of metals *N* and *F*, R_{AP} is always greater than R_P – $R_{AP} > R_P$, giving a positive MR. But we will see below, that multilayers composed of certain combinations of different *F*- and/or *N*-metals—*F*1, *F*2, and/or *N*1, *N*2—can give a negative MR ($R_{AP} < R_P$). Such an "inversion" occurs when the scattering anisotropy of *F*2 in Figure 1.4c is inverted from that of *F*1—that is, electrons in *F*2 are now weakly scattered when the *F*-layer and electron moments are parallel, giving a "shorting" in the *AP* state [1].

4.2.3 MR versus *M*

When $R_{AP} > R_P$, one expects a maximum $R = R_{AP}$ to occur when $M = 0$, and a minimum $R = R_P$, to occur when M is maximum. Both MR(H) and $M(H)$ then vary smoothly between these limits, as shown for a simple $[F/N]_n$ multilayer in Figure 4.2. But, as explained above, and shown in Figure 4.2, for magnetically uncoupled *F*-layers, the value of R at $M=0$ can depend upon the sample history.

4.2.4 CIP- and CPP-MR Geometries

GMR can be measured in two limiting geometries as shown in Figure 4.1. Current flows in the layer-planes (CIP) MR and current flows perpendicular-to-the-layer-planes (CPP) MR [19]. As already noted, GMR has also been measured with a current flow at an angle to the layer planes (CAP)-MR [2,3]. CAP-MR interpolates between CIP- and CPP-MRs and is usually more complex to analyze. We, thus, focus on the CIP- and CPP-MRs.

4.2.5 Achieving *AP* States

From the definition in Equation 4.2, to determine GMR, it is necessary to produce both *AP* and *P* states. The *P* state can always be achieved by applying H large enough to align all the *F*-layer magnetic moments in the multilayer. The problem is to achieve the *AP* state. GMR was discovered in Fe/Cr multilayers [17,18] using *AF* exchange coupling between the adjacent Fe layers to achieve the *AP* state at field $H=0$. But, for devices, such coupling is disadvantageous, since it requires a large H to switch from *AP* to *P*. Fortunately, it was soon shown that the *AP* state could be obtained without such coupling (see, e.g., Ref. [21]), as already illustrated in Figure 4.2. In the following, we describe different techniques used to reliably and controllably produce *AP* states, mostly at low H. In most of the multilayers of interest, the layers are wider rather than thicker. Due to demagnetization effects, the magnetization usually prefers to lie in the layer plane. If the layer diameters are less than microns, and the *N*-layer thicknesses are much smaller than microns, dipolar coupling will cause the adjacent *F*-layers to couple antiferromagnetically. If, instead, the layer diameters are micrometers, the dipoles at the ends of the layers will be so far out that such coupling is negligible.

1. *An mm diameter multilayer with uncoupled, identical F-layers*: If dipolar coupling in an $[F/N]_n$ multilayer is small, then making the *N*-layer thick enough to minimize exchange coupling (see Chapter 2 [22]) between the adjacent *F*-layers, will still give a GMR as shown in Figure 4.2. Figure 4.2 illustrates that the field dependences of the CIP- and CPP-MRs are usually similar.

2. *Nanowires with coupled identical layers*: In nanowires, where the *F*-layers are much wider rather than thicker (so that their magnetizations tend to lie "in-plane"), dipolar coupling will usually produce *AF* ordering of the adjacent *F*-layers. By alternating two very different thicknesses of *N*, such coupling can be used to give *AF* coupling of each pair of close *F*-layers, but little or no coupling of widely separated pairs [23].

3. *Multilayers with uncoupled, nonidentical F-layers (extension of Figure 4.1b)*: A well-defined *AP* state can be obtained by using two *F*-layers with different coercive fields. These can either be *F*-layers of different thicknesses, or grown on different underlayers [21], or involving two different *F*-layers [24]. The top panel in Figure 4.3 shows a CPP-MR example for a $[Co(6)/Cu(20)/Co(1)/Cu(20)]_8$ multilayer, with layer thicknesses in nanometer. Now, because of the very different values of H_c for the 6 nm and 1 nm thick Co layers, the multilayer reaches (and maintains for a short range of values of H) R_{AP} after saturation. In contrast to Figure 4.2, R_0 for the "initial" state is now slightly below R_{AP}. This difference suggests that R_0 in Figure 4.2 might also be slightly below the actual R_{AP}. The bottom panel in Figure 4.3 shows a CPP-MR example for a "separated" multilayer. These data will be examined in Section 4.5.9. An example of CPP-MR with two different *F*-metals, Co and Py = permalloy = $Ni_{1-x}Fe_x$ with $x \sim 0.2$, is given in Ref. [26]. The different values of H_c—$H_c(Co) \sim 150$ Oe and $H_c(Py) < 20$ Oe—yield $R_{peak} \approx R_{AP}$ as in Figure 4.3.

4. *ESBV (Figure 4.1c)*: A sharp transition and good control over the *AP* state can be achieved with an EBSV [27] of the form *AF/F/N/F*, where *AF* is an antiferromagnet (*AF*), such as FeMn, IrMn, or CoO. Here, first heating the sample to above the blocking temperature of the *AF*-layer, and then cooling to room temperature in the presence of an applied magnetic field, produces in the *AF*-layer an internal order that pins, via the exchange interaction, the magnetization of the adjacent *F*-layer to a much higher field than the coercive field of the other "free" *F*-layer. For a sample with macroscopic cross-sectional area, the "free" *F*-layer can be exchange decoupled from the pinned *F*-layer by making the *N*-layer thick enough (typically >5 nm). An example of CIP-MR data for such multilayers is shown in Figure 4.4. Similar results for CPP-MR are given in Ref. [28]. In both cases, the equality of the values of R for the $-H$- and $+H$ *P*-states shows that the *AF*-layer contributes only a constant term to R that is independent of whether the adjacent *F*-layer is aligned along or opposite to its "pinned" direction. If, however, the sample area is nanoscale, as needed for modern read heads, then two adjacent *F*-layers with magnetizations in the layer planes will couple antiferromagnetically via dipolar coupling. Real read-heads now minimize this problem by replacing the *AF/F* of the "pinned layer" by a synthetic *AF* of the form *AF/F2/Ru/F2*, where an Ru thickness ~0.6 nm gives *AF* coupling between the two *F*-layers as illustrated in Figure 4.5. Here, exchange coupling between the *AF* and adjacent *F2* sets the orientation of the synthetic *AF* and the presence of the two, oppositely magnetized *F2* layers greatly reduces the magnetic coupling with the separate "free" layer that reverses under the action of a small applied magnetic field.

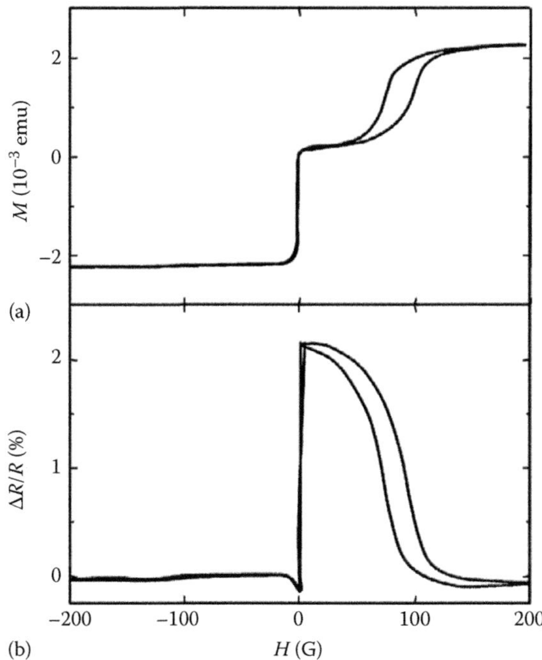

FIGURE 4.4 Magnetization curve (a) and CIP-MR (b) for an *n* EBSV multilayer Si/NiFe(15)/Cu(2.6)/NiFe(15)/FeMn(10)/Ag(2) (layer thicknesses in nm). The in-plane field *H* is parallel to the exchange anisotropy (EA) field created by the FeMn. The current flows perpendicular to this direction. (After Dieny, B. et al., *Phys. Rev. B*, 43, 1297, 1991. With permission.)

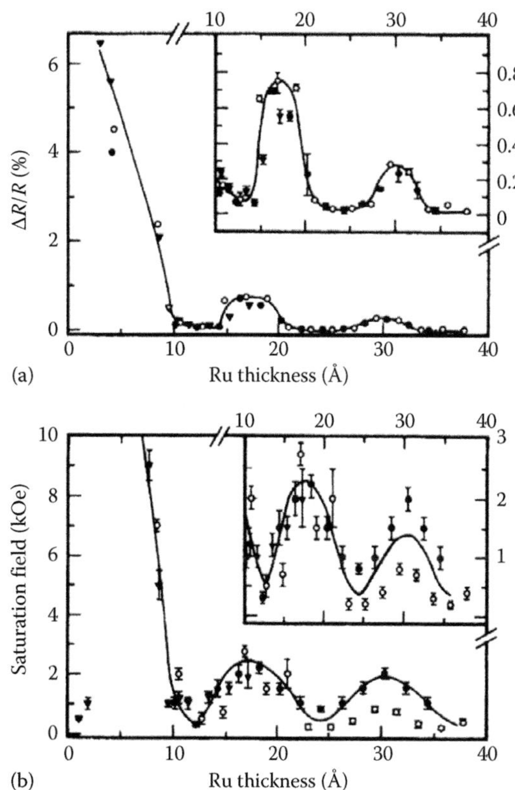

FIGURE 4.5 (a) CIP-MR and (b) saturation field, H_{sat}, versus Ru thickness, t_{Ru}, for Si(111)/Ru(10)/[Co(2)/Ru(t_{Ru})]20/Ru(5) multilayers (thicknesses in nanometer) deposited at temperatures of 313 K (filled circles), 398 K (open circles), or 473 K (crosses). (After Parkin, S. et al., *Phys. Rev. Lett.*, 64, 2304, 1990. With permission.)

4.2.6 Asymmetry Parameters

As explained in Chapters 1 and 5 [1,7], the models standardly used by experimentalists assume that electrons passing through an *F*/*N* multilayer can be separated into two current streams, one with its moments "up" and the other with its moments "down," relative to a chosen fixed axis (e.g., the direction of the initial applied field *H*). As one stream passes through a given *F*-metal (assumed monodomain), its moments can be along (↑) or opposite (↓) the magnetization of the *F*-metal. If, as the electrons flow, their moments do not "flip" so as to mix up and down currents, this assumption gives a "two-current" (2C) model [17,30–33]. In the limit of diffusive transport, within the bulk of the *F*-metal, this model has two parameters: $\rho_F^{\uparrow}$ and $\rho_F^{\downarrow}$. These parameters can be combined to give two alternative dimensionless parameters: $\alpha_F = \rho_F^{\downarrow}/\rho_F^{\uparrow}$ or $\beta_F = (\rho_F^{\downarrow} - \rho_F^{\uparrow})/(\rho_F^{\downarrow} + \rho_F^{\uparrow})$. Values of α_F for *F*-based alloys were derived a few decades ago [34] from measurements of deviations from Matthiessen's rule (DMR) in three-component *F*-based alloys. For GMR analysis, it is often more convenient to use β_F, along with an enhanced resistivity $\rho_F^{*} = (\rho_F^{\downarrow} + \rho_F^{\uparrow})/4 = \rho_F/(1-\beta_F^2)$. Here, ρ_F is the standard resistivity of the *F*-metal, which can be measured separately on films deposited in the same way as the GMR multilayers of interest. For the CPP-MR, one also defines similar quantities for the *F*/*N* interface: the asymmetry parameter $\gamma_{F/N} = (AR_{F/N}^{\downarrow} - AR_{F/N}^{\uparrow})/(AR_{F/N}^{\downarrow} + AR_{F/N}^{\uparrow})$ and the enhanced specific resistance $2AR_{F/N}^{*} = (AR_{F/N}^{\downarrow} + AR_{F/N}^{\uparrow})/2$, where *A* is the area through which the CPP-current flows, and $R_{F/N}$ is the CPP *F*/*N* interface resistance. Because the CIP-MR can rarely be written in closed form with bulk and interfacial scattering contributions separated, it is usually difficult to derive quantitative values of bulk or interfacial scattering asymmetries. In the CPP-MR, in contrast, applicability of simple two-current series-resistor (2CSR) [31–33], or more complex Valet–Fert (VF) [33] models often allow bulk and interfacial contributions and parameters to be separated. For the CPP-MR, we will list tables of values, and in several cases, compare them with values found by completely independent techniques, or else with calculated values involving no adjustable parameters.

4.2.7 Length Scales

4.2.7.1 CIP-MR

The dominant length scales in the CIP-MR are the mean free paths (λ) for scattering of electrons in the *F*-metal ($\lambda_F^{\uparrow}$ and $\lambda_F^{\downarrow}$) and the N-metal (λ_N) [1,35]. For an N-metal, λ_N is generally estimated from the metal's resistivity, ρ_N, using equation [36]:

$$\lambda_N = \frac{(\rho_B l_B)}{\rho_N} \tag{4.3}$$

Here, the parameter ($\rho_B l_B$) is a property of the N-metal of interest. Values for a number of metals, determined mostly from size-effect studies, have been collected together in Ref. [8]. For an F-metal, the situation is complicated by the presence of the two mean free paths, $\lambda_F^{\uparrow}$ and $\lambda_F^{\downarrow}$. If, for a given F-metal or alloy, one is much longer than the other, then the longer one can be estimated from Equation 4.3. In the absence of knowledge of the ratio, $\lambda_F^{\uparrow}$ and $\lambda_F^{\downarrow}$ must be determined independently. In Section 4.4.3.7, we will describe an experiment that did so.

4.2.7.2 CPP-MR

The dominant length scales in the CPP-MR are the lengths that electrons diffuse before flipping their spins, the spin-diffusion lengths l_{sf}^F in the F-metal or alloy, and l_{sf}^N in the N-metal or alloy [1,33]. In the 2CSR [31–33] and VF [33] models used to analyze most CPP-MR data, the mean free paths do not appear at all. In Section 4.5.5 we will explain how l_{sf}^F and l_{sf}^N are measured, and list some values of each. More complete listings, with values also determined by other techniques, are contained in Ref. [37].

At 4.2 K, electrons are scattered by impurities and fixed defects. Such scattering is mostly large angle, in which case any spin-flipping during the scattering does not mix spin currents [1,33]. At higher temperatures, however, electron–magnon (and perhaps also electron–phonon scattering) can mix spin currents. Then the spin-current mixing length can become important in both CIP- and CPP-MRs.

4.3 Compare CIP-MR, CPP-MR, and CAP-MR

In this section, we compare data on CIP-MR, CPP-MR, and CAP-MR.

Figure 4.2 compares values at 4.2 K for (a) CPP-MR; (b) CIP-MR; and (c) magnetization $M(H)$, for a $[Co/Ag]_{60}$ multilayer with $t_{Co} = t_{Ag} = 6$ nm; and also (d) the CPP-MR for a Co(9 nm) film between the Nb leads. The multilayer CPP-MR is several times larger than the multilayer CIP-MR, and both are much larger than the CPP-MR for a single Co film, which is <1%.

Similar data for uncoupled magnetic granules standardly require a much larger H to reach the P-state than do the CIP- or CPP-MRs for uncoupled F-layers [3,4].

Figure 4.6 compares the CPP- and CIP-MRs for (a) $[Fe(3)/Cr(1)]_{100}$ and (b) $[Co(1.2)Cu(1.1)]_{180}$ as functions of temperature T from 4.2 to 300 K. For both metal pairs, the CPP-MR is larger than the CIP-MR at all temperatures. Both MRs decrease with increasing T, due both to the increases in the denominators (R_P)

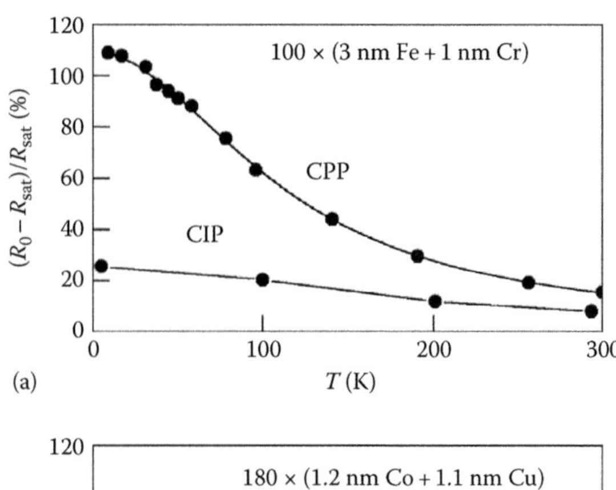

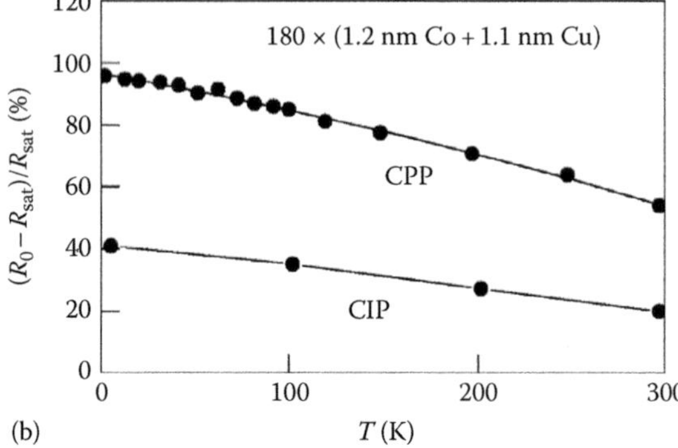

FIGURE 4.6 CPP-MR and CIP-MR versus temperature T (K) for (a) $[Fe(3)/Cr(1)]_{100}$ and (b) $[Co(1.2)/Cu(1.1)]_{180}$ multilayers (thicknesses in nanometer). (After Bass, J. and Pratt, W.P., *J. Magn. Magn. Mater.*, 200, 274, 1999; Bass, J. and Pratt, W.P., Erratum: *J. Magn. Magn. Mater.*, 296, 65, 2006; Gijs, M.A.M. et al., *Phys. Rev. Lett.*, 70, 3343, 1993; Gijs, M.A.M. et al., *J. Appl. Phys.*, 75, 6709, 1994. With permission.)

due to increased phonon scattering, as we now explain, and probably also to some mixing of spin currents (see Section 4.2.7.2). The resistivities of both the F and N layers increase substantially as T increases from cryogenic (4.2 K) to room (293 K) temperature, as scattering of electrons by phonons becomes increasingly important. As a rough guide, for sputtered or evaporated polycrystalline samples of nominally "pure" elemental metals (e.g., Cu, Ag, Co, Fe), the resistivity ratio ($RRR = \rho(293\,K)/\rho(4.2\,K)$) is typically ~2. However, RRRs as large as 4 have been seen for Cu, and RRRs closer to 1 can be found for concentrated alloys, such as Py [8,12]. If such bulk resistivities dominate R, $R(T)$ will increase by roughly a factor of 2 from 4.2 to 293 K. If, in contrast, interface scattering dominates, then the increase of R, and the variation of $\Delta R/R$, will depend upon the temperature dependence of interface scattering, which must be derived from experiment. In general, interface scattering seems to vary much less rapidly than bulk—see Refs. [40–42] and discussion of the CIP-MR(T) in Section 4.4.3.11, and of the CPP-MR(T) in Section 4.5.8.

Figure 4.7 compares CPP-MR (left) and CIP-MR (right) [43] for hybrid spin valves with different thicknesses of a $Ni_{95}Cr_5$ alloy that has negative β_{NiCr} but positive $\gamma_{NiCr/Cu}$. For thin NiCr, the positive interface dominates and both CPP- and CIP-MRs are "positive." For thick enough NiCr, the negative bulk dominates

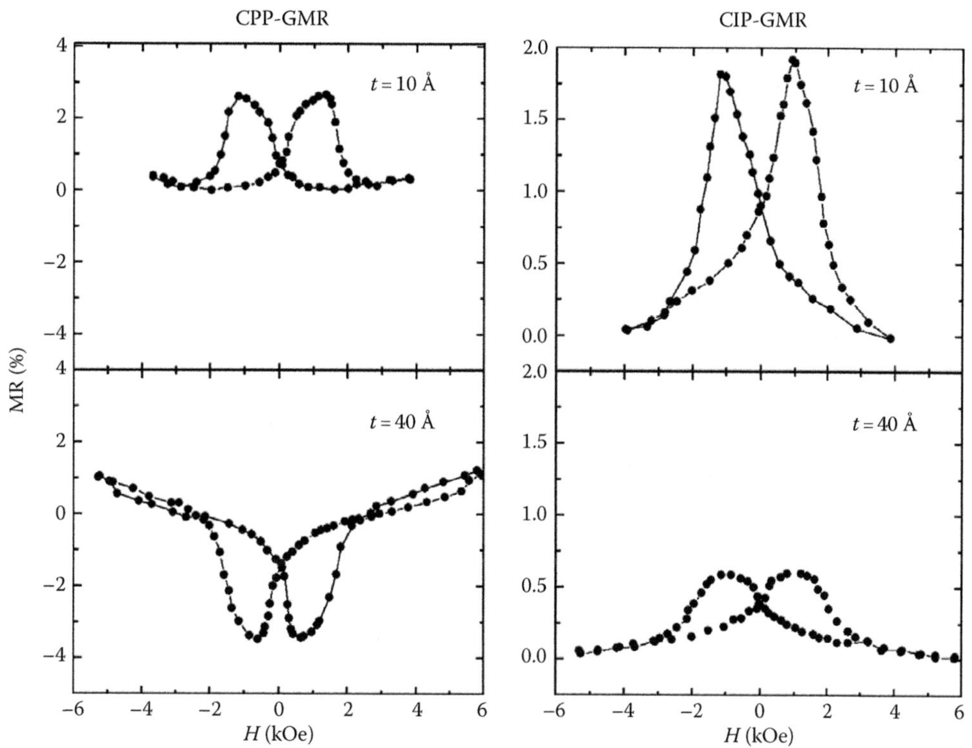

FIGURE 4.7 Comparisons of CPP-MR (left) and CIP-MR (right) for $[Ni_{95}Cr_5(t)/Cu(4\,nm)/Co(0.4\,nm)/Cu(4\,nm)]_{20}$ multilayers with $t=1$ and 4 nm. The CPP-MR inverts, but the CIP-MR does not. (After Vouille, C. et al., *Phys. Rev. B*, 60, 6710, 1999; Vouille, C. et al., *J. Appl. Phys.*, 81, 4573, 1997. With permission.)

the CPP-MR, making it negative. But the increasing bulk contribution only reduces the magnitude of the CIP-MR. These data illustrate that the relative importances of bulk and interface contributions usually differ for CIP- and CPP-MRs. For more discussion see Ref. [43].

Using V-grooves cut into an insulating surface, and then depositing layers parallel to one of the sides of the V-grooves [2,3] produced CAP data. Levy et al. [44] derived a relation between the smaller CIP, intermediate CAP, and larger CPP resistances that allow prediction of the CPP-MR from measurements of the CIP- and CAP-MRs. By evaporating highly conductive Cu or Au layers before and after the *F*/*N* multilayers of interest, Gijs et al. [40,45] converted the V-groove geometry into an almost CPP one. For an overview of these latter studies see Ref. [11].

4.4 CIP-MR

4.4.1 Measuring CIP-MR

The advantage of the CIP-MR is that a typical long (~mm), thin (<1 μm) multilayer has a resistance R large enough to be easily measured with standard laboratory equipment. But sample and lead geometries must be handled with some care to measure ΔR and R separately. In practice, most studies simply provide the GMR ratio, $MR=\Delta R/R$, for which knowledge of the detailed geometry is not essential. Said another way, if the current distribution is independent of the applied magnetic field H, the MR ratio will be independent of the applied current and will be a property of the sample at a given H. However, interpreting the MR is often not trivial, because it is affected by changes in both ΔR and R.

There are two standard ways to measure CIP-MR. The first involves a long, narrow sample with side contacts long and narrow enough to minimize the current flow into them. If the end contacts have low resistance compared to the sample, this geometry should give a uniform current and allow direct determinations of ΔR and R separately. The widely used second case involves a rectangular sample with four leads in a straight line. Unless the sample is much longer than it is wide, and the leads are well separated, the current density will be nonuniform, and separate determinations of ΔR and R require calculations to correct for this nonuniform density. As noted above, to avoid the effects of current nonuniformity, most published studies focus on $MR=\Delta R/R$.

4.4.2 Some CIP-MR Equations

A quantitative analysis of CIP-MR, in general, is difficult because the current flow through the sample is not uniform, and the CIP-MR depends upon the ratios of the thicknesses of the *F*- and *N*-layers to the mean free paths in those layers. If the *F*- and *N*-layer thicknesses are similar, the current is larger in the metal (usually the *N*-metal) with the lower resistivity.

Including contributions both within the bulk F- and N-layers and at the F/N interfaces generally involves a large number of parameters [30,35], most of which cannot easily be independently determined. Thus, analyses are mostly qualitative and highly simplified. As two examples, Ref. [30] fit data on Fe/Cr trilayers assuming a single mean free path in both the Fe and Cr and asymmetric scattering only at the interfaces, whereas [46] fit data on Py/Cu EBSVs completely neglecting scattering at the Py/Cu interfaces, but assuming five mean free paths, two in the Cu (parallel (||) and perpendicular (⊥) to the planes), and three in the Py ((||) and (⊥) for spin-up electrons and isotropic for spin down). As we will see next, data have also been fit using phenomenological equations that do not formally distinguish between bulk or interfacial scattering. Further below we will see arguments in favor of interfacial scattering often dominating the CIP-MR. For these reasons, analyses of CIP-MR data rarely yield quantitative parameters that can be compared with measurements by techniques other than CIP-MR.

Two phenomenological equations have been used to fit CIP-MR data as only t_N or t_F is varied. The physics underlying these equations is discussed in Chapter 5 [1] and Ref. [47].

For varying t_N [10,48]:

$$\frac{\Delta R}{R} = A + B\exp\left(\frac{-t_N/\lambda_N}{(1+t_N/t_0)}\right) \tag{4.4}$$

In Equation 4.4, the CIP-MR is a monotonically decreasing function of t_N. The t_N/t_0 term in the denominator gives a simple "dilution" effect [48]; the thicker t_N, the larger the fraction of current that flows through it, and the smaller the fraction scattered asymmetrically in the F-metal or at the F/N interfaces. The $\exp(t_N/\lambda_N)$ term is a mean-free-path effect; the thicker t_N, the harder for electrons scattered in one F-metal, or at its F/N interface, to get to the other F-metal before being scattered.

For varying t_F [14,49],

$$\frac{\Delta R}{R} = \frac{A\left[1-\exp\left(-t_F/\lambda_F\right)\right]}{\left(1+t_F/t_0\right)} \tag{4.5}$$

To qualitatively understand Equation 4.5, assume that the numerator corresponds to a normalized value of ΔR, and the denominator to a normalized value of R. As t_F increases, the numerator grows on scale λ_F, first linearly with t_F, but then gradually saturating, as t_F becomes much larger than λ_F. The denominator continuously grows linearly with increasing t_F, but on a scale t_0, where the ratio t_F/t_0 represents the fraction of R due to F. Coupling the growth and saturation of ΔR, with the slower but continuing growth of R, predicts an initial increase in $\Delta R/R$, a maximum, and then a slower decrease.

We will compare these equations with the data in Section 4.4.3.4.

4.4.3 Some Basic CIP-MR Questions and Answers

4.4.3.1 How Does Coupling Vary with t_N?

As described in Chapter 1 [7], GMR was discovered [17,18] in Fe/Cr samples with the Cr thickness chosen to give a strong AF coupling between the adjacent F-layers. The maximum MR then occurred for $H=0$, and the minimum MR for H above a saturation field, H_s. In Ref. [17], modest deviations of the Cr thickness from that which gave best AF coupling left the form of the MR data unchanged, but reduced its magnitude. The question naturally arose, "How does the coupling between the F-layers (e.g., Fe) across a given N-layer (e.g., Cr) vary with the thickness, t_N, of N?" Parkin et al. [29] showed, for carefully prepared samples, that the coupling and MRs oscillated together with decreasing coupling strength as t_N increased. Figure 4.5 shows an example of such behavior for F=Co and N=Ru.

4.4.3.2 Is *AF* Coupling Necessary for GMR?

The answer is "No." As noted above, all that is needed is the ability to change the magnetic order from P at high field to AP at lower field [21,50]. In Figure 4.5 we see that AF coupling (large H_s), which gives AP order at $H=0$, gives the largest CIP-MR, and that ferromagnetic coupling (small H_s), which forces the retention of approximate P order at all Hs, gives the smallest CIP-MR. From the continuous curves, we infer that zero coupling should give at least partial AP order and an intermediate CIP-MR. Figure 4.2 shows examples of data for no coupling. For devices, no coupling is preferred, so that the device can respond to small H. See the discussion of how to achieve the AP state in Section 4.2.5.

4.4.3.3 Can Other *F/N* Pairs Give MRs Competitive with Fe/Cr?

Not long after the discovery of GMR, the pure-metal pairs Co/Cu and Co/Ag were shown to give comparably large MRs. Subsequently, pairs with alloy F-metals, such as (Py=$Ni_{80}Fe_{20}$)/Cu and $Co_{91}Fe_9$/Cu were also found to give significant MRs, for Py, with much smaller H_c. For references to these studies see Ref. [10]. For devices, inserting a thin layer of Co at the Py/Cu interface combined the advantages of small H_c (for switching at low H) with high MR [51]. In contrast, a variety of other combinations of F- and N-metals gave smaller CIP-MRs (see Ref. [14, pp. 136–137]). For discussion of the band structure matching and other effects that contribute to the physics underlying these differences, see Chapter 5 [1].

4.4.3.4 How Do CIP-MRs Vary with t_F and t_N?

As the analysis of CIP-MR can involve many parameters, early investigators developed phenomenological equations to describe their data. Figures 4.8 and 4.9 show examples of multilayer data with only t_N or t_F varied. Equations 4.4 and 4.5 used for the fits are given in Section 4.4.2. The physics underlying these behaviors is discussed in Chapter 5 [1] and Ref. [47].

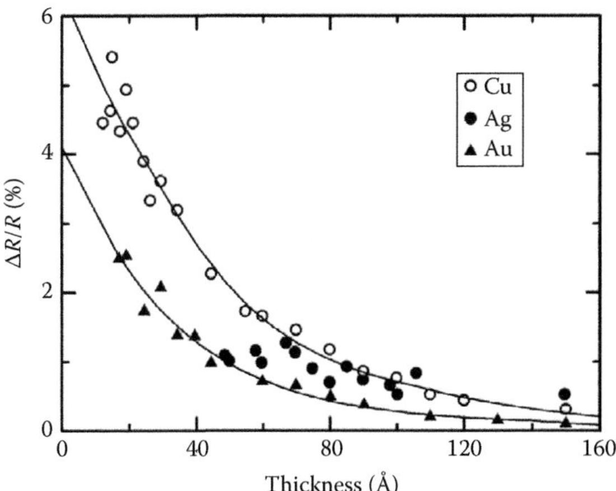

FIGURE 4.8 CIP-MR = (Δ*R*/*R*) (%) versus *N*-layer thickness for samples of the form Si/Co(7)/*N*/Py(4.7)/FeMn(7.8)/*N*(1.5) with *N* = Cu, Ag, or Au and thicknesses in nanometer. The solid curves are fits to Equation 4.4. (After Dieny, B. et al., *J. Appl. Phys.*, 69, 4774, 1991; Dieny, B. et al., *Phys. Rev. B*, 45, 806, 1992. With permission.)

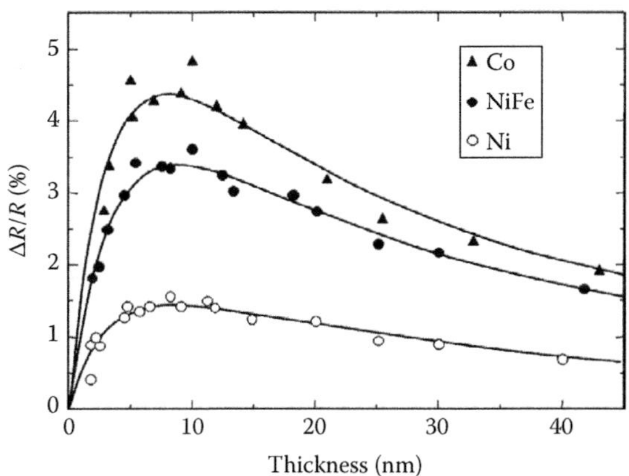

FIGURE 4.9 CIP-MR = (Δ*R*/*R*) (%) versus *F*-layer thickness for multilayers of the form Si/*F*/Cu(2.2)/Py(4.7)/FeMn(7.8)/Cu(1.5), with *F* = Ni, Py = $Ni_{0.8}Fe_{0.2}$, or Co and thicknesses in nanometer. The curves are fits to Equation 4.5. (After Dieny, B. et al., *J. Appl. Phys.*, 69, 4774, 1991; Dieny, B. et al., *Phys. Rev. B*, 45, 806, 1992. With permission.)

4.4.3.5 Is CIP-MR Dominated by Spin-Dependent Scattering in Bulk *F* or at *F*/*N* Interfaces?

Early interpretations favoring both possibilities were published [30,46]. To investigate the alternatives in more detail, investigators began to change the interfaces in different ways, of which we give some examples. In Fe/Cr multilayers, the authors of Refs. [52,53] inserted at the Fe/Cr interfaces impurities like Cr that give $\beta_F < 1$ (V, Mn) and impurities unlike Cr that give $\beta_F > 1$ (Ge, Al, Ir). For V and Mn, the CIP-MRs decreased relatively slowly with increasing impurity layer thickness, similar to the behavior for Cr. In contrast, for Ge, Al, and Ir, the CIP-MRs decreased rapidly with impurity layer thickness. The authors attributed these differences to the qualitative differences in β_F, as expected for spin-dependent interfacial scattering. Speriosu et al. [54] heated Py/Cu and Co/Cu samples to extend the intermixed regions of the interfaces. For Py/Cu, they found decreases in MR with increasing intermixing, which they attributed to growth of nonmagnetic alloys at the interfaces, which increased spin-independent scattering there. In contrast, in Fe/Cr multilayers, increased intermixing increased the CIP-MR [55,56].

Parkin [51] used EBSVs to study the effects at 293 K of inserting thin Co layers at Py/Cu interfaces or within the bulk Py layers, and of inserting thin Py layers at Co/Cu interfaces. Figure 4.10 shows that the CIP-MR grew with more Co at the Py/Cu interface until the Co contribution dominated, whereas moving the Co into the Py, or inserting Py at Co/Cu interfaces, decreased the CIP-MR. In all three cases, the scale over which the effect occurred was only 0.15–0.3 nm. He argued that these behaviors showed that interfacial scattering was dominant.

In contrast, the data for Co/Cu-based EBSVs in Figure 4.11 show that inserting fractional monolayers (MLs) of various impurities at the Co/Cu interfaces always decreased the MR, whereas inserting the same ML within the bulk of the Co, a distance *x* from the interface, could decrease, increase, or leave

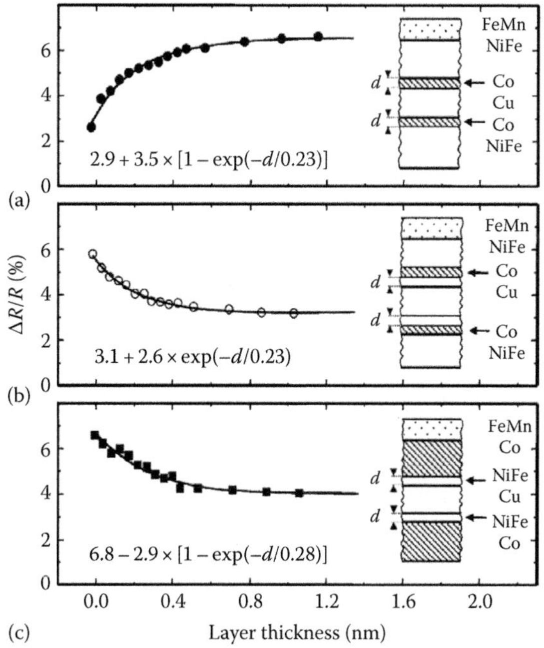

FIGURE 4.10 Dependence of room temperature CIP-MR = (Δ*R*/*R*)(%) on (a) Co interface layer thickness, *d*, in EBSVs of the form Si/Py(5.3 − *d*)/Co(*d*)/Cu(3.2)/Co(*d*)/Py(2.2 − *d*)/FeMn(9)/Cu(1); (b) distance *d* of a 0.5 nm thick Co layer from the Py/Cu interfaces in EBSVs of the form Si/Py(4.9 − *d*)/Co(0.5)/Py(*d*)/Cu(3)/Py(*d*)/Co(5)/Py(1.8 − *d*)/FeMn(9)/Cu(1); and (c) Py interface layer thickness, *d*, in EBSVs of the form Si/Co(5.7 − *d*)/Py(*d*)Cu(2.4)/Py(*d*)/Co(2.9 − *d*)/FeMn(10)/Cu(1). Py = $Ni_{0.8}Fe_{0.2s}$ is designated as NiFe, and layer thicknesses are in millimeters. (After Parkin, S.S.P., *Phys. Rev. Lett.*, 71, 1641, 1993. With permission.)

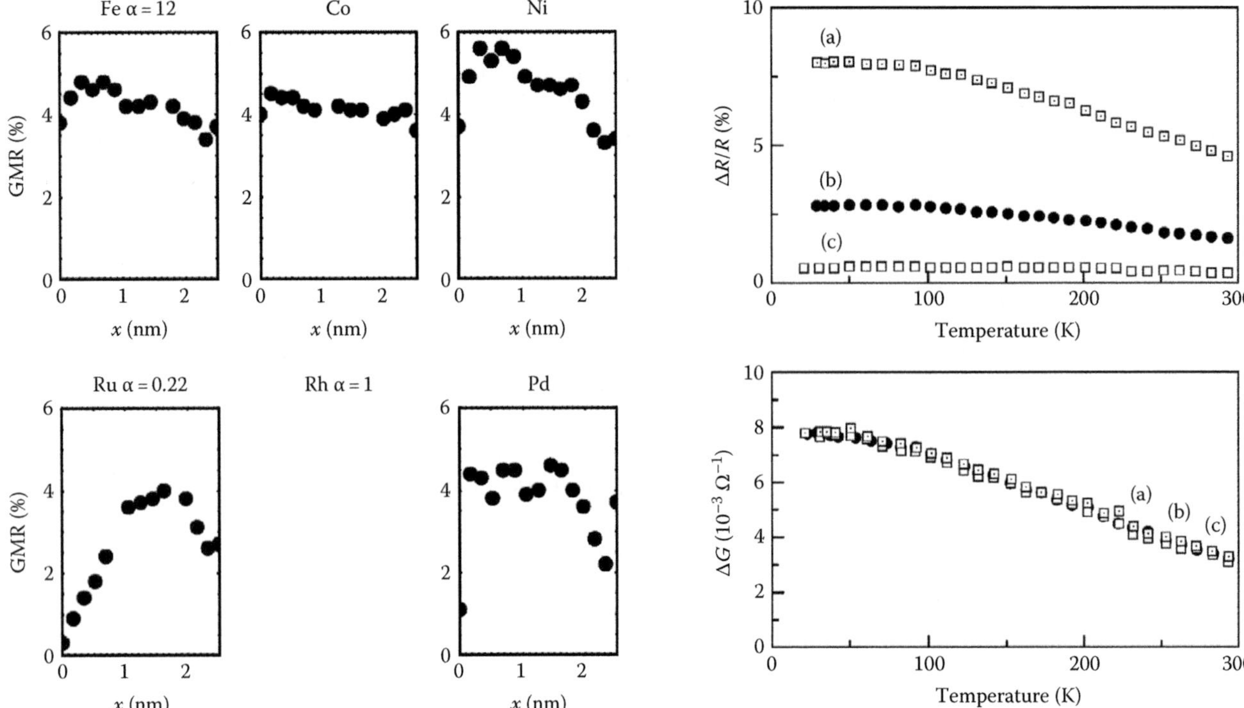

FIGURE 4.11 GMR (%) = CIP-MR versus distance x of the specified impurity from a Co/Cu interface. (After Marrows, C.H. and Hickey, B.J., *Phys. Rev. B*, 63, 220405R, 2001. With permission.)

FIGURE 4.12 Top: CIP-MR, and bottom: ΔG, versus temperature for multilayers (a), (b), and (c) with differing capping layers—see text. (After Dieny, B. et al., *Appl. Phys. Lett.*, 61, 2111, 1992. With permission.)

the MR unchanged, depending upon both the impurity and x. The authors attributed these different behaviors to deviations of the β_F for these impurities in Co from the scattering anisotropy of the undoped Co. They concluded that surface scattering was important, but that bulk scattering by impurities could also be important much further into the Co than the fraction of a nanometer seen in Ref. [51].

As illustrated in Figure 4.7, for thin F-layers, interface effects predominate, but bulk effects grow in thicker layers. The relative importance of bulk and interfacial scattering depends upon both the particular host F-metal or the alloy and the particular N-metal. If impurities are selectively inserted into a "pure" host F-metal, these effects depend on just where the impurities are inserted into the host F-metal or at the interface.

4.4.3.6 What Is the Fundamental Quantity for CIP-MR?

Dieny et al. [58] argue that the fundamental quantity for the CIP-MR is the change in sheet conductance, $\Delta G = \Delta R/(R_P R_{AP})$. Figure 4.12 shows the different behaviors of MR $= \Delta R/R$ and ΔG for three samples that differ only in the capping layer: (a) cap = Ta(5), (b) cap = Cu(10)/Ta(5), and (c) cap = Cu(30)/Ta(5), where the layer thicknesses are in nanometer. The data collapse from three different curves for MR, to a single curve for ΔG, consistent with the prediction in Ref. [58] of $\Delta G = n(\Delta G_{per}) - \Delta G_{corr}$, where "$\Delta G_{per}$ is independent of the number of periods n in the multilayer, and ΔG_{corr} corrects for boundary effects and is independent of n." Despite the advantage argued for ΔG, most researchers have stuck with the dimensionless CIP-MR, where knowing the precise sample dimensions is not crucial.

4.4.3.7 How Long Are Mean-Free-Paths in F-Metals?

As explained in Chapter 5 [1] and Section 4.2.7.1 above, the CIP-MR depends upon three mean free paths: $\lambda_F^{\uparrow}$ (majority = long) and $\lambda_F^{\downarrow}$ (minority = short), and one in the N-metal, λ_N. λ_N can be found from Equation 4.3. Gurney et al. [59] described a technique for estimating $\lambda_F^{\uparrow}$ and $\lambda_F^{\downarrow}$. For three standard F metals or alloys, they reported room temperature values of (in nanometer): $\lambda_{Py}^{\uparrow} = 4.6 \pm 0.3$, $\lambda_{Co}^{\uparrow} = 5.5 \pm 0.4$, and $\lambda_{Fe}^{\uparrow} = 1.5 \pm 0.2$, and, $\lambda_{Py}^{\downarrow} \leq 0.6$, $\lambda_{Co}^{\downarrow} \leq 0.6$, and $\lambda_{Fe}^{\downarrow} = 2.1 \pm 0.5$.

4.4.3.8 How Does Interface Roughness Affect the CIP-MR?

The published evidence suggests that the effects of interface roughness on the CIP-MR are not simple, but depend upon both the particular pair of F- and N-metals studied, and the structure of the roughness. The first model of CIP-MR [30] assumed that the MR was dominated by the scattering of electrons at interfacial roughness. Attempts to deduce the effects of such roughness soon followed. Early studies gave what appeared to be conflicting results: greater roughness was claimed to increase the CIP-MR in Fe/Cr [55,56,60], but to decrease it in Py/Cu [54].

A detailed study of Fe/Cr [61] concluded that the specific structure of the roughness was important, with the CIP-MR magnitude increasing with increasing vertical roughness amplitude, or with decreasing lateral correlation length. For a more detailed discussion, see Ref. [14].

4.4.3.9 How Do Values of β_F from CIP-MR and DMR Compare?

In the CIP-MR, the difficulty of separating bulk and interface scattering contributions makes it hard to isolate the bulk (β_F) and interface ($\gamma_{F/N}$) scattering anisotropies. We know of no quantitative comparisons for the CIP-MR. Examples have been published of the inversion of the CIP-MR [47,62,63], where the DMR values predict such inversion (see Chapters 1 and 5 [1,7]). Such examples show qualitative understanding of the physics underlying the CIP-MR.

4.4.3.10 How Does the CIP-MR Vary with Angle?

For a sample in which the magnetization of one *F*-layer is held fixed, and the other allowed to rotate, the CIP-MR is expected to grow closely as $[1 - \cos(\theta_1 - \theta_2)]$ [1,14], where $\theta_1 - \theta_2$ is the angle between the magnetizations of the two *F*-layers. Figure 4.13 shows such a variation for a Py-based EBSV with one Py layer pinned and the other rotated by a 10 Oe field. The CIP-MR data in Figure 4.13 are corrected for a modest AMR effect (shown by the curve labeled AMR). For more details, see Ref. [14].

4.4.3.11 How Does the CIP-MR Vary with Temperature?

As illustrated in Figure 4.6 for Fe/Cr and Co/Cu multilayers, the CIP-MR generally decreases monotonically with increasing temperature. Part of the decrease is due to the phonon-scattering driven increase in the bulk resistivity components of the sample resistance, R_P, which forms the denominator of the

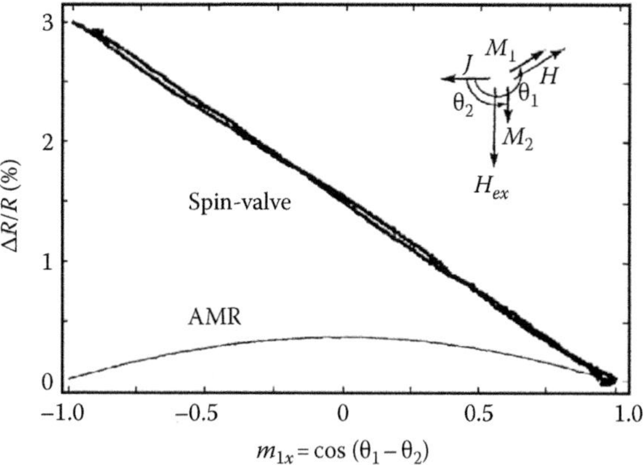

FIGURE 4.13 CIP-MR = ($\Delta R/R$) (%) (corrected for the AMR shown) versus the cosine of the angle between the magnetizations of the two NiFe layers in a Si/NiFe(6)/Cu(2.6)/NiFe(3)/FeMn(6)/Ag(2) EBSV with thicknesses in nanometer. Inset shows orientations of the current *J*, exchange field, H_{ex}, and magnetizations M_i and M_2. (After Dieny, B. et al., *Phys. Rev. B*, 43, 1297, 1991. With permission.)

GMR. As explained in Section 4.3, the bulk contributions to R_P generally increase by roughly a factor of two between 4.2 K and room temperature (293 K). For Co/Cu, doubling R_P, and holding ΔR nearly constant, would account for most of the decrease in CIP-MR in Figure 4.6. For Fe/Cr, in contrast, the decrease is more rapid, suggesting that ΔR also decreases strongly with increasing *T*. Contributors to this latter decrease could include interband transitions [1] and scattering-driven intermixing of the two spin currents [64].

4.4.3.12 Can CIP-MR Be Increased by Modifying Top and Bottom Interfaces?

In the moderate thickness EBSVs used for devices, such as read-heads, scattering from the interfaces at the bottom and top of the sample can affect the CIP-MR. Specifically, if such scattering can be made specular, holding other factors constant, one would expect to reduce the total *R*, and thus increase the CIP-MR. Several papers have reported enhanced CIP-MRs in samples with NiO and other top and bottom layers, and in samples subject to oxidation of the top layer. Most attributed these enhancements to increased specular reflection at the interfaces with such layers. A number of such studies are examined in Ref. [14], which concludes that the sources of the observed enhancements are not yet fully understood.

4.5 CPP-MR

In contrast to the CIP-MR, the CPP resistance *R* of a thin film with widths ~mm (giving $A \sim mm^2$) and thickness ~100 nm (giving "length" ~100 nm) is $\sim 10^{-8}\ \Omega$, difficult to measure and much too small for devices. Area miniaturization, with sample widths <100 nm, is now raising CPP resistances to ~Ω, leading to device-potential and device-related research. Tunneling MR, which also uses the CPP geometry, is now the device preference for read-heads and magnetic random access memory (MRAM) [65]. But because of the large *AR* product of tunneling samples, there is a likely need for alternatives as *A* decreases still further. The CPP-MR with just metals is one candidate for the next generation of read-heads. We begin this section with a discussion of the different techniques used to measure the CPP-MR, and their relative advantages and disadvantages.

4.5.1 Measuring CPP-MR

Three techniques have been used to measure the CPP-MR. They are (1) short, wide samples between superconducting cross-strips [12], (2) micron- or nano-sized pillars, and (3) electrodeposited nanowires [13].

1. The first technique [19,66–68] involves a short-(≤μm) sample, sandwiched between wide (~1 mm) crossed superconducting (*S*) strips. Such *S* leads give lead resistances due only to the *F/S* interfaces, which can be measured separately [67]. The "capacitor-like" geometry gives

uniform current flow [68]. Slater et al. [69] developed a way to measure CPP-MR using micron-sized top superconducting contacts on an otherwise mm^2 area sample. The disadvantages of superconducting leads are (1) the limitation to cryogenic temperatures (typically 4.2 K using the usual superconductor, Nb, which stays superconducting to only about 9 K [36]; and (2) the need for a high sensitivity, high precision, bridge to measure the typically very small sample resistances ($R \sim 10\,\mathrm{n\Omega}$) [66,70]. Cyrille et al. [71] avoided the need for a sensitive bridge by using one hundred ~10 μm sized samples in series. The advantages of superconducting leads are their ability to (a) achieve uniform current distributions and contacts with known and reproducible lead resistances and (b) combine a wide variety of metals with control of AP and P states via the full range of available techniques.

2. The second technique involves patterning a multilayer into micron-sized [38] or more recently, nanosized pillars (see, e.g., Ref. [72]). The advantage is the ability (a) to combine a wide variety of F and N metals and (b) to allow measurements of CPP-MR from cryogenic temperatures to room temperature. The disadvantages are difficulties in (a) achieving uniform current distributions [73], (b) reproducibly preparing nanosize samples, and (c) controlling, determining, and including contact resistances in equations of interest (simplified equations that neglect them can potentially give large uncertainties).
3. The third technique involves nanowires electrodeposited into nanopores in either plastics [74–76] or more ordered structures, such as Al_2O_3 [77]. The nanopore diameters can range from ~20 to 100 nm. The long, thin wire geometry gives a uniform current flow and large enough resistances for convenient measurement. Most published studies have the disadvantage of not knowing the number of wires. Contact resistances can be a problem, but long samples can make them unimportant. Recently, procedures have been developed to contact to just single nanowires [78,79]. As yet, only a modest number of different F/N combinations (mostly Co/Cu, Fe/Cu, and Py/Cu) have been studied with this technique [13]. Its advantages are simplicity and the ability to measure MRs from cryogenic to high temperatures. Its limitations include difficulty in making very thin layers—interfaces are usually "thicker" than with other deposition techniques; difficulty in making EBSVs; and the need for care to minimize coupling to achieve P and AP states at relatively small fields.

4.5.2 Some 2CSR and VF Equations for CPP-MR Analysis

For details of theoretical analysis of CPP-MR, see Chapter 5 [1]. We briefly describe here the simplest picture of the underlying physics and give some 2CSR [31–33] and VF [33] equations used by experimentalists to analyze CPP-MR data. The 2CSR equations hold when all layers are thinner than their spin-diffusion lengths. The more general VF equations apply to thicker layers.

4.5.2.1 2CSR Model and Equations

As noted above, if scattering of electrons transiting a multilayer does not mix the moment up and down currents, then a 2C model results. If, in addition, the spin-diffusion lengths in the F- and N-layers are much longer than the layer thicknesses, and spin-flipping at interfaces is negligible, then for free electron Fermi surfaces and diffusive transport, it was quickly shown [31,33] that the CPP-MR should be given by a 2CSR model in which the total specific resistance, AR(up) or AR(down) for each electron-moment direction is just the series sum of the specific resistances of each layer ($\rho_N^{\uparrow} t_N$, $\rho_N^{\downarrow} t_N$, $\rho_F^{\uparrow} t_F$, and $\rho_F^{\downarrow} t_F$) and the specific resistances of each interface ($AR_{F/N}^{\uparrow}$ and $AR_{F/N}^{\downarrow}$). As in Section 4.1, ↑ means that the electron moment is along the local F-moment, and ↓ means it is opposite to the local F-moment. For a simple $[F/N]_n$ multilayer, where each F- and N-layer is just repeated, in the AP state, AR(up) and AR(down) are the same, and AR_{AP} takes a simple form, just half of the common value for each. We give here some general 2CSR model equations that we employ in later analyses. These use as unknowns, the three parameters: β_F, $\gamma_{F/N}$, and $2AR_{F/N}^{*}$.

We start first with simple $[F/N]_n$ or $[F/N]_nF$ multilayers and superconducting (S) leads. Such leads have the advantage that the contact resistance, $2AR_{S/F}$, can be measured independently [12,67,80,81]. But they give the following complications. For simple $[F/N]_n$ multilayers, the contact of the top N-layer with the top S-lead causes the proximity effect to turn the top N-layer superconducting, and there are only $(2n-2)$ F/N interfaces. Conversely, for $[F/N]_nF$ multilayers, there are $n+1$ F-layers. For simplicity, we neglect these complications in the following three equations used with S-contacts [12,32]:

$$AR_{AP} = 2AR_{S/F} + n\left(\rho_N t_N + \rho_F^{*} t_F + 2AR_{F/N}^{*}\right) \tag{4.6}$$

$$A\Delta R = \frac{n^2\left(\beta_F \rho_F^{*} t_F + \gamma_{F/N} 2AR_{F/N}^{*}\right)^2}{AR_{AP}} \tag{4.7}$$

$$\sqrt{AR_{AP} A\Delta R} = n\left[\beta_F \rho_F^{*} t_F + \gamma_{F/N} 2AR_{F/N}^{*}\right] \tag{4.8}$$

We turn next to nanowires. For long nanowires, the lead resistance can be neglected, giving the following 2CSR equations useful for determining the desired parameters [41,82].

$$R_{AP} = n\left[\rho_F^{*} t_F + 2AR_{F/N}^{*} + \rho_N t_N\right] \tag{4.9}$$

$$\left[\frac{\Delta R}{R_{AP}}\right]^{-1/2} = \left[\frac{\left(\rho_F^* t_F + 2AR_{F/N}^*\right)}{\left(\beta_F \rho_F^* t_F + \gamma_{F/N} 2AR_{F/N}^*\right)}\right] + \left[\frac{\rho_N t_N}{\left(\beta_F \rho_F^* t_F + \gamma_{F/N} 2AR_{F/N}^*\right)}\right] \tag{4.10}$$

If the square root in Equation 4.10 is plotted versus t_N for two different fixed thicknesses t_F of the F-metal, then, for each thickness of t_F, the data should give a straight line, and the ordinate value where the two lines cross should be just $(\beta_F)^{-1}$ [41].

Finally, for nanopillars, the sample resistance is often not large compared to the lead resistances, which then cannot be neglected, and in many cases are not known. Determining their contribution is usually nontrivial. However, simply neglecting them might lead to significant uncertainties in derived parameters.

4.5.2.2 Valet–Fert (VF) Model and Some Equations with Finite l_{sf}

VF [33] showed that the same assumptions (free-electron Fermi surfaces and dominant diffusive scattering) that give the 2CSR model when $l_{sf}^F \gg t_F$, and $l_{sf}^N \gg t_N$, give more general equations that depend upon l_{sf}^F and l_{sf}^N when l_{sf}^F " t_F and l_{sf}^N " t_N. The most general VF equations are so complex that they require numerical analysis. But the equations sometimes simplify enough for quantitative solution. We give here examples that have been used in determining l_{sf}^F.

We start with superconducting leads and a symmetric EBSV of the form Nb/FeMn/F(t_F)/N/F(t_F)/Nb, where FeMn is an AF used to pin the magnetization of the adjacent F-layer. In the limits $t_F \gg l_{sf}^F$ (e.g., $l_{sf}^{Py} \sim 5.5$ nm [37,83]) and $t_N \ll l_{sf}^N$ (e.g., $l_{sf}^{Cu} > 200$ nm [37], VF theory gives $A\Delta R$ as a variant of Equation 4.6 where l_{sf}^F replaces t_F in both the numerator and denominator, and the denominator reduces to just the magnetically active central region between the l_{sf}^F thicknesses of the two F-layers [12]:

$$A\Delta R = \frac{\left(2\beta_F \rho_F^* l_{sf}^F + \gamma_{F/N} 2AR_{F/N}^*\right)^2}{\left[2\rho_F^* l_{sf}^F + \rho_N t_N + 2AR_{F/N}^*\right]} \tag{4.11}$$

Equation 4.11 is independent of t_F, thus predicting that $A\Delta R$ should become constant for $t_F \gg l_{sf}^F$.

We end with an equation used for nanowires. For a multilayer nanowire with a long $t_F \gg l_{sf}^F$, but $t_N \ll l_{sf}^N$, VF theory gives the equation [41,82]:

$$\left[\frac{R_P}{\Delta R}\right] = \frac{\left[(1-(\beta_F)^2\right] t_F}{2p(\beta_F)^2 l_{sf}^F} \tag{4.12}$$

Here, p is an assumed fraction of AP alignment in the nominal AP state. As illustrated by the data in Figure 4.2, a multilayer with equal thickness F-layers usually does not reach the AP state after saturation. From Equation 4.12, a plot of $[R_P/\Delta R]$ versus t_F should give a straight line with slope inversely proportional to l_{sf}^F.

4.5.3 CPP-MR Test of the 2CSR Model at 4.2 K

In Section 4.5.2 we listed some equations used to determine the CPP-MR parameters. Before describing what we know about those parameters, we present a set of confirmatory tests of the 2CSR that do not require explicit knowledge of specific parameters. Such tests have been made at 4.2 K using multilayers of Co with both Ag [32,84] and Cu [85]. We focus here on Co/Ag. As we will see in Section 4.5.5.2, Co was a fortunate choice, as l_{sf}^{Co} is unusually long.

The tests involve comparing Equations 4.6 through 4.8, for multilayer data with the N-metal Ag and a dilute $Ag_{96}Sn_4$ alloy. Alloying with 4 at% Sn gave $\rho_{AgSn} \sim 180$ nΩ m $\gg \rho_{Ag} \sim 10$ nΩ m [32], but AgSn gives only weak spin–orbit scattering [86], so that the alloy spin-diffusion length stays long. Equations 4.6 through 4.8 then make very explicit, directly testable, 2CSR model predictions about the different behaviors of the data for Co/Ag and Co/AgSn multilayers.

We rewrite Equation 4.6 for $[F/N]_n$ multilayers with the Co-thickness held fixed at $t_{Co} = 6$ nm and the total sample thickness fixed at $t_T = nt_F + nt_N$. We assume superconducting crossed leads and neglect differences between n and $n \pm 1$.

Equation 4.6 then becomes [32]:

$$AR_{AP} = 2AR_{Nb/Co} + \rho_N t_T + n\left[\left(\rho_{Co}^* - \rho_N\right)(6) + 2AR_{Co/N}^*\right] \tag{4.6'}$$

Equation 4.6′ predicts that plots of AR_{AP} versus n should give straight lines, with a much larger intercept for AgSn, but only about a third the slope (due to the negative contribution from $(\rho_F^* - \rho_N)t_F$). Figure 4.14 shows $AR_0 \approx AR_{AP}$ and AR_P versus n for $[Co(6)/Ag]_n$ and $[Co(6)/AgSn]_n Co(6)$ multilayers. As predicted, the data for Co/Ag fall along straight lines, and the intercept agrees with an independent prediction from measurements of $2AR_{Nb/Co}$ and ρ_{Ag} [32,67]. The downward curvatures of the broken lines at small n for the Co/Ag data are due to the presence of only $2n-2$ interfaces. The data for Co/AgSn have the predicted much larger intercept and smaller slope. Because of the extra Co layer, all of the n AgSn layers and $2n$ Co/AgSn interfaces contribute to AR_T.

Turning next to Equation 4.7, it stays unchanged, independent of whether t_T stays fixed, except for inserting Co for F. From Figure 4.14, because ρ_{Ag} is small, for Co/Ag, AR_{AP} is approximately linear in n, and Equation 4.7 predicts that $A\Delta R$ should grow closely linearly with n. Thus, we expect AR_P to also be linear in n, which it is. In contrast, because ρ_{AgSn} is large, for Co/AgSn and small n, AR_{AP} is almost constant. Equation 4.7 then predicts that $A\Delta R$ should grow initially as n^2. This means that AR_P should be curved, falling further below AR_{AP} as n grows. Such behavior

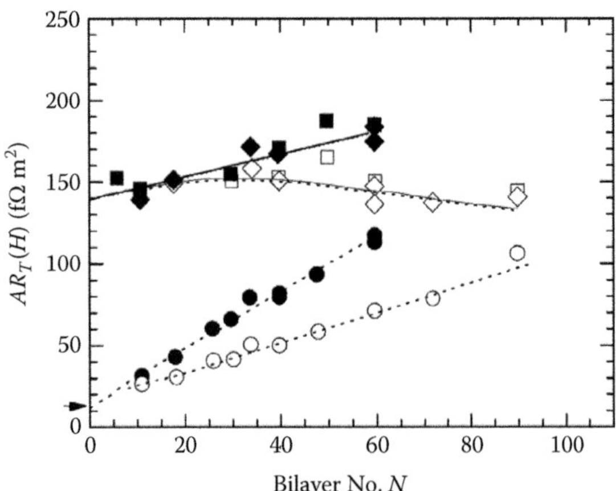

FIGURE 4.14 $AR_T(H)$ versus bilayer number N for $[\text{Co}(6)/\text{Ag}(t_{\text{Ag}})]_N$ (circles and broken lines) and $[\text{Co}(6)/\text{AgSn}(t_{\text{AgSn}})]_N\text{Co}(6)$ (squares or diamonds and solid or dashed curves) multilayers with fixed $t_{\text{Co}} = 6\,\text{nm}$ and fixed $t_T = 720\,\text{nm}$ (as closely as possible with integer N). The filled symbols are for AR_0 (approximating AR_{AP}) and the open symbols are for AR_p. The arrow to the lower left indicates the independently predicted value of the Co/Ag ordinate intercept. The dotted curves for the Co/Ag data are best fits using the 2CSR model. The solid and dotted curves for the Co/AgSn data are similar fits with two slightly different choices of parameters. The downturns of the dashed curves for Co/Ag at small N are an "end effect" due to the presence of only $2N - 1$ interfaces. (After Pratt Jr., W.P. et al., *J. Appl. Phys.*, 73, 5326, 1993. With permission.)

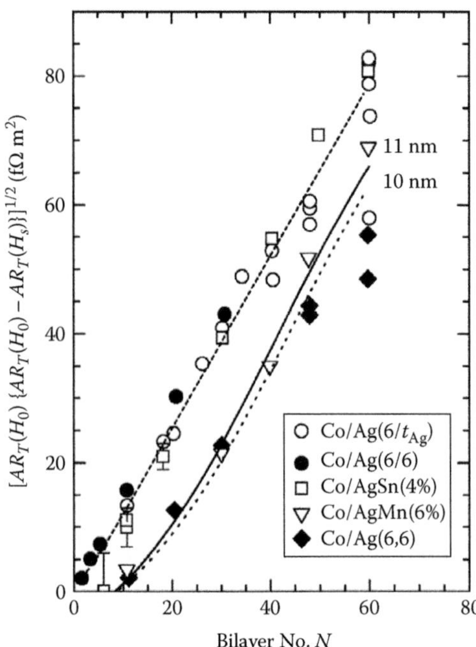

FIGURE 4.15 $\sqrt{AR_0(AR_0 - AR_P)}$ versus bilayer number N for $[\text{Co}(6\,\text{nm})/N(t_x)]_N$ multilayers with fixed total thickness $t = 720\,\text{nm}$, except for the samples indicated by filled circles, which had the form $[\text{Co}(6\,\text{nm})/\text{Ag}(6\,\text{nm})]_N$. The spin-diffusion lengths, l_{sf}^N, of Ag and AgSn are long enough so that the data stayed (to within uncertainties) on the straight line passing through the origin predicted by Equation 4.12. In contrast, the spin-diffusion lengths of AgMn and AgPt are short enough to cause deviations from this line. The solid and broken curves passing through the AgPt and AgMn data are fits using VF theory. The deduced values of l_{sf}^N are given in Table 4.3. (After Bass, J. et al., *Mater. Sci. Eng. B*, 31, 77, 1995. With permission.)

is illustrated by the dashed curve in Figure 4.14 and is shown explicitly in Refs. [32,84].

Lastly, the most important test involves revisiting Equation 4.8 for fixed $t_{\text{Co}} = 6\,\text{nm}$ to give a square root quantity predicted to grow strictly linearly with n, with zero intercept and slope that is independent of ρ_N [32]. That is, for a dilute enough AgSn alloy, so that $\gamma_{\text{Co/AgSn}} = \gamma_{\text{Co/Ag}}$ and $2AR_{\text{Co/AgSn}} = 2AR_{\text{Co/Ag}}$, the 2CSR model predicts that the data for Ag and AgSn should fall along exactly the same straight line passing through the origin. Requiring a straight line to pass through zero is a much more stringent test of a model prediction than a straight line with unknown intercept. Despite the very different data for Co/Ag and Co/AgSn in Figure 4.14, Figure 4.15 shows that the prediction is obeyed not only for the data of Figure 4.14, but also for independent data on Co/Ag samples with fixed $t_{\text{Co}} = t_{\text{Ag}} = 6\,\text{nm}$ and variable t_T [85]. The combination of data in Figures 4.15 and 4.16 is strong evidence for the 2CSR model for Co and Ag.

4.5.4 How to Determine β_F, $\gamma_{F/N}$, and $2AR^{\star}_{F/N}$

The tests just described in Section 4.5.3 support the applicability of the 2CSR model to CPP data for multilayers with F- and N-metals for which l_{sf}^F and l_{sf}^N are long. For simple $[F/N]_n$ multilayers with layer thicknesses $t_F \ll l_{sf}^F$ and $t_N < l_{sf}^N$, and where ρ_F, ρ_N, and (with superconducting leads) $2AR_{S/F}$ have been separately measured, there are only three unknown parameters: β_F, $\gamma_{F/N}$, and $2AR^{\star}_{F/N}$. Determining these parameters requires measuring AR_{AP} and AR_P on multilayers with systematic variations of t_F, t_N, and n, and analyzing the data using equations like Equations 4.6 through 4.10. A detailed study of Co/Ag with superconducting leads, using several sets of multilayers, is given in Ref. [68]. Comparable studies using nanowires are described in Refs. [41,82]. If the value of β_F for a given F-metal or alloy has already been determined in the same laboratory, then, in principle, $2AR^{\star}_{F/N}$ and $\gamma_{F/N}$ could be determined simply from either measurements of AR_{AP} and AR_P versus n for $[F/N]_n$ multilayers with fixed t_F, or measurements of AR_{AP} and $A\Delta R$ on EBSVs. In practice, uncertainty in AR_{AP} for the multilayers and fluctuations in AR_{AP} for the EBSVs (due to the usually large resistivities of thick F-metal and AF layers), limit the reliability of each procedure alone. For better results, recent determinations of $2AR^{\star}_{F/N}$ and $\gamma_{F/N}$ combined measurements on $[F/N]_n$ multilayers and EBSVs—for example, Refs. [87,88].

4.5.4.1 β_F from CPP-MR for Elemental F-Metals: Co, Fe, Ni

The values of β_F for elemental F-metals are not necessarily intrinsic. While the band structure must play a role, an important role may also be played by the scatterers in the

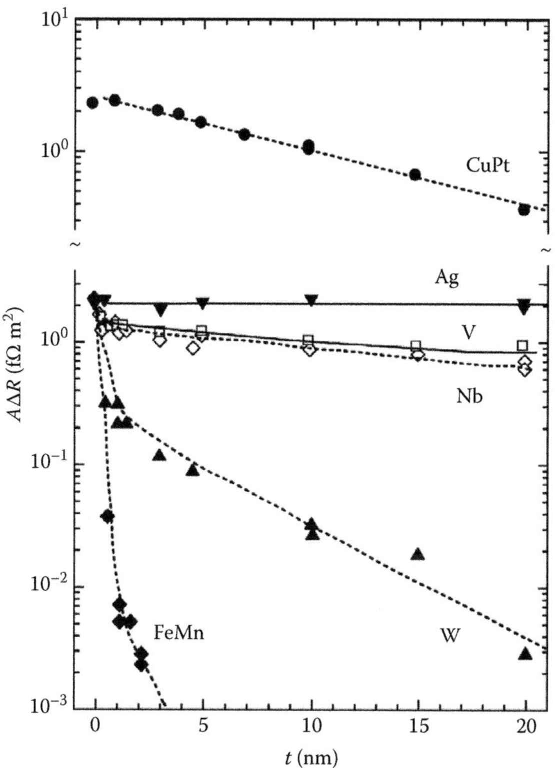

FIGURE 4.16 $A\Delta R$ (log scale) versus t_X (in nm) for single layer inserts X of CuPt, FeMn, Ag, V, Nb, and W into EBSVs of the form [FeMn/(8)/Py(24)/Cu(10)/X/Cu(10)/Py(24)]. The solid curves are fits assuming $l_{sf}^N = \infty$. The dashed curves are fits to finite values of l_{sf}^N as given in the text. (After Park, W. et al., *Phys. Rev. B*, 62, 1178, 2000. With permission.)

given F-metal. At 4.2 K, the dominant scatterers in sputtered, evaporated, or electrodeposited nominally elemental F-metals are not phonons or other electrons, but a usually unknown collection of "dirt," composed of defects (grain boundary and dislocations) and residual impurities. Even at room temperature, 293 K, the scattering from such "dirt" is typically comparable to that from phonons. Thus, one cannot count on β_F being intrinsic. To analyze new CPP-MR data involving a given F-metal, it is thus safest to have a value of β_F derived from multilayers prepared in the same way as those to be analyzed. Despite this caveat, Table 4.1 shows that the values of β_{Co}, the only F-metal measured by several different investigators, span a surprisingly narrow range, 0.38–0.48 (except for V-grooves), at low T (4.2–20 K) whether determined from the more reliable R_0 (see Figure 4.2) or from R_{peak}. Moreover, values of β_{Co} obtained by the same investigator decrease only slightly from 4.2 to 300 K [40–42]. One interpretation is that the defect, impurity, and phonon scatterers happen to give anisotropies similar to that of the band structure.

4.5.4.2 Compare β_F for F-Alloys from CPP-MR and DMR

If the concentration of a given impurity in a given host F-metal is large enough to dominate scattering, then β_F for that alloy should be intrinsic. As noted in Chapter 1 [7], in the 1970s–1980s, studies of deviations from Matthiessen's rule (DMR) in three component F-based alloys gave estimates of β_F for a range of F-alloys [34]. We examine some of these alloys.

Table 4.2 compares values of β_F derived from CPP-MR experiments with those found from studies of impurity- and temperature-driven DMRs [34]. Unfortunately, most of the CPP-MR values listed were derived with the 2CSR model, which assumes $l_{sf}^F = \infty$ in the F-alloy. Since values of l_{sf}^F in the host F-metals (see Figure 4.18) are finite, and because adding impurities should reduce l_{sf}^F (see Figure 4.18), such values are unlikely to be reliable in magnitude. We, thus, list them without bolding. All of these values agree with DMR results in sign, but disagree in magnitude. The four cases where values of β_F and l_{sf}^F were derived simultaneously are given in bold. The three that can be directly compared with DMR values agree with them reasonably well.

TABLE 4.1 β_F for the Elemental F-Metals: Co, Fe, and Ni

F-Metal	β_F	T (K)	Technique	Field	ρ_F (nΩ m)	References
Co	0.48±0.06	4.2	S	H_0	68	[68]
Co	0.46±0.05	4.2	S	H_0	58	[12,89]
Co	0.38±0.06	4.2	S	H_{peak}	58	[12,89]
Co	0.48±0.04	4.2	S	H_0	24	[12,90]
Co	0.49±0.08	20	NW	H_{peak}	400	[42]
Co	0.43±0.08	300	NW	H_{peak}	520	[42]
Co	0.36±0.04	77,300	NW	H_{peak}	160	[42,82,91]
Co	0.27±0.05	4.2	V-groove	H_0	53	[40]
Co	0.26±0.05	300	V-groove	H_0		[40]
Fe	0.78±0.05	4.2	S	H_0	38	[92]
Ni	0.14±0.02	4.2	S	Other	33	[93]

This table lists the F-metal, β_F, the temperature T (K) of the measurement, the technique used (S, superconducting leads; NW, nanowires), the estimated AP state (R_0 or R_{peak}—see Figure 4.2), the measured or estimated F residual resistivity, ρ_F, and the reference.

TABLE 4.2 β_F at 4.2 K for *F*-Based Alloys from DMR [34] and CPP-MR

F-Host	Imp.	β_F (DMR)	Reference	β_F (CPP-MR)	References
Ni	**Fe**	**+0.83±0.1**	**C&F**	**+0.65±0.1**	**[94]**
				+0.76±0.07	**[95]**
Ni	**FeCo**			**~+0.82**	**[95]**
Ni	Cu	+0.53±0.05	C&F	+0.19±0.04	[43]
Ni	**Cr**	**−0.43±0.1**	**C&F**	**−0.13±0.01**	**[43]**
Ni	**Cr**	**−0.43±0.1**	**C&F**	**−0.35±0.1**	**[96]**
Co	Cr	−0.5	C&F	−0.10±0.02	[43]
Co	Mn	−0.1	C&F	−0.03±0.03	[43]
Co	**Fe**	**+0.85±0.1**	**C&F**	**+0.65±0.05**	**[97]**
Fe	Cr	−0.6±0.1	C&F	−0.22±0.06	[43]
Fe	V	−0.78±0.05	C&F	−0.11±0.03	[43,98]

The four cases β_F and l_{sf}^F where derived simultaneously are given in bold. Listed are the host *F*-metal, the impurity, β_F from DMR and its reference, and β_F from CPP-MR and its reference.

4.5.5 Spin-Flipping within *N*- and *F*-Metals: l_{sf}^N and l_{sf}^F

In this section, we first briefly review how values of l_{sf}^N in "pure" *N*-metals and alloys, and of l_{sf}^F in "pure" *F*-metals and alloys, are found from CPP-MR measurements. We then provide examples of the results obtained. More complete information about how l_{sf}^N and l_{sf}^F are determined from both CPP-MR and alternative techniques, and extensive lists of values for each, are given in a recent review [37].

Just as β_F is "intrinsic" in an *F*-metal only when the main scatterer is known, l_{sf}^N and l_{sf}^F are "intrinsic" only when the main scatterer in the *N*- or *F*-metal (or alloys) is known. For this reason, *N*-based alloys were first used to test the CPP-MR techniques for finding l_{sf}^N. We begin with such tests, which allow quantitative comparisons with determinations from completely independent information.

4.5.5.1 Determining l_{sf}^N

Figures 4.14 and 4.15 showed that data for $[Co/Ag]_n$ and $[Co/AgSn]_n$ multilayers with fixed t_{Co} and long l_{sf}^N agreed with the 2CSR model prediction that a plot of the square root in Equation 4.8 versus *n* should give a straight line passing through the origin. If, instead, l_{sf}^N is shortened, by adding an impurity that flips electron spins by spin–orbit or spin–spin interactions, the data should fall below this line, by a larger fraction the smaller is *n*. Yang et al. [99] showed how such deviations can be used to find l_{sf}^N for dilute alloys. Figure 4.15 shows such data for AgPt (strong spin–orbit interaction [86]) and AgMn (strong spin–spin interaction [64]). Table 4.3 compares selected values from CPP-MR with electron spin-resonance (ESR) measurements of spin–orbit scattering [86] for nonmagnetic impurities, or separate calculations for the magnetic impurity Mn [64]. More comparisons are given in Ref. [37]. The good agreements between CPP and ESR results further support using VF theory for such data.

With the quantitative support of Table 4.3 for the VF theory in hand, we turn to another technique developed in Refs. [101,104]

TABLE 4.3 Comparing Alloy Values of l_{sf}^N from CPP-MR and ESR

Alloy	Tech.	l_{sf}^N (CPP)	l_{sf}^N (ESR)	References
Ag(4% Sn)	ML	≈ 39		[99,100]
Ag(6% Pt)	ML	≈ 10	≈ 7	[99]
Ag(6% Mn)	ML	≈ 11	≈ 12*	[99]
Cu(6% Pt)	ML	≈ 8	≈ 7	[99]
Cu(6% Pt)	SV	11±3	≈ 7	[101]
Cu(22.7% Ni)	ML	7.5	6.9	[102]
Cu(22.7% Ni)	SV	8.2±0.6	7.4	[103]

This table specifies the alloy, the technique used (multilayer [ML] or EBSV [SV]), and the values of l_{sf}^N in nm from CPP-MR and ESR [37]. The superscript '*' indicates that the listed value of l_{sf}^N was calculated for spin–spin scattering in Ref. [64].

to measure l_{sf}^N in nominally "pure" *N*-metals or dilute *N*-based alloys. In this case, a thickness t_N of the *N*-metal (or alloy) of interest is sputtered into the middle of an EBSV of the form: FeMn(8)/Py(24)/Cu(10 or 20)/N(t_N)/Cu(10 or 20)/Py(24), and ln ($A\Delta R$) is plotted against t_N. If no spin-flipping occurs within *N*, $A\Delta R$ should be given by Equation 4.11, except that the denominator would now also contain the additional terms $\rho_N t_N + 2AR_{N/Cu}$. As t_N grows, $A\Delta R$ should decrease as a linear function of t_N in the denominator. If, however, spin-flipping occurs in *N*, on a length scale l_{sf}^N, then $A\Delta R$ should decrease exponentially as $\exp(-t_N/l_{sf}^N)$. If this decrease dominates, a plot of $\ln(A\Delta R)$ versus t_N should be close to a straight line with a slope determined by l_{sf}^N. In practice, data such as those in Figure 4.16 are fit numerically with VF theory, including the contributions from both phenomena just described. The values of l_{sf}^N for the metals shown in Figure 4.16 are $l_{sf}^{Ag} \geq 40$ nm, $l_{sf}^{V} \geq 40$ nm, $l_{sf}^{Nb} = 25_{-5}^{+\infty}$ nm, $l_{sf}^{W} = 4.8 \pm 1$ nm, and $l_{sf}^{CuPt} = 11 \pm 3$ nm. A complete listing of values of l_{sf}^N for a variety of metals and alloys, also measured by other techniques, is given in Ref. [37].

4.5.5.2 Determining l_{sf}^F

Values of l_{sf}^F have also been determined by two techniques, one using superconducting leads and one nanowires.

4.5.5.2.1 S Leads, EBSVs

With superconducting leads, the sample is a symmetric EBSV, with two equal thicknesses t_F of the *F*-layers, one of which is pinned by the *AF* FeMn. If $l_{sf}^F = \infty$, $A\Delta R$ grows approximately linearly with t_F, as illustrated by the dashed line in Figure 4.17 for CoFe. If, in contrast, l_{sf}^F is finite, $A\Delta R$ breaks away from the $l_{sf}^F = \infty$ line approximately at $t_F = l_{sf}^F$, and then saturates at a constant value for $t_F \gg l_{sf}^F$, as predicted in Equation 4.11. Examples of data are shown in Figure 4.17 for CoFe and Co [81]. Analysis of the behavior of Co in Figure 4.17 might be complicated by spin-flipping at the Co/Cu interfaces (see Sections 4.5.6.4 and 4.5.9).

4.5.5.2.2 Nanowires

For nanowires, the samples are $[F/Cu]_n$ multilayers with *F* = Co or Py. For *F* = Co, all layer pairs were identical, and the equation

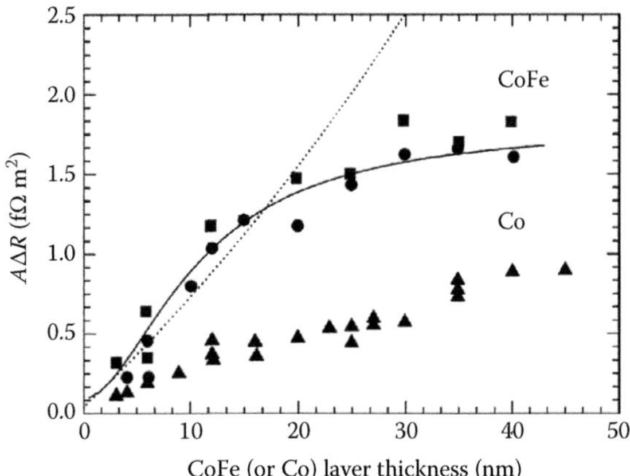

FIGURE 4.17 $A\Delta R$ versus t_F for $F = Co_{91}Fe_9$ (circles and squares) and Co (triangles) in $[F(t_F)/Cu(20)/F(t_F)/FeMn(8)]$ symmetric EBSVs. The circles and squares are CoFe runs made at different times. The dotted curve is VF theory with $\beta_{CoFe} = 0.66$ and $l_{sf}^{CoFe} = \infty$; the solid curve is VF theory with $\beta_{CoFe} = 0.66$ and $l_{sf}^{CoFe} = 12$ nm. (After Reilly, A.C. et al., *J. Magn. Magn. Mater.*, 195, L269, 1999. With permission.)

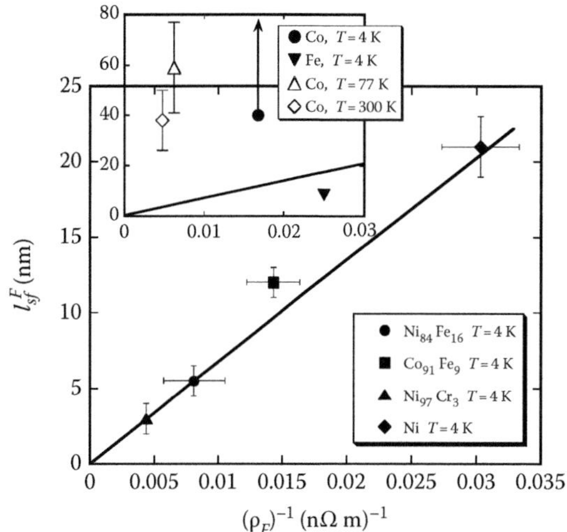

FIGURE 4.18 l_{sf}^F versus $1/\rho_F$ for several *F*-metals and *F*-based alloys. The solid line (repeated in the inset) is a straight line fit to the data in the main figure, constrained to go through zero. Note that the Ni symbol is in the main figure, but the Fe and Co symbols are in the insert. References: Co [41,97]; Fe [92]; Ni [93]; Py [83]; CoFe [81]; NiCr [96]. (After Bass, J. and Pratt, W. P., *J. Phys. Condens. Matter*, 19, 183201, 2007. With permission.)

of interest was Equation 4.12. In Ref. [82], plots of $(R_P/\Delta R)$ versus t_{Co} gave straight lines, the slopes of which were used to infer values of $l_{sf}^{Co} \sim 59 \pm 18$ nm at 77 K and $l_{sf}^{Co} \sim 38 \pm 12$ nm at 300 K [41]. The data for Co at 4.2 K in Figure 4.17 are consistent with this relatively long value at 77 K.

For $F = $ Py [13,23], the nanowires had two different *N*-layer thicknesses to better set *AP* states. Here, using Equation 4.11 with $p = 1$ gave $l_{sf}^{Py} = 4.3 \pm 1$ nm at 77 K, consistent with the value of 5.5 ± 1 at 4.2 K from Ref. [83] found with superconducting contacts.

Figure 4.18 collects together values of l_{sf}^F for a set of alloys plus Co, Ni, and Fe, all plotted versus $1/\rho_F$ [41,81,83,92,93,96,97]. This graph approximates a plot of l_{sf}^F versus λ_F, where λ_F is the effective mean free path. Note that l_{sf}^{Co} is much longer than would have been extrapolated from the alloy data; the residual resistivity of Co is likely dominated by stacking faults, which may be weak spin-flippers.

4.5.6 Interfaces

In this section, we consider first $2AR$ for interfaces, then spin-scattering anisotropy, $\gamma_{F/N}$, for *F*/*N* interfaces, then spin-flipping at interfaces, and finally potential effects of interface roughness.

4.5.6.1 2AR for Interfaces

Procedures for determining $2AR^*$ for *F*/*N* interfaces are usually complex, as outlined in Section 4.5.4. In contrast, procedures for determining $2AR$ for *N*1/*N*2 interfaces are simple enough to be described in more detail below. We will see in Table 4.4 that values of both $2AR^*$ and $2AR$ for lattice-matched ($\Delta a/a \le 1\%$–2%) pairs with the same crystal structure agree surprisingly well with calculations that involve no adjustable parameters.

An issue raised at the start of this review was whether the CPP-MR is dominated by scattering at *F*/*N* interfaces or in the bulk of the *F*-metal. Equations 4.7 and 4.10 show that their relative importance is determined by comparing the products $\beta_F\rho_F^* t_F$ for bulk and $\gamma_{F/N} 2AR_{F/N}^*$ at interfaces. Tables 4.1, 4.2, 4.4, and 4.5 give the information needed to evaluate these quantities for given layer thicknesses. Usually, for thin t_F layers, interface scattering dominates, but sometimes thick *F*-layers can make bulk scattering predominate.

4.5.6.2 Interface Specific Resistances and Comparisons with Calculations

We start with $2AR_{N1/N2}$, values of which have been determined using two techniques.

1. In the first, multilayers of fixed total thickness, t_T, are sequentially subdivided into more and more subunits with equal thicknesses $t_{N1} = t_{N2} = t_T/2n$ of *N*1 and *N*2. Such a procedure leaves the total thicknesses of *N*1 and *N*2 fixed at $t_T/2$, but adds more interfaces as t_n decreases. So long as t_n is larger than the interface thickness, t_I, the total specific resistance, AR_T, should grow linearly with *n* as [105]:

$$AR_T = 2AR_{S/Co} + 2\rho_{Co}(10) + AR_{Co/N1} + AR_{Co/N2} + \rho_{N1}\left(\frac{t_T}{2}\right) + \rho_{N2}\left(\frac{t_T}{2}\right) - AR_{N1/N2} + 2nAR_{N1/N2} \quad (4.13)$$

Here, we have assumed superconducting (*S*) leads, and 10 nm thick Co layers outside the nonmagnetic multilayer to eliminate proximity effects. Figure 4.19 [105] shows examples of such data for Au/Cu and Ag/Au. The data initially grow linearly with *n*, but eventually saturate to an approximation to a uniform alloy when t_n is less than the interface thickness [105].

TABLE 4.4 $2AR_{N1/N2}$ or $2AR^{*}_{F/N}$ at 4.2 K

Metals	$2AR$(exp)	$2AR$ (perf.)	$2AR$ (50–50)	References
Ag/Au	0.1	0.09	0.13	[105,106]
Co/Cu*	1.0	0.9	1.1	[106]
Fe/Cr*	1.6	1.7	1.5	[106,107]
Pd/Pt	0.3	0.4	0.4	[106,108]
Pd/Ir	1.0	1.1	1.1	[106]
Au/Cu	0.3	0.45	0.7	[105,109]
Ag/Cu	0.1	0.45	0.6	[105,110]
Pd/Cu	0.9	1.5	1.6	[110,111]
Pd/Ag	0.7	1.6	2.0	[110]
Pd/Au	0.5	1.7	1.9	[110]
Pt/Cu	0.9			[111]
Co/Ag*	1.2			[68]
Ni/Cu*	0.36			[93]
W/Cu	3.1			[101]
Nb/Cu	2.2			[101]
V/Cu	2.3			[101]
Co/Ru*	~1			[112]
Co/Al*	11			[88]
Fe/Al*	8			[88]
Py/Al*	9			[88]
Py/Cu*	1			[26]
Fe/Cu*	1.5			[92]
Co/Pt*	1.7			[87]
Py/Pd*	0.4			[87]
Fe/Nb*	2.7			[87]
Fe/V*	1.4.			[87]
Co/Cr*	1.0			[43]
Py/Cr	1.9			[43]

Values of $2AR_{N1/N2}$ or $2AR^{*}_{F/N}$ are in fΩ m² and are rounded to significant figures. The first five pairs are lattice matched ($\Delta a/a \leq 1\%$–2%); the other pairs are not.

* Indicates $2AR^*$. Orientations are (111) for fcc and (011) for bcc.

2. In the second, a multilayer of n layers of fixed thicknesses $t_{N1} = t_{N2}$ is inserted into the center of an EBSV [101]. Again, AR_T should grow linearly with n. But now this growth also contains a term $n(\rho_{N1}t_{N1} + \rho_{N2}t_{N2})$ that must be corrected for. This correction makes this technique more uncertain than the other. On the other hand, as described in Ref. [101] and Section 4.5.6.4 below, this technique can also be used to derive a spin-flip probability at the $N1/N2$ interface.

Values of $2AR$ (or $2AR^*$) found with these techniques are collected together in Table 4.4 for elemental F- and N-metals (but including also F=Py), along with calculations for several of the interfaces. Additional values for some F-alloys are given in Ref. [43]. Atomic planes within layers are closest-packed, corresponding to (111) for face-centered-cubic (fcc) and (011) for body-centered-cubic (bcc) layers. Of special interest are lattice-matched metal pairs, with the same crystal structure and lattice parameters, a, almost the same (i.e., $\Delta a/a \leq 1\%$–2%). Then $2AR$ or $2AR^*$ can be calculated with no adjustable parameters for a given interfacial structure [106,113–116] as described in

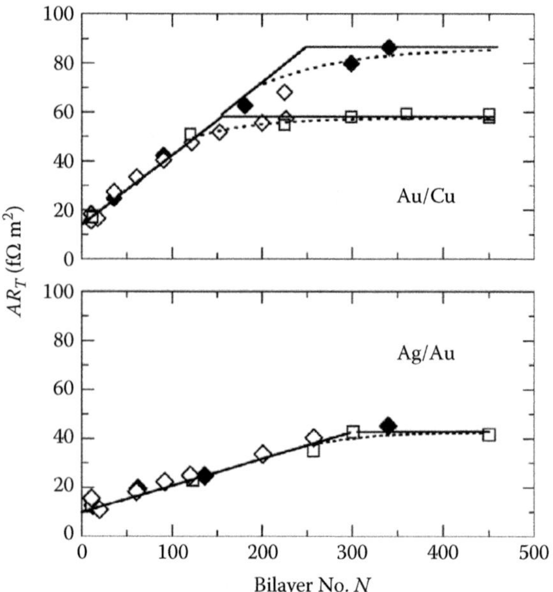

FIGURE 4.19 AR versus N for $[N1/N2]_N = [\text{Au/Cu}]_N$ and $[\text{Ag/Au}]_N$ multilayers with equal thickness ($t_{N1} = t_{N2} = t_T/N$) layers of $N1$ and $N2$, sputtered at standard (diamonds) or half (squares) rates. Open symbols are for fixed total sample thickness t_T=360 nm; filled symbols are for t_T=540 nm. (After Henry, L.L. et al., *Phys. Rev. B*, 54, 12336, 1996. With permission.)

Chapter 5 [1]. The simplest structure is a perfectly flat interface. The next simplest is n layers of a 50%–50% random alloy. Table 4.5 contains experimental and calculated values of $2AR$ (or $2AR^*$ for F/N interfaces) for both closely matched pairs (first five) and unmatched pairs (all the others). For the closely matched pairs, the experimental values fall close to (often between) the values calculated for perfect interfaces and for two MLs of a 50%–50% alloy. For the non-lattice-matched pairs, the calculations are in the right ballpark, but 50% to a factor of 4 too large. A separate test comparing experimental values and calculations of residual resistivities for dilute alloys [117] found the results to be sensitive to lattice strains, which provides a plausible explanation for much of the discrepancies between data and calculations for non-lattice-matched pairs.

4.5.6.3 $\gamma_{F/N}$ for F/N Interfaces

Table 4.5 lists values of $\gamma_{F/N}$ for F/N interfaces with F and N elemental metals, except for the well studied alloy Py. Inferred values for other F-alloys with N=Cu or Cr are given in Ref. [43]. The same calculations that gave the values of $2AR^*$ in Table 4.4, gave the following values for $\gamma_{Co/Cu}$ and $\gamma_{Fe/Cr}$:$\gamma_{Co/Cu}$(perf) = +0.60; $\gamma_{Co/Cu}$(50–50) = +0.68 and $\gamma_{Fe/Cr}$(perf) = −0.49; $\gamma_{Fe/Cr}$(50–50) = −0.35. Our choices for "best" experimental values from Table 4.5 are $\gamma_{Co/Cu} = +0.77 \pm 0.04$ and $\gamma_{Fe/Cr} = -(0.7 \pm 0.15)$. For Co/Cu, the calculated values are 80%–90% of the experimental one, with the disordered value closer to experiment. In contrast, the Fe/Cr values are only 50%–70% of the experimental one, with the ordered value closer to experiment.

TABLE 4.5 Values of $\gamma_{F/N}$

F/N	$\gamma_{F/N}$	*T* (K)	Tech.	*R*	References
Co/Cu	+0.77±0.04	4.2	S	R_0	[12]
Co/Cu	+0.85±0.15	77	NW		[82,91]
Co/Cu	+0.53±0.08	4.2	V-gr.	R_0	[40]
Co/Cu	+0.46±008	300	V-gr.	R_0	[40]
Co/Cu	+0.55±0.07*	20	NW	R_{peak}	[42]
Co/Cu	+0.40±0.10*	300	NW	R_{peak}	[42]
Co/Ag	+0.84±0.03	4.2	S	R_0	[68]
Fe/Cr	−(0.7±0.15)	4.2	S	R_0	[107]
Fe/Cu	+0.55±0.07	4.2	S	R_0	[92]
Ni/Cu	+0.29±0.15	4.2	S	Other	[93]
Py/Cu	+0.7±0.1	4.2	S	Other	[80]
Co/Ru	−0.2	4.2	S	Other	[112]
Co/Al	+0.1±0.08	4.2	S	R_0	[88]
Fe/Al	+0.05±0.02	4.2	S	R_0	[88]
Py/Al	+0.025±0.01	4.2	S	R_0	[88]
$Co_{91}Fe_9$/Al	+0.1±0.01	4.2	S	R_0	[88]
Co/Pt	+0.38±0.06	4.2	S	R_0	[87]
Py/Pd	+0.41±0.14	4.2	S	R_0	[87]
Fe/V	−(0.27±0.08)	4.2	S	R_0	[87]
Fe/Nb	−(~0.14)	4.2	S	R_0	[87]
Co/Cr	−(0.24±0.17)	4.2	S	EBSV	[43]
Py/Cr	−(0.03±0.03)	4.2	S	EBSV	[43]

This table lists the *F/N* pair, $\gamma_{F/N}$, the measuring temperature, the technique used, the *AP* state (or a note that another method was used), and the reference. The * by Co/Cu values from Ref. [42] means that we chose the larger of the two alternatives.

4.5.6.4 Spin-Flipping at Interfaces from CPP-MR

4.5.6.4.1 N1/N2 Interfaces

In Section 4.5.6.2, we described how to determine $2AR_{N1/N2}$ by measuring *AR* after inserting into the middle of the Cu layer of Py/Cu/Py EBSVs *n* layers of a multilayer of two *N*-metals $[N1/N2]_n$, each of fixed thickness (say $t_{N1}=t_{N2}=3$ nm). Measurements of $A\Delta R$ on these same EBSVs give quite different information. They allow determination of the probability of spin-flipping at the *N1/N2* interface, characterized by a parameter $\delta_{N1/N2}$ as described in Refs. [101,104]. Just as spin-flipping in an inserted bulk metal of thickness t_N led to the exponential decrease in $A\Delta R$ with increasing t_N in Figure 4.16, spin-flipping at the *F/N* interfaces in an inserted $[F(3)/N(3)]_n$ multilayer should lead to an exponential decrease in $A\Delta R$ with increasing *n*. Just as the data in Figure 4.16 with ML inserts allowed derivation of l_{sf}^N, similar data [101] for multilayer inserts allow derivation of the spin-flipping probability $\delta_{N1/N2}$. Published data are given in Ref. [37]. We list only a few representative examples: $\delta_{N1/N2}$: ~0 for Cu/Ag; 0.07±0.04 for Cu/V; 0.19±0.05 for Cu/Nb, and 0.96±0.1 for Cu/W [101].

The physics underlying the values of $\delta_{N1/N2}$ is not yet established. The values obtained are associated with interfaces that are intermixed, typically to 3–4 ML [105]. Whether similar values would be obtained for "perfect" interfaces (flat and with *N*1 and *N*2 completely separated) is yet to be determined. Theoretical guidance is needed.

4.5.6.4.2 F/N Interfaces

Unfortunately, a general procedure for determining $\delta_{F/N}$ similar to that for $\delta_{N1/N2}$, as just described, has not yet been published. Estimates of $\delta_{F/N}$ have been published for several metal pairs, as summarized in Ref. [37]. But, for the reasons given in Ref. [37], with the possible exception of $\delta_{Co/Cu} \approx 0.25$ for Co/Cu [25,112], we do not have much confidence in these numbers. [Note added in proof. Such a general procedure has now been described and used for Co/Cu, $Co_{90}Fe_{10}$/Cu, and Co/Ni (See Ref. [129]).

4.5.6.5 Interface Roughness

Most studies of CPP-MR simply take the interfaces as they are formed. As described in Section 4.5.6.2, and Table 4.4, the values of 2*AR* for several intermixed interfaces of lattice-matched metals agree surprisingly well with no-free-parameter calculations. In several cases, the calculated values are not sensitive to interfacial intermixing at the interface. But the calculations do not take account of physical roughness of the interfaces—that is, both the perfect interfaces and the boundaries of the intermixed interfaces, are assumed to be flat.

Only in sputtered Fe/Cr multilayers have attempts been made to determine explicitly effects of interfacial roughness, with conflicting results. In Ref. [71], the CPP-MR was reported to grow with increasing interface roughness, produced by increasing the sputtering pressure and inferred from observations of broadening and decreasing the height of low angle x-ray peaks. Reference [107], in contrast, reported a modest decrease in CPP-MR with roughness produced and inferred by similar techniques. The values of $2AR^{\star}_{Fe/Co}$ found in Ref. [107] fall between those calculated for perfect or two ML of intermixed 50%–50% Fe(Cr) alloy interfaces, and the calculations predict a decrease in $2AR^{\star}_{Fe/Co}$ with increasing intermixing, rather than the increase reported in Ref. [71].

4.5.7 Attempts to Enhance CPP-AR and CPP-MR

For standard multilayer nanopillars with "length" <100 nm and area $A \sim 10^{-2}\ \mu m^2$, the total resistance is typically ~1 Ω, still too small for impedance matching in devices, such as read heads. Moreover, the CPP-MRs of standard samples subject to the dimensional constraints of read heads are smaller than desired. Several techniques have been tried to address these problems. We discuss three: (1) lamination of the *F*-layer to increase the CPP-MR, (2) current confinement via nano-oxide layers to increase *AR* without decreasing the CPP-MR, and (3) search for half-metallic *F*-metals.

4.5.7.1 Laminated *F*-Layers

If the enhanced interface specific resistance, $2AR^{\star}_{F/N}$ for an *F/N* pair is large, and the scattering asymmetry, $\gamma_{F/N}$, of the *F/N* interface is larger than the asymmetry, β_F, of the *F*-layer, then a potential way to increase both $A\Delta R$ and the CPP-MR is to laminate the *F*-layers by inserting very thin *N*-layers between very thin F-layers. Very thin N-layers ensure that the thin F-layers are ferromagnetically exchange-coupled, so that they behave like a single F-layer, but with enhanced properties. Eid et al. [118] found that laminating Co with thin Cu layers ($t_{Cu} = 0.5$ nm) in Py–Co hybrid SVs with fixed total Co thickness did increase $A\Delta R$. But with more laminations, the data fell further below the linear growth expected from a simple 2CSR model. Qualitatively similar results were found in Ref. [119] with $(Co_{50}Fe_{50})$/Cu.

4.5.7.2 Current-Confinement via Nano-Oxide Layers

Another process to increase the AR product without reducing $\Delta R/R$ is inserting what is called a nano-oxide layer in the middle of the nonmagnetic (N) layer of a spin valve [120,121]. This insertion involves stopping the sample growth in the middle of the N (e.g., Cu) layer, admitting enough oxygen to produce a monolayer or so of oxide, and then continuing the sample growth. The present view is that the insulating oxide contains small conducting channels (pinholes) through which the current flows. This current confinement substantially reduces the effective cross-sectional area through which the current flows, thereby increasing R, often without significantly decreasing the CPP-MR. The focus now is on control and reproducibility of the process [121,122].

4.5.7.3 Search for Half-Metallic F-Layers

In principle, inclusion of a true half-metallic F-layer, in which current is carried by only one electron spin, should give an infinite CPP-MR. In practice, F-alloys that are predicted to be half-metallic when they contain no defects, have yet to give the hoped for CPP-MRs, perhaps because of defects leading to incompletely half-metallic behavior, along with changes in properties of such alloys at their interfaces. Initial studies of a NiMnSb-based Heusler alloy [123] gave CPP-MRs larger than the equivalent CIP-MRs, but even at 4.2 K they were only ~10%. More recently, Heusler alloys have given CPP-MRs ≥ 8% at room temperature [124,125].

4.5.8 Temperature Dependences of CPP-MR Parameters

Three studies of how CPP-MR parameters for Co/Cu multilayers vary with T reached similar conclusions. One used a V-groove geometry incorporating Cu connections to more closely approach a CPP geometry [40], and two others used nanowires [41,42]. Combining the results listed in those papers suggests that the changes with temperature of CPP-MR are due mainly to phonon-scattering driven increases of the resistivities, ρ_{Co} and ρ_{Cu}, and decreases of the spin-diffusion length, l_{sf}^{Co}, of Co. In contrast, the asymmetry parameters, β_{Co} and $\gamma_{Co/Cu}$ (see Tables 4.2 and 4.5), and the interface resistance, $2AR_{Co/Cu}$ (see Refs. [40–42]), all seem to vary only weakly with T.

4.5.9 Interleaved versus Separated Multilayers

In Section 4.5.3, we showed that the CPP-MR of multilayers with Co and various N-layers much thinner than their spin-diffusion lengths can be well described by a 2CSR model. An interesting property of this model is that the CPP-MR does not depend upon the order of the F-moments. That is, the AR should be the same for a sample with F-moments oriented ↑↓↑↓↑↓ (interleaved) or with F-moments oriented ↑↑↑↓↓↓ (separated). In contrast to this prediction, the first such comparison, involving $[Co/Cu/Py/Cu]_n$ versus $[Co/Cu]_n[Py/Cu]_n$ multilayers [126], showed clear differences, similar in form to those in Figure 4.3. These differences could be explained by the presence of a short spin-diffusion length $l_{sf}^{Py} = 5.5$ nm in Py [83], which made the 2CSR model inappropriate. A similar study involving Fe and Co showed similar differences that can also be plausibly explained by spin-flipping within the Fe-layers [127]. However, Bozec et al. [128] then reported new results, like those in Figure 4.3 for samples of the form $[Co(1)/Cu/Co(8)/Cu]_n$ and $[Co(1)/Cu]_n[Co(8)/Cu]_n$. Because l_{sf}^{Co} is long (see, e.g., Figure 4.17), these differences cannot be so easily explained. They have been attributed to (1) mean-free-path effects that require ballistic transport and specular reflection at interfaces (see Ref. [1]); (2) magnetic moment orientation effects of the adjacent F-layers (also called mean-free-path effects, but with different physics from that just noted [128]; and (3) to spin-flipping at Co/Cu interfaces [25]. For further references, and a detailed examination of this topic, see Appendix C of Ref. [37].

4.6 Summary and Conclusions

We described results of studies of CIP- and CPP-MRs with a wide range of F/N metals, and with several different kinds of samples: simple F/N multilayers; hybrid spin valves; and EBSVs. In the following, we first give our best answers to the questions that we raised in Section 4.1, and then conclude with some still unanswered questions.

4.6.1 CIP-MR

1. Interface scattering almost always dominates for small layer thicknesses. Bulk scattering can sometimes be made to dominate with larger thicknesses.
2. Data can be qualitatively understood with a semiclassical, two current model with mean free paths as characteristic lengths. Qualitative agreements of parameters with those from deviations from Matthiessen's rule studies are affirmed by observations of the expected inversion of CIP-MR. But we know of no fully quantitative comparisons with other measurements or calculations.
3. Interface structure can affect the CIP-MR, but available studies suggest that details of the effects depend upon the particular F/N pair, and the physical nature of the structure or roughness.
4. The presence of large CIP-MRs with certain F/N pairs, but not with others, argues that band structures are important. But we know of no successful quantitative agreements between data and calculations.

4.6.2 CPP-MR

1. Interface scattering almost always dominates for small layer thicknesses. Bulk scattering can sometimes be made to dominate with larger thicknesses.
2. Data can often be quantitatively understood with a semi-classical two current model with spin-diffusion lengths, l_{sf}^F and l_{sf}^N, as characteristic lengths. When $t_F \ll l_{sf}^F$ and $t_N \ll l_{sf}^N$, a 2CSR model works well. When $t_F \geq l_{sf}^F$ and/or $t_N \geq l_{sf}^N$, the more general VF model works. There is no compelling evidence for "quantum-coherence effects."
3. For dilute $F(N)$ alloys, measured values of $\beta_{F(N)}$ agree quantitatively with values determined from measurements of deviations from Matthiessen's rule in similar alloys.
4. For the two lattice-matched F/N pairs, Co/Cu and Fe/Cr, measured values of $\gamma_{F/N}$ agree semiquantitatively with the calculated ones.
5. Measured values of l_{sf}^F and l_{sf}^F agree well with values obtained by other techniques.
6. For lattice-matched metal pairs, measured values of $2AR_{N1/N2}$ and $2AR_{F/N}^{*}$ agree quantitatively with no-free-parameter calculations. These agreements suggest that band structures usually dominate $2AR$, with interfacial intermixing usually playing a secondary role.

4.6.3 CIP-MR versus CPP-MR

1. Studies of inverse CIP- and CPP-MRs strongly suggest that the relative weightings of bulk and interface contributions differ for the CIP- and CPP-MRs.

4.6.4 Some Unanswered Questions

1. Is there significant spin-flipping at any F/N interface?
2. Is the source of observed spin-flipping at $N1/N2$ interfaces due to spin–orbit effects on band structures or on interfacial alloys?
3. What physics underlies the measured values of $2AR_{S/F}$ for (superconducting Nb)/F interfaces?

References

1. E. Y. Tsymbal, D. G. Pettifor, and S. Maekawa, Giant magnetoresistance: Theory, in *Handbook of Spin Transport and Magnetism*, E. Y. Tsymbal and I. Žutić (Eds.), Taylor & Francis, Boca Raton, FL, Chapter 5 (2011).
2. T. Ono and T. Shinjo, Magnetoresistance of multilayers prepared on microstructured substrates, *J. Phys. Soc. Jpn.* **64**, 363 (1995).
3. M. A. M. Gijs, M. T. Johnson, A. Reinders et al., Perpendicular giant magnetoresistance of Co/Cu multilayers deposited under an angle on grooved substrates, *Appl. Phys. Lett.* **66**, 1839 (1995).
4. J. Q. Xiao, J. S. Jiang, and C. L. Chien, Giant magnetoresistance in nonmultilayer magnetic systems, *Phys. Rev. Lett.* **68**, 3749 (1992).
5. A. Berkowitz, J. R. Mitchell, M. J. Carey et al., Giant magnetoresistance in heterogeneous Cu–Co alloys, *Phys. Rev. Lett.* **68**, 3745 (1992).
6. T. L. Hylton, Limitations of magnetoresistive sensors based upon the giant magnetoresistive effect in granular magnetic composites, *Appl. Phys. Lett.* **62**, 2431 (1993).
7. A. Fert, Historical overview: From electron transport in magnetic materials to spintronics, in *Handbook of Spin Transport and Magnetism*, E. Y. Tsymbal and I. Žutić (Eds.), Taylor & Francis, Boca Raton, FL, Chapter 1, 2011.
8. J. Bass, Metals, electronic transport phenomena, in *Landolt-Bornstein Numerical Data and Functional Relationships in Science and Technology*, K. H. Hellwege and J. L. Olsen (Eds.), Group III, V.15a, p. 1, Springer-Verlag, Berlin, Germany, 1982.
9. P. M. Levy, Giant magnetoresistance in magnetic layered and granular materials, in *Solid State Physics Series*, H. Ehrenreich and D. Turnbull (Ed.), vol. 47, Academic Press, New York, p. 367, 1994.
10. B. Dieny, Giant magnetoresistance in spin-valve multilayers, *J. Magn. Magn. Mater.* **136**, 335 (1994).
11. M. A. M. Gijs and G. E. W. Bauer, Perpendicular giant magnetoresistance of magnetic multilayers, *Adv. Phys.* **46**, 285 (1997).
12. J. Bass and W. P. Pratt, Current-perpendicular (CPP) magnetoresistance in magnetic metallic multilayers, *J. Magn. Magn. Mater.* **200**, 274 (1999); J. Bass and W. P. Pratt, Erratum: *J. Magn. Magn. Mater.* **296**, 65 (2006).
13. A. Fert and L. Piraux, Magnetic nanowires, *J. Magn. Magn. Mater.* **200**, 338 (1999).
14. E. Y. Tsymbal and D. G. Pettifor, Perspectives of giant magnetoresistance, in *Solid State Physics*, H. Ehrenreich and F. Spaepen (Eds.), vol. 56, Academic Press, San Diego, CA, pp. 113–239, 2001.
15. W. Thomson, On the electro-dynamic qualities of metals: Effects of magnetization on the electrical conductivity of nickel and of iron, *Proc. Roy. Soc.* **8**, 546 (1857).
16. T. R. McGuire and R. I. Potter, Anisotropic magnetoresistance in ferromagnetic 3d alloys, *IEEE Trans. Magn. Magn.* **11**(4), 1018 (1975).
17. M. N. Baibich, J. M. Broto, A. Fert et al., Giant magnetoresistance of (001)Fe/(001)Cr superlattices, *Phys. Rev. Lett.* **61**, 2472 (1988).
18. G. Binasch, P. Grunberg, F. Saurenbach, and W. Zinn, Enhanced magnetoresistance in layered magnetic structures with antiferromagnetic interlayer exchange, *Phys. Rev. B* **39**, 4828 (1989).
19. W. P. Pratt, S. F. Lee, J. M. Slaughter, R. Loloee, P. A. Schroeder, and J. Bass, Perpendicular giant magnetoresistances of Ag/Co multilayers, *Phys. Rev. Lett.* **66**, 3060 (1991).
20. J. A. Borchers, J. A. Dura, J. Unguris et al., Observation of antiparallel order in weakly coupled Co/Cu multilayers, *Phys. Rev. Lett.* **82**, 2796 (1999).
21. J. Barnas, A. Fuss, R. E. Camley, P. Grunberg, and W. Zinn, Novel magnetoresistance effect in layered magnetic structures: Theory and experiment, *Phys. Rev. B* **42**, 8110 (1990).

22. B. Heinrich, Basics of nano-thin film magnetism, in *Handbook of Spin Transport and Magnetism*, E. Y. Tsymbal and I. Žutić (Eds.), Taylor & Francis, Boca Raton, FL, Chapter 2 (2011).
23. S. Dubois, L. Piraux, J. M. George, K. Ounadjela, J. L. Duvail, and A. Fert, Evidence for a short spin diffusion length in permalloy from the giant magnetoresistance of multilayered nanowires, *Phys. Rev. B* **60**, 477 (1999).
24. A. Chaiken, P. Lubitz, J. J. Krebs, G. A. Prinz, and M. Z. Harford, Low-field spin-valve magnetoresistance in Fe/Cu/Co sandwiches, *Appl. Phys. Lett.* **59**, 240 (1991).
25. K. Eid, D. Portner, J. A. Borchers et al., Absence of mean-free-path effects in CPP magnetoresistance of magnetic multilayers, *Phys. Rev. B* **65**, 054424 (2002).
26. Q. Yang, P. Holody, R. Loloee et al., Prediction and measurement of perpendicular (CPP) giant magnetoresistance of Co/Cu/$Ni_{84}Fe_{16}$/Cu multilayers, *Phys. Rev. B* **51**, 3226 (1995).
27. B. Dieny, V. S. Speriosu, S. S. P. Parkin, B. A. Gurney, D. R. Wilhoit, and D. Mauri, Giant magnetoresistance in soft ferromagnetic multilayers, *Phys. Rev. B* **43**, 1297 (1991).
28. S. D. Steenwyk, S. Y. Hsu, R. Loloee, J. Bass, and W. P. Pratt Jr., A comparison of hysteresis loops from giant magnetoresistance and magnetometry of perpendicular-current exchange-biased spin-valves, *J. Appl. Phys.* **81**, 4011 (1997).
29. S. S. P. Parkin, N. More, and K. P. Roche, Oscillations in exchange coupling and magnetoresistance in metallic superlattice structures: Co/Ru, Co/Cr, and Fe/Cr, *Phys. Rev. Lett.* **64**, 2304 (1990).
30. R. E. Camley and J. Barnas, Theory of giant magnetoresistance effects in magnetic layered structures with antiferromagnetic coupling, *Phys. Rev. Lett.* **63**, 664 (1989).
31. S. Zhang and P. M. Levy, Conductivity perpendicular to the plane of multilayered structures. *J. Appl. Phys.* **69**, 4786 (1991).
32. S. F. Lee, W. P. Pratt, Q. Yang et al., Two-channel analysis of CPP-MR data for Ag/Co and AgSn/Co multilayers, *J. Magn. Magn. Mater.* **118**, L1 (1993).
33. T. Valet and A. Fert, Theory of perpendicular magnetoresistance in magnetic multilayers, *Phys. Rev. B* **48**, 7099 (1993).
34. I. A. Campbell and A. Fert, Transport properties of ferromagnets, in *Ferromagnetic Materials*, E. P. Wolforth (Ed.), vol. 3, North-Holland, Amsterdam, the Netherlands, p. 747, 1982.
35. R. Q. Hood and L. M. Falicov, Theory of the negative magnetoresistance of ferromagnetic-normal metallic multilayers, *Phys. Rev. B* **46**, 8287 (1992); R. Q. Hood and L. M. Falicov, Effects of interfacial roughness on the magnetoresistance of magnetic metallic multilayers, *Phys. Rev. B* **49**, 368 (1994).
36. N. W. Ashcroft and N. D. Mermin, *Solid State Physics*, W. B. Saunders, Philadelphia, PA, 1976.
37. J. Bass and W. P. Pratt, Spin-diffusion lengths in metals and alloys, and spin-flipping at metal/metal interfaces: An experimentalist's critical review, *J. Phys. Condens. Matter* **19**, 183201 (2007).
38. M. A. M. Gijs, S. K. J. Lenczowski, and J. B. Giesbers, Perpendicular giant magnetoresistance of microstructured Fe/Cr magnetic multilayers from 4.2 to 300 K, *Phys. Rev. Lett.* **70**, 3343 (1993).
39. M. A. M. Gijs, J. B. Giesbers, M. T. Johnson et al., Perpendicular giant magnetoresistance of microstructures in Fe/Cr and Co/Cu multilayers, *J. Appl. Phys.* **75**, 6709 (1994).
40. W. Oepts, M. A. M. Gijs, A. Reinders, R. M. Jungblut, R. M. J. van Gansewinkel, and W. J. M. de Jonge, Perpendicular giant magnetoresistance of Co/Cu multilayers on grooved substrates: Systematic analysis of the temperature dependence of spin-dependent scattering, *Phys. Rev. B* **53**, 14024 (1996).
41. L. Piraux, S. Dubois, A. Fert, and L. Belliard, The temperature dependence of the perpendicular giant magnetoresistance in Co/Cu multilayered nanowires, *Europhys. J. B* **4**, 413 (1998).
42. B. Doudin, A. Blondel, and J. P. Ansermet, Arrays of multilayered nanowires, *J. Appl. Phys.* **79**, 6090 (1996).
43. C. Vouille, A. Barthelemy, F. Elokan Mpondo et al., Microscopic mechanisms of giant magnetoresistance, *Phys. Rev. B* **60**, 6710 (1999); C. Vouille, A. Fert, A. Barthelemy, S. Y. Hsu, R. Loloee, and P. A. Schroeder, Inverse CPP-GMR in (A/Cu/Co/Cu) multilayers (A = NiCr, FeCr, FeV) and discussion of the spin asymmetry induced by impurities, *J. Appl. Phys.* **81**, 4573 (1997).
44. P. M. Levy, S. Zhang, T. Ono, and T. Shinjo, Electrical transport in corrugated multilayered structures, *Phys. Rev. B* **52**, 16049 (1995).
45. M. A. M. Gijs, M. T. Johnson, A. Reinders et al., Perpendicular giant magnetoresistance of Co/Cu multilayers deposited under an angle on grooved substrates, *Appl. Phys. Lett.* **66**, 1839 (1995); M. A. M. Gijs, S. K. J. Lenczowski, J. B. Giesbers et al., Perpendicular giant magnetoresistance using microlithography and substrate patterning techniques, *J. Magn. Magn. Mater.* **151**, 333–340 (1995).
46. B. Dieny, Quantitative interpretation of giant magnetoresistance properties of permalloy-based spin-valve structures, *Europhys. Lett.* **17**, 261 (1992).
47. A. Barthelemy and A. Fert, Theory of the magnetoresistance in magnetic multilayers: Analytical expressions from a semiclassical approach, *Phys. Rev. B* **43**, 13124 (1991).
48. S. S. P. Parkin, Dependence of giant magnetoresistance on Cu-layer thickness in Co/Cu multilayers: A simple dilution effect, *Phys. Rev. B* **47**, 9136 (1993).
49. B. Dieny, V. S. Speriosu, S. Metin et al., Magnetotransport properties of magnetically soft spin-valve structures, *J. Appl. Phys.* **69**, 4774 (1991); B. Dieny, P. Humbert, V. S. Speriosu et al., Giant magnetoresistance of magnetically soft sandwiches: Dependence on temperature and on layer thickness, *Phys. Rev. B* **45**, 806 (1992).
50. E. Velu, C. Dupas, D. Renard, J. P. Ranard, and J. Seiden, Enhanced magnetoresistance of ultra-thin $(Au/Co)_n$ multilayers with perpendicular anisotropy, *Phys. Rev. B* **37**, 668 (1988).

51. S. S. P. Parkin, Origin of enhanced magnetoresistance of magnetic multilayers: Spin-dependent scattering from magnetic interface states, *Phys. Rev. Lett.* **71**, 1641 (1993).
52. P. Baumgart, B. A. Gurney, D. R. Wilhoit, T. Nguyen, B. Dieny, and V. S. Speriosu, The role of spin-dependent impurity scattering in Fe/Cr giant magnetoresistance multilayers, *J. Appl. Phys.* **69**, 4792 (1991).
53. B. A. Gurney, P. Baumgart, D. R. Wilhoit, D. Dieny, and V. S. Speriosu, Giant magnetoresistance of Fe/Cr multilayers: Impurity scattering model of the influence of third elements deposited at the interface, *J. Appl. Phys.* **70**, 5867 (1991).
54. V. Speriosu, J. P. Nozieres, B. A. Gurney, B. Dieny, T. C. Huang, and H. Lefakis, Role of interfacial mixing in giant magnetoresistance, *Phys. Rev. B* **47**, 11579 (1993).
55. F. Petroff, A. Barthelemy, A. Hamzic et al., Magnetoresistance of Fe/Cr superlattices, *J. Magn. Magn. Mater.* **93**, 95 (1991).
56. E. E. Fullerton, D. M. Kelly, J. Guimpel, I. K. Schuller, and Y. Bruynseraede, Roughness and giant magnetoresistance in Fe/Cr superlattices, *Phys. Rev. Lett.* **68**, 859 (1992).
57. C. H. Marrows and B. J. Hickey, Impurity scattering from δ-layers in giant magnetoresistance systems, *Phys. Rev. B* **63**, 220405R (2001).
58. B. Dieny, J. P. Nozieres, V. S. Speriosu, B. A. Gurney, and D. R. Wilhoit, Change in conductance is the fundamental measure of spin-valve magnetoresistance, *Appl. Phys. Lett.* **61**, 2111 (1992).
59. B. A. Gurney, V. S. Speriosu, J.-P. Nozieres, H. Lefakis, D. R. Wilhoit, and O. U. Need, Direct measurement of spin-dependent conduction-electron mean free paths in ferromagnetic metals, *Phys. Rev. Lett.* **71**, 4023 (1993).
60. D. M. Kelly, I. K. Schuller, V. Korenivski et al., Increases in giant magnetoresistance by ion irradiation, *Phys. Rev. B* **50**, 3481 (1994).
61. R. Schad, P. Belien, G. Verbanck et al., Giant magnetoresistance dependence on the lateral correlation length of the interface roughness in magnetic superlattices. *Phys. Rev. B* **59**, 1242 (1999).
62. J. M. George, L. G. Pereira, A. Barthelemy et al., Inverse spin-valve-type magnetoresistance in spin engineered multilayered structures, *Phys. Rev. Lett.* **72**, 408 (1994).
63. J.-P. Renard, P. Bruno, R. Megy et al., Inverse magnetoresistance in the simple spin-valve system $Fe_{1-x}V_x$/Au/Co, *Phys. Rev. B* **51**, 12821 (1995).
64. A. Fert, J. L. Duvail, and T. Valet, Spin relaxation effects in the perpendicular magnetoresistance of magnetic multilayers, *Phys. Rev. B* **52**, 6513 (1995).
65. J. Åkerman, Magnetoresistive random access memory, in *Handbook of Spin Transport and Magnetism*, E. Y. Tsymbal and I. Žutić (Eds.), Taylor & Francis, Boca Raton, FL, Chapter 35, 2012.
66. J. M. Slaughter, W. P. Pratt Jr., and P. A. Schroeder, Fabrication of layered metallic systems for perpendicular resistance measurements, *Rev. Sci. Instrum.* **60**, 127 (1989).
67. C. Fierz, S. F. Lee, J. Bass, W. P. Pratt Jr., and P. A. Schroeder, Perpendicular resistance of thin Co films in contact with superconducting Nb, *J. Phys.: Condens. Mater.* **2**, 9701 (1990).
68. S. F. Lee, Q. Yang, P. Holody et al., Current perpendicular and parallel giant magnetoresistances in Co/Ag multilayers, *Phys. Rev. B* **52**, 15426 (1995).
69. R. D. Slater, J. A. Caballero, R. Loloee, and W. P. Pratt Jr., Perpendicular-current exchange-biased spin valve structures with micron-size superconducting top contacts, *J. Appl. Phys.* **90**, 5242 (2001).
70. D. L. Edmunds, W. P. Pratt Jr., and J. R. Rowlands, 0.1 ppm four-terminal resistance bridge for use with a dilution refrigerator, *Rev. Sci. Instrum.* **51**, 1516 (1980).
71. M. C. Cyrille, S. Kim, M. E. Gomez, J. Santamaria, K. M. Krishnan, and I. K. Schuller, Enhancement of perpendicular and parallel giant magnetoresistance with the number of bilayers in Fe/Cr superlattices, *Phys. Rev. B* **62**, 3361 (2000).
72. M. AlHajDarwish, H. Kurt, S. Urazhdin et al., Controlled normal and inverse current-induced magnetization switching and magnetoresistance in magnetic nanopillars, *Phys. Rev. Lett.* **93**, 157203-1 (2004).
73. S. K. J. Lenczowski, R. J. M. van de Veerdonk, M. A. M. Gijs, J. B. Giesbers, and H. H. J. M. Janssen, Current-distribution effects in microstructures for perpendicular magnetoresistance measurements, *J. Appl. Phys.* **75**, 5154 (1994).
74. A. Blondel, J. P. Meir, B. Doudin, and J.-Ph. Ansermet, Giant magnetoresistance of nanowires of multilayers, *Appl. Phys. Lett.* **65**, 3019 (1994).
75. L. Piraux, J. M. George, J. F. Despres et al., Giant magnetoresistance in magnetic multilayered nanowires, *Appl. Phys. Lett.* 65, 2484 (1994).
76. K. Liu, K. Nagodawithana, P. C. Searson, and C. L. Chien, Perpendicular giant magnetoresistance of multilayered Co/Cu nanowires, *Phys. Rev. B* **51**, 7381 (1995).
77. P. R. Evans, G. Yi, and W. Schwarzacher, Current perpendicular to plane giant magnetoresistance of multilayered nanowires electrodeposited in anodic aluminum oxide membranes, *Appl. Phys. Lett.* **76**, 481 (2000).
78. J. E. Wegrowe, S. E. Gilbert, D. Kelly, B. Doudin, and J.-Ph. Ansermet, Anisotropic magnetoresistance as a probe of magnetization reversal in individual nano-sized nickel wires, *IEEE Trans. Magn.* **34**, 903 (1997).
79. X.-T. Tang, G.-C. Wang, and M. Shima, Layer thickness dependence of CPP giant magnetoresistance in individual Co/Ni/Cu multilayer nanowires grown by electrodeposition, *Phys. Rev. B* **75**, 134404 (2007).
80. W. P. Pratt Jr., S. D. Steenwyk, S. Y. Hsu et al., Perpendicular-current transport in exchange-biased spin-valves, *IEEE Trans. Magn.* **33**, 3505 (1997).
81. A. C. Reilly, W. Park, R. Slater et al., Perpendicular giant magnetoresistance of $Co_{91}Fe_9$/Cu exchange-biased spin-valves: Further evidence for a unified picture, *J. Magn. Magn. Mater.* **195**, L269 (1999).

82. L. Piraux, S. Dubois, and A. Fert, Perpendicular giant magnetoresistance in magnetic multilayered nanowires, *J. Magn. Magn. Mater.* **159**, L287 (1996).
83. S. Steenwyk, S. Y. Hsu, R. Loloee, J. Bass, and W. P. Pratt Jr., Perpendicular current exchange biased spin-valve evidence for a short spin diffusion length in permalloy, *J. Magn. Magn. Mater.* **170**, L1 (1997).
84. W. P. Pratt Jr., S. F. Lee, Q. Yang et al., Giant magnetoresistance with current perpendicular to the layer planes of Ag/Co and AgSn/Co, *J. Appl. Phys.* **73**, 5326 (1993).
85. J. Bass, P. A. Schroeder, W. P. Pratt et al., Studying spin-dependent scattering in magnetic multilayers by means of perpendicular (CPP) magnetoresistance measurements, *Mater. Sci. Eng. B* **31**, 77 (1995).
86. P. Monod and S. Schultz, Conduction electron spin-flip scattering by impurities in copper, *J. Phys. Paris* **43**, 393 (1982).
87. A. Sharma, J. A. Romero, N. Theodoropoulou, R. Loloee, W. P. Pratt Jr., and J. Bass, Specific resistance and scattering asymmetry of Py/Pd, Fe/V, Fe/Nb, and Co/Pt interfaces, *J. Appl. Phys.* **102**, 113916 (2007).
88. A. Sharma, N. Theodoropoulou, T. Haillard et al., Current-perpendicular to plane (CPP) magnetoresistance of ferromagnetic (F)/Al interfaces (F = Py, Co, Fe, and $Co_{91}Fe_9$) and structural studies of Co/Al and Py/Al, *Phys. Rev. B* **77**, 224438 (2008).
89. P. Holody, Perpendicular giant magnetoresistance: Study and application of spin dependent scattering in magnetic multilayers of cobalt/copper and nickel(84)iron(16)/copper, PhD thesis, Michigan State University, East Lansing, MI, 1996.
90. N. J. List, W. P. Pratt Jr., M. A. Howson, J. Xu, M. J. Walker, and D. Greig, Perpendicular resistance of Co/Cu multilayers prepared by molecular beam epitaxy, *J. Magn. Magn. Mater.* **148**, 342 (1995).
91. L. Piraux, S. Dubois, C. Marchal et al., Perpendicular magnetoresistance in Co/Cu multilayered nanowires, *J. Magn. Magn. Mater.* **156**, 317 (1996).
92. D. Bozec, Current perpendicular to the plane magnetoresistance of magnetic multilayers. Physics and astronomy. PhD thesis, Leeds University, West Yorkshire, U.K., 2000
93. C. E. Moreau, I. C. Moraru, N. O. Birge, and W. P. Pratt Jr., Measurement of spin diffusion length in sputtered Ni films using a special exchange-biased spin valve geometry, *Appl. Phys. Lett.* **90**, 012101 (2007).
94. P. Holody, W. C. Chiang, R. Loloee, J. Bass, W. P. Pratt Jr., and P. A. Schroeder, Giant magnetoresistance in copper/permalloy multilayers, *Phys. Rev. B* **58**, 12230 (1998).
95. L. Vila, W. Park, J. A. Caballero et al., Current perpendicular magnetoresistances of NiCoFe and NiFe permalloys, *J. Appl. Phys.* **87**, 8610 (2000).
96. W. Park, R. Loloee, J. A. Caballero et al., Test of unified picture of spin dependent transport in perpendicular (CPP) giant magnetoresistance and bulk alloys, *J. Appl. Phys.* **85**, 4542 (1999).
97. A. C. Reilly, W. C. Chiang, W. Park et al., Giant magnetoresistance of current-perpendicular exchange-biased spin-valves of Co/Cu, *IEEE Trans. Magn.* **34**, 939 (1998).
98. S. Y. Hsu, A. Barthelemy, P. Holody, R. Loloee, P. A. Schroeder, and A. Fert, Towards a unified picture of spin dependent transport in perpendicular giant magnetoresistance and bulk alloys, *Phys. Rev. Lett.* **78**, 2652 (1997).
99. Q. Yang, P. Holody, S. F. Lee et al., Spin diffusion length and giant magnetoresistance at low temperatures, *Phys. Rev. Lett.* **72**, 3274 (1994).
100. Q. Fowler, B. Richard, A. Sharma et al., Spin-diffusion lengths in dilute Cu(Ge) and Ag(Sn) alloys, *J. Magn. Magn. Mater.* **321**, 99 (2009).
101. W. Park, D. V. Baxter, S. Steenwyk, I. Moraru, W. P. Pratt, and J. Bass, Measurement of resistance and spin-memory loss (spin relaxation) at interfaces using sputtered current perpendicular-to-plane exchange-biased spin valves, *Phys. Rev. B* **62**, 1178 (2000).
102. S. Y. Hsu, P. Holody, R. Loloee, J. M. Rittner, W. P. Pratt, and P. A. Schroeder, Spin-diffusion Lengths of $Cu_{1-x}Ni_x$ using current perpendicular to plane magnetoresistance measurements of magnetic multilayers, *Phys. Rev. B* **54**: 9027 (1996).
103. R. Loloee, B. Baker, and W. P. Pratt Jr., Unpublished, 2006.
104. D. V. Baxter, S. D. Steenwyk, J. Bass, and W. P. Pratt Jr., Resistance and spin-direction memory loss at Nb/Cu interfaces, *J. Appl. Phys.* **85**, 4545 (1999).
105. L. L. Henry, Q. Yang, W. C. Chiang et al., Perpendicular interface resistances in sputtered Ag/Cu, Ag/Au, and Au/Cu multilayers, *Phys. Rev. B* **54**, 12336 (1996).
106. K. M. Schep, P. J. Kelly, and G. E. W. Bauer, Ballistic transport and electronic structure, *Phys. Rev. B* **57**, 8907 (1998).
107. M. S. Stiles and D. R. Penn, Calculation of spin-dependent interface resistance, *Phys. Rev. B* **61**, 3200 (2000); D. R. Penn and M. D. Stiles, Spin transport for spin diffusion lengths comparable to the mean free path, *Phys. Rev. B* **72**, 212410 (2005).
108. K. Xia, P. J. Kelly, G. E. W. Bauer, I. Turek, J. Kudrnovsky, and V. Drchal, Interface resistance of disordered magnetic multilayers, *Phys. Rev. B* **63**, 064407 (2001).
109. P. X. Xu, K. Xia, M. Zwierzycki, M. Talanana, and P. J. Kelly, Orientation-dependent transparency of metallic interfaces, *Phys. Rev. Lett.* **96**, 176602 (2006).
110. R. Acharyya, H. Y. T. Nguyen, R. Loloee et al., Specific resistance of Pd/Ir interfaces, *Appl. Phys. Lett.* **94**, 022112 (2009).
111. P. X. Xu and K. Xia. Ab-initio calculations of alloy resistivities, *Phys. Rev. B* **74**, 184206 (2006).
112. A. Zambano, K. Eid, R. Loloee, W. P. Pratt, and J. Bass, Interfacial properties of Fe/Cr multilayers in the current-perpendicular-to-plane geometry, *J. Magn. Magn. Mater.* **253**, 51 (2002).
113. S. K. Olson, R. Loloee, N. Theodoropoulou et al., Comparison of measured and calculated specific resistances of Pd/Pt interfaces, *Appl. Phys. Lett.* **87**, 252508 (2005).

114. H. Kurt, W. C. Chiang, C. Ritz, K. Eid, W. P. Pratt, and J. Bass, Spin-memory loss and CPP-magnetoresistance in sputtered multilayers with Au, *J. Appl. Phys.* **93**, 7918 (2003).
115. C. Galinon, K. Tewolde, R. Loloee et al., Pd/Ag and Pd/Au interface specific resistances and interfacial spin-flipping, *Appl. Phys. Lett.* **86**, 182502 (2005).
116. H. Kurt, R. Loloee, K. Eid, W. P. Pratt, and J. Bass, Spin-memory loss at 4.2 K in sputtered Pd, Pt, and at Pd/Cu and Pt/Cu interfaces, *Appl. Phys. Lett.* **81**, 4787 (2002).
117. K. Eid, R. Fonck, M. A. Darwish, W. P. Pratt, and J. Bass, Current-perpendicular-to-plane magnetoresistance properties of Ru and Co/Ru interfaces, *J. Appl. Phys.* **91**, 8102 (2002).
118. K. Eid, W. P. Pratt, and J. Bass, Enhancing current-perpendicular-to-plane magnetoresistance (CPP-MR) by adding interfaces within ferromagnetic layers, *J. Appl. Phys.* **93**, 3445 (2003).
119. F. Delille, A. Manchon, N. Strelkov et al., Thermal variation of current perpendicular-to-plane giant magnetoresistance in laminated and nonlaminated spin valves, *J. Appl. Phys.* **100**, 013912 (2006).
120. K. Nagasaka, Y. Seyama, L. Varga, Y. Shimizu, and A. Tanaka, Giant magnetoresistance properties of specular spin valve films in a current perpendicular to Plane structure, *J. Appl. Phys.* **89**, 6943 (2001).
121. M. Takagishi, K. Koi, M. Yoshikawa, T. Funayama, H. Iwasaki, and M. Sahashi, The applicability of CPP-GMR heads for magnetic recording, *IEEE Trans. Magn.* **38**, 2277 (2002).
122. H. Fukuzawa, H. Yuasa, and H. Iwasaki, CPP-GMR films with a current-confined-path nano-oxide layer (CPP-NOL), *J. Phys. D: Appl. Phys.* **40**, 1213 (2007).
123. J. A. Caballero, Y. D. Park, J. R. Childress et al., Magnetoresistance of NiMnSb-based multilayers and spin-valves, *J. Vac. Sci. Tech. A* **16**, 1801 (1998).
124. T. Furubayashi, Y. Takahashi, and K. Hono, Search for half metallic Heusler alloys and their application to CPP-GMR, *IEICE Tech. Rep.* **108**, MR2008-34, 25 (2008).
125. K. Kodama, T. Furubayashi, H. Sukegawa, T. M. Nakatani, K. Inomata, and K. Hono, Current perpendicular-to-plane giant magnetoresistance of a spin valve using Co_2MnSi Heusler alloy electrodes, *J. Appl. Phys.* **105**, 07E905 (2009).
126. W. C. Chiang, Q. Yang, W. P. Pratt Jr., R. Loloee, and J. Bass, Variation of multilayer magnetoresistance with ferromagnetic layer sequence: Spin-memory loss, *J. Appl. Phys.* **81**, 4570 (1997).
127. D. Bozec, M. J. Walker, B. J. Hickey, M. A. Howson, and N. Wiser, Comparative study of the magnetoresistance of MBE-grown multilayers: $[Fe/Cu/Co/Cu]_N$ and $[Fe/Cu]_N[Co/Cu]_N$, *Phys. Rev. B* **60**, 3037 (1999).
128. D. Bozec, M. A. Howson, B. J. Hickey et al., Mean free path effects on the current perpendicular to the plane magnetoresistance of magnetic multilayers, *Phys. Rev. Lett.* **85**, 1314 (2000).
129. B. Dassonneville, R. Acharyya, H. Y. T. Nguyen, R. Loloee, W. P. Pratt Jr., and J. Bass, A way to measure electron spin-flipping at F/N interfaces and application to Co/Cu, *Appl. Phys. Lett.* **96**, 022509 (2010); H. Y. T. Nguyen, R. Acharyya, E. Huey, B. Richard, R. Loloee, W. P. Pratt Jr., J. Bass, Shuai Wang and Ke Xia, Conduction electron scattering and spin-flipping at sputtered Co/Ni interfaces, *Phys. Rev. B* **82**, 220401(R) (2010); H. Y. T. Nguyen, R. Acharyya, W. P. Pratt, Jr. and J. Bass, Spin-flipping at sputtered $Co_{90}Fe_{10}$/Cu interfaces, *J. Appl. Phys.* **109**, 07C903 (2011).

5

Giant Magnetoresistance: Theory

Evgeny Y. Tsymbal
University of Nebraska

D.G. Pettifor
University of Oxford

Sadamichi Maekawa
Advanced Science Research Center, Japan Atomic Energy Agency

5.1 Origin of GMR

Giant magnetoresistance (GMR) is the change in electrical resistance of metallic layered systems when the magnetizations of the ferromagnetic layers are reoriented in relative to one another under the application of an external magnetic field [1,2]. As was discussed in Chapter 4, GMR originates from spin-dependent electronic transport intrinsic to magnetic metal systems. The discovery of GMR has stimulated significant progress in the theory of spin-dependent transport. This chapter addresses the theoretical understanding of the physical mechanisms responsible for the spin-dependent conduction in magnetic metallic multilayers and complements previous review articles devoted to theory of GMR [3–8].

GMR can be qualitatively understood using the Mott model [9]. There are two main points proposed by Mott. First, the electrical conductivity in metals can be described in terms of two largely independent conducting channels, corresponding to up- and down-spin electrons, which are distinguished according to the projection of their spins along the quantization axis. The probability of spin-flip scattering in metals is normally small as compared to the probability of scattering in which the spin is conserved. This means that the up- and down-spin electrons do not mix over long distances and, therefore, the electrical conduction occurs in parallel for the two spin channels. Second, in ferromagnetic metals the scattering rates of the up- and down-spin electrons are quite different, whatever be the nature of the scattering centers. The electric current is primarily carried by electrons from the valence *sp* bands due to their low effective mass and high mobility. The *d* bands play an important role in providing final states for the scattering of the *sp* electrons. In ferromagnets the *d* bands are exchange-split, such that the density of states (DOS) is not the same for the up- and down-spin electrons at the Fermi energy. The probability of scattering into these states is proportional to their density, such that the scattering rates are spin dependent, that is, they are different for the two conduction channels. Although this picture is too simplified in view of the strong hybridization between the *sp* and *d* states, it forms a useful basis for a qualitative understanding of the spin-dependent conduction in transition metals (TM).

Using Mott's arguments it is straightforward to explain GMR. We consider collinear magnetic configurations, as shown in Figure 5.1, and assume that scattering is strong for electrons with spin antiparallel to the magnetization direction, and is weak for electrons with spins parallel to the magnetization direction. For the parallel (P)-aligned magnetic layers, the up-spin electrons pass through the structure almost without scattering, because their spin is parallel to the magnetization. On the contrary, the down-spin electrons are scattered strongly, because their spin is antiparallel to the magnetization of the both ferromagnetic layers. Since conduction occurs in parallel for the two spin channels, the total resistivity of the multilayer is determined mainly by the highly conductive up-spin electrons and appears to be low. For the antiparallel (AP)-aligned multilayer, both the up- and down-spin electrons are scattered strongly within one of the

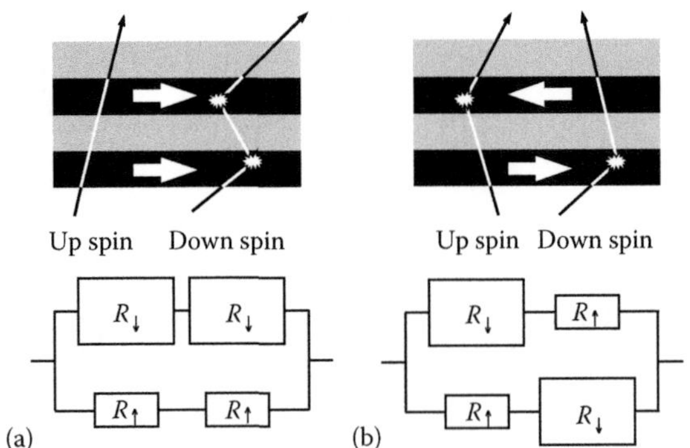

FIGURE 5.1 Schematic illustration of electron transport in a multilayer for P (a) and AP (b) magnetizations of the successive ferromagnetic layers. The magnetization directions are indicated by the arrows. The solid lines are individual electron trajectories within the two spin channels. Bottom panels show the resistor network within the two-current series resistor model. (From Tsymbal, E.Y. and Pettifor, D.G., Perspectives of giant magnetoresistance, in: *Solid State Physics*, H. Ehrenreich and F. Spaepen (Eds.), vol. 56, Academic Press, San Diego, CA, 2001, pp. 113–239. With permission.)

ferromagnetic layers, because within it, the spin is antiparallel to the magnetization direction. Therefore, in this case, the total resistivity of the multilayer is high.

The above picture of GMR explains its origin, based entirely on spin-dependent scattering, occurring in the bulk of ferromagnetic layers. In reality, a significant contribution to the spin-dependent scattering occurs due to the presence of interfaces. The spin transmission across the interface between a ferromagnetic metal and a nonmagnetic spacer layer may be very different for the up- and down-spin electrons, resulting in the strong spin dependence of the interface resistance. In this case, the mechanism of GMR, as discussed in relation to Figure 5.1, remains valid assuming that the dominant contribution to spin-dependent scattering occurs at the interfaces.

5.1.1 Series-Resistor Model

The simplest quantitative description of GMR can be obtained using the most simple resistor model. According to this model, each metallic layer is treated as an independent resistor added in series. This assumption is a better approximation for the current-perpendicular-to-the-plane (CPP) geometry, but qualitatively it can be used for the current-in-the-plane (CIP) geometry (see, e.g., Figure 5.2 in Section 5.2.3.2), assuming that the mean free path is sufficiently long and the total probability of scattering is the sum of scattering probabilities within each layer. Therefore, within a given spin channel the total resistance is the sum of resistances of each layer.

We consider a GMR multilayer shown in the top panels of Figure 5.1. Within each ferromagnetic layer, the electron spin can be either parallel or antiparallel to the magnetization direction. In the former case the electron is locally a majority-spin electron and in the latter case a minority-spin electron. The majority- and minority-spin resistivities of the ferromagnetic layer are different and are equal to $\rho_\uparrow$ and $\rho_\downarrow$, respectively. The resistance of the bilayer, which consists of the ferromagnetic layer and the spacer layer, for either of the two spin channels is equal to

$$R_{\uparrow,\downarrow} = \rho_{NM} d_{NM} + \rho_{\uparrow,\downarrow} d_{FM}, \tag{5.1}$$

where

ρ_{NM} and d_{NM} denote the resistivity and the thickness of the nonmagnetic spacer layer

d_{FM} is the thickness of the ferromagnetic layer

The equivalent network of resistors for the P and AP magnetizations are shown in the bottom panels of Figure 5.1. The total resistance of the P-aligned multilayer is given by

$$R_P = N\frac{R_\uparrow R_\downarrow}{R_\uparrow + R_\downarrow}, \tag{5.2}$$

where N is the number of the four-layer unit cells within the multilayer. The total resistance of the AP-aligned multilayer is equal to

$$R_{AP} = N\frac{R_\uparrow + R_\downarrow}{2}. \tag{5.3}$$

Thus, the GMR ratio is determined by the simple expression

$$\frac{\Delta R}{R} = \frac{R_{AP} - R_P}{R_P} = \frac{(R_\downarrow - R_\uparrow)^2}{4R_\downarrow R_\uparrow}. \tag{5.4}$$

Note that we use a definition in which GMR is normalized to the low resistance R_P, so that the GMR ratio can be larger than 100%. Using Equations 5.1 and 5.4, we find

$$\frac{\Delta R}{R} = \frac{(\alpha - 1)^2}{4\left(\alpha + pd_{NM}/d_{FM}\right)\left(1 + pd_{NM}/d_{FM}\right)}, \tag{5.5}$$

where $p = \rho_{NM}/\rho_\uparrow$ and the spin-asymmetry parameter is defined by $\alpha = \rho_\downarrow/\rho_\uparrow$. As is obvious from Equation 5.5, the magnitude of GMR is strongly dependent on α. Large spin asymmetry, that is, $\alpha \gg 1$ or $\alpha \ll 1$, is an important requirement for obtaining high values of GMR.

For a given α, the GMR increases with decreasing pd_{NM}/d_{FM}. Therefore, to obtain higher GMR, it is important to have a low resistivity of the nonmagnetic spacer layer. As a function of the spacer thickness d_{NM}, the GMR decreases monotonically and at large spacer thickness it falls off as $1/d_{NM}^2$. Although the drop in GMR with spacer thickness is also found in experiments, the actual dependence on d_{NM} is different compared to this simplified model (see Section 5.3.1).

The series-resistor model is able to account for the inverse GMR effect [10]. Equation 5.5 suggests that the resistance of the

P configuration is always smaller than the resistance of the AP configuration. In most cases this statement is correct. However, there are exceptions. For a multilayer that comprises different ferromagnetic layers, GMR is given by

$$\frac{\Delta R}{R} = \frac{(\alpha_1 - 1)(\alpha_2 - 1)}{\alpha_1(1+q) + \alpha_2(1+q^{-1})}, \tag{5.6}$$

where

α_1 and α_2 are the asymmetry parameters for the two different ferromagnetic layers, that is, $\alpha_1 = \rho_\downarrow^{(1)}/\rho_\uparrow^{(1)}$ and $\alpha_2 = \rho_\downarrow^{(2)}/\rho_\uparrow^{(2)}$

q is the ratio of the up-spin resistivities in the two ferromagnets, that is, $q = \rho_\uparrow^{(1)}/\rho_\uparrow^{(2)}$

It is clear from Equation 5.6 that in the case when the two ferromagnetic layers have different spin asymmetries in resistivity, that is, $\alpha_1 > 1$ and $\alpha_2 < 1$ or vice versa, then GMR is reversed.

The series-resistor model can be readily generalized to include spin-dependent interface resistances, by adding additional resistors in the network. This model provides a good approximation for GMR within the CPP geometry. For this reason, it has often been used to obtain values of the spin-dependent bulk and interface resistances from CPP experimental data (see Chapter 4). An accurate description of GMR in the CIP geometry requires, however, more sophisticated models, as discussed below.

5.2 Basics of Electronic Transport

5.2.1 Spin-Dependent Conduction

According to Mott's first argument, the conductivity of a metal is the sum of the independent conductivities for the up- and down-spin electrons:

$$\sigma = \sigma_\uparrow + \sigma_\downarrow. \tag{5.7}$$

Within each conduction channel the conductivity is determined by various factors. In order to illustrate their role we use the Drude formula (e.g., Ref. [11]), which can be expressed as follows:

$$\sigma_{Drude} = \frac{e^2}{h}\frac{k_F^2}{3\pi}\lambda, \tag{5.8}$$

where

σ_{Drude} is the Drude conductivity per spin

$e^2/h \approx 0.387 \times 10^{-4}\ \Omega^{-1}$ is the spin-conductance quantum

k_F is the Fermi momentum

λ is the mean free path, $\lambda = v_F\tau$

We do not display explicitly the spin indices here—it is assumed that all the above quantities are, in general, spin dependent. Although the Drude formula is valid only for free electrons, it is useful to understand qualitatively the factors affecting the spin-dependent conductivity.

As can be seen from Equation 5.8, the conductivity is proportional to the cross-sectional area of the Fermi surface $\sim k_F^2$, which characterizes the number of electrons contributing to the conduction. The mean free path depends on the Fermi velocity and the relaxation time, where the latter can be estimated from the Fermi golden rule

$$\tau^{-1} = \frac{2\pi}{\hbar}\langle V_{scat}^2\rangle n(\varepsilon_F), \tag{5.9}$$

where

$\langle V_{scat}^2\rangle$ is an average value of the scattering potential squared

$n(\varepsilon_F)$ is the density of electronic states at the Fermi energy ε_F for the appropriate spin

Although all the quantities that enter Equations 5.8 and 5.9 are, in general, spin dependent, the origin of the spin dependence is different. The Fermi momentum k_F and the Fermi velocity v_F are intrinsic properties of the metal and are entirely determined by the electronic band structure of the metal. In ferromagnetic metals these quantities are different for the up- and down-spin electrons. The DOS at the Fermi energy $n(\varepsilon_F)$ is also determined by the spin-polarized band structure.

On the contrary, the scattering potential that enters Equation 5.9 is not an intrinsic property of the metal. It is generated by scatterers such as defects, impurities, or lattice vibrations. The scattering potential and its spin dependence is determined by a particular mechanism of scattering. For example, spin-dependent scattering potentials, produced by impurities in dilute magnetic alloys, lead to the spin asymmetry of the conductivity in these alloys [12]. This has to be taken into account when treating GMR in magnetic layered systems in which ferromagnetic layers are often alloys. Spin-dependent scattering potentials might also contribute to GMR at the interfaces between ferromagnetic and nonmagnetic layers. In real magnetic multilayers, these interfaces are not ideal. Interfacial roughness and/or substitutional disorder (i.e., mixing of the adjacent metal atoms at the interface) are always present in experiments. Randomness of the atomic potentials at the interface results in enhanced interfacial scattering.

The relative importance of spin-dependent scattering potentials can, however, be diminished in real GMR structures, which are far from being perfect. Various types of defects such as grain boundaries, stacking faults, and misfit dislocations are always present in the multilayers. Because the relaxation time in Equation 5.9 is determined by the configurationally averaged value of the scattering potential squared, various types of scattering centers can make this average value spin independent or weakly spin dependent. In these circumstances, the spin-polarized band structure of the multilayer becomes decisive and usually gives the dominant contribution to the spin dependence of the mean free path and the conductivity.

5.2.2 Semiclassical Theory of Transport

The semiclassical Boltzmann theory of transport is based on a semiclassical description of electrons in metals in the presence of external fields (e.g., Ref. [11]). The central quantity of the theory is a statistical distribution function $f_{\mathbf{k}}(\mathbf{r},t)$, which is defined as the number of electrons with given position **r** and wave-vector **k** at time t. The Boltzmann transport equation is obtained by balancing the change in the distribution function caused by the applied electric field and the scattering processes that act to bring it back toward equilibrium, that is,

$$\frac{df_{\mathbf{k}}(\mathbf{r},t)}{dt} = -\dot{\mathbf{r}}\cdot\nabla_{\mathbf{r}} f - \dot{\mathbf{k}}\cdot\nabla_{\mathbf{k}} f + \left(\frac{\partial f}{\partial t}\right)_{scatt}. \tag{5.10}$$

The first term in this equation describes the electron drift due to their velocity, the second term reflects the acceleration of electrons due to the applied field, and the third term describes scattering of electrons by imperfections in the lattice. The latter can be written in terms of the probability $P_{\mathbf{kk}'}$ for an electron to be scattered between momentum **k** and **k′**.

$$\left(\frac{\partial f}{\partial t}\right)_{scatt} = \sum_{\mathbf{k}'}\left\{P_{\mathbf{kk}'}\left[1-f_{\mathbf{k}}(\mathbf{r},t)\right]f_{\mathbf{k}'}(\mathbf{r},t) - P_{\mathbf{k}'\mathbf{k}}\left[1-f_{\mathbf{k}'}(\mathbf{r},t)\right]f_{\mathbf{k}}(\mathbf{r},t)\right\}, \tag{5.11}$$

where the right-hand term describes "scattering-out" processes in which an electron from an occupied state of momentum **k** scatters into unoccupied states **k′**, and the left-hand term describes "scattering-in" processes in which electrons from occupied states of momentum **k′** scatter into an unoccupied state **k**. Assuming that the perturbation due to electric field **E** is small, it is convenient to represent the distribution function as $f_{\mathbf{k}}(\mathbf{r}) = f_{\mathbf{k}}^0 + g_{\mathbf{k}}(\mathbf{r})$, where $g_{\mathbf{k}}(\mathbf{r})$ is the deviation from the equilibrium Fermi–Dirac distribution $f_{\mathbf{k}}^0 = \left\{1+\exp[(\varepsilon_{\mathbf{k}}-\varepsilon_F)/kT]\right\}^{-1}$. In the steady state ($df/dt=0$), the linearized Boltzmann equation (5.10) can be then written as

$$\mathbf{v}_{\mathbf{k}}\cdot\nabla_{\mathbf{r}} g_{\mathbf{k}}(\mathbf{r}) - e\mathbf{E}\cdot\mathbf{v}_{\mathbf{k}}\frac{\partial f_{\mathbf{k}}^0}{\partial \varepsilon_{\mathbf{k}}} = \sum_{\mathbf{k}'} P_{\mathbf{kk}'}\left[g_{\mathbf{k}'}(\mathbf{r}) - g_{\mathbf{k}}(\mathbf{r})\right], \tag{5.12}$$

where we assumed that the electric field **E** is uniform and took into account the principle of microscopic reversibility, that is, $P_{\mathbf{kk}'} = P_{\mathbf{k}'\mathbf{k}}$. The density of electric current is given by

$$\mathbf{J}(\mathbf{r}) = -\frac{e}{\Omega}\sum_{\mathbf{k}} \mathbf{v}_{\mathbf{k}} g_{\mathbf{k}}(\mathbf{r}), \tag{5.13}$$

where Ω is the volume of the system and can be obtained from the solution of Equation 5.12. The evaluation is, however, not easy to perform because of the scattering-in term $\sum_{\mathbf{k}'} P_{\mathbf{kk}'} g_{\mathbf{k}'}(\mathbf{r})$, which links the values of $g_{\mathbf{k}}(\mathbf{r})$ at various momenta. The Boltzmann equation can be simplified using the relaxation time approximation, where the scattering-in term is neglected:

$$\mathbf{v}_{\mathbf{k}}\cdot\nabla_{\mathbf{r}} g_{\mathbf{k}}(\mathbf{r}) - e\mathbf{E}\cdot\mathbf{v}_{\mathbf{k}}\frac{\partial f_{\mathbf{k}}^0}{\partial E_{\mathbf{k}}} = -\frac{g_{\mathbf{k}}(\mathbf{r})}{\tau_k}. \tag{5.14}$$

Here, $\tau_{\mathbf{k}}$ is the relaxation time for an electron to scatter out of momentum state **k**:

$$\tau_{\mathbf{k}}^{-1} = \sum_{\mathbf{k}} P_{\mathbf{kk}'}. \tag{5.15}$$

Within the relaxation time approximation, it is straightforward to derive the expression for the conductivity tensor $\sigma^{\mu\nu}$, which is defined by

$$J^{\mu} = \sum_{\nu}\sigma^{\mu\nu}E^{\nu}, \tag{5.16}$$

where the indices μ and ν denote the Cartesian components. For a bulk homogeneous system, $\nabla_{\mathbf{r}} g(\mathbf{r},\mathbf{k}) = 0$ and it follows from Equation 5.14 that

$$g_{\mathbf{k}} = e\tau_{\mathbf{k}}\frac{\partial f_{\mathbf{k}}^0}{\partial E_{\mathbf{k}}}\mathbf{v}_{\mathbf{k}}\cdot\mathbf{E}. \tag{5.17}$$

Taking the zero-temperature limit, that is, $\partial f_n^0(\mathbf{k})/\partial\varepsilon_n(\mathbf{k}) = -\delta\left[\varepsilon_n(\mathbf{k}) - \varepsilon_F\right]$, and substituting Equation 5.17 into Equation 5.13, we obtain the well-known expression for the conductivity per single spin channel within the relaxation time approximation [11]

$$\sigma^{\mu\nu} = \frac{e^2}{\Omega}\sum_{\mathbf{k}} v_{\mathbf{k}}^{\mu} v_{\mathbf{k}}^{\nu}\tau_{\mathbf{k}}\delta(\varepsilon_{\mathbf{k}} - \varepsilon_F). \tag{5.18}$$

In the case of films and multilayers that are assumed to be homogeneous in the xy plane of the layers but inhomogeneous in the z direction perpendicular to the planes (due to the interfaces and boundaries), the distribution function $g_{\mathbf{k}}(z)$ is dependent on z, but independent of x and y. In this case, the solution of the Boltzmann equation (Equation 5.14) takes the form

$$g_{\mathbf{k}}^{\pm}(z) = e\tau_{\mathbf{k}}E\cdot\mathbf{v}_{\mathbf{k}}\frac{\partial f_{\mathbf{k}}^0}{\partial \varepsilon_{\mathbf{k}}}\left[1 + A_{\mathbf{k}}^{\pm}\exp\left(\frac{\mp z}{\tau_{\mathbf{k}}\,|v_{\mathbf{k}}^z|}\right)\right]. \tag{5.19}$$

Here, signs ± refer to whether the z component of the electron velocity is positive or negative. The coefficients $A^{\pm}$ are determined from matching the boundary conditions at the interfaces and outer boundaries in terms of reflection and transmission probabilities.

5.2.3 Quantum-Mechanical Theory of Transport

There are several different quantum-mechanical formulations of transport theory, which include those of Kubo [13], Landauer [14], and Keldysh [15]. The Kubo (linear response) formalism considers the electronic transport in a disordered metallic system as a linear response to an applied electric field [16]. The Landauer formalism describes the conductance from the point of view of the transmission of electrons through a conductor and is applicable to mesoscopic transport [17]. The Keldysh (nonequilibrium Green's function) formalism is more general in providing a description of the quantum transport in the presence of dissipative interactions [18]. Here, we outline basic principles of the Kubo and Landauer theory.

5.2.3.1 Kubo

The starting point of the Kubo formalism is the density matrix. The density matrix is the quantum-mechanical operator, which describes the statistical properties of a quantum-mechanical system. The density matrix ρ_t, satisfies the quantum-mechanical equation of motion

$$i\hbar\frac{d\rho_t}{dt}=[H_t,\rho_t], \tag{5.20}$$

where H_t is the Hamiltonian of the system. This equation describes the evolution of the system affected by a time-dependent perturbation $U(t)$ due to the applied electric field. We assume that the electric field takes the form $\mathbf{E}\exp(\eta t)$, so that it is uniform in space, is applied at $t=-\infty$, and grows adiabatically to its value $\mathbf{E}$ at $t=0$. The latter is taken into account by an infinitesimal positive η. The single-electron Hamiltonian of the system can, then, be represented by

$$H_t=H+U(t)=H+e\mathbf{E}\cdot\mathbf{r}e^{\eta t}, \tag{5.21}$$

where H is the time-independent Hamiltonian of the unperturbed system.

Equation 5.20 is the quantum-mechanical analogue of the semiclassical Equation 5.10. It describes the time evolution of the system. Initially, that is, at $t=-\infty$, the system is at equilibrium and is characterized by the unperturbed density matrix $\rho=[e^{(H-\varepsilon_F)/kT}+1]^{-1}$. Due to the applied electric field the system adiabatically follows the perturbation (5.21). We are looking for the solution of Equation 5.20 to first order with respect to $\mathbf{E}$. It is then convenient to represent the density matrix as $\rho_t=\rho+\delta\rho\exp(\eta t)$, where $\delta\rho\exp(\eta t)$ is a small deviation from the equilibrium due to the applied electric field. The linearized equation (5.20) then takes the form

$$i\hbar\left(\frac{d\,\delta\rho}{dt}+\eta\,\delta\rho\right)=[H,\delta\rho]+e\varepsilon\cdot[\mathbf{r},\rho]. \tag{5.22}$$

Now, we rewrite this operator equation in terms of the matrix elements by introducing a basis of eigenstates $|\alpha\rangle$ of the unperturbed Hamiltonian H. The equilibrium density matrix ρ has the same eigenstates as H so that $\rho_{\alpha\beta}=\delta_{\alpha\beta}f(\varepsilon_\alpha)$, where $f(\varepsilon_\alpha)=[e^{(\varepsilon_\alpha-\varepsilon_F)/kT}+1]^{-1}$ and ε_α is an eigenvalue of H. The operator $\mathbf{r}$ is nondiagonal in the $|\alpha\rangle$ representation. We represent the matrix elements of $\mathbf{r}$ in terms of the matrix elements of the velocity operator $\mathbf{v}$ using the relation $i\hbar\mathbf{v}=[\mathbf{r},H]$, where we can use H instead of H_t by neglecting high-order terms. This allows us to find the solution for the density matrix at $t=0$

$$\delta\rho_{\alpha\beta}=\frac{i\hbar e\mathbf{E}\cdot\mathbf{v}_{\alpha\beta}}{E_\beta-E_\alpha+i\eta}\frac{f(\varepsilon_\alpha)-f(\varepsilon_\beta)}{\varepsilon_\alpha-\varepsilon_\beta}. \tag{5.23}$$

For a spatially homogeneous system of volume Ω, the current-density operator is $\mathbf{J}=-e\mathbf{v}/\Omega$. We need to calculate the expectation value of $\mathbf{J}$, which is determined by $\langle\mathbf{J}\rangle=Tr(\mathbf{J}\delta\rho)$. Using the definition of the conductivity (5.16) and taking the limit $\eta\to 0$, we obtain

$$\sigma^{\mu\nu}=\frac{e^2\pi\hbar}{\Omega}\sum_{\alpha\beta}v^\mu_{\alpha\beta}v^\nu_{\beta\alpha}\delta(\varepsilon_\alpha-\varepsilon_\beta)[-f'(\varepsilon_\alpha)], \tag{5.24}$$

where we used Re $\lim_{\eta\to 0}(i/(x+i\eta))=\pi\delta(x)$ and $[f(\varepsilon_\alpha)-f(\varepsilon_\beta)]/(\varepsilon_\alpha-\varepsilon_\beta)\to f'(\varepsilon_\alpha)$. Using the identity $\delta(\varepsilon_\alpha-\varepsilon_\beta)=\int_{-\infty}^{\infty}d\varepsilon\delta(\varepsilon-\varepsilon_\alpha)\delta(\varepsilon-\varepsilon_\beta)$ and taking the zero-temperature limit at which $[-f'(\varepsilon)]=\delta(\varepsilon-\varepsilon_F)$, formula (5.24) for the conductivity can be neatly represented by

$$\sigma^{\mu\nu}=\frac{e^2\pi\hbar}{\Omega}Tr\left[v^\mu\delta(\varepsilon_F-H)v^\nu\delta(\varepsilon_F-H)\right]. \tag{5.25}$$

This expression for the conductivity, generally, depends on the particular type of disorder responsible for scattering. To obtain a result that is independent of the particular disorder configuration but depends only on average characteristics (e.g., defect density), one has to perform a configurational averaging of this expression. In this form, the above expression is known as the Kubo–Greenwood formula [19], and is often written in the following form:

$$\sigma^{\mu\nu}=\frac{e^2\hbar}{\Omega\pi}\left\langle Tr\left[v^\mu\,\mathrm{Im}G(\varepsilon_F)v^\nu\,\mathrm{Im}G(\varepsilon_F)\right]\right\rangle. \tag{5.26}$$

Here, the angular brackets stand for the configurational average and we have introduced the Green's function, $G(\varepsilon)=\lim_{\eta\to 0}(\varepsilon-H+i\eta)^{-1}$, and taken into account $\delta(\varepsilon-H)=-\mathrm{Im}G(\varepsilon)/\pi$.

By configurational averaging, we replace the system that is characterized by a random nonperiodic potential V by an effective medium, which possess translational invariance [16,20]. The configurational averaging leads to the renormalization of the Green's function so that

$$\langle G(\varepsilon)\rangle = \frac{1}{\varepsilon - H^0 - \Sigma(\varepsilon)}, \tag{5.27}$$

where

H^0 in the Hamiltonian of the unperturbed periodic system
Σ is the self-energy

Its real part shifts the energy levels of the unperturbed system, whereas the imaginary part characterizes the broadening of the levels due to the finite scattering lifetime. Im $\Sigma(\varepsilon_F)$ determines, therefore, the relaxation time τ, which has been introduced in Section 5.2.2.

Equation 5.26 requires an average over the product of two Green's functions, that is, $\sigma \propto \langle GG\rangle$. In general, performing this averaging explicitly is a complicated problem. This is the reason why the conductivity is often approximated by the product of the average of the Green's functions, that is, $\sigma \propto \langle G\rangle\langle G\rangle$. This approximation ignores the contribution from the *vertex corrections* in the linear response formalism [16] and is equivalent to neglecting the scattering-in term in the semiclassical theory (Section 5.2.2).

The Kubo–Greenwood formula (5.26) is valid for a homogeneous system, where the current and applied field are uniform. In a general case of an inhomogeneous system, the current density is determined by the nonlocal conductivity according to

$$J^{\mu}(\mathbf{r}) = \int d\mathbf{r}' \sum_{\nu} \sigma^{\mu\nu}(\mathbf{r},\mathbf{r}')E^{\nu}(\mathbf{r}'). \tag{5.28}$$

The electric field $\mathbf{E}(\mathbf{r})$ in this equation is the local electric field. For inhomogeneous magnetic systems, the local internal field is position and spin dependent.

5.2.3.2 Landauer

In the Kubo formalism, an electric current is the result of the response of the system to an applied electric field. Alternatively, current flow can be considered as a transmission process between two electrodes across a conductor. This approach was proposed by Landauer [14] and appeared to be very useful for the description of transport properties at the nanoscale [17].

A formulation of the problem involves geometry of the system, which is shown schematically in Figure 5.2. The sample under consideration, for example, a magnetic multilayer, is placed between two semiinfinite leads. The sample can be, in general, disordered but the leads are perfect. At infinity, the leads are assumed to be connected to reservoirs (not shown), which are at thermodynamic equilibrium. The electric current in the system is driven by an electrochemical potential difference between the reservoirs. Within the above formulation, the

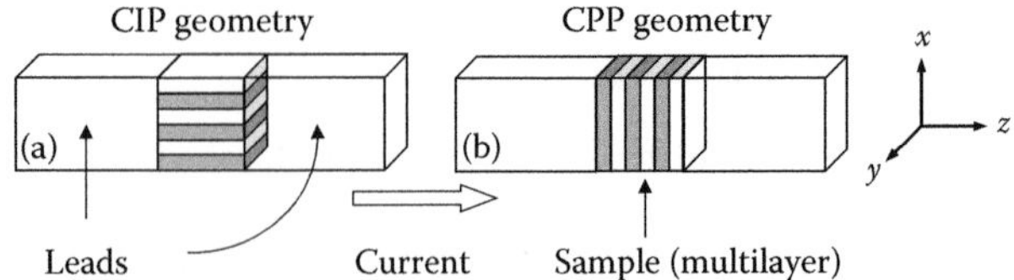

FIGURE 5.2 Geometry for the CIP (a) and CPP (b) transport calculation within the Landauer approach. (From Tsymbal, E.Y. and Pettifor, D.G., Perspectives of giant magnetoresistance, in: *Solid State Physics*, H. Ehrenreich and F. Spaepen (Eds.), vol. 56, Academic Press, San Diego, CA, 2001, pp. 113–239. With permission.)

conductance rather than the conductivity is the subject of interest. The conductance per spin is given by

$$\Gamma = \frac{e^2}{h}T(\varepsilon_F) = \frac{e^2}{h}\sum_{\mu\nu} T_{\mu\nu}(\varepsilon_F), \tag{5.29}$$

where

$T(\varepsilon_F)$ is the transmission coefficient evaluated at the Fermi energy
$T_{\mu\nu}$ is the probability for an electron to be transmitted from the state μ of the left electrode to the state ν of the right electrode

The above expression is rather general and can be applied to a system that is described by a multiband electronic structure.

In the ballistic transport regime the electrons passing through the sample are not scattered and all the scattering probabilities are $T_{\mu\nu}(\varepsilon_F) = \delta_{\mu\nu}$. In this case, the conductance is finite due to finite cross section of the sample and is entirely determined by the kinematic motion of electrons

$$\Gamma = \frac{1}{2}Ae^2 \sum_{\mathbf{k}} |\mathbf{n}\cdot\mathbf{v}_{\mathbf{k}}|\delta(\varepsilon_{\mathbf{k}} - \varepsilon_F), \tag{5.30}$$

where A is the area, and we assume that the band index is included in $\mathbf{k}$. By performing explicitly the integration in Equation 5.30 with respect to the momentum k_z along the direction of transport z, this formula can be rewritten as

$$\Gamma = \frac{e^2}{h}A\sum_{\mathbf{k}_{\|}} Q_{\mathbf{k}_{\|}} = \frac{e^2}{h}N, \tag{5.31}$$

where

$\mathbf{k}_{\|}$ is the transverse wave vector perpendicular to the current
$Q(\mathbf{k}_{\|})$ is the number of bands crossing the Fermi energy for a given $\mathbf{k}_{\|}$ and spin

According to Equation 5.31 the conductance is determined by the number of conducting channels N, which is determined by the density of the transverse modes at the Fermi energy.

In a more general case of diffusive transport regime, the Landauer formula (5.29) may be evaluated by employing Green's function formalism [17]

$$\Gamma = -\frac{e^2}{h} Tr\left[(\Sigma_L - \Sigma_L^\dagger)G(\Sigma_R - \Sigma_R^\dagger)G^\dagger\right]. \tag{5.32}$$

Here, $G(\varepsilon) = (\varepsilon - H - \Sigma_L - \Sigma_R)^{-1}$ is the Green's function of the sample coupled to the electrodes, which produce broadening and shift of the energy levels in the sample.

It has been proved that the Landauer and Kubo formalisms are equivalent [21]. Equation 5.32 or the equivalent formula within the Kubo approach can be used for calculating the conductance within a tight-binding description of the electronic structure and the recursive technique [22]. This is performed by "cutting" the system in the transverse direction (i.e., in the xy plane in Figure 5.2) and calculating the matrix elements of the Green's function between the atomic planes l and $l+1$. For this purpose, first, one must find the matrix elements of the surface Green's function for the detached semiinfinite leads. Then, the sample is grown by adding atomic layers with impurity atoms, layer by layer, onto the left lead. At every step, the Green's function matrix elements between atomic sites in the last added layer are calculated by solving numerically the Dyson equation. Once the sample has been fully grown, the last layer is bonded to the right lead in order to obtain the Green's function $G(\varepsilon_F)$ of the full system. The configurational averaging is performed numerically by generating a number of random disorder/impurity configurations.

5.3 Single-Band Models

5.3.1 Free-Electron Models

Within a free-electron model, the band structure of a magnetic multilayer or a thin film is described using a single parabolic band, which is independent of the spin direction. In this case, the expression for the conductivity per spin is simplified (e.g., Ref. [23])

$$\sigma = \frac{1}{d}\sum_i^N \left\{\frac{d_i}{\rho_i} - \frac{3}{4}\frac{\lambda_i}{\rho_i}\int_0^1 d\mu(1-\mu^2)\mu\left[A_i^+(\mu)\left(1-e^{-(d_i/\lambda_i\mu)}\right)\right.\right.$$
$$\left.\left. + A_i^-(\mu)\left(1-e^{d_i/\lambda_i\mu}\right)\right]\right\}. \tag{5.33}$$

Here, the relaxation time is assumed to be independent of **k** and the layer-dependent mean free paths $l_i = \tau_i v_F$ are introduced. In Equation 5.33, μ refers to the cosine of the momentum perpendicular to the interfaces, d is the total thickness of the multilayer, d_i and ρ_i are the thickness and the resistivity of the metal layer i, and, as before, the spin indices are omitted. The first term in this expression gives the conductivity, if the various layers were carrying the electric current in *parallel*. The second term is responsible for finite size effects. The coefficients $A^\pm$ can be found using boundary conditions.

For of a homogeneous nonmagnetic thin film, boundary conditions involve a coefficient of specular reflection p [24]. A fraction p of the electrons that are specularly reflected at the two surfaces of the film of thickness d enters the boundary conditions at $z=0$ and $z=d$

$$g^+(0,v^z) = pg^-(0,v^z), \quad g^-(d,v^z) = pg^+(d,v^z). \tag{5.34}$$

In the case $p=1$, which corresponds to perfect reflection from the boundaries, a free-electron model predicts that the conductivity of the film is identical to the conductivity of an infinite homogeneous metal. If $p<1$, a fraction of electrons, $(1-p)$, is scattered diffusively, so that $p=0$ corresponds to perfectly diffuse scattering. In the presence of diffuse scattering, the conductivity of a thin film decreases with the film thickness.

In the case of layered structures, additional boundary conditions at the interfaces are required. These boundary conditions can be imposed by assuming that the electrons are coherently transmitted with probability T, coherently reflected with probability R, or diffusely scattered with probability D at the interface. The transmission and reflection probabilities are related by the expression $D=1-T-R$. Thus, at the interface between layers i and $i+1$,

$$g_{i+1}^+ = Tg_i^+ + Rg_{i+1}^-, \quad g_i^- = Tg_{i+1}^- + Rg_i^+. \tag{5.35}$$

This model can be generalized to treat GMR in magnetic multilayers [25] by assuming that the electric current is spin dependent. Scattering in the bulk ferromagnetic layers is treated by introducing spin-dependent relaxation times. Scattering at the interfaces is taken into account by assuming spin-dependent transmission coefficients in the boundary conditions (5.35), that is, $T^\uparrow \neq T^\downarrow$. Using this model GMR can be analyzed in terms of a number of phenomenological parameters such as spin-dependent mean free paths, transmission probabilities at the interfaces, and specular reflection coefficients at the outer boundaries. This model has been extensively used for calculations of GMR and interpreting experimental data [23–28]. Below, we shall discuss the main predictions for GMR that is taken from this model.

The GMR ratio $\Delta R/R \equiv \Delta\rho/\rho_P = \Delta\sigma/\sigma_{AP}$ decreases monotonically by increasing the thickness of the nonmagnetic spacer layer d_{NM}. There are two contributions to this decrease; the first one contributes to the drop of $\Delta\sigma = \sigma_P - \sigma_{AP}$, whereas the second one results in the increase of σ_{AP}. The drop of $\Delta\sigma$ is seen in Figure 5.3a and reflects the reduction in the number of electrons, which can reach an opposite FM/NM interface before being scattered within the spacer layer. Asymptotically, the decrease of $\Delta\sigma$ is exponential with a characteristic decay length equal to λ_{NM}, that is, $\exp(-d_{NM}/\lambda_{NM})$. However, this asymptotic regime is reached only for $d_{NM} \gg \lambda_{NM}$, because according to Equation 5.33, the conductivity may be expressed as an integral

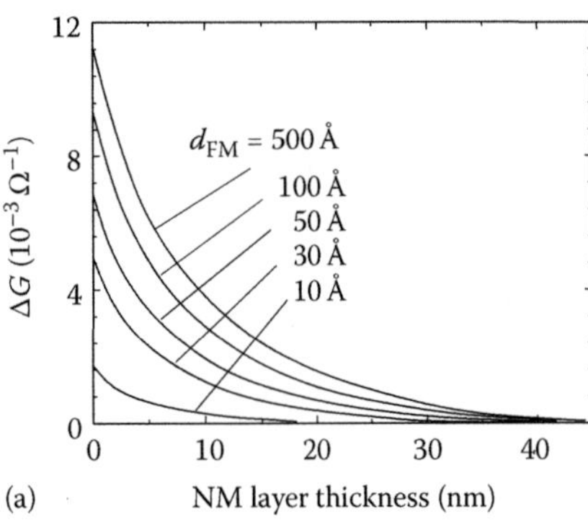

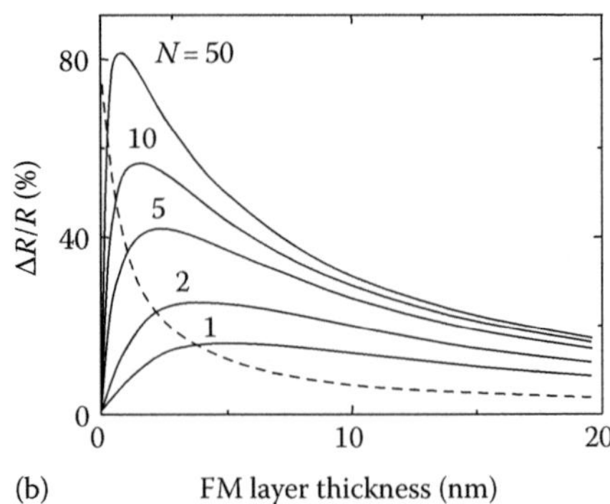

FIGURE 5.3 Dependence of GMR on layer thickness: (a) change in the sheet conductance versus thickness of the nonmagnetic spacer layer in FM(d_{FM})/NM/FM trilayer for different d_{NM}; (b) GMR versus ferromagnetic layer thickness in $(FM/NM)_N FM$ multilayers for bulk (solid lines) and interface (dashed line, $N = \infty$) scattering. (After Dieny, B., *J. Phys.: Cond. Mater.*, 4, 8009, 1992; Dieny, B., *Europhys. Lett.*, 17, 261, 1992. With permission.)

over an exponential with respect to various angles of incidence of the electrons. Another contribution to the decrease of GMR as a function of d_{NM} comes from the shunting current within the spacer layer, which leads to an increase in value of σ_{AP} approximately in a linear fashion with d_{NM}. If bulk scattering is negligible compared to the interface scattering, then the change in $\Delta\sigma$ with d_{NM} will be small. In this case, the shunting current becomes the dominant contribution and $\Delta\sigma/\sigma_{AP}$ drops as $1/d_{NM}$.

The variation of GMR as a function of the thickness of the magnetic spacer layer d_{FM} is different, depending on whether the bulk or interface spin-dependent scattering is dominant. In the case of *bulk* scattering the GMR ratio exhibits a maximum at a certain thickness. This can be seen from Figure 5.3b (solid lines), which displays GMR as a function of d_{FM} for multilayers with various number N of FM/NM bilayers. The position of the GMR maximum shifts toward lower FM layer thicknesses with increasing N. The presence of the GMR maximum and its shift with N is a direct consequence of the bulk spin-dependent scattering and the diffuse scattering at the outer boundaries. In the case of a small number of bilayers, for example, for $N = 1$, the GMR maximum is related to the long mean free path in the FM layer, so that $d_{FM} \sim \lambda_{FM}^{\uparrow}$. At this FM layer thickness the up-spin electrons are able to contribute to the conduction before being diffusely scattered at the outer boundaries. On the other hand, in the case of a large number of the FM/NM bilayers, the position of the GMR maximum is determined by the shorter of the two mean free paths, so that $d_{FM} \sim \lambda_{FM}^{\downarrow}$. This is because, for large N, the scattering at the outer boundaries becomes unimportant so that the maximum value of GMR will be obtained when the ferromagnetic layer thickness is sufficient to provide scattering of the down-spin electrons.

In the case of *interface* spin-dependent scattering, GMR decreases monotonically as a function of the FM layer thickness (see the dashed line in Figure 5.3b). The decrease of GMR reflects the fact that the increasing FM layer thickness enhances the relative contribution of the bulk spin-independent scattering in the FM layers. When d_{FM} becomes much longer than λ_{FM}, GMR is inversely proportional to d_{FM}, that is, $\propto \lambda_{FM}/d_{FM}$. This dependence follows from the fact that only those electrons that leave the FM region of thickness λ_{FM} adjacent to the interface have a sufficiently high probability of not to be scattered within this FM layer and, therefore, reach the opposite interface. The rest of the FM layer is inactive and serves only as a shunt, which reduces GMR that is inversely proportional to d_{FM}.

The semiclassical model predicts that by increasing the number of FM/NM bilayers within a multilayer, the value of GMR increases until it reaches saturation. This tendency can be seen in Figure 5.3b, according to which, at a fixed value of the FM layer thickness, the increment of the GMR growth decreases for larger N. This behavior is due to the diffuse scattering at the outer boundaries. If the longest mean free path is larger than the total thickness of the multilayer, then the diffuse outer-boundary scattering reduces the conductivity of the "good" spin channel and, hence, affects GMR negatively. GMR becomes independent of the number of bilayers when the total thickness of the multilayer is larger than the longest mean free path.

A free-electron model was used by Levy and coworkers to analyze GMR within a quantum-mechanical model, involving electron scattering from a spin-dependent potential that is produced by random-point scatterers [29,30]. By performing the configurational averaging in momentum space, they derived a simple formula for the local conductivity, which can be expressed in terms of the local spin-dependent scattering rate $\Delta(z)$ as

$$\sigma(z) = \frac{ne^2}{m\Delta(z)}, \tag{5.36}$$

where z is a coordinate perpendicular to the layers. The local scattering rate $\Delta(z)$ is determined by the appropriate average of the z-dependent scattering rates within the different layers. The total CIP conductivity of the multilayer can be found by integrating Equation 5.36 over the multilayer thickness. In the limit of the mean free path being long compared to the layer thickness, $\lambda \gg d$, the conductivity is "self-averaging" and the layering is no longer important. In this case, the model reduces to the series-resistor model (see Section 5.1.1). For the CPP geometry, it was predicted that the self-averaging occurs independent of the relationship between λ and d [4]. This result is intrinsic to a free-electron model, which ignores the variation of the potential across the multilayer. Assuming that the electronic structure is different for different layers, it leads to a solution for conductivity involving the exponential terms that decay at a rate comparable to the mean free path [31]. The real-space quantum-mechanical approach for the conductivity of free electrons affected by bulk spin-dependent scattering [32] shows that the quantum theory results in the same spin-dependent transport properties as the semiclassical approach, provided the effect of quantum interference and quantum-size effects are neglected [33].

An extension of the free-electron theory for magnetic multilayers to include the intrinsic potential within the Kronig and Penney model [34] shows that that the spin-dependent steps at the interfaces could enhance or reduce GMR, depending on the parameters characterizing the potential. It was also demonstrated that GMR could occur in structures with no spin asymmetry in the relaxation times, but with a spin-dependent potential within the multilayer. Including the effect of the intrinsic step-like potential on GMR in spin-valve structures [35,36] indicates that results are strongly influenced by quantum-size effects at small layer thicknesses and predicted oscillations in GMR as a function of the layer thicknesses. Unfortunately, these quantum-size effects in GMR, which make the quantum free-electron models conceptually different from the semiclassical free-electron models, have not yet been observed. Apart from the quantum-size effects, the variation of CIP GMR as a function of magnetic and nonmagnetic layer thickness within the quantum free-electron models qualitatively reproduces semiclassical results [30,37].

5.3.2 Tight-Binding Models

A tight-binding model describes the electronic structure of a metal in terms of localized atomic orbitals, which overlap due to the bonding between neighboring atoms [38]. The propagating (Bloch) states within this model can be built up from the atomic orbitals by solving the respective Schrödinger equation. A single-band tight-binding Hamiltonian is determined in terms of the on-site atomic energies and hopping matrix elements, which are usually assumed to be nonzero only between the nearest-neighbor lattice sites. The exchange splitting of the spin bands can be included in this model by taking different on-site atomic energies for the up- and down-spin electrons. Disorder can be introduced by assuming randomness in the on-site energies or in the hopping matrix elements. Impurities can be included by putting atoms with their on-site energies, which are different from those of the host atoms. Using such a tight-binding model helps to elucidate the microscopic origin of spin-dependent scattering, an important issue, which has not been addressed within the free-electron models described above.

In order to calculate the transport properties of a GMR multilayer, one can employ the recursive Green's function technique, as was outlined in Section 5.2.3.2. Asano et al. [39] implemented this approach for studying the effect of interface roughness and bulk disorder on GMR in a Fe/Cr magnetic multilayer within the CIP and CPP geometries. Their model assumes that Cr is nonmagnetic with on-site atomic energy equal to the minority on-site atomic energies of Fe. Due to this, the minority-spin electrons do not experience a potential step (or roughness potential) for the P alignment of magnetization. The interface roughness is introduced through substitutional randomness characterized by intermixing concentration c at the interfacial layers. The bulk disorder is modeled by a random variation of the on-site atomic energies of the Fe and Cr atoms with a uniform distribution of width γ, which is assumed to be spin independent.

As is evident from Figure 5.4a, the magnitude of CPP GMR is much larger than the magnitude of CIP GMR and they behave differently as a function of the interface roughness. The interface roughness enhances spin-dependent scattering and, therefore, CIP GMR increases with c (Figure 5.4a, squares). The presence of steps in the electronic potential at the interfaces has little influence on CIP GMR, which is close to zero in the absence of roughness. On the contrary, if a sizeable CPP GMR is found in the absence of any roughness, the latter only weakly reducing

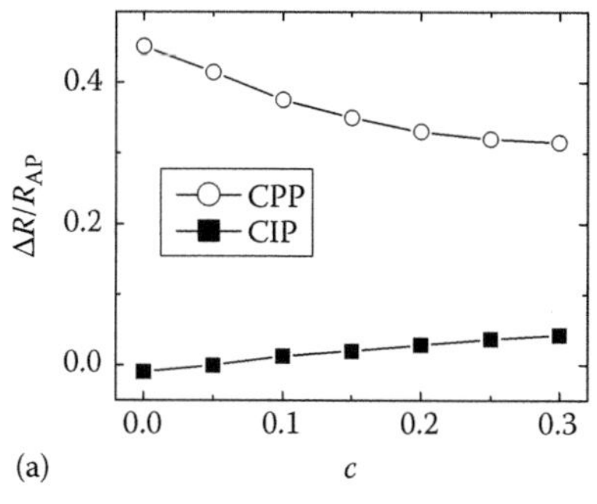

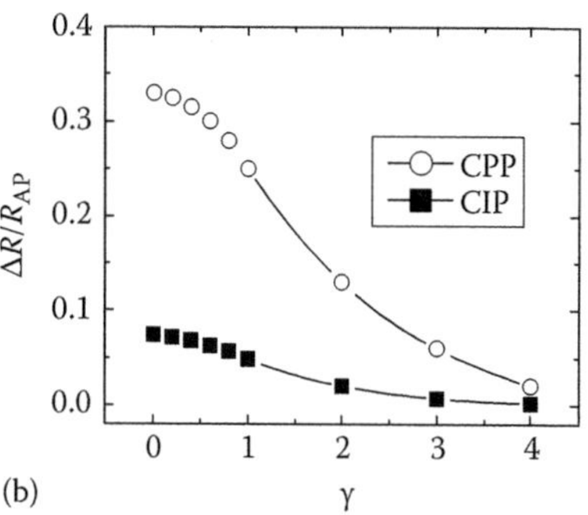

FIGURE 5.4 Calculated GMR of $(Fe/Cr)_2$ multilayer as a function of interface roughness c (a) and bulk disorder γ (b). Roughness is measured in terms of the intermixing concentration c of the two monolayers forming the interface. Disorder is introduced as a random variation in the on-site atomic energy levels of width γ. (After Asano, W.Y. et al., *Phys. Rev. B*, 48, 6192, 1993. With permission.)

GMR (Figure 5.4a, circles). This is the result of the spin-dependent potential of the multilayer, which affects differently the number of electrons contributing to the conduction of the P and AP configurations. Figure 5.4b indicates that GMR decreases with increasing bulk disorder parameter γ. This is a direct consequence of the increasing scattering in the "good" minority conduction channel, which reduces the conductance within the P configuration of the multilayer.

A similar model was implemented within coherent potential approximation (CPA) to study the combined effect of roughness and potential barriers at the interfaces on CIP and CPP GMR in magnetic multilayers [40]. Within the CPA [20,41] the disordered alloy (at the interface) is replaced by an effective medium, which is characterized by the Hamiltonian $H_{eff} = H_0 + \Sigma$, where the non-hermitian self-energy operator Σ is defined by Equation 5.27. Calculations of GMR in the weak-scattering limit of CPA [40] are found to be consistent with those predicted within the recursive technique [39]. In particular, the CPP GMR is larger than the CIP GMR, which is a consequence of the spin-dependent potential steps at the interfaces. The difference between the CPP and CIP geometries decreases strongly with decreasing step size. The interfacial roughness is favorable to CIP GMR but has a small negative effect on the CPP GMR.

The enhancement of CIP GMR due to the interface roughness is also predicted in the studies by Todorov et al. [42]. The contribution of the roughness, however, becomes quantitatively significant only in the limit of sufficiently dense interfacial steps (small correlation length of the roughness), sufficiently weak bulk scattering, and sufficiently thin magnetic layers. With increasing correlation length, GMR decreases, approaching its value without roughness. These results are consistent with the experimental data on GMR in epitaxial Fe/Cr multilayers [43].

5.4 Multiband Models

5.4.1 Role of Band Structure

In most experiments on GMR, the ferromagnetic 3*d* TM Co, Fe, and Ni, and their alloys, are used in combination with nonmagnetic spacer layers, such as Cr or the noble metals Cu, Ag, and Au. The electronic structure of these metals is characterized by a number of similar features. First, the 3*d* elements are characterized by the presence of the 4*s*, 4*p*, and 3*d* valence states, which are distinguished by their orbital momentum. The 4*s* and 4*p* states create a dispersive *sp* band, which is similar to a free-electron band. The *sp* electrons have a high velocity, a low DOS, and consequently, a long mean free path. On the contrary, the *d* bands are localized in a relatively narrow energy window and are characterized by a high DOS and a low velocity of electrons. In the interval of energy where the *sp* and *d* bands cross, they cannot be considered as independent because of the strong *sp*–*d* hybridization. These features are evident from Figure 5.5a, showing the electronic band structure of Cu and its Fermi surface [44].

In ferromagnetic 3*d* metals the *d* band is exchange-split. Due to the localized nature of the *d* electrons, two *d* electrons experience a strong Coulomb repulsion if they have AP spins and occupy the same orbital. To reduce the energy it is advantageous for the *d* electrons to have P-oriented spins, because the Pauli exclusion principle does not permit two electrons with the same spin to approach each other closely (i.e., occupy the same orbital) and, hence, the Coulomb interaction is reduced. Therefore, the Coulomb repulsion in conjunction with the Pauli principle leads to the exchange interaction and favors the formation of a spontaneous magnetization. The band structure and Fermi surfaces [44] of Co is shown in Figure 5.5b and c.

The conductivity is determined by electrons at the Fermi energy. In the case of Cu, the *d* bands are fully occupied and the Fermi level lies within the *sp* band, exhibiting a nearly

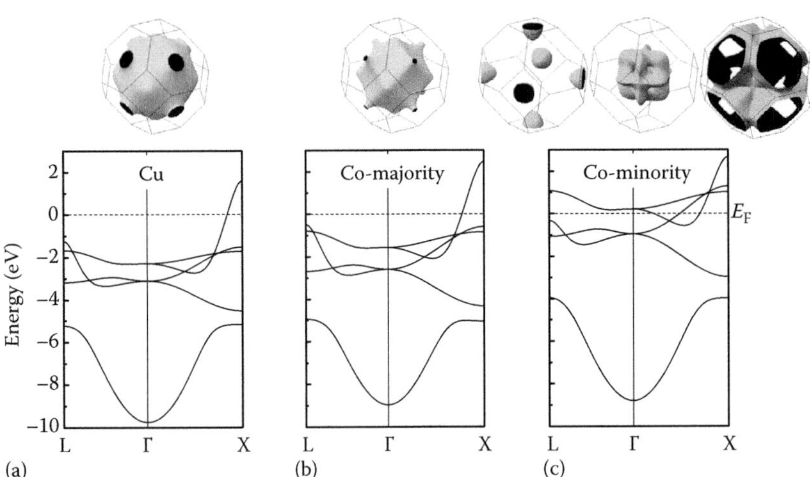

FIGURE 5.5 Band structures of Cu (a) and FCC Co for majority (b) and minority (c) spin. The horizontal dashed line indicates the Fermi energy. Top panels show the respective Fermi surfaces of the metals (note the three Fermi sheets for Co minority spin). (From Fermi Surfaces, 1997, Technical University Dresden, Institute for Theoretical Physics, http://www.physik.tu-dresden.de/~fermisur/. With permission.)

free-electron-type Fermi surface (Figure 5.5a, top panel). Due to the high velocity of the *sp* electrons, the mean free path is long and Cu is a very good conductor. This is also the case for the other noble metals Ag and Au. On the other hand, in the case of a ferromagnetic metal like Co, as a result of the exchange splitting, the majority *d* band is fully occupied, whereas the *d* minority band is only partly occupied (Figure 5.5b and c). The Fermi level lies, therefore, within the *sp* band for the majority spins and exhibit the free-electron-like Fermi surface (Figure 5.5b, top panel). For the minority spins, however, it lies within the *d* band, and the Fermi surface has three sheets (Figure 5.5c, top panels). Thus, the conductivity of majority-spin electrons is governed by the *sp* electrons and is high. On the contrary, the conductivity of the minority-spin electrons is controlled by the hybridized *spd* bands, which are not dispersive and have a high DOS. This makes the mean free path associated with these bands relatively short and the minority-spin conductivity low, despite a sizeable factor proportional to the area of the multiband Fermi surface. These arguments explain the strong spin asymmetry in the conductivity of bulk Co.

The presence of the interfaces in a magnetic multilayer adds a new important feature to spin-dependent transport. If two adjacent metals creating the interface have different band structures, the transmission probability across the interface is <1. If the interface separates ferromagnetic and nonmagnetic metals the transmission will be spin dependent due to the spin dependence of the band structure of the ferromagnetic layer. It is seen by comparing Figure 5.5a and b that the Fermi surfaces of Cu and the majority spins in Co are similar. This good band matching implies a high transmission for the majority-spin electrons across the Co/Cu interface. On the contrary, as is seen from Figure 5.5a and c, the Fermi surfaces of Cu and the minority spins in Co are very different. This band mismatch makes the transmission of the minority-spin electrons across the Co/Cu interface to be poor. Therefore, the interfaces of the Co/Cu multilayer act as spin filters. When the filters are aligned, the majority-spin electrons can pass through relatively easily. When the filters are antialigned, the electrons in both spin channels are reflected at one of the interfaces. This spin-dependent transmission is an important ingredient of the electronic transport in GMR structures (see also Section 5.4.6).

Band matching also plays an important role in the spin-dependent interface scattering due to the intermixing of atoms near the interfaces. The intermixing at the interface produces a random potential, which is strongly spin dependent. This spin dependence is a direct consequence of the good band matching for the majority spins in Co/Cu, which implies a small scattering potential, and the poor band matching for the minority spins in Co/Cu, which implies a large scattering potential. A similar behavior takes place in Fe/Cr multilayers, where a very small scattering potential (good band matching) is expected for the minority-spin electrons, but a large scattering potential (bad band matching) is expected for the majority-spin electrons. Thus, the matching or mismatching of the bands between the ferromagnetic and nonmagnetic metals results in spin-dependent scattering potentials at disordered interfaces, which can contribute to GMR.

To describe realistically the electronic structure, powerful computational methods have been developed, based on density functional theory (DFT) (see, e.g., Ref. [45]). These methods do not involve any empirical parameters and are therefore referred to as *ab-initio* or "first-principles."

5.4.2 GMR without Scattering

The effect of the band structure on GMR can be illustrated by considering *ballistic* conductance of a disorder-free system. Although this regime of conductance is not relevant to experimental conditions for GMR multilayers, it provides a useful insight into the role of the band structure. In the ballistic regime the conductance is given by Equation 5.31 and is entirely determined by the electronic band structure of the multilayer [46]. The GMR ratio, in this case, is given by

$$GMR = \frac{\sum_{\mathbf{k}} \left(v_{\mathbf{k}}^{P\uparrow}\right)^m \delta(\varepsilon_{\mathbf{k}} - \varepsilon_F) - \sum_{\mathbf{k}} \left(v_{\mathbf{k}}^{P\downarrow}\right)^m \delta(\varepsilon_{\mathbf{k}} - \varepsilon_F)}{\sum_{\mathbf{k}} \left(v_{\mathbf{k}}^{AP}\right)^m \delta(\varepsilon_{\mathbf{k}} - \varepsilon_F)}, \quad (5.37)$$

where $m = 1$ and the Fermi velocity is different for CIP and CPP transport and for P and AP magnetization.

Using this model, Schep et al. [46] calculated GMR for Co/Cu multilayers in the ballistic transport regime (Figure 5.6). They predicted that the CPP conductance increases by more than a factor of 2 when the magnetization changes from AP to P. Such a sizable difference in the conductance reflects the contribution from electrons of different orbital momenta to the current. Although the majority spins within the P configuration resemble the free-electron picture, the minority spins within the P configuration and both spins within the AP configuration are not free-electron-like. The majority-spin *s* electrons shunt the current within the P magnetizations making the conductance for this configuration much higher than that for the AP configuration, which is mainly determined by the *d* electrons. As seen from Figure 5.6, the CPP GMR displays oscillations that approach a constant value of about 100% for larger layer thicknesses. The

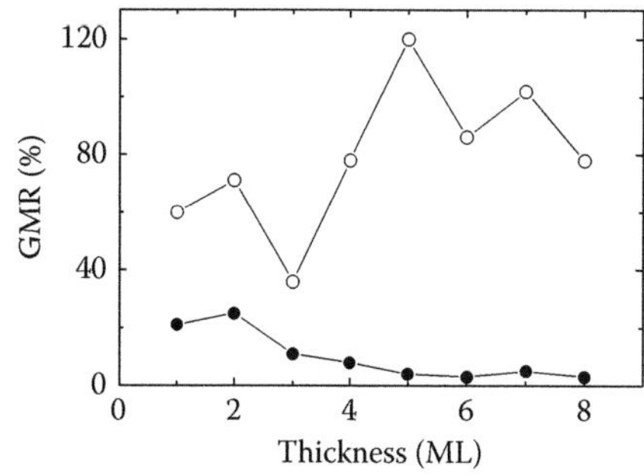

FIGURE 5.6 CIP (solid circles) and CPP (open circles) GMR for Co_n/Cu_n multilayers and as a function of number of layer thickness *n*. (After Schep, K.M. et al., *Phys. Rev. B*, 57, 8907, 1998. With permission.)

CIP GMR is much lower and decreases with layer thickness. This reflects the decreasing effect of interfaces, where the *sp*–*d* hybridization effects play a role, making the difference between P and AP configurations. Oscillations of GMR in the ballistic transport regime, which are seen in Figure 5.6, originate from quantum size effects [47]. The periods of these oscillations stems from the stationary points at the Fermi surface of Cu and the conductance cutoff due to a mismatch between the Co and Cu bands across the Co/Cu interfaces [48].

Another illustration of the importance of the band structure was given by Oguchi [49] who used a semiclassical model for conductivity described in Equation 5.18, in which he assumed that that the relaxation time τ is state and spin independent. In this case the conductivity is factorized into a scattering term τ and the electronic structure term. Since the relaxation time τ is constant, the GMR ratio is reduced to Equation 5.37 with $m=2$ and is entirely determined by the band structure. For a Co_3/Cu_3 multilayer Oguchi found GMR values of 47% for CIP and 170% for CPP. Although the assumption of a state- and spin-independent relaxation time is difficult to justify, these results demonstrate the critical role of the band structure (the Fermi velocities) for GMR.

5.4.3 Effects of Impurity Scattering and Intermixing

An important contribution to GMR comes from scattering by impurities, which may be either deliberately introduced in the multilayer or present as intrinsic defects. In addition, the intermixing of ferromagnetic and nonmagnetic materials often occurs at the interfaces (or in the bulk of the layers) and contributes to spin-dependent scattering. The description of these effects is important for the quantitative description of GMR. The essential role of these effects follows from the work by Itoh et al. [50] who used the tight-binding *d*-band model for TM to study the material dependence of GMR in Fe/TM and Co/TM multilayers (TM = Sc, Ti, V, Cr, Mn, Ru, Rh, or Pd). Their results indicate a strong dependence of spin-dependent scattering, and consequently GMR on a particular type of impurity.

The general approach to treat scattering by an impurity is known from studying bulk systems [51]. Impurities produce diffuse scattering, the probability of which is

$$P_{\mathbf{kk}'} = \frac{2\pi}{\hbar}\left|T_{\mathbf{kk}'}\right|^2 \delta(\varepsilon_{\mathbf{k}} - \varepsilon_{\mathbf{k}'}). \tag{5.38}$$

Here, T matrix describes the scattering of an electron by an impurity embedded in the system. This expression can be considered as a generalization of the Fermi golden rule (5.9) to include all orders of perturbation theory with respect to the scattering potential. The matrix elements $T_{\mathbf{kk}}$ were calculated using an application of multiple-scattering theory [52].

In magnetic systems, the probability of scattering given by Equation 5.38 depends on spin. If a particular type of impurity is the only mechanism contributing to resistivity, this spin-dependent scattering probability determines the spin asymmetry of resistivity α, which was introduced in Section 5.1.1. By summing up the microscopic transition probability over all final states, one can find the inverse relaxation time in Equation 5.15. In general, the relaxation time depends on the state **k**.

Mertig and coauthors calculated the relaxation time averaged over the Fermi surface for various impurities imbedded in GMR multilayers [53] (see also Ref. [7]). Since the averaged relaxation time is spin dependent, one can define the anisotropy ratio $\beta = \tau^{\uparrow}/\tau_{\downarrow}$, which determines the asymmetry in scattering by different spins. The results of first-principles calculations for different impurities that are placed at the interface of a Co/Cu multilayer are shown in Figure 5.7. The magnitude of the magnetic moment varies significantly from almost zero for Cu and Zn to largest in Cr, Mn, and Fe (Figure 5.7a). It is notable that for the early transition elements Sc–Mn the magnetic moment is oriented antiparallel to the host ferromagnet. As seen from Figure 5.7b, impurities with P magnetic moment show spin anisotropy of scattering $\beta > 1$, whereas most of impurities with AP magnetic moment show spin anisotropy $\beta < 1$.

Using the calculated state- and spin-dependent relaxation times within the semiclassical transport theory and assuming a low impurity concentration, one can calculate the conductivities of the multilayers and obtain values of GMR. Figure 5.7c presents results for the Co/Cu system. It is seen that the values of GMR are very large, especially for impurities characterized by large spin asymmetry β (about 3500% for Fe and 3000% for Ni). Similar results are obtained for GMR in Fe/Cr multilayers, assuming that the spin-dependent scattering is due to Cr impurities in bulk Fe [53]. In this case, the ratio of the spin-relaxation times is $\beta = 0.11$, which implies that the majority spins are

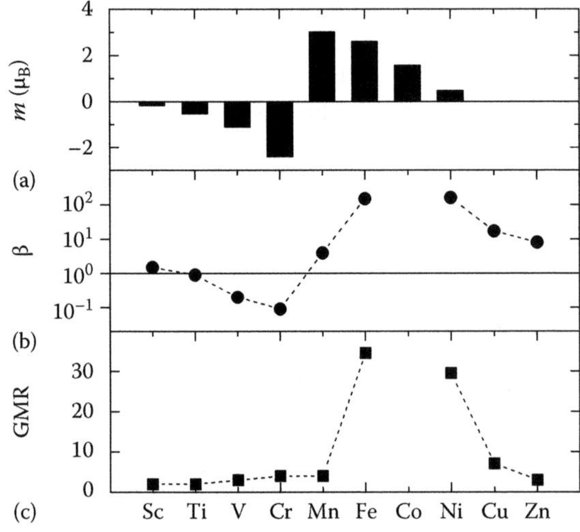

FIGURE 5.7 Magnetic moments (a), spin anisotropy of scattering (b), and CIP GMR (c) for TM impurities at the Co interfacial layer in a Co_9/Cu_7 multilayers. (After Levy, P.M. and Mertig, I., Theory of giant magnetoresistance, in: *Spin Dependent Transport in Magnetic Nanostructures*, S. Maekawa and T. Shinjo (Eds.), CRC Press, Boca Raton, FL, 2002, pp. 47–112. With permission.)

scattered much stronger at a Cr defect than the minority spins. This strong spin asymmetry in scattering leads to GMR of about 600% for CIP and of about 2500% for CPP, which are much higher than the experimental values for Fe/Cr multilayers.

Very large values of GMR are also predicted for magnetic multilayers assuming that their interfaces are interdiffused [54]. For the Co/Cu and $Fe_{20}Ni_{80}$/Cu multilayers, the values of CPP GMR are in the range of 1000%–2000%, depending on layer thicknesses. The highly overestimated GMR is due to the fact that the electronic structures of the impurities and the host atoms are nearly identical in one of the two spin channels, which leads to no scattering and shunting within this spin channel. The principle disagreement between theory and experiment originates from the neglect of scattering by intrinsic defects, which are always present in the multilayer.

More reasonable values of GMR were obtained by Blaas et al. [55], who studied intermixing at the interfaces in Co/Cu multilayers. In their model, the multilayers were embedded between two semiinfinite Cu substrates, which lead to a broadening of electronic states and circumvented shunting in the majority-spin channel. It was found that the CIP GMR increases with the number of the Co/Cu unit cells until it saturates, and that GMR is enhanced by interdiffusion. These results indicate that including additional scattering is important for obtaining more realistic values of GMR. It is not clear, however, what physical mechanism of scattering in real multilayers may produce the broadening of electronic states, as assumed in this model.

In those cases, when impurities produce strong scattering in *both* spin channels and their concentration is sufficiently large to contribute dominantly in the resistivity, a quantum mechanical theory of transport predicts values of GMR, consistent with experimental observations. This fact was nicely demonstrated by Bengone et al. [56] by the modeling of GMR in $Co_{0.8}Cr_{0.2}$/Cu/Co trilayers. It was found experimentally that this system exhibits a transition from a positive to negative GMR with increasing thickness of the ferromagnetic alloy [57]. This behavior follows from the fact that Cr impurities in Co produce stronger scattering in the majority-spin channel, which makes the spin-asymmetry parameter less than unity, $\beta < 1$ (see Figure 5.7b). If the impurity scattering in the alloy is dominant, then $\alpha = \rho_\downarrow/\rho_\uparrow < 1$, and the spin asymmetry of the $Co_{0.8}Cr_{0.2}$ resistivity is opposite to that in Co, resulting in the negative GMR (Section 5.1.1). A rigorous theoretical treatment of this phenomenon was performed [56] using the quantum-mechanical description of transport, based on the recursion method (Section 5.2.3.2) and a self-consistent electronic structure of the multilayer [58]. It was found that the origin of the negative GMR is due to the formation of a virtual bound state in the majority-spin channel of the alloy that acts as effective scattering center. With increasing Co–Cr alloy thickness the resistance of the majority-spin channel becomes high leading to an inverse GMR. Figure 5.8 shows the crossover between positive and negative GMR, which is predicted from the first-principles calculations and is consistent with the experimental data.

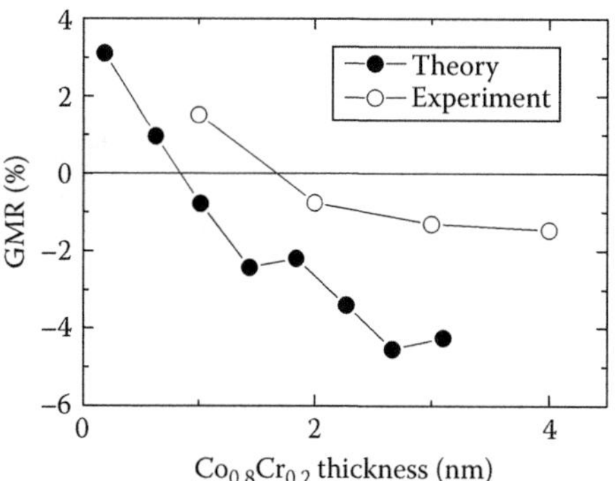

FIGURE 5.8 GMR in $Co_{0.8}Cr_{0.2}$/Cu/Co (111) multilayers: results of calculations based on density-functional theory (solid circles) and experimental data (open circles). (After Bengone, O. et al., *Phys. Rev. B*, 69, 092406 1-4, 2004. With permission.)

5.4.4 Effects of Structural Disorder

Magnetic layered systems contain intrinsic structural defects such as vacancies, stacking faults, lattice distortions, misfit dislocations, and grain boundaries. These defects make the residual resistivity larger in thin films than in corresponding bulk metals. Scattering by these defects needs to be included in a realistic model for GMR. However, simultaneous modeling of all these defects is a very complicated problem.

A simplified approach to include defect scattering in a model for GMR was developed by Tsymbal and Pettifor [59] within a multiband tight-binding theory and a quantum-mechanical approach to electronic transport. In order to describe disorder, they introduced the scattering potential V, effecting the on-site atomic energy levels randomly and characterized by mean-square displacement γ^2. The parameter γ is a parameter of the theory, which is assumed to be spin independent and is chosen to provide a realistic saturation resistivity of the multilayer. Using this model and performing configurational averaging of the Kubo–Greenwood formula (5.26), in the weak scattering limit [41] the result for conductivity can be expressed as follows:

$$\sigma^{\mu\nu} = \sum_{j\alpha} \sigma^{\mu\nu}_{j\alpha} = \sum_{j\alpha j'\alpha'} \frac{e^2}{\pi\hbar} \int \frac{d\mathbf{k}}{(2\pi)^3} \Lambda^{\mu}_{j\alpha,j'\alpha'}(\mathbf{k}) \Lambda^{\nu}_{j'\alpha',j\alpha}(\mathbf{k}), \qquad (5.39)$$

where $\Lambda^{\mu}(\mathbf{k}) = \hbar v^{\mu}(\mathbf{k}) \mathrm{Im}[E_F - H^0(\mathbf{k}) - \Sigma(\varepsilon_F)]^{-1}$ is the mean free path operator. The self-energy $\Sigma(\varepsilon_F)$ determines the partial scattering rates corresponding to cite j and orbital α:

$$\tau^{-1}_{j\alpha} = \frac{2}{\hbar} \mathrm{Im}\,\Sigma_{j\alpha,j\alpha}(\varepsilon_F) = \frac{2\pi}{\hbar} \gamma^2 n_{j\alpha}(\varepsilon_F). \qquad (5.40)$$

As seen from Equation 5.40, the scattering rates are spin dependent due to the spin dependence of the DOS $n_{j\alpha}(\varepsilon_F)$. The partial conductivities $\sigma^{\mu\nu}_{j\alpha}$ defined by Equation 5.39 can be used to determine the contributions from nonequivalent sites (layers) j and orbitals α.

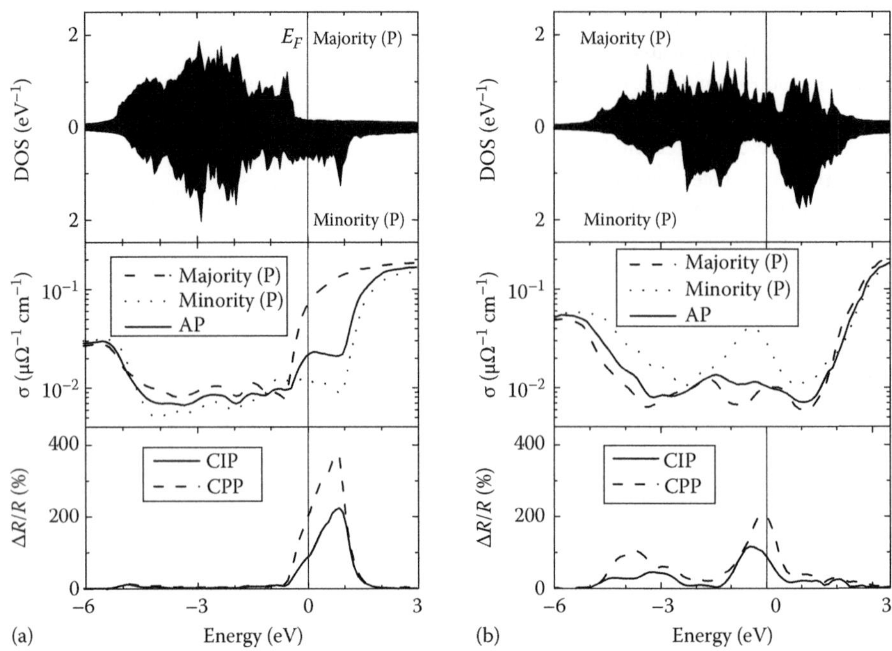

FIGURE 5.9 Density of states (top panels), conductivity (middle panels), and GMR (bottom panels) of Co_4/Cu_4 (a) and Fe_4/Cr_4 (b) multilayers as a function of energy. (After Tsymbal, E.Y. and Pettifor, D.G., *Phys. Rev. B*, 54, 15314, 1996. With permission.)

Figure 5.9a shows the DOS, conductivity, and GMR of a Co_4/Cu_4 (001) multilayer calculated as a function of energy. The DOS of the P-aligned Co/Cu multilayer is asymmetric between the majority and minority spins. This is reflected in the conductivity σ. For the energies that lie within the *d* band the *sp* electrons are slowed down by the hybridization with the *d* electrons and the conductivity is low. On top of the *d* bands σ increases rapidly. This increase is due to the "acceleration" of the *sp* electrons, which are less affected by the *sp*–*d* hybridization above the *d* bands. The top of the *d* bands for the majority spins lies at approximately 0.5 eV below the Fermi energy and for the minority spins at about 1 eV above the Fermi energy. Therefore, in the interval of the energies from −0.5 to 1 eV the difference between $\sigma^{\uparrow}$ and $\sigma^{\downarrow}$ is most pronounced. The presence of the *d* levels up to 1 eV above E_F for the AP configuration makes the conductivity per spin, in this case, similar to the conductivity of the minority spin electrons for the P configuration. This leads to the crucial difference between the σ_P and σ_{AP} and, consequently, to GMR. As seen from the bottom panel in Figure 5.9a the large values of GMR are predicted only in the interval of energies between the top of the majority-spin *d* band and the top of the minority-spin *d* band. As the Fermi level in Co/Cu multilayers lies within this interval, a sizeable value of GMR can be observed in this system.

The Fe/Cr system behaves differently. The Fermi level in the Fe_4/Cr_4 multilayer lies within the *d* bands for both spin orientations (Figure 5.9b). However, the DOS exhibits a pronounced valley for the minority spins, with the Fermi level lying almost at the bottom of this valley. This makes the minority-spin conductivity higher than the majority-spin conductivity, that is, $\alpha = \sigma^{\uparrow}/\sigma^{\downarrow} < 1$, which is opposite to bulk Co [59]. For the P-aligned multilayer $\sigma^{\downarrow}$ displays an enhancement for electron energies lying in the region of this valley (the dotted line in the middle panel of Figure 5.9b). On the contrary, $\sigma^{\uparrow}$ for the P orientation and the conductivity for the AP orientation do not change essentially in this interval of energies (the dashed and the solid lines in the middle panel of Figure 5.9b). This results in the difference between σ_P and σ_{AP} and, consequently, in GMR for the energies close to the Fermi level (the bottom panel of Figure 5.9b).

The magnitude of GMR decreases with increasing disorder. As is seen from Figure 5.10, the value of GMR in a Co_4/Cu_4 multilayer drops by an order of magnitude, that is, from about 105% to 15% as the disorder parameter γ changes from 0.4 to 1.2 eV. The

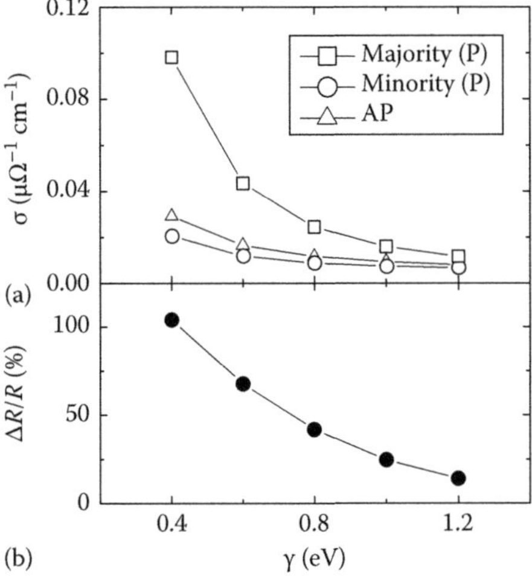

FIGURE 5.10 Conductivity (a) and GMR (b) of Co/Cu multilayer as a function of disorder parameter. (After Tsymbal, E.Y. and Pettifor, D.G., *Phys. Rev. B*, 54, 15314, 1996. With permission.)

corresponding saturation resistivity varies from 8 to 55 μΩ cm (Figure 5.10a), spanning the range of experimental values. This drop in GMR with increasing disorder has an important underlying physics. It is related to the interband transitions driven by the applied electric field, which makes the quantum mechanical description different from the semiclassical approach [59]. In metals, in the absence of disorder, these transitions may be important only in an exceedingly small region of **k**-space near points where two bands become degenerated [11]. In the presence of disorder, however, the electronic levels are broadened by the value of Im Σ. If this broadening becomes comparable with the distance between bands, even an infinitesimal small electric field can lead to the interband transitions. In this case, the semiclassical approximation breaks down.

The drop of GMR within increasing disorder (Figure 5.10b) can, therefore, be explained as follows. The conductivity of the P-aligned multilayer is mainly determined by the majority-spin electrons, which have predominantly free-electron-like *sp* character due to the position of E_F above the *d* bands. Therefore, with increasing γ, σ_P decreases approximately as $1/\gamma^2$. On the other hand, the conductivity of the AP-aligned multilayer is primarily due to the *d* electrons, because E_F falls inside the *d* band for this alignment. In this case, due to the interband transitions between the *s*, *p*, and *d* levels, the σ_{AP} decreases more slowly than $1/\gamma^2$ so that GMR drops with increasing γ. This dependence is not predicted by the semiclassical treatment of conductivity, according to which the conductivity for both the P and AP alignments is inversely proportional to γ^2 so that GMR is independent of γ.

The mechanism involving interband transitions is most likely responsible for the reduction of GMR with temperature. It is known that the dominant contribution to the temperature-dependent resistivity of metals at temperatures well below the Curie temperature originates from electron–phonon interactions. Since electron scattering processes occur on a time scale much shorter than that of lattice vibrations, the electron scattering occurs on a frozen, phonon-distorted lattice. The consequences of scattering, which is produced by the corresponding scattering potential, should be qualitatively similar to those due to disorder described above. Therefore, the reduction of GMR with increasing electron–phonon scattering is likely due to interband transitions, that is, due to the same mechanism that is responsible for the drop in GMR with increasing disorder (Figure 5.10).

5.4.5 Local Conductivity and Channeling

Useful information about local contributions to the spin-dependent current and GMR can be obtained from the layer-dependent conductivity $\sigma_j = \sum_\alpha \sigma_{j\alpha}$ (see Equation 5.39). By introducing in the model different contributions from bulk, interfacial, and boundary scattering in a spin valve, one can make conclusions about the importance of these scattering mechanisms for GMR.

Figure 5.11 shows results of calculations for a $Co_7/Cu_{10}/Co_7$ (111) trilayer, using the self-consistent electronic band structure

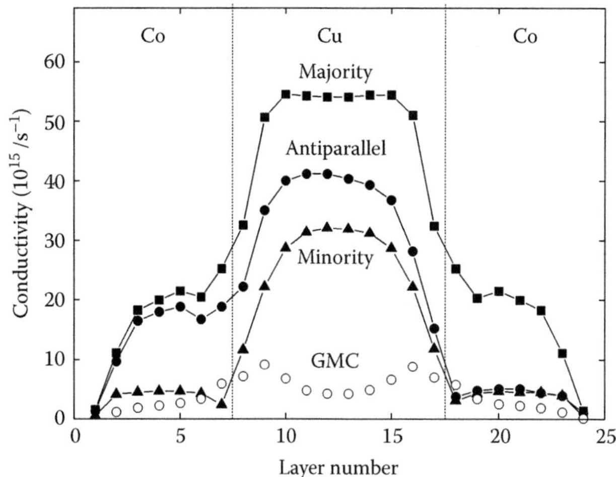

FIGURE 5.11 Layer-dependent conductivities (in units $10^{15}\,s^{-1} = 1113\,\Omega^{-1}\,cm^{-1}$) for a $Co_7/Cu_{10}/Co_7$ (111) trilayer: majority- and minority-spin contributions forP magnetization, either spin for AP magnetization, and giant magnetoconductance (GMC) defined as the difference between layer-dependent conductivities for P and AP magnetizations. (After Butler, W.H. et al., *Phys. Rev. B*, 52, 13399, 1995. With permission.)

and the Kubo–Greenwood formula for conductivity [60]. The calculations were performed by introducing phenomenological scattering rates that provide experimentally observed values of resistivity in sputtered Co and Cu films at room temperature, that is, 2.8 μΩ cm for Cu and 14.8 μΩ cm for Co. It was assumed that the scattering rate in Co is spin dependent and is proportional to the DOS for the given spin channel at the Fermi energy (similar to Equation 5.40). This led to the electron lifetime for majority carriers to be seven times that for minority carriers in the Co layers. In addition, a strong spin-dependent scattering was assumed at the interfaces, and the boundary scattering was set to reduce all the conductivities to essentially zero at the outer boundaries.

As is seen from Figure 5.11, for the P magnetization the local conductivities in the Co layer are much higher for the majority spins than for the minority spins. The local conductivities σ_{Cu} in the Cu layers are higher than those in the Co layers and are different for different spins. The latter is partly because the transport properties of electrons propagating in the Cu layer are affected strongly by the spin-dependent scattering in the adjacent Co layers. For the AP magnetization, there is an asymmetry in the local conductivities. The minority-spin conductivities in Co are almost the same for the AP configuration as for the P configuration. On the other hand, the majority-spin conductivities are reduced because the majority carriers can propagate through the Cu layer and scatter in the opposite Co layer, which has a high minority-spin DOS. This mechanism leads to local contributions to giant magnetoconductance (GMC, defined as the difference between layer-dependent conductivities for P and AP magnetizations), which comes from electrons contributing to the conductivity in the Co layers. Interestingly, a significant contribution to GMC originates from the Cu spacer layer (see

Figure 5.11, open circles). These enhanced values of GMC can be explained by electron channeling within the Cu spacer layer as a result of the potential step at the Co/Cu interface.

The presence of potential steps at the interfaces results in a contribution to GMR, which is distinct from spin-dependent scattering. This contribution originates from electron channeling, as has been demonstrated within free-electron models [34,35,61]. Channeling can occur in the CIP geometry if a large portion of the electrons in a layer is specularly reflected from both its interfaces. If the scattering rate in that layer is lower than its neighboring layers, the reflected electrons in that layer see a lower effective scattering rate and provide a larger contribution to conductivity. In magnetic multilayers, channeling contributes to GMR due to the electrons with P momenta, which are strongly reflected for one spin, but not the other [62].

Using first-principle calculations, it was shown that channeling can contribute to GMR in Co/Cu/Co spin valves [63]. The majority-spin electrons in Co have a lower Fermi momentum than electrons in Cu. This leads to the total internal reflection of those electrons whose momentum parallel to the interface exceeds the Fermi momentum of Co. These reflected electrons are, therefore, confined to the Cu layers and give a large contribution to the majority current and also to GMR. This contribution shows up, in Figure 5.11, as a contribution to GMC that is confined to the Cu layer region. The degree of channeling, which is observed in real experiments, depends strongly on the detailed nature of the interfaces between the Co and Cu layers. Interfacial roughness can reduce or eliminate the contribution of channeling to GMR.

5.4.6 Interface Transmission

The spin-dependent transmission across interfaces plays an important role in GMR and contributes to the phenomena, which have been described in the previous sections. There are several methods to evaluate the transmission probabilities across interfaces. One of them uses matching of the generalized Bloch states of the two materials creating an interface [62]. Another way is to employ the Landauer approach (Section 5.2.3.2), where the role of the scattering region is played by the interface between two semiinfinite leads [64]. Figure 5.12 shows the results of the calculation of the transmission across the Cu/Co (111) interface resolved in the 2D Brillouin zone. The majority-spin Fermi surface in Co is similar to the Fermi surface in Cu (see Figure 5.5a and b). This similarity leads to the almost complete transmission of the majority-spin electrons from Co into Cu (Figure 5.12a). The smaller size of the majority Co Fermi surface, however, results in the reflection for the states in Cu with the largest momenta parallel to the interface. These electrons contribute to channeling within the Cu layer in the Co/Cu/Co trilayer when magnetization is parallel, as was discussed in Section 5.4.5. The transmission of the minority-spin electrons is lower in magnitude and has much more complex distribution (Figure 5.12b), reflecting the presence of the three Fermi sheets in the minority-spin channel (Figure 5.5c).

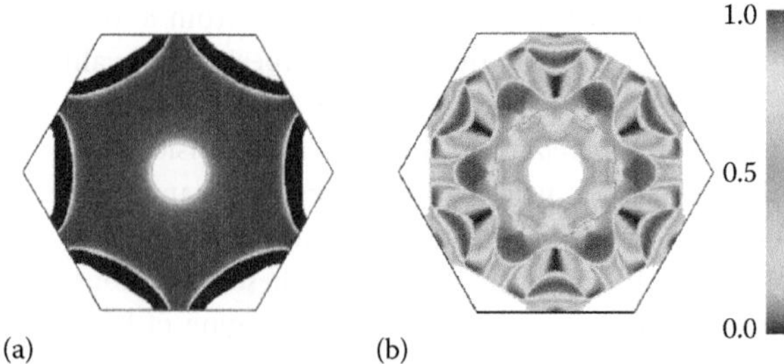

FIGURE 5.12 (See color insert.) Transmission probabilities for majority- (a) and minority- (b) spin electrons across the Co/Cu(111) interface resolved in the 2D Brillouin zone. (After Xia, K. et al., *Phys. Rev. B*, 63, 064407 1-4, 2001. With permission.)

By analyzing the spin-dependent transmission across interfaces, one can evaluate the spin-dependent interface resistance. It is known that there are two contributions to the interface resistance [65,66]: The first one comes from the potential step between the ferromagnetic and nonmagnetic metal layers and the second one results from interface disorder. The interface resistance can be calculated from first principles assuming that there is no coherent transmission between adjacent interfaces [67]. The latter approximation implies complete diffuse scattering in the bulk and results in the following expression for the spin-dependent interface resistance

$$R_{A/B} = \frac{h}{e^2}\frac{1}{T_{A/B}} - \frac{1}{2}\left(\frac{1}{\Gamma_A} + \frac{1}{\Gamma_B}\right), \quad (5.41)$$

where

Γ_A and Γ_B are ballistic conductances per spin (given by Equation 5.31) of the two metals A and B, creating the interface

$T_{A/B}$ is the spin-dependent transmission coefficient

This approach was implemented within a tight-binding linear-muffin-tin orbitals method that allows including structural disorder and comparing contributions, resulting from specular and diffuse interface scattering [68]. It was found, in particular, for Co/Cu interfaces, that including diffuse scattering somewhat affects values of the interface resistance. However, the dominant contribution comes from specular scattering, indicating the decisive role of the band structure matching at the interface [64]. This method to evaluate the spin-dependent interface resistance has extensively been used to compare the results of the calculations with experimental observations related to CPP GMR (see Chapter 4).

Normally, the interpretation of experimental data on CPP GMR assumes that the interfaces can be uniquely described using two parameters, $AR^{\star}_{FM/NM}$ and γ, independent of the layer thickness separating the interfaces. It was predicted, however, that the interface resistance depends on the distance separating them [69,70]. This effect results from the redistribution of electrons between conducting channels due to diffuse scattering and is controlled by the mean free path. It was argued that this phenomenon is responsible for the difference in CPP GMR between multilayers, where the order of thin and thick layers is changed [71]. The deviation from the series resistor model involves the ability of electrons to conserve their wave vector when propagating between the two interfaces. This effect is rather sensitive to the amount of disorder at the interfaces that produces diffuse scattering of electrons, reducing the deviation from the series-resistor model at large interface disorder [70]. The significance of the interface proximity effects for the understanding of CPP GMR data is still under debate [72].

5.5 Beyond the Two Spin-Channel Model

All the above models assume that there is no spin-flip scattering, and, hence, the electrical current can be described independently within the two spin-conduction channels. In bulk 3*d* metals the probability of spin-flip scattering processes is normally small as compared to the probability of the scattering processes in which the spin is conserved such as electron–phonon scattering. It is only at temperatures comparable to the Curie temperature that spin-flip scattering processes due to electron–magnon interactions cause mixing of the electrons within the two spin channels. This means that spin is conserved over long distances and, therefore, the electrical conduction occurs in parallel for the two spin channels. These arguments allowed us to ignore spin-flip scattering when considering the CIP geometry, because CIP GMR is exponentially dependent on the mean free path, which is normally much shorter than the spin-diffusion length.

In the CPP geometry, the situation is different. In the limit of large layer thicknesses, the series-resistor model (Section 5.1.1) predicts a constant GMR. As has been shown by Valet and Fert [73], this is not the case when a finite spin-diffusion length is taken into account. The model of Valet and Fert is based on a solution of the Boltzmann equation within a free-electron model, which, in addition to the usual electron relaxation term for the momentum, takes into account the spin-relaxation term due to spin-flip scattering. The spin relaxation balances spin accumulation, which is generated when the electrical current flows perpendicular to the interfaces due to the spin splitting of the chemical potential. The spin-flip scattering determines the spin-diffusion length—a scale that controls the distance over which Mott's two-current model holds. The model of Valet and Fert predicts that at large layer thicknesses, CPP GMR decays exponentially, the length scale of this decay being determined by the spin diffusion length.

Spin diffusion is an important process, which controls other transport phenomena, as is discussed in other chapters of this book. Several factors affect the spin diffusion length, such as the spin–orbit interaction and scattering by noncollinear magnetic moments in magnetic materials. The spin-diffusion length can vary largely from material to material, ranging from microns in normal metals like Cu or Al, to tens nanometers in elemental magnetic metals like Co and 5 nm in magnetic alloys like permalloy (see Chapter 4 for further discussion).

The microscopic description of spin diffusion should include a realistic electronic band structure of the system and treat explicitly the microscopic mechanism responsible for spin flip. For example, the spin–orbit interaction couples the electronic and spin degrees of freedom. Although the spin–orbit interaction is relatively small in the transition 3*d* metals, it can affect transport properties of bulk alloys, especially when heavy elements like gold or platinum are used as a spacer layer or introduced as impurities. Progress in developing a fully relativistic theory of transport may help to elucidate the influence of spin–orbit scattering and to calculate the spin-diffusion length in those cases when it originates from the spin–orbit interaction.

If magnetic moments in a magnetic metal are misaligned, the two spin channels are no longer independent. Spin mixing in the conductance can occur due to spin disorder at the interfaces. These misoriented moments contribute to spin-flip scattering, resulting in the reduction of GMR. A relevant process is electron–magnon scattering. At high temperatures when spin fluctuations become sizeable the electron–magnon scattering reduces the transport spin polarization. The role of this contribution at moderate temperatures is an important issue, which requires further investigations.

Acknowledgment

E.Y.T. thanks the Materials Research Institute at the Tohoku University for hospitality during his visit in May–June 2009.

References

1. M. N. Baibich, J. M. Broto, A. Fert et al., Giant magnetoresistance of (001)Fe/(001)Cr magnetic superlattices, *Phys. Rev. Lett.* **61**, 2472–2475 (1988).
2. G. Binasch, P. Grünberg, F. Saurenbach, and W. Zinn, Enhanced magnetoresistance in layered magnetic structures with antiferromagnetic interlayer exchange, *Phys. Rev. B* **39**, 4828–4830 (1989).
3. A. Fert and P. Bruno, Interlayer coupling and magnetoresistance in multilayers, in *Ultrathin Magnetic Structures II: Measurement Techniques and Novel Magnetic Properties*, A. C. Bland and B. Heinrich (Eds.), Springer, Berlin, Germany, pp. 82–116, 1994.
4. P. M. Levy, Giant magnetoresistance in magnetic layered and granular materials, in *Solid State Physics*, vol. 47, H. Ehrenreich and D. Turnbull (Eds.), Academic Press, Boston, MA, pp. 367–462, 1994.
5. M. A. M. Gijs and G. E. W. Bauer, Perpendicular giant magnetoresistance of magnetic multilayers, *Adv. Phys.* **46**, 285–445 (1997).
6. A. Barthélémy, A. Fert, and F. Petroff, Giant magnetoresistance in magnetic multilayers, in *Handbook of Magnetic Materials*, K. H. J. Buschow (Ed.), Elsevier, Amsterdam, the Netherlands, pp. 1–96, 1999.

7. P. M. Levy and I. Mertig, Theory of giant magnetoresistance, in *Spin Dependent Transport in Magnetic Nanostructures*, S. Maekawa and T. Shinjo (Eds.), CRC Press, Boca Raton, FL, pp. 47–112, 2002.
8. E. Y. Tsymbal and D. G. Pettifor, Perspectives of giant magnetoresistance, in *Solid State Physics*, H. Ehrenreich and F. Spaepen (Eds.), vol. 56, Academic Press, San Diego, CA, pp. 113–239, 2001.
9. N. F. Mott, The resistance and thermoelectric properties of the transition metals, *Proc. Roy. Soc.* **156**, 368–382 (1936); N. F. Mott, Electrons in transition metals, *Adv. Phys.* **13**, 325–422 (1964).
10. J. M. George, L. G. Pereira, A. Barthélémy et al., Inverse spin-valve-type magnetoresistance in spin engineered multilayered structures, *Phys. Rev. Lett.* **72**, 408–411 (1994).
11. N. W. Ashcroft and N. D. Mermin, *Solid State Physics*, Saunders, Philadelphia, PA, 1976.
12. I. A. Campbell and A. Fert, Transport properties of ferromagnets, in *Ferromagnetic Materials*, E. P. Wohlfarth (Ed.), North Holland, Amsterdam, the Netherlands, pp. 747–804, 1982.
13. R. Kubo, Statistical-mechanical theory of irreversible processes. I. General theory and simple applications to magnetic and conduction problems, *J. Phys. Soc. Jpn.* **12**, 570–586 (1957).
14. R. Landauer, Spatial variation of currents and fields due to localized scatterers in metallic conduction, *IBM J. Res. Dev.* **1**, 223–231 (1957); R. Landauer, Spatial variation of currents and fields due to localized scatterers in metallic conduction, *IBM J. Res. Dev.* **32**, 306–316 (1998).
15. L. V. Keldysh, Diagram technique for nonequilibrium processes, *Sov. Phys. JETP* **20**, 1018–1026 (1965).
16. G. D. Mahan, *Many-Particle Physics*, Plenum Press, New York, 1990.
17. S. Datta, *Electronic Transport in Mesoscopic Systems*, Cambridge University Press, Cambridge, U.K., 1995.
18. J. Rammer and H. Smith, Quantum field-theoretical methods in transport theory of metals, *Rev. Mod. Phys.* **58**, 323–359 (1986).
19. D. A. Greenwood, The Boltzmann equation in the theory of electrical conduction in metals, *Proc. Phys. Soc. London* **71**, 585–596 (1958).
20. H. Ehrenreich and L. M. Schwartz, The electronic structure of alloys, in *Solid State Physics*, H. Ehrenreich, F. Seitz, and D. Turnbull (Eds.), vol. 31, Academic Press, New York, pp. 149–286, 1976.
21. D. S. Fisher and P. A. Lee, Relation between conductivity and transmission matrix, *Phys. Rev. B* **23**, 6851–6854 (1981).
22. P. A. Lee and D. S. Fisher, Anderson localization in two dimensions, *Phys. Rev. Lett.* **47**, 882–885 (1981).
23. B. Dieny, Classical-theory of giant magnetoresistance in spin-valve multilayers: Influence of thicknesses, number of periods, bulk and interfacial spin-dependent scattering, *J. Phys.: Condens. Mater.* **4**, 8009–8020 (1992); B. Dieny, Quantitative interpretation of giant magnetoresistance properties of permalloy-based spinvalve structures, *Europhys. Lett.* **17**, 261–267 (1992).
24. K. Fuchs, The conductivity of thin metallic films according to the electron theory of metals, *Math. Proc. Cambridge Phil. Soc.* **34**, 100–108 (1938); E. H. Sondheimer, The mean free path of electrons in metals, *Adv. Phys.* **1**, 1–42 (1952).
25. R. E. Camley and J. Barnas, Theory of giant magnetoresistance effects in magnetic layered structures with antiferromagnetic coupling, *Phys. Rev. Lett.* **63**, 664–667 (1989).
26. J. Barnas, A. Fuss, R. E. Camley, P. Grünberg, and W. Zinn, Novel magnetoresistance effect in layered magnetic structures: Theory and experiment, *Phys. Rev. B* **42**, 8110–8120 (1990).
27. B. L. Johnson and R. E. Camley, Theory of giant magnetoresistance effects in Fe/Cr multilayers: Spin-dependent scattering from impurities, *Phys. Rev. B* **44**, 9997–10002 (1991).
28. A. Barthélémy and A. Fert, Theory of the magnetoresistance in magnetic multilayers: Analytical expressions from a semiclassical approach, *Phys. Rev. B* **43**, 13124–13129 (1991).
29. P. M. Levy, S. Zhang, and A. Fert, Electrical conductivity of magnetic multilayered structures, *Phys. Rev. Lett.* **65**, 1643–1646 (1990); S. Zhang, P. M. Levy, and A. Fert, Conductivity and magnetoresistance of magnetic multilayered structures, *Phys. Rev. B* **45**, 8689–8702 (1992).
30. H. E. Camblong and P. M. Levy, Novel results for quasiclassical linear transport in metallic multilayers, *Phys. Rev. Lett.* **69**, 2835–2838 (1992); H. E. Camblong and P. M. Levy, Magnetic multilayers: Quasiclassical transport via the Kubo formula, *J. Appl. Phys.* **73**, 5533–5535 (1993).
31. W. H. Butler, X. G. Zhang, and J. M. MacLaren, Solution to the Boltzmann equation for layered systems for current perpendicular to the planes, *J. Appl. Phys.* **87**, 5173–5175 (2000); W. H. Butler, X. G. Zhang, and J. M. MacLaren, Semiclassical theory of spin-dependent transport in magnetic multilayers. *Superconductivity* **13**, 221–238 (2000).
32. A. Vedyayev, B. Dieny, and N. Ryzhanova, Quantum theory of giant magnetoresistance of spin-valve sandwiches, *Europhys. Lett.* **19**, 329–335 (1992).
33. B. Dieny, A. Vedyayev, and N. Ryzhanova, Comparison of semi-classical and real-space quantum theories of giant magnetoresistance, *J. Magn. Magn. Mater.* **121**, 366–370 (1993).
34. B. R. Bulka and J. Barnas, Parallel magnetoresistance in magnetic multilayers, *Phys. Rev. B* **51**, 6348–6357 (1995).
35. J. Barnas and Y. Bruynseraede, Electronic transport in ultrathin magnetic multilayers, *Phys. Rev. B* **53**, 5449–5460 (1996).
36. A. Vedyayev, C. Cowache, N. Ryzhanova, and B. Dieny, Quantum effects in the giant magnetoresistance of magnetic multilayered structures, *J. Phys.: Condens. Mater.* **5**, 8289–8304 (1993); A. Vedyayev, M. Chshiev, and B. Dieny, Quantum effects in giant magnetoresistance due to interfaces in magnetic sandwiches, *J. Magn. Magn. Mater.* **184**, 145–154 (1998).
37. X.-G. Zhang and W. H. Butler, Conductivity of metallic films and multilayers, *Phys. Rev. B* **51**, 10085–10103 (1995).

38. W. A. Harrison, *Electronic Structure and the Properties of Solids*, Dover, New York, 1989; D. G. Pettifor, *Bonding and Structure of Molecules and Solids*, Clarendon Press, Oxford, 1995.
39. W. Y. Asano, A. Oguri, and S. Maekawa, Parallel and perpendicular transport in multilayered structures, *Phys. Rev. B* **48**, 6192–6198 (1993).
40. H. Itoh, J. Inoue, and S. Maekawa, Theory of giant magnetoresistance for parallel and perpendicular currents in magnetic multilayers, *Phys. Rev. B* **51**, 342–352 (1995).
41. B. Velicky, Theory of electronic transport in disordered binary alloys: Coherent-potential approximation, *Phys. Rev.* **184**, 614–627 (1969).
42. T. N. Todorov, T. E. Y. Tsymbal, and D. G. Pettifor, Giant magnetoresistance: Comparison of band-structure and interfacial-roughness contributions, *Phys. Rev. B* **54**, R12685–R12688 (1996).
43. R. Schad, P. Beliën, G. Verbanck et al., Giant magnetoresistance dependence on the lateral correlation length of the interface roughness in magnetic superlattices, *Phys. Rev. B* **59**, 1242–1248 (1999).
44. Fermi Surfaces, Technical University Dresden, Institute for Theoretical Physics, http://www.physik.tu-dresden.de/~fermisur/, 1997.
45. R. M. Martin, *Electronic Structure: Basic Theory and Practical Methods*, Cambridge University Press, New York, 2004.
46. K. M. Schep, P. J. Kelly, and G. E. W. Bauer, Giant magnetoresistance without defect scattering, *Phys. Rev. Lett.* **74**, 586–589 (1995); K. M. Schep, P. J. Kelly, and G. E. W. Bauer, Ballistic transport and electronic structure, *Phys. Rev. B* **57**, 8907–8926 (1998).
47. J. Mathon, M. Villeret, and H. Itoh, Selection rules for oscillations of the giant magnetoresistance with nonmagnetic spacer layer thickness, *Phys. Rev. B* **52**, R6983–R6986 (1995).
48. J. Mathon, A. Umerski, and M. Villeret, Oscillations with Co and Cu thickness of the current-perpendicular-to-plane giant magnetoresistance of a Co/Cu/Co(001) trilayer, *Phys. Rev. B* **55**, 14378–14386 (1997).
49. T. Oguchi, Fermi-velocity effect on magnetoresistance in Fe/transition-metal multilayers, *J. Magn. Magn. Mater.* **126**, 519–520 (1993); T. Oguchi, Magnetoresistance in magnetic multilayers: A role of Fermi velocity, *Mater. Sci. Eng.* **31**, 111–116 (1995).
50. H. Itoh, J. Inoue, and S. Maekawa, Electronic structure and transport properties in magnetic superlattices, *Phys. Rev. B* **47**, 5809–5818 (1993).
51. I. Mertig, Transport properties of dilute alloys, *Rep. Prog. Phys.* **62**, 237–276 (1999).
52. I. Mertig, R. Zeller, and P. H. Dederichs, Ab initio calculations of residual resistivities for dilute Ni alloys, *Phys. Rev. B* **47**, 16178–16185 (1993).
53. P. Zahn, I. Mertig, M. Richer, and H. Eschrig, Ab initio calculations of the giant magnetoresistance, *Phys. Rev. Lett.* **75**, 2996–2999 (1995); P. Zahn, J. Binder, I. Mertig, R. Zeller, and P. H. Dederichs, Origin of giant magnetoresistance: Bulk or interface scattering, *Phys. Rev. Lett.* **80**, 4309–4312 (1998); J. Binder, P. Zahn, and I. Mertig, Ab initio calculations of giant magnetoresistance, *J. Appl. Phys.* **87**, 5182–5184 (2000).
54. W. H. Butler, X.-G. Zhang, D. M. C. Nicholson, and J. M. MacLaren, Spin-dependent scattering and giant magnetoresistance, *J. Magn. Magn. Mater.* **151**, 354–362, 1995.
55. C. Blaas, P. Weinberger, L. Szunyogh, P. M. Levy, and C. Sommers, Ab initio calculations of magnetotransport for magnetic multilayers, *Phys. Rev. B* **60**, 492–501 (1999).
56. O. Bengone, O. Eriksson, S. Mirbt, I. Turek, J. Kudrnovsky, and V. Drchal, Origin of the negative giant magnetoresistance effect in $Co_{1-x}Cr_x$/Cu/Co (111) trilayers, *Phys. Rev. B* **69**, 092406 1-4 (2004).
57. C. Vouille, A. Barthélémy, F. Elokan Mpondo et al., Microscopic mechanisms of giant magnetoresistance, *Phys. Rev. B* **60**, 6710–6722 (1999).
58. J. Kudrnovsky, V. Drchal, C. Blaas, P. Weinberger, I. Turek, and P. Bruno, Ab initio theory of perpendicular magnetotransport in metallic multilayers, *Phys. Rev. B* **62**, 15084–15095 (2000).
59. E. Y. Tsymbal and D. G. Pettifor, The influence of spin-independent disorder on giant magnetoresistance, *J. Phys.: Condens. Matter* **8**, L569–L575 (1996); E. Y. Tsymbal and D. G. Pettifor, Effects of band structure and spin-independent disorder on conductivity and giant magnetoresistance in Co/Cu and Fe/Cr multilayers, *Phys. Rev. B* **54**, 15314–15329 (1996); E. Y. Tsymbal and D. G. Pettifor, Importance of spin-independent scattering potentials in giant magnetoresistance, *J. Appl. Phys.* **81**, 4579–4581 (1997).
60. W. H. Butler, X.-G. Zhang, D. M. C. Nicholson, and J. M. MacLaren, First-principles calculations of electrical conductivity and giant magnetoresistance of Co/Cu/Co spin valves, *Phys. Rev. B* **52**, 13399–13410 (1995).
61. R. Q. Hood and L. M. Falicov, Boltzmann-equation approach to the negative magnetoresistance of ferromagnetic–normal-metal multilayers, *Phys. Rev. B* **46**, 8287–8296 (1992); R. Q. Hood, L. M. Falicov, and D. R. Penn, Effects of interfacial roughness on the magnetoresistance of magnetic metallic multilayers, *Phys. Rev. B* **49**, 368–377 (1994).
62. M. D. Stiles, Spin-dependent interface transmission and reflection in magnetic multilayers (invited), *J. Appl. Phys.* **79**, 5805–5810 (1996); M. D. Stiles and D. R. Hamann, Ballistic electron transmission through interfaces, *Phys. Rev. B* **38**, 2021–2037 (1988); M. D. Stiles, Electron transmission through $NiSi_2$–Si interfaces, *Phys. Rev. B* **40**, 1349–1352 (1989).
63. W. H. Butler, X.-G. Zhang, D. M. C. Nicholson, T. C. Schulthess, and J. M. MacLaren, Giant magnetoresistance from an electron waveguide effect in cobalt–copper multilayers, *Phys. Rev. Lett.* **76**, 3216–3219 (1996).
64. K. Xia, M. Zwierzycki, M. Talanana, P. J. Kelly, and G. E. W. Bauer, First-principles scattering matrices for spin transport, *Phys. Rev. B* **73**, 064420 1-26 (2006).

65. J. Barnas and A. Fert, Interface resistance for perpendicular transport in layered magnetic structures, *Phys. Rev. B* **49**, 12835–12838 (1994); J. Barnas and A. Fert, Interfacial scattering and interface resistance for perpendicular transport in magnetic multilayers, *J. Magn. Magn. Mater.* **136**, 260–268 (1994).
66. S. Zhang and P. M. Levy, Interplay of the specular and diffuse scattering at interfaces of magnetic multilayers, *Phys. Rev. B* **57**, 5336–5339 (1998).
67. K. M. Schep, J. B. A. N. van Hoof, P. J. Kelly, G. E. W. Bauer, and J. E. Inglesfield, Interface resistances of magnetic multilayers, *Phys. Rev. B* **56**, 10805–10808 (1997).
68. K. Xia, P. J. Kelly, G. E. W. Bauer, I. Turek, J. Kudrnovsky, and V. Drchal, Interface resistance of disordered magnetic multilayers, *Phys. Rev. B* **63**, 064407 1-4 (2001).
69. E. Y. Tsymbal, Effect of disorder on perpendicular magnetotransport in Co/Cu multilayers, *Phys. Rev. B* **62**, R3608–R3611 (2000).
70. R. J. Baxter, D. G. Pettifor, and E. Y. Tsymbal, Interface proximity effects in current-perpendicular-to-plane magnetoresistance, *Phys. Rev. B* **71**, 024415 1–6 (2005).
71. D. Bozec, M. A. Howson, B. J. Hickey et al., Mean free path effects on the current perpendicular to the plane magnetoresistance of magnetic multilayers, *Phys. Rev. Lett.* **85**, 1314–1317 (2000).
72. W. C. Chiang, C. Ritz, K. Eid, R. Loloee, W. P. Pratt, and J. Bass, Search for mean-free-path effects in current-perpendicular-to-plane magnetoresistance, *Phys. Rev. B* **69**, 184405 1–10 (2004).
73. T. Valet and A. Fert, Theory of the perpendicular magnetoresistance in magnetic multilayers, *Phys. Rev. B* **48**, 7099–7113 (1993).

6

Spin Injection, Accumulation, and Relaxation in Metals

Mark Johnson
Naval Research Laboratory

6.1 Introduction

The main focus of this chapter are early developments of spin injection in metals and the related phenomena prior to the year 2000. More recent related developments are discussed in Chapters 31 and 32.

There are many topics that relate to the effects of carrier spin on transport in electronic systems. Among the most fundamental, spin-dependent tunneling (SDT) [1] (see also Chapter 9) has the longest history and offers a technique for measuring the spin polarization of electric current in a ferromagnet. Magnetic tunnel junctions (MTJs) [2–4] form the device family with the largest magnetoresistance ratio (MR) and with the highest promise for applications (see also Chapter 34). Giant magnetoresistance (GMR) [5,6] (see also Chapter 4) provides a descriptive understanding of spin-dependent scattering in ferromagnets and in thin-film ferromagnet/nonmagnetic (*F/N*) metal multilayers. The *Spin Injection* phenomenology [7] has a history that is second to SDT. Developed for the study of *F/N* metal interfacial spin transport and the study of spin-relaxation processes, *Spin Injection* introduced a number of ideas and techniques that have been commonly adopted. Researchers today invoke fundamental concepts such as spin injection, spin accumulation, spin-polarized currents in ferromagnet/nonmagnet structures, and pure spin currents. All these concepts were described, defined, and demonstrated in a series of experimental and theory papers in the mid-1980s [7–10].

The spin-injection experiments were the first to use a ferromagnet/nonmagnetic metal/ferromagnet (*F1/N/F2*) device structure, using a geometry that has become known as the lateral spin valve (LSV) and the first to demonstrate a spin-mediated magnetoresistance (at temperatures above 4K) that was controlled by using a magnetic field to manipulate the magnetization orientations $\vec{M}_1$ and $\vec{M}_2$ of *F*1 and *F*2 between parallel and antiparallel configurations [7]. The *F1/N/F2* device family was later expanded by the introduction of the current-in-plane (CIP) spin valve [5,11] and the current-perpendicular-to-the-plane (CPP) spin valve [12]. The spin-injection theory was the first to use unique conductances, $g_\uparrow$ and $g_\downarrow$, for up and down-spin electrons in both ferromagnetic and nonmagnetic materials in a *F/N/F* structure. Finally, the spin-injection technique has been successful at studies of spin transport in novel categories of materials, enabling measurements of spin relaxation times, τ_s, in a variety of materials systems. While transmission electron spin resonance (TESR) had been used to measure transverse spin relaxation times, T_2, of order ns or longer in bulk metals at low temperature [13], spin injection has been capable of measuring relaxation times of 0.1 ns and shorter and has been applied to metal, semiconductor, semimetal, and superconducting films. The treatment of Spin Injection in this chapter will cover these themes.

The concept that an electric current in a metal could be spin-polarized is nearly 70 years old. Mott calculated scattering rates for up-spin and down-spin itinerant electrons in the exchange-split *d*-band of transition metal ferromagnets [14] and deduced that the electrons carrying an electrical current in a ferromagnetic metal had a net spin polarization *P*,

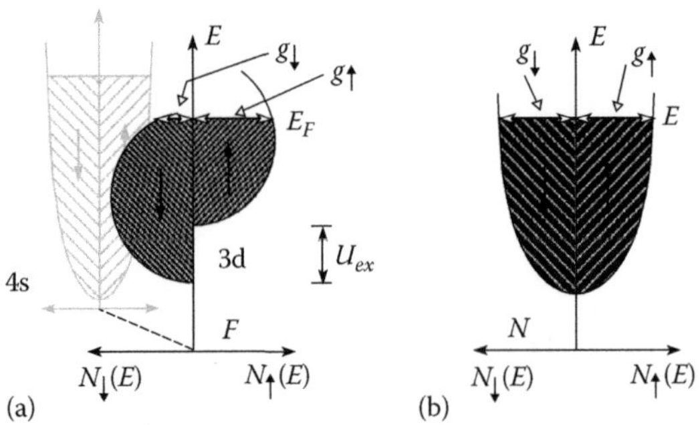

FIGURE 6.1 Density of states as a function of energy for (a) the exchange split 3d band of a transition metal ferromagnet and (b) a free electron nonmagnetic metal, $T=0$. (From Johnson, M., *Magnetoelectronics*, Copyright 2004, Figure 1.8 by permission of Oxford University Press, Elsevier, Oxford, U.K.)

$$P = \frac{(J_\uparrow - J_\downarrow)}{(J_\uparrow + J_\downarrow)} \neq 0, \tag{6.1}$$

where $J_\uparrow$ and $J_\downarrow$ are spin-dependent partial currents that sum to form the total electric current, $J=J_\uparrow+J_\downarrow$.

This concept can be described with the aid of Figure 6.1a, which shows the density of states as a function of energy, $N(E)$, for the 3d band of a transition metal ferromagnet F such as Ni, Fe, Co, or an alloy of these or similar constituents. Because of exchange splitting U_{ex}, the majority, down-spin subband is filled with more electrons than the minority, up-spin subband, and this difference accounts for the equilibrium, spontaneous magnetization of F. Another consequence of U_{ex} is that the density of states at the Fermi level, $N(E_F)$, is different for the down- and up-spin subbands, $N_\downarrow(E_F)\neq N_\uparrow(E_F)$. The density of states at E_F of the minority spin subband is larger than that of the majority spin subband in the simplified schematic representation of Figure 6.1a.

The conductivity of a metal, $g=1/\rho$ with ρ the resistivity (note that notations g and σ are both used for conductivity, and σ is used elsewhere in this book), can be related to $N(E_F)$ by an Einstein relation,

$$g = e^2 DN(E_F),$$

where
D is the diffusion constant
e is the electron charge

Mott showed that the scattering rates for up- and down-spin subbands were different, that scattering between subbands was relatively rare, and therefore that electrons in the two subbands do not intermix. It follows that a conductance g_s can be defined individually for each spin subband,

$$g_{\uparrow,\downarrow} = e^2 DN_{\uparrow,\downarrow}(E_F), \tag{6.2}$$

where D is assumed to be the same (in metals) for up- and down-spin carriers. It is clear from Figure 6.1a that $g_\uparrow\neq g_\downarrow$, and Equation 6.1 can be rewritten as

$$P = \frac{(g_\uparrow - g_\downarrow)}{(g_\uparrow + g_\downarrow)} \neq 0, \tag{6.3}$$

by noting that $J\propto g$. The conceptually important point is that the electric current in a ferromagnet is the sum of spin-up and spin-down partial currents, and these partial currents can be modeled as independent. This idea is often called the "two current model" of charge and spin transport.

A transition metal ferromagnet also has a free electron 4s band that intersects the Fermi surface, depicted in Figure 6.1a with light shading. Since an electric current in F involves all electrons within k_BT of the Fermi surface, where k_B is Boltzmann's constant and T is temperature, an accurate model of spin-polarized currents must take account of contributions from the unpolarized 4s band. It is common practice for experimental results to be discussed in a way that neglects the 4s band, and this practice will be used within the pedagogical context of this chapter. More formally, an acceptable model can be based on combining the 4s and 3d bands into a single, hybridized spin-polarized band [15]. Figure 6.1b shows a free electron density of states diagram characteristic of a nonmagnetic metal N. For a nonmagnetic metal in equilibrium, the Pauli exclusion principle demands that the number of up spins equals that of the down spins. An early contribution of spin-injection theory [9] is the observation that spin-relaxation scattering rates in nonmagnetic materials are very low, and the subsequent assertion that a conductance g_s can be defined individually for each spin subband s in a nonmagnetic material.

Many of the advances in the field of spin-polarized transport in the solid state have involved tunneling, and experimental studies began with superconducting tunneling spectroscopy about 30 years ago. As reviewed in Chapter 9, planar tunnel junctions were fabricated using thin, superconducting aluminum films, aluminum oxide tunnel barriers, and a ferromagnetic metal counterelectrode. By deconvolving the measured tunneling conductance, Tedrow and Meservey [1] could deduce the net polarization P of the tunnel current crossing the F/S interface, directly proportional to the asymmetry of the density of states of F at the Fermi level,

$$P = \frac{[N_\uparrow(E_F) - N_\downarrow(E_F)]}{[N_\uparrow(E_F) + N_\downarrow(E_F)]}. \tag{6.4}$$

These experiments provided both an experimental confirmation of Mott's assertion and a quantitative estimate of the polarization, $P\approx 10\%–45\%$ for polycrystalline films of nickel, iron, or cobalt. They also demonstrated that polarized current could tunnel across a barrier at a boundary of F and maintain its polarization. A few years later, the SDT experiments of Julliere [2] and Maekawa and Gafvert [3] introduced the sandwich MTJ device structure, $F1/I/F2$ where I is an insulating tunnel barrier that is most commonly used in technological applications. Although the first planar MTJ measurements were performed at cryogenic temperatures, spin-polarized scanning tunneling microscopy (SP STM) experiments gave very good SDT results at room temperature [16,17] and encouraged the development of planar MTJs that could be operated at room temperature [4]. These MTJs have been the dominant device for magnetic random access memory (MRAM), which is the most important integrated electronics application of the physics of spin-dependent transport (refer to Chapter 34). However, the broad approach of spin injection/detection and geometries in the LSV family continue to be important for novel applications (refer to Chapter 38) and for spin-transport studies of novel materials.

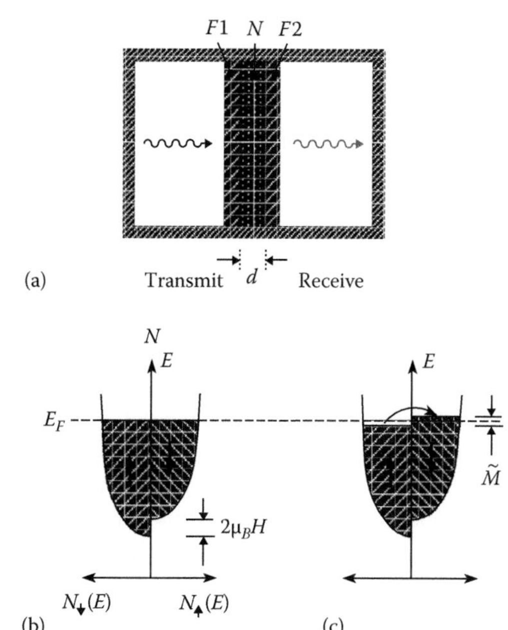

FIGURE 6.2 (a) Schematic cross-section of a TESR experiment. Resonantly tuned r_f transmit and receive cavities are separated by a nonmagnetic N metal foil sample. Transmission was enhanced when the foil was coated by thin ferromagnetic films, $F1$ and $F2$, on either side. Note that the sketch is not to scale. (b) Density of states representation of Pauli paramagnetic splitting in a N foil in the presence of external magnetic field H. (c) Resonant absorption of r_f photons creates a nonequilibrium population of spin-polarized electrons in N.

6.1.1 Transmission Electron Spin Resonance

The tunneling experiments of Tedrow and Meservey led to a fundamental question: Could a spin-polarized current, equivalently denoted as a magnetization current J_M of oriented dipoles, enter a nonmagnetic material and maintain its polarization over a penetration depth of significant length? The prevailing view at the time was based on the observation that the RKKY interaction, a coupling between dilute magnetic impurities in a nonmagnetic metal host that is mediated by conduction electrons, has a length scale of order 1 nm. Argument by analogy suggested that a spin-polarized current injected into a nonmagnetic metal N would quickly lose its polarization. The contrary view was presented by Aronov, who predicted that the penetration depth was relatively long and proposed that current J_M could be injected successfully into nonmagnetic metals [18], semiconductors [19], and superconductors [20].

At about the same time, Silsbee was studying spin diffusion in nonmagnetic metals using TESR [21]. Stimulated by Meservey's observation of spin-polarized tunnel currents, thin ferromagnetic films $F1$ and $F2$ were deposited on either side of a high-purity metal foil N. The trilayer sample was mounted in a spectrometer where the foil separated the "transmit" and "receive" cavities (Figure 6.2a). As discussed below, the transmission amplitude for samples coated with ferromagnetic films was substantially enhanced when compared to samples with no ferromagnetic coatings. This suggested that spin-polarized electrons were transported across the F/N interface and diffused into N, and these results motivated the first spin-injection experiment.

6.1.1.1 Nonequilibrium Spin Populations

The physics of spin accumulation and spin diffusion are closely related to the principles of electron spin resonance, and there are similarities in the techniques for measuring spin-diffusion lengths and spin relaxation times. Beginning with a reminder of basic definitions, the magnetic dipole moment of a conduction electron is related to its spin-angular momentum, $\vec{S}$, by

$$\vec{\mu} = \frac{-ge}{mc}\vec{S}, \tag{6.5}$$

where

- $-e$ is the electron charge
- m is the electron mass
- c is the speed of light
- $g=2.0$ for free electrons

Choosing a coordinate system that allows measurement of spin to be quantized along a convenient axis, the electrons are described as having up or down spin; the angular momentum is $s_z = s\hbar$, where $s = \pm 1/2$ is the spin-quantum number, and it follows that

$$\mu = \pm\frac{e}{2mc}\hbar = \pm\mu_B, \tag{6.6}$$

with μ_B being the Bohr magneton.

If a magnetic field H is applied along the quantization axis, each electron acquires a Zeeman energy associated with its magnetic moment, $\pm\mu_B H$. Figure 6.2b depicts the energy band of free electrons in a nonmagnetic metal in the presence of a magnetic field. The relative shift of the two spin subbands is called Pauli paramagnetism. The subband with moments oriented parallel (antiparallel) with the field H have energy lower (higher) by $\mu_B H$. The net splitting between the subbands, $\Delta E = 2\mu_B H$, is equivalent to the energy required for a transition from one spin subband to the other. The transition energy can be supplied by absorption of a photon of energy

$$\hbar\omega = \frac{e}{mc}\hbar H \quad \text{or} \quad \omega = -\gamma H = \frac{e}{mc}H, \tag{6.7}$$

where γ is the gyromagnetic ratio. As depicted in Figure 6.2c, spin-up electrons of relatively low energy absorb photons and move to higher energy spin-down states. The result is a nonequilibrium distribution of spin-polarized electrons, $\tilde{M}$, near the Fermi level. The electrochemical potential of down-spin electrons is raised by the addition of carriers from the up-spin subband, and the electrochemical potential of up-spin electrons is depleted by the loss of carriers. One can say that the down-spin subband is pumped to a nonequilibrium level by the resonant absorption of photons. This is a nonequilibrium process because the spin-subband electrochemical potentials relax to equilibrium in the absence of a flux of appropriate photons. In Figure 6.2c, the energy height of $\tilde{M}$ is less than the Pauli paramagnetic splitting because of spin relaxation. Spin-relaxation processes, described later in this chapter, are reverse transitions and act to limit and deplete the population of spin-down electrons. The resulting nonequilibrium magnetization is a balance between the spin pumping of resonant absorption and the spin depletion caused by spin relaxation. The nonequilibrium population is often called "spin accumulation," and similar populations appear in optical orientation experiments [22].

In TESR, microwave photons are supplied to a "transmit" cavity (Figure 6.2a) and absorbed within a skin depth on one side of a foil sample. The nonequilibrium population of spin-polarized electrons diffuses across the foil and, within a skin depth of the other side, spin relaxation results in reverse transitions that emit microwave photons into the "detect" cavity. An experimentally convenient frequency range for a microwave source and detector is called X-band, $\nu \approx 10\,\text{GHz}$, $\lambda \approx 3\,\text{cm}$. Using Equation 6.7, the resonant condition for free electrons is $\nu = 2.8 \times 10^6\,\text{Hz/Oe}$, and X-band TESR experiments require a resonant field of about $H = 3600\,\text{Oe}$.

6.1.1.2 Spin Relaxation and Diffusion

The transmitted spin resonance "signal" detected in the receive cavity is described as an amplitude of microwave radiation that varies as a function of magnetic field, $A(H)$, because typically the frequency is fixed and the magnetic field is varied. The signal has maximum amplitude centered at the resonant frequency "line," and the characteristic line shape of $A(H)$ can be related to the spin-transport processes that occur in the metal sample. Examples are shown in the schematic drawings of Figure 6.3. The panels on the left depict a cross section of a sample N of thickness d. The experiment is designed such that a population of nonequilibrium spin-up electrons is introduced within a skin depth on the left side of N, the external field H is perpendicular to the plane of the cross section, and the "receive" cavity is on the right side of N. The panels on the right show the line shapes $A(H)$ of resonant signals for several different conditions.

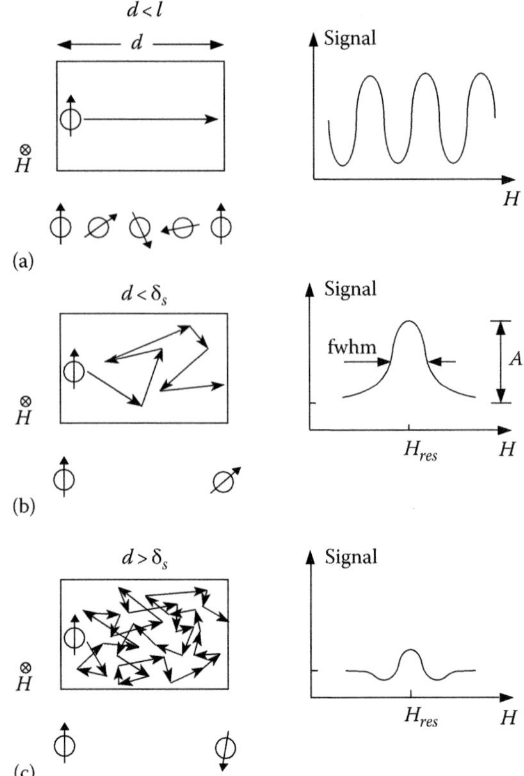

FIGURE 6.3 Resonant emission of photons detected by the receive cavity as a function of field H for several different conduction electron transport conditions across a foil with thickness d. (a) Ballistic transport results in Larmor waves. (b) Diffusive transport, $d < \delta_s$. Thin limit. (c) Diffusive transport, $d > \delta_s$. Thick limit. Left panels: in cross-section of foil, sketch shows schematic description of trajectory of a spin-polarized electron. External field H is perpendicular to the plane of the drawing (parallel to the plane of the foil). Right panels: shape of signal detected at receiver.

Experiments require low-spin-scattering rates, $1/T_2$ where T_2 is the spin relaxation time for N, and the bulk metal foil samples at low temperature have a very high conductivity. In Figure 6.3a, the conductivity is so high that the foil thickness is less than the electron mean free path, $d < \lambda$. The spin-polarized electrons introduced on one side travel with ballistic trajectories, at the Fermi velocity υ_F, across the foil. Throughout the foil, and in particular on the other side, there is a nonequilibrium population of "hot" electrons, near the Fermi surface, having electron spin with the same phase. As the ballistic electrons scatter off the surface on the other side, there is some probability that the scattering event will flip the electron spin by π radians, diminishing the nonequilibrium population and emitting a microwave photon in the process. If the field H were held at constant value, the ballistic electrons reaching the other side would not change their phase and the spin would be up. As the magnetic field is monotonically varied, the spin phase changes at a constant rate along the ballistic trajectory. The nonequilibrium population at the right edge will have spin up (down) when the magnetic field, Fermi velocity, and trajectory length correspond to phase rotations of $n\pi$, with n even (odd). The "receive" signal amplitude $A(H)$ takes the shape of repeating oscillations with undiminished amplitude, called Larmor waves [23] because the spin precession occurs at the Larmor frequency.

Larmor waves are not commonly observed for several reasons. A fundamental reason relates to the effects of spin relaxation. A variety of physical mechanisms can result in changes of the spin orientation of an electron, and the most common mechanism is discussed in Section 6.1.1.3. Spin relaxation can be discussed by introducing the mean time between the scattering events that affect the spin. For TESR, the transverse relaxation time, T_2, measures the mean time it takes for the projection of electron spin in the plane perpendicular to applied field to become random. Experiments such as direct magnetization relaxation measurements [24] are described by the longitudinal relaxation time, T_1, defined as a measure of the mean time it takes for the projection of electron spin along the axis parallel to applied field to become random. It is commonly assumed that $T_1 = T_2$ in metals [25] and the distinction between them is not important. It is also common to describe spin relaxation in metals by using a parameter τ_s, called a "spin-flip time," which can be defined as the mean time it takes for a spin of a given orientation to change its orientation by π radians. The randomization of spin orientation, represented by a phase change of $\pi/2$, is often described by τ_s, but this carries some ambiguity. In this chapter, spin relaxation is characterized by time T_2 unless otherwise noted.

In a heuristic picture, one can describe spin relaxation in the following way. In every scattering event, the collision results in a random and small change of phase, $\delta\theta$. After a large number N_s of such scattering events, the phase changes $\delta\theta$ accumulate, and the spin phase has no relation to the initial orientation; the spin orientation has been completely randomized. For a given material and a given relaxation mechanism, the characteristic number of scattering events is $N_s = 1/\alpha$. An equivalent statement is $\alpha \equiv \tau/T_2$, where τ is the mean electron scattering time given by Mathiesen's rule. Parameter α can be expected to differ for different scattering mechanisms, for example, those associated with phonon, impurity, or surface scattering. However, the crude assumption that α is roughly the same within a given material, for any mechanism, is approximately correct and permits a crude description of spin scattering and diffusion. For many metals, α is the order of 0.001, meaning that the orientation of a spin-polarized electron is randomized after about 1000 scattering events. Described in a detailed model in Section 6.3.2, each scattering event marks a "step" in the path of a conduction electron in its diffusive, random walk motion. The spin-diffusion length δ_s is defined as the average distance the electron diffuses before its spin orientation is randomized, $\delta_s = \sqrt{DT_2}$, where D is the electron diffusion constant. Using $D \sim v_F^2\tau$, δ_s can be estimated as $\delta_s \sim \sqrt{v_F^2\tau T_2} \sim \sqrt{\alpha} v_F \tau \sim 30\ell$.

In Figure 6.3b, the foil thickness is less than the spin-diffusion length, $d < \delta_s$. The nonequilibrium population $\tilde{M}$ is induced near the left surface of N. Since there is no "spin pumping" inside N, the electron distribution is in equilibrium, and, therefore, a gradient of electrochemical potential exists, which drives self-diffusion of the spin-polarized electrons across the foil. All conduction electrons leave the left surface with the same spin phase and acquire precessional phase angle at the same rate. At any given time, some of the electrons reaching the right surface have moved across the foil width in only a few steps, but others have taken many steps to reach the other side. In the limit of negligible field H, all the electrons reach the other side with nearly the original phase (e.g., spin up), and the signal is maximum. As the magnitude of H is increased, the electrons that have spent a longer time in transit have acquired a phase angle large enough that the spin orientation is random. Spin relaxation imposes a limit on the contributions of the electrons with the most phase accumulation. The result is the Lorentzian line shape depicted in the right panel of Figure 6.3b. The half width at half maximum is $H_{hw} = 1/(\gamma T_2)$.

When the foil thickness is larger than a spin-diffusion length (Figure 6.3c), the amplitude of microwave emission detected at the receive cavity is diminished. In this case, very few electrons have trajectories that go straight across the foil. By contrast, the common trajectory has many steps, and many of the electrons precess by π radians or more if they have reached the right side before spin relaxation. The resulting line shape shows a small dip that represents a net negative spin polarization, which then asymptotically approaches zero with further increases of H. These symmetric features are called side lobes of the resonant line.

The rigorous derivation of fitting functions for TESR line shapes involves solving the Bloch equations in a reference frame that is rotating at the precessional Larmor frequency, ω_L. The identical physical processes occur in spin-injection experiments, but they occur at zero frequency. The spin-injection experiment is therefore more easily described, both conceptually and rigorously, and formal solutions will be presented in Section 6.2. However, it should be clear from Figure 6.3b that the relaxation time T_2 can be deduced from analysis of TESR data. The number of results is limited because only a few metals could be prepared with appropriate thickness and conductivity.

6.1.1.3 Elliott–Yafet Spin Relaxation Mechanism

Whereas any scattering event can perturb an orbital state, tipping the spin angular momentum requires a magnetic torque. In the absence of impurities with local magnetic moments, there are few interactions capable of providing such a torque. In metals, the dominant spin-relaxation mechanism is the weakly relativistic spin–orbit interaction, derived independently by Yafet [26] and Elliott [27]. An electron with spin $\vec{s}$ moves in a periodic potential $U(r)$ with a weakly relativistic Fermi velocity v_F (Figure 6.4). In the rest frame of the electron, any scattering event that changes the electron trajectory is a perturbation of

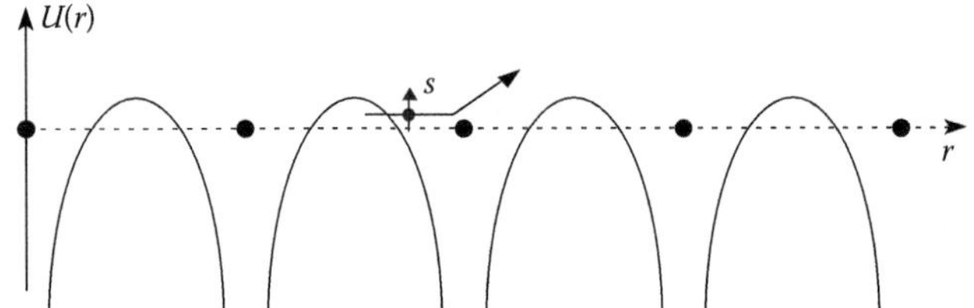

FIGURE 6.4 An electron with spin $\vec{s}$ moves in the periodic potential $U(r)$ of a metal with Fermi velocity v_F. A weakly relativistic spin–orbit interaction can tip the spin orientation. (From *Magnetoelectronics*, Johnson, M., Copyright (2004), with permission from Elsevier, Oxford, U.K., Figure 1.8.)

the periodic electric field, which transforms as a magnetic field. This effective magnetic field "pulse" can exert a torque on the spin. Formally, the effect derives from a term in the Dirac equation [26]

$$\frac{h}{4m^2c^2}\left(\nabla U \times \vec{p}\right)\cdot\vec{s}, \tag{6.8}$$

where
 m is the electron mass
 c is the speed of light

Inspecting Equation 6.8, any scattering event that changes the momentum $\vec{p}$ can couple weakly to the spin. Scattering events with impurities, phonons, grain boundaries, and surfaces can affect $\vec{s}$ with different strength. As mentioned above, a phenomenological parameter, $\alpha = \tau/T_2$, can be used to describe the spin-scattering probability per scattering event, for any given material. In Elliott–Yafet theory, the parameter $c_Y^2 \propto \tau/T_1$ is a property of a given material and is relatively large (small) for high (low) Z elements (recall $T_1 = T_2$ for metals).

The utility of introducing parameter α to describe the probability of flipping an electron spin, per scattering event and averaged over all scattering events, has been discussed. However, a counter example can be used to illustrate the limits of such simple pictures. The ratio $\alpha = \tau/T_2$ has been measured experimentally for bulk aluminum by TESR [28] and by spin injection [7,10], and the value $\alpha = 0.001$ was found by both techniques. Since Elliott–Yafet theory predicts that α should be small for Al, an element with a low value of Z, the measurement of such a relatively large value was a mystery. The contradiction was solved by detailed theoretical research [29]. The Fermi surface of polyvalent aluminum is not spherical, and most of the volume is characterized by a low spin–orbit interaction and a small spin scattering rate, $\alpha \approx c_Y^2 \approx 10^{-4}$. However, a few small pockets exist in which the spin–orbit coupling is quite large and the spin scattering rate is very rapid. Although these pockets are a small portion of the Fermi surface volume, they dominate spin scattering to the extent that the mean spin-scattering rate measured experimentally is indeed an order of magnitude smaller than the rate over most of the volume, arate consistent with Elliott–Yafet theory.

6.2 Models of Spin Injection, Accumulation, and Detection

The spin-injection phenomenology is unusual in that detailed theoretical modeling preceded the experimental results. This section begins with a review of a microscopic transport model that gave quantitatively accurate predictions of the first experimental measurements, and which continues to be used for the analysis of recent spin-injection experiments in semiconductors and mesoscopic devices. The Johnson–Silsbee thermodynamic theory is also reviewed. This treatment permitted a detailed analysis of charge and spin transport across a F/N interface and is also recognized as the earliest contribution to a new field called "spin caloritronics" [30]. A derivation of the line shape of the Hanle effect will be presented in Section 6.3.2, in the same section with experimental results.

6.2.1 Microscopic Transport Model

Pictures of nonequilibrium populations of spin-polarized electrons have been introduced in the context of TESR experiments. When Silsbee et al. [21] coated a metal foil sample with a ferromagnetic film on each surface, a large enhancement of the amplitude of the TESR signal was observed. Silsbee explained these results as an enhanced transfer of spin-polarized electrons across the $F1/N$ interface, along with enhanced transmission at the second, $N/F2$ interface. He proposed that similar transport phenomena would occur in a dc, electrically biased experiment.

Silsbee's idea [31] can be explained with a simple pedagogical geometry (Figure 6.5a) and the microscopic transport model described in Figure 6.5b. A bias current J_e driven through a single domain ferromagnetic film $F1$ and into a nonmagnetic metal sample N carries magnetization across the interface (with unit area A) and into N at the rate J_M. Because current J_M may have arbitrary direction and carry spin magnetization with arbitrary polarization orientation, in general, J_M is a second-rank tensor [8]. A convenient spin quantization axis may be chosen, and then spin polarization has a positive or negative value and $\mathbf{J}_M$ is a vector. For one-dimensional flow (e.g., Figure 6.5a), positive current I_M can be reduced to a scalar.

Referring to Figure 6.5b, there is a finite interface resistance between $F1$ and N, so that $F1$ is biased up by a small voltage V_0, $E_{F;f1,\uparrow} = E_{F;f1,\downarrow} = E_{F;n} + eV_0$. The electric current from $F1$ to N is

$$J_{e,1} = \left(\frac{1}{e}\right)\left[g_\uparrow(E_{F;f1,\uparrow} - E_{F;n}) + g_\downarrow(E_{F;f1,\downarrow} - E_{F;n})\right] = (g_\uparrow + g_\downarrow)V_0. \tag{6.9}$$

In Figure 6.5b, the up-spin conductance $g_\uparrow$ dominates. In the steady state, current and charge conservation are strict, the total charge in N must be conserved, and the increase of nonequilibrium up spins must be compensated by an equal decrease in the population of down spins. There is also a small partial current of down-spin electrons that creates a nonequilibrium population of

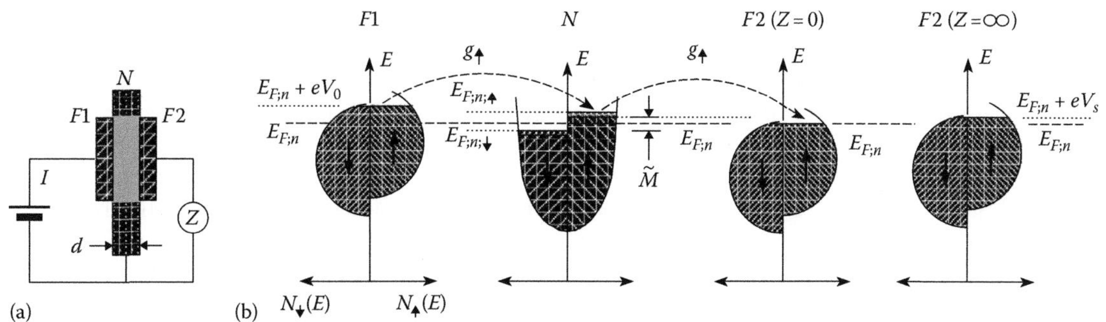

FIGURE 6.5 (a) Cross-section sketch of pedagogical geometry of spin-injection experiment. The nonmagnetic metal sample N has thickness d. Single domain ferromagnetic films serve as spin injector ($F1$) and detector $F2$. A bias current can be driven through the injector and into N. The detector circuit has an arbitrary impedance Z. (b) Density of states drawings for microscopic transport model of spin injection, accumulation, and detection. (From Johnson, M. and Silsbee, R.H., *Phys. Rev. B*, 37, 5312, 1988. With permission.)

down spins, and this is compensated by an equal decrease of up spins. The net magnetization current is

$$J_M = \left(\frac{\mu_B}{e}\right)\left[g_\uparrow(E_{F;f1,\uparrow} - E_{F;n}) - g_\downarrow(E_{F;f1,\downarrow} - E_{F;n})\right]$$
$$= \left(\frac{\mu_B}{e}\right)(g_\uparrow - g_\downarrow)V_0. \tag{6.10}$$

The ratio of J_M to J_e is

$$\frac{J_M}{J_e} = \frac{g_\uparrow - g_\downarrow}{g_\uparrow + g_\downarrow}\frac{\mu_B}{e} = \eta_1\frac{\mu_B}{e} \quad \text{or} \quad J_M = \eta_1\frac{\mu_B}{e}J_e. \tag{6.11}$$

This defines the interfacial spin-polarization coefficient η_1. The interfacial spin current is not necessarily the same as the fractional polarization p_f in the bulk of F, and the interfacial parameter, η, is similar, but not identical, to P (refer to Equation 6.1).

Silsbee identified the penetration length of polarized carriers injected into N to be identically the same as the TESR spin depth $\delta_s = \sqrt{DT_2}$. Consider a sample with thickness d (refer to Figure 6.5a) larger than an electron mean free path but smaller than a spin depth, $d < \delta_s$. In the steady state, J_M is the rate that spin magnetization is added to the sample region, and relaxation at the rate $1/T_2$ is steadily removing magnetization by spin relaxation. The nonequilibrium magnetization,

$$\tilde{M} = \frac{I_M T_2}{Ad}, \tag{6.12}$$

that results is a balance between these source and sink rates and is called *spin accumulation*. It represents a difference in spin-subband electrochemical potential in N, $\tilde{M} \propto N_{n,\uparrow}(E_{F,\uparrow}) - N_{n,\downarrow}(E_{F,\downarrow})$ and is depicted as gray shading in Figure 6.5a.

A second ferromagnetic film $F2$ can be added, fabricated to be in interfacial contact with the sample region, and connected to ground through a low-impedance current meter detector ($Z = 0$ in Figure 6.5b). A positive current $I_d \propto N_n(E_{F,n,v}) - N_n(E_{F,n})$, where $E_{F,n}$ is the average chemical potential of the two spin subbands, is driven across the $N/F2$ interface and through the current detector when the magnetizations $M1$ and $M2$ are parallel. When $M1$ and $M2$ are antiparallel, the current $I_d \propto N_n(E_{F,n,\downarrow}) - N_n(E_{F,n})$ is negative. Conceptually, this induced electric current is the converse of the injection process and is an interface effect: A gradient of spin subband electrochemical potential across the $N/F2$ interface causes an interfacial electric field that drives an electrical current, either positive or negative depending on the sign of the gradient, across the interface. This is an emf source and current conservation demands that a clockwise (counterclockwise) current must be driven in the detecting loop. The current induced across the $N/F2$ interface is [9]

$$J_{e,2} = \left(\frac{1}{e}\right)\left[g_\uparrow(E_{F;n,\uparrow} - E_{F;f2}) + g_\downarrow(E_{F;n,\downarrow} - E_{F;f2})\right]$$
$$= \frac{1}{e}\left[(E_{F;n} - E_{F;f2})(g_\uparrow + g_\downarrow) + \frac{\mu_B\tilde{M}}{\chi}(g_\uparrow - g_\downarrow)\right]. \tag{6.13}$$

Electrode $F2$ may be connected to ground through a high-impedance voltmeter detector ($Z = \infty$ in Figure 6.5a and b). It is convenient to think of $F2$ as a "floating electrode," giving no concern to experimental issues of how to configure an appropriate ground for the voltage measurement. For this case, a positive (negative) voltage V_d is developed at the $N/F2$ interface [8] when $M1$ and $M2$ are parallel (antiparallel). It will be convenient to denote detected voltage V_d as a spin-dependent voltage V_s. The voltage V_s is directly related to the interfacial, spin-subband electrochemical potential gradient described above. The high impedance ($Z \to \infty$) forces the interfacial current flow to vanish, and the rise (or fall) of electrochemical potential of $F2$ is the mechanism that stops the current flow. Setting $J_{e,2} = 0$, the voltage V_s is found to be

$$V_s = \frac{\eta_2\mu_B}{e}\frac{\tilde{M}}{\chi}, \tag{6.14}$$

where η_2 is defined in analogy with η_1 as the ratio of the difference and sum of spin-up and spin-down conductances.

The magnitude of the spin accumulation is described by Equation 6.14. A physical interpretation is easily made after multiplying both sides of the equation by e. On the left-hand side, eV_s represents the energy of a spin-polarized electron at the top of the nonequilibrium distribution. On the right-hand side, $\tilde{M}/\chi$ has units of magnetic field and represents the effective magnetic field that is associated with the spin-polarized electrons. The product $\mu_B\tilde{M}/\chi$ represents the effective Zeeman energy of the spin-polarized electron at the top of the distribution. The polarization η_2 is the fractional efficiency with which the effective Zeeman energy is measured. Equation 6.14 thus describes the effective Zeeman energy of a spin-polarized electron at the top of the nonequilibrium spin population, arising from the effective magnetic field associated with all of the spin-polarized electrons and measured with efficiency η_2.

Equations 6.12 and 6.14 can be combined to give the spin-coupled transresistance R_s that is observed in a spin-injection/detection experiment. In the pedagogical geometry of Figure 6.5a, spin-polarized electrons are injected across the $F1/N$ interface having area A and are confined to a volume Ad, defined by the thickness d of the sample. The case $d<\delta_s$ is analogous with the TESR thin limit. Using an Einstein relation to relate the electron diffusion constant D to the measured resistivity, the transresistance for the thin limit is [32]

$$R_s = \frac{\eta_1\eta_2}{\chi}\frac{\mu_B^2}{e^2}\frac{T_2}{\text{Vol}}, \tag{6.15}$$

$$R_s = \eta_1\eta_2\frac{\rho\delta_s^2}{\text{Vol}} = \eta_1\eta_2\frac{\rho\delta_s^2}{Ad}, \tag{6.16}$$

$$\Delta R = 2R_s = \frac{\eta_1\eta_2 T_2}{\chi\text{Vol}}\cdot\text{constant.} \tag{6.17}$$

Equation 6.16 is valid for a confined geometry, such as the pedagogical geometry of Figure 6.5a. Equation 6.16 is also valid for the unconfined geometry of the LSV, discussed in Section 6.3.1, when volume Vol is properly defined.

6.2.2 Attributes of Spin Accumulation Devices

Equation 6.17 is a simple form that illustrates one of the remarkable features of devices that operate with spin injection, detection, and accumulation. The resistance modulation ΔR is inversely proportional to the sample volume occupied by the spin accumulation $\tilde{M}$. At first, this is a result of the simple observation that the units of $\tilde{M}$ are magnetic moments divided by volume. It immediately follows that given a constant number of magnetic moments, shrinking the volume results in a large value of $\tilde{M}$. This simple scaling rule, however, is uniquely different from the more common kinds of devices. It is highly common that the performance of a device fails to scale with shrinking dimensions. In other words, the output level tends to diminish, rather than remain constant, as the dimensions of the device structure are reduced. It is a remarkable contrast that the output level of a spin accumulation device can be expected to increase as its dimensions are reduced. The important caveats are (1) that the spin relaxation time is not dramatically shortened and (2) the interfacial injection and detection efficiencies are not dramatically reduced. It will be shown in Section 6.3.5.2 that the simple inverse scaling rule of Equation 6.17 has been confirmed for sample volumes that vary by 10 decades.

Inverse scaling is also a characteristic that is uniquely different from GMR. In GMR, interfacial spin-dependent scattering is approximately constant and independent of sample size. The magnetoresistance ratio, MR=$\Delta R/R$, is expected to remain constant. A second important difference is that the electrical output of spin-accumulation devices is bipolar; the output voltage V_s is positive (negative) when magnetizations $M1$ and $M2$ are parallel (antiparallel). Bipolar output has been demonstrated in several experiments [7,33]. If V_s is in superposition with a nonzero baseline voltage, $|V_B|$, then V_s is relatively high (low) when $M1$ and $M2$ are parallel (antiparallel). This is opposite to the convention of GMR devices, where the resistance is relatively low (high) when $M1$ and $M2$ are parallel (antiparallel). A third difference is that spin accumulation can be measured in a nonlocal geometry, and a fourth difference is that spin accumulation can be reduced to zero by the Hanle effect and the application of a perpendicular magnetic field.

6.2.3 Johnson–Silsbee Thermodynamic Theory

By applying the methods of nonequilibrium thermodynamics to study the transport of charge, heat and nonequilibrium magnetization in discrete and continuous systems, Johnson and Silsbee extended the linear dynamic equations of thermoelectricity to describe interface thermoelectric and thermomagnetoelectric (spin-charge) effects [8]. These fully general equations can be used to study issues of charge and spin transport in F/N systems and give a deeper understanding of the phenomena.

6.2.3.1 Linear Equations of Motion

Beginning with a review of the thermoelectric system, flows of charge and heat are driven by an electric field and a gradient of temperature. Classically, the dynamic laws are derived for continuous systems and thermoelectric effects are described in terms of bulk properties of conductors, such as the absolute thermopower ε, the thermal conductivity k, and the electrical conductivity σ (recall that g and σ are synonymous and σ is often used in thermodynamic derivations). In the classical treatment of thermoelectric effects, the junctions between two conductors are considered to be in complete thermal equilibrium; no temperature or voltage difference appears across junctions that are assumed to have infinite conductance. This need not be the case, however, and Johnson–Silsbee theory developed equations describing a number of transport effects specific to the behavior of a junction, between two conductors, that has nonzero resistance. Differences of nonequilibrium magnetization $\tilde{M}$ across an interface can drive currents of spin and charge, and the classical thermoelectric system was generalized to include thermomagnetoelectric effects across metal–metal interfaces. The same generalization was then applied to bulk systems. In order to include currents of nonequilibrium spin magnetization, J_M, the magnetization potential H^* was identified as the generalized force associated with the flux J_M, $-H^* = \tilde{M}/\chi - H$. This expression allows for the possible presence of an external magnetic field H, but $H=0$ is assumed for the discussion below.

The formal approach uses an entropy production calculation, where a *flux* of interfacial current I_N (or bulk current J_N) of a thermodynamic parameter N (charge, heat, and nonequilibrium spin magnetization) is associated with a generalized force, or affinity, F_N (differences or gradients of voltage, temperature, and magnetization potential). Each flux can, in general, be driven by each of the generalized forces, so that each flux can be expanded in powers of F_N. Only the first-order terms are kept for linear response theory, and the coefficients are known as the kinetic coefficients $L_{m,n}$.

For the discrete case of two metals separated by an interface with intrinsic electrical conductance G, electronic transport across the interface is given by the following linear dynamic transport equations [8]:

$$\begin{pmatrix} I_q \\ I_Q \\ I_M \end{pmatrix} = -G \begin{pmatrix} 1 & \dfrac{k_B^2 T}{e\varepsilon} & \dfrac{\eta\mu_B}{e} \\ \dfrac{k_B^2 T^2}{e\varepsilon} & \dfrac{a k_B^2 T}{e^2} & \eta' \dfrac{\mu_B}{\varepsilon}\left[\dfrac{k_B T}{e}\right]^2 \\ \dfrac{\eta\mu_B}{e} & \eta' \dfrac{\mu_B T}{\varepsilon}\left[\dfrac{k_B}{e}\right]^2 & \xi \dfrac{\mu_B^2}{e^2} \end{pmatrix} \begin{pmatrix} \Delta V \\ \Delta T \\ \Delta - H^* \end{pmatrix}. \tag{6.18}$$

The kinetic coefficients $L'_{m,n}$ can be provided phenomenologically or estimated within a specific transport model. Coefficients $L'_{1,3} = L'_{3,1}$ have been derived from the microscopic transport model (refer to Equation 6.14). The phenomenological parameter η' is comparable with η but would be measured under conditions associated with heat flow, and ε is an energy-dependent transport parameter [8]. Finally, the $L'_{3,3}$ term describes the self-diffusion of nonequilibrium spin populations, and $\xi \approx 1$ is a very good approximation.

The same approach can be followed for a continuous medium, and electronic transport inside a bulk conductor is given by these linear dynamic transport equations [8]:

$$\begin{pmatrix} J_q \\ J_Q \\ J_M \end{pmatrix} = -\sigma \begin{pmatrix} 1 & \dfrac{a'' k_B^2 T}{eE_F} & \dfrac{p\mu_B}{e} \\ \dfrac{a'' k_B^2 T^2}{eE_F} & \dfrac{a' k_B^2 T}{e^2} & p' \dfrac{\mu_B}{E_F}\left[\dfrac{k_B T}{e}\right]^2 \\ \dfrac{p\mu_B}{e} & p' \dfrac{\mu_B T}{E_F}\left[\dfrac{k_B}{e}\right]^2 & \zeta \dfrac{\mu_B^2}{e^2} \end{pmatrix} \begin{pmatrix} \nabla V \\ \nabla T \\ \nabla - H^* \end{pmatrix}. \tag{6.19}$$

The kinetic coefficients $L_{m,n}$ are similar to those derived for interfacial transport. For example, $L_{1,3}=L_{3,1}=p_f(\mu_B/e)$ describes the flow of a magnetization current associated with an electric current, with fractional polarization p_f. Values of p_f are $p_f \approx 0.35$–0.45, according to experimental measurements [34]. In a nonmagnetic material, $p_n=0$. The fractional polarization constant p' would be associated with thermal gradients and has not yet been experimentally measured. Constants a' and a'' can be related to the thermal conductivity k and thermopower ε with a free electron model. Finally, the term $L_{3,3}=\zeta(\mu_B/e)^2$, with $\zeta=1$, describes self-diffusion of nonequilibrium spins. In most cases, gradients and differences of temperature are small, heat flow is minimal, and all terms except $L_{1,1}$, $L_{3,1}$, $L_{1,3}$, and $L_{3,3}$ are negligible.

6.2.3.2 Charge and Spin Transport across an F/N Interface

Measurements of the average fractional spin polarization in a ferromagnet have provided values for transition metal ferromagnets of $p_f=0.35$–0.6. In spin-injection experiments, however, the fractional interfacial polarization, η, typically has values of 0.04–0.25. Because of this discrepancy, details of charge and spin transport across the *F*/*N* interface may play an important role. Referring to Figure 6.6a, a ferromagnetic metal *F* and nonmagnetic material *N* are in interfacial contact, and, for simplicity, the cross-sectional area of the interface is taken to be unity and flows are isothermal. A constant electric current J_q is imposed (notation J_q is used for the convenience of a reader who seeks further details in [8,35]), and the solution for the resultant magnetization current J_M is calculated. There are three distinct regions of electron transport, the bulk of *F* and of *N*, and the *F*/*N* interface. Equation 6.19 is used to relate currents to potential gradients and to describe steady-state flows in each of the materials *F* and *N*. Equation 6.18 is used to relate interfacial currents with differences of potential across the interface. Under the assumption of no interfacial spin relaxation, boundary conditions demand that the magnetization currents of all three regions are equal at the interface ($x=0$), $J_{M,f}=J_M=J_{M,n}$ (where J_M is the interfacial magnetization current) and that the electric currents are also equal, $J_{q,f}=J_q=J_{q,n}$.

In the general case, the flow of spin-polarized current J_M into *N* generates a nonequilibrium magnetization (spin accumulation) $-H_n^* = \tilde{M}_n/\chi_n$ in *N* (Figure 6.6b). Because of spin diffusion in *N*, $-H_n^*(x)$ decreases as x increases away from the *F*/*N* interface, and the decay length is the classical spin diffusion length, $\delta_{s,n}$. The nonequilibrium spin population can also diffuse *backward*, along $-x$, going back across the *F*/*N* interface and into *F*. The nonequilibrium population $-H^*$ in *F* is not constrained to match that in *N*, $-H_f^*(x=0) \neq -H_n^*(x=0)$ (Figure 6.6b) because, for example, the susceptibilities χ_f and χ_n can be quite different. The existence of $-H_f^*$ was ignored in the simple microscopic transport model of Section 6.2.1.

The interface has some intrinsic resistance $R_i=1/G$, and the backflow of diffusing, spin-polarized electrons must be overcome by the imposed current. The backflow thereby acts as an additional, effective interface resistance (Figure 6.6c).

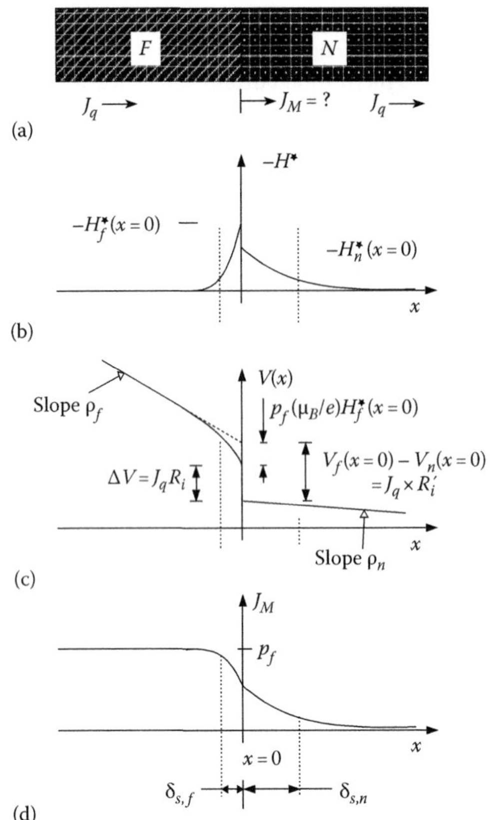

FIGURE 6.6 (a) Model for flow of charge and spin currents, J_q and J_M, at the interface between a ferromagnetic metal and nonmagnetic material, $x=0$ at the interface; (b) magnetization potential. The nonequilibrium spin population decays in F and N with characteristic lengths $\delta_{s,f}$ and $\delta_{s,n}$, respectively; (c) voltage; (d) current of spin magnetization, J_M. (From Johnson, M. and Silsbee, R.H., *Phys. Rev. B*, 35, 4959, 1987. With permission.)

The apparent resistance R_i measured at the interface, $J_q R_i' = V_f(x=0) - V_n(x=0)$, may be larger than the intrinsic resistance, R_i. The spatial extent of the backflow is described by the spin-diffusion length in F, $\delta_{s,f}$. There are few reliable experimental measurements for $\delta_{s,f}$, but an estimate for transition metal ferromagnetic films is $\delta_{s,f} \approx$ nm [36].

The backflow of polarized spins near the F/N interface effectively cancels a portion of the polarized current $J_{M,f}$. The result is that the fractional polarization of the magnetization current that reaches and crosses the interface, J_M, is reduced relative to the bulk value, $J_M < J_{M,f}$ (Figure 6.6d). As described below, this reduction of polarization is related to the difference in resistivities of the F and N materials and arises from the L_{33} and L_{33} self-diffusion terms.

After appropriate algebraic manipulations, a general form for the interfacial magnetization current is found to be [8]:

$$J_M = \frac{\eta\mu_B}{e} J_q \left[\frac{1 + G(p_f/\eta) r_f (\xi - \eta^2)/(\zeta_f - p_f^2)}{1 + G(\xi - \eta^2)\left[(r_n/\zeta_n) + r_f/(\zeta_f - p_f^2)\right]} \right], \quad (6.20)$$

where

$r_f = \delta_f \rho_f = \delta_f / \sigma_f$
$r_n = \delta_n \rho_n = \delta_n / \sigma_n$
$G = 1/R_i$

Noting that ξ, ζ_f, $\zeta_n \approx 1$, this can be rewritten as follows [8,37]:

$$J_M = \frac{\eta\mu_B}{e} J_q \left[\frac{1 + G(p_f/\eta) r_f (1 - \eta^2)/(1 - p_f^2)}{1 + G(1 - \eta^2)\left[r_n + r_f/(1 - p_f^2)\right]} \right] \quad (6.21)$$

It is important to note that spin transport is governed by the relative values of the intrinsic interface resistance $1/G$, the resistance of a length of normal material equal to a spin depth, r_n, and the resistance of a length of ferromagnetic material equal to a spin depth, r_f. Typical values of these resistances are easily estimated. A thin-transition metal ferromagnetic film has resistivity and spin depth of roughly 20 μΩ cm and 15 nm [36], respectively, giving a value $r_f \sim 10^{-11}\ \Omega\ \text{cm}^2$ that is nearly temperature independent. Similarly, a nonmagnetic metal film has resistivity and spin depth of roughly 2 μΩ cm and 0.1–1.0 μm, respectively, at low temperature, giving a value in the range $r_n \sim 2 \times 10^{-11}$ to $2 \times 10^{-10}\ \Omega\ \text{cm}^2$. A typical value for R_i can be found from the contact resistance R_c measured between a narrow Permalloy electrode and a mesoscopic Ag wire in a lithographically patterned structure, $R_c \approx 10^{-11}\ \Omega\ \text{cm}^2$ [36]. Since all of the typical characteristic values fall within a range of a factor of 10, all of the terms in Equation 6.21 are expected to be important for the general case.

It is instructive to examine several limiting cases of Equation 6.21. One case is that of high interfacial conductance, $G \to \infty$ (equivalently, low interfacial resistance, $R_i \to 0$). An appropriate experimental system is a multilayer, CPP GMR sample grown in ultra high vaccum (UHV) [12]. In this case, $R_i \approx 3 \times 10^{-12}\ \Omega\ \text{cm}^2 \ll r_f$ may justify the high conductance approximation. Equation 6.17 reduces to the simpler form [37]:

$$J_M = p_f \frac{\mu_B}{e} J_q \left[\frac{1}{1 + (r_n/r_f)(1 - p_f^2)} \right], \quad (6.22)$$

and we see that the polarization of the injected current is reduced from that in the bulk ferromagnet by the *resistance mismatch* factor $(1 + M')^{-1} = [1 + (r_n/r_f)(1 - p_f^2)]^{-1}$. Using the above estimates for r_f and r_n, the mismatch factor can be expected to be as large as $M' \sim 20$. This result is readily described. With no interface resistance, the spin accumulation in N flows back across the interface because of self-diffusion. The nonequilibrium spin populations in N and F are closely coupled, and there are two consequences. First, spin accumulation acts as a "bottleneck" for spin *and charge* flow, and there is an apparent interface resistance, even though the intrinsic resistance is assumed to be zero. Second, the backflow of polarized spins partially cancels the forward biased polarized current and reduces the fractional polarization of current crossing the interface. These effects depend directly on the transport properties of carriers in N. If spin relaxation and carrier diffusion are rapid, $r_n \leq r_f$, then spin accumulation (and the spin "bottleneck") is negligible. All of the polarized current carried in *F* is dumped into *N*, where it rapidly relaxes [8].

The case of high interfacial resistance R_i is even more important. Spin accumulation in *N* can be large, but the resistive barrier prevents back diffusion [38]. The nonequilibrium spin population in *F* remains small, and the voltage drop across the interface is almost entirely due to R_i. The interfacial magnetization current is now given by

$$J_M = \eta \frac{\mu_B}{e} J_q, \tag{6.23}$$

and the fractional polarization is dominated by the interface parameter η. The resistive barrier may be asymmetric with respect to spin. For example, spin-up (or spin-down) electrons may have preferential transmission. The reflection of polarized carriers may diminish the polarization of current in *F* as it approaches the interface and, in general, the limit $\eta \leq p_f$ is imposed.

A claim that "resistance mismatch" was a fundamental obstacle [39] that would prevent spin injection across a ferromagnetic metal/semiconductor interface was based on a calculation using the infinite interface conductance limit, Equation 6.22. However, a ferromagnetic metal/semiconductor interface is always characterized by a Schottky barrier, tunnel barrier, or low conductance Ohmic contact. In each case, the intrinsic interface resistance is high. As one example, for Ohmic contact to an InAs heterostructure [40], one has $R_i > 10^{-4} \Omega$ cm$^2 \gg r_f, r_n$, and electrical spin injection and detection [41] have been observed. For a second example, spin injection from a ferromagnetic metal across a Schottky barrier contact and into GaAs has been demonstrated, using optical detection [42]. As a third example, an MgO tunnel barrier has been shown to provide highly polarized injection into GaAs at room temperature [43].

Rashba has treated the case of high interface resistance, $R_i \gg r_f, r_n$, using a tunneling calculation to show that "resistance mismatch" effects disappear when transport across a ferromagnet/semiconductor interface is mediated by a resistive barrier [44]. The resulting form for the fractional polarization of injected current agrees with Equation 6.20 within factors of order unity.* Thus, calculations based on nonequilibrium thermodynamics and those based on tunneling formalism give the same result.

More generally, the result described by Equation 6.23 is valid for many experimental conditions independent of the detailed nature of the interface resistance. Specifically, the junction does not need to be a perfect tunnel barrier. Any resistance that creates a voltage drop that blocks the backflow of polarized spins will permit efficient spin injection across the interface.

Equations 6.18 and 6.19 for charge and spin transport can be used to derive the diffusion equations for charge and spin. These equations are more commonly recognized and are useful for solving specific problems. The derivation is straightforward [38], but the boundary conditions for spin diffusion deserve some comment. For any geometry used to study spin injection and diffusion, charge current is conserved: The charge current that enters a sample at the injector is equal to that which exits at ground. However, spin orientation is *not conserved*. If a carrier with an oriented spin is injected into the sample, there is no constraint on the orientation of the carrier removed at ground. For LSVs that have *F* injectors and detectors and are characterized by spin diffusion in *N*, the constraint $I_M = \eta(\mu_B/e)I_q$ is imposed at the *F*1/*N* injector interface, but J_M is unrelated to J_q in *N* because $p_n = 0$ in a nonmagnetic material.

6.3 Techniques and Experimental Results

The spin-injection technique enables basic research of spin transport in a variety of materials systems. Spin injection can study issues such as spin relaxation and interfacial spin polarization directly, whereas GMR techniques provide such information indirectly, at best. This section begins with a brief overview of experimental techniques and then reviews the experimental results from the original spin-injection experiments on bulk aluminum samples.

6.3.1 In-Plane Magnetoresistance Measurements

The techniques of spin injection and accumulation have been applied to several different geometries, but the nonlocal geometry used in the LSV is the most common. The microscopic transport model (Section 6.2.1) introduced a "floating electrode" *F*2 to make a voltage measurement of the spin accumulation (Figure 6.5a). In any real experiment, the voltage must be measured between *F*2 and a voltage ground that is chosen to minimize the effects of any background voltages. Figure 6.7a shows "floating electrode" *F*2 along with an appropriate choice for ground for the measurement of voltage $V_{F2} \equiv V_s$. The ground is provided at a portion of the sample that is remote from the charge flow associated with the bias current.

The *nonlocal geometry* introduced by Johnson and Silsbee [7] provides a physical embodiment of the schematic solution shown in Figure 6.7a. A quasi-one-dimensional metal wire is chosen to lie along the $\hat{x}$ axis and to have a cross-sectional area *A*. Ferromagnetic electrodes *F*1 and *F*2 cross the *N* wire near the middle (Figure 6.7b). This structure was introduced by Johnson and Silsbee [7] and is often called an LSV. The electric bias current *I* enters the sample wire at $x = 0$ and is grounded at the left end, $x = -b$. Electric current in the region $x < 0$ is uniform and is characterized by equally spaced lines of constant potential (dash/dot lines in Figure 6.7b). There is zero current flow to the right of $x = 0$, and the portion of the wire $x > 0$ is equipotential. In

* Specifically, Equation (A11) of Ref. [8] is seen to be identically the same as Equation (18) of Ref. [44] by using the following identifications between variables in the former and latter: $\eta^* = \gamma$, $p = \Delta\sigma/\sigma$, $n = \Delta\Sigma/\Sigma, G(\xi - \eta^2) = 1/r_c$, $\delta_n/(\sigma_n \zeta_n) = r_N$, and $\delta_f/[\sigma_f(\zeta_f - p_f^2)] = r_N$.

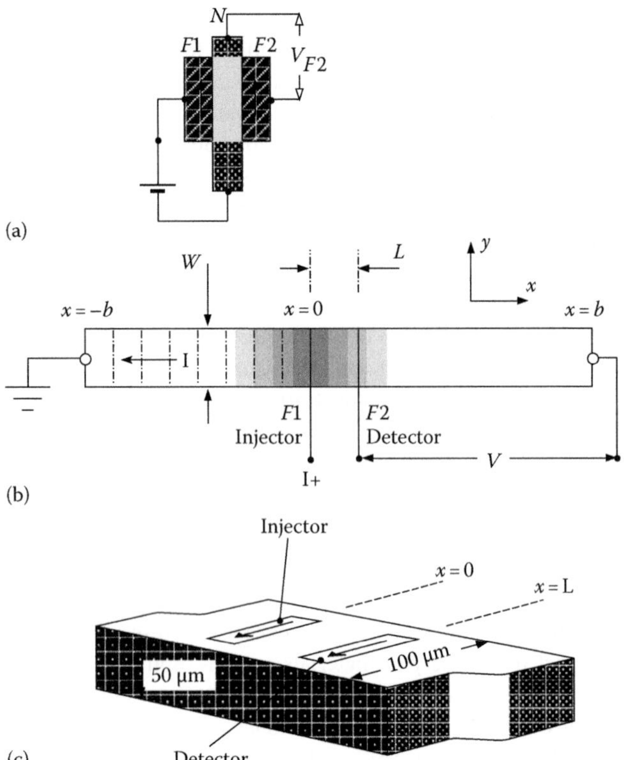

FIGURE 6.7 (a) Schematic cross-sectional sketch of the geometry of a spin-injection experiment, showing the floating detector $F2$. The measurement can be made by choosing the nonlocal ground shown. (b) Schematic top view of nonlocal, quasi-one-dimensional geometry used in original spin-injection experiment as well as in recent mesoscopic metal wire samples. Dotted lines represent equipotentials characterizing electrical current flow. Gray shading represents diffusing population of nonequilibrium spin-polarized electrons injected at $x=0$, with darker shades corresponding to higher density of polarized electrons. (c) Perspective sketch of bulk Al sample used in original spin-injection experiment. (From Johnson, M. and Silsbee, R.H., *Phys. Rev. Lett.*, 55, 1790, 1985. With permission.)

the absence of nonequilibrium effects, a voltage measurement V from the electrode $F2$ to a ground at $x=b$ would give $V=0$. The ideal geometry described here is achieved when the electrodes have negligible width, $W_{F1}, W_{F2} > W$, and have uniform contact across the width of the wire. Departures from ideality result in a small, spin-independent baseline voltage $V \neq 0$ [45].

The electric bias current also is a current of spin-polarized electrons that are injected at the $F1/N$ interface, at $x=0$, and they diffuse into N along the $\hat{x}$ axis. Diffusion is equal for $x>0$ and $x<0$, and the approximate volume occupied by the nonequilibrium spin population in this unconfined geometry is $2A\delta_s$, where the factor of 2 comes from a length of a spin depth on both the positive and negative sides of $x=0$. The density of diffusing spin-polarized electrons is depicted in Figure 6.7b by the shaded region, with darker shades representing higher density.

When the electrode at $x=L$ is also a ferromagnetic film, $F2$, a measurement of V records a spin-dependent voltage that is relatively high (low) when the magnetization orientation $\vec{M}_2$ is parallel (antiparallel) with $\vec{M}_1$. The difference between high and low values is $\Delta R = 2R_s$ (Equation 6.17). Ferromagnetic films $F1$ and $F2$ are fabricated to have different coercivities, and a small, external magnetic field H applied along the $\hat{y}$ axis is used to manipulate the magnetizations $\vec{M}_1$ and $\vec{M}_2$ between parallel and antiparallel. These measurements with in-plane magnetic fields are similar to those that were later used for studies of GMR and are often called the "magnetoresistance" or "in-plane magnetoresistance" technique.

Since $V=0$ in the absence of nonequilibrium spin effects, this measurement uniquely discriminates against any background voltages. Furthermore, the electric current density is zero to the right of $x=0$, and the current of spin-polarized electrons, detected as a spin accumulation at $x=L$, is therefore a *pure spin current*, driven by self-diffusion. Such a pure spin current was first demonstrated in the original spin-injection experiment [7].

A detector $F2$ located at a separation L from the injector, $L<\delta_s$, is in the "thin limit." Using Equation 6.17 and the volume $\text{Vol} = 2A\delta_s$, the transresistance in this geometry is [32]

$$R_s = \eta_1\eta_2 \frac{\rho\delta_s}{2A}. \tag{6.24}$$

A detector $F2$ located at a separation from injector $L<\delta_s$ is in "the thick limit," and the transresistance is [7]

$$R_s = \eta_1\eta_2 \frac{\rho\delta_s}{2A} e^{-L/\delta_s}. \tag{6.25}$$

Because of the explicit exponential dependence $e^{-L/\delta s}$ in Equation 6.25, in-plane magnetoresistance measurements can be performed on a number of samples that are identically prepared while having variable separation L between injector and detector. A plot of the spin transresistance, $R_s(L)$, directly gives a measure of the spin-diffusion length, δ_s. A separate measurement of the diffusion constant from the resistivity ρ then yields a measurement of the spin relaxation time τ_s.

6.3.2 Nonresonant TESR: Hanle Effect

Precessional dephasing of spin-polarized electrons was described in Section 6.1.1.2 for TESR. The same dynamics occur for the nonequilibrium population of spin-polarized electrons in spin-injection experiments. The experimental technique employs an external magnetic field, applied perpendicular to the plane of the spins. It is called the *Hanle* effect, after Walter Hanle performed optical pumping studies and provided an explanation for the depolarization of mercury vapor luminescence in a transverse magnetic field [46].

To derive the line shape experimentally observed in the Hanle effect, we begin with a one-dimensional random walk model that gives a description of the physical principles that is both intuitive and rigorous. The parameters of the model are chosen for the original experiment performed on a "wire" of bulk, high-purity aluminum characterized residual resistivity ratio ($T=4$ K) of $RRR=10^4$. The aluminum was rolled into a foil with thickness 50 μm, narrow wires were cut and fastened to a substrate, and then annealed. Each aluminum wire sample was about 100 μm wide and 50 μm thick [7,10] and is sketched in a perspective view in Figure 6.7c. After process, the sample was measured to have $RRR=1100$.

The discrete random walk model [7] is similar to the Lewis and Carver [47] model of TESR. Referring to Figure 6.7b and c, spin-polarized electrons are injected in steady state at $x=0$ and eventually relax by a T_2 process. Transverse motion is neglected, and we assume that each electron undergoes a random walk along the x axis. Every τ seconds it moves a discrete step of length $\lambda=v_F\tau$, where v_F is the Fermi velocity, either forward or backward. Whenever an electron is at $x=L=p$, it has a small probability of exiting from the bar through the detector. For this sample of electrons, we plot the distribution of times spent in the bar in Figure 6.8. Parameters for the experimental results presented below are $\tau=9$ ps, $\lambda=17$ μm, and $p\sim3$. In the absence of relaxation, the probability $P_{n,p}$ is just the number of ways to get to $x=p$, in n steps, which is the binomial coefficient ${}_nC_{(n+p)/2}$ divided by 2^n:

$$P_{n,p}=\binom{n}{(n+p)/2}2^{-n},\quad n+p \text{ even}, \tag{6.26}$$

$$=0,\quad n+p \text{ odd}. \tag{6.27}$$

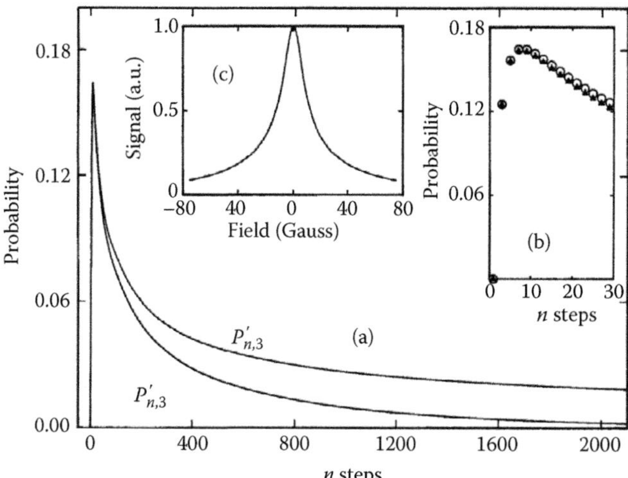

FIGURE 6.8 (a) The unnormalized probability $P_{n,3}$ that a spin-polarized electron took n steps to arrive at the detector located at $L=3$ in the absence of relaxation. $P'_{n,3}$ is the reduced probability in the presence of relaxation. (b) A detail of (a) showing the distribution of arrival times for a small number of steps. (c) Signal calculated numerically using this model. (From Johnson, M. and Silsbee, R.H., *Phys. Rev. Lett.*, 55, 1790, 1985. With permission.)

The peak probability occurs at around nine steps, as expected for a random walk. There follows a long tail that eventually falls off as $n^{-1/2}$; it takes a long time for electrons to diffuse away from the origin in a one-dimensional bar. In the presence of relaxation, there is a probability per step of $\alpha\equiv\tau/T_2$ ($\alpha=0.001$ for the results below) that the orientation of a spin is randomized by its collision, and the spin is lost to the system. The probability that a spin-polarized electron arrives at the detector after n steps without relaxation is reduced from $P_{n,p}$ by the factor $\exp(-n\tau/T_2)$ and is plotted in Figure 6.8 as $P_{n,3}$.

To determine the effect of an imposed magnetic field H parallel to $\hat{z}$, we suppose that the injected electrons are polarized along $\hat{y}$, and the detector is a spin analyzer with orientation along $\hat{y}$ as well. A moment that points at angle θ to $\hat{y}$ will register a magnetization proportional to $\cos\theta$. An external field along $\hat{z}$ will cause every spin to precess through a phase angle of $\delta\theta=\gamma H\tau$ in each step. The signal at the analyzer in a static field H is the sum of contributions of all the electrons, weighted by the probability $P_{n,p}$ that they started from the injector n steps earlier and multiplied by the precession factor $\cos(\gamma Hn\tau)$:

$$\text{Signal} \propto \sum e^{-n\tau/T_2}\,{}_nC_{(n+p)/2}2^{-n}\cos(\gamma Hn\tau). \tag{6.28}$$

The signal as a function of field, calculated numerically from Equation 6.28 with $p=3$, is plotted in Figure 6.8c. In this example, the detector is well within a spin-diffusion length of the injector. Recall from the discussion of TESR that this is the "thin limit." The moments dephase before the net magnetization has precessed, and the feature width is characterized by T_2, which determines the tail in Figure 6.8a. For larger probe separations and/or shorter mean free paths (achieved at higher temperature), the regime changes to $\delta_s>L$. This is known as the "thick limit," in which the detector preferentially samples electrons with relatively long travel times, and the Hanle signal develops side lobes. The width and shape become influenced by the arrival-time distributions as well as the decay time, T_2. In this discrete model, there is a clear physical interpretation of the signal: It is the Fourier transform of the probability $P_{n,p}$ of finding an unrelaxed moment at the detector at time t.

Although the random walk model is rigorously correct and gives good physical insight into the physics of precessional dephasing, the same physics is described by solving the Bloch equations with a diffusion term [9]. The advantage of the latter is that the solutions are readily adapted as fitting functions to experimental data. The picture described by the Bloch equations, the same description used in Section 6.1.1 for TESR, is the following. Spin-polarized electrons diffuse across a distance L from the injector to the detector. Under the influence of a perpendicular magnetic field, each spin precesses by a phase angle that is proportional to the time it takes to reach the detector. Since the electrons are moving diffusively, there is a distribution of arrival times. In the limit of zero external magnetic field, all the diffusing, polarized electrons that reach the detector have the same phase, as long as they have spent a time less than T_2 in the sample. For sufficiently large field, the spin-phase angles of the electrons reaching the detector at any one time are completely random. As T_2 is increased, there is a larger distribution of arrival times at the detector, and a very small field is needed to randomize the distribution of phases. The characteristic field H_{hw} is given by the condition that the product of the precessional frequency and T_2 is a complete precessional rotation angle of 2π, $H_{hw}=1/\gamma T_2$. It is explicit, in this inverse relationship, that long values of T_2 result in narrow lineshapes of the Hanle effect.

The dependence of the spin-coupled voltage V_s on transverse field, detected at ferromagnetic electrode $F2$ and presented for the quasi-one-dimensional sample geometry of Figure 6.7, has the form [9]

$$V_s = \frac{\eta_1\eta_2\hbar^2\pi^2 IT_2}{e^2 m^* k_F 2A\delta_s}\exp\left(\frac{-L_x f(H_\perp)}{\sqrt{DT_2}}\right)\left[F1(H_\perp),F2(H_\perp)\right], \quad (6.29)$$

where

$$F1(H_\perp) = \frac{1}{f(H_\perp)}\left[f(H_\perp)\cos[g(H_\perp)] - \frac{\gamma H_\perp T_2}{f(H_\perp)}\sin[g(H_\perp)]\right] \quad (6.30)$$

$$F2(H_\perp) = \frac{1}{f(H_\perp)}\left[\frac{\gamma H_\perp T_2}{f(H_\perp)}\cos[g(H_\perp)] + f(H_\perp)\sin[g(H_\perp)]\right] \quad (6.31)$$

$$g(H_\perp) = \frac{(d/\sqrt{DT_2})\gamma H_\perp T_2}{f(H_\perp)}, \quad f(H_\perp) = \sqrt{1+(\gamma H_\perp T_2)^2}.$$

m^* is the electron effective mass and k_F is the Fermi wavevector. Functions $F1(H_\perp)$ and $F2(H_\perp)$ determine the shape of $V_s(H_\perp)$, either "absorptive" or "dispersive" for parallel or crossed injector and detector polarizations, respectively, and examples are shown below. Experimentally, the advantage of using the Hanle effect is that a single measurement gives relaxation time T_2 from the width of the line shape, and polarization η is then the sole parameter to be fit to the amplitude.

6.3.3 Experiments: In-Plane Magnetoresistance Measurements

Figure 6.9 shows data for a sample with $L=50\,\mu$m using in-plane fields, $H=H_y$ (refer to Figure 6.7) [7,10]. As the field is swept along the easy axis of both $F1$ and $F2$ from negative to positive values, the positive voltage for $H<H_{C,1}$ is a measure of spin accumulation, R_s, with M_1 and M_2 parallel. The region $H_{C,1}<H<H_{C,2}$ represents the region where magnetizations M_1 and M_2 are reorienting between parallel and antiparallel, and the detected voltage V_s drops from positive to negative. In the region $H>H_{C,2}$, the orientations M_1 and M_2 return to parallel, and the original, positive voltage is regained. A field sweep from positive to negative values shows the same feature in the field range $-100\,\text{Oe}<H<-80\,\text{Oe}$, as expected for the hysteresis of the ferromagnetic films. Note that the voltage V_s is bipolar, positive for parallel magnetizations and negative for antiparallel magnetizations.

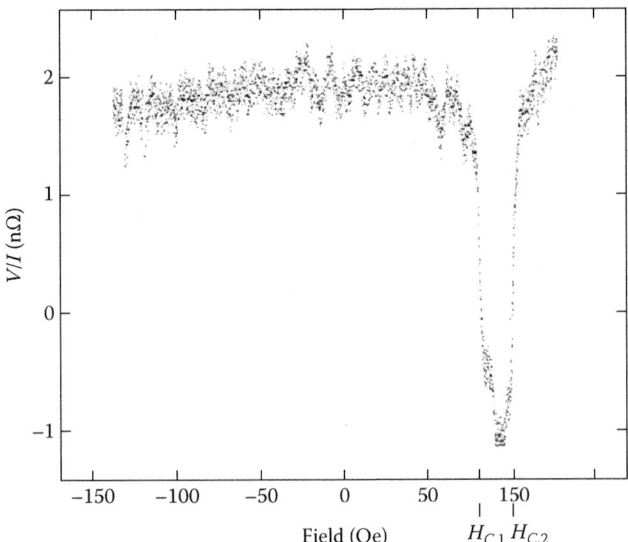

FIGURE 6.9 Magnetoresistance data presented in units of resistance as $R=V/I$. External field is applied along $\hat{y}$ axis, in the plane of $F1$ and $F2$, starting at $H_y=-130\,$Oe, sweeping up and stopping at $H_y=-130\,$Oe. The dip occurs when $\vec{M}_1$ and $\vec{M}_2$ change their relative orientation from parallel to antiparallel. A similar dip occurs for symmetric negative field when H_y is swept from positive to negative. $L=50\,\mu$m, $T=27\,$K. (From Johnson, M. and Silsbee, R.H., *Phys. Rev. Lett.*, 55, 1790, 1985. With permission.)

An important ramification of this first *spin-injection* experiment was the first demonstration of magnetization-dependent resistance modulation in a $F1/N/F2$ device structure. The ratio of the difference of resistance $\Delta R=2R_s$ to the resistance of a segment of Al wire of length L was $\Delta R/R\approx 4\%$ for thin limit samples, somewhat larger than the early spin valve results that were reported a few years later [5].

6.3.4 Experiments: Hanle Measurements

Examples of the Hanle effect from Al wire samples are shown in Figure 6.10, presented in units of resistance, $R=V/I$. In Figure 6.10a, the injector and detector magnetizations are oriented along perpendicular axes and the resulting lineshape, dispersive in appearance, is fit (solid line) using Equations 6.29 and 6.31. For these data, $L=300\,\mu$m, $T=4.3\,$K, and the deduced fitting parameters are $\eta=5.0\%$ and $T_2=10\,$ns. Note that the voltage V_s is bipolar, positive when the injected spins have precessed to be parallel with the detector and negative when precession has caused the spins to be antiparallel with the detector magnetization. Also note that when the injector

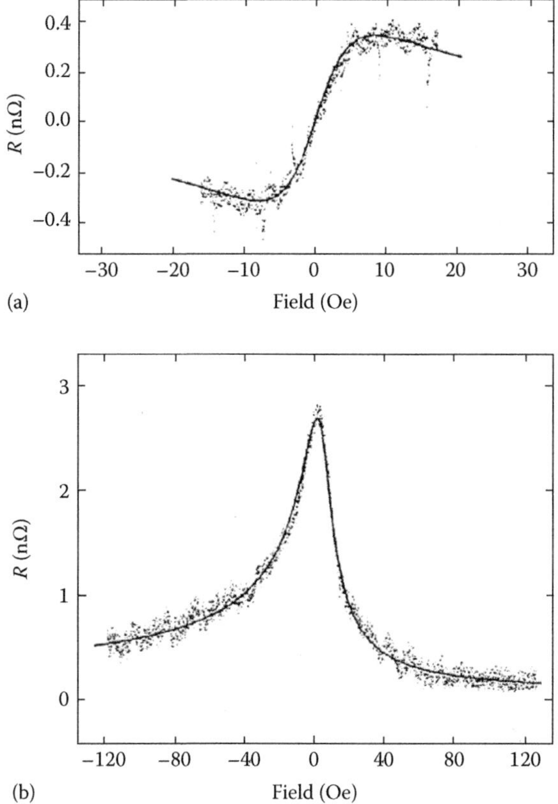

FIGURE 6.10 Examples of Hanle data from bulk Al wire sample presented in units of resistance as $R = V/I$. (a) Example of dispersive lineshape, $L = 300\,\mu\text{m}$, $T = 4.3\,\text{K}$. Solid line is fit to equations in the text: $T_2 = 10\,\text{ns}$, $\eta = 0.050$. (b) Example of Hanle data having absorptive lineshape, with small admixture of dispersive character. $L_x = 50\,\mu\text{m}$, $T = 21\,\text{K}$. Solid line is fit to equations in the text: $T_2 = 7.0\,\text{ns}$, $\eta = 0.075$. (From *Magnetoelectronics*, Johnson, M., Copyright 2004, with permission from Elsevier, Oxford, U.K., Figure 1.8.)

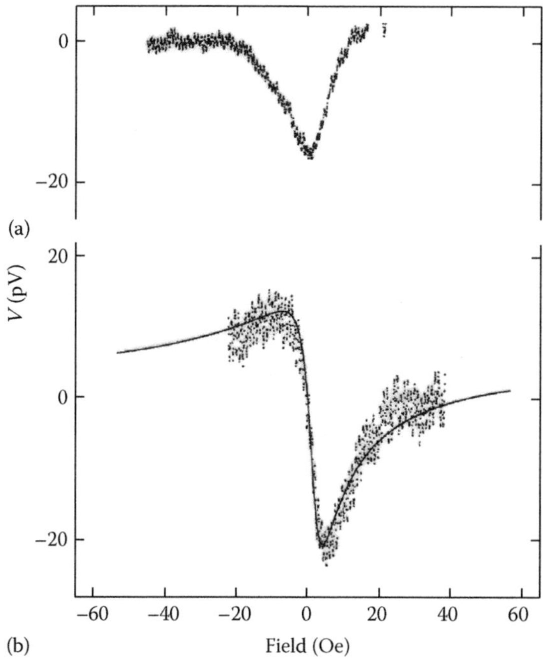

FIGURE 6.11 (a) A field sweep along $\hat{y}$ is halted between $H_{C,1}$ and $H_{C,2}$ (refer to Figure 6.9). The magnetizations $\vec{M}_1$ and $\vec{M}_2$ are now antiparallel. The field along $\hat{y}$ is reduced to zero, and a field sweep along $\hat{z}$ is performed. The resulting Hanle signal now appears with opposite sign, as a dip rather than a peak. (b) An example of the Hanle effect in a sample where the magnetizations have unusual orientations. $L = 200\,\mu\text{m}$. (With permission from Johnson, M. and Silsbee, R.H., *J. Appl. Phys.*, 63, 3934. Copyright 1988, by the American Institute of Physics.)

and detector are fabricated from the same material, we have $\eta_1 = \eta_2 = \eta$.

In general, the Hanle data are fit to a mixture of absorptive and dispersive contributions, using a linear combination of Equations 6.30 and 6.31 to represent the relative orientation between injector and detector. Figure 6.10b presents more typical data, primarily absorptive in appearance, for a sample with $L = 50\,\mu\text{m}$ and $T = 21\,\text{K}$. The deduced fitting parameters are $\eta = 7.5\%$, $T_2 = 7.0\,\text{ns}$, $C_1^2 = 0.84$, and $C_2 = -0.39$, where C_1 and C_2 are the fractional contributions of absorptive and dispersive components [10].

The absorptive Hanle effect line shape appears when the magnetization orientations of injector and detector are aligned parallel, and a dispersive shape results when the two orientations are orthogonal. When the orientations have antiparallel alignment, the Hanle effect should have an absorptive shape but with opposite sign, appearing as a dip rather than a peak. This prediction was confirmed by sweeping magnetic field along the $\hat{y}$ axis and stopping the sweep at a field $H_{C,1} < H_y < H_{C,2}$ (refer to Figure 6.7), such that the voltage is negative and the magnetization orientations are antiparallel. The field H_y is reduced to zero, and the antiparallel alignment is maintained because of hysteresis, assuming the loops of the injector and detector magnetizations to be square. The field is then swept along the $\hat{z}$ axis and the voltage is recorded. The resulting data, shown in Figure 6.11a, show a Hanle effect with a dip rather than a peak. The Hanle data shown in Figure 6.11b are unusual because the dispersive component is stronger than the absorptive contribution. The angle between magnetization axes of injector and detector was determined to be 124°, an angle beyond perpendicular. These data give an example of the range of injector and detector parameters that were observed. The deduced values of η for these first experiments on bulk samples ranged from 5% to 8%. Values of T_2 were in the range 1–10 ns for temperatures $4\,\text{K} < T < 45\,\text{K}$. These parameters will be discussed further in Section 6.3.5.1.

These experiments demonstrated that *spin injection* is a highly sensitive technique for the detection of spin-polarized electrons. The noise floor of the data represents the threshold of detection, and it corresponds to about 1 nonequilibrium spin out of 10^{12} equilibrium spins. Given the sample volume, typical data with V_s/I the order of $1\,\text{n}\Omega$ represent a population of roughly one million nonequilibrium spins, several orders of magnitude better sensitivity than TESR. Recently, the Hanle technique has been used with in-plane magnetoresistance measurements to study spin diffusion and relaxation in silicon [48] (refer also to Chapter 17).

A structure comprising one or more thin ferromagnetic films, and having resistance that varied according to the magnetization orientation of the ferromagnet was first realized using anisotropic magnetoresistance [49]. Anisotropic magnetoresistance sensors were used in read heads from 1991 to 2001 and represented a true paradigm shift for the magnetic recording industry. The first MTJ experiments [2,3] also represented devices with magnetoresistive states. The early spin-injection experiments gave a new and clear demonstration that an electron spin state could be converted to a resistance or a voltage and helped open the possibilities for the devices and applications that followed.

6.3.5 Experiments: Mesoscopic Metal Structures

One result of the broad interest in Spintronics is a renewed interest in the topic of spin injection in thin-film metal wires. The experiments typically involve mesoscopic LSVs in which the sample dimensions, including the separation L between injector and detector, are the order of the electron mean free path, ~100 nm in N. The dimensions of mesoscopic samples have the size scale required for device applications and this generates further interest. These experiments use the Johnson–Silsbee nonlocal geometry that is sketched in Figure 6.7. An atomic force microscope (AFM) image of a typical mesoscopic device is shown in Figure 6.12. The aluminum wire (N) has width w = 100–150 nm and thickness 6–10 nm. Ferromagnetic electrodes $F1$ and $F2$ are fabricated using different materials, CoFe and NiFe, respectively, which have intrinsically different coercivities. The F electrodes are fabricated using a three-angle shadow-evaporation technique. The gray wires that do not touch N result from off-angle deposition and can be ignored when understanding the device. The Al is oxidized prior to deposition of $F1$ and $F2$, forming tunnel barriers at the electrode interfaces. In Figure 6.12, L is about 200 nm.

The spin relaxation time in narrow, thin wires is typically so short that the Hanle method would require perpendicular magnetic fields sufficiently large so that the anisotropy fields of $F1$ and $F2$ are exceeded, $\vec{M}_1$ and $\vec{M}_2$ tip to orientations perpendicular to the film plane, and injected spins no longer precess because the external field is now parallel with the axis of spin polarization. Therefore, most experiments use in-plane fields to manipulate the relative magnetizations M_1 and M_2 between parallel and antiparallel orientations to measure resistance dips, $\Delta R = 2R_s$. An example of such data is shown in Figure 6.13 for the LSV shown in Figure 6.12 (T = 4 K). The slope of a semilog plot of values $\Delta R(L)$ gives the inverse of the spin-diffusion length. At low temperature, the values of δ_s varied with the thickness of the N wire, δ_s = 200 nm (thickness of 6 nm) and δ_s = 420 nm (thickness of 10 nm). Using Equation 6.25, all parameters were measured with the exception of η_1 and η_2. Since the product $\eta_1\eta_2$ is the sole-fitting parameter, there is high confidence in the deduced value of the fractional polarization of injected current. The authors find values of the geometric mean $\bar{\eta} = \sqrt{\eta_1\eta_2}$ as high as $\bar{\eta}$ = 25% (4 K) and 16.5% (295 K) [33], for tunnel barriers with impedance values of order 1–10 KΩ. These values corresponded to very large amplitudes of $\Delta R = 2R_s$, larger than 2 Ω at 4 K and larger than 1 Ω at 295 K. Such a large value of ΔR is larger than values of ΔR observed in CPP spin valves and may have ramifications for magnetoelectronic device applications.

Without discussing all of the recent work, a few interesting experiments can be reviewed. Additional studies of nonlocal effects in metallic lateral spin valves are discussed in Chapters

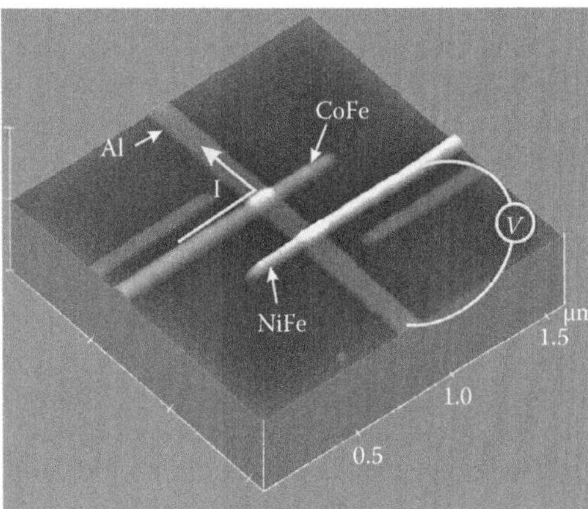

FIGURE 6.12 Scanning atomic force micrograph of mesoscopic spin-injection device in the nonlocal geometry of Figure 6.7. (From Valenzuela, S.O. and Tinkham, M., *Appl. Phys. Lett.*, 85, 5914, 2004. With permission.)

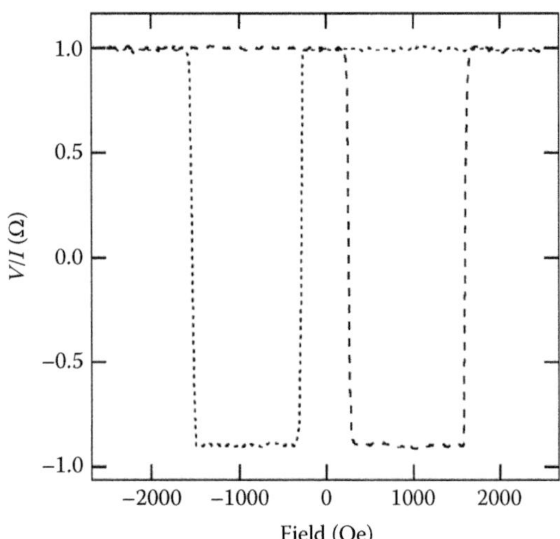

FIGURE 6.13 Example of magnetoresistance data from one of the devices of Ref. [33]. Note the width of the dips, the nearly symmetric bipolar output, and the large magnitude, ΔR = 1.9 Ω. (From Valenzuela, S.O. and Tinkham, M., *Appl. Phys. Lett.*, 85, 5914, 2004. With permission.)

31 and 32. Mesoscopic Cu wires were studied over a temperature range from 4 to 295 K [50]. The ferromagnetic electrodes were cobalt with slightly different widths, and tunnel barriers were fabricated between the *F* electrodes and the Cu wire. The Cu wires had a resistivity of 3.4 μΩ cm at 4 K. The Hanle effect was used to measure a spin relaxation time ($T = 4$ K) $T_2 = 22$ ps ($\delta_s = 550$ nm) and a fractional polarization with average value for the two electrodes of $\eta = 5.5\%$ was deduced. The ratio $\tau/T_2 = \alpha_{Cu} = 0.00066$ can be determined. These spin-injection experiments in Cu also studied the temperature dependence of spin transmission across the *F/I/N* junctions.

A second experiment has reproduced the large value of ΔR observed in thin Al films [51]. Furthermore, it directly confirmed predictions of the Johnson–Silsbee model: spin currents in *N* are not coupled to electric currents but instead are driven by self-diffusion, and spin diffusion is isotropic [7,9,38]. On a 100 nm wide, 15 nm thick Al wire, ferromagnetic detectors were placed on either side of a ferromagnetic injector, at equal distances of $|L_x| = 300$ nm. The aluminum was oxidized before deposition of the cobalt ferromagnetic electrodes so that a tunnel barrier, having an impedance of about 20 KΩ, separated the *F* and *N* layers. The injected current was grounded at one end of the wire. The voltages of the two Co detecting films were compared and found to be equal. Since one of the detectors was along the current path and the other was in the nonlocal portion of the wire, the equivalence of the detected spin-dependent voltages demonstrated that the spin accumulation diffused equally in both directions. Quantitatively, the magnitude of ΔR was about 0.15 Ω. By analyzing $\Delta R(L)$, a long spin-diffusion length of $\delta_s = 850$ nm was measured ($T = 4$ K). The Al resistivity was low, $\rho = 4\,\mu\Omega$ cm, and the authors deduced values $\eta = 12\%$ for their tunnel junctions, $T_2 = 110$ ps, and $\alpha = \tau/T_2 = 10^{-4}$. Allowing for the different thickness of the Al wire and differing values of η, the magnitude of spin accumulation is quite comparable with that observed by Valenzuela and Tinkham [33].

As a final example of a mesoscopic system, an LSV using a novel, nonplanar geometry has been introduced [52]. Called a film edge nonlocal (FENL) spin valve, the device is fabricated across the edge of a multilayered film stack so that the critical dimension *L* is determined by the thickness of one of the films in the stack. Using this geometry, length *L* can be an order of magnitude smaller than the lengths that are typical of planar LSVs [33,50,51]. The multilayer film consists of two *F* layers, *F*1 ($Ni_{81}Fe_{19}$) and *F*2 ($Co_{90}Fe_{10}$), separated and capped by insulating layers. After deposition, the multilayer stack was masked with photoresist, leaving a portion near one edge exposed. This portion was completely removed by ion milling, creating a new edge. The removal process allowed the individual layers to terminate cleanly against the new edge. A narrow copper wire (*N*) was deposited across the edge of the multilayer film, making electrical contact separately to each *F* layer. The center-to-center spacing was $L = 42 \pm 2$ nm in several prototype devices.

Magnetoresistance measurements were made with an in-plane field, using both local and nonlocal configurations. The spin-diffusion length could not be measured because *L* could not be varied over a large range of values. Using a value $\delta_s \approx 450$ nm, found in LSVs made with copper with similar resistivity [53], the FENL spin valves are characterized by $L \approx \delta_s/10 \ll \delta_s$. The observed values of $\Delta R = 2R_s$, about 2.6 mΩ at room temperature, are relatively large. The temperature dependence ΔR (295 K)/ΔR (77 K) was relatively small, which can be expected when ΔR (*T*) is weakly dependent on the exponential because its argument is very small, $L/\delta_s \ll 1$.

The greater importance of this geometry is the demonstration that spin transport is efficient through the edge of a ferromagnetic film, an observation that enables new kinds of spin valves and MTJs that have dimensions that are not limited by a lithographic feature size but rather are determined by the film thickness. Thin films can be made with thicknesses of order 1 nm, and the FENL spin valve demonstrates a technique for fabricating spintronic device structures with truly nanometer dimensions.

6.3.5.1 Experiment: Spin Relaxation Times

As discussed in Section 6.1.1.2, the simplest model of spin relaxation assigns a scattering probability, $\alpha = \tau/T_2$, per scattering event and averaged for all kinds of events. The mean scattering time τ is found from Matthiesen's rule and represents a mean time that is independent of the nature of the scattering. A more careful interpretation of the Yafet–Elliott theory of spin–orbit interactions assumes that electron scattering with impurities, grain boundaries, phonons, and sample surfaces are all different, and each is associated with a different spin-scattering probability.

The first spin-injection experiments, performed on bulk aluminum wires, included measurements of T_2 over a temperature range $4\text{ K} < T < 55\text{ K}$. The values $T_2(T)$ are shown in Figure 6.14, plotted as line width $1/\gamma T_2$ (open stars). Spin relaxation from scattering with surfaces and impurities dominate the residual width at low temperature. This width is subtracted from the data

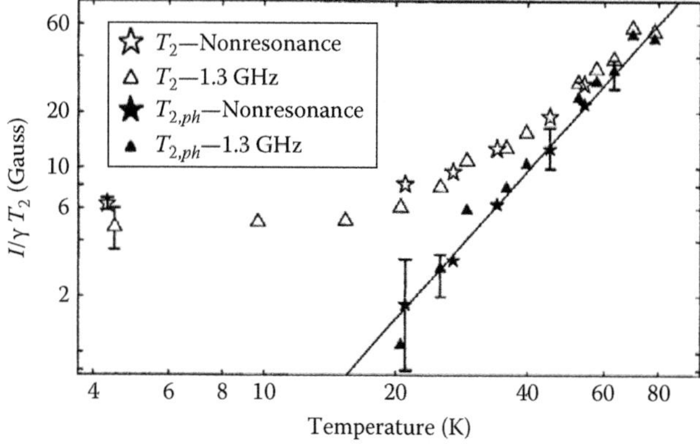

FIGURE 6.14 Comparison of relaxation times T_2 between spin-injection measurements [7] and TESR performed at 1.3 GHz [28], for bulk Al samples of comparable purity. (From Johnson, M. and Silsbee, R.H., *Phys. Rev. Lett.*, 55, 1790, 1985. With permission.)

in order to isolate the temperature dependence of spin relaxation from phonon scattering (solid stars).

Spin relaxation in aluminum foil samples of comparable purity has been studied by TESR [28]. In order to minimize the effect of g anisotropies in aluminum, measurements of T_2 from spin injection are compared with data measured at the relatively low frequency of 1.3 GHz. Values of $T2$ from resonance measurements are also plotted in Figure 6.14 (open triangles). The residual width is subtracted from the data to isolate the temperature dependence of spin relaxation from phonon scattering (closed triangles). The values of $T_{2,ph}$, adjusted by subtraction of the residual line width, are in extremely good agreement for the two different techniques. This agreement is graphic demonstration that the physics of the Hanle effect is exactly the same as the physics of TESR, with a simple adjustment for the rotating frame of reference in the latter. The low-temperature spin relaxation time in Figure 6.14 corresponds to a spin-diffusion length of about 0.4 mm.

TESR is an effective technique for measuring T_2, but it suffers from some limitations. The deduced value T_2 is inversely proportional to the line width (half width at half maximum), $\Delta H_{hw} = 1/\gamma T_2$, and long values of T_2 result in narrow line widths that are easily detected and fit. As T_2 becomes short, on the order of 1 ns or less, ΔH_{hw} becomes so large that the resonance shape can be difficult to distinguish from other variations in the baseline produced by the spectrometer. Using the spin-injection technique, the separation L between injector and detector can be varied, measurements of $V_s(L)$ can be used to determine δ_s, and a determination of the spin relaxation time T_2 then follows (τ_s and T_2 are considered to be equal, noting that T_2 is commonly used in TESR and τ_s is commonly used for describing films). Spin-injection experiments in mesoscopic samples, where L can vary in the range $50\,\text{nm} < L < \mu\text{m}$, have resulted in measurements of τ_s on the order of 1 ps, thereby extending the range of measured values of τ_s (or T_2) by three orders of magnitude.

Table 6.1 shows a variety of values of T_2 for metal samples, measured by several different techniques. The primary purpose of Table 6.1 is to give a sampling of values that are representative of the different materials classes, bulk foil, and thin film. Some remarks can also be made. Consistent with the simple model of spin relaxation described by $\alpha = \tau/T_2$, relaxation times for high resistivity films are much shorter than times for low resistivity foils. The aluminum film studied by Urech et al. [51] has surprisingly large values of δ_s and τ_s. The discrepancy of a factor of 10 in the values of α for this thin film, compared with bulk aluminum, is not understood. With this as an exception, values of α vary by one order of magnitude while values of the relaxation time vary by four orders of magnitude. Since α is expected to differ for different materials, the data in Table 6.1 give support to the simple picture of spin relaxation.

TABLE 6.1 Examples of Spin-Relaxation Times for a Variety of Metals, Measured by a Variety of Techniques

Material	T (K)	Technique	T_2 (s)	α
Cu, bulk [47]	10	TESR	1.8×10^{-8}	7.5×10^{-3}
Al, bulk [54]	4	TESR	1.7×10^{-8}	1.6×10^{-3}
Al, bulk [28]	4	Spin injection	1.4×10^{-8}	1.0×10^{-3}
Al, film [7]	4	Spin injection	1.1×10^{-10}	1.0×10^{-4}
Ag, film [36]	77	Spin injection	3.5×10^{-12}	4.5×10^{-3}
Ag, film [36]	295	Spin injection	2.6×10^{-12}	4.5×10^{-3}
Au, film [55]	295	Spin injection	4.5×10^{-11}	5.0×10^{-4}

Contributions to spin relaxation may come from scattering with impurities, phonon, or surfaces. TESR and spin-injection measurements on high conductivity, bulk samples at low temperature are dominated by surface scattering. The value $\alpha = \tau/T_2$ is an average that uses a simple Drude time for τ.

6.3.5.2 Experiment: Inverse Scaling in Lateral Spin Valves

Among the characteristics of spin injection and accumulation device structures discussed in Section 6.2.2, the idea of "inverse scaling" is perhaps the most unique. Equation 6.17 can be used for analyzing and comparing results for a variety of spin-injection experiments performed on mesoscopic samples in the limit $L < \delta_s$. Noting that the susceptibility χ may vary for different materials, Equation 6.17 can be rewritten as

$$\frac{\Delta R \cdot \text{Vol}}{\eta_1 \eta_2 T_2} \approx \text{constant}. \tag{6.32}$$

Equation 6.32 is independent of temperature, although T_2 may have intrinsic temperature dependence. Values of $\Delta R \cdot \text{Vol}/\eta_1\eta_2 T_2$ are shown in Table 6.2 for several recent experiments, along with experimentally derived parameters. All experiments were performed using LSVs, with the third volume dimension given by twice the spin-diffusion length. Furthermore, the ΔR measurements were achieved using an injector/detector spacing L that is comparable with δ_s. In order to make a thin limit ($L \ll \delta_s$) comparison that is independent of injector/detector separation L, the values of ΔR are extrapolated to a value $L \approx 0$, resulting typically in an increase by a factor e^1. These values for experiments on three different N materials are in good agreement, with a variation that is less than a factor of 2.

This inverse scaling analysis can be extended to include results from the original LSV [7,10]. Comparing values in Table 6.2 for bulk and thin-film Al samples, the volume differs by 10 orders of magnitude, but the inverse scaling parameter $\Delta R \cdot \text{Vol}/\eta_{av}^2 T_2$ agrees within about 10%. It is remarkable that inverse scaling is upheld so well over 10 decades of volume change.

A few remarks can be made about the experimentally measured values of interfacial spin polarization η. First, the values of η observed in spin-injection experiments are much smaller than the limit of $p_f = 0.4–0.6$ assumed for transition metal ferromagnetic films, and the reasons are not yet fully understood. Second, the temperature dependence of η is consistent among most experiments. The value decreases as temperature increases

TABLE 6.2 Verification of Inverse Scaling as Predicted by Equation 6.3 and Described in the Text

Material	Vol (cm^3)	η_{av}	T_2 (ps)	δ_s (μm)	L (μm)	$\Delta R'$ (Ω)	$\Delta R'\text{Vol}/\eta_{av}^2 T_2$ (Ω; cm^3/s)
Al, film	3×10^{-16}	0.17	50	0.20	0.20	2.7	5.9×10^{-4}
Cu, film	5.9×10^{-15}	0.055	22	0.55	0.43	1.2×10^{-2}	1.0×10^{-3}
Ag, film	4.5×10^{-15}	0.22	3.3	0.19	0.30	3.9×10^{-2}	1.1×10^{-3}
Al, bulk	4×10^{-6}	0.06	9000	400	50	2.0×10^{-9}	5.4×10^{-4}

Values of ΔR measured in the thick limit, $L\sim\delta_s$, are extrapolated to $L\to0$ in order to make a thin limit comparison. Thin film data from Al, Cu, and Ag are from Refs. [33,36,50], respectively, and bulk Al data are from Ref. [7].

up to 295K. This may involve spin asymmetries at the injecting interface [50], but a more complete understanding is needed. Third, the condition $R_i \ll r_f$, r_n is rarely realized, and, in those few cases, the effects of resistance mismatch are correctly treated using Equation 6.22.

To give some examples of η, for *F/I/N* tunnel junctions using aluminium oxide barriers, at low temperature, η = 5.5% was measured for *Co/Cu* samples [50], and η = 15%–25% was measured for *CoFe/Al* samples [33]. In recent work, Mgo barriers were used in Permalloy/Ag samples and η = 11% was measured [56]. The agreement is better at room temperature, η = 12% for *Co/Al* samples [51] and η = 16% for *CoFe/Al* [33]. For low-resistance *F/N* junctions, η = 25% was measured for *Co/Au* junctions on mesoscopic Au wires at 4K [55], and η = 24% was measured for *NiFe/Ag* junctions on mesoscopic Ag wires [36]. The results showing high values of η for LSVs with low interface resistance are important for another reason: They demonstrate that the polarization efficiency η and the output modulation ΔR are independent of the interface resistance and, therefore, independent of the device output impedance, Z_{out}.

6.4 Spin-Injection Experiments in Novel Materials

The spin-injection technique has been used to generalize the study of spin-dependent transport to a variety of materials systems. The principal examples are semiconductors, nanostructured materials (graphene [57]), and superconductors. The former topics are reviewed in Chapters 23 and 29, respectively. Discussions of transport effects in ferromagnet/superconductor structures [58–60] are found in several reviews [61] and mentioned in Chapter 31. This section offers a discussion of a few topics in the area of spin injection in semiconductors.

Measurement of the spin lifetimes of carriers in semiconductors has been studied optically and with ESR for many years [22,62]. Although Aronov had proposed the possibility of spin injection into semiconductors [19], interest in the topic was triggered by the proposal to fabricate a spin-injected field effect transistor (FET) [63] (refer also to Chapter 23). In the Datta–Das structure, a ferromagnetic source and drain are connected by a two-dimensional electron gas (2DEG) channel, with the source–drain distance L on the order of an electron ballistic mean free path. The magnetizations of both source and drain were oriented along the axis of the channel, the $\hat{x}$ axis. Injected electrons travel along $\hat{x}$ with a weakly relativistic Fermi velocity, $v_F/c\sim1\%$. An intrinsic electric field E_z perpendicular to the 2DEG plane transforms, in the rest frame of the carriers, as an effective magnetic field H_y^* [64]. Carriers are injected at the source with their spin axes oriented along $\hat{x}$, precess under the influence of H_y^*, and arrive at the drain with a spin phase ϕ that depends on their time of transit, $\theta\propto L/\upsilon_F$. The source–drain conductance was predicted to be proportional to the projection of the carrier spin on the magnetization orientation of the drain and therefore to be a function of θ. By applying a gate voltage to the channel, the electric field E_z, the effective field H_y^*, and θ could be varied, and an oscillatory modulation of source–drain conductance was predicted to result.

The Datta–Das device, which has become the paradigmatic device of Spintronics, was designed as a FET. The geometry is similar to the LSV, but two conceptual differences are important. First, the Johnson–Silsbee idea applied to metals involves diffusive charge and spin transport and the development of spin accumulation. By contrast, the Datta–Das device was designed to involve ballistic charge and spin transport, and no spin accumulation was expected. Second, the transport of spin-polarized carriers in the Datta–Das device is determined by internal electric fields (the Rashba effect), and these fields can be modulated by a gate voltage. This has no analogue in metal-based devices. Largely motivated by possible device applications, basic research in semiconductor Spintronics is active and addresses individual topics that relate to device development. Because these topics are discussed elsewhere in this book, this section will focus on selected issues and experimental results with which the author is associated.

A convincing demonstration of electrical spin injection and detection in a semiconducting channel was the goal of early transport experiments. In one of the earliest [41], two *F/S* junctions (where *S* refers to a semiconducting 2DEG) were fabricated on a common 2DEG channel. The transition metal ferromagnet *F* electrodes were fabricated on top of the top barrier layer of the quantum well structure. This semi-insulating barrier layer provided a large interface resistance R_i and the conductance mismatch problem was mediated (refer to Section 6.2.3.2). In the actual device, six separate channels were connected in parallel for improved signal-to-noise. Narrow channels, about 900 nm wide, were defined on an InAs single quantum well (SQW) heterostructure using optical lithography and an Ar ion mill dry etch. The chip was backfilled with SiN to planarize the surface at the level of the mesa and to cover the side edges of the 2DEG. The interelectrode spacing L was the order of magnitude of the carrier mean free path, a few microns. Magnetoresistance measurements were made using the nonlocal geometry and an in-plane field. Hysteretic dips were observed for two values of L. From the amplitude dependence of these two electrode separations, an upper bound of the spin-dependent mean free path was estimated to be $\Lambda_s = 4\,\mu\text{m}$. Results from a much more definitive experiment also used the nonlocal geometry [65], and, significantly, the measurements included observation of the Hanle effect as well as magnetoresistance with in-plane magnetic fields. This experiment is discussed in detail in Chapter 23.

Very recently, the oscillatory source–drain conductance predicted by Datta and Das has been demonstrated [66]. A channel was defined in a high-mobility InAs heterostructure, Permalloy injecting and detecting electrodes were fabricated on top, and the entire structure was covered by an insulating dielectric and a gate. The carrier mean free path l was measured independently, and the distance L between injector and detector was less than l. Measurements were made using the nonlocal geometry, significantly different from the FET configuration proposed by Datta and Das. As a function of gate voltage V_G, oscillations of the detector resistance were observed for two different values of L and for temperatures $4\,\text{K} < T < 40\,\text{K}$. The spin–orbit parameter $\alpha(V_G)$ was independently determined by Shubnikov–de Haas measurements, and the amplitude of the spin-dependent detector resistance was determined from magnetoresistance measurements with an in-plane field. The oscillating resistance was then fit to the Datta–Das model with no adjustable parameters. These results appear to give a convincing demonstration of the predictions made two decades ago.

In summary, several key issues for semiconductor spin-injection transport experiments have been resolved. First, "conductance mismatch" is never a true problem because the interface can be designed to have a resistance that is large and mediates the effect (Section 6.2.3.2). Second, the Johnson–Silsbee nonlocal geometry is used, almost universally, in order to minimize background voltages and improve signal-to-noise ratios (Section 6.3.1). Third, the Hanle effect, originally used for optical pumping experiments [46] and adapted as the signature technique of spin injection [7], is often used with semiconducting samples having relatively low spin–orbit interaction and relatively long spin relaxation times (Section 6.3.2).

Acknowledgments

The author is grateful for many important collaborations, beginning with Prof. R. H. Silsbee and extending to numerous colleagues. Specifically, the colleagues whose work is mentioned in this chapter include J. Clarke, P. R. Hammar, B. R. Bennett, M. J. Yang, J. Eom, R. Godfrey, A. McCallum and J. Chang. The author gratefully acknowledges the support of the Office of Naval Research.

References

1. P. M. Tedrow and R. Meservey, Spin-dependent tunneling into ferromagnetic nickel, *Phys. Rev. Lett.* **26**, 192 (1971).
2. M. Julliere, Tunneling between ferromagnetic films, *Phys. Lett. A* **54A**, 225 (1975).
3. S. Maekawa and U. Gafvert, Electron-tunneling between ferromagnetic-films, *IEEE Trans. Magn.* **MAG-18**, 707 (1982).
4. J. S. Moodera, L. R. Kinder, T. M. Wong, and R. Meservey, Large magnetoresistance at room temperature in ferromagnetic thin film tunnel junctions, *Phys. Rev. Lett.* **74**, 3273 (1995).
5. G. Binasch, P. Grnberg, F. Saurenbach, and W. Zinn, Enhanced magnetoresistance in layered magnetic structures with antiferromagnetic interlayer exchange, *Phys. Rev. B* **39**, 4828 (1989).
6. M. N. Baibich, J. M. Broto, A. F. Fert et al., Giant magnetoresistance of (001)Fe/(001)Cr magnetic superlattices, *Phys. Rev. Lett.* **61**, 2472 (1988).
7. M. Johnson and R. H. Silsbee, Interfacial charge-spin coupling: Injection and detection of spin magnetization in metals, *Phys. Rev. Lett.* **55**, 1790 (1985).
8. M. Johnson and R. H. Silsbee, A thermodynamic analysis of interfacial transport and of the thermomagnetoelectric system, *Phys. Rev. B* **35**, 4959 (1987).
9. M. Johnson and R. H. Silsbee, Coupling of electronic charge and spin at a ferromagnetic–paramagnetic interface, *Phys. Rev. B* **37**, 5312 (1988).
10. M. Johnson and R. H. Silsbee, The spin injection experiment, *Phys. Rev. B* **37**, 5326 (1988).
11. B. Dieny, V. S. Speriosu, B. A. Gurney, S. S. P. Parkin, D. R. Whilhoit, and D. Mauri, Giant magnetoresistance in soft ferromagnetic multilayers, *Phys. Rev. B* **43**, 1297 (1991).
12. Q. Yang, P. Holody, S.-F. Lee et al., Spin flip diffusion length and giant magnetoresistance at low temperatures, *Phys. Rev. Lett.* **72**, 3274 (1994).
13. S. Schultz, M. R. Shanabarger, and P. M. Platzman, Transmission spin resonance of coupled local-moment and conduction-electron systems, *Phys. Rev. Lett.* **19**, 749 (1967).
14. N. F. Mott, The electrical conductivity of transition metals, *Proc. Roy. Soc. A* **153**, 699 (1936).
15. M. B. Stearns, Simple explanation of tunneling spin-polarization of Fe, Co, Ni and its alloys, *J. Magn. Magn. Mater.* **5**, 167 (1977).
16. M. Johnson and J. Clarke, Spin-polarized scanning tunneling microscope: Concept, design, and preliminary results from a prototype operated in air, *J. Appl. Phys.* **67**, 6141 (1990).
17. R. Weisendanger, H. J. Guntherodt, G. Guntherodt, R. J. Gambino, and R. Ruf, Observation of vacuum tunneling of spin-polarized electrons with the scanning tunneling microscope, *Phys. Rev. Lett.* **65**, 247 (1990).

18. A. G. Aronov, Spin injection in metals and polarization of nuclei, *JETP Lett.* **24**, 32 (1976).
19. A. G. Aronov and G. E. Pikus, Spin injection into semiconductors, *Sov. Phys. Semicond.* **10**, 698 (1976).
20. A. G. Aronov, Spin injection in a superconductor, *Sov. Phys. JETP* **44**, 193 (1976).
21. R. H. Silsbee, A. Janossy, and P. Monod, Coupling between ferromagnetic and conduction-spin-resonance modes at a ferromagnetic normal-metal interface, *Phys. Rev. B* **19**, 4382 (1979).
22. F. Meier and B. P. Zakharchenya (Eds.), *Optical Orientation*, North-Holland, New York, 1984.
23. A. Janossy and P. Monod, Spin waves for single electrons in paramagnetic metals, *Phys. Rev. Lett.* **37**, 612 (1976).
24. A. Elezabi, M. R. Freeman, and M. Johnson, Direct measurement of the conduction electron spin-lattice relaxation time T_1 in gold, *Phys. Rev. Lett.* **77**, 3220 (1996).
25. C. P. Slichter, *Principles of Magnetic Resonance, Springer Series in Solid-State Sciences*, vol. 1, 3rd edn., Springer-Verlag, New York, 1990.
26. Y. Yafet, Calculation of the g factor of metallic sodium, *Phys. Rev.* **85**, 478 (1952).
27. R. J. Elliott, Theory of the effect of spin-orbit coupling on magnetic resonance in some semiconductors, *Phys. Rev.* **96**, 266 (1954).
28. D. Lubzens and S. Schultz, Observation of an anomalous frequency-dependence of conduction electron spin-resonance in Al, *Phys. Rev. Lett.* **36**, 1104 (1976).
29. J. Fabian and S. Das Sarma, Spin relaxation of conduction electrons in polyvalent metals: Theory and a realistic calculation, *Phys. Rev. Lett.* **81**, 5624 (1998).
30. M. Johnson, Spin caloritronics and the thermomagnetoelectric system, *Solid State Commun.* **150**, 543–547 (2010).
31. R. H. Silsbee, Novel method for the study of spin transport in conductors, *Bull. Magn. Reson.* **2**, 284 (1980).
32. M. Johnson, Spin polarization of gold films via transport, *J. Appl. Phys.* **75**, 6714 (1994).
33. S. O. Valenzuela and M. Tinkham, Spin-polarized tunneling in room-temperature mesoscopic spin valves, *Appl. Phys. Lett.* **85**, 5914 (2004).
34. B. Nadgorny et al., Transport spin polarization of Ni_xFe_{1-x}: Electronic kinematics and band structure, *Phys. Rev. B* **61**, R3788 (2000).
35. M. Johnson, Charge–spin coupling at a ferromagnet–nonmagnet interface, *J. Supercond.* **16**, 679 (2003).
36. R. Godfrey and M. Johnson, Spin injection in mesoscopic silver wires: Experimental test of resistance mismatch, *Phys. Rev. Lett.* **96**, 136601 (2006).
37. M. Johnson and R. H. Silsbee, Ferromagnet–nonferromagnet interface resistance, *Phys. Rev. Lett.* **60**, 377 (1988).
38. M. Johnson and J. Byers, Charge and spin diffusion in mesoscopic metal wires and at ferromagnet/nonmagnet interfaces, *Phys. Rev. B* **67**, 125112 (2003).
39. G. Schmidt, D. Ferrand, L. W. Molenkamp, A. T. Filip, and B. J. van Wees, Fundamental obstacle for electrical spin injection from a ferromagnetic metal into a diffusive semiconductor, *Phys. Rev. B* **62**, R4790 (2000).
40. P. R. Hammar, B. R. Bennett, M. J. Yang, and M. Johnson, Observation of spin polarized transport across a ferromagnet/two dimensional electron gas interface, *J. Appl. Phys.* **87**, 4665 (2000).
41. P. R. Hammar and M. Johnson, Detection of spin-polarized electrons injected into a two-dimensional electron gas, *Phys. Rev. Lett.* **88**, 066806 (2002).
42. H. J. Zhu, M. Ramsteiner, H. Kostial, M. Wassermeier, H.-P. Schonherr, and K. H. Ploog, Room temperature spin injection from Fe into GaAs, *Phys. Rev. Lett.* **87**, 016601 (2001).
43. X. Jiang, R. Wang, R. M. Shelby et al., Highly spin-polarized room-temperature tunnel injector for semiconductor spintronics using MgO (100), *Phys. Rev. Lett.* **94**, 056601 (2005).
44. E. I. Rashba, Theory of electrical spin injection: Tunnel contacts as a solution of the conductivity mismatch problem, *Phys. Rev. B* **62**, 16,267 (2000).
45. M. Johnson and R. H. Silsbee, Calculation of nonlocal baseline resistance in a quasi-one dimensional wire, *Phys. Rev. B* **76**, 153107 (2007).
46. W. Hanle, Uber magnetische beeinflussung der polarisation der resonanzfluoreszenz, *Z. Phys.* **30**, 93 (1924).
47. R. B. Lewis and T. R. Carver, Spin transmission resonance: Theory and experimental results in lithium metal, *Phys. Rev.* **155**, 309 (1967).
48. I. Appelbaum, B. Huang, and D. J. Monsma, Electronic measurement and control of spin transport in silicon, *Nature* **447**, 295 (2007).
49. R. P. Hunt, Magnetoresistive head, U.S. Patent no. 3,493,694 (1970).
50. S. Garzon, I. Žutić, and R. A. Webb, Temperature dependent asymmetry of the nonlocal spin-injection resistance: Evidence for spin non-conserving interface scattering, *Phys. Rev. Lett.* **94**, 176601 (2005).
51. M. Urech, V. Korenivski, N. Poli, and D. B. Haviland, Direct demonstration of decoupling of spin and charge currents in nanostructures, *Nano Letters* **6**, 871 (2006).
52. A. McCallum and M. Johnson, Film edge nonlocal spin valves, *Nano Letters* **9**, 2350 (2009).
53. T. Kimura, T. Sato, and Y. Otani, Temperature evolution of spin relaxation in a NiFe/Cu lateral spin valve, *Phys. Rev. Lett.* **100**, 066602 (2008).
54. S. Schultz and C. Lathan, Observation of electron spin resonance in copper, *Phys. Rev. Lett.* **15**, 148 (1965).
55. J.-H. Ku, J. Chang, H. Kim, and J. Eom, Effective spin injection in Au film from Permalloy, *Appl. Phys. Lett.* **88**, 172510 (2006).
56. Y. Fukuma, L. Wang, H. Idzuchi and Y. Otani, Enhanced spin accumulation obtained by inserting low-resistance MgO interface in metallic lateral spin values, *App. Phys. Lett.* 97, 012507 (2010).
57. N. Tombros, C. Jozsa, M. Popinciuc, H. T. Jonkman, and B. J. van Wees, Electronic spin transport and spin precession in single graphene layers at room temperature, *Nature* **448**, 571 (2007).

58. M. Johnson, Spin-coupled resistance observed in ferromagnet–superconductor–ferromagnet trilayers, *Appl. Phys. Lett.* **65**, 1460 (1994).
59. Y.-S. Shin, H.-J. Lee, and H.-W. Lee, Spin relaxation in mesoscopic superconducting Al wires, *Phys. Rev. B* **71**, 144513 (2005).
60. S. Takahashi and S. Maekawa, Spin injection and detection in magnetic nanostructures, *Phys. Rev. B* **67**, 052409 (2003).
61. A. Yu Aladyshkin, A. V. Silhanek, W. Gillijns, and V. V. Moshchalkov, Nucleation of superconductivity and vortex matter in superconductor ferromagnet hybrids, *Supercond. Sci. Technol.* **22**, 053001 (2009).
62. D. Stein, K. v. Klitzing, and G. Weimann, Electron-spin resonance on GaAs-AlxGa1-xAs heterostructures, *Phys. Rev. Lett.* **51**, 130 (1983).
63. S. Datta and B. Das, Electronic analog of the electro optic modulator, *Appl. Phys. Lett.* **56**, 665 (1990).
64. Y. A. Bychkov and E. I. Rashba, Properties of a 2D electron-gas with lifted spectral degeneracy, *JETP Lett.* **39**, 78 (1984).
65. X. Lou, C. Adelmann, S. A. Crooker et al., Electrical detection of spin transport in lateral ferromagnet-semiconductor devices, *Nat. Phys.* **3**, 197–202 (2007).
66. H. C. Koo, J. H. Kwon, J. Eom, J. Chang, S. H. Han, and M. Johnson, Gate voltage control of spin precession in a spin injected field effect transistor, *Science* **325**, 1515–1518 (2009).
67. M. Johnson, *Magnetoelectronics*, Elsevier Scientific Ltd., Oxford, U.K., Figure 1.8, 2004.
68. M. Johnson and R. H. Silsbee, Electron spin injection and detection at a ferromagnetic, *J. Appl. Phys.* **63**, 3934–3939 (1988), Figure 6.

7

Spin Torque Effects in Magnetic Systems: Experiment

Maxim Tsoi
The University of Texas at Austin

7.1 Introduction

The rapid pace of progress in the computer industry over the past 40 years has been based on the miniaturization of chips and other computer components. Further miniaturization, however, faces serious challenges, e.g., due to increasingly high power dissipation. To continue on pace the industry must go beyond incremental improvements and embrace radically new technologies. This book discusses a number of topics from a promising nanoscale technology known as *spintronics* (or magnetoelectronics) in which information is carried not only by the electron's charge, as in conventional microchips, but also by the electron's intrinsic *spin*. Changing the spin of an electron is faster and requires less power than moving it. Therefore, if a reliable way can be found to control and manipulate spins, spintronic devices could offer higher data processing speeds, lower electric consumption, and many other advantages over conventional chips including, perhaps, the ability to carry out radically new quantum computations.

This chapter is focused on spin-transfer-torque (STT) phenomenon, which refers to a novel method to actively control and manipulate magnetic moments, or spins, using an electrical current. This method offers unprecedented spatial and temporal control of spin distributions and attracts a great deal of attention because it combines interesting fundamental science with the promise of applications in a broad range of technologies. STT effect is sometimes regarded as inverse to giant magnetoresistance (GMR) effect, in that the magnetic state of a GMR structure affects its transport properties, while transport currents can alter the magnetic state via STT. This intimate connection can be traced throughout both theoretical and experimental studies of STT and makes GMR, examined in the previous chapters, the most excellent foundation for our discussion. Here we will focus on various aspects of STT experiments in metallic systems including device fabrication and STT detection techniques. In what follows we introduce the basic physics of STT effect in magnetic metallic multilayers. For additional details on STT theory in metallic systems, we refer to Chapter 8. STT in magnetic tunnel junctions and semiconductors will be discussed in Chapters 13 and 20, respectively.

This chapter is organized as follows. After introducing STT basics, we discuss a variety of STT devices from point contacts to nanopillars and nanowires. Then we focus on experimental methods to detect the effects of STT in such devices. Finally we conclude by reviewing some of potential STT applications.

7.2 STT Basic Principles

In the preceding chapters we have seen that the magnetic state of a ferromagnet can affect its electrical transport properties—the relative orientation of the magnetic moments in magnetic multilayers underlies the phenomenon of GMR [1,2]. The inverse effect—STT, in which a large electrical current density can perturb the magnetic state of a multilayer—has also been predicted [3,4] and observed [5–8] in systems with magnetic layers. Here the current transfers vector spin between the ferromagnetic layers and induces precession and/or reversal of magnetic moments in the multilayer. Altering magnetic state with spin currents is based on quantum mechanical exchange interaction and represents a novel method to control and manipulate magnetic moments on nanometer length scales as well as picosecond time scales. In the late 1970s, Berger [9] was the first to realize that the exchange interaction between conduction electron spins and localized magnetic moments in a ferromagnet should result in a torque on the moments, e.g., in a magnetic domain wall traversed by an electric current. In a series of remarkable but only recently appreciated works [9–15], Berger set the groundwork for one of the forefront and most exciting areas in magnetism research today. However, it had to wait until after the discovery of GMR, which spawned a major technological change in the information storage industry and, what is of particular interest for STT, provided a simple and efficient means to detect the effects of STT in experiments. A vast majority of such experiments rely on GMR to make STT visible, so it is natural to precede the discussion of these experiments with GMR basics. However, as both theory and experiment of GMR were already discussed in the previous chapters, here we only briefly revisit its physical concepts important for the sections to follow.

Both GMR and STT have been observed in magnetic metallic multilayers where several atomic layers of one, ferromagnetic (F), material alternate with layers of another, nonmagnetic (N), material (see Figure 7.1). To prepare such F/N multilayers a wide variety of deposition methods have been used, such as electrochemical deposition techniques [16,17] and various vacuum deposition techniques [18,19]. The latter shares mainly between two methods using either sputter deposition or molecular beam

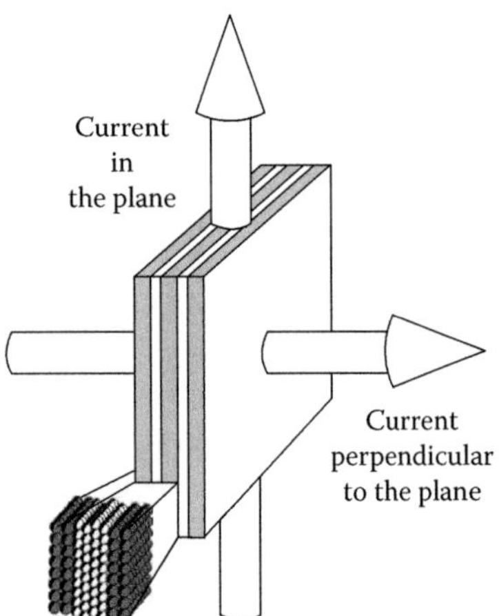

FIGURE 7.1 In a magnetic multilayer several atomic layers of magnetic material (shown in gray) alternate with layers of nonmagnetic material (shown in white). GMR occurs in two geometries: (i) when an electric current flows in the plane (CIP geometry) of the layers, or (ii) when it flows perpendicular to the layers (CPP geometry). The latter will be of particular interest for STT experiments.

epitaxy (MBE) systems. Sputter deposition involves knocking off the atoms of the material of interest from a target by particle bombardment followed by the deposition of highly energetic atoms (~2–30 eV) onto the substrate. A principal advantage of sputter deposition is the ease with which many different materials can be deposited at relatively high deposition rates. In contrast, deposition rates in MBE systems are usually much lower than for sputtering systems, but much lower energies (~0.1 eV) of the evaporated material make this technique favorable for growth of highly oriented single crystalline films.

In GMR experiments, a dramatic reduction in resistance of an F/N multilayer film is observed when the film is subject to an external magnetic field. GMR occurs in two different geometries (see Figure 7.1), namely, when the current flows in the plane of the layers (CIP geometry) or when current flows perpendicular to the layers (CPP geometry). Most of the experiments on GMR are carried out in the CIP geometry because it is quite easy to measure the fairly large resistance of a thin film (film length is typically orders of magnitude larger than its thickness). Experiments in the CPP geometry are more difficult [20] and require special techniques for precision measurements of very small resistances $\sim 10^{-7}$–$10^{-8}\,\Omega$ that is the consequence of the "short and wide" geometry of a 1 mm^2 "wide" and 1 μm "long" sample. In order to increase the resistances to easily observable values, microfabrication techniques can be used to reduce the sample's cross-sectional area [21–23]. Finally, a simple and inexpensive technique as the point-contact technique [24] may be also suitable for the purpose. The samples with reduced cross-sectional area will be of interest for STT experiments, which are mostly carried out in CPP geometry.

GMR originates from the spin-dependent scattering that occurs when electrons move across the multilayer film. The electron scattering and, therefore, the film resistivity depend on the magnetic configuration of the multilayer while the role of the external magnetic field is to change this configuration. The connection between the resistivity and magnetic configuration can be understood most simply by considering electrons traversing an $F_1/N/F_2$ trilayer where two thin film ferromagnets (F_1 and F_2) are separated by a nonmagnetic spacer layer (N). Figure 7.2 illustrates the case in the simple limit where each F can be considered a perfect spin filter; it transmits only electrons whose spin is aligned with the magnetic moment of F, while electrons with antiparallel (AP) spin are completely reflected. When electrons cross such an $F_1/N/F_2$ spin valve from left to right, the electrons transmitted through F_1 will be polarized along F_1. If F_2 is antiparallel to F_1, then all these electrons are reflected off F_2 and no current can flow across the trilayer (Figure 7.2a). The spin-valve resistance R_{AP} in this AP configuration is thus infinite. In contrast, if F_2 is parallel to F_1 (Figure 7.2b), electrons with like spins will be transmitted through both F layers and result in a finite resistance R_P in this parallel (P) configuration. The size of GMR is defined as $(R_{AP} - R_P)/R_P$ and is infinite for perfect spin-filtering. In real multilayers the filtering, i.e., transmission and reflection of electrons, is not perfect that gives finite GMR ratios.

The resulting magnetoresistance (MR) curve—resistance R versus external magnetic field—of an $F_1/N/F_2$ trilayer may look

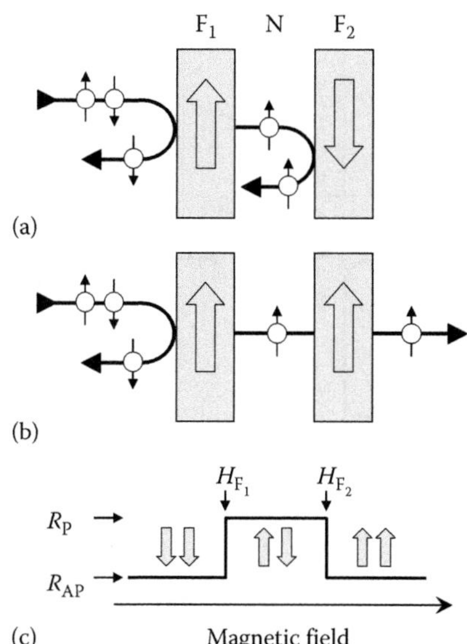

FIGURE 7.2 Qualitative picture of GMR in an $F_1/N/F_2$ spin valve. Right-going electrons are initially polarized by F_1 layer. If F_2 is antiparallel to F_1 (a), the polarized electrons are filtered by F_2 and no current flows across the spin valve. If F_2 is parallel to F_1 (b), the electrons will be transmitted through both Fs. (c) The resulting spin-valve magnetoresistance (resistance vs magnetic field) for F_1 and F_2 switching at different applied fields (H_{F_1} and H_{F_2}). The resistance is low (R_P) when F_1 and F_2 are parallel and high (R_{AP}) when they are antiparallel.

like that in Figure 7.2c. Here F_1 and F_2 were made to respond differently to the applied field H and switch at different applied fields H_{F_1} and H_{F_2}, respectively. Starting up from high negative fields where both Fs are aligned with the field, the resistance is constant at a minimum value R_P below H_{F_1} (P Fs), rises to a maximum value R_{AP} when F_1 reverses at H_{F_1} (AP Fs), and reduces back to R_P as F_2 reverses at H_{F_2} (P Fs again). The different switching fields of F_1 and F_2 can be achieved either by making them of different materials/thicknesses or by placing an antiferromagnet in contact with one of Fs to effectively "pin" its magnetization through an effect called "exchange bias."

In contrast to this simple picture of GMR where the magnetic configuration of $F_1/N/F_2$ trilayer is assumed to be affected only by external fields, the STT phenomenon is concerned with the effect of transport currents on the magnetic configuration. The basic physical picture of STT relies on conservation of angular momentum as the current propagates through the multilayer. Consider a pedagogically simple case where a conduction electron crosses an interface between a nonmagnetic metal (N) and a ferromagnet (F). We assume that initial state of the electron's spin $\boldsymbol{S}$ in N is not collinear with the F's magnetic moment $\boldsymbol{M}$. Once into F, $\boldsymbol{S}$ is subject to an exchange torque caused by $\boldsymbol{M}$, which tends to reorient $\boldsymbol{S}$. Therewith, according to Newton's third law, there should also exist a reaction torque, which acts on $\boldsymbol{M}$—STT torque. Deep into F, $\boldsymbol{S}$ is aligned with $\boldsymbol{M}$ and the change in angular momentum that occurs from its reorientation has been transferred to $\boldsymbol{M}$. Therefore the phenomenon is called spin-transfer-torque (STT). Of course the transfer of angular momentum to $\boldsymbol{M}$ by a single spin $\boldsymbol{S}$ is tiny due to $\boldsymbol{S}$ being negligibly small compared to $\boldsymbol{M}$. For a high-density current injected into F, however, the number of such spins can be very large and the effective $\boldsymbol{S}$ might become comparable to $\boldsymbol{M}$. This would result in a significant STT torque on $\boldsymbol{M}$ and highlights the need for high current densities in experimental demonstrations of STT phenomenon.

The first observation of STT in magnetic multilayers was recorded by Tsoi et al. [5] (see Figure 7.3). In this experiment

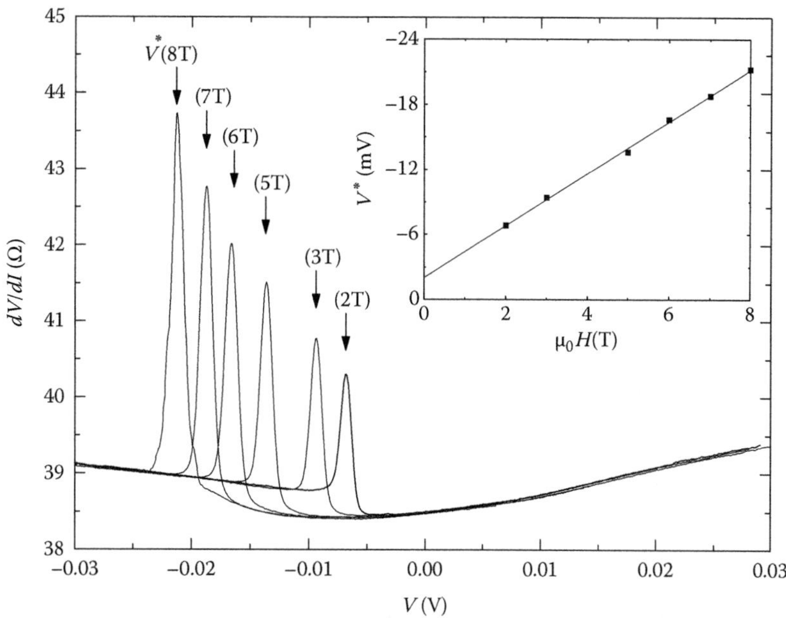

FIGURE 7.3 Differential resistance dV/dI of a mechanical point contact to Co/Cu multilayer as a function of bias voltage V (equivalent to current) for a series of magnetic fields. The peak in dV/dI indicates the onset of spin-transfer-torque excitations. The inset shows that the threshold current at the peak in dV/dI increases linearly with the applied field. (From Tsoi, M. et al., *Phys. Rev. Lett.*, 80, 4281, 1998. With permission.)

the spin-transfer-induced excitations were produced by injecting high-density electric currents into a Co/Cu magnetic multilayer through a mechanical point contact. Point contacts with sizes smaller than 10 nm are formed when a sharpened Cu metal wire (tip) is carefully brought into contact with the multilayer. The extremely small cross-sectional area of such a contact makes it possible to achieve current densities in excess of 10^{12} A/m^2. Because of its extremely small size (<10 nm), point contact is a very efficient probe of electrical transport properties in extremely small sample volumes yet inaccessible with other techniques (e.g., electron-beam-lithography patterning). The latter qualifies point contact as the smallest probe of STT today.

A typical experiment on current-driven excitation of a ferromagnet usually involves two single-domain thin film magnets separated by a nonmagnetic spacer. Here one magnet (F_1) is "hard" and used to polarize the current while the spacer (N) is thin enough for the polarized current to get through and excite the second "free" magnet (F_2). This F_1/N/F_2 trilayer structure is similar to the GMR spin valve discussed earlier. The GMR effect can thus be used to monitor the orientation of F_1 relative to F_2—GMR varies linearly with cos θ, where θ is the angle between magnetic moments of F_1 and F_2, and a phenomenological description [25] gives the trilayer resistance $R(\theta) = R_P + (R_{AP} - R_P)(1 - \cos\theta)/2$. When current flows across F_1/N/F_2, the current-induced torques act on both F_1 and F_2 layers [4], as schematically illustrated in Figure 7.4; here black arrows indicate directions of the torques. This qualitative picture of STT assumes that both F_1 and F_2 layers are perfect spin filters [26], so that electron spins aligned with magnetic moment of, e.g., F_1 layer are completely transmitted through the layer, while spins aligned antiparallel to the layer moment are completely reflected. When electron current crosses the F_1/N/F_2 trilayer from left to right (Figure 7.4b and e), electrons transmitted through F_1 will be polarized along F_1. If spin-diffusion length in N is long enough, this spin-polarized current will reach F_2 and exert a torque on F_2 in a direction so as to align F_2 with F_1. Repeating the argument for F_2, we find that electrons reflected from F_2 will be polarized antiparallel to F_2 and, hence, exert a torque on F_1 trying to align F_1 antiparallel with F_2. The net result is a pinwheel-type motion with both F_1 and F_2 rotating in the same direction (anticlockwise in Figure 7.4b), as described previously by Slonczewski [4]. The higher-order reflections between F_1 and F_2 (Figure 7.4 shows total of four) can change the magnitudes of torques on F_1 and F_2. Note that all subsequent reflections from F_2 in Figure 7.4b will produce torques opposite to the initial torque acting on F_2 and would reduce it. In contrast the subsequent reflections from F_1 produce torques in the same direction as the initial torque acting on F_1 and would increase it. However, such higher-order contributions from multiple reflections are expected to be small as the electron flux is reduced upon each reflection. When the current crosses the trilayer from right to left, the roles played by F_1 and F_2 are reversed and so are the directions of the initial torques (see Figure 7.4c and f)—the torque on F_1 is trying to align F_1 parallel with F_2, while the torque on F_2 is trying to align F_2 antiparallel with F_1. Figure 7.4e and f illustrates the situation for the AP alignment of F_1 and F_2.

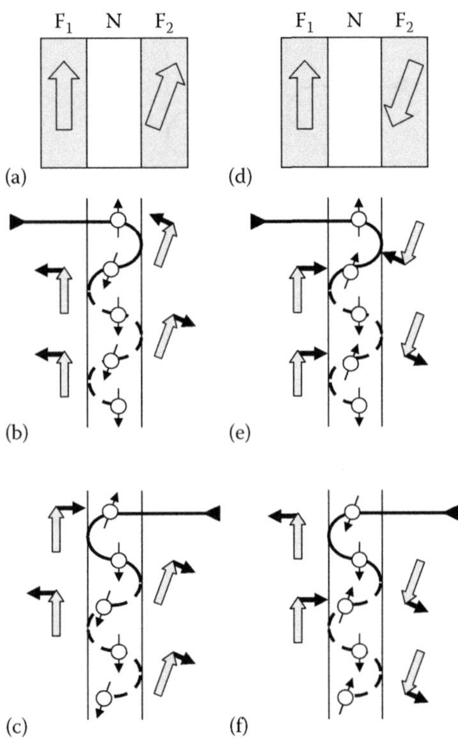

FIGURE 7.4 Qualitative picture of STT in an F_1/N/F_2 spin valve. For almost P (a) or AP (d) alignment of F_1 and F_2 the right-going electrons (b, e) are first polarized along F_1 and exert a torque on F_2 in a direction so as to align F_2 with F_1. Electrons reflected from F_2 will be polarized antiparallel to F_2 and exert a torque on F_1 trying to align F_1 antiparallel with F_2. Electrons bouncing between F_1 and F_2 (reflections up to 4th are shown) exert higher-order torques on F_1 and F_2. Higher-order contributions from multiple reflections are expected to be small as the electron flux is reduced upon each reflection. Black arrows indicate torque directions. For reverse polarity of the current flow (c, f) the roles played by F_1 and F_2 are reversed and so are the directions of the initial torques.

The above discussion implies the asymmetry of STT with respect to current direction as follows. Let's fix the orientation of the polarizer F_1; in experiments this is usually accomplished by making F_1 very thick (compared to F_2) or by pinning its orientation with an adjacent antiferromagnetic layer via the phenomenon of exchange bias. If initially F_2 is almost parallel with F_1, the right-going electrons (forward bias) will stabilize this P alignment and no STT excitation is present. When current bias is reversed, the torque on F_2 will try to rotate F_2 away from F_1 and will result in STT excitation of the system. If the initial orientation of F_2 is almost antiparallel with F_1, the forward bias will try to switch its orientation while reverse bias will stabilize the AP alignment of F_2 and F_1. This asymmetry with respect to bias (current) polarity is one of the fingerprint-like features of STT in experiments; see, for instance, Figure 7.3 where STT excitations are present only at negative bias.

STT phenomenon currently attracts a great deal of attention because it combines interesting fundamental science with

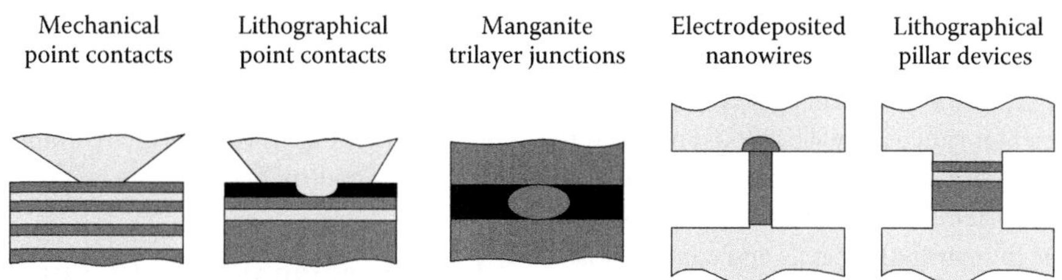

FIGURE 7.5 Overview of devices for STT observation. High current density needed for STT is achieved by forcing the electric current to flow through a very small constriction: point contact (mechanical or lithographically defined), manganite junction, nanowires, or nanopillar.

the promise of applications in a broad range of technologies. In high-speed high-density magnetic recording technology, for instance, the spin transfer torque could replace the Oersted field currently used for writing magnetic bits in storage media (e.g., in magnetic random access memory or MRAM). This may lead to a smaller and faster magnetic memory. Another possible application is based on the spin-transfer-induced precession of magnetization, which converts a dc current input into an ac voltage output. The frequency of such a precession can be tuned from a few GHz to >100 GHz by changing the applied magnetic field and/or dc current, effectively resulting in a current-controlled oscillator to be used in practical microwave circuits.

7.3 STT Devices

Since its prediction by Berger [3] and Slonczewski [4], the STT effect has been observed in a number of experiments, including those with mechanical [5,27–34] and lithographic point contacts [7,35–40], lithographically defined nanopillars [41–48], lateral structures [49–52], electrochemically grown nanowires [53–58], thin-film stripes [11,14,15,59–63], manganite junctions [6,64], tunnel junctions [65–68], and semiconductor structures [69–75]. These different methods all share one characteristic feature—they make it possible to attain extremely high current densities ($\sim 10^{11}$–10^{13} A/m^2 for metallic structures) needed to produce sufficiently large spin-transfer torques. This is achieved by forcing the electric current to flow through a very small constriction. The latter can be a mechanical point contact, lithographically defined point contact or nanopillar, or nanowire, as illustrated in Figure 7.5. In all cases, the maximum current density $j_{max} = I/A$ is defined by the current I flowing through the device and the minimum cross-sectional area A for the current flow. For typical mechanical point contacts $I \sim 1$ mA and $A \sim 100$ nm^2 gives $j_{max} \sim 10^{13}$ A/m^2. In lithographically defined structures, both I and A are typically larger, $I \sim 10$ mA and $A \sim 10{,}000$ nm^2, that gives $j_{max} \sim 10^{12}$ A/m^2. In this section we review devices used in STT experiments.

7.4 Point Contacts

Nanoscale electrical contacts are convenient tools for the realization of local injection and detection of conduction electrons in experimental studies of electron kinetics in metals [76–81]. Recently, such point contacts have received an increased amount of attention due to their ability to produce extremely high current densities needed in STT experiments. Two standard types of point contacts are mechanical and lithographic, each with their own advantages. Mechanical contacts, made by gently pressing a metal wire tip onto a sample surface, are relatively simple to make and can be used on samples of arbitrary composition and shape; whereas lithographic techniques offer much greater control of contact geometry and placement, in addition to being more stable and robust, e.g., to temperature variations.

7.4.1 Mechanical Point Contacts

Mechanical point contacts were used for the original observation of STT in metallic multilayers by Tsoi et al. [5]. In the following we discuss the mechanical setup used in Ref. [5] that provides a means to produce point contacts to a variety of samples of arbitrary composition and shape. Similar systems were used for STT observations in F/Cu multilayers with F = Co [5,27,28,82,83] and F = $Co_{90}Fe_{10}$, $Ni_{80}Fe_{20}$, $Ni_{40}Fe_{10}Cu_{50}$, or Fe [84], single F = Co layer [29–31], bulk F = Co [30], antiferromagnetic AFM = FeMn, IrMn alloys [32,85], and Co nanoparticles [33,34].

Figure 7.6 shows schematically a mechanical point contact between a sharpened metallic wire and a thin film sample. An electrical contact between the wire tip and the sample is achieved by gently pressing the wire, bent in the usual way to minimize damage to the sample, onto the sample surface. Usually, the wire is moved toward the sample surface by a standard differential screw mechanism, but a more precise piezo-control can also be used. Typical contacts made in this way have an area from ~10 to 10^4 nm^2, estimated from the measured contact resistance

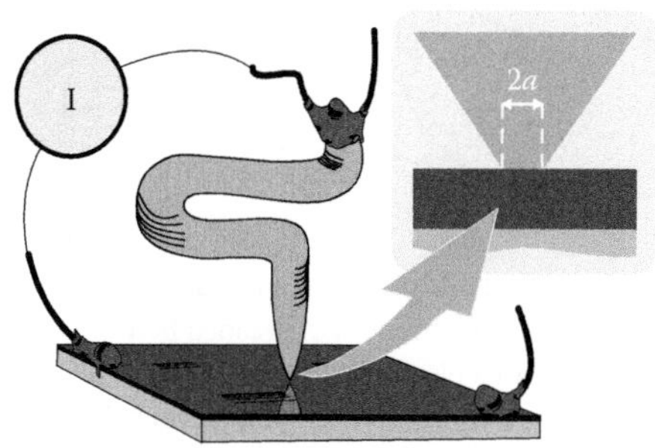

FIGURE 7.6 Schematic view of a point contact. A sharpened metal wire is gently pressed onto a sample surface. The resulting electrical contact between the wire tip and sample can be as small as $2a \sim 1$ nm.

[24,86,87] assuming a combination of Sharvin (ballistic) [76] and Maxwell (diffusive) [88] scattering. Because the contact is so small, one of the main experimental challenges is to preserve it throughout an experimental run, as any vibrations may lead to the tip displacements with respect to the sample and, thereby, to contact instability. Such systems have been used to successfully study ballistic transport in many bulk metals and compounds.

Preparation of the contact tip is similar to that used in scanning tunneling microscopy (STM). In Ref. [5] the tip was made of Ag wire (ø 100–200 μm) electrochemically etched to a sharp point. Other wire materials (Cu, Au, Nb, Fe, Co) have also been used. The tip is moved toward the sample with a standard differential screw mechanism (see Figure 7.7), which consists of a frame (1), a threaded bolt (2), and a piston (3). A room-temperature control knob (not shown) on rod (4) is used to control the rod rotation and, via fork-blade coupling (5), the threaded bolt (2). The bolt has different pitch threads on opposite bolt ends so that the bolt engages the piston (3) faster than the bolt is disengaged from the frame (1). The resulting vertical displacement of the piston is transferred to the tip (10) via a metallic ball (13). A complete rotation of the room-temperature control knob displaces the tip by 25 μm. The leaf spring (11) minimizes the tip displacements in the horizontal plane and damps unwanted vibrations. A hole in spacer (9) provides visual control of initial separation ~200 μm between the tip and the sample (8).

The system just described was used for STT observations both at helium (4.2 K) [5] and room temperature [89], and in magnetic fields up to 20 T [90]. If compared to other types of point contacts, the mechanical contacts have a number of advantages.

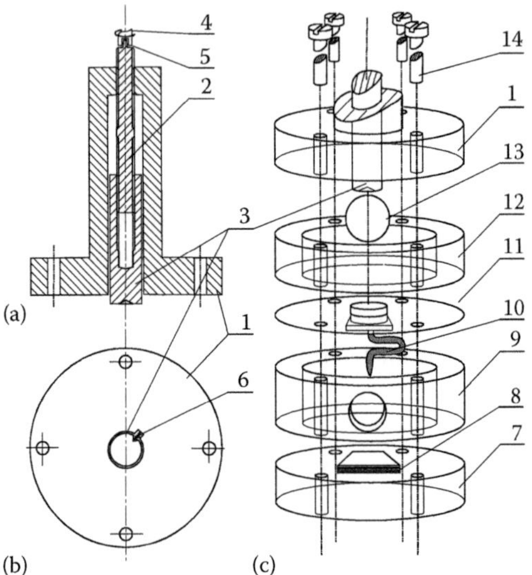

FIGURE 7.7 The mechanical point-contact setup. (a, b) A differential screw mechanism includes frame (1), threaded bolt (2), and piston (3). (c) An exploded view of a block, containing the tip (10) and sample (8). The structural elements of the block—sample table (7), leaf spring (11), and spacers (9, 12)—are tightly fixed to the frame (1) with four screws (14). The vertical displacement of piston (3), transmitted through a metallic ball (13) to the leaf spring (11), results in tip's motion up/down. A small spring (6) prevents the piston from rotating.

In addition to being inexpensive and simple to control, the mechanical system provides a unique means to produce a contact of only few nanometers in size and to change the contact within a single experimental run by simple rotation of the room-temperature control knob.

7.4.2 Lithographic Point Contacts

Although mechanical point contacts have the advantage of being small and simple to control, they also have some disadvantages. First of all, it's a poor control of the contact geometry—the contact size can be defined only indirectly, after establishing the contact, by measuring its resistance. Second, mechanical contacts are too fragile to perform a number of experiments, including temperature-dependent measurements, which fail due to different thermal expansions of tip and sample materials.

All of the above limitations can be overcome by microfabricating point contacts with the help of electron beam lithography (EBL). The microfabricated contacts provide excellent stability, e.g., under temperature cycling, and good control over the contact geometry [91]. Such lithographic point contacts were first used in STT experiments by Myers et al. [7]. The device fabrication process is schematically illustrated in Figure 7.8a–c. The process begins with a 50-nm-thick suspended membrane of silicon nitride (Si_3N_4). The membrane is patterned using EBL and reactive ion etching (RIE) to open a hole in the insulator. Reduced RIE rates in constrained geometry result in a bowl-shaped hole in the membrane and an opening 5–10 nm in diameter, which is much smaller than a 40–100 nm feature actually patterned by EBL. Next a thick Cu layer is evaporated onto the bowl-shaped side of the device and, without breaking vacuum, the metal multilayer needed for STT experiment is deposited on the opposite side of the membrane. An alternative fabrication process (Figure 7.8d–f) starts with the multilayer of interest being deposited on a substrate. The multilayer is covered by a thin insulating layer and a standard EBL patterning is used to create a hole in the insulator. Note that a layer of EBL resist (PMMA = polymethyl-methacrylate) can be used as the insulator. After in situ cleaning of the exposed metal surface, a top electrode is deposited to fill the hole and establish an electric contact to the multilayer [35]. The hole in the insulating layer can also be milled by a focused ion beam (FIB) [92].

7.5 Nanopillars

Lithographically defined nanopillars of magnetic multilayers are probably the most-studied STT devices to date [93–112]. One of the main reasons for that is the opportunity they provide to realize the geometry originally suggested by Berger [3] and Slonczewski [4] for STT observation, two ferromagnetic (F) layers separated by a nonmagnetic (N) spacer layer, and a uniform current density traversing this F/N/F stuck (see Figure 7.9). The first observation of STT phenomenon with lithographically defined nanopillars was made by Katine et al. [41]. They have used EBL to produce pillars consisting of two F = Co layers of different

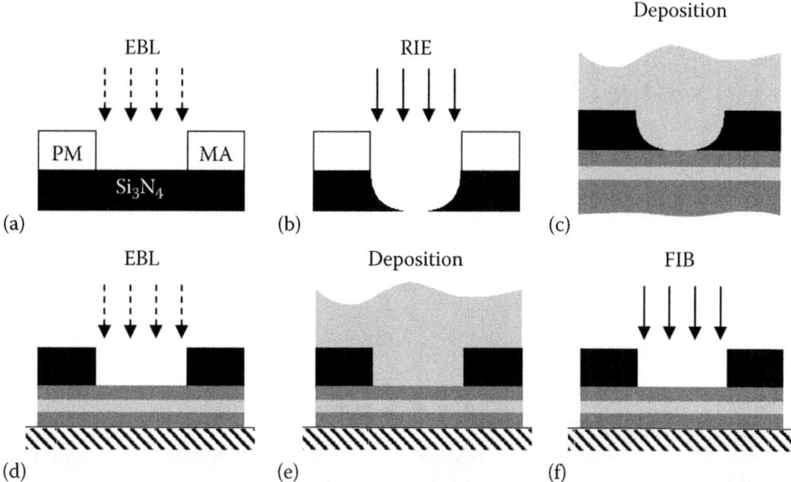

FIGURE 7.8 Illustration of the point-contact fabrication process flow. (a) PMMA resist layer is patterned by EBL. (b) RIE is used to transfer the EBL pattern to the insulating (Si_3N_4) membrane. (c) A thick Cu is evaporated on one side of the membrane, followed by a multilayer deposition on the other side. Alternatively (d) EBL can be used to create a hole in an insulating layer (black) on top of an existing multilayer, followed by a Cu electrode deposition (e). (f) The hole in the insulator can also be made by FIB.

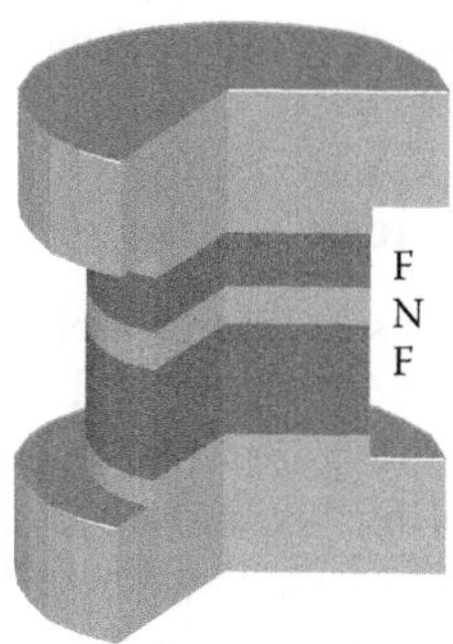

FIGURE 7.9 Schematic view of a nanopillar. Two ferromagnetic (F) layers are separated by a nonmagnetic (N) spacer. One F is typically made much thicker than the other one to minimize current-induced effects on the thick F and focus on STT in the thin F.

thicknesses separated by a N = Cu layer spacer. The lithographic process allows for greater control of device geometry while reducing cross-sectional area of this otherwise standard F/N/F spin valve to ~10^4 nm^2. The pillars can be patterned using two different generic approaches: subtractive—growth of the spin-valve film followed by patterning a hard-mask on top and etching of the film, or additive—metal deposition onto a predefined template. The etching step (typically ion milling) required for subtractive processing can degrade the magnetic behavior of nanoscale-patterned elements. On the other hand, additive process can result in less well-defined edges of the nanopillar. So either approach has to be optimized to avoid potential problems.

7.5.1 Subtractive Patterning

Fabrication of nanopillar by subtractive patterning process, as used in Ref. [41], is outlined in Figure 7.10. The fabrication begins with buffer layer/multilayer stuck/capping layer deposition onto an oxidized Si substrate. The buffer layer will later serve as the bottom electrode. Next a Cr dot is patterned on top of the capping layer by EBL: a single dot in resist (PMMA) layer is exposed and developed (Figure 7.10a); after Cr metallization and lift-off the Cr dot remains on the layer surface (Figure 7.10b). This dot serves as the mask for incoming ions during the ion-mill step (Figure 7.10b and c) and thus defines the pillar cross section. The mask and multilayer around it are completely eroded during milling while redeposition of the milled material results in a rounded pillar such as that shown in Figure 7.10c. Following the milling step, planarization with polyimide, RIE, and metallization are used to seal the pillar with insulator (Figure 7.10d), open the pillar cap (Figure 7.10e), and form the top electrode (Figure 7.10f).

7.5.2 Additive Patterning

An alternative way to fabricate nanopillars of magnetic multilayers is the additive patterning method [43] shown schematically in Figure 7.11. Here the fabrication begins with bottom electrode/insulating (SiO_2) spacer/Pt layer deposition onto an oxidized Si substrate. EBL patterning and ion milling are used to create a small hole in the Pt layer, which, in the following steps, serves as a mask and defines the pillar size. Next a wet etch is used to open up the insulating spacer, followed by deposition of the multilayer of interest. The nanopillar is formed by the portion of the multilayer, which goes through the mask and is laid down directly on the bottom electrode. Finally, without breaking vacuum, a thick metallic layer is deposited on top to make the top electrode.

To speed up the fabrication process and eliminate the EBL patterning, an FIB milling can be used to punch a hole in the Pt mask layer [113] similar to the process shown in Figure 7.8f. Another alternative may use a layer of PMMA resist as the insulating spacer and an EBL patterned hole in this layer as a template for the nanopillar [114]. This process eliminates ion mill

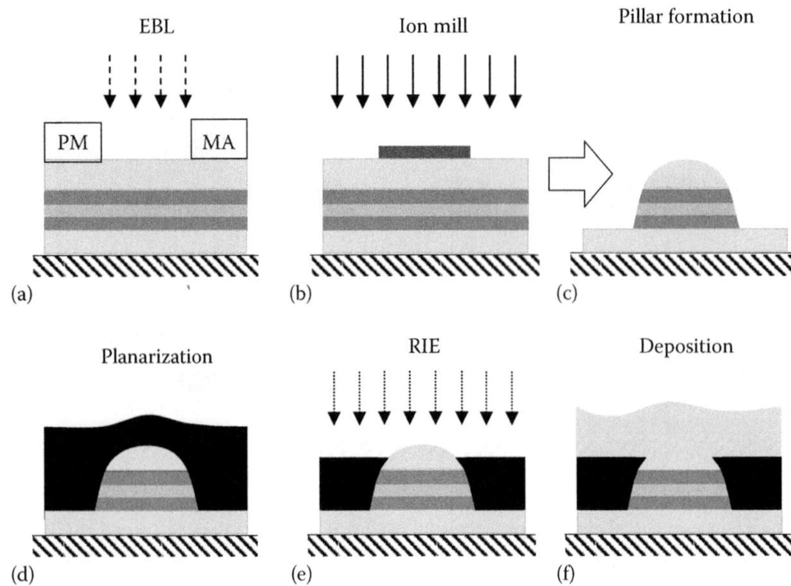

FIGURE 7.10 Illustration of nanopillar fabrication via subtractive patterning. First, buffer layer (light gray), multilayer stuck, and capping layer (light gray) are deposited on a substrate. (a) A Cr dot is patterned on the surface of multilayer stuck by EBL, Cr deposition (not shown), and lift-off process (not shown). (b–c) Ion milling defines the pillar. (d) Planarization with polyimide seals the pillar with insulator. (e) RIE opens the pillar cap. (f) The top electrode is formed by final metallization.

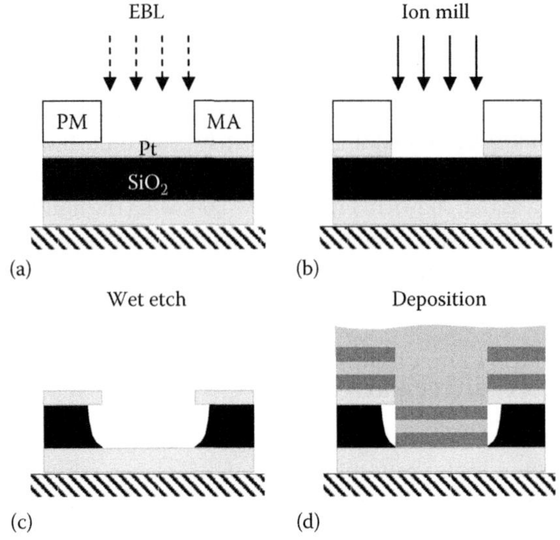

FIGURE 7.11 Additive patterning process for nanopillar fabrication. EBL (a) and ion milling (b) define a small hole in a thin Pt layer. (c) Wet etch opens up the insulating spacer. (d) Deposition of multilayer stack and top electrode.

and wet etch steps—the multilayer is deposited directly into the template hole, similar to Cu electrode deposition in Figure 7.8d and e. A shortcoming of this method lies in the absence of a controlled undercut (see Figure 7.11c) obtained during wet etch, which means that the nanopillar edges may not be well-defined due to deposition onto side-walls of the hole. Yet another method to make up a pillar exploits a three-dimensional (3D) FIB milling [115] to sculpt a nanopillar by carving away material from a bulk multilayer film.

7.5.3 Lateral Devices

The two F magnets of a standard nanopillar (Figure 7.9) can be separated laterally [49,116] in a so-called lateral spin valve (LSV) shown schematically in Figure 7.12a. In addition to providing, e.g., optical access to the N spacer to study relaxation of spin currents, this multiterminal LSV geometry provides a means for generating pure spin currents, which is impossible in vertical pillar structures of Figure 7.9 where the spin current, needed for STT excitation, always flows together with the charge current. Driving charge current between terminals 1 and 2 (see Figure 7.12a) would result in spin accumulation near the N/F_1 interface. The diffusion of the accumulated spins away from the interface generates pure spin current toward the second magnet F_2, which, in turn, generates STT on F_2 and was shown to switch it [49,50].

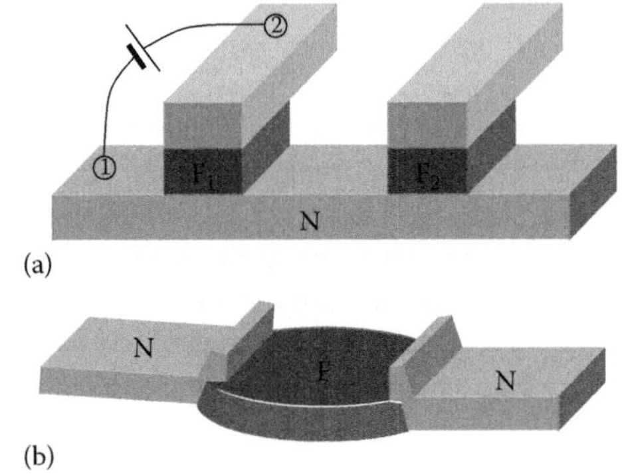

FIGURE 7.12 Lateral devices. (a) Lateral spin valve where two magnets (F_1 and F_2) are laterally separated by a nonmagnetic (N) spacer. (b) Single F element (thin film disk) contacted by two nonmagnetic electrodes.

LSV, as in Figure 7.12a, can be made by a standard multilevel lithography process [117]. Here every level of device elements is laid on a planar substrate using EBL patterning and lift-off process. Caution should be taken to ensure a good electrical contact between elements of different levels as oxidation and contamination during lithographic patterning is known to degrade the contact transparency. A short ion milling is typically used to prepare the element surfaces. To further improve the quality of N/F interfaces, magnetic F elements can be deposited directly onto the N spacer without breaking vacuum using oblique deposition techniques [50]. Figure 7.12b illustrates a lateral structure with a single F element as used for studying the current-driven dynamics and switching of magnetic vortices in an F = Py (permalloy) disk [51,52].

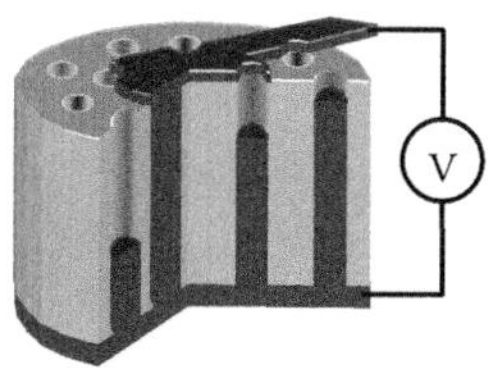

FIGURE 7.13 Electrochemically grown nanowires. A metallic film is evaporated on one (bottom) face of a track-etched membrane to serve as a cathode for electroplating. The filling of membrane pores by electroplating results in wires of nearly uniform cross section. Monitoring potential between the cathode and a patterned top electrode allows for stopping the electrodeposition when a nanowire emerges from membrane pores, attaches to the top electrode, and short-circuits the membrane.

7.6 Nanowires and Nanostripes

Currently it is challenging to make lithographically defined nanopillars with diameters smaller than 50 nm. In this section we will discuss the template-synthesis method for STT device fabrication, which potentially may be used to prepare metallic nanopillars (wires) with diameters down to a few nanometers. Larger wires, however, are still useful for STT experiments with magnetic domain walls. Here we will discuss magnetic nanostripes employed in such experiments prepared by lithographic methods and FIB patterning.

7.6.1 Electrochemically Grown Nanowires

An alternative method to produce cylindrical pillars of the desired material, known as membrane-based or template-based synthesis, was around for decades and successfully used to prepare nanowires of metals, semiconductors, conducting polymers, and other materials [118]. Here the desired material is deposited within the pores of a template membrane. For cylindrical pores with diameters in the nanometer range, e.g., as in track-etch polymeric membranes or porous aluminas, this results in a nanocylinder of the desired material (nanowire) obtained in each pore.

The template synthesis of metallic nanowires was first demonstrated by Possin [119]. He used electrochemical reduction of metal ions to deposit the appropriate metal within the pores of a template membrane. The process is schematically illustrated in Figure 7.13. Here one (bottom) face of the membrane is covered with an evaporated metallic film. The film serves as a cathode for electroplating when the membrane is immersed in a plating solution containing ions of the desired metal. The filling of membrane pores by electroplating results in wires of nearly uniform cross section and reasonable purity. The method can be applicable to any metal, which can be plated from an acid solution.

The effect of electric current on magnetic reversal in individual electrodeposited nanowires was evidenced by Wegrowe et al. [8]. They have developed a method where contact to a single wire is made in situ during electrodeposition. The contact is accomplished by continuously monitoring potential between the cathode and a patterned gold electrode on top of the membrane (see Figure 7.13) and by stopping the electrodeposition process as soon as the first nanowire emerges from the membrane pore, attaches to the top electrode, and short-circuits the membrane. Applying this method to track-etch polymeric membranes has led to STT observations in Ni nanowires [8,53,120,121], hybrid wires where a Ni half-wire is joined with a Co/Cu multilayer [53–120], nanowires of Co/Cu multilayer [120], and Co/Cu/Co spin valves embedded in Cu nanowires [54,55]. A relatively low pore density (typically $<10^9$ cm^{-2}) of polymeric membranes facilitates contacting a single nanowire, which can even be made with a sharpened Au tip [56]. However in templates with higher pore densities, like in porous aluminas with densities up to $\sim10^{12}$ cm^{-2}, it is difficult to contact a single nanowire in this way and alternative methods should be used. A nanolithography process based on electrically controlled nanoindentation of alumina templates has been used to study STT properties in nanowires of Py/Cu/Py [57] and Co/Cu/Co [58] spin valves.

7.6.2 Lithographically Defined Nanostripes

Standard lithographic methods are typically used to prepare nanowires of thin-film materials [121–147]. Figure 7.14a and b illustrates the fabrication process. A nanowire pattern is first defined, e.g., by EBL, on a PMMA resist layer spanned on an oxidized silicon wafer (Figure 7.14a). A thin-film deposition and lift-off process are then used to transfer the pattern to a thin film of the desired material (Figure 7.14b). The resulting nanowire is rather a nanostripe with cross section defined by the film thickness and the width of lithographic pattern. Scanning electron microscopy (SEM) image in Figure 7.14c shows a 100 nm wide Py stripe connecting two diamond-shaped elements used for nucleation and injection of magnetic domain walls; constrictions in the stripe are typically used to pin the injected domain walls [62].

An alternative method for fabricating similar nanostripes begins with depositing a thin continuous film on a substrate

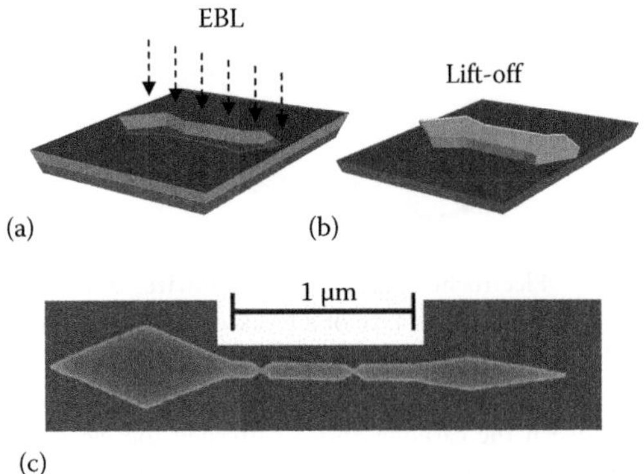

FIGURE 7.14 Lithographically defined nanowires. (a) A desired shape is patterned by EBL in a resist layer on a substrate. (b) The EBL pattern is then transferred to a thin film of the desired material using thin-film deposition and lift-off process. (c) SEM image of a Py thin-film device used in STT experiments where a 100 nm wide stripe connects two diamond shaped elements and has two narrow constrictions.

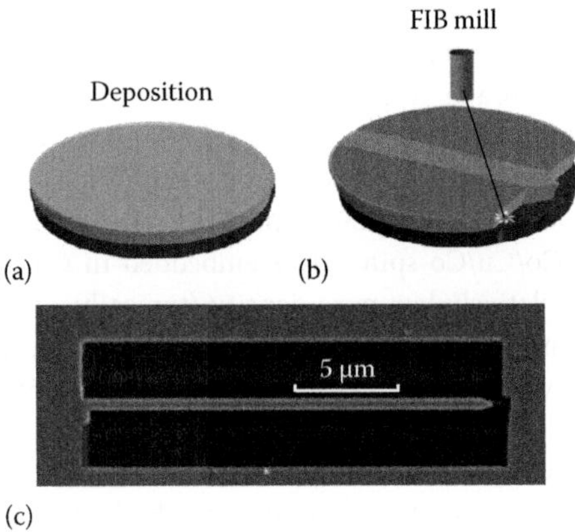

FIGURE 7.15 FIB patterned nanowires. (a) A thin film is evaporated on a substrate. (b) FIB is used to selectively remove the film material. (c) SEM image of a 600 nm wide Py stripe. Darker color indicates areas where Py film was milled away by FIB.

(Figure 7.15a), followed by a selective milling of the film by FIB [126,130,144]. Removal of selected areas of the film by FIB (Figure 7.15b) results in an arbitrary-shaped thin-film pattern, which can be a nanostripe with desired width and length. The SEM image in Figure 7.15c shows an FIB prepared 600 nm wide 20 μm long Py nanostripe. Darker color in Figure 7.15c indicates areas where Py film was milled away. One end of the remaining stripe is left connected to the extended unmilled film for domain-wall injection purposes. Although the FIB patterning process is fast and may be used to produce nanostripes as small as those prepared by EBL lithography, it has the disadvantage of damaging the film material with heavy ions. For instance, it is known that bombarding magnetic films with Ga ions may result in a magnetically dead material that in the stripe of Figure 7.15c would be concentrated near its edges [144].

7.7 Magnetic Junctions

In contrast to metallic pillars where two F elements are separated by a metallic N spacer, magnetic junctions employ an insulating spacer layer, as schematically illustrated in Figure 7.16.

7.7.1 Manganite Magnetic Trilayer Junctions

One of the first STT experiments was performed by Sun [6] with manganite magnetic trilayer junctions (MMTJs). The junctions are epitaxial LSMO60 nm/STO3 nm/LSMO40 nm thin-film trilayers grown by laser ablation on NGO substrates (LSMO = $La_{0.67}Sr_{0.33}MnO_3$, STO = $SrTiO_3$, NGO = $NdGaO_3$). After deposition, a lithographic process, similar to that used for metallic nanopillars, was used to form "pillars" of LSMO/STO/LSMO junctions where bottom LSMO is only partially patterned and serves as the bottom electrode, and a SiO_2 layer is used to isolate the top (Au) electrode from the bottom LSMO. Due to significant heat sensitivity of the junction only optical lithography, with maximum temperature reaching ~90°C during processing, was used for fabrication. The junction size is, thus, limited to ~10 μm^2.

Electron transport across the STO barrier is believed to be controlled by morphology of the junction interface. LSMO films grown by laser ablation are known to contain particulates with sizes ranging from 10 nm to more than 100 nm. The STO barrier over such LSMO particulates is nonuniform and may couple some particles to the top LSMO more conductively than others. Local conduction paths associated with the nonuniform coupling can dominate the electron transport across MMTJ and result in high densities of local current, thus making STT observable at relatively low current levels [6,64].

7.7.2 Magnetic Tunneling Junctions

Magnetic tunneling junctions (MTJs) have similar structure and fabrication process flow to that of MMTJs discussed in the previous section (see Figure 7.16), except that F_1/barrier/F_2

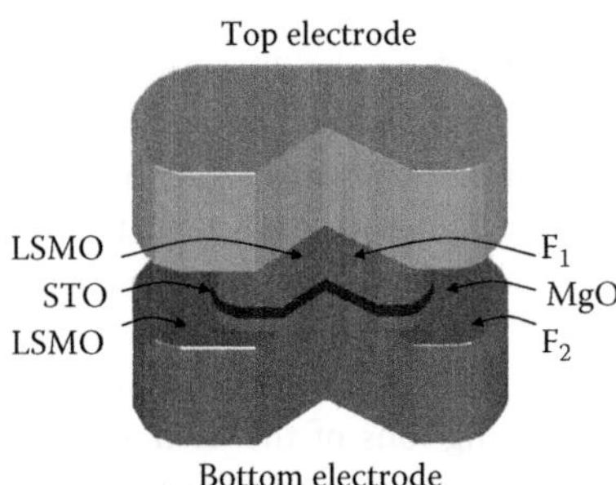

FIGURE 7.16 Magnetic junction: Two thin film magnets are separated by a nonmagnetic barrier. Top and bottom electrodes are used for electrical access to the junction. Bottom magnet can be a part of the bottom electrode. In MMTJs $F_1 = F_2$ = LSMO with STO barrier. In MTJs F_1 and F_2 are transition-metal magnets separated by AlO_x, MgO, or other tunneling barrier.

trilayer is typically prepared by sputtering with F_1 and F_2 layers made of transition metals or their alloys separated by an oxide barrier. Initial observations of STT in MTJs were done with barrier = AlO_x [65–67] later replaced by MgO [68], which is more robust to electrical breakdown and produce higher MR ratios. More details on STT in MTJs can be found in the following section on spin transport in MTJs.

Current-induced magnetization reversal has also been reported in MTJs based on ferromagnetic semiconductors [69–72]. Here the roles of two F electrodes are played by ferromagnetic III–V semiconductors, such as (Ga,Mn)As, characterized by carrier-induced ferromagnetism and strong spin–orbit interaction, while GaAs or AlAs is used as a nonmagnetic spacer layer. Such semiconductor-based trilayers are typically grown by MBE.

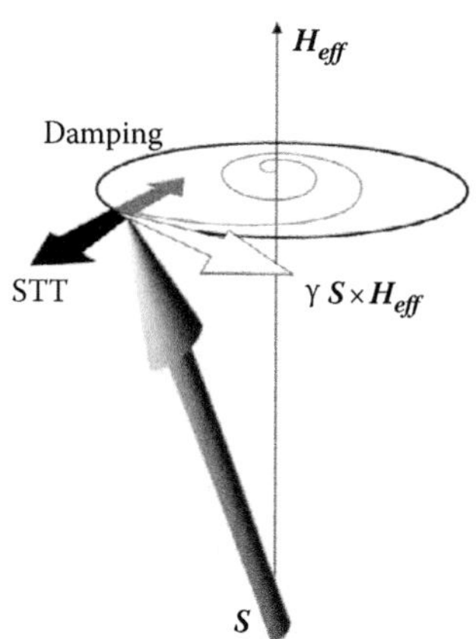

FIGURE 7.17 Torques on a magnetic moment in a magnetic field and subject to an electric current.

7.8 STT Experiment

Previous sections introduced the physics of STT and various magnetic nanosystems where STT can be observed. We have seen that a high-density electric current can result in torques on magnetic elements of a system. These torques may be used to control and manipulate the system's magnetic state. However, the resulting behavior of the system can differ significantly from case to case and depends on particular conditions of observation. For instance, in modest external magnetic fields, magnetization of a small element can be repeatedly reversed between two stable configurations; while at higher fields, where the reversal is energetically unfavorable, the moment can be set into precession at a very high frequency. In order to understand details of what happens with a magnetic moment $\boldsymbol{S}$ in a particular situation, one can use Newton's second law. For $\boldsymbol{S}$ this would be the Landau–Lifshitz–Gilbert equation where the rate of change of $\boldsymbol{S}$ is set equal to the net torque acting on $\boldsymbol{S}$:

$$\dot{S} = \gamma S \times H_{eff} - \alpha \hat{s} \times \dot{S} + I \frac{\eta}{e} \hat{s} \times \left(\hat{s}^{*} \times \hat{s} \right) \tag{7.1}$$

Here the first term on the right is the torque on a magnetic moment $\boldsymbol{S}$ in an effective magnetic field $\vec{H}_{eff}$ (including applied, demagnetizing, anisotropy, etc., fields), with γ the gyromagnetic ratio; the second term is a phenomenological damping term introduced by Gilbert, with α the Gilbert damping parameter; the third term is the STT introduced by Slonczewski [4] where small $\hat{s}$ and $\hat{s}^{*}$ are unit vectors along $\boldsymbol{S}$ and the polarizer $\boldsymbol{S}^{*}$, I is the current, e the electron charge, and η is the spin-polarization factor.

The diagram in Figure 7.17 shows the directions of the three torques from Equation 7.1. Note that depending on the polarity of applied current, the STT torque can be either in the same direction as the damping torque or opposite to it. In the former case, STT will effectively result in an increased damping for any magnitude of the applied current and suppress any possible excitations of $\boldsymbol{S}$ from its equilibrium state along $\boldsymbol{H}_{eff}$. If, however, the STT torque is opposite to the damping we can distinguish two situations. For currents below a critical current, where STT is small compared to damping, $\boldsymbol{S}$ spirals toward $\boldsymbol{H}_{eff}$ (gray trajectory in Figure 7.17). For currents larger than the critical current, STT exceeds the damping torque and causes $\boldsymbol{S}$ to spiral away from $\boldsymbol{H}_{eff}$, with a steadily increasing precession angle. The ultimate result can be, depending on angular dependence of STT and damping, either stable steady-state precession of $\boldsymbol{S}$ around $\boldsymbol{H}_{eff}$ (black trajectory in Figure 7.17) or magnetic reversal of $\boldsymbol{S}$ into a state antiparallel to $\boldsymbol{H}_{eff}$.

Almost all experimental observations of STT rely on GMR phenomenon to detect the current-induced reorientation of magnetic moments in nanodevices. Typical measurements include (i) measuring static device resistance $R = V/I$ as a function of applied dc bias current I in an applied magnetic field and (ii) measuring R versus applied field at a constant I. Figure 7.3 shows how the differential resistance dV/dI of a Cu point contact to Co/Cu magnetic multilayer varies with the bias voltage V (equivalent to I) applied across the contact. Here the multilayer magnetic moments are saturated out of the plane of the layers by a large external magnetic field (≥2 T). The onset of STT-driven magnetic precession is revealed by a peak in the differential dV/dI contact resistance. The peak in dV/dI may indicate that the transition into precession is a reversible process, and in a small range of currents one can continuously increase/decrease the angle of precession, although other scenarios, e.g., fast switching between steady-state precession and static states with current-dependent dwell time, are also possible [112].

If the applied magnetic field is small, the magnetic system can have more than one low-energy state. In the simple case of a magnetic element with uniaxial anisotropy, STT can trigger a transition between two static states, which are equally energetically favorable without current. An example of such behavior is shown in Figure 7.18. Here the current is driven across a trilayer Py20 nm/Cu12 nm/Py4.5 nm spin-valve structure patterned by EBL into a nanopillar with cross-sectional area $40 \times 120\,\text{nm}^2$ [148]. The differential resistance of the nanopillar exhibits a hysteresis as a function of an applied bias current as the magnetization of the thin (free) permalloy (Py) layer is aligned parallel or antiparallel to the thick (hard) Py layer by the current.

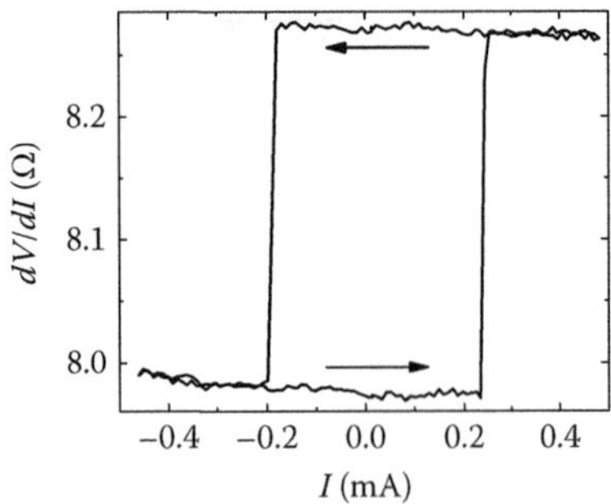

FIGURE 7.18 Spin-torque-driven magnetic switching for a Py20nm/Cu12nm/Py4.5nm spin valve with cross-sectional area $40 \times 120\,\text{nm}^2$, as the magnetization of the thin (free) magnetic layer is aligned parallel (at high negative currents) and antiparallel (at high positive currents) to the thicker magnetic layer by an applied current. Data are courtesy of Braganca and Ralph. (From Braganca, P.M. et al., *Appl. Phys. Lett.*, 87, 112507, 2005. With permission.)

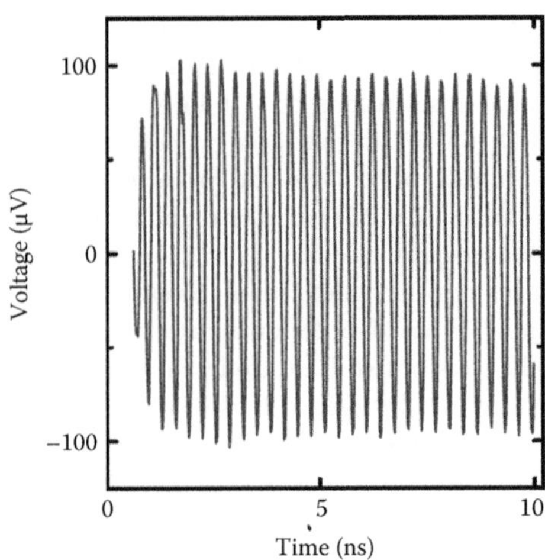

FIGURE 7.19 Oscillatory voltage generated by precessional motion of the free magnet in IrMn8nm/NiFe4nm/Cu8nm/NiFe4nm nanopillar, in response to a 335 mV dc voltage step applied to the device at $B = 630\,\text{Oe}$. Data are courtesy of Krivorotov and Ralph. (From Krivorotov, I.N. et al., *Science*, 307, 228, 2005.)

The two examples presented above (Figures 7.3 and 7.18) demonstrate how simple dc resistance measurements can be used for STT observation. Here measured resistance of a device provides indirect information about the relative orientation of magnetic elements in the device. However, measured critical currents highlighted by sharp variations in the resistance remain almost the only experimental information, which can be used for quantitative comparison with theory. Moreover the dc measurements of Figure 7.3 provide no information about fast evolution of magnetization in the device associated with high-frequency precession of magnetic moments. High-frequency techniques must be employed to provide such capabilities as discussed next.

The first experiment that gave unequivocal evidence that a dc electric current can result in high-frequency (tens of GHz) precession of magnetic moments was reported by Tsoi et al. [27]. Here an STT device, point contact, was placed in a microwave cavity of a high-frequency high-field electron spin resonance (ESR) spectrometer. This arrangement allowed performing dc transport experiments, such as those described earlier, while the contact was irradiated with high-frequency microwaves. Such irradiation induces microwave currents in the contact and was shown to be equivalent to the case where microwave currents are fed to the contact via electric leads [89]. When the frequency of external microwaves matched the precession frequency excited by dc current, an additional dc voltage was detected across the contact. By detecting this voltage, while varying the external frequency, field, and applied current, the STT-driven frequency could be mapped as a function of these parameters. Rippard et al. [149] have fed microwaves to a lithographically defined point contact via electric leads and reported observation of a similar dc response.

Finally, the high-frequency dynamics of the free-layer magnetization can be measured directly by detecting high-frequency oscillations in voltage across a spin valve under dc bias. Here the hard magnetic layer is fixed while the free layer is STT-driven into high-frequency precession relative to the hard layer. GMR results in a high-frequency modulation of the spin-valve resistance, which, in turn, leads to a high-frequency component of the voltage across the spin valve traversed by a dc current. This voltage can be directly probed with a high-frequency spectrum analyzer, as was demonstrated by Kiselev et al. [93] and Rippard et al. [35]. Moreover the voltage oscillations due to spin-torque-driven magnetic precession can be directly measured in time domain using a sampling oscilloscope [94]; an example of the high-frequency oscillations in voltage across an exchange-biased IrMn/NiFe/Cu/NiFe spin valve is shown in Figure 7.19.

The earlier examples illustrate how broadband instrumentation for measuring voltage in GMR devices may provide important, often unique, information about high-frequency magnetic dynamics driven by spin-transfer torques. However, the detailed understanding of STT is still the subject of debate and requires new experimental techniques that are capable of probing magnetization dynamics on nanometer length scales and sub-nanosecond time scales. In principle, this can be accomplished by the use of synchrotron x-rays that were recently shown [98] to probe interfacial phenomena and directly image the time-resolved response of magnetic nanostructures to sub-nanosecond magnetic field pulses (Oersted switching) and spin-polarized current pulses (STT switching). Figure 7.20 shows scanning transmission x-ray microscopy [150] images of in-plane components of magnetization $\boldsymbol{M} - M_x$ in panel (a) and M_y in panel (b), in the free CoFe layer (indicated by ellipse) of a $100 \times 150\,\text{nm}^2$ magnetic nanopillar. The images were obtained by scanning a focused (diameter ~30 nm) circularly polarized x-ray beam across the CoFe layer, with the photon energy tuned to the characteristic Co L3 resonance to provide magnetic contrast through the x-ray magnetic circular dichroism (XMCD) effect [151], and by monitoring transmission of the x-rays as a function of the position x, y with a fast avalanche detector. The $\boldsymbol{M}$-vector field of the free layer can be reconstructed from the measured M_x and M_y components, as illustrated in Figure 7.20c, and the ultrafast x-ray microscopy

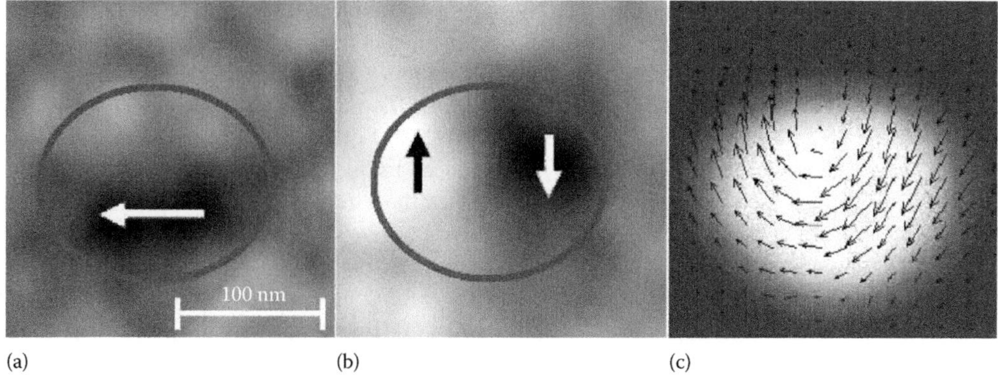

FIGURE 7.20 Scanning transmission x-ray microscopy images of M_x (a) and M_y (b) components of magnetization M combined into the vector field (c), which represents the direction of M in the plane of the CoFe free layer. Data are courtesy of Acremann and Stöhr. (From Acremann, Y. et al., *Phys. Rev. Lett.*, 96, 217202, 2006. With permission.)

technique provided a means to monitor this field as a function of time with ~100 ps resolution. The spatial resolution of the technique is set by the spot size of x-ray beam (~30 nm) and currently limits its application to spintronic devices >100 nm in size [98]. Potentially, however, technical development of the ultrafast x-ray microscopy may lead to an ultimate technique for STT studies, which can probe the ***M***-vector field on the nanometer length scale with picosecond time resolution.

7.9 Current-Induced Domain Wall Motion

Yet another manifestation of STT in metallic ferromagnets is a motion of magnetic domain walls traversed by an electric current. The original prediction of the effect dates back to 1978 when Luc Berger predicted that a spin-polarized current should apply a torque to a magnetic domain wall [9]. In a series of theoretical [9,10,12,13] and experimental [11,14,15] works, Berger set the groundwork for current-induced domain wall motion (CIDWM), which is now documented in materials ranging from superlattices with perpendicular anisotropy [127] to magnetic semiconductors [121]. But the most widely studied materials by far have been metallic ferromagnets, including Py ($Ni_{81}Fe_{19}$), CoFe, Co, because of their decades-long ubiquity in magnetic storage technology.

The CIDWM effect can be qualitatively understood on the basis of the following arguments. Consider an electric current flowing between two magnetic domains (A and B) with opposite magnetizations and, thus, traversing a 180° magnetic domain wall. The situation is somewhat similar to the case of a single N/F interface discussed earlier in conjunction with STT in magnetic multilayers. While in domain A, spins of conduction electrons are preferentially aligned with the magnetic moment of A. But once into B, the spins reverse to align with the magnetic moment of B. In reversing the electron spins, magnetic moments in the domain wall experience a torque associated with the change in angular momentum that occurs from the rotation of spins of transport electrons. This torque can move the domain wall in the direction of the electron flow, CIDWM.

Models that describe the mechanisms responsible for CIDWM fall into two classes, termed "adiabatic" [9,152,153] and "nonadiabatic" [10,152,154,155], referring essentially to the way electron spins traverse the magnetization texture of domain walls. The adiabatic torque arises when the spins of transport electrons follow the underlying magnetic landscape adiabatically. On the other hand, when spatial variations in magnetization are too rapid to be followed adiabatically by the transport spins, nonadiabatic contributions appear. The resulting magnetodynamics of domain walls can be resolved by introducing these current-induced torques into the equation of motion for magnetization (see, e.g., Equation 7.1). Most analytical work performs this operation within the framework of one-dimensional (1D) domain walls schematically illustrated in Figure 7.21a. Here we have chosen a Néel-type domain wall appropriate for thin films with large out-of-plane demagnetizing factor that promotes in-plane orientation of magnetic moments. The in-plane orientation of each spin is defined by angle θ, which varies from 0 in one magnetic domain to π in another domain over a characteristic domain-wall width Δ [156]. Experimental realization of such thin-film stripe geometry provides a workbench for testing CIDWM models.

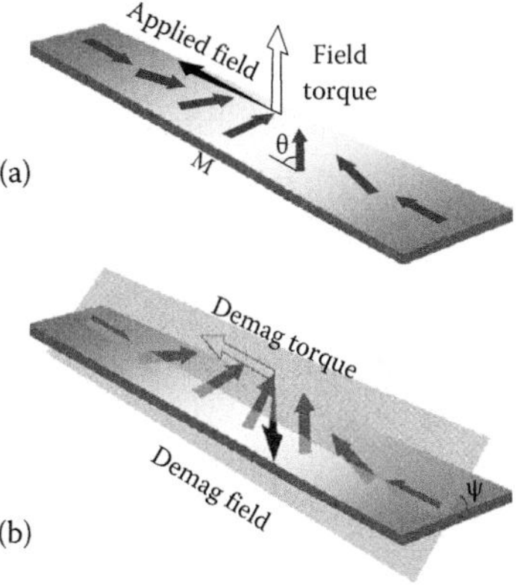

FIGURE 7.21 Schematic of (a) a 1D domain wall in a ferromagnetic nanostripe and (b) the torques involved in driving its motion.

The early experiments on CIDWM by Berger and coworkers [11,14,15] used ~3-mm-wide and 14–86 nm thick stripes of NiFe and the Faraday effect to observe magnetic domain walls. Relatively large cross sections of such samples (~10^{-10} m^2) required very large currents (up to 45 A) in order to achieve high enough current densities (~10^{11} A/m^2) needed for CIDWM; and even though the current was pulsed with pulse duration of ~2 μs, most of the samples were destroyed through electrical breakdown or local heating. Those samples that survived the pulses exhibited displacements of domain walls on the scale from 1 to 10 μm. For films thicker than 40 nm, different walls were observed to be moved by current in different directions, as expected for the effect of Oersted field produced by the current. However, all thinner (14–30 nm) samples exhibited domain-wall displacements in the direction of the electron current flow, which is consistent with STT mechanism of CIDWM most pronounced in thinner samples. Gan et al. [59] have used 100 nm thick NiFe stripes with cross sections almost two orders of magnitude smaller than those used by Berger and coworkers, but still needed currents of the order of 1 A to produce micron-scale displacements of domain walls observed with a magnetic force microscope (MFM).

With advances in nanofabrication techniques, magnetic stripes with ~10^{-15} m^2 cross sections can now be routinely produced where currents of only a few mA and below can trigger CIDWM. Moreover, unlike the initial CIDWM observations of multiple domain walls, which are the consequence of relatively large samples, CIDWM of individual magnetic domain walls is now possible. Most recent experiments [128–147] study domain-wall propagation in thin-film magnetic nanostripes prepared by EBL or FIB patterning. Such nanostripes are typically connected to extended magnetic elements, as shown in Figures 7.14 and 7.15. These extended areas are used for domain-wall nucleation and subsequent injection into the stripe where it can propagate under the influence of applied current, or magnetic field, or both. The 1D wall geometry shown in Figure 7.21a is an appropriate starting point for analysis of most experiments.

The 1D domain wall is described by two collective coordinates—the wall displacement q and its canting angle ψ (see Figure 7.21b), while (in-plane, out-of-plane) orientation of individual spins is denoted (θ, ϕ). Motion of the wall requires a torque on θ to align the wall spins along the nanostripe and drive q and ψ motion. For 1D domain wall, the equation of motion (see Equation 7.1) reduces to the following equations [152,155]:

$$\dot{q} = (2\pi \boldsymbol{M}_S)\gamma\Delta\sin 2\psi + \alpha\Delta\dot{\psi} + \eta\boldsymbol{u} \tag{7.2a}$$

$$\dot{\psi} = \gamma\boldsymbol{H}_a - \left(\frac{\alpha}{\Delta}\right)\dot{q} + \frac{(\beta\boldsymbol{u})}{\Delta} \tag{7.2b}$$

Here current density j is included via $u = -(g\mu_B p/2eM_S)j$, p polarization, and M_S magnetization; Δ is domain-wall width. The last terms of Equations 7.2a and 7.2b describe adiabatic and nonadiabatic interactions, respectively.

Without current (without last terms in Equations 7.2a and 7.2b) the domain wall is driven solely by an applied magnetic field ($\boldsymbol{H}_a$). If the field is applied along the nanostripe, the field-induced torque $\gamma\boldsymbol{M}\times\boldsymbol{H}_a$ causes the domain wall spins to cant out of the film plane, as shown in Figure 7.21b, but cannot directly drive the wall motion (cannot change θ). Instead, a demagnetizing field $\boldsymbol{H}_d$, which develops in the opposite direction as the out-of-plane component of $\boldsymbol{M}_S$, will produce the demagnetizing torque $\gamma\boldsymbol{M}\times\boldsymbol{H}_d$ that drives θ and consequent wall motion $\dot{q}$. For small applied fields, a simple translational motion of the wall occurs with a constant canting ψ. When applied magnetic field increases, so does the demagnetizing torque, the canting angle ψ, and the resulting domain-wall velocity $\dot{q}$. But the increase cannot continue indefinitely. At $\psi = \pi/4$, the demagnetizing torque, and thus $\dot{q}$, peaks. If H_a drives ψ past this limit, ψ can no longer remain stationary—a transition known as Walker breakdown [157] occurs. Here ψ advances continually and the demagnetizing torque averages to zero as it changes direction with each quarter period of ψ rotation, i.e., cannot move the wall forward. As ψ rotation becomes more and more rapid with increasing H_a, a small net damping torque $\alpha\boldsymbol{M}\times\dot{\boldsymbol{M}}$ cants the wall spins toward H_a and drives the wall forward in place of the demagnetizing torque. The Walker breakdown of individual domain walls was observed by Beach et al. [126] in NiFe nanostripes using high-bandwidth Kerr polarimetry (see Figure 7.22).

If we now turn the effect of current on (last terms in Equations 7.2a and 7.2b), one can distinguish two limiting cases—below and far above the Walker breakdown. Below the breakdown ψ is

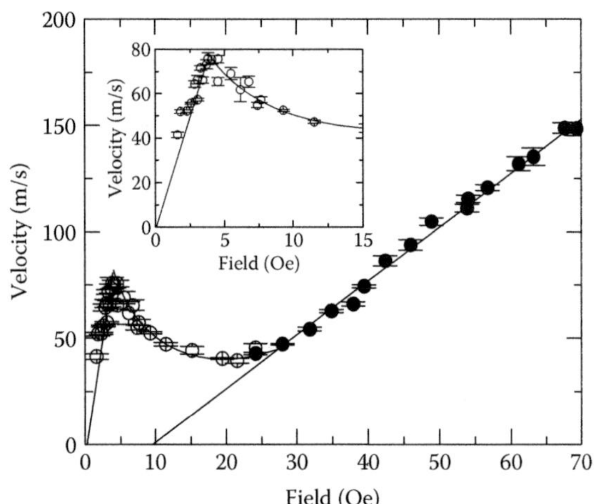

FIGURE 7.22 Average domain-wall velocity in a 600-nm-wide and 20-nm-thick NiFe nanostripe versus applied magnetic field. The inset shows the detail around the velocity peak. Straight solid lines are linear fits to the data below and far above the Walker breakdown at $H_W \cong 4$ Oe; the curved line is a visual guide. (From Beach, G.S.D. et al., *Nat. Mater.*, 4, 741, 2005. With permission.)

stationary ($\dot{\psi} = 0$) and only the nonadiabatic term can drive wall motion. From Equation 7.2a the wall velocity is

$$\dot{q} = \left(\frac{\gamma\Delta}{\alpha}\right) \boldsymbol{H}_a + \frac{(\beta \boldsymbol{u})}{\alpha} \tag{7.3}$$

and the sole effect of adiabatic torque is to shift the steady state of ψ and vary the Walker breakdown field $\boldsymbol{H}_W = 2\pi\alpha \boldsymbol{M}_S + (\alpha\eta - \beta)\boldsymbol{u}/\gamma\Delta$.

Far above the breakdown, it is only the adiabatic torque that can drive the wall motion. For $\dot{\psi} \gg 0$ and, as a result time-averaged $\langle \sin 2\psi \rangle = 0$, the wall velocity is given by

$$\dot{q} = \gamma\Delta\alpha \boldsymbol{H}_a + \eta \boldsymbol{u} \tag{7.4}$$

In experiments adiabatic and nonadiabatic interactions may thus be probed independently simply by using an applied magnetic field to select between the regimes of Equation 7.3 or Equation 7.4. Such experiments were performed in permalloy nanowires by Beach et al. [130] using ultrafast Kerr polarimetry and by Hayashi et al. [158] using real-time resistance measurements. An applied electric current causes the linear parts, at low and high H_a, of the curve in Figure 7.22 to shift vertically, as expected from Equations 7.3 and 7.4. This indicates that the equation of motion (Equation 7.2) is capable of describing the effects of field and current at least qualitatively. However, a detailed analysis of the data in Refs. [126,130] suggests that this 1D model has difficulties to account for quantitative characteristics of both field- and current-driven domain-wall dynamics on an equal footing [159]. Micromagnetic simulations that involve more complex domain-wall textures, e.g., vortex walls, provide a better agreement with experiments and indicate that meaningful comparison between experiment and theory will require models that fully account for realistic, time-variant domain-wall structures.

7.10 STT Applications

The STT method to manipulate magnetic moments by an electric current offers unprecedented spatial and temporal control of spin distributions and has a great potential to be applied in a broad range of technologies. Here we briefly discuss the perspective of STT for GHz communication applications and magnetic recording technology. More details on some of the applications can be found in Chapters 33–38.

The STT application in high-frequency technologies is based on the spin-transfer-induced precession of spins. In the previous section we have seen that precession of magnetization in GMR devices can convert a dc current input into an ac voltage output. The frequency of this output can be tuned from a few GHz to >100 GHz by changing the applied magnetic field and/or the dc current, effectively resulting in a current-controlled oscillator to be used in practical microwave circuits. Hence, the STT effect in GMR structures provides a means to engineer a nanoscale high-frequency oscillator powered and tuned by dc current. Such an oscillator could potentially have frequency characteristics spanning more than 100 GHz and perhaps into THz range. Linewidths as narrow as 2 MHz were demonstrated [160], leading to quality factors over 18,000. Potential applications for such high-frequency sources include integrated transceivers for wireless and wired applications, and chip-to-chip, as well as on-chip communications both wireless and wired. For the latter, logic circuits with a spin wave bus were proposed [161,162] as an interface between electronic circuits and integrated spintronics circuits. Here spin waves are used for information transmission and processing, and the STT effect can provide a means for efficient spin-wave generation on the nanoscale.

In high-speed, high-density magnetic recording technology, STT could replace the Oersted field currently used for writing magnetic bits in a storage media, thus leading to smaller and faster magnetic memory. For instance, the bit cell of a conventional MRAM is programmed to a "1" or "0" by switching between the P and AP states of a GMR-like storage element, e.g., spin valve. The first-generation MRAM utilizes MTJs as storage elements because of their higher MR ratios and impedance-matching constrains. However, in scaling MRAM to small dimensions, the same constrains are expected to drive a transition from MTJs to fully metallic spin-valve storage elements. The switching between "1" and "0" states (writing) relies on magnetic reversal of the MTJ's free layer achieved by passing electric currents down the "bit" and "write" lines that generates a sufficiently strong magnetic field at their intersection, i.e., for a given MTJ. However, as the spatial decay of this Oersted field is rather slow (~$1/r^2$), it may affect neighboring cells. This makes scaling of conventional MRAM to smaller dimensions challenging. So-called STT-MRAM removes the above constrains on scalability. Here the switching of the free-layer magnetic moment is achieved by STT switching when a high-density electric current is driven directly through the storage element (MTJ or spin valve). The writing is thus performed at high current levels, while the reading (measuring resistance of element) is still done at low currents. Since STT-MRAM eliminates the need for "write" and "bypass" lines, a more compact memory can be realized.

Finally, CIDWM was proposed as the basis for a new type of magnetic memory called "racetrack" [135]. In contrast to today's hard disc drives (HDDs), which rely on spinning motion of a disk to move its magnetic regions where the data are stored past a "read" head, the racetrack memory exploits the idea of moving magnetically stored data electronically. The racetrack is a ferromagnetic nanowire, with data encoded as a pattern of magnetic domains along the nanowires. An electric current can drive the entire data-pattern in the racetrack past "read" and "write" elements. If a tunnel junction to the racetrack is used as a reading element, the reading operation is achieved simply by measuring resistance of the junction, which depends on the orientation of magnetic domains in the junction area. The same junction can be used to inject a spin-polarized current into the racetrack locally, thus mediating a local reversal of magnetization via STT—the data writing. The reversal (writing) can also be triggered by magnetic fields applied locally. A high-density 3D array of the racetracks would result in a memory chip with storage density higher than that in solid-state memory devices like Flash RAM and similar to conventional HDDs, but potentially may have much higher "read/write" performance than HDD.

What is the future of STT in spintronic applications? We have seen that a number of new STT-based spintronic devices have been proposed. In the foreseeable future, however, a great deal of fundamental work remains to be done before we see commercial applications of these devices. For memory industry, this may lead to a universal memory, which would combine cost benefits of DRAM, speed of SRAM, and nonvolatility of Flash RAM. Potentially all logic operations on a chip can be carried out by manipulating spins in metallic systems instead of manipulating charges in semiconductor transistors, as in conventional microchips, and, moreover, could be combined on-chip with a universal memory. This would result in a new, scalable, radiation-resistant, and more power-efficient computers and electronic devices on which we rely in almost every aspect of our everyday life. If STT is the method of choice to control and manipulate spins in such devices, its future will be very bright.

References

1. M. N. Baibich, J. M. Broto, A. Fert et al., Giant magnetoresistance of (001)Fe/(001)Cr magnetic superlattices, *Phys. Rev. Lett.* **61**, 2472–2475 (1988).
2. G. Binasch, P. Grünberg, F. Saurenbach, and W. Zinn, Enhanced magnetoresistance in layered magnetic structures with antiferromagnetic interlayer exchange, *Phys. Rev. B* **39**, 4828–4830 (1989).
3. L. Berger, Emission of spin waves by a magnetic multilayer traversed by a current, *Phys. Rev. B* **54**, 9353–9358 (1996).
4. J. C. Slonczewski, Current-driven excitation of magnetic multilayers, *J. Magn. Magn. Mater.* **159**, L1–L7 (1996).
5. M. Tsoi, A. G. M. Jansen, J. Bass et al., Excitation of a magnetic multilayer by an electric current, *Phys. Rev. Lett.* **80**, 4281–4284 (1998).
6. J. Z. Sun, Current-driven magnetic switching in manganite trilayer junctions, *J. Magn. Magn. Mater.* **202**, 157–162 (1999).
7. E. B. Myers, D. C. Ralph, J. A. Katine, R. N. Louie, and R. A. Buhrman, Current-induced switching of domains in magnetic multilayer devices, *Science* **285**, 867–880 (1999).
8. J.-E. Wegrowe, D. Kelly, Y. Jaccard, Ph. Guittienne, and J.-Ph. Ansermet, Current-induced magnetization reversal in magnetic nanowires, *Europhys. Lett.* **45**, 626–632 (1999).
9. L. Berger, Low-field magnetoresistance and domain drag in ferromagnets, *J. Appl. Phys.* **49**, 2156–2161 (1978).
10. L. Berger, Exchange interaction between domain wall and current in very thin metallic films, *J. Appl. Phys.* **55**, 1954–1956 (1984).
11. P. P. Freitas and L. Berger, Observation of s–d exchange force between domain-walls and electric-current in very thin permalloy-films, *J. Appl. Phys.* **57**, 1266–1269 (1985).
12. L. Berger, Possible existence of Josephson effect in ferromagnets, *Phys. Rev. B* **33**, 1572–1578 (1986).
13. L. Berger, Exchange interaction between electric-current and magnetic domain-wall containing Bloch lines, *J. Appl. Phys.* **63**, 1663–1669 (1988).
14. C.-Y. Hung and L. Berger, Exchange forces between domain-wall and electric-current in permalloy-films of variable thickness, *J. Appl. Phys.* **63**, 4276–4278 (1988).
15. C.-Y. Hung, L. Berger, and C. Y. Shih, Observation of a current-induced force on Bloch lines in Ni–Fe thin-films, *J. Appl. Phys.* **67**, 5941–5943 (1990).
16. R. Weil and R. G. Barradas, *Electrocrystallization*, The Electrochemical Society, Pennington, NJ, 1981.
17. D. S. Lashmore and M. P. Dariel, Electrodeposited Cu–Ni textured superlattices, *J. Electrochem. Soc.* **135**, 1218–1221 (1988).
18. M. Ohring, *The Materials Science of Thin Films*, Academic Press, Boston, MA, 1992.
19. J. M. Slaughter, W. P. Pratt Jr., and P. A. Schroeder, Fabrication of layered metallic systems for perpendicular resistance measurements, *Rev. Sci. Instrum.* **60**, 127–131 (1989).
20. W. P. Pratt Jr., S. F. Lee, J. M. Slaughter, R. Loloee, P. A. Schroeder, and J. Bass, Perpendicular giant magnetoresistances of Ag/Co multilayers, *Phys. Rev. Lett.* **66**, 3060–3063 (1991).
21. B. Dieny, V. S. Speriosu, S. Metin et al., Magnetotransport properties of magnetically soft spin-valve structures, *J. Appl. Phys.* **69**, 4774–4779 (1991).
22. L. Piraux, J. M. George, J. F. Despres et al., Giant magnetoresistance in magnetic multilayered nanowires, *Appl. Phys. Lett.* **65**, 2484–2486 (1994).
23. A. Blondel, J. P. Meier, B. Doudin, and J.-Ph. Ansermet, Giant magnetoresistance of nanowires of multilayers, *Appl. Phys. Lett.* **65**, 3019–3021 (1994).
24. M. Tsoi, A. G. M. Jansen, and J. Bass, Search for point-contact giant magnetoresistance in Co/Cu multilayers, *J. Appl. Phys.* **81**, 5530–5532 (1997).
25. B. Dieny, V. S. Speriosu, S. S. P. Parkin, B. A. Gurney, D. R. Wilhoit, and D. Mauri, Giant magnetoresistance in soft ferromagnetic multilayers, *Phys. Rev. B* **43**, 1297–1300 (1991).
26. X. Waintal, E. B. Myers, P. W. Brouwer, and D. C. Ralph, Role of spin-dependent interface scattering in generating current-induced torques in magnetic multilayers, *Phys. Rev. B* **62**, 12317–12327 (2000).
27. M. Tsoi, A. G. M. Jansen, J. Bass, W.-C. Chiang, V. Tsoi, and P. Wyder, Generation and detection of phase-coherent current-driven magnons in magnetic multilayers, *Nature* **406**, 46–48 (2000).
28. W. H. Rippard, M. R. Pufall, and T. J. Silva, Quantitative studies of spin-momentum-transfer-induced excitations in Co/Cu multilayer films using point-contact spectroscopy, *Appl. Phys. Lett.* **82**, 1260–1262 (2003).

29. Y. Ji, C. L. Chien, and M. D. Stiles, Current-induced spin-wave excitations in a single ferromagnetic layer, *Phys. Rev. Lett.* **90**, 106601 (2003).
30. I. K. Yanson, Yu. G. Naidyuk, D. L. Bashlakov et al., Spectroscopy of phonons and spin torques in magnetic point contacts, *Phys. Rev. Lett.* **95**, 186602 (2004).
31. O. P. Balkashin, V. V. Fisun, I. K. Yanson, L. Yu. Triputen, A. Konovalenko, and V. Korenivski, Spin dynamics in point contacts to single ferromagnetic films, *Phys. Rev. B* **79**, 092419 (2009).
32. Z. Wei, A. Sharma, A. S. Nunez et al., Changing exchange bias in spin valves with an electric current, *Phys. Rev. Lett.* **98**, 116603 (2007).
33. Y. Luo, M. Esseling, M. Münzenberg, and K. Samwer, A novel spin transfer torque effect in Ag_2Co granular films, *New J. Phys.* **9**, 329 (2007).
34. X. J. Wang, H. Zou, and Y. Ji, Spin transfer torque switching of cobalt nanoparticles, *Appl. Phys. Lett.* **93**, 162501 (2008).
35. W. H. Rippard, M. R. Pufall, S. Kaka, S. E. Russek, and T. J. Silva, Direct-current induced dynamics in $Co_{90}Fe_{10}/Ni_{80}Fe_{20}$ point contacts, *Phys. Rev. Lett.* **92**, 027201 (2004).
36. S. Kaka, M. R. Pufall, W. H. Rippard, T. J. Silva, S. E. Russek, and J. A. Katine, Mutual phase-locking of microwave spin torque nano-oscillators, *Nature* **437**, 389–392 (2005).
37. F. B. Mancoff, N. D. Rizzo, B. N. Engel, and S. Tehrani, Phase-locking in double-point-contact spin-transfer devices, *Nature* **437**, 393–395 (2005).
38. O. Ozatay, N. C. Emley, P. M. Braganca et al., Spin transfer by nonuniform current injection into a nanomagnet, *Appl. Phys. Lett.* **88**, 202502 (2006).
39. Q. Mistral, M. van Kampen, G. Hrkac et al., Current-driven vortex oscillations in metallic nanocontacts, *Phys. Rev. Lett.* **100**, 257201 (2008).
40. S. Bonetti, P. Muduli, F. Mancoff, and J. Akerman, Spin torque oscillator frequency versus magnetic field angle: The prospect of operation beyond 65 GHz, *Appl. Phys. Lett.* **94**, 102507 (2009).
41. J. A. Katine, F. J. Albert, R. A. Buhrman, E. B. Myers, and D. C. Ralph, Current-driven magnetization reversal and spin-wave excitations in Co/Cu/Co Pillars, *Phys. Rev. Lett.* **84**, 3149–3152 (2000).
42. J. Grollier, V. Cros, A. Hamzic et al., Spin-polarized current induced switching in Co/Cu/Co pillars, *Appl. Phys. Lett.* **78**, 3663–3665 (2001).
43. J. Z. Sun, D. J. Monsma, D. W. Abraham, M. J. Rooks, and R. H. Koch, Batch-fabricated spin-injection magnetic switches, *Appl. Phys. Lett.* **81**, 2202–2204 (2002).
44. B. Özyilmaz, A. D. Kent, D. Monsma, J. Z. Sun, M. J. Rooks, and R. H. Koch, Current-induced magnetization reversal in high magnetic fields in Co/Cu/Co nanopillars, *Phys. Rev. Lett.* **91**, 067203 (2003).
45. S. Urazhdin, N. O. Birge, W. P. Pratt Jr., and J. Bass, Current-driven magnetic excitations in permalloy-based multilayer nanopillars, *Phys. Rev. Lett.* **91**, 146803 (2003).
46. M. Tsoi, J. Z. Sun, and S. S. P. Parkin, Current-driven excitations in symmetric magnetic nanopillars, *Phys. Rev. Lett.* **93**, 036602 (2004).
47. M. Covington, A. Rebei, G. J. Parker, and M. A. Seigler, Spin momentum transfer in current perpendicular to the plane spin valves, *Appl. Phys. Lett.* **84**, 3103–3105 (2004).
48. M. R. Pufall, W. H. Rippard, S. Kaka, S. E. Russek, and T. J. Silva, Large-angle, gigahertz-rate random telegraph switching induced by spin-momentum transfer, *Phys. Rev. B* **69**, 214409 (2004).
49. T. Kimura, Y. Otani, and J. Hamrle, Switching magnetization of a nanoscale ferromagnetic particle using nonlocal spin injection, *Phys. Rev. Lett.* **96**, 037201 (2006).
50. T. Yang, T. Kimura, and Y. Otani, Giant spin-accumulation signal and pure spin-current-induced reversible magnetization switching, *Nat. Phys.* **4**, 851–854 (2008).
51. S. Kasai, Y. Nakatani, K. Kobayashi, H. Kohno, and T. Ono, Current-driven resonant excitation of magnetic vortices, *Phys. Rev. Lett.* **97**, 107204 (2006).
52. K. Yamada, S. Kasai, Y. Nakatani et al., Electrical switching of the vortex core in a magnetic disk, *Nat. Mater.* **6**, 269–273 (2007).
53. J.-E. Wegrowe, D. Kelly, T. Truong, Ph. Guittienne, and J.-Ph. Ansermet, Magnetization reversal triggered by spin-polarized current in magnetic nanowires, *Europhys. Lett.* **56**, 748–754 (2001).
54. J.-E. Wegrowe, A. Fábián, Ph. Guittienne, X. Hoffer, D. Kelly, and J.-Ph. Ansermet, Exchange torque and spin transfer between spin polarized current and ferromagnetic layers, *Appl. Phys. Lett.* **80**, 3775–3777 (2002).
55. A. Fábián, C. Terrier, S. Serrano Guisan et al., Current-induced two-level fluctuations in pseudo-spin-valve (Co/Cu/Co) nanostructures, *Phys. Rev. Lett.* **91**, 257209 (2003).
56. N. Biziere, E. Murè, and J.-Ph. Ansermet, Microwave spin-torque excitation in a template-synthesized nanomagnet, *Phys. Rev. B* **79**, 012404 (2009).
57. L. Piraux, K. Renard, R. Guillemet et al., Template-grown NiFe/Cu/NiFe nanowires for spin transfer devices, *Nano Lett.* **7**, 2563–2567 (2007).
58. T. Blon, M. Mátéfi-Tempfli, S. Mátéfi-Tempfli et al., Spin momentum transfer effects observed in electrodeposited Co/Cu/Co nanowires, *J. Appl. Phys.* **102**, 103906 (2007).
59. L. Gan, S. H. Chung, K. H. Aschenbach, M. Dreyer, and R. D. Gomez, Pulsed-current-induced domain wall propagation in permalloy patterns observed using magnetic force microscope, *IEEE Trans. Magn.* **36**, 3047–3049 (2000).
60. J. Grollier, D. Lacour, V. Cros et al., Switching the magnetic configuration of a spin valve by current-induced domain wall motion, *J. Appl. Phys.* **92**, 4825–4827 (2002).
61. M. Klaui, C. A. F. Vaz, J. A. C. Blanda et al., Domain wall motion induced by spin polarized currents in ferromagnetic ring structures, *Appl. Phys. Lett.* **83**, 105–107 (2003).
62. M. Tsoi, R. E. Fontana, and S. S. P. Parkin, Magnetic domain wall motion triggered by an electric current, *Appl. Phys. Lett.* **83**, 2617–2619 (2003).

63. T. Kimura, Y. Otani, I. Yagi, K. Tsukagoshi, and Y. Aoyagi, Suppressed pinning field of a trapped domain wall due to direct current injection, *J. Appl. Phys.* **94**, 7226–7229 (2003).
64. J. Z. Sun, Spin-dependent transport in trilayer junctions of doped manganites, *Physica C* **350**, 215–226 (2001).
65. Y. Liu, Z. Zhang, J. Wang, P. P. Freitas, and J. L. Martins, Current-induced switching in low resistance magnetic tunnel junctions, *J. Appl. Phys.* **93**, 8385–8387 (2003).
66. Y. Huai, F. Albert, P. Nguyen, M. Pakala, and T. Valet, Observation of spin-transfer switching in deep submicron-sized and low-resistance magnetic tunnel junctions, *Appl. Phys. Lett.* **84**, 3118–3120 (2004).
67. G. D. Fuchs, N. C. Emley, I. N. Krivorotov et al., Spin-transfer effects in nanoscale magnetic tunnel junctions, *Appl. Phys. Lett.* **85**, 1205–1207 (2004).
68. A. A. Tulapurkar, Y. Suzuki, A. Fukushima et al., Spin-torque diode effect in magnetic tunnel junctions, *Nature* **438**, 339–342 (2005).
69. D. Chiba, Y. Sato, T. Kita, F. Matsukura, and H. Ohno, Current-driven magnetization reversal in a ferromagnetic semiconductor (Ga;Mn)As/GaAs/(Ga;Mn)As tunnel junction, *Phys. Rev. Lett.* **93**, 216602 (2004).
70. R. Moriya, K. Hamaya, A. Oiwa, and H. Munekata, Current-induced magnetization reversal in a (Ga,Mn)As-based magnetic tunnel junction, *Jpn. J. Appl. Phys. Part 2 Lett. Express Lett.* **43**, L825–L827 (2004).
71. R. Moriya, K. Hamaya, A. Oiwa, and H. Munekata, Magnetization reversal by electrical spin injection in ferromagnetic (Ga,Mn)As-based magnetic tunnel junctions, *J. Supercond.* **18**, 3–7 (2005).
72. P. N. Hai, S. Ohya, M. Tanaka, S. E. Barnes, and S. Maekawa, Electromotive force and huge magnetoresistance in magnetic tunnel junctions, *Nature* **458**, 489–492 (2009).
73. M. Yamanouchi, D. Chiba, F. Matsukura, and H. Ohno, Current-induced domain-wall switching in a ferromagnetic semiconductor structure, *Nature* **428**, 539–542 (2004).
74. M. Yamanouchi, D. Chiba, F. Matsukura, T. Dietl, and H. Ohno, Velocity of domain-wall motion induced by electrical current in the ferromagnetic semiconductor (Ga,Mn) As, *Phys. Rev. Lett.* **96**, 096601 (2006).
75. M. Yamanouchi, J. Ieda, F. Matsukura, S. E. Barnes, S. Maekawa, and H. Ohno, Universality classes for domain wall motion in the ferromagnetic semiconductor (Ga,Mn) As, *Science* **317**, 1726–1729 (2007).
76. Yu. V. Sharvin, A possible method for studying Fermi surfaces, *Sov. Phys. JETP* **21**, 655 (1965).
77. V. S. Tsoi, Focusing of electrons in metal by a transverse magnetic field, *JETP Lett.* **19**, 70–71 (1974).
78. I. K. Yanson, Nonlinear effects in electric–conductivity of point junctions and electron-phonon interaction in normal metals, *Zh. Eksp. Teor. Fiz.* **66**, 1035–1050 (1974).
79. A. G. M. Jansen, A. P. van Gelder, and P. Wyder, Point-contact spectroscopy in metals, *J. Phys. C* **13**, 6073–6118 (1980).
80. V. S. Tsoi, J. Bass, and P. Wyder, Studying conduction-electron/interface interactions using transverse electron focusing, *Rev. Mod. Phys.* **71**, 1641–1693 (1999).
81. Yu. G. Naidyuk and I. K. Yanson, *Point-Contact Spectroscopy*, Springer, New York, 2004.
82. M. Tsoi, V. Tsoi, J. Bass, A. G. M. Jansen, and P. Wyder, Current-driven resonances in magnetic multilayers, *Phys. Rev. Lett.* **89**, 246803 (2002).
83. M. Tsoi, V. S. Tsoi, and P. Wyder, Generation of current-driven magnons in Co/Cu multilayers with antiferromagnetic alignment of adjacent Co layers, *Phys. Rev. B* **70**, 012405 (2004).
84. M. R. Pufall, W. H. Rippard, and T. J. Silva, Materials dependence of the spin-momentum transfer efficiency and critical current in ferromagnetic metal/Cu multilayers, *Appl. Phys. Lett.* **83**, 323–325 (2003).
85. J. Basset, A. Sharma, Z. Wei, J. Bass, and M. Tsoi, Towards antiferromagnetic metal spintronics, *Proc. SPIE* **7036**, 703605 (2008).
86. G. Wexler, Size effect and non-local Boltzmann transport equation in orifice and disk geometry, *Proc. Phys. Soc. London* **89**, 927 (1966).
87. B. Nikolic and P. B. Allen, Electron transport through a circular constriction, *Phys. Rev. B* **60**, 3963–3969 (1999).
88. J. C. Maxwell, *A Treatise on Electricity and Magnetism*, Clarendon Press, Oxford, 1873.
89. Z. Wei, Spin-transfer-torque effect in ferromagnets and antiferromagnets, Thesis, The University of Texas at Austin, Austin, TX, 2008.
90. M. Tsoi, Electromagnetic wave radiation by current-driven magnons in magnetic multilayers, *J. Appl. Phys.* **91**, 6801–6805 (2002).
91. K. S. Ralls, R. A. Buhrman, and R. C. Tiberio, Fabrication of thin-film metal nanobridges, *Appl. Phys. Lett.* **55**, 2459–2461 (1989).
92. B. O'Gorman and M. Tsoi, Fabrication of point contacts by FIB patterning, *Eur. Phys. J. Appl. Phys.* **49**, 10801 (2010).
93. S. I. Kiselev, J. C. Sankey, I. N. Krivorotov et al., Microwave oscillations of a nanomagnet driven by a spin-polarized current, *Nature* **425**, 380–383 (2003).
94. I. N. Krivorotov, N. C. Emley, J. C. Sankey, S. I. Kiselev, D. C. Ralph, and R. A. Buhrman, Time-domain measurements of nanomagnet dynamics driven by spin-transfer torques, *Science* **307**, 228–231 (2005).
95. T. Devolder, A. Tulapurkar, Y. Suzuki, C. Chappert, P. Crozat, and K. Yagami, Temperature study of the spin-transfer switching speed from dc to 100 ps, *J. Appl. Phys.* **98**, 053904 (2005).
96. N. Smith, J. A. Katine, J. R. Childress, and M. J. Carey, Angular dependence of spin torque critical currents for CPP-GMR read heads, *IEEE Trans. Magn.* **41**, 2935–2940 (2005).
97. S. Mangin, D. Ravelosona, J. A. Katine, M. J. Carey, B. D. Terris, and E. E. Fullerton, Current-induced magnetization reversal in nanopillars with perpendicular anisotropy, *Nat. Mater.* **5**, 210–215 (2006).

98. Y. Acremann, J. P. Strachan, V. Chembrolu et al., Time-resolved imaging of spin transfer switching: Beyond the macrospin concept, *Phys. Rev. Lett.* **96**, 217202 (2006).
99. J. C. Sankey, P. M. Braganca, A. G. F. Garcia, I. N. Krivorotov, R. A. Buhrman, and D. C. Ralph, Spin-transfer-driven ferromagnetic resonance of individual nanomagnets, *Phys. Rev. Lett.* **96**, 227601 (2006).
100. X. Jiang, L. Gao, J. Z. Sun, and S. S. P. Parkin, Temperature dependence of current-induced magnetization switching in spin valves with a ferrimagnetic CoGd free layer, *Phys. Rev. Lett.* **97**, 217202 (2006).
101. T. Yang, A. Hirohata, M. Hara, T. Kimura, and Y. Otani, Temperature dependence of intrinsic switching current of a Co nanomagnet, *Appl. Phys. Lett.* **89**, 252505 (2006).
102. O. Boulle, V. Cros, J. Grollier et al., Shaped angular dependence of the spin-transfer torque and microwave generation without magnetic field, *Nat. Phys.* **3**, 492–497 (2007).
103. V. S. Pribiag, I. N. Krivorotov, G. D. Fuchs et al., Magnetic vortex oscillator driven by d.c. spin-polarized current, *Nat. Phys.* **3**, 498–503 (2007).
104. D. Houssameddine, U. Ebels, B. Delaët et al., Spin-torque oscillator using a perpendicular polarizer and a planar free layer, *Nat. Mater.* **6**, 447–453 (2007).
105. O. Ozatay, P. G. Gowtham, K. W. Tan et al., Sidewall oxide effects on spin-torque- and magnetic-field-induced reversal characteristics of thin-film nanomagnets, *Nat. Mater.* **7**, 567–573 (2008).
106. S. Urazhdin and S. Button, Effect of spin diffusion on spin torque in magnetic nanopillars, *Phys. Rev. B* **78**, 172403 (2008).
107. G. Finocchio, O. Ozatay, L. Torres, R. A. Buhrman, D. C. Ralph, and B. Azzerboni, Spin-torque-induced rotational dynamics of a magnetic vortex dipole, *Phys. Rev. B* **78**, 174408 (2008).
108. S. Garzon, L. Ye, R. A. Webb, T. M. Crawford, M. Covington, and S. Kaka, Coherent control of nanomagnet dynamics via ultrafast spin torque pulses, *Phys. Rev. B* **78**, 180401R (2008).
109. S. H. Florez, J. A. Katine, M. Carey, L. Folks, O. Ozatay, and B. D. Terris, Effects of radio-frequency current on spin-transfer-torque-induced dynamics, *Phys. Rev. B* **78**, 184403 (2008).
110. S. Mangin, Y. Henry, D. Ravelosona, J. A. Katine, and E. E. Fullerton, Reducing the critical current for spin-transfer switching of perpendicularly magnetized nanomagnets, *Appl. Phys. Lett.* **94**, 012502 (2009).
111. T. Seki, H. Tomita, A. A. Tulapurkar, M. Shiraishi, T. Shinjo, and Y. Suzuki, Spin-transfer-torque-induced ferromagnetic resonance for Fe/Cr/Fe layers with an antiferromagnetic coupling field, *Appl. Phys. Lett.* **94**, 212505 (2009).
112. B. O'Gorman, S. Dietze, and M. Tsoi, Current-sweep-rate dependence of spin-torque driven dynamics in magnetic nanopillars, *J. Appl. Phys.* **105**, 07D111 (2009).
113. B. Özyilmaz, G. Richter, N. Müsgens et al., Focused-ion-beam milling based nanostencil mask fabrication for spin transfer torque studies, *J. Appl. Phys.* **101**, 063920 (2007).
114. A. Parge, T. Niermann, M. Seibt, and M. Münzenberg, Nanofabrication of spin-transfer torque devices by a polymethylmethacrylate mask one step process: Giant magnetoresistance versus single layer devices, *J. Appl. Phys.* **101**, 104302 (2007).
115. M. C. Wu, A. Aziz, D. Morecroft et al., Spin transfer switching and low-field precession in exchange-biased spin valve nanopillars, *Appl. Phys. Lett.* **92**, 142501 (2008).
116. M. Johnson and R. H. Silsbee, Interfacial charge-spin coupling: Injection and detection of spin magnetization in metals, *Phys. Rev. Lett.* **55**, 1790–1793 (1985).
117. F. J. Jedema, A. T. Filip, and B. J. van Wees, Electrical spin injection and accumulation at room temperature in an all-metal mesoscopic spin valve, *Nature* **410**, 345–348 (2001).
118. C. R. Martin, Nanomaterials: A membrane-based synthetic approach, *Science* **266**, 1961–1966 (1994).
119. G. E. Possin, A method for forming very small diameter wires, *Rev. Sci. Instrum.* **41**, 772–774 (1970).
120. J.-E. Wegrowe, D. Kelly, X. Hoffer, Ph. Guittienne, and J.-Ph. Ansermet, Tailoring anisotropic magnetoresistance and giant magnetoresistance hysteresis loops with spin-polarized current injection, *J. Appl. Phys.* **89**, 7127–7129 (2001).
121. M. Yamanouchi, D. Chiba, F. Matsukura, and H. Ohno, Current-induced domain-wall switching in a ferromagnetic semiconductor structure, *Nature* **428**, 539–542 (2004).
122. E. Saitoh, H. Miyajima, T. Yamaoka, and G. Tatara, Current-induced resonance and mass determination of a single magnetic domain wall, *Nature* **432**, 203–206 (2004).
123. A. Yamaguchi, T. Ono, S. Nasu, K. Miyake, K. Mibu, and T. Shinjo, Real-space observation of current-driven domain wall motion in submicron magnetic wires, *Phys. Rev. Lett.* **92**, 077205 (2004).
124. N. Vernier, D. A. Allwood, D. Atkinson, M. D. Cooke, and R. P. Cowburn, Domain wall propagation in magnetic nanowires by spin-polarized current injection, *Europhys. Lett.* **65**, 526–532 (2004).
125. C. K. Lim, T. Devolder, C. Chappert et al., Domain wall displacement induced by subnanosecond pulsed current, *Appl. Phys. Lett.* **84**, 2820–2822 (2004).
126. G. S. D. Beach, C. Nistor, C. Knutson, M. Tsoi, and J. L. Erskine, Dynamics of field-driven domain-wall propagation in ferromagnetic nanowires, *Nat. Mater.* **4**, 741–744 (2005).
127. D. Ravelosona, D. Lacour, J. A. Katine, B. D. Terris, and C. Chappert, Nanometer scale observation of high efficiency thermally assisted current-driven domain wall depinning, *Phys. Rev. Lett.* **95**, 117203 (2005).
128. L. Thomas, M. Hayashi, X. Jiang, R. Moriya, C. Rettner, and S. S. P. Parkin, Oscillatory dependence of current-driven magnetic domain wall motion on current pulse length, *Nature* **443**, 197–200 (2006).
129. M. Laufenberg, W. Bührer, D. Bedau et al., Temperature dependence of the spin torque effect in current-induced domain wall motion, *Phys. Rev. Lett.* **97**, 046602 (2006).

130. G. S. D. Beach, C. Knutson, C. Nistor, M. Tsoi, and J. L. Erskine, Nonlinear domain-wall velocity enhancement by spin-polarized electric current, *Phys. Rev. Lett.* **97**, 057203 (2006).
131. L. Thomas, M. Hayashi, X. Jiang, R. Moriya, C. Rettner, and S. S. P. Parkin, Resonant amplification of magnetic domain-wall motion by a train of current pulses, *Science* **315**, 1553–1556 (2007).
132. M. Hayashi, L. Thomas, C. Rettner, R. Moriya, and S. S. P. Parkin, Direct observation of the coherent precession of magnetic domain walls propagating along permalloy nanowires. *Nat. Phys.* **3**, 21–25. 2007.
133. G. Meier, M. Bolte, R. Eiselt, B. Krüger, D.-H. Kim, and P. Fischer, Direct imaging of stochastic domain-wall motion driven by nanosecond current pulses, *Phys. Rev. Lett.* **98**, 187202 (2007).
134. S. Laribi, V. Cros, M. Muñoz et al., Reversible and irreversible current induced domain wall motion in CoFeB based spin valves stripes, *Appl. Phys. Lett.* **90**, 232505 (2007).
135. S. S. P. Parkin, M. Hayashi, and L. Thomas, Magnetic domain-wall racetrack memory, *Science* **320**, 190–194 (2008).
136. M. Hayashi, L. Thomas, R. Moriya, C. Rettner, and S. S. P. Parkin, Current-controlled magnetic domain-wall nanowire shift register, *Science* **320**, 209–211 (2008).
137. R. Moriya, L. Thomas, M. Hayashi, Y. B. Bazaliy, C. Rettner, and S. S. P. Parkin, Probing vortex-core dynamics using current-induced resonant excitation of a trapped domain wall, *Nat. Phys.* **4**, 368–372 (2008).
138. L. Heyne, M. Klaui, D. Backes et al., Relationship between nonadiabaticity and damping in permalloy studied by current induced spin structure transformations, *Phys. Rev. Lett.* **100**, 066603 (2008).
139. M. Bolte, G. Meier, B. Krüger et al., Time-resolved X-ray microscopy of spin-torque-induced magnetic vortex gyration, *Phys. Rev. Lett.* **100**, 176601 (2008).
140. O. Boulle, J. Kimling, P. Warnicke et al., Nonadiabatic spin transfer torque in high anisotropy magnetic nanowires with narrow domain walls, *Phys. Rev. Lett.* **101**, 216601 (2008).
141. C. Burrowes, D. Ravelosona, C. Chappert et al., Role of pinning in current driven domain wall motion in wires with perpendicular anisotropy, *Appl. Phys. Lett.* **93**, 172513 (2008).
142. T. A. Moore, I. M. Miron, G. Gaudin et al., High domain wall velocities induced by current in ultrathin Pt/Co/AlO_x wires with perpendicular magnetic anisotropy, *Appl. Phys. Lett.* **93**, 262504 (2008).
143. L. Bocklage, B. Krüger, R. Eiselt, M. Bolte, P. Fischer, and G. Meier, Time-resolved imaging of current-induced domain-wall oscillations, *Phys. Rev. B* **78**, 180405R (2008).
144. C. Knutson, Magnetic domain wall dynamics in nanoscale thin film structures, Thesis, The University of Texas at Austin, Austin, TX, 2008.
145. S. A. Yang, G. S. D. Beach, C. Knutson et al., Universal electromotive force induced by domain wall motion, *Phys. Rev. Lett.* **102**, 067201 (2009).
146. I. M. Miron, P.-J. Zermatten, G. Gaudin, S. Auffret, B. Rodmacq, and A. Schuhl, Domain wall spin torquemeter, *Phys. Rev. Lett.* **102**, 137202 (2009).
147. A. Fernández-Pacheco, J. M. De Teresa, R. Córdoba et al., Domain wall conduit behavior in cobalt nanowires grown by focused electron beam induced deposition, *Appl. Phys. Lett.* **94**, 192509 (2009).
148. P. M. Braganca, I. N. Krivorotov, O. Ozatay et al., Reducing the critical current for short-pulse spin-transfer switching of nanomagnets, *Appl. Phys. Lett.* **87**, 112507 (2005).
149. W. H. Rippard, M. R. Pufall, S. Kaka, T. J. Silva, S. E. Russek, and J. A. Katine, Injection locking and phase control of spin transfer nano-oscillators, *Phys. Rev. Lett.* **95**, 067203 (2005).
150. A. L. D. Kilcoyne, T. Tyliszczak, W. F. Steele et al., Interferometer-controlled scanning transmission X-ray microscopes at the advanced light source, *J. Synchrotron Radiat.* **10**, 125–136 (2003).
151. J. Stöhr, Y. Wu, B. D. Hermsmeier et al., Element-specific magnetic microscopy with circularly polarized X-rays, *Science* **259**, 658–661 (1993).
152. G. Tatara and H. Kohno, Theory of current-driven domain wall motion: Spin transfer versus momentum transfer, *Phys. Rev. Lett.* **92**, 086601 (2004).
153. Z. Li and S. Zhang, Domain-wall dynamics and spin-wave excitations with spin-transfer torques, *Phys. Rev. Lett.* **92**, 207203 (2004).
154. S. Zhang and Z. Li, Roles of nonequilibrium conduction electrons on the magnetization dynamics of ferromagnets, *Phys. Rev. Lett.* **93**, 127204 (2004).
155. A. Thiaville, Y. Nakatani, J. Miltat, and Y. Suzuki, Micromagnetic understanding of current-driven domain wall motion in patterned nanowires, *Europhys. Lett.* **69**, 990–996 (2005).
156. A. P. Malozemoff and J. C. Slonczewski, *Magnetic Domain Walls in Bubble Materials*, Academic Press, New York, 1979.
157. N. L. Schryer and L. R. Walker, Motion of 180 degrees domain-walls in uniform dc magnetic-fields, *J. Appl. Phys.* **45**, 5406–5421 (1974).
158. M. Hayashi, L. Thomas, Ya. B. Bazaliy et al., Influence of current on field-driven domain wall motion in permalloy nanowires from time resolved measurements of anisotropic magnetoresistance, *Phys. Rev. Lett.* **96**, 197207 (2006).
159. G. S. D. Beach, C. Knutson, M. Tsoi, and J. L. Erskine, Field- and current-driven domain wall dynamics: An experimental picture, *J. Magn. Magn. Mater.* **310**, 2038–2040 (2007).
160. W. H. Rippard, M. R. Pufall, S. Kaka, T. J. Silva, and S. E. Russek, Current-driven microwave dynamics in magnetic point contacts as a function of applied field angle, *Phys. Rev. B* **70**, 100406(R) (2004).
161. A. Khitun, D. E. Nikonov, M. Bao, K. Galatsis, and K. L. Wang, Feasibility study of logic circuits with a spin wave bus, *Nanotechnology* **18**, 465202 (2007).
162. T. Schneider, A. A. Serga, B. Leven, B. Hillebrands, R. L. Stamps, and M. P. Kostylev, Realization of spin-wave logic gates, *Appl. Phys. Lett.* **92**, 022505 (2008).

8

Spin Torque Effects in Magnetic Systems: Theory

A. Manchon
University of Arizona and King Abdullah University of Science and Technology

Shufeng Zhang
University of Arizona

Manipulating the magnetization direction of a magnetic layer without the use of an external magnetic field represents an outstanding opportunity and challenge for spintronic applications [1]. In the last decade, vast experimental data and many detailed theoretical results had been available. We do not intend to review all the development made to date. Rather we want to provide a coherent picture of this subject for various phenomena involving the interplay between the spin transport and magnetization dynamics. There are already a few excellent review papers [2–5], on this same subject and we refer some of the details to these articles. In this chapter, we address the theoretical aspects of current-induced magnetization control using the so-called spin torque (ST) effect from both the transport and magnetization dynamics points of view. Although numerous models have been developed, discussing them all is far beyond the scope of this chapter and we rather refer the reader to specific literature. After a brief overview of ST studies in Section 8.1, transport characteristics in a number of magnetic systems such as spin valves (SVs), magnetic tunnel junctions (MTJs), and magnetic domain walls (DW) are described in Section 8.2. At the end of the section, several unconventional STs will be briefly discussed. In Section 8.3, the ST-driven magnetization dynamics for these magnetic systems is studied. Finally, perspectives and conclusions of the ST effects are given in Section 8.4.

8.1 Overview of Spin Transfer Studies

Despite earlier insights by Berger [6] in magnetic DWs and Slonczewski [7] in MTJs, the concept of *spin transfer torque* (STT) has been formalized independently by these authors in 1996 [8,9] for the case of the metallic SV for quite different reasons. In Berger's picture, the magnetic instability occurs when the current-driven chemical potential splitting for the spin-up and spin-down electrons exceeds the spin wave gap (threshold). Since the chemical potential splitting scales with the spin accumulation, which is proportional to the current, the spin wave is generated at a critical current density. Slonczewski proposed that the absorption of the transverse spin current by a ferromagnetic layer is equivalent to an STT exerted on the layer and this torque can be used for rotating the magnetization of the layer. Although Berger addresses spin-wave emission and Slonczewski considers a macrospin representation of the magnetic layers, the two models are largely equivalent. With the Berger–Slonczewski model, one anticipates that a sufficiently large current applied perpendicular to the SV layers can either generate spin waves or rotate the magnetization.

Less than 2 years after its prediction, evidences of STT were observed in magnetic multilayers [10,11], metallic spin-valves [12–14], MTJs [15–17], DWs [18,19], antiferromagnets [20,21], semiconductors devices [22,23], and single ferromagnetic layers in the presence of spin–orbit coupling [24]. The details of these experimental achievements are extensively reviewed by Tsoi in the previous chapter.

These observations of ST raised an array of fundamental questions concerning the microscopic origin of the STs in these magnetic structures. Many theoretical approaches have been developed in the past 10 years. For the metallic magnetic structures with noncollinear electrodes' magnetizations, most of the theories have focused on diffusive scattering in addition to the quantum mechanical interface reflection, which induces rotation of spins for reflected and transmitted electrons [2,25–30]. In MTJs, the STs were calculated via quantum mechanical tunneling by using tight-binding model [31], free electron models [32–34], as well as first principle methods [35]. At the high bias voltage for MTJs, the inelastic tunneling becomes increasingly important and a generalized form of the transfer Hamiltonian formalism [36–38] for noncollinear structures was developed and its prediction was compared to recent experiments [39–42]. For magnetic DWs, the STs were studied semi-phenomenologically [43,44], and microscopically [45–47]. Several novel concepts, such as spin-motive forces [48–50] and spin-current mediated damping [44,51,52] have been developed. In the first part of this article, we intend to use a unified approach to formulate the ST for all these systems.

The STs generated by the current cannot be directly measured. Instead, the anticipated consequences of the ST, such as the critical current for the magnetization reversal [12,15] and the spin wave excitations [53–56] were experimentally measured. These measured results depend not only on the nature of the STs but also on other parameters, for example, the damping coefficient of the materials and detailed magnetization distributions of the film during magnetization dynamics. Furthermore, the influence of the finite temperature is usually quite complex for the current-driven magnetization dynamics and thus the connection between the theoretically calculated ST and experimentally observed dynamic motion becomes a challenging issue [57]. In fact, most of the research efforts were paid on the detailed analysis of the magnetization dynamics by assuming a specific form of the STs. Usually, either the macrospin model [58,59] or the micromagnetic simulations [60,61] are used to determine the possible magnetic phases and dynamics driven by the currents; this subject is the focus of the second half of the chapter.

8.2 Spin Torque in Magnetic Structures

8.2.1 Principle of Spin Transfer Torque

The spin transport in heterogeneous magnetic structures with collinear magnetization configuration gives rise to a longitudinal *spin density* [62] (also called *spin accumulation* in diffusive systems). When the electrodes' magnetizations become noncollinear, as in SVs or magnetic DW structures, a spin accumulation showing a component *transverse* to the background magnetization appears (see Figure 8.1a). This transverse component of the spin accumulation exerts a torque on the background magnetization via the exchange interaction, leading to magnetization reorientation, switching, or excitations [8,9]. Another way to understand the STT is through the angular momentum conservation: the net balance of spin current flowing through an enclosed region is the rate of the total angular momentum change, which is defined as an STT (see Figure 8.1b). We will show later that these two pictures are equivalent only when one neglects the spin-flip scattering and the spin–orbit coupling.

Throughout this article, we choose an *sd* model to describe the spin transport and magnetization dynamics. The *sd* model artificially separates the itinerant electrons in *sp* bands, which are responsible for the spin transport from the localized *d* bands, which determine the magnetization [63]. From the microscopic point of view, the *sd* model is excessively simplistic since the Fermi surface for the majority and minority electrons are very different in 3*d* transition metals. This simplified description introduces inaccuracies in the prediction as will be mentioned in the following text. However, the *sd* model remains a very useful and pedagogical tool that provides qualitatively valuable results in the context of STT. The Hamiltonian of the total spin system is

$$\hat{H} = \hat{H}^{d} + \hat{H}^{sp} - J_{sd}\sum_i \hat{\mathbf{S}}_i \cdot \hat{\mathbf{s}}, \tag{8.1}$$

$$\hat{H}^{d} = -\frac{g\mu_B}{2}\sum_i \hat{\mathbf{S}}_i \cdot \mathbf{H}_{\text{eff}}, \tag{8.2}$$

$$\hat{H}^{sp} = \frac{\hat{\mathbf{p}}^2}{2m} + U(\hat{\mathbf{r}}) - \frac{\hbar}{4m^2c^2}(\nabla V \times \hat{\mathbf{p}}) \cdot \hat{\mathbf{s}}, \tag{8.3}$$

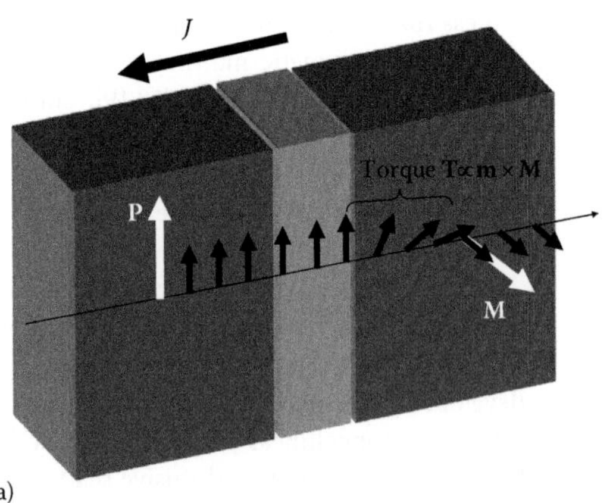

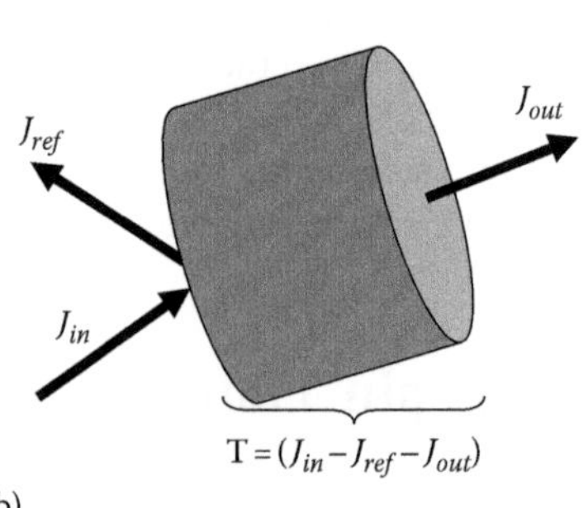

FIGURE 8.1 (a) Schematics of the spin density profile in a metallic SV, without spin diffusion: when impinging into the right layer, the spin density **m** possesses a component perpendicular to the local magnetization **M** and exerts a torque on the form $\mathbf{T} \propto \mathbf{m} \times \mathbf{M}$. (b) Spin transfer picture: the net balance of spin current density $-\nabla \cdot \mathcal{J}$ is equivalent to a spin transfer to the local magnetization, in the absence of spin diffusion.

where

$\hat{\mathbf{S}}_i(\hat{\mathbf{s}})$ is the dimensionless spin operator for the i-localized (itinerant) electron

$\mathbf{H}_{eff}$ is the effective magnetic field (including the anisotropy, dipolar, and exchange fields), which self-consistently depends on the local spin $\mathbf{S}_i$

J_{sd} is the exchange interaction between localized and itinerant electrons

The itinerant electron Hamiltonian $\hat{H}^{sp}$ includes a kinetic term, the potential profile and the spin–orbit coupling. (In (8.2), the spin–orbit coupling is taken into account within the anisotropy field.) Applying Ehrenfest's theorem, one obtains the local spin density continuity equation for a localized and an itinerant electron, respectively:

$$\frac{\partial\langle\hat{\mathbf{S}}_i\rangle}{\partial t} = -\frac{g\mu_B}{2i\hbar}\langle[\hat{\mathbf{S}}_i, \hat{\mathbf{S}}_i\cdot\mathbf{H}_{eff}]\rangle + \frac{2J_{sd}}{\hbar}\langle\hat{\mathbf{S}}_i\rangle\times\langle\hat{\mathbf{s}}\rangle, \quad (8.4)$$

$$\frac{\partial\langle\hat{\mathbf{s}}\rangle}{\partial t} = \nabla\cdot\mathcal{J}_s + \frac{1}{2m^2c^2}\langle(\nabla V\times\hat{\mathbf{p}})\times\hat{\mathbf{s}}\rangle - \frac{2J_{sd}}{\hbar}\langle\hat{\mathbf{S}}_i\rangle\times\langle\hat{\mathbf{s}}\rangle, \quad (8.5)$$

where

$\langle\cdots\rangle$ denotes quantum mechanical averaging on either the local or the itinerant electronic states with a nonequilibrium distribution function

$\mathcal{J}_s \equiv -\langle\hat{\mathbf{v}}\otimes\hat{\mathbf{s}}\rangle$ is the spin current tensor

$\gamma = |g|\mu_B/\hbar$ is the gyromagnetic ratio

In (8.4), the commutator $[\hat{\mathbf{S}}_i, \hat{\mathbf{S}}_i\cdot\mathbf{H}_{eff}]$ depends on the detailed dependence of $\mathbf{H}_{eff}$ on $\hat{\mathbf{S}}_i$. One usually approximates it by the Landau–Lifshitz–Gilbert (LLG) equation, which includes a precessional term (i.e., takes $\mathbf{H}_{eff}$ as a c-number) and a damping term. There are quite a number of recent studies on the microscopic origins of the Gilbert damping [64–66]. Here we simply assume that the magnetization motion takes the same form except the additional contribution from the second term of (8.4), that is,

$$\frac{\partial\mathbf{M}}{\partial t} = -\gamma\mathbf{M}\times\mathbf{H}_{eff} + \frac{\alpha}{M_s}\mathbf{M}\times\frac{\partial\mathbf{M}}{\partial t} - \frac{2J_{sd}}{\hbar}\mathbf{M}\times\mathbf{m}, \quad (8.6)$$

where we have defined the magnetization densities for the local and itinerant electron spins $\mathbf{M}(\mathbf{r},t)/M_s = -\langle\hat{\mathbf{S}}_i\rangle/S$ and $\mathbf{m}(\mathbf{r},t) = -\langle\hat{\mathbf{s}}\rangle$, M_s being the saturation magnetization, S is the localized spin magnitude and γ is the gyromagnetic ratio.* The first two terms are standard LLG terms and thus the only new term is due to the interaction between the magnetization $\mathbf{M}$ and the spin density $\mathbf{m}$, which is usually defined as the ST, $\mathbf{T}_{st} = -2SJ_{sd}/\hbar M_s\mathbf{M}\times\mathbf{m}$. To determine the spin density $\mathbf{m}$, a similar equation of motion for the itinerant spins is needed:

$$\frac{\partial\mathbf{m}}{\partial t} = -\nabla\cdot\mathcal{J}_s - \frac{1}{2m^2c^2}\langle(\nabla V\times\hat{\mathbf{p}})\times\hat{\mathbf{s}}\rangle - \frac{\delta\mathbf{m}}{\tau_{sf}} + \frac{2SJ_{sd}}{\hbar M_s}\mathbf{M}\times\mathbf{m}, \quad (8.7)$$

where we have phenomenologically introduced a spin relaxation time τ_{sf} (the last term) to model the spin-flip processes by impurities and magnons, $\delta\mathbf{m} = \mathbf{m} - \mathbf{m}_0$ being the nonequilibrium spin density. $\delta\mathbf{m}$ is actually the quantity responsible for the current-driven torque since the equilibrium spin accumulation $\mathbf{m}_0$ gives rise to a zero-bias interlayer exchange coupling that is usually included in the effective field $\mathbf{H}_{eff}$ for the sake of simplicity. Therefore, the nonequilibrium ST has the form: $\mathbf{T}_{st} = -2SJ_{sd}/\hbar M_s\mathbf{M}\times\delta\mathbf{m}$. The coupled equations (8.6) and (8.7) determine $\mathbf{M}$ and $\delta\mathbf{m}$ as long as one can relate the spin current $\mathcal{J}_s$ to the magnetization and one can properly evaluate the spin–orbit term. One immediately realizes that in the steady-state, $\partial\mathbf{m}/\partial t = 0$, and without the spin–orbit coupling and spin relaxation $1/\tau_{sf} = 0$, the previous equation reduces to $\mathbf{T}_{st} = -2SJ_{sd}/\hbar M_s\mathbf{M}\times\delta\mathbf{m} = -\nabla\mathcal{J}_s$, that is, the ST can be viewed as the spin current transfer (the processes illustrated by Figure 8.1a and b are equivalent).

Before we proceed to consider the details of the ST in several experimental systems, we emphasize that the ST introduced here is always transverse to the magnetization $\mathbf{M}$. In principle, there is also a longitudinal component, which may change the magnitude of the magnetization. In the ferromagnets at low temperature (compared to the Curie temperature), we assume that the magnitude of the magnetization is rigid as it is assumed in the LLG equation. Thus we only consider the two transverse components of the ST, which can be formally written as

$$\mathbf{T}_{st} = \frac{a_J}{M_s^2}\mathbf{M}\times(\mathbf{M}\times\mathbf{P}) + \frac{b_J}{M_s}\mathbf{M}\times\mathbf{P}, \quad (8.8)$$

where $\mathbf{P}$ is any fixed vector that is conveniently chosen as the magnetization direction of the pinned layer in SVs and MTJs. The first term is sometimes called the in-plane ST or simply the ST. The second is named the perpendicular ST or the effective field term. Depending on the detail of (8.7), a_J and b_J possess different characteristics (angular and bias dependencies) related with the specific system under consideration. The detailed form of the torque $\mathbf{T}_{st}$ will be addressed in the next sections.

8.2.2 Metallic Spin Valves

A metallic SV is composed of two ferromagnetic systems (ferromagnetic layers with or without antiferromagnets, laminated layers etc.) separated by a metallic spacer (see Figure 8.2a). Since the voltage drop across the metallic SV is rather small, on the order of only 1 mV even for a current density as high as 10^8 A/cm^2, the linear response theory is sufficiently accurate for the calculation of the ST. Furthermore, the spin–orbit coupling

* The electron spin and magnetic moment are antiparallel. For itinerant holes, we would have $\mathbf{m}(\mathbf{r},t) = +\langle\hat{\mathbf{s}}\rangle$.

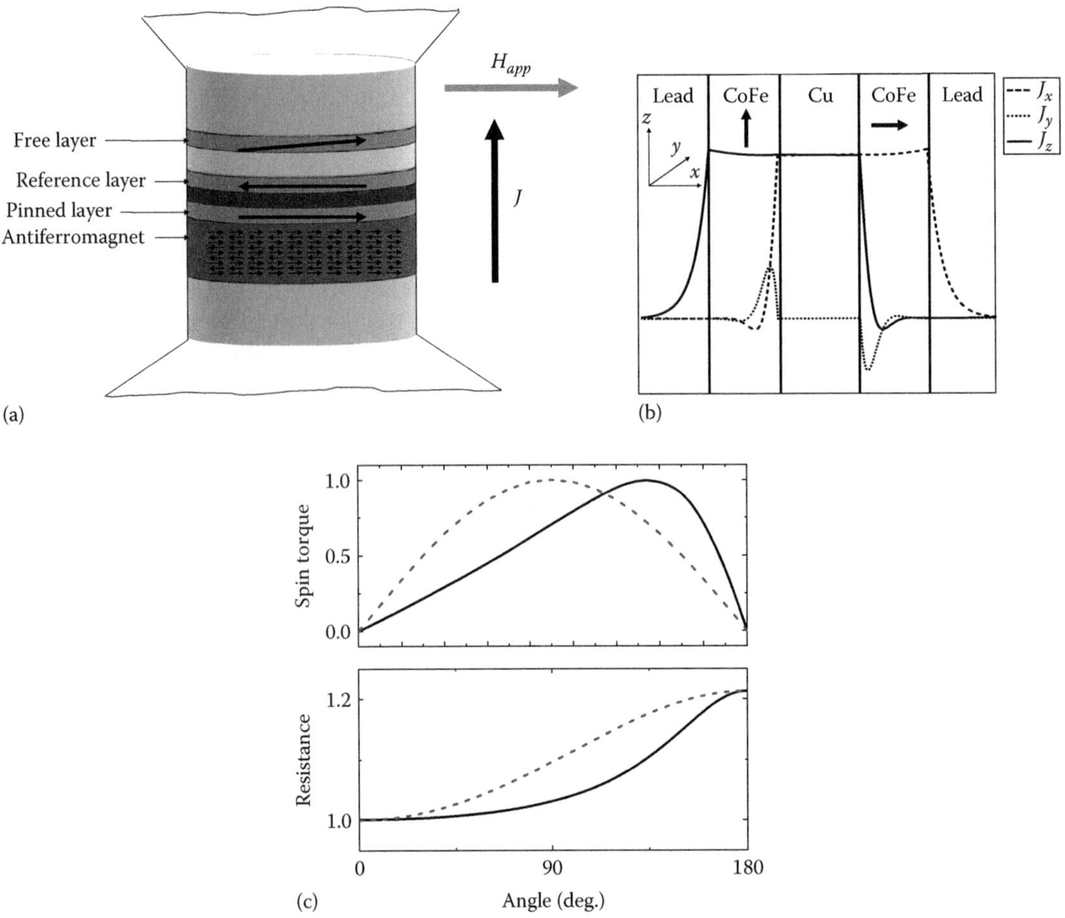

FIGURE 8.2 (a) Schematics of a common metallic SV, where the reference layer is antiferromagnetically coupled (usually through Ru) to a ferromagnetic layer pinned by an antiferromagnet (IrMn, PtMn, etc.). The spacer layer that separates the free layer from the reference layer is usually Cu, whereas the ferromagnetic layers can be Co, CoFe, NiFe, etc. (b) Calculated spatial profile of the spin current densities in a symmetric SV: Lead/Co/Cu/Co/Lead. (c) Angular dependence of the normalized spin transfer torque (top) and normalized resistance (bottom) in the same SV. The dotted lines represent the purely ballistic case (≡MTJs). The calculations are adapted from Manchon et al. [30].

is usually small compared to the *sd* interaction and thus (8.7) reduces to

$$\frac{\partial \mathbf{m}}{\partial t} = -\nabla \cdot \mathcal{J}_s - \frac{\delta \mathbf{m}}{\tau_{sf}} - \mathbf{T}_{st}. \quad (8.9)$$

In the steady-state ($\partial \mathbf{m}/\partial t = 0$), the ST is thus $\mathbf{T}_{st} = -\nabla \cdot \mathcal{J}_s - \delta \mathbf{m}/\tau_{sf}$.

In order to explicitly find the ST from the previous equation, one needs to relate the spin current tensor $\mathcal{J}_s$ to the nonequilibrium spin accumulation $\delta\mathbf{m}$. In the ballistic transport ($1/\tau_{sf} \equiv 0$), however, the current and the spin accumulation are not directly related because the spin accumulation is not well defined. In this case, one simply has $\mathbf{T}_{st} = -\nabla \cdot \mathcal{J}_s$. If one integrates the entire magnetic layer, the total ST on the layer is $\mathbf{T}_{st} = -\int_{int}^{out} d\Omega \nabla \cdot \mathcal{J}_s = \mathcal{J}_s^{in} - \mathcal{J}_s^{out}$, that is, the total ST is determined by the difference between the incoming and outgoing spin current at two sides of the layer. This simple identification makes first principle calculation possible [67–69].

The assumption of the ballistic transport is in fact a very poor approximation in the case of metallic SVs. It has been well established that diffusive spin-dependent transport is the proper description for metallic multilayers, as long as the layer thickness is large compared to the mean free path: diffusive scattering in the layers and at interfaces contributes to the spin current and magnetoresistance. For the collinear magnetization in multilayers and for the current perpendicular to the plane of the layer, the most effective model is Valet–Fert's diffusive equation [62]. For the noncollinear magnetization, the extension to Valet–Fert model is straightforward as long as one replaces the spin up and down channels by the 2×2 spinor tensor [28]. Specifically, one defines the spinor form of the current as

$$\hat{j} = \mathbf{j}_e \hat{I} + \mathcal{J}_s \cdot \hat{\sigma}, \quad (8.10)$$

where

$\hat{I}$ is the 2×2 unity matrix

$\hat{\sigma}$ is the vector of Pauli spin matrices

In the spinor form, the spin-dependent current density is expressed in the basis $(\hat{I}, \hat{\sigma})$; therefore, the charge current vector and spin current tensor are determined consistently. Within this formalism, in the presence of external electric field **E** and inhomogeneous (spin-dependent) electronic density $\hat{n}$, the spinor form of the current reads:

$$\hat{j} = \hat{C}\mathbf{E} - \hat{D}\nabla\hat{n}, \tag{8.11}$$

where
$\hat{C} = C_0(\hat{I} + \beta\hat{\sigma}\cdot\mathbf{M}/M_s)$ is the generalized conductivity
$\hat{D} = D_0(\hat{I} + \beta'\hat{\sigma}\cdot\mathbf{M}/M_s)$ is the generalized diffusion constant
$\hat{n} = n_0\hat{I} + \hat{\sigma}\cdot\mathbf{m}$ is the generalized accumulation accounting for both charge $(2n_0)$ and spin $(\mathbf{m})$ accumulations

β and β' are the spin asymmetries of the conductivity and diffusion constants, responsible for the spin polarization of the current. All these quantities are 2×2 matrices, with a similar form as in Equation 8.10. The diffusion constant and the conductivity are related via the Einstein relation $\hat{C} = e^2\hat{N}(E_F)\hat{D}$, where $\hat{N}(E_F)$ is the density of states at the Fermi energy. The spinor form of Equation 8.11 provides one scalar relation for $\mathbf{j}_e$ that must be completed with the steady-state charge density continuity equation $\nabla\cdot\mathbf{j}_e + \partial n/\partial t = \nabla\cdot\mathbf{j}_e = 0$.

Equations 8.10 and 8.11 define the spin-dependent Ohm's law. From these relations, the spin current $\mathcal{J}_s = Tr_\sigma(\hat{\sigma}\hat{j})$, can be explicitly expressed in terms of the charge current plus a diffusion term, that is, $\mathcal{J}_s = -(\beta/eM_s)\mathbf{j}_e\otimes\mathbf{M} - 2D_0\nabla\mathbf{m}$, where $e>0$ is the electric charge. By placing this spin current into Equation 8.9, we have

$$\nabla^2\mathbf{m} = -\frac{\mathbf{M}\times\mathbf{m}}{M_s\lambda_J^2} + \frac{\delta\mathbf{m}}{\lambda_{sf}^2}, \tag{8.12}$$

where
$\lambda_J = \sqrt{\hbar D_0/2SJ_{sd}}$ is the transverse decoherence length
$\lambda_{sf} = \sqrt{2D_0\tau_{sf}}$ is the spin diffusion length

The spin diffusion equation is the generalization of Valet–Fert's spin-diffusion model to the noncollinear magnetization configurations. While the longitudinal spin accumulation has a length scale determined by the spin-flip length λ_{sf}, which is on the order of 10 nm or longer for the transition ferromagnets, the transverse spin accumulation has much shorter length scale λ_J, ranging from a few angstroms to a few tens of angstroms. The solution of (8.12) can be immediately obtained:

$$\delta\mathbf{m}_\parallel = \mathbf{A}_\parallel\exp\left(\frac{x}{\lambda_{sf}}\right) + \mathbf{B}_\parallel\exp\left(\frac{-x}{\lambda_{sf}}\right), \tag{8.13}$$

$$\delta\mathbf{m}_\perp = \mathbf{A}_\perp\exp\left(\frac{\pm x}{l_+}\right) + \mathbf{B}_\perp\exp\left(\frac{\pm x}{l_-}\right), \tag{8.14}$$

where $l_\pm^{-1} = \sqrt{\lambda_{sf}^{-2} \pm i\lambda_J^{-2}}$, and the constants of integration are determined by the boundary conditions. In Figure 8.2b, we show the spin current density in the SV when the two ferromagnetic layers are perpendicular to each other. As we have seen, the longitudinal spin current (directly related to the spin accumulation) exists in the layers for a substantial distance, whereas the transverse spin current decay is much faster, on the order of λ_J. The ST $\mathbf{T}_{st} = -2SJ_{sd}/\hbar M_s\mathbf{M}\times\delta\mathbf{m}$ occurs at the interface because the transverse spin accumulation $\delta\mathbf{m}_\perp$ is confined near the interface region.

This treatment leads to two main characteristics of the STT in metallic SVs. First, it is found that the coefficients a_J and b_J in (8.8) are angular dependent due to the strong influence of the longitudinal spin accumulation on the spin transport, as shown in Figure 8.2c. This angular dependence has been used in the "wavy-structure" proposed by Boulle et al. [70] to generate zero-field steady magnetization precession. Secondly, at metallic interfaces, almost all the Fermi surface contributes to the current, so that after averaging over all the electron k-states, the effective spin density possesses only a small perpendicular component. Therefore, the out-of-plane torque is very small and usually neglected in experiment.

One concern on this diffusive approach is the validity of the diffusion equation for the transverse spin density. As we mentioned earlier, in realistic band structure, the Fermi surface for majority and minority spins are different so that the coherence length is actually very short, much shorter than in the conventional *sd* model [67]. Since the transverse decoherence length may not be much larger than the Fermi wavelength, the semiclassical approach might be inappropriate. A better treatment is to consider an extended boundary condition at the interface region where the reflection and transmission coefficients are calculated quantum mechanically. Then a transport theory that is ballistic at the interfaces but diffusive in the layers can be derived [71].

For layers much thinner than the mean free path, the diffusive approach is no more valid and the spin scattering at the interfaces become dominant. Slonczewski [27] proposed a mixed model, treating the spacer ballistically and while the rest of the system is treated diffusively [72]. Alternatively, a number of other formalisms have been developed, among which are the magneto-electronics theory by Brataas et al. [2] and the thermokinetic approach by Saslow [73] and Wegrowe et al. [74].

8.2.3 Magnetic Tunnel Junctions

In MTJs, the spacer layer that separates the two ferromagnetic layers is a thin insulator (e.g., AlO_x, MgO, etc.) in contrast to the metallic layer in SVs. This distinction makes the MTJs more promising in terms of magnetic memory applications: MTJs have a higher value of magnetoresistance, up to several hundred percent, and the tunnel resistance is much larger than in metallic SVs. Unlike the metallic SVs, the resistance and magnetoresistance are solely determined by the electronic states near the barrier. As a consequence, the modification of the electrochemical potential due to the spin accumulation build-up (≈meV) is negligible compared to the barrier height (≈eV). Then, it is convenient to describe the spin transport within the MTJ assuming a ballistic approximation. Indeed, most of the theoretical calculations have been carried out by assuming a perfect tunnel junction within the elastic tunneling formalism. When the bias voltage exceeds several hundred millivolts, however, the inelastic tunneling becomes more important. We will describe the main characteristics of the elastic and inelastic contributions to the ST in the text that follows.

8.2.3.1 Elastic Tunneling

The elastic tunneling process is a textbook example for the elementary application of quantum mechanics. Several approaches, depending on the complexity of the detailed calculation, are free electron [32–34], tight-binding [31], and ab initio methods [35]. These theories all assume the absence of spin relaxation within the barrier and the electrodes. Instead of calculating the spin accumulation, the spin current for a fixed bias voltage is computed due to the direct relation between the ST and the spin current, $\mathbf{T}_{st} = \mathcal{J}_s^{int}$. In the free electron approach, the Hamiltonian is solved layer-by-layer and the wave functions are connected using the usual continuity principle (the barrier is treated using the Airy functions [32] or Wentzel–Kramers–Brillouin (WKB) approximation [33,34].

As mentioned in the previous section, the spin transport in nanodevices is controlled by the spin accumulation within the bulk of the layers and the quantum mechanical reflection/transmission of the spin at the interfaces. Whereas the role of the longitudinal spin accumulation is negligible in MTJs (a_J and b_J in (8.8) are generally independent of the angle, as shown in Figure 8.3b), the spin rotation during the tunneling can be quite significant. Contrary to metallic SVs where the spin-current is averaged over the complete Fermi surface, in MTJs, only electrons close to the perpendicular incidence significantly contribute to the transport yielding a large effective spin rotation [34]. This effective spin rotation is at the origin of a significant out-of-plane torque ($b_J \neq 0$ in (8.8)). Furthermore, in the case of a symmetric junction with elastic tunneling, it can be shown that the components of the torque have the following form (see Figure 8.3a):

$$\mathbf{T}_{\parallel} = (a_1 V + a_2 V^2)\mathbf{M} \times (\mathbf{M} \times \mathbf{P}), \tag{8.15}$$

$$\mathbf{T}_{\perp} = (b_0 + b_2 V^2)\mathbf{M} \times \mathbf{P}. \tag{8.16}$$

These bias dependencies have been mentioned by Slonczewski [7,36] and calculated numerically by others [31–34]. Note that in MTJs, the torque is expressed as a function of the bias voltage, rather than the current density, since bias rather than current controls the tunneling transport. One of the important concerns for application is the actual bias dependence of the out-of-plane torque that significantly modifies the magnetization dynamics [39–41,75,76]. In particular, the presence of inelastic scattering as well as junction structural asymmetries alters this bias dependence [42].

8.2.3.2 Inelastic Tunneling

The inelastic scattering of tunneling electrons by impurities/defects, phonons or magnons is known to be detrimental to the tunneling magnetoresistance (TMR) in MTJs [77]. In the case of interfacial magnons and phonons, the interaction Hamiltonian reads:

$$\hat{H}_{inel} = -\frac{J_{sd}}{\sqrt{N}} \sum_{\mathbf{k},\mathbf{q}} (C^{+}_{\mathbf{k}-\mathbf{q},\uparrow} C_{\mathbf{k},\downarrow} a^{+}_{\mathbf{q}} + C^{+}_{\mathbf{k}+\mathbf{q},\downarrow} C_{\mathbf{k},\uparrow} a_{\mathbf{q}}) + \sum_{\mathbf{k},\mathbf{p},\sigma} M_{\mathbf{p}} (C^{+}_{\mathbf{k}-\mathbf{q},\sigma} C_{\mathbf{k},\sigma} b^{+}_{\mathbf{q}} + C^{+}_{\mathbf{k}+\mathbf{p},\sigma} C_{\mathbf{k},\sigma} b_{\mathbf{p}}), \tag{8.17}$$

where

$C_{\mathbf{k},\sigma}$ is the annihilation operator for a σ-electron with wave vector **k**

$a_{\mathbf{q}}(b_{\mathbf{p}})$ is the magnon (phonon) annihilation operator

In the electron–magnon interaction, an incoming electron relaxes (enhances) its energy by emitting (absorbing) a magnon (see Figure 8.3c). Since the total spin needs to be conserved, the electron flips its spin during the interaction. The electron–phonon interaction is similar except the electron conserves its spin. The influence of electron–magnon scattering on the spin transport has been studied within the transfer Hamiltonian formalism for the case of TMR [78] and ST [37,38]. The spin-dependent current density is expressed in the spinor form, as a function of the interfacial spin-dependent densities of states matrices $\hat{\rho}$ and of the transfer matrix $\hat{T}_{\mathbf{kp}}$ derived from the tunneling Hamiltonian:

$$\hat{J} = \sum_{\mathbf{k},\mathbf{p}} \hat{\rho}_{\mathbf{k}} \hat{T}_{\mathbf{kp}} \hat{\rho}_{\mathbf{p}} (\hat{T}_{\mathbf{kp}})^{+} f_L(1-f_R) - \hat{\rho}_{\mathbf{p}} \hat{T}_{\mathbf{pk}} \hat{\rho}_{\mathbf{k}} (\hat{T}_{\mathbf{pk}})^{+} f_R(1-f_L). \tag{8.18}$$

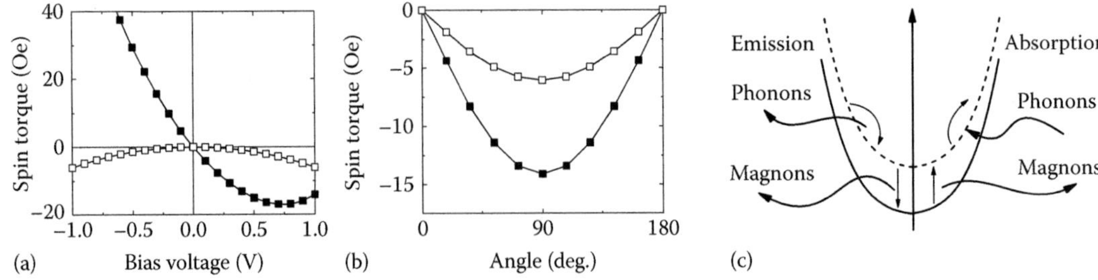

FIGURE 8.3 Free electron calculation of the bias dependence (a) and angular dependence (b) of the in-plane (plain symbols) and out-of-plane (open symbols) torques. (c) Principle of the inelastic scattering by phonons and magnons. The calculations are adapted from Manchon et al. [34].

The transfer matrix $\hat{T}_{\mathbf{kp}}$ accounts for the elastic tunneling as well as inelastic contributions [37,38]. One can show that the general form of the in-plane torque is not modified, whereas the out-of-plane torque acquires a asymmetric linear component:

$$\mathbf{T}_{\|} = (a_1(T)V + a_2V^2)\mathbf{M} \times (\mathbf{M} \times \mathbf{P}), \tag{8.19}$$

$$\mathbf{T}_{\perp} = (b_0 + b_1(T)\,|V|\, + b_2V^2)\mathbf{M} \times \mathbf{P}. \tag{8.20}$$

Note that the coefficients are now temperature-dependent and that the out-of-plane torque remains an *even* function of the bias voltage since the symmetry of the MTJ is conserved. Interestingly, a bias-dependent field-like effect has recently been observed by some authors [42,75,79]. This field effect is clearly asymmetric against bias and is responsible for *back-hopping* phenomena [42,79]. Although the interfacial magnon scattering cannot reproduce this bias asymmetry, Li et al. [75] showed that in the case of *bulk* magnons the out-of-plane torque may acquire a component on the form $\alpha\ j_e|V|$, which is therefore antisymmetric against the bias. However, other possible mechanisms (temperature effects, non-macrospin behavior, other inelastic interactions) are still under investigation.

8.2.3.3 Structural Asymmetries

In the previous treatment of the inelastic tunneling, we assumed that the symmetry of the MTJ is conserved, namely, the electrodes are equivalent and the barrier has a perfectly symmetric shape at zero bias. However, when structural asymmetries are present (the exchange splitting are different in the left and right electrodes or when the barrier displays an asymmetric potential profile at zero bias) the transport properties of the junction become asymmetric against bias. Brinkman et al. [80] and Wilczynski et al. [32] have shown that in the case of asymmetric barrier or asymmetric exchange splitting in the electrodes, the conductance has the form:

$$G(V) = G_0 + (\alpha_\phi \Delta\phi + \alpha_J \Delta J)eV + G_2V^2, \tag{8.21}$$

where $\Delta\phi = \phi_R - \phi_L$ and $\Delta J = J_{sd}^R - J_{sd}^L$ are the asymmetries in the barrier height and ferromagnetic exchange splitting. Similarly, the out-of-plane torque is no more an even function of the bias voltage and an antisymmetric linear term appears at the first order [32,33]. It can be shown that the form of the torque exerted on the *right* electrode is therefore:

$$T_{\perp} = b_0 + (\beta_\phi \Delta\phi + \beta_J \Delta J)eV + b_2V^2. \tag{8.22}$$

The calculation of coefficients $\alpha_{\phi,J}$, $\beta_{\phi,J}$ following Brinkman's method [80] at the lowest order in voltage is a standard derivation that we do not develop here.

Up until now, the free electron and tight-binding models have provided valuable results that have been generally verified experimentally. However, in the case of MgO-based MTJs the complexity of spin transport in crystalline structures limits the validity of these approaches [81]. The presence of interfacial resonant states [82] as well as the metal-induced gap states [83] are expected to deeply influence the spin transport for very thin insulators. For such systems, an *ab initio* description of the ST is necessary [35].

8.2.4 Magnetic Domain Walls

Magnetic DWs are another example where the magnetization is not collinear (see Figure 8.4a). As an electric current flows in a DW, the spin current whose spin direction is varying along with the direction of the magnetization $\mathbf{M}(\mathbf{r},t)$ is locally absorbed, producing an STT. Again, we use the diffusive transport model (Equation 8.9) and separate the equilibrium and nonequilibrium parts of the spin density and the spin current

$$\mathbf{m} = \frac{m_0}{M_s}\mathbf{M}(\mathbf{r},t) + \delta\mathbf{m}, \tag{8.23}$$

$$\mathcal{J}_s = -\frac{P}{eM_s}\mathbf{j}_e \otimes \mathbf{M}(\mathbf{r},t) + \delta\mathcal{J}_s, \tag{8.24}$$

where

$\delta\mathbf{m}$ and $\delta\mathcal{J}_s$ are the nonequilibrium spin density and spin current

m_0 is the equilibrium itinerant spin density

P is the current polarization

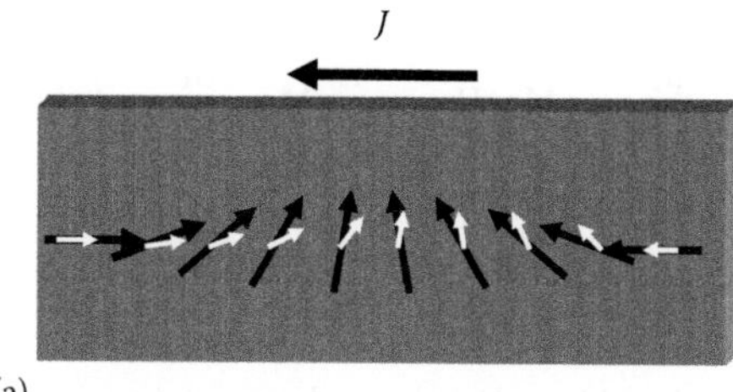

(a)

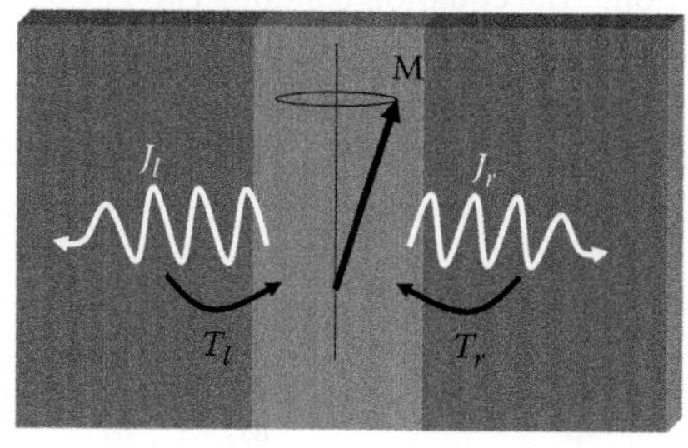

(b)

FIGURE 8.4 (a) Schematics of spin transport through a magnetic transverse DW. (b) Principle of spin pumping: the temporal oscillation of the magnetization injects a spin current in the neighboring electrodes. Its absorption by the leads gives rise to a torque exerted at the interfaces.

The semiclassical diffusive theory presented earlier relates the nonequilibrium current density to the nonequilibrium spin density, in the form: $\delta\mathcal{J}_s = -2D_0\nabla\mathbf{m}$. By inserting (8.23) and (8.24) into Equation 8.9, and discarding the time derivation of the magnetization, we have

$$2D_0\Delta\delta\mathbf{m} + 2\frac{SJ_{sd}}{\hbar M_s}\mathbf{M}(\mathbf{r},t)\times\delta\mathbf{m} - \frac{\delta\mathbf{m}}{\tau_{sf}} = -\frac{P}{eM_s}(\mathbf{j}_e\cdot\nabla)\mathbf{M}. \quad (8.25)$$

Assuming a slow variation of the magnetization in the DW, the first term in the left-hand side can be neglected and the torque $\mathbf{T}_{st} = -2SJ_{sd}/\hbar\mathbf{M}\times\delta\mathbf{m}$, exerted by the spin accumulation on the magnetic texture is

$$\mathbf{T}_{st} = -\frac{\mu_{BP}}{eM_s^3}\mathbf{M}\times[\mathbf{M}\times(\mathbf{j}_e\cdot\nabla)\mathbf{M}] - \beta\frac{\mu_{BP}}{eM_s^2}\mathbf{M}\times(\mathbf{j}_e\cdot\nabla)\mathbf{M}, \quad (8.26)$$

where $\beta = \hbar/(2SJ_{sd}\tau_{sf})$. These two terms are sometimes referred to as the *adiabatic* and *nonadiabatic* torques, respectively. Both torques vanish in the absence of DW, and interestingly, the nonadiabatic torque is proportional to the spin-flip scattering in this model. Then, in the absence of spin-flip scattering, the nonadiabatic torque vanishes and the DW only generates an adiabatic torque.

The adiabatic torque does not depend on the detail scattering mechanism and it is based on the angular momentum conservation. The nonadiabatic torque remains controversial in theories and in experiments, in particular regarding the model-dependent magnitude of β. To see the importance of the nonadiabatic torque, we write the full LLG equation including the current-driven STs,

$$\frac{\partial\mathbf{M}}{\partial t} = -\gamma\mathbf{M}\times\mathbf{H}_{eff} + \frac{\alpha}{M_s}\mathbf{M}\times\frac{\partial\mathbf{M}}{\partial t} - \frac{b_J}{M_s^2}\mathbf{M} \times\left(\mathbf{M}\times\frac{\partial\mathbf{M}}{\partial x}\right) - \beta\frac{b_J}{M_s}\mathbf{M}\times\frac{\partial\mathbf{M}}{\partial x}, \quad (8.27)$$

where $b_J = \mu_B P j_e/eM_s$ and we have chosen the current flowing in the x-direction.* The equation contains two key parameters: the damping coefficient α and the nonadiabatic parameter β; both parameters are proportional to the spin angular momentum loss. In the special case where $\alpha = \beta$, the above equation can be reduced to original LLG simply by a transformation $t' = t$ and $x' = x - b_J t$,

$$\frac{d\mathbf{M}}{dt'} = -\gamma\mathbf{M}\times\mathbf{H}_{eff} + \frac{\alpha}{M_s}\mathbf{M}\times\frac{d\mathbf{M}}{dt'}, \quad (8.28)$$

where $\mathbf{M}(x') = \mathbf{M}(x - b_J t)$. This property makes several authors to claim that $\alpha = \beta$ is valid in general [52,73,84]. However, one should not assume the Galilean invariance for the system with DWs and scattering centers. In fact, microscopic calculations can be performed for a given model Hamiltonian to separately compute α and β. It has been found that α and β are different in general, but the order of the magnitude is quite close [85–88]. Recently, other features related to the STs in the DW have been investigated; we summarize them in the following text.

8.2.4.1 Spin–Orbit Systems

Nguyen et al. [89] have shown that the mistracking between the carrier spin and the magnetization, responsible for the nonadiabatic torque, may be dramatically enhanced in dilute magnetic semiconductors. The authors showed that in such hole-mediated systems, described by Luttinger Hamiltonian, the strong intrinsic spin–orbit interaction allows for a significant reflection of the holes by the DW. This reflection increases the spin accumulation and the spin mistracking, giving rise to a larger nonadiabatic torque [90].

Obata and Tatara [91] proposed to use the specific spin–orbit interaction existing in two-dimensional electron gas with asymmetric interfaces to enhance the amplitude of the nonadiabatic torque. In such systems, the spin–orbit interaction is described by the Rashba spin–orbit coupling $H_{so} = \alpha(\sigma_x k_y - \sigma_y k_x)$ (see Section 8.2.5). Since this spin–orbit coupling has an inversion asymmetry, a nonequilibrium magnetic field can be generated when injecting a current in the 2DEG. This magnetic field is perpendicular to the current direction and lies in the 2DEG plane (see Section 8.2.5). Then, for particular shapes of the DWs, this magnetic field can induce a DW motion.

8.2.4.2 Thermal Activation and Magnons

The influence of a finite temperature and the role of spin waves have also been studied theoretically. As an alternative to the nonadiabatic torque, Duine et al. [92] showed that the inclusion of the thermal fluctuations together with the adiabatic ST term leads to a linear dependence of the DW velocity. Due to the thermal noise, the domain is not intrinsically pinned and it is then possible to induce the DW motion in the absence of nonadiabatic torque. Ohe and Kramer [93] studied the influence of electron–magnon scattering on the current-induced DW motion and showed that the spin-dependent scattering due to the spin waves contribute to a nonadiabatic torque, in agreement with a detailed analysis from Le Maho et al. [94].

8.2.4.3 Gilbert Damping and Spin-Pumping

A very closely related topic is the generation of spin current induced by the magnetization dynamics and the related enhancement of the Gilbert damping (see Figure 8.4b). The principle of spin pumping has been first proposed by Tserkovnyak et al. [47,95] in magnetic multilayers and can be viewed as the reciprocal of the spin current-driven ST: the time-dependent magnetization precession generates a spin current that leaks out of the magnetic layer to a normal metal in contact. The

* The convention is different for the spin torque in spin valves and domain wall: In the former b_J is the amplitude of the field-like torque, whereas in the latter, it is the amplitude of the adiabatic torque.

absorption of this spin current by the leads induces a torque that contributes to the magnetization damping.

The concept of the spin pumping given earlier can be equally applied to magnetic DWs [96]. For the temporal and spatial varying magnetization, the spin pumping leads to nonuniform spatial distribution of the spin current, and, thus, a ST is generated due to nonzero divergence of the spin current, that is, $\mathbf{T}_{st} = -\nabla \cdot \mathcal{J} \neq 0$. This ST has two interesting consequences: it generates a spin motive force via $V = \int \mathbf{T}_{st} \cdot d\phi$ on the conduction electrons and provides an additional damping mechanism for the local magnetization. Indeed, Barnes and Maekawa [48] and Ohe et al. [49] have obtained the explicit expression of the spin motive force for a rotating DW. Another way to understand these phenomena is by considering an interaction Hamiltonian between the itinerant electrons and the local magnetization—$-J_{sd}\hat{\mathbf{s}} \cdot \hat{\mathbf{S}}$. By assuming $\mathbf{M}(x,t)$ varies slowly in space and in time, a unitary transformation rotates the quantization axis from that parallel to $\mathbf{M}$ to a fixed axis. This rotation leads to a spin-dependent vector potential $\mathbf{A}$ and a scalar potential V_s, which in turn generate the spin-dependent electric and magnetic fields. These fields are the sources for the charge and spin currents, and therefore, the spin motive force and magnetic damping [97].

8.2.5 Unconventional Spin Torques

Up to now we considered systems with inhomogeneous magnetic texture, like DWs or (metallic and tunneling) SVs. In these two cases, the ST emerges from the spatial variation of the magnetic structure, through the divergence of the spin current driven by an external electric field (see Ref. [8.7]). In this section, we address three different types of ST, different from the "conventional" STT presented earlier. The first torque acts on a single uniform magnetic layer, due to the presence of spin–orbit coupling. The second type of torque acts in a usual SV but is driven by an external *thermal* gradient. The third torque is predicted to occur in antiferromagnets.

8.2.5.1 Spin–Orbit-Induced Torque

As shown in (8.7), the spin–orbit coupling itself can act as a source for the ST. This effect can be viewed as the reciprocal of the anisotropic magnetoresistance effect (see Figure 8.5a). Assuming a uniformly magnetized single layer ($\nabla \cdot \mathcal{J}_s = 0$) in steady-state ($\partial \mathbf{m}/\partial t = 0$) condition and neglecting the spin relaxation ($1/\tau_{sf} = 0$), the out-of-equilibrium torque on the magnetization is directly proportional to the spin density of the conduction electrons,

$$\mathbf{T}_{st} \equiv -\frac{2SJ_{sd}}{\hbar M_s}\mathbf{M} \times \mathbf{m} = -\frac{1}{2m^2c^2}\langle(\nabla V \times \hat{\mathbf{p}}) \times \hat{\mathbf{s}}\rangle. \quad (8.29)$$

Contrary to the STT presented in the previous sections, this torque does not come from the spatial variation of the magnetic texture. Considering a Hamiltonian with both Rashba and **k**-linear Dresselhauss coupling for a two-dimensional electron gas with both asymmetric interfaces and strained zinc-blende structure ((Ga,Mn)As and (In,Mn)As are typical candidates), the Hamiltonian reads [98]:

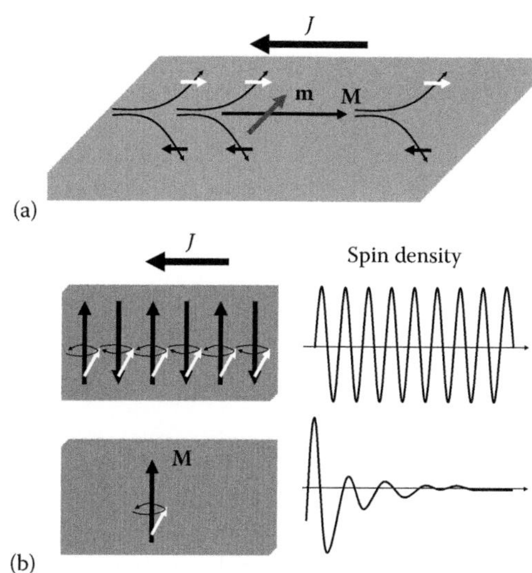

FIGURE 8.5 (a) In a single magnetic layer with spin–orbit interaction, the spin-dependent scattering of the carriers gives rise to the well-known anisotropic magnetoresistance effect. Similarly, it is responsible for the presence of a nonequilibrium spin density **m**, that can be used to reorient the magnetization **M**. (b) Principle of spin transport in an antiferromagnet following Nunez et al. physical picture: In an antiferromagnet (top), the torque exerted by the local magnetization on the itinerant spin is reversed at each layer, which provides a *staggered* nonequilibrium spin density texture (right diagram) in the absence of spin diffusion. This is in sharp contrast with the ferromagnetic case (bottom), where the spin precession remains unchanged along the ferromagnet, allowing for destructive interferences and therefore a damped oscillatory nonequilibrium spin density (right diagram).

$$\hat{H} = \frac{\hat{\mathbf{p}}}{2m} + \frac{SJ_{sd}}{M_s}\mathbf{M} \cdot \hat{\sigma} + \alpha(\sigma_x k_y - \sigma_y k_x) + \beta(\sigma_x k_x - \sigma_y k_y). \quad (8.30)$$

Assuming that the magnetization lies in the (x,y) plane $\mathbf{M}/M_s = \cos\theta\,\mathbf{x} + \sin\theta\,\mathbf{y}$, one can exactly derive the electron wave function and eigenenergy:

$$\psi_{\mathbf{k}}^s = \frac{e^{i\mathbf{k}\cdot\mathbf{r}}}{\sqrt{2A}}\begin{pmatrix} se^{i\gamma_{\mathbf{k}}} \\ 1 \end{pmatrix},$$

$$\varepsilon_{\mathbf{k}}^s = \frac{\hbar^2\mathbf{k}^2}{2m} + s\sqrt{(J_{sd}\cos\theta + \alpha k_y + \beta k_x)^2 + (-J_{sd}\sin\theta + \alpha k_x + \beta k_y)^2}, \quad (8.31)$$

where

$\tan\gamma_{\mathbf{k}} = (-J_{sd}\sin\theta + \alpha k_x + \beta k_y)/(J_{sd}\cos\theta + \alpha k_y + \beta k_x)$, $s = \pm 1$, and A is the area of the film

Interestingly, the electron is no more in a pure spin state and we can define two chiralities related to the superposition of majority and minority electron spins. To calculate the ST in the presence of the current, we use the Boltzmann equation for the two bands [98] and find

$$\mathbf{T}_{st} = j_e \frac{J_{sd}}{\varepsilon_F}\frac{m}{e\hbar}(\alpha\cos(\theta-\theta_0)+\beta\sin(\theta+\theta_0))\mathbf{z}, \tag{8.32}$$

where

$j_e=\sigma_0 E$, $\sigma_0=e^2\tau_0\varepsilon_F/\pi\hbar^2$ is the conductivity in the absence of spin–orbit coupling

θ_0 is the angle between the applied current and **x**

In the presence of Rashba and **k**-linear Dresselhaus spin–orbit coupling, the two spin chiralities are scattered in different directions, giving rise to an effective spin density perpendicular to the applied current. Due to the exchange interaction between the itinerant electrons and the magnetization, this spin-density acts like an effective magnetic field on the local magnetization.

While (8.29) is the general form of the spin-orbit interaction (SOI) coupling, the potential V includes all electric potentials in the solids. In the case of a system with a *spatial inversion symmetry* (impurities-induced SOI and Luttinger SOI, for instance) or a *rotational symmetry* (cubic Dresselhaus SOI), the **k**-integration vanishes and the ST is found to be zero at the first order. However, in the case of *structure inversion asymmetry* and *bulk inversion asymmetry with strain*, the spin–orbit interaction is linear in **k** and depends on its sign, so that $H_{so}(\mathbf{k})=-H_{so}(-\mathbf{k})$ and a nonzero torque is found at the first order in SOI.

The present SOI-induced ST is created by the intrinsic (or extrinsic, in case of impurities-induced SOI) spin–orbit coupling in the nonequilibrium condition and thus it does not involve the transfer of the conduction electron spin to the magnetization, in sharp contrast with the *STT* presented in the previous sections. Then, this torque exists for a *uniformly magnetized single layer* with the current applied in the plane of the layer, and acts as an effective field. Since the torque does not compete with the damping, it cannot excite current-driven steady magnetization precessions, contrary to the STT. Such a torque has also recently been studied by others [99,100].

8.2.5.2 Thermoelectric Torque

It is known that in a heterogeneous nonmagnetic multilayer, the application of a temperature gradient across the device generates a voltage via the "Seebeck effect": the spatial dependence of the electronic thermal distribution along the structure is equivalent to the spatial dependence of the electron chemical potential, which in turn gives rise to a bias voltage.

Recently, Hatami et al. [101] proposed a so-called *thermoelectric* ST, based on the Seebeck effect in magnetic multilayers. In a heterogeneous magnetic system with noncollinear magnetic electrodes and with a temperature gradient, the thermal distributions of majority and minority electrons are different. The thermalization of the electronic system through inelastic scattering generates a spin current (the total number of electron is conserved but not the spin polarization) that can be absorbed within the neighboring ferromagnetic layers, inducing a torque. The current spin polarization that drives this torque is no more the conventional polarization of the conductance, but the polarization of the conductance *derivative*, since the electronic thermal distribution is involved (it comes directly from the Sommerfield expansion of the generalized Ohm's law). This effect has recently been studied by Kovalev and Tserkovnyak [102] for DWs.

8.2.5.3 Spin Torque in Antiferromagnets

In the past 3 years a new type of *non-transfer* ST has been predicted to appear in antiferromagnets. In these systems, the overall magnetization is zero, due to strong antiferromagnetic interactions within the layer ($J_{ij}<0$), but locally, the band structure of the itinerant electrons is split due to the strong *sd* exchange coupling. Nunez et al. [103] proposed that the out-of-equilibrium spin density $\delta\mathbf{m}$ within such materials follows the microscopic magnetic texture of the antiferromagnet, displaying a *staggered* structure: the transport properties are essentially determined by the local magnetic properties of the antiferromagnet. As a result, the precession direction of the incoming spins varies from one monolayer to another, due to the antiparallel configuration of the local magnetization, preventing the destructive interference that usually occurs in ferromagnets, as illustrated in Figure 8.5b. Therefore, in antiferromagnets the spin memory is predicted to be conserved over a distance on the order of the spin diffusion length [104] (compared to $2\hbar\upsilon_F/J_{sd}$ in the case of conventional ferromagnets—see Section 8.2.2).

Consequently, when a spin-polarized current impinges on an antiferromagnet, as long as the incoming spins are not aligned on the local order parameter of the antiferromagnet, the spin-polarized current exerts a torque on the local magnetization. A number of *ab-initio* studies have been published on uncompensated and compensated antiferromagnets [105,106]. A few experiments have tried to observe the influence of a spin current on antiferromagnet, showing a clear modification of the exchange bias [20,21], but without explicit evidence of this torque.

8.3 Current-Induced Magnetization Dynamics

In the previous section, we showed that, depending on the structure under consideration (metallic SVs, MTJs, DWs, single magnetic layers, antiferromagnets), a torque arises, driven by an external bias (voltage or temperature). This ST modifies the dynamics of the local magnetization and provides a powerful tool to control the magnetization direction, essential for technological applications. In this section, we address the characteristics of the magnetization dynamics in the presence of ST.

In principle, the magnetization dynamics involves simultaneous time-dependent solutions for $\mathbf{M}(\mathbf{r},t)$ and $\mathbf{m}(\mathbf{r},t)$; see (8.6) and (8.7). However, the analysis of the timescale for $\mathbf{M}(\mathbf{r},t)$ and $\mathbf{m}(\mathbf{r},t)$ can reduce the problem to a steady-state condition for the conduction electrons, that is, $\partial\mathbf{m}/\partial t = 0$. To see this, one notices that it only takes about 1 fs (transport relaxation time) to establish a steady-state charge current after one applies an electric field to any heterogeneous conducting system. To establish a steady-state spin density and spin current in a magnetic system, the time-scale would be $\tau_{sf} = l_{sf}/v_F \approx 10$ fs – 1 ps. The time-scale of the transverse magnetization described by (8.6) is on the order of nanoseconds, much longer than τ_{sf} and it is an excellent approximation to assume $\partial\mathbf{m}/\partial t = 0$ if one is only interested in the nanosecond dynamics of the magnetization $\mathbf{M}(\mathbf{r},t)$. With this simplification, the spin accumulation $\mathbf{m}(\mathbf{r},t)$ is solely determined by $\mathbf{M}(\mathbf{r},t)$, that is, there is no retardation effect. In fact, all the explicit expressions of the STs, for example, (8.14) through (8.16), (8.26), and (8.34) are derived under this approximation. The starting point for the current-driven magnetization dynamics is the following equation:

$$\frac{\partial \mathbf{M}}{\partial t} = -\gamma \mathbf{M} \times \mathbf{H}_{eff} + \frac{\alpha}{M_s} \mathbf{M} \times \frac{\partial \mathbf{M}}{\partial t} + \mathbf{T}_{st}(\mathbf{M}), \qquad (8.33)$$

where the ST $\mathbf{T}_{st}$ takes a specific form for each magnetic system we want to discuss in the following section.

8.3.1 Magnetization Dynamics in Spin Valves

As we have shown in the previous sections, the general expressions of the ST in SVs and MTJs have taken the form of (8.8). Since the perpendicular torque, the second term of (8.8), can be absorbed to the effective field in (8.33), we should neglect this term and solely consider the ST

$$\mathbf{T}_{st} = \frac{a_J}{M_s^2} \mathbf{M} \times (\mathbf{M} \times \mathbf{P}). \qquad (8.34)$$

The torque coefficient a_J may depend on the angle between the magnetization directions of the pinned and free layers (e.g., in Ref. [70]). In this chapter, we limit our discussions to a constant a_J. While (8.33) can be solve by the standard micromagnetic simulations, we shall here consider analytical solutions in order to gain physical insight on the role of the ST by assuming the magnetization $\mathbf{M}$ is uniform with the layer, that is, the macro-spin model.

8.3.2 Magnetic Phase Diagrams

A useful tool to describe the dynamic properties of a magnetic layer under a spin-polarized current is the magnetic phase diagram that displays the magnetic stability regions of the layer as a function of the external magnetic field and applied bias. The geometry of the system is displayed in Figure 8.2a: the SV consists of two ferromagnetic layers separated by a (metallic or insulator) spacer. The system is a pillar along the z direction with an elliptic shape lying in the (x,y) plane (other configurations can be used experimentally as described by Tsoi in the previous chapter). The bottom (reference) layer has a magnetization direction $\mathbf{P} = +\mathbf{z}$ assumed to be fixed, and the top (free) layer has a magnetization direction $\mathbf{M}$, an easy axis $\mathbf{u} = \mathbf{x}$, and a demagnetizing field in the z direction.

8.3.2.1 Switching Threshold

Solving (8.33) requires the precise knowledge of the magnetization distribution as well as the effective field $\mathbf{H}_{eff}$. In principle, it comprises the exchange field $\mathbf{H}_{ex}$, the anisotropy field $\mathbf{H}_K$, the demagnetization field $\mathbf{H}_d$, and the external applied field $\mathbf{H}_{ext}$. The current-induced Oersted field as well as the interlayer exchange coupling (and perpendicular ST) can also be included. In the macrospin approach, the spatial variations of $\mathbf{M}$ are neglected so that the exchange field and the Oersted field are disregarded. Then, the total effective field reduces to $\mathbf{H}_{eff} = \mathbf{H}_{ext} + \mathbf{H}_{an} + \mathbf{H}_d$. The modified LLG equation for this system is

$$\frac{\partial \mathbf{M}}{\partial t} = -\gamma \mathbf{M} \times (\mathbf{H}_{ext} + M_x \mathbf{H}_K \mathbf{x} - 4\pi M_s M_z \mathbf{z}) + \frac{\alpha}{M_s} \mathbf{M} \times \frac{\partial \mathbf{M}}{\partial t} + \mathbf{T}_{st}, \qquad (8.35)$$

and $\mathbf{T}_{st}$ is given by (8.34). In spherical coordinates, $\mathbf{M} = M_s$ ($\sin\theta \cos\phi$, $\sin\theta \sin\phi$, $\cos\theta$) and (8.35) reduces to [107]:

$$\frac{1+\alpha^2}{\gamma} \dot{\theta} = h_\phi + \alpha h_\theta, \qquad (8.36)$$

$$\frac{1+\alpha^2}{\gamma} \sin\theta \dot{\phi} = \alpha h_\phi - h_\theta, \qquad (8.37)$$

where h_ϕ and h_θ are functions of (ϕ, θ) that we do not write explicitly [107]. The conventional method to obtain the stability conditions for the magnetization $\mathbf{M}$ is to study the stability of small deviations from the equilibrium $\delta\mathbf{M}$, under a spin-polarized current j_e. The equilibrium magnetization direction $\mathbf{M}_0$ is defined as $h_{\theta,\phi}(\mathbf{M}_0) = 0$, with $a_J = b_J = 0$ ($j_e = 0$). In the case where the external field is $\mathbf{H}_{ext} = H_x\mathbf{x}$, $\mathbf{M}_0 = M_s(\pm 1,0,0)$. Inserting the perturbed magnetization $\mathbf{M} = \mathbf{M}_0 + \delta\mathbf{M}$ in (8.36) and (8.37), and assuming that $|\mathbf{M}|^2 = M_s^2$, one obtains two differential equations for δM_x and δM_y, that can be solved analytically. For metallic SVs ($a_J = aj_e$ and $b_J = 0$), the stability conditions for the parallel (P) and antiparallel (AP) states are [108]:

$$j_e < \alpha \frac{(H_x + H_K + H_d/2)}{a}, \quad j_e > -\alpha \frac{(H_x - H_K - H_d/2)}{a}. \qquad (8.38)$$
$$H_x > -H_K \qquad\qquad H_x < H_K$$

As long as the conditions (8.38) are met, either P or AP or even both states are stable, as illustrated in Figure 8.6a. However, when these conditions are not met, for example, $j_e > \alpha(H_x + H_K + (H_d/2))/a$, the P and AP configurations become

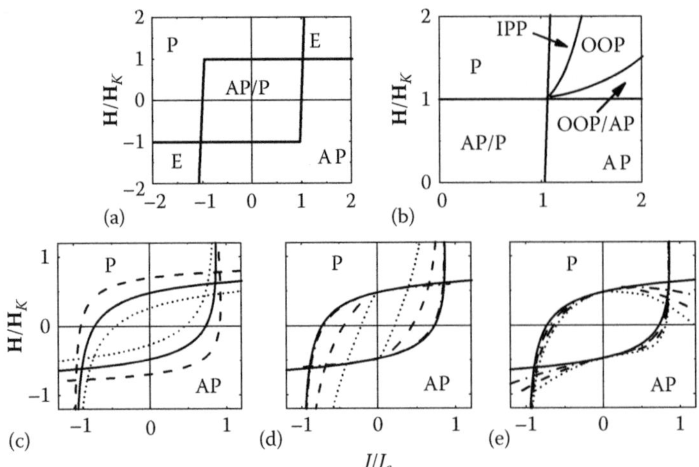

FIGURE 8.6 Stability diagrams for the free layer: (a) At zero temperature, assuming $b_J = 0$. (b) Zoom of the zero temperature diagram for positive field and current, including IPP and OPP. (c) Finite temperature ($\tau = 1$ s), assuming $b_J = 0$ (solid: 300 K, dashed: 100 K, dotted: 600 K). (d) Finite temperature ($\tau = 1$ s), with current heating and $b_J = 0$ ($c_0 = 1 \times 10^4$ K^2/V^2: dotted-dashed, $c_0 = 5 \times 10^5$ K^2/V^2: dashed, $c_0 = 5 \times 10^6$ K^2/V^2: dotted). (e) Finite temperature ($\tau = 1$ s), $b_J = b_2 V^2$ (at $V = 1$, $b_J/a_J = 0$: solid, $b_J/a_J = 0.6$: dotted-dashed, $b_J/a_J = 1$: dashed, $b_J/a_J = 2$: dotted).

unstable: the magnetization jumps to new stable states (denoted E in Figure 8.6a). These novel states are unique characteristics of the ST and are described as follows.

8.3.2.2 Steady-State Precessions

Investigating the stability diagrams out of the stability regions reveals a rich variety of magnetization regimes, including *in-plane precessions* (IPP), *out-of-plane* precessions (OPP), and *inhomogeneous* magnetic excitations. The fact that the STT term a_J can be either positive or negative, depending on the current direction, allows reaching steady precessional states in which the damping is exactly compensated by the ST.

The usual method to determine current-induced self-sustained precessional motion of the magnetization is based on the work of Bertotti et al. [59]. In the following, we disregard the role of the out-of-plane torque and focus on the influence of the in-plane torque. In the absence of both damping and in-plane torque, the magnetic system is energy conservative and the magnetization trajectories can be found straightforwardly by solving

$$\frac{\partial \mathbf{M}}{\partial t} = -\gamma \mathbf{M} \times \mathbf{H}_{eff}. \tag{8.39}$$

Equation 8.39 possesses fixed points (e.g., along the anisotropy axis) as well as two types of trajectories: IPPs and OPPs. Details of the derivation of these trajectories are found in Bertotti et al.[59]. In the following, we assume that we know the detail of such trajectories $\mathcal{L}(E)$, their energy E, and precession period $T(E)$.

The method introduced by Bertotti et al. consists in determining how the competition between the Gilbert damping and the ST stabilizes or destabilizes these unperturbed trajectories and fixed points. This perturbative method is based on Melkinov Integrals [109]:

$$\mathcal{M}(E) = \int_0^{T(E)} \frac{\partial \varepsilon}{\partial t} dt, \tag{8.40}$$

where

ε is the total energy of the perturbed magnetic system
E is the conserved energy of the unperturbed trajectory $\mathcal{L}(E)$

A nonzero integral would indicate that the magnetization **M** deviates from its initial trajectory, whereas a zero value means that the competition between the ST and the damping produces a self-sustained precessional motion of the magnetization over the unperturbed trajectory. Therefore, a self-sustained precession is described by $\mathcal{M}(E) = 0$ and is stable (unstable) under $\partial\mathcal{M}/\partial E > 0 (<0)$. Considering the complete LLG equation (8.33), the Melkinov Integral yields

$$\mathcal{M}(E) = \int_0^{T(E)} \left[-\alpha \left| \tilde{\mathbf{M}} \times \mathbf{H}_{eff} \right|^2 + a_J (\tilde{\mathbf{M}} \times \mathbf{P}) \cdot (\tilde{\mathbf{M}} \times \mathbf{H}_{eff}) \right] dt, \tag{8.41}$$

where $\tilde{\mathbf{M}}$ is determined in the *absence* of damping and ST. This integral can be understood as the work done by the damping and ST on the magnetization on a trajectory over one period. This method is then convenient to describe the critical currents for the onset of precessional states and magnetization switching, revealing the rich variety of dynamical behavior, consistent with experimental observations. For example, this method allows for the determination of the IPP and OPP states, as illustrated in Figure 8.6b.

8.3.2.3 Inclusion of Thermal Effects

Thermal fluctuations of the magnetization are known to deeply influence the magnetization dynamics [110] and the current densities required to manipulate the magnetization are expected to significantly increase the effective temperature of the sample [57,111]. Experimental evidence of thermally assisted current-induced dynamics [112–114] has led to numerous theoretical works and simulations on this topic [115–117]. Note that the thermal fluctuations are unlikely to significantly affect the ST amplitude itself since the Fermi level is usually larger than the thermal energy ($E_F \gg k_BT$).

Up to now, thermal activation processes are accounted for using the Brown model of thermal magnetic fluctuations at equilibrium [118–120]: thermal fluctuations add a Gaussian stochastic field $\mathbf{h}_r$ to the effective field $\mathbf{H}_{eff}$, with

$$\left\langle h_r^i(t)\right\rangle_h = 0,\quad \left\langle h_r^i(t)h_r^j(t')\right\rangle_h = 2D\delta_{ij}\delta(t-t'), \tag{8.42}$$

where

$\langle\cdots\rangle_h$ denotes averaging over all the realizations of the fluctuating field

D is the strength of thermal fluctuations

The fluctuation–dissipation theorem yields $D=\alpha k_BT/\gamma M_s$. In the framework of the LLG equation, $|\mathbf{M}|$ is kept constant, so that the random field $\mathbf{h}_r$ induces a random walk on the $\mathbf{M}$-sphere. The evolution of the probability density for magnetic orientation vectors $\rho(\mathbf{M},t)$ is therefore described by the continuity equation [110]:

$$\frac{\partial\rho(\mathbf{M},t)}{\partial t} + \nabla_{\mathbf{M}}\cdot\mathbf{J}_{\mathbf{M}}(\mathbf{M},t) = 0, \tag{8.43}$$

where $\mathbf{J}_{\mathbf{M}}(\mathbf{M},t)$ is the probability current density, containing a convective part (driven by the magnetization dynamics) and a diffusive part (driven by the random diffusion)*:

$$\mathbf{J}_{\mathbf{M}}(\mathbf{M},t) = \frac{\rho(\mathbf{M},t)}{M_s}\frac{\partial\mathbf{M}}{\partial t} - D\nabla\rho(\mathbf{M},t). \tag{8.44}$$

In the case of static states, Equation 8.43 is sufficient to determine the temperature-dependence of the stability diagrams. However, to account for both static and dynamics states, one has to consider the energy diffusion equation derived by Kramers [121]:

$$\frac{\partial[E\rho(\mathbf{M},t)]}{\partial t} = -\frac{D}{M_s}\frac{\partial\mathbf{M}}{\partial t}\cdot(\nabla_{\mathbf{M}}\rho\times\mathbf{M}). \tag{8.45}$$

The determination of $\rho(\mathbf{M},t)$ is done using a perturbative procedure, in the same spirit as Bertotti et al. [59]. The integration of (8.45) along an unperturbed trajectory $\mathcal{L}(E)$ yields

$$ET(E)\frac{\partial\rho(\mathbf{M},t)}{\partial t} = -\rho(\mathbf{M},t)\mathcal{M}(E) + \frac{D}{M_s}\frac{\partial\rho(\mathbf{M},t)}{\partial E}\oint_{\mathcal{L}(E)} d\mathbf{M}\cdot(\mathbf{H}_{eff}\times\mathbf{M}). \tag{8.46}$$

* Notice the similarity between (8.44) and (8.11).

In steady-state ($\partial/\partial t=0$), the probability density has the form [118–120]:

$$\rho(\mathbf{M},t) \propto \exp\left(-\frac{E_{eff}}{k_BT}\right), \tag{8.47}$$

$$E_{eff} = E - \int^{E} C(x)dx, \tag{8.48}$$

$$C(E) = \frac{a_J}{\alpha}\frac{\oint_{\mathcal{L}(E)} \mathbf{M}\times\mathbf{P}\cdot d\mathbf{M}}{\oint_{\mathcal{L}(E)} \mathbf{M}\times\mathbf{H}_{eff}\cdot d\mathbf{M}}. \tag{8.49}$$

In the case of static states, $C(E)=a_J/a_c$, where a_c is the amplitude of the ST at the critical switching voltage. In the case of an in-plane applied field $\mathbf{H}_{ext}=H_x\mathbf{x}$ with $\mathbf{H}_K=H_KM_x/M_s\mathbf{x}$, the lifetime of the magnetization within a given static state is

$$\tau^{-1} = f_0\exp\left[-\frac{E_b(1-a_J/a_c)}{k_BT}\right], \tag{8.50}$$

where

f_0 is the attempt frequency

E_b is the energy barrier, defined as $E_b=K_uV_0(1-H_{ext}/H_c)^{\beta}$

Here, K_uV_0 is the anisotropy energy and H_c is the switching field defined in (8.38). The exponent β is 2 when the external field is along the easy axis and 1.5 otherwise. The out-of-plane torque can be included as a bias-dependent magnetic field, and one replaces H_{ext} by $H_{ext}+b_J$. Equation 8.50 relates the lifetime τ of the magnetization in a given stable state. Therefore, assuming a constant τ (e.g., equals to the experimental current-pulse duration), it is possible to analytically determine the boundaries between two stable states, as done in Figure 8.6.

Finally, one has to account for the temperature heating due to Joule heating. From the usual thermal diffusion equations [122], the contribution of the Joule heating to the effective temperature of the free layer is obtained by replacing the temperature of the environment T by $\sqrt{T^2+c_0V^2}$ [111], so that the equilibrium thermal factor of the free layer $\Delta=K_uV_0/k_BT$ is replaced by $\Delta = K_uV_0/k_B\sqrt{T^2+c_0V^2}$ in the presence of an external current.

8.3.2.4 Stability Diagrams

The stability diagrams obtained from the analytical treatment given earlier are displayed in Figure 8.6. Figure 8.6a shows the stability diagram of a metallic SV at 0K ($a_J=aj_e$ and $b_J=0$). The diagram shows two stability regions where the perturbation $\delta\mathbf{M}$ is damped and the magnetization relaxes toward either P or AP configuration, and two regions where the perturbation $\delta\mathbf{M}$ diverges (regions E). These regions are detailed in Figure 8.6b and show both IPP and OPP. Interestingly, the inclusion of thermal effects modifies the boundaries of the stability diagram. Figure 8.6c shows the same diagram at 100, 300, and 600 K, assuming a lifetime (of pulse duration) $\tau = 1$ s. The critical switching voltage and field both decrease with the temperature as the result of the thermally assisted magnetization switching. When including the Joule heating of the junction (Figure 8.6d), the effective temperature varies with the bias and the critical switching voltage is then dramatically decreased.

Finally, Figure 8.6e displays the stability diagram in the case of a symmetric MTJ, assuming elastic tunneling. In this case, the out-of-plane torque is no more negligible and reads $b_J = b_2 V^2$ and $b_2 < 0$. In this case, the out-of-plane torque favors the P state, and modifies the stability diagram asymmetrically: for large positive bias and small negative external field, b_J competes with H_{ext} and a_J, then decreases the stability region AP.

8.3.2.5 Inhomogeneous Magnetic Excitations

Although the macrospin model has been shown to provide a simple and powerful tool for the interpretation of a number of experiments [58,59,108], some experimental data are not reproduced by this method, indicating the micromagnetic nature of the magnetization [53,123]. In this case, the Oersted and dipolar fields are taken into account as well as the magnetic exchange arising from the spatially varying magnetization texture. These two terms, Oersted field and magnetic exchange, are responsible for vortex nucleation and collective magnetic excitations (spin waves) that are absent in the macroscopic model and can deeply modify the magnetization dynamics.

As an example, the current-induced magnetization switching shows a much more complex behavior in micromagnetic modeling [124]. Whereas in the macrospin model, the magnetization switching occurs after fast coherent magnetization precessions, in micromagnetic modeling, it appears to occur in two steps: First, incoherent spin waves are excited by the current close to the edge of the layer; then after a few nanosecond, these spin waves extend along the edges and begin to excite the center of the layer, generating inhomogeneous magnetic domains and vortices. Eventually, the magnetization of the free layer switches. The contribution of the Oersted field is essential to distort the initial magnetic configuration and break the symmetry of the system. This type of incoherent magnetization switching has been recently directly observed by Strachan et al. [125] by time-resolved x-ray imaging.

Most of the magnetic excitation features predicted by the macrospin model are observed within the micromagnetic simulations [126,127]. In particular, IPPs and OPPs (in Figure 8.6b) are found with both approaches, even if the line width of the magnetic excitations is usually different. However, the presence of spin wave modes within the ferromagnetic layer can produce magnetic behaviors far from the macrospin predictions [61,124,127–131] up to chaotic dynamics [130,131]. These modes can be excited either by small perturbations (linear regime) or by a strong spin-polarized current (nonlinear regime), resulting in strongly inhomogeneous magnetic excitations, as observed both experimentally and theoretically.

Finally, we mention the increasing interest for phase-locked nano-oscillators. Mancoff et al. [54] and Kaka et al. [55] simultaneously demonstrated the mutual phase-locking of the precession frequency of two SVs in point-contact geometry. A number of theoretical investigations [132–135] addressed this effect within the nonlinear regime of spin waves excitations. The current-driven phase-locking of an array of nano-oscillators via an AC current has been recently achieved experimentally [136].

8.3.3 Spin Wave Dynamics and Domain Wall Motion

There are very rich dynamic properties in ferromagnetic films. Most of the theoretical studies are carried out via micromagnetic simulations where standard codes are now available. The dynamic equation in the presence of the current and the magnetic field is given by (8.27)

$$\frac{\partial \mathbf{M}}{\partial t} = -\gamma \mathbf{M} \times \mathbf{H}_{eff} + \frac{\alpha}{M_s}\mathbf{M} \times \frac{\partial \mathbf{M}}{\partial t} - \frac{b_J}{M_s^2}\mathbf{M} \times \left(\mathbf{M} \times \frac{\partial \mathbf{M}}{\partial x}\right) - \beta \frac{b_J}{M_s}\mathbf{M} \times \frac{\partial \mathbf{M}}{\partial x}. \tag{8.51}$$

As we have discussed earlier, the above model contains two crucial parameters: damping parameter α and the nonadiabaticity constant β; both parameters are material dependent and have not been well understood microscopically. In this section, we simply choose for them reasonable values for transition metals.

The study of the current-driven dynamics given by (8.51) falls into two broadly defined categories. If a DW preexists before the application of the current, theory and simulation are usually focused on the current-driven wall dynamics, including the wall distortion, transportation, and transformation from one type of wall structures to the others. For application purposes, one is often interested in the critical current density I_{c1} required for overcoming various DW pinning potentials [18,19]. Another interesting subject is the current-induced ferromagnetic instability where the initially uniformly magnetized film or nanowire becomes unstable at a second and larger critical value I_{c2} of the current.

8.3.3.1 Ferrodynamics: Destabilization of Ferromagnetism by Currents

The critical current density for the linear instability of a uniformly magnetized film can be readily obtained by using the conventional linear instability analysis [43]. Consider a small deviation of the uniform magnetization, $\tilde{\mathbf{m}}(x,t) = \mathbf{e}_x + \delta\tilde{\mathbf{m}}e^{i(\omega(k)t+kx)}$ where we have defined $\tilde{\mathbf{m}} = \mathbf{M}/M_s$ as a unit vector, $\delta\tilde{\mathbf{m}}$, k, and $\omega(k)$ are the amplitude (small vector perpendicular to $\mathbf{e}_x$), wave vector, and frequency of the spin wave, respectively. Since we assume the wire is infinitely long and the boundary effect is discarded, $\delta\tilde{\mathbf{m}}$ can be treated as a spatially independent constant. By placing $\tilde{\mathbf{m}}(x,t)$ into (8.51) and carrying out the linear instability analysis, we find

$$\begin{bmatrix} -i(\omega + b_J k) & f_1 \\ f_2 & -i(\omega + b_J k) \end{bmatrix} \begin{bmatrix} \delta\tilde{m}_y \\ \delta\tilde{m}_v \end{bmatrix} = 0, \tag{8.52}$$

where we have defined $f_1 = -\gamma(J_{ex}k^2 + H_1) - i(\alpha\omega + \beta b_J k)$ and $f_2 = \gamma(J_{ex}k^2 + H_2) + i(\alpha\omega + \beta b_J k)$, where $H_1 = H_{ext} + H_K + 4\pi M_s$ and $H_2 = H_{ext} + H_K$, and where H_{ext} and H_K are the external field and the anisotropy field along the wire; J_{ex} is the exchange constant. By calculating the determinant of the secular equation (8.52), one could find the critical current $j^c(k)$ at the onset of instabilities ($\mathcal{Im}[\omega] = 0$).

$$j^c(k) = \frac{\alpha\gamma}{kb(\alpha - \beta)} \sqrt{\left(J_{ex}k^2 + H_1\right)\left(J_{ex}k^2 + H_2\right)}, \tag{8.53}$$

where $b_J = bj_e$. By minimizing $j^c(k)$ with respect to k, that is, $\partial j^c/\partial k = 0$, we find that the minimum critical current density j_{c2} is

$$j_{c2} = \frac{\alpha\gamma\sqrt{J_{ex}}\left(\sqrt{H_1} + \sqrt{H_2}\right)}{b(\alpha - \beta)}. \tag{8.54}$$

Interestingly, only a single spin wave with frequency $\omega_c = j_{c2}(\beta b/\alpha)(H_1 H_2/J_{ex}^2)^{1/4}$ is excited at j_{c2}. One notes that j_{c2} is infinite when $\beta = \alpha$. The lack of spin wave excitations in this case can be traced back to (8.51): when one makes a Galilean transformation, $x' = x - bjt$ and $t' = t$, (8.51) becomes a usual LLG equation without the ST terms. Thus, the spin wave excitations are prohibited when $\beta = \alpha$ as expected for a Galilean invariance system.

When the current density is larger than j_{c2}, the excited spin wave frequency expands from a single value to a finite range. Equation 8.52 has two solutions for $\omega(k)$:

$$\omega_1 = -bjk + \gamma\sqrt{(J_{ex}k^2 + H_1)(J_{ex}k^2 + H_2)}, \tag{8.55}$$

$$\omega_2 = -\frac{\beta bj}{\alpha}k. \tag{8.56}$$

Equation 8.55 is known as the spin wave dispersion relation with Doppler shift [137], and Equation 8.56 is an additional solution associated with the nonadiabatic ST. To fully describe the ferrodynamic phase for $j > j_{c2}$, it is necessary to quantitatively determine the number of the spin waves excited by the current by using the micromagnetic simulation. The detailed features, such as fast spin wave propagation, critical exponents on the order parameters, and generation of localized spin packets have been reported in several references [43,119,137–139]. Although a number of experimental results have reported the current-driven domain nucleation [19,140], the relation to ferrodynamics is not clear yet.

Finally, we mention the observation of current-driven magnetic excitations in a single magnetic layer when the current is injected perpendicular to the plane of the layer [141,142]. These excitations have been associated with the STT mediated by the electron diffusion along the interfaces [143,144] (also called *self-torque*). Another interpretation could be the current-driven instabilities in the framework of the ferrodynamic effect. Up until now, no clear experimental data has been able to determine which phenomenon is more likely responsible for these excitations.

8.3.3.2 Domain Wall Motion: Transverse and Vortex Walls

The effects of the current on the DW dynamics have been extensively studied in the last several years [145–148]. To address the theoretical dynamics of the DW, two approaches are commonly used. One is the numerical simulation based on micromagnetic codes [149–153]. This approach can quantitatively address detailed dynamic features of various DWs in the presence of defects. Another approach is based on the Thiele's method [154], which approximates the DW as a rigid object that can be characterized by a few parameters such as the center of the wall, the orientation of the wall plane and the size of the vortex core [151,153]. In this section, we discuss a few elementary features in the current-driven motion.

Magnetic DWs in magnetic films have various structural forms, which are determined by geometrical and material parameters. For a magnetic wire with an "infinite" length and a finite width, there are commonly two types of DWs: transverse wall (TW—Figure 8.7a) and vortex wall (VW—Figure 8.7b). Both of them are stable but depending on the wire thickness and width, one of the walls is usually more stable than the other. However, in a certain range of the parameters, the static energies of these two walls are comparable and thus one can produce both types of walls in the same wire [145,146]. By using different initialization methods, one can create either wall [155]. When a magnetic field or an electric current is applied, both TW and VW are able to move along the wire. The dynamics of the walls is generally very complex and micromagnetic simulations are required in order to describe the details of the DW motion [149,150].

For the TW, a simple scenario assumes the DW undergoes a uniform motion along the direction of the electron current [152]. This is modeled by $\mathbf{M}(x,t) = \mathbf{M}(x - v_x t)$ and $\partial\mathbf{M}/\partial t = -v_x \partial\mathbf{M}/\partial t$, where v_x is the wall velocity. Placing this into (8.51), we readily find that

$$\gamma\mathbf{M} \times \mathbf{H}_{eff} = -\frac{v_x\alpha + \beta b_J}{M_s}\mathbf{M} \times \frac{\partial\mathbf{M}}{\partial x} + (b_J + v_x)\frac{\partial\mathbf{M}}{\partial x}. \tag{8.57}$$

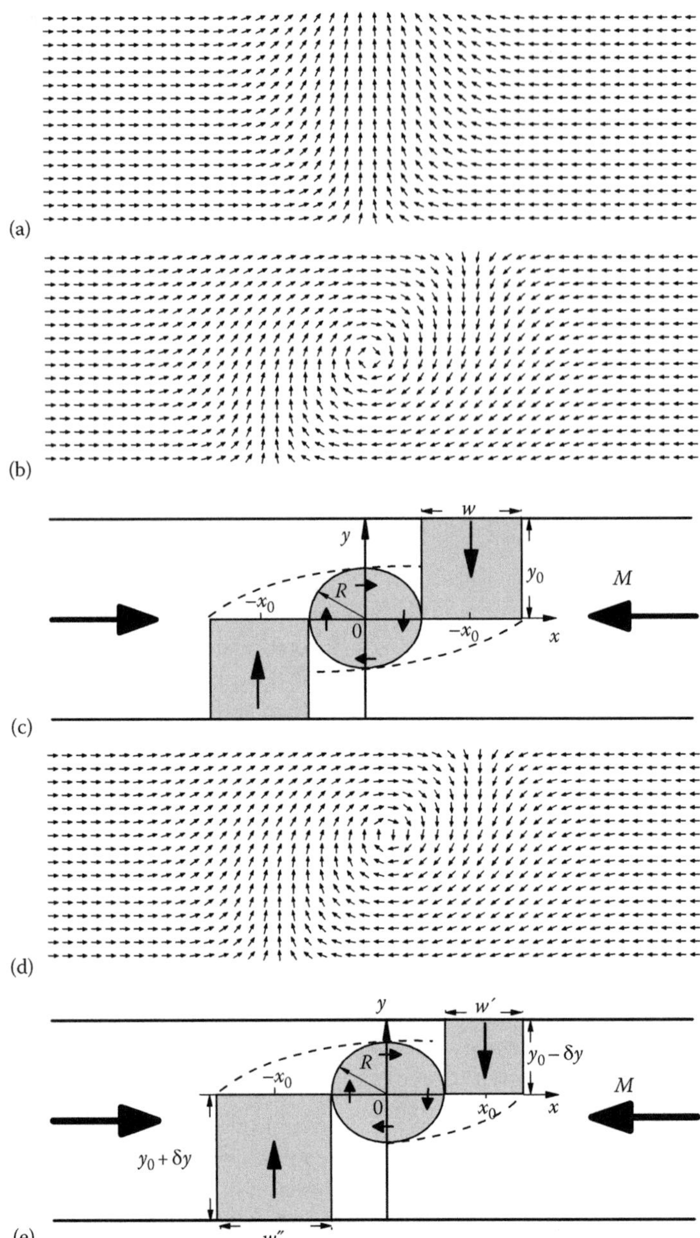

FIGURE 8.7 Magnetization patterns of (a) a TW and (b) a VW, without applied current or field. (c) Schematic model for the VW: a VW is modeled by two TWs and a vortex core, shown in gray areas. The vortex core is at the center of the wire and $\pm x_0$ are the positions of the centers of the TWs. R is the outer radius of the vortex core, w is the wall width of two TWs, and y_0 is the half-width of the wire. (d) Magnetization pattern of the VW in the presence of the current. The parameters are $b_J = 25$ m/s, $\alpha = 0.05$, and $\beta = 0.04$. (e) Schematic model for the VW. δy is the displacement of the vortex toward the edge of the wire. The wall structure is not symmetrical, and one TW is expanded while the other is compressed. These figures are a courtesy from He and Zhang [152].

After some algebra [152], the TW velocity for a rigid wall is obtained:

$$v_x = \frac{\pm\gamma H_{ext}\Delta - \beta b_J}{\alpha}, \tag{8.58}$$

where the DW width Δ is defined as $\Delta \propto \int dV \left|\partial \mathbf{M}/\partial t\right|^2$. Equation 8.58 demonstrates that the DW velocity solely depends on the external magnetic field and on the *nonadiabatic* torque βb_J. Actually, the DW width Δ weakly depends on the magnetic field and on the adiabatic torque b_J. Therefore, under an external field, the final DW velocity slightly depends on b_J.

In the case where the TW is not rigid ($\Delta = \Delta(t)$), the velocity is time dependent and we adopt Schryer and Walker's description [156], by introducing two polar angles θ and φ, where θ is the angle between the magnetization vector and the propagation direction and φ is the out-of-plane angle of the magnetization vector:

$$\varphi = \varphi(t), \quad \theta(x,t) = 2\tan^{-1}\exp\left(\frac{\left(x - \int_0^t v_x(\tau)d\tau\right)}{\Delta(t)}\right). \tag{8.59}$$

Inserting the previous forms in (8.51), and keeping only the first-order terms, we obtain:

$$\frac{d\varphi}{dt} = \gamma(H_{ext} - 2\pi\alpha M_s \sin 2\varphi) + b_J \frac{\alpha - \beta}{\Delta(t)}, \tag{8.60}$$

$$\alpha v_x(t) + \Delta(t)\frac{d\varphi}{dt} = \gamma H_{ext}\Delta(t) - \beta b_J. \tag{8.61}$$

Assuming that the wall distortions as well as the polar angle ϕ are small, a dynamical equation for the TW is obtained:

$$\frac{1}{2\pi\gamma^2\Delta(t)}\frac{dv_x(t)}{dt} + \frac{2\alpha M_s}{\gamma\Delta(t)}v_x(t) - 2M_sH_{ext} + \frac{2M_s}{\gamma\Delta(t)}\beta b_J = 0. \tag{8.62}$$

Equation 8.62 is a Newton-type dynamical equation, which shows that the TW can be treated as a quasi-particle with an effective mass $m^* = (2\pi\gamma^2\Delta(t))^{-1}$ and where the magnetization damping (second term), the external field (third term), and the spin current (fourth term) act like external forces. From (8.62), one can deduce the velocity of the TW under both external field and current:

$$v_x = \frac{\gamma H_{ext}\Delta - \beta b_J}{\alpha} + Ce^{-t/\tau}, \tag{8.63}$$

where

$\tau = (4\pi\gamma\alpha M_s)^{-1}$ is a relaxation time due to the friction

C is a constant depending on the initial conditions of the DW and that can be related to the adiabatic torque b_J

Equation 8.63 shows that the adiabatic torque causes the distortion that absorbs the adiabatic spin angular momentum. Therefore, once this distortion is complete, the net effect of the adiabatic torque vanishes and the wall stops [151,152]. Inspecting the role of the adiabatic torque in the absence of external field and nonadiabatic torque, we find that it gives rise to a component of the magnetization perpendicular to the film plane, which is balanced by the demagnetization field and there is no contribution to the steady translation motion of the wall. In contrast, the nonadiabatic torque acts like a magnetic field and induces a steady-state DW motion: the steady velocity is solely determined by the nonadiabatic torque and the external magnetic field.

For the VW, the one-dimensional wall profile fails to capture the wall structure and one needs at least to use a two-dimensional model to approximately characterize the vortex structure (Figure 8.7c). The VW structure might be schematically understood as the combination of two symmetrical TWs diagonally crossing the wire and a central vortex core connecting the two TWs. Following Thiele's method, we find that the initial VW velocities are

$$v_x|_{t=0} \simeq -b_J, \tag{8.64}$$

$$v_x|_{t=0} \simeq -\frac{1}{2}b_J p(\alpha - \beta)\text{In}\frac{R}{a_0}, \tag{8.65}$$

where

$p = \pm 1$ is the vortex polarity

a_0 is the lattice parameter (here, we disregarded the influence of the applied magnetic field)

Actually, the distorted TWs produce a reacting (or restoring) force to the vortex core during the DW motion. Then, two scenarios are possible: the reacting force is strong enough to completely halt the perpendicular wall velocity or the reacting force is unable to stop the vortex from colliding with the wall edge. In the former case, the perpendicular wall velocity eventually reaches zero and a steady-state wall velocity along the wire is achieved. The final wall velocity is precisely the same as that of the TW, that is, $v_x|_{t=\infty} = -\beta b_J/\alpha$ and $v_y|_{t=\infty} = 0$. In the latter case, the vortex core collides with the wire edge and can either vanish, move out of the wire, or be reflected, that is, the transformation of the VW to other types of walls occurs. Usually, the vortex core disappears at the edge and the initial VW becomes a TW [145–148].

However, the above conclusion is not universal. In the presence of the defects or DW pinning centers, the adiabatic torque plays critical role in determining the critical depinning current [149–153]. Furthermore, the wall velocity becomes a complicate function of the current and field, depending on the wall structure as well as the strength of currents and fields.

8.4 Conclusion

Since its first theoretical prediction, the ST mechanism has attracted an extensive attention from the Spintronics community. Understanding the interplay between the spin current and the magnetization dynamics in SVs, tunnel junctions andDW motion still represent one of the most important challenges for applications, in magnetic memories, read-heads, and oscillators.

The theory of spin-dependent transport has also made a lot of progress in the understanding of spin currents, and related phenomena in all the structures mentioned earlier. Even if the main physics is now captured within magnetoelectronics or diffusive theories, quantitative agreement between the theory and the experiment remains to be achieved, and the large interest for numerical studies (first principle calculations and micromagnetic modeling) illustrates the need for parameter-free understanding.

The conventional ST is now aiming toward applications, and recent theoretical studies have proposed different new types of STs, using thermokinetic properties, spin–orbit interaction, or specific magnetic texture of the antiferromagnets. In these last three examples, convincing experimental evidence is needed to make more progress and to propose an alternative method for information storage and magnetic field-free magnetization control.

References

1. J. A. Katine and E. E. Fullerton, Device implications of spin-transfer torques, *J. Magn. Magn. Mater.* **320**, 1217–1226 (2008).
2. A. Brataas, G. E. W. Bauer, and P. J. Kelly, Non-collinear magnetoelectronics, *Phys. Rep.* **427**, 157–255 (2006).
3. D. C. Ralph and M. D. Stiles, Spin transfer torques, *J. Magn. Magn. Mater.* **320**, 1190–1216 (2008).
4. J. Z. Sun and D. C. Ralph, Magnetoresistance and spin-transfer torque in magnetic tunnel junctions, *J. Magn. Magn. Mater.* **320**, 1227–1237 (2008).
5. A. Manchon, N. Ryzhanova, M. Chschiev et al., Spin transfer torques in magnetic tunnel junctions, in *Giant Magnetoresistance New Research*, A. D. Torres and D. A. Perez (Eds.), Nova Science Publishers, New York, pp. 63–106, 2009.
6. L. Berger, Low-field magnetoresistance and domain drag in ferromagnets, *J. Appl. Phys.* **49**, 2156–2161 (1978).
7. J. C. Slonczewski, Conductance and exchange coupling of two ferromagnets separated by a tunneling barrier, *Phys. Rev. B* **39**, 6995–7002 (1989).
8. L. Berger, Emission of spin waves by a magnetic multilayer traversed by a current, *Phys. Rev. B* **54**, 9353–9358 (1996).
9. J. C. Slonczewski, Current-driven excitation of magnetic multilayers, *J. Magn. Magn. Mater.* **159**, L1–L7 (1996).
10. M. Tsoi, A. G. M. Jansen, J. Bass et al., Excitation of a magnetic multilayer by an electric current, *Phys. Rev. Lett.* **80**, 4281–4284 (1998).
11. E. B. Myers, D. C. Ralph, J. A. Katine, R. N. Louie, and R. A. Buhrman, Current-induced switching of domains in magnetic multilayer devices, *Science* **285**, 867–870 (1999).
12. J. A. Katine, F. J. Albert, R. A. Buhrman, E. B. Myers, and D. C. Ralph, Current-driven magnetization reversal and spin-wave excitations in Co/Cu/Co pillars, *Phys. Rev. Lett.* **84**, 3149–3152 (2000).
13. A. Fert, V. Cros, J.-M. George et al., Magnetization reversal by injection and transfer of spin: Experiments and theory, *J. Magn. Magn. Mater.* **272**, 1706–1711 (2004).
14. S. Mangin, D. Ravelosona, J. A. Katine, M. J. Carey, B. D. Terris, and E. E. Fullerton, Current-induced magnetization reversal in nanopillars with perpendicular anisotropy, *Nat. Mater.* **5**, 210–215 (2006).
15. J. Z. Sun, Current-driven magnetic switching in manganite trilayer junctions, *J. Magn. Magn. Mater.* **202**, 157–162 (1999).
16. Y. Huai, F. Albert, P. Nguyen, M. Pakala, and T. Valet, Observation of spin-transfer switching in deep submicron-sized and low-resistance magnetic tunnel junctions, *Appl. Phys. Lett.* **84**, 3118–3120 (2004).
17. G. D. Fuchs, N. C. Emley, I. N. Krivorotov et al., Spin-transfer effects in nanoscale magnetic tunnel junctions, *Appl. Phys. Lett.* **85**, 1205–1207 (2004).
18. A. Yamaguchi, T. Ono, S. Nasu et al., Real-space observation of current-driven domain wall motion in submicron magnetic wires, *Phys. Rev. Lett.* **92**, 077205 (2004).
19. M. Klaui, P.-O. Jubert, R. Allenspach et al., Direct observation of domain-wall configurations transformed by spin currents, *Phys. Rev. Lett.* **95**, 026601 (2005).
20. Z. Wei, A. Sharma, A. S. Nunez et al., Changing exchange bias in spin valves with an electric current, *Phys. Rev. Lett.* **98**, 116603 (2007).
21. S. Urazhdin and N. Anthony, Effect of polarized current on the magnetic state of an antiferromagnet, *Phys. Rev. Lett.* **99**, 046602 (2007).
22. D. Chiba, Y. Sato, T. Kita, F. Matsukura, and H. Ohno, Current-driven magnetization reversal in a ferromagnetic semiconductor (Ga,Mn)As/GaAs/(Ga,Mn)As tunnel junction, *Phys. Rev. Lett.* **93**, 216602 (2004).
23. M. Yamanouchi, D. Chiba, F. Matsukura, and H. Ohno, Current-induced domain-wall switching in a ferromagnetic semiconductor structure, *Nature* **428**, 539–542 (2004).
24. A. Chernyshov, M. Overby, X. Liu, J. K. Furdyna, Y. Lyanda-Geller, and L. P. Rokhinson, Evidence for reversible control of magnetization in a ferromagnetic material by means of spin–orbit magnetic field, *Nat. Phys.* **5**, 656–659 (2009).
25. X. Waintal, E. B. Myers, P. W. Brouwer, and D. C. Ralph, Role of spin-dependent interface scattering in generating current-induced torques in magnetic multilayers, *Phys. Rev. B* **62**, 12317–12327 (2000).
26. M. D. Stiles and A. Zangwill, Anatomy of spin-transfer torque, *Phys. Rev. B* **66**, 014407 (2002).
27. J. C. Slonczewski, Currents and torques in metallic magnetic multilayers, *J. Magn. Magn. Mater.* **247**, 324–338 (2002).
28. S. Zhang, P. M. Levy, and A. Fert, Mechanisms of spin-polarized current-driven magnetization switching, *Phys. Rev. Lett.* **88**, 236601 (2002).
29. J. Barnas, A. Fert, M. Gmitra, I. Weymann, and V. K. Dugaev, From giant magnetoresistance to current-induced switching by spin transfer, *Phys. Rev. B* **72**, 024426 (2005).
30. A. Manchon, N. Ryzhanova, N. Strelkov, A. Vedyayev, and B. Dieny, Modeling spin transfer torque and magnetoresistance in magnetic multilayers, *J. Phys.: Condens. Matter* **19**, 165212 (2007).
31. I. Theodonis, N. Kioussis, A. Kalitsov, M. Chshiev, and W. H. Butler, Anomalous bias dependence of spin torque in magnetic tunnel junctions, *Phys. Rev. Lett.* **97**, 237205 (2006).

32. M. Wilczynski, J. Barnas, and R. Swirkowicz, Free-electron model of current-induced spin-transfer torque in magnetic tunnel junctions, *Phys. Rev. B* 77, 054434 (2008).
33. J. Xiao, G. E. W. Bauer, and A. Brataas, Spin-transfer torque in magnetic tunnel junctions: Scattering theory, *Phys. Rev. B* 77, 224419 (2008).
34. A. Manchon, N. Ryzhanova, N. Strelkov, A. Vedyayev, M. Chshiev, and B. Dieny, Description of current-driven torques in magnetic tunnel junctions, *J. Phys.: Condens. Matter* **20**, 145208 (2008).
35. C. Heiliger and M. D. Stiles, Ab initio studies of the spin-transfer torque in magnetic tunnel junctions, *Phys. Rev. Lett.* **100**, 186805 (2008).
36. J. C. Slonczewski, Currents, torques, and polarization factors in magnetic tunnel junctions, *Phys. Rev. B* **71**, 024411 (2005).
37. P. M. Levy and A. Fert, Spin transfer in magnetic tunnel junctions with hot electrons, *Phys. Rev. Lett.* **97**, 097205 (2006).
38. A. Manchon and S. Zhang, Influence of interfacial magnons on spin transfer torque in magnetic tunnel junctions, *Phys. Rev. B* **79**, 174401 (2009a).
39. J. C. Sankey, Y. T. Cui, J. Z. Sun, J. C. Slonczewski, R. A. Buhrman, and D. C. Ralph, Measurement of the spin-transfer-torque vector in magnetic tunnel junctions, *Nat. Phys.* **4**, 67 (2008).
40. H. Kubota, A. Fukushima, K. Yakushiji et al., Quantitative measurement of voltage dependence of spin-transfer torque in MgO-based magnetic tunnel junctions, *Nat. Phys.* 4, 37 (2008).
41. A. M. Deac, A. Fukushima, H. Kubota et al., Bias-driven high-power microwave emission from MgO-based tunnel magnetoresistance devices, *Nat. Phys.* **4**, 803 (2008).
42. S.-C. Oh, S.-Y. Park, A. Manchon et al., Bias-voltage dependence of perpendicular spin-transfer torque in asymmetric MgO-based magnetic tunnel junctions, *Nat. Phys.* (2009), doi:10.1038/nphys1427.
43. Ya. B. Bazaliy, B. A. Jones, and S.-C. Zhang, Modification of the Landau–Lifshitz equation in the presence of a spin-polarized current in colossal- and giant-magnetoresistive materials, *Phys. Rev. B* **57**, R3213–R3216 (1998).
44. S. Zhang and Z. Li, Roles of nonequilibrium conduction electrons on the magnetization dynamics of ferromagnets, *Phys. Rev. Lett.* **93**,127204 (2004).
45. G. Tatara and H. Kohno, Theory of current-driven domain wall motion: Spin transfer versus momentum transfer, *Phys. Rev. Lett.* **92**, 086601 (2004).
46. J. Xiao, A. Zangwill, and M. D. Stiles, Spin-transfer torque for continuously variable magnetization, *Phys. Rev. B* **73**, 054428 (2006).
47. Y. Tserkovnyak, H. J. Skadsem, A. Brataas, and G. E. W. Bauer, Current-induced magnetization dynamics in disordered itinerant ferromagnets, *Phys. Rev. B* **74**, 144405 (2006).
48. S. E. Barnes and S. Maekawa, Generalization of Faraday's law to include nonconservative spin forces, *Phys. Rev. Lett.* **98**, 246601 (2007).
49. J.-I. Ohe, A. Takeuchi, and G. Tatara, Charge current driven by spin dynamics in disordered Rashba spin–orbit system, *Phys. Rev. Lett.* **99**, 266603 (2007).
50. R. A. Duine, Effects of nonadiabaticity on the voltage generated by a moving domain wall, *Phys. Rev. B* **79**, 014407 (2009).
51. A. Thiaville, Y. Nakatani, J. Miltat, and Y. Suzuki, Micromagnetic understanding of current-driven domain wall motion in patterned nanowires, *Europhys. Lett.* **69**, 990 (2005).
52. S. E. Barnes and S. Maekawa, Current-spin coupling for ferromagnetic domain walls in fine wires, *Phys. Rev. Lett.* **95**, 107204 (2005).
53. S. I. Kiselev, J. C. Sankey, I. N. Krivorotov et al., Microwave oscillations of a nanomagnet driven by a spin-polarized current, *Nature* **425**, 380–383 (2003).
54. F. B. Mancoff, N. D. Rizzo, B. N. Engel, and S. Tehrani, Phase-locking in double-point-contact spin-transfer devices, *Nature* **437**, 393–395 (2005).
55. S. Kaka, M. R. Pufall, W. H. Rippard, T. J. Silva, S. E. Russek, and J. A. Katine, Mutual phase-locking of microwave spin torque nano-oscillators, *Nature* **437**, 389–392 (2005).
56. D. Houssameddine, U. Ebels, B. Delaet et al., Spin-torque oscillator using a perpendicular polarizer and a planar free layer, *Nat. Mater.* **6**, 447–454 (2007).
57. R. H. Koch, J. A. Katine, and J. Z. Sun, Time-resolved reversal of spin-transfer switching in a nanomagnet, *Phys. Rev. Lett.* **92**, 088302 (2004).
58. J. Z. Sun, Spin-current interaction with a monodomain magnetic body: A model study, *Phys. Rev. B* **62**, 570–578 (2000).
59. G. Bertotti, C. Serpico, I. D. Mayergoyz, A. Magni, M. dAquino, and R. Bonin, Magnetization switching and microwave oscillations in nanomagnets driven by spin-polarized currents, *Phys. Rev. Lett.* **94**, 127206 (2005).
60. K.-J. Lee, A. Deac, O. Redon, J.-P. Nozieres, and B. Dieny, Excitations of incoherent spin-waves due to spin-transfer torque, *Nat. Mater.* **3**, 877–881 (2004).
61. D. Berkov and N. Gorn, Transition from the macrospin to chaotic behavior by a spin-torque driven magnetization precession of a square nanoelement, *Phys. Rev. B* **71**, 052403 (2005).
62. T. Valet and A. Fert, Theory of the perpendicular magnetoresistance in magnetic multilayers, *Phys. Rev. B* **48**, 7099 (1993).
63. S. V. Vonsovsky, *Magnetism*, Wiley, New York, 1974.
64. T. L. Gilbert, A phenomenological theory of damping in ferromagnetic materials, *IEEE Trans. Magn.* **40**, 3443–3449 (2004).
65. K. Gilmore, Y. U. Idzerda, and M. D. Stiles, Identification of the dominant precession-damping mechanism in Fe, Co, and Ni by first-principles calculations, *Phys. Rev. Lett.* **99**, 027204 (2007).
66. V. Kambersky, Spin–orbital Gilbert damping in common magnetic metals, *Phys. Rev. B* **76**, 134416 (2007).

67. M. Zwierzycki, Y. Tserkovnyak, P. J. Kelly, A. Brataas, and G. E. W. Bauer, First-principles study of magnetization relaxation enhancement and spin transfer in thin magnetic films, *Phys. Rev. B.* **71**, 064420 (2005).
68. P. M. Haney, D. Waldron, R. A. Duine, A. S. Nunez, H. Guo, and A. H. MacDonald, Current-induced order parameter dynamics: Microscopic theory applied to Co/Cu/Co spin valves, *Phys. Rev. B* **76**, 024404 (2007).
69. S. Wang, Y. Xu, and K. Xia, First-principles study of spin-transfer torques in layered systems with noncollinear magnetization, *Phys. Rev. B* **77**, 184430 (2008).
70. O. Boulle, V. Cros, J. Grollier et al., Shaped angular dependence of the spin-transfer torque and microwave generation without magnetic field. *Nat. Phys.* **3**, 492–497 (2007).
71. Y.-N. Qi and S. Zhang, Crossover from diffusive to ballistic transport properties in magnetic multilayers, *Phys. Rev. B* **65**, 214407 (2002).
72. J. Xiao, A. Zangwill, and M. D. Stiles, A numerical method to solve the Boltzmann equation for a spin valve, *Eur. Phys. J. B* **59**, 415–427 (2007).
73. W. M. Saslow, Spin pumping of current in non-uniform conducting magnets, *Phys. Rev. B* **76**, 184434 (2007).
74. J.-E. Wegrowe, S. M. Santos, M.-C. Ciornei, H.-J. Drouhin, and J. M. Rubi, Magnetization reversal driven by spin injection: A diffusive spin-transfer effect, *Phys. Rev. B* **77**, 174408 (2008).
75. Z. Li, S. Zhang, Z. Diao et al., Perpendicular spin torques in magnetic tunnel junctions, *Phys. Rev. Lett.* **100**, 246602 (2008).
76. S. Petit, C. Baraduc, C. Thirion et al., Spin-torque influence on the high-frequency magnetization fluctuations in magnetic tunnel junctions, *Phys. Rev. Lett.* **98**, 077203 (2007).
77. E. Y. Tsymbal, O. N. Mryasov, and P. R. LeClair, Spin-dependent tunnelling in magnetic tunnel junctions, *J. Phys.: Condens. Matter* **15**, R109–R142 (2003).
78. S. Zhang, P. M. Levy, A. C. Marley, and S. S. P. Parkin, Quenching of magnetoresistance by hot electrons in magnetic tunnel junctions, *Phys. Rev. Lett.* **79**, 3744–3747 (1997).
79. J. Z. Sun, M. C. Gaidis, G. Hu et al., High-bias backhopping in nanosecond time-domain spin-torque switches of MgO-based magnetic tunnel junctions, *J. Appl. Phys.* **105**, 07D109 (2009).
80. W. F. Brinkman, R. C. Dynes, and J. M. Rowell, Tunneling conductance of asymmetrical barriers, *J. Appl. Phys.* **41**, 1915 (1970).
81. W. H. Butler, X.-G. Zhang, T. C. Schulthess, and J. M. MacLaren, Spin-dependent tunneling conductance of Fe|MgO|Fe sandwiches, *Phys. Rev. B* **63**, 054416 (2001).
82. C. Tiusan, J. Faure-Vincent, C. Bellouard, M. Hehn, E. Jouguelet, and A. Schuhl, Interfacial resonance state probed by spin-polarized tunneling in epitaxial Fe/MgO/Fe tunnel junctions, *Phys. Rev. Lett.* **93**, 106602 (2004).
83. P. Mavropoulos, N. Papanikolaou, and P. H. Dederichs, Complex band structure and tunneling through ferromagnet/insulator/ferromagnet junctions, *Phys. Rev. Lett.* **85**, 1088–1091 (2000).
84. M. D. Stiles, W. M. Saslow, M. J. Donahue, and A. Zangwill, Adiabatic domain wall motion and Landau–Lifshitz damping, *Phys. Rev. B* **75**, 214423 (2007).
85. H. Kohno, G. Tatara, and J. Shibata, Theory of current-driven domain wall motion: Spin transfer versus momentum transfer, *J. Phys. Soc. Jpn.* **75**, 113706 (2006).
86. R. A. Duine, A. S. Nunez, J. Sinova, and A. H. MacDonald, Functional Keldysh theory of spin torques, *Phys. Rev. B* **75**, 214420 (2007).
87. F. Piechon and A. Thiaville, Spin transfer torque in continuous textures: Semiclassical Boltzmann approach, *Phys. Rev. B* **75**, 174414 (2007).
88. I. Garate, K. Gilmore, M. D. Stiles, and A. H. MacDonald, Nonadiabatic spin-transfer torque in real materials, *Phys. Rev. B* **79**, 104416 (2009).
89. A. K. Nguyen, H. J. Skadsem, and A. Brataas, Giant current-driven domain wall mobility in (Ga,Mn)As, *Phys. Rev. Lett.* **98**, 146602 (2007).
90. D. Culcer, M. E. Lucassen, R. A. Duine, and R. Winkler, Current-induced spin torques in III–V ferromagnetic semiconductors, *Phys. Rev. B* **79**, 155208 (2009).
91. K. Obata and G. Tatara, Current-induced domain wall motion in Rashba spin–orbit system, *Phys. Rev. B* **77**, 214429 (2008).
92. R. A. Duine, A. S. Nunez, and A. H. MacDonald, Thermally assisted current-driven domain-wall motion, *Phys. Rev. Lett.* **98**, 056605 (2007).
93. J.-I. Ohe and B. Kramer, Dynamics of a domain wall and spin-wave excitations driven by a mesoscopic current, *Phys. Rev. Lett.* **96**, 027204 (2006).
94. Y. Le Maho, J.-V. Kim, and G. Tatara, Spin-wave contributions to current-induced domain wall dynamics, *Phys. Rev. B* **79**, 174404 (2009).
95. Y. Tserkovnyak, A. Brataas, G. E. W. Bauer, and B. I. Halperin, Nonlocal magnetization dynamics in ferromagnetic heterostructures, *Rev. Mod. Phys.* **77**, 1375–1421 (2005).
96. R. A. Duine, Spin pumping by a field-driven domain wall, *Phys. Rev. B* **77**, 014409 (2008).
97. S. Zhang and S. S.-L. Zhang, Generalization of the Landau-Lifshitz-Gilbert equation for conducting ferromagnets, *Phys. Rev. Lett.* **102**, 086601 (2009).
98. A. Manchon and S. Zhang, Theory of spin torque due to spin–orbit coupling, *Phys. Rev. B* **79**, 094422 (2009).
99. I. Garate and A. H. MacDonald, Influence of a transport current on magnetic anisotropy in gyrotropic ferromagnets, *Phys. Rev. B* **80**, 134403 (2009).
100. A. Matos-Abiague and R. L. Rodriguez-Suarez, Spin–orbit coupling mediated spin torque in a single ferromagnetic layer, *Phys. Rev. B* **80**, 094424 (2009).
101. M. Hatami, G. E. W. Bauer, Q. Zhang, and P. J. Kelly, Thermal spin-transfer torque in magnetoelectronic devices, *Phys. Rev. Lett.* **99**, 066603 (2007).

102. A. A. Kovalev and Y. Tserkovnyak, Thermoelectric spin transfer in textured magnets, *Phys. Rev. B* **80**, 100408 (2009).
103. A. S. Nunez, R. A. Duine, P. M. Haney, and A. H. MacDonald, Theory of spin torques and giant magnetoresistance in antiferromagnetic metals, *Phys. Rev. B* **73**, 214426 (2006).
104. R. A. Duine, P. M. Haney, A. S. Nunez, and A. H. MacDonald, Inelastic scattering in ferromagnetic and antiferromagnetic spin valves, *Phys. Rev. B* **75**, 014433 (2007).
105. P. M. Haneyand A. H. MacDonald, Current-induced torques due to compensated antiferromagnets, *Phys. Rev. Lett.* **100**, 196801 (2008).
106. Y. Xu, S. Wang, and K. Xia, Spin-transfer torques in antiferromagnetic metals from first principles, *Phys. Rev. Lett.* **100**, 226602 (2008).
107. M. D. Stiles and J. Miltat, Spin transfer torque and dynamics in spin dynamics in confined structures, in *Spin Dynamics in Confined Magnetic Structures III*, Topics in Applied Physics, vol. 101, B. Hillebrands and A. Thiaville (Eds.), Springer, Berlin, pp. 225–308, 2006.
108. J. Grollier, V. Cros, H. Jaffrès et al., Field dependence of magnetization reversal by spin transfer, *Phys. Rev. B* **67**, 174402 (2003).
109. V. K. Melkinov, On the stability of the center for time periodic perturbations, *Trans. Moscow Math. Soc.* **12**, 1–57 (1963).
110. W. F. Brown Jr., Thermal fluctuations of a single-domain particle, *Phys. Rev.* **130**, 1677–1686 (1963).
111. G. D. Fuchs, I. N. Krivorotov, P. M. Braganca et al., Adjustable spin torque in magnetic tunnel junctions with two fixed layers, *Appl. Phys. Lett.* **86**, 152509 (2005).
112. A. Fabian, C. Terrier, S. Serrano Guisan et al., Current-induced two-level fluctuations in pseudo-spin-valve (Co/Cu/Co) nanostructures, *Phys. Rev. Lett.* **91**, 257209 (2003).
113. I. N. Krivorotov, N. C. Emley, A. G. Garcia et al., Temperature dependence of spin-transfer-induced switching of nanomagnets, *Phys. Rev. Lett.* **93**, 166603 (2004).
114. W. L. Lim, N. Anthony, A. Higgins, and S. Urazhdin, Thermal dynamics in symmetric magnetic nanopillars driven by spin transfer, *Appl. Phys. Lett.* **92**, 172501 (2008).
115. S. E. Russek, S. Kaka, W. H. Rippard, M. R. Pufall, and T. J. Silva, Finite-temperature modeling of nanoscale spin-transfer oscillators, *Phys. Rev. B* **71**, 104425 (2005).
116. M. Gmitra and J. Barnas, Thermally assisted current-driven bistable precessional regimes in asymmetric spin valves, *Phys. Rev. Lett.* **99**, 097205 (2007).
117. R. Zhu and P. B. Visscher, Spin torque switching in perpendicular films at finite temperature, *J. Appl. Phys.* **103**, 07A722 (2008).
118. Z. Li and S. Zhang, Thermally assisted magnetization reversal in the presence of a spin-transfer torque, *Phys. Rev. B* **69**, 134416 (2004).
119. Z. Li, J. He, and S. Zhang, Stability of precessional states induced by spin-current, *Phys. Rev. B* **72**, 212411 (2005).
120. D. M. Apalkov and P. B. Visscher, Spin-torque switching: Fokker-Planck rate calculation, *Phys. Rev. B* **72**, 180405(R) (2005).
121. H. A. Kramers, Brownian motion in a field of force and the diffusion model of chemical reactions, *Physica (Utrecht)* **7**, 284–304 (1940).
122. R. Holm, *Electric Contacts: Theory and Applications*, Springer-Verlag, New York, 1967.
123. I. N. Krivorotov, N. C. Emley, R. A. Buhrman, and D. C. Ralph, Time-domain studies of very-large-angle magnetization dynamics excited by spin transfer torques, *Phys. Rev. B* **77**, 054440 (2008).
124. K.-J. Lee, Micromagnetic investigation of current-driven magnetic excitation in a spin valve structure, *J. Phys.: Condens. Matter* **19**, 165211 (2007).
125. J. P. Strachan, V. Chembrolu, Y. Acremann et al., Direct observation of spin-torque driven magnetization reversal through nonuniform modes, *Phys. Rev. Lett.* **100**, 247201 (2008).
126. J. Xiao, A. Zangwill, and M. D. Stiles, Macrospin models of spin transfer dynamics, *Phys. Rev. B* **72**, 014446 (2005).
127. D. V. Berkov and J. Miltat, Spin-torque driven magnetization dynamics: Micromagnetic modeling, *J. Magn. Magn. Mater.* **320**, 1238–1259 (2008).
128. K. Ito, Micromagnetic simulation on dynamics of spin transfer torque magnetization reversal, *IEEE Trans. Magn.* 2630–2632 (2005).
129. I. Firastrau, D. Gusakova, D. Houssameddine et al., Modeling of the perpendicular polarizer-planar free layer spin torque oscillator: Micromagnetic simulations, *Phys. Rev. B* **78**, 024437 (2008).
130. Z. Li, Y. C. Li, and S. Zhang, Dynamic magnetization states of a spin valve in the presence of dc and ac currents: Synchronization, modification, and chaos, *Phys. Rev. B* **74**, 054417 (2006).
131. Z. Yang, S. Zhang, and Y. C. Li, Chaotic dynamics of spin-valve oscillators, *Phys. Rev. Lett.* **99**, 134101 (2007).
132. A. N. Slavin and V. S. Tiberkevich, Nonlinear self-phase-locking effect in an array of current-driven magnetic nanocontacts, *Phys. Rev. B* **72**, 092407 (2005).
133. A. N. Slavin and V. S. Tiberkevich, Theory of mutual phase locking of spin-torque nanosized oscillators, *Phys. Rev. B* **74**, 104401 (2006).
134. J.-V. Kim, V. Tiberkevich, and A. N. Slavin, Generation linewidth of an auto-oscillator with a nonlinear frequency shift: Spin-torque nano-oscillator, *Phys. Rev. Lett.* **100**, 017207 (2008).
135. J.-V. Kim, Q. Mistral, C. Chappert, V. S. Tiberkevich, and A. N. Slavin, Line shape distortion in a nonlinear auto-oscillator near generation threshold: Application to spin-torque nano-oscillators, *Phys. Rev. Lett.* **100**, 167201 (2008).
136. B. Georges, J. Grollier, M. Darques et al., Coupling efficiency for phase locking of a spin transfer nano-oscillator to a microwave current, *Phys. Rev. Lett.* **101**, 017201 (2008).

137. J. Fernandez-Rossier, M. Braun, A. S. Nunez, and A. H. MacDonald, Influence of a uniform current on collective magnetization dynamics in a ferromagnetic metal, *Phys. Rev. B* **69**, 174412 (2004).
138. J. He and S. Zhang, Magnetic dynamic phase generated by spin currents, *Phys. Rev. B* **78**, 012414 (2008).
139. J. Shibata, G. Tatara, and H. Kohno, Effect of spin current on uniform ferromagnetism: Domain nucleation, *Phys. Rev. Lett.* **94**, 076601 (2005).
140. M. Klaui, M. Laufenberg, L. Heyne et al., Current-induced vortex nucleation and annihilation in vortex domain walls, *Appl. Phys. Lett.* **88**, 232507 (2006).
141. Y. Ji, C. L. Chien, and M. D. Stiles, Current-induced spin-wave excitations in a single ferromagnetic layer, *Phys. Rev. Lett.* **90**, 106601 (2003).
142. B. Ozyilmaz, A. D. Kent, J. Z. Sun, M. J. Rooks, and R. H. Koch, Current-induced excitations in single Cobalt ferromagnetic layer nanopillars, *Phys. Rev. Lett.* **93**, 176604 (2004).
143. M. L. Polianski and P. W. Brouwer, Current-induced transverse spin-wave instability in a thin nanomagnet, *Phys. Rev. Lett.* **92**, 026602 (2004).
144. M. D. Stiles, J. Xiao, and A. Zangwill, Phenomenological theory of current-induced magnetization precession. *Phys. Rev. B* **69**, 054408 (2004).
145. M. Hayashi, L. Thomas, C. Rettner, R. Moriya, Ya. B. Bazaliy, and S. S. P. Parkin, Current driven domain wall velocities exceeding the spin angular momentum transfer rate in permalloy nanowires, *Phys. Rev. Lett.* **98**, 037204 (2007).
146. L. Heyne, M. Klaui, D. Backes et al., Relationship between nonadiabaticity and damping in permalloy studied by current induced spin structure transformations, *Phys. Rev. Lett.* **100**, 066603 (2008).
147. S. Kasai, Y. Nakatani, K. Kobayashi, H. Kohno, and T. Ono, Current-driven resonant excitation of magnetic vortices, *Phys. Rev. Lett.* **97**, 107204 (2006).
148. R. Moriya, L. Thomas, M. Hayashi, Ya. B. Bazaliy, C. Rettner, and S. S. P. Parkin, Probing vortex-core dynamics using current-induced resonant excitation of a trapped domain wall, *Nat. Phys.* **4**, 368–372 (2008).
149. J.-H. Moon, D.-H. Kim, M. H. Jung, and K.-J. Lee, Current-induced resonant motion of a magnetic vortex core: Effect of nonadiabatic spin torque, *Phys. Rev. B* **79**, 134410 (2009).
150. P. Warnicke, Y. Nakatani, S. Kasai, and T. Ono, Long-range vortex domain wall displacement induced by an alternating current: Micromagnetic simulations, *Phys. Rev. B* **78**, 012413 (2008).
151. Z. Li and S. Zhang, Domain-wall dynamics and spin-wave excitations with spin-transfer torques, *Phys. Rev. Lett.* **92**, 207203 (2004).
152. Z. Li, J. He, and S. Zhang, Effects of spin current on ferromagnets (invited), *J. Appl. Phys.* **99**, 08Q702 (2006).
153. J. He, Z. Li, and S. Zhang, Current-driven vortex domain wall dynamics by micromagnetic simulations, *Phys. Rev. B* **73**, 184408 (2006).
154. A. A. Thiele, Steady-state motion of magnetic domains, *Phys. Rev. Lett.* **30**, 230–233 (1973).
155. R. D. McMichael and M. J. Donahue, Head to head domain wall structures in thin magnetic strips, *IEEE Trans. Magn.* **33**, 4167–4169 (1997).
156. N. L. Schryer and L. R. Walker, The motion of 180° domain walls in uniform dc magnetic fields, *J. Appl. Phys.* **45**, 5406 (1974).

9

Hot Carrier Spin Transport in Ferromagnetic Metals

Ron Jansen

National Institute of Advanced Industrial Science and Technology (AIST)

9.1 Introduction

Spin-dependent transport has been extensively studied in solid-state systems, particularly since the discovery of the giant magnetoresistance effect (see Chapters 4 and 5 in this book). In parallel, the properties of excited electrons in ferromagnetic materials have received significant attention. Such hot electrons, characterized by an energy of more than a few times the thermal energy above the Fermi energy (E_F), exhibit spin-dependent transport as well. The aim of this chapter is to describe the fundamental rules and the underlying phenomena that control the spin-dependent transport of hot carriers in ferromagnetic metals. We focus on the study of these phenomena using hybrid structures comprising ferromagnetic metals and semiconductors, where the latter are employed to create the required potential energy landscape for hot-electron transport.

Historically, the desire to understand the spin-dependent behavior of hot electrons in ferromagnetic materials derives from the widespread use of spin-polarized electrons in a variety of electron spectroscopy techniques, particularly those in the area of surface science. Here, hot electrons are used either to excite a magnetic system or to monitor the response of the system to an excitation created by other means such as irradiation with photons. Photoemission is the most well-known example, for which it was recognized that spin-dependent transport of the photoexcited hot electrons on their way out of the sample has significant impact on the spin asymmetry of the detected photoelectrons and can even lead to a sign reversal of the spin polarization [1–4]. Hence, a correct interpretation of the data for many of these techniques requires a sound knowledge of hot-electron spin transport.

A second motivation is the development of spintronic devices that utilize hot-electron spin transport. These devices, the spin-valve transistor (SVT [5–9], shown in Figure 9.1) and magnetic tunnel transistor (MTT [10–14]), are hybrid devices combining semiconductors with ferromagnetic metals and have been shown to exhibit very large magnetic field sensitivity. While it is now generally accepted that the output current level of such devices is too small for practical use, they continue to play a role as they allow detailed fundamental studies of spin transport in new regimes and in materials such as silicon [15].

A third field of relevance is that of magnetic systems out of equilibrium, such as magnetic tunnel junctions at finite applied bias voltage or ferromagnets that exhibit ultrafast magnetization dynamics induced by excitation with a femtosecond laser pulse or a spin-transfer torque. In these situations, the system is far from equilibrium, and, to a certain extent, the behavior is determined by excited carriers, not only electrons with energy above E_F, but also excited holes with energy below E_F. Both have to be taken into account, which is one of the reasons why attention has recently been paid to the study of the spin-dependent transport of hot holes in metallic ferromagnets. This is covered in a separate section of this chapter. We will not describe magnetization dynamics, as this is done in other chapters.

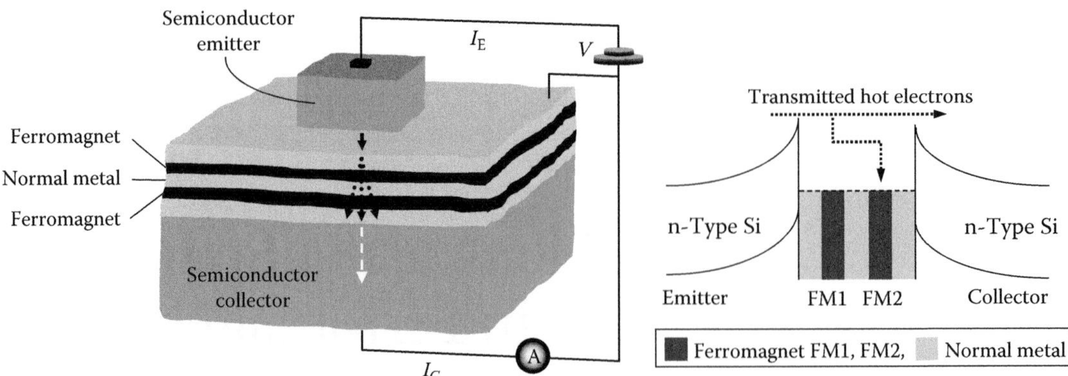

FIGURE 9.1 The spin-valve transistor. Left: Layout and electrical connections. Right: Energy band diagram.

The chapter is organized as follows. The next section more precisely defines what hot carrier spin transport entails and gives an overview of the salient features. A specific aim is to clarify the difference with transport at the Fermi energy. The third section focuses on spin transport of hot electrons. It covers the experimental techniques and devices used to examine hot-carrier spin transport and subsequently focuses on the fundamental scattering mechanisms that lead to the large spin asymmetry of the transmission. An important aim is to explain why hot-electron spin transport is poorly described if only the inelastic scattering via electron–hole pair excitation is considered. The hot electron section is, in part, based on a previously published topical review [9]. The fourth section is then devoted to the spin transport of hot holes. The fifth and concluding section gives a brief description of the application of the unique spin filtering properties of hot carriers to other areas of spintronics, including spin-polarized tunneling at finite energy, spin injection into semiconductors, and spin-sensitive scanning tunneling microscopy.

9.2 Overview and Basics of Hot Carrier Spin Transport

We define a hot electron as an electron with an energy that is higher than the Fermi energy by more than a few times the thermal energy. At room temperature, this implies all electrons with energy more than about 0.1 eV above E_F. Under normal conditions, such electrons are very few, but they can be created by injection from an external electron source through vacuum or by injection across an energy barrier in solid-state structures. In order to understand spin-dependent transport of such hot carriers, it is best to "forget" what we know about giant magnetoresistance and spin-dependent scattering at E_F because any description in terms of (spin-dependent) resistivity fails. To illustrate this, consider a thin film of ferromagnetic metal and ask what happens to a beam of initially unpolarized hot electrons when it is transmitted through the ferromagnet (FM) in the direction perpendicular to its surface. Once the hot electrons have entered the FM, their transport is driven by their kinetic energy since there is no voltage difference between the front and the rear surface of the FM thin film. A second essential difference to resistive transport is that hot electrons experience large energy losses due to inelastic scattering. At a typical energy of 0.3–1.5 eV above E_F, new scattering processes become available, including excitation of electron–hole pairs and small-wavelength (high-energy) spin waves. In this respect, it is important to keep in mind that experiments are typically done with FM films of only 3–10 nm thick, which, for carrier velocities in metals, means that it takes hot electrons less than 10 fs to traverse the layer. Hence, scattering processes associated with such short time scales play a dominant role. A third important point is that in virtually all the experiments, the beam of injected electrons does not have an isotropic distribution over states with different crystal momentum k. As a result, momentum scattering also changes the transmission. In fact, one of the greatest misunderstandings is that hot-electron transport is automatically ballistic (presumably the name of an often used technique to study hot-electron transport, "ballistic electron emission microscopy" [16,17], has contributed to this). This is not correct from a strict scientific point of view (the term ballistic refers to transport without scattering, not to the carrier energy), but also in many hot-electron experiments momentum scattering has been proven to contribute to the reduced transmission. We will come back to this and other aspects later on, but first review the salient experimental observations and their phenomenological description.

9.2.1 Hot Carrier Spin Filtering in a Ferromagnet

Spin-dependent transmission of hot electrons in ferromagnetic metals such as Co, Ni, and Fe has been extensively studied for several decades [1–4,18–20], and it is known that the scattering rate of hot electrons in these materials is spin dependent. Early work on hot-electron scattering at energies of about 5 eV employed spin-polarized photoemission from overlayer structures [2]. This and also later experiments [3,4] at energies as low as 1.5 eV showed unambiguously that the inelastic mean free path of hot electrons is spin dependent in ferromagnets. More precisely, experiments carried out thus far have always found that the inelastic mean free path of hot electrons is shorter for the minority spin electrons and larger for the majority spin electrons [1]. Scattering processes have also been investigated using transmission through free-standing magnetic thin film foils [19,20].

Consider an FM thin film sandwiched between two nonmagnetic metals, typically Au or Cu. The transmission T^{M} and T^{m} of majority (M) and minority (m) spin hot electrons is

$$T_{\mathrm{fm}}^{\mathrm{M}} = \Gamma_{\mathrm{in}}^{\mathrm{M}} \exp\left(\frac{-t_{\mathrm{fm}}}{\lambda_{\mathrm{fm}}^{\mathrm{M}}}\right) \Gamma_{\mathrm{out}}^{\mathrm{M}}, \quad (9.1)$$

$$T_{\mathrm{fm}}^{\mathrm{m}} = \Gamma_{\mathrm{in}}^{\mathrm{m}} \exp\left(\frac{-t_{\mathrm{fm}}}{\lambda_{\mathrm{fm}}^{\mathrm{m}}}\right) \Gamma_{\mathrm{out}}^{\mathrm{m}}. \quad (9.2)$$

where

t_{fm} is the film thickness

$\lambda_{\mathrm{fm}}^{\mathrm{M}}$ and $\lambda_{\mathrm{fm}}^{\mathrm{m}}$ are the hot-electron attenuation lengths for majority and minority spin

fm denotes the ferromagnet

The most important feature is that the hot-electron transmission depends exponentially on the thickness of the FM, which reflects the nonequilibrium nature of the transport. The factors Γ_{in} and Γ_{out} denote the transmission across the interface of the ferromagnet with the normal metal at each side, which can be spin dependent due to the mismatch in the bandstructure of the magnetic and nonmagnetic metals. Thus, the interfacial attenuation is generally dependent on the combination of ferromagnet and normal metal. Now, an initially unpolarized current acquires, after transmission of an FM film, a spin polarization P given by

$$P_{\mathrm{fm}} = \frac{T_{\mathrm{fm}}^{\mathrm{M}} - T_{\mathrm{fm}}^{\mathrm{m}}}{T_{\mathrm{fm}}^{\mathrm{M}} + T_{\mathrm{fm}}^{\mathrm{m}}}. \quad (9.3)$$

An important point is that λ_{fm} is not the (inelastic) mean free path. Rather, it is the length scale that describes the exponential decay of the transmission that is measured when the thickness of the FM is increased. Obviously, such attenuation can only occur if on the collector side of the layers there is some kind of barrier that decides which hot electrons are blocked and which ones are transmitted and contribute to the collected current. It is easy to see that inelastic scattering accompanied by large energy loss leaves a fraction of the injected electrons unable to surmount this collector barrier. However, we should keep in mind that the barrier, typically a Schottky barrier at a metal/semiconductor interface, acts not only as an energy barrier but also as a momentum filter, because the electronic states on either side of the (metal/semiconductor) interface overlap only partially in momentum space. As a result, the attenuation is in general caused by a combination of inelastic and elastic scattering processes. While the scattering parameters are uniquely defined for a given ferromagnet, hot-electron energy, momentum, and spin, the attenuation length is not [21] since it depends on the energy and momentum distribution of the electrons injected into the FM and on the energy and momentum selection applied by the collector barrier. For example, in the hypothetical case that a barrier is used that passes hot electrons regardless of their momentum, elastic scattering would not directly cause attenuation, and the attenuation length would be governed by inelastic processes.

9.2.2 Magnetocurrent

Let us next consider a structure with two FM layers (labeled fm1 and fm2), separated by a nonmagnetic metal spacer. One can now easily see that the total transmission of this layer stack depends on whether the magnetizations of the two FM layers are parallel (P) or antiparallel (AP). For the transmitted current collected at the rear side for P and AP magnetic state, $I_{\mathrm{C}}^{\mathrm{P}}$ and $I_{\mathrm{C}}^{\mathrm{AP}}$, respectively, we have

$$I_{\mathrm{C}}^{\mathrm{P}} \propto T_{\mathrm{fm1}}^{\mathrm{M}} T_{\mathrm{fm2}}^{\mathrm{M}} + T_{\mathrm{fm1}}^{\mathrm{m}} T_{\mathrm{fm2}}^{\mathrm{m}}, \quad (9.4)$$

$$I_{\mathrm{C}}^{\mathrm{AP}} \propto T_{\mathrm{fm1}}^{\mathrm{M}} T_{\mathrm{fm2}}^{\mathrm{m}} + T_{\mathrm{fm1}}^{\mathrm{m}} T_{\mathrm{fm2}}^{\mathrm{M}}. \quad (9.5)$$

The magnetocurrent MC is then

$$\mathrm{MC} = \frac{I_{\mathrm{C}}^{\mathrm{P}} - I_{\mathrm{C}}^{\mathrm{AP}}}{I_{\mathrm{C}}^{\mathrm{AP}}} = \frac{2P_{\mathrm{fm1}}P_{\mathrm{fm2}}}{1 - P_{\mathrm{fm1}}P_{\mathrm{fm2}}}. \quad (9.6)$$

9.2.3 Lifetime and Group Velocity

Another important general aspect that should be emphasized is that the spin asymmetry of hot carrier transport is related to, but not solely determined by, the spin asymmetry of the lifetime of the excited carriers. Transport involves scattering lengths, which are the product of a scattering time τ and the group velocity v_g of the states in which the hot carriers are propagating. Both terms may be spin dependent. In fact, it was proposed in analyzing the unexpected strong spin filtering for hot holes that the difference in the group velocity of the states in which holes of different spin propagate may be responsible [22]. This notion then also stimulated theoretical work [23] calculating the spin-dependent group velocity of hot electrons in ferromagnetic metals. In general, v_g in s,p-like states is larger than in more localized d-like states, and, indeed, the states above E_F in transition metal ferromagnets have a different character for majority and minority spin. The contribution of a spin-dependent v_g may also explain the discrepancy between the strong spin asymmetry observed in hot-electron spin transport and the much weaker spin asymmetry observed in direct measurements of the inelastic lifetime for majority and minority spin in time-resolved, two photon photoemission (2 PPE) experiments [24,25]. This issue will be discussed in more detail in the next section in connection with available experimental data.

9.3 Hot-Electron Spin Transport

9.3.1 Experimental Techniques and Devices

Different techniques have been used to study spin-dependent transport of hot carriers, including overlayer experiments in photoemission and electron beam transmission through free-standing thin film foils. Typically, the hot-electron energy is above a few electron volts. In the last decade, devices such as the SVT and MTT have been developed and used extensively to study hot carrier spin transport. These devices operate at lower energy, typically 0.3–1.5 eV and allow versatile studies under a wide range of conditions. Moreover, a unique feature of these solid-state devices is that they can also be designed to study transport of hot holes, which is impossible for spectroscopy techniques in vacuum (try creating a "beam of holes"). We therefore focus here on the SVT and MTT and the related technique of ballistic electron magnetic microscopy (BEMM) [26,27].

9.3.1.1 Spin-Valve Transistor

The spin-valve transistor (SVT) is a three-terminal device that has the typical emitter/base/collector structure of a (bipolar) transistor, but is different in that the base region is metallic and contains at least two magnetic layers separated by a normal metal spacer (see Figure 9.1). The two magnetic layers act as polarizer and analyzer of electron spins, such that the relative orientation of the magnetization of the two layers determines the transmission of the base. This was first demonstrated by Monsma et al. [5] in an SVT, showing a huge change of 390% of the collector current in an applied magnetic field at low temperature. In 1998, the first device operating at room temperature was reported, having an MC of 15% in fields of a few kOe [6]. More recently, the reproducible fabrication of SVTs that exhibit magnetocurrent of up to 400% at room temperature was achieved [7,8].

The specific structure used in Ref. [7] has silicon as the semiconductor for the emitter and collector and a metallic base that contains an $Ni_{80}Fe_{20}$/Au/Co spin valve. The fabrication technology of the SVT is based on vacuum metal bonding, which was described elsewhere [28]. Schottky barriers are formed at the interfaces between the metal base and the semi-conductors. Low-doped n-type Si (1–10 Ω cm) is used and thin layers of, for example, Pt and Au are incorporated at the emitter and collector side, respectively, resulting in the desired high-quality Schottky barriers. When a current is now established between the emitter and the base (the emitter current I_E), electrons are injected into the base perpendicular to the layers of the spin valve. Since the injected electrons have to go over the Si/Pt Schottky barrier, they enter the base as hot electrons. Their energy is determined by the emitter Schottky barrier height, which is 0.86 eV for Si/Pt, but typically between 0.5 and 1 eV, depending on the metal/semiconductor combination. As the hot electrons traverse the base, scattering changes their energy as well as momentum distribution. Electrons are only able to enter the collector if they have retained sufficient energy to overcome the energy barrier at the collector side, which is chosen to be somewhat lower than the emitter barrier. Equally important, a hot electron can only enter the collector if its momentum matches with that of one of the available states in the conduction band of the collector semiconductor. The collector current I_C depends sensitively on the scattering in the base, which is spin dependent when the base contains magnetic materials. The total scattering rate is controlled with an external applied magnetic field, which changes the relative magnetic alignment of the ferromagnetic $Ni_{80}Fe_{20}$ and Co layers in the base. This is illustrated in Figure 9.2, where I_C at room temperature is plotted as a function of magnetic field, for a transistor with an Si(100) emitter, an Si(111) collector and the following base: Pt(20 Å)/NiFe(30 Å)/Au(35 Å)/Co(30 Å)/Au(20 + 20 Å). At large applied fields, the two magnetic layers have their magnetization directions aligned parallel. This gives the largest collector current (I_C^P = 11.2 nA). When the magnetic field is reversed, the difference in switching fields of Co (22 Oe) and NiFe (5 Oe) creates a field region where the NiFe and Co magnetizations are antiparallel. In this state, the collector current is drastically reduced (I_C^{AP} = 3.3 nA). The result is a large MC of 240%. On the down side, the absolute value of the output current is rather small. Typically, the transfer ratio [29], defined as $\alpha = I_C/I_E$, is between 10^{-6} and 10^{-4}. Nevertheless, signals are large enough to allow systematic studies of hot-electron spin transport.

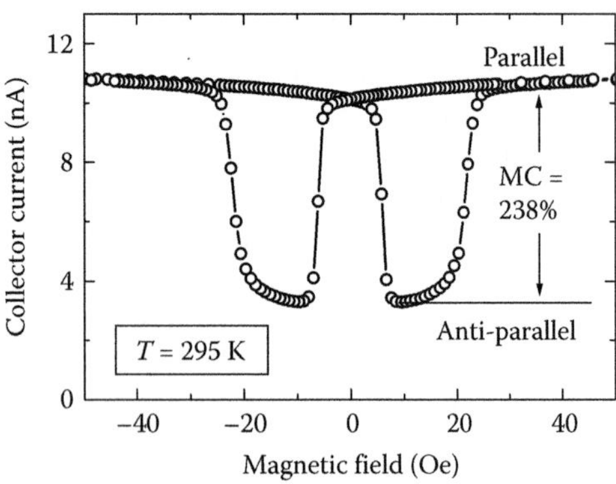

FIGURE 9.2 Collector current versus magnetic field for an SVT with structure Si/Pt/$Ni_{80}Fe_{20}$/Au/Co/Au/Si measured at T = 295 K with an emitter current of 2 mA.

9.3.1.2 Magnetic Tunnel Transistor

The magnetic tunnel transistor (MTT) uses the same principle as the SVT, only the emitter structure differs. Whereas the SVT uses a Schottky barrier, the MTT has a tunnel barrier as the emitter. The first MTT was fabricated in 1997 by Mizushima et al. [10,11]. The MTT exists in two slightly different variations, as shown in Figure 9.3. The design on the left has a thin insulating layer to separate the metal base from a nonmagnetic metal emitter electrode. When a voltage V_{EB} is applied across the tunnel barrier between emitter and base, unpolarized electrons are injected by tunneling into the metal base, arriving at an energy eV_{EB} above the Fermi level in the metal base. The rest

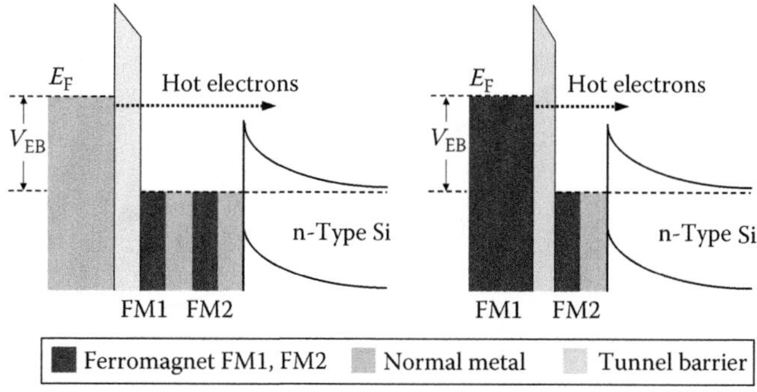

FIGURE 9.3 Energy band diagrams of magnetic tunnel transistors with (left) a spin-valve base and nonmagnetic emitter electrode and (right) a ferromagnetic emitter.

is the same as in the SVT: spin-dependent transmission of the hot electrons through the two ferromagnetic thin films in the metal base and collection across a Schottky barrier with energy and momentum selection. The design on the right in Figure 9.3 also has a tunnel barrier, but uses a ferromagnetic emitter electrode. Because the tunnel probability is spin dependent in such a structure, the injected hot electrons are already spin polarized as they enter the transistor base. Therefore, there is only one magnetic layer in the base (acting as spin analyzer). Again, the collector current is determined by spin-dependent transmission of the hot electrons through the ferromagnetic base and collection across a Schottky barrier with energy and momentum selection.

The most notable difference with the SVT is that the energy of the injected hot electrons in the MTT is given by the voltage V_{EB} applied across the tunnel barrier. The hot-electron energy is thus tunable over a certain range of energies, typically up to 2 eV, allowing spectroscopic studies. Figure 9.4 shows the typical behavior. A collector current is detected above a certain

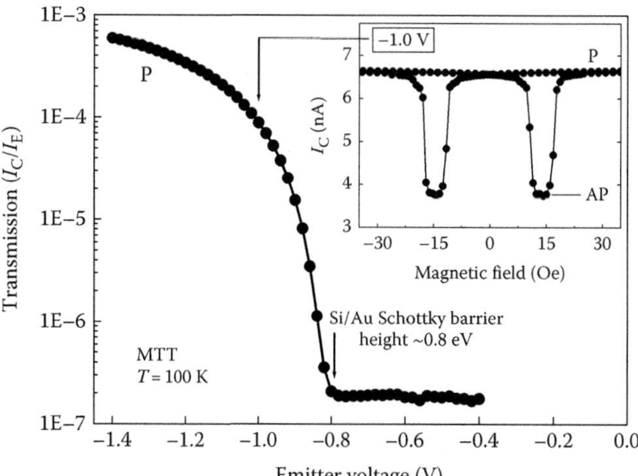

FIGURE 9.4 Hot-electron transmission of an MTT with structure $Ni_{80}Fe_{20}/Al_2O_3/Co/Au/Si$. Inset: collector current versus magnetic field at –1.0 V emitter bias. $T = 100$ K.

threshold voltage set by the collector Schottky barrier, and I_C is different for P and AP magnetic state.

The expressions for the transmission and magnetocurrent of the MTT with nonmagnetic tunnel emitter (left design of Figure 9.3) are the same as for the SVT and given by Equations 9.1 through 9.6. The situation is different for the MTT with a ferromagnetic tunnel emitter, since the emitter current now becomes spin polarized, with the polarization determined by the tunneling process. For the parallel magnetization state, the emitter tunnel current is a sum of a majority and a minority spin contribution according to

$$I_E^P \propto \delta_E^M \delta_B^M + \delta_E^m \delta_B^m, \tag{9.7}$$

where

δ_E^M and δ_E^m are the fractions of majority and minority tunneling electrons associated with the emitter/insulator interface (label E)

δ_B^M and δ_B^m are the fractions of tunnel electrons associated with the insulator/base interface (label B)

These are defined in the usual way with the help of the tunneling spin polarization P_t:

$$P_{t,E} = \frac{\delta_E^M - \delta_E^m}{\delta_E^M + \delta_E^m} \tag{9.8}$$

and

$$P_{t,B} = \frac{\delta_B^M - \delta_B^m}{\delta_B^M + \delta_B^m}. \tag{9.9}$$

To obtain the collector current, we need to multiply each of the two emitter current components in Equation 9.7 with the majority or minority spin hot-electron transmission factor appropriate for the ferromagnetic layer in the base (M2 in Figure 9.3). We then have

$$I_C^P \propto \delta_E^M \delta_B^M T_B^M + \delta_E^m \delta_B^m T_B^m, \tag{9.10}$$

where T^M and T^m are defined as before in Equations 9.3 and 9.4. For the antiparallel case, the magnetization of the base is reversed:

$$I_C^{AP} \propto \delta_E^M \delta_B^m T_B^m + \delta_E^m \delta_B^M T_B^M. \tag{9.11}$$

Using these equations, the magnetocurrent can be cast into the familiar form:

$$\text{MC} = \frac{I_C^P - I_C^{AP}}{I_C^{AP}} = \frac{2P_{t,E}P_B^*}{1 - P_{t,E}P_B^*}, \tag{9.12}$$

where we have defined a renormalized polarization P_B^* of the base as

$$P_B^* = \frac{\delta_B^M T_B^M - \delta_B^m T_B^m}{\delta_B^M T_B^M + \delta_B^m T_B^m}. \tag{9.13}$$

Note that this renormalized polarization is a combination of tunneling factors and hot-electron transmission factors and as such should not be confused with the regular tunneling spin polarization as defined in Equation 9.9. Thus, we see that the MC of an MTT with a ferromagnetic tunnel injector is determined by two completely unrelated physical processes. The polarization of the emitter current is due to spin-dependent tunneling, while the subsequent transmission of the hot electrons through the base of the MTT is given by the spin-dependent scattering of hot electrons. Although we are here primarily interested in probing hot-electron spin transport, the MTT with ferromagnetic emitter has proven to be a powerful tool to study spin-polarized tunneling at high bias. This is briefly addressed in the last section of this chapter.

9.3.1.3 Ballistic Electron Magnetic Microscopy

From a conceptual point of view, the ballistic electron magnetic microscopy (BEMM) technique [26,27] is identical to the MTT. The difference is that the emitter part does not consist of a fixed oxide tunnel barrier and metal electrode, but instead the tip of a scanning tunneling microscope (STM) is used to inject hot electrons into the magnetic metal stack across a vacuum (or air) tunnel gap. The injection and also the collection are local, and the technique can thus also be used to obtain spatially resolved information about hot-electron spin transport. Magnetic imaging with spatial resolution down to about 20 nm was indeed reported. It should be noted that the momentum distribution of the hot electrons injected by tunneling from a sharp STM tip is different compared to that of tunneling across the solid-state (oxide) tunnel barrier of the MTT. This has consequences for the attenuation length, as will be shown in one of the following sections.

9.3.2 Hot-Electron Attenuation Lengths

We first examine what the typical attenuation lengths λ_{fm}^{M} and λ_{fm}^{m} are for majority and minority spin hot electrons in transition metal ferromagnets, and whether the dominant scattering occurs within the volume of the magnetic layers or at the interfaces. Surely, volume scattering should dominate as its contribution grows exponentially with film thickness and will quickly outweigh the (constant) interface contribution. Experiments confirm this. As an example, we describe the data in Figure 9.5 for an SVT with the base structure Pt(30 Å)/$Ni_{80}Fe_{20}$(x)/Au(44 Å)/Co(30 Å)/Au(44 Å), where the thickness x of the $Ni_{80}Fe_{20}$ layer was varied [30]. The transmitted hot-electron current I_C^P and I_C^{AP} (top panel) and MC (bottom panel) at $T = 100$ K are measured as a function of x. For the parallel magnetic state of the spin valve, minority spins are strongly attenuated in both magnetic layers and only majority spins contribute. The exponential decay of I_C^P with thickness yields $\lambda_{NiFe}^{M} = 43 \pm 3$ Å for majority hot spins in $Ni_{80}Fe_{20}$ (at 0.9 eV). The interface attenuation for majority spins is extracted by comparing the I_C value of 169 nA for the device with x = 0 to the value of 78 nA obtained by extrapolation of the I_C^P data to zero thickness. The latter value is a factor of 2.2 lower, which is due to the extra attenuation created when the original Pt/Au interface is replaced by

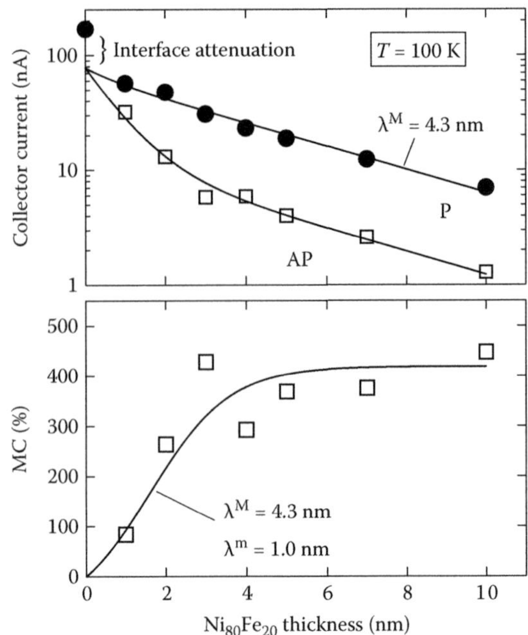

FIGURE 9.5 Collector current for P and AP magnetization (top) and MC (bottom) as a function of $Ni_{80}Fe_{20}$ thickness for an SVT with $Ni_{80}Fe_{20}$/Au/Co base.

two new Pt/$Ni_{80}Fe_{20}$ and $Ni_{80}Fe_{20}$/Au interfaces. The interface attenuation originates from a combination of the mismatch of the electronic states at both sides of the interface and elastic scattering due to interface disorder and defects. Next, the difference between I_C^P and I_C^{AP} is used [30] to obtain the minority spin attenuation length $\lambda_{NiFe}^{m} = 10 \pm 2$ Å. With these parameters, a 30 Å thick $Ni_{80}Fe_{20}$ film attenuates minority spins a factor of 10 more strongly. Volume scattering is thus by far the dominant source of spin dependence.

Hot-electron scattering has been investigated in a similar fashion for different transition metal ferromagnets using SVT [9,30], MTT [10–14,31–38], and BEMM [26,27,39–42]. The extracted attenuation lengths are summarized in Table 9.1 from which a consistent picture emerges. Minority spin attenuation lengths for hot electrons are around 10 Å, while majority spin scattering lengths are in the range of 25–80 Å. The ratio $\lambda_{fm}^{M}/\lambda_{fm}^{m}$ is between 2 and 6. Moreover, data obtained with MTT and BEMM showed that λ_{fm}^{M} decreases with increasing hot-electron energy [27,31], but much slower than expected from a simple model based on Fermi's golden rule [31].

In order to understand the origin of the spin asymmetry, we start with the conventional picture, which, while insightful, turned out to be much too simplified. It relies on the notion that in transition metal ferromagnets, the density of empty states above E_F is different and much larger for the minority spin due to the presence of empty d-states (see Figure 9.6). If inelastic scattering by electron–hole pair excitation is the dominant mechanism, a minority spin hot electron has a much larger number of states to decay into and therefore a shorter inelastic lifetime. This is qualitatively consistent with early experiments [1–4,18–20]

TABLE 9.1 Attenuation Lengths for Hot Electrons of Majority (λ^M) and Minority (λ^m) Spin in Transition Metal Ferromagnets

Ferromagnet	λ^M (Å)	λ^m (Å)	Technique	Collector	T (K)	Energy (eV)	Reference
$Ni_{81}Fe_{19}$	43	10	SVT	Si(100)	100	0.9	Vlutters et al. [30]
$Ni_{81}Fe_{19}$	32	—	SVT	Si(100)	300	0.9	Vlutters et al. [30]
Co	24–21	8	BEMM	Si(111)	300	1.0–2.0	Rippard et al. [27]
Co	34	—	BEMM	Si(100)	150	1.0–1.5	Jansen et al. [45]
Co	68	—	MTT	Si(100)	100	1.0–1.4	Jansen et al. [45]
Co	70	10	MTT	Si(100)	100	1.0–1.4	Park et al. [38]
Co	52–33	10	BEMM	Si(111)	300	1.0–2.0	Kaidatzis et al. [42]
Fe	15	4	BEMM	Si(100)	150	1.2–1.6	Banerjee et al. [40]
$Ni_{81}Fe_{19}$	79–56	13	MTT	GaAs(001)	77	1.0–1.8	van Dijken et al. [31]
$Co_{84}Fe_{16}$	58–42	9	MTT	GaAs(001)	77	1.0–1.9	van Dijken et al. [31]

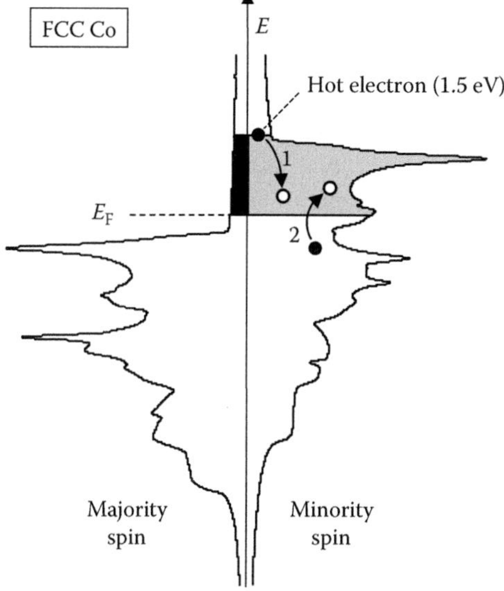

FIGURE 9.6 Density of states for majority and minority spin versus energy for FFC cobalt. Arrows indicate (1) the inelastic decay of a hot electron and (2) the accompanying excitation of an electron–hole pair. Shaded areas represent the states available for decay of a hot electron of majority spin (dark gray) and minority spin (light gray).

and also with time- and spin-resolved two photon photoemission (2 PPE) experiments [24,25]. Nevertheless, this picture is too simplified for a number of reasons:

- It predicts that no significant spin asymmetry should exist for hot holes, while experiments described in Section 9.4 clearly contradict this.
- While the phase space of states for the hot electron to decay into plays an important role, the phase space for two more particles, the excited electron and hole, has to be taken into account, with conservation of (total) energy, momentum, and spin. A recent comparison with a full ab initio calculation of the inelastic lifetime [23] showed that a convolution of three phase spaces can be a rather good approximation, but neglecting the phase space for the electron–hole pair gives clearly erroneous results.
- Inelastic scattering by excitation of spin waves is not included, but was found to play an important role. This will be addressed separately in Section 9.3.4.
- A point that was already made in Section 9.2 is that in order to relate the inelastic lifetime to the scattering lengths measured with SVT, MTT, or BEMM, one needs to consider the group velocity and its dependence on spin.
- Similarly, attenuation lengths are controlled not only by inelastic, but also by elastic scattering. This aspect will be discussed in more detail in the Section 9.7.

Significant effort has also been made in the calculation of spin-dependent quasiparticle lifetimes in ferromagnets. The most noteworthy are the sophisticated calculations reported by Zhukov et al. [23,43], in which the inelastic lifetime in Ni and Fe was evaluated including the role of spin waves, but also the group velocity was calculated. For Ni, the inelastic lifetime for hot electrons in the energy range 0.5–3 eV was found to be the dominant source of spin asymmetry, while the group velocity is similar for hot electrons of either spin. In contrast, the result for Fe was strikingly different: contrary to simple models, the inelastic lifetime showed little difference between majority and minority spin between 0.7 and 1.5 eV, but the group velocity was found to be larger for the majority spin by more than a factor of 4 (!). As a result, the inelastic mean free path has a spin asymmetry that is dominated by the group velocity, rather than the lifetime. The basic reason for this was already eluded to at the end of Section 9.2: majority spin hot electrons propagate in states that are predominantly s,p-like, while minority spin hot electrons propagate in states that are strongly hybridized and have a more d-like character. The group velocity for the latter is smaller. The same is to be expected for other transition metal ferromagnets that have large density of empty d-states above E_F for the minority spin. For Ni, the effect is much weaker, since the empty d-states extend only up to about 0.3 eV above E_F.

Including the group velocity as a source of spin asymmetry provided a much more consistent picture of hot-electron spin transport. Nevertheless, two other aspects, originally overlooked, were found to have significant impact. These aspects, elastic scattering and the interaction of hot electrons with spin waves, are discussed in detail in the next two sections.

9.3.3 Elastic Scattering of Hot Electrons

In hot-electron scattering it is obvious that inelastic scattering plays a role, as it leads to energy loss and thus makes the electrons "less hot." But, as noted, elastic scattering cannot be neglected a priori. In the metals of the base, the density of states and hence the number of k-vectors available for hot electrons is much larger than that in the semiconductor collector, whose states near the conduction band minimum are limited and have much smaller k-vector. Hence, any elastic scattering in the metal base is likely to redistribute electrons over momentum states, which are not available in the semiconductor collector, causing a drop in the transmission. This was illustrated by phenomenological model calculations in which scattering parameters were systematically varied, and the effect on (spin) transport was studied [21]. Elastic scattering in the volume of the metal layers and at the interfaces of the layers was included. Experimental support for the latter comes from a simple estimate based on exponential decay of the transmission of each metal layer. Using measured attenuation lengths (see previous section), one would expect a total transmission of about 0.01. Since observed base transfer ratios are about 2 orders of magnitude smaller, a significant fraction of the attenuation must be attributed to scattering at interfaces, which is primarily elastic. The importance of elastic scattering is supported by several experimental facts:

- The attenuation due to metal–metal interfaces in the base of the SVT was proven and quantified. An interface between $Ni_{80}Fe_{20}$ and Au or Pt typically reduces the hot-electron transmission by a factor of 2.
- The magnitude of the collector current of the SVT was found to be insensitive to the crystal orientation (100) or (111) of the Si collector. Although the density of states in the Si collector for both cases is (obviously) identical, the six pockets near the conduction band minimum of Si have a very different momentum distribution when projected onto the interface. For instance, states at zero momentum parallel to the interface are present for (100), but not for (111). The fact that the transmission is nevertheless the same indicates that the hot-electrons incident on the metal/semiconductor interface have acquired a broad distribution of momentum states after their transport through the metal base.
- In a unique experiment, the anisotropic scattering of hot electrons in a ferromagnet due to spin–orbit interaction has been observed [44]. Using an SVT with $Pt/Ni_{80}Fe_{20}/Au$ base, thus, with a single FM layer, the transmission was found to depend on whether the magnetization of the $Ni_{80}Fe_{20}$ was parallel to the hot-electron current or perpendicular to it. This anisotropic behavior is similar to the well-known anisotropic magnetoresistance (AMR) of ferromagnetic metals and caused by the spin–orbit interaction, hence directly proving that a purely elastic scattering process affects the hot-electron transmission.
- The most recent, and, perhaps, most compelling experimental observation is that the value of the hot-electron attenuation length of Co was found to depend sensitively on the momentum distribution of the electrons injected into it [45]. The experiment compared injection across two different types of tunnel barriers, namely amorphous Al_2O_3 and a vacuum tunnel barrier. The former was part of a MTT with $Ni_{80}Fe_{20}/Al_2O_3/Co/Au/Si$ structure, while the latter was part of a BEMM experiment in ultrahigh vacuum with a Au/Co/Au/Si sample and a nonmagnetic metal tip. For both, the Co/Au/Si (base/collector) part was identical and prepared in the same growth system. Nevertheless, the decay of the hot-electron transmission as a function of Co thickness is strikingly different, the attenuation length being a factor of 2 (!) shorter for injection from the STM tip in vacuum (Figure 9.7). This can

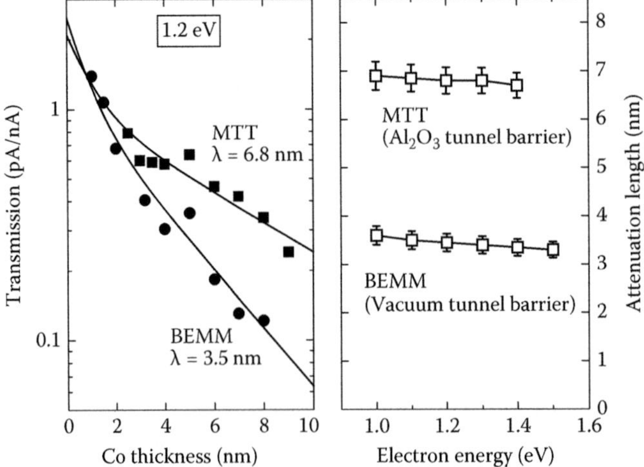

FIGURE 9.7 Hot-electron transmission versus Co thickness at 1.2 eV (left) and attenuation length versus energy (right) measured using MTT and BEMM.

be understood by the significantly broader momentum distribution for tunneling across amorphous Al_2O_3 compared to vacuum tunneling. If the tunnel barrier itself already provides significant momentum scattering and the hot electrons are injected into the Co base with an isotropic momentum distribution, then elastic scattering in the Co base has little further effect and does not contribute to the attenuation. The resulting attenuation length becomes larger and is limited by inelastic scattering only. In contrast, tunneling across a vacuum barrier, in which scattering sites are absent, produces a more forward-focused momentum distribution dominated by electrons with small momentum component parallel to the surface. In this case, elastic scattering in the Co is important and contributes to the shorter attenuation length.

Overall, it seems that the picture of purely ballistic transport is rather inappropriate and that elastic scattering plays a (perhaps) surprisingly important role in hot-electron transport.

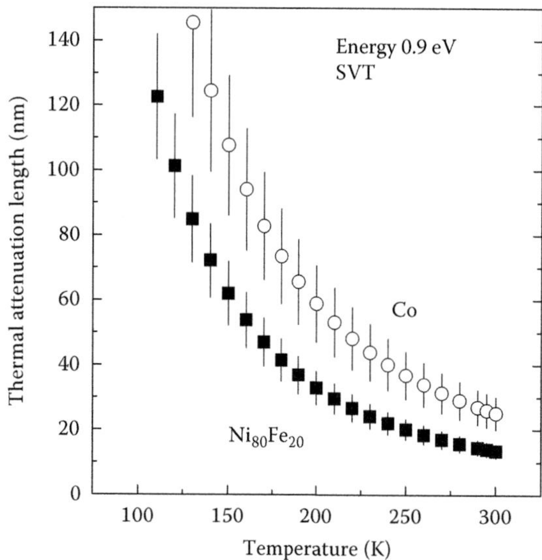

FIGURE 9.8 Attenuation length due to thermal scattering for Co and $Ni_{80}Fe_{20}$, measured with the SVT at 0.9 eV hot-electron energy.

9.3.4 Temperature and Interaction with Spin Waves

While originally neglected, there exists a body of experimental [8,30,46] and theoretical work [23,43,47,48] that suggests that scattering by spin-wave excitation contributes to the spin dependence of hot-electron transport. To describe this, we use Matthiessens' rule and write λ_{fm}^{M} and λ_{fm}^{m} in terms of the attenuation lengths for all the distinctive scattering processes:

$$\frac{1}{\lambda_{fm}^{M}} = \frac{1}{\lambda_{e-h}^{M}} + \frac{1}{\lambda_{el}^{M}} + \frac{1}{\lambda_{ph}} + \frac{1}{\lambda_{TSW\,abs}^{M}}, \tag{9.14}$$

$$\frac{1}{\lambda_{fm}^{m}} = \frac{1}{\lambda_{e-h}^{m}} + \frac{1}{\lambda_{el}^{m}} + \frac{1}{\lambda_{ph}} + \frac{1}{\lambda_{TSW\,emis}^{m}} + \frac{1}{\lambda_{SSW\,emis}^{m}}, \tag{9.15}$$

where λ_{e-h}, λ_{el}, and λ_{ph} are the attenuation lengths associated with electron–hole pair excitations, elastic scattering, and phonon scattering, respectively. Also included are attenuation lengths $\lambda_{TSW\,abs}^{M}$ and $\lambda_{TSW\,emis}^{m}$ due to absorption and emission of thermal spin waves, respectively, as well as a term $\lambda_{SSW\,emis}^{m}$ due to spontaneous emission of spin waves. Due to the conservation of angular momentum, only majority spin hot electrons can absorb spin waves, whereas (spontaneous and thermal) emission is allowed only for minority spins. Thus, the overall rate of spin-wave scattering has a spin asymmetry due to spontaneous spin-wave emission $\lambda_{SSW\,emis}^{m}$, which is allowed only for minority spins.

Let us first focus on the thermal component of the attenuation due to spin waves. From a measurement of I_C versus temperature, one can extract the attenuation length λ_{TSW} (T) due to thermal spin waves, using the procedure described in Ref. [30]. Results for $Ni_{80}Fe_{20}$ and Co are shown in Figure 9.8 for an energy of 0.9 eV. The value of λ_{TSW} (T) is much shorter than originally assumed, with room temperature values of 130 ± 20 Å for $Ni_{80}Fe_{20}$ and 270 ± 40 Å for Co. For $Ni_{80}Fe_{20}$, this is only three times larger than the majority spin attenuation length at low T. Hence, hot-electron attenuation lengths are certainly dependent on T. For instance, the addition of thermal spin-wave scattering with a length scale of 130 Å reduces the attenuation length for $Ni_{80}Fe_{20}$ from 43 Å at 100 K to a significantly lower value of $(1/43 + 1/130)^{-1} = 32$ Å at room temperature. This clearly establishes that (thermal) spin waves interact with the hot electrons and cause attenuation. To appreciate this, one should keep in mind that hot electrons traverse the FM layer in about 10 fs. That spin waves are able to interact with the hot electrons at such a short timescale was a rather astonishing finding.

The above result suggests that spin waves also contribute to the spin asymmetry of hot-electron transmission. Compared to thermal spin waves, the attenuation due to spontaneous emission of spin waves (minority spin only) is expected to be significantly stronger. This is because thermal spin waves occupy only a small fraction of the spin-wave phase space up to energies of the order of kT, while for spontaneous spin wave emission, the complete phase space up to the hot-electron energy ($\approx$0.9 eV$\gg$ kT) is available at all temperatures. If we crudely estimate the spontaneous emission rate to be an order of magnitude larger than the thermal emission rate, we obtain an attenuation length for spontaneous emission in the 10 Å range. This is close to the measured minority spin attenuation lengths in transition metal ferromagnets, and since the process cannot contribute to attenuation of majority spins, it seems that the spin asymmetry of the attenuation length may well be due to spontaneous spin-wave emission. The spin asymmetry then has its origin in the fundamental law of angular momentum conservation, instead of in the exchange split band structure (which produces the spin-dependent rate of electron–hole pair generation).

Further support for this interpretation comes from calculations of the inelastic hot-electron lifetime in which spin wave excitation is explicitly included [23,43,47,48]. Early work already showed that spin-wave excitation dominates the inelastic lifetime in transition metal ferromagnets at electron energies below ≈1 eV and is the major source of small energy losses [46]. More recently, this was put on a firm theoretical basis using full ab initio calculations using the so-called GW + T method, which treats inelastic scattering by excitation of electron–hole pairs and emission of spin waves on an equal footing [23,43]. The result is striking and shows convincingly that at low energy, the inelastic lifetime for minority spin hot electrons is strongly reduced when spin-wave emission is included, being the main source of spin asymmetry at energy below about 1 eV.

Spin waves have also been shown to affect hot-electron spin transport is a different way, namely by inducing spin-flip scattering [8]. Since a spin wave has an angular momentum of $-\hbar$, the absorption or emission of a spin wave requires the hot electron to change its spin. This causes mixing of the two spin channels and results in a decay of the MC with increasing temperature. The top panel of Figure 9.9 shows the variation of the collector current with applied magnetic field at two temperatures (80 and 290 K), for an SVT with a Co/Au/$Ni_{80}Fe_{20}$ spin-valve base. At T = 80 K, MC = 560%, while a huge effect of 350% still remains at room temperature. Figure 9.9, bottom panel, shows the decay of MC as a function of temperature. The decay of MC is attributed to thermally induced spin-flip scattering by thermal spin waves [8]. This produces spin mixing for those scattering events in which the energy and/or momentum transferred to the spin wave is such that after scattering, the hot electron still has sufficient energy and the proper momentum to enter the collector.

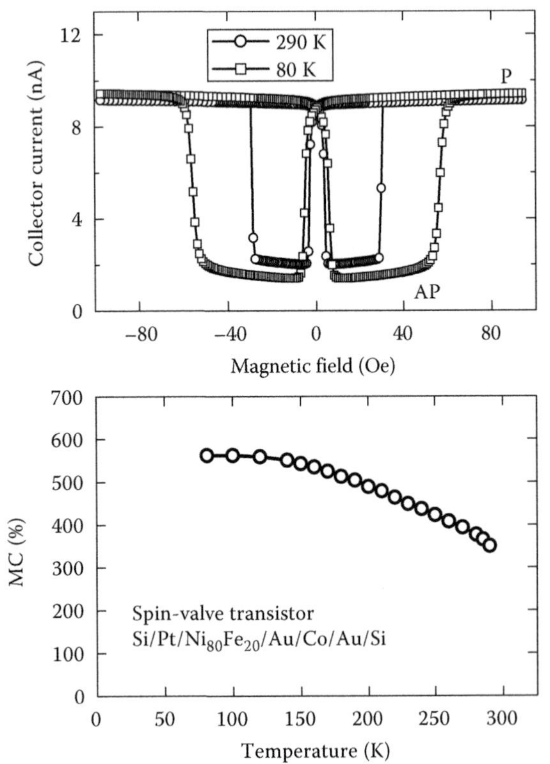

FIGURE 9.9 Top: SVT collector current versus magnetic field at 290 and 80 K. Bottom: MC versus temperature.

In summary, spin waves are found to interact with hot electrons on the 10 fs timescale, producing attenuation and spin mixing, while they are also an important source of spin asymmetry. In the next section, we will reach a similar conclusion for spin transport of hot holes.

9.4 Hot-Hole Spin Transport

Research on spin transport of nonequilibrium holes below E_F is still at its infancy. Nevertheless, some remarkable results have been reported, including the measurement of attenuation lengths of hot holes in cobalt, the observation of spin-dependent hole transmission in an $Ni_{81}Fe_{19}$/Au/Co trilayer with surprisingly large MC, and the observation of hot-hole scattering by spin waves with inverse spin asymmetry. These results are treated in this section.

9.4.1 Experimental Techniques and Devices

Spin-dependent transport of hot holes has been studied with hole versions of MTT and BEMM. The energy band diagrams are displayed in Figure 9.10. In ballistic hole magnetic microscopy (BHMM), a p-type semiconductor is used as collector. Second, the emitter bias polarity is kept positive, such that electrons tunnel from the metal surface into the STM tip. This is equivalent to holes tunneling in the opposite direction. Therefore, a current of hot holes is injected into the magnetic metal stack at energies below the metal E_F. After transmission of the metal base, these holes are collected in the valence band of the p-type semiconductor. A Schottky barrier for holes is present at the

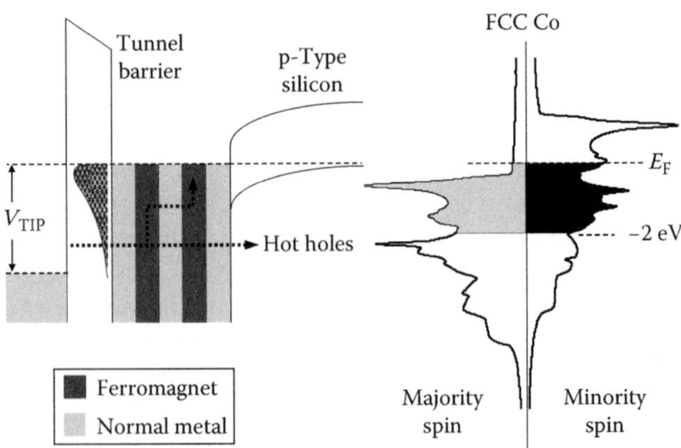

FIGURE 9.10 Left: energy band diagram for ballistic hole magnetic microscopy (BHMM). The hatched area in the tunnel barrier represents the energy distribution of the injected holes. Right: density of states for FCC cobalt, with indicated the states available for decay of a hot hole at −2 eV with majority spin (light gray) or minority spin (dark gray).

metal/semiconductor interface with typical height of 0.3 eV (for p-type Si/Au), providing the necessary energy and momentum selection. The configuration for studying hole transport in a p-type MTT is essentially the same, except that the tunnel barrier is a fixed oxide (typically Al_2O_3) and the emitter can be a ferromagnetic layer, in which case the holes injected into the metal base are spin polarized. It should be noted that for tunnel injection of holes, the energy distribution is peaked near the Fermi energy of the metal base, as illustrated in Figure 9.10. Therefore, only holes in the tail of the energy distribution have sufficient energy to be transmitted across the collector barrier. For the interpretation of the results below, this is not an essential feature. In fact, similar results have been obtained with a p-type SVT [49], where a p-type semiconductor is used as the emitter of hot holes. In this case, the energy distribution of the injected holes is determined by thermal excitation over the Schottky barrier, just as in the electron version of the SVT (see Figure 9.1). Measurements are usually done at low temperatures (80–150 K) to limit noise and offsets associated with the low Schottky barrier height.

9.4.2 Spin Filtering of Hot Holes

The first experimental observation of spin-dependent transport of holes in a ferromagnet was made in 2005 using the BHMM technique [22]. The basic result is shown in Figure 9.11. The sample structure was p-type Si/Au(70 Å)/$Ni_{81}Fe_{19}$(18 Å)/Au(70 Å)/Co(18 Å)/Au(30 Å). The curves in the top panel show the collected hole current, where the sign of the current corresponds to holes flowing from the metal stack into the Si. The current is below the detection limit for voltage under 0.8 V. The hole current detected above 0.8 V depends significantly on magnetic field. It is largest when the magnetization of the two ferromagnetic layers is aligned parallel (P) and about a factor of two smaller in the antiparallel (AP) configuration. The bottom panel shows the variation of the transmitted hole current as a function of magnetic field, taken at a constant tunnel current of 10 nA and tip bias of 2 V. The data exhibits the expected magnetic hysteresis and a clear difference in the transmitted hole current between the P and AP states. This spin-valve effect unambiguously demonstrated for the first time that the transmission of hot holes in a ferromagnetic metal is spin dependent.

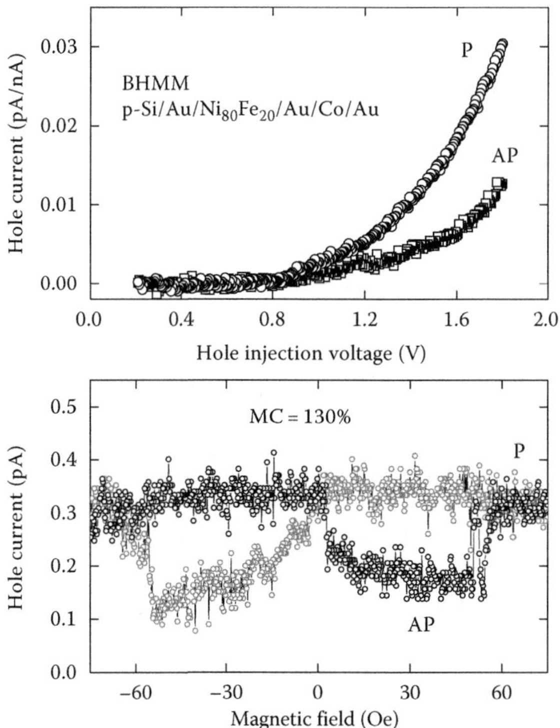

FIGURE 9.11 Top: transmitted hot-hole current versus tip bias in BHMM for P and AP state. Bottom: hole current at −2 eV versus magnetic field. $T = 150$ K.

From the change of the hole current of 0.34 pA in the P state to 0.15 pA in the AP state, we obtain a hole MC of 130% ± 30%. The difference between I^{P}_{hole} and I^{AP}_{hole} is related to the transmission of majority T^{M} and minority T^{m} spin hot holes in the ferromagnetic layers and can be expressed in a similar way as for hot electrons (Equations 9.4 and 9.5). After transmission of a ferromagnetic film, the hole current acquires a spin polarization $P = (T^{M} - T^{m})/(T^{M} + T^{m})$, and if P is the same for both ferromagnets, the measured MC corresponds to P = 63% (or −63%) for each of the 18 Å films. This gives a value of the spin asymmetry of transmission $T^{M}/T^{m} = 4.4$ (or 0.23), which is quite large indeed.

Such a large spin asymmetry was completely unexpected. It cannot be explained by the phase space available for the inelastic decay of the hot holes via electron–hole pair excitation (for hot holes this involves an electron from below E_F that drops down into the hot hole state, leaving behind a hot hole at lower energy (closer to E_F), while the energy released is used to excite an electron–hole pair). Consider the electronic structure of Co (Figure 9.10). There is no significant difference in the number of filled states below E_F for holes of different spin to decay into. Therefore, a large spin asymmetry, comparable in magnitude to that of hot electrons, was not expected. This indicates that other factors are at work and led Banerjee et al. [22] to propose that the difference in the group velocity might play a role. Until then, this had not been considered, neither for hot holes nor for hot electrons. Subsequent calculations [50] of the group velocity for holes in Co yielded values of about 1.5×10^5 m/s in the energy range −0.5 to −2 eV below E_F with a slight difference between holes in the minority and majority spin bands, but not sufficient to explain the large MC. To date, this is still unresolved.

9.4.3 Hole Attenuation Lengths and Scattering Mechanisms

The first data (Figure 9.12) on the attenuation length for hot holes was obtained using BHMM on a p-type Si/Au(70 Å)/Co/Au(30 Å) structure with Co of varying thickness [22]. The hole attenuation length was 6.5 ± 1 Å at 0.8 eV, increasing to 10 ± 1 Å at

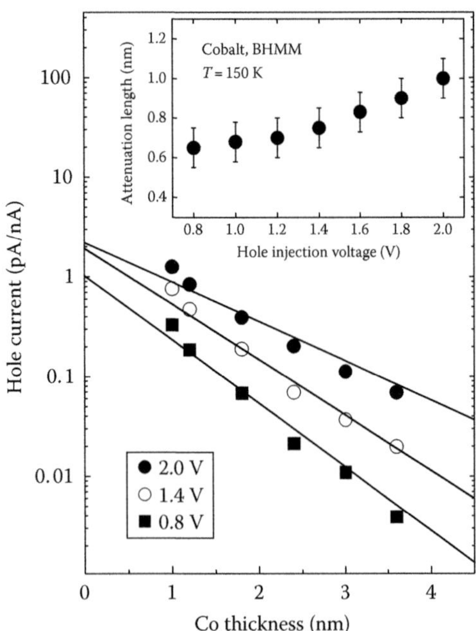

FIGURE 9.12 Transmitted hole current versus Co thickness measured in BHMM. The extracted attenuation length for different injection voltage is shown in the inset. $T = 150\,\text{K}$.

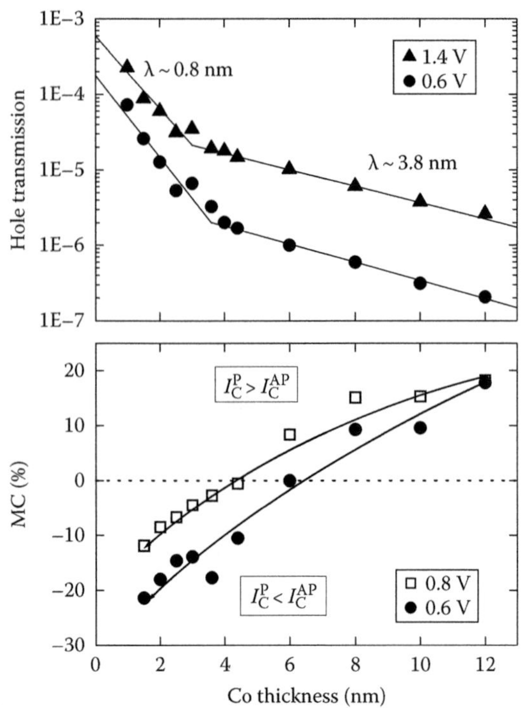

FIGURE 9.13 Top: hole transmission versus Co thickness measured in a p-type MTT. Bottom: MC versus Co thickness. The MC is positive when $I_C^P > I_C^{AP}$.

2 eV. The hole attenuation length is quite short and shorter by a factor of 2 to 6 compared to the majority spin attenuation length for hot electrons. The attenuation length for holes increases with energy, despite the fact that at higher energy there is a larger phase space of final states for holes to decay into. According to Fermi's golden rule, this would result in a shorter attenuation length at higher energy, as is observed in noble metals and for majority spin hot electrons.

In a subsequent study [51], the spin-dependent attenuation of holes was investigated using a p-type MTT with structure $Ni_{80}Fe_{20}$(5 nm)/Al_2O_3(2 nm)/Co/Au(7 nm)/p-type Si. It was concluded that nonequilibrium holes, when injected into an FM, scatter (quasi-)elastic in the first few nanometer of material due to spin-wave emission, while at larger depth into the film the attenuation becomes dominated by inelastic scattering by e–h pair excitations, but with opposite spin asymmetry. The presence of two different scattering mechanisms is apparent in the decay of the transmitted hole current as a function of the Co thickness (Figure 9.13). There is a strong decay for small thickness up to ~3 nm and a transition to a much slower decay for larger thickness. The two slopes correspond to attenuation lengths of about 0.8 and 3.8 nm. This behavior was explained in the following way [51]. Close to the injection interface, the hole current is very sensitive to elastic scattering, producing a strong decay. But as the holes proceed through the base layer, their momentum distribution gradually broadens and after a certain distance becomes isotropic. Elastic scattering, though still present, is then no longer effective in attenuating the holes, and, at large thickness, inelastic scattering with a different (longer) attenuation length dominates.

Perhaps even more remarkable is that the transition in scattering mechanism is accompanied by a sign reversal of the MC [51], as shown in Figure 9.13. In the small thickness regime, where elastic scattering dominates, a negative MC is found. This is rather unusual and means that the hole current is largest for the AP state and smallest in the P state. Thus, elastic scattering is stronger for holes in the majority spin bands, producing a shorter attenuation length compared to holes in the minority spin bands ($\lambda_M^{el} < \lambda_m^{el}$). On the other hand, for large Co thickness, where inelastic scattering dominates, a positive MC is observed. It follows that inelastic scattering has the opposite spin asymmetry ($\lambda_M^{inel} > \lambda_m^{inel}$). Control experiments ruled out interfaces as a possible source of this behavior [51].

The origin of the spin asymmetry of the inelastic scattering of holes is not yet clear. However, for the elastic scattering of hot holes with opposite spin asymmetry, a plausible explanation exists, as it is consistent with spontaneous emission of spin waves. As discussed, for hot electrons, spontaneous emission of a spin wave is forbidden for majority spin hot electrons. However, for holes, the spin asymmetry is opposite, and spin-wave emission is forbidden for hot holes in the minority spin bands. Spin-wave emission is possible when a hot hole in the majority spin band is annihilated by an electron that flips its spin from minority to majority, leaving a hole in the minority spin band. The interaction with spin waves thus predicts a shorter lifetime for holes in the majority spin bands and $\lambda_M < \lambda_m$. Thus, the observed negative MC is consistent with the (quasi-)elastic scattering of hot holes by spontaneous emission of spin waves.

9.5 Hot Carrier Spin Transport in Other Areas of Spintronics

9.5.1 Probing Spin-Polarized Tunneling and Its Energy Dependence

Besides the study of hot carrier spin transport, an important application of the MTT is to probe tunnel spin polarization. It has been shown [36–38] that the MTT allows (a) quantitative determination of the tunnel spin polarization P_t of the emitter ferromagnet/insulator interface and (b) extraction of the variation of P_t with energy of the states from which the tunnel current is drawn, above as well as below E_F. This works as follows. For an MTT with ferromagnetic emitter electrode, the MC is a function of P_t and of the hot-electron spin transport in the magnetic base, as described by Equation 9.12. The MC grows with increasing thickness of the ferromagnetic base and because $\lambda^M > \lambda^m$, at sufficiently large base thickness (~5 nm), only majority spins are transmitted ($P^*_B \sim 1$). The MC then saturates at a value of $2P_{t,E}/(1 - P_{t,E})$ and is dependent only on $P_{t,E}$. This is illustrated by the data in Figure 9.14 for a $Co/Al_2O_3/Co/Si$ transistor with varying thickness of the Co base. From the value of the MC at large base thickness, a tunnel spin polarization of 34% is obtained for electrons tunneling from the emitter Co/Al_2O_3 interface at −1 V. This technique has also been applied to obtain the variation of P_t with temperature [36] and to probe other ferromagnet/insulator combinations. For instance, it was shown [37] that for $Ni_{81}Fe_{19}/SiO_2$ interfaces, P_t is about 27%, which is only slightly smaller compared to $Ni_{81}Fe_{19}/Al_2O_3$ interfaces.

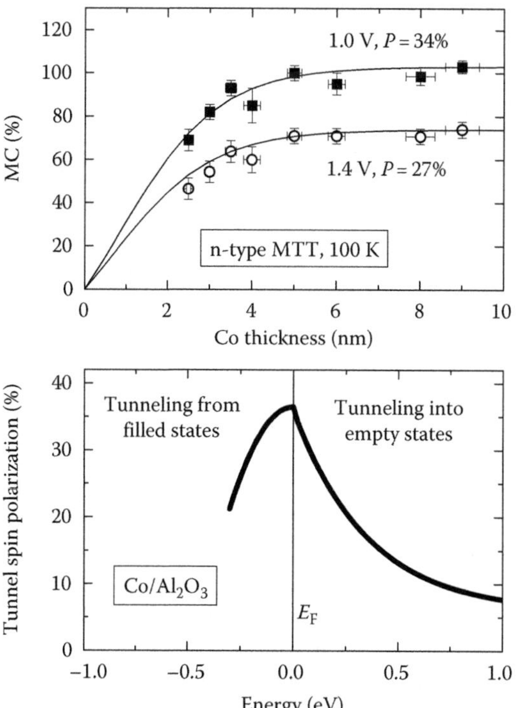

FIGURE 9.14 Top: MC versus thickness of the Co base for an n-type MTT with structure $Co/Al_2O_3/Co/Au/Si$ at two different emitter voltages. Bottom: tunnel spin polarization for Co/Al_2O_3 interfaces versus energy above and below E_F, extracted from MTT data as explained by Park et al. [38].

With typical emitter bias of −1 V, the MTT results described above prove that highly spin-polarized electrons are injected into the base by tunneling at high bias. Interestingly, it was shown [38] that the full energy variation of P_t can be obtained, via a deconvolution procedure. As an example, the tunnel spin polarization of the Co/Al_2O_3 interface versus energy is displayed in Figure 9.14. It is found that P_t is highest for states near the Fermi energy and decays with energy for states below and above E_F, and faster so for the latter. Since methods to extract the energy dependence of the tunnel spin polarization are scarce [52], this approach provides unique and valuable information.

9.5.2 Spin Injection into Semiconductors

One of the crucial ingredients of semiconductor-based spintronics is the ability to inject a highly spin-polarized current into a semiconductor. As described in Chapter 17, direct electrical spin injection from a ferromagnetic metal contact into a semiconductor, via resistive transport at the Fermi energy, was argued to be nearly impossible due to the large mismatch of resistance between metal and semiconductor [53]. It was suggested that the problem may be overcome by using some kind of spin-dependent "interface resistance," for instance, a tunnel barrier. Other suggestions were to use ballistic transport [53] or hot-electron transport [9]. In this case, the transport is not resistive but driven by the electron's kinetic energy, and the resistance mismatch between metal and semiconductor is irrelevant. Spin filtering of a current of hot electrons in a ferromagnetic thin film, just prior to entering the semiconductor, allows injection of a current with very high spin polarization (typically above 85% for FM film thickness above 3 nm). This approach was experimentally demonstrated in 2003 for GaAs structures [54]. More recently, this approach was applied by Appelbaum to inject spin-polarized hot electrons into silicon and then extended with a complementary hot-electron scheme for detection of the spins after transport through the Si [15]. This enabled unique systematic investigations of spin transport and spin coherence in Si, as described in Chapter 18.

9.5.3 Spin-Polarized Scanning Tunneling Microscopy

Spin-polarized tunneling is, when applied in a scanning tunneling microscope, capable of imaging magnetic structure at the surface of a magnetic sample with atomic resolution. However, when a magnetic probe tip is used, special schemes are needed to isolate the magnetic component of the tunnel current and distinguish it from artifacts that can be misinterpreted as magnetic contrast. Interestingly, hot-electron spin transport has recently been proposed [55] as a means to achieve this in a new technique called spin-filter STM (SF-STM). To extract and quantify the spin polarization of the tunnel current it will make use of the spin-dependent filtering of hot carriers in a semiconductor/ferromagnet heterostructure, integrated into a two-terminal STM probe tip. As spin analysis will occur within the STM tip (after tunneling), the magnetic contrast is effectively decoupled from the tunneling signal, which in parallel but independently provides the usual topographic information about the sample. Although a detailed description is beyond the present scope, the unique spin-filtering properties of hot electrons are expected to enable this new technique to provide quantitative information on the spin polarization of materials with atomic resolution.

References

1. G. Schönhense and H. C. Siegmann, Transmission of electrons through ferromagnetic material and applications to detection of electron spin polarization, *Ann. Phys.* **2**, 465 (1993).
2. D. P. Pappas, K.-P. Kämper, B. P. Miller et al., Spin-dependent electron attenuation by transmission through thin ferromagnetic films, *Phys. Rev. Lett.* **66**, 504 (1991).
3. J. C. Gröbli, D. Guarisco, S. Frank, and F. Meier, Spin-dependent transmission of polarized electrons though a ferromagnetic iron film, *Phys. Rev. B* **51**, 2945 (1995).
4. J. C. Gröbli, D. Oberli, and F. Meier, Crucial tests of spin filtering, *Phys. Rev. B* **52**, R13095 (1995).
5. D. J. Monsma, J. C. Lodder, Th. J. A. Popma, and B. Dieny, Perpendicular hot electron spin-valve effect in a new magnetic field sensor: The spin-valve transistor, *Phys. Rev. Lett.* **74**, 5260 (1995).
6. D. J. Monsma, R. Vlutters, and J. C. Lodder, Room temperature-operating spin-valve transistors formed by vacuum bonding, *Science* **281**, 407 (1998).
7. P. S. Anil Kumar, R. Jansen, O. M. J. van't Erve, R. Vlutters, P. de Haan, and J. C. Lodder, Low-field magnetocurrent above 200% in a spin-valve transistor at room temperature, *J. Magn. Magn. Mater.* **214**, L1 (2000).
8. R. Jansen, P. S. Anil Kumar, O. M. J. van't Erve, R. Vlutters, P. de Haan, and J. C. Lodder, Thermal spin-wave scattering in hot-electron magnetotransport across a spin valve, *Phys. Rev. Lett.* **85**, 3277 (2000).
9. R. Jansen, The spin-valve transistor: A review and outlook, *J. Phys. D: Appl. Phys.* **36**, R289 (2003).
10. K. Mizushima, T. Kinno, K. Tanaka, and T. Yamauchi, Energy-dependent hot electron transport across a spin-valve, *IEEE Trans. Magn.* **33**, 3500 (1997).
11. T. Kinno, K. Tanaka, and K. Mizushima, Ballistic-electron-emission spectroscopy on an Fe/Au/Fe multilayer, *Phys. Rev. B* **56**, R4391 (1997).
12. K. Mizushima, T. Kinno, K. Tanaka, and T. Yamauchi, Strong increase of the effective polarization of the tunnel current in Fe/AlOx/Al junctions with decreasing Fe layer thickness, *Phys. Rev. B* **58**, 4660 (1998).
13. T. Yamauchi and K. Mizushima, Theoretical analysis of energy-dependent hot-electron transport in a magnetic multilayer, *Phys. Rev. B* **61**, 8242 (2000).
14. R. Sato and K. Mizushima, Spin-valve transistor with an Fe/Au/Fe(001) base, *Appl. Phys. Lett.* **79**, 1157 (2001).
15. I. Appelbaum, B. Huang, and D. J. Monsma, Electronic measurement and control of spin transport in silicon, *Nature* **447**, 295 (2007).
16. W. J. Kaiser and L. D. Bell, Direct investigation of subsurface interface electronic structure by ballistic electron emission microscopy, *Phys. Rev. Lett.* **60**, 1406 (1988).
17. M. Prietsch, Ballistic-electron emission microscopy (BEEM): Studies of metal/semiconductor interfaces with nanometer resolution, *Phys. Rep.* **253**, 163 (1995).
18. R. J. Celotta, J. Unguris, and, D. T. Pierce, Hot electron spin-valve effect in coupled magnetic layers, *J. Appl. Phys.* **75**, 6452 (1994).
19. H.-J. Drouhin, G. Lampel, Y. Lassailly, A. J. van der Sluijs, and C. Marlière, Electron transmission through ultra-thin metal layers and its spin dependence for magnetic structures, *J. Magn. Magn. Mater.* **151**, 417 (1995).
20. D. Oberli, R. Burgermeister, S. Riesen, W. Weber, and H. C. Siegmann, Total scattering cross section and spin motion of low energy electrons passing through a ferromagnet, *Phys. Rev. Lett.* **81**, 4228 (1998).
21. R. Vlutters, R. Jansen, O. M. J. van't Erve et al., Modeling of spin-dependent hot-electron transport in the spin-valve transistor, *Phys. Rev. B* **65**, 024416 (2002).
22. T. Banerjee, E. Haq, M. H. Siekman, J. C. Lodder, and R. Jansen, Spin filtering of hot holes in a metallic ferromagnet, *Phys. Rev. Lett.* **94**, 027204 (2005).
23. V. P. Zhukov, E. V. Chulkov, and P. M. Echenique, Lifetimes and inelastic mean free path of low-energy excited electrons in Fe, Ni, Pt, and Au: Ab initio GW + T calculations, *Phys. Rev. B* **73**, 125105 (2006).
24. M. Aeschlimann, M. Bauer, S. Pawlik et al., Ultrafast spin-dependent electron dynamics in fcc Co, *Phys. Rev. Lett.* **79**, 5158 (1997).
25. R. Knorren, K. H. Bennemann, R. Burgermeister, and M. Aeschlimann, Dynamics of excited electrons in copper and ferromagnetic transition metals: Theory and experiment, *Phys. Rev. B* **61**, 9427 (2000).
26. W. H. Rippard and R. A. Buhrman, Ballistic electron magnetic microscopy: Imaging magnetic domains with nanometer resolution, *Appl. Phys. Lett.* **75**, 1001 (1999).
27. W. H. Rippard and R. A. Buhrman, Spin-dependent hot electron transport in Co/Cu thin films, *Phys. Rev. Lett.* **84**, 971 (2000).
28. T. Shimatsu, R. H. Mollema, D. J. Monsma, E. G. Keim, and J. C. Lodder, Metal bonding during sputter film deposition, *J. Vac. Sci. Technol.* A **16**, 2125 (1998).
29. O. M. J. van't Erve, R. Vlutters, P. S. Anil Kumar et al., Transfer ratio of the spin-valve transistor, *Appl. Phys. Lett.* **80**, 3787 (2002).
30. R. Vlutters, O. M. J. van't Erve, S. D. Kim, R. Jansen, and J. C. Lodder, Interface, volume, and thermal attenuation of hot-electron spins in $Ni_{80}Fe_{20}$ and Co, *Phys. Rev. Lett.* **88**, 027202 (2002).

31. S. van Dijken, X. Jiang, and S. S. P. Parkin, Spin-dependent hot electron transport in $Ni_{81}Fe_{19}$ and $Co_{84}Fe_{16}$ films on GaAs(001), *Phys. Rev. B* **66**, 094417 (2002).
32. S. van Dijken, X. Jiang, and S. S. P. Parkin, Comparison of magnetocurrent and transfer ratio in magnetic tunnel transistors with spin-valve bases containing Cu and Au spacer layers, *Appl. Phys. Lett.* **82**, 775 (2003).
33. S. van Dijken, X. Jiang, and S. S. P. Parkin, Giant magnetocurrent exceeding 3400% in magnetic tunnel transistors with spin-valve base layers, *Appl. Phys. Lett.* **83**, 951 (2003).
34. S. van Dijken, X. Jiang, and S. S. P. Parkin, Nonmonotonic bias voltage dependence of the magnetocurrent in GaAs-based magnetic tunnel transistors, *Phys. Rev. Lett.* **90**, 197203 (2003).
35. T. Hagler, C. Bilzer, M. Dumm, W. Kipferl, and G. Bayreuther, Semiepitaxial magnetic tunnel transistor: Effect of electron energy and temperature, *J. Appl. Phys.* **97**, 10D505 (2005).
36. B. G. Park, T. Banerjee, B. C. Min, J. S. M. Sanderink, J. C. Lodder, and R. Jansen, Temperature dependence of magnetocurrent in a magnetic tunnel transistor, *J. Appl. Phys.* **98**, 103701 (2005).
37. B. G. Park, T. Banerjee, B. C. Min, J. C. Lodder, and R. Jansen, Tunnel spin polarization of $Ni_{81}Fe_{19}/SiO_2$ probed with a magnetic tunnel transistor, *Phys. Rev. B* **73**, 172402 (2006).
38. B. G. Park, T. Banerjee, J. C. Lodder, and R. Jansen, Tunnel spin polarization versus energy for clean and doped Al_2O_3 barriers, *Phys. Rev. Lett.* **99**, 217206 (2007).
39. R. Heer, J. Smoliner, J. Bornemeier, and H. Brückl, Ballistic electron emission microscopy on spin valve structures, *Appl. Phys. Lett.* **85**, 4388 (2004).
40. T. Banerjee, J. C. Lodder, and R. Jansen, Origin of the spin-asymmetry of hot-electron transmission in Fe, *Phys. Rev. B* **76**,140407(R) (2007).
41. E. Heindl, J. Vancea, and C. H. Back, Ballistic electron magnetic microscopy on epitaxial spin valves, *Phys. Rev. B* **75**, 073307 (2007).
42. A. Kaidatzis, S. Rohart, A. Thiaville, and J. Miltat, Hot electron transport and a quantitative study of ballistic electron magnetic imaging on Co/Cu multilayers, *Phys. Rev. B* **78**, 174426 (2008).
43. V. P. Zhukov, E. V. Chulkov, and P. M. Echenique, Lifetimes of excited electrons in Fe and Ni: First-principles GW and the T-Matrix theory, *Phys. Rev. Lett.* **93**, 096401 (2004).
44. R. Jansen, S. D. Kim, R. Vlutters, O. M. J. van't Erve, and J. C. Lodder, Anisotropic spin-orbit scattering of hot-electron spins injected into ferromagnetic thin-films, *Phys. Rev. Lett.* **87**, 166601 (2001).
45. R. Jansen, T. Banerjee, B. G. Park, and J. C. Lodder, Probing momentum distributions in magnetic tunnel junction via hot electron decay, *Appl. Phys. Lett.* **90**, 192503 (2007).
46. M. Plihal, D. L. Mills, and J. Kirschner, Spin wave signature in the spin polarized electron energy loss spectrum of ultrathin Fe films: Theory and experiment, *Phys. Rev. Lett.* **82**, 2579 (1999).
47. J. Hong and D. L. Mills, Theory of the spin dependence of the inelastic mean free path of electrons in ferromagnetic metals: A model study, *Phys. Rev. B* **59**, 13840 (1999).
48. J. Hong and D. L. Mills, Spin dependence of the inelastic electron mean free path in Fe and Ni: Explicit calculations and implications, *Phys. Rev. B* **62**, 5589 (2000).
49. H. Gökcan, Magnetotransport of hot electrons and holes in the spin-valve transistor, PhD thesis, University of Twente, Enschede, the Netherlands, 2006.
50. V. P. Zhukov, E. V. Chulkov, and P. M. Echenique, Private communication.
51. B. G. Park, T. Banerjee, J. C. Lodder, and R. Jansen, Opposite spin asymmetry of elastic and inelastic scattering of non-equilibrium holes injected into a ferromagnet, *Phys. Rev. Lett.* **97**, 137205 (2006).
52. S. O. Valenzuela, D. J. Monsma, C. M. Marcus, V. Narayanamurti, and M. Tinkham, Spin polarized tunneling at finite bias, *Phys. Rev. Lett.* **94**, 196601 (2005).
53. G. Schmidt, D. Ferrand, L. W. Molenkamp, A. T. Filip, and B. J. van Wees, Fundamental obstacle for electrical spin injection from a ferromagnetic metal into a diffusive semiconductor, *Phys. Rev. B* **62**, R4790 (2000).
54. X. Jiang, R. Wang, S. van Dijken et al., Optical detection of hot-electron spin injection into GaAs from a magnetic tunnel transistor source, *Phys. Rev. Lett.* **90**, 256603 (2003).
55. I. J. Vera Marún and R. Jansen, Multiterminal semiconductor/ferromagnet probes for spin-filter scanning tunneling microscopy, *J. Appl. Phys.* **105**, 07D520 (2009).

III

Spin Transport and Magnetism in Magnetic Tunnel Junctions

10

Tunneling Magnetoresistance: Experiment (Non-MgO Magnetic Tunnel Junctions)

Patrick R. LeClair
University of Alabama

Jagadeesh S. Moodera
Massachusetts Institute of Technology

Tunneling magnetoresistance (TMR) is a consequence of spin-dependent tunneling, and in a broader sense, is another manifestation of spin-dependent transport related to giant magnetoresistance (GMR) and anisotropic magnetoresistance (AMR) effects. Spin-dependent tunneling between two ferromagnets, an outgrowth of earlier work by Meservey and Tedrow [1], was first proposed [2] and observed [3] in 1975 and was reliably demonstrated only in 1995 [4,5]. In the past decade, magnetic tunnel junctions (MTJs) have aroused considerable interest due to their suitability for applications in spin-electronic devices and indeed have found applications in hard disk read heads and magnetic random access memories (MRAMs). The diversity of the physical phenomena governing the operation of these magnetoresistive devices also makes MTJs very attractive from the fundamental physics point of view. These facts have recently stimulated tremendous activity in both the experimental and theoretical realms of investigation, with a view to understanding and manipulating the electronic, magnetic, and transport properties of MTJs.

An MTJ, in its essence, consists of two ferromagnetic metal layers separated by a thin insulating barrier layer, as shown in Figure 11.1. The insulating layer is sufficiently thin (a few nm or less) for electrons to tunnel through the barrier, provided a bias voltage is applied between the two metal electrodes across the insulator. The most important property of MTJs is that the tunneling current depends on the relative orientation of the magnetizations of the two ferromagnetic layers, which can be changed by an applied magnetic field. This phenomenon is called tunneling magnetoresistance (sometimes referred to as junction magnetoresistance). Although TMR has been known for over 30 years from the experiments of Jullière [3], only a relatively modest number of studies had been performed in this field until the mid-1990s. The technologically demanding fabrication process was partly responsible, as it was difficult to fabricate robust and reliable tunnel junctions. Also, the fact that the reported values of TMR were small (at the most a small percent at low temperatures) did not immediately trigger a considerable interest with regard to sensor/memory applications. In 1995, however, Miyazaki and Tezuka [4] and Moodera et al. [5] demonstrated reliable magnetoresistances of over 10% that triggered an explosion in research.

In this chapter, we review the history and fundamental physics of spin-dependent tunneling, aiming to provide the reader with a conceptual understanding and an overview of what factors control magnetoresistance in MTJs, and what we view as some of the crucial illustrative experiments. Starting from early experiments on spin-dependent tunneling that established the basis of the field, we subsequently overview the characteristic features of MTJs and consider recent experiments and models that highlight the role of the electronic structure of ferromagnets, the insulating layer, and the ferromagnet/insulator interfaces.

10.1 Introduction to Electron Tunneling

The quantum mechanical tunnel effect is one of the oldest quantum phenomena, dating back to the late 1920s. A general review of electron tunneling and all its nuances is well beyond the scope of this review. We will assume the reader has at least a basic familiarity with the tunneling problem from introductory quantum physics, and for a more complete introduction can refer to the excellent overview in Ref. [6] and Chapter 12 for a discussion of the theory of spin-dependent tunneling. Here, we only briefly introduce the key features of the tunneling problem as applied to metal–insulator–metal structures before embarking on a tour of spin-polarized tunneling.

Electron tunneling is a phenomenon by which an electric current may flow through a thin insulating region. A simple way to understand how tunneling is possible is to consider an electron wave incident on a potential step, as shown in Figure 10.1a. Though most of the wave is reflected, if the potential barrier is sufficiently thin, the evanescent states in the barrier region can emerge on the other side of the barrier, and the electron will have a finite probability of "tunneling" across the potential barrier. Strictly speaking, tunneling is not a purely quantum phenomenon, but merely a *wave phenomenon*, and has an optical analogy in frustrated total internal reflection. In the step potential model, the tunneling probability depends exponentially on the tunnel barrier thickness d, $P \propto e^{-\kappa d}$, where the decay constant κ in turn depends on the difference between the electron's energy and that of the potential barrier. Even in the simplest realization of the problem, two crucial length scales are apparent: one is set by the electron wave vector in the classically allowed region (k), and the second by the wave function's decay constant in the insulating region (κ), which is in turn dictated by the details of the potential landscape in the barrier region.

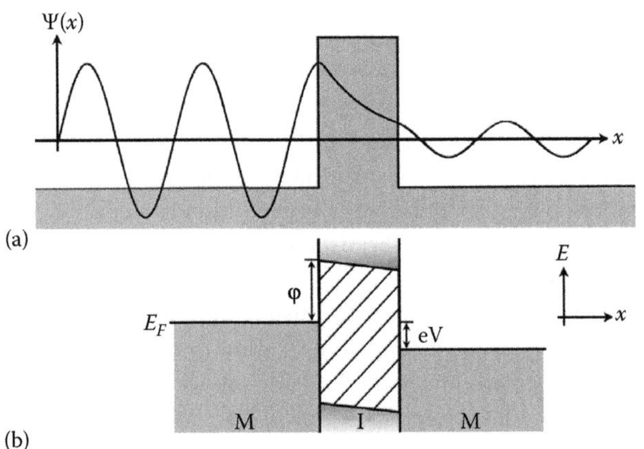

FIGURE 10.1 Tunneling in metal–insulator–metal structures. (a) Electron wave function decays exponentially in the barrier region, and for thin barriers, some intensity remains in the right side. (b) Potential diagram for an M/I/M structure with applied bias eV. Shaded areas represent filled states, open areas are empty states, and the hatched area represents the forbidden gap in the insulator.

10.1.1 Tunneling between Two Free-Electron Metals

The most straightforward realization of this problem is in a metal–insulator–metal (M/I/M) trilayer structure (Figure 10.1b), commonly called a tunnel junction, with the insulator typically provided by a metal oxide (e.g., Al_2O_3). In this case, a bias voltage (V) is applied between the two metal electrodes. Owing to the enormous difference in resistance between the insulating region and the metal electrodes, this potential difference is essentially all across the tunnel barrier, and this has the effect of raising the Fermi energy of one electrode relative to the other by eV. Thus, in Figure 10.1b, electrons near the Fermi energy in the leftmost electrode can tunnel through the insulating barrier elastically into the available states in the rightmost electrode. The two length scales in the tunneling problem now have important implications. First, the electron wavelength at the Fermi energy in a typical metal is extremely short, on the order of a few lattice spacings [7]. This means that tunneling is sensitive to the atomic-scale details of the physical and electronic structure of the metal–insulator interface. Second, the decay constant in the insulating region in a real insulator is dictated by the evanescent states from the metal electrode. The potential barrier height is thus no longer a simple step (except for illustrative purposes), but is in reality governed by the complex electronic structure of the insulator and its coupling to the metal electrodes.

In general, one must consider the electronic structure of the entire trilayer system, which ultimately makes tunneling in realistic systems a subtle problem with a wide variety of rich phenomena exhibited. We should note that our discussion in this chapter focuses almost exclusively on incoherent systems, for example, those with nonepitaxial electrodes and amorphous barriers. In incoherent systems, k conservation rules are not strict [8], and any analysis of tunneling based on ideal k conservation is generally invalid. For all but a handful of the experiments discussed in this chapter, these considerations play a negligible role, but they will be a primary motivator for the results of Chapters 11 and 12.

In spite of these complications, it is nonetheless illustrative to further consider the idealized M/I/M structure of Figure 10.1b. In the simplest phenomenological picture [6], the number of electrons tunneling from one electrode to the other is given by the product of the density of states (DOS) at a given energy in the left electrode, $\rho_l(E)$, and the DOS at the same energy in the right electrode, $\rho_r(E)$, multiplied by the matrix elements $|M|^2$ representing the probability of transmission through the barrier. One must also then multiply by the probabilities that the states in the left electrode are occupied, $f(E)$, and that the states in the right electrode are empty, $1 - f(E - eV)$, where $f(E)$ is the Fermi–Dirac function. This is simply stating the requirement that electrons on one side of the barrier must have empty states to tunnel into on the other side of the barrier. Taking the total tunnel current as the difference between left- and right-going currents, within this picture, we have

$$I(V) = I_{l\to r} - I_{r\to l}$$

$$= \int_{-\infty}^{+\infty} \rho_l(E)\cdot\rho_r(E-eV)\,|M|^2 f(E)\big[f(E-eV)-f(E)\big]dE \qquad (10.1)$$

For a simple square barrier of height φ and thickness d, the matrix elements can be evaluated exactly, and one finds $I \sim e^{-d\sqrt{\varphi}}$. Essentially, the same holds for more arbitrarily shaped barriers [9]. The effect of an applied potential difference eV is also already clear in this simple model: the potential difference will develop almost wholly across the insulating barrier, which serves to effectively lower the average barrier height by $eV/2$ for electrons tunneling in one direction, and raise it by the same amount for tunneling in the opposing direction. This leads to a rapidly (and nonlinearly) increasing current with increasing potential difference, one hallmark characteristic of tunneling behavior. Extensions to finite temperature further predict an increase in tunnel current as temperature is increased, also commonly used as one criteria for establishing the presence of tunneling [10,11].

Often, it is more common to measure the derivative of the current–voltage characteristic, the *conductance dI/dV*. In the limit of low bias voltage, where the tunnel probability matrix elements are essentially independent of energy,

$$G \equiv \frac{dI}{dV} \propto |M|^2 \int \rho_l(E)\rho_r(E-eV)\frac{df(E-eV)}{dV}dE \qquad (10.2)$$

At low temperatures, the derivative of the Fermi–Dirac distribution is similar to a delta function, zero everywhere except where its argument is zero and with a width proportional to k_BT. This gives the result that for low temperature and bias voltage, the tunnel conductance should be proportional to the product of the densities of states in the two electrodes.

10.2 Density of States in Tunneling

In his pioneering experiments, Giaever [12] investigated the current–voltage and conductance–voltage characteristics of Al/Al_2O_3/Pb normal metal–insulator–superconductor (M/I/S) tunnel junctions well below the superconducting transition temperature of the Pb electrode. By driving the Pb into its normal state through the application of a magnetic field, Giaever could perform a comparative study of the tunneling characteristics between normal metals, or between a normal metal and a superconductor. When the Pb electrode was in the normal state, the tunnel current was linear in voltage and the conductance dI/dV was essentially constant, Figure 10.2, as expected from the previous section. However, when the Pb was in the superconducting state, the tunnel current was dramatically reduced at low voltages, independent of current polarity. Most interestingly, the conductance $dI/dV(V)$, Figure 10.2, very closely resembles the Bardeen-Cooper-Schrieffer (BCS) quasiparticle "density of states" [6,12]:

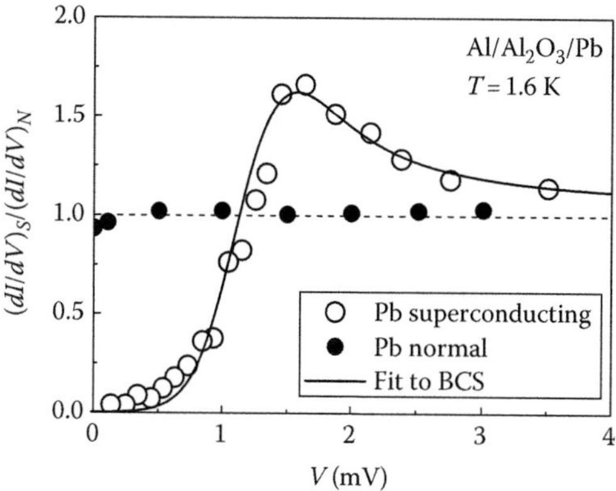

FIGURE 10.2 Conductance vs. voltage curve dI/dV (V) for an Al/Al_2O_3/Pb junction at T=1.6K when the Pb is in the normal state (closed) and superconducting state (open). The Al is in the normal state for both curves. The solid line is a fit to a thermally broadened BCS DOS. (Adapted from Giaever, I., *Phys. Rev. Lett.*, 5(4), 147, 1960.)

$$\frac{\rho_s(E)}{\rho_n(E)} = sgn(E)\,Re\left(\frac{E}{\sqrt{E^2-\Delta^2}}\right) \qquad (10.3)$$

where

ρ_n is the density of states in the normal state

Δ is the energy gap in the quasiparticle excitation spectrum

Giaever interpreted this result [12] as an indication that tunneling conductance is proportional to the DOS in the electrodes, and subsequent measurements confirmed these results shortly thereafter [13]. This result had no adequate theoretical explanation at the time. In fact, several of Giaever's coworkers were initially skeptical of his "naive" experiment to directly measure the superconducting energy gap [14]. This "naive" experiment later led to a shared Nobel Prize for Giaever in 1973.

Tunneling into superconductors is now so ubiquitous that it is difficult to appreciate how striking a result it was. A phenomenological explanation, closely following Giaever's original treatment, is now presented in most introductory solid-state physics texts. However, a proper theoretical justification of Giaever's ideas is subtle, with important consequences for MTJs, and will be treated more rigorously in Chapter 12. Despite the fact that a simple calculation starting with Equation 10.1 seems to include the DOS directly, it has been shown that [15] in a simple independent electron model, the tunneling probability matrix elements are *inversely proportional to the density of states*, and, thus, they exactly cancel the DOS factors in Equation 10.1. Thus, the DOS does *not* directly enter the expression for the tunnel current in a free-electron model [15]. Shortly after Giaever's experiments, Bardeen [16] clarified this apparent contradiction and found that one could indeed recover the result that the tunneling conductance should be proportional to the DOS in the electrodes. The crucial point is that the dependence of the tunnel conductance on the DOS does not come as simply as one might expect from Equation 10.1, as will be pointed out in more detail in Chapter 12. For MTJs, this means that although an explanation of the TMR effect (Jullière's model, Section 10.4.1) seems to come quite easily from Equation 10.1, the proper justification of such an explanation is subtle. In fact, in our view, it is only in recent years that a deeper understanding of the role of electronic structure in tunneling has arisen, and we refer the reader to Chapter 12 for further discussion.

In the decade following Giaever's groundbreaking results, more sophisticated theoretical treatments by Appelbaum and Brinkman [17] and Mezei and Zawadowski [18], as well as more recent refinements [19–23], put the role of the electrode DOS on firmer ground. The essential result is that, particularly in incoherent systems, tunneling is specifically sensitive to the *local density of states at the electrode–barrier interface*. Qualitatively, in these more sophisticated treatments, the relevant length scale over which the DOS are "sampled" in the tunneling process is dictated by the Fermi wavelength [17].

In a normal metal, this scale is on the order of a few lattice spacings. However, at distances of only a few lattice spacings from the metal–insulator interface, the wave function in the normal metal is strongly perturbed by the presence of the insulator and the details of the interface bonding. In short, in normal metal tunneling structures, one can only measure the DOS in the normal metal *in this strongly perturbed region*, rather than the bulk-like or even the surface-like behavior. Put slightly differently, although one may probe electronic structure effects by normal-metal tunneling, these effects are probed only within a few Fermi wavelengths of the electrode–barrier interface, where the DOS may differ greatly from that in the bulk. One may already imagine that because of this interfacial sensitivity, tunneling characteristics will be a property of both the electrode and barrier *together*. In fact, the junctions with the same electrodes but different barriers *do* behave differently, a phenomenon that will be discussed in Section 10.5. Generally speaking, the interfacial sensitivity of tunneling in nonsuperconducting junctions has extremely important consequences for spin-dependent tunneling [1], particularly in MTJs [24–26] (see Section 10.5).

10.3 Beginnings of Spin-Dependent Tunneling

Based on Giaever's ideas about tunneling, in the late 1960s and early 1970s, Meservey and Tedrow embarked on a series of studies to investigate the influence of spin paramagnetism on high-magnetic-field superconductivity, and in doing, so ushered in the era of spin-dependent tunneling. In a simple picture, one can already imagine how spin should play a role in determining tunneling current: if the tunnel current is proportional to the DOS, and one presumes a two-current model in a ferromagnetic metal, one should have a spin-dependent tunneling current. It was essentially this fact, combined with the unique properties of thin film Al superconductors to serve as spin detectors, which created the field of spin-dependent tunneling. These pioneering experiments are the fundamental basis for the magnetoresistance effect in MTJs, as well as many other spin-polarized tunneling phenomena discussed in later chapters. Though the full details of spin-polarized tunneling are beyond the scope of this review (and have been excellently reviewed by Meservey and Tedrow [1]), we briefly describe the concepts behind this powerful technique.

10.3.1 Spin-Polarized Tunneling Technique

The spin-polarized tunneling technique (SPT) developed by Meservey and Tedrow [1] utilizes a superconductor as one electrode in a tunneling device to probe the tunneling spin polarization of electrons in the counter-electrode within a few hundreds of microelectronvolt of the Fermi level (i.e., on the order of the superconducting energy gap Δ). If the superconductor is sufficiently thin (~4–5 nm), such that the orbital screening currents in the superconductor are largely suppressed, the effects of the electron spin interaction with the magnetic field may be observed, resulting in the Zeeman splitting of the quasiparticle states in the superconductor. However, not only must the critical field of the superconductor be large enough to have observable Zeeman splitting (Zeeman splitting larger than the thermal smearing, i.e., $3.5k_BT<\mu_BH_{ext}$) at fields below the superconductor's critical temperature, elements with low atomic number must be used to avoid significant spin scattering in the superconductor via the spin-orbit effect (spin-orbit scattering). In practice, this usually means temperatures of ~500 mK, magnetic fields of a few tesla, and typically, but not exclusively, the superconductor is pure or doped Al, since ultra-thin Al films can have critical fields >5 T, and being a light element, spin-orbit scattering is negligible.

As discussed above, when tunneling into a superconductor, the conductance $dI/dV(V)$ is proportional to the BCS [6,12], and the SPT technique achieves spin sensitivity by applying a magnetic field (~2–3 T) parallel to the film plane. If Zeeman splitting is achieved, the total quasiparticle DOS is a superposition of spin-up and spin-down DOS, each shifted in energy by $\pm\,\mu_BB$ from the zero-field curve as shown in Figure 10.3a. The total conductance without spin polarization in the normal metal thus exhibits a four-peaked structure. If the tunneling electrons are spin polarized, however, each of the separate spin channels is weighted by the relative DOS, and $dI/dV(V)$ is asymmetric (Figure 10.3b and c). The heights of the four peaks relate to P. Assuming a proportionality between the conductance and the normal metals' densities of states at E_F, ρ, the tunneling spin polarization may be *estimated* by the heights of the conductance peaks [27,28] shown in Figure 10.3, σ_{1-4}:

$$P(E_F)\equiv\frac{\rho_\uparrow-\rho_\downarrow}{\rho_\uparrow+\rho_\downarrow}\approx\frac{(\sigma_4-\sigma_2)-(\sigma_1-\sigma_3)}{(\sigma_4-\sigma_2)+(\sigma_1-\sigma_3)}\approx\frac{\sigma_2-\sigma_3}{\sigma_2+\sigma_3}\qquad(10.4)$$

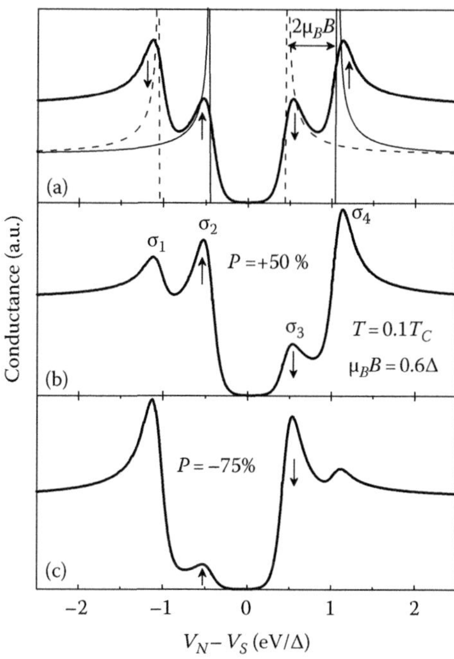

FIGURE 10.3 Conductance (dI/dV) of an S/I/N tunnel junction with a Zeeman splitting of the DOS in the superconductor due to a magnetic field B. In (a) the normal metal has zero polarization, $P = +50\%$ in (b), and $P = -75\%$ in (c). The thin, solid, and dashed lines in (a) show the spin-up and spin-down partial densities of states. The conductance curves are calculated for $T = 0.1T_C$ and $\mu_B B = 0.6\Delta$.

The curve in Figure 10.3b has a tunneling spin polarization of $P = +50\%$, while the curve in Figure 10.3c has $P = -75\%$. The difference in the conductance peak heights is only an estimate of P, and an accurate determination must account for spin-orbit scattering in the superconductor and depairing due to the orbital motion of the carriers [1] (using (10.4) tends to overestimate P). Physically, the depairing corresponds to an interaction between the condensate and unpaired electrons. Due to this interaction, electrons continuously enter and leave the condensate, leading to a finite lifetime for unpaired electrons. This effect is dramatically strengthened by the presence of a magnetic field (particularly perpendicular to the film plane), the presence of magnetic impurities, or the current-generated field when a superconductor carries a current [1]. The spin-orbit interaction represents a transfer of electrons between spin-up and spin-down subsystems. From the reference frame of a moving electron, the positively charged nuclei constitute a current, which from the electrons' perspective creates a magnetic field. When electrons are scattered, this field leads to a finite probability of a spin flip accompanying the scattering event. This effect scales as the fourth power of the atomic number, necessitating the use of superconductors comprised of light elements. Both the spin-orbit and depairing interactions represent a spin-dependent correction to the electron energy, and the DOS in the superconductor accounting for both interactions may be determined self-consistently to accurately extract a spin polarization [1].

From our point of view, a crucial advantage of the SPT technique is that a robust, well-tested, and microscopic theory of SPT has been developed over the years [1]. This, coupled with the simple, fundamental nature of the experiment makes SPT agreeable for detailed comparison with theory. Though restricted to low temperatures and bias voltages, SPT is a powerful technique for assessing novel spintronic systems. We should again point out that the DOS measured via tunneling (and thus the tunneling spin polarization) is never the "raw" DOS, but always weighted by the barrier transmission probability [29,30]. For SPT experiments, the transmission factors are essentially evaluated at E_F due to the small bias involved (>1 mV), and any energy dependence may be safely ignored. In systems where the spin dependence of the transmission factors may be ignored, such as disordered Al_2O_3, the tunneling probability is spin-independent and the situation may be somewhat simplified [27]. This is not generally true; however, for ordered MgO, for example (see Chapter 12), the transmission factors are not equal for each spin due to the "symmetry filtering" effect.

For comparison with the theoretical curves, Figure 10.4 shows the first SPT measurement by Tedrow and Meservey [27] using $Al/Al_2O_3/Ni$ junctions. While the zero-field curve is symmetric, the curves for $\mu_0 H = 3.37$ T show an obvious asymmetry, consistent with Figure 10.3c. Though the spin polarization is small in this case, $P_{Ni} = 8.0, \pm 0.2\%$ is obtained from fitting to a model incorporating spin-orbit scattering and orbital depairing in the Al [1], and the presence of a finite and *positive* tunneling spin polarization in Ni is clear. Parenthetically, we note that the recent value of tunneling spin polarization for (polycrystalline) Ni (see Table 10.1) is more than a factor of 5 higher, almost certainly due to three decades of improved deposition techniques resulting in cleaner junctions with better interfaces. Table 10.1 lists recently obtained tunneling spin-polarization values obtained for several ferromagnetic metals, after correction for spin-orbit scattering. Of particular interest is that the sign of the tunneling spin polarization for the 3*d* metals and alloys is, in all cases, positive. In fact, the only negative polarization measured with SPT to date is for $SrRuO_4$ [31].

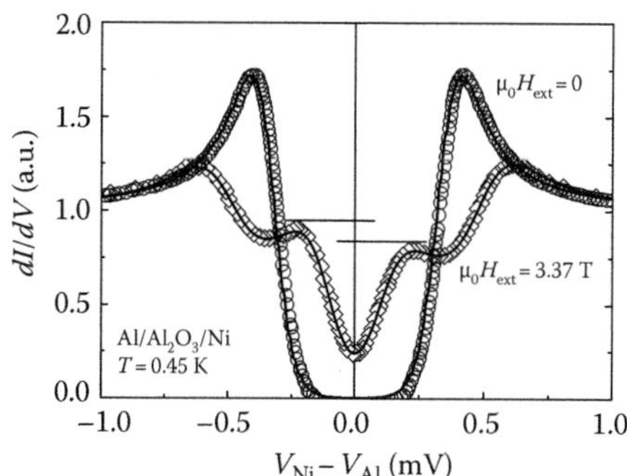

FIGURE 10.4 First observation of spin-polarized tunneling using Ni. Fitting the curves (*lines*) with a model accounting for spin-orbit scattering and orbital depairing [1] gives a spin polarization of 8.0% ± 0.2%. (Adapted from Tedrow, P.M. and Meservey, R., *Phys. Rev. Lett.*, 27, 919, 1971.)

TABLE 10.1 Tunneling Spin-Polarization Values Obtained from SPT Measurements on FM/I/Al or Al/I/FM Junctions, after Corrections for Spin Orbit Scattering and Orbital Depairing

FM	*I*	*P* (%)	References
Ni	Al_2O_3	46	[54]
Co	Al_2O_3	42	[42]
$Co_{84}Fe_{16}$	Al_2O_3	55	[42,46]
$Co_{60}Fe_{40}$	Al_2O_3	50	[46]
$Co_{50}Fe_{50}$	Al_2O_3	55	[42]
$Co_{40}Fe_{60}$	Al_2O_3	51	[46]
Fe	Al_2O_3	44–45	[42]
$Ni_{90}Fe_{10}$	Al_2O_3	36	[46]
$Ni_{80}Fe_{20}$	Al_2O_3	48	[42]
$Ni_{40}Fe_{60}$	Al_2O_3	55	[46]
$SrRuO_3$*	STO*	−9.5	[31]
$La_{0.67}Sr_{0.33}MnO_3$*	STO*	78	[53]
CrO_2*	Cr_2O_3	100	[49]
Fe	MgO*	74	[62]
$Co_{70}Fe_{30}$	MgO*	85	[62]

Crystalline electrodes or barriers are indicated by *.

Though, in retrospect, the SPT technique may seem like a natural outgrowth of work on superconducting tunneling in the late 1960s, its development was in fact something of a fortuitous accident. Meservey himself recently clarified [32] the story surrounding the development of SPT:

> What we were doing was actually trying to observe the theoretically predicted paramagnetic critical field of superconducting Al, which seemed to be attainable because of our ability to make thin enough Al films. It was expected that this result would help in understanding the critical magnetic fields of other superconductors with higher values of spin-orbit scattering.
>
> If you read [33] you will understand the situation. We had spins in mind, but had never thought of spin-splitting of the density of states. However, by the time the $x-y$ recorder had traced out the first two peaks, I was calculating what the spin splitting should be and we knew immediately what we were seeing. I think [33] should make it clear that the paramagnetic critical field was one of the few things not known about superconductors at that time and was the center of much interest. The next and obvious step was to apply the technique to study ferromagnetic metals.
>
> —**Robert Meservey [32]**

10.3.2 Interface Sensitivity, Spin-Flip Tunneling

In the preceding analysis, the determination of *P* with the SPT technique relied on spin being conserved during the tunneling process. In fact, the Zeeman splitting of the DOS may be also used directly to *prove* that spin flipping does not take place during tunneling. In an elegant experiment [1,27], Meservey and Tedrow studied spin-polarized tunneling with *two* Zeeman split superconductors, and showed that no spin flipping takes place during tunneling, a crucial result for the development of MTJs. This extension of SPT directly measures the probability of spin-altering events during the tunneling process. Briefly, if no spin flipping or filtering takes place during tunneling, the tunneling $dI/dV(V)$ should display two peaks at $V=(\Delta_1+\Delta_2)/e$, and will be unaffected by a magnetic field [1,6,27]. If the quasiparticle spin changes during the tunneling process, additional conductance peaks are expected at $V=\pm(\Delta_1+\Delta_2+2\mu H)/e$ [1,27]. The absence of these conductance peaks was used to directly verify that spin is conserved in tunneling through Al_2O_3 barriers.

In another application of this powerful technique, spin polarization measurements on ultrathin ferromagnetic films [1,34] beautifully demonstrated the anticipated interfacial sensitivity [17,18,20] in tunnel junctions with nonsuperconducting electrodes. In this case, ultrathin (0–3 nm) films of Fe or Co were deposited on a normal metal layer to form the bottom electrode of the N/F/I/S junctions, on which the Al_2O_3 tunnel barrier and superconducting Al electrode were grown (Figure 10.5). The spin polarization of electrons tunneling from these ultrathin layers was then measured as a function of average layer thickness, and compared with the polarization values for thick Fe or Co layers. For Fe and Co backed with (normal-state) Al, shown in Figure 10.5 for Co, the polarization increased rapidly with increasing thickness, showing clear spin polarization (and hence, ferromagnetism) for even a single monolayer. The most striking result is that the "bulk" tunneling spin polarization measured on thick layers is approximately reached by only 0.7–1.0 nm, or ~3–5 monolayers. Not only does this experiment clearly indicate that the onset of ferromagnetism is extremely rapid, it definitively

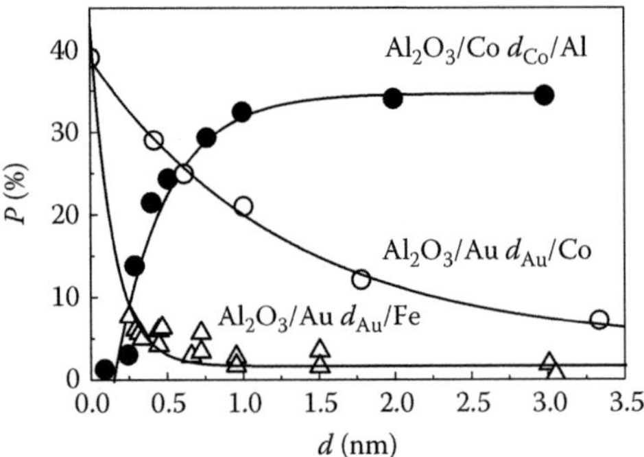

FIGURE 10.5 Closed circles: spin polarization of ultrathin layers of Co deposited on (normal-state) Al. Open circles: spin polarization of Co as a function of the thickness of an Au layer inserted at the Co–Al_2O_3 interface. Open triangles: spin polarization of Fe as a function of the thickness of an Au layer inserted at the Fe–Al_2O_3 interface. The lines are fits to an exponential decay, which yields length scales ξ of $\xi_{Co\text{-}Al}\sim 0.2$ nm, $\xi_{Au\text{-}Co}\sim 1.4$ nm, and $\xi_{Au\text{-}Fe}\sim 0.2$ nm. (Adapted from Tedrow, P.M. and Meservey, R., *Solid State Commun.*, 16, 71, 1975; Moodera, J.S. et al., *Phys. Rev. B*, 40(17), 11980, 1989; Gabureac, M.S. et al., *J. Appl. Phys.*, 103(7), 07A915, 2008.)

demonstrates that tunneling in noncoherent normal-state structures is highly interface sensitive. The 3–5 monolayers adjacent to the insulating barrier appear to dominate the transport properties, in agreement with theoretical expectations [17,18,20]. Moodera et al. [35] first probed the opposite side of the question, viz., how far away from the barrier interface may the ferromagnetic layer be to observe spin polarization? They used SPT to measure the spin polarization in $Al/Al_2O_3/Au/Fe$ junctions as a function of Au interlayer thickness. Consistent with the results of Meservey and Tedrow, they found that the polarization was rapidly quenched for the first two monolayers Au, but a small polarization persisted over even a 10 nm Au. A recent follow-up to this experiment by Gabureac et al. [36] probed the spin polarization transmitted across an interfacial Au layer between Co and Al_2O_3, finding a somewhat longer length scale (~1.4 nm). However, as with later experiments involving thin interface layers in MTJs (see Section 10.5.1), we caution that without a detailed structural analysis of the interlayers, the absolute numbers should be regarded with caution, since effects, such as intermixing and three-dimensional growth, may play an important role [26].

10.3.3 What Is Tunneling Spin Polarization?

Originally, the positive (i.e., majority) tunneling spin polarization measured for Fe, Ni, and Co was a bit of a puzzle theoretically, since these materials have a DOS that is strongly minority-dominated near the Fermi level. One early explanation was given by Stearns [37]. She recognized that within the independent electron model of tunneling [15], the transmission probability depends on the electron effective mass, which is different for different bands. More generally, the transmission probabilities in Equation 10.1 are different for states of different symmetry, and depend on the nature of the evanescent states in the insulator [22,38–41]. In Fe, Co, and Ni, the relatively localized ("*d*-like") electrons carry the vast majority of the total magnetic moment, while the itinerant ("*s*-like") electrons contribute little. However, the localized electrons necessarily have a large effective mass and would be expected to decay extremely rapidly into a vacuum barrier, compared to the mobile hybridized electrons. Since the itinerant bands have a spin polarization opposite in sign, the positive tunneling spin polarization can be rationalized in the case of a vacuum barrier, in spite of a minority-dominated total DOS. Calculations based on a tight binding model through a vacuum barrier [42] essentially reproduce this simple picture (as do different treatments by Mazin [30] and Butler et al. [38]).

However, this picture neglects the electronic structure of the insulating barrier and bonding at the ferromagnet–insulator interface. Based on the discussions above, one expects the tunnel current to be sensitive to the *interfacial* DOS at the ferromagnet–insulator barrier, which is strongly modified by the presence of interface bonding. Tsymbal and Pettifor [21] first addressed the issue, finding that the tunnel conductance strongly depends on the type of bonding present. They showed that different interface bonding between Co or Fe and the insulating barrier could result in either positive or negative tunneling spin polarization. Indeed, experimentally, both positive and negative spin polarizations have been measured for Co, depending on the type of insulating barrier used (see Section 10.5).

For coherent systems, such as fully epitaxial tunnel junctions, $k_{||}$ conservation must be considered in the tunneling process [8], and the situation is even more complex. As *ab initio* calculations have demonstrated (Chapter 12), the idea of "tunneling spin polarization" then tends to lose meaning, with the transport characteristics becoming a rather complex function of *k*, barrier thickness, and the symmetry of the Bloch states involved. In realistic systems, ordered or otherwise, the tunneling spin polarization is not simply a property of the electrode alone or an electrode–barrier combination, but is (at least) a property of the electrode–barrier interface, and generally of the electronic structure of the entire M–I–M system. We refer the readers to Chapter 12 for more details.

10.4 Magnetic Tunnel Junctions

The dependence of tunneling current on the DOS, coupled with spin conservation during tunneling, makes it reasonable to anticipate that interesting spin-polarized tunneling effects may be observed without resorting to the use of a superconducting spin detector. In 1975, Jullière [3] and Slonczewski [2] had exactly this idea in mind. Rather than use the Zeeman-split DOS in a superconductor as a spin detector, one should be able to make use of the exchange-split DOS in a second ferromagnetic electrode. In this case, tunneling between two ferromagnets, the tunnel current is expected to depend on the relative magnetization orientation of the two ferromagnetic electrodes, giving rise to a magnetoresistance—the TMR effect.

In the section we introduce the basis for the TMR effect and its characteristic dependencies, for instance, the dependence of the TMR and conductance on applied bias, temperature, and materials choice. Though some of these features may be readily understood from the previous sections (for instance, the magnitude of the TMR and its angular dependence), others are more subtle and their origins have not been definitively determined (for instance, the bias and temperature dependence). Nevertheless, we will outline the main features of MTJs and the fundamental physics behind them as they are now understood.

10.4.1 Basis for the TMR Effect

Following Meservey and Tedrow's model of spin-polarized tunneling, Jullière [3] first proposed a model for tunneling between ferromagnets, which we sketch here. First, we consider two identical ferromagnetic electrodes with parallel magnetizations, separated by an insulating barrier. Assuming spin conservation during the tunneling process (Section 10.3.1), we may use a two-current model, and tunneling may only occur between bands of the same spin orientation in either electrode. That is, tunneling must proceed from a filled-up spin band in one electrode to an empty up spin band in the other, and vice versa. In this simplest case, we assume that the tunneling *probability* is itself spin-independent, and we worry only about the spin-dependent DOS in determining the tunneling current (in Chapter 12, we see how band symmetries may effectively provide a spin-dependent tunneling probability). We also presume a simple proportionality between the tunneling conductance and the product of the spin-dependent densities of states in the two electrodes. With these assumptions, one arrives at Jullière's formula for TMR, defined as the difference in *conductance* between parallel and antiparallel magnetization states, normalized by the antiparallel conductance:

$$\text{TMR}\,|_{V=0} \equiv \frac{G_p - G_{ap}}{G_{ap}} = \frac{R_{ap} - R_p}{R_p} = \frac{2P_l P_r}{1 - P_l P_r} \quad (10.5)$$

where

$P_{l(r)}$ is the tunneling spin polarization in the left (right) ferromagnetic electrode

$G_{p(ap)}$ is the tunneling conductance dI/dV

$R_{p(ap)}$ is the tunneling resistance in the parallel (antiparallel) state

The conductances are different for parallel and antiparallel magnetization orientations, and we expect MTJs to display a magnetoresistance, so long as there is a net spin polarization at the Fermi level in both electrodes.

Once again, we must keep in mind that even when the tunneling probability is assumed to be spin- and energy-independent (which ignores any symmetry-based spin filtering, see Chapter 12), P is not simply the difference in total density states at the Fermi level, but is still determined by, for example, the effective masses of different band electrons, interface bonding, etc. We also note parenthetically that in the tunneling literature "conductance" is defined as $G \equiv dI/dV$, which is *not* the same as $I/V = 1/R$ except at the vanishing bias, since tunnel junctions are nonlinear elements.

10.4.2 Early MTJ Experiments

Before we discuss the characteristics of MTJs in more detail, we will first review some of the early experiments. In 1975, Jullière [3] reported the first observations of TMR in Fe/Ge/Co tunnel junctions. By growing the Fe and Co layers with different coercive fields, he was able to realize both parallel and antiparallel magnetization orientations. At zero bias, the maximum observed effects were ~14%, but decreased rapidly with an increasing bias voltage. Though these results were indeed groundbreaking and stimulated much future research, they were not reproduced by other workers, and their true interpretation is still the subject of debate (e.g., the observed magnetoresistance effects were not correlated with the magnetic switching of the Fe and Co layers). Later spin-polarized tunneling experiments showed very little spin conservation when tunneling through amorphous Ge or Si barriers [1]. Whatever the interpretation, Slonczewski [2] quickly realized the potential of the TMR effect, and others soon followed this research.

Maekawa and Gäfvert [43] demonstrated TMR effects in 1982, using Ni/NiO/Ni and Ni/NiO/Co junctions. For Ni/NiO/Co junctions at 4.2 K, ≈2% magnetoresistance was observed, which was, for the first time, clearly correlated with the measured hysteresis behavior of the magnetic electrodes. The effects were still small compared to those anticipated, but this work showed that the magnetoresistance effects were due to the relative magnetization alignment between the two electrodes.

In 1995, nearly 20 years after the original "discovery" of the TMR effect, Moodera et al. [5] solved most of the earlier problems with MTJs, and demonstrated >10% magnetoresistance. The same year, Miyazaki and Tezuka [4] also showed TMR effects at room temperature. These demonstrations quickly garnered a great deal of attention, and catalyzed many groups to investigate MTJs. The work by Moodera et al. [5] already established many of the basic features associated with MTJs, most notably, the bias and temperature dependence of the TMR, which are discussed in Sections 10.4.7 and 10.4.8.

10.4.3 Resistance vs. Field

Jullière's simple model for the TMR effect was discussed previously in this section, with the result that the conductance (resistance) of an MTJ *with spin polarizations of the same sign for both electrodes* is higher (lower) when the magnetizations of the two ferromagnets are parallel. Once said, one must realize experimentally both a parallel and antiparallel magnetization alignment. Perhaps, the simplest way to realize this is to use two ferromagnets with different coercive fields, for example, by using two different ferromagnets, as shown in Figure 10.6a.

More relevant for sensor or MRAM technology are exchange-biased MTJs. In this case, one of the magnetic electrodes is in direct contact with an antiferromagnetic (AFM) material (e.g., FeMn, IrMn, NiO). The presence of an exchange anisotropy at the FM/AFM interface shifts the entire magnetization-field loop of the ferromagnet away from the zero field. Typical TMR

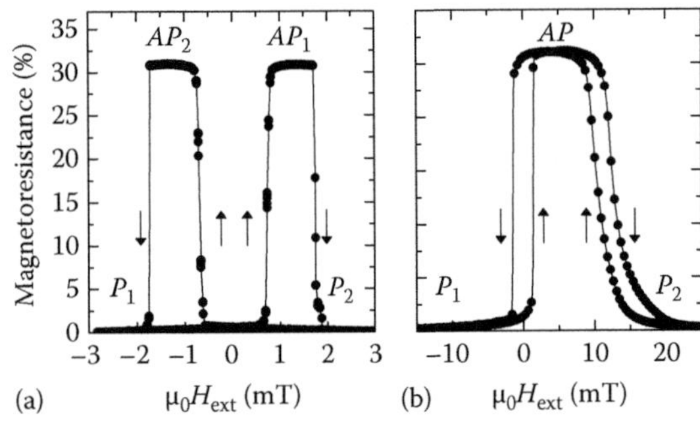

FIGURE 10.6 Magnetoresistance vs. magnetic field for (a) a hard-soft MTJ and (b) an exchange-biased MTJ, both at 10 K. Vertical arrows refer to sweep direction. Both curves are taken at $V = 0$. (Adapted from LeClair, P. et al., in Heinrich, B. and Bland, J.A.C., Eds., *Ultrathin Magnetic Structures*, Vol. III, Springer-Verlag, Berlin, Germany, 2005.)

vs. magnetic field behavior for an exchange biased system (at $V=0$) is shown in Figure 10.6. The exchange-biased MTJ is the most commonly studied configuration by far. Technologically, exchange biasing is advantageous because the resistance transition takes place near the zero magnetic field, and it generally results in greater magnetic stability [44], while from a fundamental point of view, it allows one to study MTJs with nominally identical electrodes. For more details, we refer to Chapters 2 and 4.

10.4.4 Ferromagnet Dependence

Using the measured tunneling spin polarization for Co with Al_2O_3 tunnel barriers (see Table 10.1), we may expect a TMR effect of more than 40% for Co/Al_2O_3/Co MTJs, only slightly above the observed (low temperature) value [45]. In general, the most recent spin polarization values with Al_2O_3 barriers [42,46] obtained via the SPT technique agree well with the maximum TMR values reported withAl_2O_3 barriers [46,47]. Clearly, one expects the largest TMR values for materials with the largest tunneling spin polarization. This explains a great deal of the recent interest in the so-called half-metallic ferromagnets, materials for which only one spin band is occupied at the Fermi level, resulting in a perfect 100% spin polarization [48]. Many compounds have been predicted to be half metallic; however, CrO_2 [49] and $La_{0.67}Sr_{0.33}MnO_3$ [50] are somewhat unique in that they have been shown to be essentially half-metallic by SPT and TMR experiments, respectively. While recently very large TMR effects have been recently observed in MTJs with Heusler alloy electrodes, it is unclear in most cases if the high TMR is due to a high polarization in the Heusler alloy or a spin filtering effect in the MgO tunnel barriers typically used (see Chapters 11 and 12).

Not long after the demonstrations of Moodera and Miyazaki, Lu et al. [51] and Viret et al. [52] observed TMR effects of more than 400% at low temperature utilizing $La_{0.7}Sr_{0.3}MnO_3$ (LSMO) with $SrTiO_3$ (STO), $PrBaCu_{2.8}Ga_{0.2}O_7$, or CeO_2 barriers. Using Equation 10.5, this implies a spin polarization of more than 80%, in agreement with SPT experiments [53]. More recently, Bowen et al. [50] have observed more than 1800% TMR in LSMO/STO/LSMO junctions, implying a spin polarization of 95% based on Jullière's model and essentially corroborating photoemission results showing LSMO to be half-metallic. While these TMR results, particularly those of [50], currently exceed anything observed in more traditional MTJs (including MgO-based junctions), the fact that the magnetic ordering temperatures of these oxidic ferromagnets is well below room temperature prevents practical applications, and perhaps as a result, there has been less work in this area compared to Al_2O_3- or MgO-based systems.

In more traditional systems, the search for high-spin polarization has often focused on analogies with bulk magnetization (perhaps for lack of a better rationale, owing to the complexity of P). In fact, the relationship between the total moment of the ferromagnetic electrode and the tunneling spin polarization has been a point of controversy since the advent of spin-polarized tunneling. Early experiments by Meservey and Tedrow [1] on various transition-metal alloys apparently showed a simple scaling relationship between the magnetic moment and polarization. More recent studies with improved deposition techniques and cleaner junctions (e.g., Refs. [46,54]) have found essentially no correlation. When the tunnel current is dominated by itinerant electrons, as for Al_2O_3 barriers, and when the localized electrons provide most of the magnetic moment, the relationships between P and, for example, magnetic moment are not straightforward. Measurements on the Ni–Fe [55] and Co–Fe systems showed a weak dependence of P on magnetic moment, implying no direct correlation. However, in the Co–Mn [56] system, polarization and moment were found to be directly related.

Kaiser et al. [57] studied Co–V and Co–Pt alloys, and suggested that the weaker composition dependence and higher P in Co–Pt alloys compared to Co–V alloys could be explained by different interface bonding between V and Pt to Al_2O_3, leading to different tunneling rates from different alloy sites. A higher tunneling rate from Pt sites, coupled with the tendency of Pt to polarize in a ferromagnetic matrix, leads to a weaker relative dependence of P with an increasing Pt content than the magnetization. On the other hand, Co–V alloys showed an even stronger decrease of P with an increasing V content than the magnetization, suggesting that the ease of formation of chemical bonds between alloy constituents and the barrier played a primary role in determining P.

In an interesting follow-up study, Kaiser et al. [58] studied ferrimagnetic $Co_{1-x}Gd_x$ alloys, and their work suggested that any relationship between the magnetic moment and polarization is purely coincidental. In the Co–Gd system, one has antiferromagnetically coupled sublattices of Co and Gd, the former giving rise to a much larger tunneling spin polarization, while the later provides a larger magnetic moment. At a critical composition x, the compensation point, the two sublattices give a zero net magnetic moment. However, Kaiser et al. showed that due to the far larger spin polarization of the Co sublattice, *a strongly spin-polarized current could be observed even with zero net magnetization*. Depending on the composition, both positive and negative polarizations could be observed. Using Gd/Co nanolayers, Min et al. [59] came to similar conclusions, finding that the sign of the tunneling spin polarization depended on the Gd and Co layer thicknesses, temperature, and applied voltage. After a somewhat colorful history, it seems that one is forced to retain the original conclusion that the magnetic moment and spin polarization have no obvious relationship, and further that the "spin polarization" is not a unique number even for a given metal electrode and tunnel barrier, but a property of the electronic structure of the entire system under study.

10.4.5 Crystallographic Orientation Dependence

Still, given the expected dependence of TMR on the interfacial DOS, one would anticipate a dependence on the crystallographic orientation of the electrodes. However, MTJs with

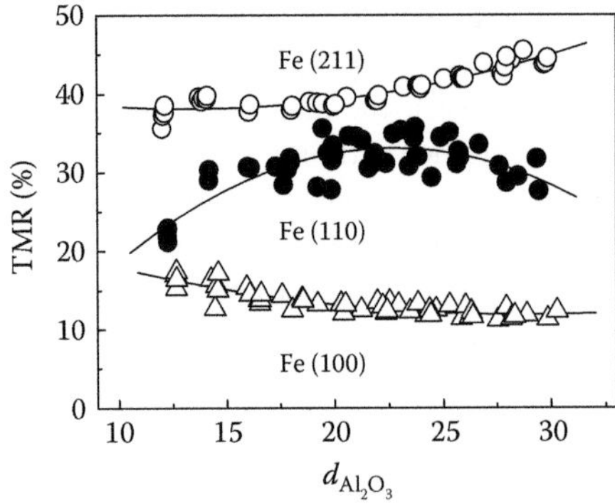

FIGURE 10.7 TMR at 2 K as a function of Al_2O_3 thickness for Fe(211), Fe(110), and Fe(100) epitaxial electrodes in Fe/Al_2O_3/CoFe junctions. Lines are a guide to the eye. (After Yuasa, S. et al., *Europhys. Lett.*, 52(3), 344, 2000.)

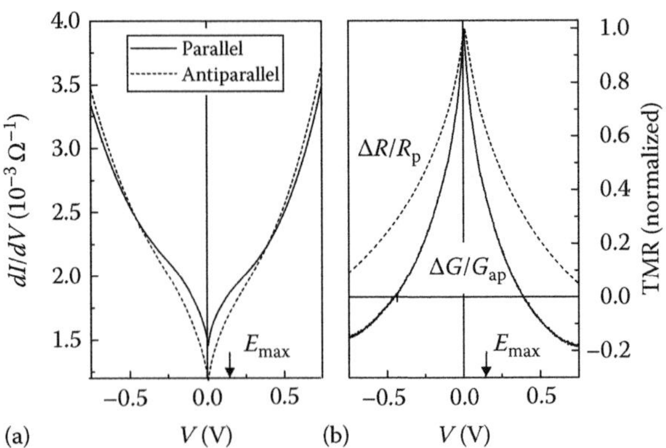

FIGURE 10.8 (a) Conductance vs. applied bias, dI/dV (V), at 5 K for parallel (*solid*) and antiparallel (*dashed*) magnetizations of a Co/Al_2O_3/Co MTJ. Vertical arrow indicates the maximum magnon energy (E_{max}) for bulk Co in a mean field approximation. (b) Bias dependence of the TMR ($\Delta R/R_p$, *dashed*), and differential TMR ($\Delta G/G_{ap}$, *solid*) for a Co/Al_2O_3/Co junction at 5 K. Note that the normal ($\Delta R/R_p$) and "differential" ($\Delta G/G_{ap}$) magnetoresistances are only equivalent at *zero bias*. (From LeClair, P. et al., *Phys. Rev. Lett.*, 88, 107201, 2002 and unpublished data from the authors.)

even a single epitaxial layer with different orientations were not prepared until 2000. Yuasa et al. [60] prepared Fe(100,110,211)/Al_2O_3/CoFe MTJs with a bottom epitaxial Fe layer to study the effect of the Fermi surface anisotropy on TMR. They observed a strong dependence of the TMR on crystallographic orientation, as shown in Figure 10.7. The largest TMR effect was observed for Fe(211) electrodes, while Fe(100) electrodes yielded a TMR effect nearly a factor of 4 lower, while Fe(110) electrodes showed an intermediate TMR value. The fact that the TMR varies so strongly with crystallographic orientation clearly points to the details of the Fe band structure [61] and momentum filtering, which perhaps foreshadows the "giant" TMR effects [62,63] first observed in Fe–MgO–Fe MTJs (see Chapters 11 and 12).

More recent results by Kim and Moodera [54] compared spin-polarized tunneling with polycrystalline and epitaxial Ni films. For Ni(111), a spin polarization of 25% was measured, despite the fact that bulk band structure calculations for Ni predict a low or even negative polarization along the (111) direction [64]. In addition, a large increase of the polarization compared to previous studies was observed for polycrystalline Ni films (P = +46%) when the Ni surface/interface was contamination free. The dramatically large value of P observed for polycrystalline Ni is in the same range as that for Fe and Co, whereas the magnetic moment is only 0.60 μB for Ni, compared to 2.2 μB for Fe. This again shows the lack of a direct correlation between P or TMR and the magnetic moment, and that the factors controlling P (e.g., the *sp*-projected DOS and the nature of the evanescent states in the barrier region [38]) are far subtler than those controlling magnetization.

10.4.6 Conductance vs. Voltage

As we have seen previously in Section 10.1, tunneling between two nonmagnetic free-electron metals should result in a conductance (dI/dV) that is quadratic in an applied bias (V). Indeed, for nonmagnetic junctions, such as Al/Al_2O_3/Al, this behavior is often observed [6]. However, in MTJs (or even junctions with one magnetic electrode), the conductance behavior deviates significantly from the expected parabolic dependence, as shown in Figure 10.8a for a Co/Al_2O_3/Co junction (with two polycrystalline electrodes), most noticeably at low voltages. Though the conductance is approximately symmetric with respect to voltage, as expected for (nominally) identical electrodes, there is a pronounced, sharp decrease below ~150 mV (the exact value is material-dependent), where the conductance is approximately linear in voltage. This sharp dip in conductance, about zero bias, is similar to so-called zero-bias anomalies observed by many groups [6], which were attributed to (magnetic) impurities in the tunnel barrier or near the electrode-barrier interfaces. However, in contrast to those zero-bias anomalies, the conductance dip has no strong temperature dependence or magnetic field dependence, persisting even in the cleanest junctions. Further, the energy scale of the anomaly is far greater than usually observed. Other mechanisms, such as the presence of metal particles in the barrier (Giaever–Zeller anomalies [6]), are similarly inconsistent. Further, the characteristic (nearly) linear conductance contribution near the zero bias occurs *only* when at least one electrode is magnetic [47,65,66]. A linear conductance contribution suggests a continuous spectrum of inelastic excitations [6], with an upper-energy cutoff of ~150 meV and a lower-energy cutoff below 1 meV, likely in the μeV range.

Moodera and Kinder [67] suggested that magnon excitations may be at least partly responsible for the conductance anomalies as well as the TMR bias dependence. A theoretical explanation based on this idea was given by Zhang et al. [68] and Bratkovsky [69] in terms of magnon excitations localized at the FM–Al_2O_3 interface. Electrons at the Fermi level of one ferromagnet tunnel across the junction reaching the second ferromagnet with an energy eV above the Fermi level in the second electrode (assuming no other inelastic tunneling processes). These "hot" electrons may then lose energy by emitting a magnon of energy $\hbar\omega \leq eV$. A similar process holds for magnon absorption, and as one might expect this leads to a decrease of the TMR with an increasing bias. Zhang et al. [68] and Bratkovsky [69] found that these processes influence the conductance characteristics, giving an additional inelastic conductance contribution $G_{magnon} \propto V$. This simply reflects the fact that a larger bias allows more magnons to be excited. The slope of the linear contribution is larger in the antiparallel case, with the difference between parallel and antiparallel slopes dictated by the spin polarization. The linearity is preserved up to a maximum voltage determined by the maximum magnon energy in the ferromagnetic electrode, which, within a mean-field approximation, corresponds to $E_{max} = 3k_B T_C/(S+1)$, where k_B is Boltzman's constant, and T_C is the Curie temperature of the ferromagnet with spin S. Considering two identical Co electrodes with the bulk T_C, $E_{max} \sim 144$ meV. Given that the interface T_C is expected to be significantly lower, this is in fairly good agreement with width of the conductance dip shown in Figure 10.8a. On the low energy scale, a cutoff arises due to finite size effects (imposing a maximum magnon wavelength) or anisotropy, both in the μeV range and far below the thermal energy at the temperature of typical experiments. For this reason, the low-energy cutoff plays a more direct role in the temperature dependence of TMR as opposed to the bias dependence.

For a detailed application of this model, we refer to Han et al. [47,65] The primary conclusions of that study were that by *combining* different MTJ characteristics (such as $dI/dV(V, T)$, $\Delta R/R_p(V, T)$), a consistent set of all model parameters could be obtained, allowing a much more stringent test of the magnon excitation model. An excellent agreement between experimental data and model calculations were demonstrated, suggesting that indeed the additional conductance contribution consistently observed in MTJs (the zero-bias "dip") can be largely, if not completely, explained by interfacial magnons excited by hot tunneling electrons.

One further question with regard to conductance–voltage behavior that frequently arises is the role of the DOS in the tunnel conductance. Given the form of Equation 10.1, which explicitly contains the DOS as a function of energy in each electrode, it is rather surprising that clear observations of DOS and band structure effects have *not* been reported until recently [70–72]. One probable reason for this is that in Al_2O_3-based junctions, one is preferentially sensitive to the itinerant DOS, which by its dispersive nature will have little energy dependence [21,73]. By comparing MTJs with fcc(111)-textured and polycrystalline Co electrodes, LeClair et al. [74] have observed features in the conductance and TMR bias dependence that were attributed to the *s*-projected DOS of fcc(111)-textured Co electrodes, essentially corroborated by theoretical results.

10.4.7 TMR vs. Voltage

Perhaps, the most surprising feature of MTJs, initially, was the dependence of the TMR effect on applied dc bias. First observed by Jullière [3], and confirmed by Moodera et al. [5], it was at initially unclear if this was an intrinsic effect, or simply due to inelastic tunneling through a nonideal interface and barrier (e.g., impurity-assisted tunneling). However, subsequent measurements on clean junctions, as well as the observation of this effect by many other groups in the years that followed, produced the conclusion that the bias dependence of the TMR is an *intrinsic* effect [66], although its magnitude may vary considerably. A customary figure-of-merit is the voltage at which the TMR ($\Delta R/R_p$) is reduced by a factor of 2, usually denoted $V_{1/2}$. Moodera et al., in their initial observation [5], found a "half voltage" of ~200 mV, while recently several groups have improved this figure to >500 mV. Moodera et al. [66] also showed that the degree of bias dependence is relatively temperature independent.

Figure 10.8b shows the TMR ($\Delta R/R_p$) vs. bias behavior for a $Co/Al_2O_3/Co$ junction at 5 K [29], showing both $\Delta R/R_p$ and $\Delta G/G_{ap}$. Note that these two definitions of the TMR are *not* identical at finite bias, since the tunnel current is a nonlinear function of voltage. To avoid confusion, we will refer to $\Delta R/R_p$ as simply TMR, following the majority of workers, and to $\Delta G/G_{ap}$ as the *differential* TMR. The differential TMR is typically observed to be approximately linear in bias up to ~0.5 V [75], becoming negative at higher biases and then tending to zero. The normal TMR shows a roughly parabolic voltage dependence at intermediate biases (~0.5 V), and tends smoothly to zero at higher biases, with a half-voltage of typically ~0.3–0.5 V. One insight into its possible origin is the fact that the differential TMR has exactly the same voltage dependence as the magnon-assisted conductance contribution [68,69] discussed in the previous section.

The aforementioned models of Zhang et al. [68] and Bratkovsky [69] were originally proposed to explain the bias dependence of the TMR. We have seen that the magnon-assisted conductance contribution is linear in voltage, but with a differing slope for parallel and antiparallel magnetization orientations. Thus, the conductance change ΔG from parallel to antiparallel orientations will also be approximately linear in voltage up to $eV = E_{max}$, roughly as observed in $\Delta G/G_{ap}$ (after further analysis to remove other conductance contributions). The fact that both the conductance–voltage and magnetoresistance–voltage agree favorably with model calculations, and further, yield realistic parameters [47,65], emphasizes that magnon excitations may play a dominant role in the bias dependence. However, several other possible origins have also been proposed and must be considered.

Davis and MacLaren [76] have proposed that the bias dependence also has an *intrinsic* component resulting from the underlying electronic structure. They considered free electron tunneling through a square potential barrier, using simple parabolic bands modeled on the itinerant electron bands in Fe determined from *ab initio* calculations. Two primary electronic structure effects were considered, viz., shifting of the Fermi level of one electrode relative to the other, and the altered barrier shape at finite bias. The shifting of the chemical potential allows new states to be accessed for tunneling, which essentially makes the spin polarization a voltage-dependent quantity, whereas the altered barrier shape allows higher-energy states to tunnel more easily. This makes the matrix elements in Equation 10.1 *spin-* and *bias-dependent*. A strong decrease of the TMR with bias is also found within their model, and also gives good agreement with experimental data with realistic parameter choices. Further, early work by Moodera et al. [5,66] as well as recent experiments utilizing composite insulating barriers [70–72] and ultrathin nonmagnetic layers at the FM/Al_2O_3 interface [24] have pointed out the importance of a bias-dependent spin polarization, as will be discussed in Sections 10.4.9 and 10.4.10. In particular, at large biases ($eV > E_{max}$) one must consider alternative mechanisms to a purely magnon-assisted bias dependence. As one example, the effects of phonon [69] and impurity [77,78] assisted tunneling on the bias dependence of the TMR have also been considered theoretically. Experimental support exists for both magnon-assisted and intrinsic band-structure mechanisms of bias dependence as well as for an impurity-assisted contribution [79,80]. The currently accepted viewpoint is that all of these mechanisms play a key role to some degree.

10.4.8 TMR vs. Temperature

It was first noticed by Shang et al. [81] that the temperature dependence of the tunnel resistance for MTJs greatly exceeds that for nonmagnetic junctions with nominally identical barriers. Typically, Al/Al_2O_3/Al junctions showed only a 5%–10% [81] change in resistance from 4.2 to 300 K, while MTJs always exhibited a 15%–25% change in resistance, as shown in Figure 10.9 for a Co/Al_2O_3/Co junction. Further, the TMR ($\Delta R/R_p$) can change as much as 25% or more from 4.2 to 300 K depending on the magnetic electrodes (also shown in Figure 10.9). Shang et al. explained these results within a simple phenomenological

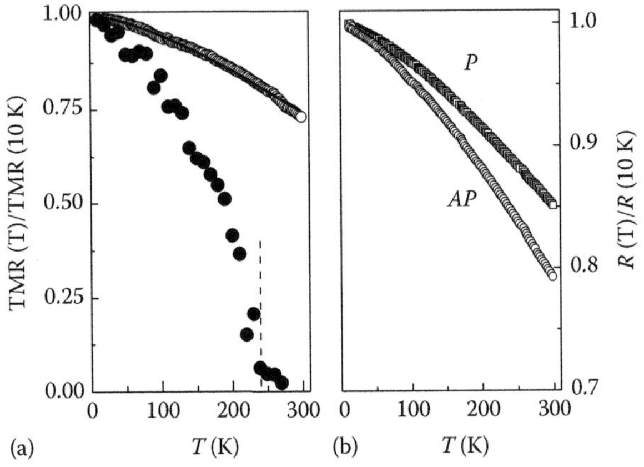

FIGURE 10.9 (a) Temperature dependence of the TMR for a Co/Al_2O_3/Co MTJ (open), along with a fit to the model of Shang et al. [81] (*line*) and a $Cu_{38}Ni_{62}$/Al_2O_3/Co junction (closed). Dashed vertical line shows the Curie temperature of $Cu_{38}Ni_{62}$ ($T_C \approx 240$ K, note $T_C = 1388$ K for fcc-Co [104]). (b) Temperature dependence of the normalized tunnel resistance for parallel (*P, squares*) and antiparallel (*AP, circles*) magnetization orientations, R_p, for a Co/Al_2O_3/Co MTJ. All curves taken at $V \approx 5$ mV. (Adapted from Davis, A.H. et al., *J. Appl. Phys.*, 89, 7567, 2001; Hindmarch, A.T. et al., *Phys. Rev. B*, 72(10), 100401, 2005.)

model, in which it was assumed that the tunneling spin polarization P decreases with increasing temperature due to spin-wave excitations, as does the surface magnetization. They thus assumed that both the tunneling spin polarization and the interface magnetization followed the same temperature dependence, the well-known Bloch $T^{3/2}$ law, that is, $M(T) = M(0)(1 - \alpha T^{3/2})$ for low T/T_C. This temperature dependence holds for surfaces as well as the bulk, though the former has a larger decay constant α [82]. Given the interfacial sensitivity of tunneling, Shang et al. [81] assumed that α was also larger than the bulk value in MTJs, and thus gave a satisfactory explanation for the temperature dependence of the TMR, as demonstrated by the fit in Figure 10.9.

MacDonald et al. [83] provided a more rigorous theoretical justification of these ideas, essentially reproducing the proportionality between $M(T)$ and $P(T)$, though the microscopic origin was slightly different than that considered by Shang et al. The aforementioned model of Zhang et al. [68] also predicts a temperature dependence of TMR, and in contrast to their explanation of the bias dependence of TMR, it is no longer the maximum magnon energy, E_{max}, which sets the relevant energy scale, but rather the lower wavelength cutoff in the spectrum, E_C. This cutoff results physically either from anisotropy, always present for the spins near the FM–I interface, or from a finite coherence length, due to, for example, a finite grain size. In this model, the zero-bias conductance scales with temperature roughly as $G(T) \approx T\ln(k_B T/E_c)$.

Davis et al. [84] also attempted to put the model of Shang et al. on firmer theoretical ground, somewhat in the spirit of the model of Zhang et al. [68], by assuming that the temperature dependence of the polarization arises from a Stoner-like collapse of the exchange splitting. As with their model of the bias dependence [76], they used free electron-like bands modeled after the *ab initio* band structure of Co. The model itinerant bands were exchange-split parabolic bands, with a spin-dependent effective mass. Recent photoemission data [85] indicate that these itinerant bands are Stoner-like, that is, the exchange splitting may depend on temperature, and collapses at T_C. Further, citing the fact that the exchange splitting is nearly proportional to $M(T)$, they used the known $M(T)$ behavior of Co to obtain the temperature-dependent exchange splitting: $\Delta E_{ex} = \beta M(T)$. Using the bulk $M(T)$ behavior of Co, Davis et al. were able to explain only about a third of the experimental drop in TMR from 0 to 300 K. This is not surprising, given that the interface T_C is expected to be significantly lower [82]. Strikingly, however, they calculated an 18% change in TMR despite the fact that the magnetization changes by only 1.5% over the same temperature range, once again indicating that in general P and M are not simply related, though their temperature dependence may share a common origin. The TMR temperature dependence over this range could be well described by assuming a lower interface ordering temperature for Co, viz., $T_C \approx 982$ K, though this is perhaps unphysically low. As with their work on the TMR bias dependence, the main message is that purely *intrinsic* band effects are able to explain much of the TMR temperature dependence, without resorting to inelastic processes. Of course, in realistic devices, the role of impurity-assisted tunneling must also be considered [79,80]. The current opinion is generally that both intrinsic (e.g., band effects) and extrinsic (inelastic processes) contributions are of importance.

More recently, Hindmarch et al. [86,87] performed a definitive study of the temperature dependence of TMR all the way through the Curie point of a $Cu_{38}Ni_{62}$ alloy, Figure 10.9a. They interpreted their results as being consistent with a Stoner-like collapse of the exchange splitting of the bands responsible for tunneling, lending support to the model of Davis et al. [84]. Further, the normalized TMR vs. temperature was found to be independent of applied bias, collapsing onto a "universal" curve. Put another way, the applied bias seems to affect only the *relative magnitude* of the TMR and spin polarization, not the functional form of the temperature dependence. While spin polarization and magnetization vs. temperature were clearly *related* in this study, a simple linear relationship was lacking. Their results further suggest that magnetization and polarization are not simply related in magnitude. Further, their results suggest that while bias- and temperature dependence may share a common origin in magnetic excitations, for all intents and purposes they may be treated independently of another. This does not seem inconsistent with theoretical arguments: the excitation of magnons by finite temperature or hot electrons is independent of one another, and from any bias dependence of the polarization. More bluntly, that these quantities have the same temperature dependence is merely a reflection of the fact that the same physical process causes both M and P to decrease with temperature, viz., magnetic disorder due to the thermal excitation of spin waves.

10.4.9 Barrier Dependence

The bulk of the early work on MTJs focused almost exclusively on Al_2O_3 tunnel barriers, for a variety of reasons. Perhaps, the most important factors are: the ease in fabricating ultrathin; pinhole-free Al_2O_3 layers; demonstrated spin conservation across Al_2O_3 barriers [1]; the first successful demonstrations of the TMR effect used Al_2O_3 barriers [4,5]. Significant effort has been made to characterize, understand, and improve the properties of alumina barriers (see, e.g., Refs. [48,49,62,82,88,89]). On the other hand, there is little doubt that a myriad of alternative barriers have indeed been explored by many groups—the lack of evidence for this in literature may simply be due to the fact that successful MTJ tunnel barriers are a rare find, and one tends not to report unsuccessful results! Nevertheless, in the last several years, several alternative barriers have been successfully employed, some with very distinct behavior compared to Al_2O_3.

As discussed previously, the magnitude, and even the sign of the tunneling spin polarization are expected to depend on not only the ferromagnetic electrode, but the barrier and nature of the electrode–barrier interface as well. In 1999, two different groups [70–72] independently observed this prediction [21]. Later theories confirmed this prediction [40,73,90,91] and demonstrated the possibility of creating novel MTJ properties by not only varying the magnetic electrode, but by varying the insulating barrier as well.

In a striking series of experiments, De Teresa et al. [71,72] also observed that the tunneling spin polarization depends explicitly on the insulating barrier used. In this case, half metallic LSMO was used as one of the electrodes with barriers of Al_2O_3, STO, $Ce_{0.69}La_{0.31}O_{1.845}$ (CLO), or a composite Al_2O_3/STO barrier. Since it is known that LSMO has only majority states at E_F, the tunneling spin polarization must be positive and close to 100% [53,89], regardless of the insulating barrier used. Indeed, SPT experiments have measured a tunneling spin polarization of $P = +78\%$, and TMR effects in LSMO-based devices have reached 1800% as noted above. Thus, an LSMO electrode can serve as a spin analyzer for the polarization of a second electrode, in the spirit of the SPT technique of Meservey and Tedrow (see Section 10.1). Moreover, a half metallic analyzer has the added advantage that the energy dependence of the polarization may be studied to some extent.

As expected, De Teresa et al. found that Co/Al_2O_3/LSMO MTJs gave a positive TMR for all biases, not surprising since both LSMO and the Co/Al_2O_3 interface are known to have positive polarizations. Surprisingly, Co/STO/LSMO junctions showed *negative* TMR values at zero bias, and further displayed a strong biasdependence as shown in Figure 10.10b. In this case, the polarization of the Co/STO interface must be *negative*, opposite that of Co/Al_2O_3 interfaces, in agreement with the d electron polarization of Co at E_F. In order to show this more conclusively, De Teresa et al. investigated Co/Al_2O_3/STO/LSMO junctions, with the expectation that since the LSMO and Co/Al_2O_3 tunneling spin polarizations are both positive, a normal positive TMR would result for all biases. As shown in the inset to Figure 10.10b, a normal positive TMR is observed for all biases, with a bias dependence that is essentially identical to standard Co/Al_2O_3/Co junctions.

De Teresa argued that the sign change of the Co tunneling spin polarization can be understood in terms of interface bonding, as discussed above (see also Chapter 12). For Co/Al_2O_3 interfaces, the nature of the bonding at the Co/Al_2O_3 interface allows the itinerant electrons in Co, which are *positively* polarized, to tunnel preferentially. For Co/STO interfaces, however, the interface bonding is sufficiently different to allow preferential tunneling of the more localized states, which are *negatively* polarized [21]. Indeed, this conjecture is borne out by recent calculations [88,90,91] for Co/STO/Co MTJs. Further experimental proof is evidenced in the bias dependence of Co/STO/LSMO junctions, shown in Figure 10.10b. At zero bias, the Co DOS is dominated by relatively localized minority states, as shown in Figure 10.10a (for the Co(100) surface), and exhibits a large negative tunneling spin polarization. For negative bias (electrons tunneling *from* the LSMO), the LSMO detector predominantly scans the DOS (convoluted with the energy-dependent tunneling probability) above the Fermi level in the Co electrode. As the bias is decreased to −0.4 V, in the antiparallel state the Fermi level of LSMO is at the same energy as the peak of the Co minority

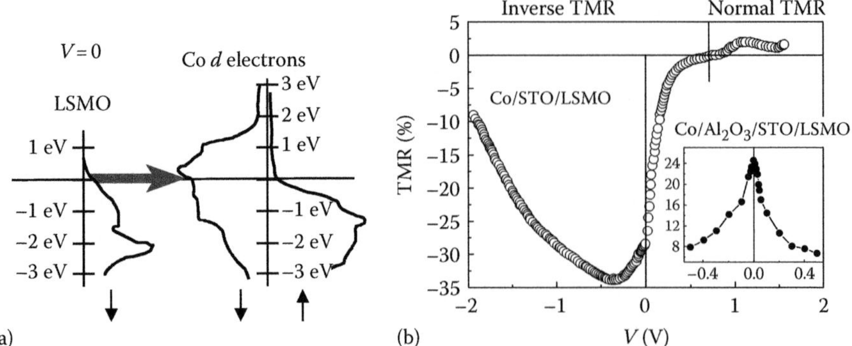

FIGURE 10.10 (a) Schematic of the spin-polarized densities of states of LSMO (as derived from photoemission) and the Co(100) surface (calculated). (b) TMR ratio versus applied bias for a Co/STO/LSMO junction at $T = 5$ K. Inverse TMR is observed for $V < 0.8$ V, while normal TMR is observed for $V > 0.8$ V, indicating that the Co/STO spin polarization is negative for $V < 0.8$ V. Inset: TMR ratio versus applied bias for a Co/Al_2O_3/STO/LSMO junction. In this case, the polarization of Co/Al_2O_3 and LSMO are both positive, and a normal (positive) TMR is seen. (Adapted from De Teresa, J.M. et al., *Phys. Rev. Lett.*, 82, 4288, 1999; De Teresa, J.M. et al., *Science*, 286, 507, 1999; Tsymbal, E.Y. et al., *J. Phys. Cond. Matt.*, 15(4), R109, 2003.)

DOS, giving a larger tunneling spin polarization and, hence, a large (negative) TMR. As the bias becomes more negative, the minority DOS progressively decreases, and the tunneling spin polarization decreases.

For positive bias (electrons tunneling *from* Co), the LSMO detector predominantly scans the DOS *below* the Fermi level in the Co electrode. As the bias is increased, the majority DOS steadily increases, until at ~0.8 V the majority and minority densities of states are equal, resulting in a zero crossing of the TMR. For a still higher bias, the majority DOS of Co continues to increase (while the minority DOS steadily decreases), resulting in a positive tunneling spin polarization and TMR. At ~1.15 V, the Fermi level of the LSMO is at the same energy as the *majority* DOS peak in Co, and the positive TMR is maximal. Thus, all the features in the bias dependence of Co/STO/LSMO junctions can be attributed to the structure of the spin-polarized Co DOS, dominated by more localized ("d-like") states. This is in contrast with experiments on Co/Al_2O_3-based tunnel junctions, which demonstrate positive polarization and structure related to the Co itinerant ("s-like") DOS [74].

Subsequent results by Sugiyama et al. [92] and Sun et al. [93] essentially corroborated these results. Experiments by Sharma et al. [70] utilizing Ta_2O_5/Al_2O_3 composite barriers have indicated that the sign of the spin polarization at Ta_2O_5 interfaces is also negative, and have proposed an explanation similar to that of De Teresa et al. However, these results have not yet been independently verified. Still, these results illustrate the rich physics behind spin-polarized tunneling in MTJs as well as the intriguing possibility of "engineering" MTJs with tailored properties, which leads us naturally to our next example.

10.4.10 Interface Dependence

As pointed out numerous times above, the tunneling current in MTJs is effectively dominated by the electronic structure in the vicinity of the ferromagnet–insulator interface. One powerful way to explore this interface sensitivity is by the insertion of ultrathin layers (often called "dusting" layers) at the electrode–barrier interfaces. As discussed in Section 10.3.2, Tedrow and Meservey [34] already used this technique in 1975 (Figure 10.5) to probe the onset of spin polarization in ultrathin ferromagnets. In one view, their experiments established that not only can a few monolayers of Fe, Co, and Ni be ferromagnetic and fully polarized, but that these few monolayers were sufficient to dominate the tunneling current. In retrospect, this was also a direct proof of the interface sensitivity of tunneling.

In 2000, LeClair et al. [26] were among early workers to extend these studies to MTJs, investigating the insertion of ultrathin Cu layers in Co/Cu/Al_2O_2/Co structures. They were able to demonstrate that much of the early controversy in MTJ studies was morphology-related. In their initial work, they grew Cu interlayers both above and below the Al_2O_3 barrier, which resulted in two different TMR decay lengths, as shown in Figure 10.11a. The longer length scale in the former case simply results from the island-like growth of Cu on Al_2O_3, and hints at the technical challenges that arise in MTJ fabrication. When Cu was grown on Co, atomic-scale smoothness resulted, and a much shorter length scale. In this latter system, LeClair et al. found that the TMR decay was well-described by an exponential decay, with a length scale of $\xi \approx 0.26$ nm, equivalent to just more than one monolayer Cu. Appelbaum and Brinkman [17] first pointed out that tunneling in(disordered) normal-metal junctions should be sensitive to the DOS within a few Fermi wavelengths of the electrode–barrier interface. In this case, the length scale for TMR decay corresponds to about $3.5/k_F$, which indeed suggests that the Fermi wavelength sets the relevant length scale in disordered systems. Perhaps more important than the quantitative results of their experiments, however, was the conclusion that a detailed characterization of the interlayers must be performed in order to elucidate intrinsic interface effects from morphological effects.

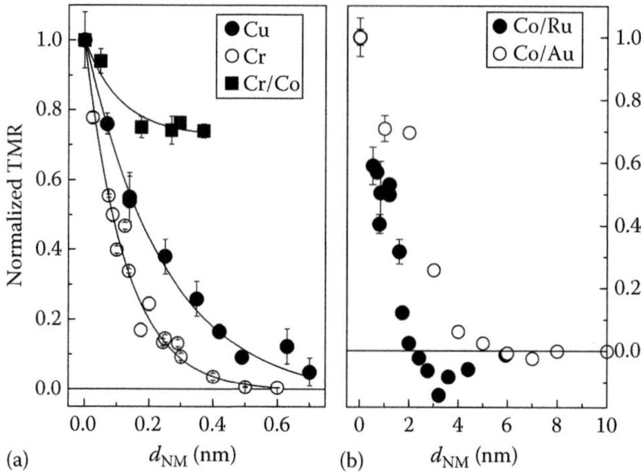

FIGURE 10.11 (a) Normalized TMR at 10 K as a function of Cu thickness for Co/Cu(d_{Cu})/Al_2O_3/Co, Co/Al_2O_3/Cu(d_{Cu})/Co, and Co/Cr(d_{Cr})/Co(d_{Co})/Al_2O_3/Co MTJs. Adding only a few monolayers of Co on Cr almost completely restores the TMR, demonstrating the interfacial sensitivity of MTJs. Lines are fits to an exponential decay. (b) Normalized TMR at 10 K as a function of Cr thickness for Co/Ru(d_{Ru})/Al_2O_3/Co and Co/Cr(d_{Au})/Al_2O_3/$Ni_{80}Fe_{20}$ tunnel junctions. (Adapted from LeClair, P., *Phys. Rev. B*, 64, 100406, 2001; LeClair, P. et al., *Phys. Rev. Lett.*, 86(6), 1066, 2001; LeClair, P. et al., *Phys. Rev. Lett.*, 84(13), 2933, 2000; Moodera, J.S. et al., *Phys. Rev. Lett.*, 83(15), 3029, 1999.)

A further demonstration of interface sensitivity was subsequently obtained by LeClair et al. [25] using ultrathin Cr layers in Co/Cr/Al_2O_3/Co MTJs. The TMR decay was again approximately exponential, and in this case was even more rapid (see Figure 10.11b) with $\xi \sim 0.1$ nm, or only about 0.5 monolayers of Cr. In these experiments, they also added an additional Co layer on top of the Cr dusting layer, that is, Co/Cr(d_{Cr})/Co/Al_2O_3/Co, shown in Figure 10.11b. Strikingly, the TMR was almost completely restored with only these 3–5 monolayers of Co. This further confirms that only the few monolayers of the electrode adjacent to the ferromagnet–insulator interface dominate MTJ properties, in very good agreement with the earlier SPT work on ultrathin magnetic layers. Subsequent work by Moodera et al. [94] and Samant and Parkin [95] come to essentially the same conclusion, establishing that at least in disordered (i.e., nonepitaxial) systems, the relevant length scale appears to be several monolayers at the most. In an elegant experiment discussed briefly above, Gabureac et al. [36] studied the effect of Au interlayers in MTJs and also probed the spin polarization directly by SPT in the spirit of Moodera et al. [35] replacing one of the ferromagnetic layers with an Al spin detector. In this direct comparison between TMR and SPT, they found very good agreement between the decay lengths of the TMR and spin polarization, though their length scales (~1–1.5 nm) were somewhat longer than in the previous work. They further inserted a 0.1 nm Au layer at a variable distance from the ferromagnet–interface, and consistent with LeClair et al. [25] found that the TMR and spin polarization recover their undoped values with length scales of ~0.6–0.7 nm.

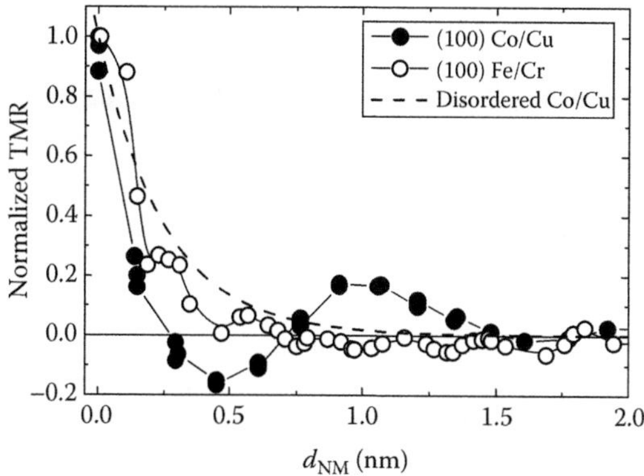

FIGURE 10.12 (closed) TMR at $T = 2$ K as a function of Cu interlayer thickness for Co(001)/Cu(001)/Al_2O_3/$Ni_{80}Fe_{20}$ junctions. The period of the oscillation observed, 1.14 nm, is in agreement with the Fermi surface of Cu. For comparison, the dashed line shows exponential decay of TMR fitted to the data of [26]. (closed) TMR at $T = 50$ K as a function of Cr interlayer thickness for Fe(100)/Cr(100)/Al_2O_3/CoFe junctions. A small oscillation of the TMR with 2 ML period is observed due to the layered AFM structure of Cr. (Adapted from Matsumoto, R. et al., *Phys. Rev. B*, 79(17), 174436, 2009; Nagahama, T. et al., *Phys. Rev. Lett.*, 95(8), 086602, 2005; Yuasa, S. et al., *Science*, 297(5579), 234, 2002.)

The initial failure to observe quantum-well states, as anticipated, has by some been deemed a sign that recent theories are not yet sophisticated enough to deal with MTJs. Although no quantum-well states were observed in the studies above, in retrospect one would not to expect to observe them except in a nearly perfect epitaxial system, as is the case for quantum-well states in metallic multilayers (see Section 10.2). Although there were observations of a sign inversion of the TMR as a function of Ru thickness in Co/Ru/Al_2O_3/Co junctions [24,96] and Au thickness in Co/Au/Al_2O_3/$Ni_{80}Fe_{20}$ junctions [97], the interpretation of these results as quantum-well states is unlikely, since none of these studies used single-crystalline electrodes in which quantum-well oscillations may reasonably be expected to occur [26,98]. This is particularly true in the case of Ru interlayers, where one explanation for the observed negative TMR is an interfacial CoRu alloy [24].

However, in 2002, Yuasa et al. [98] reported unambiguous quantum-well oscillations of the TMR using single-crystalline Cu interlayers in Co(001)/Cu(001)/Al_2O_3/$Ni_{80}Fe_{20}$ junctions. Figure 10.12 shows the TMR at 2 K as a function of the inserted Cu layer thickness. A clear (damped) oscillation of the TMR is evident, and the zero-bias period of 1.14 nm is in good agreement with the Fermi surface of Cu, as is the bias dependence of the oscillation period. Crucially, independent measurements on similarly grown Co/Cu/Co trilayers gave an oscillation of the interlayer exchange coupling with a period of 1.1 nm. This is a key distinction from earlier work with polycrystalline interface layers [26], where the polycrystalline nature of the interlayers (or interface mixing) would seem to preclude the observation of quantum-well states. The clear correlation between TMR, interlayer coupling, and the Fermi surfaces of Co and Cu suggests that the oscillation indeed arises

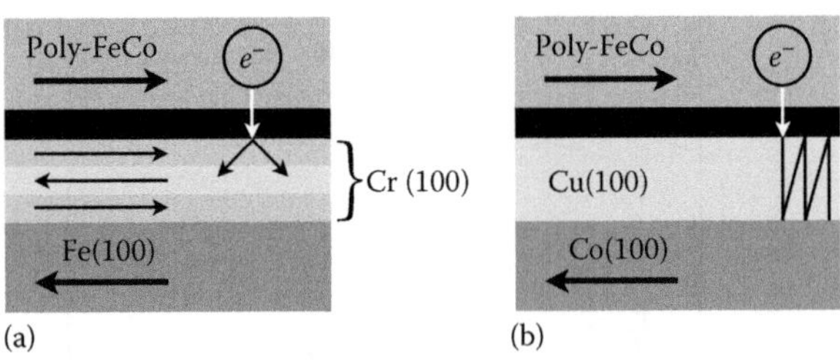

FIGURE 10.13 Dusting layers in epitaxial systems. (a) The magnetic moments in Cr(100) are magnetically aligned within each layer, with each layer aligned opposing adjacent layers. If the TMR ratio is determined by the metal monolayer closest the insulator interface, the TMR should oscillate with a period of 2 ML. If the TMR is determined by bulk electronic structure, there should be no TMR, since there is no net polarization in the Cr(100) layer. (b) If there is suitable matching of the dispersive bands throughout the structure, one spin channel will have relatively easy transmission, while the other will be preferentially reflected at the ferromagnet-dusting layer interface. If multiple scatterings occur within the dusting layer, the down-spin electrons form resonant states (QW states), giving rise to oscillations in the TMR as a function of bias and interlayer thickness. (From Nagahama, T. et al., *Phys. Rev. Lett.*, 95(8), 086602, 2005; Yuasa, S. et al., *Science*, 297(5579), 234, 2002.)

from the spin-dependent reflection at the Co/Cu interface due to the formation of spin-polarized quantum-well states within the Cu interlayer. The TMR observed in this case is relatively low, even without an inserted Cu, but the dependence of spin polarization on crystallographic orientation for fcc Co is not yet known, and the low TMR value may simply be a result of a low spin polarization for Co(100). Another curiosity is the fact that the overall decay of the TMR, on which the oscillatory behavior is superimposed, occurs at approximately the same rate as observed by LeClair et al. [26], which has still not been adequately explained. Further, the same overall quasi-exponential decay was also observed in Fe/Cr(001)-based junctions (Figure 10.12), where Nagahama et al. [99] also observed an oscillating TMR attributed to the layered AFM structure of Cr(001). In Figure 10.13, we illustrate the mechanism for this oscillatory effect in comparison to quantum-well states. That Nagahama et al. were able to observe monolayer oscillations of the TMR further confirm the interface sensitivity of tunneling: if it were the *bulk* electronic structure of the electrode that were of importance, no oscillation should be observed. The observation of quantum oscillations by Yuasa et al. [98] were at the time an encouraging first step toward the development of coherent spin-dependent tunneling devices, culminating in the "giant" TMR effects observed in Fe-MgO-based systems [62,63].

10.4.11 Curious Case of CoFeB

In 2002–2004, very large TMR effects were observed in MTJs based on the amorphous ferromagnet CoFe [100,101], as large as 113% (72% at 295 K) at low temperatures, corresponding to $P = 60\%$ in the Jullière model. At the time, this was the highest TMR observed at room temperature. More importantly, it was the highest TMR observed in a nonepitaxial junction based on 3d alloys and fabricated by industrially friendly standard sputtering techniques. Further, the bias voltage dependence of the TMR in these MTJs is remarkably weak [101]—even at 0.5 V, the TMR was still >50%, making CoFeB a very interesting system for applications.

However, what is perhaps more interesting in this case is why does the seemingly innocuous addition of B to a standard Co-Fe alloy increase the TMR (and effectively P) to such a degree? Remarkably, despite the technological importance of these materials, our understanding of their electronic and magnetic behavior is at a surprisingly low level [102]. Further, is the mechanism a general, and exploitable, means for increasing P? Mechanisms for the increased TMR have been proposed. The first assumes that the amorphous CoFeB layer grows more smoothly than standard polycrystalline layers, leading to improved barrier flatness, coverage, and quality, thus giving a larger TMR. In this model, the polarization is assumed only weakly affected by the addition of B; it is a structural effect, with the B only serving to make the CoFe amorphous. A second proposition argues that the B hybridizes in such a way with the Co and/or Fe that the itinerant majority DOS is increased near E_F. To an extent, the first argument is borne out by TEM observations [101]. However, one would not expect this to affect the bias dependence so strongly, nor the temperature dependence [101]. While both of these properties are influenced by barrier quality, the amount of TMR (T) and TMR (V) decay usually attributed to barrier imperfections is far too small to explain the differences observed. We believe this lends credence to the hybridization argument, which is borne out to an extent by examining the DOS of the Fe and Co monoborides. On the other hand, the observation of "giant" TMR effects in CoFeB/MgO/CoFeB junctions lends credence to the former hypothesis. In these structures, it is currently believed that the amorphous CoFeB electrodes allow the MgO tunnel barrier to grow in a preferred (100) orientation, promoting "symmetry filtering" [38], which leads to TMR effects an order of magnitude above those observed with amorphous Al_2O_3 barriers [103].

In any case, these results were exciting in retrospect for three reasons: first, the TMR was significantly higher than anything observed previously; second, the TMR observed was nominally higher than what one would expect from the Jullière model [101]; and third, amorphous CoFeB turned out to be crucial to the practical realization of giant TMR effects in MgO-based junctions, which will be explored in the following chapters.

10.5 Conclusions

In the relatively short time since the first demonstration of large room temperature magnetoresistance in MTJs in 1995, there has been an enormous interest in the commercial potential of MTJ devices. Some of the main proposed applications are large arrays of magnetic sensors for imaging, ultralow magnetic field sensors, and most prominently, nonvolatile MRAMs and read-head sensors for hard disks. In particular, the latter two have already been commercialized, and TMR-based read sensors have been a mainstay in hard disk drives for several years already. The great demand for improved magnetic devices has certainly played a large role in the "rebirth" of spin-polarized tunneling; a variety of these applications are covered in detail in Section 10.5. In particular, Chapters 34 and 35 have direct relevance to the present discussion.

More striking from a physics point of view, however, is that despite three decades of research in spin-polarized tunneling, it is only in the last decade that the influence of the insulating barrier and the ferromagnet–insulator interface on tunneling spin polarization has been recognized. Some of the experiments mentioned here, for example, the crystallographic orientation dependence of spin polarization, quantum-well oscillations in quasi-epitaxial junctions, and large TMR effects with CoFeB-based junctions, foreshadowed or directly led to the most recent and exciting developments in spin-polarized tunneling, namely, the "giant" TMR effects in MgO-based MTJs discussed in Chapters 11 and 12. Others, such as the spin-filtering effects discussed in Chapter 13 are natural outgrowths of the SPT technique.

References

1. R. Meservey and P. M. Tedrow, Spin-polarized electron tunneling, *Phys. Rep.* **238**, 173 (1994).
2. J. C. Slonczewski, Magnetic bubble tunnel detector, *IBM Tech. Disclosure Bull.* **19**, 2328–2336 (1976).
3. M. Jullière, Tunneling between ferromagnetic films, *Phys. Lett. A* **54**, 225 (1975).
4. T. Miyazaki and N. Tezuka, Giant magnetic tunneling effect in Fe/Al_2O_3/Fe junction, *J. Magn. Magn. Mater.* **139**, L231 (1995).
5. J. S. Moodera, L. R. Kinder, T. M. Wong, and R. Meservey, Large magnetoresistance at room temperature in ferromagnetic thin film tunnel junctions, *Phys. Rev. Lett.* **74**, 3273 (1995).
6. E. L. Wolf, *Principles of Electron Tunneling Spectroscopy*, Oxford University Press, London, U.K., 1985.
7. C. Kittel, *Introduction to Solid State Physics*, 7 edn., John Wiley & Sons, Inc., New York, 1996.
8. J. E. Dowman, M. L. A. MacVicar, and J. R. Waldram, Selection rule for tunneling from superconductors, *Phys. Rev.* **186**, 452 (1969).
9. J. G. Simmons, Generalized formula for the electric tunnel effect between similar electrodes separated by a thin insulating film, *J. Appl. Phys.* **34**, 1793 (1963).
10. J. J. Akerman, J. M. Slaughter, and I. K. Schuller, Tunneling criteria for magnetic-insulator-magnetic structures, *Appl. Phys. Lett.* **79**, 3104 (2001).
11. J. M. Rowell, *Tunneling Phenomena in Solids*, Plenum, New York, p. 273, 1969.
12. I. Giaever, Energy gap in superconductors measured by electron tunneling, *Phys. Rev. Lett.* **5**(4), 147–148 (1960).
13. I. Giaever and K. Megerle, Study of superconductors by electron tunneling, *Phys. Rev.* **122**(4), 1101–1111 (1961).
14. I. Giaever, *Nobel Lectures in Physics, 1971–1980*, World Scientific, Singapore, p. 144, 1992. See also http://nobelprize.org/nobel_prizes/physics/laureates/1973/giaever-lecture.html
15. W. A. Harrison, Tunneling from an independent-particle point of view, *Phys. Rev.* **123**, 85 (1961).
16. J. Bardeen, Tunneling from a many-particle point of view, *Phys. Rev. Lett.* **6**, 57 (1961).
17. J. A. Appelbaum and W. F. Brinkman, Theory of many-body effects in tunneling, *Phys. Rev.* **186**, 464 (1969).
18. F. Mezei and A. Zawadowski, Kinematic change in conduction-electron density of states due to impurity scattering. II. Problem of an impurity layer and tunneling anomalies, *Phys. Rev. B* **3**, 3127 (1971).
19. H. Itoh and J. Inoue, Interfacial electronic states and magnetoresistance in tunnel junctions, *Surf. Sci.* **493**, 748 (2001).
20. H. Itoh, J. Inoue, S. Maekawa, and P. Bruno, Tunnel conductance in strong disordered limit, *J. Magn. Soc. Jpn.* **23**, 52 (1999).
21. E. Y. Tsymbal and D. G. Pettifor, Modelling of spin-polarized electron tunnelling from 3d ferromagnets, *J. Phys. Condens. Matter* **9**, L411 (1997).
22. C. Uiberacker and P. M. Levy, Role of symmetry on interface states in magnetic tunnel junctions, *Phys. Rev. B* **64**, 193404 (2001).
23. S. Zhang and P. M. Levy, Models for magnetoresistance in tunnel junctions, *Eur. Phys. J. B* **10**, 599 (1999).
24. P. LeClair, B. Hoex, H. Wieldraaijer, J. T. Kohlhepp, H. J. M. Swagten, and W. J. M. de Jonge, Sign reversal of spin polarization in Co/Ru/Al_2O_3/Co magnetic tunnel junctions, *Phys. Rev. B* **64**, 100406 (2001).
25. P. LeClair, J. T. Kohlhepp, H. J. M. Swagten, and W. J. M. de Jonge, Interfacial density of states in magnetic tunnel junctions, *Phys. Rev. Lett.* **86**(6), 1066–1069 (2001).
26. P. LeClair, H. J. M. Swagten, J. T. Kohlhepp, R. J. M. van de Veerdonk, and W. J. M. de Jonge, Apparent spin polarization decay in Cu-dusted Co/Al_2O_3/Co tunnel junctions, *Phys. Rev. Lett.* **84**(13), 2933–2936 (2000).
27. P. M. Tedrow and R. Meservey, Direct observation of spin-state mixing in superconductors, *Phys. Rev. Lett.* **27**, 919 (1971).
28. P. M. Tedrow and R. Meservey, Spin-dependent tunneling into ferromagnetic nickel. *Phys. Rev. Lett.* **26**, 192 (1971).
29. P. LeClair, H. J. M. Swagten, and J. S. Moodera, Spin polarized electron tunneling, in *Ultrathin Magnetic Structures III*, B. Heinrich and J. A. C. Bland (Eds.), Springer-Verlag, Berlin, Germany, 2005.
30. I. I. Mazin, How to define and calculate the degree of spin polarization in ferromagnets, *Phys. Rev. Lett.* **83**, 1427 (1999).
31. D. C. Worledge and T. H. Geballe, Negative spin-polarization of $SrRuO_3$, *Phys. Rev. Lett.* **85**, 5182 (2000).
32. R. Meservey, Private email communications, 2009.

33. P. M. Tedrow, R. Meservey, and B. B. Schwartz, Experimental evidence for a first-order magnetic transition in thin superconducting Aluminum films, *Phys. Rev. Lett.* **24**(18), 1004–1007 (1970).
34. P. M. Tedrow and R. Meservey, Critical thickness for ferromagnetism and the range of spin-polarized electrons tunneling into Co, *Solid State Commun.* **16**, 71 (1975).
35. J. S. Moodera, M. E. Taylor, and R. Meservey, Exchange-induced spin polarization of conduction electrons in paramagnetic metals, *Phys. Rev. B* **40**(17), 11980–11982 (1989).
36. M. S. Gabureac, K. J. Dempsey, N. A. Porter, C. H. Marrows, S. Rajauria, and H. Courtois, Spin-polarized tunneling with au impurity layers, *J. Appl. Phys.* **103**(7), 07A915 (2008).
37. M. B. Stearns, Simple explanation of tunneling spin-polarization of Fe, Co, Ni and its alloys, *J. Magn. Magn. Mater.* **8**, 167 (1977).
38. W. H. Butler, X. G. Zhang, T. C. Schulthess, and J. M. MacLaren, Spin-dependent tunneling conductance of Fe|MgO|Fe sandwiches, *Phys. Rev. B* **63**, 055416 (2001).
39. J. M. MacLaren, X. G. Zhang, W. H. Butler, and X. Wang, Layer KKR approach to Bloch-wave transmission and reflection: Application to spin-dependent tunneling, *Phys. Rev. B* **59**, 5470 (1999).
40. Ph. Mavropoulos, N. Papanikolaou, and P. H. Dederichs, Complex band structure and tunneling through ferromagnet/insulator/ferromagnet junctions, *Phys. Rev. Lett.* **85**, 1088 (2000).
41. I. I. Mazin, Tunneling of Bloch electrons through vacuum barrier, *Europhys. Lett.* 55, 404 (2001).
42. J. S. Moodera, J. Nassar, and J. Mathon, Spin-dependent tunneling in ferromagnetic junctions, *J. Magn. Magn. Mater.* **200**, 248 (1999).
43. S. Maekawa and U. Gäfvert, Electron tunneling between ferromagnetic films, *IEEE Trans. Magn.* **18**, 707 (1982).
44. S. Gider, B. U. Runge, A. C. Marley, and S. S. P. Parkin, The magnetic stability of spin-dependent tunneling devices, *Science* **281**, 797 (1999).
45. P. LeClair, Fundamental aspects of spin-polarized tunneling, PhD thesis, Technische Universiteit Eindhoven, 2002. See also http://alexandria.tue.nl/extra2/200211695.pdf
46. D. J. Monsma and S. S. P. Parkin, Spin polarization of tunneling current from ferromagnet/Al_2O_3 interfaces using copper-doped aluminum superconducting films, *Appl. Phys. Lett.* **77**(5), 720–722 (2000).
47. X.-F. Han, A. C. C. Yu, M. Oogane, J. Murai, T. Daibou, and T. Miyazaki, Analyses of intrinsic magnetoelectric properties in spin-valve-type tunnel junctions with high magnetoresistance and low resistance, *Phys. Rev. B* **63**, 224404 (2001).
48. W. E. Pickett and J. S. Moodera, Half metallic ferromagnets, *Phys. Today* **5**, 39 (2001).
49. J. S. Parker, S. M. Watts, P. G. Ivanov, and P. Xiong, Spin polarization of C_rO_2 at and across an artificial barrier, *Phys. Rev. Lett.* **88**(19), 196601 (2002).
50. M. Bowen, M. Bibes, A. Barthélémy et al., Nearly total spin polarization in $La_{2/3}Sr_{1/3}MnO_3$ from tunneling experiments, *Appl. Phys. Lett.* **82**(2), 233–235 (2003).
51. Y. Lu, X. W. Li, G. Q. Gong et al., Large magnetotunneling effect at low magnetic fields in micrometer-scale epitaxial $La_{0.67}Sr_{0.33}MnO_3$ tunnel junctions, *Phys. Rev. B* **54**, R8357 (1996).
52. M. Viret, M. Drouet, J. Nassar, J. P. Contour, C. Fermon, and A. Fert, Low-field colossal magnetoresistance in manganite tunnel spin valves, *Europhys. Lett.* **39**, 545 (1997).
53. D. C. Worledge and T. H. Geballe, Spin-polarized tunneling in $La_{0.67}Sr_{0.33}MnO_3$, *Appl. Phys. Lett.* **76**, 900 (2000).
54. T. H. Kim and J. S. Moodera, Large spin polarization in epitaxial and polycrystalline Ni films, *Phys. Rev. B* **69**(2), 020403 (2004).
55. B. Nadgorny, R. J. Soulen Jr., M. S. Osofsky et al., Transport spin polarization of Ni_xFe_{1-x}: Electronic kinematics and band structure, *Phys. Rev. B* **61**, R3788 (2000).
56. T. H. Kim and J. S. Moodera, Enhanced ferromagnetism and spin-polarized tunneling studies in Co-Mn alloy films, *Phys. Rev. B* **66**(10), 104436 (2002).
57. C. Kaiser, S. van Dijken, S.-H. Yang, H. Yang, and S. S. P. Parkin, Role of tunneling matrix elements in determining the magnitude of the tunneling spin polarization of 3d transition metal ferromagnetic alloys, *Phys. Rev. Lett.* **94**(24), 247203 (2005).
58. C. Kaiser, A. F. Panchula, and S. S. P. Parkin, Finite tunneling spin polarization at the compensation point of rare-earth-metal–transition-metal alloys, *Phys. Rev. Lett.* **95**(4), 047202 (2005).
59. B. C. Min, J. C. Lodder, and R. Jansen, Sign of tunnel spin polarization of low-work-function Gd/Co nanolayers in a magnetic tunnel junction, *Phys. Rev. B* **78**(21), 212403 (2008).
60. S. Yuasa, T. Sato, E. Tamura et al., Magnetic tunnel junctions with single-crystal electrodes: A crystal anisotropy of tunnel magneto-resistance, *Europhys. Lett.* **52**(3), 344–350 (2000).
61. J. Callaway and C. S. Wang, Energy bands in ferromagnetic iron, *Phys. Rev. B* **16**, 2095 (1977).
62. S. S. P. Parkin, C. Kaiser, A. Panchula et al., Giant room-temperature magnetoresistance in single-crystal Fe/MgO/Fe magnetic tunnel junctions, *Nat. Mater.* **3**, 862–867 (2004).
63. S. Yuasa, T. Nagahama, A. Fukushima, Y. Suzuki, and K. Ando, Giant room-temperature magnetoresistance in single-crystal Fe/MgO/Fe magnetic tunnel junctions, *Nat. Mater.* **3**(12), 868–871 (2004).
64. J.-N. Chazalviel and Y. Yafet, Theory of the spin polarization of field-emitted electrons from nickel, *Phys. Rev. B* **15**, 1062–1071 (1977).
65. X.-F. Han, J. Murai, Y. Ando, H. Kubota, and T. Miyazaki, Inelastic magnon and phonon excitations in $A_{1x}Co_x$/$A_{1x}Co_x$-oxide/Al tunnel junctions, *Appl. Phys. Lett.* **78**, 2533 (2001).

66. J. S. Moodera, J. Nowak, and R. J. M. van de Veerdonk, Interface magnetism and spin wave scattering in ferromagnet–insulator–ferromagnet tunnel junctions, *Phys. Rev. Lett.* **80**, 2941 (1998).
67. J. S. Moodera and L. R. Kinder, Ferromagnetic-insulator-ferromagnetic tunneling: Spin-dependent tunneling and large magnetoresistance in trilayer junctions (invited), *J. Appl. Phys.* **79**, 4724 (1996).
68. S. Zhang, P. M. Levy, A. C. Marley, and S. S. P. Parkin, Quenching of magnetoresistance by hot electrons in magnetic tunnel junctions, *Phys. Rev. Lett.* **79**, 3744 (1997).
69. A. M. Bratkovsky, Assisted tunneling in ferromagnetic junctions and half-metallic oxides, *Appl. Phys. Lett.* **72**, 2334 (1998).
70. M. Sharma, S. X. Wang, and J. H. Nickel, Inversion of spin polarization and tunneling magnetoresistance in spin-dependent tunneling junctions, *Phys. Rev. Lett.* **82**, 616 (1999).
71. J. M. De Teresa, A. Barthélémy, A. Fert et al., Inverse tunnel magnetoresistance in Co/$SrTiO_3$/$La_{0.7}Sr_{0.3}MnO_3$: New ideas on spin-polarized tunneling, *Phys. Rev. Lett.* **82**, 4288 (1999).
72. J. M. De Teresa, A. Barthélémy, A. Fert, J. P. Contour, F. Montaigne, and P. Seneor, Role of metal-oxide interface in determining the spin polarization of magnetic tunnel junctions, *Science* **286**, 507 (1999).
73. E. Y. Tsymbal, O. N. Mryasov, and P. R. LeClair, Spin-dependent tunnelling in magnetic tunnel junctions, *J. Phys. Condens. Matter* **15**(4), R109–R142 (2003).
74. P. LeClair, J. T. Kohlhepp, C. H. van de Vin et al., Band structure and density of states effects in Co-based magnetic tunnel junctions, *Phys. Rev. Lett.* **88**, 107201 (2002).
75. P. LeClair, H. J. M. Swagten, J. T. Kohlhepp, and W. J. M. de Jonge, Tunnel conductance as a probe of spin polarization decay in Cu dusted Co/Al_2O_3/Co tunnel junctions, *Appl. Phys. Lett.* **76**, 3783 (2000).
76. A. H. Davis and J. M. MacLaren, Spin dependent tunneling at finite bias, *J. Appl. Phys.* **87**, 5224 (2000).
77. A. M. Bratkovsky, Tunneling of electrons in conventional and half-metallic systems: Towards very large magnetoresistance, *Phys. Rev. B* **56**, 2344 (1997).
78. A. Vedyayev, D. Bagrets, A. Bagrets, and B. Dieny, Resonant spin-dependent tunneling in spin-valve junctions in the presence of paramagnetic impurities, *Phys. Rev. B* **63**, 064429 (2001).
79. R. Jansen and J. S. Moodera, Influence of barrier impurities on the magnetoresistance in ferromagnetic tunnel junctions, *J. Appl. Phys.* **83**, 6682 (1998).
80. R. Jansen and J. S. Moodera, Magnetoresistance in doped magnetic tunnel junctions: Effect of spin scattering and impurity-assisted transport, *Phys. Rev. B* **61**, 9047 (2000).
81. C. H. Shang, J. Nowak, R. Jansen, and J. S. Moodera, Temperature dependence of magnetoresistance and surface magnetization in ferromagnetic tunnel junctions, *Phys. Rev. B* **58**, R2917 (1998).
82. D. T. Pierce, R. J. Celotta, J. Unguris, and H. C. Siegman, Spin-dependent elastic scattering of electrons from a ferromagnetic glass, $Ni_{40}Fe_{40}B_{20}$, *Phys. Rev. B* **26**, 2566 (1982).
83. A. H. MacDonald, T. Jungwirth, and M. Kasner, Temperature dependence of itinerant electron junction magnetoresistance, *Phys. Rev. Lett.* **81**, 705 (1998).
84. A. H. Davis, J. M. MacLaren, and P. LeClair, Inherent temperature effects in magnetic tunnel junctions, *J. Appl. Phys.* **89**, 7567 (2001).
85. T. Kreutz, T. Greber, P. Aebi, and J. Osterwalder, Temperature-dependent electronic structure of nickel metal, *Phys. Rev. B* **58**, 1300 (1998).
86. A. T. Hindmarch, C. H. Marrows, and B. J. Hickey, Tunneling magnetoresistance spectroscopy: Temperature dependent spin-polarized band structure in $Cu_{38}Ni_{62}$, *Phys. Rev. B* **72**(6), 060406 (2005).
87. A. T. Hindmarch, C. H. Marrows, and B. J. Hickey, Tunneling spin polarization in magnetic tunnel junctions near the Curie temperature, *Phys. Rev. B* **72**(10), 100401 (2005).
88. I. I. Oleinik, E. Y. Tsymbal, and D. G. Pettifor, Atomic and electronic structure of Co/$SrTiO_3$/Co magnetic tunnel junctions, *Phys. Rev. B* **65**, 020401(R) (2001).
89. J. H. Park, E. Vescovo, H. J. Kim, C. Kwon, R. Ramesh, and T. Venkatesan, Direct evidence for a half-metallic ferromagnet, *Nature* **392**, 794 (1998).
90. E. Y. Tsymbal, K. D. Belashchenko, J. P. Velev et al., Interface effects in spin-dependent tunneling, *Prog. Mater. Sci.* **52**(2–3), 401–420 (2007).
91. J. Velev, K. D. Belashchenko, D. A. Stewart, M. van Schilfgaarde, S. Jaswal, and E. Y. Tsymbal, Negative spin polarization and large tunneling magnetoresistance in epitaxial Co|$SrTiO_3$|Co magnetic tunnel junctions, *Phys. Rev. Lett.* **95**(21), 216601 (2005).
92. M. Sugiyama, J. Hayakawa, K. Itou et al., Tunneling magnetoresistance effect and transport properties of tunnel junctions using half-metal ferromagnet, *J. Magn. Soc. Jpn.* **25**, 795–798 (2001).
93. J. Z. Sun, K. P. Roche, and S. S. P. Parkin, Interface stability in hybrid metal-oxide magnetic trilayer junctions, *Phys. Rev. B* **61**(17), 11244–11247 (2000).
94. J. S. Moodera, T. H. Kim, C. Tanaka, and C. H. de Groot, Spin-polarized tunnelling, magnetoresistance and interfacial effects in ferromagnetic junctions, *Phil. Mag. B* **80**, 195 (2000).
95. M. G. Samant and S. S. P. Parkin, Magnetic tunnel junctions—Principles and applications, *Vacuum* **74**(3–4), 705–709 (2004).
96. T. Nozaki, Y. Jiang, Y. Kaneko et al., Spin-dependent quantum oscillations in magnetic tunnel junctions with Ru quantum wells, *Phys. Rev. B* **70**(17), 172401 (2004).
97. J. S. Moodera, J. Nowak, L. R. Kinder et al., Quantum well states in spin-dependent tunnel structures, *Phys. Rev. Lett.* **83**(15), 3029–3032 (1999).
98. S. Yuasa, T. Nagahama, and Y. Suzuki, Spin-polarized resonant tunneling in magnetic tunnel junctions, *Science* **297**(5579), 234–237 (2002).

99. T. Nagahama, S. Yuasa, E. Tamura, and Y. Suzuki, Spin-dependent tunneling in magnetic tunnel junctions with a layered antiferromagnetic Cr(001) spacer: Role of band structure and interface scattering, *Phys. Rev. Lett.* **95**(8), 086602 (2005).
100. H. Kano, K. Bessho, Y. Higo et al., MRAM with improved magnetic tunnel junction material, in *Magnetics Conference, 2002, INTERMAG Europe 2002, Digest of Technical Papers, 2002 IEEE International*, IEEE, page BB4, 2002.
101. D. Wang, C. Nordman, J. M. Daughton, Z. Qian, and J. Fink, 70% TMR at room temperature for SDT sandwich junctions with CoFeB as free and reference layers, *IEEE Trans. Magn.* **40**, 2269 (2004).
102. J. Hafner, M. Tegze, and Ch. Becker, Amorphous magnetism in Fe-B alloys: First-principles spin-polarized electronic-structure calculations, *Phys. Rev. B* **49**, 285–298 (1994).
103. S. Ikeda, J. Hayakawa, Y. Ashizawa et al., Tunnel magnetoresistance of 604% at 300k by suppression of Ta diffusion in CoFeB/MgO/CoFeB pseudo-spin-valves annealed at high temperature, *Appl. Phys. Lett.* **93**, 082508 (2008).
104. J. Crangle, The magnetic moments of cobalt-copper alloys, *Phil. Mag.* **46**, 499 (1955).
105. R. Matsumoto, A. Fukushima, K. Yakushiji et al., Spin-dependent tunneling in epitaxial Fe/Cr/MgO/Fe magnetic tunnel junctions with an ultrathin Cr(001) spacer layer, *Phys. Rev. B* **79**(17), 174436 (2009).

11

Tunneling Magnetoresistance: Experiment (MgO Magnetic Tunnel Junctions)

Shinji Yuasa
National Institute of Advanced Industrial Science and Technology

11.1 Introduction

11.1.1 Tunnel Magnetoresistance Effect in a Magnetic Tunnel Junction

A magnetic tunnel junction (MTJ) consists of an ultrathin insulating layer (tunnel barrier) sandwiched between two ferromagnetic (FM) metal layers (electrodes), as shown in Figure 11.1a. The resistance of MTJ depends on the relative magnetic alignment (parallel or antiparallel) of the electrodes. The tunnel resistance, R, of MTJ is lower when the magnetizations are parallel than it is when the magnetizations are antiparallel, as shown in Figure 11.1b. That is, $R_P < R_{AP}$. This change in resistance with the relative orientation of the two magnetic layers, called the tunnel magnetoresistance (TMR) effect, is one of the most important phenomena in spintronics. The size of this effect is measured by the fractional change in resistance $(R_{AP} - R_P)/R_P \times 100$ (%), which is called the magnetoresistance (MR) ratio. The MR ratio at room temperature (RT) and a low magnetic field (typically <10 mT) represent a performance index for industrial applications.

The TMR effect was first observed in 1975 by Julliere [1], who found that an Fe/Ge-O/Co MTJ exhibited an MR ratio of 14% at 4.2 K.* Although it received little attention for more than a decade because the TMR effect was not observed at RT, it attracted renewed attention after the discovery of giant magnetoresistance (GMR) in metallic magnetic multilayers in the late 1980s [2,3] (see Chapters 4 and 5 for the reviews on GMR). Because researchers recognized that the GMR effect could be applied to magnetic sensor devices such as the read heads of hard disk drives (HDDs), extensive experimental and theoretical efforts were devoted to increasing the MR ratio at RT as well as to understanding the physics of spin-dependent transport. In 1995, Miyazaki and Tezuka [4] and Moodera et al. [5] made MTJs with amorphous aluminum oxide (Al-O) tunnel barriers and 3d FM electrodes such as Fe and Co and obtained MR ratios above 10% at RT (Chapter 10). Because these MR ratios were then the highest reported for a practical ferromagnetic/nonmagnetic/ferromagnetic (FM/NM/FM) tri-layer structure called a (pseudo-) spin-valve structure, the TMR effect attracted a great deal of attention. Since 1995, extensive experimental

* In Ref. [1], the tunnel barrier material is described as Ge. However, the Ge layer should have been oxidized because it was exposed to air before the upper electrode layer was deposited.

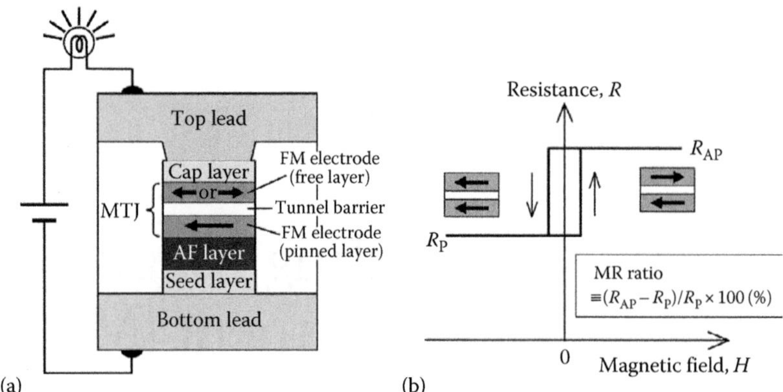

FIGURE 11.1 (a) Typical cross-sectional structure of MTJ with spin-valve structure. (b) Typical MR curve of spin-valve-type MTJ and definition of MR ratio.

efforts have been expended on MTJs with an amorphous Al-O barrier, and the FM electrode materials and conditions for fabricating the Al-O barrier have been optimized. As a result, room-temperature MR ratios have been increased to about 70%, as shown in Figure 11.2 (solid circles). These MR ratios of up to 70%, however, are still lower than those needed for many applications of spintronic devices. High-density magnetoresistive random-access-memory (MRAM) cells, for example, will need to have MR ratios that are higher than 150% at RT, and the read head in next-generation ultrahigh-density HDDs will need to have both a high MR ratio and an ultralow tunneling resistance. The MR ratios of conventional Al-O-based MTJs are simply not high enough for applications to next-generation devices.

In 2001, theoretical calculations predicted that epitaxial MTJs with a crystalline magnesium oxide (MgO) tunnel barrier would have MR ratios of over 1000%, and in 2004, huge MR ratios of about 200% were obtained at RT in MTJs with a crystalline MgO(001) barrier. The huge TMR effect in MgO-based MTJs is now called the *giant TMR effect* and is of great importance not only for device applications but also for clarifying the physics of spin-dependent tunneling.

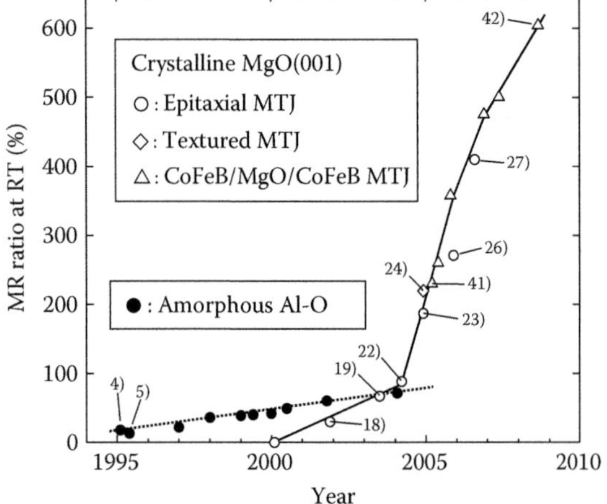

FIGURE 11.2 History of improvement in MR ratio at room temperature (RT) for MTJs with amorphous Al-O barrier (solid circles) or crystalline MgO(001) barrier (open symbols).

In this chapter, we present an overview of the huge TMR effect in MgO-based MTJs mainly from the experimental point of view. In the following section of this Introduction, we provide further background information about the TMR effect in MTJs with an amorphous Al-O barrier. In Section 11.2, we briefly explain the theoretical basis for the TMR effect in MgO-based MTJs, and in Sections 11.3 and 11.4, we summarize experimental studies done on fully epitaxial MgO-based MTJs. In Section 11.5, we explain the structure and fabrication of CoFeB/MgO/CoFeB MTJs that are important for applications, and in Section 11.7, we report recent progress in applications of the giant TMR effect and those in MgO-based MTJs.

In other chapters, other topics related to MgO-based MTJ are explained in more detail. For example, the TMR effect in Al-O-based MTJs and measurement of tunneling spin polarization (TSP) are explained in Chapter 10. The theory of the TMR effect in MgO-based MTJs is explained in Chapter 12. Spin-transfer torque in MTJs is explained in Chapter 14. The application of MTJs to magnetic sensors is explained in Chapter 34 and that to MRAMs is explained in Chapter 35.

11.1.2 Incoherent Tunneling through Amorphous Tunnel Barrier

Before going into the details of coherent tunneling through a crystalline MgO(001) barrier, we will explain the tunneling process through an amorphous tunnel barrier and the phenomenological model for the TMR effect. Because this subject is explained in detail in Sections 10.3, and 10.4, we discuss it only briefly in this chapter. Julliere proposed a simple phenomenological model in which the TMR effect is due to spin-dependent electron tunneling [1]. According to this model, the MR ratio of an MTJ could be expressed in terms of the TSPs, *P*, of the FM electrodes

$$\mathrm{MR} = \frac{2P_1P_2}{(1-P_1P_2)}, \tag{11.1}$$

where

$$P_\alpha \equiv \frac{\left[D_{\alpha\uparrow}(E_F) - D_{\alpha\downarrow}(E_F)\right]}{\left[D_{\alpha\uparrow}(E_F) + D_{\alpha\downarrow}(E_F)\right]}; \quad \alpha = 1,2. \tag{11.2}$$

Here P_α is the spin polarization of an FM electrode, and $D_{\alpha\uparrow}(E_F)$ and $D_{\alpha\downarrow}(E_F)$ correspond to the densities of states (DOSs) of the electrode at the Fermi energy (E_F) for the majority-spin and minority-spin bands. In Julliere's model, spin polarization is an intrinsic property of an electrode material. When an electrode material is NM, $P=0$. When the DOS of the electrode material is fully spin-polarized at E_F, $|P|=1$.

The spin polarization of a ferromagnet at low temperature can be directly measured using ferromagnet/Al-O/superconductor tunnel junctions (the so-called Tedrow–Meservey method; see Chapter 10) [6]. Measured this way, the spin polarizations of 3d ferromagnetic metals and alloys based on iron (Fe), nickel (Ni), and cobalt (Co) are always positive and are usually between 0 and 0.6 at low temperatures below 4.2 K [6,7]. The MR ratios estimated from Julliere's model (Equation 11.1) using these measured P values agree relatively well with the MR ratios observed experimentally in MTJs, but the theoretical values of P (Equation 11.2) obtained from band calculations do not agree with the measured spin polarizations and the MR ratios observed experimentally. Even the signs of P often differ between theoretical predictions and experimental results. This paradox in spin polarization is considered to be due to the incorrect definition of spin polarization in Equation 11.2, as explained in the following text.

Tunneling in an MTJ with an amorphous Al-O barrier is schematically illustrated in Figure 11.3a, where the top electrode layer is Fe(001), being an example of a 3d ferromagnet. Various Bloch states with different symmetries of wave functions exist in the electrode. Because the tunnel barrier is amorphous, there is no crystallographic symmetry in the tunnel barrier. Because of this non-symmetric structure, Bloch states with various symmetries can couple with evanescent states in Al-O and therefore have finite tunneling probabilities. This tunneling process can be regarded as *incoherent* tunneling. In 3d FM metals and alloys, Bloch states with Δ_1 symmetry (*spd* hybridized states) usually have a large positive spin polarization at E_F, whereas Bloch states with Δ_5 and Δ_2 symmetry (*d* states) usually have a much smaller or even negative spin polarization at E_F. Julliere's model assumes that tunneling probabilities are equal for all the Bloch states in the electrodes. This assumption corresponds to completely incoherent tunneling, in which none of the momentum or coherency of Bloch states is conserved. This assumption, however, is not valid even for an amorphous Al-O barrier. Although the spin polarization, P, defined by Equation 11.2 is negative for Co and Ni, the P experimentally observed for these materials when they are combined with an Al-O barrier is positive [6,7]. This discrepancy indicates that the tunneling probability in actual MTJs depends on the symmetry of each Bloch state. Even for amorphous barriers, the Δ_1 Bloch states that usually have a large positive P are considered to have higher tunneling probabilities than the other Bloch states [8,9]. This results in a positive TSP for the FM electrode. Because the other Bloch states that have smaller spin polarizations also contribute to the tunneling current, the TSP of the electrode is reduced below 0.6 for standard 3d FM metals and alloys. If only highly spin-polarized Δ_1 states coherently tunnel through a barrier (Figure 11.3b), a very high spin polarization of tunneling current and thus a very high MR ratio are expected. Such ideal coherent tunneling is theoretically expected in an epitaxial MTJ with a crystalline MgO(001) tunnel barrier (see the next section). It should be noted here that the actual tunneling process through the amorphous Al-O barrier is considered to be an intermediate process between the completely incoherent tunneling represented by Julliere's model and the coherent tunneling illustrated in Figure 11.3b.

Although Julliere's model is simply a phenomenological model, we risked the use of Equation 11.1 with experimentally observed P here to roughly estimate the maximum possible MR ratio for Al-O-based MTJs. Equation 11.1 with the spin polarizations measured experimentally ($0<P<0.6$) yields a maximum

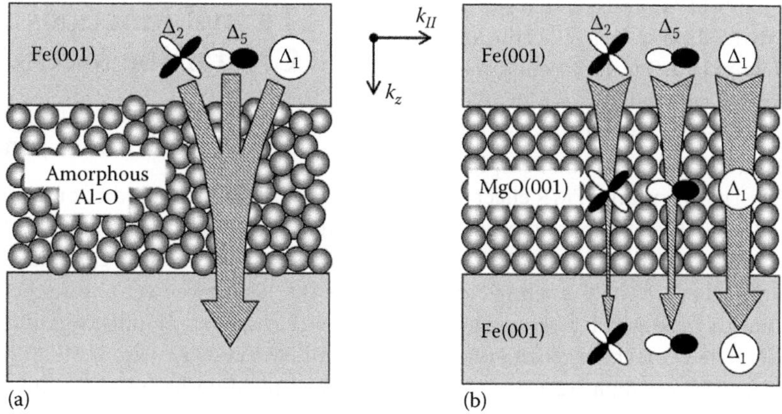

FIGURE 11.3 Schematic of electron tunneling through (a) amorphous Al-O barrier and (b) crystalline MgO(001) barrier.

MR ratio of about 100% at low temperatures. An MR ratio is reduced to about 70% at RT likely due to thermal spin fluctuations. One way to obtain an MR ratio significantly higher than 70% at RT is to use special kinds of FM materials called *half metals* as electrodes, which have full TSP ($|P|=1$) and are therefore theoretically expected to yield MTJs with huge MR ratios (up to ∞, according to Julliere's model). Some candidate half metals are CrO_2, Heusler alloys such as Co_2MnSi, Fe_3O_4, and manganese perovskite oxides such as $La_{1-x}Sr_xMnO_3$. Very high MR ratios, above several hundred percent, have been obtained at low temperatures in $La_{1-x}Sr_xMnO_3/SrTiO_3/La_{1-x}Sr_xMnO_3$ MTJs [10] and Co_2MnSi/Al-O/Co_2MnSi MTJs [11]. However, such high MR ratios have never been observed at RT for half-metal electrodes (for details, see Section 11.6). Another way to obtain a very high MR ratio is to use coherent spin-dependent tunneling in an epitaxial MTJ with a crystalline tunnel barrier such as MgO(001). That is explained in the following sections.

11.2 Theory of Coherent Tunneling through Crystalline MgO(001) Barrier

A crystalline MgO(001) barrier layer can be epitaxially grown on a bcc Fe(001) layer with a relatively small lattice mismatch of about 3% because this amount of lattice mismatch can be absorbed by lattice distortions in the Fe and MgO layers and/or by dislocations formed at the interface. Coherent tunneling transport in an epitaxial Fe(001)/MgO(001)/Fe(001) MTJ is schematically illustrated in Figure 11.3b. In ideal coherent tunneling, Fe Δ_1 states are theoretically expected to dominantly tunnel through the MgO(001) barrier using the following mechanism [12,13]. For the $k_{//}=0$ at which the tunneling probability is the highest, there are three kinds of evanescent states (tunneling states) in the band gap of MgO(001): Δ_1, Δ_5, and Δ_2. It should be noted that although free-electron theories often assume tunneling states to be plane waves, they actually have specific orbital symmetries. When the symmetries of tunneling wave functions are conserved, Fe Δ_1 Bloch states couple with MgO Δ_1 evanescent states, as schematically illustrated in Figure 11.3a. Of the three kinds of evanescent states in MgO, the Δ_1 evanescent states have the slowest decay (i.e., the longest decay length) as shown in Figure 11.3. The dominant tunneling channel for a parallel magnetic state is Fe $\Delta_1 \leftrightarrow$ MgO $\Delta_1 \leftrightarrow$ Fe Δ_1. The band dispersion of bcc Fe for the [001] ($k_{//}=0$) direction is shown in Figure 11.4a. It is seen that the Fe Δ_1 band is fully spin-polarized at E_F ($P=1$). A very large TMR effect in the epitaxial Fe(001)/MgO(001)/Fe(001) MTJ is therefore expected when Δ_1 electrons dominantly tunnel through the MgO(001) barrier. It should also be noted that a finite tunneling current flows even for antiparallel magnetic states. However, the tunneling conductance in the P state is much larger than that in the AP state, making the MR ratio very high [12,13].

It should be noted that the Δ_1 Bloch states are highly spin-polarized not only in bcc Fe(001) but also in many other bcc

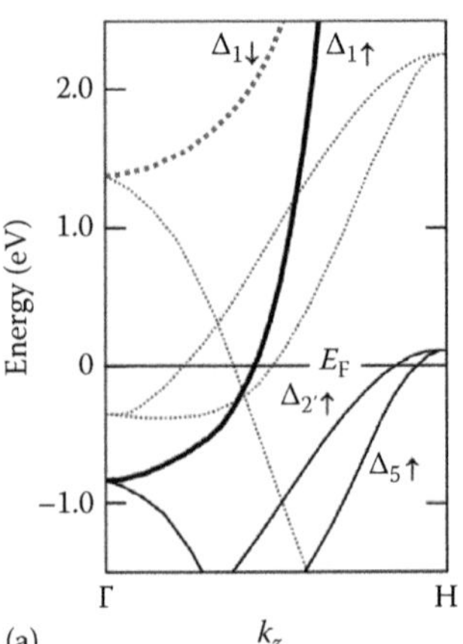

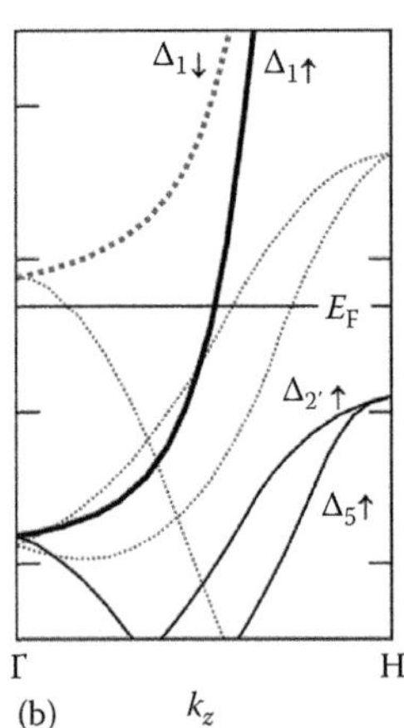

FIGURE 11.4 Band dispersions of (a) bcc Fe in [001] (Γ-H) direction and (b) bcc Co in [001] (Γ-H) direction. Black solid and gray dotted lines correspond to majority-spin and minority-spin bands. Thick black solid and gray dotted lines correspond to majority-spin and minority-spin Δ_1 bands. E_F denotes Fermi energy. (Redrawn from Yuasa, S. et al., *Appl. Phys. Lett.*, 89, 042505, 2006.)

FM metals and alloys based on Fe and Co (e.g., bcc Fe–Co, bcc CoFeB, and some of the Heusler alloys). The band dispersion of bcc Co(001) (a metastable structure) has been shown in Figure 11.4b as an example. The Δ_1 states in bcc Co, like those in bcc Fe, are fully spin-polarized at E_F. According to first-principles calculations, the TMR of a Co(001)/MgO(001)/Co(001) MTJ is even larger than that of an Fe(001)/MgO(001)/Fe(001) MTJ [14]. A very large TMR should be characteristic of MTJs with 3d-FM-alloy electrodes with a bcc(001) structure based on Fe and Co. Note also that very large TMR is theoretically expected not only for the MgO(001) barrier but also for other crystalline tunnel barriers such as ZnSe(001) [15] and $SrTiO_3$(001) [16]. Large TMR has, however, never been observed in MTJs with crystalline ZnSe(001) or $SrTiO_3$(001) barriers because of experimental difficulties in fabricating high-quality MTJs without pin holes and interdiffusion at the interfaces.

11.3 Giant TMR Effect in Magnetic Tunnel Junctions with Crystalline MgO(001) Barrier

11.3.1 Development of MgO-Based Magnetic Tunnel Junctions

Since the theoretical predictions of very large TMR effects in Fe/MgO/Fe MTJs [12,13], there have been several experimental attempts to fabricate fully epitaxial Fe(001)/MgO(001)/Fe(001) MTJs [17–19]. Bowen et al. were the first to obtain a relatively high MR ratio in Fe(001)/MgO(001)/Fe(001) MTJs (30% at RT and 60% at 30 K, Figure 11.5a) [18], but the RT MR ratios obtained in MgO-based MTJs did not exceed the highest of 70% obtained in Al-O-based MTJs. The main difficulty at the early stages of these experimental attempts was the fabrication

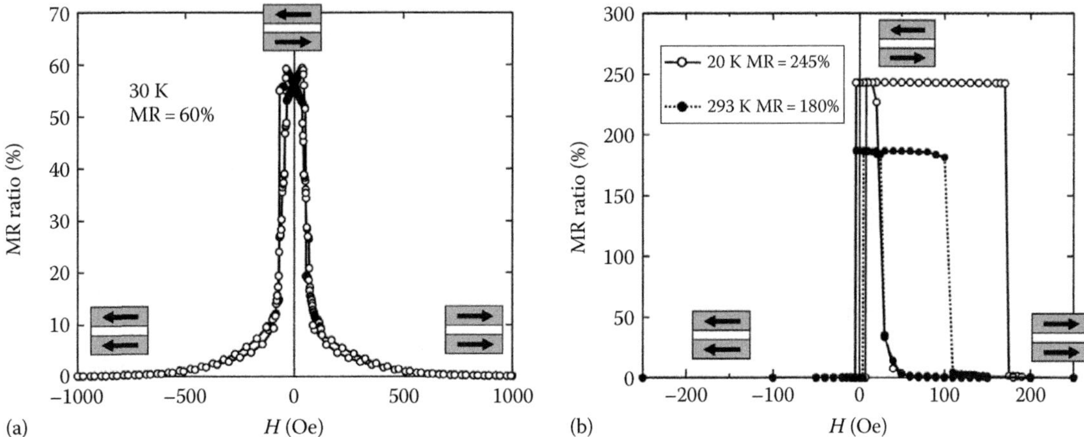

FIGURE 11.5 (a) MR curve measured at 30 K for epitaxial Fe(001)/MgO(001)/Fe(001) MTJ. (Redrawn from Bowen, M. et al., *Appl. Phys. Lett.*, 79, 1655, 2001.) (b) MR curves measured at RT and 20 K for epitaxial Fe(001)/MgO(001)/Fe(001) MTJ. (Redrawn from Yuasa, S. et al., *Nat. Mater.*, 3, 868, 2004.)

of an ideal interface structure. It was experimentally observed that Fe atoms at the Fe(001)/MgO(001) interface were easily oxidized [20]. According to first-principles calculations, tunneling processes through the ideal interface and the oxidized interface are significantly different [21]. At the ideal interface, where there are no O atoms in the first Fe monolayer at the interface, the Fe Δ_1 Bloch states effectively couple with the MgO Δ_1 evanescent states in the $k_{//} = 0$ direction, which results in the coherent tunneling of Δ_1 states and thus a very large TMR effect. At the oxidized interface, where there are excess oxygen atoms in the interfacial Fe monolayer, the Fe Δ_1 states do not effectively couple with the MgO Δ_1 states. This prevents coherent tunneling of Δ_1 states and significantly reduces the MR ratio. Coherent tunneling is thus very sensitive to the structure of barrier/electrode interfaces. Oxidation of even a monolayer at the interface significantly suppresses the TMR effect.

In 2004, Yuasa et al. fabricated high-quality fully epitaxial Fe(001)/MgO(001)/Fe(001) MTJs with a single crystal MgO(001) tunnel barrier by using molecular beam epitaxy (MBE) growth under an ultrahigh vacuum and attained RT MR ratios that were significantly higher than those of conventional Al-O-based MTJs [22,23]. There is a cross-sectional transmission electron microscope (TEM) image of an epitaxial MTJ in Figure 11.6, where single-crystal lattices of MgO(001) and Fe(001) are clearly evident. The MR curves of the epitaxial Fe(001)/MgO(001)/Fe(001) MTJ are shown in Figure 11.5b. They obtained giant MR ratios of up to 188% at RT (see also Figure 11.2) [22,23]. During the same period, Parkin et al. fabricated highly oriented poly-crystal (or textured) FeCo(001)/MgO(001)/FeCo(001) MTJs with a textured MgO(001) tunnel barrier by using sputtering deposition on an SiO_2 substrate with a tantalum nitride seed layer that was used to orient the entire MTJ stack in the (001) plane [24]. They obtained a giant MR ratio of up to 220% at RT (Figure 11.2) [24]. These were the first RT MR ratios that were higher than the previous highest

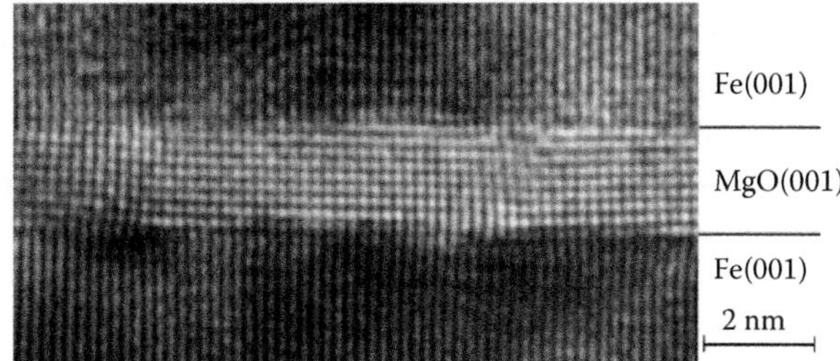

FIGURE 11.6 Cross-sectional TEM image of epitaxial Fe(001)/MgO(001)(1.8 nm)/Fe(001) MTJ. Vertical and horizontal directions correspond to MgO[001] (Fe[001]) axis and that of MgO[100] (Fe[110]). (Redrawn from Yuasa, S. et al., *Nat. Mater.*, 3, 868, 2004.)

MR ratio obtained with an Al-O-based MTJ. It should be noted that fully epitaxial MTJs and textured MTJs are basically the same from a microscopic point of view if structural defects such as grain boundaries do not strongly influence the tunneling properties. It is therefore considered that the giant MR ratios observed in the epitaxial and textured MTJs originate from the same mechanism. This very large TMR effect in MgO-based MTJs is called the *giant TMR effect*.

The key to obtaining such high MR ratios is thought to be the production of clean Fe/MgO interfaces without oxidized Fe atoms. X-ray absorption spectroscopy (XAS) and x-ray magnetic circular dichroism (XMCD) studies have revealed that none of the Fe atoms adjacent to the MgO(001) layer are oxidized, indicating that there are no oxygen atoms in the interfacial Fe monolayer [25]. Since the experimental achievements of a giant TMR effect in 2004, even higher MR ratios of up to 410% at RT have been demonstrated in epitaxial Co/MgO/Fe and Co/MgO/Co MTJs with metastable bcc Co(001) electrodes (Figure 11.2) [26,27]. The fact that the MR ratio of Co/MgO/Co MTJ is even larger than that of Fe/MgO/Fe MTJ qualitatively agrees with the theoretical prediction made by Zhang and Butler [14].

11.4 Other Phenomena Observed in Tunnel Junctions with Crystalline MgO(001) Barrier

It is difficult to investigate the detailed mechanisms of the TMR effect by using an amorphous Al-O barrier because of its structural uncertainty, and no significant progress has been made toward further understanding of the physics of the TMR effect since Julliere proposed his phenomenological model in 1975. Unlike Al-O-based MTJs, epitaxial MTJs with a single-crystal MgO(001) barrier are a model system for studying the physics of spin-dependent tunneling because of their well-defined crystalline structure with atomically flat interfaces. Other than the giant TMR effect, several interesting phenomena have been observed in epitaxial MgO-based MTJs. For example, epitaxial Fe(001)/MgO(001)/Fe(001) MTJs were observed to exhibit an oscillatory TMR effect with respect to MgO barrier thickness [23,28], and complex spin-dependent tunneling spectra [29]. It should be noted that these phenomena have never been observed in conventional MTJs with an amorphous Al-O barrier. Clarifying their underlying mechanisms should lead to a deeper understanding of the physics of spin-dependent tunneling. Of all the novel phenomena that have been observed in epitaxial MgO-based MTJs, we will discuss the interlayer exchange coupling (IEC) mediated by tunneling electrons and injection of spin-polarized electrons through the MgO(001) barrier in this section.

11.4.1 Interlayer Exchange Coupling Mediated by Tunneling Electrons

The epitaxial Fe(001)/MgO(001)/Fe(001) MTJ is also a model system used in studying IEC between two FM layers separated by an insulating NM spacer. IEC in an FM/NM/FM structure with a metallic NM spacer has been extensively studied and is well known to exhibit oscillations as a function of the spacer thickness [30]. IEC for a metallic spacer is induced by conduction electrons at E_F. Although similar IEC is also expected in an FM/NM/FM structure with an insulating NM spacer (i.e., an MTJ) [31,32], such intrinsic IEC has not been observed in MTJs with an amorphous Al-O barrier. Faure-Vincent et al. were the first to observe intrinsic antiferromagnetic (AF) IEC, on which extrinsic FM magnetostatic coupling was superposed, in an epitaxial Fe(001)/MgO(001)/Fe(001) MTJ structure, as shown in Figure 11.7a [33]. Katayama et al. obtained a more refined experimental result that exhibited little extrinsic magnetostatic coupling (Figure 11.7b) [34]. AF coupling is seen at $t_{MgO} < 0.8$ nm. With increasing t_{MgO} the IEC changes sign at $t_{MgO} = 0.8$ nm and then gradually approaches zero. Because of the atomically flat barrier/electrode interfaces (see Figure 11.6), no extrinsic magnetostatic coupling was observed. The IEC for an MgO(001) spacer is apparently mediated by spin-polarized tunneling electrons.

11.4.2 Spin Injection through Crystal MgO(001) Barrier

Spin injection into semiconductors is the key to developing semiconductor spintronic devices such as Datta–Das-type spin-dependent field-effect transistors [35] and spin-dependent metal-oxide-semiconductor field-effect transistors (spin MOSFETs) [36]. However, the direct contact of FM metal on semiconductors results in a very small spin polarization because of the so-called *conductivity mismatch* problem [37]. It was theoretically proposed that this problem can be solved by inserting a tunnel barrier between the FM metal and semiconductor [38,39] (for details, see Chapter 17). A bcc Fe–Co(001) electrode combined with a crystalline MgO(001) tunnel barrier is especially useful for injecting spin-polarized electrons into a semiconductor because the tunneling current is highly spin-polarized.

Parkin et al. [24] first injected tunneling electrons from a bcc Fe–Co(001) electrode into a superconducting counter electrode of $Al_{96}Si_4$ through the MgO(001) barrier at low temperature to

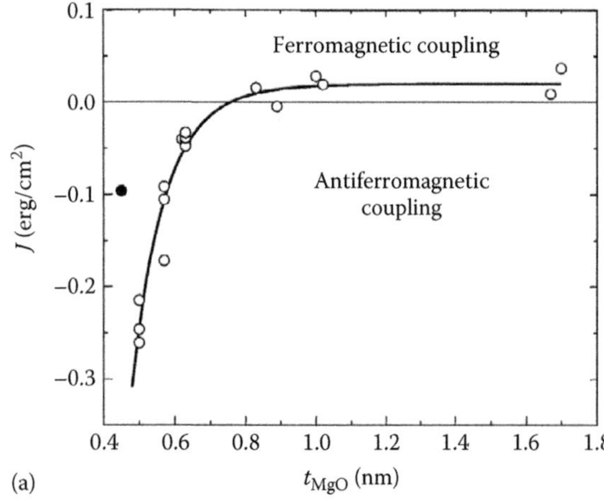

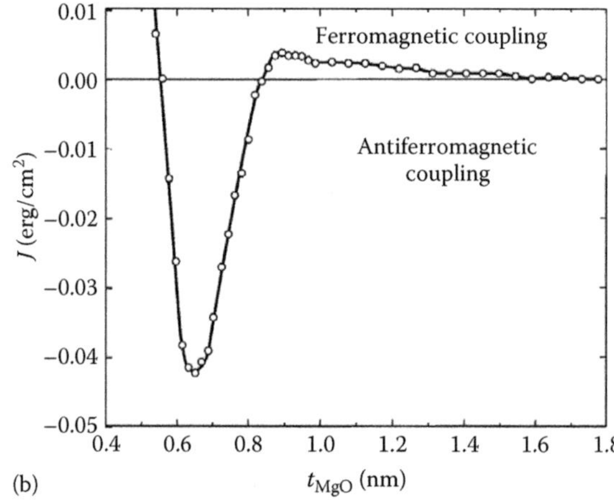

FIGURE 11.7 IEC in epitaxial Fe(001)/MgO(001)/Fe(001) at RT as measured (a) by Faure-Vincent et al. [33] and (b) by Katayama et al. [34].

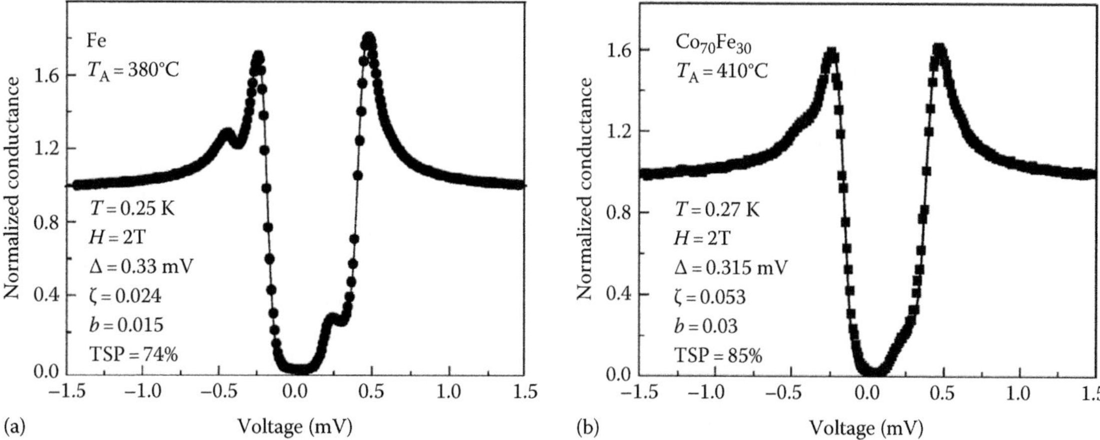

FIGURE 11.8 Measurement of TSP by using Tedrow–Meservey method [24]. Symbols denote experimental data and curves denote fitting. Superconducting counter electrode is $Al_{96}Si_4$. Tunnel barrier is crystalline MgO(001). FM electrode is (a) Fe and (b) $Co_{70}Fe_{30}$ FM. (From Parkin S. et al., *Nat. Mater.*, 3, 866, 2004. With permission.)

evaluate the magnitude of TSP P using the Tedrow–Meservey method (see Section 10.3.1) [6]. Typical results are plotted in Figure 11.8. They observed $P = 74\%$ for an Fe electrode and 85% at 0.3 K for a $Co_{70}Fe_{30}$ electrode. Such high P values have never been observed in Al-O-based tunnel junctions. Although these values do not agree precisely with the MR ratio of a CoFe/MgO/CoFe MTJ when Julliere's model is applied, the fact that a $Co_{70}Fe_{30}$ electrode has a higher P than an Fe electrode agrees qualitatively with the theory on MgO-based MTJs [14]. The high spin polarization is closely related to the high MR ratio of an MgO-based MTJ.

Jiang and Parkin et al. used the CoFe/MgO(001) structure to inject spins into a III-V semiconductor based on GaAs [40]. They fabricated the CoFe/MgO(001) spin injector on a GaAs/AlGaAs quantum well structure that acted as a light emitting diode (LED). When spin polarized current with the quantization axis perpendicular to the plane was injected into the quantum well structure, circularly polarized light was emitted. By measuring the circular polarization of the light, the spin polarization of the electrons injected into the semiconductor could be estimated. They observed a circular polarization of 57% at 100 K and 47% at 290 K. It should be noted that such a high degree of circular polarization by spin injection through an amorphous Al-O tunnel barrier has never been observed. It was thus proved that the MgO(001) tunnel barrier is very useful for obtaining spin injection into semiconductors.

11.5 CoFeB/MgO/CoFeB Magnetic Tunnel Junctions for Device Applications

11.5.1 MTJ Structure for Practical Applications

As explained in Section 11.3, epitaxial MTJs with a single-crystal MgO(001) barrier or textured MTJs with a poly-crystal MgO(001) barrier exhibit the giant TMR effect at RT. This is an attractive property for device applications such as MRAMs and the read heads of HDDs. These MTJ structures, however, cannot be applied to practical devices because of the following problem. MTJs for practical applications need to have the stacking structure shown in Figure 11.9. That is, they need to have an AF layer for exchange biasing, a synthetic ferrimagnetic (SyF) tri-layer structure that acts as a pinned layer, a tunnel barrier, and an FM layer that acts as a free layer. Ir–Mn or Pt–Mn is used as the AF layer. The SyF structure, which consists of an antiferromagnetically coupled FM/NM/FM tri-layer such as Co–Fe/Ru/Co–Fe, is exchange biased by the AF layer and acts as the pinned layer of a spin valve. This type of pinned layer structure is indispensable for device applications because of its robust exchange bias and small stray magnetic field acting on the top free layer. A reliable AF/SyF pinned layer is based on an fcc structure with (111) orientation. A fundamental problem with this structure is that an NaCl-type MgO(001) tunnel barrier and a bcc(001) FM electrode, both of which have fourfold in-plane crystallographic symmetry, cannot be grown on the fcc(111)-oriented AF/SyF structure that has threefold in-plane symmetry because of the

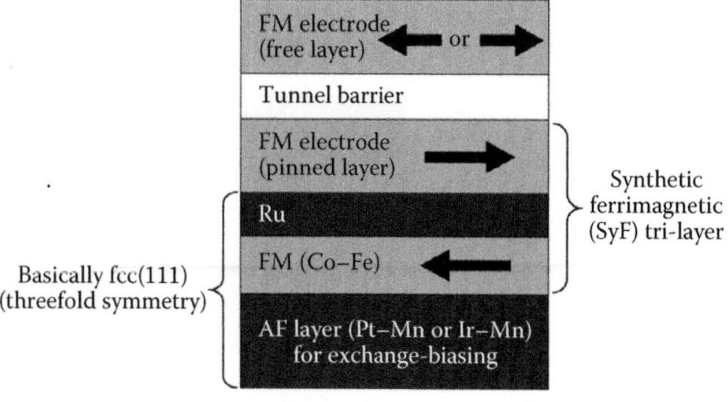

FIGURE 11.9 Cross-sectional structure of MTJ for practical applications.

mismatch in structural symmetry. We could theoretically solve this problem by developing a new pinned layer structure having fourfold in-plane symmetry. This solution, however, is not acceptable to the electronics industry because they have spent more than a decade in developing a reliable pinned layer structure. It should be noted that the reliability of practical devices is determined by the reliability of the pinned layer.

11.5.2 Giant TMR Effect in CoFeB/MgO/CoFeB MTJs

To solve this problem concerning crystal growth, Djayaprawira et al. developed a new MTJ structure, that is, CoFeB/MgO/CoFeB, by using a sputtering deposition technique [41]. There is a cross-sectional TEM image of the MTJ in an *as-grown* state in Figure 11.10a. As can be seen in the figure, the bottom and top CoFeB electrode layers have an amorphous structure in the as-grown state. Surprisingly, the MgO barrier layer grown on the amorphous CoFeB is (001)-oriented poly-crystalline. Because the bottom CoFeB electrode is amorphous, this CoFeB/MgO/CoFeB MTJ can theoretically be grown on all kinds of underlayers. For practical applications, the CoFeB/MgO/CoFeB MTJ can be grown on the standard AF/SyF pinned layer structure shown in Figure 11.10a. It is possible to grow the MgO(001) barrier with fourfold symmetry on the practical pinned layer with threefold symmetry by inserting the amorphous CoFeB bottom electrode layer between them. After annealing at 360°C, the CoFeB/MgO/CoFeB MTJ exhibited an MR ratio of 230% at RT (Figures 11.2 and 11.11) [41]. Up to now, MR ratios above 600% have been obtained at RT in CoFeB/MgO/CoFeB MTJs [42]. The giant TMR effect in CoFeB/MgO/CoFeB MTJs is closely related to the crystallization of the CoFeB electrode layers during the post-annealing process, which is explained later.

As described in Section 11.2, crystallographic symmetry in both the MgO barrier and the electrodes is essential for the coherent tunneling of Δ_1 states. Because the fully spin-polarized Δ_1 band in bcc(001) FM electrodes is essential for the giant TMR effect, amorphous CoFeB electrodes are not expected to result in a giant MR ratio. It has, however, been observed that the amorphous CoFeB electrode layers adjacent to the MgO(001) barrier layer crystallized in the bcc(001) structure during annealing above 250°C (Figure 11.10b) [43–45]. It should be noted that a $Co_{60}Fe_{20}B_{20}$ layer adjacent to an MgO(001) layer crystallizes in a bcc(001) structure although the stable structure of $Co_{60}Fe_{20}B_{20}$ is not bcc but fcc [43]. This clearly indicates that the MgO(001) layer acts as a template for crystallizing the amorphous CoFeB layers because of the good lattice matching between MgO(001) and bcc CoFeB(001). This type of crystallization process is known as *solid phase epitaxy*. The giant TMR effect observed in CoFeB/MgO/CoFeB MTJs can therefore be interpreted within the framework of the theory on epitaxial MTJs because the microscopic structure of bcc CoFeB(001)/MgO(001)/bcc CoFeB(001) MTJs is basically the same as that of epitaxial Fe(001)/MgO(001)/Fe(001) MTJs.

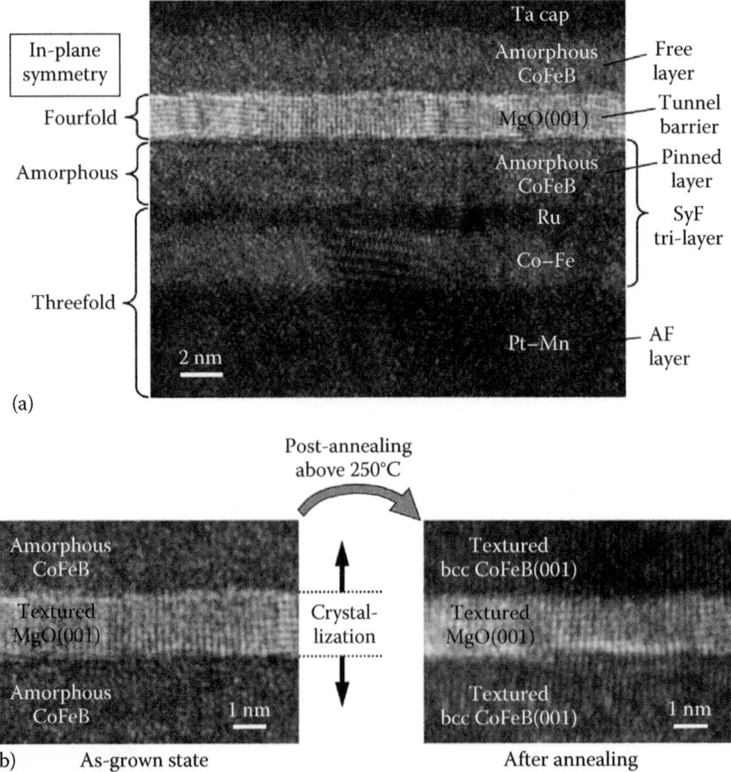

FIGURE 11.10 (a) Cross-sectional TEM image in as-grown state of CoFeB/MgO/CoFeB MTJ with practical AF/SyF structure underneath MTJ part. (b) Cross-sectional TEM image of MTJ part in as-grown state and after annealing. (Courtesy of Canon-Anelva Corp., Tokyo, Japan.)

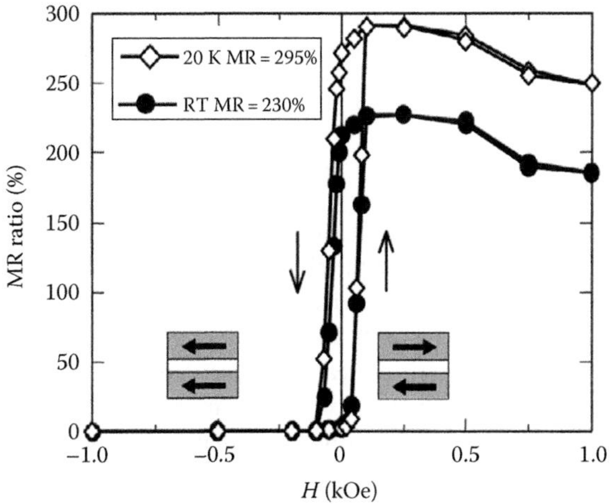

FIGURE 11.11 MR curves at RT and 20K for CoFeB/MgO/CoFeB MTJ. (Redrawn from Djayaprawira, D.D. et al., *Appl. Phys. Lett.*, 86, 092502, 2005.)

A cap layer deposited on the top CoFeB electrode in CoFeB/MgO/CoFeB MTJs was found to have a strong influence on the TMR effect in some cases. In standard CoFeB/MgO/CoFeB MTJs, a Ta or Ru cap layer is usually used (Figure 11.12a). The bottom CoFeB electrode is deposited on an Ru layer (the spacer layer of the SyF tri-layer). In this standard structure, MR ratios of over 200% are obtained at RT. Tsunekawa et al. deposited various cap-layer materials on CoFeB/MgO/CoFeB MTJs and found that the MR ratios were significantly reduced for some cap-layer materials [46]. This is because the cap layer can influence the crystallization of the top CoFeB electrode. An $Ni_{0.8}Fe_{0.2}$ (permalloy) cap layer, for example, grown on amorphous CoFeB has a textured fcc(111) structure (Figure 11.12b). When this structure is annealed, CoFeB crystallization from the interface

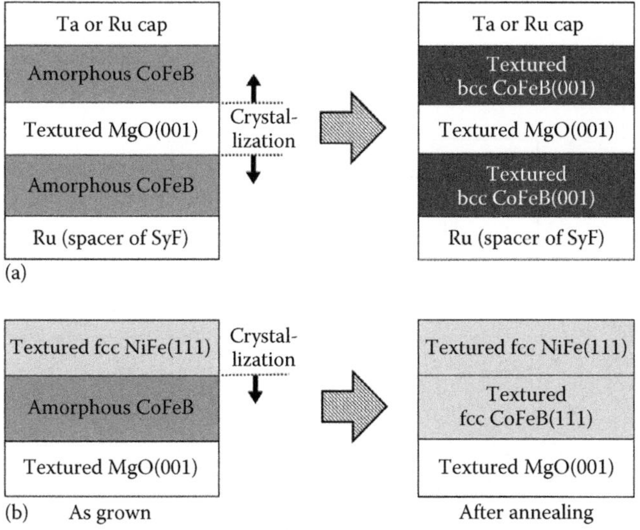

FIGURE 11.12 Schematic of structure of CoFeB/MgO/CoFeB MTJ in as-grown state and after annealing above 250°C. (a) Cap layer is Ta or Ru. (b) Cap layer is $Ni_{0.8}Fe_{0.2}$ (permalloy).

with the $Ni_{0.8}Fe_{0.2}$ layer takes place at about 200°C before crystallization from the MgO interface takes place at about 250°C. Consequently, the top CoFeB layer crystallizes in a textured fcc(111) structure (Figure 11.12b) [43]. Because the top CoFeB electrode does not have a bcc(001) structure in this case, coherent tunneling of Δ_1 electrons does not take place. This results in a significant decrease in the MR ratio. Thus, a correct choice of cap-layer material is necessary to obtain giant MR ratios in CoFeB/MgO/CoFeB MTJs.

As mentioned earlier, the CoFeB/MgO/CoFeB MTJs can be fabricated by sputtering deposition at RT followed by ex situ annealing. This makes it possible to fabricate the MTJ films on large substrates with high throughput. Djayaprawira et al. actually deposited CoFeB/MgO/CoFeB MTJ films on 8 in. diameter Si substrates by using a standard sputtering-deposition chamber to mass-produce HDD read heads and MRAMs [41]. Because this fabrication process is highly compatible with high-throughput mass-production processes, current research and development efforts devoted to producing spintronic devices such as MRAMs and HDD read heads are based on this MTJ structure. The CoFeB/MgO/CoFeB MTJs are also commonly used in basic researches on spin-torque-induced phenomena such as magnetization switching and microwave oscillation. These researches are explained more fully in Section 14.7.

11.6 MTJs with Heusler-Alloy Electrodes

Heusler alloys have an ordered bcc structure with $L2_1$-type chemical ordering. According to band calculations [47], some Heusler alloys are *half metals*, which have full TSP ($|P| = 1$) and are therefore theoretically expected to yield giant MR ratios even in Al-O-based MTJs and current-perpendicular-to-plane (CPP) GMR junctions. Experimentally, a very high MR ratio of 570% at low temperatures has been obtained in Co_2MnSi/Al-O/Co_2MnSi MTJs with an amorphous Al-O barrier [11]. However, the giant MR ratio was observed to disappear at RT as a result of a rapid decrease in the MR ratio with temperature. This giant TMR effect at low temperature must originate from the half-metallic nature of Heusler-alloy electrodes because the amorphous Al-O barrier does not have a function for orbital-symmetry filtering unlike the crystalline MgO(001) barrier.

An MTJ with an MgO(001) barrier and (001)-oriented Heusler-alloy electrodes can be fabricated experimentally because of the relatively small lattice mismatch of a few percent. High-quality epitaxial MTJs with an MgO(001) barrier and Heusler-alloy electrodes have been recently developed and have been observed to demonstrate giant MR ratios of up to about 380% at RT (about 830% at low temperatures) [48–51]. This giant TMR effect, however, is not proof of the half-metallic nature of Heusler-alloy electrodes because the MgO(001) barrier can yield a giant TMR effect if the Δ_1 Bloch states of the electrodes are fully spin-polarized at E_F. It should be noted that the Δ_1 band of half-metallic Heusler alloys is also fully spin-polarized at E_F. Note also that, as described in Sections 11.3 and 11.5, when combined with a crystalline MgO(001) barrier, even simple FM electrodes such as bcc Fe, Co, and CoFeB yield MTJs with MR ratios of up to 600% at RT (1000% at low temperatures) [23,27,41,42]. In both Al-O-based and MgO-based MTJs, RT MR ratios of Heusler-alloy electrodes have never exceeded those of Co–Fe or CoFeB electrodes since the discovery of RT TMR effect in 1995. From the engineering point of view, Heusler-alloy electrodes are much more difficult to synthesize than CoFeB electrodes.

Heusler-alloy electrodes therefore need to exhibit MR ratios that are substantially higher than those of CoFeB not only to prove their half-metallic nature but also to establish their technological importance. Theoretically, Heusler-alloy electrodes have an advantage over CoFeB especially in the ultralow-resistance range because the MgO's orbital-symmetry filtering weakens as the MgO barrier becomes ultrathin. Ultralow-resistance MTJs are crucial in magnetic-sensor applications (see Section 11.7.1). Thus, Heusler alloys still have a great deal of potential for device applications especially in the ultralow-resistance range either as the electrodes of MTJs or as the FM layers of CPP-GMR junctions.

11.7 Device Applications of MgO-Based MTJs

From the viewpoint of device applications, the history of spintronics is a history of MR at RT. The history of MR and its applications is summarized in Figure 11.13. GMR spin-valve devices have MR ratios of 5%–15% at RT and have been used in the read heads of HDDs. Al-O-based MTJs have MR ratios of 20%–70% at RT and have been used not only in HDD read heads but also in MRAM cells. MgO-based MTJs have MR ratios of 100%–600% at RT and are expected to be used in various spintronic devices such as HDD read heads, spin-transfer MRAM cells, and novel microwave devices. In this section, we briefly describe recent developments in these applications. Further details on the HDD read heads and MRAM are explained in Chapters 34 and 35.

11.7.1 Read Heads of HDDs

As explained in Section 11.5, CoFeB/MgO/CoFeB MTJs with the giant TMR effect are compatible with the mass-manufacturing processes for spintronics devices because they can be fabricated on a practical pinned-layer structure by sputtering deposition at RT followed by post-annealing. Apart from requiring giant MR ratios and compatibility with manufacturing processes, industrial applications require MTJs that have MR ratios that have little dependence on bias voltage, that are robust and have high break-down voltages, that can be manufactured uniformly and reproducibly, that have a free layer with small magnetostriction, and that have appropriate resistance-area (*RA*) products. It is now possible to satisfy all these requirements by using CoFeB/MgO/CoFeB MTJs. At the early stage of research and development, however, an ultralow *RA* product that was required for read head applications was not easy to accomplish. In this section, we explain how the ultralow *RA* product was attained and how MgO-based MTJs have been applied to the read heads of ultrahigh-density HDDs.

Because impedance matching in electronic circuits is crucial for the high-speed operation of electronic devices, the *RA* product of MTJs should be adjusted to satisfy impedance-matching conditions. MRAM applications require an *RA* in a range from a few tens of $\Omega \cdot \mu m^2$ to several $k\Omega \cdot \mu m^2$, depending on the lateral MTJ size (i.e., the areal density of an MRAM). In this *RA* range, MR ratios of over 200% at RT can easily be obtained using MgO-based MTJs. A very low *RA* product, on the other hand, is required for the read heads of high-density

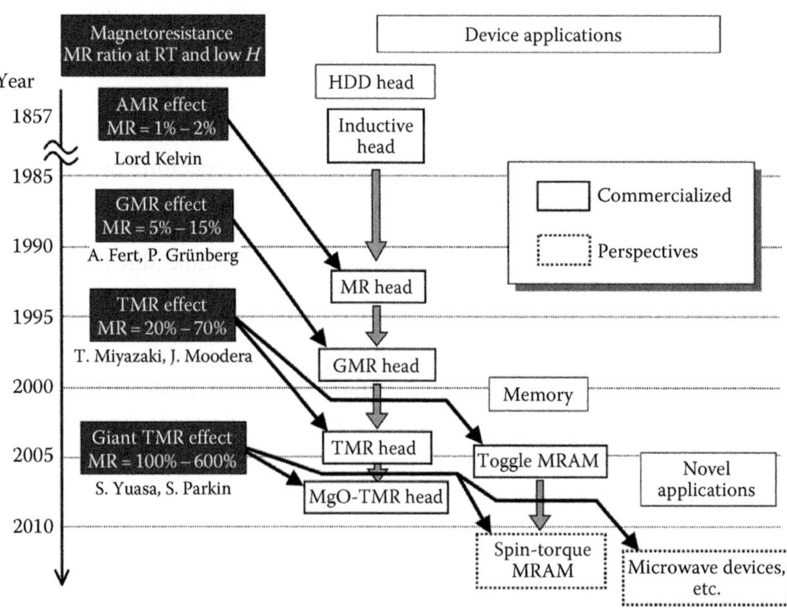

FIGURE 11.13 History and perspectives of magnetoresistive effects, MR ratios at RT and applications of MR effects in spintronic devices.

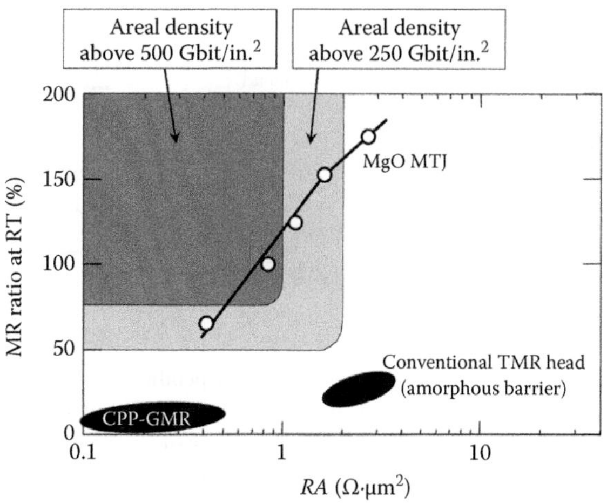

FIGURE 11.14 MR ratio at RT vs. RA product. Open circles are values for CoFeB/MgO/CoFeB MTJs. Light gray and dark gray areas are zones required for HDDs with recording densities above 250 and 500 Gbit/in.^{2}. (Redrawn from Nagamine, Y. et al., *Appl. Phys. Lett.*, 89, 162507, 2006.)

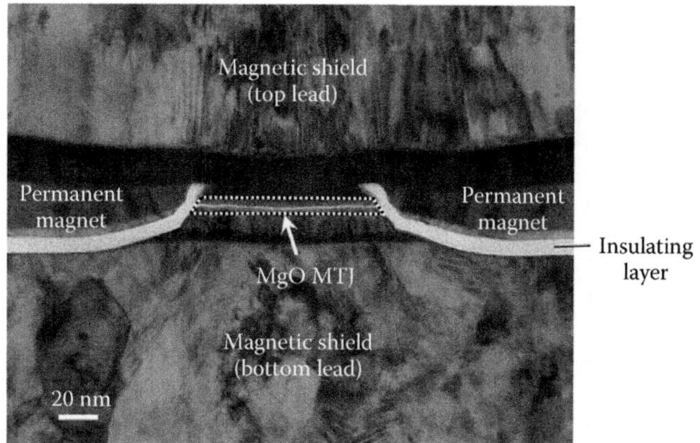

FIGURE 11.15 Cross-sectional TEM image of MgO-TMR read head for HDD with recording density of 250 Gbit/in.^{2}. (Courtesy of Fujitsu Corp., Tokyo, Japan.)

HDDs. MTJs with an amorphous Al-O or Ti-O barrier have been used in the TMR read heads for HDDs with areal recording densities of 100–150 Gbit/in.^{2} since 2004 [52]. These MTJs have low *RA* products (about 3 Ω·μm^2) and MR ratios of 20%–30% at RT. Although these properties are sufficient for recording densities of 100–150 Gbit/in.^{2}, even lower *RA* products and much higher MR ratios are needed for recording densities above 200 Gbit/in.^{2} For example, *RA* products below 2 Ω·μm^2 and MR ratios above 50% are required for areal recording densities above 250 Gbit/in.^{2} (Figure 11.14). Such ultralow *RA* products and high MR ratios have never been obtained in conventional MTJs with an amorphous Al-O or Ti-O barrier. A current perpendicular to plane (CPP) GMR device, which is one of the candidates for the next-generation of HDD read heads, has an ultralow *RA* product (below 1 Ω·μm^2), but the MR ratio of a CPP GMR device is too low (<10% for a practical spin-valve structure) for a device that is used as an HDD read head (Figure 11.14).

To reduce the *RA* product, Tsunekawa et al. made CoFeB/MgO/CoFeB MTJs with an ultrathin textured MgO(001) barrier (t_{MgO} = 1.0 nm, which corresponds to only 4–5 ML) [53,54]. Because the textured growth of MgO(001) on amorphous CoFeB deteriorates below an MgO thickness of 1.0 nm [43], it is necessary to carefully optimize the conditions for growing a 1.0-nm-thick textured MgO(001) barrier. For example, the textured growth of MgO layers was found to be sensitive to the base pressure of the sputtering chamber. Residual H_2O molecules in the chamber were found to degrade the crystalline orientation of MgO(001), reducing the MR ratio. By carefully removing residual H_2O molecules by using a Ta getter technique, Nagamine et al. were able to grow a highly textured 1.0-nm-thick MgO(001) barrier [54]. The removal of residual H_2O molecules was also found to effectively prevent the CoFeB/MgO interface from becoming oxidized [55]. The MR ratios for these CoFeB/MgO/CoFeB MTJs are indicated by the open circles in Figure 11.14. Nagamine et al. obtained both ultralow *RA* products (<1 Ω·μm^2) and high MR ratios (>100% at RT), satisfying the requirements for areal recording densities well above 500 Gbit/in.^{2}.

By using ultralow-resistance MgO-based MTJs, HDD manufacturers were able to commercialize TMR read heads for ultrahigh-density HDDs in 2007. For example, Fujitsu Corp. commercialized TMR read heads for ultrahigh-density HDDs in 2007. There is a cross-sectional TEM image of a TMR read head in Figure 11.15. MgO-TMR heads have already been used for HDDs with recording densities of 250–500 Gbit/in.^{2}. This means that the recording density of HDDs has tripled thanks to MgO-TMR heads combined with perpendicular magnetic recording media. MgO-TMR heads currently represent the mainstream technology for HDD read heads and are also expected to be applied to recording densities of up to about 1 Tbit/in.^{2}.

11.7.2 Spin-Torque MRAMs

The giant TMR effect in MgO-based MTJs is also useful in developing MRAMs. In conventional MRAMs, the writing process (i.e., magnetization reversal of a free layer) uses a magnetic field generated by pulse currents, and the readout process uses a resistance change between parallel and antiparallel magnetic states (i.e., the TMR effect). The giant TMR effect enables high-speed readout because a giant MR ratio yields a very high output signal [56]. In conventional MRAMs, however, the writing pulse currents increase when the lateral size of MTJs is reduced, which makes it difficult to develop Gbit-scale high-density MRAMs. In the new types of MRAMs called spin-transfer MRAMs or spin RAMs, on the other hand, the writing process uses the magnetization switching induced by spin-transfer torque (for details, see Chapter 14). This phenomenon, called spin-torque switching (STS) [57], is especially important in developing high-density MRAMs because the writing pulse current flowing through the

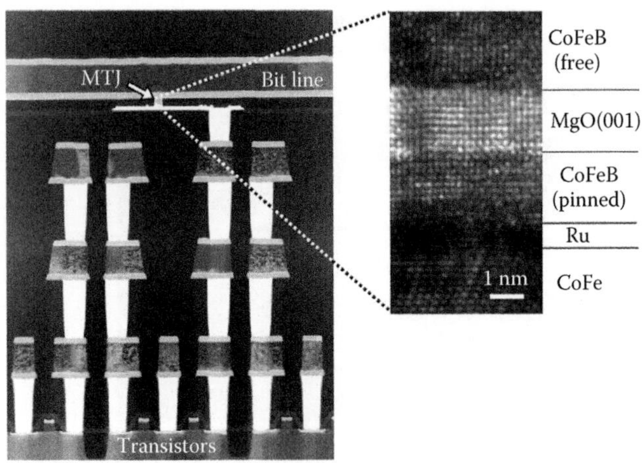

FIGURE 11.16 Cross-sectional TEM images of 4 kbit Spin-RAM using CoFeB/MgO/CoFeB MTJs. (Courtesy of Sony Corp., Tokyo, Japan.) (Redrawn from Hosomi, M. et al., A novel nonvolatile memory with spin torque transfer magnetization switching: Spin-RAM, *Technical Digest of IEEE International Electron Devices Meeting (IEDM)*: 19.1, Washington, DC, 2005.)

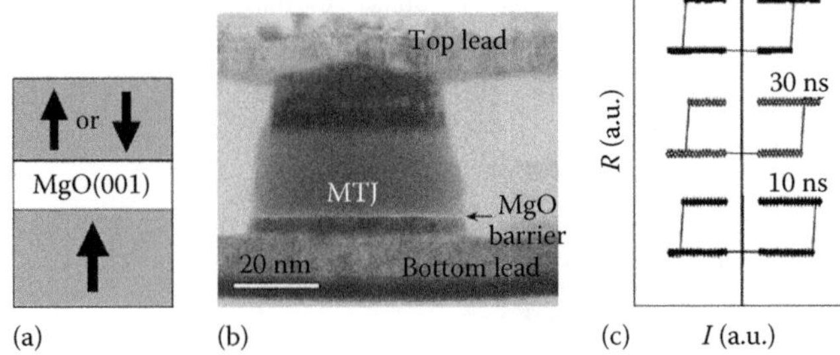

FIGURE 11.17 MgO-based MTJ with perpendicularly magnetized electrodes. (a) Schematic and (b) cross-sectional TEM image of MTJ. (c) STS at pulse durations of 10–100 ns. (Redrawn from Kishi, T. et al., Lower-current and fast switching of a perpendicular TMR for high speed and high density spin-transfer-torque MRAM, *Technical Digest of IEEE International Electron Devices Meeting (IEDM)*: 12.6, San Francisco, CA, 2008.)

MTJ can be reduced by reducing the MTJ's lateral size. STS was experimentally demonstrated first in CPP GMR devices [58] and later in Al-O-based MTJs [59]. STS in MgO-based MTJs, which is especially important for MRAMs, has been demonstrated using CoFeB/MgO/CoFeB MTJs [60–64].

Prototype spin-transfer MRAM cells based on the giant TMR effect and STS in MgO-based MTJs have been developed [64,65]. The 4 kbit spin-torque MRAM (or Spin-RAM) developed by Sony Corp. (shown in Figure 11.16), for example, provides reliable readout and write operations [64]. In CoFeB/MgO/CoFeB MTJs, the intrinsic critical current density, J_{c0}, or switching pulse-current density when the pulse duration is 1 ns, is about 2×10^6 A/cm^2 [63,64]. This J_{c0} is, however, not small enough for high-density Spin-RAMs. The J_{c0} value needs to be smaller than 1×10^6 A/cm^2 and high thermal stability for free-layer magnetization is necessary to retain data for more than 10 years to develop Gbit-scale Spin-RAMs. High thermal stability is especially difficult when the lateral size of MTJ cells is smaller than about 50 nm, which is a typical cell size in Gbit-scale Spin-RAMs.

A solution to achieving Gbit-scale Spin-RAMs is to use MTJs with perpendicularly magnetized electrodes. It is theoretically possible to simultaneously have high thermal stability in 50-nm-sized MTJ cells and a J_{c0} smaller than 1×10^6 A/cm^2 by using perpendicularly magnetized electrodes with a high perpendicular magnetic anisotropy, K_u. Toshiba Corp. recently developed a prototype spin RAM using 50 nm sized MgO-based MTJs with perpendicularly magnetized electrodes (Figure 11.17). They successfully demonstrated a small J_{c0}, thermal stability that was high enough to retain data for more than a decade, and high-speed read/write operations. The MgO-based MTJs with perpendicularly magnetized electrodes are a promising solution to accomplishing Gbit-scale Spin-RAMs.

11.7.3 Microwave Applications

MgO-based MTJs are also potentially useful for microwave device applications. Tulapurkar et al. demonstrated that a DC voltage is generated between the two electrodes when an AC current with a microwave frequency flows through a CoFeB/MgO/CoFeB MTJ with a lateral size of about 100 nm [67]. This phenomenon, called the *spin-torque diode effect* or *spin-torque ferromagnetic resonance* (*FMR*), results from a combination of the giant TMR effect and resonant precession of the free-layer magnetic moment induced by spin-transfer torque. Because of this effect, MgO-based MTJs can thus act as microwave detectors. The spin-torque diode effect is also a powerful tool for measuring the magnitude of spin-transfer torque. Using the spin-torque diode effect, Sankey et al. [68] and Kubota et al. [69] carried out quantitative measurements of the magnitude of spin-transfer torque as a function of the applied bias voltage.

An inverse of the spin-torque diode effect is the microwave emissions that occur when a DC current flows through a 100-nm-scale magnetoresistive device. Spin-transfer torque acting on the free-layer magnetic moment can induce steady precession of the free-layer moment at an FMR frequency. The steady precession of the free-layer moment in magnetoresistive device induces an AC current or AC voltage at an FMR frequency (i.e., a microwave frequency). This microwave emission was first demonstrated using a CPP GMR device with an MR ratio of about 1% at RT [70]. The microwave power from a CPP GMR device, however, is only of the order of nanowatts and is too small for practical use. However, because the microwave power is theoretically proportional to the square of the MR ratio, MgO-based MTJs with giant MR ratios are potentially able to emit high-power microwaves. Deac et al. recently obtained microwave emissions of microwatt order from a 100-nm CoFeB/MgO/CoFeB MTJ with an MR ratio of about 100% [71]. Spin-torque-induced microwave emissions based on MgO-based MTJs are expected to be the third major application of spintronics technology.

As previously described, MgO-based MTJs can be used in various spintronic devices. In these applications, output performance is roughly proportional to the MR ratio at RT. The giant TMR effect in MTJs is therefore expected not only to extend the applications of existing devices but also to help achieve novel spintronic applications (Figure 11.13).

Acknowledgments

The author would like to acknowledge the contributions made by Messrs. A. Fukushima, T. Nagahama, H. Kubota, K. Ando (AIST), D. D. Djayaprawira, K. Tsunekawa, H. Maehara, Y. Nagamine, S. Yamagata, N. Watanabe (Canon-Anelva Corp.), R. Matsumoto, A. M. Deac, and Y. Suzuki (Osaka University) for their valuable discussions, and the financial support given by the New Energy and Industrial Technology Development Organization (NEDO) and the Japan Science and Technology Corporation (JST).

References

1. M. Julliere, Tunneling between ferromagnetic films, *Phys. Lett.* **54**A, 225–226 (1975).
2. M. N. Baibich, J. M. Broto, A. Fert, F. Nguyen Van Dau, and F. Petroff, Giant magnetoresistance of (001)Fe/(001)Cr magnetic superlattices, *Phys. Rev. Lett.* **61**, 2472–2475 (1988).
3. G. Binasch, P. Grünberg, F. Saurenbach, and W. Zinn, Enhanced magnetoresistance in layered magnetic structures with antiferromagnetic interlayer exchange, *Phys. Rev. B* **39**, 4828 (1989).
4. T. Miyazaki and N. Tezuka, Giant magnetic tunneling effect in Fe/Al_2O_3/Fe junction, *J. Magn. Magn. Mater.* **139**, L231–L234 (1995).
5. J. S. Moodera, L. R. Kinder, T. M. Wong, and R. Meservey, Large magnetoresistance at room temperature in ferromagnetic thin film tunnel junctions, *Phys. Rev. Lett.* **74**, 3273–3276 (1995).
6. R. Meservey and P. M. Tedrow, Spin-polarized electron-tunneling, *Phys. Rep.* **238**, 173–243 (1994).
7. S. S. P. Parkin, X. Jiang, C. Kaiser, A. Panchula, K. Roche, and M. Samant, Magnetically engineered spintronic sensors and memory, *Proc. IEEE* **91**, 661–680 (2003).
8. S. Yuasa, T. Nagahama, and Y. Suzuki, Spin-polarized resonant tunneling in magnetic tunnel junctions, *Science* **297**, 234–237 (2002).
9. T. Nagahama, S. Yuasa, E. Tamura, and Y. Suzuki, Spin-dependent tunneling in magnetic tunnel junctions with a layered antiferromagnetic Cr(001) spacer: Role of band structure and interface scattering, *Phys. Rev. Lett.* **95**, 086602 (2005).
10. M. Bowen, M. Bibes, A. Barthélémy et al., Nearly total spin polarization in $La_2/3Sr_1/3MnO_3$ from tunneling experiments, *Appl. Phys. Lett.* **82**, 233–235 (2003).
11. Y. Sakuraba, M. Hattori, M. Oogane et al., Giant tunneling magnetoresistance in Co_2MnSi/Al-O/Co_2MnSi magnetic tunnel junctions, *Appl. Phys. Lett.* **88**, 192508 (2006).
12. W. H. Butler, X.-G. Zhang, T. C. Schulthess, and J. M. MacLaren, Spin-dependent tunneling conductance of Fe/MgO/Fe sandwiches, *Phys. Rev. B* **63**, 054416 (2001).
13. J. Mathon and A. Umersky, Theory of tunneling magnetoresistance of an epitaxial Fe/MgO/Fe(001) junction, *Phys. Rev. B* **63**, 220403R (2001).
14. X.-G. Zhang and W. H. Butler, Large magnetoresistance in bcc Co/MgO/Co and FeCo/MgO/FeCo tunnel junctions, *Phys. Rev. B* **70**, 172407 (2004).
15. P. Mavropoulos, N. Papanikolaou, and P. H. Dederichs, Complex band structure and tunneling through ferromagnet/insulator/ferromagnet junctions, *Phys. Rev. Lett.* **85**, 1088–1091 (2000).
16. J. P. Velev, K. D. Belashchenko, D. A. Stewart, M. van Schilfgaarde, S. S. Jaswal, and E. Y. Tsymbal, Negative spin polarization and large tunneling magnetoresistance in epitaxial Co vertical bar SrTiO3 vertical bar Co magnetic tunnel junctions, *Phys. Rev. Lett.* **95**, 216601 (2005).
17. W. Wulfhekel, M. Klaua, D. Ullmann et al., Single-crystal magnetotunnel junctions, *Appl. Phys. Lett.* **78**, 509–511 (2001).
18. M. Bowen, V. Cros, F. Petroff et al., Large magnetoresistance in Fe/MgO/FeCo(001) epitaxial tunnel junctions on GaAs(001), *Appl. Phys. Lett.* **79**, 1655–1657 (2001).
19. J. Faure-Vincent, C. Tiusan, E. Jouguelet et al., High tunnel magnetoresistance in epitaxial Fe/MgO/Fe tunnel junctions, *Appl. Phys. Lett.* **82**, 4507–4509 (2003).
20. H. L. Meyerheim, R. Popescu, J. Kirschner et al., Geometrical and compositional structure at metal-oxide interfaces: MgO on Fe(001), *Phys. Rev. Lett.* **87**, 076102 (2001).
21. X.-G. Zhang, W. H. Butler, and A. Bandyopadhyay, Effects of the iron-oxide layer in Fe–FeO–MgO–Fe tunneling junctions, *Phys. Rev. B* **68**, 092402 (2003).
22. S. Yuasa, A. Fukushima, T. Nagahama, K. Ando, and Y. Suzuki, High tunnel magnetoresistance at room temperature in fully epitaxial Fe/MgO/Fe tunnel junctions due to coherent spin-polarized tunneling, *Jpn. J. Appl. Phys.* **43**, L588–L590 (2004).
23. S. Yuasa, T. Nagahama, A. Fukushima, Y. Suzuki, and K. Ando, Giant room-temperature magnetoresistance in single-crystal Fe/MgO/Fe magnetic tunnel junctions, *Nat. Mater.* **3**, 868–871 (2004).
24. S. S. P. Parkin, C. Kaiser, A. Panchula et al., Giant tunnelling magnetoresistance at room temperature with MgO (100) tunnel barriers, *Nat. Mater.* **3**, 862–867 (2004).
25. K. Miyokawa, S. Saito, T. Katayama et al., X-ray absorption and x-ray magnetic circular dichroism studies of a monatomic Fe(001) layer facing a single-crystalline MgO(001) tunnel barrier, *Jpn. J. Appl. Phys.* **44**, L9–L11 (2005).
26. S. Yuasa, T. Katayama, T. Nagahama et al., Giant tunneling magnetoresistance in fully epitaxial body-centered-cubic Co/MgO/Fe magnetic tunnel junctions, *Appl. Phys. Lett.* **87**, 222508 (2005).

27. S. Yuasa, A. Fukushima, H. Kubota, Y. Suzuki, and K. Ando, Giant tunneling magnetoresistance up to 410% at room temperature in fully epitaxial Co/MgO/Co magnetic tunnel junctions with bcc Co(001) electrodes, *Appl. Phys. Lett.* **89**, 042505 (2006).
28. R. Matsumoto, A. Fukushima, T. Nagahama, Y. Suzuki, K. Ando, and S. Juasa, Oscillation of giant tunneling magnetoresistance with respect to tunneling barrier thickness in fully epitaxial Fe/MgO/Fe magnetic tunnel junctions, *Appl. Phys. Lett.* **90**, 252506 (2007).
29. Y. Ando, T. Miyakoshi, M. Oogane et al., Spin-dependent tunneling spectroscopy in single-crystal Fe/MgO/Fe tunnel junctions, *Appl. Phys. Lett.* **87**, 142502 (2005).
30. S. S. P. Parkin, N. More, and K. Roche, Oscillations in exchange coupling and magnetoresistance in metallic superlattice structures—Co/Ru, Co/Cr, and Fe/Cr, *Phys. Rev. Lett.* **64**, 2304–2307 (1990).
31. P. Bruno, Theory of interlayer magnetic coupling, *Phys. Rev. B* **52**, 411–439 (1995).
32. J. C. Slonczewski, Currents, torques, and polarization factors in magnetic tunnel junctions, *Phys. Rev. B* **71**, 024411 (2005).
33. J. Faure-Vincent, C. Tiusan, C. Bellouard et al., Interlayer magnetic coupling interactions of two ferromagnetic layers by spin polarized tunneling, *Phys. Rev. Lett.* **89**, 107206 (2002).
34. T. Katayama, S. Yuasa, J. Velev, M. Ye. Zhuravlev, S. S. Jaswal, and E. Y. Tsymbal, Interlayer exchange coupling in Fe/MgO/Fe magnetic tunnel junctions, *Appl. Phys. Lett.* **89**, 112503 (2006).
35. S. Datta and B. Das, Electronic analog of the electro-optic modulator, *Appl. Phys. Lett.* **56**, 665 (1990).
36. S. Sugahara, Spin metal-oxide-semiconductor field-effect transistors (spin MOSFETs) for integrated spin electronics, *IEE Proc. Circuits Dev. Syst.* **152**, 355–365 (2005).
37. G. Schmidt, D. Ferrand, L. W. Molenkamp, A. T. Filip, and B. J. vanWees. Fundamental obstacle for electrical spin injection from a ferromagnetic metal into a diffusive semiconductor, *Phys. Rev. B* **62**, R4790 (2000).
38. E. I. Rashba, Theory of electrical spin injection: Tunnel contacts as a solution of the conductivity mismatch problem, *Phys. Rev. B* **62**, R16267 (2000).
39. A. Fert and H. Jaffrès, Conditions for efficient spin injection from a ferromagnetic metal into a semiconductor, *Phys. Rev. B* **64**, 184420 (2001).
40. X. Jiang, R. Wang, R. M. Shelby et al., Highly spin-polarized room-temperature tunnel injector for semiconductor spintronics using MgO(100), *Phys. Rev. Lett.* **94**, 056601 (2005).
41. D. D. Djayaprawira, K. Tsunekawa, N. Motonobu et al., 230% room-temperature magnetoresistance in CoFeB/MgO/CoFeB magnetic tunnel junctions, *Appl. Phys. Lett.* **86**, 092502 (2005).
42. S. Ikeda, J. Hayakawa, Y. Ashizawa et al., Tunnel magnetoresistance of 604% at 300 K by suppression of Ta diffusion in CoFeB/MgO/CoFeB pseudo-spin-valves annealed at high temperature, *Appl. Phys. Lett.* **93**, 082508 (2008).
43. S. Yuasa, Y. Suzuki, T. Katayama, and K. Ando, Characterization of growth and crystallization processes in CoFeB/MgO/CoFeB magnetic tunnel junction structure by reflective high-energy electron diffraction, *Appl. Phys. Lett.* **87**, 242503 (2005).
44. S. Yuasa and D. D. Djayaprawira, Giant tunnel magnetoresistance in magnetic tunnel junctions with a crystalline MgO(001) barrier, *J. Phys. D: Appl. Phys.* **40**, R337–R354 (2007).
45. Y. S. Choi, K. Tsunekawa, Y. Nagamine, and D. Djayaprawira, Transmission electron microscopy study on the polycrystalline CoFeB/MgO/CoFeB based magnetic tunnel junction showing a high tunneling magnetoresistance, predicted in single crystal magnetic tunnel junction, *J. Appl. Phys.* **101**, 013907 (2007).
46. K. Tsunekawa, D. D. Djayaprawira, M. Nagai, H. Maehara, S. Yamagata, and N. Watanabe, *Digests of the IEEE International Magnetics Conference (Intermag)*: HP-08, Nagoya, Japan, 2005.
47. I. Galanakis, P. H. Dederichs, and N. Papanikolaou, Slater-Pauling behavior and origin of the half-metallicity of the full-Heusler alloys, *Phys. Rev. B* **66**, 174429 (2002).
48. N. Tezuka, N. Ikeda, S. Sugimoto, and K. Inomata, Giant tunnel magnetoresistance at room temperature for junctions using full-Heusler $Co_2FeAl_{0.5}Si_{0.5}$ electrodes, *Jpn. J. Appl. Phys.* **46**, L454–L456 (2007).
49. S. Tsunegi, Y. Sakuraba, M. Oogane, K. Takanashi, and Y. Ando, Large tunnel magnetoresistance in magnetic tunnel junctions using a Co_2MnSi Heusler alloy electrode and a MgO barrier, *Appl. Phys. Lett.* **93**, 112506 (2008).
50. N. Tezuka, N. Ikeda, F. Mitsuhashi, and S. Sugimoto, Improved tunnel magnetoresistance of magnetic tunnel junctions with Heusler $Co_2FeAl_{0.5}Si_{0.5}$ electrodes fabricated by molecular beam epitaxy, *Appl. Phys. Lett.* **94**, 162504 (2009).
51. W. H. Wang, H. Sukegawa, R. Shan, S. Mitani, and K. Inomata, Giant tunneling magnetoresistance up to 330% at room temperature in sputter deposited Co_2FeAl/MgO/CoFe magnetic tunnel junctions, *Appl. Phys. Lett.* **95**, 182502 (2009).
52. J.-G. Zhu and C.-D. Park, Magnetic tunnel junctions, *Mater. Today* **9**(11), 36–45 (2006).
53. K. Tsunekawa, D. D. Djayaprawira, M. Nagai et al., Giant tunneling magnetoresistance effect in low-resistance CoFeB/MgO(001)/CoFeB magnetic tunnel junctions for read-head applications, *Appl. Phys. Lett.* **87**, 072503 (2005).
54. Y. Nagamine, H. Maehara, K. Tsunekawa et al., Ultralow resistance-area product of $0.4\,\Omega(\mu m)^2$ and high magnetoresistance above 50% in CoFeB/MgO/CoFeB magnetic tunnel junctions, *Appl. Phys. Lett.* **89**, 162507 (2006).
55. Y. S. Choi, Y. Nagamine, K. Tsunekawa et al., Effect of Ta getter on the quality of MgO tunnel barrier in the polycrystalline CoFeB/MgO/CoFeB magnetic tunnel junction, *Appl. Phys. Lett.* **90**, 012505 (2007).

56. J. M. Slaughter, R. W. Dave, M. Durlam et al., High speed toggle MRAM with MgO-based tunnel junctions, *Technical Digest of IEEE International Electron Devices Meeting (IEDM)*: 35.7, Washington, DC, 2005.
57. J. C. Slonczewski, Current-driven excitation of magnetic multilayers, *J. Magn. Magn. Mater.* **159**, L1–L7 (1996).
58. J. A. Katine, F. J. Albert, R. A. Buhrman, E. B. Myers, and D. C. Ralph, Current-driven magnetization reversal and spin-wave excitations in Co/Cu/Co pillars, *Phys. Rev. Lett.* **84**, 3149–3152 (2000).
59. Y.-M. Huai, F. Albert, P. Nguyen, M. Pakala, and T. Valet, Observation of spin-transfer switching in deep submicron-sized and low-resistance magnetic tunnel junctions, *Appl. Phys. Lett.* **84**, 3118–3120 (2004).
60. H. Kubota, A. Fukushima, Y. Ootani et al., Evaluation of spin-transfer switching in CoFeB/MgO/CoFeB magnetic tunnel junctions, *Jpn. J. Appl. Phys.* **44**, L1237–L1240 (2005).
61. H. Kubota, A. Fukushima, Y. Ootani et al., Dependence of spin-transfer switching current on free layer thickness in Co-Fe-B/MgO/Co-Fe-B magnetic tunnel junctions, *Appl. Rev. Lett.* **89**, 032505 (2006).
62. J. Hayakawa, S. Ikeda, Y. M. Lee et al., Current-driven magnetization switching in CoFeB/MgO/CoFeB magnetic tunnel junctions, *Jpn. J. Appl. Phys.* **44**, L1267–L1270 (2005).
63. Z. Diao, D. Apalkov, M. Pakala, Y. Ding, A. Panchula, and Y. Huai, Spin transfer switching and spin polarization in magnetic tunnel junctions with MgO and AlOx barriers, *Appl. Phys. Lett.* **87**, 232502 (2005).
64. M. Hosomi, H. Yamagishi, T. Yamamoto et al., A novel nonvolatile memory with spin torque transfer magnetization switching: Spin-RAM, *Technical Digest of IEEE International Electron Devices Meeting (IEDM)*: 19.1, Washington, DC, 2005.
65. T. Kawahara, R. Takemura, K. Miura et al., 2Mb spin-transfer torque RAM (SPRAM) with bit-by-bit bidirectional current write and parallelizing-direction current read, *Technical Digest of IEEE International Solid-State Circuits Conference (ISSCC)*: 26.5, Lille, France, 2007.
66. T. Kishi, H. Yoda, T. Kai et al., Lower-current and fast switching of a perpendicular TMR for high speed and high density spin-transfer-torque MRAM, *Technical Digest of IEEE International Electron Devices Meeting (IEDM)*: 12.6, San Francisco, CA, 2008.
67. A. A. Tulapurkar, Y. Suzuki, A. Fukushima et al., Spin-torque diode effect in magnetic tunnel junctions, *Nature* **438**, 339–342 (2005).
68. J. C. Sankey, Y.-T. Cui, J. Z. Sun et al., Measurement of the spin-transfer-torque vector in magnetic tunnel junctions, *Nat. Phys.* **4**, 67–71 (2008).
69. H. Kubota, A. Fukushima, K. Yakushiji et al., Quantitative measurement of voltage dependence of spin-transfer torque in MgO-based magnetic tunnel junctions, *Nat. Phys.* **4**, 37–41 (2008).
70. S. I. Kiselev, J. C. Sankey, I. N. Krivorotov et al., Microwave oscillations of a nanomagnet driven by a spin-polarized current, *Nature* **425**, 380–383 (2003).
71. A. M. Deac, A. Fukushima, H. Kubota et al., Bias-driven high-power microwave emission from MgO-based tunnel magnetoresistance devices, *Nat. Phys.* **4**, 803–809 (2008).

12

Tunneling Magnetoresistance: Theory

Kirill D. Belashchenko
University of Nebraska

Evgeny Y. Tsymbal
University of Nebraska

12.1 Introduction

Tunneling magnetoresistance (TMR) is a change of the electrical resistance of a magnetic tunnel junction (MTJ) under the influence of an external magnetic field. An MTJ is a tunnel junction with two magnetic metals serving as electrodes; the resistance change occurs when the orientation of the electrode magnetizations is switched between parallel and antiparallel configurations (see Chapter 10). In this respect, TMR is similar to the giant magnetoresistance (GMR) effect, where the magnetic electrodes are separated by a nonmagnetic metal, rather than an insulating barrier (see Chapters 4 and 5 for the reviews on GMR).

As reviewed in the Chapters 10 and 11, TMR depends strongly on the choice of the electrode and barrier materials, as well as on the fabrication technique and quality of the interfaces. The discoveries of high and reproducible TMR values at room temperature in Al_2O_3-based MTJs (Chapter 10) and later in high-quality epitaxial Fe/MgO/Fe MTJs (Chapter 11) have stimulated a surge of interest to this phenomenon due to its many potential applications, and a variety of electrode/barrier combinations have been explored. These discoveries also stimulated significant progress in the theory of spin-dependent tunneling.

The simplest models of tunneling based on the semiclassical approximation [1–3] are often used for the analysis of the tunneling I-V curves. However, these models are not able to describe TMR because they completely ignore the nature of the electrodes and consequently cannot take into account the changes associated with the magnetization configuration. Over the years, the TMR has usually been interpreted in terms of models based either on Bardeen's view of tunneling as being driven by overlap of wavefunctions penetrating into the barrier with the electrodes [4], or on the one-dimensional picture of tunneling across a potential barrier with exchange-split free-electron bands of the ferromagnetic electrodes [5]. These models are physically transparent and, though simple, capture some important physics of TMR. However, they have no predictive power for the quantitative description of TMR because they do not incorporate an accurate band structure of MTJs. A significant progress has recently been achieved based on the band theory, which takes into account the particular material-specific form of the crystal potential and the multi-orbital electronic structure.

In this chapter, we describe the current level of the theoretical understanding of spin-dependent tunneling. Starting from the general background and simple formulations of electron tunneling, we then consider more sophisticated approaches capturing the important physics related to a realistic spin-dependent electronic band structure of MTJs. We also discuss the influence of imperfections on TMR and the phenomenon of tunneling anisotropic magnetoresistance (TAMR), which is induced by spin–orbit coupling (SOC) and is closely related to TMR. This chapter complements previous review articles that include theory of TMR [6–9].

12.2 Exploring Tunneling Magnetoresistance Using Simple Models

12.2.1 Bardeen's Description of Tunneling

In Bardeen's approach [4], the transmission probability of a tunnel barrier is assumed to be small, so that the coupling between the electrodes can be regarded as a small perturbation. At $t=0$, one considers the many-electron eigenfunctions of the unperturbed potential corresponding to the left or right electrode with the barrier extended to infinity. Since these states are coupled by tunneling, they are not eigenstates of the full Hamiltonian. The perturbation that couples the electrodes is called the transfer Hamiltonian. Using the standard time-dependent perturbation theory, one can then find the rates of tunneling transitions, that is, those involving an electron transfer between the electrodes [4], which are given by Fermi's golden rule. Multiplying by the electron charge e, one obtains for the tunneling current:

$$J = \frac{2\pi e}{\hbar}\sum_n P_n |M_{0n}|^2 \delta(E_n - E_0) \tag{12.1}$$

where

0 labels the many-body ground state that has N_L electrons in the left and N_R in the right electrode

n enumerates the excited states with $N_L \pm 1$ electrons in the left and $N_R \pm 1$ in the right electrode

M_{0n} is the tunneling matrix element

$P_n = \pm 1$ counts left-to-right and right-to-left tunneling rates with appropriate signs

Since the tunneling Hamiltonian is a one-particle operator, its matrix elements M_{0n} are nonzero only for states n in which one single quasiparticle is removed from the ground state of the left electrode and one single quasiparticle is added to the ground state of the right electrode (or vice versa). Bardeen has shown [4] that this matrix element is equal (up to the factor $-i\hbar$) to the matrix element of the current density operator between the states 0 and n taken at a plane inside the barrier.

Therefore, if we represent the many-body wavefunctions as Slater determinants of quasiparticle states occupied according to the Fermi distribution $f(\varepsilon-\mu)$ for the two electrodes (with different chemical potentials μ_L and μ_R), we find

$$J = \frac{2\pi e}{\hbar}\sum_{ij} |b_{ij}|^2 [f(\varepsilon_i - \mu_L) - f(\varepsilon_j - \mu_R)]\delta(\varepsilon_i - \varepsilon_j) \tag{12.2}$$

where

i and j label the single-quasiparticle states of the left and right electrodes, respectively

b_{ij} is the tunneling matrix element between these states (which can be calculated using Bardeen's prescription)

As a reminder, the result (Equation 12.2) for the tunneling current is not exact, because it only contains the first order of the perturbation theory, neglecting the interference of multiple scatterings from the interfaces.

12.2.2 Jullière's Model of Tunneling Magnetoresistance

Jullière's model for TMR [10] may be viewed as an application of Bardeen's approach with two assumptions. First, the electron spin is assumed to be conserved in the tunneling process. It follows then, that tunneling of up- and down-spin electrons are two independent processes. Such a two-current model has also been used to interpret the closely related phenomenon of GMR (see Chapter 5), but spin-flip corrections are sometimes needed to include the effects of SOC or thermal spin disorder. The electrons from the filled quasiparticle states of the left ferromagnetic electrode are thus accepted by unfilled states of the same spin of the right electrode. If the two ferromagnetic films are magnetized parallel, the minority spins tunnel to the minority states and the majority spins tunnel to the majority states. If, however, the two films are magnetized antiparallel, the identity of the majority- and minority-spin electrons is reversed. Second, Jullière's model assumes that the matrix elements b_{ij} are the same for all single-particle states. The corresponding factor in Equation 12.2 can then be moved outside the sum. Since the difference in the chemical potentials is equal to electron charge times the voltage drop across the barrier, at $T=0$ one then immediately obtains

$$G_P = \frac{2\pi e^2}{\hbar}|b|^2\left[\rho_L^\uparrow(\varepsilon_F)\rho_R^\uparrow(\varepsilon_F) + \rho_L^\downarrow(\varepsilon_F)\rho_R^\downarrow(\varepsilon_F)\right]$$

$$G_{AP} = \frac{2\pi e^2}{\hbar}|b|^2\left[\rho_L^\uparrow(\varepsilon_F)\rho_R^\downarrow(\varepsilon_F) + \rho_L^\downarrow(\varepsilon_F)\rho_R^\uparrow(\varepsilon_F)\right] \tag{12.3}$$

where

G_P and G_{AP} are the conductances for parallel and antiparallel electrode magnetizations

ρ_n^σ is the density of states (DOS) of the electrode n ($n=L$ or $n=R$) for spin channel σ ($\sigma=\uparrow$ for majority spin, and $\sigma=\downarrow$ for minority spin)

If we define a TMR ratio as the conductance difference between parallel and antiparallel magnetizations normalized by the antiparallel conductance, we arrive at Jullière's formula:

$$\mathrm{TMR} \equiv \frac{G_P - G_{AP}}{G_{AP}} = \frac{2P_LP_R}{1-P_LP_R} \tag{12.4}$$

which expresses the TMR via the spin polarizations of the two ferromagnetic electrodes

$$P_n = \frac{\rho_n^\uparrow - \rho_n^\downarrow}{\rho_n^\uparrow + \rho_n^\downarrow} \tag{12.5}$$

Jullière's formula (12.4) allows one to link the values of spin polarization obtained from the Tedrow–Meservey experiments [11] to TMR (see also Chapter 10). As long as both measurements are made for the same kinds of the ferromagnet/insulator interfaces, this relation usually holds reasonably well, at least qualitatively. However, any attempts to interpret Equation 12.5 literally by using bulk DOS in Equation 12.5 almost invariably fail. For example, positive spin polarizations have been measured for tunneling from Fe, Co, and Ni into Al_2O_3 [11], while the DOS at the Fermi level is much larger for minority-spin electrons in Co and Ni. This failure led to the notion of the "tunneling DOS," which is supposed to appear in Equation 12.5 instead of the bulk DOS. The idea is that there is supposedly a subset of states that tunnel easily and have similar tunneling matrix elements, while other states barely tunnel at all. In this case, it is the density of the first group of states that matters in Equation 12.4. It has been common in the literature to explain the measured TMR and positive Tedrow–Meservey spin polarization based on the notion that "*s*-electrons" dominate in the tunneling. However, one should consider this separation with caution because *s* and *d*-electrons are strongly hybridized. In particular (see, for instance, Chapter 5), band structure calculations show that transport in transition metals is dominated by 3*d*-electrons. It is often true that most of the current in the diffusive regime is carried by electrons on just one of the Fermi surface sheets that is characterized by large Fermi velocities, but meaningful predictions require detailed calculations, and also the situation may change at the interface compared to the bulk. Therefore, quantitative understanding of TMR requires calculations based on realistic electronic band structure. It is, however, helpful to consider a free-electron model as a starting point for a qualitative description of spin-dependent tunneling.

12.2.3 Free-Electron Model in One Dimension

The transmission can be easily calculated for a one-dimensional asymmetric rectangular barrier by matching the wavefunction and its derivative:

$$T = \frac{(2k_L\kappa)(2k_R\kappa)}{(k_L + k_R)^2\kappa^2 + (k_L^2 + \kappa^2)(k_R^2 + \kappa^2)\sinh^2\kappa d} \quad (12.6)$$

where

d is the barrier thickness

$\kappa = \sqrt{2m(U - E)}/\hbar$ is the decay parameter in the barrier

U is the barrier height

k_L, k_R are the wavevectors in the left and right electrode

This free-electron model can also be used for a qualitative description of a spin-polarized metal by adding an exchange splitting to the potential profile. In this case, all wavevectors in Equation 12.6 (but not κ, unless the barrier is also ferromagnetic) carry a spin index, which we omit to simplify the notation.

The conductance can be calculated exactly using the Landauer–Büttiker (LB) formula [12], which says that the conductance of a two-terminal device can be written as

$$G_\sigma = \frac{e^2}{h}\sum_{\mu\nu} T_{\mu\nu}^{\sigma} \quad (12.7)$$

where

σ is the spin index, the sum is taken over the conducting channels (μ, ν) in the two leads

$T_{\mu\nu}$ is the transmission probability from channel μ in lead L to channel ν in lead R (see Ref. [12] for details)

In our one-dimensional free-electron problem there is only one conduction channel, and the sum in Equation 12.7 contains only one term e^2T/h with T given by Equation 12.6. Let us make some general observations, which will be very useful later when we consider more realistic models.

Observation 1: The transmission probability is not equal to a product of two factors characterizing the two sides of the tunnel junction. This is because the coefficients of the propagating and evanescent waves in the wavefunction are determined from a coupled set of matching conditions; a tunneling electron can reflect off of the interfaces multiple times and interfere with itself.

Observation 2: If the barrier is sufficiently strong and tunneling weak ($\kappa d \gg 1$), then with exponential accuracy the transmission probability reduces to a product of two factors, T_L and T_R, characterizing the two interfaces:

$$T = \frac{4k_L\kappa e^{-\kappa d}}{k_L^2 + \kappa^2} \cdot \frac{4k_R\kappa e^{-\kappa d}}{k_R^2 + \kappa^2} = T_L T_R \qquad (\text{at } e^{-2\kappa d} \ll 1) \quad (12.8)$$

This approximation can be understood as follows: In the wavefunction matching at the left interface, we neglect the barrier eigenstate, which is exponentially growing to the right, whose amplitude is exponentially small at this interface. Then, T_L describes the "transmission probability" of the electron with k_L into the evanescent barrier eigenstate $e^{-\kappa x}$ (with the barrier extended to infinity). T_R is, likewise, the transmission probability of the decaying barrier eigenstate into the right electrode's outgoing eigenstate k_R. The word "transmission" may be a little confusing here because evanescent waves do not carry current, but T_L and T_R are simply analytic continuations of the transmission probability from the range of positive kinetic energies in the barrier into the range of negative kinetic energies. The reciprocity relation persists into this region (as seen from the symmetry of Equation 12.8) due to the analytic properties of the scattering amplitude. Two exponential factors in Equation 12.8 show the decay of the evanescent barrier state between the two interfaces, which may be extracted in a separate spin-independent factor. We will refer to T_L and T_R as "interface transmission probabilities."

Observation 3: Equation 12.8, which can be directly generalized to the spin-split case, resembles the formulas (12.3) used in the derivation of Jullière's formula, but T_L and T_R are by no means proportional to the electrode DOS, which in our one-dimensional case is proportional to the inverse wavevector. The relation with DOS holds only in the limit of a very low barrier height $\kappa << k_L$. We will refer to Equations 12.4 and 12.5 in which the DOS is replaced by some more appropriate quantity as the "generalized Jullière's formula" (which is also, as we will see below, not universally valid).

Observation 4: Instead of a single free electron, consider a free-electron metal in which the levels are filled according to Fermi statistics (still in one dimension). The expectation value of the density operator for one of the electrodes (with the barrier extended to infinity) over the ground state (filled Fermi sphere) is $\rho(x) = \left\langle 0 \left| \sum_i \delta(x - x_i) \right| 0 \right\rangle = \sum_k |\psi_k(x)|^2$, where $\Psi_k(x)$ are the normalized eigenfunctions, and $k \le k_F$ with k_F the Fermi wavevector. If the electrode has a large but finite length L, the values of k are quantized with an interval π/L. Converting the sum over k to an integral, and then changing the integration variable to energy ε, we find $\rho(x) = (L/\pi)\int |\psi_k(x)|^2\, dk = (L/\pi)\int v_\varepsilon^{-1} |\psi_\varepsilon(x)|^2\, d\varepsilon$, where $v_\varepsilon = d\varepsilon/dk = \hbar k/m$ is the electron velocity. On one hand, the integrand $\rho(x,\varepsilon)$ represents the energy-resolved charge density at point x, or, in other words, the DOS projected on the eigenstate of the coordinate operator at x. On the other hand, if x is placed inside the barrier, one can easily verify that the integrand is proportional to T_L from Equation 12.8. Thus, the interface transmission probability can be interpreted in terms of the electrode-induced DOS inside the barrier. This conclusion is analogous to the result of the Tersoff–Hamann theory of scanning tunneling microscopy [13]. It is the spin polarization of this quantity that determines TMR through Equations 12.4 and 12.5 in the weak tunneling limit.

Observation 5: It is obvious from wavefunction continuity that as long as the factorization (Equation 12.8) holds, the amplitude of the evanescent wave in the barrier is proportional to the amplitude of the wavefunction at the interface. Therefore, the quantities T_L or, equivalently, $\rho(x,\varepsilon)$ calculated for x in the middle of the barrier and right at the interface differ only by the spin-independent exponential decay factor. Therefore, instead of T_L and T_R in Equation 12.8, we can just as well calculate the spin polarizations of the *interface DOS* for use in Equation 12.4.

Observation 6: In the limit of weak tunneling, the Bardeen and LB approaches produce identical results. In Bardeen's approach, we need to calculate the matrix element b_{LR} of the current density operator

$$\hat{j} = \frac{i\hbar}{2m}\left[\overleftarrow{\nabla}\delta(x - x_b) - \delta(x - x_b)\overrightarrow{\nabla}\right] \tag{12.9}$$

between the states $\psi_L(x) = C_L \exp(-\kappa x)$ and $\psi_R(x) = C_R \exp[\kappa(x-d)]$, where x_b is a point inside the barrier, and the gradient operator $\nabla = d/dx$ acts in the direction shown by the arrow. The matrix element then is $b_{LR} = \langle \psi_L | \hat{j} | \psi_R \rangle = -(i\hbar/m) C_L^* C_R \kappa \exp(-\kappa d)$. The amplitudes are found from interface matching: $C_L/A_L = 2k_L/(k_L + i\kappa)$, where A_L is the amplitude of the incoming wave, whose magnitude is found from the normalization of the scattering state, $|A_L|^2 = 1/(2L_L)$, and L_L the length of the left electrode. Substituting this matrix element in Equation 12.2 and introducing the DOS of the left and right electrodes, after some cancelations we obtain the conductance identical to the LB calculation given by Equations 12.7 and 12.8.

12.2.4 New Features in Three Dimensions

Now consider an electron moving in three-dimensional space in an external potential $V(z)$, which depends only on one coordinate z and again corresponds to a rectangular tunnel junction. The Schrödinger equation for this potential is separable in Cartesian components, and the wavefunction can be represented in the following form: $\psi(\mathbf{r}) = \exp(i\mathbf{k}_{\|}\mathbf{r}_{\|})\psi(z)$, where $\mathbf{k}_{\|}$ and $\mathbf{r}_{\|}$ are the projections of $\mathbf{k}$ and $\mathbf{r}$ on the xy plane, and $\Psi(z)$ is the solution of the one-dimensional Schrödinger equation with potential $V(z)$ and energy $E_z = E - \hbar^2 k_{\|}^2/(2m)$. At $T = 0$, the electrons in each electrode fill a Fermi sphere. In the magnetic case, the Fermi wavevector depends on spin. Since the barrier decay parameter κ increases if the electron energy is decreased, it is clear that the electrons with $\mathbf{k}_{\|} = 0$ have the highest tunneling probability. In the limit of a very thick barrier, the tunneling current is therefore confined to the narrow vicinity of the point $\mathbf{k}_{\|} = 0$.

Let us estimate the size of this region for large but finite κd. Introducing κ_0 according to $\hbar^2\kappa_0^2 = 2m(U - E)$, assuming that $k_{\|}^2 \ll \kappa_{\|}^2$ in the important range of $k_{\|}$, and expanding $\kappa(k_{\|}^2)$ to first order in $k_{\|}^2$, we approximate the barrier decay factor by $\exp(-2\kappa d) \approx \exp(-2\kappa_0 d)\exp(-k_{\|}^2 d/\kappa_{\|})$. The prefactors in Equation 12.8 depend weakly on $k_{\|}$, so the tunneling is suppressed at $k_{\|}^2 d \sim \kappa_0$, which can be rewritten as $k_{\|}a \sim \sqrt{\kappa_0 d}\,(a/d)$, where a will have the meaning of the lattice parameter when we consider crystalline electrodes. Typical tunnel junctions have barriers that are a few monolayers thick, and values of $\kappa_0 d$ usually do not exceed 10 (otherwise the tunneling current is no longer measurable). We therefore see that even for relatively thick barriers $k_{\|}a$ is never very small, which means that the current is carried by electrons in a significant portion of the Brillouin zone, and not exclusively by those with $k_{\|} \approx 0$.

Each $\mathbf{k}_{\|}$ allowed by boundary conditions in the xy directions serves as an independent channel in LB formula (Equation 12.7), and the transmission probability is diagonal in the $\mathbf{k}_{\|}$ representation, because $\mathbf{k}_{\|}$ is conserved for the entire system. Thus, we find that the conductance per spin is given by

$$G = \frac{e^2}{h}\sum_{\mathbf{k}_{\|}} T(\mathbf{k}_{\|}) \tag{12.10}$$

where $T(\mathbf{k}_{\|})$ is the transmission probability obtained from the solution of the one-dimensional scattering problem given by Equation 12.6 at energy E_z. All the observations about the features of the transmission probability made in the previous section are carried over *verbatim* for tunneling at the given $\mathbf{k}_{||}$ point. However, the total current in each spin channel is now given by a sum over $\mathbf{k}_{||}$ in Equation 12.10, so that, even if the transmission probability is factorized as in Equation 12.8 in the weak tunneling limit at each $\mathbf{k}_{||}$, the entire integral can no longer be represented as a product of two factors. Therefore, the generalized Jullière's formula, in general, does not hold in three dimensions even in the weak tunneling limit.

As an illustration, let us consider the limit of a very thick barrier, when the important region of integration in Equation 12.10 is confined to the vicinity of the $\mathbf{k}_{||}=0$ point. In this case, one can neglect the variation of the prefactors in Equation 12.8 and expand the decay parameter in the exponential as we just did in the beginning of this section. Then the only $k_{||}$-dependent factor under the integral is $\exp(-k_{||}^2 d/\kappa_0)$, which is spin-independent. The generalized Jullière's formula is then restored, and the relevant quantity in Equation 12.5 for the given electrode (say, L) is the value $T_L(\mathbf{0})$. Taking these values from Equation 12.8, we obtain after a little algebra:

$$P_L = \frac{k_L^\uparrow - k_L^\downarrow}{k_L^\uparrow + k_L^\downarrow} \cdot \frac{\kappa^2 - k_L^\uparrow k_L^\downarrow}{\kappa^2 + k_L^\uparrow k_L^\downarrow} \tag{12.11}$$

and a similar expression for the right electrode. Here $k_L^\uparrow$ and $k_L^\downarrow$ are the wavevectors for spin-up and spin-down electrons in the left electrode, respectively. This result was derived by Slonczewski [5].

Equation 12.11 has a very peculiar and suggestive feature. At $\kappa \ll k_\sigma$ (very low barrier), P_L is equal to the spin polarization of the *inverse* wavevector, which is the same as the spin polarization of DOS at $\mathbf{k}_{||}=0$. We have mentioned this in *Observation 3* of the previous section. On the other hand, in the opposite high-barrier limit ($\kappa^2 \gg k^\uparrow k^\downarrow$) the spin polarization has the same magnitude but an *opposite sign*. This is because in the $\kappa \to \infty$ limit the scattering wavefunction develops a node at the interface, while in the $\kappa \to 0$ limit it has an antinode there. In the first case, the amplitude of the wavefunction at the interface is suppressed by a factor k/κ, so that the transmission probability is proportional to k. In the second case, this amplitude is just twice the incoming amplitude, and the transmission probability is proportional to k^{-1}. The spin polarization (Equation 12.5) changes sign if the underlying quantity is replaced by its inverse. This simple example [5] shows the futility of "plausible" arguments based on Jullière's formula in predicting even the sign of the spin polarization and TMR without taking into account the details of the electronic structure and interfacial scattering.

12.3 Band Theory of Tunneling

The current in a tunnel junction is directly related to transmission probabilities, which, as we have seen from previous sections, depend in a delicate way on the matching of wavefunctions at the interfaces. Solids, and especially transition metals that are usually used as electrodes in MTJs, have complicated electronic band structures; moreover, metal-oxide interfaces almost never look as a simple concatenation of bulk materials. Surfaces and interfaces develop qualitatively new features due to the reduction of the coordination number, severing of bonds, charge transfer, and metal-oxide hybridization; in reality there are also defects, intermixing, interfacial strain, and other imperfections. The TMR is sensitive to all these details.

Until 2001, the only barrier material that produced sizeable TMR at room temperature was amorphous Al_2O_3. Over the years, the fabrication techniques have been perfected, so that junctions with high TMR of about 70%–80% can now be routinely produced. Tedrow–Meservey experiments for the ferromagnet/Al_2O_3 interfaces show positive spin polarization of the tunneling current (see Chapter 10), which, however, is still not completely understood from the theoretical point of view due to the extreme complexity of the atomic structure. The situation has changed dramatically with the prediction [14,15] and successful fabrication [16,17] of epitaxial Fe(001)/MgO/Fe tunnel junctions exhibiting very high TMR values. In spite of the fact that the details of the interfacial structure are almost never perfectly controlled or even known, it has become clear that certain simple selection rules are instrumental in analyzing the TMR in epitaxial tunnel junctions. In this section, we discuss the tunneling of Bloch electrons across epitaxial interfaces.

12.3.1 Complex Band Structure of the Tunneling Barrier

Since the potential of a crystal is periodic, the Hamiltonian commutes with any lattice translation, which also commute between themselves. The Bloch theorem immediately follows, stating that the eigenstates of the crystal Hamiltonian can be taken to be the eigenstates of the lattice translation operators, namely, $\psi_i(\mathbf{r}) = \exp(i\mathbf{kr})u_i(\mathbf{r})$, where $\mathbf{k}$ is the quasi-wavevector, and $u(\mathbf{r})$ has the periodicity of the lattice. $\mathbf{k}$ is defined up to a reciprocal lattice vector and can therefore be taken inside the first Brillouin zone; the index i labels the eigenstates with the same $\mathbf{k}$, of which there is an infinite number.

In a bulk crystal, only the wavefunctions with real quasi-wavevectors $\mathbf{k}$ are allowed in the Hilbert space; those with complex $\mathbf{k}$ blow up at infinity and must be discarded. However, in the tunneling problem, the wavefunction matching across the interfaces involves the general solutions of the Schrödinger equation including all evanescent eigenstates in the barrier and the decaying evanescent states in the electrodes. These evanescent eigenstates obey the Bloch theorem, but their quasi-wavevectors are complex. The analytic properties of these states as a function of complex $\mathbf{k}$ were studied by Kohn [18] and Heine [19]. Only the states with real energies have physical significance. The set of all these states is called "complex band structure," which extends the standard band structure which includes only real-$\mathbf{k}$ solutions.

An example below illustrates this concept. For an epitaxial junction, the projection of the quasi-wavevector on the plane

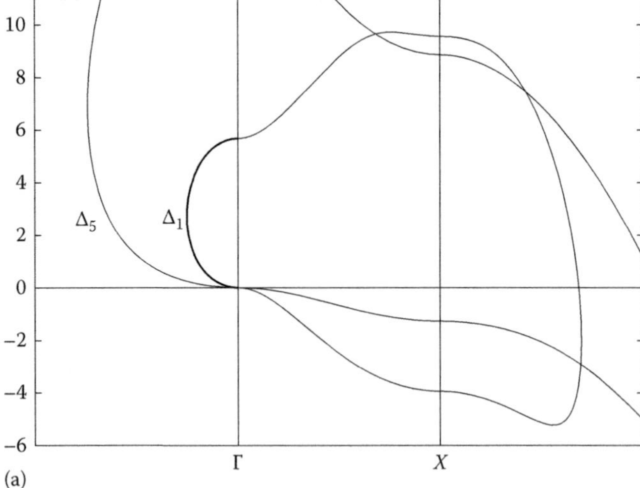

FIGURE 12.1 (See color insert.) Illustration of the complex band structure of MgO. (a) Complex band structure at $\mathbf{k}_{||}=0$ (energy in eV). The Δ_1 complex band is shown by a thick line straddling the insulating gap. For clarity, complex bands with variable real and imaginary parts, which attach to the band extrema at −5 and at 9.5 eV, are not shown. (b) Right panel: Brillouin zone plot of the lowest value of Im k_z taken at mid-gap.

parallel to the interfaces ($\mathbf{k}_{||}$) must be real. For each $\mathbf{k}_{||}$ there is an infinite number of eigenstates with complex k_z, whose imaginary part determines the decay parameter. In Figure 12.1b we show the smallest decay parameter as a function of $\mathbf{k}_{||}$ for MgO with the high-symmetry (001) interface and the energy taken in the middle of the insulating gap. Note that $\mathbf{k}_{||}$ lies in the *surface Brillouin zone* defined with respect to the two-dimensional lattice translations parallel to the interface, while Re k_z lies in the one-dimensional "Brillouin zone" defined by the translational period in the z direction. The whole idea of conserving $\mathbf{k}_{||}$ is meaningful only if the interface lies parallel to one of the low-index planes of the crystal. Points in the surface Brillouin zone are usually denoted with a bar on top of the letter; for example, the $\mathbf{k}_{||}=0$ point is denoted as $\bar{\Gamma}$.

As it was pointed out by Mavropoulos et al. [20], tunneling should be more efficient through barrier eigenstates with the smallest available decay rates. In the limit of large barrier thickness, the state with the smallest decay rate controls the tunneling current, similar to Slonczewski's result (Equation 12.11) for the free-electron case. As seen in Figure 12.1b, for MgO(001) the minimum decay rate is reached at the $\bar{\Gamma}$ point. This is similar to the case of a free-electron tunneling through a rectangular barrier, where the smallest decay parameter κ is reached at $\mathbf{k}_{||}=0$. The main difference is that a crystalline barrier has an infinite number of decaying eigenstates at each $\mathbf{k}_{||}$ within the Brillouin zone, while in the free-electron case there is only one eigenstate per $\mathbf{k}_{||}$, which is unrestricted.

Let us now set $\mathbf{k}_{||}=0$ and look at the complex bands of MgO shown in Figure 12.1a. These bands are plotted using three panels: the middle panel shows the real part of k_z, and the two side panels show its imaginary part. It is useful to use two side panels for the following reason. The complex bands can be classified in four categories [21]: (1) ordinary real bands with Im $k_z=0$, (2) imaginary bands with Re $k_z=0$, (3) imaginary bands with the maximum allowed Re $k_z=k_{max}$, and (4) complex bands for which both Im k_z and Re k_z depend on energy. Complex bands are continuous; therefore, bands of type 2 either go to infinity or connect to real bands at $k_z=0$. Likewise, bands of type 3 can connect to real bands at $k_z=k_{max}$. Bands of type 2 and 3 are therefore plotted in the left and right panels, respectively; in this way the connections to the real bands are obvious. Type 4 bands may be shown in two panels at the same time [21]; they attach to real bands at their extremal points. Two such bands, which are far from the Fermi level and irrelevant for tunneling, were omitted from Figure 12.1b to avoid clutter.

The point $k_z=0$ in Figure 12.1a corresponds to the bulk Γ point. The degenerate triplet at $E=0$ is the valence band maximum (VBM), and the singlet at $E\approx5.5$ eV is the conduction band minimum (CBM). (The band gap is underestimated due to the use of the local density approximation [LDA] in the calculation.) We see that VBM and CBM are connected by a Δ_1 band with imaginary k_z, and the decay constant is lower for the state at this loop than anywhere else in the surface Brillouin zone. This decay constant appears in the center of Figure 12.1b. Note that in the free-electron case, the "band gap" corresponds to negative kinetic energies, and the only "complex band" $\kappa(E)$ is just an inverted parabola.

The crucial point is that all bands, be they real or complex, can be classified by the irreducible representations of the group of the quasi-wavevector $\mathbf{k}$. The $\bar{\Gamma}$ point, which we found to be most important for tunneling, has a high degree of symmetry. Since the full quasi-wavevector of all states with $\mathbf{k}_{||}=0$ has only the z component, the symmetry group is that of the [001] direction in the bulk Brillouin zone, that is C_{4v} (or 4 mm in the international notation). The [001] direction in a cubic crystal is customarily denoted as Δ.

The C_{4v} group has five irreducible representations: four one-dimensional and one two-dimensional [22]. For the electronic states on the Δ line, they are usually denoted as follows: Δ_1, Δ_1', Δ_2, Δ_2', and Δ_5. This notation was introduced by Bouckaert et al. [23] and differs a little from the standard notation for the C_{4v} group [22]. Index 1 or 2 distinguishes the one-dimensional representations with even or odd symmetry with respect to the C_4 rotation; unprimed representations are even with respect to reflections in the planes parallel to the given Δ direction and normal to one of the other two cubic axes; primed representations are odd with respect to these reflections. Δ_1 is the identity representation; Δ_5 is the two-dimensional representation. If the character of the wavefunctions is described by the *s*, *p*, *d* basis, the Δ_1 bands can only contain *s*, p_z, and d_{z^2} character; Δ_5 can only have p_x, p_y, d_{xy}, and d_{yz} character; Δ_2 is $d_{x^2-y^2}$ only, and Δ_2' is d_{xy} only. These rules can be used for the assignment of symmetry labels to the electronic states. The Δ_1' wavefunction must have four node lines in the *xy* plane and is not representable by *s*, *p*, *d* functions.

In MgO, the triplet at VBM has Γ_{15} symmetry, and the singlet at CBM has Γ_1 symmetry (notation of Ref. [23]). As **k** is moved away from Γ along the Δ line, the degeneracy is partially lifted, and the Γ_{15} triplet splits in one Δ_1 band and a doubly degenerate Δ_5 band. These bands are labeled in Figure 12.1a; we see that the complex band connecting VBM and CBM has Δ_1 symmetry. (By continuity, only bands of compatible symmetry can connect to each other.)

If we move away from the $\bar{\Gamma}$ point in a general $\mathbf{k}_{||}$ direction, all symmetry operations (except the identity) are lost, and all eigenstates belong to the trivial identity representation. However, if $\mathbf{k}_{||}$ is small, the symmetry properties of the eigenstates will still strongly resemble those at the $\bar{\Gamma}$ point. For example, the matrices of the C_{4v} group operators in the basis of the new eigenstates will be almost diagonal at small $\mathbf{k}_{||}$ (except for the matrix elements in the subspace "lifted" from Δ_5). Therefore, at small $\mathbf{k}_{||}$ one can talk about, say, "Δ_1-like" or "Δ_5-like" states, according to the representation to which they are adiabatically connected by sending $\mathbf{k}_{||}$ to zero. However, this terminology loses its meaning at larger $\mathbf{k}_{||}$, especially when the different sheets of the $\kappa(\mathbf{k}_{||})$ function intersect with each other at finite $\mathbf{k}_{||}$.

12.3.2 Tunneling Spin Filtering due to Symmetry Selection Rules

If the junction is epitaxial, there is a common translational periodicity for both the electrodes and the barrier in the directions parallel to the interfaces, and $\mathbf{k}_{||}$ is a conserved quantum number for the eigenstates of the whole system. At $\mathbf{k}_{||}=0$, the scattering eigenstates may be classified by the irreducible representation of the symmetry group characterizing the junction. If we now view the whole problem in terms of wavefunction matching, it is clear that a Δ_p scattering eigenstate is obtained by matching the Δ_p states in the electrodes with the Δ_p states in the barrier. In addition to the conventional real bands, the metallic electrodes also have an infinite number of complex bands, which have a given symmetry, and all of them will be represented in all the scattering states of this symmetry (of course, only those that decay *into* the metal are allowed) [24]. However, the current is only carried by real bands, of which there are usually at most one or two.

As an example, consider an epitaxial Fe/MgO/Fe tunnel junction with both MgO and Fe grown in the [001] direction. The relevant real bands in Fe for $\mathbf{k}_{||}=0$ cross the Fermi level along the Δ line connecting the Γ and *H* points in the bulk Brillouin zone. The band structure of Fe along this line is shown in Figure 12.2. We see that there is one Δ_1 band for majority-spin electrons, and *none* for minority-spin electrons.

For a certain state to carry tunneling current across the barrier, it should have a finite amplitude of at least one real band on *each* side of the barrier. If the magnetizations of the two Fe electrodes are parallel, only the majority-spin Δ_1 eigenstate of the junction can carry the tunneling current across the barrier. However, when the electrodes are magnetized antiparallel, *none* of the Δ_1 states of the junction can carry the tunneling current. Therefore, one can expect that the tunneling conductance is larger in the parallel configuration, and hence the TMR defined by Equation 12.4 is positive. This feature of epitaxial MTJs, which was elucidated in Refs. [20,25], can be called symmetry-enforced spin filtering.

These qualitative considerations are confirmed by explicit calculations [14,15]. Figure 12.3 illustrates the spin filtering effect. The four panels show the tunneling DOS of the scattering states of the given symmetry incoming from the left electrode plotted as a function of the position in the system. Different decay rates of the states of different symmetry are apparent in the barrier region. We also see that the rightward propagating Δ_1 tunneling state exists only for majority-spin electron in the parallel configuration (upper left panel). In all other cases, such state either does not exist (upper and lower right panels) or decays exponentially into the right electrode and carries no current (lower left panel).

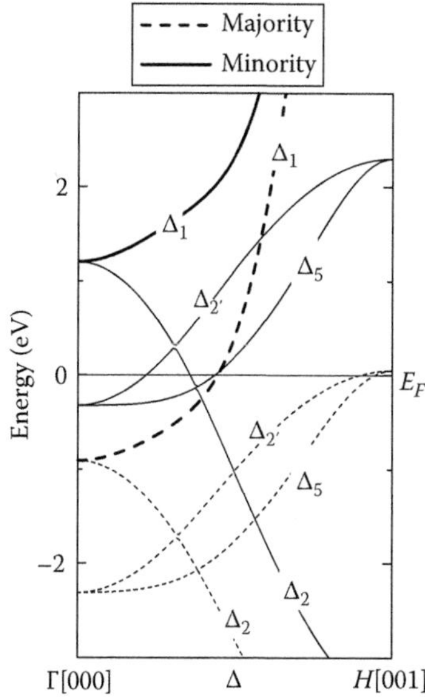

FIGURE 12.2 Spin-resolved band structure of Fe along the [001] direction. Bands are labeled by their symmetry group representation. Thick lines show the Δ_1 bands, which decay the slowest in MgO.

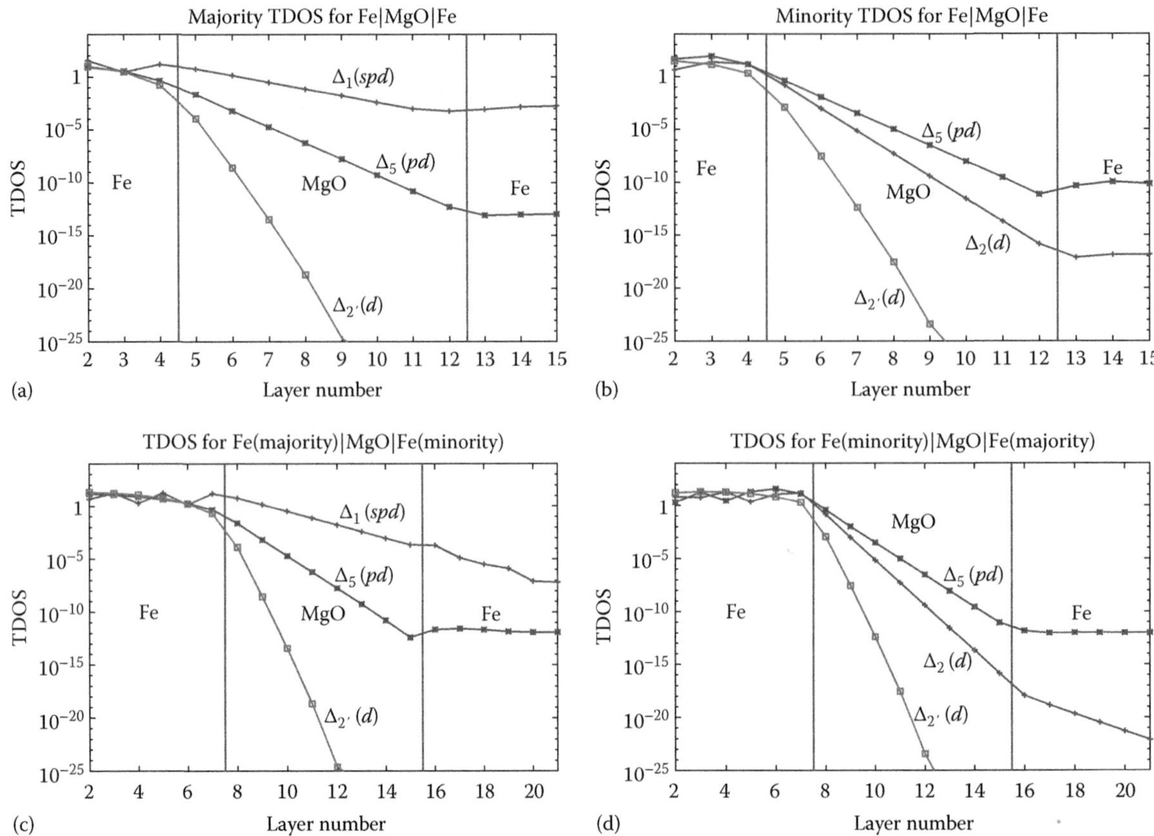

FIGURE 12.3 Tunneling DOS (TDOS) for $\mathbf{k}_{||} = 0$ for Fe(100)|MgO|Fe(100) tunnel junction for majority (a), minority (b), and antiparallel alignment of the magnetizations in the two electrodes (c,d). Each curve is labeled by the symmetry of the incident Bloch state in the left Fe electrode. (From Butler, W.H. et al., *Phys. Rev. B*, 63(5), 054416, 2001. With permission.)

Symmetry-enforced spin filtering qualitatively explains large values of TMR observed experimentally in crystalline Fe/MgO/Fe tunnel junctions [16,17,26]. In fact, symmetry-enforced spin filtering should be quite common, as long as the given junction can be grown epitaxially with low-index interfaces. In addition to bcc Fe, bcc Co and Fe–Co alloys have a Δ_1 symmetry band crossing the Fermi energy only in the majority-spin channel. Therefore, large values of TMR are expected for junctions with these electrodes and MgO barriers stacked along the [001] direction. This was confirmed by first-principles calculations [27], and high TMR was observed experimentally [28,29].

12.3.3 Factorization of Transmission

In Section 12.2.3 (see *Observation 2*), we have explained that in the one-dimensional free-electron case, when the barrier is sufficiently thick ($e^{-2\kappa d} \ll 1$), the transmission probability is factorized in a product of two interface transmission probabilities. This factorization holds at each $\mathbf{k}_{||}$ for free electrons in three dimensions. We now show that the situation for Bloch electrons is, in fact, quite similar [30]. Let us consider an epitaxial tunnel junction and treat the scattering problem for the given $\mathbf{k}_{||}$.

First, we assume that the barrier is thick enough to have a region where the potential is bulk-like. We can then draw an imaginary plane on each side of both left and right interfaces, as shown in Figure 12.4, so that we have bulk-like potential of the left electrode to the left of plane LE, bulk-like barrier potential between planes LB and RB, and bulk-like right electrode potential to the right of plane RE. Furthermore, the planes LB and RB are drawn in such a way that they can be superimposed by a bulk translation vector normal to the interface.

The general solution of the Schrödinger equation to the left of plane LE is given by a linear combination of the Bloch eigenstates of the left electrode for the given $\mathbf{k}_{||}$, and similarly for the

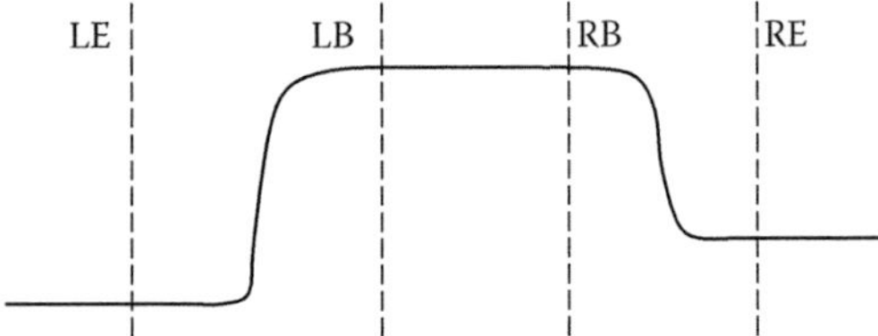

FIGURE 12.4 Schematic picture of tunneling through a barrier. The dashed lines indicate the imaginary planes showing the positions sufficiently far away from the interfaces so that the periodic potential in their vicinity is already identical to that in the bulk of the corresponding region. L and R in the labels refer to left and right; E and B mean electrode and barrier, respectively.

right electrode. Although there are an infinite number of evanescent states, we are only interested in real bands. There is an equal number of incoming and reflected waves. The general solution in the barrier (between LB and RB) is given by a linear combination of all evanescent barrier eigenstates. Similar to regular Bloch waves, the evanescent waves can be represented as $\psi(\mathbf{r}) = \exp[i\mathbf{k}_{\|}\mathbf{r}_{\|} + ik_z z]u(\mathbf{r})$, where $u(\mathbf{r})$ is periodic, and $k_z = k' + i\kappa$, where k' is the real part of k_z. We define the amplitudes of the evanescent waves in such a way that $z=0$ at plane LB (or RB for the right interface).

We now treat the problem by the standard method of combining S-matrices [12]. We will denote the barrier eigenstates decaying to the right as "outgoing," and those decaying to the left as "incoming" with respect to the left interface, and vice versa with respect to the right interface. (Outgoing eigenstates remain finite at infinity when a small positive imaginary part is added to the energy; these are the boundary conditions for the retarded Green's function.) In order to limit the number of subscripts, we restrict ourselves to the case when the barrier has a symmetry plane normal to the axis of the junction, so that the incoming and outgoing barrier eigenstates come in degenerate pairs with complex conjugate k_z. The eigenstate of the junction must satisfy the Schrödinger equation everywhere. This condition introduces a linear relation between the amplitudes A_i^L of incoming and B_i^L of outgoing waves (referenced from planes LE and LB, respectively), which is described by an S-matrix: $B_i^L = \sum_j S_{ij}^L A_j^L$ ($\bar{B}_L = \hat{S}_L \bar{A}$ in matrix notation). There is a similar relation at the right interface: $\bar{B}_R = \hat{S}_R \bar{A}_R$, where the incoming and outgoing waves are referenced from planes RE and RB.

Due to our assumption about the existence of a translation vector connecting planes LB and RB, the amplitudes of the barrier eigenstate b at these planes are related by the corresponding Bloch translation factor: $A_b^R = \exp(ik_b'd - \kappa_b d)B_b^L$, $A_b^L = \exp(ik_b'd - \kappa_b d)B_b^R$, where d is the distance between the planes LB and RB (which is a multiple of the translation vector along the z direction). We can now eliminate the amplitudes of the barrier waves from the linear relationships given by the left and right S-matrices, and find the transmission matrix connecting the amplitudes of the outgoing waves of the right electrode with those of the incoming waves of the left electrode. Following the standard convention [12], we denote the subblock of $\hat{S}_L$ connecting the outgoing LB states with incoming LE states as $\hat{t}_L$, and that connecting the outgoing LB states with incoming LB states as $\hat{r}_L$. In a similar way, for the right interface, the outgoing RE states are connected to the incoming BR states by matrix $\hat{t}_R$, and the outgoing RB states to incoming RB states by matrix $\hat{r}_R$. Elimination of the barrier amplitudes then results in the following matrix relation:

$$\hat{t} = \hat{t}_R' e^{i\hat{k}_z d/2} \left[1 - e^{i\hat{k}_z d/2} \hat{r}_L' e^{i\hat{k}_z d} \hat{r}_R e^{i\hat{k}_z d/2} \right]^{-1} e^{i\hat{k}_z d/2} \hat{t}_L \qquad (12.12)$$

where the dependence on $\mathbf{k}_{\|}$ is omitted for brevity. Here $\hat{k}_z$ is a diagonal matrix whose eigenvalues are the complex wavevectors of the barrier eigenstates with negative imaginary parts. Equation 12.12 has an obvious interpretation in terms of the Feynman paths going back and forth between the two interfaces [12]. It is exact under our assumption that the region between LB and RB is bulk-like. In fact, it is correct even in the case of a metallic spacer layer that has real k_z eigenvalues.

For a sufficiently thick barrier, the factors $\exp(-\kappa d)$ for all evanescent states become small. In this limit, the matrix in square brackets in Equation 12.12 is close to a unit matrix, and we obtain (with exponential accuracy):

$$\hat{t} = \hat{t}_R' e^{i\hat{k}_z d} \hat{t}_L \quad \text{(thick barrier, Im } k_z < 0) \qquad (12.13)$$

The generalization of the Landauer–Büttiker formula (Equation 12.10) for an epitaxial crystalline junction is

$$G = \frac{e^2}{h} \sum_{\mathbf{k}_{\|}} \sum_{ij} \left| t_{ij}(\mathbf{k}_{\|}) \right|^2 \qquad (12.14)$$

Using (12.13), for a thick barrier we obtain $G = (e^2/h) \sum_{\mathbf{k}_{\|}} T(\mathbf{k}_{\|})$, where

$$T(\mathbf{k}_{\|}) = \mathrm{Tr} \left[\hat{T}_L e^{-i\hat{k}_z^+ d}\, \hat{T}_R\, e^{i\hat{k}_z d} \right] \qquad (12.15)$$

where the Hermitian matrices $T_{bc}^L = \sum_i t_{bi} t_{ci}^* = \left(\hat{t}_L \hat{t}_L^+ \right)_{bc}$ and $T_{bc}^R = \sum_j t_{jc}' t_{jb}'^* = \left(\hat{t}_R'^+ \hat{t}_R' \right)_{bc}$ describe the matching between the propagating electrode states and the barrier eigenstates, the trace is taken over the barrier eigenstates with positive $\kappa = \mathrm{Im}\, k_z$, and we omitted the $\mathbf{k}_{\|}$ arguments. In contrast to the free-electron case where the transmission amplitude factorizes for each given $\mathbf{k}_{\|}$, we now have a double sum over the barrier eigenstates. Of course, only those barrier eigenstates that have a sufficiently small decay parameter are important. Note that $T(\mathbf{k}_{\|})$ in Equation 12.15 is automatically real, as it should be.

Using the condition of current continuity, one can derive a reciprocity relation between $\hat{t}$ and $\hat{t}'$ matrices for the same interface:

$$\hat{t}\hat{t}^+ = \hat{t}'^+ \hat{t}' \qquad (12.16)$$

which holds as long as the propagating states in the electrodes are normalized for unit flux, and the evanescent barrier eigenstates are normalized in such a way that $\sqrt{2}$ times a sum of the right and left-going evanescent states differing in the sign of κ carries unit flux. This implies that the matrix $\hat{T}$ for the given interface in Equation 12.15 does not depend on whether this interface forms the left or the right side of the system. In particular, if the junction is symmetric, we have $\hat{T}_L = \hat{T}_R$.

Equation 12.15 has an interesting consequence. If there are at least two barrier eigenstates for which $\mathrm{Re}\, k_z^b \uparrow \mathrm{Re}\, k_z^c$ (while one of them may be zero), then $T(\mathbf{k}_{\|})$ contains a part that oscillates as a function of d (and $\mathbf{k}_{\|}$) due to the interference of these eigenstates. This feature was pointed out by Butler et al. [14], who found oscillations of the transmission probability as a function of $\mathbf{k}_{\|}$ and exponentially damped oscillations as a function of d in Fe/MgO/Fe junctions. This feature is often invoked to explain the experimentally observed oscillations of the tunneling conductance (in both parallel and antiparallel configurations) and TMR as a function of barrier thickness [17,31]. However, the evanescent states in MgO have purely imaginary wavevectors in a large region of $\mathbf{k}_{\|}$ around the $\bar{\Gamma}$ point [32,33]. This implies that any conductance oscillations should decay much faster than the conductance itself, and therefore the TMR oscillations should be strongly damped. This behavior is different from the experimentally observed almost [17] or completely [31] undamped oscillations. It was argued in Ref. [34] that disorder in the barrier, counterintuitively, can restore the conductance oscillations through a mechanism involving scattering of the states at $\mathbf{k}_{\|}=0$ into the pair of interfering states at a large-$\mathbf{k}_{\|}$ "hot spot." However, this argument implicitly assumes that the phase relation between the interfering states near the hot spot depends only on the distance from the interface and not on the location of the scattering impurity, which is unphysical.

Based on the above arguments, we believe that the undamped oscillations of the tunneling conductance and TMR observed in Fe/MgO/Fe MTJs are likely unrelated to interference of evanescent states. An alternative explanation is suggested by the measurements of the anisotropy constant for MgO-based MTJs with Co/Pt multilayer electrodes, which show undamped oscillations as a function of MgO thickness with a period identical to that for TMR oscillations [35]. These oscillations are, furthermore, reduced under annealing, pointing toward an extrinsic morphology-related effect.

For a $\mathbf{k}_{\|}$ point of a general position, the evanescent barrier eigenstates are nondegenerate. Moreover, it often happens that the tunneling conductance is dominated by a region of the interface Brillouin zone where the smallest-κ evanescent state is nondegenerate and widely separated from the next-smallest κ state. This region is usually located around the $\bar{\Gamma}$ point, as occurs, for example, in Fe/MgO/Fe. In this case, one can neglect all terms in Equation 12.15 except one in which k_z in both exponential factors has the smallest imaginary part κ (to be denoted as κ_0) for the given $\mathbf{k}_{\|}$. Then, the sums are eliminated and we find

$$T(\mathbf{k}_{\|}) = T_L(\mathbf{k}_{\|})e^{-2\kappa_0(\mathbf{k}_{\|})d}T_R(\mathbf{k}_{\|}) \tag{12.17}$$

where $T_L(\mathbf{k}_{\|}) = \sum_i |t_{i0}|^2$, and likewise for T_R. Note that T_L and T_R are always real; these quantities can be called *interface transmission functions* [30].

Since we are working at fixed $\mathbf{k}_{\|}$, the problem is essentially one-dimensional, and the DOS for each propagating electrode band is inversely proportional to the projection of its group velocity on the z direction, or to the flux carried by a Bloch wave in that direction. We can therefore use arguments invoked in *Observation 4* of Section 12.3 to associate $T_L(\mathbf{k}_{\|})$ with the metal-induced $\mathbf{k}_{\|}$-resolved DOS at the plane LB in the barrier; this relation is meaningful as long as the conditions of validity of Equation 12.17 are satisfied.

12.3.4 Asymptotic Dependence on Barrier Thickness

Let us now consider the asymptotic behavior of TMR when the barrier thickness is increased. Although this limit is almost never reached in practice, it is important from the methodological point of view. We assume for definiteness that the Δ_1 state at $\mathbf{k}_{\|}=0$ has the lowest decay rate in the barrier, as it happens in MgO and in many other crystalline barriers.

If the electrodes have Δ_1 bands crossing the Fermi level in both spin channels (or if one electrode has Δ_1 bands in both spin channels, and the other electrode has them in only one spin channel), the situation is qualitatively similar to the free-electron case considered in Section 12.2.4, which led to Equations 12.11 and 12.4. At $d \to \infty$, the TMR tends to a finite constant, which is determined by Jullière's formula with the spin polarizations calculated for the interface transmission functions taken at $\mathbf{k}_{\|}=0$.

However, if both electrodes have Fermi-crossing Δ_1 bands in only one spin channel, the TMR tends to infinity at $d \to \infty$, and it is interesting to consider its asymptotic behavior at large d. We assume that the Δ_1 bands are in the "up" spin channel for both electrodes, so that the conductance in the parallel configuration is dominated by the Δ_1 band near $\mathbf{k}_{\|}=0$. The barrier decay factor can be expanded near $\mathbf{k}_{\|}=0$ as explained in Section 12.4, and we have

$$G_P = \frac{e^2}{h}e^{-2\kappa_0 d}\int T_L^{\uparrow}e^{-\alpha k_{\|}^2 d}T_R^{\uparrow}d\mathbf{k}_{\|} \sim T_L^{\uparrow}T_R^{\uparrow}\frac{e^{-2\kappa_0 d}}{\alpha d} \tag{12.18}$$

where $T_L^{\uparrow}$ and $T_R^{\uparrow}$ are taken at $\mathbf{k}_{\|}=0$, and $\alpha = d^2\kappa/d\mathbf{k}_{\|}^2$.

In the antiparallel configuration, the Δ_1 channel does not carry any current, but the Δ_1-like states with nonzero $\mathbf{k}_{\|}$ do, and it is these states that dominate in G_{AP} at large d [36]. For a general $\mathbf{k}_{\|}$, the coefficient of the Δ_1-like barrier eigenstate in the scattering state is proportional to $k_{\|}$ (which can be seen, for example, using the $\mathbf{k}\cdot\mathbf{p}$ method [37]), and the interface transmission function is proportional to $k_{\|}^2$. It is possible that tunneling through the Δ_1-like state is forbidden along some high-symmetry directions in the $\mathbf{k}_{\|}$ plane. For example, if there is a symmetry plane normal to the interface and containing $\mathbf{k}_{\|}$, all states will be classified as even or odd with respect to reflection in this plane. If the electrode Δ_1-like eigenstate is, say, odd with respect to this plane and the barrier eigenstate is even, then these states do not mix. If this is the case, then the interface transmission function will have an additional positive-definite factor depending on the direction of $\mathbf{k}_{\|}$, which we denote $f(\phi)$, that is $T(\mathbf{k}_{\|}) = Qk_{\|}^2 f(\phi)$. Thus, we have for G_{AP}:

$$G_{AP} = \frac{e^2}{h} e^{-2\kappa_0 d} \int \left[T_L^{\uparrow} Q_R^{\downarrow} f_R(\phi) + T_R^{\uparrow} Q_L^{\downarrow} f_L(\phi) \right] \times k_{\|}^2 e^{-\alpha k_{\|}^2 d} d\mathbf{k}_{\|} \sim \left[T_L^{\uparrow} Q_R^{\downarrow} \bar{f}_R + T_R^{\uparrow} Q_L^{\downarrow} \bar{f}_L \right] \frac{e^{-2\kappa_0 d}}{(\alpha d)^2} \quad (12.19)$$

Comparing Equations 12.18 and 12.19, we see that the asymptotic thickness dependence of G_P and G_{AP} is dominated by the same exponential factor, but G_{AP} has an additional factor $(\propto d)^{-1}$. Thus, the ratio G_P/G_{AP} increases linearly in d in this asymptotic regime, and the coefficient depends on the curvature of the dominant complex band at the $\mathbf{k}_{\|}$ point where k_0 reaches its minimum.

Thus, we see that the TMR increases rather slowly in the asymptotic regime, which is consistent with the fact that the important range of $\mathbf{k}_{\|}$ also shrinks slowly, as explained in Section 12.2.4. (The important range is $k_{\|} \sim (\alpha d)^{-1/2}$.) This behavior was illustrated in Ref. [32] for Fe/MgO/Fe junctions.

12.3.5 Tunneling through Interface States

Surfaces and interfaces between different materials often have spectral features that are absent in the bulk. Such features may include *surface states* (or interface states), which are localized at the surface (or interface) and are orthogonal to all bulk states [19,38,39]. Surface states appear as bands that have energy dispersion in $\mathbf{k}_{\|}$; if they cross the Fermi level, they produce Fermi contours in the $\mathbf{k}_{\|}$ plane. *Surface resonances* are similar, but they are not strictly orthogonal to bulk states. Due to coupling to bulk states, surface resonances have finite lifetime; the corresponding bands and the Fermi contours are, accordingly, broadened. Of course, it only makes sense to talk about resonant states if their lifetime is larger than the relevant inverse bandwidth. Surface resonances often form on the (001) surfaces of transition metals; in particular, the presence of minority-spin resonant states on the Fe(001) surface is well established experimentally [40].

The formation of interface resonant states is largely controlled by the atomic structure and bonding at the interface. This can already be seen within a simple tight-binding model [41], which shows that the interplay between the atomic level shifts and changes in bonding strength at the interface may result in the appearance of an interface resonance. The same parameters determine the coupling of the resonance to bulk states, and hence the width of this resonance.

Interface resonances may appear in MTJs and influence their transport properties [7,14,42]. The minority-spin interface resonances are present at the Fe(001)/MgO interface, as was found from first-principle calculations [14] and indirectly observed in tunneling experiments [43,44]. Within a two-dimensional Brillouin zone, these resonant states represent a surface band sharply peaked near the Fermi energy. Calculations show that the interface Fe(001)/MgO band is extremely sensitive to electron energy [45]. This is seen from Figure 12.5, which shows the $\mathbf{k}_{\|}$-resolved DOS for different energies. Interface (resonant) bands show up as (broadened) contours in the surface Brillouin zone. A very small change in energy of 0.02 eV leads to a significant change in the location of these bands.

In the presence of interface resonant states, the tunneling conductance is strongly enhanced for the values of $\mathbf{k}_{\|}$ lying within a linewidth of their Fermi contour. This happens simply because for these $\mathbf{k}_{\|}$ the scattering states have a large amplitude at the interface; the electron approaching the interface gets "trapped" in the resonant state for a time of the order of its inverse linewidth and therefore has a larger probability to tunnel across the barrier.

While there is no doubt that interface (resonant) states can significantly affect the tunneling conductance and TMR in MTJs, it is extremely difficult to describe their contribution quantitatively for particular junctions. First, as it is clear from the symmetry considerations of Section 12.3.2, this contribution strongly depends on the locus of the Fermi contours of the interface states, which, in turn, is usually very sensitive to the

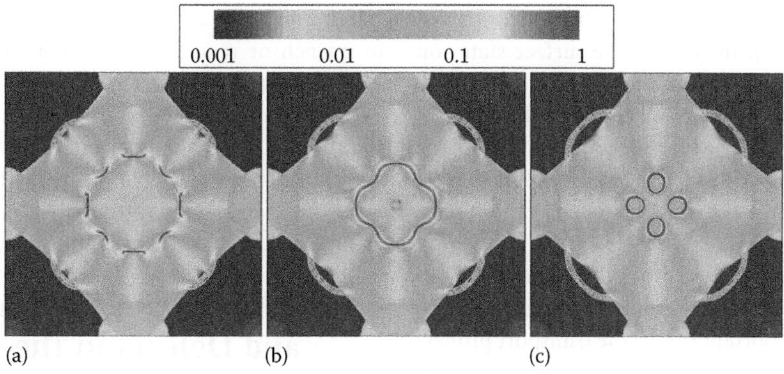

FIGURE 12.5 (See color insert.) $\mathbf{k}_{\|}$-Resolved minority-spin density of states (in arbitrary units) for the Fe(001)/MgO interface calculated for three different energies near the Fermi energy (E_F): (a) $E=E_F-0.02$ eV, (b) $E=E_F$, (c) $E=E_F+0.02$ eV. (From Belashchenko, K.D. et al., *Phys. Rev. B*, 72(14), 140404, 2005. With permission.)

atomic structure of the interface. In most cases, this structure is complicated and not known in sufficient detail. Second, even if the surface structure is known, metal/insulator interfaces are difficult to treat using density functional theory due to the well-known problems with the description of band gaps, band offsets, and correlation effects.

Apart from these serious problems, two issues related to tunneling through interface states are worth mentioning. The first one appears when the conductance of a *symmetric* MTJ is calculated theoretically. If interface resonant states exist in one of the spin channels, their dispersions at the two interfaces are identical. For the parallel magnetization of the electrodes, the interface states then appear at exactly the same contour in the $\mathbf{k}_{\|}$ plane and "match" with each other. This matching greatly enhances the conductance in the corresponding spin channel. For example, if the width of the surface resonance is γ, and the coupling between the interface states localized on the two electrodes is much smaller than γ, the $\mathbf{k}_{\|}$-integrated tunneling rate is proportional to γ^{-1} [46]. On the other hand, for *mismatched* surface resonances on the two electrodes, the integrated conductance does not depend on γ. For sharp interface resonances (i.e., those with small γ compared to bandwidth), any structural disorder, difference in the surface structure due to the different growth kinetics for the bottom and top electrodes, as well as the bias voltage will tend to destroy their matching. On the other hand, the conductance in the antiparallel configuration should not be as strongly affected by these factors. In practice, this means that the conductance calculated for a symmetric MTJ with parallel magnetization may contain a large fictitious (unobservable) contribution due to surface resonant states, which usually needs to be properly subtracted [45,46]. The suppression of this contribution was found using first-principles calculations including interface roughness or disorder [47,48].

A particularly peculiar feature occurs when the coupling between the "matched" resonant states on the two interfaces exceeds their intrinsic width due to coupling to bulk states [42]. In this case a "resonance of resonances" occurs, when in the first approximation one can neglect the coupling to the bulk and find the exact surface eigenstates to be the symmetric and antisymmetric linear combinations of the degenerate surface states on the two sides of the barrier. When the energy splitting of these states is large compared to the coupling to the bulk, the electrons transmit across the barrier with unit probability near the corresponding energies. This feature is generally regarded as an artifact of the perfectly symmetric junction, which cannot realistically be observed in practice.

Although the situation with perfectly matched interface states on the two sides of the barrier is likely unphysical, interface resonant states often play an important role in the transport properties of MTJs. Examples of first-principles calculations for specific electrode/barrier combinations can be found in the literature (see, for example, Refs. [30,45]).

The second issue is related to the neglect of all inelastic processes within the simple one-electron picture. Consider a region of $\mathbf{k}_{\|}$ in the interface Brillouin zone where there are no bulk states in one of the electrodes (like the dark blue regions in Figure 12.5). According to the single-electron Landauer–Büttiker formalism, this region does not contribute at all to the tunneling current. However, surface states may form in this region (see the pairs of arcs extending into the dark blue regions in Figure 12.5; they are revealed by adding a small imaginary part to energy). If the problem is treated using Bardeen's approach, which considers the current at $t=0$ when all states below the Fermi level are filled (including the surface states, if present), the surface states do contribute to the tunneling current. The Landauer–Büttiker approach correctly gives zero for the steady-state current, unless there is a mechanism allowing the surface states to be dynamically refilled to compensate for the tunneling outflux across the barrier. Since the rate of this refilling process should only keep up with the (slow) tunneling, it is reasonable to expect that in any practical situation it will be made possible through inelastic processes. Thus, it may be argued that Bardeen's approach correctly includes the contribution of pure surface states, while the Landauer–Büttiker approach does not. This argument was advanced by Wortmann et al. [49], who have also developed an elegant computational technique designed to implement Bardeen's formula for the tunneling current within a full-potential linear augmented plane wave (FLAPW) method. This technique is based on the embedded Green function formulation [50] for the electronic structure of an active layer sandwiched between semi-infinite electrodes, which was previously used to implement the Landauer–Büttiker calculation of the conductance in FLAPW [51]. Unfortunately, Bardeen's approach is not universal, because it is limited to the case when the tunneling coupling is weak, which allows one to treat tunneling as a first-order process in this coupling.

As it was mentioned above, bonding at the interface may be very different from the bulk, which often leads to the formation of interface resonant states. Additional complications occur due to interface reconstruction, interdiffusion, disorder, etc. Apart from the possible formation of surface states, there are, obviously, other ways in which interface bonding and atomic rearrangements can strongly influence the spin polarization and TMR. For example, excess oxidation at the interface can tend to quench or reduce the magnetic moments near the surface [30,52] and remove some of the surface spectral weight from the Fermi level. Combined with the symmetry selection rules for the surface states, this may effectively lead to the formation of additional spin-dependent tunneling barriers inside the surface layer. These changes may sometimes even lead to a reversal of the spin polarization of the tunneling current.

12.4 Influence of Interface Roughness and Defects in the Barrier

Interface imperfections and defects in the barrier may have a strong effect on spin-dependent tunneling due to the mixing of the tunneling channels, as well as resonant transmission through the localized electronic states created by these defects. In this section, we will briefly discuss the influence of point defects on the TMR within the single-electron picture.

12.4.1 Interface Roughness

Since high TMR is achieved in high-quality MTJs exhibiting symmetry-enforced spin filtering (see Section 12.3.2), defects usually should reduce the TMR. It appears that atomic-scale interface roughness or disorder is the most detrimental, because it can introduce scattering between otherwise independent tunneling channels and thereby reduce the efficiency of the symmetry-enforced spin filtering [33,48]. On the other hand, sparsely spaced steps between atomically flat terraces can be expected to be of lesser importance. The characteristic scale here is the inverse of the size of the area in $\mathbf{k}_{\parallel}$ space contributing appreciably to the tunneling current. According to Sections 12.2.4 and 12.3.4, this size is of order $\Delta r \sim \sqrt{\alpha d}$, where α is typically a few lattice parameters; Δr can be viewed as the transverse size of the tunneling wave packet. Thus, if the typical distance L between the atomically flat steps is much larger than Δr, they will make a contribution proportional to the areal density of the steps (i.e., this contribution contains a factor $\Delta r/L$). However, this contribution activates the tunneling channel with the lowest decay parameter, and therefore takes over at large barrier thickness. In any case, surface roughness limits the efficiency of spin filtering and imposes the upper bound for TMR, which saturates as a function of barrier thickness.

12.4.2 Nonresonant Defects in the Barrier

The behavior of the tunneling transparency is very different in the cases when the point defects do or do not introduce localized levels close to the Fermi level [53,54]. In this context, "close to the Fermi level" means that the distance from the Fermi level is not large compared the level's linewidth due to coupling with the electrodes. For example, a continuum of localized levels should always be present in an amorphous barrier (like Al_2O_3), while isolated point defects, such as oxygen vacancies, would be the most common in high-quality crystalline barriers.

The effect of oxygen vacancies in Fe/MgO/Fe MTJs was considered by Velev et al. [55]. Figure 12.6 shows the calculated majority-spin transmission as a function of energy for an ideal Fe/MgO/Fe MTJ (solid line) and for an MTJ with O vacancies (dashed line). For the ideal MTJ, the majority-spin transmission has a characteristic energy point (denoted by arrow in Figure 12.6) above which the transmission increases significantly due to the occurrence of the Δ_1 symmetry band in the electronic structure of Fe(001). O vacancies produce a pronounced peak in the transmission due to resonant tunneling of electrons via the O vacancy *s* state. The width of the peak depends strongly on the vacancy density and on the coupling to the electrodes determined by MgO thickness.

The transmission at resonance is not of practical interest in this case, however, because the resonant level lies at about 1 eV

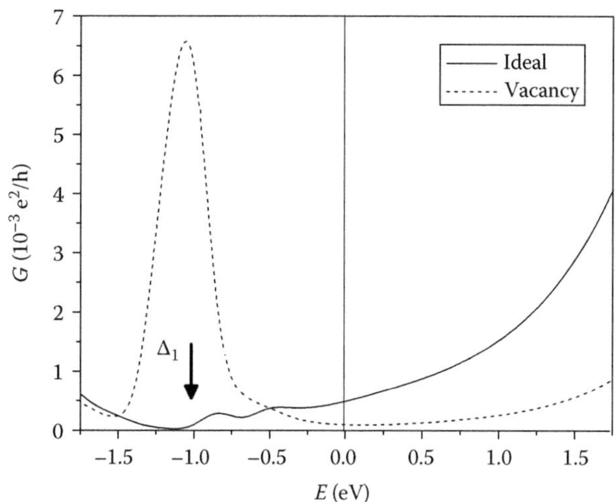

FIGURE 12.6 Majority-spin conductance per unit cell area in Fe/MgO/Fe(001) tunnel junction with five monolayers of MgO for ideal MgO and MgO with O vacancies (1/8 cell). The arrow indicates the bottom of the majority-spin Δ_1 band. The Fermi energy lies at zero. (From Velev, J.P. et al., *Appl. Phys. Lett.*, 90, 072502, 2007. With permission.)

below the Fermi energy. With correction for the self-interaction, the resonant level moves even deeper down in energy as corroborated by the experimental data [56]. Therefore, for a moderate bias, the transport mechanism is controlled by nonresonant scattering at O vacancies. It is evident from Figure 12.6 that at and above the Fermi energy the transmission of an MTJ with vacancies is reduced by a factor of 5–7 depending on energy, as compared to the transmission of a perfect MTJ. This detrimental effect of O vacancies on the majority-spin transmission is due to scattering of tunneling electrons between states with different transverse wave vectors $\mathbf{k}_{\parallel}$. In a perfect Fe/MgO/Fe junction, the tunneling probability in the majority-spin channel is dominated by the Δ_1 evanescent state at the $\bar{\Gamma}$ point ($\mathbf{k}_{\parallel} = 0$), because it has the longest attenuation length in MgO. Scattering to states with $\mathbf{k}_{\parallel} \neq 0$ reduces the transmission coefficient due to a shorter decay length of these states.

For the nonresonant case, Lifshitz and Kirpichenkov [54] developed a rigorous technique allowing one to calculate the tunneling transparency by performing the configurational averaging in a "subbarrier kinetic equation" that describes the decay of the electronic density of a particular incident wave inside the barrier. The "collision integral" in this equation appears as a power series in the defect concentration. In the first order, the kinetic equation is linear; a similar equation was used by Zhang et al. [34] to discuss the effect of nonresonant point defects on TMR in Fe/MgO/Fe tunnel junctions. The effect of point defects is to scatter electrons from one evanescent state $\mathbf{k}_{\parallel}$ to another $\mathbf{k}_{\parallel}$. (It may be more physically appealing to imagine an outgoing spherical evanescent wave emitted by the defect, on which a plane evanescent wave is incident. On each side of the defect, the evanescent spherical wave can be expanded in $\mathbf{k}_{\parallel}$ states, which decay in the corresponding direction.)

Generalizing the approach of Ref. [54], single defect scattering leads to the appearance of new $\mathbf{k}_{\parallel}$-nonconserving transmission channels, each contributing to the tunneling current as

$$T(\mathbf{k}_{\parallel},\mathbf{k}'_{\parallel},x) = ce^{-2\kappa_0 d}T_L(\mathbf{k}_{\parallel})e^{-2(\kappa-\kappa_0)x} \times |t(\mathbf{k}_{\parallel},\mathbf{k}'_{\parallel})|^2 e^{-2(\kappa'-\kappa_0)(d-x)}T_R(\mathbf{k}'_{\parallel}) \quad (12.20)$$

where

- c is the concentration of defects
- $T_L(\mathbf{k}_{\parallel})$ and $T_R(\mathbf{k}_{\parallel})$ are the surface transmission functions introduced in Section 12.3.3
- $t(\mathbf{k}_{\parallel},\mathbf{k}'_{\parallel})$ is the subbarrier scattering amplitude
- x is the coordinate of the impurity
- κ and κ' are the decay parameters of the evanescent barrier states with $\mathbf{k}_{\parallel}$ and $\mathbf{k}_{\parallel}$, respectively
- κ_0 is the smallest decay parameter for all evanescent bands

The obvious complex band indices have been omitted. It is seen that the $\mathbf{k}_{\parallel} \to \mathbf{k}'_{\parallel}$ channel can give an appreciable contribution only if $x(\kappa - \kappa_0) + (d - x)(\kappa' - \kappa_0) \leq 1$. Since the ballistic contribution is restricted to $(\kappa - \kappa_0)d \sim 1$, it is seen that the scattering with a large lateral momentum transfer (either $\mathbf{k}_{\parallel}$ or $\mathbf{k}_{\parallel}$), as well as interband scattering, make a significant contribution only if the scattering impurity is located close to one of the interfaces, that is, within a distance $(\kappa - \kappa_0)^{-1}$ from the left interface, or within $(\kappa' - \kappa_0)^{-1}$ from the right interface. Averaging over the position of the impurity in (12.20) yields, therefore, a factor $(\kappa - \kappa_0)^{-1}$, that is,

$$T(\mathbf{k}_{\parallel},\mathbf{k}'_{\parallel}) = \frac{c}{2}e^{-2\kappa_0 d}T_L(\mathbf{k}_{\parallel})\,|t(\mathbf{k}_{\parallel},\mathbf{k}'_{\parallel})|^2 \left[\frac{1}{\kappa - \kappa_0} + \frac{1}{\kappa' - \kappa_0}\right]T_R(\mathbf{k}'_{\parallel}) \quad (12.21)$$

The first (second) term in square brackets comes from impurity scattering near the left (right) interface. The behavior of $T(\mathbf{k}_{\parallel},\mathbf{k}'_{\parallel})$ in the region of $\mathbf{k}_{\parallel}$ where $(\kappa - \kappa_0)d \leq 1$ (and similar for $\mathbf{k}_{\parallel}$) needs a more detailed consideration. As shown in Ref. [54] within their simplified model, the unphysical divergence at $\mathbf{k}_{\parallel},\mathbf{k}'_{\parallel} \to 0$ is an artifact of the perturbative treatment.

Since the $\mathbf{k}_{\parallel}$-nonconserving channels produce a contribution decaying with thickness at a similar rate to the symmetry-filtered ballistic contribution, nonresonant defect scattering in the barrier leads to the saturation of TMR at a value that depends on impurity concentration (similar to the case of interface roughness discussed above), and fitting to the experimental thickness dependence of the parallel and antiparallel conductances is possible [34]. The case of very thick barriers, for which the ballistic tunneling is fully replaced by diffusive tunneling, requires a separate treatment along the lines of Ref. [54].

12.4.3 Resonant Defects in the Barrier

If localized levels in the barrier are found close to the Fermi level, and as long as the temperature is sufficiently low to preserve the wavefunction coherence across the barrier, the tunneling current is dominated by resonant channels through which an electron can tunnel with no attenuation [53,54]. If the defect concentration is sufficiently low and the barrier is thin enough, the resonant channels are formed by defects located close to the middle of the barrier, so that their coupling to both electrodes is similar. For thicker barriers or larger impurity concentrations, the resonant channels are formed either by percolating resonant paths connecting the defects lying close to a straight line perpendicular to the interfaces, or by even more complicated arrangements of defects. It is typical in this situation that the channels carrying most of the current are statistically rare, and the conductance does not behave as a self-averaging quantity. In particular, large fluctuations of the conductance, as well as irregularities in the current–voltage curves, are expected in junctions of small transverse sizes depending on the specific arrangement of defects [53,57].

Let us consider an isolated defect in the barrier, and assume that in an infinite barrier it would have one nondegenerate eigenstate ψ_0 at energy ε_0. Neglecting direct tunneling, and applying time-dependent perturbation theory in the spirit of Bardeen's approach (see Section 12.2.1), where the role of perturbation is played by the hopping matrix elements t_l and t_r between the defect level and the Bloch states in the left and right electrodes (respectively), one obtains the rate of tunneling between the states ψ_l and ψ_r in the electrodes [57]:

$$w_{lr}(\varepsilon) = \frac{2\pi}{\hbar}\frac{|t_l|^2|t_r|^2}{(\varepsilon - \varepsilon_0)^2 + (\Gamma_L + \Gamma_R)^2}\delta(\varepsilon_r - \varepsilon_l) \quad (12.22)$$

where the quantities

$$\Gamma_L = \pi\sum_l |t_l|^2\delta(\varepsilon - \varepsilon_l); \quad \Gamma_R = \pi\sum_r |t_r|^2\delta(\varepsilon - \varepsilon_r) \quad (12.23)$$

play the role of the self-energy of the impurity level due to its coupling to the electrodes. The zero-temperature conductance for the given spin channel is obtained by taking the trace:

$$G(\varepsilon) = \frac{4e^2}{h}\frac{\Gamma_L\Gamma_R}{(\varepsilon - \varepsilon_0)^2 + (\Gamma_L + \Gamma_R)^2} \quad (12.24)$$

It is seen that the conductance has a sharp maximum when the impurity level ε_0 lies close to the Fermi energy. This behavior is an example of resonant conductance.

For a planar tunnel junction, the indices l and r in the hopping matrix elements t_l and t_r denote $\mathbf{k}_{\parallel}$ and the band index. If, for each $\mathbf{k}_{\parallel}$, only one evanescent state is important, these matrix elements are related to the interface transmission functions T_L and T_R, that is,

$$|t_l|^2 \to |t_L(\mathbf{k}_{\parallel})|^2 = T_L(\mathbf{k}_{\parallel})e^{-2\kappa(\mathbf{k}_{\parallel})x}\,|V(\mathbf{k}_{\parallel})|^2 \quad (12.25)$$

where L stands for the left electrode, the electrode band index is absorbed in $\mathbf{k}_{\parallel}$, and $V(\mathbf{k}_{\parallel}) = \langle\psi_0|V|u(\mathbf{k}_{\parallel})\rangle$, where V is the perturbing defect potential, and $u(\mathbf{k}_{\|})$ the periodic part of the evanescent wave (from which the decaying Bloch exponent has been excluded). The coordinate of the impurity is denoted as x; a similar expression for t_R contains $d - x$ instead.

The conductance (Equation 12.24) at the given energy depends only on Γ_L and Γ_R, which are equal to the surface Brillouin zone integrals of Equation 12.25. The situation is very different from nonresonant tunneling. In particular, in the resonant case $\varepsilon = \varepsilon_0$, the conductance is dominated by impurities with $\Gamma_L \approx \Gamma_R$, which set up transparent channels with $G \approx e^2/h$. These impurities are located close to the *middle* of the barrier, contrary to the dominant nonresonant impurities, which lie close to the interfaces, as discussed above. Let us consider particular cases.

In the nonresonant regime, that is, at $\varepsilon_0 - \varepsilon_F \gg (\Gamma_L + \Gamma_R)$, the situation is rather similar to the direct nonresonant tunneling considered in the previous section. Whether direct or impurity-assisted nonresonant tunneling dominates depends on the relation between the matrix elements and on the distance of the impurity level from the Fermi level.

For a narrow (point) junction, the conductance of a particular sample may be dominated by a single resonant impurity. In this case, the sign of the TMR may be reversed compared to the nonresonant one. Indeed, consider the case $\varepsilon_0 - \varepsilon_F = 0$ and an asymmetric impurity position. If the impurity is closer to the left electrode, then $\Gamma_L \gg \Gamma_R$, and $G = (4e^2/h)(\Gamma_L/\Gamma_R)$. In the opposite case $\Gamma_L \ll \Gamma_R$; Γ_R we have $G = (4e^2/h)(\Gamma_L/\Gamma_R)$. In both cases the conductance is inversely proportional to the self-energy due to coupling with one of the electrodes, which results in the sign reversal of TMR [58]. This behavior was observed experimentally in narrow nanojunctions [58] and in gate-field-controlled carbon nanotubes connecting two ferromagnetic electrodes through tunnel barriers [59]. This latter effect occurs due to resonant tunneling through quantum well states in the nanotube, whose energy is changed by the gate voltage.

For a wide tunnel junction, the averaging over the random impurity positions should be performed. For simplicity, let us again take $\varepsilon_F = \varepsilon_0$. If the system consisted of independent one-dimensional threads, (12.23) would contain only one term with the x dependence from Equations 12.25, and 12.24 can then be easily integrated over x. Neglecting for simplicity the discreteness of x due to the lattice periodicity, and assuming $\kappa d \gg 1$, we find with these approximations that the conductance $G = ce^2/(h\kappa)$ does not depend on Γ_L, Γ_R at all. This means that in this case the current does not depend on the magnitude of the coupling to the bulk states, and therefore the TMR vanishes. Note that the dominant impurities are always those that maximize G in Equation 12.24; for different configurations of the electrode magnetizations and different spin channels, this maximum is achieved at different x.

The case when the impurity states are not all pinned to ε_F, but are rather distributed with a constant density may be appropriate for an amorphous barrier. In the quasi-one-dimensional case, one then has to perform an additional integration over $\varepsilon_0 > \varepsilon_F$, which results in $\langle G \rangle \propto \sqrt{T_L T_R}$ [8].

The situation for a three-dimensional system is more complicated. The conductance looks similar to Equation 12.24, but it now contains a double sum over $\mathbf{k}_{\|}$, $\mathbf{k}_{\|}$, while Γ_L, Γ_R are replaced by quantities given in Equation 12.25. If x is integrated over, the resonance for different $\mathbf{k}_{\|} \to \mathbf{k}'_{\|}$ channels is achieved at different values of x (i.e., different impurities are at resonance). Therefore, the total conductance cannot, in general, be represented as a product of two ("left" and "right") factors in the three-dimensional case.

12.5 Spin–Orbit Coupling and Tunneling Anisotropic Magnetoresistance

Until now, we have treated the transport of electrons with two different spin projections as being completely independent. This is a good approximation as long as the magnetizations of the two electrodes are collinear, the temperature is low enough so that scattering on spin fluctuations is not important, and SOC is negligible. Spin transport in noncollinear configurations is of main interest for problems involving spin torque effects, which are considered in Chapters 7, 8, and 14.

SOC may naturally be expected to affect the TMR appreciably if it can generate a significant amount of spin-flip scattering at the interfaces. In addition, in the presence of SOC, the evanescent barrier eigenstates do not have a definite spin quantum number, which weakens the selection rules leading to tunneling spin filtering. In this case, SOC should normally lead to the reduction of TMR due to the reduced spin selectivity of the tunneling process. The details of this problem were studied for a Fe/GaAs/Fe MTJ using a fully-relativistic Green's function formalism [60]. In this method, the current operator is approximated to be spin diagonal in the leads, while the spins are mixed in the "active region"; thus the Landauer–Büttiker transmission function has both spin-conserving and spin-flip components, the latter representing the effect of SOC. For Fe/GaAs/Fe, the main effect was found to be on the conductance for the antiparallel configurations, where SOC is able to open efficient spin-flip conducting channels that were otherwise forbidden; the TMR is thereby reduced.

However, under favorable conditions SOC can also lead to the appearance of an entirely new effect called tunneling anisotropic magnetoresistance in which the resistance of a tunnel junction with only *one* ferromagnetic electrode depends on the magnetization direction of that electrode. First experimental demonstrations of TAMR utilized a diluted magnetic semiconductor (Mn-doped GaAs) as the magnetic electrode [61]. In general, the TAMR effect occurs because the electronic structure of the magnetic side of the junction depends on the orientation of the magnetization with respect to the crystallographic axes, and this dependence is reflected in the tunneling conductance. A similar effect exists even in bulk resistivity (it is called anisotropic magnetoresistance), but the tunneling setup can greatly increase its magnitude, because SOC is often strongly enhanced at interfaces.

Consider the effect of SOC on the interface states, whose role in tunneling was discussed in Section 12.3.5. Suppose that the tunneling current has a large contribution from such states. Having the largest amplitude at the surface, these states are the most sensitive to enhanced interfacial SOC. (In other words, the expectation value of the SOC operator is usually larger for the interface resonant states.) The effect of SOC is to shift these states up or down in energy depending on their in-plane wave-vector $\mathbf{k}_{\|}$ and on the magnetization direction (Rashba effect) [62]. (Note that the spin degeneracy is already lifted by the exchange splitting; the SOC does not lead to any additional symmetry breaking.) In turn, these shifts deform the Fermi contour of the interface states. Since the tunneling probabilities strongly depend on $\mathbf{k}_{\|}$ (see Section 12.3.2), the tunneling current, in turn, depends on the direction of magnetization, leading to the TAMR effect [63]. A similar effect occurs in an MTJ with two magnetic electrodes, that is, the tunneling current depends on the orientation of the magnetization of both electrodes, which remain parallel to each other [64,65]. In this latter case, however, the situation is complicated by the presence of two interfaces, which may both support interface states.

One can distinguish out-of-plane and in-plane TAMR effects, with respect to the mode in which the magnetization is rotated. The out-of-plane (in-plane) TAMR refers to the resistance variation for the rotation of the magnetization in a plane normal (parallel) to the interfaces. The angular dependence of the resistance reflects the underlying (nonmagnetic) symmetry of the junction. The resistance is symmetric with respect to the change of sign of the magnetization. (This follows from the time-reversal symmetry of the Hamiltonian; it is assumed that the external field is transformed along with the magnetization.) Obviously, if the junction has a rotation axis (normal to the interfaces), the TAMR automatically has the same axis of symmetry. Further, if the junction has a reflection plane normal to the interfaces, the TAMR also has the same reflection plane. (This property follows from the consideration of the response of the axial **M** vector to the mirror reflection combined with time reversal.)

A particular case of interest is the Fe/GaAs/Au tunnel junction, whose in-plane TAMR has a C_{2v} angular symmetry [66]. The junction itself cannot have a fourfold axis, because zinc blende GaAs only has screw fourfold axes, which are destroyed in a junction setup. For example, for an As-terminated interface, the normal planes containing the GaAs bonds next to the interface are oriented in a certain unique way. For this reason, even without SOC, the electronic structure of the Fe/GaAs interface at most has a C_{2v} symmetry (see, e.g., figure 2 of Ref. [67]). Therefore, in the presence of SOC the in-plane TAMR cannot have a fourfold axis.

A further discussion of TAMR is given in Chapter 21.

12.6 Concluding Remarks

In this chapter, we have discussed the main properties of TMR, which follow from basic quantum-mechanical considerations. In many cases these considerations (along with model or first-principles calculations) are sufficient to understand the qualitative features of measured TMR and its dependence on variable parameters. However, it is worth reemphasizing that the tunneling currents are very sensitive to the electronic structure of the electrodes, barrier, and interfaces. Consequently, quantitative comparison with experiment requires that this electronic structure is known in sufficient detail. Although the epitaxial relations are known for some high-quality epitaxial interfaces (like Fe/MgO), there always remains some uncontrolled variation in such details as the degree of disorder, interdiffusion, impurity segregation, and other imperfections dependent on film growth conditions. For example, the presence of excess oxygen at the Fe/MgO interface, which likely incorporates itself in the surface Fe layer, was found to have a large effect on spin-dependent tunneling across this interface [68]. Likewise, it was found that the presence of excess oxygen at the Co/Al_2O_3 interface can lead to a reversal of the sign of TMR in $Co/Al_2O_3/Co$ MTJs [52]. The influence of surface structure on TMR is especially strong if interface resonant states are involved. Tunneling across a vacuum gap from a high-quality surface of a sufficiently simple metal, for which the spin polarization can be measured using spin-polarized STM, is an important example where the surface structure may be known in full detail.

Even if the atomic structure of the interfaces is known (or hypothesized), the standard methods used to describe the electronic structure based on density functional theory often lead to large errors in insulating band gaps, band offsets at the interfaces, as well as in the dispersion relations of the interface states. It is therefore necessary to establish whether the behavior of the calculated TMR is robust against such errors; this is, unfortunately, not always possible. The development of new methods to redress deficiencies of the local density approximation (such as the GW method) is desirable. An application of the GW method to a Fe/MgO/Fe tunnel junction was recently reported [69].

Acknowledgments

This work was supported by the National Science Foundation through the Nebraska Materials Research Science and Engineering Center (Grant No. DMR-0820521) and the Nebraska Research Initiative.

References

1. W. F. Brinkman, R. C. Dynes, and J. M. Rowell, Tunneling conductance of asymmetrical barriers, *J. Appl. Phys.* **41**(5), 1915–1921 (1970).
2. W. A. Harrison, Tunneling from an independent-particle point of view, *Phys. Rev.* **123**(1), 85–89 (1961).
3. J. G. Simmons, Generalized formula for the electric tunnel effect between similar electrodes separated by a thin insulating film, *J. Appl. Phys.* **34**(6), 1793–1803 (1963).
4. J. Bardeen, Tunnelling from a many-particle point of view, *Phys. Rev. Lett.* **6**(2), 57–59 (1961).

5. J. C. Slonczewski, Conductance and exchange coupling of two ferromagnets separated by a tunneling barrier, *Phys. Rev. B* **39**(10), 6995–7002 (1989).
6. J. S. Moodera, J. Nassar, and G. Mathon, Spin-tunneling in ferromagnetic junctions, *Ann. Rev. Mater. Sci.* **29**, 381–432 (1999).
7. E. Y. Tsymbal, K. D. Belashchenko, J. P. Velev et al., Interface effects in spin-dependent tunneling, *Prog. Mater. Sci.* **52**, 401–420 (2007).
8. E. Y. Tsymbal, O. N. Mryasov, and P. R. LeClair, Spin-dependent tunnelling in magnetic tunnel junctions, *J. Phys.: Condens. Matter* **15**, R109–R142 (2003).
9. X.-G. Zhang and W. H. Butler, Band structure, evanescent states, and transport in spin tunnel junctions, *J. Phys.: Condens. Matter* **15**, R1603–R1639 (2003).
10. M. Julliére, Tunneling between ferromagnetic films, *Phys. Lett. A* **54**(3), 225–226 (1975).
11. P. M. Tedrow and R. Meservey, Spin polarization of electrons tunneling from films of Fe, Co, Ni, and Gd. *Phys. Rev. B* **7**(1), 318–326 (1973).
12. S. Datta, *Electronic Transport in Mesoscopic Systems*, Cambridge University Press, Cambridge, U.K., 1997.
13. J. Tersoff and D. R. Hamann, Theory of the scanning tunneling microscope, *Phys. Rev. B* **31**(2), 805–813 (1985).
14. W. H. Butler, X.-G. Zhang, T. C. Schulthess, and J. M. MacLaren, Spin-dependent tunneling conductance of FejMgOjFe sandwiches, *Phys. Rev. B* **63**(5), 054416 (2001).
15. J. Mathon and A. Umerski, Theory of tunneling magnetoresistance of an epitaxial Fe/MgO/Fe(001) junction, *Phys. Rev. B* **63**(22), 220403 (2001).
16. S. S. P. Parkin, C. Kaiser, A. Panchula et al., Giant tunnelling magnetoresistance at room temperature with MgO (100) tunnel barriers, *Nat. Mater.* **3**(12), 862–867 (2004).
17. S. Yuasa, T. Nagahama, A. Fukushima, Y. Suzuki, and K. Ando, Giant room-temperature magnetoresistance in single-crystal Fe/MgO/Fe magnetic tunnel junctions, *Nat. Mater.* **3**(12), 868–871 (2004).
18. W. Kohn, Analytic properties of Bloch waves and Wannier functions, *Phys. Rev.* **115**(4), 809–821 (1959).
19. V. Heine, On the general theory of surface states and scattering of electrons in solids, *Proc. Phys. Soc.* **81**(2), 300–310 (1963).
20. Ph. Mavropoulos, N. Papanikolaou, and P. H. Dederichs, Complex band structure and tunneling through ferromagnet/insulator/ferromagnet junctions, *Phys. Rev. Lett.* **85**(5), 1088–1091 (2000).
21. Y.-C. Chang, Complex band structures of zinc-blende materials, *Phys. Rev. B* **25**(2), 605–619 (1982).
22. M. Tinkham, *Group Theory and Quantum Mechanics*, Dover, New York, 2003.
23. L. P. Bouckaert, R. Smoluchowski, and E. Wigner, Theory of Brillouin zones and symmetry properties of wave functions in crystals, *Phys. Rev.* **50**(1), 58–67 (1936).
24. G. C. Osbourn and D. L. Smith, Transmission and reflection coefficients of carriers at an abrupt GaAs-GaAlAs (100) interface, *Phys. Rev. B* **19**(4), 2124–2133 (1979).
25. J. M. MacLaren, X.-G. Zhang, W. H. Butler, and X. Wang, Layer KKR approach to Bloch-wave transmission and reflection: Application to spin-dependent tunneling, *Phys. Rev. B* **59**(8), 5470–5478 (1999).
26. J. Faure-Vincent, C. Tiusan, E. Jouguelet et al., High tunnel magnetoresistance in epitaxial Fe/MgO/Fe tunnel junctions, *Appl. Phys. Lett.* **82**(25), 4507–4509 (2003).
27. X.-G. Zhang and W. H. Butler, Large magnetoresistance in bcc Co/MgO/Co and FeCo/MgO/FeCo tunnel junctions, *Phys. Rev. B* **70**(17), 172407 (2004).
28. T. Moriyama, C. Ni, W. G. Wang, X. Zhang, and J. Q. Xiao, Tunneling magnetoresistance in (001)-oriented FeCo/MgO/FeCo magnetic tunneling junctions grown by sputtering deposition, *Appl. Phys. Lett.* **88**(22), 222503 (2006).
29. S. Yuasa, A. Fukushima, H. Kubota, Y. Suzuki, and K. Ando, Giant tunneling magnetoresistance up to 410% at room temperature in fully epitaxial Co/MgO/Co magnetic tunnel junctions with bcc Co(001) electrodes, *Appl. Phys. Lett.* **89**(4), 042505 (2006).
30. K. D. Belashchenko, E. Y. Tsymbal, M. van Schilfgaarde, D. A. Stewart, I. I. Oleynik, and S. S. Jaswal, Effect of interface bonding on spin-dependent tunneling from the oxidized Co surface, *Phys. Rev. B* **69**(17), 174408 (2004).
31. R. Matsumoto, A. Fukushima, T. Nagahama, Y. Suzuki, K. Ando, and S. Yuasa, Oscillation of giant tunneling magnetoresistance with respect to tunneling barrier thickness in fully epitaxial Fe/MgO/Fe magnetic tunnel junctions, *Appl. Phys. Lett.* **90**(25), 252506 (2007).
32. C. Heiliger, P. Zahn, B. Yu. Yavorsky, and I. Mertig, Thickness dependence of the tunneling current in the coherent limit of transport, *Phys. Rev. B* **77**(22), 224407 (2008).
33. J. Mathon and A. Umerski, Theory of tunneling magnetoresistance in a disordered Fe/MgO/Fe(001) junction, *Phys. Rev. B* **74**(14), 140404 (2006).
34. X.-G. Zhang, Y. Wang, and X. F. Han, Theory of nonspecular tunneling through magnetic tunnel junctions, *Phys. Rev. B* **77**(14), 144431 (2008).
35. C.-M. Lee, J.-M. Lee, L.-X. Ye, S.-Y. Wang, Y.-R. Wang, and T.-H. Wu, Effects of MgO barrier thickness on magnetic anisotropy energy constant in perpendicular magnetic tunnel junctions of (Co/Pd)n/MgO/(Co/Pt)n, *IEEE Trans. Magn.* **44**(11), 2558–2561 (2008).
36. I. I. Mazin, Tunneling of Bloch electrons through vacuum barrier, *Europhys. Lett.* **55**(3), 404–410 (2001).
37. W. A. Harrison, *Solid State Theory*, Dover, New York, 1980.
38. J. Bardeen, Surface states and rectification at a metal semiconductor contact, *Phys. Rev.* **71**(10), 717–727 (1947).
39. W. Shockley, On the surface states associated with a periodic potential, *Phys. Rev.* **56**(4), 317–323 (1939).
40. J. A. Stroscio, D. T. Pierce, A. Davies, R. J. Celotta, and M. Weinert, Tunneling spectroscopy of bcc (001) surface states, *Phys. Rev. Lett.* **75**(16), 2960–2963 (1995).
41. E. Y. Tsymbal and K. D. Belashchenko, Role of interface bonding in spin-dependent tunneling, *J. Appl. Phys.* **97**(10), 10C910 (2005).

42. O. Wunnicke, N. Papanikolaou, R. Zeller, P. H. Dederichs, V. Drchal, and J. Kudrnovský, Effects of resonant interface states on tunneling magnetoresistance, *Phys. Rev. B* **65**(6), 064425 (2002).
43. C. Tiusan, J. Faure-Vincent, C. Bellouard, M. Hehn, E. Jouguelet, and A. Schuhl, Interfacial resonance state probed by spin-polarized tunneling in epitaxial Fe/MgO/Fe tunnel junctions, *Phys. Rev. Lett.* **93**(10), 106602 (2004).
44. C. Tiusan, M. Sicot, J. Faure-Vincent et al., Static and dynamic aspects of spin tunnelling in crystalline magnetic tunnel junctions, *J. Phys.: Condens. Matter* **18**(3), 941–956 (2006).
45. K. D. Belashchenko, J. Velev, and E. Y. Tsymbal, Effect of interface states on spin-dependent tunneling in Fe/MgO/Fe tunnel junctions, *Phys. Rev. B* **72**(14), 140404 (2005).
46. J. P. Velev, K. D. Belashchenko, and E. Y. Tsymbal, Comment on Destructive effect of disorder and bias voltage on interface resonance transmission in symmetric tunnel junctions, *Phys. Rev. Lett.* **96**(11), 119601 (2006).
47. Y. Ke, K. Xia, and H. Guo, Disorder scattering in magnetic tunnel junctions: Theory of nonequilibrium vertex correction, *Phys. Rev. Lett.* **100**(16), 166805 (2008).
48. P. X. Xu, V. M. Karpan, K. Xia, M. Zwierzycki, I. Marushchenko, and P. J. Kelly, Influence of roughness and disorder on tunneling magnetoresistance, *Phys. Rev. B* **73**, 180402 (2006).
49. D. Wortmann, H. Ishida, and S. Blügel, Embedded Green-function formulation of tunneling conductance: Bardeen versus Landauer approaches, *Phys. Rev. B* **72**(23), 235113 (2005).
50. J. E. Inglesfield, A method of embedding, *J. Phys. C: Solid State Phys.* **14**(26), 3795–3806 (1981).
51. D. Wortmann, H. Ishida, and S. Blügel, Embedded Green-function approach to the ballistic electron transport through an interface, *Phys. Rev. B* **66**(7), 075113 (2002).
52. K. D. Belashchenko, E. Y. Tsymbal, I. I. Oleynik, and M. van Schilfgaarde, Positive spin polarization in Co/Al_2O_3/Co tunnel junctions driven by oxygen adsorption, *Phys. Rev. B* **71**, 224422 (2005).
53. I. M. Lifshitz, S. A. Gredeskul, and L. A. Pastur, *Introduction to the Theory of Disordered Systems*, Wiley, New York, 1988.
54. I. M. Lifshitz and V. Y. Kirpichenkov, On tunneling transparency of disordered systems, *Sov. Phys. JETP* **50**, 499–511 (1979).
55. J. P. Velev, K. D. Belashchenko, S. S. Jaswal, and E. Y. Tsymbal, Effect of oxygen vacancies on spin-dependent tunneling in Fe/MgO/Fe magnetic tunnel junctions, *Appl. Phys. Lett.* **90**, 072502 (2007).
56. P. G. Mather, J. C. Read, and R. A. Buhrman, Disorder, defects, and band gaps in ultrathin (001) MgO tunnel barrier layers, *Phys. Rev. B* **73**, 205412 (2006).
57. A. I. Larkin and K. A. Matveev, Current-voltage characteristics of mesoscopic semiconductor contacts, *Sov. Phys. JETP* **66**, 580–584 (1987).
58. E. Y. Tsymbal, A. Sokolov, I. F. Sabirianov, and B. Doudin, Resonant inversion of tunneling magnetoresistance, *Phys. Rev. Lett.* **90**(18), 186602 (2003).
59. S. Sahoo, T. Kontos, J. Furer et al., Electric field control of spin transport, *Nat. Phys.* **1**, 99–102 (2005).
60. V. Popescu, H. Ebert, N. Papanikolaou, R. Zeller, and P. H. Dederichs, Influence of spin-orbit coupling on the transport properties of magnetic tunnel junctions, *Phys. Rev. B* **72**(18), 184427 (2005).
61. C. Gould, C. Räuster, T. Jungwirth et al., Tunneling anisotropic magnetoresistance: A spin-valve-like tunnel magnetoresistance using a single magnetic layer, *Phys. Rev. Lett.* **93**(11), 117203 (2004).
62. O. Krupin, G. Bihlmayer, K. Starke et al., Rashba effect at magnetic metal surfaces, *Phys. Rev. B* **71**, 201403 (2005).
63. A. N. Chantis, K. D. Belashchenko, E. Y. Tsymbal, and M. van Schilfgaarde, Tunneling anisotropic magnetoresistance driven by resonant surface states: First-principles calculations on an Fe(001) surface, *Phys. Rev. Lett.* **98**(4), 046601 (2007).
64. L. Gao, X. Jiang, S.-H. Yang, J. D. Burton, E. Y. Tsymbal, and S. S. P. Parkin, Bias voltage dependence of tunneling anisotropic magnetoresistance in magnetic tunnel junctions with MgO and Al_2O_3 tunnel barriers, *Phys. Rev. Lett.* **99**(22), 226602 (2007).
65. M. N. Khan, J. Henk, and P. Bruno, Anisotropic magnetoresistance in Fe/MgO/Fe tunnel junctions, *J. Phys.: Condens. Matter* **20**(15), 155208 (2008).
66. J. Moser, A. Matos-Abiague, D. Schuh, W. Wegscheider, J. Fabian, and D. Weiss, Tunneling anisotropic magnetoresistance and spin-orbit coupling in Fe/GaAs/Au tunnel junctions, *Phys. Rev. Lett.* **99**(5), 056601 (2007).
67. A. N. Chantis, K. D. Belashchenko, D. L. Smith, E. Y. Tsymbal, M. van Schilfgaarde, and R. C. Albers, Reversal of spin polarization in Fe/GaAs (001) driven by resonant surface states: First-principles calculations, *Phys. Rev. Lett.* **99**(19), 196603 (2007).
68. X.-G. Zhang, W. H. Butler, and A. Bandyopadhyay, Effects of the iron-oxide layer in Fe-FeO-MgO-Fe tunneling junctions, *Phys. Rev. B* **68**(9), 092402 (2003).
69. S. V. Faleev, O. N. Mryasov, and M. van Schilfgaarde, Effect of electron correlations on spin polarized transport in Fe/MgO tunnel junctions, arXiv: 1010.4086.

13

Spin-Filter Tunneling

Tiffany S. Santos
Argonne National Laboratory

Jagadeesh S. Moodera
Massachusetts Institute of Technology

13.1 Introduction

Over the last two decades or more, condensed matter physics, and in particular the field of magnetism, is overwhelmingly active in the area of spintronics: the effects of spin-polarized current, its generation, injection, transport, and dynamics. The amount of information generated and the discoveries that have resulted are unprecedented, reshaping not only the direction of magnetism research but also driving technological advances to dizzying heights. Thousands of articles, countless reviews, and several books have been written that cover the field as it is evolving. The field is developing so rapidly and becoming so diverse that it is hard to keep up with. The driving force is the spin-polarized current generated from ferromagnetic elements, compounds, and alloys (metallic or semiconducting) due to their intrinsic exchange energy split conduction band. Generally, the degree of spin polarization (P) is limited to about 50% in conventional ferromagnetic elements and alloys materials. It can reach higher values in a special set of materials called half-metallic ferromagnets [1], which by design are quite difficult to achieve. In this chapter, we deal with another phenomenon called spin-filter tunneling, currently hotly pursued for its capability to provide tunable P all the way to near 100% and can be interfaced with metals or semiconductors, giving the versatility for exploring new effects.

13.2 Spin-Filter Effect

Spin polarization of the tunnel current in conventional ferromagnet/insulator/ferromagnet magnetic tunnel junctions (MTJs), or in the Meservey–Tedrow experiments on Al/Al_2O_3/ferromagnet tunnel junctions, arises from the spin imbalance of the spin-up and spin-down density of states (DOS) of the conduction electrons at the Fermi energy $[N_\uparrow(E_F) > N_\downarrow(E_F)]$ in the ferromagnet (F). It is also dependent on the F/barrier interface. An entirely different way to create a spin-polarized tunnel current is to have spin-selective tunneling through barriers that are sensitive to spin orientation. For example, EuS is a barrier material that has a different tunneling probability $|T|^2$ for the two spin directions: $|T\uparrow|^2 \gg |T\downarrow|^2$. EuS is a magnetic semiconductor having a band gap of 1.65 eV and a ferromagnetic Curie temperature, $T_C = 16.6$ K. The exchange splitting of the conduction band ($2\Delta E_{ex}$) gives rise to two barrier heights for spin-down and spin-up electrons: $\Phi_{\downarrow,\uparrow} = \Phi_0 \pm \Delta E_{ex}$ where Φ_0 is the average barrier height at a temperature above T_C. In the case of EuS, the barrier for spin-up electrons ($\Phi_\uparrow$) is lower than the one for spin-down electrons ($\Phi_\downarrow$). Since the tunneling probability depends exponentially on barrier height, the spin-up current can greatly exceed the spin-down current, yielding a highly spin-polarized current in a metal/EuS/metal tunnel junction where the metal electrodes are nonmagnetic. This phenomenon is called the spin-filter effect, shown schematically in Figure 13.1. For a given barrier thickness d, the tunnel current density J varies exponentially with the corresponding barrier height [2,3]:

$$J_{\uparrow(\downarrow)} \propto \exp(-\Phi_{\uparrow(\downarrow)}{}^{1/2} d) \tag{13.1}$$

Therefore, even with a modest separation of spin-up and spin-down barrier heights, the tunneling probability for spin-up

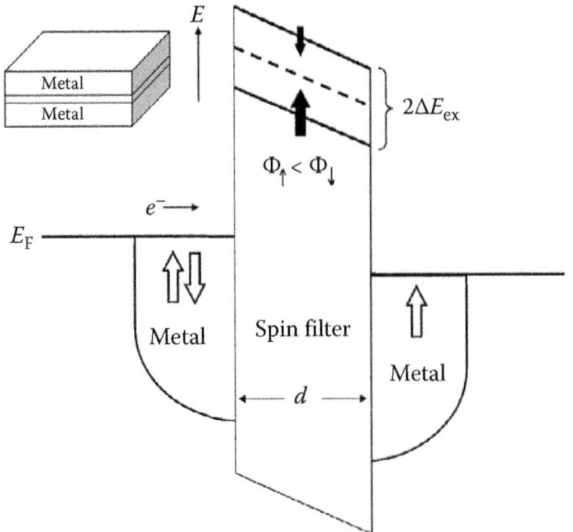

FIGURE 13.1 Schematic of the spin-filter effect. Unpolarized electrons from the nonmagnetic metal electrode tunnel through the ferromagnetic spin filter. Exchange splitting creates a lower barrier height for spin-up electrons and higher barrier height for spin-down electrons, generating a highly spin-polarized current.

electrons is much higher than for spin-down electrons, resulting in polarization (*P*) of the tunnel current:

$$P = \frac{J_\uparrow - J_\downarrow}{J_\uparrow + J_\downarrow} \tag{13.2}$$

The exchange splitting of the conduction band in europium chalcogenides (EuX, where X is O or S) is substantial, the largest being 0.54 eV for EuO, making them ideal candidates as spin-filter materials. Thus, they could completely filter out spin-down electrons, leading to total spin polarization, $P = 100\%$.

The europium chalcogenide tunnel barriers EuS [4] and EuSe [5] have demonstrated the spin-filter effect in spectacular fashion. EuS barriers have shown *P* as high as 85%, even with no magnetic field applied. EuSe, which is an antiferromagnet (AF), becomes ferromagnetic in a small applied field, and field-dependent exchange splitting of the conduction band appears. This gives rise to spin filtering where *P* in turn is field dependent: $P = 0$ in zero field and increases with *H*, reaching nearly 100% at 1 T. These results are described later in the chapter. The magnetic ordering temperatures of EuS and EuSe are 16.6 K (ferromagnetic) and 4.6 K (antiferromagnetic), respectively. Hence, they filter spins only at temperatures in the liquid helium range.

On the other hand, EuO has a higher T_C of 69.3 K and a larger exchange splitting, giving it the potential to reach greater spin-filter efficiency at higher temperatures. Doping EuO with rare earth metals is known to increase the T_C, although doping can lead to a reduced $2\Delta E_{ex}$ and metallic behavior [6]. However, demonstrating the spin-filter effect in EuO is more challenging than with EuS and EuSe, due to the difficulty in making high-quality, stoichiometric, ultrathin EuO films. The other oxide phase, Eu_2O_3, is more stable and forms more readily during film growth. Good-quality ultrathin films of EuS and EuSe are evaporated easily from a powder source of EuS and EuSe. This is not the case with EuO, as a powder source is unavailable due to its metastable form. Despite this challenge, EuO tunnel barriers have displayed large exchange splitting and near 30% polarization by direct measurement [7].

Other promising spin-filter candidates have arrived on the scene recently, namely, ferrites and perovskites. The Curie temperature of the ferrites is well above room temperature. Thus, they could potentially filter spins at a technologically useful temperature range. However, with their complex structure, the material aspects are complicated. As will be described in this chapter, some degree of spin filtering has been achieved in ferrimagnetic, insulating $NiFe_2O_4$ and $CoFe_2O_4$ [8], even at room temperature in $CoFe_2O_4$, though there is much room for improvement. Among the perovskites, some interesting tunneling results have been observed with insulating, ferromagnetic $BiMnO_3$, especially since this material is multiferroic, such that the barrier profile is modified due to both magnetic and ferroelectric order (see Chapter 14). Table 13.1 lists the properties of the known spin-filter materials.

13.3 Electronic Structure of Eu Chalcogenides

In the 1960s and 1970s, the magnetic, optical, and electronic properties of the EuX compounds in the form of bulk crystals and thick films were extensively studied. The earlier work

TABLE 13.1 Spin-Filter Materials

Material	Magnetic Behavior	T_C (K)	Moment (μ_B)	Structure *a* (nm)	E_g (eV)	$2\Delta E_{ex}$ (eV)	*P* (%)	Spin-Filter Reference
EuO	F	69.3	7	Fcc, 0.514	1.12	0.54	29	Santos and Moodera [7]
EuS	F	16.6	7	Fcc, 0.596	1.65	0.36	86	Moodera et al. [4]
EuSe	AF	4.6	7	Fcc, 0.619	1.80		100	Moodera et al. [5]
$BiMnO_3$	F	105	3.6	Perovskite			22	Gajek et al. [9]
$NiFe_2O_4$	Ferri	850	2	Spinel	1.2		22	Lüders et al. [10]
$CoFe_2O_4$	Ferri	796	3	Spinel	0.80		−25	Ramos et al. [11]

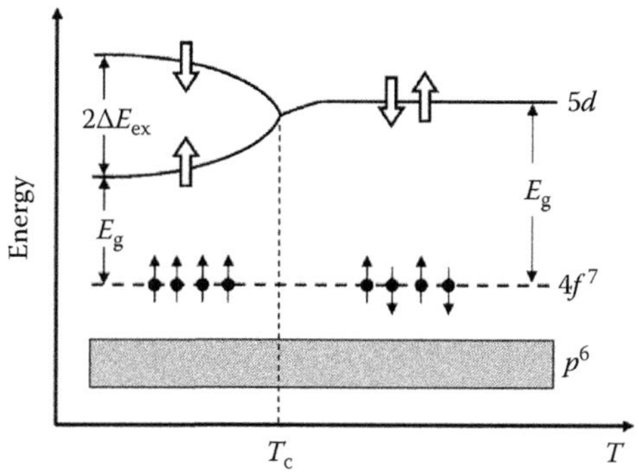

FIGURE 13.2 Simple schematic of the energy-level diagram as a function of temperature for ferromagnetic, semiconducting EuS and EuO. (Modified from Wachter, P., in *Handbook on the Physics and Chemistry of Rare Earths*, vol. 2, Eds. K. A. Schneider and L. Eyring, North-Holland Publishing Co., New York, 1979, pp. 507–574.)

has been well covered in several past reviews; see for example Refs. [12–15]. We briefly mention here some properties that are relevant for the discussion of these materials as spin filters. The europium chalcogenides are Heisenberg ferromagnets. The magnetic moment originates from the half-filled $4f$ states, giving rise to a large saturation moment of $g\mu_B J = 7\,\mu_B$ per Eu^{2+} ion. This has been experimentally observed in bulk single crystals and thin films and shown in band structure calculations [16–18]. A simple schematic of the energy gap is shown in Figure 13.2. The $4f^7$ states are localized within the energy gap between the valence band and conduction band. The energy gap between the $4f^7$ states and the $5dt_{2g}$ states at the bottom of the conduction band defines the optical band gap (E_g).

When cooled below the T_C, ferromagnetic order of the $4f^7$ spins causes Zeeman splitting of the conduction band, splitting it into two levels: the energy level for spin-up electrons is lowered by an amount ΔE_{ex}, whereas the spin-down level is raised by the same amount, thus lifting the spin degeneracy. The exchange interaction H_{exch} is given by the following Heisenberg exchange relation: $H_{exch} = 2\Delta E_{ex} = -2J_c \Sigma \vec{s} \cdot \vec{S}$, where $\vec{s}$ is the spin of a conduction electron, $\vec{S}$ is the spin of neighboring Eu^{2+} ions, and J_c is the space-dependent exchange constant between the electron spin and the ion spin [6]. This exchange splitting of the conduction band causes a reduction in E_g, and Busch et al. [19] measured this shift in the optical band gap to lower energy in a series of absorption measurements using single-crystal samples as they were cooled below T_C. In this way, they determined the magnitude of the exchange splitting: 0.36 eV for EuS and 0.54 eV for EuO. Exchange splitting of the conduction band in EuO was reported again more recently [20] in spin-resolved x-ray absorption spectroscopy measurements on a thin film of Eu-rich EuO. They concluded that the doped electrons (due to oxygen vacancies) in the exchange-split conduction band were 100% spin-polarized, forming a half-metal.

13.4 Spin Filtering in the Europium Chalcogenides

13.4.1 Early Experiments

The spin-filter effect was demonstrated for the first time in field-emission experiments with EuS by Müller et al. [21] P as high as 89% ± 7% was measured for field-emitted electrons from an EuS-coated tungsten tip at low temperature. Similar work later confirmed that electrons tunnel from the Fermi level of the tungsten into the exchange-split conduction band of the EuS [22–24]. The high P of the emitted electrons resulted from the different spin-up and spin-down barrier potentials at the W–EuS interface.

Transport experiments of EuS and EuSe films by Esaki et al. [25], performed previous to the field-emission experiments, showed indirect evidence of the spin-filter effect. The EuS and EuSe films in this study were 20 and 60 nm thick, between either Al or Au electrodes. With this trilayer structure, they observed tunneling across the potential barrier between the metal and the EuX layer, as shown in Figure 13.3a. This potential barrier was formed by the energy difference between the Fermi level of the metal electrode and the bottom of the conduction band in the EuX. When measuring the temperature dependence of the I–V behavior of these structures, a significant drop in voltage (measured with constant current) occurred below the T_C for both EuS and EuSe junctions, as shown in Figure 13.3b. This drop resulted from the exchange splitting of the conduction band, thereby lowering the barrier height. Applying a magnetic field shifted the drop to a temperature slightly higher than the T_C. However, the 20% decrease in voltage below ~10 K in zero field for EuSe is surprising given that it is an antiferromagnet at zero field and thus no exchange splitting was expected (see Section 13.4.3). Spin polarization of the current was not quantified in this early experiment, whereas it showed indirect evidence of the spin-filter effect. Similar behavior of tunnel junction resistance can be seen for all Eu chalcogenide junctions described later. A few years after Esaki's work, Thompson et al. [26] also measured an increase in conductance at low temperature due to conduction band splitting in a Schottky barrier contact between a EuS single crystal and indium metal.

Subsequently, the spin polarization resulting from spin filtering in Eu chalcogenide tunnel barriers was directly measured via the Meservey–Tedrow technique, whereby a superconducting Al electrode was used to detect the spin of the tunneling electrons, and we describe this elegant set of experiments in this section. Refer to Chapter 9 for a detailed description of this measurement technique. Tunnel junctions for these studies were fabricated in situ in a high vacuum chamber on glass substrates, with the general structure Al/EuX/M (where M = Al, Au, Ag, or Y). EuS and EuSe tunnel barriers (about 1–4 nm thick) were deposited by thermal evaporation from EuS and EuSe powder sources. EuO was grown by reactive evaporation of Eu metal in the presence of a small oxygen flow. The metal electrodes were deposited by thermal evaporation and patterned by shadow masks into a

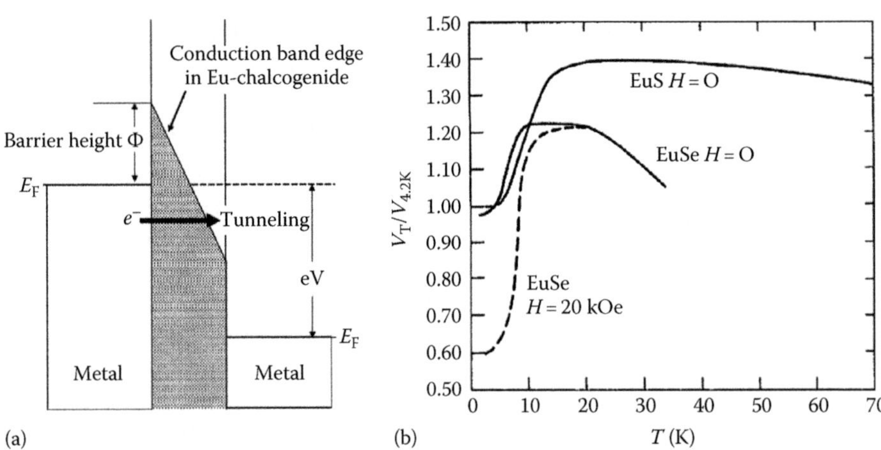

FIGURE 13.3 (a) Energy diagram for metal/EuS or EuSe/metal junction of Esaki et al. Φ is the barrier height for tunneling from the metal into the bottom of the conduction band of EuS or EuSe. (b) Normalized voltage across the junction (measured with constant current) versus temperature, with and without applying a magnetic field. (Reprinted with permission from Esaki, L. et al., *Phys. Rev. Lett.*, 19, 852, 1967. Copyright 1967 by the American Physical Society.)

cross configuration. The Al electrode, typically 4.2 nm thick, was deposited onto a liquid nitrogen–cooled substrate. Conductance was measured at 0.4 K, well below the critical temperature of the Al superconducting electrode (~2.7 K). The degree of spin polarization of the tunnel current was calculated from the dynamic tunnel conductance (dI/dV) measured within a few millivolts of the Fermi level.

The EuS, EuSe, and EuO thin films were polycrystalline, having saturation magnetization close to bulk values even for thickness down to 2 nm. For EuS, bulk T_C was maintained in this thickness range, whereas by ~1 nm thickness the T_C came down to ~10 K. Reduction of T_C with thickness for ultrathin EuO is discussed in more detail later in this section. Because the magnetic moment of Eu^{2+} is rather high (~7 μ_B), the magnetic properties of ultrathin EuX films can be easily measured using a SQUID magnetometer.

13.4.2 EuS

The initial demonstration of spin-filter tunneling using the Meservey–Tedrow technique was performed with a ferromagnetic EuS barrier [4]. Dynamic tunnel conductance versus voltage at 0.4 K for a Au/EuS/Al tunnel junction is shown in Figure 13.4. In comparison, the dI/dV curve at zero applied magnetic field for a junction with a nonferromagnetic barrier (i.e., Al_2O_3) is completely symmetric with two peaks showing the SC energy gap in Al. In the special case of this EuS barrier, the zero field conductance is asymmetric with four peaks. These four peaks correspond to the Zeeman splitting of the Al quasiparticle DOS and are more pronounced at 0.07 T. Asymmetry of the spin-up and spin-down peaks is the signature of a spin-polarized tunnel current. These dI/dV curves were fit using Maki's theory [27] for a thin film superconductor (SC) in a magnetic field, in order to extract a large polarization value $P=80\% \pm 5\%$.

The amount of Zeeman splitting is equal to $2\mu_B H_0$, where μ_B is the Bohr magneton and H_0 is the magnetic field. Notably, the amount of Zeeman splitting in the dI/dV measurement in an applied field $H_{appl}=0.07$ T actually corresponds to a much higher effective field of $H_0=3.46$ T. The internal exchange field of the ferromagnetically ordered Eu^{2+} ions in the EuS barrier, acting on the quasiparticles of the superconducting Al (discussed in detail later) causes this phenomenally enhanced Zeeman splitting. This internal field was even large enough to drive the Al to the normal state at $H_{appl}=0.15$ T, as seen in Figure 13.4, which otherwise occurs at about 5 T (e.g., without EuS; with Al_2O_3 barriers). Zeeman splitting and polarization was observed even at zero applied field, resulting from significant remanent magnetization in the EuS ultrathin layer. Because the electrodes are not

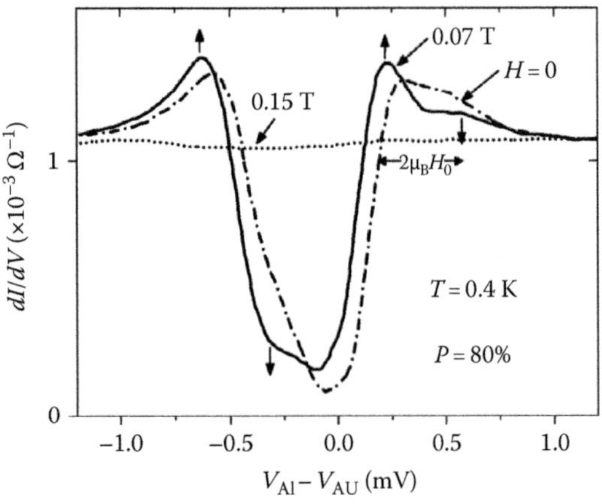

FIGURE 13.4 Dynamic conductance of a Au/EuS/Al junction at 0.4 K in the indicated applied magnetic field. The arrows indicate the spin of the Zeeman split DOS for the 0.07 T curve, showing a polarization of $P=80\% \pm 5\%$. The amount of Zeeman splitting ($2\mu_B H_0$) is indicated for the 0.07 T curve. (Modified from Moodera, J.S. et al., *Phys. Rev. Lett.*, 61, 637, 1988; Hao, X. et al., *Phys. Rev. B*, 42, 8235, 1990.)

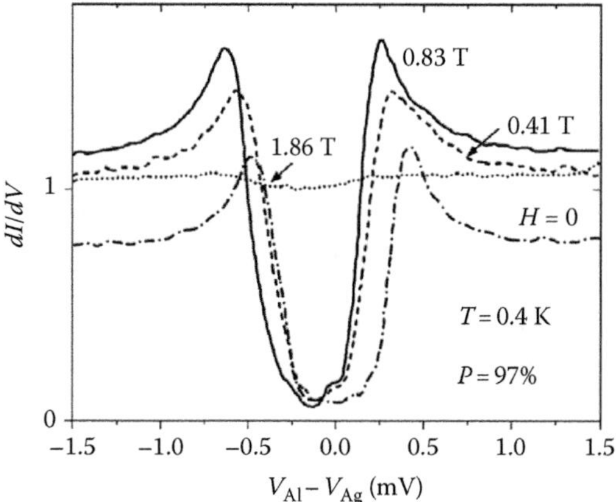

FIGURE 13.5 Conductance versus voltage for an Ag/EuSe/Al junction at 0.4 K taken at the applied fields indicated. Note the highly symmetric curve at $H=0$, whereas the peaks shift to lower voltage as H increases, corresponding to the Zeeman energy of the spin-up DOS of the Al. This indicates that the polarization is nearly 100%. (Modified from Moodera, J.S. et al., *Phys. Rev. Lett.*, 70, 853, 1993.)

ferromagnetic, and thus cannot be a source of polarized spins, this large P is generated clearly by spin filtering in the EuS barrier. Polarization as high as 85% was found for Al/3.3 nm EuS/Al SC–SC tunnel junctions [4].

The temperature dependence of the junction resistance (R_J) is another indicator of the spin-filter effect in the EuS. When the temperature decreases below the T_C of EuS, the lowering of the tunnel barrier height for spin-up electrons results in a significant decrease of R_J. This is contrary to the typical behavior for a nonmagnetic barrier, for which R_J increases continuously when cooled (by a small amount for high-quality, insulating barriers, and by a larger amount for semiconducting barriers). The amount of the drop in R_J below T_C can be used to determine the amount of exchange splitting, which is discussed later for EuO barriers.

13.4.3 EuSe

Because EuSe is an antiferromagnet at zero field, the spin-filter behavior of EuSe is quite different from EuS. EuSe has an interesting magnetic phase diagram [28] such that it transitions from the antiferromagnetic state into a ferrimagnetic state at low field or a ferromagnetic state at high field. Thus, in an applied field such that EuSe is ferromagnetic, exchange splitting develops in the EuSe conduction band, and the splitting increases as the field increases. This results in spin polarization that increases as the field increases, such that P becomes a function of H_{appl}. For the Ag/EuSe/Al tunnel junction shown in Figure 13.5 at zero field, the EuSe was in the antiferromagnetic state. Thus, no Zeeman splitting was observed and conductance was symmetric about $V=0$. As the field increased and EuSe entered the ferromagnetic state, polarization also increased. This is evident in the increasing Zeeman splitting and asymmetry as the field increased. The Zeeman splitting is much higher than the applied field, as with the EuS barriers. The increasing asymmetry means higher value of P. In fact with an optimum interface, P even reached as high as 97% ± 3%. Because polarization is near 100%, the dI/dV shows only the spin-up peaks: two peaks on either side of $V=0$, shifted left of center. There is nearly zero conductance in the spin-down DOS and thus the spin-down peaks are not seen in the dI/dV curve. This was the first demonstration of a fully polarized tunnel current in a tunnel junction. The $H_{appl}=1.86$ T curve shows that this applied field plus the large internal field in the EuSe barrier drove the Al to normal state.

Clearly displayed in Figure 13.5 is an increase in conductance as the applied field increases. In fact, this magnetoresistance is quite high, as much as 400%. This is the first observation of a resistance change of such magnitude for a tunnel junction in an applied field. This increase in conductance corresponds to a significant drop in R_J with increasing magnetic field, as shown in Figure 13.6a. This decrease in R_J resulted from the lowering of the spin-up barrier height as

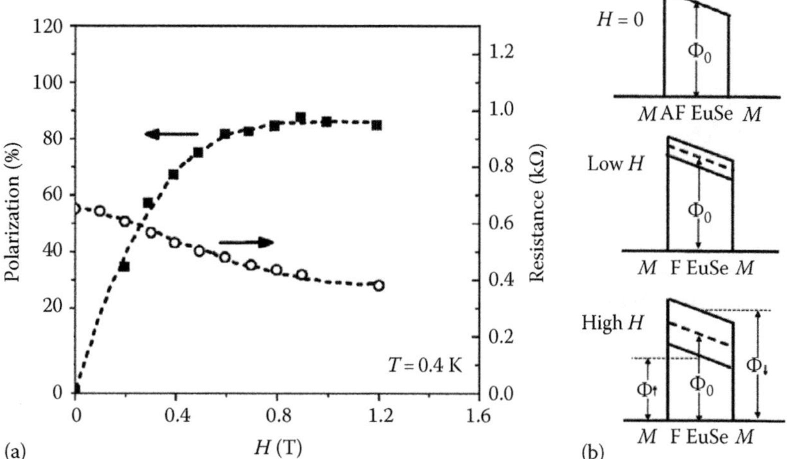

FIGURE 13.6 (a) Polarization and R_J as a function of applied field for a EuSe junction. (b) Schematic of how the exchange splitting of the EuSe conduction band develops as a field is applied. (Modified from Moodera, J.S. et al., *Phys. Rev. Lett.*, 70, 853, 1993.)

the field increased, as shown schematically in Figure 13.6b. At zero field, EuSe is in the antiferromagnetic state. When a field is applied, EuSe transitions to the ferromagnetic state and the conduction band undergoes exchange splitting, reaching a constant value at ~1 T in this junction. Accompanying the increase of exchange splitting with field is an improvement in spin-filter efficiency, resulting in increasing *P* until a constant value is reached. This dependence of polarization on applied field is unique to the EuSe spin filter. This study also confirmed that at zero field, when the EuSe is antiferromagnetic, the polarization is indeed zero.

13.4.4 EuO

Stoichiometric EuO is quite difficult to grow at the few monolayers scale necessary for tunneling, due to the formation of the more stable and nonmagnetic Eu_2O_3, which is not a spin filter. Therefore, compared to EuS and EuSe, demonstrating spin-filter tunneling with EuO barriers is much more challenging. With careful control of the Eu metal evaporation and oxygen flow, good quality EuO was grown in an Al/EuO/Y/Al tunnel junction, showing 29% polarization [7] using superconducting Al as the spin detector. Similar to EuS and EuSe barriers, EuO showed a large enhancement of Zeeman splitting, amounting to an effective magnetic field of 3.9 T in a small applied field of just 0.1 T. Thus, the exchange interaction in just 1.4 nm thick EuO was strong enough to create a large internal exchange field of 3.8 T. The exchange interaction of the EuO/Al bilayer had been observed in an earlier tunneling study as well [29]. Even though EuO has a larger $2\Delta E_{ex}$ than EuS, which should result in a higher *P*, spin-filtering efficiency of EuO is not as high. This was explained as due to the nonstoichiometry at the EuO/electrode interfaces, which is quite difficult to control during sample growth. The best polarization values were obtained using yttrium as the top electrode. Yttrium prevented overoxidation of EuO into the undesirable Eu_2O_3 phase. Confirmation of this came from an x-ray absorption spectroscopy study of the chemical nature of the EuO interface with Y, Al, and Ag capping layers [30]. The result from this study was found to agree with the tunneling result.

One must be aware that when a magnetic insulator is made thin enough to act as a tunnel barrier, it may have a lower T_C and a lower exchange splitting than the bulk. Consequently, this could limit efficient spin filtering to temperatures much less than T_C and reduce the spin-filter efficiency. A reduction in T_C with thickness was found to be the case for EuO. As shown in Figure 13.7a, T_C for thickness <3 nm is significantly reduced from bulk $T_C = 69$ K. This thickness-dependent study of the T_C was the first experimental confirmation of a theoretical calculation by Schiller and Nolting [31] showing reduced T_C for a few monolayers of EuO.

Measuring the exchange splitting in ultrathin films is more difficult. The optical absorption measurements used to measure the exchange splitting of the conduction band for crystals of EuO cannot be as easily applied to these ultrathin films. As an alterative, spin-filter tunneling was used to quantitatively determine the exchange splitting in ultrathin EuO [32]. With a tunnel junction comprised of a EuO barrier sandwiched between nonferromagnetic electrodes, the junction resistance as a function of temperature was measured when varying the temperature across the T_C of EuO. For $T > T_C$, the junction resistance increased as *T* decreased, as expected for a nonmagnetic semiconducting barrier. When the Eu^{2+} ions ferromagnetically ordered at T_C, exchange splitting lowered the spin-up tunnel barrier height, and the junction resistance decreased dramatically as the temperature was lowered further below T_C, as shown in Figure 13.7b. In accordance with an exponential dependence on barrier height, the resistance dropped by several orders of magnitude, especially for larger applied bias at which Fowler–Nordheim (FN) tunneling occurs. The barrier height Φ_0 (before splitting) at low bias

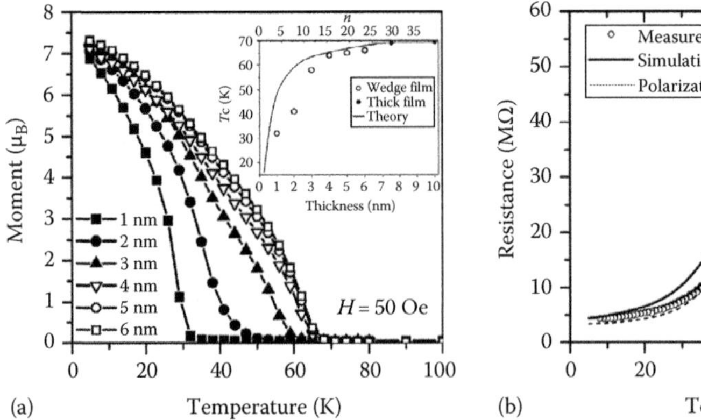

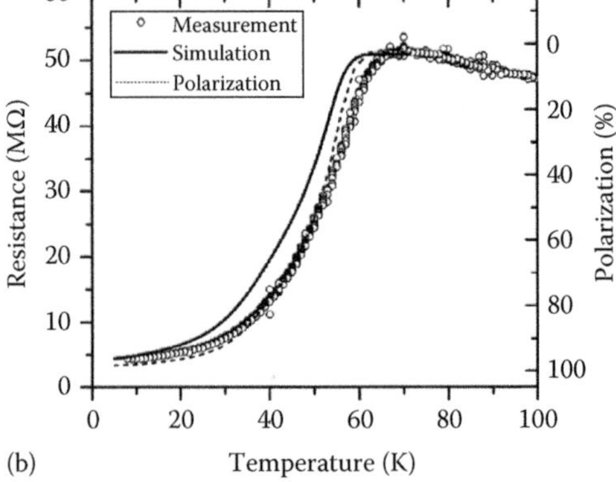

FIGURE 13.7 (a) $M(T)$ for different thicknesses of EuO, 1–6 nm, showing a lower T_C for films <4 nm. This is in agreement with a calculation by Schiller et al. [31] shown in the inset. (b) Resistance versus temperature for a Y/EuO/Al junction, plotted along with a simulation of the measurement based on the $M(T)$ curve for the 3 nm film in (a) and calculated polarization resulting from the large exchange splitting that causes the decrease in junction resistance. (Modified from Santos, T.S. et al., *Phys. Rev. Lett.*, 101, 147201, 2008.)

was found by applying the Brinkman–Dynes–Rowell [3] relation to the current–voltage characteristics of the junction for $T > T_C$, then $R_J(T)$ was used to quantify the amount of exchange splitting for $T < T_C$. Equation 13.2 was used to calculate P, assuming no spin scattering. The amount of resistance drop corresponded to an exchange splitting of 0.31 eV for the 3 nm thick EuO. Even though this is less than the bulk value, this splitting is still large enough to produce a nearly fully spin-polarized current, though polarization could not be measured directly without a spin detector. A complete study of exchange splitting as a function of thickness was not performed. However, it was demonstrated that even though thinner barriers may have smaller exchange splitting, the high spin-filter efficiency can still be maintained by the lower average barrier height of the thinner barrier.

13.4.5 Exchange Interaction at FI/SC Interfaces

A novel feature of ferromagnetic insulator/superconductor (FI/SC) interface is the large internal exchange field of the ferromagnet acting on the SC. Sarma [33] and de Gennes [34] had predicted this phenomenon over 40 years ago as an exchange interaction between the ferromagnetically ordered ions in the FI and the conduction electrons of the SC. This is dramatically displayed in the enhanced Zeeman splitting of the Al superconducting DOS when in contact with the EuS, EuSe, and EuO barriers in the spin-polarized tunneling measurements (see Figures 13.4 and 13.5). As discussed above, this internal exchange field is as much as ~4 T. For small thickness d of the superconducting film during conduction, the scattering of Al quasiparticles at the EuS/Al interface is the dominant mechanism, where they exchange interact with the ordered Eu^{2+} ions. This is analogous to a superconducting thin film in a uniform exchange field, schematically shown in the inset of Figure 13.8. The exchange field acts only on the electron spins, and not on electron motion [33,34], causing Zeeman splitting and yet having a negligible effect on orbital depairing in the SC. A result of a purely field–spin interaction is that the phase transition to normal state occurs through a first-order transition when the critical field is reached.

de Gennes predicted that for d less than the superconducting coherence length ξ, the magnitude of the field is inversely proportional to d. This was experimentally verified using an EuS/Al/Al_2O_3/Ag junction structure [35], with the EuS layer under only half of the patterned Al electrode strip. Conductance of junctions for the two configurations was compared, as shown in Figure 13.8. For the EuS/Al/Al_2O_3/Ag junction, enhanced Zeeman splitting, corresponding to a field of 1.9 T, was observed in an applied field of 0.1 T. For the Al/Al_2O_3/Ag trilayer without the EuS underlayer, a much higher conductance at zero bias and broadening of the peaks is observed with $H_{appl} = 1.93$ T, showing significant orbital depairing and no Zeeman splitting in the Al quasiparticle DOS. It was also shown in this study that the internal field drove the Al to the normal state via a first-order transition, and the internal saturation field was inversely proportional to Al film thickness. The exchange interaction only occurs for a

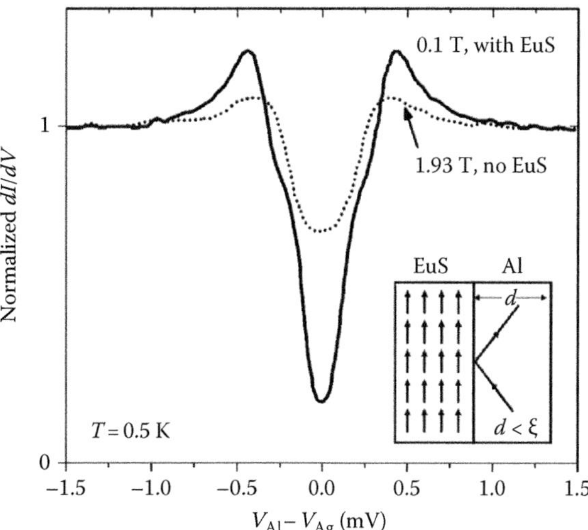

FIGURE 13.8 Comparison of the tunnel conductance for a Al/Al_2O_3/Ag junction with and without an EuS layer beneath the Al superconducting electrode. The EuS layer creates an internal field in the Al, resulting in enhanced Zeeman splitting with minimal orbital depairing in a small applied field. Inset: Schematic of the interaction between the Al quasiparticles with the ferromagnetically ordered ions in the EuS, as they scatter at the FI/SC interface. (Modified from Hao, X. et al., *Phys. Rev. Lett.*, 67, 1342, 1991.)

FI and SC in immediate proximity to each other. The effect was lost when a thin layer of Al_2O_3, even two monolayers, formed at the FI/SC interface.

13.5 Quasi-Magnetic Tunnel Junctions

The early experiments with spin-filter tunnel barriers showed the spin-filter effect by using the Meservey–Tedrow technique. In more recent studies, a quasi-magnetic tunnel junction (QMTJ) has been used to observe spin filtering, by measuring tunnel magnetoresistance (TMR) in junctions where the spin-filter barrier is sandwiched between a nonmagnetic electrode (NM) and a F electrode. If the polarization of the ferromagnet (P_F) is known, the spin-filter efficiency (P_{SF}) can be calculated from the TMR measurement by using Jullière's formula [36]:

$$\mathrm{TMR} = \frac{\Delta R}{R} = \frac{2P_F P_{SF}}{(1 - P_F P_{SF})} \tag{13.3}$$

Resistance (R_J) of the junction depends on the relative alignment of the magnetization of the spin filter and the ferromagnet, as shown in Figure 13.9. For a tunnel barrier that preferentially allows spin-up electrons to tunnel, R_J is low (high) for parallel (antiparallel) alignment. Just as in conventional MTJs, independent switching with well-separated coercivities for the two ferromagnetic layers is necessary for observing TMR and quantifying P_{SF}. In order to have a clear effect, magnetic coupling between the spin-filter barrier and the F electrode must be

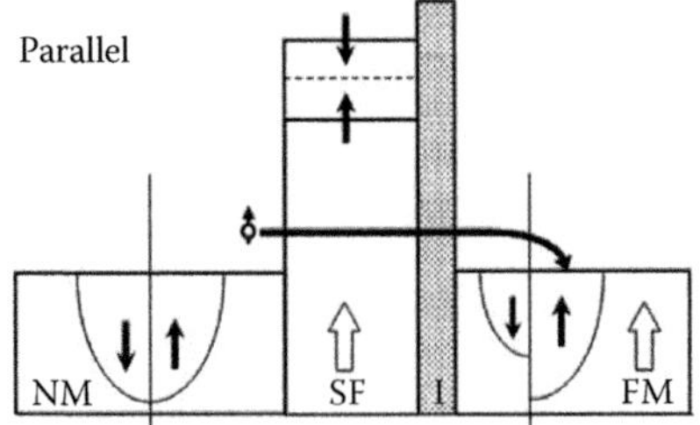

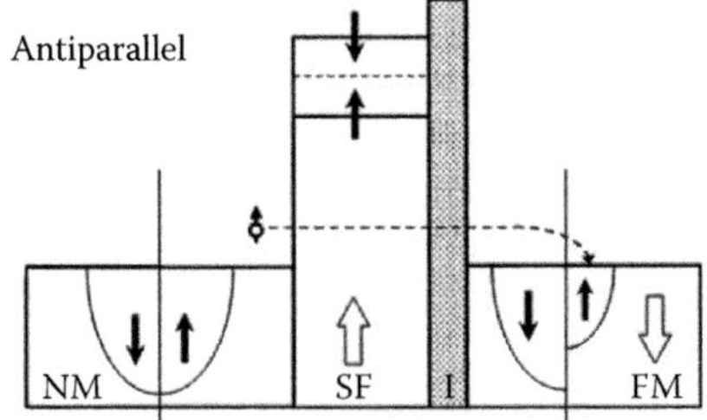

FIGURE 13.9 Left: Low resistance state (high tunnel current) for a QMTJ, with parallel alignment of magnetization of the spin-filter barrier (SF) and F electrode. Right: High resistance state (low tunnel current) for antiparallel alignment. In some junction structures, there is a very thin, nonmagnetic, insulating layer between SF and F, shown here, in order to prevent magnetic coupling between them.

prevented. For this reason, a nonmagnetic spacer layer, such as Al_2O_3 or $SrTiO_3$ is sometimes placed between the spin filter and F electrode.

The high spin polarization (~90%) found for EuS spin-filter barriers in the Meservey–Tedrow measurements described earlier motivated the incorporation of EuS in QMTJs. LeClair et al. [37] showed the results for a QMTJ. They measured MR ~ 100% at 2 K for a 5 nm EuS barrier between Al and Gd electrodes. There was considerable noise in the MR signal, attributed to instabilities in the EuS magnetization. The TMR decreased as temperature was raised closer to the T_C of EuS (16 K) and was absent at above T_C. As discussed earlier, a clear decrease of junction resistance was observed for these QMTJs when cooled below T_C, due to the lowering of the barrier height.

Much more robust and reproducible measurements of QMTJs with EuS barriers were made by inserting an AlO_x spacer layer between the EuS and a Co electrode, by Nagahama et al. [38], as shown in Figure 13.10. The AlO_x effectively decoupled the EuS and Co layers, and the antiferromagnetic CoO acted as an exchange biasing layer to pin the Co magnetization. Figure 13.10a shows the typical shape of the MR for these QMTJs, having a MR value of 33% at 4.2 K with a bias of 800 mV. The magnetic configuration is well defined and stable, with no instability in junction resistance.

In contrast to the behavior of conventional F/insulator/F MTJs, these QMTJs display a novel bias dependence of MR, in which the MR actually *increases with bias* (Figure 13.10b). This unique bias dependence for a spin-filter MTJ was predicted theoretically by Saffarzadeh [39]. The increase in MR at higher bias was attributed to FN tunneling into the fully spin-polarized conduction band of the EuS barrier. Direct tunneling via the spin-filter effect occurred in the low bias regime, and the MR decreased monotonically with bias (see Figure 13.10b inset), similar to conventional MTJs. As the applied bias was increased further, a rise in MR marked the transition from direct tunneling to FN tunneling via the spin-up conduction band for the spin-up electrons. Upon further increase of bias, a subsequent decrease in TMR occurred as the spin-down electrons also underwent FN tunneling via the higher spin-down conduction band level, resulting in a lower polarization of the tunnel current. Thus, in contrast to conventional MTJs, a sizable TMR effect in this spin-filter MTJ is not limited to low bias. As a consequence of having two tunnel barriers (EuS + AlO_x), the junction resistance is very high, especially in the low bias regime (few GΩ). Noise in the measurements at low bias (below ±400 mV in Figure 13.10b) made the measurement in this regime uncertain. Measurement of a EuS QMTJ in a low noise setup [40] confirmed high MR, up to 70% (see inset), and the

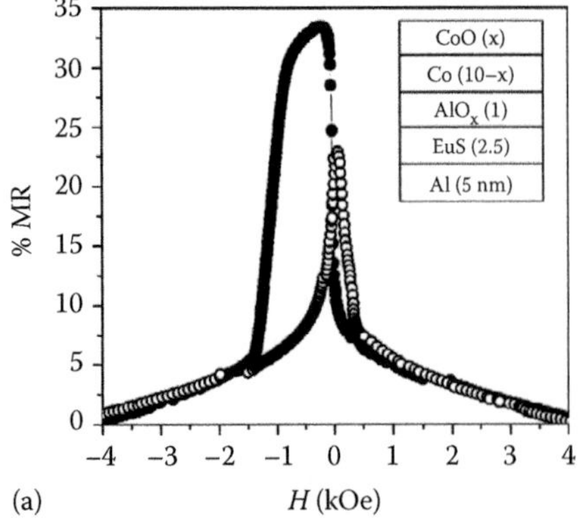

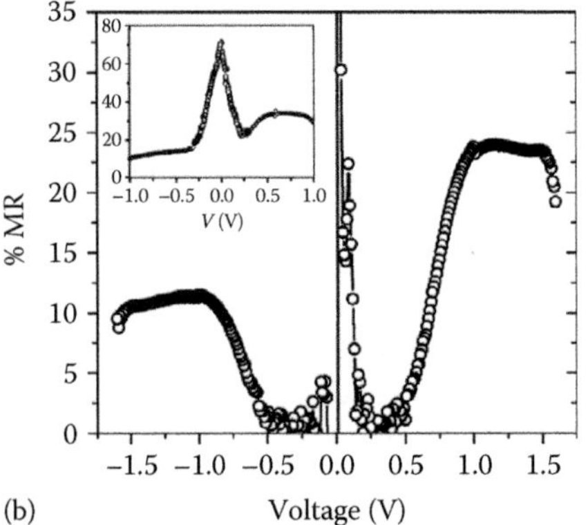

FIGURE 13.10 (a) MR of an Al/EuS/AlO_x/Co/CoO junction at 4.2 K with 800 mV bias. Filled (open) circles correspond to decreasing (increasing) field during measurement. (b) Bias dependence for a similar junction, showing an abrupt *increase* in MR at finite bias. Noise in the measurement for *V* less than ±400 mV prevented reliable data in this range. Inset: Full bias dependence measurement at 2 K in a low noise setup [40] of a similar junction with 3 nm EuS.

familiar bias dependence behavior (decreasing MR with bias) before the onset of FN tunneling.

13.6 Oxide Spin Filters

The phenomenal spin-filtering effects described up to this point were achieved using EuX compounds that were polycrystalline, although in some cases near epitaxy was seen depending on the substrate. Correspondingly, with the latter films better properties were observed. More effort is needed toward complete epitaxial EuX layer systems. Fully epitaxial structures can be prepared with a radically different material selection—perovskites or ferrites. While good structural quality has been shown for these complex oxide spin filters, on either elemental metal or oxide electrodes, the spin filtering is not yet optimized, as to show similar spectacular effects that are seen with EuX compounds. However, the performance of these oxide spin filters continues to improve, and the role of oxygen vacancies and defects comes to light. This section focuses on epitaxial spin-filter materials.

The spin-filter effect has been successfully demonstrated by Gajek et al. in the perovskite $BiMnO_3$ [9] and La-doped $BiMnO_3$ [41], which is ferromagnetic with a bulk $T_C = 105\,K$. In addition to being ferromagnetic, this material is also ferroelectric, whereby its electric dipole moment can be switched by applying an electric field. The electric polarization of the barrier modifies the barrier profile, causing a different average barrier height for the two polarization states. Thus, four stable resistance states can be obtained with this barrier having two ferroic parameters at play [42]. Multiferroic tunnel junctions is the subject of another chapter.

13.6.1 Ferrites: Leading the Way to Room Temperature Spin Filtering

All of the spin-filter materials described thus far have beautifully demonstrated the spin-filter phenomenon, but all at low temperature due to the low Curie temperature of the spin filter. Having a Curie temperature far above room temperature, the ferromagnetic, spinel ferrites are the best candidates for achieving spin filtering at room temperature. We present in this section the progress toward high-temperature spin filtering.

The spinel ferrites have the general formula XFe_2O_4 where X is a divalent transition metal cation such as Fe^{2+}, Mn^{2+}, Ni^{2+}, or Co^{2+}. With the exception of Fe_3O_4, all are insulating and ferrimagnetic with a Curie temperature above 700 K. While the conductive Fe_3O_4 has been studied as a half-metallic electrode in MTJ structures [43], its insulating relatives $NiFe_2O_4$ and $CoFe_2O_4$ have the potential to function as spin-filter tunnel barriers.

The ferrites ideally have an inverse spinel structure consisting of a face-centered cubic oxygen sublattice and a distribution of cations in both the tetrahedral (A) and octahedral (B) interstitial sites, resulting in a net moment of $2\,\mu_B$ per formula unit in $NiFe_2O_4$ and $3\,\mu_B$ in $CoFe_2O_4$. Theoretical calculations on the inverse spinel structure yield band gaps of 1.2 eV for $NiFe_2O_4$ and 0.80 eV for $CoFe_2O_4$, while the exchange splitting of the conduction band in both materials is quite large, around 1.2 eV [44]. The exchange splitting is such that the lowest energy level of the conduction band corresponds to *spin-down* states. The result should therefore be a lower tunnel barrier height for spin-down electrons, thus leading to preferential tunneling of the spin-down electrons and a *negative* spin polarization. One important factor to consider for these spinel ferrites is that they rarely exhibit a perfect inverse spinel structure. Migration of X^{2+} cations from B to A sites readily occurs [45], leading to "mixed" spinel having a moment and electronic band structure that can vary significantly from the inverse scenario. Theoretical calculations have demonstrated the effect of cation-site inversion on the electronic band structure and magnetic moment [45]. A site inversion of the Ni^{2+} and Co^{2+} cations results in a large increase in moment to $8\,\mu_B$ for $NiFe_2O_4$ and $7\,\mu_B$ for $CoFe_2O_4$. As a consequence, a largely amplified exchange splitting in the conduction band results, and therefore the possibility of even higher spin-filtering efficiencies. While theoretical predictions confirmed the high potential of $NiFe_2O_4$ and $CoFe_2O_4$ to filter spins, only a few groups have successfully fabricated tunnel junctions containing spinel ferrite barriers and measured the spin-filter effect.

13.6.1.1 $NiFe_2O_4$

Ultrathin $NiFe_2O_4$ films indeed show a dramatic enhancement of the saturation moment compared to the bulk value, indicating a large degree of cation-site inversion [46,47]. In addition, under some deposition conditions, the films are metallic, making them unsuitable as a tunnel barrier. As is the case for $CoFe_2O_4$ discussed later, the oxygen content plays a critical role in the spin-polarized transport properties of $NiFe_2O_4$, including simply the difference between insulating and metallic behavior. Using insulating $NiFe_2O_4$ barriers, the spin-filter effect was demonstrated in TMR measurements with Au and $La_{2/3}Sr_{1/3}MnO_3$ (LSMO) electrodes at low temperature [10], as shown in Figure 13.11. With one ferromagnetic electrode and one normal metal electrode, the observed TMR signal indicates that $NiFe_2O_4$ acts as a spin filter. One may apply the Jullière relation in Equation 13.3 to this TMR measurement, using $P_F = 90\%$ for $La_{2/3}Sr_{1/3}MnO_3$ (LSMO). In this way, a spin-filter efficiency $P_{SF} = 19\%$ was found for the $NiFe_2O_4$ spin filter. Further studies showed that this value was improved slightly to 22% by inserting a thin $SrTiO_3$ spacer layer in between the LSMO and $NiFe_2O_4$.

The positive sign of P_{SF} is contrary to the negative spin polarization expected from the band structure calculations for $NiFe_2O_4$ mentioned earlier. One proposed explanation attributes the positive polarization to the different effective masses for spin-up and spin-down electrons ($m^*_\uparrow < m^*_\downarrow$) [44]. Another possibility could be defects in the $NiFe_2O_4$ barrier that provide alternative tunneling pathways other than direct tunneling across the electronic band gap.

Despite the prediction of TMR at room temperature, the TMR rapidly vanishes by 150 K, well below the T_C of 850 K for $NiFe_2O_4$, as shown in Figure 13.11b. In addition, there was a systematic decrease in TMR with increasing bias for the $NiFe_2O_4$-based tunnel junctions, which is not the theoretically expected behavior

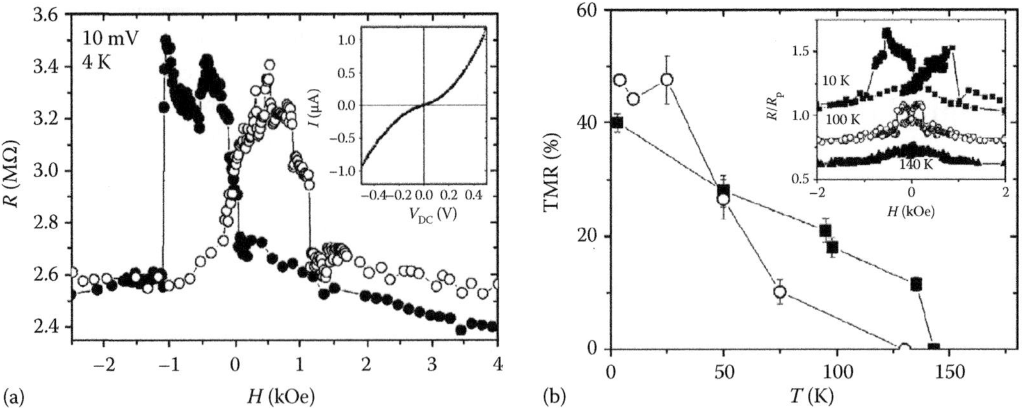

FIGURE 13.11 (a) TMR for a LSMO/$NiFe_2O_4$/Au tunnel junction at 4 K. The current–voltage curve is shown in the inset. (b) Temperature dependence of the TMR with (circles) and without (squares) $SrTiO_3$ at the LSMO/$NiFe_2O_4$ interface. (Reprinted with permission from Lüders, U. et al., *Appl. Phys. Lett.*, 88, 082505, 2006. Copyright 2006, American Institute of Physics.)

for a spin filter. The loss of TMR could be due to a decrease in coercivity for the $NiFe_2O_4$ barrier as temperature increases, or to a drop in the T_C of LSMO caused by nonstoichiometry at the LSMO/$NiFe_2O_4$ interface. Furthermore, in the case of even a small nonstoichiometry in either of the oxide films, defect states in the barrier, and/or at the interfaces can certainly lead to spin scattering events or spin-independent tunneling. These effects become more prominent at higher temperatures and bias.

13.6.1.2 $CoFe_2O_4$

Ramos et al. [48] performed a detailed study on the film growth and magnetic properties of epitaxial $CoFe_2O_4$, including $CoFe_2O_4/Fe_3O_4$ and $CoFe_2O_4$/Co heterostructures. A fully epitaxial, QMTJ structure was adopted for their TMR experiments: Pt(111)/$CoFe_2O_4$(111)/γ-Al_2O_3(111)/Co(0001) deposited by oxygen plasma-assisted molecular beam epitaxy onto a sapphire substrate. Note that the Al_2O_3 spacer layer used to decouple the $CoFe_2O_4$ barrier and Co electrode was crystalline in this case, which is rare for this type of tunnel barrier. The high quality of this fully epitaxial system, especially the interfaces, largely contributed to the successful spin transport measurements, even at room temperature. Ramos et al. [11] observed a small TMR signal of −3% at room temperature, which grew to −18% at 2 K, as shown in Figure 13.12. Using Julliere's formula with P_{Co} = 40%, this yields polarization values for $CoFe_2O_4$ of −4% and −25% respectively. This is the first demonstration of spin filtering at room temperature. Furthermore, unlike the $NiFe_2O_4$ junction, this $CoFe_2O_4$ junction showed the expected bias dependence behavior of a spin filter, with an increase of TMR at high bias.

Though the TMR signal at room temperature for this $CoFe_2O_4$ junction was small, it showed the importance of having a barrier of high structural quality for these spinel ferrites in order to see room temperature TMR and robust junctions with the expected bias dependence. It is known that antiphase boundaries in thinner films and oxygen vacancies can affect the magnetic properties and hence the spin-filtering performance. Ramos, in collaboration with the authors [49], performed tunneling experiments using the Meservey–Tedrow technique that demonstrated the influence of oxidation during film growth on the spin polarization produced by the $CoFe_2O_4$. Barriers grown in lower oxygen pressure had a stronger dependence of junction resistance on temperature compared to barriers grown in higher oxygen

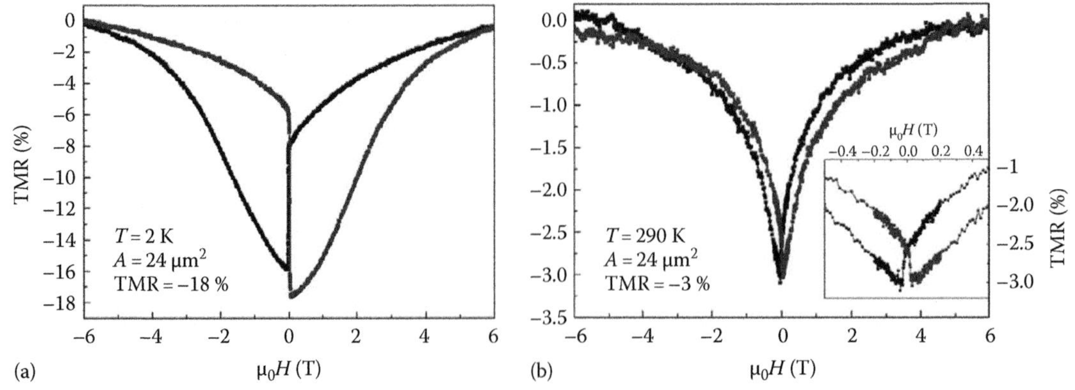

FIGURE 13.12 TMR in a 20 nm Pt/3 nm $CoFe_2O_4$/1.5 nm Al_2O_3/10 nm Co tunnel junction at 2 K (a) and at room temperature (b) with V = 200 mV. (Reprinted with permission from Ramos, A.V. et al., *Appl. Phys. Lett.*, 91, 122107, 2007. Copyright 2007, American Institute of Physics.)

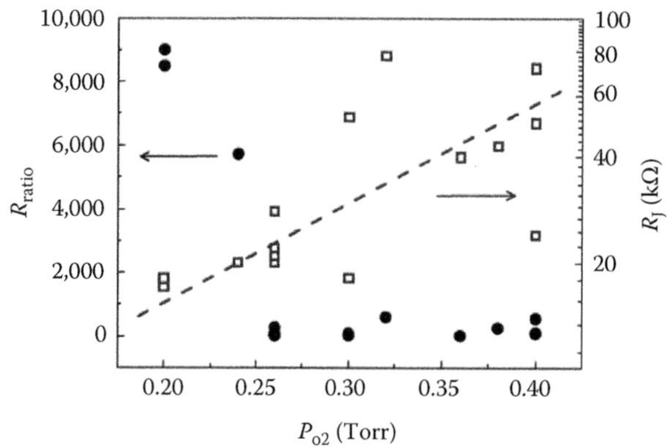

FIGURE 13.13 Junction resistance at 4 K (right axis) for junctions with $CoFe_2O_4/Al_2O_3$ barriers deposited in different oxygen pressures. R_{ratio} on the left axis is the temperature resistance ratio R_J (4 K)/R_J (300 K). The dashed line is a guide for the eyes. (From Ramos, A.V. et al., *Phys. Rev. B*, 78, 180402(R), 2008.)

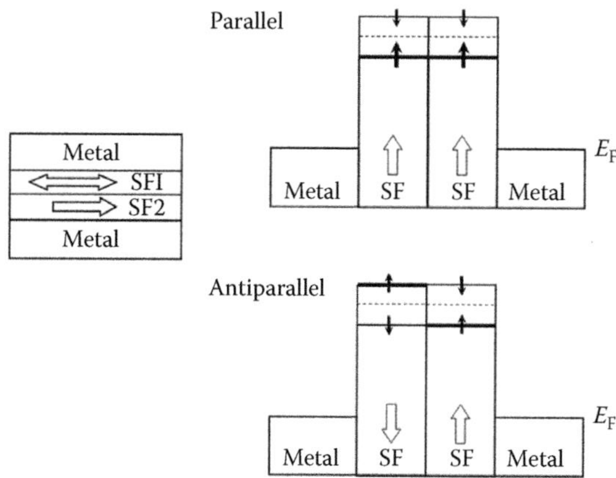

FIGURE 13.14 Schematic of the double spin-filter junction proposed by Worledge and Geballe [50]. The magnetization of one spin filter (SF1) is free to switch in a small applied field while the other (SF2) is effectively pinned. The spin-up and spin-down barrier heights for the two spin filters are shown for the parallel and antiparallel alignment. In this schematic, the two SF barriers are assumed to have identical band property.

pressure, as shown in Figure 13.13. This indicated the presence of defect states in the band gap and a lowering of the effective barrier height as a result of oxygen vacancies. Direct measurement of spin polarization in the Meservey–Tedrow experiments (with no ferromagnetic electrodes) revealed a higher spin polarization for the barrier grown in higher oxygen pressure, ranging from $P=6\%$ for oxygen pressure of 0.20 Torr to $P=26\%$ for 0.26 Torr. The sign of polarization measured in these experiments is not in agreement with the negative sign of polarization in the TMR measurements previously described. The origin of this discrepancy is unclear, but may be related to the different spin detectors or different bias regimes used in the two cases: superconducting Al spin detector and <2 mV bias regime for the Meservey–Tedrow experiments, and Co spin detector in the higher, FN tunneling regime for the TMR experiments.

It is interesting to note that the oxygen vacancies detected in these transport experiments were undetectable by extensive investigation of this material system by standard structural, magnetic, and chemical characterization techniques. This proves the sensitivity of spin-polarized tunneling to such minute defects. Even though oxygen plasma was used during film growth, enough oxygen vacancies were incorporated to degrade the spin-filter efficiency. This result emphasizes the importance and challenge of obtaining high-quality films for achieving high spin-filter efficiency with complex oxide barriers.

13.7 Double Spin-Filter Junctions

Worledge and Geballe [50] proposed another way to reach high TMR in a junction having two spin-filter barriers. In a double spin-filter junction, shown schematically in Figure 13.14, one spin filter acts as the polarizer and the other as the analyzer, and the tunnel current depends on the relative alignment of magnetization between them. When they are magnetized parallel to each other, the spin-up electrons have a lower barrier height for both spin filters, in contrast to antiparallel alignment for which the spin-up electrons preferentially tunnel through the first spin filter (the polarizer) but are blocked by the higher barrier height in the analyzer. The TMR is generated by the two magnetic barriers, without the need for a ferromagnetic electrode. In order to reach both parallel and antiparallel alignment, the two spin-filter barriers must be magnetically decoupled and have separate coercive fields.

Worledge and Geballe estimated a MR value as high at 10^5. However, one can easily anticipate the difficulties in experimentally realizing a double spin-filter junction. Because the junction resistance depends exponentially on barrier thickness (Equation 13.1), a junction with two spin-filter barriers would have very high resistance. In addition, it is likely that the two spin filters would be magnetically coupled, so that the antiparallel alignment state could be reached. Any coupling would significantly reduce the MR ratio. One could insert a nonmagnetic, insulating spacer layer between the two spin filters to decouple them. A spin-independent spacer layer would not decrease the MR ratio, but it would increase the junction resistance even further.

Miao et al. gave the first experimental demonstration of high TMR in a double spin-filter junction, using two EuS barriers and *nonmagnetic* electrodes [51]. An Al_2O_3 spacer layer was used to prevent coupling between the two EuS barriers. These junctions produced high TMR and an unconventional bias dependence, in which the TMR *increased* at high bias, reaching some peak value, then decreased as the bias was increased further, as shown in Figure 13.15. Similar to the behavior of QMTJs described earlier, this bias dependence is a result of a progression from direct tunneling for both spin orientations to FN tunneling for spin-up electrons only, to FN tunneling for both spin orientations, as the bias is increased. The asymmetry of the peaks and the peak positions for positive and negative applied bias is especially pronounced and is caused by the different thickness of the two EuS

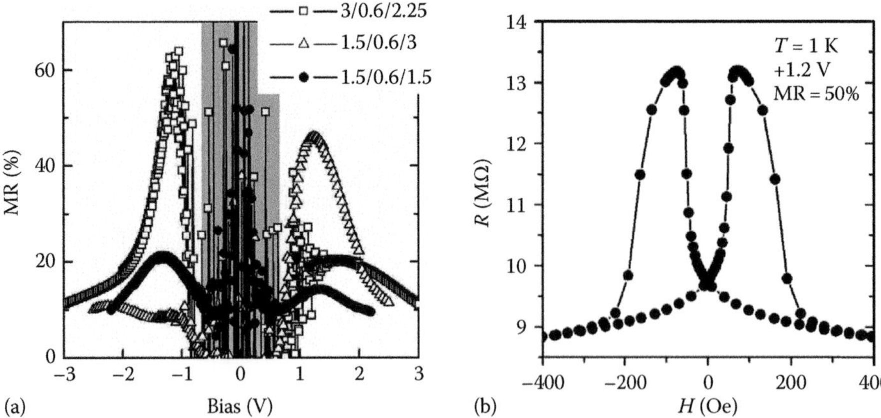

FIGURE 13.15 (a) The bias dependence measured for 10 nm Al/EuS/0.6 nm AlO_x/EuS/10 nm Al double spin-filter junction, for the various EuS thicknesses indicated. The shaded region at low bias contains noisy data due to the high impedance of the junction. (b) Magnetoresistance curve for the junction with 1.5 and 3 nm thick EuS barriers. (Modified from Miao, G.-X. et al., *Phys. Rev. Lett.*, 102, 076601, 2009.)

layers. With three barrier layers, the impedance was very high (several GΩ) so as to complicate measurement at low bias, below the onset of FN tunneling (see Figure 13.15).

For Miao et al., the trick to fabricating a double spin junction was to deposit the two EuS layers using different growth conditions. Deposition of the first EuS layer onto a liquid nitrogen-cooled substrate yielded a lower coercive field and sharper switching compared to the second EuS layer subsequently deposited at room temperature. Another important ingredient was the Al_2O_3 spacer of optimal thickness, thick enough to sufficiently decouple the two layers but not too thick to prevent single-step tunneling and/or make the junction impedance too high. Only then could the EuS layers have a large enough separation of coercive field so that a well-defined, antiparallel resistance state was reached [52].

For a double spin-filter junction, there are many parameters with which to tune the magnitude of the MR and bias dependence. Relative thickness, exchange splitting, and average barrier height all play a role in the dramatic bias dependence behavior of these junctions, as shown by the numerical calculations of Miao and Moodera based on WKB approximation [53]. As can be expected, the TMR is enhanced by increasing the thickness and exchange splitting, or by reducing the average barrier height. In qualitative agreement with experiment [51], the maximum TMR value is obtained when electrons first tunnel through the more efficient spin filter, meaning the one that is thicker or has the larger exchange splitting, whereas a lower TMR is obtained for the opposite current flow direction. This is shown in Figure 13.16, in which the relative thickness of the two layers is varied while keeping the total thickness constant. Barrier thickness is the parameter most easily varied experimentally. For optimal spin-filter efficiency, a thicker spin-filter layer is preferred [50]. Layers that are too thin are likely to have a reduced T_c and consequently a lower exchange splitting. This was shown to be the case for EuO [32], and the opposite is expected to occur in the ferrites. However, on the other hand, inherent problems associated with having a thick barrier is the increased probability of multistep

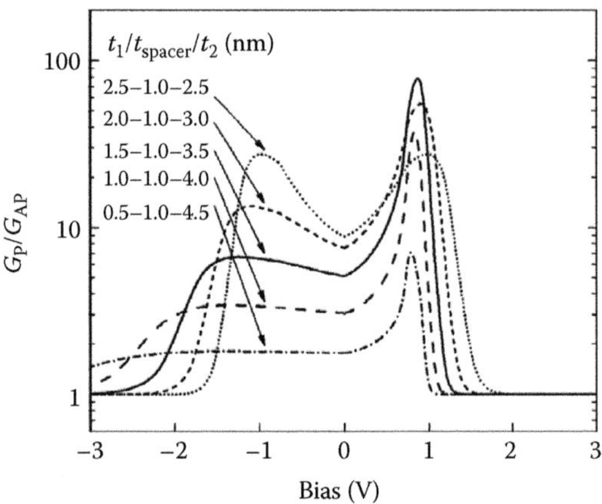

FIGURE 13.16 Calculated bias dependence for a double spin-filter junction in which the relative thicknesses of the two spin-filter layers is varied while the total thickness remains constant. (From Miao, G.-X. and Moodera, J.S., *J. Appl. Phys.*, 106, 023911, 2009.)

tunneling via defect states, which are sites for spin-flip scattering, causing a lower MR ratio than expected. Exchange splitting and average barrier heights are parameters that are determined by the material, though these parameters can be influenced by layer thickness as well [32]. Exchange splitting is more effective for a lower average barrier height.

At this time, we are not aware of any other successful demonstration of a double spin-filter junction. One potential candidate is a $CoFe_2O_4/NiFe_2O_4$ double spin filter. Given that these ferrites have the same crystal structure and similar magnetic properties, it is likely that an epitaxial, nonmagnetic spacer layer is needed.

13.8 Spin Injection into Semiconductors

Sugahara and Tanaka proposed a spin-filter transistor (SFT) device, based on the successful, experimental demonstration of spin-filter tunneling [54]. Their model device, shown in Figure 13.17, consisted of a semiconductor base separated from

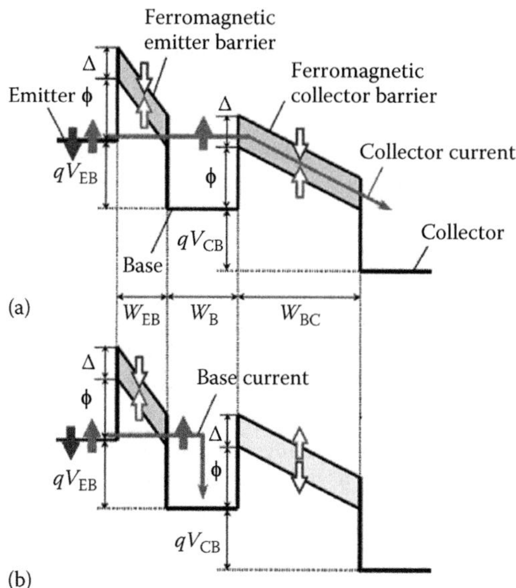

FIGURE 13.17 Schematic of the SFT proposed by Sugahara and Tanaka. (a) Parallel magnetic orientation of the emitter barrier and collector barrier, resulting in a high collector current. (b) Antiparallel magnetic orientation, resulting in no collector current. (Reprinted from *Physica E*, 21, Sugahara, S. and Tanaka, M., A novel spin transistor based on spin-filtering in ferromagnetic barriers: A spin-filter transistor, 996–1001, Fig. 1, Copyright 2004, with permission from Elsevier.)

the nonmagnetic, semiconducting emitter and collector by two spin-filter tunnel barriers. When an emitter-base bias is applied, the emitter barrier filters the hot electron spins as they tunnel into the base, allowing only spin-up electrons into the base. The collector barrier acts as an analyzer for the spin-polarized transport through the base. The width of the base must be less than the spin-flip scattering length. The relative magnetic alignment of the emitter and collector tunnel barriers determines the output of the SFT. The collector current for parallel configuration can be significantly higher than that for antiparallel configuration. Depending on the spin-filter barrier parameters, the mean free path of the hot electrons in the base layer and width of the base, Sugahara and Tanaka showed that the SFT could achieve large magneto-current ratio, current gain, and power gain.

Filip et al. [55] proposed a similar device, consisting of two EuS spin-filter tunnel barriers separated by a nonmagnetic, semiconducting quantum well. The spin injection device was modeled using the EuS/PbS system because it could be grown epitaxially as well as the high spin-filter efficiency of EuS. According to their model, when the EuS layers are aligned antiparallel, a nonequilibrium spin accumulation takes place in the PbS quantum well. No spin accumulation takes place when they are aligned parallel. Calculation showed that this could result in high magnetoresistance for a range of EuS and PbS layer thicknesses.

The above theoretical models are yet to be explored experimentally due to the degree of difficulty in integrating spin filters with semiconductors. There have been some studies in this direction: the metal/EuS Schottky contact [56] and EuS/GaAs semiconductor heterojunction [57]. For the metal/EuS Schottky contact (100 nm thick EuS layer), exchange splitting of the EuS conduction band was determined by measuring the *I–V* characteristics at temperatures above and below the T_C of EuS, and then extracting (from the forward bias *I–V*) the Schottky barrier height change as it lowers for $T < T_C$. The exchange splitting lowers the Schottky barrier height. Similarly, using forward bias *I–V* (electron injection from EuS into GaAs), the EuS/GaAs heterojunction showed an exchange splitting of 0.48 eV (much higher than bulk value). In addition, under reverse bias (injection from GaAs into EuS) a shift of the *I–V* with temperature was observed, suggesting filtering of the unpolarized electrons from GaAs. In this manner, the performance of the EuS/GaAs heterojunction as an injector and detector was probed individually. The observed shifts in the *I–V* characteristics with temperature, caused by exchange splitting of the EuS conduction band, implied spin-polarized injection and detection, although no spin detector was explicitly used.

Spin-filter tunneling is even being considered as an approach to achieving quantum computing, where the electron spin is used as a qubit [58,59] as shown schematically in Figure 13.18. The spin-filter property of the barrier provides the method for the quantum measurement. The correlation of the electron wave function from one quantum dot to the adjacent one is by tunneling through the spin-selective barrier. The energy level of the trapped electron is varied by an applied gate voltage, allowing it to tunnel or not through the barrier placed beside a quantum dot. The tunneling probability depends on the magnetic orientation of the spin filter. In this way, the spin state of the trapped electron can be read, allowing measurement of the state of each qubit individually and reliably by electrical means, rather than the extremely difficult task of magnetically measuring a single spin. Thus, spin information is converted into charge information. This is a very simple scheme, and of course, there are many

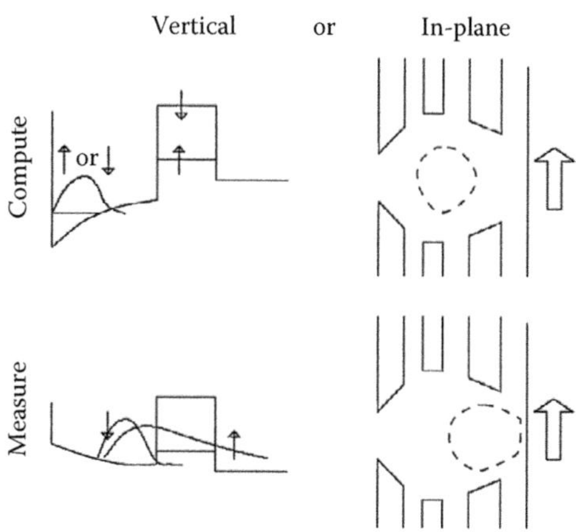

FIGURE 13.18 Proposed scheme of a spin-filter quantum measurement. Possible vertical and in-plane structures are shown. In the computing phase, the electron wave function in the quantum dot is held far from the spin filter. In the measurement phase, the gate potentials are changed so that the electron wave function is pressed up against the spin filter. (Reprinted with permission from DiVincenzo, D.P., *J. Appl. Phys.*, 85, 4785, 1999. Copyright [1999], American Institute of Physics.)

material and other parameters that must be investigated to realize it. Another group has proposed a spin filter and a spin memory using quantum dots. It is realized by the quantum dots in the Coulomb blockade regime in a magnetic field [60].

13.9 Conclusion

Although spin-filter studies mostly began in the 1970s while exploring spin-polarized photoemission and field emission, it has come of age in later years both in basic importance and application potential. Ferromagnetic Eu-chalcogenides were the only compounds serving as spin filters, limited to low temperatures, for nearly 40 years. However, recent activity in the ferrites shows promise for developing this novel property at room temperature. Understanding the defect formation in ferrites, and other complex oxide candidates, and its control need to be thoroughly investigated in order to benefit from this promising class of materials. This is true to an extent for Eu-chalcogenides as well. Very interesting physics continues to come out of the Eu-chalcogenides, including double spin filters demonstrating large magnetoresistance with further new physics anticipated from sandwiched double spin-filter layers with a spacer. With the availability of large interfacial exchange fields, exchange splitting, and quantum well states, unraveling the physics is both challenging and exciting. This can eventually lead to SFTs, opening up this field to various application possibilities. Other engrossing topics that can benefit from spin-filter tunneling would be the electrical information readout scheme for quantum computing and spin injection into semiconductors, not to mention what more new phenomena waiting to be discovered.

Acknowledgments

It is a pleasure to acknowledge the past extensive collaboration and fruitful discussions with Drs. Robert Meservey and Paul Tedrow. In recent times, we have largely benefited from the input from Drs. Guoxing Miao, Taro Nagahama, Ana Ramos, Martina Mueller, Jean-Baptiste Moussy, Raghava Panguluri, and Boris Nadgorny. The collaborations over the years with Prof. Yves Idzerda and his team as well as with Drs. Julie Borchers and Shannon Watson have greatly contributed to our understanding of the spin-filter materials and interfaces. T.S.S. is currently supported by a CNM Distinguished Postdoctoral Fellowship and the L'Oreal USA Fellowship for Women in Science. Research funding during many years of research from National Science Foundation, Office of Naval Research and KIST-MIT joint program has made all this possible.

References

1. W. E. Pickett and J. S. Moodera, Half metallic magnets, *Phys. Today* **54**, 39–44 (2001).
2. J. G. Simmons, Generalized formula for the electric tunnel effect between similar electrodes separated by a thin insulating film, *J. Appl. Phys.* **34**, 1793–1803 (1963).
3. W. F. Brinkman, R. C. Dynes, and J. M. Rowell, Tunneling conductance of asymmetrical barriers, *J. Appl. Phys.* **41**, 1915–1921 (1970).
4. J. S. Moodera, X. Hao, G. A. Gibson, and R. Meservey, Electron-spin polarization in tunnel junctions in zero applied field with ferromagnetic EuS barriers, *Phys. Rev. Lett.* **61**, 637–640 (1988); X. Hao, J. S. Moodera, and R. Meservey, Spin-filter effect of ferromagnetic europium sulfide tunnel barriers, *Phys. Rev. B* **42**, 8235–8243 (1990).
5. J. S. Moodera, R. Meservey, and X. Hao, Variation of the electron-spin polarization in EuSe tunnel junctions from zero to near 100% in a magnetic field, *Phys. Rev. Lett.* **70**, 853–856 (1993).
6. J. Schoenes and P. Wachter, Exchange optics in Gd-doped EuO, *Phys. Rev. B* **9**(7), 3097–3105 (1974).
7. T. S. Santos and J. S. Moodera, Observation of spin filtering with a ferromagnetic EuO tunnel barrier, *Phys. Rev. B* **69**, 241203(R) (2004).
8. M. Pénicaud, B. Siberchicot, C. B. Sommers, and J. Kübler, Calculated electronic band structure and magnetic moments of ferrites, *J. Magn. Magn. Mater.* **103**, 212–220 (1992).
9. M. Gajek, M. Bibes, S. Barthélémy et al., Spin filtering through ferromagnetic $BiMnO_3$ tunnel barriers, *Phys. Rev. B* **72**, 020406(R) (2005).
10. U. Lüders, M. Bibes, K. Bouzehouane et al., Spin filtering through ferrimagnetic $NiFe_2O_4$ tunnel barriers, *Appl. Phys. Lett.* **88**, 082505 (2006).
11. A. V. Ramos, M. J. Guittet, J. B. Moussy et al., Room temperature spin filtering in epitaxial cobalt-ferrite tunnel barriers, *Appl. Phys. Lett.* **91**, 122107 (2007).
12. C. Hass, Magnetic semiconductors, *CRC Crit. Rev. Solid State Sci.* **1**, 47–98 (1970).
13. P. Wachter, Europium chalcogenides: EuO, EuS, EuSe, and EuTe, in *Handbook on the Physics and Chemistry of Rare Earths*, vol. 2, K. A. Schneider and L. Eyring (Eds.), North-Holland Publishing Co., New York, pp. 507–574, 1979.
14. F. Holtzberg, S. von Molnar, and J. M. D. Coey, Rare earth magnetic semiconductors, in *Handbook on Semiconductors*, vol. 3, S. P. Keller (Ed.), North-Holland Publishing Co., New York, pp. 803–856, 1980.
15. A. Mauger and C. Godart, The magnetic, optical, and transport properties of representatives of a class of magnetic semiconductors: The europium chalcogenides, *Phys. Rep.* **141**, 51–176 (1986).
16. É. T. Kulatov, Y. A. Uspenskiĭ, and S. V. Khalilov, Electronic structure and magnetooptical properties of europium monochalcogenides EuO and EuS, *Phys. Solid State* **38**(10), 1677–1678 (1996).
17. M. Horne, P. Strange, W. M. Temmerman, Z. Szotek, A. Svane, and H. Winter, The electronic structure of europium chalcogenides and pnictides, *J. Phys.: Condens. Matter* **16**, 5061–5070 (2004).
18. D. B. Ghosh, M. De, and S. K. De, Electronic structure and magneto-optical properties of magnetic semiconductors: Europium monochalcogenides, *Phys. Rev. B* **70**, 115211 (2004).

19. G. Busch, P. Junod, and P. Wachter, Optical absorption of ferro- and antiferromagnetic europium chalcogenides, *Phys. Lett.* **12**(1), 11–12 (1964).
20. P. G. Steeneken, L. H. Tjeng, I. Elfimov et al., Exchange splitting and charge carrier spin polarization in EuO, *Phys. Rev. Lett.* **88**, 047201 (2002).
21. N. Müller, W. Eckstein, W. Heiland, and W. Zinn, Electron spin polarization in field emission from EuS-coated tungsten tips, *Phys. Rev. Lett.* **29**, 1651–1654 (1972).
22. E. Kisker, G. Baum, A. H. Mahan, W. Raith, and K. Schröder, Conduction-band tunneling and electron-spin polarization in field emission from magnetically ordered europium sulfide on tungsten, *Phys. Rev. Lett.* **36**, 982–985 (1976).
23. G. Baum, E. Kisker, A. H. Mahan, W. Raith, and B. Reihl, Field emission of monoenergetic spin-polarized electrons, *Appl. Phys.* **14**, 149–153 (1977).
24. E. Kisker, G. Baum, A. H. Mahan, W. Raith, and B. Reihl, Electron field emission from ferromagnetic europium sulfide on tungsten, *Phys. Rev. B* **18**, 2256–2275 (1978).
25. L. Esaki, P. J. Stiles, and S. von Molnar, Magneto internal field emission in junctions of magnetic insulators, *Phys. Rev. Lett.* **19**, 852–854 (1967).
26. W. A. Thompson, W. A. Holtzberg, T. R. McGuire, and G. Petrich, Tunneling study of EuS magnetization, in *Magnetism and Magnetic Materials: Proceedings of the 17th Annual Conference on Magnetism and Magnetic Materials (Chicago, 1971) (AIP Conf. Proc. No. 5)*, C. D. Graham Jr. and J. J. Rhyne (Eds.), AIP, New York, pp. 827–836, 1972.
27. K. Maki, in *Superconductivity*, vol. 2, R. D. Parks (Ed.), Marcel Dekker, New York, p. 1035, 1969; P. Fulde and K. Maki, Theory of superconductor combining magnetic impurities, *Phys. Rev.* **141**, 275 (1966).
28. R. Griessen, M. Landolt, and H. R. Ott, A new ferromagnetic phase in EuSe below 1.8 K, *Solid State Commun.* **9**, 2219–2223 (1971).
29. P. M. Tedrow, J. E. Tkaczyk, and A. Kumar, Spin-polarized electron tunneling study of an artificially layered superconductor with internal magnetic field: EuO-Al, *Phys. Rev. Lett.* **56**, 1746 (1986).
30. E. Negusse, J. Holroyd, M. Liberati et al., Effect of electrode and EuO thickness on EuO-electrode interface in tunneling spin filter, *J. Appl. Phys.* **99**, 08E507 (2006).
31. R. Schiller and W. Nolting, Prediction of a surface state and a related surface insulator-metal transition for the (100) surface of stoichiometric EuO, *Phys. Rev. Lett.* **86**, 3847 (2001).
32. T. S. Santos, J. S. Moodera, K. V. Raman et al., Determining exchange splitting in a magnetic semiconductor by spin-filter tunneling, *Phys. Rev. Lett.* **101**, 147201 (2001).
33. G. Sarma, On the influence of a uniform exchange field acting on the spins of the conduction electrons in a superconductor, *J. Phys. Chem. Solids* **24**, 1029–1032 (1963).
34. P. G. de Gennes, Coupling between ferromagnets through a superconducting layer, *Phys. Lett.* **23**(1), 10–11 (1966).
35. X. Hao, J. S. Moodera, and R. Meservey, Thin-film superconductor in an exchange field, *Phys. Rev. Lett.* **67**, 1342–1345 (1991).
36. M. Julliere, Tunneling between ferromagnetic films, *Phys. Lett.* **54A**, 225–226 (1975).
37. P. LeClair, J. K. Ha, H. J. M. Swagten, J. T. Kohlhepp, C. H. van de Vin, and W. J. M. de Jonge, Large magnetoresistance using hybrid spin filter devices, *Appl. Phys. Lett.* **80**, 625–627 (2002).
38. T. Nagahama, T. S. Santos, and J. S. Moodera, Enhanced magneto-transport at high bias in quasi-magnetic tunnel junctions with EuS spin-filter barriers, *Phys. Rev. Lett.* **99**, 016602 (2007).
39. A. Saffarzadeh, Spin-filter magnetoresistance in magnetic barrier junctions, *J. Magn. Magn. Mater.* **269**, 327–332 (2004).
40. Measurement by T. S. Santos, T. Nagahama, and J. S. Moodera, in collaboration with P. Señeor, M. Bibes, and A. Barthélémy.
41. M. Gajek, M. Bibes, M. Varela et al., $La_{2/3}Sr_{1/3}MnO_3$-$La_{0.1}Bi_{0.9}MnO_3$ heterostructures for spin filtering, *J. Appl. Phys.* **99**, 08E504 (2006).
42. M. Gajek, M. Bibes, S. Fusil et al., Tunnel junctions with multiferroic barriers, *Nat. Mater.* **6**, 296 (2007).
43. G. Hu and Y. Suzuki, Negative spin polarization of Fe_3O_4 in magnetite/magnanite-based junctions, *Phys. Rev. Lett.* **89**, 276601 (2002).
44. Z. Szotek, W. M. Temmerman, D. Ködderitzsch, A. Svane, L. Petit, and H. Winter, Electronic structures of normal and inverse spinel ferrites from first principles, *Phys. Rev. B* **74**, 174431 (2006).
45. G. Hu, J. H. Choi, C. B. Eom, V. G. Harris, and Y. Suzuki, Structural tuning of magnetic behavior in spinel-structure ferrite thin films, *Phys. Rev. B* **62**, 779 (2000).
46. U. Lüders, M. Bibes, J.-F. Bobo, M. Cantoni, R. Bertacco, and J. Fontcuberta, Enhanced magnetic moment and conductive behavior in $NiFe_2O_4$ spinel ultrathin films, *Phys. Rev. B* **71**, 134419 (2005).
47. F. Rigato, S. Estradé, J. Arbiol et al., Strain-induced stabilization of new magnetic spinel structures in epitaxial oxide heterostructures, *Mater. Sci. Eng. B* **144**, 43 (2007).
48. A. V. Ramos, J.-B. Moussy, M.-J. Guittet et al., Influence of a metallic or oxide top layer in epitaxial magnetic bilayers containing $CoFe_2O_4$(111) tunnel barriers, *Phys. Rev. B* **75**, 224421 (2007).
49. A. V. Ramos, T. S. Santos, G. X. Miao, M.-J. Guittet, J.-B. Moussy, and J. S. Moodera, Influence of oxidation on the spin-filtering properties of $CoFe_2O_4$ and the resultant spin polarization, *Phys. Rev. B* **78**, 180402(R) (2008).
50. D. C. Worledge and T. H. Geballe, Magnetoresistive double spin filter tunnel junction, *J. Appl. Phys.* **88**, 5277–5279 (2000).
51. G.-X. Miao, M. Müller, and J. S. Moodera, Magnetoresistance in double spin filter tunnel junctions with nonmagnetic electrodes and its unconventional bias dependence, *Phys. Rev. Lett.* **102**, 076601 (2009).

52. G.-X. Miao and J. S. Moodera, Controlling magnetic switching properties of EuS for constructing double spin filter magnetic tunnel junctions, *Appl. Phys. Lett.* **94**, 182504 (2009).
53. G.-X. Miao and J. S. Moodera, Numerical evaluations on the asymmetric bias dependence of magnetoresistance in double spin filter tunnel junctions, *J. Appl. Phys.* **106**, 023911 (2009).
54. S. Sugahara and M. Tanaka, A novel spin transistor based on spin-filtering in ferromagnetic barriers: A spin-filter transistor, *Physica E* **21**, 996–1001 (2004).
55. A. T. Filip, P. LeClair, C. J. P. Smits et al., Spin-injection device based on EuS magnetic tunnel barriers, *Appl. Phys. Lett.* **81**, 1815–1817 (2002).
56. C. Ren, J. Trbovic, P. Xiong, and S. von Molnár, Zeeman splitting in ferromagnetic Schottky barrier contacts based on doped EuS, *Appl. Phys. Lett.* **86**, 012501 (2005).
57. J. Trbovic, C. Ren, P. Xiong, and S. von Molnár, Spontaneous spin-filter effect across EuS/GaAs heterojunction, *Appl. Phys. Lett.* **87**, 082101 (2005).
58. D. P. DiVincenzo, Quantum computing and single-qubit measurements using the spin-filter effect, *J. Appl. Phys.* **85**, 4785 (1999).
59. C. H. Bennett and D. P. DiVincenzo, Quantum information and computation, *Nature* **404**, 247 (2000).
60. P. Recher, E. V. Sukhorukov, and D. Loss, Quantum dot as spin filter and spin memory, *Phys. Rev. Lett.* **85**, 1962–1965 (2000).

14

Spin Torques in Magnetic Tunnel Junctions

Yoshishige Suzuki
Osaka University

Hitoshi Kubota
National Institute of Advanced Industrial Science and Technology, Spintronics Research Center

14.1 Spin Injection Torque in Magnetic Tunnel Junctions

14.1.1 Magnetic Tunnel Junction for Spin Injection

As discussed in previous chapters, an electric current in a magnetic multilayer is spin polarized and yields a torque exerted on a local spin momentum. Such an electric current–induced spin torque in magnetic multilayers was first predicted theoretically [1,2] and subsequently observed experimentally in metallic nano-junctions by excitation of spin waves [3] and spin injection magnetization switching (SIMS) [4,5]. The effects of the spin torque was also claimed to be observed in a perovskite system [6]. Further, SIMS was also observed in magnetic tunnel junctions (MTJs) with an Al-O barrier [7] and an MgO barrier [8,9] and in magnetic semiconductor systems [10].

Figure 14.1 shows a schematic of an MTJ with in-plane magnetization for spin injection switching. The MTJ consists of two ferromagnetic layers (e.g., Co and Fe) separated by an insulating barrier layer (e.g., AlO and MgO). Lateral shape of an MTJ pillar with in-plane magnetization is an ellipse or a rectangle with dimensions of about 200 nm × 100 nm or less. The angular momentum in the fixed layer, $\vec{S_1}$, which is opposite to the magnetization, is fixed along the long axis of the ellipse through an exchange interaction with an antiferromagnetic layer (e.g., PtMn). Without current injection, the angular momentum in the free layer, $\vec{S_2}$, also lies along the long axis of the ellipse because of magnetostatic shape anisotropy and is either parallel (P) or antiparallel (AP) with respect to $\vec{S_1}$. To induce asymmetry between the two magnetic layers, the thickness of the free layer is often less than that of the fixed layer.

The electric conductance, G, and resistance, R, of the junction vary with the relative angle, θ, between the two magnetization directions, as already discussed in previous chapters [11],

$$G(\theta) = R^{-1}(\theta) = \frac{G_P + G_{AP}}{2} + \frac{G_P - G_{AP}}{2}\cos\theta. \tag{14.1}$$

Here, G_P $(=R_P^{-1})$ and G_{AP} $(=R_{AP}^{-1})$ are the tunnel conductances (resistance) for the parallel $(\theta = 0)$ and antiparallel $(\theta = \pi)$ configurations of magnetization, respectively. To inject enough current to switch the magnetization, the resistance-area product $((RA)_P = A/G_P$ where A is the junction area) is set to below $10\,\Omega\,\mu m^2$.

14.1.2 Spin-Transfer Torque and Torquance

When current is passed through a magnetic nano-pillar, the electrons are first polarized by the thick fixed layer (spin polarizer) and then injected into the free layer through the barrier. The spins of the injected electrons interact with the local spin

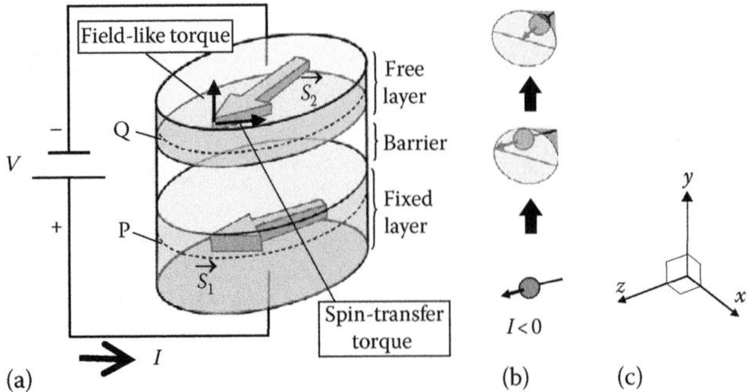

FIGURE 14.1 (a) Typical structures of magnetic nano-pillars designed for SIMS experiments. The upper magnetic layer in the bird's-eye-view images acts as a magnetically free layer, whereas the lower magnetic layer is thicker than the free layer and acts as a spin polarizer. The spins in the lower layers are usually pinned to an antiferromagnetic material, which is placed below the pinned layer (not shown), as a result of exchange interaction at the bottom interface. Therefore, the spin polarizer layer is often called the "pinned layer," "fixed layer," or "reference layer." The layer between the two ferromagnetic layers is made of insulators such as MgO. Here, the diameter of the pillar is around 100 nm. The free layer is typically a few nanometers thick. The large arrows indicate the direction of the total spin moment in each layer, and the two small arrows indicate two different spin torques. (b) The flow and precession of conduction electrons. The injected electrons from the bottom are spin-polarized by the thick ferromagnetic layer (FM1: spin polarizer) and then injected into the thin ferromagnetic layer (FM2: free layer) through NM1. The injected spins undergo precession in FM2 and lose their transverse components. The lost spin components are transferred to the local spin moment in FM2, $\vec{S}_2$ (details are provided in the text). (c) Coordinate system.

angular momentum, $\vec{S}_2$, in the free layer by exchange interaction and exert torque. If the exerted torque is large enough, the magnetization in the free layer is reversed or continuous precession is excited. As a consequence of the total spin conservation, the change in the local spin angular momentum is equal to the difference between the injected spin current and gushed spin current:

$$\left(\frac{d\vec{S}_2}{dt}\right)_{ST} = \vec{I}_1^S - \vec{I}_2^S, \tag{14.2}$$

where $\vec{I}_1^S$ and $\vec{I}_2^S$ are the spin currents at the barrier and nonmagnetic (NM) layer, respectively. Since FM2 is very thin, we neglected the spin–orbit interaction in FM2. Equation 14.2 indicates that the torque can be exerted on the local spin angular momentum as a result of spin transfer from the conduction electrons.

Slonczewski showed an intuitive way to evaluate the spin-transfer torque in MTJs [11,12] by evaluating the spin currents inside ferromagnetic layers. We assume that FM1 is sufficiently thick; therefore, at cross section P in FM1 (see Figure 14.1a and b), the conducting spins are relaxed and aligned parallel to $\vec{S}_1$. Those spin-polarized electrons are injected into FM2.

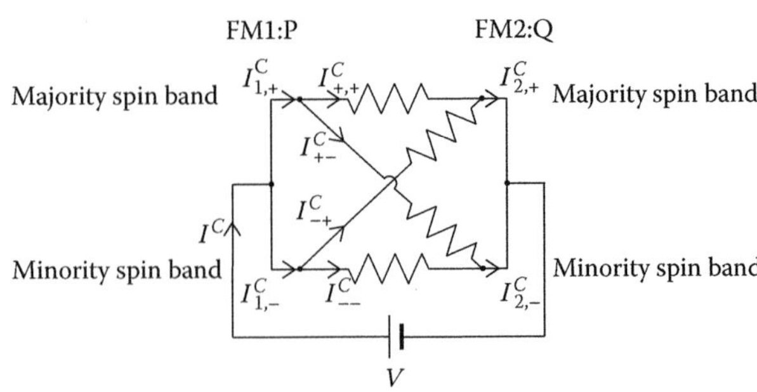

FIGURE 14.2 Circuit model of a magnetic tunnel junction. (After Slonczewski, J.C., *Phys. Rev. B*, 71, 024411(1–10), 2005. With permission.)

The injected spins are subjected to an exchange field made by the local magnetization and show precession motion. Here, we also assume that at cross section Q inside FM2, the spins of the conducting electrons have already lost their transverse spin component on average because of the decoherence of the precessions and the spins have aligned parallel to $\vec{S}_2$ on average. Therefore, the spin currents at P and Q, $\vec{I}_1^S$ and $\vec{I}_2^S$, are parallel to $\vec{S}_1$ and $\vec{S}_2$, respectively. Since the spins of the conduction electrons at *P* and Q are either the majority or minority spins of the host material, the total charge current in the MTJ can be expressed as a sum of the following four components as shown in Figure 14.2:

$$I^C = I_{++}^C + I_{+-}^C + I_{-+}^C + I_{--}^C. \tag{14.3}$$

Here, the suffixes + and − indicate the majority and minority spin channels, respectively. For example, I_{+-}^C represents a charge current flow from the FM1 minority spin band into the FM2 majority spin band. These charge currents are expressed using the conductance for each spin subchannel, $G_{\pm\pm}$.

$$\begin{cases} I_{\pm\pm}^C = VG_{\pm\pm}\cos^2\dfrac{\theta}{2}, \\ I_{\mp\pm}^C = VG_{\mp\pm}\sin^2\dfrac{\theta}{2}. \end{cases} \tag{14.4}$$

Here, *V* is the applied voltage. The angle dependence of the conductions can be derived from the fact that the spin functions in the FM1 are $|\text{maj.}\rangle = \cos(\theta/2)|\uparrow\rangle + \sin(\theta/2)|\downarrow\rangle$ for the majority spins and $|\text{min.}\rangle = \sin(\theta/2)|\uparrow\rangle - \cos(\theta/2)|\downarrow\rangle$ for the minority spins, and those in the FM2 are $|\uparrow\rangle$ and $|\downarrow\rangle$, respectively. Since the spin quantization axes at P and Q are parallel to $\vec{S}_1$ and $\vec{S}_2$, respectively, the spin currents at P and Q are obtained easily as follows:

$$\begin{cases} \vec{I}_1'^S = \dfrac{\hbar}{2}\dfrac{1}{-e}\left(I_{++}^C + I_{+-}^C - I_{-+}^C - I_{--}^C\right)\mathbf{e}_1, \\ \vec{I}_2'^S = \dfrac{\hbar}{2}\dfrac{1}{-e}\left(I_{++}^C - I_{+-}^C + I_{-+}^C - I_{--}^C\right)\mathbf{e}_2, \end{cases} \tag{14.5}$$

where unit vectors $\vec{e}_1$ and $\vec{e}_2$ are parallel to the majority spins in the FM1 and FM2, respectively. Now, we apply the total angular momentum conservation between the P and Q planes, that is,

$$\left(\frac{d}{dt}\left(\vec{S}_1+\vec{S}_2\right)\right)_{ST} = \vec{I}_1'^S - \vec{I}_2'^S. \tag{14.2'}$$

Then, after a straightforward calculation, the total current and spin-transfer torque are obtained as follows:

$$\begin{cases} I^C = \left(\dfrac{G_{++}+G_{--}+G_{+-}+G_{-+}}{2} + \dfrac{(G_{++}+G_{--})-(G_{+-}+G_{-+})}{2}\vec{e}_2\cdot\vec{e}_1\right)V \\ \quad = \left(\dfrac{G_P+G_{AP}}{2} + \dfrac{G_P-G_{AP}}{2}\cos\theta\right)V, \\ \left(\dfrac{d\vec{S}_2}{dt}\right)_{ST} = \dfrac{\hbar}{2}\dfrac{1}{-e}\left(\dfrac{G_{++}-G_{--}}{2} + \dfrac{G_{+-}-G_{-+}}{2}\right)(\vec{e}_2\times(\vec{e}_1\times\vec{e}_2))V \\ \quad = T_{ST}(\vec{e}_2\times(\vec{e}_1\times\vec{e}_2))V. \end{cases} \tag{14.6}$$

The first equation in Equation 14.6 shows $\cos\theta$ dependence of the tunnel conductance. The second equation shows $\sin\theta$ dependence of the spin torque (note that $|\vec{e}_2\times(\vec{e}_1\times\vec{e}_2)| = \sin\theta$). Slonczewski called T_{ST} the "torquance," which is an analogue of "conductance." In particular, in MTJs, the spin torques should be bias voltage dependent because $G_{\pm\pm}$ is bias voltage dependent. The direction of the spin-transfer torque is shown in Figure 14.1a for $T_{ST}<0$. This direction is the same as that in junctions utilizing current perpendicular to the plane (CPP) giant magnetoresistance (GMR), as was explained in Chapter 7.

14.1.3 Field-Like Torque in MTJs

The direction of $\vec{S}_2$ may change in two different ways. One is along the direction parallel to the spin-transfer torque, $(\vec{e}_2\times(\vec{e}_1\times\vec{e}))$. The other is the direction parallel to $(\vec{e}_1\times\vec{e}_2)$ (see Figure 14.1a). If the torque is parallel to $(\vec{e}_1\times\vec{e}_2)$, it has the same symmetry as a torque exerted by an external field. Therefore, the latter torque is called a field-like torque. It can also be called an "accumulation torque" based on one of its possible origins, or a "perpendicular torque" based on its direction in the multilayer system with in-plane magnetizations.

It has been pointed out that one of the important origins of the field-like torque in MTJs is the change in the interlayer exchange coupling through the barrier layer at a finite biasing voltage [13,14].

The field-like torque could originate from other mechanisms that are similar to those responsible for the "β-term" in magnetic nanowires. Several mechanisms, such as spin relaxation [15,16], Gilbert damping itself [17], momentum transfer [18], or a current induced ampere field have been proposed for the origin of the β-term.

14.2 Spin-Torque Diode Effect and Voltage Dependence of the Torque

14.2.1 Linearized LLG Equation and Spin-Torque Diode Effect

Spin torques have been quantitatively measured from the ferromagnetic resonance (FMR) excited by them [19–23]. To observe the spin-torque FMR, Tulapurkar et al. [19] developed a homodyne detection technique and named it as "spin-torque diode effect measurement" wherein they used the rectification function of the MTJ. To observe the spin-torque diode effect, we may apply an external field to set a specific relative angle between the free layer and fixed layer magnetizations. In Figure 14.3b, we show a case in which the free layer and fixed layer magnetizations are in-plane but perpendicular to each other. We then apply an alternative current to the junction. A negative current induces a preferential parallel configuration of the spins. Thus, the resistance of the junction becomes smaller and we observe only a small negative voltage across the junction for a given current (Figure 14.3a). A positive current of the same amplitude induces a preferential antiparallel configuration

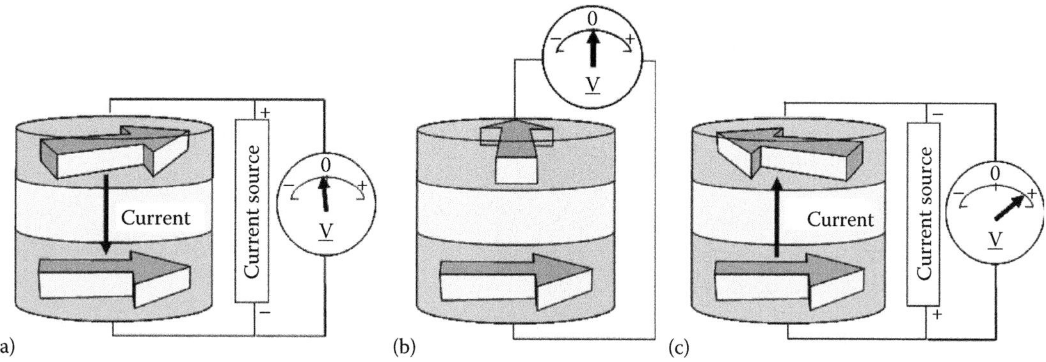

FIGURE 14.3 Schematic explanation of the spin-torque diode effect: (a) negative current, (b) null current, and (c) positive current. (After Suzuki, Y. and Kubota, H. *J. Phys. Soc. Jpn.*, 77, 031002(1–7), 2008. With permission.)

and the resistance becomes higher. We observe a larger positive voltage appearing across the junction (Figure 14.3c). As a result, we observe a positive voltage on average. This is the spin-torque diode effect. This effect can be large if the frequency of the applied current matches the FMR frequency of the free layer. In other words, this effect provides a sensitive FMR measurement technique for the nano-pillar moment excited by the spin torque and provides a quantitative measure of the spin torques.

To treat the dynamic property of the free layer spin angular momentum, here, we introduce the Landau–Lifshitz–Gilbert (LLG) equation including the spin-transfer torque and field-like torque as follows [1,12,24]:

$$\frac{d\vec{S}_2}{dt} = \gamma\vec{S}_2 \times \vec{H}_{eff} - \alpha\vec{\mathbf{e}}_2 \times \frac{d\vec{S}_2}{dt} + T_{ST}(V)V\vec{\mathbf{e}}_2 \times (\vec{\mathbf{e}}_1 \times \vec{\mathbf{e}}_2) + T_{FT}(V)V(\vec{\mathbf{e}}_2 \times \vec{\mathbf{e}}_1). \quad (14.7)$$

The first term is the effective field torque; second, Gilbert damping; third, spin-transfer torque; and fourth, field-like spin torque. $\vec{S_2} = S_2\vec{\mathbf{e}_2}$ is the total spin angular momentum of the free layer and is opposite to its magnetic moment, $\vec{M_2}$. $\vec{\mathbf{e}_2}(\vec{\mathbf{e}_1})$ is a unit vector that expresses the direction of the spin angular momentum of the free layer (fixed layer). For simplicity, we neglect the distribution of the local spin angular momentum inside the free layer and assume that the local spins within each magnetic cell are aligned in parallel and form a coherent "macrospin" [25,26]. This assumption is not strictly valid since the demagnetization field and current-induced Oersted field inside the cells are not uniform. Such non-uniformities introduce incoherent precessions of the local spins and cause domain and/or vortex formation in the cell [26–28]. Despite the predicted limitations, the macrospin model is still useful, because of both its transparency and its validity for small excitations. γ is the gyromagnetic ratio, where $\gamma < 0$ for electrons ($\gamma = -2.21 \times 10^5$ m/A·s for free electrons). The effective field, $\vec{H_{eff}}$, is the sum of the external field, demagnetization field, and anisotropy field. It should be noted that the demagnetization field and the anisotropy field depend on $\vec{\mathbf{e}_2}$. $\vec{H_{eff}}$ is derived from the magnetic energy, E_{mag}, and the total magnetic moment, M_2, of the free layer:

$$\vec{H}_{eff} = \frac{1}{\mu_0 M_2}\frac{\partial E_{mag}}{\partial \vec{\mathbf{e}}_2}, \quad (14.8)$$

where $\mu_0 = 4\pi \times 10^{-7}$ H/m is the magnetic susceptibility of vacuum.

The first term determines the precessional motion of $\vec{S_2}$. In the second term, α is the Gilbert damping factor (e.g., $\alpha > 0$, $\alpha \approx 0.007$ for Fe). V is the applied voltage, $T_{ST}(V) = \frac{\hbar}{2}\frac{1}{-e}\frac{1}{2}(G_{++} - G_{--} + G_{+-} - G_{-+})$ is the "torquance" that was defined in the second line in Equation 14.6, while $T_{FT}(V)$ is an unknown coefficient that expresses the size of the "field-like torque."

To treat the small excitation around a static equilibrium point of $\vec{S_2}$, we take small deviation around the equilibrium point as follows:

$$\begin{cases} \vec{S}_2 = \vec{S}_2^{(0)} + \cdot \vec{S}_2 = S_2\left(\vec{\mathbf{e}}_2^{(0)} + \delta\vec{\mathbf{e}}_2 e^{-i\omega t}\right), \\ \vec{H}_{eff} = \vec{H}_{eff}^{(0)} + \delta\vec{H}e^{-i\omega t}, \\ V = V_0 + \delta V e^{-i\omega t}. \end{cases} \quad (14.9)$$

At the equilibrium point, $\vec{S}_2^{(0)}$, the static effective field is $\vec{H}_{eff}^{(0)}$, the voltage is V_0, and the total static torque is equal to zero:

$$\vec{\mathbf{e}}_2^{(0)} \times \left(\gamma S_2\vec{H}_{eff}^{(0)} + T_{ST}(V_0)V_0\left(\vec{\mathbf{e}}_1 \times \vec{\mathbf{e}}_2^{(0)}\right) + T_{FT}(V_0)V_0\vec{\mathbf{e}}_1\right) = 0. \quad (14.10)$$

Employing the following spherical coordinate system:

$$\begin{cases} \delta\vec{\mathbf{e}}_2 \equiv a_\theta\vec{\mathbf{e}}_\theta + a_\phi\vec{\mathbf{e}}_\phi \equiv \begin{pmatrix} a_\theta \\ a_\phi \end{pmatrix}, \\ \vec{\mathbf{e}}_\theta \equiv -\dfrac{\vec{\mathbf{e}}_2^{(0)} \times \left(\vec{\mathbf{e}}_1 \times \vec{\mathbf{e}}_2^{(0)}\right)}{|\sin\theta_0|}, \\ \vec{\mathbf{e}}_\phi \equiv \dfrac{\vec{\mathbf{e}}_1 \times \vec{\mathbf{e}}_2^{(0)}}{|\sin\theta_0|}, \end{cases} \quad (14.11)$$

where θ_0 is the relative angle between $\vec{\mathbf{e}}_2^{(0)}$ and $\vec{\mathbf{e}_1}$, we obtain a linearized LLG equation

$$-i\omega\begin{pmatrix} a_\theta \\ a_\phi \end{pmatrix} = \left(-\alpha + i\hat{\sigma}_y\right)\left(\hat{\Omega} + \omega_{FT} - i\omega_{ST}\hat{\sigma}_y\right)\begin{pmatrix} a_\theta \\ a_\phi \end{pmatrix} - \frac{\sin\theta_0}{S_2}\begin{pmatrix} T'_{ST}(V_0) \\ T'_{FT}(V_0) \end{pmatrix}\delta V - i\hat{\sigma}_y\gamma\begin{pmatrix} \delta H_\theta \\ \delta H_\phi \end{pmatrix}, \quad (14.12)$$

where

$$\begin{cases} \hat{\Omega} \equiv \left(\omega_{i,j}\right) \equiv \gamma\left(\vec{H}_{eff}^{(0)} \cdot \vec{\mathbf{e}}_2^{(0)}\right)\delta_{ij} - \dfrac{\gamma}{\mu_0 M_2}\dfrac{\partial^2 E_{mag}}{\partial\vec{\mathbf{e}}_i\partial\vec{\mathbf{e}}_j} \quad (i,j = \theta \text{ or } \phi), \\ \omega_{ST} \equiv -\dfrac{T_{ST}(V_0)V_0}{S_2}\cos\theta_0, \quad \omega_{FT} \equiv \dfrac{T_{FT}(V_0)V_0}{S_2}\cos\theta_0, \\ T'_{ST}(V_0) \equiv \dfrac{\partial}{\partial V}(T_{ST}(V_0)V_0), \quad T'_{FT}(V_0) \equiv \dfrac{\partial}{\partial V}(T_{FT}(V_0)V_0), \end{cases} \quad (14.13)$$

and where $\sigma_y = \begin{pmatrix} 0 & -i \\ i & 0 \end{pmatrix}$ is the y-component of Pauli's matrix. The explicit expressions of $\hat{\Omega} = (\omega_{i,j})$ for the typical geometries of devices are given in Appendix.

The solution of the linearized LLG equation (14.12) is a forced oscillatory motion around the equilibrium point, driven by a small rf voltage and/or rf field with frequency ω and can be expressed as follows:

$$\begin{pmatrix} a_\theta \\ a_\phi \end{pmatrix} \cong \frac{1}{\omega^2 - \omega_0^2 + i\omega\Delta\omega} \begin{pmatrix} (\omega_{\phi\phi} + \omega_{FT}) & -(\omega_{\theta\phi} - \omega_{ST} - i\omega) \\ -(\omega_{\phi\theta} + \omega_{ST} + i\omega) & (\omega_{\theta\theta} + \omega_{FT}) \end{pmatrix} \times \left(i\sigma_y \frac{\sin\theta_0}{S_2} \begin{pmatrix} T'_{ST}(V_0) \\ T'_{FT}(V_0) \end{pmatrix} \delta V - \gamma \begin{pmatrix} \delta H_\theta \\ \delta H_\phi \end{pmatrix} \right). \quad (14.14)$$

Here, the resonance frequency, ω_0, and the full width at half maximum (FWHM) of the resonance, $\Delta\omega$, are given by the following equation:

$$-\frac{1}{2} Tr\left[\alpha\hat{\Omega} - \omega_{ST}\right] \pm i\sqrt{\det\left[\hat{\Omega} + \omega_{FT}\right]} = -\frac{\Delta\omega}{2} \pm i\omega_0. \quad (14.15)$$

Equation 14.14 shows that both the spin-transfer torque and the field-like torque can excite a uniform mode (FMR mode) in the free layer. However, the phases of the FMR excitations differ by 90°. This difference in the precessional phase is a result of the different directions of the respective torques (Figure 14.1). In addition, the width of the resonance, $\Delta\omega$, is affected only by the spin-transfer torque exerted by the direct voltage, V_0. This is the (anti)damping effect of the spin-transfer torque, which was already discussed in the previous section. The field-like torque exerted by the direct voltage, V_0, changes the resonance frequency, ω_0, in a manner similar to an external field.

When the rf voltage across the MTJ is $\delta V \cos\omega t$, Equations 14.1 and 14.14 can be used to determine the precessional motion and oscillating part of the junction resistance, $\delta R\cos(\omega t + \phi)$, which is linear in δV. The rf current through the junction can be approximated as $\delta V\, G(\theta_0)\cos\omega t$. Thus from Ohm's law, the following additional voltages appear across the junction.

$$\delta R\cos(\omega t + \phi) \times \delta V G(\theta_0)\cos\omega t = \delta R G(\theta_0)\frac{\delta V}{2}\left(\cos\phi + \cos(2\omega t + \phi)\right).$$

Here, the frequencies of the additional voltages are zero (dc) and 2ω. This implies that, under spin-torque FMR excitation, the MTJs may possess a rectification function and a mixing function. Because of these new functions, Tulapurkar et al. referred to these MTJs as spin-torque diodes and to these effects as spin-torque diode effects [19]. These are nonlinear effects that result from two linear responses, that is, the spin-torque FMR and Ohm's law.

When the MTJ is placed at the end of a waveguide, the explicit expression of the rectified dc voltage under a small bias voltage is given as follows:

$$V_{dc} \cong \frac{\eta}{4}\sin^2\theta_0 \frac{(G_P - G_{AP})}{S_2 G(\theta_0)} \times \mathrm{Re}\left[\frac{(\omega_{\theta\phi} - \omega_{ST} - i\omega)T'_{ST} + (\omega_{\phi\phi} + \omega_{FT})T'_{FT}}{\omega^2 - \omega_0^2 + i\omega\Delta\omega}\right] V_\omega^2, \quad (14.16)$$

where

V_ω is the rf voltage amplitude applied to the emission line

η is the coefficient used to correct the impedance matching between the MTJ and the waveguide with a characteristic impedance of Z_0

where

$$\eta = \left(\frac{2R(\theta_0)}{R(\theta_0) + Z_0}\right)^2. \quad (14.17)$$

If the emission line and the MTJ include some parasitic impedances (capacitance in most cases), we should employ an appropriate value of η to correct this effect [22,24].

This is a type of homodyne detection and is, thus, phase sensitive. The motion of the spin, illustrated in Figure 14.3, corresponds to that excited by the spin-transfer torque at the resonance frequency. However, the motion of the spin excited by the field-like torque shows a 90° difference in phase. As a consequence, only the resonance excited by the spin-transfer torque can rectify the rf current at the resonance frequency. In Figure 14.4, the dc voltage spectra predicted for the spin-transfer torque excitation and for the field-like torque are both shown. The spectrum excited by the spin-transfer torque exhibits a single bell-shaped peak (dashed line), whereas that excited by the field-like torque is of a dispersion type (dotted line). This very clear difference provides us with an elegant method to distinguish a spin-transfer torque from a field-like torque [19–24].

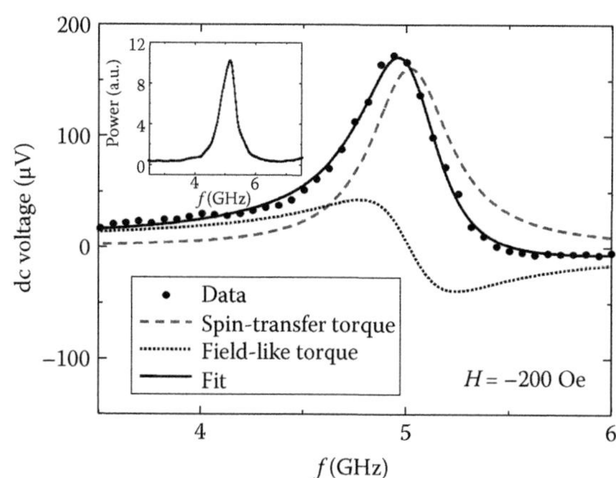

FIGURE 14.4 Spin-torque diode spectra for a CoFeB/MgO/CoFeB MTJ. Data (closed dots) are well fitted by a theoretical curve that includes contributions from both the spin-transfer torque and the field-like torque. The inset shows an rf noise spectrum obtained for the same MTJ. (After Tulapurkar, A.A. et al., *Nature*, 438, 339, 2005. With permission.)

14.2.2 Experiments and Bias Voltage Dependence of the Torque

The multilayer stack is shown in Figure 14.5a. The numbers in parentheses indicate layer thicknesses in nm units. Co–Fe/Ru/Co–Fe–B is a synthetic ferrimagnet structure, in which the magnetizations of Co–Fe and Co–Fe–B align in an antiparallel configuration. The magnetization of Co–Fe is pinned unidirectionally by an exchange-biasing field from the antiferromagnetic Ir–Mn layer. The thin MgO layer is a crystalline tunnel barrier with a (001) plane [29,30].

The thin Co–Fe–B on the MgO is a free layer whose magnetization can reverse freely by applying a magnetic field or current. The under layer consists of a Ta/Cu/Ta trilayer, which has a low resistance as a bottom lead. The multilayer was microstructured into submicron sized tunnel junctions. The cross sectional structure of the prepared MTJ is shown in Figure 14.5b. The junction with dimensions on the order of 100 nm is sandwiched between a top lead and a bottom lead. The top and bottom leads are isolated by a thick SiO_2 layer so that a current flows through only the magnetic multilayer.

The sequence of the microfabrication process is shown in Figure 14.6. The top and side views of one tunnel junction are illustrated in the process steps. The side views represent cross sections of the center part of the sample as indicated by the dotted line A–B in (Figure 14.6a). The first step was the fabrication of a bottom structure. On the surface of the multilayer, the resist pattern for the bottom lead was formed using optical lithography (Figure 14.6a). The resist pattern was transferred to the multilayer using an Ar ion etching technique and a lift-off technique (Figure 14.6b). The minimum width of the bottom lead was about 4 μm. The next step was the fabrication of a submicron sized junction and electrical contacts. On the bottom electrode, resist patterns for a small junction and large contact pads were prepared using electron beam lithography. The sizes of the junction and contact pads were 50 nm × 250 nm and 50 μm × 50 μm, respectively (Figure 14.6c). Ar ion etching was terminated at a predetermined time to stop in the Ir–Mn layer (Figure 14.6d). A thick SiO_2 layer, which was passivated on the sample surface (Figure 14.6e), was partially removed using the lift-off technique (Figure 14.6f). The final step was fabrication of a top lead. A Cr(5 nm)/Au(20 nm) double layer was sputtered. Then, the resist patterns for a top lead and two contact pads for the bottom lead were prepared using photolithography (Figure 14.6g). The shapes of the top lead and contact pads were transferred to the Cr/Au double layer using an Ar ion etching technique. Then, the remaining resist patterns were removed (Figure 14.6h). Finally, we obtained about 200 MTJs on a 20 mm × 20 mm substrate chip. The substrate chip was annealed in vacuum at 330°C under a magnetic field of 1 T. The magnetic field direction was parallel to the long axis of the junction.

The measurement setup for the spin-torque diode effect is shown in Figure 14.7. A microwave current of −15 dBm was applied to an MTJ sample through a bias-T. The microwave currents (50 MHz–15 GHz) were modulated at 10 kHz. The diode signal, that is, the ac voltage across the sample that was synchronized with the modulation, was measured using a lock-in amplifier at a dc terminal of the bias-T. A dc bias current between −1.5 and 1.5 mA was applied to the sample and the dc bias voltage was measured. By sweeping the frequency of the microwave current, the spin-torque diode spectrum (diode signal vs. microwave frequency (f)) was observed at various dc biases.

Prior to spin-torque diode measurements, the magnetoresistive properties of the sample were measured. Figure 14.8 shows magnetoresistance curves under magnetic field, H, with an angle, θ_H, of 0° (parallel to the long axis of the sample) and θ_H of 45° (tilted from the long axis of the sample) [22]. The magnetoresistance (MR) ratio defined by $(R_{AP} - R_P)/R_P \times 100$ (%) was about 150% at a low bias, where R_{AP} (R_P) denotes the tunnel resistance when the magnetizations of the Co–Fe–B layers are antiparallel (parallel) to each other, corresponding to the resistance value of point A (C) in the figure. The resistance-area product of the sample was about 2 Ω μm² at room temperature (RT). At θ_H of 0°, the resistance shows abrupt jumps at low fields because of the magnetization switching of the top Co–Fe–B free layer. At θ_H of 45°, in addition to the resistance jumps at low fields, a resistance decrease was observed at negative magnetic fields because of the rotation of the bottom Co–Fe–B magnetization. To measure the spin-torque diode effect, a finite θ_{12} is required: when θ_{12} is

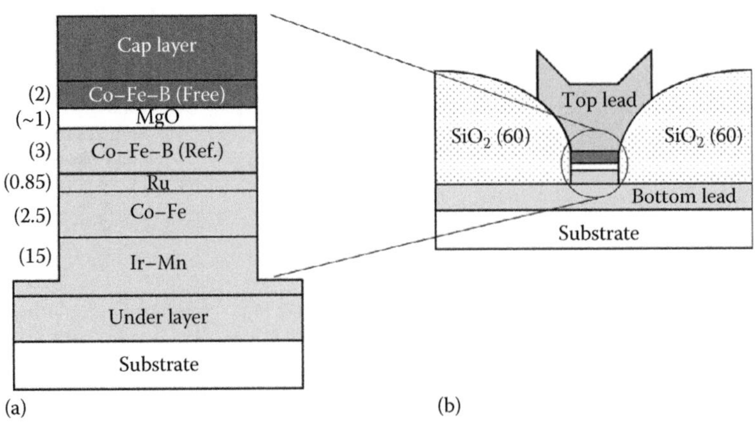

FIGURE 14.5 The multilayer stack (a) and cross sectional structure (b) of the MTJ.

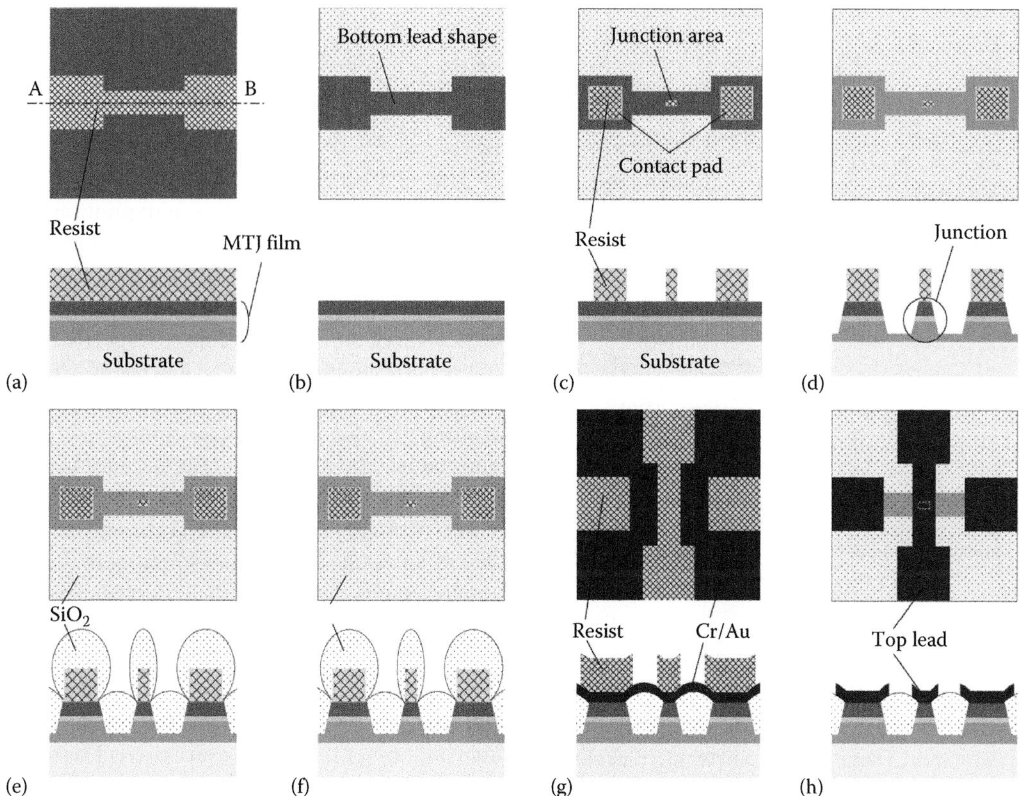

FIGURE 14.6 The sequence of steps (a–b) in the microfabrication process of the MTJs (see explanations in text).

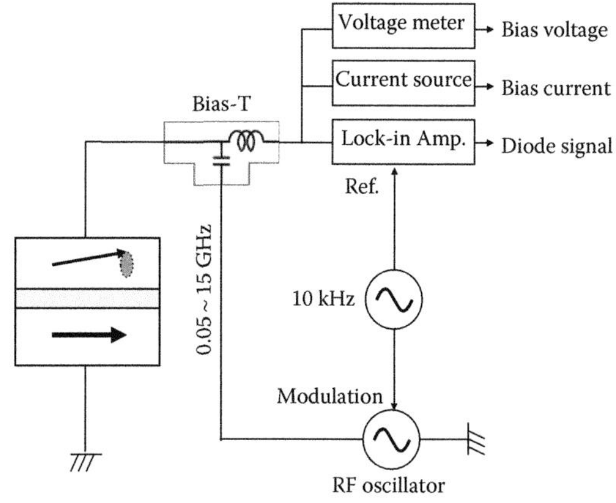

FIGURE 14.7 Schematic diagram of the setup for measuring the spin-torque diode effect.

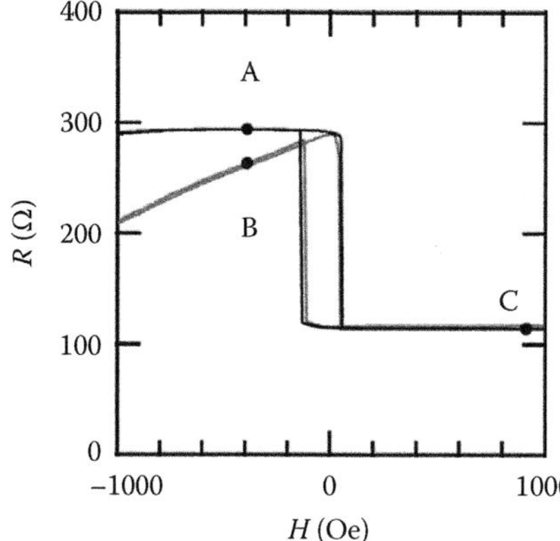

FIGURE 14.8 Magnetoresistance curves for the Ir–Mn/Co–Fe/Ru/Co-Fe-B/MgO/Co-FeB MTJ. θ_H is the angle between the magnetic fields and the long axis of the sample.

equal to 0 or π, the diode signal cancels completely. Therefore, we applied −400 Oe, in which θ_{12} is equal to 137° (point B), for the spectrum measurements. The value of θ_{12} was evaluated from Equation 14.1. To check the deviation of θ_{12} under finite biases, the bias dependence of the tunnel resistance was measured at several points. It was confirmed that the change in θ_{12} with bias was small in the range of ±400 mV.

The observed spectra are shown in Figure 14.9 [22]. At zero bias, the diode signal shows a peak at around 6.7 GHz. The peak shape is symmetric with respect to frequency. This suggests that the spin-transfer torque oscillating in phase with the microwave current was dominant. When negative (positive) biases were applied to the sample, the diode signals became larger (smaller) with the bias. The change in the peak height corresponded well with our expectation: spin precession was enhanced with a negative bias and dampened with a positive bias because of the current dependent spin-transfer torque. The peak shape also showed remarkable changes: it became asymmetric with frequency under large positive and negative biases. This asymmetry

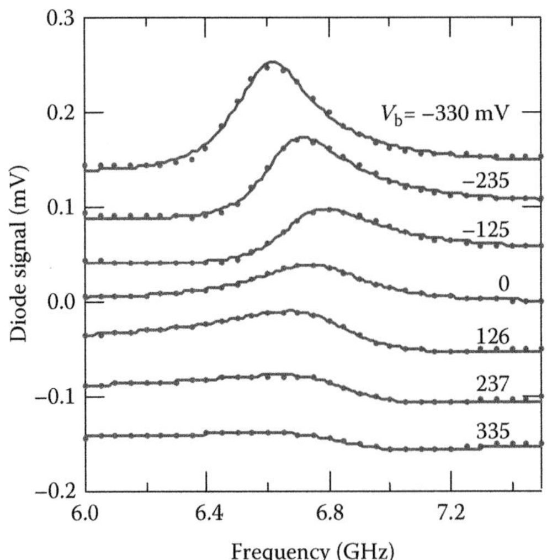

FIGURE 14.9 Spin-torque diode spectra obtained for various dc biases.

was caused by the field-like torque that oscillated out-of-phase with the microwave current. The lines in Figure 14.9 represent Equation 14.16 with appropriate parameters, which fit the experimental results very well. From the parameters, we evaluated the magnitudes of both the spin-transfer and field-like torques. Figure 14.10a shows the spin-transfer torque as a function of the bias voltage [22]. The solid circles and lines represent the experimental results and calculations, respectively. The experimental result for the spin-transfer torque shows a linear dependence around a zero bias. However, at high biases it shows a minimum of around +300 mV and a slope change of around −300 mV. This nonlinear behavior coincides with the calculated results of Theodonis et al. [14]. They calculated the bias dependence of the spin-transfer torque based on a tight binding model. In the calculation, the spin-transfer torque is expressed as

$$T_{//} = \left(\frac{1}{2}\right)[J_s(\pi) - J_s(0)]\mathbf{M}_2 \times (\mathbf{M}_1 \times \mathbf{M}_2), \qquad (14.18)$$

where

$J_s(\pi)$ ($J_s(0)$) is the spin current density for an antiparallel (parallel) magnetization configuration

$\mathbf{M}_1$ ($\mathbf{M}_2$) is the magnetization vector of the free (pinned) layer

In the calculation, $J_s(0)$ and $J_s(\pi)$ have linear and parabolic dependences on the bias voltage, respectively. The lines in the figure correspond to the calculated results with different parameter values for the minority spin onsite energy, ε, in the tight binding calculation. The linear component is large for a large ε and the parabolic component is large for a small ε. One of the calculations with ε = 2.25 eV is very close to the experimental result. Although the calculation is based on a simple model disregarding detailed electronic structures of the materials, the calculated results can explain the observed dependence well.

Similar experimental results were reported by a Cornell University group [21,23]. They reported a fairly linear bias dependence for the spin-transfer torque in a wider bias range up to about ±500 mV with a θ_{12} of close to 90°. In our analysis, the contribution of the higher order current terms was ignored but these terms may have a certain effect on the bias dependence of the torque when angle θ_{12} is far from 90°.

Figure 14.10b shows the field-like torque as a function of the bias voltage [22]. The observed field-like torque shows a quadratic bias dependence, which agrees with the calculations [14]. The calculated results again can reproduce qualitative nature and order of the intensity of the field-like torque. In this case, smaller ε values show better fit, but it is due to the over simplified model. Such a quadratic dependence of the field-like torque was also observed by other groups [21,31]. In contrast, some groups have reported that the field-like torque changes its sign when the bias

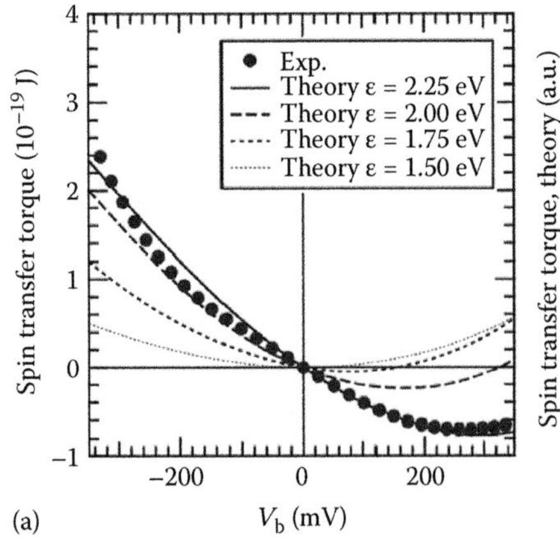

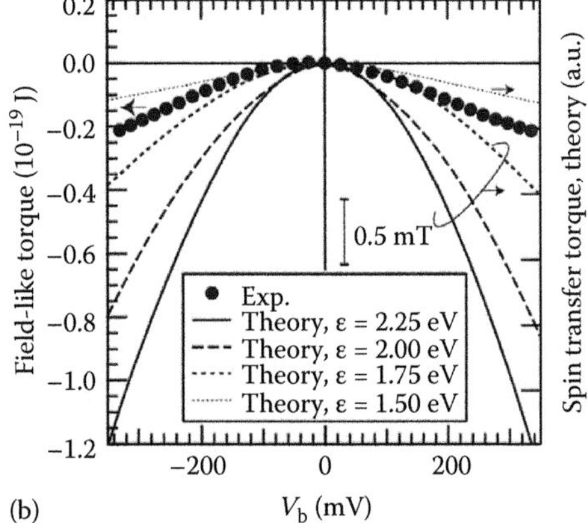

FIGURE 14.10 Bias dependence of the spin torques. (a) Bias dependence of the spin-transfer torque. (b) Bias dependence of field-like torque. Fine curves were obtained from the theoretical model calculations with different spin splitting parameters, ε. The points show the experimental results obtained for a CoFeB/MgO/CoFeB MTJ by exploiting the spin-torque diode effect. (After Theodonis, I. et al., *Phys. Rev. Lett.*, 97, 237205(1–4), 2006; Kubota, H. et al., *Nat. Phys.*, 4, 37, 2008. With permission.)

current direction is reversed. Devolder et al. analyzed an asteroid loop displacement with bias and separated the contributions of the field-like torque, heating, and spin-transfer torque. They concluded that the field-like torque has a linear dependence on the bias [32]. Li et al. also reported that the sign of the field-like torque depends on the bias direction [33]. They measured spin-transfer switching under various magnetic fields and current pulse widths. The magnitude of the observed field-like torque was proportional to the product of the current density and absolute value of the voltage. They explained the dependence qualitatively by taking into account the energy dependent inelastic tunneling. It should be noted here that because these results are affected by a charge current induced Oersted field and Joule heating, the results may vary depending on the sample geometry. By contrast, our experiment uses phase-sensitive detection technique; therefore, such artifacts do not affect the observed results. Further investigations should be performed carefully to conclude the bias dependence of the field-like torque.

14.3 Spin Injection Magnetization Switching in MTJs

14.3.1 Critical Current and Switching Time

In MTJs, SIMS may occur similar to that in CPP-GMR nano-pillars (see Chapter 7). Minor but important differences are in the angular dependences of the torque intensity and in the junction resistances. The much larger resistance in the MTJ results in much larger temperature raise during switching because of Joule heating. The critical current (instability current) at zero temperature is obtained easily by setting $\Delta\omega$ in Equation 14.15 to zero.

$$I_{c0} = \alpha S_2 \left(\frac{\omega_{\theta\theta} + \omega_{\phi\phi}}{2} \right) \frac{G(\theta)}{T_{ST}}. \tag{14.19}$$

Many efforts have been paid to reduce I_{c0}. Since above expression is essentially the same as in the CPP-GMR nano-pillars, the same techniques that were effective are also useful in MTJs. The first attempt is to reduce total spin angular momentum, S_2, in the free layer. SIMS requires effective injection of spin angular momentum that is equal to that in the free layer. Therefore, reduction in S_2 results in a reduction in I_{c0}. Reduction in the free layer thickness reduces S_2 and I_{c0}. S_2 can also be reduced by reducing the magnetization of the ferromagnetic material. Especially in the nano-pillar with in-plane magnetization, since magnetization also affects the size of the anisotropy field $(-\omega_{\phi\phi}/\gamma)$, I_{c0} is a quadratic function of the magnetization. Therefore, reduction in the magnetization is very effective [34]. Second attempt is to use double spin filter structure. This method was originally proposed by Berger [35]. By using this structure, Huai et al. observed substantial reduction of the threshold current to 2.2×10^6 A/cm^2 [36]. Third attempt is to use perpendicular magnetic anisotropy, which can reduce the size of the anisotropy field [37]. Yakata et al. pointed that a free layer with an Fe rich composition in FeCoB/MgO/FeCoB stacking possesses a perpendicular crystalline anisotropy and results in reduction of switching current [38]. The perpendicular crystalline anisotropy partially cancels demagnetization field and reduces the size of the anisotropy field. T. Nagase et al. obtained a significant reduction of the threshold current under the required thermal stability factor for the MTJ nano-pillars with the CoFeB/[Pd/Co]x2/Pd free layer and FePt/CoFeB pinned layer [39]. In this structure the films have perpendicular magnetization. With relatively small anisotropy field, small magnetization can be the origin of the reduction.

The switching time can be obtained by integrating the LLG equation with the spin-transfer torque (Equation 14.7). For an MTJ with orthogonal symmetry, the equation of motion for the opening angle of precession, θ, is derived from Equation 14.7 as follows:

$$\frac{d\theta}{dt} \cong -\alpha(-\gamma)\left(\langle H_{ani}\rangle \cos\theta - H_{ext}\right)\sin\theta - \frac{T_{ST}V}{S_2}\sin\theta + (-\gamma)H_{stochastic}(t), \tag{14.20}$$

where $H_{stochastic}$ is a stochastic field that expresses thermal agitation. In the present treatment, we neglect $H_{stochastic}$. H_{ext} is a z-direction external field. $\langle H_{ani}\rangle = (H_{ani,x} + H_{ani,y})/2$ is an averaged anisotropy field. For an in-plane magnetized film, as shown in Figure 14.1, those fields are the in-plane coercive force ($=H_{ani,x}$) and the out-of-plane demagnetization field ($=H_{ani,y}$). To obtain the previous equation, the so-called "one-cycle average" of the LLG equation was taken. The approximation is exact if $H_{ani,x} = H_{ani,y}$, that is, the uniaxial symmetry case. The switching time is obtained by integrating the equation [25]:

$$\tau_{sw} \cong \frac{I_{c0}}{I - I_{c0}} \frac{1}{\alpha(-\gamma)\langle H_{ani}\rangle} \log\left(\frac{2}{\delta\theta_0}\right), \tag{14.21}$$

where $\delta\theta_0$ is the angle between the direction of magnetization and the easy axis at the beginning of the current application. For the thermal equilibrium, the average value of $\delta\theta_0$ is estimated as

$$\delta\theta_0 = \sqrt{\frac{k_B T}{\mu_0 M_2 H_{ani,x}}} = \sqrt{\frac{1}{2\Delta}}, \tag{14.22}$$

where

k_B is the Boltzmann constant
T is the absolute temperature
Δ is the thermal stabilization factor

Koch et al. [40] and Li and Zhang [41] explained the reduction in the switching current for long pulse durations by considering the thermal excitation of the macrospin fluctuation [40] induced by the stochastic field, $H_{stochastic}(t)$. Because of the stochastic field, the direction of the free layer spin angular momentum, $\vec{\mathbf{e}}_2$, becomes also a stochastic variable. Therefore, we can only discuss its probability density, $p(\vec{\mathbf{e}}_2,t)$, by taking into account the statistical nature of $H_{stochastic}(t)$. From the fluctuation–dissipation theorem [42], $H_{stochastic}(t)$ in Equation 14.20 is related to the Gilbert damping constant:

$$\alpha \cong \frac{(-\gamma)\mu_0 M_2}{k_B T}\frac{1}{2}\int_{-\infty}^{\infty} d\tau \langle H_{stochastic}(t) H_{stochastic}(t+\tau)\rangle, \tag{14.23}$$

where ⟨⟩ expresses a thermodynamical ensemble average. Using this, the time evolution of the probability density is expressed as follows:

$$\begin{cases} \dfrac{\partial}{\partial t}p(z,t) + \dfrac{\partial}{\partial z}\left\{-\dfrac{\alpha}{S_2}(1-z^2)\left[\dfrac{dE_{eff}(z)}{dz} + k_B T\dfrac{\partial}{\partial z}\right]p(z,t)\right\} = 0, \\ E_{eff}(z) = \mu_0 M_2 H_{ext} z - \dfrac{1}{2}\mu_0 M_2\left(\dfrac{H_{ani,x} + H_{ani,y}}{2}\right)z^2 - \dfrac{z}{\alpha}T_{ST}V, \end{cases} \tag{14.24}$$

where

$E_{eff}(z)$ is the effective magnetic energy

$z = \cos\theta$ is the z-component of $\vec{\mathbf{e}}_2$

The term that includes $dE_{eff}(z)/dz$ is the contribution from the drift driven probability current and is proportional to the sum of three torques, namely, the precession, damping, and spin transfer. The term with $k_B T(\partial/\partial z)$ expresses the contribution from the thermal diffusion current of the probability density and is proportional to the gradient of the probability density. This is the Fokker–Planck equation [42,43] adapted for the LLG equation including the spin-transfer torque [44,45].

Without a charge current, Equation 14.24 has the Boltzmann distribution, $p(\vec{\mathbf{e}}_2) \propto e^{-E_{mag}(\vec{\mathbf{e}}_2)/k_B T}$, as a solution for the thermal equilibrium. If the system is out of equilibrium because of a charge current application, the probability density currents start to flow to approach a steady state.

The probabilistic distribution of the switching time, $p_{sw}(t)$, and its integration, $P_{sw}(t) = \int_0^t p_{sw}(t_1)dt_1$, in the thermal activation regime can be obtained by applying Kramers's method [43] to Equation 14.24, for a case with a high thermal barrier, which corresponds to a large thermal stability factor, small current, and small external field, as follows:

$$\begin{cases} p_{sw}(t) = \tau_{sw}^{-1} e^{-t/\tau_{sw}}, \\ P_{sw}(t) = 1 - e^{-t/\tau_{sw}}, \\ \tau_{sw}^{-1} \cong \tau_0^{-1} e^{-\Delta(1 - I/I_{c0} - H_{ext}/H_c)^2}, \end{cases} \tag{14.25}$$

where

t is the elapsed time

τ_{sw} is the average switching time

$\tau_0^{-1} = \alpha(-\gamma)\langle H_{ani}\rangle\sqrt{\Delta/\pi}$ is the attempt frequency

Δ is a thermal stability factor calculated for zero current and zero external field

The reader can find that Néel–Brown's exponential law is valid under a current injection. Li and Zhang solved the Fokker–Planck equation for in-plane magnetization with a small current and proposed a slightly different formula to also include parameter β [41]. In their paper, the term $(1-(I/I_{c0})-(H_{ext}/H_c))^2$ in Equation 14.25 is replaced with $(1-(H_{ext}/H_c))^{\beta}(1-(I/I_{c0}))$. The theoretical curve derived on the basis of the aforementioned theory fits the experimental data well when reasonable parameter values are used.

In the thermal activation regime, SIMS can occur at currents smaller than I_{c0} because of thermal activations. Moreover, the switching time is distributed widely following an exponential law for a given applied current. In another way, if we apply current pulses with constant widths but different heights (current), we will see the following switching current distribution:

$$p_{sw}(I) = \frac{2\Delta}{I_{c0}}\frac{\tau_{pulse}}{\tau_{sw}} e^{-\frac{\tau_{pulse}}{\tau_{sw}}}, \tag{14.26}$$

where

τ_{pulse} is the width of the applied pulse current

τ_{sw} is a function of the applied current (Equation 14.25)

Experimentally, I_{c0} and Δ can be determined either from the pulse width dependence of the average switching current or from the switching current distribution measured using pulses with constant width.

14.3.2 Probabilistic Switching under Long Duration Pulse Current

In a magnetic recording system such as a hard disk drive (HDD), data retention is guaranteed based on the probabilistic magnetization reversal model with a thermal activation process [46]. Spin-transfer switching is also probabilistic and governed by the thermal activation, which was discussed experimentally by Koch et al. [40] and theoretically by Li and Zhang [41]. When the current pulse duration is longer than the inverse of the attempt frequency, thermal activation plays an important role in the switching. The observed switching current I_c at a finite temperature is always reduced from the intrinsic switching current I_{c0} that corresponds to a switching current without thermal activation. In the spin dynamics, including the spin-transfer torque, I_{c0} is the current at which the effective damping constant becomes zero. Since I_c depends on the measurement condition, I_{c0} is a very important quantity for characterizing samples. The thermal stability index, Δ, is another important parameter of the switching, which is defined as $\Delta = K_u V/(k_B T)$. Here, K_u, V, k_B, and T correspond to the uniaxial magnetic anisotropy, free layer cell volume, Boltzman constant, and temperature, respectively. A higher Δ corresponds to a lower switching probability, resulting in a longer data retention time in memory applications. In the thermally activated spin-transfer switching model proposed by Li and Zhang [41], the switching probability is expressed as

$$P_{sw} = 1 - \exp\left\{-\left(\frac{t}{\tau_0}\right)\exp\left[-\Delta\left(\frac{1-I}{I_{c0}}\right)\left(\frac{1-H}{H_{c0}}\right)^2\right]\right\}, \quad (14.25')$$

where t, τ_0, I, H, and H_{c0} are the switching time, inverse of the attempt frequency, current, external magnetic field, and intrinsic coercivity, respectively. With this model, two methods are used to evaluate I_{c0} and Δ. The first method is the observation of the pulse duration dependence of the switching current [47]. The switching current is plotted as a function of $\ln(t_p/\tau_0)$, where t_p usually ranges from 10^{-1} to 10^{-5} s. By extrapolating to τ_0 (=1 ns), we can obtain I_{c0}. From the slope of the extrapolating line, we can obtain Δ. Another method is the observation of the switching current distribution at a constant pulse duration [48,49]. By measuring the switching current repeatedly, we can obtain the switching current distribution. Theoretically, this distribution is expressed as dP_{sw}/dI, which has a peak at around a mean value. By fitting the function to experimental values, we can evaluate I_{c0} and Δ.

Figure 14.11a shows the measurement setup of the switching current [50]. A lock-in amplifier is used to measure the junction resistance at a zero dc bias. The pulse generator produces sequential pulse currents as shown in Figure 14.11b, which were applied to a sample. The sample was a CoFeB/MgO/CoFeB MTJ, which had a structure similar to the one used in the previous section (Figure 14.5). Figure 14.12a shows the magnetoresistance (R–H) loop of the tunnel junction. The resistance jumps at low fields correspond to the magnetization reversal of the top 2-nm-thick CoFeB free layer. The bottom 3 nm thick CoFeB layer is a reference layer, whose magnetization does not reverse due to strong exchange biasing. When the free layer magnetization is parallel (P) to that of the reference layer, the resistance is as low as 320 Ω. When the free layer magnetization is antiparallel (AP) to that of the reference layer, the resistance is as high as 750 Ω. Figure 14.12b shows a resistance-current (R–I) loop of the sample. In this plot, the resistance was measured after each pulse current. The resistance jumps abruptly at 0.6 and −0.4 mA. The resistance change by current agrees well with that in the magnetoresistance loop as shown in Figure 14.12a. This proves that the free layer magnetization is completely switched by the current.

Using this sample, R–I loops were measured repeatedly 100 times. The switching currents were distributed from P to AP and from AP to P, as shown in Figure 14.13. The I_c's values for P to AP switching were distributed around 0.7 mA, with those for AP to P switching around −0.4 mA. The lines in the figure represent theoretical fits based on Equation 14.19. From the theoretical fit, we obtained I_{c0} and Δ values of 1.3 mA and 41 for the P to AP

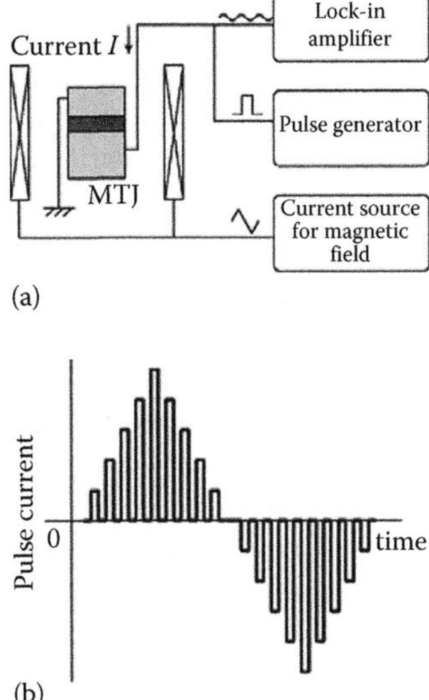

FIGURE 14.11 Schematic diagram of the setup for observing spin-transfer switching (a) and the pulse current sequence (b). (After Kubota, H. et al., *Oyo Buturi*, 78, 232, 2009. With permission.)

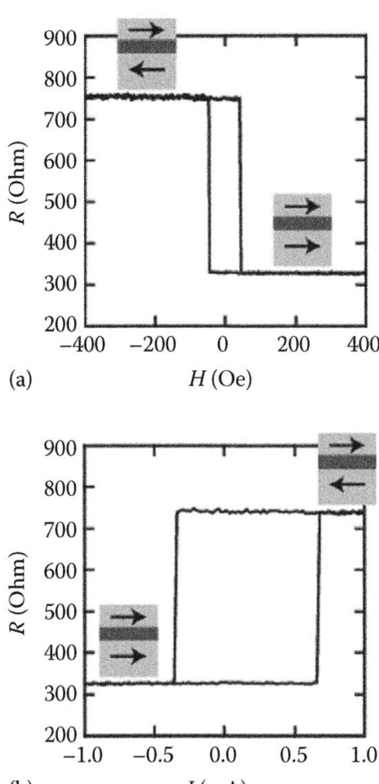

FIGURE 14.12 A R–H loop (a) and a R–I loop (b) of the CoFeB/MgO/CoFeB MTJ. (After Kubota, H. et al., *Oyo Buturi*, 78, 232, 2009. With permission.)

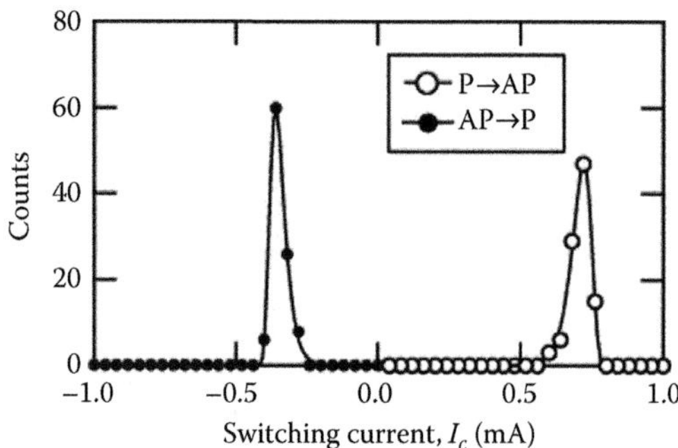

FIGURE 14.13 Distribution of switching currents for the spin-transfer switching in CoFeB/MgO/CoFeB MTJ. (After Kubota, H. et al., *Oyo Buturi*, 78, 232, 2009. With permission.)

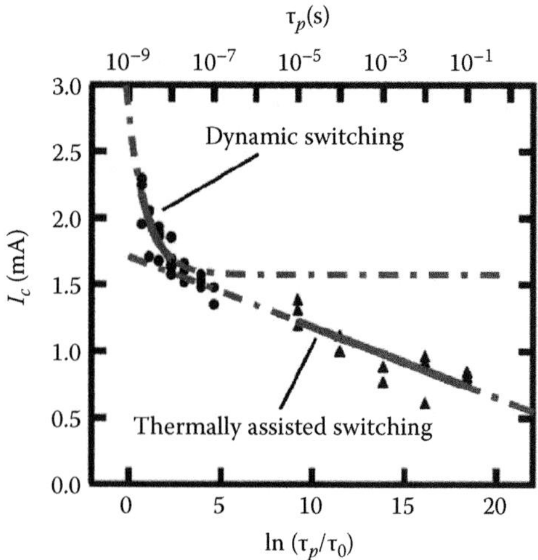

FIGURE 14.14 Pulse width (τ_p) dependence of the average critical current for the SIMS in a CoFeB/MgO/CoFeB MTJ. The junction area, free layer thickness, resistance-area product, and MR ratio are 110 nm × 330 nm, 2 nm, 17 Ω μm², and about 115%, respectively. Measurements were performed at RT using electric current pulses with durations ranging from 2 ns to 100 ms. Dotted curved line and straight line are fittings using Equations 14.21 and 14.25, respectively. (After Aoki, T. et al., *J. Appl. Phys.*, 103, 103911, 2008. With permission.)

switching and −0.8 mA and 33 for the AP to P switching, respectively. I_{c0} for AP to P is smaller than that for P to AP. This difference can be explained by the difference of $G(\theta)$ between AP and P states as shown in Equation 14.19. The difference of Δ between P and AP states probably originates in the different effective fields in the free layer cell. During *R–I* switching experiment, we adjusted the external field to the center of hysteresis in Figure 14.12a. However, due to magnetostatic coupling at cell edges or through rough interfaces, local fields would be nonuniform and the effective field can be different depending on the magnetization arrangement.

In our experience, the two earlier-mentioned methods give almost the same value for I_{c0} and Δ. The advantage of the latter method, measuring the switching current distribution, is the simplicity of the measurement setup. Since a rather long pulse can be used, impedance matching in a circuit is not very important. On the other hand, when we use the former method, the pulse duration dependence of the switching current, the short pulse measurement requires impedance matching in the circuit. Otherwise, the pulse current waveform is degraded, resulting in indistinct current values.

14.3.3 High-Speed Measurements

To analyze high-speed dynamics of SIMS in MTJs, time domain high-speed electrical observations have been performed [51–53]. An example of the pulse width dependence of the switching current for a CoFeB/MgO/CoFeB magnetic nano-pillar is shown in Figure 14.14. A clear transition from the dynamic switching regime to the thermally assisted switching regime can be seen.

Figure 14.15 shows the circuit used to observe real time switching signal (Figure 14.15a) and an example of a switching signal obtained by Tomita et al. [53]. For this measurement, a CoFeB/MgO/CoFeB nano-pillar was placed at the end of an rf wave guide. A 10 ns pulse from a pulse generator was supplied to the MTJ through a power divider. The pulse was partially reflected by the MTJ depending on its resistance states and was provided to a high-speed storage oscilloscope with a 16 GHz bandwidth after being made to pass through the power divider again. Figure 14.15b shows both the direct pulse signal and the reflected pulse signal. Since the pulse reflectivity is dependent on the sample resistance, a magnetization switching event can be observed as a step in the reflected signal, as shown in the inset of Figure 14.15b. To cancel the parasitic signals generated by pulse reflections from the pulse generator and the cable connections, the obtained signal was numerically divided by the signal obtained in the absence of switching; the resulting signals are shown in Figure 14.16 for a current slightly below I_{c0}. The signal shows a long waiting time followed by a short transition time (about 500 ps) as expected for a thermally assisted SIMS. The waiting time seems to vary considerably. Tomita et al. [53] also analyzed 1000 single-shot data and obtained the distribution of the non-switched probability, $1 - P_{sw}(t)$, which should be a simple exponential function of the elapsed time according to Néel–Brown's law. The obtained distribution was not explained by simple Néel–Brown's law but was well fitted by considering a certain initial period in which the switching probability was negligibly small, as shown in Figure 14.17. Tomita et al. [53] called this initial period a "nonreactive time (t_{NR})" and explained that this was the time required to complete a transition from the initial thermal equilibrium without current to a quasi-thermal equilibrium state for a finite current. Considering this point Equation 14.25 is rewritten as follows:

$$P_{sw}(t) = 1 - e^{-\frac{t - t_{NR}}{\tau_{sw}}}. \qquad (14.25'')$$

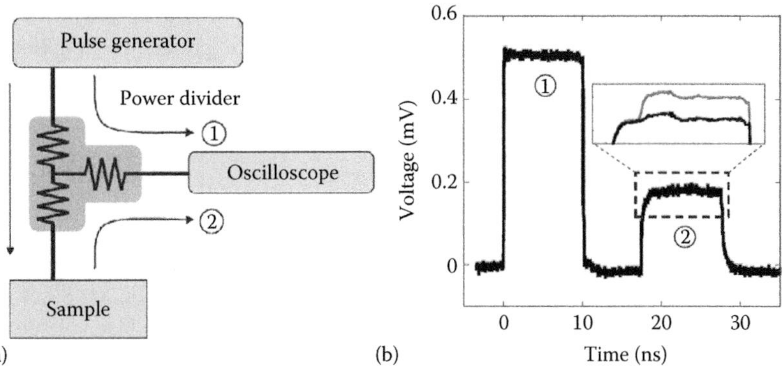

FIGURE 14.15 (a) Electric circuit and (b) an example of the obtained signal for a real-time single-shot observation of the SIMS process. Since an MTJ terminates an rf wave guide with a certain impedance mismatch, the supplied voltage pulse is partially reflected by the MTJ and observed by using a high-speed single-shot oscilloscope with a 16 GHz bandwidth. The amplitude reflectivity of the MTJ is $(R(\theta) - Z_0)/(R(\theta) + Z_0)$, where $R(\theta)$ is the MTJ resistance and Z_0 is the characteristic impedance of the wave guide (50 Ω). The magnetization switching during an application of a current pulse causes a change in the reflectivity, and is observed as a jump in the reflected voltage wave, as shown in the inset in (b). In the inset the gray (black) signal shows reflected pulse signal with(without) jump because of magnetization switch (P to AP). (After Tomita, H. et al., *Appl. Phys. Express*, 1, 061303(1–3), 2008. With permission.)

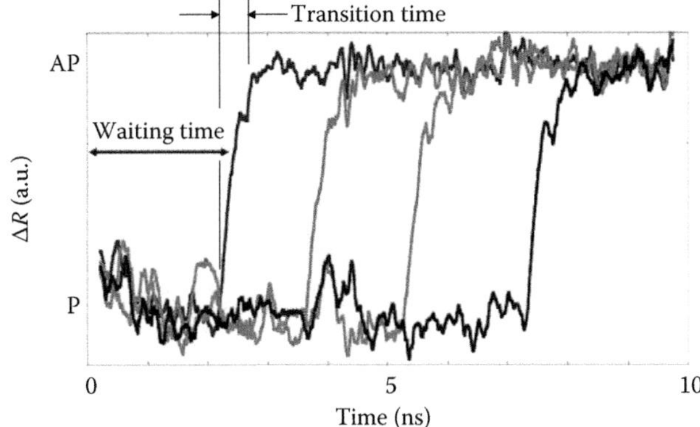

FIGURE 14.16 Real-time, single-shot observation of the SIMS process in a CoFeB/MgO/CoFeB nano-pillar. All the traces are obtained at the same current but show differences because of the probabilistic nature of switching. The current magnitude is 2.7 mA (2.4×10^7 A/cm^2), which is slightly smaller than I_{c0}. The signals are normalized by the signal obtained without switching. (After Tomita, H. et al., *Appl. Phys. Express*, 1, 061303(1–3), 2008. With permission.)

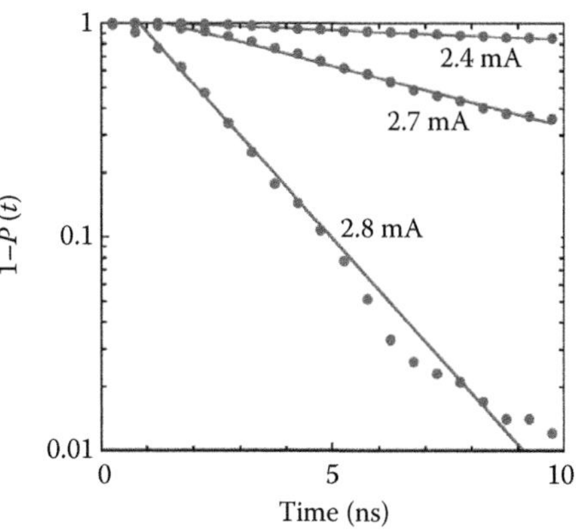

FIGURE 14.17 Non-switched probability, $1 - P_{sw}(t)$, as a function of elapsed time. The data deviate from Néel–Brown's law at the initial stages and are well fitted by the function $1 - P_{sw}(t) = e^{-(t-t_{NR})}$, where t_{NR} is the nonreactive time. (After Tomita, H. et al., *Appl. Phys. Express*, 1, 061303(1–3), 2008. With permission.)

14.4 Magnetic Noise and Spin-Transfer Auto-Oscillation in MTJs

14.4.1 Spin-Transfer Effect on the Thermally Excited FMR (Mag-Noise)

Since the magneto-resistance effect in MTJs is large, the thermal fluctuation in the magnetization in MTJs produces considerable electric noise under a dc bias. The noise is called "magnetic-noise" or "mag-noise." The noise has its peak in intensity at the FMR frequency (typically several GHz). The peak accompanies asymmetrical foots toward lower frequency side (larger) and higher frequency side (smaller). The most advanced magnetic read heads in HDDs require a wide signal band in order to assure a high data transfer rate, and they are subject to mag-noise. When the system noise is governed by the mag-noise, an enhancement in the MR ratio does not enhance the signal to noise ratio (S/N) because both the signal and the noise are proportional to the MR ratio. Therefore, understanding the mechanism of generation and control of mag-noise in MTJs is important [54–64].

For a given fluctuation of the relative angle between two spin moments, $\delta\theta(t)$, around its average angle θ_0, the output noise voltage is

$$V_{noise}(t) = \frac{R(\theta_0)}{R(\theta_0) + Z_0} I \left.\frac{dR(\theta)}{d\theta}\right|_{\theta=\theta_0} \delta\theta(t), \tag{14.27}$$

where

$R(\theta_0) = G^{-1}(\theta_0)$ is the resistance of the MTJ
I is the bias current

For a small excitation of the thermal noise, the linearized LLG equation, including the spin-torque term (Equation 14.14) can be adapted by replacing the external excitation fields δH_θ and δH_ϕ with stochastic fields $H_{stochastic,\theta}$ and $H_{stochastic,\phi}$, respectively. If the stochastic field has a very short correlation time and a white spectrum in the GHz region, $\delta\theta(t)$ and $V_{noise}(t)$ show a resonance type Fourier spectrum that is determined by the response function. From Equation 14.14, the power density spectrum (PSD) of $\delta\theta(t)$ is expressed as

$$PSD_{\delta\theta}(\omega) = \frac{(\omega_{\phi\phi} + \omega_{FT})^2 + (\omega_{\theta\varphi} - \omega_{ST})^2 + \omega^2}{\left(\omega^2 - \omega_0^2\right)^2 + (\omega\Delta\omega)^2} \times \int d\tau \langle (-\gamma) H_{stochastic}(0)(-\gamma) H_{stochastic}(\tau) \rangle. \quad (14.28)$$

The PSD of the noise voltage, $V_{noise}(t)$, is obtained by multiplying

$$\frac{1}{Z_0}\left(\frac{R(\theta_0)}{R(\theta_0) + Z_0} I \left.\frac{dR(\theta)}{d\theta}\right|_{\theta=\theta_0}\right)^2$$

with the previous equation. In the inset of Figure 14.18a, a typical "mag-noise" spectrum in Fe/MgO/Fe epitaxial MTJ is shown

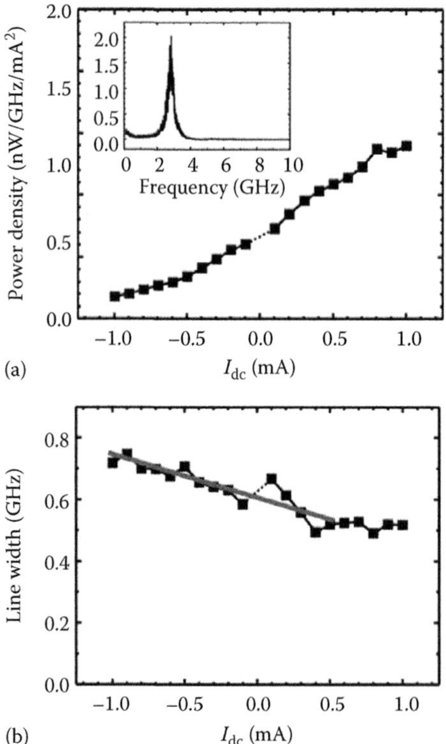

FIGURE 14.18 Rf noise properties observed in Fe/MgO/Fe single crystalline MTJs in the parallel configuration. The junction size is 220 nm × 420 nm. (a) Peak power density as a function of the biasing current. The power is normalized by I^2. (b) FWHM as a function of the biasing current. Inset: A typical noise power spectrum (I = 0.8 mA). (After Matsumoto, R. et al., *Phys. Rev. B*, 80, 174405(1–8), 2009. With permission.)

[57]. The peak frequency, $\omega_0/(2\pi)$, and the FWHM of the peak, $\Delta\omega/(2\pi)$, are expressed by Equation 14.15. In the linear regime, the width of the peak is a linear function of the bias voltage:

$$\Delta\omega = \alpha Tr\left[\hat{\Omega}\right] + 2\frac{T_{ST}V}{S_2}\cos\theta_0. \quad (14.29)$$

The peak width at a zero bias voltage is determined by the Gilbert damping factor, α. Since the injected spin-polarized current under a finite bias voltage causes a forced damping or anti-damping depending on the polarity of the current, the bias voltage contributes to an increase or decrease in the peak width depending on its sign. In Figure 14.18b, the linear dependence of $\Delta\omega/(2\pi)$ on the biasing current is shown for a parallel state. The peak height in the PSD also shows a similar dependence on the biasing current, as shown in Figure 14.18a.

According to the fluctuation-dissipation theorem (Equation 14.23), the stochastic field is related to the temperature and the damping constant. By using this relation, we can obtain the following "quasi-equi-partition theorem," which relates the integrated noise power to the "quasi-spin-temperature":

$$\begin{cases} \left\langle \delta\theta^2 \right\rangle = \dfrac{1}{2\pi}\displaystyle\int_{-\infty}^{+\infty} d\omega PSD_{\delta\theta}(\omega) = \dfrac{k_B T^*}{S_2\omega_{\theta\theta}}, \\ \Delta E_{mag,\theta} = \dfrac{\mu_0 M_2 \omega_{\theta\theta}}{2(-\gamma)}\left\langle \delta\theta^2 \right\rangle = \dfrac{1}{2}k_B T^*, \\ T^* = T\dfrac{I_{c,0}}{I_{c,0} - I}. \end{cases} \quad (14.30)$$

Here, orthogonal symmetry of the system is assumed. This equation shows that in the absence of the biasing current, the thermal energy shared with the θ–coordinate in the macro-spin system, $\Delta E_{mag,\theta}$, is equal to $k_B T/2$. This is a consequence of the equipartition theorem. Since the injection of spin-polarized current can amplify or reduce the fluctuations, the magnetic energy of the free layer can be controlled by the current. In order to express this phenomena, we introduce quasi-spin-temperature, T^*, which is defined in Equation 14.30. The magnetic energy of the system is dependent on the applied current and is equal to $k_B T^*/2$. It should be noted that in this case, the change in magnetic fluctuation does not cause a change in the S/N. Since the effects of the current on the response functions to the stochastic and signal fields are identical the S/N remains constant, as far as the current has no direct effect on the stochastic field. If there is an effect like the spin-Peltier effect, we can control the S/N by current injection.

14.4.2 Spin-Transfer Auto-Oscillation in MTJs

A large current injection may result not only in magnetization switching but also in a continuous precession of magnetization.

This spin-transfer oscillation was first observed in CPP-GMR nano-pillars and nano-contacts (see Chapter 7). The maximum output power was limited to about 1 nW because of its small MR ratio. In order to obtain higher output power, it is important to use pillars with a large MR ratio and with a resistance that matches the impedance of the wave guide (usually 50 Ω). By replacing a CPP-GMR pillar with an MTJ, we can expect a significant increase in the output power [55,56]. Deac et al. reported a large output power of 0.14 μW from a single CoFeB/MgO/CoFeB nano-pillar of cross section 70 nm × 160 nm [55]. The junction was specially designed to have a high MR ratio of about 100% and a very low resistance-area product ($RA = 4\,\Omega\,\mu m^2$) [65]. Figure 14.19 shows the output power spectra obtained from a CoFeB/MgO/CoFeB nano-pillar. With an increase in current, a steep increase in the power of the first harmonic peak at around 6 GHz can be seen.

Clearer evidence of the on-set of auto-oscillation is shown in Figure 14.20. When the injection current is less than I_{c0}, an increase in the injection current results in a linear reduction in the peak width, as in the previous case (Figures 14.18b and 14.20b). The threshold current (I_{c0}), which is indicated by an arrow in Figure 14.20b, corresponds to the current at which the peak width reduces to zero, if a leaner reduction holds until I_{c0}. In practice, when the injection current is around the threshold current, there is a sudden increase in the peak width. The peak width has its maximum value slightly below I_{c0}. Further increase in the injection current reduces the peak width and results in a sudden increase in the output power. These observations provide clear evidence of the threshold properties, which are in good agreement with the theory developed by Kim et al. [59]. The width of the spectral lines are, however, very wide when compared to the width of the spectral lines for the CPP-GMR nano-pillars and nano-contacts. This is explained by the findings of Houssameddine et al.: they observed the time-domain oscillation signals and found spontaneous mode jumps that may significantly increase the averaged line width [64]. The mechanism of the mode jump was discussed by considering the current distribution, instability mechanism, and inhomogeneities in MTJs. Further studies are required in order to realize stable oscillation.

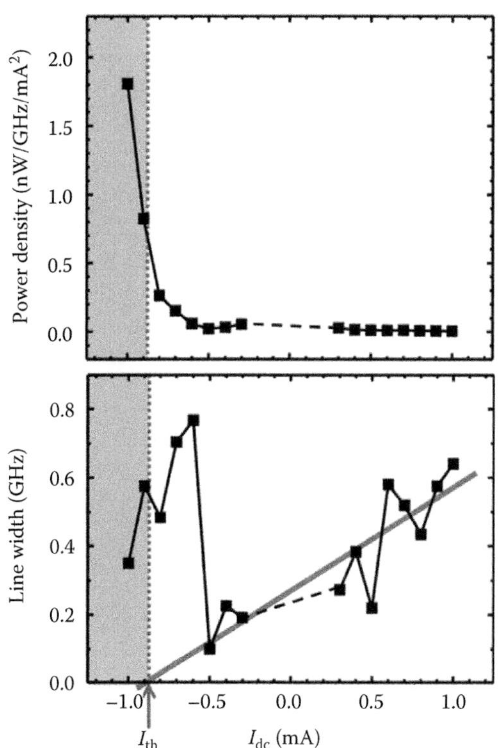

FIGURE 14.20 Rf auto-oscillation properties observed in Fe/MgO/Fe single crystalline MTJs (AP state). The junction size is 220 nm × 420 nm. (a) Peak power density as a function of the biasing current. The power is normalized by I^2. (b) FWHM as a function of the biasing current. (After Matsumoto, R. et al., *Phys. Rev. B*, 80, 174405(1–8), 2009. With permission.)

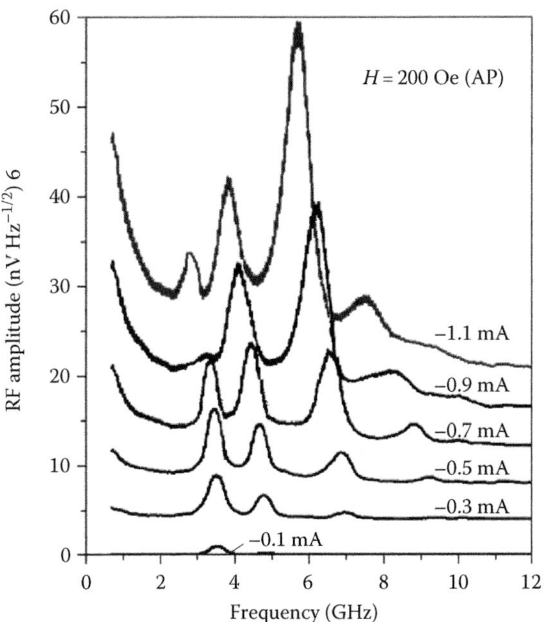

FIGURE 14.19 Rf power spectrum observed for a nano-pillar prepared from a low resistance CoFeB/MgO/CoFeB MTJ with in-plane magnetization (AP state). Large rf power emission and a steep increase in the first harmonic (around 6 GHz) intensity with an increase in the injected current were observed. (After Deac, A.M. et al., *Nat. Phys.*, 4, 803, 2008. With permission.)

Another method to obtain higher output power is to synchronize the number of oscillators. Kaka et al. [66] and Mancoff et al. [67] demonstrated the synchronization of two metal-based oscillators by an exchange coupling. For this purpose they developed two magnetic oscillators that were very close to each other (separation was about 100 nm). On the other hand, Georges et al. proposed an electric coupling among the oscillators [68]. Such a coupling would be feasible if each oscillator has sufficient power output. This is an excellent idea for enabling the use of MTJs as rf devices.

Appendix

14.A.1 Elliptical Nano-Pillar with an In-Plane Magnetization

For the in-plane magnetization case, we assume that $\vec{e}_1$ is fixed parallel to the easy axis of the free layer ($\vec{e}_1 = (0, 0, 1)$) and that an external field is applied in-plane ($\vec{H}_{ext} = H_{ext}(\sin, {}_{ext}, 0, \cos, {}_{ext})$). Here, the coordinate system in Figure 14.1 is used. Then, $\vec{e}_2^{(0)}$ is aligned almost opposite to the external-field direction, $\vec{e}_2^{(0)} = (\sin\theta_0, 0, \cos\theta_0)$. We neglect the small deviation of $\vec{e}_2^{(0)}$ caused by dc voltage application. In this case, the magnetic energy and the effective field are

$$E_{mag} = \mu_0 M\left(\vec{H}_{ext}\cdot\hat{s}_2^{(0)} + \frac{1}{2}H_{//,H}\left(\hat{e}_x\cdot\hat{s}_2^{(0)}\right)^2 + \frac{1}{2}H'_\perp\left(\hat{e}_y\cdot\hat{s}_2^{(0)}\right)^2 + \frac{1}{2}H_{//,E}\left(\hat{e}_z\cdot\hat{s}_2^{(0)}\right)^2\right),$$

$$\vec{H}_{eff}^{(0)} = \frac{1}{\mu_0 M}\frac{\partial E_{dipole\_and\_anisotropy}}{\partial\hat{s}_2^{(0)}} = \vec{H}_{ext} + \begin{pmatrix} H_{//,H} & 0 & 0 \\ 0 & H'_\perp & 0 \\ 0 & 0 & H_{//,E}\end{pmatrix}\hat{s}_2^{(0)}, \tag{14.A.1}$$

where $H_{//,H}$, $H'_\perp$, and $H_{//,E}$ are the demagnetization fields for the x-, y-, and z-directions, respectively.

From Equation 14.13, the $\hat{\Omega}$ matrix elements in ($\vec{e}_\theta, \vec{e}_\phi$) coordinate system are calculated as follows:

$$\hat{\Omega} = \gamma H_{ext}\cos(\theta_{ext} - \theta_0) - \gamma\begin{pmatrix} H_{//}\cos 2\theta_0 & 0 \\ 0 & H_\perp - H_{//}\sin^2\theta_0\end{pmatrix}, \tag{14.A.2}$$

where $H_{//} \equiv H_{//,H} - H_{//,E}$ and $H_\perp = H'_\perp - H_{//,E}$ are in-plane and out-of-plane effective anisotropy fields, respectively.

14.A.2 Out-of-Plane Magnetization Case

For a cylindrical pillar with the out-of-plane magnetization case, we assume that $\vec{e}_1$ is fixed perpendicular to the film plane ($\vec{e}_1 = (0, 0, 1)$) and that an external field is applied with angle ($\vec{H}_{ext} = H_{ext}(\sin\theta_{ext}, 0, \cos\theta_{ext})$). Here, x-axis and y-axis are taken in-plane and z-axis perpendicular to the film plane. The free-layer magnetization is assumed to have a perpendicular easy axis and to be tilted because of the in-plane external field, $\vec{e}_2^{(0)} = (\sin\theta_0, 0, \cos\theta_0)$. We again neglect the small deviation of $\vec{e}_2^{(0)}$ caused by the dc voltage application. In this case, the magnetic energy and the effective field are

$$E_{mag} = \mu_0 M\left(\vec{H}_{ext}\cdot\hat{s}_2^{(0)} + \frac{1}{2}H''_{//}\left(\left(\hat{e}_x\cdot\hat{s}_2^{(0)}\right)^2 + \left(\hat{e}_y\cdot\hat{s}_2^{(0)}\right)^2\right) + \frac{1}{2}\left(H''_\perp - H''_u\right)\left(\hat{e}_z\cdot\hat{s}_2^{(0)}\right)^2\right),$$

$$\vec{H}_{eff}^{(0)} = \frac{1}{\mu_0 M}\frac{\partial E_{dipole\_and\_anisotropy}}{\partial\hat{s}_2^{(0)}} = \vec{H}_{ext} + \begin{pmatrix} H''_{//} & 0 & 0 \\ 0 & H''_{//} & 0 \\ 0 & 0 & H''_\perp - H''_u\end{pmatrix}\hat{s}_2^{(0)}, \tag{14.A.3}$$

where $H''_{//}$, $H''_\perp$, and H_u are in-plane demagnetization, out-plane demagnetization, and uniaxial anisotropy fields, respectively.

From Equation 14.13, the $\hat{\Omega}$ matrix elements in ($\vec{e}_\theta, \vec{e}_\phi$) coordinate system are calculated as follows:

$$\hat{\Omega} = \gamma H_{ext}\cos(\theta_{ext} - \theta_0) - \gamma\begin{pmatrix} H_u\cos 2\theta_0 & 0 \\ 0 & H_u\cos^2\theta_0\end{pmatrix}, \tag{14.A.4}$$

where $H_u = H''_u + H''_{//} - H''_\perp$ is the effective uniaxial anisotropy field.

References

1. J. C. Slonczewski, Current-driven excitation of magnetic multilayers, *J. Magn. Magn. Mater.* **159**, L1–L7 (1996).
2. L. Berger, Emission of spin waves by a magnetic multilayer traversed by a current, *Phys. Rev. B* **54**, 9353–9358 (1996).
3. M. Tsoi, A. G. M. Jansen, J. Bass et al., Excitation of a magnetic multilayer by an electric current, *Phys. Rev. Lett.* **80**, 4281 (1998); **81**, 492 (1998) (Erratum).
4. E. B. Myers, D. C. Ralph, J. A. Katine, R. N. Louie, and R. A. Buhrman, Current-induced switching of domains in magnetic multilayer devices, *Science* **285**, 867–870 (1999).
5. J. A. Katine, F. J. Albert, R. A. Buhrman, E. B. Myers, and D. C. Ralph, Current-driven magnetization reversal and spin-wave excitations in Co/Cu/Co pillars, *Phys. Rev. Lett.* **84**, 3149–3152 (2000).
6. J. Z. Sun, Current-driven magnetic switching in manganite trilayer junctions, *J. Magn. Magn. Mater.* **202**, 157–162 (1999).
7. Y. Huai, F. Albert, P. Nguyen, M. Pakala, and T. Valet, Observation of spin-transfer switching in deep submicron-sized and low-resistance magnetic tunnel junctions, *Appl. Phys. Lett.* **84**, 3118–3120 (2004).
8. H. Kubota, A. Fukushima, Y. Ootani et al., Evaluation of spin-transfer switching in CoFeB/MgO/CoFeB magnetic tunnel junctions, *Jpn. J. Appl. Phys.* **44**, L1237–L1240 (2005).

9. Z. Diao, D. Apalkov, M. Pakala, Y. Ding, A. Panchula, and Y. Huai, Dependence of spin-transfer switching current on free layer thickness in Co–Fe–B/MgO/Co–Fe–B magnetic tunnel junctions, *Appl. Phys. Lett.* **87**, 232502(1–3) (2005).
10. D. Chiba, Y. Sato, T. Kita, F. Matsukura, and H. Ohno, Current-driven magnetization reversal in a ferromagnetic semiconductor (Ga,Mn)As/GaAs/(Ga,Mn)As tunnel junction, *Phys. Rev. Lett.* **93**, 216602(1–4) (2004).
11. J. C. Slonczewski, Conductance and exchange coupling of two ferromagnets separated by a tunneling barrier, *Phys. Rev. B* **39**, 6995–7002 (1989).
12. J. C. Slonczewski, Currents, torques, and polarization factors in magnetic tunnel junctions, *Phys. Rev. B* **71**, 024411(1–10) (2005).
13. D. M. Edwards, F. Federici, J. Mathon, and A. Umerski, Self-consistent theory of current-induced switching of magnetization, *Phys. Rev. B* **71**, 054407 (2005).
14. I. Theodonis, N. Kioussis, A. Kalitsov, M. Chshiev, and W. H. Butler, Anomalous bias dependence of spin torque in magnetic tunnel junctions, *Phys. Rev. Lett.* **97**, 237205(1–4) (2006).
15. S. Zhang and Z. Li, Roles of nonequilibrium conduction electrons on the magnetization dynamics of ferromagnets, *Phys. Rev. Lett.* **93**, 127204(1–4) (2004).
16. A. Thiaville, Y. Nakatani, J. Miltat, and Y. Suzuki, Micromagnetic understanding of current-driven domain wall motion in patterned nanowires, *Europhys. Lett.* **69**, 990–996 (2005).
17. S. E. Barnes and S. Maekawa, Current-spin coupling for ferromagnetic domain walls in fine wires, *Phys. Rev. Lett.* **95**, 107204(1–4) (2005).
18. G. Tatara and H. Kohno, Theory of current-driven domain wall motion: Spin transfer versus momentum transfer, *Phys. Rev. Lett.* **92**, 086601(1–4) (2004).
19. A. A. Tulapurkar, Y. Suzuki, A. Fukushima et al., Spin-torque diode effect in magnetic tunnel junctions, *Nature* **438**, 339–342 (2005).
20. J. C. Sankey, P. M. Braganca, A. G. F. Garcia, I. N. Krivorotov, R. A. Buhrman, and D. C. Ralph, Spin-transfer-driven ferromagnetic resonance of individual nanomagnets, *Phys. Rev. Lett.* **96**, 227601(1–4) (2006).
21. J. C. Sankey, Y.-T. Cui, J. Z. Sun, J. C. Slonczewski, R. A. Buhrman, and D. C. Ralph, Measurement of the spin-transfer-torque vector in magnetic tunnel junctions, *Nat. Phys.* **4**, 67–71 (2008).
22. H. Kubota, A. Fukushima, K. Yakushiji et al., Quantitative measurement of voltage dependence of spin-transfer torque in MgO-based magnetic tunnel junctions, *Nat. Phys.* **4**, 37–41 (2008).
23. C. Wang, Y. T. Cui, J. Z. Sun, J. A. Katine, R. A. Buhrman, and D. C. Ralph, Bias and angular dependence of spin-transfer torque in magnetic tunnel junctions, *Phys. Rev. B* **79**, 224416(1–10) (2009).
24. Y. Suzuki and H. Kubota, Spin-torque diode effect and its application, *J. Phys. Soc. Jpn.* **77**, 031002(1–7) (2008).
25. J. Z. Sun, Spin-current interaction with a monodomain magnetic body: A model study, *Phys. Rev. B* **62**, 570–578 (2000).
26. M. D. Stiles and J. Miltat, Spin dynamics in confined magnetic structures III, in *Topics in Applied Physics*, vol. 101, B. Hillebrands and A. Thiavilles (Eds.), Springer, Heidelberg, pp. 225–308, 2006.
27. Miltat, J., G. Albuquerque, A. Thiaville, and C. Vouille, Spin transfer into an inhomogeneous magnetization distribution, *J. Appl. Phys.* **89**, 6982–6984 (2001).
28. K.-J. Lee, A. Deac, O. Redon, J.-P. Nozières, and B. Dieny, Excitations of incoherent spin-waves due to spin-transfer torque, *Nat. Mater.* **3**, 877–881 (2004).
29. S. Yuasa, T. Nagahama, A. Fukushima, Y. Suzuki, and K. Ando, Giant room-temperature magnetoresistance in single-crystal Fe/MgO/Fe magnetic tunnel junctions, *Nat. Mater.* **3**, 868–871 (2004).
30. S. S. P. Parkin, C. Kaiser, A. Panchula et al., Giant tunnelling magnetoresistance at room temperature with MgO (100) tunnel barriers, *Nat. Mater.* **3**, 862–867 (2004).
31. C. Heiliger and M. D. Stiles, Ab initio studies of the spin-transfer torque in magnetic tunnel junctions, *Phys. Rev. Lett.* **100**, 186805(1–4) (2008).
32. T. Devolder, J. V. Kim, C. Chappert et al., Direct measurement of current-induced fieldlike torque in magnetic tunnel junctions, *J. Appl. Phys.* **105**, 113924(1–5) (2009).
33. Z. Li, S. Zhang, Z. Diao, et al., Perpendicular spin torques in magnetic tunnel junctions, *Phys. Rev. Lett.* **100**, 246602(1–4) (2008).
34. K. Yagami, A. A. Tulapurkar, A. Fukushima, and Y. Suzuki, Magnetic-field-controllable avalanche breakdown and giant magnetoresistive effects in gold/semi-insulating-GaAs Schottky diode, *Appl. Phys. Lett.* **85**, 5634–5636 (2004).
35. L. Berger, Multilayer configuration for experiments of spin precession induced by a dc current, *J. Appl. Phys.* **93**, 7693 (2003).
36. Y. Huai, M. Pakala, Z. Diao, and Y. Ding, Spin transfer switching current reduction in magnetic tunnel junction based dual spin filter structures, *Appl. Phys. Lett.* **87**, 222510 (2005).
37. S. Mangin, D. Ravelosona, J. A. Katine, M. J. Carey, B. D. Terris, and E. E. Fullerton, Current-induced magnetization reversal in nanopillars with perpendicular anisotropy, *Nat. Mater.* **5**, 210–215 (2006).
38. S. Yakata, H. Kubota, Y. Suzuki et al., Influence of perpendicular magnetic anisotropy on spin-transfer switching current in CoFeB/MgO/CoFeB magnetic tunnel junctions, *J. Appl. Phys.* **105**, 07D131(1–3) (2009).
39. T. Nagase, K. Nishiyama, M. Nakayama et al., Spin transfer torque switching in perpendicular magnetic tunnel junctions with Co based multilayer. C1.00331, *American Physics Society March Meeting*, New Orleans, LA, 2008.
40. R. H. Koch, J. A. Katine, and J. Z. Sun, Time-resolved reversal of spin-transfer switching in a nanomagnet, *Phys. Rev. Lett.* **92**, 088302 (2004).

41. Z. Li and S. Zhang, Thermally assisted magnetization reversal in the presence of a spin-transfer torque, *Phys. Rev. B* **69**, 134416(1–6) (2004).
42. W. Coffey, Yu. P. Kalmykov, and J. T. Waldron, *The Langevin Equation: With Applications to Stochastic Problems in Physics, Chemistry and Electrical Engineering*, 2nd edn., World Scientific Series in Contemporary Chemical Physics, vol. 14, World Scientific, Singapore (2004).
43. W. F. Brown, Jr., Thermal fluctuations of a single-domain particle, *Phys. Rev.* **130**, 1677–1686 (1963).
44. D. M.Apalkov and P. B. Visscher, Spin-torque switching: Fokker–Planck rate calculation, *Phys. Rev. B* **72**, 180405(R) (2005).
45. D. M. Apalkov and P. B. Visscher, Slonczewski spin-torque as negative damping: Fokker–Planck computation of energy distribution, *J. Magn. Magn. Mater.* **286**, 370–374 (2005).
46. M. P. Sharrock, Time dependence of switching fields in magnetic recording media, *J. Appl. Phys.* **76**, 6413–6418 (1994).
47. K. Yagami, A. A. Tulapurkar, A. Fukushima, and Y. Suzuki, Low-current spin-transfer switching and its thermal durability in a low-saturation-magnetization nanomagnet, *Appl. Phys. Lett.* **85**, 5634–5636 (2004).
48. M. Pakala, Y. M. Huai, T. Valet, Y. F. Ding, and Z. T. Diao, Critical current distribution in spin-transfer-switched magnetic tunnel junctions, *J. Appl. Phys.* **98**, 056107(1–3) (2005).
49. M. Morota, A. Fukushima, H. Kubota, K. Yakushiji, S. Yuasa, and K. Ando, Dependence of switching current distribution on current pulse width of current-induced magnetization switching in MgO-based magnetic tunnel junction, *J. Appl. Phys.* **103**, 07A707(1–3) (2008).
50. H. Kubota, Y. Suzuki, and S. Yuasa, Development of high-density MRAM with spin-torque writing, *Oyo Buturi* **78**, 232–235 (2009).
51. T. Aoki, Y. Ando, D. Watanabe, M. Oogane, and T. Miyazaki, Spin transfer switching in the nanosecond regime for CoFeB/MgO/CoFeB ferromagnetic tunnel junctions, *J. Appl. Phys.* **103**, 103911 (2008).
52. T. Devolder, J. Hayakawa, K. Ito et al., Single-shot time-resolved measurements of nanosecond-scale spin-transfer induced switching: Stochastic versus deterministic aspects, *Phys. Rev. Lett.* **100**, 057206(1–4) (2008).
53. H. Tomita, K. Konishi, T. Nozaki et al., Single-shot measurements of spin-transfer switching in CoFeB/MgO/CoFeB magnetic tunnel junctions, *Appl. Phys. Express* **1**, 061303(1–3) (2008).
54. S. Petit, C. Baraduc, C. Thirion et al., Spin-torque influence on the high-frequency magnetization fluctuations in magnetic tunnel junctions, *Phys. Rev. Lett.* **98**, 077203(1–4) (2007).
55. A. M. Deac, A. Fukushima, H. Kubota et al., Bias-driven high-power microwave emission from MgO-based tunnel magnetoresistance devices, *Nat. Phys.* **4**, 803–809 (2008).
56. D. Houssameddine, S. H. Florez, J. A. Katine et al., Spin transfer induced coherent microwave emission with large power from nanoscale MgO tunnel junctions, *Appl. Phys. Lett.* **93**, 022505(1–3) (2008).
57. R. Matsumoto, A. Fukushima, K. Yakushiji et al., Spin-torque-induced switching and precession in fully epitaxial Fe/MgO/Fe magnetic tunnel junctions, *Phys. Rev. B* **80**, 174405(1–8) (2009).
58. J.-V. Kim, Stochastic theory of spin-transfer oscillator linewidths, *Phys. Rev. B* **73**, 174412(1–8) (2006).
59. J.-V. Kim, Q. Mistral, C. Chappert, V. S. Tiberkevich, and A. N. Slavin, Line shape distortion in a nonlinear auto-oscillator near generation threshold: Application to spin-torque nano-oscillators, *Phys. Rev. Lett.* **100**, 167201(1–4) (2008).
60. J. V. Kim, V. Tiberkevich, and A. N. Slavin, Generation linewidth of an auto-oscillator with a nonlinear frequency shift: Spin-torque nano-oscillator, *Phys. Rev. Lett.* **100**, 017207(1–4) (2008).
61. V. S. Tiberkevich, A. N. Slavin, and J.-V. Kim, Temperature dependence of nonlinear auto-oscillator linewidths: Application to spin-torque nano-oscillators, *Phys. Rev. B* **78**, 092401(1–4) (2008).
62. A. N. Slavin and P. Kabos, Approximate theory of microwave generation in a current-driven magnetic nanocontact magnetized in an arbitrary direction, *IEEE Trans. Magn.* **41**, 1264–1273 (2005).
63. V. Tiberkevich, A. Slavin, and J.-V. Kim, Approximate theory of microwave generation in a current-driven magnetic nanocontact magnetized in an arbitrary direction, microwave power generated by a spin-torque oscillator in the presence of noise, *Appl. Phys. Lett.* **91**, 192506(1–3) (2007).
64. D. Houssameddine, U. Ebels, B. Dieny et al., Temporal coherence of MgO based magnetic tunnel junction spin torque oscillators, *Phys. Rev. Lett.* **102**, 257202(1–4) (2009).
65. Y. Nagamine, H. Maehara, K. Tsunekawa et al., Ultralow resistance-area product of 0.4 Ω (μm)^2 and high magnetoresistance above 50% in CoFeB/MgO/CoFeB magnetic tunnel junctions, *Appl. Phys. Lett.* **89**, 162507(1–3) (2006).
66. S. Kaka, M. R. Pufall, W. H. Rippard, T. J. Silva, S. E. Russek, and J. A. Katine, Mutual phase-locking of microwave spin torque nano-oscillators, *Nature* **437**, 389–392 (2005).
67. F. B. Mancoff, N. D. Rizzo, B. N. Engel, and S. Tehrani, Phase-locking in double-point-contact spin-transfer devices, *Nature* **437**, 393–395 (2005).
68. B. Georges, J. Grollier, V. Cros, and A. Fert, Impact of the electrical connection of spin transfer nano-oscillators on their synchronization: An analytical study, *Appl. Phys. Lett.* **92**, 232504(1–3) (2008).

15

Multiferroic Tunnel Junctions

Manuel Bibes
Université Paris-Sud

Agnès Barthélémy
Université Paris-Sud

Electron tunneling through vacuum or an ultrathin insulating barrier (1 to about 5 nm) is a well-known phenomenon that dates back to the development of quantum mechanics [1] and is described in all quantum mechanics textbooks. It recently attracted a lot of attention not only from a fundamental point of view but also for its device perspectives. This is exemplified by the work performed on superconducting [2] and magnetic tunnel junctions (see Chapters 10 through 12). These latter junctions are composed of a thin dielectric barrier sandwiched between two ferromagnetic electrodes. Upon changing the relative orientation of the magnetizations of the two electrodes, the conductance of the junctions changes, leading to a tunnel magnetoresistance (TMR) effect. This TMR is nowadays exploited for the realization of a new type of nonvolatile memory called magnetoresistive random access memory (MRAM) (see Chapter 35) in which the information is stored by the parallel or antiparallel orientation of the magnetizations of the electrodes.

Until now, very little work has been reported on junctions with a ferroelectric (FE) tunnel barrier despite the fact that the basic idea was reported by Esaki almost 40 years ago [3]. This may be because the growth of ultrathin films of FE materials is challenging, and also because of the collective nature of ferroelectricity that is not always conserved at the nanometer scale required for operating in the tunneling regime. Nevertheless, recent theoretical and experimental advances on perovskite FE oxide thin films demonstrate clearly that in some conditions ferroelectricity persists down to at least 1 nm (see Section 15.2). This makes the conception of ferroelectric tunnel junctions (FTJs) composed of an FE tunnel barrier and two metallic electrodes feasible (see Section 15.2). Furthermore, if the two electrodes are ferromagnetic, or if the barrier on its own is composed of a multiferroic material (see Section 15.1), the junction becomes a multiferroic tunnel junction (MFTJ) (see Sections 15.3 and 15.4), in which the interplay between ferroelectricity and magnetism not only holds great promises for applications [4] but also is a fantastic playground for physicists.

15.1 Introduction to Multiferroics

Multiferroics are a relatively rare class of multifunctional materials that simultaneously exhibit several ferroic orders among ferromagnetic, ferroelectric, and ferroelastic (ferrotoroidic ordering [5] is also sometimes included) [6,7]. Given the scarcity of compounds that present two or more strictly ferroic orders, antiferroic orders (e.g., antiferromagnetic) are usually considered. Most of the currently investigated multiferroics are generally magnetic and FE, very few showing a finite large moment (corresponding to ferro- or ferrimagnetic ordering). Practically, the vast majority of multiferroics are thus FE antiferromagnets or weak ferromagnets.

While the coexistence of several ferroic orders in the same material has been thought, since the 1970s, to be an interesting way to realize multiple-state memories [8], what looks more appealing today is the coupling that often exists between the ferroic orders. As sketched in Figure 15.1, this coupling may enable the manipulation of a given ferroic order parameter (FE polarization, magnetization, and strain) by an external stimulus different from the usual one (electric field, magnetic field, and stress, respectively). In the context of spintronics, a promising type of coupling is the so-called magnetoelectric coupling [9] that links the FE polarization and the magnetization, and potentially allows the manipulation of polarization by a magnetic field, and most importantly, of magnetization by an electric field.

Single-phase FE-magnetic multiferroics (simply referred to as "multiferroics" henceforth) are scarce and many are oxides [10,11]. The technological advances in oxide thin film growth in the 1990s naturally led researchers to favor oxide materials to explore the physics and device potential of multiferroics in

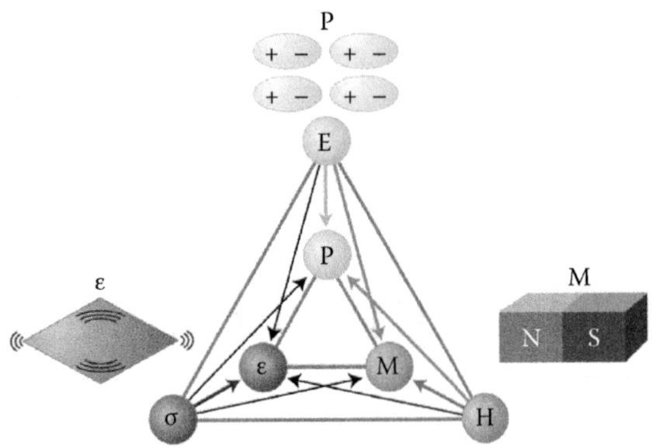

FIGURE 15.1 Phase control in ferroics and multiferroics. The electric field E, magnetic field H, and stress σ control the electric polarization P, magnetization M, and strain ε, respectively. In a ferroic material, P, M, or ε are spontaneously formed to produce ferromagnetism, ferroelectricity, or ferroelasticity, respectively. In a multiferroic, the coexistence of at least two ferroic forms of ordering leads to additional interactions. In a magnetoelectric multiferroic, a magnetic field may control P or an electric field may control M. (Reproduced from Spaldin, N.A. and Fiebig, M. *Science*, 309, 391, 2005. With permission.)

the thin film form. Figure 15.2 shows a Venn diagram classifying different types of insulating ferroic oxides according to their FE and/or magnetic properties. While many FE or ferri/ferromagnetic insulating oxides do exist, only a handful of materials are believed to possess both properties simultaneously (experimental indications, somewhat controversial, do exist for $La_{0.1}Bi_{0.9}MnO_3$ [12], $CoCr_2O_4$ [13], and Bi_2FeCrO_6 [14]), and only at low temperature. Many more are FE and antiferromagnetic, and remarkably, most of these show a magnetoelectric coupling. So far, $BiFeO_3$ is perhaps the only room temperature magnetoelectric multiferroic [15], with an FE Curie temperature of 1100 K [16], and a Néel temperature of 640 K [17]. From the inspection of Figure 15.2, it clearly appears that many FE, ferromagnetic, and multiferroic materials crystallize in a reduced number of structural families, such as the perovskite structure. Accordingly, they may be combined into epitaxial heterostructures in which their properties can be exploited independently or through couplings appearing at interfaces. In particular, so-called artificial multiferroics can be engineered by the combination of, for example, ferroelectrics with magnetic materials [18]. As we will see in Sections 15.3 and 15.4, tunnel junctions may be defined using either single-phase or artificial multiferroics.

Since the renaissance of multiferroics in 2003, much effort has been made to decipher the complex physical mechanisms at the origin of the magnetoelectric coupling. More practically, various strategies have also been deployed to achieve an electrical control of magnetization. Approaches based on single-phase multiferroics have had limited success. In $HoMnO_3$, it was shown that an electric field could modify the type of magnetic ordering and generate a large finite magnetic moment, but only at low temperature [19]. The lack of room temperature FE-ferromagnets has driven the research to bilayer systems combining either an FE-antiferromagnet and a soft ferromagnet coupled by exchange [20–23], or a piezoelectric–FE with a magnetostrictive

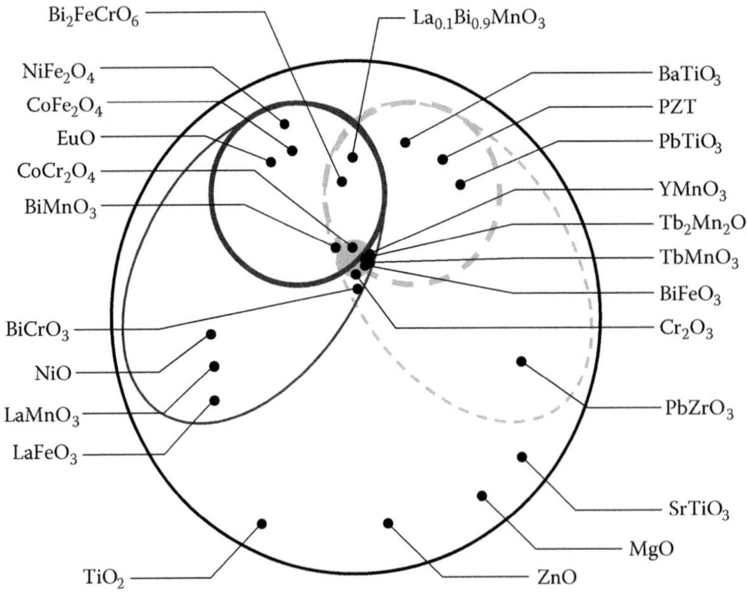

FIGURE 15.2 Classification of insulating oxides. The largest circle represents all insulating oxides among which one finds electrically polarizable materials (light gray, dotted ellipse) and magnetically polarizable materials (dark gray ellipse). Within each ellipse, the circle represents materials with a finite polarization (ferroelectrics) and/or a finite magnetization (ferro- and ferrimagnets). Depending on the definition, multiferroics correspond to the intersection between the ellipses or the circles. The small disk in the middle denotes systems exhibiting a magnetoelectric coupling. (Adapted from Béa, H. et al., *J. Phys.: Condens. Matter*, 20, 434221, 2008. With permission; inspired from Eerenstein, W. et al., *Nature*, 442, 759, 2006.)

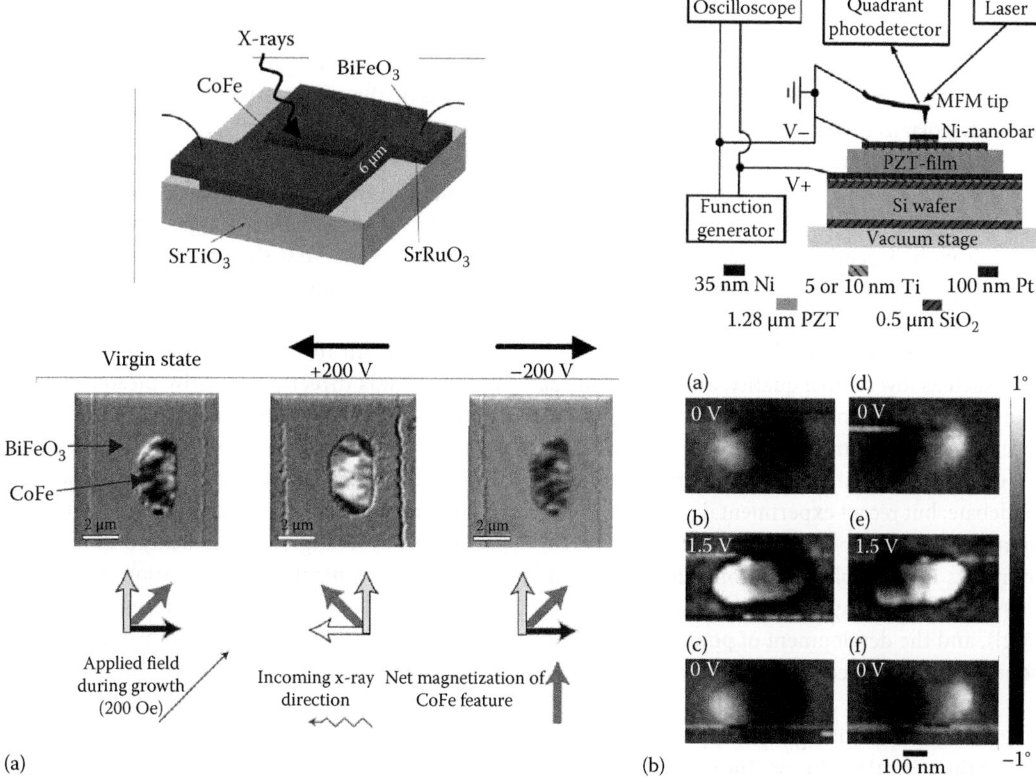

FIGURE 15.3 Magnetoelectric control of magnetization. (a) Top: Sketch of a device consisting of a micronic CoFe dot deposited onto a $BiFeO_3$ film equipped with $SrRuO_3$ planar electrodes. The ferromagnetic domain structure of the CoFe dot is probed by x-ray photoemission electron microscopy (bottom) and the domain patterned is reversibly switched by applying an in-plane voltage of ±200 V across the $BiFeO_3$. (Adapted from Chu, Y.H. et al., *Nat. Mater.* 7, 478, 2008.) (b) Top: Sketch of the magnetoelectric Ni nanobar/PZT film and associated experimental setup. Bottom: Magnetic force microscopy images of the Ni nanobar for various voltages applied across the PZT. (Reproduced from Chung, T.-K. et al., *Appl. Phys. Lett.*, 94, 132501, 2009. With permission.)

ferromagnet [24–27]. In these so-called artificial multiferroics, an "extrinsic" magnetoelectric coupling may be engineered, with coupling coefficients much larger than that of the "intrinsic" magnetoelectric coupling present in single-phase multiferroics.

Using this bilayer approach, the local domain structure of microstructured ferromagnetic pads was shown to be efficiently manipulated by an electric field. In Figure 15.3a, we present the results of Chu et al. on $BiFeO_3$/CoFe architectures [28]. The ferromagnetic domains in the CoFe are exchange-coupled to the antiferromagnetic–FE domains in the $BiFeO_3$ (BFO). Applying an in-plane voltage across the BFO modifies the antiferromagnetic–FE domain structure, and consequently, the ferromagnetic domain structure of the CoFe dot probed by XMCD-PEEM. Figure 15.3b illustrates the complementary study of Chung et al. on an architecture combining a thick piezoelectric $Pb(Zr,Ti)O_3$ (PZT) layer with Ni nanobars [26]. Due to the converse piezoelectric effect, when a voltage is applied to the PZT, it deforms and so do the other layers grown on top of it, including the Ni elements. The resulting magnetoelastic anisotropy induces changes in the ferromagnetic domain structure in Ni, as visible from magnetic force microscopy (MFM) images. Prospective developments beyond these promising results include the magnetoelectric control of the response of a spintronic device [4]. This would pave the way toward MRAMs in which a spin-based binary information would be written by the application of a voltage across a multiferroic, that is, at a very low energy cost compared to even the best state-of-the-art approaches, such as spin-transfer (see Chapters 7, 8, and 14).

En route to the grand objective of a magnetoelectrically controllable spintronics device, a lot has been learnt on the physical properties of multiferroics, especially in thin film form. It is now possible to engineer BFO films with determined multiferroic domain structures [29], to tailor their polarization and critical temperatures by epitaxial strain [30], etc. However, the behavior of multiferroic films at thicknesses compatible with tunneling (typically 1–5 nm) is not well known. In the following paragraphs, we discuss important related issues currently under investigation, namely, the physics of tunneling through a ferroelectric and the problematic depression of ferroelectricity often observed in ultrathin films.

15.2 Tunneling through Ferroelectrics

This paragraph is dedicated to FTJs that we describe both theoretically and experimentally after briefly addressing the issue of ferroelectricity at the nanoscale level (for a more complete description of ferroelectric thin films, the reader is invited to consult Refs. [31–35]).

A critical issue to realize FTJs is to be able to grow FE thin films at a thickness of a few nanometers. The influence of film thickness on the FE polarization amplitude and ferroelectric domain pattern is due to the interplay between different phenomena including intrinsic surface effects related to dipole–dipole interactions [36],* and the presence of a depolarizing field. Both effects strongly depend on the boundary conditions at the FE/electrode interface. Additional strain effects produced by external stress imposed by the underlying substrate or electrode/substrate system strongly affect the FE character as well. Extrinsic factors, such as low sample quality, also contribute as evidenced by the fact that the minimum thickness for ferroelectricity has decreased by order of magnitudes over the years. The presence or not of a minimum thickness for ferroelectricity is still a subject of debate, but recent experimental advances (notably the development of piezoresponse force microscopy (PFM) [37,38], improvements in x-ray photoelectron diffraction techniques [39,40], synchrotron x-ray studies [41], or UV Raman spectroscopy [42]), and the development of powerful first principle calculations [43,44] have recently shed light on these FE size effects.

The screening of the depolarizing field appears as the key parameter that sets the critical thickness. The effect of the depolarizing field increases as the film thickness decreases [44] and may lead to the suppression of ferroelectricity and the concomitant reduction of the tetragonality of the unit cell. Other ways to reduce this huge electrostatic energy at low thickness are by the formation of domains with FE polarization pointing in opposite directions but also by electrical conduction within the film. This depolarizing field arises from uncompensated charges appearing at the surfaces of the FE thin film. In an ideal FE capacitor, composed of perfectly conducting plates sandwiching the FE film, the screening of charges located at the electrode/FE interface exactly compensate the surface charges related to the FE polarization. However, in real FE capacitors, the screening charges extend over a finite region in the metallic electrodes resulting in the presence of finite interface dipoles and a subsequent nonzero depolarizing field. The spatial extension of this region, that is, the effective screening length, depends strongly on the material used as electrodes (e.g., 0.2–1.9 nm for $La_{2/3}Sr_{1/3}MnO_3$ [45,46], ~0.5 nm for $SrRuO_3$ [44], and <0.1 nm for Au). Through its influence on the screening length of the depolarizing field, the choice of the electrodes has important effects on the critical thickness and/or on the formation of domains. For example, using high resolution x-ray to study single-domain $PbTiO_3$ deposited on conductive Nb-doped $SrTiO_3$ substrates, Lichtensteiger et al. [47] pointed out in 2005, the systematic decrease of the *c*-axis lattice parameter with decreasing film thickness below 20 nm, reflecting the decrease of the FE polarization (see Figure 15.4a). This reduction of the polarization is attributed to the presence of a residual unscreened depolarizing field. When grown on $La_{2/3}Sr_{1/3}MnO_3$ electrodes, the same $PbTiO_3$ thin films first show a decrease of *c*/*a* with decreasing film thickness followed by a recovery of *c*/*a* at small thicknesses (Figure 15.4a). This recovery is accompanied by a change from a single domain to a polydomain configuration of the polarization, as was directly shown by piezo response atomic force microscopy (AFM) measurements (Figure 15.4b). Surprisingly, the LSMO electrodes do not seem to screen as well as Nb-doped $SrTiO_3$ electrodes despite their better bulk metallicity. More recent x-ray scattering measurements and ab initio calculations performed by Wang et al., evidenced how the chemical environment can control the polarization orientation in a $PbTiO_3$ FE thin film [49]. Interestingly, recent theoretical predictions from first principle calculations of ultrathin Au/$BaTiO_3$/Au FE capacitors predict that the covalent bonding mechanism at the interface between Ba-O-terminated films and the simple metal can lead to an overall enhancement of the driving force of the film toward a polar state rather than a suppression [50]. This result not only evidences the limitation of the simple Thomas–Fermi screening approach to understand the existence of a critical thickness, and that a microscopic analysis will generally be necessary to describe interfacial effects in FE capacitors, but also the opportunity to improve the FE properties of thin films by selecting the appropriate electrode.

Additional effects, such as polarization fatigue, have not yet been studied in ultrathin films but are well known for thin films in the 100 nm range, usually used in FE random access memories (FeRAMs). Fatigue is generally related to the migration of oxygen to the electrode–FE interface that can inhibit the switching by pinning domain walls [51], the formation of layers at electrode interfaces that effectively reduce the total electric field applied across the device, or inhibit the nucleation of oppositely polarized domains [52,53]. These mechanisms modify the dynamics of FE domains in applied electric fields and should have significant impact on the characteristics of FTJs.

Interestingly, in perovskite materials, ferroelectricity is also often highly sensitive to strain. If judiciously used, the strain induced by the substrate may lead to an enhancement of FE properties. A shift of the FE transition temperature (Tc) and an increase of the remnant polarization (Pr) with strain have been predicted [54–58] and observed [59–62]. A particularly impressive example of the power of such strain effects is the demonstration of strain-induced ferroelectricity at room temperature in $SrTiO_3$ [63]. Another remarkable illustration is the large increase of the remnant polarization (by at least 250%) as well as of the transition temperatures (by nearly 500°C) in $BaTiO_3$ thin films grown on rare-earth scandates substrates inducing a biaxial compressive strain [62]. Strain engineering may thereby

* Ferroelectricity has long been viewed as a collective phenomenon resulting from the alignment of localized dipoles (resulting in a spontaneous polarisation) within a correlation volume that extend for many ferroelectrics over a distance of the order of 10–50 nm along the polarisation axis. In this picture, the occurrence of ferroelectricity is allowed only for films thicker than this correlation length which in the case of PZT for example is 20 nm. However, with appropriate boundary conditions, recent ab-initio calculations as well as experimental results reveal that ferroelectricity can be preserved at nanometer scale.

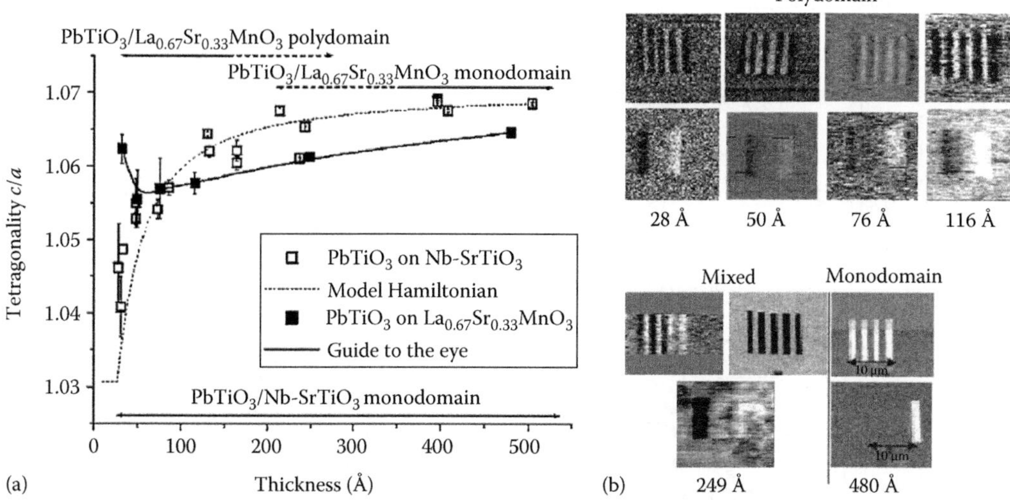

FIGURE 15.4 (a) Evolution of the *c/a* ratio with the $PbTiO_3$ film thickness grown on Nb-doped $SrTiO_3$ and $La_{2/3}Sr_{1/3}MnO_3$ buffered $SrTiO_3$. Very different behaviors are obtained due to the different screening conditions. (b) The recovery of the *c/a* ratio at low thickness is accompanied by a change from a monodomain to a polydomain configuration of the polarization, as shown by piezoresponse atomic force microscopy measurements. (Reproduced from Lichtensteiger, C. et al., *Appl. Phys. Lett.*, 90, 052907, 2007. With permission.)

counteract the intrinsic thickness effect and restore ferroelectricity at low thicknesses [64].

All ferroelectrics are nevertheless not as sensitive to strain. $BiFeO_3$, for example, in which the FE character is driven by the Bi lone pair, has been predicted to be virtually strain insensitive [65]. Experimentally, a small variation of the remnant polarization was found under biaxial strain and attributed to the polarization rotation inside the (110) plane [66]. Finally, it has to be noticed that the absence of critical thickness has been predicted in improper ferroelectrics [67].

Characterizing the FE behavior at the nanometer scale is a difficult task. Indeed, for films as thin as a few nanometers, the tunneling current impedes the characterization through standard polarization versus electric field loops (P(E) loops). Noticeable exceptions (see Figure 15.5a and b) are P(E) loop measurements realized on fully strained 3.5 nm at 77 K [68] and 5 nm at room temperature [69] in $BaTiO_3$ capacitors with $SrRuO_3$ electrodes deposited on $SrTiO_3$. As shown by the current–voltage response of the capacitor (Figure 15.5a), a large contribution of the leakage to the total current is observed in this tunneling regime. PFM appears as the technique of choice for probing ferroelectricity in ultrathin films. Figure 15.5c and d present two examples of the PFM characterization of nanometer-thick films. In the first example (Figure 15.5c), stripes with a polarization either up or down have been written on a $BiFeO_3$ thin film deposited on $La_{2/3}Sr_{1/3}MnO_3$-buffered $SrTiO_3$ using a conductive tip AFM (CT-AFM); PFM is then used to read the signal. The clear and reversible out-of-plane phase contrast is indicative of the FE character of this 2 nm $BiFeO_3$ thin film [70]. In the following example (Figure 15.5d), $BaTiO_3$ has been deposited on $La_{2/3}Sr_{1/3}MnO_3$-buffered $NdGaO_3$ that imposes a large biaxial strain in order to stabilize its FE behavior [62]. The clear PFM contrast between written stripe-shaped domains with a polarization up or down directly evidences the FE character of this 1 nm thin film. Subsequent square domains have been rewritten to assess the reversibility of the writing process [64].

Ultrathin films deposited on different substrates and electrodes, inducing different boundary conditions and different strains, have been reported to be FE, experimentally. Films as thin as 4 nm of $Pb(Zr_{0.2}Ti_{0.8})O_3$ [37], 2 nm of $BiFeO_3$ [70,71] or $(La,Bi)MnO_3$ [12], and ~1 nm of $PbTiO_3$ [41], P(VDF-TrFE) [72], and $BaTiO_3$ [64] have been reported to be FE, setting the upper limit for the effective critical thickness of these materials and allowing the exploration of their potential as tunnel barriers in FTJs and MFTJs. Tunneling through ferroelectrics is not only interesting from a fundamental point of view but it could also be of great potential for applications in the field of data storage, based on heterostructures in which the information is encoded by the direction of the FE polarization, and read nondestructively. This would represent a major advance over existing FeRAMs in which data readout is destructive, so that the initial orientation of the polarization has to be restored after reading, which is time and power consuming. Very few theoretical and experimental works have been reported on the topic of FTJs, but the recent advances in FE ultrathin films have paved the way toward more developments. Experimental and theoretical work in this nascent field* has been initiated by the group of Hermann Kohlstedt in Jülich [74–76].

From the theoretical point of view, Tsymbal and Kohlstedt have recently summarized three possible mechanisms through

* The concept of ferroelectric tunnel junction first proposed by L. Esaki in 1971 has been the object of a patent deposited by Phillips Corp. in 1996 [73].

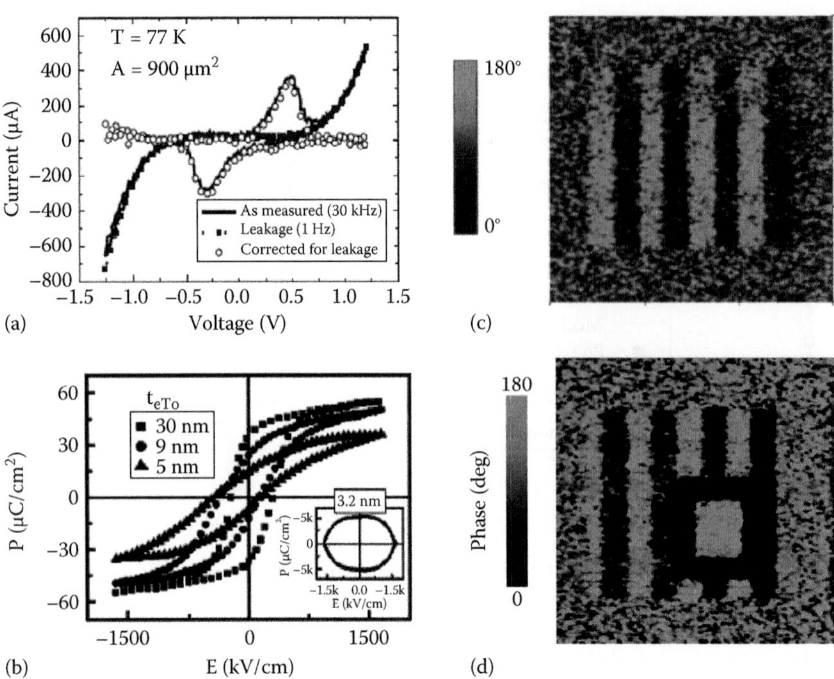

FIGURE 15.5 (a) Current versus voltage response of the 3.5 nm $BaTiO_3$ capacitor with $SrRuO_3$ electrodes measured at 77 K. The black line shows the response measured at the excitation signal frequency of 30 kHz. The measured current is the sum of three contributions: the FE displacement current caused by the switching of the spontaneous polarization and leakage current very important in the tunneling regime. The square dots denote the current response measured at a frequency of 1 Hz at which the leakage only contributes to the signal. The circles show the difference between these two current–voltage curves, which represents the leakage-compensated current response. (From Petraru, A. et al., *Appl. Phys. Lett.*, 93, 072902, 2008. With permission.) (b) Polarization versus electric field loops at room temperature for 30, 9, and 5 nm thick $BaTiO_3$ sandwiched between two $SrRuO_3$ electrodes. A clear FE loop is measured for the 5 nm thick film, whereas for the 3.2 nm one, leakage current dominates. (From Kim, Y.S. et al., *Appl. Phys. Lett.*, 86, 102907, 2005. With permission.) (c) PFM out of plane phase image of a 2 nm $BiFeO_3$ film deposited on $La_{2/3}Sr_{1/3}MnO_3$ buffered $SrTiO_3$ after poling the $BiFeO_3$ film with alternatively positive and negative DC voltage along 500 nm wide stripes. (From Béa, H. et al., *Jpn. J. Appl. Phys.*, 45, L187, 2006.) (d) PFM out-of-plane phase image of a 1 nm $BaTiO_3$ film grown on $La_{2/3}Sr_{1/3}MnO_3$ buffered $NdGaO_3$ substrate after writing 4× 0.5 μm² stripes with a polarization either up (bright) or down (dark). The polarization in these domains can be switched back and forth by applying positive or negative bias. (From Garcia, V. et al., *Nature*, 460, 81, 2009. With permission.)

which the tunnel current would be modulated by the reversal of polarization in the FE barrier [77] (see Figure 15.6). The first mechanism, described in detail in Refs. [78–80], is an electrostatic effect that yields an asymmetric deformation of the potential profile of the barrier owing to the presence of charges at the barrier–electrode interfaces that are differently screened in the two different electrodes. An interesting way to increase this effect, based on composite tunnel barriers combining an FE film with a nonpolar dielectric material, has recently been proposed [81]; the dielectric layer serves as a switch that changes its barrier height from a low to a high value when the polarization of the FE is reversed. In the second mechanism, the so-called interface effect, the interfacial density of states of the electrodes is modified according to the FE-polarization-direction-dependent position of the ions in the last atomic layer in the FE, which affects the tunnel current [82,83]. The third mechanism is related to the converse piezoelectric effect through which the tunnel barrier width would be changed upon switching the polarization direction. Since the tunnel current depends exponentially on the barrier width, a substantial modulation of the current can indeed be expected [78–80]. All these three mechanisms may contribute to the modulation of the tunneling current by the FE character of the barrier and give rise to large tunnel electroresistance (TER) phenomena. More sophisticated models based on first principle calculations predict sizable changes in the transmission probability across the Pt/$BaTiO_3$ interface as well as in the complex band structure of $BaTiO_3$ with polarization reversal, two ingredients controlling the electron tunneling rates through epitaxial tunnel junctions [82].

Experimentally, it is a challenge to unambiguously interpret the observed resistive switchings in terms of a modulation of the tunneling current by ferroelectricity. Indeed, bias-induced resistive switches have been observed in non-FE oxides [84]. A first breakthrough in FTJs was the demonstration of hysteretic I(V) curves in the $SrRuO_3$/$BaTiO_3$(6 nm)/$SrRuO_3$ [74] and $SrRuO_3$/PZT(6 nm)/$SrRuO_3$ [76] tunnel junctions with, in some cases, clear switchings from a high-resistance state to a low-resistance state giving rise to large TER phenomena (see Figure 15.7a). Nevertheless, as mentioned by the authors of these articles, the correlation between ferroelectricity and electroresistance could not be demonstrated unambiguously and additional effects, such as electromigration could not be refuted. However, a direct

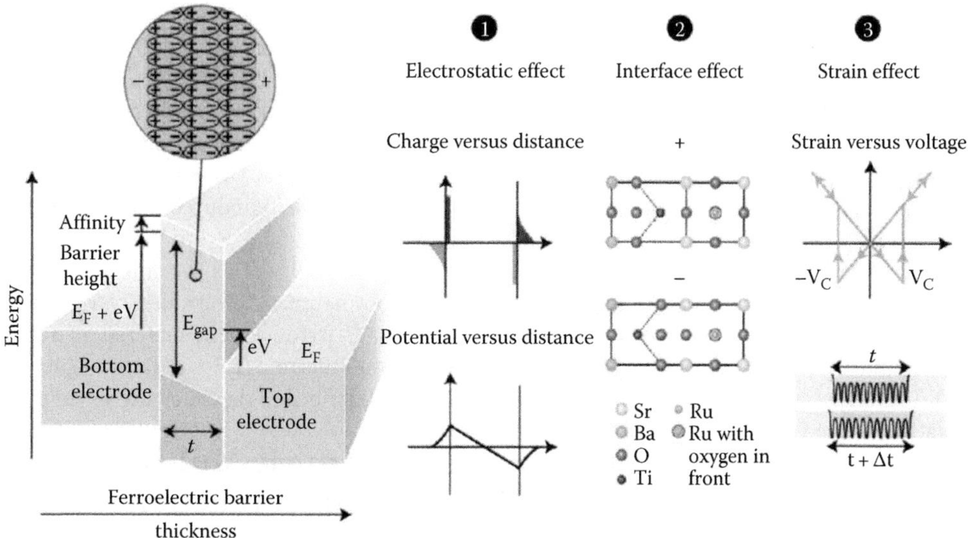

FIGURE 15.6 Schematic diagram of a tunnel junction, which consists of two electrodes separated by a nanometer-thick FE barrier layer and of the three different mechanisms that can influence the tunneling process due to the FE character of the barrier. (From Tsymbal, E.Y. and Kohlstedt, H. *Science*, 313, 181, 2006. With permission.)

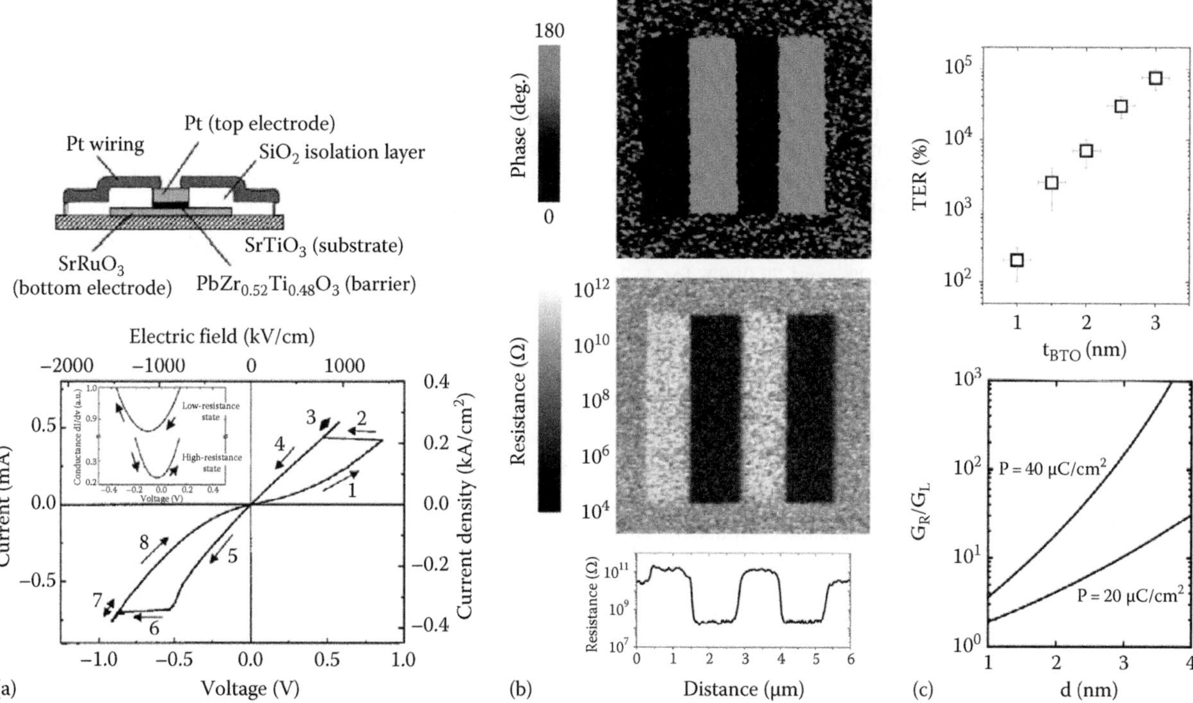

FIGURE 15.7 (a) Sketch (top) and current versus voltage behavior (bottom) of the FETJs of Rodríguez Contreras et al. [76]; the inset shows the dynamic conductance plotted versus voltage for the low- and high-resistance states. (b) Parallel PFM (top) and conductive-tip AFM (middle) mappings of a 3 nm $BaTiO_3$ film grown on $La_{2/3}Sr_{1/3}MnO_3/NdGaO_3(001)$ after poling the $BaTiO_3$ film along stripes with positive or negative DC voltage. The resistance difference between negatively and positively poled areas defines a tunnel electroresistance (TER) of ~75,000%. The bottom curve is a resistance profile across the series of stripes in the middle figure. (From Garcia, V. et al., *Nature*, 460, 81, 2009.) (c) Thickness dependence of the TER across FE $BaTiO_3$ tunnel barriers measured by Garcia et al. (top, from Garcia, V. et al., *Nature*, 460, 81, 2009. With permission) and calculated by Zhuravlev et al. (bottom, from Zhuravlev, M.Y. et al., *Phys. Rev. Lett.*, 94, 246802, 2005. With permission).

relationship between FE polarization reversal and resistance switching has been observed in more recent *in situ* scanning tunneling microscopy imaging and spectroscopy experiments, which evidence hysteretic features consistent with the FE switching process in the I(V) curves of 4–10 monolayer $BaTiO_3$ thin films deposited on $SrRuO_3$ buffered $SrTiO_3$ [85].

Recently, following a procedure defined by Kohlstedt et al. [86] to unambiguously demonstrate the ferroelectricity-related nature of the resistive switch, current–voltage curves and piezoresponse measurements performed using CT-AFM in ultrahigh vacuum were collected simultaneously on 30 nm thick $Pb(Zr_{0.2}Ti_{0.8})O_3$ films deposited on an $La_{2/3}Sr_{1/3}MnO_3$ electrode on $SrTiO_3(001)$ substrate. These authors evidenced a very large modulation of the tunneling current by a factor of about 200 and its tunability by the FE nature of the film [87]. In these experiments, electron tunneling is induced by injecting electrons from the metal tip into the oxide in the Fowler–Nordheim regime. Injection through the ferroelectrically modulated Schottky barrier at the interface was invoked to interpret the data. Additionally, a smaller modulation of the transport properties was reported using a 5 nm $BiFeO_3$ ultrathin FE tunneling barrier grown on an $La_{2/3}Sr_{1/3}MnO_3$ electrode [87].

A definite correlation between the FE character and the electroresistance phenomena in the standard tunneling regime has been recently brought out by combining PFM and CT-AFM experiments at room temperature [64]. Using PFM, Garcia et al. wrote stripes with an FE polarization pointing either up or down in highly strained $BaTiO_3$ thin films (1–3 nm) deposited on buffered $NdGaO_3$ (see the top of Figure 15.7b). Subsequently, using CT-AFM, they collected resistance maps of the poled domains (see the middle of Figure 15.7b), taking advantage of the finite tunnel current at these low barrier thicknesses. The resistance contrast between positively and negatively poled regions corresponds to TER effects of 200%, 10,000%, and 75,000% for 1, 2, and 3 nm (see the bottom of Figure 15.7b), respectively. From the thickness dependence of the TER effect (see the top of Figure 15.7c), the authors conclude that the predominant mechanism at the origin of this TER effect is mainly the modulation of the barrier profile by the FE character of the barrier as predicted by Zhuravlev et al. [79] and Kohlstedt et al. [78] (see the bottom of Figure 15.7c). The scalability down to 70 nm, corresponding to densities greater than 16 Gbit in.$^{-2}$, was also established. This has been confirmed more recently on 4.8 nm $BaTiO_3$ ultrathin films deposited on $SrRuO_3$-buffered $SrTiO_3$ substrates inducing a smaller strain [88]. The more modest TER effect may result from a smaller strain that induces a less significant polarization than in the case of Garcia et al. [64]. This demonstrates that in FTJs, the direction of the polarization can be determined without altering it by exploiting the polarization-dependent leakage current in the tunneling regime. This could be of great promise for next generation FeRAMs with nondestructive readouts [89]. Their realization in the solid-state form with top electrodes and standard row and column addressing, as well as their integration into CMOS technology will have to be demonstrated before their real potential as memory devices can be assessed [89]. Additional effects related to the piezoelectric character of the barrier, which are generally impeded in thin films due to the clamping by the substrate, could be expected if these devices are patterned with lateral dimensions of the order or below 100 nm [90]. This may enlarge [78–80] the parameters playing a key role in the ferroelectricity-induced modulation of tunneling current in FTJs.

The functionality of FTJs can be further enriched by replacing the normal metal electrodes by a ferromagnetic one or by using an FE–ferromagnetic material as tunnel barrier, thereby defining multiferroic (FE and ferromagnetic) tunnel junctions (MFTJs) described in the following sections.

15.3 Intrinsic Multiferroic Tunnel Junctions

Taking advantage of the multifunctional character of $La_{0.1}Bi_{0.9}MnO_3$, one of the only ferromagnetic–FE multiferroic compounds, Gajek et al. designed an MFTJ with a ferromagnetic–FE barrier [12]. $La_{2/3}Sr_{1/3}MnO_3$ and Au were used as electrodes. As expected from the FE nature of the $La_{0.1}Bi_{0.9}MnO_3$ compound,* a TER effect of about 20% was observed upon reversing the FE polarization of the barrier. An additional spin filtering effect and the subsequent TMR arise from the ferromagnetic character of the barrier [91,92]. This spin filtering effect, observed for both $BiMnO_3$ and $La_{0.1}Bi_{0.9}MnO_3$, is related to the fact that the barrier height is spin dependent because the bottom level of the conduction band is spin-split by exchange interaction in a ferromagnetic material [93, see also Chapter 13]. This allows the tunneling electrons to be efficiently filtered according to their spin, and results in a highly spin-polarized current, thus leading to a large tunnel magnetoresistance (TMR, defined as the difference between the junction resistance in the antiparallel (AP) and parallel (P) configurations of the $La_{2/3}Sr_{1/3}MnO_3$ and $La_{0.1}Bi_{0.9}MnO_3$ magnetizations, that is, $TMR=(R_{AP}-R_P)/R_P$) effect when one of the electrodes (in that case, $La_{2/3}Sr_{1/3}MnO_3$) is ferromagnetic. TMR effects of ~50% and ~90% have been observed for $BiMnO_3$ and $La_{0.1}Bi_{0.9}MnO_3$, respectively, corresponding to a spin-filtering efficiency of about 22% and 35%, respectively [92,94]. Using the FE and ferromagnetic character of a 2 nm $La_{0.1}Bi_{0.9}MnO_3$ tunnel barrier, Gajek et al. were consequently able to define a four-resistance state system (see Figure 15.8 [11]). It was proposed that these four resistance states are encoded by the two-order parameters in this multiferroic barrier—magnetization and polarization—through the mechanisms of spin filtering and the modulation of the barrier profile by the FE nature of the compound [12]. As expected from the ferromagnetic Curie temperature of the compound (95 K, for 10% La doping), the TMR effect disappears around 90 K, whereas the TER effect is still observed

* It has to be noticed that the ferroelectric character of $BiMnO_3$ thin films remains questionable due to the high leakage current that prevent poling of the sample. Only in $La_{0.1}Bi_{0.9}MnO_3$ films, even as thin as 2 nm, a clear ferroelectric behavior has been observed.

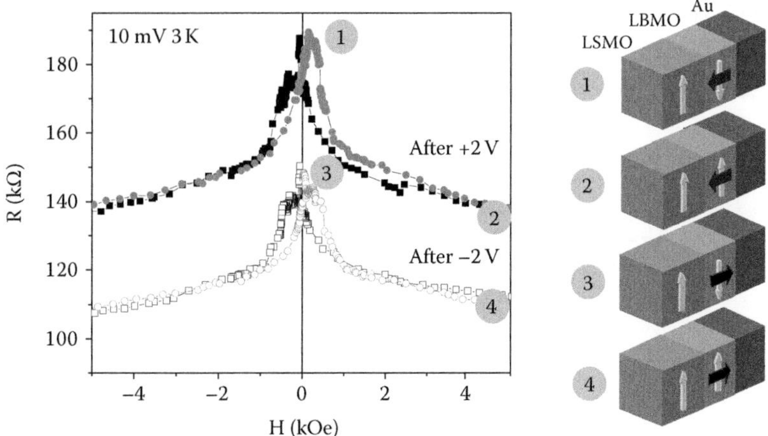

FIGURE 15.8 Variation of the resistance of an Au/LBMO(2 nm)/LSMO junction as a function of the magnetic field and previously applied bias displaying four resistance states whose magnetic and electric configurations are represented on the right-hand side. (From Velev, J.P. et al., *Phys. Rev. Lett.*, 95, 216601, 2005; Bowen, M. et al., *Phys. Rev. B* 73, 140408(R), 2006. With permission.)

at room temperature (in agreement with the estimated FE T_C of the compound, 450 K [95]).

Electron transport through these multiferroic tunneling junctions has been investigated theoretically by Ju et al. [96] taking into account the screening of the polarization charges in the metallic electrodes as well as the exchange splitting of the barrier. The TMR effect was found to depend strongly on the FE polarization of the barrier and on the contrast between the screening length of the two electrodes that are the two parameters at the origin of the TER effect in this simple model [78,79]. This could explain the quite small TMR effect observed in the $Au/La_{0.1}Bi_{0.9}MnO_3(2\,nm)/La_{2/3}Sr_{1/3}MnO_3$ multiferroic junctions for which the screening lengths of the two electrodes are very dissimilar [12]. First principle calculations of the systems, such as those performed for the $BaTiO_3$ tunnel barriers [83,97], taking into account the symmetry filtering effect affecting the transmission coefficient and the coupling of the wave function of the electrodes with the evanescent wave functions in the barrier, are however necessary to fully understand the transport properties in these epitaxial tunnel junctions. Nevertheless, these predictions [96] underline the complexity to optimize the TER and TMR effects in these multiferroic junctions based on an FE–ferromagnetic single material. Solutions can be found in extrinsic MFTJs in which the different functionalities underlying the operation of these devices—ferromagnetism and ferroelectricity—may be optimized independently (see Chapter 16).

15.4 Artificial Multiferroic Tunnel Junctions

In intrinsic MFTJs, the two-order parameters (magnetization and polarization) of the multiferroic tunnel barrier determine the transport properties. However, in artificial MFTJs, the spin-dependent and ferroelectricity-related contribution to the transport properties are, in first approximation, physically separated as the barrier is made of a nonferromagnetic FE material, and the ferromagnetic order is only present in the electrodes. These junctions are thus expected to show four resistance states due to both TMR (the same as in conventional MTJs) and to TER (as in FTJs with nonmagnetic electrodes) [98]. Additionally, artificial MFTJs allow more flexibility in the choice of materials and in the optimization of the FE and ferromagnetic properties. As will be discussed in the following paragraphs, the interplay between ferroelectricity and ferromagnetism at the interfaces between the barrier and the electrodes can also result in a giant modulation of TMR (and, therefore, of the tunnel current spin polarization) induced by FE polarization reversal.

This ferroelectricity-induced modulation of spin polarization is related to the second mechanism responsible for TER in FTJs, put forward by Tsymbal and Kohlstedt [77] (see Figure 15.6), namely, the modification of the interfacial density of states due to different orbital bonding depending on the direction of the polarization in the FE barrier. If the electrode is made of a ferromagnetic material with a spin polarized density of states, the FE -polarization-direction-dependent modulation of the density of states will be different for spin-up and spin-down, resulting in a modulation of the interfacial spin polarization and of magnetic properties depending on the electron density close to the Fermi level E_F (e.g., the magnetic moment or the magnetocrystalline anisotropy). This effect is thus reminiscent of the predicted [99,100] and observed [101,102] influence of a large external electric field on the magnetic anisotropy of ferromagnetic ultrathin layers due to changes in the spin-dependent electron density at the E_F. An important benefit of using an FE rather than applying an external electric field to induce changes in the spin-dependent density of states close to E_F is that those changes will be driven by the direction of the FE polarization and will thus be nonvolatile.

Ab initio calculations of the interfacial electronic structure of FE/ferromagnet heterostructures have been performed for

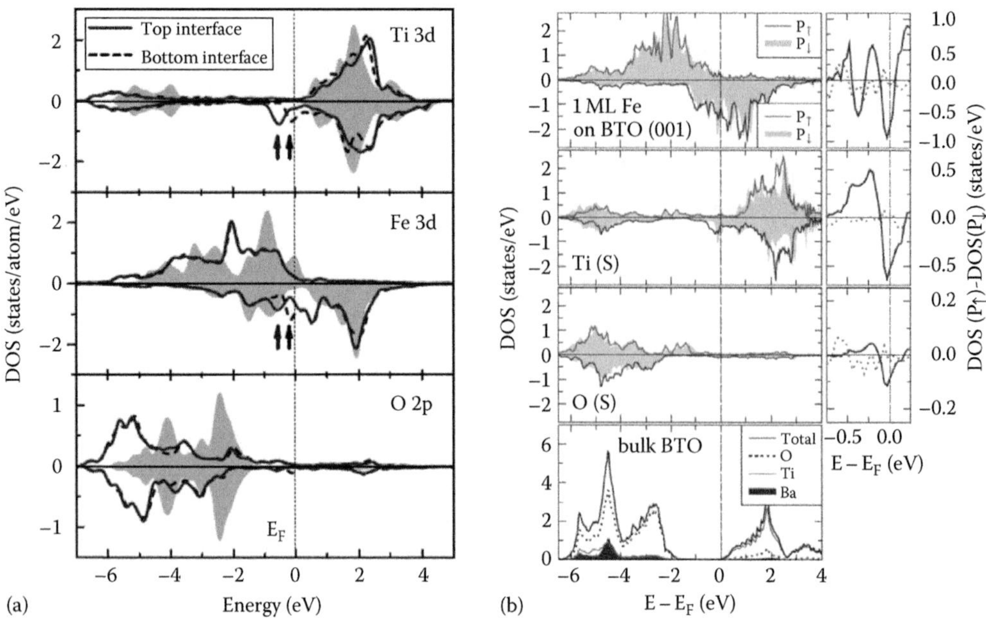

FIGURE 15.9 (a) Orbital-resolved DOS for interfacial atoms in an Fe=$BaTiO_3$ multilayer for m = 4: (top) Ti 3d, (middle) Fe 3d, and (bottom) O 2p. Majority- and minority-spin DOS are shown in the upper and lower panels, respectively. The solid and dashed curves correspond to the DOS of atoms at the top and bottom interfaces, respectively. The shaded plots are the DOS of atoms in the central monolayer of (middle) Fe or (top), (bottom) TiO_2 that can be regarded as bulk. The vertical line indicates the Fermi energy (E_F). (From Duan, C.G. et al., *Phys. Rev. Lett.*, 97, 047201, 2006. With permission.) (b) Main panels on the left: electronic structure at the surface of Fe/$BaTiO_3$(001): spin-resolved density of states (DOS) of Fe in layer S+1 (top) as well as of Ti (second from top) and O (third from top) in layer S for FE polarization pointing up (lines) and down (gray). Bottom: total and partial DOS of bulk $BaTiO_3$, with the bottom of the conduction band taken as energy reference. The small panels on the right show the spin-resolved difference of the DOS for FE polarization pointing up and down of Fe, Ti, and O (majority: dotted; minority: solid).

several systems including $BaTiO_3$/Fe [82,103–105], $PbTiO_3$/Fe [103,105], $BaTiO_3$/Co_2MnSi [106], $BaTiO_3$/$SrRuO_3$ [97], and $BaTiO_3$/Fe_3O_4 [107]. In Figure 15.9, we show typical results for the case of $BaTiO_3$/Fe. In Figure 15.9a, orbital-resolved density of states profiles for interfacial atoms are displayed [82]. Clearly, the density of states at the interface (dotted and solid curves) is modified compared to the bulk (shaded gray) in both $BaTiO_3$ and Fe, indicating some substantial redistribution of charge. Importantly, this redistribution and accordingly, the spin-dependent density of states is different depending on the direction of the FE polarization (for the solid and dotted curves, the FE polarization of the BTO is pointing toward and away from the Fe layer, respectively). For the type of interface structure considered here, the spin polarization is found more negative when the FE polarization is pointing away from the Fe. Figure 15.9b reproduces the results of calculations by Fechner et al. for one monolayer of Fe on top of $BaTiO_3$ [103]. Consistent with the prediction of Duan et al. [82] (Figure 15.9a), they also find a globally negative spin polarization for the Fe layer at the interface with $BaTiO_3$ that increases in absolute value when the FE polarization points away from the Fe. However, the precise influence of the FE polarization direction on the interfacial spin polarization is expected to depend on the details of the interface structure [108]. We emphasize that while those calculations predict relatively modest changes in the local magnetic moments induced by FE polarization reversal, the expected effect on the interfacial spin-dependent density of states at E_F is usually much larger making tunneling (with its extreme sensitivity to interfacial effects) a particularly well-suited probe of these phenomena.

Although the amplitude of the TMR may in simple cases be estimated from inspecting the spin-dependent density of states of the electrodes (using the so-called Jullière model [109]), it is now well accepted that for epitaxial systems, the complex band structure of the barrier material plays a key role (see Chapter 12). This is epitomized by the model Fe/MgO/Fe system [110,111] in which the complex band structure of MgO selectively transmits the electronic wave functions of the Δ_1 symmetry that have a very high spin polarization compared to that of the density of states (integrated over the full Brillouin zone, and thus, summing up the contribution of wave functions of all possible symmetries). Calculations of the complex electronic band structure of perovskite oxides have been carried out [112,113] and Velev et al. have even calculated the influence of ferroelectricity on that of $BaTiO_3$ [83]. More recently, full ab initio calculations of the spin-dependent transport properties of $SrRuO_3$/$BaTiO_3$/$SrRuO_3$ artificial MFTJs have been reported [97] (note that despite the use of two $SrRuO_3$ electrodes, the two barrier/electrodes interfaces are asymmetric due to different stacking sequences of the perovskite blocks). Not only are four different resistance states predicted, but also TMR ratios of different amplitude depending on the direction of the FE polarization in the $BaTiO_3$ tunnel barrier.

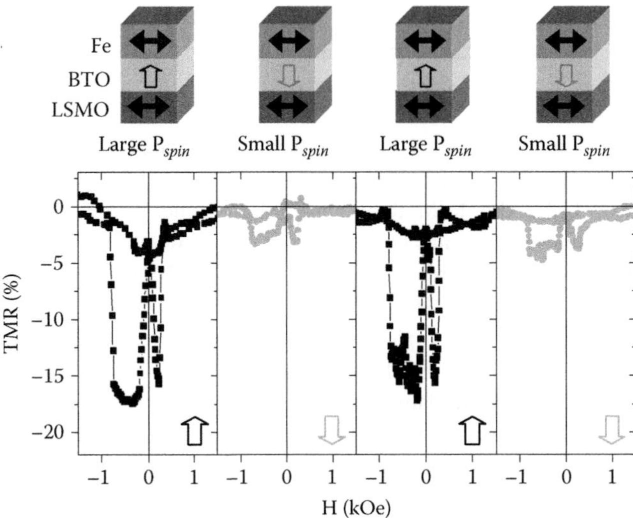

FIGURE 15.10 Tunnel magnetoresistance (TMR) versus magnetic field curves for an Fe/$BaTiO_3$(1 nm)/$La_{2/3}Sr_{1/3}MnO_3$ tunnel junction (V_{DC} = −50 mV, T = 4 K) after poling the FE $BaTiO_3$ tunnel barrier down (VP+), up (VP−), down (VP+), and up (VP−). A clear modulation of the TMR with FE polarization orientation is achieved. (From Garcia, V. et al., *Science*, 327, 1106, 2010. With permission.)

On the experimental side, there have been so far very few efforts to explore the physics and the potential of artificial MFTJs. Garcia et al. have fabricated $La_{2/3}Sr_{1/3}MnO_3$/$BaTiO_3$/Fe tunnel junctions in which the ultrathin (1 nm) $BaTiO_3$ film used as the barrier was shown to be FE at room temperature [64]. In these junctions, a negative TMR was measured, as expected (within Jullière's model [109]) from the positive spin polarization of $La_{2/3}Sr_{1/3}MnO_3$ and the calculated negative spin polarization of the density of states of Fe at the interface with $BaTiO_3$ [82,103] (see Figure 15.10). Remarkably, the amplitude of the TMR was found to vary reversibly with poling voltage pulses of ±1 V used to orient the FE polarization in the $BaTiO_3$ barrier toward or away from the Fe electrode [114]. A larger TMR, corresponding to a larger (in absolute value) spin polarization for the Fe, was found when the FE polarization was pointing toward the Fe.

While very much remains to be done in the exciting novel field of artificial MFTJs (and more generally on the electric-field and/or FE control of interfacial magnetism), these first theoretical and experimental results look promising to propose alternative solutions for the local manipulation of spin-based properties at a low energetic cost. In the future, the FE control of spin polarization predicted and observed at FE/ferromagnet interfaces may be useful to control the response of lateral spin-injection devices, such as spin transistors [98,115].

15.5 Conclusions and Perspectives

Although the research on MFTJs is still in its infancy, very important advances have been achieved in the past 3 or 4 years. The field is at the confluence of several very active areas, such as spintronics, oxide interfaces, and nanoscale ferroelectrics; and advances are thus partly fuelled by the fast progress made on these topics. From a practical perspective, MFTJs are some sort of a merger of magnetic and FTJs, with the potential not only to address issues in MRAMs (by reducing the writing energy) and in FeRAMs (by providing a simple and nondestructive readout scheme), but also to combine the best of both technologies (magnetic storage, simple electrical readout) to propose novel storage concepts. Interestingly, there is great synergy between the theoretical and experimental work in this field, with several outstanding achievements, especially in FTJs (note for instance, the remarkable agreement between the predicted [79] and observed [64] exponential increase of TER with barrier thickness). This joint effort from first-principles specialists on one side and experimentalists on the other side will be decisive in solving the numerous remaining issues, such as fatigue in FE tunnel barriers, FE nanoscale domain dynamics in the GHz regime, and interface-driven magnetoelectric effects.

Acknowledgments

The authors wish to thank V. Garcia for his critical reading of the manuscript. Partial support from the French ANR Programme Blanc ("Oxitronics") and P-Nano ("Méloic"), and the Triangle de la Physique is acknowledged.

References

1. J. Frenkel, On the electrical resistance of contacts between solid conductors, *Phys. Rev.* **36**, 1604–1618 (1930).
2. I. Giaever, Energy gap in superconductors measured by electron tunneling, *Phys. Rev. Lett.* **5**, 147–148 (1960).
3. L. Esaki, R. B. Laibowitz, and P. J. Stiles, Polar switch, *IBM Tech. Disclos. Bull.* **13**, 2161 (1971).
4. M. Bibes and A. Barthélémy, Towards a magnetoelectric memory, *Nat. Mater.* **7**, 425–426 (2008).
5. B. B. Van Aken, J.-P. Rivera, H. Schmid, and M. Fiebig, Observation of ferrotoroidic domains, *Nature* **449**, 702–705 (2007).
6. W. Eerenstein, N. D. Mathur, and J. F. Scott, Multiferroic and magnetoelectric materials, *Nature* **442**, 759–765 (2006).
7. N. A. Spaldin and M. Fiebig, The renaissance of magnetoelectric multiferroics, *Science* **309**, 391–392 (2005).
8. J. F. Scott, Multiferroic memories, *Nat. Mater.* **6**, 256–257 (2007).
9. M. Fiebig, Revival of the magnetoelectric effect, *J. Phys. D: Appl. Phys.* **38**, R123–R152 (2005).
10. M. Bibes and A. Barthélémy, Oxide spintronics, *IEEE Trans. Electron. Dev.* **54**, 1003–1023 (2007).
11. H. Béa, M. Gajek, M. Bibes, and A. Barthélémy, Spintronics with multiferroics, *J. Phys.: Condens. Matter* **20**, 434221 (2008).
12. M. Gajek, M. Bibes, S. Fusil et al., Tunnel junctions with multiferroic barriers, *Nat. Mater.* **6**, 296–302 (2007).

13. Y. Yamasaki, S. Miasaka, Y. Kanko, J.-P. He, T. Arima, and Y. Tokura, Magnetic reversal of the ferroelectric polarization in a multiferroic spinel oxide, *Phys. Rev. Lett.* **96**, 207204 (2006).
14. R. Nechache, C. Harnagea, A. Pignolet et al., Growth, structure and properties of epitaxial thin films of first-principles predicted multiferroic Bi_2FeCrO_6, *Appl. Phys. Lett.* **89**, 102902 (2006).
15. G. Catalan and J. F. Scott, Physics and applications of bismuth ferrite, *Adv. Mater.* **21**, 2463–2485 (2009).
16. J. R. Teague, R. Gerson, and W. J. James, Dielectric hysteresis in single crystal $BiFeO_3$, *Solid State Commun.* **8**, 1073–1074 (1970).
17. P. Fischer, M. Polomska, I. Sosnowska, and M. Szymanski, Temperature dependence of the crystal and magnetic structures of $BiFeO_3$, *J. Phys. C: Solid State Phys.* **13**, 1931–1940 (1980).
18. H. Zheng, J. Wang, S. E. Lofland et al., Multiferroic $BaTiO_3$–$CoFe_2O_4$ nanostructures, *Science* **303**, 661–663 (2004).
19. T. Lottermoser, T. Lonkai, U. Amann, D. Holhwein, J. Ihringer, and M. Fiebig, Magnetic phase control by an electric field, *Nature* **430**, 541–544 (2004).
20. J. Dho, X. Qi, H. Kim, J. L. MacManus-Driscoll, and M. G. Blamire, Large electric polarization and exchange bias in multiferroic $BiFeO_3$, *Adv. Mater.* **18**, 1445–1448 (2006).
21. H. Béa, M. Bibes, S. Cherifi et al., Tunnel magnetoresistance and robust room temperature exchange bias with multiferroic $BiFeO_3$ epitaxial thin films, *Appl. Phys. Lett.* **89**, 242114 (2006).
22. H. Béa, M. Bibes, F. Ott et al., Mechanisms of exchange bias with multiferroic $BiFeO_3$ epitaxial thin films, *Phys. Rev. Lett.* **100**, 017204 (2008).
23. V. Laukhin, V. Skumryev, X. Martí et al., Electric-field control of exchange bias in multiferroic epitaxial heterostructures, *Phys. Rev. Lett.* **97**, 227201 (2006).
24. W. Eerenstein, W. Wiora, J.-L. Prieto, J. F. Scott, and N. D. Mathur, Giant sharp and persistent converse magnetoelectric effects in multiferroic epitaxial heterostructures, *Nat. Mater.* **6**, 348–351 (2007).
25. C. Thiele, K. Dörr, O. Bilani, J. Rödel, and L. Schultz, Influence of strain on the magnetization and magnetoelectric effect in $La_{0.7}A_{0.3}MnO_3$/PMN-PT(001) (A=Sr,Ca), *Phys. Rev. B* **75**, 054408 (2007).
26. T.-K. Chung, S. Keller, and G. P. Carman, Electric-field-induced reversible magnetic single-domain evolution in a magnetoelectric thin film, *Appl. Phys. Lett.* **94**, 132501 (2009).
27. M. Weiler, A. Brandlmaier, S. Geprägs et al., Voltage controlled inversion of magnetic anisotropy in a ferromagnetic thin film at room temperature, *New J. Phys.* **11**, 013021 (2009).
28. Y. H. Chu, L. W. Martin, M. B. Holcomb et al., Electric-field control of local ferromagnetism using a magnetoelectric multiferroic, *Nat. Mater.* **7**, 478–482 (2008).
29. Y. H. Chu, Q. He, C.-H. Yang et al., Nanoscale control of domain architecture in $BiFeO_3$ thin films, *Nanoletters* **9**, 1726–1730 (2009).
30. C. J. Fennie and K. M. Rabe, Magnetic and electric phase control in epitaxial $EuTiO_3$ from first principles, *Phys. Rev. Lett.* **97**, 267602 (2006).
31. C. H. Ahn, K. M. Rabe, and J.-M. Triscone, Ferroelectricity at the nanoscale: Local polarization in oxide thin films and heterostructures, *Science* **303**, 488–491 (2004).
32. K. Rabe, C. H. Ahn, and J.-M. Triscone, *Physics of Ferroelectrics*, Topics in Applied Physics, vol. 105, Springer Verlag, Berlin, Heidelberg, 2007.
33. N. Setter, N. Damjanovic, L. Eng et al., Ferroelectric thin films: Review of materials, properties and applications, *J. Appl. Phys.* **100**, 051606 (2006).
34. M. Dawber, K. M. Rabe, and J. F. Scott, Physics of ferroelectric oxides, *Rev. Mod. Phys.* **77**, 1083–1130 (2005).
35. M. Dawber, N. Stucki, C. Lichtensteiger, S. Gariglio, and J.-M. Triscone, New phenomena at the interfaces of very thin ferroelectric oxides, *J. Phys.: Condens. Matter* **20**, 264015 (2008).
36. M. E. Lines and A. M. Glass, *Principles and Applications of Ferroelectrics and Related Materials*, Oxford University Press, Oxford, 1979; S. Li, J. A. Easterman, J. M. Vetrone, C. M. Foster, R. E. Newnham, and L. E. Cross, Dimension and size effects in ferroelectrics, *Jpn. J. Appl. Phys. Part 1* **36**, 5169–5174 (1977); N. Sai, A. M. Kolpak and A. M. Rappe, Ferroelectricity in ultrathin perovskite films, *Phys. Rev. B* **72**, 020101(R) (2005).
37. T. Tybell, C. H. Ahn, and J.-M. Triscone, Ferroelectricity in thin perovskite films, *Appl. Phys. Lett.* **75**, 856–858 (1999).
38. S. Kalinin and A. Gruverman, *Scanning Probe Microscopy: Electrical and Electromechanical Phenomena at the Nanoscale*, Springer, Berlin, 2006.
39. L. Despont, C. Koitzsch, F. Clerc et al., Direct evidence for ferroelectric polar distortion in ultrathin lead titanate perovskite films, *Phys. Rev. B* **73**, 094110 (2006).
40. W. F. Egelhoff Jr., X-ray photoelectron and Auger electron forward scattering: A new tool for surface crystallography, *Crit. Rev. Solid State Mater. Sci.* **16**, 213–235 (1990).
41. D. D. Fong, G. B. Stephenson, S. K. Streiffer et al., Ferroelectricity in ultrathin perovskite films, *Science* **304**, 1650–1653 (2004).
42. D. A. Tenne, A. Bruchhausen, N. D. Lanzillotti-Kimura et al., Probing nanoscale ferroelectricity by ultraviolet Raman spectroscopy, *Science* **313**, 1614–1616 (2006).
43. J. Junquera and P. Ghosez, Critical thickness for ferroelectricity in perovskite ultrathin films, *Nature* **422**, 506–509 (2003).
44. P. Ghosez and J. Junquera, First-principles modelling of ferroelectric oxides nanostructures, in *Handbook of Theoretical and Computational Nanotechnology*, M. Reith and W. Schommers (Eds.), vol. 9, American Scientific Publishers, Stevenson Ranch, CA, pp. 623–728, 2006.

45. M. Dzero, L. P. Gor'kov, and V. Z. Kresin, On magnetoconductivity of metallic manganite phases and heterostructure, *Int. J. Mod. Phys. B* **10**, 2095–2115 (2003).
46. X. Hong, A. Posadas, and C. H. Ahn, Examining the screening limit of field effect devices via the metal-insulator transition, *Appl. Phys. Lett.* **86**,142501 (2005).
47. C. Lichtensteiger, J.-M. Triscone, J. Junquera, and P. Ghosez, Ferroelectricity and tetragonality in ultrathin $PbTiO_3$ films, *Phys. Rev. Lett.* **94**, 047603 (2005).
48. C. Lichtensteiger, M. Dawber, N. Stucki et al., Monodomain to polydomain transition in ferroelectric $PbTiO_3$ thin films with $La_{0.67}Sr_{0.33}MnO_3$ electrodes, *Appl. Phys. Lett.* **90**, 052907 (2007).
49. R. V. Wang, D. D. Fong, F. Jiang et al., Reversible chemical switching of a ferroelectric film, *Phys. Rev. Lett.* **102**, 047601 (2009).
50. M. Stengel, D. Vanderbilt, and N. A. Spaldin, Enhancement of ferroelectricity at metal-oxide interfaces, *Nat. Mater.* **8**, 392–397 (2009).
51. M. Dawber and J. F. Scott, A model for fatigue in ferroelectric perovskite thin films, *Appl. Phys. Lett.* **76**, 1060–1062 (2000).
52. A. K. Tagantsev, I. Stolichnov, E. L. Colla, and N. Setter, Polarization fatigue in ferroelectric films: Basic experimental findings, phenomenological scenarios, and microscopic features, *J. Appl. Phys.* **90**, 1387–1402 (2001).
53. P. K. Larsen, G. J. M. Dormans, D. J. Taylor, and P. J. van Veldhoven, Ferroelectric properties and fatigue of $PbZr_{0.51}Ti_{0.49}O_3$ thin films of varying thickness: Blocking layer model, *J. Appl. Phys.* **76**, 2405–2413 (1994).
54. A. F. Devonshire, Theory of barium titanate, *Phil. Mag.* **42**, 1065–1079 (1951).
55. N. A. Pertsev, A. G. Zembilgotov, and A. K. Tagantsev, Effect of mechanical boundary conditions on phase diagrams of epitaxial ferroelectric thin films, *Phys. Rev. Lett.* **80**, 1988–1991 (1998).
56. Y. L. Li, S. Y. Hu, Z. K. Liu, and L. Q. Chen, Phase-field model of domain structures in ferroelectric thin films, *Appl. Phys. Lett.* **78**, 3878–3880 (2001).
57. M. Sepliarsky, S. R. Phillpot, M. G. Stachiotti, and R. L. Migoni, Ferroelectric phase transitions and dynamical behavior in $KNbO_3/KTaO_3$ superlattices by molecular-dynamics simulation, *J. Appl. Phys.* **91**, 3165–3171 (2002).
58. J. B. Neaton and K. M. Rabe, Theory of polarization enhancement in epitaxial $BaTiO_3/SrTiO_3$ superlattices, *Appl. Phys. Lett.* **82**, 1586–1588 (2003).
59. E. D. Specht, H.-M. Christen, D. P. Norton, and L. A. Boatner, X-ray diffraction measurement of the effect of layer thickness on the ferroelectric transition in epitaxial $KTaO_3/KNbO_3$ multilayers, *Phys. Rev. Lett.* **80**, 4317–4320 (1998).
60. N. Yanase, K. Abe, N. Fukushima, and T. Kawakubo, Thickness dependence of ferroelectricity in heteroepitaxial $BaTiO_3$ thin film capacitors, *Jpn. J. Appl. Phys.* **38**, 5305–5308 (1999).
61. S. K. Streiffer, J. A. Eastman, D. D. Fong et al., Observation of nanoscale 180° stripe domains in ferroelectric $PbTiO_3$ thin films, *Phys. Rev. Lett.* **89**, 067601 (2002).
62. K. J. Choi, M. Biegalski, Y. L. Li et al., Enhancement of ferroelectricity in strained $BaTiO_3$ thin films, *Science* **306**, 1005–1009 (2004).
63. H. Haeni, P. Irvin, W. Chang et al., Room-temperature ferroelectricity in strained $SrTiO_3$, *Nature* **430**, 758–761 (2004).
64. V. Garcia, S. Fusil, K. Bouzehouane et al., Giant tunnel electroresistance for non-destructive readout of ferroelectric states, *Nature* **460**, 81–84 (2009).
65. C. Ederer and N. A. Spaldin, Effect of epitaxial strain on the spontaneous polarization of thin film ferroelectrics, *Phys. Rev. Lett.* **95**, 25760 (2005).
66. H. W. Jang, S. H. Baek, D. Ortiz et al., Strain-induced polarization rotation in epitaxial (001) $BiFeO_3$ thin films, *Phys. Rev. Lett.* **101**, 107602 (2008).
67. N. Sai, C. J. Fennie, and A. A. Demkov, Absence of critical thickness in an ultrathin improper ferroelectric film, *Phys. Rev. Lett.* **102**, 107601 (2009).
68. A. Petraru, H. Kohlstedt, U. Poppe et al., Wedgelike ultrathin epitaxial $BaTiO_3$ films for studies of scaling effects in ferroelectrics, *Appl. Phys. Lett.* **93**, 072902 (2008).
69. Y. S. Kim, D. H. Kim, J. D. Kim et al., Critical thickness of ultrathin ferroelectric $BaTiO_3$ films, *Appl. Phys. Lett.* **86**, 102907 (2005).
70. H. Béa, S. Fusil, K. Bouzehouane et al., Ferroelectricity down to at least 2 nm in multiferroic $BiFeO_3$ epitaxial thin films, *Jpn. J. Appl. Phys.* **45**, L187–L189 (2006).
71. Y. H. Chu, T. Zhao, M. P. Cruz et al., Ferroelectric size effects in multiferroic $BiFeO_3$ thin films, *Appl. Phys. Lett.* **90**, 252906 (2007).
72. A. V. Bune, V. M. Fridkin, S. Ducharme et al., Two-dimensional ferroelectric films, *Nature* **391**, 874–877 (1998).
73. R. Wolf, P. W. M. Blom, and M. P. C. Krijn, U.S. Patent, Patent number 5,541,422, July 30, 1996.
74. H. Kohlstedt, N. A. Pertsev, and R. Waser, Size effects on polarization in epitaxial ferroelectric films and the concept of ferroelectric tunnel junctions including first results, *Mater. Res. Symp. Proc.* **688**, C6.5.1 (2002).
75. J. Rodríguez Contreras, J. Schubert, U. Poppe et al., Structural and ferroelectric properties of epitaxial $PbZr_{0.52}Ti_{0.48}O_3$ and $BaTiO_3$ thin films prepared on $SrRuO_3/SrTiO_3(100)$ substrates, *Mater. Res. Symp. Proc.* **688**, C8.10.1 (2002).
76. J. Rodríguez Contreras, H. Kohlstedt, U. Poppe, R. Waser, C. Buchal, and N. A. Pertsev, Resistive switching in metal-ferroelectric-metal junctions, *Appl. Phys. Lett.* **83**, 4595–4597 (2003).
77. E. Y. Tsymbal and H. Kohlstedt, Tunneling across a ferroelectric, *Science* **313**, 181–183 (2006).
78. H. Kohlstedt, N. A. Pertsev, J. Rodríguez-Contreras, and R. Waser, Theoretical current-voltage characteristics of ferroelectric tunnel junctions, *Phys. Rev. B* **72**, 125341 (2005).

79. M. Y. Zhuravlev, R. F. Sabirianov, S. S. Jaswal, and E. Y. Tsymbal, Giant electroresistance in ferroelectric tunnel junctions, *Phys. Rev. Lett.* **94**, 246802 (2005).
80. K. M. Indlekofer and H. Kohlstedt, Simulation of quantum dead-layers in nanoscale ferroelectric tunnel junctions, *Europhys. Lett.* **72**, 282–286 (2005).
81. M. Ye. Zhuravlev, Y. Wang, S. Maekawa, and E. Y. Tsymbal, Tunneling electroresistance in ferroelectric tunnel junctions with a composite barrier, *Appl. Phys. Lett.* **95**, 052902 (2009).
82. C. G. Duan, S. S. Jaswal, and E. Y. Tsymbal, Predicted magnetoelectric effect in Fe/$BaTiO_3$ multilayers: Ferroelectric control of magnetism, *Phys. Rev. Lett.* **97**, 047201 (2006).
83. J. P. Velev, C.-G. Duan, K. D. Belaschenko, S. S. Jaswal, and E. Y. Tsymbal, Effect of ferroelectricity on electron transport in Pt/$BaTiO_3$/Pt tunnel junctions, *Phys. Rev. Lett.* **98**, 137201 (2007).
84. R. Waser and M. Aono, Nanoionics-based resistive switching memories, *Nat. Mater.* **6**, 833–840 (2007).
85. S. Shin, S. V. Kalinin, E. W. Plummer, and A. P. Baddorf, Electronic transport through in-situ grown ultrathin $BaTiO_3$ films, *Appl. Phys. Lett.* **95**, 03290 (2009).
86. H. Kohlstedt, A. Petraru, K. Szot et al., Method to distinguish ferroelectric from nonferroelectric origin in case of resistive switching in ferroelectric capacitors, *Appl. Phys. Lett.* **92**, 062907 (2008).
87. P. Maksymovych, S. Jesse, P. Yu, R. Ramesh, A. P. Baddorf, and S. V. Kalini, Polarization control of electron tunneling into ferroelectric surfaces, *Science* **324**, 1421–1425 (2009).
88. A. Gruverman, D. Wu, H. Lu et al., Tunneling electroresistance in ferroelectric tunnel junctions at the nanoscale, *Nanoletters* **9**, 3539–3543 (2009).
89. P. Zubko and J.-M. Triscone, A leak of information, *Nature* **460**, 45–46 (2009).
90. S. Bühlmann, B. Dwir, J. Baborowski, and P. Muralt, Size effect in mesoscopic epitaxial ferroelectric structures: Increase of piezoelectric response with decreasing feature size, *Appl. Phys. Lett.* **80**, 3195–3197 (2002).
91. M. Gajek, M. Bibes, A. Barthélémy, M. Varela, and J. Fontcuberta, Perovskite-based heterostructures integrating ferromagnetic-insulating $La_{0.1}Bi_{0.9}MnO_3$, *J. Appl. Phys.* **97**, 103909 (2005).
92. M. Gajek, M. Bibes, A. Barthélémy et al., Spin filtering through ferromagnetic $BiMnO_3$ tunnel barriers, *Phys. Rev. B* **72**, 020406(R) (2005).
93. J. S. Moodera, T. S. Santos, and T. Nagahama, The phenomena of spin-filter tunnelling, *J. Phys. Condens. Matter* **19**, 165202 (2007).
94. M. Gajek, M. Bibes, M. Varela et al., $La_{2/3}Sr_{1/3}MnO_3$–$La_{0.1}Bi_{0.9}MnO_3$ heterostructures for spin filtering, *J. Appl. Phys.* **99**, 08E504 (2006).
95. Z. H. Chi, C. J. Xiao, S. M. Feng et al., Manifestation of ferroelectromagnetism in multiferroic $BiMnO_3$, *J. Appl. Phys.* **98**, 103519 (2005).
96. S. Ju, T. Y. Cai, G.-Y. Guo, and Z.-Y. Li, Electrically controllable spin filtering and switching in multiferroic tunneling junctions, *Phys. Rev. B* **75**, 064419 (2007).
97. J. Velev, C.-G. Duan, J. D. Burton et al., Magnetic tunnel junctions with ferroelectric barriers: Prediction of four resistance states from first principles, *Nanoletters* **9**,427–432 (2009).
98. M. Y. Zhuravlev, S. S. Jaswal, E. Y. Tsymbal, and R. F. Sabirianov, Ferroelectric switch for spin injection, *Appl. Phys. Lett.* **87**, 222114 (2005).
99. C.-G. Duan, J. P. Velev, R. F. Sabirianov et al., Surface magnetoelectric effect in ferromagnetic metal films, *Phys. Rev. Lett.* **101**, 137201 (2008).
100. K. Nakamura, R. Shimabukuro, Y. Fujiwara, T. Akiyama, and T. Ito, Giant modification of the magnetocrystalline anisotropy in transition-metal monolayers by an external electric field, *Phys. Rev. Lett.* **102**, 187201 (2009).
101. M. Weisheit, S. Fähler, A. Marty, Y. Souche, C. Poinsignon, and D. Givord, Electric-field-induced modification of magnetism in thin-film ferromagnets, *Science* **315**, 349–351 (2007).
102. T. Maruyama, Y. Shiota, T. Nozaki et al., Large voltage-induced magnetic anisotropy change in a few atomic layers of iron, *Nat. Nanotechnol.* **4**, 158–161 (2009).
103. M. Fechner, I. V. Maznichenko, S. Ostanin et al., Magnetic phase transition in two-phase multiferroics predicted from first-principles, *Phys. Rev. B* **78**, 212406 (2008).
104. J. P. Velev, C.-G. Duan, K. D. Belashchenko, S. S. Jaswal, and E. Y. Tsymbal, Effects of ferroelectricity and magnetism on electron and spin transport in Fe/$BaTiO_3$/Fe multiferroic tunnel junctions, *J. Appl. Phys.* **103**, 07A701 (2008).
105. J. Lee, N. Sai, T. Cai, Q. Niu, and A. A. Demkov, Interfacial magnetoelectric coupling in tri-component superlattices, *Phys. Rev. B* **81**, 144425 (2010).
106. K. Yamauchi, B. Sanyal, and S. Picozzi, Interface effects at a half-metal/ferroelectric junction, *Appl. Phys. Lett.* **91**, 062506 (2007).
107. M. K. Niranjan, J. P. Velev, C.-G. Duan, S. S. Jaswal, and E. Y. Tsymbal, Magnetoelectric effect at the Fe_3O_4/$BaTiO_3$ interface: A first-principles study, *Phys. Rev. B* **78**, 104405 (2008).
108. A. Gloter et al., unpublished.
109. M. Jullière, Tunneling between ferromagnetic films, *Phys. Lett. A* **54**, 225–226 (1975).
110. W. H. Butler, X.-G. Zhang, T. C. Schulthness, and J. M. MacLaren, Spin-dependent tunneling conductance of Fe|MgO|Fe sandwiches, *Phys. Rev. B* **63**, 054416 (2001).
111. J. Mathon and A. Umerski, Theory of tunneling magnetoresistance of an epitaxial Fe/MgO/Fe(001) junction, *Phys. Rev. B* **63**, 220403(R) (2001).

112. J. P. Velev, K. D. Belashchenko, D. A. Stewart, M. van Schilfgaarde, S. S. Jaswal, and E. Y. Tsymbal, Negative spin polarization and large tunneling magnetoresistance in epitaxial Co|$SrTiO_3$|Co magnetic tunnel junctions, *Phys. Rev. Lett.* **95**, 216601 (2005).
113. M. Bowen, A. Barthélémy, V. Bellini et al., Observation of Fowler-Nordheim hole tunneling across an electron tunnel junctions due to total symmetry filtering, *Phys. Rev. B* **73**, 140408(R) (2006).
114. V. Garcia, M. Bibes, L. Bocher et al., Ferroelectric control of spin polarization, *Science* **327**, 1106–1110 (2010).
115. J. Wang and Z. Y. Li, Multiple switching of spin polarization injected into a semiconductor by a multiferroic tunneling junction, *J. Appl. Phys.* **104**, 033908 (2008).

IV

Spin Transport and Magnetism in Semiconductors

16

Spin Relaxation and Spin Dynamics in Semiconductors

Jaroslav Fabian
University of Regensburg

M.W. Wu
University of Science and Technology of China

The spin of conduction electrons decays due to the combined effect of spin–orbit coupling and momentum scattering. The spin–orbit coupling couples the spin to the electron momentum that is randomized by momentum scattering off of impurities and phonons. Seen from the perspective of the electron spin, the spin–orbit coupling gives a spin precession, while momentum scattering makes this precession randomly fluctuating, both in magnitude and orientation.

The specific mechanisms for the spin relaxation of conduction electrons were proposed by Elliott [1] and Yafet [2], for conductors with a center of inversion symmetry, and by D'yakonov and Perel' [3], for conductors without an inversion center. In p-doped semiconductors, there is in play another spin relaxation mechanism, due to Bir et al. [4]. As this has a rather limited validity we do not describe it here. More details can be found in reviews [5–9].

Before we discuss the two main mechanisms, we introduce a toy model that captures the relevant physics of spin relaxation without resorting explicitly to quantum mechanics: *the electron spin in a randomly* fluctuating *magnetic field*. We will find certain universal qualitative features of the spin relaxation and dephasing in physically important situations.

The next part of this review covers the experimental as well as computational status of the field, discussing the spin relaxation in semiconductors under varying conditions such as temperature and doping density.

16.1 Toy Model: Electron Spin in a Fluctuating Magnetic Field

Consider an electron spin **S** (or the corresponding magnetic moment) in the presence of an external time-independent magnetic field $\mathbf{B}_0 = B_0\mathbf{z}$, giving rise to the Larmor precession frequency $\omega_0 = \omega_0\mathbf{z}$, and a fluctuating time-dependent field $\mathbf{B}(t)$ giving the Larmor frequency $\omega(t)$ (see Figure 16.1). We assume that the field fluctuates about zero and is correlated on the timescale of τ_c:

$$\overline{\omega(t)} = 0, \quad \overline{\omega_\alpha(t)\omega_\beta(t')} = \delta_{\alpha\beta}\overline{\omega_\alpha^2}e^{-|t-t'|/\tau_c}. \tag{16.1}$$

Here α and β denote the Cartesian coordinates and the overline denotes averaging over different random realizations $\mathbf{B}(t)$. We will see later that such fluctuating fields arise quite naturally in the context of the electron spins in solids.

The following description applies equally to the classical magnetic moment described by the vector **S** as well as to the quantum mechanical spin whose expectation value is **S**. Writing out the torque equation, $\dot{\mathbf{s}} = \omega \times \mathbf{S}$, we get the following equations of motion:

$$\dot{S}_x = -\omega_0 S_y + \omega_y(t)S_z - \omega_z(t)S_y, \tag{16.2}$$

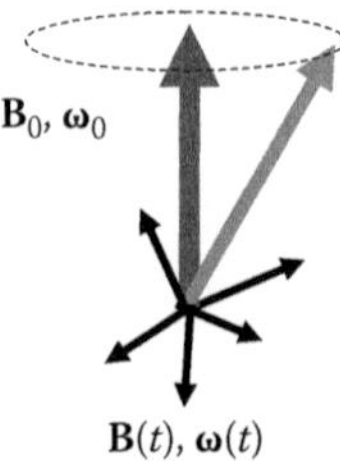

FIGURE 16.1 Electron spin precesses about the static $\mathbf{B}_0$ field along **z**. The randomly fluctuating magnetic field $\mathbf{B}(t)$ causes spin relaxation and spin dephasing.

$$\dot{S}_y = \omega_0 S_x - \omega_x(t)S_z + \omega_z(t)S_x, \tag{16.3}$$

$$\dot{S}_z = \omega_x(t)S_y - \omega_y(t)S_x. \tag{16.4}$$

These equations are valid for one specific realization of $\omega(t)$. Our goal is to find instead effective equations for the time evolution of the average spin, $\bar{s}(t)$, given the ensemble of Larmor frequencies $\omega(t)$.

It is convenient to introduce the complex "rotating" spins $S_\pm$ and Larmor frequencies $\omega_\pm$ in the (x,y) plane:

$$S_+ = S_x + iS_y, \quad S_- = S_x - iS_y, \tag{16.5}$$

$$\omega_+ = \omega_x + i\omega_y, \quad \omega_- = \omega_x - i\omega_y. \tag{16.6}$$

The inverse relations are

$$S_x = \frac{1}{2}(S_+ + S_-), \quad S_y = \frac{1}{2i}(S_+ - S_-), \tag{16.7}$$

$$\omega_x = \frac{1}{2}(\omega_+ + \omega_-), \quad \omega_y = \frac{1}{2i}(\omega_+ - \omega_-). \tag{16.8}$$

The equations of motion for the spin set (S_+, S_-, S_z) are

$$\dot{S}_+ = i\omega_0 S_+ + i\omega_z S_+ - i\omega_+ S_z, \tag{16.9}$$

$$\dot{S}_- = -i\omega_0 S_- - i\omega_z S_- + i\omega_- S_z, \tag{16.10}$$

$$\dot{S}_z = -\left(\frac{1}{2}i\right)\left(\omega_+ S_- - \omega_- S_+\right). \tag{16.11}$$

In the absence of the fluctuating fields, the spin S_+ rotates in the complex plane anticlockwise (for $\omega_0 > 0$) while S_- clockwise.

The precession about B_0 can be factored out by applying the ansatz*:

$$S_\pm = s_\pm(t)e^{\pm i\omega_0 t}. \tag{16.12}$$

Indeed, it is straightforward to find the time evolution of the set $(s_+, s_-, s_z \equiv S_z)$:

$$\dot{s}_+ = i\omega_z s_+ - i\omega_+ s_z e^{-i\omega_0 t}, \tag{16.13}$$

$$\dot{s}_- = -i\omega_z s_- + i\omega_- s_z e^{i\omega_0 t}, \tag{16.14}$$

$$\dot{s}_z = -\left(\frac{1}{2}i\right)\left(\omega_+ s_- e^{-i\omega_0 t} - \omega_- s_+ e^{i\omega_0 t}\right). \tag{16.15}$$

The penalty for transforming into this "rotating frame" is the appearance of the phase factors $\exp(\pm i\omega_0 t)$.

The solutions of Equations 16.13 through 16.15 can be written in terms of the integral equations

$$s_+(t) = s_+(0) + i\int_0^t dt'\omega_z(t')s_+(t') - i\int_0^t dt'\omega_+(t')s_z(t')e^{-i\omega_0 t'}, \tag{16.16}$$

$$s_-(t) = s_-(0) - i\int_0^t dt'\omega_z(t')s_-(t') + i\int_0^t dt'\omega_-(t')s_z(t')e^{i\omega_0 t'}, \tag{16.17}$$

$$s_z(t) = s_z(0) - \frac{1}{2i}\int_0^t dt'\left[\omega_+(t')s_-(t')e^{-i\omega_0 t'} - \omega_-(t')s_+(t')e^{i\omega_0 t'}\right]. \tag{16.18}$$

We should now substitute the above solutions back into Equations 16.13 through 16.15. The corresponding expressions become rather lengthy, so we demonstrate the procedure on the s_+ component only. We get

$$\begin{aligned}\dot{s}_+(t) &= i\omega_z(t)s_+(0) - \omega_z(t)\int_0^t dt'\omega_z(t')s_+(t') \\ &+ \omega_z(t)\int_0^t dt'\omega_+(t')s_z(t')e^{-i\omega_0 t'} - i\omega_+(t)e^{-i\omega_0 t}s_z(0) \\ &+ \frac{1}{2}e^{-i\omega_0 t}\omega_+(t)\int_0^t dt'\left[\omega_+(t')s_-(t')e^{-i\omega_0 t'} - \omega_-(t')s_+(t')e^{i\omega_0 t'}\right].\end{aligned} \tag{16.19}$$

The reader is encouraged to write the analogous equations for $\dot{s}_z$ (that for $\dot{s}_-$ is easy to write since $s_- = s_+^*$).

We now make two approximations. First, we assume that the fluctuating field is rather weak and stay in *the second* order in ω.† This allows us to factorize the averaging over the statistical realizations of the field

$$\overline{\omega(t)\omega(t')s(t')} \approx \overline{\omega(t)\omega(t')}\,\overline{s(t')} \tag{16.20}$$

* This is analogous to going to the interaction picture when dealing with a quantum mechanical problem of that type.

† More precisely, we assume that $|\omega(t)|\tau_c \ll 1$, so that the spin does not fully precess about the fluctuating field before the field makes a random change.

as the spin changes only weakly over the timescale, τ_c, of the changes of the fluctuating fields. This approximation is called the *Born approximation*, alluding to the analogy with the second-order time-dependent perturbation theory in quantum mechanics. Going beyond the Born approximation one would need to execute complicated averaging schemes of the product in Equation 16.20, since $s(t)$ in general depends on $\omega(t' \leq t)$.

As the second assumption, we consider a "coarse-grained" time evolution, meaning that we are effectively averaging $s(t)$ over the timescale of the correlation time τ_c; we are interested in times t much greater than τ_c. That allows us to approximate

$$\int_0^{t \gg \tau_c} dt' \overline{\omega(t)\omega(t')s(t')} \approx \int_0^{t \gg \tau_c} dt' \overline{\omega(t)\omega(t')}\,\overline{s(t)}, \tag{16.21}$$

since the correlation function $\overline{\omega(t)\omega(t')}$ is significant in the time interval of $|t-t'| \approx \tau_c$ only. The above approximation makes clear that the spin $s(t)$ is the representative coarse-grained (running-averaged) spin of the time interval $(t-\tau_c, t)$. Equation 16.21 is a realization of the *Markov approximation*. The physical meaning is that the spin s varies only slowly on the timescale of τ_c over which the correlation of the fluctuating fields is significant. We then need to restrict ourselves to the timescales t larger than the correlation time τ_c. In effect, we will see that in this approximation *the rate of change of the spin at a given time depends on the spin at that time, not on the previous history of the spin.*

Applying the Born–Markov approximation to Equation 16.19, we obtain for the average spin $\bar{s}_+$ the following time evolution equation*:

$$\begin{aligned}\bar{s}_+ &= i\overline{\omega_z(t)s_+(0)} - \int_0^t dt' \overline{\omega_z(t)\omega_z(t')s_+(t)} \\ &+ \int_0^t dt' \overline{\omega_z(t)\omega_+(t')}e^{-i\omega_0 t'}\overline{s_z(t)} - i\overline{\omega_+(t)}e^{-i\omega_0 t}s_z(0) + \frac{1}{2}e^{-i\omega_0 t} \\ &\times \int_0^t dt' \left[\overline{\omega_+(t)\omega_+(t')}e^{-i\omega_0 t'}\overline{s_-(t)} - \overline{\omega_+(t)\omega_-(t')}e^{i\omega_0 t'}\overline{s_+(t)}\right].\end{aligned} \tag{16.22}$$

Using the rules of Equation 16.1, Equation 16.22 simplifies to

$$\begin{aligned}\bar{s}_+ &= -\overline{\omega_z^2}\int_0^t dt' e^{-(t-t')/\tau_c}\overline{s_+(t)} + \frac{1}{2}e^{-i\omega_0 t} \\ &\times \int_0^t dt' \left[(\overline{\omega_x^2} - \overline{\omega_y^2})e^{-i\omega_0 t'}\overline{s_-(t)} - (\overline{\omega_x^2} + \overline{\omega_y^2})e^{i\omega_0 t'}\overline{s_+(t)}\right]e^{-(t-t')/\tau_c}.\end{aligned} \tag{16.23}$$

* The initial values of the spin, $s(0)$, are fixed and not affected by averaging.

Since we consider the times $t \gg \tau_c$, we can approximate

$$\int_0^t dt' e^{-(t-t')/\tau_c} \approx \int_{-\infty}^t dt' e^{-(t-t')/\tau_c} = \tau_c. \tag{16.24}$$

Similarly,

$$\int_0^t dt' e^{-(t-t')/\tau_c} e^{-i\omega_0(t \pm t')} \approx \int_{-\infty}^t dt' e^{-(t-t')/\tau_c} e^{-i\omega_0(t \pm t')} = \tau_c \frac{1 \pm i\omega_0\tau_c}{1+\omega_0^2\tau_c^2}. \tag{16.25}$$

The imaginary parts induce the precession of $s_\pm$, which is equivalent to shifting (renormalizing) the Larmor frequency ω_0. The relative change of the frequency is $(\omega\tau_c)^2$, which is assumed much smaller than one by our Born approximation. We thus keep the real parts only and obtain

$$\dot{\bar{s}}_+ = -\overline{\omega_z^2}\tau_c \bar{s}_+ + \frac{1}{2}\frac{\tau_c}{1+\omega_0^2\tau_c^2}\left[(\overline{\omega_x^2} - \overline{\omega_y^2})\bar{s}_- e^{-2i\omega_0 t} - (\overline{\omega_x^2} + \overline{\omega_y^2})\bar{s}_+\right]. \tag{16.26}$$

Using the same procedure (or simply using $s_- = s_+^*$), we would arrive for the analogous equation for s_-:

$$\dot{\bar{s}}_+ = -\overline{\omega_z^2}\tau_c \bar{s}_- + \frac{1}{2}\frac{\tau_c}{1+\omega_0^2\tau_c^2}\left[(\overline{\omega_x^2} - \overline{\omega_y^2})\bar{s}_+ e^{2i\omega_0 t} - (\overline{\omega_x^2} + \overline{\omega_y^2})\bar{s}_-\right]. \tag{16.27}$$

Similarly,

$$\dot{\bar{s}}_z = -(\overline{\omega_x^2} + \overline{\omega_y^2})\frac{\tau_c}{1+\omega_0^2\tau_c^2}\bar{s}_z. \tag{16.28}$$

For the rest of the section we omit the overline on the symbols for the spins, so that S will mean the average spin. Returning back to our rest frame of the spins rotating with frequency ω_0, we get

$$\dot{S}_+ = i\omega_0 S_+ - \overline{\omega_z^2}\tau_c \frac{1}{2}\frac{\tau_c}{1+\omega_0^2\tau_c^2}\left[(\overline{\omega_x^2} - \overline{\omega_y^2})S_- - (\overline{\omega_x^2} + \overline{\omega_y^2})S_+\right], \tag{16.29}$$

$$\dot{S}_- = i\omega_0 S_- - \overline{\omega_z^2} - \frac{1}{2}\frac{\tau_c}{1+\omega_0^2\tau_c^2}\left[(\overline{\omega_x^2} - \overline{\omega_y^2})S_+ - (\overline{\omega_x^2} + \overline{\omega_y^2})S_-\right], \tag{16.30}$$

$$\dot{S}_z = -(\overline{\omega_x^2} + \overline{\omega_y^2})\frac{\tau_c}{1+\omega_0^2\tau_c^2}S_z. \tag{16.31}$$

Finally, going back to S_x and S_y

$$\dot{S}_x = -\omega_0 S_y - \overline{\omega_z^2}\tau_c S_x - \frac{\tau_c}{1+\omega_0^2\tau_c^2}\overline{\omega_y^2}S_x, \tag{16.32}$$

$$\dot{S}_y = \omega_0 S_x - \overline{\omega_z^2}\tau_c S_y - \frac{\tau_c}{1+\omega_0^2\tau_c^2}(\overline{\omega_x^2})S_y, \tag{16.33}$$

$$\dot{S}_z = -(\overline{\omega_x^2} + \overline{\omega_y^2})\frac{\tau_c}{1+\omega_0^2\tau_c^2}S_z. \tag{16.34}$$

We can give the above equation a more conventional form by introducing two types of the spin decay times. First, we define the *spin relaxation time* T_1 by

$$\frac{1}{T_1}=\left(\overline{\omega_x^2}+\overline{\omega_y^2}\right)\frac{\tau_c}{1+\omega_0^2\tau_c^2},\tag{16.35}$$

and the spin dephasing times T_2 by

$$\frac{1}{T_{2x}}=\overline{\omega_z^2}\tau_c+\frac{\overline{\omega_y^2}\tau_c}{1+\omega_0^2\tau_c^2},\tag{16.36}$$

$$\frac{1}{T_{2y}}=\overline{\omega_z^2}\tau_c+\frac{\overline{\omega_x^2}\tau_c}{1+\omega_0^2\tau_c^2}.\tag{16.37}$$

We then write

$$\dot{S}_x=-\omega_0 S_y-\frac{S_x}{T_{2x}},\tag{16.38}$$

$$\dot{S}_y=\omega_0 S_x-\frac{S_y}{T_{2y}},\tag{16.39}$$

$$\dot{S}_z=-\frac{S_z}{T_1}.\tag{16.40}$$

Our fluctuating field is effectively at infinite temperature, at which the average value for the spin in a magnetic field is zero. A more general spin dynamics is

$$\dot{S}_x=-\omega_0 S_y-\frac{S_x}{T_{2x}},\tag{16.41}$$

$$\dot{S}_y=\omega_0 S_x-\frac{S_y}{T_{2y}},\tag{16.42}$$

$$\dot{S}_z=-\frac{S_z-S_{0z}}{T_1}.\tag{16.43}$$

where S_{0z} is the equilibrium value of the spin in the presence of the static magnetic field of the Larmor frequency ω_0 at the temperature at which the environmental fields giving rise to $\omega(t)$ are in equilibrium. The above equations are called the *Bloch equations*.

The spin components S_x and S_y, which are perpendicular to the applied static field $\mathbf{B}_0$, decay exponentially on the timescales of T_{2x} and T_{2y}, respectively. These times are termed spin *dephasing* times, as they describe the loss of the *phase* of the spin components perpendicular to the static field $\mathbf{B}_0$. They are also often called *transverse* times, for that reason. The time T_1 is termed the spin *relaxation* time, as it describes the (thermal) relaxation of the spin to the equilibrium. During the spin relaxation in a static magnetic field, the energy is exchanged with the environment. In the language of statistical physics, the relaxation process establishes the Boltzmann probability distribution for the system. Similarly, dephasing establishes the "random phases" postulate that says that there is no correlation (coherence) among the degenerate states, such as the two transverse spin orientations; in thermal equilibrium such states add incoherently.

For the sake of discussion consider an isotropic system in which

$$\overline{\omega_x^2}=\overline{\omega_y^2}=\overline{\omega_z^2}=\overline{\omega^2}.\tag{16.44}$$

If the static magnetic field is weak, $\omega_0\tau_c \ll 1$, the three times are equal:

$$T_1=T_{2x}=T_{2y}=\frac{1}{\overline{\omega^2}\tau_c}.\tag{16.45}$$

There is no difference between the spin relaxation and spin dephasing. It is at first sight surprising that the spin relaxation time is inversely proportional to the correlation time. The more random the external field appears, the less the spin decays. We will explain this fact below by the phenomenon of motional narrowing.

In the opposite limit of large Larmor frequency, $\omega_0\tau_c \gg 1$, the spin relaxation rate vanishes

$$\frac{1}{T_1}\approx 2\frac{\overline{\omega^2}}{\omega_0^2}\frac{1}{\tau_c}\to 0,\tag{16.46}$$

while the spin dephasing time is given by what is called *secular broadening*:

$$\frac{1}{T_2}\approx\overline{\omega_z^2}\tau_c.\tag{16.47}$$

If secular broadening is absent, the leading term in the dephasing time will be, as in the relaxation

$$\frac{1}{T_2}\approx\frac{\overline{\omega^2}}{\omega_0^2}\frac{1}{\tau_c}.\tag{16.48}$$

In this limit the spin dephasing rate is proportional to the correlation rate, not to the correlation time.

In many physical situations the fluctuating fields are anisotropic. Perhaps the most case is such that $\omega_z=0$ while

$$\overline{\omega_x^2}=\overline{\omega_y^2}=\overline{\omega^2}.\tag{16.49}$$

Then, from Equations 16.35 through 16.37, it follows that

$$T_2=2T_1.\tag{16.50}$$

This anisotropic spin relaxation is due to the fact that while the transverse spin can be flipped by two modes of the in-plane fluctuating field, the in-plane spins can be flipped by one mode only (the one perpendicular to the spin).

In the cases in which there is no clear distinction between T_1 and T_2, we often use the symbol

$$\tau_s = T_1 = T_2, \tag{16.51}$$

to describe the *generic spin relaxation*.

16.1.1 Motional Narrowing

The surprising fact that at low magnetic fields the spin relaxation rate is proportional to the correlation time of the fluctuating field (as opposed to its inverse) is explained by *motional narrowing*. Consider the spin transverse to an applied magnetic field and assume that the field has a single magnitude, but can randomly switch directions, between up to down, leading to a random precession of the spin clock- and anticlockwise. In effect, the spin phase executes a random walk. A single step takes the time τ_c, the correlation time of the fluctuating field. After n steps, that is, after the time $t=n\tau$, the standard deviation of the phase will be $\delta\phi = (\omega\tau_c)\sqrt{n}$, the well-known result for a random walk. We call the spin dephasing time the time it takes for $\delta\phi \approx 1$. This happens after the time $\tau_s = \tau_c/(\omega\tau_c)^2$, or $\tau_s = 1/(\omega^2\tau_c)$, which is the result we obtained earlier from the Born–Markov approximation, Equation 16.45.

16.1.2 Reversible Dephasing, Spin Ensemble, Random Walk in Inhomogeneous Fields

Our previous calculation was carried out for a single spin in a fluctuating magnetic field. After the decay of the spin components, the information about the original spin is irreversibly lost as we have no information on the actual history of the fluctuating field. Such an irreversible loss of spin is often termed spin *decoherence*. We will see that spin dephasing can be reversible. We will also see that a simple exponential decay, of the type $\exp(-t/\tau_s)$, is not a rule.

16.1.2.1 Reversible Spin Dephasing: Spin Ensemble in Spatially Random Magnetic Field

There are physically relevant cases in which the decay of spin is reversible. A typical example is an *ensemble* of localized spins, each precessing about a static local magnetic field $\mathbf{B}_0+\mathbf{B}_1$, with varying static components $\mathbf{B}_1$ (of zero average) giving rise to random precession frequencies ω_1 (see Figure 16.2). Another important example is that of the conduction electrons in noncentrosymmetric crystals, such as GaAs, in which the spin–orbit coupling acts as a momentum-dependent magnetic field. The spins of the electrons of different momenta precess with different frequencies. We are interested in the total spin as the sum of the individual spins.

Consider the external field along z direction, and take the fluctuating frequencies from the Gaussian distribution

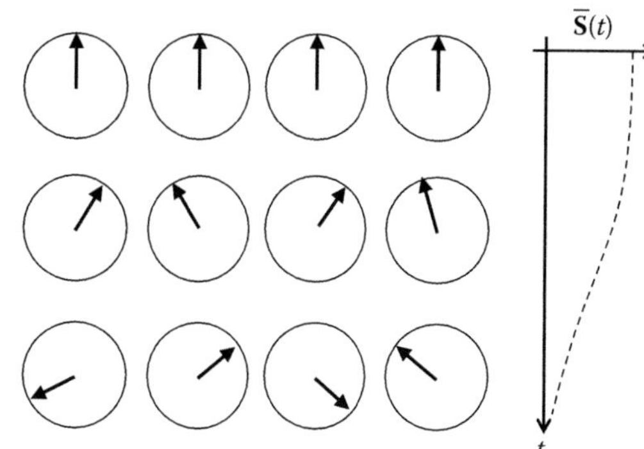

FIGURE 16.2 Electron spin precesses about the static $\mathbf{B}_0$ field along $\mathbf{z}$ perpendicular to this page. The spatially fluctuating magnetic field $B_1\mathbf{z}$ causes reversible spin dephasing.

$$P(\omega_1) = \frac{1}{\sqrt{2\pi\delta\omega^2}} e^{-\omega_1^2/2\delta\omega^2}, \tag{16.52}$$

with zero mean and $\delta\omega_1^2$ variance. Denote the in-plane spin of the electron a* by S_x^a and S_y^a. This spin precesses with the frequency $\omega_0 + \omega_1^a$, leading to the time evolution for the rotating spins:

$$S_\pm^a(t) = S_x^a(t) \pm iS_y^a(t) = S_\pm^a(0)e^{\pm i\omega_0 t}e^{\pm i\omega_1^a t}. \tag{16.53}$$

Suppose at $t=0$ all the spins are lined up, that is, $S_\pm^a(0) = S_\pm(0)$. The total spin $S_\pm(t)$ is the sum

$$S_\pm(t) = \sum_s S_\pm^a(t) = S_\pm(0)e^{\pm i\omega_0 t}\int_{-\infty}^{\infty} d\omega_1 P(\omega_1)e^{\pm i\omega_1 t}. \tag{16.54}$$

Evaluating the Gaussian integral we get

$$S_\pm(t) = S_\pm(0)e^{\pm i\omega_0 t}e^{-\delta\omega_1^2 t^2/2}. \tag{16.55}$$

The in-plane component vanishes after the time of about $1/\delta\omega_1$, but this dephasing of the spin is reversible, since each individual spin preserves the memory of the initial state. The disappearance of the spin is purely due to the statistical averaging over an ensemble in which the individual spins have, after certain time, random phases. This spin decay is not a simple exponential, but rather Gaussian.

16.1.2.2 Spin Echo

We have seen that there are irreversible and reversible effects both present in spin dephasing. It turns out that the reversible effects can be separated out by the phenomenon of the spin echo. Figure 16.3 explains this mechanism in detail. Suppose all the localized spins in our ensemble point in one direction at time $t=0$. At a later time, $t=T_\pi$, the spins dephase due to the

* This discussion applies to nuclear spins as well.

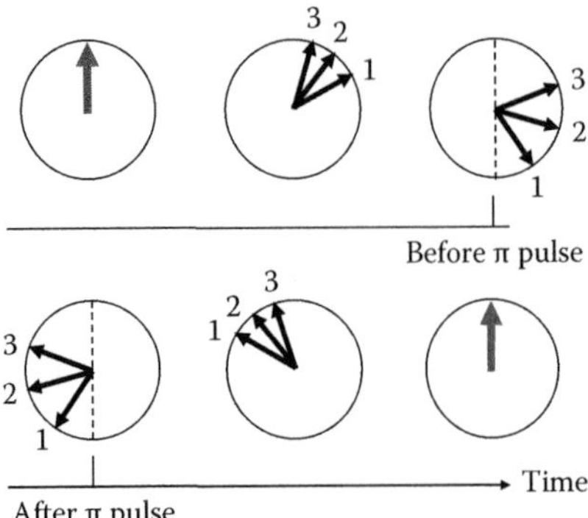

FIGURE 16.3 Initially all the spins point up. Due to random precession frequencies the spins soon point in different directions and the average spin vanishes. Applying a π-pulse rotating the spins along the vertical axis makes the fastest spin the slowest, and vice versa. After the fastest ones catch up again with the slowest, the original value of the average spin is restored. Any reduction from the original value is due to irreversible processes signal.

inhomogeneities of the precession frequencies and the total spin is small. We apply a short pulse of an external magnetic field, the so-called π pulse, that rotates the spins along the axis parallel to the original spin direction, mapping the spins as $(S_x, S_y) \to (-S_x, S_y)$. At that moment the spins will still be dephased, but the one that is the fastest is now the last, and the one that is the slowest appears as the first. At the time $t=2T_\pi$ all the spins catch on, producing a large spin signal along the original spin direction. In reality this signal will be weaker than that at $t=0$ due to the presence of irreversible processes, by $\exp(-2T_\pi/T_2)$. Important, reversible processes are not counted in T_2.

16.1.2.3 Spin Random Walk in Inhomogeneous Magnetic Field

Another interesting situation appears when we consider the possibility that the spin diffuses through a region of an inhomogeneous precession frequency.* We can model this situation on the system of spins in one dimension, each spin jumping in a time τ left or right. Suppose the precession frequencies vary in the x direction (see Figure 16.4) as

$$\omega(x) = \omega_0 + \omega' x. \tag{16.56}$$

Here ω′ is the gradient of ω. As the electron diffuses, it precesses with the precession frequency $\omega(t)=\omega[x(t)]$, given by the position of the electron $x(t)$ at time t. The time evolution for an individual spin is

$$\dot{S}_+ = \omega(t)S_+(t) = \omega(t)S_+(0) + \omega(t)\int_0^\infty dt'\omega(t')S(t'). \tag{16.57}$$

We are interested in the averaged quantities over many realizations of the random walk. We also consider the timescales much greater than the individual random walks steps τ.

* The inhomogeneity could be due to the magnetic field or to the g-factor.

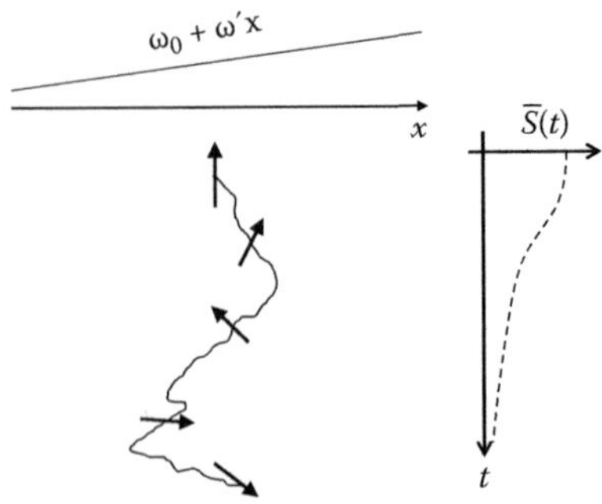

FIGURE 16.4 The electron performs a random walk. Its spin precesses by the inhomogeneous magnetic field along the x-axis. The average spin dephases to zero with time.

The accumulated phase after N steps or $t=N\tau$ time is the sum

$$\phi_N = x_1 + x_2 + \cdots + x_N = N\delta_1 + (N-1)\delta_1 + \cdots + \delta_N. \tag{16.58}$$

Here $\delta_i = \pm 1$ is a random variable representing the random step left or right. The variance of ϕ_N is

$$\sigma_N^2 = \sum_i^{N-1}(N-i)^2 \approx \frac{1}{3}N^3. \tag{16.59}$$

According to the central limit theorem, ϕ_N is distributed normally with the above variance:

$$P(\phi_N) = \frac{1}{\sqrt{2\pi\sigma_N^2}} e^{-\phi_N^2/2\sigma_N^2}. \tag{16.60}$$

We can now transform the dynamical equation into the ensemble averaging

$$S_+(t) = S_+(0)e^{i\phi(t)} = \int_{-\infty}^{\infty} d\phi_N e^{i\phi(t)} P(\phi_N) = e^{-\omega_1^2 D t^3/3}, \tag{16.61}$$

where we denoted the diffusivity by $D=\tau/2$, for the unit length step.

16.1.3 Quantum Mechanical Description

The Born–Markov approximation to the dynamics of a system coupled to an external environment can be cast in the quantum mechanical language. We refer the reader to Ref. [8] for the derivation. The physics of this derivation is the same as what we did in Section 16.1. Only the formalism is different.

We consider the system described by the Hamiltonian

$$H(t) = H_0 + V(t), \tag{16.62}$$

in which H_0 is our system per se and $V(t)$ is the time-dependent random fluctuating field of zero average and correlation time τ_c:

$$\overline{V(t)} = \overline{V(t)} = 0, \quad \overline{V(t)V(t')} \sim e^{-|t-t'|/\tau_c}. \tag{16.63}$$

The system is fully described by the density matrix ρ.

The transformation to the rotating frame is equivalent to going to the interaction picture in quantum mechanics:

$$\rho_I(t) = e^{iH_0t/\hbar}\rho e^{-iH_0t/\hbar}, \tag{16.64}$$

$$V_I(t) = e^{iH_0t/\hbar}V(t)e^{-iH_0t/\hbar}. \tag{16.65}$$

Performing the operations, as outlined in Section 16.1, for the classical model, we arrive at the effective time evolution for the density of state operator:

$$\frac{d\overline{\rho_I(t)}}{dt} = \left(\frac{1}{i\hbar}\right)^2 \int_0^{t\gg\tau_c} dt' \left[\overline{V_I(t),[V_I(t'),\overline{\rho_I(t)}]}\right]. \tag{16.66}$$

The above equation is called the *Master equation* and is the starting equation in many important problems in which a quantum system is in contact with a reservoir (Figure 16.4).

16.1.4 Spin Relaxation of Conduction Electrons

We will consider nonmagnetic conductors in zero or weak magnetic fields so that the spin dephasing and spin relaxation times are equal, $T_2 = T_1 = \tau_s$. The formula for the spin relaxation time

$$\frac{1}{\tau_s} = \omega^2\tau_c, \tag{16.67}$$

that was derived above for the spin in a fluctuating magnetic field applies in a semiquantitative sense (i.e., it gives an order of magnitude estimates and useful trends) to the conduction electron spins as well. We analyze below the D'yakonov–Perel' [3] and the Elliott–Yafet [1,2] mechanisms.

The D'yakonov–Perel' mechanism is at play in solids lacking a center of spatial inversion symmetry. The most prominent example is the semiconductor GaAs. In such solids the spin–orbit coupling is manifested as some effective magnetic field—*the spin-orbit field*—that depends on the electron momentum. Electrons in different momentum states feel different spin–orbit fields, so that the spin precesses with a given Larmor frequency, until the electron is scattered into another momentum state (see Figure 16.5). As the electron momentum changes on the timescale τ of the momentum relaxation time, the net effect of the momentum scattering on the spin is to produce random fluctuations of the Larmor frequencies. We have motional narrowing. Since these frequencies are correlated by $\tau_c = \tau$, we arrive at

$$\frac{1}{\tau_s} = \omega_{so}^2\tau, \tag{16.68}$$

for the spin relaxation time. The magnitude of ω_{so} is the measure of the strength of the spin–orbit coupling. The spin relaxation

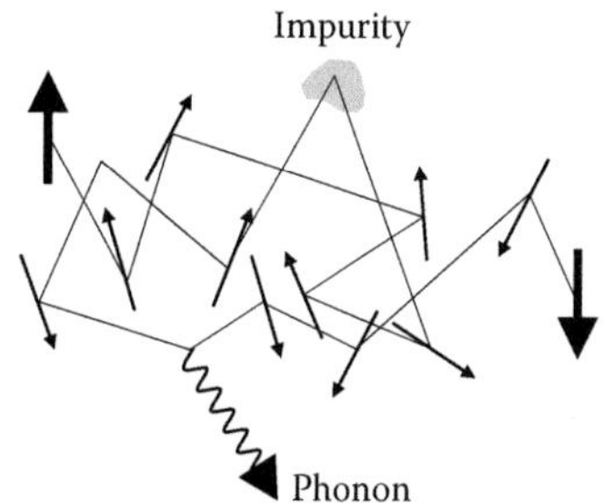

FIGURE 16.5 D'yakonov–Perel' mechanism. The electron starts with the spin up. As it moves, its spin precesses about the axis corresponding to the electron velocity. Phonons and impurities change the velocity, making the spin to precess about a different axis (and with different speed). During the scattering event the spin is preserved.

rate is directly proportional to the momentum relaxation time—the more the electron scatters, the less its spin dephases.

The Elliott-Yafet mechanism works in systems with and without a center of inversion. It relies on spin-flip momentum scattering. The spin-flip amplitudes are due to spin–orbit coupling, while the momentum scattering is due to the presence of impurities (that also contribute to the spin–orbit coupling), phonons, rough boundaries, or whatever is capable of randomizing the electron momentum. During the scattering events, the spin is preserved (see Figure 16.6). How do we account for such a scenario with our toy model? Consider an electron scattering off of an impurity with a spin flip. This spin flip can be viewed as a precession that occurs during the time of the interaction of the electron with the impurity. Let us say that the scattering takes the time of λ_F/υ_F, where υ_F is the electron velocity and λ_F is the electron wavelength at the Fermi level (considering that it is greater or at most equal to the size a of the impurity—otherwise we could equally put a/υ_F). Then, the precession angle $\varphi = \omega_{so}(\lambda_F/\upsilon_F)$, with ω_{so} denoting the spin–orbit coupling-induced precession frequency. Let us compare this with the angle of precession, ωτ, in the motional narrowing model, Equation 16.67. We see that the spin-flip can be described by the effective precession frequency $\omega = \omega_{so}(\lambda_F/\upsilon_F\tau)$. We then obtain for the spin relaxation rate

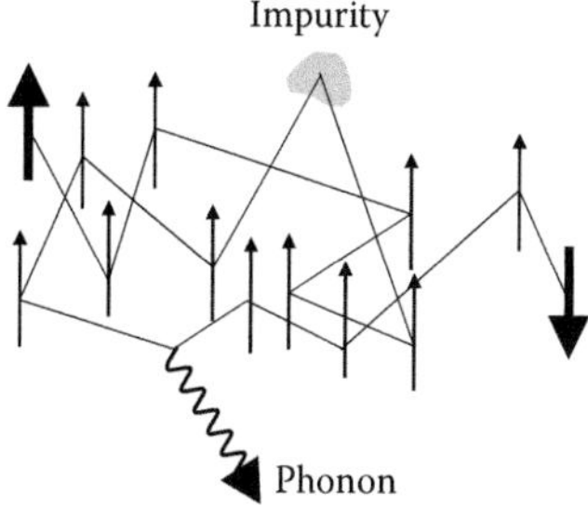

FIGURE 16.6 Elliott–Yafet mechanism. The electron starts with the spin up. As it scatters off of impurities and phonons, the spin can also flip due to spin–orbit coupling. Between the scattering the spin is preserved. After, say, a million scattering events, the spin will be down.

$$\frac{1}{\tau_s} = \omega_{so}^2 \frac{\lambda_F^2}{v_F^2 \tau} \approx \left(\frac{\varepsilon_{so}}{\varepsilon_F}\right)^2 \frac{1}{\tau}. \quad (16.69)$$

Here $\varepsilon_{so} = \hbar\omega_{so}$ and $\varepsilon = (\hbar k_F)v_F/2$ is the Fermi energy. For the Elliott–Yafet mechanism holds that the more the electron scatters, the more the spin dephases.

Below we discuss the two mechanisms in more detail, providing the formalisms for their investigation.

16.2 D'yakonov–Perel' Mechanism

D'yakonov and Perel' [3] considered solids without a center of inversion symmetry, such as GaAs or InAs. The mechanism also works for electrons at surfaces and interfaces. In such solids, the presence of spin–orbit coupling induces the spin–orbit fields, which give rise to the spin precession. Momentum relaxation then causes the random walk of the spin phases.

16.2.1 Spin–Orbit Field

In solids without a center of inversion symmetry, spin–orbit coupling splits the electron energies:

$$\varepsilon_{\mathbf{k},\uparrow} \neq \varepsilon_{\mathbf{k},\downarrow}. \quad (16.70)$$

Only the Kramers degeneracy is left, due to time reversal symmetry:

$$\varepsilon_{\mathbf{k},\uparrow} = \varepsilon_{-\mathbf{k},\downarrow}. \quad (16.71)$$

This energy splitting at a given momentum **k** is conveniently described by the spin–orbit field **Ω**, giving a Zeeman-like (but momentum-dependent) energy contribution to the electronic states, described by the additional Hamiltonian (to the usual band structure):

$$H_1 = \frac{\hbar}{2} \mathbf{W}_{\mathbf{k}} \cdot \sigma. \quad (16.72)$$

The time reversal symmetry requires that the spin–orbit field is an odd function of the momentum:

$$\mathbf{W}_{\mathbf{k}} = -\mathbf{W}_{-\mathbf{k}}. \quad (16.73)$$

The representative example of a spin–orbit field is the Bychkov–Rashba [10] (α_{BR}) and Dresselhaus [11,12] (γ_D) couplings in 2d electron gases formed at the zinc-blend heterostructures grown along [001] [8]:

$$H_1 = (\alpha_{BR} + \gamma_D)\sigma_x k_y - (\alpha_{BR} - \gamma_D)\sigma_y k_x. \quad (16.74)$$

The axes are $x = [110]$ and $y = [1\bar{1}0]$. The spin–orbit field is

$$\hbar \mathbf{W}_{\mathbf{k}} = 2\left[(\alpha_{BR} + \gamma_D)k_y, -(\alpha_{BR} - \gamma_D)k_x\right]. \quad (16.75)$$

This field has the C_{2v} symmetry, reflecting the structural symmetry of the zinc-blend interfaces (such as GaAs/GaAlAs) with the principal axes along [110] and $[1\bar{1}0]$.

16.2.2 Kinetic Equation for the Spin

Let $\mathbf{s}_{\mathbf{k}}$ be the electron spin in the momentum state **k**. The time evolution of the spin is then described by

$$\frac{\partial \mathbf{s}_{\mathbf{k}}}{\partial t} - \mathbf{W}_{\mathbf{k}} \times \mathbf{s}_{\mathbf{k}} = -\sum_{\mathbf{k}'} W_{\mathbf{k}\mathbf{k}'}\left(\mathbf{s}_{\mathbf{k}} - \mathbf{s}'_{\mathbf{k}}\right). \quad (16.76)$$

The left-hand side gives the full time derivative $d\mathbf{s}_{\mathbf{k}}/dt$, which is due to the explicit change of the spin, its direction, and the change of the spin due to the change of the momentum **k**. The right-hand side is the change of the spin at **k** due to the spin-preserving scattering from and to that state. The scattering rate between the two momentum states **k** and **k′** is $W_{\mathbf{kk}}$.

There are two timescales in the problem. First, momentum scattering, which occurs on the timescale of the momentum relaxation time τ, makes the spins in different momentum states equal. Second, this uniform spin decays on the timescale of the spin relaxation time τ_s, which we need to find. We assume that $\tau \ll \tau_s$; this assumption is well satisfied in real systems. In principle we could go directly to our model of the electron spin in a fluctuating magnetic field, with the role of the random Larmor precession playing by **Ω**, identifying $\tau_c = \tau$. However, it is instructive to see how this two time-scale problem is solved directly.

We separate the fast and slow components of the spins as follows:

$$\mathbf{s}_{\mathbf{k}} = \mathbf{s} + \mathbf{x}_{\mathbf{k}}, \quad \langle \mathbf{x}_{\mathbf{k}} \rangle = 0. \quad (16.77)$$

The symbol $\langle \cdots \rangle$ denotes averaging over different momenta. Our goal is to find the effective equation for the time evolution of **s**, which is the actual spin averaged over **k**. The fast component, $\xi_{\mathbf{k}}$, decays on the timescale of τ to the value given by the instantaneous value of **s**. Our goal is to find the effective equation for the time evolution of **s**, which is the actual spin averaged over **k**. The fast component, $\xi_{\mathbf{k}}$, decays on the timescale of τ to the quasistatic value given by the instantaneous value of **s**.

Upon substituting Equation 16.77 into the kinetic Equation 16.76 and averaging over **k**, we obtain the equation of motion for the averaged spin

$$\dot{\mathbf{s}} = \langle \mathbf{W}_{\mathbf{k}} \times \mathbf{x}_{\mathbf{k}} \rangle, \quad (16.78)$$

using that $\langle \Omega_{\mathbf{k}} = 0 \rangle$. We need to find $\xi_{\mathbf{k}}$. Since **s** is hardly changing on the timescales relevant to $\xi_{\mathbf{k}}$, we write

$$\dot{\mathbf{x}}_{\mathbf{k}} - \mathbf{W}_{\mathbf{k}} \times \mathbf{s} - \mathbf{W}_{\mathbf{k}} \times \mathbf{x}_{\mathbf{k}} = -\sum_{\mathbf{k}'} W_{\mathbf{kk}'}\left(\mathbf{x}_{\mathbf{k}} - \mathbf{x}_{\mathbf{k}'}\right). \tag{16.79}$$

We make the following assumption:

$$\Omega\tau \ll 1. \tag{16.80}$$

That is, we assume that the precession is slow on the timescale of the momentum relaxation time (see our note on the Born approximation in the toy model in Section 16.1). For simplicity, we make the relaxation time approximation to model the time evolution of the fast component $\boldsymbol{\xi}_{\mathbf{k}}$:

$$\dot{\mathbf{x}}_{\mathbf{k}} - \mathbf{W}_{\mathbf{k}} \times \mathbf{s} - \mathbf{W}_{\mathbf{k}} \times \mathbf{x}_{\mathbf{k}} = -\frac{\mathbf{x}_{\mathbf{k}}}{\tau}. \tag{16.81}$$

From the condition of the quasistatic behavior, $\partial\xi_{\mathbf{k}}/\partial t = 0$, we get up to the first order in $\Omega\tau$ the following solution for the quasistatic $\boldsymbol{\xi}_{\mathbf{k}}$:

$$\mathbf{x}_{\mathbf{k}} = \tau\left(\mathbf{W}_{\mathbf{k}} \times \mathbf{s}\right). \tag{16.82}$$

The above is a realization of coarse graining, in which we effectively average the spin evolution over the timescales of τ.

Substituting the quasistatic value of $\boldsymbol{\xi}_{\mathbf{k}}$ from Equation 16.82 to the time evolution equation for **s**, Equation 16.78, we find

$$\dot{\mathbf{s}} = \tau\left\langle \mathbf{W}_{\mathbf{k}} \times \left(\mathbf{W}_{\mathbf{k}} \times \mathbf{s}\right)\right\rangle. \tag{16.83}$$

Using the vector product identities, we finally obtain for the individual spin components α:

$$\dot{s}_{\alpha} = \left\langle \Omega_{\mathbf{k}\alpha}\Omega_{\mathbf{k}\beta}\right\rangle \tau s_{\beta} - \left\langle \mathbf{W}_{\mathbf{k}}^{2}\tau\right\rangle s_{\alpha}. \tag{16.84}$$

These equations describe the effective time evolution of the electron spin in the presence of the momentum-dependent Larmor precession $\boldsymbol{\Omega}_{\mathbf{k}}$.

For the specific model of the zinc-blend heterostructure with the C_{2v} spin–orbit field, Equation 16.75, we obtain the spin dephasing dynamics

$$\dot{s}_x = \frac{-s_x}{\tau_s}, \quad \dot{s}_y = \frac{-s_y}{\tau_s}, \quad \dot{s}_z = \frac{-s_y}{\tau_z}, \tag{16.85}$$

where

$$\frac{1}{\tau_{x,y}} = \frac{\left(\alpha_{BR} \pm \gamma_D\right)^2}{\alpha_{BR}^2 + \gamma_D^2}\frac{1}{\tau_s}, \quad \frac{1}{\tau_z} = \frac{2}{\tau_s}, \tag{16.86}$$

and

$$\frac{1}{\tau_s} = \frac{4m}{\hbar^4}\varepsilon_k\left(\alpha_{BR}^2 + \gamma_D^2\right). \tag{16.87}$$

The up (down) sign is for the s_x (s_y). The spin relaxation is anisotropic. The maximum anisotropy is for the case of equal magnitudes of the Bychkov–Rashba and Dresselhaus interactions, $\alpha_{BR} = \pm\gamma_D$. In this case one of the spin components does not decay.* The s_z component of the spin relaxes roughly twice faster than the in-plane components.

16.2.3 Persistent Spin Helix

Let us consider the case of $\alpha_{BR} = \gamma_D = \lambda/2$. According to Equation 16.86, the spin component s_x does not decay, while the decay rates of s_y and s_z are the same:

$$\dot{s}_y = \frac{-2s_y}{\tau_s}, \quad \dot{s}_z = \frac{-2s_z}{\tau_s}. \tag{16.88}$$

It turns out that a particular nonuniform superposition of s_y and s_z can exhibit no decay as well. This superposition has been termed *persistent spin helix* [13].

Let us assume that the spin is no longer uniform, so that the kinetic equation contains the spin gradient as well, due to the quasiclassical change of the electronic positions:

$$\frac{\partial \mathbf{s}_{\mathbf{k}}}{\partial t} - \mathbf{W}_{\mathbf{k}} \times \mathbf{s}_{\mathbf{k}} + \frac{\partial \mathbf{s}_{\mathbf{k}}}{\partial \mathbf{r}} \cdot \mathbf{v}_{\mathbf{k}} = -\sum_{\mathbf{k}'} W_{\mathbf{kk}'}\left(\mathbf{s}_{\mathbf{k}} - \mathbf{s}'_{\mathbf{k}}\right). \tag{16.89}$$

A running spin wave

$$\mathbf{s}_{\mathbf{k}}(\mathbf{r}) = \mathbf{s}_{\mathbf{k}} e^{i\mathbf{q}\cdot\mathbf{r}}; \quad \mathbf{s}_{\mathbf{k}} \equiv \mathbf{s}_{\mathbf{k}}(\mathbf{q}), \tag{16.90}$$

then evolves according to

$$\frac{\partial \mathbf{s}_{\mathbf{k}}}{\partial t} - \mathbf{W}_{\mathbf{k}} \times \mathbf{s}_{\mathbf{k}} + i\left(\mathbf{q}\cdot\mathbf{v}_{\mathbf{k}}\right)\mathbf{s}_{\mathbf{k}} = -\sum_{\mathbf{k}'} W_{\mathbf{kk}'}\left(\mathbf{s}_{\mathbf{k}} - \mathbf{s}'_{\mathbf{k}}\right). \tag{16.91}$$

We again separate the fast and slow spins as follows:

$$\mathbf{s}_{\mathbf{k}} = \mathbf{s} + \mathbf{x}_{\mathbf{k}}, \quad \left\langle \mathbf{x}_{\mathbf{k}}\right\rangle = 0. \tag{16.92}$$

For the dynamics of the slow part we get

$$\dot{\mathbf{s}}_{\mathbf{k}} = \left\langle \mathbf{W}_{\mathbf{k}} \times \mathbf{x}_{\mathbf{k}}\right\rangle - i\left\langle\left(\mathbf{q}\cdot\mathbf{v}_{\mathbf{k}}\right)\mathbf{x}_{\mathbf{k}}\right\rangle. \tag{16.93}$$

Proceeding as in the previous section, assuming that **s** is stationary on the timescales relevant to $\boldsymbol{\xi}$, we can write

$$\dot{\mathbf{x}}_{\mathbf{k}} - \mathbf{W}_{\mathbf{k}} \times \mathbf{s} - \mathbf{W}_{\mathbf{k}} \times \mathbf{x}_{\mathbf{k}} + i\left(\mathbf{q}\cdot\mathbf{v}_{\mathbf{k}}\right)\mathbf{s} + i\left(\mathbf{q}\cdot\mathbf{v}_{\mathbf{k}}\right)\mathbf{x}_{\mathbf{k}} = -\sum_{\mathbf{k}'} W_{\mathbf{kk}'}\left(\mathbf{x}_{\mathbf{k}} - \mathbf{x}_{\mathbf{k}'}\right). \tag{16.94}$$

* The decay of that component would be due to higher-order (such as cubic) terms in the spin–orbit fields.

We now work with the following assumptions

$$\Omega\tau \ll 1, \quad q\ell \ll 1, \tag{16.95}$$

where $\ell = g\tau$ is the mean free path. We thus assume that the precession is slow on the timescale of the momentum relaxation time, as well as (this is new here) the electronic motion is diffusive on the scale of the wavelength of the spin wave. In the momentum relaxation approximation, also considering the leading terms according to the conditions in Equation 16.95, we get

$$\mathbf{x}_{\mathbf{k}} - \mathbf{W}_{\mathbf{k}} \times \mathbf{s} + i(\mathbf{q}\cdot\mathbf{v}_{\mathbf{k}})\mathbf{s} = -\frac{\mathbf{x}}{\tau}. \tag{16.96}$$

In the steady state, corresponding to a given $\mathbf{s}(t)$, the solution is

$$\mathbf{x}_{\mathbf{k}} = \tau(\mathbf{W}_{\mathbf{k}} \times \mathbf{s}) - i\tau(\mathbf{q}\cdot\mathbf{v}_{\mathbf{k}})\mathbf{s}. \tag{16.97}$$

Substituting to the time evolution equation for $\mathbf{s}$, Equation 16.95, we find

$$\dot{\mathbf{s}} = -2i\tau\langle(\mathbf{q}\cdot\mathbf{v}_{\mathbf{k}})(\mathbf{W}_{\mathbf{k}} \times \mathbf{s})\rangle + \tau\langle \mathbf{W}_{\mathbf{k}} \times (\mathbf{W}_{\mathbf{k}} \times \mathbf{s})\rangle - \tau\langle(\mathbf{q}\cdot\mathbf{v}_{\mathbf{k}})^2\rangle\mathbf{s}. \tag{16.98}$$

Using the vector product identities and introducing the diffusivity

$$D = \langle \mathbf{v}_{\mathbf{k}\alpha}^2\rangle\tau, \tag{16.99}$$

where α denote the Cartesian coordinates (we assume an isotropic system), we finally obtain

$$\dot{s}_\alpha = -2i\tau\varepsilon_{\alpha\beta\gamma}q_\delta\langle v_{\mathbf{k}\delta}\Omega_{\mathbf{k}\beta}\rangle s_\gamma + \langle\Omega_{\mathbf{k}\alpha}\Omega_{\mathbf{k}\beta}\rangle\tau s_\beta - \langle \mathbf{W}_{\mathbf{k}}^2\tau\rangle s_\alpha - Dq^2 s_\alpha. \tag{16.100}$$

For our specific case of

$$\Omega_{\mathbf{k}} = (\lambda k_y, 0, 0), \quad \lambda = \alpha_{BR} + \beta_D, \tag{16.101}$$

we find that s_x decays only via diffusion:

$$\dot{s}_x = -Dq^2 s_x. \tag{16.102}$$

The spin dephasing is ineffective. This is an expected result.

More interesting behavior is found for the two remaining spin components, s_y and s_z. These two spin components, transverse to the spin–orbit field, are coupled

$$\dot{s}_y = +2i\frac{m\lambda}{\hbar}q_y D s_z - (\tau\Omega^2 + Dq^2)s_y, \tag{16.103}$$

$$\dot{s}_z = -2i\frac{m\lambda}{\hbar}q_y D s_y - (\tau\Omega^2 + Dq^2)s_z, \tag{16.104}$$

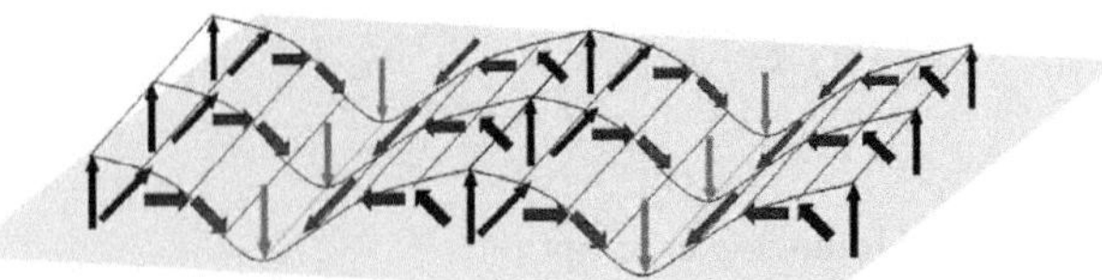

FIGURE 16.7 The persistent spin helix is a wave of circularly polarized spin. The sense of polarization, clock or counterclockwise, depends on the relative sign of the Dresselhaus and the Bychkov–Rashba spin–orbit coupling.

and we denoted $\Omega^2 \equiv \langle \mathbf{W}_{\mathbf{k}}^2\rangle$. Interestingly, this set of coupled differential equations has only decaying solutions. This is best seen by looking at the rotating spins

$$s_+ = s_y + is_z, \quad s_- = s_y - is_z, \tag{16.105}$$

whose time evolutions are uncoupled

$$\dot{s}_+ = -\left(\Omega^2\tau + Dq^2 - \frac{2m\lambda}{\hbar}q_y D\right)s_+, \tag{16.106}$$

$$\dot{s}_- = -\left(\Omega^2\tau + Dq^2 + \frac{2m\lambda}{\hbar}q_y D\right)s_-. \tag{16.107}$$

Considering that

$$\Omega^2\tau = \lambda^2\langle k_y^2\rangle\tau = \lambda^2\left(\frac{m}{\hbar}\right)^2\langle v_y^2\rangle\tau = \lambda^2\left(\frac{m}{\hbar}\right)^2 D, \tag{16.108}$$

we find that the decay of s_+ vanishes for the wave vector

$$q_y^{\mathrm{PSH}} = \frac{m\lambda}{\hbar} = \frac{m}{\hbar}(\alpha_{BR} + \beta_D). \tag{16.109}$$

The abbreviation PSH stands for the *persistent spin helix*, which is the spin wave described by s_+ at this particular wave vector (see Figure 16.7). While individually both s_y and s_z decay at a generic wave vector (and also in the uniform case, $q = 0$), the spin helix they form does not decay in the approximation of the linear spin–orbit field. The spin wave rotating in the opposite sense, s_-, on the other hand, decays. And vice versa for $q_y^{\mathrm{PSH}} = -m\lambda/\hbar$. The persistent spin helix was observed in the spin grating experiment [14].

16.3 Elliott–Yafet Mechanism

Elliott [1] was first to recognize the role of the intrinsic spin–orbit coupling—that coming from the host ions—on spin relaxation. Yafet [2] significantly extended this theory to properly treat electron–phonon spin-flip scattering. The Elliott–Yafet mechanism dominates the spin relaxation of conduction electrons in elemental metals and semiconductors, the systems with space inversion symmetry. In systems lacking this symmetry, the mechanism competes with the D'yakonov–Perel' one; the dominance of one over the other depends on the material in question and specific conditions, such as temperature and doping.

Suppose a nonequilibrium spin accumulates in a nonmagnetic degenerate conductor. The spin accumulation is given as the difference between the chemical potentials for the spin-up and the spin-down electrons. Let us denote the corresponding potentials by $\mu_\uparrow$ and $\mu_\downarrow$. The nonequilibrium electron occupation function for the spin λ is

$$f_{\lambda\mathbf{k}} \approx \frac{1}{e^{\beta(\varepsilon_\mathbf{k}-\mu_\lambda)}+1} \approx f_\mathbf{k}^0 + \left[-\frac{\partial f_\mathbf{k}^0}{\partial \varepsilon_\mathbf{k}}\right](\mu_\lambda - \varepsilon_F). \quad (16.110)$$

Here

$$f_\mathbf{k}^0 = f^0(\varepsilon_\mathbf{k}) = \frac{1}{e^{\beta(\varepsilon_\mathbf{k}-\varepsilon_F)}+1}, \quad (16.111)$$

describes the equilibrium degenerate electronic system of the Fermi energy ε_F. The electron density for the spin λ is

$$n_\lambda = \int d\varepsilon g_s(\varepsilon) f_\lambda^0(\varepsilon) \approx \frac{n}{2} + \int d\varepsilon g(\varepsilon)\left[-\frac{\partial f^0(\varepsilon)}{\partial \varepsilon}\right](\mu_\lambda - \varepsilon_F). \quad (16.112)$$

Here $g_s(\varepsilon)$ is the electronic density of states, defined per unit volume and per spin, at the Fermi level:

$$g_s = \sum_\mathbf{k}\left[-\frac{\partial f_\mathbf{k}^0}{\partial \varepsilon_\mathbf{k}}\right]. \quad (16.113)$$

The total electron density n is

$$n = 2\int d\varepsilon g(\varepsilon) f^0(\varepsilon). \quad (16.114)$$

We assume that the spin accumulation does not charge the system, that is, the charge neutrality is preserved $n_\uparrow + n_\downarrow = n$. This condition is well satisfied in metals and degenerate semiconductors that we consider. We then get

$$\mu_\uparrow + \mu_\downarrow = 2\varepsilon_F. \quad (16.115)$$

The spin density is

$$s = n_\uparrow - n_\downarrow = g_s(\mu_\uparrow - \mu_\downarrow) = g_s\mu_s, \quad (16.116)$$

where μ_s is the spin quasichemical potential, $\mu_s(\mu_\uparrow - \mu_\downarrow)$.

The spin relaxation time T_1 is defined by the decay law:

$$\frac{ds}{dt} = \frac{dn_\uparrow}{dt} - \frac{dn_\downarrow}{dt} = W_{\uparrow\downarrow} - W_{\downarrow\uparrow} = -\frac{s}{T_1} = -g_s\frac{\mu_s}{T_1}. \quad (16.117)$$

Here $W_{\uparrow\downarrow}$ is the net number of transitions per unit time from the spin $\downarrow$ to $\uparrow$. Similarly, $W_{\downarrow\uparrow}$ expresses the rate of spin flips from $\uparrow$ to $\downarrow$. In the degenerate conductors, the spin decay is directly proportional to the decay of the spin accumulation μ_s:

$$\frac{d\mu_s}{dt} = -\frac{\mu_s}{T_1}. \quad (16.118)$$

16.3.1 Electron-Impurity Scattering

We need to distinguish the cases of the impurity or host-induced spin–orbit coupling. Although the formulas for the calculation of T_1 look similar in the two cases, they are nevertheless conceptually different.

16.3.1.1 Spin–Orbit Coupling by the Impurity

If the spin–orbit coupling comes from the impurity potential (in this case the coupling is often termed *extrinsic*), the spin-flip scattering is due to that potential.

The number of transitions per unit time from the spin-up to the spin-down states is

$$W_{\uparrow\downarrow} = \sum_{\mathbf{k}n}\sum_{\mathbf{k}'n'} W_{\mathbf{k}'n'\uparrow,\mathbf{k}n\downarrow} - W_{\mathbf{k}n\downarrow,\mathbf{k}'n'\uparrow}. \quad (16.119)$$

The rate is given by the spin-flip events from down to up minus the ones from up to down. We use the Fermi golden rule to write out the individual scattering rates:

$$W_{\mathbf{k}'n'\uparrow,\mathbf{k}n\downarrow} = \frac{2\pi}{\hbar} f'_{\mathbf{k}n}\left(1 - f_{\mathbf{k}'n}\right)|U_{\mathbf{k}n\uparrow,\mathbf{k}'n'\downarrow}|^2\,\delta\left(\varepsilon_{\mathbf{k}'n'} - \varepsilon_{\mathbf{k}n}\right), \quad (16.120)$$

$$W_{\mathbf{k}n\uparrow,\mathbf{k}'n'\downarrow} = \frac{2\pi}{\hbar} f_{\mathbf{k}'n'}\left(1 - f_{\mathbf{k}n}\right)|U_{\mathbf{k}'n'\downarrow,\mathbf{k}n\uparrow}|^2\,\delta\left(\varepsilon_{\mathbf{k}'n'} - \varepsilon_{\mathbf{k}n}\right). \quad (16.121)$$

Substituting for the occupation numbers the linearized expression, Equation 16.110, and using the definition of the spin relaxation of Equation 16.117, we obtain for the spin relaxation rate the expression

$$\frac{1}{T_1} = \frac{2\pi}{\hbar}\frac{1}{g_s}\sum_{\mathbf{k}\mathbf{k}'}\left|U_{\mathbf{k}n\uparrow,\mathbf{k}'n'\downarrow}\right|^2\left[-\frac{\partial f^0(\varepsilon_\mathbf{k})}{\partial \varepsilon_\mathbf{k}}\right]\delta\left(\varepsilon_{\mathbf{k}'n'} - \varepsilon_{\mathbf{k}n}\right). \quad (16.122)$$

If we define the spin relaxation for the individual momentum state $\mathbf{k}$ as

$$\frac{1}{T_{1\mathbf{k}}} = \frac{2\pi}{\hbar}\sum_{\mathbf{k}'}\left|U_{\mathbf{k}n\uparrow,\mathbf{k}'n'\downarrow}\right|^2\left[-\frac{\partial f^0(\varepsilon_\mathbf{k})}{\partial \varepsilon_\mathbf{k}}\right]\delta\left(\varepsilon_{\mathbf{k}'n'} - \varepsilon_{\mathbf{k}n}\right), \quad (16.123)$$

which has a straightforward interpretation as the spin-flip rate by the elastic impurity scattering to all the possible states $\mathbf{k}'$, we get

$$\frac{1}{T_1} = \left\langle \frac{1}{T_{1\mathbf{k}}} \right\rangle_{\varepsilon_{\mathbf{k}} = \varepsilon_F}, \qquad (16.124)$$

as the average of the individual scattering rates over the electronic Fermi surface.

16.3.1.2 Spin–Orbit Coupling by the Host Lattice

If the spin–orbit coupling comes from the host lattice (in this case the coupling is often termed *intrinsic*), the spin-flip scattering is due to the admixture of the Pauli spin-up and spin-down states in the Bloch eigenstates.

How do the Bloch states actually look like in the presence of spin–orbit coupling? Elliott showed that the Bloch states corresponding to a generic lattice wave vector $\mathbf{k}$ and band n can be written as [1]

$$\Psi_{\mathbf{k},n\Uparrow}(\mathbf{r}) = \left[a_{\mathbf{k}n}(\mathbf{r})|\uparrow\rangle + b_{\mathbf{k}n}(\mathbf{r})|\downarrow\rangle\right]e^{i\mathbf{k}\cdot\mathbf{r}}, \qquad (16.125)$$

$$\Psi_{\mathbf{k},n\Downarrow}(\mathbf{r}) = \left[a^{*}_{-\mathbf{k}n}(\mathbf{r})|\downarrow\rangle - b^{*}_{-\mathbf{k}n}(\mathbf{r})|\downarrow\rangle\right]e^{i\mathbf{k}\cdot\mathbf{r}}. \qquad (16.126)$$

The states $|\Uparrow\rangle$ and $|\Downarrow\rangle$ are the usual Pauli spinors. We can select the two states such that $|a_{\mathbf{k}}n| \approx 1$ while $|b_{\mathbf{k}n}| \ll 1$, due to the weak spin–orbit coupling; this justifies calling the two above states "spin up" ($\Uparrow$) and "spin down" ($\Downarrow$). In fact, to "prepare" the states for the calculation of the spin relaxation, they need to satisfy

$$\langle \mathbf{k}, n\lambda | \sigma_z | \mathbf{k}, n\lambda' \rangle = \lambda\delta_{\lambda\lambda'}, \qquad (16.127)$$

with $\lambda = \Uparrow, \Downarrow$. That is, the two states should diagonalize the spin matrix S_z (or whatever spin direction one is interested in).

The Bloch states of Equations 16.125 and 16.126 allow for a spin flip even if the impurity does not induce a spin–orbit coupling. Indeed, the matrix element

$$\langle \mathbf{k}, n \Uparrow | U | \mathbf{k}, n \Downarrow \rangle \sim ab, \qquad (16.128)$$

is in general nonzero due to the spin admixture. The spin-flip probability is proportional to $|b|^2$, the spin admixture probability. This quantity is crucial in estimating the spin relaxation in the Elliott–Yafet mechanism. The spin relaxation time T_1 in this case can be calculated using the formula Equation 16.122, with

$$U_{\mathbf{k}n\uparrow,\mathbf{k}'n'\downarrow} = U_{\mathbf{k}n\Uparrow,\mathbf{k}'n'\Downarrow}, \qquad (16.129)$$

given by Equation 16.128. A useful *rule of thumb* for estimating the spin relaxation time in this case is

$$\frac{1}{T_1} \approx \frac{\langle b^2_{\mathbf{k}n} \rangle}{\tau_p}, \qquad (16.130)$$

where the averaging of the spin admixture probabilities b^2 is performed over the Fermi surface (or the relevant energy scales of the problem); τ_p is the spin-conserving momentum relaxation time. We stress that b is obtained from the states prepared according to Equation 16.127.

16.3.2 Electron–Phonon Scattering

The spin flip due to the scattering of electrons off phonons involves the intrinsic spin–orbit potential. The electron Bloch states are the ones given by Equations 16.125 and 16.126, prepared according to Equation 16.127.

The net number of transitions per unit time from the spin-up to the spin-down states is

$$W_{\uparrow\downarrow} = \sum_{\mathbf{k}n}\sum_{\mathbf{k}'n'}\sum_{\mathbf{q}\nu} W_{\mathbf{k}n\uparrow,\mathbf{q}\nu;\mathbf{k}'n'\downarrow} + W_{\mathbf{k}n\uparrow;\mathbf{k}'n'\downarrow,\mathbf{q}\nu} - W_{\mathbf{k}'n'\downarrow,\mathbf{q}\nu;\mathbf{k}n\uparrow} - W_{\mathbf{k}'n'\downarrow;\mathbf{k}n\uparrow,\mathbf{q}\nu}. \qquad (16.131)$$

We introduced the rates of the spin-flip transitions accompanied by the phonon absorption and emissions as follows. The net transition rate from the single electron state $|\mathbf{k}'n'\downarrow\rangle$ to the electron state $|\mathbf{k}n\uparrow\rangle$, while the phonon of momentum $\mathbf{q}$ and polarization ν is emitted, is

$$W_{\mathbf{k}n\uparrow,\mathbf{q}\nu;\mathbf{k}'n'\downarrow} = \frac{2\pi}{\hbar}\left|M_{\mathbf{k}n\uparrow,\mathbf{q}\nu;\mathbf{k}'n'\downarrow}\right|^2 f_{\mathbf{k}'n'\downarrow}\left(1 - f_{\mathbf{k}n\uparrow}\right)\delta\left(\varepsilon_{\mathbf{k}n} - \varepsilon_{\mathbf{k}'n'} + \hbar\omega_{\mathbf{q}\nu}\right). \qquad (16.132)$$

Similarly, the net transition rate from the single electron state $|\mathbf{k}'n'\downarrow\rangle$ to the electron state $|\mathbf{k}n\uparrow\rangle$, while the phonon of momentum $\mathbf{q}$ and polarization ν is absorbed, is

$$W_{\mathbf{k}n\uparrow;\mathbf{k}'n'\downarrow,\mathbf{q}\nu} = \frac{2\pi}{\hbar}\left|M_{\mathbf{k}n\uparrow;\mathbf{k}'n'\downarrow\mathbf{q}\nu}\right|^2 f_{\mathbf{k}'n'\downarrow}\left(1 - f_{\mathbf{k}n\uparrow}\right)\delta\left(\varepsilon_{\mathbf{k}n} - \varepsilon_{\mathbf{k}'n'} - \hbar\omega_{\mathbf{q}\nu}\right). \qquad (16.133)$$

The same way are defined the remaining two rates, $W_{\mathbf{k}'n'\downarrow,\mathbf{q}\nu;\mathbf{k}n\uparrow}$ for the spin flip from $\mathbf{k}n\uparrow$ to $\mathbf{k}'n'\downarrow$ with the phonon emission, and $W_{\mathbf{k}'n'\downarrow;\mathbf{k}n\uparrow\mathbf{q}\nu}$, for the phonon absorption:

$$W_{\mathbf{k}'n'\downarrow,\mathbf{q}\nu;\mathbf{k}n\uparrow} = \frac{2\pi}{\hbar}\left|M_{\mathbf{k}'n'\downarrow,\mathbf{q}\nu;\mathbf{k}n\uparrow}\right|^2 f_{\mathbf{k}n\uparrow}\left(1 - f_{\mathbf{k}'n'\downarrow}\right)\delta\left(\varepsilon_{\mathbf{k}'n'} - \varepsilon_{\mathbf{k}n} + \hbar\omega_{\mathbf{q}\nu}\right), \qquad (16.134)$$

$$W_{\mathbf{k}'n'\downarrow;\mathbf{k}n\uparrow\mathbf{q}\nu} = \frac{2\pi}{\hbar}\left|M_{\mathbf{k}'n'\downarrow,\mathbf{q}\nu;\mathbf{k}n\uparrow,\mathbf{q}\nu}\right|^2 f_{\mathbf{k}n\uparrow}\left(1 - f_{\mathbf{k}'n'\downarrow}\right)\delta\left(\varepsilon_{\mathbf{k}'n'} - \varepsilon_{\mathbf{k}n} - \hbar\omega_{\mathbf{q}\nu}\right). \qquad (16.135)$$

The calculation of the spin relaxation due to the electron–phonon scattering is rather involved and we cite here only the final result for degenerate conductors:

$$\frac{1}{T_1} = \frac{4\pi}{\hbar}\frac{1}{g_s}\sum_{\mathbf{k}n}\sum_{\mathbf{k}'n'}\sum_{\nu}\left(-\frac{\partial f^0_{\mathbf{k}n}}{\varepsilon_{\mathbf{k}n}}\right)\frac{\hbar}{2NM}\frac{N^2}{\omega_{\mathbf{k}'-\mathbf{k},\nu}}$$

$$\times\left|\varepsilon_{\mathbf{k}-\mathbf{k}',\nu}\cdot\langle\mathbf{k}n\Uparrow|\nabla V|\mathbf{k}'n'\Downarrow\rangle\right|^2 \tag{16.136}$$

$$\times\Big\{\left[n_{\mathbf{q}\nu} - f^0_{\mathbf{k}'n'} + 1\right]\delta\left(\varepsilon_{\mathbf{k}n} - \varepsilon_{\mathbf{k}'n'} - \hbar\omega_{\mathbf{q}\nu}\right)$$
$$+\left[n_{\mathbf{q}\nu} + f^0_{\mathbf{k}'n'}\right]\delta\left(\varepsilon_{\mathbf{k}n} - \varepsilon_{\mathbf{k}'n'} + \hbar\omega_{\mathbf{q}\nu}\right)\Big\}. \tag{16.137}$$

The electronic bands are described by the energies $\varepsilon_{\mathbf{k}n}$ of the state with momentum **k** and band index n. Phonon frequencies are $\omega_{\mathbf{q}\nu}$, for the phonon of momentum **q** and polarization ν; similarly for the phonon polarization vector $\varepsilon_{\mathbf{q}\nu}$. We further denoted by M the atomic mass, by N the number of atoms in the lattice, and by ∇V the gradient of the electron-lattice ion potential. The equilibrium phonon occupation numbers are denoted as $n_{\mathbf{q}\nu}$, given as

$$n_{\mathbf{q}\nu} = n(\omega_{\mathbf{q}\nu}) = \frac{1}{e^{\beta\hbar\omega_{\mathbf{q}\nu}} - 1}. \tag{16.138}$$

The electronic states $|\mathbf{k}n\sigma\rangle$ are normalized to the whole space.

Two types of processes contribute to the phonon-induced spin flips. First, what we call the *Elliott processes* are the Elliott-type of spin flips in which the Bloch states given by Equations 16.125 and 16.126 scatter by the scalar part of the gradient of the electron–ion potential V. Second, what we call the *Yafet processes* are the spin flips due to the gradient of the spin–orbit part of the electron–ion potential. These two processes are typically of similar order of magnitude and have to be added coherently in order to obtain T_1.

Figure 16.8 shows the experimental data of the spin relaxation in intrinsic (non-degenerate) silicon, obtained by the spin resonance [8,16] and the spin injection [17] experiments. The calculation based on the Elliott-Yafet mechanism of the phonon-induced spin flips reproduces the experiments [15].

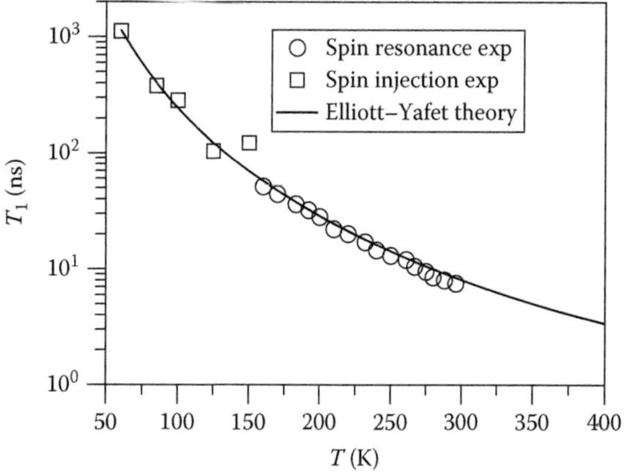

FIGURE 16.8 Spin relaxation in silicon. Phonon-induced spin relaxation in silicon results in an approximate T^3 power law [15]. Shown are the experimental data from spin resonance [16] and spin injection [17] experiments, and a calculation based on the Elliott–Yafet mechanism [15].

16.3.2.1 Yafet Relation

The expected temperature dependence of the phonon-induced spin flips in degenerate conductors, is, according to the Elliott–Yafet mechanism, $1/T_1 \sim T$ at high temperatures (starting roughly at a fraction of the Debye temperature T_D) and $1/T_1 \sim T^3$ at low temperatures, in analogy with the conventional spin-conserving electron–phonon scattering. The high-temperature dependence originates from the linear increase of the phonon occupation numbers n with increasing temperature: $n_{\mathbf{q}\nu} \sim T$, as $k_BT \gg \omega_{\mathbf{q}\nu}$. At $T > T_D$ are all the phonons excited. The low-temperature dependence of the spin-conserving scattering follows from setting the relevant phonon energy scale to k_BT. The matrix element

$$\langle\mathbf{k}n\Uparrow|\nabla V|\mathbf{k}'n'\Uparrow\rangle \sim q, \tag{16.139}$$

which gives the $1/T_1 \sim T^3$ dependence. However, Yafet showed [2] that for the spin-flip matrix element the space inversion symmetry modifies the momentum dependence to

$$\langle\mathbf{k}n\Uparrow|\nabla V|\mathbf{k}'n'\Downarrow\rangle \sim q^2, \tag{16.140}$$

so that

$$\frac{1}{T_1} \sim T^5, \tag{16.141}$$

instead of the expected T^3. Since the same temperature dependence holds for the phonon-induced electrical resistance $\rho(T)$, we can write

$$\frac{1}{T_1} \sim \rho(T), \tag{16.142}$$

known as the *Yafet relation*.

The relation Equation 16.140 holds if both the Elliott and Yafet processes are taken into account. Individually, they would lead to a linear dependence on q. The quantum mechanical interference between these two processes thus significantly reduces the spin-flip probability at low momenta q. An example is shown in Figure 16.9. The Elliott and Yafet processes would individually give much stronger spin relaxation than is observed. Their destructive interference can modify T_1 by orders of magnitude.

16.4 Results Based on Kinetic-Spin-Bloch-Equation Approach

It was shown by Wu et al. from a full microscopic kinetic-spin-Bloch-equation approach [9] that the single-particle approach is inadequate in accounting for the spin relaxation/dephasing both in the time [18–22] and in the space [23–25] domains. The momentum dependence of the effective magnetic field (the

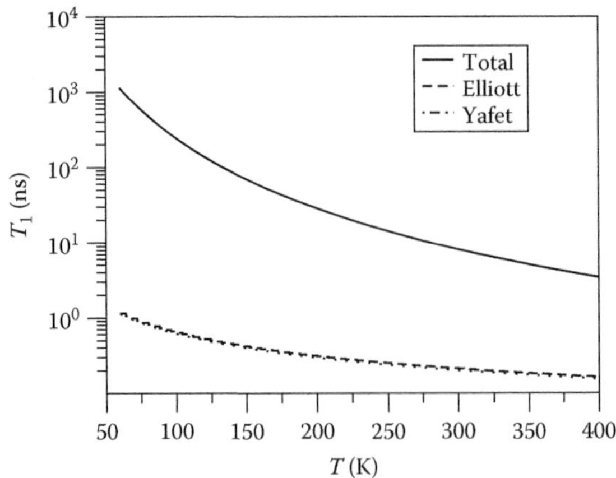

FIGURE 16.9 Interference between the Elliott and Yafet processes. Individually the Elliott and Yafet phonon-induced spin relaxation processes give spin relaxation orders of magnitude stronger than the total one. This example is from the calculation of T_1 in silicon [15].

D'yakonov–Perel' term) and the momentum dependence of the spin diffusion rate along the spacial gradient [23] or even the random spin–orbit coupling [26] all serve as inhomogeneous broadenings [19,20]. It was pointed out that in the presence of inhomogeneous broadening, any scattering, including the carrier–carrier Coulomb scattering, can cause an irreversible spin relaxation/dephasing [19]. Moreover, besides the spin relaxation/dephasing channel the scattering provides, it also gives rise to the counter effect to the inhomogeneous broadening. The scattering tends to drive carriers to a more homogeneous state and therefore suppresses the inhomogeneous broadening. Finally, this approach is valid in both strong and weak scattering limits and also can be used to study systems far away from the equilibrium, thanks to the inclusion of the Coulomb scattering.

In the following subsection, we present the main results based on kinetic-spin-Bloch-equation approach. We first briefly introduce the kinetic spin Bloch equations. Then we review the results of the spin relaxation/dephasing in the time and space domains, respectively.

16.4.1 Kinetic Spin Bloch Equations

By using the nonequilibrium Green function method with gradient expression as well as the generalized Kadanoff–Baym ansatz [27], we construct the kinetic spin Bloch equations as follows:

$$\dot{\rho}_{\mathbf{k}}(\mathbf{r},t) = \dot{\rho}_{\mathbf{k}}(\mathbf{r},t)\big|_{\text{dr}} + \dot{\rho}_{\mathbf{k}}(\mathbf{r},t)\big|_{\text{dif}} + \dot{\rho}_{\mathbf{k}}(\mathbf{r},t)\big|_{\text{coh}} + \dot{\rho}_{\mathbf{k}}(\mathbf{r},t)\big|_{\text{scat}}. \qquad (16.143)$$

Here $\rho_{\mathbf{k}}(\mathbf{r},t)$ are the density matrices of electrons with momentum $\mathbf{k}$ at position $\mathbf{r}$ and time t. The off-diagonal elements of $\rho_{\mathbf{k}}$ represent the correlations between the conduction and valence bands, different subbands (in confined structures), and different spin states. $\dot{\rho}_{\mathbf{k}}(\mathbf{r},t)|_{\text{dr}}$ are the driving terms from the external electric field. The coherent terms, in Equation 16.143, $\dot{\rho}_{\mathbf{k}}|_{\text{coh}}$ are composed of the energy spectrum, magnetic field, and effective magnetic field from the D'yakonov–Perel' term and the Coulomb Hartree-Fock terms. The diffusion terms $\dot{\rho}_{\mathbf{k}}(\mathbf{r},t)|_{\text{dif}}$ come from the spacial gradient. The scattering terms $\dot{\rho}_{\mathbf{k}}(\mathbf{r},t)|_{\text{scat}}$ include the spin-flip and spin-conserving electron–electron, electron–phonon, and electron-impurity scatterings. The spin-flip terms correspond to the Elliot–Yafet and/or Bir–Aronov–Pikus mechanisms. Detailed expressions of these terms in the kinetic spin Bloch equations depend on the band structures, doping situations, and dimensionalities [9] and can be found in the literature for different cases, such as intrinsic quantum wells [18], n-type quantum wells with [21,28–31] and without [22,32–34] electric field, p-type quantum wells [35–37], quantum wires [38,39], quantum dots [40], and bulk materials [41] in the spacial uniform case and quantum wells in spacial nonuniform case [23,24,42,43]. By numerically solving the kinetic spin Bloch equations with all the scattering explicitly included, one is able to obtain the time evolution and/or spacial distribution of the density matrices, and hence all the measurable quantities, such as mobility, diffusion constant, optical relaxation/dephasing time, spin relaxation/dephasing time, spin diffusion length, as well as hot-electron temperature, can be determined from the theory without any fitting parameters.

16.4.2 Spin Relaxation/Dephasing

In this subsection we present the main understandings of the spin relaxation/dephasing added to the literature from the kinetic-spin-Bloch-equation approach. We focus on three related issues: (i) the importance of the Coulomb interaction to the spin relaxation/dephasing; (ii) spin dynamics far away from the equilibrium; and (iii) qualitatively different behaviors from those wildly adopted in the literature.

First we address the effect of the Coulomb interaction. Based on the single-particle approach, it has been long believed that the Coulomb scattering is irrelevant to the spin relaxation/dephasing [44]. It was first pointed out by Wu and Ning [19] that in the presence of inhomogeneous broadening in spin precession, that is, the spin precession frequencies are $\mathbf{k}$-dependent, any scattering, including the spin-conserving scattering, can cause irreversible spin dephasing. This inhomogeneous broadening can come from the energy-dependent g-factor [19], the D'yakonov–Perel' term [20], the random spin–orbit coupling [26], and even the momentum dependence of the spin diffusion rate along the spacial gradient [23]. Wu and Ning first showed that with the energy-dependent g-factor as an inhomogeneous broadening, the Coulomb scattering can lead to irreversible spin dephasing [19]. In [001]-grown n-doped quantum wells, the importance of the Coulomb scattering for spin relaxation/dephasing was proved by Glazov and Ivchenko [45] by using perturbation theory and by Weng and Wu [21] from the kinetic-spin-Bloch-equation approach. In a temperature-dependent study of the spin dephasing in [001]-oriented n-doped quantum wells, Leyland et al. experimentally verified the effects of the electron–electron Coulomb scattering by closely measuring the momentum scattering rate from the mobility [46]. By showing the momentum relaxation rate obtained from the mobility cannot give the correct spin relaxation rate, they showed the difference comes from the Coulomb scattering. Later Zhou et al. even predicted a peak from the Coulomb scattering in the temperature dependence of the spin relaxation time in a high-mobility low-density n-doped (001) quantum well [29]. This was later demonstrated by Ruan et al. experimentally [47].

Figure 16.10 shows the temperature dependence of the spin relaxation time of a 7.5 nm $GaAs/Al_{0.4}Ga_{0.6}As$ quantum well at different electron and impurity densities [29]. For this small well width, only the lowest subband is needed in the calculation. It is shown in the figure that when the electron-impurity scattering is dominant, the spin relaxation time decreases with increasing temperature monotonically. This is in good agreement with the experimental findings [48] and a nice agreement of the theory and the experimental data from 20 to 300 K is given in Ref. [29]. However, it is shown that for sample with high mobility, that is, low-impurity density, when the electron density is low enough, there is a peak at low temperature. This peak, located around the Fermi temperature of electrons $T_F^e = E_F/k_B$, is identified to be solely due to the Coulomb scattering [29,49]. It disappears when the Coulomb scattering is switched off, as shown by the dashed curves in the figure. This peak also disappears at high impurity densities. It is also noted in Figure 16.10c that for electrons of high density so that T_F^e is high enough and the contribution from the electron–longitudinal optical-phonon scattering becomes marked, the peak disappears even for sample with no impurity and the spin relaxation time increases with temperature monotonically. The physics leading to the peak is due to the crossover of the Coulomb scattering from the degenerate to the non-degenerate limit. At $T < T_F^e$, electrons are in the degenerate limit and the electron–electron scattering rate $1/\tau_{ee} \propto T^2$. At $T > T_F^e$, $1/\tau_{ee} \propto T^{-1}$ [45,50]. Therefore, at low electron density so that T_F^e is low enough and the electron-acoustic phonon scattering is very weak compared with the electron–electron Coulomb scattering, the Coulomb scattering is dominant for high mobility sample. Hence, the different temperature dependence of the Coulomb scattering leads to the peak. It is noted that the peak is just a feature of the crossover from the degenerate to the non-degenerate limit. The location of the peak also depends on the strength of the inhomogeneous broadening. When the inhomogeneous broadening depends on momentum linearly, the peak tends to appear at the Fermi temperature. A similar peak was predicted in the electron spin relaxation in p-type GaAs quantum well and the hole spin relaxation in (001) strained asymmetric Si/SiGe quantum well, where the electron and hole spin relaxation times both show a peak at the hole Fermi temperature T_F^h [36,37]. When the inhomogeneous broadening depends on momentum cubically, the peak tends to shift to a lower temperature. It was predicted that a peak in the temperature dependence of the electron spin relaxation time appears at a temperature in the range of ($T_F^e/4$, $T_F^e/2$) in the intrinsic bulk GaAs [41] and a peak in the temperature dependence of the hole spin relaxation time at $T_F^h/2$ in p-type Ge/SiGe quantum well [37]. Ruan et al. demonstrated the peak experimentally in a high-mobility low-density $GaAs/Al_{0.35}Ga_{0.65}As$ heterostructure and showed a peak appears at $T_F^e/2$ in the spin relaxation time versus temperature curve [47].

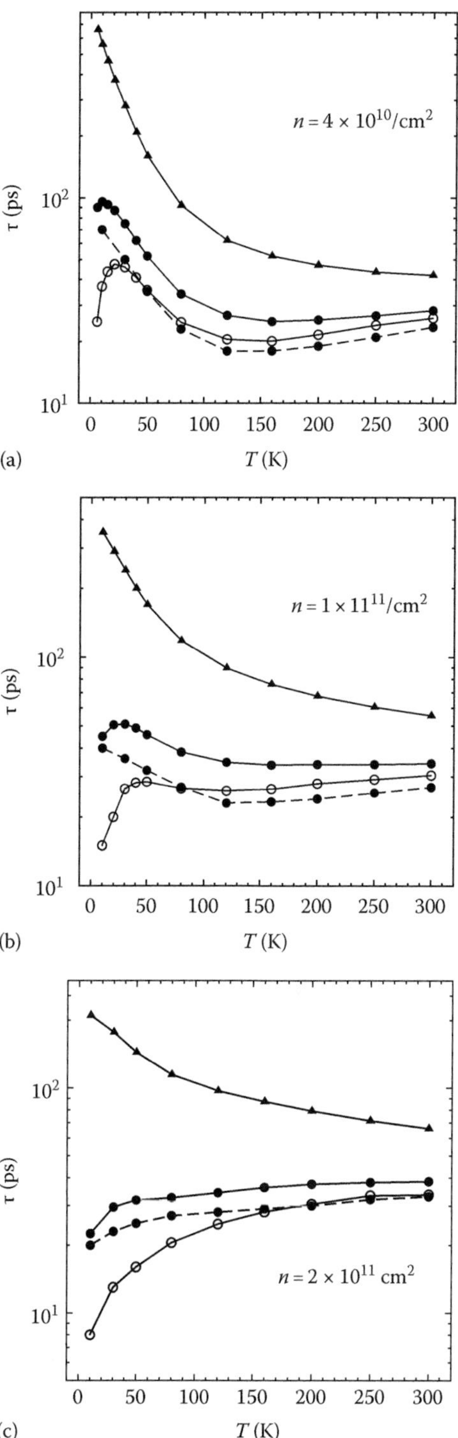

FIGURE 16.10 Spin relaxation time τ versus the temperature T with well width $a = 7.5$ nm and electron density n being (a) 4×10^{10} cm^{-2}, (b) 1×10^{11} cm^{-2}, and (c) 2×10^{11} cm^{-2}, respectively. Solid curves with triangles: impurity density $n_i = n$; solid curves with dots: $n_i = 0.1n$; solid curves with circles: $n_i = 0$; dashed curves with dots: $n_i = 0.1n$ and no Coulomb scattering. (From Zhou, J. et al., *Phys. Rev. B*, **75**, 045305, 2007.)

For larger well width, the situation may become different in the non-degenerate limit. Weng and Wu calculated the spin relaxation/dephasing for (001) GaAs quantum wells with larger well width and high mobility by including the multi-subband effect [28]. It is shown that for small/large well width so that the linear/cubic Dresselhaus term is dominant, the spin relaxation/dephasing time increases/decreases with the temperature. This is because with the increase of temperature, both the inhomogeneous broadening and the scattering get enhanced. The relative importance of these two competing effects is different when the linear/cubic term is dominant [28]. Jiang and Wu further introduced strain to change the relative importance of the linear and cubic D'yakonov–Perel' terms and showed the different temperature dependences of the spin relaxation time [51]. This prediction has been realized experimentally by Holleitner et al. where they showed that in n-type two-dimensional InGaAs channels, when the linear D'yakonov–Perel' term is suppressed, the spin relaxation time decreases with temperature monotonically [52]. Another interesting prediction related to the multi-subband effect is related to the effect of the inter-subband Coulomb scattering. From the calculation Weng and Wu found out that although the inhomogeneous broadening from the higher subband of the (001) quantum well is much larger, due to the strong inter-subband Coulomb scattering, the spin relaxation times of the lowest two subbands are identical [28]. This prediction has later been verified experimentally by Zhang et al., who studied the spin dynamics in a single-barrier heterostructure by time-resolved Kerr rotation [53]. By applying a gate voltage, they effectively manipulated the confinement of the second subband, and the measured spin relaxation times of the first and second subbands are almost identical at large gate voltage. Lü et al. showed that due to the Coulomb scattering, $T_2 = T_2^*$ in (001) GaAs quantum wells for a wide temperature and density regime [54]. It was also pointed out by Lü et al. that in the strong (weak) scattering limit, introducing the Coulomb scattering will always lead to a faster (slower) spin relaxation/dephasing [35].

Another important effect from the Coulomb interaction to the spin relaxation/dephasing comes from the Coulomb Hartree–Fock contribution in the coherent terms of the kinetic spin Bloch equations. Weng and Wu [21] first pointed out that at a high spin polarization, the Hartree–Fock term serves as an effective magnetic field along the z-axis, which blocks the spin precession. As a result, the spin relaxation/dephasing time increases dramatically with the spin polarization. They further pointed out that the spin relaxation/dephasing time decreases with temperature at high spin polarization in quantum well with small well width, which is in contrast to the situation with small spin polarizations. These predictions have been verified experimentally by Stich et al. in an n-type (001) GaAs quantum well with high mobility [55,56]. By changing the intensity of the circularly polarized lasers, Stich et al. measured the spin dephasing time in a high mobility n-type GaAs quantum well as a function of initial spin polarization. Indeed they observed an increase of the spin dephasing time with the increased spin polarization, and the theoretical calculation based on the kinetic spin Bloch equations nicely

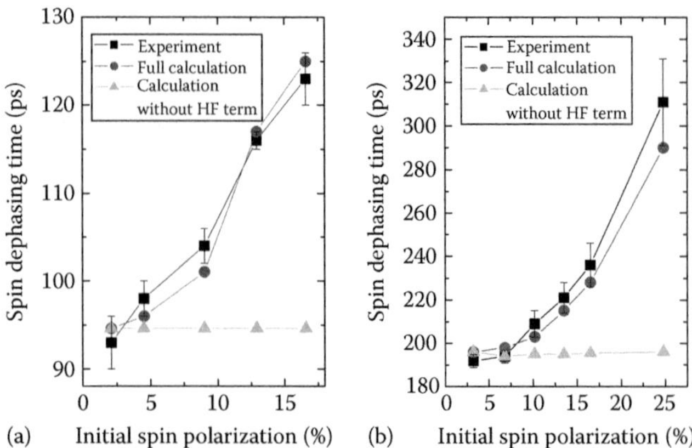

FIGURE 16.11 (a) The spin dephasing times as a function of initial spin polarization for constant, *low* excitation density and variable polarization degree of the pump beam. The measured spin dephasing times are compared to calculations with and without the Hartree–Fock (HF) term, showing its importance. (b) The spin dephasing times measured and calculated for constant, *high* excitation density and variable polarization degree. (From Stich, J. et al., *Phys. Rev. B*, 76, 205301, 2007.)

reproduced the experimental findings when the Hartree–Fock term was included [55]. It was also shown that when the Hartree–Fock term is removed, one does not see any increase of the spin dephasing time. Later, they further improved the experiment by replacing the circular-polarized laser pumping with the elliptic polarized laser pumping. By doing so, they were able to vary the spin polarization without changing the carrier density. Figure 16.11 shows the measured spin dephasing times as function of initial spin polarization under two fixed pumping intensities, together with the theoretical calculations with and without the Coulomb Hartree–Fock term. Again the spin dephasing time increases with the initial spin polarization as predicted and the theoretical calculations with the Hartree–Fock term are in good agreement with the experimental data [56]. Moreover, Stich et al. also confirmed the prediction of the temperature dependences of the spin dephasing time at low and high spin polarizations [56]. Figure 16.12a shows the measured temperature dependences of the spin dephasing time at different initial spin polarizations. As predicted, the spin dephasing time increases with increasing temperature at small spin polarization but decreases at large spin polarization. The theoretical calculations also nicely reproduced the experimental data. The hot-electron temperatures in the calculation were taken from the experiment (Figure 16.12b). The effective magnetic field from the Hartree–Fock term has been measured by Zhang et al. from the sign switch of the Kerr signal and the phase reversal of Larmor precessions with a bias voltage in a GaAs heterostructure [57]. Korn et al. [58] also estimated the average effect by applying an external magnetic field in the Faraday configuration, as shown in Figure 16.13a, for the same sample reported earlier [55,56]. They compared the spin dephasing times of both large and small spin polarizations as function of external magnetic field. Due to the effective magnetic field from the Hartree–Fock term, the spin relaxation times are

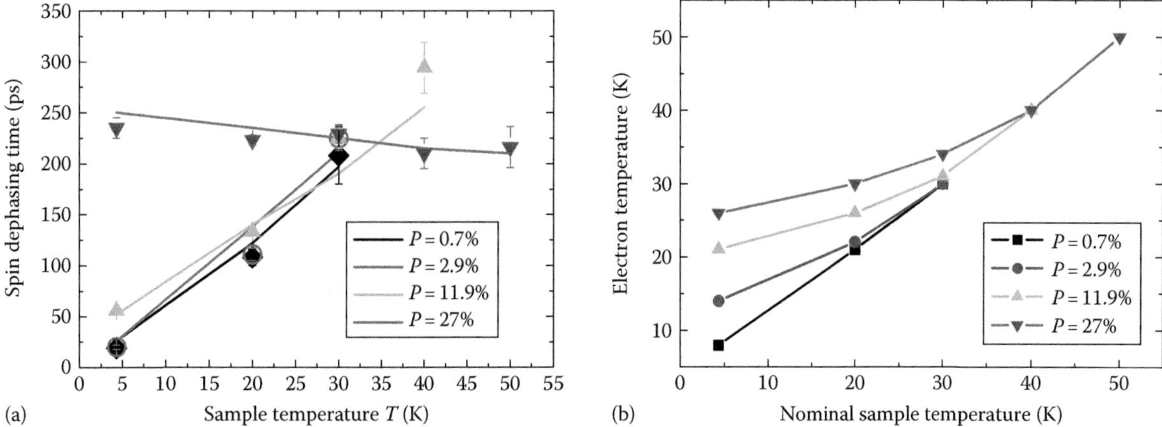

FIGURE 16.12 (a) Spin dephasing time as a function of sample temperature, for different initial spin polarizations. The measured data points are represented by solid points, while the calculated data are represented by lines of the same grayscale. (b) Electron temperature determined from intensity-dependent photoilluminance measurements as a function of the nominal sample temperature, for different pump beam fluence and initial spin polarization, under experimental conditions corresponding to the measurements shown in (a). The measured data points are represented by solid points, while the curves serve as guide to the eye. (From Stich, J. et al., *Phys. Rev. B*, 76, 205301, 2007.)

different under small external magnetic field but become identical when the magnetic field becomes large enough. From the merging point, they estimated that the mean value of the effective magnetic field is below 0.4 T. They further showed that this effective magnetic field from the Hartree–Fock term cannot be compensated by the external magnetic field, because it does not break the time-reversal symmetry and is therefore not a genuine magnetic field, as said earlier. This can be seen from Figure 16.13b that the spin relaxation time at large spin polarization shows identical external magnetic field dependences when the magnetic field is parallel or antiparallel to the growth direction.

We now turn to discuss the spin relaxation/dephasing far away from the equilibrium. In fact, the spin relaxation/dephasing of high spin polarization addressed earlier is one of the cases far away from the equilibrium. Another case is the spin dynamics in the presence of a high in-plane electric field. The spin dynamics in the presence of a high in-plane electric field was first studied by Weng et al. [22] in GaAs quantum well with only the lowest subband by solving the kinetic spin Bloch equations. To avoid the "runaway" effect [59], the electric field was calculated up to 1 kV/cm. Then, Weng and Wu further introduced the second subband into the model and the in-plane electric field was increased up to 3 kV/cm [28]. Zhang et al. included L valley and the electric field was further increased up to 7 kV/cm [32]. The effect of in-plane electric field to the spin relaxation in system with strain was investigated by Jiang and Wu [51]. Zhou et al. also investigated the electric-field effect at low lattice temperatures [29].

The in-plane electric field leads to two effects: (a) It shifts the center of mass of electrons to $\mathbf{k}_d = m^*\mathbf{v}_d = \mu\mathbf{E}$ with μ representing the mobility, which further induces an effective magnetic field via the D'yakonov–Perel' term [22]; (b) the in-plane electric field also leads to the hot-electron effect [60]. The first effect induces a spin precession even in the absence of any external magnetic field and the spin precession frequency changes with the direction of the electric field in the presence of an external magnetic field [22,32]. The second effect enhances both the inhomogeneous broadening and the scattering, two competing effects

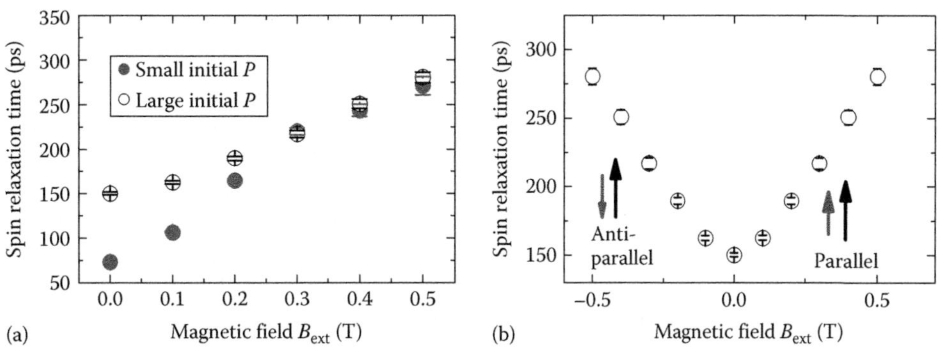

FIGURE 16.13 (a) Spin dephasing times as a function of an external magnetic field perpendicular to the quantum well plane for small and large initial spin polarization. (b) Same as (a) for large initial spin polarization and both polarities of the external magnetic field. (From Korn, T. et al., *Adv. Solid State Phys.*, 48, 143, 2009.)

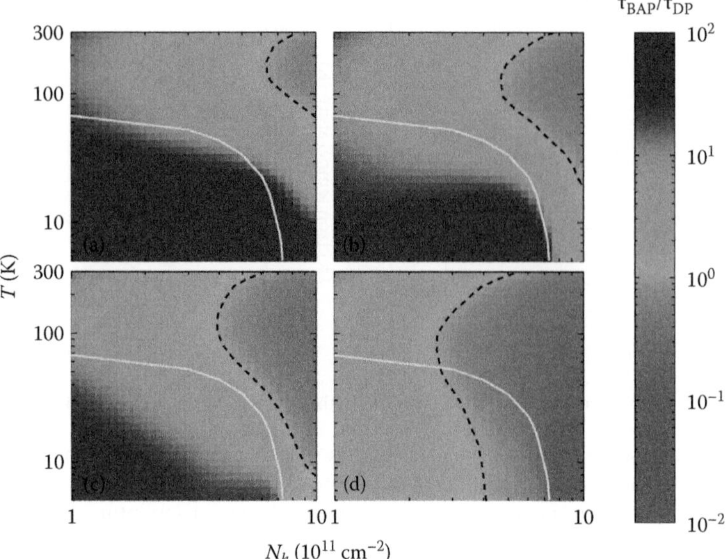

FIGURE 16.14 (See color insert.) Ratio of the spin relaxation time due to the Bir–Aronov–Pikus mechanism to that due to the D'yakonov–Perel' mechanism, τ_{BAP}/τ_{DP}, as function of temperature and hole density with (a) $N_I = 0$, $N_{ex} = 10^{11}$ cm^{-2}; (b) $N_I = 0$, $N_{ex} = 10^9$ cm^{-2}; (c) $N_I = N_h$, $N_{ex} = 10^{11}$ cm^{-2}; (d) $N_I = N_h$, $N_{ex} = 10^9$ cm^{-2}. The black dashed curves indicate the cases satisfying $\tau_{BAP}/\tau_{DP} = 1$. Note the smaller the ratio τ_{BAP}/τ_{DP} is, the more important the Bir–Aronov–Pikus mechanism becomes. The yellow solid curves indicate the cases satisfying $\partial_{\mu_h}[N_{LH^{(1)}} + N_{HH^{(2)}}]/\partial_{\mu_h} N_h = 0.1$. In the regime above the yellow curve the multi-hole-subband effect becomes significant. (Zhou, Y. et al., *New J. Phys.*, 11, 113039, 2009.)

leading to rich electric-field dependence of the spin relaxation/dephasing and thus spin manipulation [22,28,29,32,33,41,51].

Finally we address some issues of which the kinetic-spin-Bloch-equation approach gives qualitatively different predictions from those widely used in the literature. These issues include the Bir–Aronov–Pikus mechanism, the Elliot–Yafet mechanism, and some density/temperature dependences of the spin relaxation/dephasing time.

It has long been believed in the literature that for electron relaxation/dephasing, the Bir–Aronov–Pikus mechanism is dominant at low temperature in p-type samples and has important contribution to intrinsic sample with high photo-excitation [61–67]. This conclusion was made based on the single-particle Fermi golden rule. Zhou and Wu reexamined the problem using the kinetic-spin-Bloch-equation approach [30]. They pointed out that the Pauli blocking was overlooked in the Fermi golden rule approach. When electrons are in the non-degenerate limit, the results calculated from the Fermi golden rule approach are valid. However, at low temperature, electrons can be degenerate and the Pauli blocking becomes very important. As a result, the previous approaches always overestimated the importance of the Bir–Aronov–Pikus mechanism at low temperature. Moreover, the previous single-particle theories underestimated the contribution of the D'yakonov–Perel' mechanism by neglecting the Coulomb scattering. Both made the Bir–Aronov–Pikus mechanism dominate the spin relaxation/dephasing at low temperature. Later, Zhou et al. performed a thorough investigation of electron spin relaxation in p-type (001) GaAs quantum wells by varying impurity, hole and photo-excited electron densities over a wide range of values [36], under the idea that very high impurity density and very low photo-excited electron density may effectively suppress the importance of the D'yakonov–Perel' mechanism and the Pauli blocking. Then the relative importance of the Bir–Aronov–Pikus and D'yakonov–Perel' mechanisms may be reversed. This indeed happens as shown in the phase-diagram-like picture in Figure 16.14 where the relative importance of the Bir–Aronov–Pikus and D'yakonov–Perel' mechanisms is plotted as function of hole density and temperature at low and high impurity densities and photo-excitation densities. For the situation of high hole density, they even included multi-hole subbands as well as the light hole band. It is interesting to see from the figures that at relatively high photo-excitations, the Bir–Aronov–Pikus mechanism becomes more important than the D'yakonov–Perel' mechanism only at high hole densities and high temperatures (around hole Fermi temperature) when the impurity is very low (zero in Figure 16.14a). Impurities can suppress the D'yakonov–Perel' mechanism and hence enhance the relative importance of the Bir–Aronov–Pikus mechanism. As a result, the temperature regime is extended, ranging from the hole Fermi temperature to the electron Fermi temperature for high hole density. When the photo-excitation is weak so that the Pauli blocking is less important, the temperature regime where the Bir–Aronov–Pikus mechanism is important becomes wider compared to the high excitation case. In particular, if the impurity density is high enough and the photo-excitation is so low that the electron Fermi temperature is below the lowest temperature of the investigation, the Bir–Aronov–Pikus mechanism can dominate the whole temperature regime of the investigation at sufficiently high hole density, as shown in Figure 16.14d. The corresponding spin relaxation times of each mechanism under high or low-impurity and photo-excitation densities are demonstrated in Figure 16.14. They also discussed the density dependences of spin relaxation with some intriguing properties related to the high hole subbands [36]. The predicted Pauli-blocking effect in the Bir–Aronov–Pikus mechanism has been partially demonstrated experimentally by Yang et al. [68]. They showed by increasing the pumping density that the temperature dependence of the spin dephasing time deviates from the one from the Bir–Aronov–Pikus mechanism and the peaks at high excitations agree well with those predicted by Zhou and Wu [30].

Another widely accepted but incorrect conclusion is related to the Elliot–Yafet mechanism. It is widely accepted in the literature that the Elliot–Yafet mechanism dominates spin relaxation in n-type bulk III–V semiconductor at low temperature, while the D'yakonov–Perel' mechanism is important at high temperature [5,7,8,69,70]. Jiang and Wu pointed out that the previous understanding is based on the formula that can only be used in the non-degenerate limit. Moreover, the momentum relaxation rates are calculated via the approximated formula for mobility [69]. By performing an accurate calculation via the kinetic-spin-Bloch-equation approach, they showed that the Elliot–Yafet mechanism is *not* important in III–V semiconductors, including even the narrow-band InAs and InSb [41]. Therefore, the D'yakonov–Perel' mechanism is the only spin relaxation mechanism for n-type III–V semiconductors in metallic regime.

Jiang and Wu have further predicted a peak in the density dependence of the spin relaxation/dephasing time in n-type III–V semiconductors where the spin relaxation/dephasing is limited by the D'yakonov–Perel' mechanism [41]. Previously, the non-monotonic density dependence of spin lifetime was observed in low-temperature ($T \lesssim 5$ K) experiments, where the localized electrons play a crucial role and the electron system is in the insulating regime or around the metal-insulator transition point [71]. Jiang and Wu found, for the first time, that the spin lifetime in *metallic* regime is also *non-monotonic*. Moreover, they pointed out that it is a *universal* behavior for *all* bulk III–V semiconductors at *all* temperatures where the peak is located at $T_F \sim T$ with T_F being the electron Fermi temperature. The underlying physics for the non-monotonic density dependence in metallic regime can be understood as following: In the non-degenerate regime, as the distribution is the Boltzmann one, the density dependence of the inhomogeneous broadening is marginal. However, the scattering increases with the density. Consequently, the spin relaxation/dephasing time increases with the density. However, in the degenerate regime, due to the Fermi distribution, the inhomogeneous broadening increases with the density much faster than the scattering does. As a result, the spin relaxation/dephasing time decreases with the density. Similar behavior was also found in two-dimensional system [31,37] where the underlying physics is similar. The predicted peak was later observed by Krauß et al. [72], as shown in Figure 16.15 where theoretical calculation based on the kinetic spin Bloch equations nicely reproduced the experimental data by Shen [73].

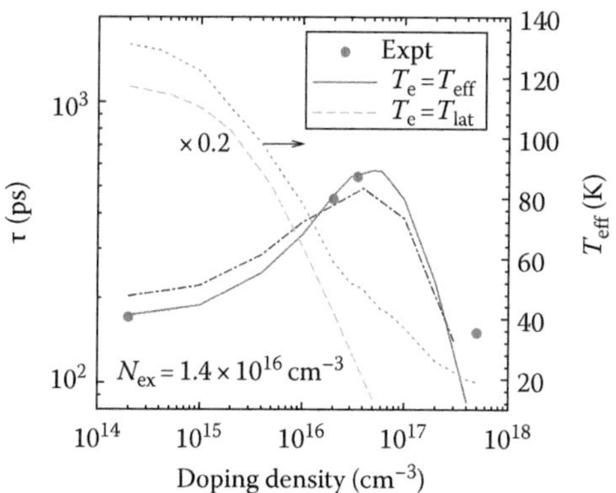

FIGURE 16.15 Electron spin relaxation times from the calculation via the kinetic-spin-Bloch-equation approach (red solid curve) and from the experiment [72] (blue •) as function of the doping density. The green dashed curve shows the results without hot-electron effect. The hot-electron temperature used in the computation is plotted as the blue dotted curve. (Note the scale is on the right-hand side of the frame.) The chain curve is the calculated spin relaxation time with a fixed hot-electron temperature 80 K, and $N_{ex} = 6 \times 10^{15}$ cm^{-3}. (From Shen, K., *Chin. Phys. Lett.*, 26, 067201, 2009).

16.4.3 Spin Diffusion/Transport

By solving the kinetic spin Bloch equations together with the Poisson equation self-consistently, one is able to obtain all the transport properties such as the mobility, charge diffusion length, and spin diffusion/injection length without any fitting parameter. It was first pointed out by Weng and Wu [23] that the drift-diffusion equation approach is inadequate in accounting for the spin diffusion/transport. It is important to include the off-diagonal term between opposite spin bands $\rho_{k\uparrow\downarrow}$ in studying the spin diffusion/transport. With this term, electron spin precesses along the diffusion and therefore $\mathbf{k}\cdot\nabla_r\rho_\mathbf{k}(\mathbf{r},t)$ in the diffusion terms $\dot{\rho}_\mathbf{k}(\mathbf{r},t)|_{dif}$ provides an additional inhomogeneous broadening. With this additional inhomogeneous broadening, any scattering, including the Coulomb scattering, can cause an irreversible spin relaxation/dephasing [23]. Unlike the spin precession in the time domain where the inhomogeneous broadening is determined by the effective magnetic field from the D'yakonov–Perel' term, $\mathbf{h}(\mathbf{k})$, in spin diffusion and transport it is determined by

$$W_\mathbf{k} = \frac{|g\mu_B\mathbf{B} + \mathbf{h}(\mathbf{k})|}{k_x}, \tag{16.144}$$

provided the diffusion is along the x-axis [42]. Here, the magnetic field is in the Voigt configuration. Therefore, even in the absence of the D'yakonov–Perel' term $\mathbf{h}(\mathbf{k})$, the magnetic field *alone* can provide an inhomogeneous broadening and leads to the spin relaxation/dephasing in spin diffusion and transport. This was first pointed out by Weng and Wu back to 2002 [23] and has been realized experimentally by Appelbaum et al. in bulk silicon [74,75], where there is no D'yakonov–Perel' spin–orbit coupling due to the center inversion symmetry. Zhang and Wu further investigated the spin diffusion and transport in symmetric Si/SiGe quantum wells [43].

When $B = 0$ but the D'yakonov–Perel' term is present, then the inhomogeneous broadening for spin diffusion and transport is determined by $\mathbf{\Omega}_{\mathbf{k}} = \mathbf{h}(\mathbf{k})/k_x$. In (001) GaAs quantum well where the D'yakonov–Perel' term is determined by the Dresselhaus term [76], the average of $\mathbf{\Omega}_{\mathbf{k}}$ reads $\langle W_{\mathbf{k}} \rangle = C\left(\langle k_y^2 \rangle - \langle k_z^2 \rangle, 0, 0\right)$ with C being a constant. For electrons in quantum well, this value is not zero. Therefore, the spacial spin oscillation due to the Dresselhaus effective magnetic field survives even at high temperature when the scattering is strong. This effect was first predicted by Weng and Wu by showing that a spin pulse can oscillate along the diffusion in the absence of the magnetic field at very high temperature [24]. Detailed studies were carried out later on this effect [25,42,77]. The spin oscillation without any applied magnetic field in the transient spin transport was later observed experimentally by Crooker and Smith in strained bulk system [78]. Differing from the two-dimensional case, in bulk, the average of $\mathbf{\Omega}_{\mathbf{k}}$ from the Dresselhaus term is zero, since $\langle W_{\mathbf{k}} \rangle = C\left(\langle k_y^2 \rangle - \langle k_z^2 \rangle, 0, 0\right) = 0$ due to the symmetry in the y- and z-directions. This is consistent with the experimental result that there is no spin oscillation for the system without stress. However, when the stress is applied, an additional spin–orbit coupling, namely, the coupling of electron spins to the strain tensor, appears, which is linear in momentum [5]. This additional spin–orbit coupling also acts as an effective magnetic field. Therefore, once the stress is applied, one can observe spacial spin oscillation even when there is no applied magnetic field [78].

Cheng and Wu further developed a new numerical scheme to calculate the spin diffusion/transport in GaAs quantum wells with very high accuracy and speed [42]. It was discovered that due to the scattering, especially the Coulomb scattering, $T_2 = T_2^*$ is valid even in the spacial domain. This prediction remains yet to be verified experimentally. Moreover, as the inhomogeneous broadening in spin diffusion is determined by $|\mathbf{h}(\mathbf{k})|/k_x$ in the absence of magnetic field, the period of the spin oscillations along the x-axis is independent of the electric field perpendicular to the growth direction of the quantum well [42], which is different from the spin precession rate in the time domain [22]. This is consistent with the experimental findings by Beck et al. [79].

Cheng et al. applied the kinetic-spin-Bloch-equation approach to study the spin transport in the presence of competing Dresselhaus and Rashba fields [80]. When the Dresselhaus and Bychkov–Rashba [10] terms are both important in semiconductor quantum well, the total effective magnetic field can be highly anisotropic and spin dynamics is also highly anisotropic in regard to the spin polarization [81]. For some special polarization direction, the spin relaxation time is extremely large [81–84]. For example, if the coefficients of the linear Dresselhaus and Bychkov–Rashba terms are equal to each other in (001) quantum well of small well width and the cubic Dresselhaus term is not important, the effective magnetic field is along the [110] direction for all electrons. For the spin components perpendicular to the [110] direction, this effective magnetic field flips the spin and leads to a finite spin relaxation/dephasing time. For spin along the [110] direction, this effective magnetic field cannot flip it. Therefore, when the spin polarization is along the [110] direction, the Dresselhaus and Bychkov–Rashba terms cannot cause any spin relaxation/dephasing. When the cubic Dresselhaus term is taken into account, the spin dephasing time for spin polarization along the [110] direction is finite but still much larger than other directions [85]. The anisotropy in the spin direction is also expected in spin diffusion and transport. When the Dresselhaus and Bychkov–Rashba terms are comparable, the spin injection length L_d for the spin polarization perpendicular to [110] direction is usually much shorter than that for the spin polarization along [110] direction. In the ideal case when there are only the linear Dresselhaus and Bychkov–Rashba terms with identical strengths, spin injection length for spin polarization parallel to the [110] direction becomes infinity [82,83]. This effect has promoted Schliemann et al. to propose the non-ballistic spin-field-effect transistor [83]. In such a transistor, a gate voltage is used to tune the strength of the Bychkov–Rashba term and therefore control the spin injection length. However, Cheng et al. pointed out that spin diffusion and transport actually involve both the spin polarization and spin transport directions [80]. The latter has long been overlooked in the literature. In the kinetic-spin-Bloch-equation approach, this direction corresponds to the spacial gradient in the diffusion term $[\dot{\rho}_{\mathbf{k}}(\mathbf{r},t)|_{\mathrm{dif}}]$ and the electric field in the drifting term $[\dot{\rho}_{\mathbf{k}}(\mathbf{r},t)|_{\mathrm{dr}}]$. The importance of the spin transport direction has not been realized until Cheng et al. pointed out that the spin transport is highly anisotropic not only in the sense of the spin polarization direction but also in the spin transport direction when the Dresselhaus and Bychkov–Rashba effective magnetic fields are comparable [80]. They even predicted that in (001) GaAs quantum well with identical linear Dresselhaus and Bychkov–Rashba coupling strengths, the spin injection along [$\bar{1}$10] or [110]* can be infinite *regardless of* the direction of the spin polarization. This can be easily seen from the inhomogeneous broadening Equation 16.144, which well defines the spin diffusion/transport properties. For the spin diffusion/transport in a (001) GaAs quantum well with identical Dresselhaus and Bychkov–Rashba strengths (the schematic is shown in Figure 16.16 with the transport

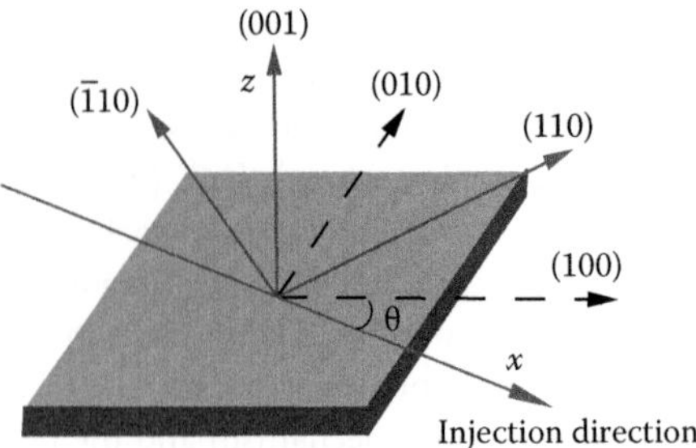

FIGURE 16.16 Schematic of the different directions considered for the spin polarizations [(110), ($\bar{1}$10), and (001)-axes] and spin diffusion/injection (x-axis). (From Cheng, J.L. et al., *Phys. Rev. B*, 75, 205328, 2007.)

* [$\bar{1}$10] or [110] depends on the relative signs of the Dresselhaus and Rashba coupling strengths.

direction chosen along the x-axis), the inhomogeneous broadening is given by [80]

$$\mathbf{W}_{\mathbf{k}} = \left\{ 2\beta\left(\sin\left(\theta - \frac{\pi}{4}\right) + \cos\left(\theta - \frac{\pi}{4}\right)\frac{k_y}{k_x}\right)\hat{\mathbf{n}}_0 + \gamma\left(\frac{k_x^2 - k_y^2}{2}\sin 2\theta + k_x k_y \cos 2\theta\right)\left(\frac{k_y}{k_x}, -1, 0\right)\right\}, \quad (16.145)$$

with θ being the angle between the spin transport direction (x-axis) and [001] crystal direction. It can be split into two parts: the zeroth-order term (on k), which is always along the same direction of $\hat{\mathbf{n}}_0$, and the second-order term, which comes from the cubic Dresselhaus term. If the cubic Dresselhaus term is omitted, the effective magnetic fields for all $\mathbf{k}$ states align along $\hat{\mathbf{n}}_0$ (crystal [110]) direction. Therefore, if the spin polarization is along $\hat{\mathbf{n}}_0$, there is no spin relaxation even in the presence of scattering since there is no spin precession. Nevertheless, it is interesting to see from Equation 16.145 that when $\theta = 3\pi/4$, that is, the spin transport is along the $[\bar{1}10]$ direction, $\Omega_{\mathbf{k}} = 2m^*\beta\hat{\mathbf{n}}_0$ is independent of $\mathbf{k}$ if the cubic Dresselhaus term is neglected. Therefore, in this special spin transport direction, there is no inhomogeneous broadening in the spin transport for *any* spin polarization. The spin injection length is therefore infinite regardless of the direction of spin polarization. This result is highly counterintuitive, considering that the spin relaxation times for the spin components perpendicular to the effective magnetic field are finite in the spacial uniform system. The surprisingly contradictory results, that is, the finite spin relaxation/dephasing time versus the infinite spin injection length, are due to the difference in the inhomogeneous broadening in spacial uniform and nonuniform systems. For genuine situation, due to the presence of the cubic term, the spin injection length is still finite and the maximum spin injection length does not happen at the identical Dresselhaus and Bychkov–Rashba coupling strengths, but shifted by a small amount due to the cubic term [80]. However, there is strong anisotropy in regard to the spin polarization and spin injection direction, as shown in Figure 16.17. This predication has not yet been realized experimentally. However, very recent experimental findings on spin helix [13,14] have provided strong evidence to support this predication [86].

Now we turn to the problem of spin grating. Transient spin grating, whose spin polarization varies periodically in real space, is excited optically by two non-collinear coherent light beams with orthogonal linear polarization [14,87–89]. Transient spin grating technique can be used to study the spin transport since it can directly probe the decay rate of nonuniform spin distributions. Spin diffusion coefficient D_s can be obtained from the transient spin grating experiments [87–89]. In the literature, the drift-diffusion model was employed to extract D_s from the experimental data. With the drift-diffusion model, the transient spin grating was predicted to decay

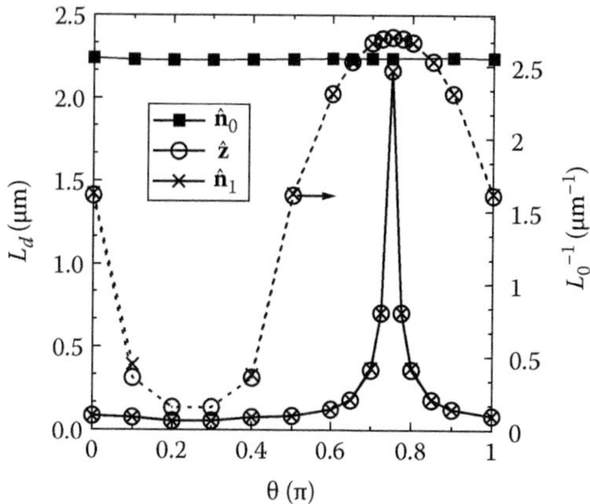

FIGURE 16.17 Spin diffusion length L_d (solid curves) and the inverse of the spin oscillation period L_0^{-1} (dashed curves) for identical Dresselhaus and Rashba coupling strengths as functions of the injection direction for different spin polarization directions $\hat{\mathbf{n}}_0$, $\hat{\mathbf{z}}$ and $\hat{\mathbf{n}}_1$ ($\hat{\mathbf{n}}_1 = \hat{\mathbf{z}} \times \hat{\mathbf{n}}_0$, i.e., crystal direction $[\bar{1}10]$) at $T = 200$ K. It is noted that the scale of the spin oscillation period is on the right-hand side of the frame. (From Cheng, J.L. et al., *Phys. Rev. B*, 75, 205328, 2007.)

exponentially with time with a decay rate of $\Gamma_q = D_s q^2 + 1/\tau_s$, where q is the wavevector of the spin grating and τ_s is the spin relaxation time [87,88]. However, this result is not accurate since it neglects the spin precession, which plays an important role in spin transport, as first pointed out by Weng and Wu [22]. Indeed, experimental results show that the decay of transient spin grating takes a double-exponential form instead of single exponential one [14,87,89]. Also the relation that relates the spin injection length with the spin diffusion coefficient D_s and the spin relaxation time τ_s, i.e., $L_s = 2\sqrt{D_s \tau_s}$ from the drift-diffusion model should be checked. In fact, if this relation is correct, the above prediction of infinite spin injection length at certain spin injection direction for any spin polarization in the presence of identical Dresselhaus and Bychkov–Rashba coupling strengths cannot be correct.

Weng et al. studied this problem from the kinetic-spin-Bloch-equation approach [90]. By first solving the kinetic spin Bloch equations analytically by including only the elastic scattering, that is, the electron-impurity scattering, they showed that the transient spin grating should decay double exponentially with two decay rates $\Gamma_\pm$. In fact, none of the rates is quadratic in q. However, the average of them reads [90]

$$\Gamma = \frac{(\Gamma_+ + \Gamma_-)}{2} = \frac{Dq^2 + 1}{\tau_s'} \quad (16.146)$$

with $1/\tau_s' = (1/\tau_s + 1/\tau_{s1})/2$, which differs from the current widely used formula by replacing the spin decay rate by the average of the in- and out-of-plane relaxation rates. The difference of these two decay rates is a linear function of the wavevector q when q is relatively large:

$$\Delta\Gamma = cq + d, \quad (16.147)$$

with c and d being two constants. The steady-state spin injection length L_s and spin precession period L_0 are then [90]

$$L_s = \frac{2D_s}{\sqrt{|c^2 - 4D_s(1/\tau_s^{'} - d)|}}, \tag{16.148}$$

$$L_0 = \frac{2D_s}{c}. \tag{16.149}$$

They further showed that the above relation Equations 16.146 and 16.147 are valid even including the inelastic electron–electron and electron-phonon scatterings by solving the full kinetic spin Bloch equations, as shown in Figure 16.18. A good agreement with the experimental data [89] of double exponential decays $\tau_\pm = \Gamma_\pm^{-1}$ is shown in Figure 16.19. Finally it was shown that the infinite spin injection length predicted by Cheng et al. in the presence of identical Dresselhaus and Bychkov–Rashba coupling strengths [80] addressed earlier can exactly be obtained from Equation 16.148 as in that special case $\tau_s^{'} = \tau_s$, $d=0$, and $c = 2\sqrt{D_s/\tau_s}$ [90]. However, $\sqrt{D_s\tau_s}$ always remains finite unless $\tau_s = \infty$. Therefore, Equations 16.146 through 16.149 give the correct way to extract the spin injection length from the spin grating measurement.

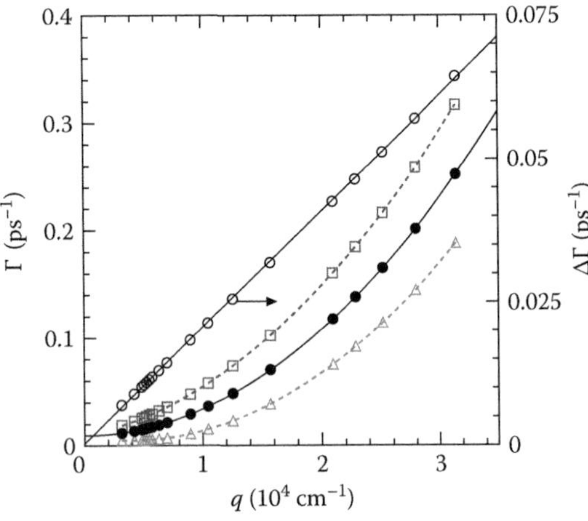

FIGURE 16.18 $\Gamma=(\Gamma_+ + \Gamma_-)/2$ and $\Delta\Gamma=(\Gamma_+ + \Gamma_-)/2$ versus q at $T=295$ K. Open boxes/triangles are the relaxation rates ($\Gamma_{+/-}$ calculated from the full kinetic spin Bloch equations. Filled/open circles represent Γ and $\Delta\Gamma$ respectively. Noted that the scale for $\Delta\Gamma$ is on the right-hand side of the frame. The solid curves are the fitting to Γ and $\Delta\Gamma$, respectively. The dashed curves are guide to eyes. (From Weng, M.Q. et al., *J. Appl. Phys.*, 103, 063714, 2008.)

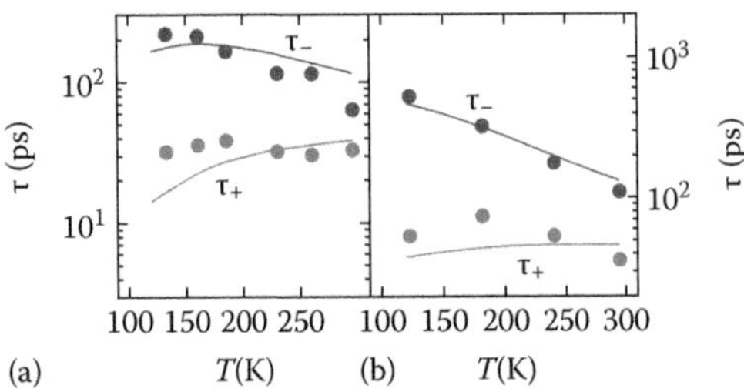

FIGURE 16.19 Spin relaxation times $\tau_\pm$ versus temperature for (a) high-mobility sample with $q=0.58\times10^4$ cm^{-1} and (b) low-mobility sample with $q=0.69\times10^4$ cm^{-1}. The dots are the experiment data from Ref. [90]. (From Weng, M.Q. et al., *J. Appl. Phys.*, 103, 063714, 2008.)

Acknowledgments

This work was supported by DFG SFB 689 and SPP1286, Natural Science Foundation of China under Grant No. 10725417, the National Basic Research Program of China under Grant No. 2006CB922005, and the Knowledge Innovation Project of Chinese Academy of Sciences.

References

1. R. J. Elliott, Theory of the effect of spin–orbit coupling on magnetic resonance in some semiconductors, *Phys. Rev.* **96**, 266 (1954).
2. Y. Yafet, F. Seitz, and D. Turnbull (Eds.), G-factors and spin-lattice relaxation of conduction electronics, *Solid State Physics*, Vol. 14, Academic, New York, p. 2, 1963.
3. M. I. D'yakonov and V. I. Perel', Spin relaxation of conduction electrons in noncentrosymmetric semiconductors, *Sov. Phys. Solid State* **13**, 3023 (1971).
4. G. L. Bir, A. G. Aronov, and G. E. Pikus, Spin relaxation of electrons due to scattering by holes, *Sov. Phys. JETP* **42**, 705 (1976).
5. F. Meier and B. P. Zakharchenya (Eds.), *Optical Orientation*, North-Holland, New York, 1984.
6. J. Fabian and S. Das Sarma, Spin relaxation of conduction electrons, *J. Vac. Sci. Technol. B* **17**, 1708 (1999).
7. I. Žutić, J. Fabian, and S. Das Sarma, Spintronics: Fundamentals and applications, *Rev. Mod. Phys.* **76**, 323 (2004).
8. J. Fabian, A. Matos-Abiague, C. Ertler, P. Stano, and I. Žutić, Semiconductor spintronics, *Acta Phys. Slov.* **57**, 565 (2007).
9. M. W. Wu, J. H. Jiang, and M. Q. Weng, Spin dynamics in semiconductors, *Phys. Rep.* **493**, 61 (2010).
10. Y. A. Bychkov and E. I. Rashba, Properties of a 2D electron gas with lifted spectral degeneracy, *JETP Lett.* **39**, 78 (1984).
11. G. Dresselhaus, Spin-orbit coupling effects in zinc blende structures, *Phys. Rev.* **100**, 580 (1955).
12. M. I. D'yakonov and V. Y. Kachorovskii, Spin relaxation of two-dimensional electrons in noncentrosymmetric semiconductors, *Sov. Phys. Semicond.* **20**, 110 (1986).
13. B. A. Bernevig, J. Orenstein, and S. C. Zhang, Exact SU(2) symmetry and persistent spin helix in a spin–orbit coupled system, *Phys. Rev. Lett.* **97**, 236601 (2007).
14. J. D. Koralek, C. P. Weber, J. Orenstein et al., Emergence of the persistent spin helix in semiconductor quantum wells, *Nature* **458**, 610 (2009).

15. J. L. Cheng, M. W. Wu, and J. Fabian, The spin relaxation of conduction electrons in silicon, *Phys. Rev. Lett.* **104**, 016601 (2010).
16. D. J. Lepine, Spin resonance of localized and delocalized electrons in phosphorus-doped silicon between 20 and 30 K, *Phys. Rev. B* **2**, 2429 (1970).
17. B. Huang and I. Appelbaum, Coherent spin transport through a 350 micron thick silicon wafer, *Phys. Rev. Lett.* **99**, 177209 (2007).
18. M. W. Wu and H. Metiu, Kinetics of spin coherence of electrons in an undoped semiconductor quantum well, *Phys. Rev. B* **61**, 2945 (2000).
19. M. W. Wu and C. Z. Ning, A novel mechanism for spin dephasing due to spin-conserving scatterings, *Eur. Phys. J. B* **18**, 373 (2000).
20. M. W. Wu, Spin dephasing induced by inhomogeneous broadening in D'yakonov–Perel' effect in a n-doped GaAs quantum well, *J. Phys. Soc. Jpn.* **70**, 2195 (2001).
21. M. Q. Weng and M. W. Wu, Spin dephasing in n-type GaAs quantum wells, *Phys. Rev. B* **68**, 075312 (2003).
22. M. Q. Weng, M. W. Wu, and L. Jiang, Hot-electron effect in spin dephasing in n-type GaAs quantum wells, *Phys. Rev. B* **69**, 245320 (2004).
23. M. Q. Weng and M. W. Wu, Longitudinal spin decoherence in spin diffusion in semiconductors, *Phys. Rev. B* **66**, 235109 (2002).
24. M. Q. Weng and M. W. Wu, Kinetic theory of spin transport in n-type semiconductor quantum wells, *J. Appl. Phys.* **93**, 410 (2003).
25. M. Q. Weng, M. W. Wu, and Q. W. Shi, Spin oscillations in transient diffusion of a spin pulse in n-type semiconductor quantum wells, *Phys. Rev. B* **69**, 125310 (2004).
26. E. Ya. Sherman, Random spin–orbit coupling and spin relaxation in symmetric quantum wells, *Appl. Phys. Lett.* **82**, 209 (2003).
27. H. Haug and A. P. Jauho, *Quantum Kinetics in Transport and Optics of Semiconductors*, Springer, Berlin, 1996.
28. M. Q. Weng and M. W. Wu, Multisubband effect in spin dephasing in semiconductor quantum wells, *Phys. Rev. B* **70**, 195318 (2004).
29. J. Zhou, J. L. Cheng, and M. W. Wu, Spin relaxation in n-type GaAs quantum wells from a fully microscopic approach, *Phys. Rev. B* **75**, 045305 (2007).
30. J. Zhou and M. W. Wu, Spin relaxation due to the Bir–Aronov–Pikus mechanism in intrinsic and p-type GaAs quantum wells from a fully microscopic approach, *Phys. Rev. B* **77**, 075318 (2008).
31. J. H. Jiang, Y. Zhou, T. Korn, C. Schüller, and M. W. Wu, Electron spin relaxation in paramagnetic Ga(Mn)As quantum wells, *Phys. Rev. B* **79**, 155201 (2009).
32. P. Zhang, J. Zhou, and M. W. Wu, Multivalley spin relaxation in the presence of high in-plane electric fields in n-type GaAs quantum wells, *Phys. Rev. B* **77**, 235323 (2008).
33. J. H. Jiang, M. W. Wu, and Y. Zhou, Kinetics of spin coherence of electrons in n-type InAs quantum wells under intense terahertz laser fields, *Phys. Rev. B* **78**, 125309 (2008).
34. P. Zhang and M. W. Wu, Effect of nonequilibrium phonons on hot-electron spin relaxation in n-type GaAs quantum wells, *Europhys. Lett.* **92**, 47009 (2010).
35. C. Lü, J. L. Cheng, and M. W. Wu, Hole spin dephasing in P-type semiconductor quantum wells, *Phys. Rev. B* **73**, 125314 (2006).
36. Y. Zhou, J. H. Jiang, and M. W. Wu, Electron spin relaxation in P-type GaAs quantum wells, *New J. Phys.* **11**, 113039 (2009).
37. P. Zhang and M. W. Wu, Hole spin relaxation in [001] strained asymmetric Si/SiGe and Ge/SiGe quantum wells, *Phys. Rev. B* **80**, 155311 (2009).
38. C. Lü, U. Zülicke, and M. W. Wu, Hole spin relaxation in P-type GaAs quantum wires investigated by numerically solving fully microscopic kinetic spin Bloch equations, *Phys. Rev. B* **78**, 165321 (2008).
39. C. Lü, H. C. Schneider, and M. W. Wu, Electron spin relaxation in n-type InAs quantum wires, *J. Appl. Phys.* **106**, 073703 (2009).
40. J. H. Jiang, Y. Y. Wang, and M. W. Wu, Reexamination of spin decoherence in semiconductor quantum dots from the equation-of-motion approach, *Phys. Rev. B* **77**, 035323 (2008).
41. J. H. Jiang and M. W. Wu, Electron-spin relaxation in bulk III–V semiconductors from a fully microscopic kinetic spin Bloch equation approach, *Phys. Rev. B* **79**, 125206 (2009).
42. J. L. Cheng and M. W. Wu, Spin diffusion/transport in n-type GaAs quantum wells, *J. Appl. Phys.* **101**, 073702 (2007).
43. P. Zhang and M. W. Wu, Spin diffusion in Si/SiGe quantum wells: Spin relaxation in the absence of D'yakonov–Perel' relaxation mechanism, *Phys. Rev. B* **79**, 075303 (2009).
44. M. E. Flatté, J. M. Bayers, and W. H. Lau, Spin dynamics in semiconductors, in *Spin Dynamics in Semiconductors*, D. D. Awsehalom, D. Loss, and N. Samarth (Eds.), Springer, Berlin, 2002.
45. M. M. Glazov and E. L. Ivchenko, Precession spin relaxation mechanism caused by frequent electron–electron collisions, *JETP Lett.* **75**, 403 (2002).
46. M. A. Brand, A. Malinowski, O. Z. Karimov et al., Precession and motional slowing of spin evolution in a high mobility two-dimensional electron gas, *Phys. Rev. Lett.* 236601 (2002); W. J. H. Leyland, R. T. Harley, M. Henini, A. J. Shields, I. Farrer, and D. A. Ritchie, Energy-dependent electron–electron scattering and spin dynamics in a two-dimensional electron gas, *Phys. Rev. B* **77**, 205321 (2008).
47. X. Z. Ruan, H. H. Luo, Y. Ji, Z. Y. Xu, and V. Umansky, Effect of electron–electron scattering on spin dephasing in a high-mobility low-density two-dimensional electron gas, *Phys. Rev. B* **77**, 193307 (2008).
48. Y. Ohno, R. Terauchi, T. Adachi, F. Matsukura, and H. Ohno, Electron spin relaxation beyond D'yakonov–Perel' interaction in GaAs/AlGaAs quantum wells, *Physica E* **6**, 817 (2000).

49. F. X. Bronold, A. Saxena, and D. L. Smith, Semiclassical kinetic theory of electron spin relaxation in semiconductors, *Phys. Rev. B* **70**, 245210 (2004).
50. G. F. Giulianni and G. Vignale, *Quantum Theory of the Electron Liquid*, Cambridge University Press, Cambridge, U.K., 2005.
51. L. Jiang and M. W. Wu, Control of spin coherence in n-type GaAs quantum wells using strain, *Phys. Rev. B* **72**, 033311 (2005).
52. A. W. Holleitner, V. Sih, R. C. Myers, A. C. Gossard, and D. D. Awschalom, Dimensionally constrained D'yakonov–Perel' spin relaxation in n-InGaAs channels: Transition from 2D to 1D, *New J. Phys.* **9**, 342 (2007).
53. F. Zhang, H. Z. Zheng, Y. Ji, J. Liu, and G. R. Li, Spin dynamics in the second subband of a quasi–two-dimensional system studied in a single-barrier heterostructure by time-resolved Kerr rotation, *Europhys. Lett.* **83**, 47007 (2008).
54. C. Lü, J. L. Cheng, M. W. Wu, and I. C. da Cunha Lima, Spin relaxation time, spin dephasing time and ensemble spin dephasing time in n-type GaAs quantum wells, *Phys. Lett. A* **365**, 501 (2007).
55. D. Stich, J. Zhou, T. Korn, R. Schulz, D. Schuh, W. Wegscheider, M. W. Wu, and C. Schüller, Effect of initial spin polarization on spin dephasing and the electron g factor in a high-mobility two-dimensional electron system, *Phys. Rev. Lett.* **98**, 176401 (2007).
56. D. Stich, J. Zhou, T. Korn et al., Dependence of spin dephasing on initial spin polarization in a high-mobility two-dimensional electron system, *Phys. Rev. B* **76**, 205301 (2007).
57. F. Zhang, H. Z. Zheng, Y. Ji, J. Liu, and G. R. Li, Electrical control of dynamic spin splitting induced by exchange interaction as revealed by time-resolved Kerr rotation in a degenerate spin-polarized electron gas, *Europhys. Lett.* **83**, 47006 (2008).
58. T. Korn, D. Stich, R. Schulz, D. Schuh, W. Wegscheider, and C. Schüller, Spin dynamics in high-mobility two-dimensional electron system, *Adv. Solid State Phys.* **48**, 143 (2009).
59. A. P. Dmitriev, V. Y. Kachorovskii, and M. S. Shur, High-field transport in a dense two-dimensional electron gas in elementary semiconductors, *J. Appl. Phys.* **89**, 3793 (2001).
60. E. M. Conwell, *High Field Transport in Semiconductors*, Pergamon, Oxford, 1972.
61. T. C. Damen, L. Vina, J. E. Cunningham, J. Shah, and L. J. Sham, Subpicosecond spin relaxation dynamics of excitons and free carriers in GaAs quantum wells, *Phys. Rev. Lett.* **67**, 3432 (1991).
62. J. Wagner, H. Schneider, D. Richards, A. Fischer, and K. Ploog, Observation of extremely long electron-spin-relaxation times in p-type δ-doped GaAs/$Al_xGa_{1-x}As$ double heterostructures, *Phys. Rev. B* **47**, 4786 (1993).
63. H. Gotoh, H. Ando, T. Sogawa, H. Kamada, T. Kagawa, and H. Iwamura, Effect of electron–hole interaction on electron spin relaxation in GaAs/AlGaAs quantum wells at room temperature, *J. Appl. Phys.* **87**, 3394 (2000).
64. T. F. Boggess, J. T. Olesberg, C. Yu, M. E. Flatté, and W. H. Lau, Room-temperature electron spin relaxation in bulk InAs, *Appl. Phys. Lett.* **77**, 1333 (2000).
65. S. Hallstein, J. D. Berger, M. Hilpert et al., Manifestation of coherent spin precession in stimulated semiconductor emission dynamics, *Phys. Rev. B* **56**, R7076 (1997).
66. P. Nemec, Y. Kerachian, H. M. van Driel, and A. L. Smirl, Spin-dependent electron many-body effects in GaAs, *Phys. Rev. B* **72**, 245202 (2005).
67. H. C. Schneider, J.-P. Wüstenberg, O. Andreyev et al., Energy-resolved electron spin dynamics at surfaces of p-doped GaAs, *Phys. Rev. B* **73**, 081302 (2006).
68. C. Yang, X. Cui, S.-Q. Shen, Z. Xu, and W. Ge, Spin relaxation in submonolayer and monolayer InAs structures grown in a GaAs matrix, *Phys. Rev. B* **80**, 035313 (2009).
69. P. H. Song and K. W. Kim, Spin relaxation of conduction electrons in bulk III–V semiconductors, *Phys. Rev. B* **66**, 035207 (2002).
70. D. D. Awschalom, D. Loss, and N. Samarth (Eds.), *Semiconductor Spintronics and Quantum Computation*, Springer-Verlag, Berlin, 2002; M. I. D'yakonov (Ed.), *Spin Physics in Semiconductors*, Springer, Berlin, 2008, and references therein.
71. R. I. Dzhioev, K. V. Kavokin, V. L. Korenev et al., Low-temperature spin relaxation in n-type GaAs, *Phys. Rev. B* **66**, 245204 (2002).
72. M. Krauß, R. Bratschitsch, Z. Chen, S. T. Cundiff, and H. C. Schneider, Ultrafast spin dynamics in optically excited bulk GaAs at low temperatures, *Phys. Rev. B* **81**, 035213 (2010).
73. K. Shen, A peak in density dependence of electron spin relaxation time in n-type bulk GaAs in the metallic regime, *Chin. Phys. Lett.* **26**, 067201 (2009).
74. I. Appelbaum, B. Huang, and D. J. Monsma, Electronic measurement and control of spin transport in silicon, *Nature* **447**, 295 (2007).
75. B. Huang, L. Zhao, D. J. Monsma, and I. Appelbaum, 35% magnetocurrent with spin transport through Si, *Appl. Phys. Lett.* **91**, 052501 (2007).
76. G. Dresselhaus, Spin–orbit coupling effects in zinc blende structures, *Phys. Rev.* **100**, 580 (1955).
77. L. Jiang, M. Q. Weng, M. W. Wu, and J. L. Cheng, Diffusion and transport of spin pulses in an n-type semiconductor quantum well, *J. Appl. Phys.* **98**, 113702 (2005).
78. S. A. Crooker and D. L. Smith, Imaging spin flows in semiconductors subject to electric, magnetic, and strain fields, *Phys. Rev. Lett.* **94**, 236601 (2005).
79. M. Beck, C. Metzner, S. Malzer, and G. H. Döhler, Spin lifetimes and strain-controlled spin precession of drifting electrons in GaAs, *Europhys. Lett.* **75**, 597 (2006).
80. J. L. Cheng, M. W. Wu, and I. C. da Cunha Lima, Anisotropic spin transport in GaAs quantum wells in the presence of competing Dresselhaus and Rashba spin–orbit coupling, *Phys. Rev. B* **75**, 205328 (2007).

81. N. S. Averkiev and L. E. Golub, Giant spin relaxation anisotropy in zinc-blende heterostructures, *Phys. Rev. B* **60**, 15582 (1999).
82. N. S. Averkiev, L. E. Golub, and M. Willander, Spin relaxation anisotropy in two-dimensional semiconductor systems, *J. Phys.: Condens. Matter* **14**, R271 (2002).
83. J. Schliemann, J. C. Egues, and D. Loss, Nonballistic spin-field-effect transistor, *Phys. Rev. Lett.* **90**, 146801 (2003).
84. R. Winkler, Spin orientation and spin precession in inversion-asymmetric quasi-two-dimensional electron systems, *Phys. Rev. B* **69**, 045317 (2004).
85. J. L. Cheng and M. W. Wu, Spin relaxation under identical Dresselhaus and Rashba coupling strengths in GaAs quantum wells, *J. Appl. Phys.* **99**, 083704 (2006).
86. K. Shen and M. W. Wu, Infinite spin diffusion length of any spin polarization along direction perpendicular to effective magnetic field from Dresselhaus and Rashba spin–orbit couplings with identical strengths in (001) GaAs quantum wells, *J. Supercond. Nov. Magn.* **22**, 715 (2009).
87. C. P. Weber, N. Gedik, J. E. Moore, J. Orenstein, J. Stephens, and D. D. Awschalom, Observation of spin Coulomb drag in a two-dimensional electron gas, *Nature* **437**, 1330 (2005).
88. A. R. Cameron, P. Rickel, and A. Miller, Spin gratings and the measurement of electron drift mobility in multiple quantum well semiconductors, *Phys. Rev. Lett.* **76**, 4793 (1996).
89. C. P. Weber, J. Orenstein, B. A. Bernevig, S.-C. Zhang, J. Stephens, and D. D. Awschalom, Nondiffusive spin dynamics in a two-dimensional electron gas, *Phys. Rev. Lett.* **98**, 076604 (2007).
90. M. Q. Weng, M. W. Wu, and H. L. Cui, Spin relaxation in n-type GaAs quantum wells with transient spin grating, *J. Appl. Phys.* **103**, 063714 (2008).

17

Electrical Spin Injection and Transport in Semiconductors

Berend T. Jonker
Naval Research Laboratory

17.1 Introduction

The field of semiconductor electronics has been based exclusively on the manipulation of charge. The remarkable advances in performance have been due in large part to size scaling, i.e., systematic reduction in device dimensions, enabling significant increases in circuit density. This trend, known as Moore's Law [1], is likely to be curtailed in the near future by practical and fundamental limits. Consequently, there is keen interest in exploring new avenues and paradigms for future technologies. *Spintronics*, or the use of carrier spin as a new degree of freedom in an electronic device, represents one of the most promising candidates for this paradigm shift [2,3]. Like charge, spin is an intrinsic fundamental property of an electron, and is one of the alternative state variables under consideration on the International Technology Roadmap for Semiconductors (ITRS) for processing information in the new ways that will be required beyond the ultimate scaling limits of the existing silicon-based complementary metal-oxide-semiconductor (CMOS) technology [4].

Semiconductor-based spintronics offers many new avenues and opportunities which are inaccessible to metal-based spintronic structures. This is due to the characteristics for which semiconductors are so well-known: the existence of a band gap that can often be tuned over a significant range in ternary compounds; the accompanying optical properties on which a vast optoelectronic industry is based; and the ability to readily control carrier concentrations and transport characteristics via doping, gate voltages, and band offsets. Coupling the new degree of freedom of carrier spin with the traditional band gap engineering of modern electronics offers opportunities for new functionality and performance for semiconductor devices (examples can be found in Chapters 38 and 39).

There are four essential requirements for implementing a semiconductor-based spintronics technology:

1. Efficient *electrical injection* of spin-polarized carriers from an appropriate contact into the semiconductor heterostructure.
2. Adequate *spin diffusion lengths and spin lifetimes* within the semiconductor host medium.
3. Effective *control and manipulation* of the spin system to provide the desired functionality.
4. Efficient *detection* of the spin system to provide the output.

Although the focus of this chapter is nominally on the first, each of these factors will necessarily be addressed to some extent, as each plays a role in any experimental research effort.

The generation of spin-polarized carrier populations in semiconductors has historically been accomplished by optical excitation, an approach referred to as "optical pumping" or "optical orientation" [5]. This has been the basis for decades of research exploring spin-dependent phenomena in semiconductor hosts, beginning with the pioneering work of Lampel [6], and enabling the development of new spectroscopic techniques which provide extraordinary insight into spin properties [7,8]. Injection of spin-polarized electrons from a scanning tunneling microscope tip in ultra-high vacuum was first reported in 1992 [9], and provides a highly localized probe, suitable for basic research. However, a broad technology based on spin manipulation in a semiconductor host requires an efficient and practical means of spin injection and detection. This is best accomplished *via a discrete electrical contact.* Such a contact defines the extent of the spin source (or detector), provides an interface between the spin system and the canonical electronic input/output parameters of voltage and current, and offers a very simple and direct means of implementing spin injection/detection compatible with existing device design and fabrication technology [10–12].

In this chapter, we will discuss the factors relevant to spin injection and transport in semiconductors, review the necessary concepts, and illustrate these by way of examples to provide both a practical guide and an overview of the current state of the art. We conclude with a brief section (Section 17.6) on electrical injection and *detection* of spin accumulation using multiterminal lateral transport geometries in silicon, where an operation well above room temperature has recently been demonstrated.

17.2 Spin Polarization by Optical Pumping

The earliest method of generating a spin-polarized population of carriers within a semiconductor was by optical excitation using circularly polarized light, often referred to as "optical pumping" [5]. A photon carries a unit of angular momentum either parallel or antiparallel to its propagation direction, and the corresponding polarization is referred to as positive (σ+) or negative (σ−) helicity, respectively. Absorption of a photon whose energy is equal to or greater than the band gap promotes an electron from the valence band (VB) to the conduction band (CB). If the light is circularly polarized, the electron and hole must share the angular momentum transferred in a manner that satisfies conservation laws [13]. In a direct gap zinc-blende semiconductor such as GaAs, the CB is *s*-like in character and twofold spin degenerate, and the electron can occupy states with values of spin angular momentum $m_j = \pm 1/2$. The VB is *p*-like in character and fourfold spin degenerate in bulk material, so that the hole can occupy states with values of angular momentum $m_j = \pm 1/2$,

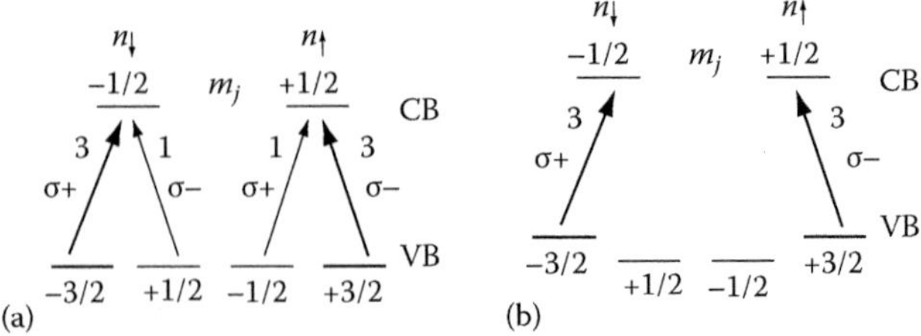

FIGURE 17.1 Optical transitions allowed due to absorption (or emission—reverse the arrows) of circularly polarized light for (a) bulk material and (b) a structure in which the light/heavy hole band degeneracy is lifted, as in a quantum well or strained material. The HH transitions are three times more probable than the LH.

±3/2, corresponding to "light hole" (LH) and "heavy hole" (HH) states, respectively (Figure 17.1a). For photons incident along the surface normal, interband transitions at the zone center ($k_{//} = 0$) satisfy the selection rule $\Delta m_j = \pm 1$, reflecting absorption of the photon's original angular momentum. The probability of a transition involving a LH or HH state is weighted by the square of the corresponding matrix element connecting it to the appropriate electron state, so that HH transitions are three times more likely than LH transitions. Thus, absorption of photons with angular momentum +1 produces three "spin-down" ($m_j = -1/2$) electrons for every one "spin-up" ($m_j = +1/2$) electron, resulting in an electron population with a spin polarization of 50% in a bulk material, where the HH and LH states are degenerate. In a quantum well (QW), either quantum confinement or strain can lift this degeneracy, as illustrated in Figure 17.1b, where the HH states are at lower energy (nearer the center of the band gap). In this case, circularly polarized light whose energy is just sufficient to excite these states (and not the LH states) can in principle produce an electron population, which is 100% spin polarized. The electron spin lifetime is much longer than that of the hole in most materials, and the holes are generally assumed to depolarize instantaneously. Spin relaxation and lifetimes are discussed in Chapter 20.

Note that the resulting electron spin is oriented parallel or antiparallel to the propagation direction of the incident photon (typically the surface normal), consistent with the angular momentum of the absorbed photon. This is an important factor to bear in mind when performing and interpreting an experiment.

17.3 Spin Polarization by Electrical Injection: General Considerations

A discrete electrical contact to create, manipulate, and detect spin currents and populations in a semiconductor host enables the development of a much broader range of technologies than that allowed by optical excitation alone. A discrete contact clearly defines the source/detector dimensions and is readily scalable, an indispensible attribute, in contrast with optical approaches which are diffraction limited. Existing electronics is based upon a voltage or current for input and output. A spin contact must, therefore, provide a suitable interface between these conventional parameters and the spin system to be utilized for information processing or sensing in the semiconductor. A magnetic material has a spin-polarized band structure and density of states (DOS), and is well suited for such a contact—the electron population at the Fermi energy, E_F, naturally exhibits a net spin polarization, which serves as a source for electrical injection under appropriate bias into an unpolarized host.

The notion that a spin current could in principle accompany a charge current was first hypothesized by Aronov and Pikus in 1976 [14], when they noted that the flow of electric current I_e from a ferromagnetic (FM) metal should exhibit a net spin polarization (magnetization) due to its spin-polarized DOS. If this charge current flows into a nonmagnetic material, it simultaneously corresponds to a flow of magnetization, since each electron also bears the fundamental unit of magnetization, μ_B, the Bohr magneton. This magnetization current I_m is simply written as

$$I_m = \eta\mu_B\left(\frac{I_e}{e}\right) \tag{17.1}$$

where

e is the electron's charge

$0 \le \eta \le 1$ is a phenomenological factor to account for the fact that the Fermi surface of the ferromagnet may contain both spin bands (or to account for possible spin scattering at the interface)

This remarkably simple and intuitive relation provided the seed from which the burgeoning field of spin injection and transport has sprung.

A contact for spin injection/detection should exhibit some specific characteristics. In general, one desires high spin-injection efficiency at low bias power extending well above room temperature for practical device applications. While it is common to think of these properties as attributes of the contact material alone, it is essential to note that the interface between the contact and the semiconductor is equally important and is often more challenging to understand. Thus, the following should be considered as desirable attributes for the combination of spin contact material and the corresponding interface:

- A Curie temperature $T_c > 400$ K, although certain specialized applications (e.g., cooled infrared detectors) may permit T_c's as low as 100 K.
- High remanent magnetization, i.e., the magnetization retained at zero field. The nonvolatile behavior essential for many applications (memory, optical isolators, reprogrammable logic) is predicated upon a remanence that is a substantial fraction of the saturation magnetization.
- The magnetization should be readily switched at modest power expense, important for applications, which require fast and routine switching (memory logic). Historically, this has meant a low coercive field ($H_c < 200$ Oe), but more recent interest is focused on reducing the critical current required to switch the free layer in a spin torque transfer element (see Chapters 7 and 8 for a discussion of spin torque effects), or utilizing an electric field to assist in magnetization reversal [15].
- A highly polarized DOS at E_F. This is often viewed to first order as the source term when discussing spin injection from any contact, and higher polarizations produce larger signal levels. Half metals such as select Heusler alloys ideally exhibit 100% spin polarization at E_F, and are of keen interest, although this polarization is rarely, if ever, achieved in practice.
- The contact/interface should have a reasonably low specific resistance to allow the useable current densities. A 100% spin-polarized current of 10^{-3} A cm^{-2} may not be as useful as a 30% spin-polarized current of 10 A cm^{-2}, particularly in generating spin accumulation in semiconductor structures (with the exception of single-electron devices). High resistances also degrade signal-to-noise ratios, compromising device characteristics such as operating frequency. The contact resistance will be a compromise involving consideration of many factors.
- The contact/interface should offer spin injection which is robust against defects. Defects at complex heterointerfaces are to be expected—their role in spin scattering is poorly understood and poses an immediate challenge for development of spin transport devices.
- The contact/interface should be thermodynamically stable, and thermally stable at temperatures typically encountered for growth, processing, and device operation.

The above list is neither exhaustive nor intended to be an absolute yardstick by which to gauge the merits of a particular contact/interface—performance and intended application define the latter. It is useful, however, to keep these items in mind as we continue to develop the spin contacts, corresponding interface, and the new functionality that semiconductor spintronic devices offer.

Two classes of magnetic materials have been utilized as contacts for electrical spin injection into semiconductors: *semiconductors* and *metals* [10,12]. FM metals are familiar to all and offer some significant advantages, but also present some challenges for efficient utilization as spin contacts—they will be discussed in later paragraphs. Magnetic semiconductors are less familiar, but have, in fact, been studied for decades. They have enjoyed a resurgence of research interest since 1990, and were successfully used before FM metals to electrically inject spin-polarized carriers into another semiconductor. An overview of recent developments in III–V and oxide magnetic semiconductors appears in Chapters 19 and 20.

17.4 Semiconductor/Semiconductor Electrical Spin Injection

17.4.1 Magnetic Semiconductors: Material Properties

Magnetic semiconductors simultaneously exhibit semiconducting properties and typically either paramagnetic or FM order. The coexistence of these properties in a single material provides fertile ground for fundamental studies, and offers exciting possibilities for a broad range of applications. The use of a magnetic semiconductor as a spin contact enables design of a spin-injecting semiconductor/semiconductor interface guided by known principles of band-gap engineering (CB and VB offsets, doping, and carrier transport) and epitaxial growth (lattice match, interface structure, and materials compatibility). Therefore, they seem ideal candidates for incorporation in semiconductor-spintronic devices. However, their magnetic properties to date fall well short of those required for practical devices—e.g., the Curie temperature is typically well below room temperature, despite much effort to increase it.

Ferromagnetic semiconductors (FMS) were intensively studied from 1950 to 1970—classic examples include the europium chalcogenides (e.g., EuO, S, and Se) and the chalcogenide spinels, in which the magnetic element forms a significant fraction of the atomic constituents (see Figure 17.2a) [16,17]. Simple transport experiments have demonstrated that the exchange splitting of the band edges below the Curie temperature could be used as a spin-dependent potential barrier, which selectively passes one spin component while blocking the other, leading to current with a net spin polarization. This "spin filter effect" was demonstrated in EuS- and EuSe-based single barrier heterostructures in 1967 by Esaki et al. [18], and recently revisited [19,20] (spin filtering is reviewed in Chapter 10). However, device applications languished due to low Curie temperatures and the inability to incorporate these materials in thin-film form with mainstream semiconductor device materials.

Recent work successfully incorporated epilayers of n-type $CdCr_2Se_4$, another classic magnetic semiconductor, with GaAs-based heterostructures by molecular beam epitaxy (MBE) [21] and demonstrated electrical spin injection [22]. $CdCr_2Se_4$ is a direct gap ($E_g = 1.3\,eV$) chalcogenide spinel, which is reasonably lattice matched to technologically important materials such as

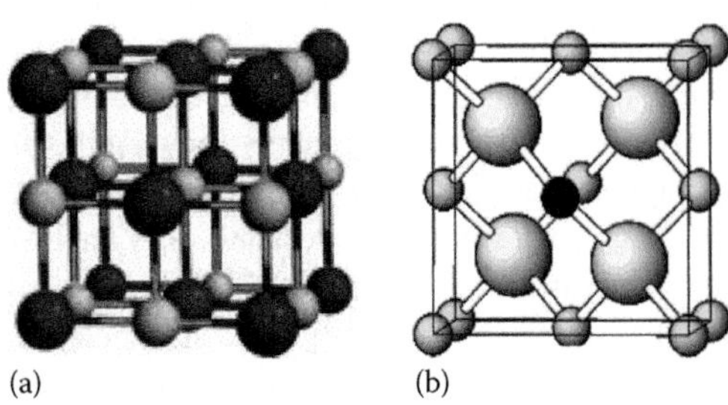

FIGURE 17.2 Lattice models of a classic magnetic semiconductor EuO (a), and a diluted magnetic semiconductor $Ga_{1-x}Mn_xAs$ (b).

Si and GaP (~1.7% tensile mismatch), and to GaAs (5.2% tensile mismatch), assuming one uses an effective lattice constant of $a_o/2$ for the $CdCr_2Se_4$. Single crystal epilayers grown on both GaP(001) and GaAs(001) substrates are n-type ($n \sim 10^{18}\,cm^{-3}$), FM with the easy magnetization axis in plane (along the GaAs(110)), and have a Curie temperature of 132 K [21]. Subsequent work successfully demonstrated electrical injection of spin-polarized electrons from $CdCr_2Se_4$ contacts into GaAs [22] using the spin-polarized light-emitting diode (spin-LED) approach, described in detail below.

Oxide MBE has been used to successfully grow epitaxial films of EuO on Si(001) [23] and GaN(0001) substrates [24]. EuO has a highly spin-polarized band structure and a large specific Faraday rotation. Andreev reflection measurements show that the electron spin polarization in these films exceeds 90%. However, no spin injection from EuO into either substrate has been reported to date. It should be noted that EuO has a Curie temperature, T_c, of only 69 K, a common shortcoming of most FMS compounds, which essentially precludes their use in practical device applications.

Pronounced magnetic behavior can be introduced into some semiconductors by incorporating a small amount (~1%–10%) of certain magnetic impurities into the lattice. These materials are referred to as "*diluted magnetic semiconductors*" (DMS), because the material is formed by diluting a host semiconductor lattice with a substitutional magnetic impurity, most typically Mn, to form an alloy (see Figure 17.2b). Strong exchange interactions between the magnetic impurity and the host carriers lead to magnetic behavior. Some of these materials are truly FM, while others are paramagnetic yet still exhibit very useful and highly flexible magnetic properties.

The paramagnetic compounds are commonly based on the II–VI or IV–VI semiconductors, where there is a high degree of solubility for the magnetic element. Classic examples include $Zn_{1-x}Mn_xSe$, $Cd_{1-x}Mn_xTe$, and $Pb_{1-x}Mn_xSe$. Comprehensive reviews may be found in Refs. [25,26]. These materials are often called "semimagnetic semiconductors" (SMS) because their paramagnetic behavior and Zeeman splitting of the CB and VB are dramatically enhanced over what might be expected from the magnetic ions alone. This enhancement originates from very large exchange interactions between the s- and p-like carriers of the CB and VB of the host, and the d electrons of the substitutional magnetic impurity. This sp-d exchange leads to a tremendous amplification of the Zeeman splitting of the band edges in an applied magnetic field, and produces amplified magneto-optical properties such as giant Faraday rotation, as well as other field dependent effects. If this Zeeman splitting is parametrized by $\Delta E = g^* u_B H$, then the effective value of the host g-factor is increased from 2, typical of most hosts to a value of $g^* \sim 100$ or greater [25]. For modest fields, the spin splitting significantly exceeds $k_B T$ at low temperature. For example, the splitting of the $m_j = +1/2$ and $-1/2$ electron states in $Zn_{0.94}Mn_{0.06}Se$ is ~10 meV at 3 T and 4.2 K, so that the CB effectively forms a completely polarized source of $m_j = -1/2$ electrons, and this spin polarization can be tuned by the external magnetic field. These properties are certainly attractive for a spin-injecting contact.

Oestreich et al. initially proposed the use of an SMS as a spin-injecting contact [27]. They used time-resolved photoluminescence (PL) to demonstrate that *optically* excited carriers became spin aligned in a $Cd_{1-x}Mn_xTe$ layer on a picosecond time scale, and that the spin-polarized electrons were transferred into an adjacent CdTe layer with little loss in spin polarization. However, this pronounced behavior is limited to low temperatures ($T < 25\,K$) and relatively highly applied magnetic fields ($H > 0.5\,T$), making these paramagnetic semiconductors less attractive for device fabrication. Nevertheless, bulk crystals have found application as magnetically tunable optical polarizers operating at room temperature [28], where the long optical path length compensates for the severe reduction in *g*-factor and magnetization at higher temperatures.

A concerted effort was made to introduce magnetic order into semiconductor compounds already recognized for device applications by alloying small amounts of Mn with GaAs and InAs to form new DMS materials. Nonequilibrium growth at relatively low substrate temperatures by MBE permits incorporation of Mn at levels well above the solubility limit of the host lattice. Pioneering work lead to the discovery of spontaneous FM order in DMS alloys such as $In_{1-x}Mn_xAs$ in 1989 [29] and $Ga_{1-x}Mn_xAs$ in 1996 [30,31]. Since Mn acts as both the magnetic element and an acceptor, they are *p*-type. After much research [32–34], these new FMS materials now exhibit Curie temperatures up to 50 K [35] and 192 K [36], respectively. Although their optical and electronic properties are not nearly as clean and controllable as their nonmagnetic hosts, these materials have been vigorously studied for their potential in future spin-dependent semiconductor device technologies. For example, $Ga_{1-x}Mn_xAs$ has been used as a source of spin-polarized holes in resonant tunneling diodes (RTD) [37,38], LED [39,40], and for current-induced magnetization switching [41]. Electric field control of FM order has been demonstrated in $In_{1-x}Mn_xAs$ [42], Mn_xGe_{1-x} [43], and $Ga_{1-x}Mn_xAs$ [44] heterostructures, demonstrating one of the highly unusual multiferroic properties of these materials and portending a host of new applications.

The theory based on a mean field Zener model has predicted that FM order should be stabilized in a wide variety of semiconductor hosts when diluted or alloyed with Mn at concentrations of ~5% and sufficiently high hole densities [45,46]. Subsequent work has indicated that other magnetic atoms should produce similar effects. This has stimulated a groundswell of research activity to synthesize new diluted FMS compounds, with the goal of achieving technologically attractive materials [47]. Unfortunately, progress has been hampered on both the theoretical and experimental fronts. Although the mean field model works well for $Ga_{1-x}Mn_xAs$ for which it was developed, it is clear that it is not appropriate for many other semiconductor hosts—the relevant mechanisms leading to potential magnetic order and the proper choice of theoretical framework depend strongly upon the position of the electronic states introduced by the magnetic atom relative to the band edges of the host semiconductor. New theoretical approaches are now being developed and appear promising [48].

The principal obstacle in experimental efforts to synthesize new diluted FMS materials has been the formation of unwanted phases due to the relatively low solubility of the magnetic atoms (Mn, Cr, Fe, Co) in most of the hosts considered (e.g., the group III-As, III-Sb, and III-N families). In many cases, nano- and microscale precipitates of known FM bulk phases form, which are exceedingly difficult to detect by the usual structural probes of x-ray diffraction and electron microscopy, but are all too readily detected with standard magnetometry measurements. A major challenge facing the experimentalist is to utilize characterization techniques which discriminate against the potential presence of such precipitates, clearly distinguish the contribution of the FMS material from that of the precipitates, and directly probe the key characteristics expected for the FMS itself. Electrical spin injection from the FMS provides one such litmus test, since it is enabled by the net carrier spin polarization intrinsic to a true FMS.

17.4.2 Magnetic Semiconductors as Spin-Injecting Contacts: Initial Transport Studies

As noted earlier, the use of a magnetic semiconductor as a contact enables design of a spin-injection semiconductor device guided by known principles of band-gap engineering and epitaxial growth. This has greatly facilitated demonstration of electrical spin injection from such contacts.

The first report of electrical injection of spin-polarized carriers from a diluted FMS, *p*-type GaMnAs, was obtained for RTD structures [37,38]. A *p*-$Ga_{0.965}Mn_{0.035}As$ emitter layer was grown by MBE on a GaAs/AlAs/GaAs-QW/AlAs/p-GaAs(001) RTD structure—the corresponding VB diagram is shown as the inset of Figure 17.3. The structure was biased to inject holes from the GaMnAs emitter, and resonant tunneling occurs at bias voltages where the emitter VB edge is aligned with one of the confined LH or HH states in the GaAs QW (labeled LH1, HH2, etc.). As the temperature is lowered below the GaMnAs Curie temperature (~70 K), the resonant peak labeled HH2 splits, and the temperature dependence of the splitting mirrors that of the GaMnAs magnetization. The authors attribute this to the spontaneous exchange splitting of the GaMnAs VB edge below T_c and resonant tunneling from the spin-polarized hole states, which accompanies the onset of FM order in the emitter. Thus, by selecting the bias voltage, one can use the RTD to preferentially transmit one hole spin state or the other. The fact that other resonant peaks did *not* split was tentatively attributed to some selection rule process, and is likely due to the orbital character/spin orientation of the GaAs QW hole states relative to the in-plane magnetization of the GaMnAs emitter contact. Although the RTD provides some qualitative indication of electrical spin injection, it is difficult to obtain more specific and very basic information such as the net spin polarization achieved.

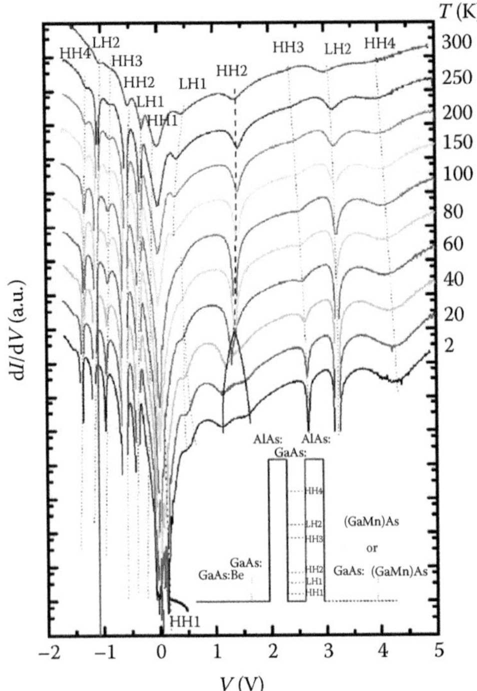

FIGURE 17.3 Derivative of current (dI/dV) vs. voltage (V) of a resonant tunneling diode with FM (Ga,Mn)As emitter. The labeling indicates the relevant resonant state in the GaAs well. When holes are injected from the (Ga,Mn)As side (positive bias), a spontaneous splitting of resonant peaks HH2 is observed below 80 K. The transition temperature of (GaMn)As is expected to be 70 K. The splitting is attributed to spin splitting of (Ga,Mn)As VB states. (From Ohno, H., *Science*, 281, 951, 1998. Reprinted with permission from AAAS.)

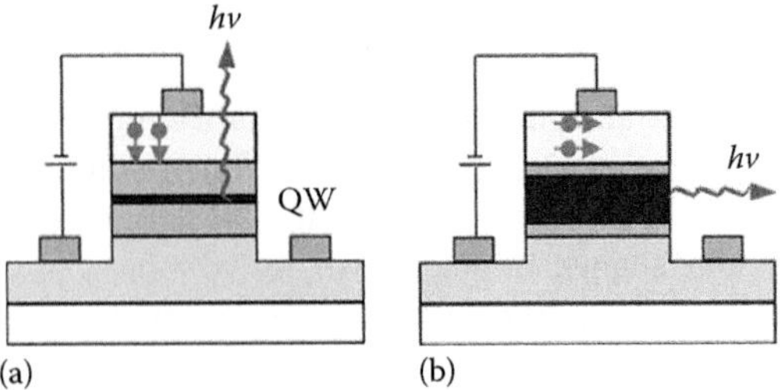

FIGURE 17.4 Schematic of spin-LEDs showing relative orientation of electron spins and light propagation direction appropriate for deducing the spin polarization of the electrons participating in the radiative recombination process for (a) surface-emitting and (b) edge-emitting geometries. The holes are assumed to be unpolarized in these examples.

17.4.3 Spin-Polarized Light-Emitting Diode: Spin-LED

In order to study and develop electrical spin injection from a discrete contact, one must first have a reliable means of both *detecting* the presence of spin-polarized carriers and of *quantifying* the spin polarization achieved. A simple LED structure takes advantage of one of the distinguishing characteristics of semiconductors—radiative recombination of carriers—and provides a powerful platform for this purpose. In a normal LED, electrons and holes recombine in the vicinity of a *p*-*n* junction or QW to produce light when a forward bias current flows. This light is unpolarized, because all carrier spin states are equally populated. However, if electrical injection produces a spin-polarized carrier population within the semiconductor, the same selection rules discussed above for optical pumping also describe the radiative recombination pathways allowed. Inspection of Figure 17.1 reveals that if injected carriers retain their spin polarization, radiative recombination results in the emission of circularly polarized light. A simple analysis based on these selection rules provides *a quantitative and model independent measure* of the spin polarization of the carriers participating.

A schematic of such a spin-LED [49] is shown in Figure 17.4 for surface and edge-emitting geometries. As an example, we consider injection of spin-polarized electrons from a magnetic contact layer, which recombine with holes supplied from the substrate. The net circular polarization P_{circ} of the light emitted is readily determined from the measured intensities of the positive and negative helicity components of the electroluminescence (EL), $I(\sigma+)$ and $I(\sigma-)$, respectively (Equation 17.2). These in turn are directly related to the occupation of the carrier states. Assuming a spin-polarized electron population and an unpolarized hole population in a bulk-like sample (i.e., all of the hole states are at the same energy and thus have the same probability of being occupied), a general expression for the degree of circular polarization in the Faraday geometry (Figure 17.4a) follows directly from Figure 17.1a. P_{circ} can be written in terms of the relative populations of the electron spin states $n_\uparrow$ ($m_j = +1/2$) and $n_\downarrow$ ($m_j = -1/2$), where $0 \le n \le 1$, and $n_\uparrow + n_\downarrow = 1$

$$\begin{aligned} P_{circ} &= \frac{[I(\sigma+) - I(\sigma-)]}{[I(\sigma+) + I(\sigma-)]} \\ &= 0.5\frac{(n_\downarrow - n_\uparrow)}{(n_\downarrow + n_\uparrow)} \\ &= 0.5P_{spin} \end{aligned} \tag{17.2}$$

The optical polarization is directly related to the electron spin polarization $P_{spin} = (n_\downarrow - n_\uparrow)/(n_\downarrow + n_\uparrow)$ at the moment of radiative recombination, and has a maximum value of 0.5 due to the bulk degeneracy of the HH and LH bands.

In a QW, the HH and LH bands are separated in energy by quantum confinement, which modifies Equation 17.2 and significantly impacts the analysis. The HH/LH band splitting is typically several meV even in shallow QWs, and is much larger than the thermal energy at low temperature (~0.36 meV at 4.2 K) so that the LH states are at higher energy and are not occupied (Figure 17.1b). For typical $Al_xGa_{1-x}As$/GaAs QW structures with Al concentration $0.03 \le x \le 0.3$ and a 15 nm QW, a simple calculation yields a value of 3–10 meV for the HH/LH splitting [50]. Thus, only the HH levels participate in the radiative recombination process at low temperature, as shown in Figure 17.1b, and P_{circ} is calculated as before

$$P_{circ} = \frac{(n_{\downarrow} - n_{\uparrow})}{(n_{\downarrow} + n_{\uparrow})} = P_{spin} \tag{17.3}$$

In this case, P_{circ} is equal to the electron spin polarization in the well, and can be as high as 1.

The use of a QW offers several distinct advantages over a *p–n* junction in this approach. The QW provides a specific spatial location within the structure where the spin polarization is measured, and hence depth resolution. Varying the distance of the QW from the injecting interface may then provide a measure of spin transport lengths. This feature was utilized by Hagele et al. to obtain a lower bound of 4 μm for spin diffusion lengths in optically pumped GaAs at 10 K [51]. In addition, the light emitted from the QW has an energy characteristic of the QW structure, and may therefore be easily distinguished from spectroscopic features arising from other areas of the structure or impurity-related emission.

Note that the carrier spin polarization, P_{spin}, determined by this procedure is the spin polarization *at the instant and location* at which radiative recombination occurs in the structure (Figure 17.5). After injection at the interface, the spin-polarized carriers (a) must first transport some distance to the point of radiative recombination (e.g., to the QW), and then (b) wait some time τ_r characteristic of the structure before radiative recombination occurs, where τ_r is the radiative lifetime. Spin relaxation is likely to occur during both of these processes—scattering events during transport lead to loss of polarization, and the spin polarization decays exponentially with time as $\exp(-t/\tau_s)$, where τ_s is the spin lifetime (see Chapter 16 for a detailed discussion of spin relaxation). It is difficult to correct for (a) to obtain the spin polarization at the point of injection, P_{inj}, without very detailed knowledge of the interface structure and relevant scattering mechanisms. However, a straightforward procedure can be used to correct for (b). Essentially, one takes a snapshot of the spin system with τ_r as the shutter speed. If τ_r was much shorter than τ_s, this would provide an accurate measure of P_{spin} in much the same way that a short exposure

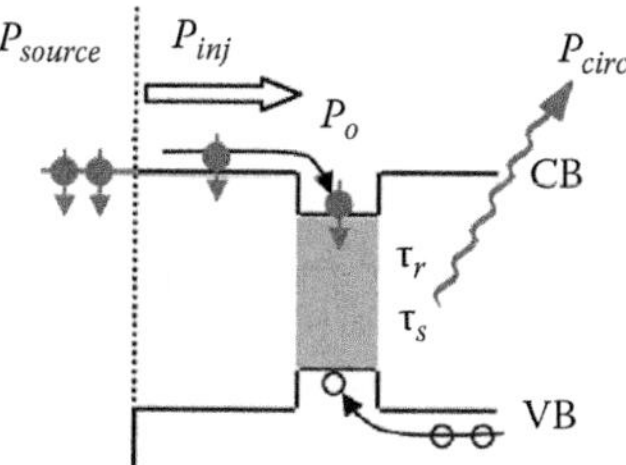

FIGURE 17.5 Band diagram illustrating lifetimes and polarizations in a spin-LED. The electron spin polarization in the electrical contact, P_{source}, results in some net injected spin polarization P_{inj} just inside the semiconductor. Through drift and diffusion, this electron spin reaches the QW and produces a spin polarization P_o. After radiative recombination, this is manifested as a circular polarization in the EL, P_{circ}. If the VB degeneracy is lifted, $P_{circ} = P_{spin}$. The initial QW spin polarization can be obtained from a rate equation analysis, $P_o = P_{spin}\,(1 + \tau_r/\tau_s)$.

time or shutter speed is used to capture action photographs with a camera. This is rarely the case, however, and the spin system decays over the time τ_r. Therefore, the result measured represents a lower bound for the carrier spin polarization achieved by electrical injection. A simple rate equation analysis may be applied to obtain a more accurate measure of the initial carrier spin polarization, P_o, that exists when the electrically injected carriers enter the region of radiative recombination (e.g., the QW). P_o is given by [52]

$$P_o = P_{spin}\left(1 + \frac{\tau_r}{\tau_s}\right) \tag{17.4}$$

where P_{spin} is the value determined experimentally as described above. This provides a first-order correction for what is effectively the instrument response function of the LED to serve as a spin detector and the efficiency of the radiative recombination process in the particular structure and material utilized.

The value of τ_r/τ_s can be determined by a simple PL measurement using near band edge circularly polarized excitation (optical pumping) [5,52]. Assuming that both LH and HH states are excited, the initial spin polarization produced by optical pumping is $P_o = 0.5$ (see Figure 17.1a). One measures the circular polarization of the PL, P_{PL}, and uses Equation 17.4 to solve for τ_r/τ_s: $0.5 = P_{PL}\ (1 + \tau_r/\tau_s)$. Typical values are $1 \leq \tau_r/\tau_s \leq 10$, so that this correction can be significant.

Thus, the spin-LED serves as a *polarization transducer*, effectively converting carrier spin polarization, which is difficult to measure by any other method, to an optical polarization, which can be easily and accurately measured using standard optical spectroscopic techniques. The existence of circularly polarized EL demonstrates successful electrical spin injection (subject to appropriate control experiments), and an analysis of the circular polarization using these fundamental selection rules provides a quantitative assessment of carrier spin polarization in the QW without resorting to a specific model. Note that this approach measures the spin polarization of the carrier *population* in the semiconductor (electron or hole) achieved by electrical injection, and not the spin polarization of the injected *current*. Detailed knowledge of the transport mechanism and spin scattering is necessary to connect the two.

A number of conditions facilitate application of the spin-LED approach:

1. The analysis of the measurements is considerably more reliable if the experimental geometry is appropriate for the optical selection rules [5,13]. The hole spin, electron spin, and the optical emission/analysis axes (or projections thereof) must be colinear to extract information on carrier spin polarization from the circular-polarized EL. The hole states in a QW have preferred orientations due to quantum confinement and reduced symmetry [53,54]: the $k = 0$ HH orbital angular momentum is oriented entirely along the growth direction (*z*-axis), whereas the $k = 0$ LH orbital angular momentum has nonzero projections in all

three directions. Analysis of the HH QW exciton requires that the injected electron spin must be along the z-axis *and* the optical measurement must also be performed along the same axis, meaning that surface-rather than edge-emitted light should be analyzed (Figure 17.4a) [55]. Thus, the Faraday rather than the Voigt geometry must be used. If the injected spins are oriented in-plane, then an edge-emission geometry (Figure 17.4b) may be utilized to analyze the EL *if* the radiative recombination region is bulk-like, or if one measures the LH exciton in a QW. An example of spin analysis for edge emission from a spin-LED is provided in Ref. [56].

2. The radiative recombination region should not be highly strained. Strain modifies the selection rules, compromising the quantitative relationship between P_{circ} and P_{spin}. Strain may produce optical polarization in the absence of carrier spin polarization, or reduce P_{circ} below that expected from the corresponding P_{spin}. While no system is perfect due to lattice mismatch, thermal expansion coefficients, and contact issues, the $Al_xGa_{1-x}As$/GaAs QW system comes very close due to the very small variation of lattice constant with Al concentration. In contrast, the lattice constant varies rapidly with In concentration and strain is almost unavoidable in the GaAs/$In_xGa_{1-x}As$ QW system. Other factors complicate the use of InGaAs in spin-LEDs—the introduction of In, with its stronger spin–orbit interaction, may reduce spin lifetimes [57], and the relatively large g-factor introduces significant magnetic field-dependent effects, which complicate interpretation of the EL polarization [58].
3. The origin of the EL must be correctly identified to determine its spatial origin within the structure, and to confirm that it derives from recombination processes for which the selection rules are valid [59,60], i.e., spin-conserving processes such as free exciton or free electron recombination. In practice, this means that the EL must be spectroscopically resolved and standard analyses applied to assist in the identification of the emission peaks. Note that bound exciton and impurity-related emission typically involve nonspin-conserving processes, and, therefore, cannot be used.
4. The VB degeneracy must be correctly identified, as described above. In a QW, the LH/HH splitting can be readily calculated for many materials [50], and should be compared to the measurement temperature to determine their relative contributions.

To briefly summarize, a quantitative and model independent determination of the initial spin polarization P_o can be obtained from a spin-LED experiment if a few straightforward conditions are met: one employs the proper experimental geometry, identifies the radiative transition producing the EL, and determines a value for τ_r/τ_s for the samples under consideration. This is all relatively easy to do in the $Al_xGa_{1-x}As$/GaAs QW system, and with some additional care in the GaAs/$In_xGa_{1-x}As$ QW system.

17.4.4 Magnetic Semiconductors as Spin-Injecting Contacts: Spin-LED Studies

17.4.4.1 $Zn_{1-x}Mn_xSe$

As an example illustrating the concepts and procedures above, we consider electrical spin injection from a diluted magnetic semiconductor contact, n-type $Zn_{0.94}Mn_{0.06}Se$, into an AlGaAs/GaAs QW/AlGaAs LED structure, as first described by Fiederling et al. [61] and Jonker et al. [62]. $Zn_{1-x}Mn_xSe$ was chosen as the contact material for a number of reasons. It can readily be doped n-type, allowing one to focus on *electron* transport, and the giant Zeeman splitting described above provides an essentially 100% spin-polarized electron population at modest applied magnetic fields, albeit at low temperature. $Zn_{1-x}Mn_xSe$ forms high quality epitaxial films on GaAs due to a close lattice match, and the CB offset can be tailored to facilitate electron flow from the $Zn_{1-x}Mn_xSe$ into the $Al_yGa_{1-y}As$/GaAs structure by suitable choice of the Mn and Al concentrations [62].

A flat band diagram of the structure is shown in Figure 17.6 illustrating the band alignments and the spin splitting of the CB and VB edges of the $Zn_{1-x}Mn_xSe$ with applied magnetic field. The ZnMnSe spin polarization may be varied simply by varying the applied field, a very useful handle for experimental studies. At sufficiently high fields, the CB spin splitting ($\Delta E = g^* u_B H \sim 10\,\text{meV}$) is much larger than the measurement temperature (~0.4 meV), so that the $Zn_{1-x}Mn_xSe$ contact is essentially 100% spin polarized. The samples were grown by MBE and processed into surface-emitting LED mesas 200–400 μm in diameter using standard photolithographic techniques. Figure 17.7 shows a schematic cross section and photograph of the final devices. The top metallization to the ZnMnSe consists of concentric Au rings to help insure uniform current distribution, leaving most of the mesa surface optically transparent. Details of the MBE growth and device fabrication may be found in Ref. [62].

The EL was measured by electrically biasing the LEDs to inject electrons from the n-ZnMnSe into the GaAs QW at

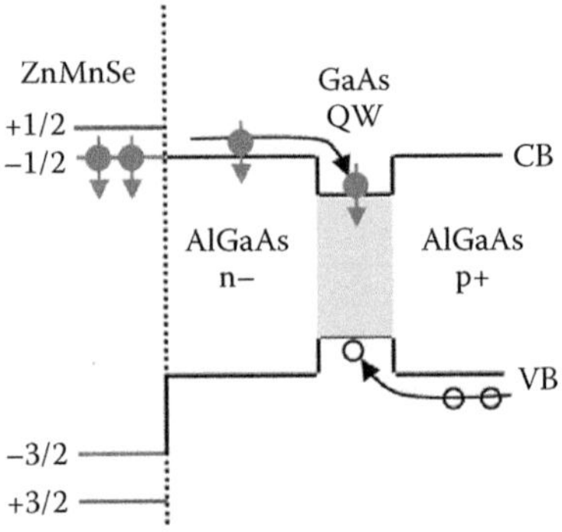

FIGURE 17.6 ZnMnSe/AlGaAs spin-LED band diagram, illustrating the giant Zeeman splitting of the ZnMnSe band edges and electrical injection of spin-polarized electrons into the GaAs QW.

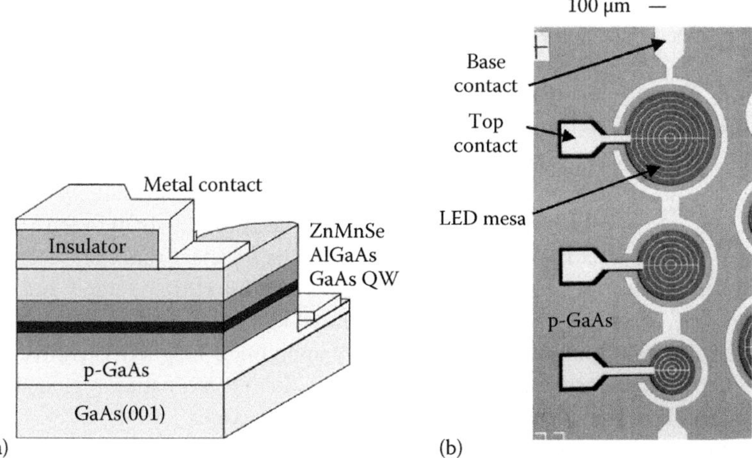

FIGURE 17.7 (a) Schematic cross section of spin-LED illustrating the structure and processing steps. (b) Photograph of completed surface-emitting devices. The active mesa areas (dark circular regions) are 200, 300, and 400 μm in diameter (From Ohno, H., *Science*, 281, 951, 1998. With permission.)

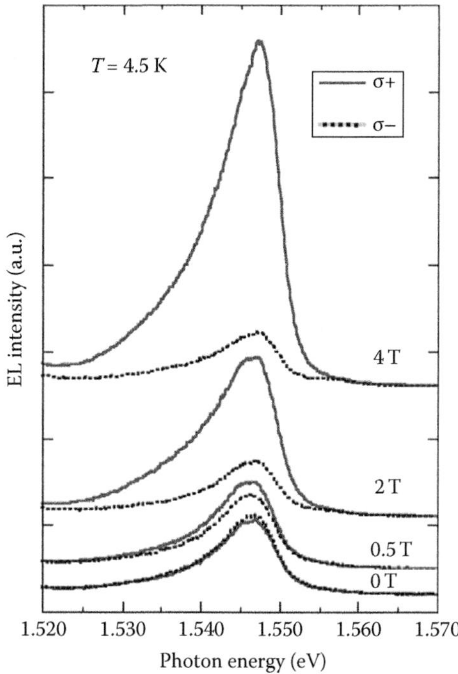

FIGURE 17.8 EL spectra from a surface-emitting spin-LED with a $Zn_{0.94}Mn_{0.06}Se$ contact for selected values of applied magnetic field, analyzed for positive (σ+) and negative (σ−) circular polarization. The magnetic field is applied along the surface normal (Faraday geometry). The spectra are dominated by the HH exciton. Typical operating parameters are 100 μA and 2.5 V.

current densities of ~0.01–1.0 A cm^{-2}. Representative EL spectra from such a structure (in this case with a multiple QW LED (3 × (20 nm $Al_{0.08}Ga_{0.92}As$/10 nm GaAs)) are shown in Figure 17.8 for selected values of the applied field. The light emitted along the surface normal and magnetic field direction (Faraday geometry) was analyzed for σ+ and σ− circular polarization and spectroscopically resolved. The energy of the emission confirms that the radiative recombination occurs in the GaAs QW via the HH ground state exciton (note that other tests are applied to confirm this identification—see Ref. [59]). As noted in the previous section (condition (1)), quantum confinement locks the HH spins along the *z*-axis (surface normal), and therefore, both the injected electron spins and the axis for optical analysis must be oriented along the *z*-axis, as well, to interpret any circular polarization observed in terms of spin polarization. Edge emission will give a spurious or null result [55].

At zero field, the σ+ and σ− components are identical, as expected, since no spin polarization yet exists in the ZnMnSe. As the magnetic field increases, the ZnMnSe bands split into spin states due to the giant Zeeman effect, and spin-polarized electrons are injected into the AlGaAs/GaAs LED structure. The corresponding spectra exhibit a large difference in intensity between the σ+ and σ− components, demonstrating that the spin-polarized electrons successfully reach the GaAs QW. The circular polarization, $P_{circ} = [I(\sigma+) - I(\sigma-)]/[I(\sigma+) + I(\sigma-)]$, increases with applied field as the ZnMnSe polarization increases, and saturates at a value of ~80% at 4 T. From the discussion leading to Equation 17.3, the spin polarization of the electron population in the GaAs QW, P_{spin}, is therefore 80%.

This value is the QW spin polarization at the time of radiative recombination, i.e., after a time ~τ_r has elapsed. A value for the initial QW spin polarization, P_o, can be obtained using Equation 17.4 to correct for the effect that the relative values of the radiative and spin lifetimes have on $I(\sigma+)$ and $I(\sigma-)$. Independent PL measurements using circularly polarized excitation provide a value of τ_r/τ_s, as described previously. The PL is analyzed for σ+ and σ− helicity, and exhibits a polarization $P_{PL} = 40\%$. Applying $0.5 = P_{PL}\ (1 + \tau_r/\tau_s)$ gives $\tau_r/\tau_s = 0.25 \pm 0.05$. Equation 17.4 then gives $P_o \sim 100\%$, demonstrating that a well-ordered ZnMnSe/AlGaAs interface is essentially transparent to spin transport. Thus, an all-electrical process using a discrete contact can produce a carrier spin polarization within the semiconductor that equals or exceeds that achieved via optical pumping.

Several control experiments rule out contributions to the circular polarization measured from spurious sources, and are an essential part of any such experiment. LEDs with nonmagnetic *n*-ZnSe contact layers show no circular polarization with magnetic field, as expected. The circular dichroism resulting from transmission through the ZnMnSe is negligible because the GaAs QW emission wavelength is very far from that corresponding to the band gap of $Zn_{0.94}Mn_{0.06}Se$. PL data from the GaAs QW excited with linearly polarized light from the same LED mesa structures used for the EL studies show little polarization, providing a very effective built-in reference for each mesa LED. Such dichroism effects could be much larger for emission energies very near the ZnMnSe band gap, where strong absorption occurs, and must be considered when designing the LED structure.

17.4.4.2 $Ga_{1-x}Mn_xAs$

Spin-LED structures have also been fabricated with *p*-type GaMnAs as the spin contact. The first report [39] utilized a 300 nm $Ga_{0.955}Mn_{0.045}As$/GaAs/10 nm $In_{0.13}Ga_{0.87}As$ QW/*n*-GaAs(001) heterostructure in which spin-polarized holes injected from the GaMnAs radiatively recombined in the strained InGaAs QW with unpolarized electrons from the substrate. Because the magnetization of the GaMnAs (and nominal spin orientation of the injected holes) was in-plane, they used an edge-emission geometry with the optical axis parallel to the magnetization. They measured the EL at the QW ground state transition, and observed a 1% polarization, which exhibited the same temperature dependence as the GaMnAs magnetization. A reliable interpretation in terms of hole spin injection and polarization, however, is compromised by the use of a strained QW as the radiative recombination region—as discussed previously, (1) the orbital angular momentum of the QW HH ground state is oriented along the growth direction (*z*-axis), orthogonal to the nominal in-plane orientation of the spins injected from the GaMnAs, and (2) strain in the QW leads to admixture of states and modifies the selection rules. Inducing an out-of-plane magnetization in the GaMnAs (e.g., by applying a magnetic field) or utilizing a structure with a bulk-like recombination region (Figure 17.4b) may have alleviated these issues.

More definitive measurements were performed on identical sample structures in 2002 using the Faraday geometry (surface emission) to address the first issue noted above [40,63]. The EL spectra at $T = 5$ K and the field dependence of the circular polarization for selected temperatures are shown in Figure 17.9. The field dependence tracks the out-of-plane (hard axis) magnetization of the GaMnAs (Figure 17.9b), and the saturation value depends upon the thickness, *d*, of the undoped GaAs spacer layer between the GaMnAs injector and the InGaAs QW. The circular polarization is 4% for $d = 70$ nm, and increases to 7% for $d = 20$ nm, attributed to the hole spin diffusion length effects in the GaAs spacer. These data indicate that the spin polarization of the QW hole population is at least 7% at the measurement temperature. The polarization decreases with increasing temperature, and disappears by 62 K—the Curie temperature of the GaMnAs injector. In Ref. [63], the use of small mesas to force

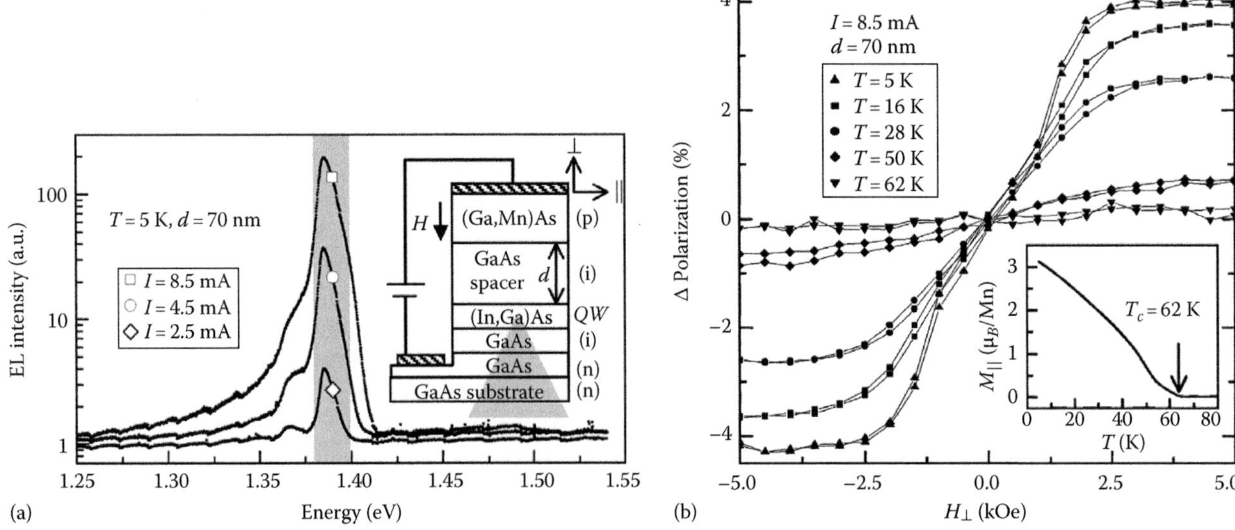

FIGURE 17.9 (a) Spectrally resolved EL intensity along the growth direction for several bias currents, *I*. Inset shows device schematic and EL collection geometries. (b) Temperature dependence of the relative changes in the energy-integrated [gray shaded area in (a)] polarization ΔP as a function of out-of-plane magnetic field. When $T < 62$ K, polarization saturates at $H_\perp \sim 2.5$ kOe. Inset shows $M(T)$, indicating that the polarization is proportional to the magnetic moment. (Reprinted with permission from Young, D.K. et al. Anisotropic electrical spin injection in FM semiconductor heterostructures. *Appl. Phys. Lett.*, 80, 1598. Copyright 2002, American Institute of Physics.)

the easy axis of the GaMnAs contact out of plane permitted the demonstration of hole spin injection without an applied magnetic field in the surface emission geometry.

17.4.4.3 Spin Scattering by Interface Defects

Spin scattering by interface defects is an important issue in heteroepitaxial systems, and can rapidly suppress the spin polarization achieved by electrical injection. A comprehensive understanding of this process remains a major challenge for future research. Initial work addressed the effect of interface nanostructure on the spin-injection efficiency at the ZnMnSe/AlGaAs interface [64]. High resolution transmission electron microscopy (TEM) was used to assess the interface morphology in spin-LED structures, and found that the most prevalent defects were stacking faults (SF) in <111> directions nucleating at or near the ZnMnSe/AlGaAs interface, where they formed line defects (Figure 17.10a). The electron spin polarization in the GaAs QW determined from the circular polarization P_{circ} of the EL exhibited an approximately linear dependence on the density of interface defects, as shown in Figure 17.10b. This correlation was explained by a model that incorporated spin–orbit (Elliot–Yafet) scattering—the model showed that the asymmetric potential of the interface defect results in strong spin-flip scattering in the forward direction. A simple expression, $P_{circ} \sim 1 - 4r_o n$, gave excellent agreement with the experimental data with no adjustable parameters, where $r_o \sim 10\,\text{nm}$ is the Thomas–Fermi screening length and n is the measured defect density. These results provided the first experimental demonstration that interface defect structure limits spin-injection efficiency in the diffusive transport regime. It is interesting to note that strong spin injection persists even when the misfit dislocation density greatly exceeds values which would be fatal for conventional devices such as III–V-based diode lasers and field effect transistors. This is especially reassuring, since interface misfit dislocations are a generic defect routinely encountered in heteroepitaxial device structures.

17.5 Ferromagnetic Metal/Semiconductor Electrical Spin Injection: Optical Detection with Spin-LED Structures

FM metals offer most of the properties desired for a practical spin-injecting contact material: reasonable spin polarizations at E_F (~40%–50%), high Curie temperatures, low coercive fields, fast switching times, and a well-developed material technology due to decades of research and development driven in large part by the magnetic storage industry. An FM contact introduces *nonvolatile* and *reprogrammable* operation in a very natural way, due to both the material's intrinsic magnetic anisotropy and the ability to tailor the magnetic characteristics in numerous ways. Spin-torque transfer switching is rapidly emerging as the mechanism of choice for manipulating the contact magnetization, and offers many advantages for spintronic applications such as embedded memory or field programmable gate arrays (see Chapters 7, 8, and 20 for an overview). This emerging technology can readily be incorporated in semiconductor spintronic devices. In addition, metallization is a standard process in any semiconductor device fabrication line, so that an FM metallization could easily be incorporated into existing processing schedules.

Initial efforts to inject spin-polarized carriers from an FM metal contact into a semiconductor were not very encouraging. Several groups reported a change in voltage or resistance on the order of 0.1%–1%, which they attributed to spin accumulation or transport in the semiconductor [65–67]. Such small effects, however, make it difficult to either unambiguously confirm spin injection or successfully implement new device concepts. In addition, some have argued that these measurements were compromised by contributions from anisotropic magnetoresistance or a local Hall effect, which can easily contribute a signal of 1%–2% [68–70]. Thus, great care must be taken in the

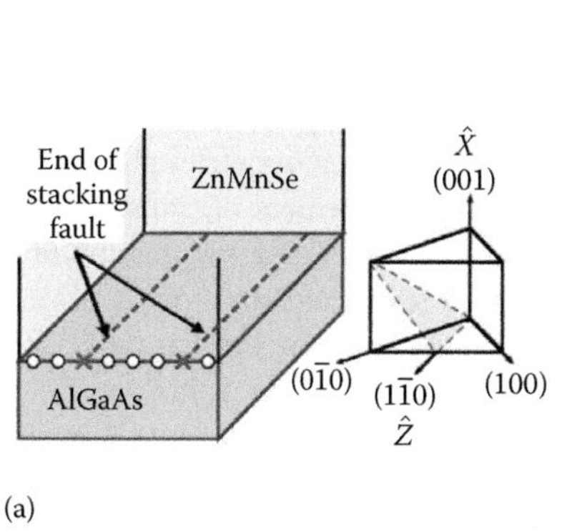

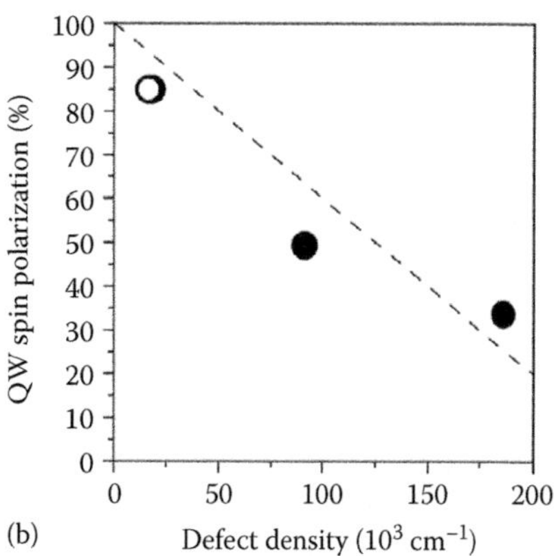

FIGURE 17.10 (a) Diagram illustrating the linear interface defects resulting from the intercept of (111)-type SF planes and the interface plane. Only one of the four possible (111)-type SF planes is shown for clarity. (b) Correlation of GaAs QW spin polarization with stacking fault density. Two data points nearly overlap at 85% polarization. The error bars are comparable to the symbol size. The dashed line is the calculated result with no adjustable parameters.

experimental design to recognize and eliminate such spurious contributions.

Zhu et al. [71] utilized optical rather than electrical detection in a spin-LED structure (the spin-LED is described in Section 17.4.3) consisting of an Fe Schottky contact to a GaAs/$In_{0.2}Ga_{0.8}As$ QW LED detector. They were unable to observe any clear difference in intensity when they analyzed the EL for positive (σ+) or negative (σ–) helicity polarization, in contrast with earlier results for a ZnMnSe magnetic semiconductor contact, as described in Section 17.4.4.1. However, by examining the high and low energy tails of the EL peak (which they attributed to LH and HH contributions, respectively) using pulsed current injection and lock-in detection techniques, with some background subtraction, they identified a signal which they attributed to electrical spin injection from the Fe contact. They concluded that a spin polarization of ~2% had been achieved in the GaAs, and found that this signal was independent of temperature from 25 to 300 K. The lack of temperature dependence raises some question as to their interpretation of this signal, since the spin lifetime is known to decrease rapidly with temperature for GaAs(001) structures [72].

Two fundamental issues are key to understanding, successfully demonstrating, and optimizing electrical spin injection from an FM metal into a semiconductor—the role of band symmetries in facilitating spin transmission across a heterointerface, and the impact of the large difference in conductivity between the metal and semiconductor.

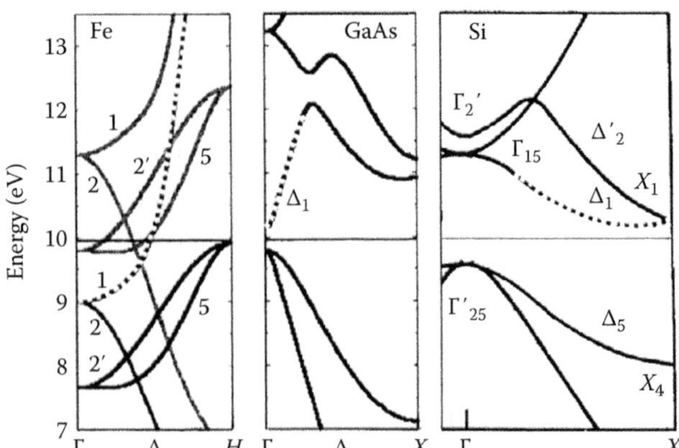

FIGURE 17.11 Band structures along (001) for Fe, GaAs and Si. The majority spin band of Fe crossing the Fermi level is of Δ_1 symmetry (labeled "1" in the plot, dashed curve), and the CB edge of both GaAs and Si (dashed curve) is also of Δ_1 symmetry. The Fe minority spin bands are of different symmetry (Δ_2, Δ_5, labeled "2" and "5"). This is expected to enhance transmission of majority spin electrons from the Fe, and suppress minority spin injection, leading to highly polarized spin currents in the semiconductor. The plots for Fe and GaAs are taken from Ref. [76]—the underestimate of the GaAs band gap is characteristic of the computational approach used. The plot for Si is derived from Ref. [78], and the energy scale is expanded relative to that used for Fe and GaAs. (Reprinted with permission from Wunnicke, O., Mavropoulos, Ph., Zeller, R., Dederichs, P.H., and Grundler, D., *Phys. Rev. B*, 65, 241306(R) and 2002; Chelikowsky, J.R. and Cohen, M.L., *Phys. Rev. B*, 14, 556 and 1976. Copyright 2002 and 1976 by the American Physical Society.)

17.5.1 Band Symmetries and Spin Transmission

The first issue can be stated rather simply: the symmetries and orbital composition of the bands participating in the transport process in the metal must be compatible with those of the semiconductor for optimum spin transmission to occur. Theoretical work has explored the electronic structure at the FM metal/semiconductor interface in an effort to elucidate the role of band structure in spin injection [73–76]. This work has emphasized the importance of matching the *symmetries* as well as the energies of the bands between metal and semiconductor to optimize spin-injection efficiency.

Some initial insight can be gained by examining the bulk band structures. Figure 17.11 shows selected portions of the band structure along (001) for Fe, GaAs [76,77], and Si [78]. For typical semiconductors of interest, including GaAs, InAs, GaP, ZnSe, and Si, the CB for the (001) surface exhibits Δ_1 symmetry near the zone center (dashed curves), with significant *s* and p_z-orbital contributions. The Fe *majority* spin band (dashed curve), which crosses E_F midway across the zone, is also of Δ_1 symmetry, with significant *s*-, p_z-, and d_z^2-orbital contributions. The *s* and p_z components have large spatial extent, and the p_z and d_z^2 orbitals point directly into the semiconductor, leading to strong overlap with the corresponding states comprising the semiconductor CB. The corresponding Fe Δ_1 *minority* spin band is 1.3 eV above E_F and, therefore, will not contribute to transport. In contrast, the Fe *minority* spin bands, which cross E_F, exhibit quite different symmetries (Δ_2, Δ_5), with orbital components that do not couple strongly to the semiconductor. Thus, the CB of the semiconductor is well matched in orbital composition/symmetry and energy to the majority spin bands of the Fe, while a poor match exists to the Fe minority spin bands.

Theoretical treatments have addressed several FM metal/semiconductor systems, and generally assume a well-ordered interface so that the electron momentum parallel to the interface plane, $k_{//}$, is conserved to simplify the calculation. For the case of spin injection from Fe into GaAs(001), calculations show that these band symmetries play a critical role, leading to a significant enhancement of transmission from the Δ_1 Fe majority spin band at E_F, and a suppression of transmission from the Fe minority spin bands (Δ_2, $\Delta_{2'}$, Δ_5) [76,79]. Consequently, majority spin electrons are preferentially transmitted from Fe into the GaAs, while minority spins are blocked, producing a significant enhancement of spin-injection efficiency. In such cases, the metal/semiconductor interface essentially serves as a *band structure spin filter* due to basic issues of band symmetry and orbital composition.

Similar arguments can be made for the case of Fe and Si(001). As seen in Figure 17.11, the bottom of the Si(001) CB is also of Δ_1 symmetry, enhancing transmission of Fe Δ_1-band majority spin electrons, and suppressing minority spin current. Calculations assuming an ideal Fe/Si(001) interface indicated that the current injected from the Fe contact should be strongly spin-polarized [80]. Similar conclusions were obtained for Fe/InAs(001) [81], and earlier for epitaxial Fe/MgO tunnel barriers by Butler et al. [82].

The band bending which occurs at the metal/semiconductor interface complicates the picture, but can be included in the calculation to some extent. Wunnicke et al. [76] and Mavropoulos [80] simulated the effect of a Schottky barrier by introducing a potential step between the Fe and the semiconductor, and found that the current remained strongly polarized.

Thus, the knowledge of the interface band structure is important for understanding and optimizing spin-injection efficiency. The *physical structure* of the interface is a key ingredient here. The perfectly abrupt interface typically assumed for calculation is unlikely to exist in practice. Intermixing leading to compound formation or disorder will alter the band symmetries and strongly impact the arguments above. Symmetry breaking due to disorder at the Fe/InAs(001) interface in the form of Fe atoms occupying In or As interface sites was shown to rapidly suppress the spin filtering effect and resultant high spin polarization in the InAs by opening more channels for minority spin transport [81].

It should be noted that compound formation per se is not necessarily detrimental, provided that bands with appropriate symmetries and/or orbital composition are preserved or created. It is indeed possible that the spin-injection efficiency at certain interfaces with poorly matched band symmetries will improve with compound formation which produces bands of more favorable orbital composition. However, disorder due to either random intermixing, poorly ordered compound formation, or defects will, in general, lead to mixing of states, compromising any state-specific transmission, and reducing the spin-injection efficiency.

17.5.2 Conductivity Mismatch: Description of the Problem

The second fundamental issue presenting an unanticipated challenge to utilizing an FM metal as a spin-injecting contact on a semiconductor is the very large difference in conductivity between the two materials. Simply stated, the ability of the semiconductor to accept carriers is independent of spin, and much smaller (lower conductivity) than that of the metal to deliver them. Consequently, equal numbers of spin-up and spin-down electrons are injected, regardless of the FM metal's initial polarization, resulting in essentially zero spin polarization in the semiconductor.

This issue, commonly referred to in the literature as the problem of "conductivity mismatch," can readily be understood in the context of the canonical two channel model typically used to describe spin-polarized current flow in FM metals [83]. In this model, spin-up or "majority spin" electrons flow in one channel, while spin-down or "minority spin" electrons flow in the other.* Using the analogy of water flow through a hose for electrical resistance, the relative conductivities of these channels in each material can be represented by the diameter of the hose, as shown in Figure 17.12a. In the metal, the diameter of both spin-up and spin-down hoses is very large (high conductivity), while in the semiconductor, the diameter is very small (low conductivity). The problem of conductivity mismatch and its impact on spin injection is thus reduced to the intuitive physical picture of attempting to effectively transfer water from a sewer pipe to a drinking straw.

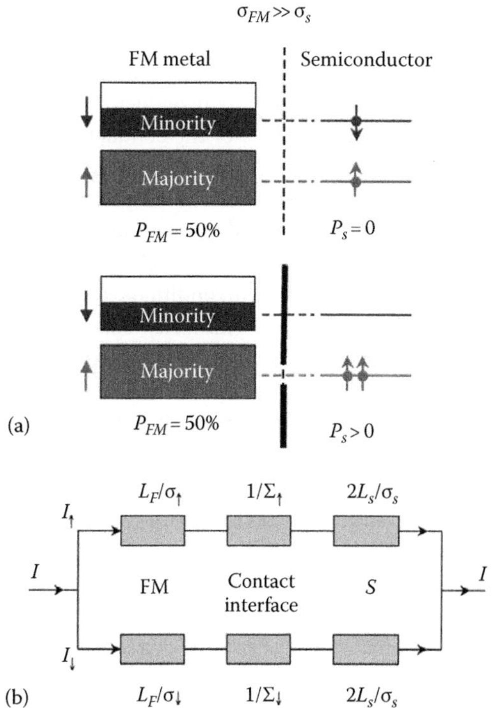

FIGURE 17.12 (a) Schematic of the two channel model of spin transport illustrating the conductivity mismatch issue, which must be considered for spin injection from an FM metal into a semiconductor. A spin-selective interface resistance provides a solution, as shown in the lower panel. (b) Equivalent resistor model (due to A.G. Petukhov). (From Jonker, B.T. et al., *MRS Bull.*, 28, 740, 2003.)

In the FM metal, the carriers are partially spin polarized (by definition), so that one of the pipes is nearly full while the other may be nearly empty. In the nonpolarized, low carrier density semiconductor, the two spin channels are of equal diameter and partially full. If one now considers the flow of water (spin-polarized current) from the sewer pipes (metal) to the drinking straws (semiconductor), it is immediately apparent that the comparatively small conductivity of the semiconductor limits current flow. While one drop of water may be transferred from the majority spin sewer pipe of the metal to the corresponding majority spin drinking straw of the semiconductor, an identical amount will flow in the minority spin channel, even though the metal pipe may be nearly empty, due to the limited conductivity

* For a discussion of this terminology, see Ref. [84].

of the semiconductor. This results in *zero* spin polarization in the semiconductor. It is apparent that this will be the case regardless of how nearly empty the metal minority spin band may be, due to the very limited ability of the semiconductor bands to accept current flow. This is the essence of the conductivity mismatch issue.

This simple picture provides rather surprising insight into efforts to transfer spin-polarized carriers between these two materials. In the diffusive transport regime, significant spin injection can occur for only two conditions: (1) the FM metal must be 100% spin polarized (minority channel completely empty), or (2) the conductivity of the FM metal and semiconductor must be closely matched. No FM metal meets either of these criteria. While half metallic materials offer 100% spin polarization in principle [85,86], defects such as antisites or interface structure rapidly suppress this value [87].

Model calculations by several groups [70,88–92] have addressed this issue and provide a quantitative treatment. These authors extended the work of van Son et al. [93] and Valet and Fert [83] to calculate the spin polarization achieved in a semiconductor due to transport of spin-polarized carriers from the ferromagnet. An equivalent resistor circuit explicitly incorporates the conductivities of the FM metal, interface, and semiconductor [11,88], as shown in Figure 17.12b, and permits discussion of spin-injection efficiency in terms of these physical parameters in the classical diffusive transport regime. The resistances representing the FM metal and semiconductor are given by L_F/σ_F and L_s/σ_s, where L and σ are the spin diffusion lengths and conductivities, respectively, with $\sigma_F = \sigma_\uparrow + \sigma_\downarrow$ summing the two FM spin channels. The interface conductivity $\Sigma = \Sigma_\uparrow + \Sigma_\downarrow$ is assumed to be spin dependent. With some algebra and effort, the spin-injection coefficient γ can be shown to be

$$\gamma = \left(\frac{I_\uparrow - I_\downarrow}{I}\right) = \frac{\left(r_F \frac{\Delta\sigma}{\sigma_F} + r_c \frac{\Delta\Sigma}{\Sigma}\right)}{(r_F + r_S + r_c)} \quad (17.5a)$$

where

$$r_F = L_F \frac{\sigma_F}{4\sigma_\uparrow\sigma_\downarrow} \quad (17.5b)$$

$$r_c = \frac{\Sigma}{4\Sigma_\uparrow\Sigma_\downarrow} \quad (17.5c)$$

$$r_s = \frac{L_s}{\sigma_s} \quad (17.5d)$$

are the effective resistances of the ferromagnet, contact interface, and semiconductor, respectively, $I_\uparrow$ and $I_\downarrow$ are the majority and minority spin current, $\Delta\sigma = \sigma_\uparrow - \sigma_\downarrow$ and $\Delta\Sigma = \Sigma_\uparrow - \Sigma_\downarrow$.

For an FM metal, $r_F/(r_F + r_s + r_c) \ll 1$, and the contribution of the first term in Equation 17.5a is negligible. The second term is significant only if $\Delta\Sigma \neq 0$ and $r_c \gg r_F$, r_s. Thus, two criteria must be satisfied for significant spin injection to occur across the interface between a typical FM metal and a semiconductor: the interface resistance r_c (1) must be spin selective, and (2) must dominate the series resistance in the near-interface region.

17.5.3 Conductivity Mismatch: Practical Solution

A tunnel barrier between the FM metal and semiconductor satisfies both these criteria, and was suggested as a potential solution to the conductivity mismatch problem by several groups [70,88–91]. The spin selectivity ($\Delta\Sigma \neq 0$) comes naturally from the spin-polarized DOS of the FM metal at E_F, which serves as the source term in a mathematical treatment of the tunneling process. The resistance of the tunnel barrier is readily controlled by its thickness, and can easily be the largest in the series resistance such that $r_c \gg r_F$, r_s. A detailed calculation was provided by Rashba [89], and a more comprehensive overview was provided recently by Fert et al. [94].

A tunnel barrier can be introduced at a metal/semiconductor interface in at least two ways: tailoring the band bending in the semiconductor, which typically leads to Schottky barrier formation, or physically inserting a discrete insulating layer such as an oxide. Examples of both are presented below for GaAs and Si.

17.5.3.1 Tailored Schottky Barrier as a Tunnel Barrier

One avenue is to take advantage of the band bending which occurs at the metal/semiconductor interface. This approach exploits a natural characteristic of the interface, and avoids the use of a discrete barrier layer and the accompanying problems with pinholes and thickness uniformity. Schottky contacts are also routine ingredients in semiconductor technology. In the case of an *n*-type semiconductor, electrons are transferred into the metal, depleting the semiconductor interfacial region and causing the CB to bend upward, forming a pseudo-triangular-shaped barrier with a quadratic falloff with distance into the semiconductor [95]. The depletion width associated with the Schottky barrier depends upon the doping level of the semiconductor, and is generally far too large to allow tunneling to occur. For example, in *n*-GaAs, the depletion width is on the order of 100 nm for $n \sim 10^{17}\,\text{cm}^{-3}$, and 40 nm for $n \sim 10^{18}\,\text{cm}^{-3}$ [96]. However, this width can be tailored by the doping profile used at the semiconductor surface [97]. Heavily doping the surface region can reduce the depletion width to a few nanometers, so that electron tunneling from the metal to the semiconductor becomes a highly probable process under reverse bias.

This approach was first utilized to achieve large electrical spin injection from Fe epilayers into AlGaAs/GaAs QW LED structures [10,98,99]. The *n*-type doping profile of the surface AlGaAs was designed by solving Poisson's equation with several criteria in mind: (a) minimize the Schottky barrier depletion width to facilitate tunneling, (b) use a minimum amount of dopant and heavily doped regions to accomplish this, since high *n*-doping is associated with stronger spin scattering and short spin lifetimes

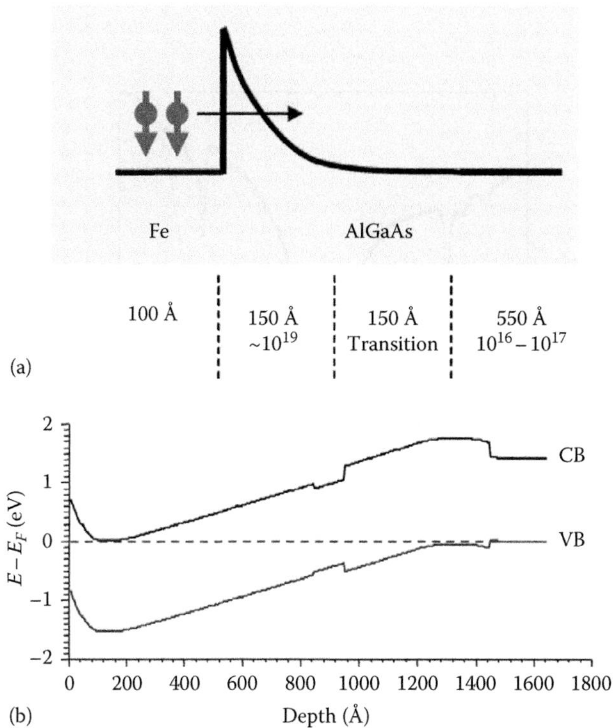

FIGURE 17.13 Design of an Fe Schottky barrier and spin-LED using a doping profile to facilitate tunneling of spin-polarized electrons from the Fe through the Schottky tunnel barrier. (a) Doping profile and resultant reduction of depletion width at the Fe/AlGaAs interface, and (b) Poisson equation solution corresponding to doping profile and AlGaAs/GaAs QW/AlGaAs structure described in the text.

[100,101], and (c) avoid formation of an electron "puddle" or accumulation region (i.e., pushing E_F above the CB edge), which may either dilute the polarization of the electrons injected from the Fe contact or contribute to spin scattering. A schematic of the doping profile and resultant barrier is shown in Figure 17.13a, and the band diagram resulting from the Poisson equation solution for the full LED structure is illustrated in Figure 17.13b. The doping of the top 150 Å of *n*-type $Al_{0.1}Ga_{0.9}As$ was chosen to be $n = 1 \times 10^{19}\,cm^{-3}$ to minimize the depletion width, followed by a 150 Å transition region, while the rest was $n = 1 \times 10^{16}\,cm^{-3}$ with a 100 Å dopant setback at the QW. The LED structure consisted of 850 Å *n*-$Al_{0.1}Ga_{0.9}As$/100 Å undoped GaAs/500 Å *p*-$Al_{0.3}Ga_{0.7}As$/*p*-GaAs buffer layer on a *p*-GaAs(001) substrate. The width of the GaAs QW was chosen to be 100 Å to insure separation of the LH and HH levels and corresponding excitonic spectral features, an important consideration for quantitative interpretation of the data, as discussed earlier. Details of the growth may be found elsewhere [98,99].

17.5.3.1.1 *Confirmation of Tunneling: Analysis of the Transport Mechanism*

To demonstrate that the tailored doping profile reduces the depletion width sufficiently to produce a tunnel barrier, it is necessary to analyze the current–voltage (*I–V*) characteristics of the Fe/AlGaAs Schottky contact and apply well-known criteria to identify the dominant transport process. Such criteria should also be applied when a discrete oxide barrier is used. The mere addition of a layer intended to serve as a nominal tunnel barrier does not insure that transport occurs by tunneling—indeed, pinholes are a chronic problem, and extensive work in the metal/insulator/metal tunnel junction community has shown that their presence cannot be ruled out easily [102–104]. Pinholes form low-resistance areas, which essentially short out the high resistance tunnel barrier layer. Application of the "Rowell criteria" for tunneling [102,105] and observation of phonon signatures and a zero bias anomaly in the low temperature conductance spectra provide clear, unambiguous tests to determine whether tunneling is the dominant transport mechanism.

There are three Rowell criteria. The first—the conductance ($G = dI/dV$) should have an exponential dependence on the thickness of the barrier—cannot be readily applied in this case due to the nonrectangular shape of the barrier and variations of the barrier width with bias. The second criterion states that the conductance should have a parabolic dependence on the voltage and can be fit with known models, e.g., a Simmons (symmetric barrier) [106] or Brinkman, Dynes, and Rowell (BDR) model (asymmetric barrier) [107]. The inset to Figure 17.14a shows *G–V* data at a variety of temperatures, and a representative fit (dashed line) using the BDR model. Parameters of this asymmetric barrier model are defined in the diagram. Fits to the data between ±100 Mv* at several different temperatures yield an average barrier thickness of $d = 29$ Å, and barrier heights of $\phi_1 = 0.46$ eV and $\phi_2 = 0.06$ eV. The large potential difference between the two sides of the barrier is physically consistent with a triangular Schottky barrier tunnel junction. Although ϕ_1 is lower than might be expected for an Fe/AlGaAs interface, image force lowering of the barrier due to the highly degenerate nature of the AlGaAs can lead to reduction of the barrier by >0.3 eV [97]. The "goodness" of the fits, energy range considered, and deviation of the fit parameters from "known" physical characteristics of the barrier are typical of similar treatments in the literature [102,108]. Therefore, one may conclude that the second Rowell criterion is satisfied.

While the first two criteria are routinely invoked as proof of tunneling, it has been argued that *neither* can reliably distinguish tunneling from contributions due to spurious effects such as pinholes [102–104]. It has been shown that the *G–V* data can be fit with reasonable parameters even when tunneling was not the dominant transport path [102]. Jönsson-Åkerman et al. have presented convincing evidence that the *third* Rowell criterion is indeed a definitive confirmation of tunneling [102]. This criterion states that the zero-bias resistance (ZBR), i.e., the slope of the *I–V* curve at zero bias, should exhibit a weak, insulating-like

* Notice that the voltage range here is much smaller than the 1–2 V applied to the LED in the spin injection experiment. In the transport measurement, the voltage drop is primarily across the tunnel barrier interface, while in the LED, there are a variety of series resistances and contact resistances, as well as the band gap of GaAs that need to be considered.

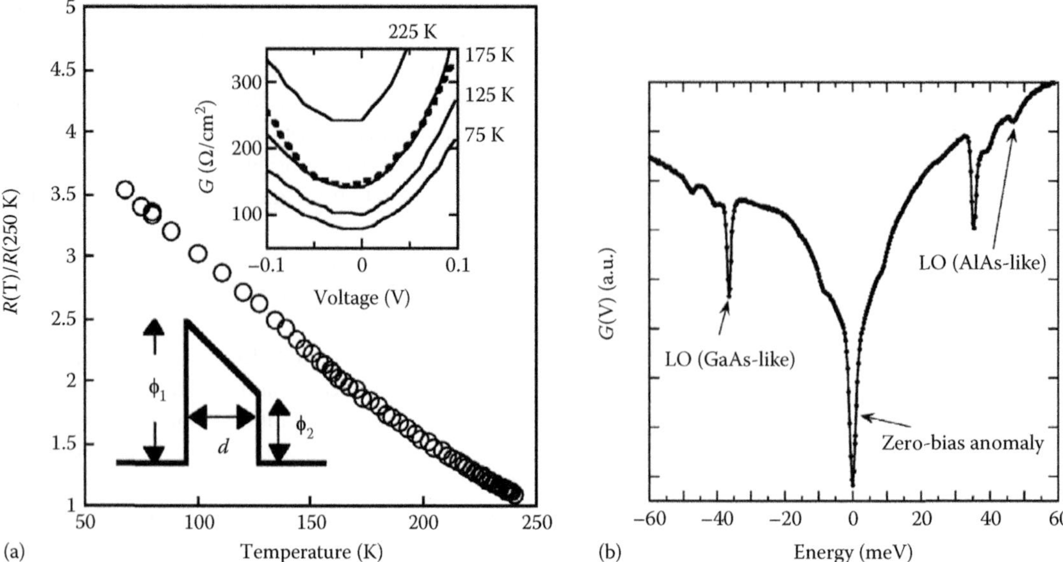

FIGURE 17.14 (a) Inset: series of conductance curves taken at different temperatures. The dotted lines are representative fits to the data. Parameters for the fitting are defined in the schematic of the tunnel barrier. Normalized ZBR as a function of temperature for an Fe/$Al_{0.1}Ga_{0.9}As$ Schottky barrier contact, showing minimal temperature dependence consistent with tunneling. (b) Conductance vs. applied voltage at 2.7 K. A zero-bias anomaly as well as phonon peaks attributed to GaAs-like and AlAs-like LO phonons are clearly visible.

temperature dependence. ZBR data are shown in Figure 17.14a as a function of temperature for the tailored Fe/AlGaAs Schottky contact, and clearly exhibit such a weak dependence over a wide range of temperature. Thus, the third Rowell criterion is also satisfied, confirming that single-step tunneling is the dominant conduction mechanism.

Further evidence for tunneling may be provided by the observation of phonon signatures and a pronounced zero bias anomaly in the conductance spectra.* For transport across an ohmic contact, electron energies can never reach a value higher than a few kT above E_F, regardless of the applied bias, and phonon modes are not observed because the injected electron energy is too low to excite them. A tunnel barrier enables injection of electrons at higher energies sufficient to excite phonons, thereby enabling a corresponding spectroscopy [109]—the observation of such features then provides further proof for tunneling.

The conductance spectrum for an Fe/$Al_{0.1}Ga_{0.9}As$ sample at 2.7 K is shown in Figure 17.14b, and exhibits two distinct features between 30 and 50 meV. In AlGaAs, two sets of longitudinal-optical (LO) phonon modes are present, one GaAs-like and the other AlAs-like, with energies of 36 and 45 meV, respectively. The observed features agree very well with these nominal values, and are labeled accordingly. In addition, the relative intensities of the GaAs and AlAs phonon interactions are positively correlated with the relative Ga:Al content [110]—the observed features exhibit an intensity ratio of ~10:1, further confirming their identity.

At the lowest temperatures, the *G–V* data have a pronounced feature at zero bias. Zero bias anomalies are generally observed in semiconductor tunneling devices [109]. Although poorly understood, they have been attributed to inelastic scattering effects arising from acoustic phonons and barrier defects, and are closely associated with tunneling. The observation of such a feature in these data provides further evidence for tunneling.

17.5.3.1.2 Examples: Spin Injection via Tailored Schottky Tunnel Barrier

The Fe/AlGaAs/GaAs/AlGaAs LED structures described above (Figure 17.13) provide conclusive evidence for efficient spin injection from an FM metal into a semiconductor, using the tunnel barrier to circumvent the conductivity mismatch [98,99]. These samples were processed to form surface emitting LEDs (Figures 17.4a and 17.7b), annealed at 200°C to improve the interface structure, and biased to inject spin-polarized electrons from the Fe through the tailored Schottky tunnel barrier and into the semiconductor, where they radiatively recombine with unpolarized holes. The EL spectra are shown in Figure 17.15a, where the light emitted along the surface normal (Faraday geometry) is analyzed for σ+ and σ− circular polarization for selected values of applied magnetic field. The spectra are dominated by the GaAs QW HH exciton, with a linewidth of 5 meV, clearly identifying the location where radiative recombination occurs. With no applied magnetic field, the Fe magnetization (easy axis) and corresponding electron spin orientation are entirely in the plane of the thin film. Although spin injection may indeed occur, it cannot be detected via surface emission because the average electron spin along the surface normal (*z*-axis) is zero, and the σ+ and σ− components are nearly coincident, as expected. The magnetic field is applied to rotate the Fe magnetization (and electron spin orientation in the Fe) out of

* See, e.g., Ref. [109].

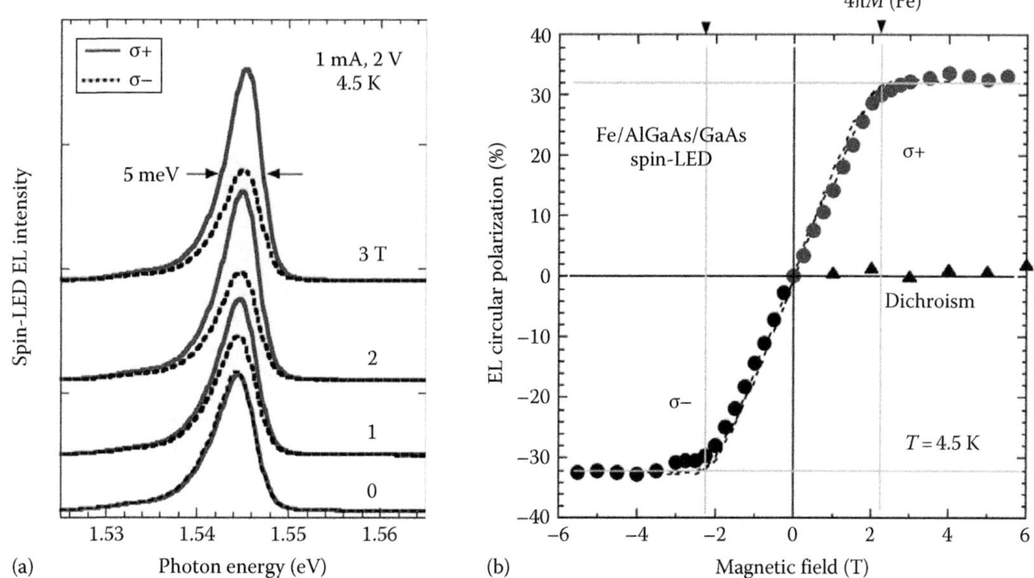

FIGURE 17.15 EL data from an Fe Schottky tunnel barrier spin-LED. (a) EL spectra for selected values of applied magnetic field, analyzed for positive and negative helicity circular polarization. (b) Magnetic field dependence of $P_{circ} = P_{spin}$. The dashed line shows the out-of-plane Fe magnetization obtained with SQUID magnetometry and scaled to fit the EL data. The triangles indicate the measured background contribution, including dichroism, using PL from an undoped reference sample.

plane, so that any net electron spin polarization can be manifested as circular polarization in the EL via the quantum selection rules. As the magnetic field increases, the component of Fe magnetization and electron spin polarization along the *z*-axis continuously increase, and the corresponding spectra exhibit a substantial difference in intensity of the σ+ and σ− components—this difference rapidly increases with field, signaling successful electrical spin injection.

The field dependence of the circular polarization $P_{circ} = [I(\sigma+) - I(\sigma-)]/[I(\sigma+) + I(\sigma-)]$, where $I(\sigma+)$ and $I(\sigma-)$ are the EL component peak intensities when analyzed as σ+ and σ−, respectively, is summarized in Figure 17.15b. P_{circ} rapidly increases with field and directly tracks the out-of-plane magnetization of the Fe film obtained by independent magnetometry measurements (dashed line). P_{circ} saturates at a value of 32% at a magnetic field value characteristic of the Fe contact, $B = 2.2$, $T = 4\pi M_{Fe}$, where the Fe magnetization is saturated out-of-plane. Thus, the electron spin orientation is preserved during injection from the Fe contact, with a net spin polarization $P_{spin} = P_{circ} = 32\%$ at the moment of radiative recombination in the GaAs QW.

Significant polarization is observed to nearly room temperature. Preliminary analysis of the temperature dependence of P_{circ} shows that it is dominated by that of the QW spin lifetimes, indicating that the injection process itself is independent of temperature [98], as expected for tunneling. Previous work has shown that the electron spin relaxation in a GaAs QW generally occurs more rapidly with increasing temperature [5,72], suppressing the measured circular polarization—the GaAs(001) QW is simply an imperfect spin detector, with a strong temperature dependence of its own. As noted earlier, the optical polarization measured depends upon the values of the spin and radiative carrier lifetimes. Both have been extensively studied, vary with temperature, and depend upon the physical parameters of the structure and the "quality" of the sample material.

One can determine the initial spin polarization P_o that exists at the moment the electrically injected electrons enter the GaAs QW by independently determining the value of the ratio τ_r/τ_s by PL measurements and applying Equation 17.4, as described previously for the case of spin injection from ZnMnSe films. These measurements yield $\tau_r/\tau_s = 0.78 \pm 0.05$, resulting in a value $P_o = P_{spin}\ (1 + \tau_r/\tau_s) = 57\%$. It is interesting to note that this value exceeds the nominal spin polarization of bulk Fe (~45%), clearly demonstrating that the bulk spin polarization of FM metals does not represent a limit to the spin polarization that can be achieved via electrical injection. Effects such as spin filtering at the Fe/AlGaAs interface and spin accumulation in the GaAs QW enable the generation of highly polarized carrier populations, and can be exploited in the design and operation of semiconductor spintronic devices.

A number of control experiments must be performed to rule out spurious effects. For example, LED structures fabricated with a nonmagnetic metal contact showed little circular polarization and very weak field dependence, eliminating contributions from Zeeman splitting in the semiconductor itself. Possible contributions to the measured P_{circ} arising from magnetic dichroism as the light emitted from the QW passes through the Fe film may be determined both analytically and directly measured. This contribution was calculated to be 0.9% using well-established models at the appropriate wavelength for the thickness of the Fe film [111]. This contribution was also directly measured by independent PL measurements on undoped Fe/AlGaAs/GaAs QW test structures. Linearly polarized laser excitation creates unpolarized electrons and holes in the AlGaAs/GaAs, which emit unpolarized light when they recombine in the GaAs QW. This light passes through the Fe film, and any circular polarization measured is therefore due to a combination of dichroism in the Fe, field-induced splittings in the QW, or other background effects. The solid triangles in Figure 17.15b summarize these measurements, and show that the sum of any such contributions is <1%. Note that the effect measured due to electron spin injection is over 30 times larger.

The sign of the polarization demonstrates that $m_j = -1/2$ electrons injected from the Fe Schottky tunnel contact dominate the radiative recombination process in the QW when the magnetic field is applied along the surface normal (Figure 17.1b). This is confirmed by the results from the ZnMnSe-based spin-LEDs, where the CB spin-splitting is known and the σ+ component also dominates for a similar field orientation. In the nomenclature of the magnetic metals community [84,112], such electrons are referred to as "majority spin" or "spin-up," even though the *spin* is antiparallel to the net magnetization. In this community, the term "spin-up" is used to describe an electron whose *moment* (rather than spin) is parallel to the magnetization. The corresponding state is at lower energy and more populated, and is therefore synonymously referred to as the "majority spin" state. Such carriers are designated here as "$n_m\uparrow$" where the subscript "m" is used to unambiguously indicate the convention used in this community [84]. Since the electron's spin and moment are antiparallel [113,114], the *spin* of a "spin-up" electron is actually antiparallel to the magnetization. Similarly, the terms "minority spin" and "spin-down" (designated $n_m\downarrow$) are used synonymously to refer to electrons whose *moment* is antiparallel to the magnetization, and therefore have a higher energy than the "majority spin" electrons.

The experimental observation that the polarization of the current injected from the Fe contact through the Schottky tunnel barrier corresponds to majority spin in Fe is consistent with the pioneering work of Meservey and Tedrow [112], and with the model proposed by Stearns [115]. While one might reasonably expect majority spin injection to dominate, recent work has shown that a number of factors are likely to contribute to the spin polarization of the tunneling current from an FM contact, including barrier thickness, band bending, and details of bonding at the interface, which effect the interface electronic structure [116–118].

The interface atomic structure and its correlation with spin-injection efficiency was addressed for the Fe/AlGaAs/GaAs system by combining the spin-LED results described above with high resolution TEM analysis of spin-LED samples and density functional theory (DFT) calculations of the Fe/AlGaAs interface [119]. The TEM analysis showed that the as-grown Fe/AlGaAs interface exhibited some degree of disorder, as shown in Figure 17.16a. The chemical order and coherence of the interface could be significantly improved by a low temperature anneal (200°C for 10 min), resulting in a highly ordered interface

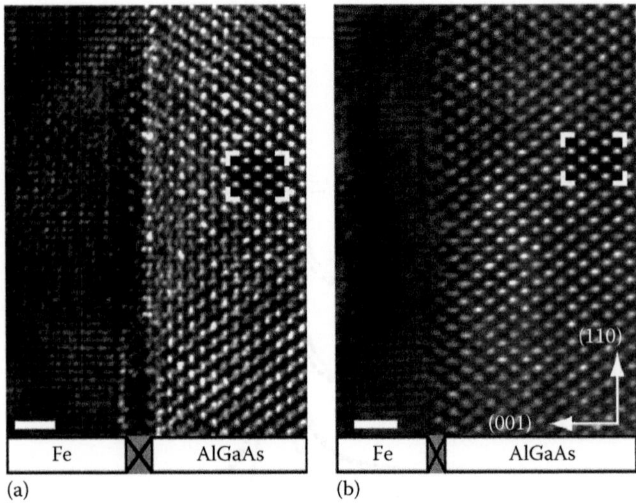

FIGURE 17.16 HRTEM images ([−110] cross section) of an Fe/AlGaAs spin-LED sample. (a) As-grown, exhibiting 18% spin polarization, and (b) following a mild postgrowth anneal, exhibiting a 26% spin polarization. Image simulations are inset (white brackets). The rectangles on the bottom of the images indicate Fe, AlGaAs, and the interfacial region (gray box with cross). Scale bars equal 1.0 nm. The contrast variation across the Fe regions is a result of changes in thickness.

(Figure 17.16b) and a 44% increase in spin-injection efficiency. The annealing temperature was well below the ~600°C growth temperature of the AlGaAs/GaAs QW structure, and therefore did not affect the characteristics of the semiconductor away from the Fe interface.

Phase and *Z*-contrast images were compared with those simulated from four interface models determined by DFT to be likely low energy candidates. Three of these models, shown in Figure 17.17 (from left to right: abrupt, partially intermixed, and fully intermixed), were previously proposed and studied theoretically [120]. The fourth model contained several monolayers of Fe_3GaAs (a known stable alloy in the Fe-Ga-As phase diagram) sandwiched as an interlayer between the GaAs and Fe. For each model, *Z*-contrast images were simulated with software [121] using the relaxed atomic coordinates determined by the DFT calculations. Based upon a mathematical comparison of line scans of the atomic intensity profiles, both parallel and perpendicular to the interface between the experimental and simulated images (Figure 17.18), the authors concluded that the interface of the annealed sample forms by intermixing of the Fe and AlGaAs occurring on a single atomic plane, resulting in an interface with alternating Fe and As atoms (see Figure 17.18b, and middle model structure in Figure 17.17).

The 44% increase in spin-injection efficiency observed experimentally was attributed to the greater tunneling efficiency for spin-polarized electrons across the chemically and structurally coherent annealed interface. As discussed previously (Section 17.5.1), band symmetries play an important role in spin injection from a metal to a semiconductor. Strong spin filtering is expected to occur at the ideal Fe/GaAs(001) interface (abrupt with no intermixing), because the Δ_1 symmetry of the bulk Fe

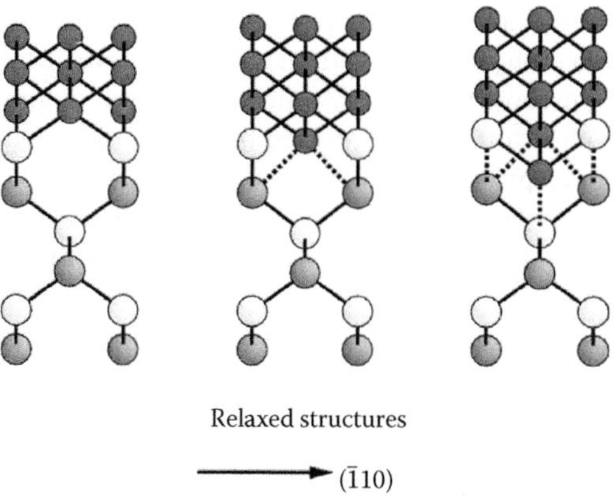

FIGURE 17.17 Low energy structures calculated by DFT after layer relaxation for the As-terminated Fe/GaAs(001) interface. Left—abrupt interface, middle—partially intermixed, with one plane consisting of alternating Fe and As atoms, and right—fully intermixed. The gray and light gray spheres represent Ga and As atoms, and the dark gray spheres represent Fe atoms. Highly strained bonds ~15%–20% longer than ideal are shown as dotted lines. (Reprinted with permission from Erwin, S.C., Lee, S.-H., and Scheffler, M., *Phys. Rev. B*, 65, 205422 and 2002. Copyright 2002 by the American Physical Society.)

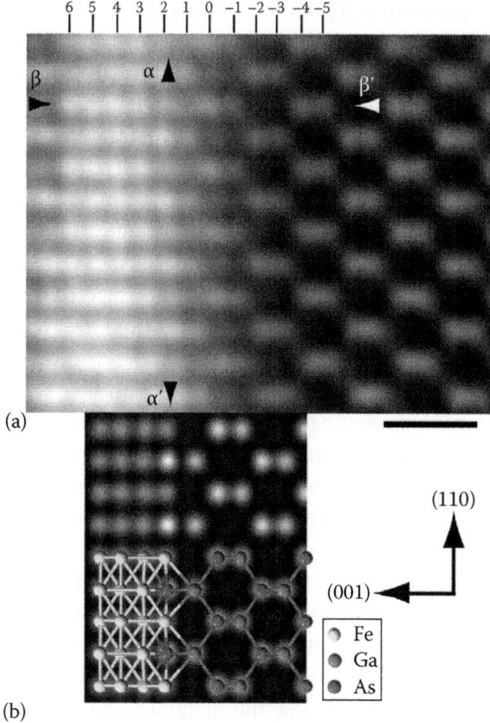

FIGURE 17.18 *Z*-contrast TEM images of the Fe/AlGaAs(001) [−110] interface from spin-LED samples. (a) Experimental image from a sample with 26% spin polarization. The arrowheads, α–α′ and β–β′, indicate the location and direction of the line profiles parallel and perpendicular to the interface, respectively, as used in the analysis. (b) **(See color insert.)** Simulated image of a partially intermixed interface and inset ball-and-stick model from which it was calculated (Fe atoms appear in yellow, As in blue, and Ga in red). The scale bar equals 0.5 nm.

majority-spin state near E_F matches that of the bulk GaAs CB states, enhancing transmission of majority spin electrons [76]. The symmetries of the Fe minority-spin bands do not match the GaAs, so that these states decay very quickly, suppressing minority spin transmission. The same analysis was applied to the partially intermixed and fully intermixed interface models reported in Ref. [120] and considered in the TEM analysis above. The results show no significant change in the Δ_1 decay rate between the abrupt and partially intermixed models [119], indicating that both should enable highly polarized spin injection. However, a significantly faster decay of the majority spin Δ_1 state into the GaAs was found for the fully intermixed model, indicating lower spin polarization of the injected carriers and consistent with the measured polarizations from the spin-LED samples.

This picture of the Fe/AlGaAs interface structure is consistent with earlier detailed studies of the initial nucleation, interface formation, and magnetic properties of Fe films grown on the As-dimer terminated GaAs(001) 2×4 and $c(4\times4)$ surfaces reported by Kneedler et al. [122–124] and Thibado et al. [125]. These authors concluded that the interface that was ultimately formed after a few monolayers of Fe deposition was planar with little intermixing, and characterized by Fe–As bonding. Excess As diffuses through the Fe film and segregates to the surface. Complementary studies further indicated that the electrical character of the Fe/GaAs interface was not dominated by As antisite defects [126,127], consistent with surface segregation of As. An overview of the growth and magnetic properties may be found in Ref. [128].

This tailored doping profile and Schottky tunnel barrier has been widely used in subsequent studies to elucidate spin injection, transport, and detection in a variety of FM metal/AlGaAs/GaAs heterostructures [129–134]. Generation of *minority spin* electrons has been reported for a limited range of bias conditions in Fe/AlGaAs/GaAs structures with interface doping profiles and structures nominally identical to those described above [129,131]. Two possible theoretical explanations have been offered. The first invokes formation of interface states which mediate minority spin transmission when the structure is biased so that electrons flow from the GaAs into the Fe (spin extraction) [118]. Note that this calculation was based on the perfectly abrupt Fe/GaAs interface model (Figure 17.17 left image) rather than the partially intermixed model (Figure 17.17 middle image) as concluded from the TEM analysis above. The specific structure of the interface will have a significant impact on the formation and character of interface states, and such calculations can provide further insight by addressing other interface structures. The second explanation invokes formation of bound states which form in the CB QW produced by the heavy doping near the semiconductor interface [117]. The relative contribution of these bound states to the spin transport process can also lead to the generation of minority spin accumulation in the GaAs. These results underscore the fact that several factors need to be considered in developing a more comprehensive understanding of spin transmission across a heterointerface. The tailored Schottky tunnel barrier approach has also been used very recently for spin injection into ZnSe [135] and Si [136].

17.5.3.2 Spin Injection via a Discrete Layer as a Tunnel Barrier

Discrete insulating layers may also be used to form a tunnel barrier between the FM metal and the semiconductor. These provide the more familiar and well-studied rectangular potential barrier, but also introduce an additional interface into the structure and raise issues of pinholes and uniformity of layer thickness. FM metal/oxide/metal structures have been extensively studied, since they form the basis for a tunneling spectroscopy used to determine the metal spin polarization [137], and for tunnel magnetoresistance devices being developed for nonvolatile memory [138–141]. Reviews of this area may be found in Chapters 10 and 11. Various oxides have been used as tunnel barriers for spin injection into a semiconductor, including Al_2O_3, SiO_2, MgO, and Ga_2O_3. Examples of each are presented below.

17.5.3.2.1 Al_2O_3/AlGaAs-GaAs

Spin-polarized tunneling has been well studied in metal/Al_2O_3/metal heterostructures. Careful measurements using a superconductor such as Al for one contact have provided quantitative information on the spin polarization of the tunneling current from many FM metals [137,142,143].

Several groups have utilized Al_2O_3 barriers in FM metal/Al_2O_3/AlGaAs-GaAs spin-LED structures with great success [144–147]. Motsnyi et al. [144] used a CoFe contact and the oblique Hanle effect to measure and analyze the EL, and reported values for the GaAs electron spin polarization P_{spin} of 14%–21% at 80 K, and 6% at 300 K. In addition, they were able to experimentally determine the radiative and spin lifetimes of the GaAs itself. When they corrected their measured values for this "instrument response function" of the GaAs detector as discussed previously, they obtained a value $P_o = 16\%$ at 300 K (this quantity is denoted by "Π" in their notation). Manago and Akinaga obtained spin polarizations of ~1% using either Fe, Co, or NiFe contacts [146]. van't Erve et al. obtained a value of $P_{spin} = 40\%$ at 5 K for samples, which utilized a lightly n-doped (10^{16} cm^{-3}) $Al_{0.1}Ga_{0.9}As$ layer adjacent to the Al_2O_3 [147]. However, they reported lower operating efficiency (higher bias voltages and currents) than for the Schottky barrier based spin-LEDs of Refs. [98,99]. This was subsequently remedied by including the same semiconductor doping profile shown in Figure 17.13 before the growth of the Al_2O_3 layer, so that high $P_{spin} = 40\%$ was achieved at $T = 5$ K and bias conditions comparable to those utilized for the Schottky tunnel barrier devices. They determined the efficiency of their GaAs spin detector by optical pumping to obtain the value $\tau_r/\tau_s = 0.75$, resulting in a value $P_o = P_{spin}(1 + \tau_r/\tau_s) = 70\%$—a value that again significantly exceeds the spin polarization of bulk Fe.

17.5.3.2.2 Al_2O_3/Si

One of the first demonstrations of spin injection into Si utilized an Al_2O_3 tunnel barrier in an Fe/Al_2O_3/Si(001) n–i–p spin-LED heterostructure [148]. While Si is clearly an attractive candidate for spintronic devices, its indirect band gap makes it much more difficult to probe optically, and led many to believe that the optical spectroscopic techniques, which had proven so productive when applied to the III–Vs, would be unable to provide much insight into spin-dependent behavior in Si. Although the indirect gap complicates application of polarized optical techniques used routinely in GaAs, it is worth noting that the first successful optical pumping experiments to induce a net electron/nuclear spin polarization were performed in Si rather than in a direct gap material by Lampel [6].

Several fundamental properties of Si make it an ideal host for spin-based functionality. Spin–orbit effects producing spin relaxation are much smaller in Si than in GaAs due to the lower atomic mass and the inversion symmetry of the crystal structure itself. The dominant naturally occurring isotope, Si^{28}, has no nuclear spin, suppressing hyperfine interactions. Consequently, spin lifetimes are relatively long in Si, as demonstrated by an extensive literature on both donor bound [149] and free electrons [150,151]. In addition, silicon's mature technology base and overwhelming dominance of the semiconductor industry make it an obvious choice for implementing spin-based functionality. Several spin-based Si devices have indeed been proposed, including transistor structures [152,153] and elements for application in quantum computation/information technology [154].

Jonker et al. electrically injected spin-polarized electrons from a thin FM Fe film through an Al_2O_3 tunnel barrier into an Si(001) n–i–p doped heterostructure, and observed circular polarization of the EL [148]. This signals that the electrons retain a net spin polarization at the time of radiative recombination, which they estimated to be ~30%, based on simple arguments of momentum conservation. This interpretation was confirmed by similar measurements on Fe/Al_2O_3/Si/AlGaAs/GaAs QW structures in which the spin-polarized electrons injected from the Fe drift under applied field from the Si across an air-exposed interface into the AlGaAs/GaAs structure and recombined in the GaAs QW. In this case, the polarized EL can be quantitatively analyzed using the standard selection rules, yielding an electron spin polarization of 10% in the GaAs. More recently, theory provided a quantitative link between the circular polarization of the Si EL and the electron spin polarization. These results are summarized in the following paragraphs.

The Si n–i–p LED structures were grown by MBE, transferred in air, and introduced to a second chamber for deposition of the Fe/Al_2O_3 tunnel contact. An n-doping level ~2×10^{18} cm^{-3} at 300 K was chosen to prevent carrier freeze-out at lower temperatures. This is well below the metal–insulator transition of 5.6×10^{18} cm^{-3}. Details of the growth and processing may be found in Ref. [148]. A schematic of the sample structure, corresponding band diagram of the spin-injecting interface, and a photograph of the processed surface-emitting LED devices are shown in Figure 17.19a. A high resolution cross-sectional TEM image of the Fe/Al_2O_3/Si contact interface region is shown in Figure 17.19b and reveals a uniform and continuous oxide tunnel barrier with relatively smooth interfaces. The Fe film is polycrystalline on the nominally amorphous Al_2O_3 layer.

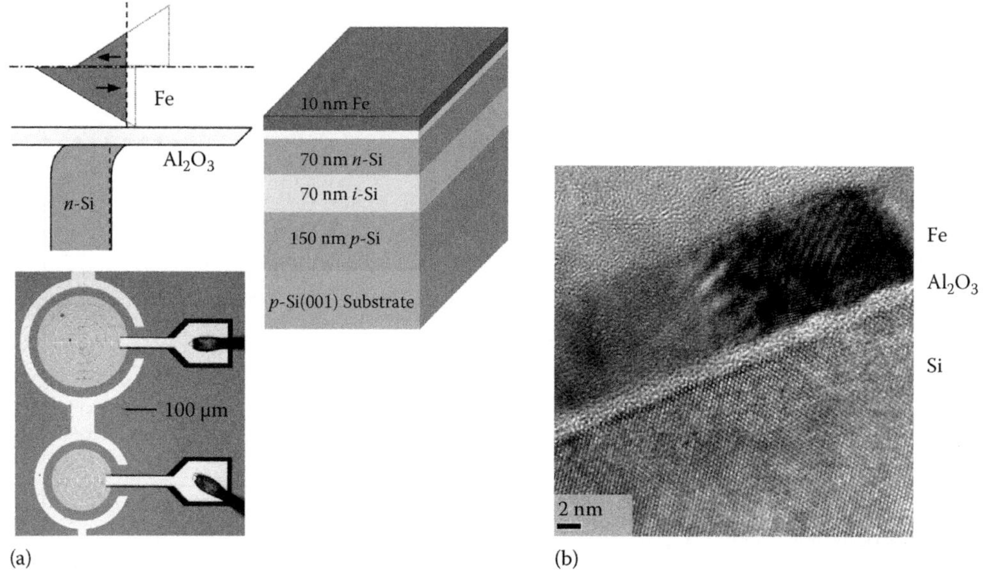

FIGURE 17.19 Fe/Al_2O_3/Si spin-LED structures. (a) Band diagram of the spin-injecting interface and schematic of the sample structure, with an optical photograph of the processed spin-LED devices. The light circular mesas are the active LED regions. (b) Cross-sectional TEM image of the spin-injecting interface region.

Typical EL spectra from a surface-emitting Fe/Al_2O_3/Si *n–i–p* spin-LED structure are shown in Figure 17.20a and b for $T = 5–80$ K, analyzed for σ+ and σ− circular polarization. At 5 K, the spectra are dominated by features arising from electron-hole recombination accompanied by transverse acoustic (TA) or transverse optical (TO) phonon emission. Subsequent analysis to determine the origin of the emission features has resulted in a revised identification of the EL peaks observed. This analysis revealed that the TO/TA peaks occur in correlated pairs, with the peaks at 1105 meV (TA_s) and 1065–1070 meV (TO_s) due to recombination in the *p*-Si substrate ($p \sim 10^{19}$ cm^{-3}), and the peaks at 1090 meV (TA_1) and 1050 meV (TO_1) likely arising from recombination in the interface region. At 80 K, the TO_s feature dominates. At zero field, no circular polarization is observed because the Fe magnetization and corresponding electron spin orientation lie in-plane and orthogonal to the light propagation direction. As discussed previously, although spin injection may occur, it cannot be detected with this orthogonal alignment. Therefore, a magnetic field is applied to rotate the Fe spin orientation out-of-plane, and the main spectral features each exhibit circular polarization, as shown by the difference between the red (σ+) and blue (σ−) curves in the 3 T spectra at $T = 5$, 50, and 80 K.

The magnetic field dependence of P_{circ} for each feature is summarized in Figure 17.20c. As the Fe magnetization (and majority electron spin orientation) rotates out of plane with increasing field, P_{circ} increases and saturates above ~2.5 T with average values of 3.7%, 3.5%, and 1.9% for the TA_s, TA_1, and TO_1 features, respectively, at 5 K, and 2.1% and 2% for the TO_s feature at 50 and 80 K. Note that, for each feature, P_{circ} tracks the magnetization of the Fe contact, shown as a solid line scaled to the data, indicating that the spin orientation of the electrons that radiatively recombine in the Si directly reflects that of the electron spin orientation in the Fe contact. The field at which the Fe magnetization saturates out of plane is characteristic of Fe ($H_{sat} = 4\pi M_{Fe} = 2.2$ T), and is unaffected by lateral patterning, which might alter the in-plane coercive field. This also argues against possible compound formation (e.g., FeSi, etc.) at the interface due to pinholes, which would lead to a different field dependence. The monotonic decrease in P_{circ} from the higher energy (TA_s) to the lowest energy feature (TO_1) as seen in Figure 17.20c is consistent with spin relaxation expected to accompany electron energy relaxation. The sign of P_{circ} indicates that Fe majority spin electrons dominate, as was the case for the tailored Schottky Fe/AlGaAs and Fe/Al_2O_3/AlGaAs contacts. Data from the requisite reference samples used to determine background effects such as dichroism show only a weak paramagnetic field dependence ~0.1%/T, ruling out spurious contributions. These data together unambiguously demonstrate that spin-polarized electrons are electrically injected from the Fe contact into the Si heterostructure.

An analysis of the EL to extract the electron spin polarization P_{spin} in the Si is complicated by the indirect gap character and the participation of phonons in the radiative recombination process. Since the electron spin angular momentum must be shared by the photons and any phonons involved in radiative recombination, the photons will carry away only some fraction of the spin angular momentum of the initial spin-polarized electron population. Thus, our experimentally measured value of P_{circ} will result in a significant underestimate of the corresponding Si electron polarization. A second significant factor affecting P_{circ} is the very long radiative lifetime typical of indirect gap materials—as noted previously, P_{spin} decays exponentially with a characteristic time τ_s before radiative recombination occurs. Despite these issues, the broader rule of conservation of momentum

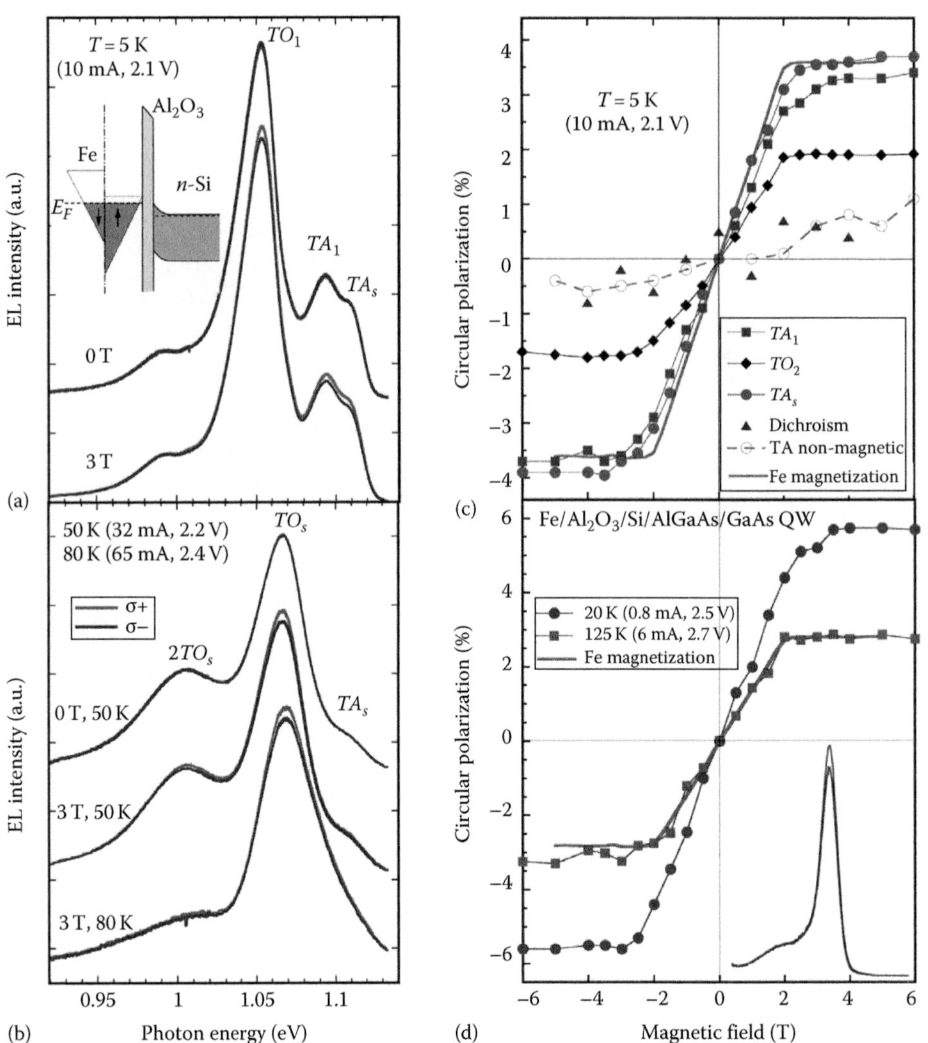

FIGURE 17.20 (See color insert.) EL spectra from surface-emitting Fe/Al_2O_3/Si *n–i–p* spin-LED structures, analyzed for positive (σ+) and negative (σ−) helicity circular polarization at temperatures of (a) 5 K, and (b) 50 and 80 K. The spectra are dominated by features arising from electron-hole recombination accompanied by transverse acoustic (*TA*) and transverse optical (*TO*) phonon emission. (c) The magnetic field dependence of the circular polarization for each feature in the Si spin-LED spectrum at 5 K. (d) Inset shows the EL spectrum from Fe/Al_2O_3/80 nm *n*-Si/*n*-$Al_{0.1}Ga_{0.9}As$/GaAs QW/*p*-$Al_{0.3}Ga_{0.7}As$ spin-LEDs at $T = 20$ K and $H = 3$ T. The dominant feature is the GaAs QW free exciton, which exhibits strong polarization. The magnetic field dependence of this circular polarization is plotted for $T = 20$ and 125 K. The out-of-plane Fe magnetization curve appears in panel (c) and (d) as a solid line for reference.

can be used to estimate a *lower bound* for the initial electron spin polarization, and the same rate equation analysis leading to Equation 17.4 applies to correct for these lifetime effects: $P_o = P_{spin}(1 + \tau_r/\tau_s) > P_{circ}(1 + \tau_r/\tau_s)$. Typical values for τ_r in doped Si at low temperature are 0.1–1 ms [155,156]. The radiative lifetime of emission features associated specifically with acoustic phonon emission was measured to be 480 μs for $1\,K < T < 5\,K$ [155]. The spin lifetime for free electrons well above the metal–insulator transition has been determined to be ~1 μs in both bulk [150] and modulation doped samples [151], and is expected to increase at lower electron densities [150]. Therefore, using $P_{circ} \sim 0.03$ (Figure 17.20c), $\tau_r = 100$ μs and $\tau_s = 10$ μs, a conservative estimate is given by $P_o > P_{circ}(1 + \tau_r/\tau_s) \sim 0.3$ or 30%, comparable to that achieved in GaAs.

This value is supported by similar experiments on Fe/Al_2O_3/80 nm *n*-Si/80 nm *n*-$Al_{0.1}Ga_{0.9}As$/10 nm GaAs QW/200 nm *p*-$Al_{0.3}Ga_{0.7}As$ heterostructures [148], in which electrons injected from the Fe into the Si drift across the Si/$Al_{0.1}Ga_{0.9}As$ interface with applied bias and radiatively recombine in the GaAs QW. In this case, the standard quantum selection rules can be rigorously applied to quantify the electron spin polarization, which eventually reaches the GaAs [157]. The EL spectrum at 20 K and 3 T is shown as insert in Figure 17.20d, and is dominated by the free exciton emission at 1.54 eV from the GaAs QW. The field dependence of the polarization of this feature, P_{circ}(GaAs), is shown in Figure 17.20d. P_{circ}(GaAs) again tracks the magnetization of the Fe contact, and saturates at values of 5.6% at 20 K and 2.8% at 125 K. Separate optical pumping measurements on the GaAs

QW provide a direct measure of τ_r/τ_s (GaAs QW) = 0.8. Thus, the spin polarization of the electrons in the GaAs is $P_{GaAs} = P_{circ}(1 + \tau_r/\tau_s) = 0.1$ or 10% at 20 K. This is consistent with our estimate above of $P_{Si} \sim 30\%$, and establishes a firm lower bound for $P_{Si} > 10\%$. It is indeed remarkable that the electrons injected from the Fe contact drift through the Si, cross the Si/AlGaAs interface and still retain a significant spin polarization, given (1) the relatively poor crystalline quality of Si epilayers on GaAs (lattice mismatch 3.9%), (2) the heterovalent interface structure, (3) the 0.3 eV CB offset (Si band lower than $Al_{0.1}Ga_{0.9}As$) [158], and (4) the fact that the sample surface was exposed to air before growth of the Si, and then again before growth of the Fe/Al_2O_3 contact.

Recent theory provides a fundamental and more rigorous interpretation of the circular polarization of the TO and TA features in the Si spin-LED EL spectra, and thereby an independent determination of the electron spin polarization achieved in the Si by electrical injection. Li and Dery [159] have shown that the circular polarization of these features is directly related to the electron spin polarization for a given doping level in the region where radiative recombination occurs. For $p = 10^{19}\,cm^{-3}$, they conclude that the TA feature should have a maximum circular polarization of 13% for a 100% spin-polarized electron population. Thus, the measured value $P_{circ} \sim 3.5\%$ for the TA_s feature (Figure 17.20c) indicates an electron spin polarization of 27%. The estimates of the electron spin polarization summarized here are in remarkably good agreement.

The EL intensity decreases rapidly with increasing temperature. It is too low to reliably analyze at temperatures higher than those shown in Figure 17.20 for emission from either the Si (80 K) or GaAs (125 K) even though clear polarization remains—a further limitation of the LED as a spin detector. Subsequent work using electrical rather than optical spin detection confirms that spin injection in the FM metal/Al_2O_3/Si system persists to room temperature, as described in the next section.

The Al_2O_3 layer serves as both a tunnel barrier and a diffusion barrier, preventing interdiffusion between the Fe and Si which is likely to occur. Spin injection is also observed in Fe/Si *n–i–p* samples when the Fe is deposited directly on the Si, where the Schottky barrier serves as a tunnel barrier [160]. However, the net spin polarization achieved is lower due to the inhomogeneous character of the interface that is formed.

Grenet et al. have also used the spin-LED approach to demonstrate spin injection into Si [161]. They employed a (Co/Pt)/Al_2O_3 tunnel barrier contact, where the Co/Pt exhibits strong perpendicular magnetic anisotropy with the remanent magnetization along the surface normal. This obviates the need for an applied magnetic field to saturate the out-of-plane magnetization, as required for the Fe/Al_2O_3 contacts described above. The entire structure consisted of (3 nm Pt/1.8 nm Co)/2 nm Al_2O_3/100 nm *n*-Si/10 nm $Si_{0.7}Ge_{0.3}$ QW/550 nm *p*-Si/p-Si(001), with 50 nm dopant setbacks at the QW. The strained $Si_{0.7}Ge_{0.3}$ QW served as the radiative recombination region, which produces a Type II band alignment with strong hole localization in the $Si_{0.7}Ge_{0.3}$. They measure a maximum (typical) circular polarization $P_{circ} = 3\%$ (1.2%) in the EL signal from the QW, which is nearly constant at 200 K. The magnetic field dependence of P_{circ} tracks the Co/Pt magnetization, confirming that the injected electrons retain their spin orientation from the Co/Pt contact.

17.5.3.2.3 SiO_2/Si

SiO_2 has been the gate dielectric of choice for generations of Si metal-oxide-semiconductor devices, because it is easy to form and provides the low interface state density necessary for device operation. It has been the cornerstone of the vast Si electronics industry, because no other semiconductor has a robust native oxide which forms such an electrically stable self-interface. As such, it is an obvious choice to use as the tunnel barrier in a Si-based spin transport device. However, there are few reports of its use as a spin tunnel barrier in any structure. Smith et al. reported a smaller-than-expected magnetoresistance of 4% at 300 K in Co/SiO_2/CoFe tunnel junctions, which they attributed to overoxidation at the metal interfaces [162]. A composite NiFe/SiO_2/Al_2O_3 tunnel barrier was used as the emitter in a magnetic tunnel transistor by Park et al. [163]. They reported a tunnel spin polarization of 27% at 100 K, which was lower than the value of 34% obtained for an NiFe/Al_2O_3 emitter in the same structure. No data were reported for a tunnel barrier consisting of SiO_2 alone.

Li et al. successfully used SiO_2 as a spin tunnel barrier on Si, and recently reported spin injection from Fe through SiO_2 into a Si *n–i–p* heterostructure, producing an electron spin polarization in the Si, $P_{Si} > 30\%$ at 5 K [164]. The samples were similar to those described above and consist of 10 nm Fe/2 nm SiO_2/70 nm *n*-Si/70 nm undoped Si/150 nm *p*-Si/*p*-Si(001) substrate, forming a spin-LED structure identical to that described above. The SiO_2 layer was formed by natural oxidation of the *n*-Si surface assisted by illumination from an ultraviolet lamp. A TEM image of the Fe/SiO_2/Si interface region appears in Figure 17.21a, and shows a reasonably uniform SiO_2 layer with well-defined interfaces. The SiO_2 appears amorphous, as expected, while the Fe is polycrystalline.

Transport measurements based on the Rowell criteria confirmed that conduction from the Fe through the SiO_2 occurred by tunneling. The conductance exhibits a parabolic dependence upon the bias voltage, as shown in Figure 17.21b for several temperatures. Good fits (solid lines) to the conductance data are obtained within an energy range of ±100 meV using the BDR model [107], which yield an effective barrier height of ~1.7 eV and a barrier thickness of ~21 Å at 10 K. The temperature dependence of the ZBR, expressed as $R_0(T)/R_0(300\,K)$, where R_0 is the slope of the *I–V* curve at zero bias, is summarized in Figure 17.21c. The ZBR exhibits a weak temperature dependence indicative of single-step tunneling rather than the exponential dependence of thermionic emission. A weak temperature dependence of the ZBR has been shown to be a reliable indicator for a pinhole-free tunnel barrier [102], as discussed earlier in this chapter.

EL spectra at $T = 5\,K$ and $B = 0$ and 3 T are shown in Figure 17.22a, analyzed for $\sigma+$ and $\sigma-$ circular polarization. The spectra are very similar to those of Figure 17.20a (Fe/Al_2O_3/Si *n–i–p* spin-LEDs), and are dominated by TA and TO

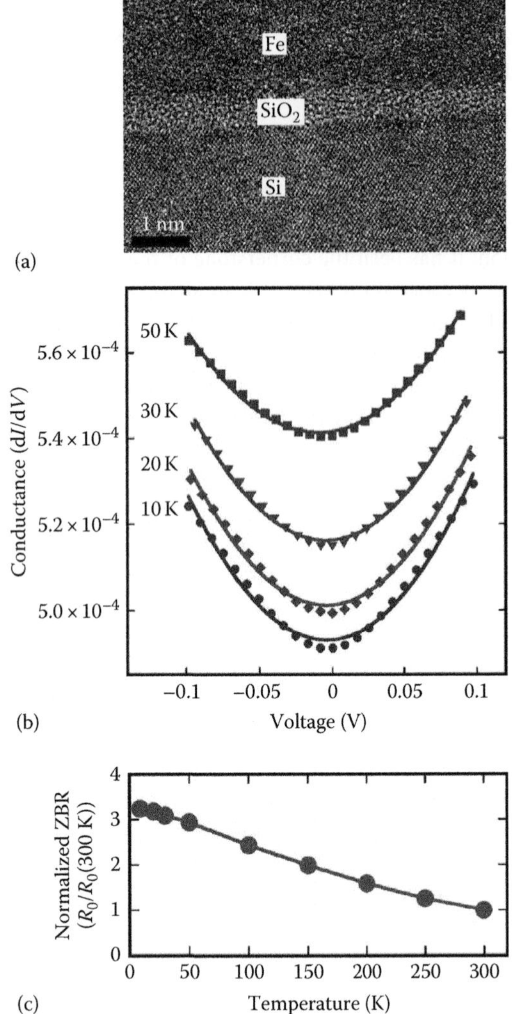

FIGURE 17.21 (a) Cross-sectional TEM image of the Fe/SiO_2/Si interface region. (b) Conductance curves for transport through an Fe/SiO_2/*n*-Si(001) contact exhibiting parabolic behavior, and fits using the BDR model. (c) Zero-bias resistance of the contact as a function of temperature. The weak temperature dependence indicates single-step tunneling through a pinhole-free SiO_2 tunnel barrier.

phonon-mediated recombination, with specific peak identifications as discussed in Section 17.5.3.2.2. Although emission is initially detected at a bias of 1.6 V, higher biases are typically used to reduce data acquisition time, as the polarization measured is relatively independent of the bias. A magnetic field along the surface normal rotates the Fe magnetization out of plane, enabling the injected electron spin polarization to be manifested as circular polarization in the EL. At $B = 3$ T, where the Fe magnetization is saturated out of plane, the spectra exhibit polarization for each of the spectral features, with $P_{circ} \sim 3\%$. The magnetic field dependence of P_{circ} for the three main peaks in the spectra is summarized in Figure 17.22b, and the behavior and discussion parallel that of Figure 17.20c—the electrons that tunnel from the Fe through the SiO_2 to eventually radiatively recombine in the Si clearly retain a net spin polarization from the Fe. Applying the same rate equation analysis, discussed above, yields a value of 30% as a lower bound for the electron spin polarization achieved in the silicon for temperatures up to 50 K. A similar value is obtained by applying the phonon-mediated recombination theory of Li and Dery [159] to the TA_s feature.

17.5.3.2.4 MgO/AlGaAs-GaAs

MgO has been widely used in metal/oxide/metal tunnel structures, and more recently in metal spin torque transfer devices (see Chapters 7, 11, and 14 for a comprehensive overview). Its rocksalt structure is closely lattice-matched to bcc-Fe, and high-quality epitaxial growth has been demonstrated. The band symmetry arguments originally applied to the Fe/MgO/Fe system by MacLaren, Butler et al. [74,75,82] show that the Fe Δ_1 majority spin band couples well to the propagating state in the MgO barrier, while the minority spin bands do not, resulting in highly spin-polarized current. Thus, Fe/MgO is an attractive candidate as a spin-injecting contact for a semiconductor such as GaAs or Si (see Figure 17.11).

Jiang et al. demonstrated robust spin injection up to room temperature with CoFe/MgO tunnel contacts on AlGaAs/GaAs QW/AlGaAs(001) spin-LED structures [165]. A cross-sectional TEM image of the tunnel contact region is shown in Figure 17.23a. The experimental procedure and analysis are identical to that described previously in Section 17.5.3.1.2 (Figure 17.15) for Fe Schottky tunnel/AlGaAs/GaAs QW/AlGaAs spin-LED structures [98,99]. The EL spectra analyzed for σ+ and σ− polarization for temperatures of 100 and 290 K are shown in Figure 17.23b and c, and exhibit significant circular polarization of 52% and 32%, respectively, after background subtraction when the Fe magnetization is saturated along the surface normal.

The authors argue that these high values for P_{circ} are due to improved spin-injection efficiency at higher temperatures than either the Fe/Al_2O_3 or Fe Schottky tunnel contacts provide. Because spin-dependent tunneling is expected to have a weak temperature dependence, a more likely explanation lies in the temperature dependence of the spin and radiative lifetimes of the GaAs QW—the *detector* efficiency rather than the spin-injection efficiency was much higher due to the particular characteristics of the GaAs (which can vary significantly from one MBE machine to another). This was explicitly investigated by Salis et al., who used time-resolved optical techniques to measure the lifetimes for these same Fe/MgO spin-LED samples and found that they were strongly temperature dependent, as shown in Figure 17.24. Using the values at 10 K of P_{circ} (P_{EL} in Figure 17.24b) = 0.47, $\tau_r = 300$ ps, and $\tau_s = 560$ ps (Figure 17.24a), from Equation 17.4, we find $P_o = P_{circ}(1 + \tau_r/\tau_s) = 0.72$. For the second sample they studied (see Figure 3 of Ref. [166]), at 10 K, the corresponding values are P_{circ} (P_{EL}) = 0.32, $\tau_r = 500$ ps, $\tau_s = 700$ ps, and $P_o = 0.55$. These values for the initial spin polarization achieved in the GaAs QW are very similar to those obtained using either the Fe/GaAs Schottky tunnel barrier ($P_o = 0.57$) or Fe/Al_2O_3 spin-injecting contacts ($P_o = 0.70$), described in Sections 17.5.3.1.2 and 17.5.3.2.1, indicating that all three tunnel barriers are of

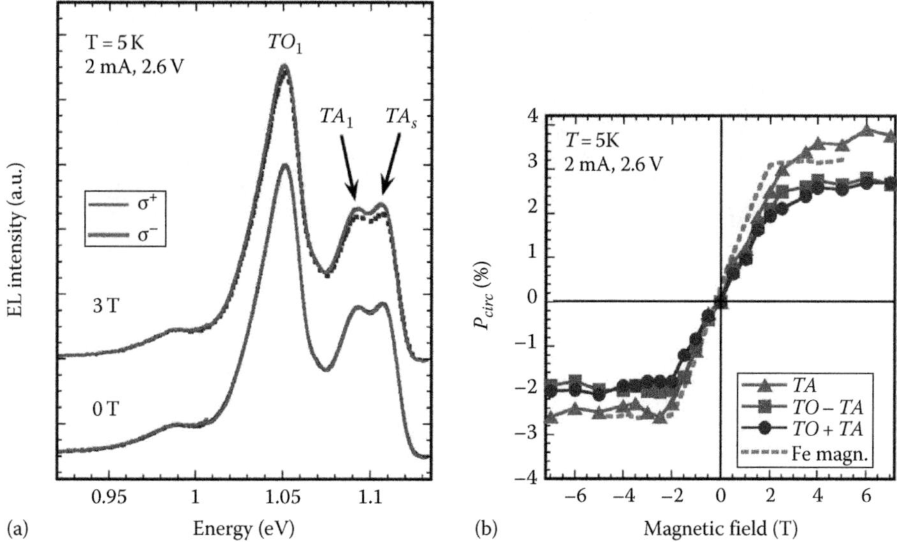

FIGURE 17.22 (a) EL spectra at 5 K from a Si *n–i–p* structure with an Fe/SiO_2 contact at zero field and 3 T, analyzed for σ+ and σ− circular polarization. (b) Magnetic field dependence of P_{circ} for the TA_s, TA_1, and TO_1 features at 5 K. P_{circ} consistently tracks the out-of-plane magnetization of the Fe (dashed line).

comparable efficiency for spin injection. These data underscore the importance of quantifying the detector response function. Nevertheless, these results conclusively show that efficient electrical spin injection persists to room temperature and presents no obstacle for future development of semiconductor spintronic devices.

Lu et al. investigated the effect of MgO barrier thickness and crystallinity on the spin-injection efficiency in CoFeB/MgO/AlGaAs/GaAs spin-LED structures [167]. They found that the injection efficiency as determined from P_{circ} increased as the MgO thickness was increased from 1.4 to 4.3 nm. They attributed this to a suppression of band bending due to hole accumulation in the AlGaAs at the MgO interface for the thicker barriers, which prevented the holes from tunneling into the FM injector. They also found that the spin-injection efficiency was higher for MgO grown at 300°C where it exhibited some degree of crystalline texture vs. amorphous films grown at room temperature, which they attributed to contributions from the band symmetry matching arguments discussed earlier in this chapter. Truong et al. combined pulsed electrical input and time-resolved optical spectroscopy to analyze the temporal response of the spin polarization, following injection with these same spin-LED structures [168]. They concluded that the build-up time of the electron spin polarization in the GaAs QW was much faster than the rise time of the EL signal, which was limited to ~700 ps due to parasitics in the device circuit.

17.5.3.2.5 GaO_x/AlGaAs-GaAs

Unlike Si, the native oxide of GaAs is difficult to grow with proper stoichiometry and is less stable than SiO_2, hence a metal-oxide-semiconductor technology, which allows inversion mode operation in a GaAs device, has never blossomed. Previous efforts reported a very low interface state density (10^{10}–10^{11} cm^{-2} eV^{-1}) at the GaO_x/GaAs interface [169], comparable to that of SiO_2/Si, making GaO_x an attractive candidate as a gate insulator. Saito et al. utilized GaO_x as the spin tunnel barrier in Fe/GaO_x/AlGaAs/GaAs QW spin-LED structures [170]. They fabricated surface emitting devices and measured the circular polarization of the EL as a function of applied magnetic field (Faraday geometry) and temperature. The experimental procedure and analysis are again identical to that described previously in Section 17.5.3.1.2 (Figure 17.15) for Fe Schottky tunnel/AlGaAs/GaAs QW structures. They find that P_{circ} tracks the out-of-plane magnetization of the Fe contact with a maximum value $P_{circ} = P_{spin} = 20\%$ at $T = 2$ K. Using a typical value for $\tau_r/\tau_s \sim 1$ from the literature, they calculate the initial spin polarization in the GaAs: $P_o = P_{spin}(1 + \tau_r/\tau_s) = 40\%$. P_{circ} decreased rapidly with temperature to a value of 2% at 300 K, as found previously for GaAs QW based spin-LEDs with both Fe Schottky and Fe/Al_2O_3 tunnel barrier contacts, and this is likely due to the temperature dependence of τ_r and τ_s [10]. Thus, the temperature dependence observed is that of the GaAs QW detector rather than that of the spin-injection process through GaO_x.

In summary, the use of a tunnel barrier allows one to circumvent the conductivity mismatch issue, enabling the use of FM metals as spin-injecting contacts into a semiconductor. FM metals introduce many desirable attributes to a device technology, including intrinsic nonvolatile and reprogrammable operating characteristics. They exhibit reasonable spin polarizations at E_F (~40%–50%), high Curie temperatures, low coercive fields, fast switching times, and an advanced materials technology, which permits one to tailor the magnetic characteristics for a particular application. An FM metal/tunnel barrier contact injects spin-polarized electrons near the semiconductor CB edge with high efficiency, where the electron energy is determined by the bias applied across the tunnel barrier (~10–100 meV). The realization

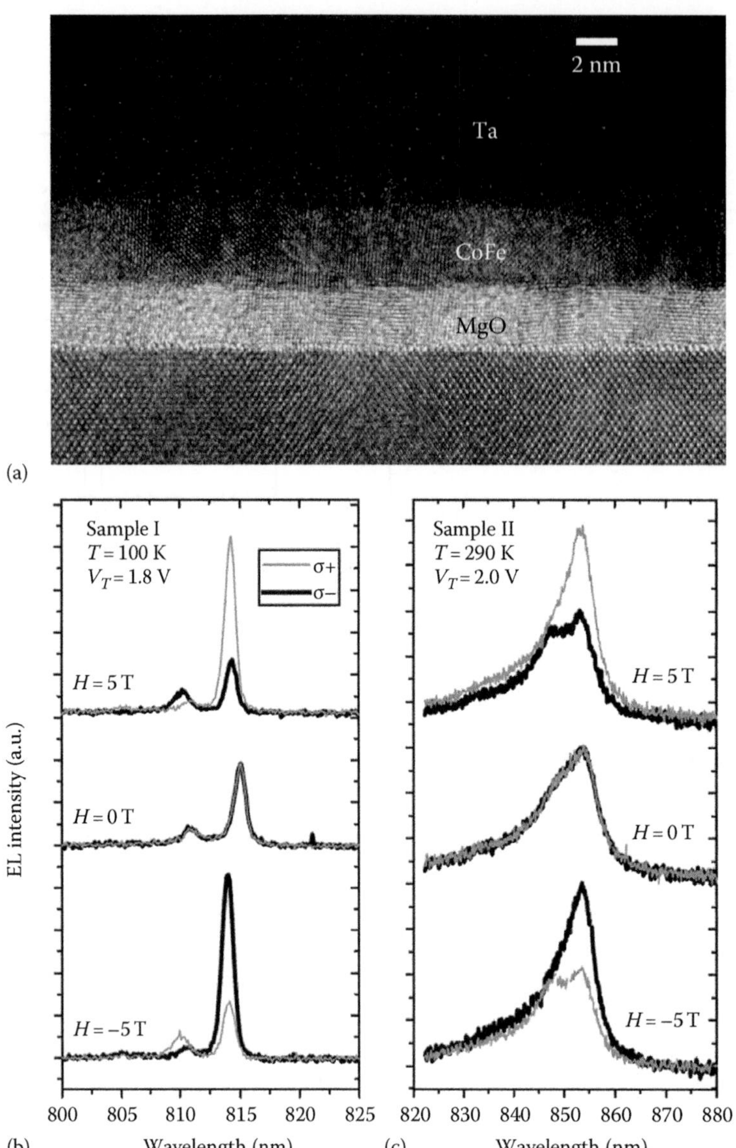

FIGURE 17.23 (a) TEM image of the CoFe/MgO/AlGaAs/GaAs spin-LED structure. Electroluminescence spectra from the spin-LEDs are shown at temperatures of (b) 100 K and (c) 290 K. (Reprinted with permission from Jiang, X., Wang, R., Shelby, R.M., Macfarlane, R.M., Bank, S.R., Harris, J.S., and Parkin, S.S.P., *Phys. Rev. Lett.*, 94, 056601. Copyright 2005 by the American Physical Society.)

of efficient electrical injection and significant spin polarization using a simple tunnel barrier compatible with "back-end" semiconductor processing should greatly facilitate progress in the development of semiconductor spintronic devices.

17.5.4 Hot Electron Spin Injection

We conclude this section by describing a completely different avenue toward electrical injection of spin-polarized carriers from an FM metal into a semiconductor, which also circumvents the conductivity mismatch issue. This is the process of hot electron injection, which builds on the development of the spin-valve transistor and magnetic tunnel transistor [171–173]. This topic is covered in detail in Chapters 9 and 18, and we include a brief discussion here for the sake of providing a more complete overview on electrical spin injection.

Electrical spin injection into a semiconductor by hot electron injection was first demonstrated by Jiang et al. using a GaAs/$In_{0.2}Ga_{0.8}As$ multiple QW structure as the spin detector, incorporating the spin-LED approach [174]. A schematic of the heterostructure is shown in Figure 17.25, and consists of a metal emitter/Al_2O_3 tunnel barrier/FM metal base deposited on GaAs, which forms the top layer of the GaAs/$In_{0.2}Ga_{0.8}As$ structure on a *p*-GaAs(001) substrate. The FM metal base layer forms a Schottky barrier contact with the GaAs. Electrons from the metal emitter (which need not be FM) tunnel through the Al_2O_3

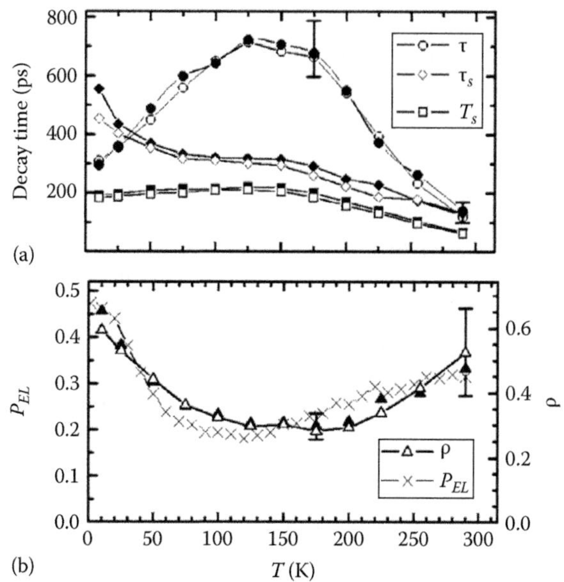

FIGURE 17.24 (a) Measured radiative (τ), spin (τ_s) and combined $\left(T_s^{-1} = \tau_s^{-1} + \tau^{-1}\right)$ lifetimes as a function of temperature for the spin-LEDs of Figure 17.23. (b) Measured circular polarization P_{EL} and spin detector efficiency $\rho = T_s/\tau_r$ as a function of temperature for $B = 0$ (open symbols) and 0.8 T (closed symbols). (Reprinted with permission from Salis, G., Wang, R., Jiang, X., Shelby, R.M., Parkin, S.S.P., Bank, S.R., and Harris, J.S., Temperature independence of the spin-injection efficiency of a MgO-based tunnel spin injector. *Appl. Phys. Lett.*, 87, 262503. Copyright 2005, American Institute of Physics.)

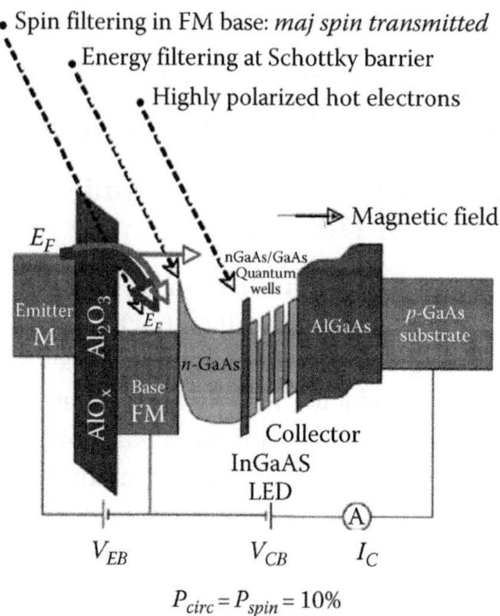

FIGURE 17.25 Schematic energy band diagram of a magnetic tunnel transistor merged with a QW LED collector. The emitter/base bias V_{EB} controls the energy of the injected hot electrons, while the collector/base bias V_{CB} can be used to adjust the band bending of the LED. The emitter consists of 5 nm $Co_{84}Fe_{16}$. The 2.2 nm thick Al_2O_3 tunnel barrier is formed by reactive sputtering of Al in the presence of oxygen. The base layer consists of 3.5 nm $Ni_{81}Fe_{19}$ and 1.5 nm $Co_{84}Fe_{16}$ with the NiFe layer adjacent to the GaAs. (Reprinted with permission from Jiang, X., Wang, R., van Dijken, S., Shelby, R., Macfarlane, R., Solomon, G.S., Harris, J., and Parkin, S.S.P., *Phys. Rev. Lett.*, 90, 256603. Copyright 2003 by the American Physical Society.)

tunnel barrier and enter the FM metal base at an energy well above E_F, determined by the emitter-base bias voltage. Electrons that are minority spin relative to the base magnetization are rapidly scattered due to their short mean free path and lose energy in the FM base, relaxing to E_F, where they are blocked from entering the semiconductor by the Schottky barrier. Although majority spin electrons have a longer mean free path, most are also scattered in the base, relax to E_F, and are blocked by the Schottky barrier. But a small fraction of the majority spin electrons do not scatter, and retain sufficient energy to surmount the Schottky barrier and enter the semiconductor, forming a spin-polarized current. These electrons thermalize to the CB and recombine with unpolarized holes in the InGaAs QWs. The circular polarization of the resultant EL provides an assessment of the spin injection.

Using the procedures described above, Jiang et al. measure a value $P_{circ} = 10\%$ at $T = 1.4$ K when the FM base magnetization is rotated out of plane by an applied field (Faraday geometry), signaling successful spin injection using this hot electron approach. The value of P_{circ} is somewhat lower than measured in other spin-LED experiments, and the authors suggest two likely reasons: (1) the injected hot electrons lose a significant amount of polarization during the process of thermalization to the bottom of the CB, and (2) after the hot electrons enter the QW region, further spin relaxation can occur in the QWs before recombination. The measured result, therefore, sets a lower bound for the injected electron spin polarization.

This hot electron injection approach was later adopted by Appelbaum et al. to demonstrate electrical spin injection into Si [175]. The structure is similar to that of Figure 17.25 (substituting Si for GaAs/AlGaAs), but they replace the InGaAs QW detector section with a second FM metal layer, which serves as a spin analyzer, functioning in the same way as described above for the FM emitter. This enables them to detect spin currents by measuring changes in the current through the device, as one changes the orientation of the magnetization of emitter and analyzer layers from parallel to antiparallel. Rather than measuring an optical polarization or a voltage produced by spin accumulation [130], this approach measures the change in current. A detailed treatment may be found in Chapter 18.

Thus, the combination of hot electron injection, spin filtering in the FM base or analyzer layer, and energy filtering by a Schottky barrier results in preferential transmission of majority spin current and a blocking of the minority spin current, enabling spin injection and analysis. A practical concern with this approach is that it is very inefficient—most of the current flows in the emitter-base circuit (or FM analyzer circuit), while a very small fraction ~10^{-4} enters the semiconductor as spin-polarized current. This is a recognized problem with the magnetic tunnel transistor [173], and is discussed in detail in Chapter 9.

17.6 Ferromagnetic Metal/Semiconductor Electrical Spin Injection: Electrical Detection

17.6.1 General Considerations and Nonlocal Detection

The preceding sections summarize rapid progress in developing a practical and technologically attractive methodology for electrical spin injection, enabling one to produce electron populations in a semiconductor with spin polarizations exceeding those of typical FM metals. It was repeatedly noted that the temperature dependence observed experimentally was dominated by the spin *detector*, e.g., the temperature dependence of the radiative and spin lifetimes in the spin-LED structures. Indeed, spin injection by tunneling is expected to depend only weakly upon temperature. Optical detection of spin polarization in a semiconductor has proven to be invaluable as a research tool, and will enable a subset of applications such as spin lasers (see Chapter 38). However, it is not practical for the broader range of electronic applications, where a voltage input/output with minimal temperature dependence is required to interface to the rest of the circuit. Thus, utilizing *an FM metal as a spin detector* has great technological appeal.

The issue of conductivity mismatch has significant implications for electrical spin detection, and places constraints on the range of contact resistances at the FM metal/semiconductor interface, which permit electrical detection of spin accumulation. This is reviewed in detail in Refs. [91,94], and in Chapter 1 by Fert. The specific cases of FM metal contacts as spin detectors on GaAs and on Si were discussed by Dery et al. [176] and Min et al. [177], respectively. The criteria for the conventional two-terminal magnetoresistance device are particularly demanding, and have yet to be experimentally realized in a conventional semiconductor such as Si or GaAs.

The concept of *nonlocal* spin detection was introduced by Johnson and Silsbee in 1985 in their study of spin accumulation in metals [178] (see Chapter 6), and developed extensively by others for spin detection in metals [179,180], GaAs [130,131], Si [181,182], and graphene [183]. It is based upon the fact that spin diffuses away from the point of generation independently of the direction of charge flow, i.e., *a pure spin current flows outside of the charge current path*. This process is similar to that of heat diffusing away from a hot contact on a surface produced, e.g., by the tip of a soldering gun. This is illustrated in Figure 17.26, where an FM metal tunnel barrier contact (labeled 3 in Figure 17.26) is used to inject a spin-polarized charge current into the semiconductor. A second FM contact (labeled 2) is placed outside of the spin-polarized charge current path defined by the bias applied between contacts 3 and 4. Contact 2 senses the spin-splitting of the chemical potential, $\Delta\mu = \mu_{up} - \mu_{down}$, produced by the spin accumulation resulting from spin diffusing away from the FM injector, contact 3. The spin accumulation is the greatest directly underneath the injector contact, and decays with increasing distance from the contact with a certain spin diffusion length, λ_{sf}. The voltage measured at the detector contact 2 is proportional to $\Delta\mu$ and the projection of the semiconductor spin polarization onto the contact magnetization. Therefore, one can use some mechanism to manipulate the semiconductor spin polarization (magnitude or orientation) to encode or process information with the pure spin system, and directly read out the result as a voltage at this second contact. There are two key concepts to be recognized: (1) *the nonlocal geometry enables one to separate and distinguish a pure spin current from a spin-polarized charge current*, and (2) *the spin accumulation produces a real voltage, which can be sensed by an FM contact*. The following section illustrates these processes, and summarizes recent work on spin injection and nonlocal electrical detection in Si. Nonlocal spin injection/detection in GaAs is covered in Chapter 23, and in graphene in Chapter 29.

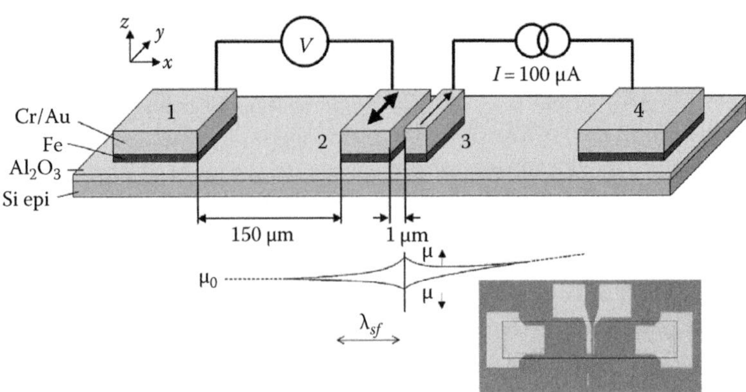

FIGURE 17.26 Schematic layout of four terminal nonlocal spin valve (NLSV) device and depiction of the spin-dependent electrochemical potential. A spin-polarized current is injected at contact 3 and changes in the electrochemical potential are measured non-locally at contact 2, placed 1 μm from contact 3. The difference in shape (100 × 24 μm) and (100 × 6 μm) for contact 2 and 3, respectively, allows for independent control of the magnetizations of the two contacts. The large reference contacts (100 × 150 μm) are spaced 150 μm (several spin flip lengths) from the injector and detector contact. The inset shows an image of the actual device.

Spin accumulation can also be generated and detected using a single FM metal tunnel barrier contact, which serves as both the spin injector and detector in a simpler three terminal geometry with two nonmagnetic reference contacts [130,184]. This approach is illustrated in Sections 17.6.3 and 17.6.4, using FM metal/oxide tunnel barrier contacts on both Si and GaAs.

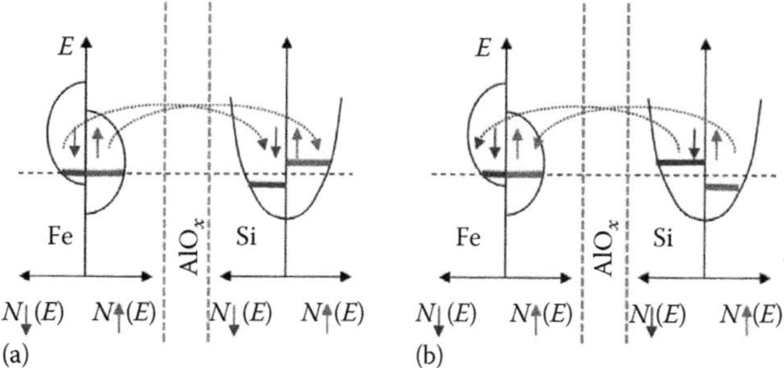

FIGURE 17.27 Schematic illustration of (a) injection and (b) extraction of spins from silicon by means of spin-dependent tunneling.

17.6.2 Nonlocal Spin Injection and Detection in Silicon: Pure Spin Currents

The nonlocal spin valve (NLSV) geometry of Figure 17.26 has been used to generate and detect pure spin current in silicon in a lateral transport device, and determine the corresponding spin lifetimes via the Hanle effect [181,182]. The silicon samples were 250 nm thick epilayers grown by MBE on insulating Si(001) substrates at 500°C. The films were n-doped to achieve an electron concentration of $2–3 \times 10^{18}\,cm^{-3}$ at room temperature, below the metal–insulator transition ($n_{MIT} \sim 5 \times 10^{18}\,cm^{-3}$), but high enough to avoid carrier freeze-out at low temperature. The samples were exposed to atmosphere upon removal from the MBE system. After a 10% HF acid etch and deionized water rinse to produce a hydrogen passivated surface, the samples were loaded into a second MBE system, heated to desorb the hydrogen, and a 1.5 nm Al_2O_3 tunnel barrier and 10 nm polycrystalline Fe film were deposited to form the spin-injecting tunnel barrier contacts, as described in Ref. [148].

Conventional photolithography and wet chemical etching were used to define multiple NLSV structures, each consisting of four Fe/Al_2O_3 contacts spaced as shown in Figure 17.26. Contacts 2 ($24 \times 100\,\mu m$) and 3 ($6 \times 100\,\mu m$) are separated by $1\,\mu m$ and serve as the spin detector and injector, respectively. The different aspect ratios provide different coercive fields with the easy axis along the long axis so that the contact magnetizations can be aligned parallel or antiparallel. Contacts 1 and 4 are reference contacts ($100 \times 150\,\mu m$) located several spin diffusion lengths away from the transport channel to insure that the spin polarization is zero at these spin grounds.

When a negative bias is applied to contact 3, a spin-polarized electron current is injected from the Fe into the Si (Figure 17.27a) to produce (a) a spin-polarized charge current, which flows to contact 4 due to the applied bias, and (b) a pure spin current, which diffuses isotropically. The electron spin is majority spin oriented in the plane of the surface due to the in-plane magnetization of the Fe injector contact (and opposite the magnetization of the injector by convention). These spin currents produce spin accumulation in the Si described by the splitting of the spin-dependent electrochemical potential, $\Delta\mu = \mu_{up} - \mu_{down}$, which decreases with distance from the point of injection, as discussed above and shown explicitly in Figure 17.28a. The pure spin

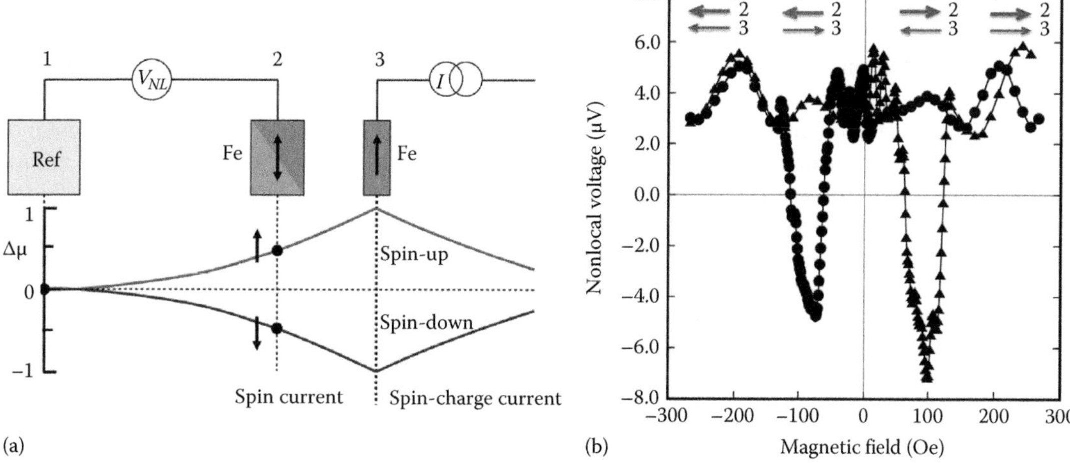

FIGURE 17.28 (a) Schematic of the spin splitting of the electrochemical potential $\Delta\mu$ corresponding to spin accumulation in the silicon produced by spin injection from contact 3 in the four terminal nonlocal spin valve geometry (reference contact 4 is not shown). Depending upon the direction of its magnetization, the nonlocal voltage V_{NL} measured at contact 2 measures either the spin-up or spin-down branch of the chemical potential relative to the reference contact. (b) Nonlocal voltage vs. in-plane magnetic field, for an injection current of $-100\,\mu A$ at 10 K in an $Fe/Al_2O_3/n$-Si nonlocal spin valve structure. Two levels corresponding to the parallel and antiparallel remanent states are clearly visible. The triangles (circles) correspond to increasing (decreasing) the in-plane magnetic field. The arrows indicate the relative orientation of the detector (2) and injector (3) contact magnetizations.

diffusion current is analyzed at contact 2, which is outside of the charge current path ("nonlocal"), where Δμ is manifested as a voltage. No charge current flows in the detection circuit defined by reference contact 1 and magnetic detector contact 2, thus excluding spurious contributions from anisotropic magnetoresistance and local Hall effects.

The nonlocal voltage V_{NL} measured at the detector contact 2 is sensitive to the relative orientation of the contact magnetization and that of the spin polarization in the Si beneath it. In Figure 17.28b, V_{NL} is plotted as a function of an applied in-plane magnetic field B_y used to switch the magnetization of the injector and detector. At large negative field, the magnetizations are parallel, and the majority spins injected into the Si are parallel to the electron spin orientation of the detector, resulting in a positive voltage at contact 2 corresponding to the difference between the point indicated on the "spin-up" curve under contact 2 in Figure 17.28a and the spin ground. As the field is increased (triangles), contact 2 with the lower coercive field switches so that the injector/detector magnetizations are antiparallel. The orientation of the spin current diffusing to contact 2 is now antiparallel to that of the detector, producing an abrupt change at $B_y \sim 50$ Oe and a plateau in the nonlocal voltage, with the value corresponding to the difference between the point indicated on the "spin-down" curve under contact 2 (Figure 17.28a) and the spin ground. As B_y is increased further, contact 3 also switches so that injector and detector contact magnetizations are again parallel, and the nonlocal voltage abruptly changes to its original value. The process is repeated as the field is swept in the opposite direction (circles, Figure 17.28b), producing an output characteristic similar to that seen in metal pseudo spin-valve structures [185], and confirming remanent (nonvolatile) parallel and antiparallel contact orientations.

When a positive bias is applied at contact 3, electrons flow from the Si into the Fe contact (Figure 17.27b). The higher conductivity of the Fe majority spin channel results in efficient majority spin current, leaving an accumulation of *minority* spin electrons in the Si and a reversal of the spin-splitting of the chemical potential. This bias configuration is often referred to as "spin extraction [186]," since majority spins are *extracted* from the semiconductor. A minority rather than majority pure spin current now diffuses from the injector to the detector contact 2, resulting in an inversion of the nonlocal voltage peaks for antiparallel alignment of injector and detector contact magnetizations, as seen in Figure 17.29. The magnitude of the nonlocal voltage is roughly linear with the magnitude of the bias current.

Thus, the orientation of the spin in the Si, and the corresponding splitting of the chemical potential, can be reversed either by switching the contact magnetization or by changing the polarity of the bias voltage on the injecting contact.

Further confirmation of spin transport and another avenue enabling manipulation of the spin in the silicon is provided by the Hanle effect [187], in which a magnetic field perpendicular to the surface (electron spin) induces precession and dephasing of the spins in the silicon, resulting in a modulation and eventual

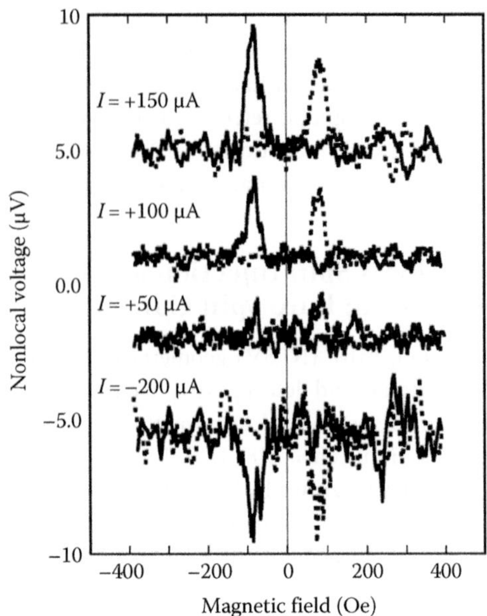

FIGURE 17.29 Nonlocal voltage vs. in-plane magnetic field at 10 K for several values of the injection current. Graphs are offset for clarity. For negative bias, electrons are injected from Fe into the Si channel, and the change in nonlocal voltage is consistent with majority spin injection. For positive bias, electrons are extracted from the silicon into the Fe contact. The majority spins are more readily extracted, resulting in the accumulation of minority spins in the silicon. A change in sign is seen for nonlocal voltage peaks for the antiparallel state, consistent with minority spin accumulation.

suppression of the detected NL MR signal. The Hanle effect is widely regarded as the gold standard of spin injection and transport, providing definitive proof of the existence of a spin-polarized electron population [130,131,175,180,181]. The Hanle effect curve at $T = 10$ K and a bias current of −100 μA (spin injection) is shown in Figure 17.30. The magnetizations of the injector and detector contacts are placed in the parallel remanent state, and a magnetic field (Hanle field, B_z) is applied perpendicular to the surface of the device. As the Hanle field increases from zero, the injected spins in the silicon precess around B_z during transit to the detector contact, resulting in both dephasing and an increasing degree of antiparallel alignment relative to the detector spin orientation, producing a corresponding decrease in the magnitude of the nonlocal signal. Variations in transit times in the transport channel due to the width of the injector and detector contacts truncate the full Hanle curve and suppress further precessional oscillations in the nonlocal voltage. If the injector bias is reversed to operate in the spin extraction mode, or the injector and detector contacts are placed in the antiparallel remanent state, the sign of the Hanle curve reverses [181,182]. Hanle curves are visible to measurement temperatures of ~150 K for the NLSV geometry and doping levels $n \sim 2 \times 10^{18}$ cm^{-3}, used in these samples.

The Hanle lineshape can be fit analytically to obtain the spin lifetime in this pure spin current. A fit to the data using an approach similar to that of Ref. [131] (and discussed in

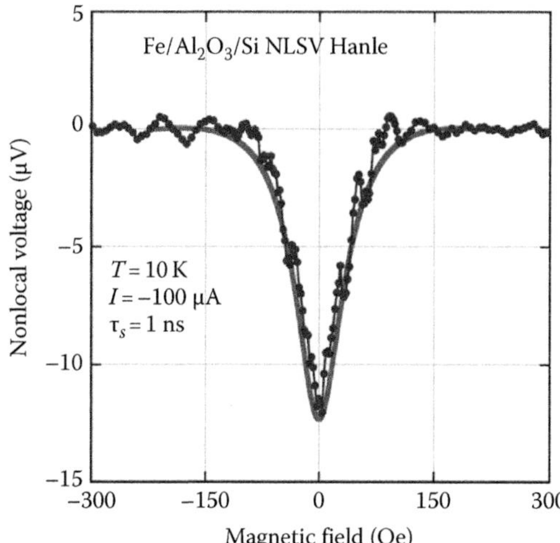

FIGURE 17.30 The Hanle effect curve at $T=10$ K and a bias current of $-100\,\mu$A (spin injection) obtained for the four terminal nonlocal spin valve contact geometry with Fe/Al_2O_3/Si contacts. The spin in the n-Si transport channel ($n \sim 2-3 \times 10^{18}\,cm^{-3}$) precesses during transit between injector and detector contacts due to an applied perpendicular magnetic field. The magnetizations of the injector and detector contacts are in the parallel remanent state. The solid line is a fit to the data using the model described in the text, and yields a spin lifetime of 1 ns and diffusion constant of 10 cm²/s.

Chapter 23) is shown by the smooth curve in Figure 17.30, and yields a spin lifetime of ~1 ns at $T=10$ K. The lifetimes for majority and minority spins (referenced to the injector contact 3) in the Si channel are comparable, as expected for a nonmagnetic semiconductor. This value is shorter than in the bulk at a comparable electron density and temperature. The relatively short spin lifetimes are attributed to the fact that this lateral transport geometry probes spin diffusion near the Si/Al_2O_3 interface, where surface scattering and interface states are likely to produce more rapid spin relaxation than in the bulk.

NLSV measurements using the geometry of Figure 17.26 were also reported recently in degenerately doped n-Si (doped well above the metal–insulator transition) by Ando et al. using Fe_3Si/Si Schottky tunnel barrier contacts [188], and by Sasaki et al. using Fe/MgO tunnel barrier contacts [189]. Both groups observed the nonlocal magnetoresistance behavior (Figure 17.28b) due to reversal of the contact magnetization. However, neither reported a Hanle signal, and thus provide no information on spin lifetimes. By measuring the dependence of the NLSV signal on injector/detector contact spacing, Sasaki et al. obtained an estimate for the spin diffusion length of 2.25 μm at $T=8$ K and a carrier density $n \sim 1 \times 10^{20}\,cm^{-3}$. They also report that a measurable NL MR signal persists to $T=120$ K.

Thus, an FM metal/Al_2O_3 tunnel barrier contact can be effectively used as both a spin injector and a spin detector for Si, providing a direct voltage readout of the Si spin polarization in a lateral transport device. With the nonlocal geometry, one can generate and detect a pure spin current that flows outside of the charge current path. The spin orientation of this pure spin current is controlled in one of three ways: (a) by switching the magnetization of the FM metal injector contact, (b) by changing the polarity of the bias on the "injector" contact, which enables the generation of either majority or minority spin populations in the Si, and provides a way to electrically manipulate the injected spin orientation without changing the magnetization of the contact itself, and (c) by inducing spin precession through application of a small perpendicular magnetic field. The contact bias and magnetization, together with spin precession, allow full control over the orientation of the spin in the silicon channel and subsequent detection as a voltage, demonstrating that information can be transmitted and processed with pure spin currents in silicon. The FM contacts provide nonvolatile functionality as well as the potential for reprogrammability. This is accomplished in a lateral transport geometry using lithographic techniques compatible with existing device geometries and fabrication methods.

17.6.3 Three Terminal Spin Injection and Detection in Si at Room Temperature: Spin-Polarized Charge Current

A simpler three terminal geometry can also be used to generate and measure spin accumulation in a semiconductor [130,184], as illustrated in Figure 17.31 for Si. In this case a single FM tunnel barrier contact serves as both the spin injector and detector, and the spin accumulation Δμ measured is that which develops directly under the contact in the presence of a spin-polarized charge current. As discussed in the next

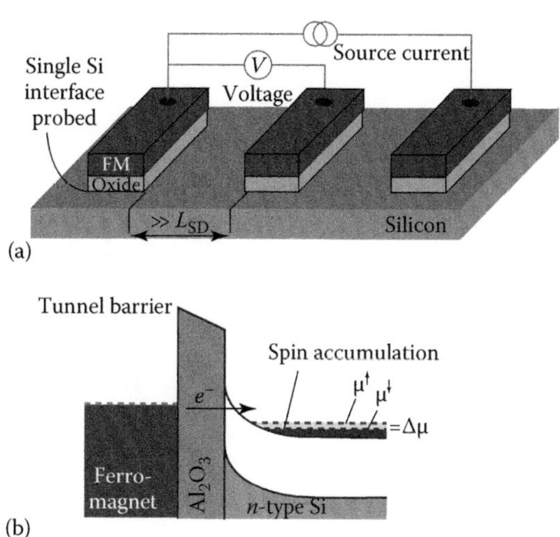

FIGURE 17.31 (a) Three terminal measurement geometry for injection and detection of spin accumulation under left contact. (b) Energy band diagram of the junction, depicting the ferromagnet, tunnel barrier, and n-type Si conduction and VBs. (Reprinted by permission from MacMillan Publishers Ltd. [*Nature*] Dash, S.P. et al., Electrical creation of spin polarization in silicon at room temperature, 462, 491, copyright 2009.)

section, *this spin accumulation does not necessarily develop in the semiconductor itself*—it may, in fact, develop in the interface states or trap states within the oxide tunnel barrier. Thus, spin injection into the semiconductor is not required to generate spin accumulation and corresponding voltage signal in this measurement geometry. The density of such states depends upon the procedures used to form the interface and oxide layer, and may vary widely. Spin accumulation in such states may indeed prove to be just as interesting and technologically relevant as in the semiconductor, but may complicate and compromise spin injection/transport/detection in the semiconductor channel.

This three terminal approach was used by Dash et al. [190] to demonstrate spin accumulation and electrical detection in heavily *n*- and *p*-doped Si at room temperature. They used FM metal/Al_2O_3/Si tunnel barrier contacts for which spin injection into Si had been successfully demonstrated [148]. In contrast with previous work, their Si was degenerately doped, with *n*-(*p*-) doping of $1.8\times10^{19}\,cm^{-3}$ ($4.8\times10^{18}\,cm^{-3}$), well above the metal–insulator transition. The Hanle effect was used to observe the spin accumulation and determine the corresponding spin lifetimes directly under the contact as a function of bias and temperature.

A cross-sectional diagram of the contact interface region illustrating the spin injection, accumulation, and subsequent precession induced by the perpendicular magnetic field B_z is shown in Figure 17.32a. At $B_z=0$, the spin accumulation $\Delta\mu$ builds and reaches a static equilibrium value. For $B_z\neq0$, the spins precess at the Larmor frequency $\omega_L=g\mu_B B_z/\hbar$, resulting in a reduction of the net spin accumulation due to precessional dephasing (Hanle effect). Here, *g* is the Lande *g*-factor, μ_B the Bohr magneton, and $\hbar$ is the reduced Planck's constant. Thus, the voltage at the detector contact decreases as the magnitude of B_z increases with an approximately Lorentzian lineshape given by $\Delta\mu(B)=\Delta\mu(0)/[1+(\omega_L\tau_s)^2)]$, and the spin lifetime τ_s is obtained from fits to this lineshape.

The Hanle data showing spin accumulation at the $Ni_{80}Fe_{20}$/Al_2O_3/Si contact are shown in Figure 17.32b and c as a function of temperature. A clear signal due to spin accumulation is observed at room temperature, and increases at lower temperatures. The magnitude of the spin accumulation $\Delta\mu$ at the tunnel interface is obtained from the magnitude of the Hanle

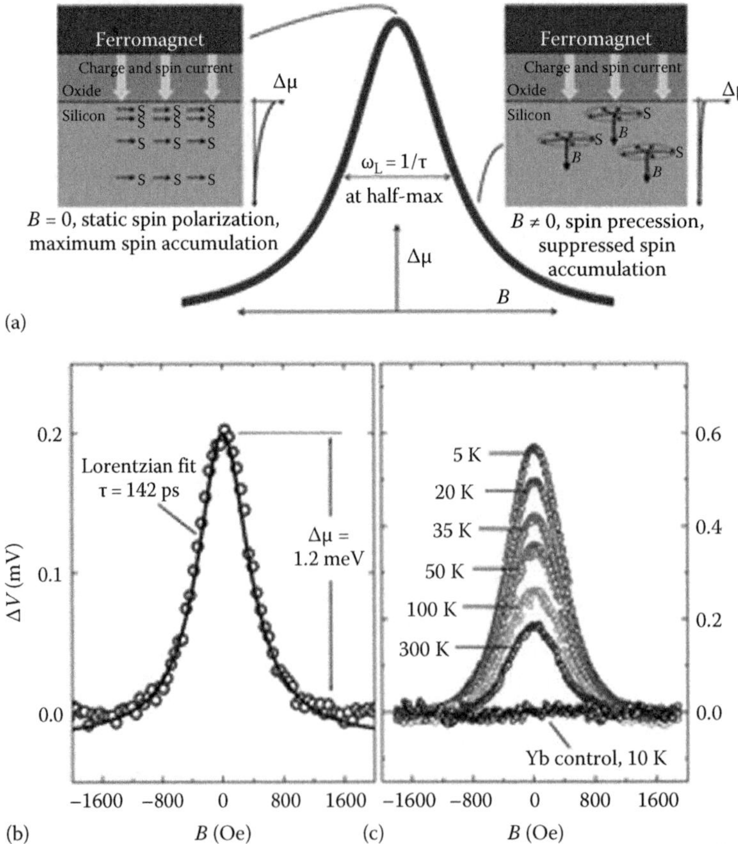

FIGURE 17.32 (a) Illustration of the Hanle effect, in which spin precesssion produced by a perpendicular magnetic field leads to a decay of the spin accumulation as measured by the voltage at the injection/detector terminal in the three terminal geometry. A plot of the voltage vs. magnetic field gives a pseudo-Lorentzian lineshape from which the spin lifetime can be determined. (b) Hanle curves shown for measurement temperatures of 5–300 K. A spin lifetime of 142 ps is obtained from the fit (solid line), and is independent of temperature. (Reprinted by permission from MacMillan Publishers Ltd. [*Nature*] Dash, S.P. et al., Electrical creation of spin polarization in silicon at room temperature, 462, 491, copyright 2009.)

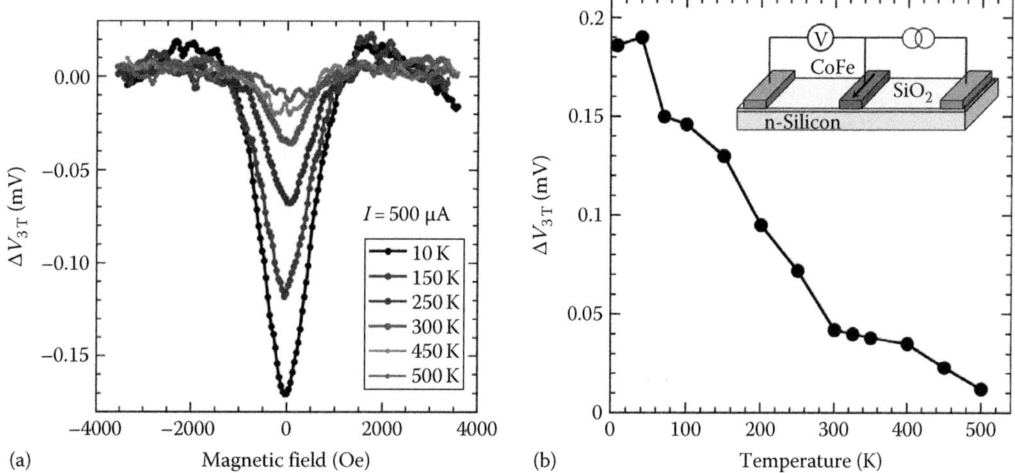

FIGURE 17.33 (a) Hanle curves obtained at various temperatures to 500 K for $Co_{0.9}Fe_{0.1}/SiO_2$ tunnel barrier contacts on Si(001). The amplitude decreases with increasing temperature. The linewidth and corresponding spin lifetime shows little variation with temperature. (b) Temperature dependence of the amplitude of the Hanle signal. The inset shows the three terminal measurement geometry.

signal $\Delta V = TSP \times \Delta\mu/2e$, where $TSP = 0.3$ is the known tunneling spin polarization [177] for $Al_2O_3/Ni_{80}Fe_{20}$ at 300 K. Thus, $\Delta\mu = 1.2$ meV at 300 K, and increases to a value of 3 meV at 5 K. Dash et al. [190] note that their measured voltage and these $\Delta\mu$ values are about two orders of magnitude larger than that expected from the model of Fert and Jaffrès for spin injection and accumulation in the semiconductor itself [91,94]. They suggest that lateral inhomogeneity of the tunnel current leads to higher localized spin-injection current densities, but an alternate explanation involving accumulation in localized interface states [184] must also be considered, as discussed below. Al_2O_3/Si interfaces formed by sputter- or electron-beam deposition of the oxide are known to exhibit large interface state densities.*

The data are well fit by a simple Lorentzian of the form described above, as shown by the solid line in Figure 17.32b. The full width at half maximum corresponds to $\omega_L = 1/\tau_s$, yielding a spin lifetime $\tau_s = 142$ ps at 300 K for the heavily doped *n*-Si with a measured electron density of $1.8 \times 10^{19}\ cm^{-3}$. The lifetime does not change significantly with temperature to 5 K. Similar results were obtained for *p*-doped Si, with the surprising result that the hole spin lifetimes deduced were a factor of two longer (~270 ps at 300 K) than the electron spin lifetimes, in marked contrast to what is typically reported for III–V semiconductors such as GaAs.

More recent work [192], has utilized $Co_{0.9}Fe_{0.1}/SiO_2$ and $Ni_{0.8}Fe_{0.2}/SiO_2$ tunnel barrier contacts on Si(001) [164] in a similar three terminal geometry, and observed spin accumulation and the Hanle effect to 500 K. This is a significant temperature milestone, since commercial and military electronic systems must operate at temperatures well above room temperature. The temperature 500 K easily exceeds the maximum temperature specification required to meet commercial (360 K), industrial (375 K), and even military (400 K) ratings. SiO_2 is highly desirable as a tunnel barrier for at least two reasons: (1) the SiO_2/Si interface is well developed and known to have a low level of interface trap states, and (2) the existing semiconductor processing infrastructure is very familiar and comfortable with this system, facilitating future transition.

The measurements were performed on silicon-on-insulator (SOI) samples with a doping levels of $n(As) = 3 \times 10^{18}\ cm^{-3}$ to $6 \times 10^{19}\ cm^{-3}$, above and below the metal–insulator transition. The SiO_2 tunnel barrier was formed in situ by plasma oxidation, followed immediately by sputter-beam deposition of a 10 nm $Co_{0.9}Fe_{0.1}$ film, which was subsequently patterned to define the magnetic contacts. Nonmagnetic reference contacts were deposited by liftoff. The Hanle data obtained are shown in Figure 17.33a. The signal amplitude is nearly 0.2 mV at low temperature, and signal is clearly visible at 500 K. The signal amplitude is in good agreement with that predicted by theory—the measured spin resistance-area product $A\Delta V_{3T}(B_z = 0)/I$ is within a factor of 2–3 of the simplest model, which assumes that the observed spin accumulation occurs in the bulk. (A is the contact area, I is the bias current, and $\Delta V_{3T}(B_z = 0)$ is the measured Hanle amplitude).

The Hanle curves exhibit the classic pseudo-Lorentzian lineshape expected, and can be fit very well with the Lorentzian model, described above. The spin lifetimes, thus, obtained are ~100 ps and essentially independent of temperature. This is similar to spin lifetimes measured in bulk Si at similar carrier densities of $n = 3.4 \times 10^{19}\ cm^{-3}$ (As), and attributed to the metallic character of the sample, where impurity dominated spin–orbit interaction dominates spin relaxation for a wide temperature range [193]. The amplitude of the signal $\Delta V_{3T}(B_z = 0)$ decreases monotonically with temperature, as shown in Figure 17.33b. This behavior is inconsistent with that of the spin lifetime,

* See, e.g., Ref. [191].

indicating that other factors, such as spin- injection/detection efficiency, may be responsible. Thus, spin precessional effects in this technologically dominant material are readily observed at the elevated temperatures necessary for utilization in commercial electronic systems.

17.6.4 Three Terminal Spin Injection and Detection in GaAs: Interpretation and the Role of Localized States

A clear interpretation of such three terminal Hanle data is still being developed. In contrast with the NLSV (four terminal) geometry, where the spin current *must flow through the semiconductor* before producing the spin accumulation enabling detection at a second contact, in the three terminal geometry it is not always clear that spin transport/accumulation occurs in the semiconductor itself.

Significant insight was provided by recent experiments of Tran et al. [184] who used the three terminal geometry and Hanle measurements with $Co/Al_2O_3/n$-GaAs(001) samples. The GaAs 50 nm thick channel was heavily n-doped (Si) at $5 \times 10^{18}\,cm^{-3}$ (well above the metal–insulator transition of $2 \times 10^{17}\,cm^{-3}$) on a semi-insulating substrate, with a 15 nm thick surface layer at $2 \times 10^{19}\,cm^{-3}$ to minimize the depletion region and facilitate spin injection. They observed a Hanle signal $\Delta V \sim 1.2$ meV at $T = 10$ K (Figure 17.34a), corresponding to a spin accumulation $\Delta\mu = 2e\,(\Delta V/\gamma) \sim 6$ meV (where $\gamma = 0.4$ is the tunnel spin polarization for Co/Al_2O_3), much larger than that expected from the predictions of the standard theory of spin injection in the diffusive regime for a single ferromagnet/semiconductor interface [91,94]. They concluded that their measured signal did *not* originate from spin accumulation and precession in the GaAs channel. Rather, states localized near the Al_2O_3/GaAs interface served as a spin accumulation layer with long spin lifetimes leading to exceptionally large signal levels, as illustrated in Figure 17.34b. These localized states participate in a sequential tunneling process, and act as a confining layer which enhances the spin accumulation. The thickness of the layer containing these states was estimated to be ~1 nm. The authors speculate that the origin of these states may be either ionized Si donors within the depletion layer or interface states at the Al_2O_3/GaAs interface. The oxide/GaAs interface is well-known to have a high density of interface states, which has thwarted the development of a metal-oxide-semiconductor transistor technology.* Thus, the voltage measured in such a three-terminal geometry will include contributions from spin accumulation in localized states associated with the tunnel barrier and corresponding interfaces, and may not demonstrate spin injection or accurately reflect spin accumulation in the semiconductor itself. Four terminal (NLSV) measurements from the same samples, in which the spin current must flow through the semiconductor to a second detector contact, should provide valuable insight and clarification.

* See, e.g., Ref. [194].

17.7 Outlook and Critical Research Issues

A great deal of progress has been made in spin injection and transport in semiconductors in the last 10 years. The contributions from many groups summarized here and elsewhere have rapidly advanced the state-of-the-art in electrical injection, detection, and spin manipulation in semiconductors of immediate technological interest such as GaAs and Si. There are several important areas that require further in-depth research.

Interfaces continue to play a critical role in at least two ways. First, they impact spin-injection efficiency from a given contact, which has been the primary topic of this chapter. Second, they impact spin lifetimes in lateral transport structures or in any thin film, nanoscale device. While spin lifetimes have been reasonably well studied in bulk Si, there is very little information available on these lifetimes near interfaces. In Si, the reduced symmetry will enable spin relaxation via mechanisms which are symmetry-forbidden in the bulk, such as D'yakonov–Perel'. For all semiconductors, the specific interface structure, defects and impurities, and band bending are expected to play a role, but the relative significance of these contributions is yet to be determined.

Contact resistance plays a major role in the relative efficiency of a spin-injecting contact, as illustrated by several model calculations and described in Sections 17.5.2 and 17.5.3. While a large value seems, at first glance, to be desirable for spin injection, this will severely limit the spin-polarized current flow through the contact into the semiconductor, and thereby limit the spin accumulation achieved. Therefore, a compromise must be reached. If one considers electrical *detection*, further constraints apply as discussed by Min et al. [177]—for the case of moderately doped Si ($n \sim 10^{16}\,cm^{-3}$), contact resistances of order 10^{-2}–$10^{-5}\ \Omega\ cm^2$ will be required. Similar considerations apply to other materials such as GaAs, although the specific values vary. These values are achievable even with tunnel barriers, but will be strongly case-specific and application-dependent.

Acknowledgments

It is a pleasure to acknowledge the many colleagues and collaborators who have made significant contributions through either experimental work, theory, or discussion: Brian Bennett, Steve Erwin, Aubrey Hanbicki, George Kioseoglou, Jim Krebs, Connie Li, Athos Petrou, Andre Petukhov, Philip Thompson, Olaf van't Erve, and many others. I wish to especially acknowledge Gary Prinz, who provided me the opportunity to enter and contribute to the field of magnetoelectronics. Various aspects of the work presented here were supported by the Office of Naval Research, DARPA and core program at the Naval Research Laboratory. This effort would not have been possible without

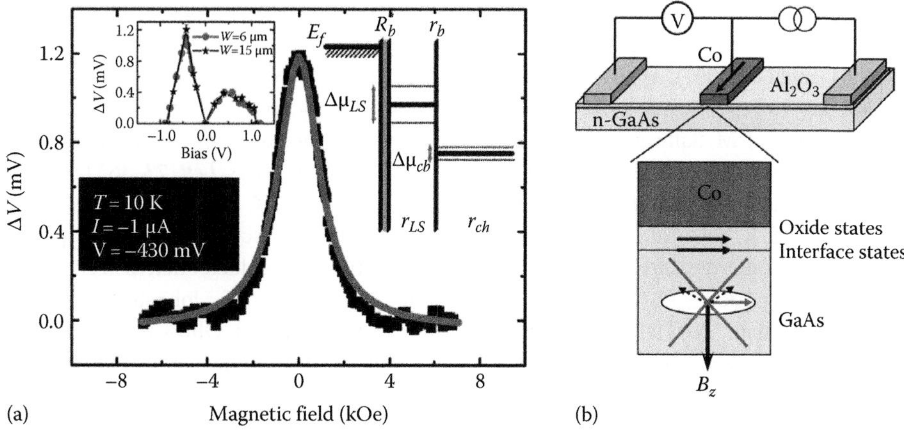

FIGURE 17.34 (a) Hanle signal obtained in a three terminal geometry for a $15 \times 196\,\mu m^2$ Co/Al_2O_3 tunnel barrier contact to GaAs. A small quadratic background has been subtracted. The solid line is a fit to the data using the model described assuming spin accumulation in interface states. The left inset shows the bias dependence of the Hanle signal, and the right inset illustrates the spin splitting of the electrochemical potential in the interfacial region. (Reprinted with permission from Tran, M., Jaffrès, H., Deranlot, C., George, J.-M., Fert, A., Miard, A., and Lemaitre, A., *Phys. Rev. Lett.*, 102, 036601. Copyright 2009 by the American Physical Society.) (b) Schematic of the measurement geometry and cross section of the region under the magnetic tunnel barrier contact, illustrating the location of interface states which the authors concluded produced the spin accumulation and strong Hanle signal measured.

the long-term basic research support of both the Office of Naval Research and NRL.

References

1. G. Moore, Cramming more components onto integrated circuits, *Electronics* **38** (April 19, 1965).
2. G. A. Prinz, Magnetoelectronics, *Science* **282**, 1660 (1998).
3. S. A. Wolf, D. D. Awschalom, R. A. Buhrman et al., Spintronics: A spin-based electronics vision for the future, *Science* **294**, 1448 (2001).
4. International Technology Roadmap for Semiconductors, 2009 edn., http://www.itrs.net. See *Executive Summary*, *Emerging Research Devices*, and *Emerging Research Materials*.
5. F. Meier and B. P. Zakharchenya (Eds.), *Optical Orientation*, Vol. 8, North-Holland, New York, 1984.
6. G. Lampel, Nuclear dynamic polarization by optical electronic saturation and optical pumping in semiconductors, *Phys. Rev. Lett.* **20**, 491 (1968).
7. S. A. Crooker, J. J. Baumberg, F. Flack, N. Samarth, and D. D. Awschalom, Terahertz spin precession and coherent transfer of angular momenta in magnetic quantum wells, *Phys. Rev. Lett.* **77**, 2814 (1996).
8. M. J. Stevens, A. L. Smirl, R. D. R. Bhat, A. Najmaie, J. E. Sipe, and H. M. van Driel, Quantum interference control of ballistic pure spin currents in semiconductors, *Phys. Rev. Lett.* **90**, 136603 (2003).
9. S. F. Alvarado and P. Renaud, Observation of spin-polarized-electron tunneling from a ferromagnet into GaAs, *Phys. Rev. Lett.* 68, 1387 (1992).
10. B. T. Jonker, Progress toward electrical injection of spin-polarized electrons into semiconductors, *Proc. IEEE* **91**, 727 (2003).
11. B. T. Jonker, S. C. Erwin, A. Petrou, and A. G. Petukhov, Electrical spin injection and transport in semiconductor spintronic devices, *MRS Bull.* **28**, 740–748 (2003).
12. B. T. Jonker and M. E. Flatte, Electrical spin injection and transport in semiconductors, in *Nanomagnetism: Ultrathin Films, Multilayers and Nanostructures*, D. L. Mills and J. A. C. Bland (Eds.), in *Contemporary Concepts of Condensed Matter Science*, E. Burstein, M. L. Cohen, D. L. Mills, and P. J. Stiles (Eds.), Elsevier, Amsterdam, the Netherlands, p. 227, 2006.
13. C. Weisbuch and B. Vinter, *Quantum Semiconductor Structures*, Academic Press, New York, p. 65, 1991.
14. A. G. Aronov and G. E. Pikus, Spin injection into semiconductors, *Fiz. Tekh. Poluprovodn.* **10**, 1177 (1976) [*Sov. Phys. Semicond.* **10**, 698 (1976)]; A. G. Aronov, *Pis'ma Zh. Eksp. Teor. Fiz.* **24**, 37 (1976) [*JETP Lett.* **24**, 32 (1976)].
15. D. Chiba, M. Sawicki, Y. Nishitani, Y. Nakatani, F. Matsukura, and H. Ohno, Magnetization vector manipulation by electric fields, *Nature* **455**, 515 (2008).
16. C. Haas, Magnetic semiconductors, *CRC Crit. Rev. Solid State Sci.* **1**, 47 (1970).
17. E. L. Nagaev, Photoinduced magnetism and conduction electrons in magnetic semiconductors, *Phys. Stat. Sol. B* **145**, 11 (1988).
18. L. Esaki, P. J. Stiles, and S. von Molnar, Magnetointernal field emission in junctions of magnetic insulators, *Phys. Rev. Lett.* **19**, 852 (1967).
19. J. S. Moodera, X. Hao, G. A. Gibson, and R. Meservey, Electron–spin polarization in tunnel junctions in zero applied field with ferromagnetic EuS barriers, *Phys. Rev. Lett.* **61**, 637 (1988).
20. J. S. Moodera, R. Meservey, and X. Hao, Variation of the electron spin polarization in EuSe tunnel junctions from zero to near 100% in a magnetic field, *Phys. Rev. Lett.* **70**, 853 (1993).

21. Y. D. Park, A. T. Hanbicki, J. E. Mattson, and B. T. Jonker, Epitaxial growth of an n-type ferromagnetic semiconductor $cdCr_2Se_4$ on GaAs (001) and GaP (001), *Appl. Phys. Lett.* **81**, 1471–1473 (2002).
22. G. Kioseoglou, A. T. Hanbicki, J. M. Sullivan et al., Electrical spin injection from an n-type ferromagnetic semiconductor into a III–V device heterostructure, *Nat. Mater.* **3**, 799 (2004).
23. J. Lettieri, V. Vaithyanathan, S. K. Eah et al., Epitaxial growth and magnetic properties of EuO on (001)Si by molecular beam epitaxy, *Appl. Phys. Lett.* **83**, 975 (2003).
24. A. Schmehl, V. Vaithyanathan, A. Herrnberger et al., Epitaxial integration of the highly spin-polarized ferromagnetic semiconductor EuO with silicon and GaN, *Nat. Mater.* **6**, 882–887 (2007).
25. J. K. Furdyna and J. Kossut, *Diluted Magnetic Semiconductors*, Semiconductors and Semimetals, Vol. 25, R. K. Willardson and A. C. Beer (series Eds.), Academic Press, New York, 1988.
26. N. Samarth and J. K. Furdyna, Diluted magnetic semiconductors, *Proc. IEEE* **78**, 990 (1990).
27. M. Oestreich, J. Hubner, D. Hagele et al., Spin injection into semiconductors, *Appl. Phys. Lett.* **74**, 1251 (1999).
28. K. Onodera, T. Masumoto, and M. Kimura, 980 nm compact optical isolators using $Cd_{1-x-y}Mn_xHg_yTe$ single crystals for high power pumping laser diodes, *Electron. Lett.* **30**, 1954 (1994).
29. H. Munekata, H. Ohno, S. von Molnar, A. Segmüller, L. L. Chang, and L. Esaki, Diluted magnetic III–V semiconductors, *Phys. Rev. Lett.* **63**, 1849 (1989).
30. J. De Boeck, R. Oesterholt, A. Van Esch et al., Nanometer-scale magnetic MnAs particles in GaAs grown by molecular beam epitaxy, *Appl. Phys. Lett.* **68**, 2744 (1996).
31. H. Ohno, A. Shen, F. Matsukura et al., (Ga, Mn)As: A new diluted magnetic semiconductor based on GaAs, *Appl. Phys. Lett.* **69**, 363 (1996).
32. S. J. Potashnik, K. C. Ku, S. H. Chun, J. J. Berry, N. Samarth, and P. Schiffer, Effects of annealing time on defect-controlled ferromagnetism in $Ga_{1-x}Mn_xAs$, *Appl. Phys. Lett.* **79**, 1495 (2001).
33. K. W. Edmonds, K. Y. Wang, R. P. Campion et al., High-Curie-temperature $Ga_{1-x}Mn_xAs$ obtained by resistance-monitored annealing, *Appl. Phys. Lett.* **81**, 4991 (2002).
34. T. Hayashi, Y. Hashimoto, S. Katsumota, and Y. Iye, Effect of low-temperature annealing on transport and magnetism of diluted magnetic semiconductor (Ga, Mn)As, *Appl. Phys. Lett.* **78**, 1691 (2001).
35. T. Slupinski, A. Oiwa, S. Yanagi, and H. Munekata, Preparation of ferromagnetic (In, Mn)As with relatively low hole concentration and Curie temperature 50 K, *J. Cryst. Growth* **237–239**, 1326 (2002).
36. M. Tanaka, Spintronics: Recent progress and tomorrow's challenges, *J. Cryst. Growth* **278**, 25 (2005); A. M. Nazmul, T. Amemiya, Y. Shuto, S. Sugahara, and M. Tanaka, High temperature ferromagnetism in GaAs-based heterostructures with Mn delta doping, *Phys. Rev. Lett.* **95**, 017201 (2005).
37. H. Ohno, Making nonmagnetic semiconductors ferromagnetic, *Science* **281**, 951 (1998).
38. H. Ohno, N. Akiba, F. Matsukura, A. Shen, K. Ohtani, and Y. Ohno, Spontaneous splitting of ferromagnetic (Ga,Mn) As valence band observed by resonant tunneling spectroscopy, *Appl. Phys. Lett.* **73**, 363 (1998).
39. Y. Ohno, D. K. Young, B. Beschoten, F. Matsukura, H. Ohno, and D. D. Awschalom, Electrical spin injection in a ferromagnetic semiconductor heterostructure, *Nature* **402**, 790 (1999).
40. D. K. Young, E. Johnston-Halperin, D. D. Awschalom, Y. Ohno, and H. Ohno, Anisotropic electrical spin injection in ferromagnetic semiconductor heterostructures, *Appl. Phys. Lett.* **80**, 1598 (2002).
41. M. Yamanouchi, D. Chiba, F. Matsukura, and H. Ohno, Current-induced domain-wall switching in a ferromagnetic semiconductor structure, *Nature* **428**, 539 (2004).
42. H. Ohno, D. Chiba, F. Matsukura et al., Electric field control of ferromagnetism, *Nature* **408**, 944 (2000).
43. Y. D. Park, A. T. Hanbicki, S. C. Erwin et al., A group-IV ferromagnetic semiconductor: Mn_xGe_{1-x}, *Science* **295**, 651 (2002).
44. A. M. Nazmul, S. Kobayashi, S. Sugahara, and M. Tanaka, Control of ferromagnetism in Mn delta doped GaAs based semiconductor heterostructures, *Physica E* **21**, 937–943 (2004).
45. T. Dietl, H. Ohno, F. Matsukura, J. Cibert, and D. Ferrand, Zener model description of ferromagnetism in zinc-blende magnetic semiconductors, *Science* **287**, 1019 (2000).
46. T. Dietl, H. Ohno, and F. Matsukura, Hole-mediated ferromagnetism in tetrahedrally coordinated semiconductors, *Phys. Rev. B* **63**, 195205 (2001).
47. A. H. MacDonald, P. Schiffer, and N. Samarth, Ferromagnetic semiconductors: Moving beyond (Ga,Mn)As, *Nat. Mater.* **4**, 195 (2005).
48. H. Raebiger, S. Lany, and Z. Zunger, Control of ferromagnetism via electron doping in In_2O_3:Cr, *Phys. Rev. Lett.* **101**, 027203 (2008).
49. B. T. Jonker, Polarized optical emission due to decay or recombination of spin-polarized injected carriers, U.S. patent 5,874,749 (filed June 23, 1993, awarded February 23, 1999 to U.S. Navy).
50. G. Bastard, *Wave Mechanics Applied to Semiconductor Heterostructures*, Halsted, New York, p. 109, 1988.
51. D. Hagele, M. Oestreich, W. W. Ruhle, N. Nestle, and K. Eberl, Spin transport in GaAs, *Appl. Phys. Lett.* **73**, 1580 (1998).
52. M. I. D'yakonov and V. I. Perel', Theory of optical spin orientation of electrons and nuclei in semiconductors, in *Optical Orientation*, F. Meier and B. P. Zakharchenya (Eds.), North-Holland, Amsterdam, the Netherlands, Chapter 2, p. 27, 1984.
53. R. W. Martin, R. J. Nicholas, G. J. Rees, S. K. Haywood, N. J. Mason, and P. J. Walker, Two-dimensional spin confinement in strained-layer quantum wells, *Phys. Rev. B* **42**, 9237 (1990).

54. S. A. Crooker, D. D. Awschalom, J. J. Baumberg, F. Flack, and N. Samarth, Optical spin resonance and transverse spin relaxation in magnetic semiconductor quantum wells, *Phys. Rev. B* **56**, 7574 (1997).
55. R. Fiederling, P. Grabs, W. Ossau, G. Schmidt, and L. W. Molenkamp, Detection of electrical spin injection by light-emitting diodes in top- and side-emission configurations, *Appl. Phys. Lett.* **82**, 2160 (2003).
56. O. M. J. van't Erve, G. Kioseoglou, A. T. Hanbicki, C. H. Li, and B. T. Jonker, Remanent electrical spin injection from Fe into AlGaAs/GaAs light emitting diodes, *Appl. Phys. Lett.* **89**, 072505 (2006).
57. A. Tackeuchi, O. Wada, and Y. Nishikawa, Electron spin relaxation in InGaAs/InP multiple quantum wells, *Appl. Phys. Lett.* **70**, 1131 (1997).
58. A. F. Isakovic, D. M. Carr, J. Strand, B. D. Schultz, C. J. Palmstrom, and P. A. Crowell, Optical pumping in ferromagnet-semiconductor heterostructures: Magneto-optics and spin transport, *Phys. Rev. B* **64**, 161304(R) (2001).
59. B. T. Jonker, A. T. Hanbicki, Y. D. Park et al., Quantifying electrical spin injection: Component-resolved electroluminescence from spin-polarized light-emitting diodes, *Appl. Phys. Lett.* **79**, 3098 (2001).
60. G. Kioseoglou, A. T. Hanbicki, B. T. Jonker, and A. Petrou, Bias-controlled hole degeneracy and implications for quantifying spin polarization, *Appl. Phys. Lett.* **87**, 122503 (2005).
61. R. Fiederling, M. Kelm, G. Reuscher et al., Injection and detection of a spin-polarized current in a light-emitting diode, *Nature* **402**, 787 (1999).
62. B. T. Jonker, Y. D. Park, B. R. Bennett, H. D. Cheong, G. Kioseoglou, and A. Petrou, Robust electrical spin injection into a semiconductor heterostructure, *Phys. Rev. B* **62**, 8180 (2000).
63. D. K. Young, J. A. Gupta, E. Johnston-Halperin, R. Epstein, Y. Kato, and D. D. Awschalom, Optical, electrical and magnetic manipulation of spins in semiconductors, *Semicond. Sci. Technol.* **17**, 275 (2002).
64. R. M. Stroud, A. T. Hanbicki, Y. D. Park et al., Reduction of spin injection efficiency by interface defect spin scattering in ZnMnse/AlGaAs-GaAs spin-polarized light emitting diodes, *Phys. Rev. Lett.* **89**, 166602 (2002).
65. P. R. Hammar, B. R. Bennett, M. J. Yang, and M. Johnson, Observation of spin injection at a ferromagnet–semiconductor interface, *Phys. Rev. Lett.* **83**, 203 (1999).
66. C.-M. Hu, J. Nitta, A. Jensen, J. B. Hansen, and H. Takayanagi, Spin-polarized transport in a two dimensional electron gas with interdigital-ferromagnetic contacts, *Phys. Rev. B* **63**, 125333 (2001).
67. S. Gardelis, C. G. Smith, C. H. W. Barnes, E. H. Linfield, and D. A. Ritchie, Spin-valve effects in a semiconductor field-effect transistor: A spintronic device, *Phys. Rev. B* **60**, 7764 (1999).
68. F. G. Monzon, H. X. Tang, and M. L. Roukes, Magnetoelectronic phenomena at a ferromagnet–semiconductor interface, *Phys. Rev. Lett.* **84**, 5022 (2000).
69. B. J. van Wees, Comment on observation of spin injection at a ferromagnet–semiconductor interface, *Phys. Rev. Lett.* **84**, 5023 (2000).
70. A. T. Filip, B. H. Hoving, F. J. Jedema, B. J. van Wees, B. Dutta, and S. Borghs, Experimental search for the electrical spin injection in a semiconductor, *Phys. Rev. B* **62**, 9996 (2000).
71. H. J. Zhu, M. Ramsteiner, H. Kostial, M. Wassermeier, H.-P. Schönherr, and K. H. Ploog, Room temperature spin injection from Fe into GaAs, *Phys. Rev. Lett.* **87**, 016601 (2001).
72. R. C. Miller, D. A. Kleinman, W. A. Nordland Jr., and A. C. Gossard, Luminescence studies of optically pumped quantum wells in GaAs-$Al_xGa_{1-x}As$ multilayer structures, *Phys. Rev. B* **22**, 863 (1980).
73. W. H. Butler, X.-G. Zhang, X. Wang, J. van Ek, and J. M. MacLaren, Electronic structure of FM/semiconductor/FM spin tunneling structures, *J. Appl. Phys.* **81**, 5518 (1997).
74. J. M. MacLaren, W. H. Butler, and X.-G. Zhang, Spin-dependent tunneling in epitaxial systems: Band dependence of conductance, *J. Appl. Phys.* **83**, 6521 (1998).
75. J. M. MacLaren, X.-G. Zhang, W. H. Butler, and X. Wang, Layer KKR approach to Bloch-wave transmission and reflection: Application to spin-dependent tunneling, *Phys. Rev. B* **59**, 5470 (1999).
76. O. Wunnicke, Ph. Mavropoulos, R. Zeller, P. H. Dederichs, and D. Grundler, Ballistic spin injection from Fe(001) into ZnSe and GaAs, *Phys. Rev. B* **65**, 241306(R) (2002).
77. P. Mavropoulos, O. Wunnicke, and P. H. Dederichs, Ballistic spin injection and detection in Fe/semiconductor/Fe junctions, *Phys. Rev. B* **66**, 024416 (2002).
78. J. R. Chelikowsky and M. L. Cohen, Nonlocal pseudopotential calculations for the electronic structure of eleven diamond and zinc-blende semiconductors, *Phys. Rev. B* **14**, 556 (1976).
79. S. Vutukuri, M. Chshiev, and W. H. Butler, Spin-dependent tunneling in FM/semiconductor/FM structures, *J. Appl. Phys.* **99**, 08K302 (2006).
80. P. Mavropoulos, Spin injection from Fe into Si(001): Ab initio calculations and role of the Si complex band structure, *Phys. Rev. B* **78**, 054446 (2008).
81. M. Zwierzycki, K. Xia, P. J. Kelly, G. E. W. Bauer, and I. Turek, Spin injection through an Fe/InAs interface, *Phys. Rev. B* **67**, 092401 (2003).
82. W. H. Butler, X.-G. Zhang, T. C. Schulthess, and J. M. MacLaren, Spin-dependent tunneling conductance of Fe/MgO/Fe sandwiches, *Phys. Rev. B* **63**, 054416 (2001).
83. T. Valet and A. Fert, Theory of the perpendicular magnetoresistance in magnetic multilayers, *Phys. Rev. B* **48**, 7099 (1993).
84. B. T. Jonker, A. T. Hanbicki, D. T. Pierce, and M. D. Stiles, Spin nomenclature for semiconductors and magnetic metals, *J. Magn. Magn. Mater.* **277**, 24 (2004).
85. R. A. de Groot, New class of materials: Half-metallic ferromagnets, *Phys. Rev. Lett.* **50**, 2024 (1983).

86. W. E. Pickett and J. S. Moodera, Half metallic magnets, *Phys. Today* **54**(5), 39 (May 2001).
87. D. Orgassa, H. Fujiwara, T. C. Schulthess, and W. H. Butler, First-principles calculation of the effect of atomic disorder on the electronic structure of the half-metallic ferromagnet NiMnSb, *Phys. Rev. B* **60**, 13237 (1999).
88. G. Schmidt, D. Ferrand, L. W. Molenkamp, A. T. Filip, and B. J. van Wees, Fundamental obstacle for electrical spin injection from a ferromagnetic metal into a diffusive semiconductor, *Phys. Rev. B* **62**, R4790–R4793 (2000).
89. E. I. Rashba, Theory of electrical spin injection: Tunnel contacts as a solution of the conductivity mismatch problem, *Phys. Rev. B* **62**, R16267–R16270 (2000).
90. D. L. Smith and R. N. Silver, Electrical spin injection into semiconductors, *Phys. Rev. B* **64**, 045323 (2001).
91. A. Fert and H. Jaffrès, Conditions for efficient spin injection from a ferromagnetic metal into a semiconductor, *Phys. Rev. B* **64**, 184420 (2001).
92. Z. G. Yu and M. Flatte, Electric-field dependent spin diffusion and spin injection into semiconductors, *Phys. Rev. B* **66**, 201202(R) (2002).
93. P. C. van Son, H. van Kempen, and P. Wyder, Boundary resistance of the ferromagnetic-nonferromagnetic metal interface, *Phys. Rev. Lett.* **58**, 2271 (1987).
94. A. Fert, J.-M. George, H. Jaffrès, and R. Mattana, Semiconductors between spin-polarized source and drain, *IEEE Trans. Electron. Dev.* **54**, 921 (2007).
95. E. H. Rhoderick and R. H. Williams, *Metal-Semiconductor Contacts*, 2nd edn., Clarendon, Oxford, U.K., 1988.
96. M. Ilegems, Properties of III–V layers, in *The Technology and Physics of Molecular Beam Epitaxy*, E. H. C. Parker (Ed.), Plenum, New York, p. 119, 1985.
97. S. M. Sze, *Physics of Semiconductor Devices*, 2nd edn., Wiley, New York, p. 294, 1981.
98. A. T. Hanbicki, B. T. Jonker, G. Itskos, G. Kioseoglou, and A. Petrou, Efficient electrical spin injection from a magnetic metal/tunnel barrier contact into a semiconductor, *Appl. Phys. Lett.* **80**, 1240 (2002).
99. A. T. Hanbicki, O. M. J. van't Erve, R. Magno, G. Kioseoglou, C. H. Li, and B. T. Jonker, Analysis of the transport process providing spin injection through an Fe/AlGaAs Schottky barrier, *Appl. Phys. Lett.* **82**, 4092 (2003).
100. J. M. Kikkawa and D. D. Awschalom, Resonant spin amplification in n-type GaAs, *Phys. Rev. Lett.* **80**, 4313 (1998).
101. R. I. Dzhioev, K. V. Kavokin, V. L. Korenev et al., Low-temperature spin relaxation in n-type GaAs, *Phys. Rev. B* **66**, 245204 (2002).
102. B. J. Jönsson-Åkerman, R. Escudero, C. Leighton, S. Kim, I. K. Schuller, and D. A. Rabson, Reliability of normal-state current-voltage characteristics as an indicator of tunnel-junction barrier quality, *Appl. Phys. Lett.* **77**, 1870 (2000).
103. D. A. Rabson, B. J. Jönsson-Åkerman, A. H. Romero et al., Pinholes may mimic tunneling, *J. Appl. Phys.* **89**, 2786 (2001).
104. U. Rüdiger, R. Calarco, U. May et al., Temperature dependent resistance of magnetic tunnel junctions as a quality proof of the barrier, *J. Appl. Phys.* **89**, 7573 (2001).
105. E. Burstein and S. Lundqvist (Eds.), *Tunneling Phenomena in Solids*, Plenum, New York, 1969.
106. J. G. Simmons, Generalized formula for the electric tunnel effect between similar electrodes separated by a thin insulating film, *J. Appl. Phys.* **34**, 1793 (1963).
107. W. F. Brinkman, R. C. Dynes, and J. M. Rowell, Tunneling conductance of asymmetrical barriers, *J. Appl. Phys.* **41**, 1915 (1970).
108. S. H. Chun, S. J. Potashnik, K. C. Ku, P. Schiffer, and N. Samarth, Spin-polarized tunneling in hybrid metal-semiconductor magnetic tunnel junctions, *Phys. Rev. B* **66**, 100408(R) (2002).
109. E. L. Wolf, *Principles of Electron Tunneling Spectroscopy*, Oxford, New York, 1989.
110. A. M. Andrews, H. W. Korb, N. Holonyak, C. B. Duke, and G. G. Kleiman, *Phys. Rev. B* **5**, 2273 (1972).
111. J. J. Krebs and G. A. Prinz, *NRL Memo Rep. 3870*, Naval Research Lab, Washington, DC, 1978.
112. R. Meservey and P. M. Tedrow, Spin-polarized electron tunneling, *Phys. Rep.* **238**, 173–234 (1999). See 200–202.
113. C. Kittel, *Introduction to Solid State Physics*, 5th edn., Wiley, New York, pp. 438–439, 1976.
114. N. W. Ashcroft and N. D. Mermin, *Solid State Physics*, Holt, Rinehart, Winston, New York, pp. 654, 661, 1976.
115. M. B. Stearns, Simple explanation of tunneling spin polarization of Fe, Co, Ni and its alloys, *J. Magn. Magn. Mater.* **5**, 167–171 (1977).
116. J. M. De Teresa, A. Barthelemy, A. Fert, J. P. Contour, F. Montaigne, and P. Seneor, Role of metal-oxide interface in determining the spin polarization of magnetic tunnel junctions, *Science* **286**, 507 (1999).
117. H. Dery and L. J. Sham, Spin extraction theory and its relevance to spintronics, *Phys. Rev. Lett.* **98**, 046602 (2007).
118. A. N. Chantis, K. D. Belashchenko, D. L. Smith, E. Y. Tsymbal, M. van Schilfgaarde, and R. C. Albers, Reversal of spin polarization in Fe/GaAs(001) driven by resonant surface states: First-principles calculations, *Phys. Rev. Lett.* **99**, 196603 (2007).
119. T. J. Zega, A. T. Hanbicki, S. C. Erwin et al., Determination of interface atomic structure and its impact on spin transport using Z-contrast microscopy and density-functional theory, *Phys. Rev. Lett.* **96**, 196101 (2006).
120. S. C. Erwin, S.-H. Lee, and M. Scheffler, First-principles study of nucleation, growth, and interface structure of Fe/GaAs, *Phys. Rev. B* **65**, 205422 (2002).
121. E. J. Kirkland, *Advanced Computing in Electron Microscopy*, Plenum, New York, 1998.
122. E. Kneedler, P. M. Thibado, B. T. Jonker et al., Epitaxial growth, structure, and composition of Fe films on GaAs(001)-2 × 4, *J. Vac. Sci. Technol. B* **14**, 3193 (1996).
123. E. M. Kneedler and B. T. Jonker, Kerr effect study of the onset of magnetization in Fe films on GaAs(001)-2 × 4, *J. Appl. Phys.* **81**, 4463 (1997).

124. E. M. Kneedler, B. T. Jonker, P. M. Thibado, R. J. Wagner, B. V. Shanabrook, and L. J. Whitman, Influence of substrate surface reconstruction on the growth and magnetic properties of Fe on GaAs(001), *Phys. Rev. B* **56**, 8163 (1997).
125. P. M. Thibado, E. Kneedler, B. T. Jonker, B. R. Bennett, B. V. Shanabrook, and L. J. Whitman, Nucleation and growth of Fe on GaAs(001)-2×4 studied by scanning tunneling microscopy, *Phys. Rev. B* **53**, 10481 (1996).
126. B. T. Jonker, O. J. Glembocki, R. T. Holm, and R. J. Wagner, Enhanced carrier lifetimes and suppression of midgap states in GaAs at a magnetic metal interface, *Phys. Rev. Lett.* **79**, 4886 (1997).
127. C. S. Gworek, P. Phatak, B. T. Jonker, E. R. Weber, and N. Newman, Pressure dependence of Cu, Ag, and Fe n-GaAs Schottky barrier heights, *Phys. Rev. B* **64**, 045322 (2001).
128. B. T. Jonker, Electrical spin injection into semiconductors, in *Ultrathin Magnetic Structures IV: Applications of Nanomagnetism*, B. Heinrich and J. A. C. Bland (Eds.), Springer, Berlin, Germany, 2005.
129. S. A. Crooker, M. Furis, X. Lou et al., Imaging spin transport in lateral ferromagnet/semiconductor structures, *Science* **309**, 2191 (2005).
130. X. Lou, C. Adelmann, M. Furis, S. A. Crooker, C. J. Palmstrøm, and P. A. Crowell, Electrical detection of spin accumulation at a ferromagnet-semiconductor interface, *Phys. Rev. Lett.* **96**, 176603 (2006).
131. X. Lou, C. Adelmann, S. A. Crooker et al., Electrical detection of spin transport in lateral ferromagnet-semiconductor devices, *Nat. Phys.* **3**, 197 (2007)
132. M. C. Hickey, C. D. Damsgaard, S. N. Holmes et al., Spin injection from Co_2MnGa into an InGaAs quantum well, *Appl. Phys. Lett.* **92**, 232101 (2008).
133. M. Holub, J. Shin, D. Saha, and P. Battacharya, Electrical spin injection and threshold reduction in a semiconductor laser, *Phys. Rev. Lett.* **98**, 146603 (2007).
134. P. Kotissek, M. Bailleul, M. Sperl et al., Cross-sectional imaging of spin injection into a semiconductor, *Nat. Phys.* **3**, 872 (2007).
135. A. T. Hanbicki, G. Kioseoglou, M. A. Holub, O. M. J. van't Erve, and B. T. Jonker, Electrical spin injection from Fe into ZnSe(001), *Appl. Phys. Lett.* **94**, 082507 (2009).
136. G. Kioseoglou, A. T. Hanbicki, R. Goswami et al., Electrical spin injection into Si: A comparison between Fe/Si Schottky and Fe/Al_2O_3 tunnel contacts, *Appl. Phys. Lett.* **94**, 122106 (2009).
137. R. Meservey and P. M. Tedrow, Spin polarized electron tunneling, *Phys. Rep.* **238**, 173–234 (1994).
138. J. S. Moodera, L. R. Kinder, T. M. Wong, and R. Meservey, Large magnetoresistance at room temperature in ferromagnetic thin film tunnel junctions, *Phys. Rev. Lett.* **74**, 3273 (1995).
139. J. Daughton, Magnetic tunneling applied to memory, *J. Appl. Phys.* **81**, 3758 (1997).
140. S. S. P. Parkin, C. Kaiser, A. Panchula et al., Giant tunneling magnetoresistance at room temperature with MgO(100) tunnel barriers, *Nat. Mater.* **3**, 862 (2004).
141. S. Yuasa, T. Nagahama, A. Fukushima, Y. Suzuki, and K. Ando, Giant room temperature magnetoresistance in single crystal Fe/MgO/Fe magnetic tunnel junctions, *Nat. Mater.* **3**, 868 (2004).
142. J. S. Moodera, J. Nowak, L. R. Kinder et al., Quantum well states in spin-dependent tunnel structures, *Phys. Rev. Lett.* **83**, 3029 (1999).
143. D. J. Monsma and S. S. P. Parkin, Spin polarization of tunneling current from ferromagnet/Al_2O_3 interfaces using copper-doped aluminum superconducting films, *Appl. Phys. Lett.* **77**, 720 (2000).
144. V. F. Motsnyi, J. De Boeck, J. Das et al., Electrical spin injection in a ferromagnet/tunnel barrier/semiconductor heterostructure, *Appl. Phys. Lett.* **81**, 265 (2002).
145. V. F. Motsnyi, P. Van Dorpe, W. Van Roy et al., Optical investigation of electrical spin injection into semiconductors, *Phys. Rev. B* **68**, 245319 (2003).
146. T. Manago and H. Akinaga, Spin-polarized light-emitting diode using metal/insulator/semiconductor structures, *Appl. Phys. Lett.* **81**, 694 (2002).
147. O. M. J. van't Erve, G. Kioseoglou, A. T. Hanbicki et al., Comparison of Fe/Schottky and Fe/Al_2O_3 tunnel barrier contacts for electrical spin injection into GaAs, *Appl. Phys. Lett.* **84**, 4334 (2004).
148. B. T. Jonker, G. Kioseoglou, A. T. Hanbicki, C. H. Li, and P. E. Thompson, Electrical spin injection into silicon from a ferromagnetic metal/tunnel barrier contact, *Nat. Phys.* **3**, 542 (2007).
149. G. Feher and E. A. Gere, Electron spin resonance experiments on donors in silicon: II. Electron spin relaxation effects, *Phys. Rev.* **114**, 1245–1256 (1959).
150. V. Zarifis and T. G. Castner, ESR linewidth behavior for barely metallic n-type silicon, *Phys. Rev. B* **36**, 6198–6201 (1987).
151. A. M. Tyryshkin, S. A. Lyon, W. Jantsch, and F. A. Schaffler, Spin manipulation of free two-dimensional electrons in Si/SiGe quantum wells, *Phys. Rev. Lett.* **94**, 126802 (2005).
152. S. Sugahara and M. Tanaka, A spin metal-oxide-semiconductor field effect transistor using half-metallic contacts for the source and drain, *Appl. Phys. Lett.* **84**, 2307–2309 (2004).
153. C. L. Dennis, C. V. Tiusan, J. F. Gregg, G. J. Ensell, and S. M. Thompson, Silicon spin diffusion transistor: Materials, physics and device characteristics, *IEEE Proc. Circuits Dev. Syst.* **152**, 340–354 (2005).
154. E. Yablonovitch, H. W. Jiang, H. Kosaka, H. D. Robinson, D. S. Rao, and T. Szkopek, Quantum repeaters based on Si/SiGe heterostructures, *Proc. IEEE* **91**, 761–780 (2003).
155. M. L. W. Thewalt, S. P. Watkins, U. O. Ziemelis, E. C. Lightowlers, and M. O. Henry, Photoluminescence lifetime, absorption and excitation spectroscopy measurements on isoelectronic bound excitons in beryllium-doped silicon, *Solid State Commun.* **44**, 573–577 (1982).
156. P. Yu and M. Cardona, *Fundamentals of Semiconductors*, Springer-Verlag, Berlin, Germany, 1996. ISBN 3-540-25470-6.

157. I. Žutić, J. Fabian, and S. C. Erwin, Spin injection and detection in silicon, *Phys. Rev. Lett.* **97**, 026602 (2006).
158. J. C. Costa, F. Williamson, T. J. Miller et al., Barrier height variation in Al/GaAs Schottky diodes with a thin silicon interfacial layer, *Appl. Phys. Lett.* **58**, 382–384 (1991).
159. P. Li and H. Dery, Theory of spin-dependent phonon-assisted optical transitions in silicon, *Phys. Rev. Lett.* **105**, 037204 (2010).
160. G. Kioseoglou, A. T. Hanbicki, R. Goswami et al., Electrical spin injection into Si: A comparison between Fe/Si Schottky and Fe/Al_2O_3 tunnel contacts, *Appl. Phys. Lett.* **94**, 122106 (2009).
161. L. Grenet, M. Jamet, P. Noé et al., Spin injection in silicon at zero magnetic field, *Appl. Phys. Lett.* **94**, 032502 (2009).
162. D. J. Smith, M. R. McCartney, C. L. Platt, and A. E. Berkowitz, Structural characterization of thin film ferromagnetic tunnel junctions, *J. Appl. Phys.* **83**, 5154 (1998).
163. B. G. Park, T. Banerjee, B. C. Min, J. C. Lodder, and R. Jansen, Tunnel spin polarization of $Ni_{80}Fe_{20}/SiO_2$ probed with a magnetic tunnel transistor, *Phys. Rev. B* **73**, 172402 (2006).
164. C. Li, G. Kioseoglou, O. M. J. van't Erve, P. E. Thompson and B. T. Jonker, Electrical spin injection into Si(001) through an SiO_2 tunnel barrier, *Appl. Phys. Lett.* **95**, 172102 (2009).
165. X. Jiang, R. Wang, R. M. Shelby et al., Highly spin polarized room temperature tunnel injector for semiconductor spintronics using MgO(100), *Phys. Rev. Lett.* **94**, 056601 (2005).
166. G. Salis, R. Wang, X. Jiang et al., Temperature independence of the spin-injection efficiency of a MgO-based tunnel spin injector, *Appl. Phys. Lett.* **87**, 262503 (2005).
167. Y. Lu, V. G. Truong, P. Renucci et al., MgO thickness dependence of spin injection efficiency in spin-light emitting diodes, *Appl. Phys. Lett.* **93**, 152102 (2008).
168. V. G. Truong, P.-H. Binh, P. Renucci et al., High speed pulsed electrical spin injection in spin-light emitting diode, *Appl. Phys. Lett.* **94**, 141109 (2009).
169. M. Passlack, M. Hong, J. P. Mannaerts, J. R. Kwo, and L. W. Tu, Recombination velocity at oxide–GaAs interfaces fabricated by in situ molecular beam epitaxy, *Appl. Phys. Lett.* **68**, 3605 (1996).
170. H. Saito, J. C. Le Breton, V. Zayets, Y. Mineno, S. Yuasa, and K. Ando, Efficient spin injection into semiconductor from an Fe/GaO_x tunnel injector, *Appl. Phys. Lett.* **96**, 012501 (2010).
171. D. J. Monsma, J. C. Lodder, Th. J. A. Popma, and B. Dieny, Perpendicular hot electron spin-valve effect in a new magnetic field sensor: The spin-valve transistor, *Phys. Rev. Lett.* **74**, 5260 (1995).
172. K. Mizushima, T. Kinno, T. Yamauchi, and K. Tanaka, Energy-dependent hot electron transport across a spin-valve, *IEEE Trans. Magn.* **33**, 3500 (1997).
173. R. Jansen, The spin-valve transistor: A review and outlook, *J. Phys. D* **36**, R289–R308 (2003).
174. X. Jiang, R. Wang, S. van Dijken et al., Optical detection of hot-electron spin injection into GaAs from a magnetic tunnel transistor source, *Phys. Rev. Lett.* **90**, 256603 (2003).
175. I. Appelbaum, B. Huang, and D. J. Monsma, Electronic measurement and control of spin transport in silicon, *Nature* **447**, 295 (2007)
176. H. Dery, Ł. Cywiński, and L. J. Sham, Lateral diffusive spin transport in layered structures, *Phys. Rev. B* **73**, 041306(R) (2006).
177. B. C. Min, K. Motohashi, J. C. Lodder, and R. Jansen, Tunable spin-tunnel contacts to silicon using low work function ferromagnets, *Nat. Mater.* **5**, 817 (2006).
178. M. Johnson and R. H. Silsbee, Interfacial charge-spin coupling: Injection and detection of spin magnetization in metals, *Phys. Rev. Lett.* **55**, 1790 (1985).
179. F. J. Jedema, A. T. Filip, and B. J. van Wees, Electrical spin injection and accumulation at room temperature in an all-metal mesoscopic spin valve, *Nature* **410**, 345 (2001).
180. F. J. Jedema, H. B. Heersche, A. T. Filip, J. J. A. Baselmans, and B. J. vanWees, Electrical detection of spin precession in a metallic mesoscopic spin valve, *Nature* **416**, 713–716 (2002).
181. O. M. J. van't Erve, A. T. Hanbicki, M. Holub et al., Electrical injection and detection of spin-polarized carriers in silicon in a lateral transport geometry, *Appl. Phys. Lett.* **91**, 212109 (2007).
182. O. M. J. van't Erve, C. Awo-Affouda, A. T. Hanbicki, C. H. Li, P. E. Thompson, and B. T. Jonker, Information processing with pure spin currents in silicon: Spin injection, extraction, modulation and detection, *IEEE Trans. Electron. Dev.* **56**, 2343 (2009).
183. N. Tombros, C. Jozsa, M. Popinciuc, H. T. Jonkman, and B. J. van Wees, Electronic spin transport and spin precession in single graphene layers at room temperature, *Nature* **448**, 571 (2007).
184. M. Tran, H. Jaffrès, C. Deranlot, J.-M. George, A. Fert, A. Miard, and A. Lemaitre, Enhancement of the spin accumulation at the interface between a spin-polarized tunnel junction and a semiconductor, *Phys. Rev. Lett.* **102**, 036601 (2009).
185. A. V. Pohm, B. A. Everitt, R. S. Beech, and J. M. Daughton, Bias field and end effects on the switching thresholds of "pseudo spin valve" memory cells, *IEEE Trans. Magn.* **33**, 3280–3282 (1997).
186. A. M. Bratkovsky and V. V. Osipov, Efficient spin extraction from nonmagnetic semiconductors near forward-biased ferromagneticsemiconductor modified junctions at low spin polarization of current, *J. Appl. Phys.* **96**(8), 4525–4529 (2004).
187. M. I. D'yakonov and V. I. Perel', in *Optical Orientation, Modern Problems in Condensed Matter Science*, F. Meier and B. P. Zakharchenya (Eds.), Vol. 8, North-Holland, Amsterdam, the Netherlands, Chapter 2, Theory or optical spin orientation of electrons and nuclei, p. 39, 1984.
188. Y. Ando, K. Hamaya, K. Kasahara et al., Electrical injection and detection of spin-polarized electrons in silicon through an Fe_3Si/Si Schottky tunnel barrier, *Appl. Phys. Lett.* **94**, 182105 (2009).

189. T. Sasaki, T. Oikawa, T. Suzuki, M. Shiraishi, Y. Suzuki, and K. Tagami, Electrical spin injection into silicon using MgO tunnel barrier, *Appl. Phys. Express* **2**, 053003 (2009).
190. S. P. Dash, S. Sharma, R. S. Patel, M. P. de Jong, and R. Jansen, Electrical creation of spin polarization in silicon at room temperature, *Nature* **462**, 491 (2009).
191. L. Manchanda, M. D. Morris, M. L. Green et al., Multi-component high-K gate dielectrics for the silicon industry, *Microelectron. Eng.* **59**, 351 (2001), and references therein.
192. C. Li, O. M. J. van't Erve, and B. T. Jonker, Electrical injection and detection of spin accumulation in silicon at 500K with magnetic metal/silicon dioxide contacts, *Nat. Commun.* **2**, 245. doi: 10.1038/ncomms1256 (2011).
193. Y. Ochiai and E. Matsuura, ESR in heavily doped n-type silicon near a metal-nonmetal transition, *Phys. Stat. Sol. (a)* **38**, 243 (1976).
194. M. Passlack, R. Droopad, Z. Yu et al., Screening of oxide/GaAs interfaces for MOSFET applications, *IEEE Trans. Electron. Dev. Lett.* **29**, 1181 (2008) and references therein.

18

Spin-Polarized Ballistic Hot-Electron Injection and Detection in Hybrid Metal–Semiconductor Devices

Ian Appelbaum
University of Maryland

18.1 Introduction

The spin-injection and detection limitations posed by conductivity and lifetime mismatch between semiconductors and ferromagnetic (FM) metals have been variously solved by lifting the constraints of ohmic transport with the insertion of a tunnel barrier [1] or the use of FM semiconductors [2,3]. However, for many semiconductors, there are other material-dependent obstacles to the observation of spin transport. The material properties of silicon (Si), for instance, pose problems such as alloy formation at metallic contacts and difficulties with the use of optical methods for spin detection due to indirect bandgap [4,5].

Here, we solve both issues related to the classical "fundamental obstacle" [6–9] and the material-dependent limitations by using a mechanism for spin-injection and detection that differs fundamentally from ohmic transport in that (1) the inelastic mean free path (mfp) is not the smallest length scale in the device and that (2) electrochemical potentials cannot be uniquely defined in thermal equilibrium. With these techniques, the physical length scale of the metallic electron injection contacts is shorter than the mfp, and conduction occurs through states far above the Fermi level (as compared to the thermal energy k_BT), far out of thermal equilibrium. Because this transport mode utilizes electrons with high kinetic energy that do not suffer appreciable inelastic scattering, it is known as "ballistic hot-electron transport." Both spin injection and detection are performed all-electronically, and interfacial structure [10] plays only a minor role, enabling observation and study of spin transport in materials such as Si, which had been previously excluded from the field.

18.2 Hot-Electron Generation and Collection

There are several techniques for generating hot electrons for injection from a metal into a semiconductor. Since metals have a very high density of electrons at and below the Fermi energy E_F, all therefore rely on an electrically rectifying barrier to eliminate transport of these thermalized electrons across the metal–semiconductor interface, which would dilute the injected hot-electron current. This barrier is ideally created by the difference in work functions of the metal and the electron affinity of the semiconductor [11–14], but, in reality, its height is determined more by the details of surface states, which lie deep in the bandgap of the semiconductor that pins the Fermi level [15]. There is of course always a leakage current due to thermionic emission over this "Schottky" barrier at nonzero temperature; because typical barrier heights are in the range of 0.6–0.8 eV [16] for the common semiconductors Si and GaAs, hot-electron collection with Schottky barriers is often performed at temperatures below ambient conditions to reduce current leakage to negligible levels.

Some of the earliest hot-electron work in metal–semiconductor devices involved the use of sub-bandgap photon absorption in the metal, which scattered electrons from a state with energy $E<E_F$ to another state with energy $E+\hbar\omega>E_F$, where $\hbar\omega$ is the photon energy. If the Schottky barrier height is $q\phi$, where q is the elementary charge, then electrons with initial energy $E>E_F+q\phi-\hbar\omega$ can cross the interface, assuming that they are injected or generated within an mfp (typically in the range 1–100 nm) of it. Because this is similar to the photoelectric effect (except that the electrons cross an internal interface), the method is known as internal photoemission (IPE). By varying the photon energy, Schottky barrier heights [17] (and even buried heterojunction band offsets [18]) can be determined in a model-independent way by measuring the turn-on threshold of the collected hot-electron current without biasing the device, and the hot-electron current can be independently tuned by changing the illumination intensity.

In a later experiment, hot electrons were generated electrically using a forward-biased Schottky diode and again collected by another (lower height) Schottky interface in a semiconductor–metal–semiconductor (SMS) device. Because this provides a narrow distribution of energies (width $\approx k_BT$), the hot-electron mfp of metal thin films could reliably be measured [19]. However, in contrast to IPE, the energy itself cannot be changed since this is fixed by the emitter Schottky barrier height [20,21].

The use of tunnel junctions (TJs, which consist of two metallic conductors separated by a thin insulator) [22–27] remedies this problem. The electrostatic potential energy qV provided by voltage bias V tunes the energy of hot electrons emitted from the cathode, and the exponential energy dependence of quantum-mechanical tunneling assures a narrow distribution. Hot electrons thus created can be collected by the Schottky barrier if $qV>q\phi$. Because of its robust insulating native oxide, aluminum (Al) is often employed during TJ fabrication; although, in principle, any insulator can be used, it has been empirically found that the best are Al_xO, MgO, and AlN [28,29]. The first proposal for a tunnel-junction hot-electron injector was by Mead, who suggested another TJ or the vacuum level as a collector [30–32]. Very soon thereafter, Spratt, Schwartz, and Kane showed that a semiconductor collector could be used to realize this device with an Au/Al_xO/Al TJ [33]. All three techniques (IPE, SMS, and TJ), along with illustrations of their corresponding hot-electron energetic distributions, are shown in Figure 18.1.

Of course, a solid-state TJ barrier insulator is not an absolute requirement; if two conductors can be reliably placed close enough, the vacuum level comprises a thin barrier of height equal to the metal work function through which electrons can tunnel [34,35]. This is the basis for scanning tunneling microscopy (STM) [36], which can be used to spectroscopically image conductive surfaces. The depth-of-probe can be extended to buried metal–semiconductor interfaces by collecting ballistic hot electrons passing through a thin metallic film with a third-terminal Schottky semiconductor collector to perform ballistic electron emission microscopy (BEEM) [37–39]. Because hot electrons can have relatively long mfps in semiconductors as well [40–43], it is worth noting that this technique also has the ability to measure

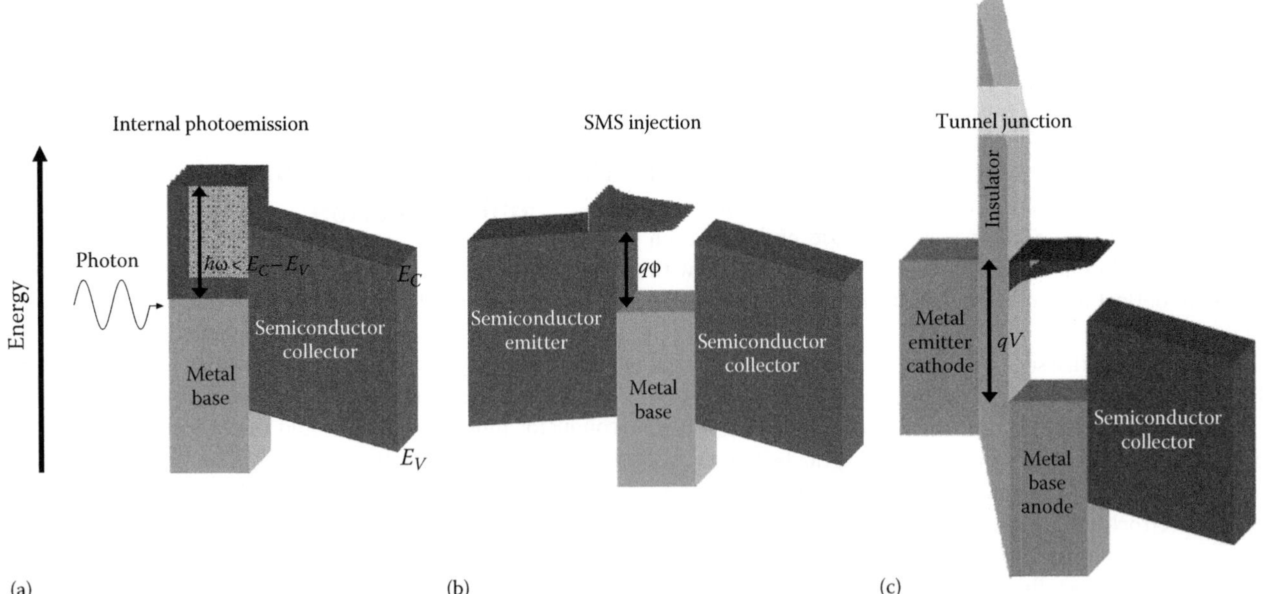

FIGURE 18.1 (See color insert.) Illustration of band diagrams for three different hot-electron generation techniques: internal photoemission, semiconductor–metal–semiconductor (SMS) metal-base transistor, and tunnel junction (TJ) emission. The energetic distributions of the hot electrons generated are roughly homogeneous up to $\hbar\omega$ above the collector Fermi energy for IPE (a), peaked at the fixed Schottky barrier height of the emitter ($q\phi$) for SMS (b), and peaked near the emitter cathode Fermi energy qV (where V is the applied voltage and q is the elementary charge) for TJ (c).

coherent transmission through semiconductor heterostructures grown close to the metal–semiconductor interface [44,45].

18.3 Spin-Polarized Hot-Electron Transport

The mfp of hot electrons in FM metal films is spin dependent: majority ("spin up") electrons have a longer mfp than minority ("spin down") electrons. Therefore, an initially unpolarized hot-electron current will become spin polarized by spin-selective scattering during ballistic transport through an FM metal film, where the polarization is given by

$$P = \frac{e^{-l/\lambda_{maj}} - e^{-l/\lambda_{min}}}{e^{-l/\lambda_{maj}} + e^{-l/\lambda_{min}}} \quad (18.1)$$

where

l is the FM film thickness
λ_{maj} is the majority spin mfp
λ_{min} is the minority spin mfp

This "ballistic spin filtering" effect can be used not only for spin polarization at injection but also for spin analysis of a hot-electron current at detection, much as an optical polarizing filter can be used for both electric field polarization and analysis of photons by changing the relative orientation of the optical axis.

A device making direct use of this effect is called the spin-valve transistor (SVT) [45–49]. It is an SMS metal-base transistor device [19] with a multilayer base consisting of at least two FM layers (spin polarizer and analyzer) whose magnetizations can be set in a parallel or antiparallel configuration; switching between these two relative orientations induces a large change in the collector current of several hundred percent because of the ballistic spin-filtering effect.

Other non-SMS SVT-related designs have been demonstrated as well. To generate hot electrons, Monsma et al. [46] also suggested the use of a magnetic TJ together with a Schottky hot-electron collector in their initial work introducing the SVT; this was then later demonstrated using solid-state devices [50,51] and with metal–vacuum–metal TJs using STM/BEEM [52]. In 2002, Van Dijken et al. [53] suggested that moving the FM polarizer to the emitter cathode of the TJ comprises a new device, which they named the magnetic tunnel transistor (MTT). (This configuration was however first proposed by Monsma in 1998, without describing it as a new device distinct from the SVT.) [54] Optical methods such as IPE with sub-bandgap illumination [55,56] and photogeneration via interband transition with super-bandgap illumination [57] have also been used to demonstrate hot-electron spin-valve effects.

Although we can speculate that the hot electrons presumably maintain some residual spin polarization after collection by the Schottky barrier [58], no spin detection after transport through the semiconductor was ever demonstrated in the above works, so spin-injection success is unknown. The first successful use of spin-polarized ballistic hot-electron injection was not reported until 2003, by Jiang et al. [59], where an MTT was used for injection into GaAs, and circular polarization analysis of band edge light emitted from InGaAs quantum wells was used for detection (a "spinLED"). However, because of the optical selection rules demanding a polar "Faraday" geometry, a large perpendicular magnetic field was required to induce a perpendicular magnetization of the magnetic layers in the MTT injector, so spin precession measurements (necessary to unambiguously confirm spin transport [60,61]) could not be done.

After injection across the interface, the hot electrons must relax to the conduction band edge before transport and recombination. Because GaAs is a noncentrosymmetric zincblende lattice that does not preserve inversion symmetry, Dresselhaus spin–orbit terms in the Hamiltonian break the spin degeneracy of the conduction band away from $\vec{k}=0$, and the D'yakonov–Perel' effect can induce strong spin scattering [62]. For this reason, injection near the conduction band edge in states close to the Brillouin zone center is strongly preferred, or application of hot-electron spin injection to a material without this scattering mechanism is desired. Si has the diamond lattice, which *does* preserve inversion symmetry and hence has no D'yakonov–Perel' mechanism, satisfying this requirement. Therefore, all subsequent work presented in this chapter will focus on spin injection, transport, and detection in Si.

18.4 Introduction to Spins in Silicon

Spin-polarized electrons in Si were first studied decades ago using resonance between Zeeman-split levels in a large magnetic field. Of interest in this field of electron spin resonance (ESR) was not only the conduction electrons but also electrons bound to donors [63–68]. The microwave absorption line positions (or electrically detected magnetic resonance (EDMR) [69,70] conductance features) give the gyromagnetic ratio $\gamma = g\mu_B/\hbar$, where g is the g-factor and μ_B is the Bohr magneton, and their linewidths give a measure of the inverse of the spin lifetime [67,71]. It was found that, due to a relative lack of spin–orbit coupling, the gyromagnetic ratio in Si corresponds to a g-factor very close to that of a bare electron in vacuum ($g \approx 2$) [72]. Furthermore, since the spin is essentially insulated from its environment, the observed lifetimes were very long, with very narrow line widths. This apparent superiority of Si over other semiconductors is due to several fortuitous intrinsic properties, which make it a relatively simple spin system [73]; Si could be considered as the "hydrogen atom" of semiconductor spintronics:

1. A low atomic number ($Z = 14$) leads to a reduced spin–orbit interaction (which scales as Z^4 in atomic systems).
2. As discussed previously, the preservation of spatial inversion symmetry of the diamond lattice leads to a spin-degenerate conduction band, eliminating D'yakonov–Perel' spin scattering in bulk Si.

3. The most abundant isotope of Si (≈92%) is ^{28}Si, which has no nuclear spin, and this abundance can be fortified with isotopic purification to very high levels [74]. Therefore, the electron-nuclear (hyperfine) coupling is weak, as compared with other semiconductors where no such nuclear-spin-free isotope exists.

This long spin lifetime (especially for electrons bound in neutral donors) was exploited to achieve spontaneous emission of tunable microwaves [75] and was suggested much later as the basis for quantum computation [76]. In fact, Si was the first material with which optical orientation (generation of spin-polarized electrons in the conduction band through interband transitions induced by circularly polarized super-bandgap photon illumination) was demonstrated [77,78]. Yet, the importance of achieving true spin transport in Si was often overlooked in recent reviews of the field [79]. This milestone was accomplished convincingly for the first time only in 2007, using the ballistic hot-electron techniques described in Section 18.5 [80].

Due to its apparent advantages over other semiconductors, many groups tried to demonstrate phenomena attributed to spin transport in Si [81–89], but this was typically done with ohmic FM-Si contacts and two-terminal magnetoresistance measurements or in transistor-type devices [90,91], which are bound to fail due to the "fundamental obstacle" for ohmic spin injection mentioned in Section 18.1 [1,7–9]. Although weak spin-valve effects are often presented, no evidence of spin precession is available, so the signals measured are ambiguous at best [60,61]. Indeed, although magnetic exchange coupling across ultrathin tunneling layers of Si was seen, not even any spin-valve magnetoresistance was observed [92].

These failures were addressed by pointing out that only in a narrow window of FM-Si resistance–area (RA) product was a large spin polarization and hence large magnetoresistance expected [93]. Subsequently, several efforts to tune the FM-Si interface resistance were made [94–98]. However, despite the ability to tune the RA product by over eight orders of magnitude and even into the anticipated high-MR window (for instance, by using a low-work-function Gd layer), no evidence that this approach has been fruitful for Si can be found, and the theory has yet been confirmed only for the case of low-temperature-grown 5 nm thick GaAs [99,100].

Subsequent to the first demonstration of spin transport in Si [80], optical detection (circular electroluminescence analysis) was shown to indicate spin injection from an FM metal and transport through several tens of nm of Si, using first a Al_xO tunnel barrier [101] and then tunneling through the Schottky barrier [102], despite the indirect bandgap and relative lack of spin–orbit interaction, which couples the photon angular momentum to the electron spin. These methods required a large perpendicular magnetic field to overcome the large in-plane shape anisotropy of the FM contact, but later a perpendicular anisotropic magnetic multilayer was shown to allow spin injection into Si at zero external magnetic field [103]. While control samples with nonmagnetic injectors do show negligible spin polarization, again no evidence of spin precession was presented.

More recently, four-terminal nonlocal measurements on Si devices have been made, for instance, with Al_xO [104] or MgO [105,106] tunnel barriers, or Schottky contacts using FM silicide injector and detector [107]. Only Ref. [106] presents convincing evidence of spin precession, although Ref. [104] does show a Hanle measurement, which unfortunately has low signal-to-noise, obscuring geometrical effects of precession expected from the oblique magnetic field configuration used [108]. Recent claims of "electrical creation of spin polarization in silicon at room temperature" [109], supported by Hanle measurements using a three-terminal geometry (without any actual evidence of transport through Si), remain ambiguous in part due to the likely influence of interface states [110].

As mentioned in Section 18.1, another way to overcome the conductivity/lifetime mismatch for spin injection is to use a carrier-mediated FM semiconductor heterointerface. The interfacial quality may have a strong effect on the injection efficiency so epitaxial growth will be necessary. Materials such as dilute magnetic semiconductors Mn-doped Si [111], Mn-doped chalcopyrites [5,112–115], or the "pure" FM semiconductor EuO [116] have all been suggested, but none as yet has been demonstrated as spin injectors for Si. It should be noted that while their intrinsic compounds are indeed semiconductors, due to the carrier-mediated nature of the ferromagnetism, it is seen only in highly (i.e., degenerately) doped and essentially metallic samples in all instances.

Several experimental [117,118] and theoretical [119] works also addressed spin-polarized electrons and holes [120,121] in Si/SiGe 2DEGs, but this is outside the scope of the present introduction; we deal here only with spin transport in 3D bulk Si, but do note that SiGe alloys could very well become useful as a way of tailoring spin transport devices to make use of local spin–orbit effects.

18.5 Ballistic Hot-Electron Injection and Detection Devices

Two types of TJs have been employed to inject spin-polarized ballistic hot electrons into Si. The first used ballistic spin filtering of initially unpolarized electrons from a nonmagnetic Al cathode by an FM anode base layer in direct contact to undoped, 10 μm thick single-crystal Si(100) [80]. Despite SVT measurements suggesting the possibility of 90% spin polarization in the metal [49,122], only 1% polarization was found after injection into and transport through the Si. It was discovered later that a nonmagnetic Cu interlayer spacer could be used to increase the polarization to approximately 37% [123], likely due to Si's tendency to readily form spin-scattering "magnetically dead" alloys (silicides) at interfaces with FM metals.

Despite the possibilities for high spin polarization with these ballistic spin filtering injector designs, the short hot-electron mfps in FM thin-film anodes causes a very small injected charge current on the order of 100 nA with an emitter electrostatic potential energy approximately 1 eV above the Schottky barrier and a contact area approximately 100 × 100 μm². Because the injected spin density and spin current are dependent on the product of spin polarization and charge current, this technique is not ideal for transport measurements. Therefore, an alternative injector utilizing an FM tunnel-junction cathode and nonmagnetic anode (which has a larger mfp) was used for approximately 10 times greater charge injection and hence larger spin signals, despite somewhat smaller potential spin polarization of approximately 15% [124,125]. These injectors can be thought of as one-half of a magnetic TJ [126], with a spin polarization proportional to the Fermi-level density-of-states spin asymmetry, rather than exponentially dependent on the spin-asymmetric mfp as is the case with ballistic spin filtering described above.

Although the injection is due to ballistic transport in the metallic contact, the mfp in the Si is only on the order of 10 nm [127], *so the vast majority of the subsequent transport to the detector over a lengthscale of tens* [80,123,124], *hundreds* [108,125], *or thousands* [128] *of microns occurs at the conduction band edge following momentum relaxation.* Typically, relatively large accelerating voltages are used so that the dominant transport mode is carrier drift; the presence of rectifying Schottky barriers on either side of the transport region assures that the resulting electric field does nothing other than determine the drift velocity of spin polarized electrons and hence the transit time [129,130]—there are no spurious (unpolarized) currents induced to flow. Furthermore, undoped Si transit layers are primarily used; otherwise, band-bending would create a confining potential and increase the transit time, potentially leading to excessive depolarization (see Section 18.6.2) [131].

The ballistic hot-electron spin detector comprises a semiconductor–FM–semiconductor structure (both Schottky interfaces), fabricated using ultrahigh vacuum (UHV) metal-film wafer bonding (a spontaneous cohesion of ultraclean metal film surfaces, which occurs at room temperature and nominal force in UHV) [47,132]. After transport through the Si, spin-polarized electrons are ejected from the conduction band over the Schottky barrier and into hot-electron states far above the Fermi energy. Again, because the mfp in FMs is larger for majority-spin (i.e., parallel to magnetization) hot electrons, the number of electrons coupling with conduction band states in an n-Si collector on the other side (which has a smaller Schottky barrier height due to contact with Cu) is dependent on the final spin polarization and the angle between spin and detector magnetization. Quantitatively, we expect the contribution to our signal from each electron with spin angle θ to be proportional to

$$\cos^2\frac{\theta}{2}e^{-l/\lambda_{maj}}+\sin^2\frac{\theta}{2}e^{-l/\lambda_{min}}=\frac{1}{2}\left[\left(e^{-l/\lambda_{maj}}-e^{-l/\lambda_{min}}\right)\cos\theta+\left(e^{-l/\lambda_{maj}}+e^{-l/\lambda_{min}}\right)\right] \quad (18.2)$$

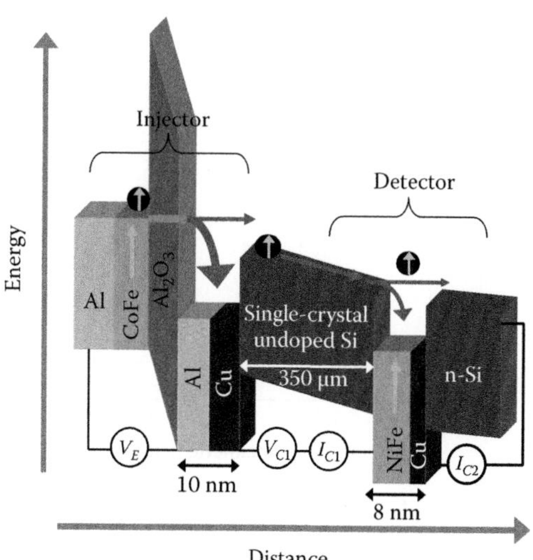

FIGURE 18.2 (See color insert.) Schematic band diagram of a four-terminal (two for TJ injection and two for FM SMS detection) ballistic hot-electron injection and detection device with a 350 μm thick Si transport layer.

Because the exponential terms are constants, this has the simple form ∝ cos θ + *const*; in the following, we disregard the constant term as it is spin independent.

The spin transport signal is thus the (reverse) current flowing across the n-Si collector Schottky interface. In essence, this device (whose band diagram is schematically illustrated in Figure 18.2) can be thought of as a split-base tunnel-emitter SVT with several hundred to thousands of microns of Si between the FM layers.

Two types of measurements are typically made: "spin valve" in a magnetic field parallel to the plane of magnetization and spin precession in a magnetic field perpendicular to the plane of magnetization. The former allows the measurement of the difference in signals between parallel (P) and antiparallel (AP) injector/detector magnetization and hence is a straightforward way of determining the conduction electron spin polarization,

$$P=\frac{I_P-I_{AP}}{I_P+I_{AP}} \quad (18.3)$$

Typical spin-valve measurement data, indicating ≈8% spin polarization after transport through 350 μm undoped Si, are shown in Figure 18.3.

Measurements in perpendicular magnetic fields reveal the average spin orientation after transit time t through the Si, due to precession at frequency $\omega=g\mu_B B/\hbar$, where B is magnetic field. If the transit time is determined only by drift, that is, $t=L/\mu E$, we expect our spin transport signal to behave ∝ cos $g\mu_B Bt/\hbar$. However, due to transit time uncertainty Δt caused by random diffusion, there is likewise an uncertainty in spin precession angle $\Delta\theta=\omega\Delta t$, which increases as the magnetic field (and hence ω) increases. When this uncertainty approaches 2π rad, the spin signal is fully suppressed by a cancellation of

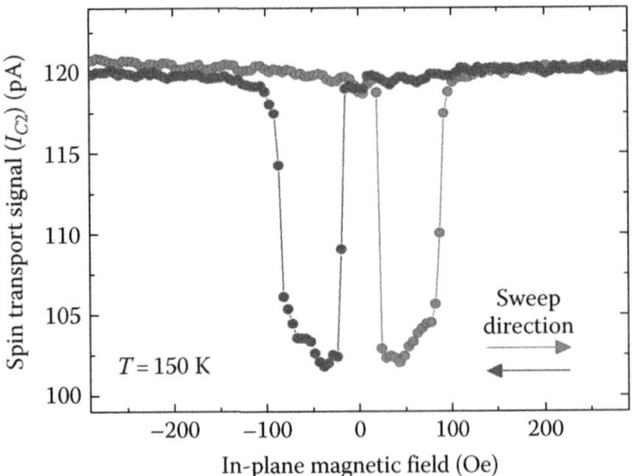

FIGURE 18.3 In-plane magnetic field measurements show the "spin-valve" effect and can be used to calculate the spin polarization after transport.

contributions from antiparallel spins, a phenomenon called spin "dephasing" or Hanle effect [133].

On a historical note, our device is essentially a solid-state analogue of experiments performed in the 1950s that were used to determine the *g*-factor of the free electron in vacuum using Mott scattering as spin polarizer and analyzer and spin precession in a solenoid [72]. In our case, we already know the *g*-factor (from, e.g., ESR lines), so our experiments in strong drift electric fields where spin dephasing is weak can be used to measure transit time with $t=h/g\mu_B B_{2\pi}$, where $B_{2\pi}$ is the magnetic field period of the observed precession oscillations, despite the fact that we make DC measurements, not time-of-flight [129,130]. Typical spin-precession data, indicating transit time of approximately 12 ns to cross 350 μm undoped Si in an electric field of ≈580 V/cm, are shown in Figure 18.4a.

One important application of this transit time information from spin precession is to correlate it to the spin polarization determined from spin-valve measurements and Equation 18.3 to extract spin lifetime. By varying the internal electric field, we change the drift velocity and hence average transit time. A reduction of polarization is seen with an increase in average transit time (as in Figure 18.5a) that we can fit well to first order using an exponential-decay model $P \propto e^{-t/\tau}$ and extract the timescale, τ [125]. In this way, we have observed spin lifetimes of approximately 1 μs at 60 K in 350 μm thick transport devices [134].* The temperature dependence of spin lifetime is compared to the $T^{-5/2}$ power law predicted by Yafet [68] and the more recent (and more complete) theory of Cheng et al. giving T^{-3} [135] in Figure 18.5b.

If we can adequately model the transport and the expected signals, we can make a spin-polarized electron version of the Haynes–Shockley experiment (originally used to measure diffusion coefficient, mobility, and lifetime of minority charge carriers) [136]. This experiment was very useful in the design of bipolar electronics devices such as junction transistors. By measuring the *spin* transport analogues of these parameters, we imagine enabling the design of useful semiconductor spintronic devices.

18.6 Modeling

Modeling the spin transport signal involves summing all the detector signal contributions of spins, which begin at the injector at the same time and with the same spin orientation [i.e., initial conditions on spin density $s(x,t=0)=\delta(x)$], where $s(x,t)$ is determined by the drift-diffusion equation

$$\frac{\partial s}{\partial t} = D\frac{\partial^2 s}{\partial x^2} - v\frac{\partial s}{\partial x} - \frac{s}{\tau} \tag{18.4}$$

where

D is the diffusion coefficient
v is the (constant) drift velocity

The solution to this partial differential equation with the $\delta(x)$ Dirac-delta initial condition is called the "Green's function," which can be used to construct the response to arbitrary spin-injection conditions, including the DC currents used so far in experiments.

Unlike open-circuit voltage spin detection [61,137,138], our ballistic hot-electron mechanism employs current sensing. This must be accounted for in the modeling, and since electrons crossing the metal–semiconductor boundary do not return via diffusion, it imposes an absorbing boundary condition on the spin transport [139]. Whereas voltage detection is sensitive to the spin density $s(x=L,t)$ and the boundaries imposed on the drift-diffusion Green's function are at infinity, here, our signal is sensitive to the spin current $J_s(x=L,t)=-D(ds/dx)|_{x=L}$, where we must impose $s(x=L,t)=0$.

18.6.1 Undoped Si

Assuming *v* is a constant independent of position *x* (as it is in undoped semiconductors), the Green's function of Equation 18.4 satisfying this boundary condition can be found straightforwardly using the method of images:

$$s(x,t) = \frac{1}{2\sqrt{\pi D t}}\left[e^{-(x-vt)^2/4Dt} - e^{Lv/D}e^{-(x-2L-vt)^2/4Dt}\right]e^{-t/\tau} \tag{18.5}$$

The corresponding spin current at the detector is then

$$J_s(x=L,t) = \frac{1}{2\sqrt{\pi D t}}\frac{L}{t}e^{-(L-vt)^2/4Dt}e^{-t/\tau} \tag{18.6}$$

* In Ref. [125], a more conservative estimate of the spin lifetime (e.g., 520 ns at 60 K) was obtained by fitting to the transit time dependence of an alternative quantity expected to be proportional to the spin polarization, rather than using Equation 18.3 directly.

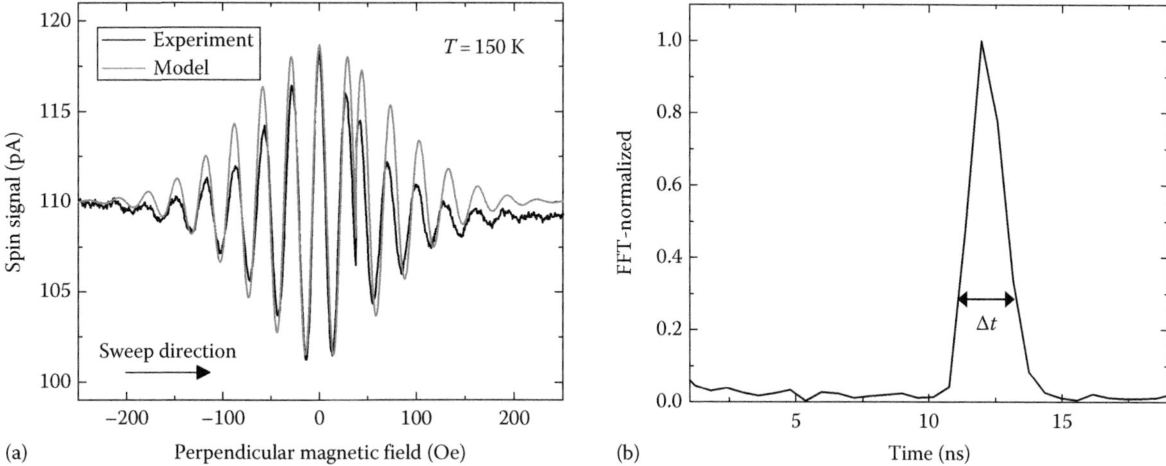

FIGURE 18.4 (a) A typical spin precession measurement shows the coherent oscillations due to drift and the suppression of signal amplitude ("dephasing") as the precession frequency rises. Our model simulates this behavior well. (b) The real part of the Fourier transform of the precession data in (a) reveals the spin current arrival distribution.

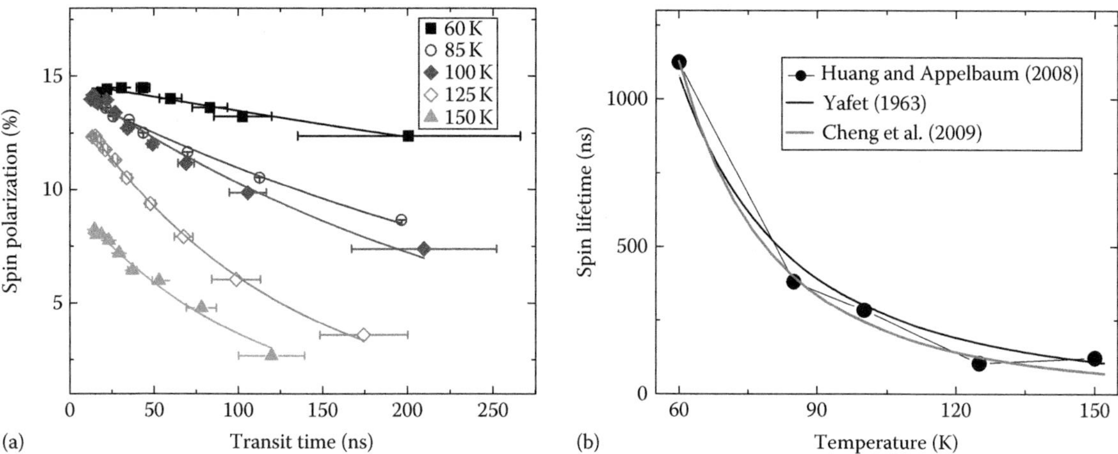

FIGURE 18.5 (a) Fitting the normalized spin signal from in-plane spin-valve measurements to an exponential decay model using transit times derived from spin precession measurements at variable internal electric field yields measurement of spin lifetimes in undoped bulk Si. (b) The experimental spin lifetime values obtained as a function of temperature are compared to Yafet's $T^{-5/2}$ power law for indirect-bandgap semiconductors [68] and Cheng et al.'s T^{-3} derived from a full band structure theory [135].

The Green's function in Equation 18.5 and the spin current derived from it in Equation 18.6 (using typical values for v, D, and L at various times t) are shown in Figure 18.6.

Note that the expression in Equation 18.6 is simply the spin density in the absence of the boundary condition [137,138] multiplied by the spin velocity L/t. Because our measurements are under conditions of strong drift fields, we have approximated this term with a constant (e.g., the average drift velocity v) in previous work [125,140,141]. To first order, this approximation involves a rigid shift of the signal peak to lower arrival times of $\approx 2D/v^2$ and a subsequent error in the measured velocity of $2D/L$. Because typical values for these variables are on the order of $v > 10^6$ cm/s, $L > 10^{-2}$ cm, and $10^2 < D < 10^3$ cm^2/s, the relative error associated with this approximation ($2D/Lv$ in both cases) is then just a few percent.

We can use this Green's function in the kernel of an integral expression to model, for instance, the expected precession signal as the weighted sum of all the cosine contributions from electrons with different arrival times ($0 < t < \infty$):

$$\int_0^\infty J_s(x = L, t)\cos\omega t\,dt \qquad (18.7)$$

Note that because J_s is causal [i.e., $J_s(t < 0) = 0$], we can extend the lower integration bound to $-\infty$ so that this expression is equivalent to the real part of the Fourier transform of $J_s(x = L, t)$. We can therefore use precession measurements as a direct probe of the spin current transit time distribution by making use of the Fourier transform, as shown in Figure 18.4b. By fitting the data

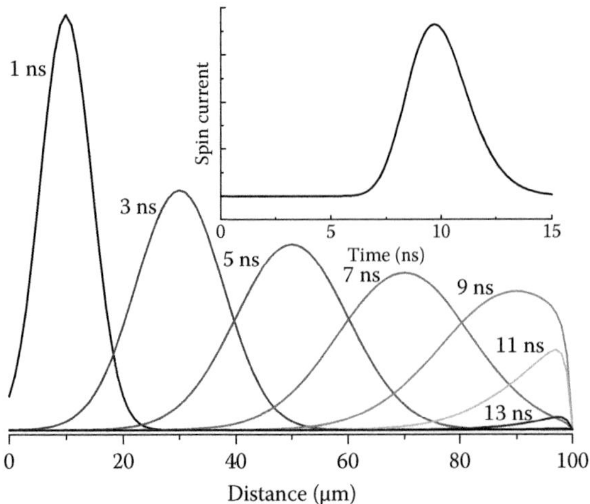

FIGURE 18.6 Evolution of a Delta function spin distribution (Green's function) injected on the left side at time increments shown in the legend, with drift velocity 10^6 cm/s, diffusion coefficient 100 cm²/s, and absorbing boundary conditions at the detector ($x = 100\,\mu$m). Inset: Simulated spin current, given by the gradient of the Green's function at the detector, as in Equation 18.6.

to the simulation, we can extract the transport variables v and D; along with the spin lifetime measurement, this comprises a spin Haynes–Shockley experiment.

Equation 18.7 is valid only for the case of a purely perpendicular magnetic field $\vec{B} = B_z\hat{z}$; [142]. In the most general case, if there is also an in-plane field component $B_y\hat{y}$ (due to misalignment or a static field) and/or if the perpendicular component of the field B_z is strong enough to overcome the in-plane magnetic anisotropy of the film, the cos ωt term must be replaced with

$$\frac{(B_z\cos\theta_1 + B_y\sin\theta_1)(B_z\cos\theta_2 + B_y\sin\theta_2)}{B_y^2 + B_z^2}\cos\omega t + \frac{(B_z\sin\theta_1 - B_y\cos\theta_1)(B_z\sin\theta_2 - B_y\cos\theta_2)}{B_y^2 + B_z^2} \qquad (18.8)$$

where, in the simplest case, the magnetization rotation angles $\theta_{1,2} = \arcsin(\tanh(B_z/\eta_{1,2}))$, and $\eta_{1,2}$ are the demagnetization fields of injector and detector FM layers (typically several Tesla) [131]. From Equation 18.8, we can show that small misalignments from the out-of-plane direction during single-axis precession measurements cause only in-plane magnetization switching of injector and detector and a negligible correction to the amplitude of precession oscillations to first order [108]. Furthermore, the expression is now composed of a coherent term proportional to cos ωt and an incoherent term independent of t. In the case of extremely strong dephasing (caused, e.g., by geometrically induced transit length uncertainty), the first term contributes a negligible amount to the integral expression in Equation 18.7, and the signal is dominated by the field and magnetization geometry dictated in the second term [128].

The measurements shown so far have used only magnetic field to control the spin precession angle θ via the spin precession frequency ω in a static electric field (and hence constant transit time t). Because $\theta = \omega t$, we can alternatively control the final spin orientation via transit time modulation with a changing electric field in a fixed magnetic field [123,140]. Measured data at a fixed perpendicular magnetic field of 200 Oe is favorably compared to simulation results in Figure 18.7a and b.

18.6.2 Doped Si

In the above, we have assumed that the drift velocity v in Equation 18.4 was independent of x. This assumption is not valid in doped semiconductors due to the presence of inhomogeneous electric fields and band-bending caused by ionized dopants. Equation 18.4 then no longer has constant coefficients, so, in general, its Green's function cannot be solved analytically, and a computational approach must be taken.

Direct simulation of an ensemble of electrons can be used to assemble a histogram of transit times, which in the limit of large numbers of electrons yields the transit-time distribution function. This "Monte Carlo" method, which incorporates the proper boundary conditions for our current-based spin detection

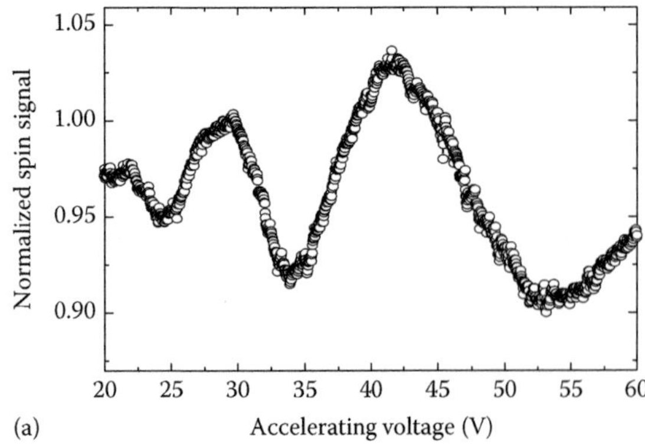

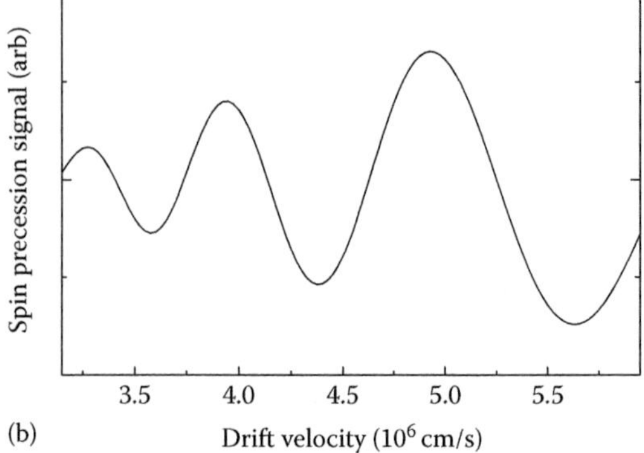

FIGURE 18.7 Spin orientation control using electric field modulation in a static precession magnetic field of 200 Oe during transport through 350 μm Si. (a) Shows experimental results, (b) panel shows simulation results.

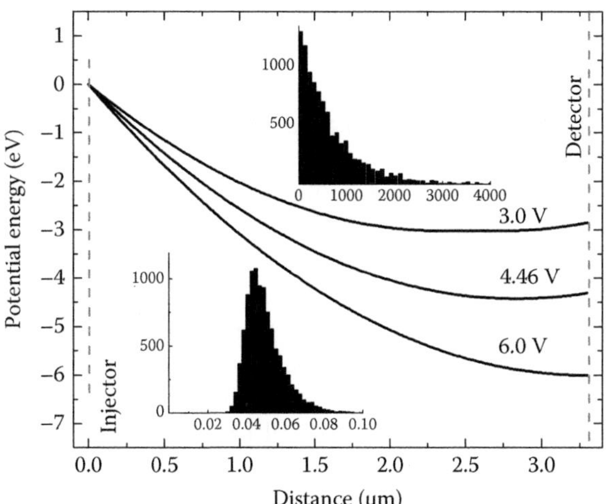

FIGURE 18.8 Depletion-approximation conduction-band diagrams of a 3.3 μm thick *n*-type (7.2×10^{14} cm^{-3}) doped Si spin-transport layer with injector–detector voltage drop of 3 V (where the transport is dominated by diffusion against the electric field at the detector side and results in the exponential-like arrival distribution with typical timescale 1000 ns shown in the top inset); 4.46 V (where the bias is enough to eliminate the electrostatically neutral region and fully deplete the transport layer); and 6 V (where the potential well has been eliminated and transport is dominated by drift in the unipolar electric field, resulting in the Gaussian-like arrival distribution with typical timescale of 0.050 ns shown in the bottom inset).

technique automatically, was used to model spin transport through doped Si [131].

The technique involves discretely stepping through time a duration δt, modeling drift with a spatial translation $v(x)\delta t$, and diffusion with a translation of $\pm\sqrt{2D\delta t}$ (where the sign is randomly chosen to model the stochastic nature of the process), until $x > L$. The drift velocity $v(x)$ can be constructed from a depletion-approximation model of the transport layer to determine the electric field profile $E(x)$ and a mobility model to get $v(E)$ [130]. Our experimental studies with *n*-type ($\approx 10^{15}$ cm^{-3} phosphorus) 3.3 μm thick Si devices indicate that despite highly "non-ohmic" spin transport where the transit time can be varied by several orders of magnitude resulting from a confining electrostatic potential, Monte-Carlo device modeling shows that spin lifetime is not measurably impacted from its value in intrinsic Si [131]. Figure 18.8 shows example band diagrams under different bias conditions (calculated in the depletion approximation) and corresponding arrival time distributions from the Monte-Carlo simulation.

18.7 Outlook

There has been significant progress in using ballistic hot-electron spin injection and detection techniques for spin transport studies in Si. However, there are limitations of these methods. For example, the small ballistic transport transfer ratio is typically no better than 10^{-3}–10^{-2}; the low injection currents and detection signals obtained will result in subunity gain and limit direct applications of these devices. In addition, our reliance on the ability of Schottky barriers to serve as hot-electron filters presently limits device operation temperatures to approximately 200 K (although materials with higher Schottky barrier heights could extend this closer to room temperature) and also limits application to only nondegenerately doped semiconductors. Carrier freeze-out in the n-Si spin detection collector at approximately 20 K introduces a fundamental low-temperature limit as well [141].

Despite these shortcomings, there are also unique capabilities afforded by these methods, such as independent control over internal electric field, injection current, and spectroscopic control over the injection energy level [143]. Unlike, for instance, optical techniques, other semiconductor materials should be equally well suited to study with these methods. The purpose is to use these devices as tools to understand spin transport properties for the design of spintronic devices just as the Haynes–Shockley experiment [136] enabled the design of electronic minority-carrier devices such as the bipolar junction transistor.

There is still much physics to be done with ballistic hot-electron spin injection and detection. Recently, we have adapted our fabrication techniques to assemble lateral spin transport devices, where in particular very long transit lengths [128] and the effects of an electrostatic gate to control the proximity to an Si/SiO_2 interface can be investigated [144]. The massive spin lifetime suppression induced by the interface that was observed in this work suggests that such devices can be used to explore the effects of broken inversion symmetry [145] and paramagnetic [70] or charged [146,147] defects on conduction electron spins at the semiconductor/insulator interface. In addition, this opportunity for sensitive characterization may be of importance to the electronics community as they continue to push MOSFET scaling toward its ultimate limits.

Hopefully, more experimental groups will develop the technology necessary to compete in this wide-open field. Theorists, too, are eagerly invited to address topics such as whether this injection technique fully circumvents the "fundamental obstacle" because of a remaining Sharvin-like effective resistance [148], or whether it introduces anomalous spin dephasing [147].

Acknowledgments

I.A. would especially like to thank Douwe Monsma for introducing him to the field and Biqin Huang, Hyuk-Jae Jang, Jing Li, and Jing Xu for fabrication, measurement, and programming efforts that have made much of this work possible.

This work has been financially supported by the Office of Naval Research, DARPA/MTO, and the National Science Foundation.

References

1. E. Rashba, Theory of electrical spin injection: Tunnel contacts as a solution of the conductivity mismatch problem, *Phys. Rev. B* **62**, R16267 (2000).

2. R. Fiederling, M. Keim, G. Reuscher et al., Injection and detection of a spin polarized current in a n-i-p light emitting diode, *Nature* **402,** 787 (1999).
3. Y. Ohno, D. K. Young, B. Beschoten, F. Matsukura, H. Ohno, and D. D. Awschalom, Electrical spin injection in a ferromagnetic semiconductor heterostructure, *Nature* **402**, 790 (1999).
4. I. Žutić, J. Fabian, and S. D. Sarma, Spintronics: Fundamentals and applications, *Rev. Mod. Phys.* **76**, 323 (2004).
5. I. Žutić, J. Fabian, and S. C. Erwin, Spin injection and detection in silicon, *Phys. Rev. Lett.* **97**, 026602 (2006).
6. M. Johnson and R. Silsbee, Thermodynamic analysis of interfacial transport and of the thermomagnetoelectric system, *Phys. Rev. B* **35**, 4959 (1987).
7. M. Johnson and R. Silsbee, Spin-injection experiment, *Phys. Rev. B* **37**, 5326 (1988).
8. G. Schmidt, D. Ferrand, L. W. Molenkamp, A. T. Filip, and B. J. van Wees, Fundamental obstacle for electrical spin injection from a ferromagnetic metal into a diffusive semiconductor, *Phys. Rev. B* **62**, R4790 (2000).
9. G. Schmidt, Concepts for spin injection into semiconductors—A review, *J. Phys. D* **38**, R107 (2005).
10. T. J. Zega, A. T. Hanbicki, S. C. Erwin et al., Determination of interface atomic structure and its impact on spin transport using Z-contrast microscopy and density-functional theory, *Phys. Rev. Lett.* **96**, 196101 (2006).
11. W. Schottky, Semiconductor theory of the barrier film, *Naturwissenschaften* **26**, 843 (1938).
12. W. Schottky, Deviations from Ohm's law in semiconductors, *Phys. Z.* **41**, 570 (1940).
13. N. F. Mott, Note on the contact between a metal and an insulator or semi-conductor, *Proc. Cambridge Philos. Soc.* **34**, 568 (1938).
14. N. F. Mott, The theory of crystal rectifiers, *Proc. R. Soc. London A* **171**, 27–38 (1939).
15. J. Tersoff, Schottky barrier heights and the continuum of gap states, *Phys. Rev. Lett.* **52**, 465–468 (1984).
16. S. Sze, *Physics of Semiconductor Devices*, 2nd edn., Wiley-Interscience, New York, 1981.
17. C. R. Crowell, W. G. Spitzer, L. E. Howarth, and E. E. LaBate, Attenuation length measurements of hot electrons in metal films, *Phys. Rev.* **127**, 2006 (1962).
18. M. Heiblum, M. I. Nathan, and M. Eizenberg, Energy band discontinuities in heterojunctions measured by internal photoemission, *Appl. Phys. Lett.* **47**, 503–505 (1985).
19. C. Crowell and S. Sze, Hot electron transport and electron tunneling in thin film structures, in *Physics of Thin Films*, G. Hass and R. Thun (Eds.), Academic Press, New York, 1967.
20. C. Crowell and S. Sze, Quantum-mechanical reflection of electrons at metal-semiconductor barriers: Electron transport in semiconductor-metal-semiconductor structures, *J. Appl. Phys.* **37**, 2683 (1966).
21. S. Sze and H. Gummel, Appraisal of semiconductor-metal-semiconductor transistor, *Solid State Electron.* **9**, 751 (1966).
22. A. Sommerfeld and H. Bethe, Electronentheorie der Metalle, in *Handbuch der Physik*, S. Flügge (Ed.), Springer, Berlin, Germany, 1933.
23. C. Duke, *Tunneling in Solids*, Academic Press, New York, 1969.
24. J. Fisher and I. Giaever, Tunneling through thin insulating layers, *J. Appl. Phys.* **32**, 172 (1961).
25. I. Giaever, Energy gap in superconductors measured by electron tunneling, *Phys. Rev. Lett.* **5**, 147–148 (1960).
26. J. G. Simmons, Generalized thermal j-v characteristic for electric tunnel effect, *J. Appl. Phys.* **35**, 2655 (1964).
27. W. F. Brinkman, R. C. Dynes, and J. M. Rowell, Tunneling conductance of asymmetrical barriers, *J. Appl. Phys.* **41**, 1915–1921 (1970).
28. J. S. Moodera, J. Nassar, and G. Mathon, Spin-tunneling in ferromagnetic junctions, *Annu. Rev. Mater. Sci.* **29**, 381–432 (1999).
29. Y. Yang, P. F. Ladwig, Y. A. Chang, L. Feng, B. B. Pant, and A. E. Schultz, Thermodynamic evaluation of the interface stability between selected metal oxides and Co, *J. Mater. Res.* **19**, 1181 (2004).
30. C. Mead, The tunnel-emission amplifier, *Proc. IRE* **48**, 359 (1960).
31. C. Mead, Operation of tunnel-emission devices, *J. Appl. Phys.* **32**, 646 (1961).
32. C. A. Mead, Transport of hot electrons in thin gold films, *Phys. Rev. Lett.* **8**, 56–57 (1962).
33. J. Spratt, R. Schwartz, and W. Kane, Hot electrons in metal films: Injection and collection, *Phys. Rev. Lett.* **6**, 341 (1961).
34. J. Frenkel, On the electrical resistance of contacts between solid conductors, *Phys. Rev.* **36**, 1604 (1930).
35. R. Young, J. Ward, and F. Scire, The topografiner: An instrument for measuring surface microtopography, *Rev. Sci. Instrum.* **43**, 999–1011 (1972).
36. G. Binnig, H. Rohrer, Ch. Gerber, and E. Weibel, Tunneling through a controllable vacuum gap, *Appl. Phys. Lett.* **40**, 178–180 (1982).
37. W. Kaiser and L. Bell, Direct investigation of subsurface interface electronic structure by ballistic-electron-emission microscopy, *Phys. Rev. Lett.* **60**, 1406 (1988).
38. L. D. Bell and W. J. Kaiser, Observation of interface band structure by ballistic-electron-emission microscopy, *Phys. Rev. Lett.* **61**, 2368–2371 (1988).
39. W. Yi, A. Stollenwerk, and V. Narayanamurti, Ballistic electron microscopy and spectroscopy of metal and semiconductor nanostructures, *Surf. Sci. Rep.* **64**, 169 (2009).
40. M. Heiblum, Tunneling hot-electron transfer amplifier: A hot-electron GaAs device with current gain, *Appl. Phys. Lett.* **47**, 1105 (1985).
41. M. Heiblum, Direct observation of ballistic transport in GaAs, *Phys. Rev. Lett.* **55**, 2200 (1985).
42. T. Daniels-Race, K. Yamasaki, W. J. Schaff, and L. F. Eastman, Observation of ballistic transport in hot-electron vertical FET spectrometer using ultrathin planar-doped barrier launcher, *Electron. Lett.* **27**, 1144 (1991).

43. J. Hayes, F. Levia, and W. Wiegmann, Hot electron spectroscopy of GaAs, *Phys. Rev. Lett.* **54**, 1570 (1985).
44. G. N. Henderson, T. K. Gaylord, E. N. Glytsis, P. N. First, and W. J. Kaiser, Ballistic electron emission testing of semiconductor heterostructures, *Solid State Commun.* **80**, 591–596 (1991).
45. T. Sajoto, J. J. O'Shea, S. Bhargava, D. Leonard, M. A. Chin, and V. Narayanamurti, Direct observation of quasi-bound states and band-structure effects in a double barrier resonant tunneling structure using ballistic electron emission microscopy, *Phys. Rev. Lett.* **74**, 3427–3430 (1995).
46. D. J. Monsma, J. C. Lodder, T. J. Popma, and B. Dieny, Perpendicular hot electron spin-valve effect in a new magnetic field sensor: The spin-valve transistor, *Phys. Rev. Lett.* **74**, 5260 (1995).
47. D. Monsma, R. Vlutters, and J. Lodder, Room temperature-operating spin-valve transistors formed by vacuum bonding, *Science* **281**, 407 (1998).
48. I. Appelbaum, K. J. Russell, D. J. Monsma et al., Luminescent spin-valve transistor, *Appl. Phys. Lett.* **83**, 4571 (2003).
49. R. Jansen, The spin-valve transistor: A review and outlook, *J. Phys. D* **36**, R289 (2003).
50. K. Mizushima, T Kinno, T. Yamauchi, and K. Tanaka, Energy-dependent hot electron transport across a spin-valve, *IEEE Trans. Magn.* **33**, 3500 (1997).
51. G. Tae, J. Eom, J. Song, and K. Kim, Spin-polarized hot electron injection into two-dimensional electron gas by magnetic tunnel transistor, *Jpn. J. Appl. Phys.* **46**, 7717 (2007).
52. W. Rippard and R. Buhrman, Spin-dependent hot electron transport in Co/Cu thin films, *Phys. Rev. Lett.* **84**, 971 (2000).
53. S. van Dijken, X. Jiang, and S. S. P. Parkin, Room temperature operation of a high output current magnetic tunnel transistor, *Appl. Phys. Lett.* **80**, 3364–3366 (2002).
54. D. Monsma, The spin-valve transistor, PhD thesis, University of Twente, the Netherlands, 1998.
55. I. Appelbaum, D. J. Monsma, K. J. Russell, V. Narayanamurti, and C. M. Marcus, Spin-valve photo-diode, *Appl. Phys. Lett.* **83**, 3737 (2003).
56. B. Huang and I. Appelbaum, Perpendicular hot-electron transport in the spin-valve photodiode, *J. Appl. Phys.* **100**, 034501 (2006).
57. B. Huang, I. Altfeder, and I. Appelbaum, Spin-valve phototransistor, *Appl. Phys. Lett.* **90**, 052503 (2007).
58. S. A. Wolf, D. D. Awschalom, R. A. Buhrman et al., Spintronics: A spin-based electronics vision for the future, *Science* **294**, 1488 (2001).
59. X. Jiang, R. Wang, S. van Dijken et al., Optical detection of hot-electron spin injection into GaAs from a magnetic tunnel transistor, *Phys. Rev. Lett.* **90**, 256603 (2003).
60. F. Monzon, H. Tang, and M. Roukes, Magnetoelectronic phenomena at a ferromagnet-semiconductor interface, *Phys. Rev. Lett.* **84**, 5022 (2000).
61. M. Johnson and R. Silsbee, Interfacial charge-spin coupling: Injection and detection of spin magnetization in metals, *Phys. Rev. Lett.* **55**, 1790 (1985).
62. X. Jiang, R. Wang, R. M. Shelby, and S. S. P. Parkin, Optical detection of hot-electron spin injection into GaAs from a magnetic tunnel transistor, *IBM Res. Dev.* **50**, 111 (2006).
63. A. M. Portis, A. F. Kip, C. Kittel, and W. H. Brattain, Electron spin resonance in a silicon semiconductor, *Phys. Rev.* **90**, 988 (1953).
64. G. Feher, Electron spin resonance experiments on donors in silicon 1. Electronic structure of donors by the electron nuclear double resonance technique, *Phys. Rev.* **114**, 1219 (1959).
65. G. Feher and E. Gere, Electron spin resonance experiments on donors in silicon 2. Electron spin relaxation effects, *Phys. Rev.* **114**, 1245 (1959).
66. D. Wilson and G. Feher, Electron spin resonance experiments on donors in silicon 3. Investigation of excited states by application of uniaxial stress and their importance in relaxation processes, *Phys. Rev.* **124**, 1068 (1959).
67. D. Lepine, Spin resonance of localized and delocalized electrons in phosphorus-doped silicon between 20 and 300 degrees K, *Phys. Rev. B* **2**, 2429 (1970).
68. Y. Yafet, G-factors and spin-lattice relaxation of conduction electrons, in *Solid State Physics-Advances in Research and Applications*, F. Seitz and D. Turnbull (Eds.), Academic Press, New York, 1963.
69. R. Ghosh and R. Silsbee, Spin-spin scattering in a silicon two-dimensional electron gas, *Phys. Rev. B* **46**, 12508 (1992).
70. M. Xiao, I. Martin, E. Yablonovitch, and H. W. Jiang, Electrical detection of the spin resonance of a single electron in a silicon field-effect transistor, *Nature* **430**, 435 (2004).
71. J. Fabian, A. Matos-Abiague, C. Ertler, P. Stano, and I. Žutić, Semiconductor spintronics, *Acta Phys. Slovaca* **57**, 565 (2007).
72. W. Louisell, R. Pidd, and H. Crane, An experimental measurement of the gyromagnetic ratio of the free electron, *Phys. Rev.* **94**, 7 (1954).
73. I. Žutić and J. Fabian, Spintronics: Silicon twists, *Nature* **447**, 269 (2007).
74. M. Cardona and M. Thewalt, Isotope effects on the optical spectra of semiconductors, *Rev. Mod. Phys.* **77**, 1173 (2005).
75. G. Feher, J. P. Gordon, E. Buehler, E. A. Gere, and C. D. Thurmond, Spontaneous emission of radiation from an electron spin system, *Phys. Rev.* **109**, 221–222 (1958).
76. B. Kane, A silicon-based nuclear spin quantum computer, *Nature* **393**, 1331 (1998).
77. G. Lampel, Nuclear dynamic polarization by optical electronic saturation and optical pumping in semiconductors, *Phys. Rev. Lett.* **20**, 491 (1968).
78. N. Bagraev, L. Vlasenko, and R. Zhitnikov, Optical orientation of Si^{29} nuclei in n-type silicon and its dependence on the pumping light intensity, *Sov. Phys. JETP* **44**, 500 (1976).
79. D. Awschalom and M. E. Flatté, Challenges for semiconductor spintronics, *Nat. Phys.* **3**, 153 (2007).

80. I. Appelbaum, B. Huang, and D. J. Monsma, Electronic measurement and control of spin transport in silicon, *Nature* **447**, 295 (2007).
81. Y. Q. Jia, R. C. Shi, and S. Y. Chou, Spin-valve effects in nickel/silicon/nickel junctions, *IEEE Trans. Magn.* **32**, 4707 (1996).
82. K. I. Lee, H. J. Lee, J. Y. Chang, S. H. Han, Y. K. Kim, and W. Y. Lee, Spin-valve effect in an FM/Si/FM junction, *J. Mater. Sci.* **16**, 131 (2005).
83. W. J. Hwang, H. J. Lee, K. I. Lee et al., Spin transport in a lateral spin-injection device with an FM/Si/FM junction, *J. Magn. Magn. Mater.* **272–276**, 1915 (2004).
84. S. Hacia, T. Last, S. F. Fischer, and U. Kunze, Study of spin-valve operation in permalloy-SiO_2-silicon nanostructures, *J. Supercond.* **16**, 187 (2003).
85. S. Hacia, T. Last, S. F. Fischer, and U. Kunze, Magnetotransport study of nanoscale permalloy-Si tunnelling structures in lateral spin-valve geometry, *J. Phys. D* **37**, 1310 (2004).
86. C. L. Dennis, C. Sirisathitkul, G. J. Ensell, J. F. Gregg, and S. M. Thompson, High current gain silicon-based spin transistor, *J. Phys. D* **36**, 81 (2003).
87. C. L. Dennis, J. F. Gregg, G. J. Ensell, and S. M. Thompson, Evidence for electrical spin tunnel injection into silicon, *J. Appl. Phys.* **100**, 043717 (2006).
88. C. L. Dennis, C. V. Tiusan, R. A. Ferreira et al., Tunnel barrier fabrication on Si and its impact on a spin transistor, *J. Magn. Magn. Mater.* **290**, 1383 (2005).
89. J. F. Gregg, I. Petej, E. Jouguelet, and C. Dennis, Spin electronics—A review, *J. Phys. D* **35**, R121 (2002).
90. S. Sugahara and M. Tanaka, A spin metal-oxide-semiconductor field-effect transistor using half-metallic-ferromagnet contacts for the source and drain, *Appl. Phys. Lett.* **84**, 2307 (2004).
91. M. Cahay and S. Bandyopadhyay, Room temperature silicon spin-based transistors, in *Device Applications of Silicon Nanocrystals and Nanostructures*, N. Koshida (Ed.), Springer, New York, 2009.
92. R. R. Gareev, L. L. Pohlmann, S. Stein, D. E. Bürgler, P. A. Grünberg, and M. Siegel, Tunneling in epitaxial Fe/Si/Fe structures with strong antiferromagnetic interlayer coupling, *J. Appl. Phys.* **93**, 8038 (2003).
93. A. Fert and H. Jaffrès, Conditions for efficient spin injection from a ferromagnetic metal into a semiconductor, *Phys. Rev. B* **64**, 184420 (2001).
94. B.-C. Min, J. C. Lodder, R. Jansen, and K. Motohashi, Cobalt-Al_2O_3-silicon tunnel contacts for electrical spin injection into silicon, *J. Appl. Phys.* **99**, 08S701 (2006).
95. B.-C. Min, K. Motohashi, C. Lodder, and R. Jansen, Tunable spin-tunnel contacts to silicon using low-work-function ferromagnets, *Nat. Mater.* **5**, 817 (2006).
96. K. Wang, J. Stehlik, and J.-Q. Wang, Tunneling characteristics across nanoscale metal ferric junction lines into doped Si, *Appl. Phys. Lett.* **92**, 152118 (2008).
97. T. Uhrmann, T. Dimopoulos, H. Brückl et al., Characterization of embedded MgO/ferromagnet contacts for spin injection in silicon, *J. Appl. Phys.* **103**, 063709 (2008).
98. T. Dimopoulos, D. Schwarz, T. Uhrmann et al., Magnetic properties of embedded ferromagnetic contacts to silicon for spin injection, *J. Phys. D* **42**, 085004 (2009).
99. R. Mattana, J.-M. George, H. Jaffrès et al., Electrical detection of spin accumulation in a *p*-type GaAs quantum well, *Phys. Rev. Lett.* **90**, 166601 (2003).
100. A. Fert, J.-M. George, H. Jaffrès, and R. Mattana, Semiconductors between spin-polarized sources and drains, *IEEE Trans. Electron. Dev.* **54**, 921–932 (2007).
101. B. T. Jonker, G. Kioseoglou, A. T. Hanbicki, C. H. Li, and P. E. Thompson, Electrical spin-injection into silicon from a ferromagnetic metal/tunnel barrier contact, *Nat. Phys.* **3**, 542 (2007).
102. G. Kioseoglou, A. T. Hanbicki, R. Goswami et al., Electrical spin injection into Si: A comparison between Fe/Si Schottky and Fe/Al_2O_3 tunnel contacts, *Appl. Phys. Lett.* **94**, 122106 (2009).
103. L. Grenet, M. Jamet, P. Noé et al., Spin injection in silicon at zero magnetic field, *Appl. Phys. Lett.* **94**, 032502 (2009).
104. O. M. J. van't Erve, A. T. Hanbicki, M. Holub, et al., Electrical injection and detection of spin-polarized carriers in silicon in a lateral transport geometry, *Appl. Phys. Lett.* **91**, 212109 (2007).
105. T. Sasaki, T. Oikawa, T. Suzuki, M. Shiraishi, Y. Suzuki, and K. Tagami, Electrical spin injection into silicon using MgO tunnel barrier, *Appl. Phys. Express* **2**, 53003 (2009).
106. T. Sasaki, T. Oikawa, T. Suzuki, M. Shiraishi, Y. Suzuki, and K. Noguchi, Temperature dependence of spin diffusion length in silicon by Hanle-type spin precession, *Appl. Phys. Lett.* **96**, 122101 (2010).
107. Y. Ando, K. Hamaya, K. Kasahara et al., Electrical injection and detection of spin-polarized electrons in silicon through an Fe_3Si/Si Schottky tunnel barrier, *Appl. Phys. Lett.* **94**, 182105 (2009).
108. J. Li, B. Huang, and I. Appelbaum, Oblique Hanle effect in semiconductor spin transport devices, *Appl. Phys. Lett.* **92**, 142507 (2008).
109. S. P. Dash, S. Sharma, R. S. Patel, M. P. de Jong, and R. Jansen, Electrical creation of spin polarization in silicon at room temperature, *Nature* **462**, 491 (2009).
110. M. Tran, H. Jaffrès, C. Deranlot et al., Enhancement of the spin accumulation at the interface between a spin-polarized tunnel junction and a semiconductor, *Phys. Rev. Lett.* **102**, 036601 (2009).
111. M. Bolduc, C. Awo-Affouda, A. Stollenwerk et al., Above room temperature ferromagnetism in Mn-ion implanted Si, *Phys. Rev. B* **71**, 033302 (2005).
112. S. C. Erwin and I. Žutić, Tailoring ferromagnetic chalcopyrites, *Nat. Mater.* **3**, 410 (2004).
113. S. Cho, S. Choi, G.-B. Cha et al., Synthesis of new pure ferromagnetic semiconductors: $MnGeP_2$ and $MnGeAs_2$, *Solid State Commun.* **129**, 609 (2004).

114. S. Cho, S. Choi, G.-B. Cha et al., Room-temperature ferromagnetism in $Zn_{1-x}Mn_xGeP_2$ semiconductors, *Phys. Rev. Lett.* **88**, 257203 (2002).
115. Y. Ishida, D. D. Sarma, K. Okazaki et al., In situ photoemission study of the room temperature ferromagnet $ZnGeP_2$:Mn, *Phys. Rev. Lett.* **91**, 107202 (2003).
116. A. Schmehl, V. Vaithyanathan, A. Herrnberger et al., Epitaxial integration of the highly spin-polarized ferromagnetic semiconductor EuO with silicon and GaN, *Nat. Mater.* **6**, 882 (2007).
117. S. D. Ganichev, S. N. Danilov, V. V. Bel'kov et al., Pure spin currents induced by spin-dependent scattering processes in SiGe quantum well structures, *Phys. Rev. B* **75**, 155317 (2007).
118. M. Friesen, P. Rugheimer, D. E. Savage et al., Practical design and simulation of silicon-based quantum-dot qubits, *Phys. Rev. B* **67**, 121301 (2003).
119. P. Zhang and M. Wu, Spin diffusion in Si/SiGe quantum wells: Spin relaxation in the absence of D'yakonov–Perel' relaxation mechanism, *Phys. Rev. B* **79**, 075303 (2009).
120. N. T. Bagraev, N. G. Galkin, W. Gehlhoff, L. E. Klyachkin, A. M. Malyarenko, and I. A. Shelykh, Spin interference in silicon three-terminal one-dimensional rings, *J. Phys. D* **18**, L567 (2006).
121. P. Zhang and M. Wu, Hole spin relaxation in [001] strained asymmetric Si/SiGe and Ge/SiGe quantum wells, *Phys. Rev. B* **80**, 155311 (2009).
122. S. Dijken, X. Jiang, and S. S. P. Parkin, Giant magnetocurrent exceeding 3400% in magnetic tunnel transistors with spin-valve base layers, *Appl. Phys. Lett.* **83**, 951 (2003).
123. B. Huang, D. J. Monsma, and I. Appelbaum, Experimental realization of a silicon spin field-effect transistor, *Appl. Phys. Lett.* **91**, 072501 (2007).
124. B. Huang, L. Zhao, D. J. Monsma, and I. Appelbaum, 35% magnetocurrent with spin transport through Si, *Appl. Phys. Lett.* **91**, 052501 (2007).
125. B. Huang, D. J. Monsma, and I. Appelbaum, Coherent spin transport through a 350 micron thick silicon wafer, *Phys. Rev. Lett.* **99**, 177209 (2007).
126. J. S. Moodera, L. R. Kinder, T. M. Wong, and R. Meservey, Large magnetoresistance at room temperature in ferromagnetic thin film tunnel junctions, *Phys. Rev. Lett.* **74**, 3273–3276 (1995).
127. L. D. Bell, S. J. Manion, M. H. Hecht, W. J. Kaiser, R. W. Fathauer, and A. M. Milliken, Characterizing hot-carrier transport in silicon heterostructures with the use of ballistic-electron-emission microscopy, *Phys. Rev. B* **48**, 5712–5715 (1993).
128. B. Huang, H.-J. Jang, and I. Appelbaum, Geometric dephasing-limited Hanle effect in long-distance lateral silicon spin transport devices, *Appl. Phys. Lett.* **93**, 162508 (2008).
129. C. Canali, C. Jacoboni, F. Nava, G. Ottaviani, and A. Alberigi-Quaranta, Electron drift velocity in silicon, *Phys. Rev. B* **12**, 2265 (1975).
130. C. Jacoboni, C. Canali, G. Ottaviani, and A. Alberigi-Quaranta, A review of some charge transport properties of silicon, *Solid State Electron.* **20**, 77–89 (1977).
131. H.-J. Jang, J. Xu, J. Li, B. Huang, and I. Appelbaum, Non-ohmic spin transport in n-type doped silicon, *Phys. Rev. B* **78**, 165329 (2008).
132. I. Altfeder, B. Huang, I. Appelbaum, and B. C. Walker, Self-assembly of epitaxial monolayers for vacuum wafer bonding, *Appl. Phys. Lett.* **89**, 223127 (2006).
133. W. Hanle, The magnetic influence on the polarization of resonance fluorescence, *Z. Phys.* **30**, 93 (1924).
134. B. Huang, Vertical transport silicon spintronic devices, PhD thesis, University of Delaware, 2008.
135. J. L. Cheng, M. W. Wu, and J. Fabian, Theory of the spin relaxation of conduction electrons in silicon, *Phys. Rev. Lett.* **104**, 016601 (2010).
136. J. Haynes and W. Shockley, The mobility and lifetime of injected holes and electrons in germanium, *Phys. Rev.* **81**, 835 (1951).
137. X. Lou et al., Electrical detection of spin transport in lateral ferromagnet-semiconductor devices, *Nat. Phys.* **3**, 197 (2007).
138. F. J. Jedema, H. B. Heersche, A. T. Filip, J. J. A. Baselmans, and B. J. van Wees, Electrical detection of spin precession in a metallic mesoscopic spin valve, *Nature* **416**, 713 (2002).
139. J. Li and I. Appelbaum, Modeling spin transport with current-sensing spin detectors, *Appl. Phys. Lett.* **95**, 152501 (2009).
140. I. Appelbaum and D. Monsma, Transit-time spin field-effect transistor, *Appl. Phys. Lett.* **90**, 262501 (2007).
141. B. Huang and I. Appelbaum, Spin dephasing in drift-dominated semiconductor spintronics devices, *Phys. Rev. B* **77**, 165331 (2008).
142. V. F. Motsnyi, P. Van Dorpe, W. Van Roy et al., Optical investigation of electrical spin injection into semiconductors, *Phys. Rev. B* **68**, 245319 (2003).
143. B. Huang, D. J. Monsma, and I. Appelbaum, Spin lifetime in silicon in the presence of parasitic electronic effects, *J. Appl. Phys.* **102**, 013901 (2007).
144. H.-J. Jang and I. Appelbaum, Spin polarized electron transport near the Si/SiO_2 interface, *Phys. Rev. Lett.* **103**, 117202 (2009).
145. M. D'yakonov and V. Perel', Spin relaxation of conduction electrons in noncentrosymmetric semiconductors, *Sov. Phys. Solid State* **13**, 3023 (1971).
146. E. Y. Sherman, Random spin–orbit coupling and spin relaxation in symmetric quantum wells, *Appl. Phys. Lett.* **82**, 209–211 (2003).
147. J. H. Shim, K. V. Raman, Y. J. Park et al., Large spin diffusion length in an amorphous organic semiconductor, *Phys. Rev. Lett.* **100**, 226603 (2008).
148. E. Rashba, Inelastic scattering approach to the theory of a magnetic tunnel transistor source, *Phys. Rev. B* **68**, 241310 (2003).

19

III–V Magnetic Semiconductors

Carsten Timm
Technische Universität Dresden

19.1 Introduction

This chapter discusses the properties of III–V diluted magnetic semiconductors (DMS) and their current theoretical understanding. By DMS we mean semiconducting compounds doped with magnetic ions and showing ferromagnetic order below some Curie temperature T_C. Due to the semiconducting properties of DMS, the carrier concentration can be changed significantly by doping and even *in situ* by the application of a gate voltage or by optical excitations. Since the carriers couple strongly to the magnetic moments, this leads to unique properties and makes DMS promising for spintronics. In particular, magnetotransport effects in DMS tunnel junctions and DMS quantum dots are discussed in Chapters 21 and 30, respectively. The strong spin–orbit coupling in most DMS is responsible for the anomalous Hall effect discussed in Chapter 25.

Of the many excellent reviews on III–V DMS we only mention a few. Dietl and Ohno's concise review of the experimental and theoretical situation as of 2001 [1] is a good starting point. The extensive review by Jungwirth et al. [2] focuses on the theory of DMS, covering both ab initio calculations and the Zener kinetic-exchange theory discussed in this chapter. This review of 2006 remains the most definitive discussion.

Dietl [3,4] reviews ferromagnetic semiconductors not limited to III–V DMS. Sato et al. [5] review ab initio calculations as well as model-based calculations and simulations for III–V and II–VI DMS. Pearton et al. [6] and Liu et al. [7] concentrate on wide-gap III–V DMS and dilute oxides. Specialized reviews discuss disorder effects in DMS [8], ferromagnetic resonance (FMR) in (Ga,Mn)As [9], carrier localization in DMS [10], optical properties of III–V DMS [11], and the anomalous Hall effect [12]. The basic picture put forward in Ref. [11] has been challenged by Jungwirth et al. [13], as discussed below. The physics of doped (nonmagnetic) semiconductors is discussed in depth in the well-known book by Shklovskii and Efros [14].

Two classes of III–V DMS have emerged: on the one hand compounds with narrow to intermediate gaps, such as antimonides and arsenides, and on the other hand wide-gap materials, such as nitrides and phosphides. The first class appears to be much better understood and will be discussed first and in more detail. At least some of the wide-gap materials are more similar to the dilute magnetic oxides discussed in Chapter 20. We will here restrict ourselves to a number of comments pertinent to III–V compounds.

19.2 III–V DMS with Narrow to Intermediate Gaps

The equilibrium solubility of Mn in these DMS lies below 1% of cations, too low for ferromagnetic order to occur. This led to efforts to achieve higher Mn concentrations by means of nonequilibrium molecular beam epitaxy (MBE) at low substrate temperatures. With this technique, ferromagnetism in (In,Mn)As with $T_C \approx 7.5\,\mathrm{K}$ was achieved in 1992 [15] and in (Ga,Mn)As with $T_C \approx 60\,\mathrm{K}$ in 1996 [16]. The Curie temperature of (Ga,Mn)As has been pushed above 100 K [17] and recently up to around 190 K [18–20] by improvements in the MBE technique and careful post-growth annealing, which allow higher Mn concentrations up to nominally $x \approx 12\%$ per cation site [18–20]. In GaAs δ-doped with Mn, that is, containing an essentially 2D Mn-rich layer, $T_C \approx 250\,\mathrm{K}$ has been reached [21]. (Ga,Mn)As with large Mn, concentration has also been produced by Mn-ion implantation of GaAs followed by pulsed-laser annealing [22]. The values of T_C (exceeding 100 K), the magnetization curves, and the temperature-dependent resistivity are comparable to annealed MBE-grown samples [22]. Meanwhile, T_C for (In,As)Mn has been increased to 90 K [23].

(Ga,Mn)As is the most extensively studied III–V DMS, not the least because the host material GaAs is technologically important and well understood. It is known that ferromagnetism in (Ga,Mn)As requires Mn concentrations $x \gtrsim 1\%$ [17,24,25]. There is a zero-temperature insulator-to-metal transition at a slightly different concentration of about $x = 1.5\%$ for well-annealed samples [24–26]. The narrow-gap antimonides (Ga,Mn)Sb and (In,Mn)Sb show lower T_C [27–29], but are otherwise similar to the arsenides.

At present, two central questions regarding (Ga,Mn)As remain partially open. It is clear that for relatively light doping, bound acceptor states form a narrow impurity band (IB) in the gap. It is also obvious that for heavy doping, this IB will become very broad, merge with the valence band (VB), and lose its identity [30]. There is an ongoing debate whether in high-T_C (Ga,Mn)As one can still identify an IB. A related but separate question is whether the states close to the Fermi energy, which are responsible for charge transport and for the carrier-mediated magnetic interaction, are localized or extended. In the following sections, we discuss the physics of (III,Mn)V DMS, progressing from light to heavy Mn doping.

19.2.1 Properties of Isolated Mn Dopants: Single-Electron Picture

Manganese in narrow and intermediate-gap III–V semiconductors forms Mn^{2+} ions at cation sites of the zinc blende lattice, as seen in electron paramagnetic resonance (EPR) [31] and FMR [32]. The lattice with substitutional Mn impurities is depicted for the example of (Ga,Mn)As in Figure 19.1a. Since Mn^{2+} replaces a 3+ ion, manganese provides one fewer electron or, equivalently, donates a hole, and thus acts as an acceptor. Since it has a negative charge relative to the cation sublattice, it can bind the hole. The neutral acceptor-hole complex has been observed in infrared (IR) spectroscopy [33,34] and EPR [35]. The hole binding energy is found to be 112.4 meV [33–35], in agreement with the thermal activation energy seen in transport [36]. Mn acceptors close to the surface have been imaged and switched between the ionized (no bound hole) and neutral states with a scanning tunneling microscope (STM) at room temperature [37].

In contrast to simple shallow acceptors, Mn^{2+} has a half-filled d-shell. Photoemission experiments [38,39] find the maximum weight of occupied d-orbitals about 45 eV below the Fermi energy and do not find evidence for a mixed-valence state so that the unoccupied d-levels also do not lie close to the Fermi energy. The absence of d-level signatures in the optical conductivity shows that they lie above the conduction-band (CB) minimum. The d-shell thus mostly acts as a local spin $S = 5/2$, consistent with EPR experiments [35]. It is useful to notionally divide the Mn impurity into a simple acceptor and the half-filled d-shell. Ignoring the d-shell for the moment, the attractive potential should lead to hydrogenic energy levels for the hole. A spherical approximation for the band structure of GaAs gives a binding energy of 25.7 meV [40,41]. Binding energies of typical hydrogenic acceptors are larger, for example, 30.6 meV for Zn^{2+} [42], since the dielectric screening of the impurity charge becomes less effective close to the impurity. This can be taken into account by a central-cell correction [41], that is, an additional short-range attractive potential.

A more complex picture emerges if we take into account that the hole also experiences the periodic potential of the host lattice. The bound acceptor states are derived from small-wavevector states close to the VB top, which mostly consist of anion p-orbitals. The point group at the impurity (cation) site in zinc blende semiconductors is the tetrahedral group T_d [43]. Neglecting spin–orbit coupling, the anion p-orbitals correspond to the Γ_5 irreducible representation of T_d, that is, they have t_2 symmetry [43]. Γ_5 is three dimensional. Together with the electronic spin $s = 1/2$, we obtain a sixfold degenerate bound acceptor state.

In the presence of spin–orbit coupling, the theory of double groups [43] shows that the level splits into a Γ_7 (e'') doublet and

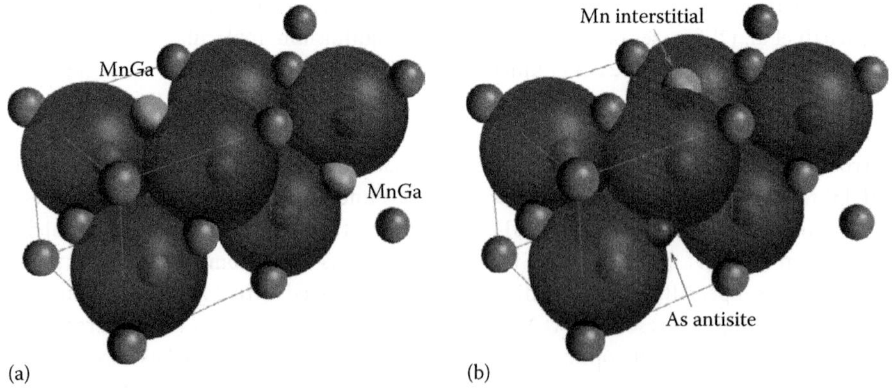

FIGURE 19.1 (See color insert.) (a) Zinc blende lattice structure of GaAs with two substitutional Mn impurities. Ga^{3+} ions are shown as small red spheres, As^{3-} as large, semitransparent blue spheres, and Mn^{2+} as small green spheres. The gray cube denotes the conventional unit cell. The ionic radii are used in the plot, which is an exaggeration, since the bonds are partially covalent. (b) Lattice structure of GaAs with an Mn^{2+} interstitial in the As-coordinated tetrahedral position and an As antisite defect (small blue sphere with the As^{5+} ionic radius).

a Γ_8 (u') quartet. The doublet can be thought of as resulting from the split-off band, which for GaAs is separated from the heavy-hole and light-hole bands by $\Delta_{SO} = 341$ meV at the Γ point. The doublet is therefore expected to form a resonance in the VB. The remaining fourfold degeneracy can be understood in terms of a total angular momentum $j = 3/2$ resulting from orbital angular momentum $l = 1$ and spin $s = 1/2$. STM experiments [26,37,44,45] find a pronounced spin-dependent spatial anisotropy of these states, in agreement with theory [37,44–47].

We now turn to the Mn d-electrons. In the T_d crystal field, the five d-orbitals split into three orbitals with t_2 character and two with e character. All five are singly occupied and the electron spins are aligned, showing that the Hund's-rule coupling is larger than the crystal-field splitting. Importantly, the t_2-type d-orbitals hybridize with the p-orbitals of the neighboring anions. This hybridization strongly affects the bound acceptor states, which are also of t_2 type if spin–orbit coupling is neglected and have a large amplitude at the neighboring sites [48,49]. The occupied d-orbitals, which all have the same spin, can only hybridize with the bound acceptor states with the same spin. Consequently, level repulsion shifts the occupied d-levels down in energy and the bound acceptor states with the same spin-up. The resulting bound acceptor states, called the *dangling-bond hybrid* [48,49], contain some d-level admixture and are deeper in the gap and thus more strongly bound than for a hydrogenic impurity [41,50]. The occupied d-levels with some admixture of anion p-orbitals are called the *crystal-field resonance* [48,49]. The resulting spin-up single-particle levels are sketched in the left part of Figure 19.2.

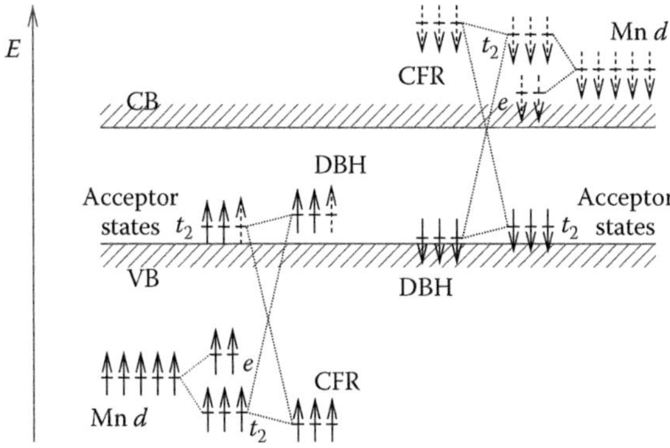

FIGURE 19.2 Sketch of single-particle states introduced by substitutional Mn acceptors in GaAs. The spin of the Mn d-shell is assumed to point upward. Spin–orbit coupling is ignored. Bold (dashed) arrows indicate occupied (unoccupied) states. The VB and CB edges are indicated. Left: level repulsion between spin-up d-states of t_2 symmetry and bound acceptor states also of t_2 symmetry leads to the occurrence of a spin-up dangling-bond hybrid (DBH) in the gap and a crystal-field resonance (CFR) deep in the VB. Right: for spin-down states, the unoccupied d-states are high above the Fermi energy and the DBH is shifted downward in energy. In the ground state, the single hole introduced by the Mn acceptor occupies one of the most strongly bound spin-up DBHs.

On the other hand, the bound acceptor state with spin opposite to the Mn d-shell hybridizes with the unoccupied t_2 orbitals and is shifted downward, as shown in the right part of Figure 19.2. This leads to spin splitting of the bound state, which depends on the orientation of the Mn spin [48,49]. In the ground state, a hole will occupy the most strongly bound state, which has the same spin as the d-electrons. Thus, an electron with this spin is missing and the total spin of the system is reduced. In this sense, the exchange coupling J_{pd} between the Mn d-shell and the VB-derived states is antiferromagnetic.

19.2.2 Properties of Isolated Mn Dopants: Many-Particle Picture

For several reasons, the single-particle picture is not sufficient for the understanding of Mn dopants [13]: First, the splitting between occupied and unoccupied Mn d-levels is large, as noted above, mostly due to the on-site Coulomb repulsion $U \approx 3.5$ eV [39]. Second, there is a large Hund's-rule coupling $J_H \approx 0.55$ eV [51]. The resulting strong correlations are not well captured by density-function theory (DFT) using standard approximations such as the local (spin) density approximation (LDA, LSDA). There has been progress in incorporating strong correlations either by phenomenologically introducing a Hubbard-U (LDA + U, LSDA + U) [52,53] or by correcting for self-interaction artifacts [54,55]. These approaches generically push Mn d-orbital weight away from the Fermi energy, leading to small Mn d-orbital weight in the bound acceptor states.

Since the hybridization between the anion p-orbitals and the Mn d-orbitals is small compared to their energy separation, it can be described perturbatively. Canonical perturbation theory [56,57] maps the d-electrons onto a local spin and a short-range potential scatterer. The latter is a typically negligible correction to the Coulomb attraction. One finds an antiferromagnetic pd-exchange interaction $J_{pd}(\mathbf{r})$ between the hole and Mn spins, which falls off on the length scale of the anion–cation separation. In reciprocal space, this leads to a $\mathbf{k}$-dependent interaction $\tilde{J}_{pd}(\mathbf{k})$ [57]. The exchange coupling quoted in the literature is the limit $J_{pd} = \tilde{J}_{pd}(\mathbf{k} \to 0)$.

The third reason for the failure of single-particle physics is that each Mn acceptor can bind a single hole, but not a second one [49]. Thus, the Coulomb repulsion U_{acc} in the bound acceptor state is at least of the same order as the binding energy. The physics of impurity states in doped semiconductors is strong-correlation physics [58]. Naive single-particle approaches predict too many bound acceptor states, seen, for example, in Figure 19.2.

The emerging picture describes an impurity state containing a single hole with total angular momentum $j = 3/2$ antiferromagnetically coupled to a local spin $S = 5/2$. In the four-dimensional subspace spanned by the $j = 3/2$ impurity states, one finds the operator identity $\mathbf{s} = \mathbf{j}/3$ [2,41], which allows one to write the exchange interaction either as $\varepsilon \mathbf{j} \cdot \mathbf{S}$ or as $3\varepsilon \mathbf{s} \cdot \mathbf{S}$. Taking the factor of three into account, one can express the exchange constant ε in terms of the exchange coupling J_{pd} and the envelope function $f(\mathbf{r})$ describing the spatial extent of the bound acceptor state [2,41],

$$\varepsilon = \frac{1}{3}\int d^3r\, J_{pd}(\mathbf{r})|\, f(\mathbf{r})\,|^2 \approx \frac{1}{3} J_{pd}\overline{|\, f(\mathbf{r})\,|^2}. \quad (19.1)$$

The approximation is valid if the envelope function changes much more slowly in space than the exchange interaction. The average is taken over the range of $J_{pd}(\mathbf{r})$. From IR spectroscopy [34], a value of $\varepsilon \approx 5$ meV is obtained. Photoemission experiments [38] independently give $J_{pd} \approx 54$ meV nm^3. (In the literature, J_{pd} is often denoted by β and $J_{pd}n_{\text{Mn}}$ by βN_0, where n_{Mn} is the Mn concentration [2]. J_{pd} has units of energy times volume.) The exchange interaction can be rewritten as

$$\varepsilon \mathbf{j}\cdot\mathbf{S} = \frac{\varepsilon}{2}[F(F+1) - j(j+1) - S(S+1)], \quad (19.2)$$

where $\mathbf{F} = \mathbf{j} + \mathbf{S}$ is the total angular momentum. The splitting between the ground state triplet ($F = 1$) and the first excited quintet ($F = 2$) is thus $2\varepsilon \approx 10$ meV [34].

19.2.3 Other Isolated Defects

III–V DMS grown by low-temperature MBE contain relatively large concentrations of defects apart from substitutional magnetic dopants. For (Ga,Mn)As, the most important ones are As ions at Ga sites (arsenic antisites) and Mn ions in interstitial positions. The positions of these two defects are shown in Figure 19.1b. Vacancy defects are present in much smaller concentrations, but can strongly enhance the mobilities of other defects [59].

Even undoped GaAs grown by low-temperature MBE contains a large concentration of As antisites [60], which act as double donors, as expected for As^{5+} cations. In (Ga,Mn)As, antisites should thus lead to partial compensation: there are fewer holes than Mn acceptors. Variation of the As:Ga ratio during growth has been found to lead to a corresponding change in the hole concentration, confirming that antisites are also present in (Ga,Mn)As [61]. Cracking As_4 clusters to As_2 during growth can substantially reduce the compensation [62], probably by reducing the concentration of antisites. The mobility of antisites in GaAs is likely small at relevant temperatures [63] so that their distribution is not affected by post-growth annealing. However, due to their Coulomb attraction, antisite donors could preferentially be incorporated close to Mn acceptors during growth [64].

During low-temperature growth of (Ga,Mn)As, some Mn^{2+} ions are incorporated in interstitial positions. Channeling Rutherford backscattering experiments [65] have provided evidence for about 17% of Mn residing in interstitial positions for as-grown samples with an Mn concentration of $x \approx 7\%$. DFT calculations find a lower formation energy of the As-coordinated tetrahedral Mn interstitial than of the Ga-coordinated one [49,66,67], as expected for a cation, but the predicted energy differences do not agree well. Channeling Rutherford backscattering cannot distinguish between these two positions.

Mn^{2+} interstitials in GaAs are double donors contributing to compensation. Since the mobility of Mn interstitials at typical growth and annealing temperatures around 250°C is substantial [65,66,68,69], they can lower their electrostatic energy by assuming positions close to substitutional Mn acceptors during annealing, not just during growth [64,70]. Mn interstitials also diffuse to the film surface during annealing, where they form a magnetically inactive layer [66,68,69,71]. This is thought to be the main origin of the observed increase of the hole concentration (i.e., decrease of compensation) during annealing [59,71,72] and of the higher T_C of thinner films [73,74]. This is nicely demonstrated by capping the surface with a thin GaAs layer, which strongly suppresses the effect of annealing [71].

Mn^{2+} interstitials also carry a local spin $S = 5/2$. For an Mn interstitial and a substitutional Mn dopant in nearest-neighbor positions, one expects a strong direct antiferromagnetic exchange interaction. X-ray magnetic circular dichroism experiments [75] support this picture. On the other hand, the question of the exchange interaction between VB holes and Mn-interstitial spins is still open, with DFT calculations giving conflicting results [67,70].

Since As antisites and Mn interstitials are double donors, the electrostatic interaction favors their formation for higher concentrations of substitutional Mn acceptors. This *self-compensation* is a limiting factor in the growth of heavily Mn-doped (Ga,Mn)As.

19.2.4 Effects of Weak Mn Doping

A nonzero concentration of Mn impurities does not have a strong effect on the exchange interaction J_{pd}, since J_{pd} is mainly due to hybridization between Mn and neighboring anion orbitals. In contrast, it does have important effects on the electronic system. For small Mn concentrations, the total potential seen by the holes consists of the sum of the Coulomb potentials of the individual acceptors. The impurity potentials fall off like $1/r$ in the absence of metallic screening, whereas the bound-state wave function falls off exponentially. Hence, the overlap of the potentials remains relevant for arbitrarily small concentrations, unlike the overlap of the bound states. This situation of very weak doping is sketched in Figure 19.3a. The acceptor binding energy in this regime is reduced, since an acceptor can be ionized by exciting the hole to roughly the *average* potential. This leads to a reduction of the binding energy proportional to $x^{1/3}$ [36,76], which is observed in Mn-doped GaAs [13].

In this regime, IR spectroscopy shows a broad mid-IR peak in the ac conductivity [33,34,77]. This peak is broad despite of the sharp energy of bound acceptor states, since the bound impurity wave functions contain many wavevector components so that transitions from large-$\mathbf{k}$ states deep within the VB to the impurity states are possible [13].

With increasing Mn doping x, the bound acceptor states start to overlap. Let us for the moment assume a periodic superlattice

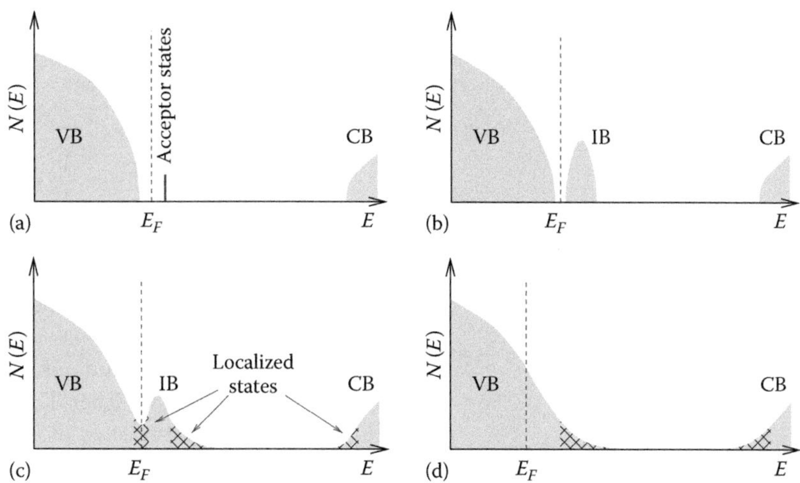

FIGURE 19.3 Density of states of III–V DMS with narrow to intermediate gaps. The Fermi energy E_F is in all cases shown assuming no compensation. (a) For very weak doping, the overlap and thus the hybridization of bound acceptor states is negligible. (b) For larger doping, but assuming a regular superlattice of dopants, a sharply delimited IB emerges. If the doping level is not too large, there is a true gap between IB and VB. For zero compensation, E_F falls into this gap. (c) A quite different picture emerges if disorder is included. The bands develop tails of localized states (shown crosshatched) and the IB becomes asymmetric. Note that it is a present unknown whether extended states exist in the center of the IB before it merges with the VB, as shown here, or not. (d) For heavy doping, VB and IB merge completely and the Fermi energy lies deep in the region of extended states.

of Mn dopants, ignoring disorder. Tunneling of electrons between the acceptor states leads to a narrow IB. Consequently, the mid-IR conductivity peak broadens and shifts slightly to lower energies [77]. The total weight of the IB increases linearly with x, whereas the tunneling matrix elements and thus the band width increase exponentially. Thus, the density of states does not become very large. This might explain why the Stoner enhancement of ferromagnetism, which is governed by the product of local Coulomb repulsion U_{acc} and density of states, remains small [78,79,87,93].

Since we have so far ignored disorder, the density of states shows sharp edges of IB and VB and a true gap between the bands. A density of states of this type is predicted by theories that effectively average over disorder [50,80,81]. The doping also affects the host VB, for example, since the impurity states take away spectral weight from the VB. The density of states in this regime is sketched in Figure 19.3b.

If we include the random positions of Mn dopants, the system is no longer periodic and, strictly speaking, band theory breaks down, whereas the density of states remains a well-defined quantity. Due to disorder, the band edges in the density of states develop tails [46,82], which always overlap, so that there is no true gap. However, it is reasonable to say that the IB still significantly affects the physics if there is a clear minimum in the density of states in the overlap region with the VB. In addition, since the tunneling matrix elements decay exponentially with separation, the density of states of the IB becomes highly asymmetric [83]. This situation is sketched in Figure 19.3c. With compensation, E_F can shift to the extended states close to the IB center.

It remains true that the IB contains one state per dopant, whereas single-particle approaches taking the angular momentum $j = 3/2$ into account find four states. Such approaches predict the Fermi energy E_F to lie deep within the IB, even for vanishing compensation, leading to a relatively large density of states at E_F [84,85]. However, taking many-particle effects into account, E_F falls into the minimum of the density of states between IB and VB, leading to very different physics. A separate IB with too many states can evidently mimic the large density of states at E_F found if the bands have merged.

19.2.5 Effects of Heavy Mn Doping

For further increased Mn doping, the IB peak in the density of states grows in weight and width. Consequently, the minimum in the density of states between IB-type and VB-type states is gradually filled in. Krstajić et al. [30] estimate that the energy of the IB center minus half its width—a rough measure of its lower edge—crosses the unperturbed VB top already for an Mn concentration below 1%.

In the heavily doped limit, the IB completely merges with the VB. Due to disorder, the states in the band tail are expected to be localized, as shown in Section 19.2.7, but the hole concentration is relatively high so that E_F lies deep in the band, unless compensation is nearly complete. Both the impurity potential and the Coulomb repulsion between the holes are screened by the itinerant holes close to E_F. In this regime, it is reasonable to start from the unperturbed VB with the chemical potential located in the band and to incorporate disorder perturbatively in the second step [2]. The density of states for this case is sketched in Figure 19.3d.

The picture emerging for heavily doped DMS—holes in a weakly perturbed VB coupled to local spins by an antiferromagnetic exchange interaction—is the Zener kinetic-exchange or *pd*-exchange model [2,30,78,86–89]. Taking the *pd*-exchange interaction to be local, the model is characterized by a Hamiltonian of the form

$$H = H_{\text{holes}} + J_{pd} \sum_{I,i} \mathbf{S}_I \cdot \mathbf{s}_i \delta(\mathbf{R}_I - \mathbf{r}_i), \tag{19.3}$$

where

$$H_{\text{holes}} = \sum_{n\mathbf{k}\sigma} \varepsilon_{n\mathbf{k}\sigma} c^\dagger_{n\mathbf{k}\sigma} c_{n\mathbf{k}\sigma} \tag{19.4}$$

describes noninteracting holes in the VB, $\mathbf{S}_I$ are the impurity spins at positions $\mathbf{R}_I$, and $\mathbf{s}_i$ are the hole spins at positions $\mathbf{r}_i$. In this basic form, the Coulomb potentials due to acceptors and donors are neglected. Since properties of the host band structure are crucial for this model, a realistic description is called for. The band structure of GaAs is sketched in Figure 19.4. In DFT, one obtains the (Kohn–Sham) band structure as a matter of course, which for clean semiconductors is in fair agreement with the real (Landau-quasiparticle) bands, except for well-understood problems such as the underestimation of the energy gap. It should be noted, however, that many DFT-based works neglect the spin–orbit coupling, which is strong in arsenides and antimonides.

There are two main model-based approaches to a realistic band structure $\varepsilon_{n\mathbf{k}\sigma}$. On the one hand, in Luttinger–Kohn $\mathbf{k}\cdot\mathbf{p}$ theory [47,90–93], one restricts the Hilbert space to the relevant bands (usually the heavy-hole, light-hole, and split-off VBs, perhaps adding the CB) and expands $\varepsilon_{n\mathbf{k}\sigma}$ in the wave vector $\mathbf{k}$ around the Γ point ($\mathbf{k}=0$). The approach incorporates spin–orbit coupling and the nonspherical $\mathbf{k}$-dependence of $\varepsilon_{n\mathbf{k}\sigma}$ and includes the correct effective masses at Γ. The bands become increasingly inaccurate away from $\mathbf{k}=0$. If only the light-hole and heavy-hole bands are taken into account and the

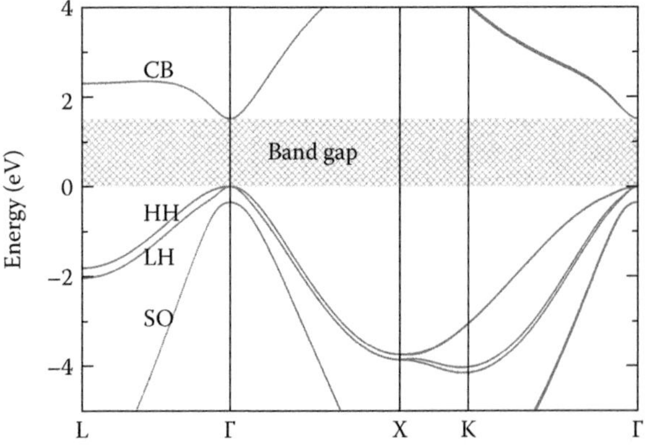

FIGURE 19.4 Sketch of the band structure of GaAs (more precisely, the result of a tight-binding calculation using parameters from Ref. [95]). The band structure of other zinc blende semiconductors is topologically identical. The conduction band (CB) and the heavy-hole (HH), light-hole (LH), and spin-off (SO) valence bands are marked.

nonspherical $\mathbf{k}$-dependence is averaged over all directions, one arrives at the spherical approximation, which is described by the first-quantized 4×4 Hamiltonian

$$H_{\text{sph}}(\mathbf{k}) = \frac{\hbar^2}{2m}\left[\left(\gamma_1 + \frac{5}{2}\gamma_2\right)k^2 - 2\gamma_2(\mathbf{k}\cdot\mathbf{j})^2\right] \tag{19.5}$$

in terms of the 4×4 total-angular-momentum operator $\mathbf{j}$ with $j=3/2$. γ_1 and γ_2 are parameters. The last term stems from spin–orbit coupling and is responsible for the splitting between the light-hole and heavy-hole bands away from the Γ point.

The other approach, Slater–Koster tight-binding theory [46,57,82,94–96], is formulated in real space by specifying on-site energies and hopping amplitudes for the relevant orbitals. Spin–orbit coupling is included [95]. These parameters are chosen so as to match the real band structure at high-symmetry points in the Brillouin zone. Therefore, the overall shape of the band structure agrees well with experiments, compare Figure 19.4, but local properties, such as effective masses, are not well described. To resolve this problem, Yildirim et al. [97] have introduced a tight-binding model that reduces to the $\mathbf{k}\cdot\mathbf{p}$ Hamiltonian for small $\mathbf{k}$.

There is a persistent controversy regarding the range of Mn concentrations for which an IB can still be identified. Model-based theories tend to predict that the IB and VB have merged at concentrations x of a few percent [2,30,50,83]. The VB picture has been very successful in explaining many experiments, for example, magnetization curves, magnetic anisotropies, magnetic stiffness, Gilbert damping, magnetotransport properties, the anomalous Hall effect, and optical properties [2,12,98].

However, a number of experiments appear to be inconsistent with this picture [2,11,98]: First, photoemission experiments [99] have found a weakly $\mathbf{k}$-dependent feature in the density of states of (Ga,Mn)As with $x\approx3.5\%$, which has been interpreted as an IB. This is not easily reconciled with the large energy of this feature, about 0.5 eV below E_F. An alternative explanation invokes localized states from the tail of the merged VB, which could strongly affect the photoemission signal [2].

Second, hot-electron photoluminescence experiments [100] are interpreted in terms of transitions of electrons from the CB into an IB. The high-energy cutoff in the spectra is a measure of the difference between the energy of photoexcited CB electrons and the Fermi energy, regardless of whether it lies in an IB or in the VB. Detailed calculations based on the VB or IB picture do not yet exist, though. Third, magnetic circular dichroism experiments [101] on (Ga,Mn)As with $x\lesssim3\%$ have been interpreted in terms of multiple IBs. The data and in particular the positive sign of the dichroism signal are claimed to be inconsistent with the VB picture [101]. However, a recent tight-binding calculation based on the VB picture [102] does find a positive signal.

Fourth, the effects of intentionally increased disorder are interpreted in terms of the IB picture. On the one hand, Stone et al. [103] find that in the quaternary DMS (Ga,Mn) (As,P) and (Ga,Mn) (As,N) already a concentration of phosphorus or nitrogen in the percent range makes the samples insulating and significantly reduces T_C. This is in agreement with estimates based on the IB picture. On the other hand, Sinnecker et al. [104] have irradiated (Ga,Mn)As with ion beams to induce disorder, finding that the transport and magnetic properties are rather robust. This is argued to indicate that transport takes place in an IB, since the VB should be much more strongly affected by the disorder [104]. Note that high and low sensitivity of the DMS against disorder are both cited as support for the IB picture [103,104], indicating that further work is needed.

Fifth, Singley et al. [105] and Burch et al. [106] have performed IR spectroscopy on (Ga,Mn)As with $x = 1.7\%–7.3\%$. Except for $x = 1.7\%$, the samples are metallic and show a broad mid-IR peak in the ac conductivity [105,106]. This peak increases in weight for increasing x and also with annealing. The peak maximum lies around 250 meV for $x = 2.8\%$ and shifts to lower energies (redshifts) for increasing x and with annealing.

The main tenet of Refs. [11,105,106] is to interpret this peak in terms of VB-to-IB transitions. The redshift with increasing x is then naturally explained in terms of increased electronic screening, which moves the IB closer to the VB, and growing IB width [11,105,106]. These features are incorporated in a recent theoretical study [85], which treats the disorder by exact diagonalization in a supercell. The authors consider $x \leq 1.5\%$, mostly outside of the controversial region. Note that the calculation is based on a single-particle picture and therefore overestimates the number of states in the IB. The calculated peak shows a slight redshift or blueshift with increasing x depending on the compensation [85].

In the VB picture, one naturally expects a blueshift, in contrast to the observation [105,106]. In this picture, the mid-IR peak is due to transitions between the light-hole and heavy-hole bands. For increasing x, the Fermi energy moves deeper into the VB, the typical $\mathbf{k}$ vectors of these transitions grow, and the energy difference between light-hole and heavy-hole bands increases. An older calculation [107] indeed predicts a blueshift. A blueshift is actually observed in GaAs heavily doped with carbon, a hydrogenic acceptor [108].

For some of the samples, the ac conductivity also shows a smaller maximum at zero frequency [105,106], which is interpreted as a Drude peak. By fitting a two-component model to the data and applying the conductivity sum rule to the Drude part alone, Burch et al. [106] extract the effective mass of the carriers. The resulting large effective mass is interpreted in terms of strong electron–electron interactions, as expected for an IB [106]. However, the analysis should be viewed with caution, since it relies on a single sample and neglects that the Drude peak can lose weight by disorder scattering [13].

Jungwirth et al. [13] challenge the IB interpretation [11,105,106]. They point out that the maximum of the mid-IR peak is at a higher energy than expected from extrapolating the weak-doping data [77] and that the VB picture can also explain a redshift of the mid-IR peak, if disorder is treated beyond the first-order Born approximation (see Section 19.2.7). The main idea is that transitions into localized states in the tail of the merged VB do not have to conserve crystal momentum and can therefore contribute with large weight at relatively large energies [13]. For increasing x, the samples become more metallic and the high-energy contribution from localized states decreases, which could lead to a redshift. Within the VB picture, Yang et al. [109] have performed calculations based on a $\mathbf{k}\cdot\mathbf{p}$ Hamiltonian and *pd*-exchange interaction with frozen spins, treating the electron–electron interaction in the Hartree approximation and the disorder by exact diagonalization for supercells. Only two parameter sets have been treated, but these are consistent with a redshift [13,109]. Systematic calculations of the x-dependence of the ac conductivity might settle this issue. Note that recent IR magneto-optical experiments have been explained within the VB picture [110]. Related ideas have been explored with regard to circular dichroism [111].

Furthermore, several strong arguments support the VB picture for Mn concentrations above a few percent. First, for carrier-mediated magnetic interactions, a minimum in the density of states as sketched in Figure 19.3c would lead to a suppression of T_C at weak compensation. Experimentally, there is no sign of this for metallic (Ga,Mn)As [2,4,18, 19,112–114].

Second, the temperature-dependent dc conductivity should show an upturn when thermal activation of holes from the Fermi energy, supposed to lie in an IB, to extended states in the VB becomes possible. Indeed, this is observed for weakly Mn-doped GaAs [13,115]. In metallic (Ga,Mn)As with $x \gtrsim 2\%$, no activated behavior is observed [13].

Third, Neumaier et al. [116] have extracted the density of states from the temperature-dependent correction to the conductivity due to the electron–electron interaction [117] for samples of various geometries with $x \approx 4\%$ and 6%. They do not find any evidence for a dip in the density of states and do find effective masses of the order of the bare mass. Their results are close to the prediction of a tight-binding model including disorder [46] and suggest merged bands and itinerant holes at E_F. Note, however, that this interpretation of the data depends quite heavily on theory. All in all, further experimental and theoretical work is needed to elucidate the doping-dependent electronic structure.

19.2.6 Carrier-Mediated Magnetic Interaction

For heavily doped (Ga,Mn)As, we have arrived at the Zener *kinetic-exchange* model [86]. This is different from the Zener *double-exchange* model [118], which is concerned with the interaction between mixed-valence ions. The two pictures are opposite limiting cases on a continuum, though. For heavy doping, superexchange between neighboring Mn dopants might also be relevant [52,79,119,120]. For Mn^{2+} in III–V DMS, superexchange is antiferromagnetic [121]. Dietl et al. [89] have argued that it is small. Within canonical perturbation theory, the superexchange interaction is indeed smaller than the Ruderman–Kittel–Kasuya–Yosida (RKKY) interaction discussed below by a factor of the order of E_F divided by the Coulomb repulsion U in the d-shell [57].

There are a number of "smoking-gun" experiments showing that the magnetic interaction is indeed carrier mediated: in field-effect transistors based on (In,Mn)As [122] and (Ga,Mn)As [112–114] as well as on GaAs δ-doped with Mn [123], the hole concentration can be changed by a gate voltage. A scheme of the device is shown in Figure 19.5. A positive (negative) voltage, which repels (attracts) the holes, leads to a reversible decrease (increase) of the magnetization and of T_C, as expected for a hole-mediated interaction. The reverse effect, that is, using a magnetic field close to T_C to change the magnetization and *thereby* the carrier concentration [124] has not yet been observed. In a nonuniform magnetic field, this should lead to a voltage drop in the direction of the field gradient [124].

Furthermore, Koshihara et al. [125] have shown that irradiation of originally paramagnetic (In,Mn)As/GaSb heterostructures with deep red to IR light induces magnetic hysteresis. The results are easily understood in terms of photoinduced holes transferred from the GaSb layer to the (In,Mn)As layer. In ferromagnetic (Ga,Mn)As, Wang et al. [126] have observed an ultrafast enhancement of ferromagnetism after irradiation. The light pulse is thought to generate additional holes in the VB, which increase the magnetic interaction.

Measurements of the anomalous Hall effect in (Ga,Mn)As [12,127] also support the picture of carrier-mediated magnetic interactions, since the magnetization inferred from the anomalous Hall effect agrees well with the one obtained from direct magnetometer measurements. The anomalous Hall effect is discussed in Chapter 25. In addition, experiments show a large and complex magnetic anisotropy in (In,Mn)As and (Ga,Mn)As films [9,128–131], which, for example, leads to reorientation transitions of the easy axis as a function of carrier concentration, temperature, etc. In view of the tiny spin–orbit splitting [132] and resulting tiny magnetic anisotropy of the Mn d^5 configuration, this would be hard to understand based on direct exchange interactions.

The kinetic-exchange model naturally leads to an RKKY interaction [133] between the Mn spins. The interaction is essentially proportional to $J_{pd}^2 \chi(\mathbf{r}_1 - \mathbf{r}_2)$, where χ is the nonlocal magnetic susceptibility of the carriers. If the Zeeman splitting of the VB is small, as is certainly the case close to T_C, it is sufficient to assume unpolarized bands. The RKKY interaction calculated for realistic bands [84,57,134–136] is quite different from the textbook expression for a parabolic band [133]. Using a tight-binding band structure of GaAs, Timm and MacDonald [57] find an RKKY interaction that is highly anisotropic both in real space and in spin space. The interaction is ferromagnetic and large at small separations and shows the characteristic Friedel oscillations with period π/k_F at larger separations, albeit with anisotropic Fermi wavenumber k_F. In compensated samples, typically several neighboring impurities lie within the first ferromagnetic maximum, as sketched in Figure 19.6a, favoring ferromagnetic order [84,87,134,135,137]. Interactions at larger distances have random signs, but are small and average out for most impurity spins. However, new high-quality samples are only weakly compensated so that the period of the Friedel oscillations is comparable to the impurity separation, as shown in Figure 19.6b. In this case, it might be important that the first minimum in the (110) direction does not reach negative, that is, antiferromagnetic, values [57], that is, in Figure 19.6b the innermost antiferromagnetic (gray) region is absent in some directions. Similar "avoided frustration" is also found in DFT calculations [119,138].

Kitchen et al. [44] have performed STM experiments for Mn ions on top of the (110) surface of Zn-doped p-type GaAs. The observed effective magnetic interactions are also highly anisotropic in real space. They agree well with tight-binding calculations using the bulk GaAs band structure [44].

The effective interaction between impurity spins has also been calculated within DFT from the total energy of aligned and anti-aligned spin pairs [119,138]. While the DFT and model-based results show similar strong anisotropy in real space, the DFT results typically lead to an overestimation of T_C, while model-based theories underestimate T_C [57]. Probably contributing to this discrepancy are the neglect of the Coulomb potential of the acceptors in the model calculations, the neglect of spin-orbit coupling in the DFT results, and the unphysically large weight of Mn d-orbitals close to the Fermi energy in DFT. Indeed, LDA + U usually leads to a reduction of the predicted T_C [52]. Disorder is also expected to play a major role, as discussed below.

FIGURE 19.5 Field-effect transistor with (In,Mn)As active region (schematically). By applying a voltage V_g to the gate electrode, which is electrically isolated from the (In,Mn)As channel, the hole concentration in the DMS can be changed, leading to a change in the magnetic interactions and thus in the magnetization. Source (S) and drain (D) electrodes can be used to measure transport properties of the gated DMS film.

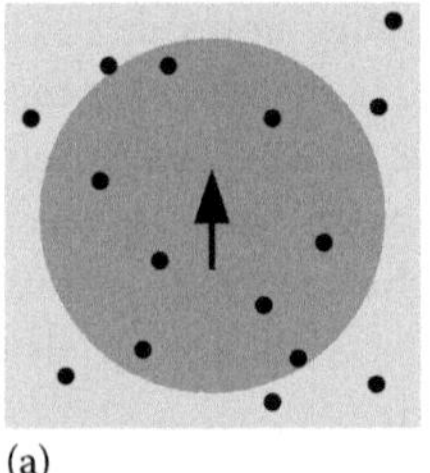

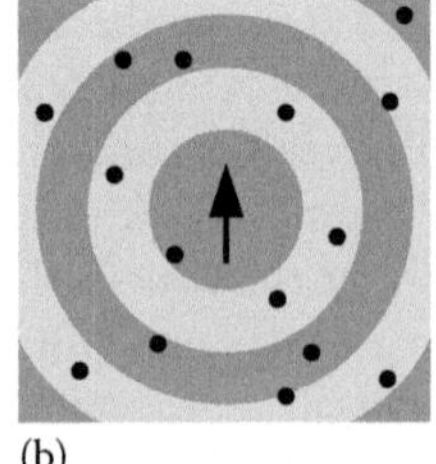

FIGURE 19.6 Sketch of regions with ferromagnetic (dark gray) and antiferromagnetic (light gray) RKKY interaction with regard to a given impurity spin, denoted by the arrow in the center. In (a) several neighboring impurity spins are ferromagnetically coupled, favoring ferromagnetic order, whereas in (b) the nearest-neighbor interaction is essentially random in sign.

The conceptually simplest treatment of the kinetic-exchange model involves two distinct mean-field-type approximations: First, the *pd*-exchange interaction is decoupled at the mean-field (MF) level, and, second, the random distribution of impurity spins is replaced by a smooth magnetization (virtual-crystal approximation [VCA]). Note that the same result for T_C is obtained by first calculating the RKKY interaction between impurity spins, and then making the VCA/MF approximations [87]. The result reads [2,78,87,137,139,140] as follows:

$$k_B T_C = \frac{S(S+1)}{3} \frac{J_{pd}^2 n_{\mathrm{Mn}} \chi_{\mathrm{Pauli}}}{g^2 \mu_B^2}, \qquad (19.6)$$

where

n_{Mn} is the concentration of Mn impurities
g is their g-factor
μ_B is the Bohr magneton
χ_{Pauli} is the Pauli susceptibility of the carriers

For a single parabolic band, one obtains $k_B T_C = S(S+1)N(E_F)J_{pd}^2 n_{\mathrm{Mn}}/6$ [137], where $N(E_F)$ is the density of states per spin direction at E_F. This leads to $T_C \propto a_0^{-6} p^{1/3} x^{4/3}$, where a_0 is the lattice constant, x is the Mn doping level related to the Mn concentration by $n_{\mathrm{Mn}} = 2.21 \times 10^{22}\,\mathrm{cm}^{-3} x$, and p is the number of carriers per impurity. For increasing x, the Curie temperature should thus increase without any saturation. An increase is indeed observed for high-quality samples [2,4,18,19,112–114]. The present model does not predict a reduction of T_C close to vanishing compensation, $p = 1$, in contrast to theories based on an IB.

Dietl et al. [89] have predicted the Curie temperatures of 5% Mn-doped III–V and II–VI DMS based on the $\mathbf{k} \cdot \mathbf{p}$ Hamiltonian. The main trend is a strong increase of T_C for lighter elements [89], which can already be understood from the parabolic band model: assuming that J_{pd} is approximately independent of the host semiconductor [89], the strong dependence of T_C on the lattice constant a_0 favors high T_C for hosts with smaller a_0. Reinforcing this trend, the effective masses and thus the density of states $N(E_F)$ are typically larger for lighter elements.

Several routes have been taken to go beyond the MF approximation. We here restrict ourselves to equilibrium properties; spin dynamics is discussed in Chapter 16. A common approach is to first map the system onto a (Mn) spin-only model [119,136,141] as discussed earlier. While Gaussian fluctuations (noninteracting spin waves) do not affect T_C, higher-order fluctuations can appreciably reduce it, as has been shown within the Tyablikov approximation [141] and within the random-phase approximation (RPA) [136]. Both approaches include disorder and achieve quantitative agreement of $T_C(x)$ with experiments for high-quality, annealed (Ga,Mn)As.

Calculations of the spin-wave stiffness find a much higher energy of spin waves for realistic band structures incorporating spin–orbit coupling compared to a parabolic band [2,88]. This is partly due to the splitting between heavy-hole and light-hole bands, which can be understood as magnetic anisotropy of the hole total angular momentum $\mathbf{j}$. The large stiffness stabilizes ferromagnetism and is consistent with FMR experiments [142]. The theoretical situation is not quite clear, though, since the spin-only model mentioned earlier [136,141] does not include spin–orbit coupling but nevertheless within the RPA yields a stiffness in good agreement with experiments [143]. In addition, a calculation going beyond RPA but neglecting disorder [144] finds a strong enhancement of the spin-wave stiffness compared to the RPA. The relative importance of spin–orbit coupling, disorder, and higher-order many-particle corrections for the spin-wave stiffness is thus not yet clear.

19.2.7 Disorder and Transport

The resistivity of high-T_C (Ga,Mn)As generically shows a finite residual value at low temperatures, sometimes with a small upturn for $T \to 0$ [17,18,22,59,72,145,146]. A finite limit of the resistivity is the defining property of a metal. (Ga,Mn)As is metallic due to the heavy doping. In addition, there is a broad and high peak in the resistivity in the vicinity of T_C [13,17,18,22, 59,72,105,145,146], also seen in (In,Mn)As [147]. In high-quality samples, its slope on the $T > T_C$ side is relatively flat [13,18,72].

DMS are highly disordered, since the magnetic dopants and the compensating defects are randomly distributed. Due to the strong interactions between charged defects, they are expected to be incorporated in correlated positions, likely as defect clusters, or to achieve such positions during annealing [64,70,79,148–154]. Random telegraph noise in the charge transport in (Ga,Mn)As [155] is interpreted in terms of the random switching of the magnetic moments of small impurity clusters.

There are two main contributions to the disorder felt by the carriers. First, the Mn acceptors attract the holes. In metallic DMS, the long-range part of this attraction is screened by itinerant holes, but screening is not very efficient due to the small hole concentration. The strong short-range attraction is not appreciably screened. The acceptor potentials, together with the repulsive potentials from Mn interstitials and antisites, form the Coulomb disorder potential. Clustering of defects of course has a large effect on this potential [64,153].

Second, the *pd*-exchange interaction leads to disorder even if the Mn spins are frozen at low temperatures. Kyrychenko and Ullrich [154] have found that dynamical spin fluctuations are important for transport. Most theoretical approaches have nevertheless assumed the Mn spins to be frozen on the timescale relevant for electronic transport. In this limit, they act as a short-range, spin-dependent contribution to the disorder potential, which is small compared to the Coulomb disorder [2]. The metallic behavior of high-quality samples suggests that disorder can be treated perturbatively in the first-order Born approximation [156,157], which leads to a finite lifetime of quasiparticle states and to the exponential decay of the electronic Green function on the length scale of the elastic mean free path l. This leads to a finite conductivity at $T = 0$. For annealed (Ga,Mn)As, the product $k_F l$, which provides a measure of the strength of disorder, is experimentally found to increase from $k_F l \approx 1$ for $x \approx 1.5\%$ to $k_F l \approx 3$–5 for $x \approx 7\%$ [13,106], consistent with a bad metal.

Going beyond the weak-disorder limit, disorder can lead to localized electronic states [10,117,157,158]: electrons in low-energy states, that is, close to band minima, tend to be trapped in the lowest depressions of the random potential landscape. The same holds for holes close to band maxima. If disorder is not too strong, only states in the band tails are localized, whereas states in the band center are extended, with *mobility edges* separating the two types. If the states in the vicinity of E_F are localized, conduction is only possible by thermal activation into extended states, leading to an Arrhenius form of the conductivity, $\sigma \propto \exp(-\Delta E/T)$, where ΔE is the difference between E_F and the nearest mobility edge [159]. In (Ga,Mn)As, E_F seems to pass through a mobility edge for $x \approx 1.5\%$ [24–26], corresponding to an insulator-to-metal transition. A higher dopant concentration is required to make the system metallic compared to GaAs with shallow dopants, since the potential of a single Mn acceptor and consequently the disorder are stronger [13].

Moca et al. [146] find that measurements of the magnetoresistivity of (Ga,Mn)As with $x \lesssim 6.7\%$ and small $k_F l \sim 1$ are reasonably well fit by expressions derived from the scaling theory of localization [158]. The comparison shows that the drop in resistivity roughly below T_C is strongly correlated with the magnetization and suggests that the upturn of the resistivity at low T [17,18,59,72,145,146] results from disorder scattering. Recent STM and scanning tunneling spectroscopy experiments [26] exhibit a rich spatial structure of the density of states at the surface of (Ga,Mn)As close to the metal–insulator transition. The correlations of the density of states at E_F, but not at other energies, decay as a power law with distance [26]. If this power-law behavior were pinned to E_F and not to the mobility edge, this would indicate that the electron–electron interaction plays a crucial role [26]. How the pronounced spatial dependence [26] can be reconciled with the homogeneous ferromagnetism observed by muon spin relaxation (μSR) [160] is not yet clear.

As noted earlier, the question of localization of states is separate from the question of the survival of the IB [10]. In particular, an insulator-to-metal transition has been observed in p-doped GaAs quantum wells with E_F in an IB [161]. Transport measurements also show metallic conduction in an IB in (Ga,Mn)As with $x \approx 0.3\%$ [162], contradicting earlier results [24,25]. A picture of localized and extended states within an IB, as sketched in Figure 19.3c, is corroborated by exact-diagonalization studies [84,85]. At present, it is unclear whether such a scenario pertains to III–V DMS.

In the regime of weak doping, where an IB exists and the states at E_F are localized, it is reasonable to start from a picture of holes hopping between bound acceptor states [84,85,149,163–166]. The bound-hole spin is exchange-coupled to one or more impurity spins, forming a hole dressed with a spin-polarization cloud, called a *bound magnetic polaron* (BMP) [164,165]. The exchange coupling to the same hole spins leads to an effective ferromagnetic coupling of the impurity spins in the BMP.

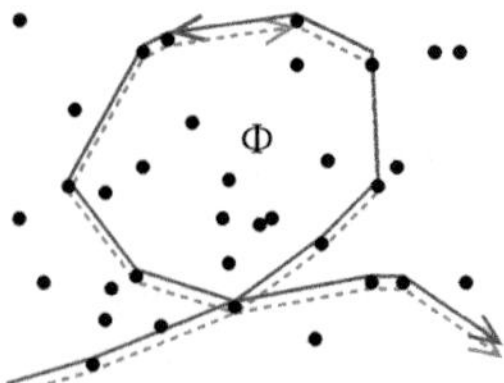

FIGURE 19.7 Scattering paths traversed in opposite directions. The black dots represent random scatterers. The constructive interference of the wave functions of carriers following the solid and dashed paths leads to weak localization. A magnetic flux Φ threading the loop leads to a phase difference between the two paths.

The confinement of carriers in DMS quantum dots increases the tendency of BMP formation (see Chapter 30).

In metallic DMS, where the strong localization discussed earlier is not relevant, disorder can still lead to subtle effects, in particular in reduced dimensions. Closed multiple-scattering paths like the ones sketched in Figure 19.7 are important here: in the absence of a magnetic field and of spin–orbit coupling, the wave functions of carriers traversing a closed loop in opposite directions accumulate the same quantum phase and therefore interfere constructively. This increases the probability for carriers to return to the same point and, consequently, decreases the conductivity. This *weak localization* [117,156,157] leads to a logarithmic increase in the resistivity for $T \to 0$, as observed in (Ga,Mn)As nanowires [167].

If a magnetic field is applied, the quantum phases accumulated for the two directions become different by an amount proportional to the magnetic flux Φ enclosed by the loop, as shown in Figure 19.7. For a single loop, the conductivity shows Aharonov–Bohm oscillations as a function of Φ. In typical nanoscopic samples, loops of different sizes contribute, which have different oscillation periods. This leads to a characteristic quasi-random dependence of the conductivity on magnetic field with universal amplitude (universal conductance fluctuations) [168]. Both the Aharonov–Bohm effect and universal conductance fluctuations have been observed in nanoscopic (In,Mn)As [147] and (Ga,Mn)As [167,169,170] devices. The reason why these effects can occur at all is that DMS are both diluted ferromagnets and bad metals, as has been discussed by Garate et al. [171]. In normal metallic ferromagnets, they are suppressed by the stronger internal magnetic field.

In larger samples, so many loops of different sizes contribute such that the net effect is the suppression of weak localization by a magnetic field, leading to negative magnetoresistance. Negative magnetoresistance has been attributed to weak localization for (Ga,Mn)As films with up to $x \approx 8\%$ Mn and nonoptimal T_C [172]. The effect was observed at low temperatures, $T \lesssim 3$ K, and in weak magnetic fields of the order of 20 mT [172].

Since spin–orbit coupling in (Ga,Mn)As is strong, its effect on weak localization has to be taken into account. For a parabolic band with strong spin–orbit coupling, one expects *positive* magnetoresistance (weak anti-localization) [157,171]. Rokhinson et al. [172] argue that their observation of *negative* magnetoresistance in (Ga,Mn)As with $x \approx 8\%$ therefore indicates that E_F is not located in the VB but in an IB. However, it is at present unclear why spin–orbit effects should be significantly smaller in an IB derived from VB states. Garate et al. [171] show that within a $\mathbf{k} \cdot \mathbf{p}$ approach the complex band structure leads to negative magnetoresistance and weak localization, as observed.

Weak anti-localization has also been invoked to explain the broad resistivity peak around T_C [173]. However, it is now thought that the peak maximum is located a few Kelvin above T_C and that the peak is mostly due to temperature-dependent changes in the band structure and in the scattering rate [2,146]. Indeed, Novák et al. [18] observe a singularity in the temperature derivative of the resistivity, $d\rho/dT$, precisely at T_C slightly below the resistivity maximum. The singularity is consistent with the Fisher–Langer theory [174] for the resistive anomaly in metallic ferromagnets. Deviations from Fisher–Langer theory are more likely to be observed if the resistivity at T_C is dominated by disorder scattering [173].

19.2.8 Disorder and Magnetism

Disorder affects the magnetic properties in two principal ways. First, the carrier-mediated magnetic interaction is sensitive to carrier scattering. Second, the Mn local moments occupy random positions in the DMS lattice. We discuss these mechanisms in turn.

The Curie temperature of (Ga,Mn)As decreases with decreasing x, but remains nonzero in the doping range $1\% \lesssim x \lesssim 1.5\%$, where the material is insulating at low temperatures. It has been suggested that an RKKY description remains valid in the presence of strong localization [3,87,89,175]. In this regime, the RKKY interaction between two typical impurity spins decays exponentially on the scale of the localization length [3,175]. Ferromagnetic order is still possible, as long as the localization length is larger than the separation between impurity spins. While the RKKY paradigm might hold in this regime, disorder effects are very strong, precluding a VCA/MF description. Myers et al. [61] have measured the dependence of T_C on the number p of holes per Mn in this regime by systematically varying the As:Ga ratio across their samples during growth. They indeed find a strong deviation from the prediction $T_C \propto p^{1/3}$ of VCA/MF theory [61].

In this regime, the complementary BMP picture may be more appropriate [84,85,149,163–166]. For strong compensation, a single BMP comprises a hole and several ferromagnetically aligned impurity spins. For larger impurity concentrations, the BMPs overlap, leading to a ferromagnetic interaction between impurity spins within the same cluster of overlapping BMPs. At zero temperature, ferromagnetic long-range order emerges if the BMPs percolate. At finite temperatures, thermal fluctuations destroy the alignment between weakly coupled clusters. Kaminski and Das Sarma [164] have calculated T_C based on this picture. A more recent theory based on variable-range hopping [166] is similar in spirit. Above the percolation transition at T_C, rare but large ordered regions have been predicted to lead to Griffiths singularities [176], which would result in characteristic scaling behavior of the susceptibility and specific heat. Such ordered regions have been observed in (In,Mn)Sb [177], but Griffiths singularities have not yet been found in III–V DMS.

Carrier scattering has another important consequence: one cannot simply grow quaternary DMS such as (Ga,Mn)(As,P) and hope to achieve properties interpolating between the end points. The alloying introduces strong disorder, which dramatically reduces the mean free path and T_C [103].

We now turn to the effects of disorder due to the random distribution of magnetic dopants. In the extreme case, certain regions of the sample may be effectively decoupled from the rest and thus not take part in the spontaneous magnetization. There is evidence that such regions are present even in high-quality samples [113], whereas a recent study using μSR as the local probe found homogeneous ferromagnetism in the metallic regime and even through the metal–insulator transition [160]. A random distribution can lead to noncollinear ground states, since the RKKY interaction changes sign as a function of separation, which leads to frustration, and is anisotropic both in real space and in spin space [8,57,84,134,135,178,179]. Based on a tight-binding band structure, the resulting reduction of the low-temperature magnetization has been estimated to be small, though [57].

In various theoretical approaches, disorder is generically found to lead to an anomalous temperature dependence of the magnetization, characterized by an upward curvature or nearly linear behavior in a broad intermediate temperature range [3,97,137,149,163,179–182]. Magnetization curves of this type are observed for insulating (Ga,Mn)As [183–185] and also for early metallic samples [127,183,184]. Note that an upward curvature can also result from a temperature-induced change of the easy axis [2]. New high-quality samples show normal, Brillouin-function-like magnetization curves, though [18].

Simultaneously treating both effects of disorder—the carrier scattering and the spatial disorder of spins—in a realistic calculation is computationally demanding. Extending earlier work [149,186,187], Yildirim et al. [97] perform large-scale Monte Carlo simulations of (Ga,Mn)As, where the local spins are treated classically and an electronic tight-binding Hamiltonian is diagonalized for each spin configuration. For reasonably large x and weak compensation, the magnetization curves are fairly Brillouin-function–like, in agreement with experiments [18]. However, to achieve realistic values of T_C, J_{pd} had to be assumed to be much smaller than usually thought [188]. One might add that the Coulomb disorder potential is neglected, although it is probably the larger contribution to the disorder seen by the holes [2]. Its inclusion would probably reduce T_C, counteracting the effect of a larger J_{pd}.

The disorder due to the random distribution of spins has also been studied in spin-only models, including magnetic fluctuations within the Tyablikov approximation [141] and the RPA [136], and disorder by exact diagonalization in a supercell. As noted earlier, the results for $T_C(x)$ are in good agreement with experiments. Disorder is crucial for this.

19.3 III–V DMS with Wide Gaps

Wide-gap III–V DMS such as (Ga,Mn)N are very interesting, since for several of these compounds Curie temperatures above room temperature have been reported [189–197]. Indeed, both the arguments based on a parabolic band given in Section 18.2.6 and more sophisticated $\mathbf{k}\cdot\mathbf{p}$ theory predict high T_C values [89]. The underlying kinetic-exchange model is unlikely to be valid here, though.

Dietl [3,4] has discussed several scenarios for apparently ferromagnetic response of DMS. For quasi-uniform DMS with merged VB and IB, the kinetic-exchange model can be applied, leading to an RKKY-type magnetic interaction. Deep Mn-derived levels in the gap likely invalidate this picture for wide-gap DMS, though.

Ferromagnetic response can also result from precipitates of known magnetic compounds [3,4]. In this case, the semiconducting and magnetic properties result from different parts of the sample. A more interesting possibility is chemical nanoscale phase separation into regions rich and poor in magnetic ions [3,4]. In contrast to precipitates, the crystal structure of the magnetically rich regions is in this case determined by the host, which can stabilize phases that are unstable in the bulk [3,4]. Nanoscale phase separation has been observed in (Ga,Mn)N [198] and (Ga,Fe)N [196] as well as in (Ga,Mn)As [199]. Note that magnetic and transport properties can still be strongly coupled.

In both cases of phase separation, an apparently ferromagnetic response is seen below a blocking temperature T_B: For $T < T_B$, the thermal energy is too small to overcome the size-dependent magnetic anisotropy energy of the magnetic regions, which therefore show frozen magnetic moments. This superparamagnetic behavior has been observed for Gd-ion-implanted GaN [200] and for (Ga,Fe)N [196]. Ney et al. [201] have pointed out that a large fraction of the apparently ferromagnetic signals could result from metastable states. Careful analysis is required to check that static, quasi-uniform ferromagnetism is actually present in a wide-gap DMS. In the following, we discuss such a still partially hypothetical state.

We concentrate on (Ga,Mn)N. Like in (Ga,Mn)As, substitutional Mn^{2+} introduces one hole, which is attracted by the Mn^{2+} core. Neglecting hybridization, one would find a hydrogenic acceptor state. However, the crucial difference between (Ga,Mn)N and (Ga,Mn)As is that in the nitride the $d^5 \to d^4$ ($Mn^{2+} \to Mn^{3+}$) transition lies in the band gap [54,55,202]. It is probably deeper than the bound acceptor state, becoming a $d^4 \to d^5$ transition. Due to the smaller energy difference, the effect of hybridization and in particular the *pd*-exchange interaction J_{pd} are larger, as seen in photoemission as well as in optical and x-ray spectroscopy for (Ga,Mn)N [203,204]. J_{pd} may even become ferromagnetic [205,206]. Hybridization and level repulsion lead to an unoccupied level with more than 50% *d*-orbital character deep in the gap and an occupied level with more than 50% VB character, which may occur as a resonance in the VB.

The strong *pd*-hybridization agrees with LDA + U and LSDA + U calculations [52,53,207], which find large Mn *d*-orbital weight at the Fermi energy in (Ga,Mn)N as opposed to very small Mn *d*-orbital admixture in (Ga,Mn)As [52,53]. Employing self-interaction corrected LSDA, Schulthess et al. [54,55] also find a large Mn *d*-orbital weight in the gap, close to E_F, and a predominantly d^4 state of the Mn dopants for both (Ga,Mn)N and (Ga,Mn)P, but not for (Ga,Mn)As. This picture is supported by optical spectroscopy [208], x-ray spectroscopy [203], and transport and magnetization measurements [209]. However, in other x-ray spectroscopy experiments [210], (Ga,Mn)P appears to be more similar to (Ga,Mn)As, suggesting that its properties are intermediate between (Ga,Mn)As and (Ga,Mn)N.

The overlap between the strongly *d*-orbital-like states in the gap is small. The IB formed for given Mn doping x is thus narrower than in (Ga,Mn)As. Furthermore, it is much deeper in the gap. Therefore, a significant overlap, let alone merging, of IB and VB is unlikely [211]. The density of states is sketched in Figure 19.8. Dynamical MF theory calculations assuming local Coulomb and exchange interactions between holes and Mn impurities support this view [50]. The kinetic-exchange picture with the Fermi energy in a weakly perturbed VB is thus not valid.

The high observed T_C [189,194,195] at concentrations of magnetic dopants below the percolation threshold suggests a carrier-mediated mechanism. The suppression of the remanent magnetization in (Ga,Mn)N below about 10 K, accompanied by an increase of the resistivity attributed to localization also supports a carrier-mediated scenario [195]. Since in the absence of compensation the narrow IB would be completely empty, a carrier-mediated mechanism requires partial compensation [54]. A high concentration of nitrogen vacancies, which are donors with one weakly bound electron [212], is typical for GaN and thus provides a likely source of compensation. Transport and x-ray spectroscopy experiments [213] on (Ga,Mn)N with $x \lesssim 1\%$, grown by metal-organic chemical vapor deposition, suggest a strong increase in the concentration of nitrogen vacancies with x (self-compensation). Due to the large *d*-orbital weight in the IB, strong compensation leads to strong charge fluctuations in the Mn *d*-shell. This mixed-valence nature of Mn might lead to interactions through the double-exchange mechanism [53,118,194,195,207,214], which can lead to high T_C.

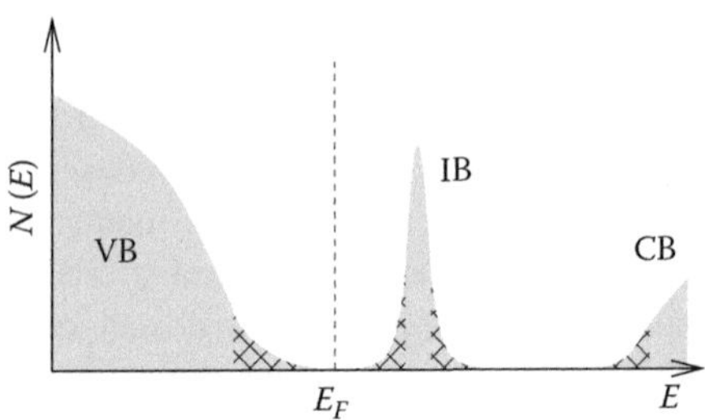

FIGURE 19.8 Sketch of the density of states for a typical wide-gap III–V DMS. Compared to Figure 19.3c, the IB is both narrower and deeper in the gap. The Fermi energy is shown for vanishing compensation.

Dhar et al. have observed ferromagnetic order in MBE-grown [191] and ion-implanted [193] GaN with very low concentrations of Gd, n_{Gd}, down to $7\times10^{5}\,cm^{-3}$. Dividing the measured magnetization by n_{Gd}, they obtain magnetic moments per Gd of up to $5000\,\mu_B$. This of course implies that most of the magnetization is *not* carried by the Gd dopants. The natural candidates are charged defects [215]. Since the concentrations of unwanted impurities are small [191], native defects, most likely nitrogen vacancies [212], are probably responsible. Electrons weakly bound to nitrogen-vacancy donors [212] would each contribute a spin 1/2. Indeed, the total magnetization is found to depend only weakly on n_{Gd} for small concentrations, followed by a crossover to a roughly linear behavior for larger n_{Gd} with a coefficient of the expected order of magnitude for the f^7 configuration of Gd^{3+} [191]. However, it is not yet understood how very weak Gd doping could induce magnetic ordering of the defect spins. It has been found that in some Gd-doped GaN samples the magnetic effects are metastable [201], perhaps due to electrons trapped by defects.

Magnetic moments of defects have also been invoked to explain magnetic signals in other semiconductors, including samples without transition-metal or rare-earth dopants [216–219]. This so-called d^0 magnetism [216], which is particularly relevant for oxides, is discussed in Chapter 20.

References

1. T. Dietl and H. Ohno, Ferromagnetism in III–V and II–VI semiconductor structures, *Physica E* **9**, 185–193 (2001).
2. T. Jungwirth, J. Sinova, J. Mašek, J. Kučera, and A. H. MacDonald, Theory of ferromagnetic (III,Mn)V semiconductors, *Rev. Mod. Phys.* **78**, 809–864 (2006).
3. T. Dietl, Origin of ferromagnetic response in diluted magnetic semiconductors and oxides, *J. Phys.: Condens. Matter* **19**, 165204 (2007).
4. T. Dietl, Origin and control of ferromagnetism in dilute magnetic semiconductors and oxides (invited), *J. Appl. Phys.* **103**, 07D111 (2008).
5. K. Sato, L. Bergqvist, J. Kudrnovský et al., First-principles theory of dilute magnetic semiconductors, *Rev. Mod. Phys.* **82**, 1633 (2010).
6. S. J. Pearton, C. R. Abernathy, M. E. Overberg et al., Wide band gap ferromagnetic semiconductors and oxides, *J. Appl. Phys.* **93**, 1–12 (2003).
7. C. Liu, F. Yun, and H. Morkoç, Ferromagnetism of ZnO and GaN: A review, *J. Mater. Sci.: Mater. Electron.* **16**, 555–597 (2005).
8. C. Timm, Disorder effects in diluted magnetic semiconductors, *J. Phys.: Condens. Matter* **15**, R1865–R1896 (2003).
9. X. Liu and J. K. Furdyna, Ferromagnetic resonance in $Ga_{1-x}Mn_xAs$ dilute magnetic semiconductors, *J. Phys.: Condens. Matter* **18**, R245–R279 (2006).
10. T. Dietl, Interplay between carrier localization and magnetism in diluted magnetic and ferromagnetic semiconductors, *J. Phys. Soc. Jpn.* **77**, 031005 (2008).
11. K. S. Burch, D. D. Awschalom, and D. N. Basov, Optical properties of III-Mn-V ferromagnetic semiconductors, *J. Magn. Magn. Mater.* **320**, 3207–3228 (2008).
12. N. Nagaosa, J. Sinova, S. Onoda, A. H. MacDonald, and N. P. Ong, Anomalous Hall effect, *Rev. Mod. Phys.* **82**, 1539–1592 (2010).
13. T. Jungwirth, J. Sinova, A. H. MacDonald et al., Character of states near the Fermi level in (Ga,Mn)As: Impurity to valence band crossover, *Phys. Rev. B* **76**, 125206 (2007).
14. B. I. Shklovskii and A. L. Efros, *Electronic Properties of Doped Semiconductors*, Springer, Berlin, Germany, 1984.
15. H. Ohno, H. Munekata, T. Penney, S. von Molnár, and L. L. Chang, Magnetotransport properties of p-type (In,Mn)As diluted magnetic III–V semiconductors, *Phys. Rev. Lett.* **68**, 2664–2667 (1992).
16. H. Ohno, A. Shen, F. Matsukura et al., (Ga,Mn)As: A new diluted magnetic semiconductor based on GaAs, *Appl. Phys. Lett.* **69**, 363–365 (1996).
17. F. Matsukura, H. Ohno, A. Shen, and Y. Sugawara, Transport properties and origin of ferromagnetism in (Ga,Mn)As, *Phys. Rev. B* **57**, R2037–R2040 (1998).
18. V. Novák, K. Olejník, J. Wunderlich et al., Curie point singularity in the temperature derivative of resistivity in (Ga,Mn)As, *Phys. Rev. Lett.* **101**, 077201 (2008).
19. M. Wang, R. P. Campion, A. W. Rushforth, K. W. Edmonds, C. T. Foxon, and B. L. Gallagher, Achieving high Curie temperature in (Ga,Mn)As, *Appl. Phys. Lett.* **93**, 132103 (2008).
20. L. Chen, S. Yan, P. F. Xu et al., Low-temperature magnetotransport behaviors of heavily Mn-doped (Ga,Mn)As films with high ferromagnetic transition temperature, *Appl. Phys. Lett.* **95**, 182505 (2009).
21. A. M. Nazmul, T. Amemiya, Y. Shuto, S. Sugahara, and M. Tanaka, High temperature ferromagnetism in GaAs-based heterostructures with Mn δ-doping, *Phys. Rev. Lett.* **95**, 017201 (2005).
22. M. A. Scarpulla, R. Farshchi, P. R. Stone et al., Electrical transport and ferromagnetism in $Ga_{1-x}Mn_x$ As synthesized by ion implantation and pulsed-laser melting, *J. Appl. Phys.* **103**, 073913 (2008).
23. T. Schallenberg and H. Munekata, Preparation of ferromagnetic (In,Mn)As with a high Curie temperature of 90 K, *Appl. Phys. Lett.* **89**, 042507 (2006).
24. S. J. Potashnik, K. C. Ku, R. Mahendiran et al., Saturated ferromagnetism and magnetization deficit in optimally annealed $Ga_{1-x}Mn_x$ As epilayers, *Phys. Rev. B* **66**, 012408 (2002).
25. R. P. Campion, K. W. Edmonds, L. X. Zhao et al., The growth of GaMnAs films by molecular beam epitaxy using arsenic dimers, *J. Cryst. Growth* **251**, 311–316 (2003).
26. A. Richardella, P. Roushan, S. Mack et al., Visualizing critical correlations near the metal-insulator transition in $Ga_{1-x}Mn_xAs$, *Science* **327**, 665–669 (2010).
27. E. Abe, F. Matsukura, H. Yasuda, Y. Ohno, and H. Ohno, Molecular beam epitaxy of III–V diluted magnetic semiconductor (Ga,Mn)Sb, *Physica E* **7**, 981–985 (2000).

28. T. Wojtowicz, G. Cywiński, W. L. Lim et al., $In_{1-x}Mn_xSb$—A narrow-gap ferromagnetic semiconductor, *Appl. Phys. Lett.* **82**, 4310–4312 (2003).
29. M. Csontos, G. Mihály, B. Jankó, T. Wojtowicz, X. Liu, and J. K. Furdyna, Pressure-induced ferromagnetism in (In,Mn) Sb dilute magnetic semiconductor, *Nat. Mater.* **4**, 447–449 (2005); G. Mihály, M. Csontos, S. Bordács et al., Anomalous Hall effect in the (In,Mn)Sb dilute magnetic semiconductor, *Phys. Rev. Lett.* **100**, 107201 (2008).
30. P. M. Krstajić, F. M. Peeters, V. A. Ivanov, V. Fleurov, and K. Kikoin, Double-exchange mechanisms for Mn-doped III–V ferromagnetic semiconductors, *Phys. Rev. B* **70**, 195215 (2004).
31. N. Almeleh and B. Goldstein, Electron paramagnetic resonance of manganese in gallium arsenide, *Phys. Rev.* **128**, 1568–1570 (1962).
32. Y. Sasaki, X. Liu, J. K. Furdyna, M. Palczewska, J. Szczytko, and A. Twardowski, Ferromagnetic resonance in GaMnAs, *J. Appl. Phys.* **91**, 7484–7486 (2002).
33. R. A. Chapman and W. G. Hutchinson, Photoexcitation and photoionization of neutral manganese acceptors in gallium arsenide, *Phys. Rev. Lett.* **18**, 443–445 (1967).
34. M. Linnarsson, E. Janzén, B. Monemar, M. Kleverman, and A. Thilderkvist, Electronic structure of the GaAs:Mn_{Ga} center, *Phys. Rev. B* **55**, 6938–6944 (1997).
35. J. Schneider, U. Kaufmann, W. Wilkening, M. Baeumler, and F. Köhl, Electronic structure of the neutral manganese acceptor in gallium arsenide, *Phys. Rev. Lett.* **59**, 240–243 (1987).
36. J. S. Blakemore, W. J. Brown, Jr., M. L. Stass, and D. A. Woodbury, Thermal activation energy of manganese acceptors in gallium arsenide as a function of impurity spacing, *J. Appl. Phys.* **44**, 3352–3354 (1973).
37. A. M. Yakunin, A. Yu. Silov, P. M. Koenraad, W. Van Roy, J. De Boeck, and J. H. Wolter, Imaging of the ($Mn^{2+}3d^5$ + hole) complex in GaAs by cross-sectional scanning tunneling microscopy, *Physica E* **21**, 947–950 (2004); A. M. Yakunin, A. Yu. Silov, P. M. Koenraad et al., Spatial structure of an individual Mn acceptor in GaAs, *Phys. Rev. Lett.* **92**, 216806 (2004).
38. J. Okabayashi, A. Kimura, O. Rader et al., Core-level photoemission study of $Ga_{1-x}Mn_xAs$, *Phys. Rev. B* **58**, R4211–R4214 (1998).
39. J. Okabayashi, A. Kimura, T. Mizokawa, A. Fujimori, T. Hayashi, and M. Tanaka, Mn 3d partial density of states in $Ga_{1-x}Mn_xAs$ studied by resonant photoemission spectroscopy, *Phys. Rev. B* **59**, R2486–R2489 (1999).
40. A. Baldereschi and N. O. Lipari, Spherical model of shallow acceptor states in semiconductors, *Phys. Rev. B* **8**, 2697–2709 (1973).
41. A. K. Bhattacharjee and C. B. à la Guillaume, Model for the Mn acceptor in GaAs, *Solid State Commun.* **113**, 17–21 (2000).
42. R. F. Kirkman, R. A. Stradling, and P. J. Lin-Chung, An infrared study of the shallow acceptor states in GaAs, *J. Phys. C* **11**, 419–433 (1978).
43. M. S. Dresselhaus, G. Dresselhaus, and A. Jorio, *Group Theory: Application to the Physics of Condensed Matter*, Springer, Berlin, Germany, 2008.
44. D. Kitchen, A. Richardella, J.-M. Tang, M. E. Flatté, and A. Yazdani, Atom-by-atom substitution of Mn in GaAs and visualization of their hole-mediated interactions, *Nature* **442**, 436–439 (2006).
45. C. Çelebi, J. K. Garleff, A. Yu. Silov et al., Surface induced asymmetry of acceptor wave functions, *Phys. Rev. Lett.* **104**, 086404 (2010).
46. J.-M. Tang and M. E. Flatté, Multiband tight-binding model of local magnetism in $Ga_{1-x}Mn_xAs$, *Phys. Rev. Lett.* **92**, 047201 (2004).
47. M. J. Schmidt, K. Pappert, C. Gould, G. Schmidt, R. Oppermann, and L. W. Molenkamp, Bound-hole states in a ferromagnetic (Ga,Mn)As environment, *Phys. Rev. B* **76**, 035204 (2007).
48. A. Zunger, Electronic structure of 3*d* transition-atom impurities in semiconductors, in *Solid State Physics*, vol. 39, H. Ehrenreich and D. Turnbull (Eds.), Academic, Orlando, FL, p. 275, 1986.
49. P. Mahadevan and A. Zunger, Ferromagnetism in Mn-doped GaAs due to substitutional-interstitial complexes, *Phys. Rev. B* **68**, 075202 (2003).
50. F. Popescu, C. Şen, E. Dagotto, and A. Moreo, Crossover from impurity to valence band in diluted magnetic semiconductors: Role of Coulomb attraction by acceptors, *Phys. Rev. B* **76**, 085206 (2007).
51. L. Craco, M. S. Laad, and E. Müller-Hartmann, Ab initio description of the diluted magnetic semiconductor $Ga_{1-x}Mn_xAs$: Ferromagnetism, electronic structure, and optical response, *Phys. Rev. B* **68**, 233310 (2003).
52. L. M. Sandratskii, P. Bruno, and J. Kudrnovský, On-site Coulomb interaction and the magnetism of (GaMn)N and (GaMn)As, *Phys. Rev. B* **69**, 195203 (2004).
53. M. Wierzbowska, D. Sánchez-Portal, and S. Sanvito, Different origins of the ferromagnetic order in (Ga,Mn)As and (Ga,Mn)N, *Phys. Rev. B* **70**, 235209 (2004).
54. T. C. Schulthess, W. M. Temmerman, Z. Szotek, W. H. Butler, and G. M. Stocks, Electronic structure and exchange coupling of Mn impurities in III–V semiconductors, *Nat. Mater.* **4**, 838–844 (2005).
55. T. C. Schulthess, W. M. Temmerman, Z. Szotek, A. Svane, and L. Petit, First-principles electronic structure of Mn-doped GaAs, GaP, and GaN semiconductors, *J. Phys.: Condens. Matter* **19**, 165207 (2007).
56. J. R. Schrieffer and P. A. Wolff, Relation between the Anderson and Kondo Hamiltonians, *Phys. Rev.* **149**, 491–492 (1966); K. A. Chao, J. Spałek, and A. M. Oleś, Canonical perturbation expansion of the Hubbard model, *Phys.*

Rev. B **18**, 3453–3464 (1978); B. E. Larson, K. C. Hass, H. Ehrenreich, and A. E. Carlsson, Theory of exchange interactions and chemical trends in diluted magnetic semiconductors, *Phys. Rev. B* **37**, 4137–4154 (1988).
57. C. Timm and A. H. MacDonald, Anisotropic exchange interactions in III–V diluted magnetic semiconductors, *Phys. Rev. B* **71**, 155206 (2005).
58. E. Nielsen and R. N. Bhatt, Search for ferromagnetism in doped semiconductors in the absence of transition metal ions, *Phys. Rev. B* **82**, 195117 (2010).
59. S. J. Potashnik, K. C. Ku, S. H. Chun, J. J. Berry, N. Samarth, and P. Schiffer, Effects of annealing time on defect-controlled ferromagnetism in $Ga_{1-x}Mn_xAs$, *Appl. Phys. Lett.* **79**, 1495–1497 (2001).
60. M. Luysberg, H. Sohn, A. Prasad et al., Electrical and Structural Properties of LT-GaAs: Influence of As/Ga flux ratio and growth temperature, *Mater. Res. Soc. Symp. Proc.* **442**, 485–490 (1997).
61. R. C. Myers, B. L. Sheu, A. W. Jackson et al., Antisite effect on hole-mediated ferromagnetism in (Ga,Mn)As, *Phys. Rev. B* **74**, 155203 (2006).
62. R. P. Campion, K. W. Edmonds, L. X. Zhao et al., High-quality GaMnAs films grown with arsenic dimers, *J. Cryst. Growth* **247**, 42–48 (2003).
63. M. Suezawa, in *International Conference on Science and Technology of Defect Control in Semiconductors*, vol. 2, K. Sumino (Ed.), North Holland, Amsterdam, 1990, p. 1043.
64. C. Timm, F. Schäfer, and F. von Oppen, Correlated defects, metal-insulator transition, and magnetic order in ferromagnetic semiconductors, *Phys. Rev. Lett.* **89**, 137201 (2002).
65. K. M. Yu, W. Walukiewicz, T. Wojtowicz et al., Effect of the location of Mn sites in ferromagnetic $Ga_{1-x}Mn_xAs$ on its Curie temperature, *Phys. Rev. B* **65**, 201303(R) (2002).
66. K. W. Edmonds, P. Bogusławski, K. Y. Wang et al., Mn interstitial diffusion in (Ga,Mn)As, *Phys. Rev. Lett.* **92**, 037201 (2004).
67. J. Mašek and F. Máca, Interstitial Mn in (Ga,Mn)As: Binding energy and exchange coupling, *Phys. Rev. B* **69**, 165212 (2004).
68. V. Holý, Z. Matěj, O. Pacherová et al., Mn incorporation in as-grown and annealed (Ga,Mn)As layers studied by x-ray diffraction and standing-wave fluorescence, *Phys. Rev. B* **74**, 245205 (2006).
69. G. S. Chang, E. Z. Kurmaev, L. D. Finkelstein et al., Post-annealing effect on the electronic structure of Mn atoms in $Ga_{1-x}Mn_xAs$ probed by resonant inelastic x-ray scattering, *J. Phys.: Condens. Matter* **19**, 076215 (2007).
70. J. Blinowski and P. Kacman, Spin interactions of interstitial Mn ions in ferromagnetic GaMnAs, *Phys. Rev. B* **67**, 121204 (2003).
71. M. B. Stone, K. C. Ku, S. J. Potashnik, B. L. Sheu, N. Samarth, and P. Schiffer, Capping-induced suppression of annealing effects on $Ga_{1-x}Mn_xAs$ epilayers, *Appl. Phys. Lett.* **83**, 4568–4570 (2003).
72. K. W. Edmonds, K. Y. Wang, R. P. Campion et al., Hall effect and hole densities in $Ga_{1-x}Mn_xAs$, *Appl. Phys. Lett.* **81**, 3010–3012 (2002).
73. B. S. Sørensen, J. Sadowski, S. E. Andresen, and P. E. Lindelof, Dependence of Curie temperature on the thickness of epitaxial (Ga,Mn)As film, *Phys. Rev. B* **66**, 233313 (2002).
74. K. C. Ku, S. J. Potashnik, R. Wang et al., Highly enhanced Curie temperature in low-temperature annealed [Ga,Mn]As epilayers, *Appl. Phys. Lett.* **82**, 2302–2304 (2003).
75. K. W. Edmonds, N. R. S. Farley, T. K. Johal et al., Ferromagnetic moment and antiferromagnetic coupling in (Ga,Mn)As thin films, *Phys. Rev. B* **71**, 064418 (2005).
76. G. L. Pearson and J. Bardeen, Electrical properties of pure silicon and silicon alloys containing boron and phosphorus, *Phys. Rev.* **75**, 865–883 (1949).
77. W. J. Brown, Jr. and J. S. Blakemore, Transport and photoelectrical properties of gallium arsenide containing deep acceptors, *J. Appl. Phys.* **43**, 2242–2246 (1972).
78. T. Jungwirth, W. A. Atkinson, B. H. Lee, and A. H. MacDonald, Interlayer coupling in ferromagnetic semiconductor superlattices, *Phys. Rev. B* **59**, 9818–9821 (1999).
79. T. Jungwirth, K. Y. Wang, J. Mašek et al., Prospects for high temperature ferromagnetism in (Ga,Mn)As semiconductors, *Phys. Rev. B* **72**, 165204 (2005).
80. A. Chattopadhyay, S. Das Sarma, and A. J. Millis, Transition temperature of ferromagnetic semiconductors: A dynamical mean field study, *Phys. Rev. Lett.* **87**, 227202 (2001).
81. C. Santos and W. Nolting, Ferromagnetism in the Kondo-lattice model, *Phys. Rev. B* **65**, 144419 (2002); J. Kienert and W. Nolting, Magnetic phase diagram of the Kondo lattice model with quantum localized spins, *Phys. Rev. B* **73**, 224405 (2006).
82. M. Turek, J. Siewert, and J. Fabian, Electronic and optical properties of ferromagnetic $Ga_{1-x}Mn_xAs$ in a multiband tight-binding approach, *Phys. Rev. B* **78**, 085211 (2008).
83. C. Timm, F. Schäfer, and F. von Oppen, Comment on "Effects of disorder on ferromagnetism in diluted magnetic semiconductors," *Phys. Rev. Lett.* **90**, 029701 (2003).
84. G. A. Fiete, G. Zaránd, and K. Damle, Effective Hamiltonian for $Ga_{1-x}Mn_xAs$ in the dilute limit, *Phys. Rev. Lett.* **91**, 097202 (2003); G. A. Fiete, G. Zaránd, K. Damle, and C. P. Moca, Disorder, spin–orbit, and interaction effects in dilute $Ga_{1-x}Mn_xAs$, *Phys. Rev. B* **72**, 045212 (2005).
85. C. P. Moca, G. Zaránd, and M. Berciu, Theory of optical conductivity for dilute $Ga_{1-x}Mn_xAs$, *Phys. Rev. B* **80**, 165202 (2009).
86. C. Zener, Interaction between the d shells in the transition metals, *Phys. Rev.* **81**, 440–444 (1951).
87. T. Dietl, A. Haury, and Y. Merle d'Aubigné, Free carrier-induced ferromagnetism in structures of diluted magnetic semiconductors, *Phys. Rev. B* **55**, R3347–R3350 (1997).

88. J. König, H.-H. Lin, and A. H. MacDonald, Theory of diluted magnetic semiconductor ferromagnetism, *Phys. Rev. Lett.* **84**, 5628–5631 (2000); J. König, T. Jungwirth, and A. H. MacDonald, Theory of magnetic properties and spin-wave dispersion for ferromagnetic (Ga,Mn)As, *Phys. Rev. B* **64**, 184423 (2001).
89. T. Dietl, H. Ohno, F. Matsukura, J. Cibert, and D. Ferrand, Zener model description of ferromagnetism in zinc-blende magnetic semiconductors, *Science* **287**, 1019–1022 (2000).
90. J. M. Luttinger and W. Kohn, Motion of electrons and holes in perturbed periodic fields, *Phys. Rev.* **97**, 869–883 (1955).
91. I. Vurgaftman, J. R. Meyer, and L. R. Ram-Mohan, Band parameters for III–V compound semiconductors and their alloys, *J. Appl. Phys.* **89**, 5815–5875 (2001).
92. M. Abolfath, T. Jungwirth, J. Brum, and A. H. MacDonald, Theory of magnetic anisotropy in $III_{1-x}Mn_xV$ ferromagnets, *Phys. Rev. B* **63**, 054418 (2001).
93. T. Dietl, H. Ohno, and F. Matsukura, Hole-mediated ferromagnetism in tetrahedrally coordinated semiconductors, *Phys. Rev. B* **63**, 195205 (2001).
94. J. C. Slater and G. F. Koster, Simplified LCAO method for the periodic potential problem, *Phys. Rev.* **94**, 1498–1524 (1954); G. F. Koster and J. C. Slater, Wave functions for impurity levels, *Phys. Rev.* **95**, 1167–1176 (1954).
95. D. J. Chadi, Spin–orbit splitting in crystalline and compositionally disordered semiconductors, *Phys. Rev. B* **16**, 790–796 (1977).
96. T. Jungwirth, J. Mašek, J. Sinova, and A. H. MacDonald, Ferromagnetic transition temperature enhancement in (Ga,Mn)As semiconductors by carbon codoping, *Phys. Rev. B* **68**, 161202 (2003).
97. Y. Yildirim, G. Alvarez, A. Moreo, and E. Dagotto, Large-scale Monte Carlo study of a realistic lattice model for $Ga_{1-x}Mn_xAs$, *Phys. Rev. Lett.* **99**, 057207 (2007).
98. H. Ohno and T. Dietl, Spin-transfer physics and the model of ferromagnetism in (Ga,Mn)As, *J. Magn. Magn. Mater.* **320**, 1293–1299 (2008).
99. J. Okabayashi, A. Kimura, O. Rader et al., Angle-resolved photoemission study of $Ga_{1-x}Mn_xAs$, *Phys. Rev. B* **64**, 125304 (2001).
100. V. F. Sapega, M. Moreno, M. Ramsteiner, L. Däweritz, and K. H. Ploog, Polarization of valence band holes in the (Ga,Mn)As diluted magnetic semiconductor, *Phys. Rev. Lett.* **94**, 137401 (2005); V. F. Sapega, M. Ramsteiner, O. Brandt, L. Däweritz, and K. H. Ploog, Hot-electron photoluminescence study of the (Ga,Mn)As diluted magnetic semiconductor, *Phys. Rev. B* **73**, 235208 (2006).
101. K. Ando, H. Saito, K. C. Agarwal, M. C. Debnath, and V. Zayets, Origin of the anomalous magnetic circular dichroism spectral shape in ferromagnetic $Ga_{1-x}Mn_xAs$: Impurity bands inside the band gap, *Phys. Rev. Lett.* **100**, 067204 (2008).
102. M. Turek, J. Siewert, and J. Fabian, Magnetic circular dichroism in $Ga_xMn_{1-x}As$: Theoretical evidence for and against an impurity band, *Phys. Rev. B* **80**, 161201 (R) (2009).
103. P. R. Stone, K. Alberi, S. K. Z. Tardif et al., Metal-insulator transition by isovalent anion substitution in $Ga_{1-x}Mn_xAs$: Implications to ferromagnetism, *Phys. Rev. Lett.* **101**, 087203 (2008).
104. E. H. C. P. Sinnecker, G. M. Penello, T. G. Rappoport et al., Ion-beam modification of the magnetic properties of $Ga_{1-x}Mn_xAs$ epilayers, *Phys. Rev. B* **81**, 245203 (2010).
105. E. J. Singley, R. Kawakami, D. D. Awschalom, and D. N. Basov, Infrared probe of itinerant ferromagnetism in $Ga_{1-x}Mn_xAs$, *Phys. Rev. Lett.* **89**, 097203 (2002); E. J. Singley, K. S. Burch, R. Kawakami, J. Stephens, D. D. Awschalom, and D. N. Basov, Electronic structure and carrier dynamics of the ferromagnetic semiconductor $Ga_{1-x}Mn_xAs$, *Phys. Rev. B* **68**, 165204 (2003).
106. K. S. Burch, D. B. Shrekenhamer, E. J. Singley et al., Impurity band conduction in a high temperature ferromagnetic semiconductor, *Phys. Rev. Lett.* **97**, 087208 (2006).
107. J. Sinova, T. Jungwirth, S.-R. E. Yang, J. Kučera, and A. H. MacDonald, Infrared conductivity of metallic (III,Mn)V ferromagnets, *Phys. Rev. B* **66**, 041202 (2002).
108. W. Songprakob, R. Zallen, D. V. Tsu, and W. K. Liu, Intervalenceband and plasmon optical absorption in heavily doped GaAs:C, *J. Appl. Phys.* **91**, 171–177 (2002).
109. S.-R. E. Yang, J. Sinova, T. Jungwirth, Y. P. Shim, and A. H. MacDonald, Non-drude optical conductivity of (III,Mn)V ferromagnetic semiconductors, *Phys. Rev. B* **67**, 045205 (2003).
110. G. Acbas, M.-H. Kim, M. Cukr et al., Electronic structure of ferromagnetic semiconductor $Ga_{1-x}Mn_xAs$ probed by subgap magneto-optical spectroscopy, *Phys. Rev. Lett.* **103**, 137201 (2009).
111. J.-M. Tang and M. E. Flatté, Magnetic circular dichroism from the impurity band in III–V diluted magnetic semiconductors, *Phys. Rev. Lett.* **101**, 157203 (2008).
112. D. Chiba, F. Matsukura, and H. Ohno, Electric-field control of ferromagnetism in (Ga,Mn)As, *Appl. Phys. Lett.* **89**, 162505 (2006).
113. M. Sawicki, D. Chiba, A. Korbecka et al., Experimental probing of the interplay between ferromagnetism and localization in (Ga, Mn)As, *Nat. Phys.* **6**, 22–25 (2010).
114. Y. Nishitani, D. Chiba, M. Endo et al., Curie temperature versus hole concentration in field-effect structures of $Ga_{1-x}Mn_xAs$, *Phys. Rev. B* **81**, 045208 (2010).
115. D. A. Woodbury and J. S. Blakemore, Impurity conduction and the metal-nonmetal transition in manganese-doped gallium arsenide, *Phys. Rev. B* **8**, 3803–3810 (1973).
116. D. Neumaier, M. Turek, U. Wurstbauer et al., All-electrical measurement of the density of states in (Ga,Mn)As, *Phys. Rev. B* **80**, 205202 (2009).
117. P. A. Lee and T. V. Ramakrishnan, Disordered electronic systems, *Rev. Mod. Phys.* **57**, 287–337 (1985).
118. C. Zener, Interaction between the d-Shells in the transition metals. II. Ferromagnetic compounds of manganese with perovskite structure, *Phys. Rev.* **82**, 403–405 (1951).
119. J. Kudrnovský, I. Turek, V. Drchal, F. Máca, P. Weinberger, and P. Bruno, Exchange interactions in III–V and group-IV

diluted magnetic semiconductors, *Phys. Rev. B* **69**, 115208 (2004).

120. G. Bouzerar, R. Bouzerar, and O. Cépas, Superexchange induced canted ferromagnetism in dilute magnets, *Phys. Rev. B* **76**, 144419 (2007).
121. J. B. Goodenough, An interpretation of the magnetic properties of the perovskite-type mixed crystals $La_{1-x}Sr_xCoO_{3-\lambda}$, *J. Phys. Chem. Solids* **6**, 287–297 (1958); J. Kanamori, Superexchange interaction and symmetry properties of electron orbitals, *J. Phys. Chem. Solids* **10**, 87–98 (1959).
122. H. Ohno, D. Chiba, F. Matsukura et al., Electric-field control of ferromagnetism, *Nature* **408**, 944–946 (2000); D. Chiba, M. Yamanouchi, F. Matsukura, and H. Ohno, Electrical Manipulation of magnetization reversal in a ferromagnetic semiconductor, *Science* **301**, 943–945 (2003).
123. A. M. Nazmul, S. Kobayashi, S. Sugahara, and M. Tanaka, Electrical and optical control of ferromagnetism in III–V semiconductor heterostructures at high temperature (~100 K), *Jpn. J. Appl. Phys. Part* **43**, L233–L236 (2004).
124. C. Timm, Charge and magnetization inhomogeneities in diluted magnetic semiconductors, *Phys. Rev. Lett.* **96**, 117201 (2006); Possible magnetic-field-induced voltage and thermopower in diluted magnetic semiconductors, *Phys. Rev. B* **74**, 014419 (2006).
125. S. Koshihara, A. Oiwa, M. Hirasawa et al., Ferromagnetic order induced by photogenerated carriers in magnetic III–V semiconductor heterostructures of (In,Mn)As/GaSb, *Phys. Rev. Lett.* **78**, 4617–4620 (1997); H. Munekata, T. Abe, S. Koshihara et al., Light-induced ferromagnetism in III–V-based diluted magnetic semiconductor heterostructures, *J. Appl. Phys.* **81**, 4862–4864 (1997).
126. J. Wang, I. Cotoros, K. M. Dani, X. Liu, J. K. Furdyna, and D. S. Chemla, Ultrafast enhancement of ferromagnetism via photoexcited holes in GaMnAs, *Phys. Rev. Lett.* **98**, 217401 (2007).
127. H. Ohno, Properties of ferromagnetic III–V semiconductors, *J. Magn. Magn. Mater.* **200**, 110–129 (1999); H. Ohno and F. Matsukura, A ferromagnetic III–V semiconductor: (Ga,Mn)As, *Solid State Commun.* **117**, 179–186 (2001).
128. X. Liu, Y. Sasaki, and J. K. Furdyna, Ferromagnetic resonance in $Ga_{1-x}Mn_xAs$: Effects of magnetic anisotropy, *Phys. Rev. B* **67**, 205204 (2003); X. Liu, W. L. Lim, L. V. Titova et al., External control of the direction of magnetization in ferromagnetic InMnAs/GaSb heterostructures, *Physica E* **20**, 370–373 (2004).
129. M. Sawicki, F. Matsukura, A. Idziaszek et al., Temperature dependent magnetic anisotropy in (Ga,Mn)As layers, *Phys. Rev. B* **70**, 245325 (2004); M. Sawicki, K.-Y. Wang, K. W. Edmonds et al., In-plane uniaxial anisotropy rotations in (Ga,Mn)As thin films, *Phys. Rev. B* **71**, 121302 (2005).
130. S. C. Masmanidis, H. X. Tang, E. B. Myers et al., Nanomechanical measurement of magnetostriction and magnetic anisotropy in (Ga,Mn)As, *Phys. Rev. Lett.* **95**, 187206 (2005).
131. K. W. Edmonds, G. van der Laan, N. R. S. Farley et al., Strain dependence of the Mn anisotropy in ferromagnetic semiconductors observed by x-ray magnetic circular dichroism, *Phys. Rev. B* **77**, 113205 (2008).
132. Y. Wan-Lun and T. Tao, Theory of the zero-field splitting of $^6S(3d^5)$-state ions in cubic crystals, *Phys. Rev. B* **49**, 3243–3252 (1994).
133. K. Yosida, *Theory of Magnetism*, Springer, Berlin, Germany, 1996.
134. G. Zaránd and B. Jankó, $Ga_{1-x}Mn_xAs$: A frustrated ferromagnet, *Phys. Rev. Lett.* **89**, 047201 (2002).
135. L. Brey and G. Gómez-Santos, Magnetic properties of GaMnAs from an effective Heisenberg Hamiltonian, *Phys. Rev. B* **68**, 115206 (2003).
136. G. Bouzerar, T. Ziman, and J. Kudrnovský, Calculating the Curie temperature reliably in diluted III–V ferromagnetic semiconductors, *Europhys. Lett.* **69**, 812–818 (2005); Compensation, interstitial defects, and ferromagnetism in diluted ferromagnetic semiconductors, *Phys. Rev. B* **72**, 125207 (2005).
137. S. Das Sarma, E. H. Hwang, and A. Kaminski, Temperature-dependent magnetization in diluted magnetic semiconductors, *Phys. Rev. B* **67**, 155201 (2003).
138. P. Mahadevan, A. Zunger, and D. D. Sarma, Unusual directional dependence of exchange energies in GaAs diluted with Mn: Is the RKKY description relevant?, *Phys. Rev. Lett.* **93**, 177201 (2004).
139. A. A. Abrikosov and L. P. Gor'kov, *Sov. Phys. JETP* **16**, 1575 (1963) [*Zh. Eksp. Teor. Fiz.* **43**, 2230 (1962)].
140. T. Jungwirth, J. König, J. Sinova, J. Kučera, and A. H. MacDonald, Curie temperature trends in (III,Mn)V ferromagnetic semiconductors, *Phys. Rev. B* **66**, 012402 (2002).
141. S. Hilbert and W. Nolting, Magnetism in (III,Mn)-V diluted magnetic semiconductors: Effective Heisenberg model, *Phys. Rev. B* **71**, 113204 (2005); Disorder in diluted spin systems, *Phys. Rev. B* **70**, 165203 (2004).
142. S. T. B. Goennenwein, T. Graf, T. Wassner et al., Spin wave resonance in $Ga_{1-x}Mn_xAs$, *Appl. Phys. Lett.* **82**, 730–732 (2003).
143. G. Bouzerar, Magnetic spin excitations in diluted ferromagnetic systems: The case of $Ga_{1-x}Mn_xAs$, *EPL* **79**, 57007 (2007).
144. M. D. Kapetanakis and I. E. Perakis, Spin dynamics in (III,Mn)V ferromagnetic semiconductors: The role of correlations, *Phys. Rev. Lett.* **101**, 097201 (2008).
145. A. Van Esch, L. Van Bockstal, J. De Boeck et al., Interplay between the magnetic and transport properties in the III–V diluted magnetic semiconductor $Ga_{1-x}Mn_xAs$, *Phys. Rev. B* **56**, 13103–13112 (1997).
146. C. P. Moca, B. L. Sheu, N. Samarth, P. Schiffer, B. Jankó, and G. Zaránd, Scaling theory of magnetoresistance and carrier localization in $Ga_{1-x}Mn_xAs$, *Phys. Rev. Lett.* **102**, 137203 (2009).
147. S. Lee, A. Trionfi, T. Schallenberg, H. Munekata, and D. Natelson, Mesoscopic conductance effects in InMnAs structures, *Appl. Phys. Lett.* **90**, 032105 (2007).

148. J. Mašek, I. Turek, V. Drchal, J. Kudrnovský, and F. Máca, Correlated doping in semiconductors: The role of donors in III–V diluted magnetic semiconductors, *Acta Phys. Pol. A* **102**, 673–678 (2002); J. Mašek, I. Turek, J. Kudrnovský, F. Máca, and V. Drchal, Compositional dependence of the formation energies of substitutional and interstitial Mn in partially compensated (Ga,Mn)As, *Acta Phys. Pol. A* **105**, 637–644 (2004).
149. G. Alvarez, M. Mayr, and E. Dagotto, Phase diagram of a model for diluted magnetic semiconductors beyond mean-field approximations, *Phys. Rev. Lett.* **89**, 277202 (2002).
150. G. Alvarez and E. Dagotto, Clustered states as a new paradigm of condensed matter physics, *J. Magn. Magn. Mater.* **272–276**, 15–20 (2004).
151. P. Mahadevan, J. M. Osorio-Guillén, and A. Zunger, Origin of transition metal clustering tendencies in GaAs based dilute magnetic semiconductors, *Appl. Phys. Lett.* **86**, 172504 (2005).
152. H. Raebiger, M. Ganchenkova, and J. von Boehm, Diffusion and clustering of substitutional Mn in (Ga,Mn)As, *Appl. Phys. Lett.* **89**, 012505 (2006).
153. F. V. Kyrychenko and C. A. Ullrich, Enhanced carrier scattering rates in dilute magnetic semiconductors with correlated impurities, *Phys. Rev. B* **75**, 045205 (2007).
154. F. V. Kyrychenko and C. A. Ullrich, Transport and optical conductivity in dilute magnetic semiconductors, *J. Phys.: Condens. Matter* **21**, 084202 (2009); Temperature-dependent resistivity of ferromagnetic $Ga_{1-x}Mn_xAs$: Interplay between impurity scattering and many-body effects, *Phys. Rev. B* **80**, 205202 (2009).
155. M. Zhu, X. Li, G. Xiang, and N. Samarth, Random telegraph noise from magnetic nanoclusters in the ferromagnetic semiconductor (Ga,Mn)As, *Phys. Rev. B* **76**, 201201(R) (2007).
156. H. Bruus and K. Flensberg, *Many-Body Quantum Theory in Condensed Matter Physics*, Oxford University Press, Oxford, U.K., 2004.
157. J. Rammer, *Quantum Transport Theory*, Westview Press, Boulder, CO, 2004.
158. E. Abrahams, P. W. Anderson, D. C. Licciardello, and T. V. Ramakrishnan, Scaling theory of localization: Absence of quantum diffusion in two dimensions, *Phys. Rev. Lett.* **42**, 673–676 (1979).
159. D. M. Basko, I. L. Aleiner, and B. L. Altshuler, Metalinsulator transition in a weakly interacting many-electron system with localized single-particle states, *Ann. Phys.* **321**, 1126–1205 (2006); On the problem of many-body localization, cond-mat/0602510 (2006).
160. S. R. Dunsiger, J. P. Carlo, T. Goko, G. Nieuwenhuys, T. Prokscha, A. Suter, E. Morenzoni et al., Spatially homogeneous ferromagnetism of (Ga, Mn)As, *Nat. Mater.* **9**, 299–303 (2010).
161. N. V. Agrinskaya, V. I. Kozub, D. V. Poloskin, A. V. Chernyaev, and D. V. Shamshur, Transition from strong to weak localization in the split-off impurity band in two-dimensional p-GaAs/AlGaAs structures, *JETP Lett.* **80**, 30–34 (2004) [*P. Zh. Eksp. Teoret. Fiz.* **80**, 36–40 (2004)]; Crossover from strong to weak localization in the split-off impurity band in two-dimensional p-GaAs/AlGaAs structures, *Phys. Stat. Sol. (C)* **3**, 329–333 (2006).
162. T. Słupiński, J. Caban, and K. Moskalik, Hole transport in impurity band and valence bands studied in moderately doped GaAs:Mn single crystals, *Acta Phys. Pol. A* **112**, 325–330 (2007).
163. M. Berciu and R. N. Bhatt, Effects of disorder on ferromagnetism in diluted magnetic semiconductors, *Phys. Rev. Lett.* **87**, 107203 (2001); Mean-field approach to ferromagnetism in (III,Mn)V diluted magnetic semiconductors at low carrier densities, *Phys. Rev. B* **69**, 045202 (2004).
164. A. Kaminski and S. Das Sarma, Polaron percolation in diluted magnetic semiconductors, *Phys. Rev. Lett.* **88**, 247202 (2002); Magnetic and transport percolation in diluted magnetic semiconductors, *Phys. Rev. B* **68**, 235210 (2003).
165. A. C. Durst, R. N. Bhatt, and P. A. Wolff, Bound magnetic polaron interactions in insulating doped diluted magnetic semiconductors, *Phys. Rev. B* **65**, 235205 (2002).
166. B. L. Sheu, R. C. Myers, J.-M. Tang et al., Onset of ferromagnetism in low-doped $Ga_{1-x}Mn_xAs$, *Phys. Rev. Lett.* **99**, 227205 (2007).
167. D. Neumaier, K. Wagner, S. Geißler et al., Weak localization in ferromagnetic (Ga,Mn)As nanostructures, *Phys. Rev. Lett.* **99**, 116803 (2007).
168. P. A. Lee, A. D. Stone, and H. Fukuyama, Universal conductance fluctuations in metals: Effects of finite temperature, interactions, and magnetic field, *Phys. Rev. B* **35**, 1039–1070 (1987).
169. K. Wagner, D. Neumaier, M. Reinwald, W. Wegscheider, and D. Weiss, Dephasing in (Ga,Mn)As nanowires and rings, *Phys. Rev. Lett.* **97**, 056803 (2006).
170. L. Vila, R. Giraud, L. Thevenard et al., Universal conductance fluctuations in epitaxial GaMnAs ferromagnets: Dephasing by structural and spin disorder, *Phys. Rev. Lett.* **98**, 027204 (2007).
171. I. Garate, J. Sinova, T. Jungwirth, and A. H. MacDonald, Theory of weak localization in ferromagnetic (Ga,Mn)As, *Phys. Rev. B* **79**, 155207 (2009).
172. L. P. Rokhinson, Y. Lyanda-Geller, Z. Ge et al., Weak localization in $Ga_{1-x}Mn_xAs$: Evidence of impurity band, transport, *Phys. Rev. B* **76**, 161201(R) (2007).
173. C. Timm, M. E. Raikh, and F. von Oppen, Disorder-induced resistive anomaly near ferromagnetic phase transitions, *Phys. Rev. Lett.* **94**, 036602 (2005).
174. M. E. Fisher and J. S. Langer, Resistive anomalies at magnetic critical points, *Phys. Rev. Lett.* **20**, 665–668 (1968).
175. J. A. Sobota, D. Tanasković, and V. Dobrosavljević, RKKY interactions in the regime of strong localization, *Phys. Rev. B* **76**, 245106 (2007).
176. V. M. Galitski, A. Kaminski, and S. Das Sarma, Griffiths phase in diluted magnetic semiconductors, *Phys. Rev. Lett.* **92**, 177203 (2004).

177. A. Geresdi, A. Halbritter, M. Csontos et al., Nanoscale spin polarization in the dilute magnetic semiconductor (In,Mn)Sb, *Phys. Rev. B* **77**, 233304 (2008).
178. J. Schliemann and A. H. MacDonald, Noncollinear ferromagnetism in (III,Mn)V semiconductors, *Phys. Rev. Lett.* **88**, 137201 (2002); J. Schliemann, Disorder-induced noncollinear ferromagnetism in models for (III,Mn)V semiconductors, *Phys. Rev. B* **67**, 045202 (2003).
179. G. A. Fiete, G. Zaránd, B. Jankó, P. Redliński, and C. P. Moca, Positional disorder, spin–orbit coupling, and frustration in $Ga_{1-x}Mn_xAs$, *Phys. Rev. B* **71**, 115202 (2005).
180. M. Mayr, G. Alvarez, and E. Dagotto, Global versus local ferromagnetism in a model for diluted magnetic semiconductors studied with Monte Carlo techniques, *Phys. Rev. B* **65**, 241202(R) (2002).
181. D. J. Priour, Jr., E. H. Hwang, and S. Das Sarma, Disordered RKKY lattice mean field theory for ferromagnetism in diluted magnetic semiconductors, *Phys. Rev. Lett.* **92**, 117201 (2004).
182. K. Aryanpour, J. Moreno, M. Jarrell, and R. S. Fishman, Magnetism in semiconductors: A dynamical mean-field study of ferromagnetism in $Ga_{1-x}Mn_xAs$, *Phys. Rev. B* **72**, 045343 (2005).
183. B. Beschoten, P. A. Crowell, I. Malajovich et al., Magnetic circular dichroism studies of carrier-induced ferromagnetism in $(Ga_{1-x}Mn_x)As$, *Phys. Rev. Lett.* **83**, 3073–3076 (1999).
184. R. Mathieu, B. S. Sørensen, J. Sadowski et al., Magnetization of ultrathin (Ga,Mn)As layers, *Phys. Rev. B* **68**, 184421 (2003).
185. W. Limmer, A. Koeder, S. Frank et al., Electronic and magnetic properties of GaMnAs: Annealing effects, *Physica E* **21**, 970–974 (2004).
186. J. Schliemann, J. König, and A. H. MacDonald, Monte Carlo study of ferromagnetism in (III,Mn)V semiconductors, *Phys. Rev. B* **64**, 165201 (2001).
187. G. Alvarez and E. Dagotto, Single-band model for diluted magnetic semiconductors: Dynamical and transport properties and relevance of clustered states, *Phys. Rev. B* **68**, 045202 (2003).
188. G. Bouzerar and R. Bouzerar, Comment on "Large-scale Monte Carlo study of a realistic lattice model for $Ga_{1-x}Mn_xAs$," *Phys. Rev. Lett.* **100**, 229701 (2008); A. Moreo and E. Dagotto, Reply, *Phys. Rev. Lett.* **100**, 229702 (2008).
189. S. Sonoda, S. Shimizu, T. Sasaki, Y. Yamamoto, and H. Hori, Molecular beam epitaxy of wurtzite (Ga,Mn)N films on sapphire(0001) showing the ferromagnetic behaviour at room temperature, *J. Cryst. Growth* **237–239**, 1358–1362 (2002); T. Sasaki, S. Sonoda, Y. Yamamoto et al., Magnetic and transport characteristics on high Curie temperature ferromagnet of Mn-doped GaN, *J. Appl. Phys.* **91**, 7911–7913 (2002).
190. N. Theodoropoulou, A. F. Hebard, M. E. Overberg et al., Unconventional carrier-mediated ferromagnetism above room temperature in ion-implanted (Ga,Mn)P:C, *Phys. Rev. Lett.* **89**, 107203 (2002).
191. S. Dhar, O. Brandt, M. Ramsteiner, V. F. Sapega, and K. H. Ploog, Colossal magnetic moment of Gd in GaN, *Phys. Rev. Lett.* **94**, 037205 (2005); S. Dhar, L. Pérez, O. Brandt et al., Gd-doped GaN: A very dilute ferromagnetic semiconductor with a Curie temperature above 300 K, *Phys. Rev. B* **72**, 245203 (2005).
192. R. Rajaram, A. Ney, G. Solomon, J. S. Harris, Jr., R. F. C. Farrow, and S. S. P. Parkin, Growth and magnetism of Cr-doped InN, *Appl. Phys. Lett.* **87**, 172511 (2005).
193. S. Dhar, T. Kammermeier, A. Ney et al., Ferromagnetism and colossal magnetic moment in Gd-focused ion-beam-implanted GaN, *Appl. Phys. Lett.* **89**, 062503 (2006).
194. S. Sonoda, I. Tanaka, H. Ikeno et al., Coexistence of Mn^{2+} and Mn^{3+} in ferromagnetic GaMnN, *J. Phys.: Condens. Matter* **18**, 4615–4621 (2006).
195. S. Yoshii, S. Sonoda, T. Yamamoto et al., Evidence for carrier-induced high-T_C ferromagnetism in Mn-doped GaN film, *EPL* **78**, 37006 (2007).
196. A. Bonanni, M. Kiecana, C. Simbrunner et al., Paramagnetic GaN:Fe and ferromagnetic (Ga,Fe)N: The relationship between structural, electronic, and magnetic properties, *Phys. Rev. B* **75**, 125210 (2007).
197. J. M. Zavada, N. Nepal, C. Ugolini et al., Optical and magnetic behavior of erbium-doped GaN epilayers grown by metal-organic chemical vapor deposition, *Appl. Phys. Lett.* **91**, 054106 (2007).
198. G. Martínez-Criado, A. Somogyi, S. Ramos et al., Mn-rich clusters in GaN: Hexagonal or cubic symmetry?, *Appl. Phys. Lett.* **86**, 131927 (2005).
199. M. Yokoyama, H. Yamaguchi, T. Ogawa, and M. Tanaka, Zinc-blende-type MnAs nanoclusters embedded in GaAs, *J. Appl. Phys.* **97**, 10D317 (2005).
200. S. Y. Han, J. Hite, G. T. Thaler et al., Effect of Gd implantation on the structural and magnetic properties of GaN and AlN, *Appl. Phys. Lett.* **88**, 042102 (2006).
201. A. Ney, R. Rajaraman, T. Kammermeier et al., Metastable magnetism and memory effects in dilute magnetic semiconductors, *J. Phys.: Condens. Matter* **20**, 285222 (2008).
202. P. Mahadevan and A. Zunger, First-principles investigation of the assumptions underlying model-Hamiltonian approaches to ferromagnetism of 3d impurities in III–V semiconductors, *Phys. Rev. B* **69**, 115211 (2004).
203. J. I. Hwang, Y. Ishida, M. Kobayashi et al., High-energy spectroscopic study of the III–V nitride-based diluted magnetic semiconductor $Ga_{1-x}Mn_xN$, *Phys. Rev. B* **72**, 085216 (2005).
204. S. Marcet, D. Ferrand, D. Halley et al., Magneto-optical spectroscopy of (Ga,Mn)N epilayers, *Phys. Rev. B* **74**, 125201 (2006).
205. T. Dietl, Hole states in wide band-gap diluted magnetic semiconductors and oxides, *Phys. Rev. B* **77**, 085208 (2008).
206. W. Pacuski, P. Kossacki, D. Ferrand et al., Observation of strong-coupling effects in a diluted magnetic semiconductor $Ga_{1-x}Fe_xN$, *Phys. Rev. Lett.* **100**, 037204 (2008).

207. B. Sanyal, O. Bengone, and S. Mirbt, Electronic structure and magnetism of Mn-doped GaN, *Phys. Rev. B* **68**, 205210 (2003).
208. T. Graf, M. Gjukic, M. S. Brandt, M. Stutzmann, and O. Ambacher, The $Mn^{3+/2+}$ acceptor level in group III nitrides, *Appl. Phys. Lett.* **81**, 5159–5161 (2002).
209. M. A. Scarpulla, B. L. Cardozo, R. Farshchi et al., Ferromagnetism in $Ga_{1-x}Mn_xP$: Evidence for inter-Mn exchange mediated by localized holes within a detached impurity band, *Phys. Rev. Lett.* **95**, 207204 (2005).
210. P. R. Stone, M. A. Scarpulla, R. Farshchi et al., Mn $L_{3,2}$ x-ray absorption and magnetic circular dichroism in ferromagnetic $Ga_{1-x}Mn_xP$, *Appl. Phys. Lett.* **89**, 012504 (2006).
211. L. Kronik, M. Jain, and J. R. Chelikowsky, Electronic structure and spin polarization of $Mn_xGa_{1-x}N$, *Phys. Rev. B* **66**, 041203 (2002).
212. J. Neugebauer and C. G. Van de Walle, Atomic geometry and electronic structure of native defects in GaN, *Phys. Rev. B* **50**, 8067–8070 (1994); P. Bogusławski, E. L. Briggs, and J. Bernholc, Native defects in gallium nitride, *Phys. Rev. B* **51**, 17255–17258 (1995); S. Limpijumnong and C. G. Van de Walle, Diffusivity of native defects in GaN, *Phys. Rev. B* **69**, 035207 (2004).
213. X. L. Yang, Z. T. Chen, C. D. Wang et al., Effects of nitrogen vacancies induced by Mn doping in (Ga,Mn)N films grown by MOCVD, *J. Phys. D: Appl. Phys.* **41**, 125002 (2008).
214. K. Sato and H. Katayama-Yoshida, First principles materials design for semiconductor spintronics, *Semicond. Sci. Technol.* **17**, 367–376 (2002).
215. G. M. Dalpian and S.-H. Wei, Electron-induced stabilization of ferromagnetism in $Ga_{1-x}Gd_xN$, *Phys. Rev. B* **72**, 115201 (2005).
216. J. M. D. Coey, M. Venkatesan, and C. B. Fitzgerald, Donor impurity band exchange in dilute ferromagnetic oxides, *Nat. Mater.* **4**, 173–179 (2005).
217. T. Dubroca, J. Hack, R. E. Hummel, and A. Angerhofer, Quasiferromagnetism in semiconductors, *Appl. Phys. Lett.* **88**, 182504 (2006).
218. D. M. Edwards and M. I. Katsnelson, High-temperature ferromagnetism of sp electrons in narrow impurity bands: Application to CaB_6, *J. Phys.: Condens. Matter* **18**, 7209–7225 (2006).
219. J. M. D. Coey, K. Wongsaprom, J. Alaria, and M. Venkatesan, Charge-transfer ferromagnetism in oxide nanoparticles, *J. Phys. D: Appl. Phys.* **41**, 134012 (2008).

20

Magnetism of Dilute Oxides

J.M.D. Coey
Trinity College, Dublin

Reports of Curie temperatures well above room temperature for wide-bandgap oxides doped with a few percent of transition-metal cations have triggered intense interest in these materials as potential magnetic semiconductors. The origin of the magnetism is debated; in some systems, the ferromagnetism can be attributed to nanoparticles of a ferromagnetic secondary phase, but in others, properties are found that are incompatible with any secondary phase, and an intrinsic origin related to structural defects is implicated. The magnetic interactions in these materials are qualitatively different from those in magnetically concentrated compounds, and the standard "m-J paradigm" of localized magnetism seems unable to explain their behavior. In this chapter, we first summarize the established view of magnetism in the oxides before reviewing data on a selection of the most widely studied materials; then we consider the physical models that have been advanced to explain the magnetism, and raise issues that still have to be addressed. As noted in Chapter 19, some of the concepts also apply to wide-gap III–V compounds.

20.1 Introduction

Transition-metal oxides have long been regarded as an important and well-understood class of magnetic materials. We have built up an excellent working knowledge of the spinel ferrites TFe_2O_4, garnets $M_3Fe_5O_{12}$, and hexagonal ferrites $MFe_{12}O_{19}$, which have widespread industrial applications, and represent about 40% of the global market for hard and soft magnets. This knowledge can be regarded as a familiar landmark in our understanding of magnetism in solids, epitomized by the award of the 1970 Nobel Prize in physics to Louis Néel.

The crystal structures of these oxides are usually based on a close-packed lattice of oxygen anions where the 3d cations mostly occupy octahedral interstices with six oxygen neighbors or tetrahedral interstices with four oxygen neighbors. The largest of the 3d cations show some preference for the octahedral sites. Rare earth or alkali ions such as Ba or Sr have a greater coordination number, and may sometimes replace an anion in the close-packed oxygen lattice. A list of the ionic radii of common 3d cations is given in Table 20.1, based on a radius of 140 pm for O^{2-}.

An ionic picture of the bonding, with formal cation valences of 2+, 3+, or 4+ and an oxygen valence of 2− leading to a $2p^6$ closed shell, is the traditional starting point for looking at the electronic structure. The number of electrons per cation is an integer, and the magnetic moments are well localized. The oxides are usually insulators or wide-gap semiconductors. They owe their colors—green for yttrium iron garnet, brown for barium hexaferrite—to electronic transitions involving the 3d levels, which normally lie in the primary bandgap E_0 between a valence band with largely oxygen 2p character and a conduction band formed from unoccupied metal 4s orbitals. The magnitude of E_0 decreases on moving along the 3d series from 8.1 eV for TiO_2 or 5.7 eV for Sc_2O_3 to 4.9 eV for Ga_2O_3 or 3.4 eV for ZnO. The measured bandgap E_g can be much less than E_0 because of the possibility of p–d, or d–s transitions.

According to Hund's rules, the 3d ions are in either an A (d^5), D (d^1, d^4, d^6, d^9), or F (d^2, d^3, d^7, d^8) orbital ground state, depending on the number of d electrons they possess. The ground-state levels are split by the crystal/ligand field interaction [1,2] and there are higher energy levels, which can be excited optically. A schematic representation of the electronic structure is given in Figure 20.1a.

It is often helpful to think of one-electron d states, even though the d shell is strongly correlated. The five d orbitals split

TABLE 20.1 Charge State, Electronic Configuration, and Ionic Radius in Octahedral Coordination for Common 3*d* Cations in Oxides

Cation	Charge State	Configuration	Ionic Radius (pm)
Sc	3^+	$3d^0$	83
Ti	$3^+/4^+$	$3d^0/3d^1$	61/69
V	2^+	$3d^3$	72
Cr	3^+	$3d^3$	64
Mn	$2^+/3^+/4^+$	$3d^5/3d^4/3d^3$	83/65/53
Fe	$2^+/3^+$	$3d^6/3d^5$	82/65
Co	$2^+/3^+$	$3d^7/3d^6$	82/61
Ni	$2^+/3^+$	$3d^8/3d^7$	78/69
Cu	2^+	$3d^9$	72

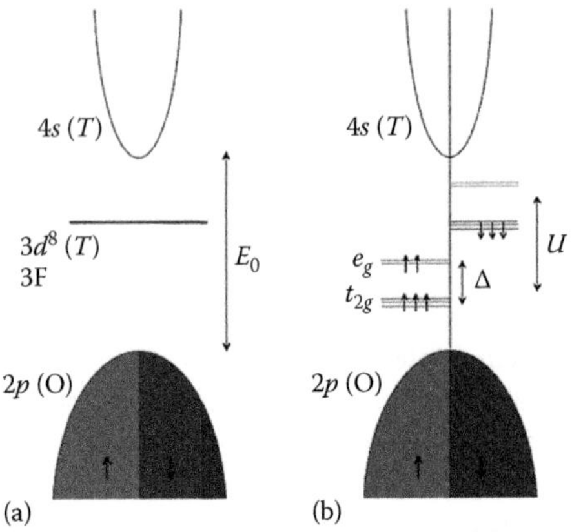

FIGURE 20.1 Electronic structure diagrams for a 3*d* metal oxide (a) shows the 3*d* electrons as a highly correlated level, (b) shows the 3*d* one-electron states in the crystal field. The example illustrated is a $3d^8$ cation with $S = 1$ in an octahedral site.

into a group of three (d_{xy}, d_{yz}, d_{zx}) and a group of two ($d_{x^2-y^2}$, d_{z^2}) in cubic symmetry—undistorted octahedral and tetrahedral sites have cubic symmetry. These are labeled as t_{2g} and e_g states in octahedral symmetry, and as t_2 and e states in tetrahedral symmetry. The e_g and t_2 overlap and hybridize strongly with the oxygen 2*p* ligand orbitals (p_x, p_y, p_x) but the t_{2g} and *e* do not. This produces the ligand field splitting of the triplet and the doublet, which reinforces the purely electrostatic crystal field effect of the oxygen charge on the *d* orbitals to give the splitting Δ. The *d*–*d* correlations are roughly taken into account in this picture by splitting the ↑ and ↓ levels by an energy *U*, as shown in Figure 20.1b. When $U > \Delta$, Hund's first rule applies to the ion, and it is in a *high-spin* state. Otherwise, when $U < \Delta$, it is in a *low-spin* state, with a spin moment determined by the orbital occupancy. The crystal field quenches the orbital moments for 3*d* ions in solids, so their magnetic moments are, to a first approximation, spin-only moments of just one Bohr magneton, μ_B, per unpaired electron. *S* is the good quantum number for the 3*d* ion. The one-electron picture works well for D-state ions, where there is a single electron or hole outside a filled or half-filled shell. It is not nearly as good for the F-state ions.

Three other points complete the beginner's toolkit for 3*d* ions in oxides. (a) Filled shells and half-filled shells are particularly stable. In other cases, a crystal field stabilization energy can be calculated for the ion. For Cr^{3+} d^3, for example, this is $3 \times (2/5)$ Δ in octahedral sites. (The crystal field splitting of the energies of the five *d* orbitals preserves their center of gravity.) (b) Jahn–Teller distortion of the octahedron or tetrahedron may stabilize the energy of the ion. The effect is strongest for a singly occupied e_g or *e* level, in other words d^4 or d^9 ions in octahedral coordination, and d^1 and d^6 ions in tetrahedral coordination. (c) There is a tendency for the *d*-levels to become more stable as we move across the series from Sc to Zn because of the increasing nuclear charge. For example, the *d* levels of Ti are close to the conduction band, whereas those of Cu may overlap the 2*p* valence band.

Electrons or holes, introduced into oxides by doping or nonstoichiometry, are often trapped. If they are at all mobile, they tend to form *polarons* with a large effective mass. Electron doping can populate states in the 4*s* conduction band or $3d^{n+1}$ states in the primary gap, but the extra electrons do not part easily from their dopant atom. Tin in In_2O_3 is an exception, and indium tin oxide (ITO) is the best example of a transparent conducting oxide. Hole doping, by oxygen substoichiometry, for example, usually creates donor states near the bottom of the conduction band, which also tend to remain localized. The insulating character of the oxides effectively resists tinkering with the stoichiometry. Transport properties are quite different from those of the classical covalent semiconductors, where electrons or holes diffuse freely in the conduction or valence bands. The best oxide semiconductors are those with the most covalent bonding, especially those with the 3*d* cations in tetrahedral coordination—ZnO and $CuAlO_2$, for example.

Electronic structure calculations based on density functional theory have become very popular. These calculations work well for metals, and can accurately predict the magnetic moments. There are difficulties in applying them to oxides, where the primary gap E_0 and the unoccupied electronic levels are difficult to reproduce.

The source of the magnetism in oxides is the metal cations, which bear a moment due to unpaired electrons in their 3*d* shells. Any 3*d* cation with $1 \le n \le 9$ has a spin moment, with very few exceptions—$3d^6$ cations in octahedral coordination and $3d^4$ cations in tetrahedral coordination may be in a nonmagnetic low-spin state. The wave functions of the 3*d* cations have exponentially decaying tails, which ensures that there is negligible overlap between cations that are not nearest neighbors. These neighbors share one or more common oxygen ligands. Hence, the magnetic superexchange coupling essentially involves a threesome of two nearest-neighbor cations and an intervening oxygen as shown schematically for the Mn^{2+}–O^{2-}–Mn^{2+} bond in Figure 20.2. The Mn has a half-filled 3*d* shell, so only minority-spin electron transfer from the oxygen to the manganese is possible. A 2*p* electron with ↓ is transferred to the Mn on the left, leaving a 2*p* hole that can be filled by an electron from the other Mn, provided

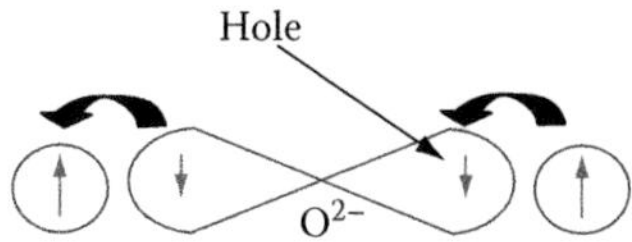

FIGURE 20.2 Superexchange interaction between ions with more than half-filled *d* shells *via* an O^{2-} anion.

it is ↓. Superexchange theory leads to a Heisenberg-type Hamiltonian

$$\mathcal{H} = -2J\mathbf{S}_i \cdot \mathbf{S}_j \tag{20.1}$$

where the exchange constant $J = J_0 \cos^2\theta$ where θ is the superexchange bond angle and $J_0 = -t^2/U'$, where t is the M–O transfer integral and U' is the on-site Coulomb interaction for adding an electron to the M ion. Typically, $t \approx 0.1$ eV and $U \approx 5$ eV. Hence, the order of magnitude of the exchange constant is 2 meV or −20 K.

The expression for the Curie temperature in mean-field theory is

$$T_C = \frac{2ZJS(S+1)}{3k} \tag{20.2}$$

Hence, if $S = 5/2$, the largest value possible in the 3*d* series, and the cation coordination number $Z = 8$, we find $T_C \approx 800$ K. In practice, the superexchange interactions are often frustrated—the antiparallel coupling of all nearest neighbors cannot be achieved for geometric reasons imposed by the topology of the lattice. Furthermore, Equation 20.3 is known to overestimate T_C by about 30%, because it takes no account of spin-wave excitations. In practice, the magnetic-ordering temperatures of oxides do not exceed 1000 K. Hematite—αFe_2O_3—has a Néel temperature of 960 K. In most cases the ordering temperatures are a few hundred Kelvin, at best, as indicated in Figure 20.3. When the oxide is diluted with nonmagnetic cations, the coordination number $Z = Z_0 x$ is reduced, and T_C varies as x above the percolation threshold as shown in Figure 20.4.

The validity of the Heisenberg model in oxides has been amply demonstrated by fitting the exchange constants to the spin-wave dispersion relations determined by inelastic neutron scattering. Complete data sets have been obtained for Fe_2O_3, Cr_2O_3, and Fe_3O_4, among others [3–5]. The values of the exchange constants range from −30 to +6 K in α-Fe_2O_3. Negative (antiferromagnetic) interactions are for superexchange bonds involving the same ions, with large bond angles. Positive (ferromagnetic) values are for bonds involving ions in different valence states, or near-90° bonds between ions in the same valence state.

During the 1960s, a wealth of information was accumulated on the nature of the exchange interactions between different cations for different superexchange bond angles. An extensive set of rules was formulated by Goodenough [6] and Kanamori [7]. These rules were simplified by Anderson [8], as follows:

1. 180° exchange between half-filled orbitals is strong and antiferromagnetic.
2. 90° exchange between half-filled orbitals is ferromagnetic, and rather weak.
3. Exchange between a half-filled orbital and an empty orbital of different symmetry is ferromagnetic and rather weak.

The stronger interactions tend to be negative (antiferromagnetic), coupling the two cation spin moments antiparallel, and leaving no net moment on the intervening oxygen. The three main families of magnetic insulators mentioned in the opening paragraph are all ferrimagnetic, with dominant nearest-neighbor antiferromagnetic exchange between ions on different, unequal sublattices.

Not all transition-metal oxides are ferrimagnetic or antiferromagnetic, and not all are insulators. The extended 3*d* orbitals of

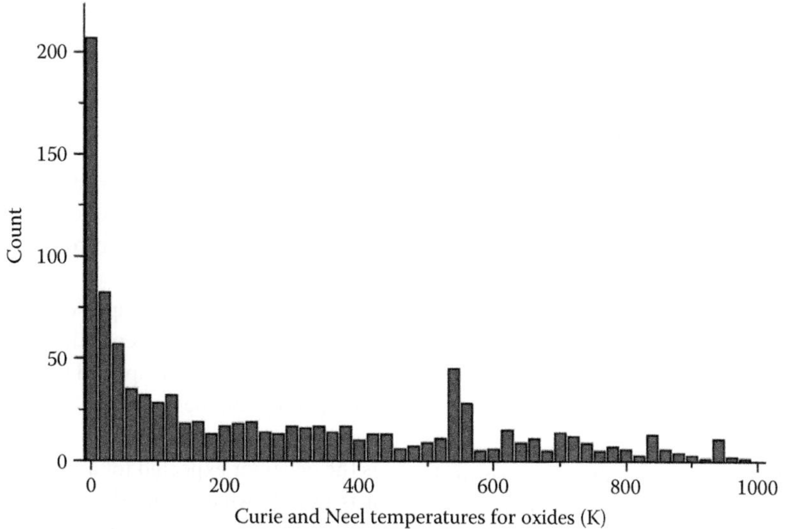

FIGURE 20.3 Histogram of the magnetic-ordering temperature of a selection of magnetic oxides.

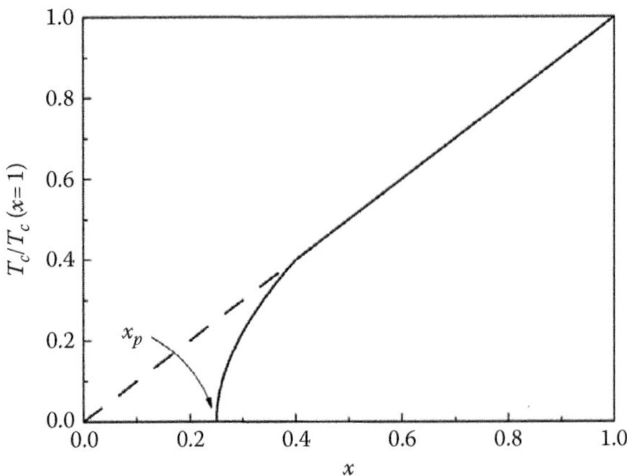

FIGURE 20.4 Sketch of the magnetic-ordering temperature in a dilute magnetic oxide with nearest-neighbor superexchange interactions. The dashed line is the prediction of mean-field theory for long-range interactions, x_p is the percolation threshold.

the elements at the beginning of the series may form band states. There exist a few ferromagnetic oxides, such as CrO_2, which is the only simple oxide that is a ferromagnetic metal ($T_C = 390\,K$).

In an effort to increase the magnetization of oxides, beginning in the 1950s, ways were sought to create strong ferromagnetic coupling. A successful approach, implemented in mixed-valence manganites, was to impose two different charge states on the manganese on the octahedral sites of the perovskite structure by including cations on the cubic sites, which had well-defined, but different charge states [9]. A famous example is $(La_{0.7}Sr_{0.3})MnO_3$. Formally, 70% of the manganese is Mn^{3+} $3d^4$, while 30% is Mn^{4+} $3d^3$. The fourth manganese d electrons can hop around among the $3d^3$ ion cores, preserving their spin memory and providing ferromagnetic coupling via a mechanism known as *double exchange* [10]. The delocalization of the d electrons is conditional on parallel alignment of the ion cores, which gives rise to the characteristic colossal magnetoresistance in these oxides near the Curie point, which can be as high as 370 K. Another route to ferromagnetism is illustrated by the double perovskite Sr_2FeMoO_6, where the spins of the iron $3d^5$ ion cores are coupled antiparallel to the spins of the electrons in a spin-polarized $4d^1$ molybdenum band, giving ferromagnetic alignment of the iron.

To summarize, concentrated magnetic oxides are well described in terms of the "m-J paradigm": there are localized magnetic moments m on the cations, and superexchange interactions couple them. The paradigm provides a good account of both the magnetic order and the spin waves. Exchange interactions are normally antiferromagnetic, and magnetic-ordering temperatures do not exceed 1000 K.

20.2 Dilute Oxides

So much for magnetic general knowledge. Our concern in this chapter is dilute magnetic oxides with general formula

$$(M_{1-x}T_x)O_n \tag{20.3}$$

where

- M is a nonmagnetic cation
- T is a transition-metal cation
- n is an integer or rational fraction
- $x < 0.1$ (10%)

Typical values of x are a few percent.

An important limit is x_p, the percolation threshold [11], where continuous nearest-neighbor paths first appear, which link M cations throughout the crystal. We are interested in magnetic bond percolation, where cations are regarded as neighbors when their spins are coupled via a superexchange bond. Below x_p there are only isolated cations, and small clusters of nearest-neighbor pairs, triplets, etc. Above x_p there is a bulk cluster, which encompasses most of the magnetic cations. Provided the exchange interactions only involve nearest neighbors, there is no possibility of long-range order below x_p, which is approximately $2/Z_0$ [12], where Z_0 is the cation coordination number. Depending on the structure, Z_0 lies between 6 and 12, which means that x_p is in the range 16%–33%. Our dilute oxides fall below the percolation threshold, and should not be expected to order magnetically. A random distribution of magnetic dopant ions is illustrated schematically in Figure 20.5a. At low concentrations, most of the dopants are isolated, with no magnetic nearest neighbors, and a Curie-law susceptibility is expected for them:

$$\chi_1 = \frac{x_1 C}{T} \tag{20.4}$$

Here x_1 is the fraction of isolated ions, and C is the Curie constant $\mu_0 N g^2 \mu_B^2 S(S+1)/3k_B$, where N is the number of cations per unit volume, and $g = 2$ for spin-only moments. As the concentration increases, there will be an increasing proportion of dimers and larger groups. The dimers (and other even-membered groups) have a Curie–Weiss susceptibility

$$\chi_2 = \frac{x_2 C}{(T-\theta)} \tag{20.5}$$

where

- x_2 is the fraction ions present as dimers
- θ is a negative constant of order −10 to −100 K, which is proportional to the antiferromagnetic exchange coupling J_0

The odd-membered groups with a net moment contribute a modified Curie-law susceptibility. With all this, deviations from a linear response to an applied field of up to 1 T, for example, are only perceptible below about 10 K. The room-temperature susceptibility is of order $10^{-3}x$ (x is defined in Equation 20.3). This is the behavior expected for a dilute magnetic oxide, and indeed it is exactly what is found in bulk crystalline materials and in well-crystallized, defect-free thin films. There are numerous examples in the literature, for example, Refs. [13–17].

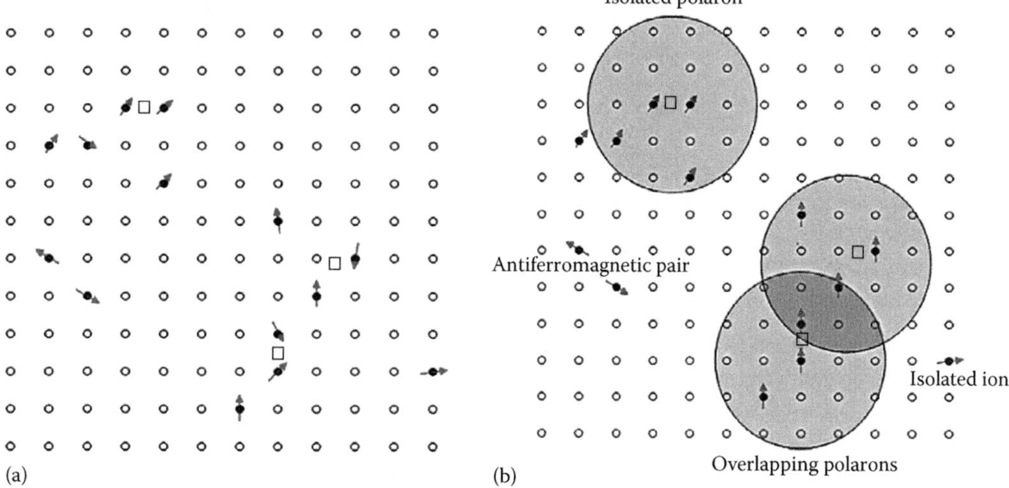

FIGURE 20.5 (a) Schematic representation of distribution of dopant ions in a DMS. (b) The same, but with donor defects (□) that create magnetic polarons where the dopant ions are coupled ferromagnetically.

The conventional picture of magnetism in insulators or magnetic semiconductors is at a loss to account for high-temperature ferromagnetism in dilute magnetic oxides. The Curie or Néel temperatures of oxides were plotted on a histogram in Figure 20.3. None exceeded the Néel point of αFe_2O_3, which is 960 K. The ordering temperatures in solid solutions vary as x or as $x^{1/2}$, depending on the nature of the interactions [18]. No order is expected at room temperature when $x < 10\%$, or in insulators when $x < x_p$. The magnetic behavior we would normally expect from the m-J paradigm is a superposition of Curie-law paramagnetism for isolated ions and Curie–Weiss behavior for the small, antiferromagnetically coupled groups of two, three, or more ions. Indeed, this is precisely what is found in well-crystallized, bulk material [15].

It therefore came as a surprise when, at the beginning of 2001, Matsumoto et al. published their claim that anatase TiO_2 doped with 7% cobalt was ferromagnetically ordered at room temperature. There followed a deluge of similar reports on other materials. The first samples were all thin films, but soon reports of high-temperature ferromagnetism were appearing for nanoparticles and nanocrystalline material as well. Table 20.2 lists some of these early reports. There are now well over 1000 publications on dilute ferromagnetic oxides, more than half of them on ZnO, and new ones appear every week. Many of these are electronic

TABLE 20.2 Early Reports of Ferromagnetic Oxide Thin Films with T_C above Room Temperature

Material	E_g (eV)	Doping	Moment (μ_B)	T_C (K)	Reference
TiO_2	3.2	V—5%	4.2	>400	Hong et al. [34]
		Co—7%	0.3	>300	Matsumoto et al. [25,26]
		Co—7%	1.4	>650	Shinde et al. [29]
		Fe—2%	2.4	300	Wang et al. [35]
SnO_2	3.5	Fe—5%	1.8	610	Coey et al. [54]
		Co—5%	7.5	650	Ogale et al. [51]
HfO_2		Co—2%	9.4	>400	Coey et al. [72]
		Fe—1%	2.2	>400	Hong et al. [73]
CeO_2		Co—3%	6.1	725	Tiwari et al. [79]
In_2O_3	2.9	Mn—5% Sn	0.8	>400	Philip et al. [88]
		Fe—5%	1.4	>600	He et al. [90]
		Cr—2%	1.5	900	Philip et al. [96]
ZnO	3.4	V—15%	0.5	>350	Saeki et al. [110]
		Mn—2.2%	0.16	>300	Sharma et al. [108]
		Fe 5%, Cu 1%	0.75	550	Han et al. [153]
		Co—10%	2.0	280–300	Ueda et al. [104]
		Ni—0.9%	0.06	>300	Radovanovic and Gamelin [68]

structure calculations for dilute magnetic oxides, with various defects, which are not the focus of this chapter.

The experimental results were astonishing for three reasons:

1. The magnetic order appears in oxide materials where the doping is far below x_p.
2. The magnetic order is ferromagnetic, whereas the super-exchange in oxides is usually antiferromagnetic.
3. The films are ferromagnetic at room temperature and above, although no dilute magnetic material had ever been found to be magnetically ordered at room temperature.

In these circumstances, the claims of ferromagnetism were regarded with deep skepticism, and the conviction that the high-temperature ferromagnetism must somehow be associated with a segregated, magnetically concentrated impurity phase, or else it was simply a measurement artifact. Otherwise, one is led to believe that the high-temperature ferromagnetism of dilute oxides is a new and significant magnetic phenomenon. Although the field is fraught with irreproducibility, and impurity phases are often implicated, such is the view of the present author.

20.3 Magnetic Measurements

Before considering the experimental evidence, some words of caution are in order. We are dealing mainly with thin films, about 100 nm thick. If $x = 1\%$, there may the equivalent of just four layers of ferromagnetic cations in the film. The ferromagnetic volume fraction of a typical $5 \times 5\,\text{mm}^2$ sample on its substrate, which is usually about 0.5 mm thick, is just two parts per million. Assuming $1\,\mu_B$ per cation, the ferromagnetic moment is approximately 1.5×10^{-8} A m^2 (1.5×10^{-5} emu). While this is comfortably above the sensitivity of a superconducting quantum interference device magnetometer (SQUID), an alternating-gradient force magnetometer (AGFM) or even a good vibrating-sample magnetometer (VSM), it does represent a *very small moment*. A 0.1 μg speck of iron or magnetite has a moment of this magnitude. These contaminants are ubiquitous, and some care is needed to be sure that what is being measured is the sample, and not magnetic contamination. Co nanoparticles are difficult to detect, but Fe and Fe_3O_4 are readily picked up by Mössbauer spectroscopy. Element-specific x-ray absorption is useful. Secondary phases are an important consideration, and are probably the complete explanation of the ferromagnetism in certain cases. Other high-temperature ferromagnets are spinel phases, where the antiferromagnetic interactions lead to ferromagnetic structures where the moment does not exceed $1.5\,\mu_B$ per magnetic cation. Low concentrations of nanoscale ferromagnetic inclusions may elude detection by x-ray diffraction, but should be detectable by careful transmission-electron microscopy. In magnetic measurements, they are expected to behave superparamagnetically, with the characteristic signature of anhysteretic magnetization curves that superpose perfectly when plotted as a function of H/T.

Recent publications have emphasized how bogus magnetic signals may arise from markers, kapton tape, silver paint, contamination from steel tweezers, unsuitable substrate holders, and other sources of external pollution [19–24]. It is essential, as part of the experimental protocol, to run blank, control samples that have been treated in exactly the same way as those containing the magnetic cations. In order to make a case that magnetic cations or unpaired electrons in the oxide film are the source of the observed magnetism, precautions in sample preparation and handling are imperative, to do justice to the sensitivity of the magnetometer.

The problems for nanocrystalline material are different. Here the moments reported are of order 2×10^{-4} A m^2 kg^{-1} (2 10^{-4} emu g^{-1}). The issue now is sample purity; the presence of ferromagnetic metals or oxides at the ppm level has to be discounted. The specific moments of Fe, Co, Ni, and Fe_3O_4 are, respectively, 217, 162, 55, and 92 A m^2 kg^{-1} (emu g^{-1} is an equivalent unit); their magnetizations are 1710, 1440, 490, and 480 kA m^{-1} (emu cc^{-1} is an equivalent unit).

For thin films deposited on substrates, the huge mismatch between the mass of the thin film sample—typically some tens of micrograms—and that of the substrate—which is about a thousand times greater—presents a challenge. Commonly used substrates, sapphire (Al_2O_3), MgO, $SrTiO_3$, $LaAlO_3$, and Si, are all diamagnetic with susceptibilities of −4.8, −3.1, −1.3, −2.7, and -1.5×10^{-9} m^3 kg^{-1}. The problem is illustrated in Figure 20.6. The substrate gives a diamagnetic signal, which is perhaps reproducible to within 1%, given the uncertainties in positioning and centering the substrate in the magnetometer used to measure the hysteresis loop. It is therefore practically impossible to determine the diamagnetic susceptibility of the undoped film. A Curie-law signal due to paramagnetic dopant ions can be readily detected at low temperatures, but it is difficult to measure any high-field slope that may be associated with the thin film.

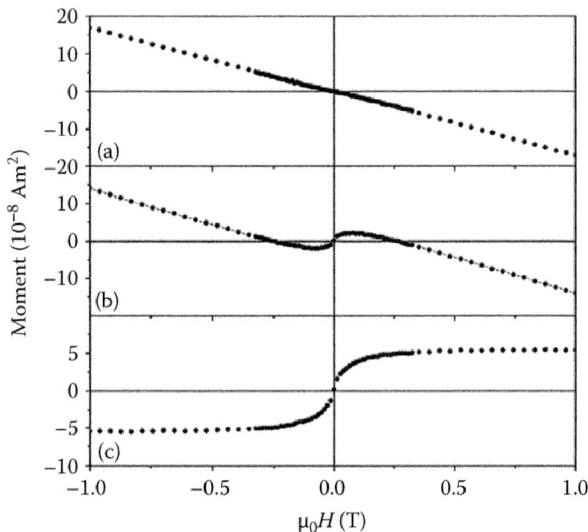

FIGURE 20.6 Data reduction for magnetization measurements on a thin film of a typical dilute magnetic oxide: (a) diamagnetic substrate, (b) ferromagnetic thin film with diamagnetic substrate, (c) ferromagnetic component isolated from (b).

The measured response for a ferromagnetic thin film is shown in Figure 20.6b. The magnetization appears to saturate in a field of order 1 T, and the high-field slope becomes that of the substrate, within experimental error. The procedure is then to suppose the magnetization is indeed saturated and subtract the high-field diamagnetic slope as the substrate correction, yielding the magnetization curve for the film shown in Figure 20.6c. Typically, these "ferromagnetic" hysteresis loops for dilute magnetic oxides exhibit little coercivity (≈10 mT), and a small remanence ratio ($M_r/M_s \approx 5\%-10\%$). Trapped flux in a SQUID, or the remanence of an iron-cored electromagnet create fields of this magnitude, so it is important to measure the field actually acting on the sample.

At first, it seemed reasonable to quote the magnetization of the samples in units of Bohr magnetons per 3*d* cation, normalizing the measured magnetic moment by the number of transition-metal cations thought to be in the sample. This unit is useful when comparing different doping concentrations and also when comparing different host materials, provided the transition-metal concentration is accurately known. However, the practice is misleading when the oxides turn out not to be dilute magnetic semiconductors (DMSs), and the dopant cations are not the primary source of the magnetic order.

We now present experimental results on the dilute oxides, which can exhibit ferromagnetic behavior when they are in thin film or nanocrystalline form. The crystal structures of host materials are presented in Figure 20.7. Some structural details are included in Table 20.3, including the percolation threshold x_p, the cation site coordination, and bandgap E_g. The main systems are now discussed in turn.

20.4 Results on Specific Oxides

20.4.1 Dioxides

20.4.1.1 TiO_2

Anatase with 7% Co-doping was the first example of high-temperature ferromagnetism in a dilute oxide film [25]. The same group reported ferromagnetism in rutile films [26]. The solubility of Co in these oxides is low, and there is the possibility of cobalt-metal clustering [27,28]. But films with 1%–2% of Co can be prepared, which are apparently free of clusters, and exhibit ferromagnetism with a moment of $\approx 1\ \mu_B$/Co up to 700 K [29]. Other evidence that Co-doped TiO_2 can be intrinsically ferromagnetic, and carrier-mediated include electric-field modulation of the magnetization [30], observation of band-edge optical dichroism [31], anomalous Hall effect [32] as well as tunnel magnetoresistance [33].

Other cation dopants reported to make TiO_2 ferromagnetic include V–Ni [34], Cr [16], Mn [35], Fe [36], and Cu [37,38]. In some highly perfect, Fe-doped films, an Fe_3O_4 secondary phase forms at the surface [39]. It has been shown that highly perfect Cr-doped TiO_2 films are not ferromagnetic, whereas films with structural defects can be [40]. The Cr is trivalent in TiO_2 films, and the magnetization of (001) anatase films grown on $LaAlO_3$ can be highly anisotropic [41], but the magnetism again

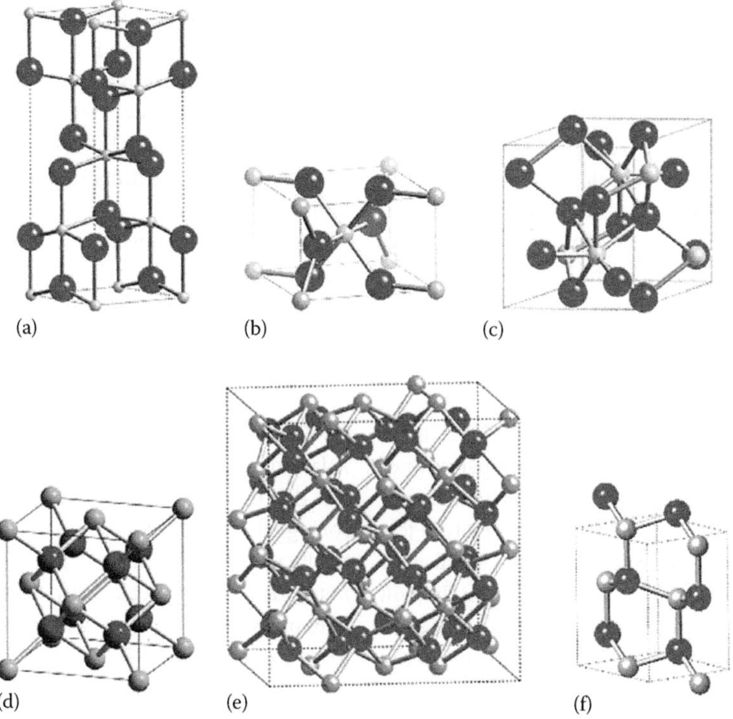

FIGURE 20.7 Crystal structures of host semiconductors: (a) TiO_2 (anatase), (b) SnO_2, TiO_2 (rutile), (c) HfO_2, (d) CeO_2 (fluorite), (e) In_2O_3 (bixbyite), and (f) ZnO (wurtzite), oxygen are the large dark ions.

TABLE 20.3 Parameters for Some Oxides

Material	Structure	ε	m^*/m	E_g (eV)	Coordination	x_p
TiO_2	Anatase	25	0.4	5.7	Octahedral	0.25
SnO_2	Rutile	26	0.4	3.2	Octahedral	0.25
HfO_2	Monoclinic				Sevenfold	
CeO_2	Fluorite				Cubic	
In_2O_3	Bixbyite				Octahedral	
ZnO	Wurtzite	9	0.3	2.9	Tetrahedral	0.18

n_c, cation density; E_g, optical bandgap; x_p, cation percolation threshold.

disappears as the structural quality improves [42]. The order of magnitude of the magnetization of the doped TiO_2 films is 10–20 kA m^{-1}.

TiO_2, unlike ZnO or SnO_2, is usually quite insulating in thin film form. Griffin et al. [43–45] have established that free carriers are not necessary for the high-temperature ferromagnetic order in Co-TiO_2, and that it does not depend critically on the uniformity of the cobalt distribution. The materials may be *dilute magnetic dielectrics*, rather than DMSs. Nevertheless, Fukumura et al. [46] concluded from a detailed study of magnetization, anomalous Hall effect and magnetic circular dichroism of epitaxial films with different cobalt contents and carrier densities that the ferromagnetism *is* related to the electron density. Many studies indicate that a critical requirement for ferromagnetism in TiO_2 films is an abundance of oxygen vacancies, O_v. These form easily in titanium oxide, as is evident from the existence of a sequence of Magneli phases Ti_nO_{2n-1}.

Giant moments of 4.2 μ_B/V were reported for 5% V-doped TiO_2 [34], and moments as large as 22.9 μ_B/Co have been reported for 0.15% Co-doping [47]. These numbers imply that the transition-metal cation cannot be the carrier of the moment. Ferromagnetism with a magnetization of ≈10–20 kA m^{-1} at temperatures up to 880 K has also been reported in 200–300 nm thick films of undoped anatase-TiO_2 deposited on $LaAlO_3$ [48,49]. Recently it has been shown by Zhou et al. [50] that a TiO_2 crystal irradiated with a dose of 5×10^{15} cm^{-2} of 2 MeV oxygen ions develops a tiny magnetization of 2 A m^{-1}. The induced defects are Ti^{3+}–O_v complexes, which are thought to be responsible for the magnetism of TiO_2.

20.4.1.2 SnO_2

First reports of high-temperature ferromagnetism in SnO_2 films came from Ogale et al. [51], who found a Curie temperature of 650 K and an extraordinary moment of 7.5 Bohr magnetons per cobalt atom for films doped with 5% Co. High-temperature ferromagnetism was subsequently observed for films doped with V [52], Cr [53], Fe [54], Mn [55], and elements from Cr–Ni. [56]; in some of these films, the moments again exceed the cation spin-only values [18]. In the case of V-doping, for example, the results depend rather critically on the choice of substrate [57,58]. Co-doped films exhibit Faraday rotation of 570° cm^{-1}, and show some sign of an anomalous Hall effect [59]. The magnetization of the 3*d* doped SnO_2 films is typically 20 kA m^{-1}.

There is a remarkable anisotropy of the magnetization when the films grow with a (101) texture, which is found in films with V, Mn, Fe, and Co-doping [56,58], and even in undoped films [60]. The effect cannot be explained by magnetocrystalline anisotropy, because the ferromagnetic moments measured in different directions do not converge in high fields. The anisotropy becomes very pronounced at low temperature, and it has been modeled in terms of orbital current loops [61].

There is much evidence that the magnetism is somehow associated with defects related to the oxygen stoichiometry. It was originally suggested that these could be F-centers—an oxygen vacancy that traps an electron [54]. The defects are not necessarily adjacent to the cation dopant, which are usually octahedrally coordinated in substitutional sites. SnO_2 forms dilute magnetic oxides not only in thin film form, but also in nanoparticles [62–64] and nanocrystalline ceramics [65,66]. A significant experiment by Archer and Gamelin [67] demonstrated that the magnetism of colloidal, 2 nm Ni-doped SnO_2 nanoparticles could be switched on and off as they aggregate under gentle thermal annealing at 100°C. The ferromagnetism is controlled by grain-boundary oxygen defects, as was shown in similar experiments on colloidal particles of Ni-doped ZnO [68] and Co-doped TiO_2 [69], which become strongly ferromagnetic when spin coated into thin films. Magnetic force microscopy of Fe-doped SnO_2 thin films shows magnetic contrast at the grain boundaries [56].

There are also reports of magnetic moments in undoped SnO_2 films [60] and nanoparticles [70]. Magnetization is of order 10 kA m^{-1} and 10 A m^{-1}, respectively. The latter authors reported very small ferromagnetic signals in a range of undoped, nominally pure oxides with a crystallite size ≤20 nm.

20.4.1.3 HfO_2

A landmark in the study of ferromagnetism in oxide thin films was the report by Venkatesan et al. [71] of high-temperature ferromagnetism in undoped HfO_2 films. These results helped shift the focus from dopants to defects, and led to coining the term *d^0 magnetism* for the weak, anhysteretic high-temperature magnetism of a wide range of materials that are not normally expected to be magnetic because they contain no 3*d* ions [18]. A subsequent investigation [72] of more than 40 HfO_2 films of different thickness, deposited on different substrates by pulsed-laser deposition, with and without doping established that moments were largest on c-cut sapphire and smallest on Si; they showed no systematic variation with film thickness (except that the thinnest films were not magnetic). Dopants, whether magnetic or nonmagnetic, tended to reduce the moment. The moments were widely scattered around an average of ~200 μ_B nm^{-2}, but they were unstable, decaying over times of order 6 months. Magnetization of the thinnest films was comparable to that of Ni (400 kA m^{-1}), and in some samples it was very anisotropic, being larger when the field was applied perpendicular to the film, than when it was applied parallel. The results were substantially confirmed in a series of papers by Hong and coworkers [49,73–75] who looked at HfO_2 undoped, or with Fe or Ni dopants. Oxygen annealing reduced the moment, which was associated with defects near the film/substrate interface. However, other studies could find no intrinsic ferromagnetic signal in similarly prepared HfO_2 films, irrespective of oxygen pressure during deposition [76,77] and it was shown by Abraham et al. [19] that similar signals could be observed in samples manipulated with stainless steel tweezers.

A study of colloidal HfO_2 nanorods 2.6 nm in diameter [78] found small magnetic moments ~100 A m^{-1}. Furthermore, a magnetization of comparable magnitude can be reversibly induced by heating pure HfO_2 powder in vacuum or air at 750°C [72].

20.4.1.4 CeO_2

The first report of ferromagnetism in CeO_2 was by Tiwari et al. [79] who found a moment of 6.1 μ_B/Co in 3% Co-doped films that increased with temperature up to 725 K; it corresponds to a magnetization of 40 kA m^{-1} and is too big to be explained by the presence of Co metal in the films. The magnetization depends on the substrate and oxygen pressure during growth [80,81]. Very much smaller magnetization has been found in CeO_2 nanoparticles that are undoped [70,82–84] or doped with Ni [85] or Ca [86]. Values range from 1 to 5000 A m^{-1}, and magnetization is greatest in undoped sub-20 nm particles [82]. The high-temperature ferromagnetism in the nanoparticles is associated with Ce^{3+}–Ce^{4+} surface pairs [84], rather than surface oxygen vacancies [83].

20.4.2 Sesquioxides

20.4.2.1 In_2O_3

Indium oxide, In_2O_3, has the cubic bixbyite structure, with indium in undistorted 8*b* and highly distorted 24*d* octahedral sites (Figure 20.7e). The structure is related to fluorite (Figure 20.7d) with a quarter of the oxygen atoms removed. The bandgap E_0 is 2.93 eV [87]. The oxide is usually oxygen-deficient, creating donor levels near the bottom of the conduction band. Electron concentrations of order 10^{24}–10^{26} m^{-3} lead to resistivities of about 10 µΩ m, a high mobility and an effective mass of $0.3m_e$. The electrons tend to accumulate at the oxide surface. Substituting some of the In by Sn is a convenient way of controlling the carrier concentration in In_2O_3. Indium tin oxide $(In_{1-x}Sn_x)_2O_3$ (ITO) with $x \approx 0.5$ is a widely used transparent *n*-type conducting oxide.

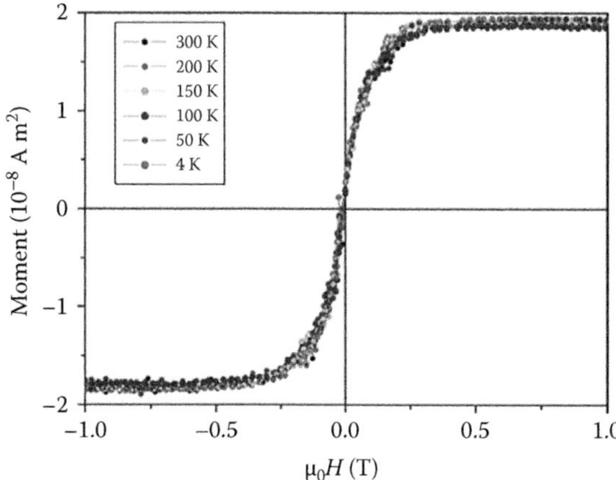

FIGURE 20.8 (See color insert.) Ferromagnetic magnetization curves of a thin film of ITO doped with 5% Mn, at different temperatures. The manganese in these films is paramagnetic. (From Venkatesan, M. et al., *Nature*, 430, 630, 2004. With permission.)

The first report of ferromagnetism in Mn-doped ITO by Philip et al. [88] showed a moment of about 1 μ_B per manganese and evidence of an anomalous Hall signal. It was followed by indications of ferromagnetism in In_2O_3, for bulk material doped with Fe and Cu [89], and for thin films doped with Fe [90,92] and other dopants [91,93,94]. Subsequent investigations of ferromagnetic Mn- and Fe-doped ITO films [95] found that the manganese behaved paramagnetically at low temperature, and was not the primary source of the magnetism. The anhysteretic magnetization curve, shown in Figure 20.8, may be regarded as typical for any ferromagnetic dilute magnetic oxide thin film. Hysteretic ferromagnetism of the iron-doped samples was due to an Fe_3O_4 secondary phase. Anhysteretic ferromagnetism was observed in Fe-doped films thought to be intrinsically ferromagnetic [92].

A study of Cr-doped In_3O_3 by Philip et al. [96] demonstrated that the ferromagnetism was related to carrier concentration, appearing at a concentration $n_e = 2 \times 10^{25}$ m^{-3}, which would correspond to the formation of an impurity band [154]. Moments of up to 1.5 μ_B/Cr were found for 2% doping. Chromium is a good dopant, because there is little prospect of ferromagnetic secondary phases; CrO_2 is ferromagnetic, with $T_C = 390$ K, but T_C was measured in this study to be ≈900 K. Domains were observed by magnetic force microscopy (MFM). Further investigations by Panguluri et al. [97] on both undoped and Cr-doped films found paramagnetic Cr coexisting with a film moment of about 500 A m^{-1}. Both materials exhibited a spin polarization of 50% in measurements of point-contact Andreev reflection. The implication is that the 3*d* dopant is inessential for the ferromagnetism, which is associated with a narrow defect-related band. This picture is consistent with the dependence of the anomalous Hall effect on carrier concentration [94].

There are also reports of ferromagnetism of similar magnitude in undoped thin films [93]. Sundaresan et al. [70] found very small moments in 12 nm nanoparticles of In_2O_3 (<5 A m^{-1}), but in another study by Bérardan et al. [98], In_2O_3 nanoparticles prepared in a variety of different conditions were found to be completely diamagnetic, and transition-metal doped samples were paramagnetic.

20.4.3 Monoxides

20.4.3.1 ZnO

Although the first report of ferromagnetism in thin films of a dilute magnetic oxide was for TiO_2, it is on ZnO that the majority of the investigations have been carried out. Zinc oxide is a promising semiconductor material with a bandgap of 3.37 eV. It crystallizes in the wurtzite structure (Figure 20.7f), where Zn occupies tetrahedral sites. The oxide has a natural tendency to be *n*-type on account of oxygen vacancies or interstitial zinc atoms. Recently, it has been possible to make nitrogen-doped *p*-type material, which opens the way to producing light-emitting diodes and laser diodes. An extensive review of the semiconducting properties of ZnO is available [99].

Various cations can replace zinc in the structure, and small dopings of all the 3*d* elements have somewhere been reported to make ZnO ferromagnetic. There is an admirable review of the voluminous literature on doped ZnO by Pan et al. [100]. The largest moments are found for Co^{2+}. The Co^{2+} ion gives a characteristic pattern of optical absorption in the bandgap, due to crystal field transitions of the ^4F ion in tetrahedral coordination. Studies of the magnetic properties of bulk material doped with Co [15,101] or Mn [102] show only the paramagnetism expected for isolated ions and small antiferromagnetically coupled nearest-neighbor clusters that arise statistically at a low doping level, as discussed above (Figure 20.5a).

The range of solid solubility of cations of the 3*d* series in ZnO films prepared by pulsed-laser deposition was established by Jin et al. [103]; solubility limits as high as 15% were found for Co, but most other cations could be introduced at the 5% level. None of these films turned out to be to be magnetic, but following Ueda et al. [104], who were the first to report high-temperature ferromagnetism in their films containing cobalt at the 10% level, there have been numerous reports of ferromagnetic moments in films doped with Co (see among others Janisch et al. [105], Prellier et al. [106], Pearton et al. [107], and references therein), as well as every other 3*d* dopant from Sc to Cu [68,108–110]. The variation of magnetic moment for a series of films nominally doped with 5% of various transition-metal ions prepared in the same conditions is shown in Figure 20.9. There are also numerous counterexamples where no room-temperature moment has been found in doped films, or an extrinsic origin is found. Negative results tend to be underreported in the literature, so the observation of ferromagnetism in ZnO and other oxide films is probably less prevalent than one might imagine. It is very sensitive to deposition conditions such as substrate temperature and oxygen pressure [99,100]. However, nanoparticles and nanorods of ZnO with cobalt and other dopants have also been found to be magnetic under certain process conditions [111], although the moments per cobalt atom are an order of magnitude less than those found for the thin films, which are typically 0.1–1 μ_B/Co.

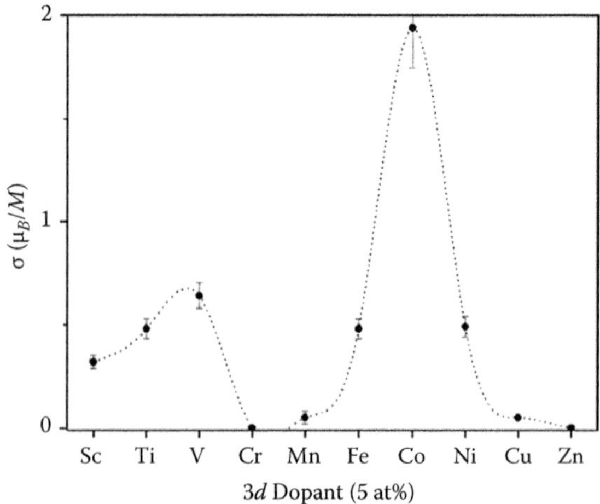

FIGURE 20.9 Magnetic moment measured at room temperature for ZnO thin films with various 3*d* dopants. (From Venkatesan, M. et al., *Appl. Phys. Lett.*, 90, 242508, 2006. With permission.)

Some progress has been made toward providing a systematic experimental account of the phenomenon. Narrow process windows have been delimited where ferromagnetism can be observed, which differ depending on the deposition method—pulsed-laser deposition, sputtering, evaporation, organometallic chemical vapor deposition, and others. With Cr and Mn, for example, it is difficult to produce ferromagnetic moments in *n*-type material, whereas *p*-type samples can exhibit the symptoms [80] (Figure 20.10). The substrate temperature required for ferromagnetism is often around 400°C, where the films are not of the best crystalline quality. This inference points to a defect-related origin of the magnetism [112]. Other evidence in this sense comes from the appearance of magnetism in Zn-doped and oxygen-deficient films [113,114], and the creation of magnetization of ~200 kA m^{-1} in the volume implanted by 100 keV Ar ions [115].

Evidence that Co-doped ZnO is an intrinsically ferromagnetic semiconductor is relatively sparse. No magnetoresistance has been observed at room temperature, despite the high T_C, nor are there clear signs of an anomalous Hall effect. Semiconductor-like magnetoresistance can, however, be observed below about 30 K [116]; there is a positive component, as expected in the classical two-band model, and a negative component due to ionized impurity scattering. There is also quite a large anisotropic magnetoresistance below 4 K (≈10%), which is a band effect in the semiconductor, not ferromagnetic AMR. Furthermore, there is band-edge magnetic dichroism with a paramagnetic dichroic response from the dilute Co^{2+} ions in tetrahedral sites in the wurtzite lattice, as well as a component with a ferromagnetic response [117,118]. The ferromagnetic dichroism has been observed for Ti, V, Mn, and Co-doping [118]. Magnetic circular dichroism experiments on single nanoparticles in the electron microscope by Zhang et al. [119] indicate that their particles are intrinsically ferromagnetic at room temperature when doped with Co but not with Fe.

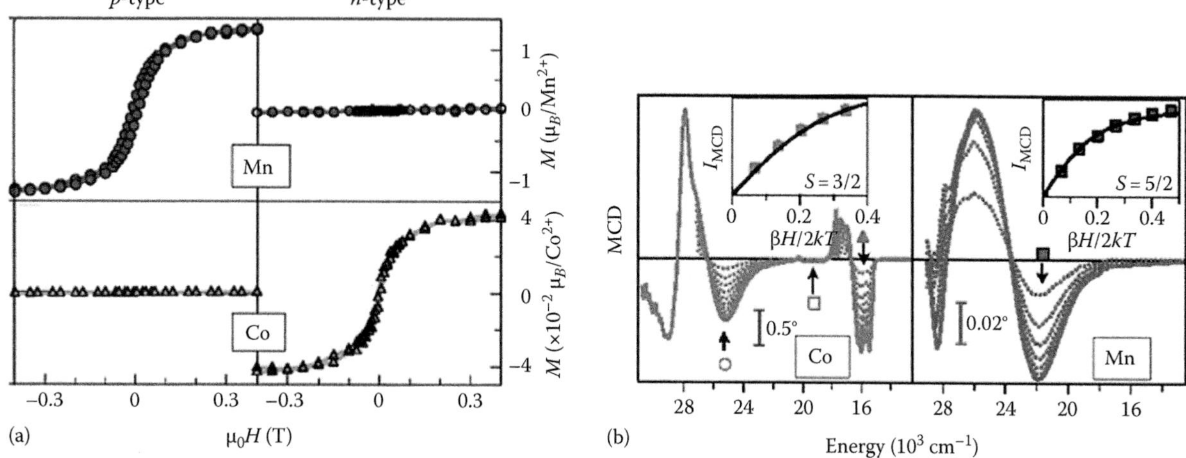

FIGURE 20.10 (a) Development of magnetism in *n*-type ZnO with Co or *p*-type ZnO with Mn. (From Kittilstved, K.R. et al., *Nat. Mater.*, 5, 291, 2006. With permission.) (b) MCD spectra and the magnetic field dependence of the intensity of the MCD signal (insets) recorded at different energies in ZnO doped with Co (left) and Mn (right).

Some reports [120,121] have related the carrier concentration to the existence of ferromagnetism, as would be expected if ZnO were a DMS. However, the interaction between the conduction electrons and the Co spins seems to be rather weak, <20 K, and in other studies the magnetic properties are little influenced by changing the carrier concentration by Al doping, for example [122]. It has been shown that the sign of the weak Mn–Mn exchange in colloidal Mn–ZnO nanocrystals can be changed from antiferromagnetic to ferromagnetic, by exciting carriers optically [123], although the strength of the interaction is only a few Kelvin. It seems that there may be a correlation between carrier concentration and ferromagnetism, but this is not a cause and effect relation.

A problem in interpreting the magnetic data for some dopants is the tendency to form ferromagnetic impurity phases, which may escape detection by x-ray diffraction. For cobalt in particular, absence of evidence cannot be taken as evidence of absence. There is a well-documented tendency for nanometric coherent cobalt precipitates to appear in the ZnO films [124–127] and single crystals [128], which exhibit ferromagnetic properties, especially at doping levels above 5%. An impurity-phase-based explanation ($Mn_{2-x}Zn_xO_3$) has been advanced also for Mn-doped material [129]. For other dopants, such as vanadium or copper, contamination by high-temperature ferromagnetic impurity phases seems improbable.

In some cases, there are features of the data, which make an impurity-phase explanation untenable. These are (a) observation of a moment per transition-metal ion that is greater than that of any possible ferromagnetic impurity phase based on the dopant ion, and which may exceed the spin-only moment of the cation, and (b) observation of an anisotropy of the magnetization, measured in different directions relative to the crystal axes of the film, which is not a feature of any known ferromagnetic phase at room temperature [71,109,130]. An example is shown in Figure 20.11. Both the features have occasionally been reported in the other thin film oxide systems we have been considering.

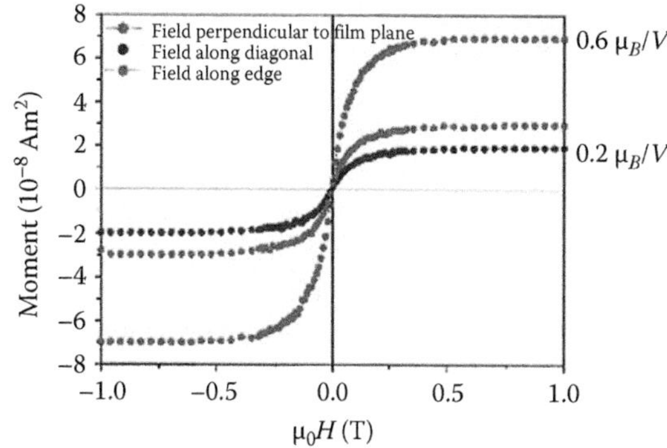

FIGURE 20.11 (See color insert.) Anisotropy of the magnetization of a thin film of $(Zn_{0.95}V_{0.05})O$.

A common feature of the ferromagnetism in ZnO and the other oxides is its tendency to decay over times of order of a few months [21,130]. This has led to the name *phantom ferromagnetism*, but *fickle ferromagnetism* may be a better term for the syndrome presented by dilute magnetic oxides.

The tunnel spin polarization of transition-metal-doped ZnO has been measured [131]. Although poorly reproducible, these results are in line with the observation of MCD at the band-edge as discussed above. The possibility to obtain and manipulate a spin current in ZnO could be of great interest, as the spin lifetime in ZnO is rather long [132].

As in the other oxides, there are reports of magnetism in undoped films [133–135], where the magnetization is greatest in the thinnest ZnO films. Undoped nanocrystals are also reported to be ferromagnetic [136], but with a very small magnetization ~1 A m^{-1}. A compilation of data from several groups by Straumal et al. [137] indicates that weak ferromagnetism is only

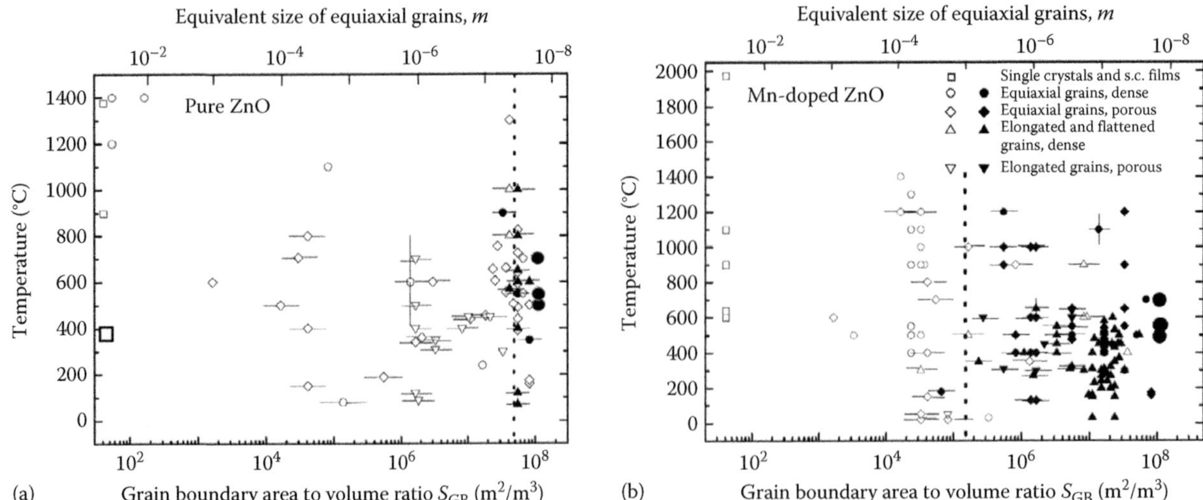

FIGURE 20.12 Plot of magnetic moment versus grain-boundary area for undoped (a) and Mn-doped (b) ZnO ceramics. (From Straumal, B.B. et al., *Phys. Rev. B*, 79, 205206, 2009. With permission.)

found in undoped material with a grain size below 60 nm and in Mn-doped ceramic samples with a grain size of less than about 1.5 μm (Figure 20.12).

There is much evidence to suggest that ferromagnetism in transition-metal-doped ZnO is unambiguously correlated with structural defects, which depend on the substrate temperature and oxygen pressure used during film deposition [23,100]; the carriers are by-products of these defects. Ney et al. [17] have demonstrated that highly perfect Co-doped epitaxial films, for example, are purely paramagnetic, with cobalt occupying zinc sites and no sign of anything other than weak antiferromagnetic Co–O–Co nearest-neighbor superexchange. Droubay et al. [138] show that Mn-doped epitaxial films are paramagnetic, with Mn substituting for Zn, and no moments on either oxygen or zinc. Barla et al. [139] showed that the cobalt in ferromagnetic Co-doped ZnO films was paramagnetic, and there was no moment on the zinc. A study by Tietze et al. [140] who investigated 5% Co-doped films produced by pulsed laser deposition (PLD), which exhibited a moment of ≈1 μ_B/Co, corresponding to a magnetization of ≈20 kA m^{-1} came to a remarkable conclusion; extensive x-ray magnetic circular dichroism (XMCD) measurements on the Co and O edges with a very good signal/noise ratio showed only the signatures of paramagnetic cobalt and oxygen. There is no sign of element-specific ferromagnetism. By a process of elimination, only oxygen vacancies remained as the possible intrinsic source of the ferromagnetism observed in the ZnO films.

These results contradict the view of Behan et al. [121], based on a systematic investigation of samples with 5% Co and carrier densities ranging from 10^{16} to 10^{21} cm^{-3} that the material is a dilute magnetic insulator at low carrier concentrations and a DMS at high concentrations, with an intermediate concentration range around 10^{19} cm^{-3} where the ferromagnetism is quenched. Thin films and tunnel junctions made of Cr-doped material, which is a magnetic insulator, show that defects near the surface rather than throughout the bulk are responsible for the magnetism [141]. The studies of many systems as a function of thickness, which indicate magnetic moments of 100–200 μ_B nm^{-2} argue in the same sense [21,130]. The analysis of the nanocrystalline ceramic samples of undoped and Co- or Mn-doped ZnO by Straumal et al. [137] suggests that the magnetism is associated with the grain boundaries, and they envisage the ferromagnetism as a grain-boundary foam, which only occupies a few percent of the volume of a the samples.

Although there is yet no consensus, the current view of transition-metal-doped ZnO is moving far away from the initial expectations that the materials are DMSs.

20.5 Discussion

The experimental picture is confused by the difficulty in reproducing many of the results, the often ephemeral nature of the magnetism, the problem of characterizing defects and interface states in thin films and nanoparticles, as well as the difficulty in detecting small amounts (≈1 wt%) of secondary phases in the films [23]. Despite the tendency for experiments where nothing unusual is found to pass unreported in the literature, there does seem to be a *prima facie* case that there is something to explain in the numerous reports of fickle ferromagnetism in dilute magnetic oxides, which cannot be attributed to sloppy experimentation. If so, it seems better to seek a general explanation of the phenomenon, rather than to rely on a different *ad hoc* explanation for each material. For this reason, we will not discuss the numerous electronic structure calculations for specific defects in the six oxides, which we have been considering, but focus instead on generic physical models.

It is now quite clear that the anomalous ferromagnetism is not found in highly perfect films or well-crystallized bulk material. Those dilute magnetic oxides behave paramagnetically down to liquid helium temperatures, in the way that was described in Section 20.1, albeit with a nonstatistical distribution of dopants

[138]. There is accumulating evidence, reviewed in Section 20.4, that the phenomenon is closely associated with defects of some sort, which are present in thin films and nanocrystals, but are absent in perfectly crystalline material [142]. The poor reproducibility of the experimental data may be due to the difficulty in precisely recapturing the process conditions that lead to a specific defect distribution and density. The samples as prepared are not in equilibrium, and they evolve with time and temperature.

At first, it was taken for granted that the magnetism was *spatially homogeneous*, and due to the *spin moments of the 3d ions*. Both these assumptions look increasingly questionable. The difficulty with any spatially homogeneous explanation based on localized $3d$ moments is that the magnetic interactions have to be ferromagnetic, long-range, and extraordinarily strong—at least one or two orders of magnitude greater than anything previously encountered in oxides (Figure 20.3). An influential view [143] was that dilute magnetic oxides were DMSs, similar to the canonical example of $(Ga_{1-x}Mn_x)As$, discussed in Chapter 19. There the manganese introduces occupied $3d$ states deep below the top of the As $4p$ valence band, and the ferromagnetism is mediated by $4p$ holes that are polarized by exchange with the $3d^5$ Mn^{2+} ion cores. Unfortunately, the model does not work for most $3d$ dopants in oxides, because the $3d$ level does not lie below the top of the oxygen $2p$ valence band. Exchange via the $4s$ electrons in the conduction band introduced by donor doping, as in Gd-doped EuO, for example [144], provides weaker coupling than exchange via the $2p$ holes.

The third possibility illustrated in Figure 20.13 is to couple the $3d$ ion cores via electrons in a spin-polarized impurity band. The impurity band in a semiconductor is made up of large Bohr-like orbitals associated with the dopant electrons. In dilute magnetic oxides, it could be derived from defect states such as F-centers: the idea is that the impurity band becomes spin-split on account of s–d interaction with the magnetic dopant, and is formed of overlapping magnetic polarons (Figure 20.5b). Calculations of T_C in the molecular field approximation [145,146] lead to the expression

$$T_C = \left[\frac{S(S+1)s^2x\delta}{3}\right]^{1/2}\frac{J_{sd}\omega_c}{k_B} \tag{20.6}$$

where

S and s are the cation core spin and the donor spin, respectively
δ is the donor or acceptor defect concentration
$\omega_c \approx 8\%$ is the cation volume fraction in the oxide

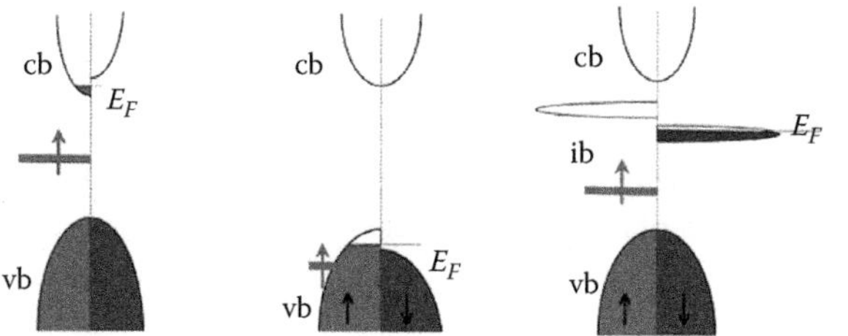

FIGURE 20.13 Schematic models for DMSs left; spin-split conduction band center; spin-split valence band and right; spin-split impurity band.

The difficulty is that, knowing the values of the parameters in the model, especially $J_{sd} \approx 1$ eV, the predicted Curie temperatures are of order 10 K, which is one or two orders of magnitude too low. The model can be modified by introducing hybridization and charge transfer from the donor orbitals to those of the dopants, but at some point in the dilute limit it has to fail, and a different approach is needed. Small concentrations of conduction electrons do provide ferromagnetic coupling via the RKKY interaction, but an estimate of the magnitude of the magnetic-ordering temperature due to this interaction is

$$T_C \approx \frac{Z_0 n^{5/3} m^* x^{1/3} \delta J_{sf}^2 S(S+1)}{(96\pi\hbar^2 n_c^{2/3} k_B)} \tag{20.7}$$

The Curie temperature at high carrier concentrations $n_c \sim 10^{21}$ cm^{-3} does not exceed a few tens of Kelvin.

One way of increasing the exchange energy density is to look for an inhomogeneous distribution of magnetic ions. At one extreme, they may form a segregated secondary phase, which should have a high T_C if it is to explain the magnetism. Phase segregation is expected when the solubility limit is exceeded, but the tolerance for dissolved ionic species may be greater in thin films or nanoparticles, which are not in thermodynamic equilibrium, than it is in the bulk. Magnetically ordered transition-metal oxides such as CoO or Fe_3O_4 have predominantly antiferromagnetic superexchange coupling, and are usually antiferromagnetic or ferrimagnetic. However, it is possible to find the reduced metal as a secondary phase, especially after deposition or thermal treatment in vacuum at high temperature. Fe, Co, and Ni are ferromagnetic metals, and the presence of segregated ferromagnetic metal or ferromagnetic oxide inclusions is undoubtedly the explanation of the ferromagnetism (with telltale hysteresis) of *some* of the dilute magnetic oxide samples. It is a pity that so much effort has been invested in studying these dopants, and especially cobalt, which is the ferromagnet with the highest Curie temperature of all. It is easier to make a case for an intrinsically new magnetic phenomenon when ferromagnetism is observed in oxides that are undoped, or doped with elements such as Sc, Ti, V, Cr, or Cu, which form no phases with high-temperature magnetic order.

The near-anhysteretic magnetization curves of dilute ferromagnetic oxides are practically temperature-independent, and emphatically do not evoke superparamagnetism, where anhysteretic magnetization curves taken at different temperatures must superpose when plotted as a function of H/T. In the dilute ferromagnetic oxides, there is almost no difference in the curves measured at 4 and 300 K (Figure 20.8). It may be that in some cases the macrospin moments of cobalt or iron nanoinclusions are coupled by electrons in the host oxide. Such high-temperature-conducting ferromagnetic nanocomposites could find possible applications in electronic devices.

Even when substituting for cations in the host oxide, the magnetic dopant ions may experience a tendency to cluster. Spontaneous chemical segregation in a solid solution is known as *spinodal decomposition*, and it often takes the form of a composition density wave with alternating regions rich and poor in one or other element. Spinodal decomposition has been suggested as a mechanism for high-temperature ferromagnetism in dilute oxides [147]. It has the virtue of segregating regions of the sample, which still looks like a solid solution, with a higher magnetic energy density than the average. Since these regions percolate as sheets throughout the material, there is no expectation of independent superparamagnetic clusters. The problem is to explain why the regions rich in transition-metal ions should be ferromagnetic, granted the propensity for 3*d* ions in oxides to couple antiferromagnetically by superexchange. There may be an unbalanced moment in a lamella related to certain antiferromagnetic arrangements of the spins, but it seems unlikely that this could be a general explanation, especially for the 3*d* ions from the beginning of the series, which have a small spin moment, and do not form any oxide with a high magnetic-ordering temperature.

The accumulating evidence that defects play a central role in the magnetism of dilute magnetic oxides does not mean that the 3*d* cations are not involved, but it suggests that they do not contribute in the way at first assumed. Beyond a threshold thickness, the moments do not scale with film thickness, but seem to depend on surface area, with values of $100–400\,\mu_B\,nm^{-2}$. Unphysically large moments have been inferred "per cation" at low doping levels in all the oxide systems, and there are the reports of d^0 magnetism in some samples of undoped films and nanoparticles. Both these are rather uncommon. It is much more usual to find that the $x=0$ end-member is nonmagnetic, and this then serves as a credibility check of the experimental protocol, establishing the 3*d* dopant's role in initiating the magnetism, and incidentally confirming that the laboratory is not irredeemably contaminated with airborne magnetic dust.

20.5.1 Models

Having discussed the messages that the experimental data seem to convey, we now summarize the generic models that have been proposed for ferromagnetism in dilute magnetic oxides. There are five of them, depending on whether the magnetic moments are localized or delocalized, and the nature of the electrons involved in the exchange.

1. The DMS model [143]. There are well-defined local moments on the dopant cations, which are coupled ferromagnetically via 2*p* holes, or by RKKY interactions with 4*s* electrons.
2. The bound magnetic polaron (BMP) model [145]. Here again, there are well-defined local moments on the dopants, but they interact with electrons associated with defects, which form an impurity band. Each defect electron occupies a large orbit and interacts with several dopant cations to form a magnetic polaron and ferromagnetism sets in at the percolation threshold for these polarons.
3. A variant of the model (BMP), [97] dispenses with the magnetic dopants, but retains the idea of localized moments, which are now associated with electrons trapped at defects. Triplet pairs could form to give $S=1$ moments, which are then coupled by electrons in an impurity or conduction band.
4. The spin-split impurity band (SIB) model [148], which is based on a local density of states associated with defects where the density of states at the Fermi level is sufficient to satisfy the Stoner criterion.
5. The charge-transfer ferromagnetism (CTF) model [149,152] is also a Stoner model with a defect-based impurity band, but there is another charge reservoir in the system that allows for the facile transfer of electrons to or from the impurity band to create a filling that leads to spontaneous spin splitting. In dilute magnetic oxides, this reservoir is associated with the dopant ions.

The first three of the models are Heisenberg models, with localized spin moments that fit the *m*-J paradigm. The first two of them require roughly uniform ferromagnetism of the sample, since the moments are borne by the 3*d* dopants that are distributed more or less uniformly throughout the material. We have seen little proof that the dopants in the dilute magnetic oxides are magnetically ordered, and growing evidence that they are actually paramagnetic. The models also require long-range ferromagnetic exchange interactions of a strength not found in any concentrated magnetically ordered material. Curie temperatures of order 10–50 K might be possible for materials with a few percent doping, but not values of order 800 K. Neither model accounts for d^0 ferromagnetism in undoped, defective oxides. The third model associates the local moments with defects. They can occur in oxides with no 3*d* ions [150]. The defects could be distributed near interfaces or grain boundaries, and so the model can account for both inhomogeneous and d^0 ferromagnetism. The drawback is that, as with any Heisenberg model, spin-wave excitations are inevitable, and there is no reason to expect high Curie temperatures.

The last two models are Stoner models. Localized magnetic moments are not involved; all that is needed is a defect-related density of states with the right structure and occupancy. The magnetism is not homogeneous, but is confined to percolating regions such as surfaces, interfaces, or grain boundaries. Furthermore, Edwards and Katsnelson [148], who developed the SIB model for the d^0 ferromagnet CaB_6, show that spin waves are suppressed for the defect-related impurity band, so Curie temperatures may be very high. The main problem with the model is its lack of flexibility. There is no specific role for the 3*d* dopants. The number of electrons in the impurity band is determined by the defect structure of the sample, and it would be something of a fluke for the electron occupancy to coincide with a peak in the density of states where the Stoner criterion is satisfied:

$$IN(E_F) > 1 \tag{20.8}$$

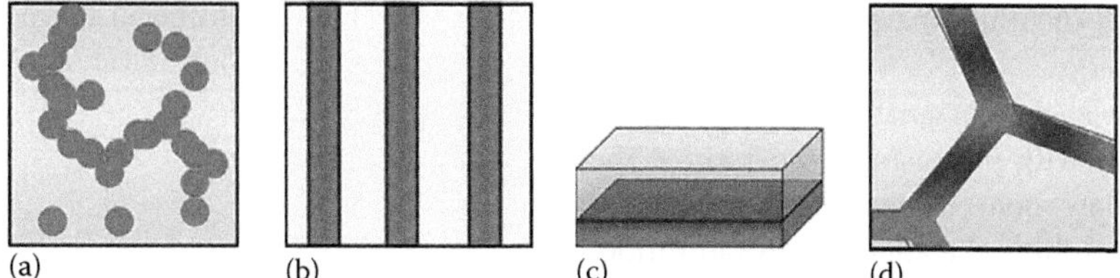

FIGURE 20.14 Inhomogeneous ferromagnetism in a dilute magnetic oxide. The ferromagnetic regions are distributed (a) at random, (b) in spinodally segregated regions, (c) at the surface/interface of a film, and (d) at grain boundaries.

where I is the Stoner exchange integral, which is of order 1 eV. The model can have ferromagnetic metallic, half-metallic, and insulating ground states.

The difficulty of having a fixed number of electrons in the impurity band is resolved in the CTF model. In this model, there is a proximate electron reservoir, which can feed or drain the impurity band as required to meet the Stoner criterion. This makes it possible for a wider range of materials to become magnetically ordered. In the case of dilute magnetic oxides, the $3d$ dopants can serve as the reservoir provided they are able to coexist in different valence states such as Ti^{3+}/Ti^{4+}, Mn^{3+}/Mn^{4+}, Fe^{2+}/Fe^{3+}, Co^{2+}/Co^{3+}, or Cu^{+}/Cu^{2+}. The point of doping with $3d$ cations is not that they carry a magnetic moment, but that their mixed valence allows them to act as the electron reservoir. The cations can remain paramagnetic. It is possible that they tend to migrate to the surfaces or grain boundaries, where the defects lie. The defect-based models can explain the inhomogeneous nature of magnetism. The magnetic regions are those that are rich in defects, which tend to concentrate at interfaces and grain boundaries (Figure 20.14).

It is a challenge to the models to account for the observations that the dilute magnetic oxides may be insulating, semiconducting, or metallic. The Heisenberg models account for a nonconducting magnetic state by supposing that the carriers that propagate the exchange are immobile; the Fermi energy in the conduction band lies below the mobility edge [121]. However, the Stoner models have the option of a half-filled SIB, which can arise from single-electron defect states. The CTF model has a rich electronic phase diagram, with insulating, half-metallic, and metallic ferromagnetic states, as well as nonmagnetic states (Figure 20.15). The undoped end-member is most often nonmagnetic, which shows that the doping is critically important. Ferromagnetic d^0 end-member oxide films remain very much an exception, rather than the rule.

Finally, there are two features of the ferromagnetism, which have received less attention than they deserve. One is the magnetization process—the fact that the ferromagnetism exhibits very little hysteresis, and the magnetization curve is practically independent of temperature. Magnetization curves like that in Figure 20.8 are found in all the systems we have been discussing. It may be asked whether a material with a magnetization curve showing no hysteresis is really a ferromagnet and where is the broken symmetry? But the magnetization curves of iron, permalloy, and other soft ferromagnets often fail to show hysteresis,

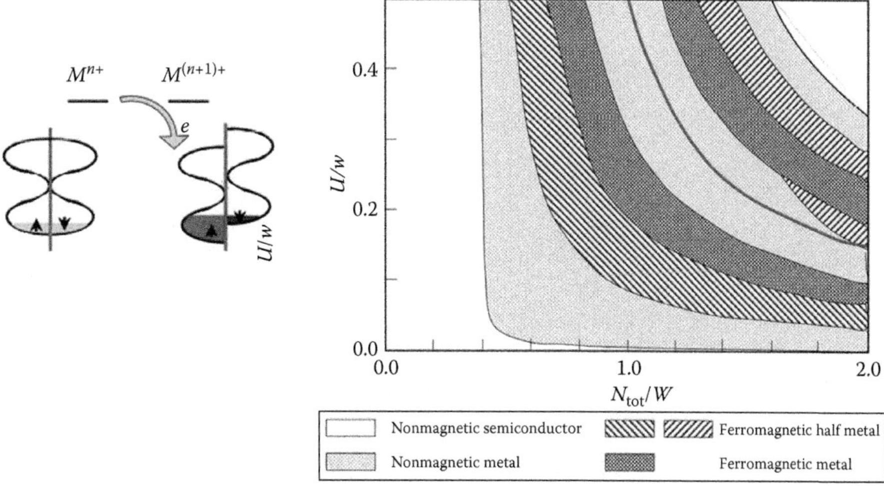

FIGURE 20.15 Phase diagram for the CTF model. Electron transfer from the $3d$ charge reservoir into the defect-based impurity band (a "double-peak" structure is chosen) can lead to spin splitting, as shown on the left. Axes are the total number of electrons in the system N_{tot} and the $3d$ coulomb energy U_d, each normalized by the impurity bandwidth w. The Stoner integral I is taken to be 0.6. In the unshaded nonmagnetic semiconductor regions the impurity band is empty or full. The curved gray line marks half-filling. In the gray regions it is partly filled, but not spin-split. In the blue regions, the Stoner criterion is satisfied, and it is ferromagnetic. (From Coey, J.M.D. et al., *New J. Phys.*, 12, 053025, 2010. With permission.)

because they form domains or other flux-closed structures to reduce their dipolar self-energy.

Hysteresis is normally related to magnetocrystalline anisotropy, which falls off rapidly with increasing temperature. The virtual lack of hysteresis below room temperature indicates that *dipolar interactions* are controlling the approach to saturation. A uniformly magnetized soft ferromagnetic film has a demagnetizing factor of zero when the magnetization is in-plane and it is unity when the magnetization is perpendicular to the plane; the magnetization curve saturates in a small field in the first case and it is linear up to saturation at $H = M_0$ in the second. The measured magnetization curves are usually not like this; they saturate in almost the same way regardless of field direction, with a magnetization curve that can be approximated by a tanh function

$$M = M_0 \tanh\left(\frac{H}{H_d}\right) \tag{20.9}$$

where $H_d \approx 70$–$130\,\text{kA m}^{-1}$. Isolated micron and submicron iron particles ($M_0 = 1720\,\text{kA m}^{-1}$) behave similarly, but with $H_d \approx 300\,\text{kA m}^{-1}$. The iron particles adopt a vortex magnetic configuration to minimize the demagnetizing energy, which progressively unwinds as the magnetization approaches saturation.

The spin-split defect-based impurity band would not be expected to exhibit magnetocrystalline anisotropy. Assuming the ferromagnetic regions are thin layers following the grain boundaries, the ferromagnetic grain-boundary foam will tend to have a wandering ferromagnetic axis that lies in the plane of the grain boundary. On applying an increasing magnetic field, the wandering axis will unwind and straighten up as saturation is approached. Modeling the process of a macrospin subjected to a randomly oriented demagnetizing field gives a magnetization curve with $H_d \approx 0.16 M_0$ [151], where M_0 is the magnetization of the ferromagnetic regions. From the plot in Figure 20.16,

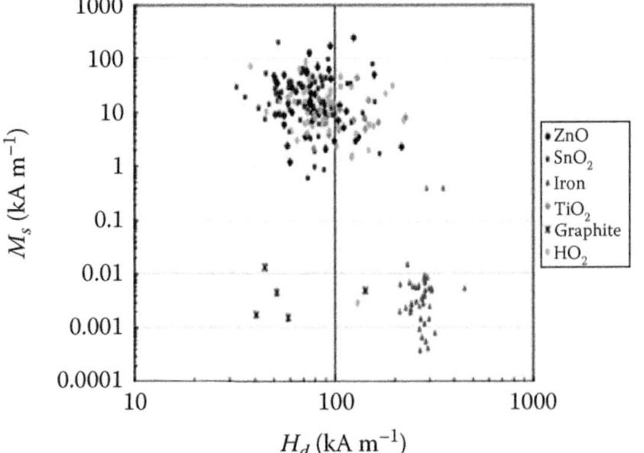

FIGURE 20.16 (See color insert.) Plot of magnetization M_s versus internal field H_d for thin films and nanoparticles of doped and undoped oxides. For films of the thin films of oxides, M_s clusters around $10\,\text{kA m}^{-1}$, whereas for nanoparticles it is three orders of magnitude lower. H_d in both cases is about $100\,\text{kA m}^{-1}$. Also included on the scatter plot are data on graphite and dispersed iron nanoparticles.

TABLE 20.4 Local Dipole Field H_d

	H_d (kA m^{-1})
TiO_2	125 (40)
SnO_2	79 (30)
HfO_2	94 (35)
ZnO	83 (30)
Graphite	68 (42)
Fe	275 (40)

based on the magnetization curves of 250 dilute ferromagnetic oxide films and the summary in Table 20.4, it can be deduced that $M_0/M_s \approx 100$, where M_s is the saturation magnetization of the sample. We infer that *the ferromagnetic volume fraction in the dilute magnetic oxides is about 1%*. When data on the nanocrystalline ceramics is analyzed in the same way, it turns out that the volume fraction there is around one part in 10^5 or 10^6.

The other point to flag is the anisotropy of the saturation magnetization of oriented films, either between parallel and perpendicular to the plane or within the plane. The effect has been reported for HfO_2 [71,72,109], TiO_2 [41], SnO_2 [56,58,60,61], and ZnO [109,130]; the anisotropy may be as large as a factor 10. The prospect of a large and anisotropic moment associated with two-dimensional grain-boundary cells is intriguing, and merits further investigation.

20.6 Conclusions

The magnetism of dilute and undoped magnetic oxides remains one of the most puzzling and potentially important open questions in magnetism today. There is sufficient evidence that the observations in many cases are not simply attributable to measurement artifacts or trivial impurities. There is definitely something to explain.

Two possible sources of the magnetism are defects and dopants. Both may be involved, but in the d^0 systems and perhaps in most of the dilute magnetic oxide systems that have been investigated, the former suffice. The nature of the magnetic order and the coupling mechanism have to be better understood, but it seems that the *m*-J paradigm of magnetism in solids is unable to encompass these materials. On present evidence, the coupling between the 3*d* dopants is weak. They are not DMSs.

A new picture of defect-based magnetism is emerging, where the high-temperature ferromagnetism derives from a spin-split defect-based impurity band. Charge transfer from the 3*d* cations can help stabilize the spin splitting. There is much to investigate in the CTF model, and to connect with specific materials systems—spin waves, micromagnetism, orbital moments, anisotropy, magnetotransport, and other spin–orbit coupling effects.

At present, progress is impeded by the lack of reproducibility of magnetic properties between and within laboratories. There is a desperate need for a dilute magnetic "fruitfly"—a system that is easy to prepare and reproduce. The widely investigated

Co-doped ZnO films produced by PLD really do not fit the bill! It is possible that dip-coated nanocrystalline ceramic films with early transition-metal dopants may be better.

Device applications can follow by design or by serendipity. Once the fruitfly is identified and better understood, the challenge will be to generate stable and controllable defect structures where we can reap the benefit of this unusual high-temperature ferromagnetism, whose occurrence is by no means limited to dilute magnetic oxides.

Acknowledgment

This work was supported by Science Foundation Ireland as part of the MANSE project.

References

1. R. G. Burns, *Mineralogical Applications of Crystal Field Theory*, 2nd edn., Cambridge University Press, Cambridge, U.K., 1993.
2. D. J. Newman and B. Ng, *Crystal Field Handbook*, Cambridge University Press, Cambridge, U.K., 290pp, 2000.
3. E. J. Samuelsen and G. Shirane, Inelastic neutron scattering investigation of spin waves and magnetic interactions in alpha-Fe_2O_3, *Phys. Stat. Sol.* **42**, 241 (1970).
4. E. J. Samuelsen, M. T. Hutchings, and G. Shirane, Inelastic neutron scattering investigation of spin waves and magnetic interactions in Cr_2O_3, *Physica* **48**, 13 (1970).
5. H. Bourdonnay, A. Bousquet, J. P. Cotton et al., Experimental determination of exchange integrals in magnetite, *J. Phys.* **32(2–3)**, C1–C1182 (1970).
6. J. Goodenough, Theory of the role of covalence in the perovskite-type manganites [La, M(II)]MnO_3, *Phys. Rev.* **100**, 564 (1955).
7. J. Kanamori, Superexchange interaction and symmetry properties of electron orbitals, *J. Phys. Chem. Solids* **10**, 87 (1959).
8. P. W. Anderson, Theory of magnetic exchange interactions: Exchange in insulators and semiconductors, *Solid State Phys.* **14**, 99 (1963).
9. J. M. D. Coey, M. Viret, and S. von Molnar, Mixed-valence manganites, *Adv. Phys.* **48**, 167–293 (1999).
10. C. Zener, Interaction between the *d*-shells in the transition metals II: Ferromagnetic compounds of manganese with the perovskite structure, *Phys. Rev.* **82**, 403–405 (1951).
11. D. Stauffer, *Introduction to Percolation Theory*, Taylor & Francis, London, U.K., 1985.
12. G. Deutscher, R. Zallen, and J. Adler (Eds.), *Percolation Structures and Processes*, Adam Hilger, Bristol, U.K., 1983.
13. P. Sati, R. Hayn, R. Kuzian et al., Magnetic anisotropy of Co^{2+} as signature of intrinsic ferromagnetism in ZnO:Co, *Phys. Rev. Lett.* **96**, 017203 (2006).
14. W. Pacuski, D. Ferrand, J. Cibert et al., Effect of the s,p-d exchange interaction on the excitons in $Zn_{1-x}Co_xO$ epilayers, *Phys. Rev. B* **73**, 035214 (2006).
15. C. N. R. Rao and F. L. Deepak, Absence of ferromagnetism in Mn- and Co-doped ZnO, *J. Mater. Chem.* **15**, 573–578 (2005).
16. S. A. Chambers and R. F. Farrow, New possibilities for ferromagnetic semiconductors, *MRS Bull.* **28**, 729–733 (2003).
17. A. Ney, K. Ollefs, S. Ye et al., Absence of intrinsic ferromagnetic interactions with isolated and paired Co dopant atoms in $Zn_{1-x}Co_xO$ with high structural perfection, *Phys. Rev. Lett.* **100**, 157201 (2008).
18. J. M. D. Coey, d^0 ferromagnetism, *Solid State Sci.* **7**, 660–667 (2005).
19. D. W. Abraham, M. M. Frank, and S. Guha, Absence of magnetism in hafnium oxide films, *Appl. Phys. Lett.* **87**, 252502 (2005).
20. Y. Belghazi, G. Schmerber, S. Colis et al., Extrinsic origin of ferromagnetism in ZnO and $Zn_{0.9}Co_{0.1}O$ magnetic semiconductor films prepared by sol-gel technique, *Appl. Phys. Lett.* **89**, 122504 (2006).
21. J. M. D. Coey, Dilute magnetic oxides, *Curr. Opin. Solid State Mater. Sci.* **10**, 83 (2007).
22. M. A. Garcia, E. Fernandez Pinel, J. de la Venta et al., Sources of experimental errors in the observation of nanoscale magnetism, *J. Appl. Phys.* **105**, 013925 (2009).
23. K. Potzger and S. Q. Zhou, Non-DMS related ferromagnetism in transition-metal doped zinc oxide, *Phys. Stat. Sol. B* **246**, 1147–1167 (2009).
24. M. Khalid, A. Setzer, M. Ziese et al., Ubiquity of ferromagnetic signals in common diamagnetic oxides, *Phys. Rev. B* **81**, 214414 (2010).
25. Y. Matsumoto, M. Murakami, T. Shono et al., Room-temperature ferromagnetism in transparent transition metal-doped titanium dioxide, *Science* **291**, 854–856 (2001).
26. Y. Matsumoto, R. Takahashi, M. Murakami et al., Ferromagnetism in Co-doped TiO_2 rutile thin films grown by laser molecular beam epitaxy, *Jpn. J. Appl. Phys.* **40**, L1204–L1206 (2001).
27. J. Y. Kim, J. H. Park, B. G. Park et al., Ferromagnetism induced by clustered Co in Co-doped anatase TiO_2 thin films, *Phys. Rev. Lett.* **90**, 017401 (2003).
28. P. A. Stampe, R. J. Kennedy, Y. Xin, and J. S. Parker, Investigation of the cobalt distribution in the room temperature ferromagnet TiO_2:Co, *J. Appl. Phys.* **93**, 7864–7866 (2003).
29. S. R. Shinde, S. B. Ogale, S. Das Sarma et al., Ferromagnetism in laser deposited anatase $Ti_{1-x}Co_xO_{2-\delta}$ films, *Phys. Rev. B* **67**, 115211 (2003).
30. T. Zhao, S. R. Shinde, S. B. Ogale et al., Electric field effect in diluted magnetic insulator anatase Co: TiO_2, *Phys. Rev. Lett.* **94**, 126601 (2005).
31. H. Toyosaki, T. Fukumura, Y. Yamada, and M. Kawasaki, Evolution of ferromagnetic circular dichroism coincident with magnetization and anomalous Hall effect in Co-doped rutile TiO_2, *Appl. Phys. Lett.* **86**, 182503 (2005).

32. H. Toyosaki, T. Fukumura, Y. Yamada et al., Anomalous Hall effect governed by electron doping in a room-temperature transparent ferromagnetic semiconductor, *Nat. Mater.* **3**, 221–224 (2004).
33. H. Toyosaki, T. Fukumura, K. Ueno et al., A ferromagnetic oxide semiconductor as spin injection electrode in a magnetic tunnel junction, *Jpn. J. Appl. Phys.* **44**, L896–L898 (2005).
34. N. H. Hong, J. Sakai, W. Prellier et al., Ferromagnetism in transition-metal doped TiO_2 thin films, *Phys. Rev. B* **70**, 195204 (2004).
35. Z. J. Wang, J. K. Tang, Y. X. Chen et al., Room-temperature ferromagnetism in manganese-doped reduced rutile titanium dioxide thin films, *J. Appl. Phys.* **95**, 7384 (2004).
36. Z. Wang, W. Wang, J. Tang et al., Extraordinary Hall effect and ferromagnetism in Fe-doped reduced rutile, *Appl. Phys. Lett.* **83**, 518–520 (2003).
37. S. Duhalde, M. F. Vignolo, F. Golmar et al., Appearance of room-temperature ferromagnetism in Cu-doped $TiO_{2-\delta}$ films, *Phys. Rev. B* **72**, 161313 (2005).
38. D. L. Hou, R. B. Zhao, H. J. Meng et al., Room-temperature ferromagnetism in Cu-doped TiO_2 thin films, *Thin Solid Films* **516**, 3223–3226 (2008).
39. Y. J. Kim, S. Thevuthasan, T. Droubay et al., Growth and properties of molecular beam epitaxially grown ferromagnetic Fe-doped TiO_2 rutile films on TiO_2(110), *Appl. Phys. Lett.* **84**, 3531–3533 (2004).
40. T. C. Kaspar, S. M. Heald, C. M. Wang et al., Negligible magnetism in excellent structural quality $Cr_xTi_{1-x}O_2$ anatase: Contrast with high-T_C ferromagnetism in structurally defective $Cr_xTi_{1-x}O_2$, *Phys. Rev. Lett.* **95**, 217203 (2005).
41. J. Osterwalder, T. Droubay, T. Kaspar et al., Growth of Cr-doped TiO_2 films in the rutile and anatase structures by oxygen plasma assisted molecular beam epitaxy, *Thin Solid Films* **484**, 289–298 (2005).
42. S. A. Chambers, Ferromagnetism in doped thin film oxide and nitride semiconductors and dielectrics, *Surf. Sci. Rep.* **61**, 345–381 (2006).
43. K. A. Griffin, A. B. Pakhomov, C. M. Wang et al., Intrinsic ferromagnetism in insulating cobalt doped anatase TiO_2, *Phys. Rev. Lett.* **94**, 157204 (2005).
44. K. A. Griffin, A. B. Pakhomov, C. M. Wang et al., Cobalt-doped anatase TiO_2: A room-temperature dilute magnetic dielectric material, *J. Appl Phys.* **97**, 10D320 (2006).
45. K. A. Griffin, M. Varela, S. J. Pennycook et al., Atomic-scale studies of cobalt distribution in Co-TiO_2 anatase TiO_2 thin films: Processing, microstructure and the origin of ferromagnetism, *J. Appl. Phys.* **97**, 10D320 (2006).
46. T. Fukumura, H. Toyosaki, K. Ueno et al., Role of charge carriers for ferromagnetism in Co-doped rutile TiO_2, *New J. Phys.* **10**, 055018 (2008).
47. A. F. Orlov, L. A. Balagurov, A. S. Konstantinova et al., Giant magnetic moments in dilute magnetic semiconductors, *J. Magn. Magn. Mater.* **320**, 895–897 (2008).
48. S. D. Yoon, Y. Chen, A. Yang et al., Oxygen-defect-induced magnetism at 880 K in semiconducting anatase $TiO_{2-\delta}$ films, *J. Phys.: Condens. Mat.* **18**, L355–L361 (2006).
49. N. H. Hong, J. Sakai, N. Proirot et al., Room-temperature ferromagnetism observed in undoped semiconducting and insulating oxide thin films, *Phys. Rev. B* **73**, 132404 (2006).
50. Z. Q. Zhou, C. Cizmar, K. Potzger et al., Origin of magnetic moments in defective TiO_2 single crystals, *Phys. Rev. B* **79**, 113201 (2009).
51. S. B. Ogale, R. J. Choudhary, J. P. Buban et al., High temperature ferromagnetism with a giant magnetic moment in transparent Co-doped $SnO_{2-\delta}$, *Phys. Rev. Lett.* **91**, 077205 (2003).
52. N. H. Hong and J. Sakai, Ferromagnetic V-doped SnO_2 thin films, *Physica B* **358**, 265–268 (2005).
53. N. H. Hong, J. Sakai, W. Prellier, and A. Hassini, Transparent Cr-doped SnO_2 thin films: Ferromagnetism beyond room temperature with a giant magnetic moment, *J. Phys.: Condens. Mat.* **17(10)**, 1697 (2005).
54. J. M. D. Coey, A. P. Douvalis, C. B. Fitzgerald, and M. Venkatesan, Ferromagnetism in Fe-doped SnO_2 thin films, *Appl. Phys. Lett.* **84**, 1332–1334 (2004).
55. K. Gopinadhan, S. C. Kashyap, D. K. Pandya et al., High temperature ferromagnetism in Mn-doped SnO_2, *J. Appl. Phys.* **102**, 113513 (2007).
56. C. B. Fitzgerald, M. Venkatesan, L. S. Dorneles et al., Magnetism in dilute magnetic oxide thin films based on SnO_2, *Phys. Rev. B* **74**, 11530 (2006).
57. N. H. Hong, J. Sakai, N. T. Huong et al., Role of defects in tuning ferromagnetism in diluted magnetic oxide thin films, *Phys. Rev. B* **72**, 045336 (2005).
58. J. Zhang, R. Skomski, L. P. Yue et al., Structure and magnetism of V-doped SnO_2 thin films: Effect of the substrate, *J. Phys.: Condens. Mat.* **19**, 256204 (2007).
59. H. S. Kim, L. Bi, G. F. Dionne et al., Structure, magnetic and optical properties, and Hall effect of Co- and Fe-doped SnO_2 films, *Phys. Rev. B* **77**, 214436 (2008).
60. N. H. Hong, N. Poirot, and J. Sakai, Ferromagnetism observed in pristine SnO_2 thin films, *Phys. Rev. B* **77**, 033205 (2008).
61. J. Zhang, R. Skomski, Y. F. Lu et al., Temperature-dependent orbital-moment anisotropy in dilute magnetic oxides, *Phys. Rev. B* **75**, 214417 (2007).
62. P. I. Archer, P. V. Radovanovic, S. M. Heald et al., Low-temperature activation and deactivation of high-curie-temperature ferromagnetism in a new diluted magnetic semiconductor: Ni^{2+}-doped SnO_2, *J. Am. Chem. Soc.* **127**, 14479 (2005).
63. C. Van Komen, A. Thurber, K. M. Reddy et al., Structure-magnetic property relationship in transition metal (M = V, Cr, Mn, Fe, Co, Ni) doped SnO_2 nanoparticles, *J. Appl. Phys.* **103**, 07D141 (2008).
64. A. Thurber, K. M. Reddy, V. Shutthanandan et al., Ferromagnetism in chemically-substituted CeO_2 nanoparticles by Ni doping, *Phys. Rev. B* **76**, 165206 (2007).

65. C. B. Fitzgerald, M. Venkatesan, L. S. Dorneles et al., SnO_2 doped with Mn, Fe or Co; room-temperature dilute magnetic semiconductors, *J. Appl. Phys.* **95**, 7390 (2004).
66. A. F. Cabrera, A. M. N. Mudarra, C. E. T. Rodriguez et al., Mechanosynthesis of Fe-doped SnO_2 nanoparticles, *Physica B* **398**, 215 (2007).
67. P. I. Archer and D. R. Gamelin, Controlled grain-boundary defect formation and its role in the high-T_c ferromagnetism on Ni^{2+}: SnO_2, *J. Appl. Phys.* **99**, 08M107 (2006).
68. P. V. Radovanovic and D. R. Gamelin, High-temperature ferromagnetism in Ni^{2+}-doped ZnO aggregates prepared from colloidal dilute magnetic semiconductor quantum dots, *Phys. Rev. Lett.* **91**, 157302 (2003).
69. J. D. Bryan, S. M. Heald, S. A. Chambers, and D. R. Gamelin, Strong room-temperature ferromagnetism in Co^{2+}-doped TiO_2 made from colloidal nanocrystals, *J. Am. Chem. Soc.* **126**, 11640–11647 (2004).
70. A. Sundaresan, R. Bhargavi, N. Rangarajan et al., Ferromagnetism as a universal feature of nanoparticles of otherwise nonmagnetic oxides, *Phys. Rev. B* **74**, 161306(R) (2006).
71. M. Venkatesan, C. B. Fitzgerald, and J. M. D. Coey, Unexpected magnetism in a dielectric oxide, *Nature* **430**, 630 (2004).
72. J. M. D. Coey, M. Venkatesan, P. Stamenov et al., Magnetism in hafnium dioxide, *Phys. Rev. B* **72**, 024450 (2005).
73. H. N. Hong, N. Poirot, and J. Sakai, Evidence for magnetism due to oxygen vacancies in Fe-doped HfO_2 thin films, *Appl. Phys. Lett.* **89**, 042503 (2006).
74. H. N. Hong, Magnetism due to defects/oxygen vacancies in HfO_2 thin films, *Phys. Stat. Sol. C* **4**, 1270–1275 (2007).
75. N. H. Hong, J. Sakai, N. Poirot, and A. Ruyter, Laser ablated Ni-doped HfO_2 thin films: Room temperature ferromagnets, *Appl. Phys. Lett.* **86**, 242505 (2005).
76. M. S. R. Rao, D. C. Kundaliya, S. B. Ogale et al., Search for ferromagnetism in undoped and cobalt-doped $HfO_{2-\partial}$, *Appl. Phys. Lett.* **88**, 142505 (2006).
77. N. Hadacek, A. Nosov, L. Ranno et al., Magnetic properties of HfO_2 thin films, *J. Phys.: Condens. Mat.* **19**, 486206 (2007).
78. E. Tirosh and G. Markovich, Control of defects and magnetic properties in colloidal HfO_2 nanorods, *Adv. Mater.* **19**, 2608–2612 (2007).
79. A. Tiwari, V. M. Bhosle, R. Ramachandran et al., Ferromagnetism in Co-doped CeO_2: Observation of a giant magnetic moment with a high Curie temperature, *Appl. Phys. Lett.* **88**, 142511 (2006).
80. Y. Q. Song, H. W. Zhang, Q. Y. Wen et al., Room-temperature ferromagnetism in Co-doped CeO_2 thin films on Si (111) substrates, *Chin. Phys. Lett.* **24**, 218–221 (2007).
81. R. Vodungbo, Y. Zheng, F. Vidal et al., Room-temperature ferromagnetism of Co-doped $CeO_{2-\delta}$ diluted magnetic oxide: Effect of oxygen and anisotropy, *Appl. Phys. Lett.* **90**, 162510 (2007).
82. Y. L. Liu, Z. Lockmann, A. Aziz et al., Size-dependent ferromagnetism in cerium oxide (CeO_2) nanostructures independent of oxygen vacancies, *J. Phys.: Condens. Mat.* **20**, 165201 (2008).
83. M. Y. Ge, H. Wang, E. Z. Liu et al., On the origin of ferromagnetism in CeO_2 nanocubes, *Appl. Phys. Lett.* **93**, 062505 (2008).
84. M. L. Li, S. H. Ge, W. Qiao et al., Relationship between the surface chemical states and magnetic properties of CeO_2 nanoparticles, *Appl. Phys. Lett.* **94**, 152511 (2009).
85. A. Thurber, K. M. Reddy, and A. Punnoose, Influence of oxygen level on structure and ferromagnetism in $Sn_{0.95}Fe_{0.05}O_2$ nanoparticles, *J. Appl. Phys.* **105**, 07E706 (2009).
86. X. B. Chen, G. S. Li, Y. G. Su et al., Synthesis and room-temperature ferromagnetism of CeO_2 nanocrystals with nonmagnetic Ca^{2+} doping, *Nanotechnology* **20**, 115600 (2009).
87. P. D. C. King, T. D. Veal, F. Fuchs et al., Band gap, electronic structure and surface electron accumulation in cubic and rhombohedral In_2O_3, *Phys. Rev. B* **79**, 205211 (2009).
88. J. Philip, N. Theodoropoulou, G. Berera et al., High-temperature ferromagnetism in manganese-doped indium–tin oxide films, *Appl. Phys. Lett.* **85(5)**, 777–779 (2004).
89. Y. K. Yoo, Q. Z. Xue, H. C. Lee et al., Bulk synthesis and high-temperature ferromagnetism of $(In_{1-x}Fe_x)_2O_{3-s}$ with Cu co-doping, *Appl. Phys. Lett.* **86**, 042506 (2005).
90. J. He, S. F. Xu, Y. K. Yoo et al., Room-temperature ferromagnetic n-type semiconductor $(In_{1-x}Fe_x)_2O_{3-\sigma}$, *Appl. Phys. Lett.* **87**, 052503 (2005).
91. G. Peleckis, X. L. Wang, and S. X. Dou, Room-temperature ferromagnetion in Mn and Fe codoped In_2O_3, *Appl. Phys. Lett.* **88**, 132507 (2006).
92. X. H. Xu, F. X. Jiang, J. Zhang et al., Magnetic properties of *n*-type Fe-doped In_2O_3 ferromagnetic thin films, *Appl. Phys. Lett.* **94**, 212510 (2009).
93. N. H. Hong, J. Sakai, N. T. Huong et al., Magnetism in transition-metal-doped In_2O_3 thin films, *J. Phys.: Condens. Mat.* **18**, 6897–6905 (2006).
94. Z. G. Yu, J. He, S. F. Xu et al., Origin of ferromagnetism in semiconducting $(In_{1-x-y}Fe_xCu_y)_2O_{3-\sigma}$, *Phys. Rev. B* **74**, 165321 (2006).
95. M. Venkatesan, R. D. Gunning, P. Stamenov et al., Room-temperature ferromagnetism in Mn- and Fe-doped indium tin oxide thin films, *J. Appl. Phys.* **103**, 07D135 (2008).
96. J. Philip, A. Punnoose, B. I. Kim et al., Carrier-controlled ferromagnetism in transparent oxide semiconductors, *Nat. Mater.* **5**, 298 (2006).
97. R. P. Panguluri, P. Kharel, C. Sudakar et al., Ferromagnetism and spin-polarized charge carriers in In_2O_3 thin films, *Phys. Rev. B* **79**, 165208 (2009).
98. D. Bérardan, E. Guilmeau, and D. Pelloquin, Intrinsic magnetic properties of In_2O_3 and transition-metal doped In_2O_3, *J. Magn. Magn. Mater.* **320**, 983–989 (2008).

99. U. Ozgur, Y. I. Alivov, C. Liu et al., A comprehensive review of ZnO materials and devices, *J. Appl. Phys.* **98**, 041301 (2005).
100. F. Pan, C. Song, X. J. Liu, Y. C. Yang, and F. Zeng, Ferromagnetism and possible application in spintronics of transition-metal doped ZnO films, *Mater. Sci. Eng. R* **62**, 1–35 (2008).
101. M. Bouloudenine, N. Viart, S. Colis et al., Antiferromagnetism in bulk $Zn_{1-x}Co_xO$ magnetic semiconductors prepared by the coprecipitation technique, *Appl. Phys. Lett.* **87**, 052501 (2005).
102. J. Alaria, P. Turek, M. Bernard et al., No ferromagnetism in Mn-doped ZnO semiconductors, *Chem. Phys. Lett.* **415**, 337 (2005).
103. Z. Jin, T. Fukumura, M. Kawasaki et al., High throughput fabrication of transition-metal-doped epitaxial ZnO thin films: A series of oxide-diluted magnetic semiconductors and their properties, *Appl. Phys. Lett.* **78(24)**, 3824–3826 (2001).
104. K. Ueda, H. Tabata, and T. Kawai, Magnetic and electric properties of transition-metal-doped ZnO films, *Appl. Phys. Lett.* **79**, 988–990 (2001).
105. R. Janisch, P. Gopal, and N. A. Spaldin, Transition metal-doped TiO_2 and ZnO-present status of the field, *J. Phys.: Condens. Mat.* **17**, R657 (2005).
106. W. Prellier, A. Fouchet, and B. Mercey, Oxide-diluted magnetic semiconductors: A review of the experimental status, *J. Phys.: Condens. Mat.* **15**, R1583 (2003).
107. S. J. Pearton, W. H. Heo, M. Ivill et al., Dilute magnetic semiconducting oxides, *Semicond. Sci. Technol.* **19**, R59 (2004).
108. P. Sharma, A. Gupta, K. V. Rao et al., Ferromagnetism above room temperature in bulk and transparent thin films of Mn-doped ZnO, *Nat. Mater.* **2**, 673 (2003).
109. M. Venkatesan, C. B. Fitzgerald, J. G. Lunney, and J. M. D. Coey, Anisotropic ferromagnetism in substituted zinc oxide, *Phys. Rev. Lett.* **93**, 177206 (2004).
110. H. Saeki, H. Tabata, and T. Kawai, Magnetic and electric properties of vanadium doped ZnO films, *Solid State Commun.* **120**, 439 (2001).
111. B. Martinez, F. Sandiumenge, L. Balcells et al., Role of the microstructure on the magnetic properties of Co-doped ZnO nanoparticles, *Appl. Phys. Lett.* **86**, 103113 (2005).
112. N. Khare, M. J. Kappers, M. Wei et al., Defect induced ferromagnetism in Co-doped ZnO, *Adv. Mater.* **18**, 1449–1452 (2006).
113. L. E. Halliburton, N. C. Giles, N. Y. Garces et al., Production of native donors in ZnO by annealing at high temperature in Zn vapor, *Appl. Phys. Lett.* **87**, 172108 (2005).
114. D. A. Schwartz and D. R. Gamelin, Reversible 300 K ferromagnetic ordering in a diluted magnetic semiconductor, *Adv. Mater.* **16**, 2115–2119 (2004).
115. R. P. Borges, R. C. da Silva, S. Magalhaes et al., Magnetism in Ar-implanted ZnO, *J. Phys.: Condens. Mat.* **19**, 476207 (2007) (see also ibid **20**, 429801).
116. P. Stamenov, M. Venkatesan, L. S. Dornales et al., Magnetoresistance of Co-doped ZnO thin films, *J. Appl. Phys.* **99**, 08M124 (2006).
117. K. R. Kittilstved, W. K. Liu, and D. R. Gamelin, Electronic origins of polarity-dependent high T_C ferromagnetism in oxide diluted magnetic semiconductors, *Nat. Mater.* **5**, 291 (2006).
118. J. R. Neal, A. J. Behan, R. M. Ibrahim et al., Room temperature magneto-optics of ferromagnetic transition-metal doped ZnO thin films, *Phys. Rev. Lett.* **96**, 107208 (2006).
119. Z. H. Zhang, X. F. Wang, J. B. Xu et al., Evidence of intrinsic ferromagnetism in individual dilute magnetic semiconducting nanostructures, *Nat. Nanotechnol.* **4**, 523–527 (2009).
120. X. H. Xu, H. J. Blythe, M. Ziese et al., Carrier-induced ferromagnetism in ZnMnAlO and ZnCoAlO thin films at room temperature, *New J. Phys.* **8**, 135 (2006).
121. A. J. Behan, H. Mokhtari, H. Blythe et al., Two magnetic regimes in doped ZnO corresponding to a dilute magnetic semiconductor and a dilute magnetic insulator, *Phys. Rev. Lett.* **100**, 047206 (2008).
122. M. Venkatesan, R. D. Gunning, P. Stamenov et al., Magnetic, magnetotransport and optical properties in Al doped $Zn_{0.95}Co_{0.05}$ O thin films, *Appl. Phys. Lett.* **90**, 242508 (2006).
123. S. T. Oschenbein, Y. Feng, K. M. Whittaker et al., Charge-controlled magnetism in colloidal doped semiconductor nanocrystals, *Nat. Nanotechnol.* **4**, 681 (2009).
124. J. H. Park, M. G. Kim, H. M. Jang et al., Co metal clustering as the origin of ferromagnetism in Co-doped ZnO thin films, *Appl. Phys. Lett.* **84(8)**, 1338 (2004).
125. L. S. Dorneles, M. Venkatesan, R. Gunning et al., Magnetic and structural properties of Co-doped ZnO thin films, *J. Magn. Magn. Mater.* **310**, 2087 (2007).
126. K. Rode, R. Mattana, A. Anane et al., Magnetism of (Zn,Co) O thin films probed by x-ray absorption spectroscopies, *Appl. Phys. Lett.* **92**, 012509 (2008).
127. M. Ivill, S. J. Pearton, S. Rawal et al., Structure and magnetism of cobalt-doped ZnO thin films, *New J. Phys.* **10**, 065002 (2008).
128. D. P. Norton, M. E. Overberg, S. J. Pearton et al., Ferromagnetism in VCo-implanted ZnO, *Appl. Phys. Lett.* **83**, 5488–5490 (2003).
129. D. C. Kundaliya, S. B. Ogale, S. E. Lofland et al., On the origin of high-temperature ferromagnetism in the low-temperature processed Mn–Zn–O system, *Nat. Mater.* **3**, 709 (2004).
130. A. Zhukova, A. Teiserskis, S. van Dijken et al., Giant moment and magnetic anisotropy in Co-doped ZnO films grown by pulse-injection metal organic chemical vapor deposition, *Appl. Phys. Lett.* **89**, 232503 (2006).
131. K. Rode, Contribution a l'etude des semiconducteurs ferromagnetiques dilues: Cas des films minces d'oxyde de zinc dope au cobalt, PhD thesis, Université Paris XI, France, 2006.
132. S. Ghosh, V. Sih, W. H. Lau et al., Room-temperature spin coherence in ZnO, *Appl. Phys. Lett.* **86**, 232507 (2005).

133. Q. Y. Xu, H. Schmidt, S. Q. Zhou et al., Room temperature ferromagnetism in ZnO films due to defects, *Appl. Phys. Lett.* **92**, 082508 (2008).
134. N. H. Hong, E. Chikoidze, and Y. Dumont, Ferromagnetism in laser-ablated ZnO and Mn-doped ZnO thin films: A comparative study from magnetization and Hall-effect measurements, *Physica B* **404**, 3978 (2009).
135. H. N. Hong, A. Barla, J. Sakai et al., Can undoped semiconducting oxides be ferromagnetic? *Phys. Stat. Sol. c* **4**(12), 4461–4466 (2007).
136. A. Sundaresan and C. N. R. Rao, Implications and consequences of ferromagnetism universally exhibited by inorganic nanoparticles, *Solid State Commun.* **140**, 1197–2000 (2009).
137. B. B. Straumal, A. A. Mazilkin, S. G. Protasova et al., Magnetization study of nanograined pure and Mn-doped ZnO films: Formation of a ferromagnetic grain-boundary foam, *Phys. Rev. B* **79**, 205206 (2009).
138. T. C. Droubay, D. J. Keavney, T. C. Kaspar et al., Correlated substitution in paramagnetic Mn^{2+}-doped ZnO epitaxial films, *Phys. Rev. B* **79**, 155203 (2009).
139. A. Barla, G. Schmerber, E. Beaurepaire et al., Paramagnetism of the Co sublattice in ferromagnetic $Zn_{1-x}Co_xO$ films, *Phys. Rev. B* **76**, 125201 (2007).
140. T. Tietze, M. Gacic, G. Schütz et al., XMCD studies on Co and Li doped ZnO magnetic semiconductors, *New J. Phys.* **10**, 055009 (2008).
141. B. K. Roberts, A. B. Pakhomov, P. Voll et al., Surface scaling of magnetism in Cr-ZnO dilute magnetic dielectric thin films, *Appl. Phys. Lett.* **92**, 162511 (2008).
142. J. M. D. Coey, High-temperature ferromagnetism in dilute magnetic oxides, *J. Appl. Phys.* **97**, 10D313 (2004).
143. T. Dietl, H. Ohno, F. Matsukura et al., Zener model description of ferromagnetism in zinc-blende magnetic semiconductors, *Science* **287**, 1019 (2000).
144. S. Methfessel and D. C. Mattis, Magnetic semiconductors, in *Handbuch der Physik*, vol. 18(1), S. Flügge (Ed.), Springer-Verlag, Berlin, 1968.
145. J. M. D. Coey, M. Venkatesan, and C. B. Fitzgerald, Donor impurity band exchange in dilute ferromagnetic oxides, *Nat. Mater.* **4**, 173–179 (2005).
146. D. J. Priour Jr. and S. Das Sarma, Clustering in disordered ferromagnets: The Curie temperature in diluted magnetic semiconductors, *Phys. Rev. B* **73**, 165203 (2006).
147. T. Dietl, Origin and control of magnetism in dilute magnetic semiconductors and oxides, *J. Appl. Phys.* **103**, 07D111 (2008).
148. D. M. Edwards and M. I. Katsnelson, High-temperature ferromagnetism of sp electrons in narrow impurity bands: Application to CaB_6, *J. Phys. C: Condens. Mat.* **18**, 7209–7225 (2006).
149. J. M. D. Coey, K. Wongsaprom, J. Alaria et al., Charge transfer ferromagnetism in oxide nanoparticles, *J. Phys. D: Appl. Phys.* **41**, 134012 (2008).
150. I. S. Elfimov, S. Yunoki, and G. A. Sawatzky, Possible path to a new class of ferromagnetic and half-metallic ferromagnetic materials, *Phys. Rev. Lett.* **89**, 216403 (2002).
151. J. M. D. Coey, J. T. Mlack, M. Venkatesan, and P. Stamenov, Magnetization process in dilute magnetic oxides, *IEEE Trans. Magn.* **46**, 2501 (2010).
152. J. M. D. Coey, R. D. Gunning, M. Venkatesan, P. Stamenov, and K. Paul, Magnetism in defect-ridden oxides, *New J. Phys.* **12**, 053025 (2010).
153. S. J. Han, J. W. Song, C. H. Yang et al., A key to room-temperature ferromagnetism in Fe-doped ZnO: Cu, *Appl. Phys. Lett.* **81(22)**, 4212–4214 (2002).
154. H. Raebiger, S. Lam, and A. Zunger, Control of ferromagnetism via electron doping in In_2O_3:Cr, *Phys. Rev. Lett.* **101**, 027203 (2008).

21

Tunneling Magnetoresistance and Spin Transfer with (Ga,Mn)As

Henri Jaffrès
Université Paris-Sud

Jean-Marie George
Université Paris-Sud

This chapter reviews some fundamental properties of tunneling transport of holes and transfer of angular momenta in the valence band of III–V heterojunctions involving the ferromagnetic semiconductor (Ga,Mn)As, one of the key materials for semiconductor spintronics [1,2]. Additional discussion of (Ga,Mn)As is provided in Chapter 18. This current research, which has roots in the work of Tanaka et al. in the early 2000s [3], is motivated by (1) the wealth of the physics involving transport of spin–orbit coupled states carrying quantized angular momenta in epitaxial heterostructures, (2) the need to use materials and tunnel devices providing an efficient spin injection [4–7] into semiconductors able to circumvent the conductivity mismatch problem [8], and (3) the possibility, as demonstrated in very recent pump-probe optical investigations [9,10], to use the high coherence of hole spin states in a semiconductor quantum dot as a quantum bit for quantum computing or spin-memory state [11]. Consequently, two prerequisites for spin manipulation—(1) creating a current of angular momenta in a semiconductor and (2) converting the latter information into an electrical signal by means of tunnel filtering—still remain challenging for today's spintronics. Despite a relatively low Curie temperature currently reaching 185 K [12,13], the p-type ferromagnetic semiconductor (Ga,Mn)As, whose electronic structure is largely described in this volume [14] and already well documented [15–20], offers the possibility of a good integration with conventional III–V species and related structures. The (Ga,Mn)As ferromagnetism properties may be, in the metallic regime beyond the metal-insulating transition (MIT) [21,22], semiquantitatively understood by the p–d Zener approach [14,20,23]. It results in particular magnetic properties qualitatively different from conventional ferromagnetic transition metals. It appears then interesting to revisit spin-dependent transport properties and spin-transfer phenomena with (Ga,Mn)As. In addition, (Ga,Mn)As exhibits a specific anisotropic character of its different valence bands, an interplay of a large spin–orbit coupling, exchange interactions, and an anisotropic strain-field [24–27], which makes it very attractive. In that sense, III–V heterostructures including (Ga,Mn)As as a ferromagnetic reservoir represent model systems to study injection and transport phenomena involving angular momenta of holes as quantum numbers.

This chapter is divided into four different sections. In Section 21.1, we will present results of tunneling magnetoresistance (TMR) obtained on (Ga,Mn)As/III–V/(Ga,Mn)As junctions [3,28–31] studied hereafter for spin injection problem. TMR results on hybrid (Ga,Mn)As/insulator(I)/transition metal junctions will be also briefly discussed [32–35]. In this section, tunneling phenomena observed on resonant (Ga,Mn)As [36,37] and (In,Ga)As [38] quantum wells (QWs) structures will be addressed to question about the coherent nature of transport in heterojunctions and related systems. We then tackle about the role of the different material properties, mainly hole filling and exchange energy, on the TMR properties and will establish phase diagrams that correlate TMR versus latter parameters. Section 21.2 will be devoted to tunneling anisotropic magnetoresistance (TAMR) effects [39–44] including, at least, one (Ga,Mn) As electrode. TAMR generally manifests itself by a significant variation of the tunneling current with respect to the magnetization direction of the ferromagnetic electrode (for a recent review on TAMR, the readers can refer to Matos-Abiague and Fabian [45,46]). In the present case, these physical phenomena originate from the strong anisotropy of the hole Fermi surfaces of the different (Ga,Mn)As subbands involved in the tunneling transport of holes. In the third part, we will describe spin-transfer torque experiments [47–50], which have been demonstrated in these systems [51–54] accounting for the conservation of the angular momenta of holes carried by the current. The last part will be devoted to the modeling of angular momenta transport in the valence band of semiconductors within the frame of the generalized $k \cdot p$ kinetic approach. Describing (Ga,Mn)As in a mean-field approach [24,55] includes the average p–d interaction between spin-up and spin-down holes in the valence band of the semiconductor parameterized by the spin-exchange parameter $\Delta_{exc} = -5/2xN_0\beta$ or, equivalently, the spin-splitting parameter $B_G = \Delta_{exc}/6$ introduced by Dietl et al. [24]. This should be added to the Kinetic or Kohn–Luttinger Hamiltonian [56]. Within the envelope function approximation, the tunneling transport of holes in the valence band of semiconductors will be described in detail according to the procedure given by Petukhov et al. [57,58] for matching hole wave functions and derivatives at each interface. This approach constitutes the generalization to magnetic systems of the boundary conditions for hole envelope wave functions and tunneling current examined about 20 years ago by Andreani et al. [59] and Wessel and Altarelli [60] for semiconducting heterojunctions.

21.1 Tunneling Magnetoresistance in the Valence Band of *p*-Type Semiconductors

21.1.1 Introduction

The quantum mechanical tunnel effect is one of the oldest quantum phenomena and still carries on to enrich our understanding of many fields in physics. The generality of the tunneling phenomenon is such that every dedicated textbook on quantum mechanics treats tunneling through a potential barrier, and the possibility for an electron to tunnel through such a barrier with a characteristic imaginary velocity and imaginary wave vector. Despite this long history and many experimental investigations, the tunneling process of electrons (or holes) continues to amaze physicists in the solid-state community. For spin-dependent tunneling between two ferromagnetic materials, first proposed and reported [61] in 1975, and subsequently observed, it has taken 2 decades to be reliably demonstrated at room temperature [62]. The large TMR effects possible in ferromagnet–insulator–ferromagnet tunnel junctions and magnetic tunnel junctions (MTJs) have led to the development of nonvolatile magnetic random access memories (MRAMs) and new generation of magnetic field sensors in hard disks, for example.

For sufficiently thin barriers, typically a few nanometers insulating layer, carriers have a certain finite probability to be detected on the other side of the potential barrier due to the coupling between the evanescent wave functions on each side of the barrier. With the condition that the density of states (DOS) of the electrodes is spin polarized in the case of ferromagnets, and in a simple picture of a tunneling current proportional to the DOS within each spin channel, the transmission probability is spin dependent, that is, depends on the relative orientation of the two magnetizations. This is the so-called TMR effect. Additionally, for tunneling of holes in the valence band of semiconductors, one can expect an influence of the large spin–orbit coupling on the transmission probability through the barrier as well as a possible deflection of the characteristic carrier wavevector leading to the so-called TAMR effects.

21.1.2 Overview of TMR Phenomena Including (Ga,Mn)As

The first generation of MTJs has integrated amorphous oxide as tunnel barrier, and spin-dependent tunneling effects were mainly described in terms of the phenomenological Jullière's model [61]. This has become to change with the synthesis of high-quality epitaxial Fe/MgO/Fe MTJs [63,64] and related junctions showing up huge TMR ratio explained in terms of spin-symmetry filtering of the electron wavefunction during their tunneling transfer [65,66]. From a fundamental point of view, the ferromagnetic semiconductor (Ga,Mn)As provided, at about the same time, new opportunities to study spin-polarized transport phenomena in semiconductors heterojunctions [3]. One of the interests in (Ga,Mn)As lies in the wealth and varieties of its electronic valence band structure seat of exchange interactions and strong spin–orbit coupling. In this vein, advantages of semiconductor tunnel junctions are fourfold: (1) III–V heterostructures can be epitaxially grown in a wide variety of tunnel devices with abrupt interfaces and with atomically controlled layer thicknesses, (2) junctions can be easily integrated with other III–V structures and devices, (3) many structural and band parameters are controllable like barrier thickness and barrier height allowing a possible engineering of the band profile, and (4) one can easily introduce quantum heterostructures much easier than any other material system.

The first large TMR ratio (>70% at low temperature) observed on III–V-based (Ga,Mn)As/AlAs/(Ga,Mn)As MTJs [3] and purely due to a spin-valve effect was obtained by Tanaka and Higo in 2001 (Figure 21.1). One of the first interesting properties of these epitaxially grown MTJs was the evidence of a rapid drop of TMR with increasing the AlAs barrier thickness demonstrating a

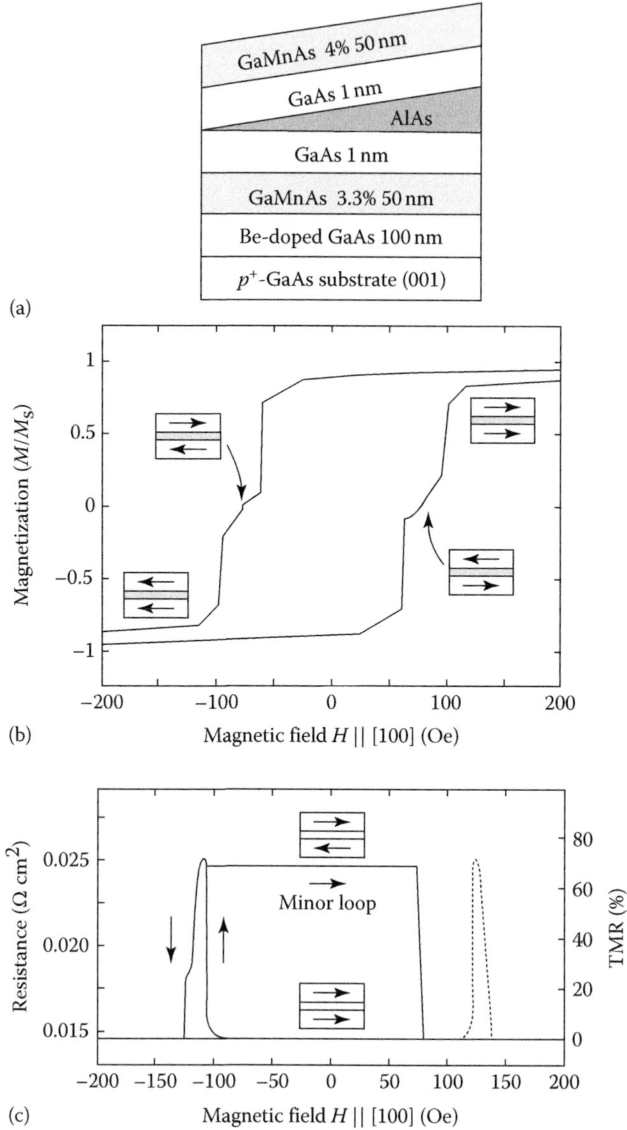

FIGURE 21.1 (a) Schematic illustration of a wedge-type ferromagnetic semiconductor trilayer heterostructure sample grown by LT-MBE. (b) Magnetization of a $Ga_{0.96}Mn_{0.04}As$/AlAs 3 nm/$Ga_{0.967}Mn_{0.033}As$ trilayer measured by SQUID at 8 K. (c) TMR curves at 8 K of the trilayer 200 μm in diameter. Bold solid and dashed curves were obtained by sweeping the magnetic field from positive to negative and negative to positive, respectively. A minor loop is shown by a thin solid curve. In both (b) and (c), the magnetic field was applied along the [154] axis in the plane. (From Tanaka, M. and Higo, Y., *Phys. Rev. Lett.*, 87, 026602, 2001. With permission.)

change of the effective tunneling spin polarization of holes with barrier thickness. This was ascribed, in the frame of spin–orbit nearest neighbor sp^3s^* tight-binding model of tunneling, to the conservation of the in-plane wavevector parallel to the interface during the tunneling transfer process due to translation invariance. Some theoretical investigations performed in the frame of the 6 × 6 multiband $k \cdot p$ theory emphasize the specific role of the spin–orbit coupling within the barrier to possibly explain such a loss of TMR with the III–V barrier thickness [67,68]. However, other calculations performed by Krstajic and Peeters gave opposite conclusions [69].

Beyond single tunnel junctions, TMR obtained on (Ga,Mn)As-based III–V double barrier structures, (Ga,Mn)As/AlAs/GaAs/AlAs/(Ga,Mn)As [70] and (Ga,Mn)As/AlAs/InGaAs/AlAs/(Ga,Mn)As [38] including (Ga,Mn)As/AlAs junctions as spin injector/detector devices demonstrated spin injection phenomena into III–V materials in current perpendicular to plane (CPP) geometry. Nevertheless, such spin-valve effects observed on double barrier structures with in-plane magnetization still question about the symmetry-state of the hole wave function injected in the QW whose natural quantization axis is the growth direction. In the case of experiments of Mattana et al. [70], the formation of hydrogenic impurity states inside the QW of quasi-spherical or cubic symmetry was suggested. Such extended envelope hole wave function of a given symmetry has been described theoretically in the $k \cdot p$ framework by Baldereschi and Lipari in the early 1970s [71–73] and discussed in Appendix A of this chapter. Such localized states could give rise to resonant tunneling effects, through Ga vacancies states [74] for instance in experiments of Mattana et al. [70]. More generally, resonant tunneling effects in the conduction band were already observed, in our group, through As antisites deep levels in MnAs/GaAs/MnAs tunnel junctions synthesized at low temperature [75,76]. Nevertheless, the oscillatory character of the TMR highlighted by Ohya et al. [38] on structures of higher epitaxial quality shows a more surprising nature about the hole symmetry-state. A part of the answer lies in the possible existence of two orthogonal mixed-spin hole states on resonant levels in the QW. This conclusion should be supported by future theoretical investigations. Several kind of III–V (Ga,Mn)As-based tunnel junctions were developed hereafter including a thin GaAs layer [28,29,53,54] or (In,Ga)As barriers [52,30] specifically grown for their low resistance area product with the view to proceed to spin-transfer experiments. The results of Xiang et al. [77] provide an intriguing observation of small in-plane magnetoresistance effects on (Ga,Mn)As/highly Be-doped GaAs/(Ga,Mn)As trilayers [77] where GaAs should play the role of barrier for holes.

In parallel to tunneling transport experiments, (Ga,Mn)As/GaAs Esaki diodes were used to demonstrate spin injection by electrical means on lateral structures with a nonlocal detection [78] and by optical technics [79–81] giving more than 80% for the degree of helicity of the light emitted from these spin-light-emitting diodes (spin-LED) [79]. Spin-dependent tight-binding Hamiltonian calculations including exchange interactions within (Ga,Mn)As was then employed to understand such a high spin injection efficiency [82]. On the other hand, Chun et al. [32] demonstrated spin-polarized tunneling effects on hybrid MnAs/AlAs/(Ga,Mn)As structures as well as Saito et al. with hybrid Fe/ZnSe/(Ga,Mn)As and Fe/GaO/(Ga,Mn)As junctions [33,35]. This shows the possibility to observe large TMR ratio with bipolar *n/i/p* structures and consequently to convert spin-polarized electrons into spin-polarized holes and vice versa.

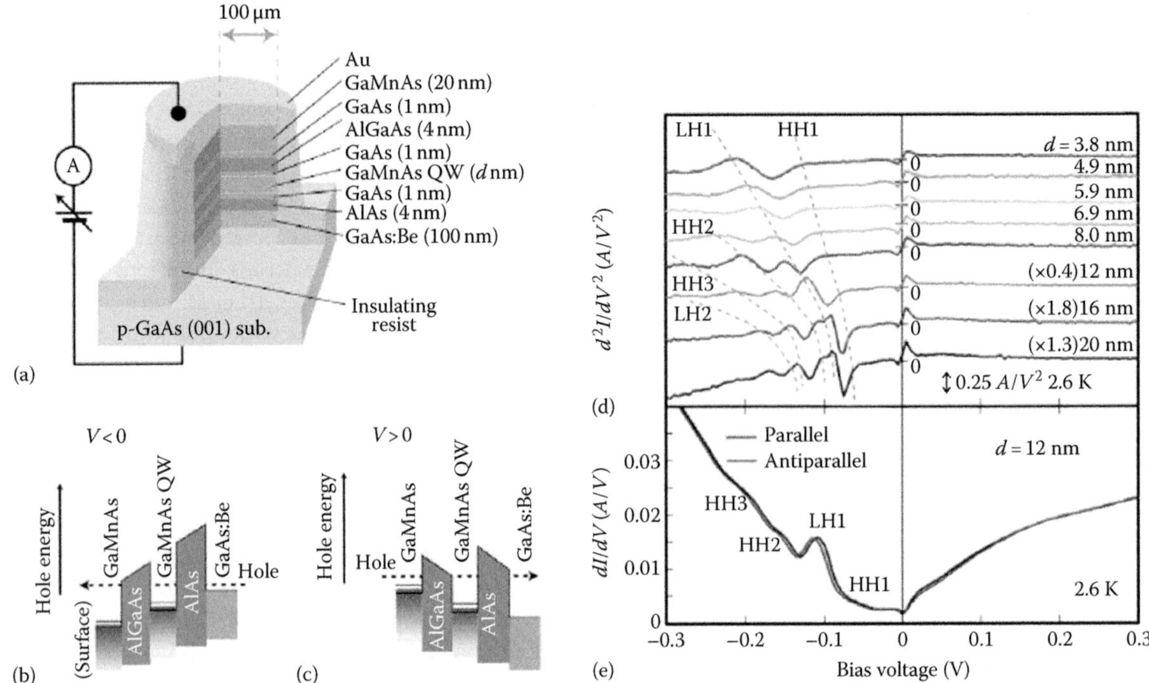

FIGURE 21.2 (a) Schematic device structure of the $Ga_{0.95}Mn_{0.05}As$ 20 nm/GaAs 1 nm/$Al_{0.5}Ga_{0.5}As$ 4 nm/GaAs 1 nm/$Ga_{0.95}Mn_{0.05}As$ *d* nm/GaAs 1 nm/AlAs 4 nm/GaAs:Be 100 nm RTD junction. (b) and (c) Schematic band diagrams of the RTD junction when the bias polarity is negative and positive, respectively. (d) $d^2I/dV^2 - V$ characteristics of these RTD junctions with various QW thicknesses *d* in parallel magnetization at 2.6 K. Numbers in the parentheses express the magnification ratio for the vertical axis. (e) $dI/dV - V$ characteristics of the junction with $d = 12$ nm at 2.6 K in parallel (blue curve) and antiparallel (red curve) magnetization. (From Ohya, S. et al., *Phys. Rev. B Condens. Matter*, 75, 155328, 2007. With permission.)

From a more fundamental point of view, could tunneling of holes through thin (Ga,Mn)As layer and through the whole III–V heterojunction be realistically described by coherent processes? An affirmative answer can be brought by referring to a recent work of Tanaka's group [36,37] who have reported on resonant TMR effects obtained on (Ga,Mn)As QW III–V heterostructures. Specifically, clear resonant tunneling effects were observed through relatively thin AlAs/(Ga,Mn)As/(Al,Ga)As resonant tunneling diode with varying QW thickness ranging from 3.8 to 20 nm (Figure 21.2). These experiments seems to support a highly coherent nature for spin-polarized holes tunneling processes through relatively thin (Ga,Mn)As QWs despite the low-temperature growth procedure (250°C) often described to be detrimental for the observation of such phenomena. Moreover, these experiments are well explained by 4 × 4 [36] and 6 × 6 [37] valence band $k \cdot p$ [56] model including p–d mean-field exchange interactions, thus supporting the validity of the coherent approach to describe tunneling transport of holes in a wide family of (Ga,Mn)As/III–V heterostructures.

21.1.3 Material and Chemical Trends in Heterojunctions

We next focus on our experimental results. Figure 21.3a and b displays TMR results acquired at low bias (1 mV) and low temperature (3 K) on $Ga_{0.93}Mn_{0.07}As$(80 nm)/$In_{0.25}Ga_{0.75}As$ (6 nm)/$Ga_{0.93}Mn_{0.07}As$ (15 nm) tunnel junctions patterned in micron-sized [30] (Figure 21.3b) (standard UV lithography) and submicron pillars [52] (Figure 21.3a) fabricated by e-beam lithography for spin-transfer experiments. Concerning the e-beam lithography process, the discussion of the fabrication procedure is given in Section 21.3 devoted to spin-transfer experiments. These samples exhibit respectively 120% (micron sized) and 150% (submicron) TMR ratio after an annealing procedure, which is among the highest values found at this temperature with (Ga,Mn)As-based MTJs [3,28,29]. The TMR is assumed to be larger for submicron junctions owing to a better homogeneity of the magnetization configuration. One can notice that the TMR ratio is only 30% for as-grown micron-sized junctions before any annealing treatment. This highlights a particular change of the (Ga,Mn)As properties. Although the magnetic properties derive from the whole magnetic layers (volume effect), whereas the tunneling process is more sensitive to interfaces, it seems however possible to draw some qualitative conclusions. A comparative study of the transport and magnetic properties of these stacks has clearly demonstrated that the rise of TMR observed after annealing conjugated to the decrease of the specific resistance × area (RA) product from 5×10^{-2} to 3×10^{-3} Ω cm² was correlated to the change of the top (Ga,Mn)As layer properties [30]. In addition, the drop of the RA product

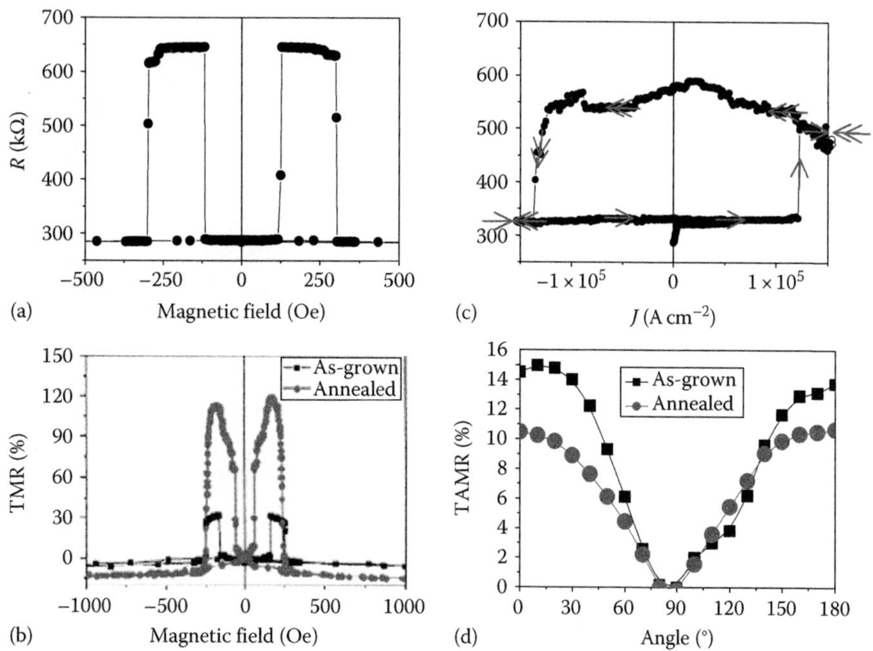

FIGURE 21.3 (a) Tunneling magnetoresistance (TMR) measured on 700×200nm nanopillar $Ga_{0.93}Mn_{0.07}As$(80nm)/$In_{0.25}Ga_{0.75}As$ (6nm)/$Ga_{0.93}Mn_{0.07}As$ (15nm) at 1mV and at 3K. (b) TMR obtained on 128 μm² junction patterned by standard UV lithography on as-grown and annealed samples. (c) Current-induced magnetization switching (CIMS) experiments on submicronic tunnel junction (a) at 3K. (d) Tunneling anisotropic magnetoresistance (TAMR) experiments measured at 3K on annealed sample (b). The reference for angle 0° is the direction along the growth axis ($\hat{m}$||[001]).

conjugated to a reduction of the coercive field (softening of the magnetic film) observed on the top (Ga,Mn)As layer [30,83–85] unambiguously stems from an increase of the hole concentration in (Ga,Mn)As due to the annealing. The consequences are (1) a change of the Fermi energy within (Ga,Mn)As, (2) a reduction of the (In,Ga)As barrier height, and (3) an increase of the exchange energy Δ_{exc} within the top (Ga,Mn)As layer. With a typical exchange integral of $-5/2xN_0\beta=-1.2$eV [86], the average spin-splitting in (Ga,Mn)As can be estimated in the mean-field approach from the saturation magnetization according to $\Delta_{exc}=6B_G=(A_F\beta M_s)/(g\mu_B)$ [24] where A_F is the hole Fermi Liquid parameter. Considering that only the top (Ga,Mn)As layer is affected by thermal treatment, it is then possible to evaluate the increase of the spin-splitting parameter, B_G, from 17 to 24meV [30] and then Δ_{exc} from 100 up to 140meV, for a characteristic Curie temperature T_C of 60K [30]. We will see in the following section that such range of values for Δ_{exc} is also compatible with TAMR experiments performed on (Ga,Mn)As-based resonant tunneling devices. More recent experiments led on epitaxially strained (Ga,Mn)As samples gave increased values of B_G from 23 up to 39meV by annealing treatment [26].

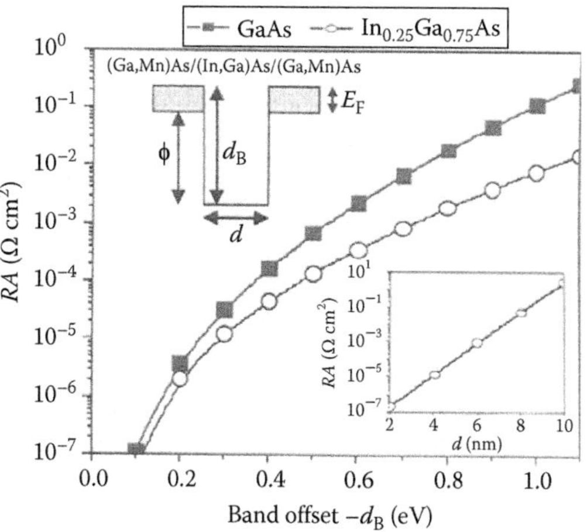

FIGURE 21.4 Calculated Resistance×Area (*RA*) product of trilayer structures versus the band offset d_B between the ferromagnetic semiconductor and a 6nm barrier of GaAs or (In,Ga)As. The physical parameters used for (Ga,Mn)As are $\Delta_{exc}=6B_G=120$meV and $p=1.5\times10^{20}$ cm^{-3}. Insets: (Bottom) *RA* product as a function of the (In,Ga)As barrier width *d*. This gives a quasiexponential variation of the *RA* product versus the barrier thickness. (Top) Valence band profile of the heterojunctions.

21.1.4 Experiments versus $k\cdot p$ Theory of Transport

In Figure 21.4, we present our results of $k\cdot p$ calculations (details are given in Appendix B) concerning specific RA products of Ga,Mn)As/$In_{0.25}Ga_{0.75}As$(6nm)/(Ga,Mn)As junction vs. the valence band offset (VBO), d_B, between (Ga,Mn)As and GaAs or (In,Ga)As. The VBO determines the effective barrier height ϕ according to $d_B=-\varepsilon_F+\phi$ (left inset of Figure 21.4). The top of paramagnetic (Ga,Mn)As valence band ($\Delta_{exc}=0$) is taken here as the reference for energy as well as for the Fermi energy. From tunneling experiments, one can estimate the barrier height between metallic (Ga,Mn)As and high-temperature (H-T) grown GaAs in the range between 100 meV [44] and 90 meV [28], whereas photoemission spectra gave a barrier height ϕ of 0.45 eV for low-temperature (L-T) grown GaAs [87]. This is in agreement with our $k \cdot p$ model giving a RA product $\simeq 10^{-3}\,\Omega\,cm^2$ for annealed samples by taking a VBO $d_B = -0.55$ eV and with experimental results of Chiba et al. [28]. Concerning (In,Ga) As, the band offset between $In_{0.25}Ga_{0.75}As$ and GaAs grown at high temperature is known to be less than 50 meV for samples grown at optimized temperatures [88].* Nevertheless, this VBO has to be taken larger, on the order of $d_B = -0.7$eV, to match with our calculations. This determines the effective barrier height at about 0.6 eV. The reason is that the real value of VBO (and barrier height) is known to depend on (1) the growth temperature that affects the nature and density of the dangling bonds at interfaces between two semiconductors [89] and (2) the local density of ionized defects in the barrier (such as As antisites). These generally lead to valence band bending susceptible to modify the average barrier height. The bottom inset of Figure 21.4 gives the exponential dependence of the RA product of (Ga,Mn)As/(In,Ga)As/(Ga,Mn)As junctions $\propto \exp(2\kappa d)$ versus the (In,Ga) As barrier width corresponding to an effective imaginary wavevector $\kappa_{InGaAs} \simeq 1.15\,nm^{-1}$. Experimental results on AlAs barriers with larger effective hole mass (Section 21.2) gave attenuation transmission on the order of $\kappa_{AlAs} \simeq 2.9\,nm^{-1}$ [3].

Figure 21.5a displays, in color code, the calculated TMR of (Ga,Mn)As/$In_{0.25}Ga_{0.75}As$(6 nm)/(Ga,Mn)As junctions versus the spin-splitting parameter B_G and Fermi energy ε_F of (Ga,Mn) As. B_G represents the exchange splitting between two consecutive hole bands at the Γ_8 symmetry point of the valence band. Also are plotted on these phase diagrams, the energy of the four first bands at the Γ_8 point as well as three different iso-hole concentration lines corresponding to 1×10^{20}, 3.5×10^{20}, and 5×10^{20} cm^{-3} as a guide for the readers. High TMR values, up to several hundred percent, can be expected either for spin-splitting parameter B_G larger than several tens of meV or for low carrier concentration, that is when only the first subband is involved in the tunneling transport (the different (Ga,Mn)As subbands $n = 1-4$ are described in Appendix A). This corresponds to a quasi half-metallic character for (Ga,Mn)As. Starting from the first subband ($n = 1$) and increasing the carrier concentration to fill the consecutive lower subbands ($n = 2,3,4$), up and down spin populations start to mix-up, leading to a significant drop of TMR. For a high carrier concentration ($n = 4$), a smaller TMR is expected, which implies difficulties to simultaneously attain higher T_C and large TMR. We specify that for low values of spin-splitting and Fermi energy, ferromagnetic phase induced by carrier delocalization may not exist (top right corner of the diagram), which is, of course, not taken into account in our $k \cdot p$ model using a basis of propagating envelope wave function. Comparing with our experiments, the estimated value of B_G gives a good qualitative agreement for TMR as illustrated by the trajectory represented in Figure 21.5a between point 1 (as-grown sample) and point 2 (after annealing). TMR ratio obtained on as-grown sample (Figure 21.3b) is well reproduced for a hole concentration in $Ga_{0.93}Mn_{0.07}As$ approaching 10^{20} cm^{-3}, in rather good agreement with the one measured for single (Ga,Mn)As layer and already reported for a Mn concentration close to 7% [21,90]. Annealing procedures are known to remove Mn interstitial atoms and increase hole concentration [20]. This should lead to a reduction of the effective barrier height, even if the position of the topmost (Ga,Mn)As valence bands is expected to move upward in the GaAs band gap due to an increase of the exchange term Δ_{exc}. The strong reduction of the RA product, together with the rise of TMR, is consistent with such assumption. Nevertheless, the hole concentration extracted here after annealing, $\sim 1.7 \times 10^{20}$ cm^{-3}, appears to be slightly smaller as compared to the value derived from Hall effect [21,90] and from magnetic anisotropy measurements [26], which gave hole concentration at least two times larger. The existence of a possible concentration gradient can be at the origin of such discrepancy, as shown by Nishitani et al. [91]. A reduction of the hole

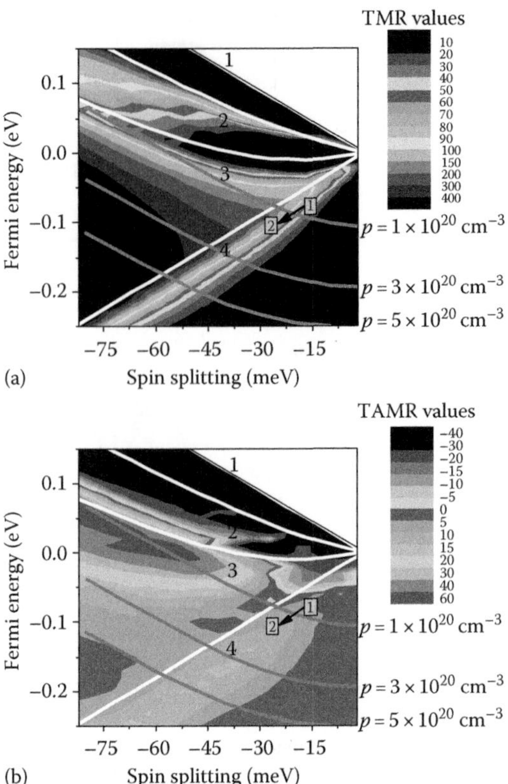

FIGURE 21.5 (See color insert.) Tunnel magnetoresistance (a) and TAMR (b) versus Fermi energy and spin-splitting for a 6 nm (In,Ga)As barrier and a band offset of $d_B = 700$ meV. White lines represent the four bands at the Γ_8 point. Gray lines indicate the Fermi energy for different hole concentrations.

* If one takes into account explicitly the stress between (Ga,Mn)As and (In,Ga)As, this shift should correspond to the valence band offset between (Ga,Mn)As and the LH component of (In,Ga)As that is preferentially transmitted through thick barriers.

concentration at the interface with the barrier is also possible due to a significant charge transfer between p-type (Ga,Mn)As and the Be-doped GaAs counter-electrode and/or the n-type (In,Ga)As (excess of As antisites) [92,93].

21.2 Tunneling Anisotropic Magnetoresistance in the Valence Band of *p*-Type Semiconductors

21.2.1 Overview of TAMR Phenomena

In the absence of spin–orbit interactions, spin and orbital quantum numbers are fully uncoupled. Consequently, no change of the tunnel conductance is expected when the magnetization of the ferromagnet is rotated in space. This assertion changes when the spin–orbit coupling is turned on. The combination of spin–orbit coupling and exchange interactions in MTJs leads to TAMR effects. This can be understood as a variation of the tunneling current versus the magnetization direction of the ferromagnetic electrode(s). TAMR phenomena can be divided into two different classes according to the following: (1) the magnetization always remains perpendicular to the tunneling current leading to in-plane TAMR effects [39,40] or (2) the magnetization rotates from in-plane to out-of-plane directions, for example, parallel to the tunneling current (out-of-plane TAMR) [30,41]. The physical basis is the same as for anisotropic magnetoresistance (AMR) effects observed in a single ferromagnetic layer except that, in magnetic junctions, the transport of spin-polarized carriers admits a tunneling character. In that sense, TAMR can be viewed as a natural extension of AMR in the tunneling regime of transport. In principle, the latter geometry is more favorable to the observation of a larger intrinsic TAMR signal due to a high selectivity in k-space by tunneling filtering. First discovered in (Ga,Mn)As-based III–V MTJs, TAMR has since been demonstrated on transition metal-based nanodevices [94–97], transition metal-based tunneling contacts [98,99], and MTJs [100–102]. TAMR is now used in theoretical proposals for resonant tunneling devices [103]. For a recent review on TAMR phenomena, the readers can refer to Matos-Abiague and Fabian [45,46].

The strong spin–orbit coupling Δ_{so} acting in the valence band of III–V compounds is very favorable to the observation of TAMR phenomena. Significant TAMR effects have been observed on (Ga,Mn)As-based tunnel junctions in the respective in-plane [39,104] and out-of-plane geometry [30,41]. This also includes huge TAMR effects obtained in GaAs samples doped with low Mn content [40] at the vicinity of the metal-to-insulator (MIT) transition (Figure 21.6). This effect originates from the occurrence of an Efros–Shklovskii gap in (Ga,Mn)As at low bias and low temperature controlled by the extent of the hole wave function on the Mn acceptor atoms within a certain depletion region [105]. Using $k \cdot p$ calculations, it has been shown that this extent could depend crucially on the direction of magnetization in (Ga,Mn)As. This was explained by the strong anisotropy of hole envelope function around a single Mn in (Ga,As) in a low Mn doping regime where anisotropic exchange interactions mediate a significant difference in the size of hole wave function between $\mathbf{r} \parallel \hat{m}$ and $\mathbf{r} \perp \hat{m}$. This argument holds also in the reciprocal space ($\hat{k}$) as emphasized in the following. In the present case, this leads to a magnetization-dependent hole integral transfer, tuning the system from an insulating phase into a metallic state by changing the magnetization direction. This anisotropy of the hole envelope wave function around a single Mn center calculated by an exchange-including tight-binding method leading to huge TAMR effects near the MIT transition is sketched in Figure 21.7 [106,107].

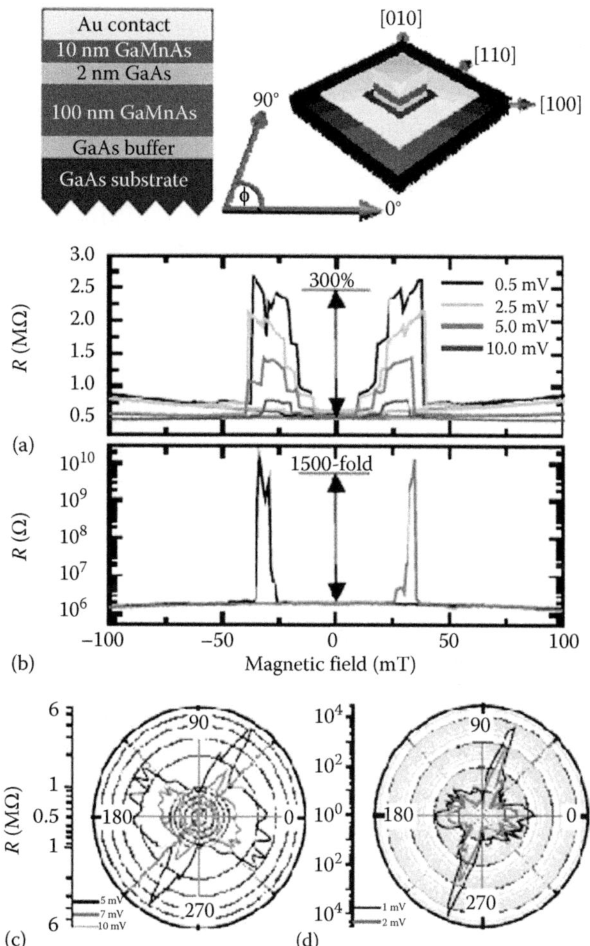

FIGURE 21.6 Amplification of the TAMR effect at low bias voltage and low temperatures. (a) TAMR along $\phi=65°$ at 4.2 K for various bias voltages. (b) Very large TAMR along $\phi=95°$ at 1.7 K and 1 mV bias. (c, d) ϕ at various bias at 1.7 K. (Top) Layer stack under study. (From Ruster, C. et al., *Phys. Rev. Lett.*, 94, 027203, 2005. With permission.)

Furthermore, an unconventional TAMR bias-dependence has been evidenced on (Ga,Mn)As/GaAs(n^+) Esaki diodes [42] exhibiting a crossover between positive and negative TAMR signals. This particular feature was also observed in our group (Figure 21.8a) in (Ga,Mn)As/AlAs junctions with opposite convention for the definition of TAMR giving rise to an opposite sign. The TAMR changes from positive values at positive bias to negative values for negative bias that is when holes are injected from the GaAs:Be electrode toward lowest hole energy subbands corresponding to $n=4$ (inset of Figure 21.8).

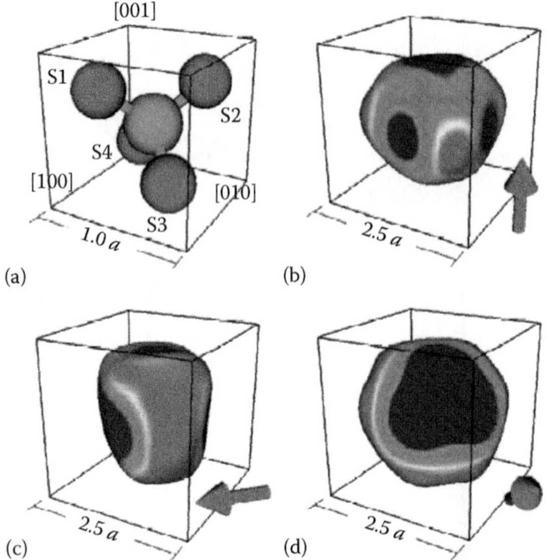

FIGURE 21.7 (a) Atomic structure close to a single Mn atom in substitutional site in GaAs host. As atoms are indicated by S1, S2, S3, and S4. Local spin density surfaces for Mn magnetization along [001] (b), [1$\bar{1}$0] (c), and [112] (d) showing the anisotropy of the hole envelope wave function. (From Tran, M. et al., *Appl. Phys. Lett.*, 95, 172101, 2009; Tang, J.M. and Flatté, M.E., *Phys. Rev. B Condens. Matter*, 72, 161315(R), 2005. With permission.)

21.2.2 TAMR with Holes

Physical phenomena explaining TAMR with holes can be understood with the following arguments:

1. The thick tunnel barriers (AlAs, GaAs, InGaAs) transmit easily any light hole (LH) projection along the zone axis (or growth direction labeled by $\hat{z}$) and switches off any heavy hole (HH) components because of their larger effective mass.
2. As emphasized below, the exchange interactions mediate a strong anisotropic LH↔HH mixing. Since holes are mainly spin polarized along their propagation direction (at least in the limit of a large spin–orbit coupling $\Delta_{so} \gg \varepsilon_F$), the LH↔HH conversion is efficient for wavevectors $\hat{k}$ perpendicular to the magnetization $\hat{m}$. This derives from the introduction of non-diagonal exchange terms in the overall Hamiltonian H_h [24].
3. Arguments 1 and 2 explain the large positive TAMR associated to the LH subbands near the Γ point ($k \| \hat{z}$): LH–HH conversion along $\hat{z}$ is forbidden for out-of-plane magnetization ($\hat{m} \| \hat{z}$) while it is possible for in-plane magnetization ($\hat{m} \| \hat{x}$). The same reasoning yields a large negative TAMR for subbands of HH symmetry near the Γ point.

Our calculations performed on (Ga,Mn)As/AlAs(3 nm)/(Ga,Mn)As tunnel junctions seem to corroborate such conclusions. In Figure 21.8c, one can observe a change of sign for

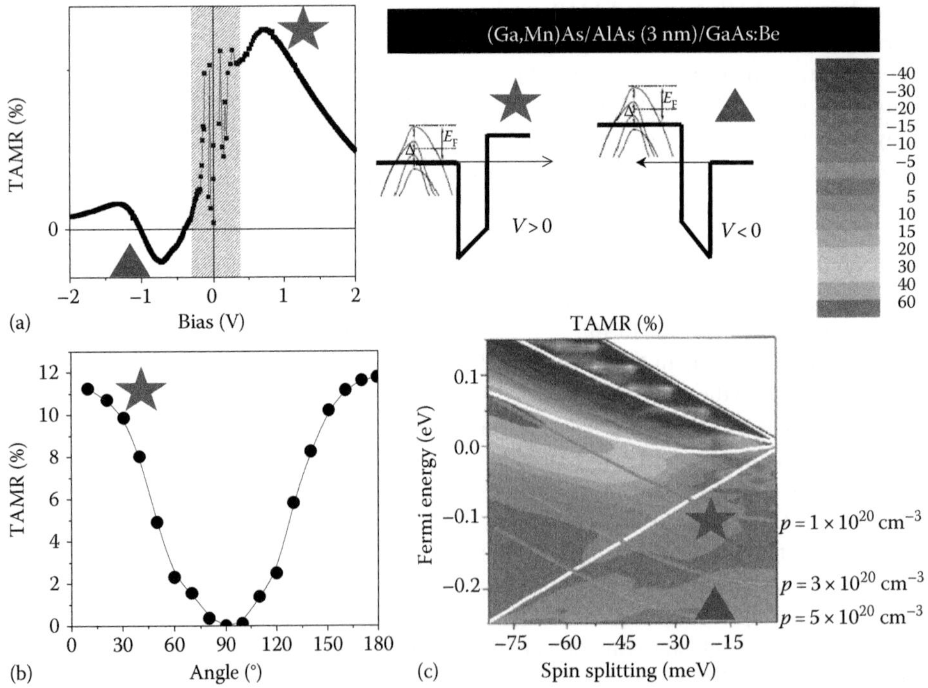

FIGURE 21.8 (See color insert.) (a) Out-of-plane TAMR versus bias acquired on a (Ga,Mn)As/AlAs(3 nm)/GaAs:Be tunnel junction. Positive biases correspond to hole injected from (Ga,Mn)As toward GaAs:Be. (b) Angular variation of the resistance acquired at a positive bias of +1 V (star) displaying a positive TAMR. The angle 0° corresponds to $\hat{m} \|$ [001] (growth axis). (c) Calculation of the corresponding TAMR with the same parameters for (Ga,Mn)As than the one used for Figure 21.6. The injection in the negative bias regime implies an hole injection towards the lowest HH band (triangle) thus giving rise to a negative TAMR.

TAMR on crossing the third subband that is when the LH subbands become dominant in the tunneling transport (star in Figure 21.8c). The first subband clearly gives a negative contribution to TAMR as well as the fourth one that has the effect to reverse the overall TAMR sign at higher energy (triangle in Figure 21.8c). This originates from the predominant HH character of these subbands, an in-plane magnetization allowing, through off diagonal components, a possible HH–LH conversion, and then a larger transmission through the barrier [44]. This argument is reversed for the second and third subbands with the results that TAMR becomes positive when the $n=2$ and 3 subbands are dominant in the tunneling transport. This was theoretically established through tight-binding treatment of TAMR [108]. Reducing the hole concentration through hydrogenation technics should give the possibility to probe this possible crossover from positive to negative TAMR [109]. Coming back to experiments on (Ga,Mn)As/(In,Ga)As/(Ga,Mn)As junctions, taking into account conjugate TMR and TAMR values obtained before and after annealing, one can roughly evaluate the projection of the corresponding signals trajectories in the [ε_F, B_G] plane followed during annealing (Figure 21.5b). A good qualitative agreement can be found even though symmetrical junctions were simulated in order to restrict the number of parameters.

21.2.3 Resonant TAMR in Double Tunnel Junctions

In order to go a little bit further in the understanding of TAMR phenomena with holes, we will now discuss resonant-TAMR experiments observed on AlAs/GaAs/AlAs resonant tunneling diodes (RTD) spin-filtered by a (Ga,Mn)As electrode [44]. In this work, large TAMR oscillations correlated to the *on/out-of* resonance of QW's quantized states have been observed, up to 60K (inset of Figure 21.9a) corresponding to the (Ga,Mn)As Curie temperature (T_C), on RTD consisting of 50nm ($Ga_{0.94}$,$Mn_{0.06}$)As/ with a 5nm undoped GaAs spacer/5nm AlAs barrier/6nm GaAs QW/5nm AlAs barrier/5nm undoped GaAs spacer/50nm Be-doped ($p=1\times10^{17}$ cm^{-3}) GaAs/50nm Be-doped ($p=1\times10^{17}$ cm$^{-3}\rightarrow 2\times10^{19}$ cm^{-3}) GaAs/p^+ GaAs substrates, from top to bottom. The whole structure was grown at an optimal temperature of 600°C except the top (Ga,Mn)As layer synthesized at 250°C to avoid the formation of MnAs precipitates. Figure 21.9a displays resonant tunnel conductance ($dI/dV-V$ curves in logarithmic scale) and TAMR versus bias acquired on the 6nm QW 16μm² mesa under a saturation field of 6kOe. Positive bias still corresponds to holes injected from the (Ga,Mn)As layer into GaAs:Be. Note that (1) the differential conductivity is broadened near the resonances due to the large hole density within the (Ga,Mn)As injector and that (2) the two first HH1 and LH1 resonances are hardly visible. Comparing with previous results of H. Ohno et al. [110], a series of negative differential conductance peaks is clearly evidenced with successive contributions of heavy (HH*n*) and light hole (LH*n*) states. TAMR data have been obtained by plotting the relative difference of the tunneling current, j_T, measured respectively for out-of plane ($j_T^{[001]}$), and in-plane (Ga,Mn)As magnetization ($j_T^{[100]}$) according to TAMR $=[j_T^{[001]}-j_T^{[100]}]/j_T^{[100]}$ ($\hat{z}=[001]$ is the growth direction).

For positive bias $V>0$, we observe interesting positive TAMR oscillations characterized by successive peaks and dips whose positions are strongly correlated to the successive resonances.

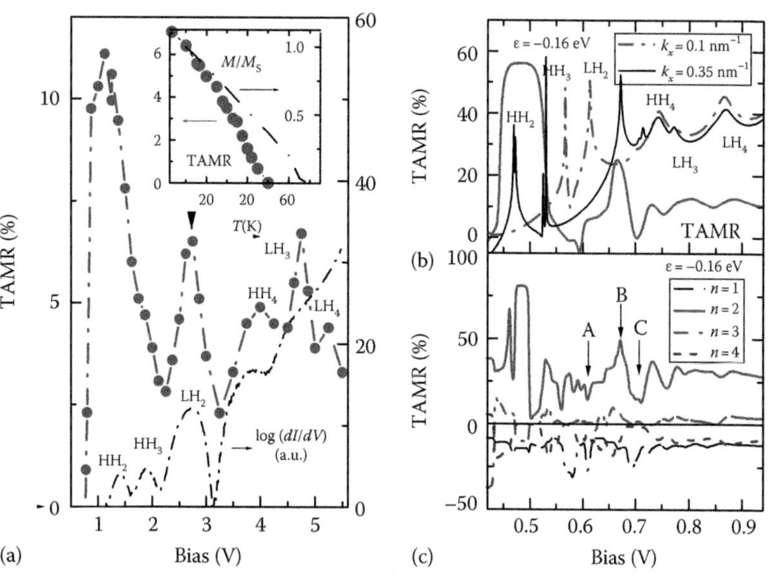

FIGURE 21.9 (a) TAMR-V signal (dash-dot red line) and $dI/dV-V$ (dashed line) acquired on the 6nm GaAs QW. The left axis displays the measured TAMR; the right one takes into account the up-renormalization (see text). Inset of (a) TAMR-T acquired at 2.75V (LH2 peak) and M-T. (b) TAMR-V calculated (straight red line) and transmission coefficient (unit not shown) calculated at an average energy ε_h −160meV and for respective small $k_{\|}$(dotted blue line) and intermediate $k_{\|}$ (dashed black line). (c) Energy and band-resolved TAMR-V calculated for $\varepsilon_h=-160$meV. (From Elsen, M. et al., *Phys. Rev. Lett.*, 99, 127203, 2007. With permission.)

However, the resonance appears at bias much larger than twice the quantized energy of the discrete levels in the QWs. This discrepancy originates from the introduction of a resistance R_{DZ} in series with the double barrier structure in the GaAs:Be electrode because of a gradient in the doping [44]. While the measured TAMR amplitude appears rather moderate (10% or less), the overall *intrinsic* TAMR signal is much larger, by a factor of ≈5, (right scale of Figure 21.9a) when one takes into account the presence of such nonmagnetic serial resistance R_{DZ}. Inset of Figure 21.9a displays the temperature dependence of the TAMR peak acquired at 2.75 V, corresponding to resonant tunneling into the LH2 state. The signal amounts to 7% at 4 K and decreases gradually with temperature to vanish at about 50 K close to the (Ga,Mn)As Curie temperature (M-T curve). This demonstrates that the oscillatory behavior of TAMR is induced by the exchange interaction within (Ga,Mn)As layer. One can thus discard (1) any magneto-tunneling effects deriving from the quantization of Landau levels on QW states as well as (2) deflection of the hole parallel wavevector by Lorentz force in the plane of the junction [111]. Moreover both mechanisms would hardly appear at such intermediate field (6 kOe). Our conclusion is that such an oscillatory TAMR behavior reflects the spin-dependent transmission coefficients of the polarized holes injected from (Ga,Mn)As, thus emphasizing the anisotropy of the (Ga,Mn)As valence band. However, this can include an extension of mechanism (2), that is, a deviation of the hole wavevector along the in-plane $\hat{y}$= [010] direction normal to the in-plane $\hat{x}$= [100] magnetization driven by spin–orbit interactions. We will come back to this very important point at the end of this section.

Based on the aforementioned arguments on differential transmission of HH and LH components, one can suggest that the oscillation character of TAMR on/out-of resonance may result from the fourth HH (Ga,Mn)As subband contribution. To this end, we have plotted in Figure 21.9b the transmission amplitude of holes of average energy $\varepsilon_h = -100$ meV (−160 meV from the top of the valence band) corresponding to $p = 2 \times 10^{20}$ cm^{-3} for two different parallel wavevectors $k_{\parallel}$, respectively, small $k_x = 0.1$ nm^{-1} (dot curve) and intermediate $k_x = 0.35$ nm^{-1} (straight curve) across the double barrier structure. $\varepsilon_h = -160$ meV is then assumed to be the elastic energy transmitted during the tunneling process. The resulting TAMR shape, displayed in Figure 21.9c, is in excellent agreement with the experimental data exhibiting the successive peaks and dips correlated with the successive positions of resonances. In addition, the magnitude of TAMR is also well reproduced after renormalization by a factor of 5 (right scale of Figure 21.9a). We observe that most of the TAMR peaks correspond to a specific resonance at intermediate $k_{\parallel}$(what we call *out-of*-resonance) whereas the TAMR dips match generally with a specific resonance at small $k_{\parallel}$(what we call *on*-resonance). We then considered the respective contributions of the different hole subbands to the TAMR. The four band-resolved TAMR signals extracted for $\varepsilon_h = -160$ meV (still normalized by the *total* tunneling current for an in-plane magnetization) are plotted in Figure 21.9c. The band indexes $n = 1,4$ *resp.* $n = 2,3$ stand for the HH *resp.* LH symmetries near the Γ point. Remarkably, the main contribution to the TAMR oscillations derives from the first LH subband ($n = 2$) exhibiting the same trends as the summed signal (Figure 21.9b) as well as the experimental data (Figure 21.9a). Let us focus on the specific LH2 resonance (Figure 21.9c) where we hereafter designate V as the effective bias applied to the double barrier structure. The second subband contribution to TAMR is small, about 10%, *on*-resonance ($V = 0.59$ eV: point A in Figure 21.10c increases up to about 60% *out-of*-resonance for intermediate $k_{\parallel}$($V = 0.68$ eV: point B), before vanishing at even larger bias and larger $k_{\parallel}$($V = 0.71$ eV: point C).

The following demonstration is based on the calculations of the band-selected tunnel transmissions mapped onto the different (Ga,Mn)As Fermi surfaces like sketched in Figures 21.10 and 21.11. These figures display the different band-selected transmission for respective in-plane (Figure 21.11: $\hat{m} \parallel \hat{x}$) and out-of-plane magnetization (Figure 21.12: $\hat{m} \parallel \hat{z}$). The transmission coefficients $\sum_{n'} \mathrm{T}_{n'}^{k_{\parallel},\hat{m}}(\varepsilon_h)$ mapped on the two-dimensional $k_{\parallel}$Fermi surfaces for the two magnetization configurations are plotted in Figures 21.11a–l and 21.12 for the four subbands. In both figures, the indices (a–d) correspond to the mapping of transmission calculated for a bias $V = 0.59$ eV, indices (e–h) for a bias $V = 0.68$ eV, and indices (i–l) for a bias $V = 0.71$ eV.

1. *On LH2 resonance* (*V* = 0.59 *eV*): *point* (*A*): In Figures 21.10 and 21.11a through d, the bright spot with a relative small radius at the center indicates that the tunnel transmission is mainly driven by small $k_{\parallel}$. The band-resolved TAMRs obtained by $k_{\parallel}$-integration approach +125% (*resp.* −38%) for $n = 2$ (*resp.* $n = 4$), whereas it gives −38% (resp. +3%) for $n = 1$, (*resp.* $n = 3$). Correspondingly, the weighted tunneling currents calculated for an in-plane magnetization are 18% and 22%, respectively, for $n = 2$ and 4 (30% for both $n = 1$ and 3). The overall TAMR is less than 10%, since *the negative HH contribution* (*n* = 1,4) *is partially compensated by the positive LH one* (*n* = 2).
2. *Out-of*-resonance *at* *V* = 0.68 *eV*: *point* (*B*): The tunneling transmission in the reciprocal space is now mainly confined within a ring of an intermediate radius from the zone axis, for example, driven by intermediate $k_{\parallel} \approx 0.35$ nm^{-1} (Figures 21.10 and 21.11e through h). Due to the finite extension of the minority spin HH Fermi surface ($n = 4$), the resonance condition can be only partially achieved for this particular band: the tunnel conductance falls off. Each weighted tunnel conductivity is then 33%, 30%, 33%, and 4% for $n = 1$–4 whereas the corresponding TAMRs equal now −30%, +180%, 20%, and −90%. The consequence is an overall increase of the TAMR up to 60%, due to the deficit of negative TAMR from the minority spin HH subband.
3. *At higher bias* *V* = 0.71 *eV*: *point* (*C*): The tunnel transmission is driven by holes of even larger $k_{\parallel}$(≈0.5 nm^{-1}) away from the zone axis $\hat{z}$ (Figures 21.10 and 21.11i through l). It results in a strong LH↔HH mixing along $\hat{z}$, which, in turn, strongly reduces selected TAMR down to −22%, +40%, 9%, and −55% for $n = 1$–4 (Figure 21.7c). The

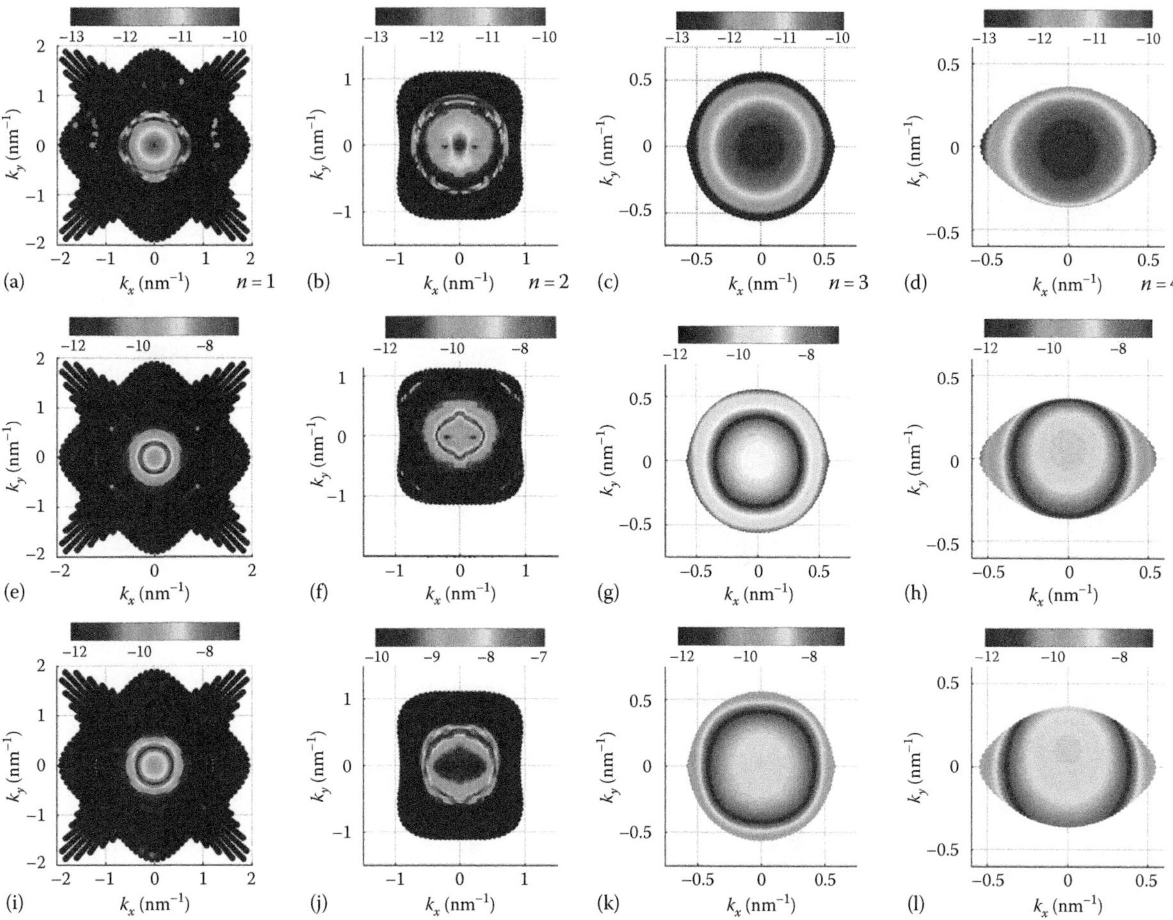

FIGURE 21.10 (See color insert.) Band-selected ($n = 1-4$) tunnel transmissions (in color logarithmic scale) mapped onto the different (Ga,Mn)As Fermi surfaces calculated for in-plane magnetization ($\hat{m} \| \hat{x}$). The different columns correspond to a specific selected band. Calculations have been performed for a bias applied to the double barrier structure equal to $V = 0.59$ eV (a through d: first line), $V = 0.68$ eV (e through h: second line), and $V = 0.71$ eV (i through l: third line) around the LH2 resonance.

weighted tunnel conductivities are 60%, 30%, 7%, and 3%, respectively. As a result, the overall TAMR is low due to the strong reduction of the second subband LH contribution (from +180% to +40%), as shown in Figures 21.10j and 21.11j. The conclusion is that the oscillatory behavior of TAMR originates from the two merging mechanisms established at the beginning of the section and based on the HH and LH characters of each subband. One can then argue that acquiring a large $k_{\|}$ *out-of*-resonance naturally mixes the LH and HH characters, resulting in a significant drop of TAMR (*from 180% to 40%*). Conversely, the argument (1) yields a large negative TAMR for subbands of HH symmetry at intermediate or small $k_{\|}$ before decreasing at larger $k_{\|}$. The ensemble of band-selected *intrinsic* TAMR and their corresponding contribution to the resonant tunneling current are reported in Tables 21.2 and 21.3 for respective *on*- and *off*-resonant transmission on LH2 peak Table 21.1.

Furthermore, the high resolution of resonance peaks obtained at negative bias $V < 0$ (when holes are injected from GaAs:Be into (Ga,Mn)As, not shown) allows us to investigate other superimposed effects related to new TAMR phenomena. Very recently, a significant resonant-TAMR signal has been highlighted in asymmetric (Ga,Mn)As-based double barrier structure at negative bias [112]. Such experiments exhibiting a magnetization-dependent shift in bias of the consecutive HHn and LHn resonances seem to originate from a significant modulation of the Fermi level of (Ga,Mn)As imposed by the anisotropic strain-field. This could be closely related to Coulomb-blockade magnetoresistive (CB-TAMR) phenomena [113] in the in-plane geometry although, in the latter case, a full understanding of CB-TAMR is still missing.

21.2.4 Does Tunneling (Spin) Hall Effect Exist?

To conclude our discussion about TAMR phenomena, we note that Figure 21.10a through l displays in general a clear non-cubic asymmetry of the tunneling coefficients for in-plane magnetization ($\hat{m} \| \hat{x}$). This results from the action

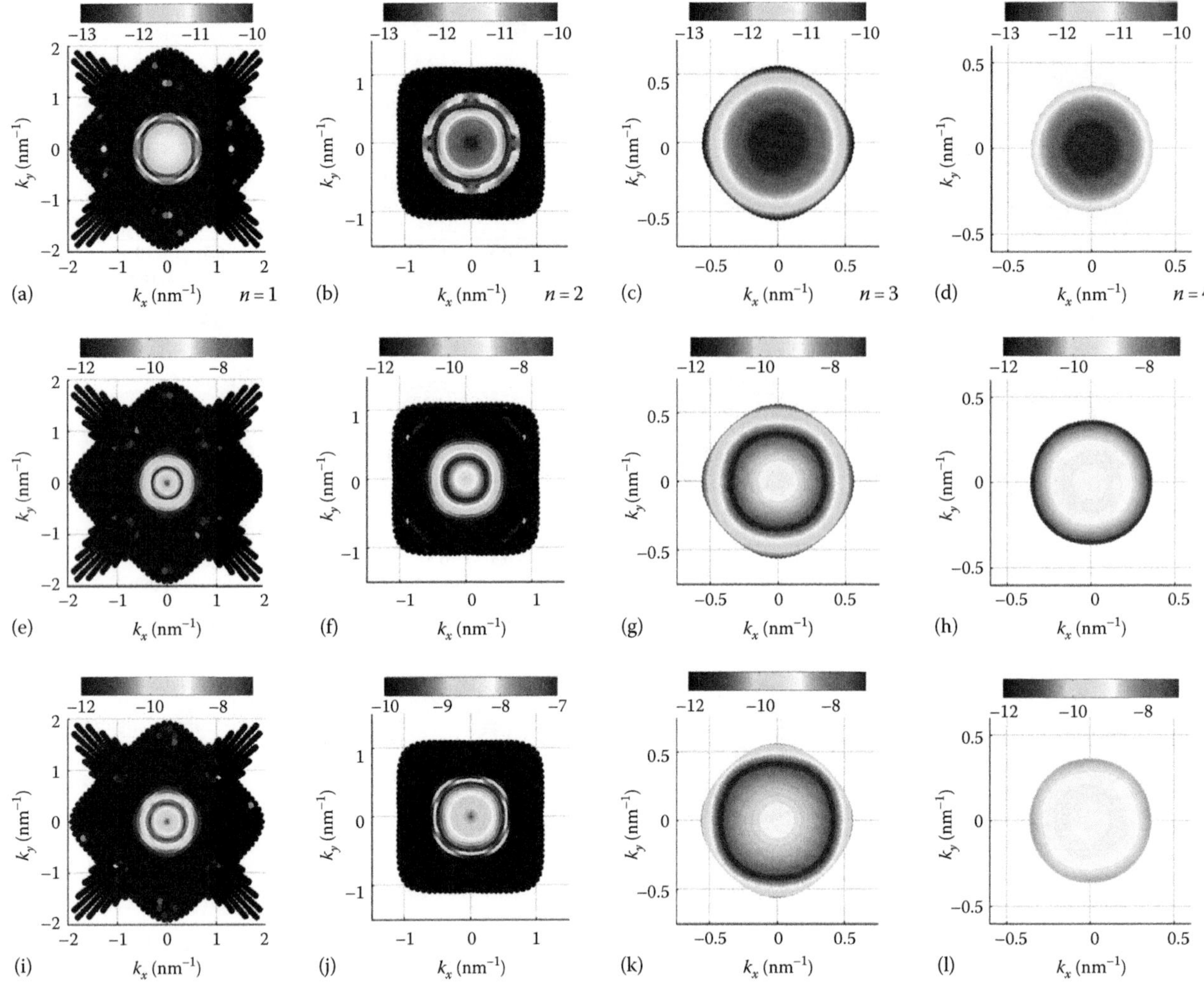

FIGURE 21.11 (See color insert.) Band-selected ($n = 1-4$) tunnel transmissions (in color logarithmic scale) mapped onto the different (Ga,Mn)As Fermi surfaces calculated for out-of-plane magnetization ($\hat{m}\|\hat{x}$). The different columns correspond to a specific selected band. Calculations have been performed for a bias applied to the double barrier structure equal to $V = 0.59$ eV (a through d: first line), $V = 0.68$ eV (e through h: second line), and $V = 0.71$ eV (i through l: third line) around the LH2 resonance.

of the spin–orbit coupling on hole motion during their tunneling transfer. This can be viewed as a generalization of the Lorentz force to magnetic systems by which a deviation of the hole wavevector along the in-plane $\hat{y} = [010]$ direction normal to the $\hat{x} = [100]$ magnetization vector is expected [38]. From the general expression of the spin–orbit coupling $\mathcal{H}_{so} = -\Delta_{so}\left(\vec{\nabla}V(\vec{r}) \times \vec{p}\right)\cdot\vec{S}$, it results that a potential discontinuity due to band offset ($\vec{\nabla}V(\vec{r}) \propto \vec{z}$) introduces a supplementary energy term $\mathcal{H}_{so} \propto -\Delta_{so}(\vec{z} \times \vec{p})\cdot\vec{S} = \Delta_{so}(\vec{S} \times \vec{z})\cdot\vec{p}$ breaking the in-plane symmetry as soon as $\vec{S}$ admits a non-vanishing in-plane component. This term is at the origin of the transmission asymmetry between $\pm k_y$ in-plane wavevectors. In the envelope function approximation, this effect is implicitly included in the boundary conditions for the wave functions and their derivate (current operator). Such an asymmetry in the transmission coefficient is at the origin of what should be termed as *tunneling (Spin) Hall effect* and whose experimental demonstration still remains a challenging task.

21.3 Spin-Transfer in the Valence Band of *p*-Type Semiconductors

21.3.1 Short Overview of Spin-Transfer Phenomena

The magnetic moment of a ferromagnetic body can be reversed or reoriented by transfer of the spin angular momentum carried by a spin-polarized current. This concept of spin transfer has been introduced by Slonczewski [47] and Berger [48], before it was confirmed by extensive experiments on pillar-shaped magnetic trilayers (for a review, the reader can refer to Stiles and J. Miltat [50] and Chapters 7 and 8. Most current-induced magnetization switching (CIMS) experiments have been performed on purely metallic trilayers [114–119], such as Co/Cu/Co, in lithographically patterned nanopillars with detection of the magnetic switching by giant magnetoresistance effects. Typical critical current density j_c required for magnetization

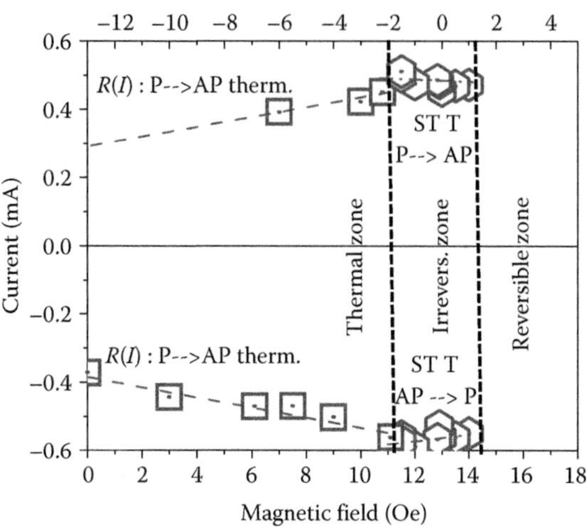

FIGURE 21.12 Phase diagram displaying the different regime of magnetization reversal versus applied field that are thermal switching, irreversible switching (CIMS), and reversible switching. The top abscise corresponds to the effective field equal to applied field added to the stray field equaling −13 Oe.

TABLE 21.1 Luttinger Parameters γ, Spin–Orbit Splitting Δ_{so} (eV), and Valence Band Offset *VBO* (eV) Used

	γ_1	γ_2	γ_3	Δ_0	*VBO*
GaAs	6.85	2.1	2.9	0.34	0
AlAs	3.25	0.64	1.21	0.275	−0.64
InAs	20.4	8.3	9.1	0.38	
$In_{0.25}Ga_{0.75}As$	10.24	8.3	4.45	0.35	−0.7 (Low-T)
(Ga,Mn)As	6.85	2.1	2.9	0.34	+0.22

TABLE 21.2 Band-Restricted TAMR and Contribution to the LH2 *On*-Resonance Tunneling Current (V = 0.59 eV)

	n = 1 (HH)	n = 2 (LH)	n = 3 (LH)	n = 4 (HH)
TAMR	−38%	+125%	+3%	−38%
Current	30%	18%	30%	22%

The kinetic energy of holes in (Ga,Mn)As is equal to 160 meV. (HH) and (LH) indicate the Heavy or Light character for holes near the Γ_8 point.

TABLE 21.3 Band-Restricted TAMR and Contribution to the LH2 *Off*-Resonance Tunneling Current (V = 0.68 eV)

	n = 1 (HH)	n = 2 (LH)	n = 3 (LH)	n = 4 (HH)
TAMR	−30%	+180%	+20%	−90%
Current	33%	30%	33%	4%

The kinetic energy of holes in (Ga,Mn)As is equal to 160 meV. (HH) and (LH) indicate the Heavy or Light character for holes near the Γ_8 point.

reversal in these systems has been on the order of 10^7 A cm^{-2} or higher. Since then, there have been several reports on CIMS on transition metal MTJs with low junction resistance at critical current density below 10^6 A cm^{-2} [120–122]. CIMS experiments on tunnel junctions bring new physical problems [123], and are also of particular interest for their promising application to the switching of the MTJ of MRAM (see Chapters 35 and 36). For a recent review of spin transfer in MgO-based MTJs, the reader can refer to the paper of Katine and Fullerton [124] and references therein. (Ga,Mn)As is known to have small magnetization of less than 0.05 T and a high spin polarization of holes, which must result in the reduction of critical current according to the Slonczewski's spin-transfer torque model. It results in a low current density, on the order of 10^5 A cm^{-2}, needed for CIMS with (Ga,Mn)As [51,52] and shows the interest in magnetic semiconductors for spin transfer. From a fundamental point of view, as emphasized by Chiba et al. [125], the specific valence band structure and spin–orbit interactions have to be taken into account and should mix the spin states of carriers. In the following, we present results of CIMS experiments obtained on our own (Ga,Mn)As MTJs whose structure and TMR properties are described in Section 21.1, comparable to the one obtained in Ohno's group [51]. We also discuss open questions related to the valence band electronic structure compared to the basis applicable to metal CPP-GMR structures.

21.3.2 Spin Transfer with Holes in (Ga,Mn)As Tunnel Devices

Concerning the fabrication of our nanoscale MTJs, a circular resist mask (height = 200 nm, diameter = 700 nm) is first defined by e-beam lithography and oxygen plasma etching. A pillar is then etched down to the conducting GaAs buffer layer by ion beam etching, and an Si_3N_4 layer is sputtered to cover the bottom electrode. The next step is the planarization of the surface by spin-coating. The nitride layer on top of the pillar is then removed by ion etching. Finally, the top of the pillar is cleaned with an oxygen plasma and top electrodes are fabricated by a lift-off process. For CIMS experiments, the application of a significant bias of about 800 mV in our devices is needed to observe spin-transfer phenomena so that this switching cannot be detected directly by a clear change of resistance. Instead, we used the following procedure. Starting, for example, from a parallel (P) configuration at V = 0, the voltage is increased step by step, and, after each step, brought back to 20 mV to compare with that found at 20 mV before the step. We can check in this way whether the magnetic configuration has been irreversibly switched. Only irreversible switchings can be detected and the reversible changes of the "steady precession regime" [50,126,127] cannot be detected. Examples of results obtained with this procedure are shown in Figure 21.3c. Comparing the MR curve at 20 mV of Figure 21.3a, one sees that the magnetic configuration is switched irreversibly from an almost parallel (P) to an almost antiparallel (AP) configuration by a positive current density (current flowing from the thin magnetic layer to the thick one), $j_{c+} = 1.23 \times 10^5$ A cm^{-2} (V_{c+} = 810 mV) at 3 K and $j_{c+} = 0.939 \times 10^5$ A cm^{-2} (V_{c+} = 680 mV) at 30 K. The configuration is then switched back to parallel by a negative current above a threshold current density of the same order ($j_{c-} = -1.37 \times 10^5$ A cm^{-2} at 3 K and $j_{c-} = -0.986 \times 10^5$ A cm^{-2} at 30 K). Opposite current directions for the P to AP and AP to P transitions are the characteristic behavior of switching by spin transfer [50,114–122]. Reversing the initial orientation of the magnetizations does not reverse the sign of the switching current, which confirms that Oersted field effects can be ruled out. By more complex cycling procedures, we have checked that, as in standard CIMS experiments, the transition is due to the magnetic switching of the thin layer. The fact that the bottom thin layer is the free layer is also consistent with the spin-transfer torque model as the total magnetic moment of the bottom layer is less than that of the thicker top layer in (Ga,Mn)As as already reported. Further experiments [52] have demonstrated that CIMS on (Ga,Mn)As-based MTJs could be obtained only in a small field window of 2–3 Oe around 13 Oe at 3 K, which may be explained by the combined effects of the Joule heating and temperature dependence of the magnetic properties. Indeed, from superconducting quantum interference device (SQUID) and TMR measurements, we know that the anisotropy and coercive field of the thin layer drop to a couple of Oe above 30 K, or, equivalently, by heating the sample with a bias voltage value of about 700 mV. This study has established a particular phase diagram (Figure 21.12) for CIMS versus the amplitude of the external magnetic field applied and has shown that the heating of the sample and the manifestation of a reversible switching regime [50,115,117,119] (which cannot be detected by the experimental protocol used) can prevent any switching of (Ga,Mn)As out of a certain narrow field window.

21.3.3 Phase Diagram for CIMS with (Ga,Mn)As

A significant difference with respect to the behavior with magnetic metals is that CIMS curves, similar to those of Figure 21.3c, can be obtained only in a small field window of 2–3 Oe around 13 Oe at 3 K. From an analysis of the shifts of minor MR cycles, we found that the dipolar field generated by the thick (Ga,Mn)As layer and acting on the thin layer is close to 13 Oe at 3 K so that $H_{app} = 13$ Oe corresponds to approximately an effective field $H_{eff} = 0$ Oe on the thin layer when the moment of the thick one is in the positive direction (case of Figure 21.3c). The behavior of Figure 21.3c in a field range of a couple of Oe around $H_{eff} = 0$ Oe can be explained by the combined effects of the Joule heating and temperature dependence of the magnetic properties. From SQUID and TMR measurements, we know that the anisotropy and coercive field of the thin layer drop to a couple of Oe above 30 K or, equivalently, by heating the sample with a bias voltage value of about 700 mV [30]. Suppose, for example, a P state with both magnetizations in the direction of positive H with $H_{eff} = 9$ Oe or, equivalently, $H_{eff} \cong -4$ Oe. If, at a current I^* smaller than the critical current for CIMS, the heating of the sample reduces the magnitude of the coercive field of the thin layer to 4 Oe or below, the effective field of −4 Oe switches the configuration to AP when the current reaches I^* or $-I^*$. This behavior, with switching to AP by positive as well as by negative currents, is what we observe for H_{app} below the window centered on $H_{app} = 13$ Oe. Suppose now the same P state with $H_{app} = 17$ Oe or, equivalently $H_{eff} = +4$ Oe. If the heating of the sample reduces the anisotropy field below 4 Oe, we enter the regime of reversible switching [50,115,117,119] that we cannot detect. Consistently, we cannot detect any switching for H_{app} above the window. We conclude that, with respect to CIMS with magnetic metals, the dramatic variation of the properties of (Ga,Mn)As with temperature, combined with the heating of the sample, introduces significant complications in CIMS experiments with (Ga,Mn)As and limits the observation of CIMS by spin transfer to a narrow field range. These arguments give rise to the phase diagram shown in Figure 21.12 and highlight the three different regions that are thermal switching, irreversible switching (CIMS), and reversible switching. We conclude that, with respect to CIMS with magnetic metals, the dramatic variation of the properties of (Ga,Mn)As with temperature, combined with the heating of the sample, introduces significant complications in CIMS experiments with (Ga,Mn)As and limits the observation of CIMS by spin transfer to a narrow field range.

21.3.4 Discussion

The order of magnitude of the switching current densities for MTJs (and 10^5 A cm^{-2}) is consistent with the model of Slonczewski [47] when one takes into account that the magnitude of the (Ga,Mn)As magnetization is significantly reduced as compared to transition metals. This argument was already put forward by Chiba et al. [51,125]. In the simplest approximation, the switching currents at zero (or low) field are expected to be given by [116,117]: $I^{P(AP)} = G^{P(AP)}\alpha tM[2\pi M + H_K]$ where $G^{P(AP)}$ is a coefficient depending on the current spin polarization, α is the Gilbert damping coefficient, t is the layer thickness, and H_K is the anisotropy field. The main difference with transition metals originates from the factor $tM[2\pi M + H_K]$. Comparing (Ga,Mn)As-based junctions with $t = 15$ nm, $M = 0.035$ T, and a negligible H_K after heating, and a typical Co/Cu/Co pillar [114–116], with $t = 2.5$ nm, $M = 1.78$ T, $H_K = 0.02$ T, we find that the factor $tM[2\pi M + H_K]$ is larger by a factor of 500 for the metallic structure. This mainly explains the difference by two orders of magnitude. Additional factors should come from the current spin polarization (probably higher with (Ga,Mn)As) and the Gilbert coefficient (probably larger by almost an order of magnitude for (Ga,Mn)As grown at low temperature) [128], but a quantitative prediction is still a challenging task. What about the sign of the switching currents in (Ga,Mn)As devices, which is the same as for standard metallic pillars [50,114,116,117]? This is consistent with what we expect from the valence bands structure of (Ga,Mn)As, whereby the majority of electron spins (and no hole p state) at the Fermi level are aligned parallel to the local Mn spins (d state) due to the negative p–d exchange interaction parameterized by the exchange integral β [24,53]. Complementary experiments [53,54] performed on double (Ga,Mn)As-based MTJs in ms-pulsed current injection regime have demonstrated the possibility to reduce the critical current densities to less than 3×10^4 A cm^{-2} that is by more than a factor of 3 as compared to single (Ga,Mn)As MTJs. This is due, in part, to the interfacial torque applied at both side of the middle (Ga,Mn)As sensitive layer and imposed by the symmetric structure.

A still pending question lies in the observation of switching phenomena by spin transfer at bias (≈800 mV) corresponding to a quasi vanishing TMR and then to a loss of the spin polarization of holes with bias. Although in standard MTJs with metallic electrodes, the decrease of the MR with the bias is generally ascribed to electron-magnon scattering (emission and annihilation of magnons) [129] and to other types of spin-dependent inelastic scatterings [130], calculations performed by Sankowski et al. [108] put forward a natural decrease of TMR with bias due to the band structure of (Ga,Mn)As. In the former case, even if the spin current is strongly affected by spin-flip effects, the transverse component of the spin-polarized tunneling current cannot be lost and should be finally transferred to the magnetic moment of the layer as proposed by Levy and Fert [131]. On the other hand, concerning the band structure itself, recent theoretical investigations on torques and TMR in MTJs performed in the frame of Keldysh formalism have demonstrated the possibility to reconcile an increase of spin-transfer efficiency and a decrease of TMR in the same device [132–134]. The same kind of calculation remains to investigate in a multiband formalism adapted to valence band of semiconducting tunnel junctions and in particular the role of respective Slonczewski's and field's contributions of the torque in (Ga,Mn)As-based MTJs. It also remains to study theoretically the efficiency of the current-induced torque exerted by general angular momenta (J) and no more spin (S) including the role of spin–orbit interactions as discussed in the next section.

21.4 Summary

To conclude, we have reviewed different properties of spin-polarized tunneling transport involving the (Ga,Mn)As diluted magnetic semiconductor (DMS). It has been shown that a relatively good agreement can be found between experimental TMR and TAMR phenomena and Dietl's theory for DMS in a molecular field approach for (Ga,Mn)As ferromagnetism. The general description of spin-polarized tunnel transport including (Ga,Mn)As is fully described using standard $k \cdot p$ theory and Laudauer–Buttiker formalism. In particular, it has been possible to understand the large modification of TMR properties as a function of annealing procedure. Indeed, it is now commonly accepted that an optimized annealing procedure, by the reduction of As interstitial at the free interface, leads in parallel to an increase of the carrier density and an increase of the spin-exchange interaction. Further experiments exploring a wider region of the phase diagram established theoretically are required to refine our understanding of the (Ga,Mn)As band structure. A more advanced technological control of the material will be needed to extract precisely (Ga,Mn)As characteristic parameters. More generally, our $k \cdot p$ framework seems to catch the essential trends of the spin-dependent tunneling phenomena involving holes emitted from (Ga,Mn)As. As an example, the sizeable TAMR effects originating from the anisotropy of the electronic band structure have been accurately reproduced. Exploring the TAMR signal on a resonant QW has revealed the intricate competition of the different (Ga,Mn)As subbands. It implies that a respective positive and negative contribution can be expected from LH and HH band. More generally, these different contributions emphasize the band curvature of (Ga,Mn)As with HH and LH characters near the Γ_8 point assuming that the LH character is favorably transmitted by the tunnel barrier.

In the last part, we have demonstrated current-induced magnetization reversal phenomena in nanoscale tunnel junction integrating (Ga,Mn)As. From the particular magnetic properties of (Ga,Mn)As, a smaller current density compared to transition metal is required to reverse the magnetization. In the future, a deep understanding of spin-transfer phenomena involving holes will imply the development of more sophisticated model like the one developed in the standard $k \cdot p$ approach. In parallel, time-resolved (or frequency-resolved) experiments performed in the reversible regime of spin transfer will be certainly helpful for a complete description of the inner mechanisms for spin transfer. Moreover, we believe that the specificity of carrier-mediated ferromagnetism of (Ga,Mn)As will give new opportunities to explore spin-dependent phenomena taking advantage of semiconductor functionalities. Despite the difficulties in path toward the room temperature operations for this class of materials, further spin-related studies with ferromagnetic semiconducting heterojunctions seem necessary to pursue fundamental physics as well as to establish promising concepts in other systems.

Appendix A: Exchange Interactions in (Ga,Mn)As—From the Mn-Isolated Atom to the Metallic Phase

From general group-theory arguments, in the effective mass approximation [71–73], nonmagnetic shallow acceptors can be described by hydrogenic states of fundamental symmetry term $1S_{3/2}$ of binding energy equal to 28 meV for GaAs. These are characterized by a total angular momentum $F = L + J = 3/2$, which is a constant of motion where L is the angular momentum of the envelope wave function. The result is that the fundamental $1S_{3/2}$ wave function is written as $\Phi(S_{3/2}) = f_0(r)|L = 0, J = 3/2, F = 3/2, F_z\rangle + g_0(r)|L = 2, J = 3/2, F = 3/2, F_z\rangle$. According to optical studies, Mn is known to form a shallow acceptor center in GaAs of $A^0 = d^5 + h$ electronic configuration characterized by a binding energy [135] of 110 meV, and an energy difference on the order of 10 (±3) meV between the $J = 1$ and $J = 2$ quantized values of the $d^5 + h$ state [135]. In this picture, the $J = S + j$ quantum number, constant of motion, is the sum of the d^5 Mn spin angular momentum $S = 5/2$ and the $j = 3/2$ hole angular momentum [136,137]. In the $S - j$ exchange coupling scheme where the exchange interaction is $J_{exc}\ S \cdot j$, the energy difference between the extrema $J = 1$ and 4 states is equal to $9J_{exc}$ whereas it gives $2J_{exc}$ between the two successive $J = 1$ and 2 states. It follows that the energy difference between the states corresponding to the spin of the bound hole parallel and antiparallel to the Mn spin can be estimated to be 45 meV. This relative small p–d exchange energy originates from the relative long extent of the bound hole wave function where the Mn d states are mostly localized within an effective Bohr radius $a_0^* \propto \varepsilon/m^* \simeq 0.8$ nm and corresponding to an effective volume of 3 nm^3 as well as an effective Mn concentration approaching $x_{loc} = 1.35\%$. A direct consequence is that the average exchange integral Δ_{exc} is expected to be enhanced with increasing the Mn content x above this threshold Mn concentration $x_{loc} = 1.35\%$.

In the metallic regime and in the **S–s** exchange coupling scheme, the average exchange interaction in (Ga,Mn)As reads $\Delta_{exc} = -5/2xN_0\beta$, N_0 is the concentration of cations and $N_0\beta = -(16/S)\left[1/(-\Delta_{eff} + U_{eff}) + 1/\Delta_{eff}\right] \times \left[1/3(pd\sigma) - 2\sqrt{3}/9\,(pd\pi)\right]^2 < 0$ is the exchange integral found by treating the p–d hybridization as a perturbation in the configuration interaction picture [138] giving rise to antiferromagnetic interactions between p and d shells. Here, S is the localized d spin, $U_{eff} = E(d^{n-1}) + E(d^{n+1}) - 2E(d^n)$ is the characteristic $3d$–$3d$ Coulomb interaction, $\Delta_{eff} = E(Ld^n) - E(d^{n-1})$ is the ligand-to-$3d$ charge transfer energy. On the other hand, $(pd\sigma)$ and $(pd\pi)$ are characteristic Slater–Koster hopping integrals. The value of $N_0\beta = -1.2$ eV ($\beta = -54$ meV nm^3) is generally admitted from core level photoemission measurements for (Ga,Mn)As with a T_c close to 60 K [86] corresponding to $x_{eff} \simeq 4\%$. Figure 21.13 displays the four-different exchange-split (Ga,Mn)As subbands calculated for a hole density $p = 1.7 \times 10^{20}$ cm^{-3} and an average exchange energy of $\Delta_{exc} = 120$ meV.

From a point of view of material properties, questions remain on the general trends of TMR versus exchange interactions, $\Delta_{exc} = -5/2xN_0\beta$, where xN_0 is the concentration of Mn atoms as well as hole band filling within (Ga,Mn)As. Also, what are the possible effects of the low-temperature (L-T) growth procedure used for the synthesis of these MTJs on the band lineup, VBO, and related barrier heights of heterostructures integrating (Ga,Mn)As? Remarkably, as shown by nonlinear I–V characteristics observed in junctions, both (Ga,As) and (In,Ga)As materials play the role of a tunnel barrier for holes injected from/into (Ga,Mn)As [28–30,139]. The same qualitative feature has been demonstrated through optical measurement of the hole chemical potential in ferromagnetic (Ga,Mn)As/GaAs heterostructures by photoexcited resonant tunneling [140]. These results favor the band-edge discontinuity due to the smaller band gap of (Ga,Mn)As, as compared to GaAs [141], and

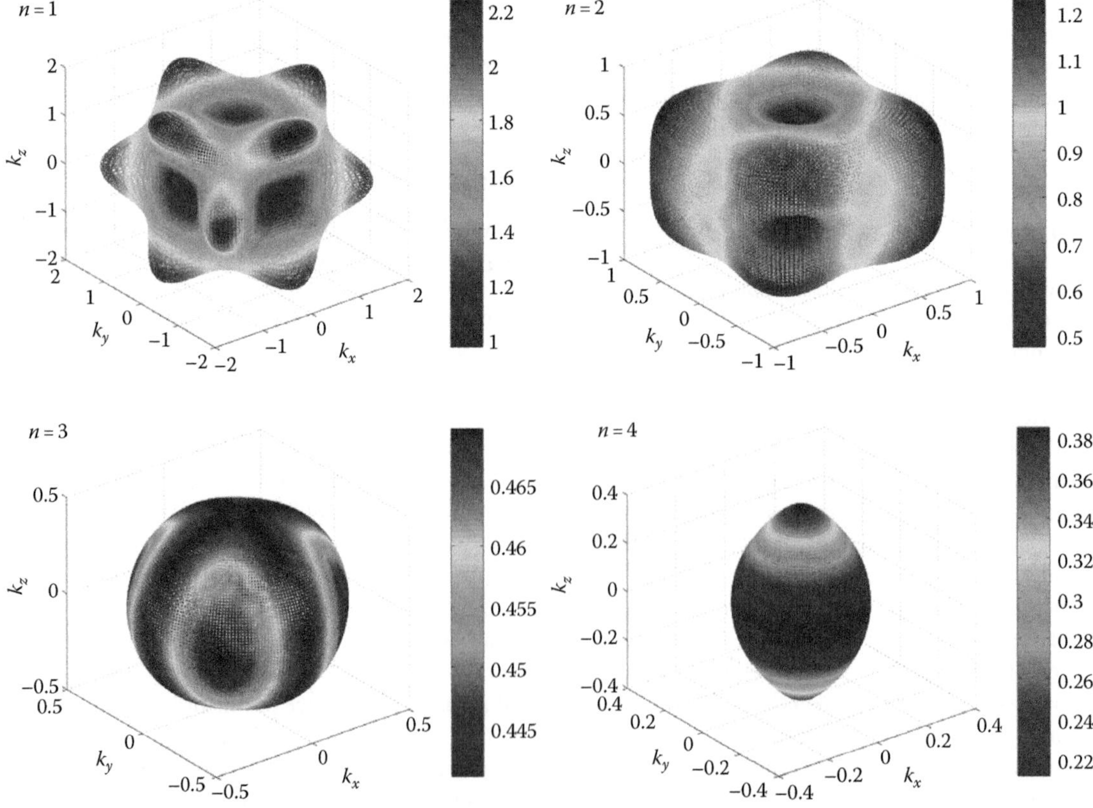

FIGURE 21.13 (See color insert.) The four different Fermi surfaces of (Ga,Mn)As calculated in the $k \cdot p$ formalism for a Fermi energy equal −135 meV counted from the top of the valence band ($p = 1 \times 10^{20}$ cm^{-3}) and an exchange interaction $\Delta_{exc} = 120$ meV (calculated without strain). The magnetization is along the [001] growth crystalline axis. The color code scales the Fermi wavevector (in nm^{-1}) along the corresponding crystalline axis. The corresponding band structure of (Ga,Mn)As in the reciprocal space can be found in Refs. [3,18]. (From Tanaka, M. and Higo, Y., *Phys. Rev. Lett.*, 87, 026602, 2001; Matsukura, F. et al., III-V ferromagnetic semiconductors, in *Handbook of Magnetic Materials*, Vol. 14, K.H.J. Buschow (Ed.), North Holland, Amsterdam, the Netherlands, pp. 1–87, 2002. With permission.)

indicate a pinning of the Fermi level deep inside the band gap of the (Ga,As) host. A part of the answer lies in the incorporation of n-type double-donor As antisites during the low-temperature growth procedure [20] that governs, in part, the pinning of the Fermi level at a higher energy position than expected near the midgap of GaAs. The second reason is due to the positive coulombic-exchange potential experienced by holes and introduced by Mn species playing the role of hydrogenic centers for holes orbiting around it [19]. This is at the origin of an impurity-band formation at smaller or intermediate doping level in the host bandgap [142], which is currently intensively debated regarding the position of the Fermi level relative to the impurity band [143]. While infrared measurements [144] as well as dichroism (MCD) [145] experimental and theoretical studies [146] seem to support the scenario of a detached impurity band, recent low-temperature conductivity measurements [147] validate the approach of a valence band picture for (Ga,Mn)As more compatible with a $k\cdot p$ treatment of its electronic properties.

Appendix B: Model for Spin-Polarized Transport in Standard $k\cdot p$ Theory

B.1 $k\cdot p$ Approach for Describing Metallic (Ga,Mn)As

This section gives some insights about the procedure to calculate transport properties of spin-polarized holes in the valence band of heterojunctions integration the (Ga,Mn)As ferromagnetic semiconductor in the frame of the $k\cdot p$ theory. This includes the contribution of the p–d exchange interaction in the molecular field approximation. For zinc-blende semiconductors, we take into account explicitly the four Γ_8 and the two Γ_7 valence subbands, for which we choose the basis functions in the form according to Dietl et al. [24]:

$$
\begin{aligned}
u_1 &= \left|J=\frac{3}{2}, j_z=\frac{3}{2}\right\rangle = \frac{1}{\sqrt{2}}(X+iY)\uparrow \\
u_2 &= \left|J=\frac{3}{2}, j_z=\frac{1}{2}\right\rangle = i\frac{1}{\sqrt{6}}\left[(X+iY)\downarrow -2Z\uparrow\right] \\
u_3 &= \left|J=\frac{3}{2}, j_z=-\frac{1}{2}\right\rangle = i\frac{1}{\sqrt{6}}\left[(X-iY)\uparrow +2Z\downarrow\right] \\
u_4 &= \left|J=\frac{3}{2}, j_z=-\frac{3}{2}\right\rangle = i\frac{1}{\sqrt{2}}(X-iY)\downarrow \\
u_5 &= \left|J=\frac{1}{2}, j_z=\frac{1}{2}\right\rangle = \frac{1}{\sqrt{3}}\left[(X+iY)\downarrow +Z\uparrow\right] \\
u_6 &= \left|J=\frac{1}{2}, j_z=-\frac{1}{2}\right\rangle = i\frac{1}{\sqrt{3}}\left[-(X-iY)\uparrow +Z\downarrow\right]
\end{aligned}
\qquad (21.B.1)
$$

where X, Y, and Z denote Kohn–Luttinger amplitudes, which for the symmetry operations of the crystal point group transform as p_x, p_y, and p_z wave functions of the hydrogen atom.

Added to the Kohn–Luttinger kinetic Hamiltonian [56], the hole Hamiltonian H_h includes a p–d exchange term introduced by the interaction between the localized Mn magnetization and the holes derived in the mean-field approximation, thus giving [24,55]:

$$
\begin{aligned}
H_h = &-(\gamma_1+4\gamma_2)k^2 + 6\gamma_2\sum_{\alpha} L_\alpha^2 k_\alpha^2 \\
&+ 6\gamma_3\sum_{\alpha\neq\beta}(L_\alpha L_\beta + L_\beta L_\alpha)k_\alpha k_\beta + \Delta_{so}\vec{L}\vec{S} + \Delta_{exc}\vec{m}\vec{S}
\end{aligned}
\qquad (21.B.2)
$$

where
- $\alpha = \{x,y,z\}$
- L_α are $l=1$ angular momentum operators
- $\vec{S}$ is the vector spin operator
- $\vec{m}$ the unit magnetization vector
- γ_i are Luttinger parameters of the host semiconductor
- Δ_{exc} represents the spin-splitting between the HHs at the Γ_8 point

The detailed form of the 6×6 Hamiltonian is given in Ref. [24]. We chose $\Delta_{exc}=120\,\text{meV}$ in agreement with the value extracted from the magnetization saturation of (Ga,Mn)As [30]. Note that the standard 4×4 $k\cdot p$ Hamiltonian projected on the Γ_8 ($J=3/2$) basis can be written in a cubic form as

$$
\begin{aligned}
H_h = &\left(\gamma_1+\frac{5}{2}\gamma_2\right)\kappa^2 I_4 - 2\gamma_2(\kappa\cdot J)^2 \\
&+ 2(\gamma_3-\gamma_2)\sum_{i<j}\left[\kappa_i\kappa_j\left(J_iJ_j+J_jJ_i\right)\right] + \Delta_{exc}\vec{m}\vec{S}
\end{aligned}
\qquad (21.B.3)
$$

in unit of $\hbar^2/2m_0$ where J_i is the 4×4 momentum angular operator in the subspace $J=3/2$ along the direction i.

B.2 Theory of Transport within the $k\cdot p$ Framework

Our calculations of the transmission coefficients are based on the multiband transfer-matrix technique developed in Refs. [57,58] and applied to the hole 6×6 valence band $k\cdot p$ Hamiltonian H_h. To calculate the transmission coefficient, we will use the transfer-matrix technique and represent the device as a stack of two-dimensional flat-band interior layers of *constant* average electrostatic potential. The 0th (L) and ($N+1$)th (R) layers are semi-infinite emitter and collector. We will take into account elastic processes only, that is, assume that the hole energy ε_h and lateral momentum $k_{\parallel}$ are conserved during tunneling. Substituting $k_z \rightarrow -i\hbar\partial/\partial z$ into the Kane Hamiltonian matrix, we can find wave function eigenvectors $\psi_{n,\alpha}$ depending on $k_{\parallel}$ in the αth flat-band region (α) according to

$$
\psi_n(z) = \sum_{\alpha=1\rightarrow6}\left[A_{n,\alpha}^{+}v(k_{n,\alpha}^{+})\exp(ik_{n,\alpha}^{+}z) + A_{n,\alpha}^{-}v(k_{n,\alpha}^{-})\exp(ik_{n,\alpha}^{-}z)\right]
\qquad (21.B.4)
$$

where $\psi_{n,\alpha}$ and $\upsilon_{n,\alpha}$, are six-component column vectors and we separate all our solutions into two subsets with $k^{+}_{n,\alpha}$ corresponding to either traveling waves carrying the probability current from left to right or to evanescent waves decaying to the right and $k^{-}_{\alpha,n}$ corresponding to their left counterparts. Since the Kramers symmetry is broken in a magnetic region, $k^{-}_{n} \neq -k^{+}_{n}$ for real k (traveling wave). However, complex k always occur in complex conjugated pairs, that is, $k^{-}_{n} = k^{+*}_{n}$. The latter condition is a consequence of the Hermitian character of the hole Hamiltonian. This condition is necessary to ensure that the current across the device is steady state. It results also that, in the $k \cdot p$ approach, the current operator in the z direction writes $\hat{j} = 1/\hbar \nabla_k [H_h]$ [57,58].

The boundary conditions to match at a single interface between two different materials are as follows:

1. The continuity of the six components of the envelope function according to $\psi^{+}_{n} + \sum_{\bar{n}} r_{n,\bar{n}} \psi^{-}_{\bar{n}} = \sum_{n'} t_{n,n'} \psi^{+}_{n'}$ where the subscript $t_{n,n'}(r_{n,\bar{n}})$ refers to the respective transmission (reflection) amplitude from *incident* (n), *reflected* ($\bar{n}$), and *transmitted* (n') waves together with
2. The continuity of the six components of the current wavevector according to $\hat{j}\psi^{+}_{n} + \sum_{\bar{n}} r_{n,\bar{n}} \hat{j}\psi^{-}_{\bar{n}} = \sum_{n'} t_{n,n'} \hat{j}\psi^{+}_{n'}$.

The boundary conditions at each interface z_α ($\alpha = 0 \ldots N$), corresponding to the continuity of the wave function and the current, can be expressed in matrix form as

$$\chi_\alpha \pi^\alpha (z_\alpha) A_\alpha = \chi_{\alpha+1} \pi^{\alpha+1} (z_{\alpha+1}) A_{\alpha+1} \tag{21.B.5}$$

where $\mathcal{X}_\alpha$ is the 12×12 matrix containing the 12 eigenvectors $\upsilon(k_{n,\alpha})$ as well as the 12 current vectors $\hat{j}_z(k_{n,\alpha})\upsilon(k_{n,\alpha})$

$$\chi_\alpha = \begin{bmatrix} \upsilon(k^{+}_{n,\alpha}) & \upsilon(k^{-}_{n,\alpha}) \\ \hat{j}_z(k^{+}_{n,\alpha})\upsilon(k^{+}_{n,\alpha}) & \hat{j}_z(k^{-}_{n,\alpha})\upsilon(k^{-}_{n,\alpha}) \end{bmatrix} \tag{21.B.6}$$

where $\pi^\alpha_{\nu,\mu} = \exp(ik_{n,\alpha} z)\delta_{\nu\mu}$ is the propagation matrix within the layer α.

We adopt the following convention for enumeration of the lead layers and all related quantities: $0 \equiv L$ (emitter) and $N+1 \equiv R$ (collector). The system of linear equations above can be solved as

$$\begin{bmatrix} A^{+}_{L} \\ A^{-}_{\bar{L}} \end{bmatrix} = \mathcal{M} \begin{bmatrix} A^{+}_{R} \\ A^{-}_{\bar{R}} \end{bmatrix} = \begin{bmatrix} \mathcal{M}_{++} & \mathcal{M}_{+-} \\ \mathcal{M}_{-+} & \mathcal{M}_{--} \end{bmatrix} \begin{bmatrix} A^{+}_{R} \\ A^{-}_{\bar{R}} \end{bmatrix} \tag{21.B.7}$$

where the 12×12 transfer matrix

$$\mathcal{M} = \chi^{-1}_{L} \left[\Pi_{n=1\ldots N} \chi_\alpha \pi^\alpha (z_{n+1} - z_n) \chi^{-1}_\alpha \right] \chi_R \pi^{N+1}(z_N) \tag{21.B.8}$$

is partitioned according to the hole propagation direction in the emitter (L) and collector (R) and where the boundary condition is given by $A_{\bar{R}} = 0$. It follows that the respective 6×6 amplitude transmission and reflection matrices write respectively $t_{n,n'} = [\mathcal{M}_{++}]^{-1}$ and $r_{n,n'} = [\mathcal{M}_{-+}] \times [\mathcal{M}_{++}]^{-1}$.

Nevertheless, one must be aware that the $k \cdot p$ theory of bulk semiconductors is based on the consideration of the symmetry of the zone-center eigenstates and a small number of interband matrix elements of the momentum: It generally ignores the problem of the intracell shape and localization of the Bloch functions. However, when the crystal potential is modified on the scale of a unit cell, as, for example, in the case of an abrupt interface, it is intuitively evident that the matrix elements of the perturbation will be strongly dependent on the local properties of the Bloch functions [148–150]. In general, the reduced point symmetry $C_{2\upsilon}$ of a zinc-blende-based (001) interface allows mixing between heavy and light hole states even under normal incidence. However, this HH–LH nondiagonal mixing for the wave function matching at one interface is calculated to be smaller than 10% for the elastic hole energy considered [148]. More refined calculations of spin-polarized hole transmission through zinc-blende (001) heterojunction would require tight-binding treatment as proposed by Sankowski et al. [108].

Assuming that the in-plane wavevector $k_{\parallel}$ is conserved, the spin-polarized transmission $t_{n,n'}$ and reflection $r_{n,n'}$ amplitudes through a specific heterojunction can be determined by a matrix numerical procedure detailed in Refs. [57,58]. The transmission coefficients $\mathbf{T}^{k_\parallel,\hat{m}}_{n,n'}$ and elastic tunneling current $j^{\hat{m}}_{T}$ follow on the shell ($\varepsilon, k_{\parallel}$) according to Laudauer–Buttiker formula:

$$\mathbf{T}^{k_\parallel,\hat{m}}_{n,n'} = [t^{k_\parallel,\hat{m}}_{n,n'}]^{*} t^{k_\parallel,\hat{m}}_{n,n'} \frac{\left\langle \psi^{k_\parallel,\hat{m}}_{n'} \mid \hat{j} \mid \psi^{k_\parallel,\hat{m}}_{n'} \right\rangle}{\left\langle \psi^{k_\parallel,\hat{m}}_{n} \mid \hat{j} \mid \psi^{k_\parallel,\hat{m}}_{n} \right\rangle} \tag{21.B.9}$$

$$j^{\hat{m}}_{T} = \frac{e}{h} \sum_{k_\parallel, n, n'} \int_{\varepsilon_F}^{\varepsilon_m} \mathbf{T}^{k_\parallel,\hat{m}}_{n,n'} (\varepsilon, eV) d\varepsilon \tag{21.B.10}$$

where ε is the hole energy considered whereas ε_m refers to the lowest absolute hole energy involved in the transport.

The TAMR $= \left[j^{[001]}_{T} - j^{[100]}_{T} \right] / \left[j^{[100]}_{T} \right]$ is obtained numerically after summation of the tunneling current (Equation 21.B.3) over n, n', and $k_{\parallel}$ and integration on the energy shell ε taking $\varepsilon_F = -180$ meV and $\varepsilon_m = -140$ meV (considering a typical energy broadening for (Ga,Mn)As of 40 meV). In the calculations performed throughout the chapter, when not explicitly written, the Fermi level ε_F of (Ga,Mn)As was fixed at −180 meV from the top of the (Ga,Mn)As valence band, thus giving a reasonable hole concentration of about 2×10^{20} cm^{-3}. Concerning the heterostructure itself, the average VBO between (Ga,Mn)As and AlAs (GaAs), varying linearly with bias, was fixed at 0.7 eV (0.22 eV) at zero bias corresponding to an effective barrier height of 0.58 eV (0.1 eV). Note that the biaxial strain Hamiltonian H_{BS} can also added in the Hamiltonian form H_h when strain and stress effect need to be considered.

B.3 General Arguments on Spin-Transfer Phenomena

Concerning the calculation of spin torque in III–V heterojunctions, one can start from the above equation giving the time derivative of the angular momenta density operator $\hat{\mu}_i = \hat{\rho}\hat{J}_i$ at a certain position inside the heterostructure where $\vec{J}_i$ is the angular momentum operator along direction i:

$$\frac{\partial \hat{\mu}_i}{\partial t} = -\nabla\left(\vec{j}\hat{J}_i\right) + \hat{\rho}\frac{\partial \hat{J}_i}{\partial t}$$
$$\frac{\partial \hat{\mu}}{\partial t} = -\nabla\left(\vec{j}\hat{J}_i\right) + \frac{\Delta_{\text{exc}}}{\hbar}\,\hat{m}\times\hat{\mu} \quad (21.B.11)$$

Summing the latter equation over the different hole states included in the transport gives out (after having considered by the time derivative of the angular moment operator is null in the limit of coherent elastic transport)

$$\left\langle \nabla\left(\vec{j}\hat{J}_i\right)\right\rangle = \frac{\Delta_{\text{exc}}}{\hbar}\,\hat{m}\times\langle\hat{\mu}\rangle \quad (21.B.12)$$

This gives the torque at a given interface (int.) after integration over the semi half-infinite structure:

$$\left\langle \vec{j}\hat{J}_i\right\rangle_{\text{int.}} = \vec{\mathcal{H}}_{\text{exc}}\times\left\langle\vec{\mu}_J\right\rangle \quad (21.B.13)$$

where

$\vec{\mathcal{H}}_{\text{exc}}$ is the average exchange field experienced by holes

μ_J is the magnetic moment vector (per unit surface) injected in the thin layer subject to the torque by the spin-polarized current

We recover here the general results established by Kalitsov et al. [151] in the case of a transport in the conduction band using the Keldysh framework.

This demonstrates the presence of two intertwined contributions $\langle\vec{j}\hat{J}_i\rangle_x$ (parallel to the layer) and $\langle\vec{j}\hat{J}_i\rangle_z$ (perpendicular to the layer) for the spin torque ($\hat{y}$ is the magnetization direction) as derived for symmetric [132,133] and asymmetric barriers [134] and demonstrated experimentally in epitaxial MgO barrier MTJs [152,153]. These two components of the transverse spin currents are also responsible for the particular angular dependence of the tunneling transmission as in the case of the conduction band [154,155]. These two components of the torque are linked to the average value of the two components of the angular momenta carried by the hole current [151]. Within the general circuit-theory formalism developed by Braatas et al. [156] and adapted to metallic trilayers, one can establish that these two contributions in MTJs to the torque can be linked to the magnetic moment in the barrier (generalization of the spin accumulation) through the real and imaginary parts of the tunnel mixing conductance $G_{\uparrow\downarrow}$. These should be calculated in a future work.

Acknowledgment

We gratefully acknowledge A. Fert for his help in the work of spin transfer, T. Dietl, T. Jungwirth, O. Krebs, A. Lemaitre, P. Voisin, and J. Sinova for fruitful discussions.

References

1. I. Žutić, J. Fabian, and S. Das Sarma, Spintronics: Fundamentals and applications, *Rev. Mod. Phys.* **76**, 323–410 (2004).
2. C. Chappert, A. Fert, and F. Nguyen Van Dau, The emergence of spin electronics in data storage, *Nat. Mater.* **6**, 813–823 (2007).
3. M. Tanaka and Y. Higo, Large tunneling magnetoresistance in GaMnAs/AlAs/GaMnAs ferromagnetic semiconductor tunnel junctions, *Phys. Rev. Lett.* **87**, 026602 (2001).
4. E. Rashba, Theory of electrical spin injection: Tunnel junction as a solution of the conductivity mismatch problem, *Phys. Rev. B Condens. Matter* **62**, R16267–R16270 (2000).
5. A. Fert and H. Jaffrès, Conditions for efficient spin injection from a ferromagnetic metal into a semiconductor, *Phys. Rev. B Condens. Matter* **64**, 184420 (2001).
6. H. Jaffrès and A. Fert, Spin injection from a ferromagnetic metal into a semiconductor, *J. Appl. Phys.* **91**, 8111–8113 (2002).
7. A. Fert, J.-M. George, H. Jaffrès, and R. Mattana, Semiconductors between spin-polarized sources and drains, *IEEE Trans. Electr. Dev.* **54**, 921–932 (2007).
8. G. Schmidt, D. Ferrand, L. W. Molenkamp et al., Fundamental obstacle for electrical spin injection from a ferromagnetic metal into a diffusive semiconductor, *Phys. Rev. B Condens. Matter* **62**, R4790–R4793 (2000).
9. B. Eble, C. Testelin, P. Desfonds, F. Bernardot et al., Hole–nuclear spin interaction in quantum dots, *Phys. Rev. Lett.* **102**, 146601 (2009).
10. D. Brunner, B. D. Gerardot, P. A. Dalgarno et al., A coherent single-hole spin in a semiconductor, *Science* **325**, 70–72 (2009).
11. A. Marent, T. Nowozin, J. Gelze, F. Luckert, and D. Bimberg, Hole-based memory operation in an InAs/GaAs quantum dot heterostructure, *Appl. Phys. Lett.* **95**, 242114 (2009).
12. V. Novak, K. Olejnyk, J. Wunderlich, M. Cukr et al., Curie point singularity in the temperature derivative of resistivity in (Ga,Mn)As, *Phys. Rev. Lett.* **101**, 077201 (2008).
13. M. Wang, R. P. Campion, A. W. Rushforth et al., Achieving high Curie temperature in (Ga,Mn)As, *Appl. Phys. Lett.* **93**, 132103 (2008).
14. C. Timm, Magnetic semiconductors: III–V semiconductors, Chapter 18 in this book and references therein.
15. H. Ohno, Properties of ferromagnetic III–V semiconductors, *J. Magn. Magn. Mater.* **200**, 110–129 (1999).
16. T. Dietl, H. Ohno, F. Matsukura, J. Cibert, and D. Ferrand, Zener model description of ferromagnetism in zinc-blende magnetic semiconductors, *Science* **287**, 1019–1022 (2000).

17. H. Ohno and F. Matsukura, A ferromagnetic III–V semiconductor: (Ga,Mn)As, *Solid State Commun.* **117**, 179–186 (2001).
18. F. Matsukura, H. Ohno, and T. Dietl, III–V ferromagnetic semiconductors, in *Handbook of Magnetic Materials*, Vol. 14, K. H. J. Buschow (Ed.), North Holland, Amsterdam, the Netherlands, pp. 1–87, 2002.
19. J. Konig, J. Schliemann, T. Jungwirth, and A. H. MacDonald, Ferromagnetism in (III,V) semiconductors, in *Electronic Structure & Magnetism of Complex Material*, D. J. Singh and D. A. Papaconstantopoulos (Eds.), Springer, Material Science, Berlin, Germany, pp. 163–211, 2002.
20. T. Jungwirth, J. Sinova, J. Masek, J. Kucera, and A. H. M. MacDonald, Theory of ferromagnetic (III,Mn)V semiconductors, *Rev. Mod. Phys.* **78**, 809 (2006).
21. M. Sawicki, D. Chiba, A. Korbecka et al., Experimental probing of the interplay between ferromagnetism and localization in (Ga,Mn)As, *Nat. Phys.* **6**, 22–55 (2010).
22. S. R. Dunsiger, J. P. Carlo, T. Goko et al., Spatially homogeneous ferromagnetism of (Ga,Mn)As, *Nat. Mater.* **9**, 299–303 (2010).
23. C. Zener, Interaction between the *d* shells in the transition metals, *Phys. Rev.* **81**, 440–444 (1951).
24. T. Dietl, H. Ohno, and F. Matsukura, Hole-mediated ferromagnetism in tetrahedrally coordinated semiconductors, *Phys. Rev. B Condens. Matter* **63**, 195205 (2001).
25. L. Thevenard, L. Largeau, O. Mauguin et al., Magnetic properties and domain structure of (Ga,Mn)As films with perpendicular anisotropy, *Phys. Rev. B Condens. Matter* **73**, 195331 (2006).
26. M. Glunk, J. Daeubler, L. Dreher et al., Magnetic anisotropy in (Ga,Mn)As: Influence of epitaxial strain and hole concentration, *Phys. Rev. B Condens. Matter* **79**, 195206 (2010).
27. W. Stafanowicz, C. Sliwa, P. Aleshkevych et al., Magnetic anisotropy of epitaxial (Ga,Mn)As on (113)*A* GaAs, *Phys. Rev. B Condens. Matter* **81**, 155203 (2010).
28. D. Chiba, F. Matsukura, and H. Ohno, Tunneling magnetoresistance in (Ga,Mn)As-based heterostructures with a GaAs barrier, *Physica E: Low Dimens. Syst. Nanostruct.* **21**, 966–969 (2004).
29. R. Mattana, M. Elsen, J.-M. George et al., Chemical profile and magnetoresistance of $Ga_{1-x}Mn_xAs$/GaAs/AlAs/GaAs/$Ga_{1-x}Mn_xAs$ tunnel junctions, *Phys. Rev. B Condens. Matter* **71**, 075206 (2005).
30. M. Elsen, H. Jaffrès, R. Mattana et al., Spin-polarized tunneling as a probe of the electronic properties of $Ga_{1-x}Mn_xAs$ heterostructures, *Phys. Rev. B Condens. Matter* **76**, 144415 (2007).
31. S. Ohya, I. Muneta, P. N. Hai, and M. Tanaka, GaMnAs-based magnetic tunnel junctions with an AlMnAs barrier, *Appl. Phys. Lett.* **95**, 242503 (2009).
32. S. H. Chun, S. J. Potashnik, K. C. Ku, P. Schiffer, and N. Samarth, Spin-polarized tunneling in hybrid metal-semiconductor magnetic tunnel junctions, *Phys. Rev. B Condens. Matter* **66**, 100408 (2002).
33. H. Saito, S. Yuasa, K. Ando, Y. Hamada, and Y. Suzuki, Spin-polarized tunneling in metal-insulator-semiconductor Fe/ZnSe/$Ga_{1-x}Mn_xAs$ magnetic tunnel diodes, *Appl. Phys. Lett.* **89**, 232502 (2006).
34. H. Saito, A. Yamamoto, S. Yuasa, and K. Ando, Spin-dependent density of states in $Ga_{1-x}Mn_xAs$ probed by tunneling spectroscopy, *Appl. Phys. Lett.* **92**, 192512 (2008).
35. H. Saito, A. Yamamoto, S. Yuasa, and K. Ando, High tunneling magnetoresistance in Fe/GaO_x/$Ga_{1-x}Mn_xAs$ with metal/insulator/semiconductor structure, *Appl. Phys. Lett.* **93**, 172515 (2008).
36. S. Ohya, P. N. Hai, Y. Mizuno, and M. Tanaka, Quantum-size effect and tunneling magnetoresistance in ferromagnetic-semiconductor quantum heterostructure, *Phys. Rev. B Condens. Matter* **75**, 155328 (2007).
37. S. Ohya, I. Muneta, P. N. Hai, and M. Tanaka, Valence-band structure of the ferromagnetic semiconductor GaMnAs studied by spin-dependent resonant tunneling spectroscopy, *Phys. Rev. Lett.* **104**, 167204 (2010).
38. S. Ohya, P. N. Hai, and M. Tanaka, Tunneling magnetoresistance in GaMnAs/AlAs/InGaAs/AlAs/GaMnAs double-barrier magnetic tunnel junctions, *Appl. Phys. Lett.* **87**, 012105 (2005).
39. C. Gould, C. Ruster, T. Jungwirth, E. Girgis et al., Tunneling anisotropic magnetoresistance: A spin-valve-like tunnel magnetoresistance using a single magnetic layer, *Phys. Rev. Lett.* **93**, 117203 (2004).
40. C. Ruster, C. Gould, T. Jungwirth et al., Very large tunneling anisotropic magnetoresistance of a (Ga,Mn)As/GaAs/(Ga,Mn)As stack, *Phys. Rev. Lett.* **94**, 027203 (2005).
41. H. Saito, S. Yuasa, and K. Ando, Origin of the tunnel anisotropic magnetoresistance in $Ga_{1-x}Mn_xAs$/ZnSe/$Ga_{1-x}Mn_xAs$ magnetic tunnel junctions of II-VI/III-V heterostructures, *Phys. Rev. Lett.* **95**, 086604 (2005).
42. R. Giraud, M. Gryglas, L. Thevenard, A. Lemaitre, and G. Faini, Voltage-controlled tunneling anisotropic magnetoresistance of a ferromagnetic p^{++}-(Ga,Mn)As/n^+-GaAs Zener-Esaki diode, *Appl. Phys. Lett.* **87**, 242505 (2005).
43. A. Giddings, M. Khalid, T. Jungwirth, J. Wunderlich et al., Large tunneling anisotropic magnetoresistance in (Ga,Mn)As nanoconstrictions, *Phys. Rev. Lett.* **94**, 127202 (2005).
44. M. Elsen, H. Jaffrès, R. Mattana, M. Tran et al., Exchange-mediated anisotropy of (Ga,Mn)As valence-band probed by resonant tunneling spectroscopy, *Phys. Rev. Lett.* **99**, 127203 (2007).
45. A. Matos-Abiague and J. Fabian, Anisotropic tunneling magnetoresistance and tunneling anisotropic magnetoresistance: Spin–orbit coupling in magnetic tunnel junctions, *Phys. Rev. B Condens. Matter* **79**, 155303 (2009).
46. A. Matos-Abiague, M. Gmitra, and J. Fabian, Angular dependence of the tunneling anisotropic magnetoresistance in magnetic tunnel junctions, *Phys. Rev. B Condens. Matter* **80**, 045312 (2009).

47. J. Slonczewski, Current-driven excitations of magnetic multilayers, *J. Magn. Magn. Mater.* **159**, L1 (1996).
48. L. Berger, Emission of spin waves by a magnetic multilayer traversed by a current, *Phys. Rev. B Condens. Matter* **54**, 9353–9358 (1996).
49. A. Fert, A. Barthelemy, J. Ben Youssef et al., Review of recent results on spin polarized tunneling and magnetic switching by spin injection, *Mater. Sci. Eng. B* **84**, 1–9 (2001).
50. M. Stiles and J. Miltat, Spin-transfer torque and dynamics, in *Spin Dynamics in Confined Magnetic Structures III*, Vol. 101, Hillebrands and A. Thiaville (Eds.), Springer, Berlin, Germany, pp. 225–308, 2006.
51. D. Chiba, Y. Sato, T. Kita, F. Matsukura, and H. Ohno, Current-driven magnetization reversal in a ferromagnetic semiconductor (Ga,Mn)As/GaAs/(Ga,Mn)As tunnel junction, *Phys. Rev. Lett.* **93**, 216602 (2004).
52. M. Elsen, O. Boulle, J.-M. George et al., Spin transfer experiments on (Ga,Mn)As/(In,Ga)As/(Ga,Mn)As tunnel junctions, *Phys. Rev.* **B73**, 035303 (2006).
53. J. Okabayashi, M. Watanabe, H. Toyao, T. Yamagushi, and J. Yoshino, Pulse-width dependence in current-driven magnetization reversal using GaMnAs-based double-barrier magnetic tunnel junction, *J. Supercond. Nov. Magn.* **20**, 443–446 (2007).
54. M. Watanabe, J. Okabayashi, H. Toyao, T. Yamagushi, and J. Yoshino, Current-driven magnetization reversal at extremely low threshold current density in (Ga,Mn)As-based double-barrier magnetic tunnel junctions, *Appl. Phys. Lett.* **92**, 082506 (2008).
55. M. Abolfath, T. Jungwirth, J. Brum, and A. H. MacDonald, Theory of magnetic anisotropy in $III_{1-x}Mn_xV$ ferromagnets, *Phys. Rev. B Condens. Matter* **63**, 054418 (2001).
56. J.-M. Luttinger and W. Kohn, Motion of electrons and holes in perturbed periodic fields, *Phys. Rev.* **97**, 869–883 (1955).
57. A. G. Petukhov, A. N. Chantis, and D. O. Demchenko, Resonant enhancement of tunneling magnetoresistance in double-barrier magnetic heterostructures, *Phys. Rev. Lett.* **89**, 107205 (2002).
58. A. G. Petukhov, D. O. Demchenko, and A. N. Chantis, Electron spin polarization in resonant interband tunneling devices, *Phys. Rev. B Condens. Matter* **68**, 125332 (2003).
59. L. C. Andreani, A. Pasquarello, and F. Bassani, Hole subbands in strained GaAs-$Ga_{1-x}Al_xAs$ quantum wells: Exact solution of the effective mass equation, *Phys. Rev. B Condens. Matter* **36**, 5887–5894 (1987).
60. R. Wessel and M. Altarelli, Resonant tunneling of holes in double-barrier heterostructures in the envelope-function approximation, *Phys. Rev. B Condens. Matter* **39**, 12802–12807 (1989).
61. M. Jullière, Tunneling between ferromagnetic films, *Phys. Lett. A* **54**, 225–226 (1975).
62. J. S. Moodera, L. R. Kinder, T. M. Wong, and R. Meservey, Large magnetoresistance at room temperature in ferromagnetic thin film tunnel junctions, *Phys. Rev. Lett.* **74**, 3273 (1995).
63. S. S. P. Parkin, C. Kaiser, A. Panchula et al., Giant tunnelling magnetoresistance at room temperature with MgO(100) tunnel barriers, *Nat. Mater.* **3**, 862–867 (2004).
64. S. Yuasa, T. Nagahama, A. Fukushima, Y. Suzuki, and K. Ando, Giant room-temperature magnetoresistance in single-crystal Fe/MgO/Fe magnetic tunnel junctions, *Nat. Mater.* **3**, 868–871 (2004).
65. W. H. Butler, X.-G. Zhang, T. C. Schulthess, and J. M. MacLaren, Spin-dependent tunneling conductance of Fe/MgO/Fe sandwiches, *Phys. Rev. B Condens. Matter* **63**, 054416 (2001).
66. J. Mathon and A. Umerski, Theory of tunneling magnetoresistance of an epitaxial Fe/MgO/Fe(001) junction, *Phys. Rev. B Condens. Matter* **63**, 220403 (2001).
67. L. Brey, C. Tejedor, and J. Fernandez-Rossier, Tunnel magnetoresistance in GaMnAs: Going beyond Jullière formula, *Appl. Phys. Lett.* **85**, 1996 (2004).
68. L. Brey, J. Fernandez-Rossier, and C. Tejedor, Spin depolarization in the transport of holes across $Ga_{1-x}Mn_xAs$/$Ga_yAl_{1-y}As$/*p*-GaAs, *Phys. Rev. B Condens. Matter* **70**, 235334 (2004).
69. P. Krstajic and F. M. Peeters, Spin-dependent tunneling in diluted magnetic semiconductor trilayer structures, *Phys. Rev. B Condens. Matter* **72**, 125350 (2005).
70. R. Mattana, J.-M. George, H. Jaffrès et al., Electrical detection of spin accumulation in a *p*-type GaAs quantum well, *Phys. Rev. Lett.* **90**, 166601 (2003).
71. N. O. Lipari and A. Baldereschi, Angular momentum theory and localized states in solids. Investigation of shallow acceptor states in semiconductors, *Phys. Rev. Lett.* **25**, 1660–1664 (1970).
72. A. Baldereschi and N. O. Lipari, Spherical model of shallow acceptor states in semiconductors, *Phys. Rev. B Condens. Matter* **8**, 2697–2709 (1973).
73. A. Baldereschi and N. O. Lipari, Cubic contributions to the spherical model of shallow acceptor states, *Phys. Rev. B Condens. Matter* **9**, 1525–1539 (1973).
74. J. M. George, H. Jaffrès, R. Mattana et al., Electrical spin detection in GaMnAs-based tunnel junctions: Theory and experiments, *Mol. Phys. Rep.* **40**, 23–33 (2004).
75. V. Garcia, H. Jaffrès, M. Eddrief, M. Marangolo, V. H. Etgens, and J.-M. George, Resonant tunneling magnetoresistance in MnAs/III-V/MnAs junctions, *Phys. Rev. B Condens. Matter* **72**, 081303(R) (2005).
76. V. Garcia, H. Jaffrès, J.-M. George, M. Marangolo, M. Eddrief, and V. H. Etgens, Spectroscopic measurement of spin-dependent resonant tunneling through a 3D disorder: The case of MnAs/GaAs/MnAs junctions, *Phys. Rev. Lett.* **97**, 246802 (2006).
77. G. Xiang, B. L. Sheu, M. Zhu, P. Schiffer, and N. Samarth, Noncollinear spin valve effect in ferromagnetic semiconductor trilayers, *Phys. Rev. B Condens. Matter* **76**, 035324 (2007).
78. M. Ciorga, A. Einwanger, U. Wurstbauer, D. Schuh, W. Wegscheider, and D. Weiss, Electrical spin injection and detection in lateral all-semiconductor devices, *Phys. Rev. B Condens. Matter* **79**, 165321 (2009).

79. P. Van Dorpe, Z. Liu, W. Van Roy et al., Very high spin polarization in GaAs by injection from a (Ga,Mn)As Zener diode, *Appl. Phys. Lett.* **84**, 3495 (2004).
80. M. Kohda, T. Kita, Y. Ohno, F. Matsukura, and H. Ohno, Bias voltage dependence of the electron spin injection studied in a three-terminal device based on a (Ga,Mn)As/n$^+$-GaAs Esaki diode, *Appl. Phys. Lett.* **89**, 012103 (2006).
81. M. Kohda, Y. Ohno, F. Matsukura, and H. Ohno, Effect of n$^+$/GaAs thickness and doping density on spin injection of (Ga,Mn)As/n$^+$-GaAs Esaki tunnel junction, *Physica E Low Dimens. Syst. Nanostruct.* **32**, 438–441 (2006).
82. P. Van Dorpe, W. Van Roy, J. De Boeck et al., Voltage-controlled spin injection in a (Ga,Mn)As/(Al,Ga)As Zener diode, *Phys. Rev. B Condens. Matter* **72**, 205322 (2005).
83. T. Wojtowicz, W. L. Lim, X. Liu et al., Correlation of Mn lattice location, free hole concentration, and Curie temperature in ferromagnetic GaMnAs, *J. Supercond. Nov. Magn.* **16**, 41–44 (2003).
84. S. J. Potashnik, K. C. Ku, R. F. Wang et al., Coercive field and magnetization deficit in $Ga_{1-x}Mn_xAs$ epilayers, *J. Appl. Phys.* **93**, 6784 (2003).
85. K. Y. Wang, M. Sawiki, K. W. Edmonds et al., Control of coercivities in (Ga,Mn)As thin films by small concentrations of MnAs nanoclusters, *Appl. Phys. Lett.* **88**, 022510 (2006).
86. J. Okabayashi, A. Kimura, O. Rader et al., Core-level photoemission study of $Ga_{1-x}Mn_xAs$, *Phys. Rev. B Condens. Matter* **58**, R4211–R4214 (1998).
87. M. Adell, J. Adell, L. Ilver, J. Kanski, and J. Sadowski, Photoemission study of the valence band offset between low temperature GaAs and (GaMn)As, *Appl. Phys. Lett.* **89**, 172509 (2006).
88. S. Tiwari and D. J. Frank, Empirical fit to band discontinuities and barrier heights in III–V alloy systems, *Appl. Phys. Lett.* **60**, 630 (1992).
89. S. Lodhaa, D. B. Janes, and N.-P. Chen, Unpinned interface Fermi-level in Schottky contacts to n-GaAs capped with low-temperature-grown GaAs; experiments and modeling using defect state distributions, *J. Appl. Phys.* **93**, 2772 (2003).
90. M. Malfait, J. Vanacken, W. V. Roy, G. Borghs, and V. Moshchalkov, Low-temperature annealing study of $Ga_{1-x}Mn_xAs$: Magnetic properties and Hall effect in pulsed magnetic fields, *J. Magn. Magn. Mater.* **290**, 1387–1390 (2005).
91. Y. Nishitani, D. Chiba, M. Endo et al., Curie temperature versus hole concentration in field-effect structures of $Ga_{1-x}Mn_xAs$, *Phys. Rev. B Condens. Matter* **81**, 045208 (2010).
92. A. Koeder, S. Frank, W. Schoch et al., Curie temperature and carrier concentration gradients in epitaxy-grown $Ga_{1-x}Mn_xAs$ layers, *Appl. Phys. Lett.* **82**, 3278 (2003).
93. K. Pappert, M. J. Schmidt, S. Humpfner et al., Magnetization-switched metal-insulator transition in a (Ga,Mn)As tunnel device, *Phys. Rev. Lett.* **97**, 186402 (2006).
94. K. I. Bolotin, F. Kuemmeth, and D. C. Ralph, Anisotropic magnetoresistance and anisotropic tunneling magnetoresistance due to quantum interference in ferromagnetic metal break junctions, *Phys. Rev. Lett.* **97**, 127202 (2006).
95. S. F. Shi, K. I. Bolotin, F. Kuemmeth, and D. C. Ralph, Temperature dependence of anisotropic magnetoresistance and atomic rearrangements in ferromagnetic metal break junctions, *Phys. Rev. B Condens. Matter* **76**, 184438 (2007).
96. M. Viret, M. Gabureac, F. Ott et al., Giant anisotropic magneto-resistance in ferromagnetic atomic contacts, *Eur. Phys. J. B Condens. Matter* **51**, 1–4 (2006).
97. A. Bernand-Mantel, P. Seneor, K. Bouzehouane et al., Anisotropic magneto-Coulomb effects and magnetic single-electron-transistor action in a single nanoparticle, *Nat. Phys.* **5**, 920–924 (2009).
98. T. Uemura, Y. Imai, M. Harada, K. Matsuda, and M. Yamamoto, Tunneling anisotropic magnetoresistance in epitaxial CoFe/n-GaAs junctions, *Appl. Phys. Lett.* **94**, 182502 (2009).
99. M. Wimmer, M. Lobenhofer, J. Moser et al., Orbital effects on tunneling anisotropic magnetoresistance in Fe/GaAs/Au junctions, *Phys. Rev. B Condens. Matter* **80**, 121301(R) (2009).
100. J. Moser, A. Matos-Abiague, D. Schuh, W. Wegscheider, J. Fabian, and D. Weiss, Tunneling anisotropic magnetoresistance and spin–orbit coupling in Fe/GaAs/Au tunnel junctions, *Phys. Rev. Lett.* **99**, 056601 (2007).
101. L. Gao, X. Jiang, S. H. Yang et al., Bias voltage dependence of tunneling anisotropic magnetoresistance in magnetic tunnel junctions with MgO and Al_2O_3 tunnel barriers, *Phys. Rev. Lett.* **99**, 226602 (2007).
102. B. G. Park, J. Wunderlich, D. A. Williams et al., Tunneling anisotropic magnetoresistance in multilayer-(Co/Pt)/AlO_x/Pt structures, *Phys. Rev. Lett.* **100**, 087204 (2008).
103. A. N. Chantis, K. D. Belashchenko, E. Y. Tsymbal, and M. van Schilfgaarde, Tunneling anisotropic magnetoresistance driven by resonant surface states: First-principles calculations on an Fe(001) surface, *Phys. Rev. Lett.* **98**, 046601 (2007).
104. M. Zhu, M. J. Wilson, P. Mitra, P. Schiffer, and N. Samarth, Quasireversible magnetoresistance in exchange-spring tunnel junctions, *Phys. Rev. B Condens. Matter* **78**, 195307 (2008).
105. M. J. Schmidt, K. Pappert, C. Gould, J. Schmidt, R. Oppermann, and L. W. Molenkamp, Bound-hole states in a ferromagnetic (Ga,Mn)As environment, *Phys. Rev. B Condens. Matter* **76**, 035204 (2007).
106. J. M. Tang and M. E. Flatté, Multiband tight-binding model of local magnetism in $Ga_{1-x}Mn_xAs$, *Phys. Rev. Lett.* **92**, 047201 (2004).
107. J. M. Tang and M. E. Flatté, Spin-orientation-dependent spatial structure of a magnetic acceptor state in a zinc-blende semiconductor, *Phys. Rev. B Condens. Matter* **72**, 161315(R) (2005).

108. P. Sankowski, P. Kacman, J. Majewski, and T. Dietl, Spin-dependent tunneling in modulated structures of (Ga,Mn)As, *Phys. Rev. B Condens. Matter* **75**, 045306 (2007).
109. L. Thevenard, L. Largeau, O. Mauguin, A. Lemaitre, and B. Theys, Tuning the ferromagnetic properties of hydrogenated GaMnAs, *Appl. Phys. Lett.* **87**, 182506 (2005).
110. H. Ohno, N. Akira, F. Matsukura et al., Spontaneous splitting of ferromagnetic (Ga,Mn)As valence band observed by resonant tunneling spectroscopy, *Appl. Phys. Lett.* **73**, 363 (1998).
111. R. K. Hayden, D. K. Maude, L. Eaves et al., Probing the hole dispersion curves of a quantum well using resonant magnetotunneling spectroscopy, *Phys. Rev. Lett.* **66**, 1749–1752 (1991).
112. M. Tran, J. Peiro, H. Jaffrès, J.-M. George, and A. Lemaitre, Magnetization-controlled conductance in (Ga,Mn)As-based resonant tunneling devices, *Appl. Phys. Lett.* **95**, 172101 (2009).
113. J. Wunderlich, T. Jungwirth, B. Kaestner et al., Coulomb blockade anisotropic magnetoresistance effect in a (Ga,Mn)As single-electron transistor, *Phys. Rev. Lett.* **97**, 077201 (2006).
114. J. A. Katine, F. J. Albert, R. A. Buhrman, E. B. Myers, and D. C. Ralph, Current-driven magnetization reversal and spin-wave excitations in Co/Cu/Co pillars, *Phys. Rev. Lett.* **84**, 3149–3152 (2000).
115. S. I. Kiselev, J. Sankey, I. Krivorotov et al., Microwave oscillations of a nanomagnet driven by a spin-polarized current, *Nature* **425**, 380–383 (2003).
116. J. Grollier, V. Cros, A. Hamzic et al., Spin-polarized current induced switching in Co/Cu/Co pillars, *Appl. Phys. Lett.* **78**, 3663 (2001).
117. J. Grollier, V. Cros, H. Jaffrès et al., Field dependence of magnetization reversal by spin transfer, *Phys. Rev. B Condens. Matter* **67**, 174402 (2003).
118. J. Sun, D. Monsma, D. Abraham, M. Rooks, and R. Koch, Batch-fabricated spin-injection magnetic switches, *Appl. Phys. Lett.* **81**, 2202 (2002).
119. S. Urazdhin, N. Birge, W. P. Pratt, and J. Bass, Switching current versus magnetoresistance in magnetic multilayer nanopillars, *Appl. Phys. Lett.* **84**, 1516 (2004).
120. Y. Huai, F. Albert, P. Nguyen, M. Pakala, and T. Valet, Observation of spin-transfer switching in deep submicron-sized and low-resistance magnetic tunnel junctions, *Appl. Phys. Lett.* **84**, 3118 (2004).
121. Y. Liu, Z. Zhang, P. P. Freitas, and J. L. Martins, Current-induced magnetization switching in magnetic tunnel junctions, *Appl. Phys. Lett.* **82**, 2871 (2003).
122. G. D. Fuchs, N. C. Emley, I. N. Krivorotov, P. M. Braganca et al., Spin-transfer effects in nanoscale magnetic tunnel junctions, *Appl. Phys. Lett.* **85**, 1205 (2005).
123. J. Slonczewski, Currents, torques, and polarization factors in magnetic tunnel junctions, *Phys. Rev. B Condens. Matter* **71**, 024411 (2005).
124. J. A. Katine and E. E. Fullerton, Device implications of spin-transfer torques, *J. Magn. Magn. Mater.* **320**, 1217–1226 (2008).
125. D. Chiba, F. Matsukura, and H. Ohno, Electrical magnetization reversal in ferromagnetic III–V semiconductors, *J. Phys. D: Appl. Phys.* **39**, R215–R225 (2006).
126. M. Tsoi, A. G. M. Jansen, J. Bass et al., Excitation of a magnetic multilayer by an electric current, *Phys. Rev. Lett.* **80**, 4281–4284 (1998).
127. M. Tsoi, A. G. M. Jansen, J. Bass, W.-C. Chiang, V. Tsoi, and P. Wyder, Generation and detection of phase-coherent current-driven magnns in magnetic multilayers, *Nature* **406**, 46–48 (2000).
128. J. Sinova, T. Jungwirth, X. Liu et al., Magnetization relaxation in (Ga,Mn)As ferromagnetic semiconductors, *Phys. Rev. B Condens. Matter* **69**, 085209 (2004).
129. S. Zhang, P. M. Levy, A. Marley, and S. S. P. Parkin, Quenching of magnetoresistance by hot electrons in magnetic tunnel junctions, *Phys. Rev. Lett.* **79**, 3744–3747 (1997).
130. A. M. Bratkovsky, Assisted tunneling in ferromagnetic junctions and half-metallic oxides, *Appl. Phys. Lett.* **72**, 2334–2336 (1998).
131. P. M. Levy and A. Fert, Spin transfer in magnetic tunnel junctions with hot electrons, *Phys. Rev. Lett.* **97**, 097205 (2006).
132. I. Theodonis, N. Kioussis, A. Kalitsov, M. Chshiev, and W. H. Butler, Anomalous bias dependence of spin torque in magnetic tunnel junctions, *Phys. Rev. Lett.* **97**, 237205 (2006).
133. A. Kalitsov, M. Chshiev, I. Theodonis, N. Kioussis, and W. H. Butler, Spin-transfer torque in magnetic tunnel junctions, *Phys. Rev. B Condens. Matter* **79**, 174416 (2009).
134. Y. H. Tang, N. Kioussis, A. Kalitsov et al., Controlling the nonequilibrium interlayer exchange coupling in asymmetric magnetic tunnel junctions, *Phys. Rev. Lett.* **103**, 057206 (2009).
135. M. Linnarsson, E. Janzén, B. Monemar, M. Kleverman, and A. Thilderkvist, Electronic structure of the GaAs:Mn_{Ga} center, *Phys. Rev. B Condens. Matter* **55**, 6938–6944 (1997).
136. J. Schneider, U. Kaufmann, W. Wilkening, M. Baeumler, and F. Kohl, Electronic structure of the neutral manganese acceptor in gallium arsenide, *Phys. Rev. Lett.* **59**, 240–243 (1987).
137. A. O. Govorov, Optical probing of the spin state of a single magnetic impurity in a self-assembled quantum dot, *Phys. Rev. B* **70**, 035321 (2004).
138. B. E. Larsson, K. C. Hass, H. Ehrenreich, and A. E. Carlsson, Theory of exchange interactions and chemical trends in diluted magnetic semiconductors, *Phys. Rev. B Condens. Matter* **37**, 4137–4154 (1988).
139. Y. Ohno, I. Arata, F. Matsukura, and H. Ohno, Valence band barrier at (Ga,Mn)As/GaAs interfaces, *Physica E Low Dimens. Syst. Nanostruct.* **13**, 521–524 (2002).
140. O. Thomas, O. Makarovsky, A. Patanè et al., Measuring the hole chemical potential in ferromagnetic $Ga_{1-x}Mn_xAs$/GaAs heterostructures by photoexcited resonant tunneling, *Appl. Phys. Lett.* **90**, 082106 (2007).

141. T. Tsuruoka, N. Tachikawa, S. Ushioda, F. Matsukura, K. Takamura, and H. Ohno, Local electronic structures of GaMnAs observed by cross-sectional scanning tunneling microscopy, *Appl. Phys. Lett.* **81**, 2800 (2002).
142. B. I. Shklovskii and A. L. Efros, *Electronic Properties of Doped Semiconductors*, Solid-State Sciences, Vol. 45, Springer-Verlag, Berlin, Germany, pp. 1–65, 1984.
143. T. Jungwirth, J. Sinova, A. H. MacDonald et al., Character of states near the Fermi level in (Ga,Mn)As: Impurity to valence band crossover, *Phys. Rev. B Condens. Matter* **76**, 125206 (2007).
144. K. S. Burch, D. B. Shrekenhamer, E. J. Singley et al., Impurity band conduction in a high temperature ferromagnetic semiconductor, *Phys. Rev. Lett.* **97**, 087208 (2006).
145. K. Ando, H. Saito, K. C. Agarwal, M. C. Debnath, and V. Zayets, Origin of the anomalous magnetic circular dichroism spectral shape in ferromagnetic $Ga_{1-x}Mn_xAs$: Impurity bands inside the band gap, *Phys. Rev. Lett.* **100**, 067204 (2008).
146. J. M. Tang and M. E. Flatté, Magnetic circular dichroism from the impurity band in III-V diluted magnetic semiconductors, *Phys. Rev. Lett.* **101**, 157203 (2008).
147. D. Neumaier, M. Turek, U. Wurstbauer et al., All-electrical measurement of the density of states in (Ga,Mn)As, *Phys. Rev. Lett.* **103**, 087203 (2009).
148. E. L. Ivchenko, A. Yu. Kaminski, and U. Rossler, Heavy-light hole mixing at zinc-blende (001) interfaces under normal incidence, *Phys. Rev. B Condens. Matter* **54**, 5852–5889 (1996).
149. O. Krebs and P. Voisin, Giant optical anisotropy of semiconductor heterostructures with no common atom and the quantum-confined pockels effect, *Phys. Rev. Lett.* **77**, 1829–1832 (1996).
150. S. Cortez, O. Krebs, and P. Voisin, Breakdown of rotational symmetry at semiconductor interfaces: A microscopic description of valence subband mixing, *Eur. Phys. J. B* **21**, 241–250 (2001).
151. A. Kalitsov, I. Theodonis, N. Kioussis, M. Chshiev, W. H. Butler, and A. Vedyayev, Spin-polarized current-induced torque in magnetic tunnel junctions, *J. Appl. Phys.* **99**, 08G501 (2006).
152. H. Kubota, A. Fukushima, K. Yakushiji et al., Quantitative measurement of voltage dependence of spin-transfer torque in MgO-based magnetic tunnel junctions, *Nat. Phys.* **4**, 37–41 (2007).
153. J. C. Sankey, Y. T. Cui, J. Z. Sun et al., Measurement of the spin-transfer-torque vector in magnetic tunnel junctions, *Nat. Phys.* **4**, 67–71 (2007).
154. J. Slonczewski, Conductance and exchange coupling of two ferromagnets separated by a tunneling barrier, *Phys. Rev. B Condens. Matter* **39**, 6995–7002 (1989).
155. H. Jaffrès, D. Lacour, F. Nguyen Van Dau et al., Angular dependence of the tunnel magnetoresistance in transition-metal-based junctions, *Phys. Rev. B Condens. Matter* **64**, 064427 (2001).
156. A. Braatas, G. E. W. Bauer, and P. J. Kelly, Non-collinear magnetoelectronics, *Phys. Rep.* **427**, 157–255 (2006).

22

Spin Transport in Organic Semiconductors

Valentin Dediu
Institute for Nanostructured Materials Studies

Luis E. Hueso
Basque Foundation for Science

Ilaria Bergenti
Institute for Nanostructured Materials Studies

22.1 Introduction

Recently, spintronics has expanded its choice of materials to the organic semiconductors (OSCs) [1–3]. OSCs offer unique properties toward their integration with spintronics, the main one being the weakness of the spin-scattering mechanisms in OSC. Spin–orbit coupling is very small in most OSCs; carbon has a low atomic number (Z) and the strength of the spin–orbit interaction (SOI) is in general proportional to Z^4 [4]. Typical SOI values in OSC are less/about a few meV [5], well below any other characteristic energy, including main vibrational modes. Small spin–orbit coupling implies that the spin polarization of the carriers could be maintained for a very long time. Indeed, spin relaxation times in excess of 10 μs have been detected by various resonance techniques [6,7], and these values compare very favorably (at least 10^3 times larger) with those obtained with high-performing inorganic semiconductors, such as GaAs [8].

The field of OSCs combines nowadays considerable achievements in research and applications with challenging expectations. Over decades, the interest toward OSCs was mainly sustained by their exceptional optical properties, as the multicolor "offer" was backed by an enormous choice of available or easily modifiable molecules as well as by significant luminescence intensity. Indeed, these optical properties promoted a real breakthrough in the optoelectronics field where, along with the achievement of a deep basic knowledge, a variety of important applications have been developed. Display products based on hybrid light-emitting diodes with organic emitters (OLEDs) have become available to consumers, and organic photovoltaic devices are challenging existing commercial applications. These accomplishments are advancing a future strong demand for high quality control and guiding circuitry based on novel hybrid organic–inorganic devices. Organic field effect transistors (OFETs), in which considerable improvements have already been achieved, together with various elements based on magnetic and electrical resistance switching are foreseen as one of the most important candidates for these newly emerging applications.

Organic molecules are also considered as a promising contender for the future "beyond CMOS devices," offering 1–10 nm-sized building blocks with rich optical, electrical, and magnetic properties. OSCs have become a requisite of various roadmaps and research agendas, although many serious difficulties have still to be circumvented in order to meet these challenges. One of the biggest obstacles for commercial applications is the resistance contact problem [9], namely, the way to insert and extract electrically driven information, interfacing fragile organic blocks with hard metallic or oxide electrodes. However, it has to be noted, on the other hand, that the hybrid nature of most organic-based devices has also positive aspects, providing routes for large and fundamental modifications of the properties of the OSC, especially in the interface region.

22.2 Organic Semiconductors: General Background

OSCs combine a strong intramolecular covalent carbon–carbon bonding with a weak van der Waals interaction between molecules. This combination of different interactions generates unusual materials whose optical properties are very similar to those of their constituent molecules, whereas their transport properties are instead firmly governed by the intermolecular interactions.

As a general rule, there are two very different classes of OSCs: small molecules and polymers. These two cases differ not only on their physical and chemical properties but also on their technological processing. Small-molecule OSCs are rigid objects of about 1 nm size, with very different shapes and symmetries, but able to form highly ordered van der Waals layers and polycrystalline films. The small-molecule film growth involves either high or ultrahigh vacuum molecular beam deposition. A few of the main representatives of this kind of materials are the classes of oligothiophenes (4T, 6T, and others), metal chelates (Alq_3, Coq_3, and others), metal-phthalocyanines (CuPc, ZnPc, and others), acenes (tetracene, pentacene, and others), rubrene, and many more. Polymers, on the other hand, are long chains formed from a given monomer, and usually feature connection of oligo-segments of various lengths leading to a random conjugation length distribution. Polymer films are strongly disordered and generally cannot be produced in UHV conditions; instead, totally different processing technologies have to be applied, including ink-jet printing, spin coating, drop casting, and other extremely cheap methods. This potentially inexpensive fabrication constitutes one of the main technological advantages of the polymers. However, these growth approaches generally preclude a real interface control, as the electrode surface is not as clean as in UHV conditions and cannot be directly characterized by such important surface techniques as x-ray photoemission spectroscopy (XPS), ultraviolet photoemission spectroscopy (UPS), x-ray magnetic dichroism (XMCD), and others.

The electrical properties of OSC are usually depicted by two main components: charge injection and charge transport. The charge injection process from a metal into an OSC is conceptually different from the case of metals/inorganic semiconductors, mainly by the vanishingly low density of intrinsic carriers—about 10^{10} cm^{-3}—in OSC. The external electrodes provide the electrical carriers responsible for the transport as the OSC molecules can easily accommodate an extra charge, although this charge modifies considerably the molecular spatial conformation and creates strong polaronic effects. Carrier injection into OSC is best described [10] in terms of thermally and field-assisted charge tunneling across the inorganic–organic interface, followed by carrier diffusion into the bulk of OSC. The carriers propagate via a random site-to-site hopping between pseudo-localized states distributed within approximately a 0.1 eV energy interval. This interval describes the spatial distribution of the localized levels and has not to be confused with the bandwidth. OSCs are mainly families of π-conjugated molecules, which combine a considerable carrier delocalization inside the molecules with a weak Van der Waals intermolecular interaction that limit considerably the carrier mobility. The electron mean free path is usually of about a molecule size, namely, about 1 nm, while standard mobilities are well below 1 cm^2/V s. Two conducting channels are usually considered active: lowest unoccupied molecular level (LUMO) for n-type and highest occupied molecular level (HOMO) for p-type carriers. However, defects and interface states have to be considered depending on the material and its structural quality. These states usually emerge via one of the most important characteristics of OSC—trapping centers. Puzzlingly, in OSCs, the impurities do not induce additional carriers like in inorganic semiconductors but mainly generate trap states. Energetically distributed shallow (about/below 0.1 eV) and deep (about 0.5 eV, but even up to 1 eV) traps are known to characterize the electrical properties (especially the mobility) of many OSCs, their role being stronger in low-order materials like Alq_3, polymers, etc.

The mobility of OSC has a strong dependence on the electric field, and in the absence of traps it is usually described by the so-called Poole–Frenkel dependence [11]

$$\mu(E) = \mu_0 \exp\left(\frac{\beta}{\sqrt{E}}\right) \tag{22.1}$$

where μ_0 is the field-independent mobility. Including trapping complicates the equations and numerical solutions for different cases have to be calculated.

In order to complete this very brief description of electrical properties of OSC, we would like to add two more concepts.

The first concept highlights the difference between the two main transport regimes present for most organic-based devices: injection-limited current (ILC) and space charge–limited current (SCLC). In the first case, the interface resistance dominates the current–voltage curves and the flowing current is basically independent on the device thickness. However, the SCLC current (J_{SCLC}) requires at least one ohmic electrode and corresponds to the case when the injected carrier density reaches the charging capacity of the material. This SCLC current is characterized by a strong thickness (d) dependence, which for a trap-free OSC is given by [11]

$$J_{SCLC} \sim \frac{\mu(E)V^2}{d^3} \tag{22.2}$$

where V is the voltage across the device and, at that given voltage, the electric field (E) value inside OSC depends as well on the thickness.

The second concept is the distinction between unipolar and bipolar transport in OSCs (the later featuring significant recombination effects). These two different transport regimes depend on the intrinsic properties of the OSC as well as on the matching of the electrode work function with either the LUMO or the HOMO levels. For a more extensive reading on the electrical properties, we suggest the excellent reviews of Coropceanu et al. [12] and Brütting et al. [11].

The optical gap in many OSCs is about 2–3 eV, which conveniently fits very well in the visible range. The simultaneous injection of electrons and holes from two independent electrodes generates Frenkel excitons with a 3:1 triplet/singlet ratio. The $S1 \rightarrow S0$ optical transition is usually the only one allowed as a consequence of low spin–orbit coupling and Franck–Condon selection. The former rules out the singlet–triplet mixing, while the latter points out once again on the strong role of the polaronic effects in OSC [13,14], which excludes the transitions where the equilibrium positions of the nuclei in both the ground and the excited states would be identical.

22.3 Electrical Operation: Injection

In the following discussion on the electrically operated organic spintronic devices (OSPDs), it is necessary to distinguish between two different cases: tunneling devices characterized by zero residence time of charges/spins in the organic barrier, and injection devices in which there is a net flow of spin-polarized carriers directly into the electronic levels of the OSC. While tunneling devices can feature large magnetoresistance (MR) values and are useful as 0–1 switching elements or magnetic sensors, injection devices are characterized by a finite lifetime of the generated spin polarization in OSC and offer such opportunities as spin manipulation and control of the exciton statistics in OSC. These options may well open the way to magnetically controlled OFETs or OLEDs.

In the description of the spin injection process into OSC, we will need to mention numerous times two materials with exceptional importance in the field: manganites, mostly $La_{0.7}Sr_{0.3}MnO_3$ (LSMO), and Alq_3 (Tris(8-hydroxyquinoline) aluminum(III)). Importantly, about half of all the OSC spintronic devices available in the literature involve either one or both of these materials. This fact has some historical grounds, as the first successful experiments dealt with these materials, but it also indicates some outstanding spintronic properties of these materials, which have not been understood and developed in detail yet.

$La_{1-x}Sr_xMnO_3$ (with $0.2 < x < 0.5$) is a ferromagnetic material with 100% spin-polarized carriers at $T \ll T_c$. The Curie temperature is around 370 K in bulk samples, while thin films have T_c around 330 K, changing with thickness and deposition method. The carrier hopping between Mn^{3+} and Mn^{4+} ions is governed by a strong on-site Hund interaction (nearly 1 eV), which makes $La_{1-x}Sr_xMnO_3$ an unusual metal, characterized by a narrow band (~1–2 eV) and small carrier density (10^{21}–10^{22} cm^{-3}), very much like the high-temperature cuprate superconductors and some organic metals.

A critical issue for the spintronic applications of manganites is the difference between their bulk and surface magnetic properties: the latter, being responsible for the injected spin-polarized current, decays much faster with temperature than the former. This leads to the fact that in the room temperature region, the spin polarization is much smaller than the expected magnetization ratio $M(T)/M(0)$ [15]. Nevertheless, special surface treatments have succeeded to improve the thin film surface properties, and samples with reproducible XMCD signals at room temperature are currently routinely obtained [16]. Well before raising of organic spintronics, LSMO was already well known for its spin injection properties and has proved to be highly successful in diverse inorganic spintronic devices, such as tunnel junctions [17] and artificial grain boundaries devices [18].

Tris(8-hydroxyquinoline aluminum(III) or Alq_3 is a popular OSC from the early stages of xerographic and optoelectronics applications [19]. This material might look somehow as a strange choice for transport-oriented applications, as Alq_3 has very low mobility values and a permanent electric dipole [20,21]. These two factors are in principle able to compromise spin transport, but in spite of the expected problems with Alq_3, strong spin-mediated effects have been repeatedly reported by different groups.

The first experimental report on injection in OSPD featured a lateral device that combined LSMO electrodes and OSC conducting channels between 100 and 500 nm long (Figure 22.1a). Sexithiophene (6T), a rigid conjugated oligomer rod, pioneer in organic thin film transistors [22], was chosen for the spin transport channel. A strong magnetoresistive response was recorded up to room temperature in 100 and 200 nm channels and was explained as a result of the conservation of the spin polarization of the injected carriers. Using the time-of-flight approach, a spin relaxation time of the order of 1 μs was found.

This work introduced OSCs as very appealing materials for long-distance spin transport, although some important questions were left open. For example, the experiment did not provide a straightforward demonstration that the MR was related to the magnetization of the electrodes, and hence to its spin polarization, as no comparison between parallel and antiparallel magnetization of electrodes was available.

Following this report, an important step forward was the fabrication of a vertical spin-valve device consisting of LSMO and Cobalt electrodes, sandwiching a thick (100–200 nm) layer of Alq_3 [23]. The spin valve showed MR up to a temperature around 200 K, and for voltages below 1 V. Unlike standard spin-valve devices, in which lower resistance corresponds to parallel electrode magnetization, the LSMO/Alq_3/Co devices showed lower resistance for an antiparallel magnetization configuration.

Starting from this work, the vertical geometry has acquired a prevailing role in organic spintronics, mainly because of its technological simplicity with respect to planar devices, requiring a substantial lithography involvement. Several groups have confirmed the so-called inverse spin-valve effect in LSMO/Alq_3/Co devices [24–28]. Nevertheless, its nature has not been yet clearly understood, although at least two different explanations have been proposed [23,26]. One of these models requires the injection of a spin-down-oriented current from the cobalt electrode [23] (i.e., the spin polarization of carriers antiparallel to the applied magnetic field), in a serious disagreement with many experimental and theoretical reports [29,30]. The second model introduces a very narrow hopping channel in OSCs, connecting at some given voltage the spin-up bands of one of the electrodes with the spin-down bands of the opposite one [26]. While in principle feasible, this model can be applied for the n-type carriers only and works mainly at voltages about 1 V or higher, that is, values at which spintronics effects in LSMO/Alq_3/Co devices generally vanish. This situation clearly indicates the limits of the overall comprehension of the main laws governing both the spin injection and transport in organic spintronics.

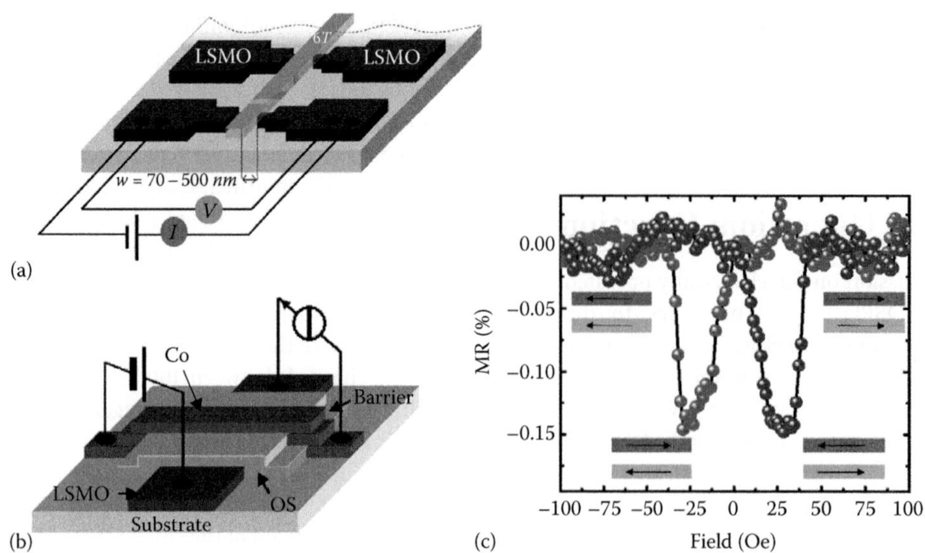

FIGURE 22.1 (a) Spin lateral device presented by Dediu et al. [32]: two LSMO electrodes obtained by electron beam lithography were connected by an organic thin film (T6). The channel length ranged between 70 and 500 nm. (b) Vertical spin-valve device consisting of LSMO electrode as bottom electrode and cobalt as the top one. The typical MR measurements are performed by applying a constant biasing voltage *V* across the electrodes while measuring the resulting current as a function of an in-plane swipping external magnetic field. (c) GMR loop of an LSMO (20 nm)/Alq_3 (100 nm)/AlO_X (2 nm)/Co (20 nm) vertical spin-valve device measured at room temperature. The antiparallel configurations of the magnetizations correspond to the low resistance state.

As explained above, Alq_3 is characterized by its very low mobility: about 10^{-5} cm^2/V s for electrons and less than 10^{-6} cm^2/V s for holes [31]; while the full relevance of the mobility has yet to be understood for OSPD, the wide use of Alq_3 is probably due to the good-quality thin films that can be grown on various ferromagnetic substrates by standard UHV evaporation.

Nevertheless, electrical detection of spin injection and transport has been actively pursued in a number of other OSCs (see Table 22.1). Thus, α-NPD (*N*,*N*-bis 1-naphtalenyl-*N*,*N*-bis phenylbenzidiane) and CVB (4, 4′-(Bis(9-ethyl-3-carbazovinylene)-1,1′-biphenyl) [24] were demonstrated to behave very similarly to Alq_3 with LSMO and Co set of magnetic electrodes: inverse spin-valve effect (around 10% at low temperatures) and fast decrease of the MR with increasing temperature, vanishing between 210 and 240 K. The same couple of electrodes was used to investigate spin injection in tetraphenylporphyrin (TTP)—the detected MR in this case was attributed, however, to tunneling effects [28]. A planar geometry with two LSMO electrodes, similar to the first publication in the field [32], provided about 30% MR at low temperature in pentacene and BTQBT [33].

Moving from low to room temperature operation in vertical devices has required an advanced control of the interface quality. Thus, room temperature MR in LSMO/Alq_3/Al_2O_3/Co devices was demonstrated [26] by improving the quality of both injecting interfaces, especially the top one, in which a 2 nm thick Al_2O_3 insulating layer was added (see Figure 22.1b and c). Similarly, about 1% room temperature MR across thick Alq_3 and CuPc layers was achieved in vertical devices with both metallic electrodes and excellent interface quality (Figure 22.2c) [34]. Importantly, the substitution of the LSMO electrode by Fe has removed the inversion of the spin-valve effect [34], indicating that inversion is strongly caused by the manganite properties [26].

All the experiments described above were performed using UHV conditions for the OSC growth. Spin-coated polymers such as regio-random and regio-regular poly(3-hexylthiophene), (RRaP3HT and RRP3HT, respectively) have also shown noticeable MR [27,35]. In these reports, the spin-valve effect was positive and a small value was recorded even at room temperature.

We can summarize the main findings obtained, thanks to the study of electrically operated injection devices. First, they unambiguously demonstrated both spin injection and spin transport in OSPDs. The available data confirms so far spin injection and transport only for voltages not exceeding approximately 1 V. Moreover, highly efficient MR are constrained in a narrower interval of about 0.1–0.2 V. Current MR measurements provide mainly qualitative information and in the absence of well-defined models for spin injection and transport in OSC, it is difficult to extract the values of the spin polarization or its decay inside the organic material. The investigation of the thickness and temperature dependences of the MR in OSPDs provides anyway important indications for the basic physics underlying spintronics effects.

Surprisingly, very important information comes from the "confusing" set of data on the thickness dependence. Except few reported sequences where an expected decrease of MR upon increasing thickness was reported, many published [28,36] as well as unpublished data (various conference discussions) indicate quite casual dependences.

A reasonable and, in our opinion, credible explanation to this comes amazingly from the analysis of the temperature

TABLE 22.1 Main Properties of OSCs Investigated in Organic Spintronics.

Class	Material	Mobility / type	Material	Mobility / type
Oligoacenes	Rubrene	$\mu = 1\ cm^2/V\ s$ *p*-type	Pentacene	$\mu = 10^{-1}\ cm^2/V\ s$ *p*-type
Tiophenes	Sexitiophene	$\mu = 10^{-1}\ cm^2/V\ s$ *p*-type	RRP3HT	$\mu = 10^{-1}\ cm^2/V\ s$ *p*-type
Triarylamines	NPD	$\mu = 10^{-5}\ cm^2/V\ s$ *p*-type	CVB	$\mu = 10^{-3}\ cm^2/V\ s$
Porphyrines	TPP	$\mu = 10^{-5}\ cm^2/V\ s$ *n*-type		
Metal complexes	CuPC	$\mu = 10^{-2}\ cm^2/V\ s$ *n*-type	Alq_3	$\mu = 10^{-5}\ cm^2/V\ s$ *n*-type

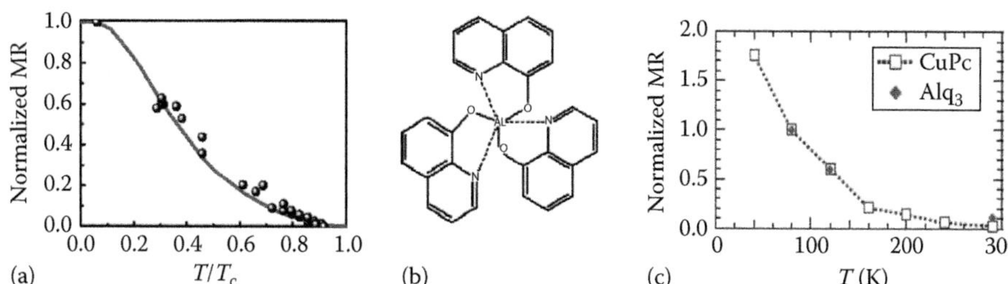

FIGURE 22.2 (a) Temperature dependence of normalized MR of vertical spin-valve devices consisting of LSMO and Co as electrodes and Alq_3 as spacer (dots) with $T_c = 325\,K$. The line represents the LSMO surface magnetization (b) Alq_3 molecular structure. (c) Temperature dependence of maximum observed MR for two junctions based on Alq_3 and CuPc, as organic spacer and Fe and Co as metallic contact. From Liu et al., the MRs for each sample are normalized to their values at $T = 80\,K$. (Reprinted with permission from Liu, Y. et al., Effects of carrier mobility and morphology in organic semiconductor spin valves (Fig. 3), *J. Appl. Phys.*, 105, 07C708, 2009. Copyright 2009, American Institute of Physics.)

behavior for the most archetypal LSMO-Alq_3-Co OSPD [24,26]. Figure 22.2a presents the normalized MR versus temperature for a device with a 100 nm thick Alq_3 layer and a 2 nm thick Al_2O_3 barrier separating Alq_3 and top Co electrode [26]. Remarkably, the data extrapolate to zero at the Curie temperature of the manganite [26] (325 K in this particular case). Moreover, the temperature dependence of MR data agree with that of surface magnetization (SM) curve for LSMO (solid line in Figure 22.2a) obtained by Park et al. [15]. The SM represents the magnetization from the top 5 Å in a standard LSMO film, as determined by spin-polarized photoemission spectroscopy [15] and it is effectively the parameter of interest for spin injection in LSMO-based devices [37]. The results summarized above are in agreement with the previous claim that the temperature dependence of MR in Alq_3 spintronic devices is governed by the manganite electrode [24].

The temperature behavior described above empowers a very important statement: the LSMO-Alq_3-Co devices behave in an "injection limited"-like regime, as the electrode spin polarization nearly dominates the temperature dependence. The spin losses at the interface significantly prevail over the transport bulk losses. By analogy with purely charge injection case, it seems coherent to expect a weak or even a random thickness dependence in such devices: For example, small fluctuations of the interface quality should influence the device behavior much more than a thickness variation. Also, we believe this regime can be expanded to other materials and interfaces, although there are no a priori reasons to suppose that all OSPDs should work in injection-limited mode.

22.4 Electrical Operation: Tunneling

While the interest in spin injection and long-distance spin transport in OSCs was growing, several groups also started to explore the possibilities of these materials as spin tunnel barriers. Spin-polarized tunneling in OSPD was initially demonstrated in quite complex geometries. Petta et al. [38] fabricated Ni-octanethiol-Ni vertical tunneling devices in a nanopore geometry with an octanethiol self-assembled monolayer. These devices showed MR up to 16% at 4.2 K, although vanishing at about 30 K. The sign of the MR was observed to switch from positive to negative for different voltage values, but also from sample to sample at the same voltage. While evidence of the spin-polarized tunneling through a monolayer of organic was given, it has also been noticed that such devices are subject to strong intrinsic and extrinsic noises.

An interesting approach for the investigation of tunneling effects is by the utilization of composite systems featuring magnetic nanoparticles embedded into organic matrix. Depending on the separation between nanoparticles, a switching from interparticle tunneling to hopping conductivity between particles occurs. Such effects have been detected in Co:C_{60} nanocomposite [39].

An important step forward for organic spin tunneling was the fabrication of simple devices via direct in situ UHV organic deposition with shadow masking. Vertical tunneling devices (Co/Al_2O_3-Alq_3/NiFe [40] and Co/Al_2O_3-rubrene/Fe [41]) made use of hybrid inorganic–organic ultrathin barriers to produce positive TMR when the electrode magnetizations were switched from parallel to antiparallel. TMR around 15% was recorded at 4.2 K [41], and importantly, both OSC showed few percent TMR at room temperatures, opening the door to possible commercial applications. Another important aspect is that Al_2O_3-Alq_3 devices had the hybrid tunnel barrier well inside the typical quantum mechanical tunneling regime (0.6–1.6 nm). However, the Al_2O_3-rubrene hybrid barriers were considerably thicker, with rubrene layers ranged from 4.6 to 15 nm. These thicknesses are well over the tunneling regime and probably imply spin injection in rubrene layers. A transition from tunneling into injection regime was in principle tackled for the first time for organic–inorganic hybrid devices [41]. At the present state of the art, the organic spintronic community is not yet ready to provide a thickness value where such a transition can take place nor is it ready to describe convincingly the system behavior at the two sides of transition. Further investigation is required in order to face these key fundamental questions. One has to keep firmly in mind nevertheless that OSCs of about 20–30 nm (e.g., Alq_3) are already good light emitters, indicating that this thickness is already well beyond tunneling limit.

The experiments reported above clearly show the possibility of achieving spin-polarized tunneling across organic or organic–inorganic hybrid barriers. Although the control of spin tunneling across organic layers is still relatively poor, it is also clear that there is a great margin for improvement, especially in the fields of interface control and engineering of single molecule studies.

22.5 Alternative Approaches

In this section, we will outline different important experiments that have been reported recently in the field of organic spintronics, but not with standard electronic spin transport in devices. These alternative experiments are extremely important, as they allow us to have access to different physical parameters than the ones restricted by simple electronic transport data. Our overview here does not pretend to be systematic, but to highlight different recent approaches that are shedding light on different aspects of spin injection and transport in OSCs.

One important piece of information recently revealed is the spin polarization profile inside an OSC. Such information has deep implications, as it would allow the research community to evaluate the true potential of organic spintronics for carrying information at long distances. The data has been obtained by two powerful nonstandard techniques: two-photon photoemission spectroscopy and muon spin rotation technique.

Two-photon photoemission spectroscopy experiments allow us to obtain the spin polarization of injected electrons in the first monolayers of organic materials grown on top of a spin-polarized ferromagnetic material [42]. Spin-polarized electrons from the metal are excited by the first energy pulse into an intermediate energy state. Some of the electrons propagate into the OSC where, with a certain probability, they can be excited by the second photon causing the photoemission. The spin-polarized injection efficiency from cobalt into the first monolayer of the copper phthalocyanine was estimated to be close to 100% [42]. An obvious drawback in these measurements is that the energy requested for the electron injection process is much higher than the energy gap in the OSC. For those energies, the spin scattering is higher than for the (lower) voltages at which electron/spin injection usually takes place, making the comparison between photoemission spectroscopy and standard electron devices far from straightforward.

In a totally independent article, electrical injection was combined with low-energy muon spin rotation to study the spin diffusion length inside Alq_3 in a standard vertical spin-valve device [43]. The stopping distribution of the spin-polarized muons was varied in the range 3–200 nm by controlling the implantation energy, while a quantitative analysis determined the spin diffusion length (l_S) and its temperature dependence. On one hand, the diffusion length reached a low temperature value of 35 nm: a lower value than the one estimated from electrical measurements (>100 nm), a discrepancy that could be nevertheless explained bearing in mind the imperfect injection efficiency and very small MR of the explored device. On the other hand, the weak temperature dependence of the l_S is in good qualitative agreement with independent MR characterizations in Alq_3-based devices [26] and also in agreement with a recently proposed mechanism for the hyperfine-driven spin scattering in OSC.

In a different subset, we could position the articles that explore the spin transport properties in OSC in relation with their electroluminescence (EL) properties. In fact, EL has been key in the advance of the OSC research since it opens the door to multiple commercial applications, such as OLEDs. Similarly, the careful study of the emitted light could give us very useful information about the electron/hole recombination processes inside the organic. In OLEDs, due to low spin–orbit coupling, the dominant EL channel involves singlet–singlet transitions [44]. In OLED, the singlet/triplet ratio is determined by quantum statistics as 1:3 when considering a similar formation probability for one singlet and three triplet states (Figure 22.3). The injection of carriers with a controlled spin state could therefore enable the amplification of the singlet (triplet) exciton density, leading to a significant increase of efficiency for the EL [45], rather than a polarization in the emitted light as in other inorganic semiconductors (such as GaAs) [46–48]. This theoretically expected increase in the emitted light attracted many researchers into organic spintronics a few years ago, but unfortunately, the efficiency amplification has not been reported in the literature [49,50]. The absence of amplification of light in spin-polarized OLED might be due to the fact that light emission can only be typically detected for applied voltages in excess of a few volts, while spin injection in organic materials has been demonstrated so far for voltages not exceeding 1 V [23,45]. While this limitation does not seem to be a fundamental obstacle, it has to be yet understood whether it can be experimentally circumvented.

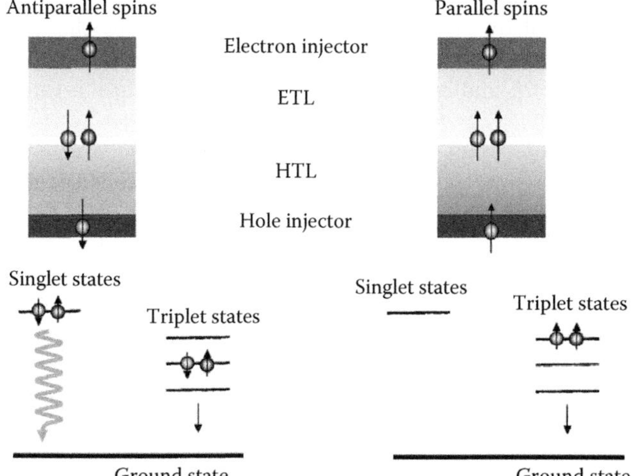

FIGURE 22.3 (See color insert.) Schematic structure of a bilayered OLED. Electrons and holes are injected by cathode and anode, respectively, into the electron-transport layer (ETL) and the hole-transport layer (HTL). Exciton formation and recombination eventually take place at the interface. Spin statistic of carrier recombination in case of spin injection. Antiparallel spin states for electrons and holes generate singlet and triplet state in ratio 1:1. Parallel spin states for electrons and holes' spin state of carriers give rise only to the triplet state.

Irrespective of the current lack of successful results in light amplification in spin-polarized OLED, they are anyway very useful for the study of the electron-spin conversion, that can be studied by electron spin resonance (ESR) [51]. In this chapter, the authors achieved a magnetically guided transfer between singlet (PP_S) and triplet (PP_T) polaron pairs caused by the spin precession. The two conversion channels give completely different inputs into the photoconductivity, and monitoring the later provides a way to control the PP dynamics. A periodical modification of the $PP_S \rightarrow PP_T$ transfer (Rabi oscillations) was observed in the conjugated polymer MEH-PPV, from which an extremely long spin coherence time above 0.5 μs [51] was extracted.

22.6 Interface Effects

In a strict sense, for the MR in a spin-valve organic device to be present, both bottom and top ferromagnetic/OSC interfaces would have to be spin selective (such is, e.g., a tunnel barrier [52]). In this context, the perception of the complex nature of the interfaces between magnetic materials and OSCs is motivating several groups to undertake detailed studies of the organic–inorganic interfaces. Important issues such as the interfacial energy levels and the degree of chemical interaction or the different interfacial states are being clarified with this kind of research. In general, spectroscopic techniques are being used to study these systems. UPS/XPS are usually the preferred methods to study these interfaces.

Two groups reported almost parallel independent research on what, in principle, should be the optimum combination of materials, a high Curie temperature ferromagnetic material (cobalt) and a high-mobility OSC (pentacene, Pc) (see Figure 22.4a). However, it is important to remember that Co, as any transition metal, is chemically very active, and this activity could result in a very complex interfacial structure. In general, interfacial dipoles of around 1 eV are observed between Co and Pc [53]; however, it is claimed that the Pc/Co interface (with the Co film deposited on top of the organic) behaves differently, showing vacuum-level alignment (no dipole) and highlighting again the importance of careful characterization of these complex devices [54].

The ubiquitous LSMO/Alq_3/Co device has been fully investigated for interface electronic properties (Figure 22.4b). Detailed XPS and UPS investigations have revealed a 0.9 eV interfacial dipole at LSMO/Alq_3 interface [55] and a 1.5 eV dipole at direct Co/Alq_3 interfaces [56]. The latter is reduced from 1.5 to 1.3 eV when a 2 nm Al_2O_3 barrier is inserted between Alq_3 and Co layer [57]. This knowledge provides a complete set of barriers across the whole device describing the current injection into LUMO and HOMO levels of OSC. On the other hand, this information is not sufficient for the spin injection efficiency, as no clear data on the spin transfer

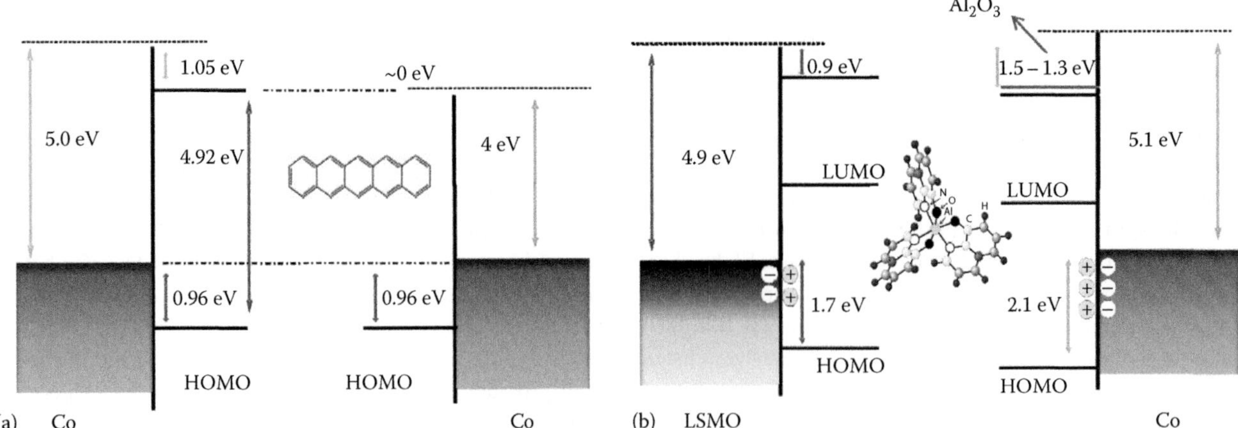

FIGURE 22.4 (a) Schematic band diagram for an LSMO/Alq_3 and Alq_3/Co structure presented by Zhang et al. [56,57]. (b) Schematic band diagram for a layered Co/pentacene/Co structure presented by Popinciuc [54].

is revealed. Thus, the necessary evolution of the interface studies consists in moving toward monitoring of the spin selectivity and injecting interfaces. For example, proximity effects at organic–inorganic interface can induce new states in the gap of OSCs with nontrivial spin polarization [58].

Recently, Grobosch et al. [59] have questioned whether interfacial energy values measured in UHV conditions can be used to explain the electronic transport behavior of actual devices measured in standard laboratory conditions.

22.7 Multifunctional Effects

A new field that is currently attracting much attention and also one in which OSC could have a significant impact is that of multifunctional or multipurpose devices. In general terms, these are devices that react either simultaneously or separately under different external stimuli, such as electric field, magnetic field, light irradiation, and pressure. In this area of research, the possibility to combine magnetic and electrical resistance switching has been realized recently [60] profiting from the unique combination of properties in OSPDs. Nonvolatile electrical memory effect in hybrid organic–inorganic materials has been extensively studied (see, e.g., the extended recent review of Scott and Bozano [61]). Under the application of an electric field, resistance changes in excess of one order of magnitude and retention times of tens of hours have been routinely detected. Following those successes, the International Technology Roadmap for Semiconductors has identified molecular-based memories among the emerging technology lines (http://www.itrs.net).

In organic spintronics, hybrid vertical devices with LSMO and Co as charge/spin-injecting electrodes show both spin-valve and nonvolatile electrical memory effects on a same chip (see Figure 22.5). This result was confirmed for two different OSCs, Alq_3 and 6T, and consists in the observation of 10% negative

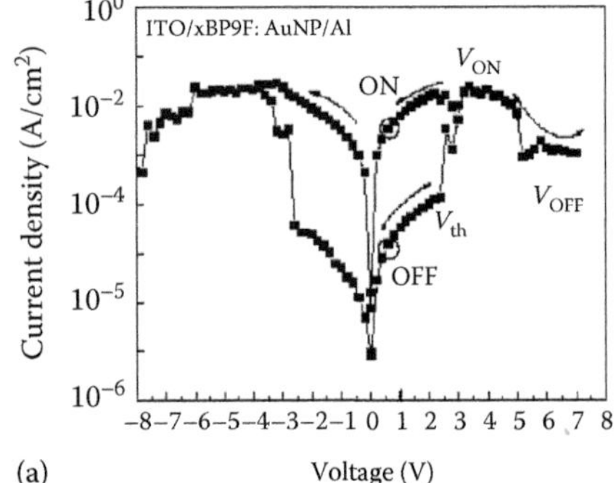

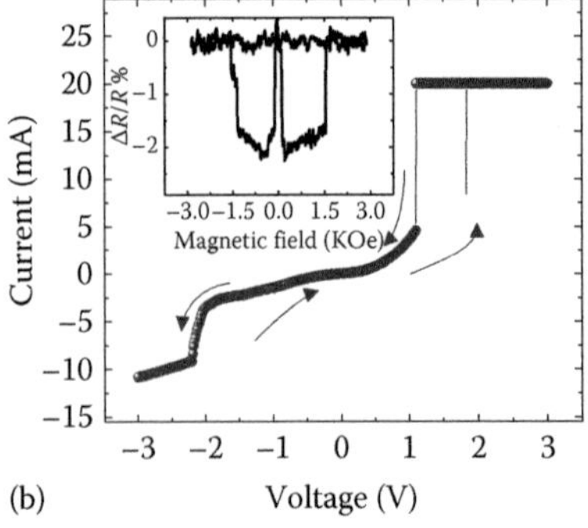

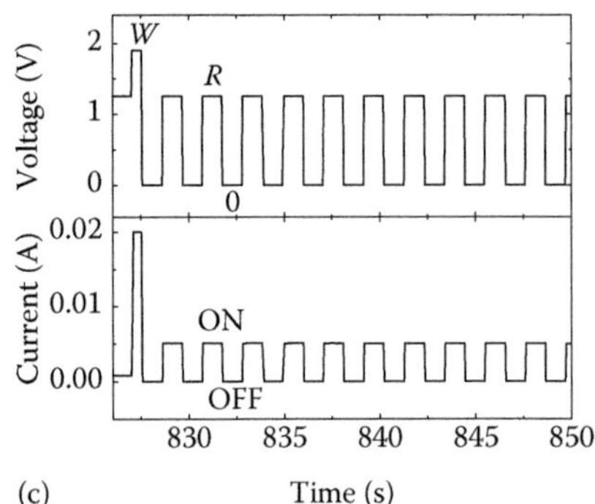

FIGURE 22.5 (a) Typical current density–voltage characteristic for a nonvolatile memory cell based on composite of metallic nanoparticles and polymers embedded in an organic semiconducting host. (Bozano, L.D., Kean, B.W., Beinhoff, M., Carter, K.R., Rice, P.M., and Scott, J.C.: Organic materials and thin-film structures for cross-point memory cells based on trapping in metallic nanoparticles (Fig. 6). *Adv. Funct. Mater.* 2005. 15. 1933–1939. Copyright Wiley-VCH Verlag GmbH & Co. KGaA. Reproduced with permission.) (b) Current–voltage room temperature characteristics in T6-based vertical device. The irreversibility at both positive and negative voltages is clearly visible and it is associated to the classical spin-valve effect presented in the inset. (c) Switching effect for the OrgSV memory cell. Writing (*W*), erasing (*E*), and reading (*R*) voltages have been used. Two very clear and reproducible ON and OFF states can be observed multiple times even if the voltage has been set to zero between different reading events.

spin-valve magnetic switching at voltages below 1 V, and 100% electrical switching at voltages above 2–3 V. Importantly, the resistance values achieved by electrical switching are not altered when the devices go back to zero voltage (nonvolatility; see Figure 22.5b and c). With an initial writing voltage pulse, the device moves to a resistance state (called 1), and this resistance state can be recovered multiple times with a reading voltage pulse (reading voltage < writing voltage) even if the voltage has been set to zero after the writing pulse. If we later apply a negative erasing pulse, the new resistance value at the reading voltage will be different (called 0) and again this value can be recovered multiple times even if the voltage is set to zero between reading pulses. This behavior, explained above, implies that a set resistance value (either 1 or 0) can be recovered even if the memory cell has been left without any voltage for a set amount of time, allowing an operation with much lower power and avoiding the need of constant refreshing of current RAM cells.

We believe this combination will become a subject of extensive investigations and can provide a basis for the generation of two bits/cell memory elements. As a future development, we envisage multifunctional devices with magnetic and electric field control of the resistance and that in addition to these responses will profit from the luminescent properties of the OSCs.

22.8 Theoretical Models

Many theoretical groups have done a considerable amount of work of different aspects of organic spintronics. Much of the current research is clearly a progress on previous works on inorganic devices; however, there are great new ideas and unique approaches that are helping the organic spintronics field to move forward. Probably, one of the most active topics is the one related to the theoretical description of tunneling devices with a single (or a few) molecule placed between two spin-polarized electrodes [62–64]. Tunneling is perhaps the less-complicated organic spintronics process to describe, and very interesting results have been obtained by applying density functional theory (DFT) combined with nonequilibrium Green's functions to calculate the transport characteristics [64,65]. Rocha et al. [64] considered two prototypical molecules, 8-alkane-dithiolate (octane-dithiolate) and 1,4-3-phenyl-dithiolate (tricene-dithiolate), sandwiched between nickel contacts. Very large MR values have been calculated for both cases: about 100% for the octane-dithiolate and up to 600% for the tricene-dithiolate molecule (Figure 22.6).

In close contact with all the experimental research dealing with the characterization of organic–inorganic interfaces, many authors have developed different theoretical approaches to understand the optimum conditions for spin injection into an OSC. A few authors have indicated that an ohmic contact would be more efficient for spin injection into OSC [66]. By contrast, it has been suggested in many other papers that spin injection into OSC can only be achieved across an interfacial tunnel or Schottky barrier [67,68]. In all the cases in which spin injection into OSC has been demonstrated, the organic–inorganic interfaces presented natural Schottky-like barriers, thus showing that experimentally this kind of interfacial contact seems more adequate for the realization of OSPD. A Su–Schrieffer–Heeger model was applied by Xie et al. [69] to study the importance of the electron–lattice interactions, considering interfaces featuring strong polaronic effects at both sides of them. The authors showed that under certain biases, spin-polarized injection can be achieved for direct manganite–OSC interface, while substitution of the manganite with a conventional (non-polaronic) ferromagnetic metal resulted in the absence of a spin-polarized flow.

If the injection process has attracted much attention, it is clear as well that the decoherence mechanisms in organic are

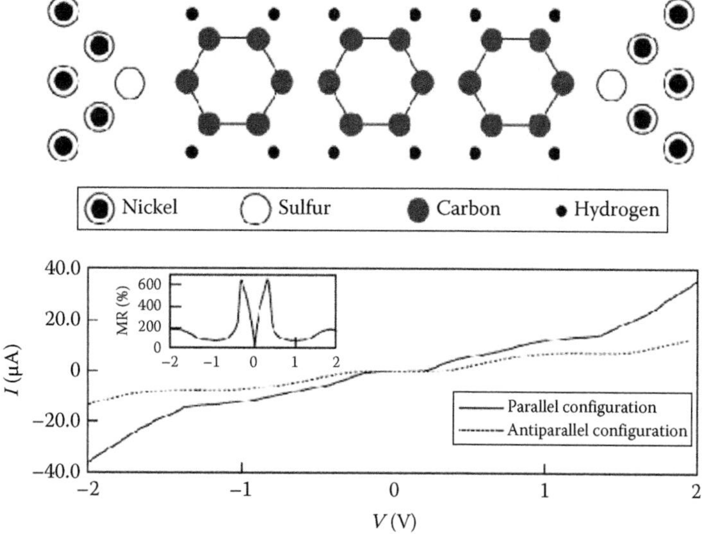

FIGURE 22.6 Schematic representation of a tricene molecule attached to a (001) Ni surface from Rocha et al. (upper figure). Theoretical calculation of the current-voltage (*I*–*V*) characteristic of a tricene–nickel spin valve. The inset shows the calculated MR ratio with values up to 600%. (From Rocha, A.R. et al., *Nat. Mater.*, 4, 335, 2005.)

being subject of intense research. In a naive approach, it is clear that spin–orbit coupling in OSC is very small, but a detailed and predictive description has been lacking. Currently, alternative mechanisms for spin decoherence are being considered. Recently, Bobbert et al. have advanced the idea that the precession of the carrier spins in the random hyperfine fields of hydrogen nuclei [70] (around 5 mT) represents the dominant source for spin scattering in OSC. The authors have derived the spin diffusion length l_S in a typical OSC and they obtain a very weak dependence of l_S on temperature, in good agreement with recent experimental data [26]. Surprisingly, it was also shown that the spin-scattering strength is much less sensitive to disorder than the carrier mobility, a result that could shed light on the apparent discrepancy between low mobility and long spin diffusion lengths in OSC. It has to be noted that spin scattering by hyperfine fields is also considered as one of the most probable mechanisms [71] responsible for the so-called organic magnetoresistance (OMAR) effect in OSC [72], that is, MR without magnetic electrodes.

22.9 Conclusions

In spite of the important achievements described above, an understanding of the basic rules governing spin injection and transport in OSC is still lacking. It is also difficult to predict now the future evolution of organic spintronics or any possible implementation into applications. Organic spintronics remains so far a fascinating scientific problem, which requires to face conceptually new physics and to develop innovative technological procedures.

In addition to this stimulating fundamental interest, the results obtained in this novel field are already encouraging. It cannot be excluded that OSC will compete with other materials for the leadership in the spintronics field or at least in some selected niche applications.

In order to link spintronics to such important organic applications like OLEDs and OFETs, it is necessary nevertheless to overcome some revealed limitations. Among these, we could cite the low voltages at which spin-polarized effects occur in electrically driven devices. While low operation voltages are required in order to reduce energy consumption, this limitation prevents the practical application of spin-polarized currents in OLEDs and OFETs, which are operated at $V > 1$ V (generally at few volts). The available data for spintronic effects in organic materials confirm so far spin injection and transport only for voltages not exceeding approximately 1 V. Expanding this working interval to about a few volts represents thus a challenging applicative goal. On the other hand, the available operation voltages do not prevent the utilization of OSPDs in sensor or memory applications.

As a final remark, we would like to point out on the immense potential of OSCs for the spintronics field. An almost infinite list of materials, easily modifiable interfaces, and cheap technologies for the growth of films on rigid and flexible substrates make us look optimistically toward the next steps of this young field.

References

1. V. A. Dediu, L. E. Hueso, I. Bergenti, and C. Taliani, Spin routes in organic semiconductors, *Nat. Mater.* **8**, 707–716 (2009).
2. F. Wang and Z. V. Vardeny, Organic spin valves: The first organic spintronics devices, *J. Mater. Chem.* **19**, 1685–1690 (2009).
3. W. J. M. Naber, S. Faez, and W. G. van der Wiel, Organic spintronics, *J. Phys. D-Appl. Phys.* **40**, R205–R228 (2007).
4. D. S. McClure, Spin-orbit interaction in aromatic molecules, *J. Chem. Phys.* **20**, 682–686 (1952).
5. D. Beljonne, Z. Shuai, G. Pourtois, and J. L. Bredas, Spin-orbit coupling and intersystem crossing in conjugated polymers: A configuration interaction description, *J. Phys. Chem. A* **105**, 3899–3907 (2001).
6. C. B. Harris, R. L. Schlupp, and H. Schuch, Optically detected electron spin locking and rotary echo trains in molecular excited states, *Phys. Rev. Lett.* **30**, 1019–1022 (1973).
7. V. I. Krinichnyi, S. D. Chemerisov, and Y. S. Lebedev, EPR and charge-transport studies of polyaniline, *Phys. Rev. B* **55**, 16233–16244 (1997).
8. R. I. Dzhioev, K. V. Kavokin, V. L. Korenev et al., Low-temperature spin relaxation in n-type GaAs, *Phys. Rev. B* **66**, 245204 (2002).
9. I. Žutić, J. Fabian, and S. Das Sarma, Spintronics: Fundamentals and applications, *Rev. Mod. Phys.* **76**, 323–410 (2004).
10. V. I. Arkhipov, E. V. Emelianova, Y. H. Tak, and H. Bassler, Charge injection into light-emitting diodes: Theory and experiment, *J. Appl. Phys.* **84**, 848–856 (1998).
11. W. Brütting, S. Berleb, and A. G. Mückl, Device physics of organic light-emitting diodes based on molecular materials, *Org. Electron.* **2**, 1–36 (2001).
12. V. Coropceanu, J. Cornil, D. A. da Silva, Y. Olivier, R. Silbey, and J. L. Bredas, Charge transport in organic semiconductors, *Chem. Rev.* **107**, 926–952 (2007).
13. T. W. Hagler, K. Pakbaz, K. F. Voss, and A. J. Heeger, Enhanced order and electronic delocalization in conjugated polymers oriented by gel processing in polyethylene, *Phys. Rev. B* **44**, 8652–8666 (1991).
14. W. P. Su, J. R. Schrieffer, and A. J. Heeger, Solitons in polyacetylene, *Phys. Rev. Lett.* **42**, 1698–1701 (1979).
15. J.-H. Park, E. Vescovo, H.-J. Kim, C. Kwon, R. Ramesh, and T. Venkatesan, Magnetic properties at surface boundary of a half-metallic ferromagnet $La_{0.7}Sr_{0.3}MnO_3$, *Phys. Rev. Lett.* **81**, 1953–1956 (1998).
16. M. P. de Jong, I. Bergenti, V. A. Dediu, M. Fahlman, M. Marsi, and C. Taliani, Evidence for Mn^{2+} ions at surfaces of $La_{0.7}Sr_{0.3}MnO_3$ thin films, *Phys. Rev. B* **71**, 014434 (2005).
17. J. M. De Teresa, A. Barthelemy, A. Fert, J. P. Contour, F. Montaigne, and P. Seneor, Role of metal-oxide interface in determining the spin polarization of magnetic tunnel junctions, *Science* **286**, 507–509 (1999).

18. J. E. Evetts, M. G. Blamire, N. D. Mathur et al., Defect-induced spin disorder and magnetoresistance in single-crystal and polycrystal rare-earth manganite thin films, *Philos. Trans. R. Soc. Lond., Ser. a-Math. Phys. Eng. Sci.* **356**, 1593–613 (1998).
19. C. W. Tang and S. A. VanSlyke, Organic electroluminescent diodes, *Appl. Phys. Lett.* **51**, 913–915 (1987).
20. A. Curioni, M. Boero, and W. Andreoni, Alq(3): ab initio calculations of its structural and electronic properties in neutral and charged states, *Chem. Phys. Lett.* **294**, 263–271 (1998).
21. A. N. Caruso, D. L. Schulz, and P. A. Dowben, Metal hybridization and electronic structure of Tris(8-hydroxyquinolato) aluminum (Alq(3)), *Chem. Phys. Lett.* **413**, 321–325 (2005).
22. G. Horowitz, D. Fichou, X. Z. Peng, Z. G. Xu, and F. Garnier, A field-effect transistor based on conjugated alpha-sexithienyl, *Solid State Commun.* **72**, 381–384 (1989).
23. Z. H. Xiong, D. Wu, Z. V. Vardeny, and J. Shi, Giant magnetoresistance in organic spin-valves, *Nature* **427**, 821–824 (2004).
24. F. J. Wang, C. G. Yang, Z. V. Vardeny, and X. G. Li, Spin response in organic spin valves based on $La_{2/3}Sr_{1/3}MnO_3$ electrodes, *Phys. Rev. B* **75**, 245324 (2007).
25. A. Riminucci, I. Bergenti, L. E. Hueso et al., Negative spin valve effects in manganite/organic based devices, arXiv:cond-mat/0701603v1.
26. V. Dediu, L. E. Hueso, I. Bergenti, et al., Room-temperature spintronic effect in Alq_3-based hybrid devices, *Phys. Rev. B* **78**, 115203 (2008).
27. S. Majumdar, H. S. Majumdar, R. Laiho, and R. Osterbacka, Comparing small molecules and polymer for future organic spin-valves, *J. Alloys Compd.* **423**, 169–171 (2006).
28. W. Xu, G. J. Szulczewski, P. LeClair et al., Tunneling magnetoresistance observed in $La_{0.67}Sr_{0.33}MnO_3$/organic molecule/Co junctions, *Appl. Phys. Lett.* **90**, 072506 (2007).
29. J. S. Moodera, L. R. Kinder, T. M. Wong, and R. Meservey, Large magnetoresistance at room-temperature in ferromagnetic thin-film tunnel-junctions, *Phys. Rev. Lett.* **74**, 3273–3276 (1995).
30. E. Y. Tsymbal, O. N. Mryasov, and P. R. LeClair, Spin-dependent tunnelling in magnetic tunnel junctions, *J. Phys. Cond. Matter* **15**, R109–R142 (2003).
31. S.-W. Liu, J.-H. Lee, C.-C. Lee, C.-T. Chen, and J.-K. Wang, Charge carrier mobility of mixed-layer organic light-emitting diodes, *Appl. Phys. Lett.* **91**, 142106 (2007).
32. V. Dediu, M. Murgia, F. C. Matacotta, C. Taliani, and S. Barbanera, Room temperature spin polarized injection in organic semiconductor, *Solid State Commun.* **122**, 181–184 (2002).
33. T. Ikegami, I. Kawayama, M. Tonouchi, S. Nakao, Y. Yamashita, and H. Tada, Planar-type spin valves based on low-molecular-weight organic materials with $La_{0.67}Sr_{0.33}MnO_3$ electrodes, *Appl. Phys. Lett.* **92**, 153304 (2008).
34. Y. Liu, T. Lee, H. E. Katz, and D. H. Reich, Effects of carrier mobility and morphology in organic semiconductor spin valves, *J. Appl. Phys.* **105**, 07C708 (2009).
35. N. A. Morley, A. Rao, D. Dhandapani, M. R. J. Gibbs, M. Grell, and T. Richardson, Room temperature organic spintronics, *J. Appl. Phys.* **103**, 07F306 (2008).
36. H. Vinzelberg, J. Schumann, D. Elefant, R. B. Gangineni, J. Thomas, and B. Büchner, Low temperature tunneling magnetoresistance on (La,Sr)MnO_3/Co junctions with organic spacer layers, *J. Appl. Phys.* **103**, 093720 (2008).
37. Y. Ogimoto, M. Izumi, A. Sawa et al., Tunneling magnetoresistance above room temperature in $La_{0.7}Sr_{0.3}MnO_3/SrTiO_3/La_{0.7}Sr_{0.3}MnO_3$ junctions, *Jpn. J. Appl. Phys. Part 2-Lett.* **42**, L369–L372 (2003).
38. J. R. Petta, S. K. Slater, and D. C. Ralph, Spin-dependent transport in molecular tunnel junctions, *Phys. Rev. Lett.* **93**, 136601 (2004).
39. S. Miwa, M. Shiraishi, S. Tanabe, M. Mizuguchi, T. Shinjo, and Y. Suzuki, Tunnel magnetoresistance of C-60-Co nanocomposites and spin-dependent transport in organic semiconductors, *Phys. Rev. B* **76**, 214414 (2007).
40. T. S. Santos, J. S. Lee, P. Migdal, I. C. Lekshmi, B. Satpati, and J. S. Moodera, Room-temperature tunnel magnetoresistance and spin-polarized tunneling through an organic semiconductor barrier, *Phys. Rev. Lett.* **98**, 016601 (2007).
41. J. H. Shim, K. V. Raman, Y. J. Park et al., Large spin diffusion length in an amorphous organic semiconductor, *Phys. Rev. Lett.* **100**, 226603 (2008).
42. M. Cinchetti, K. Heimer, J. P. Wustenberg et al., Determination of spin injection and transport in a ferromagnet/organic semiconductor heterojunction by two-photon photoemission, *Nat. Mater.* **8**, 115–119 (2009).
43. A. J. Drew, J. Hoppler, L. Schulz et al., Direct measurement of the electronic spin diffusion length in a fully functional organic spin valve by low-energy muon spin rotation, *Nat. Mater.* **8**, 109–114 (2009).
44. R. H. Friend, R. W. Gymer, A. B. Holmes et al., Electroluminescence in conjugated polymers, *Nature* **397**, 121–128 (1999).
45. I. Bergenti, V. Dediu, E. Arisi et al., Spin polarised electrodes for organic light emitting diodes, *Org. Electron.* **5**, 309–314 (2004).
46. R. Fiederling, M. Keim, G. Reuscher et al., Injection and detection of a spin-polarized current in a light-emitting diode, *Nature* **402**, 787–790 (1999).
47. V. F. Motsnyi, P. Van Dorpe, W. Van Roy et al., Optical investigation of electrical spin injection into semiconductors, *Phys. Rev. B* **68**, 245319(2003).
48. Y. Ohno, D. K. Young, B. Beschoten, F. Matsukura, H. Ohno, and D. D. Awschalom, Electrical spin injection in a ferromagnetic semiconductor heterostructure, *Nature* **402**, 790–792 (1999).
49. G. Salis, S. F. Alvarado, M. Tschudy, T. Brunschwiler, and R. Allenspach, Hysteretic electroluminescence in organic light-emitting diodes for spin injection, *Phys. Rev. B* **70**, 085203 (2004).
50. A. H. Davis and K. Bussmann, Organic luminescent devices and magnetoelectronics, *J. Appl. Phys.* **93**, 7358–7360 (2003).

51. D. R. McCamey, H. A. Seipel, S. Y. Paik et al., Spin Rabi flopping in the photocurrent of a polymer light-emitting diode, *Nat. Mat.* **7**, 723–728 (2008).
52. E. I. Rashba, Theory of electrical spin injection: Tunnel contacts as a solution of the conductivity mismatch problem, *Phys. Rev. B* **62**, R16267–R16270 (2000).
53. M. V. Tiba, W. J. M. de Jonge, B. Koopmans, and H. T. Jonkman, Morphology and electronic properties of the pentacene on cobalt interface, *J Appl. Phys.* **100**, 093707 (2006).
54. M. Popinciuc, H. T. Jonkman, and B. J. van Wees, Energy level alignment symmetry at Co/pentacene/Co interfaces, *J. Appl. Phys.* **100**, 093714 (2006).
55. Y. Q. Zhan, I. Bergenti, L. E. Hueso, and V. Dediu, Alignment of energy levels at the Alq(3)/$La_{0.7}Sr_{0.3}MnO_3$ interface for organic spintronic devices, *Phys. Rev. B* **76**, 045406 (2007).
56. Y. Q. Zhan, M. P. de Jong, F. H. Li, V. Dediu, M. Fahlman, and W. R. Salaneck, Energy level alignment and chemical interaction at Alq(3)/Co interfaces for organic spintronic devices, *Phys. Rev. B* **78**, 045208 (2008).
57. Y. Q. Zhan, X. J. Liu, E. Carlegrim et al., The role of aluminum oxide buffer layer in organic spin-valves performance, *Appl. Phys. Lett.* **94**, 053301 (2009).
58. Y. Q. Zhan and M. Fahlman, Private communication, (2009).
59. M. Grobosch, K. Dorr, R. B. Gangineni, and M. Knupfer, Energy level alignment and injection barriers at spin injection contacts between $La_{0.7}Sr_{0.3}MnO_3$ and organic semiconductors, *Appl. Phys. Lett.* **92**, 023302 (2008).
60. L. E. Hueso, I. Bergenti, A. Riminucci, Y. Q. Zhan, and V. Dediu, Multipurpose magnetic organic hybrid devices, *Adv. Mater.* **19**, 2639–2642 (2007).
61. J. C. Scott and L. D. Bozano, Nonvolatile memory elements based on organic materials, *Adv. Mater.* **19**, 1452–1463 (2007).
62. E. G. Emberly and G. Kirczenow, Molecular spintronics: Spin-dependent electron transport in molecular wires, *Chem. Phys.* **281**, 311–324 (2002).
63. R. Pati, L. Senapati, P. M. Ajayan, and S. K. Nayak, First-principles calculations of spin-polarized electron transport in a molecular wire: Molecular spin valve, *Phys. Rev. B* **68**, 100407 (R) (2003).
64. A. R. Rocha, V. M. Garcia-Suarez, S. W. Bailey, C. J. Lambert, J. Ferrer, and S. Sanvito, Towards molecular spintronics, *Nat. Mater.* **4**, 335–339 (2005).
65. D. Waldron, P. Haney, B. Larade, A. MacDonald, and H. Guo, Nonlinear spin current and magnetoresistance of molecular tunnel junctions, *Phys. Rev. Lett.* **96**, 166804 (2006).
66. J. F. Ren, J. Y. Fu, D. S. Liu, L. M. Mei, and S. J. Xie, Spin polarized injection and transport in organic polymers, *Synthetic Met.* **155**, 611–614 (2005).
67. P. P. Ruden and D. L. Smith, Theory of spin injection into conjugated organic semiconductors, *J. Appl. Phys.* **95**, 4898–4904 (2004).
68. J. H. Wei, S. J. Xie, L. M. Mei, J. Berakdar, and W. Yan, Conductance switching, hysteresis, and magnetoresistance in organic semiconductors, *Org. Electron.* **8**, 487–497 (2007).
69. S. J. Xie, K. H. Ahn, D. L. Smith, A. R. Bishop, and A. Saxena, Ground-state properties of ferromagnetic metal/conjugated polymer interfaces, *Phys. Rev. B* **67**, 125202 (2003).
70. P. A. Bobbert, W. Wagemans, F. W. A. van Oost, B. Koopmans, and M. Wohlgenannt, Theory for spin diffusion in disordered organic semiconductors, *Phys. Rev. Lett.* **102**, 156604 (2009).
71. P. A. Bobbert, T. D. Nguyen, F. W. A. van Oost, B. Koopmans, and M. Wohlgenannt, Bipolaron mechanism for organic magnetoresistance, *Phys. Rev. Lett.* **99**, 216801 (2007).
72. T. L. Francis, O. Mermer, G. Veeraraghavan, and M. Wohlgenannt, Large magnetoresistance at room temperature in semiconducting polymer sandwich devices, *New J. Phys.* **6**, 185 (2004).
73. L. D. Bozano, B. W. Kean, M. Beinhoff, K. R. Carter, P. M. Rice, and J. C. Scott, Organic materials and thin-film structures for cross-point memory cells based on trapping in metallic nanoparticles, *Adv. Funct. Mater.* **15**, 1933–1939 (2005).

23

Spin Transport in Ferromagnet/ III–V Semiconductor Heterostructures

Paul A. Crowell
University of Minnesota

Scott A. Crooker
Los Alamos National Laboratory

The processes of electrical spin injection, transport, and detection in nonmagnetic semiconductors are inherently nonequilibrium phenomena. A nonequilibrium spin polarization, which we will refer to in this chapter as *spin accumulation*, may be generated and probed by a variety of optical and transport techniques. We will focus on spin accumulation in heterostructures of ferromagnetic (FM) metals (particularly Fe) and III–V semiconductors (particularly GaAs). Much of the essential physics of these systems for both metals and semiconductors has been addressed in other chapters in this volume. The basic phenomena of spin accumulation in metals have been covered in Chapters 6 and 32. This chapter will provide a review of a sequence of experiments that have advanced the understanding of spin transport in semiconductors through their sensitivity to electron spin dynamics. To a great extent, this has been a matter of careful experimental design and execution, and a good fraction of the discussion will address tests of well-known models. The benefit of this approach has been a reliable demonstration of all-electrical devices incorporating a FM injector, semiconductor (SC) channel, and FM detector. In this context, we will show how quantitative electrical spin detection measurements can be made with these devices using both optical imaging and transport techniques, and we will conclude with some discussion of the new physics that can be addressed.

23.1 Spin Transport and Dynamics in III–V Semiconductors

The basic physics of electron spin dynamics in III–V semiconductors was established through optical pumping experiments, which were reviewed authoritatively more than two decades ago [1]. Polarized photoluminescence measurements identified many of the important spin relaxation mechanisms in GaAs, with the significant caveat that most experiments were carried out on p-type or compensated materials in order to enhance the recombination rate of spin-polarized electrons. As a result, the electron spin dynamics observed in these experiments were influenced by the presence of holes and by recombination dynamics. The advent of absorption-based techniques such as time-resolved Faraday rotation in the 1990s brought particular attention to the case of n-type GaAs doped in close proximity to the metal–insulator transition (MIT) [2]. As will be evident below, the relatively long spin lifetime (~100 ns) of electrons [2–4] in this doping range and at low temperatures provides an enormous technical advantage *if* the semiconductor can be integrated with an appropriate source and detector of spin-polarized electrons. Before considering the role of the source and detector, however, it will be useful to discuss spin transport and dynamics in GaAs in the doping range near the MIT.

The interpretation of the spin transport measurements discussed below will be based on the standard spin drift-diffusion model, which provided the framework for the original proposal of electrical spin injection into semiconductors in the 1970s [5]. A detailed discussion of the elementary physics, elucidating some of the important differences with respect to spin transport in metals, is provided in Ref. [6]. An electron spin density **S** in a semiconductor is treated as a vector field evolving according to the dynamical equation

$$\frac{\partial \mathbf{S}}{\partial t} = D\nabla_r^2\mathbf{S} + \mu(\mathbf{E}\cdot\nabla_r)\mathbf{S} + g_e\mu_B\hbar^{-1}(\mathbf{B}_{\text{eff}}\times\mathbf{S}) - \frac{\mathbf{S}}{\tau_s} + \mathbf{G}(\mathbf{r}) = 0, \tag{23.1}$$

where

D is the diffusion constant
E is the electric field
$\mathbf{B}_{\text{eff}}$ is the effective magnetic field acting on the electron spin
τ_s is the spin lifetime
G(r) is a term representing the spatial distribution of spins generated by an external source

The electron mobility is μ, and we will use $g_e = -0.44$ for the Landé g-factor of an electron in GaAs. The corresponding gyromagnetic ratio can be expressed conveniently as $\gamma = |g_e|\mu_B/\hbar = 3.8\times10^6$ s^{-1} G^{-1} or $\gamma/2\pi = 0.6$ MHz/G. Three features of semiconductors, which make the spin dynamics nontrivial, are (a) the relaxation mechanisms determining τ_s, (b) the distinct contributions to the effective field $\mathbf{B}_{\text{eff}}$ (e.g., applied fields, spin–orbit fields, or nuclear hyperfine fields), and (c) the relative importance of the drift term in the presence of internal electric fields.

In combination with ordinary diffusive transport theory, knowledge of the relevant spin lifetimes allows us to identify the critical length scales for a spin transport device. The relevant spin relaxation mechanisms are reviewed by Fabian in this volume. In the doping range of interest for us, the dominant channel of spin relaxation is the D'yakonov–Perel' (DP) mechanism [7], which corresponds to the randomization of spin during diffusive transport by precession around the instantaneous spin–orbit field. The bulk Dresselhaus spin–orbit coupling (due to the absence of inversion symmetry in GaAs) leads to a rapid decrease, proportional to n_e^{-2}, in the spin lifetime with increasing electron concentration n_e [2,4,7]. The strong dependence on carrier density reflects the k^3 dependence in the spin–orbit Hamiltonian. A maximum in the spin lifetime $\tau_s \sim 100$ ns occurs at the MIT (2×10^{16} cm^{-3}) at low temperatures, with a different set of mechanisms limiting the lifetime at lower dopings [4]. The characteristic length scales for spin transport in the semiconductor are the spin diffusion length $\lambda_s = \sqrt{D\tau_s}$ and drift length $l_d = \mu E\tau_s$. The characteristic mobility of n-GaAs at the MIT is of order 5000 cm^2/V s. Assuming a barely degenerate electron gas one finds that λ_s is on the order of several microns.

We now consider the effective fields $\mathbf{B}_{\text{eff}}$ that couple to the spin through the torque term in Equation 23.1. In this case, the relevant magnetic field scale is determined by the gyromagnetic ratio and the spin lifetime. For example, the magnetic field required in order for a spin to precess a full cycle before it decays can be estimated from $2\pi/\gamma\tau_s \sim 17$ G (for $\tau_s = 100$ ns). One immediate consequence of this relatively small field scale is that it is feasible to design a transport experiment that is extremely sensitive to the torque term in Equation 23.1. It then becomes possible to probe the effects of other "fields" that couple to spin, particularly those due to hyperfine and spin–orbit coupling. Hyperfine effects in GaAs near the MIT have been studied extensively using optical pumping [8,9], and it suffices for now to note that the effective field acting on an electron spin due to dynamically polarized nuclei can easily exceed 1 kG. As will be seen below, this hyperfine field can influence the electron spin dynamics in a transport device profoundly. In contrast, the effective spin–orbit fields in bulk GaAs are small, and their effects on spin transport and dynamics are more subtle. The reader is referred to the monograph of Winkler [10] for a complete discussion. The only case that applies to the discussion of bulk GaAs in this chapter is the effect of strain [11], which leads to spin–orbit fields that depend linearly on momentum. We will not consider heterostructures, in which spin–orbit fields linear in momentum appear due to electric fields from structural inversion asymmetry (e.g., an applied gate voltage) [12,13] or by modification of the Dresselhaus Hamiltonian by confinement [14].

For the case of a uniform electric field, the drift velocity $\mathbf{v}_d = \mu\mathbf{E}$, which for modest electric fields ($E \sim 1$ V/cm) is on the order of 10^4 cm/s. In this range, the spin drift length $\upsilon_d\tau_s$ is comparable to the spin diffusion length, and it is apparent that both the drift and diffusion terms in Equation 23.1 need to be considered. This is completely different from the situation in ordinary metals, where drift is negligible. One might guess that the drift length could become very large for transport near the MIT. In fact, however, τ_s drops precipitously due to impact ionization of donors [15] at modest electric fields above 10 V/cm. This indicates a maximum spin drift length in n-GaAs in the range of tens of microns. The physics near interfaces, where electric fields can be both large and non-uniform, is potentially much richer [6,16].

23.2 Spin Injection from Ferromagnetic Metals into Semiconductors

As the previous section makes clear, the conditions for spin transport in bulk GaAs near the MIT are rather favorable. The length scales required are readily accessible with photolithography, and the magnetic fields required to probe spin dynamics are modest. Nonetheless, significant progress on the integration of FM sources and detectors with semiconductors has occurred only since the turn of the century. The discussion here will focus on work since 2000. For a review of the status of the field at about that time, see Ref. [17].

As some stock was taken of the situation a decade ago, a few important observations emerged. First, the standard description of spin transport in metals [18–21], when applied in a straightforward manner to semiconductors [22], appeared to be consistent with the absence of spin accumulation when electrons were injected from a FM metal into a semiconductor. The essence of this argument was that the spin current flowing into the semiconductor is limited by the semiconductor's relatively large resistivity, while the low resistance of the FM metal is itself an efficient sink for any spin accumulation at the interface.

It was immediately recognized that this obstacle to spin injection, often referred to as the "conductivity mismatch problem," could be circumvented by inserting a tunnel barrier between the FM metal and the semiconductor [23–25]. In fact, this principle had already been exploited in the earliest demonstration of spin injection into GaAs using a scanning tunneling microscope with a FM tip [26]. For FM/SC devices based on $Al_xGa_{1-x}As$, in which a Schottky contact is present anyway, the only problem would appear to be engineering a barrier of the appropriate thickness and height. Two general approaches have been followed: either using a pure "tunneling Schottky" barrier [27–29], or creating an additional artificial barrier by inserting an insulator at the FM/SC interface [30–32]. For the case of a high resistance barrier, ignoring the effects of dispersion, and assuming that the spin polarization of the ferromagnet at the interface with the semiconductor is equal to the bulk value, the polarization of the tunneling current should be limited only by the spin polarization of the FM metal.

Although adopting a tunnel barrier for the injection contact was an obvious choice, it did not provide an immediate solution to the problem of electrical spin detection. At the same time, however, the community working on the problem of spin injection from magnetic *semiconductors* into their nonmagnetic counterparts developed a new version of the reliable luminescence polarization technique employed in optical-pumping experiments. By making the magnetic semiconductor one contact in a *p-i-n* junction light-emitting diode (LED), it was possible to convert a spin-polarized current of either electrons [33] or holes [34] into polarized light. This so-called "spin-LED" (originally proposed in the 1970s [35]) has since been adapted by the community working on spin injection from FM metals and is discussed in greater detail in Chapter 17. By placing the Schottky tunnel barrier in series with the *p-i-n* diode, spin-polarized electrons flowing from the metal recombine with unpolarized holes. For a quantum well, the degree of polarization of the emitted light is equal to the spin polarization of the electron component of the recombination current, provided that the component of the spin being measured is along the growth axis. Using this principle, spin injection from Fe into GaAs was first reported by Zhu et al. in 2001 [27]. A significant improvement in the efficiency of the Fe/$Al_xGa_{1-x}As$ spin-LED was then reported by the NRL group using a graded Schottky barrier [28]. In parallel, the development of spin-LEDs based on insulating tunnel barriers inserted between the ferromagnet and semiconductor was being carried out by the IMEC and IBM groups [30,31]. In all of these experiments, the extraction of the spin-polarization of the tunneling current was based on rate equation arguments that we will not discuss here [36–38]. By 2005, the Schottky spin-LED was sufficiently well-understood that it was being used for the evaluation of new potential spin injection materials, including Heusler alloys [39,40], perpendicular anisotropy materials [41,42], and others.

At the same time as spin-LEDs were being used to probe the polarization of electrons tunneling from a ferromagnet metal into a semiconductor, a different line of research was devoted to the study of spin accumulation due to unpolarized electrons incident on a FM/SC interface from the semiconductor. In 2001, Kawakami et al. [43] reported on spin accumulation at the interface between (100) GaAs and MnAs (a FM semi-metal) under optical pumping by linearly polarized light. This experiment as well as a subsequent one on Fe/GaAs [44] utilized time-resolved Faraday rotation, and hence could be carried out in simple bi-layers of the FM metal with an *n*-type GaAs epilayer. Ciuti, McGuire, and Sham interpreted this process using a tunneling approach [45,46]. The spin accumulation was enhanced in experiments in which a forward electrical bias was applied while optically pumping [47], consistent with the tunneling hypothesis. The tunneling argument implied that it should be possible to generate a spin accumulation in semiconductors under forward bias even in the absence of optical pumping, as was then observed by Stephens et al. [48] in MnAs/GaAs bilayers.

For technical reasons, the majority of the spin-LED experiments discussed above probed the component of the spin accumulation parallel to an applied magnetic field. In contrast, the experiments under optical pumping and/or a forward-bias current probed the transverse component of the accumulated spin polarization, which could be detected unambiguously through its *precession* about the applied field. The exceptions among the spin-LED experiments were those carried out by the IMEC [30,49,50] and Minnesota groups [29,36,51], which were based on the precession of spins in the semiconductor after injection from the ferromagnet. Given the relatively short spin lifetimes in quantum wells (several hundred picoseconds), significant precession could be observed in these experiments only because of the large hyperfine contributions to the effective magnetic field. Although somewhat more complicated in its implementation, a steady-state measurement that is sensitive to electron spin precession allows for a much stronger quantitative analysis of the spin dynamics. We will call experiments of this type "Hanle measurements," as they are analogs of the Hanle effect commonly observed in photoluminescence polarization under optical pumping by circularly polarized light [1].

The experiments introduced in this section demonstrated that a spin accumulation could be generated in GaAs under either forward or reverse electrical bias across a FM/SC interface. There remained, however, significant limitations. First, the effects demonstrated were local. That is, the spin polarization was measured in the semiconductor at the location of the FM source. Second, the detection techniques were strictly optical. Although there were some earlier attempts to detect optically induced spin accumulations electrically [52–54], they were not supported by the demonstration of a Hanle effect. A key point is that the tunnel barriers used in spin-LEDs, which were perfectly adequate for spin injection, were not particularly transparent to electrons at low bias voltages. The resistance-area product of a typical Schottky barrier of the design of Hanbicki et al. [28] is about 10^5–10^6 Ω μm², which is several orders of magnitude larger than the values achieved in typical magnetic tunnel junctions. This means that in simple two-terminal (FM–SC–FM) devices, the overall resistance is dominated by the interfaces. Except for the limit in which the entire semiconductor acts as a tunnel barrier [55,56], the ordinary magnetoresistance $(R_{\uparrow\uparrow} - R_{\uparrow\downarrow})/R_{\uparrow\uparrow}$ will be extremely small, as noted by Fert and Jaffrès [24]. This is in fact the other side of the two-edged sword of the "conductivity mismatch" argument. High resistivity barriers permit efficient spin injection, but the effectiveness of such a barrier for spin detection is limited by its capacity for sinking the charge current. As detailed by Fert and Jaffrès, there is a narrow range of optimal barrier resistances for two-terminal FM–SC–FM structures, but these lie away from the practical values for Fe/GaAs devices with Schottky barriers.

In the next three sections, we address how these shortcomings have been addressed in lateral Fe/GaAs devices, a process that has entailed three principal steps. The first of these was establishing that spin accumulations could be detected remotely from the FM source (or drain) contact. The experimental approach for this aspect of the project was similar to that used by Stephens et al. [48]. Scanning Kerr microscopy and local Hanle–Kerr effect studies were used to detect electron spins injected into a GaAs channel from Fe contacts under reverse bias, as expected from spin-LED experiments [57–59]. Also, spin accumulations generated by spin extraction at forward-biased contacts was demonstrated. As discussed below, these optical studies were used to directly measure the relevant spin transport parameters (τ_s, D, υ_d), and established that the drift-diffusion model of Equation 23.1 provides a satisfactory description of the spin transport.

The second step was to show that Fe/GaAs Schottky barriers, although highly resistive, could be used as electrical spin detectors [57,60]. Electrical detection of both optically injected as well as electrically injected spin was demonstrated. These latter experiments were based on the serendipitous discovery that it was possible to electrically generate *and* detect a spin accumulation at a *single* FM/SC interface. As will be evident from the discussion below, this was possible only because the measurement was sensitive to the suppression of the spin accumulation by the Hanle effect.

The third step was the implementation of a complete electrical injection and detection scheme in a lateral geometry [61]. This experiment was based on the non-local approach in which spin and charge currents are separated in space. The overall philosophy follows that developed originally by Johnson and Silsbee for metals [18,19] and implemented by many other groups in metallic nanostructures. The essential physics of the nonlocal measurement and its interpretation are covered by Johnson in another chapter in this volume, and its application in more recent experiments on metals and graphene are covered in Chapters 29 and 32. We will emphasize only those aspects that are of critical importance in semiconductors.

23.3 Fe/GaAs Heterostructures and Lateral Spin-Transport Devices

In spite of the rapid refinement of spin-LEDs and the achievement of correspondingly high spin injection efficiencies, the appropriate modification of these devices for lateral transport experiments was not immediately apparent. For some period of time, the conductivity mismatch argument and its extension by Fert and Jaffrès [24] cast a pessimistic light over efforts to fabricate a lateral FM–SC–FM spin valve. In cases where successes were reported [62], the means for extracting information about dynamics (e.g., spin lifetimes) quantitatively were not demonstrated. In particular, no Hanle effect was observed.

Our own approach to this problem evolved from a purely technical question: is it possible to image an electrically injected spin accumulation in n-GaAs in a manner similar to that demonstrated for optical-pumping experiments in the presence of an electric field [63–65]? In contrast to a spin-LED, in which electrons recombine with holes, luminescence is not a viable detection technique in this case. In contrast, Faraday or Kerr rotation is a suitable probe of spin accumulation in n-type material. Given the time and length scales for spin transport in n-GaAs samples doped near the MIT, a relatively simple magneto-optical polar Kerr effect microscope is sufficient, provided that the injected spins can be tipped perpendicular to the sample plane (thereby enabling detection by the polar Kerr effect) without perturbing the magnetization of the Fe contacts. This can be accomplished by exploiting the large uniaxial anisotropy of the epitaxial Fe/GaAs (001) interface [36]. This anisotropy, derived from the surface electronic structure, leads to a magnetic easy axis along the [011] direction and allows for small "tipping" magnetic fields to be applied along the $[01\bar{1}]$ (hard) direction without inducing significant rotation of the Fe magnetization. On the other hand, once injected into the semiconductor, an electron spin experiences a torque due to the small applied field and therefore precesses into the [100] direction, allowing for detection by polar Kerr rotation [48]. The spin accumulation, driven by lateral drift and diffusion, can then be imaged as it emerges from underneath a FM contact. (An alternative approach avoids the use of precession by imaging the spin accumulation on the cleaved edge of the sample. This has been implemented by Kotissek et al. [66].)

A cross-section of a typical FM/SC heterostructure is shown in Figure 23.1a [28]. The principal consideration in the design of these samples is the integration of an Fe/GaAs Schottky tunnel barrier with a conducting channel, while keeping the spin diffusion length long enough so that fabrication by photolithography and detection by optical Kerr microscopy are practical. The doping profile chosen for the Schottky barrier was originally perfected by Hanbicki et al. for spin LEDs [28], and the conduction band structure near the Fe/GaAs interface is shown in

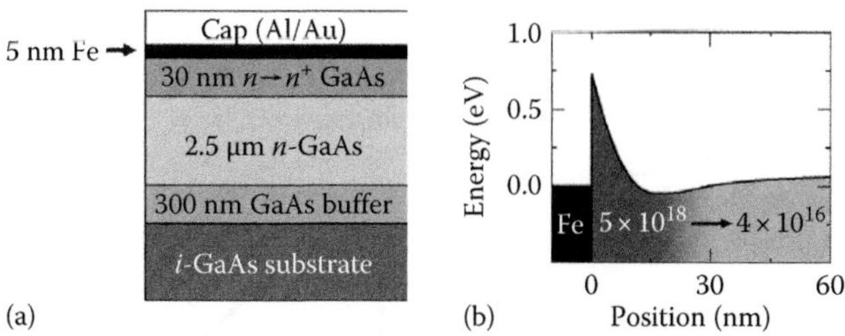

FIGURE 23.1 (a) A schematic cross-section of the heterostructures discussed in this chapter. See the text for a discussion of the dopings, which vary slightly from sample to sample. (b) Conduction band structure near the Fe/GaAs interface for the doping profile used in this work.

Figure 23.1b [28]. The heterostructures we have used in all measurements described in this chapter are grown on semi-insulating GaAs (100) wafers and typically consist of (from bottom to top) a 2.5 μm thick *n*-doped channel (Si donor concentration $n \sim 3\text{–}7 \times 10^{16}$ cm^{-3}), a 15 nm transition region in which the doping is increased from the channel doping up to 5×10^{18} cm^{-3} and an additional 15 nm thick heavily doped (5×10^{18} cm^{-3}) region. After growth of the semiconductor layers, the sample is cooled to approximately room temperature and the epitaxial Fe layer is deposited. Typically, an As-rich $c(4 \times 4)$ reconstructed surface is used. Finally, the structures are capped with Al and Au. Transverse electron microscopy (TEM) studies have established that the interface between the Fe and GaAs is atomically abrupt (within one or two monolayers) [67], and all films have the required interfacial anisotropy, with the Fe easy magnetization axis along the [011] direction. The devices are prepared using standard wet and dry etching techniques. Gold vias are deposited over SiN isolation layers to contact the FM electrodes. Micrographs of samples used in the three types of experiments discussed in this chapter are shown in Figure 23.2.

23.4 Scanning Kerr Microscopy and Local Hanle Effect Studies

23.4.1 Kerr Setup

Although any of these Fe/GaAs structures shown in Figure 23.2 can be (and have been) imaged with Kerr microscopy, the discussion of this section focuses first on the simple two-terminal device shown in Figure 23.2a. The "source" electrode is reverse-biased, meaning that electrons flow from the FM into the SC. The "drain" electrode at the opposite end of the *n*-GaAs channel is forward-biased. Figure 23.3 shows a schematic of the scanning Kerr microscopy experiment. The Fe/GaAs spin transport devices are mounted, nominally strain-free, on the vacuum cold finger of a small optical cryostat, which in turn is mounted on a x-y positioning stage. The out-of-plane component of electron spin polarization in the semiconductor, S_z, is measured at

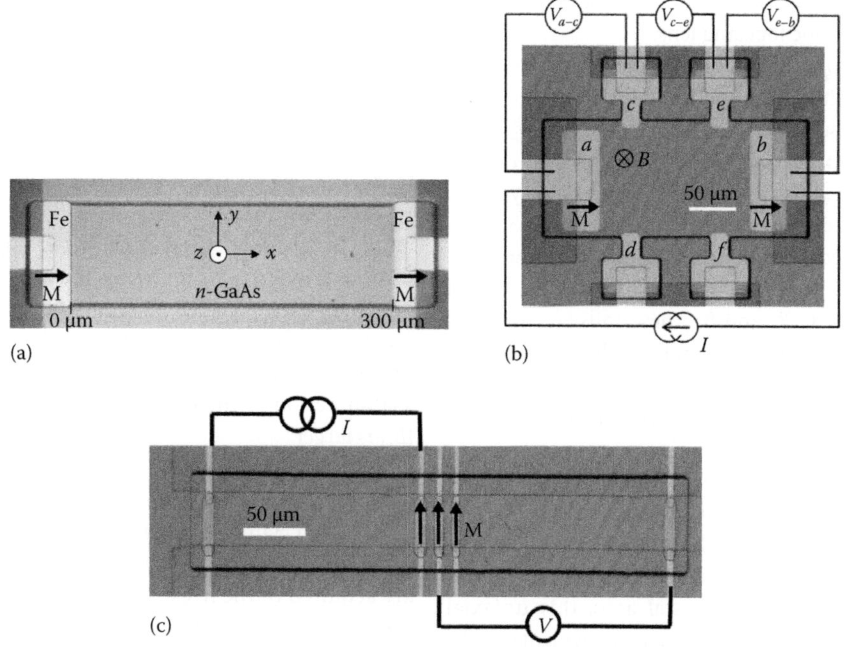

FIGURE 23.2 Micrographs of representative Fe/GaAs spin transport devices used for the experiments discussed in this chapter. (a) A simple two-terminal device used in the original Kerr imaging studies of Refs. [57,58]. (b) Hall bars used for the detection of spin accumulation at a single Fe/GaAs interface [60]. Note how the source, drain, and channel contributions to the two-terminal voltage V_{a-b} can be measured independently. (c) A typical non-local device, with a schematic of the connections to a current source and voltmeter [61]. Kerr imaging of non-local devices [59] is also discussed in the text.

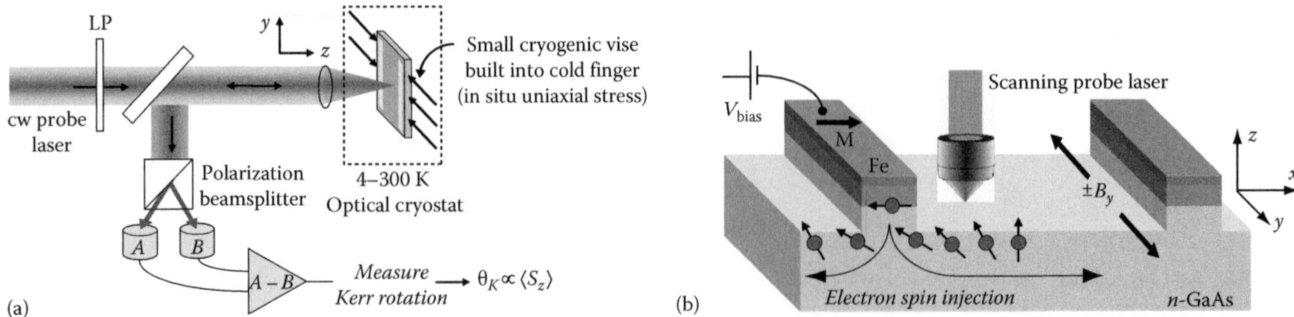

FIGURE 23.3 (a) A schematic of the scanning Kerr microscope used to image electron spin transport in Fe/GaAs devices. The measured polar Kerr rotation θ_K of the reflected probe laser beam is proportional to the out-of-plane $\hat{z}$ component of the conduction electron spin polarization, S_z. External coils (not drawn) control the applied magnetic fields B_x, B_y, and B_z. (b) A representation of an Fe/GaAs spin transport structure in cross-section. In this drawing, electrons tunnel from Fe into the n-GaAs with initial spin polarization S_0 antiparallel to the Fe magnetization **M**. A small applied transverse magnetic field $\pm B_y$ is used to precess the injected spins out-of-plane (along $\pm\hat{z}$) so that they can be measured by the probe laser via the polar Kerr effect.

a particular location via the polar Kerr rotation θ_K of a linearly polarized probe beam that is focused onto (and reflected from) the sample at normal incidence. The probe beam is derived from a narrow-band continuous-wave Ti:sapphire ring laser, which is typically tuned just below the semiconductor bandgap in order to minimize the influence of the probe beam on the resident electrons in the n-GaAs while maintaining high sensitivity to spin polarization [54,58]. The final probe focusing lens is also mounted on a positioning stage, and two-dimensional (2D) images of S_z are obtained by raster-scanning either the cryostat or (more typically) the probe focusing lens.

Electrically injected spins are studied by modulating, at kilohertz frequencies, the voltage bias applied to the Fe source/drain contacts, and then measuring the induced change in θ_K using lock-in amplifiers. As discussed above, the spins are injected along the magnetic easy axis of the Fe, which is the crystallographic [011] direction, or the $\hat{x}$-direction in Figure 23.3. They precess about the applied field **B**, and θ_K is proportional to S_z at the position of the focused laser spot. External coils control the applied tipping fields B_x, B_y, and B_z. These coils are also used together with a fixed (local) probe beam to study the relaxation and dephasing of spins at a particular spot on the sample as a continuous function of an applied transverse magnetic field—the "local Hanle effect."

The samples may also be held by a small cryogenic vise machined into the cold finger [64]. The uniaxial stress applied to the sample by the vise is uniform and can be varied in situ by a retractable actuator. For devices grown on GaAs (100) substrates and cleaved along the usual ⟨011⟩ crystal axes, this uniaxial stress along ⟨011⟩ leads to nonzero off-diagonal elements of the GaAs crystallographic strain tensor, ε_{xy}. For electrons moving in the x–y sample plane, ε_{xy} couples directly to electron spin σ and momentum **k** via a spin–orbit coupling energy $\varepsilon_{xy}(\sigma_y k_x - \sigma_x k_y)$, which has the same form [11,64] as the well-known Rashba spin–orbit Hamiltonian. The resulting effective magnetic field $\mathbf{B}_\varepsilon$ lies in-plane and is orthogonal to the electron momentum **k**.

23.4.2 Imaging Spin Injection, Local Hanle–Kerr Studies, and Modeling

In Figure 23.4 we show Kerr-rotation images and the associated local Hanle curves for electrically injected spins. Figure 23.4a shows a spin transport device having rectangular Fe/GaAs source and drain contacts at either end of a 300 μm long n-GaAs channel. We image the 80 × 80 μm region shown by the dotted square, which includes part of the Fe injection contact and the bottom edge of the n-GaAs channel. An image of the reflected probe laser power (see Figure 23.4b) clearly shows these device features. With this contact under reverse voltage bias, so that spin-polarized electrons are injected from the Fe into the channel, Figure 23.4c shows a series of Kerr-rotation images as the applied field B_y is varied from −8.4 to +8.4 G. Injected electrons, initially polarized along the $-\hat{x}$ direction, precess into the $+\hat{z}$ direction when B_y is oriented along $-\hat{y}$. When B_y inverts sign and is oriented along $+\hat{y}$, the spins are tipped into the opposite direction (along $-\hat{z}$), reversing the sign of the measured Kerr rotation. These injected electrons flow down the channel with an average drift velocity υ_d that is the same in all the images. The drifting spins precess at a rate proportional to $|B_y|$; thus, the spatial period of the observed spin precession is shorter when $|B_y|$ is larger.

When the probe laser is fixed at a point in the n-GaAs channel and $S_z \propto \theta_K$ is measured as a continuous function of B_y, we obtain local Hanle curves having the characteristic antisymmetric lineshape shown in Figure 23.4d. These local Hanle curves are acquired in the n-GaAs channel at different distances from the Fe source contact (8–120 μm), as labeled. With increasing distance, the amplitudes of the curves fall, their characteristic widths decrease, and they develop multiple oscillations associated with multiple precession cycles. The detailed structure of this family of Hanle curves contains considerable information about the dynamics of electron spin transport, including the electron spin lifetime τ_s, diffusion constant D and drift velocity

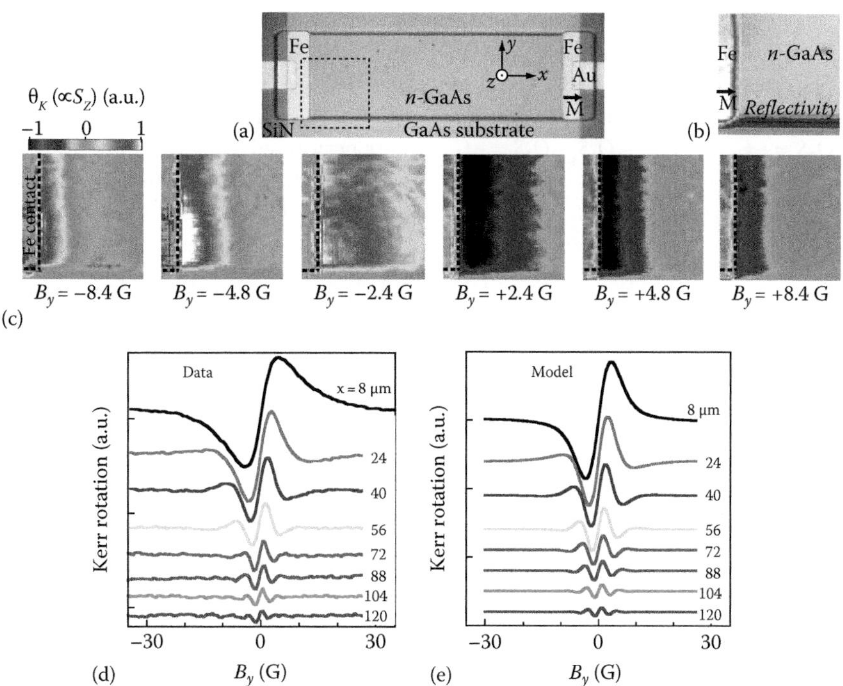

FIGURE 23.4 (See color insert) (a) A lateral spin transport device consisting of a 300 μm long *n*-GaAs channel separating Fe/GaAs source and drain contacts. The dotted square shows the 80 × 80 μm region imaged. (b) An image of the reflected probe laser power, showing device features. (c) Images of electrical spin injection and transport in the *n*-GaAs channel. The electron current $I_e = 92\,\mu A$. Electrons are injected with initial spin polarization $S_0 \parallel -\hat{x}$ and B_y is varied from −8.4 to +8.4 G (left to right), causing spins to precess out of and into the page ($\pm\hat{z}$), respectively. (d) A series of local Hanle curves (θ_K vs. B_y) acquired with the probe laser fixed at different distances (8–120 μm) from the source contact (offset for clarity). (e) Simulated local Hanle curves using a 1D spin drift-diffusion model (Equation 23.2), with $\tau_s = 125$ ns, $\upsilon_d = 2.8 \times 10^4$ cm/s, and $D = 10\,\text{cm}^2/\text{s}$.

υ_d [64,68]. The Hanle curves invert when the magnetization **M** of the Fe contacts is reversed, as expected.

To model these local Hanle data we consider a simple 1D picture of spin drift, diffusion, and precession in the GaAs channel. Spin-polarized electrons, injected with $S_0 = S_x$ (at $x = 0$ and $t = 0$), precess as they flow down the channel with a drift velocity υ_d, arriving at the point of detection x_0 at a later time $t = x_0/\upsilon_d$. At this point, $S_z = S_0 \exp(-t/\tau_s)\sin(\Omega_L t)$, where $\Omega_L = g_e \mu_B B_y/\hbar$ is the Larmor precession frequency. The actual signal is therefore computed by averaging the spin orientations of the precessing electrons over the Gaussian distribution of their arrival times (which has a half-width determined by diffusion).

$$S_z(B_y) = \int_{x_0}^{x_0+w}\int_0^{\infty} \frac{S_0}{\sqrt{4\pi Dt}} e^{-(x-\upsilon_d t)^2/4Dt} \times e^{-t/\tau_s} \sin(\Omega_L t)\,dt\,dx, \qquad (23.2)$$

where

$\upsilon_d = \mu E$ is the drift velocity
μ is the electron mobility
E is the electric field in the channel

The spatial integral accounts for the width w of the source contact. This type of averaging is the basis of the Hanle effect observed in optical-pumping experiments and in previous spin transport experiments in metals [19,69] and semiconductors [29,48,49]. Using this model, Figure 23.4e shows that good overall agreement with the local Hanle data are obtained using $D = 10\,\text{cm}^2/\text{s}$, $\upsilon_d = 2.8 \times 10^4$ cm/s, and $\tau_s = 125$ ns. The large drift velocity and spin lifetime in these devices allow access to a spatial regime far from the contacts and well beyond a spin diffusion length ($x_0 > \sqrt{D\tau_s}$), in which the average time-of-flight from the source to the point of detection, $T = (x_0 + w/2)/\upsilon_d$, determines the characteristic "age" of the measured spins. In this limit, the first peak in the data ($B_y = B_{\text{peak}}$) is the field at which electrons have precessed through one-quarter Larmor cycle, so that $T = \pi/(2\Omega_L) = \pi\hbar/(2 g_e \mu_B B_{\text{peak}})$. Thus, in Figure 23.4d, $B_{\text{peak}} = 1.05$ G when $x_0 = 88\,\mu\text{m}$, indicating that $T \sim 380$ ns and $\upsilon_d \sim 2.8 \times 10^4$ cm/s.

Simulating lateral spin flows in more complicated device geometries requires more sophisticated models of spin drift and diffusion. 2D models generally suffice for planar devices in which spin transport occurs in epilayers that are thinner than the characteristic spin diffusion lengths. For the case of applied in-plane magnetic and electric fields ($B_{x,y}$, $E_{x,y}$), the steady-state spin polarization **S(r)** can then be derived from the 2D version of the spin drift-diffusion equation. Although Equation 23.1 in 2D can be solved analytically in certain limits [59], it is often easier to compute S_z from a set of coupled spin drift-diffusion equations in 2D using numerical Fourier transform methods, particularly when including spin–orbit coupling terms [64,68]. For in-plane electric and magnetic fields, and for spin–orbit coupling due to off-diagonal strain ε_{xy} in bulk GaAs, the three equations determining the steady-state spin densities $S_{x,y,z}(\mathbf{r})$ are $O_1S_x + O_2S_z = -G_x$, $O_1S_y + O_3S_z = -G_y$ and $O_4S_z - O_2S_x - O_3S_y = -G_z$, where

$$O_1 = D\nabla_r^2 + \mu\mathbf{E}\cdot\nabla_r - C_s^2D - \frac{1}{\tau_s}, \tag{23.3}$$

$$O_2 = \frac{g_e\mu_B B_y}{\hbar} + C_s(2D\nabla_x + \mu E_x), \tag{23.4}$$

$$O_3 = \frac{-g_e\mu_B B_x}{\hbar} + C_s(2D\nabla_y + \mu E_y), \tag{23.5}$$

$$O_4 = D\nabla_r^2 + \mu\mathbf{E}\cdot\nabla_r - 2C_s^2D - \frac{1}{\tau_s}. \tag{23.6}$$

$G_{x,y,z}(\mathbf{r})$ are the generation terms, and $C_s = C_3m\varepsilon_{xy}/\hbar^2$ is the spin–orbit term that couples spin to the off-diagonal elements of the strain tensor in GaAs, ε_{xy} [11,68].

The presence of effective magnetic fields ($\mathbf{B}_{\text{eff}} = \mathbf{B}_\varepsilon$) due to spin–orbit coupling causes moving spins to precess, even in the absence of applied magnetic fields [64,70]. The effective magnetic fields are directly revealed in local Hanle measurements. For example, for electrons flowing in the $\hat{x}$ direction as shown in Figure 23.4, $\mathbf{B}_\varepsilon$ is directed along $\pm\hat{y}$. $\mathbf{B}_\varepsilon$ therefore either directly *augments* or *opposes* the applied (real) magnetic field B_y that is used to acquire the local Hanle curve. The net effect is therefore to shift the Hanle curve to the left or right, so that it is no longer antisymmetric with respect to B_y. An example of this is shown in Figure 23.5a, which shows local Hanle data acquired

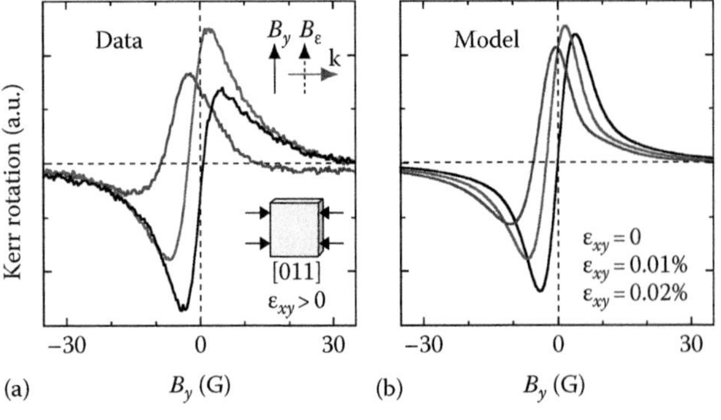

FIGURE 23.5 (a) Local Hanle effect (S_z vs. B_y) from electrically injected spins in the presence of spin–orbit coupling and effective magnetic field $\mathbf{B}_\varepsilon$ due to strain. The device is the same as shown in Figure 23.4. These data are acquired at a point 4 μm from the source contact, for three values of uniaxial stress applied along the [011] substrate axis. Curves shift to the left with increasing stress, as the effective magnetic field $\mathbf{B}_\varepsilon \parallel \hat{y}$ increases due to the induced shear strain ε_{xy}. (b) Simulated local Hanle data, for $\varepsilon_{xy} = 0, 1, 2\times10^{-4}$.

with increasing uniaxial stress applied to the device substrate along the [011] direction, which effectively turns on a Rashba-like spin–orbit interaction in the bulk n-GaAs channel of the device. The corresponding simulations shown in Figure 23.5b were performed using the numerical approach described above, and qualitatively reproduce the data when the applied strain $\varepsilon_{xy} = 0$, 1, and 2×10^{-4}.

23.4.3 Imaging Spin Injection and Extraction, Both Upstream and Downstream of the Current Path

Scanning Kerr microscopy can also be used to image the spin currents that exist when electrically injected spins diffuse outside of the charge-current path. This is demonstrated on a "non-local" Fe/GaAs spin transport device of the type shown in Figure 23.2c, and again (with contact labels and wiring diagram) in Figure 23.6a. These devices have five Fe contacts on a n-GaAs channel, with easy-axis magnetization $\mathbf{M}$ along the $\hat{y}$ direction. We focus on spin injection and spin extraction at contact 4. The dotted square in Figure 23.6a shows the 65 × 65 μm region around contact 4 that was imaged. The reflected probe power in this region (Figure 23.6b) clearly shows contact 4, the n-GaAs channel, and the edges of the SiN layer. Under bias, the electron current I_e flows *within* the imaged region when electrons flow between contacts 4 and 5. In this case the electric field in the channel, E_x, is nonzero in the imaged region and the drift velocity $v_d = \mu E_x$ may either augment or oppose electron diffusion away from contact 4, depending on the direction of I_e. Conversely, when I_e flows between contacts 4 and 1, the current path is *outside* of the imaged region. In this case, E_x is nominally zero in the imaged region regardless of I_e, and spin transport in this region should be purely diffusive.

Figure 23.6c and d shows the imaged spin polarization for the case of spin injection (that is, the Fe/GaAs Schottky contact is reverse-biased and electrons flow from Fe into n-GaAs). In this device, majority spins having spin orientation antiparallel to the Fe magnetization $\mathbf{M}$ are injected for all reverse biases. The images show two cases, in which charge current path lies within or outside of the imaged region. For clarity, contact 4 in these images is outlined by a dotted rectangle, and the black arrows indicate the direction of the electron current I_e. In Figure 23.6c, injected electrons drift slowly to the right along $+\hat{x}$. The electron current $I_e^{4\to5} = 25\,\mu\text{A}$ and the voltage across the Fe/GaAs interface is $V_{int} \simeq 53$ mV. All of these images show the *difference* between Kerr images acquired at $B_x = +2.4$ and -2.4 G. In this way, field-independent backgrounds are subtracted off, and only the signal that depends explicitly on electron spin precession remains. Figure 23.6c clearly shows a cloud of spin-polarized electrons emerging from and flowing away from contact 4, with an apparent spatial extent of order 15 μm in the n-GaAs channel. Note that the apparent extent of S_z in these images depends on $|B_x|$, as shown earlier in Figure 23.4. In this case, both drift and diffusion drive a net flow of spins to the right.

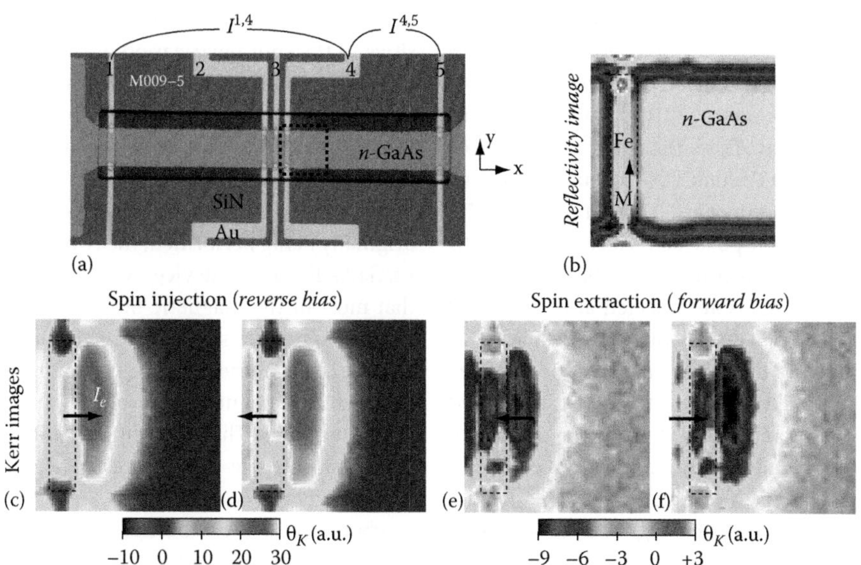

FIGURE 23.6 (See color insert.) (a) A spin transport device with five Fe/GaAs contacts (#1–5). A 65 × 65 μm region around contact 4 is imaged (dotted square). (b) The reflected probe power in this region, showing device features. (c,d) Images of the majority spin polarization that is electrically injected from contact 4 (B_x = 2.4 G). Arrows show the direction and magnitude of net electron current, I_e. In (c), spins are injected and drift/diffuse to the right ($I_e^{4\to5}$ = 25 μA). In (d), spins are injected and drift to the *left* ($I_e^{4\to1}$ = 25 μA); the spins to the right of contact 4 are diffusing outside of the current path. (e,f) Images of minority spin accumulation due to spin extraction at contact 4, both within and outside of the charge current path.

A remarkably similar image, however, is observed in Figure 23.6d for the case in which spins are injected out of contact 4 under similar bias conditions ($I_e^{4\to1}$ = 25 μA) but are drifting to the *left*, along $-\hat{x}$. In this case the current path is outside the imaged region, and E_x in the imaged region is nominally zero. The cloud of spins seen to the right of contact 4 are those spins that have diffused out from under the contact, and which are now flowing to the right due to diffusion alone. The similarity of both the magnitude and the spatial extent of S_z in Figure 23.6c and d indicates that diffusion, rather than drift, plays the major role in determining the overall transport of spins under these low injection bias conditions. These images show that robust spin currents, spatially separated from any charge currents, can be generated in the *n*-GaAs channel. It is precisely these "upstream" spin currents that we will measure using all-electrical non-local spin detection, as discussed later in the chapter.

In addition to electrical spin injection from a ferromagnet into a semiconductor, a spin polarization can also accumulate in a semiconductor when electrons flow from the semiconductor into the FM at a forward-biased Schottky barrier [48,57]. Figure 23.6e and f show the corresponding set of images for the case of spin accumulation due to spin extraction out of contact 4 in this Fe/GaAs device. At this bias, the preferential tunneling of majority electrons from *n*-GaAs into contact 4 leaves behind an accumulation of *minority*-spin electrons in the channel, as evidenced by the opposite sign of the Kerr rotation signal. This accumulated minority spin polarization diffuses out from under the contact, either into the current path (where drift now *opposes* diffusion), or outside of the current path (where spins diffuse freely).

Figure 23.6e and f show spin accumulation due to spin extraction both within and outside of the current path under low forward bias conditions ($I_e^{5\to4} = I_e^{1\to4}$ = 25 μA). In both cases, the apparent spatial extent of the accumulated S_z is similar to the case of purely diffusive spin transport. Local Hanle data confirm this, showing decay lengths of order 10 μm. Using these images and the associated local Hanle data (not shown), the important spin-transport parameters $-\tau_s$, D, and υ_d—can be accurately determined as a function of temperature and bias conditions. Knowledge of these parameters is especially useful for interpreting and understanding all-electrical measurements of spin injection and spin detection in similar devices.

23.5 Electrical Detection of Spin Accumulation

As shown by the Kerr imaging, a spin accumulation due to either spin injection or extraction forms at the Fe/GaAs interface under reverse or forward bias respectively. A rough calibration of the sensitivity of the Kerr microscope indicates that spin polarizations of 5%–10% exist within a few microns of the injection electrode. We now turn to the question of how this spin accumulation can be detected electrically.

An obvious experiment to carry out along these lines is to inject spin-polarized electrons into GaAs by optical pumping [1] and then detect the correspond change in the conductance of a nearby FM/SC interface. As noted previously, related ideas had been pursued in earlier experiments on vertical FM/SC structures [52–54]. However, because in these studies the semiconductor was optically excited through the semi-transparent FM contact, the observed signals were always comparable to background effects due to hot electrons and from magneto-absorption of the pump light by the FM contact. Moreover, no Hanle effect was observed. These shortcomings can be overcome in lateral devices by injecting spin-polarized electrons into the *n*-GaAs channel, away from the FM contact [57,71]. Doing so avoids magneto-absorption effects and allows the spins to cool before they drift and diffuse under the Fe contact. A schematic showing the implementation of this experiment is shown in Figure 23.7a, where we optically inject spin-polarized electrons, initially oriented along $\pm\hat{z}$, into the channel using circularly polarized modulated light as depicted. Spins are injected at a small spot approximately 40 μm from the edge of the forward-biased Fe drain contact. A Kerr image of these spins in zero field is shown in Figure 23.7b. Small applied magnetic fields B_y cause these spins to precess as they drift and diffuse toward the drain, tipping them into a direction parallel or antiparallel to the Fe magnetization **M**. The spin-dependent change in the conductance is then measured by lock-in techniques. Figure 23.7c shows, as a continuous function of B_y, the spin-dependent change in the voltage across the device that is explicitly due to the presence of the optically injected and spin-polarized electrons at the drain contact. The signal, which reverses when **M** is reversed, shows the same antisymmetric local-Hanle lineshape observed in the imaging of electrical injection, as expected. These data

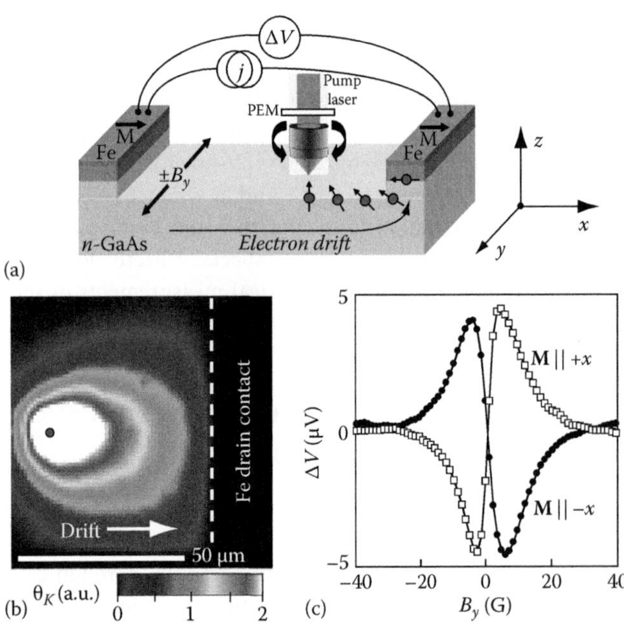

FIGURE 23.7 (a) A schematic diagram of an experiment showing electrical detection of optically injected spins. Modulated circularly polarized light produced by passing a linearly polarized pump beam (50 μW at $\lambda = 785$ nm) through a photoelastic modulator (PEM) is incident on the semiconductor channel. The spin-dependent change in device voltage is measured using lock-in techniques. (b) A Kerr image of these optically injected spins diffusing and drifting toward the drain contact. $B_y = 0$. The spins are generated in the small dot approximately 40 μm from the electrode. (c) Spin-dependent voltage change as a function of B_y, which induces precession of the optically injected spins parallel or antiparallel to **M**. Data for the two different magnetization directions of the detector are shown.

demonstrate that these Schottky tunnel barrier contacts are sensitive to the spin polarization of the electrons in the GaAs and therefore can be used as electrical spin detectors. In these studies, the spin-dependent voltages are on the order of microvolts, which represents a change of only a few parts per million of the total voltage across the device.

When the Kerr imaging experiments discussed above were originally being performed, the two-terminal resistance of these Fe/GaAs/Fe lateral devices was also measured, but we found that most of the magnetic field dependence was due to anisotropic magnetoresistance (AMR) in the Fe electrodes as well as a dependence of the effective tunneling resistance on the orientation of the magnetization of the contact with respect to the crystalline axes. The latter effect, which is now known as tunneling anisotropic magnetoresistance (TAMR) [72,73], could be as large as 1% of the device resistance. While attempting to characterize these backgrounds, we carried out measurements on devices with the electrodes several hundred microns apart, for which we reasoned that only the background AMR and TAMR effects should appear. A micrograph of one of these structures is shown in Figure 23.2b. When a small magnetic field was applied *perpendicular* to the sample, a peak in the two-terminal resistance appeared at zero magnetic field, as shown in Figure 23.8. At this field scale (~100 G), the magnetization of the contact was unchanged, and so the effect could therefore be attributed to some effect in the semiconductor. Using the additional contacts on these devices, it was also possible to measure the voltage drops across the reverse-biased source contact, the forward biased drain contact, and the channel individually. In the vast majority of cases, the peak occurred only in a measurement across the forward-biased drain, and *never* appeared in the channel voltage.

Given the correspondence with the field scales observed in the Kerr imaging experiments, we hypothesized that the signature is due to dephasing of the spin accumulation at the Fe/GaAs interface, and subsequent experiments have verified that this is the case. The magnitude of the peak is the enhancement of the voltage drop across the tunnel junction due to the presence of a spin accumulation, which is destroyed (dephased) as the magnetic field is applied. This measurement and its interpretation are discussed in more detail in Ref. [61]. We represent the Fe/GaAs interface by a 1D strip of length *w*. Spins polarized along $\hat{x}$ are generated at a point x_1 on the interface and diffuse to another point x_2, at which they are detected. The average steady-state spin accumulation is obtained by integrating the steady-state solutions of the drift diffusion equation over x_1 and x_2:

$$S_x(B_z) = \int_0^w\int_0^w\int_0^\infty \frac{S_0}{\sqrt{4\pi Dt}} e^{-(x_1 - x_2 - v_d t)^2/4Dt} \times e^{-t/\tau_s}\cos(\Omega_L t)\,dt\,dx_1\,dx_2, \quad (23.7)$$

where S_0 is the spin generation rate (per unit contact length) and the other parameters are the same as in Equation 23.2. As with

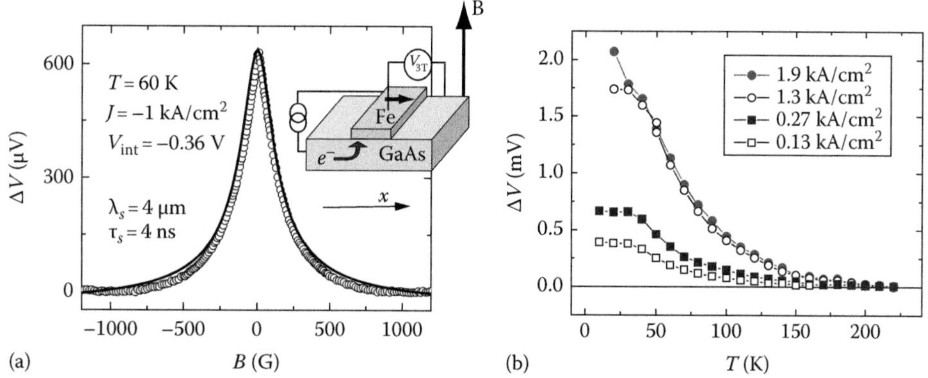

FIGURE 23.8 (a) The voltage measured across a forward-biased Fe/GaAs interface as a function of perpendicular magnetic field. A constant offset of 0.36 V has been subtracted from the data. The measurement geometry is shown in the inset. (b) The magnitude of the zero-field peak is shown as a function of temperature for several different forward bias current densities.

the imaging measurements, most of the parameters describing this Hanle curve can be obtained by independent transport or optical characterization. A typical fit is shown in Figure 23.8.

This single-contact measurement might appear to be a special case, particularly when, as demonstrated below for Fe/GaAs Schottky barriers, measurements with a separate source and detector are possible. There are some cases, however, such as that discussed by Tran et al. [74], in which this approach allows the experimentalist to work with highly resistive tunnel barriers that cannot be probed by other techniques. Outside of the formalities of the drift-diffusion model, this type of measurement has a relatively simple interpretation in terms of two time scales: a residency time τ_r between spin generation and detection as well as the spin lifetime τ_s. Some simple statements can be made by direct analogy with a traditional optical Hanle effect experiment based on polarized photoluminescence [1] and the corresponding rate equations. The magnitude of the spin accumulation at zero field is proportional to $1/(1 + \tau_r/\tau_s)$, while the width of the Hanle curve is proportional to the total relaxation rate $1/\tau_r + 1/\tau_s$. Expressed in this way, the tradeoff in the case of a resistive tunnel barrier is clear: the signal becomes small because τ_r is long, but the width of the Hanle curve can still be quite narrow (as it will depend only on τ_s). This makes a small spin accumulation easier to measure in the sense that (a) it can still be inferred from the magnetic field dependence, even if its magnitude is small; and (b) the transport properties of the bulk semiconductor are not important provided that it does not function as a strong source of spin relaxation. Recently, the single-contact approach has been applied to silicon-based devices with artificial tunnel barriers, in which observation of a Hanle effect at room temperature was reported [75].

23.6 Non-Local Detection of Spin Accumulation

We now turn to the overall goal of a device in which a spin accumulation is generated at one FM electrode, transported, and then detected with a separate electrode. If the reader consults previous reviews of transport in FM–SC heterostructures [17], a few clear themes emerge. First, backgrounds from local Hall effects, AMR, and TAMR complicate virtually any measurement in which a charge current flows through a FM detection electrode. Second, the most definitive way to establish the existence of a spin accumulation is to probe its dynamics. The first of these points can be addressed by separating the charge current in a device from the spin current. The second is addressed naturally by the observation of a Hanle effect.

The classic means by which spin and charge currents are separated is the non-local geometry of Johnson and Silsbee [18,19], which is illustrated in Figure 23.9. Spins are introduced at a FM injector by an ordinary spin-polarized charge current. The charge current is sunk at an electrode at one end of the channel, but the spin accumulation created at the injector diffuses in both directions. A second FM contact, the detector, is located on the *opposite* branch of the channel, outside the path of the charge current. The lower part of Figure 23.9 shows a spatial profile of the spin-dependent chemical potentials for the ideal case in which no electric field (and hence no charge

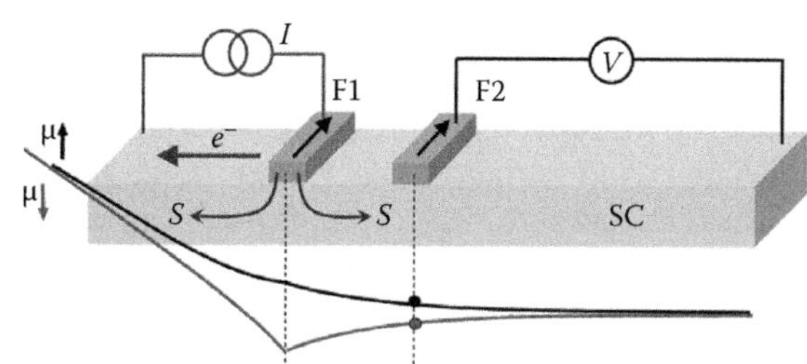

FIGURE 23.9 A schematic representation of a non-local spin transport device. F1 is the ferromagnetic injector, F2 is the detector, and SC is the semiconductor channel. Spin **S** diffuses in both directions from the injector contact. A representation of the spin-dependent chemical potentials $\mu_\uparrow$ and $\mu_\downarrow$ is shown for the ideal case in which the electric field vanishes to the right of the injector. The voltage V is measured for both magnetization states of the detector.

current) flows in the right-hand branch of the channel. The detector itself acts as a spin-dependent voltmeter, with chemical equilibrium established by the flow of a pure spin current between the FM detector and the channel. If the FM detector were a half-metal (with the Fermi level in a single spin band and therefore unity polarization), the electrochemical potentials for each of the two magnetization states ↑ and ↓ of the detector would be equal to $\mu_\uparrow$ and $\mu_\downarrow$. The voltage difference $\Delta V = (\mu_\uparrow - \mu_\downarrow)/e$ would then provide a direct measurement of the spin accumulation. In practice, the polarization P_{FM} of the FM at the Fermi level is less than one, and spin flips can occur as carriers flow back and forth across the detector interface. The latter process is accounted for by an efficiency factor η, which we can estimate from spin-LED measurements to be of order 0.5 for Fe/GaAs [37]. We make the assumption that the spin current flowing into (or out of) the detector does not impact the measurement, which corresponds to the limit of high-interface resistances. (In practice, the resistance of all of the devices discussed here is dominated by the interfaces. This is not necessarily the case in metallic devices, as discussed in Chapter 6 by Johnson.)

The spin-dependent electrochemical potential is inferred from two non-local voltage measurements $V_{\uparrow\uparrow}$ and $V_{\uparrow\downarrow}$ obtained with the detector and injector magnetizations parallel and antiparallel, respectively. The difference $\Delta V_{NL} = V_{\uparrow\downarrow} - V_{\uparrow\uparrow}$ can be related to the spin-dependent chemical potential difference $\Delta\mu$ by

$$\Delta\mu = \frac{e\Delta V_{NL}}{\eta P_{FM}}, \tag{23.8}$$

where P_{FM} is the spin polarization of the ferromagnet at the Fermi level. The electron spin polarization P_{GaAs} in the semiconductor can be obtained from $\Delta\mu$, the carrier density n and the density of states $\partial n/\partial\mu$:

$$P_{GaAs} = \frac{n_\uparrow - n_\downarrow}{n_\uparrow + n_\downarrow} = \frac{\Delta\mu}{n}\left(\frac{\partial n}{\partial\mu}\right). \tag{23.9}$$

Some indication of the overall efficiency is the *non-local resistance* $\Delta V_{NL}/I_{inj}$, which is proportional to the polarization per unit carrier flowing across the interface, which we denote P_j. We emphasize here that our samples are not truly degenerate semiconductors, and we therefore do not have direct knowledge of the density of states from the carrier density. One can, however, make a crude estimate of the order of magnitude of ΔV_{NL} if we assume a Pauli-like density of states for a degenerate semiconductor of carrier density $n \sim 3 \times 10^{16}$ cm^{-3} and an effective mass $m^* = 0.07m_e$. Given the injected polarizations from spin-LED measurements and the approximate spatial dependence $P_{GaAs}(x) \propto e^{-x/\lambda_s}$, a conservative estimate of the spin polarization at a detector several microns from the source is $P_{GaAs} \sim 0.01$. We estimate the efficiency factor $\eta \sim 0.5$ from from spin-LED measurements [37]. Using a literature value $P_{FM} = 0.4$ for the spin-polarization of iron [76], we find

$$\Delta V_{NL} \sim \frac{\eta P_{FM}\hbar^2(3\pi^2 n)^{2/3}}{3m^* e} P_{GaAs} = 10\,\mu V. \tag{23.10}$$

This is a large signal by the standards of a typical transport measurement. In our best devices, a noise floor of order 10 nV/$\sqrt{Hz}$ can be reached, although occasional devices will simply be too noisy to resolve the spin-dependent chemical potential shift. In practice, most of the difficulties in the experiments discussed below derive from the non-ideal aspects of the non-local measurement. In contrast to the cartoon representation of Figure 23.9, a significant voltage drop (up to several mV) will develop at the nonlocal detector due to spreading of the charge current from the source. This spin-independent background depends relatively strongly on magnetic field and temperature. Careful cryogenic practice, particularly temperature regulation, is necessary in order to obtain a stable background voltage throughout a measurement.

We now turn briefly to a discussion of the Hanle effect. Since the spin lifetimes and diffusion constants for these samples are known from the optical measurements discussed previously, it is possible to solve Equation 23.1 in steady state given the geometry of the source and detector. As was noted above, the typical field scale (width of a Hanle curve) is on the order of 100 G. For fields applied in the vertical direction, this will have no impact on the magnetizations of the contacts. It should therefore be possible to prepare the source and detector in either parallel or antiparallel states and obtain Hanle curves in both cases. The large interfacial anisotropy of our samples facilitates this measurement, since the Fe contacts remain uniformly magnetized at remanence.

A micrograph of a typical non-local device is shown in Figure 23.2c. Photolithography and a combination of wet and dry etching are used to pattern a large mesa and a series of FM electrodes. Two of these are located over 150 μm from the center of the device and serve as the sink for the charge current and the reference contact for the non-local voltage measurement. The bias current is generated with a current source and the non-local voltage V_{NL} is measured with a nanovoltmeter. Data for a device with channel doping 5×10^{16} cm^{-3} obtained at 60 K are shown in Figure 23.10. Figure 23.10 shows raw data obtained in the longitudinal (or "spin-valve") geometry, in which the spin-dependent non-local voltage depends only on the relative orientations of the source and detector. The expected discontinuities when the source and detector switch between parallel and antiparallel states are superimposed on a background voltage (approximately −165 μV in this case) due to the small amount of unwanted charge current spreading in the non-local geometry. This background is removed in Figure 23.10c. Data obtained in a perpendicular field (Figure 23.10b) show the Hanle effect when the source and detector are prepared in either the parallel or antiparallel states. In this case, there is a parabolic background due to the Lorentz forces acting on the carriers, but significantly the data for the two spin states merge at high fields, where the spins are completely dephased. After subtraction of the spin-independent

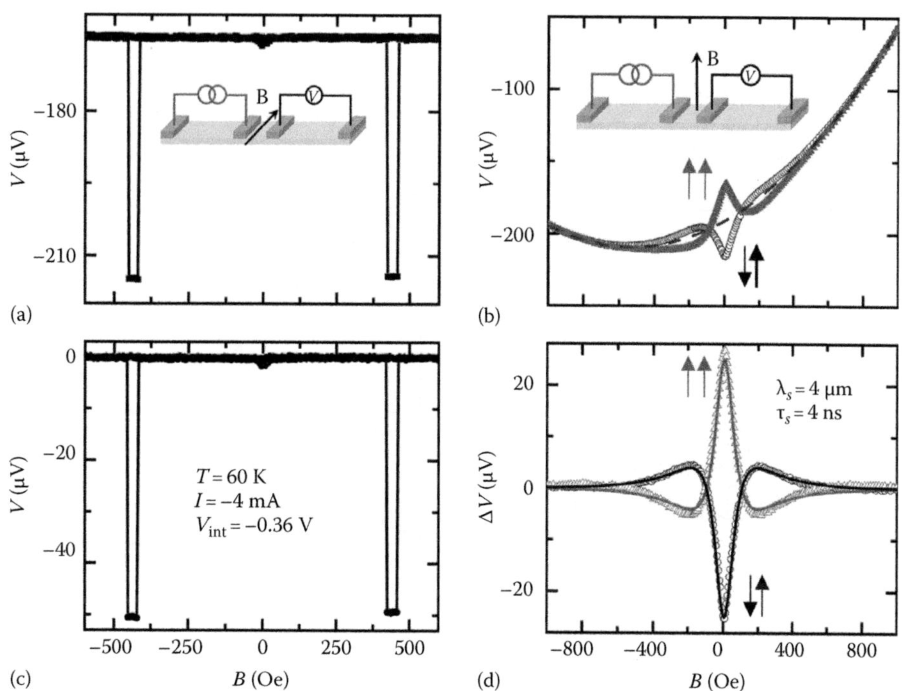

FIGURE 23.10 Non-local electrical detection of spin accumulation at $T = 60$ K under a forward bias current of 4 mA ($J = 1$ kA/cm²). The injection contact is $5 \times 80\,\mu m^2$ and the detector contact is $5 \times 50\,\mu m^2$. The center to center spacing of the contacts is 8 μm. (a) Raw data obtained in the longitudinal (or spin-valve) geometry. (b) Raw data obtained in the perpendicular (or Hanle) geometry for both parallel and antiparallel orientations of the detector and injector. The dashed line is a fit of the background obtained by a parabolic fit of the data at high fields. (c) The spin-valve data after subtraction of the spin-independent background. (d) Hanle data after subtraction of the parabolic background. The solid curves are fits to the drift-diffusion model. The lifetime $\tau_s = 4$ ns and the spin diffusion length $\lambda_s = 4\,\mu m$.

background (dashed curve in Figure 23.10b), the Hanle curves of Figure 23.10d are obtained. As expected, these two curves are equal and opposite and the difference in their peak values is equal to the magnitude of the jump in the spin-valve data. Curves showing fits to the drift-diffusion model are also shown in Figure 23.10d. In this case, we find a spin lifetime $\tau_s = 4$ ns and a spin diffusion length $\lambda_s = 4\,\mu m$.

It is also possible to determine the spin diffusion length by measuring the dependence of $\Delta V_{NL} = V_{\uparrow\downarrow} - V_{\uparrow\uparrow}$ on contact separation in multiterminal devices. By putting the detection contact in the current path (and subtracting out the additional background due to the ohmic voltage drop), it is also possible to measure the voltages $V_{\uparrow\downarrow}$ and $V_{\uparrow\uparrow}$ as well as the Hanle effect "down-stream" from the injector, where both diffusion and drift contribute to the spin transport. An example is shown in Figure 23.11, which shows ΔV_{NL} as a function of the injector/detector separation for a spin detector located outside the path of the charge current (contact separation $d > 0$), and in the current path ($d < 0$). The enhancement by drift in the latter case is clearly evident.

Demonstrations of the non-local detection of spin accumulation have been made over the last few years by other groups in Fe/GaAs [77,78] as well in other semiconductor-based systems, including MnAs/GaAs [79], Si [80], GaMnAs/GaAs [81], NiFe/InAs [82], and Co/graphene [83,84]. In many (but not all) cases, claims in these experiments have been supported by the observation of a Hanle effect, which has also been invoked extremely effectively in spin transport measurements on other spin-transport devices, particularly Si spin-valve transistors [85,86]. We now turn to some aspects of these measurements that are less well-understood.

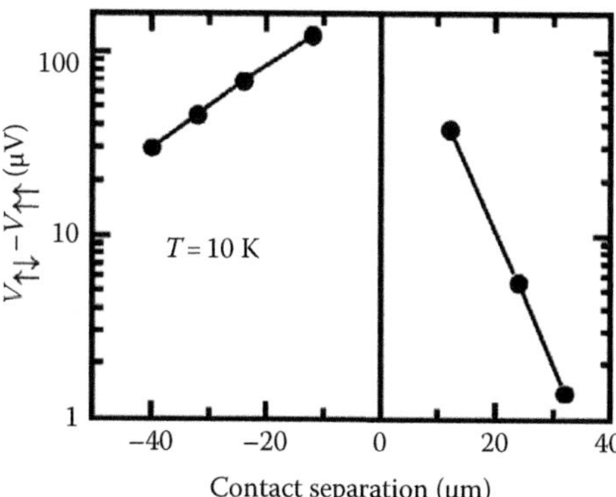

FIGURE 23.11 The spin-dependent voltage $V_{\uparrow\downarrow} - V_{\uparrow\uparrow}$ is shown as a function of the separation between the injection and detection electrodes in device with several injection/detection contacts. Positive separations correspond to the pure non-local geometry, in which the detection contact is outside the current path. For negative separations, the detector is located in the path of the charge current, and the signal is enhanced by drift.

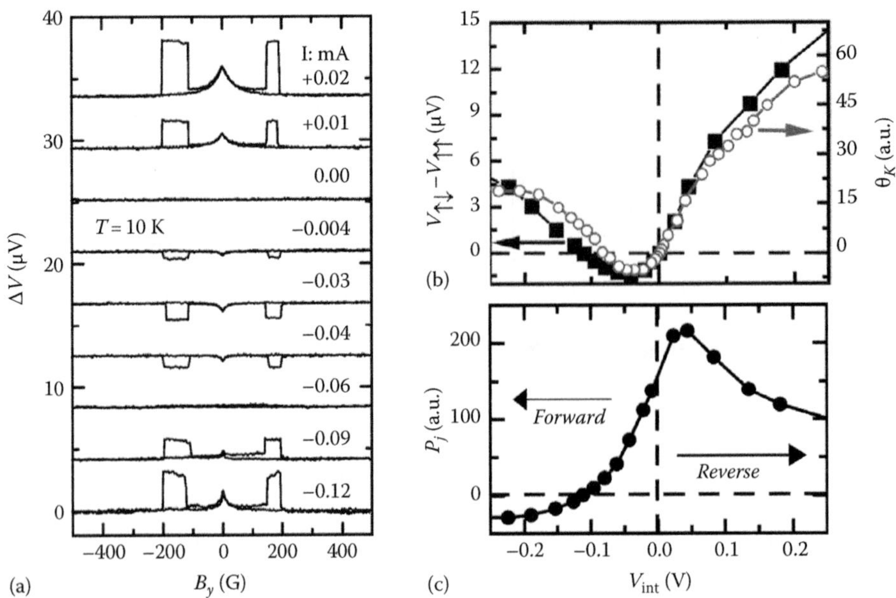

FIGURE 23.12 (a) The non-local spin-valve voltage is shown as a function of the longitudinal magnetic field for several different bias voltages. Note that the signal inverts sign *twice*. (b) The spin-dependent non-local voltage $\Delta V_{NL} = V_{\uparrow\downarrow} - V_{\uparrow\uparrow}$ (closed squares, left axis) obtained from field sweeps is shown as a function of the interfacial voltage V_{int} across the Schottky barrier injection contact. The Hanle–Kerr rotation signal θ_K (open circles, right axis) measured on this device as described in Section 23.4 is shown for the same bias conditions. (c) The polarization P_j of the interfacial tunneling current is proportional to $\Delta V_{NL}(V_{int})/I(V_{int})$, which is shown as a function of V_{int}. Note the change of sign at approximately −0.1 V.

23.6.1 Bias Dependence of the Accumulated Spin Polarization

The overall phenomenology of the non-local measurements is in general agreement with expectations based on the drift-diffusion model for spin transport in the semiconductor. We now turn to aspects of the data that reflect the important and less understood role of the Fe/GaAs interface itself. Figure 23.12a shows a series of non-local spin-valve measurements as a function of longitudinal magnetic field for different injection currents. As expected, there are jumps at the transitions between parallel and antiparallel states of the injector and detector. (The additional peak at zero field is due to hyperfine interactions and will be discussed below.) If the polarization of the tunneling current were fixed (e.g., by the bulk value of P_{Fe}), then the sign of the spin accumulation should change at zero bias. This is indeed the case, but the sign of ΔV_{NL} reverses again at a modest bias current of −60 μA (forward bias), equivalent to a voltage drop $V_{int} = -0.1$ V measured between the GaAs channel and the Fe injector. A more complete picture can be seen in Figure 23.12b, which shows ΔV_{NL} as a function of V_{int}. The same figure also shows the Kerr rotation θ_K measured in the same device (in the channel “upstream” of the injection contact) as a function of V_{int} using the methods of Section 23.4. Up to a scale factor, the two measurements agree, confirming that the sign change at −0.1 V is due to a change in the polarity of the spin accumulation. The Kerr data also determine the absolute sign of the spin accumulation, which is positive (or majority) for positive values of θ_K. Another representation of the data is provided in Figure 23.12c in which we show the effective spin polarization of the current: $P_j = (j_\uparrow - j_\downarrow)/(j_\uparrow + j_\downarrow) \propto \Delta V_{NL}(V_{int})/I(V_{int})$. In this form, it is evident that the spin polarization of the current depends very strongly on the injector bias voltage, with a peak near zero bias and a reversal in sign at −0.1 V.

The bias dependences of ΔV_{NL} and the Kerr rotation θ_K are shown for two additional samples in Figure 23.13, which illustrates that for a given sign of the injector bias voltage, the magnitude and sign of ΔV_{NL} (closed squares) can vary from heterostructure to heterostructure. In contrast, all devices grown from a single heterostructure (i.e., those based on a single growth of an epitaxial Fe/GaAs interface) will show approximately the same bias dependence.* At first glance, this wide variation among heterostructures, particularly the overall sign, appears counterintuitive. In spin-LEDs, for example, the polarization of the current at reverse bias is always observed to have the correct sign for majority spin injection [37]. A first clue as to what is going on can be found in the Kerr rotation data of Figure 23.13. As was observed for the data of Figure 23.12b, the bias dependences of θ_K and ΔV_{NL} for the heterostructure of Figure 23.13a overlap nearly exactly, but for the sample of Figure 23.13b θ_K overlaps with $-\Delta V_{NL}$. In other words, the sign of the non-local voltage for a given spin accumulation is reversed between the

* We are excluding a discussion of annealing, which can lead to a change in the sign of the non-local signal, presumably due to changes in the interfacial electronic structure. Regardless of annealing history, however, majority spin accumulation is observed for large forward and reverse bias currents in all cases.

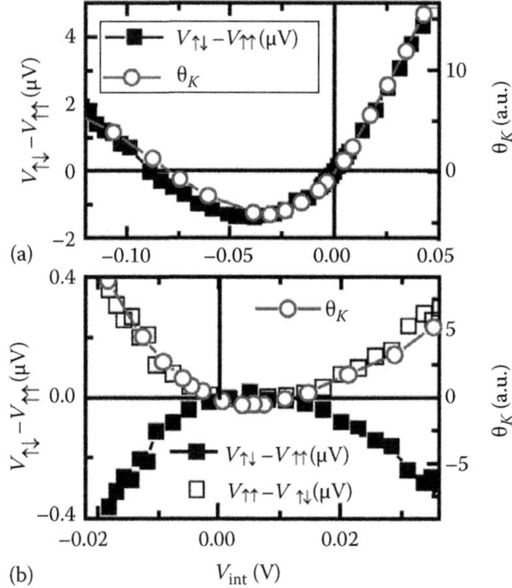

FIGURE 23.13 Bias dependences of ΔV_{NL} (closed squares, left axis) and Kerr rotation θ_K (open circles, right axis) for two different samples. As discussed in the text, the bias dependences of θ_K and ΔV_{NL} in (b) agree only after the *sign* of ΔV_{NL} is inverted (open squares).

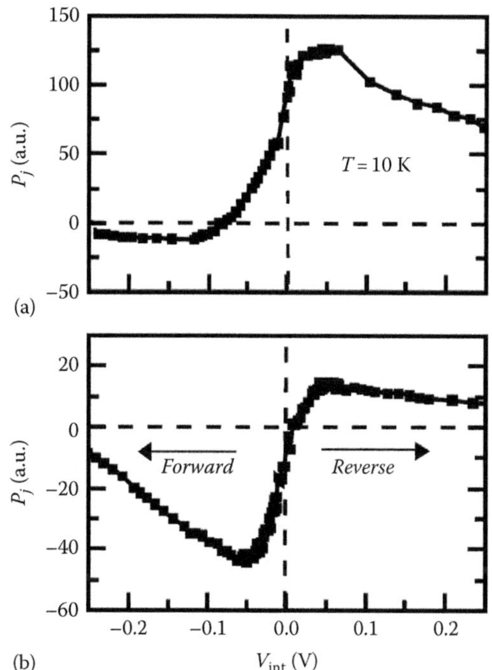

FIGURE 23.14 The spin polarization of the tunneling current flowing from the semiconductor into the ferromagnet for the two samples of Figure 23.13. Note that they have opposite sign at zero bias. Positive voltage (reverse bias) corresponds to electrons flowing from the ferromagnet into the semiconductor.

two samples. Otherwise, the ΔV_{NL} is proportional to the spin accumulation, as expected.

Focusing on the Kerr rotation, from which we can unambiguously determine the absolute sign of the spin accumulation, we always observe majority spin accumulation under large reverse bias (greater than +50 mV). In other words, a majority spin polarization flows from Fe into GaAs, as expected. Under forward bias, we observe majority spin accumulation for biases greater than (i.e., more negative than) −100 mV. This corresponds to the flow of *minority* spin from the semiconductor into the ferromagnet. In other words, the relative transmission probabilities for minority and majority spins always reverse near zero bias. This can be seen explicitly in Figure 23.14, which shows the polarization of the current $P_j \propto \Delta V_{\text{NL}}(V_{\text{int}})/I_{\text{int}}$ for the same two samples as in Figure 23.13. Based on the Kerr rotation measurements, we have chosen the sign of the polarization to be positive at large positive (reverse) bias voltages. Although the polarization of the current at zero bias can be either positive or negative, the polarization at large reverse bias is always positive, and it is always negative at large forward bias.

The absolute sign of the non-local signal, or, equivalently, the sign of P_j at zero bias, is determined by the tunneling density of states along with the exact energy at which the Fermi level is pinned at the Fe/GaAs interface. The most direct access we have to this information comes from the Kerr rotation, which measures the spin accumulation directly (to the extent that $\theta_K \propto P_{\text{GaAs}}$) and hence the quantity $\delta P_{\text{GaAs}}/\delta V_{\text{int}}$ for a small change δV_{int} in the bias. Hence, the slope of θ_K vs. V_{int} at zero bias should serve as predictor for the signs of both ΔV_{NL} (which is always measured at zero bias across the *detector*) and P_j. This is in fact the case. Although this self-consistency is reassuring, the fact remains that we cannot predict the spin polarization at zero bias *a priori*. This remains a frustrating point in the materials physics of these systems, particularly because the interfaces appear to remarkably well-controlled when evaluated by spin-LED or scanning transmission electron microscopy measurements [67,87].

While sample-specific effects may be regarded as problematic, it is important to emphasize the universal observation that P_j always reverses sign over a narrow window ±0.1 V around zero bias. For epitaxial Fe/GaAs (100), the minority spin current always dominates at large forward bias. Similar anomalous behavior (i.e., a crossover to minority-dominated tunneling) has been observed in Fe/GaAs/Fe tunnel junctions in which one interface is grown epitaxially [56]. In the case of metallic magnetic tunnel junctions, a reversal in the current polarization (and hence the sign of the tunneling magnetoresistance) has been observed in systems with crystalline barriers, as in Fe/MgO/Fe [88]. The association with crystalline barriers suggests that the effects of dispersion are important. On the other hand, the sign reversal is not necessarily observed for other ferromagnets grown epitaxially on GaAs. For example, Kotissek et al. found minority spin accumulation in $Fe_{0.32}Co_{0.68}$/GaAs structures under forward bias [66], and we find minority accumulation under forward bias in heterostructurss (otherwise identical to those under discussion here) in which Fe is replaced by epitaxial Fe_3Ga.

Addressing the bias-dependence requires moving beyond a simple Jullière-type model, in which the tunneling density of states and matrix elements are independent of energy. An extension of the Jullière model, accounting for the bulk Fe density of states, does produce a sign reversal, but only at much higher bias. The simplest approach to modifying the tunneling matrix elements, discussed by Smith and Ruden [89] accounts for the bias dependence of the Schottky barrier profile while modeling the ferromagnet as an *s*-band metal with distinct Fermi wave-vectors for majority and minority spins. Other approaches emphasize different aspects of the electronic structure of the Fe/GaAs (001) Schottky barrier [90–92]. Chantis et al. [90] focus on the tunneling density of states at the Fe/GaAs interface. By calculating the surface density of states and then integrating the tunneling current for both majority and minority bands in *k*-space, they find an enhancement of the minority current at small forward bias, leading to a crossover from majority to minority transmission. Dery and Sham [91] focus on the conduction band near the surface of the semiconductor, arguing that the minority current is enhanced by tunneling from bound states formed in the shallow potential well just inside the Schottky barrier. It is impossible to distinguish between these models based only on existing data, although Li and Dery [93] have proposed inserting a barrier (of higher band-gap material) just inside the semiconductor. This would allow for an explicit test of the Dery/Sham model.

23.6.2 Hyperfine Effects

A second set of anomalies in the non-local measurements are associated with the magnetic field dependence and are illustrated in Figure 23.15. First, in the longitudinal field sweep at low temperatures (Figure 23.15a), a peak is observed at zero field in addition to the two expected jumps when the magnetizations of the source and detector are in the anti-parallel state. (This peak can also be observed in the field sweeps of Figure 23.12.) The zero-field peak depends very strongly on the field sweep rate and vanishes if the measurement time is sufficiently long (requiring wait times over a minute at each data point) or if the temperature is sufficiently high (typically above 60 K). Various checks can be used to ascertain that the magnetizations of the contacts are indeed fixed as the field is swept through zero, and so we conclude that it is due to some change in the spin dynamics in the semiconductor. Similar observations of a "zero-field peak" have been made by Salis et al. [77] and Ciorga et al. [81].

The second anomaly, illustrated in Figure 23.15b, is the extreme sensitivity of the Hanle curves to the magnetic field orientation, an effect that becomes so strong at low temperatures that it is difficult to obtain a meaningful Hanle curve in an ordinary field sweep. As shown in Figure 23.15b, there is also a clear difference between the two different parallel configurations of the contacts. Instead of being identical, the Hanle curves for $V_{\uparrow\uparrow}$ and $V_{\downarrow\downarrow}$ are both distorted, and they appear to be mirror images with respect to inversion of the magnetic field axis. Unlike the zero-field peak in the longitudinal field sweeps, this effect persists in steady state, and the distortion as well as the apparent shift with respect to zero field increase with the degree of field misalignment as well as with the bias current. Distortion of non-local Hanle curves has also been reported by the NRL and Regensburg groups [78,81].

These effects, which are ubiquitous in Fe/GaAs devices, are due to the dynamic polarization of nuclei by electron spins localized on donor sites. This phenomenon was originally observed in optical-pumping experiments [9] and has been reviewed extensively by Fleisher and Merkulov [8]. Discussions of similar effects in GaAs-based spin-LEDs appear in Refs. [36] and [50]. From the standpoint of *electron* spin transport and dynamics, the main consequence of the dynamic nuclear polarization (DNP) is an effective mean field $\mathbf{B}_N$ which acts on the electron spins. A proper calculation of $\mathbf{B}_N$ requires a self-consistent treatment of the electron and nuclear spin dynamics. In a relatively general form:

$$\mathbf{B}_N = b_n \frac{(\mathbf{B} + b_e\mathbf{S}) \cdot \mathbf{S}(\mathbf{B} + b_e\mathbf{S})}{(\mathbf{B} + b_e\mathbf{S})^2 + \xi B_l^2}, \tag{23.11}$$

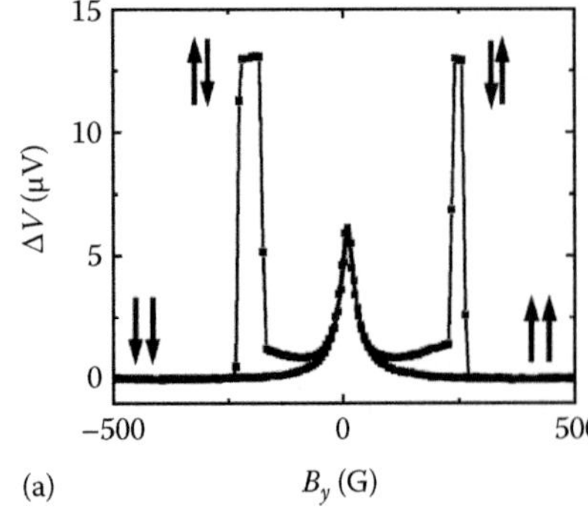

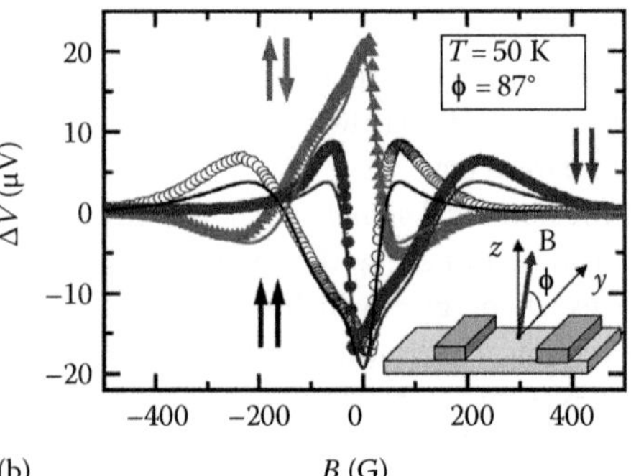

FIGURE 23.15 Effects of hyperfine interactions. (a) Data from a longitudinal field sweep (field along easy axis of contacts) at 10 K. Note the peak at zero field in addition to the switching features. (b) Hanle data at 50 K obtained with the magnetic field oriented a few degrees away from vertical. Solid lines are a fit to the drift diffusion model with a constant hyperfine field with a sign determined by the magnetization of the injector. The samples and bias conditions are different for the two panels, and so the magnitudes of the signals should not be compared.

where

b_n and b_e, which are both negative in GaAs [9], represent effective fields due to the polarized nuclei and electrons

S is the average electron spin ($|\mathbf{S}| = 1/2$ for electron spin polarization $P_{\text{GaAs}} = 1$)

B_l is the local dipolar field experienced by the nuclei

ξ parameterizes the assisting processes which allow energy to be conserved in mutual spin flips between electrons and nuclei

We have simplified the notation of Paget [9] by incorporating prefactors into b_n and b_e that cannot be determined independently. The electron spin precesses in a total field $\mathbf{B}_{\text{tot}} = \mathbf{B} + \mathbf{B}_N$, which can be much larger than the applied field **B**.

If we ignore the exchange field (Knight field) b_e acting on the nuclei, $\mathbf{B}_N$ is parallel to the applied field, but its magnitude and sign are determined by the electron spin accumulation. This is the origin of the distortion of the Hanle curves in Figure 23.15b as well as their dependence on the *sign* of the injected spins. A more extreme case is shown in Figure 23.16a, obtained for a misalignment of 15° in the direction of the applied field with respect to the sample normal. The satellite peak in the Hanle curve at ~300 G is due to the approximate cancelation of the applied field by the *z*-component of the hyperfine field. Given the dot product in Equation 23.11, the magnitude of the hyperfine field in this case is approximately 1000 G.

The sensitivity of the electron spin accumulation to $\mathbf{B}_N$ implied by Figure 23.16 can be regarded either positively or negatively depending on one's point of view, and it is important to note that it can be suppressed to a large extent by inverting the magnetization of the injector on a time scale that is fast relative to the time scale for the nuclei to become polarized, which is on the order of 1 s. This is feasible if small electromagnets are used. On the other hand, an optimist might be inclined to use the electron spin accumulation as a nuclear spin detector. For example, depolarizing a particular isotope (either ^{69}Ga, ^{71}Ga, or ^{75}As) by resonant excitation suppresses its contribution to $\mathbf{B}_N$, resulting in a corresponding shift of $\mathbf{B}_{\text{tot}}$. As a result, the Hanle curve will shift along the field axis, and the value of the spin accumulation at a fixed field will change. This means of detecting nuclear magnetic resonance is illustrated in Figure 23.16b, which shows the Hanle signal measured at an applied field of 320 G as a function of the frequency of current passing through a small coil over the sample. The primary resonances for the three main isotopes are indicated by symbols.

Other manifestations of the hyperfine interaction are less amenable to simple interpretation. The zero-field feature in the longitudinal field sweeps has been studied carefully by Salis et al. [77], who were able to model it in terms of the formalism presented here by carrying out longitudinal field sweeps in a fixed perpendicular magnetic field. Its ubiquitous presence in longitudinal field sweeps is due to the rotation of $\mathbf{B}_N$ as the applied field is swept through zero.

It is evident that nuclear spins have a profound influence on electron spin dynamics. The influence of the electronic field b_e, also known as the Knight field, is more subtle, since it acts directly only on the nuclei. An effect on the electron spin accumulation occurs only because the exchange field $b_e\mathbf{S}$ changes the magnitude and direction of the effective field in which the nuclei precess. If, however, an experiment is designed to be sensitive to the Knight field, it then allows for a measurement of the magnitude and sign of the electron spin polarization *without requiring knowledge of the density of states in the semiconductor*. This comes about because the field $b_e\mathbf{S}$, which is proportional to the electron spin polarization, modifies the magnetic field dependence of the spin accumulation signal, whereas **S** alone only impacts the magnitude.

Chan et al. [94] have solved the problem of the coupled dynamics of the electron and nuclear spin system by incorporating the effective field of Equation 23.11 into the drift-diffusion model of Equation 23.1, solving the two equations self-consistently in 1D. A model curve based on this result is shown in Figure 23.16a. The region around zero field is shown in the inset. The small double-peak near zero is due to the effective field $\mathbf{S}b_e$. A more sensitive measurement of $\mathbf{S}b_e$ described in detail in Ref. [94] has allowed for a direction quantitative measurement of the spin accumulation **S** as a function of the injector bias. The resulting spin

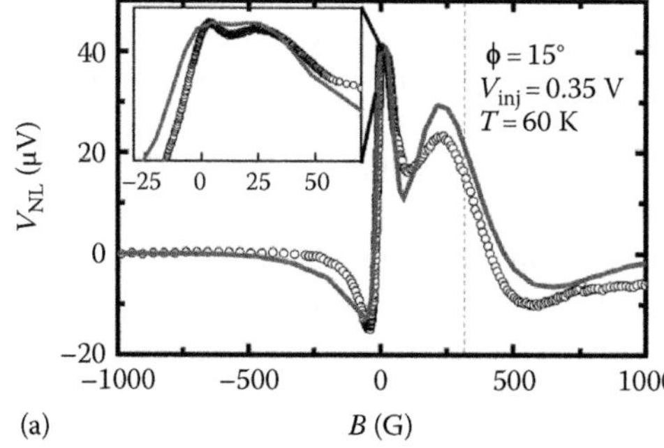

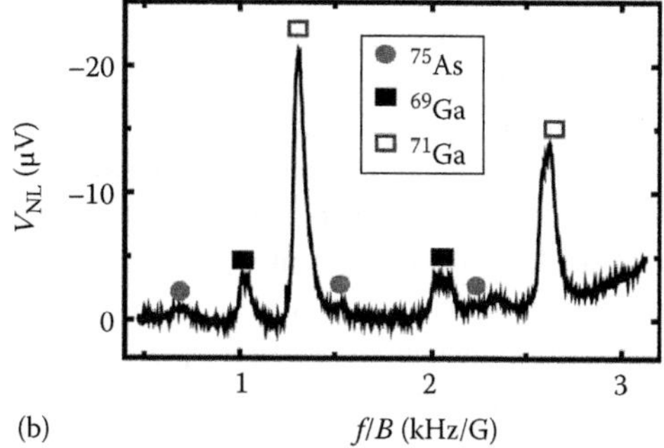

FIGURE 23.16 (a) Non-local Hanle curve obtained with the field applied at 15° from the vertical direction. An expansion of the region around zero field is shown in the inset. The solid curve is obtained from the modified drift diffusion model with a nuclear field of the form in Equation 23.11. Note the double peak in both the model and the experimental data in the inset, which is due to the non-zero Knight field b_e. (b) NMR spectrum obtained in an applied field of 320 G for the same conditions in (a) as the frequency of the current in a coil over the sample is swept. The resonances for NMR transitions of ^{75}As, ^{69}Ga, and ^{71}Ga are shown using circles, closed squares, and open squares, respectively.

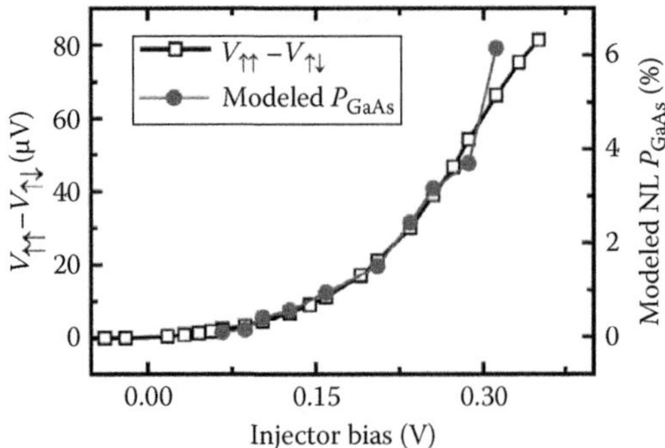

FIGURE 23.17 The non-local voltage $\Delta V_{NL} = V_{\uparrow\uparrow} - V_{\uparrow\downarrow}$ from spin-valve studies (open squares) compared with the spin polarization P_{GaAs} determined from the measurement of b_e (Knight field) as described in the text. Note the two independent y-scales.

polarization P_{GaAs} is shown in Figure 23.17 superimposed on the ordinary measurement of the spin-dependent non-local voltage $\Delta V_{NL} = V_{\uparrow\uparrow} - V_{\uparrow\downarrow}$. These two very different measurements show the same dependence on the bias voltage of the injector. As can be seen on the right-hand axis, the Knight field measurement gives an electron spin polarization of about 6% at the maximum injector bias. The sign corresponds to accumulation of majority spins, as expected based on the Kerr measurements of other structures at large forward bias. Given this polarization, the magnitude of V_{NL} is in fact quite close to the very crude estimate of Equation 23.10, although the agreement is probably serendipitous.

23.6.3 Detector Bias Dependence

Given the difficulties described above in achieving a quantitative interpretation of the injector bias dependence, it might appear premature to consider applying an additional bias voltage to the detector. A biased detector would seem to remove some of the purported advantages of the non-local approach, particularly from the standpoint of traditional objections to two-terminal measurements [17]. In the context of the above measurements, however, we now have an excellent grasp of the behavior of spin-dependent signals (and artifacts) in Fe/GaAs devices. Furthermore, in the presence of the semiconductor channel, it is easy to bias the FM source and detector electrodes individually. The actual experimental phase space is therefore much larger than considered in some of the more pessimistic assessments of two-terminal devices with high-resistivity barriers [24,95]. In fact, as demonstrated recently [71] the application of separate injector and detector bias voltages allows one to exploit the advantages of semiconductor-based spin transport devices in ways that are not possible in similar metallic devices. The potential of multiterminal FM/SC devices has only begun to be explored, and the reader is referred to Chapter 39 by Dery for a discussion of some of the possible applications.

Given the complications addressed in Section 23.6.1, we separate into two contributions the effects of an additional detector bias voltage V_d on the sensitivity of electrical spin detection [71]. The first contribution, discussed above, is simply the effect of V_d on the polarization P_j of the tunneling current that flows across the Fe/GaAs detector interface. Tuning the magnitude and/or sign of P_j with V_d will tune the magnitude and/or sign of the spin-detection sensitivity. The second contribution, which we exploit to further enhance the spin-dependent sensitivity of electrical spin detectors, originates from the electric fields in the semiconductor channel near the detector [6,16] that arise *because* the detector is biased, including the enhancement of the spin transport by drift [61,96] as well as the modification of the (spin-dependent) carrier densities near the detector interface.

We consider a situation where a bias voltage V_d is applied between the FM detector contact and the SC channel. As a measure of the detector's spin-dependent sensitivity, we then consider the voltage change ΔV_d at the detector that is induced by a remotely injected spin polarization (analogous to conventional non-local studies). If there is no electric field in the semiconductor, (e.g., if the current is purely diffusive) simple reciprocity considerations require that for a given unit of spin-polarized current, the change ΔV_d must be proportional to the change $\Delta\mu$ in the spin accumulation that occurs when the same unit of spin-polarized current is *injected* from an identical contact at injection bias V_{int}. The sensitivity of the detector at a voltage V is therefore proportional to $P_j = \Delta V_{NL}(V)/I(V)$ As noted in Section 23.6.1, this is proportional to the polarization P_j of the current flowing through the injector. From a conventional non-local experiment (in which V is the injector interface voltage), it is therefore possible to predict the electrical *detection* sensitivity of the contact *in the absence of electric field effects.*

In Ref. [71], we have addressed the reasoning of the previous paragraph by using optical pumping to generate a small spin-polarized current that is incident on a detector contact biased at a voltage V_d. The change ΔV_d in the detector voltage was measured using a standard modulation technique and compared with $\Delta V_{NL}(V)/I(V)$ for the same device. For detector bias voltages exceeding ±0.1 V, there are significant departures from the expectations of the simple reciprocity argument. In particular, the sensitivity of the detector at forward bias was observed to be dramatically enhanced. A simple model [97] showed that the enhancement is due to the modification of spin-dependent carrier densities $n_{\uparrow,\downarrow}$ (and their gradients) by the electric field at the interface.

The implication of this result is that the sensitivity of a non-local spin-valve can be tuned over a wide range by application of a bias voltage at the detector. Figure 23.18 shows that this is indeed the case. Figure 23.18a shows the magnitude of the spin-valve signal versus detector bias. We consider one of the sets of data (the black points) obtained at a *fixed* injector bias current (−1.5 mA). At zero detector bias—where non-local studies are usually performed—the spin-valve signal is quite small (<2 μV). However, when the non-local spin detector is forward-biased by −150 mV, the magnitude of the spin-valve signal increases by over an order of magnitude to approximately 20 μV. Furthermore, with the detection electrode under

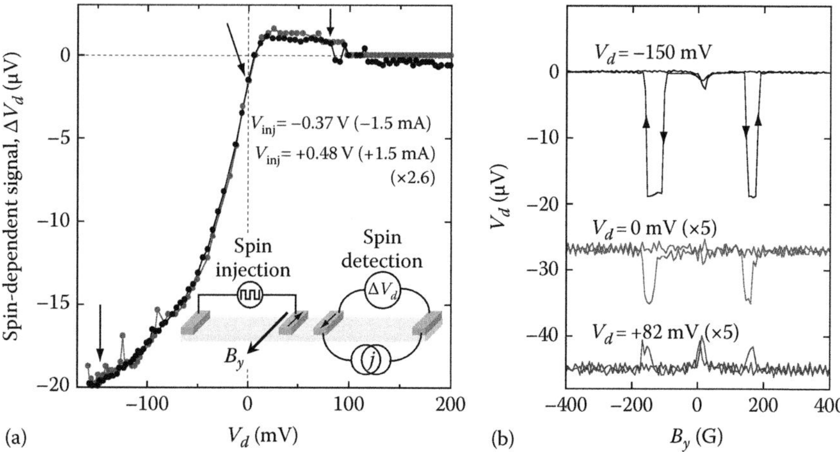

FIGURE 23.18 (a) Inset: All-electrical lateral spin-valve setup with separately biased injector and detector electrodes. Black (gray) points show the spin-valve signal ΔV_d versus detector bias V_d for a forward (reverse) biased spin injector. (b) Raw spin-valve data at three detector biases [arrows in (a)], using a fixed −1.5 mA injector bias (curves offset). Note sign switching and ten-fold enhancement of the detected signal.

a small reverse bias, the spin valve signal actually inverts sign. Note that in all three cases the injected spin density remains unchanged. Thus, electrons in GaAs with a fixed spin polarization can be made to induce either positive or negative voltage changes at a detection electrode, simply by tuning the detector bias V_d by a few tens of meV. The detector sensitivity in this voltage range is completely independent of the injector bias, as demonstrated by the second set of data (gray points) in Figure 23.18a, which were obtained with a totally different injection current of +1.5 mA. After correcting for the different spin injection efficiencies, the two sets of data collapse onto a single curve. Figure 23.18b shows spin-valve data for these three scenarios (V_d = −150, 0, and +82 mV), explicitly showing that spin-detection sensitivities are freely tunable in both sign and magnitude in all-electrical devices.

23.7 Conclusions

Before concluding, we will address spin–orbit effects very briefly, since they have served as a motivation for much of the research in this field. The classic Datta–Das spin-FET [13], in which the role of the magnetic field in a Hanle measurement is assumed by the effective spin–orbit magnetic field, requires a set of experiments, parallel to those discussed above, on a 2D electron system. As of this writing, there has been no successful demonstration of a Hanle effect for electrical spin injection into a 2D system. One might argue that this is intrinsically difficult, given that strong spin–orbit coupling is typically correlated with short lifetimes. A slightly different problem was encountered in recent work on an InAs 2D electron gas (2DEG), where gate modulation of a spin valve signal was reported in a ballistic device with an extremely short transit time, and so a Hanle effect could not be observed [82]. It should be possible, through a suitable choice of alloy composition and/or quantum well orientation [98], to engineer a FM/2DEG/FM spin valve with appropriate values of the Rashba field and spin lifetime. This would allow for observation of a Hanle effect and gate-induced modulation of spin accumulation in a single device.

Another outstanding class of problems involves the detection of spin-Hall effects (see Chapter 25) [99–101]. The first obvious candidate is the electrical detection of the spin-Hall effect in a manner analogous to the Hanle experiment of Kato et al. [100], who used Kerr microscopy to detect the spin accumulation at the edges of a Hall bar biased with an unpolarized current. In principle, this only requires placing FM Hall contacts on a semiconductor channel. Given the small values of the spin accumulation (~0.1%), however, the technical requirements are more stringent than for the experiments introduced above. Nonetheless, this experiment has recently been carried out [102], and it has allowed for a much more detailed comparison with theory than has been possible with optical probes. Detection of the *reciprocal* spin Hall effect requires the measurement of a Hall voltage induced by a spin-polarized current [103]. This has been attempted in semiconductors by using optical-pumping techniques to generate a spin-polarized current [104]. The obvious advantage of using FM contacts as sources in this case is the absence of a hole current.

We hope that the experimental philosophy shaping the research described in this chapter will be applied to these problems. It is now possible to probe spin transport and dynamics using combinations of optical and transport techniques that allow for detailed tests of theory. Through the thoughtful design of quantitative experiments, one can address the reservations that have traditionally been expressed about transport measurements on hybrid ferromagnet-semiconductor structures.

Acknowledgments

It is a pleasure to thank our collaborators who have made this work so enjoyable: Christoph Adelmann, Mun Chan, Athanasios Chantis, Madalina Furis, Eric Garlid, Qi Hu, Tsuyoshi Kondo, Xiaohua Lou, Chris Palmstrøm, Madhukar Reddy, Darryl

Smith, and Jianjie Zhang. The work discussed in this review has been supported by NSF under DMR-0804244, the Office of Naval Research, the MRSEC Program of the National Science Foundation under Award Numbers DMR-0212302 and DMR-0819885, the Los Alamos LDRD program, the National High Magnetic Field Laboratory, and the National Science Foundation NNIN program.

References

1. F. Meier and B. P. Zakharchenya, *Optical Orientation*, Amsterdam, North Holland, 1984.
2. J. M. Kikkawa and D. D. Awschalom, Resonant spin amplification in n-type GaAs, *Phys. Rev. Lett.* **80**(19), 4313–4316 (1998).
3. R. I. Dzhioev, B. P. Zakharchenya, V. L. Korenev, and M. N. Stepanova, Spin diffusion of optically oriented electrons and photon entrainment in n-gallium arsenide, *Phys. Solid State* **39**(11), 1765 (1997).
4. R. I. Dzhioev, K. V. Kavokin, V. L. Korenev et al., Low-temperature spin relaxation in n-type GaAs, *Phys. Rev. B* **66**, 245204 (2002).
5. A. G. Aronov and G. E. Pikus, Spin injection into semiconductors, *Sov. Phys. Semicond.—USSR* **10**(6), 698–700 (1976).
6. Z. G. Yu and M. E. Flatté, Spin diffusion and injection in semiconductor structures: Electric field effects, *Phys. Rev. B* **66**(23), 235302 (2002).
7. M. I. D'yakonov and V. I. Perel', Spin relaxation of conduction electrons in noncentrosymmetric semiconductors, *Sov. Phys. Solid State* **13**, 203 (1972).
8. V. G. Fleisher and I. A. Merkulov, Optical orientation of the coupled electron-nuclear spin system of a semiconductor, in *Optical Orientation*, F. Meier and B. P. Zakharchenya (Eds.), Amsterdam, North Holland, p. 173, 1984.
9. D. Paget, G. Lampel, B. Sapoval, and V. I. Safarov, Low field electron-nuclear spin coupling in gallium arsenide under optical pumping conditions, *Phys. Rev. B* **15**(12), 5780 (1977).
10. R. Winkler, *Spin-Orbit Coupling Effects in Two-Dimensional Electron and Hole Systems*, Springer-Verlag, Berlin, 2003.
11. G. E. Pikus and A. N. Titkov, Spin relaxation under optical orientation in semiconductors, in *Optical Orientation*, F. Meier and B. P. Zakharchenya (Eds.), Amsterdam, North Holland, 1984.
12. Y. A. Bychkov and E. I. Rashba, Oscillatory effects and the magnetic susceptibility of carriers in inversion layers, *J. Phys. C: Solid State Phys.* **17**(33), 6039–6045 (1984).
13. S. Datta and B. Das, Electronic analog of the electro-optic modulator, *Appl. Phys. Lett.* **56**(7), 665–667 (1990).
14. M. I. D'yakonov and V. Y. Kachorovskii, Spin relaxation of two-dimensional electrons in noncentrosymmetric semiconductors, *Sov. Phys. Semicond.—USSR* **20**, 110–112 (1986).
15. M. Furis, D. L. Smith, S. A. Crooker, and J. L. Reno, Bias-dependent electron spin lifetimes in n-GaAs and the role of donor impact ionization, *Appl. Phys. Lett.* **89**, 102102 (2006).
16. I. Žutić, J. Fabian, and S. D. Sarma, Spin injection through the depletion layer: A theory of spin-polarized p-n junctions and solar cells, *Phys. Rev. B* **64**, 121201 (2001).
17. H.X. Tang, F.G. Monzon, F.J. Jedema, A.T. Filip, B.J. Van Wees, and M. L. Roukes, Spin injection and transport in micro- and nanoscale devices, in *Semiconductor Spintronics and Quantum Computation*, D. D. Awschalom, D. Loss, and N. Samarth (Eds.), Springer-Verlag, Berlin, pp. 31–87, 2002.
18. M. Johnson and R. Silsbee, Interfacial charge-spin coupling: Injection and detection of spin magnetization in metals, *Phys. Rev. Lett.* **55**, 1790 (1985).
19. M. Johnson and R. H. Silsbee, Spin-injection experiment, *Phys. Rev. B* **37**(10), 5326 (1988).
20. P. C. van Son, H. van Kempen, and P. Wyder, Boundary resistance of the ferromagnetic-nonferromagnetic metal interface, *Phys. Rev. Lett.* **58**(21), 2271 (1987).
21. T. Valet and A. Fert, Theory of the perpendicular magnetoresistance in magnetic multilayers, *Phys. Rev. B* **48**(10), 7099–7113 (1993).
22. G. Schmidt, D. Ferrand, L. W. Molenkamp, A. T. Filip, and B. J. van Wees, Fundamental obstacle for electrical spin injection from a ferromagnetic metal into a diffusive semiconductor, *Phys. Rev. B* **62**(8), 4790–4793 (2000).
23. E. I. Rashba, Theory of electrical spin injection: Tunnel contacts as a solution of the conductivity mismatch problem, *Phys. Rev. B* **62**(24), 16267–16270 (2000).
24. A. Fert and H. Jaffrés, Conditions for efficient spin injection from a ferromagnetic metal into a semiconductor, *Phys. Rev. B* **64**(18), 184420 (2001).
25. D. L. Smith and R. N. Silver, Electrical spin injection into semiconductors, *Phys. Rev. B* **64**, 045323 (2001).
26. S. F. Alvarado and P. Renaud, Observation of spin-polarized-electron tunneling from a ferromagnet into GaAs, *Phys. Rev. Lett.* **68**(9), 1387–1390 (1992).
27. H. J. Zhu, M. Ramsteiner, H. Kostial, M. Wassermeier, H. P. Schönherr, and K. H. Ploog, Room temperature spin injection from Fe into GaAs, *Phys. Rev. Lett.* **87**, 016601 (2001).
28. A. T. Hanbicki, B. T. Jonker, G. Itskos, G. Kioseoglou, and A. Petrou, Efficient electrical injection from a magnetic metal/tunnel barrier contact into a semiconductor, *Appl. Phys. Lett.* **80**, 1240–1242 (2002).
29. J. Strand, B. D. Schultz, A. F. Isakovic, C. J. Palmstrøm, and P. A. Crowell, Dynamic nuclear polarization by electrical spin injection in ferromagnet-semiconductor heterostructures, *Phys. Rev. Lett.* **91**(3), 036602 (2003).
30. V. F. Motsnyi, J. De Boeck, J. Das et al., Electrical spin injection in a ferromagnet/tunnel barrier/semiconductor heterostructure, *Appl. Phys. Lett.* **81**, 265–267 (2002).
31. X. Jiang, R. Wang, S. van Dijken et al., Optical detection of hot-electron spin injection into GaAs from a magnetic tunnel transistor source, *Phys. Rev. Lett.* **90**, 256603 (2003).
32. X. Jiang, R. Wang, R. M. Shelby et al., Highly spin-polarized room-temperature tunnel injector for semiconductor spintronics using MgO(100), *Phys. Rev. Lett.* **94**(5), 056601 (2005).

33. R. Fiederling, M. Kelm, G. Reuscher et al., Injection and detection of a spin-polarized current in a light-emitting diode, *Nature* **402**, 787 (1999).
34. Y. Ohno, D. K. Young, B. Beschoten, F. Matsukura, H. Ohno, and D. D. Awschalom, Electrical spin injection in a ferromagnetic semiconductor heterostructure, *Nature* **402**, 790–792 (1999).
35. D. R. Scifres, B. A. Huberman, R. M. White, and R. S. Bauer, A new scheme for measuring itinerant spin polarizations, *Solid State Commun.* **13**(10), 1615–1617 (1973).
36. J. Strand, X. Lou, C. Adelmann et al., Electron spin dynamics and hyperfine interactions in Fe/$Al_{0.1}Ga_{0.9}As$/GaAs spin injection heterostructures, *Phys. Rev. B* **72**(15), 155308 (2005).
37. C. Adelmann, X. Lou, J. Strand, C. J. Palmstrøm, and P. A. Crowell, Spin injection and relaxation in ferromagnet-semiconductor heterostructures, *Phys. Rev. B* **71**(12), 121301–121301 (2005).
38. G. Salis, R. Wang, X. Jiang et al., Temperature independence of the spin-injection efficiency of a MgO-based tunnel spin injector, *Appl. Phys. Lett.* **87**(26), 262503 (2005).
39. X. Y. Dong, C. Adelmann, J. Q. Xie et al., Spin injection from the Heusler alloy Co_2MnGe into $Al_{0.1}Ga_{0.9}As$/GaAs heterostructures, *Appl. Phys. Lett.* **86**(10), 102107 (2005).
40. M. C. Hickey, C. D. Damsgaard, I. Farrer et al., Spin injection between epitaxial $Co_{2.4}Mn_{1.6}Ga$ and an InGaAs quantum well, *Appl. Phys. Lett.* **86**(25), 252106 (2005).
41. N. C. Gerhardt, S. Hovel, C. Brenner et al., Electron spin injection into GaAs from ferromagnetic contacts in remanence, *Appl. Phys. Lett.* **87**(3), 032502 (2005).
42. C. Adelmann, J. L. Hilton, B. D. Schultz et al., Spin injection from perpendicular magnetized ferromagnetic δ-MnGa into (Al,Ga)As heterostructures, *Appl. Phys. Lett.* **89**(11), 112511 (2006).
43. R. K. Kawakami, Y. Kato, M. Hanson et al., Ferromagnetic imprinting of nuclear spins in semiconductors, *Science* **294**(5540), 131–134 (2001).
44. R. J. Epstein, I. Malajovich, R. K. Kawakami et al., Spontaneous spin coherence in n-GaAs produced by ferromagnetic proximity polarization, *Phys. Rev. B* **65**, 121202–121204 (2002).
45. C. Ciuti, J. P. McGuire, and L. J. Sham, Spin polarization of semiconductor carriers by reflection off a ferromagnet, *Phys. Rev. Lett.* **89**(15), 156601–156604 (2002).
46. J. P. McGuire, C. Ciuti, and L. J. Sham, Theory of spin transport induced by ferromagnetic proximity on a two-dimensional electron gas, *Phys. Rev. B* **69**(11), 115339–115314 (2004).
47. R. J. Epstein, J. Stephens, M. Hanson et al., Voltage control of nuclear spin in ferromagnetic Schottky diodes, *Phys. Rev. B* **68**(4), 41305–41301 (2003).
48. J. Stephens, J. Berezovsky, J. P. McGuire, L. J. Sham, A. C. Gossard, and D. D. Awschalom, Spin accumulation in forward-biased MnAs/GaAs Schottky diodes, *Phys. Rev. Lett.* **93**(9), 097602–097604 (2004).
49. V. F. Motsnyi, P. Van Dorpe, W. Van Roy et al., Optical investigation of electrical spin injection into semiconductors, *Phys. Rev. B* **68**(24), 245319 (2003).
50. P. Van Dorpe, W. Van Roy, J. De Boeck, and G. Borghs, Nuclear spin orientation by electrical spin injection in an $Al_xGa_{1-x}As$/GaAs spin-polarized light-emitting diode, *Phys. Rev. B* **72**(3), 035315 (2005).
51. J. Strand, A. F. Isakovic, X. Lou, P. A. Crowell, B. D. Schultz, and C. J. Palmstrøm, Nuclear magnetic resonance in a ferromagnet-semiconductor heterostructure, *Appl. Phys. Lett.* **83**, 3335 (2003).
52. M. W. J. Prins, H. van Kempen, H. van Leuken, R. A. de Groot, W. Van Roy, and J. De Boeck, Spin-dependent transport in metal/semiconductor tunnel junctions, *J. Phys. Condens. Matter* **7**(49), 9447–9464 (1995).
53. A. Hirohata, Y. B. Xu, C. M. Guertler, J. A. C. Bland, and S. N. Holmes, Spin-polarized electron transport in ferromagnet/semiconductor hybrid structures induced by photon excitation, *Phys. Rev. B* **63**, 104425 (2001).
54. A. F. Isakovic, D. M. Carr, J. Strand, B. D. Schultz, C. J. Palmstrøm, and P. A. Crowell, Optical pumping in ferromagnet-semiconductor heterostructures: Magneto-optics and spin transport, *Phys. Rev. B* **64**(16), 161304–161301 (2001).
55. S. Kreuzer, J. Moser, W. Wegscheider, D. Weiss, M. Bichler, and D. Schuh, Spin polarized tunneling through single-crystal GaAs(001) barriers, *Appl. Phys. Lett.* **80**(24), 4582–4584 (2002).
56. J. Moser, M. Zenger, C. Gerl et al., Bias dependent inversion of tunneling magnetoresistance in Fe/GaAs/Fe tunnel junctions, *Appl. Phys. Lett.* **89**(16), 162106 (2006).
57. S. A. Crooker, M. Furis, X. Lou et al., Imaging spin transport in lateral ferromagnet/semiconductor structures, *Science* **309**(5744), 2191–2195 (2005).
58. S. A. Crooker, M. Furis, X. Lou et al., Optical and electrical spin injection and spin transport in hybrid Fe/GaAs devices, *J. Appl. Phys.* **101**(8), 081716 (2007).
59. M. Furis, D. L. Smith, S. Kos et al., Local Hanle-effect studies of spin drift and diffusion in n-GaAs epilayers and spin-transport devices, *New J. Phys.* **9**(9), 347 (2007).
60. X. Lou, C. Adelmann, M. Furis, S. A. Crooker, C. J. Palmstrøm, and P. A. Crowell, Electrical detection of spin accumulation at a ferromagnet-semiconductor interface, *Phys. Rev. Lett.* **96**(17), 176603 (2006).
61. X. Lou, C. Adelmann, S. A. Crooker et al., Electrical detection of spin transport in lateral ferromagnet-semiconductor devices, *Nat. Phys.* **3**(3), 197–202 (2007).
62. P. R. Hammar and M. Johnson, Detection of spin-polarized electrons injected into a two-dimensional electron gas, *Phys. Rev. Lett.* **88**, 066806 (2002).
63. J. M. Kikkawa and D. D. Awschalom, Lateral drag of spin coherence in GaAs, *Nature* **397**, 139 (1999).
64. S. A. Crooker and D. L. Smith, Imaging spin flows in semiconductors subject to electric, magnetic, and strain fields, *Phys. Rev. Lett.* **94**(23), 236601 (2005).

65. M. Beck, C. Metzner, S. Malzer, and G. H. Döhler, Spin lifetimes and strain-controlled spin precession of drifting electrons in GaAs, *Europhys. Lett.* **75**(4), 597–603 (2006).
66. P. Kotissek, M. Bailleul, M. Sperl et al., Cross-sectional imaging of spin injection into a semiconductor, *Nat. Phys.* **3**(12), 872–877 (2007).
67. J. M. LeBeau, Q. O. Hu, C. J. Palmstrøm, and S. Stemmer, Atomic structure of postgrowth annealed epitaxial Fe/(001) GaAs interfaces, *Appl. Phys. Lett.* **93**(12), 121909 (2008).
68. M. Hruška, Š. Kos, S. A. Crooker, A. Saxena, and D. L. Smith, Effects of strain, electric, and magnetic fields on lateral electron-spin transport in semiconductor epilayers, *Phys. Rev. B* **73**(7), 075306 (2006).
69. F. J. Jedema, H. R. Heersche, A. T. Filip, J. J. A. Baselmans, and B. J. van Wees, Electrical detection of spin precession in a metallic mesoscopic spin valve, *Nature* **416**, 713–716 (2002).
70. Y. Kato, R. C. Myers, A. C. Gossard, and D. D. Awschalom, Coherent spin manipulation without magnetic fields in strained semiconductors, *Nature* **427**, 50–53 (2004).
71. S. A. Crooker, E. S. Garlid, A. N. Chantis et al., Bias-controlled sensitivity of ferromagnet/semiconductor electrical spin detectors, *Phys. Rev. B* **80**(4), 041305 (2009).
72. C. Gould, C. Rüster, T. Jungwirth et al., Tunneling anisotropic magnetoresistance: A spin-valve-like tunnel magnetoresistance using a single magnetic layer, *Phys. Rev. Lett.* **93**(11), 117203 (2004).
73. J. Moser, A. Matos-Abiague, D. Schuh, W. Wegscheider, J. Fabian, and D. Weiss, Tunneling anisotropic magnetoresistance and spin-orbit coupling in Fe/GaAs/Au tunnel junctions, *Phys. Rev. Lett.* **99**(5), 056601 (2007).
74. M. Tran, H. Jaffrès, C. Deranlot et al., Enhancement of the spin accumulation at the interface between a spin-polarized tunnel junction and a semiconductor, *Phys. Rev. Lett.* **102**(3), 036601 (2009).
75. S. P. Dash, S. Sharma, R. S. Patel, M. P. de Jong, and R. Jansen, Electrical creation of spin polarization in silicon at room temperature, *Nature* **462**, 491–494 (2009).
76. R. J. Soulen Jr., J. M. Byers, M. S. Osofsky et al., Measuring the spin polarization of a metal with a superconducting point contact, *Science* **282**(5386), 85–88 (1998).
77. G. Salis, A. Fuhrer, and S. F. Alvarado, Signatures of dynamically polarized nuclear spins in all-electrical lateral spin transport devices, *Phys. Rev. B* **80**(11), 115332 (2009).
78. C. Awo-Affouda, O. M. J. van't Erve, G. Kioseoglou et al., Contributions to Hanle lineshapes in Fe/GaAs nonlocal spin valve transport, *Appl. Phys. Lett.* **94**(10), 102511 (2009).
79. D. Saha, M. Holub, P. Bhattacharya, and Y. C. Liao, Epitaxially grown MnAs/GaAs lateral spin valves, *Appl. Phys. Lett.* **89**(14), 142504 (2006).
80. O. M. J. van't Erve, A. T. Hanbicki, M. Holub et al., Electrical injection and detection of spin-polarized carriers in silicon in a lateral transport geometry, *Appl. Phys. Lett.* **91**(21), 212109 (2007).
81. M. Ciorga, A. Einwanger, U. Wurstbauer, D. Schuh, W. Wegscheider, and D. Weiss, Electrical spin injection and detection in lateral all-semiconductor devices, *Phys. Rev. B* **79**(16), 165321 (2009).
82. H. C. Koo, J. H. Kwon, J. Eom, J. Chang, S. H. Han, and M. Johnson, Control of spin precession in a spin-injected field effect transistor, *Science* **325**(5947), 1515–1518 (2009).
83. N. Tombros, C. Jozsa, M. Popinciuc, H. T. Jonkman, and B. J. van Wees, Electronic spin transport and spin precession in single graphene layers at room temperature, *Nature* **448**(7153), 571–574 (2007).
84. W. Han, K. Pi, W. Bao et al., Electrical detection of spin precession in single layer graphene spin valves with transparent contacts, *Appl. Phys. Lett.* **94**(22), 222109 (2009).
85. I. Appelbaum, B. Q. Huang, and D. J. Monsma, Electronic measurement and control of spin transport in silicon, *Nature* **447**(7142), 295–298 (2007).
86. B. Huang, D. J. Monsma, and I. Appelbaum, Coherent spin transport through a 350 micron thick silicon wafer, *Phys. Rev. Lett.* **99**(17), 177209 (2007).
87. T. J. Zega, A. T. Hanbicki, S. C. Erwin et al., Determination of interface atomic structure and its impact on spin transport using Z-Contrast microscopy and density-functional theory, *Phys. Rev. Lett.* **96**(19), 196101 (2006).
88. C. Tiusan, J. Faure-Vincent, C. Bellouard, M. Hehn, E. Jouguelet, and A. Schuhl, Interfacial resonance state probed by spin-polarized tunneling in epitaxial Fe/MgO/Fe tunnel junctions, *Phys. Rev. Lett.* **93**, 106602 (2004).
89. D. L. Smith and P. P. Ruden, Spin-polarized tunneling through potential barriers at ferromagnetic metal/semiconductor Schottky contacts, *Phys. Rev. B* **78**(12), 125202 (2008).
90. A. N. Chantis, K. D. Belashchenko, D. L. Smith, E. Y. Tsymbal, M. van Schilfgaarde, and R. C. Albers, Reversal of spin polarization in Fe/GaAs (001) driven by resonant surface states: First-principles calculations, *Phys. Rev. Lett.* **99**(19), 196603 (2007).
91. H. Dery and L. J. Sham, Spin extraction theory and its relevance to spintronics, *Phys. Rev. Lett.* **98**(4), 046602 (2007).
92. S. Honda, H. Itoh, J. Inoue et al., Spin polarization control through resonant states in an Fe/GaAs Schottky barrier, *Phys. Rev. B* **78**(24), 245316 (2008).
93. P. Li and H. Dery, Tunable spin junction, *Appl. Phys. Lett.* **94**(19), 192108 (2009).
94. M. K. Chan, Q. O. Hu, J. Zhang, T. Kondo, C. J. Palmstrøm, and P. A. Crowell, Hyperfine interactions and spin transport in ferromagnet-semiconductor heterostructures, *Phys. Rev. B* **80**(16), 161206 (2009).
95. R. Jansen and B. C. Min, Detection of a spin accumulation in nondegenerate semiconductors, *Phys. Rev. Lett.* **99**(24), 246604 (2007).
96. C. Jozsa, M. Popinciuc, N. Tombros, H. T. Jonkman, and B. J. Van Wees, Electronic spin drift in graphene field-effect transistors, *Phys. Rev. Lett.* **100**(23), 236603 (2008).

97. A. N. Chantis and D. L. Smith, Theory of electrical spin-detection at a ferromagnet/semiconductor interface, *Phys. Rev. B* **78**(23), 235317 (2008).
98. Y. Ohno, R. Terauchi, T. Adachi, F. Matsukura, and H. Ohno, Spin relaxation in GaAs (110) quantum wells, *Phys. Rev. Lett.* **83**, 4196 (1999).
99. M. I. D'yakonov and V. I. Perel', Possibility of orienting electron spins with current, *JETP Lett.* **13**(11), 467–469 (1971).
100. Y. K. Kato, R. C. Myers, A. C. Gossard, and D. D. Awschalom, Observation of the spin Hall effect in semiconductors, *Science* **306**(5703), 1910–1913 (2004).
101. J. Wunderlich, B. Kaestner, J. Sinova, and T. Jungwirth, Experimental observation of the spin-Hall effect in a two-dimensional spin-orbit coupled semiconductor system, *Phys. Rev. Lett.* **94**(4), 047204 (2005).
102. E. S. Garlid, Q. O. Hu, M. K. Chan, C. J. Palmstrøm, and P. A. Crowell, Electrical measurement of the direct spin hall effect in Fe/$In_xGa_{1-x}As$ heterostructures, *Phys. Rev. Lett.* **105**, 156602 (2010).
103. S. O. Valenzuela and M. Tinkham, Direct electronic measurement of the spin Hall effect, *Nature* **442**(7099), 176–179 (2006).
104. J. Wunderlich, A. C. Irvine, J. Sinova et al., Spin-injection Hall effect in a planar photovoltaic cell, *Nat. Phys.* **5**(9), 675–681 (2009).

24

Spin Polarization by Current

Sergey D. Ganichev
University of Regensburg

Maxim Trushin
University of Regensburg

John Schliemann
University of Regensburg

24.1 Introduction

Spin generation and spin currents in semiconductor structures lie at the heart of the emerging field of spintronics and are a major and still-growing direction of solid-state research. Among the plethora of concepts and ideas, current-induced spin polarization (CISP) has attracted particular interest from both experimental and theoretical point of view, for reviews see Refs. [1–9]. In nonmagnetic semiconductors or metals belonging to the gyrotropic point groups,* *dc* electric current is generically accompanied by a nonzero average nonequilibrium spatially homogeneous spin polarization and vice versa. The latter phenomenon is referred to as the spin-galvanic effect (see, e.g., Refs. [5,7,10–13]). In low-dimensional semiconductor structures, these effects are caused by asymmetric spin relaxation in systems with lifted spin degeneracy due to **k**-linear terms in the Hamiltonian, where **k** is the electron wave vector. In spite of the terminological resemblance, spin polarization by electric current fundamentally differs from the spin Hall effect [4–6,8,14–20], which refers to the generation of a pure spin current transverse to the charge current and causes spin accumulation at the sample edges. The distinctive features of the CISP are that this effect can be present in gyrotropic media only, it results in nonzero average spin polarization, and does not depend on the space coordinates. Thus, it can be measured in the whole sample under appropriate conditions. The spin Hall effect, in contrast, does not yield average spin polarization and does not require gyrotropy, at least for the extrinsic spin Hall effect.

The ability of charge current to polarize spins in gyrotropic media was predicted more than 30 years ago by Ivchenko and Pikus [27]. The effect has been considered theoretically for bulk tellurium crystals, where, almost to the same time, it has been demonstrated experimentally by Vorob'ev et al. [28]. In bulk tellurium, CISP is a consequence of the unique valence band structure of tellurium with hybridized spin-up and spin-down bands ("camel back" structure) and, in contrast to the spin polarization in quantum well (QW) structures, is not related to spin relaxation. In zinc blende structure–based QWs, this microscopic mechanism of the CISP is absent [29]. Aronov and Lyanda-Geller [30], Edelstein [31], and Vasko and Prima [32] demonstrated that spin orientation by electric current is also possible in QW systems and is caused by asymmetric spin relaxation. Two microscopic mechanisms, namely, scattering mechanism and precessional mechanism, based on Elliott–Yafet and D'yakonov-Perel' spin relaxation, respectively, were developed. The first direct experimental proofs of this effect were obtained in semiconductor QWs by Ganichev et al. in (113)-grown p-GaAs/AlGaAs QWs [33,34], Silov et al. in (001)-grown p-GaAs/AlGaAs heterojunctions [35,36], Sih et al. in (110)-grown n-GaAs/AlGaAs QWs [37], and Yang et al. in (001)-grown InGaAs/InAlAs QWs [38], as well as in strained bulk (001)-oriented InGaAs and ZnSe epilayers by Kato et al. [39,40] and Stern et al. [41], respectively. The experiments include the range of optical methods such as Faraday rotation and linear–circular dichroism in transmission of terahertz radiation, time-resolved Kerr rotation, and polarized luminescence in near-infrared up to visible spectral range. We emphasize that CISP was observed even at room temperature

* We remind that the gyrotropic point-group symmetry makes no difference between certain components of polar vectors, like electric current or electron momentum, and axial vectors, like a spin or magnetic field, and is described by the gyration tensor [7,21,22]. Gyrotropic media are characterized by the linear in light or electron wave vector **k** spatial dispersion resulting in optical activity (gyrotropy) or Rashba/Dresselhaus band spin-splitting in semiconductor structures [7,22–26], respectively. Among 21 crystal classes lacking inversion symmetry, 18 are gyrotropic, from which 11 classes are enantiomorphic (chiral) and do not possess a reflection plane or rotation–reflection axis [7,24,25]. Three nongyrotropic noncentrosymmetric classes are T_d, C_{3h}, and D_{3h}. We note that it is often, but misleading, stated that gyrotropy (optical activity) can be obtained only in noncentrosymmetric crystals having no mirror reflection plane. In fact, seven non-enantiomorphic classes groups (C_s, C_{2v}, C_{3v}, S_4, D_{2d}, C_{4v}, and C_{6v}) are gyrotropic also allowing spin orientation by the electric current.

[33,34,40]. Variation of crystallographic direction demonstrated that an in-plane electric current may result not only in the in-plane spin orientation, as in the case of (001)-grown structures [35,36,38–41], but can also orient spins normal to the 2DEG's plane in [llh]-oriented structures, for example, [113]- or [110]-grown QWs (see Refs. [33,34,37]). We also would like to note that investigating spin injection from a ferromagnetic film into a two-dimensional electron gas, Hammar et al. [42,43] used the concept of a spin orientation by current in a 2DEG (see also Refs. [44,45]) to interpret their results. Though a larger degree of spin polarization was extracted, the experiment's interpretation is complicated by other effects [46,47].

The experimental observation of CISP has given rise to extended theoretical studies of this phenomenon in various systems using various approaches and theoretical techniques. These include the semiclassical Boltzmann equation [48–52] derived from the quantum mechanical Liouville equation [53] and other diffusion-type equation describing the dynamics of spin expectation values [6,54–58]. The approach and results of Ref. [51] are reviewed in more detail later in this chapter. Other authors have performed important and fruitful studies of the same issues using various Green's function techniques of many-body physics [2,59–67]. The effect of external contacts to the system and boundaries was studied explicitly in Refs. [45,68,69], and in Refs. [70,71] the influence of oscillating electromagnetic fields was studies theoretically. The case of CISP hole systems in p-doped QWs was investigated in Refs. [64,72], and the authors of Ref. [73] performed a numerical Landauer–Keldysh study of transport and spin polarization in a four terminal geometry.

Searching for publications on CISP, one can be confused by the fact that several different names are used to describe it. Besides CISP, this phenomenon is sometimes referred to as current-induced spin accumulation (CISA), the inverse spin-galvanic effect (ISGE), the magnetoelectric effect (MEE) or kinetic magnetoelectric effect (KMEE) (this term was first used to describe the effect in nonmagnetic conductors by Levitov [74]), and electric field–mediated in-plane spin accumulation. In this chapter, we use two of these terms: CISA and the ISGE.

24.2 Model

Phenomenologically, the electron's averaged nonequilibrium spin **S** can be linked to an electric current **j** by

$$j_\lambda = \sum_\mu Q_{\lambda\mu} S_\mu, \qquad (24.1)$$

$$S_\alpha = \sum_\gamma R_{\alpha\gamma} j_\gamma, \qquad (24.2)$$

where **Q** and **R** are second-rank pseudotensors. Equation 24.1 describes the spin-galvanic effect and Equation 24.2 represents the effect inverse to the spin-galvanic effect: an electron spin

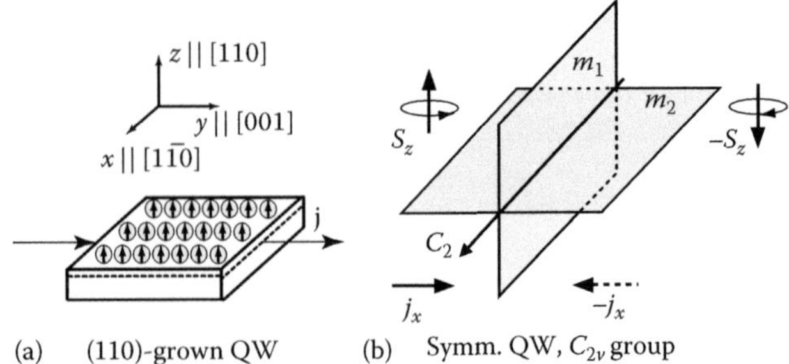

FIGURE 24.1 (a) Electric current–induced spin polarization in symmetric (110)-grown zinc blende structure–based QWs. (b) Symmetry elements of symmetrical QW grown in the z||[110] direction. Arrows in the sketch (b) show reflection of the components of polar vector j_x and axial vector S_z by the mirror reflection plane m_1. An additional reflection by the mirror reflection plane m_1 does not modify the components of an in-plane polar vector j_x as well as does not change the polarity of an out-plane axial vector S_z. Thus, the linear coupling of j_x and B_z is allowed for structures of this symmetry.

polarization induced by a *dc* electric current. We note the similarity of Equations 24.1 and 24.2 characteristic for effects due to gyrotropy: both equations linearly couple a polar vector with an axial vector. The phenomenological equation (24.2) shows that the spin polarization can only occur for those components of the in-plane components of **j**, which transform as the averaged nonequilibrium spin **S** for all symmetry operations. Thus, the relative orientation of the current direction and the average spin is completely determined by the point-group symmetry of the structure. This can most clearly be illustrated by the example of a symmetric (110)-grown zinc blende QW where an electric current along x||[1$\bar{1}$0] results in a spin orientation along the growth direction z(see Figure 24.1a). These QWs belong to the point-group symmetry C_{2v} and contain, apart from the identity and a C_2-axis, a reflection plane m_1 normal to the QW plane and x-axis and a reflection plane m_2 being parallel to the interface plane (see Figure 24.1b). The reflection in m_1 transforms the current component j_x and the average spin component S_z in the same way ($j_x \to -j_x$, $S_y \to -S_y$) (see Figure 24.1b). Also the reflection in m_2 transforms these components in an equal way, and the sign of them remains unchanged for this symmetry operation. Therefore, a linear coupling of the in-plane current and the out-plane average spin is allowed, demonstrating that a photocurrent j_x can induce the average spin polarization S_z. In asymmetric (110)-grown QWs or (113)-grown QWs, the symmetry is reduced to C_s and additionally to the z-direction, spins can be oriented in the plane of QWs.* Similar arguments demonstrate that in (001)-grown zinc blende structure–based QWs, an electric current can result in an in-plane spin orientation *only*. In this case, the direction of spins depends on the relative strengths of the structure

* Note that for the lowest symmetry C_1, which contains no symmetry elements besides identity, the relative direction between current and spin orientation becomes arbitrary.

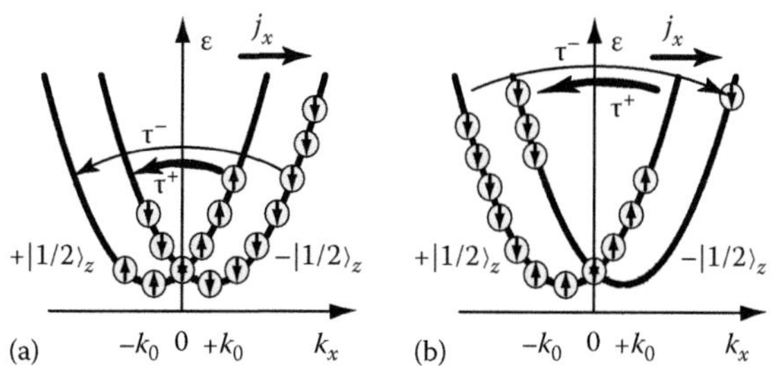

FIGURE 24.2 Current-induced spin polarization (a) and spin-galvanic effect (b) due to spin-flip scattering in symmetric (110)-grown zinc blende structure–based QWs. In this case, only the $\beta_{zx}\sigma_z k_x$ term is present in the Hamiltonian. The conduction subband is split into two parabolas with spin-up $|+1/2\rangle_z$ and spin-down $|-1/2\rangle_z$ pointing in the z-direction. In (a), biasing along the x-direction causes an asymmetry in **k**-space occupation of both parabolas. In (b), nonequilibrium spin orientation along z-direction causes an electric current in x-direction. (After Ganichev, S.D. et al., cond-mat/0403641, 2004; Ganichev, S.D. et al., *J. Magn. Magn. Mater.*, 300, 127, 2006.)

inversion asymmetry (SIA) [75] and bulk inversion asymmetry (BIA) [76] resulting in an anisotropy of the CISP.

A microscopic model of the CISP [33,34] is sketched in Figure 24.2b. To be specific, we consider an electron gas in symmetric (110)-grown zinc blende QWs. The explanation of the effect measured in structures of other crystallographic orientation or in a hole gas can be given in a similar way. In the simplest case, the electron's (or hole's) kinetic energy in a QW depends quadratically on the in-plane wave vector components k_x and k_y. In equilibrium, the spin-degenerated k_x and k_y states are symmetrically occupied up to the Fermi energy E_F. If an external electric field is applied, the charge carriers drift in the direction of the resulting force. The carriers are accelerated by the electric field and gain kinetic energy until they are scattered. A stationary state forms where the energy gain and the relaxation are balanced resulting in a nonsymmetric distribution of carriers in **k**-space yielding an electric current. The electrons acquire the average quasimomentum

$$\hbar\Delta\mathbf{k} = e\mathbf{E}\tau_{ps}, \tag{24.3}$$

where

- **E** is the electric field strength
- τ_{ps} is the momentum relaxation time for a given spin-split subband labeled by s
- e is the elementary charge

As long as spin-up and spin-down states are degenerated in **k**-space the energy bands remain equally populated and a current is not accompanied by spin orientation. In QWs made of zinc blende structure material like GaAs or strained bulk semiconductors, however, the spin degeneracy is lifted due to SIA and BIA [75,76] and dispersion reads

$$E_{ks} = \frac{\hbar^2\mathbf{k}^2}{2m^*} + \beta_{lm}\sigma_l k_m \tag{24.4}$$

with the spin–orbit pseudotensor β, the Pauli spin matrices σ_l, and effective mass m^* (note, the similarity to Equations 24.1 and 24.2). The parabolic energy band splits into two subbands of opposite spin directions shifted in **k**-space symmetrically around $\mathbf{k}=0$ with minima at $\pm k_0$. For symmetric (110)-grown zinc blende structure–based QWs, the spin–orbit interaction results in a Hamiltonian of the form $\beta_{zx}\sigma_z k_x$, the corresponding dispersion is sketched in Figure 24.2. Here, the energy band splits into two subbands of $s_z=1/2$ and $s_z=-1/2$, with minima symmetrically shifted in the **k**-space along the k_x axis from the point $k=0$ into the points $\pm k_0$, where $k_0=m^*\beta_{z'x}/\hbar^2$. Until the spin relaxation is switched off, the spin branches are equally populated and equally contribute to the current. Due to the band splitting, spin-flip relaxation processes $\pm 1/2 \rightarrow \mp 1/2$, however, become different because of the difference in quasimomentum transfer from initial to final states. In Figure 24.2a, the **k**-dependent spin-flip scattering processes are indicated by arrows of different lengths and thicknesses. As a consequence, different amounts of spin-up and spin-down carriers contribute to the spin-flip transitions causing a stationary spin orientation. In Figure 24.2a, we assume that the origin of the current-induced spin orientation is exclusively due to scattering, and hence dominated by the Elliott–Yafet spin relaxation mechanism [77] (relaxation mechanism). The other possible mechanism resulting in the current-induced spin orientation is based on the D'yakonov–Perel' spin relaxation [77] (precessional mechanism). In this case, the relaxation rate depends on the average **k**-vector [76] equal to $\bar{k}_{1/2}=-k_0+\langle k\rangle$ for the spin-up and $\bar{k}_{-1/2}=k_0+\langle k\rangle$ for the spin-down subband. Hence, also for the D'yakonov–Perel' mechanism, the spin relaxation becomes asymmetric in **k**-space and a current through the electron gas causes spin orientation.

Figure 24.2b shows that not only phenomenological equations but also microscopic models of the CISP and the spin-galvanic effect are similar. Spin orientation in the x-direction causes the unbalanced population in spin-down and spin-up subbands. As long as the carrier distribution in each subband is symmetric around the subband minimum at $k_{x\pm}$, no current flows. The current flow is caused by **k**-dependent spin-flip relaxation processes [7,10,11]. Spins oriented in the z-direction are scattered along k_x from the higher-filled spin subband $|+1/2\rangle_z$ to the less-filled spin subband $|-1/2\rangle_z$. Four quantitatively different spin-flip scattering events exist and are sketched in Figure 24.2a by bent arrows. The spin-flip scattering rate depends on the values of the wave vectors of the initial and the final states [77]. Therefore, spin-flip transitions, shown by solid arrows in Figure 24.2a, have the same rates. They preserve the symmetric distribution of carriers in the subbands and, thus, do not yield a current. However, the two scattering processes shown by broken arrows are inequivalent and generate an asymmetric carrier distribution around the subband minima in both subbands. This asymmetric population results in a current flow along the x-direction. Within this model of elastic scattering, the current is not spin-polarized since the same number of spin-up and spin-down electrons moves in the same direction with the same velocity. Like CISP, the spin-galvanic effect can also result from the precessional mechanism [7] based on the asymmetry of the D'yakonov–Perel' spin relaxation.

In order to foster the above phenomenological arguments and models, let us discuss a microscopic description of such processes as introduced in Ref. [51]. This theory is based on an analytical solution to the semiclassical Boltzmann equation, which does not include the spin relaxation time explicitly. Instead, one utilizes the so-called quasiparticle lifetime τ_0, which is defined via the scattering probability alone. This is in contrast to the momentum relaxation time τ_{ps}, which, in addition, depends on the distribution function itself and, therefore, should be self-consistently deduced from the Boltzmann equation written within the relaxation time approximation. In the framework of a simplest model dealing with the elastic delta-correlated scatterers, the quasiparticle lifetime τ_0 depends neither on carrier momentum nor spin index, and in that way it essentially simplifies our description.

Before we proceed, let us first discuss the applicability of the quasiclassical Boltzmann kinetic equation to the description of the spin polarization in QWs observed in Ref. [35–41]. There are two restrictions that are inherited by the Boltzmann equation due to its quasiclassical origin. The first one is obvious: The particle's de Broglie length must be much smaller than the mean free path. At low temperatures (as compared to the Fermi energy), the characteristic de Broglie length λ relates to the carrier concentration $n_e{\sim}m^*E_F/\pi\hbar^2$ approximately as $\lambda \sim \sqrt{2\pi/n_e}$, whereas the mean free path l can be estimated as $l \sim (\hbar/m^*)\sqrt{2\pi n_e}\tau_0$. Thus, the first restriction can be written as

$$n_e \gg \frac{m^*}{\hbar\tau_0}. \tag{24.5}$$

The second one is less obvious and somewhat more specific to our systems here, but still it is important: The smearing of the spin-split subband due to the disorder $\hbar/\tau_0$ must be much smaller than the spin-splitting energy $E_{+k}-E_{-k}$. The latter depends strongly on the quasimomentum, and therefore, at the Fermi level it is defined by the carrier concentration. As consequence, the concentration must fulfill the following inequality:

$$n_e \gg \frac{\hbar^2}{8\pi\beta^2\tau_0^2}, \tag{24.6}$$

where β is the spin–orbit coupling parameter. This restriction can also be reformulated in terms of the mean free path l and spin precession length $\lambda_s{\sim}\pi/k_0$. Namely, l must be much larger than λ_s so that an electron randomizes its spin orientation due to the spin–orbit precession between two subsequent scattering events. This restriction corresponds to the approximation that neglects the off-diagonal elements of the nonequilibrium distribution function in the spin space. Indeed, an electron spin can not only be in one of two possible spin eigen states of the free Hamiltonian but also be in an arbitrary superposition of them, the case described by the off-diagonal elements of the density matrix. The latter is not possible in classical physics where a given particle always has a definite position in the phase space. Thus, we could not directly apply Boltzmann equation in its conventional form for the description of the electron spin in 2DEGs with spin–orbit interactions as long as the inequality (24.6) is not fulfilled. Note, on the other hand, that all the quantum effects stemming from the quantum nature of the electron spin can be smeared out by a sufficient temperature larger than the spin-splitting energy $E_{+k}-E_{-k}$. Thus, the room temperature $T_{\text{room}}=25\,\text{meV}$ being much larger than the spin–orbit splitting energy of the order of 3 meV (which is relevant for InAs QWs) makes the Boltzmann equation applicable for sure. In order to verify whether the conditions (24.5) and (24.6) are indeed satisfied, one can deduce the quasiparticle lifetime τ_0 from the mobility $\mu=e\tau_0/m^*$. Then, using spin–orbit coupling parameters in the range usually found in experiments [78–80], the above inequalities turn out to be fulfilled.

In general, the Boltzmann equation describes the time evolution of the particle distribution function $f(t,\mathbf{r},\mathbf{k})$, in the coordinate $\mathbf{r}$ and momentum $\mathbf{k}$-space. To describe the electron kinetics in the presence of a small homogeneous electric field $\mathbf{E}$ in a steady state, one usually follows the standard procedure widely spread in the literature on solid-state physics. The distribution function is then represented as a sum of an equilibrium $f_0(E_{sk})$ and nonequilibrium $f_1(s,\mathbf{k})$ contributions. The first one is just a Fermi–Dirac distribution, and the second one is a time- and coordinate-independent nonequilibrium correction linear in $\mathbf{E}$. This latter contribution should be written down as a solution of the kinetic equation; however, it might also be deduced from qualitative arguments in what follows.

Let us assume that the scattering of carriers is elastic, which means, above all, that spin flip is forbidden. Then, the average momentum $\Delta\mathbf{k}$, which the carriers gain due to the electric field, can be estimated relying on the momentum relaxation time approximation from Equation 24.3. If the electric field is small (linear response) then to get the nonequilibrium term $f_1(s,\mathbf{k})$ one has to expand the Fermi–Dirac function $f_0(E_{s(\mathbf{k}-\Delta\mathbf{k})})$ into the power series for small $\Delta\mathbf{k}$ up to the term linear in $\mathbf{E}$. Recalling $\hbar\mathbf{v}=-\partial_{\Delta\mathbf{k}}E_{(s\mathbf{k}-\Delta\mathbf{k})}\big|_{\Delta\mathbf{k}=0}$, the nonequilibrium contribution $f_1(s,\mathbf{k})$ can be written as

$$f_1(s,\mathbf{k}) = -e\mathbf{E}\mathbf{v}\tau_{ps}\left[-\frac{\partial f^0(E_{sk})}{\partial E_{sk}}\right]. \tag{24.7}$$

Since $f_1(s,\mathbf{k})$ is proportional to the derivative of the steplike Fermi–Dirac distribution function, the nonequilibrium term substantially contributes to the total distribution function close to the Fermi energy only, and the sign of its contribution depends on the sign of the group velocity $\mathbf{v}$. Note that the group velocity $\mathbf{v}$ for a given energy and direction of the motion is the same for

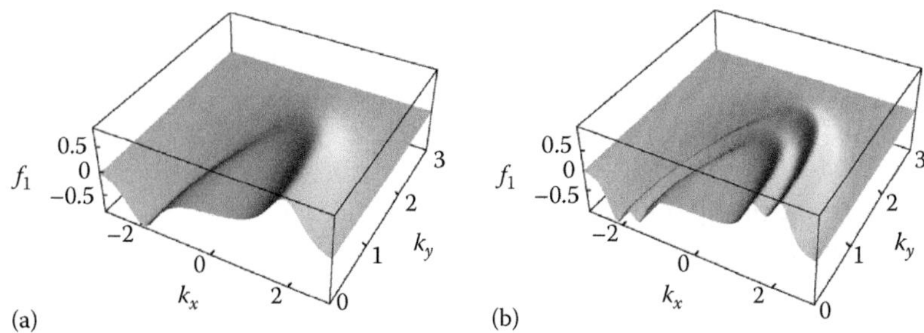

FIGURE 24.3 This plot shows schematically the nonequilibrium term of the distribution function versus quasimomentum for (a) spin-degenerate and (b) spin-split subbands. The electric field is directed along the x-axis. (a) Without spin-splitting, the nonequilibrium distribution function just provides a majority for right-moving electrons at the expense of the ones with opposite momentum. (b) The spin–orbit splitting leads to the additional electron redistribution between two spin-split subbands. As one can see from the plot, an amount of electrons belong to the outer spin-split subband is larger than for the inner one. Thus, the spins from inner and outer subbands are not compensated with each other, and spin accumulation occurs.

both spin-split subbands and, therefore, nonequilibrium addition to the distribution function would be the same for both branches as long as τ_{ps} were independent of the spin index. The latter is, however, not true, and τ_{ps} is different for two spin-split subbands, as depicted in Figure 24.2. This is the microscopic reason why the nonequilibrium correction to the distribution function gives rise to the spin polarization.

The nonequilibrium correction can also be found as an analytical solution of the Boltzmann equation [51], and has the form

$$g_1(s,\mathbf{k}) = -e\mathbf{E}\mathbf{k}\frac{\hbar\tau_0}{m^*}\left[-\frac{\partial f^0(E_{sk})}{\partial E_{sk}}\right]. \tag{24.8}$$

In contrast to Equation (24.7), Equation 24.8 contains momentum $\mathbf{k}$ and quasiparticle lifetime as a prefactor.* Since the quasiparticle lifetime does not depend on the spin index and the quasiparticle momentum calculated for a given energy is obviously different for two spin-split subbands, Equation 24.8 immediately allows us to read out the very fact that the nonequilibrium contribution depends on the spin index (see also Figure 24.3). Thus, the spins are not compensated with each other as long as the system is out of equilibrium. Therewith one can see that the efficiency of the CISP is governed by the splitting between two subbands, that is, it is directly proportional to the spin–orbit constants. One can also prove the later statements just calculating the net spin density

$$\langle S_{x,y,z}\rangle = \sum_s \int \frac{d^2k}{(2\pi)^2} S_{x,y,z}(k,s) g_1(s,\mathbf{k}), \tag{24.9}$$

with $S_{x,y,z}$ being the spin expectation values.

To conclude this section, we would like to stress that the two Equations 24.7 and 24.8 are just two ways to describe the same physical content. One can think about spin accumulation either in terms of the momentum relaxation time τ_{ps} dependent on the spin-split subband index, or one may rely on the exact solution Equation 24.8, which relates the spin accumulation to the quasi-momentum difference between electrons with the opposite spin orientations.

24.3 Anisotropy of Inverse Spin-Galvanic Effect

Similar to the spin-galvanic effect [7,78,79], the CISP can be strongly anisotropic due to the interplay of the Dresselhaus and Rashba terms. The relative strength of these terms is of general importance because it is directly linked to the manipulation of the spin-of-charge carriers in semiconductors, one of the key problems in the field of spintronics. Both Rashba and Dresselhaus couplings result in spin-splitting of the band and give rise to a variety of spin-dependent phenomena, which allow us to evaluate the magnitude of the total spin splitting of electron subbands [11,12,18,77–90]. Dresselhaus and Rashba terms can interfere in such a way that macroscopic effects vanish though the individual terms are large [77,84,85]. For example, both terms can cancel each other resulting in a vanishing spin splitting in certain $\mathbf{k}$-space directions [11,85]. This cancellation leads to the disappearance of an antilocalization [82], circular photogalvanic effect [79], magneto-gyrotropic effect [89,90], spin-galvanic effect [78] and CISP [51], the absence of spin relaxation in specific crystallographic directions [77,83], and the lack of Shubnikov–de Haas beating [84], and has also given rise to a proposal for spin field-effect transistor operating in the nonballistic regime [85].

While the interplay of Dresselhaus and Rashba spin splitting may play a role in QWs of different crystallographic orientations, here we focus on anisotropy of the inversed spin-galvanic effect in (001)-grown zinc blende structure-based QWs. For (001)-oriented QWs, linear in the wave vector part of Hamiltonian for the first subband reduces to

$$\mathcal{H}_{\mathbf{k}}^{(1)} = \alpha(\sigma_{x_0}k_{y_0} - \sigma_{y_0}k_{x_0}) + \beta(\sigma_{x_0}k_{x_0} - \sigma_{y_0}k_{y_0}), \tag{24.10}$$

* Note that the well-known textbook relation $\mathbf{v}=\hbar\mathbf{k}/m^*$ holds only in the absence of spin–orbit coupling.

where

α and β are the parameters resulting from the structure-inversion and bulk-inversion asymmetries, respectively

x_0 and y_0 are the crystallographic axes [100] and [010]

Note that, in the coordinate system with $x\|[1\bar{1}0]$ and $y\|[110]$, the matrix $\mathcal{H}_{\mathbf{k}}^{(1)}$ gets the form $\beta_{xy}\sigma_x k_y + \beta_{yx}\sigma_y k_x$ with $\beta_{xy} = \beta + \alpha$, $\beta_{yx} = \beta - \alpha$.

To study the anisotropy of the spin accumulation, one can just calculate the net spin density from Equation 24.9, where the integral over **k** can be taken easily making the substitution $\varepsilon = E(s,\mathbf{k})$ and assuming that $-\partial f^0(\varepsilon)/\partial\varepsilon = \delta(E_F - \varepsilon)$. The rest integrals over the polar angle can be taken analytically, and after some algebra we have

$$\langle \mathbf{S} \rangle = \frac{em^* \tau_0}{2\pi\hbar^3} \begin{pmatrix} \beta & \alpha \\ -\alpha & -\beta \end{pmatrix} \mathbf{E}. \tag{24.11}$$

This relation between the spin accumulation and electrical current can also be deduced by phenomenologically applying Equation 24.2.

The magnitude of the spin accumulation $\langle S \rangle = \sqrt{\langle S_x \rangle^2 + \langle S_y \rangle^2}$ depends on the relative strength of β and α and varies after

$$\langle S \rangle = \frac{eEm^* \tau_0}{2\pi\hbar^3} \sqrt{\alpha^2 + \beta^2 + 2\alpha\beta \sin(2\widehat{\mathbf{E}\mathbf{e}_x})}. \tag{24.12}$$

It is interesting to note that $\langle S \rangle$ depends on the direction of the electric field (see Figure 24.4), that is, the spin accumulation is anisotropic. This anisotropy reflects the relation between

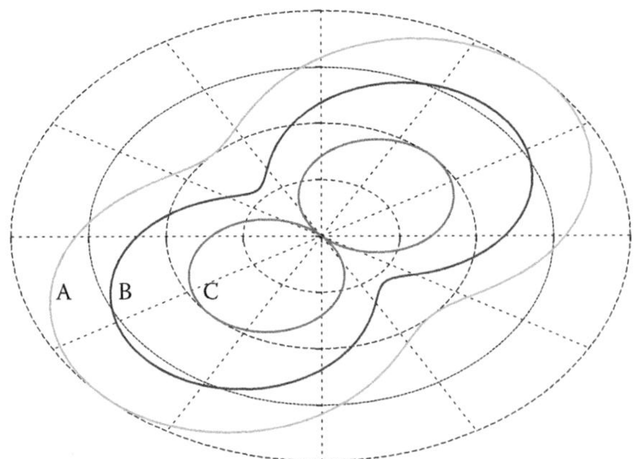

FIGURE 24.4 Anisotropy of spin accumulation (in arbitrary units) versus direction of the electric field in polar coordinates for different Rashba and Dresselhaus constants: A—$\alpha = 3\beta$, B—$\alpha = 2.15\beta$, and C—$\alpha = \beta$. The curve B corresponds to the n-type InAs-based QWs investigated in Ref. [78]. (After Trushin, M. and Schliemann, J., *Phys. Rev. B*, 75, 155323, 2007.)

Rashba and Dresselhaus spin–orbit constants. If either $\alpha = 0$ or $\beta = 0$, then the anisotropy vanishes. In the opposite case of $\alpha = \beta$ the anisotropy reaches its maximum. For the spin accumulation $\langle S_{[110]} \rangle$ and $\langle S_{[1\bar{1}0]} \rangle$ providing by the electrical current along [110] and $[1\bar{1}0]$ crystallographic directions, respectively, we obtain

$$\frac{\langle S_{[1\bar{1}0]} \rangle}{\langle S_{[110]} \rangle} = \frac{\alpha - \beta}{\alpha + \beta}. \tag{24.13}$$

The simple Drude-like relation between electrical current and field allows us to write down a useful relation between the spin accumulation and charge current density

$$\langle \mathbf{S} \rangle = \frac{m^{*2}}{2\pi\hbar^3 e n_e} \begin{pmatrix} \beta & \alpha \\ -\alpha & -\beta \end{pmatrix} \mathbf{j}. \tag{24.14}$$

Note that, in the coordinate system with $x\|[1\bar{1}0]$ and $y\|[110]$, the Hamiltonian $\mathcal{H}_{\mathbf{k}}^{(1)}$ gets the form $\beta_{xy}\sigma_x k_y + \beta_{yx}\sigma_y k_x$ with $\beta_{xy} = \beta + \alpha$, $\beta_{yx} = \beta - \alpha$, and the relation between $\langle \mathbf{S} \rangle$ and $\mathbf{j}$ looks simpler, namely,

$$\langle \mathbf{S} \rangle = \frac{m^{*2}}{2\pi\hbar^3 e n_e} \begin{pmatrix} 0 & \beta - \alpha \\ \beta + \alpha & 0 \end{pmatrix} \mathbf{j}. \tag{24.15}$$

From this equation one can see easily that the spin accumulation is strongly anisotropic if the constants α and β are close to each other. As discussed earlier, the reason of such an anisotropy can be understood either from the spin–orbit splitting different for different direction of the carrier motion or from the anisotropic spin relaxation times [83].

Equations 24.13 through 24.15 show that measuring of the spin polarization for current flowing in particular crystallographic directions one can map the spin–orbit constants and deduce their magnitudes. To find the absolute values of the spin–orbit constants, one needs to know the carrier effective mass m^* and quasiparticle lifetime τ_0. The latter can be roughly estimated from the carrier mobility. Indeed, though the band structure described by the Hamiltonian (24.10) is anisotropic, the electrical conductivity remains isotropic that can be verified directly computing the electrical current $\mathbf{j} = -e\sum_s \int (d^2k/(2\pi)^2)\mathbf{v}_s f^1(s,\mathbf{k})$. Taking this integral in the same manner as in Equation 24.9, one can see that the anisotropy of the group velocity $\mathbf{v}_s$ is compensated by the one stemming from the distribution function $f^1(s,\mathbf{k})$. Therefore, the electrical conductivity is described by the Drude-like formula $\sigma = e^2 n_e \tau_0 / m^*$ with $n_e = m^* E_F/\pi\hbar^2 + (m^*/\hbar^2)^2(\alpha^2 + \beta^2)/\pi$ being the carrier concentration. Using this Drude-like relation, one can estimate the quasiparticle lifetime as $\tau_0 \sim m^*\mu/e$. To give an example, an n-type InAs QW [78] with $\mu = 2 \times 10^4$ cm^2/(V s) and $m^* \sim 0.032 m_0$ with m_0 being the bare electron mass yields the quasiparticle lifetime $\tau_0 \sim 4 \times 10^{-13}$ s.

24.4 Concluding Remarks

We have given an overview of CISP in gyrotropic semiconductor nanostructures. Such a spin polarization as response to a charge current, also known as the magnetoelectrical effect, may be classified as the inverse of the spin-galvanic effect. Apart from reviewing the experimental status of affairs, we have provided a detailed theoretical description of both effects in terms of a phenomenological model of spin-dependent relaxation processes, and an alternative theoretical approach based on the quasiclassical Boltzmann equation. Two microscopic mechanisms of this effect, both involving **k**-linear band spin splitting, are known so far: Relaxation mechanism based on Elliott-Yafet spin relaxation and precessional mechanism due to D'yakonov–Perel' spin relaxation. We note that, recently, another mechanism of the current-induced spin polarization of different microscopic origin has been proposed, which may dominate at room temperature [91]. It is based on spin-dependent asymmetry of electron scattering and out of the scope of the present chapter. The relative direction between electric current and nonequilibrium spin for all mechanisms is determined by the crystal symmetry. In particular, we have discussed the anisotropy of the ISGE in (001)-grown zinc blende structure-based QWs in the presence of spin–orbit interaction of both the Rashba and the Dresselhaus type. Combined with the theoretical achievements derived from the Boltzmann approach, the precise measurement of this anisotropy allows in principle to determine the absolute values of the Rashba and the Dresselhaus spin–orbit coupling parameter. Moreover, these interactions can be used for the manipulation of the magnitude and direction of electron spins by changing the direction of current, thereby enabling a new degree of spin control. We focused here on the fundamental physics underlying the spin-dependent transport of carriers in semiconductors, which persists up to room temperature [33,34,40], and, therefore, may become useful in future semiconductor spintronics.

Acknowledgments

We thank E.L. Ivchenko, V.V. Bel'kov, D. Weiss, L.E. Golub, and S.A. Tarasenko for helpful discussions. This work is supported by the DFG (SFB689).

References

1. R. H. Silsbee, *J. Phys.: Condens. Matter* **16**, R179 (2004).
2. E. I. Rashba, *Phys. Rev. B* **70**, 161201(R) (2004).
3. E. I. Rashba, *J. Supercond.: Incorporat. Nov. Magn.* **18**, 137 (2005).
4. E. I. Rashba, Semiconductor spintronics: Progress and challenges, in *Future Trends in Microelectronics. Up to Nano Creek*, S. Luryi, J. M. Xu, and A. Zaslavsky (Eds.), Wiley-Interscience, Hoboken, NJ, pp. 28–40, 2007, arXiv:cond-mat/0611194.
5. R. Winkler, in Spin-dependent transport of carriers in semiconductors, in *Handbook of Magnetism and Advanced Magnetic Materials*, vol. 5, H. Kronmuller and S. Parkin (Eds.), John Wiley & Sons, New York, 2007, arXiv:cond-mat/0605390.
6. I. Adagideli, M. Scheid, M. Wimmer, G. E. W. Bauer, and K. Richter, *New J. Phys.* **9**, 382 (2007).
7. E. L. Ivchenko and S. D. Ganichev, Spin photogalvanics, in *Spin Physics in Semiconductors* M. I. D'yakonov (Ed.), Springer, Berlin, pp. 245–277, 2008.
8. Y. K. Kato and D. D. Awschalom, *J. Phys. Soc. Jpn.* **77**, 031006 (2008).
9. S. D. Ganichev, *Int. J. Mod. Phys. B* **22**, 1 (2008).
10. S. D. Ganichev et al., *Nature* **417**, 153 (2002).
11. S. D. Ganichev and W. Prettl, *J. Phys.: Condens. Matter* **15**, R935 (2003).
12. S. D. Ganichev and W. Prettl, *Intense Terahertz Excitation of Semiconductors*, Oxford University Press, Oxford, U.K., 2006.
13. E. L. Ivchenko, *Optical Spectroscopy of Semiconductor Nanostructures*, Alpha Science International, Harrow, U.K., 2005.
14. J. Schliemann, *Int. J. Mod. Phys. B* **20**, 1015 (2006).
15. H.-A. Engel, E. I. Rashba, and B. I. Halperin, in *Handbook of Magnetism and Advanced Magnetic Materials*, H. Kronmüller and S. Parkin (Eds.), John Wiley, Chichester, U.K., 2007.
16. E. M. Hankiewicz and G. Vignale, *J. Phys.: Condens. Matter* **21**, 253202 (2009).
17. M. I. D'yakonov and V. I. Perel', *Pis'ma Zh. Eksp. Teor. Fiz.* **13**, 657 (1971) [*JETP Lett.* **13**, 467 (1971)].
18. J. Fabian, A. Matos-Abiague, C. Ertler, P. Stano, and I. Žutić, *Acta Phys. Slov.* **57**, 565 (2007), arXiv:cond-mat/0711.1461.
19. J. Sinova and A. MacDonald, Theory of spin-orbit effects in semiconductors, in semiconductors and semimetals, in *Spintronics*, vol. 82, T. Dietl, D. D. Awschalom, M. Kaminska, and H. Ohno (Eds.), Elsevier, Amsterdam, 2008, pp. 45–89.
20. M. I. D'yakonov and A. V. Khaetskii, Spin Hall effect, in *Spin Physics in Semiconductors*, M. I. D'yakonov (Ed.), Springer, Berlin, 2008, pp. 211–244.
21. L. D. Landau, E. M. Lifshits, and L. P. Pitaevskii, *Electrodynamics of Continuous Media*, vol. 8, Elsevier, Amsterdam, 1984.
22. J. F. Nye, *Physical Properties of Crystals: Their Representation by Tensors and Matrices*, Oxford University Press, Oxford, U.K., 1985.
23. V. M. Agranovich and V. L. Ginzburg, *Crystal Optics with Spatial Dispersion, and Excitons*, Springer Series in Solid-State Sciences, vol. 42, Springer, Berlin, 1984.
24. V. A. Kizel', Yu. I. Krasilov, and V. I. Burkov, *Usp. Fiz. Nauk* **114**, 295 (1974) [*Sov. Phys. Usp.* **17**, 745 (1975)].
25. J. Jerphagnon and D. S. Chemla, *J. Chem. Phys.* **65**, 1522 (1976).

26. B. Koopmans, P. V. Santos, and M. Cardona, *Phys. Stat. Sol.* (*b*) **205**, 419 (1998).
27. E. L. Ivchenko and G. E. Pikus, *Pis'ma Zh. Eksp. Teor. Fiz.* **27**, 640 (1978) [*JETP Lett.* **27**, 604 (1978)].
28. L. E. Vorob'ev, E. L. Ivchenko, G. E. Pikus, I. I. Farbstein, V. A. Shalygin, and A. V. Sturbin, *Pis'ma Zh. Eksp. Teor. Fiz.* **29**, 485 (1979) [*JETP Lett.* **29**, 441 (1979)].
29. A. G. Aronov, Yu. B. Lyanda-Geller, and G. E. Pikus, *Zh. Eksp. Teor. Fiz.* **100**, 973 (1991) [*Sov. Phys. JETP* **73**, 537 (1991)].
30. A. G. Aronov and Yu. B. Lyanda-Geller, *Pis'ma Zh. Eksp. Teor. Fiz.* **50**, 398 (1989) [*JETP Lett.* **50**, 431 (1989)].
31. V. M. Edelstein, *Solid State Commun.* **73**, 233 (1990).
32. F. T. Vasko and N. A. Prima, *Fiz. Tverdogo Tela* **21**, 1734 (1979) [*Sov. Phys. Solid State* **21**, 994 (1979)].
33. S. D. Ganichev, S. N. Danilov, P. Schneider et al., cond-mat/0403641, 2004.
34. S. D. Ganichev, S. N. Danilov, P. Schneider et al., *J. Magn. Magn. Mater.* **300**, 127 (2006).
35. A. Yu. Silov, P. A. Blajnov, J. H. Wolter, R. Hey, K. H. Ploog, and N. S. Averkiev, *Appl. Phys. Lett.* **85**, 5929 (2004).
36. N. S. Averkiev and A. Yu. Silov, *Semiconductors* **39**, 1323 (2005).
37. V. Sih, R. C. Myers, Y. K. Kato, W. H. Lau, A. C. Gossard, and D. D. Awschalom, *Nat. Phys.* **1**, 31 (2005).
38. C. L. Yang, H. T. He, L. Ding et al., *Phys. Rev. Lett.* **96**, 186605 (2006).
39. Y. K. Kato, R. C. Myers, A. C. Gossard, and D. D. Awschalom, *Phys. Rev. Lett.* **93**, 176601 (2004).
40. Y. K. Kato, R. C. Myers, A. C. Gossard, and D. D. Awschalom, *Appl. Phys. Lett.* **87**, 022503 (2005).
41. N. P. Stern, S. Ghosh, G. Xiang, M. Zhu, N. Samarth, and D. D. Awschalom, *Phys. Rev. Lett.* **97**, 126603 (2006).
42. P. R. Hammar, B. R. Bennett, M. J. Yang, and M. Johnson, *Phys. Rev. Lett.* **83**, 203 (1999).
43. P. R. Hammar, B. R. Bennett, M. J. Yang, and M. Johnson, *Phys. Rev. Lett.* **84**, 5024 (2000).
44. M. Johnson, *Phys. Rev. B* **97**, 9635 (1998).
45. R. H. Silsbee, *Phys. Rev. B* **83**, 155305 (2001).
46. F. G. Monzon, H. X. Tang, and M. L. Roukes, *Phys. Rev. Lett.* **84**, 5022 (2000).
47. B. J. van Wees, *Phys. Rev. Lett.* **84**, 5023 (2000).
48. M. G. Vavilov, *Phys. Rev. B* **72**, 195327 (2005).
49. S. A. Tarasenko, *JETP Lett.* **84**, 199 (2006).
50. H.-A. Engel, E. I. Rashba, and B. I. Halperin, *Phys. Rev. Lett.* **98**, 036602 (2007).
51. M. Trushin and J. Schliemann, *Phys. Rev. B* **75**, 155323 (2007).
52. O. E. Raichev, *Phys. Rev. B* **75**, 205340 (2007).
53. D. Culcer and R. Winkler, *Phys. Rev. B* **99**, 226601 (2006).
54. A. A. Burkov, A. S. Nunez, and A. H. MacDonald, *Phys. Rev. B* **70**, 155308 (2004).
55. O. Bleibaum, *Phys. Rev. B* **73**, 035322 (2006).
56. V. L. Korenev, *Phys. Rev. B* **74**, 041308 (2006).
57. P. Kleinert and V. V. Bryksin, *Phys. Rev. B* **76**, 205326 (2007).
58. P. Kleinert and V. V. Bryksin, *Phys. Rev. B* **79**, 045317 (2009).
59. F. T. Vasko and O. E. Raichev, *Quantum Kinetic Theory and Application. Electrons, Photons, Phonons*, Springer, Berlin, pp. 662–663, 2005.
60. A. V. Chaplik, M. V. Entin, and L. I. Magarill, *Physica E* **13**, 744 (2002).
61. J. Inoue, G. E. W. Bauer, and L. W. Molenkamp, *Phys. Rev. B* **67**, 033104 (2003).
62. Y. J. Bao and S. Q. Shen, *Phys. Rev. B* **76**, 045313 (2007).
63. R. Raimondi, C. Gorini, M. Dzierzawa, and P. Schwab, *Solid State Commun.* 524 (2007).
64. C. X. Liu, B. Zhou, S. Q. Shen, and B.-F. Zhu, *Phys. Rev. B* **77**, 125345 (2008).
65. C. Gorini, P. Schwab, M. Dzierzawa, and R. Raimondi, *Phys. Rev. B* **78**, 125327 (2008).
66. M. Duckheim and D. Loss, *Phys. Rev. Lett.* **101**, 226602 (2008).
67. R. Raimondi and P. Schwab, *Euro. Lett.* **87**, 37008 (2009).
68. I. Adagideli, G. E. W. Bauer, and B. I. Halperin, *Phys. Rev. Lett.* **97**, 256601 (2006).
69. Y. Jiang and L. Hu, *Phys. Rev. B* **74**, 075302 (2006).
70. J. H. Jiang, M. W. Wu, and Y. Zhou, *Phys. Rev. B* **78**, 125309 (2008).
71. M. Pletyukhov and A. Shnirman, *Phys. Rev. B* **79**, 033303 (2009).
72. A. Zakharova, I. Lapushkin, K. Nilsson, S. T. Yen, and K. A. Chao, *Phys. Rev. B* **73**, 125337 (2007).
73. M.-H. Liu, S.-H. Chen, and C.-R. Chang, *Phys. Rev. B* **78**, 165316 (2008).
74. L. S. Levitov, Y. V. Nazarov, and G. M. Eliashberg, *Zh. Eksp. Teor. Fiz.* **88**, 229 (1985) [*Sov. Phys. JETP* **61**, 133 (1985)].
75. Y. A. Bychkov and E. I. Rashba, *Pis'ma Zh. Eksp. Teor. Fiz.* **39**, 66 (1984) [*JETP Lett.* **39**, 78 (1984)].
76. M. I. D'yakonov and V. Yu. Kachorovskii, *Fiz. Tekh. Poluprovodn.* **20**, 178 (1986) [*Sov. Phys. Semicond.* **20**, 110 (1986)].
77. N. S. Averkiev, L. E. Golub, and M. Willander, *J. Phys.: Condens. Matter* **14**, R271 (2002).
78. S. D. Ganichev, V. V. Bel'kov, L. E. Golub et al., *Phys. Rev. Lett.* **92**, 256601 (2004).
79. S. Giglberger, L. E. Golub, V. V. Bel'kov et al., *Phys. Rev. B* **75** 035327 (2007).
80. L. Meier, G. Salis, E. Gini, I. Shorubalko, and K. Ensslin, *Phys. Rev. B* **77**, 035305 (2008).
81. J. Luo, H. Munekata, F. F. Fang, and P. J. Stiles, *Phys. Rev. B* **41**, 7685 (1990).
82. W. Knap, C. Skierbiszewski, A. Zduniak et al., *Phys. Rev. B* **53**, 3912 (1996).
83. N. S. Averkiev and L. E. Golub, *Phys. Rev. B* **60**, 15582 (1999).
84. S. A. Tarasenko and N. S. Averkiev, *Pis'ma Zh. Èksp. Teor. Fiz.* **75**, 669 (2002) [*JETP Lett.* **75**, 552 (2002)]; N. S. Averkiev, M. M. Glazov, and S. A. Tarasenko, *Solid State Commun.* **133**, 543 (2005).
85. J. Schliemann, J. C. Egues, and D. Loss, *Phys. Rev. Lett.* **90**, 146801 (2003).

86. R. Winkler, *Spin-Orbit Coupling Effects in Two-Dimensional Electron and Hole Systems*, Springer Tracts in Modern Physics, vol. 191, Springer, Berlin, 2003.
87. W. Zawadzki and P. Pfeffer, *Semicond. Sci. Technol.* **19**, R1 (2004).
88. I. Žutić, J. Fabian, and S. Das Sarma, Spintronics: Fundamentals and applications. *Rev. Mod. Phys.* **76**, 323 (2004).
89. V. V. Bel'kov, P. Olbrich, S. A. Tarasenko et al., *Phys. Rev. Lett.* **100**, 176806 (2008).
90. V. Lechner, L. E. Golub, P. Olbrich et al., *Appl. Phys. Lett.* **94**, 242109 (2009).
91. S. A. Tarasenko, *Physica E* **40**, 1614 (2008).

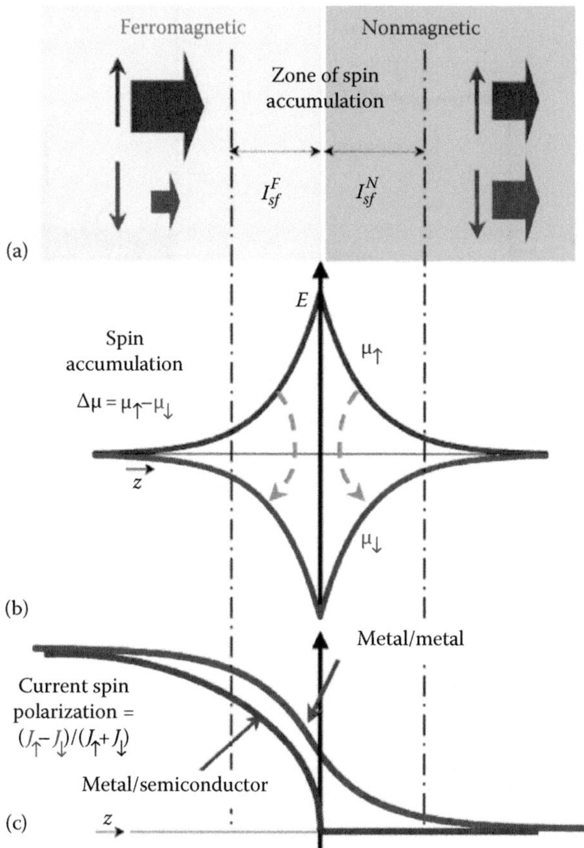

FIGURE 1.8 Schematic representation of the spin accumulation at an interface between a ferromagnetic metal and a nonmagnetic layer. (a) Incoming and outgoing spin-up and spin-down currents. (b) Splitting of the chemical potentials, $E_{F\uparrow}$ and $E_{F\downarrow}$, in the interface region with arrows symbolizing the spin-flips induced by the spin accumulation (out of equilibrium distribution), spin-flips controlling the progressive depolarization of the current. With an opposite direction of the current, the inversion of the spin accumulation and the opposite spin-flips polarize progressively the current. (c) Variation of the current spin polarization when there is an approximate balance between the spin-flips on both sides (metal/metal curve) and when the spin-flips on the right side are predominant (metal/semiconductor curve in the situation without spin-dependent interface resistance). (From Chappert, C. et al., *Nat. Mater.*, 6, 813, 2007. With permission.)

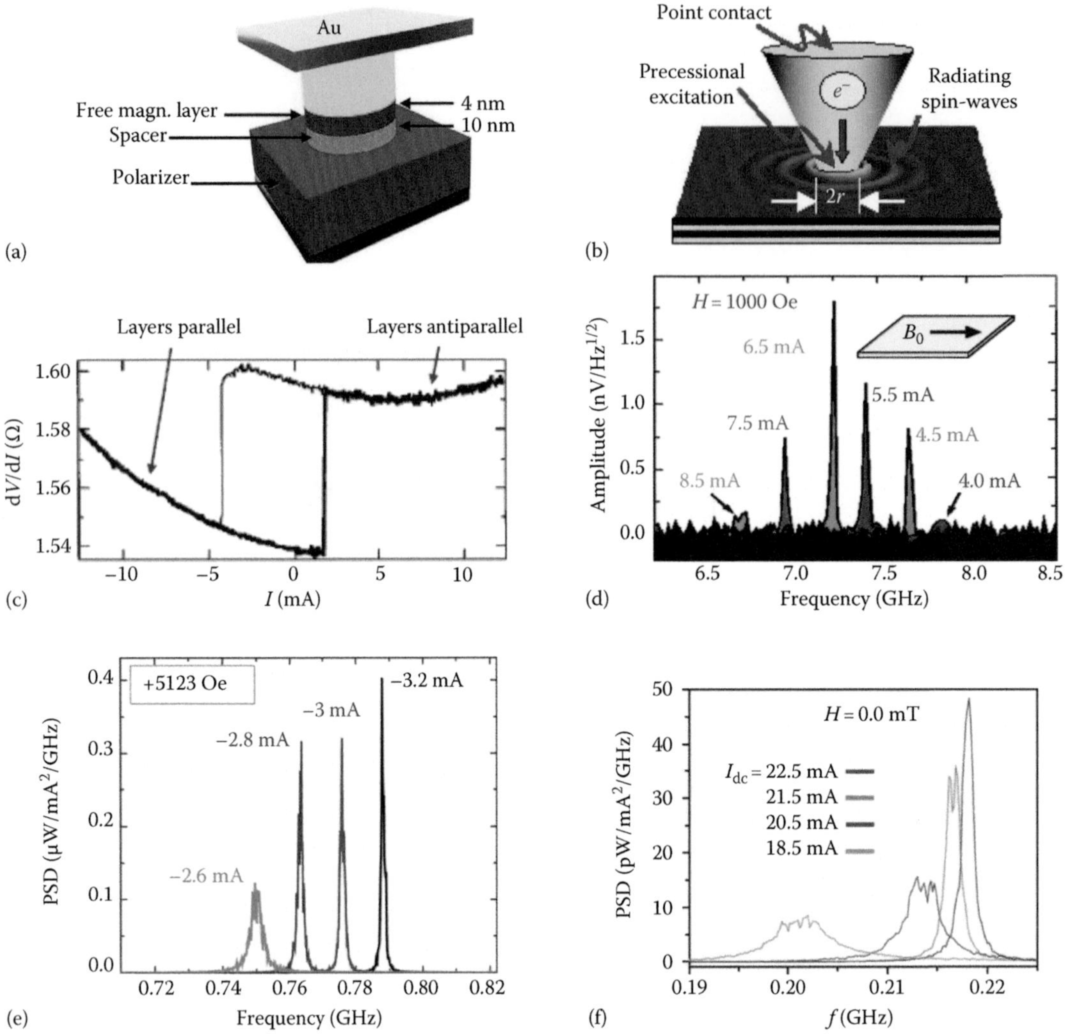

FIGURE 1.12 (a) and (b) Nanodevices for spin transfer experiments ("pillar" and nanocontacts [64]). (c) Differential resistance vs. current for a Co/Cu/Co "pillar" with current-induced switchings between parallel and antiparallel magnetic configurations [62]. (d) Microwave spectra obtained by current injection into a CoFe/Cu/NiFe trilayer through a nanocontact at different values of current [64]. (e) Microwave power spectra obtained at different values of current from vortex gyrations in the permalloy layer of a CoFeB/MgO/Permalloy MTJ [65]. (f) Progressive synchronization of the gyration of four vortices appearing at increasing current in the microwave spectrum of a device for current injection into a Co/Cu/permalloy trilayer through four nanocontacts. (From Ruotolo, A. et al., *Nat. Nanotechnol.*, 4, 528, 2009. With permission.)

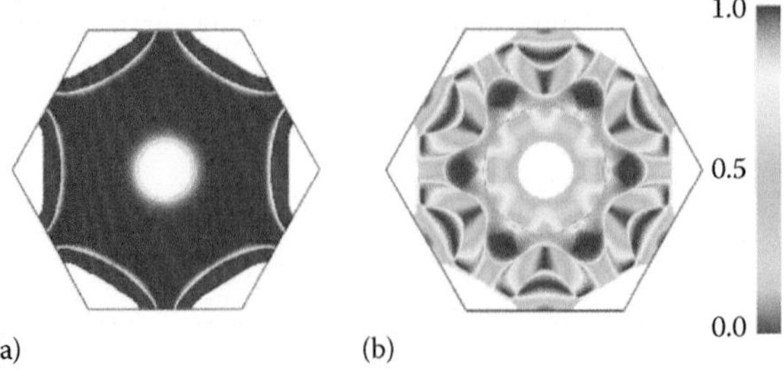

FIGURE 5.12 Transmission probabilities for majority- (a) and minority- (b) spin electrons across the Co/Cu(111) interface resolved in the 2D Brillouin zone. (After Xia, K. et al., *Phys. Rev. B*, 63, 064407 1-4, 2001. With permission.)

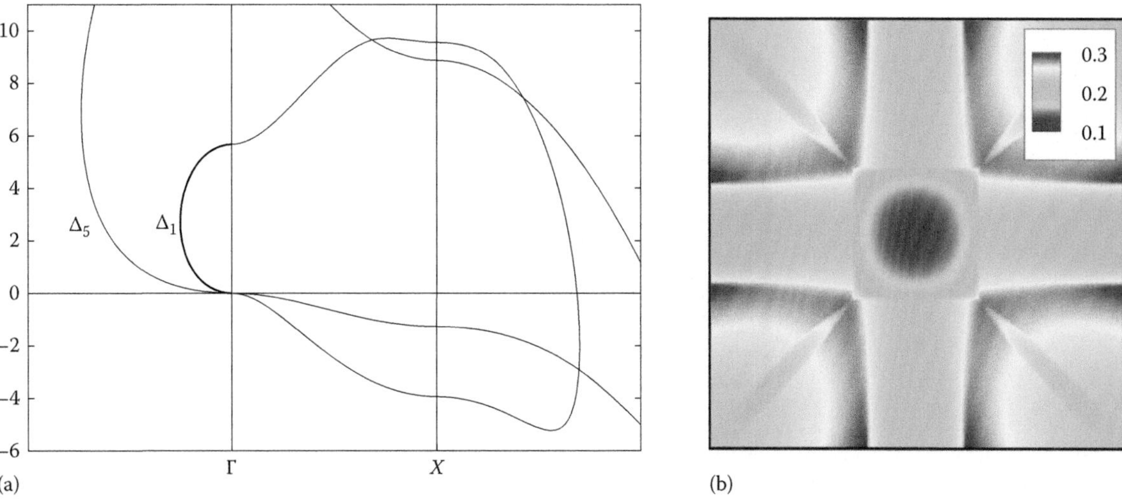

FIGURE 12.1 Illustration of the complex band structure of MgO. (a) Complex band structure at $\mathbf{k}_{\|}=0$ (energy in eV). The Δ_1 complex band is shown by a thick line straddling the insulating gap. For clarity, complex bands with variable real and imaginary parts, which attach to the band extrema at −5 and at 9.5 eV, are not shown. (b) Right panel: Brillouin zone plot of the lowest value of Im k_z taken at mid-gap.

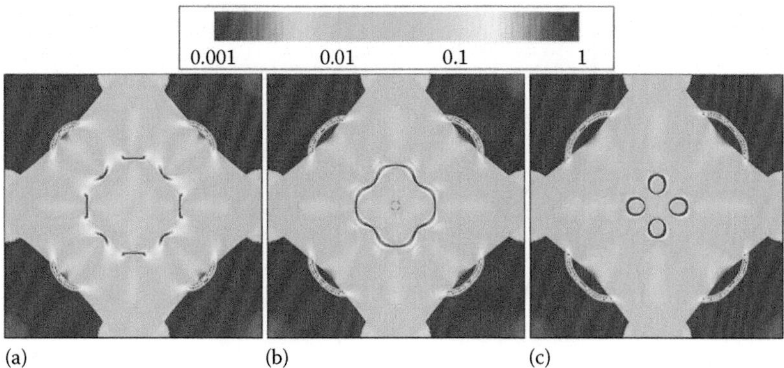

FIGURE 12.5 $\mathbf{k}_{\|}$-Resolved minority-spin density of states (in arbitrary units) for the Fe(001)/MgO interface calculated for three different energies near the Fermi energy (E_F): (a) $E=E_F-0.02$ eV, (b) $E=E_F$, (c) $E=E_F+0.02$ eV. (From Belashchenko, K.D. et al., *Phys. Rev. B*, 72(14), 140404, 2005. With permission.)

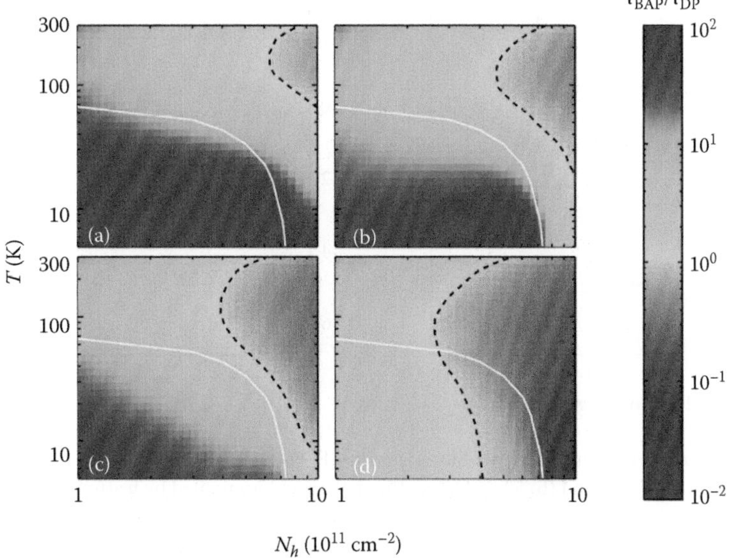

FIGURE 16.14 Ratio of the spin relaxation time due to the Bir–Aronov–Pikus mechanism to that due to the D'yakonov–Perel' mechanism, τ_{BAP}/τ_{DP}, as function of temperature and hole density with (a) $N_I=0$, $N_{ex}=10^{11}\,\text{cm}^{-2}$; (b) $N_I=0$, $N_{ex}=10^9\,\text{cm}^{-2}$; (c) $N_I=N_h$, $N_{ex}=10^{11}\,\text{cm}^{-2}$; (d) $N_I=N_h$, $N_{ex}=10^9\,\text{cm}^{-2}$. The black dashed curves indicate the cases satisfying $\tau_{BAP}/\tau_{DP}=1$. Note the smaller the ratio τ_{BAP}/τ_{DP} is, the more important the Bir–Aronov–Pikus mechanism becomes. The yellow solid curves indicate the cases satisfying $\partial_{\mu_h}[N_{LH^{(1)}}+N_{HH^{(2)}}]/\partial_{\mu_h}N_h=0.1$. In the regime above the yellow curve the multi-hole-subband effect becomes significant. (Zhou, Y. et al., *New J. Phys.*, 11, 113039, 2009.)

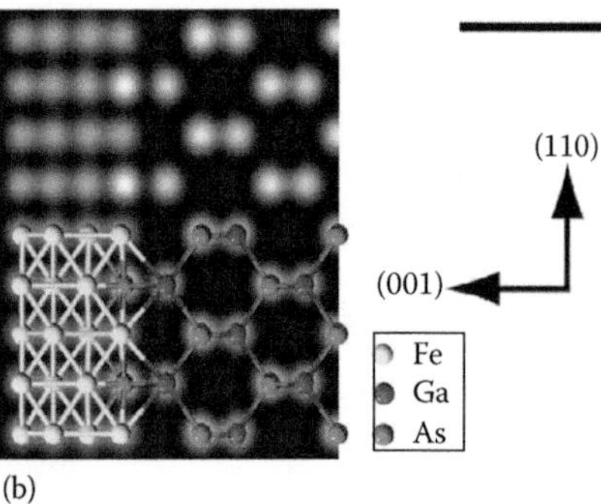

FIGURE 17.18b Simulated image of a partially intermixed interface and inset ball-and-stick model from which it was calculated (Fe atoms appear in yellow, As in blue, and Ga in red). The scale bar equals 0.5 nm.

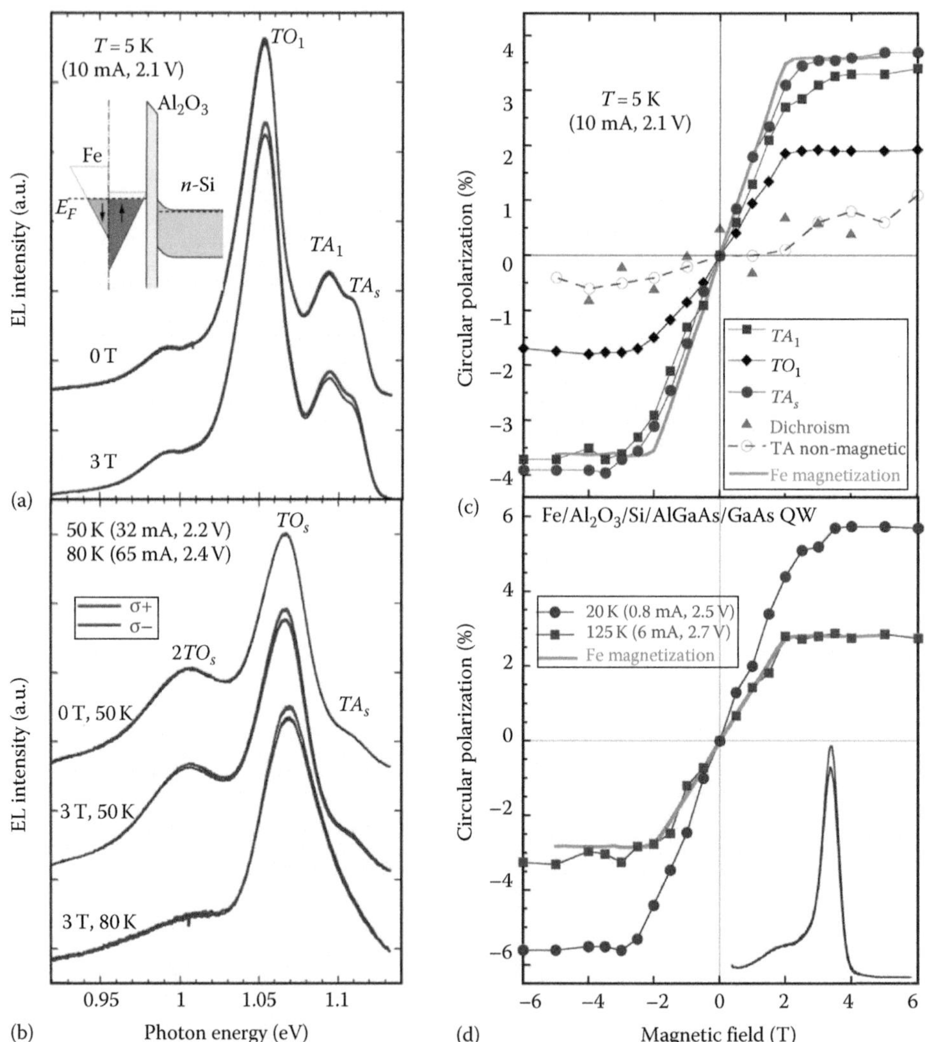

FIGURE 17.20 EL spectra from surface-emitting $Fe/Al_2O_3/Si$ *n–i–p* spin-LED structures, analyzed for positive (σ+) and negative (σ−) helicity circular polarization at temperatures of (a) 5 K, and (b) 50 and 80 K. The spectra are dominated by features arising from electron-hole recombination accompanied by transverse acoustic (*TA*) and transverse optical (*TO*) phonon emission. (c) The magnetic field dependence of the circular polarization for each feature in the Si spin-LED spectrum at 5 K. (d) Inset shows the EL spectrum from Fe/Al_2O_3/80 nm *n*-Si/*n*-$Al_{0.1}Ga_{0.9}As$/GaAs QW/*p*-$Al_{0.3}Ga_{0.7}As$ spin-LEDs at $T = 20$ K and $H = 3$ T. The dominant feature is the GaAs QW free exciton, which exhibits strong polarization. The magnetic field dependence of this circular polarization is plotted for $T = 20$ and 125 K. The out-of-plane Fe magnetization curve appears in panel (c) and (d) as a solid line for reference.

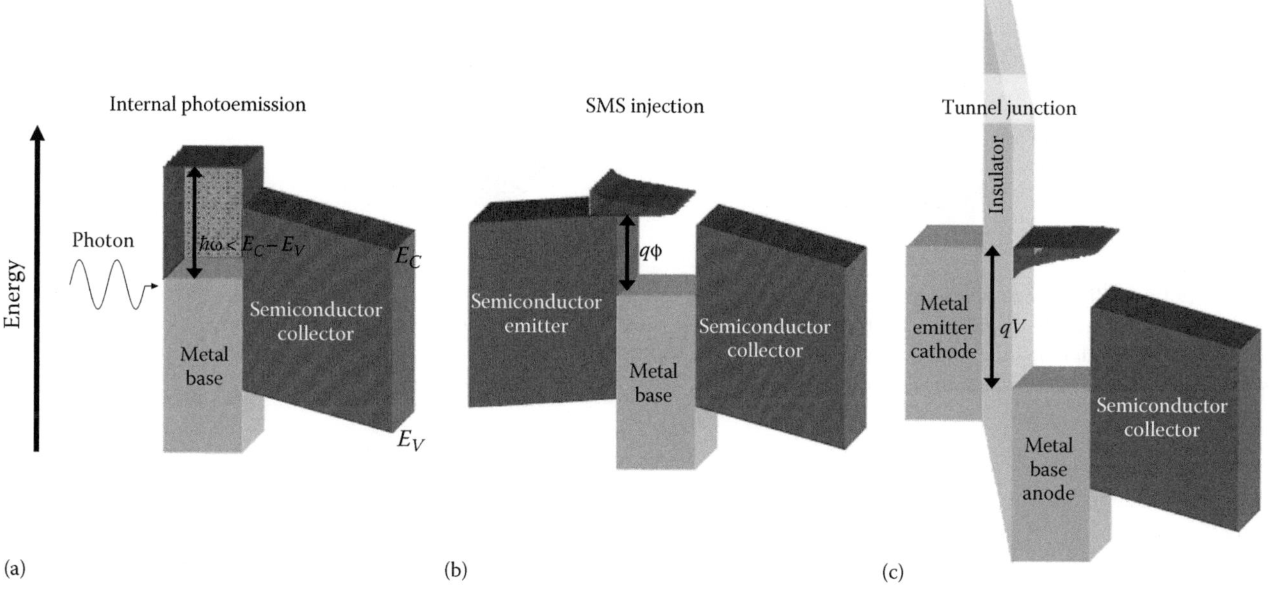

FIGURE 18.1 Illustration of band diagrams for three different hot-electron generation techniques: internal photoemission, semiconductor–metal–semiconductor (SMS) metal-base transistor, and tunnel junction (TJ) emission. The energetic distributions of the hot electrons generated are roughly homogeneous up to $\hbar\omega$ above the collector Fermi energy for IPE (a), peaked at the fixed Schottky barrier height of the emitter ($q\phi$) for SMS (b), and peaked near the emitter cathode Fermi energy qV (where V is the applied voltage and q is the elementary charge) for TJ (c).

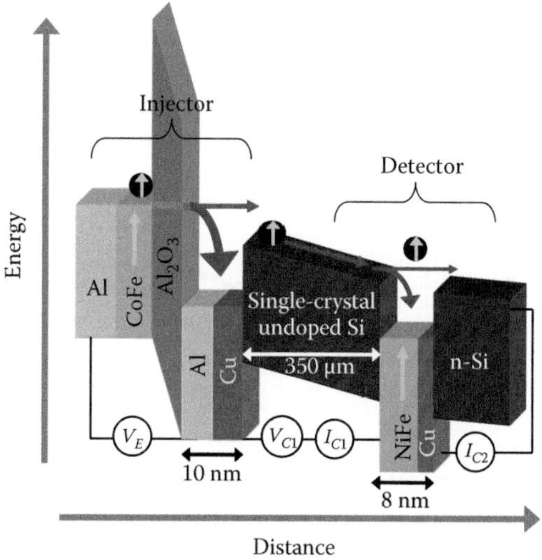

FIGURE 18.2 Schematic band diagram of a four-terminal (two for TJ injection and two for FM SMS detection) ballistic hot-electron injection and detection device with a 350 μm thick Si transport layer.

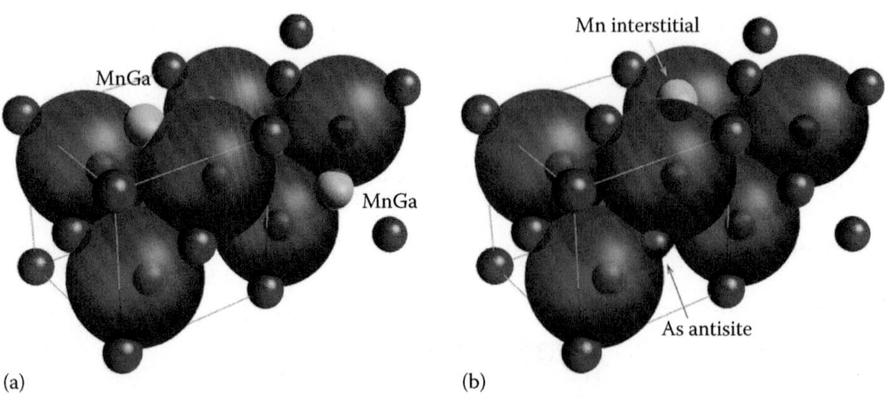

FIGURE 19.1 (a) Zinc blende lattice structure of GaAs with two substitutional Mn impurities. Ga^{3+} ions are shown as small red spheres, As^{3-} as large, semitransparent blue spheres, and Mn^{2+} as small green spheres. The gray cube denotes the conventional unit cell. The ionic radii are used in the plot, which is an exaggeration, since the bonds are partially covalent. (b) Lattice structure of GaAs with an Mn^{2+} interstitial in the As-coordinated tetrahedral position and an As antisite defect (small blue sphere with the As^{5+} ionic radius).

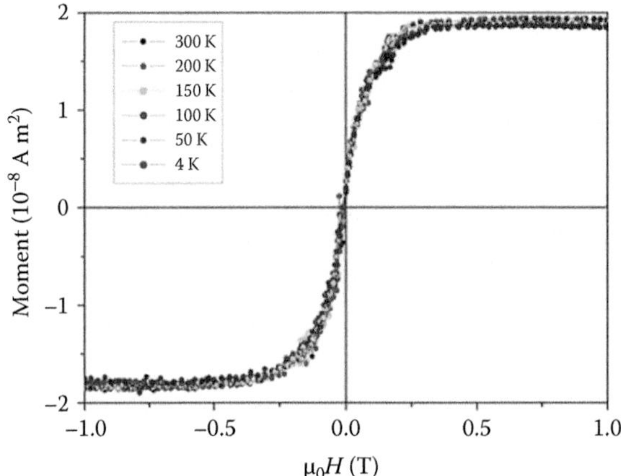

FIGURE 20.8 Ferromagnetic magnetization curves of a thin film of ITO doped with 5% Mn, at different temperatures. The manganese in these films is paramagnetic. (From Venkatesan, M. et al., *Nature*, 430, 630, 2004. With permission.)

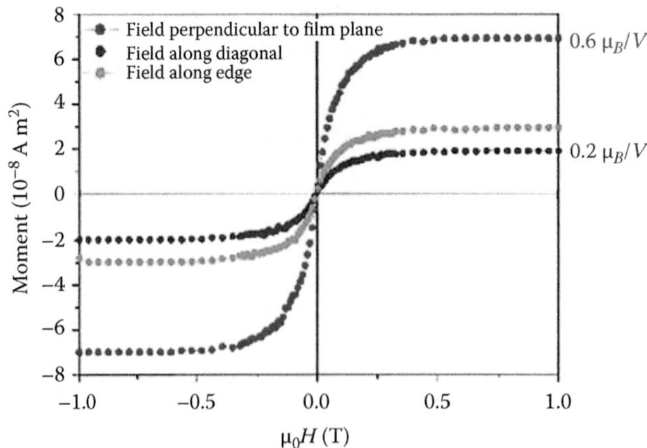

FIGURE 20.11 Anisotropy of the magnetization of a thin film of $(Zn_{0.95}V_{0.05})O$.

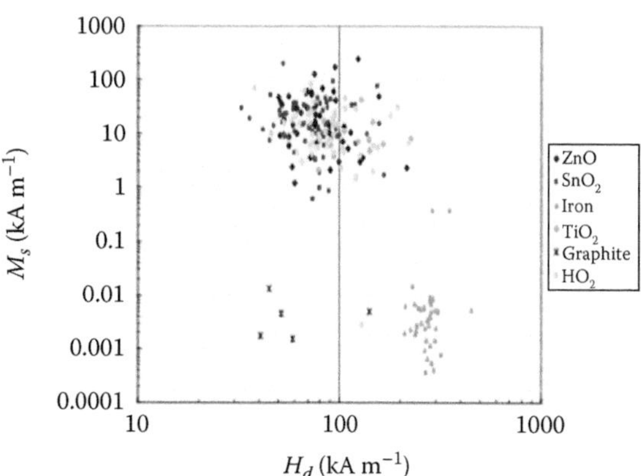

FIGURE 20.16 Plot of magnetization M_s versus internal field H_d for thin films and nanoparticles of doped and undoped oxides. For films of the thin films of oxides, M_s clusters around 10 kA m^{-1}, whereas for nanoparticles it is three orders of magnitude lower. H_d in both cases is about 100 kA m^{-1}. Also included on the scatter plot are data on graphite and dispersed iron nanoparticles.

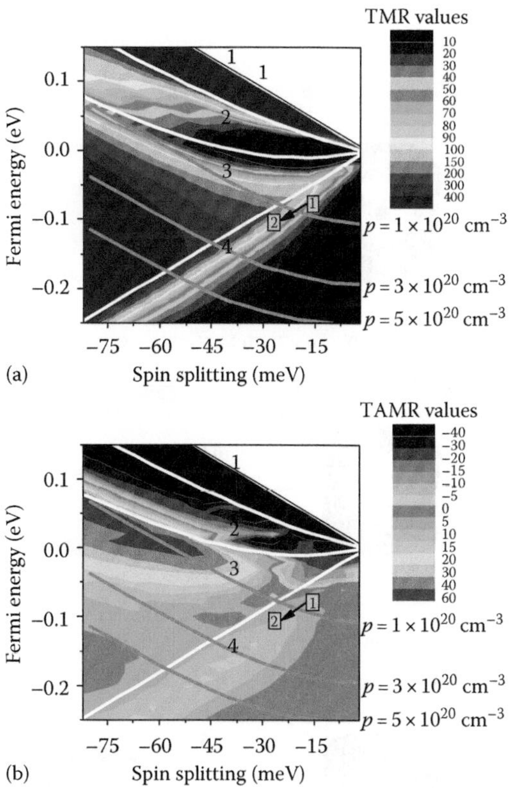

FIGURE 21.5 Tunnel magnetoresistance (a) and TAMR (b) versus Fermi energy and spin-splitting for a 6 nm (In,Ga)As barrier and a band offset of $d_B = 700$ meV. White lines represent the four bands at the Γ_8 point. Gray lines indicate the Fermi energy for different hole concentrations.

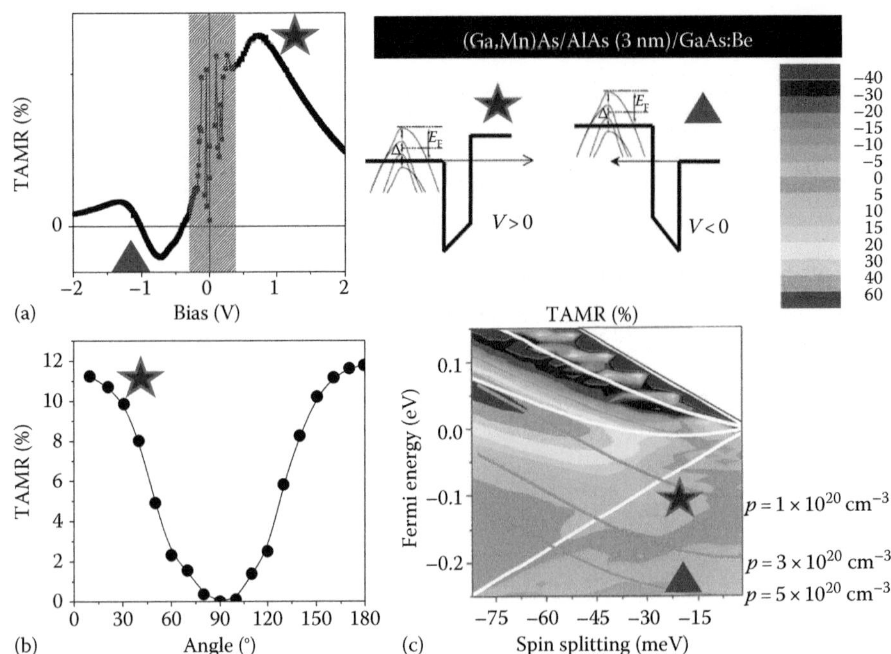

FIGURE 21.8 (a) Out-of-plane TAMR versus bias acquired on a (Ga,Mn)As/AlAs(3 nm)/GaAs:Be tunnel junction. Positive biases correspond to hole injected from (Ga,Mn)As toward GaAs:Be. (b) Angular variation of the resistance acquired at a positive bias of +1 V (star) displaying a positive TAMR. The angle 0° corresponds to $\hat{m} \parallel [001]$ (growth axis). (c) Calculation of the corresponding TAMR with the same parameters for (Ga,Mn)As than the one used for Figure 21.6. The injection in the negative bias regime implies an hole injection towards the lowest HH band (triangle) thus giving rise to a negative TAMR.

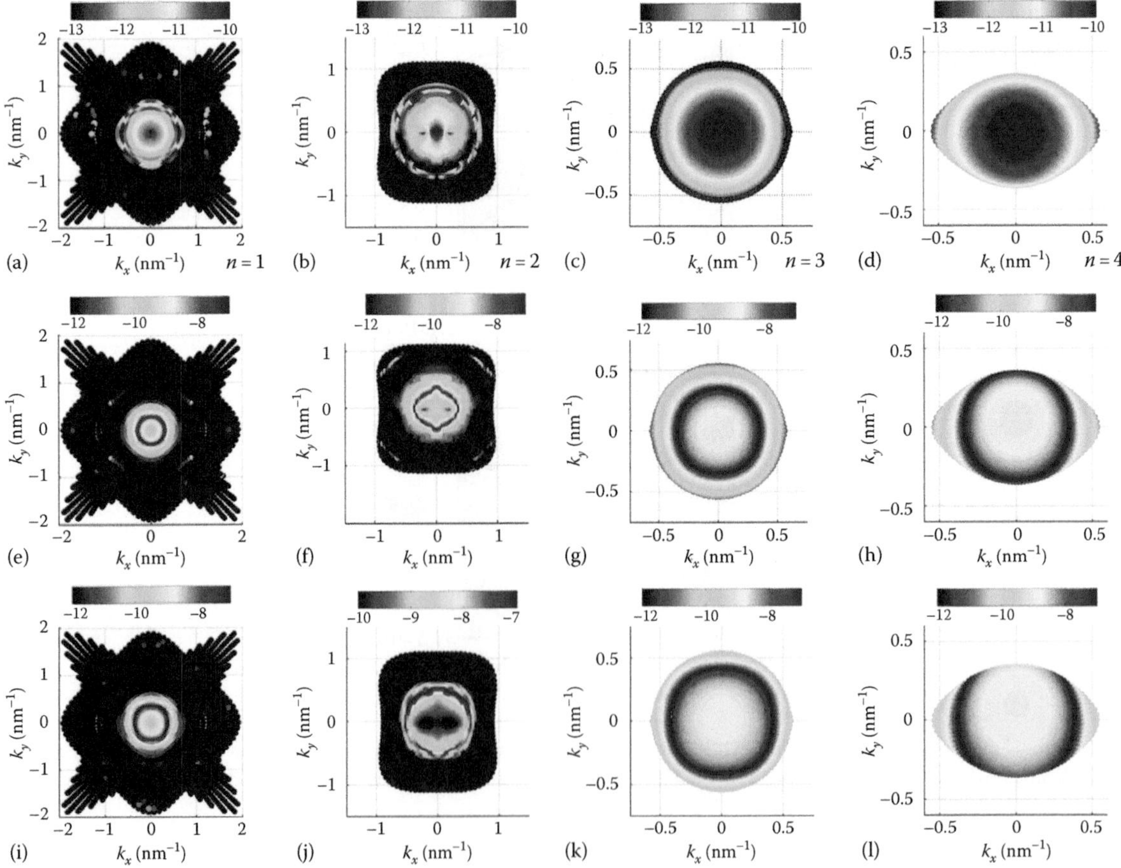

FIGURE 21.10 Band-selected ($n = 1 - 4$) tunnel transmissions (in color logarithmic scale) mapped onto the different (Ga,Mn)As Fermi surfaces calculated for in-plane magnetization ($\hat{m} \parallel \hat{x}$). The different columns correspond to a specific selected band. Calculations have been performed for a bias applied to the double barrier structure equal to $V = 0.59$ eV (a through d: first line), $V = 0.68$ eV (e through h: second line), and $V = 0.71$ eV (i through l: third line) around the LH2 resonance.

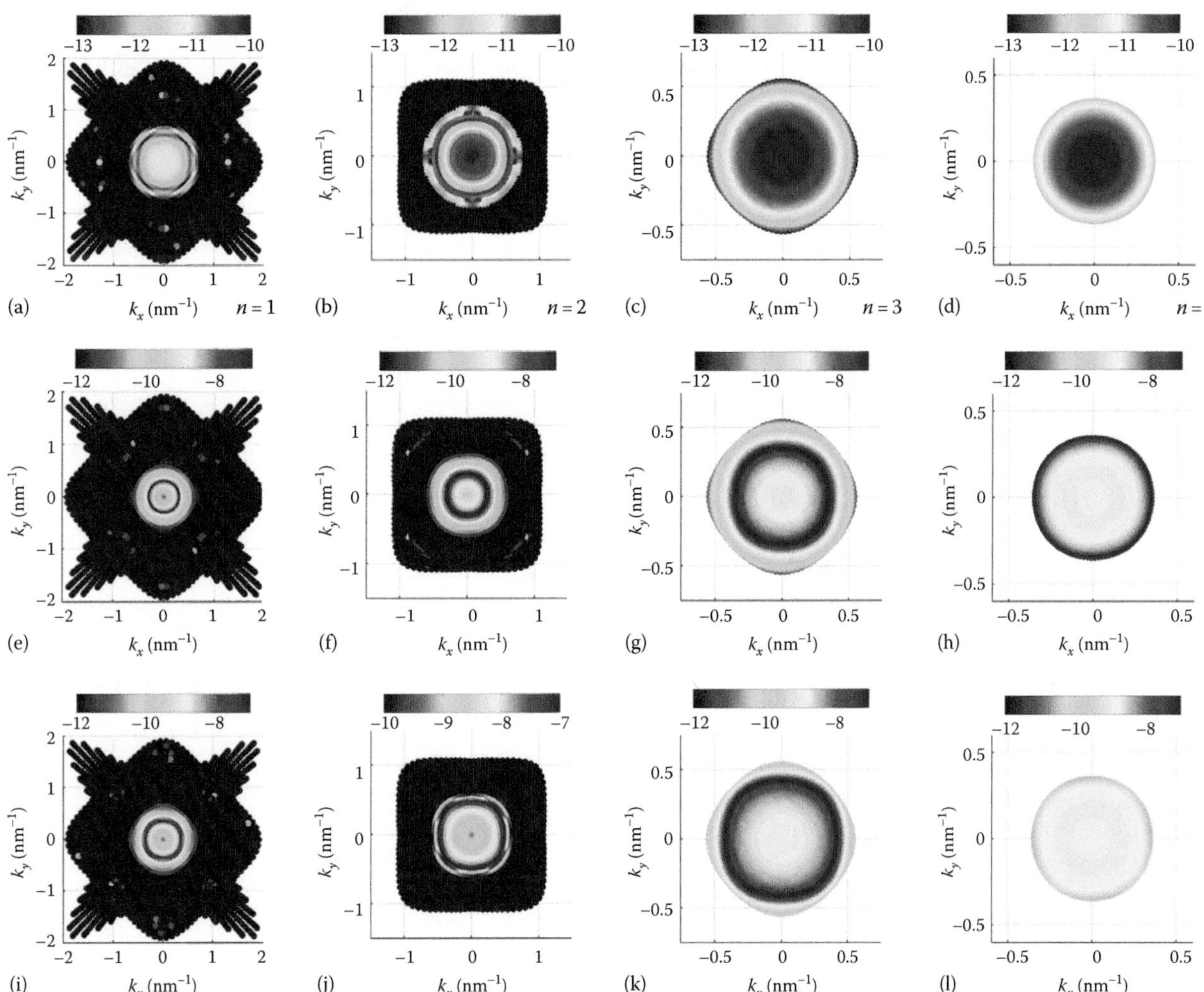

FIGURE 21.11 Band-selected ($n = 1 - 4$) tunnel transmissions (in color logarithmic scale) mapped onto the different (Ga,Mn)As Fermi surfaces calculated for out-of-plane magnetization ($\hat{m} \| \hat{x}$). The different columns correspond to a specific selected band. Calculations have been performed for a bias applied to the double barrier structure equal to $V = 0.59$ eV (a through d: first line), $V = 0.68$ eV (e through h: second line), and $V = 0.71$ eV (i through l: third line) around the LH2 resonance.

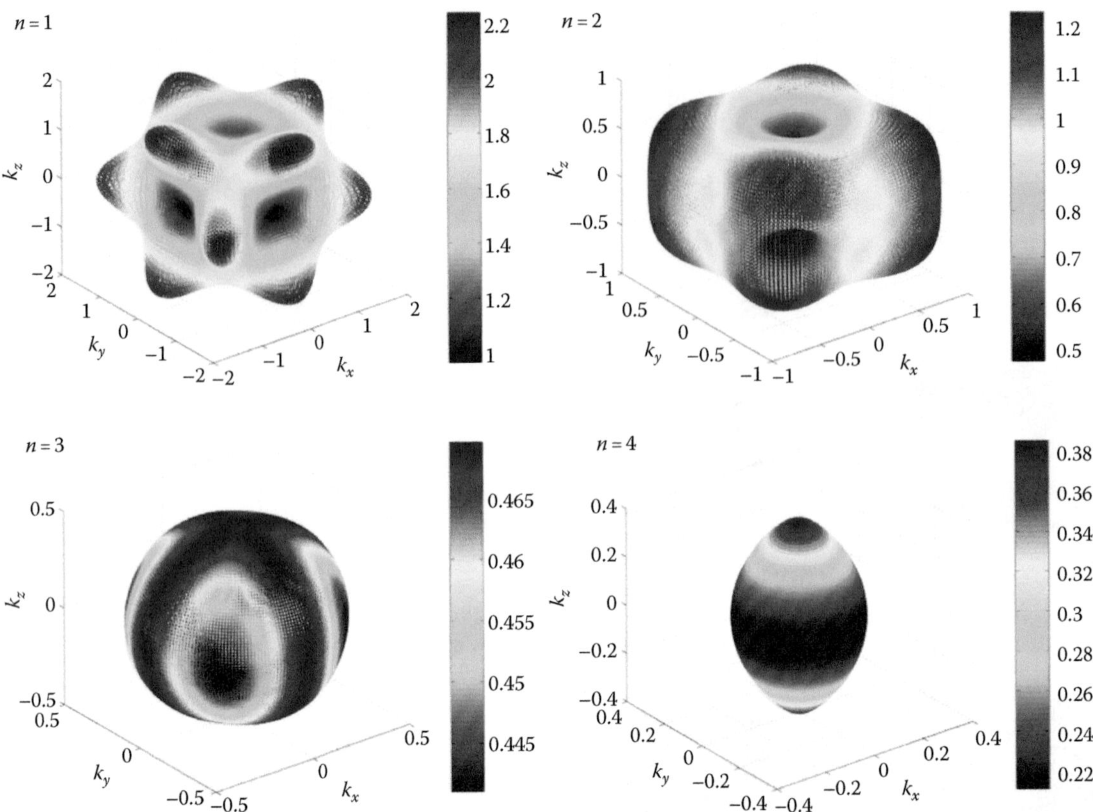

FIGURE 21.13 The four different Fermi surfaces of (Ga,Mn)As calculated in the $k \cdot p$ formalism for a Fermi energy equal −135 meV counted from the top of the valence band ($p = 1 \times 10^{20}$ cm^{-3}) and an exchange interaction $\Delta_{exc} = 120$ meV (calculated without strain). The magnetization is along the [001] growth crystalline axis. The color code scales the Fermi wavevector (in nm^{-1}) along the corresponding crystalline axis. The corresponding band structure of (Ga,Mn)As in the reciprocal space can be found in Refs. [3,18]. (From Tanaka, M. and Higo, Y., *Phys. Rev. Lett.*, 87, 026602, 2001; Matsukura, F. et al., III-V ferromagnetic semiconductors, in *Handbook of Magnetic Materials*, Vol. 14, K.H.J. Buschow (Ed.), North Holland, Amsterdam, the Netherlands, pp. 1–87, 2002. With permission.)

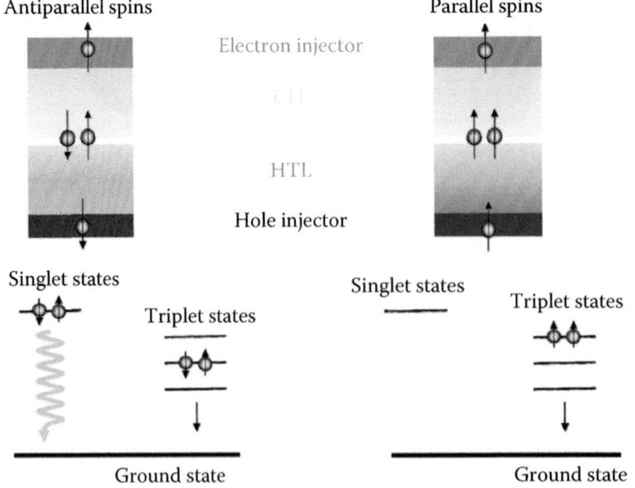

FIGURE 22.3 Schematic structure of a bilayered OLED. Electrons and holes are injected by cathode and anode, respectively, into the electron-transport layer (ETL) and the hole-transport layer (HTL). Exciton formation and recombination eventually take place at the interface. Spin statistic of carrier recombination in case of spin injection. Antiparallel spin states for electrons and holes generate singlet and triplet state in ratio 1:1. Parallel spin states for electrons and holes' spin state of carriers give rise only to the triplet state.

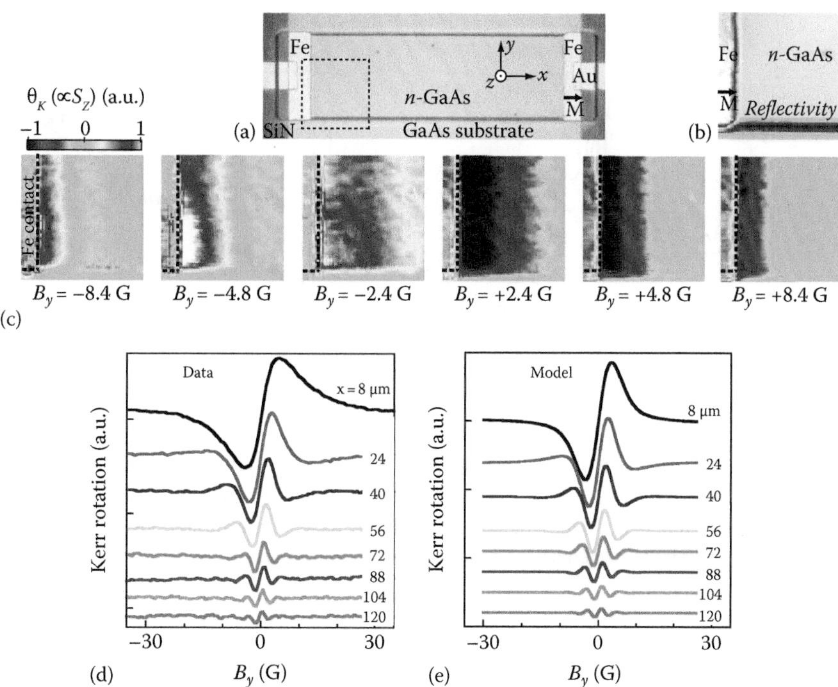

FIGURE 23.4 (a) A lateral spin transport device consisting of a 300 μm long *n*-GaAs channel separating Fe/GaAs source and drain contacts. The dotted square shows the 80 × 80 μm region imaged. (b) An image of the reflected probe laser power, showing device features. (c) Images of electrical spin injection and transport in the *n*-GaAs channel. The electron current $I_e = 92\,\mu A$. Electrons are injected with initial spin polarization $S_0 \parallel -\hat{x}$ and B_y is varied from −8.4 to +8.4 G (left to right), causing spins to precess out of and into the page ($\pm\hat{z}$), respectively. (d) A series of local Hanle curves (θ_K vs. B_y) acquired with the probe laser fixed at different distances (8–120 μm) from the source contact (offset for clarity). (e) Simulated local Hanle curves using a 1D spin drift-diffusion model (Equation 23.2), with $\tau_s = 125$ ns, $\upsilon_d = 2.8 \times 10^4$ cm/s, and $D = 10\,\text{cm}^2/\text{s}$.

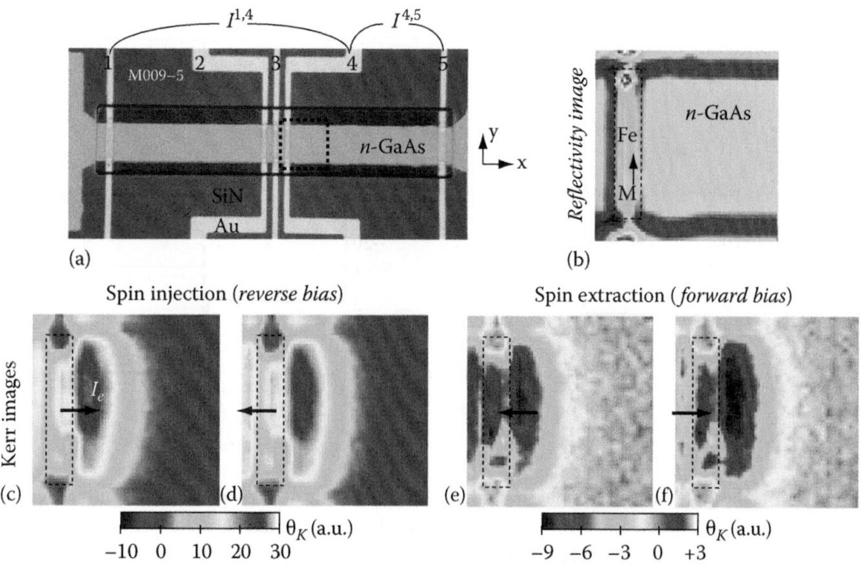

FIGURE 23.6 (a) A spin transport device with five Fe/GaAs contacts (#1–5). A 65 × 65 μm region around contact 4 is imaged (dotted square). (b) The reflected probe power in this region, showing device features. (c,d) Images of the majority spin polarization that is electrically injected from contact 4 ($B_x = 2.4$ G). Arrows show the direction and magnitude of net electron current, I_e. In (c), spins are injected and drift/diffuse to the right ($I_e^{4\to5} = 25\,\mu A$). In (d), spins are injected and drift to the *left* ($I_e^{4\to1} = 25\,\mu A$); the spins to the right of contact 4 are diffusing outside of the current path. (e,f) Images of minority spin accumulation due to spin extraction at contact 4, both within and outside of the charge current path.

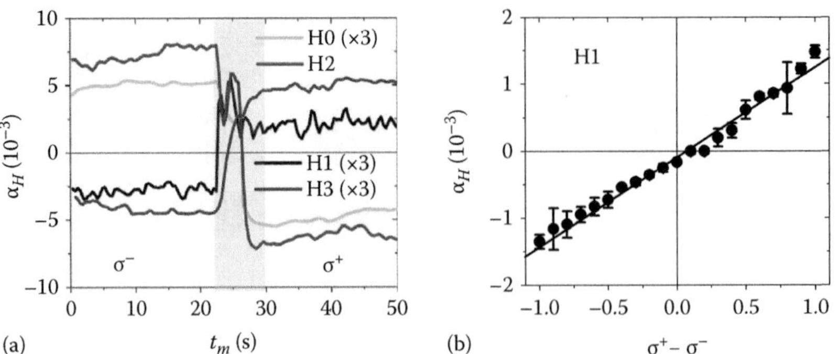

FIGURE 25.6 (a) Spin-injection Hall effect Hall probe signals are plotted for Hall crosses H0H3 in the 2DEG channel. Gray region corresponds to the manual rotation of the $\lambda/2$ wave plate, which changes the helicity of the incident light and, therefore, the spin-polarization of injected electrons. (b) Hall angles measured at n-channel Hall crosses H1 (similar results with opposite sign occur for H2). The spin-injection Hall effect angles are linear in the degree of polarization, hence behaving phenomenologically as in the anomalous Hall effect. The laser spot is focused on the p–n junction and also the bias voltage, laser wavelength, and measurement temperature are the same as in Figure 25.5. (After Wunderlich, J. et al., *Nat. Phys.*, 5, 675, 2009.)

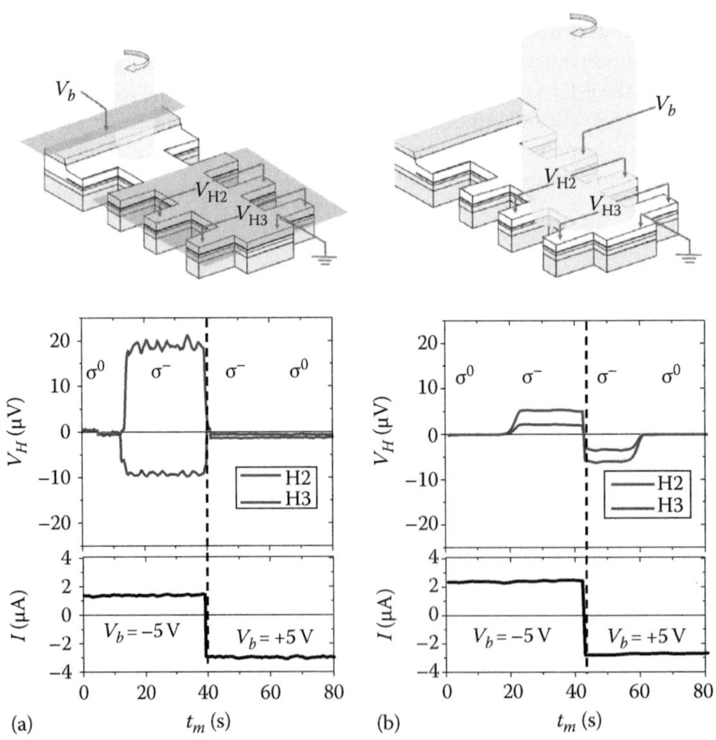

FIGURE 25.7 (a) Spin-injection Hall effect measurements in a masked sample with linearly polarized light and circularly polarized light of a fixed helicity for opposite polarities of the optical current. Schematics of each experimental setup are shown in the upper panel. Middle panel shows that spin-injection Hall effect voltages are detected only at negative bias when spin-polarized electrons move from the illuminated aperture toward the measured Hall crosses H2 and H3 in the n-channel. The optical current is plotted in the lower panel. (b) Complementary measurements to (a) in an unmasked sample with only the n-channel biased and Hall crosses H2 and H3 directly illuminated with a 10× stronger light intensity as compared to (a). Weak anomalous Hall effect signals are detected in this case, which are antisymmetric with respect to the polarity of the current. Lower panel shows the optically generated part of the current. (From Wunderlich, J. et al., *Nat. Phys.*, 5, 675, 2009.)

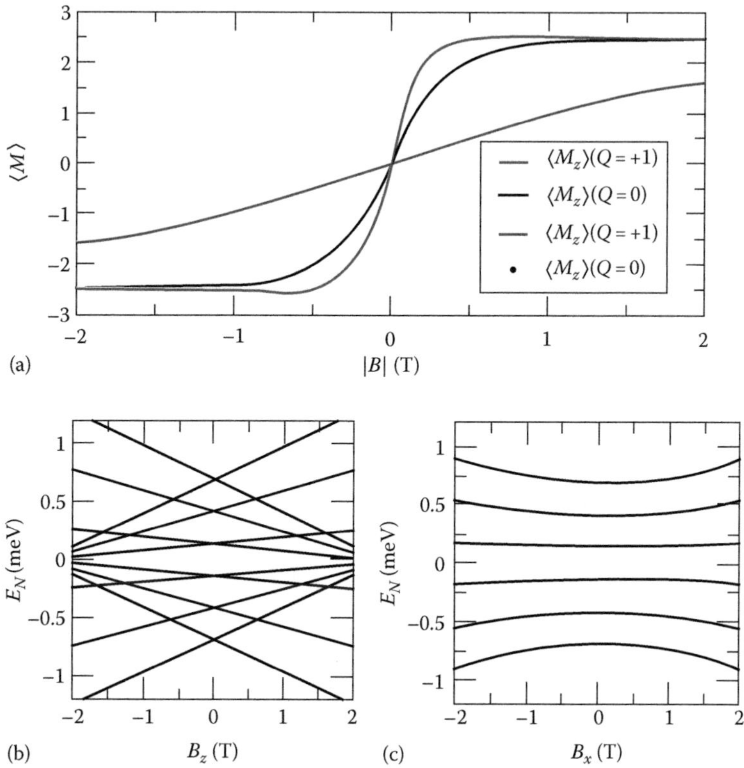

FIGURE 30.4 (a) Equilibrium magnetization of the Mn spin for a dot with $Q = 0$ and $Q = +1$ as a function of the intensity of the applied field for two orientations. (b) and (c) Energy spectra of single Mn-doped QD with one hole as a function of the magnetic field for two directions: parallel to growth axis (b) and perpendicular to the growth axis (c). (Fernández-Rossier, J. and Aguado, R.: Mn doped II-VI QDs: Artificial single molecule magnets, *Phys. Stat. Sol (c)*, 2006, 3, 3734–3739. Copyright Wiley-VCH Verlag GmbH & Co. KGaA. Reproduced with permission.)

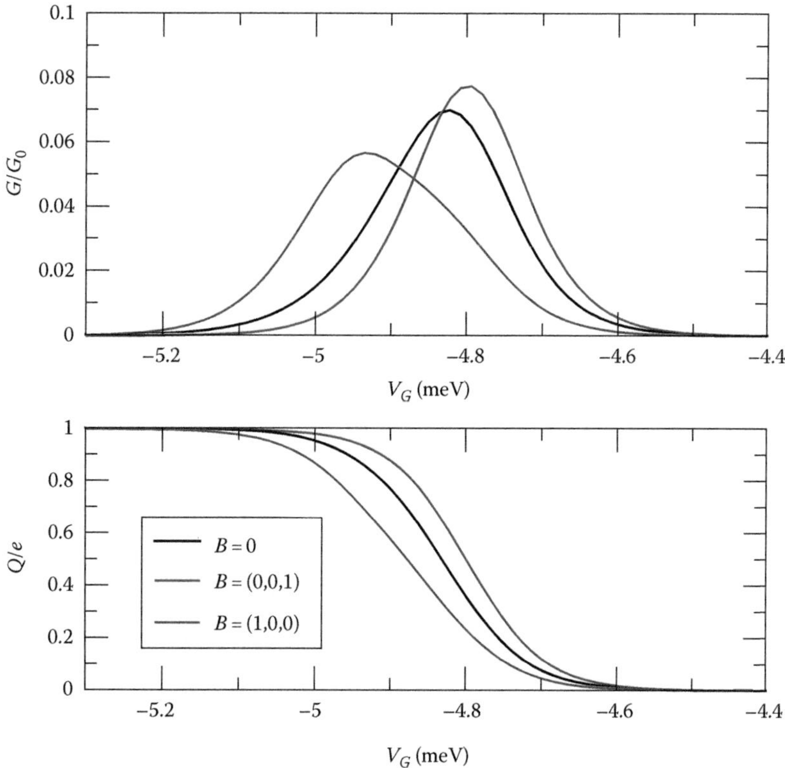

FIGURE 30.8 Conductance (upper panel) and charge (lower panel) versus gate for $\vec{B} = 0$, $\vec{B} = B_0\hat{x}$, and $\vec{B} = B_0\hat{z}$.

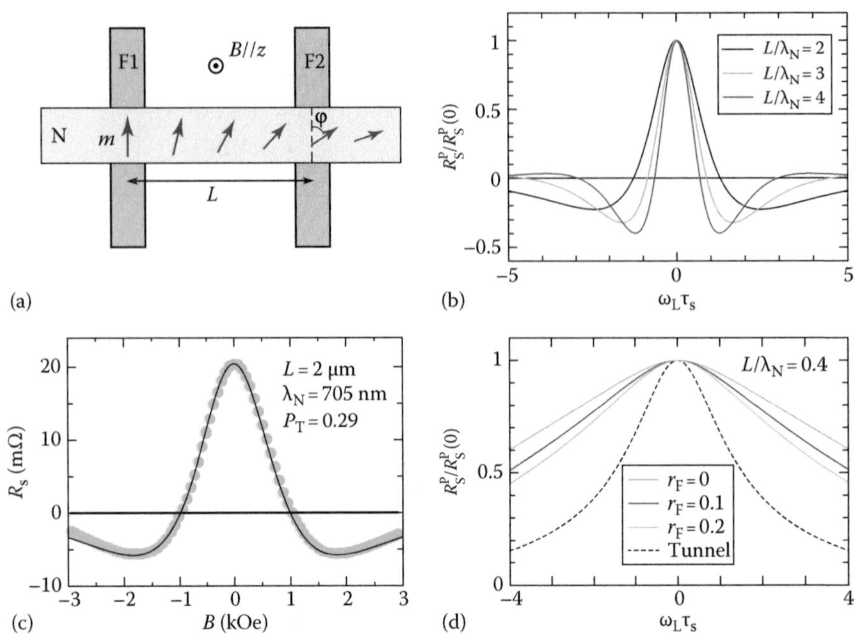

FIGURE 31.3 (a) Spin precession in which electron spins rotate an angle φ during electron travel from the injector to the detector. (b) Nonlocal resistance as a function of Larmor frequency ω_L for distances $L/\lambda_N = 2$, 3, and 4 between F1 and F2 in the parallel alignment of magnetizations in a tunnel device. (c) Nonlocal resistance as a function of perpendicular magnetic field B for $L/\lambda_N = 4.4$ in a tunnel device. The symbol (•) is the experimental data of CoFe/I/Al/I/CoFe [56]. (d) Nonlocal resistance for distance $L/\lambda_N = 0.4$ and $r_F = 2(R_F^*/R_N) = 0$, 0.1, and 0.2 in an ohmic contact device. The dashed curve corresponds to that in a tunnel device.

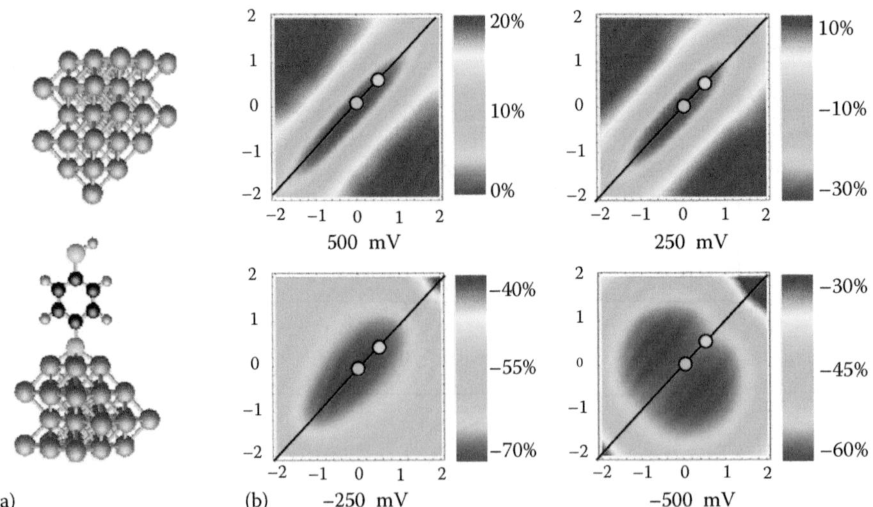

FIGURE 33.4 Spin-polarized transport through a benzene-thiol-thiolate molecule attached to an Ni surface and probed with an Ni tip. In (a), we present a sketch of the calculated system and in (b) a two-dimensional map of the spin polarization P of the current as a function of the position of the STM tip and for different voltages. The diagonal black line in the panels in (b) indicates the plane of the benzene ring, the green circle the position of S, and the blue one that of H in the SH group.

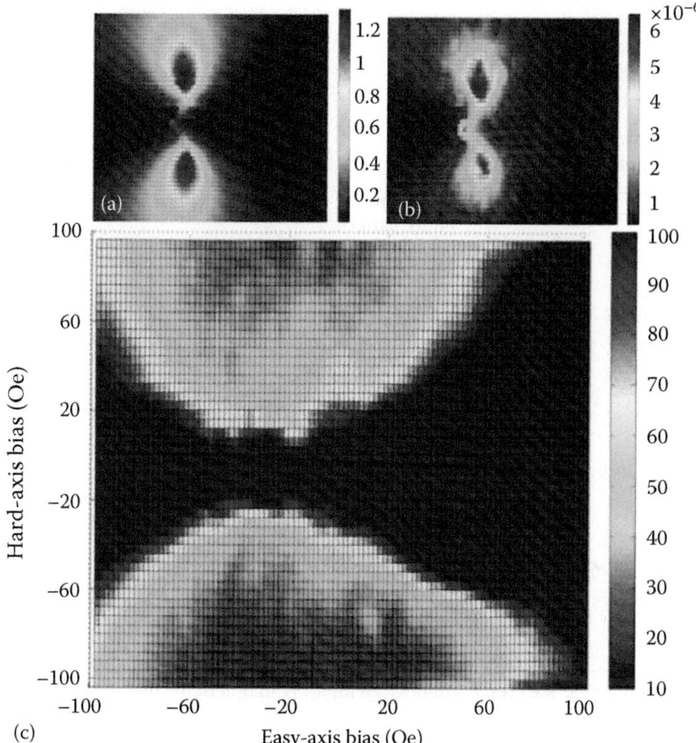

FIGURE 34.6 (a) The sensitivity map of an MgO MTJ sensor (in units of %/Oe); (b) the corresponding normalized voltage noise map in units of $Hz^{-1/2}$, both measured in a 2D orthogonal field arrangement; (c) magnetic noise map. The color bars are in units of $nT/Hz^{1/2}$ and set to 10–100 nT/$Hz^{1/2}$ (Reprinted with permission from Mazumdar, D., Shen, W., Liu, X., Schrag, B.D., Carter, M., and Xiao, G., Field sensing characteristics of magnetic tunnel junctions with (001) MgO tunnel barrier, *J. Appl. Phys.*, 103, 113911, 2008, Fig. 3. Copyright 2008, American Institute of Physics.)

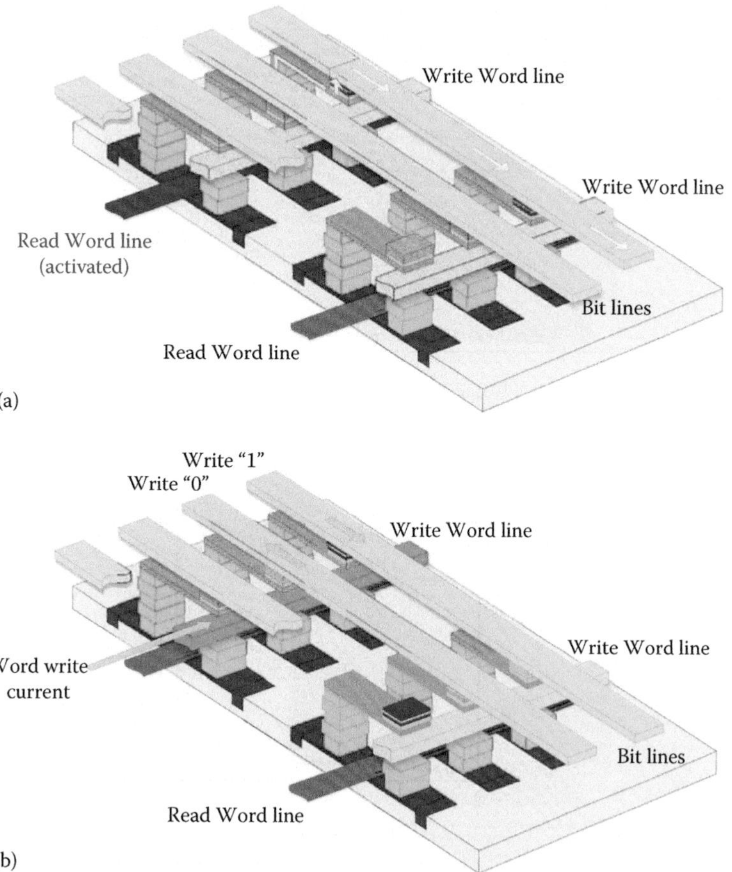

FIGURE 36.1 Diagram of a magnetic field programmable MRAM. (a) Reading a bit. Voltage is applied to the desired Read Word line (to enable the transistors in that word line) and voltage at desired bit line is measured. (b) Writing bits. Current is passed through the desired Write Word line and appropriate bit lines. Superposition of the fields generated by two currents orients the moment of the free layer in the MTJ in the desired direction. The polarity of the moment is determined by the direction of the current in the Bit line; current flowing one direction will flip it one way, and current in the opposite direction will flip it the other way. (From Parkin, S.S.P. et al., *Proc. IEEE*, 91, 661, 2003.)

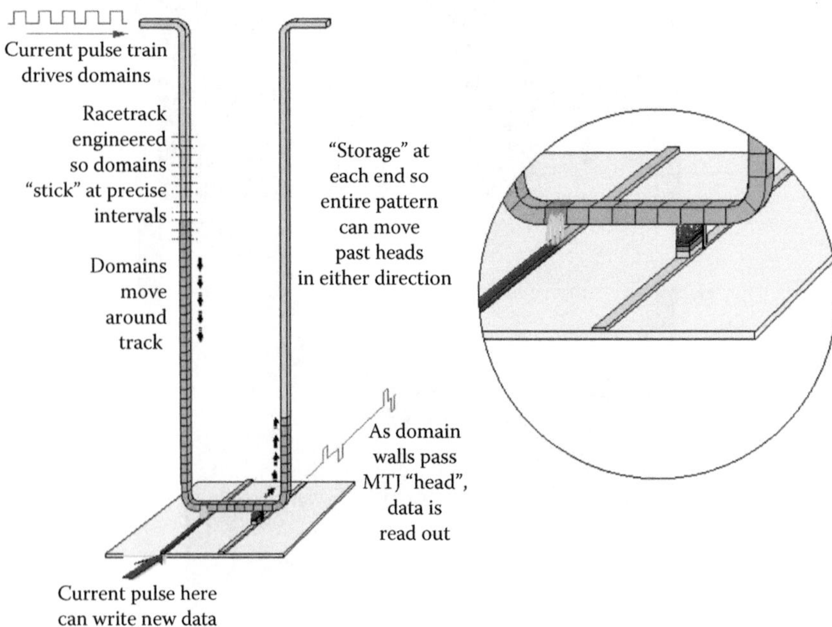
Current pulse train
drives domains
Racetrack
engineered
so domains
"stick" at precise
intervals
Domains
move
around
track
"Storage" at
each end so
entire pattern
can move
past heads
in either direction
As domain
walls pass
MTJ "head",
data is
read out
Current pulse here
can write new data

FIGURE 36.6 Vertical configuration magnetic Racetrack Memory element. A magnetic tunnel junction (MTJ) device is used for reading data in the racetrack.

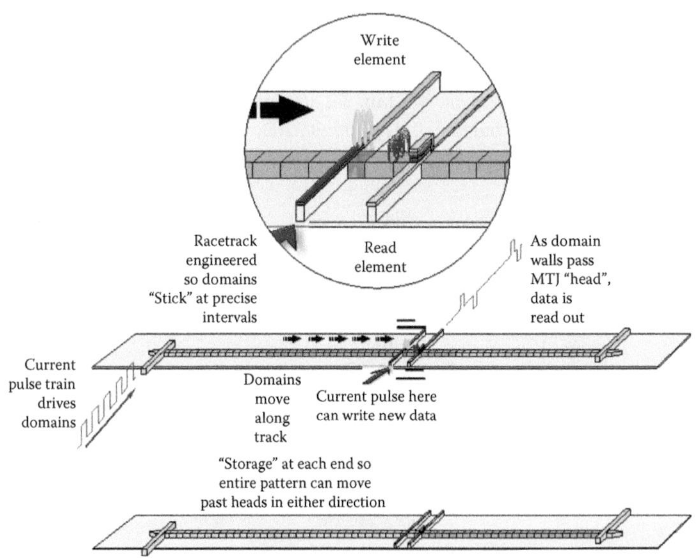
Write
element
Read
element
Racetrack
engineered
so domains
"Stick" at precise
intervals
As domain
walls pass
MTJ "head",
data is
read out
Current
pulse train
drives
domains
Domains
move
along
track
Current pulse here
can write new data
"Storage" at each end so
entire pattern can move
past heads in either direction

FIGURE 36.7 Horizontal configuration magnetic Racetrack Memory element.

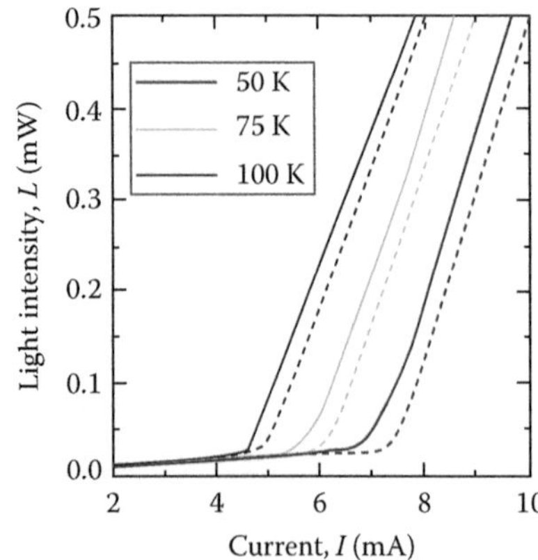
0.5
0.4
0.3
0.2
0.1
0.0
Light intensity, L (mW)
50 K
75 K
100 K
2
4
6
8
10
Current, I (mA)

FIGURE 38.7 The intensity of emitted light as function of injection proves the threshold reduction in a spin-laser using electrical spin injection. The solid (dashed) lines correspond to the injection of spin-polarized (spin-unpolarized) electrons. For every temperature, the onset of lasing is consistently lower for spin injection, which is realized with the aid of an applied magnetic field to ensure an out-of-plane component of magnetization in Fe injector (having an in-plane easy axis). (Reprinted with permission from Holub, M., Shin, J., and Bhattacharya, P., *Phys. Rev. Lett.*, 98, 146603, 2007, Fig. 3d. Copyright 2007 by the American Physical Society.)

25

Anomalous and Spin-Injection Hall Effects

Jairo Sinova
Texas A&M University

Jörg Wunderlich
Hitachi Cambridge Laboratory

Tomás Jungwirth
Institute of Physics, Academy of Sciences of the Czech Republic

25.1 Introduction

Throughout this book we have learned many aspects of spin dynamics in both metal and semiconducting systems involving the diagonal transport of spin and charge and their interactions, in some cases, with the collective ferromagnetic-order parameter driven out of equilibrium. In this chapter, we venture into another aspect of spin-charge dynamics in which an applied electric field induces a spin or a spin-charge response transverse to the field. These are the spin-dependent Hall effects.

In metallic spintronics, many studies have centered around the injection of spin and its manipulation via external magnetic fields or nonequilibrium current effects such as spin-torque effects. Within these studies, the spin–orbit (SO) coupling in the system becomes an important source of spin dephasing and, in many instances, it is important to minimize it in order to enhance the observed effects. On the other hand, in semiconductor spintronics, SO coupling comes to the forefront as a source of spin manipulation and spin generation, becoming a necessary element in physical phenomena of the spin-dependent Hall effects such as the anomalous Hall effect, spin Hall effect, inverse spin Hall effect, and spin-injection Hall effect. These spin-dependent Hall effects are of course governed by common mechanisms that we describe in the following.

In this chapter, we focus primarily on the anomalous Hall effect and the spin-injection Hall effect. The former is the primary spin-dependent Hall effect studied for many decades but one that continues to fascinate and is at the center of many theoretical and experimental studies. The latter is one of the most recent members of the spintronics Hall family [1], and its phenomenology combines many interesting physics observed in semiconducting spintronics and can be a good playground to study in detail these mechanisms and other spin-dynamics effects controlled by particular choices of SO-coupling strengths that can be tuned experimentally. The other principal spin-dependent Hall effects, the spin Hall effect and the inverse spin Hall effect, have also been intensively studied over the past 5 years and their recent progress has been presented in several reviews. Therefore, we will only provide a simple and short qualitative description of these reviews. Other chapters that are pertinent to these physics are Chapter 6 on spin-injection and detection through nonlocal measurements by Mark Johnson and Chapter 31 on spin Hall effect in metallic system by Takahashi and Maekawa.

This chapter is hence structured as follows: In the remainder of this chapter, we introduce the different spin-dependent Hall effects, which are more closely related. In the discussion of the anomalous Hall effect (Section 25.2.1), we introduce the basic mechanisms that give rise to these effects within the metallic systems (Section 25.2.2). In Section 25.2.5, we discuss the experimental phenomenology of the spin-injection Hall effect, and in Section 25.4, we discuss the theoretical aspects of this phenomenology, which incorporates anomalous Hall effect physics as well as persistent spin-helix physics. Throughout, we focus more on qualitative descriptions rather than detailed derivations and refer to the list of references for further details.

25.2 Spin-Dependent Hall Effects in Spin–Orbit-Coupled Systems

Within the family of spin-dependent Hall effects, the anomalous Hall effect, the spin Hall effect, the inverse spin Hall effect, and the spin-injection Hall effect are among the most closely related as sketched in Figure 25.1. The more extensively studied of these is the anomalous Hall effect from which many of the mechanisms have been identified. Several extensive reviews have been written over the years on these subjects. The anomalous Hall effect has been recently reviewed by Nagaosa et al. [2], in which we base part of our chapter. Within the spin Hall effect physics, several recent reviews have appeared such as the one by Schliemann [3], focusing on some aspect of the intrinsic spin Hall effect [4,5], and Engel et al. [6] and Hankiewicz and Vignale [7], focusing on the extrinsic spin Hall effect and quantum spin Hall effect. This particular field is evolving rapidly, springing off many new fields of interest such as the spin Hall insulators following the studies of the quantum spin Hall effect [8–10] and extending even to spin thermodynamic Hall effects.

25.2.1 Anomalous Hall Effect

In 1879, Edwin Hall ran a current through a gold foil and discovered that a transverse voltage was induced when the film was exposed to a perpendicular magnetic field. The ratio of this Hall voltage to the current is the Hall resistivity. For paramagnetic materials, the Hall resistivity is proportional to the applied magnetic field and Hall measurements can be shown to be equivalent to a measurement of the concentration of free carriers and determine whether they are holes or electrons. Magnetic films were found to exhibit both this ordinary Hall response and an extraordinary or anomalous Hall effect response may persist at zero magnetic fields, which is proportional to the internal magnetization perpendicular to the plane:

$$R_{Hall} = R_o H + R_s M_z, \tag{25.1}$$

where

R_{Hall} is the Hall resistance

R_o and R_s are the ordinary and anomalous Hall coefficients

M_z is the magnetization perpendicular to the plane

H is the applied magnetic field

Based on this phenomenology, the anomalous Hall effect has been an important tool to measure electrically the magnetization in many systems. The key player in this effect is the SO coupling, without which one would not observe the effect in a homogeneously magnetized system. The role of the broken time-reversal symmetry is mainly to create an asymmetric spin population that leads to a charge Hall voltage but would not produce a measurable anomalous Hall signal without the SO coupling.

The field of the anomalous Hall effect and its recent progress have been recently reviewed by Nagaosa et al. [2], focusing particularly on the new understanding of the underlying physics based on the topological nature of the effect. Let us first embark on a qualitative description of the different mechanisms.

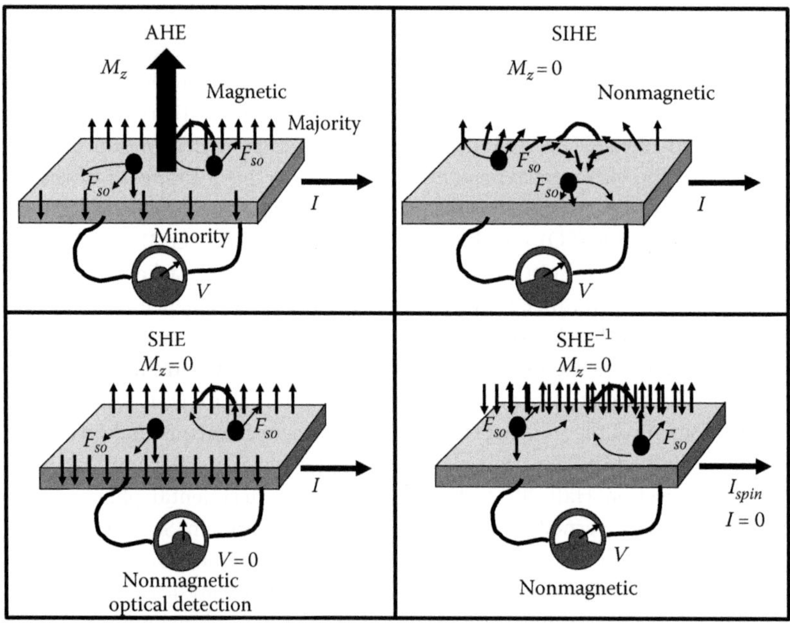

FIGURE 25.1 Schematic of the family of spin-dependent Hall effects: Anomalous Hall effect (AHE), spin-injection Hall effect (SIHE), spin Hall effect (SHE), and inverse spin Hall effect (SHE^{-1}). A common thread is their existence at zero external magnetic field (i.e., no ordinary Hall effect present), with the key difference on the nature of the currents involved and the method of detection. Within the spin-injection Hall effect, the Hall response is overlaid on a new type of spin excitation termed persistent spin-helix.

25.2.2 Mechanism of the Spin-Dependent Hall Effects

The usual descriptions of the mechanisms of the anomalous Hall effect are shown in Figure 25.2. Many of these, at least as explained in many texts, are based on semiclassical descriptions of the different scattering and deflection mechanisms. In what follows, although we are going to retain the historical labeling of each contribution, we recast their definition in a more physical and phenomenological way anticipating their correct connection to the actual microscopic physics and to a precise experimental definition of each.

A natural classification of contributions to the anomalous Hall effect and other spin-dependent Hall effects, which is guided by experiment and by the microscopic theory of metals, is to separate them according to their dependence on the Bloch state transport lifetime τ, which is equivalent to a $1/k_F l$ expansion, where k_F is the Fermi wavevector and l is the mean free path. It is relatively easy to identify contributions to the anomalous Hall conductivity, σ_{xy}^{AH}, which vary as τ^1 and as τ^0. In experiment, a similar separation can sometimes be achieved by plotting σ_{xy}^{AH} versus the longitudinal conductivity $\sigma_{xx} \propto \tau$, when τ is varied by altering disorder or varying temperature. More commonly (and equivalently), the Hall resistivity is separated into contributions proportional to ρ_{xx} and ρ_{xx}^2. Since in almost all cases $\sigma_{xx} \gg \sigma_{xy}^{AH}$, it means that an anomalous Hall resistivity, $\rho_{xy}^{AH} = \sigma_{yx}^{AH} / \left(\sigma_{xx}^2 + \sigma_{xy}^{AH^2}\right)$, proportional to ρ_{xx}^2 implies a dominating mechanism that is scattering independent, that is, $\sigma_{xy}^{AH} \approx \rho_{yx}/\rho_{xx} \sim \sigma_{xx}^0$, and an anomalous Hall resistivity proportional to ρ_{xx} implies a dominating mechanism that is very dependent on scattering, that is, $\sigma_{xy}^{AH} \approx \rho_{yx}^{AH}/\rho_{xx} \sim \sigma_{xx}^1$. The illustration of this scaling is shown in Figure 25.3a and b. This already highlights the counterintuitive nature of the anomalous Hall effect, which implies that a "dirtier" system will be dominated by scattering-independent mechanisms whereas a "cleaner" system will be dominated by impurity-scattering mechanisms.

This partitioning seemingly gives only two contributions to σ_{xy}^{AH}, one $\sim\tau$ and the other $\sim\tau^\circ$. The first contribution is known as the *skew-scattering* contribution, $\sigma_{xy}^{AH\text{-}skew}$. Note that in this parsing of anomalous Hall effect contributions, it is the dependence on τ (or σ_{xx}) that defines it, not a particular mechanism linked to a microscopic or semiclassical theory. The second contribution proportional to τ° (or independent of σ_{xx}) is further separated into two parts: *intrinsic* and *side-jump*. It has long been believed that this partitioning is only a theoretical exercise. However, although these two contributions cannot be separated experimentally by dc measurements, they *can* be separated experimentally by defining the intrinsic contribution, $\sigma_{xy}^{AH\text{-}int}$, as the extrapolation of the ac-interband Hall conductivity to zero frequency in the limit of $\tau \to \infty$, with $1/\tau \to 0$ faster than $\omega \to 0$. This then leaves a unique definition for the third and last contribution, termed side-jump, as $\sigma_{xy}^{AH\text{-}sj} \equiv \sigma_{xy}^{AH} - \sigma_{xy}^{AH\text{-}skew} - \sigma_{xy}^{AH\text{-}int}$. This possible experimental extraction of $\sigma_{xy}^{AH\text{-}sj}$ and $\sigma_{xy}^{AH\text{-}int}$ is illustrated in Figure 25.3c.

We examine these three contributions in the following. It is important to note that the above definitions have not relied on identifications of semiclassical processes such as side-jump

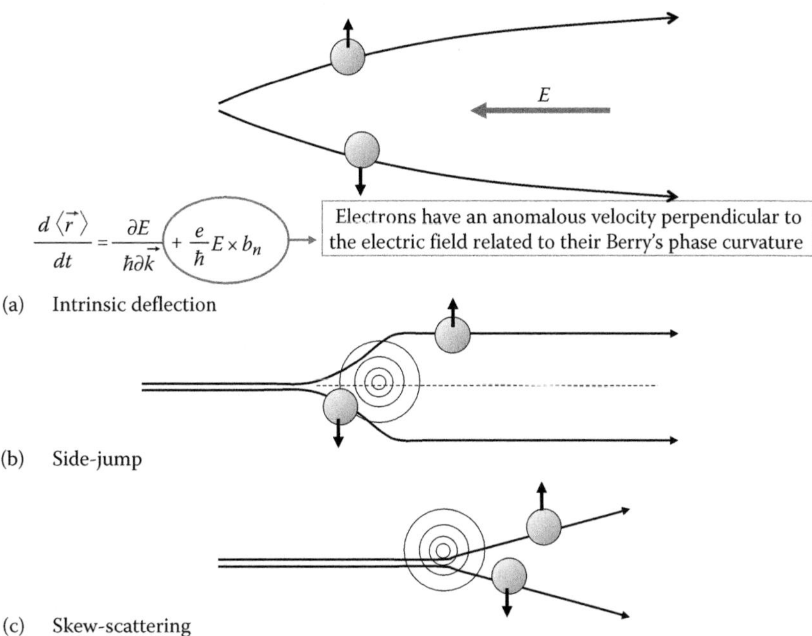

FIGURE 25.2 Schematic description of the mechanisms of the anomalous Hall effect. (a) Intrinsic deflection arises from interband coherence induced by an external electric field, giving rise to a velocity contribution perpendicular to the field direction. (b) Side-jump arises from the fact that electron velocity is deflected in opposite directions by the opposite electric fields experienced upon approaching and leaving an impurity. (c) Skew-scattering arises from asymmetric scattering due to the effective SO coupling of the electron or the impurity.

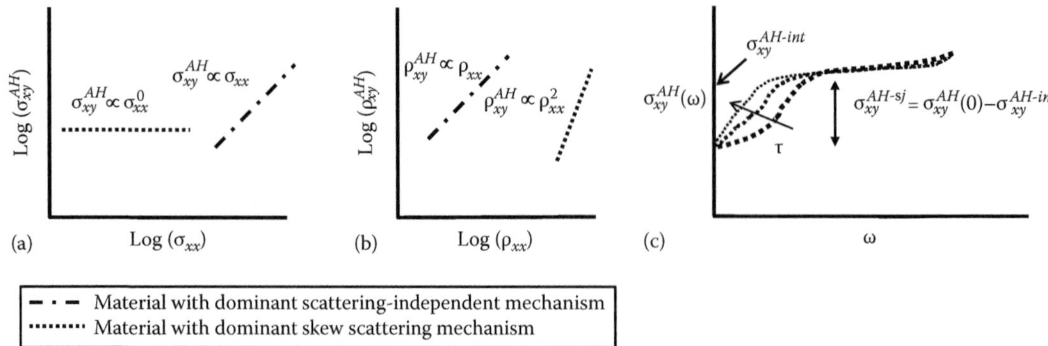

FIGURE 25.3 Scaling of anomalous Hall conductivity (a) and resistivity (b) versus diagonal conductivity and resistivity, respectively, for a material whose anomalous Hall effect is dominated by skew scattering (dot-dashes) and for a material dominated by scattering-independent mechanisms. (c) Illustration of the scaling of ac-anomalous Hall conductivity for samples with decreasing impurity concentration (increasing τ) to determine $\sigma_{xy}^{AH\text{-}int}$ and, from this, obtain $\sigma_{xy}^{AH\text{-}sj}$.

scattering [11] or skew scattering from asymmetric contributions to the semiclassical scattering rates [12] identified in earlier theories. Not surprisingly, the contributions defined earlier contain these semiclassical processes. However, it is now understood that other contributions are present in the fully generalized semiclassical theory, which were not precisely identified previously and are necessary to be fully consistent with microscopic theories [2,13,14].

25.2.2.1 Intrinsic Contribution, $s_{xy}^{AH\text{-}int}$

We have defined the intrinsic contribution microscopically as the *dc* limit of the interband *ac* conductivity, a quantity that is not zero in ferromagnets when the SO-coupling interactions are included. There is, however, a direct link to semiclassical theory in which the induced interband coherence is captured by a momentum space Berry-phase–related contribution to the anomalous velocity. We show this equivalence in the following.

Historically, this contribution to the anomalous Hall effect was first derived by Karplus and Luttinger [15] but its topological nature was not fully appreciated until recently [16–18]. The concept of intrinsic anomalous Hall effect was emphasized by Jungwirth et al. [17], motivated by the experimental importance of the anomalous Hall effect in ferromagnetic semiconductors and also by the earlier analysis of the relationship between momentum space Berry phases and anomalous velocities in semiclassical transport theory by Niu and coworkers [18,19]. The frequency-dependent interband Hall conductivity, which reduces to the intrinsic anomalous Hall conductivity in the *dc* limit, had been evaluated earlier for a number of materials by Mainkar et al. [20] and Guo and Ebert [21].

The intrinsic contribution to the conductivity is dependent only on the band structure of the perfect crystal. It can be calculated directly from the simple Kubo formula for the Hall conductivity for an ideal lattice, given the eigenstates $|n, \boldsymbol{k}\rangle$ and eigenvalues $\varepsilon_n(\boldsymbol{k})$ of a Bloch Hamiltonian H:

$$\sigma_{ij}^{AH\text{-}int} = e^2\hbar \sum_{n\neq n'} \int \frac{d\boldsymbol{k}}{(2\pi)^d}\left[f(\varepsilon_n(\boldsymbol{k})) - f(\varepsilon_{n'}(\boldsymbol{k}))\right] \times \frac{\mathrm{Im}\left[\langle n,\boldsymbol{k}|v_i(\boldsymbol{k})|n',\boldsymbol{k}\rangle\langle n',\boldsymbol{k}|v_j(\boldsymbol{k})|n,\boldsymbol{k}\rangle\right]}{(\varepsilon_n(\boldsymbol{k})-\varepsilon_{n'}(\boldsymbol{k}))^2}, \quad (25.2)$$

and the velocity operator is defined by

$$\boldsymbol{v}(\boldsymbol{k}) = \frac{1}{i\hbar}[\boldsymbol{r}, H(\boldsymbol{k})] = \frac{1}{\hbar}\nabla_k H(\boldsymbol{k}), \quad (25.3)$$

where H is the $\boldsymbol{k}$-dependent Hamiltonian for the periodic part of the Bloch functions. Note the restriction $n\neq n'$ in Equation 25.2.

What makes this contribution quite unique is that it is proportional to the integration over the Fermi sea of the Berry's curvature of each occupied band, or equivalently [22,23] to the integral of Berry phases over cuts of Fermi surface segments. This result can be derived by noting that

$$\langle n,\boldsymbol{k}|\nabla_k|n',\boldsymbol{k}\rangle = \frac{\langle n,\boldsymbol{k}|\nabla_k H(\boldsymbol{k})|n',\boldsymbol{k}\rangle}{\varepsilon_{n'}(\boldsymbol{k})-\varepsilon_n(\boldsymbol{k})}. \quad (25.4)$$

Using this expression, Equation 25.2 reduces to

$$\sigma_{ij}^{AH\text{-}int} = -\varepsilon_{ijl}\frac{e^2}{\hbar}\sum_n \int \frac{d\boldsymbol{k}}{(2\pi)^d} f(\varepsilon_n(\boldsymbol{k}))b_n^l(\boldsymbol{k}), \quad (25.5)$$

where

ε_{ijl} is the antisymmetric tensor

$\boldsymbol{a}_n(\boldsymbol{k})$ is the Berry-phase connection $\boldsymbol{a}_n(\boldsymbol{k}) = i\langle n,\boldsymbol{k}|\nabla_k|n,\boldsymbol{k}\rangle$,

$\boldsymbol{b}_n(\boldsymbol{k})$ is the Berry-phase curvature

$$\boldsymbol{b}_n(\boldsymbol{k}) = \nabla_k \times \boldsymbol{a}_n(\boldsymbol{k}) \quad (25.6)$$

corresponding to the states $\{|n,\boldsymbol{k}\rangle\}$.

This same linear response contribution to the anomalous Hall effect conductivity can be obtained from the semiclassical theory of wave-packet dynamics [18,19,24]. It can be shown that the wave-packet group velocity has an additional contribution in the presence of an electric field: $\dot{r}_c = \partial E_n(\boldsymbol{k})/\hbar\partial \boldsymbol{k} - (\boldsymbol{E}/\hbar) \times \boldsymbol{b}_n(\boldsymbol{k})$. The intrinsic Hall conductivity formula, Equation 25.5, is obtained simply by summing the second (anomalous) term over all occupied states.

One of the motivations for identifying the intrinsic contribution $\sigma_{xy}^{AH\text{-}int}$ is that it can be evaluated accurately even for relatively complex materials using first-principles electronic structure theory techniques. In many materials that have strongly SO-coupled bands, the intrinsic contribution seems to dominate the anomalous Hall effect.

25.2.2.2 Skew-Scattering Contribution, $\sigma_{xy}^{AH\text{-}skew}$

This contribution is simply the component of σ_{xy}^{AH} proportional to the Bloch state transport lifetime. It therefore tends to dominate in nearly perfect crystals. It is the only contribution to the anomalous Hall effect that appears within the confines of traditional Boltzmann transport theory in which interband coherence effects are completely neglected. Skew-scattering is due to chiral features that appear in the disorder scattering of SO-coupled ferromagnets [12,25].

Treatments of semiclassical Boltzmann transport theory found in textbooks often appeal to the principle of detailed balance, which states that the transition probability $W_{n\to m}$ from n to m is identical to the transition probability in the opposite direction ($W_{m\to n}$). Although these two transition probabilities are identical in a Fermi's golden-rule approximation, since $W_{n\to n'} = (2\pi/\hbar)\left|\langle n|V|n'\rangle\right|^2 \delta(E_n - E_{n'})$, where V is the perturbation inducing the transition, detailed balance in this microscopic sense is not generic. In the presence of SO coupling, either in the Hamiltonian of the perfect crystal or in the disorder Hamiltonian, a transition that is right-handed with respect to the magnetization direction has a different transition probability than the corresponding left-handed transition. When the transition rates are evaluated perturbatively, asymmetric chiral contributions appear first at third order. In simple models, the asymmetric chiral contribution to the transition probability is often assumed to have the form [12,25]

$$W_{\boldsymbol{k}\boldsymbol{k}'}^{A} = -\tau_A^{-1}\boldsymbol{k} \times \boldsymbol{k}' \cdot \boldsymbol{M}_s. \tag{25.7}$$

When this asymmetry is inserted into the Boltzmann equation, it leads to a current proportional to the longitudinal current driven by $\boldsymbol{E}$ and perpendicular to both $\boldsymbol{E}$ and $\boldsymbol{M}_s$. When this mechanism dominates, both the anomalous Hall conductivity σ_{xy}^{AH} and the conductivity σ_{xx} are approximately proportional to the transport lifetime τ, and the anomalous Hall resistivity $\rho^{AH\text{-}skew} = \sigma^{AH\text{-}skew}\rho_{xx}^2$ is therefore proportional to the longitudinal resistivity ρ_{xx}.

There are several specific mechanisms for skew scattering. Evaluation of the skew-scattering contribution to the Hall conductivity or resistivity requires simply that the conventional linearized Boltzmann equation be solved using a collision term with accurate transition probabilities, since these will generically include a chiral contribution. We emphasize that skew-scattering contributions to σ_{xy}^{AH} are present not only because of SO coupling in the disorder Hamiltonian, but also because of SO coupling in the perfect crystal Hamiltonian combined with purely scalar disorder. Either source of skew scattering could dominate $\sigma_{xy}^{AH\text{-}skew}$, depending on the host material and also on the type of impurities, and, as we will see, it dominates the physics of the present experiments in spin-injection Hall effect.

One important thing to note is that we have not defined earlier the skew-scattering contribution to the anomalous Hall effect as the sum of *all* the contributions arising from the asymmetric scattering rate present in the collision term of the Boltzmann transport equation. We know from microscopic theory that this asymmetry also makes an anomalous Hall effect contribution of order τ^0. There exists a contribution from this asymmetry, which is actually present in the microscopic theory treatment associated with the so-called ladder diagram corrections to the conductivity, and therefore of order τ^0. In our experimentally practical parsing of anomalous Hall effect contributions, we do not associate this contribution with skew scattering but place it under the umbrella of side-jump scattering even though it does not physically originate from any sidestep type of scattering.

25.2.2.3 Side-Jump Contribution $\sigma_{xy}^{AH\text{-}sj}$

Given the sharp definition we have provided for the intrinsic and skew-scattering contributions to the anomalous Hall effect conductivity, the equation

$$\sigma_{xy}^{AH} = \sigma_{xy}^{AH\text{-}int} + \sigma_{xy}^{AH\text{-}skew} + \sigma_{xy}^{AH\text{-}sj} \tag{25.8}$$

defines the side-jump contribution as the difference between the full Hall conductivity and the two simpler contributions. In using the term side-jump for the remaining contribution, we are appealing to the historically established taxonomy. Establishing this connection mathematically has been the most controversial aspect of anomalous Hall effect theory and the one that has taken the longest to clarify from a theory point of view. Although this classification of Hall conductivity contributions is often useful, it is not generically true that the only correction to the intrinsic and skew contributions can be physically identified with the side-jump process defined as in the earlier studies of the anomalous Hall effect [26].

The basic semiclassical argument for a side-jump contribution can be stated straightforwardly: when considering the scattering of a Gaussian wave-packet from a spherical impurity with SO interaction ($H_{SO} = (1/2m^2c^2)\ (r^{-1}\partial V/\partial r)S_zL_z$), a wave-packet with incident wavevector $\boldsymbol{k}$ will suffer a displacement transverse to $\boldsymbol{k}$ equal to $(1/6)k\hbar^2/m^2c^2$. This type of contribution was first noticed, but discarded, by Smit [25] and reintroduced by Berger [26] who argued that it was the key contribution to the anomalous Hall effect. This kind of mechanism clearly lies outside the bounds of traditional Boltzmann transport theory in which only the probabilities of transitions between Bloch states appears, and not microscopic details of the scattering processes. This contribution to the conductivity ends up being independent of τ and therefore contributes to the anomalous Hall effect at the same order as the intrinsic contribution in an expansion in powers of scattering rate. The separation between intrinsic and side-jump contributions, which cannot be distinguished by their dependence on τ, has been perhaps the most argued aspect of anomalous Hall effect theory [2].

Side-jump and intrinsic contributions have different dependences on more specific system parameters, specially in systems with complex band structures [14]. Some of the initial controversy that surrounded side-jump theories was associated with the physical meaning ascribed to quantities that were plainly gauge dependent, like the Berry's connection, which in early theories is typically identified as the definition of the side step upon scattering. Studies of simple models, for example, models of semiconductor conduction bands, also gave results in which the side-jump contribution seemed to be the same size but opposite in sign compared to the intrinsic contribution [27]. We now understand [13] that these cancelations are unlikely, except in models with a very simple band structure, for example, one with a constant Berry's curvature. It is only by a careful comparison between fully microscopic linear response calculations, such as Keldysh (nonequilibrium Green's function) or Kubo formalisms, and the systematically developed semiclassical theory that the specific contribution due to the side-jump mechanism can be separately identified with confidence [13].

A practical approach, which is followed at present for materials in which σ_{xy}^{AH} seems to be independent of σ_{xx}, is to first calculate the intrinsic contribution to the anomalous Hall effect. If this explains the observation (and it appears that it usually does), then it is deemed that the intrinsic mechanism dominates. If not, we can take some comfort from understanding on the basis of simple model results, that there can be other contributions to σ_{xy}^{AH}, which are also independent of σ_{xx} and can for the most part be identified with the side-jump mechanism. It seems extremely challenging, if not impossible, to develop a predictive theory for these contributions, partly because they require many higher-order terms in the perturbation theory that must be summed, but more fundamentally because they depend sensitively on details of the disorder in a particular material, which are normally unknown and difficult to model accurately.

25.2.3 Recent Developments on the Anomalous Hall Effect

The earlier description of the different mechanisms of the anomalous Hall effect in ferromagnetic metals has been refined over the past few years, with some of the recent research focused on unifying the different theories that disagreed. We finalize this section by pointing out some of the key recent developments, which are important to know when discussing the anomalous Hall effect:

1. When the anomalous Hall conductivity, σ_{xy}^{AH}, is independent of σ_{xx}, the anomalous Hall effect can often be understood in terms of the geometric concepts of Berry phase and Berry curvature in momentum space. This anomalous Hall effect mechanism is responsible for the intrinsic anomalous Hall effect (see following text). In this regime, the anomalous Hall current can be thought of as the unquantized version of the quantum Hall effect.
2. Three broad regimes have been identified when surveying a large body of experimental data for diverse materials [2]: (i) A high-conductivity regime ($\sigma_{xx} > 10^6\ (\Omega\ \text{cm})^{-1}$) in which a linear contribution to $\sigma_{xy}^{AH} \sim \sigma_{xx}$ due to skew scattering dominates σ_{xy}^{AH}. In this regime, the normal Hall conductivity contribution can be significant and even dominate σ_{xy}. (ii) An intrinsic or scattering-independent regime in which σ_{xy}^{AH} is roughly independent of σ_{xx} ($10^4\ (\Omega\ \text{cm})^{-1} < \sigma_{xx} < 10^6\ (\Omega\ \text{cm})^{-1}$). (iii) A bad-metal or insulating regime ($\sigma_{xx} < 10^4\ (\Omega\ \text{cm})^{-1}$) in which σ_{xy}^{AH} decreases with decreasing σ_{xx} at a rate faster than linear. This regime remains still a challenge, theoretically.
3. The band structure of the ferromagnetic materials plays a key role in these systems. In particular, the band (anti-) crossings near the Fermi energy has been identified using first-principles Berry curvature calculations as a mechanism that can lead to a large intrinsic anomalous Hall effect. The effect of scattering on these crossings is still not well understood.
4. A semiclassical treatment by a generalized Boltzmann equation taking into account the Berry curvature and coherent interband mixing effects due to band structure and disorder has been formulated. This theory provides a clearer physical picture of the anomalous Hall effect than early theories by identifying correctly all the semiclassically defined mechanisms [13,14]. This generalized semiclassical picture has been verified by comparison with controlled microscopic linear response treatments for identical models.

25.2.4 Spin Hall Effect

A natural progression of the physics of the anomalous Hall effect to systems with time-reversal symmetry and with SO coupling is the spin Hall effect. In describing the different mechanisms giving rise to anomalous Hall effect, the exchange field is only important in creating a spin population imbalance that will give rise to a finite voltage that one can measure. Hence, since the mechanisms that give rise to the anomalous Hall effect are also present in these systems, the "spin" separation will supposedly take place but it will not be directly measurable through a voltage perpendicular to the flow of the current in the sample since there is no exchange field that induces the population asymmetry between spin-up and spin-down states. However, consistent with the global time-reversal symmetry, a spin current will be generated perpendicular to an applied charge current, which in turn should lead to spin accumulations with opposite magnetization at the edges. This is the spin Hall effect first predicted over three decades ago by simply invoking the phenomenology of the anomalous Hall effect in ferromagnets and applying it to semiconductors [28].

Motivated by studies of intrinsic anomalous Hall effect in strongly SO-coupled ferromagnetic materials, where scattering contributions do not seem to dominate, the possibility of an intrinsic spin Hall efffect was put forward by Murakami et al. [4] and by Sinova et al. [5], with scattering playing a minor role. This proposal generated an extensive theoretical debate motivated by its potential as a spin injection tool in low dissipative devices. The interest in the spin Hall effect has also been dramatically enhanced by experiments by two groups reporting the first observations of the spin Hall effect in n-doped semiconductors [29,30] and in 2D hole gases (2DHG) [31]. These observations, based on optical experiments, have been followed by metal-based experiments in which both the spin Hall effect and the inverse spin Hall effect have been observed [32–36], and other experiments in semiconductors at other regimes, even at room temperature [37–40].

In the case of the inverse spin Hall effect, it is a spin current that generates a charge current in the so-called nonlocal measurements. These spin currents can be utilized in the spin-torque experiments and are being studied very actively. Given the extent of detail and still unsettled debate in this field, and the several reviews already mentioned in the introduction, we will not go further into the experiment and theory of spin Hall effect and inverse spin Hall effect. Instead, we will focus on the new phenomenology of the spin-injection Hall effect that combines several effects, is at present well understood in the diffusive regime, and is perhaps the most promising from a technology standpoint.

25.2.5 Spin-Injection Hall Effect

The spin-injection Hall effect was observed in a photovoltaic cell device that opens direct application potential in opto-spintronics, for example, as a solid-state polarimeter. Importantly, the spin-injection Hall effect can be also used to measure spin dynamics of the electrons propagating in the semiconducting channel, in particular the spin-helix effect. The spin-injection Hall effect can, therefore, be used as an efficient spin-detection tool for basic studies of SO coupling phenomena in semiconductors and can also be directly incorporated in a number of spintronic devices, including the spin-field-effect transistor of the Datta–Das type [63].

The electrical detection of spin-polarized transport in semiconductors is a key requisite for the possible incorporation of spin in semiconductor microelectronics. In the spin-injection Hall effect, a polarized current is injected (optically in the present case) and its polarization can be detected by transverse electrical signals, via the anomalous Hall effect, directly along the semiconducting channel. One key practical feature is that this is achieved without disturbing the spin-polarized current or using magnetic elements, hence, opening the door for possible spin-based sequential logic schemes in logic circuits.

Spin-polarized transport phenomena in semiconductors have been studied by a range of techniques used in ferromagnets and novel approaches developed specifically for semiconductor spintronics. The techniques include magneto-optical scanning probes [29,37,41–43], spin-polarized electroluminescence [31,44–47], and magnetoelectric measurements utilizing spin-valve effects, magnetization dependence of nonequilibrium chemical potentials, and SO coupling phenomena [32,42,43,48–54]. Two of these spintronic research developments have been particularly important for the observation of the spin-injection Hall effect. First, it has been demonstrated that electrons carrying electrical current in a ferromagnet align their spins with the local direction of magnetization and that the resulting electrical signals due to the anomalous Hall effect can be used as an alternative to magneto-optical scanning probes to measure the local magnetization [55,56]. The second is the observation of the spin Hall effect that verified that the physics of the anomalous Hall effect translates directly to semiconducting systems without broken time-reversal symmetry, that is, the spin Hall effect. Combining these two facts one can surmise that injecting spin-polarized electrical currents into nonmagnetic semiconductors should also generate a Hall effect, which, as long as the spins of the charge carriers remain coherent, yields transverse charge accumulation and is therefore detectable electrically. In Wunderlich et al. [1], the device diode is operated in the reverse regime as a photocell in order to inject spin-polarized electrical currents into the semiconductor from a spatially confined region.

The effect, as discussed in the theory section, is well understood by microscopic calculations that reflect spin dynamics induced by an internal SO field. The spin-injection Hall effect is observed up to high temperatures and the devices can also be thought of as a realization of a nonmagnetic spin-photovoltaic polarimeter, which directly converts polarization of light into transverse voltage signals (Figure 25.4).

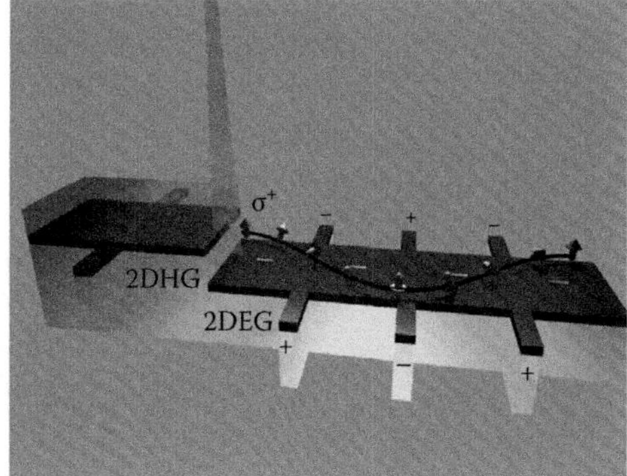

FIGURE 25.4 Sketch of the spin Hall effect device. Polarized light injects spin-polarized electron–hole pairs. The coplanar p–n junction [31] is connected in reverse bias, and the electrons, moving down the semiconductor channel in the $(1\bar{1}0)$ direction, experience a precessing motion in their spin, which is recorded via Hall probes. (From Wunderlich, J. et al., *Nat. Phys.*, 5, 675, 2009.)

25.3 Spin-Injection Hall Effect: Experiment

In this section, we first describe the experimental findings of Wunderlich et al. [1] and follow it up with the theoretical analysis in the subsequent and final section of this chapter.

Figure 25.5a shows lateral micrographs of the planar 2D electron-hole gas (2DEG-2DHG) photodiodes with the p-region and n-region patterned into unmasked 1 μm wide Hall bars (similar experiments were performed in masked samples) in the $(1\bar{1}0)$ direction. The effective width of each Hall contact is 50–100 nm and separation between two Hall crosses is 2 μm. The sketch of the device and its experimental setup is also illustrated in Figure 25.1. As in the spin Hall effect experiments [31], the semiconductor wafer consists of a modulation p-doped AlGaAs/GaAs heterojunction on top of the structure separated by 90 nm of intrinsic GaAs from an n-doped AlGaAs/GaAs heterojunction underneath. Without etching, the top heterojunction in the wafer is populated by the 2DHG while the 2DEG at the bottom heterojunction is depleted. The n-side of the coplanar p–n junction is formed by removing the p-doped surface layer from a part of the wafer through wet etching. The removal of this top heterojunction results in populating the 2DEG. At zero or reverse bias, the device is not conductive in dark due to charge depletion of the p–n junction. Counter-propagating electron and hole currents can be generated by illumination at sub-gap wavelengths. Because of the optical selection rules, the spin-polarization of injected electrons and holes is determined by the sense and degree of the circular polarization of vertically incident light. The optical spin-injection area is controlled by bias-dependent p–n junction depletion and, additionally, by the position and focus of the laser spot or by including metallic masks on top of the Hall bars [1].

Figure 25.5c shows the measurements at the Hall cross bar H2 in the n-channel at 4 K, laser wavelength of 850 nm, and for 0 and −10 V reverse bias with the laser spot fixed on top of the p–n junction and focused on approximately 1 μm in diameter. While the longitudinal resistance R_L is insensitive to the polarization, the transverse signal R_H is observed only for polarized spin injection into the electron channel and reverses sign upon reversing the polarization. With the laser spot focused on the p–n junction, the transverse voltage is only weakly bias dependent. Note that the large signals of tens of microvolts, corresponding to transverse resistances of tens of ohms, are detected outside the spin-injection area at a Hall cross separated by 3.5 μm from the p–n junction.

Figure 25.6a shows the simultaneous electrical measurements at Hall cross H0, which is wider and partially overlaps with the injection area, and at remote Hall crosses H1, H2, and H3. To confirm that the transverse signals do not result from spurious effects but originate from the polarized spin injection, the helicity of the incident light is reversed by manually rotating a λ/2

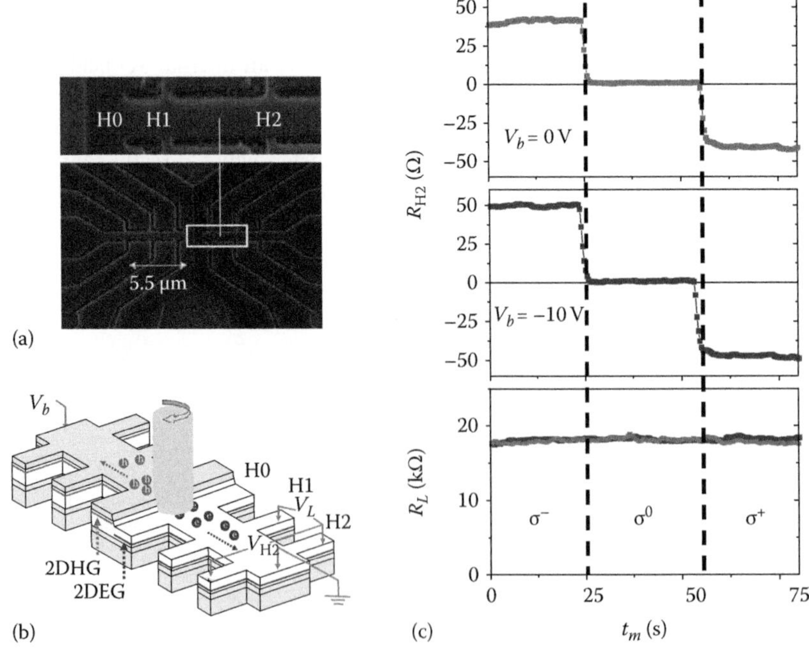

FIGURE 25.5 Devices, schematics of the experiment, and basic observation of the spin-injection Hall effect. (a) Micrograph of the coplanar p–n junction device. (b) Schematic diagram of the wafer structure of the 2DEG-2DHG p–n diode and of the spin-injection Hall effect measurement setup. (c) Steady-state spin-injection Hall effect signals changing sign for opposite helicities (σ_- and σ_+) of the incident light beam, that is, for opposite spin-polarizations of the injected electrons. R_{H2} is the second Hall bar resistance and R_L is the longitudinal resistance. For linearly polarized light (σ^0) the injected electron current is spin-unpolarized and the spin-injection Hall effect vanishes. Measurements were performed at the 2DEG Hall cross H2 at a temperature of 4 K. (From Wunderlich, J. et al., *Nat. Phys.*, 5, 675, 2009.)

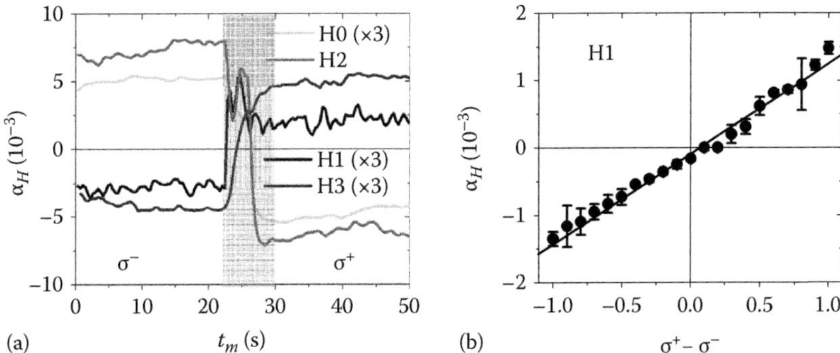

FIGURE 25.6 (See color insert.) (a) Spin-injection Hall effect Hall probe signals are plotted for Hall crosses H0H3 in the 2DEG channel. Gray region corresponds to the manual rotation of the λ/2 wave plate, which changes the helicity of the incident light and, therefore, the spin-polarization of injected electrons. (b) Hall angles measured at n-channel Hall crosses H1 (similar results with opposite sign occur for H2). The spin-injection Hall effect angles are linear in the degree of polarization, hence behaving phenomenologically as in the anomalous Hall effect. The laser spot is focused on the p–n junction and also the bias voltage, laser wavelength, and measurement temperature are the same as in Figure 25.5. (After Wunderlich, J. et al., *Nat. Phys.*, 5, 675, 2009.)

wave plate by 45°. The signals are recalculated to Hall angles, $\alpha_H = R_H/R_L$, whose magnitude 10^{-3}–10^{-2} is comparable to the anomalous Hall effect in conventional metal ferromagnets. The spatial variation of the sign of the Hall signals observed suggests a modulated spin dynamics in the length scale of microns. The scattering mean free path in this system is on the order of ~10–100 nm and the typical SO-coupling length to ~1 μm. The first length scale determines the onset of the anomalous Hall effect-like transverse charge accumulation due to skew scattering off SO-coupled impurity potentials (see theory section). The second length scale governs the spin precession about the internal SO field, which in this configuration tends to point in-plane and perpendicular to the electron momentum. In quasi 1D channels, electrons injected with out-of-plane spins would precess coherently about this internal SO field, but in wider channels (diffusive transport), the spin-coherence length is expected to be much shorter. As we will see in the next section, the coherence can also exceed micrometer scales in wider channels due to the additional Dresselhaus SO field originating from inversion asymmetry of GaAs. In Figure 25.6b the spin-injection Hall effect as a function of the degree of polarization at Hall cross H1 is shown. The linear dependence of the Hall angle on the polarization of injected electrons is analogous to the linear dependence on magnetization of the anomalous Hall effect in a ferromagnet.

Within the experiment, other important checks on the Hall symmetries were made in order to discard external effects for the observed phenomenology. Masked samples (only injection area open) show the same phenomenology (shown in Figure 25.7a) as the unmasked ones. Hence, the possibility that the observed effects arises from stray light from the focused polarized beam seems unlikely. Complete Hall symmetries of the transverse signals measured in our devices are confirmed by the experiments presented in Figure 25.7. Figure 25.7a shows data recorded in a sample with Hall crosses H2 and H3 fully covered by an insulating thin film and a metallic mask. The spin-injection Hall effect is observed only at reverse bias when electrons move from the illuminated aperture toward the n-channel Hall crosses. At forward bias, optically generated electrons are accelerated in the opposite direction and no Hall signals are detected independent of polarization.

The distinct features of the spin-injection Hall effect are highlighted by comparison to complementary data shown in Figure 25.7b. After etching the surface layers in the n-side of the p–n diode, Wunderlich et al. [1] were able to select wafers with residual sub-gap optical activity in an unmasked n-channel. The dark current generates zero Hall voltage at crosses H2 and H3 while clear (albeit weak) Hall signals are detected upon directly illuminating the crosses with intense circularly polarized light. The signals are attributed to the anomalous Hall effect since they occur inside the spin-generation area and, as expected, they are antisymmetric with respect to the current polarity. Also consistent with the anomalous Hall effect, they observe in these experiments Hall voltages with polarity depending only on current orientation and spin-polarization but with same signs on all irradiated Hall crosses measured. It contrasts the spin-injection Hall effect measurements of spin injection from the p–n junction in which the sign can alternate among the Hall crosses.

It is also interesting that when measuring within the p-region, consistently with expectations, no clear Hall signal was measured at the first p-channel Hall cross with the laser spot focused on top of the p–n junction when the measurements were taken simultaneously with the n-channel measurements. This confirms that it is the spin-decoherence of propagating holes rather than an inherent absence of the Hall effect in the 2DHG, which explains the negative result of measurements.

Wunderlich et al. [1] also found that the measurements in a sample that showed rectifying p–n junction characteristics at temperatures up to 240 K, hence, demonstrating that the spin-injection Hall effect is readily detectable at high temperatures. Together with the zero-bias operation shown in Figure 25.5c and linearity of the spin-injection Hall effect in the degree of circular

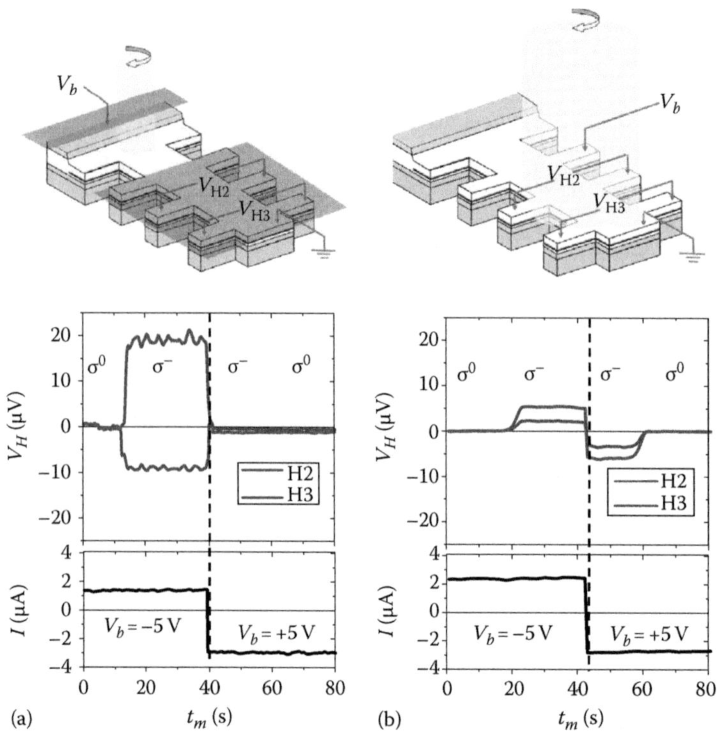

FIGURE 25.7 (See color insert.) (a) Spin-injection Hall effect measurements in a masked sample with linearly polarized light and circularly polarized light of a fixed helicity for opposite polarities of the optical current. Schematics of each experimental setup are shown in the upper panel. Middle panel shows that spin-injection Hall effect voltages are detected only at negative bias when spin-polarized electrons move from the illuminated aperture toward the measured Hall crosses H2 and H3 in the n-channel. The optical current is plotted in the lower panel. (b) Complementary measurements to (a) in an unmasked sample with only the n-channel biased and Hall crosses H2 and H3 directly illuminated with a 10× stronger light intensity as compared to (a). Weak anomalous Hall effect signals are detected in this case, which are antisymmetric with respect to the polarity of the current. Lower panel shows the optically generated part of the current. (From Wunderlich, J. et al., *Nat. Phys.*, 5, 675, 2009.)

polarization of the incident light, these characteristics represent the realization of the spin-photovoltaic effect in a nonmagnetic structure and demonstrate the utility of the device as an electrical polarimeter [52,57,58].

25.4 Theory of Spin-Injection Hall Effect

The previous section was limited to presenting the experimental phenomenology of the spin-injection Hall effect. As it may have become apparent to the reader, it is rather surprising that such large spin-coherence lengths are also associated with an apparent coherent precession across enormous length scales beyond what is expected in normal 2DEG. On the other hand, because of the particular parameters chosen in the experiments, this is a system that we can model rather precisely and, as we will see following text, it is highly anisotropic in this coherent behavior.

The theoretical approach is based on the observation that the micrometer length scale governing the spatial dependence of the nonequilibrium spin-polarization is much larger than the ~10–100 nm mean free path in our 2DEG, which governs the transport coefficients. This allows us to disregard the Hall signal on a first instance and first calculate the steady-state spin-polarization profile along the channel and only afterward consider the spin-injection Hall effect as a response to the local out-of-plane component of the polarization. Note that this would not be possible when these length scales are of similar order as they will be when mobilities increase and we exit the diffusive regime.

Our starting point is the description of GaAs near the Γ point with the effect of the valence bands taken into account through a two-band effective Hamiltonian; this can be achieved through the so-called Lowin transformation. In the presence of an electric potential $V(r)$, the corresponding 3D SO-coupling Hamiltonian reads

$$H_{3D\text{-}SO} = \left[\lambda^* \mathbf{s}\cdot(\boldsymbol{k}\times\nabla V(\boldsymbol{r}))\right] + \left[\mathcal{B}k_x\left(k_y^2 - k_z^2\right)\sigma_x + \text{cyclic permutations}\right], \quad (25.9)$$

where

σ are the Pauli spin matrices
$\boldsymbol{k}$ is the momentum of the electron
$\mathcal{B} \approx 10\,\text{eV Å}^3$
$\lambda^* = 5.3\,\text{Å}^2$ for GaAs [59,60]

Equation 25.9 together with the 2DEG confinement yields an effective 2D Rashba and Dresselhaus SO-coupled Hamiltonian [61,62],

$$H_{2DEG} = \frac{\hbar^2 k^2}{2m} + \alpha(k_y\sigma_x - k_x\sigma_y) + \beta(k_x\sigma_x - k_y\sigma_y)$$
$$+ V_{dis}(\boldsymbol{r}) + \lambda^* \mathbf{s}\cdot(\boldsymbol{k}\times\nabla V_{dis}(\boldsymbol{r})), \quad (25.10)$$

where

$m = 0.067m_e$

$\beta = -\mathcal{B}\langle k_z^2\rangle \approx -0.02\,\text{eV Å}$

$\alpha = e\lambda^* E_z \approx 0.01 - 0.03\,\text{eV Å}$ for the strength of the confining electric field

$eE_z \approx 2\text{–}5\times10^{-3}\,\text{eV/Å}$ pointing along the [001] direction

The strength of the confinement is obtained from a self-consistent Poisson-Schrödinger simulation of the conduction band profile of our GaAs/AlGaAs heterostructure [31]. Typically α can be tuned whereas β is a material-dependent parameter fixed by the choice of growth direction and, to a smaller extent, the degree of confinement.

25.4.1 Nonequillibrium Polarization Dynamics along the [1$\bar{1}$0] Channel

The realization of the original Datta–Das device concept in a purely Rashba SO-coupled system has been unsuccessful until recently due to spin-coherence issues; that is, in the required length scales in which transport is diffusive, no oscillating persistent precession states are present [63]. However, in a 2DEG where both Rashba and Dresselhaus SO coupling have similar strengths, a long-lived precessing excitation of the system has been shown to exist along a particular direction [37,62]. When α and β are equal in magnitude, the component of the spin along the [110] direction for $\alpha=-\beta$ or along the [1$\bar{1}$0] direction for $\alpha=\beta$ is a conserved quantity [61], as well as a precessing spin-wave (spin-helix) of wavelength $\lambda_{spin\text{-}helix} = \pi\hbar^2/(2m\alpha)$ in the direction perpendicular to the conserved spin component [62]. This spin-helix state has been observed through optical transient spin-grating experiments [37].

In the strong scattering regime of the structure considered, with αk_F and $\beta k_F \sim 0.5\,\text{meV}$, much smaller than the disorder scattering rate $\hbar/\tau \sim 5\,\text{meV}$, the system obeys a set of spin-charge diffusion equations [62] for arbitrary ratio of α and β:

$$\partial_t n = D\nabla^2 n + B_1\partial_x + S_{x-} - B_2\partial_{x-}S_{x+},$$

$$\partial_t S_{x+} = D\nabla^2\partial_t S_{x+} - B_2\partial_{x-}n - C_1\partial_{x+}S_z - T_1 S_{x+},$$

$$\partial_t S_{x-} = D\nabla^2\partial_t S_{x-} + B_1\partial_{x+}n - C_2\partial_{x-}S_z - T_2 S_{x-},$$

$$\partial_t S_z = D\nabla^2\partial_t S_z + C_2\partial_{x-}S_{x-} + C_1\partial_{x+}S_{x+} - (T_1+T_2)S_z,$$

where

x_+ and x_- correspond to the [110] and [1$\bar{1}$0] directions

$B_{1/2} = 2(\alpha \mp \beta)^2(\alpha \pm \beta)k_F^2\tau^2$

$T_{1/2} = (2/m)(\alpha \pm \beta)^2(k_F^2\tau/\hbar^2)$

$D = v_F^2\tau/2$

$C_{1/2}^2 = 4DT_{1/2}$

For the present device, where $\alpha \approx -\beta$, the 2DEG channel is patterned along the [1$\bar{1}$0] direction, which corresponds to the direction of the spin-helix propagation. Within this direction, the dynamics of S_{x-} and S_z couple through the diffusion equations above. Seeking steady-state solutions of the form exp $[qx_-]$ yields the transcendental equation $(Dq^2+T_2)(Dq^2+T_1+T_2) - C_2^2q^2 = 0$, which can be reduced to $q^4 + (\tilde{Q}_1^2 - 2\tilde{Q}_2^2)q^2 + \tilde{Q}_1^2\tilde{Q}_2^2 + \tilde{Q}_2^4 = 0$, where $\tilde{Q}_{1/2} \equiv \sqrt{T_{1/2}/D} = 2m|\alpha \pm \beta|/\hbar^2$. This yields a physical solution for $q = |q|\exp[i\theta]$ of

$$|q| = \left(\tilde{Q}_1^2\tilde{Q}_2^2 + \tilde{Q}_2^4\right)^{1/4} \text{ and } \theta = \frac{1}{2}\arctan\left(\frac{\sqrt{2\tilde{Q}_1^2\tilde{Q}_2^2 - \tilde{Q}_1^4/4}}{\tilde{Q}_2^2 - \tilde{Q}_1^2/2}\right). \quad (25.11)$$

The resulting damped spin precession of the out-of-plane polarization component for the parameter range of the device, where we have set $\beta = -0.02\,\text{eV Å}$ and varied α from -0.5β to -1.5β, is shown in Figure 25.8 in the main text. These results are in agreement with Monte-Carlo calculations on similar systems (modeling an InAs 2DEG) but with higher applied biases [64].

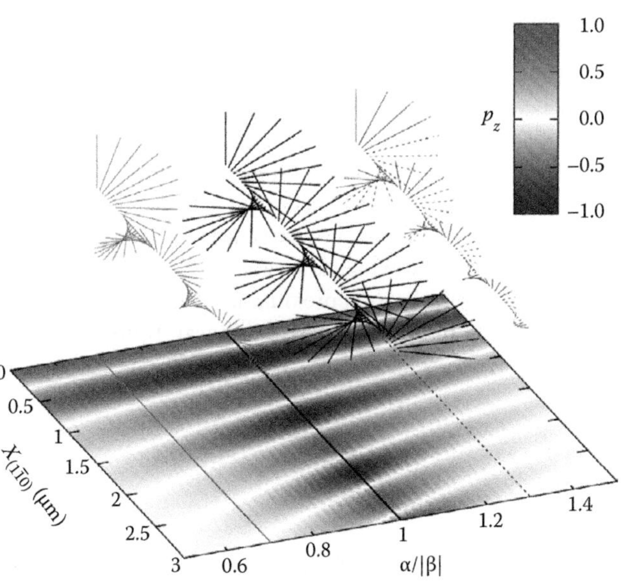

FIGURE 25.8 Steady-state solution of the spin precession for $\alpha/|\beta|$=0.7, 1.0, 1.3. The bottom panel indicates the precession of the polarization out-of-plane component for the parameter range of our device, where we have $\beta = -0.02\,\text{eV Å}$ and varied α from -0.5β to -1.5β.

In the Monte-Carlo calculations, longer decaying lengths where observed at higher biases. However, the present device is well within the linear regime with very low driving fields; this results in shorter decay length scales of the oscillations as compared to Ref. [64].

These results show good agreement with the steady-state variations (changes in the length scale of 1–2 μm) in the out-of-plane nonequilibrium polarization of our system observed indirectly through the Hall signals. We note that Monte-Carlo simulations including temperature broadening of the quasiparticle states confirm the validity of the analytical results up to the high temperatures used in the experiment.

25.4.2 Hall Effect

The Hall effect signal can be understood within the theory of the anomalous Hall effect. The contributions to the anomalous Hall effect in SO-coupled systems with nonzero polarization can be classified into two types: The first type arises from the SO-coupled quasiparticles interacting with the spin-independent disorder and the electric field, and the second type arises from the non-SO-coupled part of the quasiparticles scattering from the SO-coupled disorder potential (last term in Equation 25.10). The contributions of the first type have been recently studied in 2DEG ferromagnetic systems with Rashba SO coupling [65–68]. These have shown that, within the regime applicable to our present devices, the anomalous Hall effect contribution due to the intrinsic and side-jump mechanisms vanish even in the presence of moderate disorder. In addition, the skew-scattering contribution from this type of contribution is also small (with respect to the contribution shown in the following) and furthermore is not linear in polarization [65]. Hence, we can disregard the contributions of the first type within our devices.

This is not surprising since in our devices αk_F, $\beta k_F \ll \hbar/\tau$ and hence these contributions arising from the SO coupling of the bands are not expected to dominate. Instead, the observed signal originates from contributions of the second type, that is, from interactions with the SO-coupled part of the disorder [27,69]. Within this type, one contribution is due to the anisotropic scattering, the extrinsic skew scattering, and is obtained within the second Born approximation treatment of the collision integral in the semiclassical linear transport theory [27,69]:

$$|\sigma_{xy}|^{skew} = \frac{2\pi e^2\lambda^*}{\hbar^2} V_0 \tau n(n_\uparrow - n_\downarrow), \tag{25.12}$$

where $n = n_\uparrow + n_\downarrow$ is the density of photoexcited carriers with polarization $p_z = (n_\uparrow - n_\downarrow)/(n_\uparrow + n_\downarrow)$. Using the relation for the mobility $\mu = e\tau/m$ and the relation between n_i, V_0, and τ, $\hbar/\tau = n_i V_0^2 m/\hbar^2$ the extrinsic skew-scattering contribution to the Hall angle, $\alpha_H \equiv \rho_{xy}/\rho_{xx} \approx \sigma_{xy}/\sigma_{xx}$, can be written as

$$\begin{aligned}\alpha_H^{skew} &= 2\pi\lambda^* \sqrt{\frac{e}{\hbar n_i \mu}} n p_z (x_{[1\bar{1}0]}) \\ &= 2.44 \times 10^{-4} \frac{\lambda^*\left[\text{Å}^2\right](n_\uparrow - n_\downarrow)\left[10^{11}\,\text{cm}^{-2}\right]}{\sqrt{\mu\left[10^3\,\text{cm}^2/\text{V s}\right] n_i \left[10^{11}\,\text{cm}^{-2}\right]}} \\ &\sim 1.1 \times 10^{-3} p_z,\end{aligned} \tag{25.13}$$

where we have used $n = 2 \times 10^{11}$ cm^{-2}, p_z is the polarization, $\mu = 3 \times 10^3$ cm^2/V s, and $n_i = 2 \times 10^{11}$ cm^{-2}; the last estimate is introduced to give a lower bound to the Hall angle contribution within this model. In addition to this contribution, there also exists a side-jump scattering contribution in the SO-coupled disorder term given by

$$\alpha_H^{s-j} = \frac{2e^2\lambda^*}{\hbar\sigma_{xx}}(n_\uparrow - n_\downarrow) \tag{25.14}$$

$$= 3.0 \times 10^{-4} \frac{\lambda^*\left[\text{Å}^2\right]}{\mu\left[10^3\text{cm}^2/\text{V s}\right]} p_z \sim 5.3 \times 10^{-4} p_z. \tag{25.15}$$

As expected, this is an order of magnitude lower than the skew-scattering contribution within this system.

We can then combine this result with the results from the previous section to predict, in this diffusive regime, the resulting theoretical α_H along the $[1\bar{1}0]$ direction for the relevant range of Rashba and Dresselhaus parameters corresponding to the experimental structure [1]. This is shown in Figure 25.9. We have assumed a donor impurity density n_i on the order of the equilibrium density $n_{2DEG} = 2.5 \times 10^{11}$ cm^{-2} of the 2DEG in dark, which is an upper bound for the strength of the impurity scattering in our modulation-doped heterostructure and, therefore, a lower bound for the Hall angle. For the mobility of the injected electrons in the 2DEG channel one can consider the experimental value determined from ordinary Hall measurements without illumination, $\mu = 3 \times 10^3$ cm^2/V s. The density of photoexcited carriers of $n \approx 2 \times 10^{11}$ cm^{-2} was obtained from the measured longitudinal resistance between successive Hall probes under illumination assuming constant mobility.

The theory results shown in Figure 25.9 provide a semiquantitative account of the magnitude of the observed spin-injection Hall effect angle (~10^{-3}) and explain the linear dependence of the spin-injection Hall effect on the degree of spin-polarization of injected carriers. The calculations are also consistent with the experimentally inferred precession length on the order of a micrometer and the spin coherence exceeding micrometer length scales. We emphasize that the 2DEG in the strong disorder, weak SO-coupling regime realized in our experimental structures is a particularly favorable system for theoretically

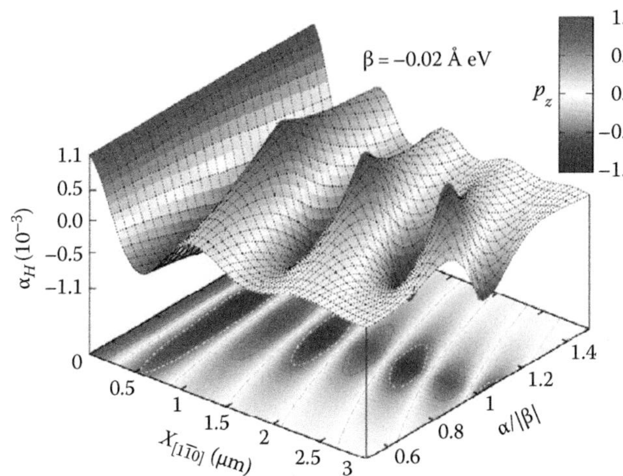

FIGURE 25.9 Microscopic theory of the spin-injection Hall effect assuming SO-coupled band structure parameters of the experimental 2DEG system. The calculated spin-precession and spin-coherence lengths and the magnitude of the Hall angles are consistent with experiment. The grayscale surface shows the proportionality between the Hall angle and the out-of-plane component of the spin-polarization. (From Wunderlich, J. et al., *Nat. Phys.*, 5, 675, 2009.)

establishing the presence of the spin-injection Hall effect. In this regime and for the simple band structure of the archetypal 2DEG, the spin-diffusion equations and the leading skew-scattering mechanism of the SO-coupling-induced Hall effect are well understood areas of the physics of quantum-relativistic spin-charge dynamics.

25.4.3 Combined Spin–Orbit and Zeeman Effects in External Magnetic Fields

The above-mentioned theoretical considerations help explain the observed phenomenology and predict other expectations, such as higher damped oscillations observed in the orthogonal direction for the given device. Another aspect that will affect these oscillations is the application of an external magnetic field whose consequences we consider next.

The spin dynamics at high temperatures and in external magnetic field can be studied theoretically by Monte-Carlo–Boltzmann simulations assuming the additional Zeeman term in the Hamiltonian, $g_e\mu_B\mathbf{B}\cdot\sigma/2$, where g_e is the g-factor of GaAs. The orbital effects of the magnetic field are included on a classical level by considering the Lorentz force acting on injected electrons between scattering events. To model the device, the system is chosen to have an $L \times W = 3\,\mu\text{m} \times 0.5\,\mu\text{m}$ rectangular 2DEG connected to a source at $x=0$ and a drain at $x=L$. The x-axis is along the $[1\bar{1}0]$ direction and the y-axis is along the [110] direction. The electron density is $n_{2DEG}=2.5\times10^{11}$ cm^{-2}. Fermi energy is $E_F=9$ eV and Fermi momentum is $k_F=0.125$ nm^{-1}. The Dresselhaus parameter $\beta=-0.02$ eV Å with the temperature of injected electrons is $T=300$ K. Note that temperature enters our calculations only through the Fermi distribution of electrons; we do not include temperature-dependent scattering effects like, for example, electron-phonon scattering. The results we obtain are only very weakly temperature dependent when comparing simulations at 1 and 300 K. Electrons are injected along the $[1\bar{1}0]$ direction, with initial spin-polarization along the [001] axis. In the simulations, only point-like impurities such that the mean free path is $L_mf=40$ nm were considered. The details of the Monte-Carlo–Boltzmann calculations can be found in Ref. [1].

Figure 25.10 shows the calculated spin-polarizations in the channel in magnetic fields oriented along the [001], $[1\bar{1}0]$, and [110] directions for the Rashba coupling parameter $\alpha=-0.5\beta$. (The effects of the magnetic field for $\alpha=-\beta$ are even less apparent than for the case of $|\alpha|\neq|\beta|$ shown in the figure.) Since the calculations do not include quantum orbital effects of the magnetic field, their validity is limited to fields $\lesssim 1$ T. In agreement with expectations for the present 2DEG with relatively large SO-coupling strength, the Zeeman term has only a minor effect on the spin-polarization profile along the channel even at ~1 T fields. For the out-of-plane and $[1\bar{1}0]$-oriented in-plane field, the Zeeman coupling causes a weak suppression of the spin coherence while for the [110]-oriented field it leads to a weak suppression of the spin coherence and a weak reduction or extension of the precession length depending on the sign of the magnetic field. The variations at magnetic fields $\lesssim 1$ T are not large enough to provide practical means for controlling the spin dynamics in the 2DEG channels. At higher magnetic fields, we expect an increasing role of the quantum orbital effects, which, due to the SO coupling, cannot be simply disentangled from the spin effects and their description requires more sophisticated theoretical modeling. Furthermore, the field dependence of the spin-current generation characteristics of the p–n junctions is expected to be significant at fields of several Tesla.

25.4.4 Prospectives of Spin-Injection Hall Effect

The spin-injection Hall effect in nonmagnetic semiconductors and the ability to tune independently the strengths of disorder and SO coupling in semiconductor structures open up new opportunities for resolving long-standing debates on the nature of spin-charge dynamics in the intriguing strong SO-coupling, weak disorder regime [66]. From the application perspective, the spin-injection Hall effect devices can be directly implemented as spin-photovoltaic cells and polarimeters, switches, invertors, and, due to the nondestructive nature of the spin-injection Hall effect, also as interconnects. We also foresee application of the spin-injection Hall effect in the Datta–Das [63] and other proposed spintronic transistor concepts [70]. An important next step toward a practical implementation in nanotechnology, as pointed out in Ref. [71], is the replacement of optical spin injection by other solid-state means of spin injection. These lightless devices utilizing the spin-injection Hall effect can be fabricated

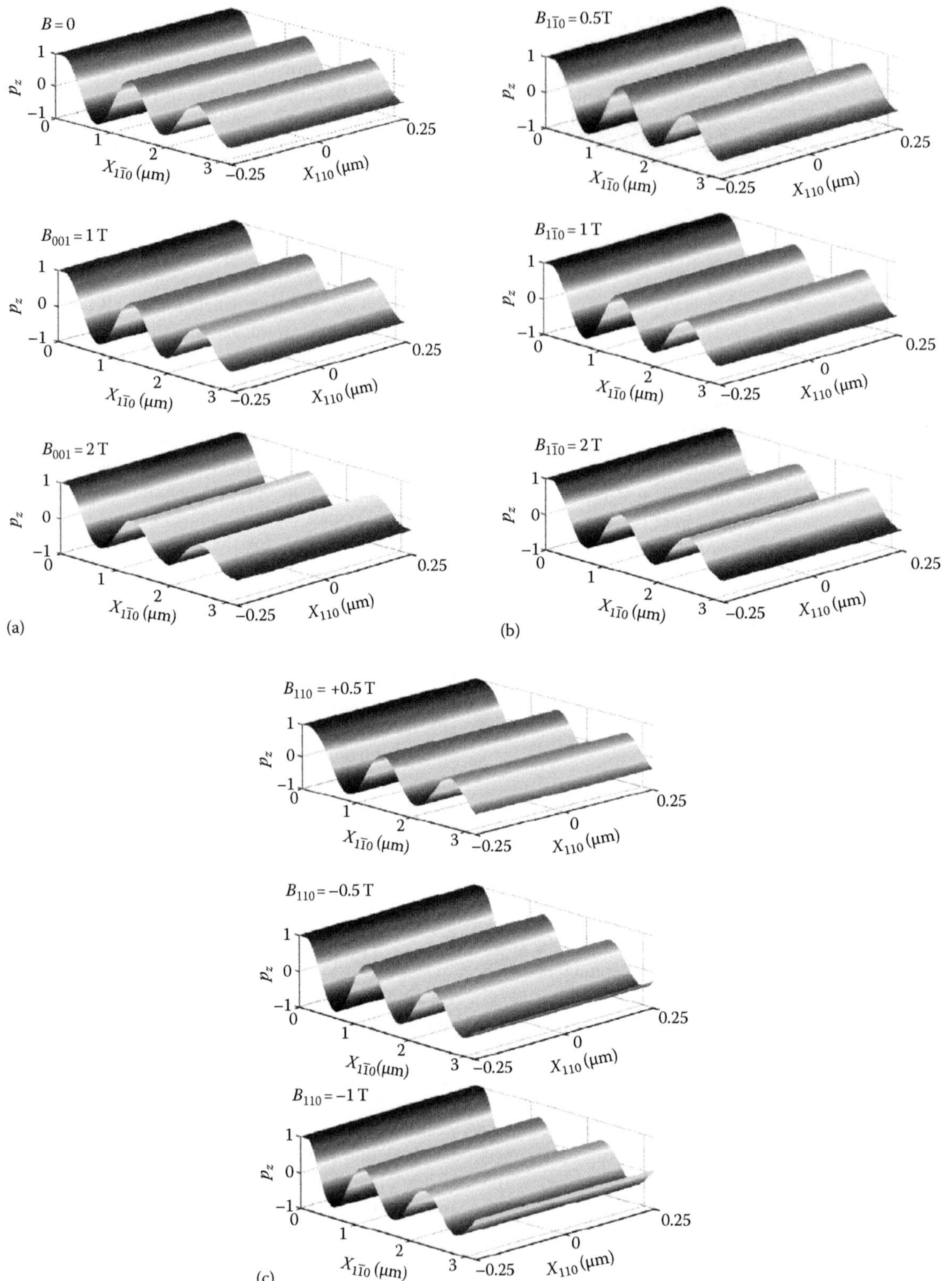

FIGURE 25.10 Monte-Carlo modeling of the out-of-plane component of the spin-polarization in the channel for $\alpha = -0.5\beta$ at 300 K. (a) Zero magnetic field and magnetic field oriented perpendicular to the plane of the 2DEG. (b) Magnetic field along the $[1\bar{1}0]$ in-plane direction (the 2DEG channel direction). (c) Magnetic field along the [110] in-plane direction (the 2DEG channel direction). Magnetic field causes a weak suppression of the spin coherence for these two field orientations. (After Wunderlich, J. et al., *Nat. Phys.*, 5, 675, 2009.)

in a broad range of materials including indirect-gap SiGe semiconductors [72]. Since the magnitude of the spin-injection Hall effect scales linearly with the SO-coupling strength, we expect ~100× weaker signals in the SiGe 2DEGs as compared to our measurements in the GaAs/AlGaAs, which is still readily detectable.

References

1. J. Wunderlich, A. C. Irvine, J. Sinova et al., Spin-injection Hall effect in a planar photovoltaic cell, *Nat. Phys.* **5**, 675 (2009).
2. N. Nagaosa, J. Sinova, S. Onoda, A. H. MacDonald, and N. P. Ong, Anomalous Hall effect, *Rev. Mod. Phys.* **82**, 1539 (2010).
3. J. Schliemann, Spin Hall effect, *Int. J. Mod. Phys. B* **20**, 1015 (2006).
4. S. Murakami, N. Nagaosa, and S. C. Zhang, Dissipationless quantum spin current at room temperature, *Science* **301**, 1348 (2003).
5. J. Sinova, D. Culcer, Q. Niu, N. A. Sinitsyn, T. Jungwirth, and A. H. MacDonald, Universal intrinsic spin Hall effect, *Phys. Rev. Lett.* **92**, 126603 (2004).
6. H.-A. Engel, E. I. Rashba, and B. I. Halperin, Theory of spin Hall effects in semiconductors, in *Handbook of Magnetism and Advanced Magnetic Materials*, H. K. S. Parkin (Ed.), John Wiley & Sons Ltd., Chichester, U.K., vol. 5, p. 2858, 2007.
7. E. M. Hankiewicz and G. Vignale, Spin-Hall effect and spin-Coulomb drag in doped semiconductors, *J. Phys.: Condens. Matter* **21**, 253202 (2009).
8. M. Koenig, H. Buhmann, L. W. Molenkamp et al., The quantum spin Hall effect: Theory and experiment, *J. Phys. Soc. Jpn.* **77**, 310 (2008).
9. B. Bernevig and S. Zhang, Quantum spin Hall effect, *Phys. Rev. Lett.* **96**, 0680 (2006).
10. C. L. Kane and E. J. Mele, Quantum spin Hall effect in graphene, *Phys. Rev. Lett.* **95**, 226801 (2005).
11. L. Berger, Side-jump mechanism for the Hall effect of ferromagnets, *Phys. Rev. B* **2**, 4559 (1970).
12. J. Smit, The spontaneous Hall effect in ferromagnetics-I, *Physica* **21**, 877 (1955).
13. N. A. Sinitsyn, A. H. MacDonald, T. Jungwirth, V. K. Dugaev, and J. Sinova, Anomalous Hall effect in a two-dimensional Dirac band: The link between the Kubo-Streda formula and the semiclassical Boltzmann equation approach, *Phys. Rev. B* **75**, 045315 (2007).
14. N. A. Sinitsyn, Semiclassical theories of the anomalous Hall effect, *J. Phys.: Condens. Matter* **20**, 023201 (2008).
15. R. Karplus and J. M. Luttinger, Hall effect in ferromagnetics, *Phys. Rev.* **95**, 1154 (1954).
16. M. Onoda and N. Nagaosa, Topological nature of anomalous Hall effect in ferromagnets, *J. Phys. Soc. Jpn.* **71**, 19 (2002).
17. T. Jungwirth, Q. Niu, and A. H. MacDonald, Anomalous Hall effect in ferromagnetic semiconductors, *Phys. Rev. Lett.* **88**, 207208 (2002).
18. G. Sundaram and Q. Niu, Wave-packet dynamics in slowly perturbed crystals: Gradient corrections and Berry-phase effects, *Phys. Rev. B* **59**, 14915 (1999).
19. M. Chang and Q. Niu, Berry phase, hyperorbits, and the Hofstadter spectrum: Semi-classical dynamics in magnetic Bloch bands, *Phys. Rev. B* **53**, 7010 (1996).
20. N. Mainkar, D. A. Browne, and J. Callaway, First-principles LCGO calculation of the magneto-optical properties of nickel and iron, *Phys. Rev. B* **53**, 3692 (1996).
21. G. Y. Guo and H. Ebert, Band-theoretical investigation of the magneto-optical Kerr effect in Fe and Co multilayers, *Phys. Rev. B* **51**, 12633 (1995).
22. F. D. M. Haldane, Berry curvature on the Fermi surface: Anomalous Hall effect as a topological Fermi-liquid property, *Phys. Rev. Lett.* **93**, 206602 (2004).
23. X. Wang, D. Vanderbilt, J. R. Yates, and I. Souza, Fermi-surface calculation of the anomalous Hall conductivity, *Phys. Rev. B* **76**, 195109 (2007).
24. M. P. Marder, *Condensed Matter Physics*, Wiley, New York, 2000.
25. J. Smit, The spontaneous Hall effect in ferromagnetics–II, *Physica* **24**, 39 (1958).
26. L. Berger, Influence of spin–orbit interaction on the transport processes in ferromagnetic nickel alloys, in the presence of a degeneracy of the 3d band, *Physica* **30**, 1141 (1964).
27. P. Nozieres and C. Lewiner, Simple theory of anomalous Hall effect in semiconductors, *J. Phys. (Paris)* **34**, 901 (1973).
28. M. I. D'yakonov and V. I. Perel', Possibility of orienting electron spins with current, *Sov. Phys. JETP* 467 (1971).
29. Y. K. Kato, R. C. Myers, A. C. Gossard, and D. D. Awschalom, Observation of the spin Hall effect in semiconductors, *Science* **306**, 1910 (2004).
30. V. Sih, R. Myers, Y. Kato, W. Lau, A. Gossard, and D. Awschalom, Spatial imaging of the spin Hall effect and current-induced polarization in two-dimensional electron gases, *Nat. Phys.* **1**, 31 (2005).
31. J. Wunderlich, B. Kaestner, J. Sinova, and T. Jungwirth, Experimental observation of the spin-Hall effect in a two dimensional spin–orbit coupled semiconductor system, *Phys. Rev. Lett.* **94**, 047204 (2005).
32. S. Valenzuela and M. Tinkham, Electrical detection of spin polarized currents: The spin-current induced Hall effect, *J. Appl. Phys.* **101**, 09B103 (2007).
33. S. Valenzuela and M. Tinkham, Direct electronic measurement of the spin Hall effect, *Nature* **442**, 176 (2006).
34. S. Valenzuela, Nonlocal spin detection, spin accumulation and the spin Hall effect, *Int. J. Mod. Phys. B* **23**, 2413 (2009).
35. E. Saitoh, M. Ueda, H. Miyajima, and G. Tatara, Conversion of spin current into charge current at room temperature: Inverse spin-Hall effect, *Appl. Phys. Lett.* **88**, 8250 (2006).

36. T. Kimura, Y. Otani, T. Sato, S. Takahashi, and S. Maekawa, Room temperature reversible spin Hall effect, *Phys. Rev. Lett.* **98**, 156601 (2007).
37. C. P. Weber, J. Orenstein, B. A. Bernevig, S.-C. Zhang, J. Stephens, and D. D. Awschalom, Non-diffusive spin dynamics in a two-dimensional electron gas, *Phys. Rev. Lett.* **98**, 076604 (2007).
38. N. P. Stern, S. Ghosh, G. Xiang, M. Zhu, N. Samarth, and D. D. Awschalom, Current-induced polarization and the spin Hall effect at room temperature, *Phys. Rev. Lett.* **97**, 126603 (2006).
39. N. P. Stern, D. W. Steuerman, S. Mack, A. C. Gossard, and D. D. Awschalom, Drift and diffusion of spins generated by the spin Hall effect, *Phys. Rev. Lett.* **91**, 6210 (2007).
40. B. Liu, J. Shi, W. Wang et al., Experimental observation of the inverse spin Hall effect at room temperature (2006), arXiv:cond-mat/0610150.
41. J. M. Kikkawa and D. D. Awschalom, Lateral drag of spin coherence in gallium arsenide, *Nature* **397**, 139 (1999).
42. S. A. Crooker, M. Furis, X. Lou, C. Adelmann, D. L. Smith, C. J. Palmstrøm, and P. A. Crowell, Imaging spin transport in lateral ferromagnet/semiconductor structures, *Science* **309**, 2191 (2005).
43. X. Lou, C. Adelmann, S. A. Crooker et al., Electrical detection of spin transport in lateral ferromagnet-semiconductor devices, *Nat. Phys.* **3**, 197 (2007).
44. R. Fiederling, M. Keim, G. Reuscher, W. Ossau, G. Schmidt, A. Waag, and L. W. Molenkamp, Injection and detection of a spin-polarized current in a light-emitting diode, *Nature* **402**, 787 (1999).
45. H. Ohno, Properties of ferromagnetic III–V semiconductors, *J. Magn. Magn. Mater.* **200**, 110 (1999).
46. H. J. Zhu, M. Ramsteiner, H. Kostial, M. Wassermeier, H. P. Schönherr, and K. H. Ploog, Room-temperature spin injection from Fe into GaAs, *Phys. Rev. Lett.* **87**, 016601 (2001).
47. X. Jiang, R. Wang, R. M. Shelby et al., Highly spin-polarized room-temperature tunnel injector for semiconductor spintronics using MgO(100), *Phys. Rev. Lett.* **94**, 056601 (2005).
48. J. N. Chazalviel, Spin-dependent Hall effect in semiconductors, *Phys. Rev. B* **11**, 3918 (1975).
49. H. Ohno, H. Munekata, T. Penney, S. von Molmar, and L. L. Chang, Magnetotransport properties of p-type (In,Mn) As diluted magnetic III–V semiconductors, *Phys. Rev. Lett.* **68**, 2664 (1992).
50. J. Cumings, L. S. Moore, H. T. Chou et al., Tunable anomalous Hall effect in a nonferromagnetic system, *Phys. Rev. Lett.* **96**, 196404 (2006).
51. M. I. Miah, Observation of the anomalous Hall effect in GaAs, *J. Phys. D: Appl. Phys.* **40**, 1659 (2007).
52. S. D. Ganichev, E. L. Ivchenko, S. N. Danilov et al., Conversion of spin into directed electric current in quantum wells, *Phys. Rev. Lett.* **86**, 4358 (2001).
53. P. R. Hammar and M. Johnson, Detection of spin-polarized electrons injected into a two-dimensional electron gas, *Phys. Rev. Lett.* **88**, 066806 (2002).
54. B. Huang, D. J. Monsma, and I. Appelbaum, Coherent spin transport through a 350 micron thick silicon wafer, *Phys. Rev. Lett.* **99**, 177209 (2007).
55. J. Wunderlich, D. Ravelosona, C. Chappert et al., Influence of geometry on domain wall propagation in a mesoscopic wire, *IEEE Trans. Magn.* **37**, 2104 (2001).
56. M. Yamanouchi, D. Chiba, F. Matsukura, and H. Ohno, Current-induced domain-wall switching in a ferromagnetic semiconductor structure, *Nature* **428**, 539 (2004).
57. I. Žutić, J. Fabian, and S. D. Sarma, Spin-polarized transport in inhomogeneous magnetic semiconductors: Theory of magnetic/nonmagnetic p–n junctions, *Phys. Rev. Lett.* **88**, 066603 (2002).
58. T. Kondo, J. ji Hayafuji, and H. Munekata, Investigation of spin voltaic effect in a pn heterojunction, *J. Appl. Phys.* **45**, L663 (2006).
59. W. Knap, C. Skierbiszewski, A. Zduniak et al., Weak antilocalization and spin precession in quantum wells, *Phys. Rev. B* **53**, 3912 (1996).
60. R. Winkler, *Spin–Orbit Coupling Effects in Two-Dimensional Electron and Hole Systems*, Spinger-Verlag, New York, 2003.
61. J. Schliemann, J. C. Egues, and D. Loss, Nonballistic spin-field-effect transistor, *Phys. Rev. Lett.* **90**, 146801 (2003).
62. B. A. Bernevig, T. L. Hughes, and S.-C. Zhang, Quantum spin Hall effect and topological phase transition in HgTe quantum wells, *Science* **314**, 1757 (2006).
63. S. Datta and B. Das, Electronic analog of the electro-optic modulator, *Appl. Phys. Lett.* **56**, 665 (1990).
64. M. Ohno and K. Yoh, Datta-Das-type spin-field-effect transistor in the nonballistic regime, *Phys. Rev. B* **77**, 045323 (2008).
65. A. A. Kovalev, K. Vyborny, and J. Sinova, Hybrid skew scattering regime of the anomalous Hall effect in Rashba systems: Unifying Keldysh, Boltzmann, and Kubo formalisms, *Phys. Rev. B* **78**, 41305 (2008).
66. M. Borunda, T. Nunner, T. Luck et al., Absence of skew scattering in two-dimensional systems: Testing the origins of the anomalous Hall effect, *Phys. Rev. Lett.* **99**, 066604 (2007).
67. T. S. Nunner, N. A. Sinitsyn, M. F. Borunda et al., Anomalous Hall effect in a two-dimensional electron gas, *Phys. Rev. B* **76**, 235312 (2007).
68. S. Onoda, N. Sugimoto, and N. Nagaosa, Quantum transport theory of anomalous electric, thermoelectric, and thermal Hall effects in ferromagnets, *Phys. Rev. B* **77**, 165103 (2008).
69. A. Crépieux and P. Bruno, Theory of the anomalous Hall effect from the Kubo formula and the Dirac equation, *Phys. Rev. B* **64**, 014416 (2001).
70. I. Žutić, J. Fabian, and S. Das Sarma, Spintronics: Fundamentals and applications, *Rev. Mod. Phys.* **76**, 323 (2004).
71. I. Žutić, Spins take sides, *Nat. Phys.* **5**, 630 (2009).
72. I. Žutić, J. Fabian, and S. C. Erwin, Spin injection and detection in silicon, *Phys. Rev. Lett.* **97**, 026602 (2006).

V

Spin Transport and Magnetism at the Nanoscale

26

Spin-Polarized Scanning Tunneling Microscopy

Matthias Bode
University of Würzburg

Fueled by the ever increasing data density in magnetic storage technology and the need for a better understanding of the physical properties of magnetic nanostructures, there exists a strong demand for high-resolution magnetically sensitive microscopy techniques. The technique with the highest available resolution is spin-polarized scanning tunneling microscopy (SP-STM), which combines the atomic-resolution capability of conventional STMs with spin sensitivity by making use of the tunneling magnetoresistance (TMR) effect between a magnetic tip and the magnetic sample surface. Beyond the investigation of ferromagnetic surfaces, thin films, and epitaxial nanostructures with unforeseen precision, it also allows the achievement of a long-standing dream, that is, the real-space imaging of atomic spins in antiferromagnetic surfaces. Furthermore, the atomic spin resolution capability of SP-STM allowed for the discovery and investigation of complex spin structures like frustrated triangular structures and helical spin spirals, which were hypothesized but could not be imaged directly before.

26.1 Introduction

In Part III, the spin transport in magnetic tunnel junctions, mostly between ferromagnetic layers separated by an insulating barrier, has been discussed. In these planar devices, the tunneling current is not localized but is—if the insulating barrier is sufficiently smooth—equally distributed over the area of the entire tunnel junction. Inspired by the extremely short decay length of electron tunneling across materials with a sufficiently high barrier, such as oxidic insulators, Binnig and Rohrer invented the STM in the early 1980s. In this technique, a sharp tip is brought close to a sample resulting in a local tunneling current across the vacuum barrier between the tip's foremost atom and the sample surface. With their atomic-resolution imaging capability, STM and numerous derived spectroscopy modes soon revolutionized surface science. These spectroscopy modes allow for the direct correlation between structural and a variety of electronic sample properties, such as the local work function [1–3], the analysis of image states [1,4], and the analysis of Shockley-like [5,6] or exchange-split surface states [7–9].

In close analogy to the development of spin-polarized electron tunneling in planar junctions, where the normal metal electrodes are replaced by ferromagnetic electrodes to achieve spin sensitivity, it was an obvious approach to apply the same principle idea to STM: When scanning a magnetic surface with a magnetic tip, the TMR effect should give rise to a characteristic variation of the tunneling current, the differential conductance, or other parameters that can then be analyzed to map the sample's spin structure down to the atomic level [10]. Although the principle appears to be straightforward, several obstacles had to be overcome: What is a suitable tip material? What is the magnetization direction at the apex of the tip? Can the stray field or exchange force of the tip manipulate the pristine domain structure of the sample? Is it possible to achieve atomic spin resolution? Can antiferromagnetic tips be used for imaging?

As so often in science, the answers to these questions, which were largely unknown 10 years ago, seem to be quite obvious in hind side. During the past two decades spin-polarized STM has been developed into a mature magnetically sensitive imaging tool. For the first time it opened the door for spin-sensitive imaging with a spatial resolution well below 10 nm. This enabled several breakthroughs as follows: the imaging of magnetic hysteresis at the nanoscale [11], the discovery of ultra-narrow domain walls [12], the direct observation of the single-domain limit in nanomagnets [13], and the analysis of the internal spin structure of magnetic vortex cores [14]. In fact, atomic spin resolution is nowadays routinely achieved allowing SP-STM the unrivaled investigation of antiferromagnetic surfaces [15–17]. This chapter will give a short overview of the existing capabilities of spin-polarized STM and how it is applied for mapping surface magnetic domains and spin structures.

26.2 Contrast Mechanism and Modes of Operation

As mentioned earlier, the contrast mechanism of spin-polarized STM bases on the TMR effect. A thorough theoretical description was developed by Wortmann et al. [18] by generalizing the Tersoff–Hamann model originally developed for conventional, that is, spin-averaged STM [19,20]. In a more recent theoretical approach by Heinze [21], the independent-orbital model is applied to approximate the surface electronic structure. In this approach, the computational demand of ab initio calculations is significantly reduced by considering the arrangement of the magnetic surface moments only while ignoring the accurate surface electronic structure. Especially for single-element films, calculated SP-STM images obtained by the independent-orbital method were found to be in excellent agreement with both more demanding theoretical data and experiments performed on periodic collinear as well as noncollinear magnetic spin structures. In both methods, ab initio calculations and the independent-orbital model, the tunneling current I at any given tip position $\vec{r}_T$ and bias voltage U is separated into a spin-averaged and a spin-dependent contribution, I_0 and I_{SP}, respectively:

$$I(\vec{r}_T,U,\theta) = I_0(\vec{r}_T,U) + I_{SP}(\vec{r}_T,U,\theta). \tag{26.1}$$

Here, the magnetization vectors of tip and sample, $\vec{m}_T$ and $\vec{m}_S$, include the angle θ. Using Bardeen's tunneling formalism, the tunneling current can be written by the following expression:

$$I(\vec{r}_T,U,\theta) = \frac{4\pi^3 C^2 \hbar^3 e}{\kappa^2 m^2}\left[n_T \tilde{n}_S(\vec{r}_T,U) + \vec{m}_T \tilde{\vec{m}}_S(\vec{r}_T,U)\right], \tag{26.2}$$

where

n_T is the non-spin-polarized local density of states (LDOS) at the tip apex

$\tilde{n}_S$ is the energy-integrated LDOS of the sample

$\vec{m}_T$ and $\tilde{\vec{m}}_S$ are the corresponding vectors of the (energy-integrated) spin-polarized (or magnetic) LDOS:

$$\tilde{\vec{m}}_S(\vec{r}_T,U) = \int_{E=E_F}^{E=eU} \vec{m}_S(\vec{r}_T,E)dE, \tag{26.3}$$

with

$$\vec{m}_S = \sum \delta(E_\mu - E)\mathbf{y}_\mu^{S\dagger}(\vec{r}_T)\sigma\mathbf{y}_\mu^S(\vec{r}_T). \tag{26.4}$$

$\mathbf{y}^S$ denotes the spinor of the sample wave function

$$\mathbf{y}_\mu^S = \begin{pmatrix} \psi_{\mu\uparrow}^S \\ \psi_{\mu\downarrow}^S \end{pmatrix} \tag{26.5}$$

and σ is Pauli's spin matrix.

The interpretation of Equation 26.2 is relatively simple and very similar to Slonczewski's model [22] for planar junctions, namely, the spin-dependent contribution to the tunneling current, I_{SP}, scales with the projection of $\tilde{\vec{m}}_S$ onto $\vec{m}_T$, or, in other words, with the cosine of the angle included between the magnetization directions of the two electrodes, $\cos\theta$.

More importantly, Wortmann et al. [18] gave a detailed theoretical description of the three modes of operation that are usually applied in SP-STM. First to mention is the constant-current mode where a feedback circuit acts on the z-component of the piezoelectric actuator in order to stabilize the tunneling current at a predetermined value during the lateral scanning process. Because of the very short decay length of electron tunneling across a vacuum barrier (typically of the order of Å), however, even a TMR effect of 100% will result in a relatively small variation of the tip height on the order of a few 10 pm (1 pm = 10^{-12} m) only [23,24]. Therefore, as will be shown in Section 26.5, the constant-current mode is only suited to study atomic-scale spin structures on extremely flat surfaces. In contrast, the separation of magnetically induced height variations from structural or electronic inhomogeneities that are not related to the surface magnetic structure is almost impossible.

The difficulty of separating topographic, electronic, and magnetic information can be overcome by adding a small modulation U_{mod} to the bias voltage U and detecting the resulting current modulation by lock-in technique [25]. Usual modulation amplitudes amount to about ten millivolts. For theoretical assessment of this so-called differential conductance, or dI/dU, mode, the simplifying assumption of a tip with a featureless electronic structure has to be made. While a spin-averaged measurement of the differential conductivity, $\mathrm{d}I/\mathrm{d}U_0(\vec{r}_T, U) \propto n_S$, gives access to the sample's non-spin-polarized local density of sample states n_S [19,26], the application of magnetic tips leads to a spin-polarized contribution to the differential conductance, which can be written as [18]

$$\frac{\mathrm{d}I}{\mathrm{d}U}(\vec{r}_T,U) = \frac{\mathrm{d}I}{\mathrm{d}U}(\vec{r}_T,U)_0 + \frac{\mathrm{d}I}{\mathrm{d}U}(\vec{r}_T,U)_{SP}$$
$$\propto n_T n_S(\vec{r}_T, E_F + eU) + \vec{m}_T \vec{m}_S(\vec{r}_T, E_F + eU). \tag{26.6}$$

The dI/dU mode detects the spin polarization within a narrow energy interval ΔE around $E_F + eU$. Although this mode has been applied to surfaces that exhibit spin-polarized surface states in proof-of-principle experiments [23,25], it is important to stress that no particular surface electronic structure is required. As long as the d*I*/d*U* signal can be measured simultaneously with the topographic information, the imbalance of spin-up and spin-down states, which is intrinsic to any magnetic material, allows for an efficient separation between topographic, electronic, and magnetic information. It is important to note, however, that the spin polarization of the magnetic LDOS, $\vec{m}_S$, not only changes in size but may also change its sign if different energy intervals ΔE are compared. For example, the surface (and also the tip) may exhibit a positive spin polarization in one energy interval but a negative spin polarization in another [25,27]. Consequently, a high d*I*/d*U* signal does not imply that the magnetization directions of both electrodes are parallel but rather that the magnetic LDOS in both electrodes have the same sign. As we will see in the next chapter, the differential conductance mode allows for a rather straightforward imaging of magnetic domains at high spatial resolution on surfaces with homogeneous electronic properties. On surfaces with an inhomogeneous electronic structure, however, the separation of spin-averaged electronic from magnetic effects is more challenging. Only by scanning at voltages where the spin asymmetry, as defined by

$$A = \frac{\mathrm{d}I/\mathrm{d}U_{\uparrow\downarrow} - \mathrm{d}I/\mathrm{d}U_{\uparrow\uparrow}}{\mathrm{d}I/\mathrm{d}U_{\uparrow\downarrow} + \mathrm{d}I/\mathrm{d}U_{\uparrow\uparrow}}, \tag{26.7}$$

vanishes and becomes maximal, the magnetic domain structure can be retrieved [28].

An alternative method for separating the magnetic contributions is based an approach originally proposed by Johnson and Clarke [29], which was later refined and made useful for the imaging of magnetic domains by Wulfhekel and Kirschner [30]. In this experimental setup, the tip magnetization $\vec{m}_T$ is periodically switched back and forth by a small coil wrapped around the tip shaft. With the modulation frequency exceeding the cutoff frequency of the feedback loop, the magnetic contribution to the tunneling current is no longer leveled out by the feedback circuit. Differentiation of Equation 26.2 with respect to $\vec{m}_T$ leads to [18]

$$\frac{\mathrm{d}I}{\mathrm{d}\vec{m}_T}(\vec{r}_T) \propto \vec{m}_S(U), \tag{26.8}$$

that is, the resulting modulation of the tunneling current in this so-called local tunneling magnetoresistance, or TMR, mode is proportional to the local magnetization of the sample at the energy corresponding to the particular bias voltage *U*. In the experiment the TMR signal is detected by lock-in technique.

In fact, the voltage dependence of spin-polarized STM is distinctly different from planar junctions. While the TMR effect of planar junctions generally decrease continuously with increasing bias voltage, pronounced band structure effects were observed in the STM geometry [31,32]. A particular showcase is the Co(0001) surface, which, at the center of the Brillouin zone, is dominated by two spin-split minority bands resulting in a virtually energy-independent polarization, as shown in Figure 26.1a. Only for small tip-sample, distances state with a significant $\vec{k}_{\|}$ but lower polarization may contribute to the tunneling current, which leads to a dip in the TMR signal at ≈200 meV (Figure 26.1b). These STM-based experiments with a vacuum barrier lead to the conclusion that the so-called zero-bias anomaly in planar tunnel junctions with insulator barriers can be attributed to defect scattering in the barrier, rather than to magnon creation or spin excitations at the interfaces [32].

Even though the TMR mode of operation effectively separates structural and spin-averaged electronic contributions from magnetic effects, data interpretation of measurements taken

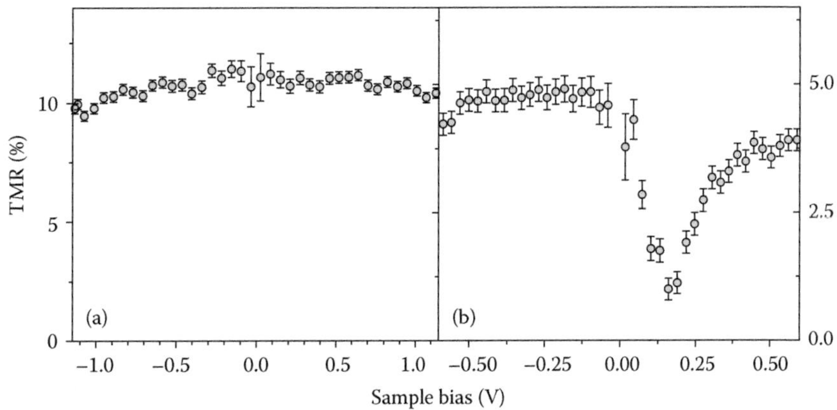

FIGURE 26.1 Bias voltage–dependent tunnel magnetoresistance δ of a clean Co(0001) surface. The data were obtained at room temperature with a bulk magnetic tip stabilized at (a) 1 V, 1 nA and (b) 100 mV, 1 nA. In contrast to experiments with planar tunnel junctions, where the TMR effect monotonously decreases with increasing bias voltage, spin-polarized tunneling across a vacuum barrier reveals pronounced band structure effects. The constant polarization is a result of the particular surface band structure of Co(0001), which is dominated by two spin-split minority bands at the center of the Brillouin zone resulting in a virtually energy-independent polarization. (Courtesy of Ding, H.F. et al., *Phys. Rev. Lett.*, 90, 116603, 2003.)

on inhomogeneous surfaces still remains challenging. As mentioned earlier, this is caused by the fact that in general, the spin polarization of the magnetic LDOS $\widetilde{m}_S$ is different for different materials and varies or even changes sign in different energy ranges.

26.3 Tip Materials

Already the initial proposal by Pierce [10] included two basic concepts for obtaining spin sensitivity, namely, the use of magnetic and of GaAs tips. While magnetic tips exhibit an intrinsic spin polarization at the Fermi level, it was planned to generate a spin-polarized density of states (DOS) in GaAs tips by illumination with circularly polarized light. The physical principle is based on the direct band-gap properties of GaAs together with the 0.34 eV spin–orbit splitting of the p-like valance band. The latter leads to a fourfold degenerate $p_{3/2}$-level and a twofold degenerate $p_{1/2}$-level. While light with an energy slightly above the energy gap $E_g = 1.52$ eV of p-GaAs can excite electron from $p_{3/2}$-level into the conduction band, the effective band gap for $p_{1/2}$-level states is too large. For circularly polarized light, the optical selection rule $\Delta m_j = m_f - m_i = \pm 1$, where $m_{f,i}$ is the angular momentum of the signal and the initial state, for σ^+-light the relative transmission probability into $m_j = -1/2$ states is three times higher than into $m_j = +1/2$ states (and vice versa for σ^{-1}-light). As a result, the theoretical limit of the electron spin polarization in photoemission is $P_{GaAs} = \pm 50\%$ [33]. One particularly compelling advantage of GaAs tips would be the complete absence of magnetic material in the tip that might potentially interact with the sample's magnetization. Even though a similar concept is widely used for the generation of spin-polarized electrons from planar GaAs wafers [33], experiments with GaAs tips on magnetic surfaces turned out to be plagued by competing spurious effects, such as the Faraday effect, which could never be reliably eliminated [34–38].

In contrast, magnetic tips were successfully used by different research groups on a large variety of sample systems in the three modes of operation described in the previous section. Two types of magnetic tips were used: Magnetic thin film tips with a static magnetization in the dI/dU mode [25] and amorphous low-remanence tips in the local TMR mode [30]. In order to obtain tips with either the out-of-plane or the in-plane magnetization direction in the TMR mode, tip shapes were engineered by making use of the shape dependence of the magnetostatic anisotropy. While a tapered magnetic wire is out-of-plane sensitive as it is preferentially magnetized along the wire axis to reduce its magnetostatic field, the preparation of in-plane sensitive bulk tip requires a more sophisticated approach. Schlickum et al. [39] obtained in-plane magnetic contrast by electrochemically etching a ring of about 1 mm diameter from a thin foil. Similar to the wire, by periodically switching the magnetization of the ring with a coil wrapped around it, the spin-dependent tunneling current between the ring and a spin-polarized sample can be measured. The magnetization direction of TMR tips can be reliably predicted as the shape anisotropy is by far the dominant anisotropy contribution. In spite of the above-mentioned advantages of the local TMR mode, such as the clear separation of magnetic contributions to the tunneling current, and some compelling results [39] this mode is not widely applied any more. Possible reasons are the relative large amount of magnetic material that is brought in very close vicinity to the tip and that—in spite of attempts to reduce the stray magnetic field by using low-remanence material—might modify the sample's domain structure and issues with magnetostriction.

Another approach for measurements performed in the dI/dU mode is based on thin film tips that are grown in situ under ultrahigh vacuum conditions. Thin film tips are usually prepared by deposition of magnetic material onto an electrochemically etched W tip. Upon etching, the tip is introduced into the vacuum system via a load lock. Early experiments indicated that films evaporated onto untreated tips are mechanically unstable and can easily be lost when approaching the sample surface or while scanning [40]. It was found that the film stability can be improved significantly by briefly heating the tip by electron bombardment prior to deposition to about 2200 K. The high-temperature ash removes contaminations, the presence of which seems to weaken the interfacial sticking and also removes potential remains of previous magnetic coatings. However, it also melts the tip apex. Thin film tips have several advantages over conventional bulk magnetic tips. First, the amount of magnetic material is drastically reduced, which helps to minimize the accidental modification of the sample's magnetic domain structure. Second, if refractory metals are used as a tip material, the nonmagnetic core of the tip can be recycled many times. As evidenced by the scanning electron microscopy images shown in Figure 26.2, the melting and subsequent cooling typically result in relatively blunt tips with a typical diameter of about 1 μm [41]. Past results show that the magnetization direction of such tips is no longer governed by the shape anisotropy due to the tapered shape of the tip but rather by material-specific parameters such as the surface and interface anisotropies of the material system, that is, the tip material and the thin film. For example, Fe-coated tips with a film thickness of about 10 atomic layers (AL) are usually in-plane sensitive. In contrast, Gd, GdFe alloy, and thin Cr film tips (2545 AL) are mostly out-of-plane sensitive. These results are largely consistent with the easy magnetization directions of equivalent films on flat W(110) substrates [42,43]. However, as different contributions to the anisotropy may also play a role, the

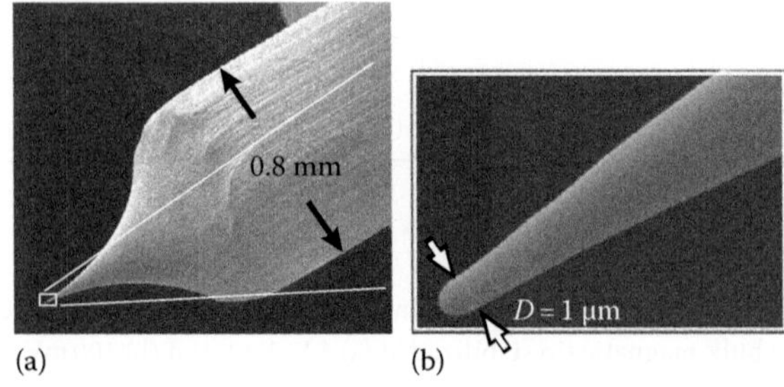

FIGURE 26.2 Scanning electron microscopy images of a flashed W tip at (a) medium and (b) higher magnification revealing a tip diameter of about 1 μm.

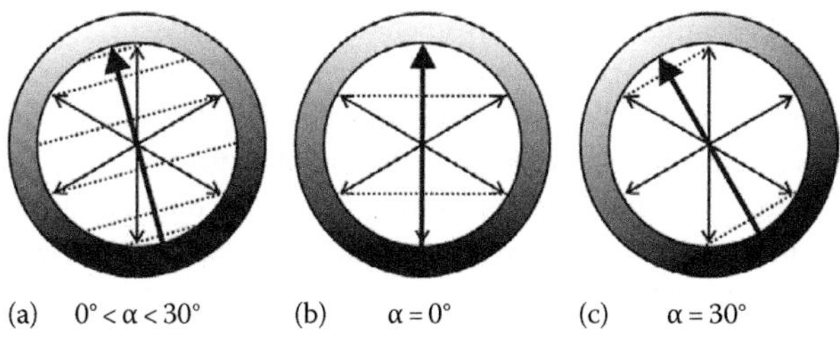

FIGURE 26.3 Schematic representation of the possible orientations of the tip magnetization $\vec{m}_t$ (thick arrow) with respect to the six possible easy axes magnetization directions $\vec{m}_s$ of Dy (thin arrows): (a) $0° < \alpha < 30°$, (b) $\alpha = 30°$, and (c) $\alpha = 0°$, resulting in six, four, and three contrast levels, respectively. α denotes the azimuthal angle between $\vec{m}_t$ and $\vec{m}_s$ of the domain with the largest dI/dU signal.

magnetization direction of thin film tips is less predictable than that of the engineered TMR tips mentioned above and canted magnetization directions are frequently observed.

As compared to the TMR mode mentioned above, the main advantage of the dI/dU mode is probably that antiferromagnetic coatings can be used. By studying the magnetic coercivity and saturation of thin iron films on W(110), Kubetzka et al. [44] could show that even thin film ferromagnetic tips lead to unavoidable distortions of the film's intrinsic domain and domain wall structure. Such problems can be avoided by the use of antiferromagnetic tips, as, for example, prepared by the evaporation of Cr onto the tip apex. Since antiferromagnets are compensated, they exhibit no stray field.

In agreement with Slonczewski's result for planar magnetic tunnel junctions [22], Equation 26.6 shows that the magnetic contrast in dI/dU maps is proportional to the projection of $\vec{m}_s$ onto $\vec{m}_t$. Since the sample system we focus on in the next section is a hcp(0001) surface, the following discussion will focus on a system with three easy axes, that is, six equivalent magnetic orientations. This opens up three principal cases of the relative orientation between the tip and the sample magnetization. For example, it may occur that the tip magnetization $\vec{m}_t$ lies along the sample's easy (Figure 26.3b) or hard axis (Figure 26.3c). In these cases, the angle between $\vec{m}_t$ and $\vec{m}_s$ of the domain with the largest dI/dU signal is $\alpha = 0°$ or $\alpha = 30°$, respectively. As symbolized by hatched lines, this leads to a reduced number of contrast levels since in each case, magnetization directions can be found that are mirror-symmetric with respect to $\vec{m}_t$ and therefore result in the same dI/dU signal, namely, we expect four contrast levels if $\alpha = 0°$ and only three contrast levels if $\alpha = 30°$.

26.4 Imaging Magnetic Surfaces

In the following, the imaging of different magnetic surfaces will be discussed, ranging from the domain structure of a ferromagnet, the imaging of so-called topological antiferromagnetism in a layered antiferromagnet, to atomic-scale spin structures of antiferromagnetically coupled nearest neighbor atoms and complex spin spirals. In the first few examples, the contrast mechanism will be revisited in detail and the strength and limitations of different modes of operation will be discussed. At the end of this section, we will see that these enhanced magnetic imaging capabilities lead to new discoveries, such as the observation of complex spin structures in ultrathin films.

26.4.1 Layered Antiferromagnets

Since the very beginning of SP-STM the Cr(001) surface has been a favorite test sample because its topography is closely linked to surface magnetic structure. Although Cr is generally known to be an antiferromagnet, theoretical investigations performed in the 1980s showed that the topmost layer of the Cr(001) surface is ferromagnetically ordered with an enhanced magnetic moment [45,46]. Yet, Blügel et al. recognized that the antiferromagnetic state of Cr prevails and that the magnetization direction of atomically at Cr(001) planes has to alternate at monoatomic step edges. To illustrate the intimate correlation between the surface topography of Cr(001) and its magnetic structure, Blügel et al. coined the term "topological antiferromagnetism" [45].

Obviously, the surfaces of topological antiferromagnets such as Cr(001) have a vanishing magnetic moment on the macroscopic scale because the moments of adjacent Cr(001) terraces cancel. As long as the surface roughness gives rise to thickness fluctuations of ≥1 AL experimental techniques that average over areas that are large compared to the terrace width show no magnetic signal. Only in some few examples Cr(001) films could be grown with sufficiently smooth interfaces such that oscillations of the TMR effect could successfully be observed by spatially averaging measurements of wedge-shaped magnetic tunnel junctions [47,48] (see also Part III of this book). Accordingly, most magnetically sensitive experimental techniques are not feasible for the investigation of topological antiferromagnetic surfaces. For the development of SP-STM and spectroscopy, however, the nanoscale one-to-one correlation between topography and magnetism of Cr(001) represents an advantage as it simplifies the save identification of magnetic contrasts.

In a pioneering experiment, Wiesendanger et al. [24] compared line sections measured on Cr(001) with nonmagnetic W and ferromagnetic CrO_2. These early experiments, which were qualitatively confirmed by Kleiber et al. (cf. Figure 2c in Ref. [23]), revealed that the apparent step height depends on the tip material. While a step height of 0.14 nm, which is consistent with the lattice constant of Cr was found with a W tip alternating apparent step heights were observed with a magnetic CrO_2 tip. This result was explained by the contribution of spin-polarized tunneling, which, due to the constant-current mode of operation, leads to magnetization direction–dependent tip-sample distance.

As mentioned in the previous section, SP-STM operated in the constant-current mode suffers from the fact that all contributions to the signal, topography, electronic, and magnetic properties, are inseparable mingled into one data channel, that is, the vertical position of the tip (z-signal). While this mode allows for high-resolution studies of homogeneous and extremely flat surfaces, such as the topological antiferromagnet Cr(001) or for atomic-resolution studies of single-element antiferromagnets

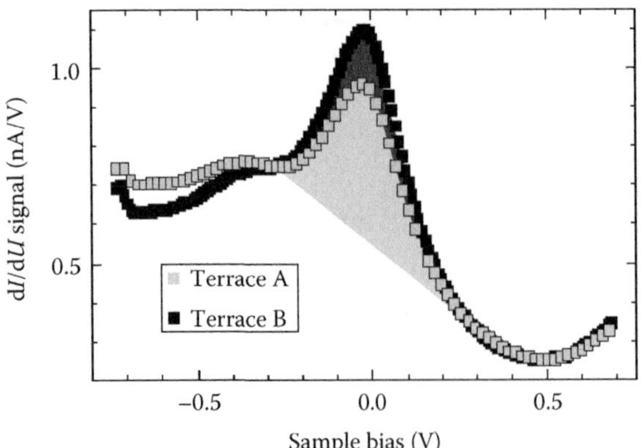

FIGURE 26.4 Tunneling spectra of Cr(001) as measured at room temperature with an Fe-coated magnetic tip on adjacent terraces. Due to the TMR effect different peak heights can be observed.

(discussed in Section 26.5), it is unsuitable on strongly corrugated or inhomogeneous surfaces.

This shortcoming can be overcome by SP-STM. The origin of the contrast can be understood from the Cr(001) tunneling spectra shown in Figure 26.4. The spectra show a peak just at the Fermi level E_F, which is known to be characteristic for Cr(001) [49] even though it is still under discussion whether the origin of this peak is a surface state [49–51] or an orbital Kondo resonance [52]. As long as nonmagnetic probe tips are employed, the spectra of all terraces, irrespective of their magnetization direction, are identical (not shown here). The spectra of Figure 26.4, however, were taken with an Fe-coated probe tip. The magnetoresistance effects between tip and sample leads to striking quantitative differences between spectra taken on adjacent terraces. While the peak position remains essentially identical, the peak intensity varies by about 10%–30%. This behavior is typical for spin-polarized tunneling with the STM: The spectrum of a given surface remains qualitatively unchanged with very similar peak positions for the two domains (as the spin-averaged electronic properties of the surface are identical), but quantitative variations of the differential conductance are observed.

While it is very time consuming to take a complete tunneling spectrum at every pixel of an image, the intensity variation of the differential conductance dI/dU can be used for mapping the sample magnetic structure at certain bias voltages. For the particular tip used in the experimental run of Figure 26.4, suitable bias voltages that can be expected to result in high magnetic contrast would be close to the peak position, that is, around 0 V, or at -0.6 ± 0.1 V. It shall be mentioned, however, that because of uncontrollable structural differences of the tip apex's atomic arrangement, the optimal bias voltage may differ from tip to tip and has to be figured out in each individual experiment.

Since the signal is obtained with lock-in technique after adding a small bias modulation (typically a few mV) with a frequency well beyond the response frequency of the feedback loop (typically several kHz) this can not only be recorded simultaneously with the topographic image but it practically separates topography and magnetic signal into two virtually independent data channel. As an example, Figure 26.5 shows such a data set taken with an in-plane sensitive, Fe-coated probe tip. The topography of Figure 26.5a is the tip's z-position as obtained in the constant-current mode. Due to the presence of several edge and screw dislocations, the sample's terrace and step structure is rather intricate and it would be essentially impossible to unravel the surface magnetic domain structure from these topographic data alone. The differential conductance dI/dU map shown in Figure 26.5b, however, immediately reveals the complex surface magnetic structure with numerous irregularities. Generally, edge dislocations, as marked by a white arrow in Figure 26.5a, do not disturb the topological antiferromagnetic order and similar to any other monatomic step edge lead to a reversal of

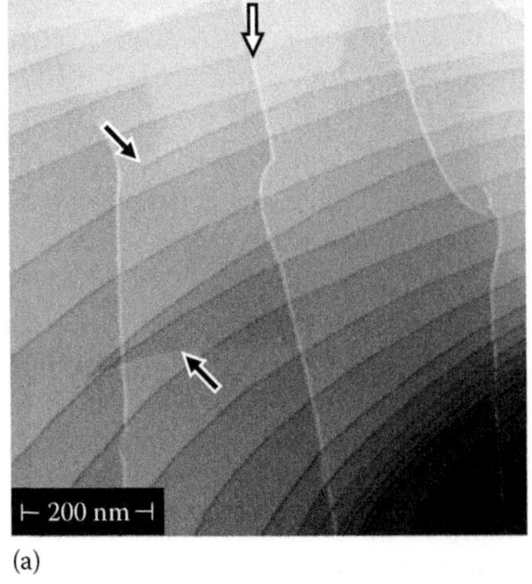

(a)

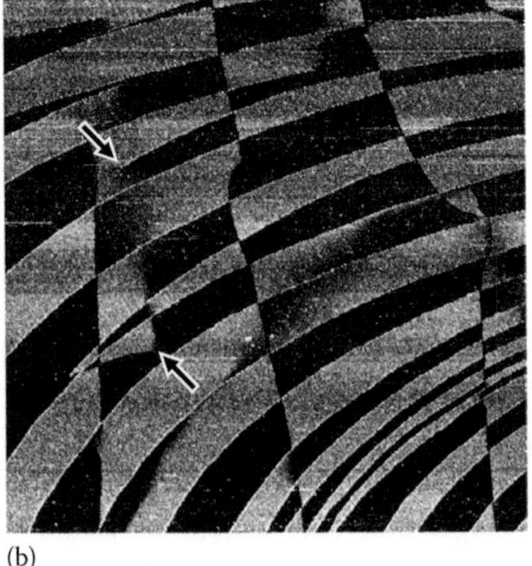

(b)

FIGURE 26.5 (a) Topography and (b) magnetic dI/dU map of a Cr(001) surface as measured with an Fe-coated probe tip at 300 K. (Adapted from Kleiber, M. et al., *Phys. Rev. Lett.*, 85, 4606, 2000.) Screw (black arrows) and edge dislocations (one example marked by a white arrow) can be recognized in (a). The tunneling parameters are $I = 3$ nA and $U = -0.7$ V. Similar to other monoatomic step edges the edge dislocations leads to a change in the magnetic contrast due to the layered antiferromagnetic structure of Cr(001). Typically, a domain wall is found between two screw dislocations, for example, between the two black arrows in (b).

the surface magnetization [53]. Screw dislocations, in contrast, appear as semi-infinite step edges, which lead to a frustration of the surface magnetic order. Since this leads to an additional step edge on one side of the sample, a perfect topological antiferromagnetic order cannot be maintained. Instead, a domain wall is generated, which extends over the surface until it is annihilated at another screw dislocation [23,53], as indicated by two black arrows in Figure 26.5b.

This example shows that surfaces with a layered and compensated magnetic structure (so-called *d*-type antiferromagnets) can be imaged with SP-STM by making use of the differential conductance d*I*/d*U* mode. In spite of a certain cross talk between the topography and the d*I*/d*U* signal, these simultaneously measured data channels allow for an effective separation of structural and magnetic surface properties on scales ranging from micro- to nanometer lateral scales.

26.4.2 Ferromagnets

The imaging of the domain structure of ferromagnetic surfaces has been in the focus of scientific interest for a long time [54]. While the magnetic exchange length of the elementary 3d transition metals Fe, Co, and Ni is sufficiently wide to be imaged by traditional magnetically sensitive microscopy techniques such as the Bitter technique, Lorentz microscopy, or magnetic force microscopy, many technologically relevant materials like rare earth metals and alloys consisting of 3d and 4f metals exhibit an exchange length well below the respective spatial resolution of the above-mentioned microscopy techniques.

One particularly illustrative example of the imaging of the domain structure of ferromagnetic surfaces by spin-polarized STM is a recent study on Dysprosium (Dy) films by Berbil-Bautista et al. [55]. Quite contrary to most other experiments, these spin-resolved STM data were not obtained with tips that were epitaxially coated with magnetic material but rather by dipping the tip into the Dy surface, which reliably resulted in a Dy cluster at the tip apex.

Figure 26.6a shows the topography of 90 ALs thick Dy(0001) film grown on W(110). The Dy (0001) surface exhibits six equivalent magnetization directions. Therefore, according to Equation 26.6, up to six differential conductance levels can be expected if an SP-STM experiment is performed with a magnetic tip.

In fact, the d*I*/d*U* map of Figure 26.6b, which was acquired simultaneously with the topographic image of Figure 26.6a shows six well-separated levels of the d*I*/d*U* signal. This is particularly obvious in the histogram of Figure 26.6b, plotted in Figure 26.6c. Because of the TMR effect, the surface areas that are magnetized almost parallel with respect to the tip magnetization gave the highest intensity of the d*I*/d*U* signal, peak i, while those that are almost antiparallel to the tip exhibit the lowest, vi. The other magnetization directions give rise to d*I*/d*U* signals in between these two extremes (ii–v). By a variational approach, this even allows for a precise determination of the tip magnetization, as shown in Figure 26.6c.

However, these six contrast levels can only be expected if the tip magnetization is sufficiently well separated from the easy ($\alpha = 0°$) and hard magnetic axes ($\alpha = 30°$) of a surface with six equivalent magnetization directions. This situation was schematically

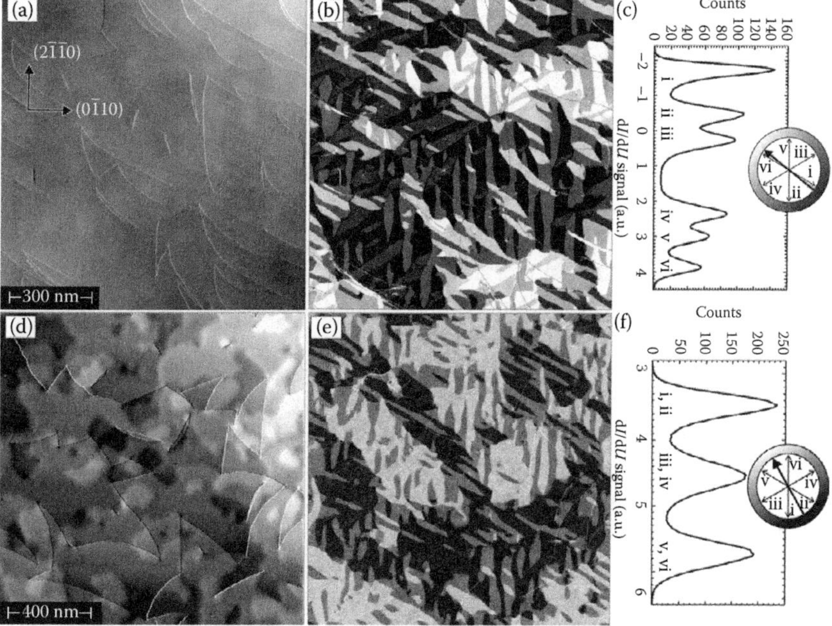

FIGURE 26.6 Example of spin-polarized STM on a ferromagnetic surface: (a) Topography and (b) spin-resolved d*I*/d*U* map of a 90 AL Dysprosium film on W(110) imaged at $T = 25$ K (tunneling parameters: $U = -1{:}0$ V, $I = 30$ nA). (c) The histogram of the d*I*/d*U* map shown in (b) reveals six contrast levels. Panels (d) and (e) show corresponding measurements performed with a different magnetic tip. Because of the particular magnetic orientation of this tip, which is just in between two domains of the Dy sample (see inset), the histogram of the d*I*/d*U* signal in (f) shows only three contrast levels.

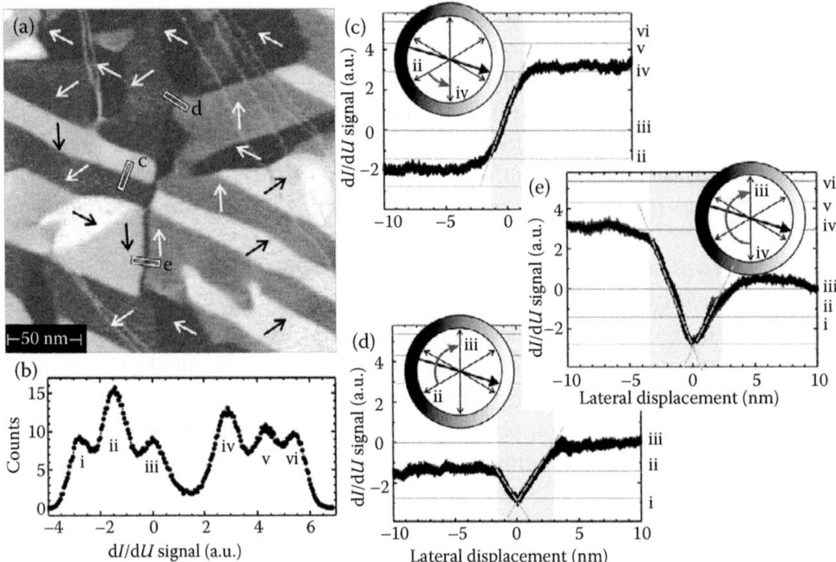

FIGURE 26.7 Domain wall profile analysis of a 90 AL Dy/W(110) film. (Adapted from Berbil-Bautista, L. et al., *Phys. Rev. B*, 76, 064411, 2007.): (a) spin-resolved dI/dU map of 90 AL Dy/W(110) showing the magnetic domain structure (T=59 K, U=−0.6 V, I=30 nA) and (b) the histogram of the data shown in (a) reveals six peaks. The directions of $\vec{m}_s$ for the domains is indicated in (a) by arrows. (c–e) Line profiles and schematic representation of the path of the magnetization for a 60°, a 120°, and a 180° domain wall labeled with c–f in (a). The profiles are between 2.5 and 6 nm wide.

represented in Figure 26.3a in the six magnetization direction (thin arrows) exhibits distinguishable projections onto the tip magnetization (thick arrow). In contrast, if the tip magnetization is close to the sample's easy or hard magnetic axis, as represented in Figure 26.3b and c, some projections fall onto in the same vector and can no longer be distinguished, resulting in three or four signal levels, respectively. One such example is shown in Figure 26.6d through f. Here the tip was accidentally magnetized along a hard magnetization axis and the six magnetization directions of Dy(0001) project onto three separate signals only.

As a first example, the high spatial resolution capability of SP-STM shall be demonstrated by an analysis of the internal spin structure of domain walls observed on Dy films. The Dy film thickness on W(110) amounts to 90 AL. The domain structure is shown in Figure 26.7a. A histogram of the dI/dU signal intensity in Figure 26.7b again reveals six contrast levels, that is, a situation similar to Figure 26.3a. Figure 26.7c through e shows line profiles of the magnetic dI/dU signal across a 60° domain wall, a 120° domain wall, and a 180° domain wall labeled c–f in Figure 26.7a, respectively. The horizontal lines in the domain wall profiles (c–f) represent the peak positions in the histogram of Figure 26.7b. While the dI/dU signal changes monotonously for the 60° domain walls, extreme dI/dU values, that is, a non-monotonous profile, are observed in the cases of the 120° domain wall and the 180° domain wall.

The width of the domain walls is between 2.5 and 6 nm, with the 60° domain wall being narrowest. It is important to note that the high spatial resolution of SP-STM not only allows for resolving the domain structure of the Dy film but also shows the internal spin structure even of domain walls that are far below the limit of resolution of other magnetically sensitive imaging techniques. In fact, as we will see below, atomic spin resolution can be achieved on appropriate surfaces.

26.5 Imaging of Antiferromagnetic Surfaces

As already discussed in the introduction of this chapter, the imaging of antiferromagnetic materials represents a particular challenge to conventional magnetic microscopy methods, such as photoelectron emission microscopy (PEEM) or Lorentz microscopy, since the sample's magnetic moments cancel if averaged beyond the atomic scale. While spatially averaging polarized neutron diffraction allowed for a quite thorough understanding of the processes in bulk samples, much less in known about antiferromagnetic thin films or even nanoparticles. Although methods like Mössbauer spectroscopy and neutron diffraction were adapted to large ensembles of presumably identical nanoparticles (for a recent review see Ref. [56]) the analysis of the properties of single particles remained out of reach.

The well-established topographic atomic-resolution capability of STM suggests that the goal of atomic spin resolution can also be achieved, but this task requires an appropriate and electrically conducting test sample. The smallest conceivable magnetic structure is a magnetically ordered surface where the magnetic moment alternates between adjacent atomic sites, that is, a so-called g-type antiferromagnet.

26.5.1 Proof of Principle

Many materials that are known to be g-type antiferromagnets in the bulk are alloys with a complex stoichiometry and cannot be stabilized at the surface. A structurally simpler sample system that can be grown epitaxially on a refractory metal is the Fe monolayer on W(001). Early spin-resolved photoemission and magneto-optical measurements surprisingly showed no sign of magnetic remanence [57,58]. Indeed, density-functional calculations performed in the local density approximation (LDA) scheme showed an antiferromagnetic ground state [59] and were recently confirmed using the improved generalized gradient approximation (GGA) scheme [17].

The topography of 1.1 AL of Fe, which was epitaxially grown on W(001), is shown in Figure 26.8a. In spite of the substantial misfit of the Fe and W lattice constants, which amounts to approximately 10%, the Fe film growth is pseudomorph, that is, the Fe atoms occupy lattice sites the lateral position of which are equivalent to substrate sites. While the first Fe layer is completely filled, only few double-layer islands with lateral dimensions of less than 10 nm can be recognized. Indeed, atomic-scale images taken with a magnetic tip reveal the atomic-scale spin structure; the magnetic checkerboard pattern of the antiferromagnetic Fe monolayer can be seen in the higher-resolution images shown in Figure 26.8b and c, which were taken in the area marked by a hatched box approximately in the center of Figure 26.8a. Figure 26.8b and c were simultaneously measured and show results obtained in the two magnetic imaging modes discussed earlier, that is, the constant-current image and the dI/dU mode, respectively. Around two adsorbates that appear as protrusions, the constant-current image clearly shows a dark and bright checkerboard pattern: due to the spin-polarized contribution to the tunneling current, the tip is retracted from the sample surface whenever the integrated tip's magnetic LDOS is parallel to the sample atom and approached above atoms the magnetization direction of which is antiparallel to the tip. The line section in Figure 26.8d shows a corrugation of about 2 pm. Qualitatively, the same pattern can be seen in the dI/dU map of Figure 26.8c whereby the signal-to-noise ratio is significantly improved as compared to the constant-current image of Figure 26.8b because of the frequency- and phase-sensitive detection of the lock-in amplifier. Comparison of the dI/dU line section in the lower panel of Figure 26.8d with the topographic section shown in upper panel also reveals that the two signals are anti-corrugated which might be caused by different signs of $\vec{\tilde{m}}_S$ and $\vec{m}_s$.

It is important to note that the observation of a checkerboard pattern in an STM experiment performed with a magnetic tip is no unambiguous proof of surface antiferromagnetism. Only if no similar superstructure is observed with a nonmagnetic tip a potential structural surface reconstruction as it might, for example, be caused by adsorbates can be excluded with sufficient certainty. The definite proof requires the reversal of the magnetization direction of the ferromagnetic tip by an external field. If an Fe-coated tip is used, a field of about 2 T is sufficient to overcome the dominant magnetostatic anisotropy energy. A vertical field of that magnitude safely forces the magnetization, which is parallel to the sample surface in the absence of external fields, into the out-of-plane direction. Since the exchange coupling in antiferromagnets is usually much stronger than the Zeeman energy, an external field of a few Tesla leaves the sample's magnetic structure essentially unaffected and only changes the vertical component of the tip magnetization.

The described reversal of the tip magnetization and its influence on the contrast is demonstrated in the example in Figure 26.9. The middle row of Figure 26.9 shows a series of images that were taken within the same $4 \times 4\,\text{nm}^2$ area. The images are aligned at the position of a native adsorbate in the center. Between successive measurements, the field direction was changed between

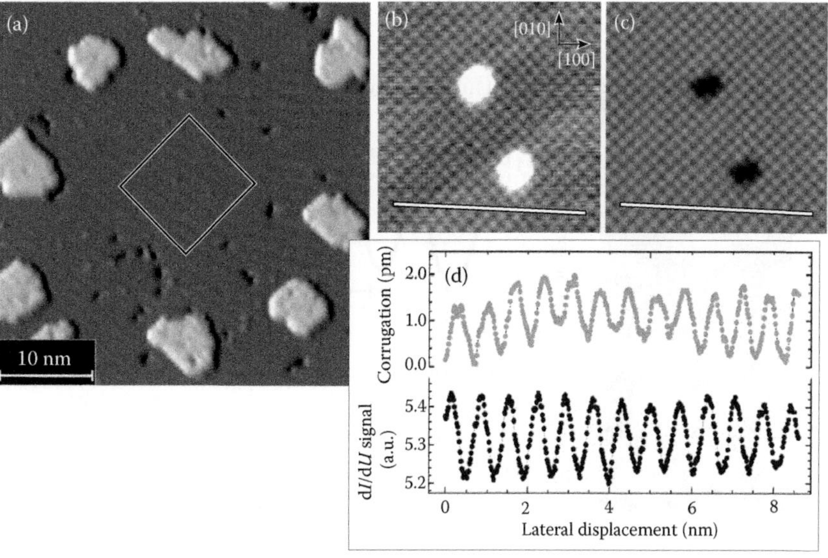

FIGURE 26.8 (a) Topography of 1.1 AL Fe on W(001). (b) The spin-resolved constant-current image and (c) a spin-resolved dI/dU map were measured simultaneously in the box of (a) and show the antiferromagnetic checkerboard pattern of the Fe monolayer on W(001) with alternating spins pointing up and down. Measurement temperature was 13 ± 1 K. Note that the two adsorbates appear as protrusions in the topography but exhibit a lower dI/dU signal strength. Line sections taken along the lines in (b) and (c) are shown in (d).

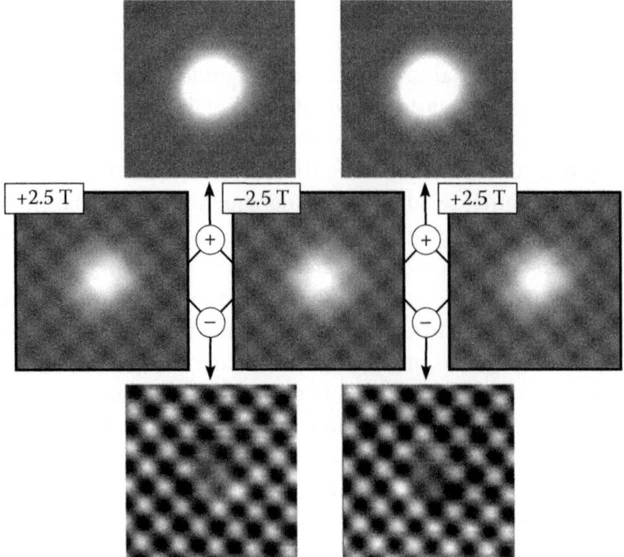

FIGURE 26.9 Constant-current images of a 4×4 nm² region of an Fe monolayer on W(001) around a native adsorbate measured with an Fe-coated tip (I=30 nA, U=−40 mV, T=13±1 K) (middle row). Opposite external fields were applied to force the tip magnetization out of the in-plane easy axis into perpendicular "up" and "down" directions. Since the exchange coupling within the Fe monolayer is much stronger than the Zeeman energy, the external field leaves the sample's magnetic structure unaffected. By changing the magnetization direction of the tip, any magnetic contrast is inverted but topographic and spin-averaged electronic contributions remain unchanged. Consequently, the sum (top row) and the difference (bottom row) of two images recorded with opposite tip magnetization directions allow the separation of non-magnetic from magnetic contributions, respectively.

+2.5 T and −2.5 T. Since the TMR effect depends on the relative orientation of tip and sample magnetization, the observed superstructure has to invert if the pattern is caused by spin-dependent tunneling. In contrast, the superstructure would remain unchanged if it were caused by structural effects. Careful inspection of the data indeed confirms the expected corrugation reversal. An easier evaluation is achieved by calculating sum and difference images of data taken at opposite magnetic fields. The top and bottom images of Figure 26.9 display such images, respectively. As expected for SP-STM the sum images show a vanishing contrast on clean areas. More importantly, the contrast is enhanced in the difference images in the bottom row. It is quite remarkable that the magnetic superstructure is even visible even at the position of the adsorbate, which indicates that the antiferromagnetic state of the Fe monolayer is largely unaffected by the adsorbate and that spin-polarized LDOS of Fe protrudes through the adsorbate into the vacuum.

26.5.2 Antiferromagnetic Domain Walls

The ability of imaging compensated spin structures with atomic resolution opens up the possibility of directly observing domain walls in antiferromagnetic materials that could only be imaged indirectly previously [60]. However, domain walls in antiferromagnets are rare and represent a metastable state. While ferromagnets reduce their total magnetic energy by domain formation, for example, by avoiding magnetic charges in the flux closure configuration, domain formation in antiferromagnets is generally associated with a higher total energy state as their spin structure is already compensated and the additional exchange energy of domain walls cannot be compensated.

Accordingly, the antiferromagnetic domain wall imaged in Figure 26.10a on the surface of a antiferromagnetic monolayer

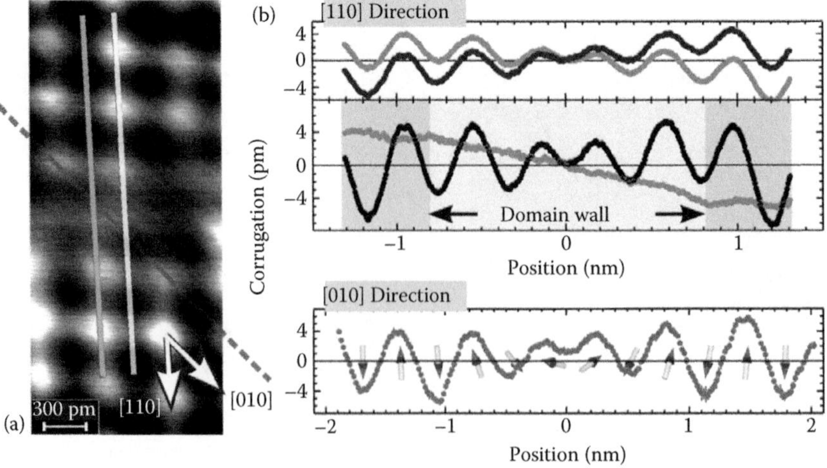

FIGURE 26.10 (a) Detailed view of a ⟨110⟩-oriented domain wall in an out-of-plane antiferromagnetic Fe monolayer on W(001) measured at T = 13 ± 1 K. (b) The line sections drawn along the middle panel shows the sum (black) and the difference (gray) of the line profiles shown in the upper panel, which were measured on adjacent atomic [110] rows. The data reveal that the wall is about 1.6 nm wide and that its out-of-plane component exhibits mirror symmetry. (Adapted from Bode, M. et al., *Nat. Mater.*, 5, 477, 2006.)

of Fe on W(001) is very short [15]. Typically the length of domain walls amounts to a few nanometers only and they are mostly trapped at structural imperfections. In the spin-polarized STM image, antiferromagnetic domain walls appear as an area with a faded magnetic contrast in the spin-sensitive image of Figure 26.10a. At the position of the domain wall, the magnetic period shifts by one atomic row, thereby separating the two domains in the lower and upper part of the image. The domain wall is very narrow as indicated by the fact that maximum contrast is fully regained only a few atomic rows away from the domain wall center. A quantitative understanding can be achieved by careful analysis of line sections taken along low-index crystallographic directions. The upper panel of Figure 26.10b shows data taken on adjacent atomic rows along the [110] direction, that is, perpendicular to the wall. The middle panel shows the sum and the difference of these lines in black and gray, respectively. The difference (gray) shows an almost constant signal of opposite sign at the left and right rim of the line section, that is, far away from the domain wall center. These regions (domains) are connected by a constant slope that extends over approximately 1.6 nm, which corresponds to a domain wall width of six to eight atomic rows. The sum (black) reveals that the average out-of-plane component of these atomic rows is mirror-symmetric. The mirror-symmetric appearance, which is also found in the line profile taken along the [010] direction shown in the lower panel of Figure 26.10b, indicates that the domain wall center is located between two atomic rows.

26.5.3 Complex Noncollinear Spin Structures

Structure-induced magnetic frustration [61], the competition of nearest and next-nearest neighbor exchange interactions, or the so-called Dzyaloshinskii–Moriya interaction [62–64] may lead to complex two- or even three-dimensional spin structures. For bulk samples, a variety of experimental techniques, the most prominent of which is polarized neutron scattering, have been used to study ground state magnetic properties and phase transitions. The adaption of those techniques to single thin films or even nanostructures is not practically achievable yet. On one hand, this is caused by the extraordinarily small scattering cross section of neutrons with atoms. Also, the flux of neutron sources could not be increased to the same extent as synchrotron x-ray sources that increased their brilliance by many orders of magnitude over the past decades. As a result, ordered complex atomic-scale spin structures remained largely unexplored experimentally.

To a certain extent spin-polarized STM can fill this gap. While being limited to conducting, relatively clean, and preferentially single-crystalline sample surfaces, its atomic spin-resolution capabilities should essentially allow for the imaging of arbitrary spin structures. Based on the independent orbital approximation Heinze predicted SP-STM images of helical and geometrically frustrated spin structures of thin films and suggested that SP-STM might even be suited for imaging "more complex noncollinear magnetic structures such as domain walls in antiferromagnets, spin-spiral states, spin glasses, or disordered states" [21].

Recent experiments have confirmed these predictions. In the following, two recent achievements will be presented: the successful imaging of the Néel state and the discovery of spin-orbit-driven chiral spin structures on surfaces. The Néel state originates from the geometrical frustration of the magnetic coupling, which cannot be matched with the crystal lattice. A theoretically well-known example is an antiferromagnet on a triangular or hexagonal lattice. While the antiferromagnetic coupling can be maintained in the dimer and linear trimer, it cannot be realized in a compact trimer or larger clusters. Theoretically a 120° arrangement of nearest neighbor spins was predicted, called the Néel state. One possible configuration is represented in Figure 26.11a. It consists of three sub-lattices,

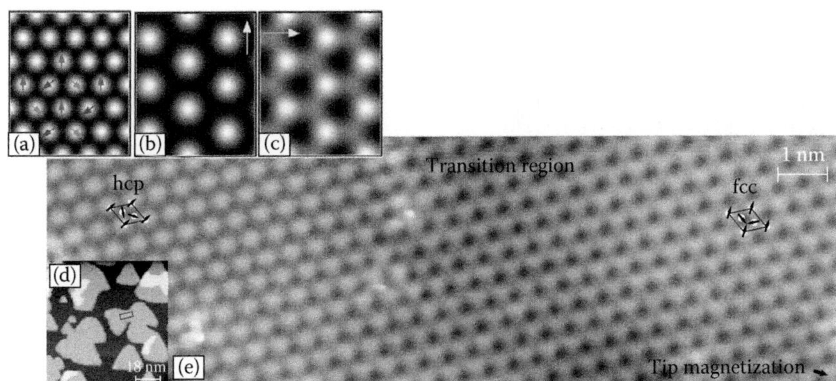

FIGURE 26.11 (a) Néel state consisting of three sub-lattices that are rotated by 120°. (b) and (c) Simulated spin-polarized STM images for two high-symmetry tip magnetization directions. (Courtesy of Heinze, S. et al., *Science*, 288, 1805, 2000.) (d) Topography of submonolayer Mn Islands are connected together. (e) Atomic-scale constant-current SP-STM image taken in the rectangular area in (d). The tunneling parameters are $U = 20$ mV and $I = 20$ nA. The right part is the hcp region, and the left part is the fcc region. In the middle, the image appears darker, which is considered as the smooth structural transition region from hcp to fcc. With the assumed tip magnetization orientation (bottom right), one possible configuration of the 120° Néel structure is indicated by arrows. (Courtesy of Gao, C.L. et al., *Phys. Rev. Lett.*, 101, 267205, 2008.)

which are rotated by 120°. The independent orbital approximation by Heinze [21] allows for the simulation of spin-polarized STM images expected for two high-symmetry tip magnetization directions (Figure 26.11b and c).

Although the Néel state has been discussed for decades, its direct observation was not possible until very recently. Following a proposal by Wortmann et al. [18], Gao et al. performed SP-STM experiments on Mn monolayer films on Ag(111) [65]. They could show that even though four principle Néel configurations exist, which are rotated by modulo a 120° spin rotation, a transition between the two patterns shown in Figure 26.11b and c would require a rotation of the tip magnetization direction. Nevertheless, Figure 26.11e shows the result of a constant-current image measured with a Cr-coated probe tip. As can be seen in the inset of Figure 26.11d, this high-resolution image was taken on a coalesced Mn island. While the left part exhibits a stacking fault with respect to the substrate, that is, hcp stacking, the right part is regularly stacked, fcc. According to Gao et al. [65] this surprising result can only be explained by 30° rotation of the Mn moments as shown by arrows in the left and right of the image. It should be noted that the tip magnetization is not known. In the dark transition region, a smooth structural change seems to be accompanied by a change of magnetic rotation.

Noncollinear spin structures may also result from relativistic effects. Namely, the Dzyaloshinskii–Moriya interaction, a spin-orbit-induced chiral contribution to the exchange, can result in very complex spin structures such as Skyrmion lattices [66] and two- or three-dimensional spin spirals [67–69]. However, the Dzyaloshinskii–Moriya interaction is irrelevant in most bulk materials as it cancels in inversion-symmetric lattices, such as bcc-, fcc-, or hcp-crystals. Only in some rare cases where the bulk inversion symmetry is broken, such as MnSi [70] or $Fe_{0.5}Co_{0.5}Si$ [71], the Dzyaloshinskii–Moriya interaction leads to chiral spin structures.

The situation is completely different, however, on surfaces, clusters, and in magnetic nanoparticles. Here the inversion symmetry is broken at the interfaces and surfaces that separate the magnetic object from the environment, that is, the substrate and/or the vacuum. Similar to collinear antiferromagnetic films and nanoparticles, these chiral spin structures are compensated and cannot be detected by spatially averaging detection methods like neutron diffraction. Only very recently some cases of ferromagnetic [68] and antiferromagnetic spin spirals [67] have been confirmed by SP-STM on monolayer films. Figure 26.12a shows a constant-current STM image of an Mn monolayer measured with an Fe-coated probe tip without an externally applied magnetic field [67]. A location with a few adsorbates has intentionally been chosen to find exactly the same position in later experiments. Dark and bright stripes can be recognized, which correspond to rows of parallel spins oriented along the [001] direction of the sample. These stripes alone would be consistent with a row-wise antiferromagnet [16]. However, the amplitude of the stripes is not constant but varies periodically. This becomes particularly obvious in the experimental line section in the bottom right panel of Figure 26.12a, which shows an apparent periodicity of about 6 nm. A more detailed analysis, however, revealed that

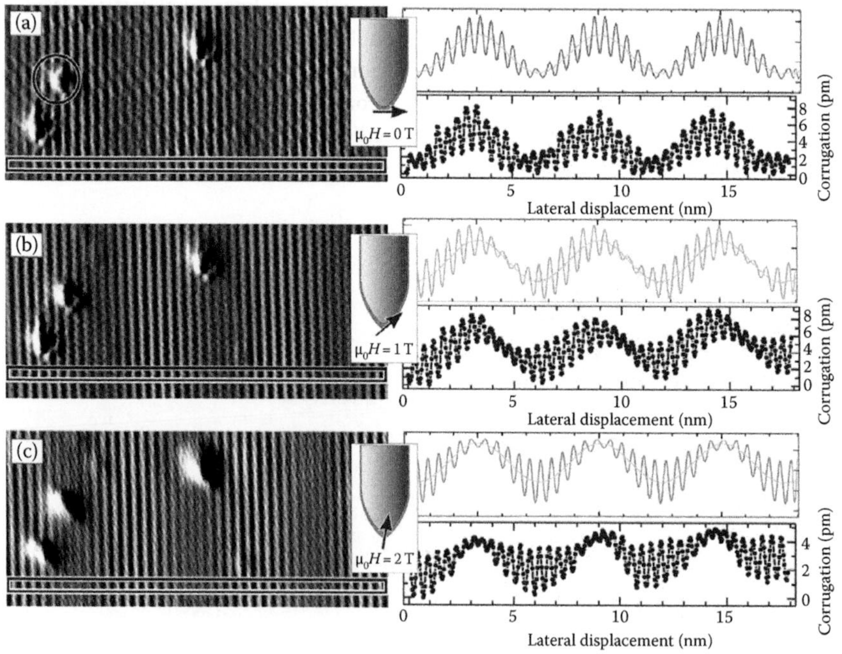

FIGURE 26.12 Magnetically sensitive constant-current images of the Mn monolayer on W(110) (left panels) and corresponding experimental (bottom right) and theoretical line sections (bottom left panels, Courtesy of Stefan Heinze). (Tunneling parameters: $I = 2$ nA, $U = 30$ mV, $T = 13 \pm 1$ K.) The data show the situation at external fields of 0 T (a), 1 T (b), and 2 T (c). (Adapted from Bode, M. et al., *Nature*, 447, 190, 2007.) As sketched in the insets, the external field rotates the tip magnetization from in-plane (a) to out-of-plane (c), shifting the position of maximum spin contrast. This proves that the Mn layer does not exhibit a spin-density wave but rather a cycloidal spin spiral.

the pattern of maxima and minima is shifted by π such that a full period extends over 12 nm.

This pattern is consistent with a spin spiral but could also be caused by a spin-density wave. Phenomenologically, the difference between these two spin structures becomes obvious if we compare the size and orientation of their local magnetic moments. While the orientation of the magnetic moment changes but the size remains constant in a spin spiral, the opposite situation can be found in a spin-density wave material; here the orientation is constant but the size oscillates. With this in mind, it is relatively straightforward to distinguish these two cases by spin-polarized STM in an external field where the magnetization direction of a ferromagnetic tip can be changed by the Zeeman energy. In a material with a spin-density wave, the minimum has to stay at the same position since it is caused by a vanishing sample's magnetic moment. In contrast, the minimum magnetic contrast in the case of a spin spiral is caused by an orthogonal orientation between tip and local sample magnetization. If the tip magnetization is rotated into the formerly orthogonal direction, the minimum of the magnetic contrast is expected to shift to a new position, which is now orthogonal to the tip. In fact, this shift can clearly be observed in Figure 26.12b and c. The upper right panels of Figure 26.12a through c show simulated spin-polarized STM line sections obtained by Heinze with the independent orbital approximation [21]. An excellent agreement between experiment and simulation can be recognized.

26.6 Outlook

The application of SP-STM and spectroscopy has contributed considerably to our modern understanding of static domain and domain wall structures in ferromagnetic, but in particular in antiferromagnetic, surfaces, thin films, and nanoparticles. Recently, some progress was also made toward imaging and manipulating the dynamic magnetic properties of magnetic nanoparticles [72–74]. By investigating the switching behavior of single nanoscale Fe islands the superparamagnetic re-magnetization rate [72] and the Arrhenius pre-factor could be correlated to the size and shape of the nanoislands [74]. At unusually low tunnel resistance values, that is, high tunneling currents, the spin polarization of the current between the magnetic SPM tip and the sample was also used for spin-torque experiments on superparamagnetic islands [73]. These experiments may lead to a better understanding of magnetization reversal processes, but further progress hinges on achieving a higher time resolution, which is currently limited by the bandwidth of the current preamplifiers because of a signal-to-noise trade-off.

Another highly exciting field is true atomic-scale magnetism. Recent spin-averaged inelastic tunneling experiments performed at temperatures below 1 K shows that so-called spin-flip spectroscopy can be used to determine the g-value [75], allows for a detailed insight into magnetic coupling behavior in engineered assemblies [76], and even enables to determine the anisotropy barrier of single magnetic atoms [77]. Also, spin-polarized STM has been used for the measurement of magnetization curves on single atoms that were deposited on a nonmagnetic substrate [78]. Although no hysteresis was observed, this experiment demonstrates the ability of spin-polarized STM to detect spin signals even from the most diluted amount of magnetic material we can imagine.

References

1. G. Binnig, K. H. Frank, H. Fuchs et al., Tunneling spectroscopy and inverse photoemission: Image and field states, *Phys. Rev. Lett.* **55**, 991–994, 1985.
2. M. Bode, R. Pascal, and R. Wiesendanger, Distance-dependent STM-study of the W(110)/C-R(15×3) surface, *Z. Phys. B* **101**, 103–107 (1996).
3. L. Olesen, M. Brandbyge, M. R. Sørensen et al., Apparent barrier height in scanning tunneling microscopy revisited, *Phys. Rev. Lett.* **76**, 1485–1488 (1996).
4. T. Jung, Y. W. Mo, and F. J. Himpsel, Identification of metals in scanning tunneling microscopy via image states, *Phys. Rev. Lett.* **74**, 1641–1644 (1995).
5. M. F. Crommie, C. P. Lutz, and D. M. Eigler, Imaging standing waves in a two dimensional electron gas, *Nature* **363**, 524–527 (1993).
6. Y. Hasegawa and Ph. Avouris, Direct observation of standing wave formation at surface steps using scanning tunneling spectroscopy, *Phys. Rev. Lett.* **71**, 1071–1074 (1993).
7. A. Bauer, A. Mühlig, D. Wegner, and G. Kaindl, Lifetime of surface states on (0001) surfaces of lanthanide metals, *Phys. Rev. B* **65**, 075421 (2002).
8. M. Bode, M. Getzlaff, A. Kubetzka, R. Pascal, O. Pietzsch, and R. Wiesendanger, Temperature-dependent exchange splitting of a surface state on a local-moment magnet: Tb(0001), *Phys. Rev. Lett.* **83**, 3017–3020 (1999).
9. D. Wegner, A. Bauer, and G. Kaindl, Magnon-broadening of exchange-split surface states on lanthanide metals, *Phys. Rev. B* **73**, 165415 (2006).
10. D. T. Pierce, Spin-polarized electron microscopy, *Phys. Scr.* **38**, 291–296 (1988).
11. O. Pietzsch, A. Kubetzka, M. Bode, and R. Wiesendanger, Observation of magnetic hysteresis at the nano-scale by spin polarized scanning tunneling spectroscopy, *Science* **292**, 2053 (2001).
12. M. Pratzer, H. J. Elmers, M Bode, O. Pietzsch, A. Kubetzka, and R. Wiesendanger, Atomic-scale magnetic domain walls in quasi-one-dimensional Fe nanostripes, *Phys. Rev. Lett.* **87**, 127201 (2001).
13. A. Yamasaki, W. Wulfhekel, R. Hertel, S. Suga, and J. Kirschner, Direct observation of the single-domain limit of Fe nano-magnets by spin-polarized STM, *Phys. Rev. Lett.* **91**, 127201 (2003).
14. A. Wachowiak, J. Wiebe, M. Bode, O. Pietzsch, M. Morgenstern, and R. Wiesendanger, Direct observation of internal spin structure of magnetic vortex cores, *Science* **298**, 577 (2002).

15. M. Bode, E. Y. Vedmedenko, K. von Bergmann et al., Atomic spin structure of antiferromagnetic domain walls, *Nat. Mater.* **5**, 477 (2006).
16. S. Heinze, M. Bode, A. Kubetzka et al., Real-space imaging of two-dimensional antiferromagnetism on the atomic scale, *Science* **288**, 1805 (2000).
17. A. Kubetzka, P. Ferriani, M. Bode et al., Revealing antiferromagnetic order of the Fe monolayer on W(001): Spin-polarized scanning tunneling microscopy and first-principles calculations, *Phys. Rev. Lett.* **94**, 087204 (2005).
18. D. Wortmann, S. Heinze, Ph. Kurz, G. Bihlmayer, and S. Blügel, Resolving complex atomic-scale spin structures by spin-polarized scanning tunneling microscopy, *Phys. Rev. Lett.* **86**, 4132 (2001).
19. J. Tersoff and D. R. Hamann, Theory and application for the scanning tunneling microscope, *Phys. Rev. Lett.* **50**, 1998 (1983).
20. J. Tersoff and D. R. Hamann, Theory of the STM, *Phys. Rev. B* **31**, 805 (1985).
21. S. Heinze, Simulation of spin-polarized scanning tunneling microscopy images of nanoscale non-collinear magnetic structures, *Appl. Phys. A* **85**, 407–414 (2006).
22. J. C. Slonczewski, Conductance and exchange coupling of two ferromagnets separated by a tunneling barrier, *Phys. Rev. B* **39**, 6995 (1989).
23. M. Kleiber, M. Bode, R. Ravlić, and R. Wiesendanger, Topology-induced spin frustrations at the Cr(001) surface studied by spin-polarized scanning tunneling spectroscopy, *Phys. Rev. Lett.* **85**, 4606 (2000).
24. R. Wiesendanger, H. J. Güntherodt, G. Güntherodt, R. J. Gambino, and R. Ruf, Observation of vacuum tunneling of spin-polarized electrons with the STM, *Phys. Rev. Lett.* **65**, 247 (1990).
25. M. Bode, M. Getzlaff, and R. Wiesendanger, Spin-polarized vacuum tunneling into the exchange-split surface state of Gd(0001), *Phys. Rev. Lett.* **81**, 4256 (1998).
26. A. Selloni, P. Carnevali, E. Tosatti, and C. D. Chen, Voltage-dependent scanning tunneling microscopy of a crystal surface: Graphite, *Phys. Rev. B* **31**, 2602 (1985).
27. R. Wiesendanger, M. Bode, and M. Getzlaff, Vacuum-tunneling magnetoresistance: The role of spin-polarized surface states, *Appl. Phys. Lett.* **75**, 124 (1999).
28. O. Pietzsch, A. Kubetzka, M. Bode, and R. Wiesendanger, Spin-polarized scanning tunneling spectroscopy of nanoscale cobalt islands on Cu(111), *Phys. Rev. Lett.* **92**, 057202 (2004).
29. M. Johnson and J. Clarke, Spin-polarized scanning tunneling microscope: Concept, design, and preliminary results from a prototype operated in air, *J. Appl. Phys.* **67**, 6141 (1990).
30. W. Wulfhekel and J. Kirschner, Spin-polarized scanning tunneling microscopy on ferromagnets, *Appl. Phys. Lett.* **75**, 1944 (1999).
31. M. Bode, M. Getzlaff, and R. Wiesendanger, Quantitative aspects of spin-polarized scanning tunneling spectroscopy of Gd(0001), *J. Vac. Sci. Technol. A* **17**, 2228 (1999).
32. H. F. Ding, W. Wulfhekel, J. Henk, P. Bruno, and J. Kirschner, Absence of zero-bias anomaly in spin-polarized vacuum tunneling in Co(0001), *Phys. Rev. Lett.* **90**, 116603 (2003).
33. D. T. Pierce and F. Meier, Photoemission of spin-polarized electrons from GaAs, *Phys. Rev. B* **13**, 5484 (1976).
34. R. Jansen, M. W. J. Prins, and H. van Kempen, Theory of spin-polarized transport in photoexcited semiconductor/ferromagnet tunnel junctions, *Phys. Rev. B* **57**, 4033 (1998).
35. R. Jansen, M. C. M. M. van der Wielen, M. W. J. Prins, D. L. Abraham, and H. van Kempen, Progress toward spin-sensitive scanning tunneling microscopy using optical orientation in GaAs, *J. Vac. Sci. Technol. B* **12**, 2133 (1994).
36. V. P. LaBella, D. W. Bullock, Z. Ding et al., Spatially resolved spin-injection probability for GaAs, *Science* **292**, 1518 (2001).
37. M. W. J. Prins, M. C. M. M. van der Wielen, R. Jansen, D. L. Abraham, and H. van Kempen, Photoamperic probes in scanning tunneling microscopy, *Appl. Phys. Lett.* **64**, 1207 (1994).
38. M. W. J. Prins, H. van Kempen, H. van Leuken, R. A. de Groot, W. van Roy, and J. de Boeck, Spin-dependent transport in metal/semiconductor tunnel junctions, *J. Phys.: Condens. Mat.* **7**, 9447 (1995).
39. U. Schlickum, W. Wulfhekel, and J. Kirschner, Spin-polarized scanning tunneling microscope for imaging the in-plane magnetization, *Appl. Phys. Lett.* **83**, 2016–2018 (2003).
40. M. Bode, R. Pascal, and R. Wiesendanger, Scanning tunneling spectroscopy of Fe/W(110) using iron covered probe tips, *J. Vac. Sci. Technol. A* **15**, 1285 (1997).
41. M. Bode, O. Pietzsch, A. Kubetzka, and R. Wiesendanger, Imaging magnetic nanostructures by spin-polarized scanning tunneling spectroscopy, *J. Electr. Spectr. Rel. Phenom.* **114–116**, 1055 (2001).
42. G. André, A. Aspelmeier, B. Schulz, M. Farle, and K. Baberschke, Temperature dependence of surface and volume anisotropy in Gd/W(110), *Surf. Sci.* **326**, 275 (1995).
43. H. J. Elmers and U. Gradmann, Magnetic anisotropies in Fe(110) films on W(110), *Appl. Phys. A* **51**, 255 (1990).
44. A. Kubetzka, M. Bode, O. Pietzsch, and R. Wiesendanger, Spin-polarized scanning tunneling microscopy with antiferromagnetic probe tips, *Phys. Rev. Lett.* **88**, 057201 (2002).
45. S. Blügel, D. Pescia, and P. H. Dederichs, Ferromagnetism versus antiferromagnetism of the Cr(001) surface, *Phys. Rev. B* **39**, 1392 (1989).
46. C. L. Fu and A. J. Freeman, Surface ferromagnetism of Cr(001), *Phys. Rev. B* **33**, 1755 (1986).
47. R. Matsumoto, A. Fukushima, K. Yakushiji et al., Spin-dependent tunneling in epitaxial Fe/Cr/MgO/Fe magnetic tunnel junctions with an ultrathin Cr(001) spacer layer, *Phys. Rev. B* **79**, 174436 (2009).
48. T. Nagahama, S. Yuasa, E. Tamura, and Y. Suzuki, Spin-dependent tunneling in magnetic tunnel junctions with a layered antiferromagnetic Cr(001) spacer: Role of band structure and interface scattering, *Phys. Rev. Lett.* **95**(8), 086602 (2005).

49. J. A. Stroscio, D. T. Pierce, A. Davies, R. J. Celotta, and M. Weinert, Tunneling spectroscopy of bcc(001) surface states, *Phys. Rev. Lett.* **75**, 2960 (1995).
50. M. Budke, T. Allmers, M. Donath, and M. Bode, Surface state vs orbital kondo resonance at Cr(001): Arguments for a surface state interpretation, *Phys. Rev. B* **77**, 233409 (2008).
51. T. Hänke, M. Bode, S. Krause, L. Berbil-Bautista, and R. Wiesendanger, Temperature dependent scanning tunneling spectroscopy of Cr(001): Orbital kondo resonance versus surface state, *Phys. Rev. B* **72**, 085453 (2005).
52. O. Y. Kolesnychenko, R. de Kort, M. I. Katsnelson, A. I. Lichtenstein, and H. van Kempen, Real-space imaging of an orbital kondo resonance on the Cr(001) surface, *Nature* **415**, 507 (2002).
53. R. Ravlić, M. Bode, A. Kubetzka, and R. Wiesendanger, Correlation of dislocation and domain structure of Cr(001) investigated by spin-polarized scanning tunneling microscopy, *Phys. Rev. B* **67**, 174411 (2003).
54. A. Hubert and R. Schäfer, *Magnetic Domains*, Springer, Berlin, Heidelberg, 1998.
55. L. Berbil-Bautista, S. Krause, M. Bode, and R. Wiesendanger, Spin-polarized scanning tunneling microscopy and spectroscopy of ferromagnetic Dy(0001)/W(110) films, *Phys. Rev. B* **76**, 064411 (2007).
56. S. Mørup, D. E. Madsen, C. Frandsen, C. R. H. Bahl, and M. F. Hansen, Experimental and theoretical studies of nanoparticles of antiferromagnetic materials, *J. Phys.: Condens. Mat.* **19**(21), 213202 (31pp) (2007).
57. J. Chen and J. L. Erskine, Surface step-induced magnetic anisotropy in thin epitaxial Fe films on W(001), *Phys. Rev. Lett.* **68**, 1212 (1992).
58. G. A. Mulhollan, R. L. Fink, J. L. Erskine, and G. K. Walters, Local spin correlations in ultrathin Fe/W(100) films, *Phys. Rev. B* **43**, 13645 (1991).
59. R. Wu and A. J. Freeman, Magnetic properties of Fe overlayers on W(001) and the effects of oxygen adsorption, *Phys. Rev. B* **45**, 7532 (1992).
60. N. B. Weber, H. Ohldag, H. Gomonaj, and F. U. Hillebrecht, Magnetostrictive domain walls in antiferromagnetic NiO, *Phys. Rev. Lett.* **91**, 237205 (2003).
61. U. Bhaumik and I. Bose, Collinear Néel-type ordering in partially frustrated lattices, *Phys. Rev. B* **58**, 73–76 (1998).
62. I. E. Dzialoshinskii, Thermodynamic theory of "weak" ferromagnetism in antiferromagnetic substances, *Sov. Phys. JETP* **5**, 1259 (1957).
63. I. E. Dzyaloshinskii, Theory of helicoidal structures in antiferromagnets. 3, *Sov. Phys. JETP* **20**, 665 (1965).
64. T. Moriya, Anisotropic superexchange interaction and weak ferromagnetism, *Phys. Rev.* **120**, 91–98 (1960).
65. C. L. Gao, W. Wulfhekel, and J. Kirschner, Revealing the 120[degree] antiferromagnetic Néel structure in real space: One monolayer Mn on Ag(111), *Phys. Rev. Lett.* **101**, 267205 (2008).
66. U. K. Rössler, A. N. Bogdanov, and C. Peiderer, Spontaneous skyrmion ground states in magnetic metals, *Nature* **442**, 797–801 (2006).
67. M. Bode, M. Heide, K. von Bergmann et al., Chiral magnetic order at surfaces driven by inversion asymmetry, *Nature* **447**, 190–193 (2007).
68. P. Ferriani, K. von Bergmann, E. Y. Vedmedenko et al., Atomic-scale spin spiral with a unique rotational sense: Mn monolayer on W(001), *Phys. Rev. Lett.* **101**, 027201 (2008).
69. M. Heide, G. Bihlmayer, and S. Blugel, Dzyaloshinskii–Moriya interaction accounting for the orientation of magnetic domains in ultrathin films: Fe/W(110), *Phys. Rev. B* **78**, 140403 (2008).
70. S. Mühlbauer, B. Binz, F. Jonietz et al., Skyrmion lattice in a chiral magnet, *Science* **323**, 915 (2009).
71. M. Uchida, Y. Onose, Y. Matsui, and Y. Tokura, Real-space observation of helical spin order, *Science* **311**, 359 (2006).
72. M. Bode, O. Pietzsch, A. Kubetzka, and R. Wiesendanger, Shape-dependent thermal switching behavior of superparamagnetic nanoislands, *Phys. Rev. Lett.* **92**, 067201 (2004).
73. S. Krause, L. Berbil-Bautista, G. Herzog, M. Bode, and R. Wiesendanger, Current induced magnetization switching with a spin-polarized scanning tunneling microscope, *Science* **317**, 1537–1540 (2007).
74. S. Krause, G. Herzog, T. Stapelfeldt et al., Magnetization reversal of nanoscale islands: How size and shape affect the Arrhenius prefactor, *Phys. Rev. Lett.* **103**, 127202 (2009).
75. A. J. Heinrich, J. A. Gupta, C. P. Lutz, and D. M. Eigler, Single-atom spin-flip spectroscopy, *Science* **306**, 466–469 (2004).
76. C. F. Hirjibehedin, C. P. Lutz, and A. J. Heinrich, Spin coupling in engineered atomic structures, *Science* **312**, 1021–1024 (2006).
77. C. F. Hirjibehedin, C.-Y. Lin, A. F. Otte et al., Large magnetic anisotropy of a single atomic spin embedded in a surface molecular network, *Science* **317**, 1199–1203 (2007).
78. F. Meier, L. Zhou, J. Wiebe, and R. Wiesendanger, Revealing magnetic interactions from single-atom magnetization curves, *Science* **320**, 82 (2008).

27

Point Contact Andreev Reflection Spectroscopy

Boris E. Nadgorny
Wayne State University

27.1 Introduction

It is generally accepted that spin polarization plays a major role in *spintronics* [1,2]. Indeed, one cannot adequately discuss spin-dependent transport without addressing the question of spin imbalance in a given system. The topics included in this chapter are chosen by association one might have with the words *spin polarization*. When we think about it, the first thing that comes to mind is probably the asymmetry in the density of states (DOS) in magnetic materials. However, spin polarization may also be related to different transport, magnetic, and optical properties of charge carriers (see Chapters 17 and 23 of this book), spin injection into paramagnetic materials (see Chapters 6, 18, and 29), transmission coefficients through interfaces and tunnel barriers (see Chapters 4, 10, and 11), and different electronic properties of surfaces and the bulk. This indicates that the concept of spin polarization is not self-evident and must be defined [3]. These definitions are naturally linked to spin measurement techniques, as we will discuss later.

The efficiency of a ubiquitous spintronics device can be dramatically improved as the spin polarization approaches 100%. For example, in tunneling magnetoresistance (TMR) devices, the efficiency $\Delta R/R \rightarrow \infty$ as the spin polarization P approaches 100%, based on the qualitative Julliere's model [4], see Figure 27.1. Obviously, this conclusion corresponds to an idealized case. For any real devices, interface scattering, spin–orbit interaction, nonzero temperatures, and other spin decoherence effects would reduce the device efficiency to finite values. Full spin polarization can be nominally realized in the extreme case of a *half-metal*, simultaneously exhibiting metallic and insulating properties at a microscopic level [5]. In half-metals, first introduced by de Groot et al. [6], only one of the energy bands (majority) crosses the Fermi level, E_F, whereas another (minority) band has a gap in the DOS. Consequently, at $T=0$, a half-metal would only be conducting in one spin channel, whereas the other spin channel would be insulating (hence *half*-metal), so that all the electrical current is carried by the charges with the same spin. However, to achieve a high degree of transport spin polarization, half-metallicity in this narrow sense is not

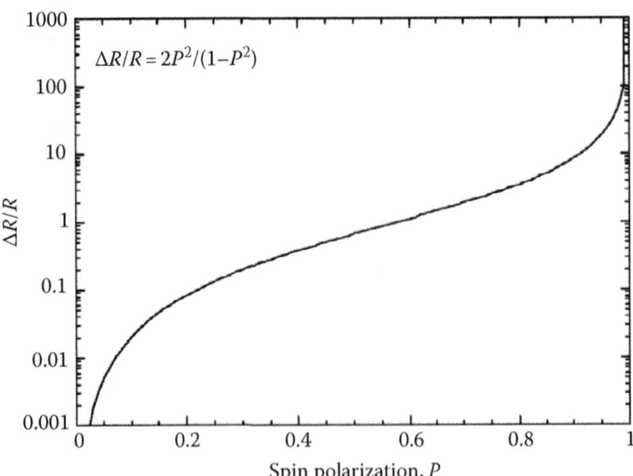

FIGURE 27.1 The efficiency of a TMR device $\Delta R/R$ as a function of spin polarization PT according to Jullier's model.

required. For example, if the charge carriers from the minority spin band are localized, the current will be due almost entirely to the majority carriers, resulting in the so-called *transport half-metallicity* [7,8]. Similarly, recent dramatic advances in interface engineering, such as the use of MgO tunnel barriers [9,10] allow one to use spin-filtering effects in magnetic tunnel junctions to obtain several hundred percent of TMR, which, to some extent, has alleviated the need for intrinsic half-metallicity.

The bulk of the chapter will be devoted to the discussion of *point contact Andreev reflection* (PCAR) spectroscopy, one of the commonly used techniques to measure spin polarization [11,12]. This technique is based on the so-called Andreev reflection (AR) process [13,14],* which determines the transport properties of a clean (transparent) normal-metal (N)–superconductor (S) interface, and on the observation that AR between a half-metal and a superconductor is forbidden for the quasiparticles with the energies below the superconducting gap [15]. The classical paper of Blonder et al. [16] gives the detailed description of AR in the case of an arbitrary interface. A different perspective on AR, based on the scattering theory formalism, may be found in the review by Beenakker [17]. These two approaches were utilized for the case of a ferromagnet (F) by Upadhyay et al. [12], and by Mazin et al. [18], among others, to create a working formalism for the analysis of the conductance data across an FS interface, allowing one to determine the values of spin polarization of the ferromagnet. AR is intimately related to the proximity effect; the interplay between these two phenomena is discussed in the reviews by Pannetier and Courtois [19], and Buzdin [20]. In recent years, AR has also become a useful tool for studying unconventional superconductivity, such as d-wave, triplet superconductivity, multiband superconductivity, and heavy fermions. We will not address these topics here; the reviews by Kashiwaya and Tanaka [21], Mackenzie and Maeno [22], and Deutscher [23] are devoted to some of these questions; the overview of AR in graphene is given by Beenakker [24].

* See the discussion on page 538.

27.2 Definitions of Spin Polarization

The problem of measuring spin polarization is twofold. While it is often assumed that the definition of spin polarization P is self-evident, similarly to magnetization, for example, which is uniquely determined by the difference between the algebraic sum of spin-up and spin-down electronic magnetic moments, this is not the case. First, the definition of P may be different for different measurement techniques and may be sensitive to experimental details. Second, while it may be straightforward to define the degree of spin polarization in the bulk of a uniform ferromagnet, the problem of spin injection, for example, require the knowledge of varying spin polarization in complex nonuniform systems, which may include various types of interfaces. As the Fermi surfaces of typical metallic ferromagnets are quite complicated, the degree of spin polarization of the injected spins may depend strongly on the crystallographic orientation and the nature of the carrier propagation across the contact [25,26]. For example, a tunnel junction may result in different values of spin polarization compared to a ballistic point contact. Various issues important for defining the concept of spin polarization were addressed by Mazin [3]. Let us start with the most intuitive case of spin imbalance: magnetization. Magnetization spin polarization can be as defined the imbalance in the electronic DOS integrated over all occupied spin states $N_\uparrow(E)$, $N_\downarrow(E)$:

$$P_M = \frac{\int N_\uparrow(E)dE - \int N_\downarrow(E)dE}{\int N_\uparrow(E)dE + \int N_\downarrow(E)dE} = \frac{n_\uparrow - n_\downarrow}{n_\uparrow + n_\downarrow} \tag{27.1}$$

This definition can also be applied to the case of spin injection into semiconductors, where the measured photoluminescence comes from carrier recombination [27]. The most common definition of spin polarization, P_N

$$P_N = \frac{N_\uparrow(E_F) - N_\downarrow(E_F)}{N_\uparrow(E_F) + N_\downarrow(E_F)} = \frac{N_\uparrow - N_\downarrow}{N_\uparrow + N_\downarrow} \tag{27.2}$$

which is defined in terms of spin-up (down) Fermi-level DOS, $N_{\uparrow(\uparrow)}(E_F)$ (we are omitting E_F here and below). $N_{\uparrow(\uparrow)}$ is obtained by integrating over the Fermi surface S_F where $v_{\uparrow(\downarrow)}$ are spin-up (spin-down) Fermi velocities.

$$N_{\uparrow(\downarrow)} = \int \frac{dS_F}{\mathbf{v}_{k\uparrow(\downarrow)}(2\pi)^3} \tag{27.3}$$

Notice that P_N may differ from P_M not only in magnitude but even in sign. This definition is useful for a comparison with spin-resolved photoemission spectroscopy, which nominally measures P_N.[†] However in the measurements related to transport

[†] Strictly speaking, this is only true in the constant matrix approximation, i.e., under the assumption that the transition probabilities of photoelectrons are spin-independent.

phenomena, different definitions are more relevant. For example, AR spectroscopy in the ballistic regime measures the so-called *current carrying* DOS $\langle Nv_x \rangle$. In a ballistic, or Sharvin point contact of an area A, an electron going through the contact gains an energy eV [28]. As the fraction of electrons with a given wave vector k reaching the contact per unit time is Nv_x, the current through the contact is

$$I_{\uparrow(\downarrow)} = e^2VA\langle Nv_x \rangle = e^2VA\int \frac{v_x dS_F}{v(2\pi)^3} = e^2VAS_{x\uparrow(\downarrow)} \tag{27.4}$$

where S_x is the projection of the Fermi surface onto the contact plane (see Sections 27.3 and 27.5 for more detail). Thus we can define the spin polarization of the current propagating through a ballistic or Sharvin contact as P_{NV}

$$P_{NV} = \frac{\langle N_\uparrow v_{x\uparrow} \rangle - \langle N_\downarrow v_{x\downarrow} \rangle}{\langle N_\uparrow v_{x\uparrow} \rangle + \langle N_\downarrow v_{x\downarrow} \rangle} \tag{27.5}$$

Let us now consider the current in a bulk ferromagnet, assuming that it is governed by the classical Boltzmann transport equation. In this case, the current density $J_{\uparrow(\downarrow)}$

$$J_{\uparrow(\downarrow)} \sim \int \frac{v^2 dS_F}{\mathbf{v}_{k_{\uparrow(\downarrow)}}(2\pi)^3}\tau_{k_{\uparrow(\downarrow)}} \int \frac{\mathbf{v}_k dS_F}{(2\pi)^3}\tau_{k_{\uparrow(\downarrow)}} \tag{27.6}$$

In general, the scattering time $\tau_{k\uparrow(\downarrow)}$ is both k and spin dependent and the current densities need to be evaluated from Equation 27.6. By neglecting the scattering anisotropy and assuming the same relaxation time τ for spin-up and spin-down carriers, we can simplify Equation 27.6 $J_{\uparrow(\downarrow)} \sim \langle N_{\uparrow(\downarrow)}v^2 \rangle \tau_{\uparrow(\downarrow)} \sim \langle N_{\uparrow(\downarrow)}v^2 \rangle \tau$, so that the spin polarization of the current can then be determined by P_{Nv^2}*

$$P_{Nv^2} = \frac{J_\uparrow - J_\downarrow}{J_\uparrow + J_\downarrow} = \frac{\langle N_\uparrow v_\uparrow^2 \rangle - \langle N_\downarrow v_\downarrow^2 \rangle}{\langle N_\uparrow v_\uparrow^2 \rangle + \langle N_\downarrow v_\downarrow^2 \rangle} \tag{27.7}$$

Equation 27.7 also applies in the case of the diffusive point contact.†

We may now ask, what is the spin polarization plotted in Figure 27.1? Contrary to some of the claims in the literature, it is not the spin polarization of the DOS, P_N. Julliere's derivation is based on Tedrow and Meservey's tunneling experiments [30], in which the current in a superconductor/oxide/ferromagnet (S/I/F) junction is measured as a function of voltage V. Assuming that the respective tunneling matrix elements $|M_{\uparrow\downarrow}(E)|$ are independent of energy for $V \ll E_F$ one can define the tunneling spin polarization P_T,

$$P_T = \frac{I_\uparrow - I_\uparrow}{I_\uparrow + I_\uparrow} = \frac{N_\uparrow |M_\downarrow(E)|^2 - N_\downarrow |M_\downarrow(E)|^2}{N_\uparrow |M_\downarrow(E)|^2 + N_\downarrow |M_\downarrow(E)|^2} \tag{27.8}$$

Thus, in the original Julliere formula, P is defined not by $N_{\uparrow\downarrow}$ but by $N_{\uparrow\downarrow}$ multiplied by $|M_{\uparrow\downarrow}(E)|^2$. In general, the two tunneling matrix elements ($|M_\downarrow(E)|$) and ($|M_\uparrow(E)|$) are not the same, as the minority and majority carriers come from different electronic sub-bands and thus may have different masses and tunneling probabilities (e.g., Refs. [31,32]). This result was further emphasized in relation to the spin polarization measurements of $La_{1-x}Sr_xMnO_3$ (LSMO) [33,34] and in the recent reviews by Butler [35] and Velev et al. [36]. There is a similarity with the giant magnetoresistance (GMR) effect, where the interface resistance is strongly spin dependent (for more detail see Chapters 4 and 5), which implies that we must separate the intrinsic spin polarization of a magnetic material from the effects of specific tunnel barriers [37–44]. A more detailed discussion of the Julliere formula is given by Belashchenko and Tsymbal in Chapter 12 and by LeClair and Moodera in Chapter 10 of this book.

For the tunneling current in a low transparency specular junction (defined by a δ-function), the spin polarization is reduced to P_{Nv^2} [45] and thus is also determined by Equation 27.7. Importantly, only for true half-metals $P_N = P_{Nv} = P_{Nv^2} = 100\%$; for all other ferromagnetic materials P_N, P_{Nv}, and P_{Nv^2} may be very different [3,7,46].

27.2.1 Spin Polarization Measurement Techniques

A number of techniques commonly used for the spin polarization measurements are *positron spin spectroscopy, spin-resolved photoemission spectroscopy, spin-polarized tunneling (Meservey–Tedrow technique), and PCAR spectroscopy.*‡

27.2.1.1 Positron Spin Spectroscopy

Positron spin spectroscopy [49] is based on the process of positron–electron annihilation, which is dependent on the electron density at the annihilation site [50]. By analyzing the emitted γ rays, the information on the electron momentum distribution

* This is a fairly rudimentary approximation, as, even in the absence of magnetic impurities, the scattering time for different spin sub-bands is dependent on the type of wave functions (e.g., s- and d-electrons) and the DOS of the respective spin sub-bands. However, it has been fairly successfully used in a number of band structure calculations, for example in the GMR calculations by Oguchi [29] (see Chapter 5).

† In contrast to the electric current, the spin current generally cannot be described by the continuity equation. Thus, the notion of current spin polarization can only be introduced in the limit of small spin–orbit interaction.

‡ Recently, a new quantitative technique, measuring P_{Nv^2}, the current-induced spin wave Doppler shift, has been introduced [47]. Qualitative measurements, designed to distinguish between half-metals and other ferromagnets based on femtosecond spin excitation, have also been realized [48].

and the electron density can be extracted, and Fermi surfaces can be reconstructed. Positrons emitted from an isotope of Na-22 are partially spin polarized [51]. Switching the magnetization direction of a magnetic material allows one to obtain a spin-resolved electronic structure, as positrons annihilate preferentially with electrons of opposite spin directions. Spin-resolved positron spectroscopy allows one to analyze the electronic structure of magnetic materials [49,52,53], enabling the measurements of P_N in a wide temperature range. While positron spin spectroscopy can be quite effective, it is difficult to apply this technique to thin films, and its impact on the field has been rather limited, due, in part, to the relative complexity of the measurements, requiring a dedicated experimental setup.

27.2.1.2 Spin-Resolved Photoemission Spectroscopy

Spin-resolved photoemission spectroscopy [54–56] is also, in principle, capable of measuring P_N (in the constant matrix approximation). While the energy resolution of this technique is inferior compared to the spin-polarized tunneling or Point contact AR techniques, it has recently been substantially improved to about 50 meV. In this technique, monochromatic photons are produced by a light source, with the whole setup evacuated to ultrahigh vacuum. The kinetic energy of the photoelectrons is analyzed by electrostatic analyzers as a function of the emission angles, the electron spin orientation, the photon energy, and the polarization [57]. The spin components of the beam are detected by a Mott detector, which contains a high Z target (typically Au). This method is quite surface sensitive, as it measures the spin of the electrons emitted from a region close to the surface of a ferromagnet on the order of 5–20 Å [58], which corresponds approximately to the mean free path of excited electrons at tens of electron volts above the Fermi energy. Thus, while photoemission can be successfully used to study surfaces of magnetic materials, in order to obtain reliable values of the bulk spin polarization, samples must be prepared in situ. Detailed spin-polarized photoemission studies are performed for various types of magnetic materials, such as Fe [59–61], as well as Fe/MgO interface [62], Co [63], Ni [64], Gd [65], Fe_3O_4 [66], LSMO [67], and CrO_2 [68], among others. *Inverse photoemission* can also be applied to the studies of magnetic materials [69].

27.2.1.3 Spin-Polarized Tunneling (Meservey–Tedrow) Technique

While the previous two techniques can work at room temperature, spin-polarized tunneling and PCAR spectroscopy are inherently low-temperature techniques as in both cases a superconductor is utilized. Spin-polarized tunneling in a planar junction geometry has high (sub-meV) energy resolution. Meservey and Tedrow [30,70,71] pioneered this technique by making ferromagnet (F)/superconductor (S) tunnel junctions, and Zeeman-splitting the superconducting single-particle excitation spectrum in very thin Al films by applying a magnetic field parallel to the plane of the junction. The resulting (dI/dV) spectrum, corrected for spin–orbit and orbital depairing effects using Maki equations [33,72], can be used to detect P_T. As it is directly related to TMR [4], spin-polarized tunneling is the technique of choice for the study of TMR devices [73]. The drawback of the technique is the constraint of fabricating a layered device with a thin ferromagnet film on top of an oxide layer. Generally, the values of tunneling spin polarization, P_T, for the same ferromagnet are sensitive to the quality of the interface [74], the degree of disorder within it [44,75], the shape and the composition of the tunnel barrier [76,77], and the type of bonding [78]. The MT technique is described in detail in Chapters 10 and 13.

27.2.1.4 Point Contact Andreev Reflection Spectroscopy

PCAR spectroscopy,* introduced by Soulen et al. [11] and Upahadya et al. [12], is probably the most flexible technique, as it places no special constraints on the types of samples that can be studied; a large number of metals, semiconductors, and magnetic oxides have been investigated (see Table 27.2 at the end of this chapter). The technique is based on AR [13,14], which facilitates a direct conversion of the quasiparticle current in a normal metal into the supercurrent, and the notion that AR is not allowed for a half-metal [15]. In contrast to spin-polarized tunneling, the PCAR technique ideally requires a clean/no-barrier interface between a ferromagnet and a superconductor. The technique often utilizes the so-called point contact, or Sharvin geometry, which may be realized by using either an adjustable tip [11], or a nanoconstriction [12]. Planar AR can also be used if the transport properties of an interface are dominated by AR [79].

27.3 Transport Properties of a Point Contact

27.3.1 Sharvin Resistance

The concept of the Sharvin resistance [28] was introduced by Sharvin who suggested to study Fermi surfaces of metals using electron focusing effect in a uniform magnetic field H_z, which is similar to the principle of a beta spectrograph. Consider a group of electrons starting from the same origin $r=0$ with the same wave vector in the direction of the field, k_z; the other two components of the wave vector may be arbitrary. As all of them have the same orbital period, they will go around an integer number of orbits in the same amount of time. Therefore, in real space, they will return to the same point at the xy plane $(x=0, y=0)$, whereas z will be determined by the total number of periods, the magnetic field, and the average velocity v_z. If a significant number of these electrons (with the same k_z) were

* While this term is often applied more broadly, e.g., to the point contact measurements of unconventional superconductors, we will limit ourselves to the spin polarization measurements.

focused at the same point, the sizable additional current would flow. This effect was demonstrated by Sharvin and Fisher [80], who fabricated two very fine (point) contacts on the opposite side of a Sn single crystal with a large mean free path, so that the electrons accelerated by the applied voltage V at one contact were focused at the other point contact, which resulted in periodic oscillations of the current when the focusing conditions were fulfilled. Later, a number of studies of focusing effects were performed in the transverse geometry, particularly in 2DEG [81–83]. These effects were successfully used to observe AR with two separate point contacts [84] and a single contact [85].

27.3.2 Ballistic (Knudsen) Regime

The main requirement for the focusing effect is that the electronic elastic mean free path l should be much larger than both the radius a and the width of the contact w, so that $l \gg a,w$. This problem is analogous to the problem of a dilute gas flow through a small hole, in the so-called Knudsen regime [86], see Figure 27.2. The degree of "ballisticity" is determined by the so-called Knudsen number $\kappa = l/a$, with $\kappa \gg 1$ defining the pure ballistic regime. Sharvin argued that in a linear regime (i.e., when eV is much smaller than E_F) as the electrons cross the area of the contact in two opposite directions they would acquire (lose) additional velocity $\delta v = eV/p_F$, where V is the voltage applied across the junction, and p_F is the Fermi momentum. The current corresponding to such a flow is $I \sim enA\delta v$, where A is the area of the contact and n is the electron density. The additional resistance is then $R_{Sh} = V/I \sim p_F/e^2nA = p_F/\pi e^2na^2$. Using the Drude expression for the resistivity $\rho = p_F/ne^2l$ and thus $R_{Sh} = \rho l/\pi a^2$. More accurately, by integrating over all angles the numerical factor of $\frac{4}{3}$ could be obtained, similar to the kinetic gas theory, where the number of molecules hitting a unit area of a wall per second is $\frac{1}{4}nv$ rather than $\frac{1}{3}nv$, which one might assume without integration. The Sharvin resistance then becomes

$$R_{Sh} = \frac{4\rho l}{3\pi a^2} = \frac{4\rho}{3\pi a}\kappa \tag{27.9}$$

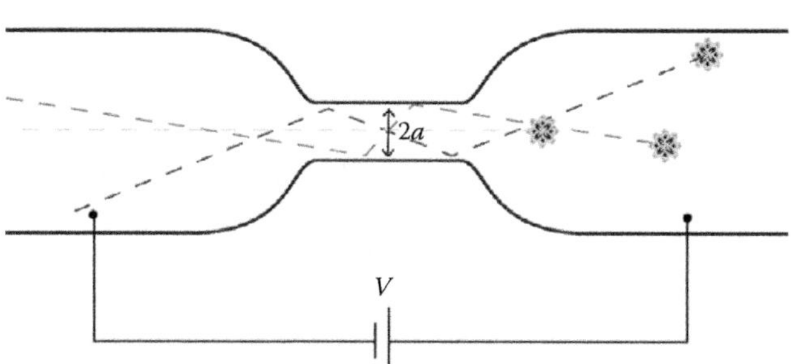

FIGURE 27.2 Ballistic (Knudsen) regime with the mean free path $l \gg a$, w.

Importantly, as in the ballistic regime, there is practically no scattering in the area of the contact, the Sharvin resistance of metals is largely independent of their materials properties, as $\rho \sim 1/l$. Indeed, in the three-dimensional (3D) geometry, the Sharvin resistance can be expressed via a number of transverse quantum modes that can propagate through the constriction $\int A d^2k/(2\pi)^2 = k_F^2 A/4\pi$. Thus, the Sharvin resistance depends only on the properties of the Fermi surface and the area of the contact:

$$R_{Sh} = \frac{4\rho l}{3\pi a^2} = \left[\frac{2e^2}{h}\right]\left(\frac{k_F^2 A}{4\pi}\right)^{-1} = R_Q\left(\frac{k_F^2 A}{4\pi}\right)^{-1} \tag{27.10}$$

where R_Q is the quantum resistance, $R_Q \simeq 25.8\,\text{k}\Omega$. While the result is given here for a spherical Fermi surface with a quadratic dispersion law, it can be easily generalized for an arbitrary Fermi surface, see Section 5.2.1. Note that the semiclassical approach works well in 3D (but not in 2D), as the electron de Broglie's wave length $\lambda_F \simeq 0.5\,\text{nm}$ is much smaller than a.

27.3.3 Diffusive (Maxwell) Regime

In the opposite limit, $l \ll a$, the diffusive transport regime with the so-called *Maxwell resistance* is realized. The problem can be solved exactly in oblate spheroidal coordinates [87]. The solution on both sides of the diaphragm (see Figure 27.3) is

$$\Phi = \pm V\left[1 - \frac{2}{\pi}\arctan(\xi)^{-1}\right] \tag{27.11}$$

where ξ is defined through cylindrical coordinates r, and z, $(r/a)^2 = (1+\xi^2)(1 - z^2/\xi^2a^2)$ where a is the radius of a circular orifice. Thus, the Maxwell resistance is inversely proportional to the diameter of the constriction $2a$

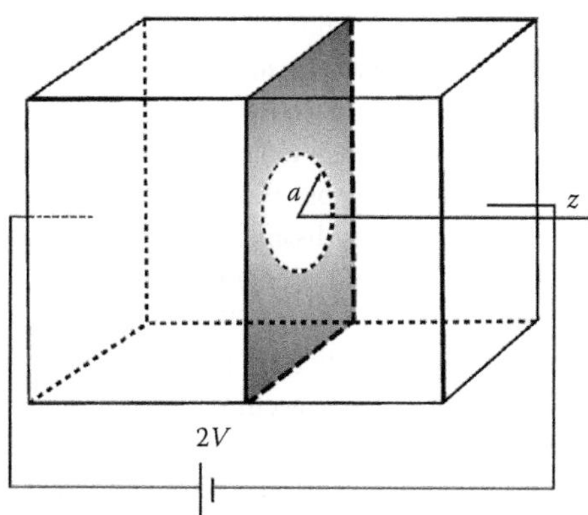

FIGURE 27.3 The Maxwell case ($l \ll a$). Electronic transport through the aperture of radius a in an insulating diaphragm separating two conducting half-spaces with the mean free path l.

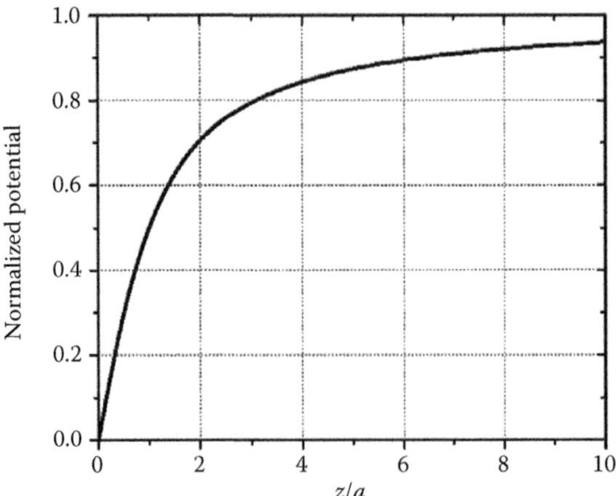

FIGURE 27.4 Voltage profile along the z axis of Figure 27.3. Note that ~75% of the total voltage drop takes place within a distance of $2a$ from the center of the orifice.

$$R_M = \frac{\rho}{2a} \quad (27.12)$$

Note that most of the voltage drop takes place within the distance of $2a$ from the orifice (see Figure 27.4).

27.3.4 Intermediate Regime

For the contact resistance R_C in the intermediate regime, the resistance is given by Wexler's formula [88]

$$R_C = \frac{4\rho l}{3\pi a^2} + \Gamma(\kappa)\frac{\rho}{2a} = R_{Sh}\left[1 + \frac{3\pi}{8}\Gamma(\kappa)\frac{a}{l}\right] = R_{Sh}\left[1 + \frac{3\pi}{8}\frac{\Gamma(\kappa)}{\kappa}\right] \quad (27.13)$$

where

$\kappa = l/a$ is the Knudsen number

$\Gamma(\kappa)$ is the geometric factor on the order of unity, which varies from $\simeq 0.7$, when $\kappa \to \infty$ (the pure ballistic case) to 1 when $\kappa = 0$ (the pure Maxwell case)

In the intermediate case (diffusive regime), the elastic mean free path l becomes smaller than a; however, the characteristic diffusion length $\Lambda = \sqrt{ll_{in}} \gtrsim a$, where l_{in} is an inelastic mean free path, which is strongly energy dependent. The larger κ is, the sharper the voltage drop across the contact becomes. In both cases, however, the energy is dissipated within the same length $\Lambda = \sqrt{ll_{in}}$, beyond the region of a potential drop [89].*

The exact solution in the same (linearized) approximation as used by Wexler has recently been found [90]. As the authors noted, Wexler's result is fairly accurate (within a few percent) compared to the exact solution. In the intermediate (quasi-ballistic) regime when $a \sim l$ it is expected that the contributions from both the Sharvin and the Maxwell resistances are approximately equal. Rather surprisingly, the naive approximation, which comes as the sum of the two limits,

$$R_C = R_{Sh}\left(1 + \frac{3\pi}{8}\frac{a}{l}\right) = R_{Sh}\left(1 + \frac{3\pi}{8}\frac{1}{\kappa}\right) \quad (27.14)$$

produces fairly accurate results, with the largest difference of only about 10% (near $\kappa = 1$) compared to the exact solution. Therefore, for all practical purposes, such as the estimates of the contact size, Equation 27.14 can be used.

In the case of materials with different but *isotropic* electrical properties on both sides of the contact, the total contact resistance R_Σ can generally be expressed as an average of the two resistances, R_{C_1} and R_{C_2} [88]:

$$R_\Sigma = \frac{1}{2}R_{C_1} + \frac{1}{2}R_{C_2} \quad (27.15)$$

This equality has some interesting implications in the case of AR, where the resistance of the superconducting side is nominally zero. In this case (at $T = 0$), the junction resistance $R_\Sigma = \frac{1}{2}R_N$ below the gap is one half of the resistance of the normal metal. Note that this is only correct in the case of a purely ballistic transport across the junction [91]. In the ideal diffusive case (disordered contact) the resistance is $R_\Sigma = R_N$ [92–94].

27.3.5 Thermal Case

In the case of $\Lambda = \sqrt{ll_{in}} \ll a$ when both elastic and inelastic mean free paths become smaller than the contact size, one can define the *thermal regime* [95]. In this case, the scattering in the area of the contact is quite strong, and, since most of the voltage drop also occurs there, this is where most of the energy is dissipated. This can, in principle, lead to a significant increase of the temperature in the area of the contact. Since l_{in} is strongly energy dependent, the maximum temperature in the area of the contact T_{max} can be much larger than the bath temperature T_{bath} [96]

$$T_{max}^2 = T_{bath}^2 + \frac{V^2}{4L} \quad (27.16)$$

where L is the Lorenz number. Thus, a contact may be in the ballistic or diffusive regime at low voltages, where one can still observe a phonon spectra, and fall into the thermal regime only at much higher voltages, as was demonstrated in the case of Ni-Ni point contacts, where the Curie temperature of Ni (≈630 K) has presumably been reached in the thermal regime [97]. Unless the contact resistance is very small (less than 1 Ω), one is unlikely to reach the thermal regime at several mV, which is typically required for the PCAR measurements with conventional superconductors.

* In both the diffusive and ballistic cases the energy relaxation is nonlocal, i.e., the Joule heat through a constriction cannot be expressed through the local electric field, only through the applied voltage as GV^2.

27.3.6 Point Contact Spectroscopy

Since the Sharvin resistance in the first approximation is independent of l, it should be largely independent of energy, except for a small (on the order of eV/E_F) variation of the DOS with the applied bias. Therefore, an ideal Sharvin contact (with $\kappa = \infty$) in this approximation is expected to be ohmic. However, this implies no scattering at all, and, therefore, is unphysical. One can show [98], that for any realistic case of finite κ, the second term in Equation 27.13 leads to a strong nonlinearity of R_C. The degree of this nonlinearity, d^2V/d^2I, is dependent upon inelastic scattering rate at different energies, which, as was first noted by Yanson, is directly related to electron–phonon interaction [99]. This pioneering work enabled the study of electron–phonon as well as other quasiparticle spectra, starting the area of *point contact spectroscopy*. For further details of this technique we refer to several extensive reviews [95,98,100].

There is a subtle relationship between the Sharvin and Maxwell parts of the resistance in Equation 27.13. On the one hand, the energy-dependent features in the point contact resistance are the result of scattering with quasiparticles of different types, and therefore are only possible in the presence of the diffusive part of the resistance. On the other hand, the more "ballistic" the contact is ($\kappa \to \infty$), the better is the energy selectivity, and thus the better these features are resolved. As the contact becomes more diffusive, the energy of the electron is no longer determined by the applied voltage, and, eventually, the d^2V/d^2I curve washes out. Therefore, the presence of the phonon spectra, is often used as a proof that the transport through a contact is ballistic or quasi-ballistic [12,85,101,102].

27.3.7 Contact Size Determination

While the presence of the phonon spectra can be used in order to confirm that one is operating in the ballistic regime, it does not help in obtaining a quantitative estimate of the contact size. The latter can be roughly estimated by solving the quadratic Equation 27.13. However, one then has to assume the bulk resistivity in the area of the contact, which may not be the case due to various defects. Therefore, it is preferable to determine the contact size by measuring the temperature dependence of both the residual resistivity and the contact resistance and by making use of the fact that the Sharvin resistance is independent of temperature. Following Akimenko et al. [103], we can differentiate Equation 27.13 to obtain

$$\frac{\partial R_C}{\partial T} = \frac{\partial}{\partial T}\left[\frac{4\rho l}{3\pi a^2} + \Gamma(\kappa)\frac{\rho}{2a}\right] = \frac{\Gamma(\kappa)}{2a}\frac{\partial \rho}{\partial T} \tag{27.17}$$

We assume the function $0.7 \le \Gamma(\kappa) \le 1$ to be 1 to find the diameter of the contact [102,103]

$$d = 2a = \frac{\partial\rho/\partial T}{\partial R_C/\partial T} \tag{27.18}$$

For the ballistic point contact of Cu and Au, for example, the following rough estimate can be used $d = 30/\sqrt{R(\Omega)}$ (nm). For a typical resistivity of a contact ranging between 1 and 100 Ω, the contact diameter varies from 30 to 3 nm [103]; more detail can be found in Naidyuk and Yanson [104].*

27.4 Andreev Reflection: Unpolarized Case

Andreev reflection, named after Alexander F. Andreev, was introduced in 1964 [13] to explain the puzzling effects of anomalously high thermal resistance in the intermediate state of type-I superconductors, which greatly exceeded the thermal resistance in the superconducting state [105]. Moreover, in contrast to the superconducting state, the thermal resistance of the intermediate state was practically independent of the electron mean free path, even in the temperature range where the heat transfer was dominated by electrons [106]. This implied that the interface played an important role, somehow preventing electrons from crossing the N/S boundary. However, a mechanism that might be responsible for this effect was not understood.

Indeed, upon arriving at an N/S interface, an electron seems to have only two obvious choices: either it goes across the interface into the superconductor, or it gets reflected back to the normal metal. A single electron with the energy below the superconducting gap Δ is not allowed to propagate in the superconductor as a quasiparticle (at $T = 0$ K), so the first option is not possible. On the other hand, if it simply gets reflected, this would imply that the net current through such a contact is zero. This is rather counterintuitive, as one would expect that if a part of the contact becomes superconducting, its resistance will go down, compared to the all-normal case (see Section 27.3.4). Moreover, if it is going to be reflected by the usual scattering process, a large barrier might be needed to reverse the Fermi momentum of the electron. Indeed, in the case of a clean interface, the only energy scale present in this problem is related to the superconducting gap, which implies [107] that the change of momentum is on the order of

$$\delta p_{Fx} \sim \frac{dp_F}{dt}\delta t \sim -F_x\delta t \sim \frac{\Delta}{\xi}\frac{\xi}{v_F} \sim \frac{\Delta}{v_F} \sim p_F\left(\frac{\Delta}{\varepsilon_F}\right) \ll p_F \tag{27.19}$$

where ξ is the coherence length. This is a classical case of an infinitely heavy ball hitting an impenetrable wall [91]. So, then, which answer will correctly describe the behavior of the electron in this case? Neither! It gets reflected—but in a highly unusual way: Its charge becomes opposite to the charge of the incoming electron and all three components of its momentum change sign, that is, it is the *hole* that gets reflected along the same trajectory

* Note that these estimates give the upper limit of the contact size, as they are based on the assumption that a single contact is established.

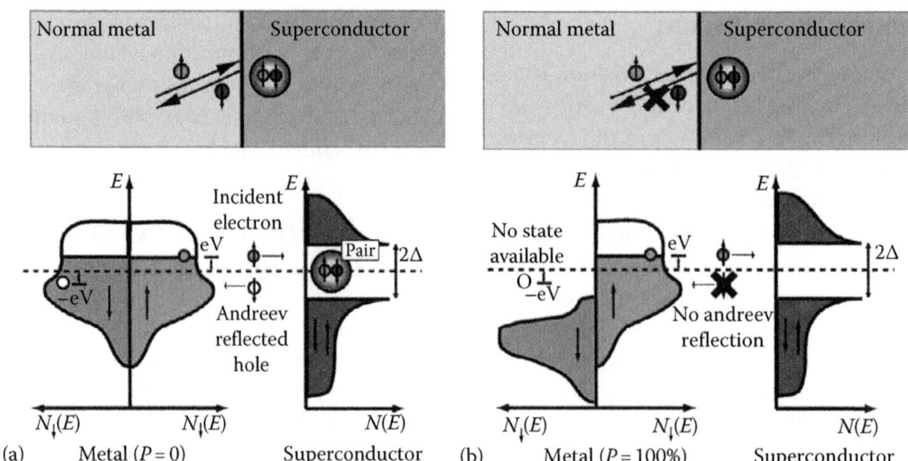

FIGURE 27.5 Schematic of quasiparticle to supercurrent conversion via Andreev reflection. (a) Andreev reflection in a normal (unpolarized) metal is unhindered as both spin up and spin down electrons are present; (b) Andreev reflection in a half-metal is suppressed, as there are no opposite spin states for the hole to get reflected to; therefore supercurrent conversion is blocked. (From Soulen, Jr. R.J. et al., *Science*, 282, 85, 1998. With permission.)

(*retro-reflected*) (see Figure 27.5a). The reflected hole also has the opposite spin.

This has a number of important consequences: (1) As the hole moving in the opposite direction is equivalent to the electron moving in the same direction, the conduction across the contact doubles (compared to the all-normal case, see Section 27.3.4). (2) As the original electron is accompanied by the electron with the opposite spin, the two of them constitute a Cooper pair that propagates inside the superconductor, thus the quasiparticle current gets converted into the supercurrent. (3) This process is elastic as the quasiparticle energy is the same for the incident electron and the reflected hole. Momentum is also conserved to the first order of Δ/E_F (usually a good approximation for conventional superconductors).

The thermal resistance of the intermediate state, calculated by Andreev using this mechanism, turned out to be in excellent agreement with the results of Zavaritskii [13]: the thermal resistance did not depend on the mean free path, and was exponentially increasing at lower temperatures, as the number of quasiparticles was decreasing proportionally. In other words, Andreev-reflected electrons, converted into Cooper pairs did not carry any heat, only quasiparticles did. This mechanism has broad implications in solid-state physics, including the proximity and quantum coherence effects in N/S systems [19,20,108–110], the symmetry of the order parameter [23,111–113], as well as in other areas of physics (e.g., Refs. [114,115]).

A different problem, rooted in the same phenomenon, was addressed by de Gennes and Saint James [116] and Saint James [14]. They considered a proximity effect of a superconductor with a normal metal, and discovered that if the thickness of a normal film is small, a bound state that was an admixture of electrons and holes is created, and the DOS in the normal metal is increased by a factor of two.* These states, which are often called *de Gennes/Saint James bound states*, have been observed experimentally, see, for example, Giazotto et al. [117].

While there has been a number of indirect experiments, indicating that AR is likely to play a major role in tunneling [118,119], the first unambiguous evidence for this effect was obtained by Krylov and Sharvin [120], who used the so-called Gantmakher (radio frequency) size effect, in which AR manifested itself by the appearance of additional lines, corresponding to trajectories of retro-reflected quasiparticles. In addition to these types of measurements [121], the most direct demonstration of AR came from using the focusing effect in the transverse geometry [84].

Let us discuss qualitatively how AR enables the process of quasiparticle to supercurrent conversion. Adopting the description of Pippard et al. [122] and Schmidt [123], we will first consider a *clean* interface† and assume that the absolute value of the energy of an electron ε_k is less than the superconducting gap $\Delta(T)$. The order parameter then raises from 0 inside the normal

* The author wishes to thank Prof. Guy Deutscher for drawing his attention to the largely unknown *Journal de Physique* paper of Saint James [14], which has never been translated into English. While the paper does not address the transport between normal and superconducting parts *per se*, it correctly identifies the origin of the DOS enhancement in the normal layer near an N/S interface, which is due to the transformation of an electron into a Cooper pair at the S interface, with a hole reflecting back to the N interface and after another reflection from N to S "annihilating" with the Cooper pair, thus leading to doubling of the DOS.

† We will define a clean or ideal interface as the one with (1) no oxide barrier and (2) with both normal and superconducting layers composed of the same material, so that no interface velocity mismatch between a superconductor and a normal metal $\gamma = v_{Fn}/v_{Fs}$ is present [16]. In practice, this can be achieved by either using a superconductor in the intermediate state, or a variable thickness structure with different critical temperatures T_{c1} and T_{c2}, working at $T_{c1} < T < T_{c2}$.

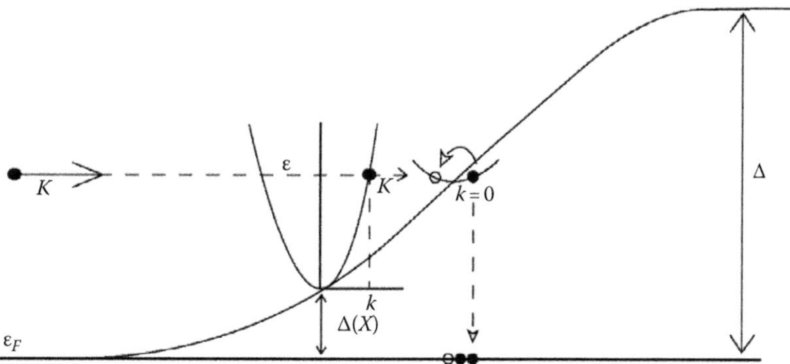

FIGURE 27.6 Quasiparticle conversion into Cooper pairs. The quasiparticle charge is gradually being transferred to the condensate. At a turning point, where the local value of the gap is equal to the quasiparticle energy, the charge of +e is transferred.

region to Δ inside the superconducting region within the coherence length ξ, see Figure 27.6. As the electron with a wave vector k moves toward the interface encountering a gradually increasing gap, it becomes a quasiparticle with the energy E_k and the dispersion law

$$E_k = [\Delta^2 + (\varepsilon_k - \varepsilon_F)^2]^{1/2} = [\Delta^2 + \hbar^2 v_F^2 (k - k_F)^2]^{1/2} \qquad (27.20)$$

which has two solutions for the momenta $k_{\pm} = k_F \pm [(E_k^2 - \Delta^2)]^{1/2}/\hbar v_F$ (where + corresponds to the electron-like particles and − to the hole-like). Accordingly, quasiparticles in a superconductor can be assigned a fractional charge [124–126].

$$q_k = \frac{\varepsilon_k}{E_k} = \frac{\varepsilon_k}{(\Delta^2 + \varepsilon_k^2)^{1/2}} \qquad (27.21)$$

As can be seen from Equation 27.21, the charge of a quasiparticle varies from the asymptotic value of −1 (pure holes) to +1 (pure electrons) (in this notation the electronic charge is + e). As the quasiparticle moves toward the interface, its charge decreases until it reaches 0 when $k = k_F$. At this point, the group velocity of the quasiparticle becomes zero; as it crosses to another branch of the spectrum, becomes negatively charged with negative group velocity, that is, it becomes the Andreev-reflected hole. Importantly, in this process, the charge is continuously transferred to the condensate, and, in the end, the total charge of $e - (-e) = 2e$ will have been transferred to a Cooper pair. This is equivalent to the incoming electron pairing up with another electron with the opposite spin and wave vector,* while the hole gets reflected back from the interface, see Figure 27.6.

* In the simplest model of a plane wave in the ballistic regime, quasiparticle momentum is a good quantum number for labeling the states. In the first approximation in (Δ/E_F), the momenta of reflected and transmitted particles can be considered equal to k. While the probability of the Andreev process is reduces at grazing angles, this effect is relatively small and can usually be neglected.

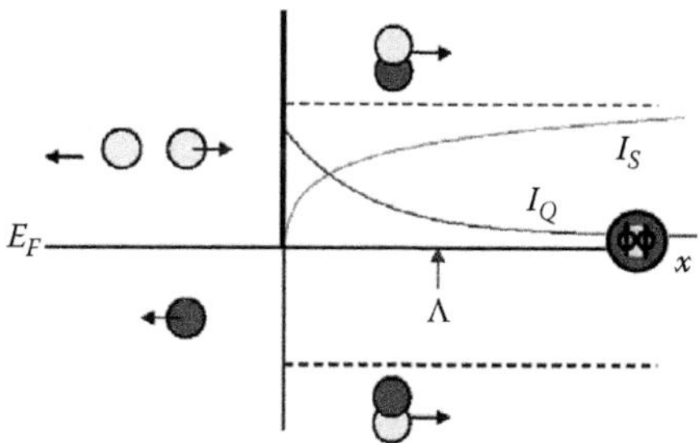

FIGURE 27.7 Quasiparticle to supercurrent conversion at a step like interface assumed in the BTK model [20]. As the quasiparticle current I_Q decays with the characteristic length Λ, the supercurrent I_S is raising at the same rate, so that the charge conservation is satisfied.

This result can be quantitatively derived for a step-like interface assumed in the BTK model $\Delta(x) = \begin{Bmatrix} \Delta, x < 0 \\ 0, x > 0 \end{Bmatrix}$ [16]. As the quasiparticle current gradually decays $J_Q = 2ev_F \exp(-x/\Lambda)$ (where $\Lambda = hv_F/2\Delta \approx \xi$, is on the order of the coherence length), the supercurrent is rising at the same rate $J_S = 2ev_F[1 - \exp(-x/\Lambda)$, so that the charge conservation is satisfied, see Figure 27.7. For clean interface, the probability of AR is

$$A(E) = \begin{Bmatrix} 1 & E < \Delta \\ \dfrac{v_0^2}{u_0^2} = \dfrac{|E| - (E^2 - \Delta^2)^{1/2}}{|E| + (E^2 - \Delta^2)^{1/2}} & E > \Delta \end{Bmatrix} \qquad (27.22)$$

Note that even above the superconducting gap AR is still possible within a narrow range of energies (which is typical for a quantum problem with sharp-edged barriers), although its probability goes down quite quickly, at higher energies, see Figure 27.8. Importantly, in contrast to the tunneling case, where the transmission probability is proportional to the DOS $N_s(E)$, the transmission probability for a clean interface is $A(E) + 1$, independent of $N_s(E)$ [16]. This property of *spectral weight nonconservation* distinguishes AR from conventional tunneling. For tunnel barriers (strong scattering) the probability of AR is rapidly

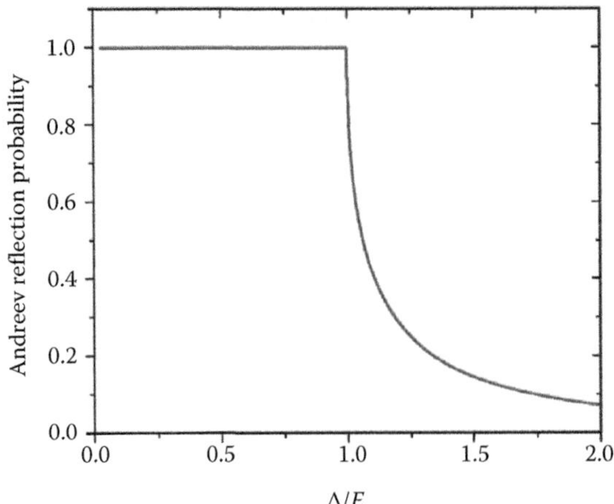

FIGURE 27.8 The probability of Andreev reflection decreases rapidly as the energy exceeds Δ. However, Andreev reflection above the superconducting gap is still possible.

approaching zero (see Section 27.4.1), so that the spectral weight is almost fully conserved.

27.4.1 Arbitrary Interface with Scattering Potential (Ballistic Limit)

While the concept of a clean or ideal interface is helpful for a qualitative understanding of AR, strictly speaking, it can only be implemented when both N and S parts are identical. However, as most real interfaces are constructed of two different materials, the quasiparticle transport is not limited to the AR process, as normal reflections are also present due to (1) a scattering potential barrier at the interface (from surface oxides, impurities, etc.) and (2) different electronic structure of the N and S components, which may result in a normal reflection even in the absence of a barrier. Let us first consider a point contact geometry in the *ballistic limit* (the diffusive limit will be discussed in Section 27.5.3.4).

The effect of a tunnel barrier modeled by a δ-function potential was first described by Demers and Griffin [127] and Griffin and Demers [128]. As the strength of the surface barrier increases, the behavior of the contact gradually changes from metallic to tunneling; both can be described by a single phenomenological dimensionless parameter $Z = H/\hbar v_F$, which is related to the strength of the δ-function potential $V(x) = H\delta(x)$ introduced in a seminal paper by Blonder, Klapwijk, and Tinkham (BTK) [16], who expanded the treatment of Demers and Griffin [127] to finite voltages. The BTK model provides a solution of the Bogoliubov–de Gennes equations in a one-dimensional (1D) case,* while at the same time assuming that the order parameter and the potential drop across the interface are the step functions. Note that such boundary conditions can only be fulfilled in a 3D case, where both the order parameter [129] and the electrical potential [130] have been shown to change on a scale comparable to a.† While, strictly speaking, the problem should be solved self-consistently to determine $\Delta(x)$ and the voltage distribution across the interface, the BTK approximation has been working surprisingly well. The parameter Z has a simple meaning: In the normal state (which can be achieved by either suppressing the superconductivity by a magnetic field, or at high bias voltages), the resistance of the junction is $R_N = R_{Sh}(1 + Z^2)$. The normalized conductance in the case of a nonzero Z is shown in Figure 27.9. The total current [16] across an N/S interface is

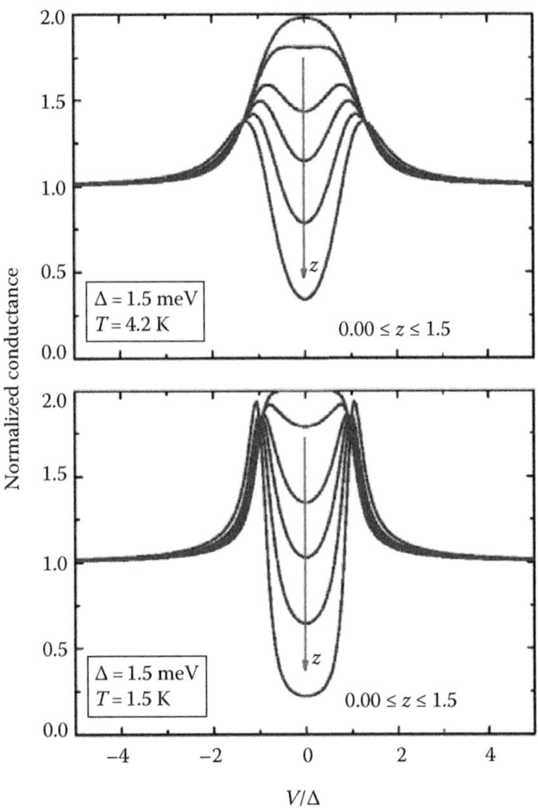

FIGURE 27.9 Normalized conductance for various values of Z (from $Z = 0$ to $Z = 1.5$) for unpolarized case, plotted using the model of Mazin et al. [18]; $\Delta = 1.5$ meV; top $T = 4.2$ K; bottom $T = 1.5$ K. Note that the conductance at zero bias is twice the normal conductance only for $Z = 0$.

$$I_{NS} = \left(\frac{1}{eR_{Sh}}\right)\int[1 + A(E) - B(E)][f(E - eV) - f(E)]dE \qquad (27.23)$$

where

$A(E)$ and $B(E)$ are the energy-dependent probabilities of Andreev and normal reflection, respectively

$f(E)$ is the Fermi function

* The main advantage of solving 1D rather than 3D equations is the computational simplicity. The only 3D case that can be exactly mapped onto a 1D problem is a hyperbolic neck problem.

† Additionally, it is assumed that the quasiparticle distribution function in the area of the contact is largely unchanged, the assumption also only justified in 3D.

As can be seen from Equation 27.23, normal reflections reduce the conductance through the N/S interface, as expected. For $Z=0$, $B=0$, and the probability of AR below the gap $A=1$ (at $T=0$ K), so that $R_{NN}=R_{Sh}=2R_{NS}$ is doubled. Thus, in this idealized case of a perfectly transparent clean interface, the NS interface conductance is completely determined by AR.

27.4.2 Impedance Mismatch

A δ-function potential used in the BTK model implies the continuity of the derivatives of the wave functions at the interface. In an idealized case, when the superconducting and normal parts are identical, the matching conditions do not affect the transport coefficients. However, when the two banks are not the same and thus have different electronic spectra, it results in the so-called *impedance mismatch* [131], which sometimes also called *Fermi velocity/Fermi wave vector* mismatch [132–136]. Such a mismatch may lead to the normal reflection from the interface, even in the absence of a barrier, thus modifying both the reflection and transmission coefficients. In the BTK approach, which adopts a free electron model, this effect can be accounted for by introducing the effective Fermi velocity $v=\sqrt{v_1 v_2}$, where v_1 and v_2 are the Fermi velocities of the two banks. Then Z is replaced by Z_{eff}:

$$Z_{eff}=\left[Z^2+\frac{(1-r)^2}{4r}\right]^{1/2} \tag{27.24}$$

where r is the ratio of the two velocities $r=v_1/v_2$. Note that the transmission coefficients, which depend on Z_{eff}, will remain the same if the two banks are interchanged, in agreement with the principle of reciprocity in quantum mechanics. Then for $Z=0$ (no "real" barrier) the minimum value of Z_{eff} will be determined by the Fermi velocity ratio r, which implies that the value of zero bias conductance will be reduced compared to the ideal value, twice the normal conductance.

While the concept of the Fermi velocity mismatch seems physically transparent, and the minimum values of Z_{eff} are in reasonable agreement with experiments in all-metal [131], as well as metal/semiconductor [137,138] contacts, this argument breaks down dramatically when one considers the case of heavy fermions, for example, in which, based on their large r (up to $\simeq 100$), one might expect $Z\approx 5$. This would have made efficient AR practically impossible, which contradicts many experiments in which Z is typically about an order of magnitude smaller (e.g., Refs. [104,113]). The main reason for this discrepancy is that the simple picture of the Fermi velocity mismatch can only be applied to a very restrictive case of the free electron model, with the electronic wave functions represented by plane waves. In a general case, when the electronic wave functions are *Bloch waves*, the *interface velocities* must be introduced to satisfy the matching conditions [139]. The concept of interface velocity is in line with the work of Deutscher and Nozieres [140], who argued that the Fermi velocity is not the one to be matched in the AR process.

27.5 Andreev Reflection: Spin-Polarized Case

27.5.1 Ideal Case: Introduction

Let us first consider qualitatively an idealized case of a clean interface. The important condition for AR process described in Section 27.4 is that for every electronic k state there is an equivalent hole state with the opposite spin. For a nonmagnetic metal this is always the case, as it has two *identical overlapping* Fermi surfaces. However, for a ferromagnet this is no longer true, as the Fermi surfaces for majority and minority electrons can be very different; so an electron from a majority sheet of the Fermi surface may not be able to find a hole state from a minority sheet to get reflected to. Therefore, in the extreme case of a half-metal ($P=100\%$), where only one type of spin is available at the Fermi level, AR would be completely suppressed [15], see Figure 27.5b. What will happen in the intermediate case, with the values of spin polarization $0<P<1$? It can be understood qualitatively if one considers the current across the interface as the sum of spin-polarized $[I_p=(I_\uparrow-I_\downarrow)]$ and unpolarized $[I_p=I-(I_\uparrow-I_\downarrow)=2I_\downarrow]$ parts [11], see Figure 27.10

$$\begin{aligned} I &= I_\uparrow + I_\downarrow = (I_p+I_u) = (I_\uparrow - I_\downarrow) + 2I_\downarrow \\ &= \frac{(I_\uparrow - I_\downarrow)}{I_\uparrow + I_\downarrow} I + \frac{2I_\downarrow}{I_\uparrow + I_\downarrow} I = PI + (1-P)I \end{aligned} \tag{27.25}$$

where $P=(I_\uparrow-I_\downarrow)/(I_\uparrow+I_\downarrow)$. The spin-polarized fraction of the current PI will not be able to Andreev-reflect, and therefore, will not contribute to the overall conductance, whereas the

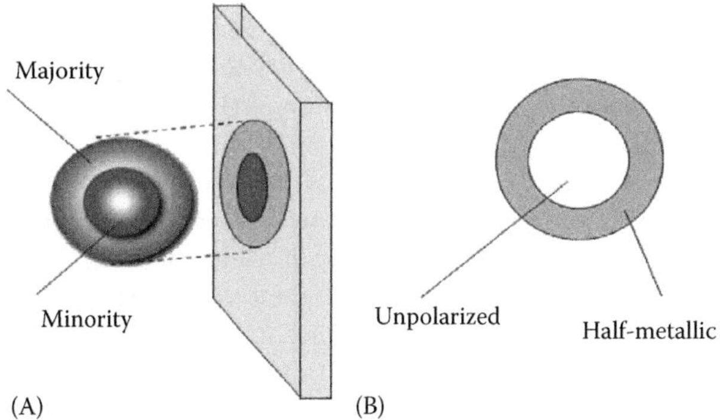

FIGURE 27.10 Current decomposition in the case of spherical Fermi surfaces for majority and minority electrons and a single band superconductor; (A) the two spin subbands projected onto the plane of a contact; (B) the number of majority and majority conductance channels (CC) in the overlapping area (white) is the same, this fraction of the current is unpolarized; the shaded area only has majority channels and thus is half-metallic (100% spin polarized) (see Section 27.5.3.1).

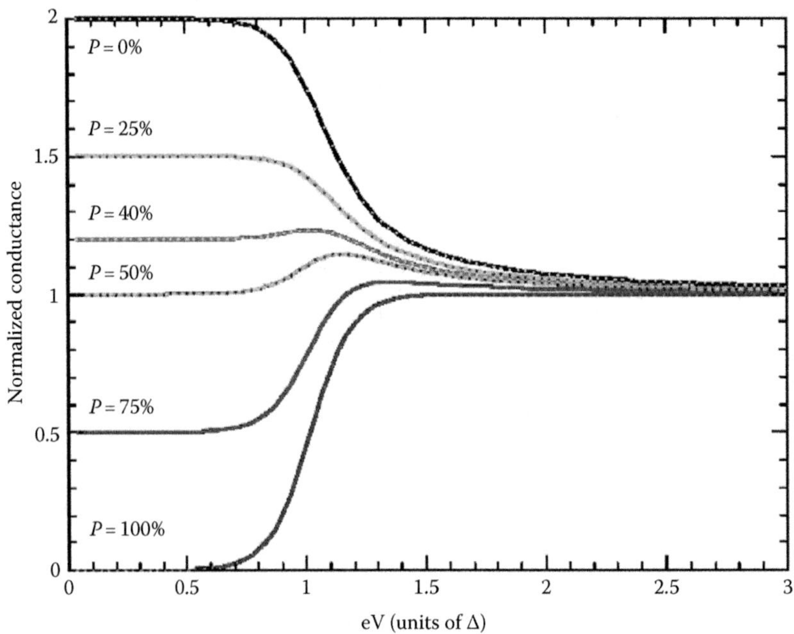

FIGURE 27.11 Normalized conductance curves for different values of spin polarization and a clean interface at $T=\Delta/10$. Every curve as uniquely determined by the value of spin polarization, P. (From Nadgorny, B. et al., *Appl. Phys. Lett.*, 80, 3973, 2002. With permission.)

unpolarized fraction $(1-P)I$ will reflect just like in a normal metal, with each electron contributing twice the normal conductance. For the two limiting cases $P=0$ (an unpolarized metal) and $P=1$ (a half-metal), the normalized conductance at zero bias (which is 2 and 0, respectively) is recovered, see Figure 27.11. Let us now take the value of $P=25\%$, for example. In this case, 25% of the current (spin-polarized fraction) going through the F/S contact will not contribute to the conductance at all (AR is suppressed), whereas for the remaining 75% of the current, the conductance will double, and thus the overall normalized conductance will be 1.5. By the same token, if the conductance is known, one should be able to solve the *inverse problem*, that is, to determine the spin polarization of a material based on the transport properties (conductance) of its contact with a superconductor. This observation becomes the foundation of the PCAR technique [11,12] which we will describe later.

Note that these simple arguments are only applicable in the case of a clean interface with $Z=0$, when there is no mixing of normal and AR (i.e., the probability of AR is 1 at $V=0$). For an arbitrary interface with nonzero Z, there is no longer a direct relationship between the conductance at zero bias and P, and one must consider the whole conductance curve $G(V)$. Judging simply by the amplitude of the conductance curve would be misleading, as demonstrated in Figure 27.12, where the same conductance at zero bias is realized for the spin-polarized case with $P=50\%$ and $Z=0$, and the unpolarized case with $P=0$, $Z=0.55$ (both contacts are assumed to be in the ballistic regime). However, as long as one can describe the conductance curve with a unique set of P and Z values, one should be able to extract the spin polarization values, see, for example, Upadhyay et al. [12] and Mazin et al. [18]. A more complicated diffusive case, first treated by Artemenko et al. [92], for $P=0$, has also been generalized to account for arbitrary values of P [18].

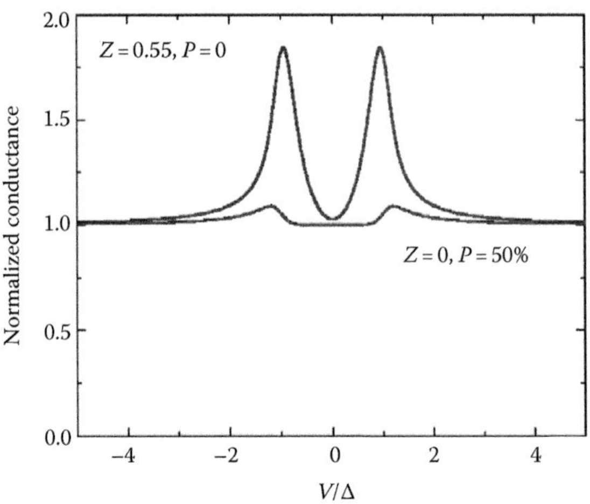

FIGURE 27.12 While the amplitudes of the two conductance curves at zero bias are the same, they correspond to different values of $P=0$ (top curve) and $P=50\%$ (bottom curve). Thus, in order to obtain the value of P for an arbitrary Z, the entire conductance curve must be analyzed.

27.5.2 Comparison of Different Models

Establishing a reliable technique for probing spin polarization based on AR requires a quantitative theory. At present, there are several approaches to formulate these theories, which can be broadly divided into the Landauer–Büttiker formalism (e.g., Ref. [142]) and the Hamiltonian formalism [143], describing tunneling in a superconductor (see, e.g., Ref. [144]). As we have mentioned in Section 27.4.1, the BTK model assumes 3D physics, which allows one to use equilibrium distribution functions, as well as sharp-edged boundary conditions, and yet solve the 1D equations. Solving the exact problem of current transfer from a metal with an arbitrary band structure into a superconductor across a magnetically active interface self-consistently may be prohibitively complicated, even with the use of Green's function techniques, which are generally quite effective in describing transport through a point contact [112,145].

A comparison of some of the models used in practice to extract the values of spin polarization from experimental data ([12,18,146] is given in a recent review [101]). The 3D model used by Upadhyay et al. [12] may be able to take into account an apparent reduction of the AR probability for a fairly clean interface with small Z values, as the angle of incidence becomes larger, due to the kinematic restrictions related to the interface mismatch [132–134,147].* On the other hand, in order to make the equations manageable in 3D, a free electron model and a Stoner model for a ferromagnet are assumed [12,101], which may be somewhat restrictive in view of the fact that the boundary conditions should be determined for the Bloch waves, rather than for the plane waves associated with the free electrons [139,149].

The 1D model of Mazin et al. [18], does not make any explicit assumptions concerning the band structure of the ferromagnet but assumes the same Z for majority and minority electrons.† By introducing an evanescent wave in the minority band, Mazin et al. [18] removed the conjecture of Strijkers et al. [146] that AR amplitude must be set to zero in the case of a half-metal,‡ which resulted in a more accurate determination of P by about 2%–4% [101,150,151]. In Section 27.5.3, we will outline the derivation of Mazin et al. [18].

27.5.3 Andreev Reflection Theory for 1D Model

A quantitative theory of AR should take into account (1) a different number of conductance channels (CC) open for AR for different spins, as was first noted by de Jong and Beenakker [15]; (2) finite interface resistance for both channels, as treated by Blonder et al. [16], for "nonmagnetic" channels, that is, the CC that exist in both spin directions; (3) band structure effects, such as arbitrary Fermi surfaces and arbitrary dispersion law, considered by Mazin [3] in the ballistic limit; (4) the effect of an evanescent Andreev hole on quasiparticle current in half-metallic CC, mentioned by Kashiwaya et al. [152] and Žutić and Valls [133]; and (5) AR for a ferromagnet in the diffusive case, addressed in part by Golubov [153], Jedema et al. [154], Fal'ko et al. [155] and Belzig et al. [156].

27.5.3.1 Ballistic Contact

We will first consider the ballistic contact, using the assumptions of the BTK model as described in Section 27.4.2.§ Conductance of such a contact is [3,149]

$$G = \frac{e^2}{\hbar}\frac{1}{2}\langle N \mid v_x \mid \rangle A \qquad (27.26)$$

where

- A is the contact area
- N is the volume density of electronic states at the Fermi level
- v is the Fermi velocity, and brackets denote Fermi surface averaging

Integrating over the states with $v_{\mathbf{k}i\sigma,x} > 0$, and summing over the band index, i, and the electron spin, σ, we obtain

$$\frac{1}{2}\langle N \mid v_x \mid \rangle = \frac{1}{(2\pi)^3}\sum_{i\sigma}\int \frac{dS_F}{\mid v_{\mathbf{k}i\sigma} \mid} v_{\mathbf{k}i\sigma,x} \qquad (27.27)$$

It is instructive to derive Equation 27.27, starting with the Landauer formula for the conductance of a single electron [17], $G_0 = e^2/h$. The overall conductance is equal to G_0 times the number of CC, N_{cc}, which is defined as the number of electrons that can pass through the contact. If the translational symmetry in the interface plane is preserved, then the quasi-momentum in this plane, $\mathbf{k}_{\parallel}$ is conserved, and N_{cc} is given by the total area of the contact times the density of the two-dimensional quasi-momentum. The latter is $S_x/(2\pi)^2$, where S_x is the area of the projection of the bulk Fermi surface onto the contact plane [158] (see Figure 29.10). Thus

$$G = \frac{e^2}{h}\frac{S_x A}{(2\pi)^2} \equiv \frac{e^2}{\hbar}\frac{1}{2}\langle N \mid v_x \mid \rangle A \qquad (27.28)$$

* In practical terms, most of the quasiparticles are approaching the interface at relatively small angles, and the difference is even less pronounced if the assumption of a clean interface is removed (the larger Z is, the more "one dimensional" the transport becomes). Overall, this effect is relatively small: a factor of 2 of the interface mismatch corresponds to about 10% reduction in transmission probability [148].

† This is a serious simplification. A more rigorous but less practical approach calls for four scattering parameters, $Z_{\uparrow\uparrow}, Z_{\downarrow\downarrow}, Z_{\uparrow\downarrow}$, and $Z_{\downarrow\uparrow}$.

‡ Recently Grein et al. [145] introduced a more general microscopic model, emphasizing the effects of a barrier shape and spin mixing, caused by magnetically active interfaces, which included the model of Mazin et al. [18] as a limiting case.

§ In fact, the assumption of a step function order parameter is more realistic for the F/S interface, as the proximity effect there is generally less pronounced, compared to the N/S case [157].

Note that the product, $\langle N|v_x|\rangle$, which is often called the current-carrying DOS [159], is very different from the DOS at the Fermi level

$$N(E_F) = \frac{1}{(2\pi)^3}\sum_{i\sigma}\int \frac{dS_F}{|v_{\mathbf{k}i\sigma}|} \quad (27.29)$$

27.5.3.2 Diffusive Contact

In the opposite limit, the contact size is much *larger* than the mean free path. The conductance is then given by the bulk conductivity

$$\sigma = \left(\frac{e^2}{\hbar}\right)\langle Nv_x^2\rangle\tau \quad (27.30)$$

where τ is the relaxation time. Ohm's law requires that the conductance $G=\sigma A/L$ where L is the length of the disordered region. This can be also reproduced within the random matrix approach of Beenakker [17], taking into account that now each CC ($\mathbf{k}_\parallel$) state, has a finite probability for an electron to get through the disordered region, $0 \le T \le 1$ and

$$G = \frac{e^2}{h}\sum_{\kappa} T_\kappa = \frac{e^2}{h}\int_\lambda^\infty \frac{d\zeta P(\zeta)}{\cosh^2(L/\zeta)} \quad (27.31)$$

where $\kappa \equiv \{\mathbf{k}_\parallel, i, \sigma\}$. T_κ is defined in terms of the probability distribution, $P(\zeta)$ of the localization lengths, ζ. The cutoff λ should be on the order of the mean free path l in fact, $\lambda=2l$ (the factor of 2 accounts for two possible directions). Ohm's law requires that G (1/L, thus the behavior of $P(\zeta)$ at large ζ must be $const/\zeta^2$. Normalization gives $const=\lambda N_{cc}$. Substituting that in Equation 27.31, we obtain

$$G = \frac{e^2}{h}\int_\lambda^\infty \frac{\lambda N_{cc} d\zeta}{\zeta^2\cosh^2(L/\zeta).} \approx \frac{e^2\lambda N_{cc}}{hL}$$

$$= \frac{e^2}{\hbar}\frac{A\lambda}{\Omega L}\sum_{i\sigma}\int \frac{dS_F}{|v_\kappa|} v_{\kappa,x} \quad (27.32)$$

where Ω is the unit cell volume

In the constant τ approximation, used in Equation 27.30, the average mean free path $l = \sum_{i\sigma}\int \frac{dS_F}{|v_\kappa|}v_{\kappa,x}^2\tau \Big/ \sum_{i\sigma}\int \frac{dS_F}{|v_\kappa|}v_{\kappa,x}$, thus $\lambda_\kappa = 2v_\kappa x\tau$. Therefore

$$\langle G\rangle_L = \frac{e^2}{\hbar}\frac{A}{\Omega L}\sum_{i\sigma}\int \frac{2dS_F}{|v_\kappa|}v_{\kappa,x}^2\tau = \frac{e^2}{\hbar}\langle Nv_x^2\rangle\frac{A}{L} = \sigma\frac{A}{L}$$

Thus in the diffusive limit the conductance is determined by $\langle Nv_x^2\rangle$, as expected.

27.5.3.3 Random Matrix Approach: Ballistic Case

Let us reproduce the BTK results using the random matrix approach. Probabilities of four processes must be considered: normal reflection, defined as the process where $\mathbf{k}_\parallel$ is conserved, while the group velocity in the direction perpendicular to the interface changes sign, AR, when $\mathbf{k}_\parallel$ changes to $-\mathbf{k}_\parallel$ and transmission into the superconductor with or without branch crossing [16]. The conductance then is

$$\langle G\rangle_{NS} = \frac{e^2}{h}\sum_{\kappa} T_S(\kappa) = \frac{e^2}{h}\sum_{\kappa}(1 + A_\kappa - B_\kappa) \quad (27.33)$$

where A and B are the probabilities of the normal and AR, respectively. "Andreev transparency," that is, the probability of the Andreev process, can be expressed in terms of the normal transparency T_N of the interface [17]. For zero bias

$$T_S = \frac{2T_N^2(\kappa)(1+\beta^2)}{\beta^2T_N^2 + [1+r_N^2]^2} = \frac{2T_N^2(1+\beta^2)}{\beta^2T_N^2 + [2-T_N]^2} \quad (27.34)$$

where

- $T_N(\boldsymbol{\kappa})$ is the normal state transparency
- $r_\kappa^2 = 1 - T_N(\kappa)$ is the corresponding normal state reflectance
- $\beta = V/\sqrt{|\Delta^2 - V^2|}$ is the coherence factor

A similar formula can be derived for $V>\Delta$. Equation 27.34 is equivalent to the BTK formulas. For a specular barrier, and neglecting the Fermi velocity mismatch at the interface, $T_N(\boldsymbol{\kappa}) = 1/[1+Z^2]$.

27.5.3.4 Random Matrix Approach: Diffusive Case

Let us apply the same approach to the diffusive AR contact, which can be viewed as a contact between the normal and the superconducting leads, separated, in addition to the interface, by a diffusive region. The size of the region is larger than the electronic mean free path [153]. In the zero-temperature and zero-bias limit, Equation 27.34 becomes

$$\langle G\rangle_{NS} = \frac{e^2}{h}\sum_{\kappa} T_A = \frac{e^2}{h}\sum_{\kappa}\frac{2\tilde{T}_\kappa^2}{(2-\tilde{T}_\kappa)^2} \quad (27.35)$$

where now the normal state transmittance for the conductance channel $\boldsymbol{\kappa}$ is given by the sequential conductor's formula

$$\tilde{T}^{-1} - 1 = (T_N^{-1} - 1) + (t^{-1} - 1) \quad (27.36)$$

where

- t is the transmittance of the diffusive region
- T_N is the barrier transparency

Using Equation 27.31 for the distribution of t's, we obtain

$$\langle G_{NS}\rangle_L = \frac{e^2}{h}\sum_{\mathbf{k}} \frac{2}{(2/T_N - 2 + 2/t_{\mathbf{k}} - 1)^2}$$

$$= \frac{e^2}{h}\frac{\lambda N_{cc}}{L}\int_0^\infty \frac{dy}{[2(1-T_N)/T_N + \cosh y]^2}$$

$$= \frac{e^2}{h}\frac{\lambda N_{cc}}{L}\frac{w\cosh w - \sinh w}{\sinh^3 w} \tag{27.37}$$

where $\cosh w = 2(1 - T_N)/T_N$. For the clean (no-barrier) interface, $T_N = 1$, $w = i\pi/2$ and this expression is reduced to Equation 27.32, thus reproducing the well-known result [17,92] that the resistance at zero bias of a diffusive Andreev contact with no additional interface barrier is the same in the superconducting and normal states.

In the following, we will derive a set of formulas for arbitrary bias, temperature, and interface resistance for both ballistic and diffusive regimes, generalizing the BTK formulas for the half-metallic CC and in the diffusive limit. These formulas are summarized in Table 27.1.

27.5.3.5 Half-Metallic Channels

Half-metallic CC correspond to the $\mathbf{k}_\parallel$ only allowed in one spin direction. We will consider an incoming plane wave and the transmitted plane wave (with and without branch crossing), assuming, for simplicity, the same wave vector for all the states

$$\psi_{in} = \begin{pmatrix}1\\0\end{pmatrix}e^{ikx};\quad \psi_{tr} = c\begin{pmatrix}u\\v\end{pmatrix}e^{ikx} + d\begin{pmatrix}v\\u\end{pmatrix}e^{-ikx}$$

Here u and v have the standard BTK meaning, for example, $u^2 = 1 - v^2 = (1+\beta)/2\beta$ at $V>\Delta$. Unlike the case of Blonder et al. [16], though, here the reflected state is a combination of a plane wave and an evanescent wave:

$$\psi_{refl} = a\begin{pmatrix}0\\1\end{pmatrix}e^{\kappa x} + b\begin{pmatrix}1\\0\end{pmatrix}e^{-ikx} \tag{27.38}$$

TABLE 27.1 Bias Dependence of the Total Interface Current in Different Regimes

	$E<\Delta$	$E>\Delta$
BNM	$\frac{2(1+\beta^2)}{\beta^2+(1+2Z^2)^2}$	$\frac{2\beta}{1+\beta+2Z^2}$
BHM	0	$\frac{4\beta}{(1+\beta)^2+4Z^2}$
DNM	$\frac{1+\beta^2}{2\beta}\mathrm{Im}[F(-i\beta)-F(i\beta)]$	$\beta F(\beta)$
DHM	0	$\beta F[(1+\beta)^2/2-1]$

BNM, ballistic nonmagnetic [16]; BHM, ballistic half-metallic; DNM, diffusive nonmagnetic; DHM, diffusive half-metallic. $F(s)$ is defined in the text.

The total current is

$$\frac{G_{HS}}{G_0} = \frac{4\beta[1+(K-2Z)^2]}{4(\beta^2-1)Z(K-Z)+[1+(K-2Z)^2][(1+\beta)^2+4Z^2]} \tag{27.39}$$

at $V>\Delta$, where $K = \kappa/k$, and zero otherwise.

As $V\to\Delta$, $G_{HS}/G_0 \to 0$, and $G_{HS}/G_0 \to G_N/G_0 = 1/(1+Z^2)$ as $V\to\infty$. This maximum of G_{HS}/G_0 at intermediate biases is due to the fact that, although the Andreev-reflected hole does not propagate and does not carry any current, the Andreev process is, however, allowed at $V>\Delta$, thus enhancing the transparency of the barrier. This effect does not exist, though, for $Z=0$ or for $K\to\infty$. To simplify the equations, we used $K\to\infty$ in the formulas given in Table 27.1. Note that the simple renormalization of the normal current at $V>\Delta$ used in Srijkers et al. [146] gives a different result: instead of $4\beta/[(1+\beta)^2+4Z^2]$ it gives $[1+\beta(1+2Z^2)]/[(1+\beta)(1+2Z^2)+2Z^4]$, which is discontinuous at $V=\Delta$.

27.5.3.6 Generalized Diffusive Case

Let us generalize the BTK formulas for the diffusive limit. For the nonmagnetic CC the calculation follows Equations 27.34 and 27.36. For zero temperature and a sub-gap bias voltage $V<\Delta(T)$

$$\langle G\rangle_{NS} = \frac{e^2}{h}\sum_{\mathbf{k}_\parallel,t} \frac{4\tilde{T}_N^2(\mathbf{k})(1+\beta^2)}{\beta^2\tilde{T}_N^2(\mathbf{k})+[2-\tilde{T}_N(\mathbf{k})]^2} \tag{27.40}$$

and

$$\tilde{T}_N^{-1} = T_N^{-1} + t^{-1} - 1 = Z^2 + t^{-1} \tag{27.41}$$

with the distribution (27.31) for t. Using this, we obtain

$$\langle G_{NS}\rangle_L = \frac{e^2}{h}\frac{\lambda N_{cc}}{L}\int_0^\infty \frac{(1+\beta^2)dy}{\beta^2+(2Z^2+\cosh y)^2} \tag{27.42}$$

Factor N_{cc} now stands for the number of CC allowed in both spin channels, λ is given by the average mean free path for the channels, and thus the total conductance is given by $\langle Nv_x^2\rangle$, averaged over these channels. For $Z=0$, this results in

$$\langle\sigma_{NS}\rangle = \frac{e^2\tau}{\Omega}\langle Nv^2\rangle_{\downarrow\uparrow}\frac{\Delta}{V}\log\left|\frac{V+\Delta}{V-\Delta}\right| \tag{27.43}$$

which diverges logarithmically at $V=\Delta$. For an arbitrary Z, the conductance is

$$\langle\sigma_{NS}\rangle = \frac{e^2\tau}{\Omega}\langle Nv^2\rangle_{\downarrow\uparrow}\frac{1+\beta^2}{2\beta}\mathrm{Im}[F(-i\beta)-F(i\beta)]$$

where

$$F(s) = \cosh^{-1}(2Z^2+s)/\sqrt{(2Z^2+s)^2-1}$$

Similarly, for $V > \Delta$

$$\langle G_{NS}\rangle_L = \frac{e^2}{h}\frac{\lambda N_{cc}}{L}\int_0^\infty \frac{2\beta dy}{\beta + (2Z^2 + \cosh y)} \tag{27.44}$$

$$\langle \sigma_{NS}\rangle = \frac{e^2\tau}{\Omega}\langle Nv^2\rangle_{\downarrow\uparrow}\beta F(\beta) \tag{27.45}$$

At $Z = 0$, this equation is reduced to*

$$\langle \sigma_{NS}\rangle = \frac{e^2\tau}{\Omega}\langle Nv^2\rangle_{\downarrow\uparrow}\frac{V}{\Delta}\log\left|\frac{V+\Delta}{V-\Delta}\right| \tag{27.46}$$

At $V \gg \Delta$ we obtain

$$\langle \sigma_N\rangle = \frac{e^2\tau}{\Omega}\langle Nv^2\rangle_{\downarrow\uparrow}\frac{\cosh^{-1}(2Z^2+1)}{Z\sqrt{Z^2+1}} \tag{27.47}$$

which should be used to normalize the conductance curve.

Finally, for the "half-metallic" CC, there is no conductance at $V < \Delta$. For $V > \Delta$

$$\langle G_{HS}\rangle_L = \frac{e^2}{h}\frac{\lambda N_{cc}}{L}\int_0^\infty \frac{2\beta dy}{(\beta+1)^2 + 2(2Z^2 - 1 + \cosh y)}$$

$$\langle \sigma_{HS}\rangle = \frac{e^2\tau}{\Omega}\langle Nv^2\rangle_{\downarrow}\beta F\left[\frac{(\beta+1)^2}{2} - 1\right]$$

where the arrow in the subscript shows that these channels are allowed only in one spin sub-band as shown in Figure 27.10. Note that in the limit of $V \gg \Delta$

$$\langle \sigma_{HS}\rangle = \frac{e^2\tau}{\Omega}\langle Nv^2\rangle_{\downarrow}\frac{\cosh^{-1}(2Z^2+1)}{2Z\sqrt{Z^2+1}} \tag{27.48}$$

which is exactly one half of the conductance of the nonmagnetic case.

The formulas corresponding to the generalized BTK equations for the finite spin polarization in both ballistic and diffusive limits are summarized in Table 27.1. The finite temperatures can be taken into account in the same way as in the BTK paper.

* Equations 27.43 and 27.46 coincide with those obtained in Artemenko et al. [92] by a different technique.

27.6 PCAR Experimental Technique

27.6.1 Experimental Geometry

In order to study the properties of an F/S interface, it is necessary to establish a stable contact between a ferromagnet and a superconductor. While planar junctions have been successfully fabricated for the spin polarization measurements [160–162], the point contact geometry is most commonly used. Among point contacts, there are two basic types: adjustable mechanical point contacts [11,131,163,164] and nanoconstrictions, fabricated in thin silicon membranes [12,101,165].

27.6.1.1 Point Contacts

In the most common geometry, a tip made out of a superconducting wire, which was either mechanically [12,166] or electrochemically polished [131,167,168] or cut with a sharp blade [169], is employed, see Figure 27.13. In principle, reversing the role of the two materials should not affect the results, as was demonstrated, for example, for Fe and Ta [16]. In practice, it is more convenient to use a superconducting tip in contact with a ferromagnetic base (the sample under study), as this allows one to study ferromagnetic films, as well as bulk samples; additionally, the effect of a magnetic field on a superconductor [170] can be minimized. The properties of a contact depend largely on the design of the approach mechanism, which may vary from a simple mechanical adjustment controlled by a micrometer [11] to more versatile systems, employing a combination of differential screws, micrometers, and piezo-driven actuators [102,171–173]. The important part of any system is the ability to maintain a stable contact, preferably within a broad temperature range. Good temperature stability is important for taking variable temperature measurements, as well as for a more accurate determination of the contact size [102,103], see Section 27.3.7. One should also be able to control the force applied to the tip, as the sample properties may be pressure sensitive. While, compared to the nanoconstriction geometry, the tip geometry is more versatile, it has a larger uncertainty in terms of the actual contact size.†

An alternative technique was used by Upadhyay et al. [12]. A silicon nitride membrane containing a tapered nanohole with a minimum diameter of 3–10 nm [174] was fabricated by chemical vapor deposition of ~50 nm thick silicon nitride, see Figure 27.14. A ferromagnet and a superconductor can then be deposited onto the opposite sides of the membrane. The important advantage of this technique is that the geometry of the contact is well defined. In addition, all the fabrication steps are typically done in situ, so that the possibility of contamination is reduced, and thus Z is usually fairly small [101]. While it may be hard to completely exclude the accumulation of defects in the area of the contact,

† Another original technique was employed by Gonnelli et al. [175] in the measurements of MgB_2 superconductors. In this technique a small amount of colloidal silver is placed on top of MgB_2, providing a stable point contact.

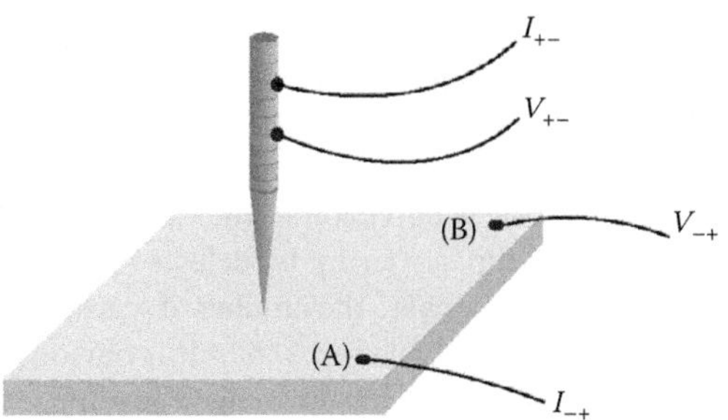

FIGURE 27.13 Schematic of a typical adjustable point contact, with two current and two voltage leads. In the most commonly used arrangement the point is the superconductor, whereas the base is the ferromagnet (leads (A) and (B)).

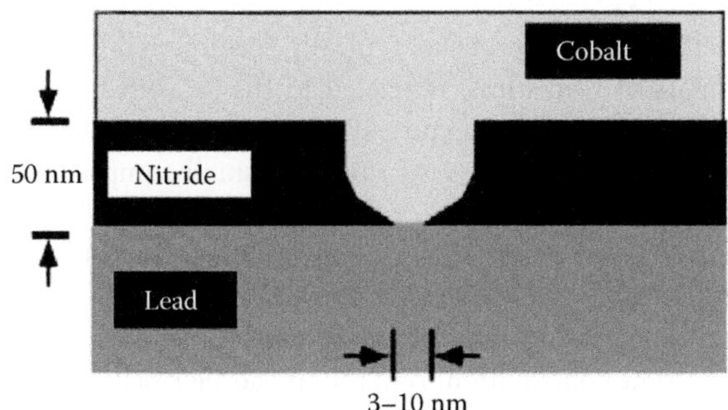

FIGURE 27.14 Schematic of a typical ferromagnet—superconductor nanocontact (nanoconstriction) device. (From Upadhyay, S.K., *Phys. Rev. Lett.*, 81, 3247, 1998. With permission.)

the presence of phonon spectra, which is routinely observed for these types of contacts, confirms that the transport is in the ballistic ($l > a$) or a quasi-ballistic ($l \approx a$) regime. Unfortunately, the fabrication of some of the most interesting new materials often requires lattice matching to a specific substrate, which limits the practical application of this method.* Ideally, the combination of an adjustable point contact with in situ sample preparation and subsequent spin polarization measurements is likely to be the most effective.

27.6.1.2 Planar Geometry

As an alternative to a point contact, a planar geometry—in which AR was first observed—can also be successfully employed for the spin polarization measurements [160–162]. An example of this geometry is shown in Figure 27.15. One of the potential disadvantages of this type of contact is the current nonuniformity—the same problem that is often encountered in tunnel junctions [178] and in CPP GMR devices [179]. In addition, the contact properties are very sensitive to surface roughness and other surface defects. As the only way to change the contact properties in this case, is to make a new sample, this method is rather labor intensive.

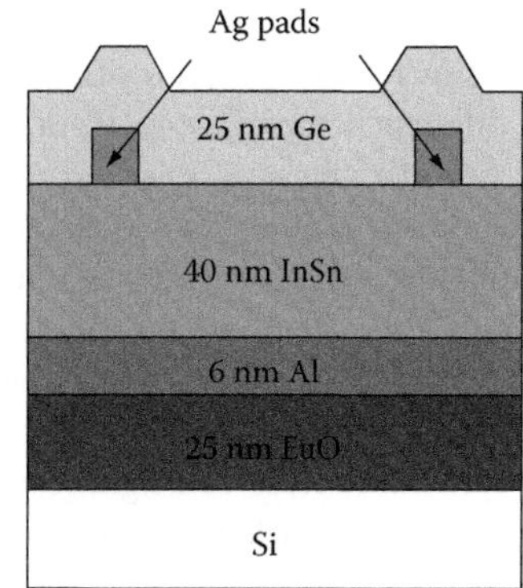

FIGURE 27.15 Example of a planar contact geometry used in the study of EuO_{1-x}. Top to bottom: 25 nm Ge cap/Ag contacts/400 nm InSn/6 nm Al/25 nm EuO_{1-x}/bulk *n*+Si substrate. The conductance is measured across the heterostructure. (From Panguluri, R.P. et al., *Phys. Rev. B*, 78, 125307, 2008. With permission.)

27.6.2 Electrical Measurements

In most cases, the point contact measurements are done in the temperature range of 0.3–4 K, using conventional superconductors, such as Nb, Pb, Ga, Ta, Sn, In, and Al, so that a typical maximum bias voltage ~10Δ is approximately 10 mV. With a typical contact resistance of 10 Ω it results in a maximum Joule heating of about 10 μW. Thus, in most cases, the standard continuous *lock-in detection* technique can be used in order to measure the differential contact resistance *dV/dI*, especially if the measurements are done in a liquid helium bath. On the other hand, for point contact spectroscopy of high-T_c superconductors with considerably larger gaps, or for contact resistances lower than 1 Ω, pulsed current techniques can be implemented to avoid heating effects, especially at lower temperatures (e.g., Ref. [180]).

27.6.3 Data Analysis

27.6.3.1 General Approach

While the actual *dI/dV* measurements are fairly straightforward, the analysis of the data is often nontrivial, as there are more fitting parameters compared to the unpolarized case. In the standard models (e.g., Refs. [12,18]) there are at least two fitting parameters, *P* and *Z*; the gap Δ is typically set to the BCS value for the respective superconductor. The fitting procedure involves normalizing the conductance curve *dI/dV* with respect to $(dI/dV)_N$, which is defined either as an asymptotic value at high bias voltages ($V \approx 10\Delta$ is usually sufficient), or upon the application of a magnetic field. A standard data analysis routine, which includes an optimization procedure minimizing the least square function, is then applied.† Additionally, a change in the value of the superconducting gap usually results in the proportional change in *P* [151]. It is desirable, therefore,

* Other nanofabrication techniques include nanoindentation using atomic force microscopy (AFM) [176], and focused ion beam (FIB) milling [177].

† It is important to keep in mind that, as in other problems of this type, this function may have multiple local minima.

to evaluate the gap independently by measuring the critical temperature of the superconducting transition and the conductance spectra at different temperatures. If, in addition to the superconducting gap, the values of quasiparticle lifetimes are varied as well, the number of adjustable parameters increases to four. For four parameters, the iteration procedure may lead to degenerate solutions, although a recipe to minimize this effect has been proposed [169]. Thus, while a typical accuracy of statistically averaged measurements of P in on the order of 2%–3%, in some cases, a systematic error can significantly exceed this value. This problem is particularly pronounced for small values of $P \simeq 10\%$, especially when additional DOS broadening is present [101,169].

27.6.3.2 Inelastic Scattering

The typical weak coupling BTK models described in Section 27.5 assume effectively infinite quasiparticle energy relaxation time $\tau_E \gg \Delta/\hbar$. However, different inelastic scattering mechanisms often contribute to the reduction of quasiparticle lifetime and thus broaden the BCS DOS. To account for this broadening, the Dynes parameter Γ can be introduced, $\Gamma = \hbar/\tau$ [181]. Accordingly, this will modify the quasiparticle DOS $N(E,\Gamma)$ [182,183]:

$$N(E,\Gamma) = \mathrm{Re}\left[\frac{E + i\Gamma}{\sqrt{(E + i\Gamma)^2 - \Delta^2}}\right] \tag{27.49}$$

which will then have to be included in the model [101]. Alternatively, one might be able to account for broadening by using the original BCS quasiparticle DOS $N(E)$, but with higher effective temperature $T_{eff} = \Gamma/\sqrt{2}$ [137]. In some cases, inelastic scattering may also be accompanied by a reduction in the superconducting gap [137,169]. Chalsani et al. [101] intentionally inserted a Pt layer between a ferromagnet and a superconductor, confirming that a modification of the standard model is necessary in order to match the experimental data when strong spin scattering is introduced.

27.6.3.3 Possible Spin Polarization Dependence on Z

Another aspect of the procedure of evaluating the values of spin polarization involves plotting P measured for different point contacts as a function of Z and extrapolating the function $P(Z)$ to $Z = 0$. As we have discussed, there may be a difference in the barrier transparency for majority and minority carriers, as well as spin-flip processes at the interface. Kant et al. [184] proposed that P could be written as

$$P_0 = \exp\left[-2\alpha\Psi Z^2\right] \tag{27.50}$$

where

P_0 is the intrinsic value of the spin-polarization

α is defined as the spin-flip scattering probability

Ψ is the ratio of the forward and backward scattering probabilities

In this model, the factor of Z^2 is derived from multiple scattering within the interface region. Z^2 is then proportional to the number of collisions and thus to the ratio of the contact diameter to the electron mean free path. Another possible interpretation of the spin polarization suppression in the case of finite barrier transparency is spin-flip scattering by defects at the interface, which in turn is proportional to the number of scattering events. This immediately leads to Equation 27.50.* It is obvious that for contacts with large Z and strong spin-flip scattering, P should eventually go to zero,† so the exponential function may have some validity. On the other hand, there is no statistically significant advantage in using Equation 27.50 compared to a quadratic or even a linear dependence [151].

27.6.3.4 Spreading Resistance

A specific technical problem, typically encountered in the measurements of thin films, is related to the so-called spreading resistance R_s ([151], see Figure 27.13). If R_s, which corresponds to the voltage drop across the sample is small compared to the differential resistance of the point contact, $R_c = dV/dI$ (i.e., well below 1 Ω—which is usually the case for bulk samples or highly conductive films), it can safely be neglected. However, if this resistance is not insignificant, the measured $(dV/dI)_{meas}$ will include some contribution from the spreading resistance R_s:

$$\left(\frac{dV}{dI}\right)_{meas} = \left(\frac{dV}{dI}\right)_{cont} + R_s \tag{27.51}$$

If this is the case, the experimentally measured conductance (which is inversely proportional to the differential resistance) does not represent the true conductance curve of the point contact, $(dI/dV)_{cont}$. Specifically, the apparent position of the zero bias conductance will change and the coherence peak will shift from $\simeq\Delta$ to higher voltages.‡ Obviously, if the effect of spreading resistance is ignored, the extracted values of P may be seriously compromised [151].

If R_s was known, it would have been easy to take R_s into account to reconstruct the actual conductance of the contact, since the current through the contact is also known. Unfortunately, it is difficult to measure the value of R_s directly, although in some cases it can be estimated. If the approximate value of R_s is known, the iteration procedure can be performed

* It should be noted [151] that this argument is only applicable for the diffusive contact, but not in the ballistic case, as stated by Kant et al. [184].

† This may be difficult to confirm experimentally, as the error in the spin polarization values also tend to increase at larger Z. In practical terms, however, it is always preferable to use low Z contacts.

‡ The origin of these distortions is understandable: Andreev reflection (or the lack thereof in the case of half-metals) results in a strongly voltage dependent conductance curve. If R_s (which is approximately Ohmic) is comparable to R_c, it will result in a "flattening" of the overall conductance curve. In the extreme case when R_s is much larger than R_c, very little voltage dependence will be observed (unless $P = 100\%$ with nominally infinite resistance at zero bias).

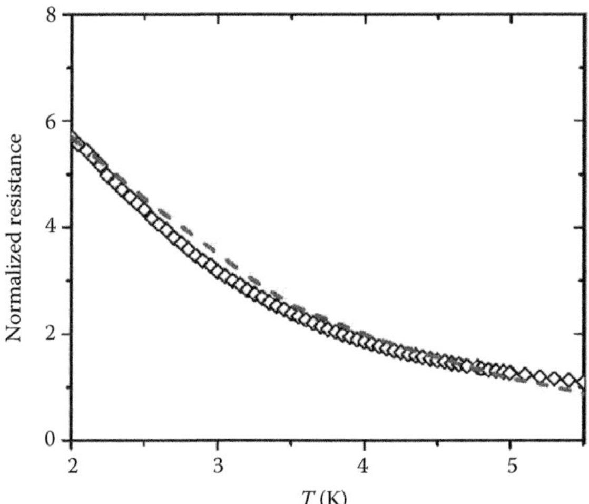

FIGURE 27.16 Numerical fit of the temperature-dependent resistivity curve of the structure shown in Figure 27.15. Dashed line is the fitting curve; open squares are the experimental data. The sharp resistivity increase below the critical temperature of InSn (~5.8 K) is due to the predominant role of Andreev reflection at the superconductor/europium oxide interface, with EuO_{1-x} nearly 90% spin polarized. (From Panguluri, R.P. et al., *Phys. Rev. B*, 78, 125307, 2008. With permission.)

as a part of the fitting routine. This procedure can be set up in two different ways: either one subtracts the contribution from R_s to obtain a "virgin" conductance curve of the contact (which is different from the originally recorded experimental data and is then renormalized), or one can fit the original curve taking into account the contribution from R_s, in which case Δ will differ from the actual superconducting gap.

Alternatively, instead of analyzing individual dI/dV curves, one might analyze a temperature-dependent $R(T)$ function. This approach should allow one to separate the contribution from R_s (which is approximately temperature independent) from a strongly temperature-dependent part related to AR [161,185], see Figure 27.16.

27.6.3.5 Comparison with Ab Initio Calculations

Let us briefly discuss ab initio calculations of the point contacts transparency and compare them with the experimental data. The calculations of Xia et al. [186], for example, were found in good agreement with the experimental results of Upadhyay et al. [12] for the nonmagnetic Cu/Pb contacts, but disagreed sharply for the magnetic Ni/Pb and Co/Pb contacts. One possible reason for this disagreement is that the δ-function approximation may be insufficient here, as it does not take into account different wave function symmetries at the interface. The other explanation is related to the problem of lattice matching across the interface, which may lead to a reduced magnetization of the interface layer [186]. The latter explanation seems to be consistent with the calculations of Taddei et al. [187], who had to use an enhanced magnetic moment at the interface to achieve a good agreement with the experimental results [101].

27.7 Spin Polarization Measurements

Since its inception in 1998, the use of the PCAR technique (including the AR measurements in a planar geometry) has increased dramatically. In addition to conventional transition metal ferromagnets, such as Fe, Co, Ni, and their alloys; many Heusler alloys; magnetic oxides, such as CrO_2, LSMO, Sr_2FeMoO_6, EuO, and EuS; dilute magnetic oxides, such as In_2O_3, ZnO; semiconductors, such as (Ga,Mn)As; and a large number of other materials have been investigated by this technique. In the following, we will briefly describe some of these results in detail; the summary of the measurements currently available in the literature is given in Table 27.2.

27.7.1 Transition Metal Ferromagnets

In addition to numerous practical applications, 3d transition metal ferromagnets and their alloys have been used as a model system to develop reliable band structure calculation techniques of different flavors. However, many aspects of this seemingly simple system are still not very well understood. For example, it is difficult to reconcile the itinerant character of these materials with the positive sign of spin polarization measured by the spin-polarized tunneling technique. While it is quite reasonable to assume that this is due to a larger tunneling matrix elements for *s*-electrons compared to *d*-electrons, which compensate for a higher minority DOS at the Fermi level, one can only marginally divide electrons in 3d metals into *s* and *d* types. Thus it may be more appropriate to discuss these effects in terms of different bands with different Fermi velocities. The PCAR measurements of various Ni_xFe_{1-x} alloys, including permalloy [164], indicate that, while the magnetic moment is dependent on the composition, the spin polarization values are almost composition independent, $P_{Nv} \simeq 45\%$–50%. These results are in surprisingly good agreement with the values of $P_T \simeq 50\%$ measured by the Meservey–Tedrow technique [188].* Similar results for permalloy have been recently obtained by the PCAR technique utilizing the point contact [191] and nanoconstriction geometries [165], and in some of the spin injection experiments (e.g., [192]). Recently introduced Doppler spin wave measurements result in $P_{Nv^2} \simeq 50\%$ at room temperature [47] and $P_{Nv^2} \simeq 75\%$ at 4 K [193], in good agreement with the band structure calculations in the diffusive limit [164].

27.7.2 Half-Metals

At present, there are many theoretical predictions of half-metallicity, far exceeding the experimental evidence for their existence. This is understandable, as band structure calculations

* Interestingly, initially *P* of Ni was measured only at 8% [70] then the value increased to 13% [189], then further increased to 23% [190], until reaching the most recent result of 50% obtained with the use of high vacuum film deposition, indicating the sensitivity of Ni and Ni-based alloys to fabrication conditions.

TABLE 27.2 Results of the Spin Polarization Measurements with the PCAR Technique

Material	Reference P(%)
Co	42 ± 2[250], 37 ± 2[251], 45 ± 2[252], 47 ± 2[253]
Fe	45 ± 2[250], 49.6 ± 1[254], 43 ± 3[252], 46 ± 3[253]
Ni	46.5 ± 1[250], 28[255], 37 ± 1[252]
CrO_2	90 ± 3.6[250], 81[314], 96 ± 4[256], 92[257], 95[258], 96 ± 1[259], 98.5[260], 96 ± 2[253]
$Ni_{0.8}Fe_{0.2}$	37 ± 5[250], 35[261]
Ni_xFe_{1-x}	44 ± 3[262]
Gd	45 ± 4[253], 52[263]
Dy	50[263]
$La_{0.7}Sr_{0.3}MnO_3$	78 ± 4[250], 60 – 90[264], 78 ± 2[265]
$La_{0.6}Sr_{0.4}MnO_3$	83 ± 2[265]
EuO	90[161], 90[267]
EuS	80[268]
EuB_6	56[269]
$Co_{1-x}Fe_xS_2$	85[270-272], 47 – 61[273]
CoS_2	56[274], 64[275]
(Ga, Mn)As	85[276], 76 – 90[277], 83 ± 17[278]
(In, Mn)Sb	52 ± 3[279], 61 ± 1[280]
(Ga, Mn)Sb	57 ± 5[281]
Sr_2FeMoO_6	70[282], 63[283]
$SrLaVMoO_6$	50[284]
In_2O_3, $Cr{:}In_2O_3$	50 ± 5[285]
$Co_{1.95}Nd_{0.05}MnSi$	59[286]
$SrRuO_3$	60 ± 2[287], 51 ± 2[288], 50[289], 60[290]
$Ni_{0.76}Al_{0.24}$	50[291]
$Co_{75}Fe_{25}$	58 ± 3[292]
Fe_4N	59[293]
$Co_xFe_{80-x}B$	65[294]
FePt	42[295]
Mn_4FeGe_3	42[296]
$Co_2Cr_xFe_{1-x}Si$	56 – 64[297]
$Co_2FeAl_xSi_{1-x}$	57 ± 1 – 60 ± 5[298]
$Co_2Cr_{1-x}Fe_xAl$	54–62[299]
$Co_2MnAl_{1-x}Sn_x$	63[300]
$Co_{1.9}Fe_{0.1}CrGa$	67 ± 3[301]
Co_2CrGa	61 ± 2[302]
Co_2MnGe	35 – 58[302], 55–60[303]
Co_2MnSi	52 – 54[302], 50 – 55[303], 54[304], 56[305], 20[258]
Co_2FeSi	49 ± 2[306]
Co_2FeGa	59[307]
NiMnSb	58 ± 2.3[250], 44[308], 45[305]
NiFeSb	52[309], 52[310]
Ni_2MnIn	28–34[261]
Mn_5Ge_3	42 ± 5[311]
Fe_3Si	45 ± 5[312]
MnAs	44 – 49[313]

Note: The numbers given in superscript are reference numbers.

are normally performed at zero temperature for perfectly ordered crystals, whereas the experiments are done at finite temperatures, and with imperfect materials. This raises an important question of stability of half-metallic behavior with respect to substitutions and magnetic disorder induced by finite temperatures. For example, just 5% of substitutional defects in NiMnSb may be sufficient to reduce its spin polarization dramatically and to close the minority gap [194]; similar effects have been predicted in PtMnSb [195] and in Sr_2FeMoO_6 [173]. Importantly for spintronics applications, the temperature scale for the decay of half-metallicity is defined by a relationship $T \ll T_C$ rather than $T \ll \Delta$ (e.g., Refs. [196–198]. In addition, there are other mechanisms that may effectively mix the spins and suppress half-metallicity, such as spin–orbit coupling [199,200] and Stoner excitations (spin flips). Another issue is a possible difference between surface and bulk electronic structure. Does half-metallicity persist all the way to the surface, where the translational symmetry is broken? The calculations in one of the layered manganite, $LaSr_2Mn_2O_7$, show that it is a half-metal in the bulk—but not at the surface [201]. On the other hand, NiMnSb is believed not to be a half-metal at the surface, but may preserve its half-metallicity at the interface with CdS [202]. These basic questions are closely related to the measurement techniques, many of which are surface sensitive. Different aspects of half-metals are reviewed, for example, by Irkhin and Katsnelson [203], Coey et al. [204], Ziese [205], Eschrig and Pickett [206] and Katsnelson et al. [207].

27.7.2.1 CrO_2

CrO_2 is seemingly the only material, where the predicted half-metallicity [208] has been confirmed experimentally by several different techniques. While the use of this material in practical applications is hindered by the presence of a thin (≃1.5 nm) Cr_2O_3 surface layer, which is an antiferromagnetic insulator, the unambiguous experimental demonstration of half-metallicity is important as a matter of principle. The early PCAR results ($P \simeq 80\%$–90% [11,209], as well as the following studies on higher quality single crystalline films (with $P > 95\%$) [163,210] agree well with the values of spin polarization obtained by other techniques, such as photoemission [68], the Meservey–Tedrow technique and the AR measurements in the planar geometry [79]. In the latter work, while the probability of AR (and thus the accuracy of the technique) is reduced rapidly as the interfacial barrier strength increases, the Meservey–Tedrow technique becomes available once the quasiparticle tunneling starts to dominate the conduction across the interface. The combined results from these measurements are shown in Figure 27.17.

27.7.3 Other Potential Half-Metals: $Fe_xCo_{1-x}S_2$, EuO, EuS, and $Sr_xFe_{1-x}MoO_6$

27.7.3.1 $Fe_xCo_{1-x}S_2$

$Fe_xCo_{1-x}S_2$ is expected to be less sensitive to the crystallographic disorder and other defects and has been predicted to be a half-metal in a wide concentrations range $0.25 < x < 0.9$ with the magnetic moment per Co of $1\,\mu_B$ [211]. Experimentally, the spin polarization seems to be quite sensitive to the composition [212] varying from $P = 56\%$ for pure CoS_2 [213] to the maximum value of approximately 85% at $x = 0.15$, even though the magnetic moment per Co was found to be practically constant ($1\,\mu_B$ within the experimental range $0.07 < x < 0.3$ [214]. While the Curie

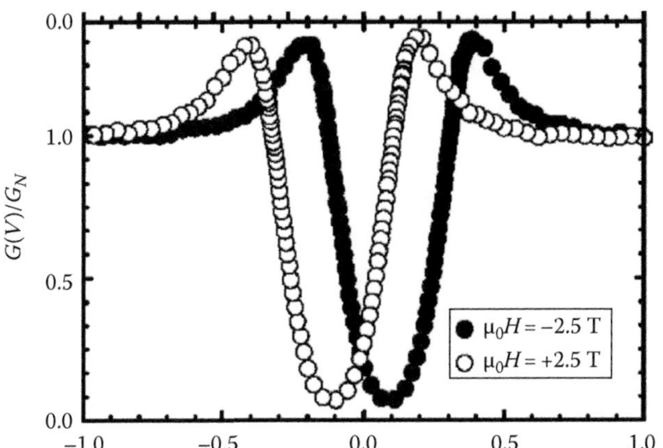

FIGURE 27.17 Spin polarization measurements of CrO_2 in a planar geometry. The conductance curve is gradually shifting upon the application of a magnetic field in the plane of the junction. No features due to the possible presence of minority spins are observed in fields as high as 2.5 T. (From Parker, J.S. et al., *Phys. Rev. Lett.*, 88, 196601, 2002. With permission.)

temperature of this system is probably not high enough for it to be used in applications, it provides an insight into the design of new half-metallic ferromagnets.

27.7.3.2 EuO/EuS

One of the few simple magnetic oxides, europium oxide, is also a viable candidate for half-metallicity. Recently, ≃90% spin polarization was measured in La-doped La_xEuO_{1-x} grown on $YAlO_3$ substrate [215], as well as in undoped EuO_{1-x} grown directly on conductively matched Si substrate with no buffer layer [161]. The latter result can potentially be used for spin injection into Si [216,217]. The PCAR measurements are consistent with spin-resolved x-ray absorption spectroscopy, showing near-complete spin polarization [218], as well as near-complete spin-filtering effect [219] (see Chapter 13, Moodera and Santos). A fairly high spin polarization ($P \simeq 80\%$) has also been measured in a planar geometry for a similar material, EuS [162].

27.7.3.3 SFMO

Sr_2FeMoO_6 (SFMO) has been predicted to be 100% spin polarized in a perfectly ordered structure [220]. It is extremely difficult to fabricate perfect SFMO samples, which has the propensity for the B-site defect formation, which, in turn, alters the half-metallic character of this compound [221]. For example, the band structure calculations for SFMO with 12.5% of anti-site defects results in $P_{Nv} \simeq 53\%$ [173].* These calculations are consistent with the PCAR measurements in single crystals with variable degree of disorder, resulting in $P \simeq 60\%–70\%$, the higher values of *P* corresponding to the smaller degree of disorder [173]. These values are in agreement with the earlier measurements in SFMO thin films [222,223]. While these results indicate that there is a correlation between the degree of disorder and *P*, it is difficult to directly interpolate the data to a perfectly ordered system, so the question of half-metallicity of SFMO remains unresolved.

* Interestingly, the calculated spin polarization in the diffusive limit for 12.5% disorder, $P_{Nv2} = 93.3\%$ is much higher, resembling the behavior of LSMO, as discussed below.

27.7.4 Transport Half-Metals

Transport half-metals were predicted [8] and experimentally observed [7] in optimally doped $La_{0.7}Sr_{0.3}MnO_3$ (LSMO), where the minority carriers were found to be approaching the metal–insulator transition limit. The name refers to materials that may not be true half-metals in terms of their DOS, but, because of their vastly different transport properties of their majority and minority carriers, carry almost 100% spin-polarized current [48,224].

27.7.4.1 LSMO

The case of LSMO has been puzzling: On the one hand, the band structure calculations predicted about 35% DOS spin polarization for LSMO [225], and higher values for P_{Nv} and P^2_{Nv}, in good agreement with the Tedrow–Meservey measurements by Worledge and Geballe [33] and the TMR results [226], with $P = 72\%$ and $P \simeq 80\%$, respectively. On the other hand, the spin-resolved photoemission measurements of the DOS [67] gave the value of 95% ± 5%, suggesting that LSMO is a conventional half-metal.† The systematic measurements of LSMO epitaxial films and bulk single crystals by the PCAR technique [7] resulted in a broad range of *P* for different samples from ≃60% to ≃90%. While a correlation was found between *P* and the residual resistivity of the samples, ρ, surprisingly, it were the highly resistive samples that had higher *P*, the opposite of what one might expect for a true half-metal (see Figure 27.18). The estimates of the contact size and the mean free path indicated that the low resistivity samples were in the ballistic regime, while the highest resistivity samples were in the diffusive regime. Using the values of the densities of states ($N_\uparrow(E_F) = 0.58$ states/eV Mn, $N_\downarrow(E_F) = 0.27$ states/eV Mn) and the Fermi velocities ($v_{F\uparrow} = 7.4 \times 10^5$ ms^{-1}, $v_{F\downarrow} = 2.2 \times 10^5$ ms^{-1}) [225], $P_{Nv} = 74\%$, and $P_{Nv^2} = 92\%$ were obtained, in a fairly good agreement with the experimental data.‡ The value of P^2_{Nv}, which also corresponds to the spin polarization of the current in bulk LSMO, implies that about 96% of the current is carried by the majority electrons, as the concept of a transport

† For a true half-metal all spin polarization measurements techniques should give the same value $P_N = P_{Nv} = P_{Nv^2} = 100\%$.

‡ One might argue though, that, it should still be possible to establish a diffusive contact even in the case of low-resistivity samples (with a long mean free path) by simply increasing the size of the contact. However, in this case, instead of a single large contact, a number of smaller contacts connected in parallel will typically be established [169]. At the same time, the extracted values of the spin polarization are practically independent of the transport regime [151].

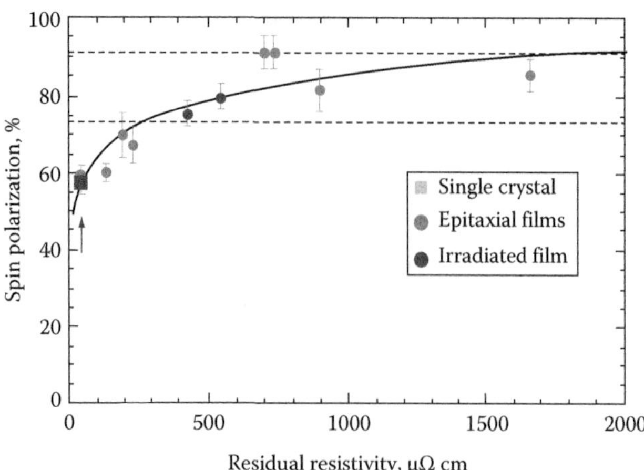

FIGURE 27.18 Spin polarization as a function of the residual resistivity of $La_{0.7}Sr_{0.3}MnO_3$ at 1.5 K. Dashed lines correspond to P_{Nv} = 74% and P_{Nv^2} = 92%. Solid line is a guide to the eye. (From Nadgorny, B. et al., *Phys. Rev. B*, 61, R3788, 2000. With permission.)

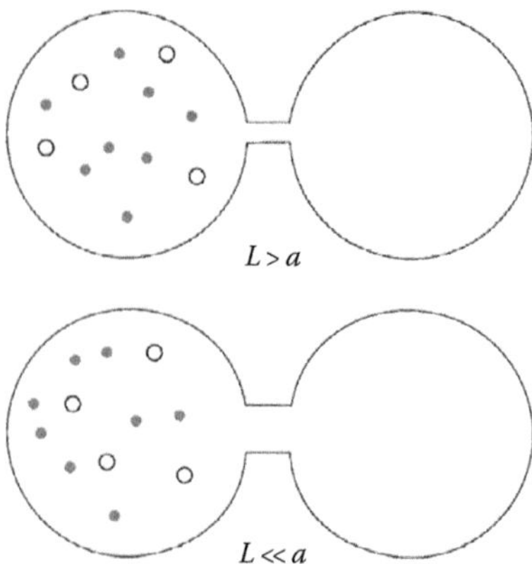

FIGURE 27.19 Geometric spin filtering. "Spin polarization" can be changed from P_{Nv} to P_{Nv^2} solely by adjusting the size of the aperture, thus changing the flow type from ballistic (Knudsen) (top) to diffusive (Maxwell) (bottom).

half-metal implies. A record high TMR ratio found in LSMO-based structures [227] corresponds to $P_T \simeq 95\%$, in close agreement with P^2_{Nv} measured by the PCAR technique. While the consensus has yet to be reached [34,207], the most recent theoretical [228] and experimental results [229] indicate the presence of minority spin states in LSMO.*

27.7.4.2 Geometric Spin Filtering

We can use the case of LSMO to draw an analogy between a spin-polarized current and a molecular flow. Consider a *toy model* (see Figure 27.19), where we have particles in one half of a vessel in equilibrium with another half. There are eight lighter particles (majority spins) and four heavier particles (minority spins). Suppose that the lighter particles are twice as fast as the heavier particles. The DOS "spin polarization" is then $P = N_1 - N_2/N_1 + N_2 = 33\%$. If the flow through the orifice is ballistic (*Knudsen regime*) the equilibrium is determined by the flux Nv and the current "spin polarization" is then higher $P_{Nv} = (N_1v_1 - N_2v_2)/(N_1v_1 + N_2v_2) = 4 - 1/4 + 1 = 60\%$. If the flow is diffusive (*Maxwell regime*), the equilibrium is determined by the pressure, which is proportional to the temperature, and, therefore, to the square of the velocity. The spin polarization in the diffusive limit becomes higher still: $P^2_{Nv} = (N_1v_1^2 - N_2v_2^2)/(N_1v_1^2 + N_2v_2^2) = 7/9 \approx 80\%$.

By increasing the size of the contact, one can transform the transport from ballistic to diffusive regimes, and thus change the degree of spin polarization by changing the contact geometry alone (see Figure 27.19), without changing any materials properties of the system [34].† This effect can be called *geometric spin filtering* in analogy with spin-filtering effect in F/I/S (Meservey–Tedrow) or F/I/F (TMR) structures, where one can suppress or enhance the transparency of the barrier for the majority and minority electrons [9,10]. Recently, a systematic decrease of P as a function of the contact size a in the nanoconstriction geometry has been reported [165]. The authors attributed the observed effect to an increased spin–orbit scattering as the contact size increases. However, the results can also be explained by a gradual transition from the ballistic ($a \approx 7$ nm) to the diffusive ($a \approx 25$ nm) regimes, as described earlier. In this case, the spin polarization of Fe would change by roughly 10% [164], in reasonable agreement with the reported results.‡

27.7.5 Dilute Magnetic Semiconductors and Magnetic Oxides

27.7.5.1 (Ga,Mn)As and Other Magnetic Semiconductors

Until recently, most of the transport studies of a semiconductor/superconductor interface, starting with the pioneering work on Nb/InGaAs junctions [232], have been performed in a two-dimensional geometry [233]. In order to extend the PCAR technique to measure the spin polarization of ferromagnetic semiconductors (FSm) several conditions must be met. First, the ability to conduct the measurements at low temperatures implies the absence of strong activation behavior with the resistivity not exceeding ≅10 mΩ cm, in which case the values of the interface mismatch [234] are typically reasonable as well. Second, the junction transparency, which is often limited by native Schottky barriers present at most Sm/S interfaces, must be high enough to ensure that AR is the dominant process below the gap. The use of high doping levels, producing the carrier concentrations

* The results from Angle Resolved Photoemission Spectroscopy (ARPES) have conclusively shown the presence of both majority and minority states in LSMO [229].

† The author wishes to acknowledge his discussions with J. Byers at the early stage of this work. A similar behavior have been predicted by ab initio calculations in cobalt nanowires [230].

‡ We would like to note that, while in the case of LSMO P_{Nv^2} is higher than P_{Nv} the opposite may be true as well; in fact the two values may even have different signs [231].

on the order of 10^{20}–10^{21} cm^{-3}, helps to alleviate these problems, dramatically reducing the width of the Schottky barrier [235]. This strategy has been successfully applied to a number of studies, in which FmS were measured together with matching nonmagnetic semiconductors with similar doping concentrations [137,138]. Fortuitously, the increase of the Curie temperature in (Ga,Mn)As is associated with higher doping levels [236] at which the PCAR technique becomes available, with $P \cong 85\%$ measured in (Ga,Mn)As thin films with the use of the planar [160] and point contact geometries [137]. These measurements also indicate the increasing importance of inelastic broadening as the semiconductor band gap increases/the mobility decreases. While practically no broadening was observed in a narrow gap semiconductor (In,Mn)Sb [138,237], it was present in (Ga,Mn)Sb [238], and even more prominently in (Ga,Mn)As [137], similarly to the effects reported for Co–Pt–Pb in the nanoconstriction geometry [101]. While the conductance broadening observed in Ga/(Ga,Mn)As was attributed primarily to the distribution of the critical temperatures in amorphous Ga used as a superconducting contact [160], additional inelastic broadening effects are also possible.

27.7.5.2 Dilute Magnetic Semiconducting Oxides

High Curie temperatures make dilute magnetic semiconducting oxides (DMSO) very attractive for applications. While the question of the nature of magnetism in these materials is still unresolved (see, e.g., the discussion by Coey [239], and in Chapter 20 of this book), it is natural to assume that at least in conducting oxides, it is mediated by delocalized carriers [240–242]. If the magnetism is carrier mediated, the spin current is spin polarized and it should be possible to use some of these materials for room temperature spin injection. Let us briefly discuss two materials systems: Cr-doped and undoped In_2O_3, in which carriers are created by oxygen vacancies [243] and Mn-doped and undoped ZnO, in which carriers are introduced by co-doping with Al [244]. It has been shown that the electrical and magnetic properties of Cr:In_2O_3 films are sensitive to the oxygen vacancy defect concentration [245]. More recent measurements of Cr-doped and undoped In_2O_3 films indicate that both are ferromagnetic with very similar magnetization, and both are almost equally spin polarized with $P \simeq 50\%$ [243], which implies that Cr is not essential for the development of magnetism in In_2O_3 with high ($\simeq 10^{21}$ cm^{-3}) concentration of oxygen vacancies.* This can be explained in the spirit of the Zener model [242], where the development of ferromagnetism is intimately related to the interaction of delocalized charge carriers and vacancies in In_2O_3. On the other hand, Al-doped ZnO with and without Mn [244] allow direct comparison between two highly conductive materials, of which only Mn-doped is ferromagnetic (with $P \simeq 55\%$, while the nonmagnetic films without Mn is unpolarized. Based on the fairly high measured values of spin polarization in both of these systems, one can argue that ferromagnetism at least in some DMSO is carrier mediated.†

27.8 Future Directions and Concluding Remarks

In addition to the conventional spin polarization measurements, the PCAR technique has recently been applied to the study of spin accumulation in a normal metal. Specifically, a spin-polarized current injected into a normal metal should be sensitive to the same AR mechanism via the spin accumulation at an N/S interface [154,242]. The observed spin accumulation will be gradually reduced as the spin diffuses from the interface, that is, as the width of the normal metal increases. Specifically, the spin injection from a Heusler alloy, $Co_2Mn_{0.5}Fe_{0.5}Si$ has been used to determine the spin diffusion length L_{sf} in Au [168], which was found to be comparable to the values of spin diffusion found by other techniques [246]. One of the potential advantages of this technique, which is yet to be realized, is the ability to perform the measurements at the nanoscale, using an STM tip, for example. Such measurements would provide a better resolution and can compliment the other spin diffusion measurement techniques described in this volume [247–249] (Chapters 6 and 23).

In summary, in a little more than 10 years since its inception in 1998, PCAR spectroscopy has become one of the most useful tools in determining the values of spin polarization. While the technique employs some very basic, perhaps oversimplified, theoretical models, they seem to work well beyond their nominal applicability range to provide the practitioners of the technique with the consistent recipe to study a wide variety of magnetic materials, see Table 27.2. The alternative in terms of theoretical development is to further advance a fully self-consistent treatment. While this approach may succeed in general, it is unlikely to completely eliminate the need for phenomenological parameters, such as mixing angles, for example, which are material dependent [145]. Moreover, considering various conceptual and technical problems, such as spin–orbit interaction, proximity and inverse proximity effects, spreading resistance, and a number of other pitfalls in the PCAR technique, it may be difficult to substantially improve the accuracy of the technique beyond the current limit of 3%–5%, even if the perfect theory is developed. As the PCAR technique is inherently a low-temperature technique, it may be hard to utilize it fully for the measurements of high Curie temperature ferromagnets, which may have to be done at room temperatures. At the same time, it is instrumental in eliminating the effects of Stoner and other temperature-driven excitations on the ground state of half-metals, for which PCAR spectroscopy may be a technique of choice.

* Unfortunately, it is difficult to fabricate an In_2O_3 sample, which would be well conducting *and* nonmagnetic, to use it as a standard in the PCAR measurements.

† While both measurements are done at liquid helium temperatures, the fact that magnetization, as well as resistivity in both systems are only weekly dependent on temperature makes it plausible (although not conclusive) that the room temperature magnetism is also carrier mediated.

Acknowledgments

I am grateful to V. Antropov, K. D. Belashchenko, N. Birge, R. A. Buhrman, A. Buzdin, G. Duetscher, A. Golubov, Yu. Lyanda-Geller, I.I. Mazin, J. S. Moodera, R. P. Panguluri, A. Petukhov, E. I. Rashba, D. Singh, B. Sinkovic, E. Tsymbal, J. Wei, S. Wolf, and I. Žutić for discussions and many useful suggestions, and to D. Dessau for sharing their experimental results prior to publication. I also wish to thank R. Panguluri and M. Faiz for their help in preparing some of the reference material and figures, and Grigori Youkov for providing most of the illustrations for this chapter.

Our work has been funded in part by DARPA, Office of Naval Research, and the National Science Foundation.

References

1. S. A. Wolf, D. D. Awschalom, R. A. Buhrman et al., Spintronics: A spin-based electronics vision for the future, *Science* **294**, 1488–1495 (2001).
2. I. Žutić, J. Fabian, and S. Das Sarma, Spintronics: Fundamentals and applications, *Rev. Mod. Phys.* **76**, 323–410 (2004).
3. I. I. Mazin, How to define and calculate the degree of spin polarization in ferromagnets, *Phys. Rev. Lett.* **83**, 1427–1430 (1999).
4. M. Julliere, Tunneling between ferromagnetic films, *Phys. Lett.* **54A**, 225–226 (1975).
5. W. E. Pickett and J. S. Moodera, Half metallic magnets, *Phys. Today* **54**, 39 (2001).
6. R. A. de Groot, F. M. Mueller, P. G. van Engen, and K. H. J. Buschow, New class of materials: Half-metallic ferromagnets, *Phys. Rev. Lett.* **50**, 2024 (1983).
7. B. Nadgorny, I. I. Mazin, M. Osofsky et al., Origin of high transport spin polarization in $La_{0.7}Sr_{0.3}MnO_3$: Direct evidence for minority spin states, *Phys. Rev. B* **63**, 184433 (2001).
8. W. E. Pickett and D. J. Singh, Chemical disorder and charge transport in ferromagnetic manganites, *Phys. Rev. B* **55**, R8642–R8645 (1997).
9. S. S. P. Parkin, C. Kaiser, A. Panchula, P. M. Rice, and B. Hughes, Giant tunneling magnetoresistance at room temperature with MgO (100) tunnel barriers, *Nat. Mater.* **3**, 862–867 (2004).
10. S. Yuasa, T. Nagahama, A. Fukushima, Y. Suzuki, and K. Ando, Giant room-temperature magnetoresistance in single-crystal Fe/MgO/Fe magnetic tunnel junctions, *Nat. Mater.* **3**, 868–871 (2004).
11. R. J. Soulen Jr., J. M. Byers, M. S. Osofsky et al., Measuring the spin polarization of a metal with a superconducting point contact, *Science* **282**, 85–88 (1998).
12. S. K. Upadhyay, A. Palanisami, R. N. Louie, and R. A. Buhrman, Probing ferromagnets with Andreev reflection, *Phys. Rev. Lett.* **81**, 3247 (1998).
13. A. F. Andreev, The thermal conductivity of the intermediate state in superconductors, *Zh. Eksp. Teor. Fiz.* **46**, 1823–1825 (1964) [*Sov. Phys. JETP* **19**, 1228–1231 (1964)].
14. D. Saint-James, Elementary excitations in the vicinity of a normal metal – superconducting metal contact, *J. Phys. (Paris)* **25**, 899 (1964).
15. M. J. M. de Jong and C. W. J. Beenakker, Andreev reflection in ferromagnet-superconductor junctions, *Phys. Rev. Lett.* **74**, 1657–1660 (1995).
16. G. E. Blonder, M. Tinkham, and T. M. Klapwijk, Transition from metallic to tunneling regimes in superconducting microconstrictions: Excess current, charge imbalance, and supercurrent conversion, *Phys. Rev. B* **25**, 4515–4532 (1982).
17. C. W. J. Beenakker, Random-matrix theory of quantum transport, *Rev. Mod. Phys.* **69**, 731–808 (1997).
18. I. I. Mazin, A. A. Golubov, and B. Nadgorny, Probing spin polarization with Andreev reflection: A theoretical basis, *J. Appl. Phys.* **89**, 7576–7578 (2001).
19. B. Pannetier and H. Courtois, Andreev reflection and proximity effect, *J. Low Temp. Phys.* **118**, 599–615 (2000).
20. A. I. Buzdin, Proximity effects in superconductor-ferromagnet heterostructures, *Rev. Mod. Phys.* **77**, 935–976 (2005).
21. S. Kashiwaya and Y. Tanaka, Tunnelling effects on surface bound states in unconventional superconductors, *Rep. Prog. Phys.* **63**, 1641 (2000).
22. A. P. Mackenzie and Y. Maeno, The superconductivity of Sr_2RuO_4 and the physics of spin-triplet pairing, *Rev. Mod. Phys.* **75**, 657 (2003).
23. G. Deutscher, Andreev-Saint James reflections: A probe of cuprate superconductors, *Rev. Mod. Phys.* **77**, 109–135 (2005).
24. C. W. J. Beenakker, Colloquium: Andreev reflection and Klein tunneling in graphene, *Rev. Mod. Phys.* **80**, 1337 (2008).
25. A. G. Aronov, Spin injection in metals and polarization of nuclei, *Pis'ma Zh. Eksp. Teor. Fiz.* **24**, 37–39 (1976) [*JETP Lett.* **24**, 32–34 (1976)].
26. A. G. Aronov and G. E. Pikus, Spin injection into semiconductors, *Fiz. Tekh. Poluprovodn.* **10**, 1177 (1976).
27. B. T. Jonker, A. T. Hanbicki, D. T. Pierce, and M. D. Stiles, Spin nomenclature for semiconductors and magnetic metals, *J. Magn. Magn. Mater.* **277**(1–2), 24–28 (2004).
28. Yu. V. Sharvin, On the possible method for studying Fermi surfaces, *Zh. Eksp. Teor. Fiz.* **48**, 984 (1965) [*Sov. Phys. JETP* **21**, 655 (1965)].
29. T. Oguchi, *Mater. Sci. Eng.* **31**, 111 (1995).
30. R. Meservey and P. M. Tedrow, Spin-polarized electron tunneling, *Phys. Rep.* **238**, 173 (1994).
31. J.-N. Chazalviel and Y. Yafet, Theory of the spin polarization of field-emitted electrons from nickel, *Phys. Rev. B* **15**, 1062 (1977).
32. J. A. Hertz and K. Aoi, Spin dependent tunneling from transition metal ferromagnets, *Phys. Rev. B* **8**, 3552 (1973).

33. D. C. Worlege and T. H. Geballe, Maki analysis of spin-polarized tunneling in an oxide ferromagnet, *Phys. Rev. B* **62**, 447–451 (2000).
34. B. Nadgorny, The case against half-metallicity in $La_{0.7}Sr_{0.3}MnO_3$, *J. Phys.: Condens. Matter* **19**, 315209 (2007).
35. W. H. Butler, Tunneling magnetoresistance from a symmetry filtering effect, *Sci. Technol. Adv. Mater.* **9**, 014106 (2008).
36. J. P. Velev, P. A. Dowben, and E. Y. Tsymbal, S. J. Jenkins, A. N. Caruso, Interface effects in spin-polarized metal/insulator layered structures, *Surf. Sci. Rep.* **63**, 400–425 (2008).
37. J. S. Moodera, J. Nowak, L. R. Kinder et al., Quantum well states in spin-dependent tunnel structures. *Phys. Rev. Lett.* **83**, 3029–3032 (1998).
38. A. M. Bratkovsky, Assisted tunneling in ferromagnetic junctions and half-metallic oxides, *Appl. Phys. Lett.* **72**, 2334 (1998).
39. A. M. Bratkovsky, Spintronic effects in metallic, semiconductor, metal–oxide and metal–semiconductor heterostructures, *Rep. Prog. Phys.* **71**, 026502 (2008).
40. W. H. Butler, X. G. Zhang, T. C. Schulthess, and J. M. MacLaren, Spin-dependent tunneling conductance of Fe|Mgo|Fe sandwiches, *Phys. Rev. B (Condens. Matter)* **63**(5), 1–12 (2001).
41. J. M. MacLaren, X.-G. Zhang, W. H. Butler, and X. Wang, Layer KKR approach to Bloch-wave transmission and reflection: Application to spin-dependent tunneling, *Phys. Rev. B* **59**, 5470–5478 (1999).
42. A. G. Petukhov, J. Niggemann, V. N. Smelyanskiy, and V. V. Osipov, 100% spin accumulation in non-half-metallic ferromagnet–semiconductor junctions, *J. Phys.: Condens. Matter* **19**, 315205 (2007).
43. J. Mathon and A. Umerski, Theory of tunneling magnetoresistance of an epitaxial Fe/MgO/Fe(001) junction, *Phys. Rev. B* **63**, 1 (2001).
44. E. Y. Tsymbal and D. G. Pettifor, Spin-polarized electron tunneling across a disordered insulator, *Phys. Rev. B* **58**, 432–437 (1998).
45. I. I. Mazin, A. A. Golubov, and A. Zaikin, "Chain Scenario" for Josephson tunneling with π shift in $YBa_2Cu_3O_7$, *Phys. Rev. Lett.* **75**, 2574 (1995).
46. C. Kaiser, S. van Dijken, S. H. Yang, H. Yang, and S. S. P. Parkin, Role of tunneling matrix elements in determining the magnitude of the tunneling spin polarization of 3d transition metal ferromagnetic alloys, *Phys. Rev. Lett.* **94**, 247203 (2005).
47. V. Vlaminck and M. Bailleul, Current-induced spin-wave Doppler shift, *Science* **322**, 410 (2008).
48. G. M. Müller, J. Walowski, M. Djordjevic, G. X. Miao, A. Gupta, A. V. Ramos, K. Gehrke et al., Spin polarization in half-metals probed by femtosecond spin excitation, *Nat. Mater.* **8**, 56 (2009).
49. K. E. H. M. Hanssen, P. E. Mijnarends, L. P. L. M. Rabou, and K. H. J. Buschow, Positron-annihilation study of the half-metallic ferromagnet NiMnSb: Experiment, *Phys. Rev. B* **42**, 1533 (1990).
50. S. Berko, and A. P. Mills, Spin distribution studies in ferromagnetic metals by polarized positron annihilation experiments, *J. Phys. (Paris), Colloq.* **32**, C1-287 (1971).
51. P. Zwart, L. P. L. M. Rabou, G. J. Langedijk et al., in *Positron Annihilation*, P. C. Jain, R. M., Singru, and K. P. Gopinathan, (Eds.), World Scientific, Singapore, p. 297, 1985.
52. S. Berko, in: *Positron Solid State Physics*, W. Brandt and A. Dupasquier (Eds.), North Holland, Amsterdam, the Netherlands, 1983.
53. K. E. H. M. Hanssen and P. E. Mijnarends, Positron-annihilation study of the half-metallic ferromagnet NiMnSb: Theory, *Phys. Rev. B* **34**, 5009 (1986).
54. S. F. Alvarado, W. Eib, H. C. Siegmann, and J. P. Remeika, Photoelectron spin polarization testing of ionic structure of 3d levels in ferrites, *Phys. Rev. Lett.* **35**, 860 (1975).
55. P. D. Johnson, Spin polarized photoemission, *Rep. Prog. Phys.* **60**, 1217 (1997).
56. R. Feder, *Polarized Electrons in Surface Physics*, World Scientific, Singapore, 1985.
57. S. Hüfner, *Photoelectron Spectroscopy*, Springer, Berlin, 1995.
58. P. D. Johnson, in *Core Level Spectroscopies for Magnetic Phenomena*, vol. 345, P. S. Bagus (Ed.), NATO ASI Series B: Physics, Plenum Press, New York, 1995.
59. B. T. Jonker, K.-H. Walker, E. Kisker, G. A. Prinz, and C. Carbone, Spin-polarized photoemission study of epitaxial Fe(001) films on Ag(001), *Phys. Rev. Lett.* **57**, 142–145 (1986).
60. E. Kisker, K. Schröder, M. Campagna, and W. Gudat, Temperature dependence of the exchange splitting of Fe by spin-resolved photoemission spectroscopy with synchrotron radiation, *Phys. Rev. Lett.* **52**, 2285 (1984).
61. D. P. Pappas, K.-P. Kämper, and H. Hopster, Reversible transition between perpendicular and in-plane magnetization in ultrathin films, *Phys. Rev. Lett.* **64**, 3179–3182 (1990).
62. L. Plucinski, Y. Zhao, B. Sinkovic, E. Vescovo, MgO+Fe(100) interface: A study of the electronic structure, *Phys. Rev. B* **75**, 214411 (2007).
63. W. Clemens, E. Vescovo, T. Kachel, C. Carbone, and W. Eberhardt, Spin-resolved photoemission study of the reaction of O_2 with fcc Co(100), *Phys. Rev. B* **46**, 4198 (1992).
64. B. Sinkovich, L. H. Tjeng, N. B. Brookes et al., Local electronic and magnetic structure of Ni below and above TC: A spin-resolved circularly polarized resonant photoemission study, *Phys. Rev. Lett.* **79**, 3510–3513 (2001).
65. H. Tang, D. Weller, T. G. Walker et al., Magnetic reconstruction of the Gd (0001) surface, *Phys. Rev. Lett.* **71**, 444–447 (1993).
66. Y. S. Dedkov, U. Rüdiger, and G. Güntherodt, Evidence for the half-metallic ferromagnetic state of Fe_3O_4 by spin-resolved photoelectron spectroscopy, *Phys. Rev. B* **65**, 064417 (2002).
67. J.-H. Park, E. Vescovo, H.-J. Kim, C. Kwon, R. Ramesh, and T. Venkatesan, Direct evidence for a half-metallic ferromagnet, *Nature* **392**, 794 (1998).

68. K. P. Kämper, W. Schmitt, G. Güntherodt, R. J. Gambino, and R. Ruf, CrO_2—A new half-metallic ferromagnet, *Phys. Rev. Lett.* **59**, 2788 (1987).
69. F. J. Himpsel, Exchange splitting of epitaxial fcc Fe/Cu(100) vs bcc Fe/Ag(100), *Phys. Rev. Lett.* **67**, 2363–2366 (1991).
70. P. M. Tedrow and R. Meservey, Spin-dependent tunneling into ferromagnetic nickel, *Phys. Rev. Lett.* **26**, 192–195 (1971).
71. P. M. Tedrow and R. Meservey, Spin polarization of electrons tunneling from films of Fe, Co, Ni, and Gd, *Phys. Rev. B* **7**, 318 (1973).
72. K. Maki, The behavior of superconducting thin films in the presence of magnetic fields and currents, *Prog. Theor. Phys.* **32**, 29 (1964).
73. J. S. Moodera, L. R. Kinder, T. M. Wong, and R. Meservey, Large magnetoresistance at room temperature in ferromagnetic thin film tunnel junctions, *Phys. Rev. Lett.* **74**, 3273 (1995).
74. R. Meservey, P. M. Tedrow, V. R. Kalvey, and D. Paraskevopoulos, Studies of ferromagnetic metals by electron spin polarized tunneling, *J. Appl. Phys.* **50**, 1935 (1979).
75. H. Itoh, S. Shibata, T. Kumazaki, J. Inoue, and S. Maekawa, Effects of interfacial randomness on tunnel conductance and magnetoresistance in ferromagnetic tunnel junctions, *J. Phys. Soc. Jpn.* **68**, 1632 (1999).
76. S. Zhang and P. M. Levy, Models for magnetoresistance in tunnel junctions, *Eur. Phys.* **B10**, 599–606 (1999).
77. C. Kaiser, A. F. Panchula, and S. S. P. Parkin, Finite tunneling spin polarization at the compensation point of rare-earth-metal–transition-metal alloys, *Phys. Rev. Lett.* **95**, 047202 (2005).
78. I. I. Oleinik, E. Yu. Tsymbal, and D. G. Pettifor, Structural and electronic properties of $Co/Al_2O_3/Co$ magnetic tunnel junction from first principles, *Phys. Rev. B* **62**(6), 3952–3959 (2000).
79. J. S. Parker, S. M. Watts, P. G. Ivanov, and P. Xiong, Spin Polarization of CrO_2 at and across an artificial barrier, *Phys. Rev. Lett.* **88**, 196601 (2002).
80. Yu. V. Sharvin, and L. M. Fisher, Observation of focused electron beams in a metal, *Pis'ma Zh. Eksp. Teor. Fiz.* **1**, 54 (1965) [*JETP Lett.* **1**, 152 (1965)].
81. C. W. J. Beenakker, H. van Houten, and B. J. van Wees, Coherent electron focusing, in: *Festkörperprobleme (Advances in Solid State Physics)*, vol. 29, U. Rössler (Ed.), Pergamon/Vieweg, Braunschweig, Germany, 1989.
82. V. S. Tsoi, Focusing of electrons in a metal by a transverse magnetic field, *Pis'ma Zh. Eksp. Teor. Fiz.* **19**, 114 (1974) [*JETP Lett.* **19**, 70 (1974)].
83. V. S. Tsoi, J. Bass, and P. Wyder, Studying conduction-electron/interface interactions using transverse electron focusing, *Rev. Mod. Phys.* **71**, 1641 (1999).
84. S. I. Bozhko, V. S. Tsoi, and S. E. Yakovlev, Observation of Andreev reflection with the use of transverse electron focusing effect, *Pis'ma Zh. Eksp. Teor. Fiz.* **36**, 123 (1982) [*JETP Lett.* **36**, 152 (1982)].
85. P. A. M. Benistant, A. P. van Gelder, H. van Kempen, and P. Wyder, Angular relation and energy dependence of Andreev reflection, *Phys. Rev. B* **32**, 3351–3353 (1985).
86. M. Knudsen, *Kinetic Theory of Gases*, Methuen, London, 1934.
87. J. C. Maxwell, *A Treatise on Electricity and Magnetism*, Dover Press, New York, 1891.
88. G. Wexler, Size effect and non-local Boltzmann transport equation in orifice and disk geometry, *Proc. Phys. Soc. Lond.* **89**, 927–941 (1966).
89. M. Rokni and Y. Levinson, Joule heat in point contacts, *Phys. Rev. B* **52**, 1882 (1995).
90. B. Nicolic, and P. Allen, Electron transport through a circular constriction, *Phys. Rev. B* **60**, 3963 (1999).
91. C. W. J. Beenakker, Why does a metal-superconductor junction have a resistance? 1999, cond-mat/9909293.
92. S. V. Artemenko, A. F. Volkov, and A. V. Zaitsev, Excess current in microbridges, *Solid State Commun.* **30**, 771–773 (1979).
93. C. W. J. Beenakker, Quantum transport in semiconductor-superconductor microjunctions, *Phys. Rev. B* **46**, 12841 (1992).
94. Y. V. Nazarov and T. H. Stoof, Diffusive conductors as Andreev interferometers, *Phys. Rev. Lett.* **76**, 823 (1996).
95. A. M. Duif, A. G. M. Jansen, and P. Wyder, Point-contact spectroscopy, *J. Phys.: Condens. Matter*, **1**, 3157 (1989).
96. R. Holm, *Electric Contacts: Theory and Application*, 4th edn., Springer-Verlag, New York, 1967.
97. A. A. Lysykh, I. K. Yanson, O. I. Shklyarevski, and Yu. G. Naydyuk, Point-contact spectroscopy of electron-phonon interaction in alloys, *Solid State Commun.* **35**, 987 (1980).
98. A. G. M. Jansen, A. P. van Gelder, and P. Wyder, Point-contact spectroscopy in metals, *J. Phys. C: Solid State Phys.* **13**, 6073 (1980).
99. I. K. Yanson, *Zh. Éksp. Teor. Fiz.* **66**, 1035 (1974) [*Sov. Phys. JETP* **39**, 506 (1974)].
100. Yu. G. Naidyuk and I. K. Yanson, *Point-Contact Spectroscopy*, Springer Series in Solid-State Sciences, vol. 145, Springer, Berlin, Germany, 2004.
101. P. Chalsani, S. K. Upadhyay, O. Ozatay, and R. A. Buhrman, Andreev reflection measurements of spin polarization, *Phys. Rev. B* **75**, 094417 (2007).
102. R. P. Panguluri, G. Tsoi, B. Nadgorny, S. H. Chun, and N. Samarth, Point contact spin spectroscopy of MnAs ferromagnetic films, *Phys. Rev. B* **68**, 201307 (2003).
103. A. I. Akimenko, A. B. Verkin, N. M. Ponomarenko, and I. K. Yanson, *Fiz. Nizk. Temp.* **4**, 1267 (1982); [*Sov. J. Low Temp. Phys.* **4**, 1267].
104. Yu. G. Naidyuk and I. K. Yanson, Point-contact spectroscopy of heavy-fermion systems, *J. Phys.: Condens. Matter* **10**, 8905 (1998).
105. K. Mendelssohn and J. L. Olsen, Anomalous heat flow in superconductors, *Proc. Phys. Soc. (Lond.)* **A63**, 2 (1950); *Phys. Rev.* **80**, 859 (1950).
106. N.V. Zavaritskii, Thermal conductivity of superconductors in the intermediate state, *JETP* **38**, 1673 (1960) [*Sov. Phys. JETP* **11**, 1207 (1960)].

107. A. A. Abrikosov, *Fundamentals of the Theory of Metals*, North-Holland, Amsterdam, the Netherlands, 1988.
108. J. M. Byers and Flatté, M. E. Probing spatial correlations with nanoscale two-contact tunneling, *Phys. Rev. Lett.* **74**, 306–309 (1995).
109. P. Cadden-Zimansky, J. Wei, and V. Chandrasekhar, Cooper-pair-mediated coherence between two normal metals, *Nat. Phys.* **5**, 393 (2009).
110. G. Deutscher and P. G. de Gennes, Proximity effects, in: *Superconductivity*, R. D. Parks (Ed.), Dekker, New York, pp. 1005–1034, 1969.
111. C. Bruder, Andreev scattering in anisotropic superconductors, *Phys. Rev. B* **41**, 4017–4032 (1990).
112. M. Eschrig, Distribution functions in nonequilibrium theory of superconductivity and Andreev spectroscopy in unconventional superconductors, *Phys. Rev. B* **61**, 9061 (2000).
113. G. Goll, *Unconventional Superconductors: Experimental Investigation of Order Parameter Theory*, Springer Tracts in Modern Physics, vol. 214, Springer, Berlin, 2006.
114. T. Jacobson, On the origin of the outgoing black hole modes, *Phys. Rev. D* **53**, 7082 (1996).
115. M. Sadzikowski and M. Tachibana, Andreev reflection at the quark-gluon-plasma-color-flavor-locking interface, *Phys. Rev. D* **66**, 045024 (2002).
116. P. G. de Gennes and D. Saint James, Elementary excitations in the vicinity of a normal metal-superconducting metal contact, *Phys. Lett.* **4**, 151–152 (1963).
117. F. Giazotto, P. Pingue, F. Beltram et al., Resonant transport in Nb/GaAs AlGaAs heterostructures: Realization of the de Gennes–Saint-James model, *Phys. Rev. Lett.* **87**, 216808 (2001).
118. J. M. Rowell and W. L. McMillan, Electron interference in a normal metal induced by superconducting contracts, *Phys. Rev. Lett.* **16**, 453–456 (1966).
119. W. J. Tomasch, Geometrical resonance in the tunneling characteristics of superconducting Pb, *Phys. Rev. Lett.* **15**, 672–675 (1965).
120. I. P. Krylov and Yu. V. Sharvin, Observation of "Andreev" reflection of electrons from the boundary between a normal and superconducting phase, using the radio-frequency size effect, *Pis'ma Zh. Eksp. Teor. Fis.* **12**, 102 (1970) [*JETP Lett.* **12**, 71 (1970).
121. S. V. Gudenko and I. P. Krylov, Radio-frequency size effect and diffusion of impurities, *Zh. Eksp. Teor. Fiz.* **86**, 2304 (1984), S. V. Gudenko and I. P. Krylov, *Sov. Phys. JETP* **86**, 2304 (1984) (Engl. Transl.).
122. A. B. Pippard, J. G. Shepherd and D. A. Tindall, Resistance of superconducting-normal interfaces, *Proc. R. Soc. Lond. A* **324**, 17–35 (1971).
123. V. V. Schmidt, in: *The Physics of Superconductors*, P. Muller and A. V. Ustinov (Eds.), Springer, Berlin, Germany, 1997.
124. M. Tinkham and John Clarke, Theory of pair-quasiparticle potential difference in nonequilibrium superconductors, *Phys. Rev. Lett.* **28**, 1366–1369 (1972).
125. C. J. Pethick and H. Smith, Relaxation and collective motion in superconductors: A two-fluid description, *Ann. Phys.* **119**, 133–169 (1979).
126. K. K. Likharev, Superconducting weak links, *Rev. Mod. Phys.* **51**, 101–159 (1979).
127. J. Demers and A. Griffin, Scattering and tunneling of electronic excitations in the intermediate state of superconductors, *Can. J. Phys.* **49**, 285 (1971).
128. A. Griffin and J. Demers, Tunneling in the normal-metal-insulator-superconductor geometry using the Bogoliubov equations of motion, *Phys. Rev. B* **4**, 2202–2208 (1971).
129. K. K. Likharev and L. A. Yakobson, *Zh. Eksp. Teor. Fiz.* **68**, 1150 (1976) [*Sov. Phys.-JETP* **41**, 570 (1976)].
130. I. O. Kulik, A. N. Omelyanchuk, and R. I. Shekhter, *Sov. J. Low Temp. Phys.* **3**, 740 (1977).
131. G. E. Blonder and M. Tinkham, Metallic to tunneling transition in Cu-Nb point contacts, *Phys. Rev. B* **27**, 112–118 (1983).
132. N. A. Mortensen, K. Flensberg, and A. P. Jauho, Angle dependence of Andreev scattering at semiconductor–superconductor interfaces, *Phys. Rev. B* **59**, 10176 (1999).
133. I. Žutić and O. T. Valls, Spin-polarized tunneling in ferromagnet/unconventional superconductor junctions, *Phys. Rev. B* **60**, 6320–6323 (1999).
134. I. Žutić, and O. T. Valls, Tunneling spectroscopy for ferromagnet/superconductor junctions, *Phys. Rev. B* **61**, 1555–1561 (2000).
135. K. Halterman and O. T. Valls, Proximity effects at ferromagnet-superconductor interfaces, *Phys. Rev. B* **65**, 014509 (2002).
136. J. Linder and A. Sudbø, Dirac fermions and conductance oscillations in s- and d-wave superconductor-graphene junctions, *Phys. Rev. Lett.* **99**, 147001 (2007).
137. R. P. Panguluri, K. C. Ku, T. Wojtowicz et al., Andreev reflection and pair-breaking effects at the superconductor/magnetic semiconductor interface, *Phys. Rev. B* **72**, 054510 (2005).
138. R. P. Panguluri, B. Nadgorny, T. Wojtowicz, W. L. Lim, X. Liu, and J. K. Furdyna, Measurement of spin polarization by Andreev reflection in ferromagnetic $In_{1-x}Mn_xSb$ epilayers, *Appl. Phys. Lett.* **84**, 4947 (2004).
139. A. A. Golubov, A. Brinkman, Y. Tanaka, I. I. Mazin, and O. V. Dolgov, Andreev spectra and subgap bound states in multiband superconductors, *Phys. Rev. Lett.* **103**, 077003 (2009).
140. G. Deutscher and P. Nozières, Cancellation of quasiparticle mass enhancement in the conductance of point contacts, *Phys. Rev. B* **50**, 13557–13562 (1994).
141. B. Nadgorny and I. I. Mazin, High efficiency nonvolatile ferromagnet/superconductor switch, *Appl. Phys. Lett.* **80**, 3973 (2002).
142. R. Landauer, Electrical resistance of disordered one-dimensional lattices, *Philos. Mag.* **21**, 863 (1970).
143. J. Bardeen, Tunnelling from a many-particle point of view, *Phys. Rev. Lett.* **6**, 57–59 (1961).
144. J. C. Cuevas, A. Martín-Rodero, and A. L. Yeyati, Hamiltonian approach to the transport properties of superconducting quantum point contacts, *Phys. Rev. B* **54**, 7366–7379 (1996).

145. R. Grein, T. Löfwander, G. Metalidis, and M. Eschrig, Theory of superconductor-ferromagnet point-contact spectra: The case of strong spin polarization, *Phys. Rev. B* **81**, 094508 (2010).
146. G. J. Strijkers, Y. Ji, F. Y. Yang, C. L. Chien, and J. M. Byers, Andreev reflections at metal/superconductor point contacts: Measurement and analysis, *Phys. Rev. B* **63**, 104510 (2001).
147. B. P. Vodopyanov and L. R. Tagirov, Andreev conductance of a ferromagnet-superconductor point contact, *Pis'ma Zh. Eksp. Teor. Fiz.* **77**, 126–131 (2003) [*JETP Lett.* **77**, 153–158.]
148. I. I. Mazin, Tunneling of Bloch electrons through vacuum barrier, *Europhys. Lett.* **55**, 404 (2001).
149. K. M. Schep, P. M. Kelly, and G. E. W. Bauer, Ballistic transport and electronic structure, *Phys. Rev. B* **57**, 8907 (1998).
150. Y. Ji, G. J. Strijkers, F. Y. Yang, and C. L. Chien, Comparison of two models for spin polarization measurements by Andreev reflection, *Phys. Rev. B* **64**, 224425 (2001).
151. G. T. Woods, R. J. Soulen Jr., I. Mazin et al., Analysis of point-contact Andreev reflection spectra in spin polarization measurements, *Phys. Rev. B* **70**, 054416 (2004).
152. S. Kashiwaya, Y. Tanaka, N. Yoshida, and M. R. Beasley, Spin current in ferromagnet-insulator-superconductor junctions, *Phys. Rev. B* **60**, 3572–3580 (1999).
153. A. A. Golubov, Interface resistance in ferromagnet/superconductor junctions, *Physica C* **327**, 46 (1999).
154. F. J. Jedema, B. J. van Wees, B. H. Hoving, A. T. Filip, and T. M. Klapwijk, Spin-accumulation-induced resistance in mesoscopic ferromagnet-superconductor junctions, *Phys. Rev. B* **60**, 16549–16552 (1999).
155. V. I. Fal'ko, C. J. Lambert, and A. F. Volkov, Andreev reflections and magnetoresistance in ferromagnet–superconductor mesoscopic structures, *Zh. Eksp. Teor. Fiz. Pisma Red.* **69**, 497–503 (1999) [*JETP Lett.* **69**, 532–538 (1999)].
156. W. Belzig, A. Brataas, Yu. V. Nazarov, and G. E. W. Bauer, Spin accumulation and Andreev reflection in a mesoscopic ferromagnetic wire, *Phys. Rev. B* **62**, 9726–9739 (2000).
157. J. Aarts, J. M. E. Geers, E. Brück, A. A. Golubov, and R. Coehoorn, Interface transparency of superconductor/ferromagnetic multilayers, *Phys. Rev. B* **56**, 2779–2787 (1997).
158. W. A. Harrison, Tunneling from an independent-particle point of view, *Phys. Rev.* **123**, 85–89 (1961).
159. S. K. Yip, Energy-resolved supercurrent between two superconductors, *Phys. Rev. B* **58**, 5803 (1998).
160. J. G. Braden, J. S. Parker, P. Xiong, S. H. Chun, and N. Samarth, Direct measurement of the spin polarization of the magnetic semiconductor (Ga,Mn)As, *Phys. Rev. Lett.* **91**, 056602 (2003).
161. R. P. Panguluri, T. S. Santos, E. Negusse et al., Half-metallicity in europium oxide conductively matched with silicon, *Phys. Rev. B* **78**, 125307 (2008).
162. C. Ren, J. Trbovic, R. L. Kallaher et al., Measurement of the spin polarization of the magnetic semiconductor EuS with zero-field and Zeeman-split Andreev reflection spectroscopy, *Phys. Rev. B* **75**, 205208 (2007).
163. A. Anguelouch, A. Gupta, G. Xiao et al., Near-complete spin polarization in atomically-smooth chromium-dioxide epitaxial films prepared using a CVD liquid precursor, *Phys. Rev. B* **64**, 180408 (2001).
164. B. Nadgorny, R. J. Soulen Jr., M. S. Osofsky et al., Transport spin polarization of Ni_xFe_{1-x}: Electronic kinematics and band structure, *Phys. Rev. B* **61**, R3788 (2000).
165. M. Stokmaier, G. Goll, D. Weissenberger, C. Sürgers, and H. v. Löhneysen, Size dependence of current spin polarization through superconductor/ferromagnet nanocontacts, *Phys. Rev. Lett.* **101**, 147005 (2008).
166. B. Nadgorny, I. I. Mazin, M. Osofsky et al., Origin of high transport spin polarization in $La_{0.7}Sr_{0.3}MnO_3$: Direct evidence for minority spin states, *Phys. Rev. B* **63**, 184433 (2001).
167. W. K. Park and L. H. Greene, Andreev reflection and order parameter symmetry in heavy-fermion superconductors: the case of $CeCoIn_5$, *J. Phys.: Condens. Matter* **21**, 103203 (2009).
168. M. Faiz, R. P. Panguluri, B. Balke et al., Direct spin diffusion measurements in metals using Andreev reflection spectroscopy, unpublished.
169. Y. Bugoslavsky, Y. Miyoshi, S. K. Clowes et al., Possibilities and limitations of point-contact spectroscopy for measurements of spin polarization, *Phys. Rev. B* **71**, 104523 (2005).
170. Y. Miyoshi, Y. Bugoslavsky, and L. F. Cohen, Andreev reflection spectroscopy of niobium point contacts in a magnetic field, *Phys. Rev. B* **72**, 012502 (2005).
171. L. Ozyuzer, J. F. Zasadzinski, and K. E. Gray, Point contact tunnelling apparatus with temperature and magnetic field control, *Cryogenics* **38**, 911 (1998).
172. K. A. Yates, L. F. Cohen, Z. A. Ren et al., Point contact Andreev reflection spectroscopy of $NdFeAsO_{0.85}$, *Supercond. Sci. Technol.* **21**, 092003 (2008).
173. R. P. Panguluri, Sheng Xu, Yutaka Moritomo, I. V. Solovyev, and B. Nadgorny, Disorder effects in half-metallic Sr_2FeMoO_6 single crystals, *Appl. Phys. Lett.* **94**, 012501 (2009).
174. K. S. Ralls, R. A. Buhrman, and R. C. Tiberio, Fabrication of thin-film metal nanobridges, *Appl. Phys. Lett.* **55**, 2459 (1989).
175. R. S. Gonnelli, D. Daghero, G. A. Ummarino et al., Direct evidence for two-band superconductivity in MgB_2 single crystals from directional point-contact spectroscopy in magnetic fields, *Phys. Rev. Lett.* **89**, 247004 (2002).
176. E. Clifford and J. M. D. Coey, Point contact Andreev reflection by nanoindentation of polymethyl methacrylate, *Appl. Phys. Lett.* **89**, 092506 (2006).
177. F. Magnus, K. A. Yates, S. K. Clowes et al., Interface properties of Pb/InAs planar structures for Andreev spectroscopy, *Appl. Phys. Lett.* **92**, 012501 (2008).
178. R. J. M. van de Veerdonk, J. Nowak, R. Meservey, J. S. Moodera, and W. J. M. de Jonge, Current distribution effects in magnetoresistive tunnel junctions, *Appl. Phys. Lett.* **71**, 2839 (1997).

179. J. Bass and W. P. Pratt Jr., Current-perpendicular (CPP) magnetoresistance in magnetic metallic multilayers, *J. Magn. Magn. Mater.* **200**, 274 (1999).
180. P. M. Rourke, M. A. Tanatar, C. S. Turel, J. Berdeklis, C. Petrovic, and J. Y. T. Wei, Spectroscopic evidence for multiple order parameter components in the heavy fermion superconductor $CeCoIn_5$, *Phys. Rev. Lett.* **94**, 107005 (2005).
181. R. C. Dynes, V. Narayanamurti, and J. P. Garno, Direct measurement of quasiparticle-lifetime broadening in a strong-coupled superconductor, *Phys. Rev. Lett.* **41**, 1509 (1978).
182. A. Plecenik, M. Grajcar, S. Benacka, P. Seidel, and A. Pfuch, Finite-quasiparticle-lifetime effects in the differential conductance of $Bi_2Sr_2CaCu_2Oy$/Au junctions, *Phys. Rev. B* **49**, 10016 (1994).
183. H. Srikanth and A. K. Raychaudhuri, Modeling tunneling data of normal metal-oxide superconductor point contact junctions, *Physica C* **190**, 229, 1992.
184. C. H. Kant, O. Kurnosikov, A. T. Filip, P. LeClair, H. J. M. Swagten, and W. J. M. de Jonge, Origin of spin-polarization decay in point-contact Andreev reflection, *Phys. Rev. B* **66**, 212403 (2002).
185. J. Aumentado and V. Chandrasekhar, Mesoscopic ferromagnet-superconductor junctions and the proximity effect, *Phys. Rev. B* **64**, 054505 (2001).
186. K. Xia, P. J. Kelly, G. E. W. Bauer, and I. Turek, Spin-dependent transparency of ferromagnet/superconductor interfaces, *Phys. Rev. Lett.* **89**, 166603 (2002).
187. F. Taddei, S. Sanvito, and C. J. Lambert, Spin-polarized transport in F/S nanojunctions, *J. Low Temp. Phys.* **124**, 305 (2001).
188. R. J. M. Veerdonk, J. S. Moodera, and W. J. M. de Jonge, *Conference of Digest of the 15th International Colloquium on Magnetic Films and Surfaces*, Queensland, August 4–8, 1997.
189. R. Meservey, P. M. Tedrow, V. R. Kalvey, and D. Paraskevopoulos, Studies of ferromagnetic metals by electron spin polarized tunneling, *J. Appl. Phys.* **50**, 1935 (1979).
190. P. M. Tedrow and R. Meservey, Spin-polarized electron tunneling, *Phys. Rep.* **238**, 173 (1994).
191. L. Bocklage, J. M. Scholtyssek, U. Merkt, and G. Meier, Spin polarization of Ni_2MnIn and $Ni_{80}Fe_{20}$ determined by point-contact spectroscopy, *J. Appl. Phys.* **101**, 09J512 (2007).
192. R. Godfrey and M. Johnson, Spin injection in mesoscopic silver wires: Experimental test of resistance mismatch, *Phys. Rev. Lett.* **96**, 136601 (2006).
193. M. Zhu, C. L. Dennis, and R. D. McMichael, Temperature dependence of magnetization drift velocity and current polarization in $Ni_{80}Fe_{20}$ by spin-wave Doppler measurements, *Phys. Rev. B* **81**, 140407(R) (2010).
194. D. Orgassa, H. Fujiwara, T. C. Schulthess, and W. H. Butler, First-principles calculation of the effect of atomic disorder on the electronic structure of the half-metallic ferromagnet NiMnSb, *Phys. Rev. B* **60**, 13237 (1999).
195. H. Ebert and G. Schütz, Theoretical and experimental study of the electronic structure of PtMnSb, *J. Appl. Phys.* **69**, 4627 (1991).
196. P. A. Dowben and R. Skomski, Are half-metallic ferromagnets half metals? *J. Appl. Phys.* **95**, 7453 (2004).
197. M. Ležaić, Ph. Mavropoulos, J. Enkovaara, G. Bihlmayer, and S. Blügel, Thermal collapse of spin polarization in half-metallic ferromagnets, *Phys. Rev. Lett.* **97**, 026404 (2006).
198. R. Skomski and P. A. Dowben, The finite-temperature densities of states for half-metallic ferromagnets, *Europhys. Lett.* **58**, 544 (2002).
199. H. Eschrig and W. E. Pickett, Density functional theory of magnetic systems revisited, *Solid State Commun.* **118**, 123 (2001).
200. P. Mavropoulos, K. Sato, R. Zeller, P. H. Dederichs, V. Popescu, and H. Ebert, Effect of the spin–orbit interaction on the band gap of half metals, *Phys. Rev. B* **69**, 054424 (2004).
201. P. K. Boer and R. A. de Groot, Electronic structure of the layered manganite $LaSr_2Mn_2O_7$, *Phys. Rev. B* **60**, 10758–10762 (1999).
202. G. A. Wijs and R. A. de Groot, Towards 100% spin-polarized charge-injection: The half-metallic NiMnSb/CdS interface, *Phys. Rev. B* **64**, 020402 (2001).
203. V. Yu. Irkhin and M. I. Katsnelson, Half-metallic ferromagnets, *Usp. Fiz. Nauk* **164**, 705 (1994) [*Phys. Usp.* **37**, 659 (1994)].
204. J. M. D. Coey, M. Viret, and S. von Molnár, Mixed-valence manganites, *Adv. Phys.* **48**, 167 (1999).
205. M. Ziese, Extrinsic magnetotransport phenomena in ferromagnetic oxides, *Rep. Prog. Phys.* **65**, 143 (2002).
206. H. Eschrig and W. E. Pickett, Half metals: From formal theory to real material issues, *J. Phys.: Condens. Matter* **19**, 315203 (2007).
207. M. I. Katsnelson, V. Yu. Irkhin, L. Chioncel, A. I. Lichtenstein, and R. A. de Groot, Half-metallic ferromagnets: From band structure to many-body effects, *Rev. Mod. Phys.* **80**, 315 (2008).
208. K. Schwarz, CrO_2 predicted as a half-metallic ferromagnet, *J. Phys. F: Met. Phys.* **16**, L211 (1986).
209. W. J. DeSisto, P. R. Broussard, T. F. Ambrose, B. E. Nadgorny, and M. S. Osofsky, Highly spin-polarized chromium dioxide thin films prepared by chemical vapor deposition from chromyl chloride, *Appl. Phys. Lett.* **76**, 3789 (2000).
210. Y. Ji, G. J. Strijkers, F. Y. Yang et al., Determination of the spin polarization of half-metallic CrO_2 by point contact Andreev reflection, *Phys. Rev. Lett.* **86**, 5585–5588 (2001).
211. I. I. Mazin, Robust half metalicity in $Fe_xCo_1-_xS_2$, *Appl. Phys. Lett.* 77, 3000 (2000).
212. C. Leighton, M. Manno, A. Cady et al., Composition controlled spin polarization in $Co_{1-x}Fe_xS_2$ alloys, *J. Phys.: Condens. Matter* **19**, 315219 (2007).
213. L. Wang, T. Y. Chen, and C. Leighton, Spin-dependent band structure effects and measurement of the spin polarization in the candidate half-metal CoS_2, *Phys. Rev. B* **69**, 094412 (2004).

214. L. Wang, K. Umemoto, R. M. Wentzcovitch et al., $Co_{1-x}Fe_xS_2$: A tunable source of highly spin-polarized electrons, *Phys. Rev. Lett.* **94**, 056602 (2005).
215. A. Schmehl, V. Vaithyanathan, A. Herrnberger et al., Epitaxial integration of the highly spin-polarized ferromagnetic semiconductor EuO with silicon and GaN, *Nature Mater.* **6**, 882 (2007).
216. I. Appelbaum, B. Huang, and D. Monsma, Electronic measurement and control of spin transport in silicon, *Nature* **447**, 295–298, 2007.
217. B. T. Jonker, G. Kioseoglou, A. T. Hanbicki, C. H. Li, and P. E. Thompson, Electrical spin-injection into silicon from a ferromagnetic metal/tunnel barrier contact, *Nat. Phys.* **3**, 542–546 (2007).
218. P. G. Steeneken, L. H. Tjeng, I. Elfimov et al., Exchange splitting and charge carrier spin polarization in EuO, *Phys. Rev. Lett.* **88**, 047201 (2002).
219. T. S. Santos, J. S. Moodera, K. V. Raman et al., Determining exchange splitting in a magnetic semiconductor by spin-filter tunneling, *Phys. Rev. Lett.* **101**, 147201 (2008).
220. K. I. Kobayashi, T. Kimura, H. Sawada, K. Terakura, and Y. Tokura, Room-temperature magnetoresistance in an oxide material with an ordered double-perovskite structure, *Nature* **395**, 357, 1998.
221. I. V. Solovyev, Electronic structure and stability of the ferrimagnetic ordering in double perovskites, *Phys. Rev. B* **65**, 144446 (2002).
222. N. Auth, G. Jakob, T. Block, and C. Felser, Spin polarization of magnetoresistive materials by point contact spectroscopy, *Phys. Rev. B* **68**, 024403 (2003).
223. Y. Bugoslavsky, Y. Miyoshi, G. K. Perkins et al., Electron diffusivities in MgB_2 from point contact spectroscopy, *Phys. Rev. B* **72**, 224506 (2005a).
224. J. M. D. Coey and S. Sanvito, Magnetic semiconductors and half-metals, *J. Phys. D: Appl. Phys.* **37**, 988 (2004).
225. W. E. Pickett and D. J. Singh, Electronic structure and half-metallic transport in the $La_1-_xCa_xMnO_3$ system, *Phys. Rev. B* **53**, 1146–1160 (1996).
226. J. Z. Sun, L. Krusin-Elbaum, P. R. Duncombe, A. Gupta, and R. B. Laibowitz, Temperature dependent, non-ohmic magnetoresistance in doped perovskite manganate trilayer junctions, *Appl. Phys. Lett.* **70**, 1769 (1997).
227. M. Bowen, M. Bibes, A. Barthélémy et al., Nearly total spin polarization in $La_2/3Sr_1/3MnO_3$ from tunneling experiments, *Appl. Phys. Lett.* **82**, 233 (2003).
228. D. I. Golosov, New correlated model of colossal magnetoresistive manganese oxides, *Phys. Rev. Lett.* **104**, 207207 (2010).
229. D. Dessau, Private communications, 2011.
230. B. Hope and A. Horsfield, Contrasting spin-polarization regimes in Co nanowires studied by density functional theory, *Phys. Rev. B* **77**, 094442 (2008).
231. B. Nadgorny, M. S. Osofsky, D. J. Singh et al., Measurements of spin polarization of epitaxial $SrRuO_3$ thin films, *Appl. Phys. Lett.* **82**, 427 (2003).
232. A. Kastalsky, A. W. Kleinsasser, L. H. Greene, R. Bhat, F. P. Milliken, and J. P. Harbison, Observation of pair currents in superconductor-semiconductor contacts, *Phys. Rev. Lett.* **67**, 3026 (1991).
233. B. J. van Wees, P. de Vries, P. Magnée, and T. M. Klapwijk, Excess conductance of superconductor-semiconductor interfaces due to phase conjugation between electrons and holes, *Phys. Rev. Lett.* **69**, 510–513 (1992).
234. I. Žutić, and S. Das Sarma, Spin-polarized transport and Andreev reflection in semiconductor/superconductor hybrid structures, *Phys. Rev. B* **60**, R16322–R16325 (1999).
235. P. A. Barnes and A. Y. Cho, Nonalloyed Ohmic contacts to n-GaAs by molecular beam epitaxy, *Appl. Phys. Lett.* **33**, 651 (1978).
236. A. H. MacDonald, P. Schiffer, and N. Samarth, Ferromagnetic semiconductors: Moving beyond (Ga,Mn)As, *Nat. Mater.* **4**, 195 (2005).
237. A. Geresdi, A. Halbritter, M. Csontos et al., Nanoscale spin polarization in the dilute semiconductor (In,Mn)Sb, *Phys. Rev. B* **77**, 233304 (2008).
238. R. P. Panguluri, B. Nadgorny, T. Wojtowicz, X. Liu, and J. K. Furdyna, Inelastic scattering and spin polarization in dilute magnetic semiconductor (Ga,Mn)Sb, *Appl. Phys. Lett.* **91**, 252502 (2007).
239. J. M. D. Coey, Dilute magnetic oxides, *Curr. Opin. Solid State Mater. Sci.* **10**, 83 (2006).
240. M. J. Calderon and S. Das Sarma, Theory of carrier mediated ferromagnetism in dilute magnetic oxides, *Ann. Phys.* **322**(11), 2618 (2007).
241. T. Dietl, Origin of ferromagnetic response in diluted magnetic semiconductors and oxides, *J. Phys.: Condens. Matter* **19**, 165204 (2007).
242. A. G. Petukhov, I. Žutić, and S. C. Erwin, Thermodynamics of carrier-mediated magnetism in semiconductors, *Phys. Rev. Lett.* **99**, 257202 (2007).
243. R. P. Panguluri, P. Kharel, C. Sudakar et al., Ferromagnetism and spin-polarized charge carriers in In_2O_3 thin films, *Phys. Rev. B* **79**, 165208 (2009).
244. K. A. Yates, A. J. Behan, J. R. Neal et al., Spin-polarized transport current in n-type codoped ZnO thin films measured by Andreev spectroscopy, *Phys. Rev. B* **80**, 245207 (2009).
245. J. Philip, A. Punnoose, B. I. Kim et al., Carrier-controlled ferromagnetism in transparent oxide semiconductors, *Nat. Mater.* **5**, 298 (2006).
246. J. Bass and W. P. Pratt Jr., Spin-diffusion lengths in metals and alloys, and spin-flipping at metal/metal interfaces: An experimentalist's critical review, *J. Phys.: Condens. Matter* **19**, 183201 (2007).
247. S. A. Crooker, M. Furis, X. Lou et al., Imaging spin transport in lateral ferromagnet/semiconductor structures, *Science* **309**, 2191–2195 (2005).
248. F. J. Jedema, A. T. Filip, and B. J. van Wees, Electrical spin injection and accumulation at room temperature in an all-metal mesoscopic spin valve, *Nature* **410**, 345–348 (2001).

249. M. Johnson and R. H. Silsbee, Interfacial charge-spin coupling: Injection and detection of spin magnetization in metals, *Phys. Rev. Lett.* **55**, 1790 (1985).
250. R. J. Soulen Jr., J. M. Byers, M. S. Osofsky et al., Measuring the spin polarization of a metal with a superconducting point contact, *Science* **282**, 85 (1998).
251. K. S. Upadhyay, A. Palanisami, R. N. Louie, and R. A. Buhrman, Probing ferromagnets with Andreev reflection, *Phys. Rev. Lett.* **81**, 3247 (1998).
252. G. J. Strijkers, Y. Ji, F. Y. Yang, C. L. Chien, and J. M. Byers Andreev reflections at metal/superconductor point contacts: Measurement and analysis *Phys. Rev. B* **63**, 104510 (2001).
253. C. H. Kant, O. Kurnosikov, A. T. Filip, P. LeClair, H. J. M. Swagten, and W. J. M. de Jonge, Origin of spin-polarization decay in point-contact Andreev reflection, *Phys. Rev. B* **66**, 212403 (2002).
254. M. Stokmaier, G. Goll, D. Weissenberger, C. Sürgers, and H. v. Löhneysen, Size dependence of current spin polarization through superconductor/ferromagnet nanocontacts, *Phys. Rev. Lett.* **101**, 147005 (2008).
255. J. Aumentado and V. Chandrasekhar, Meoscopic ferromagnet-superconductor junctions and the proximity effect, *Phys. Rev. B* **64**, 054505 (2001).
256. Y. Ji, G. J. Strijkers, F. Y. Yang et al., Determination of the spin polarization of half-metallic CrO_2 by point contact Andreev reflection, *Phys. Rev. Lett.* **86**, 5585 (2001).
257. J. S. Parker, S. M. Watts, P. G. Ivanov, and P. Xiong, Spin polarization of CrO_2 at and across an artificial barrier, *Phys. Rev. Lett.* **88**, 196601 (2002).
258. E. Clifford and J. M. D. Coey, Point contact Andreev reflection by nanoindentation of polymethyl methacrylate, *Appl. Phys. Lett.* **89**, 092506 (2006).
259. W. J. DeSisto, P. R. Broussard, T. F. Ambrose, B. E. Nadgorny, and M. S. Osofsky, Highly spin-polarized chromium dioxide thin films prepared by chemical vapor deposition from chromyl chloride, *Appl. Phys. Lett.* **76**, 3789 (2000).
260. K. A. Yates, W. R. Branford, F. Magnus et al., The spin polarization of CrO_2 revisited, *Appl. Phys. Lett.* **91**, 172504 (2007).
261. L. Bocklage, J. M. Scholtyssek, U. Merkt, and G. Meier, Spin polarization of Ni_2MnIn and $Ni_{80}Fe_{20}$ determined by point-contact spectroscopy *J. Appl. Phys.* **101**, 09J512 (2007).
262. B. Nadgorny, R. J. Soulen Jr., M. S. Osofsky et al., Tranpsort spin polarization of Ni_xFe_{1-x}: Electronic kinematics and band structure, *Phys. Rev. B* **61**, R3788 (2000).
263. J. M. Valentine and C. L. Chien, Determination of spin polarization of Gd and Dy by point-contact Andreev reflection, *J. Appl. Phys.* **99**, 08P902 (2006).
264. B. Nadgorny, I. I. Mazin, M. Osofsky et al., Origin of high transport spin polarization in $La_{0.7}Sr_{0.3}MnO_3$: Direct evidence for minority spin states, *Phys. Rev. B* **63**, 184433 (2001).
265. Y. Ji, C. L. Chien, Y. Tomioka, and Y. Tokura, Measurement of spin polarization of single crystals of $La_{0.7}Sr_{0.3}MnO_3$ and $La_{0.6}Sr_{0.4}MnO_3$, *Phys. Rev. B* **66**, 012410 (2002).
266. R. P. Panguluri, T. S. Santos, E. Negusse et al., Half-metallicity in europium oxide conductively matched with silicon, *Phys. Rev. B* **78**, 125307 (2008).
267. A. Schmehl, V. Vaithyanathan, A. Herrnberger et al., Epitaxial integration of the highly spin-polarized ferromagnetic semiconductor EuO with silicon and GaN, *Nat. Mater.* **6**, 882 (2007).
268. C. Ren, J. Trbovic, R. L. Kallaher et al., Measurement of the spin polarization of the magnetic semiconductor EuS with zero-field and Zeeman-split Andreev reflection spectroscopy, *Phys. Rev. B* **75**, 205208 (2007).
269. X. H. Zhang, S. von Molnár, Z. Fisk, and P. Xiong, Spin-dependent electronic states of the ferromagnetic semimetal EuB_6, *Phys. Rev. Lett.* **100**,167001 (2008).
270. C. Leighton, M. Manno, A. Cady et al., Composition controlled spin polarization in $Co_{1-x}Fe_xS_2$ alloys *J. Phys. Cond. Matter* **19**, 315219 (2007).
271. L. Wang, T. Y. Chen, C. L. Chien et al., Composition controlled spin polarization in $Co_{1-x}Fe_xS_2$: Electronics, magnetic, and thermodynamic properties, *Phys. Rev. B* **73**, 144402 (2006).
272. L. Wang, K. Umemoto, R. M. Wentzcovitch et al., $Co_{1-x}Fe_xS_2$: A tunable source of highly spin-polarized electrons, *Phys. Rev. Lett.* **94**, 056602 (2005).
273. S. F. Cheng, G. T. Woods, K. Bussmann et al., Growth and magnetic properties of single crystal $Fe_{1-x}Co_xS_2$ (x = 0.35-1), *J. Appl. Phys.* **93**, 6847 (2003).
274. L. Wang, T. Y. Chen, and C. Leighton, Spin-dependent band structure effects and measurement of the spin polarization in the candidate half-metal CoS_2, *Phys. Rev. B* **69**, 094412 (2004).
275. L. Wang, T. Y. Chen, C. L. Chien, and C. Leighton., Sulfur stoichiometry effects in highly spin polarized CoS_2 single crystals, *Appl. Phys. Lett.* **88**, 232509 (2006).
276. J. G. Braden, J. S. Parker, P. Xiong, S. H. Chun, and N. Samarth Dirct measurement of the spin polarization of the magnetic semiconductor (Ga,Mn)As, *Phys. Rev. Lett.* **91**, 056602 (2003).
277. T. W. Chiang, Y. H. Chiu, S. Y. Huang et al., Spectra broadening of point-contact Andreev reflection measurement on GaMnAs, *J. Appl. Phys.* **105**, 07C507 (2009).
278. R. P. Panguluri, K. C. Ku, T. Wojtowicz et al., Andreev reflection and pair-breaking effects at the superconductor/magnetic semiconductor interface, *Phys. Rev. B* **72**, 054510 (2005).
279. R. P. Panguluri, B. Nadgorny, T. Wojtowicz, W. L. Lim, X. Liu, and J. K. Furdyna, Measurement of spin polarization by Andreev reflection in ferromagnetic $In_{1-x}Mn_xSb$ epilayers, *Appl. Phys. Lett.* **84**, 4947 (2004).
280. A. Geresdi, A. Halbritter, M. Csontos et al., Nanoscale spin polarization in the dilute semiconductor (In,Mn)Sb, *Phys. Rev. B* **77**, 233304 (2008).

281. R. P. Panguluri, B. Nadgorny, T. Wojtowicz, X. Liu, and J. K. Furdyna, Inelastic scattering and spin polarization in dilute magnetic semiconductor (Ga,Mn)Sb, *Appl. Phys. Lett.* **91**, 252502 (2007).
282. R. P. Panguluri, Sheng Xu, Yutaka Moritomo, I. V. Solovyev, and B. Nadgorny, Disorder effects in half-metallic Sr_2FeMoO_6 single crystals, *Appl. Phys. Lett.* **94**, 012501 (2009).
283. W. R. Branford, S. K. Clowes, Y. V. Bugoslavsky et al., Effect of chemical substitution on the electronic properties of highly aligned thin films of $Sr_{2-x}A_xFeMoO_6$(A = Ca,Ba,La ;x = 0,0.1), *J. Appl. Phys.* **94**, 4714 (2003).
284. H. Gotoh, Y. Takeda, H. Asano, J. Zhong, A. Rajanikanth, and K. Hono, Antiferromagnetism and spin polarization in double perovskite $SrLaVMoO_6$, *Appl. Phys. Exp.* **2**, 013001 (2009).
285. R. P. Panguluri, P. Kharel, C. Sudakar et al., Ferromagnetism and spin-polarized charge carriers in In_2O_3 thin films, *Phys. Rev. B* **79**, 165208 (2009).
286. A. Rajanikanth, Y. K. Takahashi, and K. Hono, Suppression of magnon excitations in Co2MnSi Heusler alloy by Nd doping, *J. Appl. Phys.* **105**, 063916 (2009).
287. G. T. Woods, J. Sanders, S. Kolesnik et al., Measurement of the transport spin polarization of doped strontium ruthenates using point contact Andreev reflection, *J. Appl. Phys.* **104**, 083701 (2008).
288. P. Raychaudhuri, A. P. Mackenzie, J. W. Reiner, and M. R. Beasley, Transport spin polarization in $SrRuO_3$ measured through point-contact Andreev reflection, *Phys. Rev. B* **67**, 020411 (2003).
289. B. Nadgorny, M. S. Osofsky, D. J. Singh et al., Measurements of spin polarization of epitaxial $SrRuO_3$ thin films, *Appl. Phys. Lett.* **82**, 427 (2003).
290. J. Sanders, G. T. Woods, P. Poddar et al., Spin polarization measurements on polycrystalline strontium ruthenates using point-contact Andreev reflection, *J. Appl. Phys.* **97**, 10C912 (2005).
291. S. Mukhopadhyay, S. K. Dhar, and P. Raychaudhuri, Study of spin fluctuations in $Ni_{3\pm x}Al_{1\pm x}$ using point contact Andreev reflection, *Appl. Phys.* **93**, 102502 (2008).
292. S. V. Karthik, T. M. Nakatani, A. Rajanikanth, Y. K. Takahashi, and K. Hono, Spin polarization of Co-Fe alloys estimated by point contact Andreev reflection and tunneling magnetoresistance, *J. Appl. Phys.* **105**, 07C916 (2009).
293. A. Narahara, K. Ito,T. Suemasu, Y. K. Takahashi, A. Ranajikanth, and K. Hono, Spin polarization of Fe_4N thin films determined by point-contact Andreev reflection, *Appl. Phys. Lett.* **94**, 202502 (2009).
294. S. X. Huang, T. Y. Chen, and C. L. Chien, Spin polarization of amorphous CoFeB determined by point-contact Andreev reflection, *Appl. Phys. Lett.* **92**, 242509 (2008).
295. K. M. Seemann, V. Baltz, M. MacKenzie, J. N. Chapman, B. J. Hickey, and C. H. Marrows, Diffusive and ballistic current spin polarization in magnetron-sputtered L1(0) ordered epitaxial FePt, *Phys. Rev. B* **76**, 174435 (2007).
296. T. Y. Chen, C. L. Chien, and C. Petrovic, Enhance Curie temperature and spin polarization in Mn_4FeGe_3, *Appl. Phys. Lett.* **91**, 142505 (2007).
297. S. V. Karthik, A. Rajanikanth, T. M. Nakatani et al., Effect of Cr substitution for Fe on the spin polarization of $Co_2Cr_xFe_{1-x}Si$ Heusler alloys, *J. Appl. Phys.* **102**, 043903 (2007).
298. T. M. Nakatani, A. Rajanikanth, Z. Gercsi, Y. K. Takahashi, K. Inomata, and K. Hono, Structure, magnetic property, and spin polarization of $Co_2FeAl_xSi_{1-x}$ Heusler alloys, *J. Appl. Phys.* 102 033916 (2007).
299. S. V. Karthik, A. Rajanikanth, Y. K. Takahashi, T. Okhubo, and K. Hono, Spin polarization of quaternary $Co_2Cr_{1-x}Fe_xAl$ Heusler alloys, *Appl. Phys. Lett.* **89**, 052505 (2006).
300. A. Rajanikanth, D. Kande, Y. K. Takahashi, and K. Hono, High spin polarization in a two phase quaternary Heusler alloy $Co_2MnAl_{1-x}Sn_x$, *J. Appl. Phys.* **101**, 09J508 (2007).
301. T. M. Nakatani, Z. Gercsi, A. Rajanikanth, Y. K. Takahashi, and K. Hono, The effect of iron addition on the spin polarization and magnetic properties of Co_2CrGa Heusler alloy, *J. Phys. D Appl. Phys.* **41**, 225002 (2008).
302. A. Rajanikanth, Y. K. Takahashi, and K. Hono, Spin polarization of Co_2MnGe and Co_2MnSi thin films with A2 and L2(1) structures, *J. Appl. Phys.* **101**, 023901 (2007).
303. S. F. Cheng, B. Nadgorny, K. Bussmann et al., Growth and magnetic properties of single crystal Co_2Mn_X(X = Si,Ge) Heusler alloys, *IEEE Trans. Magn.* **37**, 2176 (2001).
304. L. J. Singh, Z. H. Barber, Y. Miyoshi, Y. Bugoslavsky, W. R. Branford, and L. F. Cohen, Structural, magnetic, and transport properties of thin films of the Heusler alloy Co_2MnSi, *Appl. Phys. Lett.* **84**, 2367 (2004).
305. L. Ritchie, G. Xiao, Y. Ji et al., Magnetic structural, and transport propeties of the Heusler alloys Co_2MnSi and NiMnSb, *Phys. Rev. B* **68**, 104430 (2003).
306. Z. Gercsi, A. Rajanikanth, Y. K. Takahashi et al., Spin polarization of Co_2FeSi full-Heusler alloy and tunneling magnetoresistance of its magnetic tunneling junctions, *Appl. Phys. Lett.* **89**, 082512 (2006).
307. M. Zhang et al., The magnetic and transport properties of the Co_2FeGa Heusler alloy, *J. Phys. D Appl. Phys.* **37**, 2049 (2004).
308. S. K. Clowes, Y. Miyoshi, Y. Bugoslavsky et al., Spin polarization of the transport current at the free surface of bulk NiMnSb, *Phys. Rev. B* **69**, 214425 (2004).
309. M. Zhang, Z. Liu, H. Hu et al., A new semi-Heusler ferromagnet NiFeSb: Electronic structure, magnetism and transport properties, *Solid State Commun.* **128**, 107 (2003).
310. Z. Z. Li, H. J. Tao, H. H. Wen et al., Point contact Andreev reflection measurement of spin polarization of ferromagnetic alloy NiFeSb, *Chin. Phys. Lett.* **19**, 1181 (2002).
311. R. P. Panguluri, Changgan Zeng, H. H. Weitering, J. M. Sullivan, S. C. Erwin, and B. Nadgorny, Spin polarization and electronic structure of ferromagnetic Mn_5Ge_3 epilayers, *Phys. Stat. Solidi B* **242**, R67 (2005).

312. A. Ionescu, C. A. F. Vaz, T. Trypiniotis et al., Structural, magnetic, electronic, and spin transport properties of epitaxial Fe3Si/GaAs(001), *Phys. Rev. B* **71**, 094401 (2005).
313. R. P. Panguluri, G. Tsoi, B. Nadgorny, S. H. Chun, N. Samarth, and I. I. Mazin, Point contact spin spectroscopy of ferromagnetic MnAs epitaxial films, *Phys. Rev. B* **68**, 201307(R) (2003).
314. W. J. DeSisto, P. R. Broussard, T. F. Ambrose, B. E. Nadgorny, and M. S. Osofsky, Highly spin-polarized chromium dioxide thin films prepared by chemical vapor deposition from chromyl chloride, *Appl. Phys. Lett.* **76**, 3789 (2000).

28

Ballistic Spin Transport

Bernard Doudin
Centre National de la Recherche Scientifique-University Louis Pasteur

N.T. Kemp
Centre National de la Recherche Scientifique-University Louis Pasteur

This chapter is dedicated to the study of ballistic transport in magnetic nanostructures. Ballistic means that the charges carrying the electric current go through the sample without experiencing scattering. This condition requires that sample dimensions are smaller than the carriers mean free path. This concept is therefore in immediate contrast with the diffusive transport model, extensively used to explain giant magnetoresistance (GMR) properties. The spin-dependent resistivities of the ferromagnetic layers and the interfaces largely determine the GMR properties; however, for a diminishing sample size this is not solely the case. An example of transport occurring essentially without scattering is the tunnel regime, where the conduction of the system is ensured by evanescent waves through a dielectric barrier. The related tunnel magnetoresistance (TMR) is described in details in Chapter 9, but let us recall that it relies on a spin-dependent density of states in the ferromagnetic layers (more precisely at the interface), as well as the transmission properties of the barrier.

Ballistic transport is a topic of increasing importance, owing to technical advances in top-down miniaturization capabilities. Another fundamental reason is the potential relevance for spintronic devices. The Datta and Das transistor device [1] relies on ballistic transport of the carrier, with a spin orientation modulated by external gating. In the first section, we will review the basic concepts related to ballistic transport properties, emphasizing the key characteristic lengths and dimensionalities. We will restrict our discussion to the simplest mathematical expressions and the most simplifying hypotheses. More comprehensive textbooks on mesoscopic electronics and ballistic transport in nanostructures can be found elsewhere [2,3]. We will emphasize the differences between semiconductors and metallic nanostructures. Most experiments on ballistic transport were performed on semiconducting materials, taking advantage of better control of their properties, and larger characteristic lengths. Experimental examples involving semiconductors are therefore more convincing and more detailed, and will be used as illustrations. We will then introduce spin-dependent ballistic transport, approaching its importance in the field of spin electronics. In the final section, experimental results on metallic structures will be presented.

The reader should be aware that this research topic is rather controversial, especially for metallic systems, where strong size reduction down to a single atom is needed. When studying spin-dependent properties, the application of an external magnetic field also complicates the interpretation of the results. We have tried to provide a fair summary of the results up to date, essentially emphasizing the experiments and interpretations that have mostly reached a consensus in the community.

28.1 Basic Concepts

When reducing the size of a sample, one can question how geometrically restricting the flow of carriers can affect the resistance of the sample. The best pedagogical case, having the advantage of leading to analytical solutions, is the so-called Maxwell resistance. We consider a constriction of the shape of a hyperbolic volume (Figure 28.1), for which analytical solution of the Poisson's equation provides an expression for the potential, while the current is obtained by integrating Ohm's law locally over the constriction volume. The asymptotic case of an infinite curvature of the constriction, corresponding to an orifice of radius a separating two metallic half spaces, results in a conductance given by

$$G_M \equiv \frac{1}{R_M} = 2a\sigma \quad (28.1)$$

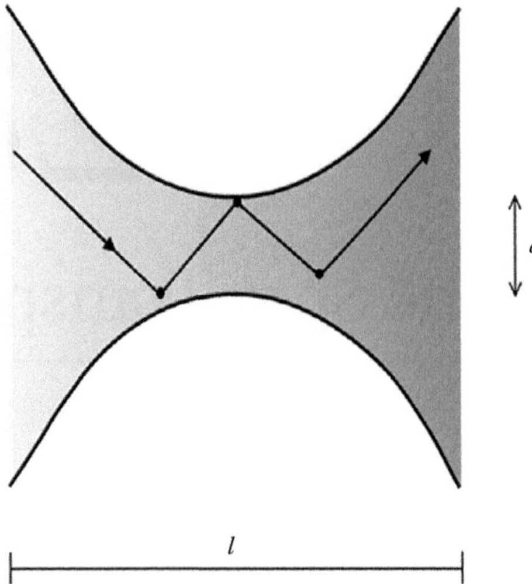

FIGURE 28.1 Sketch of a conductor with a constriction of diameter a, larger than the electronic mean free path l.

As a numerical example, if we consider an aperture of 10^{-6} m in Au ($\sigma = 4\times10^{7}$ Sm^{-1}), we obtain a resistance of the order of 10 mΩ.

28.1.1 Sharvin Model

When the constriction dimensions are reduced below the length of the electron mean free path, a diffusive model for conduction cannot apply. The flow of electrons between the left and right side of the constriction is therefore ballistic. At low temperatures, the mean free path of metallic samples of reasonable purity and crystallinity is exceeding a few hundreds of nanometers, and several microns for semiconductors. (Note: The mean free path of thin metallic films or multilayers relevant for GMR or TMR device applications does not typically exceed a few tens of nanometers.) When an electric potential is imposed between the two sides of the constriction, the restriction to the charge flow imposed by the geometry results in an effective resistance. Sharvin introduced the concept of transmission through a constriction in a material (Figure 28.2) [4], by applying perfect gas concepts to the flow of particles between two reservoirs having different total free energies. Under the hypothesis of perfect Fermi gas behavior at low temperatures and small applied voltage bias (linear approximation), the constriction along the x direction will limit the current density j by

$$j(x) = e\langle v_x\rangle \rho(\varepsilon_F)\frac{eV}{2} \tag{28.2}$$

where $\langle v_x\rangle$ is the average velocity along the x flow direction of the particles, corresponding to the direction of change of free energies. Assuming the usual expressions for $\langle v(x)\rangle$ and $\rho(\varepsilon_F)$ in terms of the free electron gas Fermi wavevector, k_F, the Sharvin conductance is therefore of the form:

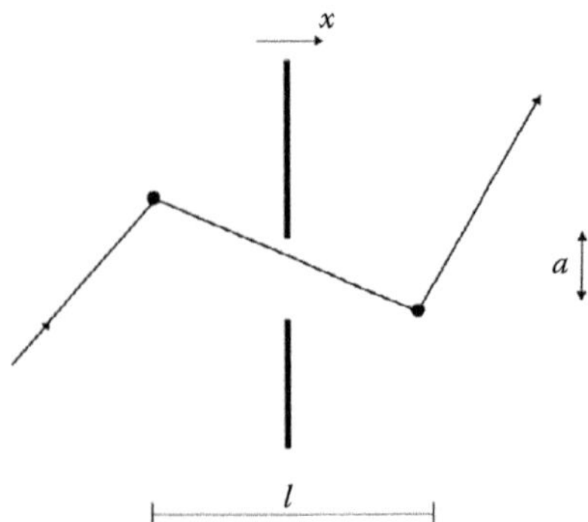

FIGURE 28.2 Sketch of a conductor with a constriction of diameter a, smaller than the mean free path l. Ohm's law does not apply, resulting in a Sharvin's conductance expressed by Equation 28.3.

$$G_S \equiv \frac{1}{R_S} = \frac{2e^2}{h}\left(k_F\frac{a}{2}\right)^2 \tag{28.3}$$

This equation reveals that locally Ohm's law does not apply, and that the conductance of a constriction does not depend on the resistivity of a material, but is related to the product of its density of states with the group velocity. This number is expected to be much less sensitive to impurities and temperature than the intrinsic resistivity of a material. If we take $k_F = 10^{10}$ m^{-1}, a 10 nm aperture will correspond to a resistance of a few ohms.

Experimental results on metallic systems using the so-called point contact geometry are well documented [5]. The Sharvin conductance can be obtained by carefully pressing a sharp needle over a metallic surface. The signature of ballistic conductance is obtained by nonlinearities in the IV curves, related to electron–phonon inelastic coupling. Anomalies in electric properties are even more striking for superconductors, where the current blocking due to the superconducting bandgap is suppressed in the ballistic regime, leading to the so-called Andreev reflection. This is of particular interest for magnetic systems, as it allows the difference between $\langle v_{\uparrow x}\rangle\rho_\uparrow(E_F)$ and $\langle v_{\downarrow x}\rangle\rho_\downarrow(E_F)$, related to majority- and minority-spin electrons, respectively, to be quantified. Extensive details about this topic can be found in Chapter 27. Such polarization of the ballistic current should, however, not be identified with the diffusive current spin polarization explaining GMR properties or the tunneling transmission spin polarization justifying the occurrence of TMR.

28.1.2 Landauer Model

Reducing further the size of a constriction requires a model taking into account the wave nature of the carriers of the electric current. If sample dimensions become comparable to or smaller than the wavelength of the carriers, the effects of quantum confinement play a role. A simple illustration is obtained if we consider the transport between the left and right side of

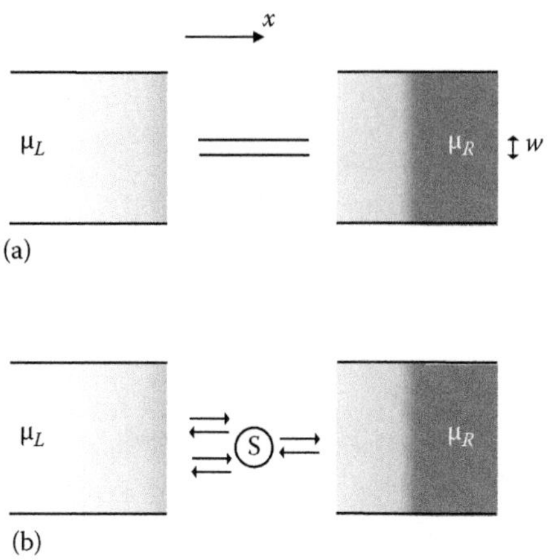

FIGURE 28.3 (a) Sketch of a conductor of width w, smaller than the conduction electrons wavelength, λ_F. In the Laudauer picture, two left and right electrons reservoirs connect the quantum conductor to the electrical circuits, with defined chemical potential μ_R and μ_L. (b) The conductor can be considered as a scattering center S, with transmission modes (arrows) connecting it to the reservoirs. The conductance relies on the number of modes and their transmission (reflection) properties. This particular example shows two modes connecting the scattering center to the left and one mode to the right.

a one-dimensional (1D) conductor (Figure 28.3). This dimensionality restriction corresponds to stating that the width of the wire connecting the two electrodes allows only one Bloch wave function to exist in the (k_y, k_z) space, implying a diameter on the order of the Fermi wavelength λ_F. As a rule of thumb, the wavelength of conduction electrons is on the order of the distance separating two conduction electron donor sites (eventually holes in semiconductor cases). For metals, of typically one conduction electron per atom, it implies that λ_F is on the order of the atomic radius, ~0.3 nm. For semiconductors of high purity, this length is significantly larger, on the order of 10 nm.

Since, in 1D, $\rho(E_F)$ is proportional to $1/\langle v(x)\rangle$, $\rho(E_F)\langle v(x)\rangle$ of Equation 28.2 remarkably simplifies to $4/h$, and the resulting conductance is of the form

$$G_0 = \frac{2e^2}{h} \tag{28.4}$$

The factor 2 comes from the assumed spin degeneracy, and the conductance G_0 is the so-called quantum of conductance, which is $\frac{1}{G_0} = \frac{h}{2e^2} \approx 12.9\,\text{k}\Omega$.

This simple outcome can be generalized by introducing the Landauer–Buttiker formalism, initially developed for modeling scattering at the mesoscale (Figure 28.3) [6]. In so-called two-terminal devices, the current flows between two infinite reservoirs, defined by their electrochemical potentials. They are separated from a central scattering region by two idealized semi-infinite leads, treated as an ensemble of 1D conductors obeying Equation 28.4. These conductors index the possible Bloch waves travelling in the leads. The interface between the leads and reservoirs are considered perfect, so that scattering only occurs between leads and the central region. A matrix formalism is used to express the linear relationship between transmission factors of waves coming from the left and right electrodes. Diagonalization of this matrix provides a series of transmission eigenvalues, T_i, corresponding to the so-called eigenchannels. At zero bias and zero temperature bias, the conductance reduces to

$$G = \frac{2e^2}{h}\sum_i T_i \tag{28.5}$$

At small bias (and negligible temperatures), the conductance of the system relates to the transmission probabilities, T_i, between the lead eigenchannels and the central scattering source taking values between 0 (perfect reflection) and 1 (perfect transmission) at the Fermi level. For the case of a 1D conductor, Equation 28.4 can be generalized to take into account the number of allowed transmission modes. This value $N(E_F)$ corresponds to the number of crossings of the 1D subbands at the Fermi level. When considering an average transmission factor $\overline{T}(E_F)$, Equation 28.5 reduces to:

$$G = \frac{2e^2}{h}\overline{T}(E_F)N(E_F) \tag{28.6}$$

In the case of perfect transmission factors, $T_i = 1$, Equation 28.6 shows that the conductance is quantized in units of G_0. Experimental confirmations of such conductance quantization were reproducibly observed in both semiconductors and metallic systems. When realizing ballistic conductors of perfect transmission factors in the ballistic quantum regime, stepwise variation of G is found. Figure 28.4 shows two examples of experimental realization of ballistic quantum conductors, obtained by electrostatically gating a 2D electron gas (2DEG), or stretching a metallic constriction, both showing so-called quantization of the conductance. The experimental results on metallic systems also confirm the necessary extreme smallness of the samples.

One should, however, note that unambiguous experimental confirmation of the ballistic quantum nature of the electronic transport can only result from a four-point measurement method, where two voltage probes should be inserted between the source and drain connections. Such an experimental tour de force has been realized for a cleaved 2DEG, using electrostatic gating to define four terminals on the resulting 1D conductor. Under the condition that the voltage probes are noninvasive (technically corresponding to voltage lines of impedance much larger than $1/G_0$), the measured voltage drop is zero when the transmission factor reaches one, confirming that no scattering event has affected the potential energy of the particles. In other words, the observed resistance of a perfect ballistic conductor, reaching several kΩ, should be interpreted as a contact

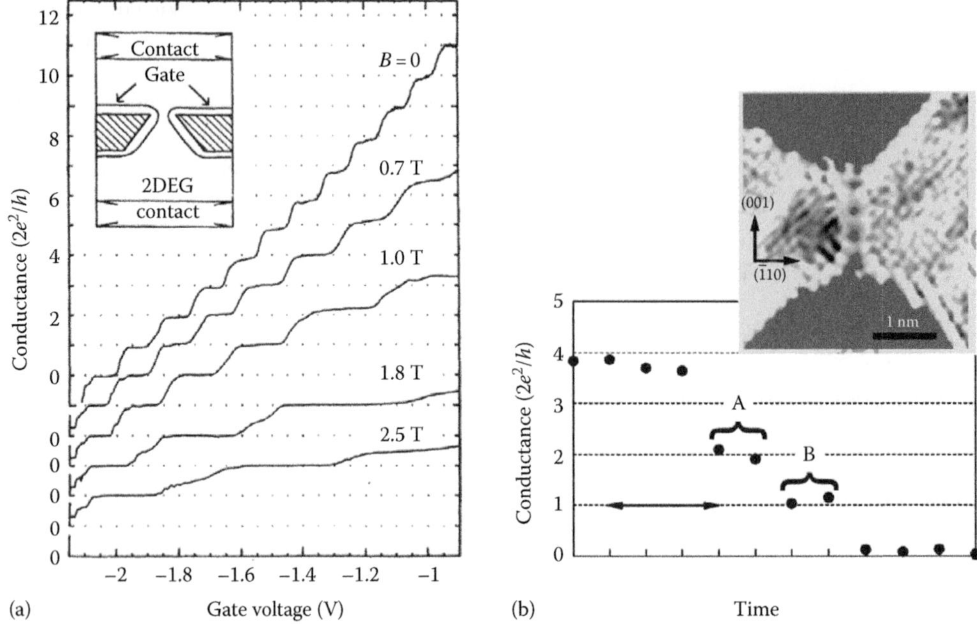

FIGURE 28.4 Experimental data showing quantization of the conductance for perfect ballistic point contacts. (a) Gated 2DEG semiconductor structure, where electrostatic gating depletes continuously the constriction between the source and drain (Reprinted with permission from van Wees, B.J., Kouwenhoven, L.P., van Houten, H., Beenakker, C.W.J., Mooij, J.E., Foxon, C.T., and Harris, J.J., *Phys. Rev. B.*, 38, 3625–3627, 1988, Fig. 1. Copyright 1988 by the American Physical Society.) (b) Metallic stretched Au linear constriction obtained by stretching Au under a transmission microscope, revealing the atomic nature of the contact. (Reprinted by permission from Macmillan Publishers Ltd. [*Nature*] Ohnishi, H., Kondo, Y., and Takayanagi, K., Quantized conductance through individual rows of suspended gold atoms, 395, 780–785, Copyright 1998.)

resistance, shared between the interfaces of the 1D conductor and the source and drain leads.

The observed two-terminal conductance in multiples of the quantum value is an idealized case, corresponding to the asymptotic transmission factors of 1 and 0. In a more general case, the conductance of a small ballistic sample requires the explicit calculation of the normal modes, and their transmission factors T_i. Detailed description of the theoretical tools and their numerical outcome are beyond the scope of this chapter, and the reader can consult more dedicated texts [2,3,7]. Two idealized geometries can be considered as the basis for discussing the number of conduction channels, their symmetry properties, and their expected transmission factors. Figure 28.5 provides a picture particularly suited to the case of metallic systems, with sample size on the order of the atomic radius. The simplest case is the pseudo-infinite 1D wire. The derivation of the allowed modes results from solid-state physics concepts applied to a 1D case, and the usual theoretical tools for calculating the propagating Bloch waves can be applied. The number of transmitting modes equals the number of crossings of the 1D dispersion curves with the Fermi level. A smooth transition from a single atom width to a width reaching the mean free path value ensures that the highest transmission factor value is obtained, that is, $T_i = 1$, if a mode is allowed to propagate in the wire. This model is useful for pedagogical purposes and for understanding which conduction channels can propagate. Calculation of the conductance is straightforward, as it corresponds to the number of allowed propagating modes.

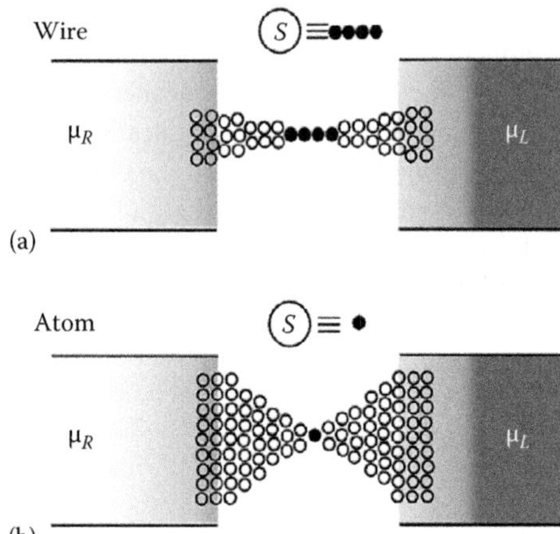

FIGURE 28.5 Sketch of the two idealized simplest pictures of atomic-sized quantum contact for metallic systems. (a) Corresponds to an idealized infinite wire, with a smooth continuous transition to the two reservoirs. (b) Corresponds to a single atom connection, with two pyramidal interconnects to the infinite reservoirs.

This image can be used for semiconductor systems, where the purity, size control, and fabrication techniques can result in well-defined 1D systems in the tens of nanometers range. The other limiting case is the single atom connection, separating two semi-infinite pyramids, which is expected to better match the experiments on metals. Transmitting channels are related to wave functions of the contact atom and the number of propagating modes can generally be matched to the number of valence electrons of the single atom contact. The two pyramid-shaped electrodes broaden the density of states and create an electronic cloud mismatch at the interface, resulting therefore in transmission factors deviating significantly from 1. Calculations of the transmission factors are, however, much less intuitive, and depend significantly on the geometrical arrangements of the pyramid atoms. One can, however, guess that atomic wave functions, being less localized (in particular *s* electrons), suffer less from the mismatch at the interface, and exhibit therefore higher transmission factors. Such a model has been particularly successful in explaining experimental results in mechanical break junctions (MBJ).

The interest of keeping in mind the two types of models is to realize the importance of the dispersion curves of the electronic structures and recall that perfect transmission or reflection are only limiting cases. One should also mention that these two models are not so much different. Following the argument of Glazman et al. [8], the adiabatic condition will allow perfect transmission and no mixing of the conduction channels when the cross section of the constriction varies slowly compared to λ_F. When contemplating on Figure. 28.5a and b, the difference between the two geometries is mostly related to checking if the adiabatic condition is fulfilled or not. For metal systems, a contact made of two well-aligned atoms could already be considered as a reasonable approximation of an infinite wire! This might explain (among other reasons) the lack of reproducibility when performing single atom contact measurements, and the necessity of ensuring statistical information on the experimental results.

28.2 Spin Transport in the Ballistic Regime

Until now, the spin of the conduction particles has not been considered, except as a doubling parameter for calculating the conductance of a system. The factor 2 in Equation 28.4 explicitly expresses the spin degeneracy. In this section, we will review several ways to lift this degeneracy, opening the possibility to create a spin-polarized flow of particles, and to expect new magnetoresistance effects. The first part of this section describes experimental realizations on semiconductors systems, opening the possibility to create spin filters and spin detectors using point contacts in a 2DEG.

28.2.1 Zeeman Splitting

Since the discovery of the quantum conductance, the use of a large magnetic field has been shown to be a usable tool for realizing point contacts in 2DEG exhibiting quantized steps of multiples of e^2/h. The Zeeman splitting energy of the form

$$\Delta E_Z = g\mu B \tag{28.7}$$

has typical orders of magnitude of 0.1 meV for 2DEG systems (using a Landé factor $g=0.4$, typical for GaAs and fields ~ few Teslas), requiring therefore experiments at temperatures below 1 K to allow conduction values of the order of $½G_0$, providing therefore a spin-filtered source of carriers. Even if larger Landé factors can be obtained on other systems, Zeeman splitting remains, however, severely limited to low temperature studies.

28.2.2 Spin Orbit Splitting

Spin polarization of quantum point contacts in 2DEG systems can be further exploited in experiments involving point contact emitters and collectors, where the cyclotron ballistic motion of charge carriers is controlled through the use of an external perpendicular magnetic induction field. A large literature on such electron-focusing experiments can be found [9]. Figure 28.6 shows the design of the experiment. Current flows through the detector when the distance L between the two point contacts matches a multiple of the cyclotron radius $r_c=hk_F/eB$, where $k_F=\sqrt{2mE_F/\hbar}$ for free-like carriers of effective mass m and charge e.

The emitter and detector can be gated to allow a controlled number of channels to be transmitted by the two point contacts. Potok et al. [10] showed that an additional Zeeman splitting can spin-polarize the emitted charges (by gating the emitter to e^2/h conductance) resulting in a modification of the B-dependence of the current passing through the detector. Figure 28.6 summarizes the experimental outcome, showing that the current through the detector can be significantly spin-polarized. It shows that the focusing condition, that is, the radius of the cyclotron trajectory is spin dependent, providing a direct illustration of spin-orbit coupling (SOC) effects.

The coupling between orbital and spin momentum of charges results from their relativistic interactions. For a single particle Hamiltonian in a crystal potential, $V(\vec{r})$, one can express:

$$H=\frac{p^2}{2m}+V(\vec{r})+\frac{\hbar}{4m^2c^2}\left[\nabla V(\vec{r})\times\vec{p}\right]\cdot\sigma \tag{28.8}$$

where the particle of mass m has a momentum $\vec{p}$ and a Pauli spin operator σ.

For the case of a 2DEG with a bulk origin of the symmetry-breaking $\nabla V(\vec{r})$, one can simplify the Hamiltonian to

$$H=\frac{1}{2m}(p_x+\gamma\sigma_x)^2+\frac{1}{2m}(p_y-\gamma\sigma_y)^2 \tag{28.9}$$

where γ is the Dresselhaus spin-orbit parameter. The resulting splitting of the dispersion curve is illustrated in Figure 28.7, for a free-carrier-like dispersion. A semiclassical treatment of the calculation of the cyclotron motion of the particle shows that the radius now depends on the projection of spin along the applied field (z-axis). One prefers to express the value of the

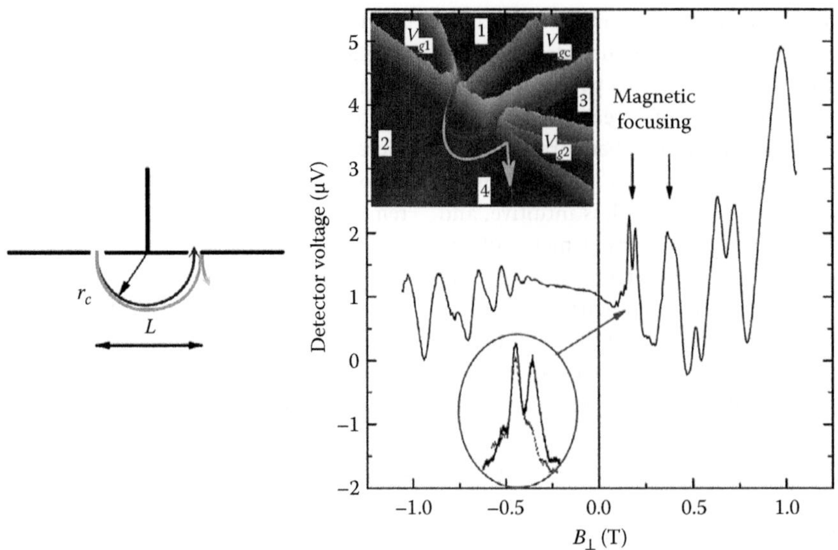

FIGURE 28.6 Spin-dependent electronic focusing experiment in GaAs 2DEG, with point contact emitters and detectors defined by top local oxidation. On the top left, the principle of the experiment is provided, with a scanning electron microscopy (SEM) picture of the corresponding device (5 × 5 mm). When varying the applied field, focusing is observed for positive field values. By applying a Zeeman splitting field of 3.5 T, the two first magnetic focusing peaks appearing for an injector of e^2/h conductance become a single peak when the injector is gated to e^2/h conductance value. Only one single spin direction (related to the black trajectory) is allowed to pass the detector point contact (From Rokhinson, L. et al., *J. Phys.: Condens. Matter,* 20, 164212, 2008, p. 2, Fig. 1a and p. 3, Fig. 3. With permission. Copyright 2008 Institute of Physics.)

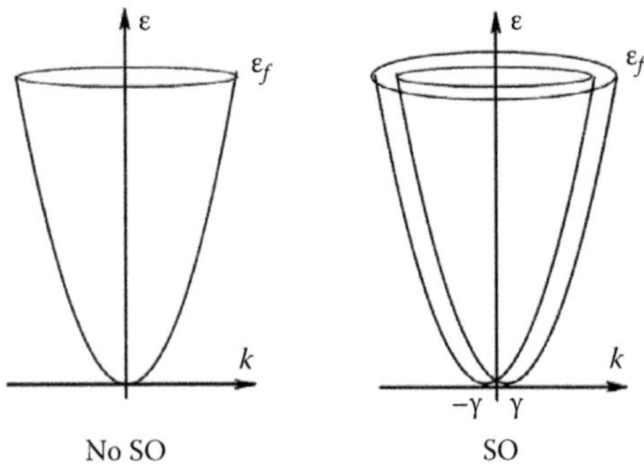

FIGURE 28.7 Dispersion curve of a free-like carrier, with the splitting due to the spin–orbit coupling of parameter γ. The two cyclotron trajectories of Figure 28.6 correspond to the top right of the figure.

applied field fulfilling the focalization condition for the two spin directions:

$$B_\perp^\pm = \frac{2(p_f \mp \gamma)}{eL} \tag{28.10}$$

where p_f relates to the momentum corresponding to focalization without SOC.

The SOC can result in splitting energies ranging from several meV to a few tens of meV in GaAs, and is therefore significantly larger than the Zeeman splitting. The observed variations in the detection currents relate to the product of the injector and detector spin selectivities, reaching several tens of percents. These devices are therefore of fundamental interest for manipulating spin populations at low temperatures, with the perspectives of performing fundamental studies. The initial experiments showed how SOC can be of use for separating the spin currents, in contrast to its detrimental effect as a source of decoherence in the spin populations. One should also emphasize that these devices allow spin electronic studies without involving magnetic elements.

28.2.3 Exchange Splitting

When considering magnetic systems, in particular transition metal materials, one can take advantage of the huge exchange energy to lift the spin degeneracy. This splitting energy, reaching eV values, effectively splits the d-type bands of transition metals giving rise to their ferromagnetic properties. This simplest argument would therefore imply that quantized conductance in multiples of e^2/h could be found in metallic nanocontacts made from transition metal systems. In such cases, Equation 28.5 is inappropriate and must be generalized for magnetic systems. In a material where magnetization defines a spin quantization axis, we can decompose the transmission factors for majority (↑) and minority (↓) spin directions.

$$G = \frac{e^2}{h}\sum_i (T_{\uparrow i} + T_{\downarrow i}) \tag{28.11}$$

If we simplify the system to a single channel with left and right transmission, it becomes possible that the exchange splitting limits the band crossing at the Fermi energy to one type of spin population only. In such a case, conductance quantization in multiples of $\frac{1}{2}G_0$ is expected. Interestingly, the ballistic conductor would play the role of a spin filter, allowing only one spin population to be transmitted, creating a perfect current spin polarization. As a result, one can expect no current transmission when the two sides of the conductor are magnetized with opposite magnetization directions (antiparallel case), and a transmission of $\frac{1}{2}G_0$ when the magnetization of the system is uniform (parallel case). The ballistic conductor would play the role of a perfect (atomic size) domain wall in the antiparallel configuration. The simplified argument given here needs more complete calculations that take into account the coherent transport through the constriction and modeling of the system in order to extract the spin-dependent transmission factors. Two illustrations are provided in Figure 28.8, both reaching a similar conclusion of strong spin-filtering effects at conductance values on the order of $\frac{1}{2}G_0$ and odd multiples of $\frac{1}{2}G_0$ that are allowed in the ferromagnetic configuration only.

Occurrence of conductance quantized in units of $\frac{1}{2}G_0$ is controversial. Experiments performed under transmission electron microscopy (TEM) observation, similar to Figure 28.4, showed half-integer conductance occurring in Co [11]. However, repetitive measurements performed on MBJ, when stretching back and forth a suspended metallic bridge lithographically defined on a flexible substrate, did not show preferred occurrence of quantized values of the conductance for transition metal contacts [12]. Many divergent experimental outcomes can be found in the literature (see review by [2,13]).

Based on the intuitive arguments of the previous section, it is not surprising that magnetic junctions of high transmission single-channel conductance cannot be obtained frequently. For example, a Fe 1D wire, without imperfections, with a diameter smoothly increasing when connected to the diffusive bank would have seven band crossings at the Fermi level, exhibiting therefore conductance that should be $7e^2/h$. Furthermore, the *s* bands are expected to exhibit limited spin splitting, which always prevents full spin polarization. When one gets away from the ideal 1D geometry, smaller transmission factors are expected, and one can expect lower measured conductance, corresponding to frequently observed values, but not guaranteeing the absence of s-type transmission. The multiple combinations of several partly conducting channels might explain the lack of preferred occurrence at simple multiples of e^2/h.

Another fundamental difficulty in making atomic spin filters is related to magnetic properties. The possibility to realize atomic-sized domain walls remains conceptually unclear. Even though magnetic energy calculations can show that such sharp domain walls can represent special cases of stability in magnetic systems of reduced dimensions, the intrinsic magnetic properties of the atoms of the structure, as well as their exchange interaction ensuring ferromagnetism, can be modified severely for atomic-sized structures. This necessitates more steps and hypothesis in the calculations, and provides more reasons to deviate from the asymptotic behavior of infinite magnetoresistance (see more details in [13]).

Experimental investigations of magnetoresistance properties also exhibited a very large spectrum of results, with lack

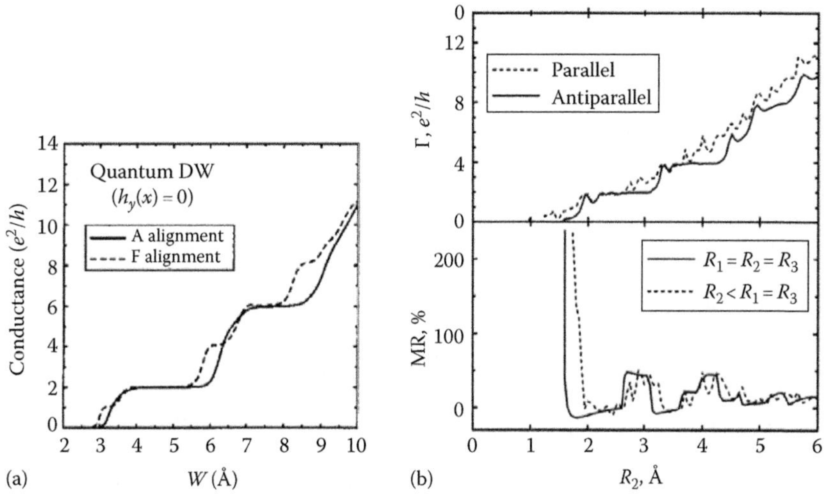

FIGURE 28.8 Free electron calculations of conductance of a ferromagnetic constriction in the quantum ballistic regime. (a) With calculated conductance as a function of the width of transition metal 1D conductor, with the conductance in the antiparallel magnetization configuration drawn in full line and conductance in the parallel configuration in dashed. (b) Same type of calculation, involving a model of a central constriction resistor R_2, separated from the electrodes by wider R_1 and R_3 resistors. (Reprinted with permission from Imamura, H., Kobayashi, N., Takahashi, S., and Maekawa, S., *Phys. Rev. Lett.*, 84, 1003–1005, 2000, Fig. 3b. Copyright 2000 by the American Physical Society.) (b) (Reprinted with permission from Zhuravlev, M.Y., Tsymbal, E.Y., Jaswal, S.S., Vedyayev, A.V., and Dieny, B., Spinblockade in ferromagnetic nanocontacts, *Appl. Phys. Lett.*, 83, 3534–3536, Figs. 2c and 2d. Copyright 2003, American Institute of Physics.)

of reproducibility. A critical issue in performing measurements of atomic-size structures under applied magnetic field is the intrinsic troublesome mechanical stability of the samples. Magnetostatic interactions, magnetoelastic properties of the bulk ferromagnetic electrodes, and magnetization-dependent strains can easily modify the distance separating the two large diffusive electrodes by several atomic distances, modifying therefore drastically the nanocontact and its conductance. There is now a significant consensus in the community that huge magnetoresistance values due to atomic domain wall structures are unlikely to happen, and that the published data revealing very large magnetoresistance values can be explained using models involving mechanical modifications of the nanocontacts under varying applied magnetic fields [14].

However, several groups reported magnetoresistance ratios of a few tens of percents within the expected range, when considering that one or a few transmitting channels can have their transmission factors turned on and off when modifying the mutual magnetic alignment of the two connecting electrodes. More interestingly, these results all showed large anisotropic magnetoresistance (AMR) effects, which is a finding that can be related to the previously discussed SOC.

28.3 Anisotropic Magnetoresistance in the Quantum Limit

28.3.1 AMR in Diffusive Transport

Magnetoresistance measurements by W. Thomson showed in 1859 that the resistance of a ferromagnetic material depends on the mutual orientation of the applied magnetic field $\vec{H}$ and current $\vec{j}$ directions. This so-called anisotropic magnetoresistance relates therefore to the experimental indication that $\rho_{//}(\vec{H} \parallel \vec{j})$ is, in most cases, larger than $\rho_{\perp}(\vec{H} \perp \vec{j})$, and follows the relation (in a cubic symmetry)

$$\rho(\theta) - \rho_{\perp} = (\rho_{//} - \rho_{\perp})\cos^2\theta \tag{28.12}$$

where θ is the angle between the current and the magnetization. This difference does not exceed a few percent at room temperature for transition metals. The mechanism by which the microscopic internal field associated with $\vec{M}$ couples to the current density in ferromagnets is the SOC interaction. It was only one century after Thomson's discovery that a scattering model, derived from Mott's two current models of conduction in transition metals, was refined to explain how SOC can lead to Equation 28.12 [15]. When the orbital motion of the d electrons is coupled to their spin and the crystal lattice, the new d electron wave functions are no longer eigenfunctions of S_z since H_{SO} mixes states of opposite spins. Hence, this limits the independence of majority and minority conduction channels of the Mott's model, and is an important source of spin mixing, detrimental to GMR properties. This mixing of states is anisotropic and results in resistivity anisotropy. Potter's analytical expression for the spin-dependent relaxation rates showed that sd scattering of minority spins (i.e., the one opposite the magnetization) is responsible for the measured AMR (with $\rho_{//} > \rho_{\perp}$). It was illustrating that the sign of the AMR depends directly on the band structure of the ferromagnetic material.

The SOC in bulk metallic systems is a rather small energetic perturbation, of typically 0.1 meV magnitude. Despite its limited magnitude, SOC is involved in explaining anisotropic magnetic properties of materials, including magneto-crystalline anisotropy, magnetostriction, and AMR. For magnetic systems of reduced dimensionalities and reaching atomic dimensions, one expects a significant enhancement of the SOC coupling, justifying the modification of magnetic properties and their anisotropy for ultrathin magnetic films, islands, and wires involving a few atoms only. The AMR properties should therefore be completely rethought when the electric transport approaches the ballistic regime and when the SOC magnitude increases.

28.3.2 Initial Experimental Results of AMR in Nanocontacts

Several groups reported magnetoresistance measurements varying significantly when the angle between magnetization and current was imposed. Figure 28.9 presents three different results obtained on three different types of samples. Results reported by Viret et al. [25] were obtained on MBJ; those reported by Keane et al. [16] were measured on nanocontacts made by electromigration breaking slowly a patterned nanoconstriction. Both experiments were performed at low temperatures. Data obtained by Yang et al. [17] used electroplated samples, where a gap, typically 100 nm wide, is closed by reducing metal ions from an electrolytic bath. This data was obtained in situ at room temperature, under variable reduction or oxidation (dissolution) electrochemical conditions of the junctions immersed in an ionic medium. The fact that such AMR-like properties were obtained on samples made by quite different methods, in different media, and at different temperatures strongly suggests that AMR on nanocontacts can be reproducibly observed with amplitudes much larger than those found on bulk samples. Therefore, there should be fundamental reasons to exacerbate AMR properties when the size of the sample is reduced.

28.3.3 AMR in the Ballistic Regime of Transport

From the previous considerations and findings, the picture of AMR of bulk materials should be completely revisited if the ballistic contribution to the sample resistance becomes dominant. A good pedagogical model can be inferred from the idealized 1D chain of atoms with perfect transmission factors and can be used to demonstrate the concept behind AMR extended to Landauer formalism (Velev et al. [26]). Ab initio calculations using the pseudopotential plane wave method and symmetry arguments have shown that the weakly dispersive bands

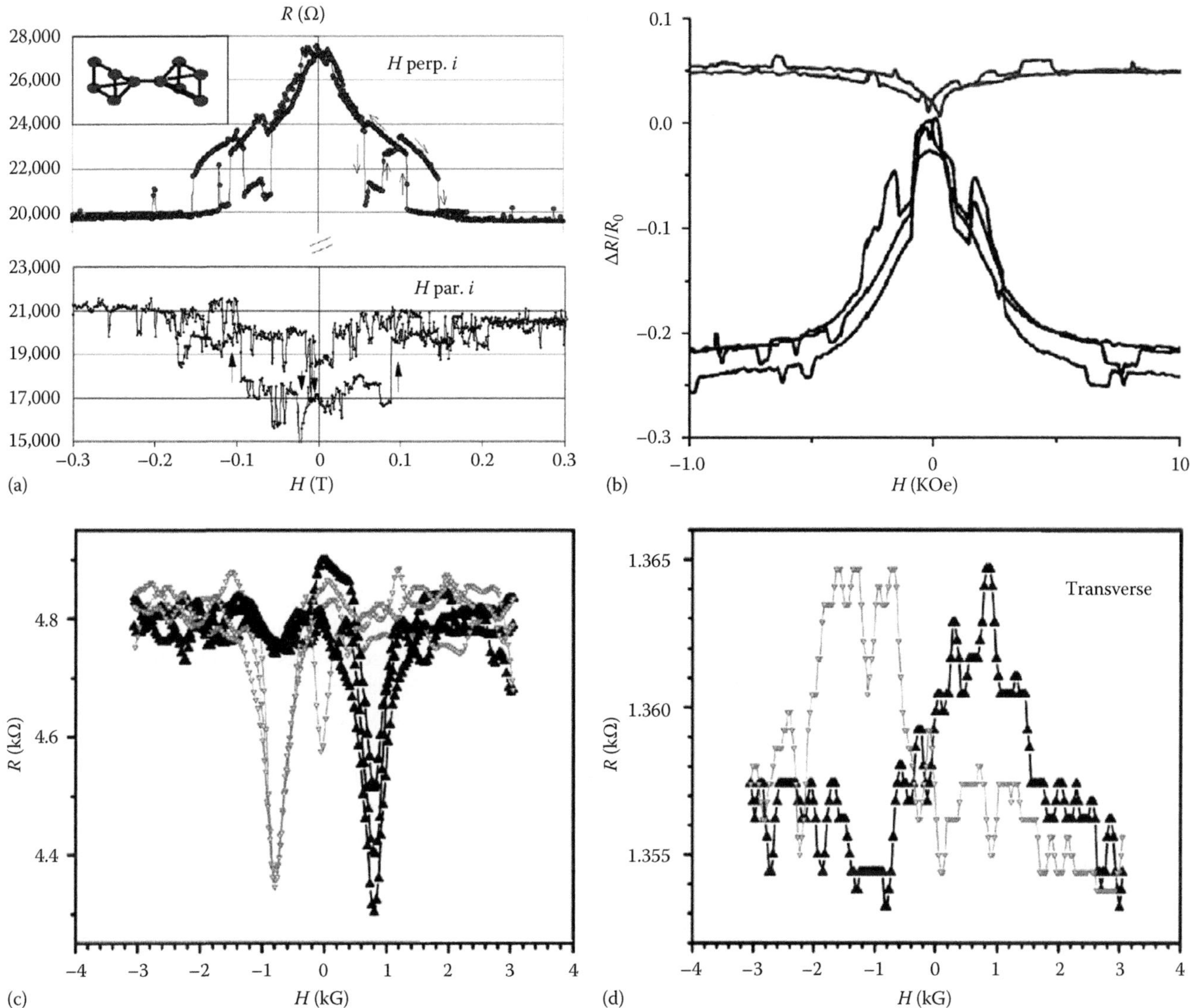

FIGURE 28.9 Angle-dependent magnetoresistance curves on Ni nanocontacts. (a) MBJ measured at 10 K. (Reprinted with permission from Viret, M., Berger, S., Gabureac, M., Ott, F., Olligs, D., Petej, I., Gregg, J.F., Fermon, G., Francinet, G., and Le Goff, G., *Phys. Rev. B*, 66, 220401, 2002, Fig. 3. Copyright 2002 by the American Physical Society.) (b) Electromigrated Ni junction at 4.2 K. (Reprinted with permission from Keane, Z.K., Yu, L.H., and Natelson, D., Magnetoresistance of atomic-scale electromigrated nickel nanocontacts, *Appl. Phys. Lett.*, 88, 062514, Fig. 3c. Copyright 2006, American Institute of Physics.) (c,d) Electroplated Ni junctions at 300 K. (Reprinted from Yang, C.S. et al., *J. Magn. Magn. Mater.*, 286, 186–189, 2005. Copyright 2006, with permission from Elsevier.)

(δ bands) arising from the coupling of nearest neighbors $3d_{xy}$ and $3d_{x^2-y^2}$ atomic orbitals are split by the spin–orbit interaction when the magnetization is parallel to the wire, and nearly degenerate when perpendicular (Figure 28.10). The energy lifting corresponds to the SOC constant reaching a magnitude of 0.1 eV, orders of magnitude larger than the bulk SOC energy. Because the Fermi energy lies close to the edge of the δ bands, one of the bands becomes expelled from the Fermi level when split by the SOC energy in the relevant geometry (M parallel to the wire). Hence, the integer number of conducting channels can be modified when the angle between magnetization and current is varied. The conductance is expected to change by e^2/h in this case, because for an infinite atomic wire the conductance is simply e^2/h per band crossing the Fermi level. For the specific example of Figure 28.10 (δ bands of a Ni monatomic wire), the conductance changes from 6 to 7 e^2/h, which is comparable in magnitude to experiments, and much larger than bulk AMR. This so-called ballistic anisotropic magnetoresistance (BAMR) is also expected to exhibit sharp angular dependence that makes the experimental signature or BAMR significantly different from bulk AMR (Figure 28.10).

It is clear that a 1D perfect wire is a highly idealized model that must be considered as a pedagogical intuitive guideline. In fact, when trying to understand transport in magnetic contacts,

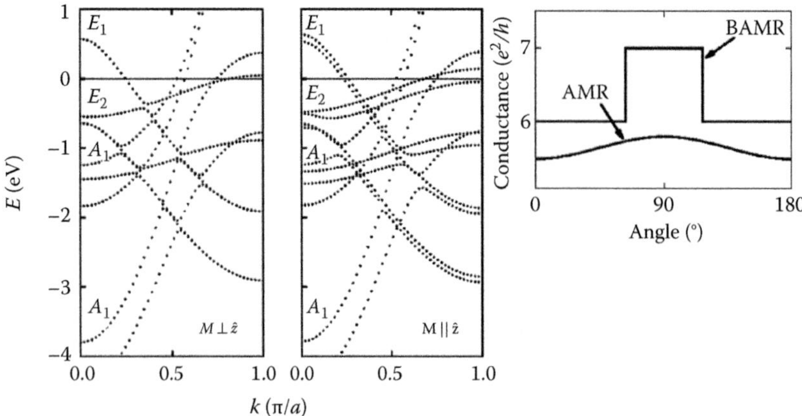

FIGURE 28.10 Calculations of the dispersion curve of a 1D Ni wire in a tight binding model, showing the splitting of the bands when magnetization and current directions are parallel. Right: comparison of angular behavior showing how AMR and BAMR can differ in magnitude and shape. (Reprinted with permission from Velev, J., Sabirianov, R.F., Jaswal, S.S., and Tsymbal, E.Y., *Phys. Rev. Lett.*, 94, 127203, 2005, Figs. 1,2. Copyright 2005 by the American Physical Society.)

one needs to model the atomic arrangement that constitutes a narrow neck and the corresponding electronic structure. A realistic modeling of the experiments must determine possible structural arrangements when making atomic contacts, compute the electronic and magnetic structures of the systems, and deduce the conductance by estimating the transmission factors of the eigenmodes of the junction. More details can be found in dedicated papers, where more information on the hypothesis for the calculations can be found [2,13].

Even though the simplicity of the infinite 1D model can be criticized, the main conclusions of Velev et al. should be kept for more sophisticated modeling. The SOC energy is increased by the atomic size of the junction, and the resulting band splitting can result in significant angular changes of the conduction, through a mechanism totally different from bulk AMR.

28.3.4 Experimental Findings of BAMR

Measuring AMR is obtained by keeping the sample magnetically saturated, and varying the angle the magnetization makes with the current. For applied magnetic fields above 1 T, the magnetic configuration of the samples is expected to be near saturation. Such studies therefore eliminate the problem of positioning an atomic-size domain wall, of importance for GMR-type studies, and investigate an unambiguous magnetic configuration. Figure 28.11a shows results on Fe contacts for MBJ measured at low temperatures, and Figure 28.11b Co electrodeposited contacts at room temperature. Both data exhibit the expected deviations for a $\cos^2 \theta$ bulk AMR variation of the resistance, with amplitude unattainable for bulk materials.

Data on Co also indicates that the sign of the effect can also be opposite. A qualitative analysis based on the arguments of Velev et al. pointed out that the dispersion curves related to the wave functions of a given nanocontact can vary significantly from one sample to the other, resulting in an angular variation of the number of band crossings that can be drastically modified from sample to sample. These findings also confirm that the sign of the BAMR has no fundamental reason to be positive, in contrast to AMR, where negative variations can only be observed in a few alloys.

Even though a set of experimental findings from different research teams appear to reach a consensus, one should mention another set of experiments, performed on MBJ, that revealed a voltage bias-dependent AMR of nanocontacts at low temperatures (Figure 28.12). They interpreted their results with quantum fluctuation arguments, and pointed out that disorder in the nanocontact region can significantly enhance this phenomenon. The remarkable sharpness of the conductance angular transitions found on Co samples (Figure 28.11) is also reminiscent of fluctuations between two (or multiple) voltage levels found in electric telegraph noise. It is known that such noise can reach large amplitudes in very small systems. For nanocontacts, the explanation of such two-level fluctuation is based on a model of atomic rearrangements at the constriction, possibly including magnetic fluctuations when considering magnetic systems. Such an occurrence can mimic BAMR data, as recently shown by Shi and Ralph on Ni samples [19]. There might be an advantage in terms of mechanical and surface stability when samples are immersed in an electrolyte and fabricated by a growth process (in opposition to a breakage process). More experiments are clearly needed to clarify the issue. The observed difference between Ni and Co is also remarkable. For example, angular measurements performed on Ni electrochemical junctions did not reveal a quantum-type behavior, but noisy features, compatible with reported occurrence of telegraph noise.

This discussion also recalls that steps in the conductance of metallic junctions are indicative of stable atomic configurations

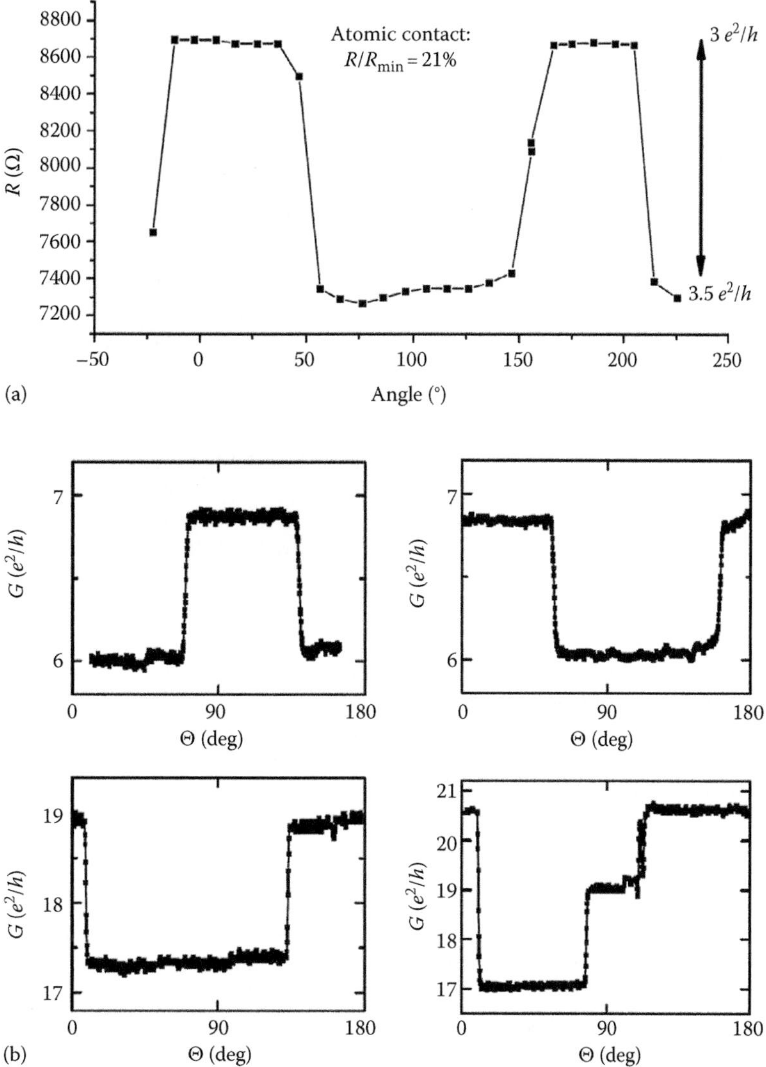

FIGURE 28.11 Angular variation of a magnetic nanocontact resistance when changing the angle between saturated magnetic state and current directions. (a) With kind permission from Springer Science+Business Media: *Eur. Phys. J. B*, Giant anisotropic magneto-resistance in ferromagnetic atomic contacts, 51, 2006, 1–5, Viret, M., Gabureac, M., Ott, F., Fermon, C., Barreteau, C., and Guirado-Lopez, R., on Fe contacts. Reproduced with permission © 2006 EDP Sciences. (b) Reprinted by permission from Macmillan Publishers Ltd. *Nat. Nanotechnol.*, Sokolov, A., Zhang, C., Tsymbal, E.Y., Redepenning, J., and Doudin, B., Quantized magnetoresistance in atomic-size contacts, 2, 522, copyright 2007, on Co contacts, showing step-like AMR of large amplitude, of behavior not reproducible from sample to sample.

at the nanocontacts, and that the mechanical stability remains a critical issue, especially for magnetic systems. More detailed information on the interplay between mechanical and electronic properties of junctions can be found in the review of Agrait et al. [2].

28.4 Conclusions

We have shown that ballistic transport can be a usable way to filter spin currents, detect preferentially one spin population, and reveal new magnetoresistance applications at the nanoscale. For metallic systems, the experiments accumulated over the past few years indicated that the mechanical stability of the samples is a key issue and limitation for applications. This chapter also emphasized how SOC is the key physical phenomenon beyond all these new properties. This coupling makes the ballistic trajectories of the two spin populations of carriers different in space, which is a property that can directly be used for semiconducting systems. This coupling also explains why large anisotropies in the resistance properties can be found. This interest complements the other aspect of SOC, namely, the limitation of spin coherence. This spin mixing effect is usually considered as negative, limiting the discrimination of the two spin populations in time or space. We think that both perspectives can be complementary, allowing therefore better and new understanding of quantum spin systems, as well as design of novel quantum spin devices.

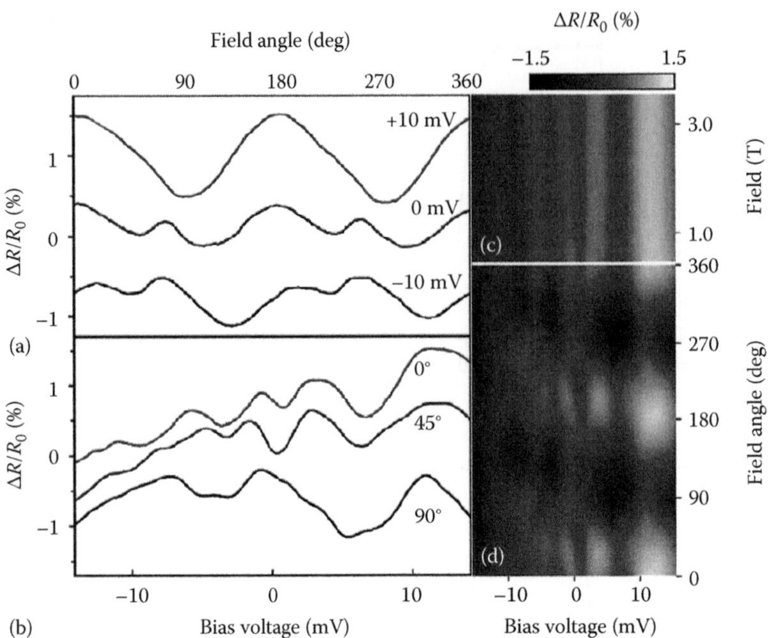

FIGURE 28.12 Variations of $R=dV/dI$ at 4.2 K in a Permalloy sample with average zero-bias resistance *of* 2.6 kΩ. (a) R versus field angle at different bias voltages ($B=0.8$ T). (b) Dependence of R on V at different fixed angles of magnetic field ($B=0.8$ T). The curves in (a) and (b) are offset vertically. (c) R as a function of V and magnetic field strength, with field directed along the x-axis. R does not have significant dependence on the magnitude of B. (d) R as a function of V and for $B=0.8$ T. (Reprinted with permission from Bolotin, K.I., Kuemmeth, F., and Ralph, D.C., *Phys. Rev. Lett.*, 97, 127202 1–127202 4, 2006, Fig. 3. Copyright 2006 by the American Physical Society.)

References

1. S. Datta and B. Das, Electronic analog of the electro-optic modulator, *Appl. Phys. Lett.* **56**, 665–667 (1990).
2. N. Agrait, A. L. Yeyati, and J. M. van Ruitenbeek, Quantum properties of atomic-sized conductors, *Phys. Rep.* **377**, 81–279 (2003).
3. Y. Imry, *Introduction to Mesoscopic Physics*, Oxford University Press, Oxford, 1997.
4. Y. V. Sharvin, A possible method for studying Fermi surfaces, *Sov. Phys. JETP* **21**, 655–656 (1965).
5. A. G. M. Jansen, A. P. van Gelder, and P. Wyder, Point-contact spectroscopy in metals, *J. Phys. C: Solid State Phys.* **13**, 6073–6118 (1980).
6. R. Landauer, Spatial variation of currents and fields due to localized scatterers in metallic conduction, *IBM J. Res. Dev.* **1**, 223–227 (1957).
7. S. Datta, *Electronic Transport in Mesoscopic Systems*, Cambridge University Press, Cambridge, U.K., 1995.
8. L. I. Glazman, G. B. Lesovik, D. E. Khmelnitskii, and R. I. Shekhter, Reflectionless quantum transport and fundamental ballistic-resistance steps in microscopic constrictions, *JETP Lett.* **48**, 238 (1988).
9. V. S. Tsoi, J. Bass, and P. Wyder, Studying conduction-electron/interface interactions using transverse electron focusing, *Rev. Mod. Phys.* **71**, 1641–1693 (1999).
10. R. M. Potok, J. A. Folk, C. M. Marcus, and V. Umansky, Detecting spin-polarized currents in ballistic nanostructures, *Phys. Rev. Lett.* **89**, 266602 (2002).
11. V. Rodrigues, J. Bettini, P. C. Silva, and D. Ugarte, Evidence for spontaneous spin-polarized transport in magnetic nanowires, *Phys. Rev. Lett.* **91**, 096801 (2003).
12. C. Untiedt, D. M. T. Dekker, D. Djukic, and J. M. van Ruitenbeek, Absence of magnetically induced fractional quantization in atomic contacts, *Phys. Rev. B* **69**, 081401 (2004).
13. J. D. Burton and E. Y. Tsymbal, Magnetoresistive phenomena in nanoscale magnetic contacts, in *Oxford Handbook of Nanoscience and Technology*, A. V. Narlikar and Y. Y. Fu (Eds.), Vol. 1, Oxford University Press, Oxford, U.K., pp. 677–718, 2010.
14. J. W. F. Egelhoff, L. Gan, H. Ettedgui, Y. Kadmon, C. J. Powell, P. J. Chen, A. J. Shapiro et al., Artifacts in ballistic magnetoresistance measurements (invited), *J. Appl. Phys.* **95**, 7554–7559 (2004).
15. R. I. Potter, Magnetoresistance anisotropy in ferromagnetic NiCu alloys, *Phys. Rev. B* **10**, 4626 (1974).
16. Z. K. Keane, L. H. Yu, and D. Natelson, Magnetoresistance of atomic-scale electromigrated nickel nanocontacts, *Appl. Phys. Lett.* **88**, 062514 (2006).
17. C. S. Yang, C. Zhang, J. Redepenning, and B. Doudin, Anisotropy magnetoresistance of quantum ballistic nickel nanocontacts, *J. Magn. Magn. Mater.* **286**, 186–189 (2005).

18. K. I. Bolotin, F. Kuemmeth, and D. C. Ralph, Anisotropic magnetoresistance and anisotropic tunneling magnetoresistance due to quantum interference in ferromagnetic metal break junctions, *Phys. Rev. Lett.* **97**, 127202 1–127202 4 (2006).
19. S. F. Shi and D. C. Ralph, Atomic motion in ferromagnetic break junctions, *Nat. Nanotechnol.* **2**, 522 (2007).
20. B. J. van Wees, H. van Houten, C. W. J. Beenakker et al., Quantized conductance of point contacts in a two-dimensional electron gas, *Phys. Rev. Lett.* **60**, 848–850 (1988).
21. H. Ohnishi, Y. Kondo, and K. Takayanagi, Quantized conductance through individual rows of suspended gold atoms, *Nature* **395**, 780–785 (1998).
22. L. Rokhinson, L. N. Pfeiffer, and K. W. West, Detection of spin polarization in quantum point contacts, *J. Phys.: Condens. Matter* **20**, 164212 (2008).
23. H. Imamura, N. Kobayashi, S. Takahashi, and S. Maekawa, Conductance quantization and magnetoresistance in magnetic point contacts, *Phys. Rev. Lett.* **84**, 1003–1005 (2000).
24. M. Y. Zhuravlev, E. Y. Tsymbal, S. S. Jaswal, A. V. Vedyayev, and B. Dieny, Spinblockade in ferromagnetic nanocontacts, *Appl. Phys. Lett.* **83**, 3534–3536 (2003).
25. M. Viret, S. Berger, M. Gabureac et al., Magnetoresistance through a single nickel atom, *Phys. Rev. B* **66**, 220401 (2002).
26. J. Velev, R. F. Sabirianov, S. S. Jaswal, and E. Y. Tsymbal, Ballistic anisotropic magnetoresistance, *Phys. Rev. Lett.* **94**, 127203 (2005).
27. M. Viret, M. Gabureac, F. Ott, C. Fermon, C. Barreteau, and R. Guirado-Lopez, Giant anisotropic magneto-resistance in ferromagnetic atomic contacts, *Eur. Phys. J. B* **51**, 1–5 (2006).
28. A. Sokolov, C. Zhang, E. Y. Tsymbal, J. Redepenning, and B. Doudin, Quantized magnetoresistance in atomic-size contacts, *Nat. Nanotechnol.* **2**, 522 (2007).
29. B. J. van Wees, L. P. Kouwenhoven, H. van Houten et al., Quantized conductance of magnetoelectric subbands in ballistic point contacts, *Phys. Rev. B.* **38**, 3625–3627 (1988).

29

Graphene Spintronics

Csaba Józsa
University of Groningen

Bart J. van Wees
University of Groningen

29.1 Introduction: Overview on Carbon Spintronics

Carbon-based electronics can be separated in two categories: The first category is the "plastic electronics," which is usually based on layers of oligomers or polymers, and carrier transport is via hopping between the localized electronic states situated on adjacent molecules. The second category is based on sp^2 conjugated pure carbon, such as fullerenes ("buckyballs"), nanotubes, graphite and, last but not least, graphene. The electronic states are in principle extended, but depending on the geometry (quantum) confinement can yield 0D, 1D, 2D, or 3D systems.

In recent years, the attention for spintronics in organic systems has grown considerably (see a review on this subject in Ref. [1]). Due to the low atomic number of carbon (Z=6), the spin–orbit interaction in organic molecules is usually very weak, leading to long spin relaxation times. In sp^2 conjugated carbon systems, however, carriers move with a high Fermi velocity of around 10^6 m/s, and it is not a priori clear that the simple argument from above can be used. The actual spin relaxation times and spin relaxation lengths are expected to depend on the structure (in particular, dimensionality) and extrinsic factors such as impurity potentials, defects, and edges.

Starting with 3D graphite, although spin injection and spin transport measurements on (bulk) graphite are lacking, electron spin resonance (ESR) has been used to obtain information on the g factor and spin relaxation times of the conduction electrons. From the narrow linewidths of only a few Gauss, spin relaxation times of 10–20 ns have been obtained [2,3].

Despite the achieved successes in observing spin transport in 1D carbon nanotubes over lengths of several micrometers [1], the situation remains unclear about the ultimate values of the spin relaxation times and spin relaxation lengths which can be reached in those systems. This is connected to the difficulty in making good quality ferromagnetic contacts and excluding spurious magnetoresistance effects, which can mask or mimic the required spin-valve behavior.

The technology to study 2D graphene as an isolated layer only became available in 2005 [4]. This triggered an avalanche of experimental and theoretical studies into (mostly) the electronic properties of graphene and graphene devices. Concerning spintronics, an obvious question is how the spin-related properties of 2D graphene compare to those of 3D graphite and 1D nanotubes. This chapter will discuss in detail the technological, experimental, and conceptual aspects of spin transport and spin dynamics in graphene field-effect transistors with ferromagnetic injector and detector contacts. The results will be linked to the available theory, and future perspectives will be discussed.

29.2 Graphene Electronics

Nature gives us elemental carbon in form of crystalline structures that are very different in many of their physical properties. The 3D variations of these (diamond, graphite) are well known

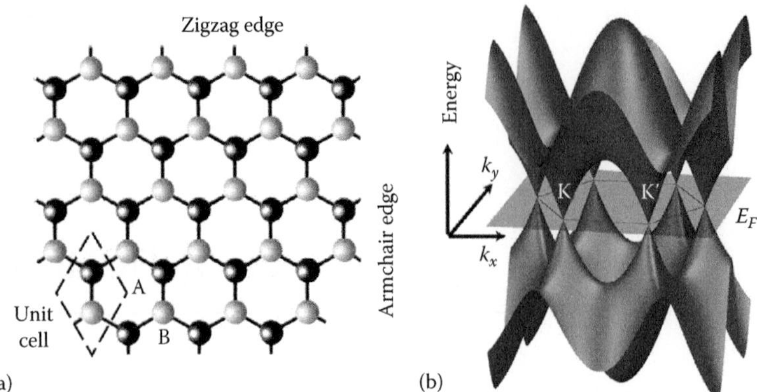

FIGURE 29.1 Graphene structure. (a) Hexagonal lattice composed of A and B sublattices. (b) Band diagram in the k-space with the valleys K, K′ and the Fermi level (image by the authors).

for a long time; 1D (nanotubes) and 0D (fullerenes) structures were discovered only some 20 years ago. One aspect that all carbon allotropes share is scientists' continuing interest in the electronic properties they display, both for fundamental research and for novel applications.

The theoretically most studied carbon allotrope is probably the 2D crystal called *graphene*, the building block of graphite and of nanotubes and fullerenes to some extent. Its hexagonal planar atomic structure (see Figure 29.1a) is a textbook example of a 2D crystal. It is simple enough to allow for band structure calculations and to be used as starting point for the more complex graphite (3D stacked graphene sheets) and carbon nanotubes (graphene sheets rolled up). In fact, its band structure was calculated already in 1947 [5] and can be described by the approximate dispersion relation in the k-space

$$E(k_x,k_y) = \pm t\sqrt{1+4\cos\left(\frac{\sqrt{3}}{2}k_xa\right)\cos\left(\frac{1}{2}k_ya\right)+4\cos^2\left(\frac{1}{2}k_ya\right)} \tag{29.1}$$

where

t represents the hopping integral

a is the lattice constant

As shown in Figure 29.1, the two atoms in the unit cell are predicted to yield two band crossings per Brillouin zone in the points marked as K and K′. The valence and conduction bands resulting from the bands of the two equivalent sublattices A and B do not overlap (as in metals) but only touch in the charge neutrality points resulting in a zero-gap semiconductor with a *linear dispersion* around K and K′ and two conical "valleys" where charge carriers can sit. The Fermi level (horizontal plane in Figure 29.1b) is, in the charge neutral case, situated exactly between the conduction and valence bands, that is, the Fermi surface consists only of the K, K′ points. On one hand the two valleys result in a new type of state for the electrons/holes in the graphene lattice, leading to a so-called valley degree of freedom. On the other hand, the linear dispersion resembles the relativistic Dirac fermions, for example, photons; charge carriers (electrons or holes) are moving with a (Fermi) velocity v_F independent of their energy and direction, obeying the Dirac equation $\hat{H}\begin{pmatrix}\Psi_A\\ \Psi_B\end{pmatrix} = E\begin{pmatrix}\Psi_A\\ \Psi_B\end{pmatrix}$ where the Hamiltonian $\hat{H} = v_F\begin{pmatrix}0 & \hat{p}_x + i\hat{p}_y\\ \hat{p}_x - i\hat{p}_y & 0\end{pmatrix} = v_F\vec{\sigma}\cdot\vec{p}$ acts on the wave functions $\Psi_{A,B}$ of the sites A and B, with $\vec{\sigma}$ the Pauli matrix and $\vec{p}$ the momentum. The relativistic behavior manifests in the high Fermi velocity ($v_F \approx 10^6$ m/s) and in the effective cyclotron (transverse) mass of the particles vanishing at the Dirac neutrality point. In the meanwhile, the longitudinal effective mass is $\pm\infty$, that is, the carriers cannot be accelerated or decelerated by applying a (e.g., electric) force.

In 2005, a simple but revolutionary technique was developed by Novoselov et al. at Manchester University to produce (by mechanical peeling of bulk graphite) single-layer graphene (SLG) flakes on Si/SiO_2 substrate on a micrometer-length scale, accessible experimentally [4].

An optical image of such single- and multilayer graphene flakes on SiO_2 (300 nm) substrate is shown in Figure 29.2a, accompanied by a scanning electron microscope image of the same sample contacted by metallic electrodes (Figure 29.2b). The first optically visible SLG flake (with the help of interference through a microscope) reported by the Manchester group started an avalanche of electrical transport experiments in this unique material. The existence of the valley-like dispersion relation was observed, and a minimum in the conductance of the graphene was measured (see Ref. [4] for single layer, respectively somewhat earlier, Ref. [6] for few layer graphene). This is illustrated in Figure 29.3 in a simple resistivity measurement on graphene using metallic contacts; the charge neutrality point is separating the metallic electron and hole conduction regimes.

The Fermi level can be raised or lowered by electrostatic gating to charge the graphene with electrons or holes at a carrier density n, creating a field-effect transistor type device. Charge carriers proved to have very high mobility μ above 10^4 cm^2/V s at room temperature (an order of magnitude larger than in Si). This

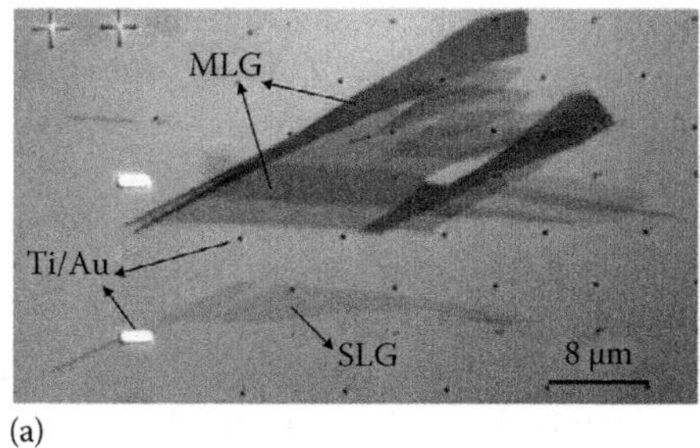

(a)

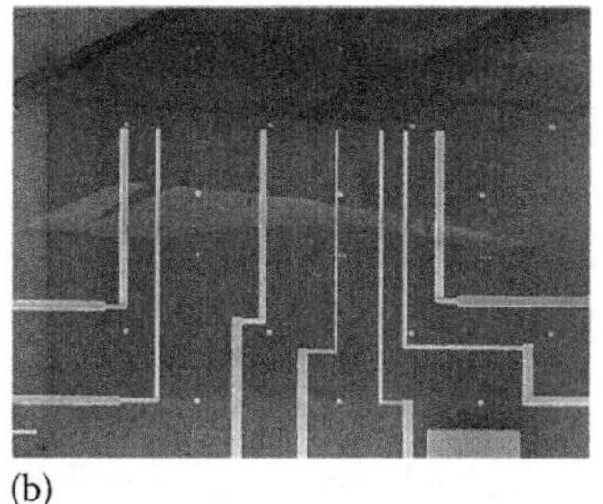
(b)

FIGURE 29.2 Single- and multilayer graphene (SLG, MLG) flakes on a Si/SiO_2 substrate fabricated at the University of Groningen. Optical microscope (a) shows the flakes and an array of Ti/Au markers. Same SLG flake contacted by cobalt electrodes (b), scanning electron microscope (SEM) image.

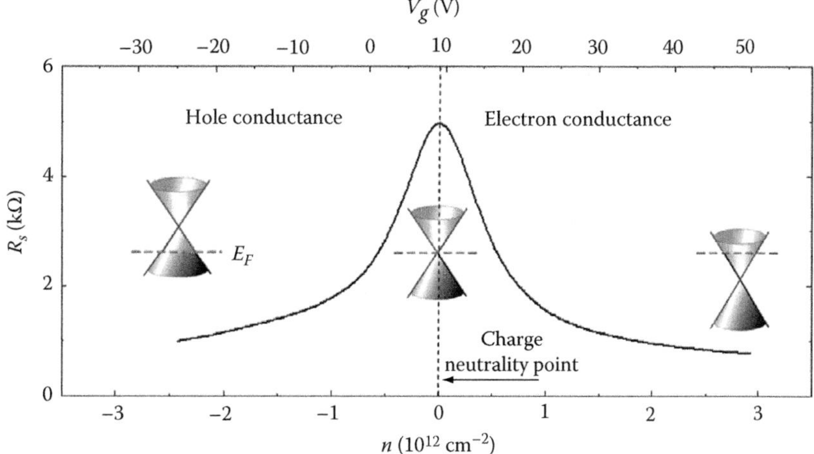

FIGURE 29.3 Sheet resistivity (square resistance) of a graphene flake measured against backgate voltage V_g (top scale). The induced charge carrier density n is shown on the bottom scale.

value, extracted from measurements of the conductivity σ with the formula $\sigma = en\mu$, was recently further increased to 2×10^5 cm^2/V s by suspending the graphene [7–9] thus removing scattering caused by the substrate and its charged impurities [10]. Using the semiclassical relation between mobility and scattering mean free path $\sigma = en\mu = 2e^2 l\sqrt{\pi n}/h$ [11], this yields a mean free path $l \approx 50$ nm for non-suspended samples and in the range of a micrometer for the suspended ones. Further, a quantized Hall effect was detected at 4.2 K and at room temperature; its anomalous nature (due to the existence of a zero-energy Landau level) is a direct proof of the Dirac fermionic nature of the charge carriers [12]. The absence of weak localization effect was also reported [13] and the existence of percolating electron–hole puddles in the charge neutrality regime was proposed that suppress the localization and limit the conductivity. The puddles were recently measured to yield a finite local carrier density of a few 10^{11} cm^{-2} using scanning single electron transistor technique [14] and scanning electron spectroscopy [15,16]. Bipolar supercurrent transport was also realized through graphene via the Josephson effect [17] using superconducting contacts. Relativity combined with quantum electronics of the charge carriers in an easy-to-produce 2D conductor makes the graphene a unique material to do fundamental research on. It presents richer physics than 1D (semi)conductors, for example, carbon nanotubes, but does this without the band complexities of a 3D system.

The fascination with graphene does not stop, however, at condensed matter physics. For instance, the movement of the Dirac fermions in curved graphene sheets seems to conveniently resemble the behavior of relativistic quantum particles in curved spaces, offering a realistic and unprecedented experimental environment for cosmologists [18].

From the applications point of view, the convenient position of the Fermi level, the selectivity of the carrier nature (electrons versus holes), very high room temperature mobilities, electric current densities above 10^8 A/cm^2 [8], and a conductance tunable by electrostatic gating are favorable features not to be met in metallic or semiconductor thin layers, and give us a hint on future electronic devices. Carbon replacing silicon in the micro- and nanoelectronics industry and single atomic thin devices is one of the (optimistic) predictions on what could be achieved in the next 50 years [19]. Additionally, the valve-type device proposed by Rycerz et al. [20] based on the selectively occupied two-valley states in graphene could lead to a new type of electronics called "valleytronics" by the authors. The very robust structure at elevated temperatures and the resistance of carbon to corrosive environments are also an important advantage. The resistivity of graphene being highly sensitive to adsorbed molecules (via charge transfer or dipole interactions), the possibility of gas- and single molecule sensing applications is being explored as well [21,22]. The reversible hydrogenation of graphene (transforming the material into *graphane*, a 2D insulator, see, e.g., Ref. [23]) or attaching other atoms and molecules is being considered as a way to create new quasi-2D crystals with designed electronic properties.

The bottleneck in using SLG on an industrial scale is, however, the mass fabrication of high (electronic) quality flakes. The "scotch tape method" (mechanical exfoliation of bulk graphite) has enough yield for doing fundamental research on, yet it is obviously inadequate for anything more. In the last years, an immense progress is to be seen from worldwide chemistry and surface science research labs, different ways are explored for graphene production on insulating, metallic, and arbitrary surfaces. One direction that shows promising results is to chemically exfoliate graphite by oxidizing it, and then disperse the obtained graphene oxide sheets in solvents that can be used to coat by Langmuir–Schaefer [24] or Langmuir–Blodgett [25–27] technique large surfaces with single or multilayer flakes. This is followed by chemical reduction to graphene and the advantage is the control over the thickness, the coverage ratio, and the choice of substrates; however, the electronic quality of the obtained layers (low carrier mobilities, low conductance) still needs improvement. Other researchers look into the direction of epitaxial graphene growth on silicon carbide via silicon sublimation [28,29] and chemical vapor deposition on metallic single- and polycrystalline surfaces (e.g., copper or nickel) with good lattice matching followed by chemical removal of the substrate and transfer to arbitrary surfaces [30–32]. Note that small-scale graphene growth on metals and on SiC has a long history; for a review of the early work, see Ref. [33].

29.3 Electron Spin Polarization in Single-Layer Graphene

We will discuss here the effects related to spin transport in graphene. For further background, we refer to Chapters 6 and 32 on metallic spintronics and the non-local measurement technique (this will be adopted for the particular case of graphene spin valves as we will explain it later). Relevant are also Chapters 16–18 that treat the spin injection and relaxation processes in(to) semiconductors as well as Chapter 23 on non-local effects in semiconductors. Finally, Chapter 22 on organic and Chapter 33 on molecular spintronics are recommended for understanding spintronics in graphitic devices.

Assuming the existence of a localized spin accumulation (spin imbalance) $\vec{n}_s$, proportional to the induced magnetization, in a non-spin-polarized sea of charge carriers in a graphene flake at position $x,y,z=0$ and at time $t=0$, created by one or another experimental method (e.g., electrical injection from a ferromagnetic contact), we can describe its spatiotemporal evolution in the diffusive limit applying basic concepts of spin transport [34]. First of all, the graphene will be considered 2D. A spin diffusion characterized by the diffusion coefficient D_s will lead to the spread of spin imbalance in the entire flake; this coefficient can be considered isotropic in the plane. Further, spin–orbit interactions will limit the lifetime of the spin imbalance (and by this, also the length scale on which the imbalance can diffuse); the scattering of spins on the local spin–orbit fields they experience is characterized by the spin relaxation time (or spin flip time) τ that includes both the spin lifetime and the dephasing time commonly denoted as $T1$ and $T2$, respectively. The characteristic length scale over which the spin imbalance can still be observed is the spin diffusion length (or spin flip length) $\lambda=\sqrt{D_s\tau}$ and it is isotropic in the (x,y) plane for an ideal graphene crystal. The presence of an external magnetic field $\vec{B}$ is accounted for by considering spin precession due to the interaction between the spins diffusing in the graphene plane and the perpendicular component of the magnetic field B_z, an interaction quantified by the gyromagnetic ratio $\gamma=g\mu_B/\hbar$ where the Landé factor is considered to be close to its free-electron value $g\approx 2$ [35]. Finally, an in-plane electric field $\vec{E}$ will lead to drift of the charge carriers of mobility μ in the graphene crystal with a velocity $\vec{v}_D=\mu\vec{E}$ parallel (holes) or antiparallel (electrons) with the field direction. This effect could not be observed in metallic spin transport, and it will add an asymmetric term to the diffusive motion of the spins as we will discuss it in detail in Section 29.5.4. (Note that this consideration is valid only for the high-density conduction regime. At the Dirac neutrality point, the mobility and, therefore the drift velocity are ill-defined.)

To summarize the above, we can write the Bloch equation for spin accumulation completed with the drift term under an electric field

$$\frac{\partial\vec{n}_s}{\partial t}D\nabla^2\vec{n}_s-\frac{\vec{n}_s}{\tau}+\left(\gamma\vec{B}\times\vec{n}_s\right)+\vec{v}_D\nabla\vec{n}_s \tag{29.2}$$

Here, $\vec{n}_s=(n_x,n_y,n_z)$ describes the spin accumulation (spin imbalance) in three directions in terms of densities for each spin. In experiments, the spin electrochemical potential $\vec{\mu}_S$ is what can be measured using, for example, ferromagnetic electrodes (see later in Section 29.5), and the two are correlated via the density of states at the Fermi energy obeying $\vec{\mu}_S=\nu(E_F)\vec{n}_s$.

Practically, what we will need further is to calculate the spin imbalance at a certain distance L from the injection point along direction x (experimentally, this will be the injector–detector distance), that is, to find the solution to the 1D Bloch equation in the presence or absence of magnetic and electric fields. In the simplest case of spin diffusion without external fields and a constant rate of spin injection (yielding a time-independent Bloch equation), the solution is

$$n_s(x)=A\exp\left(\frac{x}{\lambda}\right)+B\exp\left(\frac{-x}{\lambda}\right) \tag{29.3}$$

a symmetric exponential decay of the spin imbalance on both sides of the injection point. This is not specific to graphene and it was measured (in metallic spin valves) and explained in Johnson and Silsbee [36,37] and Jedema et al. [38]. In this chapter, we will only address the method to detect the spin injection and transport in graphene and model the results by the specific solutions of Equation 29.2 in multiterminal non-local geometries using ferromagnetic electrodes. The results will allow extracting parameters related both to the injection/detection of spins (the spin polarizations that play a role) and the spin transport in the graphene (the diffusion coefficient and the spin flip time and length). This will allow understanding the mechanisms behind the spin relaxation in graphene and give us a hint whether graphene-based spintronics is a realistic concept in modern nanotechnology.

29.4 Building Graphene-Based Multiterminal Spin Valves

Spin transport in graphene was measured at several research labs [39–47] with slightly different approach and results. One aspect that is common to all experiments reported so far is the electrical injection/detection scheme based on ferromagnetic electrodes. The first step is to obtain graphene flakes, single- or multilayers, on an insulating substrate. This is followed by lithography steps, metal deposition, and liftoff to create the electrodes on a scale of a micrometer or smaller, a well-known technology in nanofabrication. A scanning electron microscope (SEM) image of a lateral SLG spin-valve device is shown in Figure 29.4a.

29.4.1 Preparing the Substrate

Prior to graphene production, a substrate (typically a strongly doped Si wafer covered with 300–500 nm thermal oxide) is prepared. The oxide on the back of the wafer (if any) is removed and a thick metallic layer, for example, 100 nm Ti/Au is deposited that can be used as electrode to the Si. The Si serves this way as a backgate for electrostatic charging of the graphene with holes or electrons. Further, an array of 30–50 nm thick Ti/Au markers is deposited on the silicon oxide by means of lithography and liftoff (see bright square in Figure 29.4a; the method is explained in the next paragraphs) that will help locating the graphene flakes later on. These steps result in light contamination of the oxide surface, therefore a cleaning can be performed by reactive etching in oxygen plasma [48], annealing in argon/hydrogen gas flow at 200°C–400°C [49], ultrasonication, and/or wet chemical cleaning.

29.4.2 Graphene Production in the Lab

In terms of electrical transport measurements, the most studied SLG is that obtained by the exfoliation technique. A simple scotch tape is used to peel off layers from a bulk graphite source. This can be natural graphite, highly oriented pyrolytic graphite (HOPG) or Kish graphite. The slightly different electronic properties of the graphene flakes from different sources are most probably related to the contaminations present in the bulk graphite and the crystalline quality of the graphene flakes that form the bulk. There is no consensus yet whether spin lifetimes and diffusion depend on the type of graphite used. Once a clean, fresh graphite surface is obtained, this is pressed against the prepared substrate. Upon removal of the scotch tape (by peeling off or dissolving its glue in an organic solvent), thin graphite flakes are left behind on the substrate, of lateral dimensions of micrometers up to a millimeter (see Figure 29.2a for an example). These are weakly attached to the oxide surface by van der Waals interaction, the strength of the interaction depending mainly on the cleanness and hydrophobicity of the surface. The thickness of the flakes (i.e., the number of graphene layers stacked) varies from 0.3–0.5 nm to μm, therefore a rigorous inspection is needed to select the single-layer samples. Two important steps are involved here. First, the number of single to few layer flakes is relatively high compared to thicker graphite pieces. The reason behind this is most probably due to the interplay of adhesion and cohesion forces [50]. Nevertheless, the flakes are only weakly attached to the substrates; detaching of large flakes (of above 100 μm² area) when dipped into solvents with high surface tension is a common problem and leads to loss of samples during

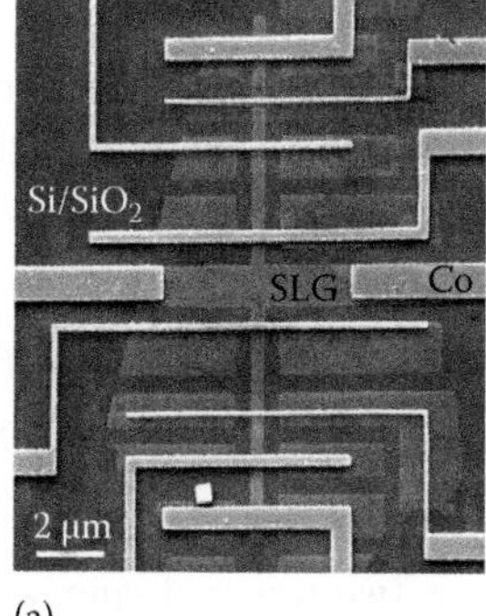

(a)

(b) Without Al_2O_3

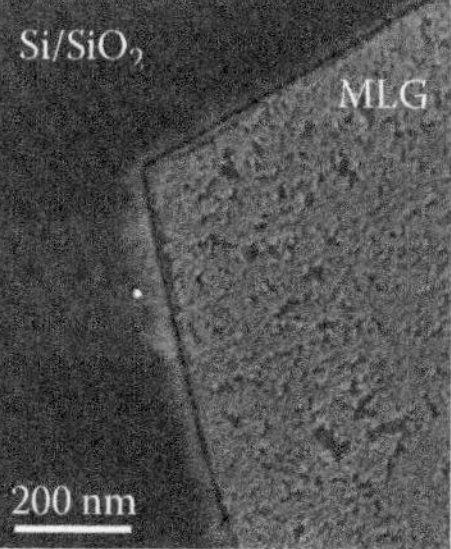

With Al_2O_3

FIGURE 29.4 SEM images: (a) Cross-shaped graphene device etched with oxygen plasma from a large polygonal flake and contacted by Co electrodes. The etch channels are visible. (b) Comparison of graphene and SiO_2 surfaces bare (left) and with the aluminum oxide cover layer (right). The two images were made using the same SEM parameters on the same sample in two different areas. (Images by the authors.)

further device fabrication steps [51]. The second surprising fact is, that a single layer of carbon atoms can be made visible under an optical microscope. For certain light wavelengths, an interference occurs between the graphene and the opaque Si substrate, provided that the SiO_2 layer is of the right thickness. The effect decreases the reflection of light from the graphene/SiO_2/Si surface giving a relatively high contrast with the regions where no graphene is present. A detailed quantitative explanation for a 285 nm thick oxide is given by Blake et al. [52] and, for a range of oxide thicknesses by Tombros [53]. The preselecting can thus be done by selecting the least visible graphene flakes on the surface. The technique can be quantified by recording 8 bit grayscale digital optical images of the surface that contains flakes of different reflectivity, measuring the brightness value on the flakes Lg as well as on the substrate in the immediate vicinity of the flakes Ls, and calculating a contrast value $c = (Ls - Lg)/Ls$. Groups of discrete contrast levels can be identified this way starting from a few percent and increasing with increased thickness of the flakes. Care must be taken that the images are recorded with the same parameters (microscope aperture and magnification, exposure, wavelength of light and contrast). Following this preselecting, typical flakes from the groups are inspected by Raman spectroscopy to determine the number of layers based on the shift and width of the 2D and G peaks [54], allowing finally for a calibration of the optical microscope selection. Another method to show that a flake is SLG is to measure the half-integer quantum Hall effect that indicates the presence of Dirac fermions in the 2D lattice; however, this is cumbersome to do for all samples. Additionally, atomic force microscope (AFM) can be used to scan the surface of the graphene flakes; this serves the purpose of accurate localization of the flake and verification of its integrity and cleanness. AFM is also used to measure the thickness of the flake determining this way the number of layers, a method much debated since adsorbed contamination (e.g., water molecules) on or under the graphene flake can give erroneous thickness values for single layers [55]. Ideally, an SLG measures 0.3–0.5 nm thick.

The shape and lateral dimensions of the graphene flake can be controlled if necessary by plasma etching down to a 50 nm scale (see in Refs. [48,56–58]) or etching on a sub-10 nm scale with thermally activated nickel nanoparticles [59]. Recently, controlled rupture of graphene involving heating–cooling cycles was also demonstrated [60] as a method to obtain graphene quantum dots of approximately 10 nm diameter. For spintronics experiments, the RF plasma etching method was employed, using argon, oxygen, or a mixture of the two gases. At 10 microbar base pressure, a flow of approximately 10 sccm of gas is ignited to form a plasma that etches graphene with a typical rate of around 0.1 nm/s. To protect parts of the graphene flake, an electron beam lithography (EBL) step is involved as follows. A thin (~100 nm) polymer resist layer is spun on the Si wafer with the graphene on it [e.g., poly(methyl methacrylate), PMMA, dissolved in *n*-amyl-acetate] and baked to form a continuous protective layer. The regions to be etched are first exposed by focused electron beam and the damaged polymer is removed using methyl isobutyl ketone solvent ("development" step). This opens a gap in the protective polymer layer allowing for the plasma to remove the parts of the graphene with an accuracy of about 50 nm. Finally, the rest of the polymer is removed by a hot acetone bath. In Figure 29.4a, an SLG etched with oxygen plasma into a cross shape is visible; the original shape of the flake can be seen from the leftover pieces (gray) on the dark Si/SiO_2 substrate.

29.4.3 Fabrication of Electrical Contacts

Next, a thin aluminum oxide (Al_2O_3) layer is created that covers the graphene flake completely or partially. The need for this will be explained in Section 29.5.2. This is done by thermal evaporation of 0.6 nm thick aluminum in an ultrahigh vacuum system with the substrate being liquid nitrogen cooled. The aluminum is then oxidized for 30–60 min in an oxygen atmosphere of about 100 mbar. The resulting oxide layer shows a roughness of 0.5–0.7 nm on the graphene surface, determined by AFM; the granular nature can be seen upon inspection with scanning electron microscope (SEM). In Figure 29.4b, two high magnification SEM images are shown, made with identical parameters on the same sample. The image on the left shows a region where no Al was deposited (due to shadowing in the evaporation system); the one on the right in comparison, illustrates the presence of the aluminum oxide both on the multilayer graphene (MLG) and on the silicon oxide.

The sequence of graphene etching and creating the aluminum oxide layer can be interchanged; this seems to decrease the chance of the graphene flake detaching from the substrate in further processing steps.

Finally, the ferromagnetic contacts are created using another step of EBL and liftoff. This process is similar to the etching described above, except the development step is followed by thermal or electron beam evaporation of a 25–40 nm thick Co layer onto the sample. The areas where the polymer is removed by development will be the contact areas where the Co grows directly on the graphene/Al_2O_3. The rest of the metallic layer is removed in acetone together with the unexposed polymer ("liftoff" step). The resolution of the technique allows creating contacts of lateral sizes below 50 nm with similar positioning accuracy. Optical lithography can also be used to speed up the fabrication process; however, its resolution is limited to a (few) hundred nanometers by the wavelength of light used and the precision of the optical mask.

29.5 Spin Transport Experiments in Lateral Graphene Spin Valves

The simplest lateral device geometry to detect spin transport electrically is a two-terminal spin valve where the graphene is contacted by two ferromagnetic electrodes [34]. The electrodes are fabricated to have different coercive fields (e.g., by shape anisotropy that varies with the width) and therefore their magnetization can be oriented parallel or antiparallel with an

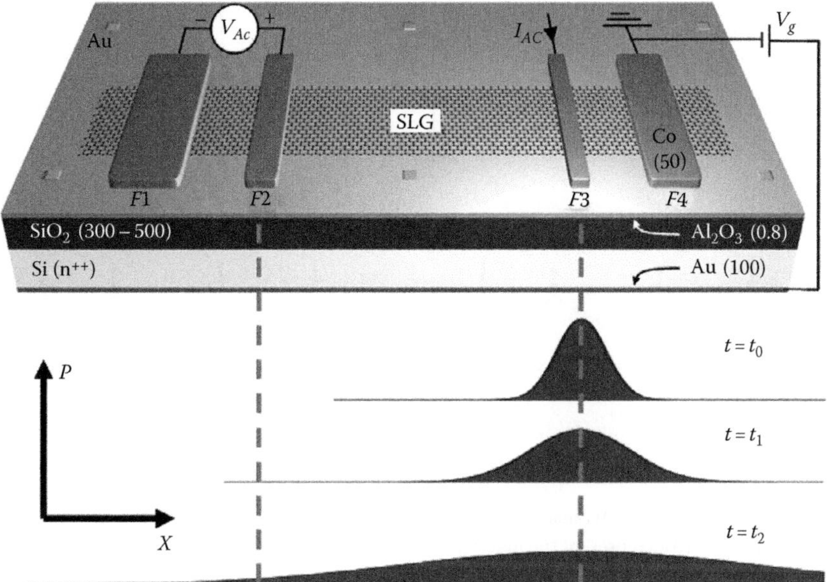

FIGURE 29.5 Non-local four-terminal spin injection geometry. *P* represents the injected spin polarization. See text for details.

external in-plane magnetic field. Such devices were built and studied by Hill et al. [39], using transparent permalloy contacts on the graphene. Magnetoresistances of a few hundred ohms were observed over an injector–detector distance of 200 nm, yet no clear distinction could be made between parallel and antiparallel configurations and the presence of spurious effects such as anisotropic magnetoresistance or Hall effect cannot be excluded.

Separating the charge current path from the voltage detection circuit and relying solely on the diffusion of spins through the device will circumvent the above problems. This was successfully performed in a four-terminal non-local geometry on metallic spin valves [36,38], semiconductors [61], carbon nanotubes [62], and also on graphene [40,44]. The non-local spin injection–detection circuit and mechanism is explained with the help of Figure 29.5.

A spin-polarized AC charge current I_{AC} is sent from the ferromagnetic electrode into the graphene layer through contact F3 and extracted on the far right edge of the device through F4, creating a spin accumulation *P* in the graphene region around F3. The spin accumulation diffuses through the graphene on both sides of the injector F3, not being tied to the charge current path. Placing a pair of ferromagnetic voltage probes F1, F2 on the left-hand side where F2 is close to the injector F3 (at a distance $L < \lambda$, the spin diffusion length), this will be sensitive to an accumulation of spins parallel to its magnetic orientation. Since F1 is further, it will detect a lower spin accumulation if any, resulting in a voltage difference $\Delta V = V_+ - V_-$ measured between F2 and F1 without any charge current passing through this region. Switching the magnetization direction of the injector F3 or detector F2 by sweeping a magnetic field oriented parallel/antiparallel with the electrodes' long axes (axis *y* in our geometry) will change the measured voltage from positive (parallel magnetizations) to negative (antiparallel magnetizations) values. A non-local resistance can be defined as $R_{non\text{-}local} = (V_+ - V_-)/I_{AC}$. Additionally, if the outer electrodes F1, F4 are also ferromagnetic and within the distance of a few spin diffusion lengths, they will also contribute to the measured voltage by injecting or detecting spin polarization. The idea is illustrated in Figure 29.6 using the spin electrochemical potential $\mu_S = \mu_\uparrow - \mu_\downarrow$; the potentials are drawn separately for the spin-up ($\mu_\uparrow$) and spin-down ($\mu_\downarrow$) channels against the *x*-axis, normalized to the average.

In Figure 29.6a, we see the starting situation with all ferromagnetic electrodes oriented parallel, in the "up" direction. (A spin-valve measurement corresponding to the situation we will describe here is presented in Figure 29.7b, however the first two steps—the switching of electrodes F4 and F1—are indiscernible in the measurement.) At moment t_0, the electrode F3 injects a spin-polarized current with the majority spin oriented "up," raising therefore the electrochemical potential for the spin-up channel (solid line) and creating a spin-up accumulation of electrons in the graphene (with the corresponding deficit of spin-down electrons) that decays exponentially along *x*. In the same time, electrode F4 extracts spin-up carriers, that is, creates a spin-down accumulation, rising μ for the spin-down channel (dotted line) and reducing the overall spin-up accumulation in the graphene. At distances $L \gg \lambda$ from the injectors, the spin-up and spin-down electrochemical potentials are overlapping (spin imbalance is decayed, $\mu_S \approx 0$). The detectors F2 and F1 are both sensitive to the spin-up channel (see dots) but they probe this at different distances from the injectors, therefore a positive voltage difference ΔV and a positive non-local resistance will be measured between them.

Switching the magnetization of the injector F4 (Figure 29.6b) changes the spin imbalance in the device: F4 will extract now spin-down carriers, meaning the spin-up accumulation generated by F3 will be enhanced. As a result, the $\mu_\uparrow$ curve will be slightly

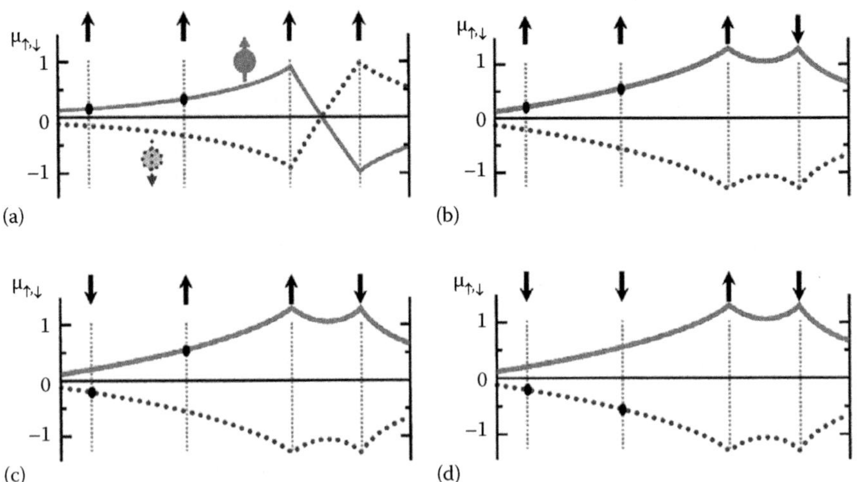

FIGURE 29.6 Electrochemical potential plotted against the x-axis for the spin-up (solid line) and spin-down channel (dashed line) in a four-terminal spin-valve device. The vertical arrows represent the magnetic orientation of the four ferromagnetic electrodes. These are switched one by one through steps (a–d).

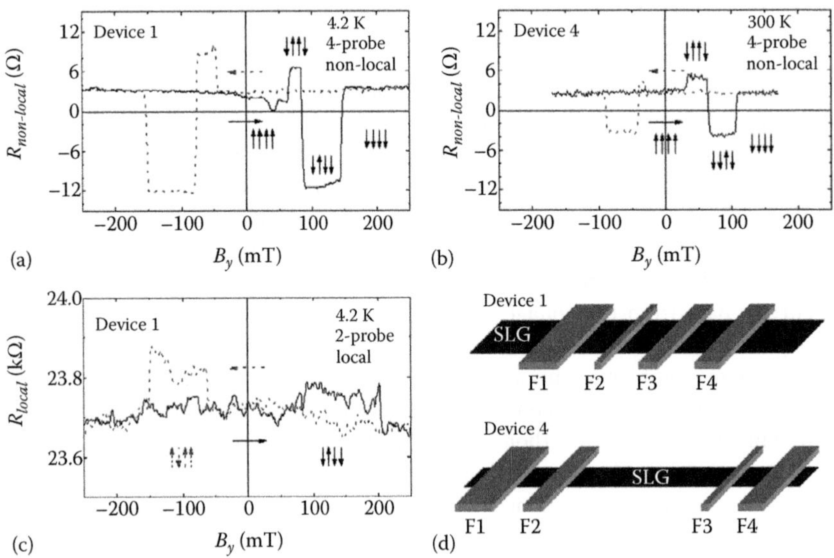

FIGURE 29.7 Spin-valve measurements on SLG in the four-terminal non-local geometry at low temperature (a) and room temperature (b) and in the two-terminal local geometry at low temperature (c). The magnetic field is swept as indicated by the horizontal arrows; the vertical arrows indicate the parallel/antiparallel combination of the electrodes' magnetizations. On (d) a sketch of the devices is drawn.

lifted up and the difference detected by F2 and F1 will increase. Switching the magnetization of the outer detector F1 will result in the situation depicted in Figure 29.6c: F1 will probe the spin-down channel, therefore the voltage difference and corresponding non-local resistance will further increase. The next step is reorienting the inner voltage probe F2's magnetization to "down" (panel d). This will set the inner injector/detector magnetizations antiparallel. Both detectors are sensitive to the spin-down channel now, yielding $\Delta V < 0$ (the non-local resistance drops to negative values). The last step is to switch the magnetization of F3 "down"; this gives an all-parallel situation equivalent to the starting configuration with $\mu_\uparrow$ and $\mu_\downarrow$ interchanged (not shown). Therefore, the measured signal returns to the (positive) starting value.

The magnetic switching order of the four contacts is not necessarily the one that we discussed here. In such experiments, it is governed by the width of the ferromagnetic electrodes as introduced earlier for the two-terminal case. The steps in the non-local resistance value, however, can always be traced back to the switching of individual electrodes through simple schematic representations as in Figure 29.6.

29.5.1 Injection/Detection of Spin-Polarized Currents

The first set of non-local spin-valve measurements done on graphene-based spin valves at low and room temperatures was presented in 2007 by Tombros et al. and reproduced since by several groups. The examples we show here are measured in the metallic conduction regime (carrier densities in the 10^{12} cm^{-2} range); however, it was shown [40,44,48,58] that the spin valves behave very similarly around the Dirac neutrality point with only a decrease of the signal amplitude. The behavior of spin injection and transport parameters in a broad density range will be discussed in Section 29.6. In Figure 29.7a, we show a spin-valve measurement at 4.2 K on a 1.4 μm wide graphene flake with uniform electrode spacing of 330 nm. A bipolar behavior and three resistance steps can be clearly identified corresponding to magnetic switches of three electrodes. The magnetic field is swept in the direction indicated by the horizontal arrows; the switching order of the electrodes is shown by the four vertical arrows for each resistance value. Non-local differential resistances up to 20 Ω were detected using AC currents of 100 nA–5 μA. Compared to the local two-terminal measurements on the same sample (Figure 29.7c) the non-local signal is considerably smaller, however the signal-to-noise ratio is much better due to the absence of a large background resistance (the resistance of the contacts and the graphene flake itself are always included in two-terminal measurements).

A similar non-local spin-valve measurement is shown in Figure 29.7b (same scale as a); this data is obtained, however, at room temperature on a different device with 3 μm spacing between F2 and F3. (The devices are sensitive to temperature changes and therefore it proves to be very difficult to measure the same device in a broad temperature range.) The shape of the $R_{non\text{-}local}$ versus B_y curve looks similar; again, three resistance steps are clearly distinguishable. Compared to the low-temperature measurements, we see a slight increase in the noise and a reduced non-local signal due also to the larger injector–detector spacing. The observation of the resistance steps for a device where the electrode spacing is in the 3–5 μm range indicates a spin diffusion length of several micrometers. At room temperature, this value is very high, almost an order of magnitude above metallic spin valves, and carries the potential of spintronic applications where it is crucial to conserve spin information over micrometers (see Part V of this book and especially Chapter 39: "Spin Logic Devices" by H. Dery).

A 3D model for the non-local resistivity measurements was given by Jedema et al. [63] for metallic spin valves. We modify this description for use in 2D on our graphene spin-valve devices and obtain the formula

$$R_{non\text{-}local} = \frac{P^2\lambda}{2W\sigma}\exp\left(-\frac{L}{\lambda}\right) \tag{29.4}$$

where

- W is the width of the graphene
- σ is its conductivity ($\sigma = 1/R_S$, to be determined from resistivity measurements similar to Figure 29.3)
- L is the distance between the central injector–detector electrodes
- P is the spin polarization of the contacts

The exponential dependence on L originates from the exponential decay of the spin imbalance as we move away from the injection point (Equation 29.3). The non-local resistance is also called *spin signal*; the spin signal difference between the parallel- and antiparallel geometries is the *spin-valve signal* and, ideally this is twice as large since the spin-valve measurement should be symmetric around zero. One way to determine the spin diffusion length λ is to verify the length dependence of the non-local signal. Several spin-valve measurements are needed on a set of samples with increasing injector–detector distances L while the rest of the parameters in Equation 29.4 are kept the same. This was attempted [40,46,48] yet it is very difficult to have the polarization P unchanged from sample to sample, even from contact to contact. The results indicate a spin diffusion length between 1 and 2 μm. Measuring the graphene's conductance σ separately, we can calculate a contact spin polarization $P \approx 10\%$ (both for injector and detector, therefore it appears squared in Equation 29.4). For a precise determination of the spin diffusion length and other spin transport parameters, one can use a more convenient technique based on the Hanle precession of the injected spins; this is related to the diffusive spin transport in graphene and thus it will be described later in Section 29.5.3.

The value of P is related to the nature of the ferromagnetic electrodes (the intrinsic spin polarization of the material) as well as to the nature of the contact between the electrode and graphene, and it is a measure of how efficiently one can inject a spin polarization into the graphene. In the following paragraphs, we will present a method to manipulate the spin injection efficiency into graphene.

29.5.2 Efficiency of Spin Injection

In the diffusive transport regime, direct electrical spin injection from a ferromagnetic metal into high-impedance materials like graphene is hindered by the conductance mismatch problem. The concept of conductivity mismatch is discussed in Schmidt et al. [64], Rashba [65], and Fert and Jaffrès [66]. In simple terms, the spin accumulation created at a clean ferromagnet/graphene interface favors flowing back into the low-impedance metal injector instead of diffusing through the high-impedance graphene, therefore the efficiency of the injection is strongly reduced in such experiments [39,41]. A 100% spin polarized injector would eliminate this problem; however, the intrinsic spin polarization of ferromagnetic metals is typically below 50% as measured in tunneling experiments by Tedrow and Meservey [67]. Another solution to improve efficiency is to reduce the contact area to ultimately a point contact [68]; this was implemented by Han et al. yielding clean, measurable non-local spin signals in the milliohm range and a polarization value $P \approx 1\%$ [46]. Yet another solution is the use of a tunnel barrier between the ferromagnet and graphene, increasing the spin-dependent interface resistance and by this combating the conductivity mismatch [65,66]. Such a technique was successfully employed for spin injection into n-doped GaAs by means of a spin-sensitive Schottky tunnel barrier [61]. The role of the barrier in the graphene spin-valve devices we present here is taken by the 0.8 nm thick aluminum oxide layer that was mentioned earlier in Section 29.4.3. The high contact resistance limits the back-diffusion of spins into the injectors and allows the spin accumulation to pass underneath the opaque contacts, however the barrier should be transparent enough to result in measurable (<1 MΩ) resistances. As we have calculated, this gives contact spin polarizations around 10%. Experimentally, it was found that contact resistances from 50 to 100 kΩ result in the highest efficiencies of spin injection, up to 18% [48,69]. Recently successful experiments were reported [105 on SCG, respectively 106 and 109 on BLG] by using oxides other than Al_2O_3 (e.g., MgO, TiO_2-seeded MgO) that yield even higher spin polarizations at the interface reaching 26%–30% and, consequently, record levels of non-local spin valve signals up to 130 Ohms.

A model for spin injection/detection in non-local ferromagnet–barrier–graphene–barrier–ferromagnet structures is constructed [48] that accounts for the effect of the conductivity mismatch in case of semitransparent contacts. It is shown that the magnitude of the spin accumulation created in graphene is drastically reduced if the contact resistance is smaller than the graphene resistance calculated over one spin relaxation length; the formula for the non-local resistance (29.4) is modified to

$$R_{non\text{-}local} = \pm \frac{2P^2\lambda}{W\sigma} \frac{(R/\lambda)^2 \exp(-L/\lambda)}{(1+2R/\lambda)^2 - \exp(-2L/\lambda)} \quad (29.5)$$

with the parameter $R = \sigma R_C W$ representing the spin relaxation due to the finite contact resistances R_C of the electrodes. Additionally to the prediction of a reduced non-local spin-valve signal, the model also shows how its exponential dependence on L from formula (29.4) becomes a $1/L$ type dependence for transparent contacts ($R < 10^{-6}$ m) and injector–detector distances $L \leq \lambda$.

The 0.8 nm thick oxide layer we employ is not a uniform tunnel barrier but an insulating layer with some relatively transparent regions on the nanometer scale. This is indicated, besides the AFM and SEM studies, both by the spread in the contact resistances within one sample, as well as by the I–V characteristics that do not show a tunneling behavior [48]. However, the model yielding Equation 29.5 is valid both for tunnel barriers and for pinholes in the oxide.

The interface injection/detection efficiency can be manipulated in a finalized spintronic device by electrical bias on the contacts. A study of the interplay of mechanisms behind electrical spin *detection* in lateral Fe/GaAs structures (again, using Schottky tunnel barriers) with biased *detector* contacts was recently reported [70]. The authors demonstrated the bias dependence of the tunneling spin polarization (interface effect) and of the spin transport in the GaAs before the spins reach the detection point (bulk effect). Simultaneously, a set of non-local experiments were done on graphene spin valves by Józsa et al. (with the 0.8 nm Al_2O_3 interface layer) where the *injector* electrodes were DC biased showing a change in the room temperature spin *injection* efficiency from 18% to 31% [69]. This resulted in non-local spin-valve signals up to 85 Ω. When a reverse DC bias was applied, the same device showed zero or even negative differential values for R_{SV} (the spin-valve signal reversed). A selection of these measurements is plotted in Figure 29.8a, for a variety of current biases between −5 and +5 μA. The bipolar non-local spin-valve effect is visible on all curves, and the resistance steps are clear enough to identify the switching of the two inner

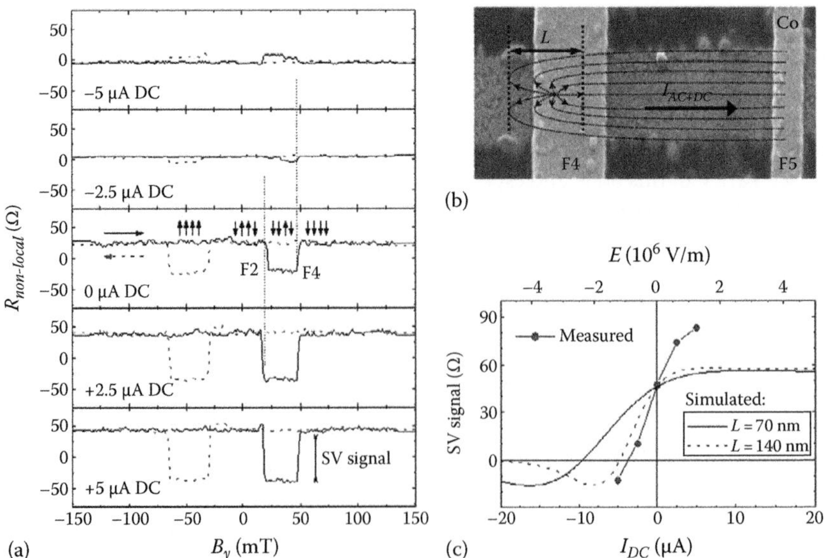

FIGURE 29.8 Spin-valve measurements with DC-biased current injector electrodes F4, F5 (a), the bias current being indicated for each measurement. The contacts are F1, F2, F4, F5; for these measurements F3 is not in the circuit. Schematic drawing of the local drift effect (b) using a SEM image of a sample. The change in the spin-valve signal is plotted on (c) against the DC bias values (bottom scale), together with simulated curves for two different values of L against the electric field (top scale). See text for details.

ferromagnetic electrodes F2, F4 from parallel to antiparallel orientation and back.

The spin-valve signals (parallel minus antiparallel) measured by the voltage probes F1, F2 are plotted in Figure 29.8c against the applied DC bias. A steep rising curve is formed, with saturation starting both for positive and negative high biases; however further increase of the current bias was not possible due to danger of permanent damage on the graphene/oxide/cobalt interface.

The physics behind the measured biasing effect can be understood considering the relatively transparent injection regions in the oxide layer. In graphene directly around these injection points, a strong current spreading is present, on a characteristic length scale **L** (see Figure 29.8b). Here, the local electric field is large, and the carriers will undergo a strong local drift. The carriers drifting away from the injection point (for a positive DC bias and p-type graphene) reduce the backflow of the spins and thus facilitate a constant injection of high spin polarization through the interface. The upper limit of the effect is a measurement of the intrinsic spin polarization P possible to inject from the ferromagnetic electrodes, when conductivity mismatch is eliminated; the signal is reduced of course by the further diffusion and relaxation toward the contacts. On the other hand, an opposite electric field polarity (or carrier type) will result in a carrier drift toward the injection point keeping the local spin polarization high, reducing the efficiency of injection. A strong negative bias enhances the conductivity mismatch problem to a point when the drift starts to dominate the spin transport. In AC measurements, this yields a *negative differential resistance* (lower spin injection for stronger electric fields), and thus the reversed spin-valve signals in Figure 29.8 (top left). Finally, above a threshold electric field value, the strong drift effect is expected to prevent any spin imbalance to reach the detectors reducing the spin-valve signal to zero.

The complicated 2D current spreading around the injection point is modeled in Józsa et al. [69] by an approximation with a 1D drift-diffusion of the spins on the drift length scale **L** along axis x. The spin current density flowing through the region due to diffusion and drift can be written as

$$j_s(x) = -D\frac{dn_s(x)}{dx} + v_D n_s(x) \tag{29.6}$$

using the same diffusion and drift parameters as in Equation 29.2. The coupling at the edges of the drift region yields boundary conditions that on one hand involve the injection parameters such as contact resistance and spin polarization of the current source; on the other hand, assure the spin current continuity into the diffusion region, toward the detector electrodes. The model predicts a spin-valve signal on the detectors as plotted in Figure 29.8 right, solid line (drift region with **L** = 70 nm as parameter) and dashed line (**L** = 140 nm). The horizontal axis of the graph (the electric field, top scale) is connected to the measurement horizontal scale (the DC bias current, bottom scale) by the linear relationship $E = I_{DC}R_S/w$ where the graphene resistivity R_S is known but the total width of the injection regions w is not. The vertical scales are the same. Comparing quantitatively the simulations and the measurements, we conclude that electric fields in the $\pm 10^6$ V/m range must be present in the drift region, yielding a drift velocity up to 0.25×10^6 m/s. The width of the injection regions $w < 5$ nm is in this case consistent with the values based on the structure of the oxide layer.

The local drift diffusion in the graphene region around the injection points is similar to the spin drift-diffusion phenomenon on a larger scale, observed in experiments where an electric field is present in the whole graphene layer. This will be discussed in Section 29.5.4.

29.5.3 Hanle Spin Precession

In the simple four-terminal non-local geometry, injection of the spin imbalance is followed by diffusion and relaxation through the graphene layer. The spin diffusion coefficient D_s and the spin flip time τ are linked via the spin diffusion length $\lambda = \sqrt{D_s\tau}$ and they cannot be determined individually from simple spin-valve experiments (see Equations 29.4 and 29.5). An experiment that allows to do so is the Hanle spin precession experiment [71]; additionally, it also represents a solid proof for spin injection and transport in lateral geometries [61]. The Hanle spin precession is simply the Larmor precession of the injected spin imbalance in a lateral spin valve, under the influence of a (small) perpendicular external magnetic field B_z. An illustration is given on the Figure 29.9 insets. For magnetic fields that are well below the demagnetization field of the injector/detector electrodes F2, F3(left inset) the spin injection is still predominantly in-plane and the spin imbalance, while diffusing through the graphene plane, undergoes a precession and dephasing that is governed by the cross-product term in the Bloch equation (29.2). The detector signal will, therefore, depend on the starting situation (parallel or antiparallel injector–detector electrode magnetizations), the strength of B_z, the injection efficiency (see previous section), the spin dephasing time τ, the spin diffusion coefficient D_s, and the injector–detector distance L. As the magnetic field is increased, the precession angle achieved over the distance L will increase and the signal will present an oscillatory behavior that is dampened due to dephasing and spin relaxation. Further increasing the magnetic field will cant the magnetizations of the electrodes out of the sample plane and, finally align them with the z-axis. In this situation (see right inset in Figure 29.9), the injected spin polarization is out of plane along z; therefore no precession will occur and the detector signal will saturate at a positive value that depends on the *out-of-plane* spin relaxation time $\tau_\perp$.

Two room temperature Hanle precession measurements (non-local resistance versus orthogonal magnetic field) are plotted in Figure 29.9 in a graphene spin valve with $L = 3.1\,\mu$m and carrier density $n \approx 5 \times 10^{11}$ cm^{-2} (holes), one for parallel configuration of the F2, F3 magnetizations (full circles) and another one for antiparallel (open circles). The signal difference between the two curves at $B_z = 0$ corresponds to the spin-valve signal measured with in-plane magnetic field. The parallel configuration

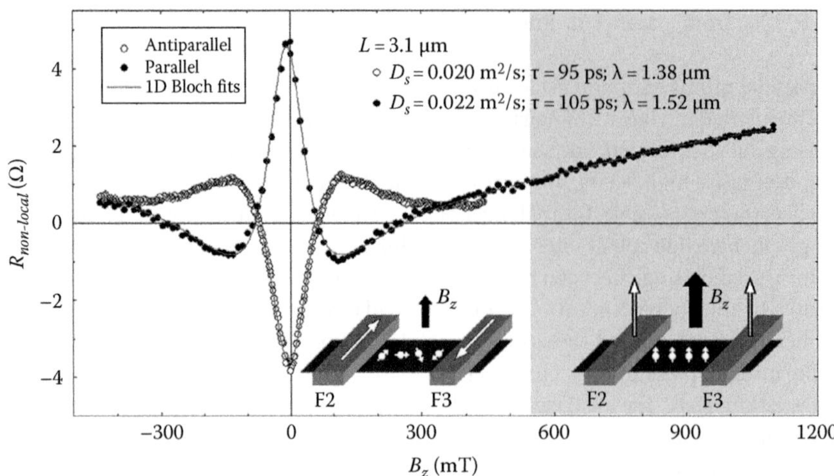

FIGURE 29.9 Hanle precession measurements for low and high magnetic fields perpendicular to the sample plane for $L = 3.1\,\mu$m injector–detector distance. The geometry is indicated in the insets. The parallel and antiparallel configuration measurements are fitted with the Bloch equation (29.2) to extract the spin diffusion coefficient D_s, relaxation time τ, and diffusion length λ.

measurement is continued for high magnetic fields up to +1.1 T (the measurement is symmetric around the y-axis) and the onset of the signal saturation can be seen.

The curves are fitted with the solutions to the Bloch equation (29.2) (solid lines in Figure 29.9) and the spin transport parameters are extracted. Aside of a small difference between the curves in the parallel and antiparallel geometry, the fit yields a spin relaxation time $\tau \approx 100\,$ps, a diffusion coefficient $D_S \approx 0.02\,\text{m}^2/\text{s}$ and a spin diffusion length $\lambda \approx 1.4\,\mu$m. The value, confirmed by measurements on many similar graphene samples [40,46–48,69], corresponds to the spin diffusion length range (1–2 μm) extracted from measurements of the spin-valve signal against L we have mentioned earlier. This also indicates that the longitudinal and transverse electron spin relaxation times are the same in graphene as expected ($T_1 = T_2 \equiv \tau$, see Ref. [72]).

An extensive spin precession study was done in the high magnetic field range [73] to determine whether the spin-valve signal saturation value at high field is the same as the zero-field value, that is, is the spin relaxation isotropic in graphene (as it is in metals, see Ref. [38]) or not. For a 2D system where spin–orbit fields in plane (Rashba or Dresselhaus type spin–orbit interactions) dominate the spin relaxation a strong anisotropy is expected up to 50% [72]. The experiments on graphene indicate that $\tau_\perp \approx 0.8\tau_\parallel$; the dominating spin relaxation mechanism seems to be of Elliot–Yafet type. More conclusive studies on this subject will be presented in Section 29.6.

An interesting effect was predicted for spin precession in a clean graphene sheet at the Dirac neutrality point: the equivalent of the Hartman effect (apparent superluminality) in optics. The dynamics of carriers at the neutrality point is "pseudodiffusive," and the authors calculate that the transmission time of spins from injector to detector (distance L) is $\tau = 0.87L/v$, that is, the electrons propagate with an apparent velocity larger than v [74].

29.5.4 Drift Diffusion under an Electric Field

Spin drift-diffusion effects in all-electrical n-GaAs-based devices were measured by Lou et al. [75], where the conduction was limited to electrons. Unlike in regular semiconductors, in graphene electrostatic gating allows for switching from hole to electron conduction while keeping the carrier mobility, diffusion constant, Fermi velocity, electric conductivity, and other parameters approximately unchanged (see Figure 29.3). The effect of spin drift was studied in metallic (hole- and electron-doped, $n \approx \pm 3.5 \times 10^{12}\,\text{cm}^{-2}$) graphene as well as in the vicinity of the charge neutrality point under an applied DC electric field in lateral spin valves [76]. The experiments were regular, four-terminal non-local spin-valve measurements using an AC spin injection and detection scheme as in Figure 29.5. The device however was extended with a fifth electric contact on the graphene (placed between the inner injector–detector contacts) allowing for a DC electric current and thus a DC electric field to be present in the injector–detector region (see Figure 29.10a). Note that there is a major difference between these experiments and the ones presented in the previous section where the AC injection contacts were DC biased to enhance injection efficiency. Here, the spin injection process is undisturbed, only the *transport* of the created spin imbalance is manipulated using the electric field.

Due to the high sheet resistance (≈1.5 kΩ in the metallic regime, 4.5 kΩ at the charge neutrality point), sending a 40 μA maximum DC current through the graphene created a 68 kV/m electric field (206 kV/m at the neutrality point). The resulting carrier drift velocity $|v_D| \leq 1.7 \times 10^4$ m/s was sufficient to change the diffusion of the spin imbalance predicted by Equation 29.2 and increase/decrease the spin-valve signals by 50%. These experiments are not possible in metals since achieving a considerable drift velocity in high conductance materials would necessitate unrealistic electrical fields.

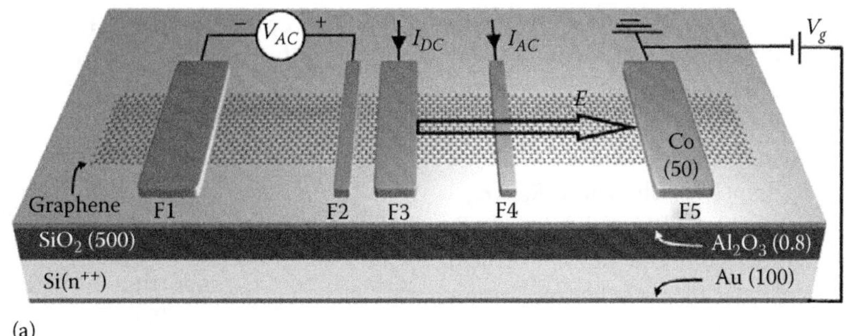

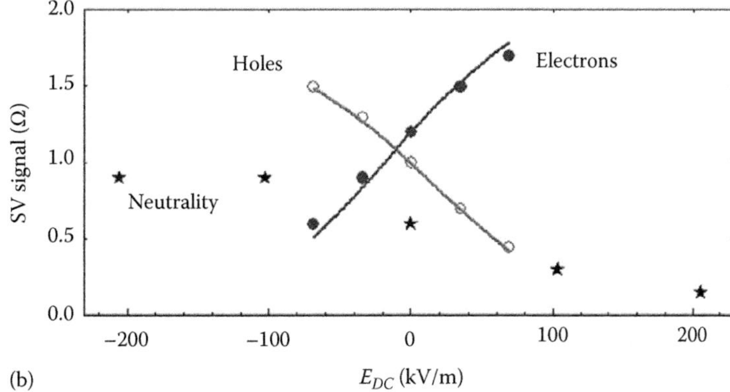

FIGURE 29.10 Sample geometry and contact scheme for the spin drift-diffusion experiments (a), with the additional electrode F3 employed to create a DC electric field E. The measured spin-valve signal is plotted against E in (b) for electrons and holes (density $|n| = 3.5 \times 10^{12}$ cm^{-2}) and for the neutrality region. The solid lines are obtained via the drift-diffusion model.

Figure 29.10b shows the spin-valve signals R_{SV} as extracted from the multitude of regular non-local measurements, plotted against the applied DC electric field. For high carrier densities, the increase in the spin-valve signal is clearly visible when the electric field is increased to positive (electrons, full circles) or negative values (holes, open circles). At the Dirac neutrality point ($n \approx 0$, stars), the conduction on the large scale happens neither via electrons nor via holes, meaning no spin drift should take place and the spin-valve signal should not depend on the magnitude of E. There is, however, a slight decreasing tendency of the signal for increasingly positive DC fields, similar to the hole conduction regime, indicating that the measurement was done not exactly at the neutrality point.

A quantitative interpretation of the data is possible by adopting the drift-diffusion model introduced by Yu and Flatté [77] for nondegenerate semiconductors. (The degenerate case is discussed in [78].) The starting equation is, again, Equation 29.2 for steady state, less the precession component:

$$D\nabla^2\vec{n}_s - \frac{\vec{n}_s}{\tau} + \vec{v}_D\nabla\vec{n}_s = 0 \tag{29.7}$$

and we will look at the transport of the spin imbalance n in the x-direction. Similar to the spin diffusion length λ, we define a spin drift length $\lambda_D = D/v_D$. The symmetric diffusion (around the injection point $x = 0$) and asymmetric drift effects add up to form a spin transport characterized by a pair of length scales, the *upstream* and *downstream* spin drift-diffusion lengths:

$$\frac{1}{\lambda_\pm} = \pm\frac{1}{2}\frac{1}{\lambda_D} + \sqrt{\frac{1}{4}\frac{1}{\lambda_D^2} + \frac{1}{\lambda^2}} \tag{29.8}$$

The general solution to the spin imbalance Equation 29.7 in the direction x is

$$n_s(x) = A\exp\left(+\frac{x}{\lambda_+}\right) + B\exp\left(-\frac{x}{\lambda_-}\right) \tag{29.9}$$

where A and B are determined by the boundary conditions. The dependence of the spin signal versus the applied DC field can, therefore, be written in the form of

$$R_{SV} = R_0\exp\left(-\frac{L}{\lambda_\pm}\right) \tag{29.10}$$

where

R_0 is the spin signal for $x = 0$

$\lambda_\pm$ carries the electric field dependence through Equation 29.8

L is the distance between F3 and F4 (the region where spin transport is of drift-diffusion type)

With the parameters calculated for the graphene spin-valve sample in case ($n \approx 3.5 \times 10^{12}\,\text{cm}^{-2}$, $\mu = 2500\,\text{cm}^2/\text{V s}$, $L = 1.5\,\mu\text{m}$, $D = 0.02\,\text{m}^2/\text{s}$, $\lambda = 2\,\mu\text{m}$), the up/downstream lengths are $\lambda_+ = 1.3\,\mu\text{m}$, $\lambda_- = 3\,\mu\text{m}$ for $E \approx \pm 34\,\text{kV/m}$ and $\lambda_+ = 0.9\,\mu\text{m}$, $\lambda_- = 4.4\,\mu\text{m}$ for $E \approx \pm 68\,\text{kV/m}$. At zero electric field $\lambda_\pm \equiv \lambda$ allows to estimate the value of $R_0 = R_{SV}\exp(L/\lambda)$. The curve R_{SV} (E_{DC}) calculated with Equation 29.10 is plotted as solid line both for electrons and holes in Figure 29.10b and shows excellent agreement with the experimental data. There are no free parameters in the calculation and, in fact, these measurements could be used to determine the spin diffusion length λ if it was unknown.

At the Dirac neutrality point, the above model does not apply since the Drude model for calculating the carrier mobility is not valid. In general, the electronic DOS in graphene is

$$\nu(\varepsilon) = \frac{g_S g_V 2\pi|\varepsilon|}{h^2 v_F^2} \tag{29.11}$$

with g_S and g_V the spin- and valley degeneracies. An expression for the drift velocity can be calculated as $v_D = (\sqrt{\pi} v_F e\tau_d)/(\hbar\sqrt{2n})E$ (see Refs. [35,53]). If the momentum scattering time τ_d is independent of the carrier density n, then at sufficiently small densities the drift velocity drastically increases and reaches the Fermi velocity but at $n = 0$ it becomes zero. However, for a $\tau_d \propto \sqrt{n}$ relationship, v_D remains constant at small densities and, again it becomes zero at $n = 0$. It is still an open theoretical question, which of the two situations holds for graphene.

29.6 Spin Relaxation Processes

29.6.1 Theoretical Considerations

We have seen that experiments done so far on graphene spin valves reveal room temperature spin relaxation times τ of the order of 100–200 ps. Combined with a high carrier mobility characteristic to graphene, this gives an already promising spin diffusion length of 1–2 μm. However, it is still below the theoretically predicted intrinsic limit (~20 μm as estimated by Huertas-Hernando [79]) and the relaxation times measured so far are too low to be attractive for spin qubit type applications in quantum computation (as proposed in Trauzettel et al. [80]; here the authors calculate a spin decoherence time of 20 μs in graphene quantum dots limited by hyperfine interactions with the nuclear spin). Some calculations also consider scattering from the substrate impurities and remote surface phonons and yet predict very long spin relaxation times up to a microsecond [81]. Most models for spin relaxation in graphene are based on a clean, defect-free graphene crystal with ripples [81–87], where the spin–orbit interactions are expected to be weak (carbon being low atomic number) and the hyperfine interactions [80,88] insignificant since the occurrence of ^{13}C is only 1%. An explanation for the fast spin relaxation must lie therefore in extrinsic mechanisms such as scattering on impurity potential, defects, boundaries, phonons; this was addressed recently by Castro Neto and Guinea [89] considering an sp^3 hybridization of carbon due to hydrogen impurities, leading to a spin–orbit coupling enhanced by three orders of magnitude.

29.6.2 Experiments: Elliot–Yafet versus D'yakonov–Perel' Mechanism

There are two possible mechanisms discussed in literature [90,91] that can be responsible for spin relaxation, scaling differently on the momentum relaxation. In case of the Elliot–Yafet mechanism (spin flip occurs with a finite probability at each momentum scattering center, Figure 29.11a) the spin scattering time τ is proportional to τ_d, the momentum scattering time. The D'yakonov–Perel' mechanism (spins precess under the influence of local spin–orbit fields in between scattering events, Figure 29.11b) is on the other hand, characterized by a $\tau \propto \tau_d^{-1}$ relationship. To identify the scattering mechanism and find the ultimate limit on spin relaxation, one can thus investigate the link between spin transport and the electronic quality of the graphene, in particular the charge diffusion coefficient $D_c = 0.5 v_F l$ (or the carrier mobility μ). Ideally, the experiments would be done on a set of graphene spin-valve devices that display a wide spread in their charge transport properties. However, it is experimentally challenging to fabricate consistently good ferromagnetic contacts to the graphene for such a set of samples where parameters like the spin injection efficiency are the same for all contacts.

Another route is to do the experiments on individual devices tuning the charge carrier density from metallic regime down to the lowest values and comparing the behavior of the spin transport to the changes in the charge diffusion coefficient.

At an energy E sufficiently far from the Dirac neutrality point (metallic regime), the charge diffusion coefficient D_c can be calculated from the measured conductivity σ using the Einstein relation $\sigma = e^2\nu(E_F)D_c$. Here, $\nu(E_F)$ is the density of states (DOS) at the Fermi level E_F given by Equation 29.11 from which by integration we can obtain the density $n(E_F) = (g_S g_V \pi E_F^2)/(h^2 v_F^2)$. This yields

$$D_c = \frac{h v_F}{2e^2\sqrt{g_S g_V \pi}} \frac{\sigma}{\sqrt{n}} \tag{29.12}$$

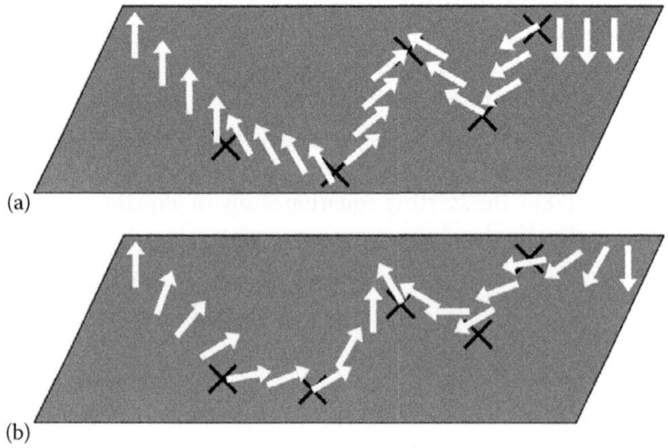

FIGURE 29.11 Illustration of the Elliot–Yafet (a) and D'yakonov–Perel' (b) type spin relaxation mechanisms.

For densities $|n| < 0.5 \times 10^{12}$ cm^{-2}, this formula gives unphysical values and results in a singularity at the Dirac neutrality point. This comes from the unrealistic assumption of vanishing carrier density and DOS at the neutrality point. To correct for it, one has to account for a broadened DOS $\nu^*(E)$ due to finite temperature, electron–hole puddles, and finite lifetime of electronic states. This can be done by adding a Gaussian broadening energy γ in form of

$$\nu^*(E) = \frac{1}{\sqrt{2\pi}\gamma} \int_{-\infty}^{\infty} \exp\left(-\frac{(\varepsilon - E)^2}{2\gamma^2}\right) \nu(\varepsilon) d\varepsilon \qquad (29.13)$$

After integration, we obtain an expression for the broadened DOS that can be used in Equation 29.12 to calculate a modified charge diffusion coefficient D_c^*.

A set of resistivity, spin valve, and Hanle precession measurements was done in a broad range of charge carrier densities to compare spin- and charge transport in SLG [58]. In Figure 29.12a through e, we plot the extracted transport parameters against the applied backgate voltage (top scale) and carrier density (bottom scale). The conductivity (inverse of the measured sheet resistivity) in panel a shows the typical electron and hole conductance regimes separated by the charge neutrality point. Due to trapped charges in the SiO_2 and/or other contaminants on the sample, the neutrality point is usually shifted by a few (tens of) volts from the zero setting of the backgate voltage. This curve is used to calculate and plot (Figure 29.12c) the charge diffusion coefficient using the unbroadened formula (29.12) (D_c, dashed line) and the broadened version (D_c^*, solid line). A broadening energy of $\gamma \approx 75$ meV was used corresponding to a density variation of $\Delta n \approx \pm\, 0.7 \times 10^{12}$ cm^{-2}. The broadening is consistent with the literature values [15] attributed to electron–hole puddles in graphene on a SiO_2 substrate, and takes care of avoiding the singularity at the neutrality point. In the same panel, the spin diffusion coefficient (D_s, circles) is also plotted, as extracted from Hanle precession measurements fitted with the solutions to the Bloch equation (similar to Figure 29.9). The spin- and charge diffusion coefficients are practically the same, although they are obtained from completely different types of experiments. This indicates a minor role of Coulomb-type electron–electron interactions in scattering processes [92,93] in agreement with the recent results of Li et al. [94] where e–e interactions are detected in high mobility, suspended graphene flakes only.

Figure 29.12b shows the spin-valve signal extracted from spin-valve measurements (solid line, continuously scanning V_g) and from the Hanle precession measurements with $B_z = 0$ (circles), which of course should be the same. We can see a threefold decrease of the signal when the carrier density goes from metallic to neutrality. Since in Equation 29.4 the signal is proportional to the inverse of the conductivity, the strong loss of spin-valve signal must come from a spin diffusion length λ that is considerably lower for low densities.

Figure 29.12d shows the behavior of the spin relaxation time τ against carrier density extracted from the Hanle experiments.

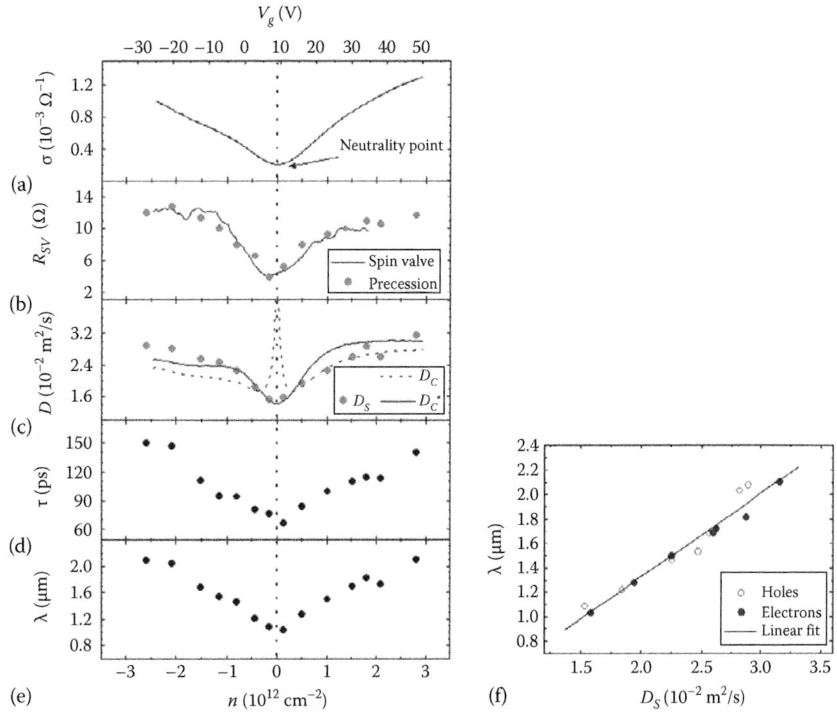

FIGURE 29.12 Charge and spin transport parameters against gate voltage (top scale) and carrier density (bottom scale): conductivity as inverse of sheet resistivity (a), spin-valve signal extracted from spin valve and Hanle precession measurements (b), spin- and charge-diffusion coefficients (c), spin relaxation time (d), spin relaxation length (e). A linear scaling is revealed when the spin relaxation length is plotted against the diffusion coefficient (f).

Indeed, combining τ with D_s the spin diffusion coefficient from panel c using $\lambda = \sqrt{D_s\tau}$ yields a spin diffusion length that decreases by more than 50% when the carrier density is reduced to the minimum (see panel e). On a device with spin injector–detector spacing $L = 3.1\,\mu m$, this is also quantitatively consistent with the behavior of the spin-valve signal displayed in panel b.

Finally, let us look at the scaling of the spin diffusion length on the (spin or charge) diffusion coefficient. Plotting λ against D_s (Figure 29.12f) shows a clear linear dependence for both the electron and the hole conduction regime. The spin scattering time is thus directly proportional to the diffusion coefficient, that is, to the momentum scattering time. This is a clear confirmation of a dominating Elliot–Yafet type spin relaxation mechanism, in agreement with the spin relaxation anisotropy studies presented in Section 29.5.3.

Improving the electronic characteristics of the graphene device is, therefore, expected both to enhance the charge transport and prolong the spin scattering time. Assuming the Elliot–Yafet mechanism is still dominating at high carrier mobilities, one can extrapolate the behavior shown in Figure 29.12f to samples displaying charge carrier mobilities in the range of 2×10^5 cm/V s reported recently, to predict a possible room temperature spin diffusion length up to 100 μm.

29.6.3 Extrinsic Factors for Spin Relaxation

A number of extrinsic factors were investigated by Popinciuc et al. [48] that could produce the impurity potential for a strong spin (and charge) scattering. One of these is the edges of the graphene flake that, especially in case of samples etched with plasma, are rough and may be chemically modified thus could reduce spin diffusion length. A set of spin valve and Hanle experiments on samples with graphene width from 0.1 to 2 μm show a considerable spread in the spin relaxation times (40–200 ps), however no clear scaling on the width is found.

Another effect that could play a role is the relaxation through the ferromagnetic contacts (i.e., the conductivity mismatch problem); this is contributing to the spin relaxation only for transparent ferromagnet–graphene interfaces, and it can be eliminated by using an oxide barrier of appropriate height as we discussed it earlier.

Yet another factor that was excluded as spin relaxation channel is the oxide layer deposited on the graphene. One could imagine a conductor that consists of a single layer of atoms to be strongly affected in its spin (and charge) transport properties by an insulating layer deposited on the top of it, of similar thickness. However, samples were made where the oxide, instead of covering the full graphene flake, was deposited only under the ferromagnetic electrodes (Al_2O_3 [48]; MgO [46]) and no improvement in the spin relaxation times was observed.

One aspect that was so far not studied systematically is how the substrate and contaminants between the substrate and graphene affects spin relaxation. It was shown experimentally that the charge transport is greatly enhanced by (partially) removing the substrate (wet chemical etching of SiO_2) and adsorbed contaminants (annealing with high current densities). Particularly, the carrier mobility is enhanced by a factor of 20 [7–9]. A SEM image of a graphene "bridge" is shown in Figure 29.13 using Cr/Au contacts. Conform to a dominating Elliot–Yafet spin relaxation mechanism such an increase in charge transport should also come with an improved spin diffusion. Currently, efforts are made to measure spin transport in suspended flakes; however, the fabrication of these devices with ferromagnetic contacts and oxide barriers is technically much more challenging due mainly to the chemical corrosiveness of the materials.

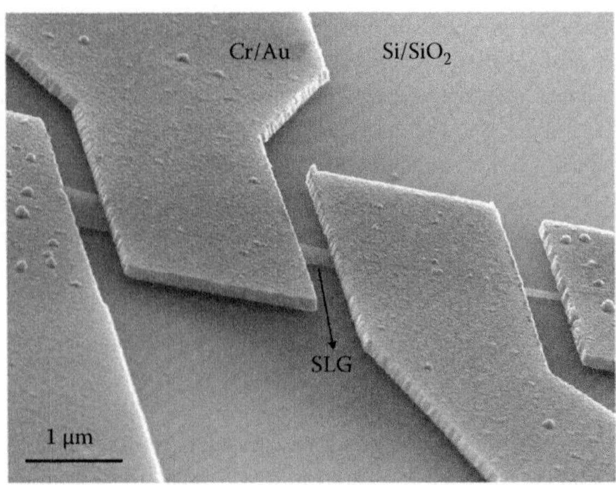

FIGURE 29.13 SEM image of an SLG flake suspended above a Si/SiO_2 (220 nm) substrate, contacted by Cr/Au (5/125 nm) electrodes. (Photo courtesy of J. van den Berg.)

29.7 New Horizons

Interesting spin transport phenomena can occur at graphene *p–n* junctions. Such junctions were already made for charge transport experiments using top- and side-gate electrodes that *n*- or *p*-dope a limited region of a large graphene flake, yet so far no spintronic measurements were reported. Žutić et al. predicted effects in GaAs-based *p–n* junctions such as spin amplification (spin density can increase away from the point of spin injection) and spin capacitance (the amount of spin stored in the *p–n* junctions changes with bias) [95]. Graphene offers the possibility to study these effects with the ease of creating *p–n* junctions that do not present a depletion region at the interface. One can imagine that an electric field across the graphene flake will cause spin drift toward the *p–n* junction (electrons and holes having opposite drift velocity for the same electric field), leading to increased spin accumulations at the junction.

Different levels of controlled hydrogenization of graphene, leading to graphane, offer the possibility to study systematically the influence of defects on the spin relaxation. Ultimately, in the strongly disordered regime (variable range hopping), there is an interesting prediction: the magnetoresistance should be exponentially sensitive to an applied magnetic field. This happens because in the localized state a magnetic field reduces the scattering by aligning the electron spin [96,97].

The idea of a "perfect spin filtering" was proposed by Karpan et al. [98] based on the calculations of the Fermi surfaces for a graphene (or graphite) crystal contacted by (111) fcc or (0001) hcp nickel or cobalt electrodes. Although difficult to realize with the current graphene production technology, such a spin-valve device is predicted to show infinite magnetoresistance due to the fact that the in-plane lattice constants of the materials match almost perfectly while their electronic structures overlap in the reciprocal space for one spin direction only.

Intrinsic ferromagnetism and half-metallic behavior of graphene nanoribbons was predicted to rise from the character of the edges (zigzag versus armchair edge, see Figure 12.1a) that is worthwhile exploring [99]. Furthermore, an extrinsic ferromagnetism is also feasible to be observed, when defects or magnetic adatoms lead to long-range ordered coupling among spins [100,101]. Ferromagnetic insulators, for example, europium oxide deposited on graphene can induce ferromagnetic correlations as well, through a proximity effect predicted by Haugen et al. [102].

Spin transport inside double- and few-layer-graphene devices on regular substrates, due to the screening effect of the outer layers and/or the onset of the D'yakonov–Perel' scattering mechanism, show improvement when compared to SLG. Recent experiments indicate already a systematic two-fold increase in the relaxation time for few-layer-graphene samples compared to single layers [103] and even stronger incertain bilayer samples [106,107]. Spin relaxation times up to 6.2 ns (at 20 K temperature) and a peculiar temperature dependence of the spin transport parameters was reported in the bilayer graphene samples described in Ref. [107].

Finally, local studies of spin transmission in *perpendicular* geometry in graphite-based spin valves (NiFe/graphite/Co) were done recently using ballistic electron magnetic microscopy [104]. No measurable loss of spin information was detected when highly (75%) spin-polarized electrons were injected vertically into 17 nm thick graphite nanostructures of 300–500 nm lateral dimensions. The energy/momentum scattering length has been found to be greater than 250 nm at 1.8 eV above the Fermi level, suggesting that the spin relaxation length in graphite would be in the micrometer range at this energy.

29.8 2010 Nobel Prize in Physics

After finishing this chapter, it was announced that the 2010 Nobel Prize in Physics would be awarded jointly to Andre Geim and Konstantin Novoselov (the University of Manchester, United Kingdom) "for groundbreaking experiments regarding the two-dimensional material graphene." The exceptionally short period (a mere 5 years) from their first electrical measurements on single-layer graphene flakes to winning the most prestigious scientific prize is typical of the rapid progress that this new material itself has made. Whether it will continue to progress as rapidly, whether it will find its way to applications, or whether it will forever remain in research labs has yet to be seen. In any case, it has generated immense experimental and theoretical interest and has made us realize that a simple hexagonal carbon lattice can still surprise us in so many different ways.

Acknowledgments

We are grateful for the crucial contribution of our collaborators Siemon Bakker, Bernard Wolfs, Johan Holstein, Nikolaos Tombros, Mihaita Popinciuc, Harry Jonkman, Paul Zomer, Thomas Maassen, Alina Veligura, Shinichi Tanabe and for the inspiring discussions with Roland Kawakami and Antonio Castro Neto. We acknowledge the financial support of the Zernike Institute for Advanced Materials, MSC[plus], NWO, FOM, and NanoNed.

References

1. W. J. M. Naber, S. Faez, and W. G. van der Wiel, Organic spintronics, *J. Phys. D: Appl. Phys.* **40**, R205 (2007).
2. G. Wagoner, Spin resonance of charge carriers in graphite, *Phys. Rev.* **118**, 647 (1960).
3. K. Matsubara, T. Tsuzuku, and K. Sugihara, Electron spin resonance in graphite, *Phys. Rev. B* **44**, 11845 (1991).
4. K. S. Novoselov, D. Jiang, F. Schedin et al., Two-dimensional atomic crystals, *Proc. Natl. Acad. Sci. USA* **102**(30), 10451 (2005).
5. P. R. Wallace, The band theory of graphite, *Phys. Rev.* **71**, 622 (1947).
6. K. S. Novoselov, A. K. Geim, S. V. Morozov et al., Electric field effect in atomically thin carbon films, *Science* **306**, 666 (2004).
7. X. Du, I. Skachko, A. Barker, and E. Y. Andrei, Approaching ballistic transport in suspended graphene, *Nat. Nanotechnol.* **3**, 491 (2008).
8. K. Bolotin, K. Sikes, Z. Jiang et al., Ultrahigh electron mobility in suspended graphene, *Solid State Commun.* **146**, 351 (2008).
9. S. V. Morozov, K. S. Novoselov, M. I. Katsnelson et al., Giant intrinsic carrier mobilities in graphene and its bilayer, *Phys. Rev. Lett.* **100**, 016602 (2008).
10. J. H. Chen, C. Jang, S. Xiao, M. Ishigami, and M. S. Fuhrer, Intrinsic and extrinsic performance limits of graphene devices on SiO_2, *Nat. Nanotechnol.* **3**, 206 (2008).
11. E. H. Hwang, S. Adam, and S. Das Sarma, Carrier transport in two-dimensional graphene layers, *Phys. Rev. Lett.* **98**, 186806 (2007).
12. Y. Zhang, J. Tan, H. L. Stormer, and P. Kim, Experimental observation of the quantum Hall effect and Berry's phase in graphene, *Nature* **438**, 201 (2005).
13. K. S. Novoselov, A. K. Geim, S. V. Morozov et al., Two-dimensional gas of massless Dirac fermions in graphene, *Nature* **438**, 197 (2005).
14. J. Martin, N. Akerman, G. Ulbricht et al., Observation of electron–hole puddles in graphene using a scanning single-electron transistor, *Nat. Phys.* **4**, 144 (2008).

15. A. Deshpande, W. Bao, F. Miao, C. N. Lau, and B. J. LeRoy, Spatially resolved spectroscopy of monolayer graphene on SiO_2. *Phys. Rev. B* **79**, 205411 (2009).
16. Y. Zhang, V. W. Brar, C. Girit, A. Zettl, and M. F. Crommie, Origin of spatial charge inhomogeneity in graphene, *Nat. Phys.* **5**, 722–726 (2009).
17. H. Heersche, P. Jarillo-Herrero, J. B. Oostinga, L. M. K. Vandersypen, and A. F. Morpurgo, Bipolar supercurrent in graphene, *Nature* **446**, 56 (2007).
18. A. Cortijo and M. A. H. Vozmediano, Effects of topological defects and local curvature on the electronic properties of planar graphene, *Nucl. Phys. B* **763**, 293 (2007).
19. J. van den Brink, From strength to strength, *Nat. Nanotechnol.* **2**, 199 (2007).
20. A. Rycerz, J. Tworzydlo, and C. W. J. Beenakker, Valley filter and valley valve in graphene, *Nat. Phys.* **3**, 172 (2007).
21. F. Schedin, A. K. Geim, S. V. Morozov et al., Detection of individual gas molecules adsorbed on graphene, *Nat. Mater.* **6**, 652 (2007).
22. T. O. Wehling, K. S. Novoselov, S. V. Morozov et al., Molecular doping of graphene, *Nano Lett.* **8**, 173 (2008).
23. D. C. Elias, R. R. Nair, T. M. G. Mohiuddin et al., Control of graphene's properties by reversible hydrogenation: Evidence for graphene, *Science* **323**, 610 (2009).
24. R. Y. N. Gengler, A. Veligura, A. Enotiadis et al., Large-yield preparation of high-electronic-quality graphene by a Langmuir-Schaefer approach, *Small* **6**(1), 35–39 (2010).
25. C. Gmez-Navarro, R. T. Weitz, A. M. Bittner et al., Electronic transport properties of individual chemically reduced graphene oxide sheets, *Nano Lett.* **7**, 3499 (2007).
26. X. L. Li, G. Y. Zhang, X. D. Bai et al., Highly conducting graphene sheets and Langmuir–Blodgett films, *Nat. Nanotechnol.* **3**, 538 (2008).
27. H. A. Becerril, J. Mao, Z. Liu, R. M. Stoltenberg, Z. Bao, and Y. Chen, Evaluation of solution-processed reduced graphene oxide films as transparent conductors, *ACS Nano* **2**, 463 (2008).
28. C. Berger, Z. Song, X. Li et al., Electronic confinement and coherence in patterned epitaxial graphene, *Science* **312**, 1191 (2006).
29. J. Hass, R. Feng, T. Li et al., Highly ordered graphene for two dimensional electronics, *Appl. Phys. Lett.* **89**, 143106 (2006).
30. X. Li, W. Cai, J. An et al., Large-area synthesis of high-quality and uniform graphene films on copper foils, *Science* **324**, 1312 (2009).
31. K. S. Kim, Y. Zhao, H. Jang et al., Large-scale pattern growth of graphene films for stretchable transparent electrodes, *Nature* **457**, 706 (2009).
32. A. Reina, X. Jia, J. Ho et al., Large Area, few-layer graphene films on arbitrary substrates by chemical vapor deposition, *Nano Lett.* **9**(1), 30–35 (2009).
33. C. Oshima and A. Nagashima, Ultra-thin epitaxial films of graphite and hexagonal boron nitride on solid surfaces, *J. Phys.: Condens. Matter* **9** (1997).
34. I. Žutić and M. Fuhrer, Spintronics: A path to spin logic, *Nat. Phys.* **1**, 85 (2005).
35. T. Ando, Screening effect and impurity scattering in monolayer graphene, *J. Phys. Soc. Jpn.* **75**, 074716 (2006).
36. M. Johnson, and R. H. Silsbee, Interfacial charge-spin coupling: Injection and detection of spin magnetization in metals, *Phys. Rev. Lett.* **55**, 1790 (1985).
37. M. Johnson and R. H. Silsbee, Thermodynamic analysis of interfacial transport and of the thermomagnetoelectric system, *Phys. Rev. B* **35**, 4959 (1987).
38. F. J. Jedema, A. T. Filip, and B. J. van Wees, Electrical spin injection and accumulation at room temperature in an all-metal mesoscopic spin valve, *Nature* **410**, 345 (2001).
39. E. W. Hill, A. K. Geim, K. S. Novoselov, F. Schedin, and P. Blake, Graphene spin valve devices, *IEEE Trans. Magn.* **42**(10), 2694 (2006).
40. N. Tombros, C. Józsa, M. Popinciuc, H. T. Jonkman, and B. J. van Wees, Electronic spin transport and spin precession in single graphene layers at room temperature, *Nature* **448**, 571 (2007).
41. S. Cho, Y.-F. Chen, and M. S. Fuhrer, Gate-tunable graphene spin valve, *Appl. Phys. Lett.* **91**, 123105 (2007).
42. M. Nishioka and A. M. Goldman, Spin transport through multilayer graphene, *Appl. Phys. Lett.* **90**, 252505 (2007).
43. M. Ohishi, M. Shiraishi, R. Nouchi, T. Nozaki, T. Shinjo, and Y. Suzuki, Spin injection into a graphene thin film at room temperature, *Jpn. J. Appl. Phys. Part 2* **46**(25), L605 (2007).
44. W. H. Wang, K. Pi, Y. Li et al., Magnetotransport properties of mesoscopic graphite spin valves, *Phys. Rev. B* **77**, 020402(R) (2008).
45. H. Goto, A. Kanda, T. Sato et al., Gate control of spin transport in multilayer graphene, *Appl. Phys. Lett.* **92**, 212110 (2008).
46. W. Han, K. Pi, W. Bao et al., Electrical detection of spin precession in single layer graphene spin valves with transparent contacts, *Appl. Phys. Lett.* **94**, 222109 (2009).
47. M. Shiraishi, M. Ohishi, R. Nouchi et al., Robustness of spin polarization in graphene-based spin valves, *Adv. Funct. Mater.* **19**, 3711 (2009).
48. M. Popinciuc, C. Józsa, P. J. Zomer et al., Electronic spin transport in graphene field effect transistors, *Phys. Rev. B* **80**, 214427 (2009).
49. M. Ishigami, J. H. Chen, W. G. Cullen, M. S. Fuhrer, and E. D. Williams, Atomic structure of graphene on SiO_2. *Nano Lett.* **7**(6), 1643 (2007).
50. D. Li, W. Windl, and N. P. Padture, Toward site-specific stamping of graphene, *Adv. Mater.* **21**, 1243 (2009).
51. X. Xie, L. Ju, X. Feng et al., Controlled fabrication of high-quality carbon nanoscrolls from monolayer grapheme, *Nano Lett.* **9**(7), 2565 (2009).
52. P. Blake, E. W. Hill, A. H. C. Neto et al., Making graphene visible, *Appl. Phys. Lett.* **91**, 063124 (2007).
53. N. Tombros, Electron spin transport in graphene and carbon nanotubes, PhD dissertation, University of Groningen, the Netherlands, 2008.

54. C. Stampfer, J. Güttinger, F. Molitor, D. Graf, T. Ihn, and K. Ensslin, Tunable Coulomb blockade in nanostructured graphene, *Appl. Phys. Lett.* **92**, 012102 (2008).
55. P. Nemes-Incze, Z. Osváth, K. Kamarás, and L. P. Biró, Anomalies in thickness measurements of graphene and few layer graphite crystals by tapping mode atomic force microscopy, *Carbon* **46**, 1435 (2008).
56. M. Y. Han, B. Ozyilmaz, Y. Zhang, and P. Kim, Energy band-gap engineering of graphene nanoribbons, *Phys. Rev. Lett.* **98**, 206805 (2007).
57. F. Molitor, J. Güttinger, C. Stampfer, D. Graf, T. Ihn, and K. Ensslin, Local gating of a graphene Hall bar by graphene side gates, *Phys. Rev. B* **76**, 245426 (2007).
58. C. Józsal, T. Maassen, M. Popinciuc et al., Linear scaling between momentum and spin scattering in graphene, *Phys. Rev. B* **80**, 241403 (2009).
59. L. C. Campos, V. R. Manfrinato, J. D. Sanchez-Yamagishi, J. Kong, and P. Jarillo-Herrero, Anisotropic etching and nanoribbon formation in single-layer graphene, *Nano Lett.* **9**(7), 2600 (2009).
60. J. Moser and A. Bachtold, Fabrication of large addition energy quantum dots in graphene, *Appl. Phys. Lett.* **95**, 173506 (2009).
61. X. Lou, C. Adelmann, S. A. Crooker et al., Electrical detection of spin transport in lateral ferromagnet–semiconductor devices, *Nat. Phys.* **3**, 197 (2007).
62. N. Tombros, S. J. van der Molen, and B. J. van Wees, Separating spin and charge transport in single-wall carbon nanotubes, *Phys. Rev. B* **73**, 233403 (2006).
63. F. J. Jedema, H. B. Heersche, A. T. Filip, J. J. A. Baselmans, and B. J. van Wees, Electrical detection of spin precession in a metallic mesoscopic spin valve, *Nature* **416**, 713 (2002).
64. G. Schmidt, D. Ferrand, L. W. Molenkamp, A. T. Filip, and B. J. van Wees, Fundamental obstacle for electrical spin injection from a ferromagnetic metal into a diffusive semiconductor, *Phys. Rev. B* **62**, 4790(R) (2000).
65. E. I. Rashba, Theory of electrical spin injection: Tunnel contacts as a solution of the conductivity mismatch problem, *Phys. Rev. B* **62**, 16267(R) (2000).
66. A. Fert and H. Jaffrès, Conditions for efficient spin injection from a ferromagnetic metal into a semiconductor, *Phys. Rev. B* **64**, 184420 (2001).
67. P. M. Tedrow and R. Meservey, Spin polarization of electrons tunneling from films of Fe, Co, Ni, and Gd, *Phys. Rev. B* **7**, 318 (1973).
68. T. Kimura, Y. Otani, and J. Hamrle, Enhancement of spin accumulation in a nonmagnetic layer by reducing junction size, *Phys. Rev. B* **73**, 132405 (2006).
69. C. Józsa, M. Popinciuc, N. Tombros, H. T. Jonkman, and B. J. van Wees, Controlling the efficiency of spin injection into graphene by carrier drift, *Phys. Rev. B* **79**, 081402(R) (2009).
70. S. A. Crooker, E. Garlid, A. N. Chantis et al., Bias-controlled sensitivity of ferromagnet/semiconductor electrical spin detectors, *Phys. Rev. B* **80**, 041305(R), 2009.
71. M. Johnson and R. H. Silsbee, Coupling of electronic charge and spin at a ferromagnetic-paramagnetic metal interface, *Phys. Rev. B* **37**, 5312 (1988).
72. J. Fabian, A. Matos-Abiague, C. Ertler, P. Stano, and I. Žutić, Semiconductor spintronics, *Acta Phys. Slov.* **57**, 565 (2007).
73. N. Tombros, S. Tanabe, A. Veligura et al., Anisotropic spin relaxation in graphene, *Phys. Rev. Lett.* **101**, 046601 (2008).
74. R. A. Sepkhanov, M. V. Medvedyeva, and C. W. J. Beenakker, Hartman effect and spin precession in graphene, *Phys. Rev. B* **80**, 245433 (2009).
75. X. Lou, C. Adelmann, M. Furis et al., Electrical detection of spin accumulation at a ferromagnet-semiconductor interface, *Phys. Rev. Lett.* **96**, 176603 (2006).
76. C. Józsa, M. Popinciuc, N. Tombros, H. T. Jonkman, and B. J. van Wees, Electronic spin drift in graphene field-effect transistors, *Phys. Rev. Lett.* **100**, 236603 (2008).
77. Z. G. Yu and M. E. Flatté, Electric-field dependent spin diffusion and spin injection into semiconductors, *Phys. Rev. B* **66**, 201202(R) (2002); ibid 235302.
78. I. D'Amico, Spin injection and electric-field effect in degenerate semiconductors, *Phys. Rev. B* **69**, 165305 (2004).
79. D. Huertas-Hernando, F. Guinea, and A. Brataas, Spin–orbit-mediated spin relaxation in graphene, *Phys. Rev. Lett.* **103**, 146801 (2009).
80. B. Trauzettel, D. V. Bulaev, D. Loss, and G. Burkard, Spin qubits in graphene quantum dots, *Nat. Phys.* **3**, 192 (2007).
81. C. Ertler, S. Konschuh, M. Gmitra, and J. Fabian, Electron spin relaxation in graphene: The role of the substrate, *Phys. Rev. B* **80**, 041405(R) (2009).
82. C. L. Kane and E. J. Mele, Quantum spin hall effect in graphene, *Phys. Rev. Lett.* **95**, 226801 (2005).
83. H. Min, J. E. Hill, N. A. Sinitsyn et al., Intrinsic and Rashba spin–orbit interactions in graphene sheets, *Phys. Rev. B* **74**, 165310 (2006).
84. D. Huertas-Hernando, F. Guinea, and A. Brataas, Spin–orbit coupling in curved graphene, fullerenes, nanotubes, and nanotube caps, *Phys. Rev. B* **74**, 155426 (2006).
85. Y. Yao, F. Ye, X.-L. Qi, S.-C. Zhang, and Z. Fang, Spin–orbit gap of graphene: First-principles calculations, *Phys. Rev. B* **75**, 041401(R) (2007).
86. K.-H. Ding, G. Zhou, and Z.-G. Zhu, Magnetotransport of Dirac fermions in graphene in the presence of spin–orbit interactions, *J. Phys.: Condens. Matter* **20**, 345228 (2008).
87. M. Gmitra, S. Konschuh, C. Ertler, C. Ambrosch-Draxl, and J. Fabian, Band-structure topologies of graphene: Spin–orbit coupling effects from first principles, *Phys. Rev. B* **80**, 235431 (2009).
88. O. V. Yazyev, Hyperfine interactions in graphene and related carbon nanostructures, *Nano Lett.* **8**(4), 1011 (2008).
89. A. H. Castro Neto and F. Guinea, Impurity-induced spin–orbit coupling in graphene, *Phys. Rev. Lett.* **103**, 026804 (2009).

90. I. Žutić, J. Fabian, and S. Das Sarma, Spintronics: Fundamentals and applications, *Rev. Mod. Phys.* **76**, 323 (2004).
91. J. Fabian and S. D. Sarma, Band structure effects in the spin relaxation of conduction electrons, *J. Appl. Phys.* **85**, 5075 (1999).
92. Y. Barlas, T. Pereg-Barnea, M. Polini, R. Asgari, and A. H. MacDonald, Chirality and correlations in graphene, *Phys. Rev. Lett.* **98**, 236601 (2007).
93. E. H. Hwang, Ben Yu-Kuang Hu, and S. Das Sarma, Density dependent exchange contribution to $\partial\mu/\partial n$ and compressibility in graphene, *Phys. Rev. Lett.* **99**, 226801 (2007).
94. G. Li, A. Luican, and E. Y. Andrei, Scanning tunneling spectroscopy of graphene on graphite, *Phys. Rev. Lett.* **102**, 176804 (2009).
95. I. Žutić, J. Fabian, and S. Das Sarma, Spin injection through the depletion layer: A theory of spin-polarized p-n junctions and solar cells, *Phys. Rev. B* **64**, 121201(R) (2001).
96. T. G. Rappoport, B. Uchoa, and A. H. Castro Neto, Magnetism and magnetotransport in disordered graphene, *Phys. Rev. B* **80**, 245408 (2009).
97. A. H. Castro Neto, Private communications, 2009.
98. V. M. Karpan, G. Giovannetti, P. A. Khomyakov et al., Graphite and graphene as perfect spin filters, *Phys. Rev. Lett.* **99**, 176602 (2007).
99. Y.-W. Son, M. L. Cohen, and S. G. Louie, Half-metallic graphene nanoribbons, *Nature* **444**, 347 (2006).
100. N. M. R. Peres, F. Guinea, and A. H. Castro Neto, Coulomb interactions and ferromagnetism in pure and doped graphene, *Phys. Rev. B* **72**, 174406 (2005).
101. M. A. H. Vozmediano, M. P. López-Sancho, T. Stauber, and F. Guinea, Local defects and ferromagnetism in graphene layers, *Phys. Rev. B* **72**, 155121 (2005).
102. H. Haugen, D. Huertas-Hernando, and A. Brataas, Spin transport in proximity-induced ferromagnetic graphene, *Phys. Rev. B* **77**, 115406 (2008).
103. T. Maassen, F. K. Dejene, M. H. D. Guimarães et al., Comparison between charge and spin transport in few-layer graphene, *Phys. Rev. B* **83**, 115410 (2011).
104. T. Banerjee, W. G. van der Wiel, and R. Jansen, Spin injection and perpendicular spin transport in graphite nanostructures, *Phys. Rev. B* **81**, 214409 (2010).
105. K. W. Han, K. Pi, K. M. McCreary et al., Tunneling spin injection into single layer graphene, *Phys. Rev. Lett.* **105**, 167202 (2010).
106. T.-Y. Yang, J. Balakrishnan, F. Volmer et al., Observation of long spin relaxation times in bilayer graphene at room temperature, arXiv:1012.1156v1 [cond-mat.mes-hall] (2010).
107. K. W. Han, and R. K. Kawakami, Long spin lifetimes in graphene spin valves, arXiv:1012.3435v1 [cond-mat.mtrl-sci] (2010).

30 Magnetism and Transport in Diluted Magnetic Semiconductor Quantum Dots

Joaquín Fernández-Rossier
Universidad de Alicante

Ramon Aguado
Instituto de Ciencia de Materiales de Madrid (ICMM), CSIC

30.1 Introduction

30.1.1 Motivation

The electronic spin already plays an important role in transport in quantum dots (QDs) made of nonmagnetic materials, like semiconductor, carbon nanotubes, and metallic grains. The standard experimental setup shown in Figure 30.1 consists of a central region, the QD, weakly coupled to two conducting electrodes, the source and drain. The charge of the dot is controlled by the electric field supplied by the gate voltage. We only consider situations where the charging energy of the dot, inversely proportional to its size, is larger than the temperature. In this Coulomb Blockade regime, the charge in the dot is quantized and whenever the highest energy electron is unpaired, the injection of an additional electron in the dot is only possible when its spin is opposite to that of the residing dot, because of the Pauli principle. At low temperature, these correlations result in the formation of the Kondo singlet, for which nontrivial spin correlations are established between the electron in the dot and those in the electrodes. Nontrivial spin effects in QD can also happen when the highest energy electrons reside a partially full degenerate shell. Other chapters of this book show how transport in magnetic materials attracts huge attention both for fundamental and applied perspectives. The influence of the magnetic state of the system over the electrical conduction gives rise to a variety of MR effects, like AMR in bulk materials, giant MR (GMR) in multilayers, tunneling MR (TMR) and AMR (TAMR) in tunnel junctions with magnetic electrodes, and so on. Conversely, the transport electrons can also affect the magnetic state of the system, through the so-called spin-transfer or spin-torque effect. Since both transport in magnetic materials and transport in nonmagnetic QDs results in very interesting spin-related phenomena, we expect that transport in dots made of magnetic materials leads to new fascinating physics. In this chapter, we focus on transport through QDs made of a fascinating class of magnetic materials, the so-called diluted magnetic semiconductors.

30.1.2 Diluted Magnetic Semiconductors

As discussed in Chapter 18, conventional semiconductor materials, like GaAs and CdTe, doped with transition metal magnetic atoms like Mn, are known as diluted magnetic semiconductors. These materials are widely studied [1–5], because they hold the promise of combining in the same sample the useful properties

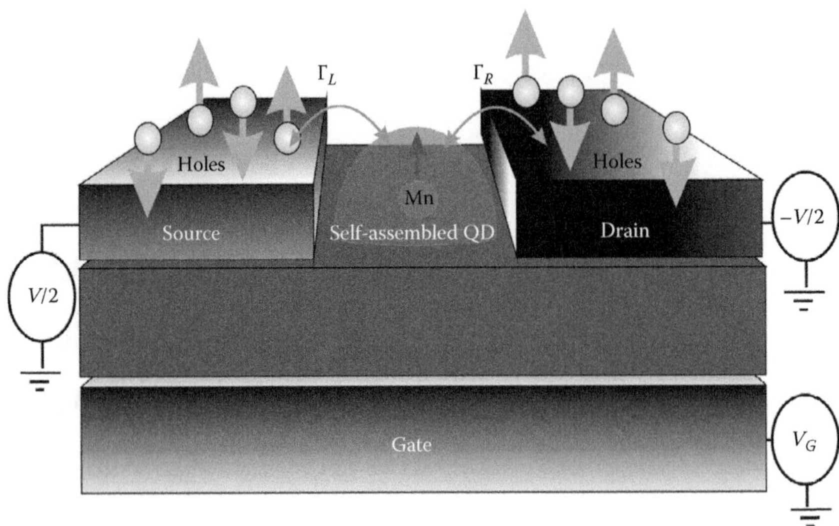

FIGURE 30.1 Schematics of the setup discussed in this chapter for studying transport in a DMS QD. The system is based on a CdTe QD doped with a single Mn^{+2} ion and coupled to metallic reservoirs (tunneling rates Γ_L and Γ_R and chemical potentials $\mu_{L/R} = \pm V/2$). V_G is the gate voltage.

of semiconducting compounds, like electrically tunable conductivity, with those of magnetic materials, like permanent magnetism. Whereas all DMS host magnetic atoms with local spins, only some of them present ferromagnetic order.

The range, strength, and sign of exchange interactions between magnetic impurities in diluted magnetic semiconductors (DMS) depend on the density and nature of the states at the Fermi level [3]. Mn-doped semiconductors of the families (II,Mn)-VI and (III,Mn)-V order ferromagnetically in the presence of carriers that mediate indirect exchange interactions between Mn. In the case of (III,Mn)V compounds like GaAsMn, Mn acts as an acceptor supplying holes that mediate ferromagnetic indirect exchange interactions responsible of the ferromagnetism below a transition temperature, which depends on both Mn and hole densities [2] and can reach 160K [6]. In contrast, Mn does not supply itinerant carriers in (II,Mn)VI compounds. These materials do not order ferromagnetically [1] unless further doping with acceptors provide holes that produce ferromagnetism below a carrier density-dependent Curie temperature (T_c) of approximately 2K in modulation-doped structures [7]. In the case of II–VI, doping with electrons is also possible, but the Curie temperature is expected to be much lower than p-doped to both smaller exchange coupling and effective mass.

DMS are also very attractive because low-dimensional heterostructures can be made with them using molecular beam epitaxy techniques that have been so successful in nonmagnetic semiconductors. Thus, quantum wells [7], delta-doped layers [8], tunnel barriers [9–11], quantum point contacts [10], and QDs (see below) have been fabricated and exciting new physical phenomena, like TAMR [11,77], have been discovered. A paradigmatic example of the potential of these artificial structures is the reversible electrical control of the Curie temperature (T_c) [12,13], the coercive field [14,15], and the magnetic moment orientation [15], demonstrated experimentally in gated layers of both II–VI and III–V DMS. For instance, the Curie temperature of InAsMn can be reversibly changed from 22 to 23K and back [12]. The small relative change of T_c is due to the small relative change in carrier density $\delta p \ll p$, because the density of holes before gating is very high in (III,Mn)-V materials.

In order to enhance the electric control of the magnetic properties, it seems desirable to reduce the density of carriers or to increase the density of electrically injected carriers. The injection of N carriers in a QD of (II,Mn)-VI affords both $p=0$ and a large $\delta p = N/V$ provided that the volume V of the dot is sufficiently small. Additionally, the number of controlled spins can be reduced down to 1 [16], making it possible to control the spin of a single Mn atom by electrical injection of a single carrier, as demonstrated by Leger et al. [17].

30.1.3 DMS Quantum Dots: Experimental State of the Art

In this section, we summarize the state of the art in the fabrication and probing of DMS QDs. According to the fabrication technique, the dots fall into three categories, ordered in increasing size: colloidal nanocrystals [18], self-assembled QD [16,19–22], and electrically defined islands [23]. In the overwhelming majority of cases, these systems have been probed optically in photoluminescence (PL) experiments under applied magnetic field. One of the outstanding features in these experiments is the strong shift of the PL spectrum under the application of a magnetic field. This is the so-called giant Zeeman splitting [1] and arises from the combination of two physical phenomena: the Mn local spin align parallel to the applied field and, because of the exchange coupling to the electron and hole spin, the Mn average magnetization that acts as an effective magnetic field on the exciton spin, which results in Zeeman splittings as large as 100meV. The exchange interaction can also give rise to the formation of the so-called magnetic polaron, that is, the spontaneous alignment at zero magnetic field of the spins of the Mn atoms parallel to the spin of an unpaired carrier or exciton. In bulk, the formation of magnetic polarons results from the competition between exchange energy, which favors the localization of the carrier, and kinetic energy, which opposes to it. QDs favor the localization of the carrier and facilitate the formation of magnetic polarons at larger temperature [19,24,25]. This is a particular manifestation of a general rule: QDs increase the exchange coupling between itinerant carriers and Mn local spins.

The fabrication of QDs doped with a single Mn atom has been one of the most striking results in this field. This was achieved first with self-assembled CdTe QD [5] and later with InGaAs [22]. In both cases, the single QD PL presents a characteristic pattern with a well-defined number of narrow peaks (six in CdTe, five in InAs) that evolve under the application of magnetic fields, permitting to associate them to the various spin states of the Mn spin. The addition of a single carrier into the dot results in a dramatic change of the PL in the case of CdTe [17]. The model Hamiltonians presented below are able to provide good account of these nontrivial PL spectra both in CdTe [26] and InGaAs QDs [27]. Thus, single Mn-doped semiconductor QDs provide an ideal test bench for our theoretical understanding of the various spin couplings in DMS. Additionally, this system has made it possible to measure the spin relaxation of a single Mn atom [28] as well as polarize its spin using optical pumping at zero applied field [29].

Thus, QDs have made it possible to measure and manipulate the spin of a single Mn atom through their interaction with optically injected carriers. This naturally leads to one of the central themes of this chapter: how would transport electrons and Mn spin affect each other in a hypothetical experiment in which a DMS QD plays the role of a central island in a single-electron transistor (SET)? Theory already provides some answers to that question, as we discuss below [30–32]. We first review the experimental progress in that direction.

To the best of our knowledge, two different transport experiments in DMS QDs have been reported. The group of Mollenkamp has been able to fabricate planar tunnel junctions with self-assembled CdSe QDs and Mn atoms in the barriers [21]. This device features resonant tunneling through the dots. The evolution of the resonant energies as a function of the applied magnetic field shows zero-field spin splitting, which might arise from the formation of a magnetic polaron in the dot. However, in this experimental configuration it is not possible to control the number of carriers in the dot with a gate voltage. The second system was an electrically defined GaMnAs island [23]. In this case, the island is in the Coulomb Blockade regime, but there is no evidence of quantum confinement. This peculiar single-hole transistor behaves very differently as the direction of an applied magnetic field is changed. This Coulomb Blockade AMR arises from the strong magnetic anisotropy of GaMnAs island, related to the spin-orbit (SO) of holes [5,33].

The fabrication of a SET based on a self-assembled QD with magnetic impurities still remains a challenge. Whereas gating self-assembled dots is relatively simple, and has been achieved both in single Mn-doped CdTe [17] and InGaAs [22], the main technological obstacle is the coupling of the QD to the conducting electrodes but very rapid progress is being achieved in that regard. The group of professor Arakawa has been able to fabricate SETs based on self-assembled InAs QDs both with ferromagnetic [34] and superconducting electrodes.

Single-hole transistors have been fabricated using CdSe nanocrystals [35]. This provides another route to have DMS SETs in the quantum confinement regime. Whether or not it is possible to dope colloidal nanocrystals in general, and with Mn atoms in particular, has been the subject of a longstanding controversy that has been recently settled with clear experimental signatures that Mn is successfully incorporated in these systems [18,36], like observation of magnetic circular dichroism and exciton magnetic polarons in Mn-doped CdSe colloidal nanocrystals. Finally, nanowire-based QDs offer another promising route toward the fabrication of DMS SET.

The rest of this chapter is organized as follows. In Section 30.3, we describe the electronic structure of DMS QDs. In Section 30.4, we discuss sequential transport in QDs doped with a single magnetic atom, a system that in spite of its simplicity shows the rich interplay between magnetism and transport in this system.

30.2 Spin Interactions in DMS

In this section, we briefly review the spin interactions relevant to Mn-doped QDs. We first consider a II–VI material and then discuss briefly the differences in the case of a III–V. As a substitutional impurity of the group II atom, Mn is an isoelectronic impurity. The electrons in the half-filled *d* shell can be considered as localized in the Mn atom and, in agreement with the Hunds rule, they form a spin $S=5/2$. In the model, electrons and holes in the confined levels of the QD are described in the *kp* effective mass approximation. We consider QDs small enough as to have a sizable quantization of their energy spectrum. As in the case of bulk, the electrons and holes are exchanged coupled to the Mn spins through a local interaction [26,32,37–40]:

$$\mathcal{H}_{e-Mn} + \mathcal{H}_{h-Mn} = \sum_I \vec{M}_I \cdot \left(\alpha \vec{S}_e(\vec{r}_I) + \beta \vec{S}_e(\vec{r}_I)\right), \qquad (30.1)$$

where

$\vec{M}_I$ is the spin 5/2 operator associated at the Mn atom located in $\vec{r}_I$

$\vec{S}_e(\vec{r}_I)$ and $\vec{S}_h(\vec{r}_I)$ are the spin densities of conduction band electrons and valence band holes, respectively

α and β are the corresponding exchange coupling constants

which have dimensions of energy times volume [1].

These constants differ in sign and magnitude because the conduction band is made mainly of *s* orbitals of group II whereas the valence band is mainly composed of *p* orbitals of group VI atom. In the case of the conduction band, the exchange is direct and ferromagnetic whereas in the case of the valence band exchange is dominated by kinetic exchange that arises from *pd* hybridization and results in antiferromagnetic Mn–hole coupling.

Even in the absence of carriers, the Mn spins interact with each other via an antiferromagnetic super-exchange that decays exponentially with distance. This term is important when two Mn spins are located as first neighbor unit cells, that is, when they are separated by a single atom. The Hamiltonian for this coupling reads [1]:

$$\mathcal{H}_{AF} = \sum_{I,I'} J(\vec{r}_I - \vec{r}_{I'})\vec{M}_I \cdot \vec{M}_{I'}. \tag{30.2}$$

The function $J(\vec{r}_I - \vec{r}_{I'})$ decays very quickly as the interatomic distance increases. For first neighbor coupling, the exchange interaction is of the order of 1 meV.

The spin 5/2 of the Mn feels the presence of the crystal environment of a zinc blende lattice through the combination of the different electrostatic energy felt by the different *d* orbtials and the SO interaction. This results in a single-ion Hamiltonian whose symmetry depends dramatically on the symmetry of the crystal environment. For instance, in the case of bulk the single-ion Hamiltonian for Mn in zinc blende is respectful with the cubic symmetry of the lattice and reads [41]:

$$\mathcal{H}_{cubic} = a\left(M_x^4 + M_y^4 + M_z^4\right). \tag{30.3}$$

The otherwise six degenerate spin levels of the Mn split in a doublet and a quartet. The splitting is proportional to α and is so small (below 10 μeV) that can only be measured with electron paramagnetic resonance (EPR). The small value *a* of is due to the fact that Mn^{2+} has a half full shell, the orbital moment is zero, and SO interaction is small.

In the case of strained quantum wells, closer to the situation expected for CdTe QDs, the single-ion Hamiltonian has the growth direction as a preferred easy axis, keeping the in-plane symmetry in the layer [42]. In the case of QDs, the in-plane symmetry might be broken so that the single-ion Hamiltonian reads:

$$\mathcal{H}_{strain} = DM_z^2 + E\left(M_x^2 - M_y^2\right). \tag{30.4}$$

In the case of the layers, $E=0$ and the reported value of D are distributed around a mean value of 10 μeV [42], much smaller than the temperature at which experiments are done. Thus, even if $M_z = \pm 1/2$ is the ground state, all the six levels are almost equally occupied at 4 K. In the case of the dots, E is probably nonzero and the sign of D could change, but their values are not known and they probably condition the single Mn dynamics observed experimentally [28,29]. Finally, in EPR experiments the hyperfine coupling between the nuclear spin (which is also 5/2) and the electronic spin is clearly observed [42] and might play some role in the Mn spin dynamics. In the remainder of this chapter, the effect single-ion anisotropy and hyperfine coupling will be neglected. Whereas this approximation is very good to describe the main features of the system, the single-ion terms are relevant when it comes to describing single-ion dynamics [29].

30.3 Electron-Doped (II,Mn),V Quantum Dot

We now consider a Mn-doped QD, where conduction band electrons are electrically injected [32,37–39,43–46]. We denote the single particle-confined wave functions and their corresponding energies by $\psi_n(\vec{r})$ and $\varepsilon_{n,\sigma} = \varepsilon_n + g\mu B\frac{\sigma}{2}B$, respectively. This notation implies that the spin quantization axis is parallel to the applied magnetic field B and also that the very small SO coupling is neglected. The second quantization operator that creates an electron with spin σ in the state n is denoted by $f_{n\sigma}^{\dagger}$:

$$\mathcal{H}_{0e} + \mathcal{H}_{e-Mn} = \sum_{n,\sigma} \varepsilon_{n,\sigma} f_{n\sigma}^{\dagger} f_{n\sigma} + \alpha \sum_{I,n,n'} \vec{M}_I \cdot \frac{\vec{\tau}_{\sigma,\sigma'}}{2} \psi_n^*(\vec{r}_I)\psi_{n'}(\vec{r}_I), \tag{30.5}$$

where $\vec{\tau}$ are the spin 1/2 Pauli matrices. Attending to the level index n, the exchange interaction has both intra and inter sublevel terms. Attending to the spin index, the exchange interaction has both spin-conserving and spin-flip terms.

In CdTe, the exchange coupling constant α takes values of −30 eV Å^3. The negative sign implies ferromagnetic coupling. The strength of the carrier–Mn exchange depends both on α, a property of the material, and on the amplitude of spin density on the Mn location, a sample property. In order to build some intuition about the latter, we make use of a very simple model for the confinement potential: a hard wall cuboid with dimensions L_x, L_y, and L_z. The wave functions for an electron in the non-degenerate effective mass model reads:

$$\psi_{nx,ny,nz}(x,y,z) = \sqrt{\frac{8}{L_x L_y L_z}} \sin\left(\frac{\pi n_x x}{L_x}\right) \sin\left(\frac{\pi n_y y}{L_y}\right) \sin\left(\frac{\pi n_z z}{L_z}\right), \tag{30.6}$$

where *nx*, *ny*, and *nz* are the positive integer numbers. For a dot of volume *V*, the strength of the maximal value of interaction between Mn–electron coupling can be estimated as α/V. In a self-assembled dot of dimensions $5\times5\times2$ nm, this yields 0.6 meV. The maximal Zeeman splitting of the electron in a dot with those dimensions and N atoms of Mn would be $0.6\times N$ meV.

When more than one carrier is present in the dot, the effect of Coulomb interactions needs to be considered (see for instance Refs. [38,39,44,45,47]). In most instances, the Coulomb interaction results only in a shift of the many-body levels that does not change their total spin. The only exception to this rule occurs in QDs with a partially filled degenerate shell [38,39]. In that situation, Coulomb repulsion favors electronic ground states with large spin even in the absence of magnetic impurities. Degenerate shells occur in dots with cylindrical or spherical symmetry, as well as in the cuboid model above, provided that at least two of the three lateral dimensions are identical. We now analyze the properties of the Mn–e Hamiltonian for situations of increasing complexity. The fact that both the number of electrons and the number of Mn atoms can be varied in the dot justifies the detailed account of the properties that follows.

30.3.1 One Electron

We consider first a QD with only one Mn atom, like the ones studied by Besombes, Mariette, and collaborators, and with two Mn atoms. In Figure 30.2, we show the spectra of a dot with dimensions ($L_x = 14$, $L_y = 12$, $L_z = 3$) nm with one and two Mn atoms, respectively, and with one and two electrons. The spectrum features three groups of levels corresponding to the three lowest energy single particle–confined levels. The fine structure within those groups is associated with spin interactions and is of the order of $\Delta \simeq \alpha/V$. The distinctive feature of these zero-dimensional systems is $\delta \gg \Delta$.

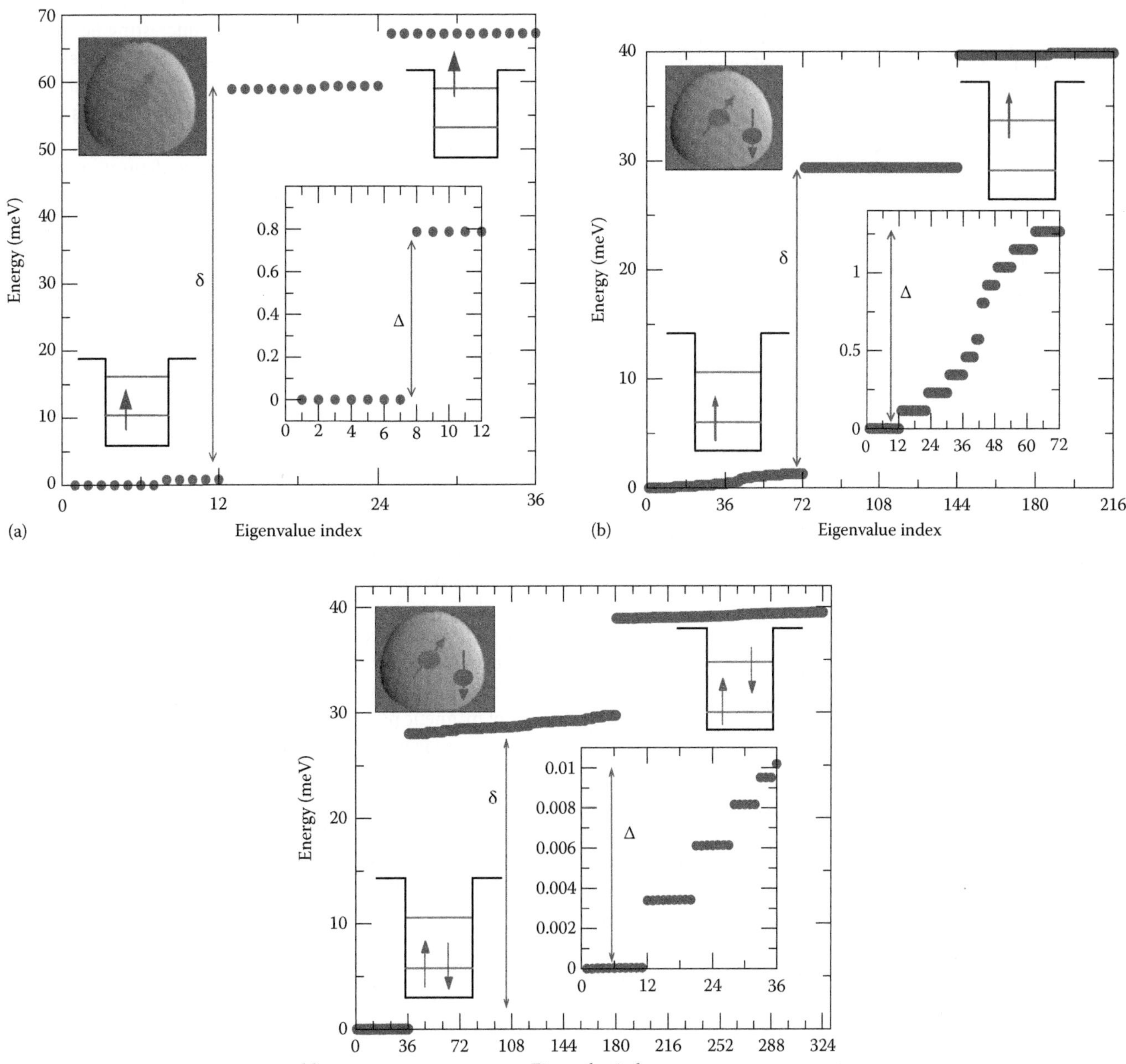

FIGURE 30.2 Spectra of QD dot doped with one Mn and one electron (left), one electron and two Mn (center), and two electron and two Mn (right). Insets: low-energy sector of the spectra.

In the dot with one Mn (Figure 30.2), the low-energy group is split into a ground state manifold with seven states and an excited state manifold with five states, as seen in the inset. Since the Hamiltonian commutes with the total spin operator, we can label the states according to the total spin S, with multiplicity $2S+1$. Thus, we can assign to the ground state manifold of the dots with 1 a total spin $S=6/2=3$, and the excited state in the low-energy group a spin of $S=2$. Notice that $S=3$ is the maximal total spin that the Mn spin 5/2 and the electron spin $S=1/2$ can form.

The low-energy groups seen on the left and central panels of Figure 30.2 correspond basically to the electron in the lowest energy confined state exchanged coupled to the Mn spin(s) through an effective Heisenberg model, because the inter-level terms of the Hamiltonian have a very small effect when $\delta \gg \Delta$ (and levels are not doubly occupied). When the higher energy single-particle levels are neglected, the effective Hamiltonian can be simplified to the effective Heisenberg model

$$\mathcal{H}_{1e} = \sum_I j_e(I)\vec{M}_I \cdot \vec{S}_e, \tag{30.7}$$

where

$j_e(I) = \alpha\,|\,\psi_0(\vec{r}_I)\,|^2$

$\vec{S}_e$ is the spin 1/2 operator of the electron

In the case of a single Mn atom, this is the Heisenberg Hamiltonian for two spins, which can be solved exactly taking advantage of the fact that the total spin and the Heisenberg coupling are related: $S_T^2 = (\vec{S}_1 + \vec{S}_2)^2 = \vec{S}_1^2 + \vec{S}_2^2 + 2\vec{S}_1 \cdot \vec{S}_2$. The square of the spin operators $\vec{S}_i^2$ are $S_i(S_i+1)$ times the unit matrix, whereas the total spin can take the values between $S_T = S_1 + S_2$ and $S_T = |S_1 + S_2|$. In this case, the allowed values are $S_T = 3$ and $S_T = 2$. Thus, in the case of the Mn spin 5/2 and the electron spin 1/2, we can write $J\vec{S}_1 \cdot \vec{S}_2 = \frac{J}{2}\left(S_T(S_T+1) - \frac{3}{4} - \frac{35}{4}\right)$, with eigenvalues $(10J/8)$ for $S=3$ and $-14J/8$ for $S=2$. The splitting is $3J$. Importantly, this intra-level splitting Δ scales linearly with the exchange coupling constant α.

In the dot with two Mn, taken sufficiently apart so that their direct coupling J_{AF} is zero, the low-energy group has 72 states, whose fine structure is shown in the inset. The ground state manifold within these 72 states is a group with 12 states that can be identified with a total spin $S=11/2$, the maximal possible spin that two Mn atoms and one electron can have. Thus the electron favors a large spin ground state, inducing ferromagnetic spin correlation between the two Mn spins. This is a clear example of indirect or carrier-mediated exchange coupling. The strength of the coupling can be estimated from the energy splitting of the low-energy group of the spectrum, Δ, which is of the order of 1 meV. Importantly, it also scales linearly with α. The splitting Δ sets the temperature scale, below which the Mn spins are not independent. This analysis implies also that one electron will induce ferromagnetic correlations between more than two Mn. This suggests that the addition of a single carrier to the QD changes significantly its spin properties. Without carriers, the Mn spins do not interact significantly, unless they are first neighbors.

30.3.2 Two Electrons and More

We now discuss how the ground state spin of the dot evolves, as additional carriers are injected into the dot by means of a gate voltage (as opposed to the injection of nonequilibrium transport carriers). When several Mn spins interact with a single electron, they tend to orient their spins parallel to that of the electron, which can take two orientations. The addition of a second electron in the same single-particle level results in the formation of an electron spin singlet that should have very little influence on the Mn spins. That is exactly what is obtained from the numerical diagonalization of the Hamiltonian. In the case of the dot with two Mn and two electrons, the ground state spin is $S=5/2$, the same as in the case of a dot without extra carriers. The case with two Mn and two electrons is more interesting because, in the absence of carriers and interactions, the spin of this system can take values between $S=0$ and $S=5$, all of them with the same energy. The presence of the two electrons breaks this degeneracy through virtual inter-level excitations (see inset of right panel in Figure 30.2). The lowest energy configuration has the two electrons with total spin zero in the lowest energy level. This configuration is coupled to higher energy configurations with two unpaired electrons, one of them in the excited QD level. This coupling enables spin density fluctuations of the two electron droplets, involving as intermediate or virtual state an excitation of one of the electrons to an excited QD level. As we show in the following, these processes are responsible for indirect exchange interaction between different Mn atoms whose strength scales as Δ^2/δ, which is much smaller than the interaction mediated by an unpaired electron of the order Δ. This is the QD version of RKKY interaction and, as in the bulk case, it can be either positive or negative, depending on the position of the two Mn atoms in the QD.

The addition of a third electron leaves the electron with two paired electrons in the low-energy level and one unpaired electron in the second energy level [37]. Not surprisingly, the behavior of such a dot is very similar to that with a single electron. As a general rule, there are two types of behaviors. When there are unpaired electrons, there are ferromagnetic correlations like those of magnetic polarons, enhanced compared to bulk because confinement favors the localization of the itinerant electron. In the absence of unpaired electrons, the carrier-mediated interactions are much weaker than in the polaron case and can be either ferromagnetic or antiferromagnetic. In the next section, we provide an analytical background for these statements, which so far are backed up by numerics.

30.3.3 Many Impurities: Classical Approximation

Here we provide an analytical argument [37] that permits to go beyond the numerical approach of the previous section, which is limited to a few magnetic atoms. The generalization is made possible by an approximation: we consider the Mn spins as classical variables. By so doing, the size of the Hilbert space is reduced by a factor 6N, where N is the number of Mn spins. The partition function of the dot is written, in the canonical ensemble, a

$$Z_N = \int d\vec{M}_1 \cdots d\vec{M}_N e^{-\beta \mathcal{H}_{AF}(\vec{M}_I)} \sum_{\bullet} e^{-\beta E_{\bullet,N}(\vec{M}_I)}, \quad (30.8)$$

where • labels a Slater determinant eigenstate of the fermionic Hamiltonian with eigen-energy, which depends parametrically on the spin arrangement. The integrals sum over all the possible classical spin configurations. We make further progress by doing some approximations. First, in the case of $\delta \gg k_B T$ we can keep the lowest energy configuration in the sum. Second, in the case of unpaired electron configuration the dominant contribution to the multielectronic energy comes from the exchange coupling of the unpaired to the classical spins. This term is formally identical to the energy of an electron interacting with an effective magnetic field $g\mu_B \vec{B}_{eff} = \sum_I j_e(I)\vec{M}_I$, where $j_e(I) \equiv \alpha \,|\, \psi_n(\vec{r}_I) \,|^2$, where n is the QD level occupied by the unpaired electron. By aligning its spin parallel to this effective magnetic field, the unpaired electron lowers its energy by an amount

$$g\mu_B \,|\, \vec{B}_{eff} \,| = \sqrt{\sum_{I,I'} j_e(I) j_e(I') \vec{M}_I \cdot \vec{M}_{I'}}. \quad (30.9)$$

It is apparent that this term favors collinear Mn spin arrangements and it scales linearly with the Mn–electron interaction constant α. Interestingly, the resulting effective model is the square root of a long range Heisenberg model.

In the case of paired ground state configurations, the contribution above has opposite signs for the two electrons and vanishes. Thus, the leading order contribution comes from interlevel exchange. Since this term is small, we can obtain it by perturbation theory, using an unperturbed Hamiltonian without exchange coupling. The sum of the second-order perturbation theory corrections of the occupied one-particle levels reads:

$$\varepsilon_e = \sum_{I,I'} \left[\frac{J_c^2}{2} \sum_{n,n'} \gamma_{n,n'}(I,I') \frac{f_n - f_{n'}}{\varepsilon_n^0 - \varepsilon_{n'}^0} \right] \vec{M}_I \cdot \vec{M}_{I'}, \quad (30.10)$$

where $\gamma_{n,n'}(I,I') \equiv \phi_n^*(\vec{x}_I)\phi_{n'}(\vec{x}_I)\phi_n^*(\vec{x}_{I'})\phi_{n'}(\vec{x}_I)$ and $f_n = 0,1$ are the occupation of the unperturbed dot in the ground state electronic configuration.

Importantly, $\gamma_{n,n'}(I,I')$ can be both positive or negative, so that the effective coupling can be either ferromagnetic or antiferromagnetic, depending on the location of the Mn atoms and the number of electrons in the dot. This full shell coupling is similar to bulk RKKY interaction and is much smaller than the unpaired electron or polaronic contribution.

In summary, the addition/removal of a single carrier can change quite drastically the effective interactions between the Mn spins. Dots with an unpaired electron tend to form a magnetic polaron. Dots with closed shell have residual carrier-induced interactions that can be either ferromagnetic or antiferromagnetic. Thus, they should behave like nanometric sized spin glasses at very low temperatures, and like standard paramagnets above a few Kelvin. In both situations, the very small size of the SO interaction for both half full d shell (Mn^{2+}) and conduction band electrons results in a very small magnetic anisotropy for the types of states. As we see in the next section, the situation is radically different in the case of Mn interacting with QD holes.

30.4 Hole-Doped (II,Mn)-VI Quantum Dot

In the case of holes, both the degeneracy of the valence band and the very strong SO interaction play an important role [1,33]. In bulk, the top of the valence band is formed by two degenerate bands, LH and HHs, well above in energy from the split-off band. In the case of CdTe, this splitting is larger than 0.8 eV reflecting the strength of SO interaction of Te. Within the effective mass approximation, the states of these bands can be described as linear combinations of the states at the Γ point, which have the same symmetry as the atomic states and are strongly affected by SO interaction. The light and heavy wave functions transform like the eigenstates of the total angular momentum J formed from the sum of a $l = 1$ and a spin $S = 1/2$ object. The eigenstates of the $J = 3/2$ sector are written as linear combinations of the product basis in which S_z and l_z are good quantum numbers [48]:

$$\begin{aligned}
\left| J = \frac{3}{2},\ J_z = \frac{+3}{2} \right\rangle &= \left| l_z = 1, \uparrow \right\rangle, \\
\left| J = \frac{3}{2},\ J_z = \frac{-1}{2} \right\rangle &= \frac{1}{\sqrt{3}} \left| l_z = -1, \uparrow \right\rangle + \sqrt{\frac{2}{3}} \left| l_z = 0, \downarrow \right\rangle, \\
\left| J = \frac{3}{2},\ J_z = \frac{+1}{2} \right\rangle &= \frac{1}{\sqrt{3}} \left| l_z = 1, \downarrow \right\rangle + \sqrt{\frac{2}{3}} \left| l_z = 0, \uparrow \right\rangle, \\
\left| J = \frac{3}{2},\ J_z = \frac{-3}{2} \right\rangle &= \left| l_z = -1, \downarrow \right\rangle.
\end{aligned} \quad (30.11)$$

It is apparent that spin is not a good quantum number in these basis states and it will not be a good quantum number for holes in general. In the QD, confinement and strain split the LHs and HHs. In most instances, the lowest energy hole states (i.e., the highest energy states in the valence band) are made almost entirely of HH states.

In order to describe the single-particle states for the QD holes, we use the same hard wall cuboid approximation as in the case of electrons. But in this case, the resulting effective mass problem leads to six coupled differential equations, due both to SO interaction and the degeneracy of the top of the valence band. This model has been proposed by Kyrychenko and Kossut [49] and is used extensively in the study of single Mn-doped QDs [26,50]. For given dimensions of the dot, L_x, L_y, and L_z, and given principal quantum numbers for the envelope function, n_x, n_y, and n_z, the six differential equations are mapped into a range of six matrix problems, given by the Kohn-Luttinger *kp* Hamiltonian. The $k=0$ eigenstates of this Hamiltonian transform like the eigenstates of $J=3/2$ ($J_z=\pm3/2$, $\pm1/2$) and $J=1/2$ ($J_z=\pm1/2$), the two values of the angular momentum that can be formed with $l=1$, the orbital momentum of *p* orbitals, and $S=1/2$, the spin of the electrons. The finite $\vec{k}$ states are linear combinations of these states, and therefore J and J_z are no longer good quantum numbers. In the approach of Kyrychencko and Kossut, the confined QD states are obtained by diagonalization of the bulk Kohn Luttinger Hamiltonian where k_a^2 is replaced by $(\pi n_a/L_a)^2$, where $a=x,y,z$, and the terms linear in $\vec{k}$ vanish. In this model, the lowest energy bands are made mostly of LHs and HHs (states within the $J=3/2$ manifold). In bulk, the wave vector determines the degree of mixing between light ($J=3/2$, $J_z=\pm1/2$) and heavy ($J=3/2$, $J_z=\pm3/2$) holes. In the QD, effective wave vector of the state is determined by the dimensions of the dot. Within this model, the ground state of a QD with $L_z \ll L_x, L_y$ is mostly made of HHs, and the degree of mixing to light holes (LHs) depends on the aspect ratio $r=L_x/L_y$. For $r=1$, the ground state is purely made of HHs.

30.4.1 One Hole

We are now in position to study the problem of QD holes interacting with Mn spin in a QD. We denote the confined QD hole states by the label ν, which combines orbital and spin degrees of freedom. The Hamiltonian for holes exchanged coupled to Mn atoms in a QD reads:

$$\mathcal{H}_{0h} + \mathcal{H}_{h-Mn} = \sum_{\nu} \varepsilon_\nu f_\nu^\dagger f_\nu + \beta \sum_{I,\nu,\nu'} \vec{M}_I \cdot \left\langle \nu \left| \frac{\vec{\tau}_{\sigma,\sigma'}}{2} \delta(\vec{r} - \vec{r}_I) \right| \nu' \right\rangle. \tag{30.12}$$

The first case we consider is a QD doped with one hole, which can occupy two states that we denote ⇑ and ⇓. The competition between SO and exchange becomes apparent in the evaluation of the spin matrix elements in the subspace formed by the lowest energy doublet. If there is absolutely no LH–HH mixing, the spin-flip process by which the HH with spin +1/2 goes to −1/2 increasing in one unit the spin of a Mn spin is forbidden, because the orbital part of initial and final HH states are orthogonal:

$$\left\langle \frac{3}{2}, \frac{-3}{2} \right| \tau^- \left| \frac{3}{2}, \frac{+3}{2} \right\rangle = \langle l=1, m=-1 | l=1, m=+1 \rangle \times \langle \downarrow | \tau^- | \uparrow \rangle = 0. \tag{30.13}$$

In contrast, the spin-conserving channel is totally open. Thus, Mn-HH spin exchange results in and Ising coupling:

$$\mathcal{H}_{\text{Ising}} = \frac{\beta}{2}\left(f_\Uparrow^\dagger f_\Uparrow - f_\Downarrow^\dagger f_\Downarrow\right) \sum_I |\psi_0(\vec{r}_I)|^2 M_I^z. \tag{30.14}$$

This Hamiltonian describes the Mn spins of the dot Ising coupled to the spin of the hole. The quantization axis is given by the growth direction that determines the shape of the dot. We define the effective coupling $j_h(I) \equiv \beta|\psi_0(\vec{r}_I)|^2$, which is positive because the hole–Mn coupling is antiferromagnetic. In the case of a CdTe dot doped with a single Mn, the spectrum of this Hamiltonian is given by six doublets $E(M_z,\pm) = \pm\frac{j_h}{2} M_z$, where M_z is the Mn spin projection along the growth axis and can take six values between −5/2 and +5/2 (see Figure 30.3. This spectrum can be rationalized into two Zeeman split families associated with the two possible orientations of the hole spin. The ground state doublet is given by the states (⇑,−5/2) and (⇓,+5/2).

This apparently simplistic model is validated by the experimentally observed PL of CdTe QDs doped with one Mn [16] that features six peaks. Basically, the dominant exciton Mn coupling is given by the hole–Mn coupling since we have $\beta \simeq 4\alpha$ [1]. Hence, the Mn is Ising coupled to the exciton and both the Mn and the exciton spin are conserved. Thus, the energy of the exciton depends on which of the six Mn spin projections along the growth axis. Therefore, in the presence of the HH the spin of the Mn acquires a strong anisotropy with the growth direction as the easy axis. This anisotropy arises from the strong SO of the hole, rather than from the very weak single-ion anisotropy of the Mn atom. Thus, the QD affords the unique opportunity of controlling the anisotropy of the Mn atom(s) by injection of a single hole [17,32,51].

In general, the QD–confined states have some mixing between HHs and LHs, determined by the shape of the dot and the strain. This mixing opens a little the spin-flip channel, resulting in strongly anisotropic Heisenberg coupling:

$$\mathcal{H}_{XXY} = \sum_I j_h(I)\left[\left(f_\Uparrow^\dagger f_\Uparrow - f_\Downarrow^\dagger f_\Downarrow\right)M_I^z + \frac{\varepsilon}{2}\left(f_\Uparrow^\dagger f_\Downarrow M_I^{(-)} + f_\Downarrow^\dagger f_\Uparrow M_I^{(+)}\right)\right], \tag{30.15}$$

where ε is a dimensionless factor linearly proportional to the overlap of both ⇑ and ⇓ states with the LH states ($J=3/2$, $J_z=\pm1/2$). As long as the anisotropy parameter ε is different from 1 (which would retrieve the isotropic Heisenberg coupling), the spin of the holes and the spin of the Mn are not good quantum numbers. Interestingly, the PL spectra of some single Mn doped can only be modeled if we take a finite value of ε, typically smaller than 0.2 [26]. The fact that different dots have different ε is understood within the theory above because of the variation of shapes, sizes, and strains from dot to dot.

30.4.2 Two Holes and More

As we did in the case of conduction band electrons, we consider now the effect of increasing the number of holes in the dot one

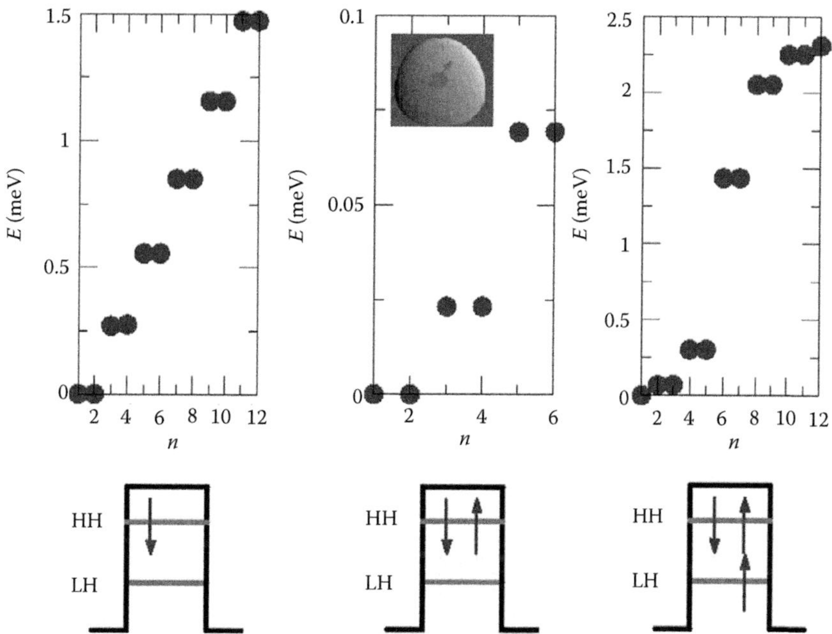

FIGURE 30.3 Spectra of single Mn QD dot doped with one, two, and three holes.

by one. On top of the even–odd effect found in the case of electrons, we also can expect strong variations in the anisotropy of the hole–Mn coupling depending on the light or heavy nature of the holes. We restrict the study to dots doped with one Mn, which are already interesting. We start with the case of two holes. The single-particle energy spacing is smaller for holes than for electrons, both because the larger effective mass and, in the case of CdTe/ZnTe, because the energy barriers are smaller for the valence band. On the other hand, exchange interaction with the Mn is larger for holes than for electrons. However, it is still true that the inter-level spacing is much larger than the exchange energy. Thus, the ground state with two holes is mostly made of configurations where the two holes occupy the lowest energy doublet and behave almost like a spin-zero object, like in the case of the dot with two electrons. Thus, the ground state is similar to that of the free Mn spin. But in contrast with the electrons, the total spin S is not a good quantum number and inter-level coupling has some observable effects on the spectrum of the system. In particular, the otherwise degenerate six states of the quasi-free Mn spin acquire a fine structure, as seen in the figure, which can be fitted according to the equation $D_{2h}M_z^2$, where $D_{2h} \propto \frac{j_h^2}{\delta}$. This fine structure splitting has been observed in the PL spectra of biexciton to exciton transitions [52], which is slightly different from the exciton to ground PL. The size of the splitting is 0.12 meV, the same order of magnitude than the one in Figure 30.3 and significantly larger than the single-ion anisotropy.

The case of a QD with three holes is also different from the case with three electrons. The ground state wave function is mostly composed configurations with two holes in the lowest energy doublet and one unpaired hole in the LH level that dominates the behavior of the system. The low-energy sector spectrum of the system, shown in the right panel of Figure 30.3, can be rationalized as follows. There are two main groups of states: five low-energy states and seven high energy states, which, in the absence of SO interaction, would merge into an $S=2$ low-energy quintuplet and an $S=3$ high energy manifold that reveals the antiferromagnetic (AF) Mn–hole coupling. The splitting inside the groups arises from the anisotropic exchange interaction that results from SO coupling.

30.5 Electrically Tunable Nanomagnet

The results of the previous sections could be summarized as follows: it is possible to reversibly tune the magnetic properties of a given QD-doped Mn by changing the charge state of the dot Q. Focusing in the case of single Mn-doped dots, electron doping permits to have two types of states, free Mn with $S=5/2$ for $Q=0,-2,-4$, in closed shell configurations and, in open shell configurations with one unpaired electron ($Q=-1,-3,-5$), a "polaronic" state where the Mn and the unpaired electron spin have strong ferromagnetic spin correlations behaving like a spin $S=3$ object. On top of this open versus closed shell dichotomy, holes add two additional ingredients. First, the $2S+1=6$ degeneracy of closed shell configurations is split due to inter-level exchange that produces a fine structure splitting identical to that of single-ion anisotropy. Second, in configurations with one unpaired hole the symmetry properties of the magnetic state are very different depending on the LH-HH character of the highest-occupied confined level.

The fact that the low-energy spectra of the single Mn-doped QD can be changed reversibly for different charge states, as shown in Figure 30.3, has lead us to propose this system as an *electrically tunable nanomagnet* [32], and is related to the idea of voltage control of the DMS QD [37,44,53]. Hole-doped single Mn

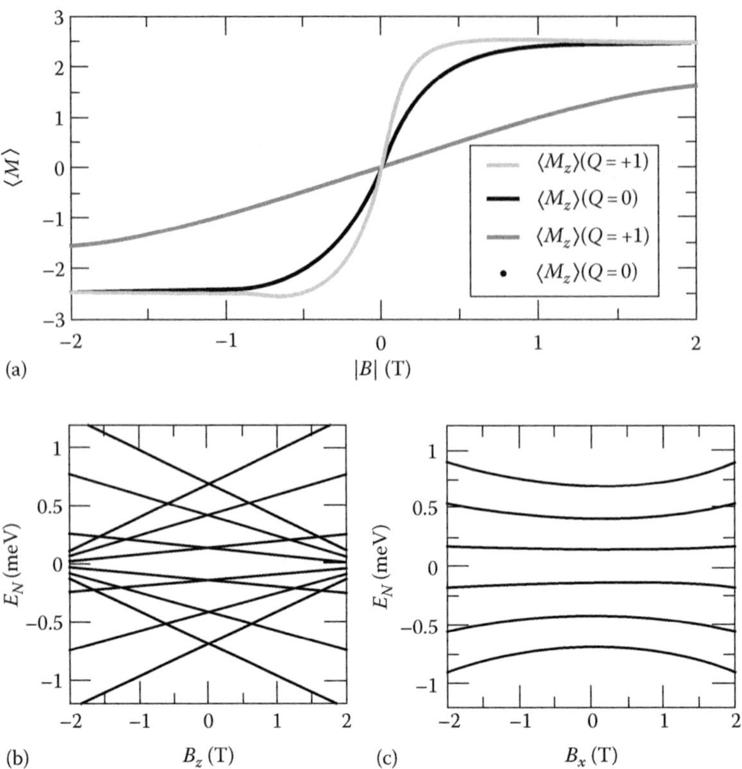

FIGURE 30.4 (See color insert.) (a) Equilibrium magnetization of the Mn spin for a dot with $Q=0$ and $Q=+1$ as a function of the intensity of the applied field for two orientations. (b) and (c) Energy spectra of single Mn-doped QD with one hole as a function of the magnetic field for two directions: parallel to growth axis (b) and perpendicular to the growth axis (c). (Fernández-Rossier, J. and Aguado, R.: Mn doped II-VI QDs: Artificial single molecule magnets, *Phys. Stat. Sol (c)*, 2006, 3, 3734–3739. Copyright Wiley-VCH Verlag GmbH & Co. KGaA. Reproduced with permission.)

QDs definitely have one of the two main properties of a nanomagnet, namely, a finite magnetic moment. Whether or not the system displays remanence and coercivity, like in the case of single molecule magnets, remains to be determined experimentally. Theoretically, remanence and coercivity require two ingredients: magnetic anisotropy and blocking of the spin relaxation. Single Mn-doped CdTe QD with one resident hole presents a very large magnetic anisotropy: the response of the system to an applied magnetic field along the growth (z) axis (Farady configuration) is very different from the one with the field applied in plane (Voigt configuration). Experimentally this is inferred from the very different Voigt and Faraday magneto-PL spectra [26,54]. In Figure 30.4, we plot the average magnetization of the Mn spin for a dot with $Q=0$ and $Q=+1$ as a function of the intensity of the applied field for two orientations. The magnetization is calculated using the equilibrium ensemble. In the $Q=0$ case, there is a very small difference, caused by the very weak crystalline anisotropy of the Mn ion. In contrast, the magnetic anisotropy caused by the hole in the $Q=+1$ case is much larger. It can be understood by inspection of the evolution of the 12 low-energy states of the QD with one Mn and one hole as a function the applied field. In the Faraday case, the field splits the levels with well-defined M_z and S_z. In the Voigt case, there is a competition between Zeeman coupling and Ising coupling that decreases the susceptibility.

30.6 Mn in III–V Quantum Dots

As opposed to the case of Mn-doped II–VI QDs, for which there is a wide range of experimental work, Mn-doped III–V QDs are much less studied. In this section, we briefly comment on the microscopic model adequate to understand the PL spectra of a single Mn-doped (In,Ga)As QD reported by Kudelsky et al. [22].

In stark contrast with the II–VI, substitutional Mn in bulk GaAs and InAs behaves like a shallow acceptor. EPR [55] and photoemission [56] experiments indicate that Mn retains the five d electrons when doping concentrations are small. Hence, Mn keeps an oxidation state of +2 resulting in an effective charge of −1 that repels the electrons nearby. The Mn impurity remains charge neutral, at the scale of a few unit cells, by binding a hole. The binding energy of the hole is of 110 meV in Ga(Mn)As and 28 meV in In(Mn)As. The acceptor hole state has a radius of 1 nm and has been probed by scanning tunneling microscope (STM) experiments both in GaAs and InAs [57]. If the size of the dot is much larger than the acceptor hole state and the Mn is far from the dot interface, we can expect that the dot has a weak influence on the acceptor state. This view is consistent with the PL experiments [22,27].

In bulk, the acceptor hole has a fourfold degeneracy inherited from the top of the valence band, which is lifted by quantum confinement and/or strain [58]. Because of the strong SO interaction, it is convenient to treat the acceptor hole as a spin $J=3/2$ object, exchanged coupled to the Mn spin $M=5/2$. It is also convenient to think of the acceptor states as if there was an unpaired electron with quantum numbers $J=3/2$, $J_z=\pm3/2$, $\pm1/2$ interacting antiferromagnetically with the Mn spin. The operators acting upon this object are the four by four $J=3/2$ angular momentum matrices, $\vec{J}$. In the spherical approximation, the Mn–acceptor hole spin coupling reads [58]: $\mathcal{H}_{G0}=\varepsilon\vec{M}\cdot\vec{J}$, where $\varepsilon=+5$ meV is the antiferromagnetic coupling between the hole and the Mn, $\vec{M}$ are the $S=5/2$ spin matrices of the Mn, and $\vec{J}$ are the $J=3/2$ matrices corresponding to the total angular momentum of the valence band states. The Hamiltonian $\mathcal{H}_{G0}$ is readily diagonalized on the basis of the total spin $F=M+J$, the spin of the Mn plus hole complex. F can take integer values between 1 and 4. The eigenvalues are $E(F)=\frac{\varepsilon F(F+1)}{2}+E_0$. Since the coupling is antiferromagnetic, the ground state has $F=1$, separated from the $F=2$ states by a sizable energy barrier of 2ε. Hence, as long as the Mn–hole complex is not distorted by perturbations that couple different F manifolds, it is a good approximation to think of it as a composite object with total spin $F=1$, whose wave functions read $|F=1,F_z=\pm1,0\rangle=\sum_{M_z,J_z}C_{M,J_z}(F,F_z)|M_z,J_z\rangle$. We see that because of the very strong exchange interaction, the spins of the Mn and the acceptor hole are strongly correlated and they are not good quantum numbers separately.

The cubic symmetry of the ideal crystal and the presence of quantum confinement and strain result in additional terms in the Hamiltonian that need to be summed to $\mathcal{H}_{G0}$. Both QD confinement and strain can result in a splitting of the LH and HH bands that can be modeled a $-DJ_z^2$ term. This term would be present in thin film layers with strain and still preserves rotational invariance in the xy plane. The presence of the QD potential will break this in-plane symmetry. To lowest order, this can be modeled by an additional term in the Hamiltonian, $E(J_x^2-J_y^2)$ [59]. It is assumed that these perturbations act on the acceptor state only and not on the Mn d electrons. This is justified since the hole is spread over tens of unit cells, whereas the Mn d states are confined in a single unit cell. Hence, the following model for the ground state Hamiltonian results: $\mathcal{H}_G=\varepsilon\vec{M}\cdot\vec{J}-DJ_z^2+E(J_x^2-J_y^2)$ Since $\varepsilon\gg D\gg E$, this Hamiltonian can be projected on the $F=1$ lowest energy manifold, resulting in the Hamiltonian [27,59]:

$$\mathcal{H}_G=-D_F F_z^2+E_F(F_x^2-F_y^2)+g_F\mu_B F_z B, \tag{30.16}$$

where

$$D_F=\tfrac{3}{10}D$$
$$E_F=\tfrac{3}{10}E$$
$$g_F=\tfrac{7}{4}g_M-\tfrac{3}{4}g_J$$

The D_F term splits the $F=1$ triplet into a singlet $|1,0\rangle$ and a doublet $|1,\pm1\rangle$. The E_F term mixes the two states in the ±1 doublet resulting in bonding and antibonding states along the Y and X axis, respectively.

A III–V QD doped with a Mn atom could undergo two types of low-energy charge fluctuation: First, removal of the acceptor hole that would liberate the spin $S=5/2$ of the Mn from its strong coupling with the acceptor hole and, second, addition of a hole. From the interpretation [27] of the PL experiments [22], we know that the second hole goes into a QD state, rather than to doubly occupy the acceptor hole. The experimental results are accounted for if the interaction between the QD hole and the $F=1$ composite spin of the Mn–acceptor complex is modeled as an Ising coupling.

30.7 Sequential Transport in Mn-Doped Quantum Dots

30.7.1 Introduction

The experimental demonstration of electrical control of DMS QDs down to the limit of a single Mn atom [17] paves the way to electric transport studies in these systems. Although such challenging experiments have not hitherto been performed, transport through individual self-assembled nonmagnetic QDs has been achieved. The main technical difficulty in performing such measurements through individual self-assembled QDs, that is, the attachment of electrodes, has been overcome by developing specific techniques such as planar SETs [60], gated submicrometer mesa devices [61,62], and nanolithographically defined metallic nanogap electrodes [63]. Recent successes in making electrical contact to a single self-assembled QD beyond the latter case (normal metallic contacts) include the coupling to ferromagnetic leads [78] and to superconducting leads [64]. Quite remarkably, this technique even allows to reach the strong tunneling limit, where the Kondo effect for both normal [65] and superconducting regimes [66] has been measured. All this experimental progress suggests that our theoretical proposal of electrical transport measurements with Mn-doped self-assembled QDs should be within reach in the next few years. Before explaining in detail transport through Mn-doped QDs, we summarize, for completeness, the main concepts and ideas of sequential transport in nonmagnetic QDs [67].

Arguably, the most relevant transport phenomenon in QDs is Coulomb Blockade. When tunneling occurs, the charge on the QD changes by the quantized amount e. This event changes the electrostatic energy of the QD by the charging energy $E_C=e^2/C$. Charging effects become important when E_C exceeds the thermal energy k_BT. Another important requirement is that the tunneling resistance of the barrier $R_t=1/(e^2\Gamma)$, with Γ being the typical level width due to tunneling, should be much larger than the resistance quantum $h/e^2=25.813\,\text{k}\Omega$, such that quantum fluctuations of charge due to tunneling are strongly suppressed. At lowest order in tunneling, transport occurs if the electrochemical potential of the QD occupied by N electrons $\mu_{dot}(N)$ lies between the electrochemical potentials of the reservoirs $\mu_L=E_F+V_L$ and $\mu_R=E_F+V_R$, where $V=\mu_L-\mu_R=V_L-V_R$ is the applied bias voltage. The electrochemical potential of the dot is, by definition, the minimum energy needed to add the Nth electron to the dot:

$\mu_{dot}(N) = U(N) - U(N-1)$, where $U(N)$ is the total ground state energy for N electrons on the dot at zero temperature. When at fixed gate voltage the number of electrons changes by one, the change in electrochemical potential is given by the addition energy $\mu_{dot}(N+1) - \mu_{dot}(N) = \Delta\varepsilon + E_C$, which is large for small capacitances and/or large energy splittings $\Delta\varepsilon$ between confined states. Typical energy scales for both $\Delta\varepsilon$ and E_C in self-assembled QDs are in the range of a few tens of meV [63] such that Coulomb Blockade is observed for temperatures below $T \approx 5$ K. Note that if two electrons are added to the same spin-degenerate level $\Delta\varepsilon = 0$n and their energy difference is given by the charging energy only. A great deal of theoretical papers consider this situation in which transport is supposed to occur through only one spin-degenerate state. When $E_C \gg k_B T$, the charging energy dominates transport. If $\mu_{dot}(N+1) > \mu_L$, then $\mu_R > \mu_{dot}(N)$ transport is blocked and the QD is in the Coulomb Blockade regime. The Coulomb Blockade can be removed by changing the gate voltage, to align $\mu_{dot}(N+1)$ between the chemical potentials of the reservoirs such that an electron can tunnel from the left reservoir to the dot and from the dot to the right reservoir, which causes the electrochemical potential to drop back to $\mu_{dot}(N)$. A new electron can now enter the dot such that the cycle $N \Rightarrow N+1 \Rightarrow N$ is repeated. This process is called single-electron tunneling and QDs exhibiting this physics are often called SETs. By changing the gate voltage, the linear conductance oscillates between zero (Coulomb Blockade) and nonzero. In the regions of zero conductance, the number of electrons inside the QD is fixed. At finite bias voltages, the large regions of nearly zero conductance form characteristic Coulomb diamonds in differential conductance plots. Excited states (e.g., electronic, vibrational, or spin degrees of freedom) appear as visible lines in the diamonds making this kind of transport measurements a useful form of level spectroscopy.

As the tunneling to the reservoirs becomes larger, namely, as the resistance of the barriers approaches the quantum of resistance, higher-order tunneling events become relevant. In this situation, quantum fluctuations dominate transport and electrons are allowed to tunnel via intermediate virtual states where first-order tunneling would be suppressed. Thus, the intrinsic width of the energy levels of the QD does not only include contributions from direct elastic tunneling but also tunneling via virtual states. These higher-order tunneling events are referred to as co-tunneling processes. These higher-order tunneling events lead to spectacular effects when the spin of the electrons is also involved. Importantly, a QD with a net spin coupled to electron reservoirs resembles a magnetic impurity coupled to itinerant electrons in a metal and, thus, can exhibit Kondo effect. In this regime, quantum fluctuations induce an effective exchange coupling $J \approx \frac{\Gamma E_C}{|\Delta\varepsilon||\Delta\varepsilon + E_C|}$ between the QD spin and the reservoir ones. This exchange results in a complete screening of the QD spin and, accordingly, the ground state becomes a singlet between confined and itinerant carriers. This mechanism gives rise to a new many-body resonance in the density of states of the QD, around the Fermi energy of the reservoirs. The most remarkable manifestation of this new "Kondo" resonance is the transition from near-zero conductance due to Coulomb Blockade to perfect transmission as the temperature is lowered well below the so-called Kondo temperature $T_K \sim e^{\frac{-\pi|\Delta\varepsilon||\Delta\varepsilon+E_C|}{2\Gamma E_C}}$. Clear experimental signatures of Kondo physics have been reported in various types of QDs as well as other systems that can be understood in terms of a SET setup. The latter includes molecules, break-junctions, etc. For a short review on the Kondo effect in QDs see Ref. [68].

30.7.2 Model

All the above physics is captured by the Hamiltonian $H = H_{QD} + H_{res} + H_{tunn}$, which describes a QD coupled to reservoirs. $H_{QD} = \sum_n \varepsilon_n \mathbf{f}_n^\dagger \mathbf{f}_n + H_C + H_{int}$ is the Hamiltonian of the isolated QD. The first term defines the energy of confined carriers within the QD. Coulomb effects are described within the constant interaction model $H_C = (\hat{Q} - Q_{gate})^2/2C$. $\hat{Q} = e\hat{N}$ is the extra charge in the QD (where $\hat{N} = \sum_n \mathbf{f}_n^\dagger \mathbf{f}_n - N_0$ is the number of extra carriers on top of the background N_0), $C = C_L + C_R + C_{gate}$ is the total capacitance, and $Q_{gate} = V_L C_L + V_R C_R + V_{gate} C_{gate}$ is the total induced gate charge. Finally, H_{int} contains extra interactions not included in the above charging model. In our case, this would correspond to the Hamiltonian of confined carriers interacting with the Mn that we have described in previous sections. The Hamiltonian of the metallic electrodes is $H_{res} = \sum_{k\alpha \in L,R} \varepsilon_{k\alpha} \mathbf{c}_{k\alpha}^\dagger \mathbf{c}_{k\alpha}$, with $\mathbf{c}_{k\alpha}^\dagger$ the creation operator of an electron/hole in the left/right reservoir (α is a channel index that takes into account spin, orbital degeneracies, etc.). The coupling between the QD and the reservoirs reads $H_T = \sum_n \sum_{k\alpha \in L,R} V_{k\alpha,n} \mathbf{c}_{k\alpha}^\dagger \mathbf{f}_n + V_{k\alpha,n}^\star \mathbf{f}_n^\dagger \mathbf{c}_{k\alpha}$, which is the standard tunneling Hamiltonian describing electron/hole tunneling in/out of the QD.

The transport properties of a QD whose dynamics is governed by the Hamiltonian in Equation 30.8.2 cannot be solved exactly and one has to adopt some approximation scheme. Most of the calculations employed for the description of transport through an interacting region fall into two different families: nonequilibrium (Keldysh) Green's functions techniques and quantum master equations (QMEs). Trying to clarify the advantages/disadvantages of each approach, and their connection, we briefly present them in the next two sections. Those readers who only want to acquire new information about the transport properties of magnetic QDs are encouraged to safely skip these two technical sections and directly go to Section 30.8.5.

30.7.3 Exact Expression for the Current Using the Keldysh Method

In the first approach, more "condensed-matter" oriented, one writes physical averages with the help of single-particle nonequilibrium Keldysh Green's functions. Formally, this approach is very powerful as it allows to derive an exact expression for the current through the interacting region [69]. Using

$$J_L = -e\langle \dot{N}_L \rangle = -\frac{ie}{\hbar}\langle [H,N_L] \rangle = \frac{ie}{\hbar}\sum_{\substack{k,\alpha\in L\\ n}}\left[V_{k\alpha,n}\langle \mathbf{c}^\dagger_{k\alpha}\mathbf{f}_n\rangle - V^*_{k\alpha,n}\langle \mathbf{f}^\dagger_n \mathbf{c}_{k\alpha}\rangle\right] \quad (30.17)$$

and assuming that the leads are non-interacting, the current can be written as

$$J_L = \frac{ie}{\hbar}\sum_{\substack{\alpha\in L\\ n,m}}\int \frac{d\varepsilon}{2\pi}\nu_\alpha(\varepsilon)V_{\alpha,n}(\varepsilon)V^*_{\alpha,m}(\varepsilon)\{f_L(\varepsilon)G^>_{nm}(\varepsilon) - [1-f_L(\varepsilon)]G^<_{nm}(\varepsilon)\}, \quad (30.18)$$

where $\nu_\alpha(\varepsilon) = \Sigma_k \delta(\varepsilon - \varepsilon_{k\alpha})$ is the density of states in channel α and $f(\varepsilon)$ is the fermi function. $G_{nm}(\varepsilon)$ is the Fourier transform of the so-called lesser (greater) Green's functions defined as $G^<(t,t') = \frac{i}{\hbar}\langle \mathbf{f}^\dagger_m(t')\mathbf{f}_n(t)\rangle$ and $G^>(t,t') = -\frac{i}{\hbar}\langle \mathbf{f}_n(t)\mathbf{f}^\dagger_m(t')\rangle$, respectively. Then it is useful to define the level-broadening functions as $[\Gamma^L(\varepsilon)]_{mn} = 2\pi\sum_{\alpha\in L}\nu_\alpha(\varepsilon)V_{\alpha,n}(\varepsilon)V^*_{\alpha,m}(\varepsilon)$. When these expressions are substituted to Equation 30.18, the current from left (right) contact to central region becomes

$$J_{L(R)} = \frac{ie}{\hbar}\int\frac{d\varepsilon}{2\pi}\mathrm{Tr}\{\Gamma^{L(R)}(\varepsilon)(f_{L(R)}(\varepsilon)\mathbf{G}^>(\varepsilon) + [1-f_{L(R)}(\varepsilon)]\mathbf{G}^<(\varepsilon))\}, \quad (30.19)$$

where the bold notations denote matrices. $\mathbf{G}(\varepsilon)$ is the many-body density of states for the addition (removal) of an extra particle of energy ε, which suggests the intuitive interpretation of the current in Equation 30.19 as "current-in" minus "current-out." In fact, using the definition of the full many-body spectral function as $\mathbf{A} = i[\mathbf{G}^>(\varepsilon) - \mathbf{G}^<(\varepsilon)]$, one can write the total current $J = (J_L - J_R)/2$ as a generalized Landauer equation

$$J = \frac{e}{\hbar}\int\frac{d\varepsilon}{2\pi}[f_L(\varepsilon) - f_R(\varepsilon)]\mathrm{Tr}\left\{\frac{\Gamma^L(\varepsilon)\Gamma^R(\varepsilon)}{\Gamma^L(\varepsilon)+\Gamma^R(\varepsilon)}(\mathbf{A}(\varepsilon))\right\}. \quad (30.20)$$

The term in brackets can be interpreted as a many-body transmission coefficient. Equation 30.20 contains broadening to arbitrary order and is, of course, equivalent to the scattering matrix method in the absence of interactions. However, its usefulness depends on whether one is able to make some suitable perturbative expansion for the interacting Green's functions. This Keldysh method, combined with ab initio quantum chemistry calculations, is extensively used in molecular electronics transport calculations (see Chapter 32). Another useful route, fully equivalent to real-time Green's functions, is the so-called imaginary-time formulation for steady-state nonequilibrium [70], which has its roots in Hershfield's nonequilibrium steady-state density matrix [71]. This formalism has the advantage of having the mathematical structure of equilibrium theory and, in principle, can be easily used in combination with established equilibrium techniques.

30.7.4 Quantum Master Approach

The second approach, more "quantum-optics" oriented, focuses on the reduced dynamics of the QD density operator after the reservoirs are integrated. In this approach, interaction effects, such as magnetic exchange, are fully taken into account from the beginning (the density matrix is written in the exact eigenbasis that diagonalizes the isolated problem). The drawback of this approach is that it is very difficult to include broadening of many-body states beyond low orders in tunneling. This results in a QME that describes the approximate dissipative dynamics of exact many-body states.

QMEs can be derived under many different approximations and assumptions. These include various Markovian schemes, such as the standard Bloch–Redfield approach, and more sophisticated superoperator methods such as the Nakajima-Zwanzig (NZ) projection operator technique or diagrammatic expansions in Liouville space, just to mention a few. A comprehensive review of all these techniques can be found in Ref. [72].

The dynamics of the density operator in the interaction picture is governed by the Liouville-von Neumann equation $\frac{d\hat{\rho}^I}{dt} = -i[H_T, \hat{\rho}^I]$. Integrating this equation with the initial condition $\hat{\rho}(t_0) = \hat{\rho}^{QD}(t_0)\otimes\hat{\rho}^{res}_{eq}$, which assumes a product state at initial times between the QD density matrix and the reservoirs at equilibrium, tracing over reservoir degrees of freedom, and iterating once yields $\frac{d\hat{\rho}^I(t)}{dt} = -\int_{t_0}^{t} dt'\mathrm{Tr}_{res}\{[H_T(t),[H_T(t'),\hat{\rho}^I(t')]]\}$. If we now assume a product state at all times $\hat{\rho}(t) = \hat{\rho}^{QD}(t)\otimes\hat{\rho}^{res}_{eq}$ (Born approximation), we end up with the equation

$$\frac{d\hat{\rho}^I(t)}{dt} = -\int_{t_0}^{t} dt'\mathrm{Tr}_{res}\{[H_T(t),[H_T(t'),\hat{\rho}^{QD,I}(t')\otimes\hat{\rho}^{res}_{eq}]]\}. \quad (30.21)$$

Equation 30.21, which is still non-Markovian, is the starting point of most weak coupling QMEs describing transport through an interacting region. In order to convert Equation 30.21 into a QME local in time, one makes one further approximation $\hat{\rho}^{QD,I}(t') \approx \hat{\rho}^{QD,I}(t)$ such that Equation 30.21 becomes

$$\frac{d\hat{\rho}^I(t)}{dt} = -\int_{t_0}^{t} dt'\mathrm{Tr}_{res}\{[H_T(t),[H_T(t'),\hat{\rho}^{QD,I}(t)\otimes\hat{\rho}^{res}_{eq}]]\}. \quad (30.22)$$

This equation is called the Redfield equation. The final form of the Markovian QME is obtained by making $\tau = t - t'$ and $t_0 \to -\infty$, such that one gets

$$\frac{d\hat{\rho}^I(t)}{dt} = -\int_0^\infty d\tau \mathrm{Tr}_{res}\{[H_T(t),[H_T(t-\tau),\, \hat{\rho}^{QD,I}(t)\otimes\hat{\rho}_{eq}^{res}]]\}. \quad (30.23)$$

Equation 30.23 is usually termed the Born–Markov QME and can also be obtained using the NZ superoperator technique at lowest order.

By projecting Equation 30.23 onto the eigenstates $|N\rangle$ that diagonalizes the many-body Hamiltonian H_{QD}, one obtains

$$\frac{d\rho_{NM}^I(t)}{dt} = \sum_{KL} e^{i(\omega_N-\omega_M-\omega_K+\omega_L)t} R_{NMKL}\rho_{KL}^I(t), \quad (30.24)$$

where $R_{NMKL} = -\delta_{L,M}\sum_R \Gamma_{NRRK}^+ - \delta_{N,K}\sum_R \Gamma_{LRRM}^- + \Gamma_{LMNK}^+ + \Gamma_{LMNK}^-$ is the so-called Redfield tensor. All the information about the reservoirs is encapsulated in the rates

$$\Gamma_{LMNK}^+ = \int_0^\infty d\tau e^{-i(\omega_N-\omega_K)\tau}\mathrm{Tr}_{res}\{\langle L|H_T(\tau)|M\rangle\langle N|H_T|K\rangle\hat{\rho}_{eq}^{res}\},$$

$$\Gamma_{LMNK}^- = \int_0^\infty d\tau e^{-i(\omega_L-\omega_M)\tau}\mathrm{Tr}_{res}\{\langle L|H_T|M\rangle\langle N|H_T(\tau)|K\rangle\hat{\rho}_{eq}^{res}\}, \quad (30.25)$$

which contain integrals of the form $\int_0^\infty d\tau e^{iz^-} = \pi\delta(z) + iP\frac{1}{z}$. The real parts that are proportional to delta functions give rise to dissipative effects whereas the imaginary principal parts induce shifts in the spectrum known as Stark and Lamb shifts in quantum optics.

One further rotating wave approximation, which involves averaging over rapidly oscillating terms $e^{i(\omega N-\omega M-\omega K+\omega L)t}\approx 1$ leads to the so-called secular approximation where only terms such that $(\omega_N-\omega_M-\omega_K+\omega_L)=0$ are considered in the summations of Equation 30.24. This approximation guarantees that the QME has a Lindblad form, namely, that it preserves the positivity of the density matrix.

After all the above approximations, the QME can be written as

$$\frac{d\rho(t)}{dt} = \mathbf{L}\rho(t), \quad (30.26)$$

where $\rho(t)$ is a vector containing both populations and coherences. The latter are important because of the intrinsic many-body degeneracies of the QD spectra. The matrix $\mathbf{L}$ contains tunneling rates of the form

$$\Gamma_{NM}^+ = \sum_{\alpha\in L,R}\Gamma_\alpha f_\alpha(E_N-E_M)\sum_n |\langle N|\mathbf{f}_n^\dagger|M\rangle|^2,$$
$$\Gamma_{NM}^- = \sum_{\alpha\in L,R}\Gamma_\alpha[1-f_\alpha(E_N-E_M)]\sum_n |\langle N|\mathbf{f}_n|M\rangle|^2. \quad (30.27)$$

The notation $\Gamma_{NM}^\pm$ implies that states M with charge Q are connected with states N with charge $Q+1$ (or $Q-1$) via the matrix elements $\langle N|\mathbf{f}_n^\dagger|M\rangle$ (or $\langle N|\mathbf{f}_n|M\rangle$). In our case, these matrix elements depend on the spin properties of the isolated system and lead to remarkable transport properties, as we will discuss in the following sections. The tunneling rates also depend on the Fermi functions $f_{L/}R$ of the left/right reservoir. As before, the tunneling coupling is parameterized as $\Gamma^\alpha(\varepsilon)=2\pi\sum_{k,\alpha\in L,R}|V_{k,\alpha}|^2\delta(\varepsilon-\varepsilon_{k\alpha})=2\pi\sum_{\alpha\in L,R}\nu_\alpha(\varepsilon)|V_\alpha(\varepsilon)|^2$, which is usually assumed to be constant. The steady-state density matrix is obtained from the solution of $\mathbf{L}\rho^{stat}=0$ and is used to compute average quantities such as charge, magnetization, and current. The most general expression for the latter is given by

$$J_{L/R} = \frac{e}{\hbar}\Gamma_{L/R}\sum_{KLM}\sum_n \{f_{L/R}(E_K-E_L)\langle L|\mathbf{f}_n|M\rangle\langle M|\mathbf{f}_n^\dagger|K\rangle$$
$$-[1-f_{L/R}(E_K-E_M)]\langle L|\mathbf{f}_n^\dagger|M\rangle\langle M|\mathbf{f}_n|K\rangle\}\rho_{KL}^{stat}. \quad (30.28)$$

The key observation to relate the above expression for the current obtained within the QME approach and the one in Equation 30.19 that we obtained using the Keldysh formalism is to note that Equation 30.19 already contains an explicit $\Gamma_{L/R}$ in front of the Green's functions. Therefore, in order to obtain the current to lowest (sequential) order we have to use Green's functions to *zeroth order in the tunneling coupling.* Using $G^<(t) = \frac{i}{\hbar}\langle \mathbf{f}_m^\dagger \mathbf{f}_n(t)\rangle = \frac{i}{\hbar}\mathrm{Tr}\left\{\mathbf{f}_m^\dagger\mathbf{f}_n(t)\hat{\rho}^{QD}\right\} = \frac{i}{\hbar}\sum_{MKL}\langle K|\hat{\rho}^{QD}|L\rangle\langle L|\mathbf{f}_m^\dagger|M\rangle\langle M|\mathbf{f}_n(t)|K\rangle$, the zeroth order is obtained by using Heisenberg operators with respect to the *isolated* many-body Hamiltonian, namely, $\mathbf{f}_n(t) = e^{iH_{QD}t}\mathbf{f}_n e^{iH_{QD}t}$, such that the lesser Green's function is $G^<(\varepsilon) = \frac{2\pi i}{\hbar}\sum_{MKL}\delta(\varepsilon-E_K+E_M)\langle K|\hat{\rho}^{QD}|L\rangle\langle L|\mathbf{f}_m^\dagger|M\rangle\langle M|\mathbf{f}_n(t)|K\rangle$. Substituting this expression in Equation 30.19, one immediately finds Equation 30.28.

30.7.5 Transport in DMS Quantum Dots

In the following, we show results for two dots of CdTe with $L_z=60$ Å, $L_x=80$ Å and different $L_y=80$ Å (dot A) and $L_y=75$ Å (dot B), both doped with one Mn atom. As explained in Section 30.8.1, the steady state of a standard SET is uniquely characterized by external voltages. In the linear response regime, a new charge is accommodated in the dot at precise values of the gate voltage, when the electrochemical potential of the dot falls within the bias window, for example, $\mu_L > \mu_{dot}(N+1) > \mu_R$>. Importantly, for DMS QDs the charge and the conductance of the SET *depend also on the quantum state of the Mn spin.* This is demonstrated in Figure 30.5, where we show the charge and linear conductance of a single dot A as the gate injects either one electron (left) or one hole (right). In both cases, the initial gate is chosen so that only the $Q=0$ states are occupied. This initial condition is described

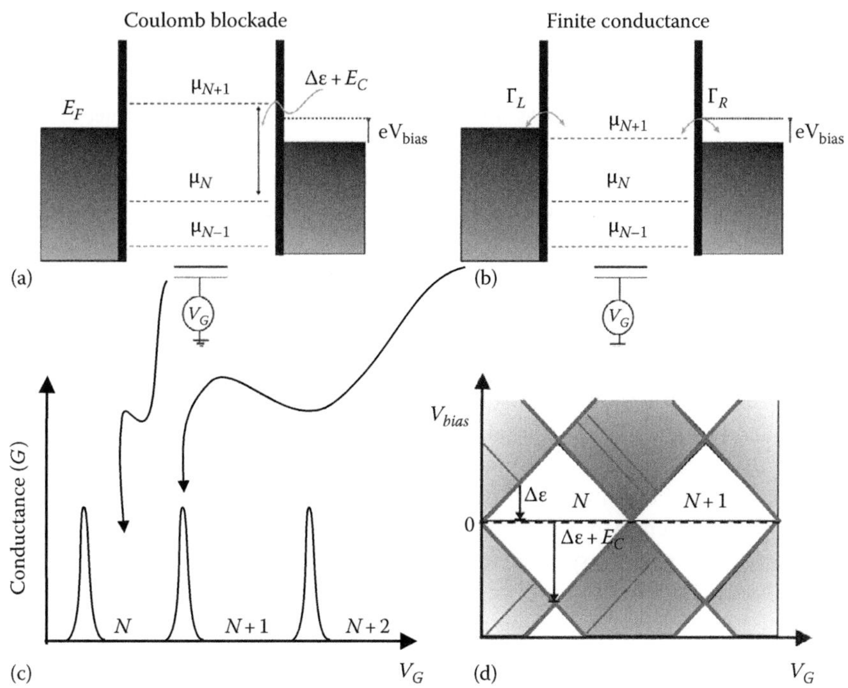

FIGURE 30.5 Schematic representation of the energy configuration of QDs in the (a) Coulomb Blockade regime and (b) transport configurations; (c) schematics of the linear conductance oscillations through such a device; (d) at finite bias voltages, the large regions of nearly zero conductance form characteristic Coulomb diamonds in differential conductance plots. Excited states appear as dashed lines.

by a thermal equilibrium density matrix (DM) with six equally populated Mn spin states, $M_z = \pm\frac{5}{2}, \pm\frac{3}{2}, \pm\frac{1}{2}$. We increase the gate and solve the master equation to obtain the steady-state density matrix, which is used as an initial condition for the next run with higher V_G. Importantly, in some cases the steady state is not equal to the thermal density matrix that has dramatic consequences for linear transport. This is the case of the positively charged dot A for which the transverse spin interaction is zero, which results in anomalous linear conductance curves that exhibit a three-peak structure in regions of gate voltage corresponding to the charge degeneracy region. This is in stark contrast with a normal QD that would exhibit instead one single Coulomb Blockade peak. Key for an understanding of this unusual Coulomb Blockade is the fact that M_z does not relax during transport: In the absence of holes, the spin of the Mn ion is free and therefore all the six projections are degenerate. Sweeping the gate voltage toward the charge degeneracy region, the only allowed transitions are those that conserve M_z. This results in three possible charge degeneracy points corresponding to the transition $Q=0$ ($Q=1$ at different values of M_z. For example, the first Coulomb Blockade peak corresponds to charge degeneracy between $|-5/2,0\rangle$ and $|-5/2,\Uparrow_h\rangle$ (or $|5/2,0\rangle$ and $|5/2,\Downarrow_h\rangle$). Importantly, at this gate voltage, the density matrix has populations in both charge sectors, namely, states with $\Theta=+1$, $|M_z|=5/2$ and states with $\Theta=0$, $|M_z|\neq 5/2$ *coexist*. This is true provided that a quasi steady-state limit for populations (and current) can be reached at time scales smaller than the spin relaxation time, see below. Increasing further V_G, states with $|M_z|=3/2$ become degenerate and a new peak in the conductance occurs (reflecting the population of the $\Theta=+1$, $|M_z|=5/2$ states). The third peak corresponds to the transition with $|M_z|=1/2$. Therefore, as a result of this mechanism Coulomb Blockade is spin dependent.

The discharge simulation is done analogously. If the initial V_G is such that there is one hole in the QD, the Ising interaction removes the degeneracy among states with different $|M_G|$. The ground state of the $Q=+1$ sector is now the lowest energy doublet with $|M_z|=5/2$, which is the only state occupied in thermal equilibrium. Clearly, this initial state is different from the steady state obtained during the charging procedure, which produces a hysteretic linear conductance versus gate voltage curve: As the gate is ramped to discharge the dot, a single peak in the conductance is obtained, corresponding to the resonance condition with the $Q=0$, $|M_z|=5/2$ states. In contrast, the charge and discharge curves for the conduction band case are equal to each other and display a single peak in the linear conductance curve, as in a standard SET. Of course, the difference between electrons and holes arises from the presence of spin-flip terms in the Hamiltonian, which sets a new time scale in the system: the Mn spin relaxation time, T_1. In the case of dot A (Ising coupling), T_1 is infinite for holes, which makes the steady state different from the thermal state. In the case of electrons (left panels), T_1 is comparable to the charge relaxation time ($\Gamma_{L,R}^{-1}$) so that steady and thermal DM are identical. In real positively charged dots, T_1 may be long but not infinite. Two independent mechanisms, missing in the simulations shown in the right panels of Figure 30.5, yield a finite T_1. For neutral dots, the Mn T_1 due to super-exchange with other *Mn* spins scales exponentially with the Mn density [73]. For bulk $Cd_{0.995}Mn_{0.005}Te$, we have $T_1 = 10^{-3}$S, which is a lower limit estimate for T_1 of the QD with a single Mn. The second mechanism is the transverse spin interaction, which is proportional to the LH HH mixing and is small according to recent experiments [17]. We have simulated a QD with finite T_1. If we integrate the master equation for $\Gamma^{-1} \ll t \ll T_1$, the $G_0(V_G)$ curve displays two peaks. In contrast, if we integrate the master equation for $\Gamma^{-1} \ll T_1 \ll t$, the $G_0(V_G)$ has a single peak, in agreement with the physical picture of the previous paragraph.

The finite bias conductance of the device also depends strongly on the charge state (=magnetic state) of the dot. In general, current flows whenever the addition of a fermion is permitted by energy conservation and spin selection rules. This provides a link between the dI/dV curve and the spectral function of the fermion in the dot. Thus, the dI/dV curve yields information about the elementary excitation of the Mn spin coupled to the carriers. In the case of a dot doped with one Mn whose charge fluctuates between 0 and 1 carrier, the relevant spectral function can be derived from Hamiltonian (30.15) changing ε and J_h. For instance, for conduction band electrons we would have $\varepsilon = 1$ and for pure HHs $\varepsilon = 0$. To illustrate how the tunneling spectra might be used to infer the degree of anisotropy, we calculate the dI/dV_B curves corresponding to the effective model of Equation 30.15 and study systematically the above changes as a function of the parameter ε (Figure 30.6). As we have explained, if valence band mixing is nonzero, $|\pm 3/2\rangle$ HHs couple with $|\mp 1/2\rangle$ LHs. This mixing allows simultaneous spin flips between the hole and the Mn spins, which results in split states that are bonding and antibonding combinations of $|M_z = -1/2, \Uparrow\rangle$ and $|M_z = +1/2, \Downarrow\rangle$. This splitting can be extracted directly from the dI/dV curves as we illustrate in Figure 30.6. For $\varepsilon = 0$, and V_G such that the conductance at zero bias is finite, we obtain six equally spaced peaks corresponding to the Ising spectrum of a QD with a single Mn and doped with a hole (solid curve). Increasing ε, a clear splitting shows up centered at $V_B = 2$ (dashed and dotted lines). This situation, $\varepsilon \leq 0.1$ corresponds to realistic parameters (see the discussion of Section 30.5.1). For larger ε, the splitting and the energy shifts are larger, which results in overlapping peaks that are not evenly spaced (dashed and dotted curve). For $\varepsilon = 1$, we recover the fully isotropic case, which, as we have discussed in Section 30.4.1, presents two peaks with a zero-field splitting corresponding to the energy difference between a spin $S = 2$ ground state and the $S = 3$ excited state septuplet. The differential conductance in this case is therefore a direct measure of the spectrum of Figure 30.2.

Interestingly, the strongly anisotropic character of the exchange can be exploited at finite magnetic fields to produce a strong AMR that can be tuned both via gate and bias voltages. This concept is illustrated in Figure 30.7. The upper plots in Figure 30.7 show the linear conductance as a function of gate voltage for three situations: $B = 0$, $|\vec{B}| = (B,0,0)$ (Faraday), and $\vec{B} = (B,0,0)$ (Voigt), whereas the lower plot shows the average charge of the dot. We see how both the charging and G curves depend on $|\vec{B}|$, since they are different from zero and finite field and also on the orientation of $\vec{B}$. The change in conductance, at fixed gate, as the orientation of the applied field can be very large. The origin of the anisotropy lies in the different spectra for the dot doped with one hole (Figure 30.4). Since the $Q = 0$ spectrum is quite isotropic, the addition energy becomes anisotropic. Interestingly, compared to the $B = 0$ case, the charging energy is smaller in the Faraday configuration and larger in the Voigt configuration. In the Faraday case, exchange and Zeeman cooperate whereas in the Voigt configuration they compete. The resulting AMR effect is maximized in regions where the charge in one of the configurations begins to saturate. This large AMR, predicted for a device with single magnetic dopant, could have practical applications (Figure 30.8).

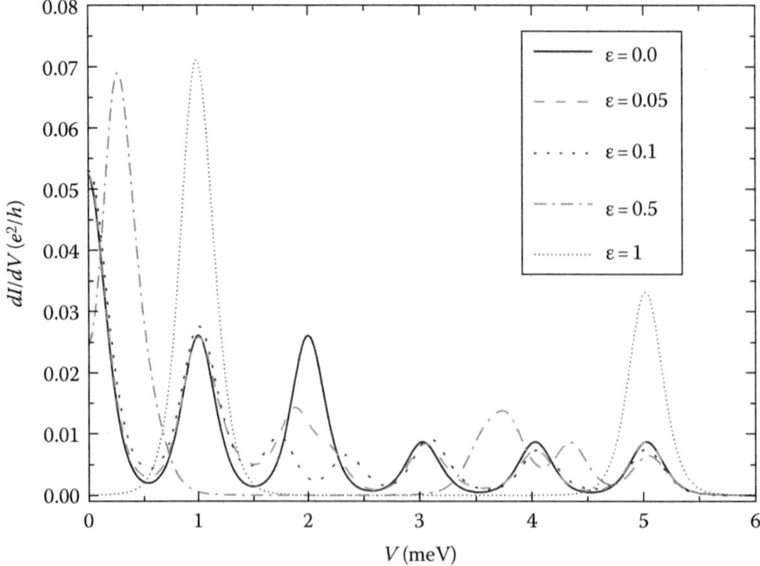

FIGURE 30.6 Differential conductance as a function of bias for holes A (a,b,c) and different anisotropy parameters. V_G chosen to yield finite $V = 0$ conductance.

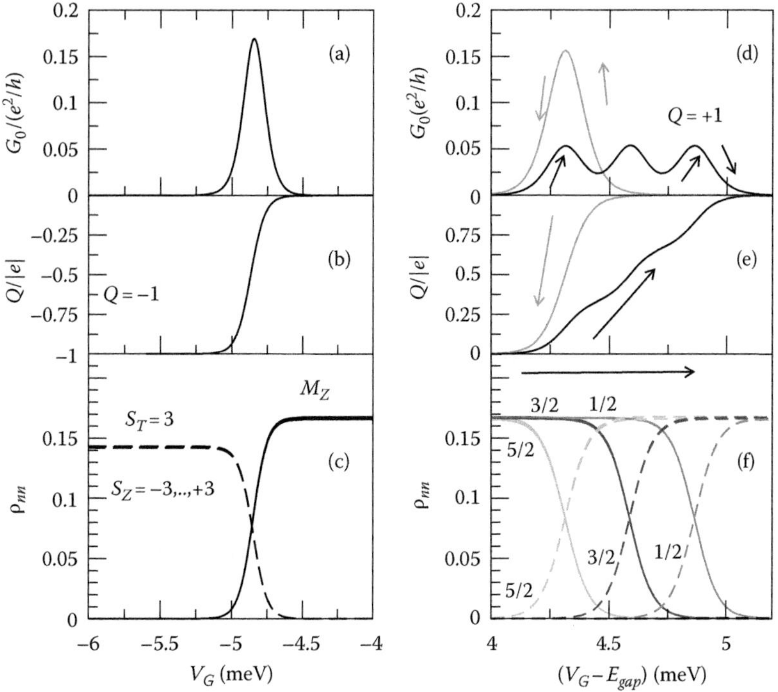

FIGURE 30.7 $G_0(V_G)$ (upper panels), charge (middle panels), and diagonal terms of the ρ (lower panels) for QD *A* as a function of V_G around the $Q = -1 \leftrightarrow 0$ transition (left) and $Q = +1 \leftrightarrow 0$ (right). Lower panels: solid (dashed) lines correspond to $Q = 0$ ($|Q| = 1$) states. Results obtained with $\Gamma_L = \Gamma_R = 0.01$ meV and $k_BT = 0.05$ meV. (From Fernandez-Rossier, J. and Aguado, R., *Phys. Rev. Lett.*, 98, 106805, 2007. Copyright [2007] by the American Physical Society.)

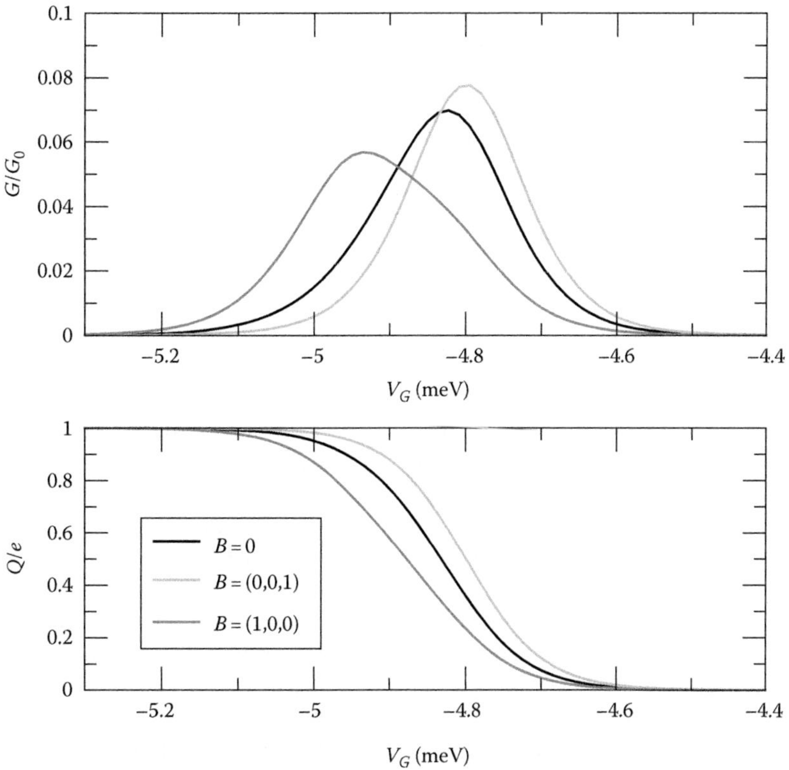

FIGURE 30.8 (See color insert.) Conductance (upper panel) and charge (lower panel) versus gate for $\vec{B} = 0$, $\vec{B} = B_0\hat{x}$, and $\vec{B} = B_0\hat{z}$.

30.8 Outlook

Semiconductor QDs with a few Mn atoms combine two very different length scales, the radius of the confined carrier and the atomic scale where the *d* electrons lie. In this unique system, the magnetic properties of the Mn atoms are thus dominated by a single or a few carriers whose wave functions extend over thousands of atoms nearby, in stark contrast with magnetic atoms in insulating crystals, where only the first neighbor atoms affect the single-ion anisotropy. Conversely, the spectrum of a semiconductor QD with thousands of atoms is dramatically modified by the presence of a single magnetic dopant, as observed experimentally [16,17,22]. Thus, probing the dot with excitons or in transport permits to access the spin state of a single or a few magnetic atoms. Here, we have reviewed the theoretical understanding of the electronic structure and transport of this promising system.

Acknowledgments

Ramon Aguado acknowledges funding from FIS2009-08744 and Joaquín Fernández acknowledges funding from MAT07-67845, FIS2010-21883-C02-01.

References

1. J. K. Furdyna, Diluted magnetic semiconductors, *J. Appl. Phys.* **64**, R29–R64 (1988).
2. H. Ohno, Making non magnetic semiconductors ferromagnetic, *Science* **281**, 951–956 (1998).
3. P. Kacman, Spin interactions in diluted magnetic semiconductors and magnetic semiconductor structures, *Semicond. Sci. Technol.* **16**, R25–R39 (2001).
4. T. Dietl, Ferromagnetic semiconductors, *Semicond. Sci. Technol.* **17**, 377–392 (2002).
5. T. Jungwirth, J. Sinova, J. Mašek, J. Kučera, and A. H. MacDonald, Theory of ferromagnetic (III,Mn)V semiconductors, *Rev. Mod. Phys.* **78**, 809 (2006).
6. K. W. Edmonds, P. Bogusławski, K. Y. Wang et al., Mn interstitial difussion in (Ga,MnAs), *Phys. Rev. Lett.* **92**, 037201 (1–4) (2004).
7. A. Haury, A. Wasiela, A. Arnoult et al., Observation of a ferromagnetic transition induced by two-dimensional hole gas in modulation-doped CdMnTe quantum wells, *Phys. Rev. Lett.* **79**, 511–514 (1997).
8. R. K. Kawakami, E. Johnston-Halperin, L. F. Chen et al., (Ga,Mn)As as a digital ferromagnetic heterostructure, *Appl. Phys. Lett.* **77**, 2379–2381 (2000).
9. M. Tanaka and Y. Higo, Large tunneling magnetoresistance in GaMnAs/AlAs/GaMnAs ferromagnetic semiconductor tunnel junctions, *Phys. Rev. Lett.* **87**, 026602 (1–4) (2001).
10. C. Rüster, T. Borzenko, C. Gould et al., Very large magnetoresistance in lateral ferromagnetic (Ga,Mn)As wires with nanoconstrictions, *Phys. Rev. Lett.* **91**, 216602 (1–4) (2003).
11. C. Gould, C. Rüster, T. Jungwirth et al., Tunneling anisotropic magnetoresistance: A spin-valve-like tunnel magnetoresistance using a single magnetic layer, *Phys. Rev. Lett.* **93**, 117203 (4pp) (2004).
12. H. Ohno, D. Chiba, F. Matsukura et al., Electric-field control of ferromagnetism, *Nature* **408**, 944946 (2000).
13. H. Boukari, P. Kossacki, M. Bertolini et al., Light and electric field control of ferromagnetism in magnetic quantum structures, *Phys. Rev. Lett.* **88**, 207204 (4pp) (2002).
14. D. Chiba, K. Takamura, F. Matsukura, and H. Ohno, Effect of low-temperature annealing on (Ga,Mn)As trilayer structures, *Appl. Phys. Lett.* **82**, 3020 (2003).
15. D. Chiba, M. Yamanouchi, F. Matsukura, and H. Ohno, Electrical manipulation of magnetization reversal in ferromagnetic semiconductor, *Science* **301**, 943–945 (2003).
16. L. Besombes, Y. Léger, L. Maingault, D. Ferrand, and H. Mariette, Probing the spin state of a single magnetic ion in an individual quantum dot, *Phys. Rev. Lett.* **93**, 207403 (4pp) (2004).
17. Y. Léger, L. Besombes, J. Fernández-Rossier, L. Maingault, and H. Mariette, Electrical control of a single Mn atom in a quantum dot, *Phys. Rev. Lett.* **97**, 107401 (4pp) (2006).
18. R. Beaulac, P. Archer, S. T. Ochsenbein, and D. Gamelin, Mn^{2+}-doped CdSe quantum dots: New inorganic materials for spin-electronics and spin-photonics, *Adv. Func. Mater.* **18**, 3873–3891 (2008).
19. A. A. Maksimov, G. Bacher, A. McDonald et al., Magnetic polarons in a single diluted magnetic semiconductor quantum dot, *Phys. Rev. B* **62**, R7767–7770 (2000).
20. G. Bacher, A. A. Maksimov, H. Schömig et al., Monitoring statistical magnetic fluctuations on the nanometer scale, *Phys. Rev. Lett.* **89**, 127201 (1–4) (2002).
21. C. Gould, A. Slobodskyy, D. Supp et al., Remanent zero field spin splitting of self-assembled quantum dots in a paramagnetic host, *Phys. Rev. Lett.* **97**, 017202 (4pp) (2006).
22. A. Kudelski, A. Lemaître, A. Miard et al., Optically probing the fine structure of a single Mn atom in an InAs quantum dot, *Phys. Rev. Lett.* **99**, 247209 (1–4) (2007).
23. J. Wunderlich, T. Jungwirth, B. Kaestner et al., Coulomb blockade anisotropic magnetoresistance effect in a (Ga,Mn) As single-electron transistor, *Phys. Rev. Lett.* **97**, 077201 (1–4) (2006).
24. T. Gurung, S. Mackowski, H. E. Jackson, and L. M. Smith, Optical studies of zero-field magnetization of CdMnTe quantum dots: Influence of average size and composition of quantum dots, *J. Appl. Phys.* **96**, 7407–7413 (2004).
25. I. R. Sellers, R. Oszwałdowski, V. R. Whiteside et al., Robust magnetic polarons in type-II (Zn,Mn)Te/ZnSe magnetic quantum dots, *Phys. Rev. B* **82**, 195320 (2010).
26. J. Fernández-Rossier, Single exciton spectroscopy in semimagnetic quantum dots, *Phys. Rev. B* **73**, 045301 (6pp) (2006).
27. J. van Brie, P. M. Koenraad, and J. Fernández-Rossier, Single-exciton spectroscopy of single Mn doped InAs quantum dots, *Phys. Rev. B* **78**, 165414 (2008).

28. L. Besombes, Y. Leger, J. Bernos et al., Optical probing of spin fluctuations of a single paramagnetic Mn atom in a semiconductor quantum dot, *Phys. Rev. B* **78**, 125324 (9pp) (2008).
29. C. Le Gall, L. Besombes, H. Boukari, R. Kolodka, J. Cibert, and H. Mariette, Optical spin orientation of a single manganese atom in a semiconductor quantum dot using quasiresonant photoexcitation, *Phys. Rev. Lett.* **102**, 127402 (1–4) (2009).
30. Al. L. Efros, E. I. Rashba, and M. Rosen, Paramagnetic ion-doped nanocrystal as a voltage-controlled spin filter, *Phys. Rev. Lett.* **87**, 206601 (4pp) (2001).
31. F. Qu and P. Vasilopoulos, Spin transport across a quantum dot doped with a magnetic ion, *Appl. Phys. Lett.* **89**, 122512 (3pp) (2006).
32. J. Fernández-Rossier and R. Aguado, Single electron transport in electrically tunable nanomagnets, *Phys. Rev. Lett.* **98**, 106805 (4pp) (2007).
33. T. Dietl, H. Ohno, F. Matsukura, J. Cibert, and D. Ferrand, Zener model description of ferromagnetism in Zinc-Blende magnetic semiconductors, *Science* **287**, 1019–1022 (2000).
34. K. Hamaya, Spin-related current suppression in a semiconductor quantum dot spin-diode structure, *Phys. Rev. Lett.* **102**, 236806 (4pp) (2009).
35. D. L. Klein, R. Roth, A. K. L. Lim, A. P. Alivisatos, and P. L. McEuen, A single-electron transistor made from a cadmium selenide, *Nature* **389**, 699 (1997).
36. S. T. Ochsenbein, Y. Feng, K. M. Whitaker et al., Charge-controlled magnetism in colloidal doped semiconductor nanocrystals, *Nat. Nanotechnol.* **4**, 681–687 (2009).
37. J. Fernández-Rossier and L. Brey, Ferromagnetism mediated by few electrons in semimagnetic quantum dots, *Phys. Rev. Lett.* **93**, 1172001 (4pp) (2004).
38. F. Qu and P. Hawrylak, Magnetic exchange interactions in quantum dots containing electrons and magnetic ions, *Phys. Rev. Lett.* **95**, 217206 (1–4) (2005).
39. F. Qu and P. Hawrylak, Theory of electron mediated Mn-Mn interactions in quantum dots, *Phys. Rev. Lett.* **96**, 157201 (1–4) (2006).
40. A. O. Govorov, Optical and electron properties of quantum dots with magnetic impurities, *C. R. Phys.* **9**, 857–873 (2008).
41. M. Blume and R. Orbach, Spin-lattice relaxation of S-state ions: Mn 2+ in a cubic environment, *Phys. Rev.* **127**, 1587 (1962).
42. M. Qazzaz, Electron paramagnetic resonance of Mn2+ in strained-layer semiconductor superlattices, *Solid State Commun.* **96**, 405–410 (1995).
43. A. O. Govorov and A. V. Kalameitsev, Optical properties of a semiconductor quantum dot with a single magnetic impurity: Photoinduced spin orientation, *Phys. Rev. B* **71**, 035338 (2005).
44. J. I. Climente, M. Korkusiński, P. Hawrylak, and J. Planelles, Voltage control of the magnetic properties of charged semiconductor quantum dots containing magnetic ions, *Phys. Rev. B* **71**, 125321 (2005).
45. R. Abolfath, P. Hawrylak, and I. Žutić, Tailoring magnetism in quantum dots, *Phys. Rev. Lett.* **98**, 207203 (1–4) (2007).
46. R. Abolfath, A. G. Petukhov, and I. Žutić, Piezomagnetic quantum dots, *Phys. Rev. Lett.* **101**, 207202 (1–4) (2008).
47. N. T. Nguyen and F. M. Peeters, Correlated many-electron states in a quantum dot containing a single magnetic impurity, *Phys. Rev. B* **76**, 045315 (2007).
48. M. Cardona and P. Y. Yu, *Fundamentals of Semiconductors*, Springer, Berlin, Heidelberg, 1999.
49. F. V. Kyrychenko and J. Kossut, Diluted magnetic semiconductor quantum dots: An extreme sensitivity of the hole Zeeman splitting on the aspect ratio of the confining potential, *Phys. Rev. B* **70**, 205317 (6pp) (2004).
50. D. E. Reiter, T. M. Kuhn, and V. M. Axt, All-optical spin manipulation of a single manganese atom in a quantum dot, *Phys. Rev. Lett.* **102**, 177403 (2009).
51. J. Fernández-Rossier and R. Aguado, Mn doped II–VI quantum dots: Artificial single molecule magnets, *Phys. Stat. Sol (c)* **3**, 3734 (2006).
52. L. Besombes, Y. Leger, L. Maingault, D. Ferrand, and H. Mariette, Carrier-induced spin splitting of an individual magnetic atom embedded in a quantum dot, *Phys. Rev. B* **71**, 161307 (2005).
53. A. O. Govorov, Voltage-tunable ferromagnetism in semimagnetic quantum dots with few particles: Magnetic polarons and electrical capacitance, *Phys. Rev. B* **72**, 075359 (2005).
54. Y. Léger, L. Besombes, L. Maingault, D. Ferrand, and H. Mariette, Hole spin anisotropy in single Mn-doped quantum dots, *Phys. Rev. B* **72**, 241309 (2005).
55. J. Szczytko, A. Twardowski, M. Palczewska, R. Jabłoński, J. Furdyna, and H. Munekata, Electron paramagnetic resonance of Mn in $In_{1-x}Mn_xAs$ epilayers, *Phys. Rev. B* **63**, 085315 (2001).
56. J. Okabayashi, T. Mizokawa, D. D. Sarma et al., Electronic structure of $In_{1-x}Mn_xAs$ studied by photoemission spectroscopy: Comparison with $Ga_{1-x}Mn_xAs$, *Phys. Rev. B* **65**, 161203(R) (4pp) (2002).
57. A. M. Yakunin, A. Yu. Silov, P. M. Koenraad et al., Spatial structure of an individual Mn acceptor in GaAs, *Phys. Rev. Lett.* **92**, 216806 (4pp) (2004).
58. A. K. Bhattarcharjee and C. Benoit a la Guillaume, Model for the Mn acceptor in GaAs, *Solid State Commun.* **113**, 117 (2000).
59. A. O. Govorov, Optical probing of the spin state of a single magnetic impurity in a self-assembled quantum dot, *Phys. Rev. B* **70**, 035321 (2004).
60. K. H. Schmidt, M. Versen, U. Kunze, D. Reuter, and A. D. Wieck, Electron transport through a single InAs quantum dot, *Phys. Rev. B* **62**, 15879 (2000).
61. T. Ota, K. Ono, M. Stopa et al., Single-dot spectroscopy via elastic single-electron tunneling through a pair of coupled quantum dots, *Phys. Rev. Lett.* **93**, 066801 (2004).
62. T. Ota, M. Rontani, S. Tarucha et al., Few-electron molecular states and their transitions in a single InAs quantum dot molecule, *Phys. Rev. Lett.* **95**, 236801 (2005).

63. M. Jung, K. Hirakawa, Y. Kawaguchi, S. Komiyama, S. Ishida, and Y. Arakawa, Lateral electron transport through single self-assembled InAs quantum dots, *Appl. Phys. Lett.* **86**, 033106 (2005).
64. K. Shibata, C. Buizert, A. Oiwa, K. Hirakawa, and S. Tarucha, Lateral electron tunneling through single self-assembled InAs quantum dots coupled to superconducting nanogap electrodes, *Appl. Phys. Lett.* **91**, 112102 (2007).
65. Y. Igarashi, M. Jung, M. Yamamoto et al., Spin-half Kondo effect in a single self-assembled InAs quantum dot with and without an applied magnetic field, *Phys. Rev. B* **76**, 081303R (2007).
66. C. Buizert, A. Oiwa, K. Shibata, K. Hirakawa, and S. Tarucha, Kondo universal scaling for a quantum dot coupled to superconducting leads, *Phys. Rev. Lett.* **99**, 136806 (4pp) (2007).
67. L. P. Kouwenhoven, C. M. Marcus, P. L. McEuen, S. Tarucha, R. M. Westervelt, and N. S. Wingreen, Electron transport in quantum dots, in *Mesoscopic Electron Transport*, L. L. Sohn, L. P. Kouwenhoven, and G. Schon (Eds.), Kluwer, the Netherlands, pp. 105–214, 1997.
68. L. P. Kouwenhoven and L. Glazman, Revival of the Kondo effect, *Phys. World* 14, 33–38 (2001).
69. Y. Meir and N. S. Wingreen, Landauer formula for the current through an interacting electron region, *Phys. Rev. Lett.* **68**, 2512 (1992).
70. J. E. Han and R. J. Heary, Imaginary-time formulation of steady-state nonequilibrium: Application to strongly correlated transport, *Phys. Rev. Lett.* **99**, 236808 (4pp) (2007).
71. S. Hershfield, Reformulation of steady state nonequilibrium quantum statistical mechanics, *Phys. Rev. Lett.* **70**, 2134 (4pp) (1993).
72. C. Timm, Tunneling through molecules and quantum dots: Master-equation approaches, *Phys. Rev. B* **77**, 195416 (2008).
73. J. Lambe and Ch. Kikuchi, Paramagnetic resonance of CdTe: Mn and CdS: Mn, *Phys. Rev.* **119**, 1256–1260 (1960).
74. D. Chiba, M. Sawicki, Y. Nishitani, Y. Nakatani, F. Matsukura, and H. Ohno, Magnetization vector manipulation by electric fields, *Nature* **455**, 515–518 (2008).
75. H. Munekata, H. Ohno, S. von Molnar, Armin Segmüller, L. L. Chang, and L. Esaki, Diluted magnetic III–V semiconductors, *Phys. Rev. Lett.* **63**, 1849–1852 (1989).
76. H. Ohno, A. Shen, F. Matsukura et al., (Ga,Mn)As: A new ferromagnetic semiconductor based on GaAs, *Appl. Phys. Lett.* **69**, 363–365 (1996).
77. C. Gould, C. Rüster, T. Jungwirth, E. Girgis, G. M. Schott, R. Giraud, K. Brunner, G. Schmidt, and L. W. Molenkamp, Tunneling anisotropic magnetoresistance: A spin-valve-like tunnel magnetoresistance using a single magnetic layer, *Phys. Rev. Lett.* **93**, 117203 (2004).
78. K. Hamaya, M. Kitabatake, K. Shibata, M. Kawamura, K. Hirakawa, T. Machida, T. Taniyama, S. Ishida, Y. Arakawa. Kondo effect in a semiconductor quantum dot coupled to ferromagnetic electrodes, *Appl. Phys. Lett.* **91**, 232105 (2007).

31

Spin Transport in Hybrid Nanostructures

Saburo Takahashi
Tohoku University

Sadamichi Maekawa
Japan Atomic Energy Agency

The spin-dependent transport in hybrid nanostructures is currently of great interest, particularly in the emergence of new phenomena as well as the potential applications to spintronic devices [1–4]. Recent experimental and theoretical studies have demonstrated that the spin-polarized carriers injected from a ferromagnet (F) into a nonmagnetic material (N), such as a normal conducting metal, semiconductor, and superconductor, create nonequilibrium spin accumulation and spin current over the spin-diffusion length in the range from nanometers to micrometers. Efficient spin injection and detection and creation of large spin current, spin accumulation, and spin transfer are key factors in utilizing the spin degrees of freedom of carriers as a new functionality in spintronic devices [5,6].

In this chapter, we discuss the basic aspects for the spin transport in hybrid nanostructures containing normal conducting metals and transition-metal ferromagnets by focusing on the spin current and spin accumulation in a nonlocal spin device of F1/N/F2 structure, where F1 is a spin injector and F2 a spin detector connected to N. We derive basic spin-dependent transport equations for the electrochemical potentials (ECPs) of up-spin and down-spin electrons, and apply them to the F1/N/F2 structure with arbitrary interface resistance ranging from a metallic contact to a tunneling regime. The injection of spin-polarized electrons and the detection of spin current and accumulation depend strongly on the nature of the junction interface (metallic contact or tunnel barrier) in the structure. By analyzing the spin-dependent transport in the structure, we show that the relative magnitude of the electrode resistance to the interface resistance plays a crucial role for the spin transport in the F1/N/F2 structure. When a tunnel barrier is inserted into the junction interfaces, spin injection and detection are most effective. When the N/F2 interface is a metallic contact, the injected spin current from F1 is strongly absorbed by F2 (spin sink) owing to small spin resistance of ferromagnets. The spin absorption effect plays a vital role in nonlocal spin manipulation. We present the spin Hall effect (SHE) in nonmagnetic metals caused by spin–orbit scattering of conducting electrons, which enables the interconversion between the spin (charge) current and charge (spin) current using a nonlocal spin device. Spin injection into a ferromagnetic insulator (FI) is briefly mentioned.

31.1 Spin Injection and Detection

Johnson and Silsbee [7,8] first reported that spins are injected from a ferromagnet into an Al film and diffuse over the spin-diffusion length of the order of 1 μm (or even several hundred μm for pure Al). This rather long spin-diffusion length led to the proposal of a spin injection and detection technique in three terminal devices of F1/N/F2 structure (F1 is an injector and F2 a detector) [9], in which the output voltage at F2 depends on the relative orientation of the magnetizations of F1 and F2. In 2001, Jedema et al. performed spin injection and detection experiments with a nonlocal measurement in a

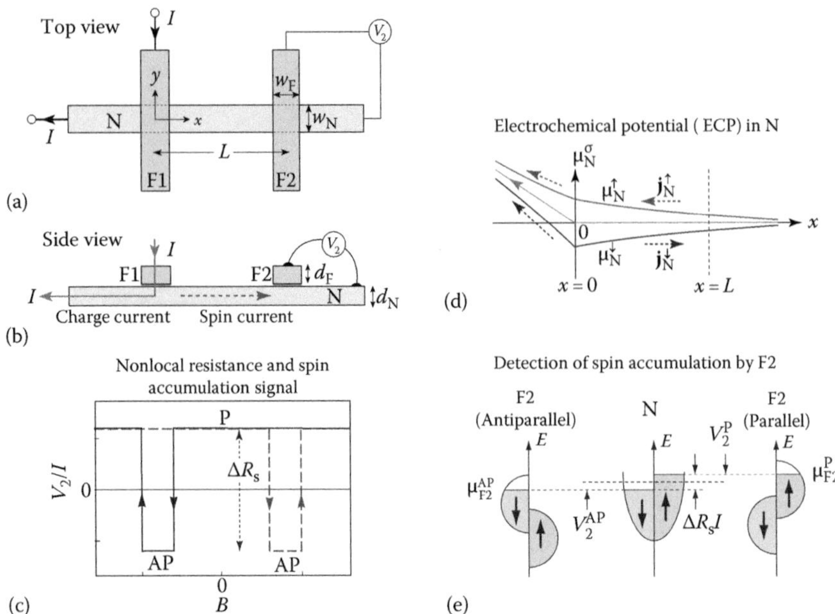

FIGURE 31.1 Nonlocal spin injection and detection device of F1/N/F2: (a) top view and (b) side view. Current I is sent from F1 to the left end of N. The spin accumulation at $x = L$ is detected by F2 by measuring voltage V_2. (c) Nonlocal resistance V_2/I as a function of in-plane magnetic field B, where P and AP represent the parallel and antiparallel orientations of magnetizations in F1 and F2. (d) Spatial variation of the ECP for up-spin and down-spin electrons in N. (e) Densities of states diagram for detection of spin accumulation in N (center) by F2 in the parallel (left) and antiparallel (right) orientations. (From Takahashi, S. and Maekawa, S., *J. Phys. Soc. Jpn.*, 77, 031009, 2008. With permission.)

lateral structure of permalloy/copper/permalloy (Py/Cu/Py) and observed a clear spin-accumulation signal at room temperature [10]. Subsequently, they measured a large spin accumulation signal in a cobalt/aluminum/cobalt (Co/I/Al/I/Co) structure with tunnel barriers ($I = Al_2O_3$) [11]. Nonlocal spin injection and detection experiments have been conducted by many groups using normal metals [12–23], graphenes [24–28], superconductors [29–31], semiconductors [32], and carbon nanotubes [33].

A basic structure of spin injection and detection device, which consists of a nonmagnetic metal N connected to the ferromagnets of the injector F1 and detector F2, is shown in Figure 31.1a and b. F1 and F2 are ferromagnetic electrodes with width w_F and thickness d_F, which are separated by distance L. N is a normal-metal electrode with width w_N and thickness d_N. The magnetizations of F1 and F2 are oriented either parallel or antiparallel. In this device, spin-polarized electrons are injected into N from F1 by sending the bias current I from F1 to the left side of N, and the spin accumulation is detected by F2 at distance L from F1, by measuring the voltage V_2 between F2 and N. Since any voltage source is absent in the right side of the device, there is no charge current in the electrodes that lie in the right side of F1. By contrast, owing to the diffusive nature of polarized spins, the injected spins are equally diffused in the left and right directions in N, creating spin current and accumulation on both sides. In this way, the spin and charge transports are separated and the region with purely spin degrees of freedom is realized, which is of great advantage of the nonlocal measurement.

31.2 Spin Transport in Nonlocal Spin Devices

We describe the nonlocal spin transport in a nanostructured F1/N/F2 device and reveal how the spin transport depends on interface resistance, electrode resistance, spin polarization, and spin-diffusion length, and obtain the conditions for efficient spin injection, spin accumulation, and spin current in the device.

31.2.1 Basic Formulation

In the presence of electric field **E** and gradient of carrier density n_σ in a conductor, the electrical current density $\mathbf{j}_\sigma$ for electrons with spin σ ($\sigma = \uparrow,\downarrow$) is given by $\mathbf{j}_\uparrow = \sigma_\uparrow \mathbf{E} - eD_\uparrow \nabla n_\uparrow$ and $\mathbf{j}_\downarrow = \sigma_\downarrow \mathbf{E} - eD_\downarrow \nabla n_\downarrow$, where σ_σ is the electrical conductivity, D_σ is the diffusion constant, and e is the electron charge ($e < 0$). Using $\nabla n_\sigma = N_\sigma \nabla \mu_c^\sigma$, where N_σ is the density of states of spin-σ electrons and μ_c^σ is the chemical potential, and the Einstein relation $\sigma_\sigma = e^2 N_\sigma D_\sigma$, we have the spin-dependent currents

$$\mathbf{j}_\uparrow = -\left(\frac{\sigma_\uparrow}{e}\right)\nabla\mu_\uparrow, \quad \mathbf{j}_\downarrow = -\left(\frac{\sigma_\downarrow}{e}\right)\nabla\mu_\downarrow, \tag{31.1}$$

where $\mu_\sigma = \mu_c^\sigma + e\phi$ is the ECP and ϕ is the electric potential ($\boldsymbol{E} = -\nabla\phi$). The continuity equations for charge and spin in the steady state are

$$\nabla \cdot (\mathbf{j}_\uparrow + \mathbf{j}_\downarrow) = 0, \quad \nabla \cdot (\mathbf{j}_\uparrow - \mathbf{j}_\downarrow) = -2e\frac{\delta n_\uparrow}{\tau_{\uparrow\downarrow}} + 2e\frac{\delta n_\downarrow}{\tau_{\uparrow\downarrow}}, \quad (31.2)$$

where

$\delta n_\sigma = n_\sigma - \bar{n}_\sigma$ is the deviation of carrier density n_σ from equilibrium one $\bar{n}_\sigma$

$\tau_{\sigma\sigma'}$ is the scattering time of an electron from spin state σ to σ'

Making use of the continuity equations and detailed balance $N_\uparrow/\tau_{\uparrow\downarrow} = N_\downarrow/\tau_{\downarrow\uparrow}$, which ensures the absence of spin imbalance in equilibrium, we obtain the basic equations for ECP that describe the charge and spin transport in each electrode [34–39]:

$$\nabla^2(\sigma_\uparrow\mu_\uparrow + \sigma_\downarrow\mu_\downarrow) = 0, \quad \nabla^2(\mu_\uparrow - \mu_\downarrow) = \frac{1}{\lambda^2}(\mu_\uparrow - \mu_\downarrow), \quad (31.3)$$

where

$\lambda = \sqrt{D\tau_s}$ is the spin-diffusion length

D and τ_s are the effective diffusion constant and spin relaxation time given by $D^{-1} = (N_\uparrow D_\downarrow^{-1} + N_\downarrow D_\uparrow^{-1})/(N_\uparrow + N_\downarrow)$ and $\tau_s^{-1} = (\tau_{\uparrow\downarrow}^{-1} + \tau_{\downarrow\uparrow}^{-1})$, respectively [37]

In the N electrode, the physical quantities are spin independent, for example, the electrical conductivity is $\sigma_N^\uparrow = \sigma_N^\downarrow = \frac{1}{2}\sigma_N$, while in the F-electrode, they are spin dependent, for example, $\sigma_F^\uparrow \neq \sigma_F^\downarrow (\sigma_F = \sigma_F^\uparrow + \sigma_F^\downarrow)$. The spin-diffusion lengths (λ_F) of transition-metal ferromagnets and alloys, such as NiFe (Py) and CoFe, are on the nanometer scale [40], whereas the spin-diffusion lengths (λ_N) of nonmagnetic metals, such as Cu, Ag, and Al, are on the micrometer scale. The large difference between λ_F and λ_N plays a crucial role in the spin-dependent transport in nonlocal devices.

The interfacial current across the junctions is described by using the CPP-GMR (current-perpendicular-plane giant magnetoresistance) theory developed by Valet and Fert [35]. In the presence of spin-dependent interface resistance R_i^σ at junction i ($i = 1,2$), the ECP changes discontinuously at the interface when the current flows across the junction. The spin-dependent interfacial current I_1^σ (I_2^σ) from F1 (F2) to N is given by the ECP difference at the interface [35–37]:

$$I_1^\sigma = \frac{1}{eR_1^\sigma}(\mu_{F1}^\sigma - \mu_N^\sigma), \quad I_2^\sigma = \frac{1}{eR_2^\sigma}(\mu_{F2}^\sigma - \mu_N^\sigma), \quad (31.4)$$

where the distribution of the current is assumed to be uniform over the interface. The total charge and spin currents across the ith interface are $I_i = I_i^\uparrow + I_i^\downarrow$ and $I_i^s = I_i^\uparrow - I_i^\downarrow$. These interfacial currents can be applicable from tunnel to transparent junction. In a transparent junction ($R_i^\sigma \to 0$), the ECP of each spin channel is continuous at the interface, which puts a constraint for spin accumulation on the N side, because the spin accumulation on F is very small due to the short spin-diffusion length. In a tunnel junction, spin accumulation on the N side is free from such constraint owing to a large discontinuous change in the ECP at the junction.

In a real device, the distribution of the current across the interface depends on the relative magnitude of the interface resistance to the electrode resistance [41]. When the interface resistance is much larger than the electrode resistance as in tunnel junctions, the current distribution is uniform over the contact area [42], which validates the assumption of uniform interface current assumed in Equation 31.4. However, when the interface resistance is comparable to or smaller than the electrode resistance as in metallic-contact junctions, the interface current may have inhomogeneous distribution with higher current density around a corner of the contact [43]. In this case, the effective contact area through which the current passes is smaller than the actual contact area $A_J = w_N w_F$ of the junction.

When current I is sent from F1 to the left side of N ($I_1 = I$), the solution of Equations 31.3 in N takes the form

$$\mu_N^\sigma = \bar{\mu}_N + \sigma(a_1 e^{-|x|/\lambda_N} - a_2 e^{-|x-L|/\lambda_N}). \quad (31.5)$$

Here, the first term describes the charge transport and takes $\bar{\mu}_N = -[eI/(\sigma_N A_N)]x$ ($A_N = d_N w_N$) for $x < 0$ and $\bar{\mu}_N = 0$ (ground level of ECP) for $x > 0$, and the second term is the shift in ECP of up-spin ($\sigma = +$) and down-spin ($\sigma = -$) electrons, where the a_1 term represents the spin accumulation due to spin injection from F1, whereas the a_2 term is the spin depletion due to spin absorption by F2. Note that, in the region of $x > 0$, the charge current ($j_N = j_N^\uparrow + j_N^\downarrow$) is absent and only the spin current ($j_N^s = j_N^\uparrow - j_N^\downarrow$) flows (Figure 31.1d), implying the pure spin current is created in this region.

In the F1 and F2 electrodes, the thicknesses are much larger than the spin-diffusion length ($d_F \gg \lambda_F$), as in the case of Py or CoFe, so that the solutions close to the interfaces may take the forms of vertical transport along the z direction:

$$\mu_{F1}^\sigma = \bar{\mu}_{F1} + \sigma b_1\left(\frac{\sigma_F}{\sigma_F^\sigma}\right)e^{-z/\lambda_F}, \quad \mu_{F2}^\sigma = \bar{\mu}_{F2} - \sigma b_2\left(\frac{\sigma_F}{\sigma_F^\sigma}\right)e^{-z/\lambda_F}, \quad (31.6)$$

where

$\bar{\mu}_{F1} = -[eI/(\sigma_F A_J)]z + eV_1$ describes the charge current flow in F1

$\bar{\mu}_{F2} = eV_2$ has a constant potential with no charge current in F2

V_1 and V_2 are the voltage drops across junctions 1 and 2

The coefficients a_i, b_i, and voltage V_i in Equations 31.5 and 31.6 are determined by the matching conditions that the spin and charge currents are continuous at the interfaces of junctions 1 and 2, and are given in Appendix 31.A. The voltages, V_2^P and V_2^{AP}, detected by F2 in the parallel (P) and antiparallel (AP) alignments of magnetizations (Figure 31.1e) are used to calculate the nonlocal resistances V_2^P/I and V_2^{AP}/I, the difference of which is the spin-accumulation signal ΔR_s (Figure 31.1c).

31.2.2 Spin Accumulation Signal

The spin-accumulation signal $\Delta R_s = (V_2^P - V_2^{AP})/I$ in the nonlocal spin injection detection device is given by [38]

$$\Delta R_s = R_N \frac{[P_1(2R_1^\star/R_N) + p_F(2R_F^\star/R_N)][P_2(2R_2^\star/R_N) + p_F(2R_F^\star/R_N)]e^{-L/\lambda_N}}{[1+(2R_1^\star/R_N)+(2R_F^\star/R_N)][1+(2R_2^\star/R_N)+(2R_F^\star/R_N)] - e^{-2L/\lambda_N}}, \quad (31.7)$$

where $R_N = (\rho_N\lambda_N)/A_N$ is the resistance of N electrode with resistivity ρ_N, length λ_N, and cross-sectional area $A_N = w_N d_N$, and

$$R_i^\star = \frac{R_i}{(1-P_i^2)}, \quad R_F^\star = \frac{R_F}{(1-p_F^2)}, \quad (31.8)$$

are, respectively, the renormalized versions of interface resistance R_i $(1/R_i = 1/R_i^\uparrow + 1/R_i^\downarrow)$ at junction i and that of resistance $R_F = (\rho_F\lambda_F)/A_J$ with resistivity ρ_F of F-electrode with resistivity ρ_F, length λ_F, and contact area $A_J = w_N w_F$ of the junctions. Hereafter, we call R_N and $R_F^\star$ the *spin resistance* of N and F, respectively. In Equation 31.8, $P_i = |R_i^\uparrow - R_i^\downarrow|/(R_i^\uparrow + R_i^\downarrow)$ is the interfacial current spin polarization of junction i and $p_F = |\rho_F^\uparrow - \rho_F^\downarrow|/(\rho_F^\uparrow + \rho_F^\downarrow)$ is the spin polarization of F1 and F2, where $\rho_F^\sigma = 1/\sigma_F^\sigma$ is the spin-dependent resistivity of F. In metallic-contact junctions, P_i and P_F are in the range of around 40%–70%, as determined from GMR experiments [44] and point-contact Andreev-reflection experiments [45]. In tunnel junctions, P_i is in the range of around 30%–55% for alumina (Al_2O_3) tunnel barriers [46–48] and very large (up to ~85%) for crystakkube MgO barriers [49,50], as determined from superconducting tunneling spectroscopy experiments.

The spin-accumulation signal ΔR_s strongly depends on the relative magnitude between the junction resistances (R_1, R_2) and the electrode resistances (R_F, R_N). Since R_F is much smaller than $R_N(R_N \gg R_F)$, as in a device with Cu and Py, Equation 31.7 is simplified as follows. When junctions 1 and 2 are both tunnel junctions $(R_1^\star, R_2^\star \gg R_N \gg R_F^\star)$ [9,11]

$$\frac{\Delta R_s}{R_N} = P_T^2 e^{-L/\lambda_N}, \quad (31.9)$$

where $P_T = P_1 = P_2$ is the tunnel spin polarization. When junction 1 is a tunnel junction and junction 2 is a transparent junction $(R_1^\star \gg R_N \gg R_F^\star \gg R_2)$ [38]

$$\frac{\Delta R_s}{R_N} = 2p_F P_T \left(\frac{R_F^\star}{R_N}\right) e^{-L/\lambda_N}. \quad (31.10)$$

When junctions 1 and 2 are both transparent junctions $(R_N \gg R_F \gg R_1^\star, R_2^\star)$ [10,36,37,51]

$$\frac{\Delta R_s}{R_N} = 4p_F^2 \left(\frac{R_F^\star}{R_N}\right)^2 \frac{e^{-L/\lambda_N}}{[1+2(R_F^\star/R_N)]^2 - e^{-2L/\lambda_N}} \approx \frac{2p_F^2}{\sinh(L/\lambda_N)}\left(\frac{R_F^\star}{R_N}\right)^2. \quad (31.11)$$

Note that ΔR_s in the above limiting cases is independent of interface resistance. In the intermediate regime $(R_F \ll R_i^\star \ll R_N)$, however, ΔR_s depends on the interface resistance as

$$\frac{\Delta R_s}{R_N} = \frac{2P_1P_2}{\sinh(L/\lambda_N)}\left(\frac{R_1^\star R_2^\star}{R_N^2}\right). \quad (31.12)$$

Figure 31.2a shows the spin-accumulation signal ΔR_s for $R_F/R_N = 0.03$, $p_F = 0.7$, and $P_1 = P_2 = 0.4$. We see that ΔR_s increases by 1 order of magnitude by replacing a metallic contact with a tunnel barrier, since the resistance mismatch, which is represented by $(R_F/R_N) \ll 1$, is removed by replacing a metallic contact with a tunnel junction. Note that the mismatch originates from a large difference in the spin-diffusion lengths between N and F$(\lambda_F \ll \lambda_N)$. When a nonmagnetic semiconductor is used for N, the resistance mismatch arises from the resistivity mismatch $(\rho_N \gg \rho_F)$ [51–53].

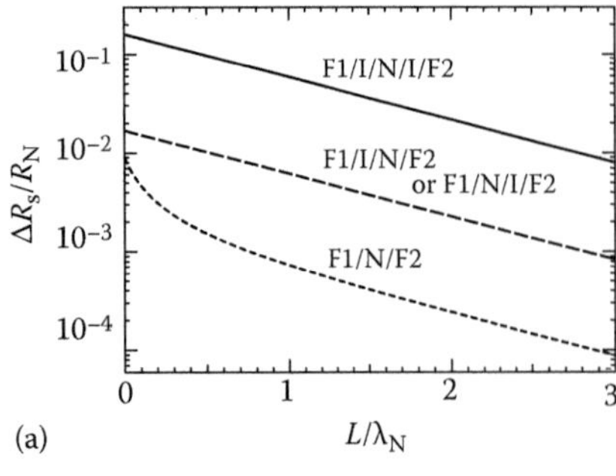

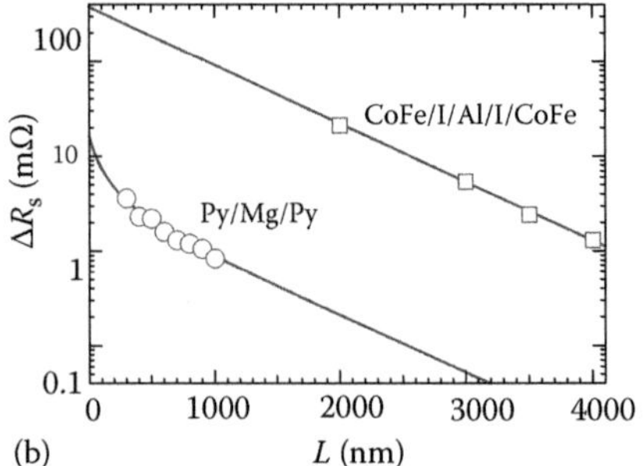

FIGURE 31.2 (a) Spin-accumulation signal ΔR_s as a function of distance L between F1 and F2. (b) Spin-accumulation signal ΔR_s as a function of distance L in tunnel and metallic-contact devices. The symbols (□) are the experimental data of CoFe/I/Al/I/CoFe tunnel devices [54] at 4.2 K, and (○) are those of Py/Mg/Py metallic-contact devices [23] at 10 K. (From Takahashi, S. and Maekawa, S., *Phys. Rev. B*, 67, 052409, 2003. With permission.)

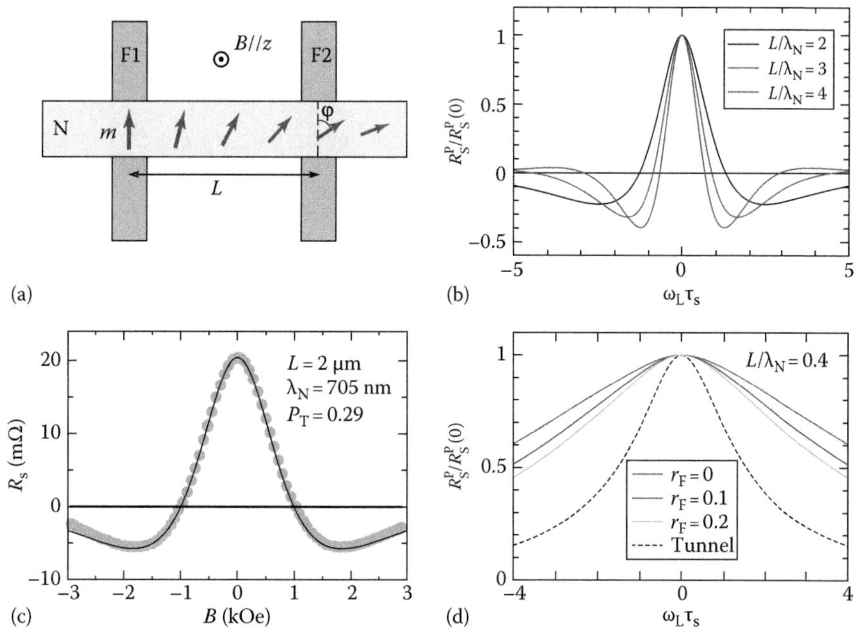

FIGURE 31.3 (See color insert.) (a) Spin precession in which electron spins rotate an angle φ during electron travel from the injector to the detector. (b) Nonlocal resistance as a function of Larmor frequency ω_L for distances $L/\lambda_N = 2$, 3, and 4 between F1 and F2 in the parallel alignment of magnetizations in a tunnel device. (c) Nonlocal resistance as a function of perpendicular magnetic field B for $L/\lambda_N = 4.4$ in a tunnel device. The symbol (●) is the experimental data of CoFe/I/Al/I/CoFe [56]. (d) Nonlocal resistance for distance $L/\lambda_N = 0.4$ and $r_F = 2(R_F^*/R_N) = 0$, 0.1, and 0.2 in an ohmic contact device. The dashed curve corresponds to that in a tunnel device.

Figure 31.2b shows the experimental data of ΔR_s as a function of distance L in devices of CoFe/I/Al/I/CoFe [54] and Py/Mg/Py [23]. In the tunnel device of CoFe/I/Al/I/CoFe ($I = Al_2O_3$), fitting Equation 31.9 to the data of Valenzuela and Tinkham [54] leads to $\lambda_N = 705$ nm at 4.2 K, $P_T = 0.3$, and $R_N = 4\,\Omega$.* The relation $\lambda_N^2 = D\tau_s$ with $\lambda_N = 705$ nm and $D = 1/[2e^2N(0)\rho_N] \sim 45\,\text{cm}^2/\text{s}$ leads to $\tau_s = 82$ ps at 4.2 K, which is consistent with the value of the spin–orbit parameter $b = \hbar/(3\tau_s\Delta_{Al}) \sim 0.01$ obtained by superconducting tunneling spectroscopy [48,49]. In the metallic-contact device of Py/Mg/Py, fitting Equation 31.11 to the data of Idzuchi et al. leads to $\lambda_N = 720$ nm, $R_N = 1.7\,\Omega$,† $P_F = 0.41$, $R_F = 0.1\,\Omega$,‡ and $\tau_s = 58$ ps at 10 K.

In the tunneling regime, the spin splitting of ECP at position x in N is given by

$$2\delta\mu_N(x) = P_T e R_N I e^{-|x|/\lambda_N}. \qquad (31.13)$$

For a device with $P_T = 0.3$, $R_N = 4\,\Omega$, and $I = 100\,\mu\text{A}$, the values of $\delta\mu_N(x)$ at $x = 0$ and λ_N are about 60 and 20 μeV, respectively, which are much smaller than the superconducting gap $\Delta \sim 200\,\mu\text{eV}$ of Al films. In a device with a superconductor (Al), it has been predicted that the spin-accumulation signal ΔR_s is dramatically enhanced due to the increase of the spin splitting of ECP in the superconducting state [38], and has been observed experimentally [29,30].

If a half-metallic ferromagnet ($p_F = 1$, $R_F^* \to \infty$) is used for F1 and F2, we have the largest signal $\Delta R_s \approx R_N e^{-L/\lambda_N}$ without tunnel barriers, which is the advantage of using a half-metallic ferromagnet with 100% spin polarization.

31.2.3 Spin Precession and Dephasing (Hanle Effect)

When a magnetic field $\mathbf{B} = (0,0,B)$ is applied perpendicular to the device plane, the injected spins in the N electrode precess around the z-axis parallel to $\mathbf{B}$, as shown in Figure 31.3a. This precession changes the direction of accumulated spins by angle $\varphi = \omega_L t$, where $\omega_L = \gamma_N B$ is the Larmor frequency and $\gamma_N = 2\mu_B/\hbar$ is the gyromagnetic ratio of conduction electrons [11,54]. The motion of the magnetization $\mathbf{m}_N$ due to the spin accumulation under the magnetic field is governed by the Bloch–Torrey equation [55]

$$\frac{\partial \mathbf{m}_N(x,t)}{\partial t} = -\gamma_N \mathbf{m}_N(x,t) \times \mathbf{B} - \frac{\mathbf{m}_N(x,t)}{\tau_s} + D_N \nabla^2 \mathbf{m}_N(x,t). \qquad (31.14)$$

In weak magnetic fields, the out-of-plane component m_z of magnetization is small, and is disregarded for simplicity. If a spin accumulation polarized in the y direction is created at position $x = 0$ and time $t = 0$ in a delta-function form $m_y(x,0) = C\delta(x)$, C being a constant, Equation 31.14 gives a time evolution of

* $\rho_N = 5.88\,\mu\Omega$ cm, $w_N = 400$ nm, and $d_N = 25$ nm for Al [54].
† $\rho_N = 4\,\mu\Omega$ cm, $w_N = 170$ nm, and $d_N = 100$ nm for Mg [23].
‡ $\rho_N = 35\,\mu\Omega$ cm, $w_F = 130$ nm, and $d_F = 20$ nm for Py [23].

magnetization as $\tilde{m}(x,t) = C\wp(x,t)e^{-t/\tau_s}e^{i\omega_L t}$, where $\tilde{m} = m_y + im_x$ is a complex representation of magnetization and $\wp(x,t) = (4\pi D_N t)^{-1/2}e^{-x^2/4D_N t}$ is the diffusion kernel in one dimension.

In a nonlocal F1/N/F2 device with tunnel barriers, spins are steadily injected from F1 to N to create the stationary magnetization at position x:

$$\tilde{m}(x) = \frac{\mu_B}{e}\frac{P_T I}{A_N}\int_0^\infty dt\,\wp(x,t)e^{-t/\tau_s}\exp(i\omega_L t) = \frac{\mu_B}{e}\frac{P_T I}{2D_N A_N}\lambda_\omega \exp\left(-\frac{|x|}{\lambda_\omega}\right), \tag{31.15}$$

where $\lambda_\omega = \lambda_N/\sqrt{1+i\omega_L\tau_s}$. Since the detected voltage by F2 at $x=L$ is given by $P_T m_y(L)/(\mu_B N(0))$, the nonlocal resistances R_s^P and R_s^{AP} in the parallel and antiparallel orientations become

$$R_s^P = -R_s^{AP} = \frac{1}{2}P_T^2 R_N^\omega \frac{\mathrm{Re}[\bar{\lambda}_\omega e^{-L/\lambda_\omega}]}{\mathrm{Re}[\bar{\lambda}_\omega]}, \tag{31.16}$$

where $R_N^\omega = \mathrm{Re}[\bar{\lambda}_\omega]R_N$ is the renormalized version of spin resistance due to spin precession and $\bar{\lambda}_\omega = \lambda_\omega/\lambda_N$. Figure 31.3b shows the normalized nonlocal resistance for distances $L/\lambda_N = 2$, 3, and 4 between F1 and F2 in the parallel alignment of the magnetizations in a nonlocal spin device with tunnel barriers. As seen from the figure, R_s^P shows a damped oscillation due to the spin precession and dephasing (Hanle effect). Figure 31.3c shows the experimental data of R_s^P as a function of perpendicular magnetic field B for $L = 2\,\mu$m in the parallel alignment of magnetizations in CoFe/I/Al/I/CoFe at 4.2 K [56]. Fitting Equation 31.16 to the data together with $\lambda_N = 705$ nm and $R_N = 4\,\Omega$ gives the values of $\tau_s = 52$ ps and $P_T = 0.29$.

When F1 and F2 have ohmic contacts with N, the nonlocal resistances are obtained as

$$R_s^P = -R_s^{AP} = 2p_F^2 R_N^\omega \left(\frac{R_F^*}{R_N^\omega}\right)^2 \times \frac{(\mathrm{Re}[\bar{\lambda}_\omega e^{-L/\lambda_\omega}]/\mathrm{Re}[\bar{\lambda}_\omega])}{[1+2(R_F^*/R_N^\omega)]^2 - (\mathrm{Re}[\bar{\lambda}_\omega e^{-L/\lambda_\omega}]/\mathrm{Re}[\bar{\lambda}_\omega])^2}, \tag{31.17}$$

which reduces to Equation 31.11 in the absence of the perpendicular magnetic field. Figure 31.3d shows the normalized R_s^P for distance $L/\lambda_N = 0.5$ between F1 and F2 in the case of ohmic contacts. With increasing field, R_s^P in the ohmic contact device decays much more slowly than in the tunnel device, which may reflect the strong absorption of the spin current by F2 in the ohmic contact device as discussed in Section 31.2.4. The results indicate that the Hanle effect depends strongly on whether the junctions are tunnel or ohmic contact junctions, in qualitative agreement with recent experiments [57].

31.2.4 Nonlocal Spin-Current Injection and Manipulation

We next discuss how the spin current is nonlocally injected from F1 to F2 (Figure 31.4a) through N, because of an interest in magnetization switching [58–60] caused by the absorption of pure spin current by F2 [61–63].

The magnitude and distribution of the spin accumulation and spin current in a nonlocal device strongly depends on the relative magnitudes between the interface resistances (R_i) and the electrode spin resistances (R_F, R_N). Figure 31.4b shows the spatial variation of spin accumulation $\delta\mu_N$ in the N electrode in

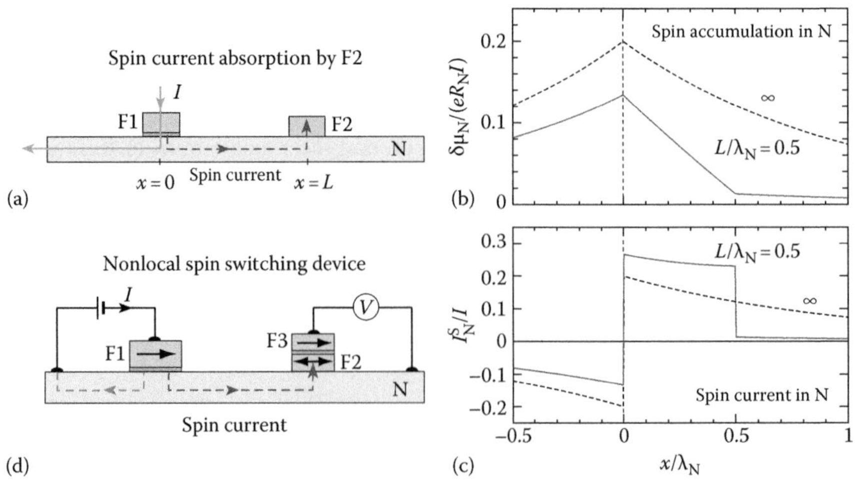

FIGURE 31.4 (a) Nonlocal spin current injection device of F1/I/N/F2, where junction 1 is a tunnel junction and 2 is a metallic contact. Spatial variations of (b) spin accumulation $\delta\mu_N$ and (c) spin current I_N^s in N for $L/\lambda_N = 0.5$ and ∞. The parameter values are the same as those in Figure 31.2. The discontinuous change of spin current at $L/\lambda_N = 0.5$ corresponds to the spin current flowing out of N through the N/F2 interface. (d) Spin-switching device utilizing nonlocal spin-current injection from F1 into F2 [61]. The magnetization direction in F2 is detected by F3. (From Takahashi, S. and Maekawa, S., *Sci. Technol. Adv. Mater.*, 9, 014105, 2008. With permission.)

the F1/I/N/F2 structure, in which the first junction is a tunnel junction and the second junction is an ohmic contact junction. In the absence of F2, the spin accumulation has a symmetric distribution around F1, as shown by the dashed curves of $L/\lambda_N = \infty$. By contrast, when F2 is placed at distance $L/\lambda_N = 0.5$, the spin accumulation is strongly suppressed by the ohmic contact of F2, leaving a small amount of spin accumulation on the right side of F2, as shown by the solid curve. This behavior is caused by the absorption of spins by F2 with small spin resistance R_F^*, and has been observed in a nonlocal device with three Py electrodes [14,15]. Note that the curve between F1 and F2 ($0 < x < L$) has a steep slope than that of the dashed curve, indicating an enhanced spin current I_N^s between F1 and F2 compared with that in the absence of F2, as seen in Figure 31.4c. The large discontinuous drop of I_N^s at $x = L$ indicates that most of the spin current flows out of N into F2 through the N/F2 interface, leaving very small spin current on the right side of F2 ($x > L$). This result clearly demonstrates strong absorption of the spin current by ferromagnets.

The spin current I_2^s across the N/F2 interface is calculated as [38,62]

$$I_2^s = \frac{2[P_1(R_1^*/R_N) + p_F(R_F^*/R_N)]e^{-L/\lambda_N} I}{[1 + 2(R_1^*/R_N) + 2(R_F^*/R_N)][1 + 2(R_2^*/R_N) + 2(R_F^*/R_N)] - e^{-2L/\lambda_N}}, \tag{31.18}$$

which leads to the spin-current injection from N into F2 being the largest in the case where the first junction is a tunnel junction and the second junction is an ohmic contact with F2 working as a strong spin absorber, such as Py or CoFe. In this case ($R_2^* \ll R_F^* \ll R_N \ll R_1^*$), the spin-current injected into F2 becomes

$$I_2^s \approx P_T I e^{-L/\lambda_N}. \tag{31.19}$$

When the both junctions are ohmic contact junctions ($R_i^* \ll R_F^* \ll R_N$), the spin-current injected into F2 becomes

$$I_2^s \approx \frac{p_F I}{\sinh(L/\lambda_N)} \left(\frac{R_F^*}{R_N} \right). \tag{31.20}$$

When a small F2 island or disk is placed on N with the contact area of $(50\,\text{nm})^2$ at distance L within the spin-diffusion length λ_N, the injected pure spin-current density into F2 becomes of the order of $I_2^s \sim 10^6 - 10^7\ \text{A/cm}^2$ for $I = 1\,\text{mA}$ and $P_T = 0.3$, suggesting that a large pure spin current can be injected, and hence, the spin-angular momentum is efficiently transferred from F1 to a small F2. This result provides a method for manipulating the orientation of magnetization due to spin transfer torque in nonlocal spin devices. Recently, the magnetization switching by the nonlocal pure spin current injection has been observed in lateral spin-valve Py/Cu/Py devices [63,64].

31.3 Spin Hall Effect

The anomalous Hall effect (AHE) originates from the relativistic interaction between the spin and orbital motion of electrons (spin–orbit interaction) in metals or semiconductors. Conduction electrons are scattered by local potentials created by impurities or defects in a crystal. The spin–orbit interaction at local potentials causes a spin-asymmetric scattering of conduction electrons [65]. In ferromagnetic materials, the electrical current is carried by up-spin (majority) and down-spin (minority) electrons, in which the flow of up-spin electrons are slightly deflected in a transverse direction while the down-spin (minority) electrons in the opposite direction, resulting in the electron flow in the direction perpendicular to both the applied electric field and the magnetization directions. Since up-spin (majority) and down-spin (minority) electrons are strongly imbalanced in ferromagnets, both spin and charge currents are generated in the transverse direction by AHE, the latter of which are probed as the Hall voltage.

Nonlocal spin injection in nanostructured devices provides a new opportunity for observing AHE in *nonmagnetic* conductors, the so-called spin Hall effect [66–71]. When a pure spin current without accompanying charge current is created in N via nonlocal spin injection, the up- and down-spin currents flowing in opposite directions are deflected in the same direction to induce a charge current in the transverse direction, resulting in a charge accumulation on the edges of N. Inversely, when an unpolarized charge current flows in N as a result of an applied electric field, the up- and down-spin currents flowing in the same direction are deflected in the opposite direction to induce a spin current in the transverse direction, resulting in a spin accumulation near the edges of N. As a consequence, the spin (charge) degrees of freedom are converted to charge (spin) degrees of freedom because of spin–orbit scattering in nonmagnetic conductors. Recently, SHE has been observed using nonlocal spin injection in metal-based nanostructured devices [54,72–75], which paves the way for future spin electronic applications. In addition to these *extrinsic* SHEs, *intrinsic* SHEs have been intensively studied in semiconductors that do not require impurities or defects [76–81].

In the following, we consider the effect of spin–orbit scattering on the spin and charge transports in nonmagnetic metals (N) such as Cu, Al, and Ag, and discuss SHE by taking into account the *side jump* (SJ) and skew *scattering* (SS) mechanisms [65,82–87], and derive formulas for the SHE induced by spin–orbit scattering in nonmagnetic metals [69].

31.3.1 Spin and Charge Currents Induced by SHE

The spin–orbit interaction in the presence of nonmagnetic impurities in a metal is derived as follows [88]. The impurity potential $V(\mathbf{r})$ gives rise to an additional electric field $\mathbf{E} = -(1/e)\nabla V(\mathbf{r})$. When an electron passes through the field with velocity $\hat{\mathbf{p}}/m = (\hbar/i)\nabla/m$, the electron feels an effective magnetic field $\mathbf{B}_{\text{eff}} = -(1/mc)\hat{\mathbf{p}}\times\mathbf{E}$, which leads to the spin–orbit coupling $V_{\text{so}} = \mu_B \boldsymbol{\sigma}\cdot\mathbf{B}_{\text{eff}} = \eta_{\text{so}}\boldsymbol{\sigma}\cdot[\nabla V(\mathbf{r})\times\nabla/i]$, where $\boldsymbol{\sigma}$ is the Pauli spin operator and η_{so} is the spin–orbit coupling parameter. Though the value of $\eta_{\text{so}} = (\hbar/2mc)^2$ in the free-electron model is too small to account for SHE as well as AHE observed in experiments, the value of η_{so} in real metals may be enhanced by several orders of magnitude for Bloch electrons [65]. In the following, η_{so} is treated as a phenomenological parameter. The total impurity potential $U(\mathbf{r})$ is the sum of the ordinary impurity potential and the spin–orbit potential: $U(\mathbf{r}) = V(\mathbf{r}) + V_{\text{so}}(\mathbf{r})$.

The one-electron Hamiltonian in the presence of the impurity potential $U(\mathbf{r})$ is given by

$$\mathcal{H} = \sum_{\mathbf{k},\sigma} \xi_k a^\dagger_{\mathbf{k}\sigma} a_{\mathbf{k}\sigma} + \sum_{\mathbf{k},\mathbf{k}'}\sum_{\sigma,\sigma'} \langle \mathbf{k}'\sigma'|U|\mathbf{k}\sigma\rangle a^\dagger_{\mathbf{k}'\sigma'} a_{\mathbf{k}\sigma}. \tag{31.21}$$

Here, the first term is the kinetic energy of conduction electrons with energies $\xi_k = (\hbar k)^2/2m - \mu_c$, and the second term describes the scattering of conduction electrons between states $|\mathbf{k}\sigma\rangle$ with momentum $\mathbf{k}$ and spin σ with the scattering amplitude

$$\langle \mathbf{k}'\sigma'|U|\mathbf{k}\sigma\rangle = \left(\frac{V_{\text{imp}}}{V}\right)\left[\delta_{\sigma'\sigma} + i\eta_{\text{so}}\boldsymbol{\sigma}_{\sigma'\sigma}\cdot(\mathbf{k}'\times\mathbf{k})\right]\sum_i e^{i(\mathbf{k}-\mathbf{k}')\cdot\mathbf{r}_i}, \tag{31.22}$$

where the first and second terms represent the matrix elements of $V(\mathbf{r})$ and $V_{\text{so}}(\mathbf{r})$, respectively, V_{imp} is the strength of impurity potential, V is the volume, σ is the Pauli matrix, and r_i is impurity position.

The velocity $\mathbf{v}^\sigma_{\mathbf{k}}$ of an electron in the presence of spin–orbit potential is calculated as follows. By taking the matrix element $\mathbf{v}^\sigma_{\mathbf{k}} = \langle \mathbf{k}^+\sigma|\hat{\mathbf{v}}|\mathbf{k}^+\sigma\rangle$ of the velocity operator $\hat{\mathbf{v}} = d\mathbf{r}/dt = [\mathbf{r}, H]/(i\hbar)$ [89] between the scattering state $|\mathbf{k}^+\sigma\rangle$ of an electron with momentum $\mathbf{k}$ and spin σ in the Born approximation, we obtain $\mathbf{v}^\sigma_{\mathbf{k}} = \boldsymbol{v}_{\mathbf{k}} + \boldsymbol{\omega}^\sigma_{\mathbf{k}}$ with the usual velocity $\boldsymbol{v}_{\mathbf{k}} = \hbar\mathbf{k}/m$ and *anomalous velocity* $\boldsymbol{\omega}^\sigma_{\mathbf{k}} = \alpha^{\text{SJ}}_{\text{SH}}(\boldsymbol{\sigma}_{\sigma\sigma}\times\hbar\mathbf{k}/m)$, where $\alpha^{\text{SJ}}_{\text{SH}}$ is the so-called spin Hall angle representing the strength of side jump

$$\alpha^{\text{SJ}}_{\text{SH}} = \frac{\hbar\bar{\eta}_{\text{so}}}{2\mu_c\tau^0_{\text{tr}}} = \frac{\bar{\eta}_{\text{so}}}{k_F l}, \tag{31.23}$$

where
- $\tau^0_{\text{tr}} = 1/[(2\pi/\hbar)n_{\text{imp}}N(0)V^2_{\text{imp}}]$ is the scattering time due to impurities
- n_{imp} is the impurity concentration
- $\bar{\eta}_{\text{so}} = k_F^2\eta_{\text{so}}$ is the dimensionless spin–orbit coupling parameter
- k_F is the Fermi momentum
- l is the mean-free path

Introducing the current operator $\hat{\mathbf{j}}_\sigma = e\sum_{\mathbf{k}}(\boldsymbol{v}_{\mathbf{k}} + \boldsymbol{\omega}^\sigma_{\mathbf{k}})a^\dagger_{\mathbf{k}\sigma}a_{\mathbf{k}\sigma}$ for conduction electrons with spin σ, the total charge current $\mathbf{J}_q = \mathbf{J}_\uparrow + \mathbf{J}_\downarrow$ and the total spin current $\mathbf{J}_s = \mathbf{J}_\uparrow - \mathbf{J}_\downarrow$ are expressed as

$$\mathbf{J}_q = \mathbf{J}'_q + \alpha^{\text{SJ}}_{\text{SH}}[\hat{\mathbf{z}}\times\mathbf{J}'_s], \quad \mathbf{J}_s = \mathbf{J}'_s + \alpha^{\text{SJ}}_{\text{SH}}[\hat{\mathbf{z}}\times\mathbf{J}'_q], \tag{31.24}$$

where

$$\mathbf{J}'_q = e\sum_{\mathbf{k}} \boldsymbol{v}_{\mathbf{k}}(f_{\mathbf{k}\uparrow} + f_{\mathbf{k}\downarrow})$$

$$\mathbf{J}'_s = e\sum_{\mathbf{k}} \boldsymbol{v}_{\mathbf{k}}(f_{\mathbf{k}\uparrow} - f_{\mathbf{k}\downarrow})$$

$f_{\mathbf{k}\sigma} = \langle a^\dagger_{\mathbf{k}\sigma}a_{\mathbf{k}\sigma}\rangle$ is the distribution function of an electron with energy ξ_k and spin σ

$\hat{\mathbf{z}}$ is the polarization axis ($\boldsymbol{\sigma}_{\sigma\sigma} = \sigma\hat{\mathbf{z}}$)

The second terms in Equations 31.24 are the charge and spin Hall currents due to side jump. In addition to the side jump contribution, there is the skew scattering contribution that originates from the modification of the distribution function due to anisotropic scattering by the spin–orbit interaction.

The distribution function $f_{\mathbf{k}\sigma}$ is calculated based on the Boltzmann transport equation in the steady state:

$$\boldsymbol{v}_{\mathbf{k}}\cdot\nabla f_{\mathbf{k}\sigma} + \frac{e\mathbf{E}}{\hbar}\cdot\nabla_{\mathbf{k}} f_{\mathbf{k}\sigma} = \left(\frac{\partial f_{\mathbf{k}\sigma}}{\partial t}\right)_{\text{scatt}}, \tag{31.25}$$

where
- $\boldsymbol{v}_{\mathbf{k}} = \hbar\mathbf{k}/m$
- $\mathbf{E}$ is the external electric field

and the collision term due to impurity scattering is written as

$$\left(\frac{\partial f_{\mathbf{k}\sigma}}{\partial t}\right)_{\text{scatt}} = \sum_{\mathbf{k}'\sigma'}[P^{\sigma\sigma'}_{\mathbf{k}\mathbf{k}'}f_{\mathbf{k}'\sigma'} - P^{\sigma'\sigma}_{\mathbf{k}'\mathbf{k}}f_{\mathbf{k}\sigma}], \tag{31.26}$$

where the first term in the brackets is the scattering-in term ($\mathbf{k}'\sigma'\to\mathbf{k}\sigma$) and the second term is the scattering-out term ($\mathbf{k}\sigma\to\mathbf{k}'\sigma'$), $P^{\sigma'\sigma}_{\mathbf{k}'\mathbf{k}} = (2\pi/\hbar)n_{\text{imp}}|\langle\mathbf{k}'\sigma'|\hat{T}|\mathbf{k}\sigma\rangle|^2\delta(\xi_{\mathbf{k}} - \xi_{\mathbf{k}'})$ is the scattering probability from state $|\mathbf{k}\sigma\rangle$ to state $|\mathbf{k}'\sigma'\rangle$, and $\hat{T}$ is the scattering matrix whose matrix elements are calculated within the second-order Born approximation. Thus, we find that the scattering probability has the symmetric and asymmetric contributions:

$$P^{\sigma'\sigma(1)}_{\mathbf{k}'\mathbf{k}} = \frac{2\pi}{\hbar}\frac{n_{\text{imp}}}{V}V^2_{\text{imp}}(\delta_{\sigma\sigma'} + \eta^2_{\text{so}}|(\mathbf{k}'\times\mathbf{k})\cdot\boldsymbol{\sigma}_{\sigma\sigma'}|^2)\delta(\xi_{\mathbf{k}'} - \xi_{\mathbf{k}}), \tag{31.27}$$

$$P^{\sigma'\sigma(2)}_{\mathbf{k}'\mathbf{k}} = -\frac{(2\pi)^2}{\hbar}\eta_{\text{so}}\frac{n_{\text{imp}}}{V}V^3_{\text{imp}}N(0)\delta_{\sigma\sigma'}[(\mathbf{k}'\times\mathbf{k})\cdot\boldsymbol{\sigma}_{\sigma\sigma}]\delta(\xi_{\mathbf{k}'} - \xi_{\mathbf{k}}). \tag{31.28}$$

In solving the Boltzmann equation, it is convenient to separate $f_{\mathbf{k}\sigma}$ into three parts [90] as

$$f_{\mathbf{k}\sigma} = f^0_{k\sigma} + g^{(1)}_{\mathbf{k}\sigma} + g^{(2)}_{\mathbf{k}\sigma}, \qquad (31.29)$$

where

$f^0_{k\sigma}$ is a nondirectional distribution function defined by the average of $f_{\mathbf{k}\sigma}$ with respect to the solid angle $\Omega_{\mathbf{k}}$ of $\mathbf{k}$: $f^0_{k\sigma} = \int f_{\mathbf{k}\sigma} d\Omega_{\mathbf{k}}/(4\pi)$

$g^{(1)}_{\mathbf{k}\sigma}$ and $g^{(2)}_{\mathbf{k}\sigma}$ are both directional distribution functions, that is, $\int g^{(i)}_{\mathbf{k}\sigma} d\Omega_{\mathbf{k}} = 0$, and are related with the symmetric and asymmetric contributions, respectively

We first determine $g^{(1)}_{\mathbf{k}\sigma}$ from the Boltzmann equation with the collision term

$$\sum_{\mathbf{k}'\sigma'} [P^{\sigma\sigma'(1)}_{\mathbf{k}\mathbf{k}'} f_{\mathbf{k}'\sigma'} - P^{\sigma'\sigma(1)}_{\mathbf{k}'\mathbf{k}} f_{\mathbf{k}\sigma}] = -\frac{g^{(1)}_{\mathbf{k}\sigma}}{\tau_{\mathrm{tr}}} - \frac{f^0_{k\sigma} - f^0_{k-\sigma}}{\tau_{\mathrm{sf}}(\theta)}, \qquad (31.30)$$

where

τ_{tr} is the transport relaxation time

$\tau_{\mathrm{sf}}(\theta)$ is the spin-flip relaxation time, which are given by $1/\tau_{\mathrm{tr}} = (1/\tau^0_{\mathrm{tr}})(1+2\bar{\eta}^2_{\mathrm{so}}/3)$ and $1/\tau_{\mathrm{sf}}(\theta) = (2\bar{\eta}^2_{\mathrm{so}}/3\tau^0_{\mathrm{tr}}) \times (1+\cos^2\theta)$ angle θ between $\mathbf{k}$ and the z-axis

Then, the Boltzmann equation (31.25) with the collision term (31.30) is [68,91]

$$\mathbf{v}_{\mathbf{k}} \cdot \nabla f_{\mathbf{k}\sigma} + \frac{e\mathbf{E}}{\hbar} \cdot \nabla_{\mathbf{k}} f_{\mathbf{k}\sigma} = -\frac{g^{(1)}_{\mathbf{k}\sigma}}{\tau_{\mathrm{tr}}} - \frac{f^0_{k\sigma} - f^0_{k-\sigma}}{\tau_{\mathrm{sf}}(\theta)}, \qquad (31.31)$$

where the first term on the rhs describes the momentum relaxation due to impurity scattering and the second term the spin relaxation due to spin-flip scattering. Since $\tau_{\mathrm{tr}} \ll \tau_{\mathrm{sf}}$, the momentum relaxation occurs first, followed by slow spin relaxation.

The distribution function $f^0_{k\sigma}$ describes the spin accumulation by the shift $\sigma\delta\mu_c(\mathbf{r})$ in the chemical potential from the equilibrium one μ_c, and may be expanded as

$$f^0_{k\sigma} \approx f_0(\xi_{\mathbf{k}}) + \sigma\left(-\frac{\partial f_0}{\partial \xi_{\mathbf{k}}}\right)\delta\mu_c(\mathbf{r}), \qquad (31.32)$$

where $f_0(\xi_{\mathbf{k}})$ is the Fermi distribution function. Using $f^0_{k\sigma}$ for $f_{\mathbf{k}\sigma}$ in Equation 31.31 and $(-\partial f_0/\partial\xi_{\mathbf{k}}) \approx \delta(\xi_{\mathbf{k}})$, we obtain

$$g^{(1)}_{\mathbf{k}\sigma} \approx -\tau_{\mathrm{tr}}\left(-\frac{\partial f_0}{\partial \xi_{\mathbf{k}}}\right)\mathbf{v}_{\mathbf{k}} \cdot \nabla\mu^{\sigma}_{\mathrm{N}}(\mathbf{r}), \qquad (31.33)$$

where

$\mu^{\sigma}_{\mathrm{N}}(\mathbf{r}) = \mu_c + \sigma\delta\mu_c + e\phi$ is the ECP

ϕ is the electric potential ($\mathbf{E} = -\nabla\phi$)

By substituting Equations 31.32 and 31.33 into the Boltzmann equation (31.31) and summing over $\mathbf{k}$, one obtains the spin-diffusion equation

$$\nabla^2(\mu^{\uparrow}_{\mathrm{N}} - \mu^{\downarrow}_{\mathrm{N}}) = \lambda^{-2}_{\mathrm{N}}(\mu^{\uparrow}_{\mathrm{N}} - \mu^{\downarrow}_{\mathrm{N}}), \qquad (31.34)$$

with $\lambda_{\mathrm{N}} = \sqrt{D\tau_{\mathrm{sf}}/2}$, $D = (1/3)\tau_{\mathrm{tr}}v^2_{\mathrm{F}}$, and $\langle\tau^{-1}_{\mathrm{sf}} = \tau^{-1}_{\mathrm{sf}}(\theta)\rangle_{\mathrm{av}} = (4/9)\bar{\eta}^2_{\mathrm{so}}/\tau^0_{\mathrm{tr}}$. Note that $\tau_{\mathrm{sf}}=2\tau_{\mathrm{s}}$. The spin-flip relaxation time τ_{sf} is related to the transport relaxation τ_{tr} time through the spin–orbit coupling parameter η_{so} as $\tau_{\mathrm{tr}}/\tau_{\mathrm{sf}} = (4/9)\bar{\eta}^2_{\mathrm{so}}/[1+(2/3)\bar{\eta}^2_{\mathrm{so}}]$.

The distribution function $g^{(2)}_{\mathbf{k}\sigma}$ contributed from *skew scattering* is determined by the asymmetric terms in the Boltzmann equation $\sum_{\mathbf{k}'\sigma'} [-P^{\sigma'\sigma(1)}_{\mathbf{k}'\mathbf{k}} g^{(2)}_{\mathbf{k}\sigma} + P^{\sigma'\sigma(2)}_{\mathbf{k}'\mathbf{k}} g^{(1)}_{\mathbf{k}'\sigma'}] = 0$ with Equations 31.27, 31.28, and 31.33, yielding

$$g^{(2)}_{\mathbf{k}\sigma} = \alpha^{\mathrm{SS}}_{\mathrm{SH}}\tau_{\mathrm{tr}}\left(-\frac{\partial f_0}{\partial \xi_{\mathbf{k}}}\right)(\sigma_{\sigma\sigma} \times \mathbf{v}_{\mathbf{k}}) \cdot \nabla\mu^{\sigma}_{\mathrm{N}}(\mathbf{r}), \qquad (31.35)$$

where $\alpha^{\mathrm{SS}}_{\mathrm{SH}}$ is the so-called spin Hall angle representing the strength of skew scattering

$$\alpha^{\mathrm{SS}}_{\mathrm{SH}} = -\left(\frac{2\pi}{3}\right)\bar{\eta}_{\mathrm{so}} N(0) V_{\mathrm{imp}}. \qquad (31.36)$$

Using the distribution function $f_{\mathbf{k}\sigma}$ obtained above, we can calculate the first terms in Equation 31.24 as $\mathbf{J}'_s = \mathbf{j}_s + \alpha^{\mathrm{SS}}_{\mathrm{SH}}(\hat{\mathbf{z}} \times \mathbf{j}_q)$ and $\mathbf{J}'_q = \mathbf{j}_q + \alpha^{\mathrm{SS}}_{\mathrm{SH}}(\hat{\mathbf{z}} \times \mathbf{j}_s)$, where the first terms are the longitudinal spin and charge currents:

$$\mathbf{j}_s = -\left(\frac{\sigma_{\mathrm{N}}}{e}\right)\nabla\delta\mu_{\mathrm{N}}, \quad \mathbf{j}_q = \sigma_{\mathrm{N}}\mathbf{E}, \qquad (31.37)$$

where

$\sigma_{\mathrm{N}} = 2e^2N(0)D$ is the electrical conductivity

$\delta\mu_{\mathrm{N}} = \frac{1}{2}(\mu^{\uparrow}_{\mathrm{N}} - \mu^{\downarrow}_{\mathrm{N}})$ is the chemical potential shift, and the second terms are the Hall spin and charge currents induced by the charge and spin currents, respectively

Therefore, the total spin and charge currents in Equation 31.24 are written as

$$\mathbf{J}_q = \mathbf{j}_q + \alpha_{\mathrm{SH}}(\hat{\mathbf{z}} \times \mathbf{j}_s), \quad \mathbf{J}_s = \mathbf{j}_s + \alpha_{\mathrm{SH}}(\hat{\mathbf{z}} \times \mathbf{j}_q), \qquad (31.38)$$

where $\alpha_{\mathrm{SH}} = \alpha^{\mathrm{SJ}}_{\mathrm{SH}} + \alpha^{\mathrm{SS}}_{\mathrm{SH}}$ with $\alpha^{\mathrm{SJ}}_{\mathrm{SH}} = \bar{\eta}_{\mathrm{so}}/(k_{\mathrm{F}}l)$ and $\alpha^{\mathrm{SS}}_{\mathrm{SH}} = -(2\pi/3)\bar{\eta}_{\mathrm{so}}N(0)V_{\mathrm{imp}}$. Equations 31.38 indicate that the spin current $\mathbf{j}_s$ induces the transverse charge current $\mathbf{j}^{\mathrm{SH}}_q = \alpha_{\mathrm{SH}}(\hat{\mathbf{z}} \times \mathbf{j}_s)$, while the charge current $\mathbf{j}_q$ induces the transverse spin current $\mathbf{j}^{\mathrm{SH}}_s = \alpha_{\mathrm{SH}}(\hat{\mathbf{z}} \times \mathbf{j}_q)$, as shown in Figure 31.5.

The spin Hall conductivity σ_{SH} is given by the sum of the side-jump contribution $\sigma^{\mathrm{SJ}}_{\mathrm{SH}} = \alpha^{\mathrm{SJ}}_{\mathrm{SH}}\sigma_{\mathrm{N}}$ and the skew scattering contribution $\sigma^{\mathrm{SS}}_{\mathrm{SH}} = \alpha^{\mathrm{SS}}_{\mathrm{SH}}\sigma_{\mathrm{N}}$. We note that the SJ conductivity has the form $\sigma^{\mathrm{SJ}}_{\mathrm{SH}} = (e^2/\hbar)\eta_{\mathrm{so}}n_e$, n_e being the carrier density, is independent of

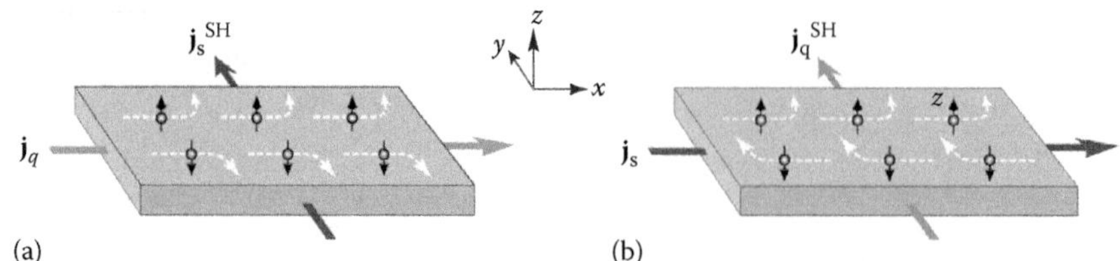

FIGURE 31.5 (a) Spin Hall effect (SHE) in which the charge current $\mathbf{j}_q$ along the x direction induces the spin current $\mathbf{j}_s^{SH}$ in the y direction with the polarization parallel to the z-axis. (b) Inverse SHE in which the spin current $\mathbf{j}_s$ flowing along the x direction with the polarization parallel to the z-axis induces the charge current $\mathbf{j}_q^{SH}$ in the y direction. (From Takahashi, S. and Maekawa, S., *Sci. Technol. Adv. Mater.*, 9, 014105, 2008. With permission.)

the impurity concentration, while the SS conductivity depends on the strength and distribution of impurities. When the impurities have a narrow distribution of potentials with definite sign (either positive or negative), as in doped impurities, the SS contribution is dominant for SHE, whereas when the impurity potentials are distributed with positive and negative sign such that the average of V_{imp} over the impurity distribution vanishes ($\langle V_{imp}\rangle \approx 0$), then SJ contribution is dominant. The spin Hall resistivity $\rho_{SH} \approx \sigma_{SH}/\sigma_N^2$ has linear and quadratic terms in ρ_N representing the contributions from side jump and skew scatterings, respectively: $\rho_{SH} \approx a_{SS}\rho_N + b_{SJ}\rho_N^2$, where $a_{SS} = -(2\pi/3)\bar{\eta}_{so}N(0)V_{imp}$ and $b_{SJ} = (2/3\pi)\bar{\eta}_{so}(e^2/h)k_F$.

31.3.2 Spin–Orbit Coupling Parameter

The electrical resistivity and the spin-diffusion length are fundamental quantities for the charge and spin transports. By multiplying the resistivity and the spin-diffusion length, we obtain $\rho_N\lambda_N = (\sqrt{3}\pi R_K/2k_F^2)(\tau_{sf}/2\tau_{tr})^{1/2}$, where $R_K = h/e^2 \approx 25.8\,\mathrm{k}\Omega$ is the quantum resistance. Since the spin–orbit coupling parameter is given by $\bar{\eta}_{so} = (3/2)(\tau_{tr}/\tau_{sf})^{1/2}$ for $\eta_{so}^2 \ll 1$, we have the relation [62,70,92]:

$$\bar{\eta}_{so} \approx \frac{3R_K}{k_F^2\rho_N\lambda_N}. \tag{31.39}$$

Equation 31.39 implies that the spin–orbit coupling parameter $\bar{\eta}_{so}$ is expressed in terms of the product $\rho_N\lambda_N$, providing a useful method of evaluating the spin–orbit coupling in nonmagnetic metals. Using the experimental data of ρ_N and λ_N for various metals in Equation 31.39, we obtain the values of the spin–orbit coupling parameter $\bar{\eta}_{so}$ in those metals, as listed in Table 31.1. We note that $\bar{\eta}_{so}$ is small for Al (light metal), large for Pt (heavy metal), and intermediate for Cu, Ag, and Au. In the case of Al, the spin–orbit coupling parameters for different samples are very close to each other, despite the scattered values of λ_N and ρ_N between them. The values of $\bar{\eta}_{so}$ estimated from the spin injection method are several orders of magnitude larger than the value of $\bar{\eta}_{so} = (\hbar k_F/2mc)^2$ in the free-electron model.

The spin Hall angle due to side-jump contribution (31.23) has the simple form $\alpha_{SH}^{SJ} = (3/8)^{1/2}/(k_F\lambda_N)$ with the aid of the relation $l/\lambda_N = (6\tau_{tr}/\tau_{sf})^{1/2} = (8/3)^{1/2}\bar{\eta}_{so}$. This formula is useful to estimate

TABLE 31.1 Spin–Orbit Coupling Parameter $\bar{\eta}_{so}$ for Al, Mg, Cu, Ag, Au, and Pt

	λ_N (nm)	ρ_N ($\mu\Omega$ cm)	τ_{sf}/τ_{tr}	$\bar{\eta}_{so}$	Reference
Al (4.2 K)	650	5.90	5.6×10^4	0.0063	[11]
Al (4.2 K)	455	9.53	7.2×10^4	0.0056	[54]
Al (4.2 K)	705	5.88	6.5×10^4	0.0059	[54]
Mg (10 K)	720	4.00	1.2×10^4	0.014	[23]
Cu (4.2 K)	1000	1.43	2.8×10^3	0.028	[10]
Cu (4.2 K)	1500	1.00	3.1×10^3	0.027	[94]
Cu (4.2 K)	546	3.44	4.9×10^3	0.021	[18]
Cu (<4 K)	520	3.5	4.6×10^3	0.022	[95]
Ag (4.2 K)	162	4.00	3.7×10^2	0.080	[19]
Ag (4.2 K)	195	3.50	4.1×10^2	0.076	[19]
Ag (<4 K)	750	3.0	5.3×10^3	0.021	[95]
Ag (77 K)	3000	1.1	9.2×10^3	0.015	[22]
Au (4.2 K)	168	4.00	3.8×10^2	0.078	[20]
Au (<4 K)	58.5	3.3	31.4	0.27	[95]
Au (10 K)	63	1.36	6.2	0.60	[16]
Pt (4.2 K)	14	4.2	1.4	1.27	[96]
Pt (5 K)	14	12.8	13	0.42	[73]

Here, the Fermi momenta, $k_F = 1.75\times10^8$ cm^{-1} (Al), 1.36×10^8 cm^{-1} (Mg, Cu), 1.20×10^8 cm^{-1} (Ag), and 1.21×10^8 cm^{-1} (Au) from [93] are taken, and 1×10^8 cm^{-1} for Pt is assumed.

α_{SH}^{SJ} and σ_{SH}^{SJ} from the measured values of λ_N and σ_N. Using the data in Table 31.1, one can obtain the magnitude of α_{SH}^{SJ} and σ_{SH}^{SJ}. For example, $\alpha_{SH}^{SJ} = 5-8\times10^{-5}$ and $\sigma_{SH}^{SJ} = 8-9\ (\Omega\ \mathrm{cm})^{-1}$ for Al [54]. For Pt, $\alpha_{SH}^{SJ} = 4.4\times10^{-3}$ and $\sigma_{SH}^{SJ} = 340\ (\Omega\ \mathrm{cm})^{-1}$ [73], which are much larger than those of Al, since Pt is a heavy metal element with large $\bar{\eta}_{so}$ and short λ_N.

31.3.3 Nonlocal Spin Hall Effect

In nonlocal spin injection devices, a pure spin current is created in a nonmagnetic metal electrode. It is fundamentally important to verify the existence of the spin current flowing in a nonmagnetic metal. A most simple and direct verification of the spin current is made by using a nonlocal spin Hall device shown in Figure 31.6a [62,70,92,97]. In this device, the magnetization of the ferromagnet (F) is in the z direction perpendicular to the plane. Spin injection is made by applying the current I from F to the left end of N, while the Hall voltage (V_{SH}) is measured by

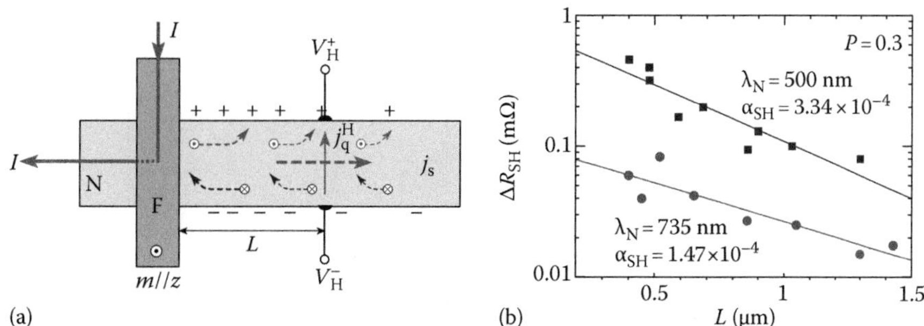

FIGURE 31.6 (a) Nonlocal spin Hall device. The magnetization of F is pointed perpendicular to the plane. The injected pure spin current generates the spin Hall voltage $V_{SH} = V_{SH}^+ - V_{SH}^-$ in the transverse direction, which is measured by the Hall bar at distance L. (b) Nonlocal spin Hall resistance ΔR_{SH} as a function of distance L between the spin injector and the Hall bar for two different devices in high magnetic fields perpendicular to the device plane. The symbols (●, ■) are the experimental data of CoFe/I/Al at 4.2 K. (From Valenzuela, S.O. and Tinkham, M., *Nature*, 442, 176, 2006. With permission.)

the Hall bar at distance L, where the pure spin current $\mathbf{j}_s = (j_s,0,0)$ flows in the x direction. Thus it follows from Equations 31.38 that $\mathbf{J}_s = \mathbf{j}_s$ and

$$\mathbf{J}_q = \sigma_N \mathbf{E} + \alpha_{SH}(\hat{\mathbf{z}} \times \mathbf{j}_s), \tag{31.40}$$

where the second term is the Hall current induced by the spin current. In an open circuit condition in the transverse direction, the ohmic current builds up in the transverse direction as opposed to the Hall current such that the y component of $\mathbf{J}_q$ in Equation 31.40 vanishes, resulting in the relation between the Hall electric field E_y and the spin current j_s, $E_y = -\alpha_{SH}\rho_N j_s$, which is integrated over the width w_N of N to yield the Hall voltage

$$V_{SH} = \alpha_{SH} w_N \rho_N j_s, \tag{31.41}$$

indicating that the induced Hall voltage is proportional to the spin current. The spin current at $x = L$ is given by

$$j_s \approx \frac{1}{2} P_{\text{eff}} \left(\frac{I}{A_N} \right) e^{\ L/\lambda_N}, \tag{31.42}$$

where P_{eff} is the effective spin polarization that takes the tunnel spin polarization $P_{\text{eff}} = P_T$ for a tunnel junction and $P_{\text{eff}} = p_F(R_F^*/R_N)$ for a metallic-contact junction. Therefore, the nonlocal Hall resistance $R_{SH} = V_{SH}/I$ becomes [62,70,92,97]

$$R_{SH} = \frac{1}{2} P_{\text{eff}} \alpha_{SH} \frac{\rho_N}{d_N} e^{-L/\lambda_N}. \tag{31.43}$$

For typical values of device parameters ($P_{\text{eff}} \sim 0.3$, $d_N \sim 10$ nm, and $\rho_N \sim 5\,\mu\Omega$ cm), and $\alpha_{SH} \sim 0.1$–0.0001 for $\bar{\eta}_{so} = 0.5 - 0.005$ (Table 31.1), $k_F l \sim 100$, and $V_{\text{imp}} N(0) \sim 0.1$–$0.01$, the expected value of R_{SH} at $L = \lambda_N/2$ is of the order of 0.05–5 mΩ, indicating that SHE is measurable using nonlocal Hall devices. Using finite element method in three dimensions, a more realistic and quantitative calculation is possible to investigate the SHE and reveal the spatial distribution of spin and charge currents in a nonlocal device by taking into account the device structure and geometry [98].

Recently, the SHE was observed by using nonlocal spin injection devices: CoFe/I/Al ($I = Al_2O_3$) under high magnetic fields perpendicular to the device plane by Valenzuela and Tinkham [54,99], Py/Cu/Pt using strong spin absorption by Pt by Kimura et al. [72,73], in FePt/Au using a perpendicularly magnetized FePt by Seki et al. [74], and by using ferromagnetic resonance (FMR) in Py/Pt bilayer films by technique by Saitoh et al. [75,100].

Figure 31.6b shows the experimental results of CoFe/Al [54] for the nonlocal spin Hall resistance $\Delta R_{SH} = R_{SH}(B_s) - R_{SH}(-B_s)$ as a function of distance L between the spin injector CoFe and the Hall bar in high perpendicular magnetic fields ($\pm B_s$) to saturate the magnetization of CoFe in the perpendicular direction, producing perpendicularly polarized spin current in Al. In the device with the Al thickness of 12 nm fitting Equation 31.43 to the data leads to $\lambda_N = 500$ nm, $\alpha_{SH} = 3.34 \times 10^{-4}$, and $\sigma_{SH} = 43\ (\Omega\ \text{cm})^{-1}$ for $P_T = 0.3$ and $\rho_N/d_N = 8\,\Omega$.* In the device with the Al thickness of 25 nm, fitting Equation 31.43 to the data leads to $\lambda_N = 735$ nm $\alpha_{SH} = 1.5 \times 10^{-4}$, and $\sigma_{SH} = 25\ (\Omega\ \text{cm})^{-1}$ for $P_T = 0.3$, and $R_N = 2.4\,\Omega$. These estimated values for the spin Hall angle and the spin Hall conductivity are rather satisfactory agreement with those presented in Section 31.3.2.

It has been pointed out that a large SHE is caused by the skew scattering mechanism in nonmagnetic metal with magnetic impurities [101]. Recently, based on a first principle band structure calculation [102] and a quantum Monte Carlo simulation [103] for Fe impurities in a Au host metal, a novel type of Kondo effect due to strong electron correlation at iron impurities tremendously enhances the spin–orbit interaction of the order of the hybridization energy (~eV), leading to a large spin Hall angle comparable to that observed in experiments of giant SHE [74].

* $\rho_N = 9.53\,\mu\Omega$ cm, $w_N = 400$ nm, and $d_N = 12$ nm for Al [54].

31.4 Spin Injection into a Ferromagnetic Insulator

Recent observation that a spin current in a Pt wire is injected into an insulator magnet $Y_3Fe_5O_{12}$ and an electric signal is detected by another Pt wire [104] has stimulated new research interests in spintronics. Magnetic insulators are unique in that they are electrically inactive with frozen charge degrees of freedom, but magnetically active due to the spins of localized electrons. The low-lying excitation of spin wave (magnon) carries spin-angular momentum. The spin current carried by spin waves is called a spin-wave (magnon) spin current. Utilizing purely magnetic excitation without charge excitation is crucial for developing energy-saving spintronic devices.

In this section, we demonstrate that the spin exchange interaction between local moments and conduction electrons at the interface of normal metal and FI plays a vital role for the spin current across the interface. Making use of linear response theory and fluctuation-dissipation theorem, we derive a formula for the spin current through the interface, and discuss the conversion efficiency of the spin currents.

31.4.1 Spin Current through an FI/N Interface

We consider a junction that consists of a normal metal (N) and an FI, as shown in Figure 31.1. The magnetization of FI is oriented in the z direction. For simplicity, we assume a spin accumulation in N, in which the up-spin and down-spin chemical potentials are split by $\delta\mu_N = \mu_N^{\uparrow} - \mu_N^{\downarrow}$ and the resulting magnetization of N is either parallel or antiparallel to the magnetization of FI, depending on the sign of $\delta\mu_N$. This situation is realized by spin injection from other ferromagnets connected to N or by the SHE by applying a current in the N layer.

When conduction electrons in N are incident on FI, the electrons are reflected back at the F/N interface, since electrons are prohibited to enter FI due to the large energy gap at the Fermi energy. At the scattering, there is a spin-flip process in which an electron reverses its spin, thereby emitting or absorbing a magnon in F. The spin-flip scattering with magnon excitation gives rise to transfer of spin-angular momentum between FI and N. When the up-spin chemical potential is higher than the down-spin chemical potential (see Figure 31.7), the spin-flip process from up-spin to down-spin state dominates over the reversed process, so that the spin current flows from N to FI across the FI/N interface.

We describe the electron–magnon interaction at the interface by using the s–d exchange interaction between local moments and conduction electrons at the interface:

$$H_{sd} = \frac{J_{sd}}{N_N} \sum_{n=1}^{N_I} \int d^3r \mathbf{S}_n \cdot \psi_{\sigma'}^{\dagger}(\mathbf{r}_n) \sigma_{\sigma'\sigma} \psi_{\sigma}(\mathbf{r}_n), \qquad (31.44)$$

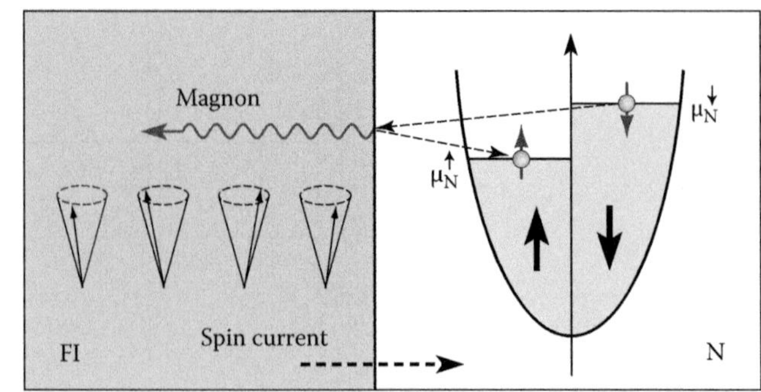

FIGURE 31.7 Schematic diagram of a junction with an FI and a normal metal (N) with spin accumulation $\delta\mu_N = \mu_N^{\uparrow} - \mu_N^{\downarrow}$. Magnon emission associated with scattering of an up-spin electron to a down-spin electron by the exchange interaction at the interface.

where

$\mathbf{S}_n$ are local moments at position $\mathbf{r}_n$ on the interface

$\psi_{\sigma}^{\dagger}(\mathbf{r}_n)$ and $\psi_{\sigma}(\mathbf{r}_n)$ are the annihilation operator of an incident electron with spin σ at position $\mathbf{r}_n$, respectively

σ is the Pauli spin operator

N_N is the number of electrons in N

N_I is the number of local moments that interact with conduction electrons at the interface

Using the Fourier transformation $\psi_{\sigma}(\mathbf{r}) = \sum_{\mathbf{k}} c_{\mathbf{k}\sigma} e^{i\mathbf{k}\cdot\mathbf{r}}$ and $\mathbf{S}_n = N_S^{-1/2} \sum_{\mathbf{q}} \mathbf{S}_{\mathbf{q}} e^{i\mathbf{q}\cdot\mathbf{r}_n}$, where $c_{\mathbf{k}\sigma}$ is the annihilation operator of an incident electron with momentum $\mathbf{k}$ and spin σ, and N_S is the number of localized moments in FI, Equation 31.44 is written as

$$H_{sd} = J_{\text{eff}} \sum_{\mathbf{k},\mathbf{k}',\mathbf{q}} [S_{\mathbf{q}}^{-} c_{\mathbf{k}'\uparrow}^{\dagger} c_{\mathbf{k}\downarrow} + S_{\mathbf{q}}^{+} c_{\mathbf{k}'\downarrow}^{\dagger} c_{\mathbf{k}\uparrow} + S_{\mathbf{q}}^{z} (c_{\mathbf{k}'\uparrow}^{\dagger} c_{\mathbf{k}\uparrow} - c_{\mathbf{k}'\downarrow}^{\dagger} c_{\mathbf{k}\downarrow})] \rho_{\mathbf{q}-\mathbf{k}'+\mathbf{k}}, \qquad (31.45)$$

where

$J_{\text{eff}} = J_{sd}/(N_N N_S^{1/2})$

$S_{\mathbf{q}}^{\pm} = S_{\mathbf{q}}^{x} \pm i S_{\mathbf{q}}^{y}$

$\rho_{\mathbf{p}-\mathbf{k}'+\mathbf{k}} = \sum_{n=1}^{N_I} e^{i(\mathbf{p}-\mathbf{k}'+\mathbf{k})\cdot\mathbf{r}_n}$.

The spin current through the interface is calculated from $I_{\text{FI/N}}^{S} = e(d/dt)\langle N_e^s \rangle$, where $N_e^s = (N_e^{\uparrow} - N_e^{\downarrow})$ and $N_e^{\sigma} = \sum_{\mathbf{k}} c_{\mathbf{k}s}^{\dagger} c_{\mathbf{k}s}$ is the number operator of electrons with spin σ. The average value of $(d/dt)\langle N_e^s \rangle$ is obtained by following the procedure to derive the Kubo formula [105], yielding the spin current across the interface

$$I_{\text{FI/N}}^{S} = \frac{2e}{\hbar^2} N_I J_{\text{eff}}^2 \sum_{\mathbf{q},\mathbf{q}'} \int_{-\infty}^{\infty} dt e^{i\delta\mu_N t/\hbar} [C_{\text{FI}}^{-+}(\mathbf{q},t) C_{\text{N}}^{+-}(\mathbf{q}',-t) - C_{\text{FI}}^{+-}(\mathbf{q},t) C_{\text{N}}^{-+}(\mathbf{q}',-t)], \qquad (31.46)$$

with the spin correlation functions

$$C_{\mathrm{FI}}^{-+}(\mathbf{q},t)=\left\langle S_{-\mathbf{q}}^{-}(t)S_{\mathbf{q}}^{+}(0)\right\rangle,\quad C_{\mathrm{FI}}^{+-}(\mathbf{q},t)=\left\langle S_{\mathbf{q}}^{+}(0)S_{-\mathbf{q}}^{-}(t)\right\rangle, \tag{31.47}$$

$$C_{\mathrm{N}}^{-+}(\mathbf{q},t)=\left\langle \sigma_{-\mathbf{q}}^{-}(t)\sigma_{\mathbf{q}}^{+}(0)\right\rangle,\quad C_{\mathrm{N}}^{+-}(\mathbf{q},t)=\left\langle \sigma_{\mathbf{q}}^{+}(0)\sigma_{-\mathbf{q}}^{-}(t)\right\rangle, \tag{31.48}$$

where $\sigma_{\mathbf{p}}^{+}=\sum_{\mathbf{k}} c_{\mathbf{k}\uparrow}^{\dagger}c_{\mathbf{k}+\mathbf{p}\downarrow}$ and $\sigma_{\mathbf{p}}^{-}=\sum_{\mathbf{k}} c_{\mathbf{k}\downarrow}^{\dagger}c_{\mathbf{k}+\mathbf{p}\uparrow}$. Using the fluctuation-dissipation theorem, which relates the spin correlation function to the imaginary part of the spin susceptibility, we obtain the spin current across the interface [104,106]

$$I_{\mathrm{FI/N}}^{S}\approx\frac{8\pi e}{\hbar}N_{\mathrm{I}}\langle S_z\rangle[J_{\mathrm{sd}}v_e N(0)]^2\frac{1}{N_S}\sum_{\mathbf{q}}(\omega_q+\delta\mu_{\mathrm{N}})$$
$$\times[n(\omega_q+\delta\mu_{\mathrm{N}})-n(\omega_q)], \tag{31.49}$$

where

$N(0)$ is the density of states of conduction electrons at the Fermi level

v_e is the volume per unit cell

$n(\omega_q)$ is the Bose distribution function

Note that $I_{\mathrm{FI/N}}^{S}$ vanishes for $\delta\mu_{\mathrm{N}}\to 0$ as expected.

To examine how the spin accumulation is converted to the magnon spin current and how the spin current depends on the magnon energy and temperature, we consider a simple model in which all the magnons have the same energy ω_0, for which Equation 31.49 becomes

$$I_{\mathrm{FI/N}}^{S}=\frac{4\pi e}{\hbar}N_{\mathrm{I}}\langle S_z\rangle[J_{\mathrm{sd}}v_e N(0)]^2(\omega_0+\delta\mu_{\mathrm{N}})$$
$$\times\left[\coth\left(\frac{\omega_0+\delta\mu_{\mathrm{N}}}{2k_{\mathrm{B}}T}\right)-\coth\left(\frac{\omega_0}{2k_{\mathrm{B}}T}\right)\right]. \tag{31.50}$$

Figure 31.8 shows the normalized spin current $I_{\mathrm{FI/N}}^{S}/I_{\mathrm{FI/N}}^{S0}$ versus spin accumulation $\delta\mu_{\mathrm{N}}$ for several values of

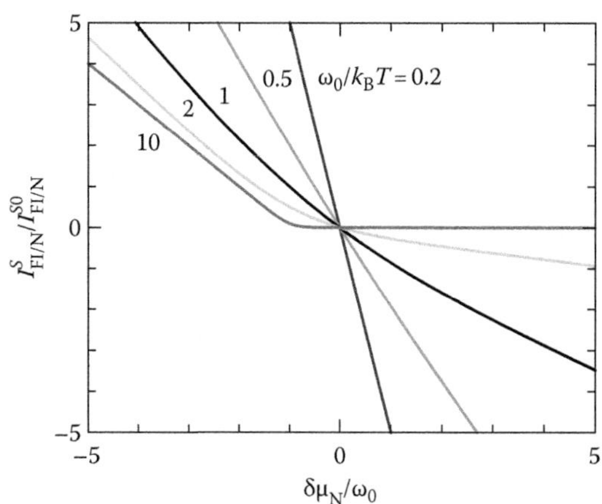

FIGURE 31.8 Spin current $I_{\mathrm{FI/N}}^{S}$ as a function of spin accumulation $\delta\mu_{\mathrm{N}}$ for the ratio $\omega_0/k_{\mathrm{B}}T$ of magnon energy to temperature.

ω_0/k_BT, where $I_{\mathrm{FI/N}}^{S0}=(4\pi/\hbar)N_{\mathrm{I}}\langle S_z\rangle[J_{\mathrm{sd}}v_eN(0)]^2\omega_0$. In the high-temperature regime ($\omega_0/k_BT\ll 1$), $I_{\mathrm{FI/N}}^{S}$ increases linearly as $I_{\mathrm{FI/N}}^{S}/I_{\mathrm{FI/N}}^{S0}\approx-(2k_BT/\omega_0)(\delta\mu_{\mathrm{N}}/\omega_0)$. In the low-temperature regime ($\omega_0/k_BT\gg 1$), $I_{\mathrm{FI/N}}^{S}$ exhibits a strong asymmetry in the $I_{\mathrm{FI/N}}^{S}$–$\delta\mu_{\mathrm{N}}$ curve; $I_{\mathrm{FI/N}}^{S}\approx 0$ for $\delta\mu_{\mathrm{N}}>-\omega_0$ and increases as $I_{\mathrm{FI/N}}^{S}\approx-I_{\mathrm{FI/N}}^{S0}\left[(\delta\mu_{\mathrm{N}}/\omega_0)+1\right]$ for $\delta\mu_{\mathrm{N}}<-\omega_0$. This result shows that the FI/N junction with an FI has a rectification effect for the spin current.

31.4.2 Spin Injection into FI via Spin Hall Effect

When an external current **I** is applied in the y direction in the plane of N, the spin current is induced by the SHE in N [5,62,100]

$$\mathbf{j}_s(x)=-\frac{\sigma_{\mathrm{N}}}{2e}\nabla\delta\mu_{\mathrm{N}}(x)+\alpha_{\mathrm{SH}}\,\mathbf{s}_s\times\left[\frac{\mathbf{I}}{A_{\mathrm{N}}}\right],\quad(0<x<d_{\mathrm{N}}), \tag{31.51}$$

where

σ_{N} is the electrical conductivity, $\delta\mu_{\mathrm{N}}(x)=\mu_{\mathrm{N}}^{\uparrow}(x)-\mu_{\mathrm{N}}^{\downarrow}(x)$

α_{SH} is the spin Hall angle

$\boldsymbol{\sigma}_s$ is the polarization vector of spin current induced by SHE

$\mathbf{I}/A_{\mathrm{N}}$ is the applied current density

A_{N} is the cross-sectional area of N

We look for a solution for the spin accumulation of the form $\delta\mu_{\mathrm{N}}(x)=ae^{-x/\lambda_{\mathrm{N}}}+be^{(x-d_{\mathrm{N}})/\lambda_{\mathrm{N}}}$ $(0<x<d_{\mathrm{N}})$, where λ_{N} is the spin-diffusion length, and a and b are determined by the boundary conditions, $j_s(0)=j_{\mathrm{FI/N}}^{s}$ and $j_s(d_{\mathrm{N}})=0$, with interface spin-current density $j_{\mathrm{FI/N}}^{s}=I_{\mathrm{FI/N}}^{S}/A_{\mathrm{J}}$ and junction area A_{J}. The resulting spin accumulation and spin current are

$$\delta\mu_{\mathrm{N}}(x)=$$
$$2e\frac{\lambda_{\mathrm{N}}}{\sigma_{\mathrm{N}}}\left[\frac{\cosh[(x-d_{\mathrm{N}})/\lambda_{\mathrm{N}}]}{\sinh(d_{\mathrm{N}}/\lambda_{\mathrm{N}})}j_{\mathrm{FI/N}}^{s}-\alpha_{\mathrm{SH}}\frac{\sinh[(x-d_{\mathrm{N}}/2)/\lambda_{\mathrm{N}}]}{\cosh(d_{\mathrm{N}}/2\lambda_{\mathrm{N}})}j\right], \tag{31.52}$$

$$j_s(x)=-\frac{\sinh[(x-d_{\mathrm{N}})/\lambda_{\mathrm{N}}]}{\sinh(d_{\mathrm{N}}/\lambda_{\mathrm{N}})}j_{\mathrm{FI/N}}^{s}-\alpha_{\mathrm{SH}}\left[1-\frac{\cosh[(x-d_{\mathrm{N}}/2)/\lambda_{\mathrm{N}}]}{\cosh(d_{\mathrm{N}}/2\lambda_{\mathrm{N}})}\right]j, \tag{31.53}$$

where $j=I/A_{\mathrm{N}}$ is the applied current density.

At the high-temperature regime, the interface spin current has the linear dependence on spin voltage $\delta\mu_{\mathrm{N}}(0)$ as

$$I_{\mathrm{FI/N}}^{s}=-\frac{G_{\uparrow\downarrow}}{2e}\delta\mu_{\mathrm{N}}(0), \tag{31.54}$$

where $G_{\uparrow\downarrow}$ is the interface spin conductance. In the model of all magnons with the same energy ω_0 as

$$G_{\uparrow\downarrow} = \frac{16\pi^2}{R_K}\frac{N_I\langle S_z\rangle[J_{sd}v_e N(0)]^2}{(\omega_0/k_BT)}, \tag{31.55}$$

where $R_K = h/e^2$ ($R_K \approx 25.8\,\mathrm{k\Omega}$) is the quantum resistance. Using Equation 31.52 for $\delta\mu_N(0)$ in Equation 31.54, we have the interface spin-current density

$$j^s_{FI/N} = -\alpha_{SH}\frac{(G_{\uparrow\downarrow}/G^s_N)\tanh(d_N/2\lambda_N)}{1+(G_{\uparrow\downarrow}/G^s_N)\coth(d_N/\lambda_N)}j, \tag{31.56}$$

where $G^s_N = A_J/(\rho_N\lambda_N)$ is the spin conductance of N.

As seen in Equation 31.56, the efficiency of spin injection is governed by the conductane ratio ($G_{\uparrow\downarrow}/G^s_N$). Using the parameter values $\rho_N = 20\,\mu\Omega\,\mathrm{cm}$, $\lambda_N = 10\,\mathrm{nm}$, and $d_N = 10\,\mathrm{nm}$ for a Pt film, and $J_{sd}v_e N(0) \sim (J_{sd}/\varepsilon_F)$, $\langle S_z\rangle = 5/2$, $a_S = 0.6\,\mathrm{nm}$, $T = 300\,\mathrm{K}$, and $(\hbar\omega_0/k_BT) = 0.5$ for an ferromagnetic insulator, we have $g_{\uparrow\downarrow} = G_{\uparrow\downarrow}/[(2e^2/h)A_J] \sim 2.2\times10^{17}(J_{sd}/\varepsilon_F)^2\ \mathrm{cm}^{-2}$. For $(J_{sd}/\varepsilon_F) \sim 0.01$, $g_{\uparrow\downarrow} \sim 2.2\times10^{13}\ \mathrm{cm}^{-2}$, which is about two orders of magnitude smaller than those in metallic junctions [104,107].

31.5 Summary

In this chapter, we have discussed the spin-dependent transport phenomena caused by spin injection from ferromagnets into normal metals in hybrid nanostructure devices and clarified the conditions for efficient spin injection, accumulation, and transport in these devices. In particular, the nonlocal spin injection creates a flow of "pure" spin current in nonmagnetic conductors (N), so that we have the opportunity to observe novel spin-current-induced phenomena, such as the SHE in N and nonlocal spin manipulation by pure spin-current injection. The observation of SHE provides a direct verification of the existence of spin current flowing in N. In a reversible way, the electrical current creates the spin current and spin accumulation by the SHE, which provides a source for spin generation without the use of magnetic materials. In addition, the nonlocal spin injection and absorption makes it possible to realize the nonlocal spin manipulation, by which the magnetization direction of a small ferromagnet connected to N can be switched by the spin transfer torque due to nonlocal spin-current absorption. Nonlocal lateral spin devices have advantages in flexibility of layout device geometry and the relative ease of fabricating multiterminal devices with different functionalities. Magnetic insulators with purely magnetic excitations will be useful for developing energy-saving spintronic devices. The development of nonlocal spin devices as well as the exploration of new phenomena open a new avenue in the research of spintronics.

31.A Appendix: Electrochemical Potentials in F1/N/F2

The coefficients (a_i, b_i) and voltages V_i ($i = 1,2$) in the ECPs of the N and F electrodes are determined by the matching conditions that the charge and spin currents are continuous at the interfaces of F1/N/F2. The coefficients a_1 and a_2 in Equation 31.5 in the N electrode are

$$a_1 = \frac{(P_1r_1 + p_Fr_F)(1+2r_2+2r_F)}{(1+2r_1+2r_F)(1+2r_2+2r_F) - e^{-2L/\lambda_N}}R_NeI, \tag{31.A.57}$$

$$a_2 = \frac{(P_1r_1 + p_Fr_F)e^{-L/\lambda_N}}{(1+2r_1+2r_F)(1+2r_2+2r_F) - e^{-2L/\lambda_N}}R_NeI, \tag{31.A.58}$$

where r_i ($i = 1,2$) and r_F are the resistances normalized to R_N: $r_i = R^\star_i/R_N$ with $R^\star_i = R_i/(1-P_i^2)$ and $r_F = R^\star_F/R_N$ with $R^\star_F = R_F/(1-p_F^2)$. The spin-accumulation voltage V_2 detected by F2 is

$$V_2 = \pm\frac{2(P_1r_1 + p_Fr_F)(P_2r_2 + p_Fr_F)e^{-L/\lambda_N}}{(1+2r_1+2r_F)(1+2r_2+2r_F) - e^{-2L/\lambda_N}}R_NI, \tag{31.A.59}$$

where "+" and "–" correspond to the parallel and antiparallel alignments of magnetizations, respectively. The coefficients b_1 and b_2 in ECP of Equation 31.6 in F1 and F2 are given in terms of a_1 and a_2 as

$$b_1 = \frac{1}{2}p_FR_FeI - \left(\frac{R_F}{R_N}\right)a_1,\quad b_2 = \left(\frac{R_F}{R_N}\right)a_2. \tag{31.A.60}$$

Thus the interfacial spin currents across junctions 1 and 2 are

$$I^s_1 = \left(\frac{2}{eR_N}\right)a_1,\quad I^s_2 = -\left(\frac{2}{eR_N}\right)a_2. \tag{31.A.61}$$

Acknowledgments

The authors would like to thank M. Ichimura, R. Sugano, J. Ieda, J. Ohe, H. Adachi, Bo Gu, H. Imamura, T. Kimura, Y. Otani, T. Seki, S. Mitani, K. Takanashi, K. Ando, K. Uchida, Y. Kajiwara, and E. Saitoh for helpful discussions and their collaboration. This work was supported by MEXT, CREST-JST, and the Next Generation Supercomputer Project of MEXT, Japan.

References

1. S. Maekawa and T. Shinjo (Eds.), *Spin Dependent Transport in Magnetic Nanostructures*, Taylor & Francis, Boca Raton, FL, 2002.
2. S. Maekawa (Ed.), *Concepts in Spin Electronics*, Oxford University Press, New York, 2006.
3. H. Kronmuller and S. Parkin (Eds.), *Handbook of Magnetism and Advanced Magnetic Materials*, Vols. 1–5, John Wiley & Sons, New York, 2007.
4. I. Žutić, J. Fabian, and S. Das Sarma, Spintronics: Fundamentals and applications, *Rev. Mod. Phys.* **76**, 323 (2004).

5. S. Takahashi and S. Maekawa, Spin current in metals and superconductors, *J. Phys. Soc. Jpn.* **77**, 031009 (2008).
6. S. Takahashi and S. Maekawa, Spin current, spin accumulation and spin Hall effect, *Sci. Technol. Adv. Mater.* **9**, 014105 (2008).
7. M. Johnson and R. H. Silsbee, Interfacial charge-spin coupling: Injection and detection of spin magnetization in metals, *Phys. Rev. Lett.* **55**, 1790 (1985).
8. M. Johnson and R. H. Silsbee, Spin-injection experiment, *Phys. Rev. B* **37**, 5326 (1988).
9. M. Johnson, Spin accumulation in gold films, *Phys. Rev. Lett.* **70**, 2142 (1993).
10. F. J. Jedema, A. T. Filip, and B. J. van Wees, Electrical spin injection and accumulation at room temperature in an all-metal mesoscopic spin valve, *Nature* **410**, 345 (2001).
11. F. J. Jedema, H. B. Heersche, A. T. Filip, J. J. A. Baselmans, and B. J. van Wees, Electrical detection of spin precession in a metallic mesoscopic spin valve, *Nature* **416**, 713 (2002).
12. M. Zaffalon and B. J. van Wees, Zero-dimensional spin accumulation and spin dynamics in a mesoscopic metal island, *Phys. Rev. Lett.* **91**, 186601 (2003).
13. M. Urech, J. Johansson, V. Korenivski, and D. B. Haviland, Spin injection in ferromagnet-superconductor/normal-ferromagnet structures, *J. Magn. Magn. Mater.* **272–276**, e1469 (2004).
14. T. Kimura, J. Hamrle, Y. Otani, K. Tsukagoshi, and Y. Aoyagi, Spin-dependent boundary resistance in the lateral spin-valve structure, *Appl. Phys. Lett.* **85**, 3501 (2004).
15. T. Kimura, J. Hamrle, and Y. Otani, Estimation of spin-diffusion length from the magnitude of spin-current absorption: Multiterminal ferromagnetic/nonferromagnetic hybrid structures, *Phys. Rev. B* **72**, 14461 (2005).
16. Y. Ji, A. Hoffmann, J. S. Jiang, and S. D. Bader, Spin injection, diffusion, and detection in lateral spin-valves, *Appl. Phys. Lett.* **85**, 6218 (2004).
17. K. Miura, T. Ono, S. Nasu, T. Okuno, K. Mibu, and T. Shinjo, Electrical spin injection in $Ni_{81}Fe_{19}/Al/Ni_{81}Fe_{19}$ with double tunnel junctions, *J. Magn. Magn. Mater.* **286**, 142 (2005).
18. S. Garzon, I. Žutić, and R. A. Webb, Temperature-dependent asymmetry of the nonlocal spin-injection resistance: Evidence for spin nonconserving interface scattering, *Phys. Rev. Lett.* **94**, 176601 (2005).
19. R. Godfrey and M. Johnson, Spin injection in mesoscopic silver wires: Experimental test of resistance mismatch, *Phys. Rev. Lett.* **96**, 136601 (2006).
20. J. H. Ku, J. Chang, H. Kim, and J. Eom, Effective spin injection in Au film from Permalloy, *Appl. Phys. Lett.* **88**, 172510 (2006).
21. T. Kimura, Y. Otani, and P. M. Levy, Electrical control of the direction of spin accumulation, *Phys. Rev. Lett.* **99**, 166601 (2007).
22. T. Kimura and Y. Otani, Large spin accumulation in a permalloy-silver lateral spin valve, *Phys. Rev. Lett.* **99**, 196604 (2007).
23. H. Idzuchi, Y. Fukuma, L. Wang, and Y. Otani, Spin diffusion characteristics in magnesium nanowires, *Appl. Phys. Exp.* **3**, 063002 (2010).
24. N. Tombros, C. Jozsa, M. Popinciuc, H. T. Jonkman, and B. J. van Wees, Electronic spin transport and spin precession in single graphene layers at room temperature, *Nature* **448**, 571 (2007).
25. M. Ohishi, M. Shiraishi, R. Nouchi, T. Nozaki, T. Shinjo, and Y. Suzuki, Spin injection into a graphene thin film at room temperature, *Jpn. J. Appl. Phys.* **46**, L605 (2007).
26. W. Han, W. H. Wang, K. Pi et al., Electron-hole asymmetry of spin injection and transport in single-layer graphene, *Phys. Rev. Lett.* **102**, 137205 (2009).
27. K. Pi, W. Han, K. M. McCreary, A. G. Swartz, Y. Li, and R. K. Kawakami, Manipulation of spin transport in graphene by surface chemical doping, *Phys. Rev. Lett.* **104**, 187201 (2009).
28. M. Popinciucl, C. Jozsa, P. J. Zomer et al., Electronic spin transport in graphene field-effect transistors, *Phys. Rev. B* **80**, 214427 (2009).
29. M. Urech, J. Johansson, N. Poli, V. Korenivski, and D. B. Haviland, Enhanced spin accumulation in a superconductor, *J. Appl. Phys.* **99**, 08M513 (2006).
30. K. Miura, S. Kasai, K. Kobayashi, and T. Ono, Non local spin detection in ferromagnet/superconductor/ferromagnet spin-valve device with double-tunnel junctions, *Jpn. J. Appl. Phys.* **45**, 2888 (2006).
31. K. Ohnishi, T. Kimura, and Y. Otani, Nonlocal injection of spin current into a superconducting Nb wire, *Appl. Phys. Lett.* **96**, 192509 (2010).
32. X. Lou, C. Adelmann, S. A. Crooker et al., Electrical detection of spin transport in lateral ferromagnet.semiconductor devices, *Nature Phys.* **3**, 197 (2007).
33. C. Feuillet-Palma, T. Delattre, P. Morfin et al., Conserved spin and orbital phase along carbon nanotubes connected with multiple ferromagnetic contacts, *Phys. Rev. B* **81**, 115414 (2010).
34. P. C. van Son, H. van Kempen, and P. Wyder, Boundary resistance of the ferromagnetic-nonferromagnetic metal interface, *Phys. Rev. Lett.* **58**, 2271 (1987).
35. T. Valet and A. Fert, Theory of the perpendicular magnetoresistance in magnetic multilayers, *Phys. Rev. B* **48**, 7099 (1993).
36. A. Fert and S. F. Lee, Theory of the bipolar spin switch, *Phys. Rev. B* **53**, 6554 (1996).
37. S. Hershfield and H. L. Zhao, Charge and spin transport through a metallic ferromagnetic-paramagnetic-ferromagnetic junction, *Phys. Rev. B* **56**, 3296 (1997).
38. S. Takahashi and S. Maekawa, Spin injection and detection in magnetic nanostructures, *Phys. Rev. B* **67**, 052409 (2003).
39. M. Johnson and J. Byers, Charge and spin diffusion in mesoscopic metal wires and at ferromagnet/nonmagnet interfaces, *Phys. Rev. B* **67**, 125112 (2003).
40. J. Bass and W. P. Pratt Jr., Spin-diffusion lengths in metals and alloys, and spin-flipping at metal/metal interfaces: an experimentalist's critical review, *J. Phys. Condens. Matter* **19**, 183201 (2007).

41. R. J. M. van de Veerdonk, J. Nowak, R. Meservey, J. S. Moodera, and W. J. M. de Jonge, Current distribution effects in magnetoresistive tunnel junctions, *Appl. Phys. Lett.* **71**, 2839 (1997).
42. M. Ichimura, S. Takahashi, K. Ito, and S. Maekawa, Geometrical effect on spin current in magnetic nanostructures, *J. Appl. Phys.* **95**, 7255 (2004).
43. J. Hamrle, T. Kimura, T. Yang, and Y. Otani, Current distribution inside Py/Cu lateral spin-valve devices, *Phys. Rev. B*, 094402 (2005).
44. J. Bass and W. P. Pratt Jr., Version 7/21/01 current-perpendicular-to-plane (CPP) magnetoresistance, *Physica B* **71**, 1 (2002).
45. R. J. Soulen Jr., J. M. Byers, M. S. Osofsky, B. Nadgorny, T. Ambrose, S. F. Cheng, P. R. Broussard et al., *Science* **282**, 85 (1998).
46. R. Meservey and P. M. Tedrow, Spin-polarized electron tunneling, *Phys. Rep.* **238**, 173 (1994).
47. J. S. Moodera and G. Mathon, Spin polarized tunneling in ferromagnetic junctions, *J. Magn. Magn. Mater.* **200**, 248 (1999).
48. D. J. Monsma and S. S. P. Parkin, Spin polarization of tunneling current from ferromagnet/Al_2O_3 interfaces using copper-doped aluminum superconducting films, *Appl. Phys. Lett.* **77**, 720 (2000).
49. S. S. P. Parkin, C. Kaiser, A. Panchula et al., Giant tunnelling magnetoresistance at room temperature with MgO (100) tunnel barriers, *Nat. Mater.* **3**, 862 (2004).
50. S. Yuasa, T. Nagahama, A. Fukushima, Y. Suzuki, and K. Ando, Giant room-temperature magnetoresistance in single-crystal Fe/MgO/Fe magnetic tunnel junctions, *Nat. Mater.* **3**, 868 (2004).
51. E. I. Rashba, Theory of electrical spin injection: Tunnel contacts as a solution of the conductivity mismatch problem, *Phys. Rev. B* **62**, 16267(R) (2000).
52. A. Fert and H. Jaffrès, Conditions for efficient spin injection from a ferromagnetic metal into a semiconductor, *Phys. Rev. B* **64**, 184420 (2001).
53. G. Schmidt, G. Richter, P. Grabs, C. Gould, D. Ferrand, and L. W. Molenkamp, Fundamental obstacle for electrical spin injection from a ferromagnetic metal into a diffusive semiconductor, *Phys. Rev. B* **62**, 4790(R) (2000).
54. S. O. Valenzuela and M. Tinkham, Direct electronic measurement of the spin Hall effect, *Nature* **442**, 176 (2006).
55. M. Johnson and R. H. Silsbee, Coupling of electronic charge and spin at a ferromagnetic-paramagnetic metal interface, *Phys. Rev. B* **37**, 5312 (1988).
56. S. O. Valenzuela, Nonlocal electronic spin detection, spin accumulation and the spin Hall effect, *Int. J. Mod. Phys. B* **23**, 2413 (2009).
57. H. Nanami, T. Kimura, and Y. Otani (unpublished).
58. J. C. Slonczewski, Current-driven excitation of magnetic multilayers, *J. Magn. Magn. Mater.* **159**, L1 (1996).
59. L. Berger, Emission of spin waves by a magnetic multilayer traversed by a current, *Phys. Rev. B* **54**, 9353 (1996).
60. J. A. Katine, F. J. Albert, R. A. Buhrman, E. B. Myers, and D. C. Ralph, Current-driven magnetization reversal and spin-wave excitations in Co/Cu/Co pillars, *Phys. Rev. Lett.* **84**, 3149 (2000).
61. S. Maekawa, K. Inomata, and S. Takahashi, Japan Patent No. 3818276 (June 23, 2006).
62. S. Takahashi and S. Maekawa, Spin injection and transport in magnetic nanostructures, *Physica C* **437–438**, 309 (2006).
63. T. Kimura, Y. Otani, and J. Hamrle, Switching magnetization of a nanoscale ferromagnetic particle using nonlocal spin injection, *Phys. Rev. Lett.* **96**, 037201 (2006).
64. T. Yang, T. Kimura, and Y. Otani, Giant spin-accumulation signal and pure spin-current-induced reversible magnetization switching, *Nat. Phys.* **4**, 851 (2008).
65. C. L. Chien and C. R. Westgate (Eds.), *The Hall Effect and its Applications*, Plenum, New York, 1980.
66. M. I. D'yakonov and V. I. Perel', Current induced spin orientation of electrons in semiconductors, *Phys. Lett. A* **35**, 459 (1971).
67. J. E. Hirsch, Spin Hall effect, *Phys. Rev. Lett.* **83**, 1834 (1999).
68. S. Zhang, Spin Hall effect in the presence of spin diffusion, *Phys. Rev. Lett.* **85**, 393 (2001).
69. S. Takahashi and S. Maekawa, Hall Effect induced by a spin-polarized current in superconductors, *Phys. Rev. Lett.* **88**, 116601 (2002).
70. S. Takahashi, H. Imamura, and S. Maekawa, Spin injection and spin accumulation in hybrid nanostructures, in *Concept in Spin Electronics*, S. Maekawa (Ed.), Oxford University Press, New York, 2006.
71. R. V. Shchelushkin and A. Brataas, Spin Hall effects in diffusive normal metals, *Phys. Rev. B* **71**, 045123 (2005).
72. T. Kimura, Y. Otani, T. Sato, S. Takahashi, and S. Maekawa, Room-temperature reversible spin Hall effect, *Phys. Rev. Lett.* **98**, 156601 (2007).
73. L. Vila, T. Kimura, and Y. Otani, Room-temperature reversible spin Hall effect, *Phys. Rev. Lett.* **99**, 226604 (2007).
74. T. Seki, Y. Hasegawa, S. Mitani, S. Takahashi, H. Imamura, S. Maekawa, J. Nitta, and K. Takanashi, Giant spin Hall effect in perpendicularly spin-polarized FePt/Au devices, *Nat. Mater.* **7**, 125 (2008).
75. E. Saitoh, M. Ueda, H. Miyajima, and G. Tatara, Conversion of spin current into charge current at room temperature: Inverse spin-Hall effect, *Appl. Phys. Lett.* **88**, 182509 (2006).
76. R. Karplus and J. M. Luttinger, Hall effect in ferromagnetics, *Phys. Rev.* **95**, 1154 (1954).
77. S. Murakami, N. Nagaosa, and S.-C. Zhang, Dissipationless quantum spin current at room temperature, *Science* **301**, 1348 (2003).
78. J. Sinova, D. Culcer, Q. Niu, N. A. Sinitsyn, T. Jungwirth, and A. H. MacDonald, Universal intrinsic spin Hall effect, *Phys. Rev. Lett.* **92**, 126603 (2002).
79. J. Inoue, G. E. W. Bauer, and L. W. Molenkamp, Suppression of the persistent spin Hall current by defect scattering, *Phys. Rev. B* **70**, 041303 (2004).

80. Y. K. Kato, R. C. Myers, A. C. Gossard, and D. D. Awschalom, Observation of the spin Hall effect in semiconductors, *Science* **306**, 1910 (2004).
81. J. Wunderlich, B. Kaestner, J. Sinova, and T. Jungwirth, Experimental observation of the spin-Hall effect in a two-dimensional spin-orbit coupled semiconductor system, *Phys. Rev. Lett.* **94**, 047204 (2005).
82. J. Smit, The spontaneous Hall effect in ferromagnetics II, *Physica* **24**, 39 (1958).
83. L. Berger, Side-jump mechanism for the Hall effect of ferromagnets, *Phys. Rev. B* **2**, 4559 (1970).
84. A. Crépieux and P. Bruno, Theory of the anomalous Hall effect from the Kubo formula and the Dirac equation, *Phys. Rev. B* **64**, 14416 (2001).
85. W.-K. Tse and S. Das Sarma, Spin Hall effect in doped semiconductor structures, *Phys. Rev. Lett.* **96**, 56601 (2006).
86. H. A. Engel, E. I. Rashba, and B. I. Halperin, Theory of spin Hall effects in semiconductors, in *Handbook of Magnetism and Advanced Magnetic Materials*, Vol. 5, H. Kronmüller and S. Parkin (Eds.), Wiley, New York, 2006.
87. N. A. Sinitsyn, Semiclassical theories of the anomalous Hall effect, *J. Phys.: Condens. Matter* **20**, 023201 (2008).
88. J. J. Sakurai and S. F. Tuan, *Modern Quantum Mechanics*, Addison-Wesley, Boston, MA, 1985.
89. S. K. Lyo and T. Holstein, Side-jump mechanism for ferromagnetic Hall effect, *Phys. Rev. Lett.* **29**, 423 (1972).
90. J. Kondo, Anomalous Hall effect and magnetoresistance of ferromagnetic metals, *Prog. Theor. Phys.* **27**, 772 (1964).
91. J.-P. Ansermet, Perpendicular transport of spin-polarized electrons through magnetic nanostructures, *J. Phys.: Condens. Matter* **10**, 6027 (1998).
92. S. Takahashi and S. Maekawa, Nonlocal spin Hall effect and spin-orbit interaction in nonmagnetic metals, *J. Magn. Magn. Mater.* **310**, 2067 (2007).
93. N. W. Ashcroft and D. Mermin, *Solid State Physics*, Saunders College, Philadelphia, PA, 1976.
94. T. Kimura, J. Hamrle, and Y. Otani, *J. Magn. Soc. Jpn.* **29**, 192 (2005).
95. A. B. Gougam, F. Pierre, H. Pothier, D. Esteve, and N. O. Birge, Comparison of energy and phase relaxation in metallic wires, *J. Low Temp. Phys.* **118**, 447 (2000).
96. H. Kurt, R. Loloee, K. Eid, W. P. Pratt Jr., and J. Bass, Spin-memory loss at 4.2 K in sputtered Pd and Pt and at Pd/Cu and Pt/Cu interfaces, *J. Appl. Phys.* **81**, 4787 (2002).
97. S. Takahashi and S. Maekawa, Effects of spin injection and accumulation, in *Spin Electronics–Basic and Applications*, K. Inomata (Ed.), CMC Publishing Co., Ltd., Tokyo, p. 28, 2004.
98. R. Sugano, M. Ichimura, S. Takahashi, and S. Maekawa, Three dimensional simulations of spin Hall effect in magnetic nanostructures, *J. Appl. Phys.* **103**, 07A715 (2008).
99. S. O. Valenzuela and M. Tinkham, *J. Appl. Phys.* **101**, 09B103 (2007).
100. K. Ando, S. Takahashi, K. Harii et al., Electric manipulation of spin relaxation using the spin Hall effect, *Phys. Rev. Lett.* **101**, 036601 (2008).
101. A. Fert and O. Jaoul, Left-right asymmetry in the scattering of electrons by magnetic impurities, and a Hall effect, *Phys. Rev. Lett.* **28**, 303 (1972).
102. G.-Y. Guo, S. Maekawa, and N. Nagaosa, Enhanced spin Hall effect by resonantskew scattering in the orbital-dependent Kondo effect, *Phys. Rev. Lett.* **102**, 036401 (2009).
103. B. Gu, J.-Y. Gan, N. Bulut et al., Quantum Renormalization of the spin Hall effect, *Phys. Rev. Lett.* **105**, 086401 (2010).
104. Y. Kajiwara, K. Harii, S. Takahashi et al., Transmission of electrical signals by spin-wave interconversion in a magnetic insulator, *Nature* **464**, 262 (2010).
105. G. D. Mahan, *Many Particle Physics*, Kluwer Academic/Plenum Publisher, New York, 2000.
106. S. Takahashi, E. Saitoh, and S. Maekawa, Spin current through a normal-metal/insulating-ferromagnet junction, *J. Phys.: Conf. Ser.* **200**, 062030 (2010).
107. Y. Tserkovnyak, A. Brataas, G. E. Bauer, and B. I. Halperin, Nonlocal magnetization dynamics in ferromagnetic heterostructures, *Rev. Mod. Phys.* **77**, 1375 (2005).

32

Nonlocal Spin Valves in Metallic Nanostructures

Yoshichika Otani
The University of Tokyo and RIKEN ASI Quantum Nano-Scale Magnetics Laboratory

Takashi Kimura
Kyushu University

32.1 Introduction

The main focus of this chapter are recent experimental developments of nonlocal spin injection in metallic systems and the related phenomena after the year 2000. Recent related theoretical work is discussed in Chapter 31 while the early developments of spin injection in metals are discussed in Chapter 6. The pioneering work on spin injection in metals, was established in 1985 by Johnson and Silsbee [1,2]. Their measurements were performed in pure "bulk wire" of aluminum with permalloy leads used for injection and detection up to ~50 K and yielded ~pV signals. By revisiting these ideas in micro-fabricated lateral ferromagnet/nonmagnet/ferromagnet (F/N/F) structures Jedema et al. [3,4] demonstrated room temperature spin injection and detection with much larger signals. All of these experiments have spurred very intensive research efforts to investigate diffusive spin transport in lateral devices with important implications for the emerging field of spintronics [5].

In Section 32.2, we describe the basic aspect of spin injection, spin accumulation, and spin current in lateral devices comprising of nonmagnetic and ferromagnetic metals. We introduce the concept of the spin resistance, that is, a measure of impedance against spin relaxation; thereby lateral spin valve structures can be expressed by equivalent spin-resistance circuits in analogy with cascaded transmission lines. This certainly facilitates designing the lateral spin valves exhibiting efficient spin injection and detection. In Section 32.3, we put our focus on the effect of the device structure, including the junction size, on the magnitude of the spin accumulation stored in the nonmagnet and also the magnitude of the spin current flowing in it. This implies that the spin injection efficiency can be tuned by connecting materials with appropriate structures and spin resistances. Thereby the optimized spin injection enables us to demonstrate pure spin current–induced phenomena such as magnetization switching, and *direct* and *inverse* spin Hall effects (SHEs).

32.2 Spin Injection, Accumulation, and Spin Current

Spin injection is a key technology to give rise to spin-dependent transport phenomena, which leads to spin accumulation over the spin-diffusion length in the vicinity of the interface of the junction consisting of ferromagnetic (F) and nonmagnetic (N) wires as in Figure 32.1, showing a spatial distribution of the spin-dependent electrochemical potential in the transparent single F/N junction device together with a schematic illustration.

When the spin-polarized electrons are injected through the interface from F to N as indicated by the dashed arrow in the illustration in Figure 32.1, the electron spins accumulate in both F and N metals across the interface over the spin-diffusion length λ_{sf}^F for F and λ_{sf}^N for N. This spin accumulation results in the spin current, which can be described by considering up ($\uparrow$) and down ($\downarrow$) diffusive spin channels based on the two-current model [6–8] as follows. The spin-dependent current density $j_{\uparrow,\downarrow}$ in both F and N is given by

$$j_{\uparrow,\downarrow} = \sigma_{\uparrow,\downarrow}E + eD_{\uparrow,\downarrow}\nabla\delta n_{\uparrow,\downarrow}. \qquad (32.1)$$

Here the first term corresponds to the ordinary Ohm's low, so-called drift current for each spin channel with the spin-dependent conductivities $\sigma_{\uparrow,\downarrow}$, and the second term describes the diffusive contribution of the accumulated spin density $\delta n_{\uparrow,\downarrow}$ with the spin-dependent diffusion constant $D_{\uparrow\downarrow}$, satisfying Einstein's

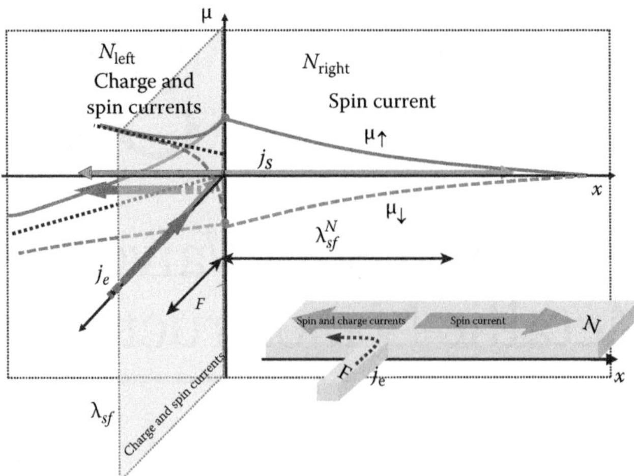

FIGURE 32.1 Spatial distribution of the spin-dependent electrochemical potential in the transparent single *F*/*N* junction device together with its schematic illustration.

relation $D_{\uparrow,\downarrow}=\sigma_{\uparrow,\downarrow}/e^2N_{\uparrow,\downarrow}$ with the density of states $N_{\uparrow,\downarrow}$ at the Fermi level for each spin. The relation between the spin-dependent current density and the electrochemical potential can thus be expressed as

$$j_{\uparrow,\downarrow}=\frac{\sigma_{\uparrow,\downarrow}}{e}\nabla\mu_{\uparrow,\downarrow}, \quad (32.2)$$

where $\mu_{\uparrow,\downarrow}=e\phi+\delta n_{\uparrow,\downarrow}/N_{\uparrow,\downarrow}(E_F)$ with ϕ the electric potential. One should note here that the total electric current density is given by $j_e\equiv j_\uparrow+j_\downarrow=\nabla(\mu_\uparrow\sigma_\uparrow+\mu_\downarrow\sigma_\downarrow)/e$ while the spin current density is $j_S\equiv j_\uparrow-j_\downarrow=\nabla(\mu_\uparrow\sigma_\uparrow-\mu_\downarrow\sigma_\downarrow)/e$, of which spin polarization P is defined as $P=j_S/j_e=(j_\uparrow-j_\downarrow)/(j_\uparrow+j_\downarrow)$. These relations imply that both the drift and diffusive contributions take place in the left-hand side of *N* with respect to the junction whereas only the diffusive contribution does in the right-hand side; that is, only the spin current flows in the right-hand side in *N* but no charge current (Figure 32.1). It is important to note here that this spin current is called "pure spin current" to distinguish from the spin (polarized) current flowing in the left-hand side and is only present in the length scale of the spin-diffusion length λ_{sf} along the *x*-axis. This implies that the spin current decays through spin relaxation process via momentum scattering by phonons or impurities in the presence of the spin–orbit coupling that mixes spin up $|\uparrow\rangle$ and down $|\downarrow\rangle$ states [9]. The diffusive spin current is proportional to the slope of the spin accumulation $\Delta\mu(=\mu_\uparrow-\mu_\downarrow)$, which is well described by the diffusion equation

$$\nabla^2(\mu_\uparrow-\mu_\downarrow)=\frac{(\mu_\uparrow-\mu_\downarrow)}{\lambda_{sf}^2}, \quad (32.3)$$

where $\lambda_{sf}=\sqrt{D\tau_{sf}}$ with $D=D_\uparrow D_\downarrow(N_\uparrow+N_\downarrow)/(N_\uparrow D_\uparrow+N_\downarrow D_\downarrow)$ and τ_{sf} the spin relaxation time. By applying the low conservation and continuity for charge current, one obtains general solutions for both up and down spin channels

$$\mu_{\uparrow,\downarrow}=a+bx\pm\frac{c}{\sigma_{\uparrow,\downarrow}}\exp\left(-\frac{x}{\lambda_{sf}}\right)\pm\frac{d}{\sigma_{\uparrow,\downarrow}}\exp\left(\frac{x}{\lambda_{sf}}\right), \quad (32.4)$$

with coefficients *a*, *b*, *c*, and *d*. These coefficients are determined by the boundary conditions at the junctions. In the solution, the first linear term $a+bx$ corresponds to a linear voltage drop due to the charge current with Ohmic resistance; thus the spin accumulation $\Delta\mu(=\mu_\uparrow-\mu_\downarrow)$ is given as a sum of the third and the fourth terms in the above equation.

The experimentally accessible quantity, the induced spin accumulation voltage, is thus given by

$$\Delta V_S=\frac{\Delta\mu}{e}=V_+\exp\left(-\frac{|x|}{\lambda_{sf}}\right)\pm V_-\exp\left(\frac{|x|}{\lambda_{sf}}\right). \quad (32.5)$$

It should be remarked here that the coefficients V_+ and V_- in Equation 32.5 are attributable to the weight of the tunnel-like character and that of the ohmic-like character of the junction.

From Equations 32.2 and 32.5, the spin current $J_S(\equiv j_SS_N$ in the *N* wire can be obtained as

$$J_S=\frac{1}{R_S^N}\left(V_+\exp\left(-\frac{x}{\lambda_{sf}^N}\right)\pm V_-\exp\left(\frac{x}{\lambda_{sf}^N}\right)\right), \quad (32.6)$$

where $R_S^N=\lambda_{sf}^N/\sigma_NS_N$ with the conductivity σ_N and the cross-sectional area S_N of the *N* wire.

Similar considerations for the *F* wire also yield the same expression for the spin current J_S in Equation 32.6 with $R_S^F=\lambda_{sf}^F/(1-P^2)\sigma_FS_F$ instead of $R_S^N=\lambda_{sf}^N/\sigma_NS_N$, where σ_F and S_F are, respectively, the conductivity and the junction size of the *F*/*N* interface. One should notice that Equations 32.5 and 32.6 mimic the electrical transmission line of the characteristic impedance R_S^N with attenuation constant of $1/\lambda_{sf}^N$ [10,11]. We call R_S^N and R_S^F the spin resistance, indicating a measure of impedance against spin relaxation. Therefore, the *F*/*N* junction in Figure 32.1 can be regarded as an equivalent spin-resistance circuit (Figure 32.2a), where Equations 32.5 and 32.6 hold in each subsection of the circuit. The spin accumulation decays exponentially along two *N* arms and one *F* arm. By taking into account these three decaying paths, the voltage V_S^i induced by the spin current J_S is

$$\Delta V_S^I=J_S\frac{R_S^NR_S^F}{R_S^N+2R_S^F}. \quad (32.7)$$

In the following section, we discuss the lateral spin valve based on the earlier discussion.

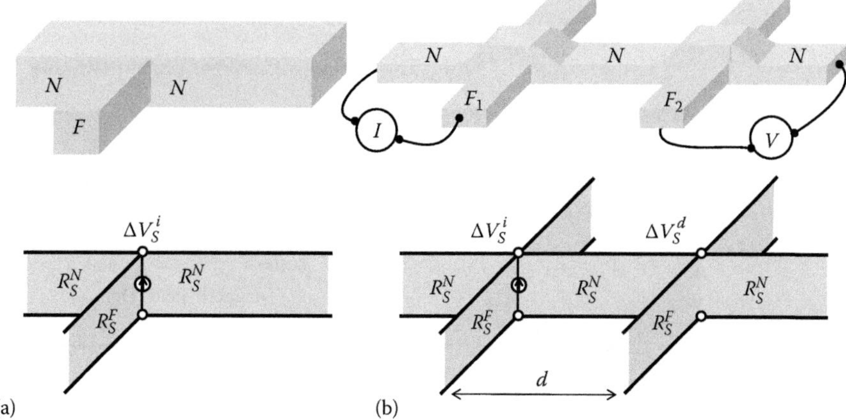

FIGURE 32.2 Schematic illustrations of (a) single *F/N* junction and (b) lateral spin valve together with equivalent spin-resistance circuits.

32.3 Nonlocal Spin Injection Experiments

32.3.1 Nonlocal Lateral Spin Valve

The nonlocal spin valve (NLSV) measurement is usually performed to determine the spin splitting $\Delta\mu$ ($=\Delta eV_S$) in the chemical potential by using the lateral spin valve structure typically consisting of two ferromagnetic wires $F_{1,2}$ bridged by a nonmagnetic wire *N*, as schematically shown in Figure 32.2b. There are two families of lateral spin valves, one is fully based on the ohmic junctions and the other is based on the tunnel junctions. Valet et al. [8] first introduced the spin-dependent interface resistance in the *F/N* junction to describe its tunnel- or ohmic-like character, that is, discontinuity appearing in the spin-dependent electrochemical potential $\mu_{\uparrow,\downarrow}$ when the current passes through it. Following their definition, the tunnel junction here means the discontinuity in $\mu_{\uparrow,\downarrow}$ much larger than the spin splitting $\Delta\mu$ ($=\Delta eV_S$), while the ohmic junction means no discontinuity in $\mu_{\uparrow,\downarrow}$. Here we limit our discussion mostly to the ohmic junctions. A series of reports on the ohmic lateral spin valves performed in the 2000s can be found in Refs. [10,12–14]. Some of representative experimental demonstrations using tunnel junctions are shown in Refs. [3,15–18]. The F_1 wire is the spin injector through which the spin-polarized current is injected into the *N* wire. The spin accumulation builds up in the vicinity of the F_1/N junction as discussed in the previous section. The spatially decaying spin accumulation then gives rise to a spin current flowing from the F_1/N interface to the F_2/N interface, and relaxing into the F_2 detector. Since $\sigma_\uparrow$ and $\sigma_\downarrow$ are different in ferromagnet, measurable nonequilibrium voltage ΔV_P or ΔV_{AP} is produced according to the relative magnetization orientation of F_1 and F_2 either parallel (*P*) or antiparallel (*AP*), as in Figure 32.3. This situation can also be well described by the equivalent spin-resistance circuit in Figure 32.2b. The total spin resistance of the device R_S^{tot} can be calculated similarly to that of the single *F/N* junction as [10]

$$R_S^{tot} = \left(\frac{1}{R_S^N} + \frac{1}{R_S^F} + \frac{1}{R_S^i}\right)^{-1} = \frac{QR_S^N\left[Q\exp(d/\lambda_{sf}) + 4\sinh(d/\lambda_{sf})\right]}{2\left[Q(4+Q)\exp(d/\lambda_{sf}) + 4\sinh(d/\lambda_{sf})\right]}, \quad (32.8)$$

where

R_S^i is the injector spin resistance consisting of *N* and *F* wires
Q is the spin resistance ratio R_S^F/R_S^N
d is the center-to-center spacing between the injector and detector wires

Hence the spin accumulation voltage V_S^d at the detector junction is given by $\Delta V_S^d = TJ_S R_S^{tot}$, where $T = Q/\left(Q\exp(d/\lambda_{sf}) + 4\sinh(d/\lambda_{sf})\right)$. Finally we obtain the

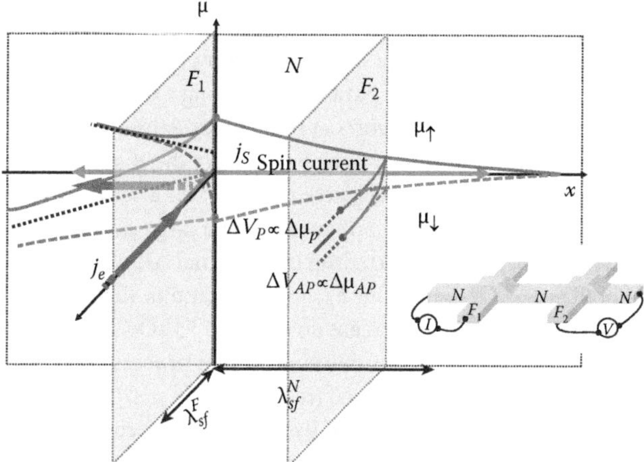

FIGURE 32.3 Spatial distribution of the spin-dependent electrochemical potential in the lateral spin valve device and a schematic illustration of the device structure.

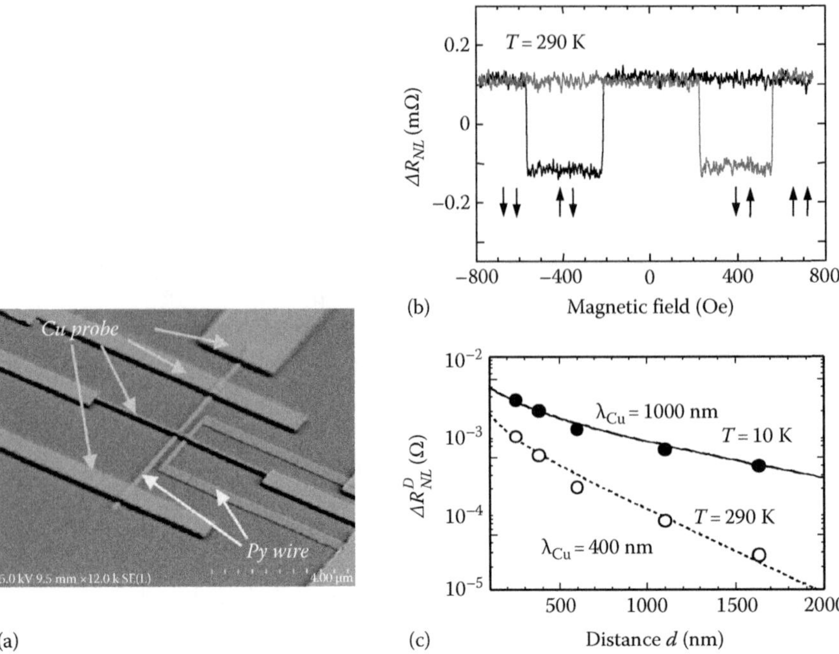

FIGURE 32.4 (a) SEM of a typical lateral spin valve consisting of Py wires bridged by a Cu wire, (b) the NLSV signal at room temperature, and (c) the overall change R_{NL}^D as a function of separating distance at various temperatures.

overall change R_{NL}^d in NLSV signal in ohm between P and AP states as

$$\Delta R_{NL}^D \equiv \frac{P\Delta V_S^d}{J_e} \approx \frac{2P^2Q^2R_S^N}{2Q\exp\left(d/\lambda_{sf}\right)+\sinh\left(d/\lambda_{sf}\right)}. \quad (32.9)$$

Figure 32.4a and b shows a scanning electron micrograph (SEM) of a typical lateral spin valve consisting of $F_{1,2}$ Permalloy (Py: $Ni_{19}Fe_{79}$) wires bridged by a N (Cu) wire and its NLSV signal ΔR_{NL} measured at room temperature as a function of applied magnetic field. The devices are usually fabricated by means of lift-off or shadow evaporation methods combined with e-beam lithography [19]. Due to the shape anisotropy, the magnetization directions of $F_{1,2}$ wires are parallel to the long axes. The F_2 wire having domain wall reservoirs at both edges exhibits a smaller switching field than the straight F_1 wire. As the magnetic field is swept from positive to negative (blue curve) after having set both magnetizations in P configuration at a large positive magnetic field of 1.0 kOe, a sharp drop in the signal ΔR_{NL} is observed at about −0.2 kOe when the $F_{1,2}$ magnetizations turn into AP configuration. Further decrease down to −0.55 kOe recovers the high ΔR_{NL} value in P configuration. A similar process takes place when the field sweep is reversed (red curve).

The spin-diffusion lengths of Py and Cu wires can be evaluated by performing above NLSV measurements as a function of the separation d. Fitting the measured overall change R_{NL}^D versus d to Equation 32.9 (Figure 32.4c) yields the spin-diffusion length of Cu as λ_{sf}^{Cu} = 1000 nm at 10 K and λ_{sf}^{Cu} = 400 nm at 290 K. Careful attention has to be paid when the cross-sectional areas S_N and S_F for Cu and Py wires are estimated. As mentioned earlier λ_{sf}^{Cu} is of the order of several hundred nanometers, whereas λ_{sf}^{Py} a few nanometers [18]. This means that the spin current flows homogeneously along the Cu wire but not in the Py wire where the spin current only exists in the vicinity of the junction. Therefore S_N for Cu should be the cross-sectional area of the wire, while S_F for Py should be the junction area. This tendency is clearly assured by the numerical calculation of spin current distribution [20].

32.3.2 Junction Size Dependence

As discussed earlier, the cross-sectional area for the spin current is represented by the effective junction size for the case of a ferromagnet such as Py with a very small λ_{sf}^{Py} of about a few nanometers. This implies that the spin accumulation in the vicinity of the Py/Cu junction is readily suppressed by the progressive spin relaxation in Py. This fact has to be well considered in order to realize efficient spin injection and detection. One of the ways to suppress the spin relaxation in the vicinity of the interface is to increase its spin resistance, equivalently meaning that the junction size has to be diminished. This was experimentally demonstrated as in Figure 32.5, showing drastic improvement in the NLSV signal R_{NL}^D as the junction size S_F is minimized [21]. If one seeks the best detection efficiency the point contact should be the one as well as the tunnel junction. However the current density for spin injection cannot be maximized for both cases. It should be noted here again that the NLSV signal greatly varies not only with the injector detector spacing d but also with the junction size S. Alternative solution for injection and detection

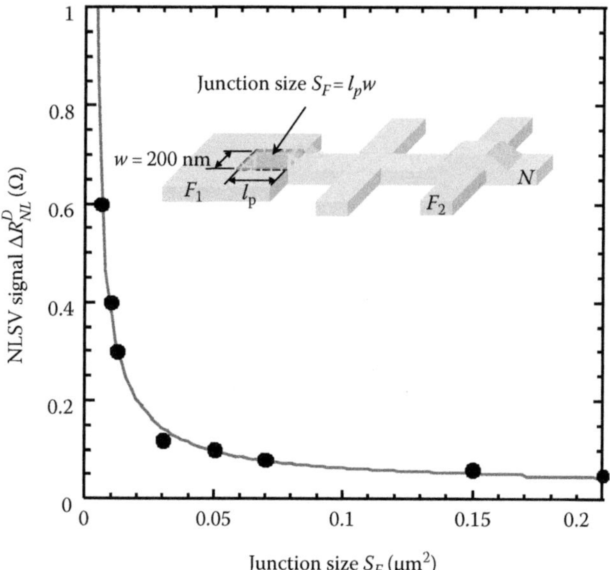

FIGURE 32.5 Overall NLSV signal R^D_{NL} as a function of the junction size S. The red curve is the best fit to the experimental data using the spin-resistance circuit model.

efficiency could rely on the interface engineering. By introducing the spin-dependent interface layer, the efficiency can be significantly optimized [22].

32.3.3 Spin Absorption (Sink) Effect

We now apply our spin-resistance circuit model to a triple junction shown schematically in Figure 32.6. In this case, the spin accumulation voltage V^d_S induced between the F_2 and N wires is expected to be influenced by the M wire inserted right in the middle at $d/2$ with the spin resistance of R^M_S and the spin-diffusion length of λ^M_{sf}, providing an additional spin relaxation volume. The magnitude of R^D_{NL} with M, $R^{D\text{with}}_{NL}$ can be calculated by using the same formalism given in Section 32.3.1

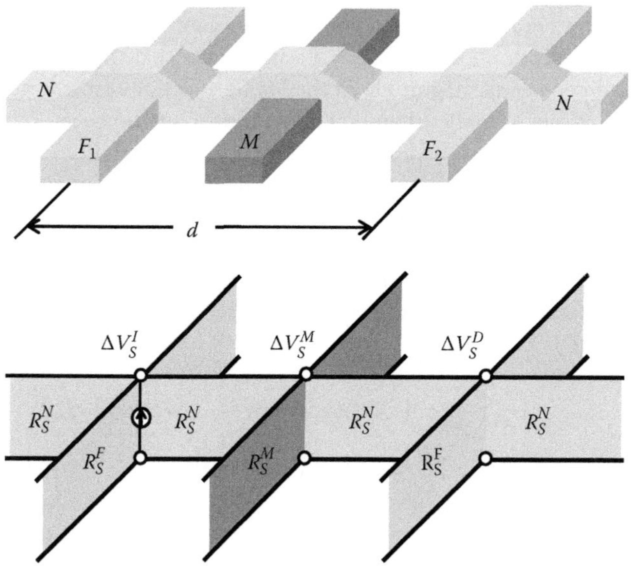

FIGURE 32.6 Schematic illustration of a triple junction device consisting of two F/N and a middle M/N junctions (top) and an equivalent spin-resistance circuit (bottom).

$$\Delta R^{D\text{with}}_{NL} \approx \frac{4P^2Q^2Q_M R^N_S}{\cosh(d/\lambda_N) - 1 + 2Q_M \sinh(d/\lambda_N) + 2Qe^{d/\lambda_N} - 2Q}, \tag{32.10}$$

where

Q_M is defined as R^M_S / R^N_S
d is the spacing between F_1 and F_2

Figure 32.7 shows typical lateral spin valve structures consisting of Py $F_{1,2}$ wires bridged by a Cu N wire with and without an inserted Pt M wire and corresponding NLSV signals. The overall change in NLSV signal $R^{D\text{without}}_{NL}$ amounts to 0.6 mΩ. However, the signal $R^{D\text{with}}_{NL}$ is considerably lowered to about 0.05 mΩ by the inserted Pt wire. This indeed indicates that the Pt wire absorbs most of the spins accumulated in between the Py injector and detector wires; in other words, the Pt wire acts as a spin sink.

Putting into Equation 32.10 the values of the spin resistances, $R^{\text{Cu}}_S = 2.60\,\Omega$ and $R^{\text{Py}}_S = 0.34\,\Omega$ separately determined by the double junction experiments, and the experimental value of $R^{D\text{with}}_{NL}$ for Pt, we obtain the value of $R^{\text{Pt}}_S = 0.15\,\Omega$ and then the spin-diffusion length $\lambda^{\text{Pt}}_{sf} = 10\,\text{nm}$. In the similar manner, the spin resistance and the spin-diffusion length for Au are determined, respectively, as $R^{\text{Au}}_S = 0.62\,\Omega$ and $\lambda^{\text{Au}}_{sf} = 60\,\text{nm}$. In this way, the value of the spin resistance R^M_S of the inserted wire can be directly determined from the measured value of $R^{D\text{with}}_{NL}$, thereby the spin-diffusion length λ^M_{sf} can be calculated. Figure 32.8 shows the R^M_S versus $R^{D\text{with}}_{NL}$ plot for spin valves with $d = 500$ and 650 nm calculated by using Equation 32.10. This plot is cross-checking that the values of the spin resistances obtained by using the triple junction are equivalent to those determined by using the double junction. The values obtained with the inserted Cu and Py wires are, respectively, 2.67 and 0.33 Ω, which are indeed in good agreement with those determined by using the double junction. Important to note is that the spin-diffusion length can be simply determined from the ratio of $R^{D\text{with}}_{NL}$ to $R^{D\text{without}}_{NL}$ without varying the injector–detector spacing d, and that all the experimental data points lie on the same curve only when the spacing d and the junction size S are the same, as can be seen in Figure 32.8.

One should also note here is that the spin current prefers to flow into the additionally connected wire when the spin resistance R^M_S is much smaller than R^N_S, meaning that this can be used as a technique to inject a pure spin current into a target material with a small R^M_S. We will show some of the examples such as the pure spin current induced switching, and SHEs in the Sections 32.4 and 32.5.

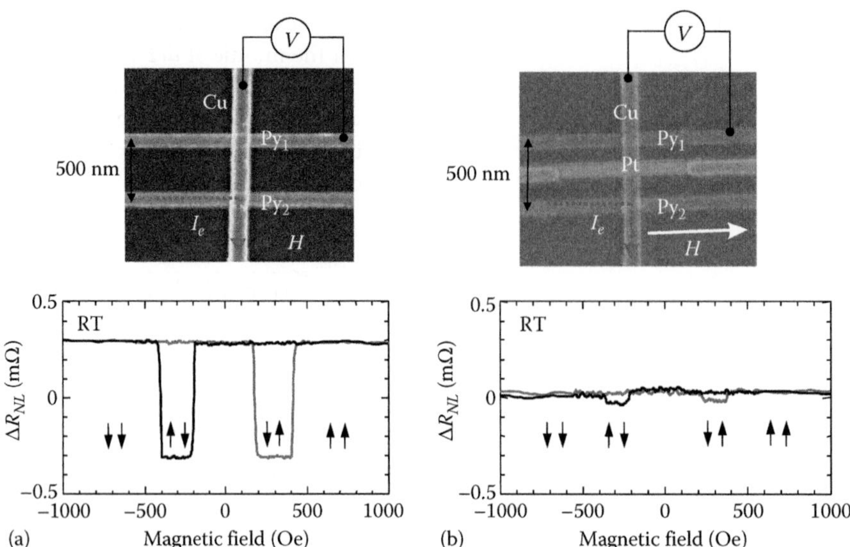

FIGURE 32.7 Typical lateral spin valves consisting of Py wires bridged by a Cu wire with (a) and without (b) an inserted Pt wire and corresponding NLSV valve signal as a function of external field measured at room temperature.

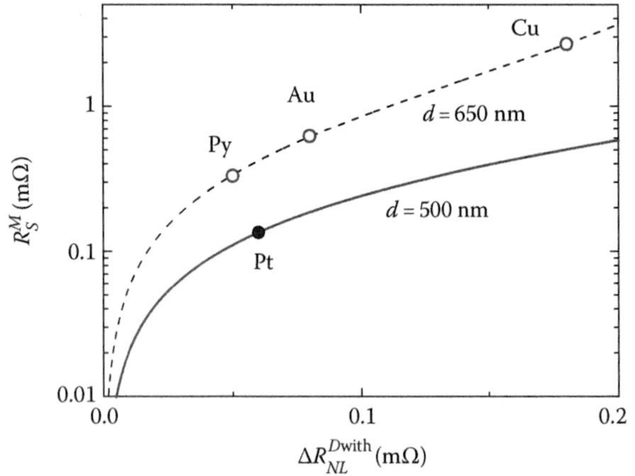

FIGURE 32.8 Calculated spin resistance as a function of the overall change R_{NL}^D for the triple junction devices with $d = 500$ and 650 nm. All measured values of R_{NL}^D for the devices with middle junctions of Py, Au, and Cu lie on the same calculated curve (dashed curve) when the value of d is the same. The calculated solid curve corresponds to the device with $d = 500$ nm with R_{NL}^D for the device with a Pt middle junction.

32.3.4 Influence of the Ohmic Contact on the Hanle Effect

The Hanle effect and NLSV measurements are regarded as complementary means to justify the presence of the spin transport in lateral spin valve systems. The Hanle effect is the manifestation of dephasing process in precessional motion of diffusively traveling collective electronic spins. In the measurement, the external field is usually applied perpendicular to the axis of the accumulated spins to induce the precessional motion about the field [23]. The averaged orientation of spins detected at the position of the detector thus results in a modulation in the NLSV signal as a function of the external field [4].

However, the spin accumulation is strongly affected by the ohmic contact as we have discussed in the previous section; the dwell time τ_D of diffusing electronic spins from the injector to the detector may be dependent on the nature of the contact. Figure 32.9a shows an SEM image of a typical lateral spin valve structure used for the Hanle effect measurement and the obtained Hanle curves at 10 K. The results demonstrate that the Hanle effect appears differently according to the nature of the injector and detector junctions. The overall change in NLSV signal R_{NL}^D for the tunnel junction is found about an order of magnitude larger than that for the ohmic junction as seen in Figure 32.9b, and this tendency can also be expected from Equation 32.5. Figure 32.9c also shows that the normalized dephasing magnitude in the Hanle effect of ohmic junction is much smaller than that of the tunnel junction.

Here we discuss, in a simplistic manner, the possibility where the dwell time τ_D could be influenced by the detector in ohmic contact with the nonmagnetic wire.

First, let us consider n electronic spins diffusing one dimensionally through the unit area with the diffusion constant D, the averaged velocity $\langle v\rangle$ may be expressed as follows by using Equation 32.5

$$\langle v\rangle = -D\frac{1}{n}\frac{\partial n}{\partial x} = \frac{D}{\lambda_{sf}}\cdot\frac{(1+\alpha)\exp\left(-x/\lambda_{sf}^N\right)-(1-\alpha)\exp\left(x/\lambda_{sf}^N\right)}{(1+\alpha)\exp\left(-x/\lambda_{sf}^N\right)+(1-\alpha)\exp\left(x/\lambda_{sf}^N\right)}, \tag{32.11}$$

where α is the ratio of the contributing weights of the tunnel- and ohmic-like junctions defined as $\alpha = (V_+ - V_-)/(V_+ + V_-)$ using coefficients V_+ and V_- in Equation 32.5. We can then deduce the dwell time τ_D using α

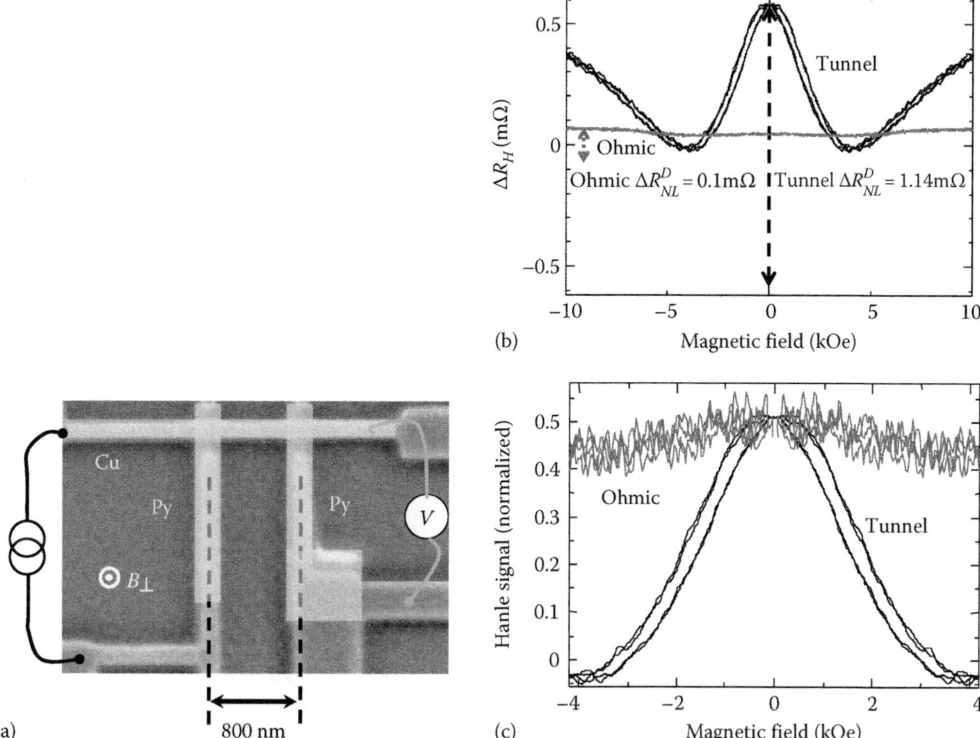

FIGURE 32.9 (a) SEM of a lateral spin valve used for the Hanle effect measurements, consisting of Py wires bridged by a Cu wire. (b) Comparison of the Hanle curve observed in a Py/Cu ohmic spin valve (red) and that observed in a Py/Cu tunnel spin valve (black). In both cases, 25 nm thick Py wires lie beneath a 100 nm thick Cu wire. The magnetizations of Py wires were first set in the parallel state by in-plane magnetic field and then the out-of-plane perpendicular magnetic field was applied for the Hanle measurements. The Hanle signal ΔR_H decreases with applied field, reflecting the dephasing process of collective spin precessional motion. The upturn around 4 kOe corresponds to the magnetization rotating out of the plane. (c) The Hanle signal normalized by the overall change R^D_{NL}, showing the dephasing magnitude for the tunnel spin valve is nearly an order of magnitude larger than that for the ohmic spin valve.

$$\tau_D = \int_0^d \frac{dx}{v} = \frac{\lambda_{sf} d}{D}\left(1 + \frac{\lambda_{sf}}{d}\ln\left(\frac{2\alpha}{(1+\alpha)-(1-\alpha)\exp(2d/\lambda_{sf})}\right)\right). \tag{32.12}$$

The dwell time τ_D is $\lambda_{sf}d/D$ for the tunnel junction with $\alpha = 1$, whereas for the transparent ohmic junction, α is given by

$$\alpha = \frac{R_S^F \cosh(d/\lambda_{sf})}{R_S^N \sinh(d/\lambda_{sf}) + R_S^F \cosh(d/\lambda_{sf})}. \tag{32.13}$$

By combining Equations 32.12 and 32.13, the dwell time τ_D can be approximated as

$$\tau_D \approx \frac{\lambda_{sf} d}{D}\left(1 - \ln\left(\frac{\cosh(d/\lambda_{sf}) + \sinh(d/\lambda_{sf})}{\cosh(d/\lambda_{sf})}\right)\right). \tag{32.14}$$

This clearly indicates that τ_D for the tunnel junction is longer than that for the ohmic junction, supporting the experimental fact that the Hanle effect in the ohmic lateral spin valve is smaller than that observed in the tunnel one.

32.4 Pure Spin Current–Induced Magnetization Reversal

Spin transfer torques arise whenever the flow of spin-angular momentum is throwing a portion of its angular momentum while passing through a ferromagnet. This happens typically in $F_1/N/F_2$ tri-layered nano-pillars where a spin-polarized current generated by spin filtering of the first ferromagnetic F_1 layer enters the second ferromagnetic F_2 layer through the N layer, and its polarization rotates to follow the magnetization direction of the F_2. This is equivalent to the situation where a pure spin current is absorbed by a magnetic wire whose magnetic moment is not collinear with the polarization direction of the pure spin current.

Kimura et al. [24,25] demonstrated for the first time the pure spin current–induced magnetization switching by using the structure similar to the lateral spin valve in Figure 32.2b but with the F_2 wire replaced by a small Py nanoscale particle to which the pure spin current was injected via spin absorption process. The above-mentioned behavior can be understood that a flow of up-spin electrons penetrate diffusively into a magnetic element and transfers their transverse component of spin angular momentum, while an equal number of down-spin electrons with an averaged spin component collinear with the magnet come out with generating no spin torque.

More recently, Yang et al. [26] succeeded in improving the spin injection and detection efficiency by a factor of 10 by employing in situ shadow evaporation technique for fabricating a lateral spin valve consisting of two separated Py (20 nm)/Au nano-pillars structured on a Cu wire, as shown in Figures 32.10b, d and 32.11a. Remarkable is that the well-defined lateral giant magneto resistance effect as well as the NLSV effect are observed (Figure 32.10a and c) and the overall change in ΔR_L of 32 mΩ is about twice as much as that in the NLSV signal ΔR_{NL} of 18.5 mΩ. In addition, these values are almost an order of magnitude larger than the reported values [3,10,20] and also in good agreement with the theoretical values calculated by using Equation 32.9. These facts assure that clean interfaces of both spin injector and detector contribute well to produce the giant NLSV signal.

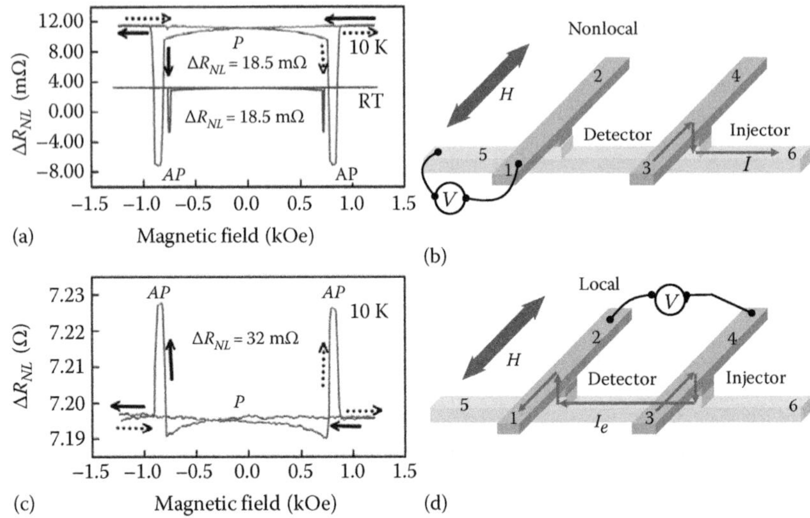

FIGURE 32.10 (a) NLSV ΔR_{NL} as a function of magnetic field together with (b) a schematic device configuration. (c) Local spin valve signal ΔR_L as a function of magnetic field together with (d) a schematic device configuration.

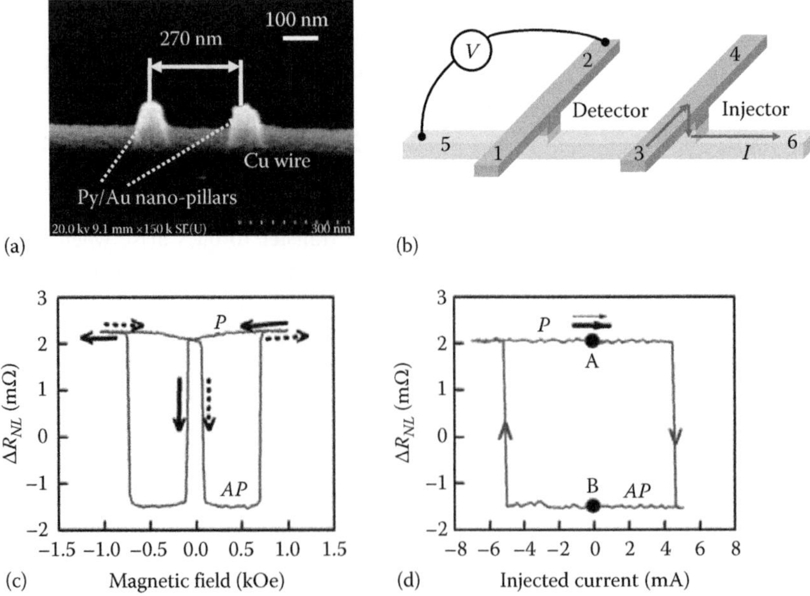

FIGURE 32.11 Nonlocal spin injection results for a sample with 4 nm thick Py detector nano-magnet. (a) SEM of two Py/Au nano-pillars bridged by a Cu wire. (b) A schematic device configuration. (c) NLSV signal as a function of magnetic field. (d) The NLSV signal as a function of injected dc current. The loop starts at the initial parallel (*P*) state A. The switching from *P* to antiparallel (*AP*) states occurs when the injected dc current reaches 5 mA. The reverse switching from *AP* to *P* states takes place when the current polarity is reversed.

For the purpose of realizing efficient pure spin current-induced switching, the detector Py/Au nano-pillar thickness was reduced to 4 nm to match the spin-diffusion length so that the spin current necessary for the switching can be minimized. The value of the NLSV signal ΔR_{NL} for the device is about 4 mΩ—smaller than that in Figure 32.10a but still larger than previously reported values. This device enables to switch reversibly the magnetization of a Py particle $75 \times 170 \times 4\,\text{nm}^3$ in size at the local spin current density in the range $2\text{–}6 \times 10^{10}\,\text{A/m}^2$, which is reasonable compared with the conventional spin injection experiments using nano-pillars [27,28].

32.5 Inverse and Direct Spin Hall Effects

As discussed in Section 32.4, spin accumulation, absorption, or spin pumping [19] enables us to inject a pure spin current into a nonmagnetic material. If the *N* material possesses the scattering centers such as impurities causing spin–orbit interaction, the injected spins are scattered asymmetrically according to their orientation [19,29–31]. Thereby the spin current is converted to the charge current flowing perpendicular to the spin current. The direction of the induced charge current J_e is given by the vector product $\boldsymbol{s} \times \boldsymbol{J}_S$, where $\boldsymbol{s}$ is the spin direction, yielding the flow of J_e normal to both J_S and $\boldsymbol{s}$. This behavior is called *inverse spin Hall effect*. In reverse, a charge current flowing in the nonmagnetic material can be converted to a spin current flowing perpendicular to the charge current and up and down electronic spins accumulate along the opposing edges of the *N* material. This is called *(direct) spin Hall effect* (SHE), which has been observed in semiconductors [32,33] and also metals [19].

For the *inverse* SHE measurements, the spin absorption technique explained in Section 32.3.3 was used for injecting a pure spin current into a Pt wire. The device structure of the SHE experiments is as shown in the inset of Figure 32.12a, typically consisting of a large Py injector pad 30 nm thick connected to an 80 nm wide Pt wire with thickness of about 5 nm via a 100 nm wide 80 nm thick Cu cross. The charge current is then injected from the Py pad into the Cu cross and drained into one of the two arms at the center of the cross. Induced spin current is diffusively transferred and absorbed into the Pt wire perpendicular to the device plane. When the spin direction $\boldsymbol{s}$ is oriented perpendicular to the Pt wire, the induced charge current J_e is directed along the wire and its magnitude is detected as the Hall voltage V_C appearing between the edges of the Pt wire. The Hall signal in ohm $\Delta V_C/J_e$ varies as a function of the external field, reflecting the magnetization process of the Py pad as shown in Figure 32.12a, since the average spin direction is defined by the magnitude of the magnetization vector.

The same device can also be used for *direct* SHE measurements. The large Py pad is now used as a detector instead of injector. It should be remarked that in the above experiment, the Pt wire is used as a spin current absorber, but the Pt wire now acts as a spin current source, which may provide a new means to produce the spin current for spintronics applications. The induced spin accumulation at the surface of the Pt wire is transferred through the Cu wire and detected by the Py pad as the spin accumulation signal in ohm $\Delta V_S/J_e$, as in Figure 32.12b, which varies almost the same as $\Delta V_C/J_e$ in Figure 32.12a corresponding to Onsager's reciprocal relation between spin and charge currents. This is equivalent to say that the spin current-induced SH conductivity σ_{SHE} is equal

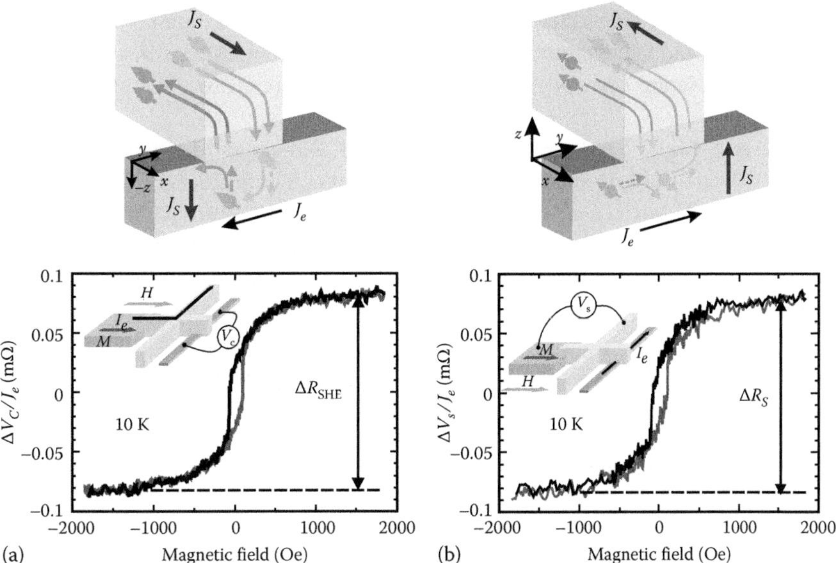

FIGURE 32.12 (a) Schematic explanation of the inverse SHE, conversion process in the Cu (light orange)/Pt (orange) junction from the spin current J_S to the charge current J_e together with the measured Hall voltage as a function of magnetic field. The inset schematically shows a device configuration. (b) Schematic explanation of the direct SHE, reversed conversion process from J_e to J_S, together with the spin accumulation signal as a function of magnetic field.

to the charge current–induced SH conductivity σ'_{SHE}. The SH conductivity can be expressed by using the experimentally determined overall change ΔR_{SHE} in $\Delta V_C/J_e$ [34]

$$\sigma_{SHE} = w\sigma^2\left(\frac{J_C}{J_S}\right)\Delta R_{SHE}, \tag{32.15}$$

where J_C/J_S is the ratio of the injected charge current to the spin current contributing to the SHE, the magnitude of which can be estimated by Equation 32.16 deduced from Equation 32.10 with the spin resistance of Pt R_S^{Pt} determined by the spin absorption experiment discussed in Section 32.3.3

$$\frac{J_C}{J_S} \approx \frac{2\left(R_S^{Py} + R_S^{Pt}\right)\cosh\left(d/\lambda_{sf}\right) + \left(R_S^{Cu} + 2R_S^{Py}\right)\sinh\left(d/\lambda_{sf}\right)}{P_{Py}R_S^{Py}}. \tag{32.16}$$

Equations 32.15 and 32.16 yield the value of σ_{SHE} of 2.4×10^4 $(\Omega m)^{-1}$ and the SH angle, that is, the ratio of the SH conductivity to the electrical conductivity of Pt $\alpha_{SHE}=\sigma_{SHE}/\sigma$ of 3.7×10^{-3} at room temperature [17]. These values were the largest at room temperature until the giant SHE was reported in 2008 [31].

References

1. M. Johnson and R. H. Silsbee, Interfacial charge-spin coupling: Injection and detection of spin magnetization in metals, *Phys. Rev. Lett.* **55**, 1790 (1985).
2. M. Johnson and R. H. Silsbee, Coupling of electronic charge and spin at a ferromagnetic-paramagnetic metal interface, *Phys. Rev. B* **37**, 5312–5325 (1988).
3. F. J. Jedema, A. T. Filip, and B. J. van Wees, Electrical spin injection and accumulation at room temperature in an all-metal mesoscopic spin valve, *Nature (London)* **410**, 345 (2001).
4. F. J. Jedema, H. B. Heershe, A. T. Filip, J. J. A. Baselmans, and B. J. van Wees, Electrical detection of spin precession in a metallic mesoscopic spin valve, *Nature (London)* **416**, 713 (2002).
5. A. Brataas, Y. Tserkovnyak, G. Bauer, and B. Halperin, Spin battery operated by ferromagnetic resonance, *Phys. Rev. B* **66**, 060404(R) (2002).
6. N. F. Mott, The electrical conductivity of transition metals, *Proc. Roy. Soc.* **153**, 699 (1936).
7. P. C. van Son, H. van Kempen, and P. Wyder, Boundary resistance of the ferromagnetic-nonferromagnetic metal interface, *Phys. Rev. Lett.* **58**, 2271 (1987).
8. T. Valet and A. Fert, Theory of the perpendicular magnetoresistance in magnetic multilayers, *Phys. Rev. B* **48**, 7099 (1993).
9. R. J. Elliott, Theory of the effect of spin-orbit coupling on magnetic resonance in some semiconductors, *Phys. Rev.* **96**, 266 (1954).
10. T. Kimura, J. Hamrle, and Y. Otani, Estimation of spin-diffusion length from the magnitude of spin-current absorption: Multiterminal ferromagnetic/nonferromagnetic hybrid structures, *Phys. Rev. B* **72**, 014461 (2005).
11. S. Ramo, J. R. Winnery, and T. V. Duzer, *Fields and Waves in Communication Electronics*, 3rd edn., Wiley, New York, p. 247, 1994.
12. Y. Ji, A. Hoffmann, J. S. Jiang, and S. D. Bader, Spin injection, diffusion, and detection in lateral spin-valves, *Appl. Phys. Lett.* **85**, 6218 (2004).
13. T. Kimura, J. Hamrle, Y. Otani, K. Tsukagoshi, and Y. Aoyagi, Spin-dependent boundary resistance in the lateral spin-valve structure, *Appl. Phys. Lett.* **85**, 3501 (2004).
14. F. Casanova, A. Sharoni, M. Erekhinsky, and I. K. Schuller, Control of spin injection by direct current in lateral spin valves. *Phys. Rev. B* **79**, 184415 (2009).
15. S. O. Valenzuela and M. Tinkham, Spin-polarized tunneling in room-temperature mesoscopic spin valves, *Appl. Phys. Lett.* **85**, 5914 (2004).
16. S. Garzon, I. Žutić, and R. A. Webb, Temperature-dependent asymmetry of the nonlocal spin-injection resistance: Evidence for spin nonconserving interface scattering, *Phys. Rev. Lett.* **94**, 176601 (2005).
17. N. Poli, M. Urech, V. Korenivski, and D. B. Haviland, Spin-flip scattering at Al surfaces, *J. Appl. Phys.* **99**, 08H701 (2006).
18. J. Bass and W. P. Pratt Jr., Spin-diffusion lengths in metals and alloys, and spin-flipping at metal/metal interfaces: An experimentalist's critical review, *J. Phys.: Condens. Matter* **19**, 183201 (2007).
19. T. Kimura, Y. Otani, T. Sato, S. Takahashi, and S. Maekawa, Room temperature reversible spin Hall effect, *Phys. Rev. Lett.* **98**, 156601 (2007); Erratum: *Phys. Rev. Lett.* **98**, 249901 (2007).
20. J. Hamrle, T. Kimura, Y. Otani, K. Tsukagoshi, and Y. Aoyagi, Three-dimensional distribution of spin-polarized current inside (Cu/Co) pillar structures, *Phys. Rev. B* **71**, 094402 (2005).
21. T. Kimura and Y. Otani, Spin transport in lateral ferromagnetic/nonmagnetic hybrid structures, *J. Phys.: Condens. Matter* **19**, 16521 (2007).
22. Y. Fukuma, H. Wang, H. Idzuchi, and Y. Otani, Enhanced spin accumulation obtained by inserting low-resistance MgO interface in metallic lateral spin valves, *Appl. Phys. Lett.* **97**, 012507 (2010).
23. M. Johnson and R. H. Silsbee, Spin-injection experiment, *Phys. Rev. B* **37**, 5326–5335 (1988).
24. T. Kimura, Y. Otani, and J. Hamrle, Enhancement of spin accumulation in a nonmagnetic layer by reducing junction size, *Phys. Rev. B* **73**, 132405 (2006).
25. T. Kimura, Y. Otani, and J. Hamrle, Switching magnetization of a nanoscale ferromagnetic particle using nonlocal spin injection, *Phys. Rev. Lett.* **96**, 037201 (2006).
26. T. Yang, T. Kimura, and Y. Otani, Giant spin-accumulation signal and pure spin-current-induced reversible magnetization switching, *Nat. Phys.* **4**, 851 (2008).

27. I. N. Krivorotov et al., Time-domain measurements of nanomagnet dynamics driven by spin-transfer torques, *Science* **307**, 228–231 (2005).
28. H. Kurt, R. Loloee, W. P. Pratt, Jr., and J. Bass, Current-induced magnetization switching in permalloy-based nanopillars with Cu, Ag, and Au, *J. Appl. Phys.* **97**, 10C706 (2005).
29. E. Saitoh, M. Ueda, H. Miyajima, and G. Tatara, Conversion of spin current into charge current at room temperature: Inverse spin-Hall effect, *Appl. Phys. Lett.* **88**, 182509 (2006).
30. S. O. Valenzuela and M. Tinkham, Direct electronic measurement of the spin Hall effect, *Nature* **442**, 176 (2006).
31. T. Seki, Y. Hasegawa, S. Mitani, S. Takahashi, H. Imamura, S. Maekawa, J. Nitta, and K. Takanashi, Giant spin Hall effect in perpendicularly spin-polarized FePt/Au devices, *Nat. Mater.* **7**, 125–129 (2008).
32. Y. Kato, R. C. Myers, A. C. Gossard, and D. D. Awschalom, Observation of the spin Hall effect in semiconductors, *Science* **306**, 1910–1913 (2004).
33. J. Wunderlich, B. Kaestner, J. Sinova, and T. Jungwirth, Experimental observation of the spin-Hall effect in a two-dimensional spin-orbit coupled semiconductor system, *Phys. Rev. Lett.* **94**, 047204 (2005).
34. S. Takahashi, H. Imamura, and S. Maekawa, in *Concepts in Spin Electronics*, S. Maekawa (Ed.), Oxford University Press, Oxford, pp. 343–370, 2006.

33

Molecular Spintronics

Stefano Sanvito
Trinity College

33.1 Introduction

While spintronics is consolidating as the leading technology in the magnetic data storage industry and is emerging into both the sensors and the random access memories arena, the interest of the scientific community has grown into investigating new materials different from conventional metals and semiconductors. Ferromagnetic insulators presenting a spontaneous electric polarization and magnetic semiconductors have been both proposed as electronic barriers for magnetic tunnel junctions (MTJ) [1,2] raising the expectation for multifunctional devices. Likewise, the search for new materials and material combinations to be used for generating highly spin-polarized currents, but at the same time having a small magnetic moment, has witnessed a substantial boost because of their potential for delivering applications based on spin-transfer torque [3,4].

Among these new materials, organic molecules seem to occupy a rather unique position. These, in fact, not only can be prepared in a practically infinite range of types and combinations at a usually modest cost, but also span an equally large horizon in terms of electronic properties and functionalities. For instance, electron transport in carbon-based materials goes from virtually scattering-free band conduction in graphene and C nanotubes to hopping-assisted conduction in molecular semiconductors, with the conductivity of these latter that can be tuned over 15 orders of magnitude [5]. As they stand, organic molecules appear as an incredibly versatile playground for exploring new spintronics concepts and/or for implementing existing ones.

As a first application, molecules have been used as nonmagnetic spacers in a conventional spin-valve architecture (two magnetic electrodes separated by a nonmagnetic material) with both planar [6] and vertical geometries [7]. The first experiments have been followed by many others over the years (for a recent and complete review see Ref. [8]) and, despite the present lack of a satisfactory rigorous metrology [9], a few indisputable facts have now emerged about the spin-scattering properties of organic molecules. These are summarized in Figure 33.1, where we report the spin relaxation time, τ_S (the average time that an electron travels before its spin direction is changed), as a function of the spin diffusion length, l_S (the average distance travelled in a time τ_S), for different materials.

Although the figure is only illustrative, since it compares data taken at different temperatures and reports values for τ_S and l_S measured/inferred in different ways, the clear message is that organic materials occupy the top left corner of the τ–l diagram, that is, they are characterized by long relaxation times but short relaxation lengths. Long τ_S's in organics are well known to the electron paramagnetic resonance community [19], and they are rooted in the lack of an efficient mechanism for relaxing spins in organics. In fact, spin–orbit interaction is generally weak (it scales as Z^4 with Z the atomic number) and hyperfine coupling is hampered by the lack of nuclei with nuclear spins in the upper part of the periodic table and by the fact that the molecular orbitals relevant for electron transport are generally delocalized [20]. Likewise also, short l_S's can be easily accounted for, since the typical mobility, μ, of an organic semiconductor is orders of magnitude smaller than that of standard semiconductors. For instance, μ in rubrene (the most conductive among the organic semiconductors) is just $10\,cm^2\,V^{-1}\,s^{-1}$ compared to $450\,cm^2\,V^{-1}\,s^{-1}$ for p-type Si.

One may then ask, what can organic molecules do for spintronics? Of course a natural answer is to think about applications where it is necessary to keep the spin coherence for long times, but not to drag the electron spins very far from the point of injection. Quantum information technology appears as one prospect [21] and other new concepts are now emerging [22]. A second possibility is to conceive devices where the organic layer

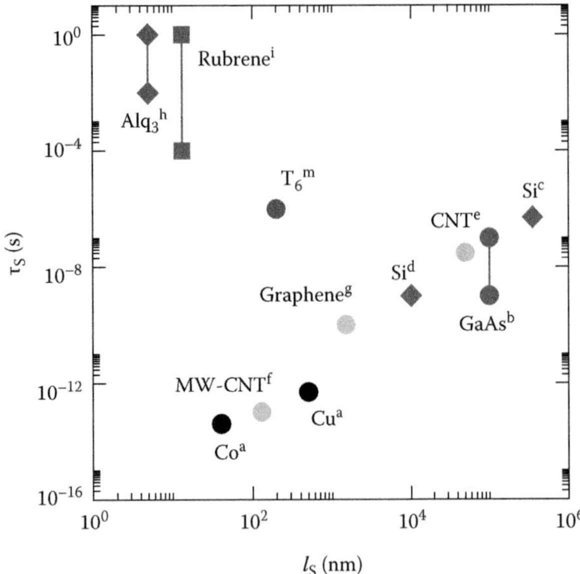

FIGURE 33.1 Spin relaxation time, τ_S, against spin diffusion length, l_S, for different classes of materials. Note that nonmagnetic organic materials occupy the upper left corner, that is, they are characterized by long τ_S and short l_S. The picture is adapted from Ref. [9]. Data in the figure correspond to the following references: a [10], b [11], c [12], d [13], e [14], f [15], g [16], h [7,17], i [18], and m [6]. (From Macmillan Publisher Ltd. [*Nat. Mater.*]; (Szulczewski, G., Sanvito, S., and Coeys, J.M.D., A spin of their own, 8, 693–695), Copyright 2009.)

is shorter than its rather short spin-diffusion length so that spin information is not lost in the transfer between the injector and the collector. Crucial to this second strategy is the design of structures where spins can be manipulated while diffusing in the organic layer. A third strategy, which departs quite drastically from conventional spintronics, is that of focusing on single-molecule transport, that is, that of looking at situations where the active part of the device is an individual molecule [23,24]. This is the main focus of the present chapter. For those readers interested in spin transport in extended organic semiconductors, we suggest a review that recently appeared in the literature [8] and Chapter 22 of this book. Furthermore, here we will primarily look at small molecules and will not consider carbon-only inorganic macromolecules, such as carbon nanotubes, graphene, and graphene nanoribbons. Also in this case, we refer the reader to the recent literature [25].

There are three main conceptual device structures that one can envision with single molecules. First, one can consider metallic magnetic electrodes sandwiching nonmagnetic molecules [23]. This is the molecular version of conventional giant magnetoresistance (GMR) or tunneling magnetoresistance (TMR) spin valves, and one can achieve one of the other depending on the conducting nature of the molecule. The important observation here is that the bonding structure of the molecule to the magnetic surface can drastically alter the spin selectivity of the interface, so that one can convert bond engineering in spin engineering. The second class of structures is that where the molecule itself carries a magnetic degree of freedom, that is, it is a single molecular magnet (SMM) [26]. These can be used both in combination with magnetic electrodes or with nonmagnetic ones, and the idea is to use the SMM to manipulate or store information. Such structures are often characterized by complex many body effects such as Coulomb and spin blockade, as well as by a rich inelastic spectrum of magnetic nature. Finally, the last class of devices is populated by molecules that possess nontrivial internal degrees of freedom (magnetic, vibrational, etc.) and that can be switched between different metastable configurations characterized by different transport properties. This is the case of spin-crossover [27] or optically switchable molecules [28].

Of course this is not a rigid classification and several proposals for combining different classes have been already made (for a complete review see Ref. [29]). For instance, one can fabricate spin valves with carbon nanotube spacers and then graft magnetic molecules on the tubes. Even more ambitious is the prospect of having devices where all the elements (electrodes and spacers) are organics, although such an idea at the moment seems more remote because of the lack, with a few exceptions [30], of room temperature organic magnets.

33.2 Establishing a Simple Conceptual Framework

Electron transport in single molecules is rather different from standard diffusion in metals and semiconductors and here we briefly review the main concepts. Consider the schematic diagram of Figure 33.2 [31]. Here, a generic molecular orbital of wave function ψ_α and single-particle level ε_α is coupled to two featureless (constant density of states—DOS) electrodes via the hopping parameters $\gamma_\alpha^m(m = \mathrm{L,R})$. The transmission rates from/to the two electrodes are $\gamma_\alpha^m/\hbar$ so that the state lifetime is $\frac{1}{\hbar}(\gamma_\alpha^{\mathrm{L}} + \gamma_\alpha^{\mathrm{L}})$.

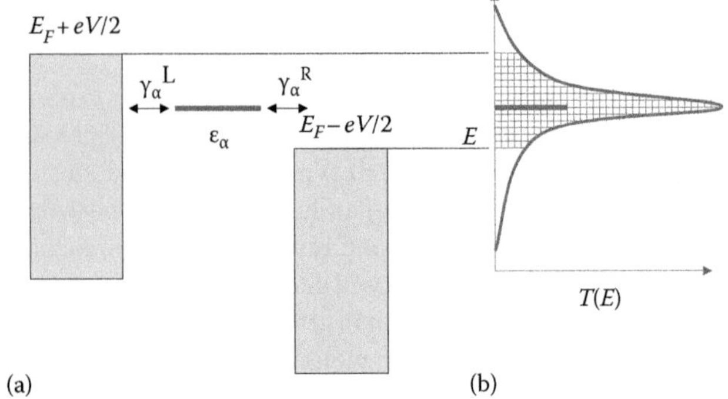

FIGURE 33.2 Schematic electronic transport through the single-particle level ε_α associated with a molecular orbital ψ_α. (a) The molecular orbital is electronically coupled to two current/voltage electrodes via the hopping parameters γ_α^m (m = L, R), so that electrons can move from an occupied state in the left electrode to an empty state in the right electrode. The electrodes are kept at a potential difference of V by a battery, which effectively shifts the position of their Fermi level, E_F. (b) The energy-integral of the resulting transmission coefficient, $T(E)$, is the current.

When ψ_α is weakly coupled to the electrodes, the associated DOS, $D_\alpha(E)$, is approximated by

$$D_\alpha(E) = \frac{1}{2\pi}\frac{\gamma_\alpha}{(E-\varepsilon_\alpha)^2+(\gamma_\alpha/2)^2}, \quad (33.1)$$

where $\gamma_\alpha = \gamma_\alpha^{\mathrm{L}} + \gamma_\alpha^{\mathrm{R}}$. Note that the inelastic electronic interaction with the molecule internal degrees of freedom (phonon, spin waves, etc.) can be incorporated in the description by adding an additional broadening, $\gamma_\alpha^{\mathrm{in}}$, whose details depend on the microscopic nature of the interaction and on how the inelastic excitations decay into the electrodes. Generally, this requires the solution of additional equations of motion, but more simply $\gamma_\alpha^{\mathrm{in}}$ can be interpreted as a "virtual" electrode with no net electron flux [32].

The actual position of the single-particle energy level ε_α is a function of the molecule charging state. At a simple mean field level, we can assume that ε_α scales linearly with the orbital occupation [33]. This is given by

$$n_\alpha = \int_{-\infty}^{+\infty} \mathrm{d}E\, D_\alpha(E)\frac{\gamma_\alpha^{\mathrm{L}} f^{\mathrm{L}}(E)+\gamma_\alpha^{\mathrm{R}} f^{\mathrm{R}}(E)}{\gamma_\alpha}, \quad (33.2)$$

where f^{L} (f^{R}) is the Fermi function evaluated at $E-\mu_{\mathrm{L}}(E-\mu_{\mathrm{R}})$ with μ_{L} (μ_{R}) the chemical potential of the left- (right-) hand side electrode ($\mu_{\mathrm{L}}-\mu_{\mathrm{R}}=eV$). The transmission coefficient associated to the molecular level ψ_α is

$$T_\alpha(E) = \frac{\gamma_\alpha^{\mathrm{L}}\gamma_\alpha^{\mathrm{R}}}{(E-\varepsilon_\alpha)^2+(\gamma_\alpha/2)^2}, \quad (33.3)$$

that is simply a resonance centered around ε_α. Finally, the current carried by ψ_α can be calculated by simply integrating $T_\alpha(E)$ within the bias window, that is, it is zero if ε_α is outside the energy interval $\mu_{\mathrm{L}}-\mu_{\mathrm{R}}$ (ψ_α is either completely empty or fully occupied), while for ε_α in the bias window we have

$$I_\alpha = 2\pi\frac{e}{h}\frac{\gamma_\alpha(1-r_\alpha^2)}{4}, \quad (33.4)$$

where

h is the Planck constant

e is the electron charge

and we have introduced the asymmetry parameter $r_\alpha = (\gamma_\alpha^{\mathrm{L}}-\gamma_\alpha^{\mathrm{R}})/(\gamma_\alpha^{\mathrm{L}}+\gamma_\alpha^{\mathrm{R}})$.

The model presented is based on the assumption that the steady state current is essentially a balance between the electron flux into and out from the molecule. If ψ_α is more strongly coupled to one of the electrodes, as in a scanning tunnel microscopy (STM) geometry, then $|r_\alpha|\sim 1$ and ε_α will be pinned to the chemical potential of that electrode. As a consequence, $T(E)$ will have a peak at $E=\varepsilon_\alpha$ moving as a function of bias together with the chemical potential that pins ψ_α and in general the current will be small. In contrast, if $\gamma_\alpha^{\mathrm{L}}\sim\gamma_\alpha^{\mathrm{R}}\left(r_\alpha\sim 0\right)$, then the dynamics of the energy level is given only by its charging properties and the current will be maximized.

Finally, it is important to mention that when no molecular levels are inside the bias window, current can still flow due to tunneling. In this case, an envelope function/effective mass picture of the tunneling process is inadequate and one simply writes the transmission coefficient as $T(E)=T_0(E)\mathrm{e}^{-\beta(E)L}$ with T_0 being a prefactor depending on the bonding structure between the molecule and the electrodes and the energy-dependent decay coefficient $-\beta(E)$ being a molecular specific quantity (L is the molecule length). In this tunneling limit, one is usually interested in linear response transport only so that the conductance is $G=\left(2e^2/h\right)T_0(E_F)e^{-\beta(E_F)L}$.

Spin can modify this description in two ways. First, the electrodes can be spin-polarized so that both their DOS and the electronic coupling to ψ_α become spin dependent. This implies $\gamma_\alpha^m \to \gamma_\alpha^{m\sigma}$, where we have introduced the spin index σ ($\sigma=\uparrow,\downarrow$). Second, the molecular level itself can be spin-polarized (for instance, if the molecule is magnetic and/or if it is spin-polarized by proximity to the electrodes), that is, $\varepsilon_\alpha \to \varepsilon_\alpha^\sigma$. Of course, both the effects can be present, so that one in general may have $\gamma_\alpha^m \to \gamma_\alpha^{m\sigma\sigma'}$, with the first spin index associated to the electrodes and the second to the molecule. In general, it is a good approximation to neglect spin-flip events so that $\sigma=\sigma'$, and all the transport quantity depends on a single spin index. A more complete description of electron and spin transport in molecules and of how this can be calculated from quantitative ab initio methods can be found in the more specialized literature [34,35].

33.3 Magnetic Electrodes and Nonmagnetic Molecules

The simpler and first device architecture investigated both experimentally and theoretically consists in a nonmagnetic molecule attached to magnetic electrodes. This is the single-molecule version of the spin valve, but it is considerably more experimentally challenging than its inorganic counterpart. In addition to the usual difficulties associated to molecular electronics (mechanical stability of the contacts, control over the bonding geometry, etc.), the use of transition metal electrodes adds the complication of the high reactivity of the surface. This effectively excludes the use of a large class of experimental protocols based on breaking junctions in wet conditions, which was proved rather successful for nonmagnetic noble metal electrodes [36]. This means that chemically stable interfaces must be made, indeed a difficult challenge.

In an early pioneering work, nanoconstrictions were constructed incorporating self-assembled monolayers of octanethiol (an insulating molecule) [37] and Ni electrodes. Remarkably, a low temperature magnetoresistance was found, although both the amplitude and the sign were extremely dependent on the specific sample, with most of the devices being either too resistive or not displaying any magnetoresistive signal at all. This rang an alarm bell toward the difficulties with maintaining clean interfaces and avoiding unwanted surface chemical reaction (oxidation of the transition metal for instance). To date, in fact, no other single-molecule MTJ device has been made, and the only other experiment available for spin-polarized transport in single molecules in a spin-valve architecture is a pump-probe optical experiment on quantum dots connected by organic linkers [38].

A better strategy appears to be that of operating with molecules that "fly," that is, molecules that can be evaporated over a magnetic surface in ultrahigh vacuum conditions and can be probed by STM (either polarized or not). Along these lines, a somehow less direct evidence that spin-polarized electrons can be injected across molecules is provided by two-photon photoemission experiments for copper phthalocyanine molecules deposited over Co surfaces [39], although no detailed microscopic information on how the various molecular features affecting the transport can be provided. Despite the fact that the experimental activity in this area is still rather limited (a much larger experimental effort is currently focused on organic tunnel junctions, where the organic layer is several nanometers in size), single-molecule spin transport has a huge potential, if not for making real devices, at least for developing an understanding of the details of spin injection across an organic/inorganic interface. Let us briefly discuss how large spin polarization (and TMR) can be achieved in molecular junctions.

First, consider the case of tunneling. Here the situation, at a first glance, may look similar to that of standard MTJs, where the symmetry of the complex band structure of the insulating barrier and its matching to the Fermi surface of the magnetic electrodes determines the spin polarization of the current [40]. However, when dealing with single molecules there are three main differences. First, the problem of electron transport through single molecules does not have translational invariance, so that attributing a symmetry to the evanescent wave function in the barrier is meaningless. Second, because of the lack of translational invariance, one expects the transmission to be maximized for electrons with very large longitudinal momentum, that is, at around the Γ point in the 2D Brillouin zone of the electrodes in the plane perpendicular to the molecule. Finally, one expects that the bonding structure at the interface and the molecular orbital character of the molecule may also affect drastically the transmission (T_0 may depend strongly on bonding structure). Overall, however, one expects the spin polarization of the current to be related to the spin polarization of a reduced DOS [42], constructed from those energy levels with a large longitudinal momentum. Since this is never too large in transition metals, we do not expect spectacularly large MR in the tunneling limit. First principle calculations for alkanethiol on Ni confirm this conjecture [23,42].

A different situation is attainable when the transport is resonant across a specific molecular level ψ_α. The main mechanism is illustrated in Figure 33.3. In this case, electrons tunnel from the electrodes to a delocalized molecular orbital, which in turns provides high transmission, that is, it supports a conductance reaching up to the quantum conductance. The total transmission of the junction thus is

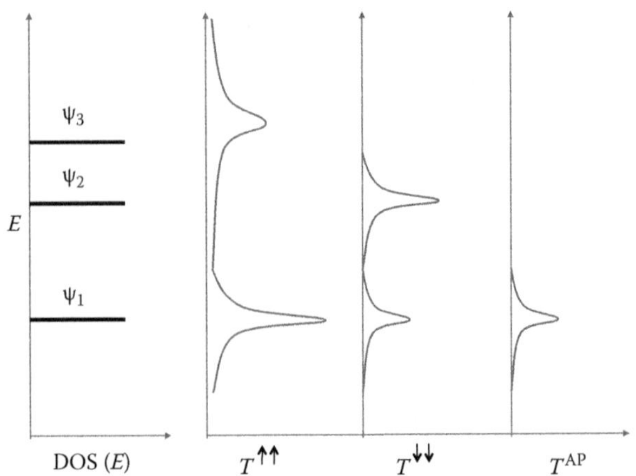

FIGURE 33.3 Illustration of the concept of orbital selectivity applied to a molecular spin valve. Assume a molecule has three molecular levels relevant for the electron transport within the bias window of interest, ψ_1, ψ_2, and ψ_3. One of them, ψ_1, is electronically coupled to both majority and minority spin electrons in the electrodes, while ψ_2 and ψ_3 are only coupled to minority and majority spins, respectively. In this case, $T^{\uparrow\uparrow}$ ($T^{\downarrow\downarrow}$) will present resonances associated to ψ_1 and ψ_3 (ψ_1 and ψ_2). In contrast, T^{AP} will be nonvanishing only where both $T^{\uparrow\uparrow}$ and $T^{\downarrow\downarrow}$ do not vanish, i.e., for ψ_1.

$$T_\alpha^{\sigma\sigma'}(E) = \frac{\gamma_\alpha^{L\sigma}\gamma_\alpha^{R\sigma'}}{(E-\varepsilon_\alpha)^2 + (\gamma_\alpha/2)^2}, \tag{33.5}$$

with $\gamma_\alpha = \gamma_\alpha^{L\sigma} + \gamma_\alpha^{R\sigma'}$. Since for transport across small organic molecules spin-flip events can be neglected, we obtain two simple expressions for the total transmission across the molecular orbital ψ_α, when the magnetization vectors of the electrodes are either parallel (P, $\sigma=\sigma'$) or antiparallel (AP, $\sigma=-\sigma'$,) to each other. These are, respectively,

$$T_\alpha^{P} = T_\alpha^{\uparrow\uparrow} + T_\alpha^{\downarrow\downarrow} \quad \text{and} \quad T_\alpha^{AP} = 2\sqrt{T_\alpha^{\uparrow\uparrow}T_\alpha^{\downarrow\downarrow}}. \tag{33.6}$$

Equation 33.6 establishes the interesting result that the transmission in the AP configuration is the convolution of the transmission for the two spin channels in the P one [23], as illustrated in Figure 33.3. As such, large GMR can be obtained in situations where the transmission in one of the spin channels is strongly suppressed.

We now look at what determines the γ's in the case of resonant tunneling across a molecular orbital. This is rooted in the concept of *orbital selectivity*. Let us assume that the molecular orbital relevant for the transport, ψ_α, has a given well-defined symmetry. It is then expected that only those states in the electrodes that share the same symmetry with ψ_α can efficiently couple to that molecular orbital [43]. As an example, let us consider the situation where the molecule is attached in a straight position to an adatom (along the z direction), so that only the bonding structure between the molecule and the adatom determines the transport. Let us also assume that the relevant molecular orbital for the transport has a π symmetry parallel to the plane of the electrodes, pointing in the x direction. Then, there are only a handful of atomic orbitals that can electronically couple to ψ_α, for instance p_x and d_{xz}, while for the other, for instance s or p_y, the electron hopping is suppressed by symmetry (in standard tight-binding theory, the matrix elements between orbitals not sharing the same angular momentum about the bond axis vanish). Therefore, if for energies $E\sim\varepsilon_\alpha$ no symmetry-allowed orbitals of the electrode are present, then $\gamma_\alpha(E\sim\varepsilon_\alpha)\sim 0$ and the transmission will be highly suppressed. In contrast, if symmetry-allowed orbitals are present, then the transmission will be large. Note that for this argument, we have in practice introduced an energy dependence of the hopping parameter $\gamma=\gamma(E)$, which originates from the nontrivial DOS of the electrodes.

Several demonstrations that orbital selectivity can drastically affect the device transmission and hence the I–V can be found in the literature for nonmagnetic electrodes [44,45]. In the case of magnetic electrodes, such a property can be used to enhance the spin polarization of the current. In general, in fact, magnetic materials have different Fermi surfaces for spins of different sign [46]. In particular, the orbital content at E_F (which atomic orbitals contribute to the DOS at E_F) can be quite different. This means that one can translate the orbital selectivity of the atomic bond into a spin selectivity via an appropriate choice of the magnetic material for the electrodes, the molecule, and the bonding/anchoring structure.

An example of orbital selectivity is provided in Figure 33.4 (see also Ref. [47]), where we plot the spin polarization of the current through a benzene-thiol-thiolate molecule attached to an Ni surface and probed with an Ni STM tip. This has been calculated with by using the nonequilibrium Green's function method implemented with density functional theory electronic structure [34]. The spin polarization of the current at a bias V is defined for the parallel alignment of the magnetization of the substrate and the tip as

$$P(V)=\frac{I^{\uparrow}-I^{\downarrow}}{I^{\uparrow}+I^{\downarrow}}, \qquad (33.7)$$

where $I^{\uparrow}$ ($I^{\downarrow}$) is the majority (minority) electrons contribution to the current. From the figure, one can immediately observe two important features. First, $P(V)$ is rather a complicate function of the applied bias, reflecting the fact that different molecular orbitals contribute to the current differently. Second, and most important for the concept of orbital selectivity, the spin polarization appears to be sensitive to the exact position of the STM tip with respect to the molecule. Thus, for instance at the positive voltage of 250 meV, P is positive if the current is collected from the π orbitals of the benzene, but it turns negative when collected from the S atom of the SH group.

33.4 Magnetic Molecule Devices

The second class of devices is made by contacting molecules comprising magnetic centers (usually transition metal ions) with either nonmagnetic or magnetic electrodes. This is a rather large class, which includes both STM-studied single spins on surfaces and more complex two- and three-terminal devices incorporating single-molecule magnets. Furthermore, depending on the interaction between the molecule and the substrate and on the nature of the substrate itself, a variety of phenomena can be

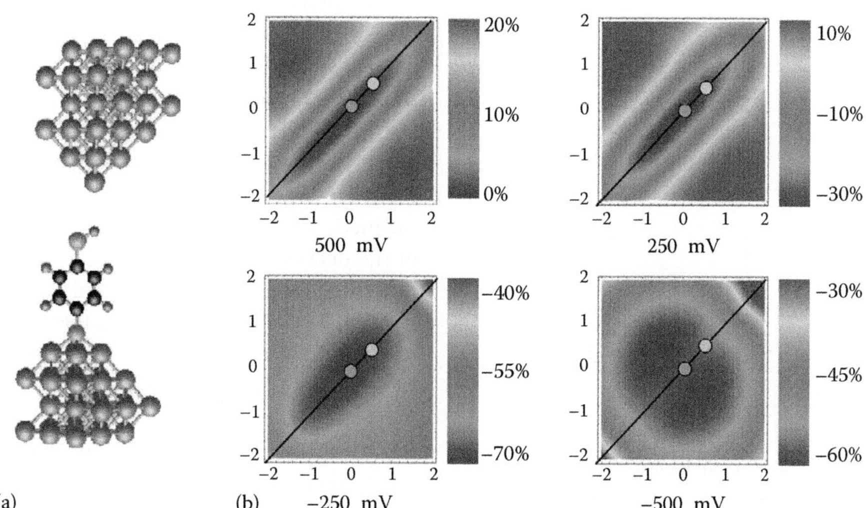

FIGURE 33.4 (See color insert.) Spin-polarized transport through a benzene-thiol-thiolate molecule attached to an Ni surface and probed with an Ni tip. In (a), we present a sketch of the calculated system and in (b) a two-dimensional map of the spin polarization P of the current as a function of the position of the STM tip and for different voltages. The diagonal black line in the panels in (b) indicates the plane of the benzene ring, the green circle the position of S, and the blue one that of H in the SH group.

observed, ranging from Coulomb and spin blockade, to Kondo resonant transport, to magnetic inelastic tunnel spectroscopy, to exchange coupling. In what follows, we will briefly review the main concepts that have emerged in the last few years.

33.4.1 STM Experiments

As for the case of nonmagnetic molecules, STM-type measurements appear simpler in terms of device fabrication than two- and three-terminal devices. In fact, there is a good choice of standard planar molecules incorporating magnetic centers (porphyrins, phthalocyanines, salens, etc.) that can be evaporated in ultrahigh vacuum conditions, that is, they are fully compatible with STM. In addition to the ease of fabrication, STM measurements for single spins can provide detailed information on the magnetic excitation spectrum and on how this is affected by magnetic coupling, anisotropy, and magnetic fields. In particular, the enabling experimental technique is spin-flip inelastic tunnel spectroscopy (SF-IETS) [48].

SF-IETS is based on a principle common to all inelastic spectroscopy techniques: a change in the electric conductance occurs every time the bias voltage reaches the critical value characteristic of a given excitation. The main concept is illustrated in Figure 33.5 and it is valid for molecules possessing internal degrees of freedom of some kind. An electron in the left electrode at a given energy E has a probability $T(E)$ of being transmitted. Let us assume that there is a molecular orbital of energy ε_α pinned at the highest of the two chemical potentials. In this situation, the transmission coefficient will show a resonance at $E=\varepsilon_\alpha$. As the bias is increased, a second transport channel becomes available. This is produced by the emission of a molecular excitation of energy $\hbar\omega$. If the energy of the final electronic state (after emission) $E=\varepsilon_\alpha-\hbar\omega$ is lower than the chemical potential, μ_2, of the right (collector) electrode, then the electrons cannot be absorbed since the final state in the electrode is occupied. However, if $\varepsilon_\alpha-\hbar\omega > \mu_2$, then the final state in the electrode is empty and the electron can be absorbed. This second inelastic channel contributes to the current and generates a satellite peak in $T(E)$ at $\varepsilon_\alpha-\hbar\omega$.

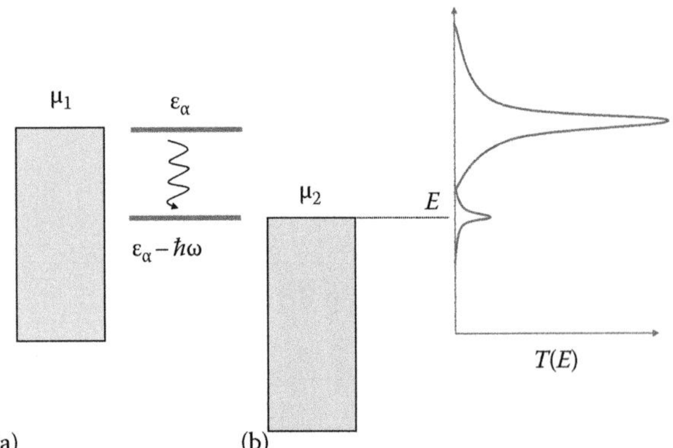

FIGURE 33.5 Graphical illustration of an inelastic scattering process. (a) An electron at energy ε_α emits an excitation of energy $\hbar\omega$ and lowers down its electronic energy to $\varepsilon_\alpha-\hbar\omega$. If the chemical potential μ_2 of the collector electrode is lower than $\varepsilon_\alpha-\hbar\omega$, such an inelastically scattered electron can be absorbed by the leads and will contribute to the current. (b) Such a scattering event produces a shoulder in the transmission coefficient at $E=\varepsilon_\alpha-\hbar\omega$.

Whether or not the opening of an inelastic channel increases the current depends strongly on the details of the junction. In the case the molecular excitations are phonons; propensity rules have been established correlating the symmetry of the various vibrational modes and of their coupling to the electrodes to the transmission [49,50]. As a rule of thumb, in junctions where the transmission is approaching the quantum limit one expects inelastic scattering to reduce the conductance, since the contributions to the current given by the additional inelastic channels are surpassed by an enhancement of the elastic backscattering [51]. In contrast, if the elastic transmission is low, inelastic channels are likely to increase the current. In the first case, one will observe a drop in the differential conductance as a function of bias and a deep in the $d^2I/dV^2(V)$, while the second case corresponds to an enhancement of $dI/dV(V)$ and a peak in $d^2I/dV^2(V)$. The details of the IETS spectrum are, however, quite complicate to predict, since phonon dissipation after emission can play an important role [52].

In the case where molecular excitations have spin origin, that is, they involve the flipping of the spin direction of the moving electrons, the technique takes the name of spin-flip inelastic tunnel spectroscopy [48]. This presents some additional features to the phonons case, since for spin transition selections rules must be satisfied while it is not the case for phonons. This indeed makes the theory more informative. Other main differences with respect to the phonon IETS is that the energy scale of the magnetic excitations is typically in the sub-meV range, that is, approximately one order of magnitude lower than that of the molecular vibrations. As such, SF-IETS requires severely low temperatures.

SF-IETS was pioneered at IBM with a number of experiments all conducted on magnetic ions deposited over metallic surfaces covered by a thin insulating film [48]. Landmark results include the measurement of the magnetic coupling [53] and of the magnetic anisotropy [54] of single atom chains. Most recently, the same technique has been employed for planar magnetic molecules deposited on surfaces. The investigations of the details of the super-exchange mechanism in Co phthalocyanine atomic layers deposited on Pb [55] and of the charging state of the same molecules [56] are some examples of this research. It is important to note in this context that SF-IETS works best in the limit of weak electronic coupling between the magnetic center and the electrodes. In the opposite situation, the transport, at least at low bias, is dominated by the Kondo effect. This should deserve a review in itself [57]; here we wish to mention that the Kondo effect in magnetic molecules is well established and can be tuned by STM manipulation of the molecule itself [58]. Also, it is important to remark that a quantitative parameter-free theory of the Kondo effect has begun to emerge [59].

Notably, SF-IETS does not require magnetic electrodes. However, the possibility to perform STM measurements by using spin-polarized currents opens a completely new prospect for the investigation of magnetism at the nano- and atomic scale. This is the growing area of spin-polarized STM [60], where both the probing STM tip and the sample to measure are magnetic (and in some cases also the substrate). A typical measurement then proceeds essentially as a spatially resolved TMR experiment, where the tunneling current is determined by the mutual orientation of the magnetization vectors of the tip and of the sample (in first approximation by their spin-polarized density of states). These, however, are rather delicate experiments, in particular if one aims at atomic scale spatial resolution. For instance, antiferromagnetic tips must be used to avoid stray fields that can affect the magnetic order of the sample to measure, that is, to avoid that the STM data depend on the tip-to-sample interaction. Furthermore, since the decay of the wave function in vacuum is different for different orbital components, one expects that the spatial mapping of the spin polarization may depend on the tip-to-sample separation.

In any case, the amount and quality of information that can be extracted from spin-polarized STM (SP-STM) have probably no equal at present in the field of atomic-scale spintronics. For instance, the fact that different energy (molecular) levels have a different spin polarization allows us to map accurately the spin-polarized spectrum of a nanoscaled object. Although no experiments are available at present for molecules, such a capability makes SP-STM a potentially major tool for the understanding of spin injection into molecular materials. Most of the applications of atomic-scale SP-STM to date have been focused on atoms on surfaces. For instance, the detection of hysteresis signals from Co adatoms of Pt [61] has allowed the investigation of Ruderman-Kittel-Kasuya-Yosida (RKKY)-interaction at the atomic level. Similarly, SP-STM experiments for O adsorbates on Fe have proved the presence of spin-polarized standing waves in thin magnetic films [62]. In a near future, we can envision new experiments where SF-IETS is combined with SP-STM to investigate in detail the transitions between different spin states and spin excitations.

33.4.2 Single Molecular Magnets Devices

An alternative to studying planar magnetic molecules by STM is that of constructing two- and three-terminal junctions incorporating magnetic molecules. This of course will be the ultimate geometry for fully functional devices, although many hurdles connected to nanofabrications and characterization must be overcome. One over many is the fact that despite many classes of magnetic molecules can be synthesized [26], most of them are extremely fragile away from solution [63,64]. This means that most likely the molecule entering a device is not the same as the one that was designed for the device. Rapid progress, however, has been made and in 2009 it was demonstrated that molecules belonging to the Fe_4 family can survive on surfaces and preserve both their spin state and most of the anisotropy [65].

Assuming a device can be made, the next important question is how this should operate, that is, what is the physical principle under which the device should function? The first option that comes to mind is that of using spin-valve-like architecture where one magnetic electrode serves as polarizer and the molecule itself as analyzer. This concept, however, lacks a fundamental ingredient; the mutual orientation of the magnetic moment of the molecule with that of the electrode should be stable. Unfortunately, the magneto-crystalline anisotropy in magnetic molecules is usually small (typical energy barriers are in the few Kelvin range), mostly because only a handful of atoms carry the magnetic moment. As a consequence, the molecule is likely to behave as a paramagnetic object at any reasonable temperature and the resulting current not to be spin-polarized. For the same reason, the idea of using magnetic molecules as spin polarizers [66] at the moment appears difficult to realize. An interesting recent proposal along this direction is that of using magnetic molecules, where the anisotropy results in little or no magnetic moment but a non-zero toroidal moment [67]. Such molecules are predicted to be able to switch the spin polarization of an injected current [68] and still be protected from dipolar interaction (one of the main sources of spin relaxation, at least in molecular crystals).

A second and in our opinion more promising strategy for making devices out of single-molecule magnets is that of relying on the exchange interaction instead of the anisotropy. This essentially means to use the spin state of the molecule as physical property to read, write, and manipulate. The question then becomes, can one address (read, write, and manipulate) the spin state of the molecule without the need of maintaining the spin-quantization axis fixed in time, that is, without the need of strong magneto-crystalline anisotropy? In this section, we will consider the problem of reading the magnetic state, while we will discuss the strategy for writing and manipulating in the last section.

The idea is to be able to convert spin information into molecular orbital information to be detected by an electrical current. In practice, one wants to demonstrate that the different spin states of a molecule present different frontier molecular orbitals or different coupling to the electrodes, both features that might affect the electric current. Again, because of the unlikely possibility of fixing the molecule anisotropy axis, the ideal situation would be to be able to read the molecule electrically without the need of using a spin-polarized current, that is, by using nonmagnetic electrodes.

Such an idea was recently explored theoretically [69] and it was demonstrated that in a two-terminal device incorporating an Mn_{12} molecule, one can distinguish between the $S=10$ ground state (GS) and a spin-flip state (SFS) obtained by flipping the magnetic moment of both a Mn^{3+} and a Mn^{4+} ion (the total spin projection of this configuration is 9). Most importantly, this is done by a single non-spin-polarized electrical readout, made possible because of magnetic state–specific molecular orbital re-hybridization under bias. In order to understand this concept, let us first observe that the two spin configurations investigated have a rather similar DOS. The main difference between the GS and the SFS at around E_F is that a molecular orbital singlet belonging to the Mn_{12} highest occupied molecular orbital (HOMO) multiplet has different spin orientation, namely, it is a majority state for the GS and a minority one for the SFS. One then expects such a difference not to be relevant if nonmagnetic electrodes are used, since different spin orientations cannot be distinguished by a non-spin-polarized current. This, however, is no longer true at finite bias as demonstrated by the calculated *I–V* curve shown in Figure 33.6.

The main feature of the *I–V* is a number of pronounced negative differential resistances (NDRs) specific for the given spin state. These arise because of the dynamic electron coupling of the various molecular orbital to the electrodes under bias, that is, $\gamma_\alpha^m = \gamma_\alpha^m(E,V)$. Such a dependence of γ_α^m over bias is a consequence of the polarization in an electric field of the wave function, as also illustrated in Figure 33.6. In fact, as the wave function polarizes under the effect of an electric field, its overlap with the electrodes' extended states changes and consequently γ_α^m also changes. If one now considers the Equation 33.4, it is clear that a change in the symmetry of the electronic coupling (change in r_α) may reduce the current carried by a given molecular orbital, leading to an NDR. The situation, however, is slightly more complicated because the molecular orbital polarization and then $\gamma_\alpha(E,V)$ is associated to the re-hybridization of molecular levels closely spaced in energy. This enables us to read the magnetic state of the molecule. In fact, the molecular orbital that gets spin-flipped when going from the GS to the SFS participates to re-hybridization only in the GS, since in absence of spin flip no hybridization can occur for states of different spins. Therefore, the re-hybridization of the HOMO multiplet in an electric field is different for the GS and the SFS, since the multiplicity of the HOMO for a given spin is different. This means that the response of a spin state to an electric field is driven by the dielectric response of its molecular orbitals, that is, spin information are translated in orbital information.

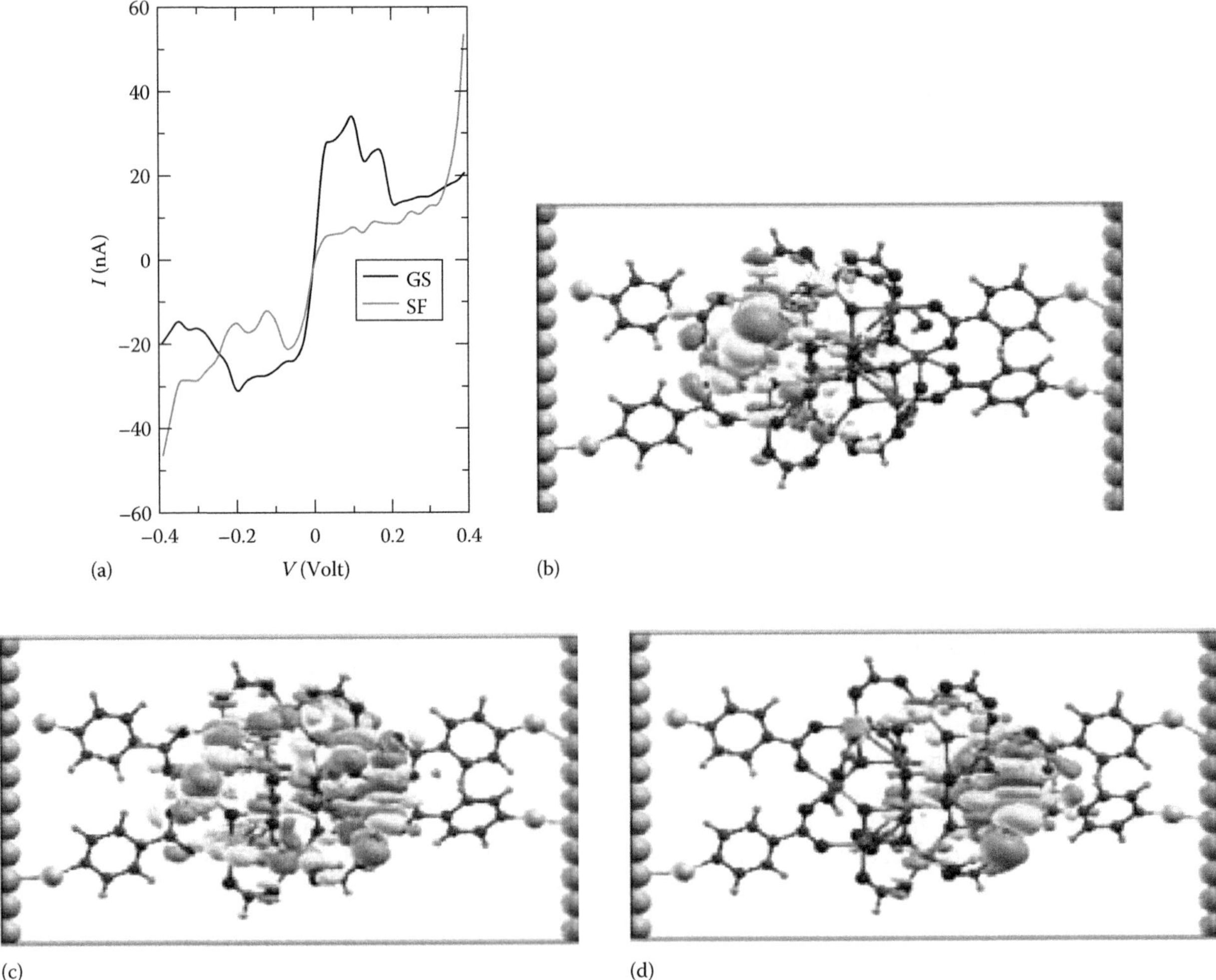

FIGURE 33.6 Transport properties of an Mn_{12} two-probe device. The *I–V* curves for both the GS and the SFS are presented in (a). The three small plots show isosurfaces of the HOMO wave function for the GS at different bias: (b) 300 mV, (c) = 0, and (d) −300 mV. Note the drastic polarization of the wave function in the electric field.

A mechanism similar to that described here can explain, at least in a qualitative way, some of the low-energy features (NDRs satellite to Coulomb blockade not involving different charging states) of the *I–V* curves measured in recent experiments on an Mn_{12} incorporated in a three-terminal device [70,71]. Alternative explanations ascribe such features to selection rule–forbidden transitions between different spin states [72,73]. Certainly more work is needed in this area, in particular in the characterization and in general in gathering more data for the various devices. In summary, we wish to remark that the demonstration that different spin states of a molecule can be distinguished electrically has been provided both experimentally and theoretically. Whether or not one will be able to assign with certainty a given spin state to a specific fingerprint in the transport and whether this will be controllable still remains an open question.

33.5 Manipulation and Control

Having demonstrated that the spin state of a molecule can be detected from a transport measurement, the next question is how to read and eventually manipulate such a state. This essentially means how to manipulate the exchange interaction in a controllable way. Note that this rather fundamental aspect underpins not only the potential for new logic devices and sensors, but also the use of magnetic molecules as elements for quantum computation [21,74]. Demonstrations that the spin state of a molecule can be changed by an external stimulus are abundant. In fact, there exists an entire class of molecules named spin crossover, whose magnetic GS can be altered by light, temperature, or pressure [27]. Typically, such molecules display a high-spin to low-spin transition, which is driven by a change in the geometric configuration between the magnetic center and the surrounding ligands, with the change initiated by the external stimulus. Importantly, spin-crossover activity has been demonstrated only in molecular crystals and to date there is little evidence that the same can happen at the single-molecule level on a surface.

A second strategy for altering the magnetic properties of a molecule is that of acting chemically. For instance, it was recently demonstrated that the magnetic exchange between two Cr_7Ni rings can be modified by changing the molecular linker coupling the rings [75]. Thus, a linker containing a single $S=1/2$ Cu^{2+} ion couples the rings antiferromagnetically, while no coupling is detected when the linker is made from a $S=0$ Cu_2 complex. An intriguing aspect of this protocol is that the electronic structure of the constituent magnetic elements (the Cr_7Ni rings) is little affected by the linker, so that the freedom in the design of the desired compound is rather large. For example, one can envision the use of active linkers, that is, of linkers whose electronic structure can be changed by a local probe such as a gate. Examples of these are the one-electron reducible $[Ru^{2+}Ru^{3+}(OCCMe_3)_4]^+$ linker still for Cr_7Ni rings [76] and $[PMo_{12}O_{40}(VO)_2]q-$ molecules [76]. Finally, it is also worth mentioning that the chemical strategy has also been successfully employed for manipulating the magnetic anisotropy of planar magnetic molecules on surfaces [77].

The important advantage of using active linkers in order to manipulate the exchange interaction between magnetic ions is that the manipulation may be driven by an electric stimulus. A second possibility for manipulating electrically the exchange interaction is based on the electrostatic spin-crossover effect (ESCE), first proposed for high electron density molecular wires [78] and then for polar molecules [79]. In general, when an electric field, $\vec{E}$, is applied to a quantum mechanical system, its energy will change (Stark effect). Such a change is proportional to the first order to $\vec{p}\cdot\vec{E}$, with $\vec{p}$ the permanent electric dipole, and to the second order to $\frac{1}{2}E_i\alpha_{ij}E_j$, where α_{ij} is the polarizability tensor. The idea behind the ESCE is that the high-spin and the low-spin state of a molecule in general have different polarizabilities and, in the case of polar molecules, also different permanent electric dipoles. For this reason, the electric field-driven energy shift depends on the magnetic state of the molecule, so that one can speculate that there exist particular conditions where a high-spin to low-spin crossover is possible. The fundamental question is how large is the crossover field.

If one considers molecules with inversion symmetry, then the first order contribution to the Stark effect vanishes and the difference in the polarizability alone induces the crossover [78]. In order to obtain realistic crossover electric fields, E_{cross}, one should consider molecules with a large spin contrast in the polarizability (i.e., the polarizability of different spin states must be largely different). This translates into a large charge density and a small HOMO-lowest unoccupied molecular orbital (LUMO) gap, and brings two serious drawbacks. First, one needs an extremely challenging chemistry to produce the desired molecules (the ones proposed in Ref. [78] were 10π benzene) and it is not clear whether such a chemical route will ever be available. Second and most importantly, the small HOMO-LUMO gap prevents the onset of a large electric field in a real device, so that even if 10π benzene are made, one will probably never be able to produce an electric field intense enough to switch the molecule.

The use of polar molecules circumvents these problems. The crucial idea is that the permanent electric dipole can effectively "bias" the crossover field to smaller fields. In fact, the energy change as a function of field, Δ_{GS}, is a parabolic function, centered at $E=0$ for nonpolar molecules (see Figure 33.7). Then, E_{cross} is determined by the interception between the parabolas associated to the different spin states (a spin singlet and a spin triplet in Figure 33.7). The addition of a linear term shifts the center of the parabolas bringing their interception closer to $E=0$ at least for one of the two field polarities. Hence, the molecule electric dipole effectively introduces a bias field, E_{bias}, which reduces the external electric field needed for the crossover. This appears to be a much more promising strategy since one can first engineer the magnetic molecule and then the specific electric dipole, as demonstrated for acetylene-bridged di-Cobaltocene($CoCp_2$) molecules functionalized with different substituents [79].

Such a field-induced crossover has important consequences for the electric transport. If one of these molecules is sandwiched in a two-terminal device in such a way that there is a

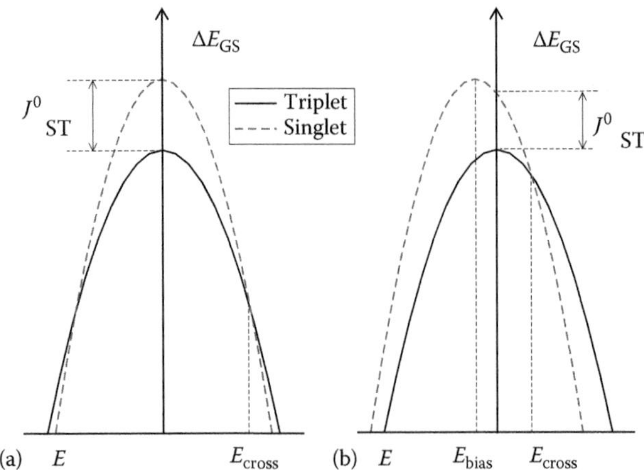

FIGURE 33.7 Stark energy gain, ΔE_{GS}, for the singlet and triplet state of a magnetic molecule as a function of the applied electric field, E. In panel (a), we represent a molecule with inversion symmetry, while in (b) one where the symmetry is broken by an electric dipole. Note that the shift of the $\Delta E_{GS}(E)$ parabola by E_{bias} generates a shift of the crossover field to lower fields. J^0_{ST} is the exchange interaction energy at zero field.

potential drop between the magnetic centers, then the strength of the magnetic coupling will change with bias. Interestingly, for biases corresponding to the crossover field the different spin states becomes degenerate, that is, on average the molecule will spend an identical amount of time in any one of them. This essentially means that there is no magnetic energy scale at E_{cross} and the current is predicted to become temperature independent [80]. Intriguingly, the electric control of a high-spin to low-spin transition has been recently demonstrated experimentally in a single Mn^{2+} ion coordinated by two tetrapyridine ligands in a three-terminal device geometry [81].

We wish to close this section by mentioning that some theoretical work has been devolved to understand how the magnetization direction of a magnetic molecule can be manipulated with an external current, by essentially translating the concept of spin-transfer torque to the molecular magnets world [82]. Importantly it was shown that a spin-polarized current can switch the magnetic moment of the molecule despite the molecule-intrinsic spin relaxation.

33.6 Conclusion and Outlook

Although the first demonstration of spin transport in an organic material is not even a decade old [6], the field has already evolved very quickly and branched out in several subfields. Here, we have examined only one of them, namely, that of single-molecule spintronics. This appears to be particularly fascinating since it explores the interaction between current electrons and spins at the most fundamental level. It is also of great technological appeal since new multifunctional devices can be envisioned to replace existing inorganic ones or to implement completely different new concepts. A particularly welcome evolution of molecular spintronics is that of addressing molecules with internal degrees of freedom, in particular magnetic molecules. This seems to be a natural evolution that can bring important consequences on other frontier areas of science such as quantum logic.

However, as the fields mature so do the challenges. At present, there is still no clear way to produce stable and reproducible single-molecule devices. This is a common problem to standard molecular electronics but the situation is even more problematic when the molecules are magnetic, because of their fragility on surfaces. Likewise, magnetic electrodes are more reactive than noble metals and by large incompatible with wet chemical deposition. Then, it is important to develop new useful molecules that are compatible with vacuum deposition, or that can be deposited on magnetic surfaces without oxidation. These need to be characterized thoroughly when in the device. For instance, one should be able to establish and control the bonding structure with the electrodes. Furthermore, it is desirable to engineer new molecules that can be switched between different magnetic states by external stimuli compatible with standard device architecture such as an electrostatic gate.

Finally, we believe it is crucial to find new applications for molecular spin devices. Simply translating concepts from inorganic spintronics do not seem to be a winning strategy. Molecules are not just small, but they are also quantum objects. Therefore, exploiting quantum properties appear like a natural choice. At the same time, it is important to realize that the electronic properties of molecules are always the result of a delicate interplay between different interactions. These can be used for constructing multifunctional devices where more than one property is addressed at the same time. We believe that this is a field that has a lot to offer in the future. In particular, the construction of functional, reproducible, and useful devices may act as a catalytic center for focusing otherwise disconnected disciplines. It is from this synergy of different fields that we expect new fascinating science and revolutionary applications to emerge.

Acknowledgment

This work is sponsored by Science Foundation of Ireland, CRANN and the EU under the FP7 program.

References

1. J. S. Moodera, T. S. Santos, and T. Nagahama, The phenomena of spin-filter tunneling, *J. Phys. Condens. Matter* **19**, 165202 (2007).
2. M. Gajek, M. Bibes, S. Fusil et al., Tunnel junctions with multiferroic barriers, *Nat. Mater.* **6**, 296–302 (2006).
3. J. C. Slonczewski, Current-driven excitation of magnetic multilayers, *J. Magn. Magn. Mater.* **159**, L1–L7 (1996).
4. L. Berger, Emission of spin-waves by a magnetic multilayer traversed by a current, *Phys. Rev. B* **54**, 9353–9358 (1996).
5. C. K. Chiang, C. R. Fincher Jr., Y. W. Park et al., Electrical conductivity in doped polyacetylene, *Phys. Rev. Lett.* **39**, 1098–1101 (1977).

6. V. Dediu, M. Murgia, F. C. Matacotta, C. Taliani, and S. Barbanera, Room temperature spin polarized injection in organic semiconductor, *Solid State Commun.* **122**, 181–184 (2002).
7. Z. H. Xiong, D. Wu, Z. Valy Vardeny, and J. Shi, Giant magnetoresistance in organic spin-valves, *Nature (London)* **427**, 821–824 (2004).
8. V. A. Dediu, L. E. Hueso, I. Bergenti, and C. Taliani, Spin routes in organic semiconductors, *Nat. Mater.* **8**, 707–716 (2009).
9. G. Szulczewski, S. Sanvito, and J. M. D. Coeys, A spin of their own, *Nat. Mater.* **8**, 693–695, (2009).
10. J. Bass and W. P. Pratt Jr., Spin-diffusion lengths in metals and alloys, and spin-flipping at metal/metal interfaces: An experimentalist's critical review, *J. Phys.: Condens. Matter* **19**, 183201 (2007).
11. J. Kikkawa and D. D. Awschalom, Lateral drag of spin coherence in gallium arsenide, *Nature (London)* **397**, 139–141 (1999).
12. B. Huang, D. Monsma, and I. Appelbaum, Coherent spin transport through a 350 micron thick silicon wafer, *Phys. Rev. Lett.* **99**, 177209 (2007).
13. I. Appelbaum, B. Huang, and D. Monsma, Electronic measurement and control of spin transport in silicon, *Nature (London)* **447**, 295–298 (2007).
14. L. E. Hueso, J. M. Pruneda, V. Ferrari et al., Transformation of spin information into large electrical signals using carbon nanotubes, *Nature (London)* **445**, 410413 (2007).
15. K. Tsukagoshi, B. W. Alphenaar, and H. Ago, Coherent transport of electron spin in a ferromagnetically contacted carbon nanotube, *Nature (London)* **401**, 572574 (1999).
16. N. Tombros, C. Jozsa, M. Popinciuc, H. T. Jonkman, and B. J. van Wees, Electronic spin transport and spin precession in single graphene layers at room temperature, *Nature (London)* **448**, 571574 (2007).
17. S. Pramanik, C.-G. Stefanita, S. Patibandla et al., Observation of extremely long spin relaxation times in an organic nanowire spin valve, *Nat. Nanotechnol.* **2**, 216219 (2007).
18. J. H. Shim, K. V. Raman, Y. J. Park et al., Large spin diffusion length in an amorphous organic semiconductor, *Phys. Rev. Lett.* **100**, 226603 (2008).
19. V. I. Krinichnyi, 2-mm Waveband electron paramagnetic resonance spectroscopy of conducting polymers, *Synth. Met.* **108**, 173–222 (2000).
20. S. Sanvito and A. R. Rocha, Molecular-spintronics: The art of driving spin through molecules, *J. Comput. Theor. Nanosci.* **3**, 624–642 (2006).
21. M. Affronte, Molecular nanomagnets for information technologies, *J. Mater. Chem.* **19**, 1731–1737 (2009).
22. H. Agarwal, S. Pramanik, and S. Bandyopadhyay, Single spin universal Boolean logic gate, *New J. Phys.* **10**, 015001 (2008).
23. A. R. Rocha, V. M. Garcia Suarez, S. W. Bailey, C. J. Lambert, J. Ferrer, and S. Sanvito, Towards molecular spintronics, *Nat. Mater.* **4**, 335–339 (2005).
24. S. Sanvito, Injecting and controlling spins in organic materials, *J. Mater. Chem.* **17**, 44554459 (2007).
25. W. J. M. Naber, S. Faez, and W. G. van der Wiel, Organic spintronics, *J. Phys. D: Appl. Phys.* **40**, R205–R228 (2007).
26. D. Gatteschi, R. Sessoli, and J. Villain, *Molecular Nanomagnets*, Oxford University Press, Oxford, U.K., 2006.
27. P. Gütlich and H. A. Goodwin (Eds.), *Spin Crossover in Transition Metal Compounds*, Springer, New York, 2004.
28. M. Alemani, M. V. Peters, S. Hecht, K.-H. Rieder, F. Moresco, and L. Grill, Electric field-induced isomerization of azobenzene by STM, *J. Am. Chem. Soc.* **128**, 14446–14447 (2006).
29. L. Bogani and W. Wernsdorfer, Molecular spintronics using single-molecule magnets, *Nat. Mater.* **7**, 179–186 (2008).
30. K. I. Pokhodnya, A. J. Epstein, and J. S. Miller, Thin-film $V[TCNE]_x$ magnets, *Adv. Mater.* **12**, 410–413 (2000).
31. S. Datta, Electrical resistance: An atomistic view, *Nanotechnology* **15**, S433–S451 (2004).
32. M. Büttiker, Role of quantum coherence in series resistors, *Phys. Rev. B* **33**, 3020–3026 (1986).
33. C. Toher, A. Filippetti, S. Sanvito, and K. Burke, Self-interaction errors in density-functional calculations of electronic transport, *Phys. Rev. Lett.* **95**, 146402 (2005).
34. A. R. Rocha, V. M. Garcá-Suárez, S. Bailey, C. J. Lambert, J. Ferrer, and S. Sanvito, Spin and molecular electronics in atomically generated orbital landscapes, *Phys. Rev. B* **73**, 085414 (2006).
35. I. Rungger and S. Sanvito, Algorithm for the construction of self-energies for electronic transport calculations based on singularity elimination and singular value decomposition, *Phys. Rev. B* **78**, 035407 (2008).
36. N. J. Tao, Electron transport in molecular junctions, *Nat. Nanotechnol.* **1**, 173–181 (2006).
37. J. R. Petta, S. K. Slater, and D. C. Ralph, Spin-dependent transport in molecular tunnel junctions, *Phys. Rev. Lett.* **93**, 136601 (2004).
38. M. Ouyang and D. D. Awschalom, Coherent spin transfer between molecularly bridged quantum dots, *Science* **301**, 1074–1078 (2003).
39. M. Cinchetti, K. Heimer, J.-P. Wüstenberg et al., Determination of spin injection and transport in a ferromagnet/organic semiconductor heterojunction by two-photon photoemission, *Nat. Mater.* **8**, 115–119 (2009).
40. W. H. Butler, X.-G. Zhang, T. C. Schulthess, and J. M. MacLaren, Spin-dependent tunneling conductance of Fe–MgO–Fe sandwiches, *Phys. Rev. B* **63**, 054416 (2001).
41. M. Julliere, Tunneling between ferromagnetic films, *Phys. Lett. A* **54**, 225–226 (1975).
42. Z. Ning, Y. Zhu, J. Wang, and H. Guo, Quantitative analysis of nonequilibrium spin injection into molecular tunnel junctions, *Phys. Rev. Lett.* **100**, 056803 (2008).
43. A. P. Sutton, *Electronic Structure of Materials*, Oxford University Press, Oxford, U.K., 1993.

44. J. Ning, Z. K. Qian, R. Li, S. M. Hou, A. R. Rocha, and S. Sanvito, Effect of the continuity of the π-conjugation on the conductance of ruthenium-octene-ruthenium molecular junctions, *J. Chem. Phys.* **126**, 174706 (2007).
45. S. M. Hou, Y. Q. Chen, X. Shen et al., High transmission in ruthenium-benzene-ruthenium molecular junctions, *Chem. Phys.* **354**, 106–111 (2008).
46. T.-S. Choy, J. Naset, J. Chen, S. Hershfield, and C. Stanton, A database of Fermi surface in virtual reality modeling language (vrml), *Bull. Am. Phys. Soc.* **45**, L36–L42 (2000).
47. A. R. Rocha and S. Sanvito, Resonant magnetoresistance in organic spin valves, J. *Appl. Phys.* **101**, 09B102 (2007).
48. A. J. Heinrich, J. A. Gupta, C. P. Lutz, and D. M. Eigler, Single-atom spin-flip spectroscopy, *Science* **306**, 466–469 (2004).
49. A. Troisi and M. A. Ratner, Molecular transport junctions: Propensity rules for inelastic electron tunneling spectra, *Nano Lett.* **6**, 17841788 (2006).
50. M. Paulsson, T. Frederiksen, H. Ueba, N. Lorente, and M. Brandbyge, Unified description of inelastic propensity rules for electron transport through nanoscale junctions, *Phys. Rev. Lett.* **100**, 226604 (2008).
51. N. Jean and S. Sanvito, Inelastic transport in molecular spin valves, *Phys. Rev. B* **73**, 094433 (2006).
52. H. Nakamura, K. Yamashita, A. R. Rocha, and S. Sanvito, Efficient ab initio method for inelastic transport in nanoscale devices: Analysis of inelastic electron tunneling spectroscopy, *Phys. Rev. B* **78**, 235420 (2008).
53. C. F. Hirjibehedin, C. P. Lutz, and A. J. Heinrich, Spin coupling in engineered atomic structures, *Science* **312**, 1021–1024 (2006).
54. C. F. Hirjibehedin, C.-Y. Lin, A. F. Otte et al., Large magnetic anisotropy of a single atomic spin embedded in a surface molecular network, *Science* **317**, 1199–1203 (2007).
55. X. Chen, Y.-S. Fu, S.-H. Ji et al., Probing superexchange interaction in molecular magnets by spin-flip spectroscopy and microscopy, *Phys. Rev. Lett.* **101**, 197208 (2008).
56. Y.-S. Fu, T. Zhang, S.-H. Ji et al., Identifying charge states of molecules with spin-flip spectroscopy, *Phys. Rev. Lett.* **103**, 257202 (2009).
57. G. D. Scott and D. Natelson, Kondo resonances in molecular devices, *ACS Nano* **4**, 3560–3579 (2010).
58. A. Zhao, Q. Li, L. Chen et al., Controlling the kondo effect of an adsorbed magnetic ion through its chemical bonding, *Science* **309**, 1542–1544 (2005).
59. P. Lucignano, R. Mazzarello, A. Smogunov, M. Fabrizio, and E. Tosatti, Kondo conductance in atomic nanocontact from first principles, *Nat. Mater.* **8**, 563–567 (2009).
60. R. Wiesendanger, Spin mapping at the nanoscale and atomic scale, *Rev. Mod. Phys.* **81**, 1495–1550 (2009).
61. F. Meier, L. Zhou, J. Wiebe, and R. Wiesendanger, The magnetization of individual adatoms, *Science* **320**, 8286 (2008).
62. K. von Bergmann, M. Bode, A. Kubetzka, M. Heide, S. Blügel, and R. Wiesendanger, Spin-polarized electron scattering at single oxygen adsorbates on a magnetic surface, *Phys. Rev. Lett.* **92**, 046801 (2004).
63. S. Voss, M. Fonin, U. Rüdiger, M. Burgert, U. Groth, and Yu. S. Dedkov, Electronic structure of Mn_{12} derivatives on the clean and functionalized Au surface, *Phys. Rev. B* **75**, 045102 (2007).
64. M. Mannini, P. Sainctavit, R. Sessoli, C. Cartier dit Moulin, F. Pineider, M.-A. Arrio, A. Cornia, and D. Gatteschi, XAS and XMCD investigation of Mn_{12} monolayers on gold, *Chem. Eur. J.* **14**, 7530–7535 (2008).
65. M. Mannini, F. Pineider, P. Sainctavit et al., Magnetic memory of a single-molecule quantum magnet wired to a gold surface, *Nat. Mater.* **8**, 194–197 (2009).
66. S. Barraza-Lopez, K. Park, V. García-Suárez, and J. Ferrer, First-principles study of electron transport through the single-molecule magnet Mn_{12}, *Phys. Rev. Lett.* **102**, 246801 (2009).
67. A. Soncini and L. F. Chibotaru, Toroidal magnetic states in molecular wheels: Interplay between isotropic exchange interactions and local magnetic anisotropy, *Phys. Rev. B* **77**, 220406(R) (2008).
68. A. Soncini and L. F. Chibotaru, Molecular spintronics using noncollinear magnetic molecules, *Phys. Rev. B* **81**, 132403 (2010).
69. C. D. Pemmaraju, I. Rungger, and S. Sanvito, Ab initio calculation of the bias-dependent transport properties of Mn_{12} molecules, *Phys. Rev. B* **80**, 104422 (2009).
70. H. B. Heersche, Z. de Groot, J. A. Folk et al., Electron transport through single Mn_{12} molecular magnets, *Phys. Rev. Lett.* **96**, 206801 (2006).
71. M.-H. Jo, J. E. Grose, K. Baheti, M. M. Deshmukh, J. J. Sokol, E. M. Rumberger, D. N. Hendrickson, J. R. Long, H. Park, and D. C. Ralph, Signatures of molecular magnetism in single-molecule transport spectroscopy, *Nano Lett.* **6**, 2014–2020 (2006).
72. C. Romeike, M. R. Wegewijs, and H. Schoeller, Spin quantum tunneling in single molecular magnets: Fingerprints in transport spectroscopy of current and noise, *Phys. Rev. Lett.* **96**, 196805 (2006).
73. C. Romeike, M. R. Wegewijs, M. Ruben, W. Wenzel, and H. Schoeller, Charge-switchable molecular magnet and spin blockade of tunneling, *Phys. Rev. B* **75**, 064404 (2007).
74. M. N. Leuenberger and D. Loss, Quantum computing in molecular magnets, *Nature* **410**, 789–793 (2001).
75. G. A. Timco, S. Carretta, F. Troiani et al., Engineering the coupling between molecular spin qubits by coordination chemistry, *Nat. Nanotechnol.* **4**, 173–178 (2009).
76. J. Lehmann, A. Gaita-Ariño, E. Coronado, and D. Loss, Spin qubits with electrically gated polyoxometalate molecules, *Nat. Nanotechnol.* **2**, 312–317 (2007).
77. P. Gambardella, S. Stepanow, A. Dmitriev et al., Supramolecular control of the magnetic anisotropy in two-dimensional high-spin Fe arrays at a metal interface, *Nat. Mater.* **8**, 189–193 (2009).
78. M. Diefenbach and K. S. Kim, Towards molecular magnetic switching with an electric bias, *Angew. Chem. Int. Ed.* **46**, 77847787 (2007).

79. N. Baadji, M. Piacenza, T. Tugsuz, F. Della Sala, G. Maruccio, and S. Sanvito, Electrostatic spin crossover effect in polar magnetic molecules, *Nat. Mater.* **8**, 813–817 (2009).
80. S. K. Shukla and S. Sanvito, Electron transport across electrically switchable magnetic molecules, *Phys. Rev. B* **80**, 184429 (2009).
81. E. A. Osorio, K. Moth-Poulsen, H. S. J. van der Zant et al., Electrical manipulation of spin states in a single electrostatically gated transition-metal complex, *Nano Lett.* **10**, 105–110 (2009).
82. M. Misiorny and J. Barnaś, Effects of intrinsic spin-relaxation in molecular magnets on current-induced magnetic switching, *Phys. Rev. B* **77**, 172414 (2008).

VI

Applications

34

Magnetoresistive Sensors Based on Magnetic Tunneling Junctions

Gang Xiao
Brown University

34.1 Introduction

The age of electronically based semiconductor devices has been with us for six decades. With more and more electronic devices being packed into smaller and smaller spaces, the limit of physical space may prevent further expansion in the direction the microelectronics industry is currently going. However, a new breed of electronics, dubbed "spintronics" [1,2],* may alleviate the challenge. Instead of relying on the electron's charge to manipulate electron motion or to store information, spintronics devices would further rely on the electron's spin or its magnetic moment. In spintronics, electrons in a device can be easily manipulated by internal and external magnetic fields or spin-polarized electrical currents. The advantage of spin-based electronics or spintronics is that they are nonvolatile [1] compared to charge-based electronics, and quantum mechanical computing based on spintronics could achieve speeds unheard of with conventional computing. Spintronics will also lead to the emergence of ultrasensitive magnetic sensors, creating new and exciting applications.

Nearly every physical object generates magnetic fields, some strong and some extremely weak. A human heart generates picotesla-scale (10^{-12} T) magnetic pulses, revealing critical cardiac information. A spinning disk inside a hard drive emits magnetic signals with frequency approaching 1 GHz, making the information age possible. The Earth's magnetic field can be a useful navigation tool, particularly where global positioning systems (GPS) are not accessible (e.g., underground and deep sea). Magnetic sensors have been used pervasively in industrial and consumer products [3]. Ultrasensitive magnetic sensors find increasing utility in a number of emerging applications [3]. Magnetocardiography (MCG) [4] uses magnetic sensors to measure the weak electrical signals from the beating heart, allowing the diagnostics of cardiac functions. Magnetoencephalography (MEG) [5], on the other hand, is the magnetic measurement of the electrical activities in the brain. The information obtained from MEG can be used to pinpoint problem regions in the brain of a patient to minimize the invasiveness of a brain surgery. Ultrasensitive magnetic sensors used in MCG and MEG are expensive superconducting quantum interference devices (SQUIDs), which require low-temperature operation.

Thin-film solid-state magnetic sensors enjoy the advantages of small physical size and low cost, though they lack the sensitivity of SQUIDs. Significant progress has been made to improve the sensitivity of thin-film sensors. In data storage application, magnetic sensors have allowed computers to store massive amount of information cheaply and reliably [6]. Magnetic sensors are quintessential in the emerging technology of magnetic random access memory (MRAM) [1].† For semiconductor manufacturers, magnetic sensors have made the diagnostics of integrated circuits (ICs) possible by sensing the high-frequency magnetic fields emitted from transistors and interconnects [7–10]. Magnetic sensors are able to detect cracks below the surface of an aircraft or engine turbine by sensing electromagnetic waves originated from the defective area. Ultrasensitive magnetic sensors also have applications in a number of defense-related applications,

* See Chapter 1 of this book.

† See Chapter 35 of this book.

such as in the detection of submerged submarines or as part of a distributed sensor network [11] ("smart dust") with the eventual goal of monitoring a battlefield remotely. Magnetic sensors are an enabling technology in many areas of science and engineering.

In this chapter, the development of magnetic sensors based on magnetic tunneling junctions (MTJs) is reviewed. Older sensing technologies suffer from varying deficiencies such as large size, poor sensitivity, and slow speed. MTJ is a disruptive sensing device. It is a tiny "dot" on silicon, invisible to the naked eye, and can be integrated with other thin-film devices such as an IC. The physics of magnetic quantum tunneling [12] dictates that MTJ can function up to multiple GHz. MTJ is truly a nature's gift containing treasure trove of rich physics and promises of applications. So much so that shortly after its discovery, MTJ has taken a large market share in the read/write heads of the data storage industry. The large magnitude of the magnetoresistance (MR) effect in MTJ has created an excellent opportunity for the creation of a new class of high-performance sensor devices.

34.1.1 Magnetic Sensor Types

It is helpful to put MTJ sensors in the context of other available magnetic sensors. There are two broad categories of magnetic sensors: one based on thin-film technology and the other either not based on thin-film technology or requiring special operating conditions [13–16]. Under each category, there are numerous types of magnetic sensors. The two most common in the latter category are SQUIDs and fluxgates. SQUIDs require cryogenic system to be functional, while fluxgates are expensive to manufacture. Both suffer the disadvantages of high prices, large power consumption, and relatively large physical sizes. SQUIDs are extremely sensitive, capable of sensing field strength on the order of femtotesla. Low-temperature SQUIDs work at 4.2 K, whereas high-temperature SQUIDs typically function at 77 K. Each SQUID can cost thousands of dollars or more. Fluxgates consist of two coils of wires wrapped around a soft magnetic cylindrical core. By driving these coils in a particular way and using the resulting magnetic field to saturate the core, the strength of the external magnetic field can be accurately measured. Fluxgates are sensitive, capable of detecting field on the order of 10–100 $pT/Hz^{1/2}$ over a limited dynamic range. Typical cost of a fluxgate is about $1000.

Sensors that can be fabricated based on standard thin-film methods can be made in large quantities and are generally priced at less than $10 each. Sensors in this category are less sensitive than SQUIDs and fluxgates. However, they enjoy the advantages of small size and low cost. They consume less power than SQUIDs and fluxgates. Thin-film-based sensors include Hall-effect (HE), anisotropic magnetoresistance (AMR), giant magnetoresistance (GMR), giant magnetoinductance (GMI), and MTJ sensors.

HE sensors [17] are based on the physics of Lorentz force on an electron inside a magnetic field. They have a simple structure and are easy to manufacture. A typical Hall sensor is very linear within a wide dynamic range, has low sensitivity and moderate to high power consumption, and is among the least expensive thin-film-based sensors. Hall sensors account for 98.5% of all magnetic sensors sold worldwide [3].

AMR sensors [13] are based on the physics of electron spin–orbit coupling inside a soft ferromagnetic metal. They use a small resistivity change (~2%–3%) experienced by certain magnetic materials in the presence of an applied field. AMR sensors are moderately priced, have good sensitivity, and low dynamic range.

GMR [6,18,19]* sensors are based on the spin-dependent scattering of electrons by the interfaces inside a ferromagnetic and metallic multilayer, first discovered in 1988 [18,19]. The maximum useful MR in commercial GMR sensors is ~10%–20%. GMR sensors are more sensitive than Hall sensors and cost about the same as AMR sensors.

GMI sensors [20] are based on a large change in impedance of an amorphous metal film inside a magnetic field. It is a relatively new sensor technology. GMI sensors require complicated electronics to detect the magnetic field. GMI sensor development is still in an early stage, and the sensors do not seem to provide a compelling advantage over MTJ sensor technology.

34.1.2 Magnetic Tunneling Junction Sensors

MTJ sensors represent the latest magnetic sensor technology with several intrinsic advantages. Because these devices rely on tunneling processes, they have a magnetoresistive response, which is fundamentally larger than those of AMR and GMR sensors. The use of a properly fabricated MgO tunnel barrier increases the device MR to over 200%. This leads to a 10-fold increase in device voltage response. In addition, because they are devices where the current flows perpendicular to the film plane, their resistance can be varied by five to six orders of magnitude by simply varying the thickness of the insulating barrier layer. These allow MTJ sensors to have a larger response, higher sensitivity, and lower power consumption than HE, AMR, and GMR sensors.

When subjected to an external magnetic field, certain materials undergo a change in their electrical resistivity. The relative resistance change is called magnetoresistance (MR), defined as

$$\frac{\Delta R}{R} = \frac{R(H) - R(0)}{R_S} \tag{34.1}$$

where $R(H)$, $R(0)$, and R_S are the resistance values, at field H, at zero field, and at saturation field, respectively. The MR property has been used in magnetoresistive sensors. The field of spintronics was given a huge boost in the late 1980s, when a GMR effect

* See Chapters 4 and 5 of this book.

was discovered in layered structures [18,19]. The commercially developed GMR spin valve, a trilayer structure (NiFe/Cu/Co), first commercialized by IBM around 1997, has been widely used as the sensing element in the read/write heads of computer hard disks [6]. The two scientists who discovered the GMR effect received the 2007 Nobel Prize in physics [21].

The discovery of GMR sparked the search for new devices that could exploit the spin of the electron. The year 1995 saw the introduction of a new type of MR called tunneling magnetoresistance (TMR) [22,23].* TMR occurs in MTJ trilayer structures composed of two ferromagnetic layers separated by a thin insulating barrier (FM/I/FM) [22,23]. The resistance of an MTJ is a function of the tunneling probability, which depends on the relative orientations of the magnetization vectors (**M**) of the magnetic layers [12]. When the **M** vectors are parallel, there is a maximal match between the numbers of occupied electron states in one electrode and available states in the other. Thus, the electron tunneling probability is maximized and the tunneling resistance (R) is at a minimum. In the antiparallel configuration, electron tunneling occurs between majority states in one electrode and minority states in the other. This mismatch reduces tunneling, causing R to increase. The maximum MR of an MTJ is given by the Julliere model [24]

$$\mathrm{MR} = \frac{R_{\uparrow\downarrow} - R_{\uparrow\uparrow}}{R_{\uparrow\uparrow}} = \frac{2P_1P_2}{1 - P_1P_2} \tag{34.2}$$

where

$R_{\uparrow\downarrow}$ ($R_{\uparrow\uparrow}$) is the resistance in the antiparallel (parallel) configuration

P_1 and P_2 are the spin polarizations of the ferromagnets (two electrodes in an MTJ)

In 2001, theoretical work predicted the possibility of extremely large MR values in certain new trilayer structures, which used MgO as the barrier material [25,26].† Due to a coherent tunneling effect, it was calculated that a properly prepared MTJ structure using an MgO barrier might exhibit up to 5000% MR. In 2004, two groups independently reported record MR ratios of over 100% in MgO-based MTJ devices [27,28].‡

In the case of Al_2O_3-based junctions, the only conditions required for the barrier were that it be sufficiently nonconductive. In contrast, MgO-based junctions rely on the quantum mechanical properties of the MgO layer for the realization of ultrahigh MR. Because of this, creating successful tunnel junctions using MgO as the barrier layer is significantly more complicated. One major reason for this is that the coherent tunneling processes required for ultrahigh MR can only occur if the MgO layer has a certain crystal orientation. Achieving this crystal orientation requires tight control of deposition and annealing parameters, and only a handful of groups worldwide have reported success in fabricating MgO-based tunnel junction devices.

Figure 34.1 is a schematic of a typical MTJ layer structure. The two FM electrodes are fabricated from CoFeB, and the insulating barrier is composed of MgO. The remaining layers in the structure are chosen to enhance the material and magnetic characteristics of the device. Typically, in order to achieve a linear, bipolar operation, one of the two magnetic electrodes (the "pinned layer") in each sensor has its magnetization fixed by the exchange-biasing phenomenon. In Figure 34.1, the bottom FM layer is the pinned layer whose magnetization is fixed by the adjacent IrMn antiferromagnetic layer. The other magnetic layer, the top FM layer in Figure 34.1, is designed to "freely" respond to an external magnetic field. This magnetic top layer is often called the "free layer." The resulting structure has an electrical resistance that varies linearly as a function of the magnetic field strength over a substantial field range. These devices can be fabricated using conventional semiconductor methods, making them potentially very cheap in large quantities. In addition, because of their two-terminal nature, they are customizable and easy to use.

MTJ magnetic sensors have been commercialized as read head in hard disk drives and are the most important application of magnetic sensors [29,30]. The first MTJ heads based on TiO tunnel barrier were introduced by Seagate in 2004, soon followed by AlO-based MTJ heads by TDK and other manufacturers. Since 2007, these first-generation MTJ heads have been replaced by MgO-based MTJ heads.

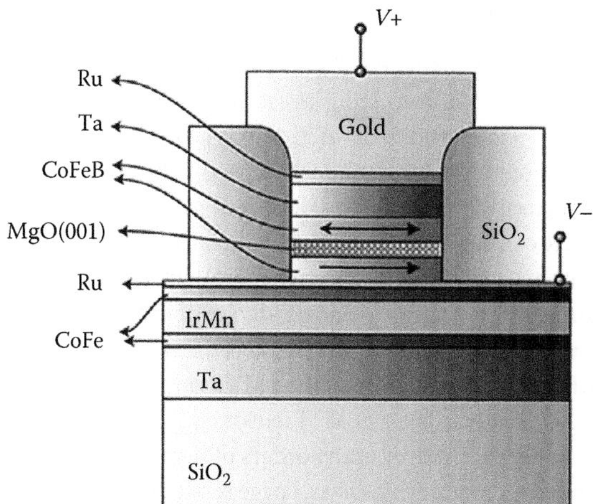

FIGURE 34.1 Schematic of a typical MTJ multilayer structure. The CoFeB layer on top of MgO barrier is the "free layer," and the CoFeB layer below the MgO is the pinned layer whose magnetization is fixed by the antiferromagnetic IrMn layer. In a sensor, the magnetization of the free layer is set to be orthogonal to that of the pinned layer at zero external field. In a memory element, the two magnetization vectors are designed to be parallel as shown.

* See Chapter 9 of this book.
† See Chapter 11 of this book.
‡ See Chapter 10 of this book.

34.2 Development of Magnetic Tunneling Junction Sensors

34.2.1 Basic Performance Parameters of MTJ Sensors

The performance parameters used to characterize any magnetic sensor are also applicable to MTJ sensors, although the latter have some unique parameters. What distinguishes MTJ sensors from other types of thin-film-based sensors is the parameter of "sensitivity," which is the largest for MTJ sensors. For an MTJ resistor sensor, the sensitivity is defined as the resistance change (%) per unit field Oe (%/Oe) at zero field. For an MTJ bridge sensor, sensitivity takes the form of mV/V/Oe (voltage output per one volt of applied voltage per Oe). Both numbers can be converted from one to another.

The sensitivity of an MTJ sensor depends on the externally applied voltage (bias) to the sensor, as a high bias voltage will lead to a lower MR due to tunneling physics. Often, an MTJ sensor consists of many MTJ elements connected in series, so that the total applied voltage is divided to only a small voltage to a single element. This allows the MTJ sensor to operate under a large applied voltage. This method also enhances the protection of MTJ sensors against electrostatic discharge (ESD).

The "detectability" of a magnetic sensor is the intrinsic magnetic noise that is dependent on frequency. For low-frequency applications, the detectability has the units of $nT/Hz^{1/2}$ at $f = 1$ Hz. Typical MTJ sensors have detectability of 1–10 nT, and the goal in the research community is to lower it to a few pT.

A linear and bipolar magnetic sensor is true only within a field "dynamic range" (±Oe). In general, the push for high sensitivity leads to the narrowing of dynamic range. A magnetic sensor should have as low a "hysteresis" as possible. Hysteresis is in general measured as a percentage of the dynamic range.

The "thermal coefficients" are resulted from the temperature dependence of the tunneling resistance (thermal coefficient of resistance, TCR) and that of the magnetoresistance (thermal coefficient of magnetoresistance [TCMR]). The effect of TCR can be compensated by using the design of a bridge, but the effect of TCMR cannot. One needs to rely on electronics and calibration data to take TCMR into consideration. At very high temperatures, MTJs can break down due to interlayer diffusion or decaying magnetization of the free layer. It is useful to know the operating temperature range of an MTJ sensor.

One of the advantages of MTJ sensors is that their "resistance" can be easily varied over many orders of magnitude by adjusting the tunneling barrier thickness. Large resistance is important for applications requiring low power consumption (mobile devices), whereas low resistance benefits high frequency (MRAM, read/write head) or low-noise applications.

Another advantage of MTJ sensors is that they can be made into very small "physical size." The size is defined by the area or volume of the free layer, which can have dimensions as small as tens of nanometers. As the resistance is concentrated near the barrier, the lead resistances can often be ignored. Nanoscaled MTJ sensors benefit the applications of scanning magnetic microscopy [7–9,31]. Here, a small size means a high spatial resolution in field measurement.

MTJ sensors enjoy a broad "frequency response." An MTJ sensor is always a good sensor for dc field. Intrinsically, the energy scale of magnetic tunneling is set by the Fermi energy, and large MR effect remains intact in broad frequency range. A well-designed MTJ sensor can be ultrafast in response. There are two frequency limits. One is the ferromagnetic resonance frequency of the free layer, which can be as high as tens of GHz. The other is the *RC* time constant of the MTJ sensor. Properly designed, the upper limit of an MTJ sensor is many GHz.

34.2.2 Physical Model of MTJ Sensor

An MTJ stack is a multilayer deposited on thermally oxidized silicon substrates using dc and rf sputtering. A representative layer structure has the sequence of (thickness in nm) 5 Ta/30 Ru/5 Ta/2 CoFe/12 IrMn/2 CoFe/1 Ru/3 CoFeB/2 MgO/3 CoFeB/2 Ta/ 2 Ru (see Figure 34.1). The 12 IrMn/2 CoFe/1 Ru/3 CoFeB layer assembly just below the MgO barrier is a synthetic antiferromagnetic (SAF) pinned layer. The net magnetization vector of the SAF layer is fixed in the plane of the layer, and only weekly perturbed by an external field. The CoFeB layer above MgO barrier is patterned into an oval shape having a certain aspect ratios and dimensions from tens of nanometers to tens of micrometers.

To make a sensor, the pinning direction of the pinned layer is set to be perpendicular to the easy axis of the free layer. The external field to be sensed is applied along the hard axis of the free layer, causing the magnetization vector of the free layer to rotate a small angle from the quiescent state. As a consequence, the MTJ sensor experiences a change in resistance, which is proportional to the external field in a bipolar fashion. In an ideal case, the free layer should behave like a single domain "particle" with a uniaxial magnetic anisotropy having a very small anisotropy constant and a very small magnetic hysteresis. The free layer should be magnetically "decoupled" from the pinned layer. Also, the pinning force should be very large so that the pinning direction will not be affected by the external field. The objective in the development of MTJ sensors is to achieve conditions as close to the ideal sensor as possible by performing research on magnetic materials, layer structures and lateral geometries, and fabrication processing.

34.2.3 Magnetoresistance Ratio and Saturation Field: Keys to Large Sensitivity

For an ideal sensor, the field sensitivity is given by

$$S = \frac{R_{\uparrow\downarrow} - R_{\uparrow\uparrow}}{R_{\uparrow\downarrow} + R_{\uparrow\uparrow}} \cdot \frac{1}{B_{sat}} \qquad (34.3)$$

where B_{sat} is the magnetic saturation field of the free layer. Note that the MR ratio, $2(R_{\uparrow\downarrow} - R_{\uparrow\uparrow})/(R_{\uparrow\downarrow} + R_{\uparrow\uparrow})$, is different from the commonly used MR ratio in the literature, $(R_{\uparrow\downarrow} - R_{\uparrow\uparrow})/R_{\uparrow\uparrow}$. According to Equation 34.3, the key to increase the field sensitivity is to obtain the largest possible MR and to reduce B_{sat}, which can be achieved by using available methods in micromagnetics. Obtaining large MR requires the discovery of new materials and new physics.

Since the discovery of MTJs at room temperature in 1995, there have been continuing efforts in developing MTJ magnetic sensors. The gradual increase in MR has benefited sensor's performance. Earlier on, most research was focused on MTJs with Al_2O_3 insulating tunneling barrier. Typically, an Al sputtering target was used and deposited in an oxidizing environment of Ar and oxygen plasma. The amorphous Al_2O_3 barrier thus formed is dense and of high quality. Between 1995 and 2004, the MR ratio steadily increased from about 10% to 70% at room temperature [22,23,32]. Figure 34.2 shows some sensor parameters of a series of *FeNi/*Al_2O_3*/FeNi* MTJs after a systematic thermal annealing treatment [33]. After an optimal annealing at 170°C, the maximum MR reaches about 35% with a maximal field sensitivity of 5%/Oe. The junction area (free layer) is $100 \times 150\,\mu m^2$. In a CoFe/Al_2O_3/NiFeCo junction with a total MR of 20%, a sensitivity of 3%/Oe was obtained by applying a biasing field along the hard axis of the free layer [34].

The Al_2O_3-based MTJs typically have polycrystalline structure and the Al_2O_3 barrier is amorphous. Electrons with different wavefunction symmetries are allowed to tunnel, and the overall spin polarization is limited. To increase MR, it was concluded theoretically that single crystal MTJ structures can enhance MR substantially [25,26]. This is because tunneling efficiency depends on wavefunction symmetry. Some particular bands having high spin polarization can lead to very high MR if these electrons become the dominant contributors to tunneling in certain crystalline orientation. Soon, the experiments on epitaxial [001] Fe/MgO/Fe and FeCoB/MgO/FeCoB MTJs confirmed the theoretical conclusions [27,28,35]. Today, MR has reached over 600% to as high as 1056% in this type of [001] MgO-based MTJs [36,37], which have become the primary MTJs for sensor development. With high MR ratios, the sensitivity of MTJ sensors has also been steadily increased.

Equally important, reducing the saturation field B_{sat} will also increase the sensitivity, according to Equation 34.3. Fe and CoFeB are not very soft ferromagnets, but they are required to form the MTJ free layer because of their special band structures. However, one can deposit a thicker ferromagnetic soft film on top of the CoFeB free layer to make a compound free layer. B_{sat} can be reduced because it is now dominated by the thicker "soft" film. Recently, Egelhoff et al. [38] have used a 100 nm of $Ni_{77}Fe_{14}Cu_5Mo_4$ as the thicker "soft" film on top of the 5 nm CoFeB free layer of a working MgO-based MTJ. The low-field hysteresis loop was found to be linear, nonhysteretical, and

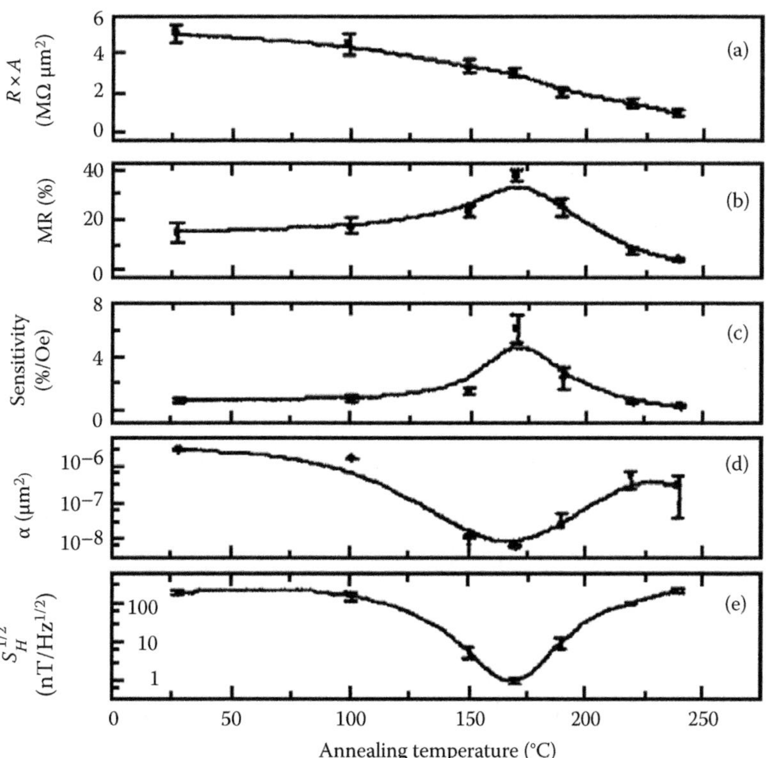

FIGURE 34.2 Annealing temperature dependence of (a) junction resistance, (b) TMR ratio, (c) sensitivity, (d) normalized $1/f$ noise, and (e) magnetic field sensitivity. (Reprinted with permission from Liu, X. and Xiao, G., Thermal annealing effects on low-frequency noise and transfer behavior in magnetic tunnel junction sensors, *J. Appl. Phys.*, 94, 6218, 2003, Fig. 2. Copyright 2003, American Institute of Physics.)

extrapolates to saturation at 0.8 Oe. The resulting sensitivity is about 70%/Oe, which is the highest sensitivity ever reported for any MTJ structure on the wafer level. Processing the MTJ structure into smaller sensor sizes will increase the B_{sat} somewhat due to shape anisotropy. Still, this extremely high sensitivity sets a limit for the MTJ sensors.

There is another approach to increase the sensitivity by incorporating a flux concentrator near the free layer. A flux concentrator is a soft ferromagnetic film with a high magnetic susceptibility and with certain geometry. By focusing the external magnetic flux toward the free layer, the magnetic field can be increased by a "gain" factor, and so can be the sensitivity. Chaves et al. demonstrated this approach by adding a 500 nm thick $Co_{93}Zr_3Nb_4$ flux concentrators with different shapes to CoFeB/MgO/CoFeB MTJ sensors [39,40]. They were able to increase the sensitivity substantially, with some hysteresis even under a hard-axis biasing field. One disadvantage of this method is that depositing and patterning the flux concentrators increases manufacturing cost and the physical size of the final sensor products. For this reason, sensors with flux concentrators are not preferred in applications where high spatial resolution is required, such as in scanning magnetic microscopy.

34.2.4 Noise Characteristics and Field Detectability

The field detectability of a magnetic sensor is determined by the intrinsic noise of the MTJ. Noise characterization can reveal physical behaviors that may be difficult to detect in other experiments. For both sensing and MRAM applications, it is important to understand noise in relation to signal levels. Earlier on, Ingvarsson et al. [41,42] measured the bias and magnetic field dependence of voltage noise in micron-scale Al_2O_3-based MTJs. At low frequency (f), they found the $1/f$ noise has two sources: one due to the charge traps in the tunnel barrier and the other due to quasi-equilibrium thermal magnetization fluctuation in the ferromagnetic electrode. The $1/f$ noise was shown to be proportional to the slope dR/dH, tracking the dc susceptibility for the top free layer [42]. This finding provides a good quantitative test of the validity of the fluctuation–dissipation theorem. It was concluded that to curtain the low-frequency noise in MTJs, the thermal stability of the magnetization must be enhanced by subduing the activated motion of domain walls. Nowak et al. [43] studied the low-frequency noise in similar junctions with large areas and attributed the $1/f$ noise only to charge trapping processes.

The field detectability is dependent on two quantities: the signal, expressed typically in change of output voltage per unit of magnetic field (V/Oe), and the noise, expressed as the noise voltage per unit frequency (V/Hz$^{1/2}$). The field detectability of any sensor is simply given by the ratio of these two outputs.

To obtain good detectability, one needs to minimize the noise of a sensor. Many types of noise inhibit the field-sensing capability of magnetic sensors. $1/f$ noise, intrinsic to sensors and amplifiers, is typically dominant at low frequencies. Beyond a knee frequency, white frequency noises dominate. Johnson–Nyquist noise is intrinsic and dependent on sensor resistance and temperature; shot noise is caused by current fluctuations and scales with current. In addition, sensors can have Barkhausen noise caused by domain wall motion. Mathematically, a typical noise spectrum in MTJs can be described as

$$S_V(f) = \alpha\frac{V^2}{Af} + 4kRT + 2eVR \tag{34.4}$$

where

- V and R are the voltage across the sensor and the sensor resistance, respectively
- A is the junction area
- α is a constant (Hooge parameter) based on material parameters
- k is the Boltzmann constant
- e is the electron charge

The first term represents a sensor's intrinsic $1/f$ noise and the second and third terms are the Johnson and shot noises, respectively. Once the voltage noise is experimentally measured, the magnetic-field noise (detectability) can be calculated from [44]

$$S_H = \left(\frac{\Delta H}{\Delta V}\right)^2 S_V = \frac{1}{\left(\frac{1}{R}\cdot\frac{\Delta R}{\Delta H}\right)^2}\cdot\frac{S_V}{V^2} \tag{34.5}$$

The term $(1/R)\cdot(\Delta R/\Delta H)$ is simply the sensitivity of a sensor (measured in relative change of resistance per Oe).

Ren et al. [44] studied in detail the low-frequency noise in $FeNi/Al_2O_3/FeNi$ MTJ sensors. They confirmed that the $1/f$ noise component scales with sensor bias voltage according to Equation 34.4, as shown in Figure 34.3. They further confirmed that the $1/f$ noise is dominated by magnetic noise. A strong correlation between the $1/f$ noise and the sensitivity $(1/R)\cdot(\Delta R/\Delta H)$ was observed. Figure 34.4 shows the transfer curve (R vs. H) and the associated sensitivity variable as a function of H. It also shows that $1/f$ noise level at $f = 1$ Hz as a function H. At fields where the sensor is most sensitive, so is the $1/f$. Figure 34.5 shows the normalized noise value versus the sensitivity. A linear correlation was observed. This was explained by the thermally activated magnetization fluctuations in the free layer. It can be concluded that a large MR ratio or sensitivity would increase $1/f$ magnetic noise, and, therefore, would not enhance field detectability as described by Equation 34.5. The only way to lower the $1/f$ magnetic noise is to reduce the magnetization fluctuations. The slope in the linear correlation in Figure 34.5 is a measure of the strength of magnetization fluctuations in the free layer. The smaller the slope, the lower the $1/f$ magnetic noise. The noise component at $(1/R)\cdot(\Delta R/\Delta H) = 0$ (the intercept on the vertical axis in Figure 34.5) is resulted from the nonmagnetic

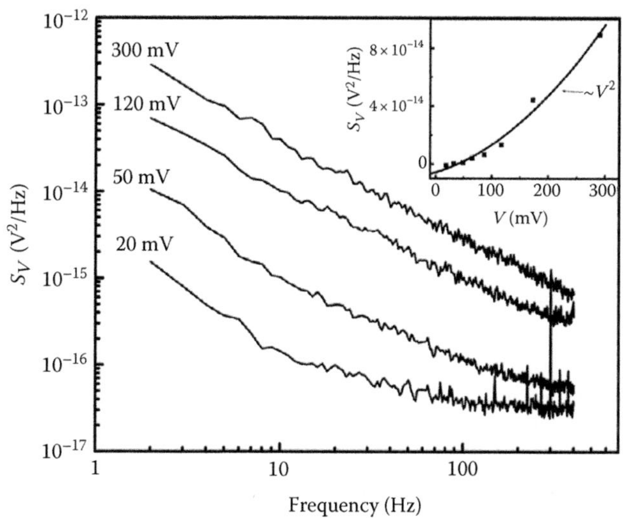

FIGURE 34.3 Voltage power spectral density S_V in frequency domain under different bias voltage V on an MTJ sensor. Inset is S_V measured at frequency f = 10 Hz as a function of bias voltage. (Reprinted with permission from Ren, C., Liu, X., Schrag, B., and Xiao, G., *Phys. Rev. B*, 69, 104405, 2004. Copyright 2004 by the American Physical Society.)

contribution to 1/*f* noise. This component is attributed to defects and localized charge traps in the barrier.

Liu and Xiao [33] found that proper thermal annealing on *FeNi*/Al_2O_3/*FeNi* MTJs reduces 1/*f* magnetic noise as shown in Figure 34.2. By varying the annealing temperature up to 250°C, they found that at 170°C, MR doubles to 35% from 17% without annealing. Most importantly, the field detectability is improved by two orders of magnitude, from over 100 nT/Hz$^{1/2}$ to 1 nT/Hz$^{1/2}$ at 1 Hz. Liou et al. [45] have also found that high-temperature annealing in hydrogen gas can reduce the 1/*f* noise by about 50% in Al_2O_3-based MTJ sensors.

The 1/*f* characteristics in MgO-based MTJ sensors are similar to that observed in the Al_2O_3-based MTJ sensors. The voltage noise was again found to be linear in the sensitivity of a sensor [46]. Even if the sensitivity is increased, the increased noise will cancel out any gains in sensor performance.

The figure of 1 nT/Hz$^{1/2}$ at f = 1 Hz for MTJ sensors is similar to the detectability of commercial AMR and GMR sensors. One major difference is that commercial AMR and GMR sensors are equipped with flux concentrators, but not in the MTJ sensors studied. Stutzke et al. [47] surveyed the low-frequency noise on commercial sensors. Most of the GMR and AMR sensors with flux concentrators have a noise figure of 3–10 nT/Hz$^{1/2}$ at 1 Hz. The best AMR sensor has a noise level of 0.3 nT/Hz$^{1/2}$ at 1 Hz. This further demonstrates that a better sensitivity in MTJ sensors than in GMR and AMR sensors does not mean better low-frequency noise level. The only way to reduce the low-frequency noise in magnetoresistive sensors is to enhance thermal magnetic stability of the free layer. It is beneficial to use a ferromagnetic material with a large magnetic exchange constant, and to configure a stable and uniform single domain structure free of minor domains and excessive domain wall motions. It is also useful to make high-quality tunnel barrier

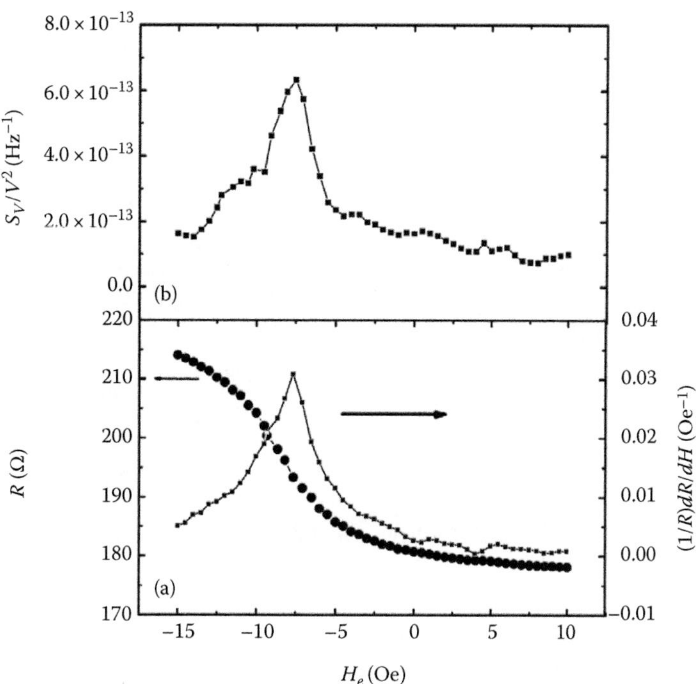

FIGURE 34.4 The voltage-normalized noise in field spectra (a) and the corresponding $R(H)$ curve and its derivative (b) for a 36 × 24 μm² MTJ with biasing field of 8 Oe. The noise value is taken from extrapolating low-frequency noise to 1 Hz. (Reprinted with permission from Ren, C., Liu, X., Schrag, B., and Xiao, G., *Phys. Rev. B*, 69, 104405, 2004. Copyright 2004 by the American Physical Society.)

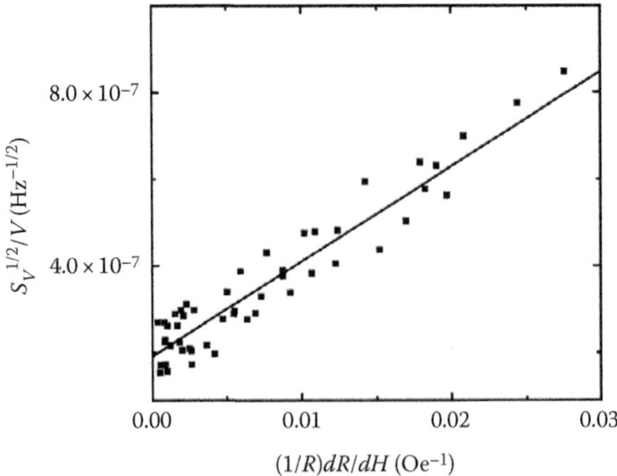

FIGURE 34.5 The voltage noise $S_V^{1/2}/V$ versus the sensitivity (derivative of MR) in an MTJ sensor. (Reprinted with permission from Ren, C., Liu, X., Schrag, B., and Xiao, G., *Phys. Rev. B*, 69, 104405, 2004. Copyright 2004 by the American Physical Society.)

free of detects and charge trappings to reduce the nonmagnetic $1/f$ noise level.

Sensitivity and $1/f$ noise are strongly dependent on biasing magnetic fields applied along the easy and hard axis of the free layer. Mazumdar et al. [48] have developed a characterization procedure to map the sensitivity, $1/f$ voltage noise and field noise, in an orthogonal magnetic field arrangement (±100 Oe in both axes), as shown in Figure 34.6, on CoFeB/MgO/CoFeB MTJ sensors. The normalized voltage noise map (in units of V/Hz$^{1/2}$ at 100 Hz) of Figure 34.6b resembles the sensitivity map (in units of %/Oe) of Figure 34.6a, implying the scaling of noise and sensitivity and the dominance of magnetization fluctuation. Using Equation 34.5, the magnetic field noise map is generated as in Figure 34.6c. It is seen that magnetic field noise can be reduced by applying a hard-axis biasing field. One can achieve this by depositing a thin-film magnet structure around the free layer of an MTJ sensor.

By incorporating a hard-axis field bias and a flux concentrator, Chaves et al. [39,40] demonstrated a noise level of about 0.25 nT/Hz$^{1/2}$ at 1 Hz and 5 pT/Hz$^{1/2}$ at 500 kHz in CoFeB/MgO/CoFeB MTJ sensors. Without the flux concentrator, the noise level

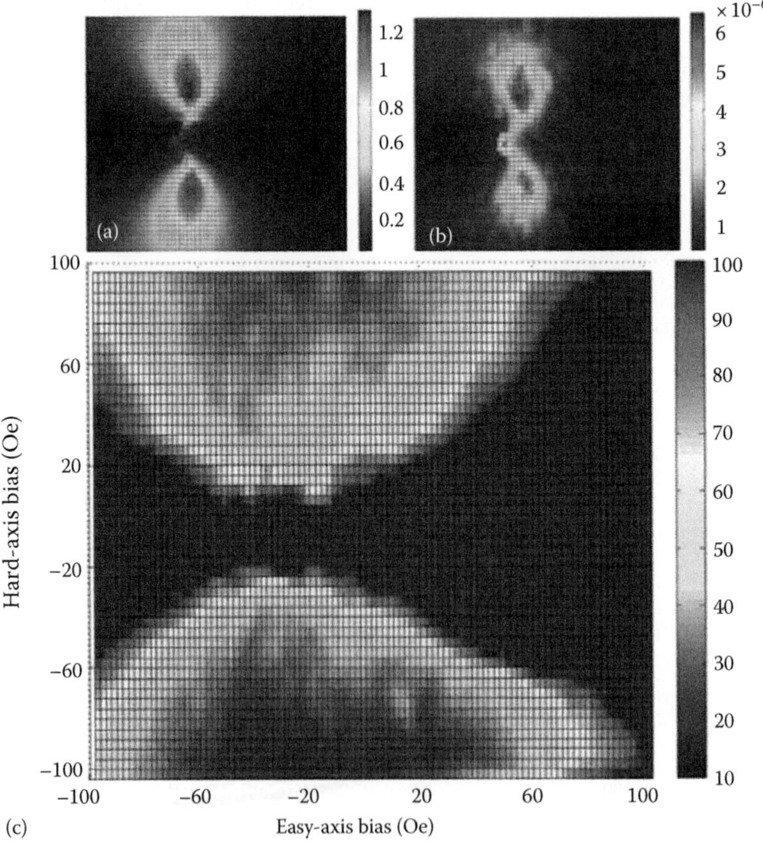

FIGURE 34.6 (See color insert.) (a) The sensitivity map of an MgO MTJ sensor (in units of %/Oe); (b) the corresponding normalized voltage noise map in units of Hz$^{-1/2}$, both measured in a 2D orthogonal field arrangement; (c) magnetic noise map. The color bars are in units of nT/Hz$^{1/2}$ and set to 10–100 nT/Hz$^{1/2}$ (Reprinted with permission from Mazumdar, D., Shen, W., Liu, X., Schrag, B.D., Carter, M., and Xiao, G., Field sensing characteristics of magnetic tunnel junctions with (001) MgO tunnel barrier, *J. Appl. Phys.*, 103, 113911, 2008, Fig. 3. Copyright 2008, American Institute of Physics.)

increased to 16 nT/Hz$^{1/2}$ at 1 Hz for a very large sensor [49] with a total area of 10 mm^2. As shown in Equation 34.4, sensor's $1/f$ noise scales inversely with junction area. Using large areas, flux concentrators and on-chip magnets present challenges to cost-effective manufacturing. More research is needed to develop better materials and sensor configurations as means to reducing the intrinsic noise level of MTJ sensors.

At frequencies above the knee frequency where $1/f$ noise is no longer dominant, Johnson and shot (white) noise set the limit of field detection (see Equation 34.4). To reduce the Johnson noise, a low resistance, for example, tens of ohms, is preferred. Johnson noise does not depend on voltage bias to a sensor. The shot noise, on the other hand, does depend on the voltage bias. Therefore, increasing the voltage bias that generates a larger signal does not lead to increased signal-to-noise ratio. Klaassen et al. [50] measured the electrical white noise of MTJ sensors used for read heads in data storage and with resistances ranging from a few ohms to over 100 Ω. They found that at low bias voltages $\ll 2\,kT/e$, the noise is dominated by Johnson noise, and $\gg 2\,kT/e$, noise is dominated by shot noise for high-quality ("healthy") junctions.

Tserkovnyak and Brataas [51] developed a semiclassical theory of shot noise in both GMR and MTJ physical systems, and calculated the Fano factor that is the ratio of real shot noise to the full shot noise defined in Equation 34.4. In the cases of ballistic, diffusive, and tunnel junctions, Fano factor is 0, 1/3, and 1, respectively. Furthermore, the Fano factor depends on the relative orientation between the magnetization vectors of the free and the pinned layer. Guerrero et al. [52] measured the shot noise of $Co/Al_2O_3/FeNi$ MTJs and found the Fano factor is about 0.7. They attribute the reduced Fano factor as evidence of sequential tunneling through nonmagnetic and paramagnetic impurity levels inside the tunnel barrier. Understanding of how the Fano factor depends on materials and junction quality will benefit applications of MTJ sensors in the intermediate frequency range.

For small MR sensors (GMR or MTJ), there exists another kind of white noise in addition to Equation 34.4. This white resistance noise, often called the mag-noise [53,54], is from the thermal magnetization fluctuations (spin waves) in the free and the pinned layers. Unlike electronic thermal noise (Johnson), the mag-noise scales with field sensitivity and inversely with the volumes of the ferromagnetic electrodes. In MR read heads used in hard disk drives, mag-noise can become the dominant white noise because of sensors' small sizes. For MTJ sensors with large junction area (hundreds of micrometers), mag-noise is less of an issue. Finite element micromagnetic simulations [55] show that the mag-noise, which increases linearly with temperature, can be suppressed significantly by an easy-axis bias field.

The current goal in the development of MR sensors is to increase field detectability from roughly 1 nT to 1 pT in sensors that are compact, ultrasensitive, not power hungry, and inexpensive to manufacture [56]. Judging from the progress in the last 10 years, such a goal is within reach in the future, but challenges abound. Research is ongoing by focusing "intrinsically" and "extrinsically" on the MTJs. The "intrinsic" way is to improve barrier and junction quality, and to search for the "good" ferromagnetic materials and the optimal micromagnetic design and device processing. One of the "extrinsic" ways is to incorporate MTJ sensors with microelectromechanical system (MEMS) actuators [57–59]. Using MEMS to create small amplitude, high-frequency oscillation of the sensor-sample separation, $1/f$ noise can be suppressed. Such an approach tends to increase the manufacturing and integration cost.

34.2.5 Sensor Linearization and Magnetic Hysteresis

In an ideal magnetic sensor, the transfer curve (output vs. field) should be linear within its intended field dynamic range and magnetic hysteresis in the transfer curve should be vanishingly small. Figure 34.7 shows a transfer curve of a commercial MgO-based MTJ sensor over a dynamic range of ±1 G. The sensitivity is 0.71%/G and the sensor is very linear (the small amount of

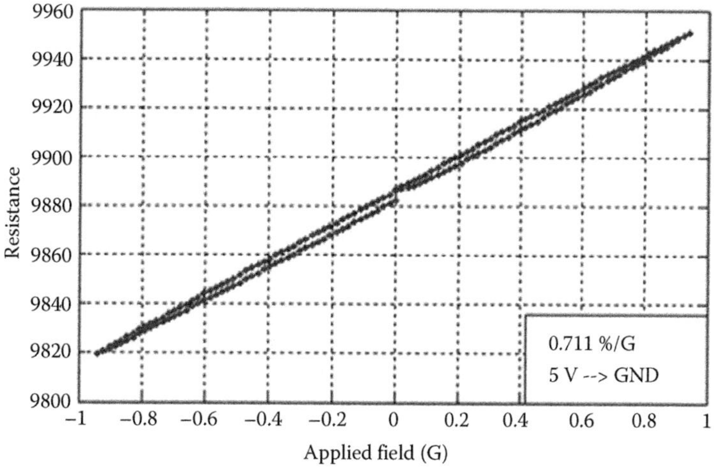

FIGURE 34.7 Transfer curve of an MgO MTJ sensor (resistance sensitivity is 0.711%/G).

hysteresis in this measurement is actually an artifact of the electromagnets, which supply the field).

To achieve linearization, MTJs should be designed such that the free and pinned layer magnetizations are at a 90° angle in the absence of applied field. The pinned layer magnetization direction is set by annealing MTJ wafers in a uniform strong magnetic field (> a few kG). After cooling back to room temperature, the AFM/FM exchange coupling keeps the pinned layer magnetization direction firmly fixed. To make the free layer magnetization direction orthogonal to that of the pinned layer, sensor designers would typically take advantage of the magnetic shape anisotropy and pattern the free layer into an oval shape. The longer (easy) axis is perpendicular to pinned layer magnetization direction. The larger the eccentricity, the larger the field dynamic range, and the lower the field sensitivity. This method, called "the orthogonal sensor," uses the intrinsic uniaxial anisotropy of the free layer to align its orientation in the desired direction, without the need for external magnetic fields. The intrinsic uniaxial anisotropy is not limited to shape, but also can be resulted from magnetocrystalline, magnetostrictive, and other types of anisotropy. Figure 34.8 shows the schematic transfer curve for an "orthogonal sensor." Lu et al. [60] have used the shape anisotropy to control the transfer curves in MTJs.

In reality, making an "orthogonal sensor" is challenging. If junction area is large, the free layer often breaks into a multidomain state, making magnetization rotation in unison impossible. Such a sensor would become nonlinear and hysteretic. On the other hand, if the junction area is small, the edge roughness and other structural defects may hinder the free rotation of local magnetizations. This would cause Barkhausen jumps in the transfer curve. To eliminate the hysteresis and Barkhausen jumps, sensor designers would add a constant magnetic field that forces the free layer magnetization into the desired state. This can be done by an external field or a local field via a thin-film magnet or current-carrying stripline. Such a sensor can be called the "offset sensor," whose schematic transfer curve is shown in Figure 34.9. GMR and MTJ read heads in hard drive disks are examples of the offset sensor.

Liu et al. [61] have studied the method of the "offset sensor" on *FeNi*/Al_2O_3/*FeNi* MTJs. The junctions have the shape of

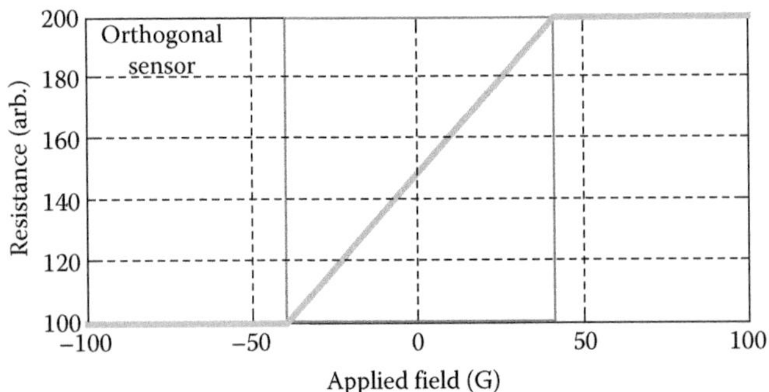

FIGURE 34.8 Schematic transfer curve for an "orthogonal sensor."

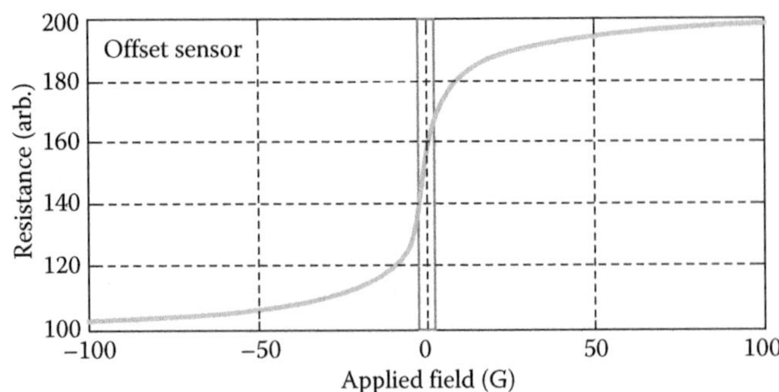

FIGURE 34.9 Schematic transfer curve for an "offset sensor."

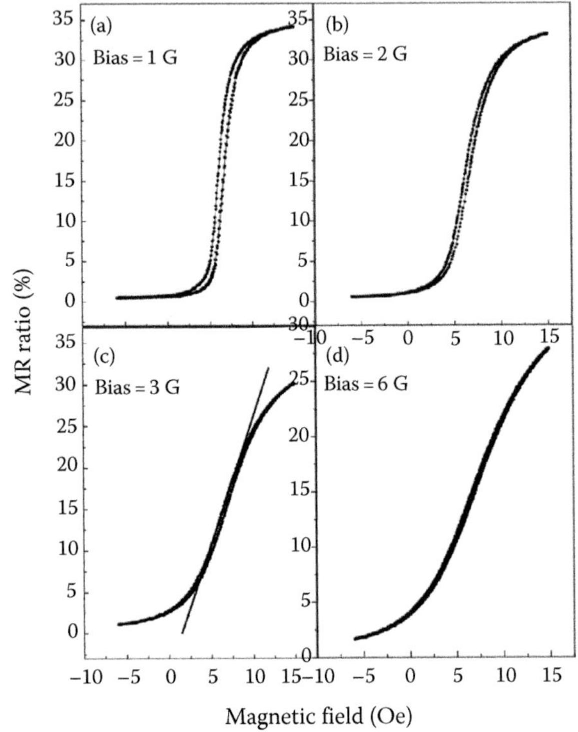

FIGURE 34.10 Transfer curves under different biasing fields (a through d) that are orthogonal to the sensed field direction. (Reprinted with permission from Liu, X., Ren, C., and Xiao, G., Magnetic tunnel junction field sensors with hard-axis bias field, *J. Appl. Phys.*, 92, 4722, 2002, Fig. 2. Copyright 2002, American Institute of Physics.)

rectangles with dimensions of $100 \times 150\,\mu m^2$. Figure 34.10 shows the transfer curves under different biasing fields, which are orthogonal to the sensing field direction. At 3 Oe, the hysteresis nearly disappears and sensor is linearized. The collapse of coercivity is further demonstrated in Figure 34.11. Pong et al. [62] have also demonstrated hysteresis collapse for linear response by sensor rotation and applying an external field on $Co_{50}Fe_{50}/Al_2O_3/Co_{50}Fe_{50}/Ni_{77}Fe_{14}Cu_5Mo_4$ MTJ wafers.

Recently, a new method to linearize an MTJ sensor was found [63,64]. The hysteresis property in CoFeB/MgO/CoFeB is sensitively dependent on the thickness of the free layer. Linear and hysteresis-free transfer curves are obtained simply

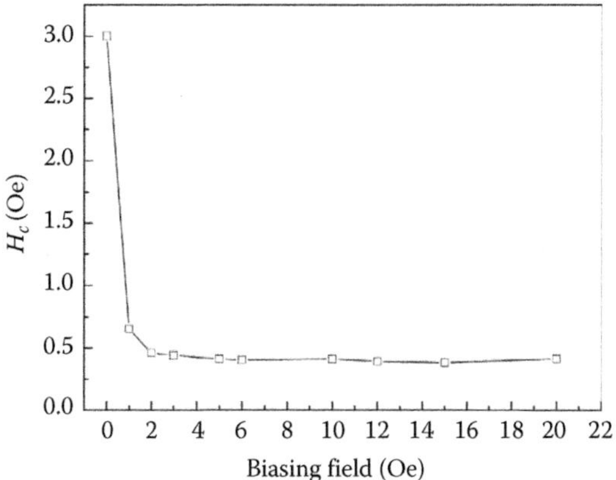

FIGURE 34.11 Coercivity of an MTJ sensor versus biasing field. Coercivity collapses under a biasing field of 2 Oe. (Reprinted with permission from Liu, X., Ren, C., and Xiao, G., Magnetic tunnel junction field sensors with hard-axis bias field, *J. Appl. Phys.*, 92, 4722, 2002, Fig. 2. Copyright 2002, American Institute of Physics.)

by reducing the free layer thickness to below a critical thickness of 1.5 nm (see Figure 34.12). Detailed analysis on the transfer curves shows that they agree well with Langevin equation describing superparamagnetism [64]. In this magnetic state, the free layer enjoys zero hysteresis and still a large magnetic susceptibility. It is not clear whether sensor performance will be strongly dependent on temperature, as is the relaxation rate of a superparamagnet.

34.2.6 Interlayer Coupling and Sensor Offset

Because of the close proximity, there exists a magnetic interlayer coupling between the free layer and the pinned layer. Such a coupling generates an internal field along the sensing axis of the free layer. As a result, the center of the transfer curve is shifted by this internal coupling field (offset field), as shown in Figure 34.13. Since the highest sensitivity typically occurs near the center of the transfer curve, this sensor offset will lower sensor's sensitivity and reduce its dynamic range. Two effects produce the extraneous magnetic fields in the plane of the free layer [65–68]: Néel "orange-peel" coupling (H_N) due to interface roughness, and magnetostatic coupling (H_M) due to uncompensated poles on the edges of the pinned layer. Figure 34.14 shows schematically the origins of these two fields.

Assuming a sinusoidal roughness profile, the Néel field [65,66] is given by

$$H_N = \frac{\pi}{\sqrt{2}}\left(\frac{h^2}{\lambda t_F}\right) M_S \exp\left(-\frac{2\pi\sqrt{2}t_s}{\lambda}\right) \tag{34.6}$$

where

h and λ are the amplitude and wavelength of the roughness profile
t_F and t_s are the thickness of the free layer and the barrier
M_S is the magnetization of the free layer

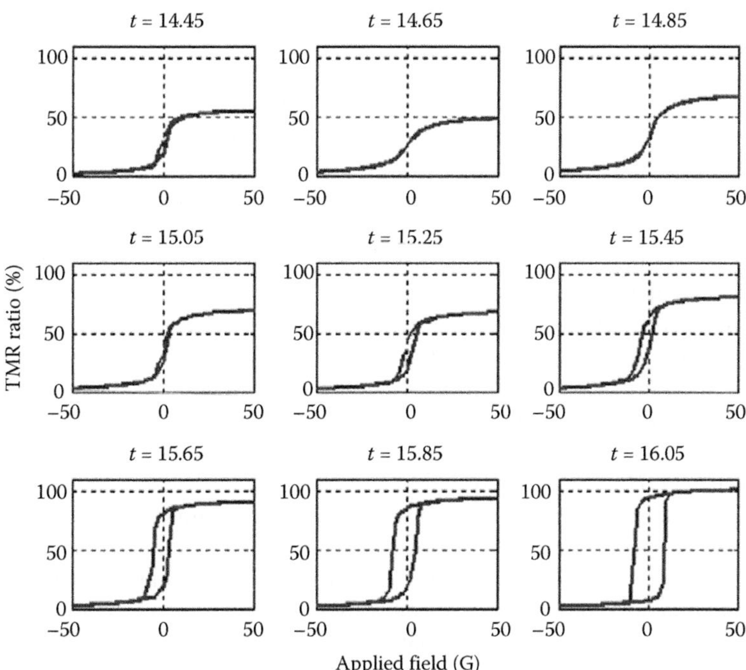

FIGURE 34.12 Representative transfer curves (resistance versus field) for a number of MgO MTJs with increasing free layer (CoFeB) thickness, ranging from 14.45 to 16.05 Å. Below the critical thickness of 15 Å, the free layer becomes superparamagnetic at room temperature, and the MTJ device behaves like a hysteresis-free linear sensor. (Reprinted with permission from Shen, W., Schrag, B., Girdhar, A., Carter, M., Sang, H., and Xiao, G., *Phys. Rev. B*, 79, 014418, 2009. Copyright 2009 by the American Physical Society.)

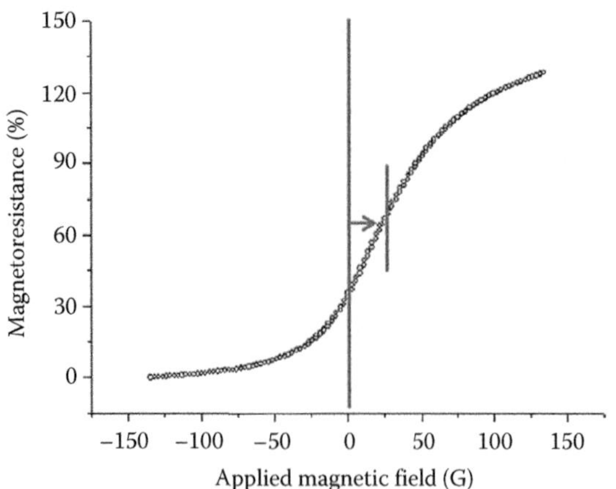

FIGURE 34.13 The center of a transfer curve of an MgO MTJ sensor is shifted by an offset field due to internal coupling field between the free layer and the pinned layer.

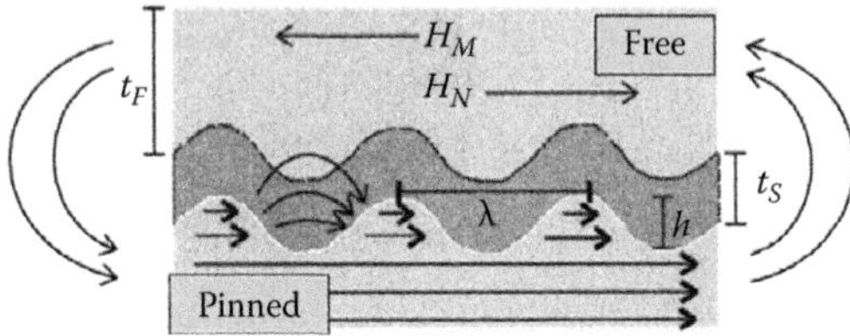

FIGURE 34.14 Internal magnetic coupling between the free layer and the pinned layer. H_N is Néel "orange-peel" coupling due to interface roughness, and H_M is magnetostatic coupling due to uncompensated magnetic poles on the edges of the pinned layer. (Reprinted with permission from Schrag, B.D., Anguelouch, A., Ingvarsson, S. et al., Neel "orange-peel" coupling in magnetic tunneling junction devices, *Appl. Phys. Lett.*, 77, 2373, 2000, Fig. 2. Copyright 2000, American Institute of Physics.)

Relation (34.6) has been confirmed in various studies on MTJs based on the measurement of H_N as a function of t_F and t_s and TEM imaging of the roughness profile [65–68]. According to Equation 34.6, the most effective way to reduce the Néel field is to reduce the interface roughness (reducing h and increasing λ). Shen et al. [69] have developed deposition process to reduce the interface roughness. They found that low (Ar) pressure sputtering can lead to interface smoothness in MgO-based MTJs. Also, when possible, using a thinner free layer and/or a thicker barrier will lower the Néel field.

The magnetostatic coupling field, H_M, can be expressed by $H_M = B/L$, with B as an adjustable constant dependent on junction shape and L as the dimension of the pinned layer along the magnetization direction [65–67]. Studies on multiple MTJ samples [55–57] have shown that for an L of ~20 μm, H_M is on the order of 1–2 Oe (barrier thickness 1–1.5 nm) for permalloy pinned layers. For smaller junctions, H_M can exceed 10 Oe or more. In order to eliminate H_M, an antiferromagnetic synthetic pinned layer is often adopted. In this case, the free layer consists of two magnetic layers antiferromagnetically coupled through a thin Ru layer [46]. The total magnetization of the pinned layer is, therefore, compensated and incapable of generating H_M.

34.2.7 Thermal Stability and Temperature Dependence

There are applications requiring magnetic sensors to function in a broad temperature range or at high temperatures. Thermal stability is an important issue. For example, magnetic sensors used in oil and gas well drilling that often encounters up to 200°C harsh conditions. Sensors used for monitoring engines face similar situations. In MTJs, high temperatures can cause interlayer diffusion and degradation of the barrier, affecting sensor performance [46,70]. Furthermore, temperature can lead to varying sensor resistance and field sensitivity. It is, therefore, necessary to characterize the temperature coefficients of both the resistance and the sensitivity.

Mazumdar and coworkers [46] have investigated the thermal stability and temperature dependence in CoFeB/MgO/CoFeB magnetic sensors. They used IrMn as the antiferromagnet for pinned layer, which is widely used due to its high thermal stability. They also used an antiferromagnetically coupled synthetic trilayer structure (CoFe/Ru/CoFeB) on the top of IrMn film as the pinned layer. The large separation between the InMn and the MgO barrier reduces potential diffusion of Mn ions, which can degrade the barrier quality at high temperatures. Figure 34.15a shows a sensor's transfer curves at various temperatures. The MTJ sensor remains functioning at 375°C. The temperature dependence of sensor resistance at zero field is approximately linear with a TCR ($TCR = (1/R)(\Delta R/\Delta T)$) of -7.1×10^{-4} °C^{-1}. Figure 34.15b shows the temperature dependence of the field sensitivity at zero field ($S = (1/R)(\Delta R/\Delta H)$) and detectability. Between room temperature and 275°C, S varies between 1.0%/Oe and 1.14%/Oe, remaining relatively constant over such a wide temperature range.

These results demonstrate that properly designed and processed MgO-based MTJ sensors enjoy a high degree of thermal stability with low TCR and stable field sensitivity. There are a few factors contributing to the good thermal properties. First of all, MgO is a robust tunnel barrier. Second, the physics of magnetic tunneling persists to high temperature. Finally, the ferromagnetic and antiferromagnetic materials used in the MTJs have high magnetic ordering temperatures.

In the bridge configuration, if all four arms are made of MTJ resistors, the effect of TCR is cancelled. Since all arms have the same TCR, the bridge will remain balanced in a varying temperature environment. The thermal coefficient of sensitivity, on the other hand, cannot be canceled in a bridge. Fortunately, the sensitivity for the MgO-based MTJ sensors can be made relatively insensitive to temperature.

34.2.8 Frequency Response

The evolution of data storage systems requires more sensitive and faster magnetic sensors. The noninvasive diagnostics of semiconductor ICs based on magnetic current density imaging

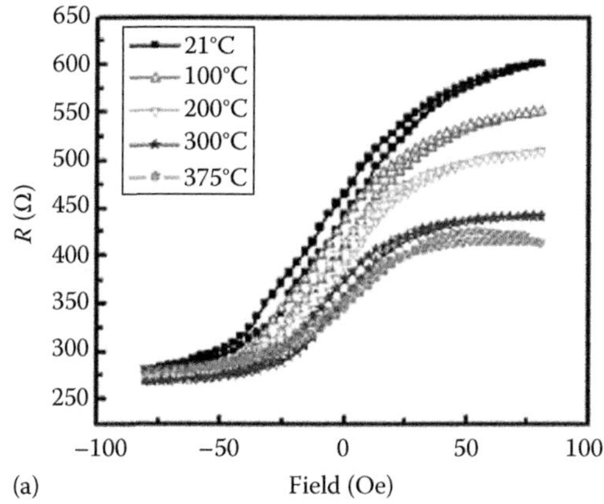

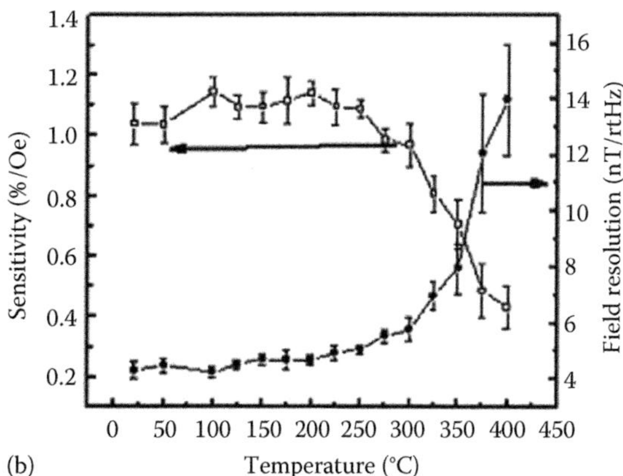

FIGURE 34.15 (a) Temperature dependence of the transfer curves of an MgO MTJ sensor. (b) Sensitivity (open squares) and estimated magnetic field resolution (solid circles) of the same MTJ sensor as a function of temperature. (Reprinted with permission from Mazumdar, D., Liu, X., Schrag, B.D., Shen, W., Carter, M., and Xiao, G., Thermal stability, sensitivity, and noise characteristics of MgO-based magnetic tunnel junctions invited, *J. Appl. Phys.*, 101, 09B502, 2007, Fig. 5. Copyright 2007, American Institute of Physics.)

has been made possible by the availability of high-performance MTJ sensors. In these and other applications, it is important that these magnetic devices can respond to signals with frequencies exceeding 1 GHz. The speed of modern electronics has been driven into the deep sub-nanosecond regime, while the dynamics of today's magnetic materials remains slower.

The physics of magnetic tunneling does allow ultrafast MTJ sensors to work at extremely high frequencies. This is because the energy scale of tunneling electrons is on the orders of eV. However, there are many factors that influence the frequency response of MTJ sensors: (1) the effective resistance-capacitance (*RC*) time constant, (2) the ferromagnetic resonance frequency of the free layer, and (3) the magnetic damping characteristics of the free layer.

MTJ sensors can be described by using a simplified circuit model for the sensor [71–73], including a lead resistance term (R_L) in series with the traditional parallel *RC* circuit (Figure 34.16). In this case, the maximum operating frequency f_{RC} of the sensor (the −3 dB point) is given by $f_{RC} = 1/2\pi RC$. Figure 34.17 shows the impedance data (real and imaginary) of an MgO-based MTJ sensor. The lines are fittings based on the simplified circuit model, with the fitting results of resistance and capacitance overlaid in the figure. In this sensor, the junction resistance is 215 Ω and the capacitance is 20.3 pF. This particular sensor works from dc to 25 MHz. With proper choice of resistance-area (RA) product and junction capacitance, it is possible to produce sensors with $R \sim 20\,\Omega$ and $C \sim 4$ pF, for an ultimate f_{RC} of ~2 GHz.

If the effective *R* and *C* of the packaged device are reduced to small enough values, the limiting factor on the speed of the MTJ sensor becomes the magnetic dynamics of the free layer. In order to respond to a change in the external magnetic field, the magnetic orientation of the free layer must change, and this process happens at a speed dictated by the Landau–Lifshitz–Gilbert (LLG) equation:

$$\frac{\partial \vec{M}}{\partial t} = -|\gamma|\vec{M} \times \vec{H}_{\text{eff}} - \frac{\alpha}{M_S}\vec{M} \times \frac{\partial \vec{M}}{\partial t} \tag{34.7}$$

where
- M is the magnetization vector
- H_{eff} is the effective magnetic field
- M_S is the saturation magnetization
- γ is the gyromagnetic ratio
- α (damping constant) is a measure of the amount of damping experienced by M

The LLG equation indicates that if a magnetic layer without damping ($\alpha \to 0$) experiences a slight perturbation, such that M and H_{eff} are not quite parallel, then M will simply precess about its lowest energy state indefinitely. α plays a key role in the high-frequency response; if α is too small, it will take a long time for M to settle into its new orientation, if too large, the transition between states may also become undesirably slow. α is dependent on the composition of the free layer. By changing the dc field, the magnitude of H_{eff} can be varied, which will in turn vary the precession frequency. Sensors normally work up to the precession frequency, which can be tuned above a few GHz. More studies are needed to understand the magnetic dynamics in MTJ sensors.

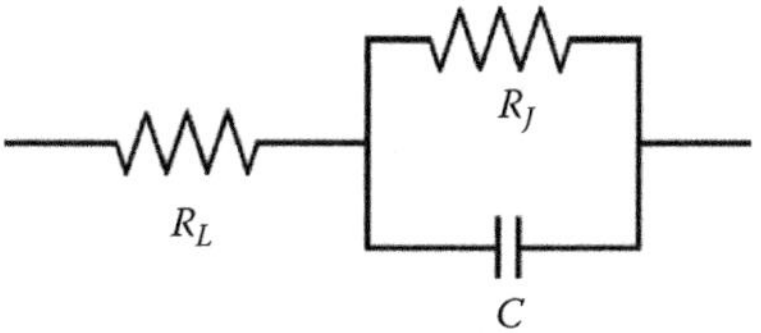

FIGURE 34.16 Effective MTJ circuit model.

34.2.9 Sensor Characterizations

Modern MTJ devices, as magnetic sensors or used for nonvolatile magnetic memory cells, have complicated structures with performance dependent on more than a dozen material and

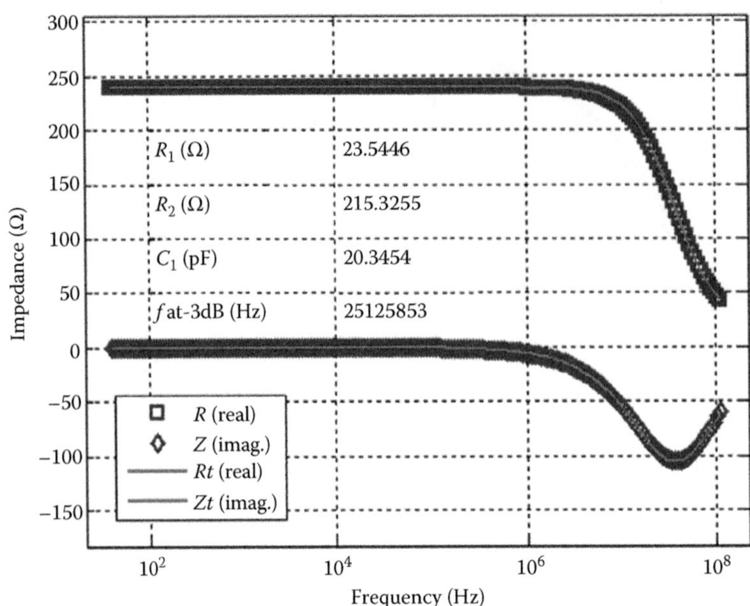

FIGURE 34.17 Fitting results (based on the *RC* circuit in Figure 34.16) for an MTJ sensor, with the extracted parameters of R_1 (lead resistance), R_2 (junction resistance), and C_1 (junction capacitance). Note that the top curve is the real part of the impedance, and the bottom curve is the imaginary part of the impedance.

structural parameters. Accurate characterization of the governing properties of MTJs is critical to understanding the physics and the performance in these devices. Since the discovery of large MR in MTJs, some novel measuring techniques have been introduced, improving the efficiency of measurement.

In MTJ research, MR and RA product are two of the most basic parameters. In the past, obtaining the two parameters took tedious process. One must pattern the MTJ wafers into vertical structures and attach electrical contacts. Worledge and Trouilloud [74] invented the technique of current-in-plane tunneling (CIPT), which allowed the measurement of MR and RA of unpatterned MTJ wafers. In this technique, a series of micro-point probes are gently placed on the top surface of a wafer. By measuring the four-point resistance at various probe spacings, the RA can be calculated using a formula that is an exact analytical solution of the CIPT conceptual model. By repeating measurement of RA as a function of magnetic field, the MR can be obtained. The CIPT method is purely a wafer-level characterization tool, and highly efficient and fast. It is a very useful tool in designing and testing the materials and structures used in MTJs.

Researchers at Brown University and Micro Magnetics, Inc. have developed the techniques of "sensitivity map" [48], "magnetic noise map" [48], and "circle transfer curves" [75] in characterizing MTJ devices. In the method of "sensitivity map," an MTJ device's field sensitivity, $S=(1/R)(\Delta R/\Delta H)$, is measured automatically in a two dimensional (2D) magnetic field (bipolar) over a large range of $\pm Bx$ and $\pm By$. A map of the S parameter is then generated in the 2D field space. An example of S map [48] is shown in Figure 34.6a. With the map, one can locate the most sensitive region for a particular sensor design. In most cases, the most sensitive region is not centered around the ambient state of $Bx=By=0$. By using hard- and easy-axis field biasings (on-chip magnets or striplines), the best sensitivity can be achieved.

In the method of "magnetic noise map" [48], an MTJ device's voltage noise at a given frequency (e.g., 1 Hz) is measured over a 2D magnetic field space as in the sensitivity map. A high-resolution map consists of hundreds of data points. Automation in data taking and analysis is necessary to generate the map. Once the voltage noise map is obtained, the "magnetic noise map" can be plotted by combining the voltage noise map and the sensitivity map, according to Equation 34.5. In sensor development, the magnetic noise map and sensitivity map will help designer to select the optimal operating regions for a magnetic sensor. Figure 34.6b and c shows the examples of voltage noise map and magnetic noise map.

The method of "circle transfer curves" [75] is an efficient technique to measure a plethora of important parameters, such as magnetic anisotropy strength, magnetostatic coupling, antiferromagnetic exchange coupling, and the absolute magnetic orientation of both the free and the pinned layers across a whole wafer. Figure 34.18 shows some of parameters of an MTJ multilayer that can be determined using this method. Traditional magnetometry methods can only measure bulk films or devices with relatively large volumes. The "circle transfer curves" method can handle devices at any size. In this technique, the MTJ resistance (or conductance) is measured as a function of the applied field angle with a fixed field strength (*R* vs. angle). Two pairs of orthogonal Helmholtz coils can be used to generate a rotating magnetic field. Only three circle transfer curves at different fixed fields are needed to extract all the desired information about

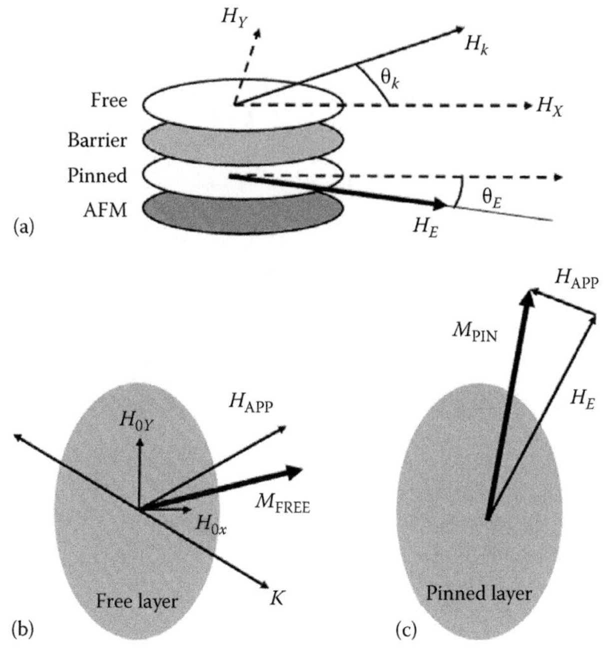

FIGURE 34.18 (a) Schematic of the MTJ multilayer showing the relevant physical quantities, which dictate the junction behavior. (b) Simplified model of the free layer, where the magnetization is determined by the two offset field components (H_{0X} and H_{0Y}), the sample's uniaxial anisotropy, and the external applied field (H_{APP}). (c) Model of the pinned layer magnetization, which is assumed to be the vector sum of two forces: the applied field and the exchange biasing field (H_E). (Reprinted with permission from Safron, N.S., Schrag, B.D., Liu, X., Shen, W., Mazumdar, D., Carter, M.J., and Xiao, G., Magnetic characterization of magnetic tunnel junction devices using circle transfer curves, *J. Appl. Phys.*, 103, 033507, 2008, Fig. 2. Copyright 2008, American Institute of Physics.)

the junction (see detailed analysis description in Ref. [75]). The "circle transfer curves" method is fast and can be used to measure the variation or uniformity parameters across a whole wafer with patterned MTJ devices.

Figure 34.19 shows three experimental circle curves (conductance vs. angle) of a representative MTJ device [75]. The deviation from a pure cosine function is due to the free layer anisotropy, which is more pronounced to the applied field strength (e.g., 40 Oe). As the field strength increases, the circle curves will be more cosine-like. The solid lines are best-fit curves based on the "circle transfer curves" model, from which many parameters (see Figure 34.18) can be derived. This technique has been used to map six parameters of MTJs across a whole wafer.

Figure 34.20 shows the distribution of junction anisotropy angle and pinned layer direction across a wafer [75]. This particular wafer consists of hundreds of different MTJ sensors. Figure 34.21 shows the distribution of junction anisotropy angle and magnitude, and pinned layer direction for a wafer consists of hundreds of MTJ memory cells. These results demonstrate that circle curves can be used as a new and complementary tool to other methods for researchers in designing better MTJ sensors and devices.

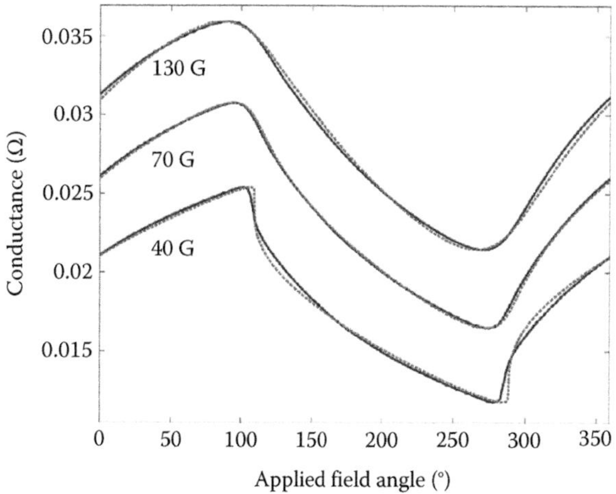

FIGURE 34.19 Experimental data (solid lines) and theoretical fitting results (dashed lines) for a set of three circle curves (taken at 40, 70, and 130 G) taken on a representative MTJ element. All fits are made using a single set of junction parameters. (Reprinted with permission from Safron, N.S., Schrag, B.D., Liu, X., Shen, W., Mazumdar, D., Carter, M.J., and Xiao, G., Magnetic characterization of magnetic tunnel junction devices using circle transfer curves, *J. Appl. Phys.*, 103, 033507, 2008, Fig. 2. Copyright 2008, American Institute of Physics.)

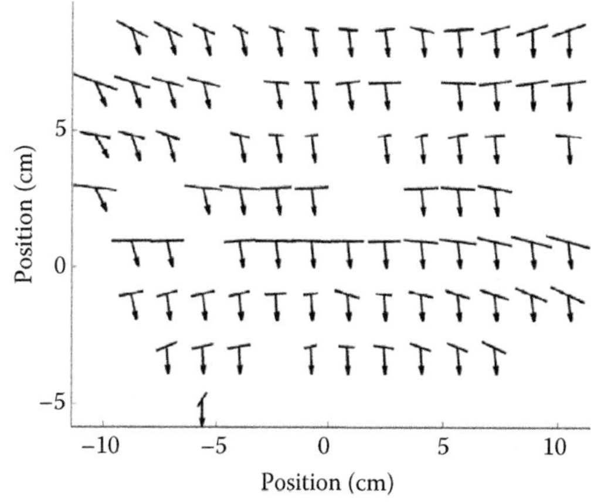

FIGURE 34.20 Plot showing the distribution of junction anisotropy angle and pinned layer direction as a function of MTJ wafer position. The arrows indicate the pinned layer orientation, while the length and orientation of the solid lines indicate the strength and direction of the sample anisotropy, respectively. (Reprinted with permission from Safron, N.S., Schrag, B.D., Liu, X., Shen, W., Mazumdar, D., Carter, M.J., and Xiao, G., Magnetic characterization of magnetic tunnel junction devices using circle transfer curves, *J. Appl. Phys.*, 103, 033507, 2008, Fig. 2. Copyright 2008, American Institute of Physics.)

34.2.10 Applications: Examples

34.2.10.1 Magnetic Current Density Microscope

The advent in MTJ sensor development has facilitated applications in different fields. Here, two examples are presented. The first example is a magnetic current density microscope. Magnetic imaging techniques have found numerous practical

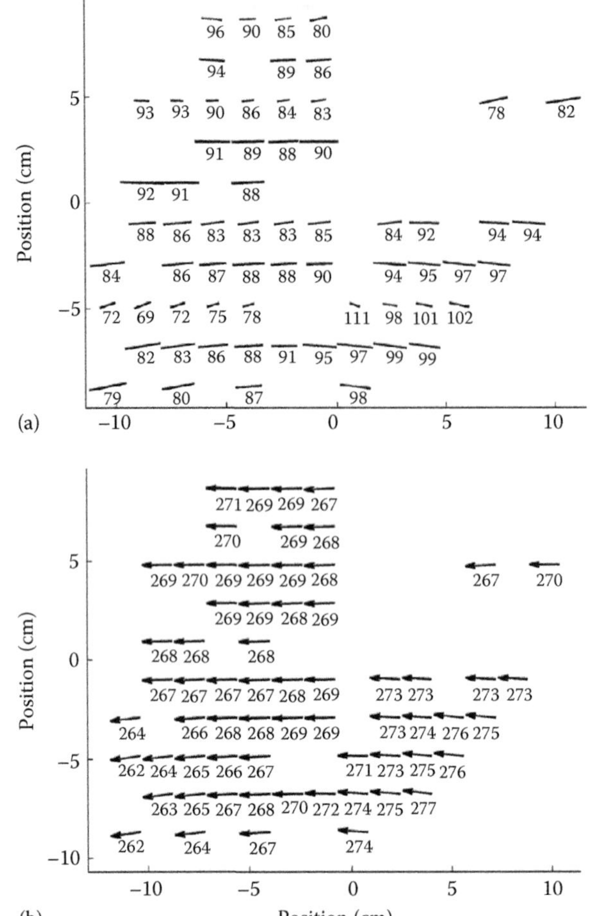

FIGURE 34.21 (a) Plot of the positional distribution of junction anisotropy angle and magnitude for an MTJ wafer. The numbers below each line show the anisotropy angle in degrees, while the length of each line is proportional to the anisotropy strength. (b) Distribution of the extracted pinned layer direction for the same sample as a function of wafer position. (Reprinted with permission from Safron, N.S., Schrag, B.D., Liu, X., Shen, W., Mazumdar, D., Carter, M.J., and Xiao, G., Magnetic characterization of magnetic tunnel junction devices using circle transfer curves, *J. Appl. Phys.*, 103, 033507, 2008, Fig. 2. Copyright 2008, American Institute of Physics.)

applications in many areas of science and technology. In addition, recent advances in spintronics have created new and powerful magnetic devices. GMR sensors have become universally adopted in the computer hard disk drive market. MTJ sensors have several significant advantages over GMR technology. Scientists have built on these sensing technologies to develop a new spintronics metrology imaging tool for mapping current densities in thin-film conductors and semiconductor devices [7–10]. The tool is noninvasive, operates under ambient conditions, and is capable of spatial resolution better than 100 nm. The MTJ sensor developed for this tool allows engineers to measure currents as small as a microampere in deeply embedded microstructures. The technique has been applied to fault isolation and failure analysis in ICs. Figure 34.22 displays a current density map in a failed IC device using a scanning MTJ microscope.

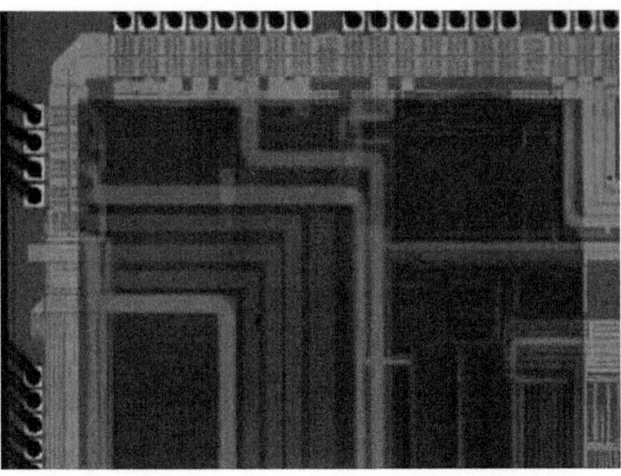

FIGURE 34.22 Map of current density distribution in a failed IC. The map is obtained using a scanning MTJ magnetic microscope. The current density image is automatically overlaid over a high-resolution optical image of the device.

The embedded structures in ICs emit a microscopic magnetic field. Imaging this field can reveal the spatial variation of current density and the physical layout of a circuit. The whole IC structure is transparent to magnetic field. The need for such a microscope is compelling and immediate in the semiconductor industry. The IC engineers can use the microscope to survey the operation of tiny components and to pinpoint defects down to the smallest units such as transistors and copper interconnects.

There are several competing technologies available to image field today. Magnetic force microscopy (MFM) [76], scanning electron microscopy with polarization analysis (SEMPA) [77], and magneto-optical microscopy [78] are capable of imaging magnetic structures. They are not, however, suitable for imaging fields created by electrical currents and, in particular, those from high-frequency currents. In some cases, the sample preparation is destructive, resulting in sample loss for further analysis. The scanning SQUID microscope [79] is poor in resolution with a lower limit of about 5 μm, and requires cryogenic cooling. A Hall probe can operate under ambient conditions, but its sensitivity is low. Ultrasensitive MTJ sensors are superior to other existing magnetic sensors for the magnetic current density microscopy [7–10]. This imaging technology is quite similar to medical x-ray imaging. In a CT scan, x-rays gather information about a body and inverse-calculate the body's internal structure. The current density microscope gathers the magnetic field information in a circuit and inverse-calculates its current distribution.

The scanning MTJ microscope is also a sensitive magnetic microscope that can directly image local magnetic field in absolute scale. The traditional scanning magnetic microscope using magnetic force only senses the local field gradient. The scanning MTJ microscope has also been used in the research of "atom chips" used in Bose–Einstein condensate [80].

34.2.10.2 Spintronics Immunoassay

Scientists have researched on a spintronics-based on-chip immunoassay system [81–86]* for the identification of biomolecules (bacteria, virus, protein, and DNA). Using MTJ sensor arrays (see Figure 34.23), it is possible to create a powerful immunoassay system that is fast, sensitive, specific, and portable. By utilizing magnetic nanoparticles (NPs) to tag the DNA, viruses, bacteria, or cancer cells, it is possible to detect the concentration of these agents in a biological solution using magnetic sensing, as shown in Figure 34.24.

Initially, magnetic NPs that bind only to specific molecules are introduced into the sample to be tested. If the molecules, viruses, DNA, or other cells that the NPs are designed for are present, they will tightly bind to the NPs. Then, the biological sample will be introduced to the surface of an array of magnetic sensors coated with "probe" molecules (antigens and complementary DNA). These probe molecules will bind tightly to certain specific agent, and agents that do not bind to the probe molecules will be washed away. The sensors can then detect the presence of these NPs by sensing their magnetic field. The strength of the signal is proportional to the concentration of the agent. Potentially, the biomagnetic sensor is sensitive enough to detect a single biomolecule.

This new immunoassay method has some very important implications. Virtually, any biomolecule can potentially be detected using this "lab-on-a-chip" technology. All tests and experiments could potentially be conducted on this single device using microfluidic circuitry, such as valves, tubes, mixers, pumps, and channels on a very small scale. This could make an impact on medical diagnoses as a noninvasive, quick, portable, and relatively inexpensive way to detect health problems.

Both GMR and MTJ sensors have been used in spintronics immunoassay [81–84]. However, in addition to a larger MR and sensitivity than GMR, MTJ-based devices boast a number of features that make them a superior sensor candidate for the application. First of all, the junction resistance (10^{-1}–10^{8} Ω) can be engineered to suit a specific application. Second, the junction area can be reduced to nanometer scale while still maintaining large MR. This is particularly beneficial to NP detection. Third, an MTJ requires very little power to operate due to the small sense currents. These advantages may allow the integration of the MTJ-based immunoassay system into silicon circuitry and coupling it with microfluidics to create a lab-on-a-chip analysis system.

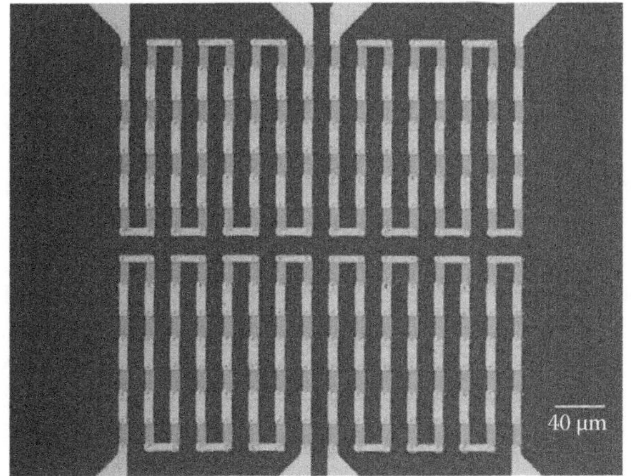

FIGURE 34.23 MTJ sensor array for spintronics immunoassay application.

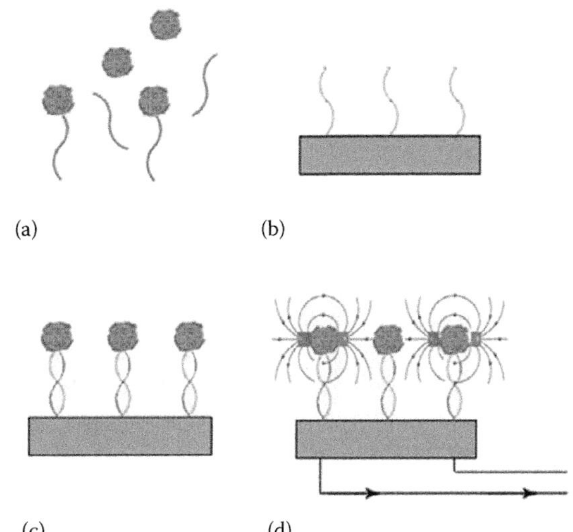

FIGURE 34.24 Using MTJ sensors to detect magnetic field of NPs bound to specific biomolecules (a–d).

34.3 Conclusions

Thin-film magnetic field sensors have had a significant impact in many different technological areas, ranging from data storage to magnetic microscopy. The development of magnetic sensor has benefited from the emerging technical discipline called "spintronics," which takes advantages of the dual spin and charge properties of electrons to form a new generation of electronic/magnetic devices. In particular, MTJ devices have become an important class of mass-market spintronics components in the information storage market and magnetic sensor market. The large MR in MgO-based MTJ sensors has been the key to the increase of field sensitivity. MTJ sensors are suitable to ultrafast sensing as in read/write heads, and applications requiring low power consumption. Continuing research is needed to reduce the intrinsic magnetic noise by using new magnetic materials and designing innovative device structures. New experimental characterization tools, such as CIPT and circle transfer curves, can benefit the research in MTJ sensors.

There are a few current objectives in MTJ sensor development. Significant impact on applications will occur if the field detectability can be enhanced to 1 pT as low frequencies (e.g., 1 Hz). The manufacturing cost can be reduced by improving fabrication process for making uniform sensors across a large wafer.

* See Chapter 36 of this book.

The potential discovery of even larger MR can always lead to improved sensor performance. More efforts are also needed to develop new applications that the MTJ sensors are capable for. The prospect for MTJ sensors in particular and spintronics in general is bright.

Acknowledgments

The author acknowledges funding by the National Science Foundation under Grant No. DMR-0907353 and by JHU MRSEC (NSF) under Grant No. DMR-0520491. The author wishes to thank W.F. Shen, B.D. Schrag, M.J. Carter, W. Egelhoff, X.Y. Liu, D. Mazumdar, X.J. Zou, W.Z. Zhang, N.S. Safron, and Hai Sang for collaboration and discussion.

References

1. S. A. Wolf, A. V. Chtchelkanova, and D. M. Treger, Spintronics: A retrospective and perspective, *IBM Journal of Research and Development* **50**, 101–110 (2006), doi: 10.1147/rd.501.0101.
2. S. A. Wolf, Spintronics: A spin-based electronics vision for the future, *Science* **294**, 1488–1495 (2001).
3. Frost & Sullivan, *World Magnetic Sensor Components and Modules/Sub-systems Markets*, February 2006.
4. H. Koch, Recent advances in magnetocardiography, *J. Electrocardiol.* **37**, 117–122 (2004).
5. D. Cohen and E. Halgren, Magnetoencephalography, in *Encyclopedia of Neuroscience*, 3rd edn., G. Adelman and B. Smith (Eds.), Elsevier, New York, 2004.
6. C. Chappert, A. Fert, and F. N. Van Dau, The emergence of spin electronics in data storage, *Nat. Mater.* **6**, 813–823 (2007).
7. B. D. Schrag, X. Liu, W. Shen, and G. Xiao, Current density mapping and pinhole imaging in magnetic tunnel junctions via scanning magnetic microscopy, *Appl. Phys. Lett.* **84**, 2937 (2004).
8. B. Schrag, X. Liu, M. J. Carter, and G. Xiao, Scanning magnetoresistive microscopy for die-level sub-micron current density mapping, in *Proceedings of the 29th International Symposium for Testing and Failure Analysis* (*ISTFA*), November 2–6, Santa Clara, CA, Awarded the Outstanding Paper Award at ISTFA (2003).
9. B. D. Schrag, M. J. Carter, X. Liu, J. Hoftun, and G. Xiao, Magnetic current imaging with magnetic tunnel junction sensors: Case study and analysis, in *Proceedings of the 2006 International Symposium for Testing and Failure Analysis*, Austin, TX.
10. B. D. Schrag, X. Liu, J. Hoftun, P. Klinger, T. Levin, and D. Vallett, Quantitative analysis and depth measurement via magnetic field imaging, *Electron. Device Fail. Anal.* **74**, 24–31 (2005).
11. K. Sohraby, *Wireless Sensor Networks: Technology, Protocols, and Applications*, Wiley-Interscience, Hoboken, NJ, 2007.
12. E. Y. Tsymbal, O. N. Mryasov, and P. R. LeClair, Spin-dependent tunnelling in magnetic tunnel junctions, *J. Phys.: Condens. Matter* **15**, R109–R142 (2003).
13. R. Popovic, The future of magnetic sensors, *Sens. Actuators A: Phys.* **56**, 39–55 (1996).
14. J. Lenz, A review of magnetic sensors, *Proc. IEEE* **78**, 973–989 (1990).
15. J. Lenz and S. Edelstein, Magnetic sensors and their applications, *IEEE Sens. J.* **6**, 631–649 (2006).
16. P. Ripka, Sensors based on bulk soft magnetic materials: Advances and challenges, *J. Magn. Magn. Mater.* **320**, 2466–2473 (2008).
17. E. Ramsden, *Hall-Effect Sensors: Theory and Applications*, Elsevier/Newnes, Amsterdam, Boston, 2006.
18. M. Baibich, J. Broto, A. Fert et al., Giant magnetoresistance of (001)Fe/(001)Cr magnetic superlattices, *Phys. Rev. Lett.* **61**, 2472–2475 (1988).
19. G. Binasch, P. Grünberg, F. Saurenbach, and W. Zinn, Enhanced magnetoresistance in layered magnetic structures with antiferromagnetic interlayer exchange, *Phys. Rev. B* **39**, 4828–4830 (1989).
20. M. Kuzminski, K. Nesteruk, and H. Lachowicz, Magnetic field meter based on giant magnetoimpedance effect, *Sens. Actuators A: Phys.* **141**, 68–75 (2008).
21. The Nobel Prize in Physics 2007, http://nobelprize.org/nobel_prizes/physics/laureates/2007/phyadv07.pdf (accessed on October 9, 2007).
22. J. Moodera, L. Kinder, T. Wong, and R. Meservey, Large magnetoresistance at room temperature in ferromagnetic thin film tunnel junctions, *Phys. Rev. Lett.* **74**, 3273–3276 (1995).
23. T. Miyazaki and N. Tezuka, Giant magnetic tunneling effect in Fe/Al_2O_3/Fe junction, *J. Magn. Magn. Mater.* **139**, L231–L234 (1995).
24. M. Julliere, Tunneling between ferromagnetic films, *Phys. Lett. A* **54**, 225–226 (1975).
25. W. Butler, X. Zhang, T. Schulthess, and J. MacLaren, Spin-dependent tunneling conductance of Fe|MgO|Fe sandwiches, *Phys. Rev. B* **63**, 054416 (1–12) (2001).
26. J. Mathon and A. Umerski, Theory of tunneling magnetoresistance of an epitaxial Fe/MgO/Fe(001) junction, *Phys. Rev. B* **63**, 220403(R) (1–4) (2001).
27. S. S. P. Parkin, C. Kaiser, A. Panchula et al., Giant tunnelling magnetoresistance at room temperature with MgO (100) tunnel barriers, *Nat. Mater.* **3**, 862–867 (2004).
28. S. Yuasa, T. Nagahama, A. Fukushima, Y. Suzuki, and K. Ando, Giant room-temperature magnetoresistance in single-crystal Fe/MgO/Fe magnetic tunnel junctions, *Nat. Mater.* **3**, 868–871 (2004).
29. J. Zhu and C. Park, Magnetic tunnel junctions, *Mater. Today* **9**, 36–45 (2006).
30. S. Yuasa and D. D. Djayaprawira, Giant tunnel magnetoresistance in magnetic tunnel junctions with a crystalline MgO(001) barrier, *J. Phys. D: Appl. Phys.* **40**, R337–R354 (2007).

31. B. D. Schrag and G. Xiao, Submicron electrical current density imaging of embedded microstructures, *Appl. Phys. Lett.* **82**, 3272 (2003).
32. D. Wang, C. Nordman, J. Daughton et al., 70% TMR at room temperature for SDT sandwich junctions with CoFeB as free and reference layers, *IEEE Trans. Magn.* **40**, 2269–2271 (2004).
33. X. Liu and G. Xiao, Thermal annealing effects on low-frequency noise and transfer behavior in magnetic tunnel junction sensors, *J. Appl. Phys.* **94**, 6218 (2003).
34. M. Tondra, J. M. Daughton, D. Wang, R. S. Beech, A. Fink, and J. A. Taylor, Picotesla field sensor design using spin-dependent tunneling devices, *J. Appl. Phys.* **83**, 6688 (1998).
35. D. D. Djayaprawira, K. Tsunekawa, M. Nagai et al., 230% room-temperature magnetoresistance in CoFeB/MgO/CoFeB magnetic tunnel junctions, *Appl. Phys. Lett.* **86**, 092502 (2005).
36. S. Ikeda, J. Hayakawa, Y. Ashizawa et al., Tunnel magnetoresistance of 604% at 300 K by suppression of Ta diffusion in CoFeB/MgO/CoFeB pseudo-spin-valves annealed at high temperature, *Appl. Phys. Lett.* **93**, 082508 (2008).
37. L. Jiang, H. Naganuma, M. Oogane, and Y. Ando, Large tunnel magnetoresistance of 1056% at room temperature in MgO based double barrier magnetic tunnel junction, *Appl. Phys. Exp.* **2**, 083002 (2009).
38. W. Egelhoff, Jr., V. Hoink, J. Lau, W. Shen, B. Schrag, and G. Xiao, Magnetic tunnel junctions with large tunneling magnetoresistance and small saturation fields, *J. Appl. Phys.* **107**, 09C705 (2010).
39. R. C. Chaves, P. P. Freitas, B. Ocker, and W. Maass, MgO based picotesla field sensors, *J. Appl. Phys.* **103**, 07E931 (2008).
40. R. C. Chaves, P. P. Freitas, B. Ocker, and W. Maass, Low frequency picotesla field detection using hybrid MgO based tunnel sensors, *Appl. Phys. Lett.* **91**, 102504 (2007).
41. S. Ingvarsson, G. Xiao, R. A. Wanner et al., Electronic noise in magnetic tunnel junctions, *J. Appl. Phys.* **85**, 5270 (1999).
42. S. Ingvarsson, G. Xiao, S. Parkin, W. Gallagher, G. Grinstein, and R. Koch, Low-frequency magnetic noise in micron-scale magnetic tunnel junctions, *Phys. Rev. Lett.* **85**, 3289–3292 (2000).
43. E. R. Nowak, M. B. Weissman, and S. S. P. Parkin, Electrical noise in hysteretic ferromagnet–insulator–ferromagnet tunnel junctions, *Appl. Phys. Lett.* **74**, 600 (1999).
44. C. Ren, X. Liu, B. Schrag, and G. Xiao, Low-frequency magnetic noise in magnetic tunnel junctions, *Phys. Rev. B* **69**, 104405 (1–5) (2004).
45. S. H. Liou, R. Zhang, S. E. Russek, L. Yuan, S. T. Halloran, and D. P. Pappas, Dependence of noise in magnetic tunnel junction sensors on annealing field and temperature, *J. Appl. Phys.* **103**, 07E920 (2008).
46. D. Mazumdar, X. Liu, B. D. Schrag, W. Shen, M. Carter, and G. Xiao, Thermal stability, sensitivity, and noise characteristics of MgO-based magnetic tunnel junctions (invited), *J. Appl. Phys.* **101**, 09B502 (2007).
47. N. A. Stutzke, S. E. Russek, D. P. Pappas, and M. Tondra, Low-frequency noise measurements on commercial magnetoresistive magnetic field sensors, *J. Appl. Phys.* **97**, 10Q107 (2005).
48. D. Mazumdar, W. Shen, X. Liu, B. D. Schrag, M. Carter, and G. Xiao, Field sensing characteristics of magnetic tunnel junctions with (001) MgO tunnel barrier, *J. Appl. Phys.* **103**, 113911 (2008).
49. R. Guerrero, M. Pannetier-Lecoeur, C. Fermon, S. Cardoso, R. Ferreira, and P. P. Freitas, Low frequency noise in arrays of magnetic tunnel junctions connected in series and parallel, *J. Appl. Phys.* **105**, 113922 (2009).
50. K. Klaassen, X. Xing, and J. van Peppen, Signal and noise aspects of magnetic tunnel junction sensors for data storage, *IEEE Trans. Magn.* **40**, 195–202 (2004).
51. Y. Tserkovnyak and A. Brataas, Shot noise in ferromagnet–normal metal systems, *Phys. Rev. B* **64**, 214402 (1–12) (2001).
52. R. Guerrero, F. Aliev, Y. Tserkovnyak, T. Santos, and J. Moodera, Shot noise in magnetic tunnel junctions: Evidence for sequential tunneling, *Phys. Rev. Lett.* **97**, 266602 (1–4) (2006).
53. N. Smith and P. Arnett, White-noise magnetization fluctuations in magnetoresistive heads, *Appl. Phys. Lett.* **78**, 1448 (2001).
54. N. Smith, Modeling of thermal magnetization fluctuations in thin-film magnetic devices, *J. Appl. Phys.* **90**, 5768 (2001).
55. V. D. Tsiantos, T. Schrefl, W. Scholz, and J. Fidler, Thermal magnetization noise in submicrometer spin valve sensors, *J. Appl. Phys.* **93**, 8576 (2003).
56. W. Egelhoff Jr., P. Pong, J. Unguris et al., Critical challenges for picoTesla magnetic-tunnel-junction sensors, *Sens. Actuators A: Phys.* **155**, 217–225 (2009).
57. G. M. Jaramillo, M. Chan, and D. A. Horsley, Integration of MEMS actuators with magnetic tunnel junction sensors, *2007 IEEE Sensors*, Atlanta, GA, pp. 1380–1383, 2007.
58. A. S. Edelstein, G. A. Fischer, M. Pedersen, E. R. Nowak, S. F. Cheng, and C. A. Nordman, Progress toward a thousandfold reduction in 1/f noise in magnetic sensors using an ac microelectromechanical system flux concentrator (invited), *J. Appl. Phys.* **99**, 08B317 (2006).
59. A. Guedes, S. B. Patil, S. Cardoso, V. Chu, J. P. Conde, and P. P. Freitas, Hybrid magnetoresistive/microelectromechanical devices for static field modulation and sensor 1/f noise cancellation, *J. Appl. Phys.* **103**, 07E924 (2008).
60. Y. Lu, R. A. Altman, A. Marley et al., Shape-anisotropy-controlled magnetoresistive response in magnetic tunnel junctions, *Appl. Phys. Lett.* **70**, 2610 (1997).
61. X. Liu, C. Ren, and G. Xiao, Magnetic tunnel junction field sensors with hard-axis bias field, *J. Appl. Phys.* **92**, 4722 (2002).
62. P. W. T. Pong, B. Schrag, A. J. Shapiro, R. D. McMichael, and W. F. Egelhoff, Hysteresis loop collapse for linear response in magnetic-tunnel-junction sensors, *J. Appl. Phys.* **105**, 07E723 (2009).

63. Y. Jang, C. Nam, J. Y. Kim, B. K. Cho, Y. J. Cho, and T. W. Kim, Magnetic field sensing scheme using CoFeB/MgO/CoFeB tunneling junction with superparamagnetic CoFeB layer, *Appl. Phys. Lett.* **89**, 163119 (2006).
64. W. Shen, B. Schrag, A. Girdhar, M. Carter, H. Sang, and G. Xiao, Effects of superparamagnetism in MgO based magnetic tunnel junctions, *Phys. Rev. B* **79**, 014418 (1–4) (2009).
65. A. Anguelouch, B. D. Schrag, G. Xiao et al., Two-dimensional magnetic switching of micron-size films in magnetic tunnel junctions, *Appl. Phys. Lett.* **76**, 622 (2000).
66. B. D. Schrag, A. Anguelouch, S. Ingvarsson et al., Néel "orange-peel" coupling in magnetic tunneling junction devices, *Appl. Phys. Lett.* **77**, 2373 (2000).
67. B. D. Schrag, A. Anguelouch, G. Xiao et al., Magnetization reversal and interlayer coupling in magnetic tunneling junctions, *J. Appl. Phys.* **87**, 4682 (2000).
68. L. Li, X. Liu, and G. Xiao, Microstructures of magnetic tunneling junctions, *J. Appl. Phys.* **93**, 467 (2003).
69. W. Shen, D. Mazumdar, X. Zou, X. Liu, B. D. Schrag, and G. Xiao, Effect of film roughness in MgO-based magnetic tunnel junctions, *Appl. Phys. Lett.* **88**, 182508 (2006).
70. X. Liu, D. Mazumdar, W. Shen, B. D. Schrag, and G. Xiao, Thermal stability of magnetic tunneling junctions with MgO barriers for high temperature spintronics, *Appl. Phys. Lett.* **89**, 023504 (2006).
71. G. Landry, Y. Dong, J. Du, X. Xiang, and J. Q. Xiao, Interfacial capacitance effects in magnetic tunneling junctions, *Appl. Phys. Lett.* **78**, 501 (2001).
72. C. Zhang, X. Zhang, P. Krstić, H. Cheng, W. Butler, and J. MacLaren, Electronic structure and spin-dependent tunneling conductance under a finite bias, *Phys. Rev. B* **69**, 134406 (1–12) (2004).
73. P. Padhan, P. LeClair, A. Gupta, K. Tsunekawa, and D. D. Djayaprawira, Frequency-dependent magnetoresistance and magnetocapacitance properties of magnetic tunnel junctions with MgO tunnel barrier, *Appl. Phys. Lett.* **90**, 142105 (2007).
74. D. C. Worledge and P. L. Trouilloud, Magnetoresistance measurement of unpatterned magnetic tunnel junction wafers by current-in-plane tunneling, *Appl. Phys. Lett.* **83**, 84 (2003).
75. N. S. Safron, B. D. Schrag, X. Liu et al., Magnetic characterization of magnetic tunnel junction devices using circle transfer curves, *J. Appl. Phys.* **103**, 033507 (2008).
76. D. Bonnell, *Scanning Probe Microscopy and Spectroscopy: Theory, Techniques, and Applications*, Wiley-VCH, New York, 2001.
77. M. R. Scheinfein, J. Unguris, M. H. Kelley, D. T. Pierce, and R. J. Celotta, Scanning electron microscopy with polarization analysis (SEMPA), *Rev. Sci. Instrum.* **61**, 2501 (1990).
78. M. R. Freeman, Advances in magnetic microscopy, *Science* **294**, 1484–1488 (2001).
79. S. Chatraphorn, E. F. Fleet, F. C. Wellstood, L. A. Knauss, and T. M. Eiles, Scanning SQUID microscopy of integrated circuits, *Appl. Phys. Lett.* **76**, 2304 (2000).
80. M. Volk, S. Whitlock, C. H. Wolff, B. V. Hall, and A. I. Sidorov, Scanning magnetoresistance microscopy of atom chips, *Rev. Sci. Instrum.* **79**, 023702 (2008).
81. W. Shen, X. Liu, D. Mazumdar, and G. Xiao, In situ detection of single micron-sized magnetic beads using magnetic tunnel junction sensors, *Appl. Phys. Lett.* **86**, 253901 (2005).
82. W. Shen, B. D. Schrag, M. J. Carter et al., Detection of DNA labeled with magnetic nanoparticles using MgO-based magnetic tunnel junction sensors, *J. Appl. Phys.* **103**, 07A306 (2008).
83. W. Shen, B. D. Schrag, M. J. Carter, and G. Xiao, Quantitative detection of DNA labeled with magnetic nanoparticles using arrays of MgO-based magnetic tunnel junction sensors, *Appl. Phys. Lett.* **93**, 033903 (2008).
84. P. P. Freitas, R. Ferreira, S. Cardoso, and F. Cardoso, Magnetoresistive sensors, *J. Phys.: Condens. Matter* **19**, 165221 (2007).
85. D. Baselt, A biosensor based on magnetoresistance technology, *Biosens. Bioelectron.* **13**, 731–739 (1998).
86. L. Ejsing, M. F. Hansen, A. K. Menon, H. A. Ferreira, D. L. Graham, and P. P. Freitas, Planar Hall effect sensor for magnetic micro- and nanobead detection, *Appl. Phys. Lett.* **84**, 4729 (2004).

35

Magnetoresistive Random Access Memory

Johan Åkerman
University of Gothenburg

35.1 Introduction

35.1.1 Background

While long-term information storage has been dominated by magnetic technologies such as hard drives and tape drives for the last 50 years, the time when magnetic core memories dominated random access memories (RAM) is long gone (for reviews on early memory technologies see, e.g., Refs. [1,2]). When static RAM (SRAM) and dynamic RAM (DRAM), both based on thin-film semiconductor technology, were introduced in the 1970s, they rapidly overtook magnetic core memory thanks to their superiority in immediate miniaturization, continued scalability, and ever-decreasing production cost per memory bit (see, e.g., Ref. [3]). Attempts at developing alternative magnetic memory technologies to compete with semiconductor memories, such as thin-film magnetic bubble memory, did not succeed commercially.

In the mid-1990s, with the looming success of giant magnetoresistance (GMR)-based spin valves as read heads in hard drives, Stuart Wolf at Defense Advanced Research Projects Agency (DARPA) initiated a program targeting the development of magnetoresistive RAM (MRAM). At about the same time, significant tunneling magnetoresistance (TMR) was demonstrated at room temperature in magnetic tunnel junctions (MTJs) [4,5]. Together, the available funding and the breakthroughs in spin valve and MTJ technologies spurred an intense research effort, which led to the birth of an entire new field of spintronics—short for spin electronics.

On July 10, 2006, a little over 10 years after the initial DARPA program, the first commercial MRAM was introduced into the market by Freescale Semiconductor [6]. Freescale then spun off its MRAM business into a separate company, Everspin Technologies, which diversified the introductory 4 Mb MRAM into similar products having different storage capacities and form factors.

The commercial use of MRAM is steadily increasing and next-generation MRAM technologies are being actively researched and developed in many laboratories and companies around the world. In this chapter, I have chosen to focus on so-called Toggle MRAM, developed by Motorola, Freescale Semiconductor, and Everspin Technologies, since it remains the only commercial MRAM product available in the marketplace (for a description of MRAM researched and developed by IBM see, e.g., Ref. [7]). Toggle MRAM may, however, soon be followed by spin transfer torque MRAM (STT-MRAM), which offers much higher storage density and greater scalability. The potential and challenges of STT-MRAM are very briefly discussed at the end of this chapter.

35.1.2 MRAM Properties

MRAM combines a set of properties not simultaneously available in any other memory technology: nonvolatility, low-voltage

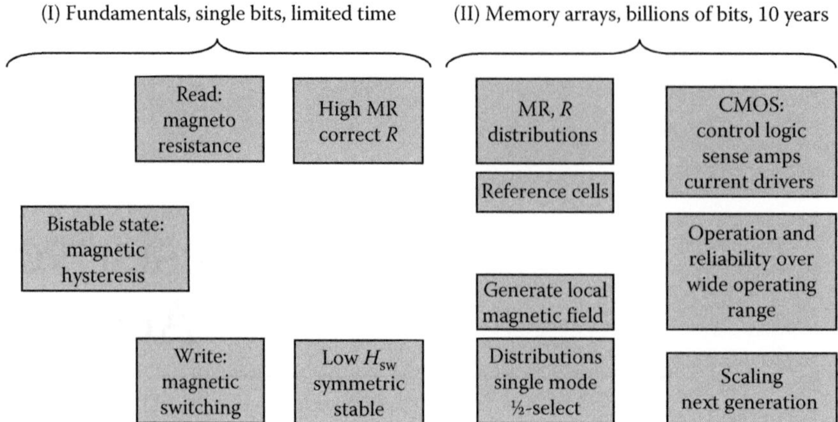

FIGURE 35.1 Different aspects of MRAM that have to be developed for a fully functional memory. While all fundamental aspects were addressed early on, and often by academic research groups, significant challenges arise when full arrays and operations over 10 years must be addressed for an actual product.

operation, unlimited read and write endurance, high-speed read and write operation, excellent reliability, and radiation hardness (for earlier reviews on MRAM see, e.g., Refs. [7–10]. For example, SRAM has excellent read and write speeds, is readily integrated into embedded applications, but lacks nonvolatility. DRAM combines high read and write speeds with a minimal cell size. However, its use of a leaky storage capacitor requires continuous refresh cycles that leads to high power consumption, lack of nonvolatility, and an increased process complexity in embedded applications. Flash memory is nonvolatile and can be fabricated with very high storage capacity. However, in many applications where high capacity is not a target, its slow write mode and limited write endurance put it at a disadvantage. While ferroelectric RAM (FeRAM) is also nonvolatile and offers better write performance than Flash, it is based on materials that dramatically increase the complexity of process integration. As a consequence of these different limitations, most systems combine a number of memory types to achieve optimal operation. MRAM on the other hand combines the most critical properties and, thanks to this universal nature, can eliminate or minimize the need for multiple memories in many applications where maximized storage capacity is not the main driver. For MRAM to also address applications where high storage capacity is needed, it must shrink its cell size significantly, and it is now generally believed that this can only be achieved to a sufficient degree by replacing Toggle MRAM with STT-MRAM where bit cell programming is no longer carried out by magnetic fields in proximity to the MTJ, but by spin-polarized current through the device.

35.1.3 Developing a Memory Technology

For the last 20 years, the solid-state memory market has been entirely dominated by SRAM, DRAM, and Flash. While the core technologies of these dominant memory types have remained largely unchanged, they have benefited from systematic evolutionary innovation toward denser, faster, and cheaper memories. To introduce an entirely new and competing memory technology is quite a daunting task since there is less established processing and test experience to draw from. While it is quite possible to demonstrate a novel memory technology on a laboratory scale, with a focus on single-bit operation at optimum conditions, it is a completely different challenge to develop a fully functioning memory with reliable operation for tens of years over a wide range of operating conditions.

Figure 35.1 shows a schematic of the different aspects of MRAM that have to be developed in order to deliver a fully operational memory into the marketplace. It is divided into two important and distinctly different parts: (I) the fundamentals of single-bit operation and (II) the realities of array operation. One may very well define and design an operating scheme that works in a limited parameter space, but with the many unknowns and imperfections of the real world, it does not take long before unexpected failure mechanisms dominate over functional operation. The devil is in the distributions: distributions of fundamental intrinsic properties, distributions of extrinsic properties, and distributions of usage conditions. While statistical analysis of accelerated tests of the fundamental properties can be used to extrapolate to actual operating conditions, only a fully tested memory array can teach us about the extrinsic failure mechanisms that will dominate the behavior in the field.

35.2 Single-Bit Fundamentals

As one embarks on the road to a new memory technology, one first has to find or design some sort of hysteresis, that is, some material or system property with memory. In SRAM, one typically uses six transistors in a flip-flop architecture, while DRAM and Flash rely on stored electric charge. In MRAM, the obvious choice is the magnetic hysteresis associated with the spontaneous magnetization of ferromagnetically ordered materials (see, e.g., Ref. [11]). Magnetic anisotropy, the property that makes the energy of a magnetic system depend on the angle of the magnetization, can be used to create two well-defined energy minima for the magnetization, which can then be interpreted as two digital states: 0 and 1. In MRAM, one of the ferromagnetic layers—the so-called free layer—has two such well-defined stable states, which represent the 0 and 1 of the MRAM bit cell, whereas the other ferromagnetic layer—the so-called fixed layer—is kept in a single well-defined magnetic state and serves as a reference to the free layer magnetization direction.

One then needs a way to detect both which state the free layer magnetization occupies at any given time (read the memory bit) and a way to prepare the free layer magnetization in either of the states (write the memory bit). In MRAM, the resistance of the bit cell MTJ depends strongly on the relative orientation of the free and fixed layer magnetizations, through the TMR effect, and is used to detect the direction of the free layer magnetization with respect to a fixed layer magnetization (see Chapters 9 through 11 for a thorough description of MTJs and TMR). External magnetic fields are then used to switch the free layer orientation between its two stable states.

These are all the prerequisites for single-bit MRAM operation. In the following sections, I will describe the fundamental properties governing single-bit operation, the additional challenges of full array operation, the long-term reliability of MRAM, and finally the emerging MRAM applications.

35.2.1 The 1T-1MTJ Memory Cell

In the so-called 1-transistor-1-magnetic-tunnel-junction (1T-1MTJ) MRAM bit cell [12] used in all commercial MRAM, a single MTJ is connected to ground via a single, minimum-sized, n-type field effect transistor (FET) (Figure 35.2). The bottom electrode of the MTJ is connected to the drain of its transistor, and its top electrode is connected to the "bit line." The gate of the transistor is connected to a "word line" that runs perpendicular to the bit line. During read operation, the word line turns on all transistors along its length and the bit line applies a bias of about 0.3 V to all MTJs along its length. As a result, current will only flow through the MTJ where the word and bit lines intersect, and by measuring the current, the resistance state of the MTJ can be determined. While many other cells are "half-selected," that is, their transistor is either open or their MTJ top electrode is biased, these bit cells do not contribute to the resulting current fed into the read-out circuitry (sense amplifier).

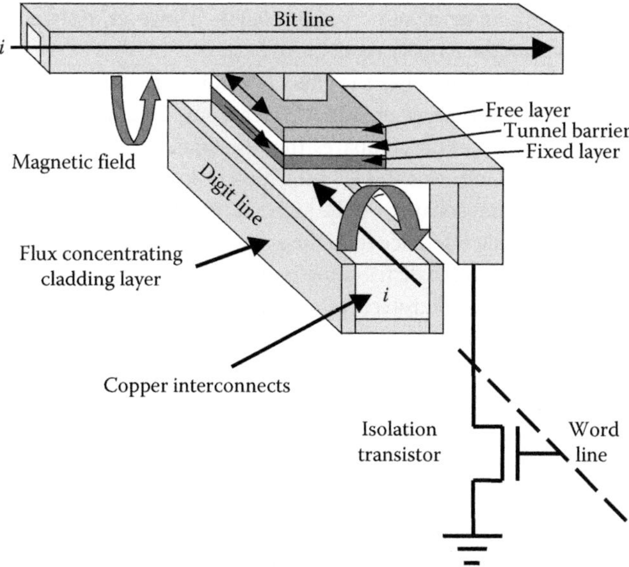

FIGURE 35.2 MRAM memory cell with one MTJ, one read-select transistor, and two write lines (bit line and digit line) for write selectivity. Each write line has flux concentrating cladding on three of the four sides of the Cu interconnect to enhance and focus the magnetic field created during writing.

A similar selection scheme is used for programming (writing) of the bit cell. Two sets of write lines run perpendicular to each other and only where two energized write lines intersect will a bit cell get written by the combined magnetic field pulses from both write lines (red bit in Figure 35.3).

However, in contrast to the read-out selection, where half-selected bits do not interfere with the read-out, the bit cells that are half-selected during writing may inadvertently switch (gray bits in Figure 35.3). Until the invention of toggle switching (see section below), this half-select problem dominated write failures in full arrays and initially threatened the success of the entire MRAM program.

35.2.2 Two Stable States

In MRAM, digital information is stored in the direction of the free layer magnetization of the MTJ. It is therefore critical that the free layer has two well-defined stable states and that switching between these two states is not only reproducible but identical between MTJs. Magnetic anisotropy in the free layer creates an energy barrier to magnetization reversal, E_b, separating the two states. One typically tries to minimize the material anisotropy, that is, magneto-crystalline and magnetostrictive anisotropies, since their direction and variation is difficult to control and leads to wide switching distributions. Magnetostriction is held at zero since strain can be quite large, arising from the different processing steps and different materials used, and furthermore may change substantially with temperature due to different thermal expansion coefficients for different materials.

Shape anisotropy, on the other hand, is primarily used to control the switching properties and critically depends on the quality of the lithographic process used in defining the bit. The elongated elliptical shape of the free layer defines an easy (long) and hard (short) axis. Either of the two stable magnetization states points along the easy axis (Figure 35.3). The barrier to reversal, E_b, is at a maximum when no magnetic field is applied. Regardless of direction, a magnetic field will always reduce the reversal barrier. Any applied field can be decomposed into an easy (H_{easy}) and hard (H_{hard}) axis field component where H_{hard} acts to reduce the energy barrier and H_{easy} not only reduces the barrier but also makes it asymmetric so as to favor one of the two stable states. At small enough dimensions, when the free layer is a single magnetic domain, one may expect magnetization

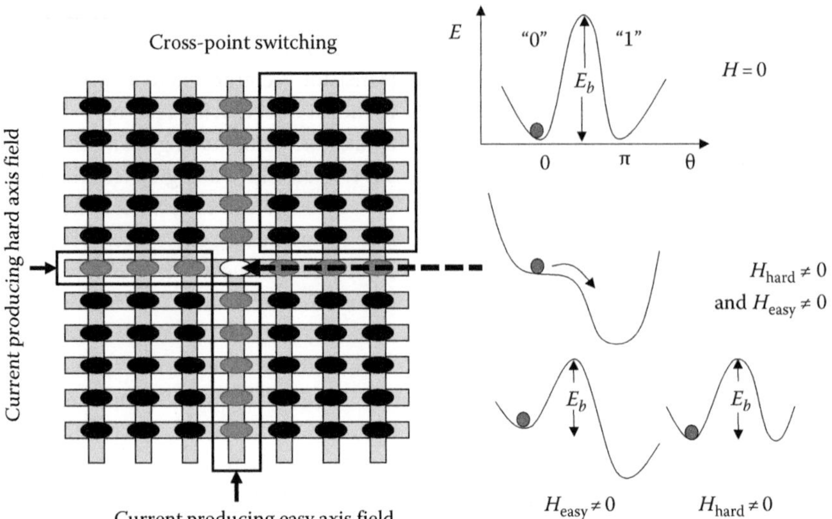

FIGURE 35.3 Conventional MRAM array showing cells in three fundamentally different states during writing: the fully selected bit (white), the half-selected bits (gray), and the unselected bits (black). To the right are shown the corresponding energy barriers.

reversal to proceed by coherent rotation. As in the case of an ellipsoidal particle, such a reversal can be described by a Stoner–Wohlfarth (SW) switching astroid [13]. However, in the planar geometry, the free layer is only elliptical viewed top down, but square in cross section and thus not ellipsoidal, which would be required for true single domain behavior. Actual switching data of elliptical free layers hence deviates somewhat from the ideal SW behavior.

35.3 Memory Arrays

35.3.1 4 Mb Array

The first commercially available MRAM, a 4 Mb MR2A16A chip from Freescale Semiconductor (today Everspin Technologies), contained 4,194,304 MTJs (not counting redundancy and error correction bits) divided into 16 I/O areas, each containing 262,144 MTJs. Both intrinsic material and lithography variations and extrinsic defects from, for example, dust and particulates, play major roles in determining the final operating windows for both read and write. When moving from single bits to millions of bits, the focus hence shifts from the intrinsic fundamental properties to the *distributions* of these properties. The final memory design must provide enough read and write *margin* and, in addition, sufficient redundancy and error correction bits, to ensure error-free operation throughout the specified lifetime of the memory.

35.3.2 Resistance Distributions

For read-out, a reference generator circuit produces a signal at the midpoint between the high and low state of the bits. Several MTJ bits produce the midpoint value, which hence follows (i) resistance variation across the wafer, (ii) changes from wafer to wafer, and (iii) changes due to temperature effects. Besides variations in operating conditions and the resistance variation across a single wafer and between different wafers, there also exist resistance variations, or distributions, within a given array. Good array uniformity is critical for correct read-out in the midpoint reference scheme. It is important to realize that the resistance distribution within an array is much larger than what one would expect from measured gradients across a wafer. Since memory arrays are only a few millimeters across, a 5% wafer-level resistance uniformity would result in bit resistance variations within an array of only about 0.05%. However, resistance sigma within an array is about 1% at best.

Figure 35.4a illustrates the effect of random resistance variations within an array on MRAM read-out. During the read operation, the circuit compares the resistance of the bit to the reference to determine if it is in the high or low state. In the low state, bits with resistance on the high side of the distribution will fall closer to the reference than the average bit. Similarly, the bits on the low side of the distribution in the high resistance state will also be closer to the reference cell. To have working megabyte memories, the circuit must be able to correctly read the state of these bits in the tails of the distributions. The circuit is designed to operate with a signal reduced by 6σ, resulting in a usable resistance change for the circuit ΔR_{use} as illustrated in Figure 35.4a. A larger ΔR_{use} allows a more aggressive circuit design with faster access time. The intrinsic quality of the MTJ barrier material plays a major role in determining the resistance distributions. Rougher barrier interfaces lead to larger local resistance variations across the MTJ surface and therefore larger final bit-to-bit variation. Since the resistance depends exponentially on barrier thickness, even small changes in roughness can cause large variations in resistance. For large roughness, conduction is largely

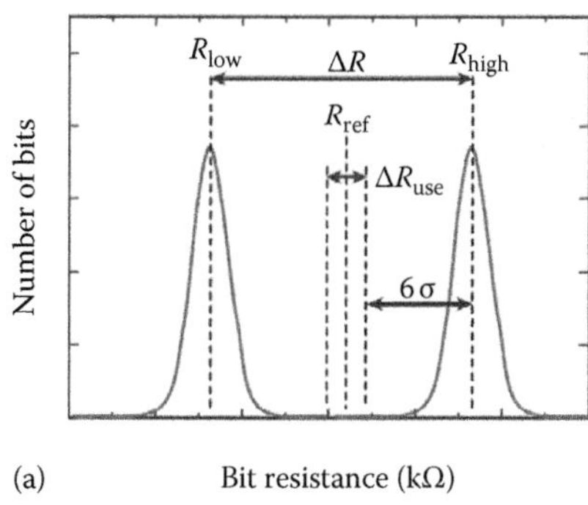

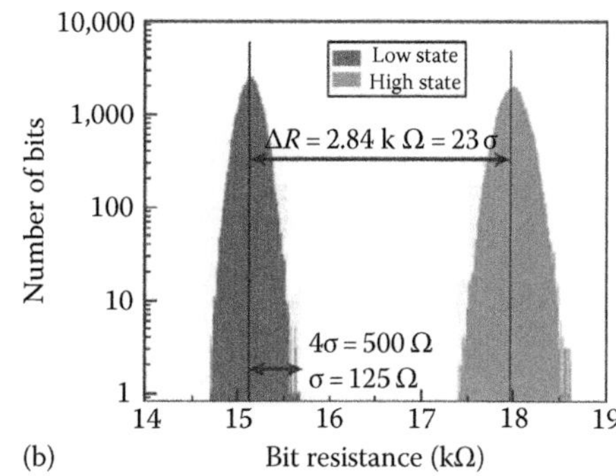

FIGURE 35.4 (a) Schematic representation of the effect of Gaussian MTJ resistance distributions resulting in the concept of an effective read margin when the full distributions are taken into account. (b) Actual MTJ resistance distributions in the high and low resistance state, respectively, showing sufficient read margin for operation. (From Engel, B. et al., *IEEE Trans. Magn.*, 41,132, 2005. With permission.)

dominated by a limited number of so-called hot-spots where most of the current is flowing [14–16]. Any variation in bit area, due to lithography or etch variations, will also cause variations in bit resistance. Figure 35.4b shows measured distributions for an array made of MTJ material with optimized array uniformity showing 23σ margin. Here, improvement in the MTJ material led to the large separation and improved uniformity.

35.3.3 Switching Distributions and the Half-Select Problem

As discussed above, material and shape anisotropy are the dominant factors governing magnetization reversal. They also determine the magnitude of the (easy axis) switching field (H_{sw}) for a given applied hard axis field. Consequently, any variation in material quality or bit shape will inevitably lead to bit-to-bit variations of the switching astroid. While both the easy and hard axis properties are affected, one typically quantifies the bit-to-bit variation as a distribution, ΔH_{sw}, of easy axis switching fields at a constant hard axis field. In this statistical description, H_{sw} denotes the average switching field of the ensemble of bits. While the bit-to-bit switching behavior may appear highly reproducible in a small laboratory experiment, the situation is quite different in an array that can consist of over four million bits. A broad ΔH_{sw} can severely limit the operating window in realistic devices. In addition, as the bit size is reduced, the relative switching distribution $\sigma_{sw} = \Delta H_{sw}/H_{sw}$ increases, reflecting the increasing difficulty in maintaining a consistent bit shape at smaller dimensions.

The switching distribution has two important consequences. First, to program all the bits in a 4Mb array with the same switching field throughout the chip, the applied field needs to be larger than the bit with the highest switching field. With over four million bits, and assuming a Gaussian distribution of switching fields, this is equivalent to a minimum field of $H_{sw} + 6\Delta H_{sw}$. However, as a consequence of the grid layout of the write lines, for every bit being programmed (the fully selected bit) with a combination of a hard axis field and an easy axis field of $H_{sw} + 6\Delta H_{sw}$, more than a thousand other bits (so-called half-selected bits) along the two write lines either experience the applied hard axis field alone, or $H_{sw} + 6\Delta H_{sw}$ without a hard axis field. Consequently, the easy axis switching field used throughout the array must also be kept below a certain *maximum* value, related to the bit with the *lowest* switching field when no hard axis field is applied. If this distribution is ignored, half-selected bits may inadvertently switch and lead to writing errors, so-called half-select disturbs. For error-free programming, conventional MRAM (in contrast to Toggle MRAM described in the next section) must keep the switching distribution and the half-select disturb distribution well-separated. Between these two distributions, there is a resulting operating window where all bits can be programmed without errors or disturbs. A schematic of the operating window superposed on the switching astroid is shown in Figure 35.5a.

Already in the early period of MRAM development, it became apparent that the half-select problem amounted to a formidable challenge. In addition to the distributions described above, the problem is exacerbated by other sources of variations such as CMOS transistor variations, (i.e., variations in the actual switching current provided by each write line driver), variations in the write line resistances, and variations in the magnetic cladding properties along the write line. The final blow to conventional MRAM switching is dealt when the true nature of the probabilistic single-bit switching is analyzed and extrapolations to 10 years of operation are considered. Due to thermal energy, the half-select disturb field for a single bit has an intrinsic temperature-dependent distribution. The more disturb attempts each bit is given, the more one probes the tail of this intrinsic distribution, and the further away from the average disturb field one has to operate in order to ensure zero disturbs.

An example of a measured switching astroid of a 256kb conventional MRAM demonstrator is shown in Figure 35.5b. While an optimum region does exist, this region is still not error free. A fully functioning operating region could not be strictly defined for this demonstrator.

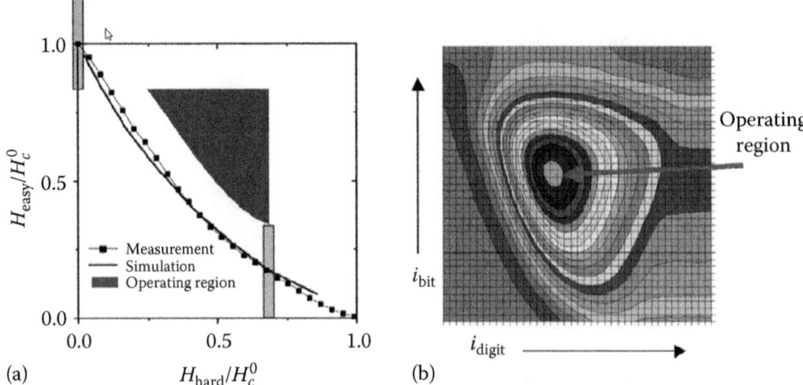

FIGURE 35.5 (a) Constructed write operating region of conventional MRAM. The solid points are quasi-static switching measurements and the solid line is the expected switching curve from micromagnetic simulation. The rectangles represent the switching field distribution within an entire array, which effectively limits the operating region (triangular area). (b) Actual measurement of the write operating region of a conventional MRAM array showing insufficient write margin for error-free operation. (From Engel, B. et al., *IEEE Trans. Magn.*, 41,132, 2005. With permission.)

35.3.4 Savtchenko Switching and Toggle MRAM

At the turn of the millennium, the half-select challenge increasingly appeared as an insurmountable brick wall. While the continued material and lithography improvements had indeed led up to important demonstrations of functional MRAM arrays, even the best distributions were nowhere near the required separations. Predictions of continued incremental improvement did not look promising and the chances that MRAM would become feasible looked quite slim. As we now know, commercial conventional MRAM never saw the light of day.

The half-select problem was instead solved through an ingenious invention by the late Motorola Staff Scientist Leonid Savtchenko. The invention includes a novel free layer structure and bit orientation, along with the use of a particular set of timed current pulses, which, in a *coup de grace*, completely removes the limitations that plagued conventional switching [17]. The mechanism proposed by Savtchenko completely removes half-select disturbs. The single free layer is replaced by a synthetic antiferromagnetic (SAF) free layer, that is, a trilayer structure where two magnetically soft ferromagnetic layers couple antiferromagnetically through a nonmagnetic spacer, for example, Ru. This is shown schematically in Figure 35.6a.

The response of the SAF to external fields bears little resemblance to the traditional switching astroid. Instead of the magnetic layers in the SAF aligning with the applied field, the generic response of the SAF is a 90° orientation of both magnetic moments with respect to the applied field, a so-called spin-flop orientation (Figure 35.6b). If each layer has a finite anisotropy field, a certain critical spin-flop field H_{sw} must first be overcome. In this spin-flopped state, each moment will cant slightly in the direction of the applied field, that is, the SAF layers will scissor an angle δ away from perfect anti-alignment. Through this scissoring mechanism, the SAF can lower its total magnetic energy, and the scissoring angle depends on the competition between

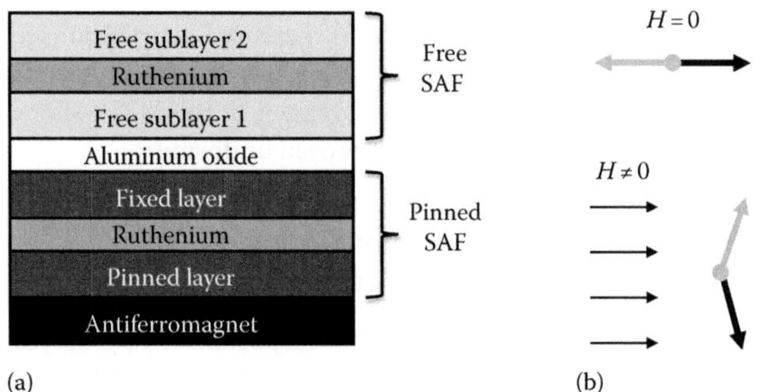

FIGURE 35.6 (a) Schematic of a Toggle MRAM MTJ with an SAF free layer. (b) The spin-flop response of an SAF free layer when an external field is applied.

the Zeeman energy due to the applied field and the antiferromagnetic exchange energy through the spacer layer [18].

To derive H_{sw}, we first analyze the case where neither of the magnetic layers has any anisotropy field, H_k (Figure 35.7a). In an applied field, H, the angular-dependent energy of the SAF can be expressed as

$$E(\theta,\delta) = -K_S\cos(2\delta) - 2M_SH\sin(\theta)\sin(\delta) \quad (35.1)$$

where

M_S is the saturation magnetization of the magnetic layers

K_S is the antiferromagnetic exchange anisotropy between the layers

Minimizing with respect to δ, one finds the energy as a function of field angle

$$E(\theta) = -\frac{M_S^2H^2}{2K_S}\sin^2(\theta) + C \quad (35.2)$$

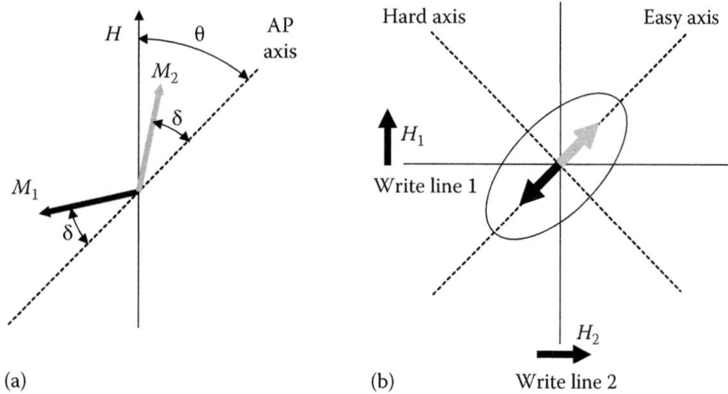

FIGURE 35.7 (a) Definitions of the angles used in the derivation of the spin-flop field. (b) Schematic Toggle MRAM bit having its easy axis oriented at 45° with respect to write lines.

where *C* is a constant independent of θ. The applied field hence acts as an anisotropy *orthogonal* to its own direction. If we now add H_k of each layer and analyze the case when *H* is applied along the same axis as H_k, the angular dependence of the energy changes to

$$E(\theta) = M_S H_k \left(1 - \frac{H^2}{H_k H_{\text{sat}}}\right) \sin^2(\theta) \tag{35.3}$$

where we have defined the saturation field of the SAF as $H_{\text{sat}} = 2K_S/M_S$. For a spin-flop reorientation to occur, the expression in the parenthesis of Equation 35.3 must be negative, that is, the critical flop field is the arithmetic mean of the anisotropy field and the saturation field:

$$H_{\text{sw}} = \sqrt{H_k H_{\text{sat}}} \tag{35.4}$$

However, overcoming the flop field is not sufficient for well-defined switching. If the field is again reduced below H_{sw}, the SAF may go back to either of its two stable zero-field states. Actual programming is hence implemented using a bit orientation at 45° with respect to the two write lines (Figure 35.7b) in combination with tailored field pulses.

The programming pulse sequence and the resulting behavior of the moment of the two magnetic layers, represented by arrows, are depicted in Figure 35.8. When write line 1 turns on (t_1), the magnetic field from this single line will prepare the SAF in an initial scissored state, with its net moment trying to align with the applied field. With write line 1 still on, the second write line turns on (t_2) and the resulting fields add up vectorially to overcome the critical flop field, H_{sw}. While the SAF is now in its fully spin-flopped state, no actual switching has yet taken place. To finally switch the bit, the first write line turns off (t_3), resulting in a final rotation of the SAF toward a scissored state with the net SAF moment largely parallel with the applied field

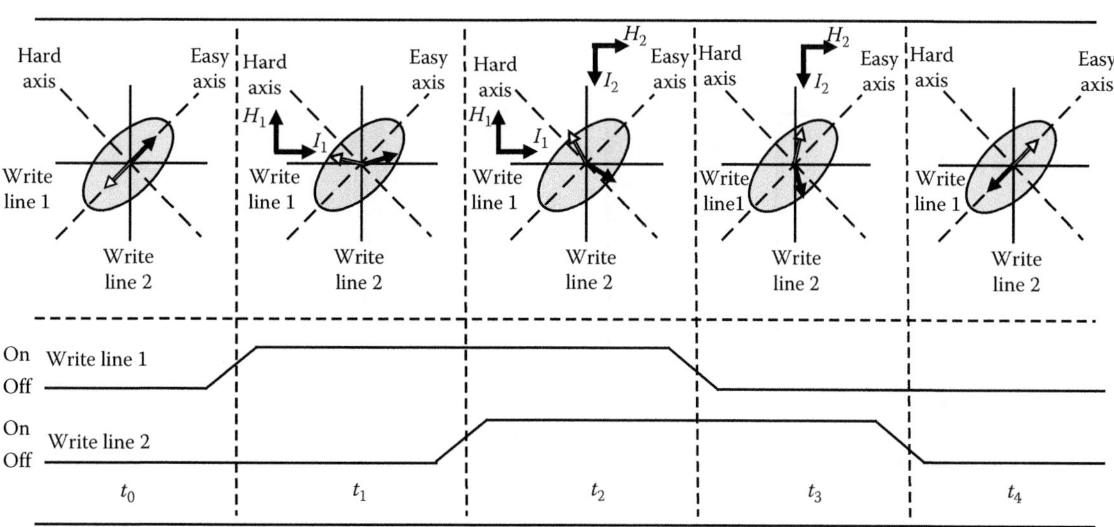

FIGURE 35.8 Programming of a Toggle MRAM cell. Write pulses are applied sequentially to rotate the free layer SAF 180° regardless of its initial state.

from write line 2. In the final step, the switching is completed by turning off write line 2 (t_4). During this field pulse sequence (t_0–t_4), the two individual layers have performed a 180° rotation. Due to symmetry, if the process is repeated with the same polarity pulses, the two layers of the SAF will again rotate 180° in the same manner. The programming scheme is hence of a toggling type, that is, it does not switch into a predetermined state; it instead toggles from one state to the other regardless of the initial state. This toggling process is a result of the carefully balanced nature of the SAF and any significant imbalance of the two moments will render it ineffective. The toggling scheme also has three direct consequences for the programming circuit: (i) it removes the need for bipolar current drivers for the easy axis field, which then simplifies the design and allows for the use of smaller transistors, (ii) it requires a decision write scheme, where the bit state is first read and only toggled if the new data differs from the existing data, and (iii) it effectively reduces the average programming power consumption by 50% since for half of the programming events, the decision circuitry will simply not energize the write lines.

However, the dominating advantage of toggle MRAM is its greatly enhanced selectivity. If only a single write line current is applied (half-selected bits) as in t_1, the 45 field angle cannot switch the state. In fact, the single-line field *raises* the switching energy barrier of those bits so that they are stabilized against reversal during the field pulse. This is in stark contrast to the conventional approach, where all of the half-selected bits have their switching energy reduced and are therefore more susceptible to thermally excited switching events. The phase relationship of the combined pulses is therefore required for switching and results in several orders of magnitude improvement of the bit selectivity in a full memory array.

Similar to the switching astroid of a conventional MRAM bit, the switching response curve for a Toggle MRAM bit can be constructed as a function of the two write lines, as shown in Figure 35.9a. The upper right and lower left quadrants are for same polarity field pulses and will allow toggling. The white regions are either insufficient field magnitudes or improper field directions for toggling. The black regions are above the toggling transition for all bits and result in 100% toggling. The gray regions represent the bit-to-bit distribution of switching thresholds that must be overcome for reliable switching of all bits in an array. Figure 35.9b is a switching versus current map for an entire 4 Mb memory. The test pattern was a checkerboard/inverse checkerboard of alternating “1” and “0” that was written at each current and read back to verify. In the region below the switching threshold, no bits changed state and hence there were no disturbs from half-selects. A large operating region is observed above the threshold consistent with the single-bit characteristics. The contours in the transition region just at the threshold are a measure of the bit-to-bit switching distribution.

35.4 Reliability

One of the key properties of MRAM is its very high reliability. The promise that a programmed “1” or “0” indeed gets stored as a “1” or “0,” and that this stored information remains stable for over 10 or 20 years under all use conditions is of paramount importance. Additionally, the information must also be read back correctly at all times and no wear-out may occur as a function of the number of reads or writes. While this may sound like quite natural demands on a memory, reliability issues are often overlooked during the early stages of development of a new technology. All well-established memory technologies have a long history of detailed reliability studies and all failure mechanisms are analyzed and modeled in order to ensure zero bit failures at usage conditions. If a new memory technology cannot demonstrate that all possible failure mechanisms have been studied, analyzed, and modeled, it is unlikely that it will gain any market acceptance. Reliability is hence part of the main function of a memory and the technology must be designed for good reliability at as early a stage as possible.

In MRAM, reliability can be divided into well-known CMOS reliability issues and new failure mechanisms related to the

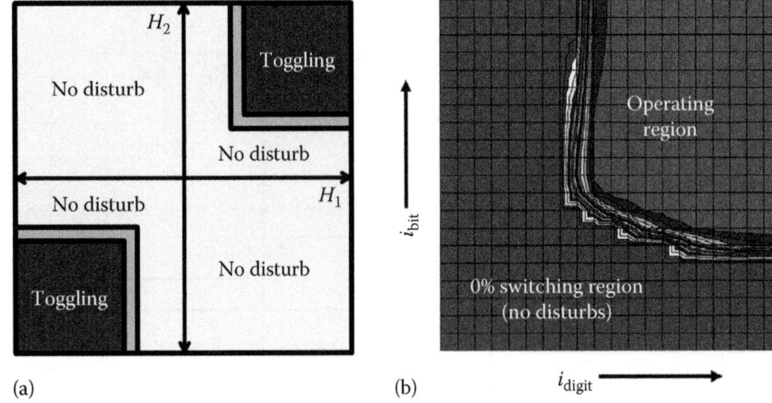

FIGURE 35.9 (a) Ideal switching map of Toggle MRAM showing two square toggling regions when both pulses have the same polarity and both pulses have sufficient magnitude. (b) An actual switching map (quadrant 1) of an entire 4 Mb array showing a large operating region for 10 ns write pulses. (From Engel, B. et al., *IEEE Trans. Magn.*, 41,132, 2005. With permission.)

MRAM module itself. The possible MRAM failure mechanisms can in turn be divided into MTJ reliability, write line reliability, (since the magnetic cladding is an entirely new part of the device), and long-term effects of temperature on bit stability and write currents.

35.4.1 MTJ Reliability

The two main types of MTJ failures are time-dependent dielectric breakdown (TDDB) and time-dependent resistance drift (TDRD), both induced by voltage stress to the junction. At constant voltage stress, dielectric breakdown is detected as an abrupt increase of the MTJ current due to a short forming through the tunneling barrier [19]. Since the metallic conduction through the short is mostly spin independent [20], the increase in current is accompanied by a loss of essentially all magnetoresistance. From the memory's point of view, a shorted MTJ results in a stuck-at-zero bit failure.

Dielectric breakdown can be successfully studied in single MTJs under ramped voltage stress, and yields a rapid assessment of the average quality of the MTJ barrier and provides rapid feedback for material and process improvements [21,22]. However, for accurate modeling of TDDB and for extrapolations to usage conditions and long lifetimes, it is necessary to study much larger arrays of actual memory bits to get enough statistics and to make sure that the correct failure mechanisms are indeed studied. For this purpose, custom-designed 1 kb memory reliability circuits, with the same MTJ as in the full 4 Mb array but with additional voltage read-out points, were designed to allow for accurate measurements of the applied bias stress over the bit as well as the bit resistance during stress [9]. Figure 35.10a shows typical TDDB data obtained from this array stressed at three different temperatures (125°C, 150°C, and 175°C) and five different nominal bias levels (1.1, 1.15, 1.2, 1.25, and 1.3 V). The actual bias level during stress is continuously measured and the exact values are used in the TDDB analysis. Each stress cell had a total of 32 bits with all bits driven to failure for optimal statistics. As expected for dielectric breakdown, the data can be well fitted (straight lines) by a model based on Weibull lifetime distributions, linear bias voltage acceleration (linear-E model), and an Arrhenius-type thermal activation.

In contrast to other solid-state memories, the MRAM bit experiences no applied bias stress during data retention, since the information is stored magnetically at zero voltage. As a consequence, only 16 memory bits are stressed simultaneously at any given time during the lifetime of the memory and the typical stress duty cycle of each bit is *extremely small.* While the exact customer usage is unknown, all possible usage falls somewhere between perfectly uniform memory usage and reading/writing the same 16 bit word for 10 years. As shown in Åkerman et al. [9], the rather high value of β (the slope of the straight lines in Figure 35.10a) makes reading the same word over and over again about two orders of magnitude more stressful than the case of uniform usage. Extrapolations to 100 ppm fallout at 95% single-sided confidence are shown in Figure 35.10b. The strong bias dependence and the low operating bias of 0.3 V makes the intrinsic reliability about four orders of magnitude better than required for 10 years of operation. Alternatively, one can argue that there is substantial design margin up to 0.6 V should one choose to increase the signal for faster read operation.

The MTJ is also subject to TDRD, which in contrast to TDDB shows a highly deterministic and uniform behavior over time in that the resistance of all bits drifts about the same for the same stress conditions. In Figure 35.11a, we show resistance drift at 120°C for bias levels ranging from 0.6 to 1.2 V. Since increasing the bias stress only shifts the curves along the log t axis, the time axis can be rescaled ($t \Rightarrow t^*$) to make all curves overlap (i.e., plot R vs. the scaled t^* instead of the original t). The single master curve, shown in Figure 35.11b, indicates that bias stress only accelerates the rate of resistance drift without introducing new sources of drift. In the inset of Figure 35.11b, it is shown that t^* versus bias voltage follows a linear-E model (the activation energy is reduced linearly with the electric field E) and that an extrapolation to operating bias (0.25 V in this case) yields

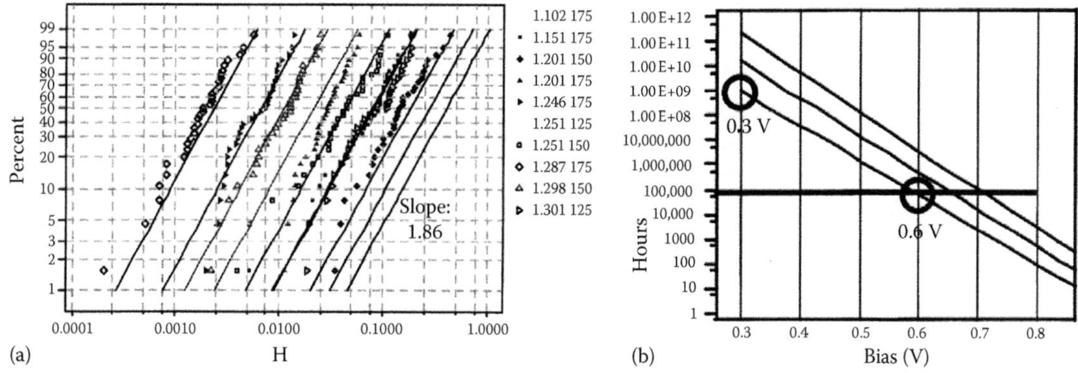

FIGURE 35.10 (a) Time-dependent dielectric breakdown data from a dedicated 1 kb MRAM reliability array. In the legend to the right of the plot, the first number indicates the applied voltage stress and the second number indicates the temperature. (b) Extrapolation to 100 ppm failures with 95% single-sided confidence. The horizontal line indicates 10 years of operation and determines the upper limit of MTJ bias (0.6 V) for reliable operation. (From Åkerman, J. et al., *IEEE Trans. Dev. Mat. Rel.*, 4, 428, 2006. With permission.)

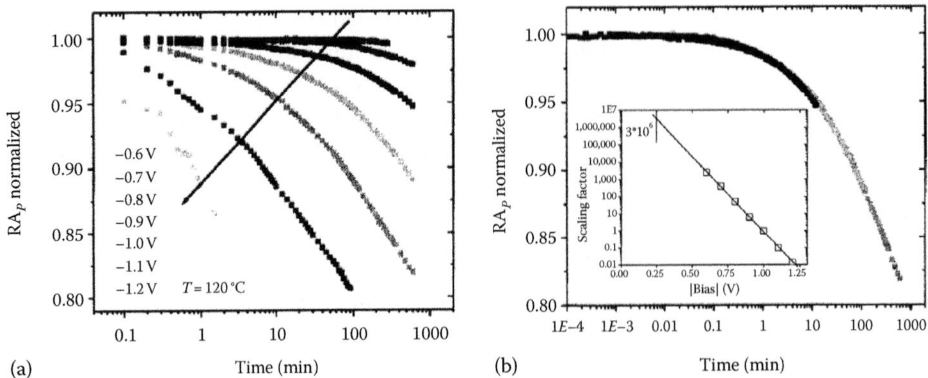

FIGURE 35.11 (a) Normalized time-dependent resistance drift for different values of the applied bias at an ambient temperature of 120°C. (b) The same data after a scaling procedure allowing for extrapolation to lower bias and longer times as well as a master curve for resistance drift. (From Åkerman, J. et al., *IEEE Trans. Dev. Mat. Rel.*, 4, 428, 2006. With permission.)

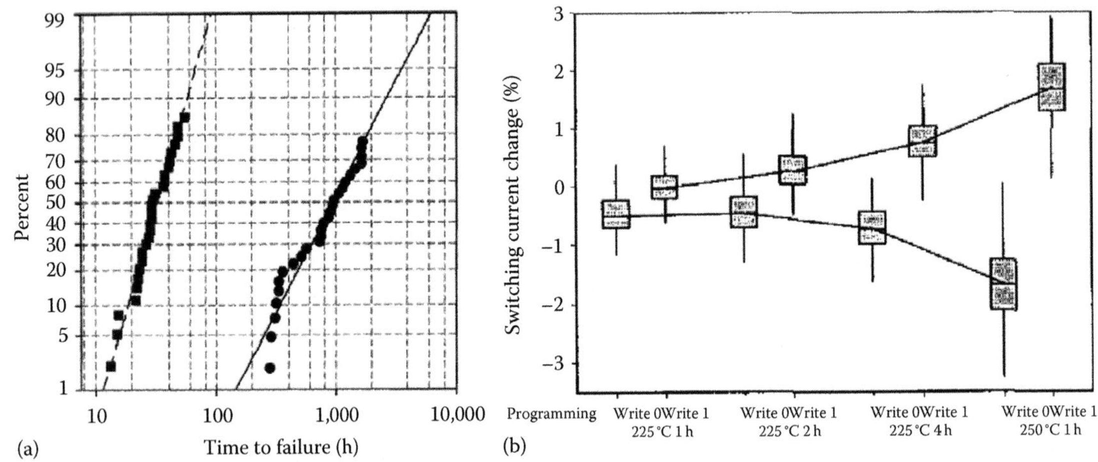

FIGURE 35.12 (a) Distribution of write line failure times due to electromigration. The left data set shows the result from standard Cu interconnects and the right data set shows an improvement when the magnetic cladding is added to the write lines. (From Åkerman, J. et al., *IEEE Trans. Dev. Mat. Rel.*, 4, 428, 2006. With permission.) (b) Change in the switching current after heating the patterned Toggle MRAM bit to different temperatures and for different times. (From Åkerman, J. et al., *IEEE Proceedings 43rd International Reliability Physics Symposium*, 163, 2005. With permission.)

a scaling factor of 3×10^6. Ten years of stress at 0.25 V is hence equivalent to 1 min and 45 s of stress at 1.0 V and the corresponding resistance drift can be directly read out from the graph. For example, 10 years of normal operation at 120°C and 100% (25%) duty cycle would result in R drift of −2.2% (−1.0%). Just as for TDDB, the worst case for resistance drift is single word usage. At continuous operation, the actual bias stress is only applied for about 50% of the cycle time. As long as the circuit can handle roughly 2% resistance drift, no read-out failures are expected due to this mechanism for 10 years at 120°C.

35.4.2 Write Line Reliability

Sufficient metal interconnect reliability is a standard requirement for any CMOS process and the required MRAM write currents and operating conditions can be designed within the known reliability performance. However, the addition of magnetic cladding onto three out of four surfaces of the metal interconnects is a significant unknown parameter and must be evaluated for reliability. Figure 35.12a shows a Weibull plot of the electromigration failure of standard Cu interconnects and Cu interconnects with cladding. As can be seen, the magnetic cladding actually improves the median of the fail time as much as 40 times, which is similar to other reports where a hard film cladding can improve the electromigration of Cu significantly [23,24].

35.4.3 Temperature Effects

Temperature has two important effects on MRAM performance: (i) instantaneous changes, which are typically reversible when the temperature is reduced and (ii) long-term irreversible changes after exposure to high temperature for extended times. Among the first type, we find changes in read signal and write current distributions due to the temperature dependence of the intrinsic magnetic properties of the magnetic layers, and the resistance and magnetoresistance of the MTJ [25]. The second, and irreversible, type is typically due to interdiffusion within the full MTJ stack and can become a reliability issue if not modeled and designed for.

For example, the toggling scheme described above relies on both the detailed moment balance between the two NiFe layers in the SAF free layer, and the exchange coupling mediated by the Ru layer. Therefore, factors determining the reliability of the write currents are the thermal endurance of the antiferromagnetic exchange coupling and the thermal endurance of the relative moment balance between the two constituent magnetic layers. This can be studied both in continuous films and in patterned structures to distinguish between intrinsic diffusive effects in the stack and increased diffusion from the sides of the patterned structures. Figure 35.12b shows the percent change in write currents versus several high temperature anneals in 150 4 Mb arrays. While the average of the program current does not change, consistent with no expected change in the antiferromagnetic exchange coupling, there is a significant drift in the individual program currents. This asymmetry in write-0 and write-1 current is a sign of an increased imbalance in the free layer SAF. Since bulk wafers show no change in the free layer moment balance at these temperatures, the observed changes are related to the bit processing. The rate of change is approximately five times greater at 250°C compared to 225°C, which suggests less than 1% change in switching current after 10 years of operation at 110°C, which is well within the design limits.

35.5 STT-MRAM

The greatest weakness of MRAM is its large cell size. Using 180 nm CMOS technology, 4 Mb Toggle MRAM currently has a cell size of 1.25 μm². Scaling toward 90 nm CMOS has been successfully demonstrated [26] and reduces the cell size to 0.29 μm². Continued scaling to 65 nm CMOS is expected to further reduce the Toggle MRAM cell size to 0.16 μm². Unfortunately, this is still about 10 times larger than the smallest cell size of commercial Flash technology, which is currently using 32 nm technology and is readying 25 nm technology for production.

By replacing the required external magnetic field to write the MRAM bit with the direct internal effects of the spin-polarized current passing through the bit, a so-called STT-MRAM bit cell can be designed (Figure 35.13) without any need for write lines (see Chapters 7, 8, and 13 for a detailed description of STT). The resulting bit cell is both smaller and less complex than the Toggle MRAM cell. The expected cell size using 65 nm CMOS is about 0.04 μm², which is similar to Flash at the 65 nm node and is expected to scale as favorably as Flash to smaller dimensions. STT-MRAM hence has the potential of becoming a true universal memory, where the good functional properties present in Toggle MRAM are combined with a high storage density and good scalability.

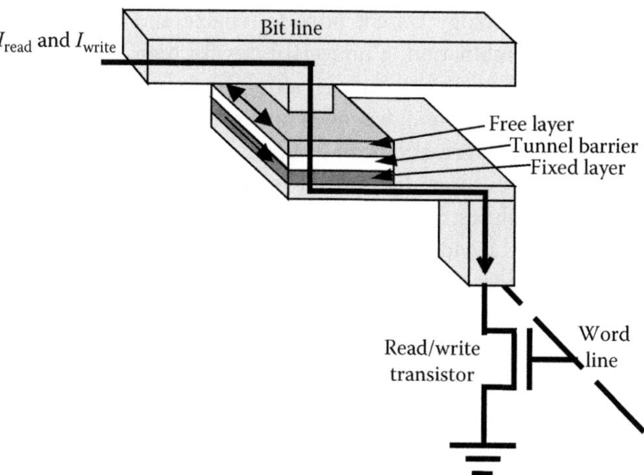

FIGURE 35.13 STT-MRAM memory cell with one MTJ and one read/write-select transistor.

Before STT-MRAM can become commercially viable, a number of challenges must be addressed. The most difficult challenge has been the reduction of the programming current while maintaining thermal stability. While a programming current of 1 mA might look like a dramatic improvement compared to the much larger programming current currently used in Toggle MRAM, it is actually not compatible with the anticipated cell size, as the CMOS transistor, which is required for each bit cell, has an upper limit to how much current it can sustain. The most promising solution, at the time of writing, appears to be perpendicularly polarized STT-MRAM where both the free and fixed layer magnetizations are oriented perpendicularly to the film plane (in other words aligned with the direction of the current through the device). This geometry allows for both thermal stability and programming currents as low as 9 μA, which is well below the estimated maximum sustainable CMOS transistor current of 20 μA, at the dimensions needed for Gbit STT-MRAM [27].

Once the single-bit operation has been solved in STT-MRAM, all possible distributions, and how they affect read and write performance, must be addressed and understood. In STT-MRAM one must control, or manage, *three* independent distributions that all are related to the current through the device: read resistance, write current, and MTJ breakdown voltage distributions. For example, a particular challenge will be to minimize the write current while also making sure that one does not accidentally write a bit during reading. Such "read disturbs" might turn out to be as detrimental to STT-MRAM as half-select disturbs were to conventional MRAM. These considerations have already spurred innovation into alternative read-out schemes using negative differential resistance, which may significantly limit the read currents [28].

35.6 Applications

With its unique combination of key memory properties, MRAM has particular potential in applications where traditional memory technology cannot provide reliable and secure solutions. The combination of nonvolatility with high write speeds and very high reliability makes it attractive for cache memories in redundant array of independent disks (RAID) storage, where critical cache information can be updated at the same speed as with SRAM, but with the added benefit of being retained during power failures. Similar continuous backup applications are applicable in a wide range of industrial applications such as communication, transportation, military, and avionic systems, where critical information needs to be retained at all times and with the highest demand on reliability. The very wide operating range, both at low and high temperatures, also makes MRAM attractive for highly reliable usage under demanding conditions such as in automotives and in space.

As a specific example, MRAM is currently employed in high volumes in Siemens's Simatic Multipanel MP 277 and MP 377 human machine interface (HMI). Such industrial automation systems, which are able to retain software-programmable logic controller process data without battery-backup, are in high demand. As a clear demonstration of the very high reliability of MRAM, Siemens has, over the course of 2 years and over 100,000 shipped multipanels, not recorded a single memory failure.

MRAM is also employed in Emerson Network Power's single-board computers used in industrial, medical, military, aerospace, and telecom platforms. In these applications, MRAM allows the capture of system data logs that are not impacted by memory wear-out or power failure. In particular, the high write speed and high reliability also allows each unit to receive frequent new programs and data updates with guaranteed data retention over 20 years even at very high temperatures.

Finally, Airbus recently announced that they will use MRAM to replace SRAM and Flash in the flight control computer on their new Airbus A350 XWB civil aircraft, currently under development.

35.7 Conclusions

The technology behind commercially available Toggle MRAM has been described, as well as the challenges and solutions associated with its development over the last 15 years. The use of MRAM is currently spreading into many different application spaces, in particular where its combination of nonvolatility, fast read/write, unlimited endurance, and high reliability is used to replace a combination of other memory technologies.

References

1. J. A. Rajchman, Computer memories—A survey of the state-of-the-art, *Proc. IRE* **49**,104 (1961).
2. L. A. Russel, R. M. Whalen, and H. O. Leilich, Ferrite memory systems, *IEEE Trans. Magn.* **4**, 134 (1968).
3. W. M. Regitz and J. A. Karp, Three-transistor-cell 1024-Bit 500-ns MOS RAM, *IEEE J. Sol. State Circuits* **SC-5**(5), 181–186 (1970).
4. T. Miyazaki and N. Tezuka, Giant magnetic tunneling effect in $Fe/Al_2O_3/Fe$ junction, *J. Magn. Magn. Mater.* **139**, L231 (1995).
5. J. S. Moodera, L. R. Kinder, T. M. Wong, and R. Meservey, Large magnetoresistance at room temperature in ferromagnetic thin film tunnel junctions, *Phys. Rev. Lett.* **74**, 3273–3276 (1995).
6. Freescale Semiconductor Inc. 2006, Freescale Leads Industry in Commercializing MRAM Technology, 4 Mbit MRAM Memory Product Now in Volume Production, Press Release on July 10, 2006. http://media.freescale.com/phoenix.zhtml?c = 196520&p = irol-newsArticle&ID = 880030
7. W. J. Gallagher and S. S. P. Parkin, Development of the magnetic tunnel junction MRAM at IBM: From first junctions to a 16-Mb demonstrator chip, *IBM J. Res. Dev.* **50**, 5–24 (2006).
8. J. Daughton, Magnetoresistive random access memories, in *Magnetoelectronics*, M. Johnson (Ed.), Elsevier, San Diego, CA, pp. 231–272 (2004).
9. J. Åkerman, M. DeHerrera, M. Durlam et al., Magnetic tunnel junction based magnetoresistive random access memory, in *Magnetoelectronics*, M. Johnson (Ed.), Elsevier, San Diego, CA, pp. 231–272 (2004).
10. J.-G. Zhu, Magnetoresistive random access memory: The path to competitiveness and scalability, *Proc. IEEE* **96**(11), 1786–1798 (2008).
11. R. O'Handley, *Modern Magnetic Materials: Principles and Applications*, Wiley-Interscience, New York, 1999.
12. P. K. Naji, M. Durlam, S. Tehrani, J. Calder, and M. F. DeHerrera, A 256 kb 3.0 V 1T1MTJ nonvolatile magnetoresistive RAM, *ISSCC Digest of Technical Papers*, February 2001, pp. 122–123.
13. E. C. Stoner and E. P. Wohlfarth, A mechanism of magnetic hysteresis in heterogeneous alloys, *Phil. Trans. Roy. Soc. A* **240**, 599 (1948).
14. V. Da Costa, C. Tiusan, T. Dimopoulos, and K. Ounadjela, Tunneling phenomena as a probe to investigate atomic scale fluctuations in metal/oxide/metal magnetic tunnel junctions, *Phys. Rev. Lett.* **85**, 876 (2000).
15. D. Rabson, B. J. Jönsson-Åkerman, A. Romero et al., Pinholes may mimic tunneling, *J. Appl. Phys.* **89**, 2786 (2001).
16. C. W. Miller, Z. Li, J. Åkerman, and I. K. Schuller, Impact of interfacial roughness on tunneling conductance and extracted barrier parameters, *Appl. Phys. Lett.* **90**, 043513 (2007).
17. B. Engel, J. Åkerman, B. Butcher et al., A 4 Mb toggle MRAM based on a novel bit and switching method, *IEEE Trans. Magn.* **41**, 132–136 (2005).
18. J. Zhu, Spin valve and dual spin valve head with synthetic antiferromagnets, *IEEE Trans. Magn.* **35**, 655–660 (1999).
19. W. Oepts, H. J. Verhagen, W. J. M. de Jonge, and R. Coehoorn, Dielectric breakdown of ferromagnetic tunnel junctions, *J. Appl. Phys.* **73**, 2363–2365 (1998).
20. J. J. Åkerman, R. Escudero, C. Leighton et al., Criteria for ferromagnetic–insulator–ferromagnetic tunneling, *J. Magn. Magn. Mater.* **240**, 86–91 (2002).

21. J. Åkerman, M. DeHerrera, J. M. Slaughter et al., Intrinsic reliability of AlO_x-based magnetic tunnel junctions, *IEEE Trans. Magn.* **42**, 2661–2663 (2006).
22. J. Åkerman, P. Brown, M. DeHerrera et al., Demonstrated reliability of 4-Mb MRAM, *IEEE Trans. Dev. Mat. Rel.* **4**, 428–435 (2006).
23. C.-K. Hu, L. Gignac, E. Liniger et al., Comparison of Cu electromigration lifetime in Cu interconnects coated with various caps, *Appl. Phys. Lett.* **83**, 869–871 (2003).
24. D. A. Gajewski, B. Feil, T. Meixner, M. Lien, and J. Walls, Electromigration performance enhancement of Cu interconnects with PVD Ta cap, *IEEE International Reliability Physics Symposium Proceedings*, Chandler, AZ, 2004, pp. 627–628.
25. J. J. Åkerman, I. V. Roshchin, J. M. Slaughter, R. W. Dave, and I. K. Schuller, Origin of temperature dependence in tunneling magnetoresistance, *Europhys. Lett.* **63**, 104–110 (2003).
26. M. Durlam, T. Andre, P. Brown et al., 90 nm toggle MRAM array with 0.29 μm 2 cells, in *2005 Symposium on VLSI Technology*, 10B-2, 2005.
27. H. Yoda, T. Kishi, T. Nagase, et al., High efficient spin torque transfer writing on perpendicular magnetic tunnel junctions for high density MRAMs, *Invited Presentation AA-02 at 11th Joint MMM-Intermag Conference*, Washington, DC, 2010.
28. D. Halupka, K. Tsunoda, C. Yoshida et al., Negative-resistance read and write schemes for STT-MRAM in 0.13 μm CMOS, *Presentation 14.1 at ISSCC 2010*, San Francisco, CA.

36

Emerging Spintronic Memories

Stuart Parkin
IBM Almaden Research Center

Masamitsu Hayashi
National Institute for Materials Science

Luc Thomas
IBM Almaden Research Center

Xin Jiang
IBM Almaden Research Center

Rai Moriya
University of Tokyo

William Gallagher
IBM Thomas J. Watson Research Center

36.1 Introduction

The electronics industry is composed of three major components: computing (logic and memory), data storage, and communications. One of the most important limitations of computing devices is the ever-increasing power consumption required by higher processing speeds. One means of reducing the power needed is the development of multi-core processors. Such processors, however, demand much larger memories so that the proportion of a silicon chip dedicated to memory will increase from typical values today of ~60%–70% to perhaps 90% in a few years' time.* The size and consequent cost of such chips could be significantly reduced by the development of a very high density and high-performance memory, if this can be done itself at a sufficiently low cost.

Traditional computing systems use a complex hierarchy of volatile and nonvolatile (NV) memory technologies (largely to manipulate data) and data-storage technologies (largely to store data) in order to achieve a reasonable balance of performance, cost, and endurance. This leads to considerable complexity in operating systems and software applications, which adds cost and decreases reliability. Magnetic disk drives are the cheapest form of non-archival data storage with a cost about 10–100 times lower per bit than solid-state memory technologies. Disk drives are capable of inexpensively storing huge amounts of data: typical capacities exceed several hundred gigabytes (GBs) today. However, hard disk drives (HDDs) have a limited mean time to failure. A hard drive includes a fixed read/write head and a moving media upon which data are written. Devices with moving parts always are intrinsically unreliable with the possibility of mechanical failure and complete loss of stored data.

The cheapest solid-state memory is FLASH, which has begun to displace disk drives in consumer applications, for example, for more reliable, lighter, and compact laptops, and, perhaps paradoxically, for high-performance storage systems, where large numbers of input–output requests are called for [2]. FLASH memory relies on a thin layer of polysilicon that is disposed in oxide on a transistor's on–off control gate. This layer of polysilicon is a floating gate, isolated by the oxide from the control gate and the transistor channel. FLASH memory is, by far, the cheapest form of solid-state memory today because it is a very dense memory, that is, the cell is relatively simple in structure, and each memory cell uses only a single transistor and therefore occupies a very small area of the silicon wafer. In some FLASH memories, more than one bit, perhaps two or four bits, may be stored per transistor although this usually leads to degraded performance and lifetime. Since the polysilicon slightly deteriorates after each write cycle, due to the very high voltages and associated electric fields required, FLASH memory cells begin to lose data after approximately 100,000 to 1 million write cycles

* See, for example, Ref. [1].

for single bit memories, and after as little as 10,000 write cycles for multibit memories. While this is adequate for many applications, particularly where the number of write operations is limited, FLASH memory cells would begin to fail rapidly if used constantly to write new data, such as in a computer's main memory. Furthermore, the data access time for FLASH memory is much too long for most computationally intensive applications.

Spintronic memories are under intensive exploration as one of the next-generation storage technologies to replace HDDs and current solid-state memories [3]. In this section, the basic concepts, underlying physics, and the state of the technology of these emerging spintronic memories are briefly reviewed.

36.2 Magnetic Random Access Memory

Magnetic random access memory (MRAM) stores data as the direction of magnetization of a ferromagnetic material [4,5]. One approach to MRAM uses a magnetic tunneling junction (MTJ) as the memory cell. The basic structure of the MTJ is a sandwich of two layers of ferromagnetic or ferrimagnetic material separated by a thin insulating material (see chapters in Part III). In a typical MTJ, the magnetization direction of one magnetic layer (the "free layer") responds to a small magnetic field, whereas the other magnetic layer (the "reference layer") has a fixed magnetization direction. The resistance of the MTJ depends on the relative magnetization directions of the free and reference layer moments and is typically a minimum (maximum) when the two layers are parallel (antiparallel) to each other. MTJs for MRAM applications are engineered so that the magnetization of the free and reference layers lies along a single direction—the easy axis of magnetization—which is usually determined by the magnetic shape anisotropy of the MTJ nanodevice, although the intrinsic magnetic crystalline anisotropy of the magnetic materials themselves can also be used. The latter is obviously critical for MTJ elements that are perpendicularly magnetized. In this way, data in the form of "ones" and "zeros" can be stored within the direction of magnetization of the free layer of the MTJ.

MRAM cell concepts based on MTJs were first proposed by IBM and others in 1995 shortly after high tunneling magneto resistance was first observed at room temperature [4]. More than a decade later, in 2006, the first commercially available MRAM chips came onto the market from Freescale Semiconductor. The first IBM and the Freescale (now the EverSpin corporation) MRAM chips [6], just 16 Mb or smaller in size, can be considered as "first-generation" MRAMs, in which the state of the MTJ memory cells is written by local magnetic fields. Localized magnetic fields at a given MTJ element are used to switch the direction of the magnetization of the free layer. A cross bar array of metal wires is used to generate local magnetic fields. In Figure 36.1, an array of MTJ memory elements are shown connected in parallel to a lower set of "word" lines and an upper set of "bit" lines. The "word" lines are oriented orthogonal to the bit lines. Currents are passed along the metal bit and word lines to generate magnetic fields. The strength of these magnetic fields is directly proportional to the current along the wires, and the field is generated uniformly along the wire. Thus, all the MTJs connected to one word line (or similarly one bit line) will be subjected to the same magnetic field. The field (current) must be chosen so that, even though it will disturb the magnetic state of each MTJ along that word or bit line, it is not large enough to change the magnetic state of the MTJ. After the current is removed, the magnetic state of the MTJ will return to its previous state. To write one MTJ element, a current is passed along both the bit line and the write line between which that MTJ is situated. The vector combination of the orthogonal magnetic fields from the bit and word lines is designed to exceed the threshold switching field for that element so that its state can be set to either a "one" or a "zero." Typically, the polarity of either the bit line or word line current is changed to set the MTJ element to one of its two possible states.

Figure 36.2 shows a top down optical micrograph of the first ever MRAM chip, which was fabricated by IBM [7]. This chip, which consists of a 1 kb CMOS twin cell MRAM array, was fabricated using IBM's CMOS-6SF technology, which is an advanced process technology with a front-end-of-line (FEOL) minimum feature size of 0.25 μm, and a 0.4 μm minimum feature size for the second level of metallization and above. This first chip used aluminum wiring. At the time this chip was fabricated in 1999, 200 mm wafer diameter MRAM processing equipment was not available within IBM, so following the fabrication of metal via above the metal 2 layer, the wafers were diced into 1 in.^{2} pieces. The MTJ and the final wiring layer were then fabricated using research-scale deposition and patterning systems. Using these devices ~3 ns read and write access times were demonstrated [7]. This excellent performance resulted from the tunability of the resistance of magnetic tunnel junctions (MTJs) over many orders of magnitude, so allowing optimization of their resistance with regard to that of the CMOS circuits [4]. Note that, as shown in Figure 36.1, there is no connection between the word line and the MTJ device. Rather, an additional metal lateral contact line is used to connect the lower electrode of the MTJ to the output of a transistor, whose state (whether on or off) is controlled by a read word line connected to the gate of the transistor. Thus, each MTJ device is connected to a transistor for the purposes of eliminating current that would otherwise leak through the MTJs in the cross-point array during reading [3].

Encouraged by these results, IBM undertook the development of full 200 mm-wafer scale MRAM technology, initially in a joint development alliance with Infineon, based on 180 nm CMOS technology [5]. Figure 36.3 shows an image of a 16 Mbit MRAM product demonstrator chip that was developed as an outgrowth of that effort, while Figure 36.4 shows a TEM (transmission electron microscopy) cross section from within the cell array [8]. The chip was designed as a 1 M × 16 b SRAM (i.e., as 1 million words each containing 16 bits) with an asynchronous interface and a 1.42 μm^2 cell size and had a read and write performance of 30 ns. For more details, see Refs. [8,9].

Whilst the concept of field writing of MTJs seems simple and straightforward, it has several drawbacks. Most importantly,

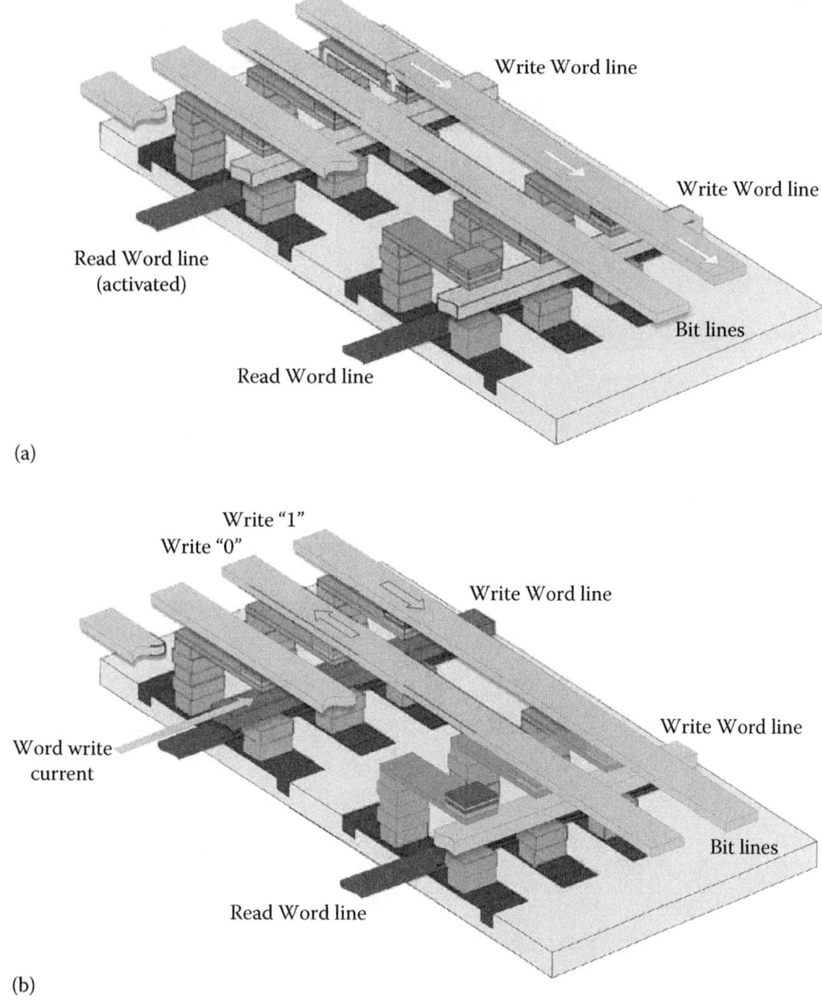

FIGURE 36.1 (See color insert.) Diagram of a magnetic field programmable MRAM. (a) Reading a bit. Voltage is applied to the desired Read Word line (to enable the transistors in that word line) and voltage at desired bit line is measured. (b) Writing bits. Current is passed through the desired Write Word line and appropriate bit lines. Superposition of the fields generated by two currents orients the moment of the free layer in the MTJ in the desired direction. The polarity of the moment is determined by the direction of the current in the Bit line; current flowing one direction will flip it one way, and current in the opposite direction will flip it the other way. (From Parkin, S.S.P. et al., *Proc. IEEE*, 91, 661, 2003.)

this method of writing scales very poorly as the MTJ devices are shrunk in size to allow for higher capacity MRAM chips. In particular, the strength of the magnetic fields generated by the word and bit line currents shrink in proportion to with the radius of the write and bit lines, for the same current density, so that ever larger current densities must be used to generate the same field strength, eventually resulting in problems such as electromigration. Moreover, due to the superparamagnetic effect, thermal fluctuations can cause the magnetic moment of the free layer to overcome the magnetic anisotropy energy barrier and so reverse the free layer's magnetization direction unless the energy barrier is large enough. As the MTJ element shrinks in size, the switching field will necessarily have to be increased in order that sufficiently large magnetic shape anisotropy energy barriers are maintained to meet the typical 10 year memory lifetimes required for a NV memory chip. The larger switching fields correspond to higher currents needed along the word and bit lines. Such high currents not only consume more power but require larger transistors to generate these currents.

Another drawback of field writing of the MTJ elements is that the elements that are half-selected (i.e., those elements that see only one field from either the write or bit lines) are more susceptible to thermal fluctuations and may be disturbed during the writing process of the selected cell. Finally, the magnetic state of the half-selected cells may be subject to the phenomenon of creep whereby their magnetic state may be slightly altered each time they are half-selected and, after an accumulation of half-select processes, may change their magnetic state [10].

To overcome some of the drawbacks of field writing (Stoner–Wohlfarth writing), a "toggle" mode writing scheme was proposed [11], which uses a free layer composed of an artificial antiferromagnet (AAF) formed from two magnetic layers coupled antiferromagnetically through a thin layer of ruthenium

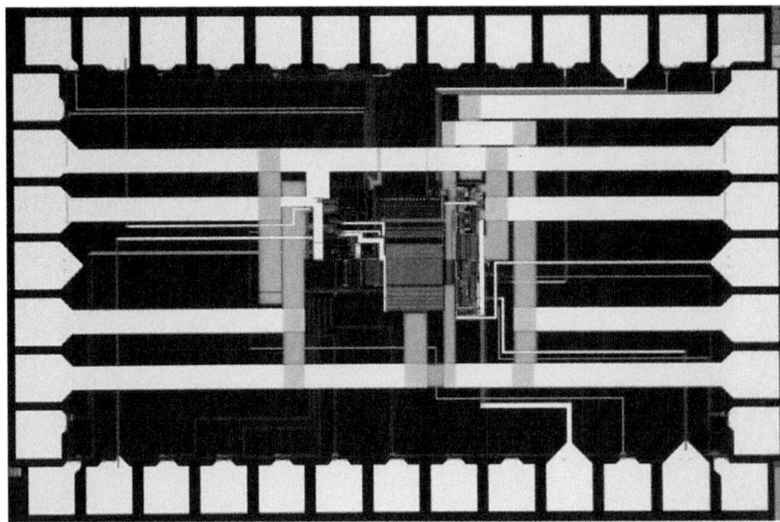

FIGURE 36.2 Top down optical micrograph of a 1 Kb CMOS MRAM array, the first MRAM chip (using field writing) fabricated at IBM in 2000.

FIGURE 36.3 Image of a 16 Mbit conventional MRAM (field writing) product demonstrator chip developed using a full 200 mm-wafer scale 180 nm CMOS technology (2004).

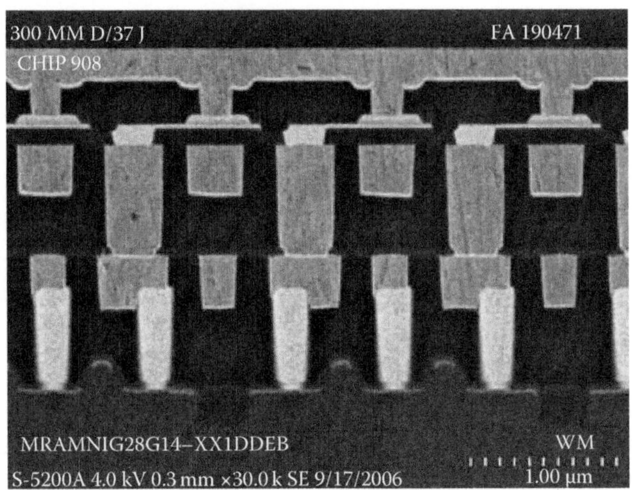

FIGURE 36.4 TEM (transmission electron microscopy) cross section image from within the cell array of a 16 Mbit conventional MRAM product demonstrator chip (shown in Figure 36.3).

[3,12]. This scheme is described in detail in Chapter 35. The main advantage of this scheme is not a lowering of the writing currents or fields needed but rather that there is a much larger range of combinations of bit line and write line fields in which the MTJ element can be written without upsetting the half-selected bits. Indeed, paradoxically, the half-selected bits actually become more stable to thermal upsets during the toggle-writing process [11], that is, the energy barrier to thermally excited switching of the half-selected bits increases during the toggle mode writing of the selected bits on the same word or bit lines.

Due to the high fields and associated high word and bit line currents needed for both Stoner–Wohlfarth and toggle mode writing MRAM, these schemes do not support scaling of MRAM to densities that could make MRAM more than just a niche memory technology. However, an alternative writing method, which is based on recent developments in magnetic tunnel junction materials and on a new spintronics phenomenon, is potentially highly attractive. This scheme, spin transfer torque (STT) writing uses a current that is passed directly through the MTJ device to "write" data bits. A current that is passed between the pinned layer and the free layer is spin polarized [13] and consequently carries spin angular momentum [14,15]. When the spin angular momentum is transferred from the tunneling electrons to the magnetic moments of the free layer atoms, these moments are subject to a torque. This spin transfer-induced torque results in excitation, rotation, and eventual reversal of the free layer magnetization direction if the torque is large enough and sufficient spin angular momentum is transferred. This will occur when a large enough current flows for a sufficiently long time (see Chapter 8). Depending on the direction of the flow of spin-polarized electrons, the free layer magnetization can be rotated toward or away from the direction of magnetization of the pinned layer. Thus, by passing a sufficiently large current through the MTJ element in one direction or the other, the element can be written to a "zero" or a "one." Spin transfer switching of the magnetization has been experimentally observed in a wide variety of MTJ nanopillars since its first observation in manganite-based tunnel junctions [16]. STT writing is the dominant writing mechanism when the lateral dimensions of the MTJ nanopillars are below ~$100 \times 100\,\text{nm}^2$ (since the field from the current increases with the radius of the device (for constant current density) these fields may influence the writing of the bit and can dominate the spin torque for larger area devices). Since the threshold current for STT writing increases approximately with the magnetic moment of the free layer, reducing the size of the MTJ element (for the same thickness) reduces the threshold writing current accordingly. Thus, the current needed for STT writing scales with the area of the MTJ nanoelement, which is very attractive (assuming that there remains sufficient stability against thermal fluctuations, for example, in perpendicularly magnetized nanoelements with large perpendicular magnetic anisotropy).

There are several important issues associated with STT writing. Perhaps, of paramount importance, is that the current for reading must be well below that needed for writing so that inadvertent upsets of the state of the MTJ elements is avoided. Since there will inevitably be a distribution of writing threshold currents for different devices, due to various factors such as variations in (1) the size and shape of the nanoelement, (2) the thickness of the magnetic layers within the free layer, (3) the spin polarization of the tunneling current due to barrier thickness inhomogeneities (the polarization of the current typically increases with thickness for thinner barriers due to improvements in the insulating characteristics of the barrier or due to increased spin filtering but typically decreases for much thicker barriers due to impurity-mediated tunneling), and (4) stray fields from neighboring elements, etc., this requirement is quite stringent. Narrowing the distribution of threshold switching currents is very important. Lowering the writing current is important for several reasons. Firstly, lowering the power used to operate the MRAM chip requires the lowest possible writing current. Secondly, to build the densest possible MRAM chip, the writing current must be provided by a minimum area transistor so that the area of a single memory cell can be its smallest possible size (for a single bit cell) of ~$4f^2$, where f is the minimum length possible in that CMOS technology node. When $f \sim 90\,\text{nm}$, then this means that the writing current must be below ~$100\,\mu\text{A}$. Currents well below this level have recently been demonstrated in perpendicularly magnetized MTJ elements [17]. Thirdly, the writing current must be such that the electric field across the MTJ tunnel barrier is well below the breakdown electric field of the dielectric material from which the barrier is formed. Since dielectric breakdown can occur gradually over many write operations even when the applied voltage is well below the breakdown voltage for a single write operation, this is a very stringent requirement.

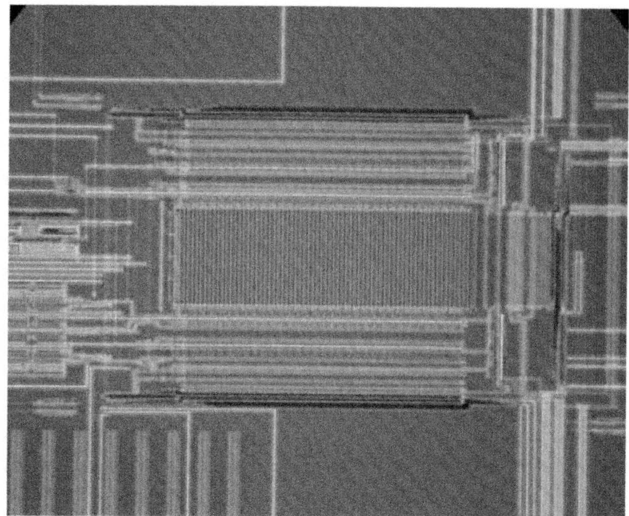

FIGURE 36.5 Image of a 4 kbit experimental array of STT-MRAM chip fabricated in a 200 mm, 180 nm ground rule CMOS technology (2007).

Figure 36.5 shows an image of a 4 kbit experimental MTJ array fabricated in a 200 mm, 180 nm ground rule CMOS technology as described in Ref. [18]. The current focus of the development of STT-MRAM at IBM is to simultaneously satisfy tight distribution requirements for write threshold control while maintaining high breakdown voltages and low disturbs from read operations [19].

36.3 Racetrack Memory

36.3.1 Concept

Racetrack Memory is a storage-class memory technology proposed by IBM in 2002 [20,21]. Like magnetic HDDs, Racetrack Memory stores data within a persistent pattern of magnetic domains that are accessed dynamically for writing, reading, or modification (namely changing the state of a domain wall [DW] bit), but unlike HDDs it uses no moving mechanical parts. The data pattern is stored as a series of magnetic DWs in a microscopic ferromagnetic wire (the racetrack), and accessed by using current pulses to move the entire pattern, intact, along the wire to reading and writing elements integrated within the device. Each storage element of the device will comprise a writing element, a magnetic racetrack, and a reading element. As the data are stored as magnetic domains, it has the same NV nature as data stored in a magnetic disk drive.

A very important feature of the Racetrack Memory is that the series of magnetic DWs in the racetrack are moved around the racetrack not by magnetic field but rather by using the spintronic concept of "spin momentum transfer" [14,15]. A spin-polarized current carries spin angular momentum. When this spin angular momentum is delivered to a magnetic DW, the DW can be moved [22–27]. In distinct contrast to their motion under the influence of magnetic field, neighboring DWs in a magnetic nanowire are moved by current in the same direction along the nanowire, defined by the direction of current (or rather, the flow of spin-polarized electrons). Magnetic field will drive neighboring DWs in opposite directions and so will cause the DWs to annihilate one another. Thus, using current, a sequence of DWs can be moved along a magnetic nanowire, in one direction or the other, depending on the direction of the current flow, one of the key concepts of Racetrack Memory.

Most conventional solid-state memories require one or more transistors or switches per data bit so that the density and thereby the cost of such memories is determined, to a large extent, by the size of a single transistor, the number of transistors needed per memory cell, and the cell layout and architecture. Moreover, since CMOS is by far the most well-established microelectronic technology, and is clearly the cheapest of today's semiconductor technologies, due to the massive scale of silicon foundries, any competitive solid-state technology will need to take advantage of and be based on CMOS. The concept of the Racetrack Memory is that many bits, as many as 10–100 or more, can be stored *per transistor*, thereby dramatically increasing the storage density.

The greatest improvement in density is achieved in the Racetrack Memory concept shown in Figure 36.6 in which the magnetic racetrack is arranged perpendicular to the surface of a silicon wafer (in which CMOS circuits are deployed). In this "vertical" or 3D Racetrack Memory (VRM), the magnetic wire extends vertically above the read and write elements in a "U"-shaped racetrack. The DWs in which the data are encoded are written and read in the bottom of the U of the racetrack in Figure 36.6 where the read and write elements are located. In this 3D design, a very high data storage density is achieved per unit area of the silicon wafer by storing large numbers of bits in the vertical dimension of the device.

As illustrated in Figure 36.6, the Vertical Racetrack Memory may include storage zones (or reservoirs) in the magnetic racetrack so the intact DW pattern may be driven back and forth repeatedly as needed. Thus, in order to move any particular DW in the series of DWs in a single racetrack to these elements, the racetrack needs to be about two times longer than the sequence of DWs. One can consider the racetrack to be composed of a "storage" region and a "reservoir" region at either end of the racetrack. These regions will dynamically occupy different portions of the racetrack as the DWs are moved back and forth during the operation of the memory. Other unidirectional designs are possible, in which data are written at one end of the wire and read at the other, by temporarily storing the data in CMOS memory elements.

Whilst the VRM has the potential to create a solid-state storage device with the capacity of a magnetic HDD, that is, about 100 times greater than that of a solid-state memory in the same area of silicon, and at about the same cost per bit as a HDD, there are considerable technical challenges in building the vertical racetrack. An alternative Racetrack Memory device is shown in Figure 36.7. In this horizontal or 2D version of the Racetrack Memory (HRM), data are stored in a long, straight racetrack element parallel to the surface of the silicon wafer. The device, as illustrated, has centrally located read and write elements, and the data pattern may be shifted back and forth across them as necessary, by varying the polarity of the current pulses. The horizontal racetrack is not as dense as the vertical racetrack but, nevertheless, has the potential to be as dense or even denser than

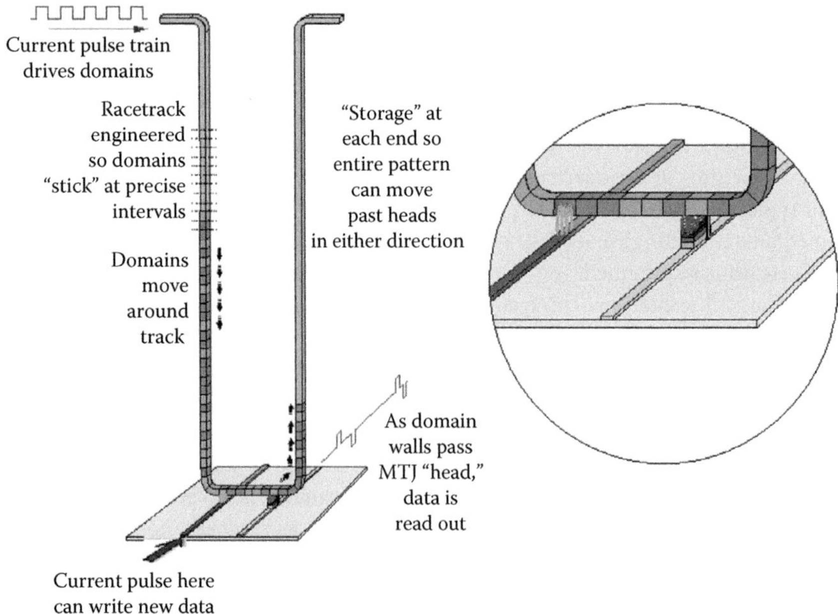
Current pulse train
drives domains
Racetrack
engineered
so domains
"stick" at precise
intervals
Domains
move
around
track
"Storage" at
each end so
entire pattern
can move
past heads
in either direction
As domain
walls pass
MTJ "head,"
data is
read out
Current pulse here
can write new data

FIGURE 36.6 (See color insert.) Vertical configuration magnetic Racetrack Memory element. A magnetic tunnel junction (MTJ) device is used for reading data in the racetrack.

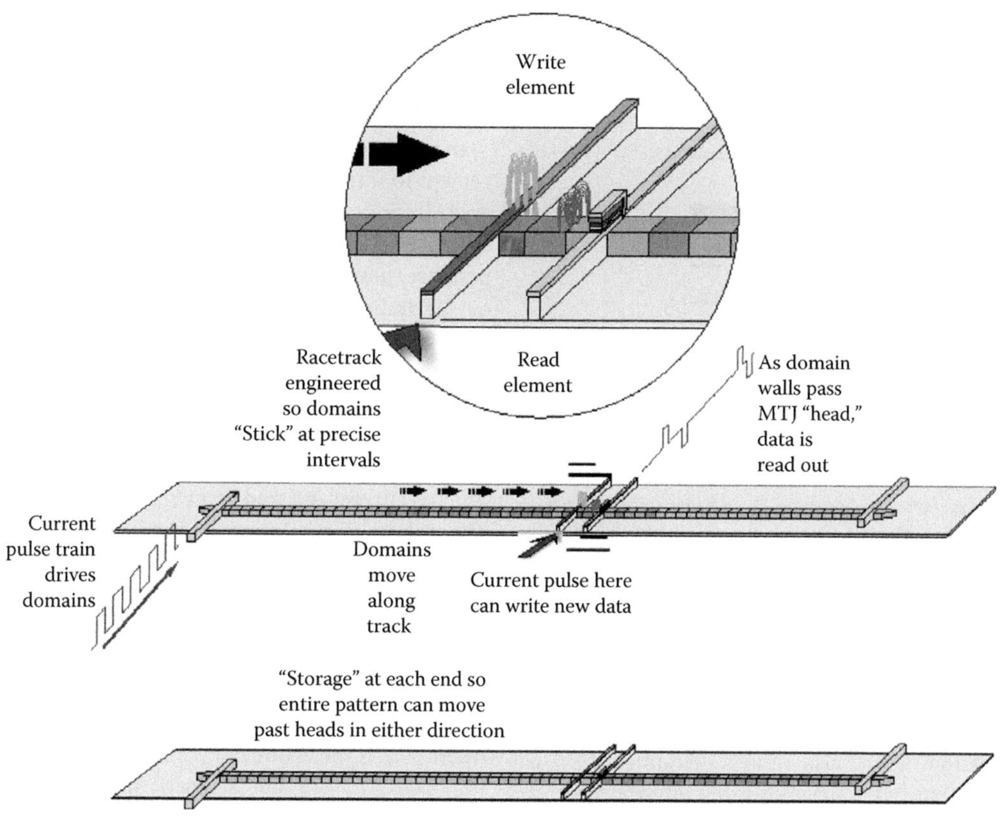
Write element
Racetrack engineered so domains "Stick" at precise intervals
Read element
As domain walls pass MTJ "head," data is read out
Current pulse train drives domains
Domains move along track
Current pulse here can write new data
"Storage" at each end so entire pattern can move past heads in either direction

FIGURE 36.7 (See color insert.) Horizontal configuration magnetic Racetrack Memory element.

FLASH memories but with much greater reading and writing performance. Moreover, even the HRM has the potential to be much denser than STT-MRAM.

Racetrack Memory encompasses the most recent advances in the field of metal spintronics. The spin-valve read head enabled a 1000-fold increase in the storage capacity of the HDD in the decade from 1997 to 2007; since about 2007 the MTJ has supplanted the spin valve in nearly all HDD recording heads because of its higher signal to noise capabilities [3]. MTJs also form the basis of modern MRAM, in which the magnetic moment of one electrode is used to store a data bit, as discussed in Section 36.2. Whereas MRAM uses a single MTJ element to store and read 1 bit, and HDDs use a single spin-valve or MTJ sensing element to read the 1,000 GB of data in a modern drive, the racetrack uses one sensing device to read ~10 to 100 bits.

The actual performance of the HRM depends, in detail, on the characteristics of the magnetic DWs in the racetrack, how closely the DWs can be spaced along a racetrack, and the velocity with which the DWs can be driven around the racetrack with current. Addressing these challenges is the central focus of current research and is briefly discussed in the following sections.

36.3.2 Domain Walls in Magnetic Nanowires

The physical width of DWs in the Racetrack Memory depends both on the magnetic properties of the material used and on the racetrack dimensions. Two very different situations occur depending on whether the racetrack is made of magnetically "soft" or "hard" materials. These two classes of materials can be characterized by the ratio of the magnetic anisotropy energy, determined, for example, by the crystal structure or surface and interface contributions, to the magnetostatic energy, which depends on the shape of the racetrack and its magnetization. This ratio is often called the quality factor $Q = K/2\pi M_S^2$, where K is the anisotropy constant and M_S the saturation magnetization of the material [28].

Soft materials, defined by $Q \ll 1$, are typically made of alloys of Fe, Co, and Ni. The most commonly used alloy is $Fe_{19}Ni_{81}$ (permalloy). For such materials, the shape anisotropy dominates and the magnetization is preferably aligned along the racetrack. Domains magnetized in opposite directions point either toward one another or away from one another, and they are separated by head-to-head and tail-to-tail DWs, respectively. Both the internal structure of the DWs (that is, the detailed magnetization profile) and their width depend on the size and shape of the track [29,30]. Figure 36.8a and b show representative DW structures observed in permalloy nanowires with widths of several hundred nanometers and thicknesses of a few tens of nanometers. In these wires, two different structures can be stabilized, the so-called transverse and vortex DWs. Of these which DW structure is most stable depends on the product of the track width and its thickness. Transverse walls are favored for thin and/or narrow tracks, whereas vortex walls have the lowest energy for wider and/or thicker tracks. However, both types of walls can be

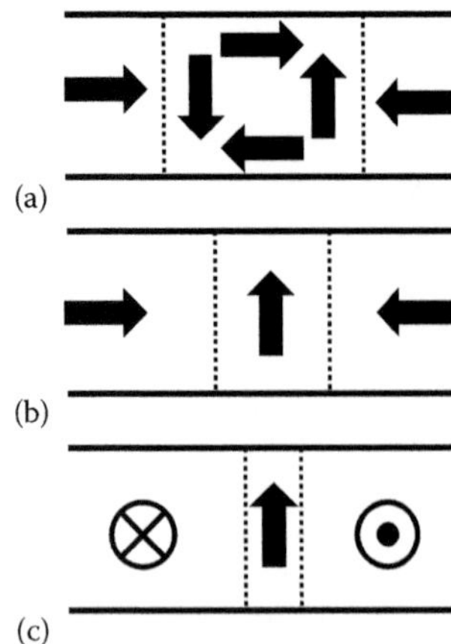

FIGURE 36.8 Schematic illustration of three possible domain wall (DW) structures for racetrack memory (a) Head to Head (HH) vortex, (b) HH transverse, and (c) Bloch domain wall.

observed for a fairly wide range of wire widths and thicknesses, and, in addition, the wall structure can be strongly influenced by the details of the mechanism by which a DW is injected into the nanowire [31–33].

Vortex and transverse walls have different widths, and their current-driven dynamics are also quite different, as will be discussed later. For both DW structures, the physical width w scales with the track width f and for tracks made of permalloy, it can be approximated by the expressions: for transverse walls, $w_{TW} \sim f$, and for vortex walls, $w_{VW} \sim 3\pi/4f \sim 2.35f$ [30,34]. These length scales define the upper bound of the linear density of the Racetrack Memory, that is, the maximum number of DWs per unit length, in the case of soft materials. The actual minimum spacing between DWs in racetracks without pinning sites will rather be determined by the magnetostatic interactions between neighboring DWs.

In the case of hard materials, with $Q \gg 1$, the magnetization orientations of the domains are aligned along the anisotropy axis. The most promising class of hard materials has strong perpendicular magnetic anisotropy (PMA). For this class of materials, for example Co/Pt [35] or Co/Ni [36] multilayers with ultrathin magnetic layers, the magnetization of the domains points out of the plane of the racetrack. The properties of the DWs separating domains (Bloch DW, Figure 36.8c or New DW) are essentially determined by the properties of the material (exchange and anisotropy energies) and are much less strongly influenced by the racetrack dimensions. The physical DW width can be extremely small, for example ~5–15 nm in Co/Pt multilayers, which makes these DWs potentially very attractive for dense Racetrack Memories.

36.3.3 Thermal Stability of Pinned Domain Walls

In Racetrack Memory, the data pattern is stored as a series of magnetic DWs. Such a pattern of DWs may be realized, for instance, by introducing uniformly spaced pinning sites along the racetrack. These pinning sites will also act to prevent the DWs from drifting out of position, perhaps propelled by sub-threshold stray currents or magnetic fields or because the controlling current pulses are not exactly of the correct intensity and duration.

In order to estimate the required pinning strength, it is convenient to introduce the concept of thermal fluctuations. In the electronics industry, it is common to define a number in units of k_BT, where k_B is the Boltzmann constant and T is room temperature, to represent the stability of the data bit. For example, 60 k_BT indicates that the data bit will hold its value for more than 10 years at room temperature. Thermal fluctuations and the superparamagnetic limit are key limitations to the areal density of HDDs, for which bits are stored in many weakly coupled magnetic grains. These are also critical issues for STT-MRAM, because the storage layer must be extremely thin (a few nm) to achieve current-induced switching for small current densities. The scaling of the Racetrack Memory is much more favorable, because DWs can be much larger, and therefore are less susceptible to thermal fluctuations. The energy barrier ΔE of a DW localized at a pinning site can be readily derived from micromagnetic simulations. The energy barrier ΔE depends in a rather complex way on the DW structure, as well as the detailed shape and depth of the pinning site, but can be described by $\Delta E = KV_{DW}$, where V_{DW} is the volume of the DW and K is an effective anisotropy constant. For soft magnetic materials such as permalloy, K (typically on the order of $\sim 5 \times 10^5$ to 5×10^6 erg/cm^3) is determined by the shape anisotropy of the racetrack. Since the DW width is proportional to the width of the racetrack f, the volume of the DW is simply given by $V_{DW} = \kappa t f^2$, where t is the racetrack thickness and κ is a numerical factor that depends on the DW structure ($\kappa = 1$ and 2.35 for transverse and vortex DWs, respectively [30,34]). The energy barrier gives rise to a pinning field H_p, which is simply given by $H_p = K/M_S$ (assuming a parabolic pinning potential well). Thus, the thermal stability target $\Delta E = nk_BT$ at room temperature is simply given by the track dimensions and magnetization and the pinning field: $\Delta E = nk_BT = \kappa t f^2 M_S H_p$. The pinning required for a typical target value of $n = 60$ can then be readily calculated for different DW structures, as a function of the track thickness.

36.3.4 Current Needed to Move Domain Walls

The critical current density for DW motion J_C is key to the performance of Racetrack Memory, since it is the main source of power dissipation during the DW shift operation. Experimentally, J_C is ~10^8 A/cm^2 for racetracks made of permalloy or closely related materials. There are reports of lower values for multilayered structures (spin valves) and nanowires with perpendicular anisotropy (Co/Pt, Co/Ni) but few data are available and more research is needed to explore the wide range of potential ferromagnetic materials.

It is crucial that a solid understanding of the mechanisms of current-driven DW motion be available so as to guide experiments toward reducing the critical current density. Let us first

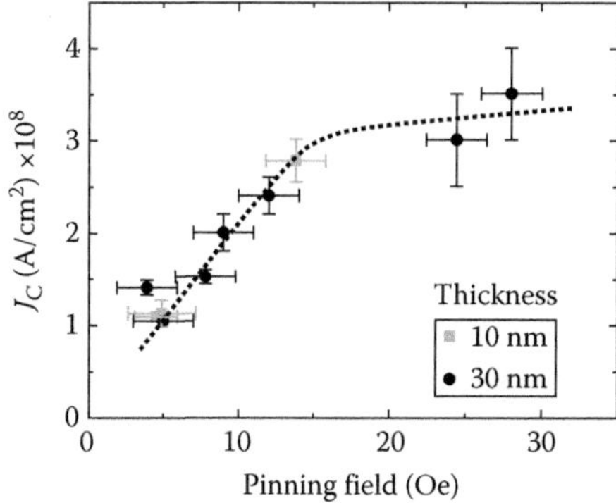

FIGURE 36.9 Threshold current density (J_C) needed to move a vortex domain wall plotted as a function of the pinning strength, represented by the pinning field. Wire widths and thicknesses are varied.

compare DW motion driven by current and magnetic field. The field analog of the critical current density J_C is the propagation (or coercive) field H_p, below which no motion occurs. The origin of this propagation field is extrinsic: for an ideal racetrack, without any defects or roughness, DWs would move for any nonzero applied field, albeit very slowly. Thus, nonzero values of the propagation field are directly related to defects, which provide local pinning sites for the DW. In fact, H_p is proportional to the pinning strength, which can be tuned, for example, by fabricating notches with variable depths.

The relationship between critical current and pinning strength for vortex DWs is shown in Figure 36.9 [21]. These data suggest the existence of two different regimes. For relatively weak pinning (below ~15 Oe), the critical current density scales linearly with the pinning field. For the lowest pinning strength (~5 Oe), the critical current is on the order of 10^8 A/cm^2. For stronger pinning (>15 Oe), the critical current appears to saturate and becomes independent of pinning strength. In this regime, however, DW motion requires very high current densities resulting in severe Joule heating, and experimental data are thus difficult to interpret.

According to microscopic theoretical models of the interaction between current and the DW, there are two different contributions by which the current interacts with a DW. The first contribution is the same as that which gives rise to switching of magnetic moments in nanopillars made from giant magnetoresistive spin valves or magnetic tunnel junctions as used in STT-MRAM [37,38]. This contribution is usually referred to as the "adiabatic spin torque." The efficiency of the adiabatic spin torque is proportional to the rate of spin angular momentum transfer u (unit of velocity, m/s), which depends on the current density J, the spin polarization of the current P, and the saturation magnetization of the racetrack M_S as follows: $u = \mu_B JP/eM_S$ (with e the electron charge and μ_B the Bohr magneton). Quantitatively, for tracks made of permalloy ($M_S = 800$ emu/cm^3) and assuming a spin polarization $P = 0.5$, $u = 100$ m/s corresponds to a current density $J \sim 2.8 \times 10^8$ A/cm^2. Note that the quantity u is equal to the maximum DW velocity when driven by adiabatic spin torque alone.

The second contribution from the current is often called the "nonadiabatic spin torque" or "β-term" and acts like a magnetic field localized at the DW [23–27]. The magnitude of this spin torque is proportional to βu. It is assumed that the dimensionless constant β is proportional to the degree of spin-dependent scattering that takes place at the DW, that is, the DW resistance. Typically, the DW resistance increases when the DW width is reduced, thereby increasing the magnitude of β and making the DW move with less current. Note that the DW velocity is on the order of $(\beta/\alpha)u$ for a nonzero β-term, where α is the Gilbert damping parameter that represents magnetization relaxation.

From theoretical models, it has been predicted that the existence of the nonadiabatic torque has dramatic consequences on the current-driven DW dynamics, that is, the critical current becomes extrinsic and scales with the pinning strength. Interestingly, the critical current shown in Figure 36.9 seems to scale with the pinning strength in the low pinning regime, suggesting a purely extrinsic origin (a similar dependence was observed in multilayers with perpendicular anisotropy [39]). However, we cannot rule out the existence of a nonzero critical current for very low pinning (<5 Oe). Further work is needed to achieve such an extremely low pinning.

To date, the critical current to move a transverse DW in zero field has not yet been unambiguously determined. It appears, as is consistent with theoretical models, that the threshold current for the motion of a transverse DW is higher than that for a vortex DW. However, it also is predicted that in nanowires of appropriate dimensions (e.g., with a uniform, e.g., a square or circular, cross section) transverse DWs can be moved much more easily, and can overcome strong pinning fields.

Finally, values of the critical currents needed to move DWs in magnetic nanowires are listed in the table below for various materials. The pinning strength, typically the coercivity of the wire, is also shown in the table.

Material	$J_C \times 10^8$ (A/cm^2)	H_P (Oe)	Notes
NiFe [21,40]	~0.6–1 × 10^8	~5–13	Weak pinning
NiFe [21]	~3 × 10^8	~30	Strong pinning
CoFeB [41]	~1 × 10^8	~17	
CoFeB [42]	~4 × 10^6	N/A[a]	Co/Cu/CoFeB multilayers
Co/Ni [39]	~5 × 10^6	~150	Multilayers, spin valve (pma)
Co/Ni [43,44]	~5 × 10^7	~100	Multilayers (pma)
CoCrPt [45]	~1 × 10^8	~500	pma
GaMnAs [46]	~8 × 10^4	~50	Temperature <80 K (pma)
$SrRuO_3$ [47]	~1 × 10^5–1 × 10^6	~50–500	Temperature <140 K (pma)
Co [48]	<9.8 × 10^6	N/A[a]	Pt|0.3 nm Co|Pt (pma)
Co [49,50]	~1–1.5 × 10^8	N/A[a]	Pt|0.6 nm Co|AlOx (pma)
TbFeCo [51]	~4–6 × 10^6	1000–2200	$Tb_{30}Fe_{58}Co_{12}$ (pma)

[a] Pinning field of the wire not specified in these reports.

36.3.5 Reading and Writing Elements

One type of reading device consists of a magnetic element on top of the track and separated from it by a tunnel barrier, thereby forming a magnetic tunnel junction (MTJ) (see Figure 36.10). Here, within the reading device, the free layer is formed from the magnetic domain of the racetrack. The tunnel barrier and the reference layer are deposited on top of the racetrack. The magnetization of the reference layer is fixed by exchange coupling to an adjacent antiferromagnetic film such as IrMn. In the reading cycle, the bit to be read is shifted so that it is within the reading device. Since the resistance of the MTJ depends on the magnetization direction of the bit with respect to that of the reading element, the bit can be read out by measuring the MTJ resistance. (Note that the reading element could also be formed under the racetrack or to one side of the racetrack.)

One method for writing DW bits is by generating localized magnetic fields in the vicinity of the racetrack to reverse the magnetization direction in a small length of the racetrack and thereby writing a magnetic DW. The magnetic field needs to be localized so that it does not perturb the neighboring bits. The simplest method to generate localized magnetic fields is to pass a current pulse along a conducting wire positioned perpendicular to and close to the racetrack. The width of the conducting wire and the distance between it and the racetrack determine how far along the racetrack does the Oersted magnetic field spread. The size of the magnetic field depends on the geometry of the conducting wire; for a given current density, the field magnitude increases with the size of the conducting wire, however the power needed to generate the current scales with the size as well.

A simple model can be used to estimate the magnitude of the magnetic field needed to nucleate a DW, that is, to write a bit. In this model, the magnetic moments under the conducting line are assumed to rotate coherently from one direction to the other via a configuration where the magnetization points perpendicular to the nanowire's long axis. In order to cause magnetization rotation, the Zeeman energy from the magnetic field must overcome the magnetostatic energy barrier when the magnetization lies perpendicular to the nanowire's long direction. The size of the nucleation field therefore depends on the nanowire dimensions as well as its saturation magnetization. Typically, a few nanoseconds (>1–2 ns) is long enough to cause DW nucleation when the local field is larger than the threshold field for nucleation.

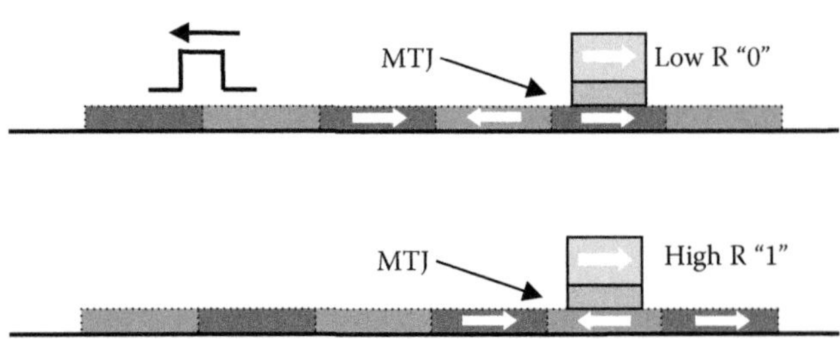

FIGURE 36.10 Reading domain wall bits using a magnetic tunnel junction sensing device.

36.3.6 Shift Register Demonstration

There have been several studies recently that have reported the demonstration of a magnetic shift register using the STT-induced motion of DWs in magnetic nanowires. The first such demonstration is shown in Figure 36.11a through c in which a 3 bit DW shift register memory was demonstrated in a permalloy nanowire [52]. The nanowire is made of a 10 nm thick permalloy film, patterned into the form of a 200 nm wide, 6 μm long nanowire using electron beam lithography. The shift register is a 3 bit unidirectional serial-in, serial-out memory in which the data are coded as the magnetization directions of individual domains (Figure 36.11a). A left (right) pointing domain represents a 0 (1). The data bit is written into the leftmost domain in section B1, is shifted by two domains, and then read from the state of the rightmost domain in section B3. Here, instead of directly reading the state of the rightmost domain, the magnetization configuration of each of the three domains is inferred from the nanowire's resistance. Owing to the relatively large anisotropic magnetoresistance of permalloy, the wire resistance provides information on the number of DWs stored in the nanowire and thus the magnetic configuration. The writing is carried out by applying a current pulse along the left contact line (vertical line in Figure 36.11a), which generates a localized magnetic field and thus can inject a DW into the nanowire. Shifting of the data bits (that is, the domains and the DWs) is carried out by injecting a nanosecond long current pulse (~1.5×10^8 A/cm^2) into the nanowire. The electron flow direction associated with the current pulse points to the right and so moves the DWs to the right via STT.

An example of the shift register operation is shown in Figure 36.11b. Here, a data sequence of 010111 is written into the register. The top panel of Figure 36.11b shows the evolution of the nanowire resistance, and Figure 36.11c shows the corresponding magnetization configuration. In Figure 36.11b, the shaded regions correspond to a shift operation whereas the light regions correspond to a write operation. As the writing/shifting operation is carried out, the resistance of the nanowire changes according to that expected from the change in the number of DWs present in the nanowire, as shown in Figure 36.11c. The bottom panel of Figure 36.11b shows the logic table of each bit. From the logic table, one can find how the data sequence (i.e., 010111) is transferred along the nanowire after each write/shift operation.

The input sequence is accurately transferred to the output after two write/shift operations. In this example, the cycle time to write and shift one bit is ~30 ns. This is determined by the write time (here ~3–4 ns) and the time to shift the series of DWs by one domain length. This time is determined by the velocity of the DWs, which here is ~150 m/s (current density is ~1.5×10^8 A/cm^2) [31].

In this first report, up to three DWs were moved with current in 6 μm long nanowires (subsequently, six walls were

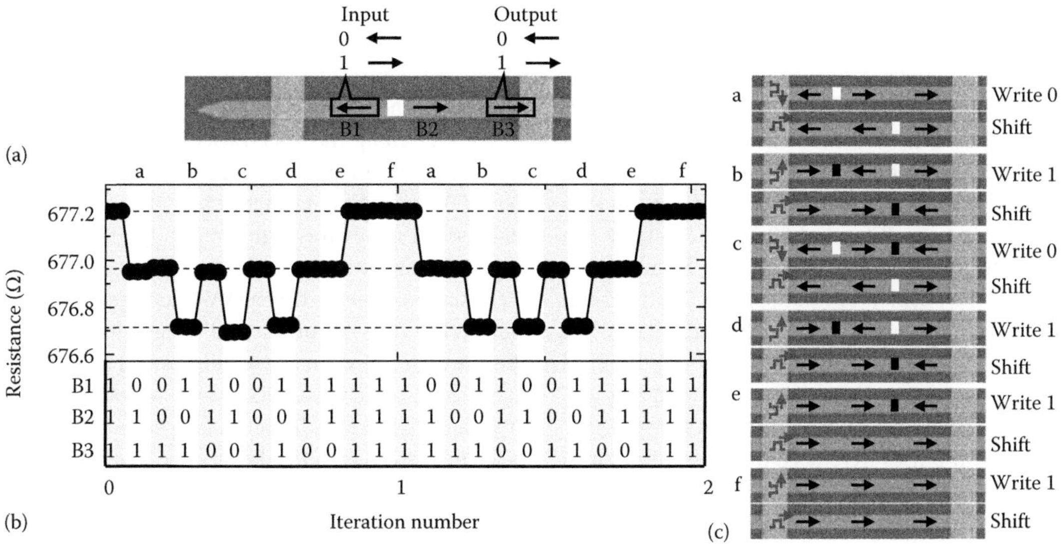

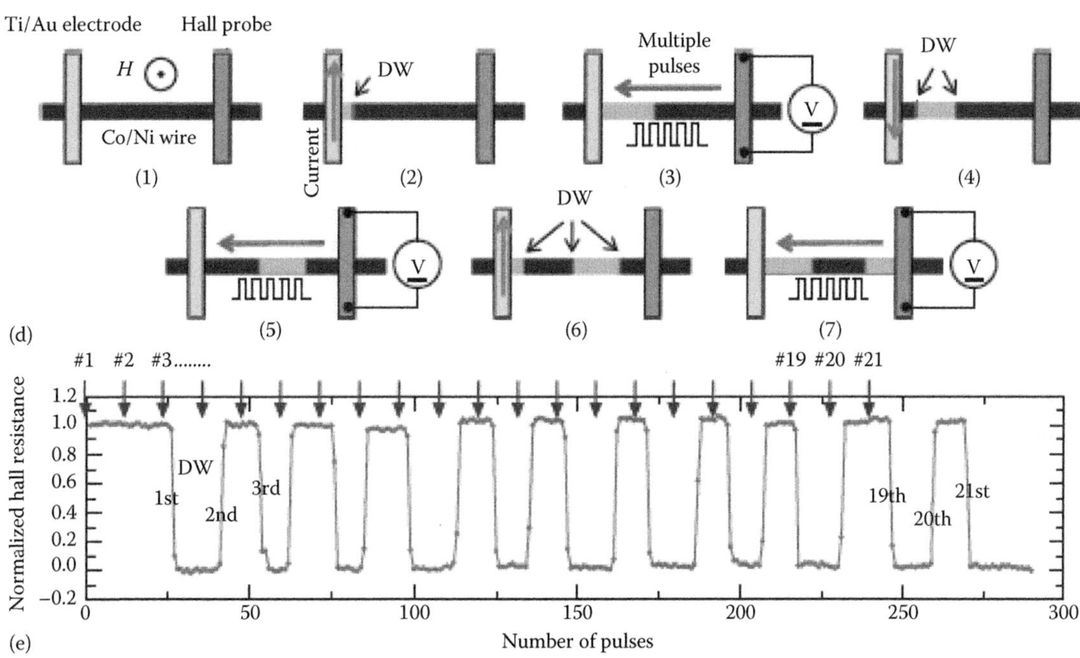

FIGURE 36.11 (a–c) A three bit unidirectional magnetic domain wall shift register using permalloy nanowires. (a) Data are encoded by the magnetization direction of three domains in the nanowire. (b) Nanowire resistance variation when a current pulse sequence is used to write and shift domain walls along the register. The sequence, "010111" is written and shifted two times in succession. The table shows the corresponding evolution of the states of the three bits during these operations. (c) Schematic illustration of the magnetic configuration of the nanowire during the shift register operation. (d,e) A unidirectional magnetic domain wall shift register using Co/Ni multilayered perpendicularly magnetized nanowires. (d) (1)–(7) Schematic illustrations of measurement procedure to create and shift multiple domain walls. (e) Normalized Hall resistance as a function of number of pulses when multiple domain walls are introduced into the wire and shifted toward the same direction. The arrows indicate the timing of the domain wall creation. (From Chiba, D. et al., *Appl. Phys. Expr.*, 3, 073004, 2010. With permission.)

reported to move using 20 μm long nanowires). However, this report also illustrated the importance of pinning for closely packing the DWs for high areal density memories. That is, due to the magnetostatic interaction between the DWs, the spacing between them is limited to a minimum distance of a few microns when the pinning field is on the order of 5–10 Oe. As discussed above, stronger pinning leads to higher threshold current densities.

These first experiments, which demonstrated the motion of a series of DWs at high speed using nanosecond current pulses, proved the fundamental viability of the shift register Racetrack Memory concept.

An example of a more recent study demonstrating a current-driven DW shift register operation is shown in Figure 36.11d through e [53]. In this study, a perpendicularly magnetized material is used. As discussed in Chapter 8, theoretical models suggest that the threshold current for the motion of DWs in perpendicularly magnetized nanowires should be lower than for in-plane magnetized nanowires. Here, 500 nm wide, 16 μm long nanowires, formed from multilayered $[0.3\,Co|0.9\,Ni]_4|0.3$ Co films (thicknesses in nm) deposited on undoped Si substrates, were used to control the position of multiple DWs with current pulses. A tantalum Hall bar was attached perpendicular to the wire to detect the DW passing under the Hall bar after each shifting and writing operation. The shifting operation was carried out with current pulses whose current density is $\sim 1.3 \times 10^8$ A/cm^2. Multiple current pulses were used to conduct one shifting operation and the total length of the pulse was ~132 ns. The wall velocity at this current density is reported to be ~38 m/s. Figure 36.11d shows the magnetization configuration during the writing/shifting operation, and Figure 36.11e shows the corresponding resistance change of the Hall bar. From the Hall bar resistance change, it can be inferred that a total of 21 DWs have been injected and shifted using the current pulses. Note that the pinning field here is reported to be on the order of ~100 Oe. At this pinning strength, dense packing of the DWs can be expected. However, due to the spread of the localized field during the writing operation, the separation distance between neighboring DWs was limited to only a few microns in these experiments.

Similar experiments have also been conducted for other perpendicularly magnetized materials, for example, nanowires made from ultrathin Co films (5 Ta|2.5 Pt|0.3 Co|1.5 Pt films [thicknesses in nm]) [48]. In these experiments the shifting of the DWs resulted from a thermally activated process of DW creep. In this process, less current density is required to move the DWs compared to the above-mentioned cases. However, since the wall motion takes place via a thermally activated hopping mechanism, the velocity is very low and thus requires longer current pulses. Shifting of multiple DWs was observed using a magneto optical Kerr effect (MOKE) microscope. Successful operation of the shifting was observed when 50 ms long current pulses with a current density of $\sim 9.8 \times 10^6$ A/cm^2 were used.

In all of these experiments, the current densities needed to move DWs along the Racetracks are similar for the same DW velocity. Even though the minimum current densities needed to move DWs may have been lower for the case of perpendicularly magnetized nanowires compared to the first experiments with in-plane magnetized permalloy [21,52] and related alloys [54], the observed current-driven DW velocities were also much lower [48,53]. It is thus important to distinguish the threshold current above which DWs can be initially moved from the current density needed to achieve a fixed DW velocity. The latter appears to be approximately the same for perpendicularly magnetized as for in-plane magnetized nanowires but the threshold current appears to be lower for the former.

One way to reduce the current density needed to manipulate DWs is to use a DW resonance excitation effect [15,55,56]. A DW trapped in a pinning potential can be considered as a classical particle trapped in a parabolic potential, which is analogous to a harmonic oscillator. When a current pulse, whose length is matched to the resonance period of the parabolic potential is applied, the motion of the wall within the potential can be amplified, thus allowing subthreshold depinning (threshold here corresponds to the threshold current density when long pulses or direct current is used) [57,58]. Reduction of the threshold current density by more than a factor of five has been demonstrated by using matched bipolar current pulses [34]. Engineering the pinning potential profile is needed to take full advantage of this interesting effect.

36.3.7 Single Domain Wall Racetrack Memory

Recently, Fukami et al. have developed what one might describe as a single DW Racetrack Memory (or otherwise a cross-point DW-motion MRAM using current-induced motion of a single DW per bit) [59] (Figure 36.12). Each memory cell is composed

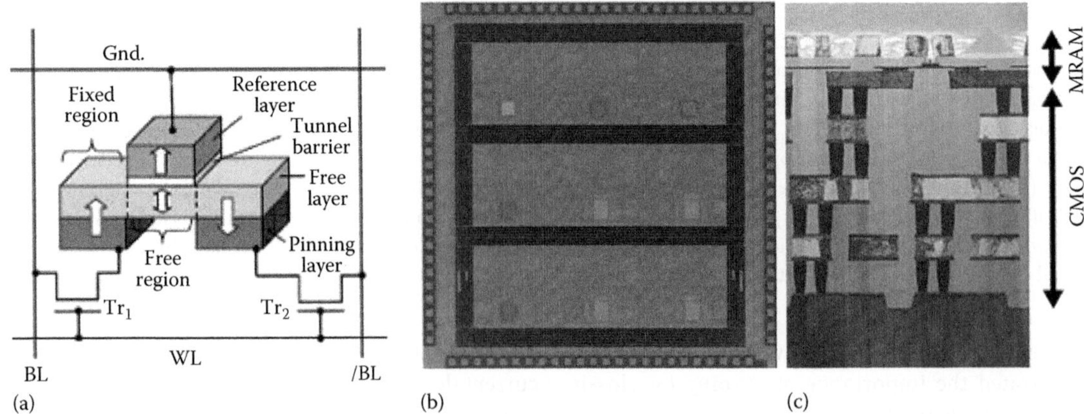

FIGURE 36.12 (a) Cell structure of a DW-motion MRAM. The cell has two transistors and one magnetic tunnel junction. (b) Photograph of a 4k-array and (c) cross-sectional TEM image of the cells. (From Fukami, S. et al., in *Symposium on VLSI Technology Digest of Technical Papers*, June 16–18, 2009, Honolulu, HI, pp. 230–231, 2009. With permission.)

of one MTJ and two transistors. The "free layer" is an elongated ferromagnetic layer, and a tunnel barrier/ferromagnetic layer element is placed on top of the free layer to probe the magnetization direction of the free layer via tunneling resistance measurements. The ends of the free layer's magnetization are pinned by adjacent "pinned layers." The magnetization directions of the pinned layers are opposite from each other, thus forming one DW in the free layer. A perpendicularly magnetized Co/Ni laminated film was used for the free layer in the demonstration reported.

The target market of the DW-motion MRAM is high-speed memory macros embedded in next-generation system on a chip (SoC) devices. Compared to conventional MRAM, including STT-MRAMs, the main advantage of DW-motion MRAM is the separation of the writing and readout circuits, and thus the larger margin for the readout levels. Owing to its reading and writing method, DW-motion MRAM does not suffer disturbance of the stored information during its reading operation and can thus has no cross talk among the cells during write operations. It has been reported that DW-motion MRAM can operate with less than 0.2 mA write currents and a 2 ns writing time with sufficient thermal stability [59].

36.3.8 Outlook

The development of a high-performance, NV MRAM began in IBM with the proposal in 1995, and the subsequent first demonstration in 1999, of a fully functional 1 kbit MTJ array [4,7]. Since this early demonstration that even MTJs with only 20% TMR and a Stoner–Wolhfarth field writing mechanism would allow for a high-speed memory (~3 ns), it is now generally accepted that to build a useful MRAM chip, magnetic tunnel junction memory elements, rather than elements based on metallic spin-valve devices, are needed due to both the very high tunneling magnetoresistance of MTJs and their much higher resistance as compared to spin-valve devices. The magnitude of the TMR signal from an MTJ has evolved enormously from 1975 with the initial discovery of the TMR phenomenon at low temperature and very low bias voltages [13] to the observation of modest TMR effects at room temperature in 1995 [60,61] in devices using conventional alumina tunnel barriers, to giant TMR effects in MTJs with MgO tunnel barriers sputter deposited on amorphous substrates in 2004 [62], to TMR values of more than 600% at room temperature in similar MgO tunnel barrier–based MTJs [63]. Although the only MRAM chips that are available today for commercial applications are based on conventional magnetic field writing, several prototype chips have been developed that use spin torque transfer (STT) writing. This method overcomes the innate problems of scaling field writing to small dimensions but, to date, the current densities needed for high-speed STT writing using practical MTJ devices, are still too high to enable high density, and therefore, price-competitive MRAM chips. Moreover, the MTJ to MTJ distributions of the read currents within a sub-array on an MRAM chip must be sufficiently narrow that the highest current to read any device must not overlap with the lowest current needed to write any device within the same array. Similarly, the highest write current must not exceed the breakdown voltage of any device in the same sub-array. Furthermore, these requirements hold not just for a single read or write operation but for as many of these operations as will take place in the lifetime of the MRAM chip (typically 10 years). The magnitude of the reading and writing currents must be carefully designed so that these processes are truly independent of one another.*

The Racetrack Memory is conceptually very different from MRAM. It overcomes the challenge of reading and writing an MTJ memory element without dielectric breakdown of the MTJ tunnel barrier by using separate devices for reading and writing. This is possible in Racetrack Memory, even while Racetrack is innately a much denser memory than MRAM. This is because Racetrack, unlike MRAM and many other resistive random access memories, stores many bits per memory cell in the form of a series of DWs along magnetic nanowires. The possibility of moving such a series of DWs to and thro along nanowires by using spin torque transfer in the form of nanosecond long current pulses has recently been demonstrated [52]. Thus, the fundamental new concept underlying Racetrack Memory has been validated.

Clearly, Racetrack Memory is in a nascent stage of development compared to MRAM but, in the opinion of the authors, has much greater potential. To achieve its potential, Racetrack Memory faces many exciting and interesting challenges. These include (1) the reliable generation of DWs with sufficiently similar mobilities, for example, by creating DWs with the same structure, (2) the repeated motion of these DWs back and forth along a Racetrack without their transformation into different DW structures, (3) the manipulation of a series of DWs without their annihilation unless planned, (4) the engineering and control of the magnetostatic fringing fields between DWs within and between Racetracks so that these do not impede the motion of the DWs, and (5) the enablement of sufficiently low current densities to move DWs along the nanowire so that the nanowires are not much heated, DW structures are not transformed, and the shift energy is minimized. Other challenges include the development of novel schemes for writing DWs at sufficiently low energy per operation and various integration challenges, including the integration of writing, and especially reading devices that do not impede the motion of the DWs along the Racetracks, the development of processes for creating uniform DW pinning sites, and the development of processes for building vertical Racetracks.

Notwithstanding these challenges, Racetrack Memory is likely to be only one of many new memory and storage and even logic technologies that go beyond the constraints embodied in conventional 2D memory and logic devices. The cost and

* If one is willing to compromise on cell size, it is possible to use a spin-valve injector with a tunnel junction detector in a three-terminal STT-MRAM device, as demonstrated by Sun et al. [64]. This device also would be capable of very fast writing.

performance of most conventional memory devices, including MRAM, are controlled by that of the access device, namely the transistor, to which each device is tethered. Racetrack Memory fundamentally overcomes such limitations by using the latest developments in spintronic materials and phenomena to ultimately allow for the storage of data bits in 3D above the 2D array of transistors that will be used to create, manipulate, and read this data.

Acknowledgments

The authors are grateful to many research colleagues, especially at IBM, who contributed to the development of concepts and experimental results described herein. IBM is also very appreciative of funding from DARPA with which the early work at IBM on MRAM was carried out. IBM also thanks the DMEA for partial support of work on MRAM in later stages of MRAM research and development at IBM. IBM also thanks ITRI for their support of Racetrack Memory research in a joint development alliance between IBM and ITRI from 2009 to 2011.

References

1. The International Technology Roadmap for Semiconductors, www.itrs.net (accessed on April 19, 2000.)
2. G. Zhang, L. Chiu, C. Dickey et al., in *26th IEEE Symposium on Massive Storage Systems and Technologies* (*MSST 2010*), Incline Village, Nevada, May 3–7, 2010.
3. S. S. P. Parkin, X. Jiang, C. Kaiser, A. Panchula, K. Roche, and M. Samant, Magnetically engineered spintronic sensors and memory, *Proc. IEEE* **91**, 661–680 (2003).
4. S. S. P. Parkin, K. P. Roche, M. G. Samant et al., Exchange-biased magnetic tunnel junctions and application to non-volatile magnetic random access memory, *J. Appl. Phys.* **85**, 5828–5833 (1999).
5. W. J. Gallagher and S. S. P. Parkin, Development of the magnetic tunnel junction MRAM at IBM: From first junctions to a 16-Mb MRAM demonstrator chip, *IBM J. Res. Dev.* **50**, 5–23 (2006).
6. J. M. Slaughter, Materials for magnetoresistive random access memory, *Ann. Rev. Mater. Res.* **39**, 277–296 (2009).
7. R. Scheuerlein, W. Gallagher, S. Parkin, A. Lee, S. Ray, R. Robertazzi, and W. Reohr, A 10ns read and write time non-volatile memory array using a magnetic tunnel junction and FET switch in each cell, *Digest of Technical Papers, 2000 IEEE International Solid State Circuits Conference*, February 8, 2000, pp. 128–129, 2000.
8. J. DeBrosse, D. Gogl, A. Bette et al., A high-speed 128-kb MRAM core for future universal memory applications, *IEEE J. Solid State Circuits* **39**, 678–683 (2004).
9. T. M. Maffitt, J. K. DeBrosse, J. A. Gabric et al., Design considerations for MRAM, *IBM J. Res. Dev.* **50**, 25–39 (2006).
10. S. Gider, B.-U. Runge, A. C. Marley, and S. S. P. Parkin, The magnetic stability of spin-dependent tunneling devices, *Science* **281**, 797 (1998).
11. B. N. Engel, J. Akerman, B. Butcher et al., A 4-Mbit toggle MRAM based on a novel bit and switching method, *IEEE Trans. Magn.* **41**, 132–136 (2005).
12. S. S. P. Parkin and D. Mauri, Spin-engineering: direct determination of the RKKY far field range function in ruthenium, *Phys. Rev. B* **44**, 7131 (1991).
13. M. Julliere, Tunneling between ferromagnetic films, *Phys. Lett.* **54A**, 225–226 (1975).
14. J. Slonczewski, Current driven excitation of magnetic multilayers, *J. Magn. Magn. Mater.* **159**, L1 (1996).
15. L. Berger, Emission of spin waves by a magnetic multilayer traversed by a current, *Phys. Rev. B* **54**, 9353 (1996).
16. J. Z. Sun, Current-driven magnetic switching in manganite trilayer junctions, *J. Magn. Magn. Mater.* **202**, 157 (1999).
17. T. Kishi, H. Yoda, T. Kai et al., Lower-current and fast switching of a perpendicular TMR for high speed and high density spin-transfer-torque MRAM, *IEDM Tech. Dig.*, 309–312 (2008).
18. S. Assefa, J. Nowakl, J. Z. Sun et al., Fabrication and characterization of MgO-based magnetic tunnel junctions for spin momentum transfer switching, *J. Appl. Phys.* **102**, 063901–063904 (2007), doi:10.1063/1.2781321.
19. R. Beach, T. Min, C. Horng et al., A statistical study of magnetic tunnel junctions for high-density spin torque transfer-MRAM (STT-MRAM), *IEDM Tech. Dig.*, pp. 305–308 (2008); D. C. Worledge, G. Hu, P. L. Trouilloud et al., Switching distributions and write reliability of perpendicular spin torque MRAM, *IEEE International Electron Devices Meeting (IEDM)*, San Francisco, CA, pp. 12.5.1–12.5.4, 2010.
20. S. S. P. Parkin, US6834005: Shiftable magnetic shift register and method of using the same U.S. patent 6834005, 2004.
21. S. S. P. Parkin, M. Hayashi, and L. Thomas, Magnetic domain-wall racetrack memory, *Science* **320**, 190–194 (2008), doi:10.1126/science.1145799.
22. L. Berger, Exchange interaction between ferromagnetic domain wall and electric current in very thin metallic films, *J. Appl. Phys.* **55**, 1954 (1984).
23. G. Tatara and H. Kohno, Theory of current-driven domain wall motion: Spin transfer versus momentum transfer, *Phys. Rev. Lett.* **92**, 086601 (2004).
24. S. Zhang and Z. Li, Roles of nonequilibrium conduction electrons on the magnetization dynamics of ferromagnets, *Phys. Rev. Lett.* **93**, 127204 (2004).
25. A. Thiaville, Y. Nakatani, J. Miltat, and Y. Suzuki, Micromagnetic understanding of current-driven domain wall motion in patterned nanowires, *Europhys. Lett.* **69**, 990–996 (2005).
26. S. E. Barnes and S. Maekawa, Current-spin coupling for ferromagnetic domain walls in fine wires, *Phys. Rev. Lett.* **95**, 107204 (2005).
27. M. D. Stiles, W. M. Saslow, M. J. Donahue, and A. Zangwill, Adiabatic domain wall motion and Landau-Lifshitz damping, *Phys. Rev. B* **75**, 214423–214426 (2007).

28. A. Hubert and R. Schäfer, *Magnetic Domains: The Analysis of Magnetic Microstructures*, Springer, Berlin, Germany, 1998.
29. R. D. McMichael and M. J. Donahue, Head to head domain wall structures in thin magnetic strips, *IEEE Trans. Magn.* **33**, 4167–4169 (1997).
30. Y. Nakatani, A. Thiaville, and J. Miltat, Head-to-head domain walls in soft nano-strips: A refined phase diagram, *J. Magn. Magn. Mater.* **290–291**, 750–753 (2005).
31. M. Hayashi, L. Thomas, C. Rettner, R. Moriya, Y. B. Bazaliy, and S. S. P. Parkin, Current driven domain wall velocities exceeding the spin angular momentum transfer rate in permalloy nanowires, *Phys. Rev. Lett.* **98**, 037204 (2007).
32. M. Hayashi, L. Thomas, C. Rettner, R. Moriya, and S. S. P. Parkin, Direct observation of the coherent precession of magnetic domain walls propagating along permalloy nanowires, *Nat. Phys.* **3**, 21–25 (2007).
33. M. Kläui C. A. F. Vaz, J. A. C. Bland et al., Head-to-head domain-wall phase diagram in mesoscopic ring magnets, *Appl. Phys. Lett.* **85**, 5637–5639 (2004).
34. L. Thomas and S. S. P. Parkin, in *Handbook of Magnetism and Advanced Magnetic Materials*, Vol. 2, H. Kronmüller and S. S. P. Parkin (Eds.), John Wiley & Sons Ltd., Chichester, U.K., pp. 942–982, 2007.
35. D. Ravelosona, D. Lacour, J. A. Katine, B. D. Terris, and C. Chappert, Nanometer scale observation of high efficiency thermally assisted current-driven domain wall depinning, *Phys. Rev. Lett.* **95**, 117203 (2005).
36. S. Mangin, D. Ravelosona, J. A. Katine, M. J. Carey, B. D. Terris, and E. E. Fullerton, Current-induced magnetization reversal in nanopillars with perpendicular anisotropy, *Nat. Mater.* **5**, 210–215 (2006).
37. J. C. Sankey, Y.-T. Cui, J. Z. Sun, J. C. Slonczewski, R. A. Buhrman, and D. C. Ralph, Measurement of the spin-transfer-torque vector in magnetic tunnel junctions, *Nat. Phys.* **4**, 67–71 (2008).
38. H. Kubota, A. Fukushima, K. Yakushiji et al., Quantitative measurement of voltage dependence of spin-transfer torque in MgO-based magnetic tunnel junctions, *Nat. Phys.* **4**, 37–41 (2008).
39. D. Ravelosona, S. Mangin, J. A. Katine, E. E. Fullerton, and B. D. Terris, Threshold currents to move domain walls in films with perpendicular anisotropy, *Appl. Phys. Lett.* **90**, 072508–072503 (2007).
40. A. Yamaguchi, T. Ono, S. Nasu, K. Miyake, K. Mibu, and T. Shinjo, Real-space observation of current-driven domain wall motion in submicron magnetic wires, *Phys. Rev. Lett.* **92**, 077205 (2004).
41. L. Heyne, M. Kläui, D. Backes et al., Relationship between nonadiabaticity and damping in permalloy studied by current induced spin structure transformations, *Phys. Rev. Lett.* **100**, 066603–066604 (2008).
42. S. Laribi, V. Cros, M. Muñoz et al., Reversible and irreversible current induced domain wall motion in CoFeB based spin valves stripes, *Appl. Phys. Lett.* **90**, 232505–232503 (2007).
43. H. Tanigawa, T. Suzuki, Y. Nakatani et al., Domain wall motion induced by electric current in a perpendicularly magnetized Co/Ni nano-wire, *Appl. Phys. Express* **2**, 053002 (2009).
44. S. Fukami, Y. Nakatani, T. Suzuki, K. Nagahara, N. Ohshima, and N. Ishiwata, Relation between critical current of domain wall motion and wire dimension in perpendicularly magnetized Co/Ni nanowires, *Appl. Phys. Lett.* **95**, 232504–232503 (2009), doi:10.1063/1.3271827.
45. H. Tanigawa, K. Kondou, T. Koyama et al., Current-driven domain wall motion in CoCrPt wires with perpendicular magnetic anisotropy, *Appl. Phys. Express* **1**, 011301 (2008).
46. M. Yamanouchi, D. Chiba, F. Matsukura, and H. Ohno, Current-induced domain-wall switching in a ferromagnetic semiconductor structure, *Nature* **428**, 539 (2004).
47. M. Feigenson, J. W. Reiner, and L. Klein, Efficient current-induced domain-wall displacement in $SrRuO_3$, *Phys. Rev. Lett.* **98**, 247204 (2007).
48. K.-J. Kim, J.-C. Lee, S.-J. Yun et al., Electric control of multiple domain walls in Pt/Co/Pt nanotrack with perpendicular magnetic anisotropy, *Appl. Phys. Express* **3**, 083001 (2010).
49. T. A. Moore, I. M. Miron, G. Gaudin et al., High domain wall velocities induced by current in ultrathin Pt/Co/AlOx wires with perpendicular magnetic anisotropy, *Appl. Phys. Lett.* **93**, 262504 (2008).
50. I. M. Miron, P.-J. Zermatten, G. Gaudin, S. Auffret, B. Rodmacq, and A. Schuh, Domain wall spin torquemeter, *Phys. Rev. Lett.* **102**, 137202–137204 (2009).
51. S. Li, H. Nakamura, T. Kanazawa, X. Liu, and A. Morisako, Current-induced domain wall motion in TbFeCo wires with perpendicular magnetic anisotropy, *IEEE Trans. Magn.* **46**, 1695–1698 (2010).
52. M. Hayashi, L. Thomas, R. Moriya, C. Rettner, and S. S. P. Parkin, Current-controlled magnetic domain-wall nanowire shift register, *Science* **320**, 209–211 (2008).
53. D. Chiba, G.Yamada, T. Koyama et al., Control of multiple magnetic domain walls by current in a Co/Ni nano-wire, *Appl. Phys. Expr.* **3**, 073004 (2010).
54. R. Moriya, M. Hayashi, L. Thomas, C. Rettner, and S. S. P. Parkin, Dependence of field driven domain wall velocity on cross-sectional area in $Ni_{65}Fe_{20}Co_{15}$ nanowires, *Appl. Phys. Lett.* **97**, 142506 (2010).
55. E. Saitoh, H. Miyajima, T. Yamaoka, and G. Tatara, Current-induced resonance and mass determination of a single magnetic domain wall, *Nature* **432**, 203–206 (2004).
56. R. Moriya, L. Thomas, M. Hayashi, Y. B. Bazaliy, C. Rettner, and S. S. P. Parkin, Probing vortex-core dynamics using current-induced resonant excitation of a trapped domain wall, *Nat. Phys.* **4**, 368–372 (2008).
57. L. Thomas, M. Hayashi, X. Jiang, R. Moriya, C. Rettner, and S. S. P. Parkin, Oscillatory dependence of current-driven magnetic domain wall motion on current pulse length, *Nature* **443**, 197–200 (2006).

58. T. Nozaki, H. Maekawa, M. Mizuguchi et al., Substantial reduction in the depinning field of vortex domain walls triggered by spin-transfer induced resonance, *Appl. Phys. Lett.* **91**, 082502–082503 (2007).
59. S. Fukami, T. Suzuki, K. Nagahara et al., Low-current perpendicular domain wall motion cell for scalable high-speed MRAM, in *Symposium on VLSI Technology Digest of Technical Papers*, June 16–18, 2009, Honolulu, HI, pp. 230–231, 2009.
60. T. Miyazaki and N. Tezuka, Giant magnetic tunneling effect in Fe/Al_2O_3/Fe junction, *J. Magn. Magn. Mater.* **139**, L231–L234 (1995).
61. J. S. Moodera, L. R. Kinder, T. M. Wong, and R. Meservey, Large magnetoresistance at room temperature in ferromagnetic thin film tunnel junctions, *Phys. Rev. Lett.* **74**, 3273–3276 (1995).
62. S. S. P. Parkin, C. Kaiser, A. Panchula et al., Giant tunneling magnetoresistance at room temperature with MgO (100) tunnel barriers, *Nat. Mater.* **3**, 862–867 (2004).
63. S. Ikeda, J. Hayakawa, Y. Ashizawa et al., Tunnel magnetoresistance of 604% at 300 K by suppression of Ta diffusion in CoFeB/MgO/CoFeB pseudo-spin-valves annealed at high temperature, *Appl. Phys. Lett.* **93**, 082508 (2008).
64. J. Z. Sun, M. C. Gaidis, E. J. Oâ Sullivan et al., A three-terminal spin-torque-driven magnetic switch, *Appl. Phys. Lett.* **95**, 083506 (2009).

37

GMR Spin-Valve Biosensors

Drew A. Hall
Stanford University

Richard S. Gaster
Stanford University

Shan X. Wang
Stanford University

37.1 Overview of Biosensing

Magnetic biosensors, particularly giant magnetoresistive (GMR) biosensors based on spin-valve sensors, used to detect surface binding reactions of biological molecules labeled with magnetic particles possess the potential to compete with fluorescent-based biosensors, which currently dominate the application space in both research and clinical settings. Magnetic biosensing offers several key advantages over conventional optical techniques and other competing sensing modalities. The samples (blood, urine, serum, etc.) naturally lack any detectable magnetic content, providing a sensing platform with a very low background. Additionally, magnetic tags do not suffer from problems that have plagued fluorescent labels such as label-bleaching and autofluorescence. Second, the sensors can be arrayed and multiplexed to perform complex protein or nucleic acid analysis in a single assay without resorting to optical scanning. Finally, the sensors are compatible with standard silicon integrated circuit (IC) technology allowing them to be manufactured cheaply with integrated electronic readout, in mass quantities, and to be deployed in a one-time use, disposable format. The technology is scalable and capable of being integrated, making it appealing for point-of-care (POC) testing applications.

Recent work in our group further shows that magnetic biosensors are able to simultaneously measure multiple binding reactions in a wash-free assay in real time. Furthermore, the sensors and the magnetic tags can be manufactured on the nanoscale, making the detector comparable in size to the biomolecules being sensed. This unique set of properties makes them practical and useful in both clinical and research settings.

37.2 History of Magnetic Biosensing

To our knowledge, the first use of magnetic nanoparticles as labels in immunoassays reported in literature was in 1997 by a group of German researchers [1]. The measurement utilized a superconducting quantum interference device (SQUID) to detect binding events of antibodies labeled with magnetic tags. While successful, the operating conditions required liquid helium cooling and a magnetically shielded room, limiting the practicality of the SQUID-based biosensors. In 1998, Baselt first demonstrated detection of magnetic nanoparticles using GMR sensors with GMR multilayers [2]. GMR sensors have the advantage of room-temperature operation and simpler instrumentation, making them more attractive, particularly for portable applications. However, proper design of GMR biosensors, magnetic tags, biochemistry, and electronics are necessary for high sensitivity bioassays, and prove to be much more sophisticated than initially perceived. This book chapter will highlight some of these advances.

All spintronic sensors, such as anisotropic magnetoresistive (AMR), GMR multilayers, GMR spin-valves, and magnetic tunnel junctions (MTJ), share a common principle of operation where a free magnetic layer (or layers) responds to a change in

TABLE 37.1 Comparison of GMR/MTJ Biosensors from Various Groups

Institution and Site	Principal Investigators	Magnetic Particles	Sensor Technology
NRL, Washington	Whitman, LJ	Dynal M280 2.8 μm	GMR, multi-segment 1.6 × 8000 μm, 42 kΩ
NVE, Eden Prairie	Tondra, M		
IST	Ferreira, HA	Nanomag-D 250 nm	SV, single-segment 2.5 × 100 μm, 1 kΩ
Lisbon, Portugal	Freitas, PP		
University of Bielefeld, Germany	Reiss, G Brueckl, H	Bangs CM01N 350 nm	GMR spiral 1 × 1800 μm, 12 kΩ
Stanford University, Stanford	Wang, SX Pourmand, N	Miltenyi MACS 40 nm	SV, Multi-segment 1.5 × 2800 μm, 45 kΩ
Brown University, Providence	Xiao, G	Dynal M280 2.8 μm	MTJ, ellipse patch 2 × 6 μm, 142 Ω
Phillips Research, the Netherlands	Prins, M	Ademtech 300 nm	GMR, gradiometer 3 × 100 μm, 250 Ω est.

Source: With kind permission from Springer Science+Business Media: *Microarrays*, MagArray biochips for protein and DNA detection with magnetic nanotags: Design, experiment, and signal-to-noise ratio, 2009, pp. 299–314, Osterfeld, S.J. and Wang, S.X.

the local magnetic field and causes a change in the resistance of the sensor. Inductive sensors, on the other hand, are only capable of monitoring the rate of temporal change in the magnetic field. The rapid development of magnetic biosensing in this decade was the result of leveraging the existing sensing technology utilized in hard disk drives. In the 1990s, hard drive manufacturers replaced the inductive pick up coil in the read head with an AMR sensor, which in turn was replaced with a spin-valve (and subsequently an MTJ) for higher sensitivity and scaling requirements in magnetic recording [3]. Since the initial conception of magnetic biosensors in the late 1990s, several groups have continued the work on research and development of magnetic biosensing technology [4–11]. Table 37.1, while not exhaustive, summarizes some of the research groups and commercial corporations currently investigating magnetoresistive biosensors. It is evident from the diversity in the table that the field is still progressing and has yet to converge on an optimal design. Furthermore, there are several research groups currently investigating using MTJs as biosensors [12,13]. Presently, the main issue limiting the adoption of MTJs is that with the large area of the devices, a single pin-hole defect renders the device unusable. Whether MTJs will take over the biological sensing applications, as they did with hard disk drives, remains to be seen.

37.3 Biology and Chemistry

An important innovation to utilizing GMR sensors as biosensors is the effective, reliable, and reproducible integration of the biology and chemistry with the inorganic sensor surface. By immobilizing recognition molecules complementary to biological molecules, such as proteins or nucleic acids, it is possible to detect binding events of proteins or nucleic acids to the recognition molecules in real time. In this chapter, we will discuss protein detection; however, nucleic acid–based detection is very similar [7]. Historically, IgG antibodies have been the recognition molecule of choice for protein-based detection assays as they are highly specific to one unique region of one protein, known as an epitope of the protein. This antibody–protein interaction is typically referred to as a lock-and-key system due to the high specificity of the interaction. It is this specificity that allows researchers to selectively capture one specific type of protein in a vast background of proteins.

The overall structure of IgG antibodies is remarkably similar whether it is reactive to a cancer tumor maker or an infectious pathogen (Figure 37.1). Every IgG antibody contains two long heavy chains and two short light chains held together via inter-chain disulfide bonds and electrostatic interactions. Both the light chains and the heavy chains can be divided into two distinct regions: a constant region at the base of each chain and a variable region at the tip. The constant region at the base of the heavy chains is typically referred to as the Fc region (which will be mentioned later). At the tip of each chain is the highly variable region known as the Fab region, which contains the antigen-binding sites. This highly variable region is where the specificity of the protein–antibody interaction takes place and the lock-and-key interface occurs. By choosing an antibody that reacts with a specific biomarker of interest, one can selectively capture that biomarker of choice over a sensor surface to facilitate protein detection.

37.3.1 Sandwich Assay

The most effective method of detecting proteins in a solution is via a "sandwich assay." Typically known for its use in the enzyme-linked immunosorbent assay (ELISA), the sandwich assay involves the formation of a three-layered structure where two antibodies (aptamers, diabodies, Fab fragments, etc., can also be used; however, antibodies are by far the most common recognition molecule used for protein detection) form a

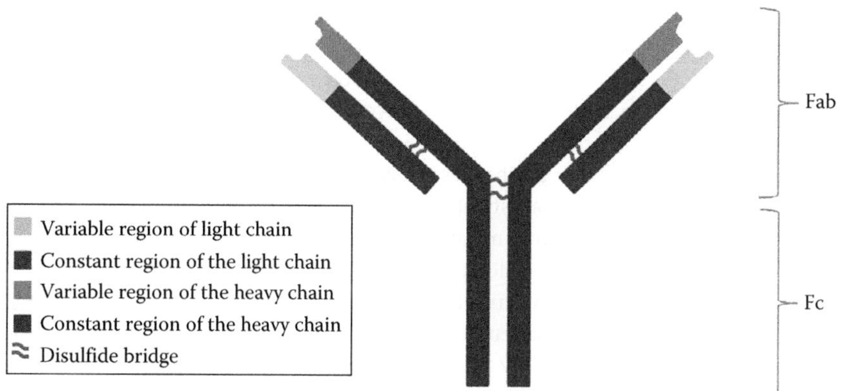

FIGURE 37.1 Structure of an IgG antibody.

sandwich around a protein (also called the analyte) of interest (Figure 37.2a). One of the antibodies, typically referred to as the capture antibody, is directly immobilized on the sensor surface. In order to make the sandwich assay highly specific, a monoclonal capture antibody is traditionally used. A solution of monoclonal antibodies means that every antibody in the solution has the exact same Fab region and therefore will bind to only one epitope on one protein. The capture antibody makes up the foundation of the sandwich assay and acts to selectively capture specific proteins of interest directly over the sensor surface.

The second antibody, known as the detection antibody, is delivered in solution, and binds to a second epitope on the captured protein of interest. The detection antibody is typically polyclonal and premodified with a reactive chemistry, enabling facile attachment of the detection antibody to the tag of interest. A polyclonal antibody solution is one in which all the antibodies react with the same protein; however, they may bind to different epitopes on that protein. Therefore, the Fab region is not identical across all the antibodies in a polyclonal solution. Typically, the antibody is modified with biotin and the tag is modified with streptavidin since the biotin–streptavidin interaction is one of the strongest non-covalent receptor–ligand interactions in biochemistry (association constant $K_a \approx 10^{14}$–10^{15} M^{-1}). In the ELISA, the tag of interest is typically fluorescent or enzymatic. However, when using GMR biosensors, the tag of interest is magnetic. Therefore, the more protein that is present in the system, the more detection antibody binds and the more magnetic tag binds. As the number of magnetic tags increases over the sensor, the magnetoresistance in the underlying GMR sensor changes proportionally, producing larger signals. In this way, quantitative protein detection is possible with this assay.

37.3.2 Reverse Phase Assay

A "reverse phase" assay can be used to detect proteins of interest in many patients' blood samples simultaneously by reorganizing the traditional sandwich assay. In the reverse phase assay, instead of functionalizing a capture antibody onto the sensor surface, patient samples containing proteins of interest such as cell lysates are immobilized onto the GMR sensor array, at least one sensor per sample. Then a solution containing antibodies complementary to the protein of interest is introduced, and will bind to the immobilized protein of interest over the GMR sensor. Subsequently, a second antibody modified with biotin will be added, which reacts with the Fc portion of the first antibody. Since the second antibody is biotinylated, it can then bind to magnetic tags coated with streptavidin in a similar fashion to the detection antibody and magnetic tag interaction in the traditional sandwich assay (Figure 37.2b). Note that the scheme in Figure 37.2b can also be accomplished with one biotinylated detection antibody if it directly binds to the protein of interest.

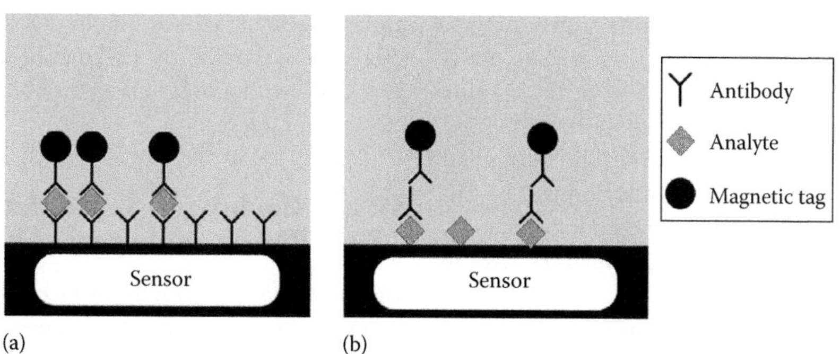

FIGURE 37.2 Schematic of (a) a sandwich assay and (b) a reverse phase assay for protein detection or for antibody detection.

In addition, if one separates the protein of interest from the sample prior to sensor immobilization, it is possible to increase capture protein density.

It is also noteworthy that the schematic in Figure 37.2b can be further modified to detect antibodies in the patient's blood that reacts against a specific protein of interest. These antibodies can be either auto-reactive antibodies (antibodies against self-causing autoimmune disorders) or antibodies conferring immunity (antibodies against infectious agents). If a physician detects antibodies against the HIV capsid protein in a patient sample, for example, then the physician can confirm that the patient has been infected with the virus without detecting the HIV viral protein itself. In this type of assay, a specific "capture protein" is selectively immobilized onto the GMR sensor. Then a sample containing antibodies complementary to the protein of interest is introduced; the patient's antibodies will bind to the immobilized protein of interest over the GMR sensor. Subsequently, a second antibody modified with biotin will be added, which reacts to the Fc portion of the first antibody (analyte). Since the second antibody is biotinylated, it can then act to bind magnetic tags coated with streptavidin.

37.3.3 Capture Antibody or Capture Protein Immobilization

The capture antibody or capture protein can be mounted onto the sensor surface via a variety of different methods. Ideally, the antibody or protein of interest will be attached to the sensor surface while retaining the native conformation and activity. The immobilization can be performed by nonspecific physical absorption where electrostatic forces between the antibody and thin film polymer on the substrate dictate the binding. For example, a highly cationic polymer can be used to nonspecifically attract negatively charged proteins or antibodies to the sensor surface. Polymers such as polyethyleneimine have been shown to provide both an effective and simple means of physical absorption–based antibody immobilization [10].

Alternatively, a covalent binding chemistry can be employed, where new molecular bonds are formed, physically connecting the recognition molecule (e.g., antibody, nucleic acid, or protein in the case of the reverse phase assay) to the sensor surface. This method of antibody immobilization is typically more effective and reproducible than direct physical absorption, however, more challenging to implement and more time consuming to run. Fancy tricks like orienting antibodies in a specific direction via selectively binding to the Fc fragment of the capture antibody or using protein A or protein G immobilization is also possible. Due to space limitations, only a few of the most common methods of covalent antibody immobilization will be discussed here; however, numerous effective methodologies exist. For more details on bioconjugation of sensor surfaces refer to "Bioconjugate Techniques" [14].

Antibodies are made out of four strings of amino acids joined via disulfide bonds to form the Y-shaped structure presented earlier. Each amino acid in the string has a specific functional group dangling. Many of these functional groups have reactive moieties that can be used to form covalent bonds for surface immobilization. Silanes, polymers, thiols, cross-linkers, etc., have all been shown to be effective methods of covalent antibody immobilization. Accordingly, choosing which chemical modification scheme to use may depend on the amino acid sequence of the antibody. All antibodies, however, consistently have an amino-terminal end, a carboxy-terminal end, and disulfide bonds connecting the strings of amino acids in the antibody. Accordingly, for every receptor–ligand combination, there are a number of effective ways one can fasten the recognition molecule of choice to the sensor surface.

Perhaps the most common method of antibody immobilization is via the formation of covalent bonds from thiol groups on antibodies to gold surfaces. Since all antibodies contain interchain disulfide bonds, it is possible to use a selective reducing agent such as 2-mercaptoethylamine or dithiothreitol to reduce only the disulfide bonds connecting the heavy chains of IgG antibodies. This will reproducibly create "half" antibodies with exposed sulfhydryl groups at the hinge region of the antibody for gold-thiol conjugation. Alternatively, if one wants to avoid reducing the antibody into half antibodies, Traut's Reagent can be used to thiolate primary amines on IgG antibodies.

Another common method of antibody immobilization is via use of primary amines on antibodies. Amine-reactive chemical reactions are commonly used because they are rapid and typically provide stable amide bonds between the antibody and polymer on the sensor surface. In addition, virtually all antibodies are equipped with primary amines making the chemistry generalizable. Therefore, virtually any antibody can be conjugated to surface immobilized polymers, silanes, or self-assembled monolayers (SAMs).

Arguably, the most critical aspect to all surface functionalization is the surface uniformity. Since GMR sensors, if coated with ultrathin passivation layers, are proximity-based biosensors [10], it is essential to ensure that the tag in the sandwich assay remains a reproducible distance from the sensor surface across all the sensors within a chip and from chip to chip. Therefore, every component of the sandwich assay must remain a reproducible distance from the sensor. If there are surface nonuniformities such as microbubbles or particulate debris, then the sensor-to-sensor reproducibility will be severely compromised. In addition, the closer the particle is to the sensor surface, the higher the signal will be. So, in order to obtain maximum sensitivity in the device, the polymer layers should be made as thin as possible without sacrificing the ability to bind the recognition molecule of interest.

37.3.4 Blocking of Sensor Surface to Reduce Nonspecific Binding

The main goal in antibody surface functionalization is to make the sensor surface as reactive as possible, but only to one unique capture antibody or capture protein. This will facilitate the maximum amount of capture agent binding as possible. Once the capture antibody or capture protein is immobilized on the sensor surface, however, the highly reactive surface must be neutralized or blocked to prevent any future nonspecific binding of protein, detection antibody, or tag. In all biosensors, such antifouling chemistry is essential to reproducible and reliable protein detection. The most common method of blocking a sensor surface is to saturate the surface with a common, high-abundance protein, such as albumin or casein. These high-concentration proteins will bind to the remaining reactive groups on the thin film and therefore prevent any future proteins or antibodies from binding to the sensor surface in the absence of the capture antibody or capture protein of interest. Therefore, upon addition of sample in the subsequent step of the sandwich assay, only proteins complementary to the selected capture antibodies of interest will bind, and all the noncomplementary proteins will be washed away.

37.4 Magnetic Labels

In magnetic biosensors, the magnetic label is a very important component of the system. Unfortunately, there are several factors that constrain the choice of label, such as the size and the functionalization. Larger ferromagnetic particles are initially attractive since they exhibit a large magnetic moment and will thus be easier to detect. Using large micron-sized magnetic tags, Li et al. have demonstrated detection down to a single particle [11]. Although exceptionally sensitive protein detection and magnetic manipulation is possible with large magnetic particles, more reproducible protein detection, with minimal nonspecific binding, and higher stability is possible when using smaller tags. Furthermore, large micron-sized labels cannot be used for basic science applications where the kinetics of the binding reactions is studied because the labels significantly degrade the mobility of the antibodies. For these reasons, the focus in our group has shifted toward smaller, nanometer-sized magnetic particles. By using magnetic tags on the order of 50 nm (46 ± 13 nm), the particles are both more stable in solution (they do not settle over the sensor) and diffuse more rapidly to the detection antibody. At this size scale, the particles can be superparamagnetic, which helps to reduce aggregation and chaining. Companies such as Miltenyi Biotec offer clusters of 10 nm Fe_2O_3 superparamagnetic particles embedded in a dextran shell, which are prefunctionalized with streptavadin or other reactive groups. Custom-made clusters of superparamagnetic nanoparticles and monodisperse iron oxide nanoparticles [15] are also promising for magnetic labels.

To overcome the difficulties associated with increasing the size of the superparamagnetic particles (aggregation, size distribution, and coercivity) for larger moment, there has been active research investigating magnetic nanoparticles with tunable magnetic properties [16,17]. Nanoparticles created by nanoimprint lithography containing a synthetic antiferromagnetic (SAF) film stack have large saturation magnetic moments, are monodisperse (at a dimension of 100 nm or less), and do not aggregate (in the absence of applied field). These particles decouple the size restriction associated with superparamagnetic particles and as a result, the magnetic moment can be substantially larger under externally applied fields, leading to higher signals. Furthermore, the properties of these particles can be tuned by adjusting the thickness of the spacer layer or magnetic layers, making them highly attractive for biological sensing, separation applications, and imaging.

37.5 GMR Sensors

To quantify the amount of biomolecules present, the GMR sensors transduce the stray field from the magnetic tags into an electrical signal. The stray field, however, is extremely small because the moment of the labels is also small. This requires the sensor to be extremely sensitive to minute magnetic field changes, making GMR spin-valve sensors preferable to AMR sensors or GMR multilayer sensors. GMR multilayer sensors, consisting of antiferromagnetically coupled magnetic layers, tend to have lower field sensitivity than spin-valves because of higher saturation fields. Spin-valves are well suited for biosensing application due to their high sensitivity, high linearity, small hysteresis, and compact feature size. The design of spin-valve biosensors was previously reviewed [18]. Figure 37.3 shows an illustration of spin-valves used as biosensors. The sensors are passivated with an ultrathin oxide to prevent the biological and chemical solutions used to functionalize the surface from causing corrosion. A thicker passivation layer is used to create a reference sensor that will be unable to sense magnetic tags and therefore allow us to monitor systematic fluctuations in the electronics (Figure 37.3c). In addition, a control sensor is typically coated with proteins, such as bovine serum albumin (BSA) or noncomplementary antibodies in order to monitor nonspecific binding in the assay (Figure 37.3b).

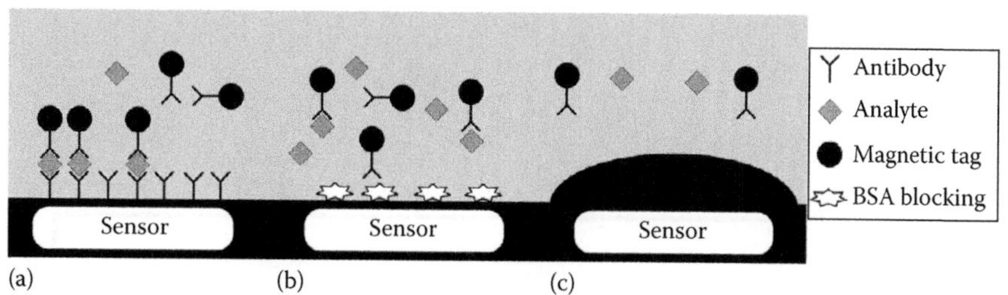

FIGURE 37.3 A sandwich assay with capture antibody, antigen, detection antibody, and magnetic particle. BSA blocking is shown on the control sensor. (a) A biologically active sensor, (b) an unfunctionalized control sensor, and (c) a passivated reference sensor.

Unlike memory applications where the desired output of the spin-valve is binary, a linear analog signal is desired for quantitative detection [19]. The sensors are typically biased such that the pinned magnetization (M_p) is fixed in the transverse direction and the free magnetization (M_F) can rotate freely in the plane. In this configuration, the resistance can be expressed as shown in Equation 37.1 (this will be derived in the next section) where ΔR_{max} is the magnetoresistance and θ_f is the angle of the free layer with respect to the x-axis.

$$R = R_0 + \frac{1}{2}\Delta R_{max}\sin(\theta_f) \qquad (37.1)$$

For small θ, the resistance versus field transfer curve has a linear dependence. In order to establish this bias condition, an external magnetic field (H_b) is applied to align the free layer along the x-axis. Alternatively, the sensor can be designed such that the shape anisotropy is sufficient to bias the sensor without the need for an external magnetic field.

37.5.1 Signal per Magnetic Label

Before designing the electronics for a spin-valve biosensor, we need to calculate the signal range (resistance change per label). The Stoner–Wohlfarth (SW) model, originally proposed to analyze the magnetic hysteresis of a single-domain particle, was extended to model spin-valves because the thin ferromagnetic layers in spin-valves are often in a single-domain state [20]. In general, all of the ferromagnetic layers in a spin-valve need to be considered for the total energy expression in the SW model. However, in cases where the strong exchange bias fields effectively pin the magnetization of the pinned layers under small external fields, the free layer alone can be used for the energy expression. As a rough approximation, the average magnetic field caused by a magnetic label positioned in the center of the sensor can be approximated as a dipole, given in Equation 37.2, where the vector $\boldsymbol{r}$ points from the center of the tag to the free layer, which has a thickness t [21] (Figure 37.4).

$$\bar{H} = \frac{1}{lwt}\int_{-l/2}^{l/2}\int_{-w/2}^{w/2}\int_{-t/2}^{t/2}\frac{3(\boldsymbol{m}\cdot\boldsymbol{r})\boldsymbol{r}}{r^s} - \frac{\boldsymbol{m}}{r^3}\,dx\,dy\,dz \qquad (37.2)$$

The moment of the magnetic particle with a bias (H_b, applied along the x-axis) and tickling field (H_t, applied along the y-axis) is shown in Equation 37.3.

$$\boldsymbol{m} = \frac{4}{3}\pi r_b^3\chi(H_b\hat{x} + H_t\hat{y}) \qquad (37.3)$$

Since the thickness of the free layer is typically significantly smaller than the other dimensions, the field induced in the spin-valve in the x and y directions can be approximated by Equation 37.4.

$$\overline{H_x} = -\frac{8m_x}{(l^2+4d^2)\sqrt{l^2+w^2+4d^2}}$$
$$\overline{H_y} = -\frac{8m_y}{(w^2+4d^2)\sqrt{l^2+w^2+4d^2}} \qquad (37.4)$$

The total energy of the free layer is taken as the sum of energies arising from anisotropy and the external fields including the bias, tickling, and dipole fields. The angle θ_f represents the rotation of the free layer magnetization. An additional term, H_{other}, is included to account for transverse fields due to other sources including interlayer coupling as well as fringing fields from the

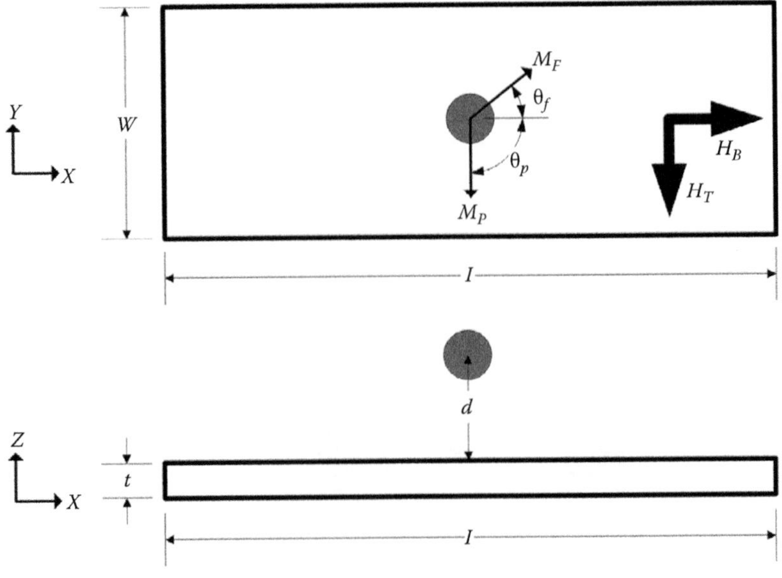

FIGURE 37.4 (Top) Top-down view and (bottom) side view of spin-valve sensor and magnetic tag. Usually $\theta_p = -90°$.

sense current and pinned layers. For simplicity, these can be considered to be negligible in most practical cases.

$$E = \frac{1}{2} H_K M_F \sin^2\theta_f - \left(H_b + H_x\right) M_F \cos\theta_f - \left(H_t + H_{\text{other}} + H_y\right) M_F \sin\theta_f \quad (37.5)$$

The total anisotropy field (H_K) is the sum of the uniaxial anisotropy field (H_U) and demagnetizing field (H_D) at the center of the free layer where the dipole field strength is at its maximum.

$$H_D = \left(N_x - N_y\right) M_F, \quad H_K = H_U + H_D \quad (37.6)$$

The demagnetizing factors used to calculate H_D are approximated in the expressions given by Li et al. [22].

$$N_x = \frac{8tw}{l\sqrt{l^2 + w^2}}, \quad N_y = \frac{8tl}{w\sqrt{l^2 + w^2}} \quad (37.7)$$

The first derivative of the energy expression is set to zero to minimize the total energy and solved for θ_f.

$$\frac{dE}{d\theta_f} = M_F \left[H_K \sin\theta_f \cos\theta_f + \left(H_b + H_x\right)\sin\theta_f - \left(H_t + H_{\text{other}} + H_y\right)\cos\theta_f\right] = 0 \quad (37.8)$$

Solving Equation 37.8 and approximating it with the Taylor expansion, (valid for $\theta \leq 25°$), we arrive at the relatively simple expression Equation 37.9.

$$\sin\theta_f = \left\langle \frac{M_Y}{M_F} \right\rangle \approx \tanh\left(\frac{H_t + H_{\text{other}} + H_Y}{H_K + H_b + H_X}\right) \quad (37.9)$$

In all practical settings, the magnetization of the pinned layer is fixed in the transverse direction such that $\theta_p = -\pi/2$, thus the resistance of the spin-valve is given as previously stated (Equation 37.1).

The derivation presented is sufficient for a rough approximation; however, several assumptions were made along the way to make the equations tractable. More accurate results can be obtained by using a finite element model simulator such as Ansoft's Maxwell simulator, or a micromagnetic simulator such as the open-source object-oriented micromagnetic simulation framework (OOMMF). Using the simulators, additional parameters can be studied, such as the effect of particle location or particle-to-particle interactions.

37.5.2 Characterization of GMR Spin-Valve Sensors

In order to convey the design process for using a spin-valve as a biosensor, we will start with a representative spin-valve and design the interface electronics. Figure 37.5a shows the transfer curve (resistance as a function of field) of a spin-valve by sweeping the field strength along the hard axis. The sensor has a minimum resistance of 2190 Ω in the parallel state and a maximum resistance of 2465 Ω in the antiparallel state, corresponding to a magnetoresistance ratio (MR) of 12%. The sensitivity of the sensor, calculated by differentiating the transfer curve, is shown in Figure 37.5b, and is maximum when no field is applied, tapering off as the field strength is increased.

To make the sensor realistic, it will also exhibit realistic noise characteristics. At low frequencies, the noise is dominated by the flicker noise, noise power that has an inverse proportionality to frequency (1/*f*) dependence and is typically believed to originate from slow defect motions and magnetic domain rotations [23]. The corner frequencies (where the 1/*f* noise intersects the thermal noise floor) are 900 Hz, 2.25 kHz, and 10 kHz for currents of 100, 250, and 500 μA, respectively. These values are consistent with typical values found in literature and actual devices [24]. The sensor dimensions are such that sufficient shape anisotropy exists to not require an external magnetic bias field.

The flicker noise of the sensor makes it prohibitive to operate at DC for sensitive detection. To escape the additional noise

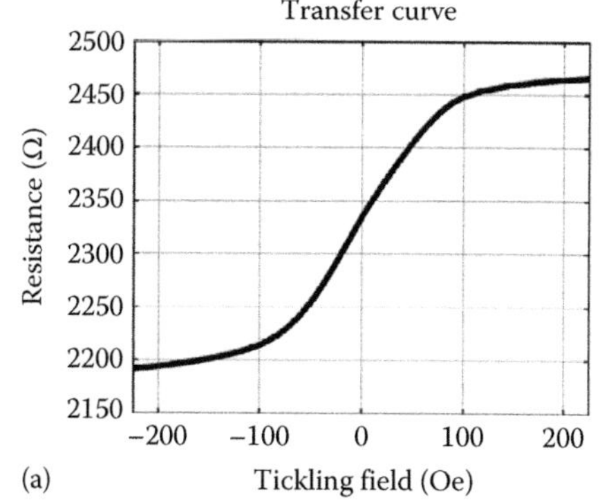

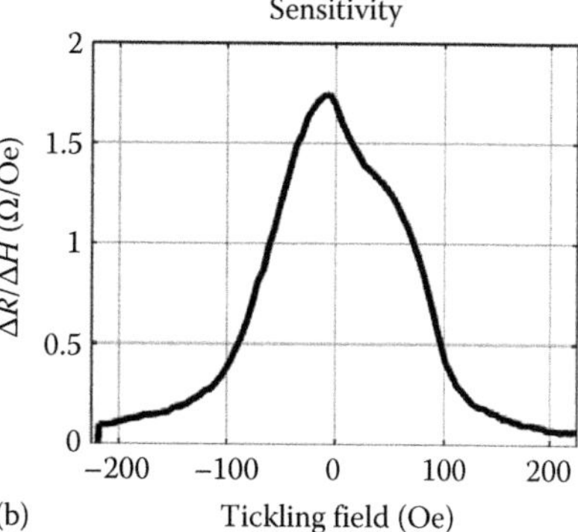

FIGURE 37.5 (a) Transfer curve of a representative spin-valve sensor. (b) Sensitivity of spin-valve in (a).

penalty incurred when operating at low frequencies, the signal is modulated up to a higher frequency with the use of an alternating external magnetic field, referred to as the tickling field, applied in the transverse direction (along the y-axis). The tickling field can also serve to magnetize the magnetic particles if they are superparamagnetic or SAF. Modulating the signal to a higher frequency is a common technique used in sensing, often called "chopping" in optical systems. Many research groups [4,8] have opted to integrate the ability to generate the tickling field on the same die as the sensors. While there are many advantages to integrating the coil, namely, precise alignment capabilities (due to photolithography), compactness, and the ability to actively attract the particles to the sensor surface, they often suffer from poor field uniformity [4].

37.6 Sensor Interface Electronics

For a sensor interface, the circuit should be transparent to the sensor, following the adage "Do No Harm" [25]. The sensor parameters for our spin-valve sensor are summarized in Table 37.2. As a case study, two front-ends will be designed to meet different design objectives: a small sensor array (8 sensors) meant for portable applications and a larger array (64 sensors) for benchtop applications.

Before diving into the two designs, there are system-level considerations that can be made to further improve the signal-to-noise ratio (SNR) in both designs. The resistance change per particle is very small (often micro-ohms to milliohms), and can be easily swamped out in the presence of temperature variations and noise. It was previously stated that by modulating the magnetic field, it is possible to modulate the signal to a frequency beyond the flicker noise of the sensor. However, there is a subtle problem with only modulating the field: the sensors have leads and from Faraday's law, a current will be induced in these leads from the time-varying electromagnetic field (EMF). This parasitic coupling occurs at the same frequency as the signal from the particles. However, if the excitation current (or voltage) applied to the sensor is also modulated at a different frequency, the output of the sensor will appear like a mixer [4,26]. In this scheme, both the resistance of the device and the current through the device are a function of time (Equation 37.10).

TABLE 37.2 Summary of Representative Spin-Valve Sensor Parameters

Sensor Property	Value
Nominal resistance	2.5 kΩ
Magnetoresistance	12%
Breakdown voltage	>0.5 V
Optimum tickling field with particles	30 Oe
Thermal noise	6.4 nV/√Hz
	900 Hz @ 100 μA
Flicker noise corner frequency	2.25 kHz @ 250 μA
	10 kHz @ 500 μA

$$\begin{aligned} R(t) &= R_0 + \Delta R_0 \cos(2\pi f_f t) \\ I(t) &= I_0 \cos(2\pi f_c t) \\ V_{GMR}(t) &= R(t) \times I(t) \\ V_{GMR}(t) &= I_0 R_0 \cos(2\pi f_c t) + \frac{I_0 \Delta R_0}{2} \cos\left(2\pi\left(f_c - f_f\right)t\right) \\ &\quad + \frac{I_0 \Delta R_0}{2} \cos\left(2\pi\left(f_c + f_f\right)t\right) \end{aligned} \tag{37.10}$$

The spectrum at the output of the sensor (Figure 37.6) will have four spectral components: a parasitic tone from the EMF (at f_f, the coil frequency, omitted in the derivation of Equation 37.10 for simplicity), a carrier tone (at f_c, the frequency of the modulated excitation), and two side tones at $(f_c \pm f_f)$. This modulation scheme separates the resistive and the magnetoresistive components of the spin-valve while only the magnetoresistive component changes under the influence of the magnetic labels. The change in amplitude of the side tone is linearly proportional to the number of magnetic labels from which the concentration of protein or nucleic acid can be inferred. This double modulation scheme improves the SNR of the side tone. In practical situations, the overall SNR of the biosensing system should be dominated by the nonspecific binding and not the electronic noise.

37.6.1 Eight-Channel Front-End Design

For the small sensor array case study, the target application is a portable biosensor [27]. The design objectives are to minimize the power consumption and component count without substantially sacrificing the sensitivity. This design will have a minimal analog-to-digital converter (ADC) back-end and very little computational power, necessitating that the demodulation be done in the analog domain. Furthermore, to minimize the form factor of the device, the excitation coil is integrated with the

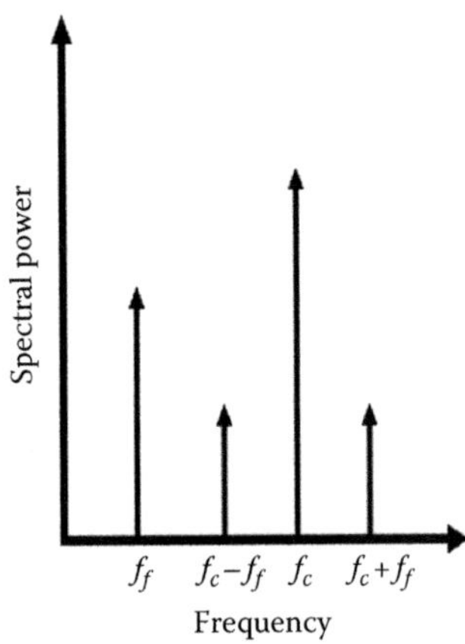

FIGURE 37.6 Spectrum at the output of the sensor using double modulation scheme.

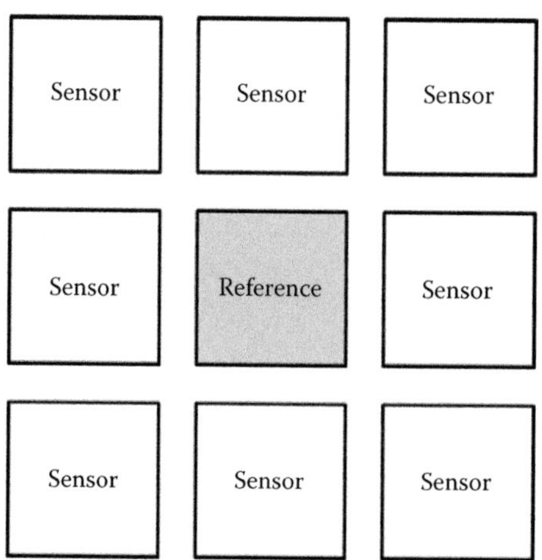

FIGURE 37.7 Sensor layout with reference sensor for matching purposes.

sensors implemented as metal traces on the same die. As such, the coil frequency can operate at a frequency much beyond the flicker noise corner frequency of the sensor. A popular method for high-precision resistance measurement is the Wheatstone bridge. It is constructed of two spin-valve sensors and two programmable resistors. One of the sensors is biologically active and the other is a reference sensor (shared among the other eight sensors) that is magnetically isolated (from magnetic nanotags, typically with a thick passivation layer) but physically near the active sensor so that it experiences the same sensing environment (such as temperature changes, magnetic bias, and tickling field gradients, etc.). The layout in Figure 37.7 shows how the reference sensor can be placed to maximize the matching between all of the biologically active sensors to minimize process variations. Furthermore, dummy sensors can be added around the outside of the sensors to further improve matching if necessary.

To reduce the component count and hence form factor, the bridge is time domain multiplexed (TDM) where only one sensor is read at any given time. The differential output of the bridge cancels the common mode signal experienced by both the active sensor and the reference sensor (such as temperature swings). In order to measure the small resistance change of a magnetic label, a high-gain instrumentation amplifier is used to amplify the small differential output of the bridge. A high-pass RC filter blocks the DC component of the signal and adjusts the common mode of the input signal to an appropriate voltage for the instrumentation amplifier. When considering the choice of amplifier, the input-referred noise and the flicker noise corner frequency are important parameters and should be chosen such that the sensor dominates the noise budget. The output of the amplifier is demodulated with a mixer to minimize the ADC requirements. A simple RC low-pass filter removes all but the lower side tone from the spectrum. Our circuit architecture is shown in Figure 37.8. The Wheatstone bridge and high-pass RC filter were used by several research groups [2,28].

The architecture presented in the preceding text is simple to design, yet very effective for small arrays and applications where the read-out time does not need to be fast. Balancing the bridge requires tuning the two programmable resistors to compensate for mismatch in the sensors due to process variations. If the bridge is not balanced, the offset can cause the high-gain differential amplifier to saturate. For small arrays, this tuning step is not a problem; however, it becomes cumbersome for larger arrays limiting the utility of this architecture in some applications.

37.6.2 64-Channel Front-End Design

The larger array of 64 sensors is meant to be used in a research laboratory or a clinical setting where it is a bench top instrument [29,30]. In this setting, power and component count are not limiting factors. Instead, achieving the best sensitivity, reproducibility, and fastest readout is the design objective. Furthermore, the computational power is not limited since it will be performed on a modern computer rather than a microprocessor as for the portable biosensor. Due to the additional computational power and the desire to read out the array as quickly as possible, this design will employ frequency domain multiplexing (FDM) in addition to TDM. By using FDM, multiple sensors can be read out simultaneously, reducing the time needed to readout the entire array. The FDM is implemented by exciting sensors with different carrier frequencies ($\omega_1, \omega_2, \ldots, \omega_n$) [26]. To implement the multiplexing scheme, this design (Figure 37.9) will use a transimpedance amplifier (TIA). The input of the amplifier is a virtual ground providing an easy node to sum the currents from each of the sensors. The demodulation of the signals will be performed in the digital domain after the signals have been digitized.

Unlike the previous example that utilized a bridge interface, this interface does not cancel the carrier signal. To remedy this, a replica path is added for carrier suppression where the sensors are replaced with resistors. The output of this path is just the carrier tones without the side tones (the signals of interest).

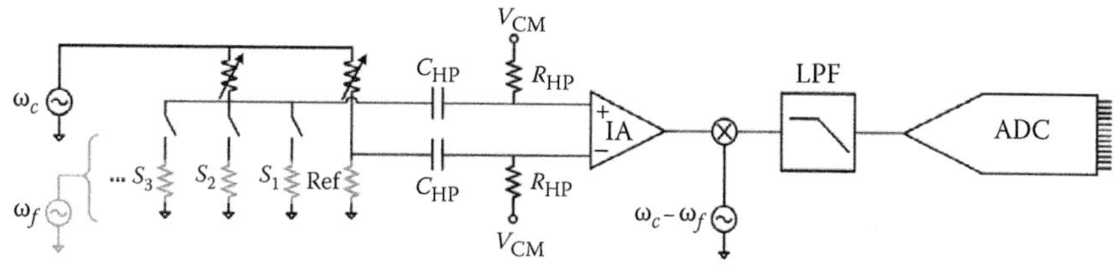

FIGURE 37.8 Circuit architecture for portable biosensor application using time domain multiplexing and double modulation.

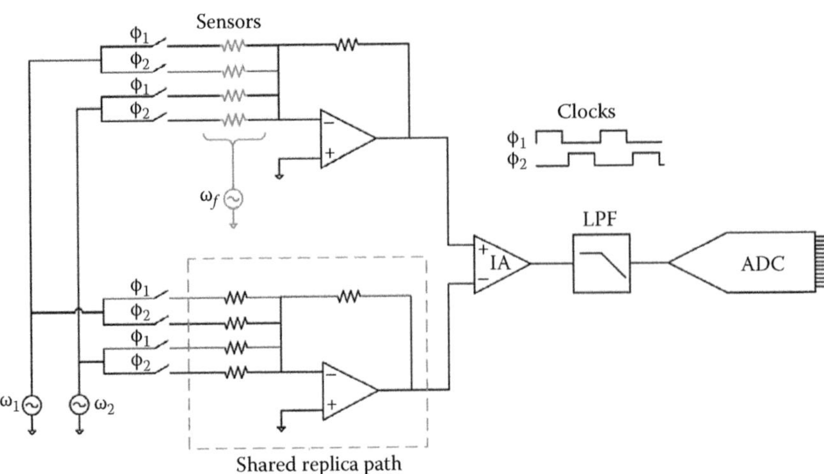

FIGURE 37.9 Circuit architecture utilizing TIA with time domain multiplexing, frequency domain multiplexing, and double modulation.

The carrier tones of the sensors are suppressed or canceled completely by subtracting the output of the replica path. This carrier suppression technique helps to reduce the dynamic range requirement of the back-end ADC.

Since the reference sensor and the biologically active sensors are acquired independently (the difference was acquired in the previous example), this interface provides an additional degree of flexibility since the reference sensor can be chosen after the experiment is completed. There is also an opportunity to use correction algorithms and digital signal processing to improve many of the nonideal characteristics of the sensors. For example, the temperature dependence of the sensors can be corrected for in software [31,32] without requiring heater elements to precisely regulate the temperature [33,34]. Additionally, correction algorithms can be designed to enhance the reproducibility of the platform, filter out common mode signals, and correct for process variations.

37.7 Real-Time Protein Assays

One of the key issues in using GMR sensors as biosensors is effectively integrating the biology with the electronics. By immobilizing recognition molecules complementary to analytes of interest directly over individually addressable GMR sensors, biological molecules of interest can be captured over the sensor surface and detected with the use of magnetic tags. Using magnetic tags and GMR biosensors, real-time protein detection is possible with a broad linear dynamic range and with high multiplex capability.

The ability to monitor protein binding events in real time provides several advantages over two-point measurements used in the ELISA. Upon incubation of the magnetic nanoparticle tags, it is possible to monitor the change in magnetic content over the sensor surface over time. This provides the user with the ability to distinguish proper nanoparticle binding kinetics, where the signal follows a smooth and continuous absorption pattern (Figure 37.10a), from irregular events that have either nonspecific binding or errors in the electronics and sensor (Figure 37.10b–d). If a sensor corrodes during the experiment or there is atypical nonspecific binding, for example, the binding signals will look abnormal and the user will be able to recognize that an error in the system has occurred and, in many cases, what went wrong (Figure 37.10). In the ELISA two-point measurements, however, an abnormally high or low data point cannot be rationalized in this manner as there is no information as to what took place during the reaction. Therefore, eliminating that abnormal data point from the test set may be controversial.

In addition, some reactions take longer than others to reach saturation. In a two-point measurement of an unknown sample, one must blindly choose when to take the final point measurement and hope that the reaction is complete or at least reproducible across all experiments. By monitoring the sensors in real time, however, it is possible to observe saturation in the signal (Figure 37.11). For low protein concentration experiments, saturation may be achieved in a minute or two, whereas in the ELISA, the technician must wait the full hour or two (depending on the protocol) prior to any analysis.

In order to translate the final saturation signal from real-time measurements into a quantifiable protein concentration, a calibration curve must first be generated (Figure 37.12). The calibration curve is created by spiking known concentrations of protein into the medium of interest (phosphate buffered saline (PBS), serum, urine, saliva, etc.) at a range of different concentrations. Therefore, by comparing the signal generated from an unknown sample to the predetermined signals on the calibration curve, it is possible to approximate the concentration of the unknown sample based on the trend observed with the known samples. In addition, the limit of detection, the linear dynamic range of analyte concentrations, and the range of linearity can be determined by plotting the calibration curve. Since each antibody–antigen interaction is unique, every new combination of antibodies and antigens warrants a new calibration curve.

According to the calibration curve in Figure 37.12, GMR biosensors are capable of detection down to fM concentrations without biological amplification or sample preconcentration.

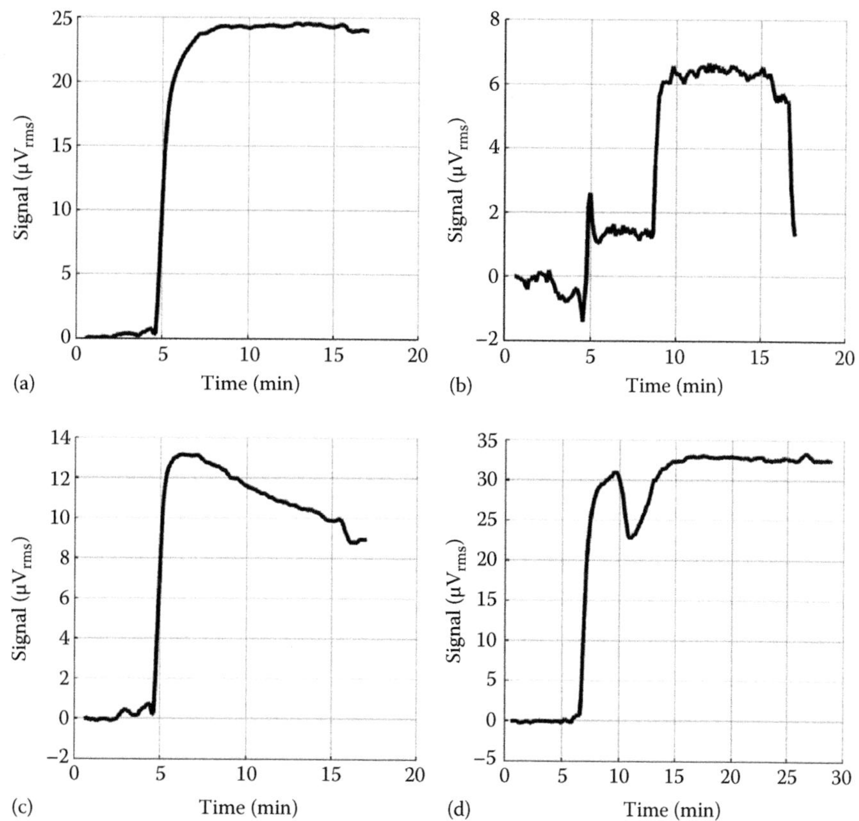

FIGURE 37.10 Normal binding curve (a) in comparison with abnormal binding curves (b–d).

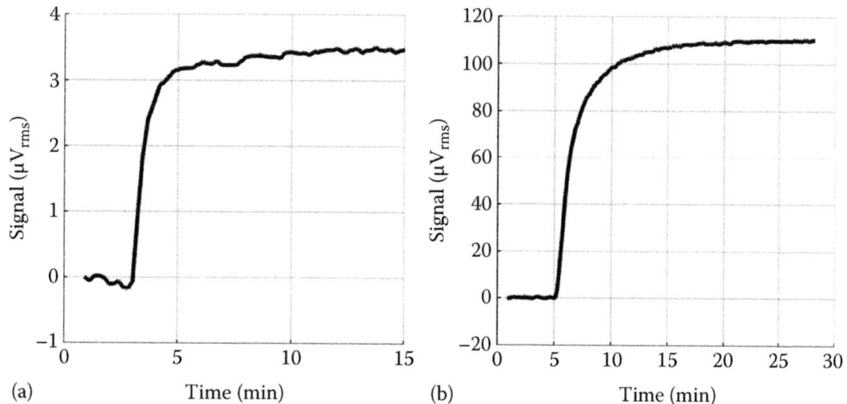

FIGURE 37.11 Example of a binding curve with (a) low concentration and (b) high concentration.

Since the calibration curve is linear (on a log–log plot) down to this concentration regime, highly sensitive and quantitative protein detection is possible when using GMR biosensors. In addition, it is important to note that the background signal (where BSA or a noncomplementary antibody is immobilized over the sensor surface instead of a capture antibody) remains flat throughout the entirety of the experiment (Figure 37.13), proving negligible magnetic tag settling and minimal nonspecific binding. It is customary to define the lower limit of detection to be the average signal on the control sensors (functionalized with noncomplementary antibody) plus three standard deviations.

In addition, since GMR sensors can be developed into sensor arrays, multiplex protein detection is also possible in a single assay. Accordingly, arrays of up to 64 unique sensors have been demonstrated [10] where an array of sensors is simultaneously monitored with several unique capture antibodies in real time. In Figure 37.13, a triplex protein detection experiment is presented where three different capture antibodies were functionalized over different sensors in the array. In this assay, biomarker

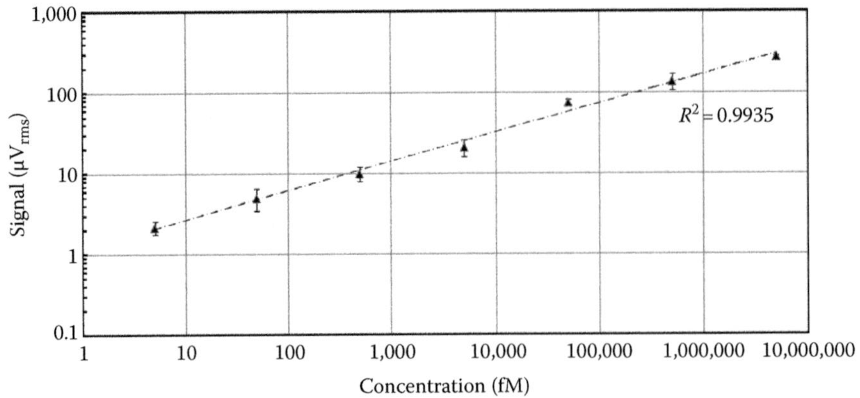

FIGURE 37.12 Calibration curve to translate signal into concentration for a protein.

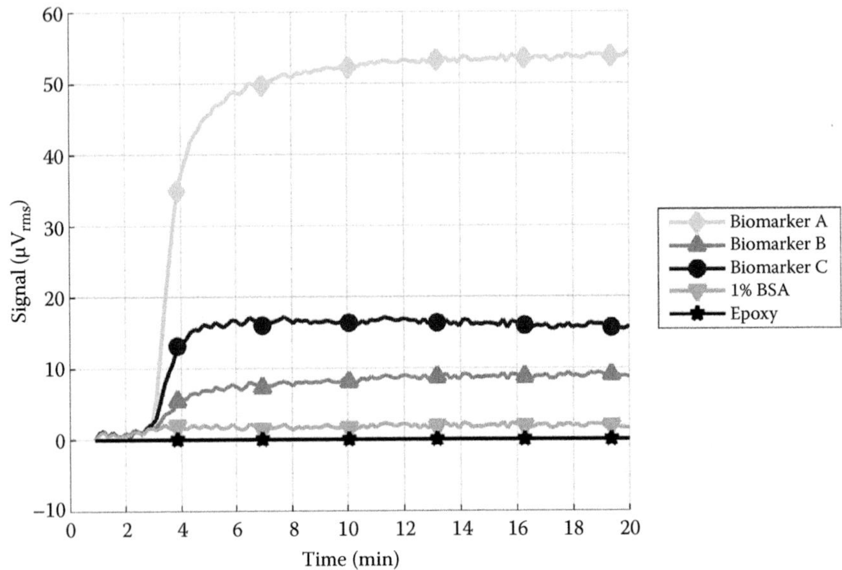

FIGURE 37.13 Triplex experiment illustrating real-time binding curves of three different biomarkers at different concentrations and two negative controls.

A is present at the highest concentration and accordingly provided the largest signal while biomarker B and biomarker C were present at successively lower concentrations revealing lower signals. In addition, the BSA control and epoxy-coated control sensors provide consistently flat signals.

37.8 Design Trade-Offs and Optimization

In adapting GMR spin-valves for biosensing applications, there are many design parameters with complex trade-offs and interdependencies. It is not surprising that a single optimal design does not exist for all applications. The following will briefly discuss a few of the design issues such as breakdown voltage, magnetic field strengths, resistance, noise, and sensor area as well as the dependencies among them.

Since the magnetic field at the sensor surface from a magnetic label is roughly proportional to the distance cubed, it is highly beneficial to minimize the separation between the label and the sensor. Some of the distance is fixed by the biological material and the surface functionalization, but the distance is typically dominated by the passivation layer on top of the sensor. Reducing the thickness of the passivation layer comes at the cost of reduced breakdown voltage and increased capacitive coupling. Furthermore, beyond a certain thickness, there are diminishing returns. A reduced breakdown voltage may limit the sense current and ultimately reduce the sensitivity of the device, negating any gain obtained from the reduced thickness.

The second set of design trade-offs concerns the sensor geometry, resistance, and noise of the sensor. The resistance is determined by the geometry, the length (L) and the width (W) of the sensor, along with the sheet resistivity (ρ) of the film stack. From an electronics perspective, the resistance of the sensor should not be too low because it would be difficult to design a front-end circuit with lower noise. Similarly, the resistance should not be too large for noise reasons. Sensors with narrow widths are more sensitive, particularly when used in conjunction with small labels. There are consequences if the width is reduced too much, since particles that land on the edge contribute an opposite signal from those that land on the center [18], and narrower sensors are more difficult to manufacture in a reproducible fashion. To quantify the relationship between the sensor resistance and the noise, Equation 37.11 shows the thermal (Johnson) noise power present in all conductors and its relationship to the geometry of the sensor:

$$\overline{V_{\text{th}}^2} = 4k_B TR\Delta f = 4k_B T \frac{L}{W}\rho\Delta f \tag{37.11}$$

The flicker noise power can be approximated from Hooge's empirical equation (Equation 37.12) where α is a dimensionless constant and N is the number of charge carriers in the conductive layers with a thickness t_c and a charge carrier density of D_c [35].

$$\overline{V_{1/f}^2} = \frac{\alpha}{Nf}(IR)^2\Delta f = \frac{\alpha}{LWt_c D_c f}\left(I\frac{L}{W}\rho\right)^2 \Delta f \tag{37.12}$$

The geometrical trade-offs can be combined by examining the SNR. There are two possible regions: when the noise is dominated by flicker noise (Equation 37.13) and when the noise is primarily thermal noise. At modest field modulation frequencies (~200 Hz), the former case typically holds. The resistance change per particle is inversely proportional to W^β where β is an empirically fitted parameter (2.536 in Li's measurements) and independent of the length [18]. When the sensor is operating in the flicker noise region, reducing the sensor area generally increases SNR.

$$\begin{aligned}\text{SNR} &= \frac{I\Delta R_{pp}}{\sqrt{(\alpha/Nf)(IR)^2\Delta f}} \propto \frac{I(1/W^\beta)}{\sqrt{(\alpha/LWf)I^2(L/W)^2\Delta f}} \\ &\propto \frac{1}{\sqrt{LW^{\beta-3/2}}} \propto \frac{1}{\sqrt{LW^{1.036}}}\end{aligned} \tag{37.13}$$

However, from a kinetics perspective, a large sensor coverage area is beneficial to reduce the time to analyte capture and stochastic noise in low-concentration measurements and improves the reproducibility. Unfortunately, for a single stripe spin-valve, the coverage area is directly proportional to the resistance. Shapes other than stripes are also possible; however, curved or round shapes may hinder coherent domain wall rotations and reduce the sensor linearity. Furthermore, the shape of the sensor can also be used to favor a specific magnetization by providing shape anisotropy. With sufficient shape anisotropy, the biasing field may not even be necessary. The geometry of the sensor is a complex trade-off between the noise, sensitivity, and coverage area.

Finally, the amplitude of the tickling field is a parameter that can be used to optimize the system performance. As the amplitude of the tickling field (typically significantly lower than the saturation field) increases, the moment of a superparamagnetic or SAF label also increases. Counteracting this, the sensitivity of the sensor decreases as the field is increased (Figure 37.5b). By multiplying these two curves and differentiating, we can find the maximum signal per particle of the entire system for a given sensor and label pair. For this particular combination of sensor and magnetic nanoparticle, we calculate that each particle induces a 4.6 μΩ change in the resistance. Detection down to as few as a few thousand 50 nm particles has been previously demonstrated [29].

37.9 Summary

In conclusion, by properly designing GMR spin-valve biosensors, labeling detection antibodies with appropriate and colloidally stable magnetic nanoparticle tags, and integrating effective and reliable biochemistry and electronics, we have shown that GMR biosensors are capable of exceptionally sensitive and selective multiplex protein detection in a single reaction well. The combination of small sample volumes, high sensitivity, and multiplex capability establishes GMR biosensor arrays as one of the most compelling biosensor platforms for both basic science research and clinical medicine applications. We anticipate that there are many other exciting applications for GMR sensors beyond magnetic recording.

After the acceptance of this book chapter, the authors have published two related papers on GMR spin-value biosensors [36,37], which the readers may find helpful to follow the latest applications in point of care diagnostics and protein-protein interaction measurements.

Acknowledgments

This work would not have been possible without the contributions from the members and collaborators of the Wang Group. In particular, the authors wish to thank Sebastian J. Osterfeld, Heng Yu, Guanxiong Li, Shu-Jen Han, Liang Xu, Robert L. White, Nader Pourmand, Ronald Davis, Robert J. Wilson, Boris Murmann, and Sanjiv S. Gambhir for their underlying work and valuable discussions. This work was supported in part by National Cancer Institute Grants 1U54CA119367 and N44CM-2009-00011, National Science Foundation (NSF) Grant ECCS-0801385-000, Defense Threat Reduction Agency Grant HDTRA1-07-1-0030-P00005, the Defense Advanced Research Projects Agency/Navy Grant N00014-02-1-0807, and the National Semiconductor Corporation. R.S.G. acknowledges financial support from Stanford Medical School MSTP program and an NSF graduate research fellowship.

References

1. R. Kotitz, H. Matz, L. Trahms et al., SQUID based remanence measurements for immunoassays, *IEEE Trans. Appl. Supercond.* **7**(2), 3678–3681 (1997).
2. D. R. Baselt, G. U. Lee, M. Natesan, S. W. Metzger, P. E. Sheehan, and R. J. Colton, A biosensor based on magnetoresistance technology, *Biosens. Bioelectron.* **13**(7,8), 731–739, doi:10.1016/S0956-5663(98)00037-2 (1998).
3. S. X. Wang and A. M. Taratorin, *Magnetic Information Storage Technology*, Academic Press, New York, April 28, 1999.
4. B. M. de Boer, J. A. H. M. Kahlman, T. P. G. H. Jansen, H. Duric, and J. Veen, An integrated and sensitive detection platform for magneto-resistive biosensors, *Biosens. Bioelectron.* **22**(9,10), 2366–2370, doi:10.1016/j.bios.2006.09.020 (2007).
5. S. P. Mulvaney, K. M. Myers, P. E. Sheehan, and L. J. Whitman, Attomolar protein detection in complex sample matrices with semi-homogeneous fluidic force discrimination assays, *Biosens. Bioelectron.* **24**(5), 1109–1115, doi:10.1016/j.bios.2008.06.010 (2009).
6. J. Schotter, P. B. Kamp, A. Becker, A. Pühler, G. Reiss, and H. Brückl, Comparison of a prototype magnetoresistive biosensor to standard fluorescent DNA detection, *Biosens. Bioelectron.* **19**(10), 1149–1156, doi:10.1016/j.bios.2003.11.007 (2004).
7. L. Xu, H. Yu, M. S. Akhras et al., Giant magnetoresistive biochip for DNA detection and HPV genotyping, *Biosens. Bioelectron.* **24**(1), 99–103, doi:10.1016/j.bios.2008.03.030 (2008).
8. M. Koets, T. van der Wijk, J. T. W. M. van Eemeren, A. van Amerongen, and M. W. J. Prins, Rapid DNA multianalyte immunoassay on a magneto-resistance biosensor, *Biosens. Bioelectron.* **24**(7), 1893–1898, doi:10.1016/j.bios.2008.09.023 (2009).
9. D. L. Graham, H. A. Ferreira, and P. P. Freitas, Magnetoresistive-based biosensors and biochips, *Trends Biotechnol.* **22**(9), 455–462, doi:10.1016/j.tibtech.2004.06.006 (2004).
10. S. J. Osterfeld, H. Yu, R. S. Gaster et al., Multiplex protein assays based on real-time magnetic nanotag sensing, *Proc. Natl. Acad. Sci.* **105**(52), 20637–20640, doi:10.1073/pnas.0810822105 (2008).
11. G. Li, V. Joshi, R. L. White et al., Detection of single micron-sized magnetic bead and magnetic nanoparticles using spin valve sensors for biological applications, *J. Appl. Phys.* **93**(10), 7557–7559, doi:10.1063/1.1540176 (2003).
12. W. Shen, X. Liu, D. Mazumdar, and G. Xiao, In situ detection of single micron-sized magnetic beads using magnetic tunnel junction sensors, *Appl. Phys. Lett.* **86**(25), 253901–253903, doi:10.1063/1.1952582 (2005).
13. T. M. Almeida, M.S. Piedade, J. Germano et al., Measurements and modelling of a magnetoresistive biosensor, in *Biomedical Circuits and Systems Conferences 2006* (*BioCAS 2006*) *IEEE*, pp. 41–44, London, U.K.
14. G. T. Hermanson, *Bioconjugate Techniques*, Academic Press, San Diego, CA, 2008.
15. S. Sun, H. Zeng, D. B. Robinson et al., Monodisperse MFe2O4 (M = Fe, Co, Mn) Nanoparticles, *J. Am. Chem. Soc.* **126**(1), 273–279, doi:10.1021/ja0380852 (2004).
16. W. Hu, R. J. Wilson, C. M. Earhart, A. L. Koh, R. Sinclair, and S. X. Wang, Synthetic antiferromagnetic nanoparticles with tunable susceptibilities, in *Proceedings of the 53rd Annual Conference on Magnetism and Magnetic Materials 2009*, 105:07B508-3, Austin, TX, AIP, April 1, doi:10.1063/1.3072028.
17. W. Hu, R. J. Wilson, A. L. Koh et al., High-moment antiferromagnetic nanoparticles with tunable magnetic properties, *Adv. Mater.* **20**(8), 1479–1483, doi:10.1002/adma.200703077 (2008).
18. S. X. Wang and G. Li, Advances in giant magnetoresistance biosensors with magnetic nanoparticle tags: Review and outlook, *IEEE Trans. Magn.* **44**, 1687–1702 (2008).
19. G. Li, S. Sun, R. J. Wilson, R. L. White, N. Pourmand, and S. X. Wang, Spin valve sensors for ultrasensitive detection of superparamagnetic nanoparticles for biological applications, *Sens. Actuators A: Phys.* **126**(1), 98–106, doi:10.1016/j.sna.2005.10.001 (2006).
20. M. Labrune, J. C. S. Kools, and A. Thiaville, Magnetization rotation in spin-valve multilayers, *J. Magn. Magn. Mater.* **171**(July 1), 1–2 (1997).
21. G. Li and S. X. Wang, Analytical and micromagnetic modeling for detection of a single magnetic microbead or nanobead by spin valve sensors, *IEEE Trans. Magn.* **39**(5), 3313–3315 (2003).
22. Li, G., S. X. Wang, and S. Sun, Model and experiment of detecting multiple magnetic nanoparticles as biomolecular labels by spin valve sensors, *IEEE Trans. Magn.* **40**(4), 3000–3002 (2004).
23. H. Wan, M. M. Bohlinger, M. Jenson, and A. Hurst, Comparison of flicker noise in single layer, AMR and GMR sandwich magnetic film devices, *IEEE Trans. Magn.* **33**(5), 3409–3411 (1997).
24. R. J. M. van de Veerdonk, P. J. L. Belien, K. M. Schep et al., 1/f noise in anisotropic and giant magnetoresistive elements, *J. Appl. Phys.* **82**(12), 6152–6164, doi:10.1063/1.366533 (1997).
25. K. A. A. Makinwa, M. A. P. Pertijs, J. C. van der Meer, and J. H. Huijsing, Smart sensor design: The art of compensation and cancellation, in *33rd European Solid State Circuits Conference 2007* (*ESSCIRC 2007*), pp. 76–82, Munich, Germany.
26. S.-J. Han, L. Xu, H. Yu et al., 2006. CMOS integrated DNA microarray based on GMR sensors, in *International Electron Devices Meeting 2006* (*IEDM'06*), pp. 1–4, doi:10.1109/IEDM.2006.346887, San Francisco, CA.
27. D. A. Hall, R. S. Gaster, S. X. Wang, and B. Murmann, Portable biomarker detection with magnetic nanotags, *Proceedings of 2010 IEEE International Symposium on Circuits and Systems* (*ISCAS*). June (2010).

28. C. R. Tamanaha, S. P. Mulvaney, J. C. Rife, and L. J. Whitman, Magnetic labeling, detection, and system integration, *Biosens. Bioelectron.* **24**(1), 1–13, doi:10.1016/j.bios.2008.02.009 (2008).
29. D. A. Hall, R. S. Gaster, T. Lin et al., GMR biosensor arrays: A system perspective, *Biosens. Bioelectron.* **25**, 2051–2057 (2010).
30. R. S. Gaster, D. A. Hall, C. H. Nielsen et al., Matrix-insensitive protein assays push the limits of biosensors in medicine, *Nat. Med.* **15**(11), 1327–1332, doi:10.1038/nm.2032 (2009).
31. D. A. Hall, R. S. Gaster, S. J. Osterfeld, B. Murmann, and S. X. Wang, GMR biosensor arrays: Correction techniques for reproducibility and enhanced sensitivity, *Biosens. Bioelectron.* **25**, 2177–2181 (2010).
32. S. J. Osterfeld and S. X. Wang, MagArray biochips for protein and DNA detection with magnetic nanotags: Design, experiment, and signal-to-noise ratio, in *Microarrays*, K. Dill, R. Liu, and R. Grodzinsky (Eds.), pp. 299–314, Springer, New York, 2009.
33. S.-J. Han, H. Yu, B. Murmann, N. Pourmand, and S. X. Wang, A high-density magnetoresistive biosensor array with drift-compensation mechanism, in *IEEE International Solid-State Circuits Conference 2007 (ISSCC 2007), Digest of Technical Papers*, pp. 168–594, San Francisco, CA.
34. H. Wang, Y. Chen, A. Hassibi, A. Scherer, and A. Hajimiri, A frequency-shift CMOS magnetic biosensor array with single-bead sensitivity and no external magnet, in *IEEE International Solid-State Circuits Conference (ISSCC 2009)—Digest of Technical Papers*, pp. 438–439,439a, doi:10.1109/ISSCC.2009.4977496, San Francisco, CA.
35. F. N. Hooge, 1/f noise is no surface effect, *Phys. Lett. A* **29**(3), 139–140, doi:10.1016/0375-9601(69)90076-0 (1969).
36. R.S. Gaster, D.A. Hall, and S.X. Wang, NanoLAB: An ultra-portable, handheld diagnostic laboratory for global health, *Lab on a Chip*, 11(5), 950–956 (2011).
37. R.S. Gaster, D.A. Hall, and S.X. Wang, Autoassembly protein arrays for analyzing antibody cross-reactivity, *Nano Letters*, in press (2011).

38

Semiconductor Spin-Lasers

Igor Žutić
University at Buffalo

Rafał Oszwałdowski
University at Buffalo

Jeongsu Lee
University at Buffalo

Christian Gøthgen
University at Buffalo

38.1 Introduction

The word laser is an abbreviation for light amplification by stimulated emission of radiation. However, considering its modern applications, it is perhaps better viewed as a light-generating device, rather than a light amplifier [1]. The operating principles of lasers can be traced back to the prediction of stimulated emission by Albert Einstein in 1917 [2]. One of the key laser characteristics is dependence of the emitted light on pumping or injection. We will distinguish conventional lasers and spin-lasers depending if this pumping/injection introduces spin-unpolarized or spin-polarized carriers, respectively. Typically, in conventional lasers two regimes are distinguished. For low pumping there is no stimulated emission, laser operates as an ordinary light source: the emitted photons are incoherent. However, the higher pumping regime with stimulated emission and coherent light makes the laser such a unique light source. The intensity of pumping/injection above which there is a phase-coherent emission of light is called the lasing threshold. Important advances in semiconductor lasers can be associated with the reduction of threshold current density for the onset of lasing. It could be shown that such a reduction would not only decrease the power consumption of lasers but can also significantly enhance their dynamic performance [4]. A schematic illustration in Figure 38.1 shows the historical evolution of threshold reduction in semiconductor lasers, realized by gradually replacing a bulk-like active region (known also as the gain medium, where the lasing action takes place) with structures of reduced dimensionality: quantum wells (QWs), quantum wires, and quantum dots (QDs).

There is also an alternative approach to reduce the lasing threshold using spin-polarized carriers generated by circularly polarized light or electrical spin injection. The principles of optical spin injection have been extensively studied for over 40 years [5–8]. Their application to QWs, which are typically used as the active region of spin-lasers, is discussed in Chapter 17. The angular momentum of absorbed circularly polarized light is transferred to the semiconductor. Electron orbital momenta are directly oriented by light and through spin–orbit coupling electron spins become polarized. While initially holes are also polarized, their spin polarization is lost almost instantaneously, owing to much stronger spin–orbit coupling in the valence band. The spin–orbit coupling, as the leading mechanism for the spin relaxation, is discussed in Chapter 16. For practical applications, it would even be more desirable to use electrical spin injection and simplify the integration with conventional electronics. We omit any detailed discussion of the electrical spin injection that has been extensively addressed, for example, in Chapters 6, 17, 18, 22, 23, 29, 31, 32, 39. Figure 38.2 illustrates theoretical prediction for the threshold reduction for completely spin-polarized injection. The key parameter to assess such a reduction is the spin relaxation time, the timescale it takes for the spin imbalance to be lost. This time strongly depends on the specific material and the geometry used for the active region of a laser. For a very short spin relaxation time, spin-lasers have no threshold reduction as compared to their conventional (spin-unpolarized) counterparts.

Both conventional lasers and spin-lasers share three main elements: the active (gain) region, the resonant cavity (resonator), and the pump, which injects (optically or electrically) energy/carriers. A typical vertical geometry, for so-called vertical cavity surface emitting lasers (VCSELs) [10] used in spin-lasers is illustrated in Figure 38.3. It is similar to the Faraday geometry of spin light emitting diodes (spin-LEDs) [7], which leads to well-defined

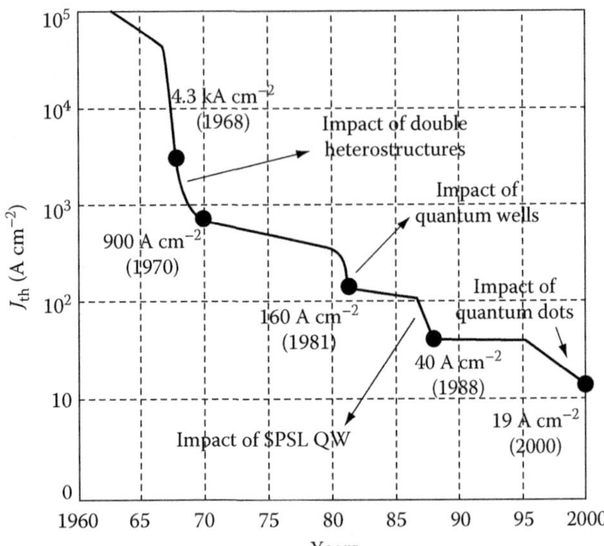

FIGURE 38.1 The evolution of threshold current for semiconductor lasers. (From Alferov, Z.I., *Rev. Mod. Phys.*, 73, 767, 2001, Fig. 9. With permission. Copyright 2001 by the American Physical Society.)

optical selection rules, further discussed in Chapter 17. Some of the spin-lasers are readily available since they are based on the commercially available semiconductor lasers to which subsequently a source of circularly polarized light is added. For electrical spin injection, one could employ ferromagnetic metals at the position of the bottom metallic contact indicated in Figure 38.3. The spin-lasers offer practical paths to room temperature spin-controlled devices that are not limited to widely employed magnetoresistive effects (another possibility is discussed in Chapter 39). Some of the applications include reconfigurable interconnects, high-speed modulators, and secure communications. In the steady-state operation, spin-lasers have demonstrated threshold reduction as compared to their conventional counterparts [11–14]. It was predicted theoretically [15] that this threshold reduction leads directly to improved dynamic operation of spin-lasers including an enhanced bandwidth and better switching properties. The spin polarization acts as a new "knob" for controlling the laser's operation. The change in spin polarization not only modifies the lasing threshold, but it can also be effectively used to turn the laser *on* and *off*, even at the fixed injection [11,15]. To facilitate the operation of spin-lasers, it is important to consider materials and geometries in which (1) the net spin polarization can be sustained over a relatively long time and (2) there is a large net (and preferably externally controllable) spin polarization of injected carriers at room temperature.

In this chapter, we first provide a description spin-unpolarized lasers that can be readily generalized to spin-lasers, which are subsequently analyzed both in their steady-state and dynamical operations. While most of our discussions pertain to QW spin-lasers, which have also been predominantly used experimentally, we also provide an outlook for the emerging QD spin-lasers.

38.2 Conventional Semiconductor Lasers

38.2.1 Bucket Model

A simple description of conventional (spin-unpolarized) lasers can be considered as the special limit of our model of spin-lasers. This simplification allows us to develop a transparent approach and analytical results for laser operation as well as to systematically include the effects of spin polarization and spin relaxation.

One of the most useful analogies to illustrate the laser operation is provided by the bucket model, previously only considered

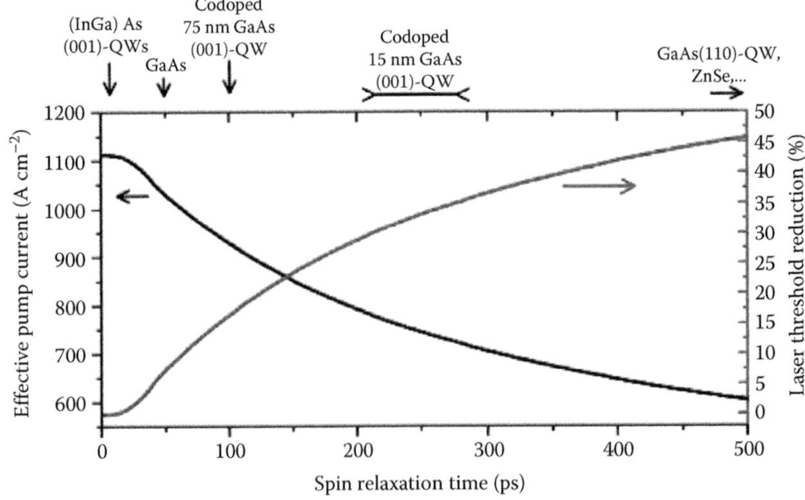

FIGURE 38.2 The laser threshold at room temperature (left axis) and the reduction of the laser threshold (right axis) as a function of spin relaxation time. The spin relaxation time of several semiconductors is depicted on the top axis. (From *Superlattices Microstruct.*, 37, Oestreich, M., Rudolph, J., Winkler, R., and Hägele, D., Design considerations for semiconductor spin lasers, 306–312 (200s). Copyright 2005, Fig. 3, with permission from Elsevier.)

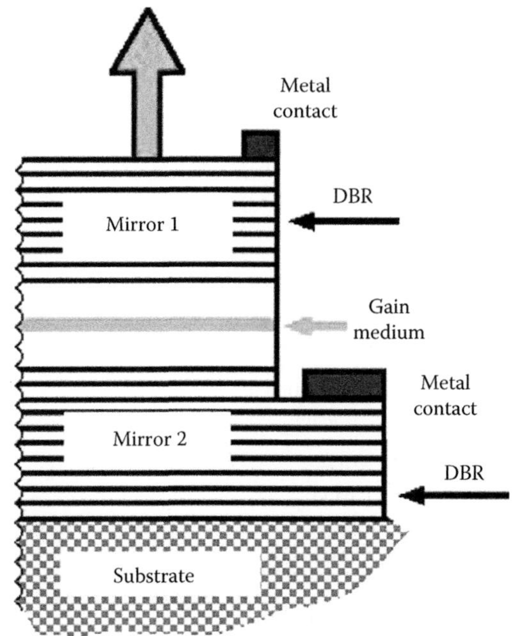

FIGURE 38.3 VCSEL. The resonant cavity is formed by a pair of parallel highly reflective mirrors made of distributed Bragg reflectors (DBRs) a layered structure with varying refractive index, and the gain medium (active region), typically consisting of QWs or QDs. Electrical injection is realized using two metal contacts and the direction of the emitted light is perpendicular to the top surface and the gain medium.

for conventional lasers [16], we generalize it for spin-lasers in Section 38.3.2 [17]. This model, depicted in Figure 38.4, allows us to establish an intuitive connection to a standard description of a conventional laser operation using rate equations (REs) [4,18,19]. For example, the water level represents electron and hole densities, which will be denoted by n and p. The faucet represents the injection of charge carriers (electrons and holes) into the lasers, while the outgoing water corresponds to the photon density, S. We will only be considering light, represented by S, which is supported by the resonant cavity. The recombination and the gain function, which corresponds to the stimulated emission in the active region, will be denoted by $R(n,p)$ and $g(n,p,S)$.

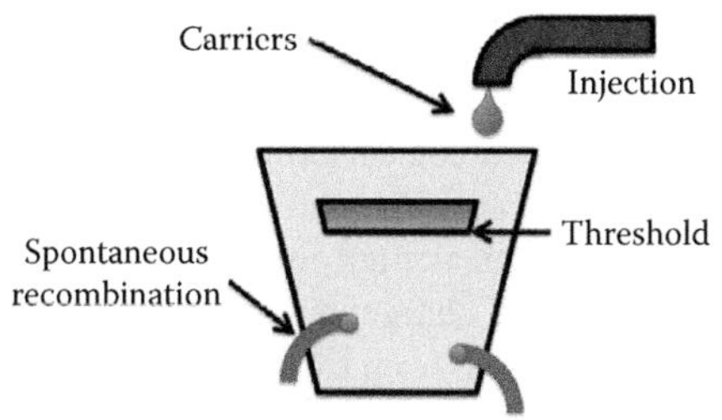

FIGURE 38.4 Bucket model of a conventional laser. The faucet represents the injection or pumping. The small leaks in the bucket represent spontaneous recombination processes and the large opening near the top of the bucket represents the lasing threshold. At low injection, in the regime I, the water level is low and the some of this water will trickle out of the leaks. A modest increase in the injection will raise the water level within the bucket, but the amount of escaping water will remain small. At high injection, regime II, the water will reach the large opening near the top and from there it will gush out. Additional water will lead only to a small change in the water level but the output will increase rapidly as compared with regime I.

The bucket model can be effectively used to illustrate the two characteristic regimes of the laser operation. At the low injection, which we call regime I, there is only a negligible output light. The operation of a laser is similar to a LED, discussed also in Chapter 17. The spontaneous recombination is responsible for the emitted light. At higher injection, when the water starts to gush out of the large slit in Figure 38.4, the lasing threshold is reached. At the threshold injection, J_T, stimulated emission starts and the emitted light intensity increases significantly. Regime II, for $J > J_T$, corresponds to fully lasing operation in which the stimulated recombination is the dominant mechanism of light emission.

As illustrated in Figure 38.3, the gain medium does not occupy the entire resonant cavity; the ratio of gain medium volume to the total cavity volume is given by the dimensionless constant Γ. The fraction of the spontaneous recombination, which produces light that couples to the resonance cavity, is modeled by a constant β. In typical VCSELs, small $\beta \propto 10^{-4}$–10^{-3} implies that there would be a negligible intensity of light below the lasing threshold. The optical losses from different sources, except from the stimulated absorption, are modeled by the photon lifetime τ_{ph}.

38.2.2 Characteristic Processes and Rate Equations

From the description of the bucket model in Figure 38.4 it is possible to establish a connection with the REs given in terms of carrier and photon densities

$$\frac{d}{dt}\alpha = J_\alpha - g(n,p,S) - R_{sp}(n,p), \tag{38.1}$$

$$\frac{d}{dt}S = \Gamma g(n,p,S) + \Gamma\beta R_{sp}(n,p) - \frac{S}{\tau_{ph}}, \tag{38.2}$$

where $\alpha = n, p$, and we consider the densities that are uniform in the gain medium. The expressions for gain and recombination are generally nonlinear functions of the corresponding densities. In Equations 38.1 and 38.2 the contributions for gain and recombination have opposite sign for carrier and photon densities, the processes that reduce the total number of carriers also increase the number of photons. The prefactor Γ introduced in Section 38.2.1, signifies that the changes of carriers and photons take place in different effective volumes (the gain medium occupies a small fraction of the resonant cavity). In order to further

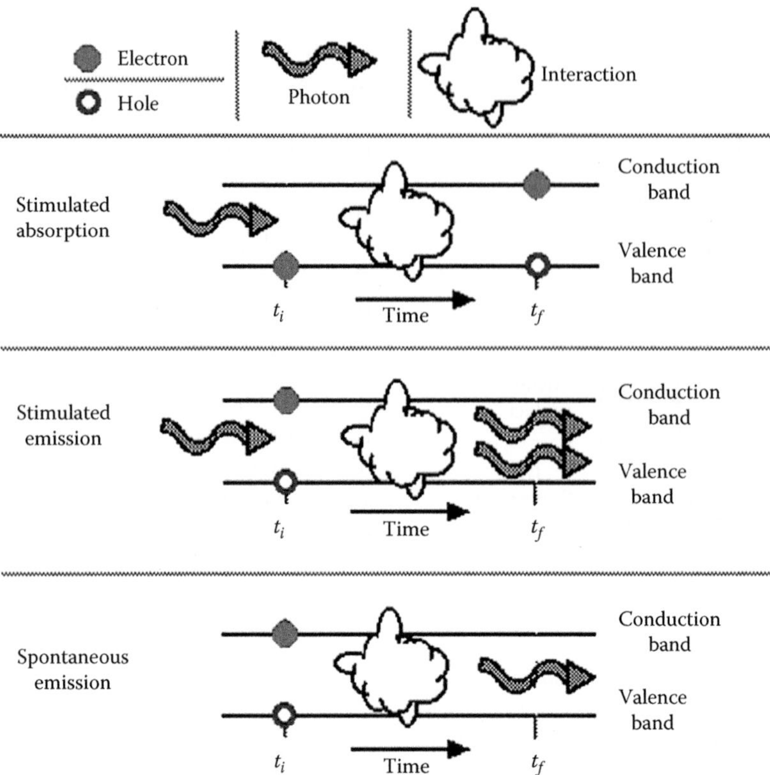

FIGURE 38.5 Schematic description of emission and absorption processes. Stimulated absorption: An incident photon strikes an electron in the valence band, promoting the electron to the conduction band leaving a hole in the valence band and annihilating the photon. Stimulated emission: An incident photon triggers a conduction band electron to relax into an empty state, hole, in the valance band releasing an additional photon that has the same energy, phase, and momentum as the incoming photon. Spontaneous emission: A conduction band electron relaxes into an empty valence band state, hole, releasing a photon in the process.

specify the REs (38.1) and (38.2), we depict relevant processes in Figure 38.5. For laser operation, characteristic carrier densities exceed the corresponding equilibrium values which when combined with the charge neutrality ($n=p$) allows for further simplifications. In the spontaneous recombination (the bottom of Figure 38.5), we distinguish two simple forms: quadratic recombination (QR), $R_{sp}=Bnp/2$, where B is a temperature-dependent constant [20], and linear recombination (LR), $R_{sp}=n/\tau_r$, where τ_r is the recombination time. LR is a particularly good approximation at high n. QR and LR can be viewed as different limits of the general expression for $R_{sp}(n, p)$. For the moderate injection considered, the Auger recombination, which involves three carriers, is typically negligible. However, for higher densities this process (expected to be $\propto n^3$) becomes quadratic [21], just like the QR considered here.

The gain function can be interpreted as the difference between stimulated emission and absorption, illustrated in Figure 38.5. A specific carrier density at which this difference vanishes is known as the transparency density, n_{tran}. It is customary [16] to linearize the gain function around n_{tran},

$$g(n,S) \approx g_0(n-n_{\text{tran}})S, \tag{38.3}$$

where g_0 is the gain coefficient. Since experimentally S does not increase indefinitely with increasing injection, the gain saturation is usually incorporated as

$$g(n,S) \equiv g_0 \frac{n-n_{\text{tran}}}{1+\varepsilon S} S, \tag{38.4}$$

where ε is the phenomenological gain compression (saturation) factor that models the influence of spectral hole burning and carrier heating, relevant at high-injection levels ($J \gg J_T$) [22]. Taken together with the charge neutrality ($n=p$ and $J_n=J_p\equiv J$), the holes can be decoupled from the REs

$$\frac{d}{dt}n = J - g_0 \frac{n-n_{\text{tran}}}{1+\varepsilon S} S - R_{sp}(n), \tag{38.5}$$

$$\frac{d}{dt}S = \Gamma g_0 \frac{n-n_{\text{tran}}}{1+\varepsilon S} S + \Gamma\beta R_{sp}(n) - \frac{S}{\tau_{ph}}. \tag{38.6}$$

These REs are widely used for conventional lasers [4]. Their generalization is typically the starting point to describe spin-lasers.

38.3 Spin-Lasers: Steady State

38.3.1 Experimental Implementation

Most of the spin-lasers are realized as VCSELs (recall Figure 38.3) with the active region consisting of III–V QWs. Optically or electrically injected/pumped spin-lasers rely on the principles of spin injection. Advances in increasing the efficiency of spin injection, pursued as an effort to improve various spintronic devices, therefore directly aid the enhanced operation of spin-lasers. For example, using a MgO tunnel barrier (discussed further in Chapter 11) that dramatically increases both the magnetoresistance in spin-valves and spin injection in semiconductors [23–25], could also be a promising direction for the next generation of spin-lasers.

The first demonstration of the threshold reduction, shown in Figure 38.6, utilized optical spin injection in (In,Ga)As QW-based spin-laser. Slow changes in the injected polarization at the fixed input power clearly lead to the change in the intensity of the emitted light (Figure 38.6b). These results suggest a new path to modulate the laser operation with a fast polarization modulation, explored further in Section 38.4. While this early work was limited to 6K operation temperature, subsequent efforts have also shown room temperature optically injected spin-lasers [12].

Alternatively, electrical spin injection employing ferromagnetic contacts has been used to realize the threshold reduction, albeit not yet at the room temperature [13,26,27]. Figure 38.7 shows the emitted light intensity in an electrically pumped spin-laser using Fe/AlGaAs Schottky tunnel barrier as the spin injector and (In,Ga)As QW as the gain medium.

These impressive experimental advances offer new paradigms in spintronic devices that would not be limited to the magnetoresistive effects. However, the theoretical understanding of spin-lasers has been far from complete, even in the steady-state operation. This has been somewhat surprising since the experimental results were supplemented with seemingly a well-established RE description, known already at the textbook level in the spin-unpolarized case and considered in Section 38.2.2. However, numerical results using such REs contain a large number of material parameters, making it difficult to systematically elucidate how the spin-dependent effects influence the device operation. As we shall see, even what is the maximum threshold reduction was not correctly understood, warranting a careful development of the spin-polarized REs in this section, followed by systematic study of their analytical solutions in different operating regimes of the spin-lasers.

38.3.2 Spin-Polarized Bucket Model

To gain an insight into the operation of the spin-lasers, we first develop a simple description by generalizing the bucket model to the spin-polarized case [17]. We introduce various spin-resolved quantities to model different projections of the spin or helicities of light. The total electron/hole density can be written as the sum of the spin up (+) and the spin down (−) electron/hole densities, $n=n_+ + n_-$ and $p=p_+ + p_-$. Analogously, we write the total photon density as the sum of the positive (+) and negative (−) helicities, $S=S^+ + S^-$. Spin injection (optical or electrical) $J=J_+ + J_-$ implies an unequal decomposition $J_+ \neq J_-$ for the two spin projections. As shown in Figure 38.8, they can be represented with two separate water faucets. For example, we can think of hot- and cold-water analogy, corresponding to spin-up and spin-down carriers. The two halves of the bucket are only coupled through a narrow tube,

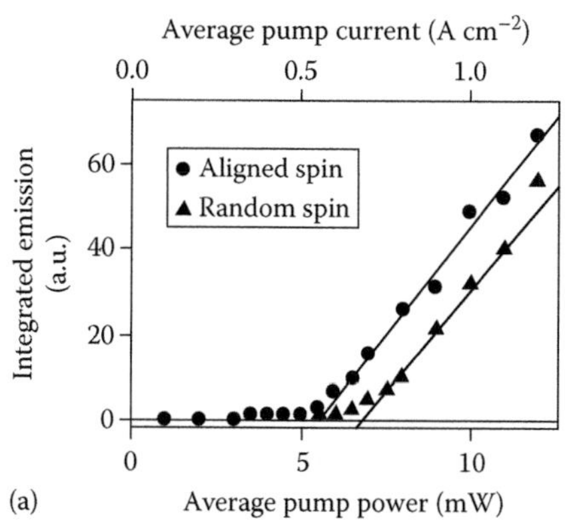

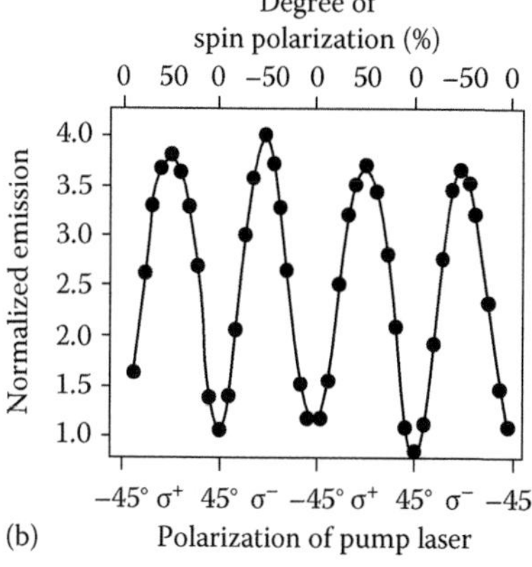

FIGURE 38.6 Threshold reduction in a spin-laser using optical spin injection. (a) The time-integrated intensity after pulsed optical injection is plotted for the 50% spin-aligned electrons (circles) and randomly oriented spins (triangles). The device yields a clear reduction of the laser threshold for spin-polarized injection. (b) Change in the VCSEL emission power by continuously changed pump polarization under 6.5mW constant pump power. The emission maxima coincide with circular pump polarization and thus with the maximum spin polarization of the excited electrons. Linearly polarized pumping creates random spin orientation causing a drop below threshold, which demonstrates that the laser emission depends not only on the pump power but also on the spin orientation. (From Rudolph, J., Hägele, D., Gibbs, H.M., Khitrova, G., and Oestreich, M., Laser threshold reduction in a spintronic device, *Appl. Phys. Lett.*, 82, 4516, 2003, Fig. 2. With permission. Copyright 2003, American Institute of Physics.)

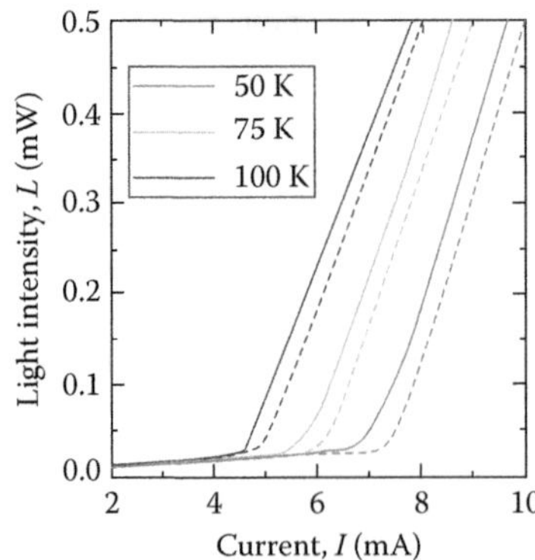

FIGURE 38.7 (See color insert.) The intensity of emitted light as function of injection proves the threshold reduction in a spin-laser using electrical spin injection. The solid (dashed) lines correspond to the injection of spin-polarized (spin-unpolarized) electrons. For every temperature, the onset of lasing is consistently lower for spin injection, which is realized with the aid of an applied magnetic field to ensure an out-of-plane component of magnetization in Fe injector (having an in-plane easy axis). (Reprinted with permission from Holub, M., Shin, J., and Bhattacharya, P., *Phys. Rev. Lett.*, 98, 146603, 2007, Fig. 3d. Copyright 2007, American Physical Society.)

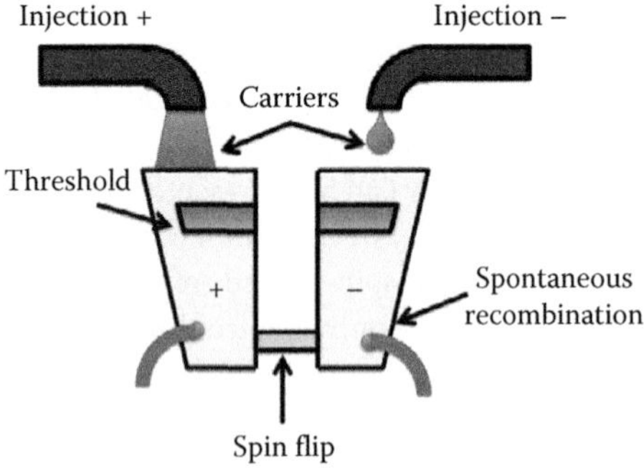

FIGURE 38.8 Spin-polarized bucket model. The two halves of the bucket, representing two spin populations are separately filled and only connected through a narrow tube to model the spin relaxation. The small holes represent carrier loss through spontaneous recombination and the large opening near the top denotes the lasing threshold. (From Gøthgen, C., Steady-state analysis of semiconductor spin-lasers, PhD thesis, University at Buffalo, New York, 2010.)

implying that their content is not completely separated. This situation corresponds to the process of spin relaxation that mixes spin up and down populations.

It is instructive to consider two limits for the coupling of the two halves of the bucket. If the connecting tube has a negligibly small diameter, the two populations will not mix, corresponding to the infinite spin relaxation time. In the opposite limit, if the connecting tube is very wide, the two unequal populations will equilibrate immediately. This leads to the negligible carrier polarization, as expected in the limit of vanishing spin relaxation time, for which the spin-lasers and conventional lasers behave the same.

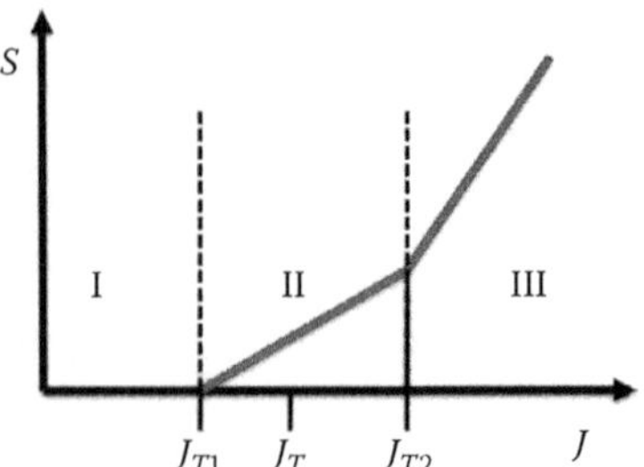

FIGURE 38.9 Total photon density, S, from a spin-polarized laser as a function of injection, J, according to the spin-polarized bucket model. At low pumping, regime I, both spin-up and spin-down carriers are in the spin LED regime thus with negligible emission. In regime II, the majority spin is lasing while the minority spin is still in the LED regime, thus all of the light is from recombination of carriers with majority spin. In regime III, both spin up and spin down are lasing. J_{T1} is the injection at which the majority spin begins to lase, and J_{T2} indicates the injection at which the minority spin begins to lase. The kink of the total photon density S at J_{T2} is from the minority light that begins to contribute to the total photon density J_T is the spin-unpolarized lasing threshold.

Based on this very simple model, we can develop an intuitive understanding for the operation of spin-lasers. With the two unequal faucets, we expect two different lasing thresholds. One-half of the bucket that is more effectively filled will overflow sooner, signaling that the lasing will first occur due to (predominantly) one type of (majority spin) carrier. Only later, when the other half becomes sufficiently filled, the lasing is expected for the other (minority) carriers. We can then identify three different operating regimes for spin-lasers. (1) Spin LED regime: in the limit of low injection, below the lasing of majority carriers, the laser resembles spin LED for both polarizations of light. There is no stimulated emission and the emitted light intensity is low. (2) Mixed or spin-filtering regime: above the threshold for majority carriers and below the threshold of minority carrier, only one spin population is responsible for the lasing while the other still operates as a spin LED. (3) Fully lasing regime: for injection above the value of minority lasing threshold, both spin populations will contribute to lasing.

Qualitatively, this picture of three operating regimes in a spin-laser can be sketched as shown in Figure 38.9. In the approximation of $\beta = 0$ (we examine its accuracy in Figure 38.10) the regime I, that of spin LED, yields negligible emission. In regime II, we expect what can be called spin-filtering effect, even for a modest polarization of the injected carriers (the two slightly unequal intensities of the faucets); there would be completely polarized light (of only one helicity, since the minority spin carries do not lase). In the regime III, there is a kink in the emitted light showing the onset of lasing due to the minority carriers. From a closer look at the Figure 38.7, we could infer a similar kink at the light intensity of ~0.1 mW, more pronounced at $T = 50$ and 75 K. This kink is completely absent in the absence of spin injection (dashed lines) further corroborating trends that can be inferred from the spin-polarized bucket model that provides a useful guidance for the experiments on spin-lasers [11–13].

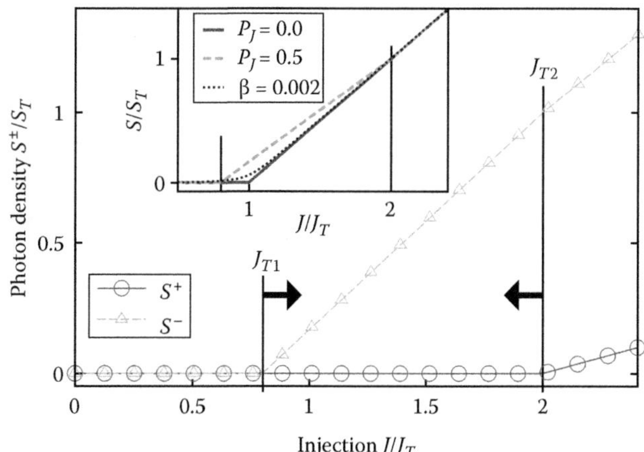

FIGURE 38.10 Photon densities of the left-(S^-) and right-(S^+) circularly polarized light as a function of electron current J with polarization $P_J = 0.5$, infinite electron spin relaxation time, and LR. J is normalized to the unpolarized threshold current J_T and $S^{\pm}$ to $S_T = J_T\Gamma\tau_{ph} = S(2J_T)$. The vertical lines indicate J_{T1} and J_{T2} thresholds, the arrows show their change when P_J is reduced. Inset: total photon density ($S = S^+ + S^-$) for unpolarized laser ($P_J = 0$) and two spontaneous-emission coupling coefficients ($\beta = 0, 0.002$), as well as a spin-laser with $P_J = 0.5$, $\beta = 0$. (From Gøthgen, C., Oszwałdowski, R., Petrou, A., and Žutić, I., Analytical model of spin-polarized semiconductor lasers, *Appl. Phys. Lett.*, 93, 042513, 2008, Fig. 1. With permission. Copyright 2008, American Institute of Physics.)

However, it is also important to elucidate some of the inherent limitations in using this spin-polarized bucket model. While we are not aware of any prior work on spin-polarized bucket model, a similar sophistication of the spin-laser description was used by other researchers to infer what is intuitively expected. In particular, since one-half of the bucket holds only a half as much of the water as the whole bucket, the limit of completely spin-polarized injection (only one faucet contributes) should give the maximum possible threshold reduction—one half of the spin-unpolarized injection, recall Figure 38.8. Our studies [28] reveal that this maximum can actually be exceeded [29]. The reason for the discrepancy is also apparent from the consideration of the spin-polarized bucket model—complete symmetry between electrons and holes is assumed, extending also to their spin-polarized properties. This questionable approximation can be already traced back to an influential work on spin-effects in lasers [30]. We consider spin-up and spin-down carriers without distinguishing if they are electrons or holes. Rather than attempting to formulate a more complex bucket model, we next develop the appropriate REs that will give us a systematic description of spin-lasers.

38.3.3 Spin-Polarized Rate Equations

When generalizing REs (38.5) and (38.6), for spin-lasers it is crucial to consistently include the effect of holes. Typically, the spin relaxation time of holes is much shorter than for electrons [7], $\tau_s^p \ll \tau_s^n$, implying that the holes can be considered unpolarized with $p_{\pm} = p/2$. In previous work [11,13] this effect was introduced only in the recombination term. Below we show its importance for the gain term by revealing that the electron densities $n_{\pm}$ are not separately clamped (pinned).

We generalize the gain (recall Equation 38.4) in the spin-polarized case as $g(n,S) \to g_{\pm}(n_{\pm}, p_{\pm}, S) = g_0(n_{\pm} + p_{\pm} - n_{\text{tran}})S^{\mp}/(1 + \varepsilon S)$. This form, even for $\varepsilon = 0$, differs from the previously employed gain expressions [11,13], as it explicitly contains the hole density, but it coincides with the rigorously derived gain expression from semiconductor Bloch equations [20]. The gain compression term is an approximate description of, mainly, spectral hole burning and carrier heating [31]. For simplicity, we also use εS for spin-lasers.

The charge neutrality $p_{\pm} = p/2 = n/2$, allows us to recover the spin-unpolarized limit for the gain expression, as well as to decouple the REs for electrons from those for holes. The spin-polarized REs for electrons thus become

$$\frac{dn_{\pm}}{dt} = J_{\pm} - g_{\pm}(n_{\pm}, S)S^{\mp} - \frac{n_{\pm} - n_{\mp}}{\tau_s^n} - R_{sp}^{\pm}, \tag{38.7}$$

$$\frac{dS^{\pm}}{dt} = \Gamma g_{\mp}(n_{\mp}, S)S^{\pm} + \beta\Gamma R_{sp}^{\mp} - \frac{S^{\pm}}{\tau_{ph}}, \tag{38.8}$$

where the notation is extended from Equations 38.5 and 38.6 for the conventional lasers. An extra term represents the (electron) spin relaxation $(n_{\pm} - n_{\mp})/\tau_s^n$. Here the recombination $R_{sp}^{\pm}$ generalizes the previous QR and LR forms [32] $R_{sp}^{\pm} = Bn_{\pm}p_{\pm} = Bn_{\pm}n/2$ and $R_{sp}^{\pm} = n_{\pm}/\tau_r$.

To describe the solutions of Equations 38.7 and 38.8, we introduce polarizations of injected electron current

$$P_J = \frac{J_+ - J_-}{J}, \tag{38.9}$$

as well as of electron and photon densities $P_n = (n_+ - n_-)/n$, $P_s = (S^+ - S^-)/S$. Electrical injection in intrinsic QWs, using Fe or FeCo, allows for [7,25,33–35] $|P_n| \sim 0.3$–0.7 with similar values for $|P_J|$, while $|P_n| \to 1$ is attainable optically at room temperature [7].* In the unpolarized limit $J_+ = J_-$, $n_+ = n_-$, $S^+ = S^-$ and $P_J = P_n = P_S \equiv 0$. As in the recent experiments [13], we consider a continuous-wave operation and look for steady-state solutions of Equations 38.7 and 38.8. Guided by the experimental range [11,13] of $\beta \sim 10^{-5}$–10^{-3}, we mostly focus on the limit $\beta = 0$, for which all the operating regimes of the spin-lasers can be simply described. Additionally, we consider $\varepsilon = 0$, relevant for moderate pumping intensities and parameters from Ref. [13]. While the most of our results are analytical and readily adapted for different laser materials and geometries, in Table 38.1 we summarize the characteristic parameter values we used guided by the experiments [11–13].

* We mostly consider $P_J \geq 0$ while the results for $P_J < 0$ can be deduced easily.

TABLE 38.1 Parameters Used to Model Spin-Lasers

Quantity		Value
$S^{\pm}$	Photon density	$\sim 10^{13}$ cm^{-3}
$(n, p)_{\pm}$	Carrier density	$\sim 10^{17}$ cm^{-3}
n_{tran}	Transparency density	4×10^{17} cm^{-3}
$J_{\pm}$	Carrier injection rate	$\sim 10^{27}$ cm^{-3} s^{-1}
P_J	Degree of injection polarization	0, 0.5, 0.9
g_0	Gain coefficient	$\sim 10^{-4}$ cm^3 s^{-1}
ε	Gain compression factor	$0-2\times 10^{-18}$ cm^3
Γ	Optical confinement factor	0.029
β	Spontaneous emission factor	0, 10^{-4}, 2×10^{-3}
τ_r	Recombination time	200 ps
τ_{ph}	Photon lifetime	1 ps
τ_s^n	Electron spin relaxation time	200 ps– ∞
$w = \dfrac{\tau_r}{\tau_s^n}$	Normalized spin relaxation rate	0, 1, 5

Source: Reprinted with permission from Lee, J., Falls, W., Oszwałdowski, R., and Žutić, I., Spin modulation in semiconductor lasers, *Appl. Phys. Lett.*, 97, 041116, 2010, Table I. Copyright 2010, American Institute of Physics.

38.3.4 Light-Injection Characteristics

We first show analytical results for an *unpolarized* laser and LR in the inset of Figure 38.10. Injection current density is normalized to the unpolarized threshold value, $J_T = N_T/\tau_r$, with N_T denoting the total electron density at (and above) the threshold, $N_T = n(J \geq J_T) = (\Gamma g_0 \tau_{ph})^{-1} + n_{tran}$, while photon density is normalized to $S_T = J_T \Gamma \tau_{ph}$. A small difference in $S(J)$ between the vanishing (solid line) and finite β (dotted line, $\beta = 0.002$ overestimates the experimental values [11,13]) shows the accuracy of $\beta = 0$ approximation. As inferred from the bucket model, the unpolarized case has two regimes: LED-like, with negligible stimulated emission, for $J < J_T$, and a fully lasing regime for $J > J_T$.

With finite P_J and $\tau_s^n \to \infty$, as expected from the spin-polarized bucket model (recall Figures 38.8 and 38.9), we reveal a more complicated behavior shown in Figure 38.10, main panel. The two threshold currents, J_{T1} and J_{T2} ($J_{T1} \leq J_T \leq J_{T2}$, the equalities hold only when $P_J = 0$), delimit three regimes of a spin-laser. (i) For $J < J_{T1}$ it operates as a spin-LED [7,33]. (ii) For $J_{T1} \leq J_T \leq J_{T2}$, there is mixed operation: lasing only with left-circularly polarized light S^- (we assume $J_+ > J_-$), which can be deduced from the spin-dependent gain in Equations 38.7 and 38.8, while S^+ is still in a spin-LED regime ($S^+ \to 0$ for $\beta \to 0$). (iii) For $J \geq J_{T2}$, it is fully lasing with both $S^{\pm} > 0$.

RE description of spin-unpolarized lasers for $\varepsilon = 0$ reveals that the carrier densities are clamped above J_T, this effect corresponds to pinning of the water level in the bucket, Figure 38.4. We find a related effect in the steady-state spin-lasers when $\varepsilon = \beta = 0$. For $J > J_{T1}$, Equation 38.8 for S^- can be divided by S^-, showing that the quantity $n_+ + p_+ = n_+ + n/2$ (we recall the charge neutrality condition and $\tau_s^p \to 0$) is clamped at N_T. If $J_{T1} < J < J_{T2}$ then neither n_+ nor n_- are separately clamped. If $J > J_{T2}$, then $n_- + n/2 = N_T$ must hold in addition to the previous condition $n_+ + n/2 = N_T$. These two conditions yield $n_{\pm} = N_T/2$, independent of P_J. Thus, above J_{T2}, the sum of Equation 38.8 for S^- and S^+ reduces to the usual unpolarized equation, and S is independent of P_J (inset of Figure 38.10). However, if $P_J \neq 0$ then we still find $S^- \neq S^+$ (Figure 38.10, $J > J_{T2}$).

Most of the results discussed above do not change qualitatively for a finite τ_s^n or QR. The general features, such as the existence of three operating regimes, remain the same. Gain compression will change J_{T2}, but even for $\varepsilon = 2\times 10^{-16}$ cm^3 (~4× larger than in Ref. [13]), the relative increase will be only ~1%. However, the photon and carrier densities (for a given J) as well as the threshold J_{T1} depend quantitatively on the spin-flip rate and the recombination form. We investigate this dependence in Figure 38.11, which shows the evolution of photon and carrier polarizations with the injection current. We consider both LR and QR forms and express our results using the ratio of the radiative recombination and spin relaxation times: $w = \tau_r/\tau_s^n$. For QR the unpolarized threshold is $J_T = BN_T^2/2$, and we define $\tau_r = N_T/J_T$ by analogy with the LR case.

When $0 < J < J_{T1}$, then $P_n(J)$ depends on the recombination form; P_n is constant for LR, while for QR it grows monotonically [6,36]. Only for a very long spin relaxation time, $w \to 0$, $P_n(J < J_{T1}) \to P_J$. When $\beta = 0$ there is no amplified spontaneous emission, so that $P_S \equiv 0$. When $J_{T1} < J < J_{T2}$, $P_S = -1$ for any w, recall Figure 38.10. With only a partially polarized electron current, the spin-laser emits fully circularly polarized light, analogous to the spin-filtering effect in magnetic materials. In the same regime, P_n decreases with J because, while n_- grows, n_+ must drop in order to maintain $n_+ + n/2 = (3n_+ + n_-)/2 = N_T$.

For $P_J \neq 0$, we can define the (normalized) spin-filtering interval

$$d = \frac{J_{T2} - J_{T1}}{J_T}, \tag{38.10}$$

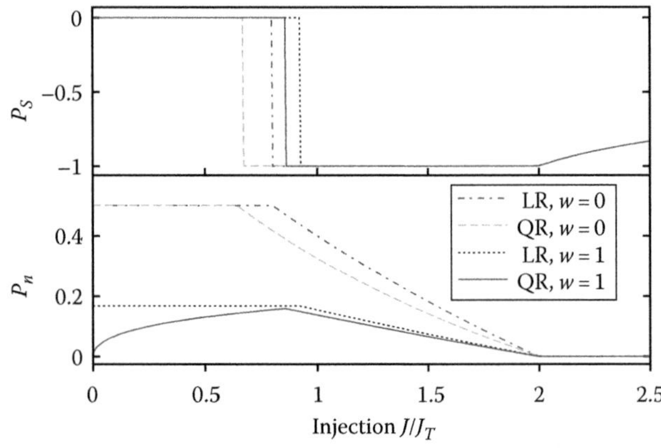

FIGURE 38.11 Electron (P_n) and photon (P_S) density polarizations as a function of injection current J, with polarization $P_J = 0.5$ and different ratios of recombination and electron spin relaxation time, w. LR and QR recombination are shown. J is normalized to the unpolarized threshold J_T. (From Gøthgen, C., Oszwałdowski, R., Petrou, A., and Žutić, I., Analytical model of spin-polarized semiconductor lasers, *Appl. Phys. Lett.*, 93, 042513, 2008, Fig. 2. With permission. Copyright 2008, American Institute of Physics.)

which is widest for $\tau_s^n \to \infty$, when we obtain

$$\text{LR}: \quad d = \frac{3|P_J|}{2-|P_J|-|P_J|^2}, \tag{38.11}$$

$$\text{QR}: \quad d = \frac{|P_J|(8+|P_J|)}{4-3|P_J|^2-|P_J|^3}, \tag{38.12}$$

showing that d diverges for $P_J=1$ and vanishes if the injection is spin-unpolarized. When τ_s^n decreases, the lower threshold J_{T1} grows toward its unpolarized counterpart J_T and d contracts, as seen in Figure 38.11. We find that $J_{T2}=J_T/(1-P_J)$, independent of τ_s^n. This should hold up to very small values of τ_s^n ($w\sim 10^2$). In the extreme case of even smaller τ_s^n, for $J_{T1}<J<J_{T2}$ the majority and minority spin densities become almost equal so that $\beta\sim 10^{-4}$ can effectively drive the device to the fully lasing regime. For typical values [11,13] of $w\sim 5$, setting $\beta=0$ is an accurate assumption. At J_{T2}, we recall $n_+=n_-(=N_T/2)$ so that the spin-relaxation term in Equation 38.7 for n_- is zero. The equation reduces to $J_- \equiv (1-P_J)J/2 = R_{sp}^-$, explaining why J_{T2} depends only on P_J but not on w. For the same reason, P_S and P_n are independent of w when $J\geq J_{T2}$. The quantity P_S increases ($|P_S|$ decreases) with J as $P_S=-P_JJ/(J-J_T)$ to the asymptotic value $P_S=-P_J$. Thus, P_J can be inferred from P_S only at sufficiently high injection. In all three regimes, defined by Figure 38.10, we find that $0\leq P_n<P_J$, (only for $\tau_s^n\to\infty$ and $J<J_{T1}$, $P_n\to P_J$). For $J>J_{T2}$ we find $P_n=0$. Therefore P_n is not enhanced in the spin-laser's active region.

38.3.5 Threshold Reduction

Experiments in Figures 38.6 and 38.7 have demonstrated that injecting spin-polarized carriers reduces the threshold current in a laser, that is, $J_{T1}(P_J\neq 0)<J_T$. We quantify this threshold-current reduction [11,13] as

$$r = \frac{J_T - J_{T1}}{J_T}, \tag{38.13}$$

which can be obtained analytically from REs (38.7) and (38.8) to yield

$$\text{LR}: \quad r = \frac{|P_J|}{2+|P_J|+4w}, \tag{38.14}$$

$$\text{QR}: \quad r = 1 - \frac{\left[1-2w+\sqrt{(1+2w)^2+4|P_J|w}\right]^2}{(2+|P_J|)^2}. \tag{38.15}$$

We find $r>0$, for any $P_J\neq 0$, $\tau_s^n>0$, and both LR and QR. As expected, $r\to 0$ for $\tau_s^n\to 0$, independent of the recombination form, since we have assumed $\tau_s^p=0$. We illustrate $r(P_J)$ for both LR and QR in Figure 38.12 and choose $w=1,5$, close to the typical values $\tau_r\sim 1$ ns and $\tau_s\sim 100$ ps for QWs at room temperature [11]. From Figure 38.12 or Equations 38.14 and 38.15 one can see that when $\tau_s^p=0$, $r(P_J,w)$ for LR is always smaller than for QR. For $|P_J|=1$ and $\tau_s^n\to\infty$ (but still $\tau_s^p=0$) we obtain the maximum threshold reductions

$$\text{LR}: \quad r = \frac{1}{3}, \quad \text{QR}: \quad r = \frac{5}{9}. \tag{38.16}$$

Remarkably, for QR the maximum r is larger than previously assumed possible $r=1/2$ [11,13] (recall also Figure 38.2), even

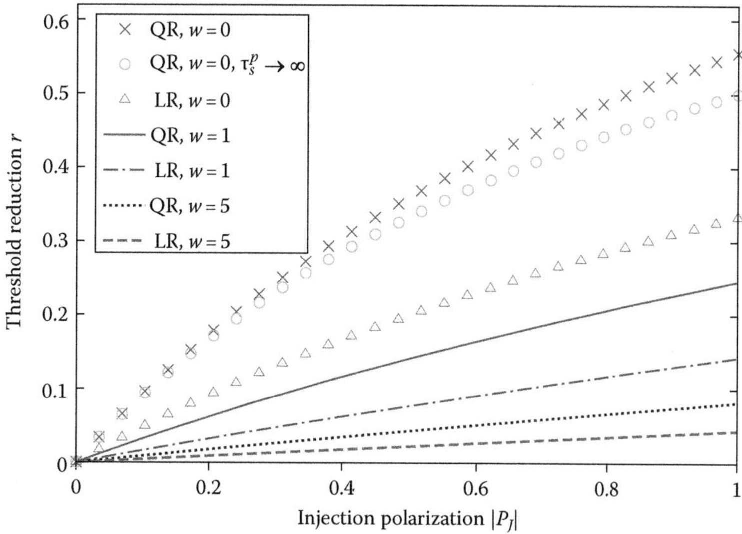

FIGURE 38.12 Threshold current reduction r as a function of injection polarization P_J. Shown are both LR and QR for various w, the ratio of recombination time and electron spin relaxation time. The hole spin relaxation time $\tau_s^p=0$, except for the circles that show the previously assumed "ideal" case, in which $\tau_s^p\to\infty$. (From Gøthgen, C., Oszwałdowski, R., Petrou, A., and Žutić, I., Analytical model of spin-polarized semiconductor lasers, *Appl. Phys. Lett.*, 93, 042513, 2008, Fig. 3. With permission. Copyright 2008, American Institute of Physics.)

though holes are completely unpolarized ($\tau_s^p = 0$). This can be explained as follows. First we calculate n from Equation 38.8 for S^-. The threshold density is reached when $n_+ + n/2 = N_T$, which gives $n = (2/3)N_T$ for $P_J = 1$, that is, for $n_+ = n$. Thus, the threshold *electron density* is reduced by only 1/3. Assuming $S^- = 0$ at the threshold, the threshold current from Equation 38.7 is given by $J_+ = Bn_+n/2$. If $P_J = 1$ then $J = J_+ = Bn^2/2 = (2/3)^2 J_T$, which yields the 5/9 *current* reduction, a direct consequence of the recombination $\propto n^2$ (note that for LR both the N_T and J_T are reduced only by 1/3).

For comparison, we calculate r for QR in a priori "ideal" case, where $n_\pm = p_\pm$ for any injection J and *both* $\tau_s^n, \tau_s^p \to \infty$. Therefore, the gain and QR terms in Equation 38.7 for n_+ (n_-) and in Equation 38.8 for S^- (S^+) must be modified by replacing $n/2$ with n_+ (n_-). Surprisingly, we obtain $r(|P_J| = 1) = 1/2$, a smaller reduction than in the case of unpolarized holes. A simple calculation yields $n = N_T/2$, that is, the reduction of the threshold *electron density* is larger than for $\tau_s^p = 0$. However, the interband recombination R_{sp}^+ is more efficient, because none of the holes undergo spin-flip. Thus, we obtain $J_+ = J_{T1} = Bn_+^2 = BN_T^2/4 = J_T/2$. For $\tau_s^p = 0$ and $P_J = 1$, half of the holes are inaccessible for recombination with the fully spin-polarized electrons. This lowers the injection necessary to overcome the recombination losses and the interplay of stimulated and spontaneous recombination leads to a smaller r in the "ideal" case.

38.4 Spin-Lasers: Dynamic Operation

The most attractive properties of conventional lasers usually lie in their dynamical performance [4]. Here we explore intriguing opportunities offered by spin-polarized modulation and study them with the REs description. Each of the key quantities, X (such as, J, S, n, and P_J), can be decomposed into a steady-state X_0 and a modulated part $\delta X(t)$, $X = X_0 + \delta X(t)$.

We focus on the amplitude and polarization modulation (*AM*, *PM*). *AM* for a steady-state polarization implies $J_+ \neq J_-$ (unless $P_J = 0$),

$$\text{AM:} \quad J = J_0 + \delta J \cos(\omega t), \quad P_J = P_{J0}, \tag{38.17}$$

where ω is the angular modulation frequency. *AM* is illustrated in Figure 38.10.

Similar to the steady-state analysis in Section 38.3, $P_J \neq 0$ leads to unequal threshold currents J_{T1} and J_{T2} (for $S^\mp$, majority and minority photons). For the injection $J_{T1} < J_T < J_{T2}$, we expect a modulation of fully polarized light, even for a partially polarized injection. Such a modulation can be contrasted with *PM*, which also has $J_+ \neq J_-$, but J remains constant*

$$\text{PM:} \quad J = J_0, \quad P_J = P_{J0} + \delta P_J \cos(\omega t). \tag{38.18}$$

A recent progress in electrically tunable P_J [37,38] suggests that fast *PM* could be realized in future spin-lasers. Currently, optically injected lasers with controllable circular polarization are more promising for implementing *PM* [11,12,39–41]. Slow *PM* using a Soleil–Babinet polarization retarder at a fixed $J(J_{T1} < J < J_{T2})$, leads to 400% modulation of laser emission [11]. Fast ($\omega/2\pi \sim 40$ GHz) *PM* can be implemented using, for example, a coherent electron spin precession in a transverse magnetic field for a pulsed laser emission with alternating circular polarization [42], or a mode conversion in a ridge waveguide electro-optic modulator [43]. As we show, the advantages of spin-lasers are also expected for $J \neq J_0$, that is, for *AM*. We present our results for LR while the advantages for QR can be even larger, as can be inferred from Figure 38.12.

38.4.1 Quasistatic Regime

In the quasi-static regime ($\omega \ll 1/\tau_{ph}, 1/\tau_r$), we can use the steady-state results from Section 38.3 to obtain *AM* (*PM*) with $J_0(P_{J0})$ substituted by $J(t)$ ($P_J(t)$) and the injection $J_\pm$ will be in phase with the response $n_\pm$ and $S^\mp$. For typical parameters in Table 38.1, we confirmed numerically that this regime is valid up to $\omega/2\pi \sim 10$ MHz. The steady-state results ($\varepsilon = \beta = 0$) for the two threshold currents (see Figures 38.10 and 38.13)

$$J_{T1} = \frac{J_T}{1 + |P_J|/2(1+2w)}, \quad J_{T2} = \frac{J_T}{1 - |P_J|}, \tag{38.19}$$

remain directly applicable for *AM* and *PM*. To reduce J_{T1} and improve dynamic performance, small ratio of the recombination and spin relaxation times $w \sim 0.1$ ($\tau_s^n > 0.4$ ns, $\tau_r \sim 40$ ps at 300 K)

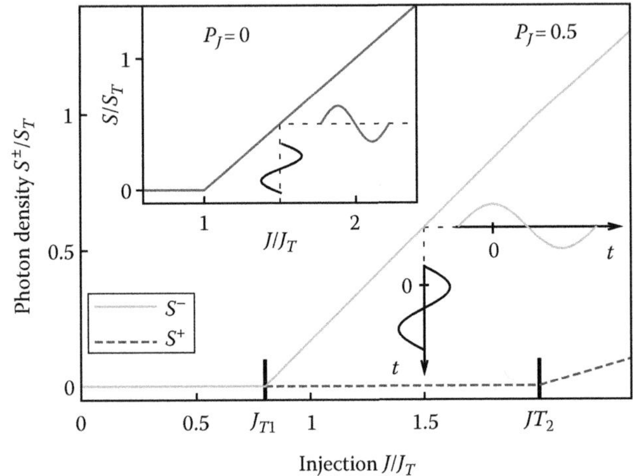

FIGURE 38.13 *AM*. Circularly polarized photon densities, $S^\pm$, as a function of injection J with polarization $P_J = P_{J0} = 0.5$, for infinite electron spin relaxation time. Vertical and horizontal harmonic curves show the modulation of (input) current and the resulting modulation of the (output) light. *AM* of a partially polarized $J < J_{T2}$ leads to the modulation of the fully polarized output light (no S^+ component). Inset: For *AM* in a spin-unpolarized laser [4], S^+ and S^- undergo identical modulations. (From Lee, J., Falls, W., Oszwałdowski, R., and Žutić, I., Spin modulation in semiconductor lasers, *Appl. Phys. Lett.*, 97, 041116, 2010, Fig. 1. With permission. Copyright 2010, American Institute of Physics.)

* This is not essential but simplifies our analytical results.

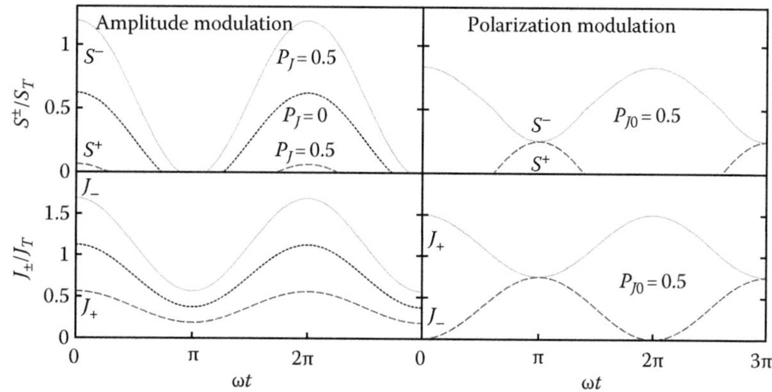

FIGURE 38.14 Time-dependence of the photon densities and injection in the quasi-static regime for large *AM* and *PM*. The results show coupling of $J_{\pm}$ with $S_{\mp}$ (Equations 38.7 and 38.8), given at $J_0 = 1.5J_T$, $w = 0$ ($\tau_s'' \to \infty$), and $\beta = \varepsilon = 0$. Left *AM*: $\delta J/J_0 = 0.5$ and $P_{J0} = 0.5$ (Equation 38.17), solid and broken lines. Reference results are given by dotted lines for $P_J = 0$: $S/2 = S^- = S^+$ (upper panel) and $J/2 = J_+ = J_-$ (lower panel). For $J > J_{T2}$, S^+ is in phase with S^-. Right *PM*: $\delta P_J = 0.5$ and $P_{J0} = 0.5$ (Equation 38.18). For $J > J_{T2}$, S^+ has an opposite phase to S^-. (From Lee, J., Falls, W., Oszwałdowski, R., and Žutić, I., Spin modulation in semiconductor lasers, *Appl. Phys. Lett.*, 97, 041116, 2010, Fig. 2. With permission. Copyright 2010, American Institute of Physics.)

can be realized in (110) GaAs-based QW spin-lasers [44–46], while $w \sim 1$ in (100) GaAs QW [7,12,45]. As in the steady state, J_{T1} and J_{T2} in Equation 38.19 delimit the three regimes of a spin-laser. For large *AM* and *PM*, these regimes determine $S^{\pm}(t)$, normalized to S_T, as shown in Figure 38.14. *AM* in the upper panel reveals $S^{\pm}$ near $\omega t = 0, 2\pi$ that corresponds to $J > J_{T2}$ (a fully lasing regime) and a constant $n_+ = n_-$. With time evolution, the laser enters the mixed regime $J_{T1} < J < J_{T2}$ (only lasing with S^-, Figure 38.13). If $P_J \geq 0$, for both *AM* and *PM* the photon densities for $w = 0$ in Figure 38.14, as well as for an arbitrary w, can be expressed as

$$\frac{S^-}{S_T} = \frac{2+4w}{3+4w}\left[\frac{J}{J_T}\left(1+\frac{P_J}{2+4w}\right)-1\right], \quad \frac{S^+}{S_T} = 0, \qquad (38.20)$$

where J and P_J are given by either Equation 38.17 or 38.18. For $P_J \leq 0$, the expressions for S^+ and S^- in Equation 38.20 are interchanged. Finally, near $\omega t = \pi, 3\pi$, no emitted $S^{\pm}$ implies $J < J_{T1}$ (the spin-LED regime) for *AM*. In the *PM* case (right panels) the injection is chosen such that $J > J_{T1}$ (S^- is present). The fully lasing regime, near $\omega t = \pi, 3\pi$, corresponds to constant $n^+ = n^-$, even as $P_J(t)$ varies.

38.4.2 Small Signal Analysis

The quasi-static approach allowed us to consider analytically large signal modulation (both for *AM* and *PM*), usually only studied numerically. We next turn to the complementary approach for laser dynamics, that is, small signal analysis (SSA), limited to a small modulation ($|\delta J/J_0| \ll 1$ for *AM* and $|\delta P_J| \ll 1$, $|P_{J0} \pm \delta P_J| \leq 1$ for *PM* [47]),* but valid for all frequencies. We confirmed that the two approaches coincide in the common region of validity. Our SSA for spin-lasers proceeds as in conventional lasers [4,48]. The decomposition $X = X_0 + \delta X(t)$ for $J_{\pm}$, $n_{\pm}$, and $S^{\pm}$ is substituted in Equations 38.7 and 38.8. The modulation terms can be written as $\delta X(t) = Re[\delta X(\omega)e^{-i\omega t}]$. We then analytically calculate $\delta n_{\pm}(\omega)$, $\delta S^{\pm}(\omega)$ and the appropriate generalized frequency response functions $R_{\pm}(\omega) = |\delta S^{\mp}(\omega)/\delta J_{\pm}(\omega)|$, which reduce to the conventional form [4] $R(\omega) = |\delta S(\omega)/\delta J(\omega)|$, in the $P_J = 0$ limit. From SSA we obtain the resonance in the modulation response $R_{\pm}(\omega)$, also known as "relaxation oscillation frequency," $\omega_R/2\pi$. For $P_{J0} = 0$ and $J > J_{T2}$, we find

$$\omega_R^{AM} \approx \sqrt{2}\omega_R^{PM} \approx \left[\Gamma g_0\left(\frac{N_T}{\tau_r}\right)\left(\frac{J_0}{J_T}-1\right)\right]^{1/2} = \left(\frac{g_0 S_0}{\tau_{ph}}\right)^{1/2}, \qquad (38.21)$$

where ω_R^{AM} recovers the standard result [4]. For $P_{J0} \neq 0$ and $J_{T1} < J < J_{T2}$

$$\omega_R^{AM} = \omega_R^{PM} \approx \left[\Gamma g_0\left(\frac{N_T}{\tau_r}\right)\left[\left(1+\frac{|P_{J0}|}{2}\right)\frac{J_0}{J_T}-1\right]\right]^{1/2} = \left(\frac{3g_0 S_0^-}{2\tau_{ph}}\right)^{1/2}. \qquad (38.22)$$

Such ω_R correspond to a peak in the frequency response and can be used to estimate its bandwidth [4], a frequency where the normalized response $R(\omega/2\pi)/R(0)$ decreases to −3 dB.

We can now look for possible advantages in the dynamic operation of spin-lasers, as compared to their conventional counterparts. From Equation 38.22 we infer that the spin-polarized

* These constraints could be relaxed with the generation of pure spin currents implying $P_J > 1$, demonstrated optically and electrically, as discussed in Ref. [7].

injection increases ω_R and thus increases the laser bandwidth—an important figure of merit [4].

In Figure 38.15, these trends are visible in the normalized frequency response. Results for $P_{J0}=0$ (using finite ε and β [12,13]) show that our analytical approximations for $\varepsilon=\beta=0$ are an accurate description of *AM* and *PM*, at moderate pumping power. The increase of ω_R and the bandwidth with P_{J0}, for *AM* and *PM*, can be understood as the dynamic manifestation of threshold reduction with increasing P_{J0}. With $\omega_R \propto (S_0^-)^{1/2}$ (Equation 38.22), the situation is analogous to the conventional lasers: ω_R and the bandwidth both increase with the square root of the output power [4,48] ($S_0^+ = 0$ for $J_{T1}<J<J_{T2}$). An important advantage of spin-lasers is that the increase in S_0^- can be achieved even at *constant* input power (i.e., $J-J_T$), simply by increasing P_{J0}. In the limit of slow modulation, this behavior can already be seen from Figure 38.6b. Additionally, a larger P_{J0} allows for a larger J_0 (maintaining $J_{T1}<J_0<J_{T2}$), which can further enhance the bandwidth, as seen in Equation 38.22. For $P_{J0}=0.9$, J_0 can be up to 10 J_T, from Equation 38.19.

We next examine the effects of finite τ_s^n, shown for $w=1$ and $P_{J0}=0.5$. *AM* results follow a plausible trend: ω_R and the bandwidth monotonically decrease and eventually attain "conventional" values for $\tau_s^n \to 0$ ($w \to \infty$). The situation is rather different for *PM*: seemingly detrimental spin relaxation *enhances* the bandwidth and the peak in the frequency response, as compared to the long τ_s^n limit ($w=0$). A shorter τ_s^n will reduce P_J and thus the amplitude of modulated light. Since $\delta S^-(0)$ decreases faster with w than $\delta S^-(\omega>0)$, we find an increase in the normalized response function, shown in Figure 38.15. The increase in the bandwidth comes at the cost of a reduced modulation signal.

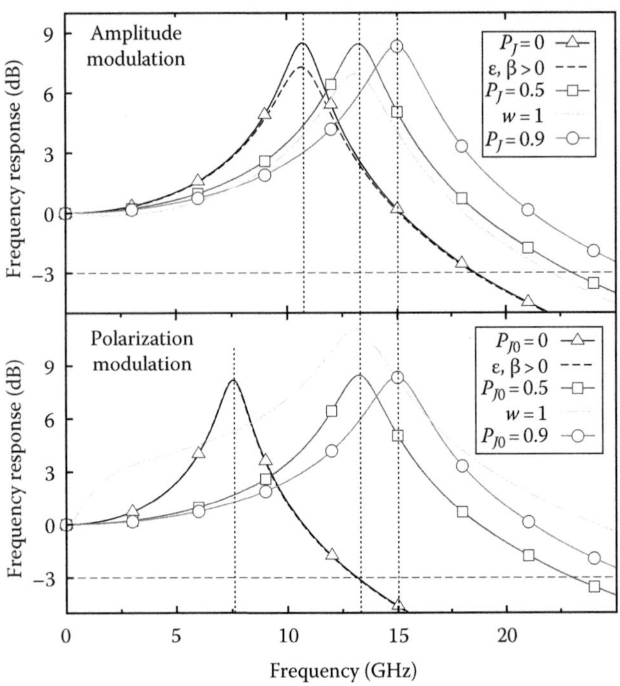

FIGURE 38.15 SSA of *AM* ($\delta J/J_0=0.01$) and *PM* ($\delta P_J=0.01$), at $J_0=1.9J_T$. The frequency response function $|\delta S^-(\omega)\delta J_+(\omega)|$ is normalized to the corresponding $\omega=0$ value. Results are shown in the limit of $w=0$, and $\beta=\varepsilon=0$, except broken lines for $w=1$, $\beta=10^{-4}$, and $\varepsilon=2\times10^{-18}$ cm^3. The frequency response at −3 dB value gives the bandwidth of the laser [4]. The vertical lines denote approximate peak positions evaluated at $w=\varepsilon=\beta=0$ from Equation 38.22, except for *PM* at $P_{J0}=0$ evaluated from Equation 38.21. When $P_{J0}=0$ there is a small difference between $\varepsilon=\beta=0$ and finite ε, β results for *AM*, the difference is nearly invisible for *PM*. (From Lee, J., Falls, W., Oszwałdowski, R., and Žutić, I., Spin modulation in semiconductor lasers, *Appl. Phys. Lett.*, 97, 041116, 2010, Fig. 3. With permission. Copyright 2010, American Institute of Physics.)

The above considered trends allow us to infer some other possible advantages of *PM* at fixed injection. In the quasi-static approximation, for $J>J_{T2}$, constant $n_+=n_-=N_T/2$ implies that *PM* would be feasible at a reduced chirp (α-factor), since $\delta n(t)$, which is a chirp source in *AM*, is eliminated [48]. These findings pertinent to the limit of low frequency and $\varepsilon=\beta=0$, can be combined with SSA in Figure 38.15, revealing substantially smaller ε, β effects for *PM* than *AM*, to suggest that the reduced chirp and therefore desirable switching properties of spin-lasers can be expected for a broad range of parameters.

38.5 Outlook and Conclusions

We have seen that the operation of spin-lasers strongly depends on robust spin injection and long spin relaxation times. While the robust optical spin injection has already been demonstrated at room temperature [49], the electrical spin injection in spin-lasers requires substantial improvements. For example, in (In,Ga)As QW spin-lasers it was realized up to ~100 K using Fe Schottky contacts [13] and in (In,Ga)As QDs it reached ~230 K, using MnAs injector [26,27,50]. Recent experiments showing room temperature spin injection in MnAs-based spin LEDs [51] are encouraging that MnAs can soon also enable room temperature injection in spin-lasers. Alternatively, MgO-based tunnel contacts, discussed in Chapter 11, provide suitable candidates to implement the room temperature spin injection. For the surface-emitting spin-lasers, magnetic injectors with out-of-plane remanent magnetization (along the emitted light, shown in Figure 38.3) would be desirable. In such a geometry, the spin-laser operation is possible without the need to apply an external magnetic field, since the optical selection rules lead to circularly polarized light [6,7].

While there is a significant progress in achieving longer spin relaxation times by varying the material and geometry of QW-based active region (recall Figure 38.2), a viable alternative could be to instead use QD-based active regions. A strong confinement in QDs can suppress the effect of spin–orbit coupling, a leading mechanism for the spin relaxation of carriers. The previous interpretation of experimental work on QD-based spin-lasers [26,27,50] has relied on QW-based REs, similar to the one we have reviewed in the previous sections. Since there are substantial differences between QW- and QD-based spin-lasers, we provide a brief illustration of some features pertinent for QDs, as well mention some materials opportunities, beyond already published work on (In,Ga)As QDs. Figure 38.16 shows

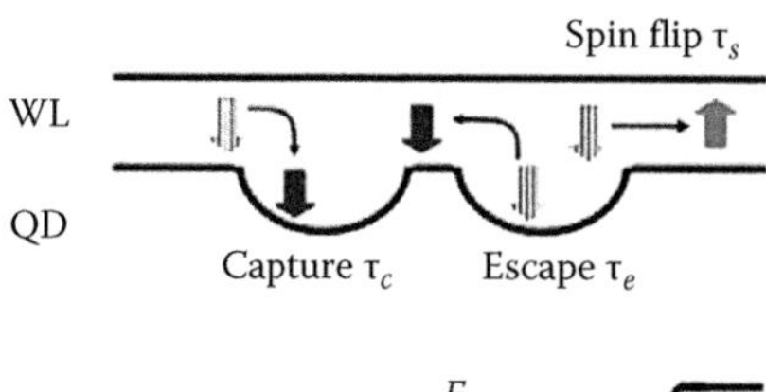

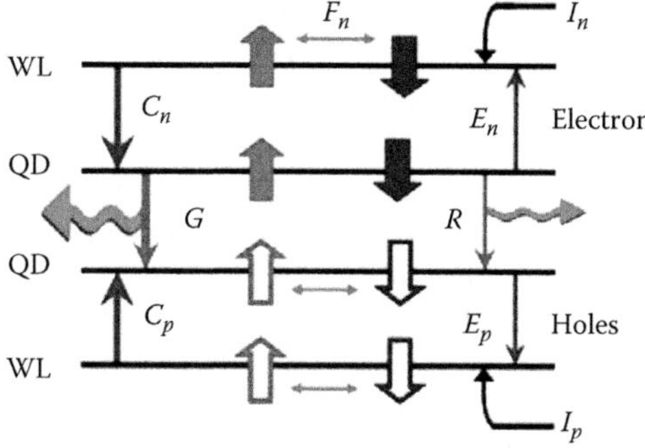

FIGURE 38.16 Characteristic processes in our model of a QD spin-laser. QD: quantum dot, WL: wetting layer. Upper panel: Thick arrows denote electron spin direction in processes labeled by their corresponding times. Lower panel: Thick vertical arrows show the carrier spin; filled for electrons, empty for holes. Curved arrows show carrier injection *I*. Thin arrows depict capture *C*, escape *E*, spin relaxation *F*, stimulated *G*, and spontaneous *R* recombination (thickness indicates relative rates). The subscripts *n* and *p* represent the electron and hole contributions, respectively. Wavy arrows depict photon emission. (From Oszwałdowski, R., Gøthgen, C., and Žutić, I., *Phys. Rev. B*, 82, 085316, 2010, Fig. 1. Copyright 2010 by the American Physical Society.)

a level structure and various processes represented in our RE model for QD-based spin-lasers [52]. The QDs capture carriers from energy levels of a two-dimensional QW-like wetting layer (WL), which acts as a reservoir of carriers. In QD spin-lasers, it is more convenient to use carrier and photon occupancies, rather than their densities, used for QWs. To correctly describe consequences of the small density of QD states, as well as saturation of the WL states at high injection, it is important that we include Pauli blocking factors of the form $1-f$, where f is the corresponding occupancy of the WL and QD states [52]. These terms, omitted in the previous work on QD-based spin-lasers, impede carrier transfer to states close to saturation. They are particularly important in description of the limited QD occupancies. Furthermore, QD- and QW-based lasers have qualitatively different densities of the initial and final states of lasing transitions. The threshold reduction in QD lasers may be hindered by the small density of QD states, if the gain or the photon cavity lifetime is too small. Unlike in QWs, there is a minimum value for the gain in QDs in order to have a lasing at all.

Long spin relaxation times have been reported in CdSe/ZnSe, an example of a II–VI structure self-organized QDs [53]. Detailed predictive studies of the spin relaxation mechanisms, as discussed in Chapter 16, could guide the future efforts in designing QD spin-lasers. It would also be interesting to consider active regions with magnetic doping, where the spin degeneracy of the lasing transition may be lifted. II–VI materials doped with Mn are a promising direction, since QD lasers based on II–VI structures have already been considered [54]. The problem of the Mn internal transition, which reduces the intensity of band-to-band transitions can be addressed by using ZnSe/(Zn,Mn)Te epitaxial QDs [55], characterized by a relatively low fundamental transition energy.

A very interesting emerging field is lasers based on colloidal semiconductor QDs typically II–VI, such as CdS, CdSe, ZnSe, and ZnTe [56,57]. These nanostructures are easily synthesized, offer a large tunability of transition energies and a long spin-coherence time [58]. However, some colloidal QD structures suffer from the very fast (~100 ps) nonradiative Auger recombination that hinders population inversion and is therefore detrimental for optical gain. This effect can be avoided by using the so-called type-II band alignment in which spatial separation of electrons from holes significantly suppresses the Auger recombination [59]. Just like their self-assembled counterparts, discussed further in Chapter 30, colloidal QDs can also be doped magnetically [60].

In the preceding sections, we have discussed the connection of steady state and dynamic operation of spin-lasers. For a versatile dynamic operation of spin-lasers, it would be important to achieve effective spin modulation schemes. Currently, optical methods appear easier to implement. However, the future advances in spin-lasers may depend on progress in magnetic memories and data storage. Answers to the key questions in these areas, about ultrafast magnetization dynamics and timescales for magnetization reversal [61–63], may also determine the switching speed limit in the modulation of electrically pumped spin-lasers. As can already be seen from the topics of spin injection and spin relaxation, the future studies of spin-lasers, and the prospect for their commercial viability, will be increasingly stimulated by many developments in spin transport and magnetism, discussed throughout this book. On the other hand, by providing practical paths to novel room temperature spin-based devices, we expect that the continued studies of spin-lasers will also offer further motivation to advance the understanding of spin transport and magnetism.

Acknowledgments

This work has been supported by the U.S. ONR, N0000140610123, AFOSR-DCT FA9550-09-1-0493, NSF-ECCS CAREER 054782, NSF-ECCS 1102092, and DOE-BES DE-SC0004890. We thank H. Dery, M. Ostereich, A. Petrou, G. Strasser, P. Bhattacharya, and X. Zhang for their valuable discussions.

References

1. V. M. Ustinov, A. E. Zhukov, A. Yu. Egorov, and N. A. Maleev, *Quantum Dot Lasers*, Oxford University, New York, 2003.
2. A. M. Prochorov, Nobel Lecture, http://nobelprize.org/nobel_prizes/physics/laureates/1964/prokhorov-lecture.html (accessed on March 17, 2011).

3. Z. I. Alferov, Nobel lecture: The double heterostructure concept and its applications in physics, electronics, and technology, *Rev. Mod. Phys.* **73**, 767–782 (2001).
4. S. L. Chuang, *Physics of Optoelectronic Devices*, 2nd edn., Wiley, New York, 2009.
5. G. Lampel, Nuclear dynamic polarization by optical electronic saturation and optical pumping in semiconductors, *Phys. Rev. Lett.* **20**, 491–493 (1967).
6. *Optical Orientation*, F. Meier and B. P. Zakharchenya, (Eds.), North-Holland, New York, 1984.
7. I. Žutić, J. Fabian, and S. Das Sarma, Spintronics: Fundamentals and applications, *Rev. Mod. Phys.* **76**, 323–410 (2004).
8. J. Fabian, A. Mathos-Abiague, C. Ertler, P. Stano, and I. Žutić, Semiconductor spintronics, *Acta Phys. Slov.* **57**, 565–907 (2007).
9. M. Oestreich, J. Rudolph, R. Winkler, and D. Hägele, Design considerations for semiconductor spin lasers, *Superlattices Microstruct.* **37**, 306–312 (2005).
10. S. F. Yu, *Analysis and Design of Vertical Cavity Surface Emitting Lasers*, Wiley, New York, 2003.
11. J. Rudolph, D. Hägele, H. M. Gibbs, G. Khitrova, and M. Oestreich, Laser threshold reduction in a spintronic device, *Appl. Phys. Lett.* **82**, 4516–4518 (2003).
12. J. Rudolph, S. Döhrmann, D. Hägele, M. Oestreich, and W. Stolz, Room-temperature threshold reduction in vertical-cavity surface-emitting lasers by injection of spin-polarized carriers, *Appl. Phys. Lett.* **87**, 241117 (2005).
13. M. Holub, J. Shin, and P. Bhattacharya, Electrical spin injection and threshold reduction in a semiconductor laser, *Phys. Rev. Lett.* **98**, 146603 (2007).
14. M. Holub and P. Bhattacharya, Spin-polarized light-emitting diodes and lasers, *J. Phys. D: Appl. Phys.***40**, R179–R203 (2007).
15. J. Lee, W. Falls, R. Oszwałdowski, and I. Žutić, Spin modulation in semiconductor lasers, *Appl. Phys. Lett.* **97**, 041116 (2010).
16. M. A. Parker, *Physics of Optoelectronics*, CRC Press, New York, 2005.
17. C. Gøthgen, Steady-state analysis of semiconductor spin-lasers, PhD thesis, University at Buffalo, New York, 2010.
18. H. Dery and G. Eisenstein, Self-consistent rate equations of self-assembly quantum wire lasers, *IEEE J. Quantum Electron.* **40**, 1398–1409 (2004).
19. H. Dery and G. Eisenstein, The impact of energy band diagram and inhomogeneous broadening on the optical differential gain in nanostructure lasers, *IEEE J. Quantum Electron.* **41**, 26–35 (2005).
20. W. W. Chow and S. W. Koch, *Semiconductor-Laser Fundamentals: Physics of the Gain Materials*, Springer, New York, 1999.
21. J. Hader, J. V. Moloney, and S. W. Koch, Beyond the ABC: Carrier recombination in semiconductor lasers, *Proc. SPIE* **6115**, 61151T (2006).
22. H.-H. Lee and W. Casperson, Gain and saturation in semiconductor lasers, *Opt. Quant. Electron.* **25**, 369–390 (1993).
23. S. S. P. Parkin, C. Kaiser, A. Panchula et al., Room temperature tunneling magnetoresistance in magnetic tunnel junctions with MgO(100) tunnel barriers, *Nat. Mater.* **3**, 862–867 (2004).
24. S. Yuasa, T. Nagahama, A. Fukushira, Y. Suzuki, and K. Ando, Giant room-temperature magnetoresistance in single-crystal Fe/MgO/Fe magnetic tunnel junctions, *Nat. Mater.* **3**, 868–871 (2004).
25. G. Salis, R. Wang, X. Jiang et al., Temperature independence of the spin-injection efficiency of a MgO-based tunnel spin injector, *Appl. Phys. Lett.* **87**, 262503 (2005).
26. D. Basu, D. Saha, C. C. Wu, M. Holub, Z. Mi, and P. Bhattacharya, Electrically injected InAs/GaAs quantum dot spin laser operating at 200 K, *Appl. Phys. Lett.* **102**, 091119 (2008).
27. D. Basu, D. Saha, and P. Bhattacharya, Optical polarization modulation and gain anisotropy in an electrically injected spin laser, *Phys. Rev. Lett.* **102**, 093904 (2009).
28. C. Gøthgen, R. Oszwałdowski, A. Petrou, and I. Žutić, Analytical model of spin-polarized semiconductor lasers, *Appl. Phys. Lett.* **93**, 042513 (2008).
29. I. Vurgaftman, M. Holub, B. T. Jonker, and J. R. Mayer, Estimating threshold reduction for spin-injected semiconductor lasers, *Appl. Phys. Lett.* **93**, 031102 (2008).
30. M. San Miguel, Q. Feng, and J. V. Moloney, Light-polarization dynamics in surface-emitting semiconductor lasers, *Phys. Rev. A* **52**, 1728–1739 (1995).
31. J. Huang and L. W. Casperson, Gain and saturation in semiconductor lasers, *Opt. Quant. Electron.* **25**, 369–390 (1993).
32. I. Žutić, J. Fabian, and S. C. Erwin, Spin injection and detection in silicon, *Phys. Rev. Lett.* **97**, 026602 (2006).
33. A. T. Hanbicki, B. T. Jonker, G. Itskos, G. Kioseoglou, and A. Petrou, Detection of spin-polarized electrons injected into two-dimensional electron gas, *Appl. Phys. Lett.* **81**, 2131–2133 (2002).
34. A. T. Hanbicki, O. M. van't Erve, R. Magno et al., Analysis of the transport process providing spin injection through an Fe/AlGaAs Schottky barrier, *Appl. Phys. Lett.* **82**, 4092–4094 (2003).
35. T. J. Zega, A. T. Hanbicki, S. C. Erwin et al., Determination of interface atomic structure and its impact on spin transport using Z-contrast microscopy and density-functional theory, *Phys. Rev. Lett.* **96**, 196101 (2006).
36. I. Žutić, J. Fabian, and S. Das Sarma, Spin injection through the depletion layer: A theory of spin-polarized p-n junctions and solar cells, *Phys. Rev. B* **64**, 121201(R) (2001).
37. P. Li and H. Dery, Tunable spin junction, *Appl. Phys. Lett.* **94**, 192108 (2009).
38. S. A. Crooker, E. S. Garlid, A. N. Chantis et al., Bias-controlled sensitivity of ferromagnet/semiconductor electrical spin detectors, *Phys. Rev. B* **80**, 041305 (2009).

39. S. Hövel, N. C. Gerhardt, C. Brenner et al., Spin-controlled LEDs and VCSELs, *Phys. Status Solid (a)* **204**, 500–507 (2007).
40. A. Bischoff, N. C. Gerhardt, M. R. Hofmann, T. Ackemann, A. Kroner, and R. Michalzik, Optical spin manipulation of electrically pumped vertical-cavity surface-emitting lasers, *Appl. Phys. Lett.* **92**, 041118 (2008).
41. H. Ando, T. Sogawa, and H. Gotoh, Photon-spin controlled lasing oscillations in surface-emitting lasers, *Appl. Phys. Lett.* **73**, 566–568 (1998).
42. S. Hallstein, J. D. Berger, M. Hilpert et al., Manifestation of coherent spin precession in stimulated semiconductor emission dynamics, *Phys. Rev. B* **56**, R7076–R7079 (1997).
43. J. D. Bull, N. A. F. Jaeger, H. Kato, M. Fairburn, A. Reid, and P. Ghanipour, 40 GHz electro-optic polarization modulator for fiber optic communication systems, in *Photonic North 2004: Optical Components and Devices*, J. C. Armitage, S. Fafard, R. A. Lessard, and G. A. Lampropoulos, (Eds.), *Proc. SPIE* **5577**, 133–143, Bellingham, WA.
44. Y. Ohno, R. Terauchi, T. Adachi, F. Matsukara, and H. Ohno, Spin relaxation in GaAs (110) quantum wells, *Phys. Rev. Lett.* **83**, 4196–4199 (1999).
45. H. Fujino, S. Koh, S. Iba, T. Fujimoto, and H. Kawaguchi, Circularly polarized lasing in a (110)-oriented quantum well vertical-cavity surface-emitting laser under optical spin injection, *Appl. Phys. Lett.* **94**, 131108 (2009).
46. K. Ikeda, T. Fujimoto, H. Fujino, T. Katayama, S. Koh, and H. Kawaguchi, Switching of lasing circular polarizations in (110)-VCSEL, *IEEE Phot. Tech. Lett.* **21**, 1350–1352 (2009).
47. M. J. Stevens, A. L. Smirl, R. D. R. Bhat, A. Najmaie, J. E. Sipe, and H. M. van Driel, Quantum interference control of ballistic pure spin currents in semiconductors, *Phys. Rev. Lett.* **90**, 136603 (2003).
48. A. Yariv, *Optical Electronics in Modern Communications*, 5th edn., Oxford University, New Yok, 1997.
49. S. Iba, S. Koh, K. Ikeda, and H. Kawaguchi, Room temperature circularly polarized lasing in an optically spin injected vertical-cavity surface-emitting laser with (110) GaAs quantum wells. *Appl. Phys. Lett.* **98**, 081113 (2011).
50. D. Saha, D. Basu, and P. Bhattacharya, High-frequency dynamics of spin-polarized carriers and photons in a laser. *Phys. Rev. B* **82**, 205309 (2010).
51. E. D. Fraser, S. Hegde, L. Schweidenback et al., Efficient electron spin injection in MnAs-based spin-light-emitting-diodes up to room temperature, *Appl. Phys. Lett.* **97**, 041103 (2010).
52. R. Oszwałdowski, C. Gøthgen, and I. Žutić, Theory of quantum dot spin lasers, *Phys. Rev. B* **82**, 085316 (2010).
53. A. Klochikhin, A. Reznitsky, S. Permogorov et al., Exciton states and spin relaxation in CdSe/ZnSe self-organized quantum dots, *Semicond. Sci. Technol.* **23**, 114010 (2008).
54. T. Passow, M. Klude, C. Kruse et al., *Adv. Solid State Phys.* **42**, 13–26 (2002).
55. I. R. Sellers, R. Oszwałdowski, V. Whiteside et al., Robust magnetic polarons in type II (Zn,Mn)Te/ZnSe magnetic quantum dots, *Phys. Rev. B* **82**, 195320 (2010).
56. V. I. Klimov, Spectral and dynamical properties of multi-excitons in semiconductor nanocrystals, *Ann. Rev. Phys. Chem.* **58**, 635–673 (2007).
57. G. D. Scholes, Controlling the optical properties of inorganic nanoparticles, *Adv. Funct. Mater.* **18**, 1157–1172 (2008).
58. N. P. Stern, M. Poggio, M. H. Bartl, E. L. Hu, G. D. Stucky, and D. D. Awschalom, Spin dynamics in electrochemically charged CdSe quantum dots, *Phys. Rev. B* **72**, 161303 (2005).
59. V. I. Klimov, S. A. Ivanov, J. Nanda et al., Single-exciton optical gain in semiconductor nanocrystals, *Nature London* **447**, 441–446 (2007).
60. R. Beaulac, P. I. Archer, S. T. Ochsenbein, and D. R. Gamelin, Mn^2+-doped CdSe quantum dots: New inorganic materials for spin-electronics and spin-photonics, *Adv. Funct. Mater.* **18**, 3873–3891 (2008).
61. S. Garzon, L. Ye, R. A. Webb, T. M. Crawford, M. Covington, and S. Kaka, Coherent control of nanomagnet via ultrafast spin torque pulses, *Phys. Rev. B* **78**, 180401(R) (2008).
62. A. V. Kimel, A. Kirilyuk, P. A. Usachev, R. V. Pisarev, A. M. Balbashov, and Th. Rasing, Ultrafast non-thermal control of magnetization by instantaneous photomagnetic pulses, *Nature London* **435**, 655–657 (2005).
63. C. Stanciu, F. Hansteen, A. V. Kimel, A. Kirilyuk, A. Tsukamoto, A. Itoh, and Th. Rasing, All-optical magnetic recording with circularly polarized light. *Phys. Rev. Lett.* **99**, 047601 (2007).

39

Spin Logic Devices

Hanan Dery
University of Rochester

39.1 Introduction

Designs using thin metallic layers, based on either giant magnetoresistive effect [1,2] or tunneling magnetoresistance [3], have already found application in hard-drive read heads, in magnetic random access memories and magnetic field sensors (see Chapters 6 and 36). These designs employ a magnetoresistive spin-valve effect: The current passing through them depends on the magnetization configuration of two terminals. Over the last two decades, the advances in magnetic storage have enabled a 1000-fold increase in the capacity of computer hard drives using metal-based spin valves. In spite of this remarkable success, there are only a few seminal attempts to propose logic gates in all-metallic magnetic systems. Cowburn and Welland [4] have implemented a room temperature magnetic quantum cellular automata network of submicrometer magnetic dots interacting via magnetostatic interactions. To trigger a logic operation, an applied oscillating magnetic field generates a magnetic soliton that carries information through the network. The logic output is then encoded in the resulting magnetic configuration. Similar proposals have used either domain walls to propagate information or shape anisotropy [5,6]. A different all-metallic magneto-logic paradigm relies on magnetic tunneling junctions as building blocks [7–9]. The output of a logic operation is decoded from the current amplitude flowing through a combination of these junctions. Similar to the case of magnetic random access memories, the logic operands are encoded using multiple bit (current) lines. This in turn changes the resistance of the junction, and the added effect from various junctions denotes an output of a logic operation.

The main advantage of magneto-logic designs lies in their scaling potential. For comparison, using the 40 nm technology node in complementary metal oxide semiconductor, the lateral size of a typical two input NAND gate with a moderate fan-out capability is about 0.5 μm^2 [10,11]. In this view, magneto-logic gate (MLG) implementations can provide about two orders of magnitude improvement [4,12]. Another important benefit of prospective MLGs is their intrinsic run-time reprogrammability; the logic functionality of a MLG can be changed at the switching speed of the operands at its input [9,12]. Conventional implementations, on the other hand, rely on field-programmable gate arrays at which the circuit is programmed via a flash memory or an electrically erasable programmable read-only-memory. The speed gap between the logic circuitry and these memories is such that these applications are practically reprogrammed only during bootup time. Moreover, the limited writing endurance of these memories questions their feasibility in run-time reprogrammable applications. Currently, the main drawback of proposed magneto-logic schemes is that the output of a logic operation cannot be carried into the input of another logic operation fast enough without having to go through a conversion step. We are still lacking with efficient propagation or conversion schemes to cascade logic operations.

The research in the emerging field of semiconductor spintronics has brought theoretical proposals of various devices exploiting the spin degree of freedom [13–22]. Remarkable progress has been made in understanding the basic physics of semiconductor spintronics on account of the advances in diluted ferromagnetic semiconductors [23–25], in discoveries of new spin phenomena [26,27], and in spin injection into semiconductors from ferromagnetic metals [23,28–39]. Unlike metals, semiconductors have a relatively low carrier density that can be drastically changed either by sample preparation using inhomogeneous doping or by applied stimuli using gate voltage or light irradiation. This renders their usage in digital logic for information processing. Yet, no viable means of using semiconductor spintronics has been presented and the introduction of spin-based semiconductor devices has been less successful, hampering the unification with traditional electronics (e.g., the integration of magnetic memory into a semiconductor chip). In this chapter, we review paths to break this impasse by using spin accumulation in nonlocal geometries [40–42] as the basis for spin-based logic operations in semiconductors [12].

In the following section, we briefly review spin-dependent transport through a Schottky barrier and in a bulk semiconductor channel (see Chapters 9, 17, 18, 23 for experimental efforts). Then, we analyze the steady-state spin-accumulation profile in two-terminal devices and in lateral multi-terminal devices. We then incorporate a time-dependent analysis in which a technique to generate digital signals is introduced. In addition, the working principles of semiconductor magneto-logic reprogrammable gates are presented and we use these devices to illustrate potential applications. In the final section, we also study semiconductor devices with magnetic contacts whose magnetization configuration is noncollinear.

39.2 Spin Transport in Semiconductors

Theoretical analysis of spin injection from metals into semiconductors shows that junctions with large resistance are necessary for the current to be polarized [43–45]. More precisely, since the spin-depth conductance of the semiconductor $G_{sc}=\sigma/L$ (with conductivity σ and spin-diffusion length L) is much smaller than its metal counterpart $G_m=\sigma_m/L_m$, for spin injection to occur the junction conductance G has to fulfill $G \leq G_{sc}$. On the other hand, with an ohmic contact ($G \gg G_{sc}$), the spin accumulation in the semiconductor channel vanishes since the ferromagnetic contact acts as a powerful spin sink. This "conductivity mismatch" effect was first analyzed by Johnson and Silsbee [46]. The spin injection constraint, $G \leq G_{sc}$, is easily achieved by an insulator barrier or by the naturally formed Schottky barrier at the interface between a metal and a semiconductor [47,48]. The electrochemical potential is discontinuous at the junction in the case of a thin tunneling barrier. This discontinuity alleviates the conductivity mismatch between the semiconductor and the ferromagnetic metal. The spin polarization of the current across the junction is then determined by the spin selectivity of the barrier, $\Delta G=G_+-G_-$, where G_s are the conductances for spin $s=\pm$ with the spin quantization axis directing along the magnetization axis of the ferromagnet. Other than non-negligible spin selectivity, an efficient spin injection occurs when $G \sim G_{sc}$ for which the ratio between the spin-accumulation density and the total electron density reaches its maximum. On the other hand, an over resistive barrier such that $G \ll G_{sc}$ results in a small spin accumulation [45,49,50]. In this view, a Schottky barrier is more suitable than a thin insulator spacer due to its reduced barrier height.

A Schottky barrier between a metal and a semiconductor is created by redistribution of charges in the space charge layer [48]. The barrier width is denoted by d and the barrier height measured from the Fermi level of the metal is denoted by ϕ_B. For a uniformly n-doped semiconductor the barrier shape is approximately parabolic,

$$\phi(x)=\frac{e^2 n_{sb} x^2}{2\varepsilon_{sc}}, \qquad (39.1)$$

where

n_{sb} is the doping concentration in the Schottky barrier region
e is the elementary charge
ε_{sc} is the static permittivity of the semiconductor (typically about 10^{-12} F/cm)

The barrier is defined by the depleted region between the interface ($\phi(d)=\phi_B$) and the edge of the semiconductor flat-band region ($\phi(0)=0$). In a typical metal/semiconductor interface where $\phi_B=0.7$ eV and $n_{sb}\sim 10^{17}$ cm^{-3}, the barrier width is $d \sim 100$ nm. At lower concentrations the tunneling processes are completely irrelevant, and the current is due to purely classical thermionic emission, which depends solely on the barrier height, not on its width or shape. Even if this current is spin polarized, its total density is too small to create an appreciable spin accumulation. For greater concentrations, the tunneling dominates the transport through the junction where one usually achieves an appreciable current density already at low bias conditions when thin barriers with $d \leq 10$ nm are employed ($n_{sb} \geq 10^{19}$ cm^{-3}). In order to achieve such thin barriers with the bulk of the semiconductor having a much lower carrier density, a strongly inhomogeneous doping profile has to be used near the interface [31,51]. The need for a much lower carrier density in the bulk region, $n_0 \ll n_{sb}$, stems from the previously mentioned condition of optimal spin-accumulation density when, $G \sim \sigma/L$. Since the barrier conductance, G, is rather low even when $d \sim 10$ nm, this "spin-impedance" matching is met in bulk regions that are moderately doped and have a relatively long spin-diffusion length. This can be achieved when $n_0 < 10^{17}$ cm^{-3} at which the bulk semiconductor conductivity is relatively small and the spin-diffusion length is relatively large. The latter is valid since the magnitudes of both the Elliot–Yafet and of the D'yakonov–Perel' mechanisms, which dominate the spin relaxation in n-type semiconductors, are reduced when the doping levels are nondegenerate [52]. As a result of the doping inhomogeneity between the bulk and the barrier regions, a potential well is likely to be created in between these regions [53–55]. Even with a careful doping design at which there is no well in equilibrium, at forward bias when less electrons need to be depleted from the semiconductor, the well creation is inevitable. The spin-related effects of this potential well may contribute to the spin accumulation in the bulk semiconductor region [56,57].

39.2.1 Spin Injection

An important quantity in the description of spin transport is a spin-dependent electrochemical potential $\mu_s(\mathbf{x})$. It is defined as

$$\mu_s(\mathbf{x})=\mu_s^c(\mathbf{x})-e\phi(\mathbf{x}), \qquad (39.2)$$

where

^c_s is the chemical potential for spin s
ϕ is the electrostatic potential
$e>0$ is the elementary charge

The spin splitting of the electrochemical potential, $\Delta\mu = \mu_+ - \mu_-$, corresponds to the presence of nonequilibrium spin density (spin accumulation). In a nonmagnetic material $\Delta\mu \neq 0$ means $\Delta n = n_+ - n_- \neq 0$, where n_s is the density of electrons with spin s. In ferromagnet/semiconductor systems, the large difference of conductivities, $\sigma_{fm} \gg \sigma_{sc}$, allows us to disregard both spatial dependence and spin dependence of μ_s in the ferromagnet, and to use a single value of chemical potential μ_{fm}. In what follows, we will simplify our analysis by assuming quasi-neutrality in the bulk semiconductor region ($n_+ + n_- = n_0$), and by assuming a linear proportionality between μ_s and the spin-accumulation density, $\delta n_s = n_s - n_0/2$. The linear proportionality is valid when $|\delta n_s| \ll n_0$ and quasi-neutrality holds if the dielectric response time of the semiconductor, $\sigma_{sc}/\varepsilon_{sc}$, is much shorter than the spin relaxation time and the transit time in the semiconductor channel [58,59]. Using these assumptions, we can define the bias voltage, V, applied across a ferromagnet/semiconductor junction as the difference between μ_{fm} and the average electrochemical potential in the semiconductor flat-band region, $\mu_{fm} - \mu_0 = -eV$ where $\mu_0 = (\mu_+ + \mu_-)/2$. This follows the conventional notation in which electrons flow from (into) the semiconductor in forward (reverse) bias, $V>0$ ($V<0$).

The current density due to tunneling of free electrons is calculated by the Landauer–Büttiker formalism assuming coherent transport across the barrier regions [47],

$$j_{s,t} = \frac{em_{sc}}{4\pi^2\hbar^3}\int_0^\infty dE[f_{fm}(E) - f_{sc,s}(E)]\int_0^E dE_t T_s(E - E_t, eV), \quad (39.3)$$

where

f_{fm} is the Fermi–Dirac distribution at the ferromagnetic metal (with our assumption of a single μ_{fm})
$f_{sc,s}$ is the Fermi–Dirac distribution at the bulk semiconductor region

Its spin dependence is introduced via μ_s. The zero reference energy is the flat conduction band in the bulk semiconductor region. E and E_t are, respectively, the total electron's energy and the electron's energy due to transverse motion. m_{sc} is the electron effective mass in the semiconductor region. The spin-dependent transmission coefficient, T_s, is a function of the kinetic energy of the impinging electron, $E - E_t$, and of the potential drop across the barrier, eV. The transmission magnitude is governed by the exact shape of the barrier. In order to quantify the spin-dependent transmission coefficient, a simple effective mass band structure is invoked. This model is sufficient for simulating the spin polarization of the tunneling current across the Schottky barrier and it is also capable of reproducing the spin polarization seen in Fe/GaAs experiments [28,29,33]. In this view, we assume that the spin-based tunneling stems from the spin-dependent kinetic energies in the ferromagnetic side of the junction and we neglect the electronic structure of the interfacial atomic layer [36,60–63]. To readily calculate the second integral in Equation 39.3, specular transmission is assumed, which conserves the transverse wave vectors due to a motion in parallel to the junction's interface, $k_{\parallel}^{fm,s} = k_{\parallel}^{sc}$. Neglecting small corrections due to spin–orbit coupling and using the effective mass approximation, then the electron's velocity and wave vector are related via $v = \hbar k/m$, and the transmission is calculated from particle and current conservation; continuity of the wave function and its first derivative divided by the effective mass. In a rectangular barrier with width d, the transmission is analytical and is given by

$$T_s(E - E_t, eV) \simeq \frac{16 v_{fm,s} v_{sc} e^{-2\kappa_b d}}{v_b^2 + v_{fm,s}^2 + v_{sc}^2 + v_{fm,s}^2 v_{sc}^2 / v_b^2}. \quad (39.4)$$

The validity of this expression is restricted to the small transmission ($T \ll 1$). All velocities are along the normal of the interface where the spin selectivity of the transmission is solely due to the spin-dependent velocities in the ferromagnetic contact, $v_{fm,s}$. The velocities in the semiconductor, v_{sc}, and in the barrier v_b are spin independent. The latter is an "effective tunneling velocity" given by

$$v_b = \frac{\hbar\kappa_b}{m_{sc}} = \sqrt{\frac{2}{m_{sc}}(\phi_B - eV - E + E_t)}. \quad (39.5)$$

In semiconductors with small electron's mass, $m_{sc} \ll m_0$, when the bias is either reverse or moderately forward, then the effective barrier velocity, v_b, dominates the denominator of Equation 39.4 and the ratio between the spin-up and spin-down transmissions is nearly identical to the ratio of their Fermi velocities in the ferromagnetic contact. In iron, this ratio is about $v_+/v_- \sim 2.5$ [64]. This picture is changed for electrons that are photoexcited in the barrier region so that their tunneling energy is high above the quasi-Fermi levels of both sides of the junction. In this case, the Fermi velocities in the ferromagnetic contact become more dominant in the denominator of Equation 39.4 and the spin polarization can switch its sign [65]. In triangular or parabolic barriers, the transmission is mainly affected by the change of κ_b and v_b [20]. In addition to the fact that spin-dependent properties of the interfacial atomic layers were neglected, the validity of this calculation is restricted to a finite bias range around equilibrium. The reverse bias range is limited by the transport across a wider depletion region with enhanced electric field [66,67] and by the enhanced spin relaxation of injected hot electrons [68]. In forward bias, the restriction to a small bias range is twofold. First, electrons tunnel to states above the Fermi level in the ferromagnetic side where new bands may exist. Second, the barrier conductance increases exponentially with the forward bias and therefore the restriction for $G \leq G_{sc}$ is eventually violated.

The description of transport is complete when we relate the spin-dependent electrochemical potentials in the semiconductor channel with their boundary conditions at the semiconductor/ferromagnetic junctions. To simplify the analysis, non-degenerate semiconductors and low electric fields are assumed. The electrochemical potentials are given by Yu and Flatté [69]

$$\mu_s = k_B T \ln\left(\frac{n_0 + 2\delta n_s}{n_0}\right) - e\phi \simeq 2k_B T\frac{\delta n_s}{n_0} - e\phi, \qquad (39.6)$$

where the second expression is the linear approximation valid for small spin-accumulation density compared with the total electron density. The spatial form of the electrochemical potential in a one-dimensional semiconductor channel follows the spin-diffusion equation [43–45]

$$\nabla^2\mu_s = \frac{\partial\mu_s}{\partial t} + \frac{\mu_s - \mu_{-s}}{2L^2}, \qquad (39.7)$$

where the spin-diffusion length, $L = \sqrt{D\tau_s}$, is defined in terms of the diffusion constant D and the spin relaxation time, τ_s. In this formalism, we employ an adiabatic approximation with respect to sub-picosecond timescale processes: the momentum scattering and dielectric relaxation [58]. The time-dependent analysis will be discussed in Section 39.4. In steady-state conditions ($\partial/\partial t = 0$), the general solution in one dimension is given by

$$\mu_s(x) = A + Bx \pm (Ce^{x/L} + Me^{-x/L}), \qquad (39.8)$$

where the A,B,C, and M are constants to be determined from the boundary conditions. The spin-s current is related to the slope of the electrochemical potential by

$$\mathbf{j}_s = \frac{\sigma_s}{e}\nabla\mu_s = \sigma_s\mathbf{E} + eD\nabla n_s, \qquad (39.9)$$

where σ_s is the conductivity for spin s. In the linear regime of small spin accumulation in nondegenerate semiconductors, the conductivity and diffusion constants are all spin independent to the first-order approximation,

$$\sigma_s = \frac{\sigma}{2} = e\nu\frac{n_0}{2}, \quad D = \nu\frac{k_B T}{e}, \qquad (39.10)$$

where ν is the spin-independent mobility. Therefore, the only way for the semiconductor to support spin accumulation is by creating a net spin density, $n_+ \neq n_-$. As a result, the linear and exponential terms in Equation 39.8 are related, respectively, to the drift and diffusion terms in Equation 39.9. In order to keep a compact analysis, we match the spin-dependent bulk current (Equation 39.9) at its boundary with the spin-dependent tunneling current (Equation 39.3) by linearizing the latter around equilibrium

$$j_{s,t} \approx \frac{G_s}{e}(\mu_{fm} - \mu_s). \qquad (39.11)$$

G_s is the barrier conductance at low bias (slope of the J–V curve near zero bias). Its spin dependence is governed by the integrated transmission coefficient. As discussed above, the ratio G_+/G_- is equal to the ratio of the velocities of carriers with different spins in the ferromagnetic contact.

39.3 Two-Terminal Devices

We are now in a position to discuss the magnetoresistance (MR) effect in a biased ferromagnet/semiconductor/ferromagnet system. Figure 39.1a sketches a one-dimensional spin valve where the semiconductor channel is sandwiched between two ferromagnetic contacts. The width of the channel is ω and it is assumed to be shorter than the spin-diffusion length so that we can linearize the exponential terms in Equation 39.8. Due to the small bias between the ferromagnetic contacts, the total barrier conductances are the same on both sides of the channel. For parallel (P) magnetization configuration, the spin-dependent conductances are the same in the right and left barriers $G_+^R = G_+^L$ and $G_-^R = G_-^L$ whereas in the antiparallel (AP) configuration we have, $G_+^R = G_-^L$ and $G_-^R = G_+^L$. Using these conditions and the former assumption of $G < \sigma/L$, Equations 39.8 through 39.11 provide a solution for the electrochemical potential in the semiconductor channel

$$\frac{2\mu_\pm^P(x)}{eV} \approx 1 + \frac{Gx}{\sigma} \pm \frac{F}{2}\frac{Gx}{\sigma}, \qquad (39.12)$$

$$\frac{2\mu_\pm^{AP}(x)}{eV} \approx 1 + \frac{Gx(1-r)}{\sigma} \mp \frac{F}{2}\frac{Gw}{\sigma}\left(\frac{2L^2}{w^2} + \frac{x^2}{w^2}\right), \qquad (39.13)$$

where

V is the applied voltage

$F = (G_+ - G_-)/(G_+ + G_-)$ is the finesse (spin selectivity) of the barrier

r denotes the MR effect of this device

The MR is defined by the relative change of the total current across the device between the parallel and antiparallel configurations

$$r \equiv \frac{I_P - I_{AP}}{I_P} \simeq F^2\left(1 + \frac{\sigma w}{2L^2 G}\right)^{-1}. \qquad (39.14)$$

This expression is valid when both ω/L and LG/σ are small fractions (say ≤1/3). One can notice that the MR effect is larger with improved matching between the barrier conductance,

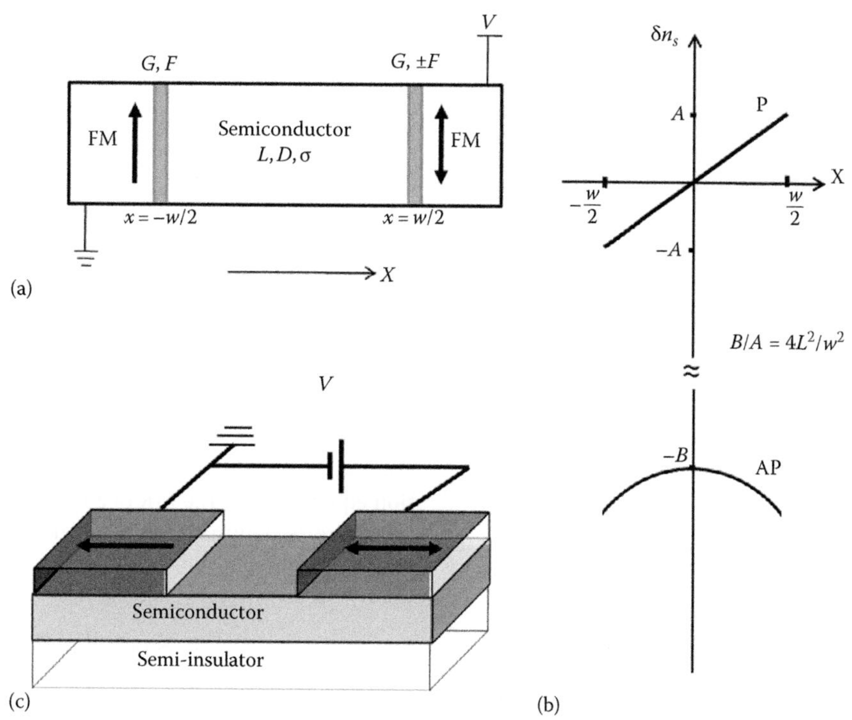

FIGURE 39.1 (a) A one-dimensional biased spin valve. (b) The spin-accumulation profile in the semiconductor channel in P and AP configurations. (c) A realistic lateral spin-valve geometry. The signal is improved when the semiconductor channel is etched away at the edges.

G, and the spin-depth factor of the semiconductor, σ/L. Nonetheless, the magnetoresistive effect is rather small when applying typical parameters. For example, consider a $\omega = 200$ nm of 10^{16} cm^{-3} n-type GaAs region in between two iron contacts at room temperature. This doping corresponds to $\sigma = 10$ $(\Omega\ \text{cm})^{-1}$ and $L = 1$ μm [52]. The barrier parameters are taken from a typical Fe/GaAs spin light emitting diode [30]. The total barrier conductance is $G = 1000\,\Omega$ cm^{-2} and its finesse, measured from the emitted circular polarization, can reach $F \sim 1/3$ [30,34]. This value is also comparable to our estimated value of $(v_{fm,+} - v_{fm,-})/(v_{fm,+} + v_{fm,-})$ in iron. Plugging these parameters results in ~1% magnetoresistive effect. Contrary to such a small effect, the spin-accumulation profile of the two magnetization configurations differs substantially. The separation between the spin-dependent electrochemical potentials is given by

$$\Delta\mu^P(x) \approx F \cdot eV \cdot \frac{Gx}{2\sigma}, \tag{39.15}$$

$$\Delta\mu^{AP}(x) \approx F \cdot eV \cdot \frac{G\omega}{4\sigma}\left(\frac{4L^2}{w^2} + \frac{2x^2}{w^2}\right), \tag{39.16}$$

where $-\omega/2 \le x \le \omega/2 < L$. One can see that the ratio between the spin-accumulation densities can easily take values of 100 or more

$$\frac{\Delta\mu^{AP}}{\Delta\mu^P} = \frac{\Delta n^{AP}}{\Delta n^P} \ge \left(\frac{2L}{w}\right)^2. \tag{39.17}$$

Figure 39.1b shows the spin-accumulation profile in both magnetization configurations. The large difference in spin-accumulation profiles does not affect the spin-dependent currents due to the fact that only the base levels of the spin-accumulation profiles differ by a factor of $4L^2/w^2$, whereas their slopes are on the same order of magnitude. To understand the difference between the spin-accumulation profiles, we can picture the following. In a parallel configuration, the reversed bias contact injects electrons whose net spin polarization is identical to the preferred extracted spin by the forward-biased contact. Since there is no bottleneck, the spin-accumulation density in the channel is very small. Figure 39.1b also shows that in the parallel configuration, the spin-accumulation changes its sign across the semiconductor channel. The reason is again simple; next to the injector, there is an accumulation of spin-up electrons that matches the spin of the majority population in the ferromagnetic contact. Next to the extractor, there is a depletion of spin-up electrons since they can easily exit the channel whereas the spin-down electrons find it "harder" to tunnel into the ferromagnetic contact. This scenario changes in the antiparallel configuration where the injector preferentially injects spin-up electrons whereas the extractor preferentially extracts spin-down electrons. This bottleneck results in a spin-accumulation profile at which the spin polarization in the channel keeps its sign and its magnitude increases dramatically.

39.3.1 Lateral Geometry

Similar to the structure of conventional field effect transistors, micro- and nano- lithography allow us to fabricate lateral spin-valve devices [33,39,42]. In this view, we need to adjust the previous analysis in which we have used a simplified one-dimensional geometry to describe the general solution of the spin-diffusion equation (Equation 39.7). To analyze the spin-diffusive transport in realistic lateral devices, as shown in Figure 39.1c, we consider an effective one-dimensional transport equation derived from the three-dimensional spin-diffusion equation. This effective one-dimensional diffusion equation can accurately describe the spin diffusion in a layer of material of thickness h smaller than the spin-diffusion length L, and covered by contacts with junction conductances G smaller than the conductance σ/h of the underlying semiconductor layer. In a structure like the one shown in Figure 39.1c, we calculate the spin diffusion by introducing the layer-averaged electrochemical potential $\xi_s(x)$ in the semiconductor channel

$$\xi_s = \frac{1}{h}\int_0^h dy \mu_s(x,y). \quad (39.18)$$

In regions of the semiconductor channel that are not covered by ferromagnetic contacts, this layer-averaged electrochemical potential is identical to $\mu_s(x)$ and we can proceed with the analysis after Equation 39.8. However, under the contacts when we integrate out the y dependence from Equation 39.7 we obtain the approximate equation [49]

$$\frac{\partial^2 \xi_s}{\partial x^2} = \frac{\xi_s - \xi_{-s}}{2L^2} + \frac{2G_s}{\sigma h}(\xi_s - \mu_{fm}), \quad (39.19)$$

where μ_{fm} is the electrochemical potential in the ferromagnetic contact and the second term was derived by incorporating the boundary condition from Equation 39.11. As before, this equation is valid when assuming small electric fields and small spin accumulations (so that $\sigma_s \approx \sigma/2$). The general solution of this equation is given by

$$\xi_\pm(x) = -eV + (1 \pm \lambda)[pe^{\lambda_s x} + qe^{-\lambda_s x}] + (\lambda \mp 1)[re^{\lambda_c x} + se^{-\lambda_c x}], \quad (39.20)$$

where
- p, q, r, and s are constants to be determined by boundary conditions
- $-eV$ is the potential level in the ferromagnetic contact

The other parameters are

$$\lambda = \cot\left[\frac{1}{2}\tan^{-1}\beta\right], \quad (39.21)$$

$$\lambda^2_{(s,c)} = \frac{\left[\alpha + 1 \pm \sqrt{1+\beta^2}\right]}{(2L^2)},$$

$$(\alpha,\beta) = \frac{2L^2(G_+ \pm G_-)}{(\sigma h)}, \quad (39.22)$$

where the first of each pair of symbols (s,c) or (α,β) takes the upper sign. Consider the case that spin selectivity is robust so that α and β are comparable. If $\alpha \ll 1$, then $\lambda_c \ll 1/L$ and $\lambda_s \sim 1/L$. The s-mode is limited by the spin-diffusion constant, and it corresponds to spin accumulation ($\lambda \gg 1$ in this case). If $\alpha \gg 1$, then both eigenvalues are nearly independent of L, and neither of the eigenvectors is a pure spin mode $\lambda \simeq 1$; the inhomogeneity of injection dominates the spatial dependence. To get a unique solution of the problem, the electrochemical potentials and their slopes are matched between covered and uncovered sections of the semiconductor channel (Equations 39.8 and 39.20). At the outermost edges of the device the spin currents are zero ($\partial\xi_\pm/dx = 0$). For Fe/GaAs structures with $G_s \sim 10^3\ \Omega^{-1}\ \text{cm}^{-2}$, this approximate formalism gives results indistinguishable from exact numerical calculations of a two-dimensional spin-diffusion equation [49]. The MR of a lateral spin-valve device whose resistance is dominated by the tunneling contacts is slightly changed compared with Equation 39.14

$$r \simeq F^2\left(1 + \frac{\mathcal{V}}{\mathcal{A}} \cdot \frac{\sigma}{2L^2G}\right)^{-1}, \quad (39.23)$$

where
- $\mathcal{V}$ is the volume of the channel
- A is the area of the contacts, which are assumed to have the same total conductance and area for simplicity

The simple two terminal geometry of a spin valve captures the essence of a logic exclusive-OR (XOR) operation if we consider the spin-accumulation magnitude as the output and the magnetization configuration as the input. By defining binary "0" and "1" as the two possible magnetization directions of the easy axis of the contacts, then the possible parallel magnetization configuration inputs of a spin valve are "00" or "11" and the possible antiparallel magnetization configurations are "01" or "10." In the next section, we elaborate on logic and memory device schemes whose outputs display clearly distinct current signals. This is achieved by utilizing the spin-accumulation profiles rather than their slopes. We will show a universal, reprogrammable MLG that relies on addition of XOR operations.

39.4 Multi-Terminal Devices

In order to achieve a more efficient electrical expression of spin accumulation, one has to move beyond a passive two-terminal spin-valve device, and consider a multi-terminal system in which additional external stimuli can control the magnetoresistive effects. In the following text, we review several proposals of devices consisting of more than two ferromagnetic terminals connected to a semiconductor channel. Similar to the all-metallic nonlocal spin valve, the common feature of these devices is the use of a ferromagnetic contact to sense the spin accumulation in the nonmagnetic channel [40–42]. Figure 39.2a shows a device that consists of three ferromagnetic tunneling contacts

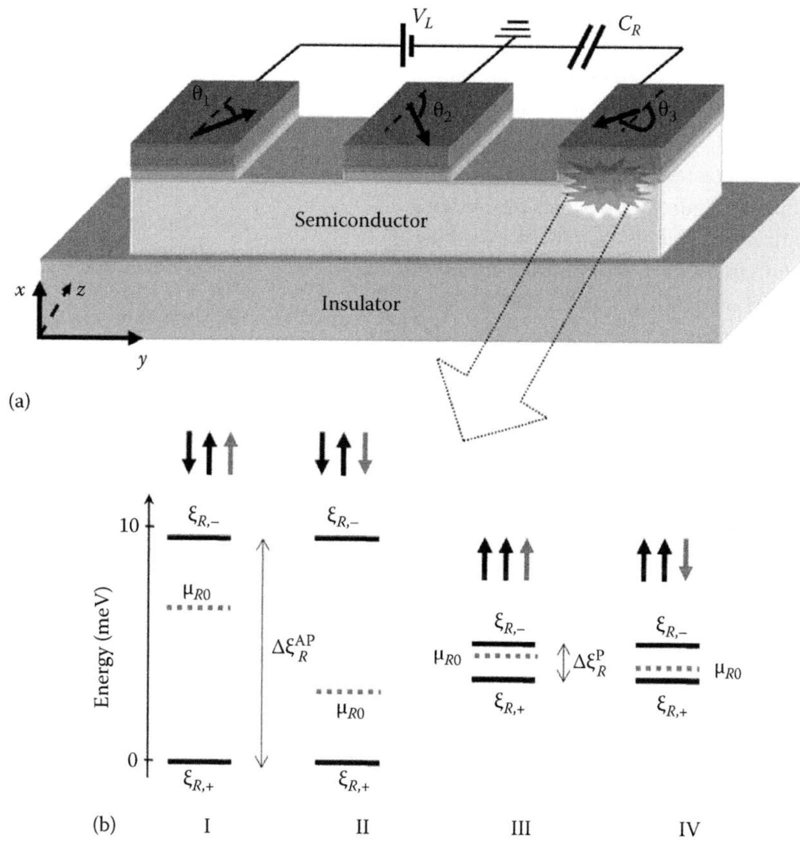

FIGURE 39.2 (a) A three-terminal nonlocal spin valve attached to a capacitor that guarantees a zero steady-state current across the right terminal. The magnetization directions are denoted by the angles θ_i (from the z-axis in the yz plane). The left and middle contacts are biased and their configuration sets the information. The floating contact (attached to the capacitor) is for reading the stored information (see text). (b) Positions of the electrochemical potentials for various magnetization configurations when $\theta_i=0$ or π. The solid lines denote the levels in the channel beneath the R contact. The dashed lines correspond to the level inside the R contact. There are four possible magnetization configurations if the middle contact is pinned in the up direction.

deposited on top of a paramagnetic semiconductor channel. The outer edges of the semiconductor channel are removed in order to confine the spin accumulation under and between the contacts [49]. In steady state, the current passes only between the left (L) and middle (M) contacts, driven by the voltage V_L. The right (R) contact is connected to a capacitor (C), which in steady state enforces zero charge current across this terminal. In Section 39.4 we will consider general noncollinear cases. Here, we focus on the collinear case in which all of the magnetization directions are either parallel or antiparallel to the $+z$ direction ($\theta_I=0$ or π). The steady-state value of the (single) electrochemical potential level inside the right contact is thus self-adjusted according to

$$J_R = 0 = G_{R,+}(\mu_{RO} - \xi_{R,+}) + G_{R,-}(\mu_{RO} - \xi_{R,-}), \tag{39.24}$$

and thus

$$\mu_{RO} = \xi_{R,0} + \frac{F_R}{2}\Delta\xi_R, \tag{39.25}$$

where

$\xi_{R,0}=(\xi_{R,+}+\xi_{R,-})/2$
$\Delta\xi_R=\xi_{R,+}-\xi_{R,-}$
$F_R=(G_{R,+}-G_{R,-})/(G_{R,+}+G_{R,-})$

$\xi_{R,\pm}$ denote the averaged spin-dependent electrochemical potentials in the semiconductor channel beneath the right contact. The positions of the energy levels of $\xi_{R,+}$ and $\xi_{R,-}$ are not affected by the parameters of the right contact (as long as it is a tunneling contact). Due to the zero steady-state current through the right contact, the creation of spin accumulation under this contact ($\Delta\xi_R$) is driven solely by the diffusion of spin-polarized electrons from the left part of the device where charge current flows between the spin selective middle and left contacts. Specifically, we can get only two possible sets of values for $\xi_{R,+}$ and $\xi_{R,-}$, which correspond to the spin-accumulation pattern of a parallel or an antiparallel magnetization configuration of the left and middle contacts. In the parallel configuration the spacing, $\Delta\xi_R^P$, is considerably smaller than in the antiparallel configuration, $\Delta\xi_R^{AP}$. In contrast to $\xi_{R,\pm}$, the value of the self-adjusting electrochemical potential inside the right contact, μ_{RO}, depends on the magnetization direction of the right contact. As can be understood from Equation 39.25, the value of μ_{R0} is closer to $\xi_{R,+}$ if $F_R > 0$, which is the case when the magnetization of the right contact is parallel to the magnetization of the middle contact. Similarly, μ_{R0} is closer to $\xi_{R,-}$ if $F_R > 0$, which is the case when the magnetization of the right contact is antiparallel to the magnetization of the middle contact. Figure 39.2b shows an energy diagram of the electrochemical potential levels for the possible magnetization collinear configurations with a pinned middle contact. Diagrams I and II (III and IV) denote an antiparallel (parallel) magnetization configurations of the middle and left contacts. The solid lines denote the energy levels of $\xi_{R,\pm}$ and the dashed lines denote the possible energy levels of μ_{R0}. Given that the middle contact is pinned, there are four possible magnetization configurations in total (denoted by three arrows above each diagram). The energy scales in this graph correspond to a semiconductor channel with a spin-diffusion length of $L = 1\,\mu m$ and conductivity $\sigma = 3\ (\Omega\ cm)^{-1}$. The three barriers are identical, $G = 10^4\,\Omega^{-1}\,cm^{-2}$ and $F = \pm 1/3$ where the sign depends on whether the magnetization direction is "1" or "0." The bias is $V_L = 0.1$ V and the lateral dimensions are $w_L = w_M = w_R = 400$ nm (contact width), $w_d = 200$ nm (contact separation), $h = 100$ nm (channel thickness). Finally, it is mentioned that the spin-dependent currents $j_{R,+}$ and $j_{R,-}$ are constantly flowing across the right junction even in steady-state conditions. Their sum (charge current) is zero and they flow in opposite directions since μ_{R0} lies in between $\xi_{R,+}$ and $\xi_{R,-}$: electrons with a spin direction that matches the accumulated spin in the channel flow into the ferromagnetic contact and the same amount of electrons with opposite spin direction flow into the semiconductor. Inside the ferromagnetic material, these spin currents decay to zero within a few nanometers due to the short spin-diffusion length of these materials [70]. That is, spin currents do not (and are not required to) reach the external capacitor in Figure 39.2a.

39.4.1 Time-Dependent Analysis

The device shown in Figure 39.2a can also be used for dynamical readout of magnetization alignment or to electrically measure the magnetization dynamics [21,71]. Consider, for example, a 2π rotation of the right magnet in the yz plane (starting from the $+z$ direction) where the magnetization direction of the middle and left contacts are fixed at parallel or antiparallel configurations ($\theta_1 = (0 \text{ or } \pi)$, $\theta_2 = 0$). After half of the rotation cycle, the spin-dependent conductances, $G_{R,+}$ and $G_{R,-}$, exchange roles and by the end of the cycle (2π), they recover their original values. As will be later explained, this rotation is described by $F_R(t) = F_0 \cos(\omega_r t)$. To explain the resulting transient current across the right magnet during the magnetization dynamics, we will first consider only the external capacitor and ignore the intrinsic Schottky barrier capacitance. If the external capacitor charging time is fast enough with respect to ω_r^{-1}, then the electrochemical potential of the right contact is instantaneously being readjusted during the rotation time. Its value guarantees that at steady state $J_R = 0$ and therefore μ_{R0} follows Equations 39.24 and 39.25. Since μ_{R0} adiabatically follows $F_R(t)$ the peak current value is proportional to

$$J_p \propto C\omega_r F_0(\Delta\xi_R) \ll GF_0(\Delta\xi_R),$$

where C is the external capacitance per unit area and the inequality corresponds to the fast charging time of a small capacitor $C/G \ll 1/\omega_r$. If, on the other hand, the external capacitance is very large then μ_{R0} is only slightly affected during the dynamics and the current across the right contact is simply

$$J_R(t) = G_{R,+}(t)(\mu_{R0} - \xi_{R,+}) + G_{R,-}(t)(\mu_{R0} - \xi_{R,-}),$$

where $G_{R,\pm}(t) = G_R(F_R(t) \pm 1)/2$. In this case, the peak of the transient current across the right contact is $GF_0(\Delta\xi_R)$ provided that initially ($t \leq 0$) the current is zero. Thus, if $C\omega_r/G \gg 1$ then the peak current is independent of C. In prospective integrated circuit applications, the size of C is limited by the available circuit area and in what follows we will work in the case that $C\omega_r/G \sim 1$. The figure of merit of this operation is that the amplitude of the resulting current oscillation, $J_R(t)$, is much smaller when the magnetization configuration of the left and middle contacts is parallel compared to the antiparallel configuration ($\Delta\xi_R^P \ll \Delta\xi_R^{AP}$). Using the small capacitance analysis, in the antiparallel configuration this perturbation refers to a swing of μ_{R0} in between its two possible values as shown by the dashed lines of diagram I and II in Figure 39.2b. In the parallel configuration, the swing is between the dashed lines of diagrams III and IV. If we rotate the magnetization of the left rather than of the right ferromagnetic contact then the measurement of the accompanying transient current ($J_R(t)$) also allows for electrical readout of magnetization dynamics at the left contact. Here the potential perturbation refers to the swing of $\xi_{R,\pm}$ between its two possible values as shown by the solid lines of diagrams I or II and those of diagrams III or IV.

In order to change the magnetization of any of the ferromagnetic contacts, the planar structure should be augmented by a set of current-carrying lines known from magnetic random access memories [72]. In these devices, there are wide metallic strips ("bit" and "word" lines) running above and below the magnets that are to be addressed. The upper and lower wires are at a 90° angle, so that every magnet is (when looking from above) located at an intersection of two wires. In order to switch its magnetization, the current pulses are passed through the appropriate bit and word lines. These currents generate transient local magnetic fields (through Ampere's law) that can switch the magnetization of the addressed terminal. The two lines are necessary for addressing purposes in a matrix consisting of lateral multi-terminal devices. Only the magnet located at the intersection of two lines is switched. The other magnets over (under) which the activated current-carrying line passes are unaffected, as the current magnitudes are such that a magnetic field from a single line is unable to switch the magnetization. Furthermore, using the two lines (giving two perpendicular magnetic fields that add up) is advantageous since the presence of magnetic field components noncollinear with the magnet's easy axis aids the fast switching [73].

In order to model the transient behavior, we add the time dependence to the formalism of lateral spin diffusion. The relevant characteristic timescales are the magnetization rotation time and the capacitor charging/discharging time, *C/G*, where we assume that the contacts are the most resistive elements in the circuit. As discussed before, *C/G* is chosen to be on the order of the magnetization dynamics whose fastest timescale is on the order of 0.1 ns [74,75]. For example, for $G \sim 10^4\ \Omega^{-1}\ \text{cm}^{-2}$ and a capacitor area of $1\ \mu\text{m}^2$ the external capacitance is $C = 40\,\text{fF}$. This conductance value maximizes the spin-accumulation density ($G \sim \sigma/L$) in a semiconductor channel whose conductivity is $1 \sim \Omega^{-1}\ \text{cm}^{-1}$ and whose spin-diffusion length is about $\sim 1\ \mu\text{m}$ (e.g., a nondegenerate n-type GaAs at room temperature). We mention, that if we were to use all-metallic versions of this device, then the typical spin accumulation in a paramagnetic metal corresponds to $\Delta\mu$ of less than $1\ \mu\text{eV}$ [41,42]. With such a small voltage swing on the capacitor for the transient current, $C(\partial\Delta u/\partial t)$, to be measurable one has to use a nearly macroscopic capacitor, which rules out an application in integrated circuits. Again, the small carrier density in a semiconductor (allowing for $\Delta\mu \sim 10\,\text{meV}$) is indispensable.

The time-dependent transport in the semiconductor channel beneath a ferromagnetic contact is described via the layer-averaged electrochemical potentials, $\xi_\pm$. By repeating the steady-state analysis that led to Equation 39.19 from Equation 39.7, we can write the time-dependent diffusion equation for the spin splitting of the electrochemical potentials $\Delta\xi = \xi_+ - \xi_-$,

$$\frac{\partial\Delta\xi}{\partial t} = D\frac{\partial^2\Delta\xi}{\partial x^2} + \frac{\beta(t)}{\tau_{sr}}(\mu^m - \xi) - \frac{\alpha}{2\tau_{sr}}\Delta\xi - \frac{\Delta\xi}{\tau_s}. \tag{39.26}$$

The α and β are dimensionless parameters given by Equation 39.22. The dynamics of magnetization is parameterized by $\beta(t) \sim \Delta G(t)$, which characterizes the contact polarization only along the *z*-axis. If we deal with a coherent precession of magnetization, then this is an approximation. In general, one should take into account the noncollinearity of spins and the magnets during the rotation time. However, for tunneling barriers the nontrivial effects of this noncollinearity are expected to be small and the only thing that matters is the average polarization along the *z* direction [76,77]. Thus, we can model the influence of the contact with magnetization making an angle $\theta(t)$ with the *z*-axis by assuming that $\beta(t) \sim \Delta G \cos\theta(t)$. On the other hand, if the magnetization reversal is incoherent (e.g., proceeding by nucleation of domains with opposite magnetization [78]), the parameter $\beta(t)$ describes an area average of spin selectivity of magnetically inhomogeneous contact, and it is naturally proportional to the *z* component of the contact's magnetization.

The dielectric relaxation time, $\tau_d = \sigma/\varepsilon_{sc}$ is much faster than the timescale of magnetization dynamics and spin diffusion (about 100 fs for nondegenerate semiconductor with $n_0 = 10^{16}\ \text{cm}^{-3}$). Thus, quasi-neutrality in the channel is assumed at all times, $\delta n_+(t) + \delta n_-(t) = 0$. In the linear regime under consideration (when $\Delta\xi < k_BT$) the average electrochemical potential $\xi = (\xi_+ + \xi_-)/2$ is equal to $-e\phi$. At every moment of time, ξ fulfills the Laplace equation with boundary conditions given by currents at the interfaces. In the time-dependent case, these currents include also a displacement current connected with charging of the barrier capacitance C_B. A Schottky barrier is a dipole layer, and its capacitance can have a strong effect on dynamics of currents on timescales of interest here. Taking this displacement current into account, the boundary condition for the spin currents are

$$j_s = \frac{G_s}{e}(\mu_{fm}(t) - \xi_s(t)) + \frac{c_B}{2e}\frac{\partial}{\partial t}(\mu_{fm}(t) - \xi(t)), \tag{39.27}$$

where c_B is the barrier capacitance per unit area and $\xi = (\xi_+ + \xi_-)/2$. The first term on the right-hand side is the time-dependent tunneling current (Equation 39.3), and the second term represents the carriers that flow toward the barrier, but do not tunnel through it. Instead, they stay in the semiconductor close to the barrier, making the depletion region slightly thinner or wider. The charge involved in this process is negligible compared to the charge already swept out from the semiconductor, so we can keep c_B constant. For small spin splitting (so that the conductivities $\sigma_+ \simeq \sigma_-$) the same amount of carriers of each spin is going to be brought from the channel into the barrier, and the displacement current is the same for each spin in Equation 39.27. For layer-averaged ξ we get then

$$\frac{\partial^2\xi}{\partial x^2} = -\frac{\alpha}{2L^2}(\mu_{fm} - \xi) + \frac{\beta(t)}{4L^2}\Delta\xi - \frac{c_B}{\sigma h}\frac{\partial}{\partial t}(\mu_{fm} - \xi). \tag{39.28}$$

The description of $\Delta\xi$ and ξ in semiconductor regions that are not covered by ferromagnetic contacts follows Equations 39.26 and 39.28 with $\alpha = \beta = 0$. The total solution is achieved by the conservation of carriers and currents throughout the semiconductor channel. Mathematically, this requires a continuity of $\Delta\xi$ and ξ and their first *x*-derivatives at all times at the boundaries between covered and uncovered regions. The magnetization dynamics of the *i*th contact translates into time-dependence of β_i, driving the spin diffusion in Equation 39.26 and the semiconductor electric potential in Equation 39.28. From ξ_s we calculate the current $I_R(t)$ charging the capacitor *C*. The electrochemical potential of the *R* terminal $\mu_R = -eV_R$ changes according to $dV_R/dt = I_R/C$. Examples of calculations for two possible modes of operation (sensing the *L* dynamics and reading out the *L/M* alignment) are shown in Figure 39.3.

39.4.2 Reprogrammable Magneto-Logic Gate

The same physical principle of operation can be harnessed to achieve a more complicated functionality. In Figure 39.4a we

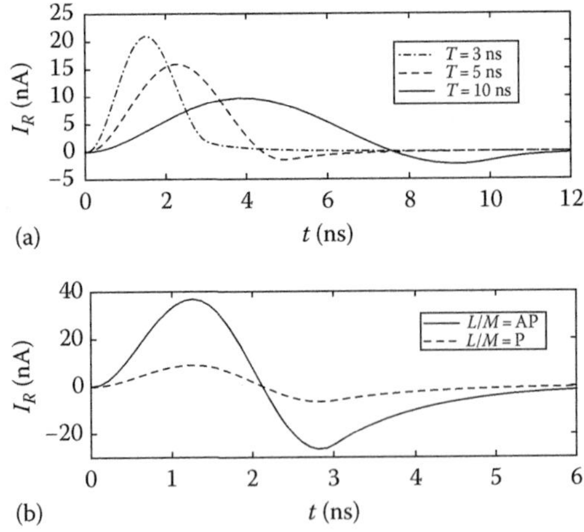

FIGURE 39.3 (a) R current signal for reversal of L magnetization occurring on timescale of 3, 5, and 10 ns starting from AP alignment of L relative to M magnet. (b) R current signal for 2π rotation of R magnet for P and AP alignments of L and M magnets. The period of rotation is 3 ns. The conductance of the barriers $G = 10^4\ \Omega^1\ \text{cm}^{-2}$. The area of a junction is $1\ \mu\text{m}^2$, and the barrier thickness is taken to be 10 nm, resulting in junction resistance $R_B = 10\ \text{k}\Omega$ and capacitance $C_B = 10$ fF. The external capacitance is $C = 40$ fF. The channel is GaAs at room temperature, with carrier density $n_0 = 10^{16} = 10^{16}\ \text{cm}^{-3}$.

present a scheme of a five-terminal system [12] in which the electric sensing of spin accumulation is used to perform a logic operation, that is, two bits of input are converted into a binary output signal. This is a reprogrammable magneto-logic gate (MLG). As mentioned in the introduction, spintronics logic gates have been proposed in purely metallic systems, but this is the first proposal that employs semiconductors as active elements of the system.

The system presented in Figure 39.4a works in the following way. The charge currents are flowing between two pairs of terminals (X and A, Y and B), between which the bias V_{dd} is applied. Depending on the alignment of these pairs of magnets, different patterns of spin accumulation are created in the channel: if both X/A and Y/B are AP, the spin accumulation underneath M is large; if only one pair of contacts is AP, the spin accumulation is approximately two times smaller; and if both pairs are P, there is very small $\Delta\xi$ beneath the M terminal. The M contact is used to directly express the differences in the average spin accumulation beneath it.

The logic inputs are encoded by magnetization directions of A, B, X, and Y terminals. We will concentrate on the case in which A and B magnetization directions are preset, defining the logic function of the gate. This reprogrammability is an important feature of magnetization-based logic [9,12]. X and Y are then the logic operands, and the output is generated when the M magnet is rotated by 2π, triggering a transient $I_M(t)$ current of amplitude proportional to the spin accumulation in the middle of the channel. Let us focus on the example of the NAND gate, as any other logic function can be realized by using a finite number of such gates. For the NAND operation, A and B magnets are set parallel to each other in a direction defining the logic "1." The amplitude of the $I_M(t)$ oscillation is two times larger for X = "0" and Y = "0" compared to the case when one of them is "0" and the other is "1," and for "11" case the current is negligible. This is shown in Figure 39.4b. The transient current can be captured by an external electronic circuit, and then used to control a suitable

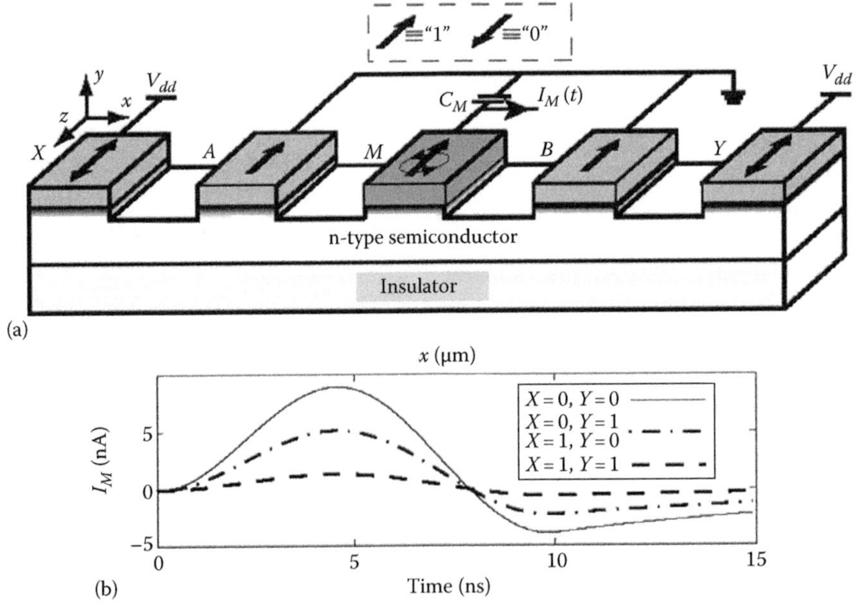

FIGURE 39.4 (a) A five-terminal MLG. The logic inputs "0" and "1" are encoded by magnetization direction of the A, B, X, Y terminals (see text for details). As shown here, the gate is set to work as a NAND operation between X and Y (A and B are fixed to "1" values). In the read-out phase the magnetization of the middle (M) terminal is rotated by 2π, or the back-gate voltage V_G is increased. (b) The $I_M(t)$ transient current triggered by rotation of the M magnetization by 2π. The solid line corresponds to the logic output "1." (From Dery, H. et al., *Nature*, 447, 573, 2007. With permission.)

write operation applied to the magnetic contact of another gate. For technical details of this electrical conversion circuit, see the supplementary material of Dery et al. [12]. In the next section, we will consider this conversion circuitry as a black box and will present a common logic application that may benefit by using these devices beyond their trivial nonvolatility merit.

39.4.3 Content-Addressable Memories

There is a significant gap between the number of devices (and thus parallelism) available in each new generation of technology and our ability to express parallelism in programs. Despite improvements in the software and hardware support, parallel programming remains challenging for the computing community at large. Clearly, in addition to traditional general-purpose parallel programming, we need other mechanisms to help bridge the gap. One important alternative mechanism is a content-addressable memory (CAM) [79]. A CAM is an example of an associative memory in which the search data (the key) is simultaneously compared with an array of data stored in memory cells to find matches. For example, in a CAM we would ask for the list of addresses at which the word "John" is stored. On the other hand, in the standard computer memory (Random Access Memory) we wish to know what is the information that is stored at a specified address (one piece of data at a time). Of these two cases, only the associative search provides a very significant amount of *data parallelism* and as such, CAMs found many uses in areas such as routing, searching, network security, and in general-purpose microprocessors (e.g., table lookaside buffers or load-store queues). In practical circuit designs, however, broadcasting the key presents technical challenges. Long buses with high capacitive loads are needed to simultaneously deliver the key to a large number of memory cells. This results in long delays and high power consumption and partly contributes to the size limitation in many applications of CAMs. An example of the importance of CAMs and their limitations is evident in the *Load-Store Queue*, which is a crucial microarchitectural structure in modern high-end microprocessors. It allows the processor to speculatively execute instructions out of original, program-specified order for high performance and yet guarantee their sequential semantics. The larger the queue, the more instruction-level parallelism the processor can exploit. The state-of-the-art designs use two queues to track loads and stores separately and yet can only allow about 20–30 entries in each queue [80,81]. Compared to the size of other microarchitectural structures (e.g., register files), these are small sizes and even these are sometimes obtained via nontrivial architectural speculations so as not to violate timing budget [82]. Scaling up these queues without affecting cycle time is a significant challenge in the design of microprocessors.

Using the MLGs, spin-based circuits have a unique advantage to support a large fan-out and make large CAM structures straightforward to implement and scale. The key factor is that transmitted information is determined by the direction of a current that magnetizes the ferromagnetic contacts. The amount of current needed for the magnetization is smaller than that needed to charge the distributed intrinsic capacitance of many memory cells along very long buses (together with parasitic capacitances). Figure 39.5 shows a bit line circuitry of a spin-based CAM. The current direction in the bit-line encodes the search information bit in one of the operand contacts of each of the MLGs under (or above) the line. The adjacent contact holds the encoded bit-information of the gate. If the magnetization directions are dissimilar/similar, then we have a mismatch/match and the spin accumulation in the semiconductor channel is high/low. As such, a XOR operation between the search and information bits

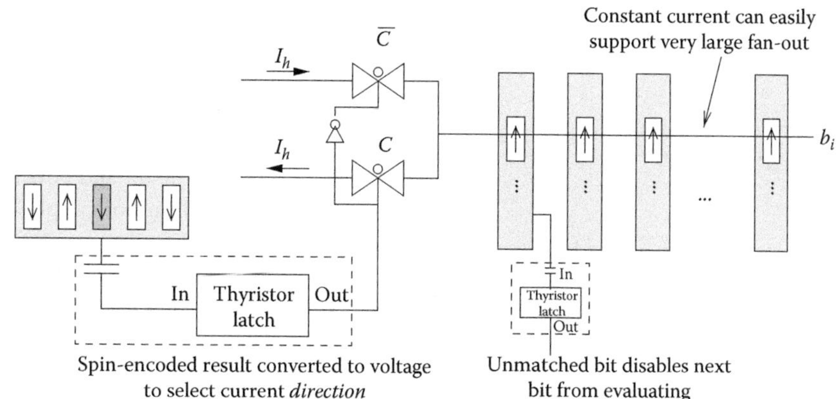

FIGURE 39.5 Schematics of an associative search in a proposed spin-based (CAM). The thyristor latch translates the current signal out of a five-terminal, MLG into a digital voltage level (left side). This in turn controls the direction of the current in the search bit-line (denoted by b_i) according to the pass-gates and the NOT gate in between. The search is performed simultaneously in all the other five-terminal, MLGs under/above the bit line without the need to charge/discharge the intrinsic capacitance of any of these gates. Once a match is identified between the searched bit and the encoded information of a given gate (XOR operation), then the accompanied thyristor latch will enable the search in the MLG at the next bit line (a similar structure to the b_i line). This procedure allows the search to be pipelined and conserves power (the search is disabled at the $i+1$ bit if the search of the i-bit failed in a given word). The details of the thyristor latch are explained in Dery et al. [12] and its supplementary material.

can tell the matching result and this can be read by perturbing the magnetization of an additional contact next to the operand contacts [12]. In addition, the half-selection problem of magnetic random access memories [72], which inhibits their downscaling is irrelevant in the magnetic CAM since the search is made simultaneously in all of memory cells (full selection of all bits).

The availability of energy-efficient *scalable* CAM designs will have a tremendous impact on general-purpose as well as application-specific processors. Possibly the most immediate and far-reaching impact is the capability to build on-chip caches with extremely high associativity. Highly associative caches can significantly cut down cache miss rate and thus reduce the need for slow, energy-intensive off-chip memory accesses.

39.4.4 Non-Collinear Configurations

To this point, we have treated the spin polarization in the channel, $p=(n_+-n_-)/n_0$, as a scalar, which implicitly relies on the assumption that the net spin has a fixed direction throughout the semiconductor channel. This description is valid in collinear systems at which the magnetization directions in all of the ferromagnetic elements share a common (easy) axis. In a more general noncollinear configuration, the boundary conditions impose a change in the direction of the spin polarization during the transport in the channel [76]. For a general coordinate system in spin space, the spin-dependent electron density is described by a 2×2 matrix

$$\hat{\mathbf{n}}(\mathbf{r},t)=\frac{n_0}{2}\left(\hat{I}+\mathbf{p}(\mathbf{r},t)\cdot\hat{\sigma}\right), \tag{39.29}$$

where

$\hat{\sigma}$ is the Pauli matrix vector

$\mathbf{p}$ has the magnitude p along the + spin direction

Using this notation and repeating the analysis, one can derive the components of the spin-diffusion equation [59]

$$\frac{1}{D}\frac{\partial p_i}{\partial t}+\frac{p_i}{\ell_{sf}^2}=\sum_j\left(\frac{\partial^2 p_i}{\partial x_j^2}+\frac{E_j}{V_T}\frac{\partial p_i}{\partial x_j}\right), \tag{39.30}$$

where i (j) enumerate the x, y, and z coordinates in spin (real) space. The components of the charge current density (vector) and of the spin current density (second-rank tensor) are

$$J_j=\sigma_0 E_j, \tag{39.31}$$

$$\mathcal{J}_{i,j}=\sigma_0\left(V_T\frac{\partial p_i}{\partial x_j}+E_j p_i\right). \tag{39.32}$$

The description of transport is complete when the spin-polarized currents across the SC/FM junctions are expressed in terms of the spin polarization vector at the semiconductor side of the junction, $\mathbf{p}(\mathbf{r}_j,t)$. We follow the notation by Brataas et al. [76,77], and write the population distribution matrices on both sides of the junction

$$\hat{f}_{sc}(\varepsilon)=\frac{1}{2}e^{(\mu_0-\varepsilon)/k_BT}\left(\hat{I}+\mathbf{p}\cdot\hat{\sigma}\right), \tag{39.33}$$

$$\hat{f}_{fm}=\frac{1}{2}\left(1+e^{(\varepsilon-\mu_0+qV)/k_BT}\right)^{-1}\hat{I}, \tag{39.34}$$

where ε denotes the energy. As mentioned before, due to the conductivity mismatch the ferromagnetic side is a reservoir with a constant chemical potential, μ_0-qV, where $\mu_0=\mu_c-q\phi(\mathbf{r}_j,t)$ is evaluated at the semiconductor side of the junction and V is the voltage drop across the SC/FM junction. The applied bias voltage is related to the electrical field via $\mathbf{E}=-\nabla\phi$. The population distribution matrices assume a nondegenerate semiconductor for which the spin-dependent electrochemical potentials, $\mu_\pm$, has the following relation with the net spin polarization $p=|\mathbf{p}|$,

$$\frac{\mu_\pm(\mathbf{r},t)}{k_BT}=\frac{\mu_0}{k_BT}+\ln\left(1\pm p(\mathbf{r},t)\right). \tag{39.35}$$

For compact notation, the boundary conditions of each SC/FM junction are written in a spin coordinate system at which the z-axis is collinear with the magnetization direction of the corresponding ferromagnetic contact. With this simplification, the reflection matrices are diagonal,

$$\hat{r}_{sc}(\varepsilon_\perp,V)=\begin{pmatrix}r_\uparrow & 0\\ 0 & r_\downarrow\end{pmatrix},\quad \hat{r}_{fm}(\varepsilon_\perp,V)=\begin{pmatrix}\tilde{r}_\uparrow & 0\\ 0 & \tilde{r}_\downarrow\end{pmatrix}. \tag{39.36}$$

The reflection coefficients in the left (right) matrix are of electrons from the semiconductor (ferromagnetic) side of the junction. For a given material system, these coefficients vary with the voltage drop and with the longitudinal energy, $\varepsilon_\perp$, which denotes the impinging energy of electrons due to their motion toward the SC/FM interface. The up and down arrows denote, respectively, the majority and minority spin directions in the ferromagnetic contact where we have set the $+z$ direction parallel to the majority direction. By using the Landauer–Büttiker formalism, the tunneling current across the SC/FM junction is given by [14,65,76,77]

$$\begin{aligned}\hat{J}(V)&=\int_0^\infty d\varepsilon\,\hat{J}(\varepsilon)\\&=\frac{q}{\hbar}\int_0^\infty d\varepsilon\int_0^{k_\varepsilon}\frac{d^2k_\parallel}{(2\pi)^2}\left\{\left[\hat{f}_{fm}-\hat{r}_{fm}\hat{f}_{fm}\hat{r}_{fm}^\dagger\right]-\left[\hat{f}_{sc}-\hat{r}_{sc}\hat{f}_{sc}\hat{r}_{sc}^\dagger\right]\right\}.\end{aligned} \tag{39.37}$$

The first (second) term in square brackets is related to the transmitted current from the ferromagnet (semiconductor) due to electrons whose total and longitudinal energies are ε and $\varepsilon_\perp$, respectively. The zero energy refers to the bottom of the semiconductor conduction band. The inner integration is carried over transverse wave vectors due to a motion in parallel to the SC/FM interface and its upper integration limit, k_ε, denotes the wave vector amplitude of an electron with energy ε. In the chosen spin coordinate system, the transmitted spin current from the ferromagnetic side is nonzero only along the z direction. Its spin-up and spin-down components are proportional, respectively, to $(1-|r_\downarrow|^2)$ and $(1-|r_\uparrow|^2)$, where we have rendered the fact that $|r_{\uparrow(\downarrow)}|^2=|\tilde{r}_{\uparrow(\downarrow)}|^2$. The transmitted current from the semiconductor side, on the other hand, includes off-diagonal mixed terms that are proportional to $r_\uparrow r_\downarrow^*$. This is the case when **p** and **z** are neither parallel nor antiparallel due to the flow of electrons from/into ferromagnetic contacts, which has noncollinear magnetization directions and that are located within about a spin-diffusion length.

The boundary conditions across a SC/FM junction were derived using the assumption that the spin z-axis is collinear with the majority spin direction in the FM. However, in order to consider all (noncollinear) ferromagnetic terminals and the semiconductor channel as one system, one should use a single spin reference coordinate system. Specifically, we transform the general expression in Equation 39.37 into this new "contact-independent" coordinate system. We use $\tilde{x}$, $\tilde{y}$ and $\tilde{z}$ to represent the contact-dependent coordinates (for which the reflection matrices are diagonal) and we reserve x, y, and z for coordinates in the contact-independent system. The angle of the majority spin direction in the ith contact with respect to $+z$ is denoted by θ_i (in the yz plane). The relation between the spin current densities in these two frames is given by

$$\begin{pmatrix} \mathcal{J}_{x,\alpha} \\ \mathcal{J}_{y,\alpha} \\ \mathcal{J}_{z,\alpha} \end{pmatrix} = \begin{pmatrix} 1 & 0 & 0 \\ 0 & \cos\theta_i & \sin\theta_i \\ 0 & -\sin\theta_i & \cos\theta_i \end{pmatrix} \begin{pmatrix} \mathcal{J}_{\tilde{x},\alpha} \\ \mathcal{J}_{\tilde{y},\alpha} \\ \mathcal{J}_{\tilde{z},\alpha} \end{pmatrix}. \quad (39.38)$$

The charge current density does not depend on the spin space coordinate. Similarly, the spin polarization vector transformation follows

$$\begin{pmatrix} p_{\tilde{x}} \\ p_{\tilde{y}} \\ p_{\tilde{z}} \end{pmatrix} = \begin{pmatrix} 1 & 0 & 0 \\ 0 & \cos\theta_i & -\sin\theta_i \\ 0 & \sin\theta_i & \cos\theta_i \end{pmatrix} \begin{pmatrix} p_x \\ p_y \\ p_z \end{pmatrix}. \quad (39.39)$$

Figure 39.6 shows the transient currents through the right contact of Figure 39.2a during its magnetization dynamics (2π rotation within 3 ns). The four curves denote various fixed magnetization directions of the left contact, $\theta_1=(0, \pi/2, \pi$ or $3\pi/2)$, whereas the middle contact is pinned ($\theta_2=0$). We have included interface capacitance $c_b=1\,\mu\text{F/cm}^2$ for the total charge current. These results also include the escape current mechanism due to localization of electrons at the interface region (the details of this effect are explained in Song and Dery [59]). The external capacitor is $C=4$ fF and the applied bias is 0.1 V. The Schottky barrier is assumed to be highly doped (2×10^{19} cm^{-3}) so that the total barrier conductance is about $G\approx2\times10^4\ \Omega^{-1}\ \text{cm}^{-2}$. By detecting the signal shape, one can use such nonvolatile magnetic transistors to hold more information than in binary devices.

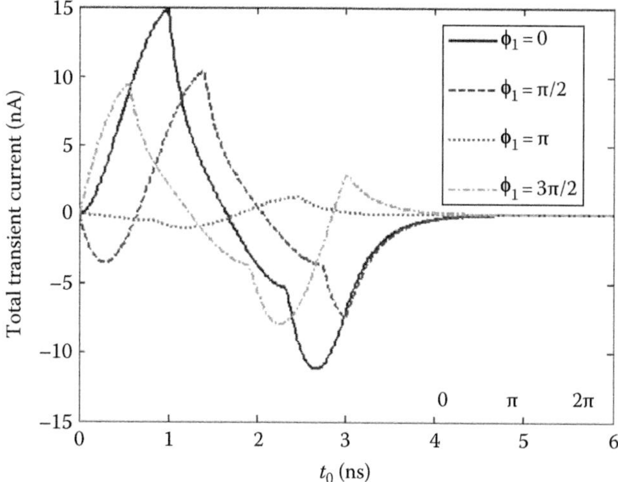

FIGURE 39.6 Transient currents of the right semifloating contact by rotating its magnetization a single clockwise rotation from $+z$ within 3 ns. The curves denote different fixed magnetization directions of the left contact whereas the magnetization direction of the middle contact is pinned ($\theta_2=0$). (From Song Y. and Dery H., *Phys. Rev. B*, 81, 045321, 2010. With permission.)

39.5 Summary

We have reviewed the physics of spin injection and studied the presence of spin accumulation in the semiconductor at small bias conditions. We have described how spin-dependent properties could be used to electrically sense the spin accumulation in a semiconductor. The multi-terminal logic and memory devices that we have described rely only on spin selectivity of the junctions and the presence of spin accumulation. These devices can work at room temperature and may improve common logic circuitry such as CAMs. Recent experimental advances in graphene [83] have also motivated us to reveal additional advantages of graphene/ferromagnet magneto-logic gates [84]. In addition, by exploiting noncollinear magnetization configuration schemes we have showed the increased storage capabilities of these devices.

Acknowledgments

I would like to thank Łukasz Cywiński, Michael Huang, and Yang Song for their active contribution in writing this chapter. I would also like to thank Lu Sham, Parin Dalal, Berry Jonker, Ian Appelbaum, Paul Crowell for many helpful discussions. This work was supported by AFOSR DCT program (contract number FA9550-09-1-0493) and by NSF contract No. ECCS-0824075.

References

1. M. N. Baibich, J. M. Broto, A. Fert et al., Giant magnetoresistance of (001)Fe/(001)Cr magnetic superlattices, *Phys. Rev. Lett.* **61**, 2472–2475 (1988).
2. G. Binasch, P. Grünberg, F. Saurenbach, and W. Zinn, Enhanced magnetoresistance in layered magnetic structures with antiferromagnetic interlayer exchange, *Phys. Rev. B* **39**, R4828–R4830 (1989).
3. J. S. Moodera, L. R. Kinder, T. M. Wong, and R. Meservey, Large magnetoresistance at room temperature in ferromagnetic thin film tunnel junctions, *Phys. Rev. Lett.* **74**, 3273–3276 (1995).
4. R. P. Cowburn and M. E. Welland, Room temperature magnetic quantum cellular automata, *Science* **287**, 1466–1468 (2000).
5. D. A. Allwood, G. Xiong, C. C. Faulkner, D. Atkinson, D. Petit, and R. P. Cowburn, Magnetic domain-wall logic, *Science* **309**, 1688–1692 (2005).
6. A. Imre, G. Csaba, L. Ji, A. Orlov, G. H. Bernstein, and W. Porod, Majority logic gate for magnetic quantum-dot cellular automata, *Science* **311**, 205–208 (2006).
7. A. T. Hanbicki, R. Magno, S.-F. Cheng, Y. D. Park, A. S. Bracker, and B. T. Jonker, Nonvolatile reprogrammable logic elements using hybrid resonant tunneling diodegiant magnetoresistance circuits, *Appl. Phys. Lett.* **79**, 1190–1192 (2001).
8. R. Richter, L. Bär, J. Wecker, and G. Reiss, Nonvolatile field programmable spin-logic for reconfigurable computing, *Appl. Phys. Lett.* **80**, 1291–1293 (2002).
9. A. Ney, C. Pampuch, R. Koch, and K. H. Ploog, Programmable computing with a single magnetoresistive element, *Nature* **425**, 485–487 (2003).
10. Taiwan Semiconductors 2009. http://www.tsmc.com/download/english/a05_literature/02_40_and_45_Nanameter_Process_Technology.pdf
11. IBM 2009. http://www-01.ibm.com/chips/techlib/techlib.nsf/productfamilies/Custom_Chip_Solution:_ASIC
12. H. Dery, P. Dalal, Ł. Cywiński, and L. J. Sham, Spin-based logic in semiconductors for reconfigurable large-scale circuits, *Nature* **447**, 573–576 (2007).
13. S. Datta and B. Das, Electronic analog of the electro-optic modulator, *Appl. Phys. Lett.* **56**, 665–667 (1990).
14. C. Ciuti, J. P. McGuire, and L. J. Sham, Spin-dependent properties of a two-dimensional electron gas with ferromagnetic gates, *Appl. Phys. Lett.* **81**, 4781–4783 (2002).
15. J. Schliemann, J. C. Egues, and D. Loss, Nonballistic spin-field-effect transistor, *Phys. Rev. Lett.* **90**, 146801 (2003).
16. I. Žutić, J. Fabian, and S. Das Sarma, Spintronics: Fundamentals and applications, *Rev. Mod. Phys.* **76**, 323 (2004).
17. J. Fabian, I. Žutić, and S. Das Sarma, Magnetic bipolar transistor, *Appl. Phys. Lett.* **84**, 85–87 (2004).
18. J. Fabian and I. Žutić, Spin-polarized current amplification and spin injection in magnetic bipolar transistors, *Phys. Rev. B* **69**, 115314 (2004).
19. J. P. McGuire, C. Ciuti, and L. J. Sham, Theory of spin transport induced by ferromagnetic proximity on a two-dimensional electron gas, *Phys. Rev. B* **69**, 115339 (2004).
20. V. V. Osipov and A. M. Bratkovsky, Efficient nonlinear room-temperature spin injection from ferromagnets into semiconductors through a modified Schottky barrier, *Phys. Rev. B* **70**, 205312, (2004).
21. Ł. Cywiński, H. Dery, and L. J. Sham, Electric readout of magnetization dynamics in a ferromagnet-semiconductor system, *Appl. Phys. Lett.* **89**, 042105 (2006).
22. I. Žutić, J. Fabian, and S. C. Erwin, Bipolar spintronics: From spin injection to spin-controlled logic, *J. Phys.: Condens. Mat.* **19**, 165219 (2007).
23. Y. Ohno, D. K. Yound, B. Beschoten, F. Matsukura, H. Ohno, and D. D. Awschalom, Electrical spin injection in a ferromagnetic semiconductor heterostructure, *Nature* **402**, 790–792 (1999).
24. T. Dietl, H. Ohno, F. Matsukura, J. Cibert, and D. Ferrand, Zener model description of ferromagnetism in zinc-blende magnetic semiconductors, *Science* **287**, 1019–1022 (2000).
25. S. J. Potashnik, K. C. Ku, S. H. Chun, J. J. Berry, N. Samarth, and P. Schiffer, Effects of annealing time on defect-controlled ferromagnetism in $Ga_{1-x}Mn_xAs$, *Appl. Phys. Lett.* **79**, 1495–1497 (2002).
26. Y. K. Kato, R. C. Myers, A. C. Gossard, and D. D. Awschalom, Observation of the spin Hall effect in semiconductors, *Science* **306**, 1910–1913 (2004).
27. J. Wunderlich, B. Kaestner, J. Sinova, and T. Jungwirth, Experimental observation of the spin-Hall effect in a two-dimensional spin-orbit coupled semiconductor system, *Phys. Rev. Lett.* **94**, 047204 (2005).
28. H. J. Zhu, M. Ramsteiner, H. Kostial, M. Wassermeier, H.-P. Schönherr, and K. H. Ploog, Room-temperature spin injection from Fe into GaAs, *Phys. Rev. Lett.* **87**, 016601 (2001).
29. A. T. Hanbicki, B. T. Jonker, G. Itskos, G. Kioseoglou, and A. Petrou, Efficient electrical spin injection from a magnetic metal/tunnel barrier contact into a semiconductor, *Appl. Phys. Lett.* **80**, 1240–1242 (2002).
30. A. T. Hanbicki, O. M. J. van't Erve, R. Magno et al., Analysis of the transport process providing spin injection through an Fe/AlGaAs Schottky barrier, *Appl. Phys. Lett.* **82**, 4092–4094 (2003).
31. B. T. Jonker, Progress toward electrical injection of spin-polarized electrons into semiconductors, *Proc. IEEE* **91**, 727–740 (2003).
32. B. T. Jonker, G. Kioseoglou, A. T. Hanbicki, C. H. Li, and P. E. Thompson, Electrical spin-injection into silicon from a ferromagnetic metal/tunnel barrier contact, *Nat. Phys.* **3**, 542–546 (2007).
33. S. A. Crooker, M. Furis, X. Lou et al., Imaging spin transport in lateral ferromagnet/semiconductor structures, *Science* **309**, 2191–2195 (2005).
34. X. Jiang, R. Wang, R. M. Shelby et al., Highly spin-polarized room-temperature tunnel injector for semiconductor

spintronics using MgO(100), *Phys. Rev. Lett.* **94**, 056601 (2005).
35. B.-C. Min, K. Motohashi, C. Lodder, and R. Jansen, Tunable spin-tunnel contacts to silicon using low-work-function ferromagnets, *Nat. Mater.* **5**, 817–822 (2006).
36. T. J. Zega, A. T. Hanbicki, S. C. Erwin et al., Determination of interface atomic structure and its impact on spin transport using Z-contrast microscopy and density-functional theory, *Phys. Rev. Lett.* **96**, 196101 (2006).
37. I. Appelbaum, B. Huang, and D. J. Monsma, Electronic measurement and control of spin transport in silicon, *Nature* **447**, 295–298 (2007).
38. X. Lou, C. Adelmann, S. A. Crooker, et al., Electrical detection of spin transport in lateral ferromagnet–semiconductor devices, *Nat. Phys.* **3**, 197–202 (2007).
39. D. Saha, M. Holub, and P. Bhattacharya, Amplification of spin-current polarization, *Appl. Phys. Lett.* **91**, 072513 (2007).
40. M. Johnson, Bipolar spin switch, *Science* **260**, 320 (1993).
41. F. J. Jedema, A. T. Filip, and B. J. van Wees, Electrical spin injection and accumulation at room temperature in an all-metal mesoscopic spin valve, *Nature* **410**, 345–348 (2001).
42. Y. Ji, A. Hoffman, J. E. Pearson, and S. D. Bader, Enhanced spin injection polarization in Co/Cu/Co nonlocal lateral spin valves, *Appl. Phys. Lett.* **88**, 052509, (2006).
43. G. Schmidt, D. Ferrand, L. W. Molenkamp, A. T. Filip, and B. J. van Wees, Fundamental obstacle for electrical spin injection from a ferromagnetic metal into a diffusive semiconductor, *Phys. Rev. B* **62**, R4790–R4793 (2000).
44. E. I. Rashba, Theory of electrical spin injection: Tunnel contacts as a solution of the conductivity mismatch problem, *Phys. Rev. B* **62**, R16267–16270 (2000).
45. A. Fert and A. Jaffrès, Conditions for efficient spin injection from a ferromagnetic metal into a semiconductor, *Phys. Rev. B* **64**, 184420 (2001).
46. M. Johnson and R. H. Silsbee, Thermodynamic analysis of interfacial transport and of the thermomagnetoelectric system, *Phys. Rev. B* **35**, 4959–4972 (1987).
47. R. Stratton, Tunneling in schottky barrier rectifiers, in *Tunneling Phenomena in Solids*, E. Burstein and S. Lundqvist (Eds.), Plenum Press, New York, pp. 105–126, 1969.
48. S. M. Sze, *Physics of Semiconductor Devices*, John Wiley, New York, 1981.
49. H. Dery, Ł. Cywiński, and L. J. Sham, Lateral diffusive spin transport in layered structures, *Phys. Rev. B* **73**, 041306 (2006).
50. A. Fert, J.-M. George, H. Jaffrès, and R. Mattana, Semiconductors between spin-polarized sources and drains, *IEEE Trans. Electron. Dev.* **54**, 921–932 (2007).
51. C. Adelmann, J. Q. Xie, C. J. Palmstrøm, J. Strand, J. Wang, and P. A. Crowell, Effects of doping profile and post-growth annealing on spin injection from Fe into (Al,Ga)As heterostructures, *J. Vac. Sci. Technol. B* **23**,1747–1751 (2005).
52. G. E. Pikus and A. N. Titkov, Spin relaxation, in *Optical Orientation*, F. Meier and B. P. Zakharchenya (Eds.), North Holland, New York, pp. 73–131,1984.
53. M. Zachau, F. Koch, K. H. Ploog, P. Roentgen, and H. Beneking, Schottky-barrier tunneling spectroscopy for the electronic subbands of a δ-doping layer, *Solid State Commun.* **59**, 591–594 (1986).
54. J. M. Geraldo et al., The effect of the planar doping on the electrical transport properties at the Al:n-GaAs(100) interface: Ultrahigh effective doping, *J. Appl. Phys.* **73**, 820–824 (1993).
55. V. I. Shashkin, A. V. Murel, V. M. Daniltsev, and O. I. Khrykin, Control of charge transport mode in the Schottky barrier by d-doping: Calculation and experiment for Al/GaAs, *Semiconductors* **36**, 505–510 (2002).
56. H. Dery and L. J. Sham, Spin extraction theory and its relevance to spintronics, *Phys. Rev. Lett.* **98**, 046602 (2007).
57. P. Li and H. Dery, Tunable spin junction, *Appl. Phys. Lett.* **94**, 192108 (2009).
58. R. A. Smith, *Semiconductors*, Cambridge University Press, Cambridge, U.K., 1978.
59. Y. Song and H. Dery, Spin transport theory in ferromagnet/semiconductor systems with noncollinear magnetization configurations, *Phys. Rev. B* **81**, 045321 (2010).
60. W. H. Butler, X.-G. Zhang, X. Wang, J. van Ek, and J. M. MacLaren, Electronic structure of FM—semiconductor—FM spin tunneling structures, *J. Appl. Phys.* **81**, 5518–5520 (1997).
61. O. Wunnicke, Ph. Mavropoulos, R. Zeller, P. H. Dederichs, and D. Grundler, Ballistic spin injection from Fe(001) into ZnSe and GaAs, *Phys. Rev. B* **65**, 241306(R) (2002).
62. M. Zwierzycki et al., Spin injection through an Fe/InAs interface, *Phys. Rev. B* **67**, 092401 (2003).
63. A. N. Chantis, K. D. Belashchenko, D. L. Smith, E. Y. Tsymbal, M. van Schilfgaarde, and R. C. Albers, Reversal of spin polarization in Fe/GaAs (001) driven by resonant surface states: First-principles calculations, *Phys. Rev. Lett.* **99**, 196603 (2007).
64. J. C. Slonczewski, Conductance and exchange coupling of two ferromagnets separated by a tunneling barrier, *Phys. Rev. B* **39**, 6995–7002 (1989).
65. C. Ciuti, J. P. McGuire, and L. J. Sham, Spin polarization of semiconductor carriers by reflection off a ferromagnet, *Phys. Rev. Lett.* **89**, 156601 (2002).
66. J. D. Albrecht and D. L. Smith, Spin-polarized electron transport at ferromagnet semiconductor Schottky contacts, *Phys. Rev. B*, **68**, 035340 (2002).
67. S. Saikin, A drift-diffusion model for spin-polarized transport in a two-dimensional non-degenerate electron gas controlled by spin–orbit interaction, *J. Phys.: Condens. Mat.* **16**, 5071–5082 (2004).
68. S. Saikin, M. Shen, and M.-C. Cheng, Spin dynamics in a compound semiconductor spintronic structure with a Schottky barrier, *J. Phys.: Condens. Mat.* **18**, 1535–1544 (2006).
69. Z. G. Yu and M. E. Flatté, Spin diffusion and injection in semiconductor structures: Electric field effects, *Phys. Rev. B* **66**, 235302 (2002).

70. B. A. Gurney, V. S. Speriosu, J. P. Nozieres, H. Lefakis, D. R. Wilhoit, and O. U. Need, Direct measurement of spin-dependent conduction-electron mean free paths in ferromagnetic metals, *Phys. Rev. Lett.* **71**, 4023–4026 (1993).
71. Ł. Cywiński, H. Dery, P. Dalal, and L. J. Sham, Electrical expression of spin accumulation in ferromagnet/semiconductor structures, *Mod. Phys. Lett. B* **21**, 1509–1520 (2007).
72. S. Tehrani, B. Engel, J. M. Slaughter et al., Recent developments in magnetic tunnel junction MRAM, *IEEE Trans. Magn.* **36**, 2752–2757 (2000).
73. B. C. Choi and M. R. Freeman, Nonequilibrium spin dynamics in laterally defined magnetic structures, in *Ultrathin Magnetic Structures III*, B. Heinrich, and J. A. C. Bland (Eds.), Springer-Verlag, Berlin, pp. 211–232, 2004.
74. Th. Gerrits, H. A. M. van den Berg, J. Hohlfeld, L. Bär, and Th. Rasing, Ultrafast precessional magnetization reversal by picosecond magnetic field pulse shaping, *Nature* **418**, 509–512 (2002).
75. W. Schumacher, C. Chappert, P. Crozat, R. C. Sousa, P. P. Freitas, and M. Bauer, Coherent suppression of magnetic ringing in microscopic spin valve elements, *Appl. Phys. Lett.* **80**, 3781–3783 (2002).
76. A. Brataas, Y. V. Nazarov, and G. E. W. Bauer, Finite-element theory of transport in ferromagnet normal metal systems, *Phys. Rev. Lett.* **84**, 2481–2484 (2000).
77. A. Brataas, Y. V. Nazarov, and G. E. W. Bauer, Spin-transport in multi-terminal normal metal-ferromagnet systems with non-collinear magnetizations, *Eur. Phys. J. B* **22**, 99–110 (2001).
78. B. C. Choi, M. Belov, W. K. Hiebert, G. E. Ballentine, and M. R. Freeman, Ultrafast magnetization reversal dynamics investigated by time domain imaging, *Phys. Rev. Lett.* **86**, 728–731 (2001).
79. T. Kohonen, *Content-Addressable Memories*, Springer-Verlag, Heidelberg, 1980.
80. Compaq Computer Corporation, *Alpha 21264/EV6 Microprocessor Hardware Reference Manual*, September 2000, Order number: DS-0027B-TE.
81. J. Tendler, J. Dodson, J. Fields, H. Le, and B. Sinharoy, POWER4 system microarchitecture, *IBM J. Res. Dev.* **46**, 5 (2002).
82. D. Boggs, A. Baktha, J. Hawkins et al., The microarchitecture of the Intel pentium 4 processor on 90 nm technology, *Intel Technol. J.* **8**, 1 (2004).
83. W. Han, K. Pi, K. M. McCreary et al., Tunneling spin injection into single layer graphene, *Phys. Rev. Lett.* **105**, 167202 (2010).
84. H. Dery, H. Wu, B. Ciftcioglu et al., Spintronic nanoelectronics based on magneto-logic gates, arXiv:1101.1497, preprint.

Index

F

G

L

M

N

O

P

Q

R

S

Printed and bound by CPI Group (UK) Ltd, Croydon, CR0 4YY
24/10/2024
01778288-0018